# ONLINE STUDY & HOMEWORK MATERIALS

**I M P O R T A N T :** Following are instructions to access online resources to support your McGraw-Hill textbook

## The URL associated with your text is:
### http://www.mhhe.com/saladin

**Option 1:** ARIS LOGIN. Your instructor may use ARIS as a homework and assessment tool. If so, you must register in order to ensure your assignments are recorded into your instructor's gradebook.

**Option 2:** If your instructor is NOT using ARIS as a homework and assessment tool, you are welcome to access the material on the site without registering. Simply go to the URL listed above. You are free to access these materials for your own self-study.

## ARIS LOGIN. *To Register you need:*

**1.** **Section Code:** Provided by your instructor.

**2.** **Registration Code:** Provided in the gray scratch-off area below.

**3.** **URL:** Go to the URL listed at the top of this card and follow the directions for creating an ARIS account.

D0169175

### Scratch off for registration code
This registration code can be used by one individual and is not transferable.

**I M P O R T A N T :** The registration code printed above can only be used once to create a unique student account. Students do not need a registration code to access the content of the site. Students choosing "ARIS Login" must login each time they visit the site in order for their grades to be saved to their instructor's gradebook.

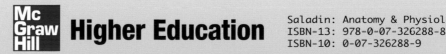 **Higher Education**

Saladin: Anatomy & Physiology, 4E
ISBN-13: 978-0-07-326288-8
ISBN-10: 0-07-326288-9

# ANATOMY & PHYSIOLOGY

## The Unity of Form and Function

## Kenneth S. Saladin

*Georgia College and State University*

*fourth edition*

# Higher Education

ANATOMY AND PHYSIOLOGY: THE UNITY OF FORM AND FUNCTION, FOURTH EDITION

1 2 3 4 5 6 7 8 9 0 VNH/VNH 0 9 8 7 6


ISBN-13   978–0–07–287506–5
ISBN-10   0–07–287506–2


Publisher: *Michelle Watnick*
Senior Developmental Editor: *Kristine A. Queck*
Director of Development: *Kristine Tibbetts*
Executive Marketing Manager: *James F. Connely*
Marketing Manager: *Lynn M. Kalb*
Senior Project Manager: *Mary E. Powers*
Senior Production Supervisor: *Laura Fuller*
Senior Media Project Manager: *Tammy Juran*
Lead Media Producer: *John J. Theobald*
Senior Designer: *David W. Hash*
Interior Designer: *Kaye Farmer*
Cover Illustration: *Joanne Brummett*
Cover Photo: *Johnny A. Ready/Getty Images*
Senior Photo Research Coordinator: *John C. Leland*
Photo Research: *Mary Reeg*
Supplement Producer: *Tracy L. Konrardy*
Compositor: *Precision Graphics*
Typeface: *10/12 Melior*
Printer: *Von Hoffmann Corporation*

The credits section for this book begins on page C-1 and is considered an extension of the copyright page.

**Library of Congress Cataloging-in-Publication Data**

Saladin, Kenneth S.
   Anatomy and physiology : the unity of form and function / Kenneth S. Saladin. — 4th ed.
      p.   cm.
   Includes index.
   ISBN 978–0–07–287506–5  —  ISBN 0–07–287506–2 (hard copy : alk. paper)
   1. Human physiology. 2. Human anatomy. I. Title. II. Title: Anatomy and physiology.

QP34.5.S23      2007
612–dc22                                    2005034469
                                            CIP


www.mhhe.com

**KEN SALADIN** has taught since 1977 at Georgia College and State University, a public liberal arts university in Milledgeville, Georgia. He earned his B.S. in zoology at Michigan State University and Ph.D. in parasitology at Florida State University. In addition to human anatomy and physiology, his courses have included histology, parasitology, animal behavior, sociobiology, introductory biology, general zoology, biological etymology, and a study abroad course in the Galápagos Islands. Nine times over the years, outstanding students inducted into Phi Kappa Phi have tapped Ken for recognition as their most significant undergraduate mentor. He received the university's Excellence in Research and Publication Award for the first edition of this book, and was named Distinguished Professor in 2001.

Ken is an active member of the Human Anatomy and Physiology Society, the Society for Integrative and Comparative Biology, the American Association of Anatomists, and the American Association for the Advancement of Science. He served as a developmental reviewer and wrote supplements for several other McGraw-Hill anatomy and physiology textbooks for a number of years before beginning this book.

Ken's avocational interests include philanthropic support of the Big Brothers/Big Sisters program for single-parent children, the Charles Darwin Research Station in the Galápagos, and several student scholarships, and he occasionally acts in college theatrical productions. Ken is married to Diane Saladin, a registered nurse. Their son, Emory, is a student in architectural design and their daughter, Nicole, is a graduate student in marine conservation ecology.

*Ken Saladin* (left) *and Les Greene.*

**DEDICATION**

*All professions depend on good teachers. I dedicate this edition to one of the very best*

*LES GREENE*

*my first biology teacher and a friend ever since.*

# BRIEF Contents

# Contents

# Preface

My students come to anatomy and physiology full of idealism, hoping to work in one of the health professions from nursing or therapy to fitness training and health education. They quickly discover that it is a tremendous task just to master the prerequisites for the clinical phase of their training. One of the greatest challenges they face is to understand the structural complexity of the human body and its intricate functional mechanisms.

My profession is to help them along the way by making this overwhelming amount of information manageable and, ideally, even stimulating. My students have long inspired me to spare no effort in presenting human form and function in a lucid, well-organized, and interesting way in my classes. Nearly 15 years ago, I accepted an invitation to put my approach on paper, with the result you are holding in your hands. I was pleasantly surprised at how quickly the first edition of this book rose to success— a gratifying validation of the way I had written and illustrated it and framed it in helpful pedagogic devices.

As enjoyable as this success has been, I've never been fully satisfied with the book. I view perfection as an asymptote, a standard to be approached more and more closely but perhaps never quite attained. The ink never dries on a new edition before I'm already compiling a list of how the next one could be better. Here is the fourth avatar, still reaching for that asymptote.

## Audience

This book is meant for a two-semester course in combined human A&P, primarily for beginning college students who need it as a prerequisite for admission to a clinical curriculum. I assume no prior college coursework; the first five chapters provide all the basic concepts of chemistry and cell biology that a student needs in order to understand the subsequent chapters on the human organ systems. The introductory chapters also serve as a refresher for those returning to college after raising a family or pursuing another career.

I also keep in mind that many A&P students, being very early in their college careers, are still developing the intellectual skills and study habits necessary for success in a health science curriculum, and that there also are many whose original language was other than English. I think of them as I choose my words and craft the structure of a sentence or paragraph. I try to avoid talking over the heads of beginning college students, but also to avoid "dumbing down" the text to an insulting level. I'm wary of idioms that may be obscure to international readers. I sprinkle the narrative with review activities such as self-testing questions; self-teaching prompts such as palpations and simple experiments the reader can do sitting at a desk; and learning aids such as pronunciation guides and insights into the roots and origins of medical terms. I try to conceive of enlightening ways to illustrate each major idea, often in ways that no other textbook has illustrated them before. To enliven the facts of science, I offer analogies, clinical insights, historical notes, biographical vignettes, and other seasoning that will make the book not only enjoyable to students, but interesting even to experienced instructors who may have "heard it all before," yet may not have heard of some of the facts and perspectives offered in these pages.

## What Sets This Book Apart?

Honored as I felt at being invited to write a textbook, I deliberated for more than a year before I accepted. There were many other A&P books already on the market, some of them very good. I didn't want to write a "me too" book, a mere imitation of the others. If I were going to write at all, it would be to provide something that other textbooks were not—a point of view, a clarity of exposition, new illustrative concepts, or content more accurate than many of the oft-repeated, but false, textbook clichés. I stake my claim to originality in the following aspects of this book.

### ORGANIZATION

Some chapters and topics are presented in a nontraditional sequence that I feel is more instructive than the conventional order. This has been well received, so I've made no changes in it since the previous edition. For those who have not used the book before, a brief explanation is in order.

## Heredity

The most fundamental principles of heredity are presented in the last few pages of chapter 4 rather than at the back of the book. I do not see how students can adequately understand such genetic traits and conditions as cystic fibrosis, color blindness, blood types, hemophilia, cancer genes, or sickle cell disease, if they do not first understand such concepts as dominant and recessive alleles, genotype and phenotype, sex linkage, and pleiotropy. It has always seemed to me a mistake to put these at the back of the book, as others do, in a chapter that many instructors never reach as we try to cover such a big subject in such a short semester. The most monumental development in medical science now taking place is genomic medicine, and textbooks can no longer defensibly make genetics take a back seat to all the rest of A&P. The next few years seem certain to demand an earlier and stronger presentation of genomics to health-care students.

## Muscle Anatomy and Physiology

I treat the functional morphology of the skeleton, joints, and muscles in three consecutive chapters, 8 through 10. Therefore, when students learn muscle origins and insertions from this book, it is only two chapters after the names of the relevant bone features; when they learn muscle actions, it is in the first chapter after learning the terms for the joint movements. This means presenting muscle anatomy and physiology in an order opposite from the one that other A&P books follow. It also brings another advantage: the physiology of muscle and nerve cells is treated in two consecutive chapters (11 and 12), which are thus closely integrated in their treatment of synapses, neurotransmitters, and membrane potentials.

## The Urinary System

Most textbooks place this system near the end of the book because of its anatomical and developmental relationships with the reproductive system. I feel, however, that its physiological ties to the circulatory and respiratory systems are much more important. The lungs and kidneys collaborate to regulate the pH of the blood; the kidneys have more impact than any other organ on blood pressure; and students should study glomerular filtration and tubular reabsorption before they have long forgotten the basic principles of capillary fluid exchange. I cannot see reason for the anatomical association between the urinary tract and reproductive system to override these important physiological ties to the car-diovascular and respiratory systems. Therefore, except for a necessary digression on lymphatics and immunity, I follow the circulatory system (chapters 18–20) almost immediately with the respiratory and urinary systems (chapters 22–24).

## WRITING STYLE

Next to scientific accuracy, the most important quality of an effective textbook is writing style. Over my years of teaching from other college textbooks, I've seen styles to admire and emulate, and others to avoid. Some are very correct and formal, but students find them aloof; such books do not "speak" to them, they say. Beginning college students get a stronger sense of engagement in the discipline from a more personal approach on the part of an author. Other writers go for such a chummy style that they appear a bit too cute and leave a student feeling patronized. My inclination has always been a middle course: a tone that is semiconversational, neither stuffy nor condescending; one that uses an occasional colloquialism without straying into baffling idioms; one that uses everyday events and innovative analogies to enable students to visualize and relate to a process; one that favors simple language and syntax over convoluted paragraphs and a needlessly multisyllabic, graduate school vocabulary.

No writer can be an objective judge of his own effectiveness, but I feel I've hit my mark when students write to me from other colleges (where I do not determine their grades) saying they enjoy the book because "it feels as if the author is talking to me," or because they can understand it so much more easily than other A&P books they've used. And when reviewers make comments like the following, I'm encouraged to preserve and enhance this tone from edition to edition:

> *This book is the clearest and most direct read in the field. His strong narrative style holds the students' interest and is an important strong area in his writing. He tends to avoid pointless and egocentric digressions that spoil many other anatomy and physiology texts. I find myself drawn into reading his text more as I would a decent novel. One particularly difficult area is immunology: so many texts are scattered and confused that I often wonder if that author understands the topic. Saladin, on the other hand, is clear and straightforward on that topic.*
> —*David L. Evans, Pennsylvania College of Technology*

> *Students who have used other textbooks rave about this one. This text has been written to simplify things for students, not to mystify them.*
> —*Lawrence N. Killian, Clark State Community College*

*Compared to [other textbooks I have used], this is the absolute best. I love the way Saladin writes and I think the students do as well. His powers of explanation are sometimes extraordinary. His analogies and interesting snippets scattered throughout the book are fascinating and greatly facilitate absorption of the material. I don't think that any other text is as "user friendly" to the students.*
*—Nikki Privacky, Palm Beach Community College*

## ILLUSTRATIONS

Captivating art and photography were especially important in stimulating my interest in the life sciences when I was a boy, and they are no less important in reinforcing a college student's interest in human structure and function. I brought several original illustrative concepts to this project, as well as a vision of the kind of art and photographs that an outstanding A&P book should have. McGraw-Hill spared no expense to commission an art package in a league of its own. In this edition, the scientific and medical illustrators at Precision Graphics have produced another quantum leap forward. I'm confident that a prospective user can compare the art in this book with the corresponding figures of others and clearly see the advantage of this one for both instructor and student. Reviewers agree:

*The artwork in Saladin is head and shoulders above all the other anatomy and physiology books on the market today….the others have not been able to come close.*
*—Robert Moldenhauer, St. Clair County Community College*

*I must say I was completely blown away by this text. The graphics in [a leading text I've been using] don't come close to the graphics in Saladin, which have an extraordinary 3-D quality.*
*—Bill Schutt, Long Island University*

*One of the major strengths of the Saladin text, one that prompted me to adopt it, was the quality and quantity of the illustrations. In my view, this text is a hands-down winner in this area.*
*—Richard A. Symmons, California State University at Hayward*

## AN EVOLUTIONARY FLAVOR

The human body can never be fully understood without a sense of how and why it came to be as it is. Since the mid-1990s, medical literature has shown an increasing interest in evolutionary medicine. Several books have recently been published on it, as have many articles in the leading medical journals. Even the first chapter of *Gray's Anatomy* is devoted mainly to the relevance of evolution for human anatomy, and Gray's reinforces this with numerous evolutionary insights from cover to cover. Yet most A&P textbooks continue to ignore it.

I briefly introduce the concept of natural selection as applied to humans in chapter 1. Later chapters have nine boxed essays (Insights) on evolutionary medicine, and many evolutionary remarks in the main body of the narrative. Students will find novel and intriguing ways of looking at such topics as mitochondria (p. 121), body hair (p. 200), skeletal anatomy (p. 284), body odors (p. 597), the taste for sweets (p. 1006), the nephron loop (p. 914), lactose intolerance (p. 985), menopause (p. 1077), and senescence (p. 1134).

*I must say that I strongly support the [evolutionary] approach. The more students hear about evolution, selection and adaptation, the better. These three variables are crucial for understanding physiology…. It is a nice idea to provide them with the underpinnings of the terms adaptation and selection.*
*—Ralph F. Fregosi, University of Arizona*

*I am particularly impressed with Saladin's inclusion of evolution and the history of medicine. I often try to interject this information into my lectures, and find most modern texts lacking in these areas. Kudos to you!*
*—Mary E. Dawson, Kingsboro Community College*

## HISTORICAL INSIGHTS— HUMANIZING HUMAN A&P

I found long ago that students especially enjoy lectures in which I remark on the personal dramas that enliven the history of medicine, so I've incorporated that into my writing as well. In these pages, I share such favorite stories as William Beaumont's digestive experiments on Alexis St. Martin, "the man with the hole in his stomach" (p. 995); Crawford Long's discovery of the surgical benefit of a popular party drug, ether (p. 630); Phineas Gage's dramatic personality changes following his remarkable brain injury (p. 539); and the testy relationship between the two men who shared a Nobel Prize for the discovery of insulin, Frederick Banting and J. J. R. MacLeod (p. 672). There's drama and poignancy in the struggles and unkind ironies of the careers of Rosalind Franklin (p. 130), Marie Curie (p. 57), and Charles Drew (p. 693); we can take inspiration from the way that Santiago Ramón y Cajal (p. 453), William

Harvey (p. 754), and Rita Levi-Montalcini (p. 454) triumphed over privation, ridicule, and even ethnic persecution.

Some say they simply don't have time in their A&P courses to cover any evolution or scientific history. Neither my book nor my semesters are longer than anyone else's, yet I find, as my students seem to do, that an occasional evolutionary remark or historical drama make the teaching and learning of A&P a lot more fun. More than a few distinguished scientists and clinical practitioners say they found their inspiration in reading of the lives of their predecessors.

## What's New?

What distinguishes this edition from the previous one? It certainly is not just the old book in a new cover. My personal list of changes runs to 95 pages of 10-point type, so obviously only the most important ones can be listed here.

### ILLUSTRATIONS

The most extensive change in this edition is a thorough revision of its 809 illustrations. The editor and I put a great deal of thought into how every piece of art might be improved, and McGraw-Hill engaged a team of talented scientific and medical illustrators at Precision Graphics, in Champaign, Illinois, to enhance, revamp, or replace almost every item of line art. The improvements are too numerous to list more than a few, but users of the previous edition will find conspicuous improvements in such figures as the plasma membrane (p. 95), endochondral ossification (p. 223), gross anatomy of the spinal cord (p. 483), the eye (p. 615), capillary histology (pp. 758 and 759), neutrophil diapedesis (p. 826), the cardiac myocyte (p. 729), neural control of respiration (p. 868), the nephron (p. 903), and histology of the teeth (p. 961) and liver (p. 976).

The illustrators' flair for human portraiture has greatly humanized and beautified such figures as wound healing (p. 181), the paranasal sinuses (p. 248), the hyoid (p. 257), the facial nerve (p. 553), and others. More realistic figure drawings also now grace the "icons" that we use to identify the bodily location of a close-up view, a dissection, or a section—for example at pages 47, 350, and 898.

Several illustrative concepts are entirely new to this edition: intervertebral disc herniation (p. 262), axes of joint rotation (p. 300), the knee menisci (p. 313), knee injuries (p. 314), CNS myelination (p. 451), the relationship of the Schwann cell to unmyelinated nerve fibers (p. 452), cross-sectional anatomy of the pons and medulla (p. 525), the sleep cycle (p. 537), the hepatic sinusoid (p. 976), development of T and B cells (p. 830), the

endometrial cycle (p. 1084), and the histology of intramembranous ossification (p. 222), the thymus (p. 816), and red bone marrow (p. 814).

Numerous figures have been enlarged and their color schemes brightened. In laying out the pages, we took great pains to ensure that in nearly every case, an illustration is placed on the same page as the text that first references and describes it, or on the facing page, so there can be a minimum of page turning between reading about a figure and looking at it.

In many cases, I have moved brief descriptions such as "Mediastinal surface, right lung" from the figure legends into the body of the figure itself (p. 862). This shortens the legends so there is less need to look up and down repeatedly between the legend and art.

I have also brought more uniformity to the "process figures": those which present a series of events. Yellow-highlighted numbers indicate the steps in a process and correspond to a numbered descriptive list of the events, placed within the figure (see p. xv) or at least on the same page (as on pp. 457 and 658). In multipart figures that show different perspectives on the same subject, such as views at levels of increasing detail, we box the area of interest and connect the figure elements with arrows (as on p. 816). Such changes give the art a stronger sense of action and a clearer sense of relationship among the figure elements.

The photographs also came under close scrutiny. Several cadaver photographs have been replaced with new ones that depict better specimens, cleaner dissections, or sharper photography (abdominal muscles, p. 345; leg muscles, p. 375; lymph nodes, p. 817; kidney, p. 899). Several other replacement photos are more stunning or instructive than the old ones (some of the joint movements, pp. 302–307; ovarian follicle, p. 635; spina bifida, p. 485; kwashiorkor, p. 684; capillary bed, p. 753; glomerular podocytes, p. 905; thalidomide effects, p. 1124). A few photos of subjects entirely new to this edition include platelet structure (p. 703), twin fetuses (p. 1110), and some of the fetal development series (pp. 1120–1121).

### REWRITES AND UPDATES

One of my most important tasks as author is to keep the science current, which I do through more than a dozen journal subscriptions, updating my library of medical books, attending conferences, monitoring and participating in online discussions of A&P, and receiving very helpful feedback from reviewers and users of the book. I have rewritten numerous passages to update the scientific content, correct errors, and improve readability. Such changes are too numerous to list in entirely, but the most significant ones in each chapter are as follows.

**Chapter 1, Major Themes of Anatomy and Physiology.** A more multicultural look at early medical history; corrections in the chronology of microscopy; update on the evolution of modern *Homo sapiens.*

**Chapter 2, The Chemistry of Life.** New discussion of the role of van der Waals forces in protein folding and molecular associations.

**Chapter 3, Cellular Form and Function.** Rewritten sections on osmosis and peroxisomes.

**Chapter 4, Genetics and Cellular Function.** Rewritten sections on chromatin coiling, mitosis, and cancer. Updates on changing concepts of the gene, alternative RNA splicing and protein diversity, genomic medicine, and the Human Genome Project.

**Chapter 5, Histology.** New sections on stem cells and tissue engineering.

**Chapter 6, The Integumentary System.** Reorganized section on the skin; new content on epidermal stem cells and dendritic cells; updates on epidermal tight junctions, eumelanin and pheomelanin, and the malignant melanoma oncogene.

**Chapter 7, Bone Tissue.** Reorganized sections on bone function, general bone structure, and the composite nature of the bone matrix. Expanded treatment of intramembranous ossification. Sections on endochondral ossification, bone growth, and bone remodeling rewritten for clarity. Wolff's law introduced. Scientific updates on osteocyte function, RANKL (formerly osteoclast-stimulating factor), and the diagnosis and treatment of osteoporosis.

**Chapter 8, The Skeletal System.** New content on the mechanical function of interosseous membranes and rotation of the embryonic limbs as a basis for understanding adult limb orientation. All bone features introduced in chapter 10 as muscle attachments are now explained here.

**Chapter 9, Joints.** Substitution of functional discussion for anatomical detail; deletion of jaw and ankle anatomy, replaced by an extensive rewrite on joint biomechanics to provide a better background for study in kinesiology and physical therapy. Reorganized section on types of synovial joint movements, with introduction of kinesiologic terminology.

**Chapter 10, The Muscular System.** Narratives on muscle anatomy condensed and moved into the tables. All muscle origins and insertions, previously from multiple sources, now conformed to *Gray's Anatomy.* Some minor and variable muscles deleted (procerus, plantaris, and deep and superficial perineal) and table entries on some other muscles divided or expanded. More everyday examples used for muscle actions.

**Chapter 11, Muscular Tissue.** Rewrites and updates on the elastic components of muscle, dystrophin and other accessory muscle proteins, energy metabolism of muscle, muscle strength, morphology of cardiac myocytes, and muscular dystrophy.

**Chapter 12, Nervous Tissue.** Enhanced discussion of astrocyte function, nerve regeneration, and the labeled line code. New historical notes on Camillo Golgi and Ramón y Cajal, and essay on Rita Levi-Montalcini.

**Chapter 13, The Spinal Cord, Spinal Nerves, and Somatic Reflexes.** Updates on spina bifida and folic acid, muscle spindle physiology, and shingles. Rewrites on the spinoreticular, tectospinal, and vestibulospinal tracts, and explanation of their related brainstem nuclei. Reorganized tables of spinal nerve plexuses with simplified lists of their muscle innervations.

**Chapter 14, The Brain and Cranial Nerves.** New concepts of cerebellar function, and other rewrites on pain modulation, sleep, and functions of the mammillary nuclei, thalamus, limbic system, basal nuclei, sensory cortex, and orbitofrontal cortex. A new format for the cranial nerve tables.

**Chapter 15, The Autonomic Nervous System and Visceral Reflexes.** Rewrites on the splanchnic nerve route, glossopharyngeal nerve, nitric oxide, cholinergic receptors, and autonomic brainstem nuclei.

**Chapter 16, Sense Organs.** Rewrites on the general properties of receptors; lamellated corpuscles; spinal gating of pain; and projection pathways for pain, olfaction, and vestibular function. Updates on olfactory sensory transduction and central processing, retinal circuitry, and inhalation anesthesia. Enhanced treatments of the auditory ossicles and physiology of hearing. Simplified treatment of retinal rod and ganglion cell function.

**Chapter 17, The Endocrine System.** Rewrites on the adrenal catecholamines and synthesis of thyroid hormones. Update on hepatic hormones including hepcidin.

**Chapter 18, The Circulatory System: Blood.** Hemopoiesis reorganized, with only general principles presented early, and details of erythropoiesis now presented in the section on RBCs, leukopoiesis with the WBCs, and thrombopoiesis with the platelets. Rewrites on plasma, blood osmolarity, blood antigens and antibodies, leukocyte functions, and platelet structure. New Insight on the complete blood count.

**Chapter 19, The Circulatory System: The Heart.** Rewrites on cardiac anatomy, cardiac myocyte morphology, angina and heart attack, blood pressure and flow, chronotropic chemicals, and myocardial contractility. Introduction of clinical terms for the coronary blood vessels.

**Chapter 20, The Circulatory System: Blood Vessels and Circulation.** Rewrites on blood vessel histology; capillary types; blood pressure, resistance, and flow relationships; and capillary fluid exchange.

**Chapter 21, The Lymphatic and Immune Systems.** An extensively rewritten chapter with new treatments of the classic three lines of defense, leukocyte functions, interferons, the complement system, the membrane attack complex, immune surveillance, perforins and granzymes, inflammation, penicillin allergy, lymphocyte development, T cell classes, antibody diversity, and AIDS chemotherapy. Addition of the structure of red bone marrow as a lymphatic organ, and a new Insight on lymph nodes and cancer. Updated and simplified cytokine terminology and deletion of some obsolete cytokine nomenclature.

**Chapter 22, The Respiratory System.** Expanded treatment of respiratory muscles and pulmonary ventilation, brainstem respiratory centers, and neural control of breathing. Rewrites on respiratory gross anatomy, restrictive and obstructive lung disorders, pressure and airflow relationships, and pulmonary surfactant.

**Chapter 23, The Urinary System.** Update on hormonal control of nephron function; enhanced discussion of ureter and bladder histology; rewrite on the neural control of micturition; and deletion of renal diabetes.

**Chapter 24, Water, Electrolyte, and Acid–Base Balance.** Rewrite on disorders of acid–base homeostasis.

**Chapter 25, The Digestive System.** Rewrites on the enteric nervous system; histology of the jejunum, ileum, and colon; tongue musculature; emesis; lactose intolerance; and micelle formation. Update on bacterial flora. Addition of gluten-sensitive enteropathy (sprue) to the table of digestive disorders.

**Chapter 26, Nutrition and Metabolism.** Extensive rewrite and update on appetite control, hunger and satiety hormones, and obesity. New Insight on hepatitis and cirrhosis. Update on thermoregulation and heat stroke.

**Chapter 27, The Male Reproductive System.** Reorganized discussion of gross anatomy for better sequencing of illustrations. Rewrites on cryptorchidism, descent of the testes, testicular thermoregulation; neural control of coital physiology; the role of voluntary muscles in erection; and treatment of erectile dysfunction.

**Chapter 28, The Female Reproductive System.** Updates on the staging of Pap smears; the role of leptin in menarche and menstruation; the relationship of oocyte count to the onset of menopause; the grandmother hypothesis for menopause; the role of plasmin in ovulation; preeclampsia; and contraception. Improved descriptions of the mesovarium, ovarian blood vessels, and cyclic histology of the uterus. Extensive rewrite on the ovarian cycle, with modified terminology and deletion of popular textbook truisms that are untrue.

**Chapter 29, Human Development.** Extensive rewrite of the entire section on prenatal development, introducing developmental trimesters and embryonic folding. Improved treatment of embryonic germ layers and gastrulation. Rewrite on the trisomies and other aneuploid conditions.

## ISSUES OF TERMINOLOGY

Aside from these chapter-specific changes, some anatomical terminology throughout the book has been updated in conformity to the *Terminologia Anatomica (TA)*, although I depart from the TA in cases where it seems that it would create more confusion than enlightenment for the beginning student. As in the previous edition, I follow the general trend of the TA and the American Medical Association in eliminating the possessive form for medical terms (such as Down's syndrome and Peyer's patches); using English terms when they have become accepted as alternatives to the Latin (such as *pilocrector muscle* in place of *arrector pili*); and using descriptive terms in place of eponyms (such as *pancreatic islets* in place of *islets of Langerhans*). The older Latin and eponymous terms are given only as parenthetical synonyms where new terms are first introduced.

## Guided Tour

Students and instructors can become acquainted with the key features of this book by browsing through the Guided Tour starting on the next page. These pages constitute a visual exposition of the book's art program and the pedagogical framework around which each chapter is organized.

## Suggestions Always Welcome

I invite my colleagues and students everywhere to continue offering their valuable and stimulating feedback as I plan the next edition.

Ken Saladin
Dept. of Biology
Georgia College & State University
Milledgeville, GA 31061 (USA)
478-445-0816
ken.saladin@gcsu.edu

# Guided Tour

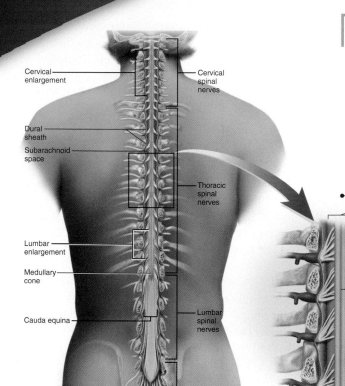

(a)

(b)

## NEW AND VIVID ILLUSTRATIONS

Saladin's all-new illustration program is unsurpassed among A&P texts! Dynamic illustrations harmonize with Saladin's clear and engaging writing style to create an A&P textbook that is visually captivating and fun to read. Colorful, precise anatomical illustrations lend a realistic view of body structures and three-dimensional details help students envision the cellular-level events of physiological processes.

### BRIGHT, BOLD COLORS

*A bright color palette provides contrast for easy distinction of structures.*

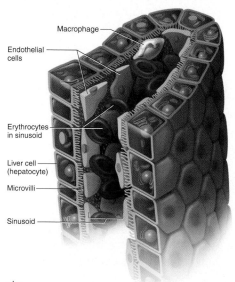

### THREE-DIMENSIONAL DETAIL

*Rich textures and shading provide the visual depth and dimension that bring structures to life.*

### CONSISTENT COLOR PALETTE

*Colors used to indicate specific structures are applied consistently for a cohesive art program from cover to cover.*

# STEPPED-OUT PROCESS FIGURES

Saladin presents physiological concepts in an easy-to-follow stepwise format. Numbered steps within figures trace the sequence of events, and brief explanations describe what is happening in each step. Combining process descriptions with artwork creates a self-contained snapshot that summarizes concepts in a convenient and consistent format.

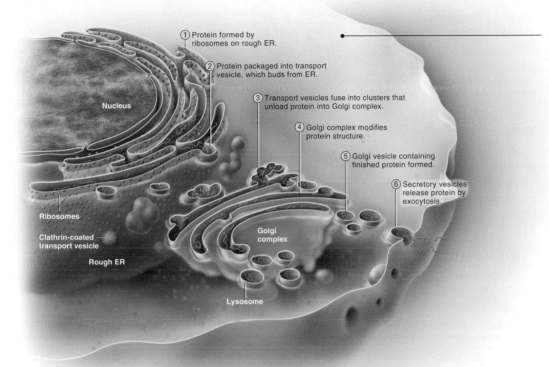

① Protein formed by ribosomes on rough ER.

② Protein packaged into transport vesicle, which buds from ER.

③ Transport vesicles fuse into clusters that unload protein into Golgi complex.

④ Golgi complex modifies protein structure.

⑤ Golgi vesicle containing finished protein formed.

⑥ Secretory vesicles release protein by exocytosis.

Nucleus

Ribosomes

Clathrin-coated transport vesicle

Rough ER

Golgi complex

Lysosome

## STEP-BY-STEP FORMAT

*Numbered steps embedded in the artwork guide students through complex processes*

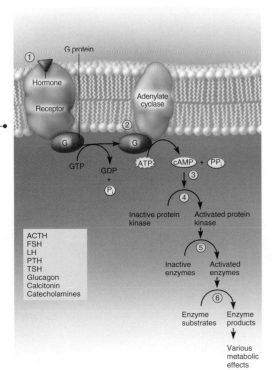

① Hormone receptor binding activates a G protein.

② G protein activates adenylate cyclase.

③ Adenylate cyclase produces cAMP.

④ cAMP activates protein kinases.

⑤ Protein kinases phosphorylate enzymes. This activates some enzymes and deactivates others.

⑥ Activated enzymes catalyze metabolic reactions with a wide range of possible effects on the cell.

G protein
Hormone
Receptor
Adenylate cyclase
GTP    GDP    +    Pi
ATP    cAMP    +    PPi
Inactive protein kinase    Activated protein kinase
Inactive enzymes    Activated enzymes
Enzyme substrates    Enzyme products
Various metabolic effects

ACTH
FSH
LH
PTH
TSH
Glucagon
Calcitonin
Catecholamines

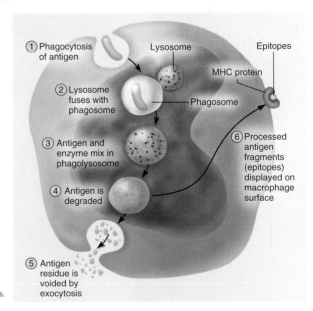

① Phagocytosis of antigen

② Lysosome fuses with phagosome

③ Antigen and enzyme mix in phagolysosome

④ Antigen is degraded

⑤ Antigen residue is voided by exocytosis

⑥ Processed antigen fragments (epitopes) displayed on macrophage surface

Lysosome    Epitopes
MHC protein
Phagosome

# Art Program

## MESSAGE-DRIVEN LAYOUTS

Creating effective, educational artwork that conveys a clear message to students requires careful consideration of the relationship of the parts to the whole, as well as a little creativity. Saladin's figures are organized in meaningful displays in which individual figure parts interact with one another to create big-picture explanations. Figures are often accented with depictions of everyday items or situations to frame concepts within familiar contexts.

### ENGAGING FIGURE PRESENTATIONS

*Figures are arranged in cohesive layouts that emphasize relationships among figure parts. For example, sharing labels between photos and drawings allows for easy comparison of structure appearance between the two mediums.*

### REAL-LIFE CONTEXT

*Framing new material within familiar contexts makes learning relevant to life. Many figures incorporate visual cues or artistic analogies that draw upon everyday experiences to facilitate understanding. This artistic technique is paralleled throughout the book by Saladin's analogy-rich writing style.*

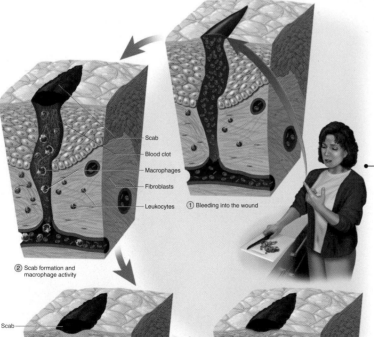

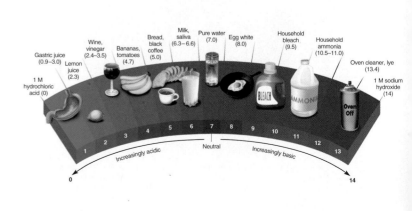

# INFORMATION-RICH VISUALS

Interpreting anatomical views can be difficult, so Saladin's figures have been designed with students in mind. Special helps like view indicators, orientation icons, and clear figure navigation paths make the focus of each figure readily apparent. Descriptive information previously confined to figure legends has been embedded directly into the artwork to make figures understandable at first glance.

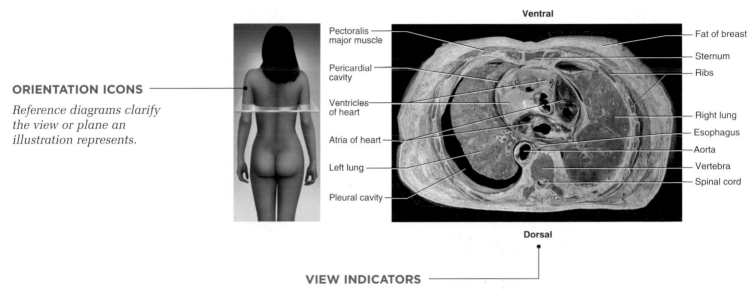

## ORIENTATION ICONS

*Reference diagrams clarify the view or plane an illustration represents.*

## VIEW INDICATORS

*Descriptive labeling indicating figure views facilitates easy interpretation of what is shown.*

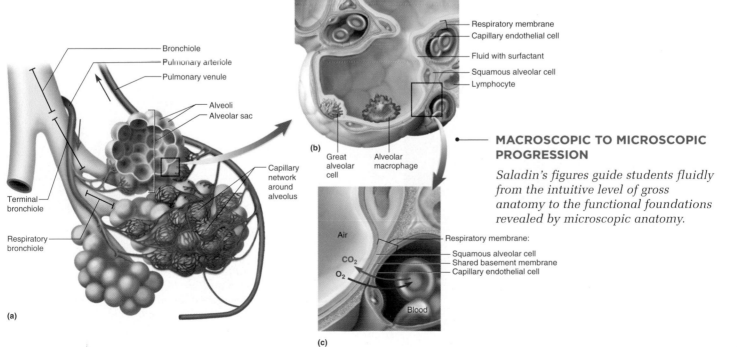

## MACROSCOPIC TO MICROSCOPIC PROGRESSION

*Saladin's figures guide students fluidly from the intuitive level of gross anatomy to the functional foundations revealed by microscopic anatomy.*

# Art Program

## ATLAS-QUALITY PHOTOGRAPHS

Photographs capture the truest appearance of gross and microscopic anatomy, and familiarize students with actual structures they may encounter in laboratory activities. Saladin's stunning collection of cadaver dissection images and light, TEM, and SEM photomicrographs balances the simplified clarity of illustrations with the realism of photos.

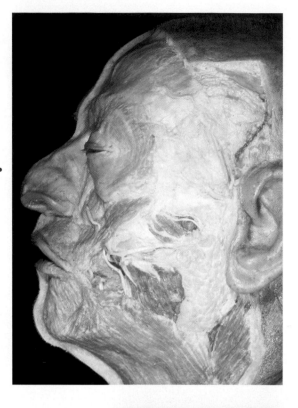

### CADAVER DISSECTIONS

*High-quality photographs of expertly dissected cadaver specimens capture the texture and detail of real human structures, and emphasize their anatomical relationships.*

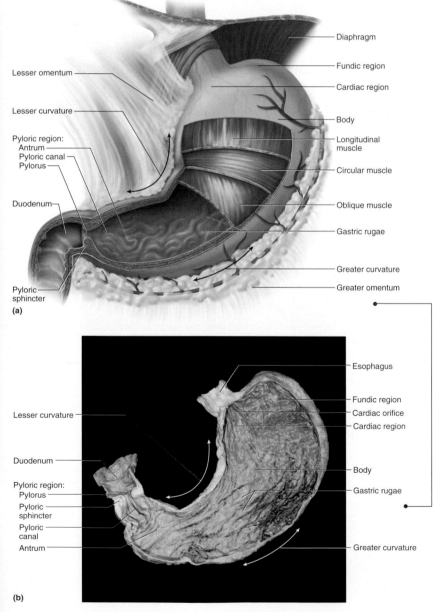

Lesser omentum

Lesser curvature

Pyloric region:
Antrum
Pyloric canal
Pylorus

Duodenum

Pyloric sphincter

(a)

Diaphragm
Fundic region
Cardiac region
Body
Longitudinal muscle
Circular muscle
Oblique muscle
Gastric rugae
Greater curvature
Greater omentum

Lesser curvature

Duodenum

Pyloric region:
Pylorus
Pyloric sphincter
Pyloric canal
Antrum

Esophagus

Fundic region
Cardiac orifice
Cardiac region

Body

Gastric rugae

Greater curvature

(b)

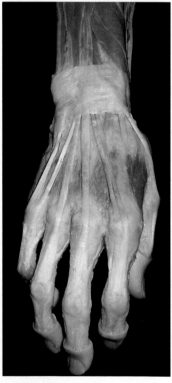

### COMPLEMENTARY VIEWS

*Drawings paired with photographs enhance visualization of structures. Labeling of art and photo mirror each other whenever possible, making it easy to correlate structures between views.*

# Art Program

## MICROGRAPHS

*A carefully researched collection of LM, SEM, and TEM photomicrographs reveal the intricate detail of microscopic structures.*

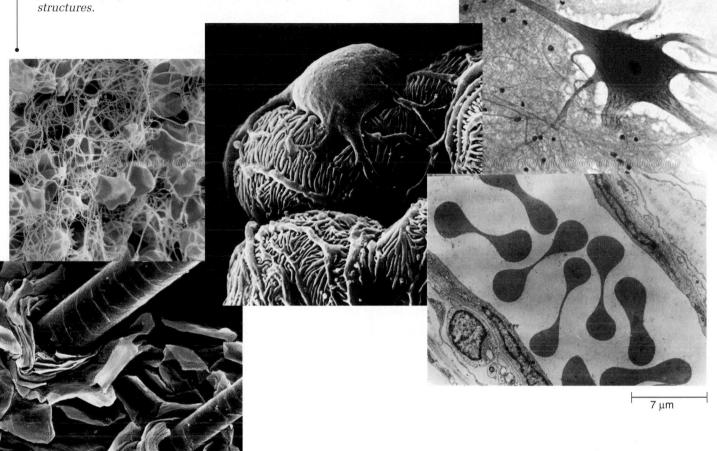

7 μm

0.1 mm

## SCALE BARS

*Scale bars are used whenever possible to provide a reference point for estimating the sizes of structures shown in micrographs.*

## PHOTOMICROGRAPHS CORRELATED WITH ILLUSTRATIONS

*Photomicrographs are often paired with illustrations to give students the best of both perspectives: the realism of photos and the explanatory clarity of drawings.*

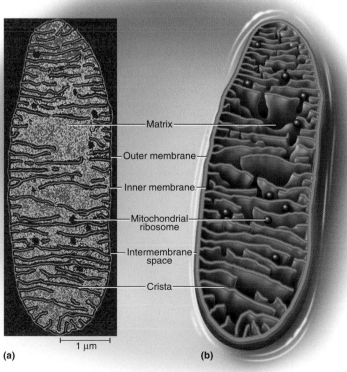

Matrix

Outer membrane

Inner membrane

Mitochondrial ribosome

Intermembrane space

Crista

1 μm

(a)  (b)

# Learning System

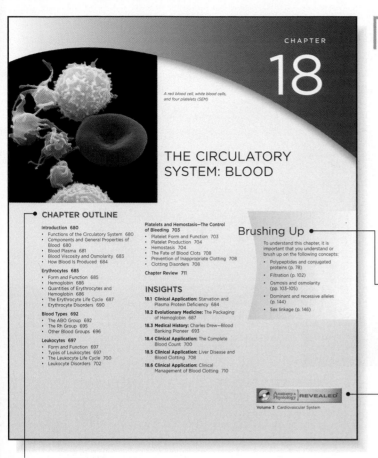

A red blood cell, white blood cells, and four platelets (SEM)

## THE CIRCULATORY SYSTEM: BLOOD

### Brushing Up

To understand this chapter, it is important that you understand or brush up on the following concepts:
- Polypeptides and conjugated proteins (p. 78)
- Filtration (p. 102)
- Osmosis and osmolarity (pp. 103–105)
- Dominant and recessive alleles (p. 144)
- Sex linkage (p. 146)

Anatomy & Physiology REVEALED

Volume 3 Cardiovascular System

## SYSTEMATIC PEDAGOGY

Saladin structures each chapter around a consistent and unique framework of pedagogic devices. Whatever the subject matter of a chapter, students can develop a consistent learning strategy. Orienting features such as chapter outlines and learning objectives help students organize study time and set goals, while self-testing questions in various formats and difficulty levels challenge students to recall terms and facts, to describe concepts, to analyze and apply ideas, and to relate concepts across chapters. Each chapter is seasoned with boxed discussions of clinical or scientific relevance that demonstrate application of concepts and add interest.

## BRUSHING UP

*A page-referenced list of previously covered concepts integral to understanding topics explained in the chapter at hand prompts students to review this material and reminds them that all organ systems are conceptually related.*

## ANATOMY & PHYSIOLOGY REVEALED

*Chapters covering systems presented in McGraw-Hill's* Anatomy & Physiology Revealed *series include an icon indicating which volume of this software coincides with the chapter.*

## CHAPTER OUTLINE

*A chapter outline provides a quick overview of the chapter contents and organization.*

## LEARNING OBJECTIVES AND BEFORE YOU GO ON

*Saladin divides each chapter typically into five or six short, digestible segments of just a few pages each, with a list of learning objectives at the beginning and a list of Before You Go On content review questions at the end of each one. This enables students to set tangible goals for short study periods and to assess their progress before moving on.*

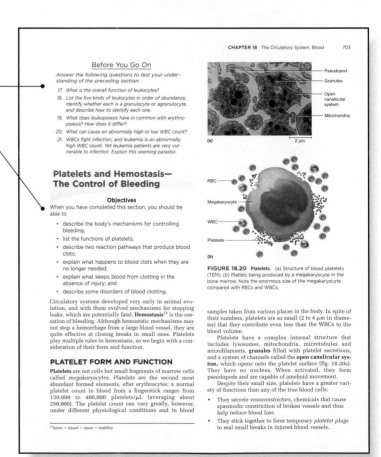

### Before You Go On

*Answer the following questions to test your understanding of the preceding section.*

17. What is the overall function of leukocytes?
18. List the five kinds of leukocytes in order of abundance, identify whether each is a granulocyte or agranulocyte, and describe how to identify each one.
19. What does leukopoiesis have in common with erythropoiesis? How does it differ?
20. What can cause an abnormally high or low WBC count?
21. WBCs fight infection, and leukemia is an abnormally high WBC count. Yet leukemia patients are very vulnerable to infection. Explain this seeming paradox.

### Platelets and Hemostasis—The Control of Bleeding

#### Objectives

When you have completed this section, you should be able to

- describe the body's mechanisms for controlling bleeding;
- list the functions of platelets;
- describe two reaction pathways that produce blood clots;
- explain what happens to blood clots when they are no longer needed;
- explain what keeps blood from clotting in the absence of injury; and
- describe some disorders of blood clotting.

Circulatory systems developed very early in animal evolution, and with them evolved mechanisms for stopping leaks, which are potentially fatal. **Hemostasis**[23] is the cessation of bleeding. Although hemostatic mechanisms may not stop a hemorrhage from a large blood vessel, they are quite effective at closing breaks in small ones. Platelets play multiple roles in hemostasis, so we begin with a consideration of their form and function.

#### PLATELET FORM AND FUNCTION

**Platelets** are not cells but small fragments of marrow cells called *megakaryocytes*. Platelets are the second most abundant formed elements, after erythrocytes; a normal platelet count in blood from a fingerstick ranges from 130,000 to 400,000 platelets/μL (averaging about 250,000). The platelet count can vary greatly, however, under different physiological conditions and in blood

[23]*hemo* = blood + *stasis* = stability

**FIGURE 18.20 Platelets.** (a) Structure of blood platelets (TEM). (b) Platelets being produced by a megakaryocyte in the bone marrow. Note the enormous size of the megakaryocyte compared with RBCs and WBCs.

samples taken from various places in the body. In spite of their numbers, platelets are so small (2 to 4 μm in diameter) that they contribute even less than the WBCs to the blood volume.

Platelets have a complex internal structure that includes lysosomes, mitochondria, microtubules and microfilaments, **granules** filled with platelet secretions, and a system of channels called the **open canalicular system,** which opens onto the platelet surface (fig. 18.20a). They have no nucleus. When activated, they form pseudopods and are capable of ameboid movement.

Despite their small size, platelets have a greater variety of functions than any of the true blood cells:

- They secrete *vasoconstrictors*, chemicals that cause spasmodic constriction of broken vessels and thus help reduce blood loss.
- They stick together to form temporary *platelet plugs* to seal small breaks in injured blood vessels.

444    **PART THREE** Integration and Control

2. **Conductivity.** Neurons respond to stimuli by producing electrical signals that are quickly conducted to other cells at distant locations.

3. **Secretion.** When the electrical signal reaches the end of a nerve fiber, the neuron secretes a chemical *neurotransmitter* that crosses the gap and stimulates the next cell.

> **Think About It**
> What basic physiological properties do a nerve cell and a skeletal muscle fiber have in common? Name a physiological property of each that the other one lacks.

### FUNCTIONAL CLASSES

There are three general classes of neurons (fig. 12.3) corresponding to the three major aspects of nervous system function listed earlier:

1. **Sensory (afferent) neurons** are specialized to detect stimuli such as light, heat, pressure, and chemicals, and transmit information about them to the CNS.

Such neurons begin in almost every organ of the body and end in the CNS; the word *afferent* refers to signal conduction *toward* the CNS. Some receptors, such as those for pain and smell, are themselves neurons. In other cases, such as taste and hearing, the receptor is a separate cell that communicates directly with a sensory neuron.

2. **Interneurons**[6] (association neurons) lie entirely within the CNS. They receive signals from many other neurons and carry out the integrative function of the nervous system—that is, they process, store, and retrieve information and "make decisions" that determine how the body responds to stimuli. About 90% of our neurons are interneurons. The word *interneuron* refers to the fact that they lie *between*, and interconnect, the incoming sensory pathways and the outgoing motor pathways of the CNS.

3. **Motor (efferent) neurons** send signals predominantly to muscle and gland cells, the effectors that carry out the body's responses to stimuli. These neurons are called *motor* neurons because most of them lead to muscle cells, and *efferent* neurons to signify the signal conduction *away from* the CNS.

### STRUCTURE OF A NEURON

There are several varieties of neurons, as we shall see, but a good starting point for discussing neuron structure is a motor neuron of the spinal cord (fig. 12.4). The control center of the neuron is its **soma**,[7] also called the **cell body** or **perikaryon**[8] (PERR-ih-CARE-ee-on). It has a single, centrally located nucleus with a large nucleolus. The cytoplasm contains mitochondria, lysosomes, a Golgi complex, numerous inclusions, and an extensive rough endoplasmic reticulum and cytoskeleton. The cytoskeleton consists of a dense mesh of microtubules and **neurofibrils** (bundles of actin filaments) which compartmentalize the rough ER into dark-staining regions called **Nissl**[9] **bodies** (fig. 12.4c, d). Nissl bodies are unique to neurons and a helpful clue to identifying them in tissue sections with mixed cell types. Mature neurons have no centrioles and apparently undergo no further mitosis after adolescence; however, they are unusually long-lived cells, capable of functioning for over a hundred years. Even into old age there are unspecialized stem cells in the CNS that can divide and develop into new neurons (see Insight 4.2, p. 139).

The major cytoplasmic inclusions in a neuron are glycogen granules, lipid droplets, melanin, and a golden brown pigment called *lipofuscin*[10] (LIP-oh-FEW-sin), produced when lysosomes digest worn-out organelles and

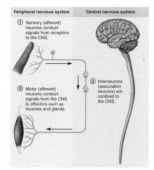

**FIGURE 12.3 Functional Classes of Neurons.** Sensory (afferent) neurons carry signals to the central nervous system (CNS), interneurons are contained entirely within the CNS and carry signals from one neuron to another, and motor (efferent) neurons carry signals from the CNS to muscles and glands.

[6] *inter* = between
[7] *soma* = body
[8] *peri* = around + *karyo* = nucleus
[9] Franz Nissl (1860–1919), German neuropathologist
[10] *lipo* = fat, lipid + *fusc* = dusky, brown

---

## THINK ABOUT IT

*Success in anatomy and physiology requires far more than memorization. More important is students' insight and ability to apply what they remember to new cases and problems. Strategically distributed throughout each chapter, Think About It questions encourage stopping and thinking more deeply about the meaning or broader significance of a concept.*

## VOCABULARY AIDS

*A&P students must assimilate a large working vocabulary. This is far easier and more meaningful if they can pronounce words correctly and if they understand the roots that compose them.*

*Pronunciation guides are given parenthetically when new words are introduced, using a "pro-NUN-see-AY-shun" format that is easy to interpret.*

*New terms are accompanied by footnotes that identify their roots and origins, and a lexicon of about 400 most commonly used word roots and affixes is printed on the inside back cover.*

## FIGURE QUESTIONS

*On average, five figures per chapter include a thought question beneath the figure legend. These questions prompt students to analyze the artwork and make connections between what they've read in the text and what they see in the figures. Answers are provided in Appendix B.*

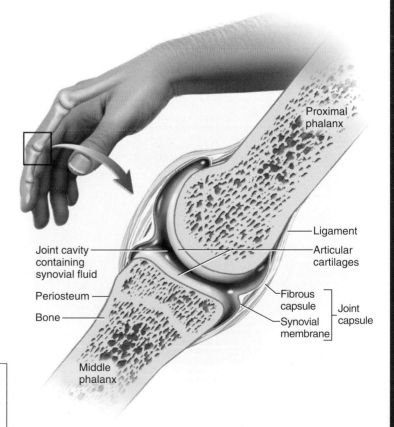

**FIGURE 9.5 Structure of a Simple Synovial Joint.**
❯ *Why is a meniscus unnecessary in an interphalangeal joint?*

# Learning System

## CLINICAL RELEVANCE

*Anatomy & Physiology* is fundamentally a textbook presenting the basic science of the human body. Yet students want to know the relevance of this science—how it relates to their career aims. Clinical examples help students see how the science relates to their long-term goals, but also provide windows of insight for understanding normal structure and function. For instance, cystic fibrosis serves to drive home the importance of membrane ion pumps, and brittle bone disease teaches the importance of collagen in osseous tissue. Saladin's Clinical Application Insights and pathology tables offer a wealth of clinical content to demonstrate concept applications and provide relevant background for students pursuing health-related careers.

### PATHOLOGY TABLES

*For each organ system, Saladin presents a table that briefly describes several well-known dysfunctions and comprehensively lists the pages where students can find comments on other disorders of that system.*

| TABLE 9.1 | Some Common Joint Disorders |
|---|---|
| Arthritis | Broad term embracing more than 100 types of joint rheumatism. |
| Bursitis | Inflammation of a bursa, usually due to overexertion of a joint. |
| Dislocation | Displacement of a bone from its normal position at a joint, usually accompanied by a sprain of the adjoining connective tissues. Most common at the fingers, thumb, shoulder, and knee. |
| Gout | A hereditary disease, most common in men, in which uric acid crystals accumulate in the joints and irritate the articular cartilage and synovial membrane. Causes gouty arthritis, with swelling, pain, tissue degeneration, and sometimes fusion of the joint. Most commonly affects the great toe. |
| Rheumatism | Broad term for any pain in the supportive and locomotory organs of the body, including bones, ligaments, tendons, and muscles. |
| Sprain | Torn ligament or tendon, sometimes with damage to a meniscus or other cartilage. |
| Strain | Painful overstretching of a tendon or muscle without serious tissue damage. Often results from inadequate warm-up before exercise. |
| Synovitis | Inflammation of a joint capsule, often as a complication of a sprain. |
| Tendinitis | A form of bursitis in which a tendon sheath is inflamed. |

***Disorders described elsewhere***

| | | |
|---|---|---|
| Hip dislocation p. 310 | Osteoarthritis p. 315 | Rotator cuff injury p. 382 |
| Knee injuries p. 314 | Rheumatoid arthritis p. 315 | Shoulder dislocation p. 308 |

***Disorders described elsewhere***

Hip dislocation p. 310    Osteoarthritis p. 315

Knee injuries p. 314    Rheumatoid arth

---

**INSIGHT 9.3**  Clinical Application

### Knee Injuries and Arthroscopic Surgery

Although the knee can bear a lot of weight, it is highly vulnerable to rotational and horizontal stress, especially when the knee is flexed (as in skiing or running) and receives a blow from behind or from the lateral side. The most common injuries are to a meniscus or the anterior cruciate ligament (ACL) (fig. 9.29). Knee injuries heal slowly because ligaments and tendons have a very scant blood supply and cartilage has no blood vessels at all.

The diagnosis and surgical treatment of knee injuries has been greatly improved by *arthroscopy,* a procedure in which the interior of a joint is viewed with a pencil-thin instrument, the *arthroscope,* inserted through a small incision. The arthroscope has a light source, a lens, and fiber optics that allow a viewer to see into the cavity, take photographs or videotapes of the joint, and withdraw samples of synovial fluid. Saline is often introduced through one incision to expand the joint and provide a clearer view of its structures. If surgery is required, additional small incisions can be made for the surgical instruments and the procedures can be observed through the arthroscope or on a monitor. Arthroscopic surgery produces much less tissue damage than conventional surgery and enables patients to recover more quickly.

Orthopedic surgeons now often replace a damaged ACL with a graft from the patellar ligament or a hamstring tendon. The surgeon "harvests" a strip from the middle of the patient's ligament (or tendon), drills a hole into the femur and tibia within the joint cavity, threads the ligament through the holes, and fastens it with screws. The grafted ligament is more taut and "competent" than the damaged ACL. It becomes ingrown with blood vessels and serves as a substrate for the deposition of more collagen, which further strengthens it in time. Following arthroscopic ACL reconstruction, a patient typically must use crutches for 7 to 10 days and undergo supervised physical therapy for 6 to 10 weeks, followed by self-directed exercise therapy. Healing is completed in about 9 months.

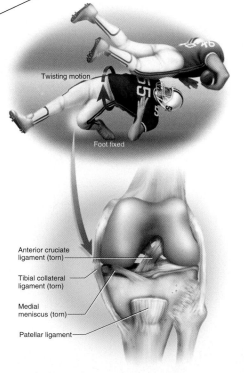

Twisting motion

Foot fixed

Anterior cruciate ligament (torn)

Tibial collateral ligament (torn)

Medial meniscus (torn)

Patellar ligament

**FIGURE 9.29**  Some Common Knee Injuries.

### CLINICAL APPLICATIONS

*Each chapter has three to five Insight boxes, over 80% of which are clinical in nature. These essays illuminate the clinical relevance of a concept and give insight on disease as it relates to normal structure and function.*

# CONCEPT CONNECTIONS

Awareness of how chapter topics relate to other sciences gives students a more holistic view of the study of anatomy and physiology. Saladin's informative Insight sidebars offer clinical, historical, and evolutionary perspectives that tie into the chapter content. Making connections between body systems is another important component of a well-rounded knowledge of A&P. Connective Issues pages underscore interactions between systems in a concise format.

## MEDICAL HISTORY

*Inspiring historical and biographical vignettes give students a more humanistic perspective of the field and a chance to consider chapter topics in other contexts.*

**INSIGHT 18.3** Medical History

### Charles Drew—Blood Banking Pioneer

Charles Drew (fig. 18.11) was a scientist who lived and died in the grip of irony. After receiving his M.D. from McGill University of ... first black person to pursue ... of Science in Medicine, for ... od-banking at Columbia ... f a new blood bank at ... 9 and organized numer-

... incing physicians to use ... ttlefield and other emer- ... ld be stored for only a ... compatible blood types. ... was less likely to cause

... sued a directive forbid- ... o blood in military blood ... resigned his position. He ... Howard University in ... f staff at Freedmen's ... s young black physicians ... d into the medical com- ... ciation, however, firmly ... Drew himself.

... ree colleagues set out to ... an annual free clinic in ... the wheel and was crit- ... Doctors at the nearest

hospital administered blood and attempted unsuccessfully to revive him. For all the lives he saved through his pioneering work in blood transfusion, Drew himself bled to death at the age of 45.

**FIGURE 18.11** Charles Drew (1904–50).

**INSIGHT 29.1** Clinical Application

### Twins

There are two ways in which twins are produced (and, by extension, other multiple births). About two-thirds of twins are *dizygotic (DZ)*—produced when two eggs are ovulated and fertilized by separate sperm. They are no more or less genetically similar than any other siblings and may be of different sexes. Multiple ovulation can also result in triplets, quadruplets, or even greater numbers of offspring. DZ twins implant separately on the uterine wall and each forms its own placenta, although their placentas may fuse if they implant close together (fig. 29.3).

*Monozygotic (MZ) twins* are produced when a single egg is fertilized and the cell mass (embryoblast) later divides into two. MZ twins are genetically identical, or nearly so, and are therefore of the same sex and nearly identical appearance. In most cases, they share the same placenta. Identical triplets and quadruplets occasionally result from the splitting of a single embryoblast.

Reproductive biologists are beginning to question whether MZ twins are truly genetically identical. They have suggested that blastomeres may undergo mutation in the course of DNA replication, and the splitting of the embryoblast may represent an attempt of each cell mass to reject the other one as genetically different and presumably foreign.

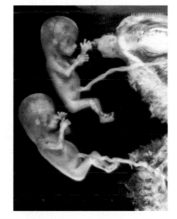

**FIGURE 29.3** Dizygotic Twins with Separate Placentas.

**INSIGHT 26.2** Evolutionary Medicine

### Evolution of the Sweet Tooth

Our craving for sugar doubtlessly originated in our prehistoric ancestors. Not only did they have to work much harder to survive than we do, but high-calorie foods were scarce in their environment and people were at constant risk of starvation. Those who were highly motivated to seek and consume sugary, high-calorie foods passed their "sweet tooth" on to us, their descendents—along with a similarly adaptive appetite for other rare but vital nutrients, namely fat and salt. The tastes that were essential to our ancestors' survival can now be a disadvantage in a culture where salty, fatty, and sugary foods are all too easy to obtain and the food industry is eager to capitalize on these tastes.

## EVOLUTIONARY MEDICINE

*Considering the evolutionary adaptations highlighted in these Insight sidebars gives students a sense of how and why the human body came to be as it is.*

## CONNECTIVE ISSUES

### Interactions Between the DIGESTIVE SYSTEM and Other Organ Systems

■ indicates ways in which this system affects other systems
■ indicates ways in which other systems affect this system

**INTEGUMENTARY SYSTEM**

Skin helps synthesize calcitriol, needed for calcium and phosphorus absorption by small intestine

**SKELETAL SYSTEM**

Small intestine adjusts calcium absorption in proportion to the needs of the skeletal system

Provides protective enclosure for some digestive organs, support for the teeth, and movements of mastication

**MUSCULAR SYSTEM**

Liver disposes of lactic acid generated by muscles

Essential for chewing, swallowing, and defecation; muscles protect lower GI organs

**NERVOUS SYSTEM**

Enteric and autonomic nervous systems regulate GI motility and secretion; somatic nervous system controls chewing, swallowing, and defecation; sense organs involved in food selection; hypothalamus contains centers for hunger, thirst, and satiety

**ENDOCRINE SYSTEM**

Liver degrades hormones; enteroendocrine cells produce many hormones

Hormones regulate GI secretion, and processing of nutrients

**CIRCULATORY SYSTEM**

GI tract absorbs fluid needed to maintain blood volume; liver degrades heme from dead RBCs, secretes clotting factors, albumin, and other plasma proteins; and regulates blood glucose and iron levels

**ALL SYSTEMS**

The digestive system provides all other systems with nutrients in a form usable for cellular metabolism and building of tissues

**CIRCULATORY SYSTEM (cont.)**

Blood transports hormones that regulate GI activity; absorbs and distributes nutrients; vasomotion alters capillary filtration and salivation

**LYMPHATIC/IMMUNE SYSTEMS**

GI mucosa is a site of lymphocyte production; acid, lysozyme, and other digestive enzymes provide nonspecific defense against pathogens; infant intestine absorbs maternal IgA to confer passive immunity on infant

Lymphatic capillaries (lacteals) absorb digested lipids; immune cells protect GI tract from infection

**RESPIRATORY SYSTEM**

Pressure of digestive organs against diaphragm aids in expiration when abdominal muscles contract

Provides O₂, removes CO₂; Valsalva maneuver aids defecation

**URINARY SYSTEM**

Intestines complement kidneys in water and electrolyte reabsorption; liver synthesizes urea and kidneys excrete it

Excretes bile pigments and other products of liver metabolism; completes the synthesis of calcitriol, needed for intestinal absorption of calcium and phosphorus

**REPRODUCTIVE SYSTEM**

Provides nutrients for fetal growth

Developing fetus crowds digestive organs; may cause constipation and heartburn during pregnancy

## CONNECTIVE ISSUES

*Found at the ends of system-specific chapters, these pages summarize ways in which a system influences all of the others of the body, and how it is influenced by them in turn. Reviewing these connections reinforces the interrelationships among organ systems and discourages the idea that a system can be forgotten after a test is over.*

# Learning System

## END-OF-CHAPTER REVIEW

A carefully devised set of learning aids at the end of each chapter helps students review the chapter content, evaluate their grasp of key concepts, and utilize what they have learned. Reading the chapter summary and completing the exercises is a great way to assess learning.

**REVIEW OF KEY CONCEPTS**

*Briefly restates the key points of the chapter, with page references back to major sections.*

**TESTING YOUR RECALL**

*Multiple choice and short answer questions allow students to check their knowledge. Answers are provided in Appendix B.*

**TRUE OR FALSE**

*True or False questions require students to explain why the false statements are untrue, thus challenging them to think more deeply into the material and to appreciate and express subtle points. Answers are provided in Appendix B.*

**TESTING YOUR COMPREHENSION**

*These questions go beyond memorization to require a deeper level of analysis and clinical application. Scenarios from sources such as* Morbidity and Mortality Weekly Reports *prompt students to apply the chapter's basic science to real-life case histories.*

**WEBSITE REMINDER**

*Each chapter ends with a reminder to visit the Saladin website for quizzes and activities.*

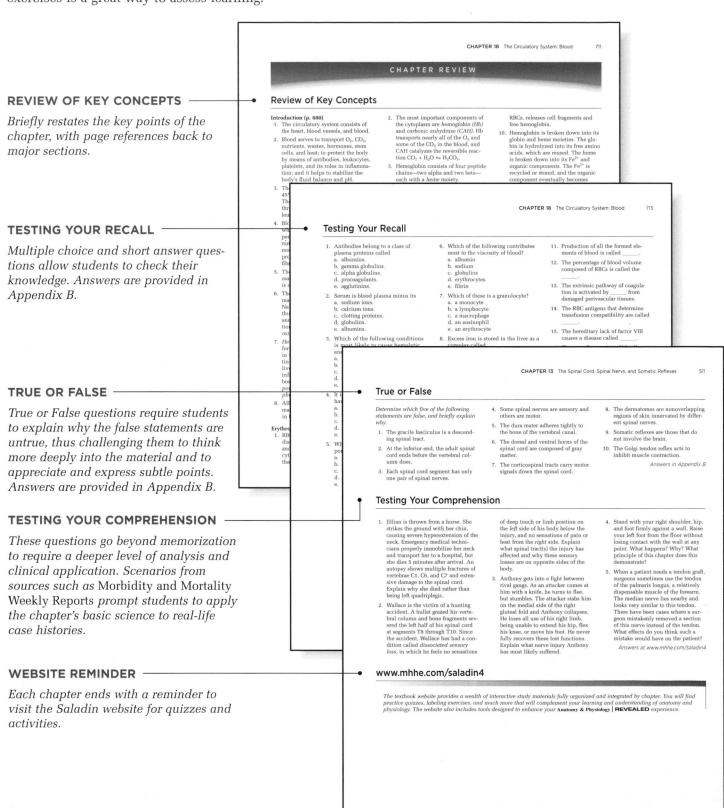

## Instructor Supplements

Instructors can obtain teaching aids to accompany this textbook by visiting www.mhhe.com/catalogs, calling 800-338-3987, or contacting a local McGraw-Hill sales representative.

### DIGITAL CONTENT MANAGER

This easy-to-use multimedia resource allows instructors to utilize artwork from the textbook in multiple formats to create customized classroom presentations, visually based tests and quizzes, dynamic course website content, and attractive printed support materials. Available in DVD and CD-ROM versions, and compatible with both Mac and Windows, the Digital Content Manager supplies the following assets grouped by chapter:

- **Art**—Full-color digital files of all illustrations from the book, plus the same art saved in unlabeled and grayscale versions, can be readily incorporated into lecture presentations, exams, or custom-made classroom materials.

- **Photos**—Digital files of photographs from the book can be reproduced for multiple classroom uses.

- **Tables**—Every table that appears in the text is provided in electronic form.

- **Text Edit Art**—Every illustration is placed in a PowerPoint presentation that allows the user to edit, move, resize, and delete labels and leader lines as desired to create custom-labeled images for presentations and/or tests.

- **Animations**—Over 150 animations explaining topics discussed in the text can be imported into classroom presentations or online course materials. Animation controls allow instructors to pause, mute, or rewind for ultimate flexibility during lecture.

- **Active Art**—Illustrations converted to the Active Art format are fully editable within PowerPoint. With Active Art, instructors can edit art colors, customize labels, or even select and copy elements of a figure to build new illustrations.

- **PowerPoint Lecture Outlines**—Ready-made presentations that combine art and lecture notes are provided for each chapter of the textbook. These outlines can be used as supplied, or can be tailored to reflect preferred lecture topics and sequences.

- **PowerPoint Slides**—For instructors who prefer to create lectures from scratch, all illustrations, photos, and tables are pre-inserted by chapter into blank PowerPoint slides for convenience. Simply select the desired slides and add your own notes.

### ARIS FOR *ANATOMY & PHYSIOLOGY*

McGraw-Hill's ARIS for *Anatomy & Physiology*, found at www.mhhe.com/saladin4, is a complete electronic homework and course management system. Free on adoption of *Anatomy & Physiology*, instructors can create and share course materials and assignments with colleagues with a few clicks of the mouse. Instructors can edit questions, import their own content, and create announcements and due dates for assignments. ARIS has automatic grading and reporting of easy-to-assign homework, quizzing, and testing. Once a student is registered in the course, all student activity within McGraw-Hill's ARIS is automatically recorded and available to the instructor through a fully integrated grade book that can be downloaded to Excel.

### INSTRUCTOR'S TESTING AND RESOURCE CD-ROM

This cross-platform CD delivers the following supplemental materials.

- **Test Bank**—A comprehensive bank of test questions is provided as simple Word files, and also within a computerized test bank powered by McGraw-Hill's flexible EZ Test program. EZ Test allows instructors to search for questions by topic, format, or difficulty level; edit existing questions or add new ones; and create multiple versions of a test. Any test can be exported for use with course management systems such as WebCT, BlackBoard, or PageOut. EZ Test Online is a new service that offers a place to easily administer EZ Test exams and quizzes online.

- **Instructor's Manual**—This handy guide prepared by David Evans, Pennsylvania College of Technology, includes discussion topics, learning strategies, thought questions, and other helpful materials specific to the textbook.

- **Clinical Applications Manual**—This manual expands on *Anatomy & Physiology*'s clinical themes, introduces new clinical topics, and provides test questions and case studies to develop students' abilities to apply knowledge to realistic situations. A print version is available for students.

## TRANSPARENCIES

A set of more than 1,000 transparency overheads includes all illustrations and many photos from the book. Unlabeled images of key figures are provided.

## LABORATORY MANUAL

The *Anatomy & Physiology Laboratory Manual* by Eric Wise of Santa Barbara City College is expressly written to coincide with the chapters of Saladin's *Anatomy & Physiology*. This lab manual includes clear explanations of physiology experiments and computer simulations that serve as alternatives to frog experimentation. Each lab also includes an extensive set of review questions. New Chapter Summary Data sheets found at the end of appropriate exercises allow students to record answers from throughout the exercise on a single summary sheet.

## eINSTRUCTION

McGraw-Hill has partnered with eInstruction to bring the revolutionary Classroom Performance System (CPS) to the classroom. An instructor using this interactive system can administer questions electronically during class while students respond via hand-held remote control keypads. Individual responses are logged into a grade book, and aggregated responses can be displayed in graphical form to provide immediate feedback on whether students understand a lecture topic or if more clarification is needed. CPS promotes student participation, class productivity, and individual student accountability. A downloadable set of textbook-specific questions, formatted for both CPS and PowerPoint, is available via the ARIS for *Anatomy & Physiology* at www.mhhe.com/saladin4.

## COURSE DELIVERY SYSTEMS

With help from our partners WebCT, Blackboard, Top-Class, eCollege, and other course management systems, professors can take complete control over their course content. Course cartridges containing content from the ARIS textbook website, online testing, and powerful student tracking features are readily available for use within these platforms.

# Student Supplements

Students can order supplemental study materials by visiting www.books.mcgraw-hill.com, calling 800-262-4729, or contacting their campus bookstore.

## ANATOMY & PHYSIOLOGY REVEALED

This amazing multimedia tool is designed to help students learn and review human anatomy using cadaver specimens. Detailed cadaver photographs blended together with a state-of-the-art layering technique provide a uniquely interactive dissection experience. This easy-to-use program features the following sections:

- **Dissection**—Peel away layers of the human body to reveal structures beneath the surface. Structures can be pinned and labeled, just like in a real dissection lab. Each labeled structure is accompanied by detailed information and an audio pronunciation. Dissection images can be captured and saved.

- **Animation**—Compelling animations demonstrate muscle actions, clarify anatomical relationships, or explain difficult concepts.

- **Imaging**—Labeled X-ray, MRI, and CT images familiarize students with the appearance of key anatomical structures as seen through different medical imaging techniques.

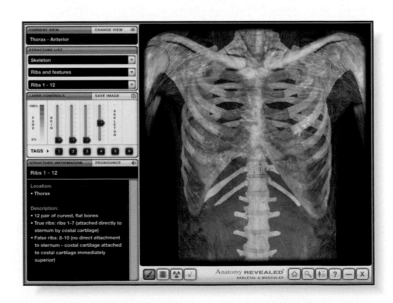

- **Self-Test**—Challenging exercises let students test their ability to identify anatomical structures in a timed practical exam format or traditional multiple choice. A results page provides analysis of test scores plus links back to all incorrectly identified structures for review.

- **Anatomy Terms**—This visual glossary of general terms includes directional and regional terms, as well as planes and terms of movement.

## TEXTBOOK WEBSITE

McGraw-Hill's ARIS (Assessment, Review, and Instruction System) for *Anatomy & Physiology* at www.mhhe.com/saladin4 offers access to a vast array of premium online content to fortify the learning experience.

- **Text-Specific Study Tools**—The Saladin ARIS site features quizzes, interactive learning games, and study tools tailored to coincide with each chapter of the text.

- **Course Assignments and Announcements**—Students of instructors choosing to utilize McGraw-Hill's ARIS tools for course administration will receive a course code to log into their specific course for assignments. Consult the ARIS login card at the front of this book for more information.

- **Essential Study Partner**—A collection of interactive study modules that contains animations, learning activities, and quizzes designed to help students grasp complex concepts.

- **Online Tutoring**—A 24-hour tutorial service moderated by qualified instructors. Help with difficult concepts is only an email away.

## STUDENT STUDY GUIDE

This comprehensive study guide written by Jacque Homan, South Plains College, in collaboration with Ken Saladin, contains vocabulary-building and content-testing exercises, labeling exercises, and practice exams.

## SOLVE SALADIN

*Solve Saladin* by Greg Reeder, Broward Community College, is a complete series of crossword puzzles compiled using the specific A&P vocabulary and textual information from each chapter of Saladin's fourth edition. Students who have used this supplement report that they enjoy learning from these fun crosswords and look forward to doing them. Used as self-testing items or just to reinforce the material, *Solve Saladin* is a unique and fun way to learn A&P.

## MEDIAPHYS 3.0

This interactive CD-ROM offers detailed explanations, high-quality illustrations, and animations to provide students with a thorough introduction to the world of physiology—giving them a virtual tour of physiological processes. MediaPhys is filled with interactive activities and quizzes to help reinforce physiology concepts that are often difficult to understand.

# Acknowledgments

Anyone who has written a textbook on this scale, especially with such lavish illustrations and sophisticated supplemental products, can attest to what a large and well-orchestrated team effort it requires. Such a book can never be the work of only one person or even a few. I am deeply indebted to the team at McGraw-Hill Higher Education who have shown continued faith in my writing and invested so generously in it.

Above all, I am profoundly indebted to my partner in this project, Developmental Editor Kristine Queck, who made the book so much better than I could have made it on my own. I have benefited greatly from Kris's vision for excellence, attention to detail, knowledge of the subject matter, and strong commitment to uncompromising quality in the art, photography, and design of this book. Kris not only gave her full support to my ideas for improvement, but pushed for even higher standards of design and illustration than I might have had the temerity to demand on my own. She was very persistent in getting the art error-free and the pages laid out for easiest cross-referencing between the narrative and the related illustrations. Her attention to this project was so constant and unflagging that I could almost have believed I was McGraw-Hill's only author.

The appearance of the book also owes a great deal to designer David Hash, photo research coordinator John Leland and photo researcher Mary Reeg, and our team of illustrators, headed by Joanne Brummett at Precision Graphics. A good copy editor makes one a better writer, and my copy editor, Linda Davoli, spared me from many embarrassing oversights, clumsy transitions, and lapses in grammar and composition. Bringing up the rear, as is the indexer's lot in life, Barbara Littlewood has skillfully assimilated innumerable details from A to Z in a way that will greatly enhance the reference value of this book to the student.

Coordinating the efforts of such a team and keeping everything running smoothly is an enormous task in itself, requiring almost supernatural power over the laws of entropy. For this, we are all indebted to Kristine Tibbetts and Mary Powers. As Director of Development, Kris ironed out the procedural details that facilitated the work of other team members and lent a helping hand with supplement development while Kris Queck and I were heavily immersed in art corrections and paging. As Project Manager, Mary coordinated all aspects of the project to keep us moving in concert in getting the book to press.

Of course none of this would have happened without the material resources so generously provided by McGraw-Hill Higher Education, and I am ever grateful for the faith they have placed in me by investing in the book's development. Publisher Michelle Watnick was my valued advocate in securing the resources to bring such high quality to this book and its supplements. All of this hinges, of course, on the success of past editions in finding a respected place in this highly competitive market, and that rests not only on those of us who produce the book but on those who market it so creatively and effectively: Executive Marketing Manager Jim Connely, Marketing Manager Lynn Kalb, and the legions of sales managers and sales representatives who have constantly promoted awareness of the book and its advantages among my colleagues in the teaching corps. I am lucky indeed to be with a publisher that has the means, ingenuity, and dedication to promote my books so effectively.

These days, a competitive textbook must be accompanied by a great diversity of supplemental products for students and instructors. The success of the flagship book therefore depends on the many talented scientists and media developers who produce those products. Thus I am further indebted to my teammates David Evans for the Instructor's Manual, Eric Wise for the Laboratory Manual, Jacqueline Homan for the Study Guide, Greg Reeder for *Solve Saladin,* and Sharon Simpson, Claudia Stanescu, and Theresa Dehne for updating several other supplements.

Here at home, I want to thank university photographer Tim Vacula for joining me once again in producing kinesiologic photographs and multiple-exposure effects for chapter 9, and two new student volunteers, Brittany Jones and Dallas White, for modeling for those photos.

Continual improvements in the accuracy and currency of this textbook are due in great part to the many reviewers who provided reams (and more reams!) of constructive reviews of the third edition, the fourth-edition manuscript, and the new art. Their specialized knowledge of the various nooks and crannies of human structure and function greatly helped to bring me up to date, to identify and assimilate more reliable sources of information, and to find more effective ways of expressing an idea for the student reader. These reviewers are identified below. Also valuable to me have been the many students around the world who have written, phoned, or dropped into my office with their observations, compliments, and suggestions for improvement. I think of them, above all, when I strive for clearer ways to describe or illustrate an idea.

# Reviewers and Advisors

## TEXTBOOK REVIEWERS

Patricia Adumanu Ahanotu
*Georgia Perimeter College*

Pegge Alciatore
*University of
Louisiana–Lafayette*

John V. Aliff
*Georgia Perimeter College*

Victor Alvarez
*Delaware Technical and
Community College*

Dennis I. Anderson
*Oklahoma City Community
College*

Gail Baker
*LaGuardia Community College*

Mary Lou Bareither
*University of Illinois at Chicago*

Sharon Barnewall
*Columbus State Community
College*

Randall Barre
*Chattanooga State Technical
Community College*

Tobie Bogart
*Jefferson State Community
College*

Sara W. Brenizer
*Shelton State Community
College*

Bob Broyles
*Butler County Community
College*

Jocelyn Cash
*Central Piedmont Community
College*

Roger D. Choate
*Oklahoma City Community
College*

Ana Christiansen
*Lamar University*

Barbara Cogdell
*University of Glasgow*

Pamela Anderson Cole
*Shelton State Community
College*

W. Wade Cooper
*Shelton State Community
College*

David T. Corey
*Midlands Technical College*

James Crowder
*Brookdale Community
College*

Mary E. Dawson
*Kingsborough Community
College*

Clementine A. deAngelis
*Tarrant County College*

Theresa A. Dehne
*Chippewa Valley Technical
College*

Sandra Denton
*Santa Fe Community College*

Charles J. Dick
*Pasco-Hernando Community
College*

Richard Doolin
*Daytona Beach Community
College*

David L. Evans
*Pennsylvania College
of Technology*

Paul Florence
*Jefferson Community College*

Clifford Fontenot
*Southeastern Louisiana
University*

Pamela B. Fouche
*Walters State Community
College*

Ralph F. Fregosi
*University of Arizona*

Purti P. Gadkari
*Wharton County Junior
College*

Louis Giacinti
*Milwaukee Area Technical
College*

Lauren S. Gollahon
*Texas Tech University*

Ron Hackney
*Volunteer State Community
College*

Josie Harder
*University of Bradford*

Susan Hines
*University of Akron*

Regina Neal Hoffman
*Midlands Technical College*

Sobrasua E. M. Ibim
*Georgia Perimeter College*

Mark Jaffe
*Nova Southeastern University*

Walter Jahn
*Orange County Community
College*

Edward W. Johnson
*Central Oregon Community
College*

Jody E. Johnson
*Arapahoe Community College*

Ian Kay
*Manchester Metropolitan
University*

Robert Kidd
*University of Western Sydney*

Lawrence N. Killian
*Clark State Community
College*

Beverly P. Kirk
*Northeast Mississippi
Community College*

Michael S. Kopenits
*Amarillo College*

Mohamed Lakrim
*Kingsborough Community
College*

Kristin Lenertz
*Black Hawk College*

Jerri K. Lindsey
*Tarrant County College*

Sue Longenbaker
*Columbus State Community
College*

Joan Lukich
*University of Akron*

Allan L. Markezich
*Black Hawk College*

Geri Mayer
*Florida Atlantic University*

Judith Megaw
*Indian River Community
College*

Ralph R. Meyer
*University of Cincinnati*

Melissa A. Mills
*Anoka-Ramsey Community
College*

Sarah L. Milton
*Florida Atlantic University*

Robert Moldenhauer
*St. Clair County Community
College*

Elek Molnar
*University of Bristol*

Alfredo Munoz
*University of Texas
at Brownsville*

Lance Myler
*SUNY Canton College
of Technology*

Ebere U. Nduka
*Medgar Evers College*

Margaret Nordlie
*University of Mary*

Margaret (Betsy) Ott
*Tyler Junior College*

James F. Palmer
*Southeastern Louisiana
University*

Vanessa Passler
*Wallace Community College*

Mark Paternostro
*Pennsylvania College
of Technology*

Andrew J. Penniman
*Georgia Perimeter College*

Robert Pope
*Miami-Dade College*

Nikki Privacky
*Palm Beach Community
College*

Kimberly Raun
*Wharton County Junior
College*

Ellen Ott-Reeves
*Blinn College*

Jennifer Regan
*University of Southern
Mississippi*

Sean M. Roe
*Queens University Belfast*

Barbara B. Rundell
*College of DuPage*

Ronald L. Salisbury
*University of Akron*

Beverly A. Schieltz
*Wright State University*

Melvin Schmidt
*McNeese State University*

Matthew J. Smith
*Pacific Lutheran University*

Robert R. Speed
*Wallace Community College*

Anthony J. Stancampiano
*Oklahoma City Community
College*

Claudia Stanescu
*University of Arizona*

Robert D. Stark
*California State
University–Bakersfield*

Brett W. Strong
*Palm Beach Community
College*

Bonnie J. Tarricone
*Ivy Tech State College*

Christine Terry
*Georgia Perimeter College*

Mary Elizabeth Torrano
*American River College*

Rafael Torres
*San Antonio College*

Randall L. Tracy
*Worcester State College*

Kathy White
*St. Phillip's College*

Shirley A. Williams
*Delaware Technical and
Community College*

Judith M. Wolff
*Delaware Technical and
Community College*

Jennifer Wortham
*University of Evansville*

Charles Wright
*Community College of
Baltimore County–Essex*

Richard Doolin
*Daytona Beach Community
College*

Christopher Dugan
*Red Rocks Community
College*

Marie L. Gabbard
*Boise State University*

Jean Helgeson
*Collin County Community
College*

Catherine J. Hurlbut
*Florida Community College
at Jacksonville*

Cindy L. A. Jones
*Colorado Community
Colleges Online*

Duncan S. MacKenzie
*Texas A & M University*

Ronald Markle
*Northern Arizona University*

Marlene M. Martinez
*American River College*

Charles W. Miller
*Colorado State University*

Melissa Ann Mills
*Anoka-Ramsey Community
College*

Margaret (Betsy) Ott
*Tyler Junior College*

Vanessa Passler
*Wallace Community College*

Karen Payne
*Chattanooga State Technical
Community College*

Nikki Privacky
*Palm Beach Community
College*

Van Wheat
*South Texas College*

Mark L. Wygoda
*McNeese State University*

## CONSULTANT BOARD MEMBERS

Pierre Deviche
*Arizona State University*

Duncan S. MacKenzie
*Texas A&M University*

Karen MacKenzie
*Greenville Technical College*

Nancy G. Morris
*Volunteer State Community
College*

Barbara B. Rundell
*College of DuPage*

Peter P. Susan
*Trident Technical College*

Rafael Torres
*San Antonio College*

Matthew A. Williamson
*Georgia Southern University*

## ART REVIEWERS

Patricia Adumanu Ahanotu
*Georgia Perimeter College*

Shylaja Akkaraju
*Bronx Community College
of CUNY*

John V. Aliff
*Georgia Perimeter College*

Shawn Bjerke
*Minnesota State Community
and Technical College*

Robert Blum
*Lehigh Carbon Community
College*

John D. Buntin
*University of
Wisconsin–Milwaukee*

Jocelyn Cash
*Central Piedmont Community
College*

Roger D. Choate
*Oklahoma City Community
College*

Pamela Anderson Cole
*Shelton State Community
College*

James Constantine
*Bristol Community College*

W. Wade Cooper
*Shelton State Community
College*

Mary E. Dawson
*Kingsborough Community
College*

## FOCUS GROUP PARTICIPANTS

Albert A. Baccari, Jr.
*Montgomery County
Community College*

Julie R. Baugh
*Community College
of Baltimore County*

Theresa A. Bissell
*Ivy Tech State College*

Richard M. Blaney
*Brevard Community College*

Perry Carter
*Midlands Technical College*

Ana Christensen
*Lamar University*

Vincent A. Cobb
*Middle Tennessee State
University*

Steve Cole
*Rochester Community
and Technical College*

Ethel Cornforth
*San Jacinto College South*

Jorge D. Cortese
*Durham Technical
Community College*

Smruti A. Desai
*Cy-Fair College*

Steve Dutton
*Amarillo College*

Dale Fatzer
*University of New Orleans*

Sharon Feaster
*Hinds Community College*

Ann M. Findley
*University of Louisiana
at Monroe*

Mym Fowler
*Tidewater Community College*

Mary Fox
*University of Cincinnati*

Purti P. Gadkari
*Wharton County Junior College*

Maureen N. Gannon
*Bronx Community College, CUNY*

Peter Germroth
*Hillsborough Community College*

Roberto B. Gonzales
*Northwest Vista College*

Chaya Gopalan
*St. Louis Community College*

Suzanne Gould
*Ivy Tech State College*

Terrence C. Harrison
*Arapahoe Community College*

James Horwitz
*Palm Beach Community College*

Michael S. Kopenits
*Amarillo College*

W. J. Loughry
*Valdosta State University*

Judy Megaw
*Indian River Community College*

Carl F. McAllister
*Georgia Perimeter College*

Janet McMillen
*Prince George's Community College*

Russell E. Moore
*Bluegrass Community and Technical College*

Ebere U. Nduka
*Medgar Evers College, CUNY*

Robyn O'Kane
*LaGuardia Community College, CUNY*

Tim Roye
*San Jacinto College–South*

Karen Payne
*Chattanooga State Technical Community College*

Bradley A. Sarchet
*Manatee Community College*

Colleen S. Sinclair
*Towson University*

Cinnamon L. VanPutte
*Southwestern Illinois College*

Mary Elizabeth Torrano
*American River College*

Charles J. Venglarik
*Jefferson State Community College*

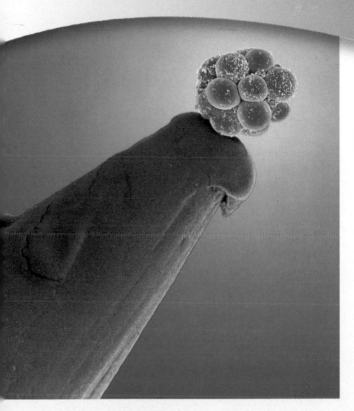

*A new life begins—a human embryo on the point of a pin*

# MAJOR THEMES OF ANATOMY AND PHYSIOLOGY

## CHAPTER OUTLINE

## INSIGHTS

No branch of science hits as close to home as the science of our own bodies. We're grateful for the dependability of our hearts; we're awed by the capabilities of muscles and joints displayed by Olympic athletes; and we ponder with philosophers the ancient mysteries of mind and emotion. We want to know how our body works, and when it malfunctions, we want to know what is happening and what we can do about it. Even the most ancient writings of civilization include medical documents that attest to humanity's timeless drive to know itself. You are embarking on a subject that is as old as civilization, yet one that grows by thousands of scientific publications every week.

This book is an introduction to human structure and function, the biology of the human body. It is meant primarily to give you a foundation for advanced study in health care, exercise physiology, pathology, and other fields related to health and fitness. Beyond that purpose, however, it can also provide you with a deeply satisfying sense of self-understanding.

As rewarding and engrossing as this subject is, the human body is highly complex and understanding it requires us to comprehend a great deal of detail. The details will be more manageable if we relate them to a few broad, unifying concepts. The aim of this chapter, therefore, is to introduce such concepts and put the rest of the book into perspective. We consider the historical development of anatomy and physiology, the thought processes that led to the knowledge in this book, the meaning of human life, and a central concept of physiology called *homeostasis*.

# The Scope of Anatomy and Physiology

**Anatomy** is the study of structure, and **physiology** is the study of function. These approaches are complementary and never entirely separable. When we study a structure, we want to know, What does it do? Physiology lends meaning to anatomy and, conversely, anatomy is what makes physiology possible. This *unity of form and function* is an important point to bear in mind as you study the body. Many examples of it will be apparent throughout the book—some of them pointed out for you, and others you will notice for yourself.

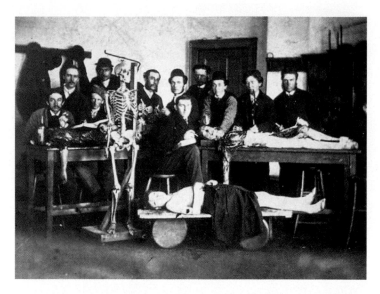

**FIGURE 1.1**  Early Medical Students in the Gross Anatomy Laboratory with Three Cadavers.

## ANATOMY—THE STUDY OF FORM

The simplest way to study human anatomy is the observation of surface structure, for example in performing a physical examination or making a clinical diagnosis from surface appearance. But a deeper understanding of the body depends on **dissection**—the careful cutting and separation of tissues to reveal their relationships. Both *anatomy*[1] and *dissection*[2] literally mean "cutting apart"; dissecting used to be called "anatomizing." The dissection of a dead human body, or **cadaver,**[3] is an essential part of the training of many health science students (fig. 1.1). Many insights into human structure are obtained from *comparative anatomy*—the study of more than one species in order to learn generalizations and evolutionary trends. Students of anatomy often begin by dissecting other animals with which we share a common ancestry and many structural similarities.

Dissection, of course, is not the method of choice when studying a living person! Physical examinations involve not only looking at the body for signs of normalcy or disease but also touching and listening to it. **Palpation**[4] is feeling structure with the fingertips, such as palpating a swollen lymph node or taking a pulse. **Auscultation**[5] (AWS-cul-TAY-shun) is listening to the natural sounds made by the body, such as heart and lung sounds. In **percussion,** the examiner taps on the body and listens to the sound for signs of abnormalities such as pockets of fluid or air.

---

[1]*ana* = apart + *tom* = cut
[2]*dis* = apart + *sect* = cut
[3]*cadere* = to fall or die
[4]*palp* = touch, feel
[5]*auscult* = listen

Structure that can be seen with the naked eye, whether by surface observation or dissection, is called **gross anatomy.** Ultimately, though, the functions of the body result from its individual cells. To see those, we usually take tissue specimens, thinly slice and stain them, and observe them under the microscope. This approach is called **histology**[6] (microscopic anatomy). *Histopathology* is the microscopic examination of tissues for signs of disease. *Ultrastructure* refers to fine details, down to the molecular level, revealed by the electron microscope.

## PHYSIOLOGY—THE STUDY OF FUNCTION

Physiology[7] uses the methods of experimental science discussed later. It has many subdisciplines such as *neurophysiology* (physiology of the nervous system), *endocrinology* (physiology of hormones), and *pathophysiology* (mechanisms of disease). Partly because of limitations on experimentation with humans, much of what we know about bodily function has been gained through *comparative physiology,* the study of how different species have solved problems of life such as water balance, respiration, and reproduction. Comparative physiology is also the basis for the development of new drugs and medical procedures. For example, a cardiac surgeon cannot practice on humans without first succeeding in animal surgery, and a vaccine cannot be used on human subjects until it has been demonstrated through animal research that it confers significant benefits without unacceptable risks.

# The Origins of Biomedical Science

### Objectives

When you have completed this section, you should be able to

- give examples of how modern biomedical science emerged from an era of superstition and authoritarianism; and
- describe the contributions of some key people who helped to bring about this transformation.

Health science has progressed far more in the last 50 years than in the 2,500 years before that, but the field did not spring up overnight. It is built upon centuries of thought

and controversy, triumph and defeat. We cannot fully appreciate its present state without understanding its past—people who had the curiosity to try new things, the vision to look at human form and function in new ways, and the courage to question authority.

## THE BEGINNINGS OF MEDICINE

As early as 3,000 years ago, physicians in Mesopotamia and Egypt treated patients with herbal drugs, salts, physical therapy, and faith healing. The "father of medicine," however, is usually considered to be the Greek physician *Hippocrates* (c. 460–c. 375 BCE). He and his followers established a code of ethics for physicians, the Hippocratic Oath, that is still recited in modern form by many graduating medical students. Hippocrates urged physicians to stop attributing disease to the activities of gods and demons and to seek their natural causes, which could afford the only rational basis for therapy. *Aristotle* (384–322 BCE) believed that diseases and other natural events could have either supernatural causes, which he called *theologi,* or natural ones, which he called *physici* or *physiologi.* We derive such terms as *physician* and *physiology* from the latter. Until the nineteenth century, physicians were called "doctors of physic." In his anatomy book, *Of the Parts of Animals,* Aristotle tried to identify unifying themes in nature. Among other points, he argued that complex structures are built from a smaller variety of simple components—a perspective that we will find useful later in this chapter.

> **Think About It**
>
> *When you have completed this chapter, discuss the relevance of Aristotle's philosophy to our current thinking about human structure.*

*Claudius Galen* (129–c. 199), physician to the Roman gladiators, wrote the most noteworthy medical textbook of the ancient era—a book that was worshiped to excess by medical professors for centuries to follow. Cadaver dissection was banned in Galen's time because of some atrocities that preceded him, including dissection of living slaves and prisoners merely to satisfy an anatomist's curiosity or to give a public demonstration. Galen was limited to learning anatomy from what he observed in treating gladiators' wounds and by dissecting pigs, monkeys, and other animals. Galen saw science as a process of discovery, not as a body of fact to be taken on faith. He warned that even his own books could be wrong, and advised his followers to trust their own observations more than they trusted any book. Unfortunately, his advice was not heeded. For nearly 1,500 years, medical professors dogmatically taught what they read in Aristotle and Galen, and few dared to question the authority of these "ancient masters."

---

[6]*histo* = tissue + *logy* = study of
[7]*physio* = nature + *logy* = study of

# THE BIRTH OF MODERN MEDICINE

In the Middle Ages, the state of medical science varied greatly from one religious culture to another. Science was severely repressed in the Christian culture of Europe until about the sixteenth century, although some of the most famous medical schools of Europe were founded during this era. Their professors, however, taught medicine primarily as a dogmatic commentary on Galen and Aristotle, not as a field of original research. Medieval medical illustrations were crude representations of the body intended more to decorate a page than to depict the body realistically. Some were astrological charts that showed which sign of the zodiac was thought to influence each organ of the body (fig. 1.2). From such pseudoscience came the word *influenza,* Italian for *influence.*

Free inquiry was less inhibited in Jewish and Muslim culture during this time. Jewish physicians were the most esteemed practitioners of their art, and none so famous as *Moses ben Maimon* (1135–1204), known in Christendom as *Maimonides.* Born in Spain, he fled to Egypt at age 24 to escape antisemitic persecution. There he served the rest of his life as physician to the eldest son of the sultan, Saladin. A highly admired rabbi, Maimonides wrote voluminously on Jewish law and theology, but also wrote 10 influential medical books and numerous treatises on specific diseases.

Among Muslims, probably the most highly regarded medical scholar was *Ibn Sina* (980–1037), known in the West as *Avicenna* or "the Galen of Islam." He studied Galen and Aristotle, combined their findings with original discoveries, and questioned authority when the evidence demanded it. Muslim medicine soon became superior to western medicine, and Avicenna's textbook, *The Canon of Medicine,* became the leading authority in European medical schools until the sixteenth century.

Chinese medicine had little influence on western thought and practice until relatively recently; the medical arts evolved in China quite independently of European medicine. Later chapters of this book describe some of the medical and anatomical insights of ancient China and India.

Modern western medicine began around the sixteenth century in the innovative minds of such people as the anatomist *Andreas Vesalius* and the physiologist *William Harvey.* Vesalius (1514–64) taught anatomy in Italy. In his time, the Catholic Church relaxed its prohibition against cadaver dissection, primarily to allow autopsies in cases of suspicious death. Dissection gradually found its way into the training of medical students throughout Europe. It was an unpleasant business, however, and most professors considered it beneath their dignity. In those days before refrigeration or embalming, the odor from the decaying cadaver was unbearable. Dissections were conducted outdoors in a nonstop 4-day race against decay. Bleary medical students had to fight the urge to vomit, lest they incur the wrath of an overbearing professor. Professors typically

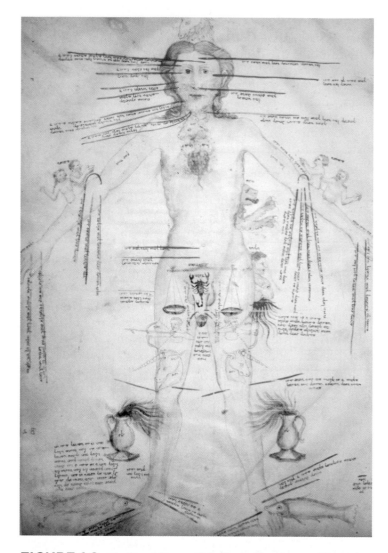

**FIGURE 1.2   Zodiacal Man.**   This illustration from a fifteenth-century medical manuscript reflects the medieval belief in the influence of astrology on parts of the body.

sat in an elevated chair, the cathedra, reading dryly from Galen or Aristotle while a lower-ranking *barber-surgeon* removed putrefying organs from the cadaver and held them up for the students to see. Barbering and surgery were considered to be "kindred arts of the knife"; today's barber poles date from this era, their red and white stripes symbolizing blood and bandages.

Vesalius broke with tradition by coming down from the cathedra and doing the dissections himself. He was quick to point out that much of the anatomy in Galen's books was wrong, and he was the first to publish accurate illustrations for teaching anatomy (fig. 1.3). When others began to plagiarize his illustrations, Vesalius published the first atlas of anatomy, *De Humani Corporis Fabrica (On the Structure of the Human Body),* in 1543. This book began a rich tradition of medical illustration that has been handed down to us

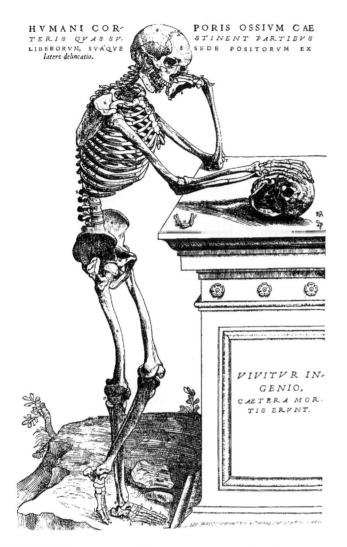

**FIGURE 1.3** **The Art of Vesalius.** Andreas Vesalius revolutionized medical illustration with the comparatively realistic art prepared for his 1543 book, *De Humani Corporis Fabrica.*

through such milestones as *Gray's Anatomy* (1856) and the vividly illustrated atlases and textbooks of today.

Anatomy preceded physiology and was a necessary foundation for it. What Vesalius was to anatomy, the Englishman *William Harvey* (1578–1657) was to physiology. Harvey is remembered especially for his studies of blood circulation and a little book he published in 1628, known by its abbreviated title *De Motu Cordis (On the Motion of the Heart).* He and Michael Servetus (1511–53) were the first western scientists to realize that blood must circulate continuously around the body, from the heart to the other organs and back to the heart again. This flew in the face of Galen's belief that the liver converted food to blood, the heart pumped blood through the veins to all other organs, and those organs consumed it. Harvey's colleagues, wedded to the ideas of Galen, ridiculed him for his theory, though we now know he was correct (see p. 754). Despite persecu-

tion and setbacks, Harvey lived to a ripe old age, served as physician to the kings of England, and later did important work in embryology. Most importantly, Harvey's contributions represent the birth of experimental physiology—the method that generated most of the information in this book.

Modern medicine also owes an enormous debt to two inventors from this era, Robert Hooke and Antony van Leeuwenhoek, who extended the vision of biologists to the cellular level.

*Robert Hooke* (1635–1703), an Englishman, designed scientific instruments of various kinds and made many improvements in the compound microscope. This is a tube with a lens at each end—an *objective lens* near the specimen, which produces an initial magnified image, and an *ocular lens (eyepiece)* near the observer's eye, which magnifies the first image still further. Although crude compound microscopes had existed since 1595, Hooke improved the optics and invented several of the helpful features found in microscopes today—a stage to hold the specimen, an illuminator, and coarse and fine focus controls. His microscopes magnified only about 30 times, but with them, he was the first to see and name cells. In 1663, he observed thin shavings of cork and observed that they "consisted of a great many little boxes," which he called *cellulae* (little cells) after the cubicles of a monastery (fig. 1.4). He later observed thin slices of fresh wood and saw living cells "filled with juices." Hooke became particularly interested in microscopic examination of such material as insects, plant tissues, and animal parts. He published the first comprehensive book of microscopy, *Micrographia,* in 1665.

*Antony van Leeuwenhoek* (an-TOE-nee vahn LAY-wen-hook) (1632–1723), a Dutch textile merchant, invented a *simple* (single-lens) *microscope,* originally for the purpose of examining the weave of fabrics. His microscope was a beadlike lens mounted in a metal plate equipped with a movable specimen clip. Owing to Leeuwenhoek's superior lens-grinding skill, his microscopes magnified about 200 times. Out of curiosity, he examined a drop of lake water and was astonished to find a variety of microorganisms—"little animalcules," he called them, "very prettily a-swimming." He went on to observe practically everything he could get his hands on, including blood cells, blood capillaries, sperm, and muscular tissue. Leeuwenhoek began submitting his observations to the Royal Society of London in 1673. He was praised at first, and his observations were eagerly read by scientists, but enthusiasm for the microscope did not last. By the end of the seventeenth century, it was treated as a mere toy for the upper classes, as amusing and meaningless as a kaleidoscope. Leeuwenhoek and Hooke had even become the brunt of satire. But probably no one in history had looked at nature in such a revolutionary way. By taking biology to the cellular level, the two men had laid an entirely new foundation for the modern medicine to follow centuries later.

The Hooke and Leeuwenhoek microscopes produced poor images with blurry edges *(spherical aberration)* and

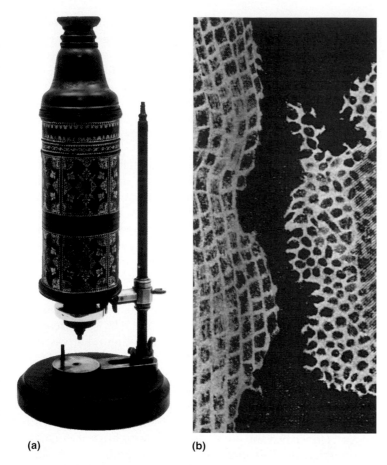

**(a)**                                              **(b)**

**FIGURE 1.4  Hooke's Compound Microscope.** (a) The compound microscope had a lens at each end of a tubular body. (b) Hooke's drawing of cork cells, showing the thick cell walls characteristic of plants.

rainbow-colored distortions *(chromatic aberration).* These problems had to be solved before the microscope could be widely used as a biological tool. In nineteenth-century Germany, *Carl Zeiss* (1816–88) and his business partner, physicist *Ernst Abbe* (1840–1905), greatly improved the compound microscope, adding the condenser and developing superior optics. With improved microscopes, biologists began eagerly examining a wider variety of specimens. By 1839, botanist *Matthias Schleiden* (1804–81) and zoologist *Theodor Schwann* (1810–82) concluded that all organisms were composed of cells. Although it took another century for this idea to be generally accepted, it became the first tenet of the **cell theory,** added to by later biologists and summarized in chapter 3. The cell theory was perhaps the most important breakthrough in biomedical history; all functions of the body are now interpreted as the effects of cellular activity.

The philosophical foundation for modern medicine was largely established by the time of Leeuwenhoek, Hooke, and Harvey, but clinical practice was still in a dismal state. Few doctors attended medical school or received any formal education in basic science or human anatomy. Physicians tended to be ignorant, ineffective, and pompous. Their practice was heavily based on expelling imaginary toxins from the body by bleeding their patients or inducing vomiting, sweating, or diarrhea. They performed operations with dirty hands and instruments, spreading lethal infections from one patient to another. Fractured limbs often became gangrenous and had to be amputated, and there was no anesthesia to lessen the pain. Disease was still widely attributed to demons and witches, and many people felt they would be interfering with God's will if they tried to treat it.

## LIVING IN A REVOLUTION

This short history brings us only to the threshold of modern biomedical science; it stops short of such momentous discoveries as the germ theory of disease, the mechanisms of heredity, and the structure of DNA. In the twentieth century, basic biology and biochemistry have given us a much deeper understanding of how the body works. Technological advances such as medical imaging (see Insight 1.5, p. 23) have enhanced our diagnostic ability and life-support strategies. We have witnessed monumental developments in chemotherapy, immunization, anesthesia, surgery, organ transplants, and human genetics. By the close of the twentieth century, we had discovered the chemical "base sequence" of every human gene and begun using gene therapy to treat children born with diseases recently considered incurable. As future historians look back on the turn of this century, they may exult about the Genetic Revolution in which you are now living.

Several discoveries of the nineteenth and twentieth centuries, and the men and women behind them, are covered in short historical sketches in later chapters. Yet, the stories told in this chapter are different in a significant way. The people discussed here were pioneers in establishing the scientific way of thinking. They helped to replace superstition with an appreciation of natural law. They bridged the chasm between mystery and medication. Without this intellectual revolution, those who followed could not have conceived of the right questions to ask, much less a method for answering them.

### Before You Go On

*Answer the following questions to test your understanding of the preceding section:*

1. *In what way did the followers of Galen disregard his advice? How does Galen's advice apply to you and this book?*

2. *Describe two ways in which Vesalius improved medical education and set standards that remain relevant today.*

3. *How is our concept of human form and function today affected by inventors from Leeuwenhoek to Zeiss?*

# Scientific Method

### Objectives

When you have completed this section, you should be able to

- describe the inductive and hypothetico-deductive methods of obtaining scientific knowledge;
- describe some aspects of experimental design that help to ensure objective and reliable results; and
- explain what is meant by *hypothesis, fact, law,* and *theory* in science.

Prior to the seventeenth century, science was done in a haphazard way by a small number of isolated individuals. The philosophers *Francis Bacon* (1561–1626) in England and *René Descartes* (1596–1650) in France envisioned science as a far greater, systematic enterprise with enormous possibilities for human health and welfare. They detested those who endlessly debated ancient philosophy without creating anything new. Bacon argued against biased thinking and for more objectivity in science. He outlined a systematic way of seeking similarities, differences, and trends in nature and drawing useful generalizations from observable facts. You will see echoes of Bacon's philosophy in the discussion of scientific method that follows.

Though the followers of Bacon and Descartes argued bitterly with one another, both men wanted science to become a public, cooperative enterprise, supported by governments and conducted by an international community of scholars rather than a few isolated amateurs. Inspired by their vision, the French and English governments established academies of science that still flourish today. Bacon and Descartes are credited with putting science on the path to modernity, not by discovering anything new in nature or inventing any techniques—for neither man was a scientist—but by inventing new habits of scientific thought.

When we say "scientific," we mean that such thinking is based on assumptions and methods that yield reliable, objective, testable information about nature. The assumptions of science are ideas that have proven fruitful in the past—for example, the idea that natural phenomena have natural causes and nature is therefore predictable and understandable. The methods of science are highly variable. **Scientific method** refers less to observational procedures than to certain habits of disciplined creativity, careful observation, logical thinking, and honest analysis of one's observations and conclusions. It is especially important in health science to understand these habits. This field is littered with more fads and frauds than any other. We are called upon constantly to judge which claims are trustworthy and which are bogus. To make such judgments depends on an appreciation of how scientists think, how they set standards for truth, and why their claims are more reliable than others.

## THE INDUCTIVE METHOD

The **inductive method,** first prescribed by Bacon, is a process of making numerous observations until one feels confident in drawing generalizations and predictions from them. What we know of anatomy is a product of the inductive method. We describe the normal structure of the body based on observations of many bodies.

This raises the issue of what is considered proof in science. We can never prove a claim beyond all possible refutation. We can, however, consider a statement as proven *beyond reasonable doubt* if it was arrived at by reliable methods of observation, tested and confirmed repeatedly, and not falsified by any credible observation. In science, all truth is tentative; there is no room for dogma. We must always be prepared to abandon yesterday's truth if tomorrow's facts disprove it.

## THE HYPOTHETICO-DEDUCTIVE METHOD

Most physiological knowledge was obtained by the **hypothetico-deductive method.** An investigator begins by asking a question and formulating a **hypothesis**—an educated speculation or possible answer to the question. A good hypothesis must be (1) consistent with what is already known and (2) capable of being tested and possibly falsified by evidence. **Falsifiability** means that if we claim something is scientifically true, we must be able to specify what evidence it would take to prove it wrong. If nothing could possibly prove it wrong, then it is not scientific.

⌐ **Think About It**

*The ancients thought that gods or invisible demons caused epilepsy. Today, epileptic seizures are attributed to bursts of abnormal electrical activity in nerve cells of the brain. Explain why one of these claims is falsifiable (and thus scientific), while the other claim is not.*

The purpose of a hypothesis is to suggest a method for answering a question. From the hypothesis, a researcher makes a deduction, typically in the form of an "if-then" prediction: *If* my hypothesis on epilepsy is correct and I record the brain waves of patients during seizures, *then* I should observe abnormal bursts of activity. A properly conducted experiment yields observations that either support a hypothesis or require the scientist to modify or abandon it, formulate a better hypothesis, and test that one. Hypothesis testing operates in cycles of conjecture and disproof until one is found that is supported by the evidence.

# EXPERIMENTAL DESIGN

Doing an experiment properly involves several important considerations. What shall I measure and how can I measure it? What effects should I watch for and which ones should I ignore? How can I be sure that my results are due to the factors (variables) that I manipulate and not due to something else? When working on human subjects, how can I prevent the subject's expectations or state of mind from influencing the results? Most importantly, how can I eliminate my own biases and be sure that even the most skeptical critics will have as much confidence in my conclusions as I do? Several elements of experimental design address these issues:

- **Sample size.** The number of subjects (animals or people) used in a study is the sample size. An adequate sample size controls for chance events and individual variations in response and thus enables us to place more confidence in the outcome. For example, would you rather trust your health to a drug that was tested on 5 people or one tested on 5,000?

- **Controls.** Biomedical experiments require comparison between treated and untreated individuals so that we can judge whether the treatment has any effect. A **control group** consists of subjects that are as much like the **treatment group** as possible except with respect to the variable being tested. For example, there is evidence that garlic lowers blood cholesterol levels. In one study, a group of people with high cholesterol was given 800 mg of garlic powder daily for 4 months and exhibited an average 12% reduction in cholesterol. Was this a significant reduction, and was it due to the garlic? It is impossible to say without comparison to a control group of similar people who received no treatment. In this study, the control group averaged only a 3% reduction in cholesterol, so garlic *seems* to have made a difference.

- **Psychosomatic effects.** Psychosomatic effects (effects of the subject's state of mind on his or her physiology) can have an undesirable effect on experimental results if we do not control for them. In drug research, it is therefore customary to give the control group a **placebo** (pla-SEE-bo)—a substance with no significant physiological effect on the body. If we were testing a drug, for example, we could give the treatment group the drug and the control group identical-looking starch tablets. Neither group must know which tablets it is receiving. If the two groups showed significantly different effects, we could feel confident that it did not result from a knowledge of what they were taking.

- **Experimenter bias.** In the competitive, high-stakes world of medical research, experimenters may want certain results so much that their biases, even subconscious ones, can affect their interpre-

tation of the data. One way to control for this is the **double-blind method.** In this procedure, neither the subject to whom a treatment is given nor the person giving it and recording the results knows whether that subject is receiving the experimental treatment or placebo. A researcher might prepare identical-looking tablets, some with the drug and some with placebo, label them with code numbers, and distribute them to participating physicians. The physicians themselves do not know whether they are administering drug or placebo, so they cannot give the subjects even accidental hints of which substance they are taking. When the data are collected, the researcher can correlate them with the composition of the tablets and determine whether the drug had more effect than the placebo.

- **Statistical testing.** If you tossed a coin 100 times, you would expect it to come up about 50 heads and 50 tails. If it actually came up 48:52, you would probably attribute this to random error rather than bias in the coin. But what if it came up 40:60? At what point would you begin to suspect bias? This type of problem is faced routinely in research—how great a difference must there be between control and experimental groups before we feel confident that the treatment really had an effect? What if a treatment group exhibited a 12% reduction in cholesterol level and the placebo group a 10% reduction? Would this be enough to conclude that the treatment was effective? Scientists are well grounded in **statistical tests** that can be applied to the data. Perhaps you have heard of the chi-square test, the *t* test, or analysis of variance, for example. A typical outcome of a statistical test might be expressed, "We can be 99.5% sure that the difference between group A and group B was due to the experimental treatment and not to random variation."

# PEER REVIEW

When a scientist applies for funds to support a research project or submits results for publication, the application or manuscript is submitted to **peer review**—a critical evaluation by other experts in that field. Even after a report is published, if the results are important or unconventional, other scientists may attempt to reproduce them to see if the author was correct. At every stage from planning to postpublication, scientists are therefore subject to intense scrutiny by their colleagues. Peer review is one mechanism for ensuring honesty, objectivity, and quality in science.

# FACTS, LAWS, AND THEORIES

The most important product of scientific research is understanding how nature works—whether it be the nature of a pond to an ecologist or the nature of a liver cell to a physiologist. We express our understanding as *facts, laws,* and *theories* of nature. It is important to appreciate the differences among these.

A scientific **fact** is information that can be independently verified by any trained person—for example, the fact that an iron deficiency leads to anemia. A **law of nature** is a generalization about the predictable ways in which matter and energy behave. It is the result of inductive reasoning based on repeated, confirmed observations. Some laws are expressed as concise verbal statements, such as the *first law of thermodynamics:* Energy can be converted from one form to another but cannot be created or destroyed. Others are expressed as mathematical formulae, such as *Boyle's law,* which states that under specified conditions, the volume of a gas *(V)* is inversely proportional to its temperature *(T)*: $V \propto 1/T$. (This law is used in respiratory physiology, chapter 22.)

A **theory** is an explanatory statement, or set of statements, derived from facts, laws, and confirmed hypotheses. Some theories have names, such as the *cell theory,* the *fluid-mosaic theory* of cell membranes, and the *sliding filament theory* of muscle contraction. Most, however, remain unnamed. The purpose of a theory is not only to concisely summarize what we already know but, moreover, to suggest directions for further study and to help predict what the findings should be if the theory is correct.

*Law* and *theory* mean something different in science than they do to most people. In common usage, a law is a rule created and enforced by people; we must obey it or risk a penalty. A law of nature, however, is a description; laws do not *govern* the universe, they *describe* it. Laypeople tend to use the word *theory* for what a scientist would call a hypothesis—for example, "I have a theory why my car won't start." The difference in meaning causes significant confusion when it leads people to think that a scientific theory (such as the theory of evolution) is merely a guess or conjecture, instead of recognizing it as a summary of conclusions drawn from a large body of observed facts. The concepts of gravity and electrons are theories, too, but this does not mean they are merely speculations.

## Think About It

*Was the cell theory proposed by Schleiden and Schwann more a product of the hypothetico-deductive method or of the inductive method? Explain your answer.*

## Before You Go On

*Answer the following questions to test your understanding of the preceding section:*

4. *Describe the general process involved in the inductive method.*

5. *Describe some sources of potential bias in biomedical research. What are some ways of minimizing such bias?*

6. *Is there more information in an individual scientific fact or in a theory? Explain.*

# Human Origins and Adaptations

### Objectives
When you have completed this section, you should be able to

- explain why evolution is relevant to understanding human form and function;
- define *evolution* and *natural selection;*
- describe some human characteristics that can be attributed to the tree-dwelling habits of earlier primates; and
- describe some human characteristics that evolved later in connection with upright walking.

If any two theories have the broadest implications for understanding the human body, they are probably the cell theory and the theory of natural selection. *Natural selection,* an explanation of how species originate and change through time, was the brainchild of *Charles Darwin* (1809–82)—probably the most influential biologist who ever lived. His book, *On the Origin of Species by Means of Natural Selection* (1859), has been called "the book that shook the world." In presenting the first well-supported theory of evolution, *On the Origin of Species* not only caused the restructuring of all of biology but also profoundly changed the prevailing view of our origin, nature, and place in the universe.

*On the Origin of Species* scarcely touched upon human biology, but its unmistakable implications for humans created an intense storm of controversy that continues even today. In *The Descent of Man* (1871), Darwin directly addressed the issue of human evolution and emphasized features of anatomy and behavior that reveal our relationship to other animals. No understanding of human form and function is complete without an understanding of our evolutionary history.

## EVOLUTION, SELECTION, AND ADAPTATION

**Evolution** simply means change in the genetic composition of a population of organisms. Examples include the evolution of bacterial resistance to antibiotics, the appearance of new strains of the AIDS virus, and the emergence of new species of organisms. The theory of **natural selection** is essentially this: Some individuals within a species have hereditary advantages over their competitors—for example, better camouflage, disease resistance, or ability to attract mates—that enable them to produce more offspring. They pass these advantages on to their offspring, and such characteristics therefore become more and more common in successive generations. This brings about the genetic change in a population that constitutes evolution.

Natural forces that promote the reproductive success of some individuals more than others are called **selection pressures.** They include such things as climate, predators, disease, competition, and the availability of food. **Adaptations** are features of an organism's anatomy, physiology, and behavior that have evolved in response to these selection pressures and enable the organism to cope with the challenges of its environment. We will consider shortly some selection pressures and adaptations that were important to human evolution.

Darwin could scarcely have predicted the overwhelming mass of genetic, molecular, fossil, and other evidence of human evolution that would accumulate in the twentieth century and further substantiate his theory. A technique called DNA hybridization, for example, suggests a difference of only 1.6% in DNA structure between humans and chimpanzees. Chimpanzees and gorillas differ by 2.3%. DNA structure suggests that a chimpanzee's closest living relative is not the gorilla or any other ape—it is us.

Several aspects of our anatomy make little sense without an awareness that the human body has a history (see Insight 1.1). Our evolutionary relationship to other species is also important in choosing animals for biomedical research. If there were no issues of cost, availability, or ethics, we might test drugs on our nearest living relatives, the chimpanzees, before approving them for human use. Their genetics, anatomy, and physiology are most similar to ours, and their reactions to drugs therefore afford the best prediction of how the human body would react. On the other hand, if we had no kinship with any other species, the selection of a test species would be arbitrary; we might as well use frogs or snails. In reality, we compromise. Rats and mice are used extensively for research because they are fellow mammals with a physiology similar to ours, but they present fewer of the aforementioned issues than chimpanzees or other mammals do. An animal species or strain selected for research on a particular problem is called a *model*—for example, a mouse model for leukemia.

## PRIMATE ADAPTATIONS

We belong to an order of mammals called the Primates, which also includes the monkeys and apes. Some of our anatomical and physiological features can be traced to the earliest primates, descended from certain squirrel-sized, insect-eating, African mammals (insectivores) that took up life in the trees 55 to 60 million years ago. This **arboreal**[8] (treetop) habitat probably afforded greater safety

---

**INSIGHT 1.1**    Evolutionary Medicine

### Vestiges of Human Evolution

One of the classic lines of evidence for evolution, debated even before Darwin was born, is *vestigial organs*. These structures are the remnants of organs that apparently were better developed and more functional in the ancestors of a species. They now serve little or no purpose or, in some cases, have been converted to new functions.

Our bodies, for example, are covered with millions of hairs, each equipped with a useless little *piloerector muscle*. In other mammals, these muscles fluff the hair and conserve heat. In humans, they merely produce goosebumps. Above each ear, we have three *auricularis muscles*. In other mammals, they move the ears to receive sounds better, but most people cannot contract them at all. As Darwin said, it makes no sense that humans would have such structures were it not for the fact that we came from ancestors in which they were functional.

---

from predators, less competition, and a rich food supply of leaves, fruit, insects, and lizards. But the forest canopy is a challenging world, with dim and dappled sunlight, swaying branches, and prey darting about in the dense foliage. Any new feature that enabled arboreal animals to move about more easily in the treetops would have been strongly favored by natural selection. Thus, the shoulder became more mobile and enabled primates to reach out in any direction (even overhead, which few other mammals can do). The thumbs became **opposable**—they could cross the palm to touch the fingertips—and enabled primates to hold small objects and manipulate them more precisely than other mammals can. Opposable thumbs made the hands **prehensile**[9]—able to grasp branches by encircling them with the thumb and fingers (fig. 1.5). The thumb is so important that it receives highest priority in the repair of hand injuries. If the thumb can be saved, the hand can be reasonably functional; if it is lost, hand functions are severely diminished.

The eyes of primates moved to a more forward-facing position (fig. 1.6), which allowed for **stereoscopic**[10] vision (depth perception). This adaptation provided better hand-eye coordination in catching and manipulating prey, with the added advantage of making it easier to judge distances accurately in leaping from tree to tree. Color vision, rare among mammals, is also a primate hallmark. Primates eat mainly fruit and leaves. The ability to distinguish subtle shades of orange and red enables them to distinguish ripe, sugary fruits from unripe ones. Distinguishing subtle

---

[8]*arbor* = tree + *eal* = pertaining to
[9]*prehens* = to seize
[10]*stereo* = solid + *scop* = vision

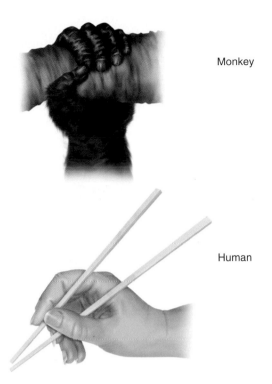

Monkey

Human

**FIGURE 1.5 Opposable Thumbs.** The opposable thumb makes the primate hand prehensile, able to encircle and grasp objects.

**FIGURE 1.6 Primitive Tool Use in a Primate.** Chimpanzees exhibit the prehensile hands and forward-facing eyes typical of primates. Such traits endow primates with stereoscopic vision (depth perception) and good hand-eye coordination, two supremely important factors in human evolution.

shades of green helps them to differentiate between tender young leaves and tough, more toxic older foliage.

Various fruits ripen at different times and in widely separated places in the tropical forest. This requires a good memory of what will be available, when, and how to get there. Larger brains may have evolved in response to the challenge of efficient food finding and, in turn, laid the foundation for more sophisticated social organization.

None of this is meant to imply that humans evolved from monkeys or apes—a common misconception about evolution that no biologist believes. Observations of monkeys and apes, however, provide insight into how primates adapt to the arboreal habitat and how certain human adaptations probably originated.

## WALKING UPRIGHT

About 4 to 5 million years ago, much of the African forest was replaced by savanna (grassland). Some primates adapted to living on the savanna, but this was a dangerous place with more predators and less protection. Just as squirrels and monkeys stand briefly on their hind legs to look around for danger, so would these early ground-dwellers. Being able to stand up not only helps an animal stay alert but also frees the forelimbs for purposes other than walking. Chimpanzees sometimes walk upright to carry food or weapons (sticks and rocks), and it is reasonable to suppose that our early ancestors did so too. They could also carry their infants.

These advantages are so great that they favored skeletal modifications that made **bipedalism**[11]—standing and walking on two legs—easier. The anatomy of the human pelvis, femur, knee, great toe, foot arches, spinal column, skull, arms, and many muscles became adapted for bipedal locomotion, as did many aspects of human family life and society. As the skeleton and muscles became adapted for bipedalism, brain volume increased dramatically (table 1.1). It must have become increasingly difficult for a fully developed, large-brained infant to pass through the mother's pelvic outlet at birth. This may explain why humans are born in a relatively immature, helpless state compared with other mammals, before their nervous systems have matured and the bones of the skull have fused.

---

[11]$bi$ = two + $ped$ = foot

| TABLE 1.1 | Brain Volumes of the Hominidae | |
|---|---|---|
| **Genus or Species** | **Time of Origin (millions of years ago)** | **Brain Volume (milliliters)** |
| *Australopithecus* | 3.9–4.2 | 400 |
| *Homo habilis* | 2.5 | 650 |
| *Homo erectus* | 1.1 | 1,100 |
| *Homo sapiens* | 0.2 | 1,350 |

The oldest bipedal primates (family Hominidae) are classified in the genus *Australopithecus* (aus-TRAL-oh-PITH-eh-cus). About 2.5 million years ago, *Australopithecus* gave rise to *Homo habilis,* the earliest member of our own genus. *Homo habilis* differed from *Australopithecus* in height, brain volume, some details of skull anatomy, and tool-making ability. It was probably the first primate able to speak. *Homo habilis* gave rise to *Homo erectus* about 1.1 million years ago.

Our own species, *Homo sapiens,* has been notoriously difficult to define. Some authorities apply this name to various forms of "archaic *Homo*" dated as far back as 600,000 years, whereas others limit it to anatomically modern humans no more than 200,000 years old (fig. 1.7). Several other species of *Homo* between *Homo erectus* and modern *Homo sapiens* have been named in recent decades; their naming, classification, and relationships are still a matter of considerable debate.

This brief account barely begins to explain how human anatomy, physiology, and behavior have been shaped by ancient selection pressures. Later chapters further demonstrate that the evolutionary perspective pro-

vides a meaningful understanding of why humans are the way we are. Evolution is the basis for comparative anatomy and physiology, which have been so fruitful for the understanding of human biology. If we were not related to any other species, those sciences would be pointless. The emerging science of **evolutionary (darwinian) medicine** traces some of our diseases and imperfections to our evolutionary past.

## Before You Go On

*Answer the following questions to test your understanding of the preceding section:*

7. *Define* adaptation *and* selection pressure. *Why are these concepts important in understanding human anatomy and physiology?*

8. *Select any two human characteristics and explain how they may have originated in primate adaptations to an arboreal habitat.*

9. *Select two other human characteristics and explain how they may have resulted from adaptation to a grassland habitat.*

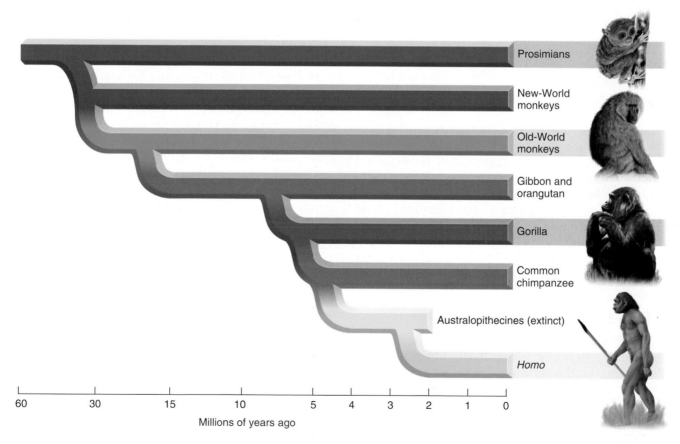

**FIGURE 1.7  The Place of Humans in Primate Evolution.**  Figures at the right show some representative primates. The branch points in this family tree show the approximate times that different lines diverged from a common ancestor. Note that the time scale is not uniform; recent events are expanded for clarity.

▶ *Which is more closely related to humans—a gorilla or a monkey? How long ago did the last common ancestor of chimpanzees and humans exist?*

# Human Structure

## Objectives

When you have completed this section, you should be able to

- list the levels of human structure from the most complex to the simplest;
- discuss the value of both reductionistic and holistic viewpoints to understanding human form and function; and
- discuss the clinical significance of anatomical variation among humans.

Earlier in this chapter, we observed that human anatomy is studied by a variety of techniques—dissection, palpation, and so forth. In addition, anatomy is studied at several levels of detail, from the whole body down to the molecular level.

## THE HIERARCHY OF COMPLEXITY

Consider for the moment an analogy to human structure: The English language, like the human body, is very complex, yet an endless array of ideas can be conveyed with a limited number of words. All words in English are, in turn, composed of various combinations of just 26 letters. Between an essay and an alphabet are successively simpler levels of organization: paragraphs, sentences, words, and syllables. We can say that language exhibits a hierarchy of complexity, with letters, syllables, words, and so forth being successive levels of the hierarchy. Humans have an analogous hierarchy of complexity, as follows (fig. 1.8):

The organism is composed of organ systems,

  organ systems are composed of organs,

    organs are composed of tissues,

      tissues are composed of cells,

        cells are composed (in part) of organelles,

          organelles are composed of molecules, and

            molecules are composed of atoms.

The **organism** is a single, complete individual.

An **organ system** is a group of organs with a unique collective function, such as circulation, respiration, or digestion. The human body has 11 organ systems, illustrated in atlas A immediately following this chapter: the integumentary, skeletal, muscular, nervous, endocrine, circulatory, lymphatic, respiratory, urinary, digestive, and reproductive systems. Usually, the organs of one system are physically interconnected, such as the kidneys, ureters, urinary bladder, and urethra, which compose the urinary system. Beginning with chapter 6, this book is organized around the organ systems.

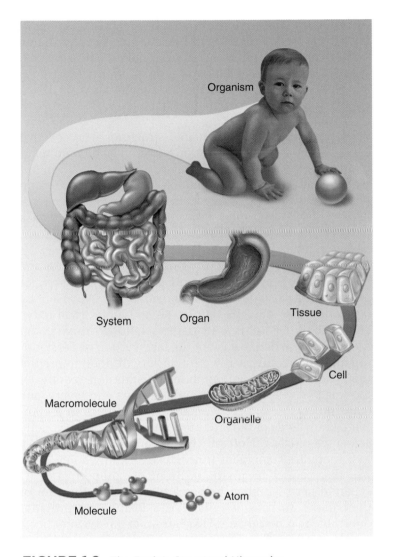

**FIGURE 1.8**  **The Body's Structural Hierarchy.**

An **organ** is a structure composed of two or more tissue types that work together to carry out a particular function. Organs have definite anatomical boundaries and are visibly distinguishable from adjacent structures. Most organs and higher levels of structure are within the domain of gross anatomy. However, there are organs within organs—the large organs visible to the naked eye often contain smaller organs visible only with the microscope. The skin, for example, is the body's largest organ. Included within it are thousands of smaller organs: each hair, nail, gland, nerve, and blood vessel of the skin is an organ in itself.

A **tissue** is a mass of similar cells and cell products that forms a discrete region of an organ and performs a specific function. The body is composed of only four primary classes of tissue: epithelial, connective, nervous, and muscular tissues. Histology, the study of tissues, is the subject of chapter 5.

**Cells** are the smallest units of an organism that carry out all the basic functions of life; nothing simpler than a cell is considered alive. A cell is enclosed in a *plasma membrane* composed of lipids and proteins. Most cells have one nucleus, an organelle that contains its DNA. *Cytology,* the study of cells and organelles, is the subject of chapters 3 and 4.

**Organelles**[12] are microscopic structures in a cell that carry out its individual functions. Examples include mitochondria, centrioles, and lysosomes.

Organelles and other cellular components are composed of **molecules.** The largest molecules, such as proteins, fats, and DNA, are called *macromolecules.* A molecule is a particle composed of at least two **atoms,** the smallest particles with unique chemical identities.

The theory that a large, complex system such as the human body can be understood by studying its simpler components is called **reductionism.** First espoused by Aristotle, this has proven to be a highly productive approach; indeed, it is essential to scientific thinking. Yet the reductionistic view is not the last word in understanding human life. Just as it would be very difficult to predict the workings of an automobile transmission merely by looking at a pile of its disassembled gears and levers, one could never predict the human personality from a complete knowledge of the circuitry of the brain or the genetic sequence of DNA. **Holism**[13] is the complementary theory that there are "emergent properties" of the whole organism that cannot be predicted from the properties of its separate parts—human beings are more than the sum of their parts. To be most effective, a health-care provider does not treat merely a disease or an organ system, but a whole person. A patient's perceptions, emotional responses to life, and confidence in the nurse, therapist, or physician profoundly affect the outcome of treatment. In fact, these psychological factors often play a greater role in a patient's recovery than the physical treatments administered.

## ANATOMICAL VARIATION

Anatomists, surgeons, and students must be constantly aware of how much one body can differ from another. A quick look around any classroom is enough to show that no two humans are exactly alike; on close inspection, even identical twins exhibit differences. Yet anatomy atlases and textbooks can easily give the impression that everyone's internal anatomy is the same. This simply is not true. Books such as this one can only teach you the most common structure—the anatomy seen in about 70%

or more of people. Someone who thinks that all human bodies are the same internally would make a very confused medical student or an incompetent surgeon.

Some people lack certain organs. For example, most of us have a *palmaris longus* muscle in the forearm and a *plantaris* muscle in the leg, but these are absent from some people. Most of us have five lumbar vertebrae (bones of the lower spine), but some people have six and some have four. Most of us have one spleen and two kidneys, but some have two spleens or only one kidney. Most kidneys are supplied by a single *renal artery,* but some have two renal arteries. Figure 1.9 shows some common variations in human anatomy, and Insight 1.2 describes a particularly dramatic and clinically important variation.

---

### INSIGHT 1.2    Clinical Application

#### Situs Inversus and Other Unusual Anatomy

In most people, the spleen, pancreas, sigmoid colon, and most of the heart are on the left, while the appendix, gallbladder, and most of the liver are on the right. The normal arrangement of these and other internal organs is called *situs* (SITE-us) *solitus.* About 1 in 8,000 people, however, are born with an abnormality called *situs inversus*—the organs of the thoracic and abdominal cavities are reversed between right and left. A selective right-left reversal of the heart is called *dextrocardia.* In *situs perversus,* a single organ occupies an atypical position—for example, a kidney located low in the pelvic cavity instead of high in the abdominal cavity.

Conditions such as dextrocardia in the absence of complete situs inversus can cause serious medical problems. Complete situs inversus, however, usually causes no functional problems because all of the viscera, though reversed, maintain their normal relationships to one another. Situs inversus is often discovered in the fetus by sonography, but many people remain unaware of their condition for decades until it is discovered by medical imaging, on physical examination, or in surgery. You can easily imagine the importance of such conditions in diagnosing appendicitis, performing gallbladder surgery, interpreting an X ray, or auscultating the heart valves.

---

⌐ **Think About It**

*People who are allergic to aspirin or penicillin often wear Medic Alert bracelets or necklaces that note this fact in case they need emergency medical treatment and are unable to communicate. Why would it be important for a person with situs inversus to have this noted on a Medic Alert bracelet?*

---

[12]*elle* = little
[13]*holo* = whole, entire

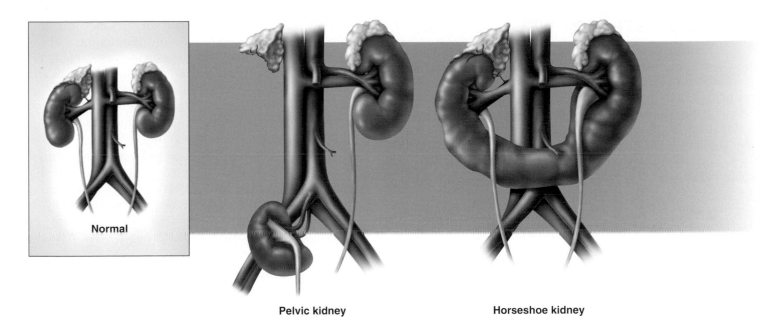

Normal          Pelvic kidney          Horseshoe kidney

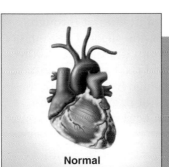

Normal

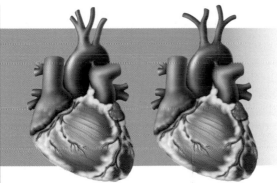

Variations in branches of the aorta

**FIGURE 1.9** Variation in Anatomy of the Kidneys and Major Arteries Near the Heart.

## Before You Go On

*Answer the following questions to test your understanding of the preceding section:*

10. *In the hierarchy of human structure, what is the level between organ system and tissue? Between cell and molecule?*

11. *How are tissues relevant to the definition of an organ?*

12. *Why is reductionism a necessary but not sufficient point of view for fully understanding a patient's illness?*

13. *Why should medical students observe multiple cadavers and not be satisfied to dissect only one?*

# Human Function

### Objectives

When you have completed this section, you should be able to

- state the characteristics that distinguish living organisms from nonliving objects;

- explain the importance of defining a reference man and woman;

- define *homeostasis* and explain why this concept is central to physiology;

- define *negative feedback,* give an example of it, and explain its importance to homeostasis; and

- define *positive feedback* and give examples of its beneficial and harmful effects.

# CHARACTERISTICS OF LIFE

Why do we consider a growing child to be alive, but not a growing crystal? Is abortion the taking of a human life? If so, what about a contraceptive foam that kills only sperm? As a patient is dying, at what point does it become ethical to disconnect life-support equipment and remove organs for donation? If these organs are alive, as they must be to serve someone else, then why isn't the donor considered alive? Such questions have no easy answers, but they demand a concept of what life is—a concept that may differ with one's biological, medical, or legal perspective.

From a biological viewpoint, life is not a single property. It is a collection of properties that help to distinguish living from nonliving things:

- **Organization.** Living things exhibit a far higher level of organization than the nonliving world around them. They expend a great deal of energy to maintain order, and a breakdown in this order is accompanied by disease and often death.

- **Cellular composition.** Living matter is always compartmentalized into one or more cells.

- **Metabolism** and **excretion.** Living things take in molecules from the environment and chemically change them into molecules that form their own structures, control their physiology, or provide them with energy. **Metabolism**[14] is the sum of all this internal chemical change. It consists of two classes of reactions: *anabolism,*[15] in which relatively complex molecules are synthesized from simpler ones (for example, protein synthesis), and *catabolism,*[16] in which relatively complex molecules are broken down into simpler ones (for example, protein digestion). Metabolism inevitably produces chemical wastes, some of which are toxic if they accumulate. Metabolism therefore requires **excretion,** the separation of wastes from the tissues and their elimination from the body. There is a constant turnover of molecules in the body; few of the molecules now in your body have been there for more than a year. It is food for thought that although you sense a continuity of personality and experience from your childhood to the present, nearly all of your body has been replaced within the past year.

- **Responsiveness** and **movement.** The ability of organisms to sense and react to **stimuli** (changes in their environment) is called *responsiveness, irritability,* or *excitability.* It occurs at all levels from the single cell to the entire body, and it characterizes all living things from bacteria to you. Responsiveness is especially obvious in animals because of nerve and muscle cells that exhibit high sensitivity to environmental stimuli, rapid transmission of information, and quick reactions. Most living organisms are capable of self-propelled movement from place to place, and all organisms and cells are at least capable of moving substances internally, such as moving food along the digestive tract or moving molecules and organelles from place to place within a cell.

- **Homeostasis.** Although the environment around an organism changes, the organism maintains relatively stable internal conditions. This ability to maintain internal stability, called *homeostasis,* is explored in more depth shortly.

- **Development.** Development is any change in form or function over the lifetime of the organism. In most organisms, it involves two major processes: (1) **differentiation,** the transformation of cells with no specialized function into cells that are committed to a particular task, and (2) **growth,** an increase in size. Some nonliving things grow, but not in the way your body does. If you let a saturated sugar solution evaporate, crystals will grow from it, but not through a change in the composition of the sugar. They merely add more sugar molecules from the solution to the crystal surface. The growth of the body, by contrast, occurs through chemical change (metabolism); for the most part, your body is not composed of the molecules you ate but of molecules made by chemically altering your food.

- **Reproduction.** All living organisms can produce copies of themselves, thus passing their genes on to new, younger containers—their offspring.

- **Evolution.** All living species exhibit genetic change from generation to generation and therefore evolve. This occurs because *mutations* (changes in DNA structure) are inevitable and because environmental selection pressures endow some individuals with greater reproductive success than others. Unlike the other characteristics of life, evolution is a characteristic seen only in the population as a whole. No single individual evolves over the course of its life.

Clinical and legal criteria of life differ from these biological criteria. A person who has shown no brain waves for 24 hours, and has no reflexes, respiration, or heartbeat other than what is provided by artificial life support, can be declared legally dead. At such time, however, most of the body is still biologically alive and its organs may be useful for transplant.

# PHYSIOLOGICAL VARIATION

Earlier we considered the clinical importance of variations in human anatomy, but physiology is even more variable. Physiological variables differ with sex, age, weight, diet,

---

[14]*metabol* = change + *ism* = process
[15]*ana* = up
[16]*cata* = down

degree of physical activity, and environment, among other things. Failure to consider such variation leads to medical mistakes such as overmedication of the elderly or medicating women on the basis of research that was done on men. If an introductory textbook states a typical human heart rate, blood pressure, red blood cell count, or body temperature, it is generally assumed that such values are for a healthy young adult unless otherwise stated. A point of reference for such general values is the reference man and reference woman. The **reference man** is defined as a healthy male 22 years old, weighing 70 kg (154 lb), living at a mean ambient (surrounding) temperature of 20°C, engaging in light physical activity, and consuming 2,800 kilocalories (kcal) per day. The **reference woman** is the same except for a weight of 58 kg (128 lb) and an intake of 2,000 kcal/day.

## HOMEOSTASIS AND NEGATIVE FEEDBACK

**Homeostasis**[17] (ho-me-oh-STAY-sis) is one of the theories that will arise most frequently in this book as we study mechanisms of health and disease. The human body has a remarkable capacity for self-restoration. Hippocrates commented that it usually returns to a state of equilibrium by itself, and people recover from most illnesses even without the help of a physician. This tendency results from homeostasis, the ability to detect change and activate mechanisms that oppose it.

French physiologist *Claude Bernard* (1813–78) observed that the internal conditions of the body remain quite stable even when external conditions vary greatly. For example, whether it is freezing cold or swelteringly hot outdoors, the internal temperature of your body stays within a range of about 36° to 37°C (97°–99°F). American physiologist *Walter Cannon* (1871–1945) coined the term *homeostasis* for this tendency to maintain internal stability. Homeostasis has been one of the most enlightening concepts in physiology. Physiology is largely a group of mechanisms for maintaining homeostasis, and the loss of homeostatic control tends to cause illness or death. Pathophysiology is essentially the study of unstable conditions that result when our homeostatic controls go awry.

Do not, however, overestimate the degree of internal stability. Internal conditions are not absolutely constant but fluctuate within a limited range, such as the range of body temperatures noted earlier. The internal state of the body is best described as a **dynamic equilibrium** (balanced change), in which there is a certain **set point** or average value for a given variable (such as, 37°C for body temperature) and conditions fluctuate slightly around this point.

The fundamental mechanism that keeps a variable close to its set point is **negative feedback**—a process in which the body senses a change and activates mechanisms that negate or reverse it. By maintaining stability, negative feedback is the key mechanism for maintaining health.

These principles can be understood by comparison to a home heating system (fig. 1.10). Suppose it is a cold winter day and you have set your thermostat for 20°C (68°F)—the set point. If the room becomes too cold, a temperature-sensitive switch in the thermostat turns on the furnace. The temperature rises until it is slightly above the set point, and then the switch breaks the circuit and turns off the furnace. This is a negative-feedback

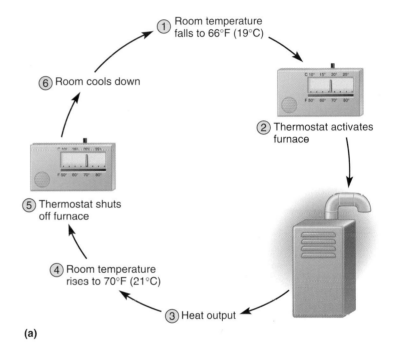

**(a)**

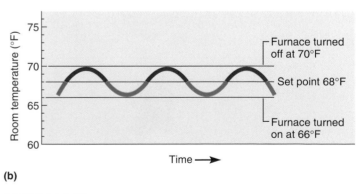

**(b)**

**FIGURE 1.10 Negative Feedback in a Home Heating System.** (a) The negative-feedback loop that maintains room temperature. (b) Fluctuation of room temperature around the thermostatic set point.

▶ *What component of the heating system acts as the sensor? What component acts as the effector?*

---

[17]*homeo* = the same + *stas* = to place, stand, stay

process that reverses the falling temperature and restores it to something close to the set point. When the furnace turns off, the temperature slowly drops again until the switch is reactivated—thus, the furnace cycles on and off all day. The room temperature does not stay at exactly 20°C but *fluctuates* a few degrees either way—the system maintains a state of dynamic equilibrium in which the temperature averages 20°C and deviates from the set point by only a few degrees. Because feedback mechanisms alter the original changes that triggered them (temperature, for example), they are often called **feedback loops.**

Body temperature is also regulated by a "thermostat"—a group of nerve cells in the base of the brain that monitors the temperature of the blood. If you become overheated, the thermostat triggers heat-losing mechanisms (fig. 1.11). One of these is **vasodilation** (VAY-zo-dy-LAY-shun), the widening of blood vessels. When blood vessels of the skin dilate, warm blood flows closer to the body surface and loses heat to the surrounding air. If this is not enough to return your temperature to normal, sweating occurs; the evaporation of water from the skin has a powerful cooling effect (see Insight 1.3). Conversely, if it is cold outside and your body temperature drops much below 37°C, these nerve cells activate heat-conserving mechanisms. The first to be activated is **vasoconstriction,** a narrowing of the blood vessels in the

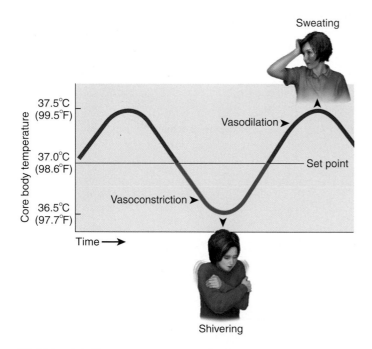

FIGURE 1.11 **Negative Feedback in Human Thermoregulation.** Negative feedback keeps the human body temperature homeostatically regulated within about 0.5°C of a 37°C set point. Sweating and cutaneous vasodilation lower the body temperature; shivering and cutaneous vasoconstriction raise it.

▶ *Why does vasodilation reduce the body temperature?*

skin, which serves to retain warm blood deeper in your body and reduce heat loss. If this is not enough, the brain activates shivering—muscle tremors that generate heat.

To take another example, a rise in blood pressure is sensed by stretch receptors in the wall of the heart and the major arteries above it. These receptors send nerve signals to a *cardiac center* in the brainstem. The cardiac center integrates this input with other information and sends nerve signals back to the heart to slow it and lower the blood pressure. Thus we can see that homeostasis is maintained by self-correcting negative-feedback loops. Many more examples are found throughout this book.

It is common, although not universal, for feedback loops to include three components: a receptor, an integrator, and an effector. The **receptor** is a structure that senses a change in the body, such as the stretch receptors that monitor blood pressure. The **integrating (control) center,** such as the cardiac center of the brain, is a mechanism that processes this information, relates it to other available information (for example, comparing what the blood pressure is with what it should be), and "makes a decision" about what the appropriate response should be. The **effector,** in this case the heart, is the structure that carries out the response that restores homeostasis. The response, such as a lowering of the blood pressure, is then sensed by the receptor, and the feedback loop is complete.

## POSITIVE FEEDBACK AND RAPID CHANGE

**Positive feedback** is a self-amplifying cycle in which a physiological change leads to even greater change in the same direction, rather than producing the corrective effects of negative feedback. Positive feedback is often a normal way of producing rapid change. When a woman is giving birth, for example, the head of the baby pushes against her cervix (the neck of the uterus) and stimulates its nerve endings (fig. 1.12). Nerve signals travel to the brain, which, in turn, stimulates the pituitary gland to secrete the hormone oxytocin. Oxytocin travels in the blood and stimulates the uterus to contract. This pushes the baby downward, stimulating the cervix still more and causing the positive-feedback loop to be repeated. Labor contractions therefore become more and more intense until the baby is expelled. Other cases of beneficial positive feedback are seen later in the book; for example, in blood clotting, protein digestion, and the generation of nerve signals.

Frequently, however, positive feedback is a harmful or even life-threatening process. This is because its self-amplifying nature can quickly change the internal state of the body to something far from its homeostatic set point. Consider a high fever, for example. A fever triggered by infection is beneficial up to a point, but if the body temperature rises much above 42°C (108°F), it may create a

dangerous positive-feedback loop. This high temperature raises the metabolic rate, which makes the body produce heat faster than it can get rid of it. Thus, temperature rises still further, increasing the metabolic rate and heat production still more. This "vicious circle" becomes fatal at approximately 45°C (113°F). Thus, positive-feedback loops often create dangerously out-of-control situations that require emergency medical treatment.

## Before You Go On

*Answer the following questions to test your understanding of the preceding section:*

14. List four biological criteria of life and one clinical criterion. Explain how a person could be clinically dead but biologically alive.

15. What is meant by *dynamic equilibrium?* Why would it be wrong to say homeostasis prevents internal change?

16. Explain why stabilizing mechanisms are called negative feedback.

17. Explain why positive feedback is more likely than negative feedback to disturb homeostasis.

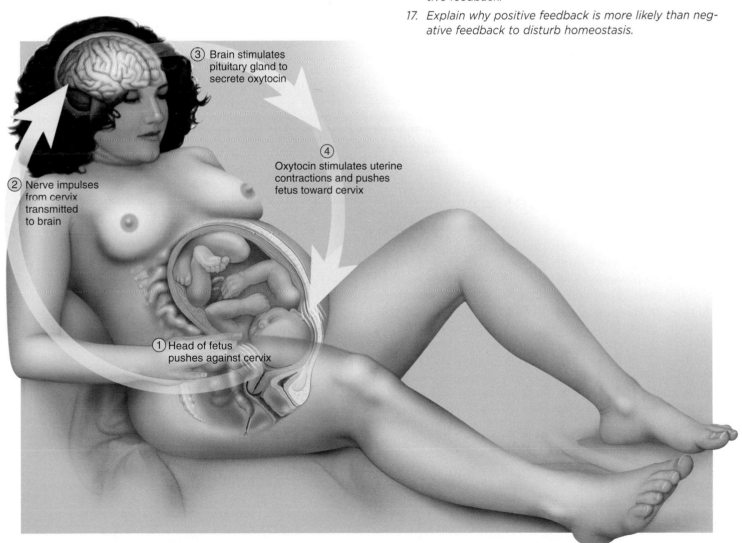

③ Brain stimulates pituitary gland to secrete oxytocin

④ Oxytocin stimulates uterine contractions and pushes fetus toward cervix

② Nerve impulses from cervix transmitted to brain

① Head of fetus pushes against cervix

**FIGURE 1.12  Positive Feedback in Childbirth.**

# The Language of Medicine

**Objectives**

When you have completed this section, you should be able to

- explain why modern anatomical terminology is so heavily based on Greek and Latin;
- recognize eponyms when you see them;
- describe the efforts to achieve an internationally uniform anatomical terminology;
- break medical terms down into their basic word elements;
- state some reasons why the literal meaning of a word may not lend insight into its definition;
- relate singular noun forms to their plural forms; and
- discuss why precise spelling is important in anatomy and physiology.

One of the greatest challenges faced by students of anatomy and physiology is the vocabulary. In this book, you will encounter such Latin terms as *corpus callosum* (a brain structure), *ligamentum arteriosum* (a small fibrous band near the heart), and *extensor carpi radialis longus* (a forearm muscle). You may wonder why structures aren't named in "just plain English," and how you will ever remember such formidable names. This section will give you some answers to these questions and some useful tips on mastering anatomical terminology.

## THE HISTORY OF ANATOMICAL TERMINOLOGY

The major features of human gross anatomy have standard international names prescribed by a book titled the *Terminologia Anatomica* (TA). The TA was codified in 1998 by an international body of anatomists, the Federative Committee on Anatomical Terminology, and approved by professional associations of anatomists in more than 50 countries.

About 90% of today's medical terms are formed from just 1,200 Greek and Latin roots. Scientific investigation began in ancient Greece and soon spread to Rome. The Greeks and Romans coined many of the words still used in human anatomy today: *uterus, prostate, cerebellum, diaphragm, sacrum, amnion,* and others. In the Renaissance, the fast pace of anatomical discovery required a profusion of new terms to describe things. Anatomists in different countries began giving different names to the same structures. Adding to the confusion, they often named new structures and diseases in honor of their esteemed teachers and predecessors, giving us such

nondescriptive terms as the *crypts of Lieberkühn* and *duct of Santorini.* Terms coined from the names of people, called **eponyms,**[18] afford little clue as to what a structure or condition is.

In hopes of resolving this growing confusion, anatomists began meeting as early as 1895 to devise a uniform international terminology. After several false starts, they agreed on a list of terms titled the *Nomina Anatomica* (NA), which rejected all eponyms and gave each structure a unique Latin name to be used worldwide. Even if you were to look at an anatomy atlas in Japanese or Arabic, the illustrations may be labeled with the same Latin terms as in an English-language atlas. The NA served for many decades until recently replaced by the TA, which prescribes both Latin names and accepted English equivalents. The terminology in this book conforms to the TA except where undue confusion would result from abandoning widely used, yet unofficial terms.

## ANALYZING MEDICAL TERMS

The task of learning medical terminology seems overwhelming at first, but there is a simple trick to becoming more comfortable with the technical language of medicine. People who find scientific terms confusing and difficult to pronounce, spell, and remember usually feel more confident once they realize the logic of how terms are composed. A term such as *hyponatremia* is less forbidding once we recognize that it is composed of three common word elements: *hypo-* (below normal), *natr-* (sodium), and *-emia* (blood condition). Thus, hyponatremia is a deficiency of sodium in the blood. Those word elements appear over and over in many other medical terms: *hypothermia, natriuretic, anemia,* and so on. Once you learn the meanings of *hypo-, natri-,* and *-emia,* you already have the tools at least to partially understand hundreds of other biomedical terms. Inside the back cover, you will find a lexicon of the 400 word elements most commonly footnoted in this book.

Scientific terms are typically composed of one or more of the following elements:

- At least one *root (stem)* that bears the core meaning of the word. In *cardiology,* for example, the root is *cardi-* (heart). Many words have two or more roots. In *cytochrome,* the roots are *cyt-* (cell) and *chrom-* (color).
- *Combining vowels* that are often inserted to join roots and make the word easier to pronounce. In

---

[18]*epo* = after, related to + *nym* = name

*cytochrome,* for example, the first *o* is a combining vowel. Although *o* is the most common combining vowel, all vowels of the alphabet are used in this way, such as *a* in *ligament,* *e* in *vitreous,* the first *i* in *spermicidal,* *u* in *ovulation,* and *y* in *tachycardia.* Some words, such as *intervertebral,* have no combining vowels. A combination of a root and combining vowel is called a *combining form:* for example, *ost* (bone) + *e* (a combining vowel) make the combining form *oste-,* as in *osteology.*

- A *prefix* may be present to modify the core meaning of the word. For example, *gastric* (pertaining to the stomach or to the belly of a muscle) takes on a variety of new meanings when prefixes are added to it: *epigastric* (above the stomach), *hypogastric* (below the stomach), *endogastric* (within the stomach), and *digastric* (a muscle with two bellies).

- A *suffix* may be added to the end of a word to modify its core meaning. For example, *microscope, microscopy, microscopic,* and *microscopist* have different meanings because of their suffixes alone. Often two or more suffixes, or a root and suffix, occur together so often that they are treated jointly as a *compound suffix;* for example, *log* (study) + *y* (process) form the compound suffix *-logy* (the study of).

To summarize these basic principles, consider the word *gastroenterology,* a branch of medicine dealing with the stomach and small intestine. It breaks down into gastro/entero/logy:

*gastro* = a combining form meaning "stomach"

*entero* = a combining form meaning "small intestine"

*logy* = a compound suffix meaning "the study of"

"Dissecting" words in this way and paying attention to the word-origin footnotes throughout this book will help you become more comfortable with the language of anatomy. Knowing how a word breaks down and knowing the meaning of its elements make it far easier to pronounce a word, spell it, and remember its definition. There are a few unfortunate exceptions, however. The path from original meaning to current usage has often become obscured by history (see Insight 1.4). The foregoing approach also is no help with eponyms or **acronyms**—words composed of the first letter, or first few letters, of a series of words. For example, *calmodulin,* a calcium-binding protein found in many cells, is cobbled together from a few letters of the three words, *cal*cium *modul*ating prote*in.*

## SINGULAR AND PLURAL FORMS

A point of confusion for many beginning students is how to recognize the plural forms of medical terms. Few peo-

### INSIGHT 1.4  Medical History

## Obscure Word Origins

The literal translation of a word doesn't always provide great insight into its modern meaning. The history of language is full of twists and turns that are fascinating in their own right and say much about the history of human culture, but they can create confusion for students.

For example, the *amnion* is a transparent sac that forms around the developing fetus. The word is derived from *amnos,* from the Greek for "lamb." From this origin, *amnos* came to mean a bowl for catching the blood of sacrificial lambs, and from there the word found its way into biomedical usage for the membrane that emerges (quite bloody) as part of the afterbirth. The *acetabulum,* the socket of the hip joint, literally means "vinegar cup." Apparently the hip socket reminded an anatomist of the little cups used to serve vinegar as a condiment on dining tables in ancient Rome. The word *testicles* literally means "little witnesses." The history of medical language has several amusing conjectures as to why this word was chosen to name the male gonads.

| TABLE 1.2 | Singular and Plural Forms of Some Noun Terminals | |
|---|---|---|
| **Singular Ending** | **Plural Ending** | **Examples** |
| -a | -ae | axilla, axillae |
| -ax | -aces | thorax, thoraces |
| -en | -ina | lumen, lumina |
| -ex | -ices | cortex, cortices |
| -is | -es | diagnosis, diagnoses |
| -is | -ides | epididymis, epididymides |
| -ix | -ices | appendix, appendices |
| -ma | -mata | carcinoma, carcinomata |
| -on | -a | ganglion, ganglia |
| -um | -a | septum, septa |
| -us | -era | viscus, viscera |
| -us | -i | villus, villi |
| -us | -ora | corpus, corpora |
| -x | -ges | phalanx, phalanges |
| -y | -ies | ovary, ovaries |
| -yx | -ices | calyx, calices |

ple would fail to recognize that *ovaries* is the plural of *ovary,* but the connection is harder to make in other cases: for example, the plural of *cortex* is *cortices* (COR-ti-sees), the plural of *corpus* is *corpora,* and the plural of *epididymis* is *epididymides* (EP-ih-DID-ih-MID-eze). Table 1.2 will help you make the connection between common singular and plural noun terminals.

## THE IMPORTANCE OF PRECISION

A final word of advice for your study of anatomy and physiology: Be precise in your use of terms. It may seem trivial if you misspell *trapezius* as *trapezium,* but in doing so, you would be changing the name of a back muscle to the name of a wrist bone. Similarly, changing *occipitalis* to *occipital* or *zygomaticus* to *zygomatic* changes other muscle names to bone names. A "little" error such as misspelling *ileum* as *ilium* changes the name of the final portion of the small intestine to the name of the hip bone. Changing *malleus* to *malleolus* changes the name of a middle-ear bone to the name of a bony protuberance of your ankle. *Elephantiasis* is a disease that produces an elephant-like thickening of the limbs and skin. Many people misspell this *elephantitis;* if such a word existed, it would mean inflammation of an elephant.

The health professions demand the utmost attention to detail and precision—people's lives may one day be in your hands. The habit of carefulness must extend to your use of language as well. Many patients have died because of miscommunication in the hospital.

### Before You Go On

*Answer the following questions to test your understanding of the preceding section:*

18. *Explain why modern anatomical terminology is so heavily based on Greek and Latin.*

19. *Distinguish between an eponym and an acronym, and explain why both of these present difficulties for interpreting anatomical terms.*

20. *Break each of the following words down into its roots, prefixes, and suffixes, and state their meanings, following the example of* gastroenterology *analyzed earlier:* pericardium, appendectomy, subcutaneous, phonocardiogram, otorhinolaryngology. *Consult the list of word elements in Appendix C for help.*

21. *Write the singular form of each of the following words:* pleurae, gyri, nomina, ganglia, fissures. *Write the plural form of each of the following:* villus, tibia, encephalitis, cervix, stoma.

## Review of Major Themes

To close this chapter, let's distill a few major points from it. These themes can provide you with a sense of perspective that will make the rest of the book more meaningful and not just a collection of disconnected facts. These are some key unifying principles behind all study of human anatomy and physiology:

- **Cell theory.** *All structure and function result from the activity of cells.* Every physiological concept in this book ultimately must be understood from the standpoint of how cells function. Even anatomy is a result of cellular function. If cells are damaged or destroyed, we see the results in disease symptoms of the whole person.

- **Homeostasis.** *The purpose of most normal physiology is to maintain stable conditions within the body.* Human physiology is essentially a group of mechanisms that produce stable internal conditions favorable to cellular function. Any serious departure from these conditions can be harmful or fatal to cells.

- **Evolution.** *The human body is a product of evolution.* Like every other living species, we have been molded by millions of years of natural selection to function in a changing environment. Many aspects of human anatomy and physiology reflect our ancestors' adaptations to their environment. Human form and function cannot be fully understood except in light of our evolutionary history.

- **Hierarchy of structure.** *Human structure can be viewed as a series of levels of complexity.* Each level is composed of a smaller number of simpler subunits than the level above it. These subunits are arranged in different ways to form diverse structures of higher complexity. For example, all the body's organs are made of just four primary classes of tissue, and the thousands of proteins are made of various combinations of just 20 amino acids. Understanding these simpler components is the key to understanding higher levels of structure.

- **Unity of form and function.** *Form and function complement each other; physiology cannot be divorced from anatomy.* This unity holds true even down to the molecular level. Our very molecules, such as DNA and proteins, are structured in ways that enable them to carry out their functions. Slight changes in molecular structure can destroy their activity and threaten life.

**Think About It**

*Architect Louis Henri Sullivan coined the phrase, "Form ever follows function." What do you think he meant by this? Discuss how this idea could be applied to the human body and cite a specific example of human anatomy to support it.*

## INSIGHT 1.5    Clinical Application

### Medical Imaging

The development of techniques for looking into the body without having to do exploratory surgery has greatly accelerated progress in medicine. A few of these techniques are described here.

### Radiography

*X rays,* a form of high-energy radiation, were discovered by William Roentgen in 1885. X rays can penetrate soft tissues of the body and darken photographic film on the other side.

They are absorbed, however, by dense tissues such as bone, teeth, tumors, and tuberculosis nodules, which leave the film lighter in these areas (fig. 1.13a). The process of examining the body with X rays is called *radiography.* The term *X ray* also applies to a photograph *(radiograph)* made by this method. Radiography is commonly used in dentistry, mammography, diagnosis of fractures, and examination of the chest. Hollow organs can be visualized by filling them with a *radiopaque* substance that absorbs X rays. Barium sulfate is given orally for examination of the esophagus, stomach, and small intestine or

*continued*

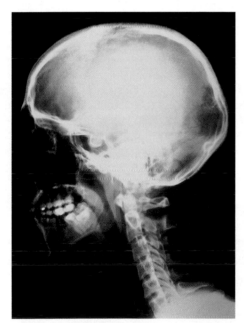

**(a) X ray (radiograph)**

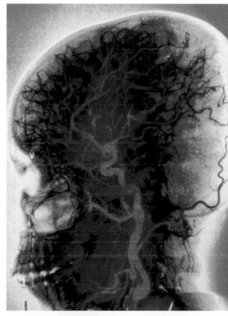

**(b) Cerebral angiogram**

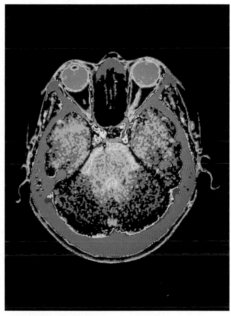

**(c) Computed tomographic (CT) scan**

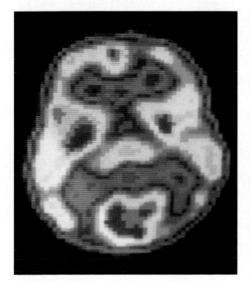

**(d) Positron emission tomographic (PET) scan**

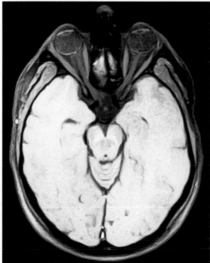

**(e) Magnetic resonance image (MRI)**

**FIGURE 1.13** Radiologic Images of the Head. (a) X ray (radiograph) showing the bones and teeth. (b) An angiogram of the cerebral blood vessels. The arteries are enhanced with false color. (c) A CT scan. The eyes and skin are shown in blue, bone in pink, and the brain in green. (d) A PET scan of the brain of an unmedicated schizophrenic patient. Red areas indicate regions of high metabolic rate. In this patient, the visual center of the brain at the rear of the head (bottom of photo) was especially active during the scan. (e) An MRI scan at the level of the eyes. The optic nerves appear in red and the muscles that move the eyes appear in green.

by enema for examination of the large intestine. Other substances are given by injection for *angiography,* the examination of blood vessels (fig. 1.13b). Some disadvantages of radiography are that images of overlapping organs can be confusing and slight differences in tissue density are not easily detected. Nevertheless, radiography still accounts for over half of all clinical imaging. Until the 1960s, it was the only method widely available.

## Computed Tomography (CT)

The *CT scan,* formerly called a computerized axial tomographic[19] (CAT) scan, is a more sophisticated application of X rays developed in 1972. The patient is moved through a ring-shaped machine that emits low-intensity X rays on one side and receives them with a detector on the opposite side. A computer analyzes signals from the detector and produces an image of a "slice" of the body about as thin as a coin (fig. 1.13c). The computer can "stack" a series of these images to construct a three-dimensional image of the body. CT scanning has the advantage of imaging thin sections of the body, so there is little overlap of organs and the image is much sharper than a conventional X ray. CT scanning is useful for identifying tumors, aneurysms, cerebral hemorrhages, kidney stones, and other abnormalities. It has virtually eliminated exploratory surgery.

## Positron Emission Tomography (PET)

The *PET scan,* developed in the 1970s, is used to assess the metabolic state of a tissue and to distinguish which tissues are most active at a given moment (fig. 1.13d). The procedure begins with an injection of radioactively labeled glucose, which emits positrons (electronlike particles with a positive charge). When a positron and electron meet, they annihilate each other and give off a pair of gamma rays that can be detected by sensors and analyzed by computer. The computer displays a color image that shows which tissues were using the most glucose at the moment. In cardiology, PET scans can show the extent of damaged heart tissue. Since it consumes little or no glucose, the damaged tissue appears dark. The PET scan is an example of *nuclear medicine—* the use of radioactive isotopes to treat disease or to form diagnostic images of the body.

## Magnetic Resonance Imaging (MRI)

*MRI* was developed in the 1970s as a technique superior to CT scanning for visualizing soft tissues. The patient lies within a cylindrical chamber surrounded by a large electromagnet that creates a magnetic field 3,000 to 60,000 times as strong as the earth's. Hydrogen atoms in the tissues align themselves with the magnetic field. The patient is then bombarded with radio waves, which cause the hydrogen atoms to absorb additional energy and align in a different direction. When the radio waves are turned off, the hydrogen atoms realign themselves to the magnetic field, giving off their excess energy at different rates that depend on the type of tissue. A computer analyzes the emitted energy to produce an image of the body. MRI can "see" clearly through the skull and spinal column to produce images of the

nervous tissue (fig. 1.13e). Moreover, it is better than CT for distinguishing between soft tissues such as the white and gray matter of the nervous system. MRI also eliminates exposure to harmful X rays. *Functional MRI (fMRI)* is a variation of this technique that visualizes moment-to-moment changes in tissue function. fMRI scans of the brain, for example, show shifting patterns of activity as the brain applies itself to a specific task. fMRI has lately replaced the PET scan as the most important method for visualizing brain function. The use of fMRI in brain imaging is further discussed in chapter 14.

## Sonography

*Sonography*[20] is the second oldest and second most widely used method of imaging. It is an outgrowth of sonar technology developed in World War II. A handheld device held firmly to the skin produces high-frequency ultrasound waves and receives the signals that echo back from internal organs. Although sonography was first used medically in the 1950s, images of significant clinical value had to wait until computer technology had developed enough to analyze differences in the way tissues reflect ultrasound. Sonography is not very useful for examining bones or lungs, but it is the method of choice in obstetrics, where the image *(sonogram)* can be used to locate the placenta and evaluate fetal age, position, and development. Sonography avoids the harmful effects of X rays, and the equipment is inexpensive and portable. Its primary disadvantage is that it does not produce a very sharp image (fig. 1.14).

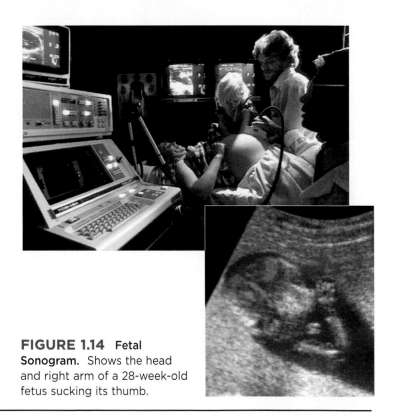

**FIGURE 1.14** Fetal **Sonogram.** Shows the head and right arm of a 28-week-old fetus sucking its thumb.

---

[19]*tomo* = section, cut, slice + *graphic* = pertaining to a recording

[20]*sono* = sound + *graphy* = recording process

## CHAPTER REVIEW

# Review of Key Concepts

**The Scope of Anatomy and Physiology (p. 2)**

1. Human *anatomy,* or structure, is studied at gross and microscopic (histological) levels.

2. The methods of anatomy include dissection, palpation, and imaging techniques such as X rays, sonography, and CT, PET, and MRI scans.

3. Human *physiology,* or function, is studied by experimental methods, and often by comparison with other species.

**The Origins of Biomedical Science (p. 3)**

1. Hippocrates and Aristotle first put medicine on a scientific basis by distinguishing natural causes from the supernatural.

2. Galen wrote the first notable medical textbook, which dominated western medicine for 1,500 years.

3. In the Middle Ages, medical science was preserved and advanced primarily in Muslim and Jewish culture, by such physicians as Avicenna and Maimonides.

4. In the sixteenth century, revolutionary work in anatomy by Vesalius and in physiology by Harvey created a foundation for modern medicine.

5. Improvements in the microscope by Hooke, Leeuwenhoek, and later Zeiss and Abbe opened the door to understanding anatomy, physiology, and disease at a cellular level.

**Scientific Method (p. 7)**

1. Philosophers Bacon and Descartes first established a systematic scientific way of thought in the seventeenth century.

2. The *inductive method,* common in anatomy, consists of generalizing about nature from numerous observations.

3. The *hypothetico-deductive method,* common in physiology, consists of formulating hypotheses and testing them by carefully crafted observational strategies.

4. The objectivity of medical science depends on experimental designs that include an adequate sample size, experimental controls such as placebos and the double-blind method, statistical analysis of the significance of the data, and peer review by other experts.

5. Science generates facts, laws, and theories. Theories are summations of our present knowledge of a natural phenomena and are the basis of much of our study in anatomy and physiology.

**Human Origins and Adaptations (p. 9)**

1. Human form and function have been shaped by millions of years of *natural selection.*

2. Many aspects of anatomy and physiology today, such as stereoscopic vision and opposable thumbs, are *adaptations* to the environments in which our prehistoric ancestors lived, including the arboreal and grassland habitats of Africa.

3. *Evolutionary medicine* is the analysis of human form, function, and disease in light of the evolutionary history of the human body.

**Human Structure (p. 13)**

1. Human structure is organized around a hierarchy of complexity. Levels of human complexity from most complex to simplest are *organism, organ systems, organs, tissues, cells, organelles, molecules,* and *atoms.*

2. Introductory textbooks teach only the most common human structure, but there are many variations in both internal and external anatomy.

**Human Function (p. 15)**

1. Life can be defined only as a collection of properties including *organization, cellular composition, metabolism, excretion, responsiveness, movement, homeostasis, development, reproduction,* and *evolution.*

2. For clinical purposes, life and legal death are differentiated on the basis of brain waves, reflexes, respiration, and heartbeat.

3. Humans vary greatly in their physiology. Data given in introductory and general textbooks are typically based on a young adult *reference male* and *reference female.*

4. An important unifying theory in physiology is *homeostasis,* mechanisms of maintaining internal constancy in spite of environmental change. Homeostasis keeps such variables as blood pressure and body temperature within a narrow range of an average called the *set point.*

5. Homeostasis is maintained by self-correcting chain reactions called *negative feedback.* This often involves detection of a change by a *receptor,* processing of this information by an *integrating center,* and reversal of the change by an *effector.*

6. *Positive feedback* is a self-amplifying chain of events that tends to produce rapid change in the body. It can be valuable in such cases as childbirth and blood clotting, but is often a cause of dysfunction and death.

**The Language of Medicine (p. 20)**

1. Anatomists the world over adhere to a lexicon of standard international terms called the *Terminologia Anatomica* (TA). Anatomy students must learn many Latin or English TA terms.

2. Biomedical terms can usually be simplified by breaking them down into familiar roots, prefixes, and suffixes. The habit of analyzing words in this way can greatly ease the difficulty of learning biomedical vocabulary, and is aided by footnotes throughout this book.

3. Precision in medical language is highly important. What may seem to be trivial spelling errors can radically change the meaning of a word, potentially causing dangerous medical errors.

# Testing Your Recall

1. Structure that can be observed with the naked eye is called
   a. gross anatomy.
   b. ultrastructure.
   c. microscopic anatomy.
   d. macroscopic anatomy.
   e. cytology.

2. The word root *homeo-* means
   a. tissue.
   b. metabolism.
   c. change.
   d. human.
   e. same.

3. The simplest structures considered to be alive are
   a. organisms.
   b. organs.
   c. tissues.
   d. cells.
   e. organelles.

4. Which of the following people revolutionized the teaching of gross anatomy?
   a. Vesalius
   b. Aristotle
   c. Hippocrates
   d. Leeuwenhoek
   e. Cannon

5. Which of the following embodies the greatest amount of scientific information?
   a. a fact
   b. a law of nature
   c. a theory
   d. a deduction
   e. a hypothesis

6. An informed, uncertain, but testable conjecture is
   a. a natural law.
   b. a scientific theory.
   c. a hypothesis.
   d. a deduction.
   e. a scientific fact.

7. A self-amplifying chain of physiological events is called
   a. positive feedback.
   b. negative feedback.
   c. dynamic constancy.
   d. homeostasis.
   e. metabolism.

8. Which of the following is *not* a human organ system?
   a. integumentary
   b. muscular
   c. epithelial
   d. nervous
   e. endocrine

9. _____ means studying anatomy by touch.
   a. Gross anatomy
   b. Auscultation
   c. Osculation
   d. Palpation
   e. Percussion

10. The prefix *hetero-* means
    a. same.
    b. different.
    c. both.
    d. solid.
    e. below.

11. Cutting and separating tissues to reveal structural relationships is called _____.

12. _____ invented many components of the compound microscope and named the cell.

13. By the process of _____, a scientist predicts what the result of a certain experiment will be if his or her hypothesis is correct.

14. Physiological effects of a person's mental state are called _____ effects.

15. The tendency of the body to maintain stable internal conditions is called _____.

16. Blood pH averages 7.4 but fluctuates from 7.35 to 7.45. A pH of 7.4 can therefore be considered the _____ for this variable.

17. Self-corrective mechanisms in physiology are called _____ loops.

18. A/an _____ is the simplest body structure to be composed of two or more types of tissue.

19. Depth perception, or the ability to form three-dimensional images, is also called _____ vision.

20. Our hands are said to be _____ because they can encircle an object such as a branch or tool. The presence of an _____ thumb is important to this ability.

*Answers in Appendix B*

# True or False

*Determine which five of the following statements are false, and briefly explain why.*

1. The technique for listening to the sounds of the heart valves is auscultation.

2. The inventions of Carl Zeiss and Ernst Abbe are necessary to the work of a modern histopathologist.

3. Abnormal skin color or dryness could be one piece of diagnostic information gained by auscultation.

4. There are more organelles than cells in the body.

5. The word *scuba,* derived from the words *self-contained underwater breathing apparatus,* is an acronym.

6. Leeuwenhoek was a biologist who invented the simple microscope in order to examine organisms in lake water.

7. A scientific theory is just a speculation until someone finds the evidence to prove it.

8. In a typical clinical research study, volunteer patients are in the treatment group and the physicians and scientists who run the study constitute the control group.

9. The great mobility of the primate shoulder joint is an adaptation to the arboreal habitat.

10. Negative feedback usually has a negative (harmful) effect on the body.

*Answers in Appendix B*

# Testing Your Comprehension

1. What aspect of William Harvey's view of blood circulation could be considered a scientific hypothesis? What would you predict from that hypothesis? What observation could you carry out today to test this hypothesis?

2. Which of the characteristics of living things are possessed by an automobile? What bearing does this have on our definition of life?

3. About 1 out of every 120 live-born infants has a structural defect in the heart such as a hole between two heart chambers. Such infants often suffer pulmonary congestion and heart failure, and about one-third of them die as a result. Which of the major themes in this chapter does this illustrate? Explain your answer.

4. How might human anatomy be different today if the forerunners of humans had never inhabited the forest canopy?

5. Suppose you have been doing heavy yard work on a hot day and sweating profusely. You become very thirsty, so you drink a tall glass of lemonade. Explain how your thirst relates to the concept of homeostasis. Which type of feedback—positive or negative—does this illustrate?

Answers at www.mhhe.com/saladin4

# www.mhhe.com/saladin4

*The textbook website provides a wealth of interactive study materials fully organized and integrated by chapter. You will find practice quizzes, labeling exercises, and much more that will complement your learning and understanding of anatomy and physiology. The website also includes tools designed to enhance your* **Anatomy & Physiology | REVEALED** *experience.*

# Atlas A

# GENERAL ORIENTATION TO HUMAN ANATOMY

Anatomy & Physiology | REVEALED®

**Volume 1** Skeletal and Muscular Systems

# Anatomical Position

**Anatomical position** is a stance in which a person stands erect with the feet flat on the floor and close together, arms at the sides, and the palms, face, and eyes facing forward (fig. A.1). This position provides a precise and standard frame of reference for anatomical description and dissection. Without such a frame of reference, to say that a structure such as the sternum, thymus, or aorta is "above the heart" would be vague, since it would depend on whether the subject was standing, lying face down, or lying face up. From the perspective of anatomical position, however, we can describe the thymus as *superior* to the heart, the sternum as *anterior* or *ventral* to the heart, and the aorta as *posterior* or *dorsal* to it. These descriptions remain valid regardless of the subject's position.

Unless stated otherwise, assume that all anatomical descriptions refer to anatomical position. Bear in mind that if a subject is facing you, the subject's left will be on your right and vice versa. In most anatomical illustrations, for example, the left atrium of the heart appears toward the right side of the page, and although the appendix is located in the right lower quadrant of the abdomen, it appears on the left side of most illustrations.

The forearm is said to be **supine** when the palms face up or forward and **prone** when they face down or rearward (fig. A.2). The difference is particularly important to descriptions of anatomy of this region. In the supine position, the two forearm bones (radius and ulna) are parallel and the radius is lateral to the ulna. In the prone position, the radius and ulna cross; the radius is lateral to the ulna at the elbow but medial to it at the wrist. Descriptions of nerves, muscles, blood vessels, and other structures of the arm assume that the arm is supine. (*Supine* also means lying face up and *prone* also means lying face down.)

**FIGURE A.1  Anatomical Position.** The feet are flat on the floor and close together, the arms are held downward and supine, and the face is directed forward.

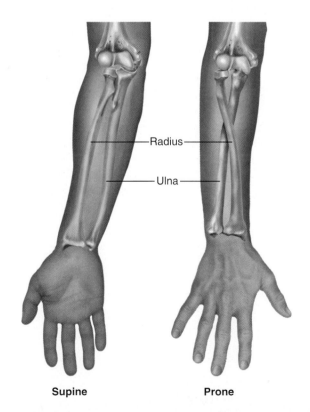

Supine                    Prone

**FIGURE A.2  Positions of the Forearm.** When the forearm is supine, the palm faces forward; when prone, it faces rearward. Note the differences in the relationship of the radius to the ulna.

# Anatomical Planes

Many views of the body are based on real or imaginary "slices" called *sections* or *planes.* "Section" implies an actual cut or slice to reveal internal anatomy, whereas "plane" implies an imaginary flat surface passing through the body. The three major anatomical planes are *sagittal, frontal,* and *transverse* (fig. A.3).

A **sagittal**[1] (SADJ-ih-tul) **plane** passes vertically through the body or an organ and divides it into right and left portions. The sagittal plane that divides the body or organ into equal halves is also called the **median (midsagittal) plane.** The head and pelvic organs are commonly illustrated on the median plane (fig. A.4a).

A **frontal (coronal) plane** also extends vertically, but it is perpendicular to the sagittal plane and divides the body into anterior (front) and posterior (back) portions. A frontal section of the head, for example, would divide it into one portion bearing the face and another bearing the back of the

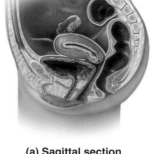

**(a) Sagittal section**

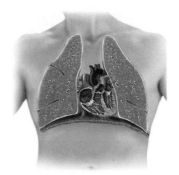

**(b) Frontal section**

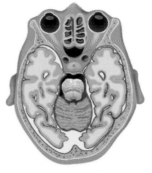

**(c) Transverse section**

**FIGURE A.4   Views of the Body in the Three Primary Anatomical Planes.**   (a) Sagittal section of the pelvic region. (b) Frontal section of the thoracic region. (c) Transverse section of the head at the level of the eyes.

head. Contents of the thoracic and abdominal cavities are most commonly shown in the frontal section (fig. A.4b).

A **transverse (horizontal) plane** passes across the body or an organ perpendicular to its long axis (fig. A.4c); therefore, it divides the body or organ into superior (upper) and inferior (lower) portions. CT scans are typically transverse sections (see fig. 1.13c, page 23).

# Directional Terms

Table A.1 summarizes frequently used terms that describe the position of one structure relative to another. Intermediate directions are often indicated by combinations of these terms. For example, one structure may be described as *dorsolateral* to another (toward the back and side).

Because of the bipedal, upright stance of humans, some directional terms have different meanings for humans than they do for other animals. *Anterior,* for example, denotes the region of the body that leads the way in normal locomotion. For a four-legged animal such as a cat, this is the head end of the body; for a human, however, it is the area of the chest and abdomen. Thus, *anterior* has the same meaning as *ventral* for a human but

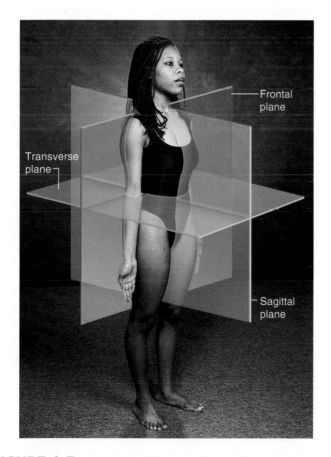

**FIGURE A.3   Anatomical Planes of Reference.**

▶ *What is another name for the particular sagittal plane shown here?*

Frontal plane

Transverse plane

Sagittal plane

---

[1]*sagitta* = arrow

| TABLE A.1 | Directional Terms in Human Anatomy | |
|---|---|---|
| **Term** | **Meaning** | **Examples of Usage** |
| Ventral | Toward the front* or belly | The aorta is *ventral* to the vertebral column. |
| Dorsal | Toward the back or spine | The vertebral column is *dorsal* to the aorta. |
| Anterior | Toward the ventral side* | The sternum is *anterior* to the heart. |
| Posterior | Toward the dorsal side* | The esophagus is *posterior* to the trachea. |
| Superior | Above | The heart is *superior* to the diaphragm. |
| Inferior | Below | The liver is *inferior* to the diaphragm. |
| Medial | Toward the median plane | The heart is *medial* to the lungs. |
| Lateral | Away from the median plane | The eyes are *lateral* to the nose. |
| Proximal | Closer to the point of attachment or origin | The elbow is *proximal* to the wrist. |
| Distal | Farther from the point of attachment or origin | The fingernails are at the *distal* ends of the fingers. |
| Superficial | Closer to the body surface | The skin is *superficial* to the muscles. |
| Deep | Farther from the body surface | The bones are *deep* to the muscles. |

*In humans only; definition differs for other animals.

not for a cat. *Posterior* denotes the region of the body that comes last in normal locomotion—the tail end of a cat but the dorsal side (back) of a human. These differences must be kept in mind when dissecting other animals for comparison to human anatomy. On the hands and feet, the dorsal surface is the one that bears the nails.

# Body Regions

Knowledge of the external anatomy and landmarks of the body is important in performing a physical examination and many other clinical procedures. For purposes of study, the body is divided into two major regions called the *axial* and *appendicular* regions. Smaller areas within the major regions are described in the following paragraphs and illustrated in figure A.5.

## AXIAL REGION

The **axial region** consists of the **head, neck** *(cervical[2] region),* and **trunk.** The trunk is further divided into the **thoracic region** above the diaphragm and the **abdominal region** below it.

One way of referring to the locations of abdominal structures is to divide the region into quadrants. Two perpendicular lines intersecting at the umbilicus (navel) divide the abdomen into a **right upper quadrant (RUQ), right lower quadrant (RLQ), left upper quadrant (LUQ),** and **left lower quadrant (LLQ)** (fig. A.6a, b). The quadrant scheme is often used to describe the site of an abdominal pain or abnormality.

The abdomen also can be divided into nine regions defined by four lines that intersect like a tic-tac-toe grid (fig. A.6c, d). Each vertical line is called a **midclavicular line** because it passes through the midpoint of the clavicle (collarbone). The superior horizontal line is called the **subcostal[3] line** because it connects the inferior borders of the lowest costal cartilages (cartilage connecting the tenth rib on each side to the inferior end of the sternum). The inferior horizontal line is called the **intertubercular[4] line** because it passes from left to right between the tubercles *(anterior superior spines)* of the pelvis—two points of bone located about where the front pockets open on most pants. The three lateral regions of this grid, from upper to lower, are the **hypochondriac,[5] lateral (lumbar),** and **inguinal[6] (iliac) regions.** The three medial regions from upper to lower are the **epigastric,[7] umbilical,** and **hypogastric (pubic)** regions.

## APPENDICULAR REGION

The **appendicular** (AP-en-DIC-you-lur) **region** of the body consists of the appendages (also called *limbs* or *extremities*): the **upper limbs** and the **lower limbs.** The upper limb includes the **brachium** (BRAY-kee-um) (arm), **antebrachium[8]** (AN-teh-BRAY-kee-um) (forearm), **carpus** (wrist), **manus** (hand), and **digits** (fingers). The lower limb includes the **thigh, crus** (leg), **tarsus** (ankle), **pes** (foot), and **digits** (toes).

In strict anatomical terms, "arm" refers only to that part of the upper limb between the shoulder and elbow. "Leg" refers only to that part of the lower limb between the knee and ankle.

---

[2]*cervic* = neck

[3]*sub* = below + *cost* = rib
[4]*inter* = between + *tubercul* = little swelling
[5]*hypo* = below + *chondr* = cartilage
[6]*inguin* = groin
[7]*epi* = above, over + *gastr* = stomach
[8]*ante* = fore, before + *brachi* = arm

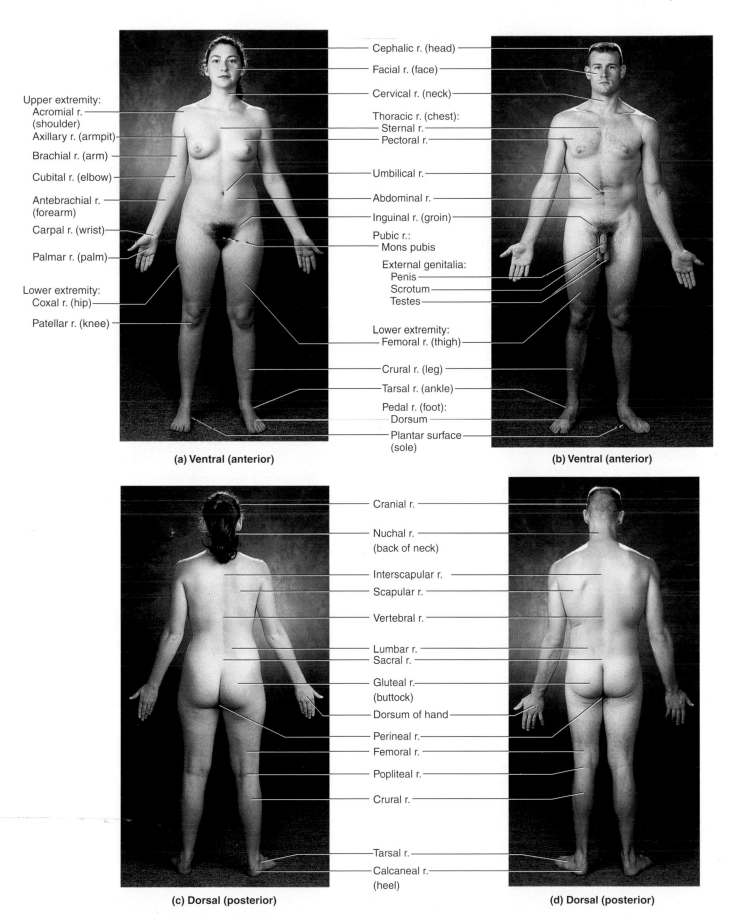

Upper extremity:
Acromial r. (shoulder)
Axillary r. (armpit)
Brachial r. (arm)
Cubital r. (elbow)
Antebrachial r. (forearm)
Carpal r. (wrist)
Palmar r. (palm)

Lower extremity:
Coxal r. (hip)
Patellar r. (knee)

Cephalic r. (head)
Facial r. (face)
Cervical r. (neck)
Thoracic r. (chest):
Sternal r.
Pectoral r.
Umbilical r.
Abdominal r.
Inguinal r. (groin)
Pubic r.:
Mons pubis
External genitalia:
Penis
Scrotum
Testes
Lower extremity:
Femoral r. (thigh)
Crural r. (leg)
Tarsal r. (ankle)
Pedal r. (foot):
Dorsum
Plantar surface (sole)

**(a) Ventral (anterior)**

**(b) Ventral (anterior)**

Cranial r.
Nuchal r. (back of neck)
Interscapular r.
Scapular r.
Vertebral r.
Lumbar r.
Sacral r.
Gluteal r. (buttock)
Dorsum of hand
Perineal r.
Femoral r.
Popliteal r.
Crural r.
Tarsal r.
Calcaneal r. (heel)

**(c) Dorsal (posterior)**

**(d) Dorsal (posterior)**

**FIGURE A.5**  The Adult Female and Male Bodies.

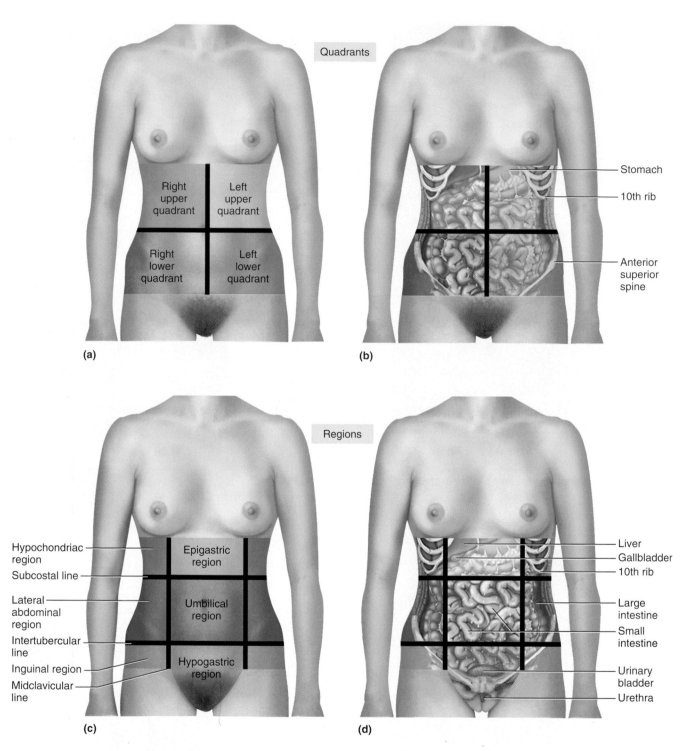

Quadrants

Right upper quadrant

Left upper quadrant

Right lower quadrant

Left lower quadrant

(a)

Stomach

10th rib

Anterior superior spine

(b)

Regions

Hypochondriac region

Subcostal line

Lateral abdominal region

Intertubercular line

Inguinal region

Midclavicular line

Epigastric region

Umbilical region

Hypogastric region

(c)

Liver

Gallbladder

10th rib

Large intestine

Small intestine

Urinary bladder

Urethra

(d)

**FIGURE A.6**   **Four Quadrants and Nine Regions of the Abdomen.**   (a) External division into four quadrants. (b) Internal anatomy correlated with the four quadrants. (c) External division into nine regions. (d) Internal anatomy correlated with the nine regions.

# Body Cavities and Membranes

The body is internally divided into two major **body cavities,** dorsal and ventral (fig. A.7). The organs within them are called the **viscera** (VISS-er-uh) (singular, *viscus*[9]). Various membranes line the cavities, cover the viscera, and hold the viscera in place (table A.2).

## DORSAL BODY CAVITY

The **dorsal body cavity** has two subdivisions: (1) the **cranial** (CRAY-nee-ul) **cavity,** which is enclosed by the cranium (braincase) and contains the brain, and (2) the **vertebral canal,** which is enclosed by the vertebral column (backbone) and contains the spinal cord. The dorsal body cavity is lined by three membrane layers called the **meninges** (meh-NIN-jeez). Among other functions, the meninges protect the delicate nervous tissue from the hard protective bone that encloses it.

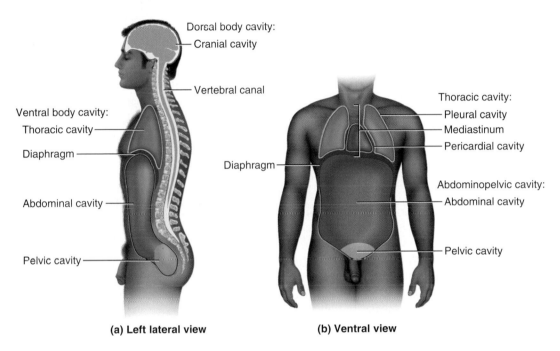

**(a) Left lateral view**    **(b) Ventral view**

**FIGURE A.7**  The Major Body Cavities.

| TABLE A.2 | Body Cavities and Membranes | |
|---|---|---|
| **Name of Cavity** | **Associated Viscera** | **Membranous Lining** |
| *Dorsal Body Cavity* | | |
| Cranial cavity | Brain | Meninges |
| Vertebral canal | Spinal cord | Meninges |
| *Ventral Body Cavity* | | |
| Thoracic cavity | | |
| Pleural cavities (2) | Lungs | Pleurae |
| Pericardial cavity | Heart | Pericardium |
| Abdominopelvic cavity | | |
| Abdominal cavity | Digestive organs, spleen, kidneys | Peritoneum |
| Pelvic cavity | Bladder, rectum, reproductive organs | Peritoneum |

[9]*viscus* = body organ

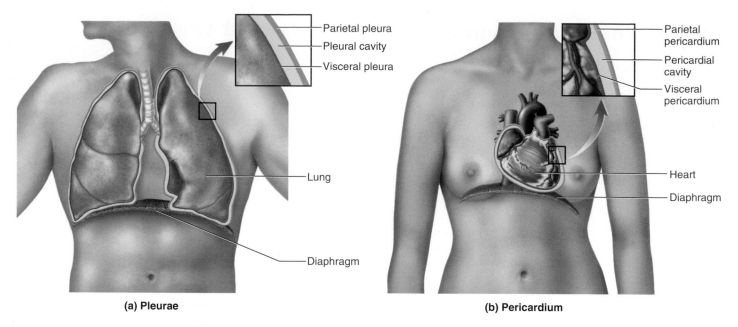

Parietal pleura
Pleural cavity
Visceral pleura

Parietal pericardium
Pericardial cavity
Visceral pericardium

Lung

Heart
Diaphragm

Diaphragm

**(a) Pleurae**          **(b) Pericardium**

**FIGURE A.8**   Parietal and Visceral Layers of Double-Walled Membranes.

# VENTRAL BODY CAVITY

During embryonic development, a space called the **coelom** (SEE-loam) forms within the trunk and eventually gives rise to the **ventral body cavity.** This cavity later becomes partitioned by a muscular sheet, the **diaphragm,** into a superior **thoracic cavity** and an inferior **abdominopelvic cavity.** The thoracic and abdominopelvic cavities are lined with thin **serous membranes.** These membranes secrete a lubricating film of moisture similar to blood serum (hence the name *serous*).

## Thoracic Cavity

The thoracic cavity is divided into right, left, and median portions by a partition called the **mediastinum**[10] (ME-dee-ass-TY-num) (fig. A.7). The right and left sides contain the lungs and are lined by a two-layered membrane called the **pleura**[11] (PLOOR-uh) (fig. A.8a). The outer layer, or **parietal**[12] (pa-RY-eh-tul) **pleura,** lies against the inside of the rib cage; the inner layer, or **visceral** (VISS-er-ul) **pleura,** forms the external surface of the lung. The narrow, moist space between the visceral and parietal pleurae is called the **pleural cavity** (see fig. A.19). It is lubricated by a slippery **pleural fluid.**

The median portion, or mediastinum, is occupied by the esophagus and trachea, a gland called the thymus, and the heart and major blood vessels connected to it. The heart is enclosed by a two-layered membrane called the

pericardium.[13] The **visceral pericardium** forms the heart surface while the **parietal pericardium** is separated from it by a space called the **pericardial cavity** (fig. A.8b). This space is lubricated by **pericardial fluid.**

## Abdominopelvic Cavity

The abdominopelvic cavity consists of the **abdominal cavity** above the brim of the pelvis and the **pelvic cavity** below the brim (see fig. A.16). The abdominal cavity contains most of the digestive organs as well as the kidneys and ureters. The pelvic cavity is markedly narrower and its lower end tilts posteriorly (see fig. A.7a). It contains the distal part of the large intestine, the urinary bladder and urethra, and the reproductive organs.

The abdominopelvic cavity contains a moist serous membrane called the **peritoneum**[14] (PERR-ih-toe-NEE-um). The **parietal peritoneum** lines the walls of the cavity, while the **visceral peritoneum** covers the external surfaces of most digestive organs. The **peritoneal cavity** is the space between the parietal and visceral layers. It is lubricated by **peritoneal fluid.**

Some organs of the abdominal cavity lie between the peritoneum and dorsal body wall (outside of the peritoneal cavity), so they are said to have a **retroperitoneal**[15] position (fig. A.9). These include the kidneys, ureters, adrenal glands, most of the pancreas, and abdominal portions of two major blood vessels—the aorta and inferior vena cava (see fig. A.15).

[10]*mediastinum* = in the middle
[11]*pleur* = rib, side
[12]*pariet* = wall

[13]*peri* = around + *cardi* = heart
[14]*peri* = around + *tone* = stretched
[15]*retro* = behind

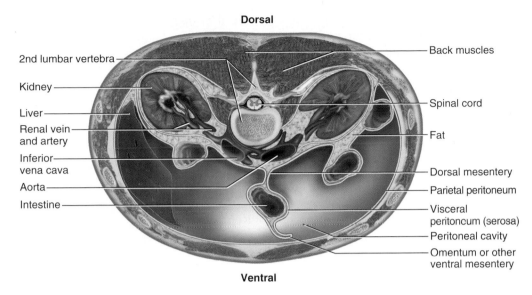

**Dorsal**

2nd lumbar vertebra

Kidney

Liver

Renal vein and artery

Inferior vena cava

Aorta

Intestine

Back muscles

Spinal cord

Fat

Dorsal mesentery

Parietal peritoneum

Visceral peritoneum (serosa)

Peritoneal cavity

Omentum or other ventral mesentery

**Ventral**

**FIGURE A.9  Transverse Section Through the Abdomen.**  Shows the peritoneum, peritoneal cavity (with most viscera omitted), and some retroperitoneal organs.

The intestines are suspended from the dorsal abdominal wall by a translucent membrane called the **mesentery**[16] (MESS-en-tare-ee), a continuation of the peritoneum. The membrane then wraps around the intestines and some other viscera, forming a membrane called the **serosa** (seer-OH-sa) on their outer surfaces (fig. A.10). The mesentery of the large intestine is called the **mesocolon.** The visceral peritoneum consists of the mesenteries and serosae.

A fatty membrane called the **greater omentum**[17] hangs like an apron from the inferolateral margin of the stomach and overlies the intestines (figs. A.10 and A.13). It is unattached at its inferior border and can be lifted to reveal the intestines. A smaller **lesser omentum** extends from the superomedial border of the stomach to the liver.

## POTENTIAL SPACES

In some places, the body has **potential spaces,** so named because under normal conditions, two membranes are pressed firmly together and there is no actual space between them. The membranes are not physically attached, however, and under unusual conditions, they may separate and create a space filled with fluid or other matter. Thus there is normally no actual space, but only a potential for membranes to separate and create one.

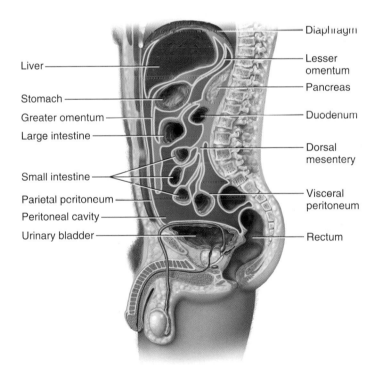

Diaphragm

Liver

Lesser omentum

Pancreas

Stomach

Greater omentum

Duodenum

Large intestine

Dorsal mesentery

Small intestine

Parietal peritoneum

Peritoneal cavity

Visceral peritoneum

Urinary bladder

Rectum

**FIGURE A.10  Serous Membranes of the Abdominal Cavity.** Sagittal section, left lateral view.

▶ *Is the urinary bladder in the peritoneal cavity?*

---

[16]*mes* = in the middle + *enter* = intestine
[17]*omentum* = covering

The pleural cavity is one example. Normally the parietal and visceral pleurae are pressed together without a gap between them, but under pathological conditions, air or serous fluid can accumulate between the membranes and open up a space. The internal cavity *(lumen)* of the uterus is another. In a nonpregnant uterus, the mucous membranes of opposite walls are pressed together so that there is no open space in the organ. In pregnancy, of course, a growing fetus occupies this space and pushes the mucous membranes apart.

# Organ Systems

The human body has 11 **organ systems** (fig. A.11) and an immune system, which is better described as a population of cells than as an organ system. These systems are classified in the following list by their principal functions, but this is an unavoidably flawed classification. Some organs belong to two or more systems—for example, the male urethra is part of both the urinary and reproductive systems; the pharynx is part of the respiratory and digestive systems; and the mammary glands can be considered part of the integumentary and female reproductive systems.

### Protection, Support, and Movement
Integumentary system
Skeletal system
Muscular system

### Internal Communication and Integration
Nervous system
Endocrine system

### Fluid Transport
Circulatory system
Lymphatic system

### Defense
Immune system

### Input and Output
Respiratory system
Urinary system
Digestive system

### Reproduction
Reproductive system

Some medical terms combine the names of two systems—for example, the *skeletomuscular (musculoskeletal) system, cardiopulmonary system,* and *urogenital (genitourinary) system.* These terms serve to call attention to the close anatomical or physiological relationships between two systems, but these are not literally individual organ systems.

# A Visual Survey of the Body

Figures A.12 through A.16 provide an overview of the anatomy of the trunk and internal organs of the thoracic and abdominopelvic cavities. Figures A.17 through A.22 are photographs of the cadaver showing the major organs of the dorsal and ventral body cavities.

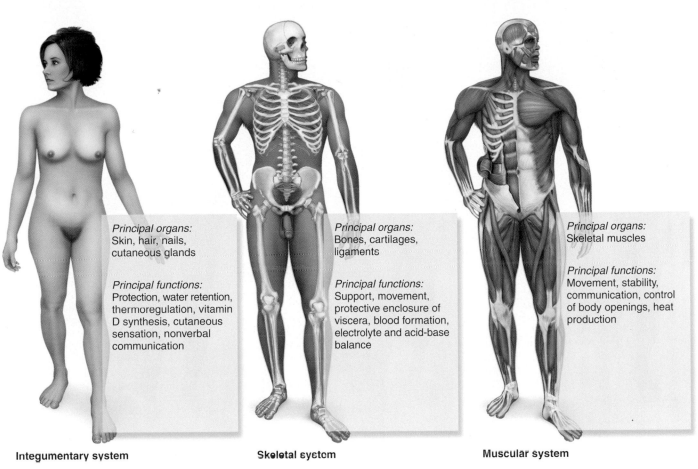

**Integumentary system**

Principal organs:
Skin, hair, nails, cutaneous glands

Principal functions:
Protection, water retention, thermoregulation, vitamin D synthesis, cutaneous sensation, nonverbal communication

**Skeletal system**

Principal organs:
Bones, cartilages, ligaments

Principal functions:
Support, movement, protective enclosure of viscera, blood formation, electrolyte and acid-base balance

**Muscular system**

Principal organs:
Skeletal muscles

Principal functions:
Movement, stability, communication, control of body openings, heat production

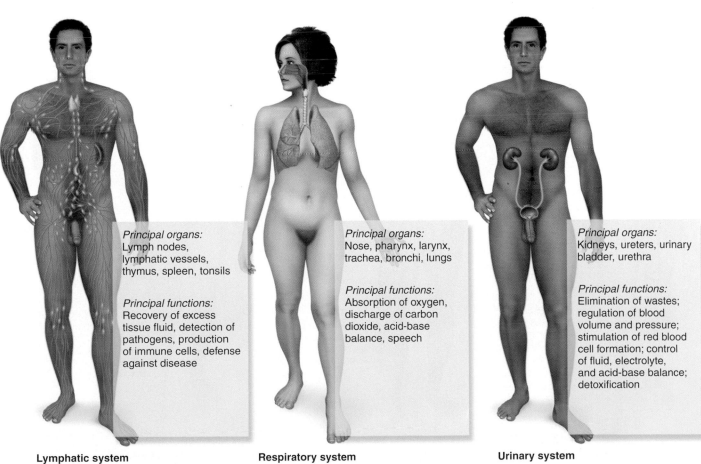

**Lymphatic system**

Principal organs:
Lymph nodes, lymphatic vessels, thymus, spleen, tonsils

Principal functions:
Recovery of excess tissue fluid, detection of pathogens, production of immune cells, defense against disease

**Respiratory system**

Principal organs:
Nose, pharynx, larynx, trachea, bronchi, lungs

Principal functions:
Absorption of oxygen, discharge of carbon dioxide, acid-base balance, speech

**Urinary system**

Principal organs:
Kidneys, ureters, urinary bladder, urethra

Principal functions:
Elimination of wastes; regulation of blood volume and pressure; stimulation of red blood cell formation; control of fluid, electrolyte, and acid-base balance; detoxification

**FIGURE A.11** The Human Organ Systems.

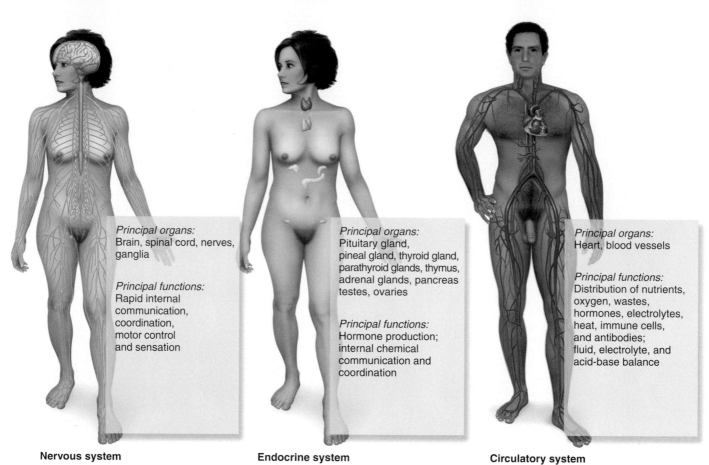

**Nervous system**

*Principal organs:*
Brain, spinal cord, nerves, ganglia

*Principal functions:*
Rapid internal communication, coordination, motor control and sensation

**Endocrine system**

*Principal organs:*
Pituitary gland, pineal gland, thyroid gland, parathyroid glands, thymus, adrenal glands, pancreas testes, ovaries

*Principal functions:*
Hormone production; internal chemical communication and coordination

**Circulatory system**

*Principal organs:*
Heart, blood vessels

*Principal functions:*
Distribution of nutrients, oxygen, wastes, hormones, electrolytes, heat, immune cells, and antibodies; fluid, electrolyte, and acid-base balance

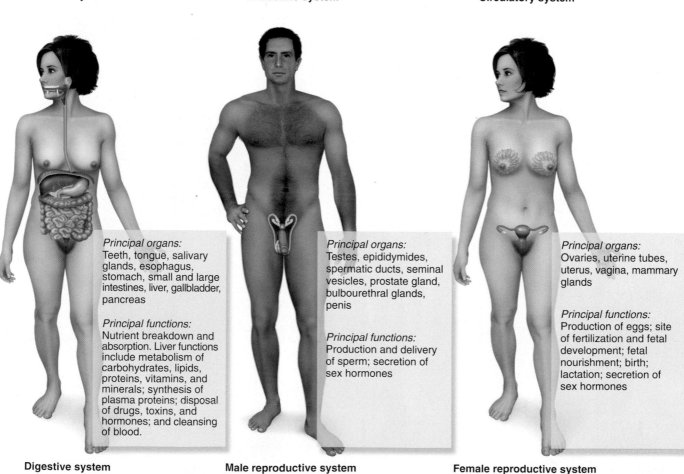

**Digestive system**

*Principal organs:*
Teeth, tongue, salivary glands, esophagus, stomach, small and large intestines, liver, gallbladder, pancreas

*Principal functions:*
Nutrient breakdown and absorption. Liver functions include metabolism of carbohydrates, lipids, proteins, vitamins, and minerals; synthesis of plasma proteins; disposal of drugs, toxins, and hormones; and cleansing of blood.

**Male reproductive system**

*Principal organs:*
Testes, epididymides, spermatic ducts, seminal vesicles, prostate gland, bulbourethral glands, penis

*Principal functions:*
Production and delivery of sperm; secretion of sex hormones

**Female reproductive system**

*Principal organs:*
Ovaries, uterine tubes, uterus, vagina, mammary glands

*Principal functions:*
Production of eggs; site of fertilization and fetal development; fetal nourishment; birth; lactation; secretion of sex hormones

**FIGURE A.11**  The Human Organ Systems *(continued)*.

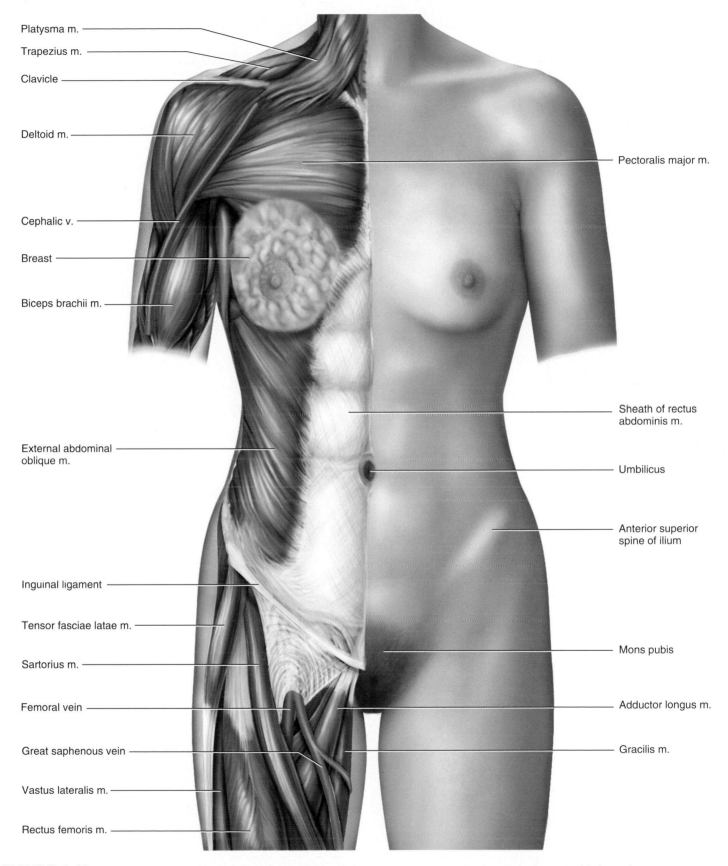

Platysma m.

Trapezius m.

Clavicle

Deltoid m.

Cephalic v.

Breast

Biceps brachii m.

External abdominal
oblique m.

Inguinal ligament

Tensor fasciae latae m.

Sartorius m.

Femoral vein

Great saphenous vein

Vastus lateralis m.

Rectus femoris m.

Pectoralis major m.

Sheath of rectus
abdominis m.

Umbilicus

Anterior superior
spine of ilium

Mons pubis

Adductor longus m.

Gracilis m.

**FIGURE A.12** **Superficial Anatomy of the Trunk (Female).** Surface anatomy is shown on the anatomical left, and structures immediately deep to the skin on the right (*m.* = muscle; *v.* = vein).

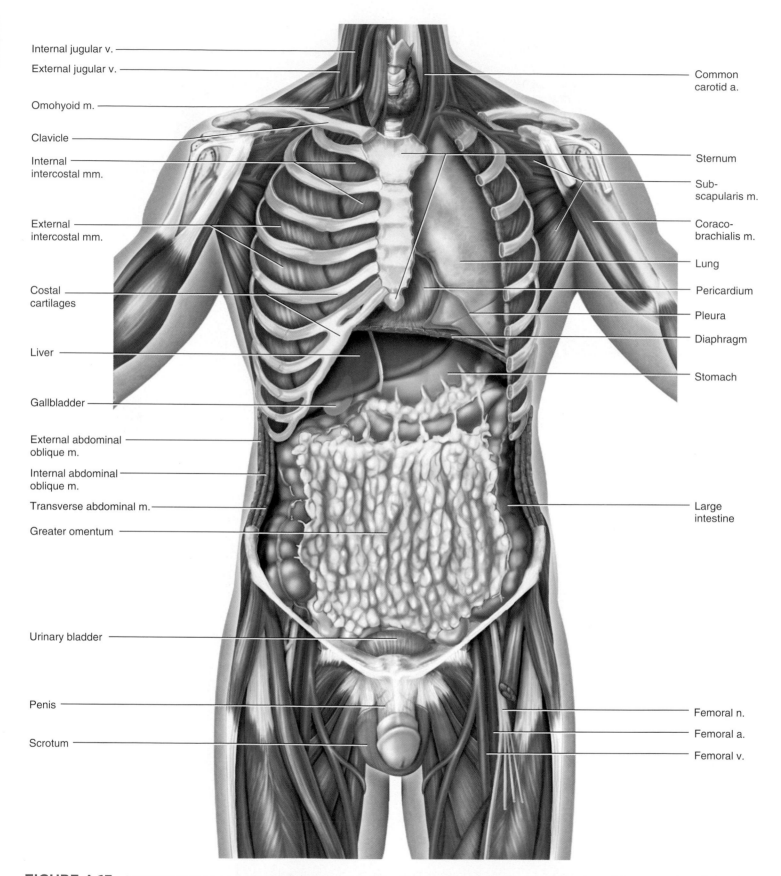

Internal jugular v.

External jugular v.

Omohyoid m.

Clavicle

Internal
intercostal mm.

External
intercostal mm.

Costal
cartilages

Liver

Gallbladder

External abdominal
oblique m.

Internal abdominal
oblique m.

Transverse abdominal m.

Greater omentum

Urinary bladder

Penis

Scrotum

Common
carotid a.

Sternum

Sub-
scapularis m.

Coraco-
brachialis m.

Lung

Pericardium

Pleura

Diaphragm

Stomach

Large
intestine

Femoral n.

Femoral a.

Femoral v.

**FIGURE A.13**   **Anatomy at the Level of the Rib Cage and Greater Omentum (Male).**   The ventral body wall is removed, and the ribs, intercostal muscles, and pleura are removed from the anatomical left (*a.* = artery; *v.* = vein; *m.* = muscle; *mm.* = muscles; *n.* = nerve).

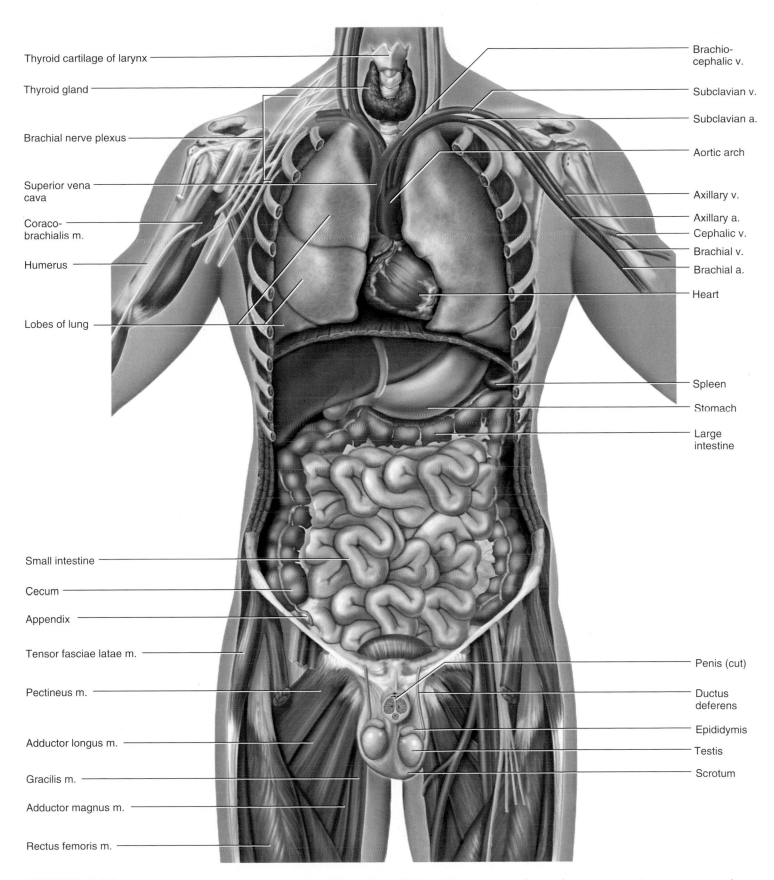

Thyroid cartilage of larynx

Thyroid gland

Brachial nerve plexus

Superior vena cava

Coraco-brachialis m.

Humerus

Lobes of lung

Small intestine

Cecum

Appendix

Tensor fasciae latae m.

Pectineus m.

Adductor longus m.

Gracilis m.

Adductor magnus m.

Rectus femoris m.

Brachio-cephalic v.

Subclavian v.

Subclavian a.

Aortic arch

Axillary v.

Axillary a.

Cephalic v.

Brachial v.

Brachial a.

Heart

Spleen

Stomach

Large intestine

Penis (cut)

Ductus deferens

Epididymis

Testis

Scrotum

**FIGURE A.14**  **Anatomy at the Level of the Lungs and Intestines (Male).**    The sternum, ribs, and greater omentum are removed (*a.* = artery; *v.* = vein; *m.* = muscle).

▶ *Name several viscera that are protected by the rib cage.*

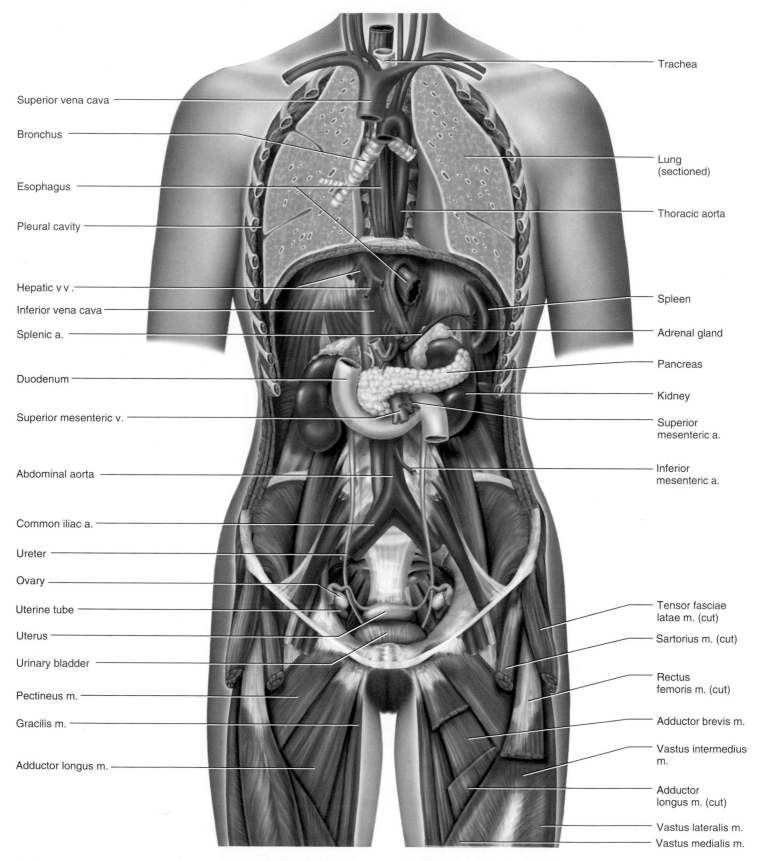

**FIGURE A.15**  **Anatomy at the Level of the Retroperitoneal Viscera (Female).**  The heart is removed, the lungs are frontally sectioned, and the viscera of the peritoneal cavity and the peritoneum itself are removed (*a.* = artery; *v.* = vein; *vv.* = veins; *m.* = muscle).

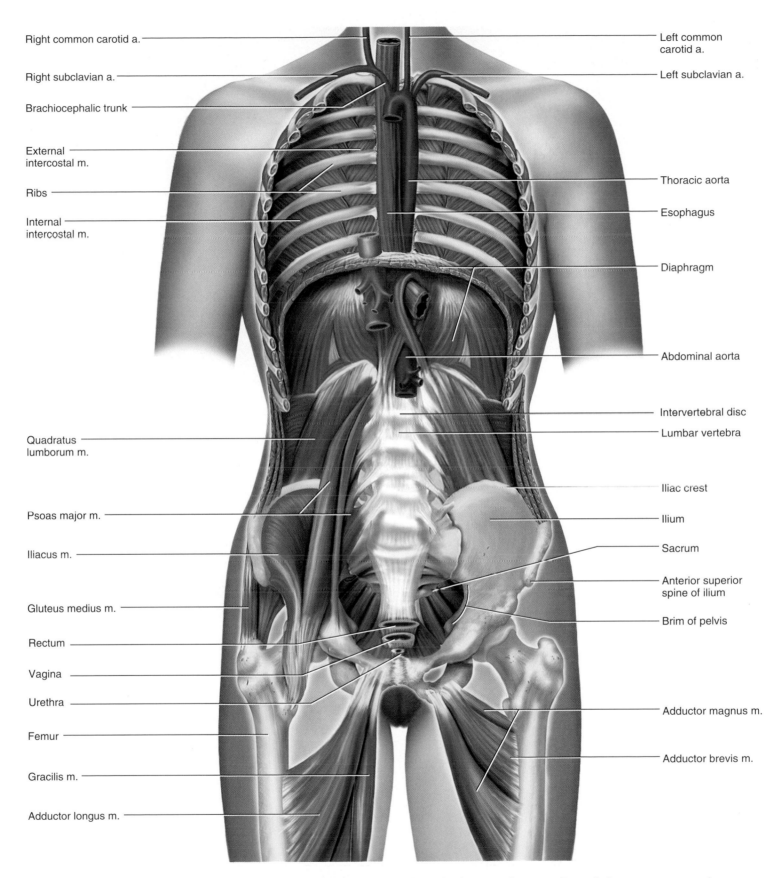

Right common carotid a.

Right subclavian a.

Brachiocephalic trunk

External intercostal m.

Ribs

Internal intercostal m.

Quadratus lumborum m.

Psoas major m.

Iliacus m.

Gluteus medius m.

Rectum

Vagina

Urethra

Femur

Gracilis m.

Adductor longus m.

Left common carotid a.

Left subclavian a.

Thoracic aorta

Esophagus

Diaphragm

Abdominal aorta

Intervertebral disc

Lumbar vertebra

Iliac crest

Ilium

Sacrum

Anterior superior spine of ilium

Brim of pelvis

Adductor magnus m.

Adductor brevis m.

**FIGURE A.16   Anatomy at the Level of the Dorsal Body Wall (Female).** The lungs and retroperitoneal viscera are removed (*a.* = artery; *m.* = muscle).

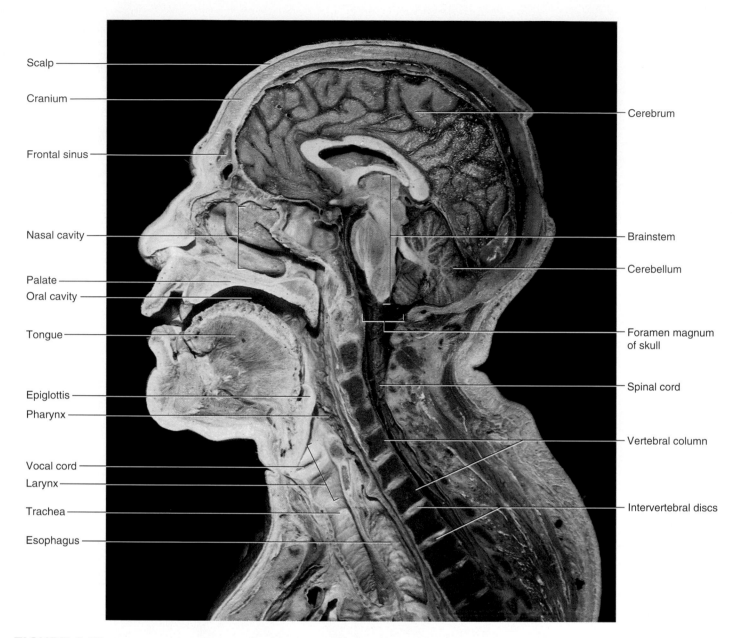

Scalp

Cranium

Frontal sinus

Nasal cavity

Palate

Oral cavity

Tongue

Epiglottis

Pharynx

Vocal cord

Larynx

Trachea

Esophagus

Cerebrum

Brainstem

Cerebellum

Foramen magnum of skull

Spinal cord

Vertebral column

Intervertebral discs

**FIGURE A.17**   **Median Section of the Head.**   Shows contents of the cranial, nasal, and buccal cavities.

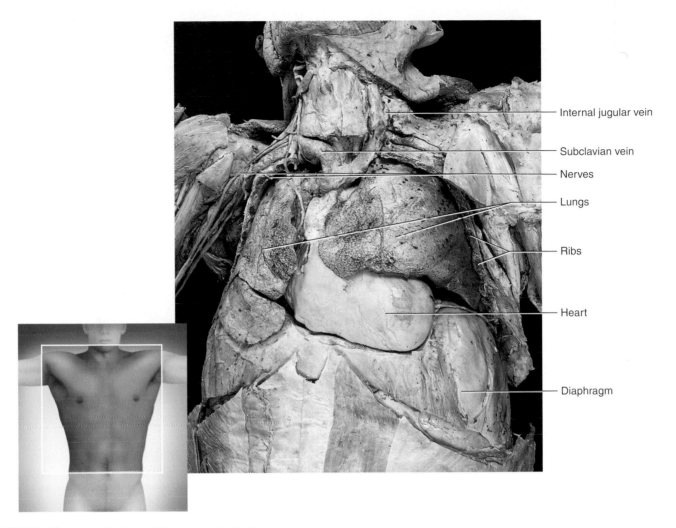

**FIGURE A.18**   Frontal View of the Thoracic Cavity.

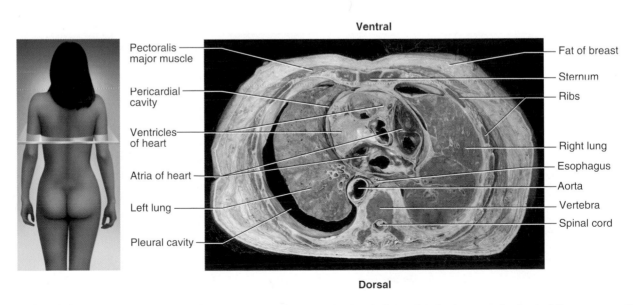

**FIGURE A.19**   **Transverse Section of the Thorax.**   Section taken at the level shown by the inset and oriented the same as the reader's body.

▶ *In this section, which term best describes the position of the aorta relative to the heart: posterior, lateral, inferior, or proximal?*

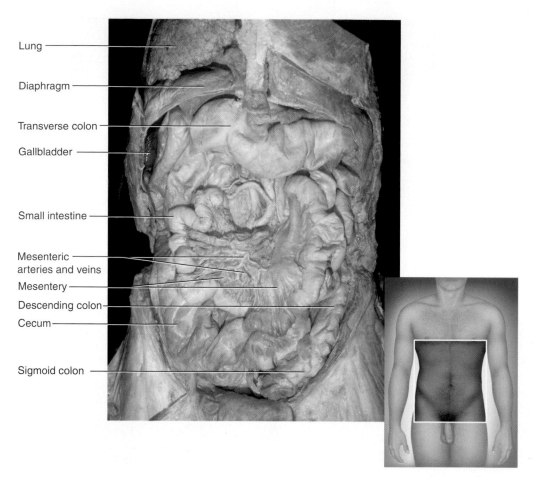

Lung

Diaphragm

Transverse colon

Gallbladder

Small intestine

Mesenteric
arteries and veins

Mesentery

Descending colon

Cecum

Sigmoid colon

**FIGURE A.20**   Frontal View of the Abdominal Cavity.

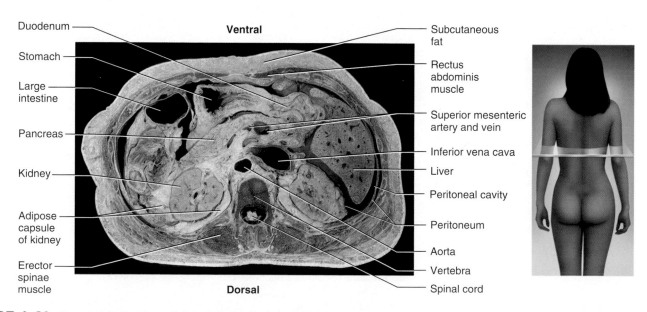

Duodenum

Stomach

Large
intestine

Pancreas

Kidney

Adipose
capsule
of kidney

Erector
spinae
muscle

**Ventral**

**Dorsal**

Subcutaneous
fat

Rectus
abdominis
muscle

Superior mesenteric
artery and vein

Inferior vena cava

Liver

Peritoneal cavity

Peritoneum

Aorta

Vertebra

Spinal cord

**FIGURE A.21**   **Transverse Section of the Abdomen.**   Section taken at the level shown by the inset and oriented the same as the reader's body.

▶ *What tissue in this photograph is immediately superficial to the rectus abdominis muscle?*

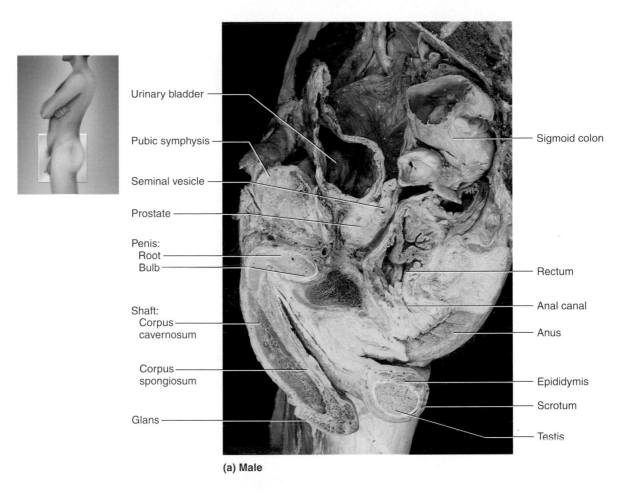

Urinary bladder

Pubic symphysis

Seminal vesicle

Prostate

Penis:
Root
Bulb

Shaft:
Corpus cavernosum

Corpus spongiosum

Glans

Sigmoid colon

Rectum

Anal canal

Anus

Epididymis

Scrotum

Testis

**(a) Male**

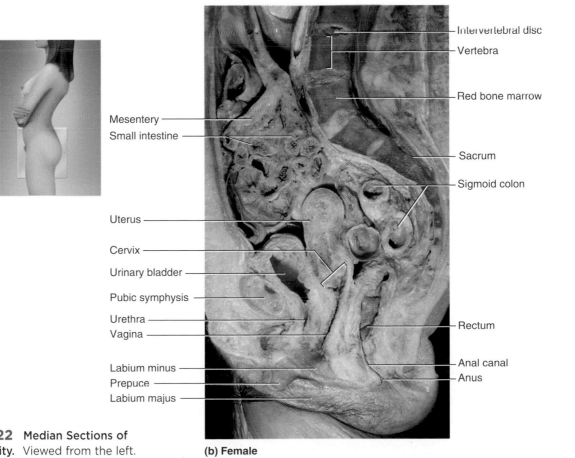

Mesentery

Small intestine

Uterus

Cervix

Urinary bladder

Pubic symphysis

Urethra

Vagina

Labium minus

Prepuce

Labium majus

Intervertebral disc

Vertebra

Red bone marrow

Sacrum

Sigmoid colon

Rectum

Anal canal

Anus

**FIGURE A.22** Median Sections of
the Pelvic Cavity. Viewed from the left.

**(b) Female**

## CHAPTER REVIEW

# Review of Key Concepts

**Anatomical Position (p. 30)**

1. Human anatomy is described with reference to a standard *anatomical position,* which avoids the ambiguity of terms that depend on the position of the body.

**Anatomical Planes (p. 31)**

1. Internal structure is often depicted along one of three mutually perpendicular planes through the body: the *sagittal, frontal,* and *transverse planes.*

**Directional Terms (p. 31)**

1. The position of one structure relative to another is often described by such pairs of terms as *superior–inferior, medial–lateral, proximal–distal,* and others (table A.1).

**Body Regions (p. 32)**

1. The body is divided into a central *axial region* (head, neck, trunk) and *appendicular region* (limbs).

2. The abdomen can be divided into either four quadrants or nine regions for describing the locations of struc-

tures, symptoms, or abnormal conditions (fig. A.6).

3. Each limb is divided into five regions from proximal to distal.

**Body Cavities and Membranes (p. 35)**

1. The body is internally divided into a *dorsal* and *ventral* body cavity. The organs in these cavities are called the *viscera.*

2. The body cavities are lined with serous membranes: the *meninges* around the brain and spinal cord, *pleurae* around the lungs, *pericardium* around the heart, and *peritoneum* in the abdominal cavity.

3. The last three of these membranes have outer and inner *parietal* and *visceral* layers, respectively, with lubricating fluid between the layers (*pleural, pericardial,* and *peritoneal fluid*).

4. *Retroperitoneal* organs such as the kidneys and pancreas lie between the peritoneum and body wall rather than within the peritoneal cavity.

5. The peritoneum continues as a *mesentery* that suspends the intestines and other organs from the dorsal body wall, a *serosa* over the surface of some abdominal organs, and two *omenta* attached to the stomach.

**Organ Systems (p. 38)**

1. The body has 11 organ systems: the *integumentary, skeletal,* and *muscular* systems for protection, support, and movement; the *nervous* and *endocrine* systems for internal communication; the *circulatory* and *lymphatic* systems for fluid transport; the *respiratory, urinary,* and *digestive* systems for input and output; and the *reproductive* system for producing offspring.

2. The body also has an immune system for protection from disease, but this is not an organ system; it is a collection of cells that populate all the organ systems.

# Testing Your Recall

1. Which of the following is *not* an essential part of anatomical position?
   a. eyes facing forward
   b. feet flat on the floor
   c. forearms supine
   d. mouth closed
   e. arms down to the sides

2. A ring-shaped section of the small intestine would be a _____ section.
   a. sagittal
   b. coronal
   c. transverse
   d. frontal
   e. median

3. The tarsal region is _____ to the popliteal region.
   a. medial
   b. superficial
   c. superior
   d. dorsal
   e. distal

4. The greater omentum is _____ to the small intestine.
   a. posterior
   b. parietal
   c. deep
   d. superficial
   e. proximal

5. A _____ line passes through the sternum, umbilicus, and mons pubis.
   a. central
   b. proximal
   c. midclavicular
   d. midsagittal
   e. intertubercular

6. The _____ region is immediately medial to the coxal region.
   a. inguinal
   b. hypochondriac
   c. umbilical
   d. popliteal
   e. cubital

7. Which of the following regions is not part of the upper limb?
   a. plantar
   b. carpal
   c. cubital
   d. brachial
   e. palmar

8. Which of these organs is within the peritoneal cavity?
   a. urinary bladder
   b. kidneys
   c. heart
   d. small intestine
   e. brain

9. In which area do you think pain from the gallbladder would be felt?
   a. umbilical region
   b. right upper quadrant
   c. hypogastric region
   d. left hypochondriac region
   e. left lower quadrant

10. Which organ system regulates blood volume, controls acid–base balance, and stimulates red blood cell production?
    a. digestive system
    b. lymphatic system
    c. nervous system
    d. urinary system
    e. circulatory system

11. The forearm is said to be _____ when the palms are facing forward.

12. The superficial layer of the pleura is called the _____ pleura.

13. The right and left pleural cavities are separated by a thick wall called the _____.

14. The back of the neck is the _____ region.

15. The manus is more commonly known as the _____ and the pes is more commonly known as the _____.

16. The dorsal body cavity is lined by membranes called the _____.

17. Organs that lie within the abdominal cavity but not within the peritoneal cavity are said to have a _____ position.

18. The sternal region is _____ to the pectoral region.

19. The pelvic cavity can be described as _____ to the abdominal cavity in position.

20. The anterior pit of the elbow is the _____ region, and the corresponding (but posterior) pit of the knee is the _____ fossa.

*Answers in Appendix B*

## True or False?

*Determine which five of the following statements are false, and briefly explain why.*

1. A single sagittal section of the body can pass through one lung but not through both.

2. It would be possible to see both eyes in one frontal section of the head.

3. The knee is both superior and proximal to the tarsal region.

4. The diaphragm is ventral to the lungs.

5. The esophagus is in the dorsal body cavity.

6. The liver is in the lateral abdominal region.

7. The heart is in the mediastinum.

8. Both kidneys could be shown in a single coronal section of the body.

9. The peritoneum lines the inside of the stomach and intestines.

10. The sigmoid colon is in the lower right quadrant of the abdomen.

*Answers in Appendix B*

## Testing Your Comprehension

1. Identify which anatomical plane—sagittal, frontal, or transverse—is the only one that could *not* show (a) both the brain and tongue, (b) both eyes, (c) both the hypogastric and gluteal regions, (d) both kidneys, (e) both the sternum and vertebral column, and (f) both the heart and uterus.

2. Laypeople often misunderstand anatomical terminology. What do you think people really mean when they say they have "planter's warts"?

3. Name one structure or anatomical feature that could be found in each of the following locations relative to the ribs: medial, lateral, superior, inferior, deep, superficial, posterior, and anterior. Try not to use the same example twice.

4. Based on the illustrations in this atlas, identify an internal organ that is (a) in the upper left quadrant and retroperitoneal, (b) in the lower right quadrant of the peritoneal cavity, (c) in the hypogastric region, (d) in the right hypochondriac region, and (e) in the pectoral region.

5. Why do you think people with imaginary illnesses came to be called hypochondriacs?

*Answers at www.mhhe.com/saladin4*

## www.mhhe.com/saladin4

*The textbook website provides a wealth of interactive study materials fully organized and integrated by chapter. You will find practice quizzes, labeling exercises, and much more that will complement your learning and understanding of anatomy and physiology. The website also includes tools designed to enhance your* **Anatomy & Physiology | REVEALED** *experience.*

*Cholesterol crystals seen through a
polarizing microscope*

# THE CHEMISTRY OF LIFE

## CHAPTER OUTLINE

## INSIGHTS

Why is too much sodium or cholesterol harmful? Why does an iron deficiency cause anemia and an iodine deficiency cause a goiter? Why does a pH imbalance make some drugs less effective? Why do some pregnant women suffer convulsions after several days of vomiting? How can radiation cause cancer as well as cure it?

None of these questions can be answered, nor would the rest of this book be intelligible, without understanding the chemistry of life. A little knowledge of chemistry can help you choose a healthy diet, use medications more wisely, avoid worthless health fads and frauds, and explain treatments and procedures to your patients or clients. Thus, we begin our study of the human body with basic chemistry, the simplest level of the body's structural organization.

We will progress from general chemistry to **biochemistry,** study of the molecules that compose living organisms—especially those unique to living things, such as carbohydrates, fats, proteins, and nucleic acids. Most people have at least heard of these—it is common knowledge that we need proteins, fats, carbohydrates, vitamins, and minerals in our diet, that we should avoid consuming too much saturated fat and cholesterol. But most people have only a vague concept of what these molecules are, much less how they function in the body. Such knowledge is very helpful in matters of personal fitness and patient education and is essential to the comprehension of the rest of this book.

# Atoms, Ions, and Molecules

### Objectives

When you have completed this section, you should be able to

- recognize elements of the human body from their chemical symbols;

- distinguish between chemical elements and compounds;

- state the functions of minerals in the body;

- explain the basis for radioactivity and the types and hazards of ionizing radiation;

- distinguish between ions, electrolytes, and free radicals; and

- define the types of chemical bonds.

## THE CHEMICAL ELEMENTS

A chemical **element** is the simplest form of matter to have unique chemical properties. Water, for example, has unique properties, but it can be broken down into two elements, hydrogen and oxygen, that have unique chemical properties of their own. If we carry this process any further, however, we find that hydrogen and oxygen are made of protons, neutrons, and electrons—and none of these are unique. A proton of gold is identical to a proton of oxygen. Hydrogen and oxygen are the simplest chemically unique components of water and are thus elements.

Each element is identified by an *atomic number,* the number of protons in its nucleus. The atomic number of carbon is 6 and that of oxygen is 8, for example. The periodic table of the elements (see appendix A) arranges the elements in order by their atomic numbers. The elements are represented by one- or two-letter symbols, usually based on their English names: C for carbon, Mg for magnesium, Cl for chlorine, and so forth. A few symbols are based on Latin names, such as K for potassium *(kalium),* Na for sodium *(natrium),* and Fe for iron *(ferrum).*

There are 91 naturally occurring elements on earth, 24 of which play normal physiological roles in humans. Table 2.1 groups these 24 according to their abundance in the body. Six of them account for 98.5% of the body's weight: oxygen, carbon, hydrogen, nitrogen, calcium, and phosphorus. The next 0.8% consists of another 6 elements: sulfur, potassium, sodium, chlorine, magnesium, and iron. The remaining 12 account for 0.7% of body weight, and no one of them accounts for more than 0.02%; thus they are known as **trace elements.** Despite their minute quantities, trace elements play vital roles in physiology. Other elements without natural physiological roles can contaminate the body and severely disrupt its functions, as in heavy-metal poisoning with lead or mercury.

Several of these elements are classified as **minerals**—inorganic elements that are extracted from the soil by plants and passed up the food chain to humans and other organisms. Minerals constitute about 4% of the human body by weight. Nearly three-quarters of this is Ca and P; the rest is mainly Cl, Mg, K, Na, and S. Minerals contribute significantly to body structure. The bones and teeth consist partly of crystals of calcium, phosphate, magnesium, fluoride, and sulfate ions. Many proteins include sulfur, and phosphorus is a major component of nucleic acids, ATP, and cell membranes. Minerals also enable enzymes and other organic molecules to function. Iodine is a component of thyroid hormone; iron is a component of hemoglobin; and some enzymes function only when manganese, zinc, copper, or other minerals are bound to them. The electrolytes needed for nerve and muscle function are mineral salts. The biological roles of minerals are discussed in more detail in chapters 24 and 26.

| TABLE 2.1 | Elements of the Human Body | |
|---|---|---|
| **Name** | **Symbol** | **Percentage of Body Weight** |
| *Major Elements (total 98.5%)* | | |
| Oxygen | O | 65.0 |
| Carbon | C | 18.0 |
| Hydrogen | H | 10.0 |
| Nitrogen | N | 3.0 |
| Calcium | Ca | 1.5 |
| Phosphorus | P | 1.0 |
| *Lesser Elements (total 0.8%)* | | |
| Sulfur | S | 0.25 |
| Potassium | K | 0.20 |
| Sodium | Na | 0.15 |
| Chlorine | Cl | 0.15 |
| Magnesium | Mg | 0.05 |
| Iron | Fe | 0.006 |

*Trace Elements (total 0.7%)*

| | | | |
|---|---|---|---|
| Chromium | Cr | Molybdenum | Mo |
| Cobalt | Co | Selenium | Se |
| Copper | Cu | Silicon | Si |
| Fluorine | F | Tin | Sn |
| Iodine | I | Vanadium | V |
| Manganese | Mn | Zinc | Zn |

## ATOMIC STRUCTURE

In the fifth century BCE, the Greek philosopher Democritus reasoned that we can cut matter such as a gold nugget into smaller and smaller pieces, but there must ultimately be particles so small that nothing could cut them. He called these imaginary particles atoms[1] ("indivisible"). Atoms were only a philosophical concept until 1803, when English chemist John Dalton began to develop an atomic theory based on experimental evidence. In 1913, Danish physicist Niels Bohr proposed a model of atomic structure similar to planets orbiting the sun (figs. 2.1 and 2.2). Although this *planetary model* is too simple to account for many of the properties of atoms, it remains useful for elementary purposes.

At the center of an atom is the *nucleus,* composed of protons and neutrons. **Protons** ($p^+$) have a single positive charge and **neutrons** ($n^0$) have no charge. Each proton or neutron weighs approximately 1 *atomic mass unit (amu),* defined as one-twelfth the mass of an atom of carbon-12. The *atomic mass* of an element is approximately equal to its total number of protons and neutrons.

Around the nucleus are one or more concentric clouds of **electrons** ($e^-$), tiny particles with a single negative charge and very low mass. It takes 1,836 electrons to equal 1 amu, so for most purposes we can disregard their mass. A person who weighs 64 kg (140 lb) contains less than 24 g (1 oz) of electrons. This hardly means that we can ignore electrons, however. They determine the chemical properties of an atom, thereby governing what molecules can exist and what chemical reactions can occur. The number of electrons equals the number of protons, so their charges cancel each other and an atom is electrically neutral.

[1] *a* = not + *tom* = cut

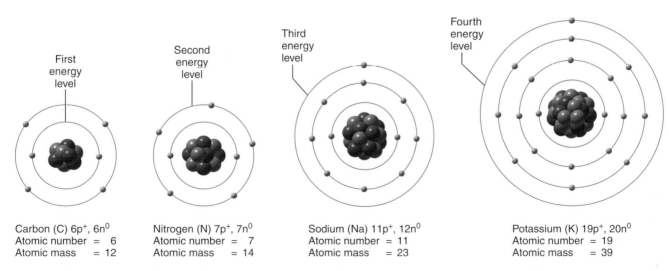

**FIGURE 2.1  Bohr Planetary Models of Four Representative Elements.**  Note the filling of electron shells as atomic number increases. ($p^+$ = protons; $n^0$ = neutrons)

▶ *Will potassium have a greater tendency to give up an electron or to take one away from another atom?*

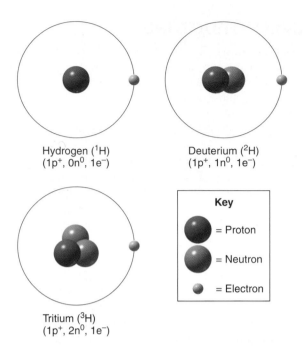

**FIGURE 2.2  Isotopes of Hydrogen.**  The three isotopes differ only in the number of neutrons present. ($p^+$ = protons; $n^0$ = neutrons; $e^-$ = electrons)

Electrons swarm about the nucleus in concentric regions called *electron shells (energy levels).* The more energy an electron has, the farther away from the nucleus its orbit lies. Each shell holds a limited number of electrons (see fig. 2.1). The one closest to the nucleus holds a maximum of 2 electrons, the second one holds a maximum of 8, and the third holds a maximum of 18. The outermost shell never holds more than 8 electrons, but a shell can acquire more electrons after another one, farther out, begins to fill. Thus, the third shell will hold 18 electrons only in atoms with four or more shells. The elements known to date have up to seven electron shells, but those ordinarily involved in human physiology do not exceed four.

The electrons of the outermost shell, called **valence electrons,** determine the chemical bonding properties of an atom. An atom tends to bond with other atoms that will fill its outer shell and produce a stable number of valence electrons. A hydrogen atom, with only one electron shell and one electron (see fig. 2.2), tends to react with other atoms that provide another electron and fill this shell with a stable number of two electrons. All other atoms react in ways that produce eight electrons in the valence shell. This tendency is called the *octet rule (rule of eights).*

## ISOTOPES AND RADIOACTIVITY

Dalton believed that every atom of an element was identical. We now know, however, that all elements have varieties called **isotopes,**[2] which differ from one another only

---

[2]*iso* = same + *top* = place (same position in the periodic table)

in number of neutrons and therefore in atomic mass. Most hydrogen atoms, for example, have only one proton; this isotope is symbolized $^1$H. Hydrogen has two other isotopes: *deuterium* ($^2$H) with one proton and one neutron, and *tritium* ($^3$H) with one proton and two neutrons (fig. 2.2). Over 99% of carbon atoms have an atomic mass of 12 ($6p^+$, $6n^0$) and are called carbon-12 ($^{12}$C), but a small percentage of carbon atoms are $^{13}$C, with seven neutrons, and $^{14}$C, with eight. All isotopes of a given element behave the same chemically. Deuterium ($^2$H), for example, reacts with oxygen the same way $^1$H does to produce water.

The *atomic weight* of an element accounts for the fact that an element is a mixture of isotopes. If all carbon were $^{12}$C, the atomic weight of carbon would be the same as its atomic mass, 12.000. But since a sample of carbon also contains small amounts of the heavier isotopes $^{13}$C and $^{14}$C, the atomic weight is slightly higher, 12.011.

Although different isotopes of an element exhibit identical chemical behavior, they differ in physical behavior. Many of them are unstable and *decay* (break down) to more stable isotopes by giving off radiation. Unstable isotopes are therefore called **radioisotopes,** and the process of decay is called **radioactivity** (see Insight 2.1). Every element has at least one radioisotope. Oxygen, for example, has three stable isotopes and five radioisotopes. All of us contain radioisotopes such as $^{14}$C and $^{40}$K—that is, we are all mildly radioactive!

Many forms of radiation, such as light and radio waves, have low energy and are harmless. High-energy radiation, however, ejects electrons from atoms, converting atoms to ions; thus it is called **ionizing radiation.** It destroys molecules and produces dangerous free radicals and ions in human tissues. In high doses, ionizing radiation is quickly fatal. In lower doses, it can be *mutagenic* (causing mutations in DNA) and *carcinogenic* (triggering cancer as a result of mutation).

Examples of ionizing radiation include ultraviolet rays, X rays, and three kinds of radiation produced by nuclear decay: *alpha* ($\alpha$) *particles, beta* ($\beta$) *particles,* and *gamma* ($\gamma$) *rays.* An alpha particle is composed of two protons and two neutrons (equivalent to a helium nucleus), and a beta particle is a free electron. Alpha particles are too large to penetrate the skin, and beta particles can penetrate only a few millimeters. They are relatively harmless when emitted by sources outside the body, but they are very dangerous when emitted by radioisotopes that have gotten into the body. Strontium-90 ($^{90}$Sr), for example, has been released by nuclear accidents and the atmospheric testing of nuclear weapons. It settles onto pastures and contaminates cow's milk. In the body, it behaves chemically like calcium, becoming incorporated into the bones, where it emits beta particles for years. Uranium and plutonium emit electromagnetic gamma rays, which have high energy and penetrating power. Gamma rays are very dangerous even when emitted by sources outside the body.

INSIGHT 2.1 Medical History

### Radiation and Madame Curie

In 1896, French scientist Henri Becquerel (1852–1908) discovered that uranium darkened photographic plates through several thick layers of paper. Marie Curie (1867–1934) and Pierre Curie (1859–1906), her husband, discovered that polonium and radium did likewise. Marie Curie coined the term *radioactivity* for the emission of energy by these elements. Becquerel and the Curies shared a Nobel Prize in 1903 for this discovery.

Marie Curie (fig. 2.3) was not only the first woman in the world to receive a Nobel Prize but also the first woman in France even to receive a Ph.D. She received a second Nobel Prize in 1911 for further work in radiation. Curie crusaded to train women for careers in science, and in World War I, she and her daughter, Irène Joliot-Curie (1897–1956), trained physicians in the use of X-ray machines. Curie pioneered radiation therapy for breast and uterine cancer.

In the wake of such discoveries, radium was regarded as a wonder drug. Unaware of its danger, people drank radium tonics and flocked to health spas to bathe in radium-enriched waters. Marie herself suffered extensive damage to her hands from handling radioactive minerals and died of radiation poisoning at age 67. The following year, Irène and her husband, Frédéric Joliot (1900–1958), were awarded a Nobel Prize for work in artificial radioactivity and synthetic radioisotopes. Apparently also a martyr to her science, Irène died of leukemia, possibly induced by radiation exposure.

**FIGURE 2.3 Marie Curie (1867–1934).** This portrait was made in 1911, when Curie received her second Nobel Prize.

Each radioisotope has a characteristic **physical half-life,** the time required for 50% of its atoms to decay to a more stable state. One gram of $^{90}Sr$, for example, would be half gone in 28 years. In 56 years, there would still be 0.25 g left, in 84 years 0.125 g, and so forth. Many radioisotopes are much longer-lived. The half-life of $^{40}K$, for example, is 1.3 billion years. Nuclear power plants produce hundreds of radioisotopes that will be intensely radioactive for at least 10,000 years—longer than the life of any disposal container yet conceived. The **biological half-life** of a radioisotope is the time required for half of it to disappear from the body. This is a function of both physical decay and physiological clearance from the body. Cesium-137, for example, has a physical half-life of 30 years but a biological half-life of only 17 days. Chemically, it behaves like potassium; it is quite mobile and rapidly excreted by the kidneys.

There are several ways to measure the intensity of ionizing radiation, the amount absorbed by the body, and its biological effects. To understand the units of measurement requires a grounding in physics beyond the scope of this book, but the standard international (SI) unit of radiation exposure is the *sievert*[3] (Sv), which takes into account the type and intensity of radiation and its biological effect. Doses of 5 Sv or more are usually fatal. The average American receives about 3.6 millisieverts (mSv) per year in *background radiation* from natural sources and another 0.6 mSv from artificial sources. The most significant natural source is *radon,* a gas that is produced by the decay of uranium in the earth and that may accumulate in buildings to unhealthy levels. Artificial sources include medical X rays, radiation therapy, and consumer products such as color televisions, smoke detectors, and luminous watch dials. Such voluntary exposure must be considered from the standpoint of its risk-to-benefit ratio. The benefits of a smoke detector or mammogram far outweigh the risk from the low levels of radiation involved. Radiation therapists and radiologists face a greater risk than their patients, however, and astronauts and airline flight crews receive more than average exposure. U.S. federal standards set a limit of 50 mSv/year as acceptable occupational exposure to ionizing radiation.

## IONS, ELECTROLYTES, AND FREE RADICALS

**Ions** are charged particles with unequal numbers of protons and electrons. Elements with one to three valence electrons tend to give them up, and those with four to

[3]Rolf Maximillian Sievert (1896–1966), Swedish radiologist

seven electrons tend to gain more. If an atom of the first kind is exposed to an atom of the second, electrons may transfer from one to the other and turn both of them into ions. This process is called *ionization*. The particle that gains electrons acquires a negative charge and is called an **anion** (AN-eye-on). The one that loses electrons acquires a positive charge (because it then has a surplus of protons) and is called a **cation** (CAT-eye-on).

Consider, for example, what happens when sodium and chlorine meet (fig. 2.4). Sodium has three electron shells with a total of 11 electrons: 2 in the first shell, 8 in the second, and 1 in the third. If it gives up the electron in the third shell, its second shell becomes the valence shell and has the stable configuration of 8 electrons. Chlorine has 17 electrons: 2 in the first shell, 8 in the second, and 7 in the third. If it can gain one more electron, it can fill the third shell with 8 electrons and become stable. Sodium and chlorine seem "made for each other"—one needs to lose an electron and the other needs to gain one. This is just what they do. When they interact, an electron

transfers from sodium to chlorine. Now, sodium has 11 protons in its nucleus but only 10 electrons. This imbalance gives it a positive charge, so we symbolize the sodium ion $Na^+$. Chlorine has been changed to the chloride ion with a surplus negative charge, symbolized $Cl^-$.

Some elements exist in two or more ionized forms. Iron, for example, has ferrous ($Fe^{2+}$) and ferric ($Fe^{3+}$) ions. Note that some ions have a single positive or negative charge, while others have charges of $\pm 2$ or $\pm 3$ because they gain or lose more than one electron. The charge on an ion is called its *valence*. Ions are not always single atoms that have become charged; some are groups of atoms—phosphate ($PO_4^{3-}$) and bicarbonate ($HCO_3^-$) ions, for example.

Ions with opposite charges are attracted to each other and tend to follow each other through the body. Thus, when $Na^+$ is excreted in the urine, $Cl^-$ tends to follow it. The attraction of cations and anions to each other is important in maintaining the excitability of muscle and nerve cells, as we shall see in chapters 11 and 12.

**Electrolytes** are salts that ionize in water and form solutions capable of conducting electricity (table 2.2). We can detect electrical activity of the muscles, heart, and brain with electrodes on the skin because electrolytes in the body fluids conduct electrical currents from these organs to the skin surface. Electrolytes are important for their chemical reactivity (as when calcium phosphate becomes incorporated into bone), osmotic effects (influence on water content and distribution in the body), and electrical effects (which are essential to nerve and muscle function). Electrolyte balance is one of the most important considerations in patient care. Electrolyte imbalances have effects ranging from muscle cramps and brittle bones to coma and cardiac arrest.

**Free radicals** are chemical particles with an odd number of electrons. For example, oxygen normally exists as a stable molecule composed of two oxygen atoms, $O_2$; but if an additional electron is added, it becomes a free radical called the *superoxide anion,* $O_2^-\bullet$. Free radicals are represented with a dot to symbolize the odd electron.

Free radicals are produced by some normal metabolic reactions of the body (such as the ATP-producing oxida-

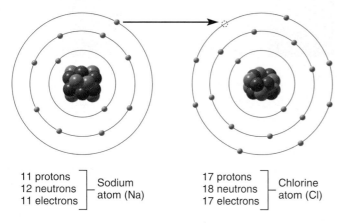

|  |  |  |  |
|---|---|---|---|
| 11 protons<br>12 neutrons<br>11 electrons | Sodium<br>atom (Na) | 17 protons<br>18 neutrons<br>17 electrons | Chlorine<br>atom (Cl) |

① Transfer of an electron from a sodium atom to a chlorine atom

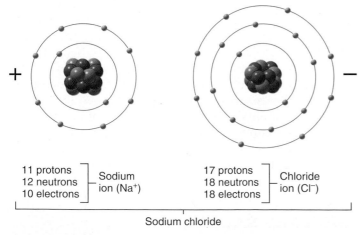

|  |  |  |  |
|---|---|---|---|
| 11 protons<br>12 neutrons<br>10 electrons | Sodium<br>ion (Na⁺) | 17 protons<br>18 neutrons<br>18 electrons | Chloride<br>ion (Cl⁻) |

Sodium chloride

② The charged sodium ion ($Na^+$) and chloride ion ($Cl^-$) that result

**FIGURE 2.4**  Ionization.

| TABLE 2.2 | Major Electrolytes and the Ions Released by Their Dissociation | | |
|---|---|---|---|
| **Electrolyte** | | **Cation** | **Anion** |
| Calcium chloride ($CaCl_2$) | → | $Ca^{2+}$ | 2 $Cl^-$ |
| Disodium phosphate ($Na_2HPO_4$) | → | 2 $Na^+$ | $HPO_4^{2-}$ |
| Magnesium chloride ($MgCl_2$) | → | $Mg^{2+}$ | 2 $Cl^-$ |
| Potassium chloride ($KCl$) | → | $K^+$ | $Cl^-$ |
| Sodium bicarbonate ($NaHCO_3$) | → | $Na^+$ | $HCO_3^-$ |
| Sodium chloride ($NaCl$) | → | $Na^+$ | $Cl^-$ |

tion reactions in mitochondria, and a reaction that some white blood cells use to kill bacteria), by radiation (such as ultraviolet radiation and X rays), and by chemicals (such as carbon tetrachloride, a cleaning solvent, and nitrites, present as preservatives in some wine, meat, and other foods). They are short-lived and combine quickly with molecules such as fats, proteins, and DNA, converting them into free radicals and triggering chain reactions that destroy still more molecules. Among the damages caused by free radicals are some forms of cancer and myocardial infarction, the death of heart tissue. One theory of aging is that it results in part from lifelong cellular damage by free radicals.

Because free radicals are so common and destructive, we have multiple mechanisms for neutralizing them. An **antioxidant** is a chemical that neutralizes free radicals. The body produces an enzyme called *superoxide dismutase (SOD)*, for example, that converts superoxide into oxygen and hydrogen peroxide. Selenium, vitamin E (α-tocopherol), vitamin C (ascorbic acid), and carotenoids (such as β-carotene) are some antioxidants obtained from the diet. Dietary deficiencies of antioxidants have been associated with increased incidence of heart attacks, sterility, muscular dystrophy, and other disorders.

## MOLECULES AND CHEMICAL BONDS

**Molecules** are chemical particles composed of two or more atoms united by a covalent chemical bond (the sharing of electrons). The atoms may be identical, as in nitrogen ($N_2$), or different, as in glucose ($C_6H_{12}O_6$). Molecules composed of two or more different elements are called **compounds.** Oxygen ($O_2$) and carbon dioxide ($CO_2$) are both molecules because both consist of at least two atoms, but only $CO_2$ is a compound, because it has atoms of two different elements.

Molecules can be represented by *molecular formulae,* as shown here, that identify their constituent elements and show how many atoms of each are present. Molecules with identical molecular formulae but different arrangements of their atoms are called **isomers**[4] of each other. For example, both ethanol (grain alcohol) and ethyl ether have the molecular formula $C_2H_6O$, but they are certainly not interchangeable! To show the difference between them, we use *structural formulae* that show the location of each atom (fig. 2.5).

The **molecular weight** (MW) of a compound is the sum of the atomic weights of its atoms. Rounding the atomic mass units (amu) to whole numbers, we can calculate the approximate MW of glucose ($C_6H_{12}O_6$), for example, as

| 6 | C atoms × 12 amu each | = | 72 amu |
|---|---|---|---|
| 12 | H atoms × 1 amu each | = | 12 amu |
| 6 | O atoms × 16 amu each | = | 96 amu |
| | Molecular weight (MW) | = | 180 amu |

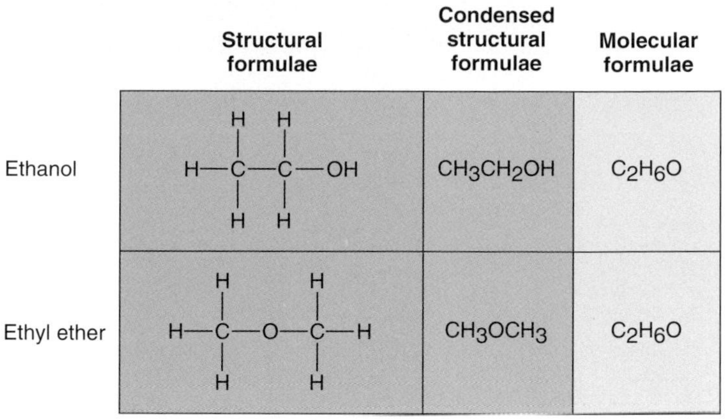

**FIGURE 2.5** Structural Isomers, Ethanol and Ethyl Ether. The molecular formulae are identical, but the structures and chemical properties are different.

Molecular weight is needed to compute some measures of concentration, as we shall see later.

A molecule is held together, and molecules are attracted to one another, by forces called **chemical bonds.** The bonds of greatest physiological interest are *ionic bonds, covalent bonds, hydrogen bonds,* and *van der Waals forces* (table 2.3).

An **ionic bond** is the attraction of a cation to an anion. Sodium ($Na^+$) and chloride ($Cl^-$) ions, for example, are attracted to each other and form the compound sodium chloride (NaCl), common table salt. Ionic compounds can be composed of more than two ions. Calcium has two valence electrons. It can become stable by donating one electron to one chlorine atom and the other electron to another chlorine, thus producing a calcium ion ($Ca^{2+}$) and two chloride ions. The result is calcium chloride, $CaCl_2$. Ionic bonds are weak and easily dissociate (break up) in the presence of something more attractive, such as water. The ionic bonds of NaCl break down easily as salt dissolves in water, because both $Na^+$ and $Cl^-$ are more attracted to water molecules than they are to each other.

**Think About It**

*Do you think ionic bonds are common in the human body? Explain your answer.*

**Covalent bonds** form by the sharing of electrons. For example, two hydrogen atoms share valence electrons to form a hydrogen molecule, $H_2$ (fig. 2.6a). The two electrons, one donated by each atom, swarm around both nuclei in a dumbbell-shaped cloud. A *single covalent bond* is the sharing of a single pair of electrons. It is symbolized by a single line between atomic symbols, for example H—H. A *double covalent bond* is the sharing of two pairs of electrons. In carbon dioxide, for example, a central carbon atom shares two electron pairs with each

| TABLE 2.3 | Types of Chemical Bonds |
|---|---|
| **Bond Type** | **Definition and Remarks** |
| Ionic bond | Relatively weak attraction between an anion and a cation. Easily disrupted in water, as when salt dissolves. |
| Covalent bond | Sharing of one or more pairs of electrons between nuclei. |
| Single covalent | Sharing of one electron pair. |
| Double covalent | Sharing of two electron pairs. Often occurs between carbon atoms, between carbon and oxygen, and between carbon and nitrogen. |
| Nonpolar covalent | Covalent bond in which electrons are equally attracted to both nuclei. May be single or double. Strongest type of chemical bond. |
| Polar covalent | Covalent bond in which electrons are more attracted to one nucleus than to the other, resulting in slightly positive and negative regions in one molecule. May be single or double. |
| Hydrogen bond | Weak attraction between polarized molecules or between polarized regions of the same molecule. Important in the three-dimensional folding and coiling of large molecules. Easily disrupted by temperature and pH changes. |
| Van der Waals force | Weak, brief attraction due to random disturbances in the electron clouds of adjacent atoms. Weakest of all bonds. |

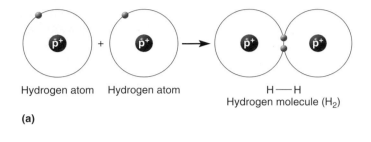

Hydrogen atom　　Hydrogen atom

H——H
Hydrogen molecule ($H_2$)

**(a)**

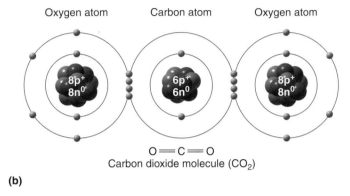

Oxygen atom　　　Carbon atom　　　Oxygen atom

O══C══O
Carbon dioxide molecule ($CO_2$)

**(b)**

**FIGURE 2.6　Covalent Bonding.** (a) Two hydrogen atoms share a single pair of electrons to form a hydrogen molecule. (b) A carbon dioxide molecule, in which a carbon atom shares two pairs of electrons with each oxygen atom, forming double covalent bonds.

▶ *How is the octet rule illustrated by the $CO_2$ molecule?*

oxygen atom. Such bonds are symbolized by two lines—for example, O══C══O (fig. 2.6b).

When shared electrons spend approximately equal time around each nucleus, they form a *nonpolar covalent bond* (fig. 2.7a), the strongest of all chemical bonds. Carbon atoms bond to each other with nonpolar covalent bonds. If shared electrons spend significantly more time orbiting one nucleus than they do the other, they lend their negative charge to the region where they spend the most time, and they form a *polar covalent bond* (fig. 2.7b). When hydrogen bonds with oxygen, for example, the electrons are more attracted to the oxygen nucleus and orbit it more than they do the hydrogen. This makes the oxygen region of the molecule slightly negative and the hydrogen regions slightly positive. The Greek delta (δ) is used to symbolize a charge less than that of one electron or proton. A slightly negative region of a molecule is represented δ− and a slightly positive region is represented δ+.

A **hydrogen bond** is a weak attraction between a slightly positive hydrogen atom in one molecule and a slightly negative oxygen or nitrogen atom in another. Water molecules, for example, are weakly attracted to each other by hydrogen bonds (fig. 2.8). Hydrogen bonds also form between different regions of the same mole-

cule, especially in very large molecules such as proteins and DNA. They cause such molecules to fold or coil into precise three-dimensional shapes. Hydrogen bonds are represented by dotted or broken lines between atoms: —C══O···H—N—. Hydrogen bonds are relatively weak, but they are enormously important to physiology.

**Van der Waals**[5] forces are weak, brief attractions between neutral atoms. When electrons orbit an atom's nucleus, they do not maintain a uniform distribution but show random fluctuations in density. If the electrons briefly crowd toward one side of an atom, they render that side slightly negative and the other side slightly positive for a moment. If another atom is close enough to this one, the second atom responds with disturbances in its own electron cloud. Oppositely charged regions of the two atoms then attract each other for a very short instant in time.

A single van der Waals attraction is only about 1% as strong as a covalent bond, but when two surfaces or large molecules meet, the van der Waals attractions between large numbers of atoms can create a very strong attraction. This is how plastic wrap clings to food and dishes, flies

[5]Johannes Diderik van der Waals (1837–1923), Dutch physicist

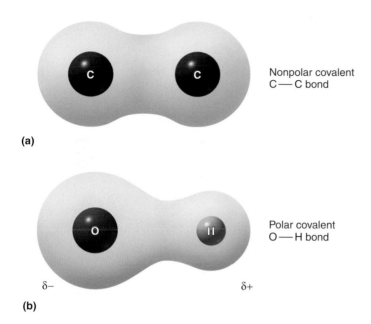

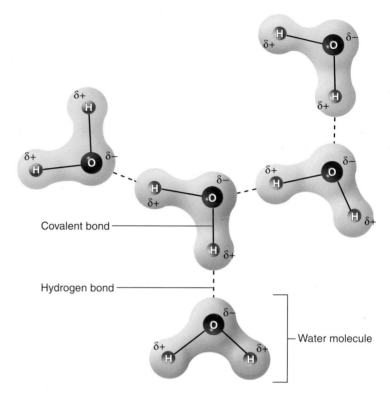

**FIGURE 2.7** **Nonpolar and Polar Covalent Bonds.**
(a) A nonpolar covalent bond between two carbon atoms, formed by electrons that spend an equal amount of time around each nucleus, as represented by the symmetric blue cloud.
(b) A polar covalent bond, in which electrons orbit one nucleus significantly more than the other, as represented by the asymmetric cloud. This results in a slight negative charge ($\delta-$) in the region where the electrons spend most of their time, and a slight positive charge ($\delta+$) at the other pole.

**FIGURE 2.8** **Hydrogen Bonding of Water.** The polar covalent bonds of water molecules enable each oxygen to form a hydrogen bond with a hydrogen of a neighboring molecule. Thus, the water molecules are weakly attracted to one another.
▶ *Why would this behavior raise the boiling point of water above that of a nonpolar liquid?*

and spiders walk across a ceiling, and even a 100-g lizard, the Tokay gecko, can run up a windowpane. Van der Waals forces also have a significant effect on the boiling points of liquids. In human structure, they are especially important in protein folding, the binding of proteins to each other and to other molecules such as hormones, and the association of lipid molecules with each other. Some of these molecular behaviors are described later in this chapter.

## Before You Go On

*Answer the following questions to test your understanding of the preceding section:*

1. *Consider iron (Fe), hydrogen gas ($H_2$), and ammonia ($NH_3$). Which of these is or are atoms? Which of them is or are molecules? Which of them is or are compounds? Explain each answer.*

2. *Why is the biological half-life of a radioisotope shorter than its physical half-life?*

3. *Where do free radicals come from? What harm do they do? What protections from free radicals exist?*

4. *How does an ionic bond differ from a covalent bond?*

5. *What is a hydrogen bond? Why do hydrogen bonds depend on the existence of polar covalent bonds?*

# Water and Mixtures

### Objectives
When you have completed this section, you should be able to

- define *mixture* and distinguish between mixtures and compounds;

- describe the biologically important properties of water;

- show how three kinds of mixtures differ from each other;

- discuss some ways in which the concentration of a solution can be expressed, and explain why different expressions of concentration are used for different purposes; and

- define *acid* and *base* and interpret the pH scale.

Our body fluids are complex mixtures of chemicals. A **mixture** consists of substances that are physically blended but not chemically combined. Each substance retains its own chemical properties. To contrast a mixture with a compound, consider sodium chloride again.

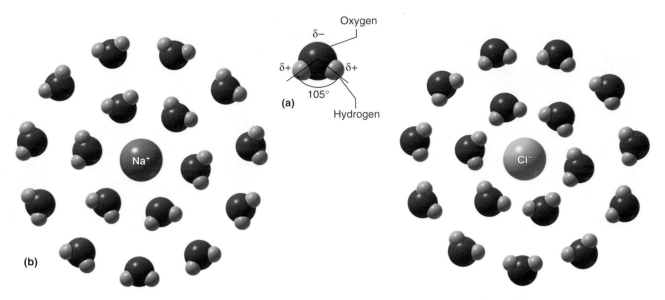

**FIGURE 2.9    Water and Hydration Spheres.**    (a) A water molecule showing its bond angle and polarity. (b) Water molecules aggregate around a sodium ion with their negatively charged oxygen poles facing the Na⁺ and aggregate around a chloride ion with their positively charged hydrogen poles facing the Cl⁻.

Sodium is a lightweight metal that bursts into flame if exposed to water, and chlorine is a yellow-green poisonous gas that was used for chemical warfare in World War I. When these elements chemically react, they form common table salt. Clearly, the compound has properties much different from the properties of its elements. But if you were to put a little salt on your watermelon, the watermelon would taste salty and sweet because the sugar of the melon and the salt you added would merely form a mixture in which each compound retained its individual properties.

## WATER

Most mixtures in our bodies consist of chemicals dissolved or suspended in water. Water constitutes 50% to 75% of your body weight, depending on age, sex, fat content, and other factors. Its structure, simple as it is, has profound biological effects. Two aspects of its structure are particularly important: (1) its atoms are joined by polar covalent bonds, and (2) the molecule is V-shaped, with a 105° bond angle (fig. 2.9a). This makes the molecule as a whole polar, with a slight negative charge ($\delta-$) on the oxygen and a slight positive charge ($\delta+$) on each hydrogen. Like little magnets, water molecules are attracted to one another by hydrogen bonds (see fig. 2.8). This gives water a set of properties that account for its ability to support life: *solvency, cohesion, adhesion, chemical reactivity,* and *thermal stability.*

*Solvency* is the ability to dissolve other chemicals. Water is sometimes called the *universal solvent* because it dissolves a broader range of substances than any other liquid. Substances that dissolve in water, such as sugar, are said to be **hydrophilic**[6] (HY-dro-FILL-ic); the relatively few substances that do not, such as fats, are **hydrophobic**[7] (HY-dro-FOE-bic). Virtually all metabolic reactions depend on the solvency of water. Biological molecules must be dissolved in water to move freely, come together, and react. The solvency of water also makes it the body's primary means of transporting substances from place to place.

To be soluble in water, a molecule must be polarized or charged so that its charges can interact with those of water. When NaCl is dropped into water, for example, the ionic bonds between Na⁺ and Cl⁻ are overpowered by the attraction of each ion to water molecules. Water molecules form a cluster, or *hydration sphere,* around each sodium ion with the $O^{\delta-}$ pole of each water molecule facing the sodium ion. They also form a hydration sphere around each chloride ion, with the $H^{\delta+}$ poles facing it. This isolates the sodium ions from the chloride ions and keeps them dissolved (fig. 2.9b).

*Adhesion* is the tendency of one substance to cling to another, whereas *cohesion* is the tendency of molecules of the same substance to cling to each other. Water adheres to the body's tissues and forms a lubricating film on membranes such as the pleura and pericardium. This helps reduce friction as the lungs and heart contract and expand and rub against these membranes. Water also is a very cohesive liquid because of its hydrogen bonds. This is why, when you spill water on the floor, it forms a puddle and evaporates slowly. By contrast, if you spill a nonpolar substance such as liquid nitrogen, it dances about and evaporates in seconds, like a drop of water in a hot dry skillet.

[6]*hydro* = water + *philic* = loving, attracted to
[7]*phobic* = fearing, avoiding

This is because nitrogen molecules have no attraction for each other, so the little bit of heat provided by the floor is enough to disperse them into the air. The cohesion of water is especially evident at its surface, where it forms an elastic layer called the *surface film* held together by a force called *surface tension.* This force causes water to hang in drops from a leaky faucet and travel in rivulets down a window.

The *chemical reactivity* of water is its ability to participate in chemical reactions. Not only does water ionize many other chemicals such as acids and salts, but water itself ionizes into $H^+$ and $OH^-$. These ions can be incorporated into other molecules, or released from them, in the course of chemical reactions such as *hydrolysis* and *dehydration synthesis,* described later in this chapter.

The *thermal stability* of water helps to stabilize the internal temperature of the body. It results from the high *heat capacity* of water—the amount of heat required to raise the temperature of 1 g of a substance by 1°C. The base unit of heat is the **calorie**[8] (cal)—1 cal is the amount of heat that raises the temperature of 1 g of water 1°C. The same amount of heat would raise the temperature of a nonpolar substance such as nitrogen about four times as much. The difference stems from the presence or absence of hydrogen bonding. To increase in temperature, the molecules of a substance must move around more actively. The hydrogen bonds of water molecules inhibit their movement, so water can absorb a given amount of heat without changing temperature (molecular motion) as much.

The high heat capacity of water also makes it a very effective coolant. When it changes from a liquid to a vapor, water carries a large amount of heat with it. One milliliter of perspiration evaporating from the skin removes about 500 cal of heat from the body. This effect is very apparent when you are sweaty and stand in front of a fan.

Think About It

*Why are heat and temperature not the same thing?*

## SOLUTIONS, COLLOIDS, AND SUSPENSIONS

Mixtures of other substances in water can be classified as *solutions, colloids,* and *suspensions.*

A **solution** consists of particles of matter called the **solute** mixed with a more abundant substance (usually water) called the **solvent.** The solute can be a gas, solid, or liquid—as in a solution of oxygen, sodium chloride, or alcohol in water, respectively. Solutions are defined by the following properties:

- The solute particles are under 1 nanometer (nm) in size. The solute and solvent therefore cannot be visually distinguished from each other, even with a microscope.

- Such small particles do not scatter light noticeably, so solutions are usually transparent (fig. 2.10a).
- The solute particles will pass through most selectively permeable membranes, such as dialysis tubing and cell membranes.
- The solute does not separate from the solvent when the solution is allowed to stand.

The most common **colloids**[9] in the body are mixtures of protein and water, such as the albumin in blood plasma. Many colloids can change from liquid to gel states—gelatin desserts, agar culture media, and the fluids within and between our cells, for example. Colloids are defined by the following physical properties:

- The colloidal particles range from 1 to 100 nm in size.
- Particles this large scatter light, so colloids are usually cloudy (fig. 2.10b).
- The particles are too large to pass through most selectively permeable membranes.
- The particles are still small enough, however, to remain permanently mixed with the solvent when the mixture stands.

The blood cells and water in our blood plasma exemplify a **suspension.** Suspensions are defined by the following properties:

- The suspended particles exceed 100 nm in size.
- Such large particles render suspensions cloudy or opaque.
- The particles are too large to penetrate selectively permeable membranes.
- The particles are too heavy to remain permanently suspended, so suspensions separate on standing. If allowed to stand, blood cells settle to the bottom of a tube, for example (fig. 2.10c, d).

An **emulsion** is a suspension of one liquid in another, such as oil-and-vinegar salad dressing. The fat in breast milk is an emulsion, as are medications such as Kaopectate and milk of magnesia.

A single mixture can fit into more than one of these categories. Blood is a perfect example—it is a solution of sodium chloride, a colloid of protein, and a suspension of cells. Milk is a solution of calcium, a colloid of protein, and an emulsion of fat. Table 2.4 summarizes the types of mixtures and provides additional examples.

## MEASURES OF CONCENTRATION

Solutions are often described in terms of their concentration—how much solute is present in a given volume of solution. Concentration is expressed in different ways for different purposes, some of which are explained here. The table of

[8]*calor* = heat

[9]*collo* = glue + *oid* = like, resembling

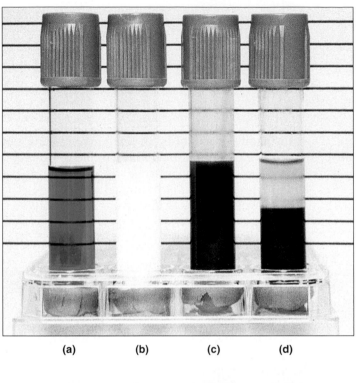

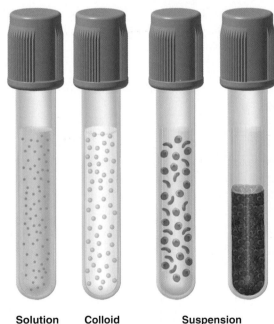

Solution    Colloid    Suspension

**FIGURE 2.10**    **A Solution, a Colloid, and a Suspension.**
*Top row:* Photographs of a representative solution, colloid, and suspension. *Bottom row:* Depiction of the particle sizes in each mixture. (a) In a copper sulfate solution, the solute particles are so small they remain permanently mixed and the solution is transparent. (b) In milk, the protein molecules are small enough to remain permanently mixed, but large enough to scatter light, so the mixture is opaque. (c) In blood, the red blood cells scatter light and make the mixture opaque; furthermore (d), they are too large to remain evenly mixed, so they settle to the bottom as in this blood specimen that stood overnight.

symbols and measures in appendix C may be helpful as you study this section.

## Weight per Volume

A simple way to express concentration is the weight of solute in a given volume of solution. For example, intravenous (I.V.) saline typically contains 8.5 grams of NaCl per liter of solution (8.5 g/L). For many biological purposes, however, we deal with smaller quantities such as milligrams per deciliter (mg/dL; 1 dL = 100 mL). For example, a typical serum cholesterol concentration may be 200 mg/dL, also expressed 200 mg/100 mL or 200 milligram-percent (mg-%).

## Percentages

Percentage concentrations are also simple to compute, but it is necessary to specify whether the percentage refers to the weight or the volume of solute in a given volume of solution. For example, if we begin with 5 g of dextrose (an isomer of glucose) and add enough water to make 100 mL of solution, the resulting concentration will be 5% weight per volume (w/v). A common intravenous fluid is D5W, which stands for 5% w/v dextrose in distilled water. If the solute is a liquid, such as ethanol, percentages refer to volume of solute per volume of solution. Thus, 70 mL of ethanol diluted with water to 100 mL of solution produces 70% volume per volume (70% v/v) ethanol.

## Molarity

Percent concentrations are easy to prepare, but that unit of measurement is inadequate for many purposes. The physiological effect of a chemical depends on how many molecules of it are present in a given volume, not the weight of the chemical. Five percent glucose, for example, contains almost twice as many glucose molecules as the same volume of 5% sucrose (fig. 2.11a). Each solution contains 50 g of sugar per liter, but glucose has a molecular weight (MW) of 180 and sucrose has a MW of 342. Since each molecule of glucose is lighter, 50 g of glucose contains more molecules than 50 g of sucrose.

To produce solutions with a known number of molecules per volume, we must factor in the molecular weight. If we know the MW and weigh out that many grams of the substance, we have a quantity known as its gram molecular weight, or 1 *mole.* One mole of glucose is 180 g and 1 mole of sucrose is 342 g. Each quantity contains the same number of molecules of the respective sugar—a number known as Avogadro's[10] number, $6.023 \times 10^{23}$. Such a large number is hard to imagine. If each molecule were the size of a pea, $6.023 \times 10^{23}$ molecules would cover 60 earth-sized planets 3 m (10 ft) deep!

---

[10]Amedeo Avogadro (1776–1856), Italian chemist

| TABLE 2.4 | Types of Mixtures | | |
|---|---|---|---|
| | **Solution** | **Colloid** | **Suspension** |
| *Particle Size* | < 1 nm | 1-100 nm | > 100 nm |
| *Appearance* | Clear | Often cloudy | Cloudy-opaque |
| *Will particles settle out?* | No | No | Yes |
| *Will particles pass through a selectively permeable membrane?* | Yes | No | No |
| *Examples* | Glucose in blood<br>$O_2$ in water<br>Saline solutions<br>Sugar in coffee | Proteins in blood<br>Intracellular fluid<br>Milk protein<br>Gelatin | Blood cells<br>Cornstarch in water<br>Fats in blood<br>Kaopectate |

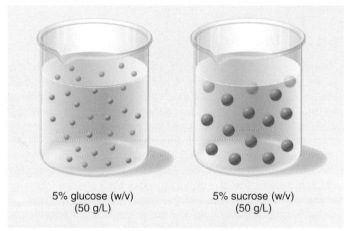

5% glucose (w/v) (50 g/L)   5% sucrose (w/v) (50 g/L)

**(a) Solutions of equal percentage concentration**

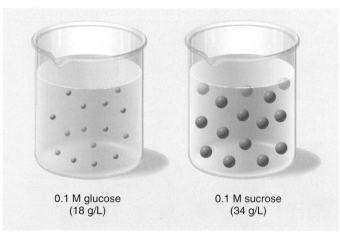

0.1 M glucose (18 g/L)   0.1 M sucrose (34 g/L)

**(b) Solutions of equal molar concentration**

**FIGURE 2.11 Comparison of Percentage and Molar Concentrations.** (a) Solutions with the same percentage concentrations can differ greatly in the number of molecules per volume because of differences in molecular weights of the solutes. Fifty grams of sucrose has about half as many molecules as 50 g of glucose, for example. (b) Solutions with the same molarity have the same number of molecules per volume because molarity takes differences in molecular weight into account.

**Molarity** (M) is the number of moles of solute per liter of solution. A *one-molar* (1.0 M) solution of glucose contains 180 g/L, and 1.0 M solution of sucrose contains 342 g/L. Both have the same number of solute molecules in a given volume (fig. 2.11b). Body fluids and laboratory solutions usually are less concentrated than 1 M, so biologists and clinicians more often work with *millimolar* (mM) and *micromolar* (μM) concentrations—$10^{-3}$ and $10^{-6}$ M, respectively.

## Electrolyte Concentrations

Electrolytes are important for their chemical, physical (osmotic), and electrical effects on the body. Their electrical effects, which determine such things as nerve, heart, and muscle actions, depend not only on their concentration but also on their electrical charge. A calcium ion ($Ca^{2+}$) has twice the electrical effect of a sodium ion ($Na^+$), for example, because it carries twice the charge. When we measure electrolyte concentrations, we must therefore take the charges into account.

One *equivalent* (Eq) of an electrolyte is the amount that would electrically neutralize 1 mole of hydrogen ions ($H^+$) or hydroxide ions ($OH^-$). For example, 1 mole (58.4 g) of NaCl yields 1 mole, or 1 Eq, of $Na^+$ in solution. Thus, an NaCl solution of 58.4 g/L contains 1 equivalent of $Na^+$ per liter (1 Eq/L). One mole (98 g) of sulfuric acid ($H_2SO_4$) yields 2 moles of positive charges ($H^+$). Thus, 98 g of sulfuric acid per liter would be a solution of 2 Eq/L.

The electrolytes in our body fluids have concentrations less than 1 Eq/L, so we more often express their concentrations in **milliequivalents per liter** (mEq/L). If you know the millimolar concentration of an electrolyte, you can easily convert this to mEq/L by multiplying it by the valence of the ion:

$$1 \text{ mM } Na^+ = 1 \text{ mEq/L}$$
$$1 \text{ mM } Ca^{2+} = 2 \text{ mEq/L}$$
$$1 \text{ mM } Fe^{3+} = 3 \text{ mEq/L}$$

## ACIDS, BASES, AND pH

Most people have some sense of what acids and bases are. Advertisements are full of references to excess stomach acid and pH-balanced shampoo. We know that drain cleaner (a strong base) and battery acid can cause serious chemical burns. But what exactly do "acidic" and "basic" mean, and how can they be quantified?

An **acid** is any *proton donor,* a molecule that releases a proton ($H^+$) in water. A **base** is a proton acceptor. Since hydroxide ions ($OH^-$) accept $H^+$, many bases are substances that release hydroxide ions—sodium hydroxide ($NaOH$), for example. A base does not have to be a hydroxide donor, however. Ammonia ($NH_3$) is also a base. It does not release hydroxide ions, but it readily accepts hydrogen ions to become the ammonium ion ($NH_4^+$).

Acidity is expressed in terms of **pH,** a measure derived from the molarity of $H^+$. Molarity is represented by square brackets, so the molarity of $H^+$ is symbolized $[H^+]$. pH is the negative logarithm of hydrogen ion molarity—that is, $pH = -\log [H^+]$. In pure water, 1 in 10 million molecules ionizes into hydrogen and hydroxide ions: $H_2O \leftrightharpoons H^+ + OH^-$. Pure water has a neutral pH because it contains equal amounts of $H^+$ and $OH^-$. Since 1 in 10 million molecules ionize, the molarity of $H^+$ and the pH of water are

$$[H^+] = 0.0000001 \text{ molar} = 10^{-7} \text{ M}$$
$$\log [H^+] = -7$$
$$pH = -\log [H^+] = 7$$

The pH scale (fig. 2.12) was invented in 1909 by Danish biochemist and brewer Sören Sörensen to measure the acidity of beer. The scale extends from 0.0 to 14.0. A solution with a pH of 7.0 is **neutral;** solutions with pH below 7 are **acidic;** and solutions with pH above 7 are **basic** (alkaline).

The lower the pH value, the more hydrogen ions a solution has and the more acidic it is. Since the pH scale is logarithmic, a change of one whole number on the scale represents a 10-fold change in $H^+$ concentration. In other words, a solution with pH 4 is 10 times as acidic as one with pH 5 and 100 times as acidic as one with pH 6.

Slight disturbances of pH can seriously disrupt physiological functions and alter drug actions (see Insight 2.2), so it is important that the body carefully control its pH. Blood, for example, normally has a pH ranging from 7.35 to 7.45. Deviations from this range cause tremors, fainting, paralysis, or even death. Chemical solutions that resist changes in pH are called **buffers.** Buffers and pH regulation are considered in detail in chapter 24.

---

### INSIGHT 2.2    Clinical Application

#### pH and Drug Action

The pH of our body fluids has a direct bearing on how we react to drugs. Depending on pH, drugs such as aspirin, phenobarbital, and penicillin can exist in charged (ionized) or uncharged forms. Whether a drug is charged or not can determine whether it will pass through cell membranes. When aspirin is in the acidic environment of the stomach, for example, it is uncharged and passes easily through the stomach lining into the bloodstream. Here it encounters a basic pH, whereupon it ionizes. In this state, it is unable to pass back through the membrane, so it accumulates in the blood. This effect, called *ion trapping* or *pH partitioning,* can be controlled to help clear poisons from the body. The pH of the urine, for example, can be manipulated so that poisons become trapped there and thus rapidly excreted from the body.

---

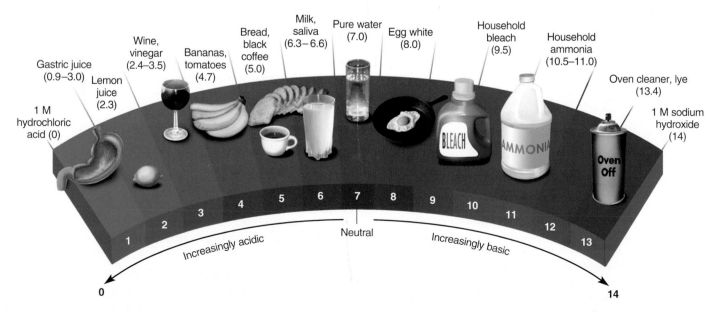

**FIGURE 2.12    The pH scale.** The pH is shown within the colored bar. $H^+$ molarity increases tenfold for every step down the scale.

*A pH of 7.20 is slightly alkaline, yet a blood pH of 7.20 is called acidosis. Why do you think it is called this?*

## Before You Go On

*Answer the following questions to test your understanding of the preceding section:*

6. *What is the difference between a mixture and a compound?*

7. *What are hydrophilic and hydrophobic substances? Give an example of each.*

8. *Why would the cohesion and thermal stability of water be less if water did not have polar covalent bonds?*

9. *How do solutions, colloids, and suspensions differ from each other? Give an example of each in the human body.*

10. *What is one advantage of percentage over molarity as a measure of solute concentration? What is one advantage of molarity over percentage?*

11. *If solution A had a $H^+$ concentration of $10^{-8}$ M, what would be its pH? If solution B had 1,000 times this $H^+$ concentration, what would be its pH? Would solution A be acidic or basic? What about solution B?*

# Energy and Chemical Reactions

### Objectives

When you have completed this section, you should be able to:

- define *energy* and *work,* and describe some types of energy;

- understand how chemical reactions are symbolized by chemical equations;

- list and define the fundamental types of chemical reactions;

- identify the factors that govern the speed and direction of a reaction;

- define *metabolism* and its two subdivisions; and

- define *oxidation* and *reduction* and relate these to changes in the energy content of a molecule.

## ENERGY AND WORK

**Energy** is the capacity to do work. To do **work** means to move something, whether it is a muscle or a molecule. Some examples of physiological work are breaking chemical bonds, building molecules, pumping blood, and contracting skeletal muscles. All of the body's activities are forms of work.

Energy is broadly classified as *potential* or *kinetic energy. Potential energy* is energy contained in an object because of its position or internal state, but which is not doing work at the time. *Kinetic energy* is energy of motion, energy that is doing work. It is observed in skeletomuscular movements, the flow of ions into a cell, and vibration of the eardrum, for example. The water behind a dam has potential energy because of its position. Let the water flow through, and it exhibits kinetic energy that can be tapped for generating electricity. Like water behind a dam, ions concentrated on one side of a cell membrane have potential energy that can be released by opening gates in the membrane. As the ions flow through the gates, their kinetic energy can be tapped to create a nerve signal or make the heart beat.

Within the two broad categories of potential and kinetic energy, several forms of energy are relevant to human physiology. *Chemical energy* is potential energy stored in the bonds of molecules. Chemical reactions release this energy and make it available for physiological work. *Heat* is the kinetic energy of molecular motion. The temperature of a substance is a measure of rate of this motion, and adding heat to a substance increases this rate. *Electromagnetic energy* is the kinetic energy of moving "packets" of radiation called *photons.* The most familiar form of electromagnetic energy is light. *Electrical energy* has both potential and kinetic forms. It is potential energy when charged particles have accumulated at a point such as a battery terminal or on one side of a cell membrane; it becomes kinetic energy when these particles begin to move and create an electrical current—for example, when electrons move through your household wiring or sodium ions move through a cell membrane.

**Free energy** is the potential energy available in a system to do useful work. In human physiology, the most relevant free energy is the energy stored in the chemical bonds of organic molecules.

## CLASSES OF CHEMICAL REACTIONS

A **chemical reaction** is a process in which a covalent or ionic bond is formed or broken. The course of a chemical reaction is symbolized by a **chemical equation** that typically shows the *reactants* on the left, the *products* on the right, and an arrow pointing from the reactants to the products. For example, consider this common occurrence: If you open a bottle of wine and let it stand for several days, it turns sour. Wine "turns to vinegar" because oxygen gets into the bottle and reacts with ethanol to produce acetic acid and water. Acetic acid gives the tart flavor to vinegar and spoiled wine. The equation for this reaction is

$$CH_3CH_2OH \quad + \quad O_2 \quad \rightarrow \quad CH_3COOH \quad + \quad H_2O$$
Ethanol  Oxygen  Acetic acid  Water

Ethanol and oxygen are the reactants, and acetic acid and water are the products of this reaction. Not all reactions are shown with the arrow pointing from left to right. In complex biochemical equations, reaction chains are often written vertically or even in circles.

Chemical reactions can be classified as *decomposition, synthesis,* or *exchange reactions.* In **decomposition**

**reactions,** a large molecule breaks down into two or more smaller ones (fig. 2.13a); symbolically, AB → A + B. When you eat a potato, for example, digestive enzymes decompose its starch into thousands of glucose molecules, and most cells further decompose glucose to water and carbon dioxide. Starch, a very large molecule, ultimately yields about 36,000 molecules of $H_2O$ and $CO_2$.

**Synthesis reactions** are just the opposite—two or more small molecules combine to form a larger one; symbolically, A + B → AB (fig. 2.13b). When the body synthesizes proteins, for example, it combines several hundred amino acids into one protein molecule.

In **exchange reactions,** two molecules exchange atoms or groups of atoms; AB + CD → AC + BD (fig. 2.13c). For example, when stomach acid (HCl) enters the small intestine, the pancreas secretes sodium bicarbonate ($NaHCO_3$) to neutralize it. The reaction between the two is $NaHCO_3 + HCl → NaCl + H_2CO_3$. We could say the sodium atom has exchanged its bicarbonate group ($—HCO_3$) for a chlorine atom.

**Reversible reactions** can go in either direction under different circumstances and are represented with paired arrows. For example, carbon dioxide combines with water to produce carbonic acid, which in turn decomposes into bicarbonate ions and hydrogen ions:

$$CO_2 + H_2O \rightleftharpoons H_2CO_3 \rightleftharpoons HCO_3^- + H^+$$

| Carbon dioxide | Water | Carbonic acid | Bicarbonate ion | Hydrogen ion |

This reaction appears in this book more often than any other, especially as we discuss respiratory, urinary, and digestive physiology.

The direction in which a reversible reaction goes is determined by the relative abundance of substances on each side of the equation. If there is a surplus of $CO_2$, this reaction proceeds to the right and produces bicarbonate and hydrogen ions. If bicarbonate and hydrogen ions are present in excess, the reaction proceeds to the left and generates $CO_2$ and $H_2O$. Reversible reactions follow the **law of mass action:** they proceed from the side with the greater quantity of reactants to the side with the lesser quantity. This law will help to explain processes discussed in later chapters, such as why hemoglobin binds oxygen in the lungs yet releases it to muscle tissue.

In the absence of upsetting influences, reversible reactions exist in a state of **equilibrium,** in which the ratio of products to reactants is stable. The carbonic acid reaction, for example, normally maintains a 20:1 ratio of bicarbonate ions to carbonic acid molecules. This equilibrium can be upset, however, by a surplus of hydrogen ions, which drive the reaction to the left, or adding carbon dioxide and driving it to the right.

## REACTION RATES

The basis for chemical reactions is molecular motion and collisions. All molecules are in constant motion, and reactions occur when mutually reactive molecules collide

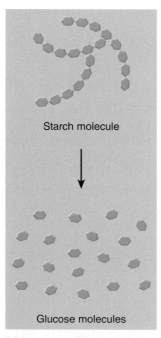

Starch molecule

Glucose molecules

**(a) Decomposition reaction**

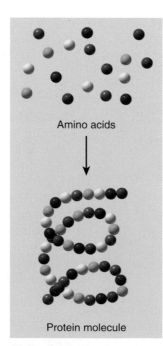

Amino acids

Protein molecule

**(b) Synthesis reaction**

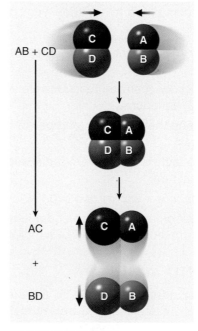

AB + CD

AC

+

BD

**(c) Exchange reaction**

**FIGURE 2.13** **Decomposition, Synthesis, and Exchange Reactions.**   (a) In a decomposition reaction, large molecules are broken down into simpler ones. (b) In a synthesis reaction, smaller molecules are joined to form larger ones. (c) In an exchange reaction, two molecules exchange atoms.

▶ *To which of these categories does the digestion of food belong?*

with sufficient force and the right orientation. The rate of a reaction depends on the nature of the reactants and on the frequency and force of these collisions. Some factors that affect reaction rates are:

- **Concentration.** Reaction rates increase when the reactants are more concentrated. This is because the molecules are more crowded and collide more frequently.

- **Temperature.** Reaction rate increases as the temperature rises. This is because heat causes molecules to move more rapidly and collide with greater force and frequency.

- **Catalysts** (CAT-uh-lists). These are substances that temporarily bind to reactants, hold them in a favorable position to react with each other, and may change the shapes of reactants in ways that make them more likely to react. By reducing the element of chance in molecular collisions, a catalyst speeds up a reaction. It then releases the products and is available to repeat the process with more reactants. The catalyst itself is not permanently consumed or changed by the reaction. The most important biological catalysts are enzymes, discussed later in this chapter.

## METABOLISM, OXIDATION, AND REDUCTION

All the chemical reactions in the body are collectively called **metabolism.** Metabolism has two divisions: *catabolism* and *anabolism.* Catabolism[11] (ca-TAB-oh-lizm) consists of energy-releasing decomposition reactions. Such reactions break covalent bonds, produce smaller molecules from larger ones, and release energy that can be used for other physiological work. Energy-releasing reactions are called *exergonic*[12] reactions. If you hold a beaker of water in your hand and pour sulfuric acid into it, for example, the beaker will get so hot you may have to put it down. If you break down energy-storage molecules to run a race, you too will get hot. In both cases, the heat signifies that exergonic reactions are occurring.

**Anabolism**[13] (ah-NAB-oh-lizm) consists of energy-storing synthesis reactions, such as the production of protein or fat. Reactions that require an energy input, such as these, are called *endergonic*[14] reactions. Anabolism is driven by the energy that catabolism releases, so endergonic and exergonic processes, anabolism and catabolism, are inseparably linked.

**Oxidation** is any chemical reaction in which a molecule gives up electrons and releases energy. A molecule is *oxidized* by this process, and whatever molecule takes the electrons from it is an **oxidizing agent** (electron acceptor). The term *oxidation* stems from the fact that oxygen is often involved as the electron acceptor. Thus, we can sometimes recognize an oxidation reaction from the fact that oxygen has been added to a molecule. The rusting of iron, for example, is a slow oxidation process in which oxygen is added to iron to form iron oxide ($Fe_2O_3$). Many oxidation reactions, however, do not involve oxygen at all. For example, when yeast ferments glucose to alcohol, no oxygen is required; indeed, the alcohol *contains less oxygen* than the sugar originally did, but it is *more oxidized* than the sugar:

$$C_6H_{12}O_6 \rightarrow 2\ CH_3CH_2OH + 2\ CO_2$$
Glucose        Ethanol        Carbon dioxide

**Reduction** is a chemical reaction in which a molecule gains electrons and energy. When a molecule accepts electrons, it is said to be *reduced;* a molecule that donates electrons to another is therefore called a **reducing agent** (electron donor). The oxidation of one molecule is always accompanied by the reduction of another, so these electron transfers are known as *oxidation–reduction (redox) reactions.*

It is not necessary that *only* electrons be transferred in a redox reaction. Often, the electrons are transferred in the form of hydrogen atoms. The fact that a proton (the hydrogen nucleus) is also transferred is immaterial to whether we consider a reaction oxidation or reduction.

Table 2.5 summarizes these energy-transfer reactions. We can symbolize oxidation and reduction as follows, letting A and B symbolize arbitrary molecules and $e^-$ represents one or more electrons:

$$Ae^- + B \rightarrow A + Be^-$$
High-energy  Low-energy  Low-energy  High-energy
reduced      oxidized    oxidized    reduced
state        state       state       state

$Ae^-$ is a reducing agent because it reduces B, and B is an oxidizing agent because it oxidizes $Ae^-$.

### Before You Go On

*Answer the following questions to test your understanding of the preceding section:*

12. *Define energy. Distinguish potential energy from kinetic energy.*

13. *Define metabolism, catabolism, and anabolism.*

14. *What does oxidation mean? What does reduction mean? Which of them is endergonic and which is exergonic?*

15. *When sodium chloride forms, which element—sodium or chlorine—is oxidized? Which one is reduced?*

[11]*cata* = down, to break down
[12]*ex, exo* = out + *erg* = work
[13]*ana* = up, to build up
[14]*end* = in

| TABLE 2.5 | Energy-Transfer Reactions in the Human Body |
|---|---|
| *Exergonic Reactions* | Reactions in which there is a net release of energy. The products have less total free energy than the reactants did. |
| Oxidation | An exergonic reaction in which electrons are removed from a reactant. Electrons may be removed one or two at a time and may be removed in the form of hydrogen atoms (H or $H_2$). The product is then said to be oxidized. |
| Decomposition | A reaction such as digestion and cell respiration, in which larger molecules are broken down into smaller ones. |
| Catabolism | The sum of all decomposition reactions in the body. |
| *Endergonic Reactions* | Reactions in which there is a net input of energy. The products have more total free energy than the reactants did. |
| Reduction | An endergonic reaction in which electrons are donated to a reactant. The product is then said to be reduced. |
| Synthesis | A reaction such as protein and glycogen synthesis, in which two or more smaller molecules are combined into a larger one. |
| Anabolism | The sum of all synthesis reactions in the body. |

# Organic Compounds

### Objectives

When you have completed this section, you should be able to

- explain why carbon is especially well suited to serve as the structural foundation of many biological molecules;
- identify some common functional groups of organic molecules from their formulae;
- discuss the relevance of polymers to biology and explain how they are formed and broken by dehydration synthesis and hydrolysis;
- discuss the types and functions of carbohydrates;
- discuss the types and functions of lipids;
- discuss protein structure and function;
- explain how enzymes function;
- describe the structure, production, and function of ATP;
- identify other nucleotide types and their functions;
- identify the principal types of nucleic acids.

## CARBON COMPOUNDS AND FUNCTIONAL GROUPS

*Organic chemistry* is the study of compounds of carbon. By 1900, biochemists had classified the organic molecules of life into four primary categories: *carbohydrates, lipids, proteins,* and *nucleic acids.* We examine the first three in this chapter but describe the details of nucleic acids, which are concerned with genetics, in chapter 4.

Carbon is an especially versatile atom that serves as the basis of a wide variety of structures. It has four valence electrons, so it bonds with other atoms that can provide it with four more to complete its valence shell. Carbon atoms readily bond with each other and can form long chains, branched molecules, and rings—an enormous variety of **carbon backbones** for organic molecules. Carbon also forms covalent bonds with hydrogen, oxygen, nitrogen, sulfur, and other elements.

Carbon backbones carry a variety of **functional groups**—small clusters of atoms that determine many of the properties of an organic molecule. For example, organic acids bear a **carboxyl** (car-BOC-sil) **group,** and ATP is named for its three **phosphate groups.** Other common functional groups include **hydroxyl, methyl,** and **amino groups** (fig. 2.14).

## MONOMERS AND POLYMERS

Since carbon can form long chains, some organic molecules are gigantic *macromolecules* with molecular weights that range from the thousands (as in starch and proteins) to the millions (as in DNA). Most macromolecules are **polymers**[15]—molecules made of a repetitive series of identical or similar subunits called **monomers** (MON-oh-murs). Starch, for example, is a polymer of about 3,000 glucose monomers. In starch, the monomers are identical, whereas in other polymers they have a basic structural similarity but differ in detail. DNA, for example, is made of 4 different kinds of monomers (nucleotides), and proteins are made of 20 kinds (amino acids).

The joining of monomers to form a polymer is called *polymerization.* Living cells achieve this by means of a reaction called **dehydration synthesis** (condensation) (fig. 2.15a). A hydroxyl (−OH) group is removed from one monomer and a hydrogen (−H) from another, producing water as a by-product. The two monomers become joined by a covalent bond, forming a *dimer.* This is repeated for each monomer added to the chain, potentially leading to a chain long enough to be considered a polymer.

---

[15]*poly* = many + *mer* = part

| Name and Symbol | Structure | Occurs in |
|---|---|---|
| Hydroxyl (—OH) | 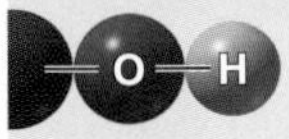 | Sugars, alcohols |
| Methyl (—CH₃) | 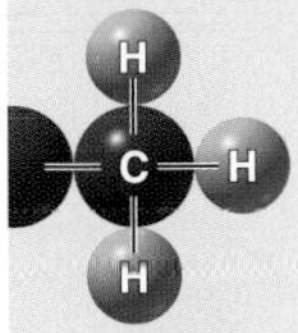 | Fats, oils, steroids, amino acids |
| Carboxyl (—COOH) | 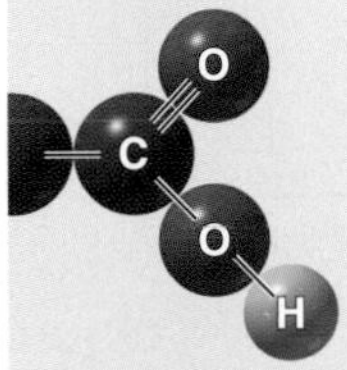 | Amino acids, sugars, proteins |
| Amino (—NH₂) | 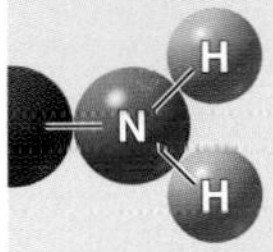 | Amino acids, proteins |
| Phosphate (—H₂PO₄) | 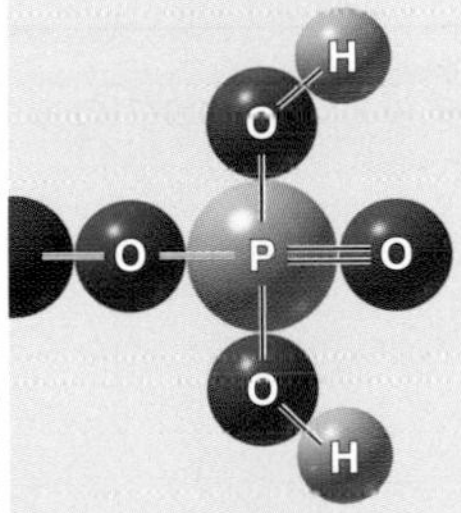 | Nucleic acids, ATP |

**FIGURE 2.14**  Functional Groups of Organic Molecules.

The opposite of dehydration synthesis is **hydrolysis**[16] (fig. 2.15b). In hydrolysis, a water molecule ionizes into $OH^-$ and $H^+$. A covalent bond linking one monomer to another is broken, the $OH^-$ is added to one monomer, and the $H^+$ is added to the other one. All digestion consists of hydrolysis reactions.

## CARBOHYDRATES

A **carbohydrate**[17] is a hydrophilic organic molecule with the general formula $(CH_2O)_n$, where $n$ represents the number of carbon atoms. In glucose, for example, $n = 6$ and the formula is $C_6H_{12}O_6$. As the generic formula shows, carbohydrates have a 2:1 ratio of hydrogen to oxygen.

> **Think About It**
>
> *Why is carbohydrate an appropriate name for this class of compounds? Relate this name to the general formula of carbohydrates.*

The names of individual carbohydrates are often built on the word root *sacchar-* or the suffix *-ose,* both of which mean "sugar" or "sweet." The most familiar carbohydrates are the sugars and starches.

The simplest carbohydrates are monomers called **monosaccharides**[18] (MON-oh-SAC-uh-rides), or simple sugars. The three of primary importance are **glucose, fructose,** and **galactose,** all with the molecular formula $C_6H_{12}O_6$; they are isomers of each other (fig. 2.16). We obtain these sugars mainly by the digestion of more complex carbohydrates. Glucose is the "blood sugar" that provides energy to most of our cells. Two other monosaccharides, ribose and deoxyribose, are important components of DNA and RNA.

---

[16]*hydro* = water + *lysis* = splitting apart
[17]*carbo* = carbon + *hydr* = water
[18]*mono* = one + *sacchar* = sugar

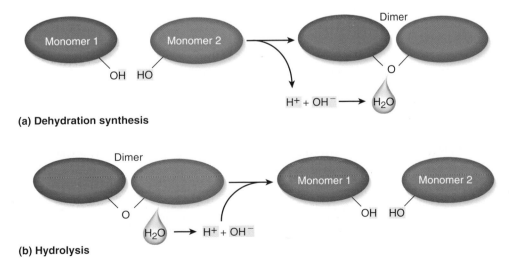

**(a) Dehydration synthesis**

**(b) Hydrolysis**

**FIGURE 2.15**  Synthesis and Hydrolysis Reactions. (a) In dehydration synthesis, a hydrogen atom is removed from one monomer and a hydroxyl group is removed from another. These combine to form water as a by-product. The monomers become joined by a covalent bond to form a dimer. (b) In hydrolysis, a covalent bond between two monomers is broken. Water donates a hydrogen atom to one monomer and a hydroxyl group to the other.

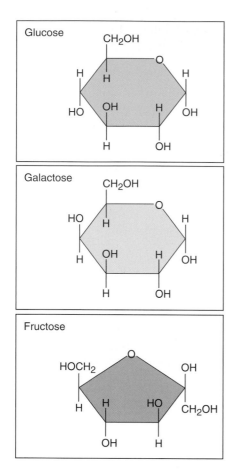

Glucose

Galactose

Fructose

**FIGURE 2.16   The Three Major Monosaccharides.**   All three have the molecular formula $C_6H_{12}O_6$. Each angle in the rings represents a carbon atom except the one where oxygen is shown. This is a conventional way of representing carbon in the structural formulae of organic compounds.

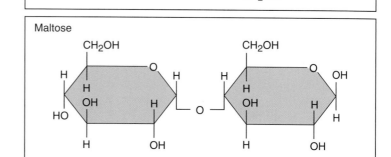

Sucrose

Lactose

Maltose

**FIGURE 2.17   The Three Major Disaccharides.**

**Disaccharides** are sugars composed of two monosaccharides. The three of greatest importance are **sucrose** (made of glucose + fructose), **lactose** (glucose + galactose), and **maltose** (glucose + glucose) (fig. 2.17). Sucrose is produced by sugarcane and sugar beets and used as common table sugar. Lactose is milk sugar. Maltose is a product of starch digestion and is present in a few foods such as germinating wheat and malt beverages.

**Polysaccharides** (POL-ee-SAC-uh-rides) are long chains of glucose. Some polysaccharides have molecular weights of 500,000 or more (compared with 180 for a single glucose). Three polysaccharides of interest to human physiology are glycogen, starch, and cellulose. Animals, including ourselves, make glycogen, while starch and cellulose are plant products.

**Glycogen**[19] is an energy-storage polysaccharide made by cells of the liver, muscles, brain, uterus, and vagina. It

is a long branched glucose polymer (fig. 2.18). The liver produces glycogen after a meal, when the blood glucose level is high, and then breaks it down between meals to maintain blood glucose levels when there is no food intake. Muscle stores glycogen for its own energy needs, and the uterus uses it in pregnancy to nourish the embryo.

**Starch** is the corresponding energy-storage polysaccharide of plants. They store it when sunlight and nutrients are available and draw from it when photosynthesis is not possible (for example, at night and in winter, when a plant has shed its leaves). Starch is the only significant digestible polysaccharide in the human diet.

**Cellulose** is a structural polysaccharide that gives strength to the cell walls of plants. It is the principal component of wood, cotton, and paper. It consists of a few thousand glucose monomers joined together, with every other monomer "upside down" relative to the next. (The $CH_2OH$ groups all face in the same direction in glycogen and starch, but alternate between facing up and down in cellulose.) Cellulose is the most abundant organic com-

---

[19]*glyco* = sugar + *gen* = producing

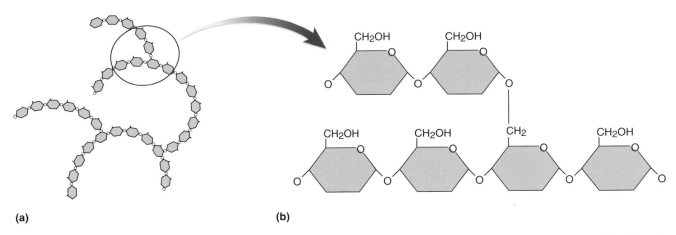

**FIGURE 2.18   Glycogen.** This is the only polysaccharide found in human tissues. (a) Part of a glycogen molecule showing the chain of glucose monomers and branching pattern. (b) Detail of the molecule at a branchpoint.

pound on earth and it is a common component of the diets of humans and other animals—yet we have no enzymes to digest it and thus derive no energy or nutrition from it. Nevertheless, it is important as dietary "fiber," "bulk," or "roughage." It swells with water in the digestive tract and helps move other materials through the intestine.

Carbohydrates are, above all, a source of energy that can be quickly mobilized. All digested carbohydrate is ultimately converted to glucose, and glucose is oxidized to make ATP, a high-energy compound discussed later. But carbohydrates have other functions as well (table 2.6). They are often **conjugated**[20] with (covalently bound to) proteins and lipids. Many of the lipid and protein molecules at the external surface of the cell membrane have chains of up to 12 sugars attached to them, thus forming **glycolipids** and **glycoproteins,** respectively. Among other functions, glycoproteins are a major component of mucus, which traps particles in the respiratory system, resists infection, and protects the digestive tract from its own acid and enzymes.

**Proteoglycans** (once called mucopolysaccharides) are macromolecules in which the carbohydrate component is dominant and a peptide or protein forms a smaller component. Proteoglycans form gels that help hold cells and tissues together, form a gelatinous filler in the umbilical cord and eye, lubricate the joints of the skeletal system, and account for the tough rubbery texture of cartilage. Their structure and functions are further considered in chapter 5.

When discussing conjugated macromolecules it is convenient to refer to each chemically different component as a **moiety**[21] (MOY-eh-tee). Proteoglycans have a protein moiety and a carbohydrate moiety, for example.

[20]*con* = together + *jug* = join
[21]*moiet* = half

| TABLE 2.6 | Carbohydrate Functions |
|---|---|
| **Type** | **Function** |
| *Monosaccharides* | |
| Glucose | Blood sugar—energy source for most cells |
| Galactose | Converted to glucose and metabolized |
| Fructose | Fruit sugar—converted to glucose and metabolized |
| *Disaccharides* | |
| Sucrose | Cane sugar—digested to glucose and fructose |
| Lactose | Milk sugar—digested to glucose and galactose; important in infant nutrition |
| Maltose | Malt sugar—product of starch digestion, further digested to glucose |
| *Polysaccharides* | |
| Cellulose | Structural polysaccharide of plants; dietary fiber |
| Starch | Energy storage in plant cells |
| Glycogen | Energy storage in animal cells (liver, muscle, brain, uterus, vagina) |
| *Conjugated Carbohydrates* | |
| Glycoprotein | Component of the cell surface coat and mucus, among other roles |
| Glycolipid | Component of the cell surface coat |
| Proteoglycan | Cell adhesion; lubrication; supportive filler of some tissues and organs |

## LIPIDS

A **lipid** is a hydrophobic organic molecule, usually composed only of carbon, hydrogen, and oxygen, with a high ratio of hydrogen to oxygen. A fat called *tristearin* (tri-STEE-uh-rin), for example, has the molecular formula $C_{57}H_{110}O_6$—more than 18 hydrogens for every oxygen.

Lipids are less oxidized than carbohydrates, and thus have more calories per gram. Beyond these criteria, it is difficult to generalize about lipids; they are much more variable in structure than the other macromolecules we are considering. We consider here the five primary types of lipids in humans—*fatty acids, triglycerides, phospholipids, eicosanoids,* and *steroids* (table 2.7).

A **fatty acid** is a chain of usually 4 to 24 carbon atoms with a carboxyl group at one end and a methyl group at the other. Fatty acids and the fats made from them are classified as *saturated* or *unsaturated.* A **saturated fatty acid** such as palmitic acid has as much hydrogen as it can carry. No more could be added without exceeding four covalent bonds per carbon atom; thus it is "saturated" with hydrogen. In **unsaturated fatty acids** such as linoleic acid, however, some carbon atoms are joined by double covalent bonds (fig. 2.19). Each of these could potentially share one pair of electrons with another hydrogen atom instead of the adjacent carbon, so hydrogen could be added to this molecule. **Polyunsaturated fatty acids** are those with many C=C bonds. Most fatty acids can be synthesized by the human body, but a few, called **essential fatty acids,** must be obtained from the diet because we cannot synthesize them (see chapter 26).

A **triglyceride** (try-GLISS-ur-ide) is a molecule consisting of three fatty acids covalently bonded to a three-carbon alcohol called **glycerol;** triglycerides are more correctly, although less widely, also known as *triacylglycerols.* Each bond between a fatty acid and glycerol is formed by dehydration synthesis (fig. 2.19). Once joined to glycerol, a fatty acid can no longer donate a proton to solution and is therefore no longer an acid. For this reason, triglycerides are also called *neutral fats.* Triglycerides are broken down by hydrolysis reactions, which split each of these bonds apart by the addition of water.

Triglycerides that are liquid at room temperature are also called *oils,* but the difference between a fat and oil is fairly arbitrary. Coconut oil, for example, is solid at room temperature. Animal fats are usually made of saturated fatty acids, so they are called *saturated fats.* They are solid at room or body temperature. Most plant triglycerides are *polyunsaturated fats,* which generally remain liquid at room temperature. Examples include peanut, olive, corn, and linseed oils. Saturated fats contribute more to cardiovascular disease than unsaturated fats, and for this reason it is healthier to cook with vegetable oils than with lard, bacon fat, or butter.

The primary function of fat is energy storage, but when concentrated in *adipose tissue,* it also provides thermal insulation and acts as a shock-absorbing cushion for vital organs (see chapter 5).

**Phospholipids** are similar to neutral fats except that, in place of one fatty acid, they have a phosphate group which, in turn, is linked to other functional groups. Lecithin is a common phospholipid in which the phosphate is bonded to a nitrogenous group called *choline* (CO-leen) (fig. 2.20). Phospholipids have a dual nature. The two fatty acid "tails" of the molecule are hydrophobic, but the phosphate "head" is hydrophilic. Thus, phospholipids are said to be **amphiphilic**[22] (AM-fih-FIL-ic). Together, the head and the two tails of a phospholipid give it a shape like a clothespin. The most important function of phospholipids is to serve as the structural foundation of cell membranes (see chapter 3).

**Eicosanoids**[23] (eye-CO-sah-noyds) are 20-carbon compounds derived from a fatty acid called *arachidonic* (ah-RACK-ih-DON-ic) *acid.* They function primarily as hormonelike chemical signals between cells. The most functionally diverse eicosanoids are the **prostaglandins,** in which five of the carbon atoms are arranged in a ring (fig. 2.21). They were originally found in the secretions of bovine prostate glands, hence their name, but they are now known to be produced in almost all tissues. They play a variety of signaling roles in inflammation, blood clotting, hormone action, labor contractions, control of blood vessel diameter, and other processes (see chapter 17).

A **steroid** is a lipid with 17 of its carbon atoms arranged in four rings (fig. 2.22). **Cholesterol** is the "parent" steroid from which the other steroids are synthesized. The others include cortisol, progesterone, estrogens, testosterone, and bile acids. These differ from each other in the location of C=C bonds within the rings and in the functional groups attached to the rings.

| TABLE 2.7 | Some Lipid Functions |
|---|---|
| **Type** | **Function** |
| Bile acids | Steroids that aid in fat digestion and nutrient absorption |
| Cholesterol | Component of cell membranes; precursor of other steroids |
| Eicosanoids | Chemical messengers between cells |
| Fat-soluble vitamins | Involved in a variety of functions including blood clotting, wound healing, vision, and calcium absorption |
| Fatty acids | Precursor of triglycerides; source of energy |
| Phospholipids | Major component of cell membranes; aid in fat digestion |
| Steroid hormones | Chemical messengers between cells |
| Triglycerides | Energy storage; thermal insulation; filling space; binding organs together; cushioning organs |

---

[22]*amphi* = both + *philic* = loving
[23]*eicosa* = 20

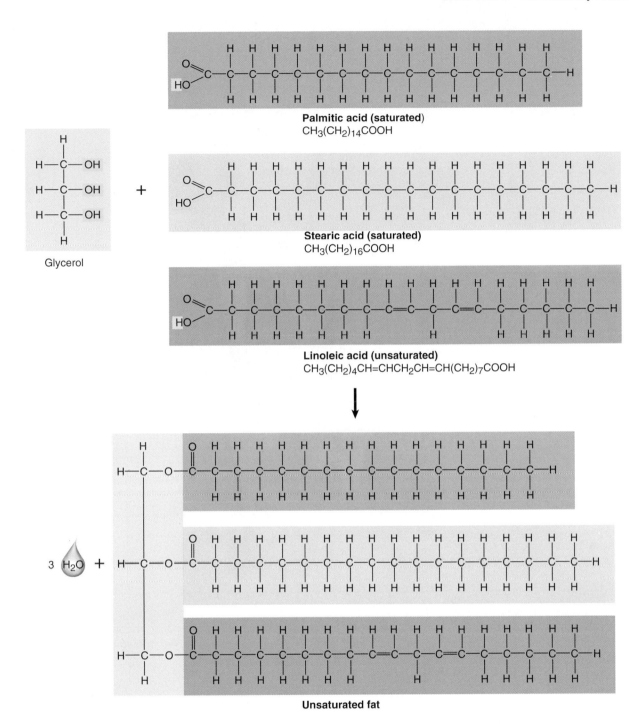

**FIGURE 2.19  Triglyceride (Fat) Synthesis.** Note the difference between saturated and unsaturated fatty acids and the production of 3 $H_2O$ as a by-product of this dehydration synthesis reaction.

Cholesterol is synthesized only by animals (especially in liver cells) and is not present in vegetable oils or other plant products. The average adult contains over 200 g (half a pound) of cholesterol. Cholesterol has a bad reputation as a factor in cardiovascular disease (see Insight 2.3), and it is true that hereditary and dietary factors can elevate blood cholesterol to dangerously high levels. Nevertheless, cholesterol is a natural product of the body. Only about 15% of our cholesterol comes from the diet; the other 85% is internally synthesized. In addition to being the precursor of the other steroids, cholesterol is an important component of cell membranes and is required for proper nervous system function.

The primary lipids and their functions are summarized in table 2.7.

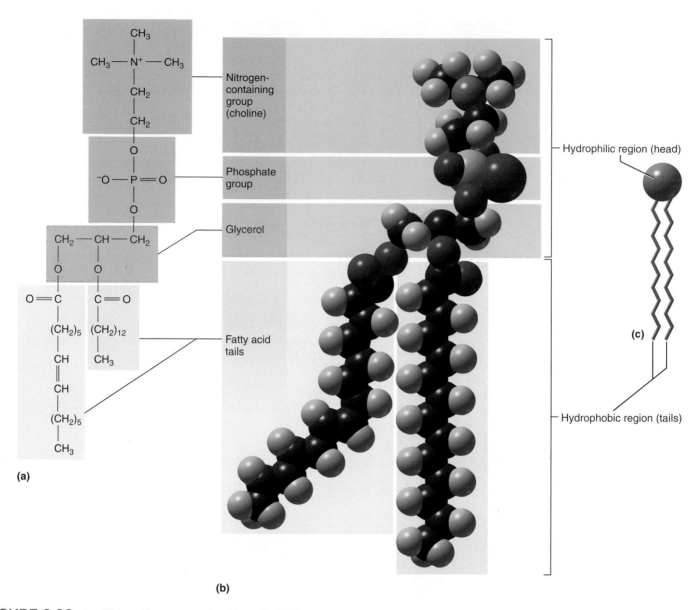

**FIGURE 2.20  Lecithin, a Representative Phospholipid.**  (a) Structural formula. (b) A space-filling model that gives some idea of the actual shape of the molecule. (c) A simplified representation of the phospholipid molecule used in diagrams of cell membranes.

## INSIGHT 2.3    Clinical Application

### "Good" and "Bad" Cholesterol

There is only one kind of cholesterol, and it does far more good than harm. When the popular press refers to "good" and "bad" cholesterol, it is actually referring to droplets in the blood called *lipoproteins,* which are a complex of cholesterol, fat, phospholipids, and protein. So-called bad cholesterol refers to low-density lipoprotein (LDL), which has a high ratio of lipid to protein and contributes to cardiovascular disease. So-called good cholesterol refers to high-density lipoprotein (HDL), which has a lower ratio of lipid to protein and may help to prevent cardiovascular disease. Even when food products are advertised as cholesterol-free, they may be high in saturated fat, which stimulates the body to produce more cholesterol. Palmitic acid seems to be the greatest culprit in stimulating elevated cholesterol levels, while linoleic acid has a cholesterol-lowering effect. Both are shown in figure 2.19. Cardiovascular disease is further discussed at the end of chapter 19, and LDLs and HDLs are more fully explained in chapter 26.

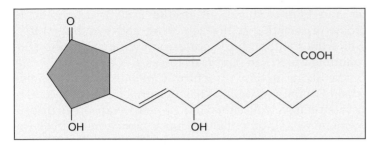

**FIGURE 2.21** A Prostaglandin. This is a modified fatty acid with five of its carbon atoms arranged in a ring.

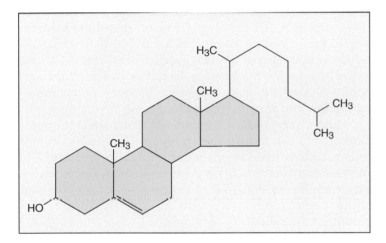

**FIGURE 2.22** Cholesterol. All steroids have this basic four-ringed structure, with variations in the functional groups and locations of double bonds within the rings.

# PROTEINS

The word *protein* is derived from the Greek word *proteios,* meaning "of first importance." Proteins are the most versatile molecules in the body, and many discussions in this book will draw on your understanding of protein structure and behavior.

## Amino Acids and Peptides

A **protein** is a polymer of **amino acids.** An amino acid has a central carbon atom with an amino (—NH₂) and a carboxyl (—COOH) group bound to it (fig. 2.23a). The 20 amino acids used to make proteins are identical except for a third functional group called the *radical* (R group) attached to the central carbon. In the simplest amino acid, glycine, R is merely a hydrogen atom, while in the largest amino acids it includes rings of carbon. Some R groups are hydrophilic and some are hydrophobic. Being composed of many amino acids, proteins as a whole are therefore often amphiphilic. The 20 amino acids involved in proteins are listed in table 2.8 along with their abbreviations.

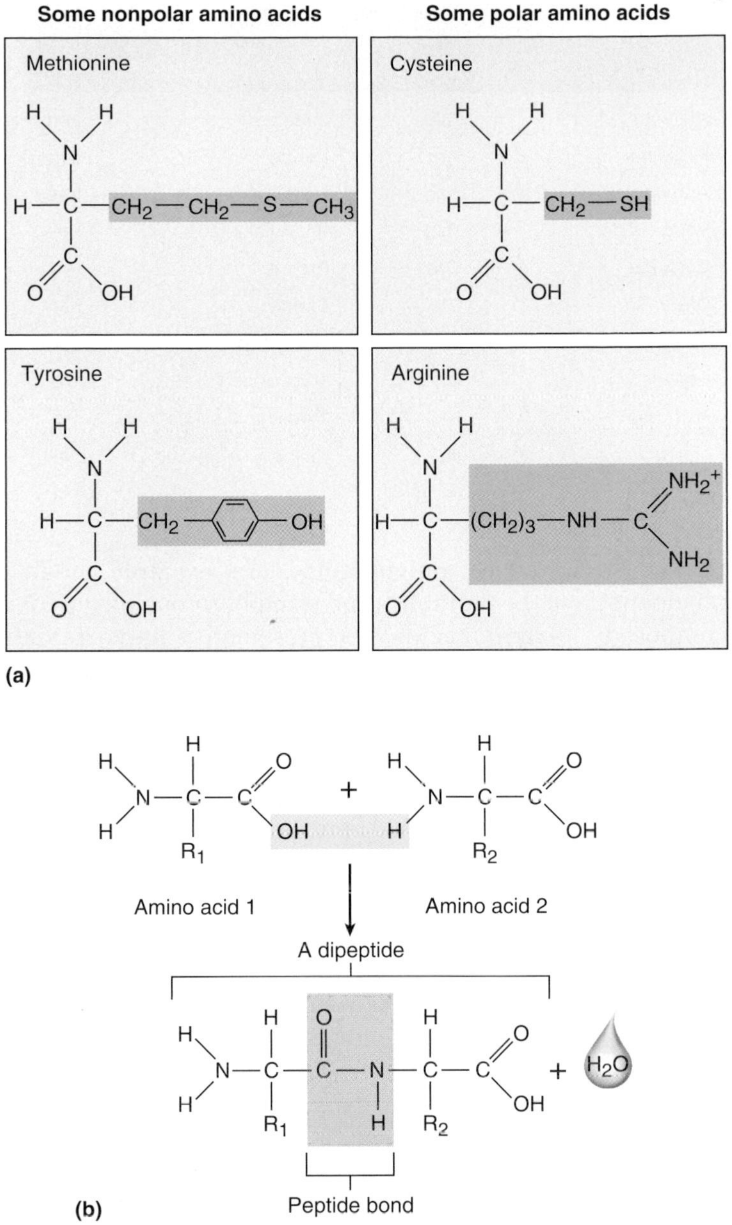

**FIGURE 2.23** Amino Acids and Peptides. (a) Four representative amino acids. Note that they differ only in the R group, shaded in pink. (b) The joining of two amino acids by a peptide bond, forming a dipeptide. Side groups $R_1$ and $R_2$ could be the groups indicated in pink in part (a), among other possibilities.

A **peptide** is any molecule composed of two or more amino acids joined by **peptide bonds.** A peptide bond, formed by dehydration synthesis, joins the amino group of one amino acid to the carboxyl group of the next (fig. 2.23b). Peptides are named for the number of amino acids they have—for example, dipeptides have two and tripeptides have three. Chains of fewer than 10 or 15 amino acids are called **oligopeptides,**[24] and chains larger

[24]*oligo* = a few

| TABLE 2.8 | The 20 Amino Acids and Their Abbreviations | | |
|---|---|---|---|
| Alanine | Ala | Leucine | Leu |
| Arginine | Arg | Lysine | Lys |
| Asparagine | Asn | Methionine | Met |
| Aspartic acid | Asp | Phenylalanine | Phe |
| Cysteine | Cys | Proline | Pro |
| Glutamine | Gln | Serine | Ser |
| Glutamic acid | Glu | Threonine | Thr |
| Glycine | Gly | Tryptophan | Trp |
| Histidine | His | Tyrosine | Tyr |
| Isoleucine | Ile | Valine | Val |

than that are called **polypeptides.** An example of an oligopeptide is the childbirth-inducing hormone oxytocin, composed of 9 amino acids. A representative polypeptide is adrenocorticotropic hormone (ACTH), which is 39 amino acids long. A protein is a polypeptide of 50 amino acids or more. A typical amino acid has a molecular weight of about 80 amu, and the molecular weights of the smallest proteins are around 4,000 to 8,000 amu. The average protein weighs in at about 30,000 amu, and some of them have molecular weights in the hundreds of thousands.

## Protein Structure

Proteins have complex coiled and folded structures that are critically important to the roles they play. Even slight changes in their **conformation** (three-dimensional shape) can destroy protein function. Protein molecules have three to four levels of complexity, from primary through quaternary structure (fig. 2.24).

**Primary structure** is the protein's sequence of amino acids. Their order is encoded in the genes (see chapter 4).

**Secondary structure** is a coiled or folded shape held together by hydrogen bonds between the slightly negative C=O group of one peptide bond and the slightly positive N—H group of another some distance away. The most common secondary structures are a springlike shape called the **alpha (α) helix** and a pleated, ribbonlike shape, the **beta (β) sheet** (or β-pleated sheet). Many proteins have multiple alpha-helical and beta-pleated regions joined by short segments with a less orderly geometry. A single protein molecule may fold back on itself and have two or more beta-pleated regions linked to one another by hydrogen bonds. Separate, parallel protein chains also may be hydrogen-bonded to each other through their beta-pleated regions.

**Tertiary**[25] (TUR-she-air-ee) **structure** is formed by the further bending and folding of proteins into various globu-

lar and fibrous shapes. It results from hydrophobic R groups associating with each other and avoiding water, while the hydrophilic R groups are attracted to the surrounding water. Van der Waals forces play a significant role in stabilizing tertiary structure. *Globular proteins,* somewhat resembling a wadded ball of yarn, have a compact tertiary structure well suited for proteins embedded in cell membranes and proteins that must move around freely in the body fluids, such as enzymes and antibodies. *Fibrous proteins* such as myosin, keratin, and collagen are slender filaments better suited for such roles as muscle contraction and providing strength to skin, hair, and tendons.

The amino acid cysteine (Cys), whose R group is —CH$_2$—SH (see fig. 2.23), often stabilizes a protein's tertiary structure by forming covalent **disulfide bridges.** When two cysteines align with each other, each can release a hydrogen atom, leaving the sulfur atoms to form a disulfide (—S—S—) bridge. Disulfide bridges hold separate polypeptide chains together in such molecules as antibodies and insulin (fig. 2.25).

**Quaternary**[26] (QUA-tur-nare-ee) **structure** is the association of two or more polypeptide chains by noncovalent forces such as ionic bonds and hydrophilic-hydrophobic interactions. It occurs in only some proteins. Hemoglobin, for example, consists of four polypeptides: two identical alpha chains and two identical, slightly longer beta chains (see fig. 2.24).

One of the most important properties of proteins is their ability to change conformation, especially tertiary structure. This can be triggered by such influences as voltage changes on a cell membrane during the action of nerve cells, the binding of a hormone to a protein, or the dissociation of a molecule from a protein. Subtle, reversible changes in conformation are important to processes such as enzyme function, muscle contraction, and the opening and closing of pores in cell membranes. **Denaturation** is a more drastic conformational change in response to conditions such as extreme heat or pH. It is seen, for example, when you cook an egg and the egg white protein (albumen) turns from clear and runny to opaque and stiff. Denaturation is sometimes reversible, but often it permanently destroys protein function.

*Conjugated proteins* have a non-amino acid moiety called a **prosthetic**[27] **group** covalently bound to them. Hemoglobin, for example, not only has the four polypeptide chains described earlier, but each chain also has a complex iron-containing ring called a *heme* moiety attached to it (see fig. 2.24). Hemoglobin cannot transport oxygen unless this group is present. In glycoproteins, as described earlier, the carbohydrate moiety is a prosthetic group.

---

[25]*tert* = third

[26]*quater* = fourth
[27]*prosthe* = appendage, addition

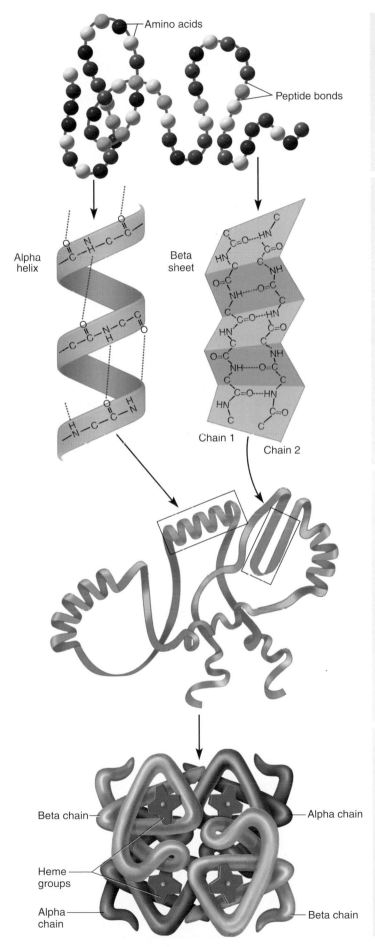

**Primary structure**

Sequence of amino acids joined by peptide bonds

**Secondary structure**

Alpha helix or beta sheet formed by hydrogen bonding

**Tertiary structure**

Folding and coiling due to interactions among R groups and between R groups and surrounding water

**Quaternary structure**

Association of two or more polypeptide chains with each other

## Protein Functions

Proteins have more diverse functions than other macromolecules. These include:

- **Structure.** *Keratin,* a tough structural protein, gives strength to the nails, hair, and skin surface. Deeper layers of the skin, as well as bones, cartilage, and teeth, contain an abundance of the durable protein *collagen.*

- **Communication.** Some hormones and other cell-to-cell signals are proteins, as are the receptors to which the signal molecules bind in the receiving cell. A hormone or other molecule that reversibly binds to a protein is called a **ligand**[28] (LIG-and).

- **Membrane transport.** Some proteins form channels in cell membranes that govern what passes through the membranes and when. Others act as carriers that briefly bind to solute particles and transport them to the other side of the membrane. Among their other roles, such proteins turn nerve and muscle activity on and off.

- **Catalysis.** Most metabolic pathways of the body are controlled by enzymes, which are globular proteins that function as catalysts.

- **Recognition and protection.** The role of glycoproteins in immune recognition was mentioned earlier. Antibodies and other proteins attack and neutralize organisms that invade the body. Clotting proteins protect the body against blood loss.

[28]*lig* = to bind

**FIGURE 2.24 Four Levels of Protein Structure.** The molecule shown for quaternary structure is hemoglobin, which is composed of four polypeptide chains. The heme groups are iron-containing nonprotein moieties.

**FIGURE 2.25   Primary Structure of Insulin.**   Insulin is composed of two polypeptide chains joined by disulfide bridges.

- **Movement.** Movement is fundamental to all life, from the intracellular transport of molecules to the galloping of a racehorse. Proteins, with their special ability to change shape repeatedly, are the basis for all such movement. Some proteins are called *molecular motors* for this reason.

- **Cell adhesion.** Proteins bind cells to each other, which enables sperm to fertilize eggs, enables immune cells to bind to enemy cancer cells, and keeps tissues from falling apart.

## ENZYMES AND METABOLISM

**Enzymes** are proteins that function as biological catalysts. They permit biochemical reactions to occur rapidly at normal body temperatures. Enzymes were initially given somewhat arbitrary names, some of which are still with us, such as *pepsin* and *trypsin*. The modern system of naming enzymes, however, is more uniform and informative. It identifies the substance the

enzyme acts upon, called its **substrate;** sometimes refers to the enzyme's action; and adds the suffix *-ase.* Thus, *amylase* digests starch (*amyl-* = starch) and *carbonic anhydrase* removes water (*anhydr-*) from carbonic acid. Enzyme names may be further modified to distinguish different forms of the same enzyme found in different tissues (Insight 2.4).

To appreciate the effect of an enzyme, think of what happens when paper burns. Paper is composed mainly of glucose (in the form of cellulose). The burning of glucose can be represented by the equation

$$C_6H_{12}O_6 + 6\ O_2 \rightarrow 6\ CO_2 + 6\ H_2O$$

Paper does not spontaneously burst into flame because few of its molecules have enough kinetic energy to react. Lighting the paper with a match, however, raises the kinetic energy enough to initiate combustion (rapid oxidation). The energy needed to get the reaction started, supplied by the match, is called the **activation energy** (fig. 2.26a).

In the body, we carry out the same reaction and oxidize glucose to water and carbon dioxide to extract its energy. We could not tolerate the heat of combustion in our bodies, however, so we must oxidize glucose in a more controlled way at a biologically feasible and safe temperature. Enzymes make this happen by lowering the activation energy—that is, by reducing the barrier to glucose oxidation (fig. 2.26b)—and by releasing the energy in small steps rather than a single burst of heat.

---

**INSIGHT 2.4**   Clinical Application

### The Diagnostic Use of Isoenzymes

A given enzyme may exist in slightly different forms, called *isoenzymes,* in different cells. Isoenzymes catalyze the same chemical reactions but have enough structural differences that they can be distinguished by standard laboratory techniques. This is useful in the diagnosis of disease. When organs are diseased, some of their cells break down and release specific isoenzymes that can be detected in the blood. Normally, these isoenzymes would not be present in the blood or would have very low concentrations. An elevation in blood levels can help pinpoint what cells in the body have been damaged.

For example, *creatine kinase* (CK) occurs in different forms in different cells. An elevated serum level of CK-1 indicates a breakdown of skeletal muscle and is one of the signs of muscular dystrophy. An elevated CK-2 level indicates heart disease, because this isoenzyme comes only from cardiac muscle. There are five isoenzymes of *lactate dehydrogenase* (LDH). High serum levels of LDH-1 may indicate a tumor of the ovaries or testes, while LDH-5 may indicate liver disease or muscular dystrophy. Different isoenzymes of *phosphatase* in the blood may indicate bone or prostate disease.

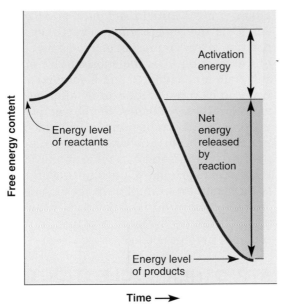

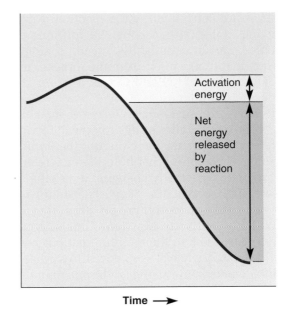

**Time ⟶**

**(a) Reaction occurring without a catalyst**

**Time ⟶**

**(b) Reaction occuring with a catalyst**

**FIGURE 2.26** **Effect of an Enzyme on Activation Energy.** (a) Without catalysts, some chemical reactions proceed slowly because of the high activation energy needed to get molecules to react. (b) A catalyst facilitates molecular interaction, thus lowering the activation energy and making the reaction proceed more rapidly.

▶ *Does an enzyme release more energy from its substrate than an uncatalyzed reaction would release?*

## Enzyme Structure and Action

Figure 2.27 illustrates the action of an enzyme, using the example of *sucrase,* an enzyme that breaks sucrose down to glucose and fructose. This process occurs in three principal steps:

① A substrate molecule approaches a pocket on the enzyme surface called the **active site.** Amino acid side groups in this region of the enzyme are arranged so as to bind functional groups on the substrate molecule. Many enzymes have two active sites, enabling them to bind two different substrates so that they can react with each other.

② The substrate binds to the enzyme, forming an **enzyme-substrate complex.** The fit between a particular enzyme and its substrate is often compared to a lock and key. Just as only one key fits a particular lock, sucrose is the only substrate that fits the active site of sucrase. Sucrase cannot digest other disaccharides such as maltose or lactose. This selectivity is called **enzyme–substrate specificity.**

③ Sucrase breaks the covalent bond between the two sugars of sucrose, adding $H^+$ and $OH^-$ groups from water (see fig. 2.15b; not illustrated in fig. 2.27). This hydrolyzes sucrose to two monosaccharides, glucose and fructose, which are then released by the enzyme as its **reaction products.** The enzyme remains unchanged and is ready to repeat the process if another sucrose is available.

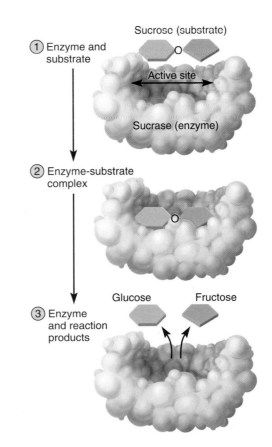

**FIGURE 2.27** **The Three Steps of an Enzymatic Reaction.**

Since an enzyme is not consumed by the reaction it catalyzes, one enzyme molecule can consume millions of substrate molecules, and at astonishing speed. A single molecule of carbonic anhydrase, for example, breaks carbonic acid ($H_2CO_3$) down to $H_2O$ and $CO_2$ at a rate of 36 million molecules per minute.

Factors that change the shape of an enzyme—notably temperature and pH—tend to alter or destroy the ability of the enzyme to bind its substrate. They disrupt the hydrogen bonds and other weak forces that hold the enzyme in its proper conformation, essentially changing the shape of the "lock" (active site) so that the "key" (substrate) no longer fits. Enzymes vary in optimum pH according to where in the body they normally function. Thus salivary amylase, which digests starch in the mouth, functions best at pH 7 and is inactivated when it is exposed to stomach acid; pepsin, which works in the acidic environment of the stomach, functions best around pH 2; and trypsin, a digestive enzyme that works in the alkaline environment of the small intestine, has an optimum pH of 9.5. Our internal body temperature is nearly the same everywhere, however, and all human enzymes have a temperature optimum (that is, they produce their fastest reaction rates) near 37°C.

## Think About It

*Why is homeostasis important for enzyme function?*

## Cofactors

Many enzymes cannot function without nonprotein partners called **cofactors**—for example, iron, copper, zinc, magnesium, or calcium ions. By binding to an enzyme, a cofactor may stimulate it to fold into a shape that activates its active site. **Coenzymes** are organic cofactors usually derived from niacin, riboflavin, and other water-soluble vitamins. They accept electrons from an enzyme in one metabolic pathway and transfer them to an enzyme in another. For example, cells partially oxidize glucose through a pathway called *glycolysis*. A coenzyme called $NAD^+$,[29] derived from niacin, shuttles electrons from this pathway to another one called *aerobic respiration*, which uses energy from the electrons to make ATP (fig. 2.28). If $NAD^+$ is unavailable, the glycolysis pathway shuts down.

## Metabolic Pathways

A **metabolic pathway** is a chain of reactions with each step usually catalyzed by a different enzyme. A simple metabolic pathway can be symbolized

$$\begin{matrix} \alpha & \beta & \gamma \\ A \rightarrow & B \rightarrow & C \rightarrow D \end{matrix}$$

where A is the initial *reactant*, B and C are *intermediates,* and D is the *end product.* The Greek letters above the reaction arrows represent enzymes that catalyze each step of the reaction. A is the substrate for enzyme $\alpha$, B is the substrate for enzyme $\beta$, and C for enzyme $\gamma$. Such a pathway can be turned on or off by altering the conformation of any of these enzymes, thereby activating or deactivating them. This can be done by such means as the binding or dissociation of a cofactor, or by an end product of the pathway binding to an enzyme at an earlier step (product D binding to enzyme $\alpha$ and shutting off the reaction chain at that step, for example). In these and other ways, cells are able to turn on metabolic pathways when their end products are needed and shut them down when the end products are not needed.

# ATP, OTHER NUCLEOTIDES, AND NUCLEIC ACIDS

**Nucleotides** are organic compounds with three principal components: a single or double carbon–nitrogen ring called a *nitrogenous base,* a monosaccharide, and one or more phosphate groups. One of the best-known nucleotides is ATP (fig. 2.29a), in which the nitrogenous base is a double ring called *adenine,* the sugar is *ribose,* and there are three phosphate groups.

## Adenosine Triphosphate

**Adenosine triphosphate (ATP)** is the body's most important energy-transfer molecule. It briefly stores energy gained from exergonic reactions such as glucose oxidation and releases it within seconds for physiological work such as polymerization reactions, muscle contraction, and pumping ions through cell membranes. The second and third phosphate groups of ATP are attached to the rest

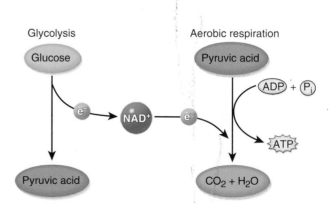

**FIGURE 2.28   The Action of a Coenzyme.** A coenzyme such as $NAD^+$ acts as a shuttle that picks up electrons from one metabolic pathway (in this case, glycolysis) and delivers them to another (in this case, aerobic respiration). ADP = adenosine diphosphate; ATP = adenosine triphosphate; $P_i$ = inorganic phosphate.

---

[29]nicotinamide adenine dinucleotide

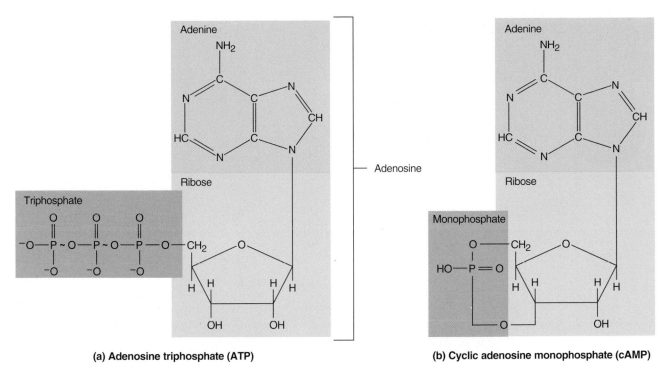

(a) Adenosine triphosphate (ATP)

(b) Cyclic adenosine monophosphate (cAMP)

**FIGURE 2.29**  Adenosine Triphosphate (ATP) and Cyclic Adenosine Monophosphate (cAMP).  The last two P~O bonds in ATP, indicated by wavy lines, are high-energy bonds.

of the molecule by high-energy covalent bonds traditionally indicated by a wavy line (~) in the molecular formula. Since phosphate groups are negatively charged, they repel each other. It requires a high-energy bond to overcome that repellent force and hold them together—especially to add the third phosphate group to a chain that already has two negatively charged phosphates. Most energy transfers to and from ATP involve adding or removing that third phosphate.

Enzymes called **adenosine triphosphatases (ATPases)** are specialized to hydrolyze the third high-energy phosphate bond, producing adenosine diphosphate (ADP) and an inorganic phosphate group ($P_i$). This reaction releases 7.3 kilocalories (kcal) of energy for every mole (505 g) of ATP. Most of this energy escapes as heat, but we live on the portion of it that does useful work. We can summarize this as follows:

$$ATP + H_2O \xrightarrow{\text{ATPase}} ADP + P_i + Energy \begin{smallmatrix} \nearrow \text{Heat} \\ \\ \searrow \text{Work} \end{smallmatrix}$$

The free phosphate groups released by ATP hydrolysis are often added to enzymes or other molecules to activate them. This addition of $P_i$, called **phosphorylation,** is carried out by enzymes called **kinases** (phosphokinases). The phosphorylation of an enzyme is sometimes the "switch" that turns a metabolic pathway on or off.

ATP is a short-lived molecule, usually consumed within 60 seconds of its formation. The entire amount in the body would support life for less than 1 minute if it were not continually replenished. At a moderate rate of physical activity, a full day's supply of ATP would weigh twice as much as you do. Even if you never got out of bed, you would need about 45 kg (99 lb) of ATP to stay alive for a day. The reason cyanide is so lethal is that it halts ATP synthesis.

ATP synthesis is explained in detail in chapter 26, but you will find it necessary to become familiar with the general idea of it before you reach that chapter—especially in understanding muscle physiology (chapter 11). Much of the energy for ATP synthesis comes from glucose oxidation (fig. 2.30). The first stage in glucose oxidation (fig. 2.31) is the reaction pathway known as **glycolysis** (gly-COLL-ih-sis). This literally means "sugar splitting," and indeed its major effect is to split the six-carbon glucose molecule into two three-carbon molecules of *pyruvic acid*. A little ATP is produced in this stage (a net yield of 2 ATP per glucose), but most of the chemical energy of the glucose is still in the pyruvic acid.

What happens to pyruvic acid depends on whether oxygen is available. If not, pyruvic acid is converted to lactic acid by a pathway called **anaerobic**[30] (AN-err-OH-bic) **fermentation.** This pathway has two noteworthy

---

[30]*an* = without + *aer* = air + *obic* = pertaining to life

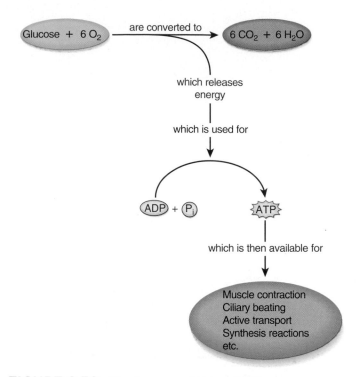

**FIGURE 2.30**  The Source and Uses of ATP.

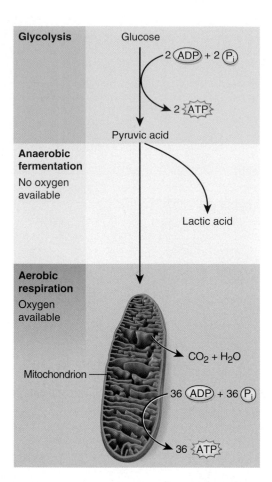

**FIGURE 2.31**  **ATP Production.**  Glycolysis produces pyruvic acid and a net gain of two ATPs. In the absence of oxygen, anaerobic fermentation is necessary to keep glycolysis running and producing a small amount of ATP. In the presence of oxygen, aerobic respiration occurs in the mitochondria and produces a much greater amount of ATP.

disadvantages: First, it does not extract any more energy from pyruvic acid; second, the lactic acid it produces is toxic, so most cells can use anaerobic fermentation only as a temporary measure. The only advantage to this pathway is that it enables glycolysis to continue (for reasons explained in chapter 26) and thus enables a cell to continue producing a small amount of ATP.

If oxygen is available, a more efficient pathway called **aerobic respiration** occurs. This breaks pyruvic acid down to carbon dioxide and water and generates up to 36 more molecules of ATP for each of the original glucose molecules. The reactions of aerobic respiration are carried out in the cell's *mitochondria* (described in chapter 3), so mitochondria are regarded as a cell's principal "ATP factories."

## Other Nucleotides

**Guanosine** (GWAH-no-seen) **triphosphate (GTP)** is another nucleotide involved in energy transfers. In some reactions, it donates phosphate groups to other molecules. For example, it can donate its third phosphate group to ADP to regenerate ATP.

**Cyclic adenosine monophosphate (cAMP)** (see fig. 2.29b) is a nucleotide formed by the removal of both the second and third phosphate groups from ATP. In some cases, when a hormone or other chemical signal ("first messenger") binds to a cell surface, it triggers an internal reaction that converts ATP to cAMP. The cAMP then acts as a "second messenger" to activate metabolic effects within the cell.

## Nucleic Acids

**Nucleic** (new-CLAY-ic) **acids** are polymers of nucleotides. The largest of them, **deoxyribonucleic acid (DNA),** is typically 100 million to 1 billion nucleotides long. It constitutes our genes, gives instructions for synthesizing all of the body's proteins, and transfers hereditary information from cell to cell when cells divide and from generation to generation when organisms reproduce. Three forms of **ribonucleic acid (RNA),** which range from 70 to 10,000 nucleotides long, carry out those instructions and synthesize the proteins, assembling amino acids in the right order to produce each protein "described" by the DNA. The detailed structure of DNA and RNA and the mechanisms of protein synthesis and heredity are described in chapter 4.

## Before You Go On

*Answer the following questions to test your understanding of the preceding section:*

16. Which reaction—dehydration synthesis or hydrolysis—converts a polymer to its monomers? Which one converts monomers to a polymer? Explain your answer.

17. What is the chemical name of blood sugar? What carbohydrate is polymerized to form starch and glycogen?

18. What is the main chemical similarity between carbohydrates and lipids? What are the main differences between them?

19. Explain the statement, All proteins are polypeptides but not all polypeptides are proteins.

20. Which is more likely to be changed by heating a protein, its primary structure or its tertiary structure? Explain.

21. Use the lock-and-key analogy to explain why excessively acidic body fluids (acidosis) could destroy enzyme function.

22. How does ATP change structure in the process of releasing energy?

23. What advantage and disadvantage does anaerobic fermentation have compared with aerobic respiration?

24. How is DNA related to nucleotides?

---

**INSIGHT 2.5**    Clinical Application

## Anabolic-Androgenic Steroids

The sex hormone testosterone stimulates muscular growth and aggressive behavior, especially in males. In Nazi Germany, testosterone was given to SS troops in an effort to make them more aggressive, but with no proven success. In the 1950s, pharmaceutical companies developed compounds related to testosterone, called *anabolic-androgenic*[31] *steroids,* to treat anemia, breast cancer, osteoporosis, and some muscle diseases, and to prevent the shrinkage of muscles in immobilized patients. By the early 1960s, athletes were using anabolic–androgenic steroids to stimulate muscle growth, accelerate the repair of tissues damaged in training or competition, and stimulate the aggressiveness needed to excel in some contact sports such as football and boxing.

The doses used by athletes, however, are 10 to 1,000 times higher than the doses prescribed for medical purposes, and they can have devastating effects on one's health. They raise cholesterol levels, which promotes fatty degeneration of the arteries (*atherosclerosis*). This can lead to coronary artery disease, heart and kidney failure, and stroke. Deteriorating blood circulation also sometimes results in gangrene, which may require amputation of the extremities. As the liver attempts to dispose of the high concentration of steroids, liver cancer and other liver diseases may ensue. In addition, steroids suppress the immune system, so the user is more subject to infection and cancer. They cause a premature end to bone elongation, so people who use anabolic steroids in adolescence may never attain normal adult height.

Anabolic–androgenic steroids have the same effect on the brain as natural testosterone. Thus, when steroid levels are high, the brain and pituitary gland stop producing the hormones that stimulate sperm production and testosterone secretion. In men,

this leads to atrophy of the testes, impotence (inability to achieve or maintain an erection), low sperm count, and infertility. Ironically, anabolic–androgenic steroids have feminizing effects on men and masculinizing effects on women. Men may develop enlarged breasts (*gynecomastia*), while in some female users the breasts and uterus atrophy, the clitoris enlarges, and ovulation and menstruation become irregular. Female users may develop excessive facial and body hair and a deeper voice, and both sexes show an increased tendency toward baldness.

Especially in men, steroid abuse can be linked to severe emotional disorders. The steroids themselves stimulate heightened aggressiveness and unpredictable mood swings, so the abuser may vacillate between depression and violence. It surely doesn't help matters that impotence, shrinkage of the testes, infertility, and enlargement of the breasts are so incongruous with the self-image of a male athlete who abuses steroids.

Partly because of the well-documented adverse health effects, the use of anabolic–androgenic steroids has been condemned by American Medical Association and American College of Sports Medicine and banned by the International Olympic Committee, National Football League, Major League Baseball Players' Association, and National Collegiate Athletic Association. But in spite of such warnings and bans, many athletes continue to use steroids and related performance-enhancing drugs, which remain available through unscrupulous coaches, physicians, Internet sources, and foreign mail-order suppliers. By some estimates, as many as 80% of weight lifters, 30% of college and professional athletes, and 20% of male high school athletes now use anabolic–androgenic steroids. The National Institutes of Health finds increasing usage among high school students in recent years, and increasing denial that anabolic–androgenic steroids present a significant health hazard.

---

[31]*andro* = male + *genic* = producing

## CHAPTER REVIEW

# Review of Key Concepts

### Atoms, Ions, and Molecules (p. 54)

1. The simplest form of matter with unique chemical properties is the *element.* Twenty-four elements play normal physiological roles in humans. Those called *trace elements* are needed in only tiny amounts.

2. An *atom* consists of a central positively charged nucleus of protons and usually neutrons, orbited by a usually multilayered cloud of negatively charged electrons.

3. The outermost electrons, called *valence electrons,* determine the chemical behavior of an element.

4. *Isotopes* are variations of an element that differ only in the number of neutrons. Some are unstable *radioisotopes,* which give off radioactivity as they change to a more stable isotope.

5. *Ions* are particles with one or more excess electrons or protons, and thus a negative charge *(anions)* or positive charge *(cations).* Oppositely charged ions are attracted to each other and tend to follow each other in the body.

6. *Electrolytes* are salts that ionize in water to form solutions that conduct electricity. These include salts of sodium, potassium, chlorine, phosphate, bicarbonate, and other elements (table 2.2).

7. *Free radicals* are highly reactive particles with an odd number of electrons. They have very destructive effects on cells and may contribute to aging and cancer, but the body has *antioxidant* chemicals that provide some protection from them.

8. A *molecule* consists of two or more atoms joined by chemical bonds. If the elements are nonidentical, the molecule is a *compound.*

9. *Isomers* are molecules with the same number and kinds of elements, but different arrangements of them and different chemical properties.

10. A molecule's *molecular weight* is the sum of the atomic weights of its elements.

11. Molecules are held together by *ionic, covalent,* or *hydrogen bonds, van der Waals forces,* or a combination of these.

### Water and Mixtures (p. 61)

1. The polarity and bond angle of water result in the hydrogen bonding of water molecules to one another. Hydrogen bonding is responsible for the diverse biologically important properties of water.

2. A *mixture* is a combination of substances that are physically blended but not chemically combined. Most mixtures in the body are a combination of water and various solutes.

3. Mixtures can be classified as *solutions, colloids,* or *suspensions* based on the size of their particles.

4. The concentration of a mixture is expressed for differing purposes as weight per volume, percentage, molarity, or (for electrolytes) milliequivalents per liter.

5. The concentration of hydrogen ions in a solution is expressed as pH, with a range from 1 to 14. A pH of 7 is neutral (with equal quantities of $H^+$ and $OH^-$), a pH < 7.0 is acidic, and a pH > 7.0 is basic.

6. *Buffers* are chemical solutions that resist changes in pH when acid or base is added to them.

### Energy and Chemical Reactions (p. 67)

1. *Energy,* the capacity to do work, exists in *potential* and *kinetic* forms.

2. Chemical reactions can be *decomposition, synthesis,* or *exchange* reactions.

3. Some chemical reactions are *reversible;* their direction depends on the relative amounts of reactants and products present. Such reactions tend to achieve an *equilibrium* state unless disrupted by the addition of new reactants or removal of products.

4. The rate of a chemical reaction is influenced by concentration, temperature, and catalysts.

5. *Metabolism* is the sum of all chemical reactions in the body. It consists of *catabolism* (breakdown of larger molecules into smaller ones) and *anabolism* (synthesis of larger molecules).

6. *Oxidation* is the removal of electrons from a molecule; *reduction* is the addition of electrons.

### Organic Compounds (p. 70)

1. *Organic* molecules contain carbon. They often have carbon atoms arranged in a *backbone* with attached *functional groups* (carboxyl and amino groups, for example) that determine the chemical behavior of the molecule.

2. Many biologically important molecules are *polymers*—large molecules composed of a chain of identical or similar subunits called *monomers.*

3. The joining of monomers to form a polymer, called *polymerization,* is achieved by a *dehydration synthesis* reaction that removes water from the reactants. Polymers are broken up into monomers by *hydrolysis* reactions, which consume water to add —H and —OH to the molecules.

4. *Carbohydrates* are organic molecules of carbon and a 2:1 ratio of H:O. The major carbohydrates are the *monosaccharides* (glucose, fructose, galactose), *disaccharides* (sucrose, lactose, maltose), and *polysaccharides* (starch, cellulose, glycogen).

5. Carbohydrates are good sources of quickly mobilized energy but also play structural and other roles (table 2.6).

6. *Lipids* are hydrophobic compounds of carbon and a high ratio of H:O. Major classes of lipids are *fatty acids, triglycerides, phospholipids, eicosanoids,* and *steroids.*

7. Lipids serve for energy storage, as chemical signals, and as structural components of cells, among other roles (table 2.7).

8. *Proteins* are polymers of amino acids.

9. An *amino acid* is a small organic molecule with an amino (—NH$_2$) and carboxyl (—COOH) group. They can join together through *peptide bonds* to form *peptides* from two to thousands of amino acids long. Proteins are generally regarded as peptides of 50 or more amino acids.

10. Proteins have four levels of structure: *primary* (amino acid sequence), *secondary* (an alpha helix or beta sheet), *tertiary* (further bending and folding), and sometimes *quaternary* (attraction of two or more polypeptide chains to each other).

11. Protein function depends strongly on three-dimensional shape, or *conformation. Denaturation* is a destructive change in conformation, usually caused by temperature or pH changes.

12. *Conjugated* proteins require a nonprotein component such as a carbohydrate or an inorganic ion attached to them in order to function. The nonprotein moiety is called the *prosthetic group.*

13. Proteins have a wide range of structural, communication, transport, catalytic, and other functions.

14. *Enzymes* are proteins that function as biological catalysts. The substances they act upon are called their *substrates,* and bind to an enzyme at specific locations called *active sites.*

15. Enzymes accelerate chemical reactions by lowering their *activation energy.*

16. An enzyme generally reacts only with specific substrates that fit its active site.

17. Enzymatic reactions are often linked together to form *metabolic pathways.*

18. *Adenosine triphosphate (ATP)* is a universal energy-transfer molecule composed of adenine, ribose, and three phosphate groups. It is essential to many physiological processes, and life ends in seconds in the absence of ATP.

19. Small amounts of ATP are generated by glycolysis linked, in the absence of oxygen, to anaerobic fermentation. Much larger amounts are generated when oxygen is available and glycolysis is linked to aerobic respiration.

20. The *nucleic acids,* DNA and RNA, are polymers of ATP-like nucleotides. They are responsible for heredity and the control of protein synthesis.

## Testing Your Recall

1. A substance that _____ is considered to be a chemical compound.
   a. contains at least two different elements
   b. contains at least two atoms
   c. has a chemical bond
   d. has a stable valence shell
   e. has covalent bonds

2. An ionic bond is formed when
   a. two anions meet.
   b. two cations meet.
   c. an anion meets a cation.
   d. electrons are unequally shared between nuclei.
   e. electrons transfer completely from one atom to another.

3. The ionization of a sodium atom to produce Na$^+$ is an example of
   a. oxidation.
   b. reduction.
   c. catabolism.
   d. anabolism.
   e. a decomposition reaction.

4. The weakest and most temporary chemical bonds are
   a. polar covalent bonds.
   b. nonpolar covalent bonds.
   c. hydrogen bonds.
   d. ionic bonds.
   e. double covalent bonds.

5. A substance capable of dissolving freely in water is
   a. hydrophilic.
   b. hydrophobic.
   c. hydrolyzed.
   d. hydrated.
   e. amphiphilic.

6. A carboxyl group is symbolized
   a. —OH.
   b. —NH$_2$.
   c. —CH$_3$.
   d. —CH$_2$OH.
   e. —COOH.

7. The only polysaccharide synthesized in the human body is
   a. cellulose.
   b. glycogen.
   c. cholesterol.
   d. starch.
   e. prostaglandin.

8. The arrangement of a polypeptide into a fibrous or globular shape is called its
   a. primary structure.
   b. secondary structure.
   c. tertiary structure.
   d. quaternary structure.
   e. conjugated structure.

9. Which of the following functions is more characteristic of carbohydrates than of proteins?
   a. contraction
   b. energy storage
   c. catalyzing reactions
   d. immune defense
   e. intercellular communication

10. The feature that most distinguishes a lipid from a carbohydrate is that a lipid has
    a. more phosphate.
    b. more sulfur.
    c. a lower ratio of carbon to oxygen.
    d. a lower ratio of oxygen to hydrogen.
    e. a greater molecular weight.

11. When an atom gives up an electron and acquires a positive charge, it is called a/an _____.

12. Dietary antioxidants are important because they neutralize _____.

13. Any substance that increases the rate of a reaction without being consumed by it is a/an _____. In the human body, _____ serve this function.

14. All the synthesis reactions in the body form a division of metabolism called _____.

15. A chemical reaction that produces water as a by-product is called _____.

16. The suffix _____ denotes a sugar, while the suffix _____ denotes an enzyme.

17. The amphiphilic lipids of cell membranes are called _____.

18. A chemical named _____ is derived from ATP and widely employed as a "second messenger" in cellular signaling.

19. When oxygen is unavailable, cells employ a metabolic pathway called _____ to produce ATP.

20. A substance acted upon and changed by an enzyme is called the enzyme's _____.

*Answers in Appendix B*

## True or False?

*Determine which five of the following statements are false, and briefly explain why.*

1. The monomers of a polysaccharide are called amino acids.

2. An emulsion is a mixture of two liquids that separate from each other on standing.

3. Two molecules with the same atoms arranged in a different order are called isotopes.

4. If a pair of shared electrons are more attracted to one nucleus than to the other, they form a polar covalent bond.

5. Amino acids are joined by a unique type of bond called a peptide bond.

6. A saturated fat is defined as a fat to which no more carbon can be added.

7. Organic compounds get their unique chemical characteristics more from their functional groups than from their carbon backbones.

8. The higher the temperature is, the faster an enzyme works.

9. Two percent sucrose and 2% sodium bicarbonate have the same number of molecules per liter of solution.

10. A solution of pH 8 has one-tenth the hydrogen ion concentration of a solution with pH 7.

*Answers in Appendix B*

## Testing Your Comprehension

1. Suppose a pregnant woman with severe morning sickness has been vomiting steadily for several days. How will her loss of stomach acid affect the pH of her body fluids? Explain.

2. Suppose a person with a severe anxiety attack hyperventilates and exhales $CO_2$ faster than his body produces it. Consider the carbonic acid reaction on page 68 and explain what effect this hyperventilation will have on his blood pH. (*Hint:* Remember the law of mass action.)

3. In one form of nuclear decay, a neutron breaks down into a proton and electron and emits a gamma ray. Is this an endergonic or exergonic reaction, or neither? Is it an anabolic or catabolic reaction, or neither? Explain both answers.

4. How would the body's metabolic rate be affected if there were no such thing as enzymes? Explain.

5. Some metabolic conditions such as diabetes mellitus cause disturbances in the acid–base balance of the body, which gives the body fluids an abnormally low pH. Explain how this could affect enzyme–substrate reactions and metabolic pathways in the body.

*Answers at www.mhhe.com/saladin4*

## www.mhhe.com/saladin4

*The textbook website provides a wealth of interactive study materials fully organized and integrated by chapter. You will find practice quizzes, labeling exercises, and much more that will complement your learning and understanding of anatomy and physiology. The website also includes tools designed to enhance your* **Anatomy & Physiology** | **REVEALED** *experience.*

*Mitochondria (blue) and rough endoplasmic reticulum (orange) in a pancreatic cell (SEM)*

# CELLULAR FORM AND FUNCTION

## CHAPTER OUTLINE

## INSIGHTS

## Brushing Up

To understand this chapter, it is important that you understand or brush up on the following concepts:

- Glycolipids and glycoproteins (p. 73)
- Phospholipids and their amphiphilic nature (p. 74)
- The relationship of protein function to tertiary structure (p. 78)
- Protein functions (p. 79)

All organisms, from the simplest to the most complex, are composed of cells—whether the single cell of a bacterium or the trillions of cells that constitute the human body. These cells are responsible for all structural and functional properties of a living organism. **Cytology,**[1] the study of cell structure and function, is therefore indispensable to any true understanding of the workings of the human body, the mechanisms of disease, and the rationale of therapy. Thus, this chapter and the next one introduce the basic cell biology of the human body, and subsequent chapters expand upon this information as we examine the specialized cellular functions of specific organs.

# Concepts of Cellular Structure

### Objectives
When you have completed this section, you should be able to

- discuss the development and modern tenets of the cell theory;
- describe cell shapes from their descriptive terms;
- state the size range of human cells and discuss factors that limit cell size;
- discuss the way that developments in microscopy have changed our view of cell structure; and
- outline the major components of a cell.

## DEVELOPMENT OF THE CELL THEORY

As you may recall from chapter 1, Robert Hooke had observed only the empty cell walls of cork when he first named the cell in 1665. Later, he studied thin slices of fresh wood and saw cells "filled with juices"—a fluid later named *protoplasm.*[2] Two centuries later, Theodor Schwann studied a wide range of animal tissues and concluded that all animals are made of cells.

Schwann and other biologists originally believed that cells came from nonliving body fluid that somehow congealed and acquired a membrane and nucleus. This idea of *spontaneous generation*—that living things arise from nonliving matter—was rooted in the scientific thought of the times. For centuries, it seemed to be simple common

sense that decaying meat turns into maggots, stored grain into rodents, and mud into frogs. Schwann and his contemporaries merely extended this idea to cells. The idea of spontaneous generation wasn't discredited until some classic experiments by French microbiologist Louis Pasteur in 1859. By the end of the nineteenth century, it was established beyond all reasonable doubt that cells arise only from other cells.

The development of biochemistry from the late nineteenth to the twentieth century made it further apparent that all physiological processes of the body are based on cellular activity and that the cells of all species exhibit remarkable biochemical unity. Thus emerged the generalizations that constitute the modern cell theory:

1. All organisms are composed of cells and cell products.
2. The cell is the simplest structural and functional unit of life. There are no smaller subdivisions of a cell or organism that, in themselves, are alive.
3. An organism's structure and all of its functions are ultimately due to the activities of its cells.
4. Cells come only from preexisting cells, not from nonliving matter. All life, therefore, traces its ancestry to the same original cells.
5. Because of this common ancestry, the cells of all species have many fundamental similarities in their chemical composition and metabolic mechanisms.

## CELL SHAPES AND SIZES

There are about 200 types of cells in the human body, and they vary greatly in shape (fig. 3.1). **Squamous**[3] (SQUAY-mus) cells are thin, flat, and often have angular contours when viewed from above. Such cells line the esophagus and cover the skin. **Polygonal**[4] cells have irregularly angular shapes with four, five, or more sides. Some nerve cells have multiple extensions that give them a starlike, or **stellate,**[5] shape. **Cuboidal**[6] cells are squarish and approximately as tall as they are wide; liver cells are a good example. **Columnar** cells, such as those lining the intestines, are markedly taller than wide. Egg cells and fat cells are **spheroid** to **ovoid** (round to oval). Red blood cells are **discoid** (disc-shaped). Smooth muscle cells are **fusiform**[7] (FEW-zih-form)—thick in the middle and tapered toward the ends. Skeletal muscle cells are described as **fibrous** because of their threadlike shape.

Most human cells range from 10 to 15 micrometers ($\mu$m) in diameter. (See appendix C for units of measurement.) The human egg cell, an exceptionally large 100 $\mu$m in diameter, is barely visible to the naked eye. The longest human cells

[1]*cyto* = cell + *logy* = study of
[2]*proto* = first + *plasm* = formed

[3]*squam* = scale + *ous* = characterized by
[4]*poly* = many + *gon* = angles
[5]*stell* = star + *ate* = characterized by
[6]*cub* = cube + *oidal* = like, resembling
[7]*fusi* = spindle + *form* = shape

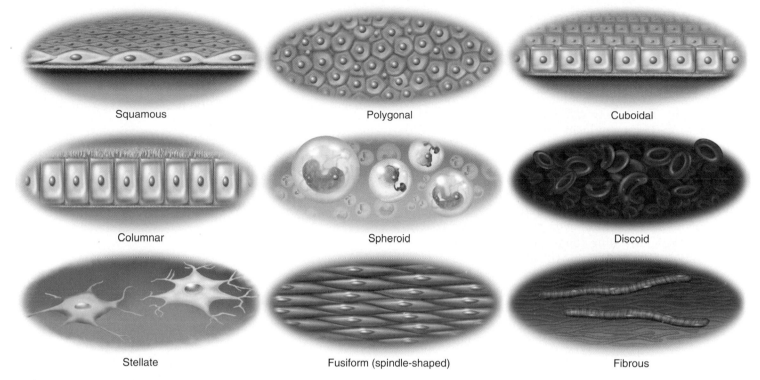

**FIGURE 3.1** Common Cell Shapes.

Squamous

Polygonal

Cuboidal

Columnar

Spheroid

Discoid

Stellate

Fusiform (spindle-shaped)

Fibrous

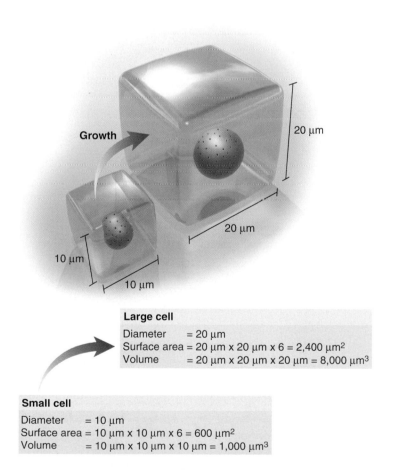

Growth

20 μm

20 μm

10 μm

10 μm

**Large cell**

Diameter    = 20 μm
Surface area = 20 μm x 20 μm x 6 = 2,400 μm²
Volume      = 20 μm x 20 μm x 20 μm = 8,000 μm³

**Small cell**

Diameter    = 10 μm
Surface area = 10 μm x 10 μm x 6 = 600 μm²
Volume      = 10 μm x 10 μm x 10 μm = 1,000 μm³

**Effect of cell growth:**

Diameter ($D$) increased by a factor of 2
Surface area increased by a factor of 4 (= $D^2$)
Volume increased by a factor of 8 (= $D^3$)

are nerve cells (sometimes over a meter long) and muscle cells (up to 30 cm long), but both are too slender to be seen with the naked eye.

There is a limit to how large a cell can be, partly due to the relationship between its volume and surface area. The surface area of a cell is proportional to the square of its diameter, while volume is proportional to the cube of its diameter. Thus, for a given increase in diameter, cell volume increases much faster than surface area. Picture a cuboidal cell 10 μm on each side (fig. 3.2). It would have a surface area of 600 μm² (10 μm × 10 μm × 6 sides) and a volume of 1,000 μm³ (10 × 10 × 10 μm). Now, suppose it grew by another 10 μm on each side. Its new surface area would be 20 μm × 20 μm × 6 = 2,400 μm², and its volume would be 20 × 20 × 20 μm = 8,000 μm³. The 20-μm cell has eight times as much protoplasm needing nourishment and waste removal, but only four times as much membrane surface through which wastes and nutrients can be exchanged. A cell that is too big cannot support itself.

**Think About It**

*Can you conceive of some other reasons for an organ to consist of many small cells rather than fewer larger ones?*

**FIGURE 3.2** **The Relationship Between Cell Surface Area and Volume.** As a cell doubles in diameter, its volume increases eightfold, but its surface area increases only fourfold. A cell that is too large may have too little plasma membrane to serve the metabolic needs of its increased volume of cytoplasm.

# GENERAL CELL STRUCTURE

In Schwann's time, little was known about cells except that they were enclosed in a membrane and contained a nucleus. The fluid between the nucleus and surface membrane, called **cytoplasm,** was thought to be little more than a gelatinous mixture of chemicals. The **transmission electron microscope (TEM),** invented in the mid-twentieth century, radically changed this concept. Using a beam of electrons in place of light, the TEM enabled biologists to see a cell's *ultrastructure* (fig. 3.3), a fine degree of detail extending even to the molecular level. The most important thing about a good microscope is not magnification but **resolution**—the ability to reveal detail. Any image can be photographed and enlarged as much as we wish, but if enlargement fails to reveal any more useful detail, it is *empty magnification*. A big fuzzy image is not nearly as informative as one that is small and sharp. The TEM reveals far more detail than the light micro-scope (LM), even at the same magnification (fig. 3.4). A later invention, the **scanning electron microscope (SEM),** produces dramatic three-dimensional images at high magnification and resolution (see fig. 3.11a), but can only view surface features.

Table 3.1 gives the sizes of some cells and subcellular objects relative to the resolution of the naked eye, light microscope, and TEM. You can see why the very existence of cells was unsuspected until the light microscope was invented, and why little was known about their internal components until the TEM became available.

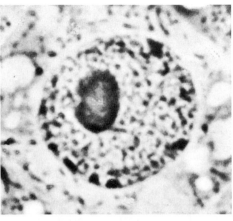

Light microscope (LM)

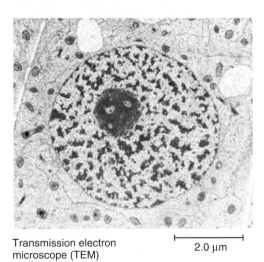

Transmission electron microscope (TEM)                    2.0 μm

**FIGURE 3.4  Magnification Versus Resolution.**
These cells were photographed at the same magnification (about ×750) through a light microscope and a transmission electron microscope.

Plasma membrane

Nucleus

Nuclear envelope

Mitochondria

Golgi vesicle

Golgi complex

Ribosomes

2.0 μm

**FIGURE 3.3**  Ultrastructure of a White Blood Cell (TEM).

| TABLE 3.1 | Sizes of Biological Structures in Relation to the Resolution of the Eye, Light Microscope, and Transmission Electron Microscope |
|---|---|
| **Object** | **Size** |
| **Visible to the naked eye (resolution 70–100 μm)** | |
| Human egg, diameter | 100 μm |
| **Visible with the light microscope (resolution 200 nm)** | |
| Most human cells, diameter | 10–15 μm |
| Cilia, length | 7–10 μm |
| Mitochondria, width × length | 0.2 × 4 μm |
| Bacteria (*Escherichia coli*), length | 1–3 μm |
| Microvilli, length | 1–2 μm |
| Lysosomes, diameter | 0.5 μm = 500 nm |
| **Visible with the transmission electron microscope (resolution 0.5 nm)** | |
| Nuclear pores, diameter | 30–100 nm |
| Centriole, diameter × length | 20 × 50 nm |
| Polio virus, diameter | 30 nm |
| Ribosomes, diameter | 15 nm |
| Globular proteins, diameter | 5–10 nm |
| Plasma membrane, thickness | 7.5 nm |
| DNA molecule, diameter | 2.0 nm |
| Plasma membrane channels, diameter | 0.8 nm |

Figure 3.5 shows some major constituents of a typical cell. The cell is surrounded by a **plasma (cell) membrane** made of proteins and lipids. The composition and functions of this membrane can differ significantly from one region of a cell to another, especially among the basal, lateral, and apical (upper) surfaces of cells like the one pictured.

The cytoplasm is crowded with fibers, tubules, passageways, and compartments (see photographs on pp. 89 and 1001). It includes several kinds of **organelles** and a supportive framework called the **cytoskeleton**—all of which we will study in this chapter. A cell may have 10 billion protein molecules, including potent enzymes with the potential to destroy the cell if they are not contained and isolated from other cellular components. You can imagine the enormous problem of keeping track of all this material, directing molecules to the correct destinations,

and maintaining order against the incessant trend toward disorder. Cells maintain order partly by compartmentalizing their contents in the organelles. The organelles and cytoskeleton are embedded in a clear gel called the **cytosol** or **intracellular fluid (ICF).** The fluid outside the cell is **extracellular fluid (ECF).**

## Before You Go On

*Answer the following questions to test your understanding of the preceding section:*

1. *What are the basic principles of the cell theory?*
2. *What does it mean to say a cell is squamous, stellate, columnar, or fusiform?*
3. *Why can cells not grow to unlimited size?*
4. *What is the difference between cytoplasm and cytosol?*
5. *Define intracellular fluid (ICF) and extracellular fluid (ECF).*

# The Cell Surface

### Objectives

When you have completed this section, you should be able to

- describe the structure of a plasma membrane;
- explain the functions of the lipid, protein, and carbohydrate components of the plasma membrane;
- describe a second-messenger system and discuss its importance in human physiology;
- describe the composition and functions of the glyco calyx that coats cell surfaces; and
- describe the structure and functions of microvilli, cilia, and flagella.

Throughout this book, you will find that many of the most physiologically important processes occur at the surface of a cell—such events as immune responses, the binding of egg and sperm, cell-to-cell signaling by hormones, and the detection of tastes and smells, for example. A substantial part of this chapter is therefore concerned with the cell surface. Before we venture into the interior of the cell, we will examine the structure of the plasma membrane, surface features such as cilia and microvilli, and methods of transport through the membrane.

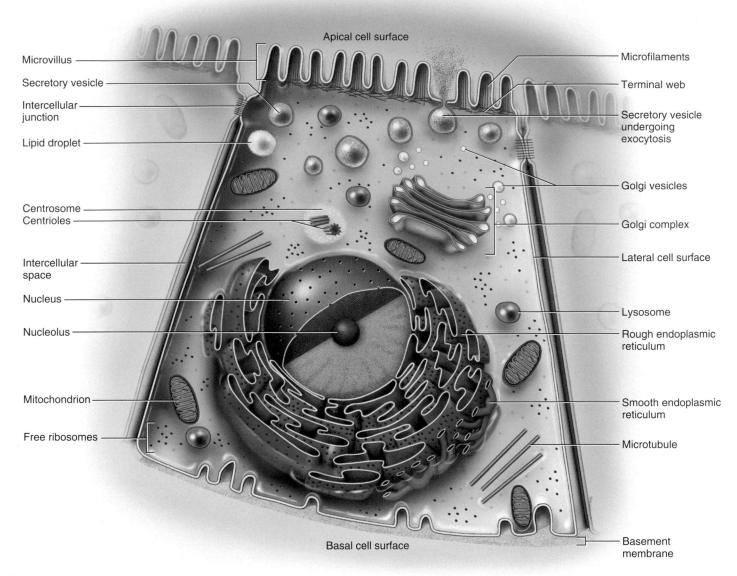

Microvillus

Secretory vesicle

Intercellular junction

Lipid droplet

Centrosome
Centrioles

Intercellular space

Nucleus

Nucleolus

Mitochondrion

Free ribosomes

Apical cell surface

Microfilaments

Terminal web

Secretory vesicle undergoing exocytosis

Golgi vesicles

Golgi complex

Lateral cell surface

Lysosome

Rough endoplasmic reticulum

Smooth endoplasmic reticulum

Microtubule

Basal cell surface

Basement membrane

**FIGURE 3.5**  Structure of a Representative Cell.

## THE PLASMA MEMBRANE

The electron microscope reveals that the cell and many of the organelles within it are bordered by a *unit membrane,* which appears as a pair of dark parallel lines with a total thickness of about 7.5 nm (fig. 3.6a). The *plasma membrane* is the unit membrane at the cell surface. It defines the boundaries of the cell, governs its interactions with other cells, and controls the passage of materials into and out of the cell. The side that faces the cytoplasm is the *intracellular face* of the membrane, and the side that faces outward is the *extracellular face.*

## Membrane Lipids

Figure 3.6b shows our current concept of the molecular structure of the plasma membrane—an oily film of lipids with diverse proteins embedded in it. Typically about 98% of the molecules in the membrane are lipids, and about 75% of the lipids are phospholipids. These amphiphilic molecules arrange themselves into a bilayer, with their hydrophilic phosphate-containing heads facing the water on each side of the membrane and their hydrophobic tails directed toward the center of the membrane, avoiding the water. The phospholipids drift laterally from place to place, spin on their axes, and flex their tails. These movements keep the membrane fluid.

*What would happen if the plasma membrane were made primarily of a hydrophilic substance such as carbohydrate? Which of the major themes at the end of chapter 1 does this point best exemplify?*

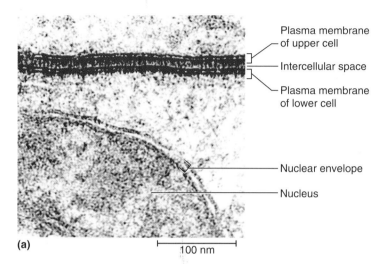

Plasma membrane of upper cell

Intercellular space

Plasma membrane of lower cell

Nuclear envelope

Nucleus

**(a)**

100 nm

Cholesterol molecules, found amid the fatty acid tails, constitute about 20% of the membrane lipids. By interacting with the phospholipids and holding them still, cholesterol can stiffen the membrane (make it less fluid) in spots. Higher concentrations of cholesterol, however, can increase membrane fluidity by preventing the phospholipids from becoming packed closely together.

The remaining 5% of the membrane lipids are glycolipids—phospholipids with short carbohydrate chains on the extracellular face of the membrane. They help to form the *glycocalyx,* a carbohydrate coating on the cell surface with multiple functions described shortly.

## Membrane Proteins

Although proteins are only about 2% of the molecules of the plasma membrane, they are larger than lipids and constitute about 50% of the membrane weight. Some of them, called **transmembrane proteins,** pass through the membrane. They have hydrophilic regions in contact with the cytoplasm and extracellular fluid, and hydrophobic

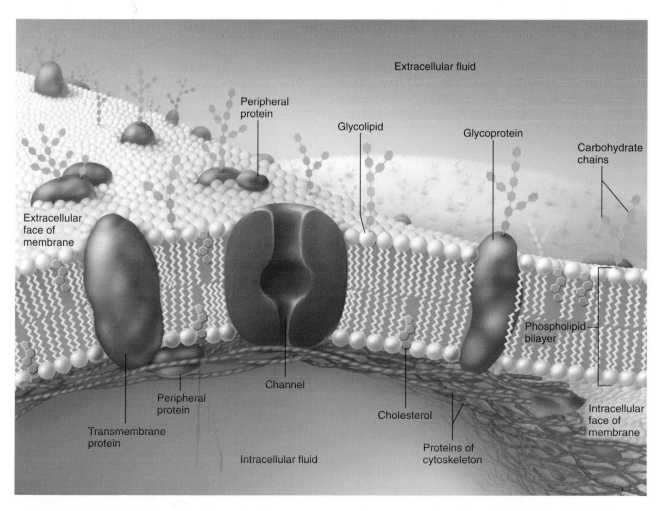

Extracellular fluid

Peripheral protein

Glycolipid

Glycoprotein

Carbohydrate chains

Extracellular face of membrane

Phospholipid bilayer

Channel

Peripheral protein

Transmembrane protein

Cholesterol

Intracellular face of membrane

Proteins of cytoskeleton

Intracellular fluid

**(b)**

**FIGURE 3.6** **The Plasma Membrane.** (a) Plasma membranes of two adjacent cells (TEM). (b) Molecular structure of the plasma membrane.

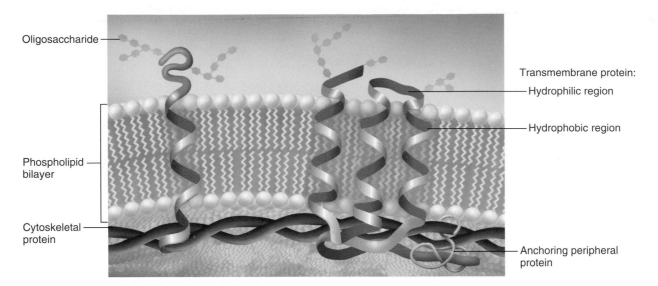

Oligosaccharide

Transmembrane protein:
Hydrophilic region
Hydrophobic region

Phospholipid bilayer

Cytoskeletal protein

Anchoring peripheral protein

**FIGURE 3.7   Transmembrane Proteins.**  A transmembrane protein has hydrophobic regions embedded in the phospholipid bilayer and hydrophilic regions projecting into the intracellular and extracellular fluids. The protein may cross the membrane once (left) or multiple times (right). The intracellular regions are often anchored to the cytoskeleton by peripheral proteins.

regions that pass back and forth through the lipid of the membrane (fig. 3.7). Most transmembrane proteins are glycoproteins, which are conjugated with oligosaccharides on the extracellular side of the membrane. Many of the transmembrane proteins drift about freely in the phospholipid film, like ice cubes floating in a bowl of water. Others are anchored to the *cytoskeleton*—an intracellular system of tubules and filaments discussed later. **Peripheral proteins** do not protrude into the phospholipid layer but adhere to one face of the membrane. A peripheral protein is typically associated with a transmembrane protein and tethered to the cytoskeleton.

The functions of membrane proteins include the following:

- **Receptors** (fig. 3.8a). The chemical signals by which cells communicate with each other (epinephrine, for example) often cannot enter the target cell, but bind to surface proteins called receptors. Receptors are usually specific for one particular messenger, much like an enzyme that is specific for one substrate.

- **Second-messenger systems.** When a messenger binds to a surface receptor, it may trigger changes within the cell that produce a second messenger in the cytoplasm. This process involves both transmembrane proteins (the receptors) and peripheral proteins. Second-messenger systems are discussed shortly in more detail.

- **Enzymes** (fig. 3.8b). Enzymes in the plasma membranes of cells carry out the final stages of starch and protein digestion in the small intestine, help produce second messengers, and break down hormones and other signaling molecules whose job is done, thus stopping them from excessively stimulating a cell.

- **Channel proteins** (fig. 3.8c). Channel proteins are transmembrane proteins with pores that allow passage of water and hydrophilic solutes through the membrane. Some channels are always open, while others are **gates** that open and close under different circumstances, thus determining when solutes can pass through (fig. 3.8d). These gates open or close in response to three types of stimuli: **ligand-regulated gates** respond to chemical messengers, **voltage-regulated gates** to changes in electrical potential (voltage) across the plasma membrane, and **mechanically regulated gates** to physical stress on a cell, such as stretch and pressure. By controlling the movement of electrolytes through the plasma membrane, gated channels play an important role in the timing of nerve signals and muscle contraction (Insight 3.1).

- **Carriers** (see figs. 3.18 and 3.19). Carriers are transmembrane proteins that bind to glucose, electrolytes, and other solutes and transfer them to the other side of the membrane. Some carriers, called **pumps,** consume ATP in the process.

- **Cell-identity markers** (fig. 3.8e). Glycoproteins contribute to the *glycocalyx,* a carbohydrate surface coating discussed shortly. Among other functions, this acts like an "identification tag" that enables our bodies to tell which cells belong to it and which are foreign invaders.

- **Cell-adhesion molecules** (fig. 3.8f). Cells adhere to one another and to extracellular material through certain membrane proteins called cell-adhesion molecules (CAMs). With few exceptions (such as blood cells and metastasizing cancer cells), cells do not grow or survive normally unless they are mechani-

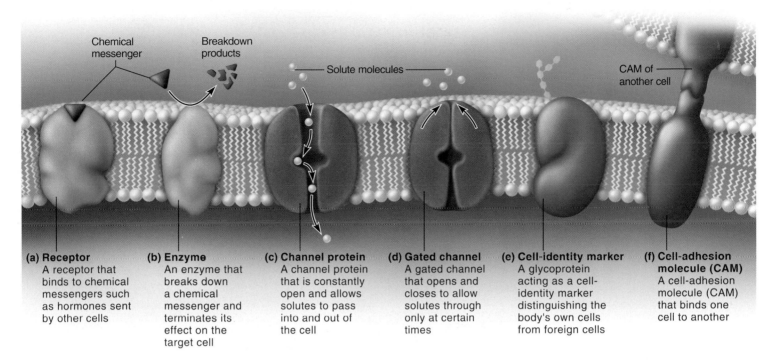

(a) **Receptor**
A receptor that binds to chemical messengers such as hormones sent by other cells

(b) **Enzyme**
An enzyme that breaks down a chemical messenger and terminates its effect on the target cell

(c) **Channel protein**
A channel protein that is constantly open and allows solutes to pass into and out of the cell

(d) **Gated channel**
A gated channel that opens and closes to allow solutes through only at certain times

(e) **Cell-identity marker**
A glycoprotein acting as a cell-identity marker distinguishing the body's own cells from foreign cells

(f) **Cell-adhesion molecule (CAM)**
A cell-adhesion molecule (CAM) that binds one cell to another

**FIGURE 3.8** Some Functions of Membrane Proteins.

cally linked to the extracellular material. Special events such as sperm–egg binding and the binding of an immune cell to a cancer cell also require CAMs.

## Second Messengers

**Second messengers** are of such importance that they require a closer look. You will find this information essential for your later understanding of hormone and neurotransmitter action. Let's consider how the hormone epinephrine stimulates a cell. Epinephrine, the "first messenger," cannot pass through plasma membranes, so it binds to a surface receptor. The receptor is linked on the intracellular side to a peripheral protein called a **G protein** (fig. 3.9). G proteins are named for the ATP-like chemical, guanosine triphosphate (GTP), from which they get their energy. When activated by the receptor, a G protein relays the signal to another membrane protein, **adenylate cyclase** (ah-DEN-ih-late SY-clase). Adenylate cyclase removes two phosphate groups from ATP and converts it to cyclic AMP (cAMP), the second messenger. Cyclic AMP then activates enzymes called **kinases** (KY-nace-es) in the cytosol. Kinases add phosphate groups to other cellular enzymes. This activates some enzymes and deactivates others, but either way, it triggers a great variety of physiological changes within the cell.

G proteins play such an enormous range of roles in physiology and disease that Martin Rodbell and Alfred Gilman received a 1994 Nobel Prize for discovering them. Up to 60% of currently used drugs work by altering the activity of G proteins.

---

**INSIGHT 3.1** Clinical Application

### Calcium Channel Blockers

The walls of the arteries contain smooth muscle that contracts or relaxes to change their diameter. These changes modify the blood flow and strongly influence blood pressure. Blood pressure rises when the arteries constrict and falls when they relax and dilate. Excessive, widespread vasoconstriction can cause hypertension (high blood pressure), and vasoconstriction in the coronary blood vessels of the heart can cause pain (angina) due to inadequate blood flow to the cardiac muscle. In order to contract, a smooth muscle cell must open calcium channels in its plasma membrane and allow calcium to enter from the extracellular fluid. Drugs called *calcium channel blockers* prevent calcium channels from opening. Thus they help to relax the arteries, relieve angina, and lower blood pressure.

## THE GLYCOCALYX

External to the plasma membrane, all animal cells have a fuzzy coat called the **glycocalyx**[8] (GLY-co-CAY-licks) (fig. 3.10), which consists of the carbohydrate moieties of membrane glycolipids and glycoproteins. It is chemically unique in everyone but identical twins, and acts like an identification tag that enables the body to distinguish its own healthy cells from transplanted tissues, invading

---

[8]*glyco* = sugar + *calyx* = cup, vessel

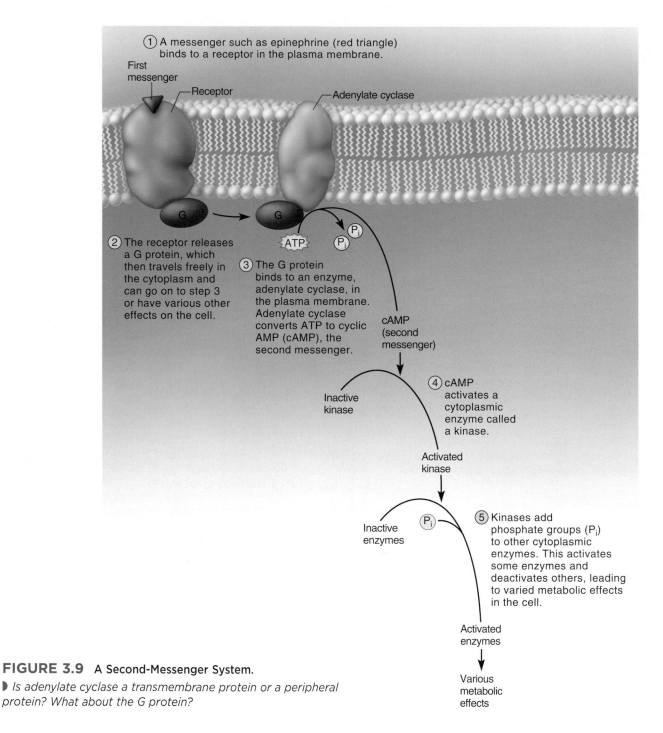

① A messenger such as epinephrine (red triangle) binds to a receptor in the plasma membrane.

First messenger

Receptor

Adenylate cyclase

G

G

ATP

$P_i$

$P_i$

② The receptor releases a G protein, which then travels freely in the cytoplasm and can go on to step 3 or have various other effects on the cell.

③ The G protein binds to an enzyme, adenylate cyclase, in the plasma membrane. Adenylate cyclase converts ATP to cyclic AMP (cAMP), the second messenger.

cAMP (second messenger)

Inactive kinase

④ cAMP activates a cytoplasmic enzyme called a kinase.

Activated kinase

Inactive enzymes

$P_i$

⑤ Kinases add phosphate groups ($P_i$) to other cytoplasmic enzymes. This activates some enzymes and deactivates others, leading to varied metabolic effects in the cell.

Activated enzymes

Various metabolic effects

**FIGURE 3.9**  A Second-Messenger System.

▶ *Is adenylate cyclase a transmembrane protein or a peripheral protein? What about the G protein?*

organisms, and diseased cells. Human blood types and transfusion compatibility are determined by glycoproteins. The glycocalyx includes the cell-adhesion molecules that enable cells to adhere to each other and guide the movement of cells in embryonic development. The functions of the glycocalyx are summarized in table 3.2.

## MICROVILLI, CILIA, AND FLAGELLA

Many cells have surface extensions called *microvilli*, *cilia*, and *flagella*. These aid in absorption, movement, and sensory processes.

## Microvilli

**Microvilli**[9] (MY-cro-VIL-eye; singular, *microvillus*) are extensions of the plasma membrane that serve primarily to increase a cell's surface area (figs. 3.10 and 3.11a, c). They are best developed in cells specialized for absorption, such as the epithelial cells of the intestines and kidney tubules. They give such cells 15 to 40 times as much absorptive surface area as they would have if their apical

_____

[9]*micro* = small + *villi* = hairs

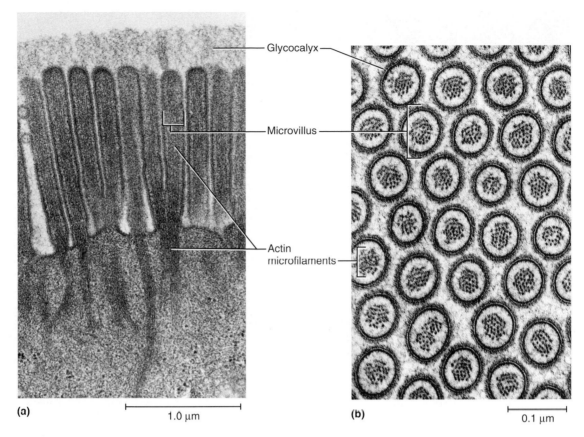

FIGURE 3.10 **Microvilli and the Glycocalyx (TEM).** The microvilli are anchored by microfilaments of actin, which occupy the core of each microvillus and project into the cytoplasm. (a) Longitudinal section, perpendicular to the cell surface. (b) Cross section.

| TABLE 3.2 | Functions of the Glycocalyx |
|---|---|
| Protection | Cushions the plasma membrane and protects it from physical and chemical injury |
| Immunity to infection | Enables the immune system to recognize and selectively attack foreign organism |
| Defense against cancer | Changes in the glycocalyx of cancerous cells enable the immune system to recognize and destroy them |
| Transplant compatibility | Forms the basis for compatibility of blood transfusions, tissue grafts, and organ transplants |
| Cell adhesion | Binds cells together so that tissues do not fall apart |
| Fertilization | Enables sperm to recognize and bind to eggs |
| Embryonic development | Guides embryonic cells to their destinations in the body |

surfaces were flat. On many cells, microvilli are little more than tiny bumps on the plasma membrane. On cells of the taste buds and inner ear, they are well developed but serve sensory rather than absorptive functions.

Individual microvilli cannot be distinguished very well with the light microscope because they are only 1 to 2 μm long. On some cells, they are very dense and appear as a fringe called the **brush border** at the apical cell surface. With the scanning electron microscope, they resemble a deep-pile carpet. With the transmission electron microscope, microvilli typically look like finger-shaped projections of the cell surface. They show little internal structure, but some have a bundle of stiff filaments of a protein called *actin*. Actin filaments attach to the inside of the plasma membrane at the tip of the microvillus, and at its base they extend a little way into the cell and anchor the microvillus to a protein mesh called the **terminal web.** When tugged by another protein in the cytoplasm, actin can shorten a microvillus to milk its absorbed contents downward into the cell.

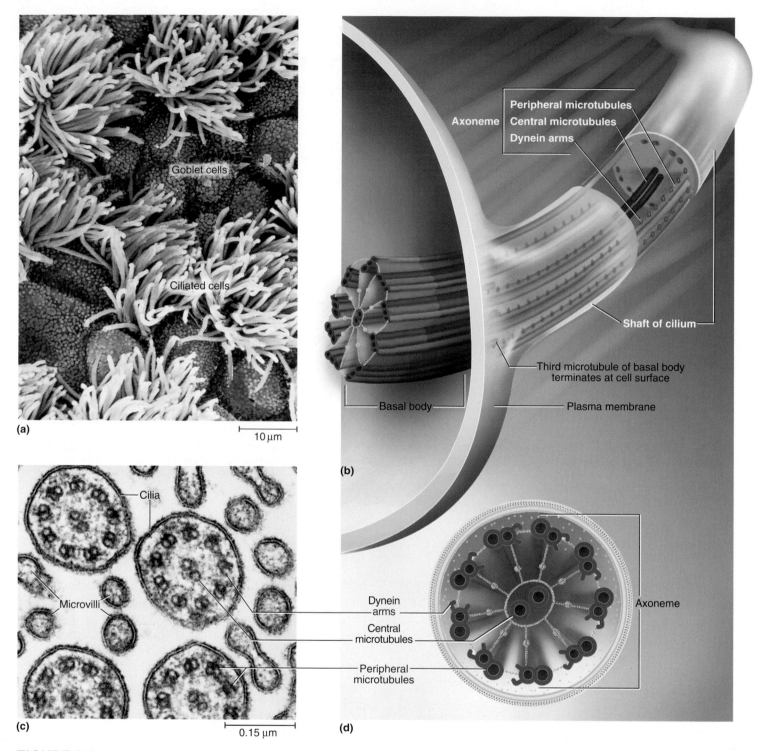

**(a)**

10 μm

**(b)**

Peripheral microtubules
Central microtubules
Dynein arms
Axoneme

Shaft of cilium

Third microtubule of basal body
terminates at cell surface

Basal body                Plasma membrane

**(c)**

Cilia

Microvilli

0.15 μm

**(d)**

Dynein
arms

Central
microtubules

Peripheral
microtubules

Axoneme

**FIGURE 3.11  Cilia.**  (a) Inner surface of the trachea (SEM). Several nonciliated, mucus-secreting goblet cells are visible among the ciliated cells. The goblet cells have short microvilli on their surface. (b) Three-dimensional structure of a cilium. (c) Cross section of a few cilia and microvilli. (d) Cross-sectional structure of a cilium. Note the relative sizes of cilia and microvilli in parts (a) and (c).

## Cilia

Cilia (SIL-ee-uh; singular, *cilium*[10]) (fig. 3.11) are hairlike processes about 7 to 10 μm long. Nearly every human cell

has a single, nonmotile *primary cilium* a few micrometers long. Its function in many cases is still a mystery, but some of them are sensory. In the inner ear, they play a role in the sense of balance; in the retina of the eye, they are highly elaborate and form the light-absorbing part of the receptor cells; and they are thought to monitor fluid flow through

---

[10]*cilium* = eyelash

the kidney tubules. In some cases they open calcium gates in the plasma membrane. Sensory cells in the nose have multiple nonmotile cilia which bind odor molecules.

Motile cilia are less widespread, occurring mainly in the respiratory tract and the uterine (fallopian) tubes. There may be 50 to 200 of these cilia on the surface of one cell. They beat in waves that sweep across the surface of an epithelium, always in the same direction (fig. 3.12), propelling mucus, an egg cell, or an embryo. Each cilium bends stiffly forward and produces a *power stroke* that pushes along the mucus or other matter. Shortly after a cilium begins its power stroke, the one just ahead of it begins, and the next and the next—collectively producing a wavelike motion. After a cilium completes its power stroke, it is pulled limply back by a *recovery stroke* that restores it to the upright position, ready to flex again.

## Think About It

*How would the movement of mucus in the respiratory tract be affected if cilia were equally stiff on both their power and recovery strokes?*

Cilia could not beat freely if they were embedded in sticky mucus (Insight 3.2). Instead, they beat within a saline (saltwater) layer at the cell surface. *Chloride pumps* in the apical plasma membrane produce this layer by pumping Cl$^-$ into the extracellular fluid. Sodium ions follow by electrical attraction and water follows by osmosis. Mucus essentially floats on the surface of this layer and is pushed along by the tips of the cilia.

The structural basis for ciliary movement is a core called the **axoneme**[11] (ACK-so-neem), which consists of

---

11 *axo* – axis + *neme* = thread

### Cystic Fibrosis

The significance of chloride pumps becomes especially evident in *cystic fibrosis (CF)*, a hereditary disease affecting especially white children of European descent. CF is usually caused by a defect in which cells make chloride pumps but fail to install them in the plasma membrane. Consequently, there is an inadequate saline layer on the cell surface and the mucus is dehydrated and overly sticky. This thick mucus plugs the ducts of the pancreas and prevents it from secreting digestive enzymes into the small intestine, so digestion and nutrition are compromised. In the respiratory tract, the mucus clogs the cilia and prevents them from beating freely. The respiratory tract becomes congested with thick mucus, often leading to chronic infection and pulmonary collapse. The mean life expectancy of people with CF is about 30 years.

an array of thin protein cylinders called *microtubules.* There are two central microtubules surrounded by a ring of nine microtubule pairs—an arrangement called the *9 + 2 structure* (fig. 3.11d). The central microtubules stop at the cell surface, but the peripheral microtubules continue a short distance into the cell as part of a **basal body** that anchors the cilium. In each pair of peripheral microtubules, one tubule has two little **dynein**[12] (DINE-een) **arms.** Dynein, a motor protein, uses energy from ATP to "crawl" up the adjacent pair of microtubules. When microtubules on the front of the cilium crawl up the microtubules behind them, the cilium bends toward the front.

---

12 *dyn* – power, energy + *in* = protein

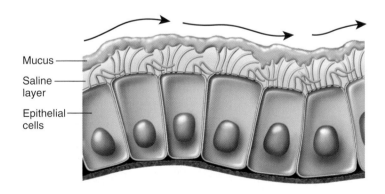

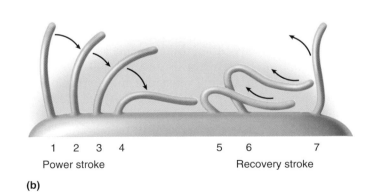

**(a)**    **(b)**

**FIGURE 3.12   Ciliary Action.**   (a) Cilia of an epithelium moving mucus along a surface layer of saline. (b) Power and recovery strokes of an individual cilium.

## Flagella

A **flagellum**[13] (fla-JEL-um) is a whiplike structure much longer than a cilium, but with an identical axoneme. The only functional flagellum in humans is the tail of a sperm cell.

### Before You Go On

*Answer the following questions to test your understanding of the preceding section:*

6. *How does the structure of a plasma membrane depend on the amphiphilic nature of phospholipids?*

7. *Distinguish between transmembrane and peripheral proteins.*

8. *Explain the differences between a receptor, pump, and cell-adhesion molecule.*

9. *How does a gate differ from other channel proteins? What three factors open and close membrane gates?*

10. *What roles do cAMP, adenylate cyclase, and kinases play in cellular function?*

11. *Identify several reasons why the glycocalyx is important to human survival.*

12. *How do microvilli and cilia differ in structure and function?*

# Membrane Transport

### Objectives

When you have completed this section, you should be able to

- explain what is meant by a selectively permeable membrane;

- describe the various mechanisms for transporting material through the plasma membrane; and

- define *osmolarity* and *tonicity* and explain their importance.

The plasma membrane is both a barrier and gateway between the cytoplasm and ECF. It is **selectively permeable**—it allows some things through, such as nutrients and wastes, but usually prevents other things, such as proteins and phosphates, from entering or leaving the cell.

The methods of moving substances into or out of a cell can be classified in two overlapping ways: as *passive* or *active* mechanisms and as *carrier-mediated* or not. Passive mechanisms require no energy (ATP) expenditure by the cell. In most cases, the random molecular motion of the particles themselves provides the energy. Passive mechanisms include filtration and diffusion (including a special

---

[13]*flagellum* = whip

case of diffusion, osmosis). Active mechanisms, however, consume ATP. These include active transport and vesicular transport. Carrier-mediated mechanisms use a membrane protein to transport substances from one side of the membrane to the other. We will first consider the mechanisms that are not carrier-mediated (filtration and simple diffusion) and then the carrier-mediated mechanisms (facilitated diffusion and active transport).

## FILTRATION

**Filtration** is a process in which particles are driven through a selectively permeable membrane by **hydrostatic pressure,** the force exerted on a membrane by water. A coffee filter provides an everyday example. The weight of the water drives water and dissolved matter through the filter, while the filter holds back larger particles (the coffee grounds). In physiology, the most important case of filtration is seen in the blood capillaries, where blood pressure forces fluid through gaps in the capillary wall (fig. 3.13). This is how water, salts, nutrients, and other solutes are transferred from the bloodstream to the tissue fluid and how the kidneys filter wastes from the blood. Capillaries hold back larger particles such as blood cells and proteins.

## SIMPLE DIFFUSION

**Simple diffusion** is the net movement of particles from a place of high concentration to a place of lower concentration as a result of their constant, spontaneous motion. It can be observed by dropping a dye crystal in a dish of still water. As the crystal dissolves, it forms a colored zone in

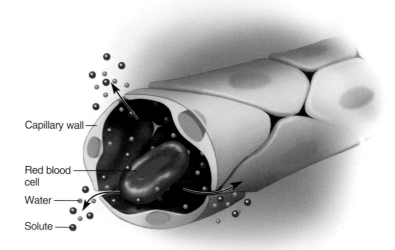

Capillary wall

Red blood cell

Water

Solute

**FIGURE 3.13   Filtration Through the Wall of a Blood Capillary.**   Water and small solutes pass through gaps between cells, while blood cells and other large particles are held back.

the water that gets larger and larger with time (fig. 3.14). The dye molecules exhibit net movement from the point of origin, where their concentration is high, toward the edges of the dish, where their concentration is low. When the concentration of a substance differs from one point to another, we say that it exhibits a **concentration gradient.** Particle movement from a region of high concentration toward a region of lower concentration is said to go *down,* or *with,* the gradient, and movement in the other direction is said to go *up,* or *against,* the gradient.

Diffusion occurs readily in air or water and has no need of a membrane. However, if there is a membrane in the path of the diffusing molecules, and if it is permeable to that substance, the molecules will pass from one side of the membrane to the other. This is how oxygen passes from the air we inhale into the bloodstream. Dialysis treatment for kidney disease patients is based on diffusion of solutes through artificial *dialysis membranes.*

Diffusion rates are very important to cell survival because they determine how quickly a cell can acquire nutrients or rid itself of wastes. Some factors that affect the rate of diffusion through a membrane are as follows:

- **Temperature.** Diffusion is driven by the kinetic energy of the particles, and temperature is a measure of that kinetic energy. The warmer a substance is, the more rapidly its particles diffuse. This is why sugar diffuses more quickly through hot tea than through iced tea.

- **Molecular weight.** Heavy molecules such as proteins move more sluggishly and diffuse more slowly than light particles such as electrolytes and gases. Small molecules also pass through membrane pores more easily than large ones.

- **"Steepness" of the concentration gradient.** The steepness of a gradient refers to the concentration difference between two points. Particles diffuse more rapidly if there is a greater concentration difference between two points. For example, we can increase the rate of oxygen diffusion into a patient's blood by using an oxygen mask, thus increasing the difference in oxygen concentration between the air and blood.

- **Membrane surface area.** As noted earlier, the apical surface of cells specialized for absorption (for example, in the small intestine) is often extensively folded into microvilli. This makes more membrane available for particles to diffuse through.

- **Membrane permeability.** Diffusion through a membrane depends on how permeable it is to the particles. For example, potassium ions diffuse more rapidly than sodium ions through a plasma membrane. Nonpolar, hydrophobic, lipid-soluble substances such as oxygen, nitric oxide, alcohol, and steroids diffuse through the phospholipid regions of a plasma membrane. Water and small charged, hydrophilic solutes such as electrolytes do not mix with lipids, but diffuse primarily through channel proteins in the membrane. Cells can adjust their permeability to such a substance by adding channel proteins to the membrane or taking them away. Kidney tubules, for example, do this as a way of controlling the amount of water eliminated from the body.

## OSMOSIS

**Osmosis**[14] (oz-MO-sis) is the diffusion of water down its concentration gradient, through a selectively permeable membrane that separates solutions with different concentrations of solutes. It occurs through nonliving membranes such as cellophane and dialysis membranes, and through the plasma membranes of cells.

Significant amounts of water diffuse even through the hydrophobic, phospholipid regions of a plasma membrane, but it diffuses more easily through the water-filled channels of transmembrane proteins. Many cells have channel proteins called **aquaporins,** which are specialized for the passage of water. Such cells can increase the rate of osmosis by installing more of these channels in the membrane, or can decrease the rate by removing them. Certain cells of the kidney, for example, regulate the rate of urinary water loss from the body by adding or removing aquaporins.

A cell can exchange a tremendous amount of water by osmosis. In red blood cells, for example, the amount of

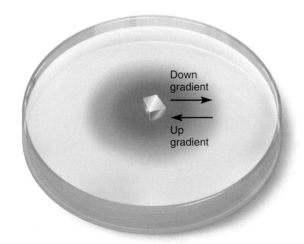

**FIGURE 3.14  Diffusion and Concentration Gradients.**
Dye molecules diffusing away from a crystal dissolving in water. The direction from high concentration (near the crystal) to low concentration is described as "down the concentration gradient"; the opposite direction is described as "up the concentration gradient."

Down gradient

Up gradient

---

[14]*osm* = push, thrust + *osis* = condition, process

water passing through the plasma membrane every second is a hundred times the volume of the cell.

To understand the direction of net water movement in osmosis, it is important to bear in mind that the higher the concentration of dissolved matter, the lower the concentration of water; the solutes take up some of the space that would otherwise be occupied by water molecules. Therefore, the direction of osmosis is from a more dilute solution (where there is more water) to a more concentrated one (where there is less water); the majority of water molecules move from a side where they are more abundant, through a membrane, to the side where they are less abundant.

In figure 3.15a, for example, we see a chamber divided by a selectively permeable membrane. Side A contains a solution of large particles that cannot pass through the membrane pores—a *nonpermeating* solute

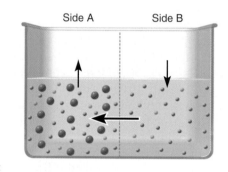

**(a) Start**

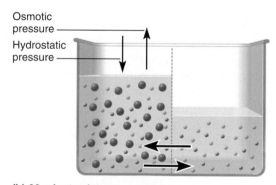

**(b) 30 minutes later**

**FIGURE 3.15   Osmosis.** The dashed line represents a selectively permeable membrane dividing the chamber in half. The large particles on side A represent any solute, such as albumin, too large to pass through the membrane. The small particles are water molecules. (a) Water diffuses from side B, where it is relatively concentrated, to side A, where it is less concentrated. Fluid level rises in side A and falls in side B. (b) Net diffusion stops when the weight (hydrostatic pressure) of the fluid in side A balances the osmotic pressure. At this point, water passes at equal rates from A to B by filtration and from B to A by osmosis. The two processes are then in equilibrium.

such as albumen (egg white protein). Side B contains distilled water. Water is more concentrated on side B than on side A, since albumen occupies some of the space on side A. Water therefore diffuses down its concentration gradient from B to A (fig. 3.15b). The rate and net direction of diffusion depend on how many molecules strike the membrane on each side in a given time period. In the illustrated system, the number of membrane encounters per second, and number of molecules passing through the membrane, is greater on side B where water molecules are more abundant.

Under these conditions, the water level on side B would fall and the level on side A would rise. It might seem as if this would go on indefinitely until side B dried up. This would not happen, however, because as water accumulated on side A, it would become heavier and exert more hydrostatic pressure on that side of the membrane. This would cause some filtration of water from side A back to B. At some point, the rate of filtration would equal the rate of osmosis, water would pass through the membrane equally in both directions, and net osmosis would slow down and stop. At this point, an equilibrium (balance between opposing forces) would exist. The hydrostatic pressure on side A that would stop osmosis is called **osmotic pressure.** The more solute there is on side A, the greater its osmotic pressure.

> ⌐ **Think About It**
>
> *If the albumen concentration on side A were half what it was in the original experiment, would the fluid on that side reach a higher or lower level than before? Explain.*

The equilibrium between osmosis and filtration will be an important consideration when we study fluid exchange through blood capillaries in chapter 20. Blood plasma also contains albumins. In the preceding discussion, side A is analogous to the bloodstream and side B is analogous to the tissue fluid surrounding the capillaries (although tissue fluid is not distilled water). Water leaves the capillaries by filtration, but this is approximately balanced by water moving back into the capillaries by osmosis.

## OSMOLARITY AND TONICITY

The osmotic concentration of body fluids has such a great effect on cellular function that it is important to understand the units in which it is measured. One **osmole** is 1 mole of dissolved particles. If a solute does not ionize in water, then 1 mole of the solute yields 1 osmole (osm) of dissolved particles. A solution of 1 molar (1 M) glucose, for example, is also 1 osm/L. If a solute does ionize, it yields two or more dissolved particles in solution. A 1 M solution of NaCl, for example, contains 1 mole of sodium ions and 1 mole of chloride ions per liter. Both ions affect osmosis and must be separately counted in a measure of osmotic concentration. Thus, 1 M NaCl = 2 osm/L. Calcium chloride ($CaCl_2$) would yield three ions if it dissociated completely (one $Ca^{2+}$ and two $Cl^-$), so 1 M $CaCl_2$ = 3 osm/L.

**Osmolality** is the number of osmoles of solute *per kilogram of water,* and **osmolarity** is the number of osmoles *per liter of solution.* Most clinical calculations are based on osmolarity, since it is easier to measure the volume of a solution than the weight of water it contains. At the concentrations of human body fluids, there is less than 1% difference between osmolality and osmolarity, and the two terms are nearly interchangeable. All body fluids and many clinical solutions are mixtures of many chemicals. The osmolarity of such a solution is the total osmotic concentration of all of its dissolved particles.

A concentration of 1 osm/L is substantially higher than we find in most body fluids, so physiological concentrations are usually expressed in terms of **milliosmoles per liter** (mOsm/L) (1 mOsm/L = $10^{-3}$ osm/L). Blood plasma, tissue fluid, and intracellular fluid measure about 300 mOsm/L.

**Tonicity** is the ability of a solution to affect the fluid volume and pressure in a cell. If a solute cannot pass through a plasma membrane, but remains more concentrated on one side of the membrane than on the other, it causes osmosis. A **hypotonic**[15] solution has a lower concentration of nonpermeating solutes than the intracellular fluid (ICF). Cells in a hypotonic solution absorb water, swell, and may burst *(lyse)* (fig. 3.16a). Distilled water is the extreme example; given to a person intravenously, it would lyse the blood cells. A **hypertonic**[16] solution is one with a higher concentration of nonpermeating solutes than the ICF. It causes cells to lose water and shrivel *(crenate)* (fig. 3.16c). Such cells may die of torn membranes and cytoplasmic loss. In **isotonic**[17] solutions, the total concentration of nonpermeating solutes is the same as in the ICF—hence, isotonic solutions cause no change in cell volume or shape (fig. 3.16b).

It is essential for cells to be in a state of osmotic equilibrium with the fluid around them, and this requires that the ECF have the same concentration of nonpermeating solutes as the ICF. Intravenous fluids given to patients are usually isotonic solutions, but hypertonic or hypotonic fluids are given for special purposes. A 0.9% solution of NaCl, called *normal saline,* is isotonic to human blood cells.

It is important to note that osmolarity and tonicity are not the same. Urea, for example, is a small organic molecule that easily penetrates plasma membranes. If cells are placed in 300 mOsm/L urea, urea diffuses into them (down its concentration gradient), water follows by osmosis, and the cells swell and burst. Thus, 300 mOsm/L urea is not isotonic to the cells. Sodium chloride, by contrast,

---

[15]*hypo* = less + *ton* = tension

[16]*hyper* = more + *ton* = tension
[17]*iso* = equal + *ton* = tension

**(a) Hypotonic**

**(b) Isotonic**

**(c) Hypertonic**

**FIGURE 3.16 Effects of Tonicity on Red Blood Cells (RBCs).** (a) In a hypotonic medium such as distilled water, RBCs absorb water, swell, and may burst. (b) In an isotonic medium such as 0.9% NaCl, RBCs gain and lose water at equal rates and maintain their normal, concave disc shape. (c) In a hypertonic medium such as 2% NaCl, RBCs lose more water than they gain and become shrunken and spiky (crenated).

penetrates plasma membranes poorly. In 300 mOsm/L NaCl, there is little change in cell volume; this solution is isotonic to cells.

## CARRIER-MEDIATED TRANSPORT

The processes of membrane transport described up to this point do not necessarily require a cell membrane; they can occur just as well through artificial membranes. Now, however, we come to processes for which a cell membrane is essential, because they employ transport proteins to get through the membrane. Thus, the next two processes are cases of **carrier-mediated transport.**

The carriers act like enzymes in some ways: the solute is a ligand that binds to a specific receptor site on the carrier, like a substrate binding to the active site of an enzyme. The carrier exhibits **specificity** for a certain ligand, just as an enzyme does for its substrate. A glucose carrier, for example, cannot transport fructose. Carriers also exhibit **saturation;** as the solute concentration rises, its rate of transport through a membrane increases, but only up to a point. When every carrier is occupied, adding more solute cannot make the process go any faster. The carriers are saturated—no more are available to handle the increased demand, and transport levels off at a rate called the **transport maximum** ($T_m$) (fig. 3.17). As we'll see later in the book, the $T_m$ explains why glucose appears in the urine of people with diabetes mellitus. An important difference between a membrane carrier and an enzyme is that carriers do not chemically change their lig-

ands; they simply pick them up on one side of the membrane and release them, unchanged, on the other.

There are three kinds of carriers: uniports, symports, and antiports. A **uniport**[18] carries only one solute at a time. For example, most cells pump out calcium by means of a uniport, maintaining a low intracellular calcium concentration so that calcium salts don't crystallize in their cytoplasm. A **symport**[19] carries two or more solutes through a membrane simultaneously in the same direction; this process is called **cotransport.**[20] As an example, absorptive cells of the small intestine and kidneys take up sodium and glucose simultaneously by means of a symport. An **antiport**[21] carries two or more solutes in opposite directions; this process is called **countertransport.** Cells everywhere have an antiport called the *sodium–potassium pump* that continually removes $Na^+$ from the cell and brings in $K^+$.

These carriers employ two mechanisms of transport called facilitated diffusion and active transport. (Any carrier type—uniport, symport, or antiport—can use either of these transport mechanisms.) **Facilitated**[22] **diffusion** is the carrier-mediated transport of a solute through a membrane *down its concentration gradient.* It is a passive transport process; that is, it does not consume ATP. It transports solutes such as glucose that cannot pass through the membrane unaided. The solute attaches to a binding site on the carrier, then the carrier changes conformation and releases the solute on the other side of the membrane (fig. 3.18).

**Active transport** is the carrier-mediated transport of a solute through a membrane *up its concentration gradient,* using energy provided by ATP. The calcium pumps mentioned previously use active transport. Even though $Ca^2$ is already more concentrated in the ECF than within the cell, these carriers pump still more of it out of the cell. Active transport also enables cells to absorb amino acids that are already more concentrated in the cytoplasm than in the ECF.

A prominent example of active transport is the **sodium–potassium ($Na^+$–$K^+$) pump,** also known as *$Na^+$–$K^+$ ATPase* because the carrier is an enzyme that hydrolyzes ATP. The $Na^+$–$K^+$ pump binds three $Na^+$ simultaneously on the cytoplasmic side of the membrane, releases these to the ECF, binds two $K^+$ simultaneously from the ECF, and releases these into the cell (fig. 3.19). Each cycle of the pump consumes one ATP and exchanges three $Na^+$ for two $K^+$. This keeps the $K^+$ concentration higher and the $Na^+$ concentration lower within the cell than in the ECF. These ions continually leak through the membrane, and the $Na^+$–$K^+$ pump compensates like bailing out a leaky boat.

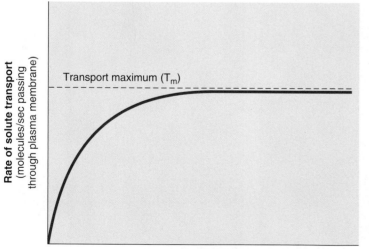

**FIGURE 3.17** Carrier Saturation and Transport Maximum. Up to a point, increasing the solute concentration increases the rate of transport through a membrane. At the transport maximum ($T_m$), however, all carrier proteins are busy and cannot transport the solute any faster, even if more solute is added.

---

[18]*uni* = one + *port* = carry
[19]*sym* = together + *port* = carry
[20]*co* = together + *trans* = across + *port* = carry
[21]*anti* = opposite + *port* = carry
[22]*facil* = easy

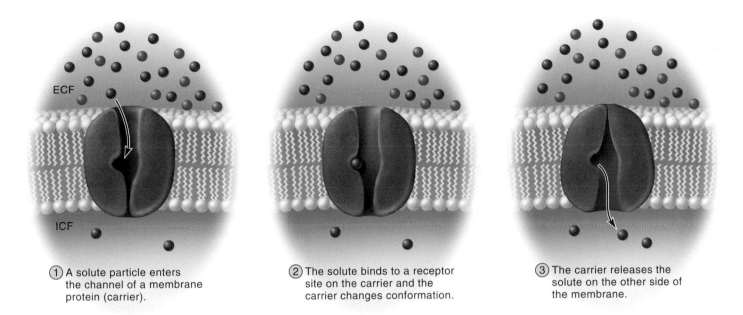

① A solute particle enters the channel of a membrane protein (carrier).

② The solute binds to a receptor site on the carrier and the carrier changes conformation.

③ The carrier releases the solute on the other side of the membrane.

**FIGURE 3.18 Facilitated Diffusion.** Note that the solute moves down its concentration gradient.

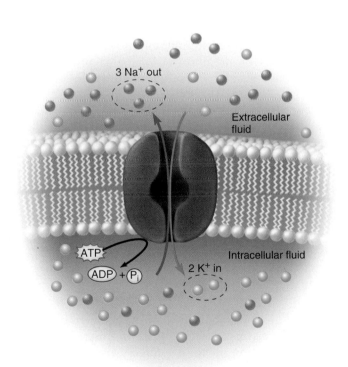

**FIGURE 3.19 The Sodium–Potassium Pump (Na⁺-K⁺ ATPase).** In each cycle of action, this membrane carrier removes three sodium ions ($Na^+$) from the cell, brings two potassium ions ($K^+$) into the cell, and hydrolyzes one molecule of ATP.

▶ *Why would the Na⁺–K⁺ pump, but not osmosis, cease to function after a cell dies?*

Lest you question the importance of the $Na^+$–$K^+$ pump, about half of the calories you consume each day are used for this alone. Beyond compensating for a leaky plasma membrane, the $Na^+$–$K^+$ pump has at least four functions:

1. **Regulation of cell volume.** Certain anions are confined to the cell and cannot penetrate the plasma membrane. These "fixed anions," such as proteins and phosphates, attract and retain cations. If there were nothing to correct for it, the retention of these ions would cause osmotic swelling and possibly lysis of the cell. Cellular swelling, however, stimulates the $Na^+$–$K^+$ pumps. Since each cycle of the pump removes one ion more than it brings in, the pumps are part of a negative feedback loop that reduces ion concentration, osmolarity, and cellular swelling.

2. **Secondary active transport.** The $Na^+$–$K^+$ pump maintains a steep concentration gradient of $Na^+$ and $K^+$ between one side of the membrane and the other. Like water behind a dam that can be tapped to generate electricity, this gradient has a high potential energy that can drive other processes. Since $Na^+$ has a high concentration outside the cell, it tends to diffuse back in. Some cells exploit this to move other solutes into the cell. In kidney tubules, for example, the cells have $Na^+$–$K^+$ pumps in the basal membrane that remove $Na^+$ from the cytoplasm and maintain a low intracellular $Na^+$ concentration. In the apical membrane, the cells have a facilitated diffusion carrier, the **sodium–glucose transport protein** (SGLT), which simultaneously binds $Na^+$ and glucose and carries

both into the cell at once (fig. 3.20). By exploiting the tendency of $Na^+$ to diffuse down its concentration gradient into these cells, the SGLT absorbs glucose and prevents it from being wasted in the urine. The SGLT in itself does not consume ATP, but it does depend on the ATP-consuming $Na^+$–$K^+$ pumps at the base of the cell. We say that glucose is absorbed by *secondary active transport,* as opposed to the *primary active transport* carried out by the $Na^+$–$K^+$ pump.

3. **Heat production.** When the weather turns chilly, we not only turn up the furnace in our home but also the "furnace" in our body. Thyroid hormone stimulates cells to produce more $Na^+$–$K^+$ pumps. As these pumps consume ATP, they release heat, compensating for the body heat we lose to the cold air around us.

4. **Maintenance of a membrane potential.** All living cells have an electrical charge difference called the *resting membrane potential* across the plasma membrane. Like the two poles of a battery, the inside of the membrane is negatively charged and the outside is positively charged. This difference stems from the unequal distribution of ions on the two sides of the membrane, maintained by the $Na^+$–$K^+$ pump. The membrane potential is essential to the function of nerve and muscle cells, as we will study in later chapters.

> **Think About It**
>
> *An important characteristic of proteins is their ability to change conformation in response to the binding or dissociation of a ligand (chapter 2). Explain how this characteristic is essential to carrier-mediated transport.*

## VESICULAR TRANSPORT

So far, we have considered processes that move from one to a few ions or molecules at a time through the plasma membrane. **Vesicular transport** processes, by contrast, move large particles, droplets of fluid, or numerous molecules at once through the membrane, contained in bubblelike *vesicles* of membrane. Vesicular processes that bring matter into a cell are called **endocytosis**[23] (EN-doe-sy-TOE-sis) and those that release material from a cell are called **exocytosis**[24] (EC-so-sy-TOE-sis).

There are three forms of endocytosis: phagocytosis, pinocytosis, and receptor-mediated endocytosis. **Phagocytosis**[25] (FAG-oh-sy-TOE-sis), or "cell eating," is the process of engulfing particles such as bacteria, dust, and cellular debris—particles large enough to be seen with a microscope. Neutrophils (a class of white blood cells), for example, protect the body from infection by phagocytizing and killing bacteria. A neutrophil spends most of its life crawling about in the connective tissues by means of blunt footlike extensions called **pseudopods**[26] (SOO-doe-pods). When a neutrophil encounters a bacterium, it surrounds it with its pseudopods and traps it in a **phagosome**[27]—a vesicle in the cytoplasm surrounded by a unit membrane (fig. 3.21). A lysosome merges with the vacuole, converting it to a *phagolysosome,* and contributes enzymes that destroy the invader. Several other kinds of phagocytic cells are described in chapter 21. In general, phagocytosis is a way of keeping the tissues free of debris and infectious microorganisms. Some cells called *macrophages* (literally "big eaters") phagocytize the equivalent of 25% of their own volume per hour.

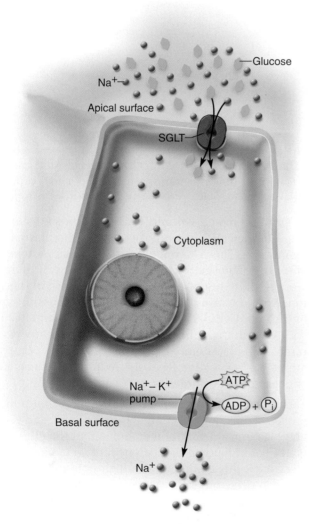

**FIGURE 3.20  Secondary Active Transport.** At the basal surface of the cell, an $Na^+$-$K^+$ pump removes sodium ions ($Na^+$) from the cytoplasm, maintaining a low sodium concentration within the cell. At the apical surface, $Na^+$ enters the cell by facilitated diffusion, following its concentration gradient. It can gain entry to the cell only by binding to a carrier, the sodium–glucose transport protein (SGLT), which simultaneously binds and transports glucose. The SGLT does not consume ATP, but does depend on the ATP-consuming pump at the base of the cell.

---

[23]*endo* = into + *cyt* = cell + *osis* = process
[24]*exo* = out of + *cyt* = cell + *osis* = process
[25]*phago* = eating + *cyt* = cell + *osis* = process
[26]*pseudo* = false + *pod* = foot
[27]*phago* = eaten + *some* = body

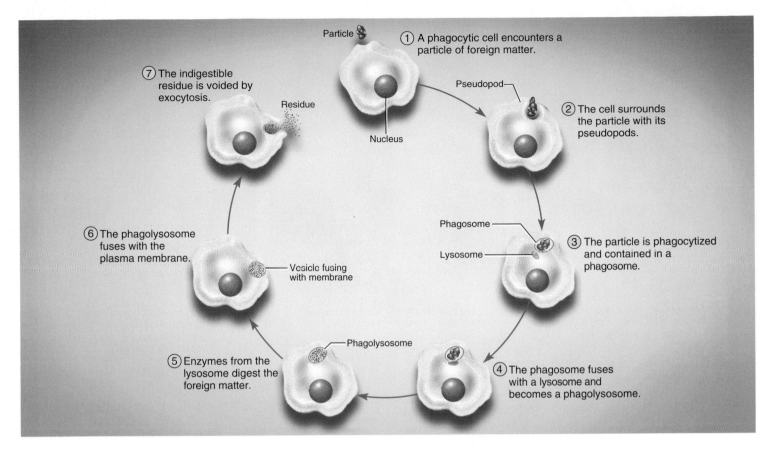

**FIGURE 3.21** Phagocytosis, Intracellular Digestion, and Exocytosis.

**Pinocytosis**[28] (PIN-oh-sy-TOE-sis), or "cell drinking," is the process of taking in droplets of ECF containing molecules of some use to the cell. While phagocytosis occurs in only a few specialized cells, pinocytosis occurs in all human cells. The process begins as the plasma membrane becomes dimpled, or caved in, at points. These pits soon separate from the surface membrane and form small membrane-bounded **pinocytotic vesicles** in the cytoplasm. The vesicles contain droplets of the ECF with whatever molecules happened to be there.

**Receptor-mediated endocytosis** (fig. 3.22) is a more selective form of either phagocytosis or pinocytosis. It enables a cell to take in specific molecules from the ECF with a minimum of unnecessary fluid. Particles in the ECF bind to specific receptors on the plasma membrane. The receptors then cluster together and the membrane sinks in at this point, creating a pit coated with a peripheral membrane protein called *clathrin*.[29] The pit soon pinches off to form a *clathrin-coated vesicle* in the cytoplasm. Clathrin may serve as an "address label" on the coated vesicle that directs it to an appropriate destination in the cell, or it may inform other structures in the cell what to do with the vesicle.

One example of receptor-mediated endocytosis is the uptake of *low-density lipoproteins (LDLs)*—protein-coated droplets of cholesterol and other lipids in the blood (see chapter 26). The thin endothelial cells that line our blood vessels have LDL receptors on their surfaces and absorb LDLs in clathrin-coated vesicles. Inside the cell, the LDL is freed from the vesicle and metabolized, and the membrane with its receptors is recycled to the cell surface. Much of what we know about receptor-mediated endocytosis comes from studies of a hereditary disease called familial hypercholesterolemia, which dramatically illustrates the significance of this process to our cardiovascular health (Insight 3.3).

Endothelial cells also imbibe insulin by receptor-mediated endocytosis. Insulin is too large a molecule to pass through channels in the plasma membrane, yet it must somehow get out of the blood and reach the surrounding

---

**INSIGHT 3.3** Clinical Application

### Familial Hypercholesterolemia

The significance of LDL receptors and receptor-mediated endocytosis is dramatically illustrated by a hereditary disease called *familial hypercholesterolemia*.[30] People with this disease have an abnormally low number of LDL receptors. Their cells therefore absorb less cholesterol than normal, and the cholesterol remains in the blood. Their blood cholesterol levels may be as high as 1,200 mg/dL, compared with a normal level of about 200 mg/dL. People who inherit the gene from both parents typically have heart attacks before the age of 20 (sometimes even in infancy) and seldom survive beyond the age of 30.

---

[28]*pino* = drinking + *cyt* = cell + *osis* = process
[29]*clathr* = lattice + *in* = protein

[30]*familial* = running in the family; *hyper* = above normal + cholesterol + *emia* = blood condition

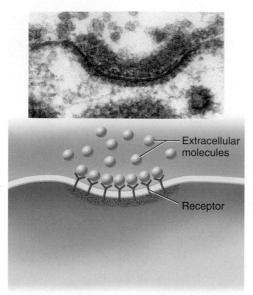 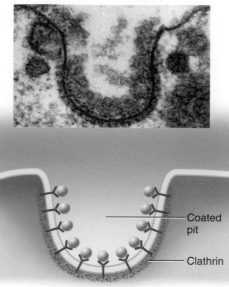 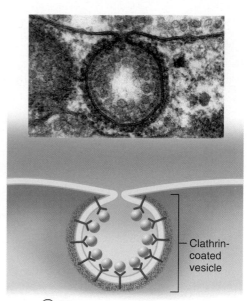

Extracellular
molecules

Receptor

Coated
pit

Clathrin

Clathrin-
coated
vesicle

① Extracellular molecules bind to receptors on plasma membrane; receptors cluster together.

② Plasma membrane sinks inward, forms clathrin-coated pit.

③ Pit separates from plasma membrane, forms clathrin-coated vesicle containing concentrated molecules from ECF.

**FIGURE 3.22** Receptor-Mediated Endocytosis.

cells if it is to have any effect. Endothelial cells take up insulin by receptor-mediated endocytosis, transport the vesicles across the cell, and release the insulin on the other side, where tissue cells await it. Such transport of a substance across a cell (capture on one side and release on the other side) is called **transcytosis**[31] (fig. 3.23). This process is especially active in muscle capillaries and transfers a significant amount of blood albumin into the tissue fluid.

Receptor-mediated endocytosis is not always to our benefit; hepatitis, polio, and AIDS viruses trick our cells into engulfing them by receptor-mediated endocytosis.

Exocytosis (fig. 3.24) is the process of discharging material from a cell. It occurs, for example, when endothelial cells release insulin to the tissue fluid, mammary gland cells secrete milk, other gland cells release hormones, and sperm cells release enzymes for penetrating an egg. It bears a superficial resemblance to endocytosis in reverse. A secretory vesicle in the cell migrates to the surface and "docks" on peripheral proteins of the plasma membrane. These proteins pull the membrane inward and create a dimple that eventually fuses with the vesicle and allows it to release its contents.

The question might occur to you, If endocytosis continually takes away bits of plasma membrane to form intracellular vesicles, why doesn't the membrane grow smaller and smaller? Another purpose of exocytosis, however, is to replace plasma membrane that has been removed by endocytosis or become damaged or worn out. Plasma membrane is continually recycled from the cell surface into the cytoplasm and back to the surface.

Table 3.3 summarizes the mechanisms of transport we have discussed.

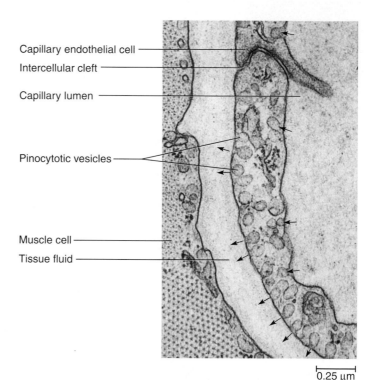

Capillary endothelial cell

Intercellular cleft

Capillary lumen

Pinocytotic vesicles

Muscle cell

Tissue fluid

0.25 μm

**FIGURE 3.23** **Transcytosis.** An endothelial cell of a capillary imbibes droplets of blood plasma at sites indicated by arrows along the right. This forms pinocytotic vesicles, which the cell transports to the other side. Here, it releases the contents by exocytosis at sites indicated by arrows along the left side of the cell. This process is especially active in muscle capillaries and transfers a significant amount of blood albumin into the tissue fluid.

▶ Why isn't transcytosis listed as a separate means of membrane transport, in addition to pinocytosis and the others?

[31]*trans* = across + *cyt* = cell + *osis* = process

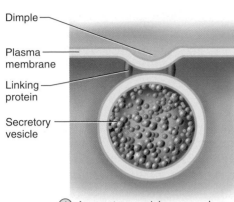

Dimple

Plasma membrane

Linking protein

Secretory vesicle

**(a)**

① A secretory vesicle approaches the plasma membrane and docks on it by means of linking proteins. The plasma membrane caves in at that point to meet the vesicle.

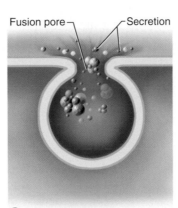

Fusion pore — Secretion

② The plasma membrane and vesicle unite to form a fusion pore through which the vesicle contents are released.

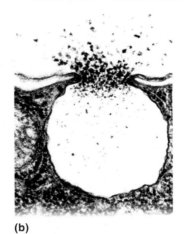

**(b)**

**FIGURE 3.24 Exocytosis.** (a) Stages of exocytosis. (b) Electron micrograph of exocytosis.

| TABLE 3.3 | Methods of Membrane Transport |
|---|---|
| *Transport Without Carriers* | Movement of material without the aid of carrier proteins |
| Filtration | Movement of water and solutes through a selectively permeable membrane as a result of hydrostatic pressure |
| Simple Diffusion | Diffusion of particles through water or air or through a living or artificial membrane, down their concentration gradient, without the aid of membrane carriers |
| Osmosis | Simple diffusion of water through a selectively permeable membrane |
| *Carrier-Mediated Transport* | Movement of material through a cell membrane with the aid of carrier proteins |
| Facilitated Diffusion | Transport of particles through a selectively permeable membrane, down their concentration gradient, by a carrier that does not directly consume ATP |
| Active Transport | Transport of particles through a selectively permeable membrane, up their concentration gradient, with the aid of a carrier that consumes ATP |
| Primary Active Transport | Direct transport of solute particles by an ATP-using membrane pump |
| Secondary Active Transport | Transport of solute particles by a carrier that does not in itself use ATP but depends on concentration gradients produced by primary active transport |
| Cotransport | Transport of two solutes simultaneously in the same direction through a membrane by either facilitated diffusion or active transport |
| Countertransport | Transport of two different solutes in opposite directions through a membrane by either facilitated diffusion or active transport |
| Uniport | A carrier that transports only one solute, using either facilitated diffusion or active transport |
| Symport | A carrier that performs cotransport |
| Antiport | A carrier that performs countertransport |
| *Vesicular (Bulk) Transport* | Movement of fluid and particles through a plasma membrane by way of vesicles of plasma membrane; consumes ATP |
| Endocytosis | Vesicular transport of particles into a cell |
| Phagocytosis | Process of engulfing large particles by means of pseudopods; "cell eating" |
| Pinocytosis | Process of imbibing droplets of extracellular fluid in which the plasma membrane sinks in and pinches off small vesicles containing droplets of fluid |
| Receptor-Mediated Endocytosis | Phagocytosis or pinocytosis in which specific solute particles bind to receptors on the plasma membrane, and are then taken into the cell in clathrin-coated vesicles with a minimal amount of fluid |
| Exocytosis | Process of eliminating material from a cell by means of a vesicle approaching the cell surface, fusing with the plasma membrane, and expelling its contents; used to release cell secretions, replace worn-out plasma membrane, and replace membrane that has been internalized by endocytosis |

*Answer the following questions to test your under-standing of the preceding section:*

13. What is the importance of filtration to human physiology?

14. What does it mean to say a solute moves down its concentration gradient?

15. How does osmosis help to maintain blood volume?

16. Define osmolarity and tonicity, and explain the differ-ence between them.

17. Define hypotonic, isotonic, and hypertonic, and explain why these concepts are important in clinical practice.

18. What do facilitated diffusion and active transport have in common? How are they different?

19. How does the $Na^+$-$K^+$ pump exchange sodium ions for potassium ions across the plasma membrane? What are some purposes served by this pump?

20. How does phagocytosis differ from pinocytosis?

21. Describe the process of exocytosis. What are some of its purposes?

# The Cytoplasm

### Objectives

When you have completed this section, you should be able to

- list the main organelles of a cell, describe their struc-ture, and explain their functions;

- describe the cytoskeleton and its functions; and

- give some examples of cell inclusions and explain how inclusions differ from organelles.

We now probe more deeply into the cell to study the structures in the cytoplasm. These are classified into three groups—*organelles, cytoskeleton,* and *inclusions*—all embedded in the clear, gelatinous cytosol.

## ORGANELLES

Organelles are internal structures of a cell that carry out specialized metabolic tasks. Some are surrounded by one or two layers of unit membrane and are therefore referred to as *membranous organelles.* These are the nucleus, mitochondria, lysosomes, peroxisomes, endoplasmic reticulum, and Golgi complex. Organelles that are not sur-rounded by membranes include the ribosomes, centro-some, centrioles, and basal bodies.

## The Nucleus

The **nucleus** is the largest organelle and usually the only one visible with the light microscope. It is usually sphe-

roid to elliptical in shape and typically about 5 μm in diameter. Most cells have a single nucleus, but there are exceptions. Mature red blood cells have none; they are **anuclear.** A few cell types are **multinucleate,** having 2 to 50 nuclei. Examples include some liver cells, skeletal muscle cells, and certain bone-dissolving and platelet-producing cells.

With the TEM, the nucleus can be distinguished by the *two* unit membranes surrounding it, which together form the **nuclear envelope** (fig. 3.25). The envelope is per-forated with **nuclear pores,** about 30 to 100 nm in diame-ter, formed by a ring of proteins. These proteins regulate molecular traffic through the envelope and act like a rivet to hold the two unit membranes together. Hundreds of molecules pass through the nuclear pores every minute. Coming into the nucleus are raw materials for DNA and RNA synthesis, enzymes that are made in the cytoplasm but function in the nucleus, and hormones that activate certain genes. Going the other way, RNA is made in the nucleus but leaves to perform its job in the cytoplasm.

The material in the nucleus is called **nucleoplasm.** This includes **chromatin**[32] (CRO-muh-tin)—fine thread-like matter composed of DNA and protein—and one or more dark-staining masses called **nucleoli** (singular, *nucleolus*), where ribosomes are produced. The genetic function of the nucleus is described in chapter 4.

## Endoplasmic Reticulum

**Endoplasmic reticulum** (ER) literally means "little network within the cytoplasm." It is a system of interconnected channels called **cisternae**[33] (sis-TUR-nee) enclosed by a unit membrane (fig. 3.26). In areas called **rough endoplas-mic reticulum,** the network is composed of parallel, flat-tened sacs covered with granules called *ribosomes.* The rough ER is continuous with the outer membrane of the nuclear envelope, and adjacent cisternae are often con-nected by perpendicular bridges. In areas called **smooth endoplasmic reticulum,** the membrane lacks ribosomes, the cisternae are more tubular in shape, and they branch more extensively. The cisternae of the smooth ER are thought to be continuous with those of the rough ER, so the two are functionally different parts of the same network.

The ER synthesizes steroids and other lipids, detoxi-fies alcohol and other drugs, and manufactures all of the membranes of the cell. Rough ER produces the phospho-lipids and proteins of the plasma membrane, and synthe-sizes proteins that are either packaged in other organelles such as lysosomes or secreted from the cell. Rough ER is most abundant in cells that synthesize large amounts of protein, such as antibody-producing cells and cells of the digestive glands. This role is discussed further in chapter 4.

---

[32]*chromat* = color

[33]*cistern* = reservoir

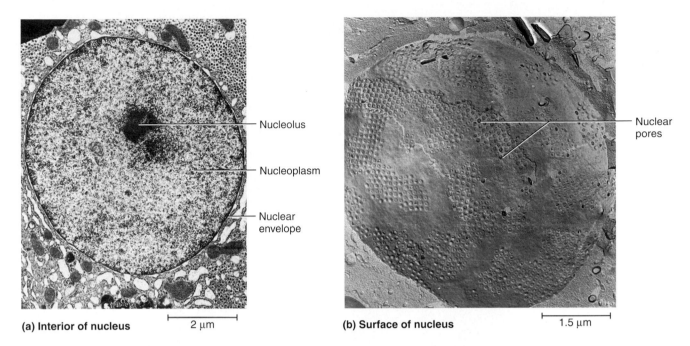

**(a) Interior of nucleus**     2 μm

    — Nucleolus

    — Nucleoplasm

    — Nuclear envelope

**(b) Surface of nucleus**     1.5 μm

    — Nuclear pores

**FIGURE 3.25** **The Nucleus.** These photomicrographs were made by different TEM methods to show the internal structure of the nucleus and surface of the nuclear envelope.

▶ *Why do these nuclear pores have to be larger in diameter than the channels in the cell's plasma membrane? (See table 3.1.)*

Most cells have only a scanty smooth ER, but it is relatively abundant in cells that engage extensively in detoxification, such as liver and kidney cells. Long-term abuse of alcohol, barbiturates, and other drugs leads to tolerance partly because the smooth ER proliferates and detoxifies the drugs more quickly. Smooth ER is also abundant in cells of the testes and ovaries that synthesize steroid hormones. Skeletal muscle and cardiac muscle contain extensive networks of smooth ER that store calcium and release it to trigger muscle contraction.

## Ribosomes

**Ribosomes** are small granules of protein and RNA found in the nucleoli, in the cytosol, and on the outer surfaces of the rough ER and nuclear envelope. They "read" coded genetic messages (messenger RNA) and assemble amino acids into proteins specified by the code. This process is detailed in chapter 4.

## Golgi Complex

The **Golgi**[34] (GOAL-jee) **complex** is a small system of cisternae that synthesize carbohydrates and put the finishing touches on protein and glycoprotein synthesis. The complex resembles a stack of pita bread. Typically, it consists of about six cisternae, slightly separated from each other; each cisterna is a flattened, slightly curved sac with swollen edges (fig. 3.27). The Golgi complex receives the newly synthesized proteins from the rough ER. It sorts them, cuts and splices some of them, adds carbohydrate moieties to some, and finally packages the proteins in membrane-bounded **Golgi vesicles.** These vesicles bud off the swollen rim of a cisterna and are seen in abundance in the neighborhood of the Golgi complex. Some vesicles become *lysosomes*, the organelle discussed next; some migrate to the plasma membrane and fuse with it, contributing fresh protein and phospholipid to the membrane; and some become **secretory vesicles** that store a cell product, such as breast milk or digestive enzymes, for later release. The role of the Golgi complex in protein synthesis and secretion is detailed in chapter 4.

## Lysosomes

A **lysosome**[35] (LY-so-some) (fig. 3.28a) is a package of enzymes bounded by a single unit membrane. Although often round or oval, lysosomes are extremely variable in shape. When viewed with the TEM, they often exhibit dark gray contents devoid of structure, but sometimes show crystals or parallel layers of protein. At least 50

---

[34]Camillo Golgi (1843–1926), Italian histologist

[35]*lyso* = loosen, dissolve + *some* = body

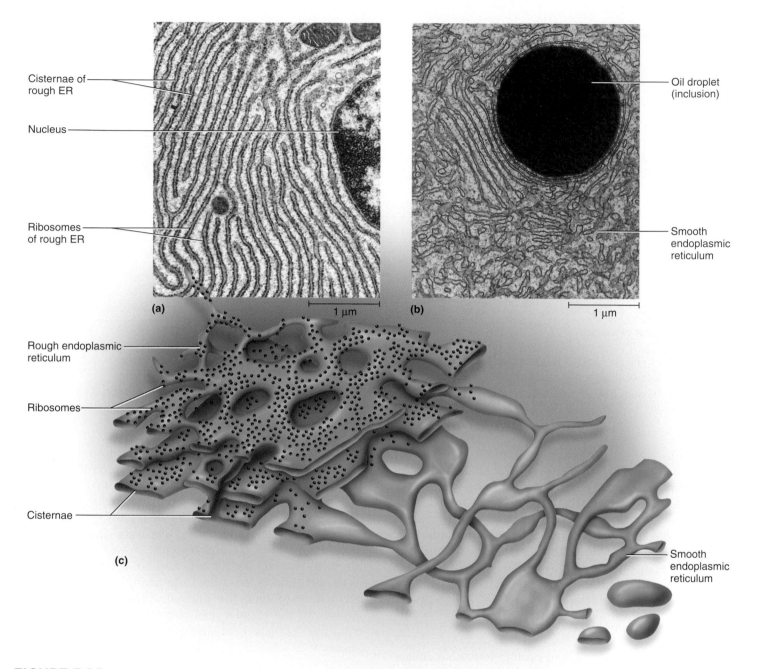

Cisternae of rough ER

Nucleus

Ribosomes of rough ER

**(a)**                    1 μm

Oil droplet (inclusion)

Smooth endoplasmic reticulum

**(b)**                    1 μm

Rough endoplasmic reticulum

Ribosomes

Cisternae

**(c)**

Smooth endoplasmic reticulum

**FIGURE 3.26  Endoplasmic Reticulum (ER).**  (a) Rough ER. (b) Smooth ER and an inclusion (oil droplet). (c) Structure of the endoplasmic reticulum, with rough and smooth regions.

lysosomal enzymes have been identified. They hydrolyze proteins, nucleic acids, complex carbohydrates, phospholipids, and other substrates. In the liver, lysosomes break down stored glycogen to release glucose into the bloodstream. White blood cells use their lysosomes to digest phagocytized bacteria. Lysosomes also digest and dispose of worn-out mitochondria and other organelles; this process is called **autophagy**[36] (aw-TOFF-uh-jee). Some cells are meant to do a certain job and then die. The

uterus, for example, weighs about 900 g at full-term pregnancy and shrinks to 60 g within 5 or 6 weeks after birth. This shrinkage is due to **autolysis,**[37] the digestion of surplus cells by their own lysosomal enzymes. Such *programmed cell death* is further discussed in chapter 5.

## Peroxisomes

**Peroxisomes** (fig. 3.28b) resemble lysosomes but contain different enzymes and are not produced by the Golgi

[36]*auto* = self + *phagy* = eating

[37]*auto* = self + *lysis* = dissolving

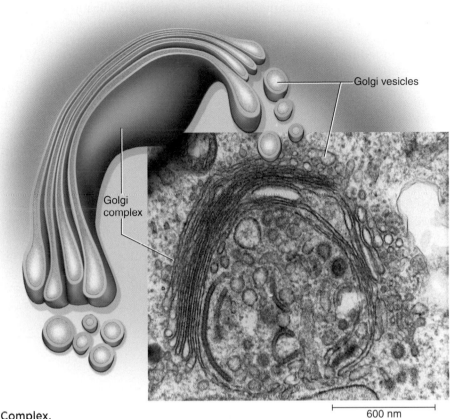

Golgi vesicles

Golgi complex

**FIGURE 3.27** The Golgi Complex.

600 nm

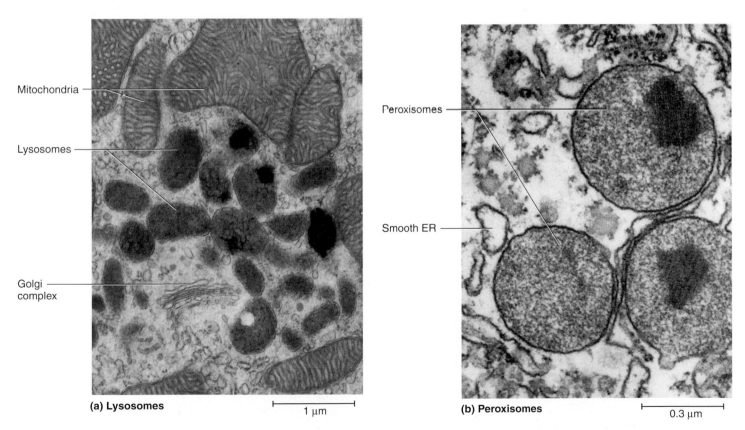

Mitochondria

Lysosomes

Golgi complex

**(a) Lysosomes**

1 μm

Peroxisomes

Smooth ER

**(b) Peroxisomes**

0.3 μm

**FIGURE 3.28** Lysosomes and Peroxisomes.

complex. Their general function is to use molecular oxygen ($O_2$) to oxidize organic molecules. These reactions produce hydrogen peroxide ($H_2O_2$), hence the name of the organelle. $H_2O_2$ is then used to oxidize other molecules, and the excess is broken down to water and oxygen by an enzyme called *catalase.*

Peroxisomes occur in nearly all cells but are especially abundant in liver and kidney cells. They neutralize free radicals and detoxify alcohol, other drugs, and a variety of blood-borne toxins. Peroxisomes also decompose fatty acids into two-carbon acetyl groups, which the mitochondria then use as an energy source for ATP synthesis.

## Mitochondria

**Mitochondria**[38] (MY-toe-CON-dree-uh) (fig. 3.29) are organelles specialized for synthesizing ATP. They have a variety of shapes: spheroid, rod-shaped, bean-shaped, or threadlike. Like the nucleus, a mitochondrion is surrounded by a double unit membrane. The inner mem-

brane usually has folds called **cristae**[39] (CRIS-tee), which project like shelves across the organelle. The space between the cristae, called the **matrix,** contains ribosomes, enzymes used in ATP synthesis, and a small, circular DNA molecule called *mitochondrial DNA* (mtDNA). Mitochondria are the "powerhouses" of the cell. Energy is not *made* here, but it is extracted from organic compounds and transferred to ATP, primarily by enzymes located on the cristae. The role of mitochondria in ATP synthesis is explained in detail in chapter 26, and some evolutionary and clinical aspects of mitochondria are discussed at the end of this chapter (Insight 3.4).

## Centrioles

A **centriole** (SEN-tree-ole) is a short cylindrical assembly of microtubules, arranged in nine groups of three microtubules each (fig. 3.30). Two centrioles lie perpendicular to each other within a small clear area of cytoplasm called the **centrosome**[40] (see fig. 3.5). They play a role in cell division

---

[38]*mito* = thread + *chondr* = grain

[39]*crista* = crest
[40]*centro* = central + *some* = body

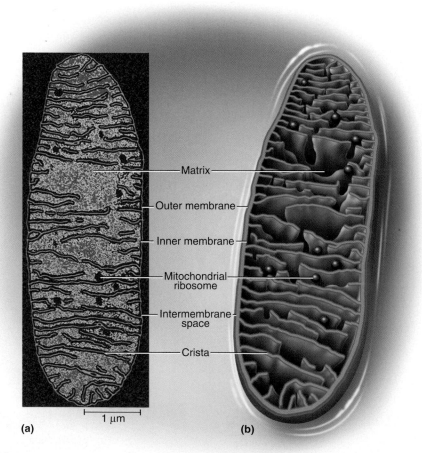

1 μm

(a)                                             (b)

**FIGURE 3.29**  A Mitochondrion.

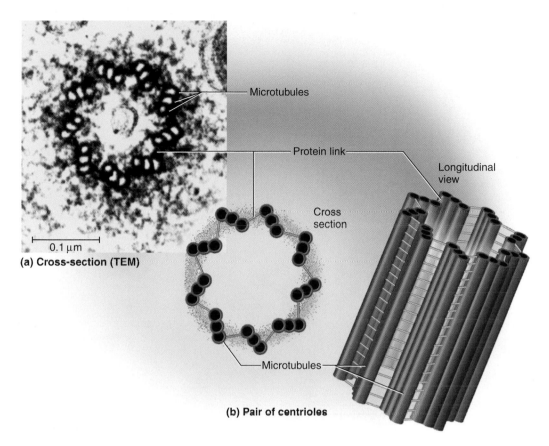

Microtubules

Protein link

Longitudinal view

Cross section

0.1 μm

**(a) Cross-section (TEM)**

Microtubules

**(b) Pair of centrioles**

**FIGURE 3.30   Centrioles.**   (a) Electron micrograph of a centriole as seen in cross section. (b) A pair of perpendicular centrioles.
▶ *How does a centriole resemble the axoneme of a cilium? How does it differ?*

described in chapter 4. Each basal body of a flagellum or cilium is a single centriole oriented perpendicular to the plasma membrane. Basal bodies originate in a *centriolar organizing center* and migrate to the plasma membrane. Two microtubules of each triplet then elongate to form the nine pairs of peripheral microtubules of the axoneme. A cilium can grow to its full length in less than an hour.

## THE CYTOSKELETON

The **cytoskeleton** is a collection of protein filaments and cylinders that determine the shape of a cell, lend it structural support, organize its contents, move substances through the cell, and contribute to movements of the cell as a whole. It can form a very dense supportive scaffold in the cytoplasm (fig. 3.31). It is connected to transmembrane proteins of the plasma membrane, and they in turn are connected to protein fibers external to the cell, so there is a strong structural continuity from extracellular material to the cytoplasm. Cytoskeletal elements may even connect to chromosomes in the nucleus, enabling physical tension on a cell to move nuclear contents and mechanically stimulate genetic function.

The cytoskeleton is composed of *microfilaments, intermediate filaments,* and *microtubules.* **Microfilaments**

are about 6 nm thick and are made of the protein *actin.* They form a network on the cytoplasmic side of the plasma membrane called the **membrane skeleton.** The phospholipids of the plasma membrane spread out over the membrane skeleton like butter on a slice of bread. It is thought that the phospholipids would break up into little droplets without this support. The roles of actin in supporting microvilli and producing cell movements were discussed earlier. Through its role in cell motility, actin plays a crucial role in embryonic development, muscle contraction, immune function, wound healing, cancer metastasis, and other processes that involve cell migration.

**Intermediate filaments** (8–10 nm in diameter) are thicker and stiffer than microfilaments. They resist stresses placed on a cell and participate in junctions that attach some cells to their neighbors. In epidermal cells, they are made of the tough protein *keratin* and occupy most of the cytoplasm. They are responsible for the strength of hair and fingernails. Intermediate filaments also line the inside of the nuclear envelope and form a cage that encloses the DNA.

A **microtubule** (25 nm in diameter) is a cylinder made of 13 parallel strands called *protofilaments.* Each protofilament is a long chain of globular proteins called *tubulin*

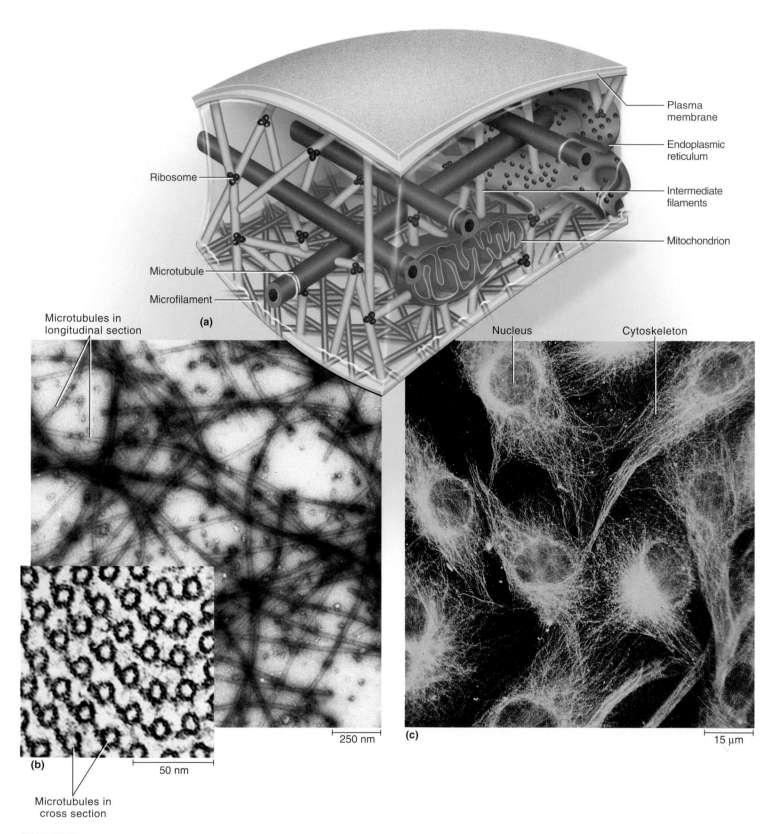

Plasma membrane

Endoplasmic reticulum

Ribosome

Intermediate filaments

Mitochondrion

Microtubule

Microfilament

**(a)**

Microtubules in longitudinal section

Nucleus          Cytoskeleton

250 nm          **(c)**          15 µm

Microtubules in cross section

**(b)**          50 nm

**FIGURE 3.31    The Cytoskeleton.**    (a) Diagram of the cytoskeleton. (b) Microtubules in a cell of the testis, shown in longitudinal and cross sections. (c) The fibrous cytoskeleton of a cell, labeled with fluorescent antibodies and photographed through a fluorescence microscope.

(fig. 3.32). Microtubules radiate from the centrosome and hold organelles in place, form bundles that maintain cell shape and rigidity, and act somewhat like railroad tracks to guide organelles and molecules to specific destinations in a cell. They form the axonemes of cilia and flagella and are responsible for their beating movements. They also form the mitotic spindle that guides chromosome movement during cell division. Microtubules are not permanent structures. They come and go moment by moment as tubulin molecules assemble into a tubule and then suddenly break apart again to be used somewhere else in the cell. The double and triple sets of microtubules in cilia, flagella, basal bodies, and centrioles, however, are more stable.

## INCLUSIONS

**Inclusions** are of two kinds: stored cellular products such as glycogen granules, pigments, and fat droplets (see fig. 3.26b), and foreign bodies such as dust particles, viruses, and intracellular bacteria. Inclusions are never enclosed in a unit membrane, and unlike the organelles and cytoskeleton, they are not essential to cell survival.

The major features of a cell are summarized in table 3.4.

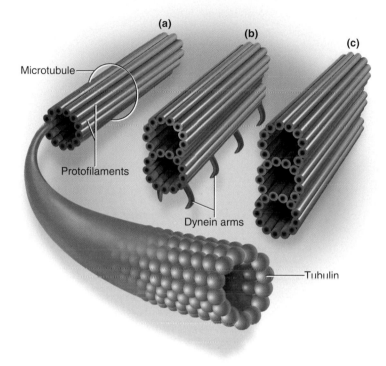

**FIGURE 3.32** **Microtubules.** (a) A microtubule is composed of 13 protofilaments. Each protofilament is a spiral chain of globular proteins called tubulin. (b) One of the nine microtubule pairs that form the axonemes of cilia and flagella. (c) One of the nine microtubule triplets that form a centriole.

### Before You Go On

*Answer the following questions to test your understanding of the preceding section:*

22. *Distinguish between organelles and inclusions. State two examples of each.*

23. *Briefly state how each of the following cell components can be recognized in electron micrographs: the nucleus, a mitochondrion, a lysosome, and a centriole. What is the primary function of each?*

24. *What three organelles are involved in protein synthesis?*

25. *In what ways do rough and smooth endoplasmic reticulum differ?*

26. *Define* centriole, microtubule, cytoskeleton, *and* axoneme. *How are these structures related to one another?*

| TABLE 3.4 | Summary of Organelles and Other Cellular Structures | |
|---|---|---|
| **Structure** | **Appearance to TEM** | **Function** |
| Plasma membrane (figs. 3.3 and 3.6) | Two dark lines at cell surface, separated by narrow light space | Prevents escape of cell contents; regulates exchange of materials between cytoplasm and extracellular fluid; involved in intercellular communication |
| Microvilli (fig. 3.10) | Short, densely-spaced, hairlike processes or scattered bumps on cell surface; interior featureless or with bundle of microfilaments | Increase absorptive surface area; some sensory roles (hearing, equilibrium, taste) |
| Cilia (fig. 3.11) | Long hairlike projections of apical cell surface; axoneme with 9 + 2 array of microtubules | Move substances along cell surface; some sensory roles (hearing, equilibrium, smell, vision) |
| Flagellum | Long, single, whiplike process with axoneme | Sperm motility |
| Nucleus (figs. 3.3 and 3.25) | Largest organelle in most cells, surrounded by double unit membrane with nuclear pores | Genetic control center of cell; directs protein synthesis |
| Rough ER (fig. 3.26a) | Extensive sheets of parallel unit membranes with ribosomes on outer surface | Protein synthesis and manufacture of cellular membranes |
| Smooth ER (fig. 3.26b) | Branching network of tubules with smooth surface (no ribosomes); usually broken into numerous small segments in TEM photos | Lipid synthesis, detoxification, calcium storage |
| Ribosomes (fig. 3.26a) | Small dark granules free in cytosol or on surface of rough ER | Interpret the genetic code and synthesize polypeptides |
| Golgi complex (fig. 3.27) | Several closely-spaced, parallel cisternae with thick edges, usually near nucleus, often with many Golgi vesicles nearby | Receives and modifies newly synthesized polypeptides, synthesizes carbohydrates, adds carbohydrates to glycoproteins; packages cell products into Golgi vesicles |
| Golgi vesicles (fig. 3.27) | Round to irregular sacs near Golgi complex, usually with light, featureless contents | Become secretory vesicles and carry cell products to apical surface for exocytosis, or become lysosomes |
| Lysosomes (fig. 3.28a) | Round to oval sacs with single unit membrane, often a dark featureless interior but sometimes with protein layers or crystals | Contain enzymes for intracellular digestion, autophagy, programmed cell death, and glucose mobilization |
| Peroxisomes (fig. 3.28b) | Similar to lysosomes; often lighter in color | Contain enzymes for detoxification of free radicals, alcohol, and other drugs; oxidize fatty acids |
| Mitochondria (fig. 3.29) | Round, rod-shaped, bean-shaped, or threadlike structures with double unit membrane and shelflike infoldings called cristae | ATP synthesis |
| Centrioles (fig. 3.30) | Short cylindrical bodies, each composed of a circle of nine triplets of microtubules | Form mitotic spindle during cell division; unpaired centrioles form basal bodies of cilia and flagella |
| Centrosome (fig. 3.5) | Clear area near nucleus containing a pair of centrioles | Organizing center for formation of microtubules of cytoskeleton and mitotic spindle |
| Basal body (fig. 3.11b) | Unpaired centriole at the base of a cilium or flagellum | Point of origin, growth, and anchorage of a cilium or flagellum; produces axoneme |
| Microfilaments (figs. 3.10 and 3.31) | Thin protein filaments (6 nm diameter), often in parallel bundles or dense networks in cytoplasm | Support microvilli; involved in muscle contraction and other cell motility, endocytosis, and cell division |
| Intermediate filaments (fig. 3.31a) | Thicker protein filaments (8–10 nm diameter) extending throughout cytoplasm or concentrated at cell-to-cell junctions | Give shape and physical support to cell; anchor cells to each other and to extracellular material; compartmentalize cell contents |
| Microtubules (figs. 3.31 and 3.32) | Hollow protein cylinders (25 nm diameter) | Form axonemes of cilia and flagella, centrioles, basal bodies, and mitotic spindles; enable motility of cell parts; direct organelles and macromolecules to their destinations within a cell |
| Inclusions (fig. 3.26b) | Highly variable—fat droplets, glycogen granules, protein crystals, dust, bacteria, viruses; never enclosed in unit membranes | Storage products or other products of cellular metabolism, or foreign matter retained in cytoplasm |

**INSIGHT 3.4**     Evolutionary Medicine

## Mitochondria—Evolution and Clinical Significance

It is virtually certain that mitochondria evolved from bacteria that invaded another primitive cell, survived in its cytoplasm, and became permanent residents. Certain modern bacteria called *ricketsii* live in the cytoplasm of other cells, showing that this mode of life is feasible. The two unit membranes around the mitochondrion suggest that the original bacterium provided the inner membrane, and the host cell's phagosome provided the outer membrane when the bacterium was phagocytized.

Several comparisons show the apparent relationship of mitochondria to bacteria. Their ribosomes are more like bacterial ribosomes than those of eukaryotic (nucleated) cells. Mitochondrial DNA (mtDNA) is a small, circular molecule that resembles the circular DNA of other bacteria, not the linear DNA of the cell nucleus. It replicates independently of nuclear DNA. mtDNA codes for some of the enzymes employed in ATP synthesis. It consists of 16,569 *base pairs* (explained in chapter 4), comprising 37 genes, compared with over a billion base pairs and about 35,000 genes in nuclear DNA.

When a sperm fertilizes an egg, any mitochondria introduced by the sperm are usually destroyed and only those provided by the egg are passed on to the developing embryo. Therefore, mitochondrial DNA is inherited almost exclusively through the mother. While nuclear DNA is reshuffled in every generation by sexual reproduction, mtDNA remains unchanged except by random mutation. Biologists and anthropologists have used mtDNA as a "molecular clock" to trace evolutionary lineages in humans and other species. mtDNA has also been used as evidence in criminal law and to identify the remains of soldiers killed in combat. mtDNA was used recently to identify the remains of the famed bandit Jesse James, who was killed in 1882. Anthropologists have gained evidence, although still controversial, that of all the women who lived in Africa 200,000 years ago, only one has any descendents still living today. This "mitochondrial Eve" is ancestor to us all.

mtDNA is very exposed to damage from free radicals normally generated in mitochondria by aerobic respiration. Yet unlike nuclear DNA, mtDNA has no effective mechanism for repairing damage. Therefore, it mutates about 10 times as rapidly as nuclear DNA. Some of these mutations are responsible for various rare hereditary diseases. Tissues and organs with the highest energy demands are the most vulnerable to mitochondrial dysfunctions—nervous tissue, the heart, the kidneys, and skeletal muscles, for example.

*Mitochondrial myopathy* is a degenerative muscle disease in which the muscle displays "ragged red fibers," cells with abnormal mitochondria that stain red with a particular histological stain. *Mitochondrial encephalomyopathy, lactic acidosis and strokelike episodes* (MELAS) is a mitochondrial disease involving seizures, paralysis, dementia, muscle deterioration, and a toxic accumulation of lactic acid in the blood. *Leber hereditary optic neuropathy* (LHON) is a form of blindness that usually appears in young adulthood as a result of damage to the optic nerve. *Kearns-Sayre syndrome* (KSS) involves paralysis of the eye muscles, degeneration of the retina, heart disease, hearing loss, diabetes, and kidney failure. Damage to mtDNA has also been implicated as a possible factor in Alzheimer disease, Huntington disease, and other degenerative diseases of old age.

# CHAPTER REVIEW

## Review of Key Concepts

### Concepts of Cellular Structure (p. 90)

1. Cytology is the study of cellular structure and function.

2. All human structure and function is the result of cellular activity.

3. Cell shapes are described as squamous, polygonal, stellate, cuboidal, columnar, spheroid, ovoid, discoid, fusiform, and fibrous.

4. Most human cells are 10 to 15 μm in diameter. Cell size is limited in part by the ratio of surface area to volume.

5. A cell is enclosed in a *plasma membrane* and contains usually one nucleus.

6. The *cytoplasm* is everything between the plasma membrane and nucleus. It consists of a clear fluid, the *cytosol* or *intracellular fluid* (ICF), and embedded organelles and other structures. Fluid external to the cell is *extracellular fluid* (ECF).

### The Cell Surface (p. 93)

1. The plasma membrane is made of lipid and protein.

2. The most abundant lipid molecules in the membrane are phospholipids, which form a bilayer with their hydrophilic heads facing the ICF and ECF. Other membrane lipids include cholesterol and glycolipids.

3. Membrane proteins are called *transmembrane proteins* if they are embedded in the lipid bilayer and extend all the way through it, and *peripheral proteins* if they only cling to the intracellular face of the lipid bilayer.

4. Membrane proteins serve as receptors, second-messenger systems, enzymes, channels, carriers, molecular motors, cell-identity markers, and cell-adhesion molecules.

5. Channel proteins are called *gates* if they can open and close. Gates are called *ligand-regulated, voltage-regulated,* or *mechanically regulated* depending on whether they open and close in response to chemicals, voltage changes across the membrane, or mechanical stress.

6. Second-messenger systems are systems for generating an internal cellular signal in response to an external one. One of the best-known examples results in the formation of a second messenger, cyclic AMP (cAMP), within the cell when certain extracellular signaling molecules bind to a membrane receptor.

7. All cells are covered with a *glycocalyx,* a layer of carbohydrate molecules bound to membrane lipids and proteins. The glycocalyx functions in immunity and other forms of protection, cell adhesion, fertilization, and embryonic development, among other roles.

8. *Microvilli* are tiny surface extensions of the plasma membrane that increase a cell's surface area. They are especially well developed on absorptive cells, as in the kidney and small intestine.

9. *Cilia* are longer, hairlike surface extensions with a central axoneme, composed of a 9 + 2 arrangement of microtubules. Some cilia are stationary and sensory in function, and some are motile and propel substances across epithelial surfaces.

10. A *flagellum* is a long, solitary, whiplike extension of the cell surface. The only functional flagellum in humans is the sperm tail.

### Membrane Transport (p. 102)

1. The plasma membrane is *selectively permeable*—it allows some substances to pass through it but prevents others from entering or leaving a cell. There are several methods of passage through a plasma membrane.

2. *Filtration* is the movement of fluid through a membrane under a physical force such as blood pressure, while the membrane holds back relatively large particles.

3. *Simple diffusion* is the spontaneous net movement of particles from a place of high concentration to a place of low concentration, such as respiratory gases moving between the pulmonary air sacs and the blood. The speed of diffusion depends on temperature, molecular weight, concentration differences, and the surface area and permeability of the membrane.

4. *Osmosis* is the diffusion of water through a selectively permeable membrane from the more watery to the less watery side. Channel proteins called *aquaporins* allow passage of water through plasma membranes.

5. The speed of osmosis depends on the relative concentrations, on the two sides of a membrane, of solute molecules that cannot penetrate the membrane. *Osmotic pressure,* the physical force that would be required to stop osmosis, is proportional to the concentration of nonpermeating solutes on the side to which water is moving.

6. An *osmole* is one mole of dissolved particles in a solution. *Osmolarity* is the number of osmoles of solute per liter of solution. The osmolarity of body fluids is usually expressed in milliosmoles per liter (mOsm/L).

7. *Tonicity* is the ability of a solution to affect the fluid volume and pressure in a cell. A solution is *hypotonic, isotonic,* or *hypertonic* to a cell if it contains, respectively, a lower, equal, or greater concentration of nonpermeating solutes than the cell cytoplasm does. Cells swell and burst in hypotonic solutions and shrivel in hypertonic solutions.

8. *Carrier-mediated transport* employs membrane proteins to move solutes through a membrane. A given carrier is usually specific for a particular solute.

9. Membrane carriers can become *saturated* with solute molecules and then unable to work any faster. The maximum rate of transport is the *transport maximum* (Tm).

10. A *uniport* is a carrier that transports only one solute at a time; a *symport*

carries two or more solutes through the membrane in the same direction (a process called *cotransport*); and an *antiport* carries two or more solutes in opposite directions (a process called *countertransport*).

11. *Facilitated diffusion* is a form of carrier-mediated transport that moves solutes through a membrane down a concentration gradient, without an expenditure of ATP.

12. *Active transport* is a form of carrier-mediated transport that moves solutes through a membrane up (against) a concentration gradient, with the expenditure of ATP.

13. The Na$^+$–K$^+$ pump is an antiport that moves Na$^+$ out of a cell and K$^+$ into it. It serves for control of cell volume, secondary active transport, heat production, and maintenance of an electrical membrane potential.

14. *Vesicular transport* is the movement of substances in bulk through a membrane in membrane-enclosed vesicles.

15. *Endocytosis* is any form of vesicular transport that brings material into a cell, including *phagocytosis, pinocytosis,* and *receptor-mediated endocytosis*.

16. *Exocytosis* is a form of vesicular transport that discharges material from a cell. It functions in the release of cell products and in replacement of plasma membrane removed by endocytosis.

**The Cytoplasm (p. 112)**

1. The cytoplasm is composed of a clear gelatinous cytosol in which are embedded organelles, the cytoskeleton, and inclusions (table 3.4).

2. *Organelles* are internal structures in the cytoplasm that carry out specialized tasks for a cell.

3. *Membranous* organelles are enclosed in one or two layers of unit membrane similar to the plasma membrane. These include the *nucleus, endoplasmic reticulum* (which has rough and smooth portions), *ribosomes,* the *Golgi complex, lysosomes, peroxisomes,* and *mitochondria*. The *centrioles* and *ribosomes* are non-membranous organelles.

4. The *cytoskeleton* is a supportive framework of protein filaments and tubules in a cell. It gives a cell its shape, organizes the cytoplasmic contents, and functions in movements of cell contents and the cell as a whole. It is composed of *microfilaments* of the protein *actin; intermediate filaments* of *keratin* or other proteins; and cylindrical *microtubules* of the protein *tubulin*.

5. *Inclusions* are either stored cellular products such as glycogen, pigments, and fat, or foreign bodies such as bacteria, viruses, and dust. Inclusions are not vital to cell survival.

# Testing Your Recall

1. The clear, structureless gel in a cell is its
   a. nucleoplasm.
   b. protoplasm.
   c. cytoplasm.
   d. neoplasm.
   e. cytosol.

2. The Na$^+$–K$^+$ pump is
   a. a peripheral protein.
   b. a transmembrane protein.
   c. a G protein.
   d. a glycolipid.
   e. a phospholipid.

3. Which of the following processes could occur *only* in the plasma membrane of a living cell?
   a. facilitated diffusion
   b. simple diffusion
   c. filtration
   d. active transport
   e. osmosis

4. Cells specialized for absorption of matter from the ECF are likely to show an abundance of
   a. lysosomes.
   b. microvilli.
   c. mitochondria.
   d. secretory vesicles.
   e. ribosomes.

5. Osmosis is a special case of
   a. pinocytosis.
   b. carrier-mediated transport.
   c. active transport.
   d. facilitated diffusion.
   e. simple diffusion.

6. Membrane carriers resemble enzymes except for the fact that carriers
   a. are not proteins.
   b. do not have binding sites.
   c. are not selective for particular ligands.
   d. change conformation when they bind a ligand.
   e. do not chemically change their ligands.

7. The cotransport of glucose derives energy from
   a. an Na$^+$ concentration gradient.
   b. the glucose being transported.
   c. a Ca$^{2+}$ gradient.
   d. the membrane voltage.
   e. body heat.

8. The function of cAMP in a cell is
   a. to activate a G protein.
   b. to remove phosphate groups from ATP.
   c. to activate kinases.
   d. to bind to the first messenger.
   e. to add phosphate groups to enzymes.

9. Most cellular membranes are made by
   a. the nucleus.
   b. the cytoskeleton.
   c. enzymes in the peroxisomes.
   d. the endoplasmic reticulum.
   e. replication of existing membranes.

10. Matter can leave a cell by any of the following means *except*
    a. active transport.
    b. pinocytosis.
    c. an antiport.
    d. simple diffusion.
    e. exocytosis.

11. Most human cells are 10 to 15 _____ in diameter.

12. When a hormone cannot enter a cell, it activates the formation of a/an _____ inside the cell.

13. _____ gates in the plasma membrane open or close in response to changes in the electrical charge difference across the membrane.

14. The force exerted on a membrane by water is called _____.

15. A concentrated solution that causes a cell to shrink is _____ to the cell.

16. Fusion of a secretory vesicle with the plasma membrane, and release of the vesicle's contents, is called _____.

17. Two organelles that are surrounded by a double unit membrane are the _____ and the _____.

18. Liver cells can detoxify alcohol with two organelles, the _____ and _____.

19. An ion gate in the plasma membrane that opens or closes when a chemical binds to it is called a/an _____.

20. The space enclosed by the unit membrane of the Golgi complex and endoplasmic reticulum is called the

_____.

*Answers in Appendix B*

## True or False

*Determine which five of the following statements are false, and briefly explain why.*

1. If a cell were poisoned so it could not make ATP, osmosis through its membrane would cease.

2. Material can move either into a cell or out by means of active transport.

3. A cell's second messengers serve mainly to transport solutes through the membrane.

4. The Golgi complex makes lysosomes but not peroxisomes.

5. Some membrane channels are peripheral proteins.

6. The plasma membrane consists primarily of protein molecules.

7. The brush border of a cell is composed of cilia.

8. Human cells swell or shrink in any solution other than an isotonic one.

9. Osmosis is not limited by the transport maximum ($T_m$).

10. It is very unlikely for a cell to have more centrosomes than ribosomes.

*Answers in Appendix B*

## Testing Your Comprehension

1. If someone bought a saltwater fish in a pet shop and put it in a freshwater aquarium at home, what would happen to the fish's cells? What would happen if someone put a freshwater fish in a saltwater aquarium? Explain.

2. A farmer's hand and forearm are badly crushed in a hay bailer. When examined at the hospital, his blood potassium level is found to be abnormal. Would you expect it to be

higher or lower than normal? Explain.

3. Many children worldwide suffer from a severe deficiency of dietary protein. As a result, they have very low levels of blood albumin. How do you think this affects the water content and volume of their blood? Explain.

4. It is often said that mitochondria make energy for a cell. Why is this statement false?

5. Kartagener syndrome is a hereditary disease in which dynein arms are lacking from the axonemes of cilia and flagella. Predict the effect of Kartagener syndrome on a man's ability to father a child. Predict its effect on his respiratory health. Explain both answers.

*Answers at www.mhhe.com/saladin4*

## www.mhhe.com/saladin4

*The textbook website provides a wealth of interactive study materials fully organized and integrated by chapter. You will find practice quizzes, labeling exercises, and much more that will complement your learning and understanding of anatomy and physiology. The website also includes tools designed to enhance your* **Anatomy & Physiology | REVEALED** *experience.*

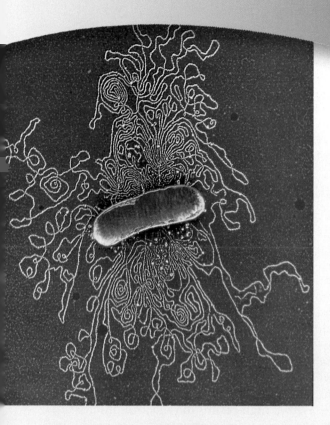

*A single DNA molecule spilling from a ruptured bacterial cell (TEM)*

# GENETICS AND CELLULAR FUNCTION

## CHAPTER OUTLINE

## INSIGHTS

## Brushing Up

To understand this chapter, it is important that you understand or brush up on the following concepts:

- Levels of protein structure (p. 78)
- Functions of proteins (p. 79)
- Exocytosis (p. 110)
- Ribosomes, rough endoplasmic reticulum, and Golgi complex (pp. 112–113)
- Centrioles and microtubules (pp. 116, 117)

Several chapters in this book describe hereditary traits such as blood type and hair color, and genetic disorders such as color blindness, cystic fibrosis, diabetes mellitus, and hemophilia. To be prepared to understand such conditions, it is necessary to have some understanding of DNA and genes. This chapter is intended to provide that preparation.

Heredity has been a matter of human interest dating even to biblical writings, but a scientific understanding of how traits are passed from parent to offspring began with the Austrian monk Gregor Mendel (1822–84) and his famous experiments on garden peas. In the early twentieth century, biologists first observed chromosomes with the microscope and began to appreciate the importance of Mendel's work. From that simple but insightful beginning, genetics has grown into a highly diverse science with several subdisciplines, and arguably the most dynamic of all natural sciences in these early years of the twenty-first century. **Mendelian genetics** deals with parent–offspring and larger family relationships to discern and predict patterns of inheritance within a family line. **Cytogenics** uses the techniques of cytology and microscopy to study the chromosomes and their relationship to hereditary traits. **Molecular genetics** uses the techniques of biochemistry to study the structure and function of DNA. **Genomic medicine** comprehensively studies the entire DNA endowment of an individual (the *genome*), how it influences health and disease, and how it can be manipulated to treat or cure diseases. We will examine all four of these perspectives in this chapter.

# Genes and Nucleic Acids

### Objectives

When you have completed this section, you should be able to

- describe how DNA is organized in the nucleus; and
- compare the structures and functions of DNA and RNA.

With improvements in the microscope, biologists of the late nineteenth century saw that the nucleus divides just before the cell does, and they came to suspect that the nucleus was the center of heredity and cellular control. This led them to search the nucleus for the biochemical secrets of heredity. Swiss biochemist Johann Friedrich Miescher (1844–95) studied the nuclei of white blood cells extracted from pus in used hospital bandages, and later the nuclei of salmon sperm, since both cell types offered large nuclei with minimal amounts of contaminating cytoplasm. In 1869, he discovered an acidic, phosphorous-rich substance he named *nuclein.* He correctly believed this to be the cell's hereditary matter, although he was never able to convince other scientists of this. We now call this substance **deoxyribonucleic acid (DNA)** and know it to be the repository of our genes.

## ORGANIZATION OF THE CHROMATIN

As we saw in chapter 3, the nucleus of a cell contains dense masses called **nucleoli** surrounded by fine thread-like material called **chromatin.** The chromatin consists of usually 46 long filaments called **chromosomes,** composed of DNA and protein. There is a stupendous amount of DNA in the nucleus—about 2 m. It is a prodigious feat to pack this much DNA into a tiny nucleus, and in such an orderly fashion that it does not become tangled, broken, and damaged beyond use. We begin our exploration of DNA with a consideration of how this is achieved.

In nondividing cells, the chromatin is so finely dispersed that it usually cannot be seen with the light microscope. With a high-resolution electron microscope, however, it has a granular appearance, like beads on a string (fig. 4.1a). The beaded string is divided into segments called **nucleosomes,** each composed of a *core particle* ("the bead") and a short segment of *linker DNA* that leads to the next core particle. There are about 30 million nucleosomes in the nucleus. The core particle is a disc-shaped cluster of eight proteins called **histones,** with the DNA wound around it for 1.65 turns like a ribbon around a spool (fig. 4.1b). Winding the DNA around the core particles makes the chromatin thread more than five times as thick (11 nm) and one-third shorter than the DNA alone.

But even at this degree of compaction, a single chromosome would cross the entire nucleus hundreds of times. There are higher orders of structure that make the chromosome still more compact. First, the nucleosomes are arranged in a zigzag pattern, folding the chromatin like an accordion. This produces a strand 30 nm wide, but still 100 times as long as the nuclear diameter. Then, the 30-nm strand is thrown into complex, irregular loops and coils that make the chromosome 300 nm thick and only 1/1,000 the length of the DNA molecule. This is the state of the DNA in a nondividing cell. This is not a static structure, but changes from moment to moment according to the genetic activity of the cell.

When a cell is preparing to divide, it makes an exact copy of all its DNA by a process described later, and each chromosome is then composed of two parallel filaments called *sister chromatids.* In the early stage of cell division, these chromatids coil some more until each one becomes another 10 times shorter and about 700 nm wide—so the chromatid is now 1/10,000 the length of the DNA. Only

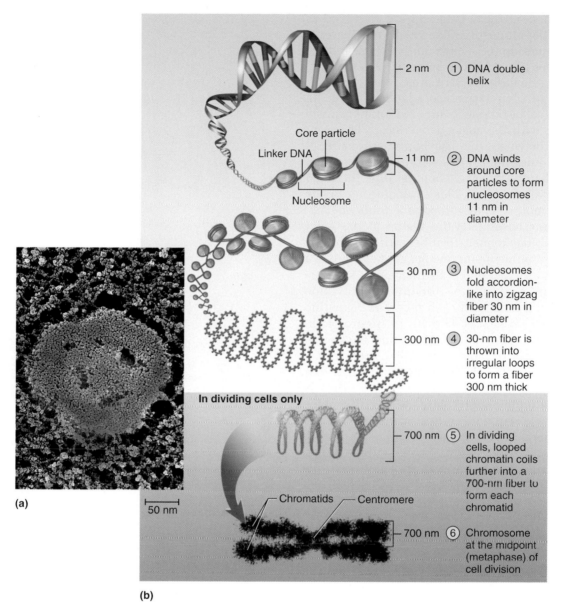

**FIGURE 4.1  Chromatin Structure.** (a) Nuclear contents of a germ cell from an 8-week-old human embryo (colorized SEM). The center mass is the nucleolus. It is surrounded by granular fibers of chromatin. Each granule is a nucleosome. (b) The coiling of chromatin and its relationship to the nucleosomes.

now are the chromosomes thick enough to be seen with a light microscope. This compaction not only allows the 2 m of DNA to be packed into the nucleus, but also enables the two sister chromatids to be pulled apart and carried to separate daughter cells without damage to the DNA.

## DNA STRUCTURE AND FUNCTION

The DNA molecule has a uniform diameter of 2 nm, but varies greatly in length from the smallest to the largest chromosomes. With 2 nm of DNA divided among the 46 chromosomes, the average DNA molecule is about 43 mm

(almost 2 in.) long. To put this in perspective, imagine that an average DNA molecule were scaled up to the diameter of a telephone pole (about 20 cm, or 8 in.). At this diameter, a proportionate pole would rise about 4,400 km (2,700 mi) into space—far higher than the orbits of space shuttles (320–390 km) and the Hubble Telescope (600 km).

At the molecular level, DNA and other nucleic acids are polymers of **nucleotides** (NEW-clee-oh-tides). A nucleotide consists of a sugar, a phosphate group, and a single- or double-ringed **nitrogenous** (ny-TRODJ-eh-nus) **base.** Three bases—**cytosine (C), thymine (T),** and **uracil (U)**—

have a single carbon–nitrogen ring and are classified as *pyrimidines* (py-RIM-ih-deens). The other two bases—**adenine (A)** and **guanine (G)**—have double rings and are classified as *purines* (fig. 4.2). The bases of DNA are C, T, A, and G, whereas the bases of RNA are C, U, A, and G.

The structure of DNA resembles a ladder (fig. 4.3a). Each sidepiece is a backbone composed of phosphate groups alternating with the sugar *deoxyribose.* The step-like connections between the backbones are pairs of nitrogenous bases. Imagine this as a soft rubber ladder that you can twist, so that the two backbones become entwined to resemble a spiral staircase. This is analogous to the shape of the DNA molecule, described as a *double helix.*

The nitrogenous bases face the inside of the helix and hold the two backbones together with hydrogen bonds. Across from a purine on one backbone, there is a pyrimidine on the other. A given purine cannot arbitrarily bind to just any pyrimidine. Adenine and thymine form two hydrogen bonds with each other, and guanine and cytosine form three, as shown in figure 4.3b. Therefore, wherever there is an A on one backbone, there is a T across from it, and every C is paired with a G. A–T and C–G are called the **base pairs.** The fact that one strand governs the base sequence of the other is called the **law of complementary base pairing.** It enables us to predict the base sequence of one strand if we know the sequence of the complementary strand. The pairing of each small, single-ringed pyrimidine with a large, double-ringed purine gives the DNA molecule its uniform 2-nm width.

## Think About It

*What would be the base sequence of the DNA strand across from ATTGACTCG? If a DNA molecule were known to be 20% adenine, predict its percentage of cytosine and explain your answer.*

The essential function of DNA is to code for the proteins synthesized by a cell. About 2% of the DNA is composed of genes, while the other 98% consists of noncoding DNA, which plays various roles in chromosome structure and regulation of gene activity, and some of which may have no function at all. Some of the noncoding DNA is sometimes thought of as "junk DNA" to suggest that it might be merely harmless debris accumulated by mutation over eons of evolutionary time.

Humans are estimated to have about 30,000 to 35,000 genes. A **gene** may be defined in various ways. It is an information-containing segment of DNA that codes for a protein or for a group of closely related proteins. A gene determines the characteristics of a species and the hereditary characteristics that distinguish each individual. A gene is thus the unit of heredity that is passed from parent to offspring. These definitions do not contradict each other, but point out different aspects of gene function.

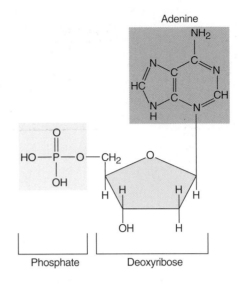
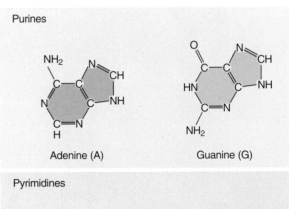

(a)

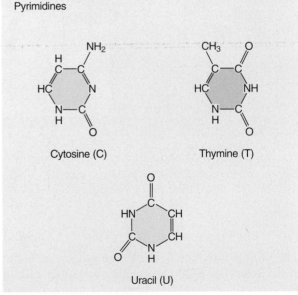

(b)

**FIGURE 4.2   Nucleotides and Nitrogenous Bases.**
(a) The structure of a nucleotide, one of the monomers of DNA and RNA. In RNA, the sugar is ribose. (b) The five nitrogenous bases found in DNA and RNA.

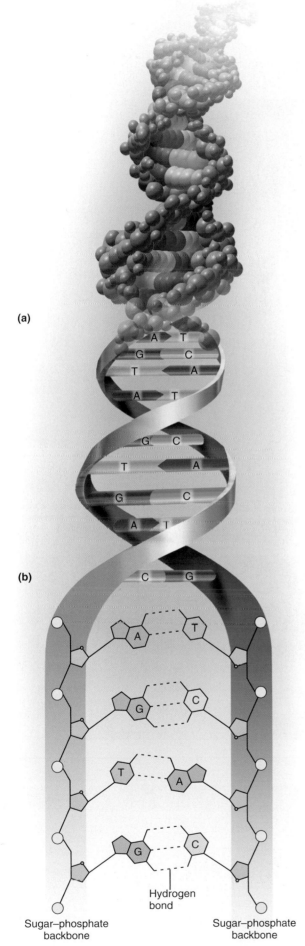

**(a)**

**(b)**

**(c)**

Sugar–phosphate backbone

Hydrogen bond

Sugar–phosphate backbone

## RNA STRUCTURE AND FUNCTION

DNA directs the synthesis of proteins by means of its smaller cousins, the ribonucleic acids (RNAs). There are three types of RNA: *messenger RNA* (mRNA), *ribosomal RNA* (rRNA), and *transfer RNA* (tRNA). Their individual roles are described shortly. For now we consider what they have in common and how they differ from DNA (table 4.1). The most significant difference is that RNA is much smaller, ranging from about 70 to 90 bases in tRNA to slightly over 10,000 bases in the largest mRNA. DNA, by contrast, averages more than 100 million base pairs long. Also, while DNA is a double helix, RNA consists of only one nucleotide chain, not held together by complementary base pairs except in certain regions of tRNA where the molecule folds back on itself. The sugar in RNA is ribose instead of deoxyribose, and one of the pyrimidines of DNA, thymine, is replaced by uracil (U) in RNA (see fig. 4.2).

The essential function of RNA is to interpret the code in DNA and direct the synthesis of proteins. RNA works mainly in the cytoplasm, while DNA remains safely behind in the nucleus, "giving orders" from there. This process is described in the next section of this chapter.

### Before You Go On

*Answer the following questions to test your understanding of the preceding section:*

1. What is the difference between DNA and chromatin?

2. What are the three components of a nucleotide? Which component varies from one nucleotide to another in DNA?

3. What two factors govern the pattern of base pairing in DNA?

4. Summarize the differences between DNA and RNA.

**FIGURE 4.3** **DNA Structure.** (a) A molecular space-filling model of DNA giving some impression of its molecular geometry. (b) The "twisted ladder" structure. The two sugar–phosphate backbones twine around each other while complementary bases (colored bars) face each other on the inside of the double helix. (c) A small segment of DNA showing the composition of the backbone and complementary pairing of the nitrogenous bases.

▶ *How would the uniform 2-nm diameter of DNA be affected if two purines or two pyrimidines could pair with each other?*

## INSIGHT 4.1     Medical History

### Discovery of the Double Helix

The components of DNA were known by 1900—the sugar, phosphate, and bases—but the technology did not exist then to determine how they were put together. The credit for that discovery went mainly to James Watson and Francis Crick in 1953 (fig. 4.4). The events surrounding their discovery of the double helix represent one of the most dramatic stories of modern science—the subject of many books and a movie. When Watson and Crick came to share a laboratory at Cambridge University in 1951, both had barely begun their careers. Watson, age 23, had just completed his Ph.D. in the United States, and Crick, 11 years older, was a doctoral candidate. Yet the two were about to become the most famous molecular biologists of the twentieth century, and the discovery that won them such acclaim came without a single laboratory experiment of their own.

Others were fervently at work on DNA, including Rosalind Franklin and Maurice Wilkins at King's College in London. Using a technique called X-ray diffraction, Franklin had determined that DNA had a repetitious helical structure with sugar and phosphate on the outside of the helix. Without her permission, Wilkins showed one of Franklin's best X-ray photographs to Watson. Watson said, "The instant I saw the picture my mouth fell open and my pulse began to race." It provided a flash of insight that allowed the Watson and Crick team to beat Franklin to the goal. They were quickly able to piece together a scale model from cardboard and sheet metal that fully accounted for the known geometry of DNA. They rushed a paper into print in 1953 describing the double helix, barely mentioning the importance of Franklin's 2 years of painstaking X-ray diffraction work in unlocking the mystery of life's most important molecule.

For this discovery, Watson, Crick, and Wilkins shared the Nobel Prize in Physiology or Medicine in 1962. Nobel Prizes are awarded only to the living, and in the final irony of her career, Rosalind Franklin had died in 1958, at the age of 37, of a cancer possibly induced by the X rays that were her window on DNA architecture.

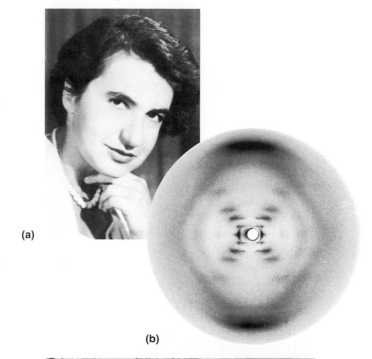

**(a)**

**(b)**

**(c)**

**FIGURE 4.4  Discoverers of the Double Helix.**  (a) Rosalind Franklin (1920–58), whose painstaking X-ray diffraction photographs revealed important information about the molecular geometry of DNA. (b) One of Franklin's X-ray photographs. (c) James Watson (1928–  ) *(left)* and Francis Crick (1916–2004) with their model of the double helix.

| TABLE 4.1 | Comparison of DNA and RNA | |
|---|---|---|
| **Feature** | **DNA** | **RNA** |
| Sugar | Deoxyribose | Ribose |
| Types of nitrogenous bases | A, T, C, G | A, U, C, G |
| Number of nitrogenous bases | Averages $10^8$ base pairs | 70–10,000 unpaired bases |
| Number of nucleotide chains | Two (double helix) | One |
| Site of action | Functions in nucleus; cannot leave | Leaves nucleus; functions in cytoplasm |
| Function | Codes for synthesis of RNA and protein | Carries out instructions in DNA; assembles proteins |

# Protein Synthesis and Secretion

### Objectives

When you have completed this section, you should be able to

- define *genetic code* and describe how DNA codes for protein structure;

- describe the process of assembling amino acids to form a protein;

- explain what happens to a protein after its amino acid sequence has been synthesized; and

- explain how DNA indirectly regulates the synthesis of nonprotein molecules.

Everything a cell does ultimately results from the action of its proteins; DNA directs the synthesis of those proteins. Cells, of course, synthesize many other substances as well—glycogen, fat, phospholipids, steroids, pigments, and so on. There are no genes for these cell products, but

their synthesis depends on enzymes that are coded for by the genes. For example, even though a cell of the testis has no genes for testosterone, testosterone synthesis is indirectly under genetic control (fig. 4.5). Since testosterone strongly influences such behaviors as aggression and sexual drive (in both sexes), we can see that genes also make a significant contribution to behavior. In this section, we examine how protein synthesis results from the instructions given in the genes.

## PREVIEW

Before studying the details of protein synthesis, it will be helpful to consider the big picture. In brief, DNA contains a genetic code that specifies which proteins a cell can make. Nearly all the body's cells except the sex cells contain identical genes, but different genes are activated in different cells; for example, the genes for digestive enzymes are active in stomach cells but not in muscle cells. When a gene is activated, a molecule of **messenger RNA (mRNA),** a sort of mirror-image copy of the gene, is made. Most mRNA migrates from the nucleus to the cytoplasm, where its code is "read" by a ribosome. Ribosomes

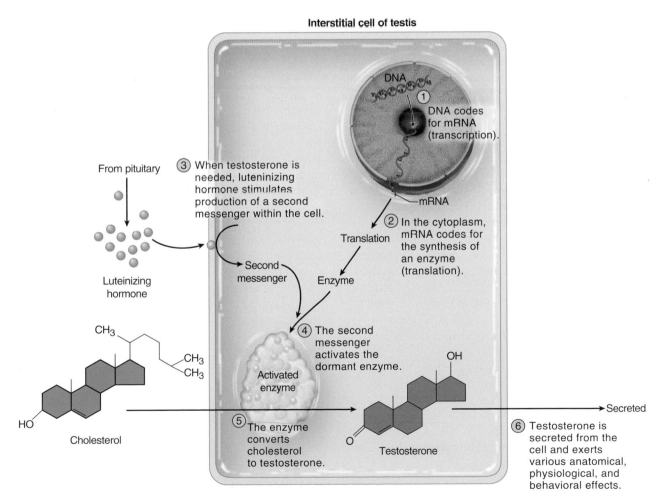

**FIGURE 4.5  Indirect Control of Testosterone Synthesis by DNA.** There is no gene for testosterone, but DNA regulates its production by coding for enzymes that convert cholesterol to testosterone.

are composed of **ribosomal RNA (rRNA)** and enzymes. **Transfer RNA (tRNA)** delivers amino acids to the ribosome, and the ribosome chooses from among these to assemble amino acids in the order directed by the mRNA.

In summary, you can think of the process of protein synthesis as DNA→mRNA→protein, with each arrow reading as "codes for the production of." The step from DNA to mRNA is called *transcription,* and the step from mRNA to protein is called *translation.* Transcription occurs in the nucleus, where the DNA is, and most translation occurs in the cytoplasm. Recent research has shown, however, that 10% to 15% of proteins are synthesized in the nucleus, with both steps occurring there.

## THE GENETIC CODE

The body makes more than 2 million different proteins, all from the same 20 amino acids and all encoded by genes made of just 4 nucleotides—A, T, C, G—a striking illustration of how a great variety of complex structures can be made from a small variety of simpler components. The **genetic code** is a system that enables these 4 nucleotides to code for the amino acid sequences of all proteins.

It is not unusual for simple codes to represent complex information. Computers store and transmit complex information, including pictures and sounds, in a binary code with only the symbols 1 and 0. It is not surprising, then, that a mere 20 amino acids can be represented by a code of 4 nucleotides; all that is required is to combine these symbols in varied ways. It requires more than 2 nucleotides to code for each amino acid, because A, U, C, and G can combine in only 16 different pairs (AA, AU, AC, AG, UA, UU, etc.). The minimum code to symbolize 20 amino acids is 3 nucleotides per amino acid, and indeed this is the case in DNA. A sequence of 3 DNA nucleotides that stands for 1 amino acid is called a **base triplet.** The "mirror image" sequence in mRNA is called a **codon.** The genetic code is expressed in terms of codons.

Table 4.2 shows a few representative triplets and codons along with the amino acids they represent. You can see from this listing that two or more codons can represent the same amino acid. The reason for this is easy to explain mathematically. Four symbols *(N)* taken three at a time *(x)* can be combined in $N^x$ different ways; that is, there are $4^3$ = 64 possible codons available to represent the 20 amino acids. Only 61 of these code for amino acids. The other 3— UAG, UGA, and UAA—are called **stop codons;** they signal "end of message," like the period at the end of a sentence. A stop codon enables the cell's protein-synthesizing machinery to sense that it has reached the end of the gene for a particular protein. The codon AUG plays two roles: It serves as a code for methionine and as a **start codon.** This dual function is explained shortly.

## TRANSCRIPTION

Most protein synthesis occurs in the cytoplasm, but DNA is too large to leave the nucleus. It is necessary, therefore, to

| TABLE 4.2 | Examples of the Genetic Code | | |
|---|---|---|---|
| **Base Triplet of DNA** | **Codon of mRNA** | **Name of Amino Acid** | **Abbreviation for Amino Acid** |
| CCT | GGA | Glycine | Gly |
| CCA | GGU | Glycine | Gly |
| CCC | GGG | Glycine | Gly |
| CTC | GAG | Glutamic acid | Glu |
| CGC | GCG | Alanine | Ala |
| CGT | GCA | Alanine | Ala |
| TGG | ACC | Threonine | Thr |
| TGC | ACG | Threonine | Thr |
| GTA | CAU | Histidine | His |
| TAC | AUG | Methionine | Met |

make a small RNA copy that can migrate through a nuclear pore into the cytoplasm. Just as we might transcribe (copy) a document, **transcription** in genetics means the process of copying genetic instructions from DNA to RNA. It is triggered by chemical messengers from the cytoplasm that enter the nucleus and bind to the chromatin at the site of the relevant gene. An enzyme called **RNA polymerase** (po-LIM-ur-ase) then binds to the DNA at this point and begins making RNA. Certain base sequences (often TATATA or TATAAA) inform the polymerase where to begin.

RNA polymerase opens up the DNA helix about 17 base pairs at a time. It transcribes the bases from the *template strand* of the DNA and makes a corresponding RNA. Where it finds a C on the DNA, it adds a G to the RNA; where it finds an A, it adds a U; and so forth. The enzyme then rewinds the DNA helix behind it. Another RNA polymerase may follow closely behind the first one; thus, a gene may be transcribed by several polymerase molecules at once, and numerous copies of the same RNA are made. At the end of the gene is a base sequence that serves as a terminator, which signals the polymerase to release the RNA and separate from the DNA.

The RNA produced by transcription is an "immature" form called *pre-mRNA.* This molecule contains "sense" portions called *exons* that will be translated into a protein and "nonsense" portions called *introns* that must be removed before translation. Enzymes remove the introns and splice the exons together into a functional mRNA molecule. Through a mechanism called *alternative splicing,* one gene can code for more than one protein. Suppose the gene produces a pre-mRNA containing six exons separated by noncoding introns. As shown in figure 4.6, these exons can be spliced together in various combinations to yield codes for two or more proteins.

## TRANSLATION

Just as we might translate a work from Spanish into English, genetic **translation** converts the language of nucleotides into the language of amino acids (fig. 4.7).

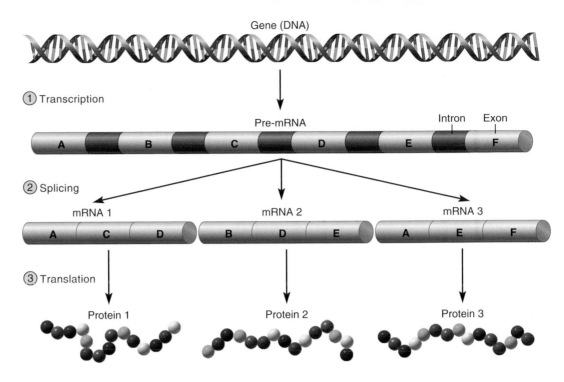

**FIGURE 4.6 Alternative Splicing of mRNA.** By splicing together different combinations of exons from a single pre-mRNA, a cell can generate multiple proteins from a single gene.

Gene (DNA)

① Transcription

Pre-mRNA

A B C D E F

Intron    Exon

② Splicing

mRNA 1          mRNA 2          mRNA 3

A C D          B D E          A E F

③ Translation

Protein 1        Protein 2        Protein 3

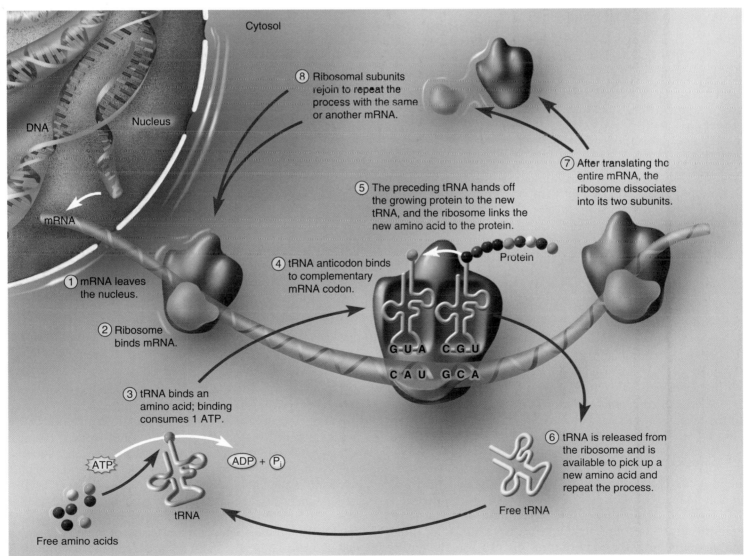

Cytosol

⑧ Ribosomal subunits rejoin to repeat the process with the same or another mRNA.

DNA        Nucleus

⑤ The preceding tRNA hands off the growing protein to the new tRNA, and the ribosome links the new amino acid to the protein.

⑦ After translating the entire mRNA, the ribosome dissociates into its two subunits.

mRNA

④ tRNA anticodon binds to complementary mRNA codon.

Protein

① mRNA leaves the nucleus.

② Ribosome binds mRNA.

G U A   C G U

C A U   G C A

③ tRNA binds an amino acid; binding consumes 1 ATP.

⑥ tRNA is released from the ribosome and is available to pick up a new amino acid and repeat the process.

ATP        ADP + Pᵢ

tRNA

Free tRNA

Free amino acids

**FIGURE 4.7 Translation of mRNA.**

❱ *Why would translation not work if ribosomes could bind only one tRNA at a time?*

This job is done by ribosomes, which are found mainly in the cytosol and on the rough ER and nuclear envelope. A ribosome consists of two granular subunits, large and small, each made of several rRNA and enzyme molecules.

The mRNA molecule begins with a *leader sequence* of bases that are not translated to protein but serve as a binding site for the ribosome. The small ribosomal subunit binds to it, the large subunit joins the complex, and the ribosome begins pulling the mRNA through it like a ribbon, reading bases as it goes. When it reaches the start codon, AUG, it begins making protein. Since AUG codes for methionine, all proteins begin with methionine when first synthesized, although this may be removed later.

Translation requires the participation of 61 types of transfer RNA (tRNA), one for each codon (except stop codons). Transfer RNA is a small RNA molecule that turns back and coils on itself to form a cloverleaf shape, which is then twisted into an angular L shape (fig. 4.8). One end of the L includes three nucleotides called an **anticodon,** and the other end has a binding site specific for one amino acid. Each tRNA picks up an amino acid from a pool of free amino acids in the cytosol. One ATP molecule is used to bind the amino acid to this site and provide the energy that is used later to join that amino acid to the growing protein. Thus, protein synthesis consumes one ATP for each peptide bond formed.

When the small ribosomal subunit reads a codon such as GCA, it must find an activated tRNA with the corresponding anticodon; in this case, CGU. This particular tRNA would have the amino acid alanine at its other end. The ribosome binds and holds this tRNA and then reads the next codon—say GGU. Here, it would bind a tRNA with anticodon CCA, which carries glycine.

The large ribosomal subunit contains an enzyme that forms peptide bonds, and now that alanine and glycine are side by side, it links them together. The first tRNA is no longer needed, so it is released from the ribosome. The second tRNA is used, temporarily, to anchor the growing peptide to the ribosome. Now, the ribosome reads the third codon—say ACG. It finds the tRNA with the anticodon UGC, which carries the amino acid threonine. The large subunit adds threonine to the growing chain, now three amino acids long. By repetition of this process, the entire protein is assembled. Eventually, the ribosome reaches a stop codon and is finished translating this mRNA. The

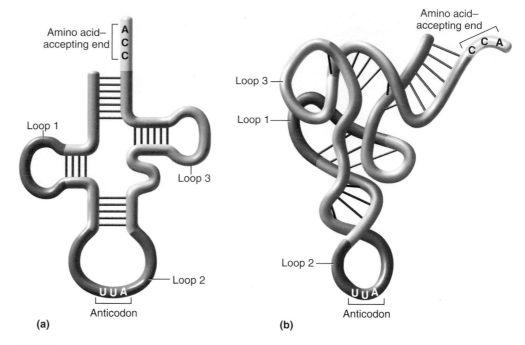

**FIGURE 4.8**  **Transfer RNA (tRNA).**  (a) tRNA has an amino acid–accepting end that binds to one specific amino acid, and an anticodon that binds to a complementary codon of mRNA. (b) The three-dimensional shape of a tRNA molecule. Black bars indicate regions of base pairing and helical structure.

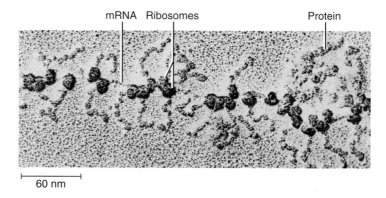

**FIGURE 4.9**  Several Ribosomes Attached to a Single mRNA Molecule, Forming a Polyribosome.

polypeptide is turned loose, and the ribosome dissociates into its two subunits.

One ribosome can assemble a protein of 400 amino acids in about 20 seconds, but it does not work at the task alone. After the mRNA leader sequence passes through one ribosome, a neighboring ribosome takes it up and begins translating the mRNA before the first ribosome has finished. One mRNA often holds 10 or 20 ribosomes together in a cluster called a **polyribosome** (fig. 4.9). Not only is each mRNA translated by all these ribosomes at once, but a cell may have 300,000 identical mRNA molecules undergoing simultaneous translation. Thus, a cell may produce over 150,000 protein molecules per second—a remarkably productive protein factory! As much as 25% of the dry weight of liver cells, which are highly active in protein synthesis, is composed of ribosomes.

Many proteins, when first synthesized, begin with a chain of amino acids called the **signal peptide.** Like a molecular address label, the signal peptide determines the protein's destination—for example, whether it will be sent to the rough endoplasmic reticulum, a peroxisome, or a mitochondrion. (Proteins used in the cytosol lack signal peptides.) Some diseases result from errors in the signal peptide, causing a protein to be sent to the wrong address, such as going to a mitochondrion when it should have gone to a peroxisome, or causing it to be secreted from a cell when it should have been stored in a lysosome. Günter Blobel of Rockefeller University received the 1999 Nobel Prize for Physiology or Medicine for discovering signal peptides in the 1970s.

Figure 4.10 summarizes transcription and translation and shows how a nucleotide sequence translates to a hypothetical peptide of 6 amino acids. A protein 500 amino acids long would have to be represented, at a minimum, by a sequence of 1,503 nucleotides (3 for each amino acid, plus a stop codon).

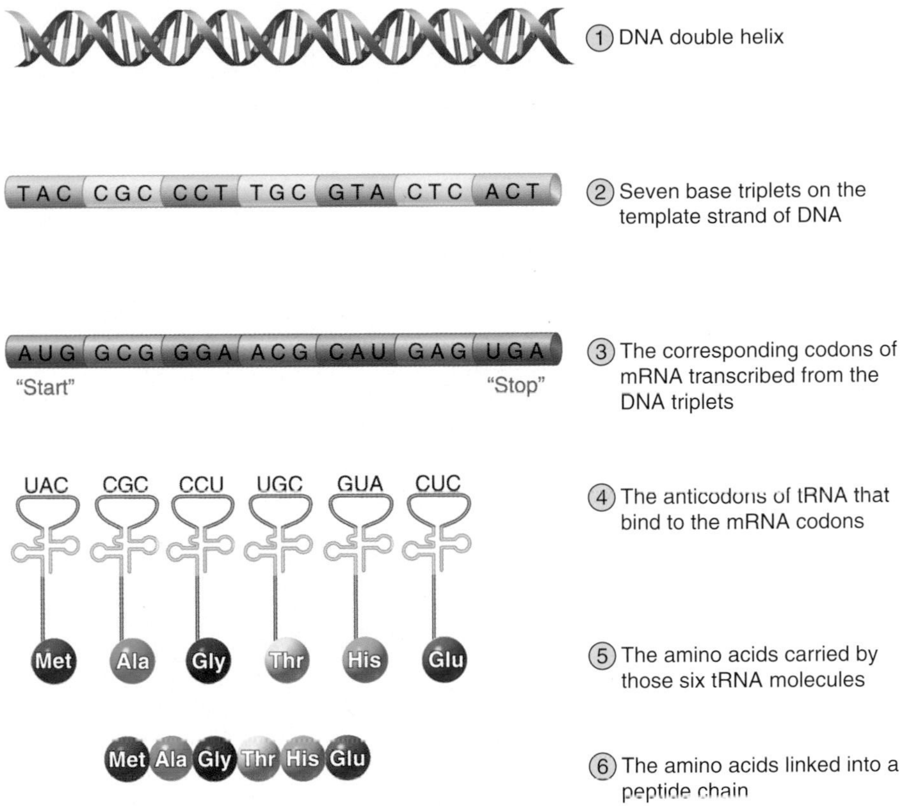

FIGURE 4.10  Relationship of a DNA Base Sequence to Peptide Structure.

## CHAPERONES AND PROTEIN STRUCTURE

The amino acid sequence of a protein (primary structure) is only the beginning; the end of translation is not the end of protein synthesis. The protein now coils or folds into its secondary and tertiary structures and, in some cases, associates with other polypeptide chains (quaternary structure) or conjugates with a nonprotein moiety, such as a vitamin or carbohydrate. It is essential that these processes not begin prematurely while the amino acid sequence is being assembled, since the correct final shape may depend on amino acids that have not been added yet. Therefore, as new proteins are assembled by ribosomes, they are sometimes picked up by older proteins called **chaperones.** A chaperone prevents a new protein from folding prematurely and assists in its proper folding once the amino acid sequence has been completed. It may also escort a newly synthesized protein to the correct destination in a cell, such as the plasma membrane, and help to prevent improper associations between different proteins. As in the colloquial sense of the word, a chaperone is an older protein that escorts and regulates the behavior of the "youngsters." Some chaperones are also called *stress proteins* or *heat-shock proteins* because they are produced in response to heat or other stress on a cell and help damaged proteins fold back into their correct functional shapes.

## POSTTRANSLATIONAL MODIFICATION AND SECRETION

If a protein is going to be used in the cytosol (for example, the enzymes of glycolysis), it is likely to be made by free ribosomes in the cytosol. If it is going to be packaged into a lysosome or secreted from the cell, however, its signal peptide causes the entire polyribosome to migrate to the rough ER and dock on its surface. Assembly of the amino acid chain is then completed on the rough ER and the protein is sent to the Golgi complex for final modification. Thus, we turn to the functions of these organelles in the modification, packaging, and secretion of a protein. Compare the following description to figure 4.11.

1. As a protein is assembled by ribosomes on the rough ER, its signal peptide threads itself through a pore in the ER membrane and drags the rest of the protein into the cisterna. Enzymes in the cisterna remove the signal peptide and modify the new protein in a variety of ways—removing some amino acid segments, folding the protein and stabilizing it with disulfide bridges, adding carbohydrates, and so forth. Such changes are called **posttranslational modification.** Insulin, for example, is first synthesized as a protein 83 amino acids long. In posttranslational modification, the chain folds back on itself, three disulfide

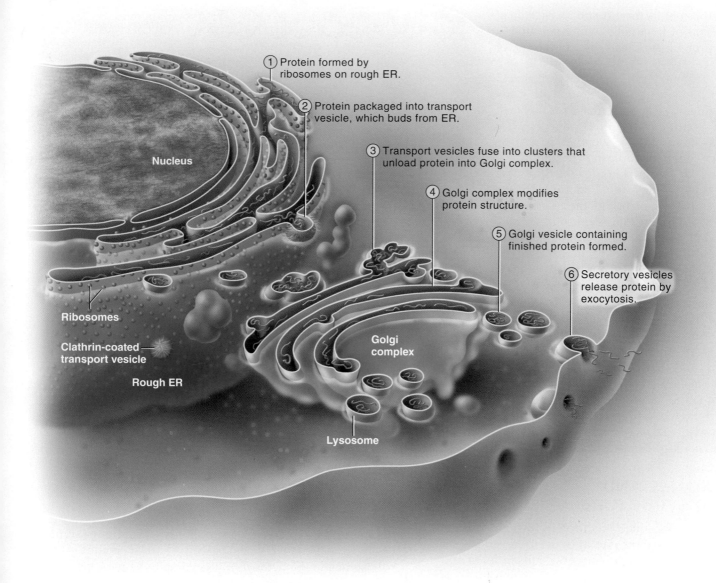

① Protein formed by ribosomes on rough ER.

② Protein packaged into transport vesicle, which buds from ER.

③ Transport vesicles fuse into clusters that unload protein into Golgi complex.

④ Golgi complex modifies protein structure.

⑤ Golgi vesicle containing finished protein formed.

⑥ Secretory vesicles release protein by exocytosis.

Nucleus

Ribosomes

Clathrin-coated transport vesicle

Rough ER

Golgi complex

Lysosome

**FIGURE 4.11    Protein Packaging and Secretion.** Some proteins are synthesized by ribosomes on the rough ER and carried in transport vesicles to the nearest cisterna of the Golgi complex. The Golgi complex modifies the structure of the protein, transferring it from one cisterna to the next, and finally packages it in Golgi vesicles. Some Golgi vesicles may remain within the cell and become lysosomes, while others migrate to the plasma membrane and release the cell product by exocytosis.

bridges are formed, and 32 amino acids are removed from the middle of the protein. The final insulin molecule is therefore made of two chains of 21 and 30 amino acids held together by disulfide bridges (see fig. 2.25, p. 80).

2. When the rough ER is finished with a protein, it pinches off clathrin-coated **transport vesicles.** Clathrin apparently helps to select the proteins to be transported in the vesicles, and as a basketlike cage, it helps to mold the forming vesicles. Soon after the vesicles detach from the ER, they fuse in irregularly shaped clusters that carry their cargo to the nearest cisterna of the Golgi complex.

3. Upon contact with the Golgi complex, the cluster fuses with it and unloads its protein cargo into the Golgi cisterna.

4. The Golgi complex further modifies the protein, often by adding carbohydrate chains. This is where the *glycoproteins* mentioned in chapter 2 are assembled. During its processing, the maturing protein is passed from one Golgi cisterna to the next. This may involve transport vesicles, but there are still competing and inconclusive hypotheses about how the transfer of protein is achieved.

5. The final Golgi cisterna, farthest from the ER, buds off new coated **Golgi vesicles** containing the finished

protein, or may simply break up into vesicles, to be replaced by a younger cisterna behind it.

6. Some of these vesicles become lysosomes, while others become **secretory vesicles** that migrate to the plasma membrane and fuse with it, releasing the cell product by exocytosis. This is how a cell of a salivary gland, for example, secretes mucus and digestive enzymes.

Table 4.3 summarizes the destinations and functions of several kinds of protein, some of which are packaged and secreted by the foregoing process and others of which are used internally by the cell that synthesizes them.

### Before You Go On

*Answer the following questions to test your understanding of the preceding section:*

5. *Define* genetic code, codon, *and* anticodon.

6. *Describe the genetic role of RNA polymerase.*

7. *Describe the genetic role of ribosomes and tRNA.*

8. *Why are chaperones important in ensuring correct tertiary protein structure?*

9. *What roles do the rough ER and Golgi complex play in protein production?*

# DNA Replication and the Cell Cycle

### Objectives

When you have completed this section, you should be able to

- describe how DNA is replicated;
- discuss the consequences of replication errors;
- describe the life history of a cell, including the events of mitosis; and
- explain how the timing of cell division is regulated.

Before a cell divides, it must duplicate its DNA so it can give a complete copy of all of its genes to each daughter cell. Since DNA controls all cellular function, this replication process must be very exact. We now examine how it is accomplished and consider the consequences of mistakes.

## DNA REPLICATION

The law of complementary base pairing shows that we can predict the base sequence of one DNA strand if we know the sequence of the other. More importantly, it enables a cell to reproduce one strand based on information in the other. This immediately occurred to Watson and Crick

| TABLE 4.3 | Some Destinations and Functions of Newly Synthesized Proteins |
|---|---|
| **Destination or Function** | **Proteins (examples)** |
| Deposited as a structural protein within cells | Actin of cytoskeleton Keratin of epidermis |
| Used in the cytosol as a metabolic enzyme | ATPase Kinases |
| Returned to the nucleus for use in nuclear metabolism | Histones of chromatin RNA polymerase |
| Packaged in lysosomes for autophagy, intracellular digestion, and other functions | Numerous lysosomal enzymes |
| Delivered to other organelles for intracellular use | Catalase of peroxisomes Mitochondrial enzymes |
| Delivered to plasma membrane to serve transport and other functions | Hormone receptors Sodium–potassium pumps |
| Secreted by exocytosis for extracellular functions | Digestive enzymes Casein of breast milk |

when they discovered the structure of DNA. Watson was hesitant to make such a grandiose claim in their first publication, but Crick implored, "Well, we've got to say *something!* Otherwise people will think these two unknown chaps are so dumb they don't even realize the implications of their own work!" Thus, the last sentence of their first paper modestly stated, "It has not escaped our notice that the specific pairing we have postulated . . . immediately suggests a possible copying mechanism for the genetic material." Five weeks later they published a second paper pressing this point more vigorously.

The basic idea of DNA replication is evident from its base pairing, but the way in which DNA is organized in the chromatin introduces some complications that were not apparent when Watson and Crick first wrote. The fundamental steps of the replication process are as follows:

1. The double helix unwinds from the histones.

2. Like a zipper, an enzyme called **DNA helicase** opens up a short segment of the helix, exposing its nitrogenous bases. The point where one strand of DNA is "unzipped" and separates from its complementary strand is called a *replication fork* (fig. 4.12a).

3. An enzyme called **DNA polymerase** moves along the opened strands, reads the exposed bases, and like a matchmaker, arranges "marriages" with complementary free nucleotides in the nucleoplasm. If the polymerase finds the sequence TCG, for example, it assembles AGC across from it. One polymerase molecule moves away from the replication fork replicating one strand of the opened DNA, and another

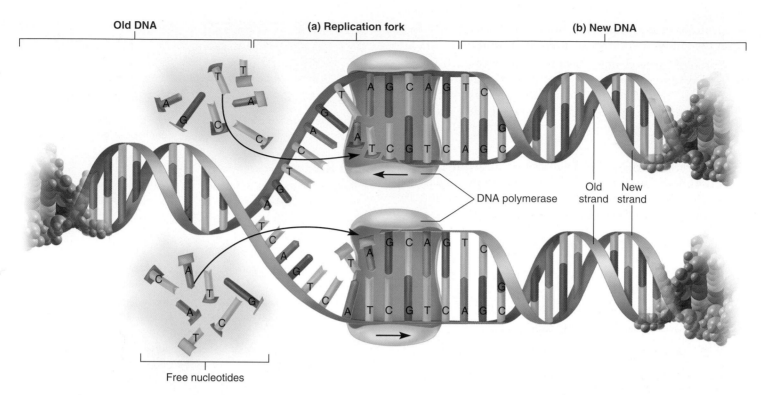

**FIGURE 4.12** **Semiconservative DNA Replication.** (a) At the replication fork, DNA helicase (not shown) unwinds the double helix and exposes the bases. DNA polymerases begin assembling new bases across from the existing ones, moving away from the replication fork on one strand and toward it on the other. (b) The result is two DNA double helices, each composed of one strand of the original DNA and one newly synthesized strand.

polymerase molecule moves in the opposite direction, replicating the other strand. Thus, from the old DNA molecule, two new ones are made. Each new DNA consists of one new helix synthesized from free nucleotides and one helix conserved from the parent DNA (fig. 4.12b). The process is therefore called **semiconservative replication.**

4. While DNA is synthesized in the nucleus, new histones are synthesized in the cytoplasm. Millions of histones are transported into the nucleus within a few minutes after DNA replication, and each new DNA helix wraps around them to make new nucleosomes.

Despite the complexity of this process, each DNA polymerase works at an impressive rate of about 100 base pairs per second. Even at this rate, however, it would take weeks for one polymerase molecule to replicate even one chromosome. But in reality, thousands of polymerase molecules work simultaneously on each DNA molecule, and all 46 chromosomes are replicated in a mere 6 to 8 hours.

## ERRORS AND MUTATIONS

DNA polymerase is fast and accurate, but it makes mistakes. For example, it might read A and place a C across from it where it should have placed a T. In *Escherichia coli,* a bacterial species in which DNA replication has been most thoroughly studied, about three errors occur for every 100,000 bases copied. At this rate of error, every generation of cells would have about 1,000 faulty proteins, coded for by DNA that had been miscopied. To help prevent such catastrophic damage to the organism, the DNA is continuously scanned for errors. After DNA polymerase has replicated a strand, a smaller polymerase comes along, "proofreads" it, and makes corrections where needed—for example, removing C and replacing it with T. This improves the accuracy of replication to one error per billion bases—only one faulty protein for every 10 cell divisions (in *E. coli*).

Changes in DNA structure, called **mutations,**[1] can result from replication errors or environmental factors. Uncorrected mutations can be passed on to the descendants of that cell, but some of them have no adverse effect. One reason is that a new base sequence sometimes codes for the same thing as the old one. For example, TGG and TGC both code for threonine (see table 4.2), so a mutation from G to C in the third place would not change protein structure. Another reason is that a change in protein structure is not always critical to its function. For example, humans and horses differ in 25 of the 146 amino acids that make up their β-hemoglobin, yet the hemoglo-

---

[1]*muta* = change

bin is fully functional in both species. Some mutations, however, may kill a cell, turn it cancerous, or cause genetic defects in future generations. When a mutation changes the sixth amino acid of β-hemoglobin from glutamic acid to valine, for example, the result is a crippling disorder called sickle cell disease. Clearly some amino acid substitutions are more critical than others, and this affects the severity of a mutation.

## THE CELL CYCLE

Most cells periodically divide into two daughter cells, so a cell has a life cycle extending from one division to the next. This **cell cycle** (fig. 4.13) is divided into four main phases: $G_1$, S, $G_2$, and M.

**$G_1$** is the **first gap phase,** an interval between cell division and DNA replication. During this time, a cell synthesizes proteins, grows, and carries out its preordained tasks for the body. Almost all of the discussion in this book relates to what cells do in the $G_1$ phase. Cells in $G_1$ also accumulate the materials needed to replicate their DNA in the next phase. In cultured cells called fibroblasts, which divide every 18 to 24 hours, $G_1$ lasts 8 to 10 hours.

**S** is the **synthesis phase,** in which a cell makes a duplicate copy of its centrioles and all of its DNA. The two identical sets of DNA molecules are then available to be divided up between daughter cells at the next cell division. This phase takes 6 to 8 hours in cultured fibroblasts.

**$G_2$,** the **second gap phase,** is a relatively brief interval (4–6 hours) between DNA replication and cell division. In $G_2$, a cell finishes replicating its centrioles and synthesizes enzymes that control cell division.

**M** is the **mitotic phase,** in which a cell replicates its nucleus and then pinches in two to form two new daughter cells. In cultured fibroblasts, the M phase takes 1 to 2 hours. The details of this phase are considered in the next section. Phases $G_1$, S, and $G_2$ are collectively called **interphase**—the time between M phases.

The length of the cell cycle varies greatly from one cell type to another. Stomach and skin cells divide rapidly, bone and cartilage cells slowly, and skeletal muscle cells and nerve cells not at all (Insight 4.2). Some cells leave the cell cycle for a "rest" and cease to divide for days, years, or the rest of one's life. Such cells are said to be in the **$G_0$ (G-zero) phase.** The balance between cells that are actively cycling and those standing by in $G_0$ is an important factor in determining the number of cells in the body. An inability to stop cycling and enter $G_0$ is characteristic of cancer cells (see Insight 4.4 at the end of the chapter).

**Think About It**

*What is the maximum number of DNA molecules ever contained in a cell over the course of its life cycle? (Assume the cell has only one nucleus.)*

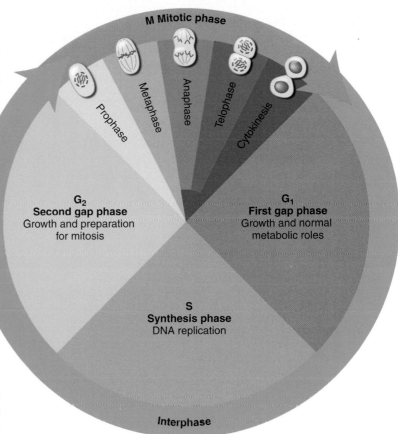

**FIGURE 4.13**  The Cell Cycle.

---

**INSIGHT 4.2**     Clinical Application

## Can We Replace Brain Cells?

Until recently, neurons (nerve cells) of the brain were thought to be irreplaceable; when they died, we thought, they were gone forever. We believed, indeed, that there was good reason for this. Motor skills and memories are encoded in intricate neural circuits, and the growth of new neurons might disrupt those circuits. Now we are not so sure.

A chemical called BrDU (bromodeoxyuridine) can be used to trace the birth of new cells, because it becomes incorporated into their DNA. BrDU is too toxic to use ordinarily in human research. However, in cancer patients, BrDU is sometimes used to monitor the growth of tumors. Peter Eriksson, at Göteborg University in Sweden, obtained permission from the families of cancer victims to examine the brain tissue of BrDU-treated patients who had died. In the hippocampus, a region of the brain concerned with memory, he and collaborator Fred Gage found as many as 200 new neurons per cubic millimeter of tissue, and estimated that up to 1,000 new neurons may be born per day even in people in their 50s to 70s. These new neurons apparently arise not by mitosis of mature neurons (which are believed to be incapable of mitosis), but from a reserve pool of embryonic stem cells. It remains unknown whether new neurons are produced late in life in other regions of the brain.

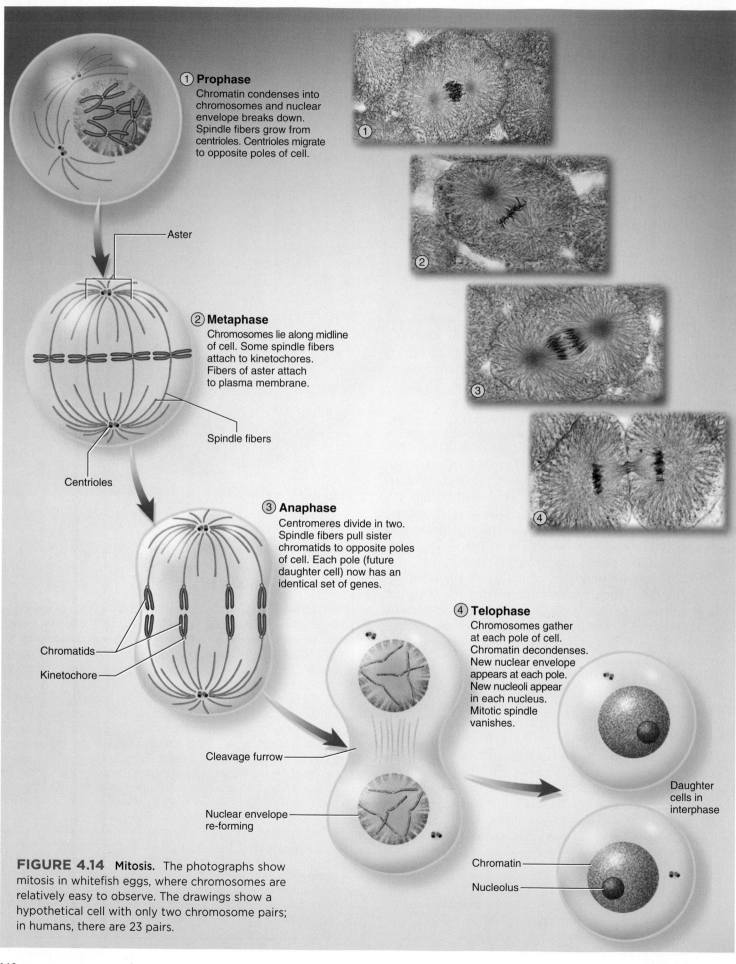

① **Prophase**
Chromatin condenses into chromosomes and nuclear envelope breaks down. Spindle fibers grow from centrioles. Centrioles migrate to opposite poles of cell.

Aster

② **Metaphase**
Chromosomes lie along midline of cell. Some spindle fibers attach to kinetochores. Fibers of aster attach to plasma membrane.

Spindle fibers

Centrioles

③ **Anaphase**
Centromeres divide in two. Spindle fibers pull sister chromatids to opposite poles of cell. Each pole (future daughter cell) now has an identical set of genes.

④ **Telophase**
Chromosomes gather at each pole of cell. Chromatin decondenses. New nuclear envelope appears at each pole. New nucleoli appear in each nucleus. Mitotic spindle vanishes.

Chromatids

Kinetochore

Cleavage furrow

Nuclear envelope re-forming

Daughter cells in interphase

Chromatin

Nucleolus

**FIGURE 4.14 Mitosis.** The photographs show mitosis in whitefish eggs, where chromosomes are relatively easy to observe. The drawings show a hypothetical cell with only two chromosome pairs; in humans, there are 23 pairs.

# MITOSIS

Cells divide by two mechanisms called mitosis and meiosis. Meiosis, however, is restricted to one purpose, the production of eggs and sperm, and is therefore treated in chapter 27 on reproduction. **Mitosis** serves all the other functions of cell division:

- development of an individual, composed of some 40 trillion cells, from a one-celled fertilized egg;

- growth of all tissues and organs after birth;

- replacement of cells that die; and

- repair of damaged tissues.

Four phases of mitosis are recognizable: *prophase, metaphase, anaphase,* and *telophase* (fig. 4.14).

① **Prophase**[2]    At the outset of mitosis, the chromosomes shorten and thicken, eventually coiling into compact rods that are easier to distribute to daughter cells than the long, delicate chromatin of interphase. At this stage, a chromosome consists of two genetically identical bodies called **chromatids,** joined together at a pinched spot called the **centromere.** There are 46 chromosomes, two chromatids per chromosome, and one molecule of DNA in each chromatid. The nuclear envelope disintegrates during prophase and releases the chromosomes into the cytosol. The centrioles begin to sprout elongated microtubules called **spindle fibers,** which push the centrioles apart as they grow. Eventually, a pair of centrioles come to lie at each pole of the cell. Some spindle fibers grow toward the chromosomes and become attached to a platelike protein complex called the **kinetochore**[3] (kih-NEE-toe-core) on each side of the centromere (fig. 4.15). The spindle fibers then tug the chromosomes back and forth until they line up along the midline of the cell.

② **Metaphase**[4]    The chromosomes are aligned on the cell equator, oscillating slightly and awaiting a signal that stimulates each of them to split in two at the centromere. The spindle fibers now form a football-shaped array called the **mitotic spindle.** Long microtubules reach out from each centriole to the chromosomes, and shorter microtubules form a star-like *aster,*[5] which anchors the assembly to the inside of the plasma membrane at each end of the cell.

③ **Anaphase**[6]    This phase begins with activation of an enzyme that cleaves the two sister chromatids from each other at the centromere. Each chromatid is now regarded as a separate, single-stranded *daughter chromosome.* One daughter chromosome migrates to

**(a)**    **(b)**    700 nm

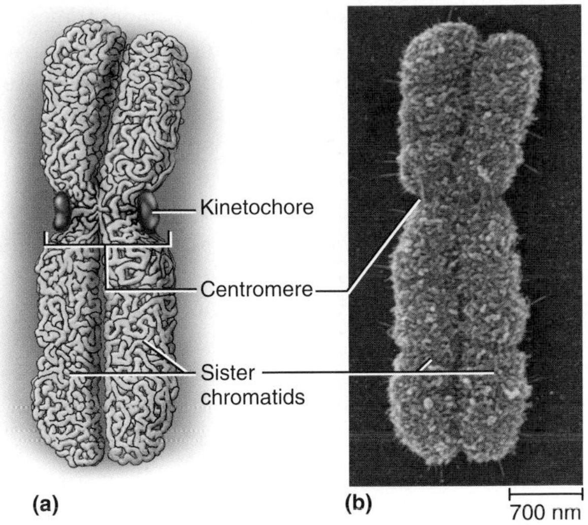

**FIGURE 4.15**  Chromosome Structure at Metaphase. (a) Drawing of a metaphase chromosome. (b) Scanning electron micrograph.

each pole of the cell, with its centromere leading the way and the arms trailing behind. Migration is achieved by means of motor proteins in the kinetochore crawling along the spindle fiber as the fiber itself is "chewed up" and disassembled at the chromosomal end. Since sister chromatids are genetically identical, and since each daughter cell receives one chromatid from each chromosome, the daughter cells of mitosis are genetically identical.

④ **Telophase**[7] The chromatids cluster on each side of the cell. The rough ER produces a new nuclear envelope around each cluster, and the chromatids begin to uncoil and return to the thinly dispersed chromatin form. The mitotic spindle breaks up and vanishes. Each new nucleus forms nucleoli, indicating it has already begun making RNA and preparing for protein synthesis.

Telophase is the end of nuclear division but overlaps with **cytokinesis**[8] (SY-toe-kih-NEE-sis), division of the cytoplasm into two cells. Early traces of cytokinesis are visible even at anaphase. It is achieved by the motor protein myosin pulling on microfilaments of actin in the membrane skeleton. This creates a crease called the *cleavage furrow* around the equator of the cell, and the cell eventually pinches in two. Interphase has now begun for these new cells.

# TIMING OF CELL DIVISION

One of the most important questions in biology is what signals cells when to divide and when to stop. The activation and inhibition of cell division are subjects of intense

---

[2]*pro* = first
[3]*kineto* = motion + *chore* = place
[4]*meta* = next in a series
[5]*aster* = star
[6]*ana* = apart

[7]*telo* = end, final
[8]*cyto* = cell + *kinesis* = action, motion

research for obvious reasons such as management of cancer and tissue repair. Cells divide when (1) they grow large enough to have enough cytoplasm to distribute to their two daughter cells; (2) they have replicated their DNA, so they can give each daughter cell a duplicate set of genes; (3) they receive an adequate supply of nutrients; (4) they are stimulated by **growth factors,** chemical signals secreted by blood platelets, kidney cells, and other sources; or (5) neighboring cells die, opening up space in a tissue to be occupied by new cells. Cells stop dividing when nutrients or growth factors are withdrawn or when they snugly contact neighboring cells. The cessation of cell division in response to contact with other cells is called **contact inhibition.**

### Before You Go On

*Answer the following questions to test your understanding of the preceding section:*

10. *Describe the genetic roles of DNA helicase and DNA polymerase. Contrast the function of DNA polymerase with that of RNA polymerase.*

11. *Explain why DNA replication is called* semiconservative.

12. *Define* mutation. *Explain why some mutations are harmless and others can be lethal.*

13. *List the stages of the cell cycle and summarize what occurs in each one.*

14. *Describe the structure of a chromosome at metaphase.*

# Chromosomes and Heredity

### Objectives

When you have completed this section, you should be able to

- describe the paired arrangement of chromosomes in the human karyotype;

- define *allele* and discuss how alleles affect the traits of an individual; and

- discuss the interaction of heredity and environment in producing individual traits.

**Heredity** is the transmission of genetic characteristics from parent to offspring. In the following discussion, we will examine a few basic principles of normal heredity, thus establishing a basis for understanding hereditary traits in later chapters. Hereditary defects are described in chapter 29 along with nonhereditary birth defects.

## THE KARYOTYPE

A **karyotype** (fig. 4.16) is a chart of the chromosomes isolated from a cell at metaphase, arranged in order by size and structure. It reveals that most human cells, with the exception of germ cells (described shortly), contain 23 pairs of similar-looking chromosomes (except for X and Y chromosomes). The two chromosomes in each pair are called **homologous**[9] (ho-MOLL-uh-gus) **chromosomes.** One is inherited from the mother and one from the father. Two chromosomes, designated X and Y, are called **sex chromosomes** and the other 22 pairs are called **autosomes.** A female normally has a homologous pair of X chromosomes, whereas a male has one X chromosome and a much smaller Y chromosome.

⌐ **Think About It**

*Why would a cell in metaphase be more useful than a cell in interphase for producing a karyotype?*

The paired state of the homologous chromosomes results from the fact that a sperm cell bearing 23 chromosomes fertilizes an egg, which also has 23. Sperm and egg cells, and the cells on their way to becoming sperm and eggs, are called **germ cells.** All other cells of the body are called **somatic cells.** Somatic cells are described as **diploid**[10] because their chromosomes are in homologous pairs, whereas germ cells beyond a certain stage of development are **haploid,**[11] meaning they contain half as many chromosomes as the somatic cells. In meiosis (see chapter 27), homologous chromosomes become *segregated* from each other into separate daughter cells leading to the haploid sex cells. At fertilization, one set of *paternal* (sperm) chromosomes unites with one set of *maternal* (egg) chromosomes, restoring the diploid number to the fertilized egg and the somatic cells that arise from it. Although the two chromosomes of a homologous pair appear to be identical, they come from different parents and therefore are not *genetically* identical.

## THE GENOME

The **genome** is all the DNA, both coding and noncoding, occurring in one haploid set of chromosomes of an individual. It consists of about 3.1 billion nucleotide pairs. Owing to a massive multinational undertaking called the **Human Genome Project (HGP),** biologists now know the base sequence (A, T, C, G) of more that 99% of the entire genome. The only unknown portions are some short, dense regions of the chromosomes inaccessible to present technology, but apparently containing very few genes. Running from 1990 to 2003, the HGP involved 20 laboratories in 6 countries and cost $2.7 billion—the only biological project ever to be funded at a level comparable to a military weapons system.

The HGP was very controversial at first. Critics questioned whether it would be worth that much money, considering that 98% of the genome does not code for

---

[9]*homo* = same + *log* = relation
[10]*diplo* = double
[11]*haplo* = half

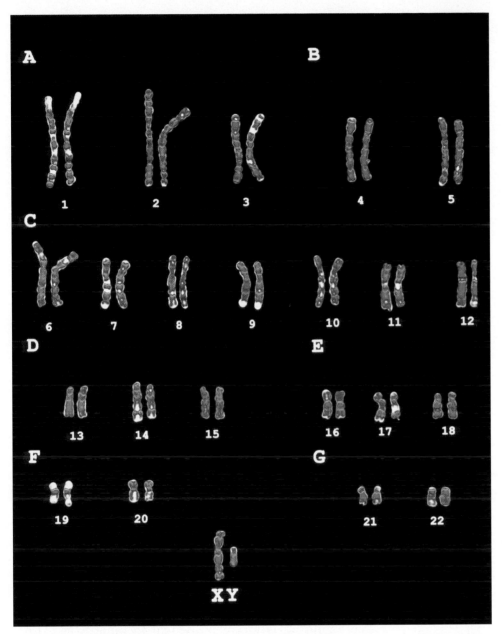

**FIGURE 4.16** **Karyotype of a Normal Human Male.** This is a false-color micrograph of chromosomes stained to accentuate their banding patterns. The two chromosomes of each homologous pair exhibit similar size, shape, and banding.

▶ *How would this karyotype differ if it were from a female?*

the fundamental structure of the DNA double helix.

What have been the results? Sequencing the human genome has been hailed as a technological accomplishment as momentous as splitting the atom and landing explorers on the moon. It has opened a new field of biology called **genomics,** the comprehensive study of the whole genome and how its genes and noncoding DNA interact to affect the structure and function of the whole organism. Genomics goes well beyond classical mendelian genetics, which focuses on single genes and their effects. Some things we have learned from the HGP are:

- *Homo sapiens* has only about 30,000 to 35,000 genes, not the 100,000 formerly believed.

- These genes generate millions of different proteins, so gone is the old idea of one gene for each protein. In its place is the revolutionary realization that a single gene can code for many different proteins through alternative splicing (see p. 132) and other means.

- The genes average about 3,000 bases long, but range up to 2.4 million bases.

- All humans, worldwide, are at least 99.99% genetically identical, but even the .01% variation means that there are over 3 million base pairs in which we vary; various combinations of these **single-nucleotide polymorphisms**[12] account for all human genetic variation.

- Some chromosomes are gene-rich (17, 19, and 22) and some are gene-poor (4, 8, 13, 18, 21, and Y).

anything and, consequently, only 2% of that budget would go to revealing the sequences of actual genes. They also questioned whether the money would be better spent on AIDS research and other pressing public health needs. But after much debate, the project went ahead. As sequencing technology improved, the 20 laboratories reached a point where they were collectively sequencing 1,000 base pairs per second, 24 hours a day, 7 days a week. The HGP was completed 2 years early and $300 million under budget. The human genome was published just 11 days before the fiftieth anniversary of Watson and Crick's publication of

- Before the HGP, we knew the chromosomal locations of fewer than 100 disease-producing mutations; we now know more than 1,400 and this number is sure to rise as the genome is further analyzed.

The most direct benefit of genomics is likely to be **genomic medicine,** a branch of diagnosis and therapy only now being born (Insight 4.3).

---

[12]*poly* = multiple + *morph* = form

INSIGHT 4.3    Clinical Application

## Genomic Medicine

*Genomic medicine* is the application of our knowledge of the genome to the prediction, diagnosis, and treatment of disease. It is relevant to disorders as diverse as cancer, Parkinson disease, Alzheimer disease, schizophrenia, obesity, and even a person's susceptibility to nonhereditary diseases such as AIDS and tuberculosis.

As the technology of gene sequencing continues to improve, geneticists expect that we may soon be able to sequence any person's entire genome for less than $1,000. Why would we want to? Because knowing one's genome could dramatically change clinical care. It may allow clinicians to forecast the risk of disease and to predict its course; mutations in a single gene can affect the severity of such diseases as hemophilia, muscular dystrophy, and cystic fibrosis. Genomics should also allow for earlier detection of diseases and for earlier, more effective clinical intervention. Drugs that are safe for most people can have serious side effects in others, owing to genetic variations in drug metabolism. Genomics may therefore provide a basis for choosing the safest or most effective drug and for adjusting dosages for different patients on the basis of their genetic makeup.

Knowing the sites of disease-producing mutations expands the potential for *gene-substitution therapy.* This is a procedure in which cells such as bone marrow stem cells are removed from a patient with a genetic disorder, supplied with a normal gene in place of the defective one, and reintroduced to the body. The hope is that these genetically modified cells will proliferate and provide the patient with a gene product that he or she was lacking—perhaps insulin for a patient with diabetes or a blood clotting factor for a patient with hemophilia. Although still fraught with great technical difficulties and some tragic setbacks, gene substitution therapy has successfully cured some people of life-threatening disorders such as *familial hypercholesterolemia* (see p. 109) and *severe combined immunodeficiency disease* (see p. 842).

Genomics is, however, introducing new problems in medical ethics and law. Should your genome be a private matter between you and your physician? Or should an insurance company be entitled to know your genome before issuing health or life insurance to you, so they can know your risk of contracting a catastrophic illness, adjust the cost of your coverage, or even deny coverage? Should a prospective employer have the right to know your genome before offering employment? These are areas in which biology, politics, and law converge to shape public policy.

## GENES AND ALLELES

Each chromosome carries many genes. The location of a particular gene on a chromosome is called its **locus.** Homologous chromosomes have the same gene at the same locus, although they may carry different forms of that gene, called **alleles**[13] (ah-LEELS), which produce alternative forms of a particular trait. Frequently, one allele is **dominant** and the other one **recessive.** If at least one chromosome carries the dominant allele, the corresponding trait is usually detectable in the individual. A dominant allele masks the effect of any recessive allele that may be present. Recessive alleles are therefore expressed only when present on both of the homologous chromosomes—that is, when the individual has no dominant allele at that locus. Typically, but not always, dominant alleles code for a normal, functional protein and recessive alleles for a nonfunctional variant of the protein.

The shape of the outer ear presents an example of dominant and recessive genetic effects. When the ears are developing in a fetus, a "death signal" is often activated in cells that attach the earlobe to the side of the head. These cells die, causing the earlobe to separate from the head. A person will then have "detached earlobes." This occurs in people who have either one or two copies of a dominant allele which we will denote *D.* If both homologous chromosomes have the recessive version of this gene, *d,* the cell suicide program is not activated, and the earlobes remain attached (fig. 4.17a). (It is customary to represent a dominant allele with a capital letter and a recessive allele with its lowercase equivalent.)

Individuals with two identical alleles, such as *DD* or *dd,* are said to be **homozygous**[14] (HO-mo-ZY-gus) for that trait. If the homologous chromosomes have different alleles for that gene *(Dd),* the individual is **heterozygous**[15] (HET-er-oh-ZY-gus). The alleles that an individual possesses for a particular trait constitute the **genotype** (JEE-no-type). A detectable trait such as attached or detached earlobes, resulting either from the genotype or from environmental influences, is called the **phenotype**[16] (FEE-no-type).

We say that an allele is *expressed* if it shows in the phenotype of an individual. Earlobe allele *d* is expressed only when it is present in a homozygous state *(dd);* allele *D* is expressed whether it is homozygous *(DD)* or heterozygous *(Dd).* The only way most recessive alleles can be expressed is for an individual to inherit them from both parents.

Recessive traits can "skip" one or more generations. A diagram called a *Punnett square* (fig. 4.17b) shows how two heterozygous parents with detached earlobes can produce a child with attached lobes. Across the top are the two genetically possible types of eggs the mother could produce, and on the left side are the possible types of sperm from the father. The four cells of the square show the genotypes and phenotypes that would result from each possible combination of sperm and egg. You can see that three of the possible combinations would produce a child with detached lobes (genotypes *DD* and *Dd*), but one combination *(dd)* would produce a child with attached

---

[13]*allo* = different

[14]*homo* = same + *zygo* = union, joined
[15]*hetero* = *different*
[16]*pheno* = showing, evident

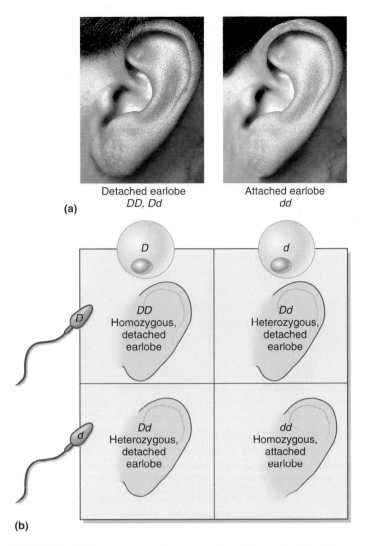

Detached earlobe
*DD, Dd*

Attached earlobe
*dd*

**(a)**

**(b)**

**FIGURE 4.17** Genetics of Attached and Detached Earlobes.
(a) Detached earlobes occur if even one allele of the pair is dominant *(D)*. Attached earlobes occur only when both alleles are recessive *(d)*. (b) A Punnett square shows why such a trait can "skip a generation." Both parents in this case have heterozygous genotypes *(Dd)* and detached earlobes. An egg from the mother can carry either allele *D* or *d* (top), as can a sperm from the father (left). This gives their offspring a one-in-four chance of having attached earlobes *(dd)*.

lobes. Therefore, the attached-lobe trait skipped the parental generation in this case but could be expressed in their child.

This phenomenon becomes more significant when parents are heterozygous **carriers** of hereditary diseases such as cystic fibrosis—individuals who carry a recessive allele and may pass it on, but do not phenotypically express it in themselves. For some hereditary diseases, tests are available to detect carriers and allow couples to weigh their risk of having children with genetic disorders. *Genetic counselors* perform genetic testing or refer

clients for tests, advise couples on the probability of transmitting genetic diseases, and assist people in coping with genetic disease.

⌐ **Think About It**

*Would it be possible for a woman with attached earlobes to have children with detached lobes? Use a Punnett square and one or more hypothetical genotypes for the father to demonstrate your point.*

## MULTIPLE ALLELES, CODOMINANCE, AND INCOMPLETE DOMINANCE

Some genes exist in more than two allelic forms—that is, there are **multiple alleles** within the collective genetic makeup, or **gene pool,** of the population as a whole. For example, there are over 100 alleles responsible for cystic fibrosis, and there are 3 alleles for ABO blood types. Two of the ABO blood type alleles are dominant and symbolized with a capital $I$ (for *immunoglobulin*) and a superscript: $I^A$ and $I^B$. There is one recessive allele, symbolized with a lowercase *i*. Which two alleles one inherits determines the blood type, as follows:

| Genotype | Phenotype |
|----------|-----------|
| $I^A I^A$ | Type A |
| $I^A i$ | Type A |
| $I^B I^B$ | Type B |
| $I^B i$ | Type B |
| $I^A I^B$ | Type AB |
| $ii$ | Type O |

⌐ **Think About It**

*Why can't one person have all three of the ABO alleles?*

Some alleles are equally dominant, or **codominant.** When both of them are present, both are phenotypically expressed. For example, a person who inherits allele $I^A$ from one parent and $I^B$ from the other has blood type AB. These alleles code for enzymes that produce the surface glycoproteins of red blood cells. Type AB means that both A and B glycoproteins are present, and type O means that neither of them is present.

Other alleles exhibit **incomplete dominance.** When two different alleles are present, the phenotype is intermediate between the traits that each allele would produce alone. Familial hypercholesterolemia, the disease discussed in Insight 3.3 (p. 109), is a good example. Individuals with two abnormal alleles die of heart attacks in childhood, those with only one abnormal allele typically die as young adults, and those with two normal alleles have normal life expectancies. Thus, the heterozygous individuals suffer an effect between the two extremes.

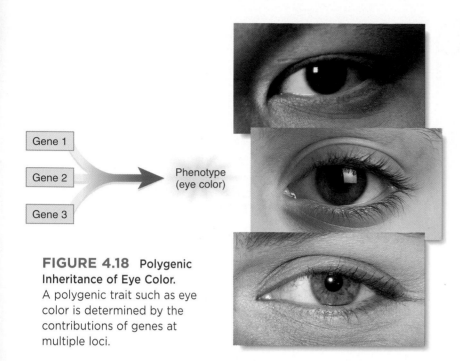

Gene 1
Gene 2 → Phenotype (eye color)
Gene 3

**FIGURE 4.18** Polygenic Inheritance of Eye Color. A polygenic trait such as eye color is determined by the contributions of genes at multiple loci.

# POLYGENIC INHERITANCE AND PLEIOTROPY

**Polygenic (multiple-gene) inheritance** (fig. 4.18) is a phenomenon in which genes at two or more loci, or even on different chromosomes, contribute to a single phenotypic trait. Human eye and skin colors are normal polygenic traits, for example. They result from the combined expression of all the genes for each trait. Several diseases are also thought to stem from polygenic inheritance, including some forms of alcoholism, mental illness, cancer, and heart disease.

**Pleiotropy** (ply-OT-roe-pee) is a phenomenon in which one gene produces multiple phenotypic effects. For example, about 1 in 200,000 people have a genetic disorder called *alkaptonuria,* caused by a mutation of chromosome 3. The mutation blocks the normal breakdown of an amino acid named tyrosine, resulting in the accumulation of an intermediate breakdown product, homogenistic acid, in the body fluids and connective tissues. When homogenistic acid oxidizes, it binds to collagen and turns the tissues gray to bluish black, and for unknown reasons, it causes degeneration of cartilages and other connective tissues. Among the multiple phenotypic effects of this disorder (fig.4.19) are darkening of the skin; darkening and degeneration of the cartilages in places such as the ears, knees, and intervertebral discs; darkening of the urine when it stands long enough for the homogentisic acid to oxidize; discoloration of the teeth and the whites of the eyes; arthritis of the shoulders, hips, and knees; and damage to the heart valves, prostate gland, and other internal organs. Another well-known example of pleiotropy is *sickle cell disease,* detailed in chapter 18.

# SEX LINKAGE

**Sex-linked traits** are carried on the X or Y chromosome and therefore tend to be inherited by one sex more than the other. Men are more likely than women to have red–green color blindness or hemophilia, for example, because the allele for each is recessive and located on the X chromosome *(X-linked).* Women have two X chromosomes. If a woman inherits the recessive hemophilia allele *(h)* on one of her X chromosomes, there is still a good chance that her other X chromosome will carry a dominant allele *(H). H* codes for normal blood-clotting proteins, so her blood clots normally. Men, on the other hand, have only one X chromosome and normally express any recessive allele found there (fig. 4.20). Ironically, even though this hemophilia is far more common among men than women, a man can inherit it only from his mother. Why? Because only his mother contributes an X chromosome to him. If he inherits *h* on his mother's X chromosome, he will have hemophilia. He has no "second chance" to inherit a normal allele on a second X chromosome. A woman, however, gets an X chromosome from both parents. Even if one parent transmits the recessive allele to her, the chances are high that she will inherit a normal allele from her other parent. She would have to have the extraordinarily bad luck to inherit it from both parents in order for her to have a trait such as hemophilia or red–green color blindness.

The X chromosome is thought to carry about 260 genes, most of which have nothing to do with determining an individual's sex. There are so few functional genes on the Y chromosome—concerned mainly with development of the testes—that virtually all sex-linked traits are associated with the X chromosome.

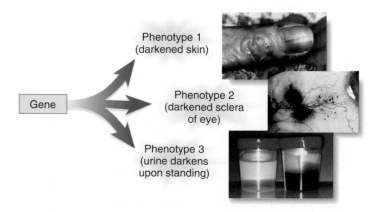

Phenotype 1 (darkened skin)

Gene → Phenotype 2 (darkened sclera of eye)

Phenotype 3 (urine darkens upon standing)

**FIGURE 4.19** Pleitropy in Alkaptonuria. Multiple phenotypic effects can result from a single gene mutation. In alkaptonuria, a single mutation leads to darkening of the skin, dark patches in the sclera ("white") of the eye, and darkening of the urine as it stands and oxidizes.

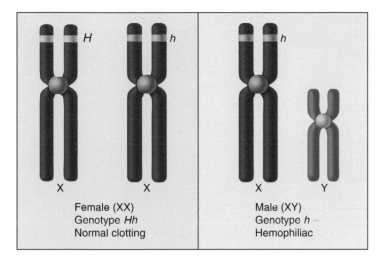

**FIGURE 4.20** **Sex-Linked Inheritance of Hemophilia.**
Left: A female who inherits a recessive allele *(h)* for hemophilia from one parent may not exhibit the trait, because she is likely to inherit the dominant allele *(H)* for a normal blood-clotting protein from her other parent. Right: A male who inherits *h* from his mother will exhibit hemophilia, because the Y chromosome inherited from his father does not have a gene locus for the clotting protein, and therefore has no ability to mask the effect of *h*.

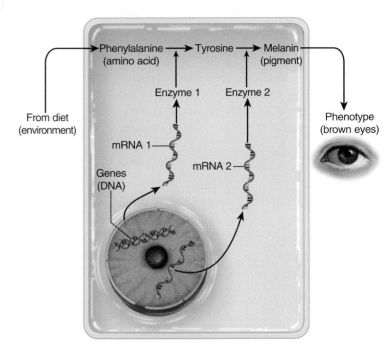

**FIGURE 4.21** **The Roles of Environment and Heredity in Producing a Phenotype.** Brown eye color requires phenylalanine from the diet (environment) and two genetically coded (hereditary) enzymes to convert phenylalanine to melanin, the eye pigment.

## PENETRANCE AND ENVIRONMENTAL EFFECTS

A given genotype does not inevitably produce the expected phenotype. For example, a dominant allele causes *poly-dactyly,*[17] the presence of extra fingers or toes. We might predict that since it is dominant, anyone who inherited the allele would exhibit this trait. Most do, but others known to have the allele have the normal number of digits. Penetrance is the percentage of a population with a given genotype that actually exhibits the predicted phenotype. If 80% of people with the polydactyly allele actually exhibit extra digits, the allele has 80% penetrance.

Another reason the connection between genotype and phenotype is not inevitable is that environmental factors play an important role in the expression of all genes. At the very least, all gene expression depends on nutrition (fig. 4.21). Children born with the hereditary disease *phenylketonuria* (FEN-il-KEE-toe-NEW-ree-uh) (PKU), for example, become retarded if they eat a normal diet. However, if PKU is detected early, a child can be placed on a diet low in phenylalanine (an amino acid) and achieve normal mental development.

No gene can produce a phenotypic effect without nutritional and other environmental input, and no nutrients can produce a body or specific phenotype without genetic instructions that tell cells what to do with them. Just as you need both a recipe and ingredients to make a cake, it takes both heredity and environment to make a phenotype.

---

[17]*poly* = many + *dactyl* = fingers, toes

## DOMINANT AND RECESSIVE ALLELES AT THE POPULATION LEVEL

It is a common misconception that dominant alleles must be more common in the gene pool than recessive alleles. The truth is that dominance and recessiveness have little to do with how common an allele is. For example, type O is the most common ABO blood type in North America, but it is caused by the recessive allele *i*. Blood type AB, caused by the two dominant ABO alleles, is the rarest. Polydactyly, caused by a dominant allele, also is rare in the population.

Definitions of some basic genetic terms are summarized in table 4.4.

### Before You Go On

*Answer the following questions to test your understanding of the preceding section:*

15. *Why must the carrier of a genetic disease be heterozygous?*

16. *State at least three reasons why a person's phenotype can't always be determined from the genotype.*

17. *A man can inherit color blindness only from his mother, whereas a woman must inherit it from both her father and mother to show the trait. Explain this apparent paradox.*

18. *Cover the left side of table 4.4 with a blank strip of paper, look at the definitions, and fill in the term to which each definition refers. Check your spelling.*

| TABLE 4.4 | Basic Terminology of Genetics |
|---|---|
| **Term** | **Definition** |
| Gene | A segment of DNA that codes for a protein or group of related proteins |
| Genome | All DNA occurring in one haploid set of chromosomes |
| Gene pool | All alleles present in a population |
| Homologous chromosomes | Two physically identical chromosomes with the same gene loci but not necessarily the same alleles; one is of maternal origin and the other paternal |
| Sex chromosomes | Two chromosomes (X and Y) that determine a person's sex |
| Autosomes | All chromosomes except the sex chromosomes; occur in 22 homologous pairs |
| Locus | The site on a chromosome where a particular gene is located |
| Allele | Any of the alternative forms that a particular gene can take |
| Genotype | The alleles that a person possesses for a particular trait |
| Phenotype | A detectable trait, such as eye color or blood type |
| Recessive allele | An allele that is not phenotypically expressed in the presence of a dominant allele; represented with a lowercase letter |
| Dominant allele | An allele that is phenotypically expressed even in the presence of any other allele; represented with a capital letter |
| Homozygous | Having identical alleles for a given gene |
| Heterozygous | Having two different alleles for a given gene |
| Carrier | A person who carries a recessive allele but does not phenotypically express it |
| Codominance | A condition in which two alleles are both fully expressed when present in the same individual |
| Incomplete dominance | A condition in which two alleles are both expressed when present in the same individual, and the phenotype is intermediate between those that each allele would produce alone |
| Polygenic inheritance | A condition in which a single phenotype results from the combined action of genes at two or more different loci, as in eye color |
| Pleiotropy | A condition in which a single gene produces multiple phenotypic effects, as in sickle cell disease |
| Sex linkage | Inheritance of a gene on the X or Y chromosome, so that the associated phenotype is expressed more in one sex than in the other |
| Penetrance | The percentage of individuals with a given genotype who actually exhibit the phenotype predicted from it |

## INSIGHT 4.4    Clinical Application

### Cancer

Proper tissue development depends on a balance between cell division and cell death. When cells multiply faster than they die, they can produce abnormal growths called *neoplasms,*[18] or tumors. The study of tumors is called *oncology.*[19]

When we have a tumor biopsied and anxiously await the result, the good news would be that it is *benign*[20] (be-NINE). This means that it is slow growing, it is surrounded by a fibrous connective tissue capsule, and its cells do not spread to other organs. Benign tumors cannot be ignored, however. Even slow-growing benign tumors can kill by compressing brain tissue, nerves, blood vessels, or airways.

The bad news would be that a tumor is *malignant*[21] (muh-LIG-nent). These are the tumors called cancer. Malignant tumors grow rapidly and their cells easily break loose and spread to other organs by way of the blood and lymph. When they lodge in other localities, such cells seed the growth of new tumors. This spreading and proliferation of tumors is called *metastasis*[22] (meh-TASS-tuh-sis); what starts as a colon cancer, for example, can metastasize to the liver, lungs, brain, and other organs. The ease with which these cells metastasize results from two factors: cell adhesion mechanisms break down in malignant cells, and malignant tumors have no fibrous capsule to prevent their spread. About 90% of cancer deaths result from metastasis rather than from the primary (original) tumor.

Malignant tumors, composed as they are of rapidly growing tissue, have an enormous demand for oxygen and nutrients. They meet their need by stimulating *angiogenesis,*[23] the growth of new blood vessels into the tumor. Such vessels often form a dense tangled mass. The word *cancer*[24] was coined by the Greek physician Hippocrates, who examined such a tangle of blood vessels in a breast tumor and felt that they reminded him of the outstretched legs of a crab.

Cancer is classified according to the cells or tissues in which the tumor originates:

| **Name** | **Origin** |
|---|---|
| Carcinoma | Epithelial cells |
| Melanoma | Pigment-producing skin cells (melanocytes) |

---

[18]*neo* = new + *plasm* = growth, formation
[19]*onco* = tumor + *logy* = study of
[20]*benign* = mild, gentle
[21]*mal* = bad, evil

[22]*meta* = beyond + *stasis* = being stationary
[23]*angio* = vessel + *genesis* = production
[24]*cancer* = crab

| Sarcoma | Bone, other connective tissue, or muscle |
| Leukemia | Blood-forming tissues |
| Lymphoma | Lymph nodes |

## Causes of Cancer

The World Health Organization estimates that 60% to 70% of cancer is caused by environmental agents called *carcinogens*[25] (car-SIN-oh-jens). These fall into three categories:

1. Chemicals such as cigarette tar, nitrites and other food preservatives, and numerous industrial chemicals.

2. Radiation such as gamma rays, alpha particles, beta particles, and ultraviolet radiation.

3. Viruses such as type 2 herpes simplex (implicated in some cases of uterine cancer), hepatitis C (implicated in some liver cancer), and human papillomavirus, HPV (implicated in cervical cancer).

Carcinogens are *mutagens*[26] (MEW-tuh-jens)—they trigger gene mutations. We have several defenses against mutagens: (1) scavenger cells may remove them before they cause genetic damage; (2) peroxisomes neutralize nitrites, free radicals, and other carcinogenic oxidizing agents; and (3) nuclear enzymes detect and repair damaged DNA. If these mechanisms fail, or if they are overworked by heavy exposure to mutagens, a cell may die of genetic damage, it may be recognized and destroyed by the immune system before it can multiply, or it may multiply and produce a tumor.

Even then, we are not left defenseless. Certain white blood cells called *natural killer* (NK) *cells* continually patrol the body, detecting and destroying malignant cells. This form of protection is called *immune surveillance*. We also gain protection from *monocytes* (another class of white blood cells) and *macrophages* (phagocytic cells derived from monocytes). They secrete a protein called *tumor necrosis factor* (TNF), which targets cancer cells, induces their death, and thus shrinks tumors.

## Growth Factors and Cancer Genes

Cancer researchers have linked many forms of cancer to abnormal growth factors or growth factor receptors. Most cells cannot divide unless a growth factor binds to a receptor on their surface. When it does so, the receptor activates cell-division enzymes in the cell. This stimulates a cell to leave the $G_0$ phase, undergo mitosis, and develop *(differentiate)* into various kinds of mature, functional cells.

Two types of genes have been identified as responsible for malignant tumors: oncogenes and tumor suppressor genes. *Oncogenes* are mutated, "misbehaving" forms of normal genes called proto-oncogenes. Healthy proto-oncogenes code for growth factors or growth-factor receptors, whereas mutated oncogenes cause malfunctions in the growth-factor mechanism. An oncogene called *sis*, for example, causes excessive secretion of growth factors that stimulate tumor angiogenesis. An oncogene known as *ras*, responsible for about one-quarter of human cancers, codes for abnormal growth-factor receptors. These receptors act like a switch stuck in the on position, sending constant cell-division signals even when there is no growth factor bound to them. Many cases of breast and ovarian cancer are caused by an oncogene called *erbB2*.

*Tumor suppressor (TS) genes* inhibit the development of cancer. They may act by opposing the action of oncogenes, promoting DNA repair, or controlling the normal histological organization of tissues, which is notably lacking in malignancies. A TS gene called *p16* acts by inhibiting one of the enzymes that drives the cell cycle. Thus, if oncogenes are like the "accelerator" of the cell cycle, then TS genes are like the "brake." Like other genes, TS genes occur in *homologous pairs*, and even one normal TS gene in a pair is usually sufficient to suppress cancer. Damage to both members of a pair, however, removes normal controls over cell division and tends to trigger cancer. The first TS gene discovered was the *Rb gene*, which causes *retinoblastoma*, an eye cancer of infants. Retinoblastoma occurs only if both copies of the *Rb* gene are damaged. In the case of a TS gene called *p53*, however, damage to just one copy is enough to trigger cancer. Gene *p53* is a large gene vulnerable to many cancer-causing mutations; it is involved in leukemia and in colon, lung, breast, liver, brain, and esophageal tumors.

Cancer often occurs only when several mutations have accumulated at different gene sites. Colon cancer, for example, requires damage to at least three TS genes on chromosomes 5, 17, and 18 and activation of an oncogene on chromosome 12. It takes time for so many mutations to accumulate, and this is one reason why colon cancer afflicts elderly people more than the young. In addition, the longer we live, the more carcinogens we are exposed to, the less efficient our DNA and tissue repair mechanisms become, and the less effective our immune system is at recognizing and destroying malignant cells.

## The Lethal Effects of Cancer

Cancer is almost always fatal if it is not treated. Malignant tumors can kill in several ways:

- Cancer displaces normal tissue, so the function of the affected organ deteriorates. Lung cancer, for example, can destroy so much tissue that oxygenation of the blood becomes inadequate to support life.

- Tumors can invade blood vessels, causing fatal hemorrhages.

- Tumors can block vital passageways. The growth of a tumor can put pressure on a bronchus of the lung, obstructing air flow and causing pulmonary collapse, or it can compress a major blood vessel, reducing the delivery of blood to a vital organ or its return to the heart.

- Tumors have a high metabolic rate and compete with healthy tissues for nutrition. Other organs of the body may even break down their own proteins to nourish the tumor. This leads to general weakness, fatigue, emaciation, and susceptibility to infections. Some forms of cancer cause *cachexia* (ka-KEX-ee-ah), an extreme wasting away of muscular and adipose tissue that cannot be corrected with nutritional therapy. In leukemia, bone marrow stem cells are so heavily diverted into producing an excess of white blood cells that they fail to produce adequate red blood cells and platelets. Thus the patient suffers anemia and bleeding. In addition, the white cells are immature and unable to fight infections, so leukemia patients also suffer from *opportunistic infection*—becoming overwhelmed by infections that a healthy person's immune system could easily ward off.

---

[25]*carcino* = cancer + *gen* = producing
[26]*muta* = change + *gen* = producing

## CHAPTER REVIEW

# Review of Key Concepts

**Genes and Nucleic Acids (p. 126)**

1. The *chromatin* in a cell nucleus is composed of DNA and protein. It is elaborately coiled to prevent damage to the DNA.

2. Nucleic acids are polymers of nucleotides. A nucleotide is composed of a sugar, a phosphate group, and a nitrogenous base.

3. Cytosine (C), thymine (T), and uracil (U) are single-ringed nitrogenous bases called *pyrimidines*. Adenine (A) and guanine (G) are double-ringed bases called *purines.*

4. The DNA molecule is like a twisted ladder, with backbones of sugar (deoxyribose) and phosphate, and "rungs" of paired bases in the middle. A base pair is always A–T or C–G (DNA contains no uracil).

5. DNA codes for the amino acid sequences of proteins. A *gene* is a segment of DNA that codes for one protein or group of proteins.

6. Three types of RNA—mRNA, rRNA, and tRNA—carry out protein synthesis.

7. RNA is much smaller than DNA and consists of just one nucleotide chain. Except in some regions of tRNA, its bases are unpaired. RNA contains ribose in place of deoxyribose, and uracil in place of thymine.

**Protein Synthesis and Secretion (p. 131)**

1. DNA directly controls protein structure and indirectly controls the synthesis of other molecules by coding for the enzymes that make them.

2. Each sequence of three bases in DNA is represented by a complementary three-base *codon* in mRNA. The codons include 61 that code for amino acids and 3 *stop codons* that code for the end of a gene. The *genetic code* is the correspondence between the mRNA codons and the 20 amino acids that they represent.

3. Protein synthesis begins with *transcription,* in which DNA uncoils at the site of a gene and RNA polymerase makes an mRNA mirror-image copy of the gene. mRNA usually leaves the nucleus and binds to a ribosome in the cytoplasm.

4. Protein synthesis continues with *translation,* in which a ribosome binds mRNA, reads the coded message, and assembles the corresponding protein.

5. In the ribosome, rRNA reads the code. tRNA molecules transport amino acids to the ribosome and contribute them to the growing peptide chain.

6. Older proteins called *chaperones* often bind new proteins, guide their folding into correct secondary and tertiary structure or their conjugation with nonprotein moieties, and escort them to their destinations in a cell.

7. Proteins destined for use in the cytosol are usually made by free ribosomes in the cytoplasm. Proteins destined to be packaged in lysosomes or secretory vesicles enter the rough ER and are modified here and in the Golgi complex. Such alterations are called *posttranslational modification.*

8. If a protein is made for use elsewhere in the body, the Golgi complex packages it into a secretory vesicle. The vesicle migrates to the cell surface and releases the product by exocytosis.

**DNA Replication and the Cell Cycle (p. 137)**

1. Since every cell division divides a cell's DNA between two daughter cells, the DNA must be replicated before the next division.

2. The enzyme *DNA helicase* prepares DNA for replication by opening up the double helix at several points and exposing the nitrogenous bases.

3. *DNA polymerase* reads the base sequence on each chain of DNA and synthesizes the complementary chain. Thus, the two helices of DNA separate from each other and each acquires a new, complementary helix to become a new, double-helical DNA molecule.

4. Most replication errors are detected and corrected by a second "proofreading" molecule of DNA polymerase. Undetected errors persist as mutations in the genome. Some mutations are harmless, but others can cause cell death or diseases such as cancer.

5. Cells have a life cycle, from division to division, of four phases: $G_1$ S, $G_2$, and M. $G_1$ through $G_2$ are collectively called *interphase* (the period between cell divisions) and M is *mitosis.*

6. Mitosis is responsible for embryonic development, tissue growth, and replacement of old, injured, or dead cells. It occurs in four stages—*prophase, metaphase, anaphase,* and *telophase*—followed by *cytokinesis,* the division of the cytoplasm into two cells.

7. Normal tissue structure depends on a balance between cell division and cell death. Cell division is stimulated by *growth factors* and suppressed by *contact inhibition.*

**Chromosomes and Heredity (p. 142)**

1. *Heredity* is the transmission of genetic characteristics from parent to offspring.

2. *Germ cells* are developing and mature eggs and sperm. They have 23 unpaired chromosomes and are thus called *haploid* cells.

3. All other cells of the body are called *somatic cells* and are *diploid,* having 46 paired chromosomes. These pairs are shown in the *karyotype,* a chart of metaphase chromosomes arranged in pairs and by size.

4. The *genome* is all the DNA in one haploid set of chromosomes of one individual.

5. The Human Genome Project has successfully sequenced the 3.1 billion base pairs of the human genome,

revealed much about the structure of the genome, and sparked the development of the field of genomic medicine.

6. Many genes occur in alternative forms called *dominant* and *recessive alleles.* Individuals with two identical alleles at a given locus are *homozygous,* and those with non-identical alleles are *heterozygous.*

7. The *genotype* is an individual's combination of genes for a particular trait; the *phenotype* is the visible trait that results from genetic and environmental influences.

8. Recessive traits can skip a generation if masked by a dominant allele. Heterozygous *carriers* lack the trait but transmit the hidden allele to the next generation.

9. For some genes, there are three or more alleles in the population, as in the ABO blood type alleles.

10. Some alleles interact in such a way that both are fully expressed *(codominance)* or an intermediate trait is produced *(incomplete dominance).*

11. In polygenic inheritance, genes at several loci collaborate to produce a sin-

gle trait such as skin or eye color. In other cases, a single gene locus can produce multiple phenotypic effects, as in alkaptonuria and sickle cell disease.

12. All traits result from a combination of genetic and environmental influences, so the environment affects whether a given genotype is expressed.

13. Whether an allele is dominant or recessive has no relationship to whether it is more or less common in the population.

## Testing Your Recall

1. Production of more than one phenotypic trait by a single gene is called
   a. pleiotropy.
   b. genetic determinism.
   c. codominance.
   d. penetrance.
   e. genetic recombination.

2. When a ribosome reads a codon on mRNA, it must bind to the _____ of a corresponding tRNA.
   a. start codon
   b. stop codon
   c. intron
   d. exon
   e. anticodon

3. The normal functions of a liver cell—synthesizing proteins, detoxifying wastes, storing glycogen, and so forth—are done during its
   a. anaphase.
   b. telophase.
   c. $G_1$ phase.
   d. $G_2$ phase.
   e. synthesis phase.

4. Two genetically identical strands of a metaphase chromosome, joined at the centromere, are its
   a. kinetochores.
   b. centrioles.
   c. sister chromatids.
   d. homologous chromatids.
   e. nucleosomes.

5. Which of the following is *not* found in DNA?
   a. thymine
   b. phosphate
   c. cytosine
   d. deoxyribose
   e. uracil

6. Genetic transcription is performed by
   a. ribosomes.
   b. RNA polymerase.
   c. DNA polymerase.
   d. helicase.
   e. chaperones.

7. A chaperone comes into play in
   a. the folding of a new protein into its tertiary structure.
   b. keeping DNA organized within the nucleus.
   c. escorting sister chromatids to opposite daughter cells during mitosis.
   d. repairing DNA that has been damaged by mutagens.
   e. preventing malignant cells from metastasizing.

8. An allele that is not phenotypically expressed in the presence of an alternative allele of the same gene is said to be
   a. codominant.
   b. lacking penetrance.
   c. heterozygous.
   d. recessive.
   e. subordinate.

9. Semiconservative replication occurs during
   a. transcription.
   b. translation.
   c. posttranslational modification.
   d. the S phase of the cell cycle.
   e. mitosis.

10. Mutagens sometimes cause no harm to cells for all of the following reasons *except*
    a. some mutagens are natural, harmless products of the cell itself.

    b. peroxisomes detoxify some mutagens before they can do any harm.
    c. the body's DNA repair mechanisms detect and correct genetic damage.
    d. change in a codon does not always change the amino acid encoded by it.
    e. some mutations change protein structure in ways that are not critical to normal function.

11. The cytoplasmic division at the end of mitosis is called _____.

12. The alternative forms in which a single gene can occur are called _____.

13. The pattern of nitrogenous bases that represents the 20 amino acids of a protein is called the _____.

14. Several ribosomes attached to one mRNA, which they are all transcribing, form a cluster called a/an _____.

15. The enzyme that produces pre-mRNA from the instructions in DNA is _____.

16. All the DNA in a haploid set of chromosomes is called a person's _____.

17. At prophase, a cell has _____ chromosomes, _____ chromatids, and _____ molecules of DNA.

18. The cytoplasmic granule of RNA and protein that reads the message in mRNA is a _____.

19. Cells are stimulated to divide by chemical signals called _____.

20. All chromosomes except the sex chromosomes are called _____.

*Answers in Appendix B*

# True or False

*Determine which five of the following statements are false, and briefly explain why.*

1. Proteins destined to be exported from a cell are made by ribosomes on the surface of the Golgi complex.

2. Each of a cell's products—such as steroids, carbohydrates, and phospholipids—is encoded by a separate gene.

3. A molecule of RNA would weigh about half as much as a segment of DNA of the same length.

4. Each amino acid of a protein is represented by a three-base sequence in DNA.

5. One gene can code for more than one protein.

6. The law of complementary base pairing describes the way the bases in an mRNA codon pair up with the bases of a tRNA anticodon during translation.

7. Most of the DNA in a human cell does not code for any proteins.

8. Some mutations are harmless.

9. Males have only one sex chromosome whereas females have two.

10. A gene can be transcribed by only one RNA polymerase at a time.

*Answers in Appendix B*

# Testing Your Comprehension

1. Why would the supercoiled, condensed form of chromosomes seen in metaphase not be suitable for the $G_1$ phase of the cell cycle? Why would the finely dispersed chromatin of the $G_1$ phase not be suitable for mitosis?

2. Suppose the DNA double helix had a backbone of alternating nitrogenous bases and phosphates, with the deoxyribose components facing each other across the middle of the helix. Why couldn't such a molecule function as a genetic code?

3. Given the information in this chapter, present an argument that evolution is not merely possible but inevitable. (*Hint:* Review the definition of evolution in chapter 1.)

4. What would be the minimum length (approximate number of bases) of an mRNA that coded for a protein 300 amino acids long?

5. Until recently, textbooks taught the concept of "one gene, one protein"—that a gene is a segment of DNA that codes for just one protein, and every protein is represented by a separate gene. Discuss the evidence that shows that this can no longer be regarded as true.

*Answers at www.mhhe.com/saladin4*

# www.mhhe.com/saladin4

*The textbook website provides a wealth of interactive study materials fully organized and integrated by chapter. You will find practice quizzes, labeling exercises, and much more that will complement your learning and understanding of anatomy and physiology. The website also includes tools designed to enhance your* **Anatomy & Physiology | REVEALED** *experience.*

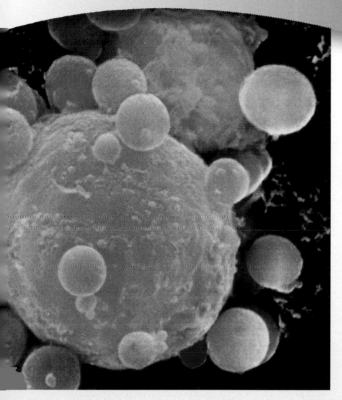

*A cancer cell (pink) undergoing apoptosis (cell suicide) under attack by an immune cell (orange) (SEM)*

# HISTOLOGY

## Brushing Up

To understand this chapter, it is important that you understand or brush up on the following concepts:

- Body cavities and membranes (p. 35)
- Glycoproteins and proteoglycans (p. 73)
- Terminology of cell shapes (p. 90)
- Secretory vesicles and exocytosis (p. 110)

With its 50 trillion cells and thousands of organs, the human body may seem to be a structure of forbidding complexity. Fortunately for our health, longevity, and self-understanding, the biologists of past generations were not discouraged by this complexity, but discovered patterns that made it more understandable. One of these patterns is the fact that these trillions of cells belong to only 200 different types or so, and these cells are organized into tissues that fall into just four broad categories: *epithelial, connective, nervous,* and *muscular tissue.*

An organ is a structure with discrete boundaries that is composed of two or more of these tissue types (frequently all four). Organs derive their function not from their cells alone but from how the cells are organized into tissues. Cells are specialized for certain tasks: muscle contraction, defense, enzyme secretion, and so forth. No one cell type has the mechanisms to carry out all of the body's vital functions. Cells work together at certain tasks and form tissues that carry out a particular function, such as nerve signaling or nutrient digestion.

The study of tissues and how they are arranged into organs is called **histology,**[1] or **microscopic anatomy.** That is the subject of this chapter. Here we study the four tissue classes; the variations within each class; how to recognize tissue types microscopically and relate their microscopic anatomy to their function; how tissues are arranged to form an organ; and how tissues change as they grow, shrink, or change from one tissue type to another over the life of the individual. Histology bridges the gap between the *cytology* of the preceding chapters and the *organ system* approach of the chapters that follow.

[1]*histo* = tissue + *logy* = study of

# The Study of Tissues

## Objectives
When you have completed this section, you should be able to

- name the four primary classes into which all adult tissues are classified;
- name the three embryonic germ layers and some adult tissues derived from each; and
- visualize the three-dimensional shape of a structure from a two-dimensional tissue section.

## THE PRIMARY TISSUE CLASSES

A **tissue** is a group of similar cells and cell products that arise from the same region of the embryo and work together to perform a specific structural or physiological role in an organ. The four *primary tissues* are epithelial, connective, nervous, and muscular tissue (table 5.1). These tissues differ from one another in the types and functions of their cells, the characteristics of the **matrix** (extracellular material) that surrounds the cells, and the relative amount of space occupied by cells versus matrix. In muscle and epithelium, the cells are so close together that the matrix is scarcely visible, while in connective tissues, the matrix usually occupies much more space than the cells do.

The matrix is composed of fibrous proteins and, usually, a clear gel variously known as **ground substance, tissue fluid, extracellular fluid (ECF), interstitial[2] fluid,** or **tissue gel.** In cartilage and bone, it can be rubbery or stony in consistency. The ground substance contains water, gases, minerals, nutrients, wastes, and other chemicals.

In summary, a tissue is composed of cells and matrix, and the matrix is composed of fibers and ground substance.

[2]*inter* = between + *stit* = to stand

| TABLE 5.1 | The Four Primary Tissue Classes | |
|---|---|---|
| **Type** | **Definition** | **Representative Locations** |
| Epithelial | Tissue composed of layers of closely spaced cells that cover organ surfaces, form glands, and serve for protection, secretion, and absorption | Epidermis<br>Inner lining of digestive tract<br>Liver and other glands |
| Connective | Tissue with more matrix than cell volume, often specialized to support, bind together, and protect organs | Tendons and ligaments<br>Cartilage and bone<br>Blood |
| Nervous | Tissue containing excitable cells specialized for rapid transmission of coded information to other cells | Brain<br>Spinal cord<br>Nerves |
| Muscular | Tissue composed of elongated, excitable cells specialized for contraction | Skeletal muscles<br>Heart (cardiac muscle)<br>Walls of viscera (smooth muscle) |

## EMBRYONIC TISSUES

Human development begins with a single cell, the fertilized egg, which soon divides to produce scores of identical, smaller cells. The first tissues appear when these cells start to organize themselves into layers—first two, and soon three strata called the **primary germ layers,** which give rise to all of the body's mature tissues. The three layers are called *ectoderm, mesoderm,* and *endoderm.* The **ectoderm**[3] is an outer layer that gives rise to the epidermis and nervous system. The inner layer, the **endoderm,**[4] gives rise to the mucous membranes of the digestive and respiratory tracts and to the digestive glands, among other things. Between these two is the **mesoderm,**[5] a layer of more loosely organized cells. Mesoderm eventually turns to a gelatinous tissue called **mesenchyme,** composed of fine, wispy collagen (protein) fibers and branching cells called *fibroblasts* embedded in a gelatinous ground substance. Mesenchyme closely resembles the connective tissue layer in figure 5.11a. It gives rise to muscle, bone, and blood, among other tissues. Most organs are composed of tissues derived from two or more primary germ layers.

[3]*ecto* = outer + *derm* = skin
[4]*endo* = inner + *derm* = skin
[5]*meso* = middle + *derm* = skin

The rest of this chapter concerns the "mature" tissues that exist from infancy through adulthood.

## INTERPRETING TISSUE SECTIONS

Histologists use a variety of techniques for preserving, sectioning (slicing), and staining tissues to show their structural details as clearly as possible. Tissue specimens are preserved in a **fixative**—a chemical such as formalin that prevents decay. After fixation, most tissues are cut into very thin slices called **histological sections.** These sections are typically only one or two cells thick, to allow the light of a microscope to pass through and to reduce the confusion of the image that would result from many layers of overlapping cells. They are mounted on slides and artificially colored with histological **stains** to bring out detail. If they were not stained, most tissues would appear very pale gray. With stains that bind to different components of a tissue, however, you may see pink cytoplasm, violet nuclei, and blue, green, or golden brown protein fibers, depending on the stain used.

Sectioning a tissue reduces a three-dimensional structure to a two-dimensional slice. You must keep this in mind and try to translate the microscopic image into a mental image of the whole structure. Like the boiled egg and elbow macaroni in figure 5.1, an object may look quite different when it is cut at various levels, or *planes of section.* A coiled tube, such as a gland of the uterus

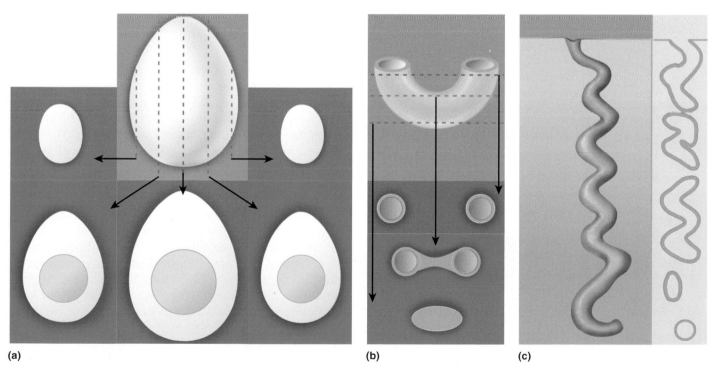

(a)  (b)  (c)

**FIGURE 5.1 Three-Dimensional Interpretation of Two-Dimensional Images.** (a) A boiled egg. Note that grazing sections (upper left and right) would miss the yolk, just as a tissue section may miss a nucleus or other structure. (b) Elbow macaroni, which resembles many curved ducts and tubules. A section far from the bend would give the impression of two separate tubules; a section near the bend would show two interconnected lumina (cavities); and a section still farther down could miss the lumen completely. (c) A coiled gland in three dimensions and as it would look in a vertical tissue section.

(fig. 5.1c), is often broken up into multiple portions since it meanders in and out of the plane of section. An experienced viewer, however, would recognize that the separated pieces are parts of a single tube winding its way to the organ surface. Note that a grazing slice through a boiled egg might miss the yolk, just as a tissue section might miss the nucleus of a cell or an egg in the ovary, even though these structures were present.

Many anatomical structures are significantly longer in one direction than another—the humerus and esophagus, for example. A tissue cut in the long direction is called a **longitudinal section (l.s.),** and one cut perpendicular to this is a **cross section (c.s.** or **x.s.),** or **transverse section (t.s.).** A section cut at an angle between a longitudinal and cross section is an **oblique section.** Figure 5.2 shows how certain organs look when sectioned on each of these planes.

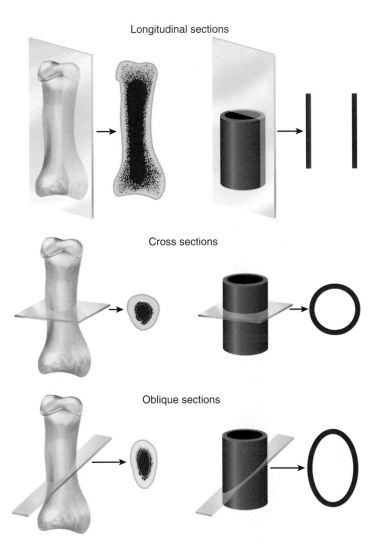

Longitudinal sections

Cross sections

Oblique sections

**FIGURE 5.2  Longitudinal, Cross, and Oblique Sections.**
Note the effect of the plane of section on the two-dimensional appearance of elongated structures such as bones and blood vessels.

▶ *Would you classify the egg sections in the previous figure as longitudinal, cross, or oblique sections? How would the egg look if sectioned in the other two planes?*

## Before You Go On

*Answer the following questions to test your understanding of the preceding section:*

1. *Classify each of the following into one of the four primary tissue classes: the skin surface, fat, the spinal cord, most heart tissue, bone, tendons, blood, and the inner lining of the stomach.*

2. *What are tissues composed of in addition to cells?*

3. *What embryonic germ layer gives rise to nervous tissue? To the liver? To muscle?*

4. *What is the term for a thin, stained slice of tissue mounted on a microscope slide?*

# Epithelial Tissue

### Objectives

When you have completed this section, you should be able to

- describe the properties that distinguish epithelium from other tissue classes;

- list and classify eight types of epithelium, distinguish them from each other, and state where each type can be found in the body;

- explain how the structural differences between epithelia relate to their functional differences; and

- visually recognize each epithelial type from specimens or photographs.

**Epithelial**[6] **tissue** consists of a flat sheet of closely adhering cells, one or more cells thick, with the upper surface usually exposed to the environment or to an internal space in the body. Epithelium covers the body surface, lines body cavities, forms the external and internal linings of many organs, and constitutes most gland tissue. The extracellular material is so thin it is not visible with the light microscope, and epithelia allow no room for blood vessels. They do, however, almost always lie on a layer of loose connective tissue and depend on its blood vessels for nourishment and waste removal.

Between an epithelium and the underlying connective tissue is a layer called the **basement membrane,** usually too thin to be visible with the light microscope. It contains collagen, adhesive glycoproteins called *laminin* and *fibronectin,* and a large protein-carbohydrate complex called *heparan sulfate.* It gradually blends with collagenous and reticular fibers on the connective tissue side. The basement membrane serves to anchor an epithelium to the connective tissue below it. The surface of an epithelial cell that faces the basement membrane is its

---

[6]*epi* = upon + *theli* = nipple, female

*basal surface,* and the one that faces away from the basement membrane is the *apical surface.*

Epithelia are classified into two broad categories—*simple* and *stratified*—with four types in each category. In a simple epithelium, every cell touches the basement membrane, whereas in a stratified epithelium, some cells rest on top of other cells and do not contact the basement membrane (fig. 5.3). *Pseudostratified columnar* epithelium is a special type of epithelium described in the next paragraph.

## SIMPLE EPITHELIA

Generally, a **simple epithelium** has only one layer of cells, although this is a somewhat debatable point in the *pseudostratified* type. Three types of simple epithelia are named for the shapes of their cells: **simple squamous**[7] (thin scaly cells), **simple cuboidal** (square or round cells), and **simple columnar** (tall narrow cells). In the fourth type, **pseudostratified columnar,** not all cells reach the free surface; the shorter cells are covered over by the taller ones. This epithelium looks stratified in most tissue sections, but careful examination, especially with the electron microscope, shows that every cell reaches the basement membrane. Simple columnar and pseudostratified columnar epithelia often produce protective mucous coatings over the mucous membranes. The mucus is secreted by wineglass-shaped **goblet cells.**

Table 5.2 illustrates and summarizes the structural and functional differences among these four types.

[7]*squam = scale*

## STRATIFIED EPITHELIA

**Stratified epithelia** range from 2 to 20 or more layers of cells, with some cells resting directly on others and only the deepest layer resting on the basement membrane. Three of the stratified epithelia are named for the shapes of their surface cells: **stratified squamous, stratified cuboidal,** and **stratified columnar epithelia.** The deeper cells, however, may be of a different shape than the surface cells. The fourth type, **transitional epithelium,** was named when it was thought to represent a transitional stage between stratified squamous and stratified columnar epithelium. This is now known to be untrue, but the name has persisted.

Stratified columnar epithelium is rare—seen only in places where two other epithelial types meet, as in limited regions of the pharynx, larynx, anal canal, and male urethra. We will not consider this type any further. The other three types are illustrated and summarized in table 5.3.

The most widespread epithelium in the body is stratified squamous epithelium, which deserves further discussion. Its deepest layer of cells are cuboidal to columnar stem cells, which undergo continual mitosis. Their daughter cells push toward the surface and become flatter (more *squamous,* or scalelike) as they migrate farther upward, until they finally die and flake off. Their separation from the surface is called **exfoliation,** or

*(text continued on p. 162)*

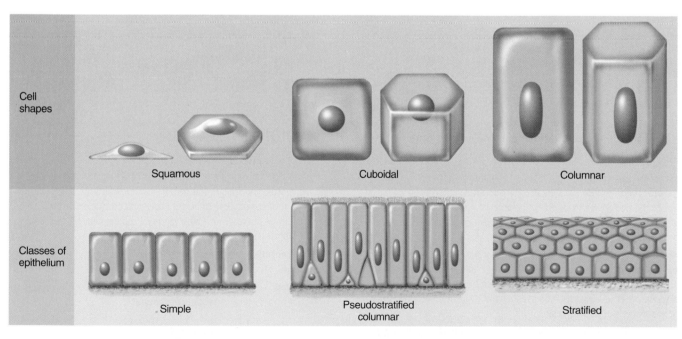

**FIGURE 5.3  Cell Shapes and Epithelial Types.** Pseudostratified columnar epithelium is a special type of simple epithelium that gives a false appearance of multiple cell layers.

| TABLE 5.2 | Simple Epithelia |
|-----------|------------------|

**Simple Squamous Epithelium**

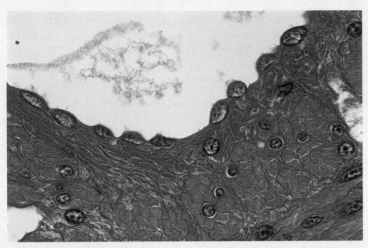

(a)

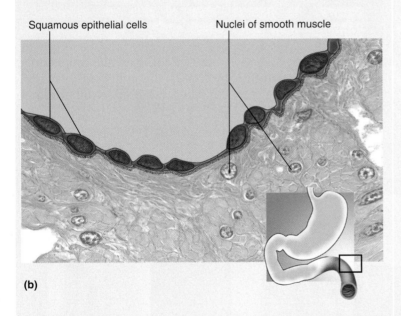

Squamous epithelial cells    Nuclei of smooth muscle

(b)

**FIGURE 5.4**    External Surface (Serosa) of the Small Intestine.

**Microscopic appearance:** Single layer of thin cells, shaped like fried eggs with bulge where nucleus is located; nucleus flattened in the plane of the cell, like an egg yolk; cytoplasm may be so thin it is hard to see in tissue sections; in surface view, cells have angular contours and nuclei appear round

**Representative locations:** Air sacs (alveoli) of lungs; glomerular capsules of kidneys; some kidney tubules; inner lining (endothelium) of heart and blood vessels; serous membranes of stomach, intestines, and some other viscera; surface mesothelium of pleurae, pericardium, peritoneum, and mesenteries

**Functions:** Allows rapid diffusion or transport of substances through membrane; secretes lubricating serous fluid

**Simple Cuboidal Epithelium**

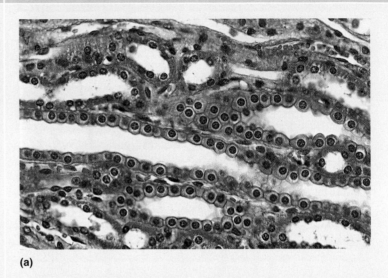

(a)

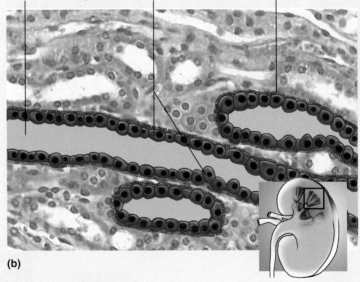

Lumen of kidney tubule    Cuboidal epithelial cells    Basement membrane

(b)

**FIGURE 5.5**    Kidney Tubules.

**Microscopic appearance:** Single layer of square or round cells; in glands, cells often pyramidal and arranged like segments of an orange around a central space; spherical, centrally placed nuclei; often with a brush border of microvilli in some kidney tubules; ciliated in bronchioles of lung

**Representative locations:** Liver, thyroid, mammary, salivary, and other glands; most kidney tubules; bronchioles

**Functions:** Absorption and secretion; production of protective mucous coat; movement of respiratory mucus

## Simple Columnar Epithelium

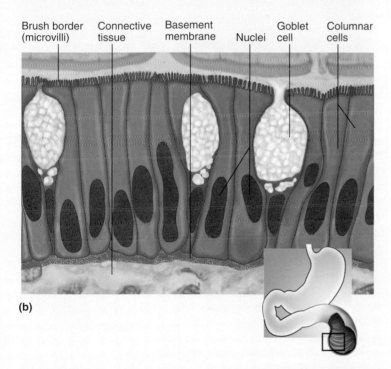

(a)

Brush border (microvilli) | Connective tissue | Basement membrane | Nuclei | Goblet cell | Columnar cells

(b)

**FIGURE 5.6**  Internal Surface (Mucosa) of the Small Intestine.

**Microscopic appearance:** Single layer of tall, narrow cells; oval or sausage-shaped nuclei, vertically oriented, usually in basal half of cell; apical portion of cell often shows secretory vesicles visible with TEM; often shows a brush border of microvilli; ciliated in some organs; may possess goblet cells

**Representative locations:** Inner lining of stomach, intestines, gallbladder, uterus, and uterine tubes; some kidney tubules

**Functions:** Absorption; secretion of mucus and other products; movement of egg and embryo in uterine tube

## Pseudostratified Columnar Epithelium

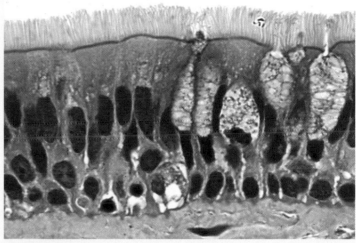

(a)

Cilia | Basement membrane | Basal cells | Goblet cell

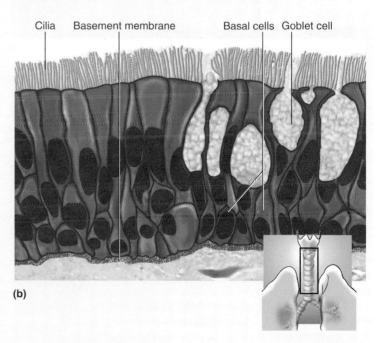

(b)

**FIGURE 5.7**  Mucosa of the Trachea.

**Microscopic appearance:** Looks multilayered; some cells do not reach free surface, but all cells reach basement membrane; nuclei at several levels in deeper half of epithelium; often with goblet cells; often ciliated

**Representative locations:** Respiratory tract from nasal cavity to bronchi; portions of male urethra

**Functions:** Secretes and propels mucus

## TABLE 5.3    Stratified Epithelia

| Stratified Squamous Epithelium—Keratinized | Stratified Squamous Epithelium—Nonkeratinized |
|---|---|

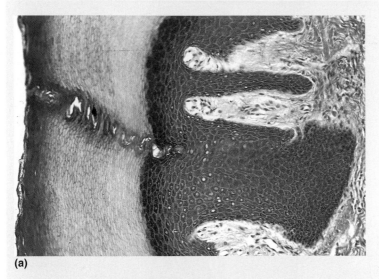

(a)

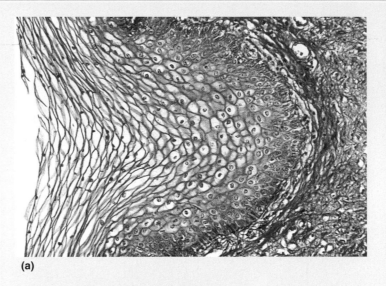

(a)

Dead squamous cells    Living epithelial cells    Dense irregular connective tissue

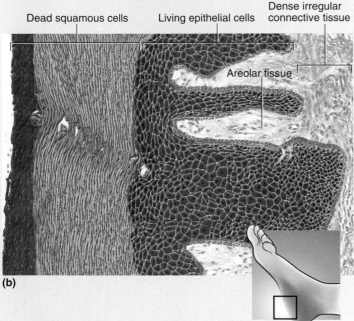

Areolar tissue

(b)

Living epithelial cells    Connective tissue

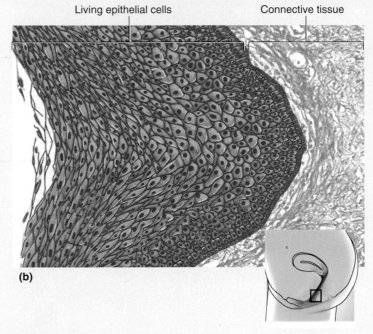

(b)

**FIGURE 5.8**  Skin from the Sole of the Foot.

**FIGURE 5.9**  Mucosa of the Vagina.

**Microscopic appearance:** Multiple cell layers with cells becoming increasingly flat and scaly toward surface; surface covered with a layer of compact dead cells without nuclei; basal cells may be cuboidal to columnar

**Representative locations:** Epidermis; palms and soles are especially heavily keratinized

**Functions:** Resists abrasion; retards water loss through skin; resists penetration by pathogenic organisms

**Microscopic appearance:** Same as keratinized epithelium but without the surface layer of dead cells

**Representative locations:** Tongue, oral mucosa, esophagus, anal canal, vagina

**Functions:** Resists abrasion and penetration by pathogenic organisms

## Stratified Cuboidal Epithelium

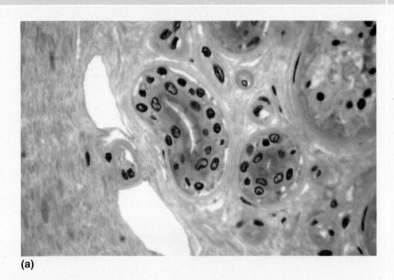

(a)

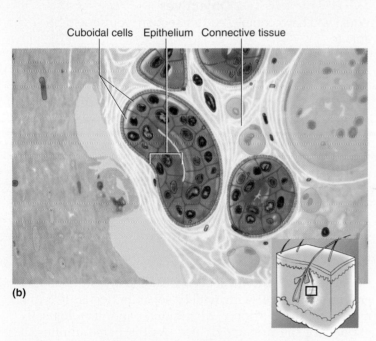

Cuboidal cells    Epithelium    Connective tissue

(b)

**FIGURE 5.10**  Duct of a Sweat Gland.

**Microscopic appearance:** Two or more layers of cells; surface cells square or round

**Representative locations:** Sweat gland ducts; egg-producing vesicles (follicles) of ovaries; sperm-producing ducts (seminiferous tubules) of testis

**Functions:** Contributes to sweat secretion; secretes ovarian hormones; produces sperm

## Transitional Epithelium

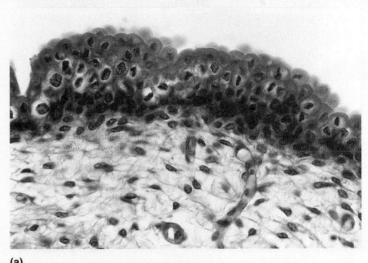

(a)

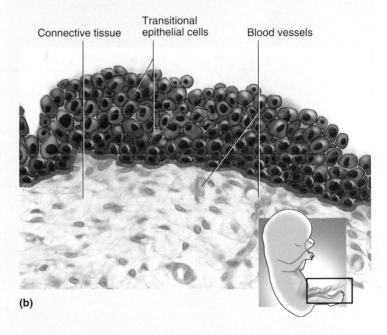

Connective tissue    Transitional epithelial cells    Blood vessels

(b)

**FIGURE 5.11**  Allantoic Duct of Umbilical Cord.

**Microscopic appearance:** Somewhat resembles stratified squamous epithelium, but surface cells are rounded, not flattened, and often bulge at surface; typically five or six cells thick when relaxed and two or three cells thick when stretched; cells may be flatter and thinner when epithelium is stretched (as in a distended bladder); some cells have two nuclei

**Representative locations:** Urinary tract—part of kidney, ureter, bladder, part of urethra; allantoic duct in umbilical cord

**Functions:** Stretches to allow filling of urinary tract

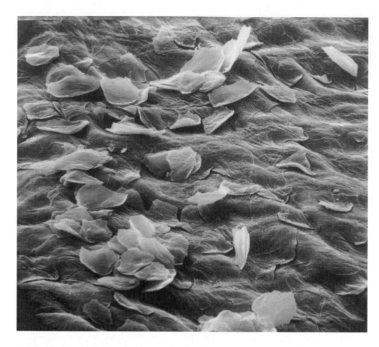

**FIGURE 5.12  Exfoliation of Squamous Cells from the Mucosal Surface of the Vagina.** [From R. G. Kessel and R. H. Kardon, *Tissues and Organs: A Text-Atlas of Scanning Electron Microscopy* (W. H. Freeman, 1979).]

▶ *Aside from the gums and vagina, name another epithelium in the body that would look like this to the scanning electron microscope.*

desquamation (fig. 5.12); the study of exfoliated cells is called *exfoliate cytology.* You can easily study exfoliated cells by scraping your gums with a toothpick, smearing this material on a slide, and staining it. A similar procedure is used in the *Pap smear,* an examination of exfoliated cells from the cervix for signs of uterine cancer (see chapter 28, fig. 28.5, for normal and cancerous Pap smears).

Stratified squamous epithelia are of two kinds: keratinized and nonkeratinized. A **keratinized** epithelium, found on the skin surface (epidermis), is covered with a layer of compact, dead squamous cells. These cells are packed with the durable protein keratin and coated with water repellent. The skin surface is therefore relatively dry, it retards water loss from the body, and it resists penetration by disease organisms. (Keratin is also the protein of which animal horns are made, hence its name.[8]) The tongue, esophagus, vagina, and a few other internal membranes are covered with the **nonkeratinized** type, which lacks the surface layer of dead cells. This type provides a surface that is, again, abrasion-resistant, but also moist and slippery. These characteristics are well suited to resist stress produced by the chewing and swallowing of food and by sexual intercourse and childbirth.

---

[8]*kerat* = horn

# Connective Tissue

## Objectives

When you have completed this section, you should be able to

- describe the properties that most connective tissues have in common;
- discuss the types of cells found in connective tissue;
- explain what the matrix of a connective tissue is and describe its components;
- name 10 types of connective tissue, describe their cellular components and matrix, and explain what distinguishes them from each other; and
- visually recognize each connective tissue type from specimens or photographs.

## OVERVIEW

**Connective tissue** typically consists mostly of fibers and ground substance, with widely separated cells. It is the most abundant, widely distributed, and histologically variable of the primary tissues. As the name implies, it often serves to connect organs to each other—for example, the way a tendon connects muscle to bone—or serves in other ways to support, bind, and protect organs. This category includes fibrous tissue, fat, cartilage, bone, and blood. The mesenchyme, described earlier in this chapter, is a form of embryonic connective tissue.

The functions of connective tissue include the following:

- **Binding of organs.** Tendons bind muscle to bone, ligaments bind one bone to another, fat holds the kidneys and eyes in place, and fibrous tissue binds the skin to underlying muscle.

- **Support.** Bones support the body, and cartilage supports the ears, nose, trachea, and bronchi.

- **Physical protection.** The cranium, ribs, and sternum protect delicate organs such as the brain, lungs, and heart; fatty cushions around the kidneys and eyes protect these organs.
- **Immune protection.** Connective tissue cells attack foreign invaders, and connective tissue fiber forms a "battlefield" under the skin and mucous membranes where immune cells can be quickly mobilized against disease agents.
- **Movement.** Bones provide the lever system for body movement, cartilages are involved in movement of the vocal cords, and cartilages on bone surfaces ease joint movements.
- **Storage.** Fat is the body's major energy reserve; bone is a reservoir of calcium and phosphorus that can be drawn upon when needed.
- **Heat production.** Brown fat generates heat in infants and children.
- **Transport.** Blood transports gases, nutrients, wastes, hormones, and blood cells.

## FIBROUS CONNECTIVE TISSUE

Fibrous connective tissues are the most diverse type of connective tissue. They are also called *fibroconnective tissue* or *connective tissue proper.* Nearly all connective tissues contain fibers, but the tissues considered here are classified together because the fibers are so conspicuous. Fibers are, of course, just one component of the tissue, which also includes cells and ground substance. Before examining specific types of fibrous connective tissue, let's examine these components.

### Components of Fibrous Connective Tissue

**Cells** The cells of fibrous connective tissue include the following types:

- **Fibroblasts.**[9] These are large, flat cells that often appear tapered at the ends and show slender, wispy branches. They produce the fibers and ground substance that form the matrix of the tissue. Fibroblasts that have finished this task and become inactive are called *fibrocytes* by some histologists.
- **Macrophages.**[10] These are large phagocytic cells that wander through the connective tissues, where they engulf and destroy bacteria, other foreign particles, and dead or dying cells of our own body. They also activate the immune system when they sense foreign matter called *antigens.* They arise from certain white blood cells called *monocytes* or from the same stem cells that produce monocytes.

- **Leukocytes,**[11] or **white blood cells (WBCs).** WBCs travel briefly in the bloodstream, then crawl out through the capillary walls and spend most of their time in the connective tissues. Most of them are *neutrophils,* which wander about in search of bacteria. Our mucous membranes often exhibit dense patches of tiny WBCs called *lymphocytes,* which react against bacteria, toxins, and other foreign agents.
- **Plasma cells.** Certain lymphocytes turn into plasma cells when they detect foreign agents. The plasma cells then synthesize disease-fighting proteins called *antibodies.* Plasma cells are rarely seen except in the walls of the intestines and in inflamed tissue.
- **Mast cells.** These cells, found especially alongside blood vessels, secrete a chemical called *heparin* that inhibits blood clotting, and one called *histamine* that increases blood flow by dilating blood vessels.
- **Adipocytes** (AD-ih-po-sites), or **fat cells.** These are large rounded cells filled mainly with a droplet of triglyceride, which forces the nucleus and cytoplasm to occupy only a thin layer just beneath the plasma membrane. They appear in small clusters in some fibrous connective tissues. When they dominate an area, the tissue is called *adipose tissue.*

**Fibers**   Three types of protein fibers are found in fibrous connective tissues:

- **Collagenous** (col-LADJ-eh-nus) **fibers.** These fibers, made of collagen, are tough and flexible and resist stretching. Collagen is about 25% of the body's protein, the most abundant type. It is the base of such animal products as gelatin, leather, and glue.[12] In fresh tissue, collagenous fibers have a glistening white appearance, as seen in tendons and some cuts of meat (fig. 5.13); thus, they are often called *white fibers.* In tissue sections, collagen forms coarse, wavy bundles, often dyed pink, blue, or green by the most common histological stains. Tendons, ligaments, and the deep layer of the skin (the dermis) are made mainly of collagen. Less visibly, collagen pervades the matrix of cartilage and bone.
- **Reticular**[13] **fibers.** These are thin collagen fibers coated with glycoprotein. They form a spongelike framework for such organs as the spleen and lymph nodes.
- **Elastic fibers.** These are thinner than collagenous fibers, and they branch and rejoin each other along their course. They are made of a protein called **elastin,** whose coiled structure allows it to stretch and recoil like a rubber band. Elastic fibers account for the ability of the skin, lungs, and arteries to spring back after they are stretched. (Elasticity is not

---

[9]*fibro* = fiber + *blast* = producing
[10]*macro* = big + *phage* = eater

[11]*leuko* = white + *cyte* = cell
[12]*colla* = glue + *gen* = producing
[13]*ret* = network + *icul* = little

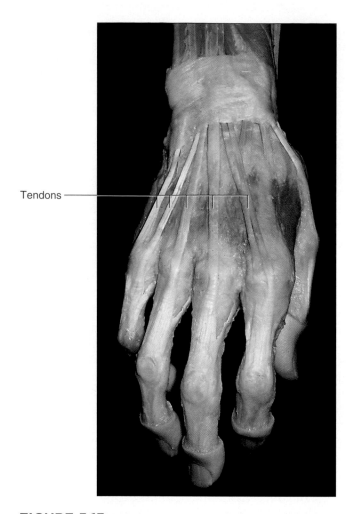

Tendons

**FIGURE 5.13  Tendons.** The white glistening appearance results from the collagen of which tendons are composed. The bracelet-like band across the wrist is also composed of collagen.

the ability to stretch, but the tendency to recoil when tension is released.) Fresh elastic fibers are yellowish and therefore often called *yellow fibers.*

*Ground Substance*    Amid the cells and fibers in some tissue sections, there appears to be a lot of empty space. In life, this space is occupied by the featureless **ground substance.** Ground substance usually has a gelatinous to rubbery consistency resulting from three classes of large molecules: glycosaminoglycans, proteoglycans, and adhesive glycoproteins. It absorbs compressive forces and, like the styrofoam packing in a shipping carton, protects the more delicate cells from mechanical injury.

A **glycosaminoglycan (GAG)** (gly-COSE-ah-MEE-no-GLY-can) is a long polysaccharide composed of unusual disaccharides called *amino sugars* and *uronic acid.* GAGs are negatively charged and thus tend to attract sodium and potassium ions, which in turn causes the GAGs to absorb and hold water. Thus, GAGs play an important role in regulating the water and electrolyte balance of tissues. The most

abundant GAG is **chondroitin** (con-DRO-ih-tin) **sulfate.** It is abundant in blood vessels and bones and is responsible for the relative stiffness of cartilage. Some other GAGs that you will read of elsewhere in this book are *heparin* (an anticoagulant) and *hyaluronic* (HY-uh-loo-RON-ic) *acid.* The latter is a gigantic molecule up to 20 µm long, as large as most cells. It is a viscous, slippery substance that forms a very effective lubricant in the joints and constitutes much of the jellylike *vitreous humor* of the eyeball.

A **proteoglycan** is another gigantic molecule. It is shaped somewhat like a test tube brush, with a central core of protein and bristlelike outgrowths composed of GAGs. The entire proteoglycan may be attached to hyaluronic acid, thus forming an enormous molecular complex. Proteoglycans form thick colloids similar to those of gravy, pudding, gelatin, and glue. This gel slows the spread of pathogenic organisms through the tissues. Some proteoglycans are embedded in the plasma membranes of cells, attached to the cytoskeleton on the inside and to other extracellular molecules on the outside. Thus, they create a strong structural bond between cells and extracellular macromolecules and help to hold tissues together.

**Adhesive glycoproteins** are protein–carbohydrate complexes that bind plasma membrane proteins to collagen and proteoglycans outside the cell. They bind all the components of a tissue together and mark pathways that guide migrating embryonic cells to their destinations in a tissue.

## Types of Fibrous Connective Tissue

Fibrous connective tissue is divided into two broad categories according to the relative abundance of fiber: *loose* and *dense connective tissue.* In **loose connective tissue,** much of the space is occupied by ground substance, which is dissolved out of the tissue during histological fixation and leaves empty space in prepared tissue sections. The loose connective tissues we will discuss are *areolar, reticular,* and *adipose tissue* (table 5.4). In **dense connective tissue,** fiber occupies more space than the cells and ground substance, and appears closely packed in tissue sections. The two dense connective tissues we will discuss are *dense regular* and *dense irregular connective tissue* (table 5.5).

**Areolar**[14] (AIR-ee-OH-lur) **tissue** exhibits loosely organized fibers, abundant blood vessels, and a lot of seemingly empty space. It possesses all six of the aforementioned cell types. Its fibers run in random directions and are mostly collagenous, but elastic and reticular fibers are also present. Areolar tissue is highly variable in appearance. In many serous membranes, it looks like figure 5.14, but in the skin and mucous membranes, it is more compact (see fig. 5.8) and sometimes difficult to distinguish from dense irregular connective tissue. Some advice on how to tell them apart is given after the discussion of dense irregular connective tissue.

---

[14]*areola* = little space

## TABLE 5.4  Loose Connective Tissues

| Areolar Tissue | Reticular Tissue | Adipose Tissue |
|---|---|---|

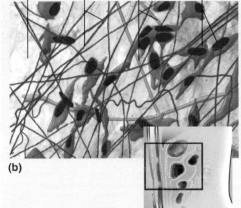

(a)

Ground substance    Elastic fibers    Collagenous fibers    Fibroblasts

(b)

**FIGURE 5.14** Spread of the Mesentery.

**Microscopic appearance:** Loose arrangement of collagenous and elastic fibers; scattered cells of various types; abundant ground substance; numerous blood vessels

**Representative locations:** Underlying nearly all epithelia; surrounding blood vessels, nerves, esophagus, and trachea; fascia between muscles; mesenteries; visceral layers of pericardium and pleura

**Functions:** Loosely binds epithelia to deeper tissues; allows passage of nerves and blood vessels through other tissues; provides an arena for immune defense; blood vessels provide nutrients and waste removal for overlying epithelia

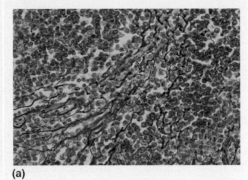

(a)

Reticular fibers
Leukocytes

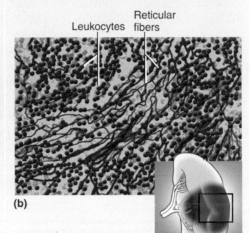

(b)

**FIGURE 5.15** Spleen.

**Microscopic appearance:** Loose network of reticular fibers and cells, infiltrated with numerous lymphocytes and other blood cells

**Representative locations:** Lymph nodes, spleen, thymus, bone marrow

**Functions:** Supportive stroma (framework) for lymphatic organs

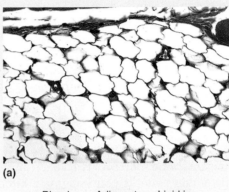

(a)

Blood vessel    Adipocyte nucleus    Lipid in adipocyte

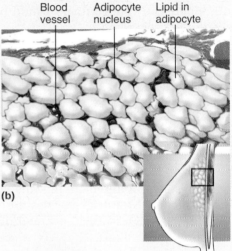

(b)

**FIGURE 5.16** Adipose Tissue.

**Microscopic appearance:** Dominated by adipocytes—large, empty-looking cells with thin margins; tissue sections often very pale because of scarcity of stained cytoplasm; adipocytes shrunken; nucleus pressed against plasma membrane; blood vessels often present

**Representative locations:** Subcutaneous fat beneath skin; breast; heart surface; mesenteries; surrounding organs such as kidneys and eyes

**Functions:** Energy storage; thermal insulation; heat production by brown fat; protective cushion for some organs; filling space, shaping body

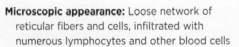

Areolar tissue is found in tissue sections from almost every part of the body. It surrounds blood vessels and nerves and penetrates with them even into the small spaces of muscles, tendons, and other tissues. Nearly every epithelium rests on a layer of areolar tissue, whose blood vessels provide the epithelium with nutrition, waste removal, and a ready supply of infection-fighting leukocytes in times of need. Because of the abundance of open, fluid-filled space, leukocytes can move about freely in areolar tissue and can easily find and destroy pathogens.

**Reticular tissue** is a mesh of reticular fibers and fibroblasts. It forms the structural framework (stroma) of such organs and tissues as the lymph nodes, spleen, thymus, and bone marrow. The space amid the fibers is filled with blood cells. If you imagine a kitchen sponge soaked with blood, the sponge fibers would be analogous to the reticular tissue stroma.

**Adipose tissue,** or **fat,** is tissue in which adipocytes are the dominant cell type. Adipocytes may also occur singly or in small clusters in areolar tissue. Adipocytes usually range from 70 to 120 μm in diameter, but they may be five times as large in obese people. The space between adipocytes is occupied by areolar tissue, reticular tissue, and blood capillaries.

Fat is the body's primary energy reservoir. The quantity of stored triglyceride and the number of adipocytes are quite stable in a person, but this doesn't mean stored fat is stagnant. New triglycerides are constantly synthesized and stored as others are hydrolyzed and released into circulation. Thus, there is a constant turnover of stored triglyceride, with an equilibrium between synthesis and hydrolysis, energy storage and energy use. Adipose tissue also provides thermal insulation, and it contributes to body contours such as the female breasts and hips. Most adipose tissue is a type called *white fat,* but fetuses, infants, and children also have a heat-generating tissue called *brown fat,* which accounts for up to 6% of an infant's weight. Brown fat gets its color from an unusual abundance of blood vessels and certain enzymes in its mitochondria. It stores lipid in the form of multiple droplets rather than one large one. Brown fat has numerous mitochondria, but their oxidation pathway is not linked to ATP synthesis. Therefore, when these cells oxidize fats, they release all of the energy as heat. Hibernating animals accumulate brown fat in preparation for winter.

## Think About It

*Why would infants and children have more need for brown fat than adults do? (Hint: Smaller bodies have a higher ratio of surface area to volume than larger bodies do.)*

**Dense regular connective tissue** is named for two properties: (1) the collagen fibers are closely packed and leave relatively little open space, and (2) the fibers are parallel to each other. It is found especially in tendons and ligaments. The parallel arrangement of fibers is an adaptation to the fact that tendons and ligaments are pulled in predictable directions. With some minor exceptions such as blood vessels and sensory nerve fibers, the only cells in this tissue are fibroblasts, visible by their slender, violet-staining nuclei squeezed between bundles of collagen. This type of tissue has few blood vessels, so injured tendons and ligaments are slow to heal.

The vocal cords, suspensory ligament of the penis, and some ligaments of the vertebral column are made of a type of dense regular connective tissue called **yellow elastic tissue.** In addition to the densely packed collagen fibers, it exhibits branching elastic fibers and more fibroblasts. The fibroblasts have larger, more conspicuous nuclei than seen in most dense regular connective tissue.

Elastic tissue also takes the form of wavy sheets in the walls of the large and medium arteries. When the heart pumps blood into the arteries, these sheets enable them to expand and relieve some of the pressure on smaller vessels downstream. When the heart relaxes, the arterial wall springs back and keeps the blood pressure from dropping too low between heartbeats. The importance of this elastic tissue becomes especially clear when there is not enough of it—for example, in Marfan syndrome (Insight 5.1)—or when it is stiffened by arteriosclerosis (see chapter 19).

## INSIGHT 5.1    Clinical Application

### Marfan Syndrome—A Connective Tissue Disease

*Marfan*[15] *syndrome* is a hereditary defect in elastin fibers, usually resulting from a mutation in the gene for *fibrillin,* a glycoprotein that forms the structural scaffold for elastin. Clinical signs of Marfan syndrome include hyperextensible joints, hernias of the groin, and vision problems resulting from abnormally elongated eyes and deformed lenses. People with Marfan syndrome typically show unusually tall stature, long limbs, spidery fingers, abnormal spinal curvature, and a protruding "pigeon breast." More serious problems are weakened heart valves and arterial walls. The aorta, where blood pressure is highest, is sometimes enormously dilated close to the heart and may rupture. Marfan syndrome is present in about 1 out of 20,000 live births, and most victims die by their mid-30s. Some authorities speculate that Abraham Lincoln's tall, gangly physique and spindly fingers were signs of Marfan syndrome, which may have ended his life prematurely had he not been assassinated. A number of star athletes have died at a young age of Marfan syndrome, including Olympic volleyball champion Flo Hyman, who died of a ruptured aorta during a game in Japan in 1986, at the age of 31.

[15]Antoine Bernard-Jean Marfan (1858–1942), French physician

| TABLE 5.5 | Dense Connective Tissues |
|---|---|

| **Dense Regular Connective Tissue** | **Dense Irregular Connective Tissue** |
|---|---|

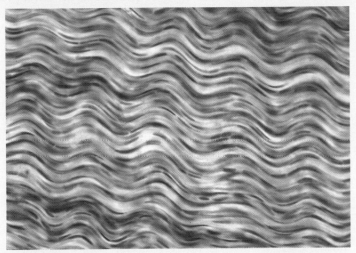

(a)

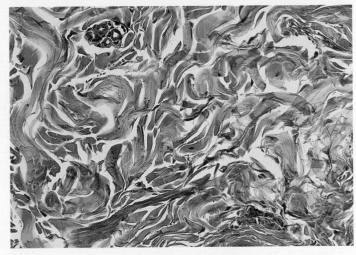

(a)

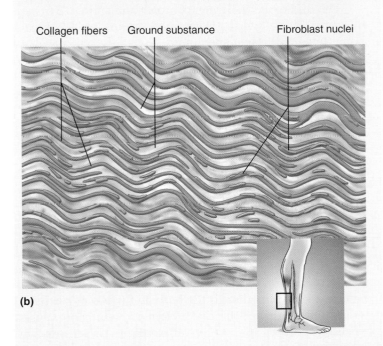

Collagen fibers   Ground substance   Fibroblast nuclei

(b)

**FIGURE 5.17**   Tendon.

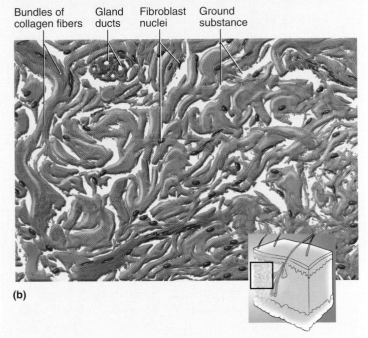

Bundles of collagen fibers   Gland ducts   Fibroblast nuclei   Ground substance

(b)

**FIGURE 5.18**   Dermis of the Skin.

**Microscopic appearance:** Densely packed, parallel, often wavy collagen fibers; slender fibroblast nuclei compressed between collagen bundles; scanty open space (ground substance); scarcity of blood vessels

**Representative locations:** Tendons and ligaments

**Functions:** Ligaments tightly bind bones together; resist stress; tendons attach muscle to bone and transfer muscular tension to bones

**Microscopic appearance:** Densely packed collagen fibers running in random directions; scanty open space (ground substance); few visible cells; scarcity of blood vessels

**Representative locations:** Deeper portion of dermis of skin; capsules around viscera such as liver, kidney, spleen; fibrous sheaths around cartilages and bones

**Functions:** Durable, hard to tear; withstands stresses applied in unpredictable directions

**Dense irregular connective tissue** also has thick bundles of collagen and relatively little room for cells and ground substance, but the collagen bundles run in random directions. This arrangement enables the tissue to resist unpredictable stresses. This tissue constitutes most of the dermis, where it binds the skin to the underlying muscle and connective tissue. It forms a protective capsule around organs such as the kidneys, testes, and spleen and a tough fibrous sheath around the bones, nerves, and most cartilages.

It is sometimes difficult to judge whether a tissue is areolar or dense irregular. In the dermis, for example, these tissues occur side by side, and the transition from one to the other is not at all obvious (see fig. 5.8). A relatively large amount of clear space suggests areolar tissue, and thicker bundles of collagen with relatively little clear space suggests dense irregular tissue.

## CARTILAGE

**Cartilage** (table 5.6) is a supportive connective tissue with a flexible rubbery matrix. It gives shape to the external ear, the tip of the nose, and the larynx (voicebox)—the most easily palpated cartilages in the body. Cells called **chondroblasts**[16] (CON-dro-blasts) secrete the matrix and surround themselves with it until they become trapped in little cavities called **lacunae**[17] (la-CUE-nee). Once enclosed in lacunae, the cells are called **chondrocytes** (CON-drosites). Cartilage is usually free of blood vessels except when transforming into bone; thus nutrition and waste removal depend on solute diffusion through the stiff matrix. Because this is a slow process, chondrocytes have low rates of metabolism and cell division, and injured cartilage heals slowly. The matrix is rich in chondroitin sulfate and contains collagen fibers that range in thickness from invisibly fine to conspicuously coarse. Differences in the fibers provide a basis for classifying cartilage into three types: *hyaline cartilage, elastic cartilage,* and *fibrocartilage.*

**Hyaline**[18] (HY-uh-lin) **cartilage** is named for its clear, glassy microscopic appearance, which stems from the usually invisible fineness of its collagen fibers. **Elastic cartilage** is named for its conspicuous elastic fibers, and **fibrocartilage** for its coarse, readily visible bundles of collagen. Elastic cartilage and most hyaline cartilage are surrounded by a sheath of dense irregular connective tissue called the **perichondrium.**[19] A reserve population of chondroblasts between the perichondrium and cartilage contributes to cartilage growth throughout life. There is no perichondrium around fibrocartilage.

You can feel the texture of hyaline cartilage by palpating the tip of your nose, your "Adam's apple" at the front of the larynx (voicebox), and periodic rings of cartilage around the trachea (windpipe) just below the larynx. Hyaline cartilage is easily seen in many grocery items—it is the "gristle" at the ends of pork ribs, on chicken leg and breast bones, and at the joints of pigs' feet, for example. Elastic cartilage gives shape to the external ear. You can get some idea of its springy resilience by folding your ear down and releasing it.

## BONE

The term *bone* has two meanings: an organ of the body such as the femur and mandible, composed of multiple tissue types, and bone tissue, or **osseous tissue,** which makes up most of the mass of bones. There are two forms of osseous tissue: (1) **Spongy bone** fills the heads of the long bones. Although it is calcified and hard, its delicate slivers and plates give it a spongy appearance. (2) **Compact (dense) bone** is a denser calcified tissue with no spaces visible to the naked eye. It forms the external surfaces of all bones, so spongy bone, when present, is always covered by compact bone.

The differences between compact and spongy bone are described in chapter 7. Here, we examine only compact bone (table 5.7). Most specimens you study will probably be chips of dead, dried bone ground to microscopic thinness. In such preparations, the cells are absent but spaces reveal their former locations. Most compact bone is arranged in cylinders of tissue that surround **central (haversian**[20] or **osteonic) canals,** which run longitudinally through the shafts of long bones such as the femur. Blood vessels and nerves travel through the central canals in life. The bone matrix is deposited in **concentric lamellae,** onionlike layers around each central canal. A central canal and its surrounding lamellae are called an **osteon.** Tiny lacunae between the lamellae are occupied in life by mature bone cells, or **osteocytes.**[21] Delicate canals called **canaliculi** radiate from each lacuna to its neighbors and allow the osteocytes to contact each other. The bone as a whole is covered with a tough fibrous **periosteum** (PERR-ee-OSS-tee-um) similar to the perichondrium of cartilage.

About a third of the dry weight of bone is composed of collagen fibers and chondroitin sulfate; two-thirds consists of minerals (mainly calcium salts) deposited around the collagen fibers.

## BLOOD

**Blood** (table 5.8) is a fluid connective tissue that travels through tubular vessels. Its primary function is to transport cells and dissolved matter from place to place. Blood consists of a ground substance called **plasma** and of cells and cell fragments collectively called **formed elements. Erythrocytes**[22] (eh-RITH-ro-sites), or red blood cells, are

---

[16]*chondro* = cartilage, gristle + *blast* = forming
[17]*lacuna* = lake, cavity
[18]*hyal* = glass
[19]*peri* = around + *chondri* = cartilage

[20]Clopton Havers (1650–1702), English anatomist
[21]*osteo* = bone + *cyte* = cell
[22]*erythro* = red + *cyte* = cell

| TABLE 5.6 | Types of Cartilage |
|---|---|

| Hyaline Cartilage | Elastic Cartilage | Fibrocartilage |
|---|---|---|

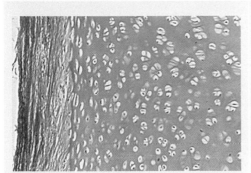

(a)

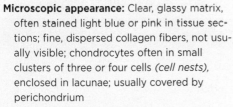

Perichondrium    Chondrocytes    Matrix

(b)

**FIGURE 5.19** Fetal Skeleton.

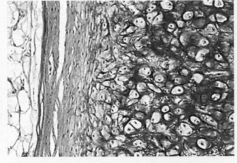

(a)

Elastic
fibers
Perichondrium    Chondrocytes

(b)

**FIGURE 5.20** External Ear.

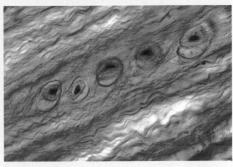

(a)

Chondrocytes    Collagen fibers    Lacuna

(b)

**FIGURE 5.21** Intervertebral Disc.

**Microscopic appearance:** Clear, glassy matrix, often stained light blue or pink in tissue sections; fine, dispersed collagen fibers, not usually visible; chondrocytes often in small clusters of three or four cells (cell nests), enclosed in lacunae; usually covered by perichondrium

**Representative locations:** Forms a thin *articular cartilage,* lacking perichondrium, over the ends of bones at movable joints; a *costal cartilage* attaches the end of a rib to the breastbone; forms supportive rings and plates around trachea and bronchi; forms a boxlike enclosure around the larynx; forms much of the fetal skeleton

**Functions:** Eases joint movements; holds airway open during respiration; moves vocal cords during speech; a precursor of bone in the fetal skeleton and the growth zones of long bones of children

**Microscopic appearance:** Elastic fibers form weblike mesh amid lacunae; always covered by perichondrium

**Representative locations:** External ear; epiglottis

**Functions:** Provides flexible, elastic support

**Microscopic appearance:** Parallel collagen fibers similar to those of tendon; rows of chondrocytes in lacunae between collagen fibers; never has a perichondrium

**Representative locations:** Pubic symphysis (anterior joint between two halves of pelvic girdle); intervertebral discs, which separate bones of vertebral column; menisci, or pads of shock-absorbing cartilage, in knee joint; at points where tendons insert on bones near articular hyaline cartilage

**Functions:** Resists compression and absorbs shock in some joints; often a transitional tissue between dense connective tissue and hyaline cartilage (for example, at some tendon–bone junctions)

## TABLE 5.7 | Bone

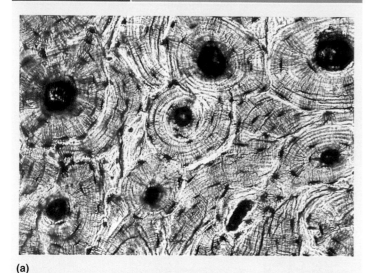

(a)

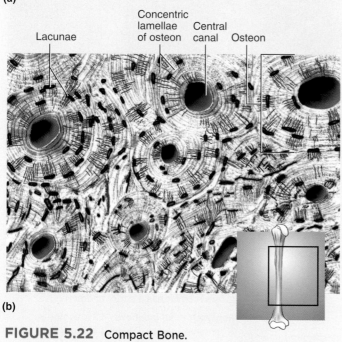

Lacunae · Concentric lamellae of osteon · Central canal · Osteon

(b)

**FIGURE 5.22**  Compact Bone.

**Microscopic appearance (compact bone):** Calcified matrix arranged in concentric lamellae around central canals; osteocytes in lacunae between adjacent lamellae; lacunae interconnected by delicate canaliculi

**Representative locations:** Skeleton

**Functions:** Physical support of body; leverage for muscle action; protective enclosure of viscera; reservoir of calcium and phosphorus

## TABLE 5.8 | Blood

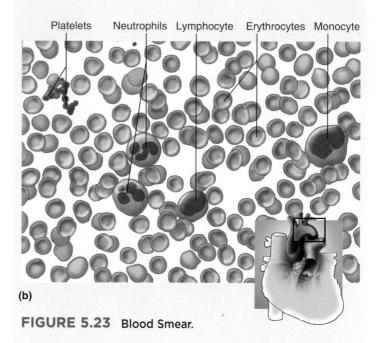

(a)

Platelets · Neutrophils · Lymphocyte · Erythrocytes · Monocyte

(b)

**FIGURE 5.23**  Blood Smear.

**Microscopic appearance:** Erythrocytes appear as pale pink discs with light centers and no nuclei; leukocytes are slightly larger, are much fewer, and have variously shaped nuclei, usually stained violet; platelets are cell fragments with no nuclei, about one-quarter the diameter of erythrocytes

**Representative locations:** Contained in heart and blood vessels

**Functions:** Transports gases, nutrients, wastes, chemical signals, and heat throughout body; provides defensive leukocytes; contains clotting agents to minimize bleeding; platelets secrete growth factors that promote tissue maintenance and repair

the most abundant formed elements. In stained blood films, they look like pink discs with a thin, pale center. They have no nuclei. Erythrocytes transport oxygen and carbon dioxide. **Leukocytes,** or white blood cells, serve various roles in defense against infection and other diseases. They travel from one organ to another in the bloodstream and lymph but spend most of their lives in the connective tissues. Leukocytes are somewhat larger than erythrocytes and have conspicuous nuclei, which usually appear violet in stained preparations. There are five kinds, distinguished partly by variations in nuclear shape: *neutrophils, eosinophils, basophils, lymphocytes,* and *monocytes.* Their individual characteristics are considered in detail in chapter 18. **Platelets** are small cell fragments scattered amid the blood cells. They are involved in clotting and other mechanisms for minimizing blood loss, and in secreting growth factors that promote blood vessel growth and maintenance.

## Before You Go On

*Answer the following questions to test your understanding of the preceding section:*

9. *What features do most or all connective tissues have in common to set this class apart from nervous, muscular, and epithelial tissue?*

10. *List the cell and fiber types found in fibrous connective tissues and state their functional differences.*

11. *What substances account for the gelatinous consistency of connective tissue ground substance?*

12. *What is areolar tissue? How can it be distinguished from any other kind of connective tissue?*

13. *Discuss the difference between dense regular and dense irregular connective tissue as an example of the relationship between form and function.*

14. *Describe some similarities, differences, and functional relationships between hyaline cartilage and bone.*

15. *What are the three basic kinds of formed elements in blood, and what are their respective functions?*

# Nervous and Muscular Tissue— Excitable Tissues

## Objectives

When you have completed this section, you should be able to

- explain what distinguishes excitable tissues from other tissues;
- name the cell types that compose nervous tissue;
- identify the major parts of a nerve cell;
- visually recognize nervous tissue from specimens or photographs;

- name the three kinds of muscular tissue and describe the differences between them; and
- visually identify any type of muscular tissue from specimens or photographs.

Excitability is a characteristic of all living cells, but it is developed to its highest degree in nervous and muscular tissue, which are therefore described as **excitable tissues.** The basis for their excitation is an electrical charge difference (voltage) called the **membrane potential,** which occurs across the plasma membranes of all cells. Nervous and muscular tissues respond quickly to outside stimuli by means of changes in membrane potential. In nerve cells, these changes result in the rapid transmission of signals to other cells. In muscle cells, they result in contraction, or shortening of the cell.

## NERVOUS TISSUE

**Nervous tissue** (table 5.9) consists of **neurons** (NOOR-ons), or nerve cells, and a much greater number of **neuroglia** (noo-ROG-lee-uh), or **glial** (GLEE-ul) **cells,** which protect and assist the neurons. Neurons are specialized to detect stimuli, respond quickly, and transmit coded information rapidly to other cells. Each neuron has a prominent **soma,** or cell body, that houses the nucleus and most other organelles. This is the cell's center of genetic control and protein synthesis. Somas are usually round, ovoid, or stellate in shape. Extending from the soma, there are usually multiple short, branched processes called **dendrites,**[23] which receive signals from other cells and transmit messages to the soma, and a single, much longer **axon,** or **nerve fiber,** which sends outgoing signals to other cells. Some axons are more than a meter long and extend from the brainstem to the foot. Nervous tissue is found in the brain, spinal cord, nerves, and ganglia, which are knotlike swellings in nerves. Local variations in the structure of nervous tissue are described in chapters 12 to 16.

## MUSCULAR TISSUE

**Muscular tissue** consists of elongated cells that are specialized to contract in response to stimulation; thus, its primary job is to exert physical force on other tissues and organs—for example, when a skeletal muscle pulls on a bone, the heart contracts and expels blood, or the bladder contracts and expels urine. Not only do movements of the body and its limbs depend on muscle, but so do such processes as digestion, waste elimination, breathing, speech, and blood circulation. The muscles are also an important source of body heat. The word *muscle* means "little mouse," apparently referring to the appearance of rippling muscles under the skin.

---

[23]*dendr* = tree + *ite* = little

| TABLE 5.9 | Nervous Tissue |
|---|---|

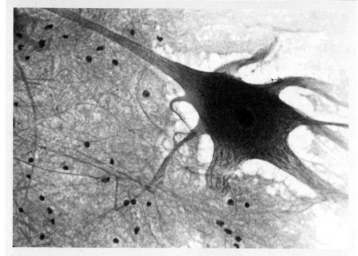

(a)

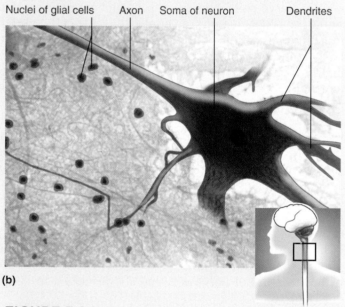

Nuclei of glial cells    Axon    Soma of neuron    Dendrites

(b)

**FIGURE 5.24**  Spinal Cord Smear.

**Microscopic appearance:** Most sections show a few large neurons, usually with rounded or stellate cell bodies (somas) and fibrous processes (axon and dendrites) extending from the somas; neurons are surrounded by a greater number of much smaller glial cells, which lack dendrites and axons

**Representative locations:** Brain, spinal cord, nerves, ganglia

**Function:** Internal communication

There are three histological types of muscle—*skeletal, cardiac,* and *smooth*—which differ in appearance, physiology, and function (table 5.10). **Skeletal muscle** consists of long, cylindrical cells called **muscle fibers.** Most of it is attached to bones, but there are exceptions in the tongue, upper esophagus, some facial muscles, and some **sphinc-**

ter[24] (SFINK-tur) muscles (ringlike or cufflike muscles that open and close body passages). Each cell contains multiple nuclei adjacent to the plasma membrane. Skeletal muscle is described as *striated* and *voluntary.* The first term refers to alternating light and dark bands, or **striations** (stry-AY-shuns), created by the overlapping pattern of cytoplasmic protein filaments that cause muscle contraction. The second term, *voluntary,* refers to the fact that we usually have conscious control over skeletal muscle.

**Cardiac muscle** is limited to the heart. It too is striated, but it differs from skeletal muscle in its other features. Its cells are much shorter, so they are commonly called **myocytes**[25] rather than fibers. The myocytes contain only one nucleus, which is located near the center and often surrounded by a light-staining region of glycogen. Cardiac myocytes are joined end to end by junctions called **intercalated**[26] (in-TUR-kuh-LAY-ted) **discs.** Electrical connections at these junctions enable a wave of excitation to travel rapidly from cell to cell, and mechanical connections keep the myocytes from pulling apart when the heart contracts. The electrical junctions allow all the myocytes of a heart chamber to be stimulated, and contract, almost simultaneously. Intercalated discs appear as dark transverse lines separating each myocyte from the next. They may be only faintly visible, however, unless the tissue has been specially stained for them. Cardiac muscle is considered *involuntary* because it is not usually under conscious control; it contracts even if all nerve connections to it are severed.

**Smooth muscle** lacks striations and is involuntary. Smooth muscle cells are fusiform (thick in the middle and tapered at the ends) and relatively short. They have only one, centrally placed nucleus. Small amounts of smooth muscle are found in the iris of the eye and in the skin, but most of it, called **visceral muscle,** forms layers in the walls of the digestive, respiratory, and urinary tracts, blood vessels, the uterus, and other viscera. In locations such as the esophagus and small intestine, smooth muscle forms adjacent layers, with the cells of one layer encircling the organ and the cells of the other layer running longitudinally. When the circular smooth muscle contracts, it may propel contents such as food through the organ. When the longitudinal layer contracts, it makes the organ shorter and thicker. By regulating the diameter of blood vessels, smooth muscle is very important in controlling blood pressure and flow. Both smooth and skeletal muscle form sphincters that control the emptying of the bladder and rectum.

⌐ **Think About It**

*How does the meaning of the word* fiber *differ in the following uses: muscle fiber, nerve fiber, and connective tissue fiber?*

[24]*sphinc* = squeeze, bind tightly
[25]*myo* = muscle + *cyte* = cell
[26]*inter* = between + *calated* = inserted

| TABLE 5.10 | Muscular Tissue |
|---|---|

| Skeletal Muscle | Cardiac Muscle | Smooth Muscle |
|---|---|---|

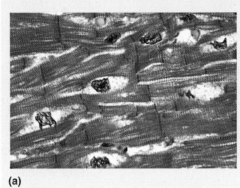

(a)

Nuclei    Striations    Muscle fiber

(b)

**FIGURE 5.25** Skeletal Muscle.

**Microscopic appearance:** Long, cylindrical, unbranched cells (fibers), relatively parallel in longitudinal tissue sections; striations; multiple nuclei per cell, near plasma membrane

**Representative locations:** Skeletal muscles, mostly attached to bones but also in the tongue, esophagus, and voluntary sphincters of the lips, eyelids, urethra, and anus

**Functions:** Body movements, facial expression, posture, breathing, speech, swallowing, control of urination and defecation, and assistance in childbirth; under voluntary control

---

(a)

Intercalated discs   Striations   Glycogen

(b)

**FIGURE 5.26** Wall of Heart.

**Microscopic appearance:** Short cells (myocytes) with notched or slightly branched ends; less parallel appearance in tissue sections; striations; intercalated discs; one nucleus per cell, centrally located and often surrounded by a light zone

**Representative locations:** Heart

**Functions:** Pumping of blood; under involuntary control

---

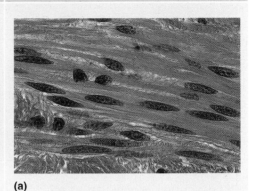

(a)

Nuclei    Muscle cells

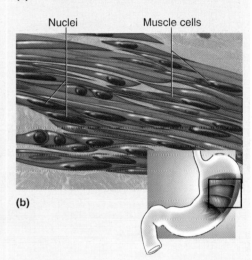

(b)

**FIGURE 5.27** Wall of Stomach.

**Microscopic appearance:** Short fusiform cells overlapping each other; nonstriated; one nucleus per cell, centrally located

**Representative locations:** Usually found as sheets of tissue in walls of viscera; also in iris and associated with hair follicles; involuntary sphincters of urethra and anus

**Functions:** Swallowing; contractions of stomach and intestines; expulsion of feces and urine; labor contractions; control of blood pressure and flow; control of respiratory airflow; control of pupillary diameter; erection of hairs; under involuntary control

## Before You Go On

*Answer the following questions to test your understanding of the preceding section:*

16. *What do nervous and muscular tissue have in common? What is the primary function of each?*

17. *What kinds of cells compose nervous tissue, and how can they be distinguished from each other?*

18. *Name the three kinds of muscular tissue, describe how to distinguish them from each other in microscopic appearance, and state a location and function for each.*

# Intercellular Junctions, Glands, and Membranes

### Objectives

When you have completed this section, you should be able to

- describe the junctions that hold cells and tissues together;
- describe or define different types of glands;
- describe the typical anatomy of a gland;
- name and compare different modes of glandular secretion;
- describe the way tissues are organized to form the body's membranes; and
- name and describe the major types of membranes in the body.

## INTERCELLULAR JUNCTIONS

Most cells, with the exception of blood and metastatic cancer cells, must be anchored to each other and to the matrix if they are to grow and divide normally. The connections between one cell and another are called **intercellular junctions.** These attachments enable the cells to resist stress and communicate with each other. Without them, cardiac muscle cells would pull apart when they contracted, and every swallow of food would scrape away the lining of your esophagus. The principal types of intercellular junctions are shown in figure 5.28.

## Tight Junctions

A **tight junction** completely encircles an epithelial cell near its apex and joins it tightly to the neighboring cells, like the plastic harness on a six-pack of soda cans. Proteins in the membranes of two adjacent cells form a zipperlike pattern of complementary grooves and ridges.

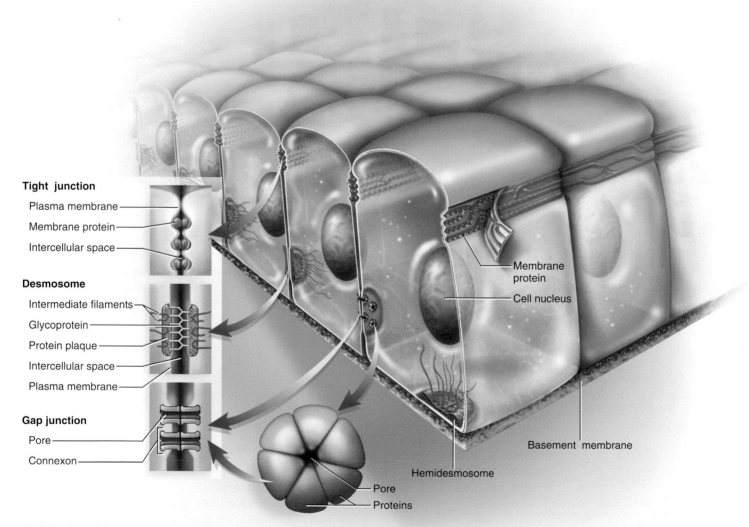

**Tight junction**
- Plasma membrane
- Membrane protein
- Intercellular space

**Desmosome**
- Intermediate filaments
- Glycoprotein
- Protein plaque
- Intercellular space
- Plasma membrane

**Gap junction**
- Pore
- Connexon

- Membrane protein
- Cell nucleus
- Basement membrane
- Hemidesmosome
- Pore
- Proteins

**FIGURE 5.28** Types of Intercellular Junctions.

▶ *Which of these junctions allows material to pass from one cell directly into the next?*

This seals off the intercellular space and makes it difficult for some substances to pass between the cells. In the stomach and intestines, tight junctions prevent digestive juices from seeping between epithelial cells and digesting the underlying connective tissue. They also help to prevent intestinal bacteria from invading the tissues, and they ensure that most digested nutrients pass *through* the epithelial cells and not *between* them.

## Desmosomes

If a tight junction is like a zipper, a **desmosome**[27] (DEZ-mo-some) is more like the snap on a pair of jeans. It is a patch that holds cells together and enables a tissue to resist mechanical stress, but does not totally encircle a cell. Desmosomes are common in the epidermis, cardiac muscle, and cervix of the uterus. The neighboring cells are separated by a small gap, which is spanned by a fine mesh of glycoprotein filaments. These filaments terminate in a thickened protein plaque at the surface of each cell. On the cytoplasmic side of each plaque, intermediate filaments from the cytoskeleton approach and penetrate the plaque, turn like a J, and return a short distance back into the cytoplasm. Each cell contributes half of the complete desmosome. The basal cells of epithelial tissue have *hemidesmosomes*—half-desmosomes that anchor them to the underlying basement membrane.

> **Think About It**
>
> *Why would desmosomes not be suitable as the only intercellular junctions in the epithelium of the stomach?*

## Gap (Communicating) Junctions

A **gap junction** is formed by a ringlike *connexon,* which consists of six transmembrane proteins surrounding a water-filled pore. Ions, glucose, amino acids, and other small solutes can pass directly from the cytoplasm of one cell into the next through these pores. In the embryo, nutrients pass from cell to cell through gap junctions until the circulatory system forms and takes over the role of nutrient distribution. Gap junctions are found in the intercalated discs of cardiac muscle and between the cells of most smooth muscle. The flow of ions through these junctions allows electrical excitation to pass directly from cell to cell so that the cells contract in near-unison. Gap junctions are absent from skeletal muscle.

## GLANDS

A **gland** is a cell or organ that secretes substances for use elsewhere in the body or releases them for elimination from the body. The gland product may be something synthesized by the gland cells (such as digestive enzymes) or

### Pemphigus Vulgaris—A Result of Defective Desmosomes

The immune system normally produces defensive *antibodies* that selectively attack foreign substances and leave the normal tissues of our bodies alone. But in a family of disorders called *autoimmune diseases,* antibodies fail to distinguish our own cells and tissues from foreign ones. Such misguided antibodies, called *autoantibodies,* thus launch destructive attacks on our own bodies. (Autoimmune diseases are discussed in more detail on p. 842.) One such disease is *pemphigus vulgaris*[28] (PEM-fih-gus vul-GAIR-iss), a disorder in which autoantibodies attack the proteins of the desmosomes in the skin and mucous membranes. This breaks down the attachments between epithelial cells and causes widespread blistering of the skin and oral mucosa, loss of tissue fluid, and sometimes death. The condition can be controlled with drugs that suppress the immune system, but such drugs reduce the patient's immune defenses against other diseases.

something removed from the tissues and modified by the gland (such as urine). Glands are composed predominantly of epithelial tissue.

## Endocrine and Exocrine Glands

Glands are broadly classified as endocrine or exocrine. They originate as invaginations of a surface epithelium. In **exocrine**[29] (EC-so-crin) **glands,** they usually maintain their contact with the surface by way of a **duct,** an epithelial tube that conveys their secretion to the surface. The secretion may be released to the body surface, as in the case of sweat, mammary, and tear glands. More often, however, it is released into the cavity (lumen) of another organ such as the mouth or intestine; this is the case with salivary glands, the liver, and the pancreas. **Endocrine**[30] (EN-doe-crin) **glands** lose their contact with the surface and have no ducts. They do, however, have a high density of blood capillaries and secrete their products directly into the blood. The secretions of endocrine glands, called *hormones,* function as chemical messengers to stimulate cells elsewhere in the body. Examples include the pituitary, thyroid, and adrenal glands. Endocrine glands are the subject of chapter 17 and are not considered further here.

The exocrine–endocrine distinction is not always clear. The liver is an exocrine gland that secretes one of its products, bile, through a system of ducts, but secretes hormones, albumin, and other products directly into the bloodstream. Several glands, such as the pancreas, testis, ovary, and kidney, have both exocrine and endocrine components. Nearly all of the viscera have at least some cells that secrete hormones, even though most of these organs are not usually thought of as glands (for example, the brain and heart).

**Unicellular glands** are secretory cells found in an epithelium that is predominantly nonsecretory. They can be endocrine or exocrine. For example, the respiratory tract, which is lined mainly by ciliated cells, also has a liberal scattering of nonciliated, mucus-secreting goblet cells, which are exocrine (see figs. 5.6 and 5.7). The stomach and small intestine have scattered endocrine cells, which secrete hormones that regulate digestion.

## Exocrine Gland Structure

Figure 5.29 shows a generalized multicellular exocrine gland—a structural arrangement found in such organs as the mammary gland, pancreas, and salivary glands. Most glands are enclosed in a fibrous **capsule.** The capsule often gives off extensions called **septa,** or **trabeculae** (trah-BEC-you-lee), that divide the interior of the gland into compartments called **lobes,** which are visible to the naked eye. Finer connective tissue septa may further subdivide each lobe into microscopic **lobules** (LOB-yools). Blood vessels, nerves, and the gland's own ducts generally travel through these septa. The connective tissue framework of the gland, called its **stroma,** supports and

organizes the glandular tissue. The cells that perform the tasks of synthesis and secretion are collectively called the **parenchyma** (pa-REN-kih-muh). This is typically simple cuboidal or simple columnar epithelium.

Exocrine glands are classified as **simple** if they have a single unbranched duct and **compound** if they have a branched duct. If the duct and secretory portion are of uniform diameter, the gland is called **tubular.** If the secretory cells form a dilated sac, the gland is called **acinar** and the sac is an **acinus**[31] (ASS-ih-nus), or **alveolus**[32] (AL-vee-OH-lus). A gland with secretory cells in both the tubular and acinar portions is called a **tubuloacinar gland** (fig. 5.30).

## Types of Secretions

Glands are classified not only by their structure but also by the nature of their secretions. **Serous** (SEER-us) **glands** produce relatively thin, watery fluids such as perspiration, milk, tears, and digestive juices. **Mucous glands,** found in the tongue and roof of the mouth among other places, secrete a glycoprotein called *mucin* (MEW-sin). After it is secreted, mucin absorbs water and forms the sticky product *mucus.* Goblet cells are unicellular mucous glands. (Note that *mucus,* the secretion, is spelled differently from *mucous,* the adjective form of the word.) **Mixed glands,** such as the two pairs of salivary glands in the chin, contain both serous and mucous cells and produce a mixture of the two types of secretions. **Cytogenic**[33] **glands** release whole cells. The only examples of these are the testes and ovaries, which produce sperm and egg cells.

## Methods of Secretion

Glands are classified as merocrine or holocrine depending on how they produce their secretions. **Merocrine**[34] (MERR-oh-crin) **glands,** also called **eccrine**[35] (EC-rin) **glands,** have vesicles that release their secretion by exocytosis, as described in chapter 3 (fig. 5.31a). These include the tear glands, pancreas, gastric glands, and many others. In **holocrine**[36] **glands,** cells accumulate a product and then the entire cell disintegrates, so the secretion is a mixture of cell fragments and the substance the cell had synthesized prior to its disintegration (fig. 5.31b). The oil-producing glands of the scalp are an example. Holocrine secretions tend to be thicker than merocrine secretions.

Some glands, such as the axillary (armpit) sweat glands and mammary glands, are named **apocrine**[37] **glands** from a former belief that the secretion was composed of

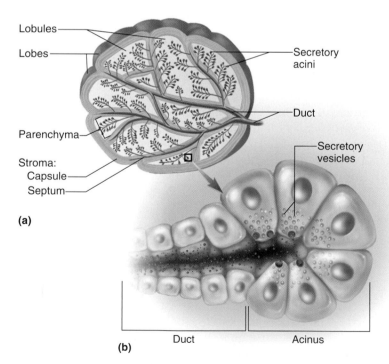

Lobules
Lobes
Parenchyma
Stroma:
Capsule
Septum
**(a)**

Secretory acini
Duct
Secretory vesicles

**(b)**    Duct    Acinus

**FIGURE 5.29    General Structure of an Exocrine Gland.**
(a) The gland duct branches repeatedly, following the connective tissue septa, until its finest divisions end on saccular acini of secretory cells. (b) Detail of an acinus and the beginning of a duct.

---

[31]*acinus* = berry
[32]*alveol* = cavity, pit
[33]*cyto* = cell + *genic* = producing
[34]*mero* = part + *crin* = to separate, secrete
[35]*ec* = *ex* = out + *crin* = to separate, secrete
[36]*holo* = whole, entire + *crin* = to separate, secrete
[37]*apo* = from, off, away + *crin* = to separate, secrete

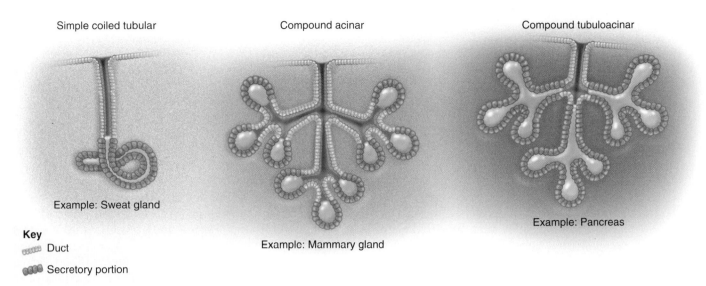

Simple coiled tubular

Example: Sweat gland

Compound acinar

Example: Mammary gland

Compound tubuloacinar

Example: Pancreas

**Key**
🔲 Duct
🔲 Secretory portion

**FIGURE 5.30 Some Types of Exocrine Glands.** Glands are simple if their ducts do not branch and compound if they do; they are tubular if they have a uniform diameter, acinar if their secretory cells are limited to saccular acini, and tubuloacinar if they have secretory cells in both the acinar and tubular regions.

▶ *Predict and sketch the appearance of a simple acinar gland.*

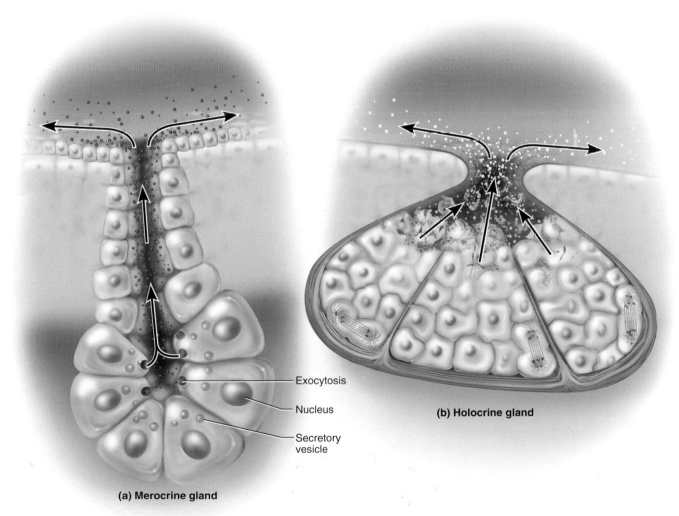

Exocytosis

Nucleus

Secretory vesicle

**(a) Merocrine gland**

**(b) Holocrine gland**

**FIGURE 5.31 Modes of Exocrine Secretion.** (a) A merocrine gland, which secretes its product by means of exocytosis at the apical surfaces of the secretory cells. (b) A holocrine gland, whose secretion is composed of disintegrated secretory cells.

▶ *Which of these glands would require a higher rate of mitosis in its parenchymal cells?*

bits of apical cytoplasm that broke away from the cell surface. Closer study showed this to be untrue; these glands are primarily merocrine in their mode of secretion. These glands are nevertheless different from other merocrine glands in function and histological appearance, and they are still referred to as apocrine glands even though their mode of secretion is not unique.

# MEMBRANES

In atlas A, the major cavities of the body were described, as well as some of the membranes that line them and cover their viscera. We now consider some histological aspects of the major body membranes.

The largest membrane of the body is the **cutaneous** (cue-TAY-nee-us) **membrane**—or more simply, the skin (detailed in chapter 6). It consists of a stratified squamous epithelium (epidermis) resting on a layer of connective tissue (dermis). Unlike the other membranes to be considered, it is relatively dry. It resists dehydration of the body and provides an inhospitable environment for the growth of infectious organisms.

The two principal kinds of internal membranes are mucous and serous membranes. A **mucous membrane** (mucosa) (fig. 5.32) lines passageways that open to the exterior environment: the digestive, respiratory, urinary, and reproductive tracts. A mucosa consists of two to three layers: (1) an epithelium, (2) an areolar connective tissue layer called the **lamina propria**[38] (LAM-ih-nuh PRO-pree-uh),

---

[38]*lamina* = layer + *propria* = of one's own

and sometimes (3) a layer of smooth muscle called the **muscularis** (MUSK-you-LAIR-iss) **mucosae.** Mucous membranes have absorptive, secretory, and protective functions. They are often covered with mucus secreted by goblet cells, multicellular mucous glands, or both. The mucus traps bacteria and foreign particles, which keeps them from invading the tissues and aids in their removal from the body. The epithelium of a mucous membrane may also include absorptive, ciliated, and other types of cells.

A **serous membrane** (serosa) is composed of a simple squamous epithelium resting on a thin layer of areolar connective tissue. Serous membranes produce watery **serous fluid,** which arises from the blood and derives its name from the fact that it is similar to blood serum in composition. Serous membranes line the insides of some body cavities and form a smooth outer surface on some of the viscera, such as the digestive tract. The pleurae, pericardium, and peritoneum described in atlas A are serous membranes.

The circulatory system is lined with a simple squamous epithelium called **endothelium,** derived from mesoderm. The endothelium rests on a thin layer of areolar tissue, which often rests in turn on an elastic sheet. Collectively, these tissues make up a membrane called the *tunica interna* of the blood vessels and *endocardium* of the heart. The simple squamous epithelium that lines the pleural, pericardial, and peritoneal cavities is called **mesothelium.**

Some joints of the skeletal system are lined by fibrous **synovial** (sih-NO-vee-ul) **membranes,** made only of connective tissue. These membranes span the gap from one bone to the next and secrete slippery *synovial fluid* into the joint.

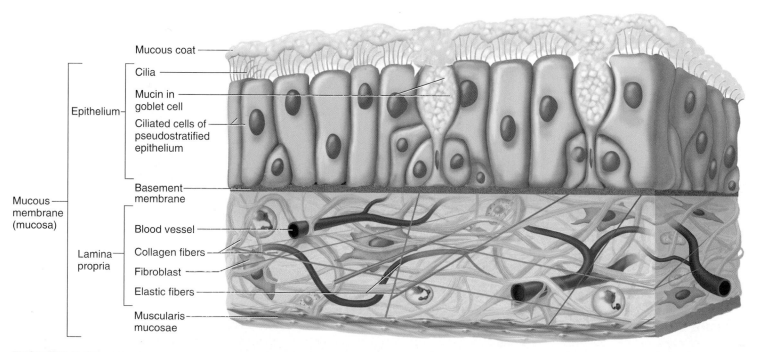

**FIGURE 5.32**  Histology of a Mucous Membrane.

## Before You Go On

*Answer the following questions to test your understanding of the preceding section:*

19. Compare the structure of tight junctions and gap junctions. Relate their structural differences to their functional differences.

20. Distinguish between a simple gland and a compound gland, and give an example of each. Distinguish between a tubular gland and an acinar gland, and give an example of each.

21. Contrast the merocrine and holocrine methods of secretion, and name a gland product produced by each method.

22. Describe the differences between a mucous and a serous membrane.

23. Name the layers of a mucous membrane, and state which of the four primary tissue classes composes each layer.

# Tissue Growth, Development, Repair, and Death

### Objectives

When you have completed this section, you should be able to

- name and describe the modes of tissue growth;
- define *adult* and *embryonic stem cells* and their varied degrees of developmental plasticity;
- name and describe the ways that a tissue can change from one type to another;
- name and describe the modes and causes of tissue shrinkage and death; and
- name and describe the ways the body repairs damaged tissues.

## TISSUE GROWTH

Tissues grow either because their cells increase in number or because the existing cells grow larger. Most embryonic and childhood growth occurs by **hyperplasia**[39] (HY-pur-PLAY-zhuh), tissue growth through cell multiplication. Exercised muscles grow, however, through **hypertrophy**[40] (hy-PUR-truh-fee), the enlargement of preexisting cells. **Neoplasia**[41] (NEE-oh-PLAY-zhuh) is the development of a tumor (neoplasm)—whether benign or malignant—composed of abnormal, nonfunctional tissue.

## CHANGES IN TISSUE TYPE

You have studied the form and function of more than two dozen discrete types of human tissue in this chapter. You should not leave this subject, however, with the impression that once these tissue types are established in the adult, they never change. Tissues are, in fact, capable of changing from one type to another within certain limits. Most obviously, unspecialized tissues of the embryo develop into more diverse and specialized types of mature tissue—mesenchyme to muscle, for example. This development of a more specialized form and function is called **differentiation.**

Epithelia often exhibit **metaplasia,**[42] a change from one type of mature tissue to another. For example, the vagina of a young girl is lined with a simple cuboidal epithelium. At puberty, it changes to a stratified squamous epithelium, better adapted to the future demands of intercourse and childbirth. The nasal cavity is lined with ciliated pseudostratified columnar epithelium. However, if we block one nostril and breathe through the other one for several days, the epithelium in the unblocked passage changes to stratified squamous. In smokers, the pseudostratified columnar epithelium of the bronchi may transform into a stratified squamous epithelium.

> ### Think About It
>
> *What functions of a pseudostratified columnar epithelium could not be served by a stratified squamous epithelium? In light of this, what might be some consequences of bronchial metaplasia in heavy smokers?*

## STEM CELLS

The growth and differentiation of tissues depends upon a supply of reserve **stem cells.** These are undifferentiated cells that are not yet performing any specialized function, but that have the potential to differentiate into one or more types of mature functional cells, such as liver, brain, cartilage, or skin cells. Such cells have various degrees of **developmental plasticity,** or diversity of mature cell types to which they can give rise.

There are two types of stem cells: *embryonic* and *adult.* **Embryonic stem cells** compose the early human embryo—for example, the cells in the photograph on page 1. In the early stages of development, these are called **totipotent** stem cells, because they have the potential to develop into any type of fully differentiated human cell—not only cells of the later embryonic, fetal, or adult body, but also cells of the temporary structures of pregnancy, such as the placenta and amniotic sac. Totipotency is unlimited developmental plasticity. About 4 days after fertilization, the developing embryo enters the *blastocyst* stage. The blastocyst is a hollow ball with an *outer cell*

---

[39]*hyper* = excessive + *plas* = growth
[40]*hyper* = excessive + *trophy* = nourishment
[41]*neo* = new + *plas* = form, growth

[42]*meta* = change + *plas* = form, growth

*mass* that helps form the placenta and other accessory organs of pregnancy, and an *inner cell mass* that becomes the embryo itself (see fig. 29.4). Cells of the inner cell mass are called **pluripotent** stem cells; they can still develop into any cell type of the embryo, but not into the accessory organs of pregnancy. Thus their developmental plasticity is already somewhat limited.

**Adult stem cells** occur in small numbers in mature organs and tissues throughout a person's life. Typically an adult stem cell divides mitotically; one of its daughter cells remains a stem cell and the other one differentiates into a mature specialized cell. The latter cell may replace another that has grown old and died, contribute to the development of growing organs (as in a child), or help to repair damaged tissue. Some adult stem cells are **multipotent**—able to develop into two or more different cell lines, but not just any type of body cell. Certain multipotent bone marrow stem cells, for example, can give rise to red blood cells, five kinds of white blood cells, and platelet-producing cells. **Unipotent** stem cells have the most limited plasticity, as they can produce only one mature cell type. Examples include the cells that give rise to sperm, eggs, keratinocytes (the majority cell type of the epidermis), and olfactory cells (the sensory cells of smell).

Both embryonic and adult stem cells have enormous potential for therapy, but stem cell research has been embroiled in great political controversy in the past several years. Insight 5.4 addresses their clinical potential and the ethical and political issues surrounding stem cell research.

## TISSUE REPAIR

Damaged tissues can be repaired in two ways: *regeneration* or *fibrosis.* **Regeneration** is the replacement of dead or damaged cells by the same type of cells as before. Regeneration restores normal function to the organ. Most skin injuries (cuts, scrapes, and minor burns) heal by regeneration. The liver also regenerates remarkably well. **Fibrosis** is the replacement of damaged tissue with scar tissue, composed mainly of collagen produced by fibroblasts. Scar tissue helps to hold an organ together, but it does not restore normal function. Examples include the healing of severe cuts and burns, the healing of muscle injuries, and scarring of the lungs in tuberculosis.

Figure 5.33 illustrates the following stages in the healing of a cut in the skin, where both regeneration and fibrosis are involved:

① Severed blood vessels bleed into the cut. Mast cells and cells damaged by the cut release histamine, which dilates blood vessels, increases blood flow to the area, and makes blood capillaries more permeable. Blood plasma seeps into the wound, carrying antibodies, clotting proteins, and blood cells.

② A blood clot forms in the tissue, loosely knitting the edges of the cut together and interfering with the spread of pathogens from the site of injury into healthy tissues. The surface of the blood clot dries and hardens in the air, forming a scab that temporarily seals the wound and blocks infection. Beneath the scab, macrophages begin to phagocytize and digest tissue debris.

③ New blood capillaries sprout from nearby vessels and grow into the wound. The deeper portions of the clot become infiltrated by capillaries and fibroblasts and transform into a soft mass called **granulation tissue.** Macrophages remove the blood clot while fibroblasts deposit new collagen to replace it. This *fibroblastic (reconstructive) phase* of repair begins 3 to 4 days after the injury and lasts up to 2 weeks.

④ Surface epithelial cells around the wound multiply and migrate into the wounded area, beneath the scab. The scab loosens and eventually falls off, and the epithelium grows thicker. Thus, the epithelium *regenerates* while the underlying connective tissue undergoes *fibrosis,* or scarring. Capillaries withdraw from the area as fibrosis progresses. The scar tissue may or may not show through the epithelium, depending on the severity of the wound. The wound may exhibit a depressed area at first, but this is often filled in by continued fibrosis and remodeling from below, until the scar becomes unnoticeable. This *remodeling (maturation) phase* of tissue repair begins several weeks after injury and may last as long as 2 years.

## TISSUE SHRINKAGE AND DEATH

**Atrophy**[43] (AT-roh-fee) is the shrinkage of a tissue through a loss in cell size or number. It results from both normal aging *(senile atrophy)* and lack of use of an organ *(disuse atrophy).* Muscles that are not exercised exhibit disuse atrophy as their cells become smaller. This was a serious problem for the first astronauts who participated in prolonged microgravity space flights. Upon return to normal gravity, they were sometimes too weak from muscular atrophy to walk. Space stations and shuttles now include exercise equipment to maintain the crews' muscular condition. Disuse atrophy also occurs when a limb is immobilized in a cast or by paralysis.

**Necrosis**[44] (neh-CRO-sis) is the premature, pathological death of tissue due to trauma, toxins, infection, and so forth. **Gangrene** is tissue necrosis resulting from an insufficient blood supply. *Gas gangrene* is necrosis of a wound resulting from infection with certain bacteria. **Infarction** is the sudden death of tissue, such as heart muscle

---

[43]*a* = without + *trophy* = nourishment
[44]*necr* = death + *osis* = process

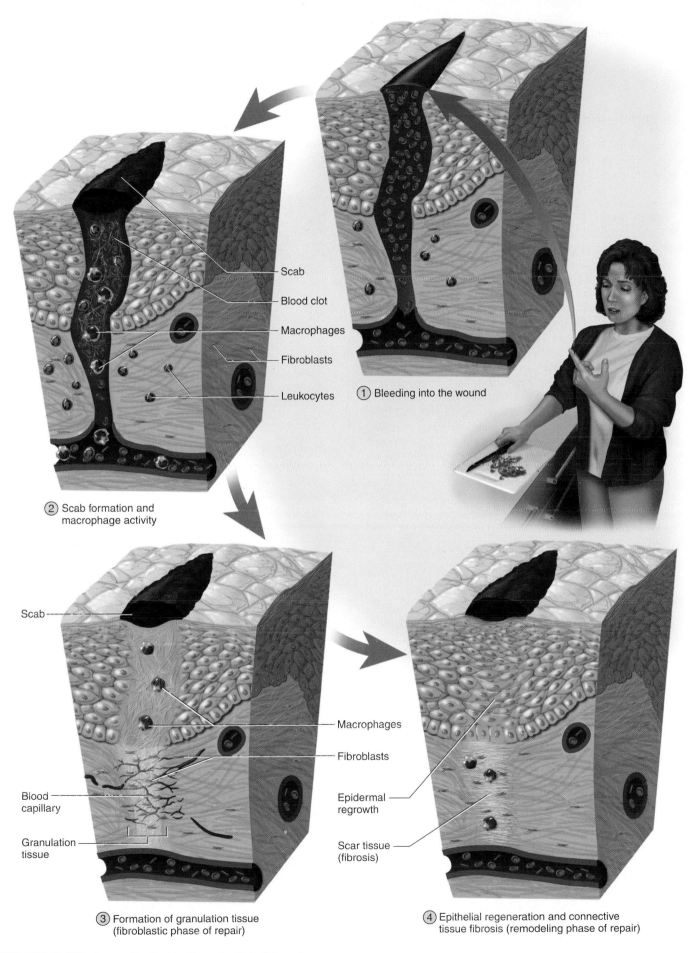

Scab

Blood clot

Macrophages

Fibroblasts

Leukocytes

① Bleeding into the wound

② Scab formation and macrophage activity

Scab

Blood capillary

Granulation tissue

③ Formation of granulation tissue (fibroblastic phase of repair)

Macrophages

Fibroblasts

Epidermal regrowth

Scar tissue (fibrosis)

④ Epithelial regeneration and connective tissue fibrosis (remodeling phase of repair)

**FIGURE 5.33** Stages in the Healing of a Skin Wound.

## INSIGHT 5.3    Clinical Application

### Tissue Engineering

Tissue repair is not only a natural process but also a lively area of research in biotechnology. **Tissue engineering** is the artificial production of tissues and organs in the laboratory for implantation in the human body. The process commonly begins with building a scaffold (supportive framework) of collagen or biodegradable polyester, sometimes in the shape of a desired organ such as a blood vessel or an ear. The scaffold is seeded with human cells and put in a "bioreactor" to grow. The bioreactor supplies nutrients, oxygen, and growth factors. It may be an artificial chamber, or the body of a human patient or laboratory animal. When a lab-grown tissue reaches a certain point, it is implanted into the patient.

Tissue-engineered skin grafts are already on the market (see Insight 6.5). Scientists are not yet close to anything as complex as a lab-grown heart, but some are working on components such as valves, coronary arteries, patches of cardiac tissue, and whole heart chambers. Others have grown liver, bone, ureter, tendon, intestinal, and breast tissue in the laboratory. Charles Vacanti of the University of Massachusetts and Linda Griffith-Cima of the Massachusetts Institute of Technology have grown a "human" outer ear on the back of a mouse (fig. 5.34). They seeded a polymer scaffold with human cartilage cells and grew it in an immunodeficient mouse unable to reject the human tissue. Vacanti and Griffith-Cima see potential in growing ears and noses for cosmetic treatment of children with birth defects or who have suffered disfiguring injuries from playground fights, accident, or animal bites.

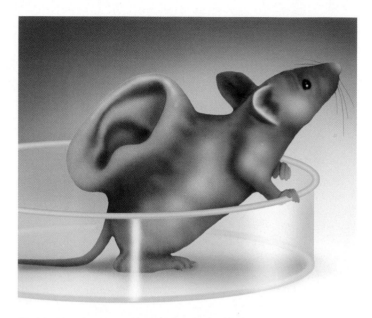

**FIGURE 5.34    Tissue Engineering.** Scientists have grown an external ear from human tissue on the back of an immunodeficient mouse. The ear can be removed without killing the mouse. In the future, such artificial organs might be used to improve the facial appearance of patients with missing parts.

---

*(myocardial infarction),* which occurs when its blood supply is cut off. A **decubitus ulcer** (bed sore) is tissue necrosis that occurs when immobilized persons, such as those confined to a hospital bed or wheelchair, are unable to move, and continual pressure on the skin cuts off blood flow to an area.

Cells dying by necrosis usually swell, exhibit *blebbing* (bubbling) of their plasma membranes, and then rupture. The cell contents released into the tissues trigger an inflammatory response in which macrophages phagocytize the cellular debris.

**Apoptosis**[45] (AP-oh-TOE-sis), or **programmed cell death,** is the normal death of cells that have completed their function and best serve the body by dying and getting out of the way. Cells undergoing apoptosis shrink and are quickly phagocytized by macrophages and other cells. The cell contents never escape, so there is no inflammatory response. Although billions of cells die every hour by apoptosis, they are engulfed so quickly that they are almost never seen except within macrophages. For this reason, apoptosis was not discovered until recently.

Apparently every cell has a built-in "suicide program" that enables the body to dispose of it when necessary. In some cases, an extracellular suicide signal binds to a receptor protein in the plasma membrane called *Fas.* Fas then activates intracellular enzymes that destroy the cell, including an *endonuclease* that chops up its DNA and a *protease* that destroys cellular proteins. In other cases, cells seem to undergo apoptosis automatically if they stop receiving growth factors from other cells. For example, in embryonic development we produce about twice as many neurons as we need. Those that make connections with target cells survive, while the excess neurons die for lack of *nerve growth factor.* Apoptosis also "dissolves" the webbing between the fingers and toes during embryonic development; it frees the earlobe from the side of the head in people with the genotype for detached earlobes (see chapter 4); and it causes shrinkage of the uterus after pregnancy and of the breasts after lactation ceases.

---

[45]*apo* = away + *ptosis* = falling

## The Stem Cell Controversy

Stem cell research has been one of the most politically controversial areas of biological science at the dawn of this century. At least 18 countries have recently debated or enacted laws to regulate it. Politicians, scientists, bioethicists, philosophers, and theologians have joined in the debate; legions of lay citizens have voiced their opinions in newspaper editorial pages; and stem cells have been a contentious issue in United States presidential politics. In 2001, President George W. Bush signed into law an act that prohibited federal funding for research on any embryonic stem (ES) cell lines created after that date, on the grounds that he regarded the harvesting of the cells to be the taking of a human life. Although the administration claimed that this made 64 preexisting ES cell lines available to investigators, most experts in the field said that only 5 or 6 of them were suitable for research. Senator John Kerry, campaigning for president in 2004, pledged to restore funding to this area of research if elected. He was supported by a number of well-known public figures. Actors Christopher Reeve and Michael J. Fox expressed hopes that stem cell research could one day help people like them who suffered, respectively, from spinal cord injuries and Parkinson disease. Nancy and Ron Reagan, the widow and son of the late President Reagan, also expressed hope that federal support for embryonic stem cell research would be permitted to move forward, since it might prove effective in treating Alzheimer disease—the neurodegenerative disorder that took President Reagan's life. President Bush won reelection, and the ban on federal funding for ES cell research remained in effect. Even before the presidential election was decided, however, California voters approved a bill to provide state funds for ES cell research, amounting to $300 million per year for 10 years, far exceeding even Senator Kerry's proposition.

Not surprisingly, biologists see stem cells as a possible treatment for diseases that result from the loss of functional cells. Skin and bone marrow stem cells have been used in therapy for many years. Scientists hope that with a little coaxing, stem cells might replace cardiac muscle damaged by heart attack; restore function to an injured spinal cord; cure parkinsonism by replacing lost brain cells; or cure diabetes mellitus by replacing lost insulin-secreting cells. But *adult stem (AS) cells* have limited developmental potential and probably cannot make all the cell types needed to treat a broad range of degenerative diseases. In addition, they are present in very small numbers and are difficult to harvest and culture in the quantities needed for therapy.

ES cells, however, may hold greater potential. New laboratory methods have made them easier to culture than AS cells and have greatly accelerated stem cell research in recent years. In animal studies, ES cells have already proven effective for treating degenerative disease. In rats, for example, neurons have been produced from ES cells, implanted into the animal, and shown to reverse the signs of Parkinson disease.

The road to therapy with ES cells remains full of technical, ethical, and legal speed bumps. Will ES cells be rejected by the recipient's immune system? Can the ES cells or the growth media in which they are cultured introduce viruses or other pathogens into the recipient? How can the ES cells be made to lodge and grow in the right place in the patient's body? Could they grow into tumors instead of healthy tissue? Can ES cell therapy ever be economical enough to be affordable to any but the very rich? Scientists can scarcely begin to tackle these problems, however, unless and until a bioethical question is resolved: Can we balance the benefits of stem cell therapy against the destruction of early human embryos from which the ES cells are harvested?

Where do these embryos come from? Most are donated by couples using *in vitro fertilization (IVF)* to conceive a child. IVF entails collecting numerous eggs from the prospective mother, fertilizing them in glassware with the father's sperm, letting them develop into embryos (technically, pre-embryos) of about 8 to 16 cells, and then transplanting *some* of these into the mother's uterus (see Insight 29.4). To overcome the low odds of success, excess embryos are always produced and some are always left over. The excess embryos are often destroyed, but many couples choose instead to donate them for research that may ultimately benefit other patients. It would seem sensible to use the embryos for beneficial purposes rather than to simply destroy and discard them. Opponents of stem cell research argue, however, that potential medical benefits cannot justify the destruction of a human embryo. Understandably, this has aroused an intense debate that is likely to restrain stem cell research for some time to come.

## Before You Go On

*Answer the following questions to test your understanding of the preceding section:*

24. Distinguish between differentiation and metaplasia.

25. Tissues can grow through an increase in cell size or cell number. What are the respective terms for these two kinds of growth?

26. Distinguish between atrophy, necrosis, and apoptosis, and describe a circumstance under which each of these forms of tissue loss may occur.

27. Distinguish between regeneration and fibrosis. Which process restores normal cellular function? What good is the other process if it does not restore function?

## CHAPTER REVIEW

# Review of Key Concepts

### The Study of Tissues (p. 154)

1. *Histology (microscopic anatomy)* is the study of tissues.

2. The body is composed of four primary tissues: epithelial, connective, nervous, and muscular tissue.

3. Tissues are composed of *cells* and *matrix (extracellular material)*. The matrix is composed of *fibers* and *ground substance*.

4. Mature tissues develop from three *primary germ layers* of the embryo: *ectoderm, mesoderm,* and *endoderm.*

5. Most tissues are studied as thin slices called *histological sections* colored with *stains* to show detail. Histological sections of elongated structures can be *longitudinal, cross,* or *oblique* sections.

### Epithelial Tissue (p. 156)

1. *Epithelia* are sheets of cells that cover organ surfaces and form glands.

2. Epithelia are composed of one or more layers of closely adhering cells, and lack blood vessels.

3. Epithelia are connected to the underlying connective tissue by a thin *basement membrane.*

4. In a *simple epithelium,* all cells contact the basement membrane. The four kinds of simple epithelium are *simple squamous* (with flat cells), *simple cuboidal* (with cubical to round cells), *simple columnar* (with tall narrow cells), and *pseudostratified columnar* (in which there are basal cells that do not reach the free surface, creating an appearance of stratification) (table 5.2).

5. In a *stratified epithelium,* the cells are multilayered and some rest on top of others, without touching the basement membrane. The four types of stratified epithelium are *stratified squamous, stratified cuboidal, stratified columnar,* and *transitional* (table 5.3).

6. Stratified squamous epithelium has two forms: *keratinized,* in which the surface cells are dead and packed with keratin, and *nonkeratinized,* in which the surface cells are living. The former constitutes the epidermis and the latter is found in internal passages such as the esophagus.

### Connective Tissue (p. 162)

1. *Connective tissue* consists mostly of fibers and ground substance, with widely separated cells.

2. Connective tissue binds, supports, and protects organs, and plays diverse roles in immunity, movement, transport, energy storage, and other processes.

3. *Fibrous connective tissue* has especially conspicuous fibers, which are of three kinds: *collagenous, reticular,* and *elastic.*

4. The cells of fibrous connective tissue include *fibroblasts, macrophages, leukocytes, plasma cells, mast cells,* and *adipocytes.*

5. The ground substance of fibrous connective tissue usually has a gelatinous consistency due to glycosaminoglycans, proteoglycans, and adhesive glycoproteins.

6. Fibrous connective tissue includes *areolar, reticular, adipose, dense irregular,* and *dense regular* types (tables 5.4 and 5.5).

7. *Cartilage* is a connective tissue with a rubbery matrix. Its principal cells are *chondrocytes,* housed in cavities called *lacunae.* The three types of cartilage are *hyaline cartilage, elastic cartilage,* and *fibrocartilage* (table 5.6).

8. *Bone (osseous tissue)* has a stony calcified matrix. The two types of bone are *spongy* and *compact bone.*

9. The principal cells of bone are *osteocytes,* housed in lacunae. Much of the matrix of compact bone is deposited in cylindrical layers around a *central canal* occupied by blood vessels and nerves (table 5.7).

10. *Blood* is a fluid connective tissue composed of *erythrocytes, leukocytes,* and *platelets* in a liquid matrix, the *plasma* (table 5.8).

### Nervous and Muscular Tissue— Excitable Tissues (p. 171)

1. Nervous and muscular tissue are called *excitable* tissues because they show quick electrical responses to stimuli.

2. *Nervous tissue* is composed of *neurons (nerve cells)* and supporting *glial cells* (table 5.9).

3. Neurons have a *cell body (soma)* and usually one *axon* and *multiple dendrites.*

4. *Muscular tissue* is specialized to contract and move other tissues.

5. There are three kinds of muscle: *skeletal, cardiac,* and *smooth* (table 5.10).

### Intercellular Junctions, Glands, and Membranes (p. 174)

1. Intercellular junctions attach cells to each other.

2. Zipperlike *tight junctions* seal off the space between cells; snap- or weldlike *desmosomes* connect cells at patches rather than continuous zones of attachment; and *gap junctions* have pores that allow substances to pass directly from cell to cell.

3. *Glands* are organs that release secretions for use in the body or for waste elimination.

4. *Exocrine glands* release their secretions through a duct onto the surface of an organ. *Endocrine glands* lack ducts and release their secretions *(hormones)* into the bloodstream.

5. The connective tissue framework of a gland is called its *stroma,* and includes a capsule and internal septa. The secretory part is the *parenchyma* and is composed of epithelial secretory cells and ducts.

6. *Simple* glands have a single unbranched duct; *compound* glands have branched ducts. *Tubular* glands have ductile and secretory portions of uniform diameter; *acinar* glands

have dilated sacs *(acini)* of secretory cells at the end of a duct.

7. *Serous* glands secrete thin runny fluids; *mucous* glands secrete viscous mucus; *mixed* glands secrete both; and *cytogenic* glands produce cells (eggs and sperm) as their products.

8. *Merocrine* gland cells release their secretion by exocytosis; *holocrine* gland cells break down to become the secretion; *apocrine* glands are specialized glands with a merocrine mode of secretion but different histological appearance.

9. Membranes of the body include the relatively dry *cutaneous membrane* (skin), moist *serous membranes* covered with serous fluid; and *mucous* membranes that secrete mucus. Blood vessels are lined with a membrane called the *endothelium;* the ventral body cavity is lined with a membrane called *mesothelium;* and some joints are lined with *synovial membranes.*

**Tissue Growth, Development, Repair, and Death (p. 179)**

1. Organs grow through tissue *hyperplasia* (cell multiplication), *hypertrophy* (cell enlargement), or *neoplasia* (abnormal growth of tumors).

2. Stem cells are undifferentiated cells that have the *developmental plasticity* to develop into multiple mature cell types. They range from *unipotent* stem cells that can develop into only one lineage, to *totipotent* stem cells that can produce any mature cell type. Embryonic stem cells, which come from embryos up to a week old, are *pluripotent;* adult stem cells are either *multipotent* or *unipotent.* Stem cell research is aimed at using stem cells to repair damaged tissues.

3. *Differentiation* is the development of a mature specialized tissue from an unspecialized one. *Metaplasia* is the normal conversion of one mature tissue type into another.

4. Two kinds of tissue repair are *regeneration* (which restores the preexisting tissue type and function) and *fibrosis* (which replaces the previous tissue with fibrous scar tissue).

5. Organs shrink through tissue *atrophy* (shrinkage due to aging or disuse).

6. Two kinds of tissue death are *necrosis* (pathological death of tissues from such causes as trauma, infection, toxins, and oxygen deprivation) and *apoptosis* (normal, programmed death of cells that have completed their function).

# Testing Your Recall

1. Transitional epithelium is found in
   a. the urinary system.
   b. the respiratory system.
   c. the digestive system.
   d. the reproductive system.
   e. all of the above.

2. The external surface of the stomach is covered by
   a. a mucosa.
   b. a serosa.
   c. the parietal peritoneum.
   d. a lamina propria.
   e. a basement membrane.

3. Which of these is a primary germ layer?
   a. epidermis
   b. mucosa
   c. ectoderm
   d. endothelium
   e. epithelium

4. A seminiferous tubule of the testis is lined with ____ epithelium.
   a. simple cuboidal
   b. pseudostratified columnar ciliated
   c. stratified squamous
   d. transitional
   e. stratified cuboidal

5. ____ prevent fluids from seeping between epithelial cells.
   a. Glycosaminoglycans
   b. Hemidesmosomes
   c. Tight junctions
   d. Communicating junctions
   e. Basement membranes

6. A fixative serves to
   a. stop tissue decay.
   b. improve contrast.
   c. repair a damaged tissue.
   d. bind epithelial cells together.
   e. bind cardiac myocytes together.

7. The collagen of areolar tissue is produced by
   a. macrophages.
   b. fibroblasts.
   c. mast cells.
   d. leukocytes.
   e. chondrocytes.

8. Tendons are composed of _____ connective tissue.
   a. skeletal
   b. areolar
   c. dense irregular
   d. yellow elastic
   e. dense regular

9. The shape of the external ear is due to
   a. skeletal muscle.
   b. elastic cartilage.
   c. fibrocartilage.
   d. articular cartilage.
   e. hyaline cartilage.

10. The most abundant formed element(s) of blood is/are
    a. plasma.
    b. erythrocytes.
    c. platelets.
    d. leukocytes.
    e. proteins.

11. Any form of pathological tissue death is called _____.

12. The simple squamous epithelium that lines the peritoneal cavity is called _____.

13. Osteocytes and chondrocytes occupy little cavities called _____.

14. Muscle cells and axons are often called _____ because of their shape.

15. Tendons and ligaments are made mainly of the protein _____.

16. The only type of muscle that lacks gap junctions is _____.

17. An epithelium rests on a layer called the _____ between its deepest cells and the underlying connective tissue.

18. Fibers and ground substance make up the _____ of a connective tissue.

19. A _____ adult stem cell can differentiate into two or more mature cell types.

20. Any epithelium in which every cell touches the basement membrane is called a/an _____ epithelium.

*Answers in Appendix B*

# True or False

*Determine which five of the following statements are false, and briefly explain why.*

1. The esophagus is protected from abrasion by a keratinized stratified squamous epithelium.

2. All cells of a pseudostratified columnar epithelium contact the basement membrane.

3. Not all skeletal muscle is attached to bones.

4. The stroma of a gland does not secrete anything.

5. In all connective tissues, the matrix occupies more space than the cells do.

6. Adipocytes are limited to adipose tissue.

7. Tight junctions function primarily to prevent cells from pulling apart.

8. Metaplasia is a normal, healthy tissue transformation but neoplasia is not.

9. Nerve and muscle cells are not the body's only electrically excitable cells.

10. Cartilage is always covered by a fibrous perichondrium.

*Answers in Appendix B*

# Testing Your Comprehension

1. A woman in labor is often told to push. In doing so, is she consciously contracting her uterus to expel the baby? Justify your answer based on the muscular composition of the uterus.

2. A major tenet of the cell theory is that all bodily structure and function is based on cells. The structural properties of bone, cartilage, and tendons, however, are due more to their extracellular material than to their cells. Is this an exception to the cell theory? Why or why not?

3. When cartilage is compressed, water is squeezed out of it, and when pressure is taken off, water flows back into the matrix. This being the case, why do you think cartilage at weight-bearing joints such as the knees can degenerate from lack of exercise?

4. The epithelium of the respiratory tract is mostly of the pseudostratified columnar ciliated type, but in the alveoli—the tiny air sacs where oxygen and carbon dioxide are exchanged between the blood and inhaled air—the epithelium is simple squamous. Explain the functional significance of this histological difference. That is, why don't the alveoli have the same kind of epithelium as the rest of the respiratory tract?

5. Which do you think would heal faster, cartilage or bone? Stratified squamous or simple columnar epithelium? Why?

*Answers at www.mhhe.com/saladin4*

# www.mhhe.com/saladin4

*The textbook website provides a wealth of interactive study materials fully organized and integrated by chapter. You will find practice quizzes, labeling exercises, and much more that will complement your learning and understanding of anatomy and physiology. The website also includes tools designed to enhance your* **Anatomy & Physiology | REVEALED** *experience.*

*Hair follicles and sebaceous (oil) glands*

# THE INTEGUMENTARY SYSTEM

## CHAPTER OUTLINE

## INSIGHTS

## Brushing Up

To understand this chapter, it is important that you understand or brush up on the following concepts:

The skin is also know as the **integument,**[1] whereas the **integumentary system** consists of the skin and its *accessory organs*—the hair, nails, and cutaneous glands. We pay more attention to this organ system than to any other. It is, after all, the most visible one, and its appearance strongly affects our social interaction. Few people venture out of the house without first looking in a mirror to see if their skin and hair are presentable. Social considerations aside, the integumentary system is important to one's self-image, and a positive self-image is important to the attitudes that promote overall good health. Care of the integumentary system is thus a particularly important part of the total plan of patient care.

The scientific study and medical treatment of the integumentary system is called **dermatology.**[2] Inspection of the skin, hair and nails is a significant part of a physical examination. The integumentary system provides clues not only to its own health, but also to deeper disorders such as liver cancer, anemia, and heart failure. The skin also is the most vulnerable of our organs, exposed to radiation, trauma, infection, and injurious chemicals. Consequently, it needs and receives more medical attention than any other organ system.

# The Skin and Subcutaneous Tissue

### Objectives

When you have completed this section, you should be able to

- list the functions of the skin and relate them to its structure;
- describe the histological structure of the epidermis, dermis, and subcutaneous tissue;
- describe the normal and pathological colors that the skin can have and explain their causes; and
- describe the common markings of the skin.

The **skin** is the body's largest organ. In adults, it covers an area of 1.5 to 2.0 m$^2$ and accounts for about 15% of the body weight. The skin consists of two layers: a stratified squamous epithelium called the *epidermis* and a deeper connective tissue layer called the *dermis* (fig. 6.1). Below the dermis is another connective tissue layer, the *hypo-*

[1]*integument* = covering
[2]*dermat* = skin + *logy* = study of

*dermis,* which is not part of the skin but is customarily studied in conjunction with it.

Most of the skin is 1 to 2 mm thick, but it ranges from less than 0.5 mm on the eyelids to 6 mm between the shoulder blades. The difference is due mainly to variation in the thickness of the dermis, although skin is classified as thick or thin based on the relative thickness of the epidermis alone. **Thick skin** covers the palms, soles, and corresponding surfaces of the fingers and toes. Its epidermis alone is about 0.5 mm thick, due to a very thick surface layer of dead cells called the *stratum corneum* (fig. 6.2). Thick skin has sweat glands but no hair follicles or sebaceous (oil) glands. The rest of the body is covered with **thin skin,** which has an epidermis about 0.1 mm thick, with a thin stratum corneum (see fig. 6.5). It possesses hair follicles, sebaceous glands, and sweat glands.

## FUNCTIONS OF THE SKIN

The skin is much more than a container for the body. It has a variety of important functions that go well beyond appearance, as we shall see here.

1. **Resistance to trauma and infection.** The skin bears the most physical injuries to the body, but it resists and recovers from trauma better than other organs do. The epidermal cells are packed with the tough protein **keratin** and linked by strong desmosomes that give this epithelium its durability. Few infectious diseases can penetrate the intact skin. Bacteria and fungi colonize the skin surface, but their numbers are kept in check by the relative dryness and slight acidity (pH 4-6) of the surface. This protective, acidic film is called the *acid mantle.*

2. **Other barrier functions.** The skin is important as a barrier to water. It prevents the body from absorbing excess water when you are swimming or bathing, but even more importantly, it prevents the body from losing excess water. The epidermis is also a barrier to ultraviolet (UV) radiation, blocking much of this cancer-causing radiation from reaching deeper tissue layers; and it is a barrier to many potentially harmful chemicals. It is, however, permeable to several drugs and poisons (Insight 6.1).

3. **Vitamin D synthesis.** The skin carries out the first step in the synthesis of vitamin D, which is needed for bone development and maintenance. The liver and kidneys complete the process.

4. **Sensation.** The skin is our most extensive sense organ. It is equipped with a variety of nerve endings that react to heat, cold, touch, texture, pressure, vibration, and tissue injury (see chapter 16). These sensory receptors are especially abundant on the face, palms, fingers, soles, nipples, and genitals. There are relatively few on the back and in skin overlying joints such as the knees and elbows.

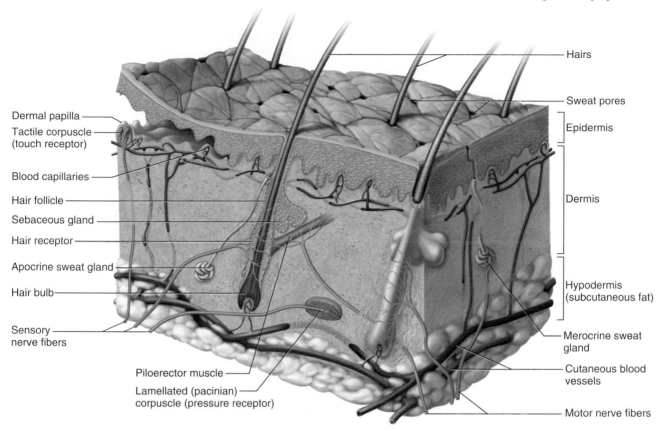

Hairs

Sweat pores

Epidermis

Dermis

Hypodermis (subcutaneous fat)

Merocrine sweat gland

Cutaneous blood vessels

Motor nerve fibers

Dermal papilla

Tactile corpuscle (touch receptor)

Blood capillaries

Hair follicle

Sebaceous gland

Hair receptor

Apocrine sweat gland

Hair bulb

Sensory nerve fibers

Piloerector muscle

Lamellated (pacinian) corpuscle (pressure receptor)

**FIGURE 6.1** Structure of the Skin and Subcutaneous Tissue.

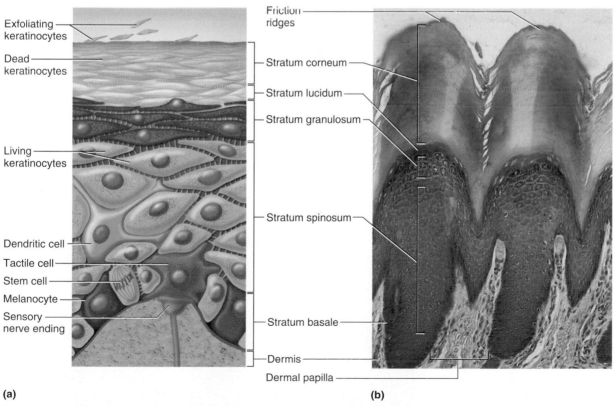

Exfoliating keratinocytes

Dead keratinocytes

Living keratinocytes

Dendritic cell

Tactile cell

Stem cell

Melanocyte

Sensory nerve ending

Friction ridges

Stratum corneum

Stratum lucidum

Stratum granulosum

Stratum spinosum

Stratum basale

Dermis

Dermal papilla

**(a)**

**(b)**

**FIGURE 6.2** **Layers and Cell Types of the Epidermis.** (a) Drawing of epidermal layers and cell types. (b) Photograph of thick skin from the fingertip showing two of the surface friction ridges responsible for the fingerprints.

## Transdermal Absorption

The ability of the skin to absorb chemicals makes it possible to administer several medicines as ointments or lotions, or by means of adhesive patches that release the medicine steadily through a membrane. For example, inflammation can be treated with a hydrocortisone ointment, nitroglycerine patches are used to relieve heart pain, nicotine patches are used to help overcome tobacco addiction, and other medicated patches are used to control high blood pressure and motion sickness.

Unfortunately, the skin can also be a route for absorption of poisons. These include toxic alkaloids from poison ivy and other plants; metals such as mercury, arsenic, and lead; and solvents such as carbon tetrachloride (a cleaning solvent), acetone (nail polish remover), paint thinner, and pesticides. Some of these can cause brain damage, liver failure, or kidney failure, which is good reason for using protective gloves when handling such substances.

5. **Thermoregulation.** In response to chilling, the skin helps to retain heat. The dermis has nerve endings called **thermoreceptors** that transmit signals to the brain, and the brain sends signals back to the dermal blood vessels. Vasoconstriction, or narrowing of these blood vessels, reduces the flow of blood close to the skin surface and thus reduces heat loss. When one is overheated, vasodilation, or widening of the dermal blood vessels, increases cutaneous blood flow and increases heat loss. If this is not enough to restore normal temperature, the brain also triggers sweating.

6. **Social functions.** The skin is an important means of nonverbal communication. Humans, like most other primates, have much more expressive faces than most mammals. Complex skeletal muscles insert on dermal collagen fibers and pull on the skin to create subtle and varied facial expressions (fig. 6.3). The general appearance of the skin, hair, and nails is also important to social acceptance and to a person's self-image and emotional state—whether the ravages of adolescent acne, the presence of a birthmark or scar, or just a "bad hair day."

## THE EPIDERMIS

The **epidermis**[3] is a keratinized stratified squamous epithelium, as described in chapter 5. That is, its surface consists of dead cells packed with the tough protein keratin. Like other epithelia, the epidermis lacks blood vessels and depends on the diffusion of nutrients from the underlying connective tissue. It has sparse nerve endings for touch and pain, but most sensations of the skin are due to nerve endings in the dermis.

### Cells of the Epidermis

The epidermis is composed of five types of cells (see fig. 6.2):

1. **Stem cells** are undifferentiated cells that undergo mitosis and give rise to the keratinocytes described next. They are found only in the deepest layer of the epidermis, called the *stratum basale* (described later).

2. **Keratinocytes** (keh-RAT-ih-no-sites) are the great majority of epidermal cells. They are named for their role in synthesizing keratin. In ordinary histological specimens, nearly all of the epidermal cells you see are keratinocytes.

---

[3]*epi* = above, upon + *derm* = skin

**FIGURE 6.3**  **Importance of the Skin in Nonverbal Expression.**  Primates differ from other mammals in having very expressive faces due to facial muscles that insert on the collagen fibers of the dermis and move the skin.

3. **Melanocytes** also occur only in the stratum basale, amid the stem cells and deepest keratinocytes. They synthesize the brown to black pigment *melanin.* They have branching processes that spread among the keratinocytes and continually shed melanin-containing fragments from their tips. The keratinocytes phagocytize these fragments and accumulate melanin granules on the "sunny side" of the nucleus. Like a parasol, the pigment shields the DNA from ultraviolet radiation. People of all races have about equal numbers of melanocytes. Differences in skin color result from differences in the rate of melanin synthesis and how clumped or spread out the melanin is. In light skin, the melanin is less abundant and is relatively clumped near the keratinocyte nucleus, imparting less color to the cells.

4. **Tactile (Merkel[4]) cells,** relatively few in number, are receptors for the sense of touch. They, too, are found in the basal layer of the epidermis and are associated with an underlying dermal nerve fiber. The tactile cell and its nerve fiber are collectively called a *tactile (Merkel) disc.*

5. **Dendritic[5] (Langerhans[6]) cells** are found in two layers of the epidermis called the *stratum spinosum* and *stratum granulosum* (described in the next section). They are macrophages that originate in the bone marrow but migrate to the epidermis and epithelia of the oral cavity, esophagus, and vagina. The epidermis has as many as 800 dendritic cells per square millimeter. They stand guard against toxins, microbes, and other pathogens that penetrate into the skin. When they detect such invaders, they alert the immune system so the body can defend itself.

## Layers of the Epidermis

The epidermis consists of four to five layers of cells (five in thick skin) (see fig. 6.2). The following description progresses from deep to superficial, and from the youngest to the oldest keratinocytes.

1. The **stratum basale** (bah-SAY-lee) consists mainly of a single layer of cuboidal to low columnar stem cells and keratinocytes resting on the basement membrane. Scattered among these are the melanocytes and tactile cells. As stem cells of the stratum basale undergo mitosis, they give rise to keratinocytes that migrate toward the skin surface and replace lost epidermal cells. The life history of these cells is described in the next section.

2. The **stratum spinosum** (spy-NO-sum) consists of several layers of keratinocytes. In most skin, this is the

thickest stratum, but in thick skin it is usuallly exceeded by the stratum corneum. The deepest cells of the strata spinosum remain capable of mitosis, but as they are pushed farther upward, they cease dividing. Instead, they produce more and more keratin filaments, which cause the cells to flatten. Therefore, the higher up you look in the stratum spinosum, the flatter the cells appear. Dendritic cells are also found throughout the stratum spinosum, but are not usually visible in tissue sections.

The stratum spinosum is named for an artificial appearance *(artifact)* created by the histological fixation of tissue specimens. Keratinocytes are firmly attached to each other by numerous desmosomes, which partly account for the toughness of the epidermis. Histological fixatives shrink the keratinocytes and cause them to pull away from each other, but they remain attached by the desmosomes—like two people holding hands while they step farther away from each other. The desmosomes thus create bridges from cell to cell, giving each cell a spiny appearance from which we derive the word *spinosum.*

Epidermal keratinocytes are also bound to each other by tight junctions, which make an essential contribution to water retention by the skin. This is further discussed in the next section.

3. The **stratum granulosum** consists of three to five layers of flat keratinocytes—more in thick skin than in thin skin. The keratinocytes of this layer contain coarse, dark-staining *keratohyalin granules* that give the layer its name. The functional significance of these granules will be explained shortly.

4. The **stratum lucidum**[7] (LOO-sih-dum) is a thin translucent zone superficial to the stratum granulosum, seen only in thick skin. Here, the keratinocytes are densely packed with *eleidin* (ee-LEE-ih-din), an intermediate in the synthesis of keratin. The cells have no nuclei or other organelles. Because organelles are absent and eleidin does not stain well, this zone has a pale, featureless appearance with indistinct cell boundaries.

5. The **stratum corneum** consists of up to 30 layers of dead, scaly, keratinized cells that form a durable surface layer. This layer is especially resistant to abrasion, penetration, and water loss.

## The Life History of a Keratinocyte

Dead cells constantly flake off the skin surface. They float around as tiny white specks in the air, settling on household surfaces and forming much of the house dust that accumulates there (Insight 6.2). Because we constantly lose these epidermal cells, they must be continually replaced.

---

[4]Friedrich Sigmund Merkel (1845–1919), German anatomist
[5]*dendr* = tree, branch
[6]Paul Langerhans (1847–88), German anatomist

[7]*lucid* = light, clear

## INSIGHT 6.2    Clinical Application

### Dead Skin and Dust Mites

In the beams of late afternoon sun that shine aslant through a window, you may see tiny white specks floating through the air. Most of these are flakes of dander; the dust on top of your bookshelves is largely a film of dead human skin. Composed of protein, this dust in turn supports molds and other microscopic organisms that feed on the skin cells and each other. One of these organisms is the house dust mite, *Dermatophagoides*[8] (der-MAT-oh-fah-GOY-deez) (fig. 6.4). (What wonders may be found in humble places!)

*Dermatophagoides* thrives abundantly in pillows, mattresses, and upholstery—warm, humid places that are liberally sprinkled with edible flakes of keratin. No home is without these mites, and it is impossible to entirely exterminate them. What was once regarded as "house dust allergy" has been identified as an allergy to the inhaled feces of these mites.

0.1 mm

**FIGURE 6.4**    The House Dust Mite, *Dermatophagoides.*

Keratinocytes are produced deep in the epidermis by the mitosis of stem cells in the stratum basale. Some of the deepest keratinocytes in the stratum spinosum also multiply and increase their number. Mitosis requires an abundant supply of oxygen and nutrients, which these deep epidermal cells can acquire from the blood vessels in the nearby dermis. Once the epidermal cells migrate more than two or three cells away from the dermis, their mitosis ceases. Mitosis is seldom seen in prepared slides of the skin, because it occurs mainly at night, and most histological specimens are taken during the day.

As new keratinocytes are formed, they push the older ones toward the surface. Over the course of 30 to 40 days a keratinocyte makes its way to the skin surface and then flakes off. This migration is slower in old age and faster in skin that has been injured or stressed. Injured epidermis regenerates more rapidly than any other tissue in the body. Mechanical stress from manual labor or tight shoes accelerates keratinocyte multiplication and results in *calluses* or *corns,* thick accumulations of dead keratinocytes on the hands or feet.

As keratinocytes are shoved upward by the dividing cells below, their cytoskeleton proliferates, the cells grow flatter, and they produce lipid-filled **membrane-coating vesicles.** In the stratum granulosum, three important developments occur: (1) The keratinocytes undergo apoptosis (programmed cell death). (2) The keratohylin granules release a substance that binds to the intermediate filaments of the cytoskeleton and converts them to keratin. (3) The membrane-coating vesicles release a lipid mixture that spreads out over the cell surface and waterproofs it.

An *epidermal water barrier* forms between the stratum granulosum and the stratum spinosum. It consists of the lipids secreted by the keratinocytes, tight junctions between the keratinocytes, and a thick layer of insoluble protein on the inner surfaces of the keratinocyte plasma membranes. The epidermal water barrier is crucial to retaining water in the body and preventing dehydration. Cells above the barrier quickly die because the barrier cuts them off from the supply of nutrients below. Thus, the stratum corneum consists of compact layers of dead keratinocytes and keratinocyte fragments. Dead keratinocytes soon *exfoliate* (fall away) from the epidermal surface as tiny specks called **dander.** *Dandruff* is composed of clumps of dander stuck together by sebum (oil).

## THE DERMIS

Beneath the epidermis is a connective tissue layer, the **dermis.** It ranges from 0.2 mm thick in the eyelids to about 4 mm thick in the palms and soles. It is composed mainly of collagen but also contains elastic and reticular fibers, fibroblasts, and the other cells typical of fibrous connective tissue (described in chapter 5). It is well supplied with blood vessels, sweat glands, sebaceous glands, and nerve endings. The hair follicles and nail roots are embedded in the dermis. Smooth muscles (piloerector muscles) associated with hair follicles contract in response to such stimuli as cold, fear, and touch. This makes the hair stand on end, causes "goosebumps," and

---

[8]*dermato* = skin + *phag* = eat

wrinkles the skin in areas such as the scrotum and areola. In the face, skeletal muscles attach to dermal collagen fibers and produce such expressions as a smile, a wrinkle of the forehead, and the lifting of an eyebrow.

The boundary between the epidermis and the dermis is histologically conspicuous and usually wavy. The upward waves are fingerlike extensions of the dermis called **dermal papillae**[9] (see fig. 6.2), and the downward waves are extensions of the epidermis called **epidermal ridges.** The dermal and epidermal boundaries thus interlock like corrugated cardboard, an arrangement that resists slippage of the epidermis across the dermis. If you look closely at your hand and wrist, you will see delicate furrows that divide the skin into tiny rectangular to rhomboid areas. The dermal papillae produce the raised areas between the furrows. On the fingertips, this wavy boundary forms the *friction ridges* that leave fingerprints on the things we touch. In highly sensitive areas such as the lips and genitals, tall dermal papillae allow nerve fibers to reach close to the surface.

> **Think About It**
>
> *Dermal papillae are relatively high and numerous in palmar and plantar skin but low and few in number in the skin of the face and abdomen. What do you think is the functional significance of this difference?*

There are two zones of dermis called the papillary and reticular layers (fig. 6.5). The **papillary** (PAP-ih-lerr-ee) **layer** is a thin zone of areolar tissue in and near the dermal papillae. This loosely organized tissue allows for mobility of leukocytes and other defenses against organisms introduced through breaks in the epidermis. This layer is especially rich in small blood vessels.

The **reticular**[10] **layer** of the dermis is deeper and much thicker. It consists of dense irregular connective tissue. The boundary between the papillary and reticular layers is often vague. In the reticular layer, the collagen forms thicker bundles with less room for ground substance, and there are often small clusters of adipocytes. Stretching of

---

[9]*pap* = nipple + *illa* = little

[10]*reti* = network + *cul* = little

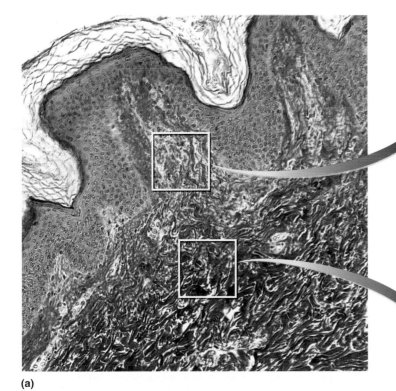

(a)

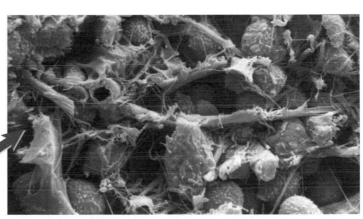

**(b) Papillary layer of dermis**

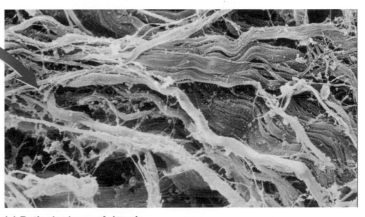

**(c) Reticular layer of dermis**

**FIGURE 6.5  Layers of the Dermis.**  (a) Light micrograph of axillary skin, with the collagen stained blue. (b) The papillary layer, made of loose (areolar) tissue, forms the dermal papillae. (c) The reticular layer, made of dense irregular connective tissue, forms the deeper four-fifths of the dermis.  [Parts (b) and (c) from R. G. Kessel and R. H. Kardon, *Tissues and Organs: A Text-Atlas of Scanning Electron Microscopy* (W. H. Freeman, 1979).]

the skin in obesity and pregnancy can tear the collagen fibers and produce *striae* (STRY-ee), or stretch marks. These occur especially in areas most stretched by weight gain: the thighs, buttocks, abdomen, and breasts.

## THE HYPODERMIS

Beneath the skin is a layer called the **hypodermis,**[11] **subcutaneous tissue,** or **superficial fascia**[12] (FASH-ee-uh). The boundary between the dermis and hypodermis is indistinct, but the hypodermis generally has more areolar and adipose tissue. The hypodermis binds the skin to the underlying tissues and pads the body. Drugs are introduced into the hypodermis by injection because the subcutaneous tissue is highly vascular and absorbs them quickly.

**Subcutaneous fat** is hypodermis composed predominantly of adipose tissue. This fat serves as an energy reservoir and thermal insulation. It is not uniformly distributed; for example, it is virtually absent from the scalp but relatively abundant in the breasts, abdomen, hips, and thighs. The subcutaneous fat averages about 8% thicker in women than in men, and varies with age. Infants and elderly people have less subcutaneous fat than other people and are therefore more sensitive to cold.

Table 6.1 summarizes the layers of the skin and hypodermis.

## SKIN COLOR

The most significant factor in skin color is **melanin.** This is produced by the melanocytes but accumulates in the keratinocytes of the stratum basale and stratum spinosum (fig. 6.6). There are two forms of melanin: a brownish black **eumelanin**[13] and a reddish yellow sulfur-containing pigment, **pheomelanin.**[14] People of different races have essentially the same number of melanocytes, but in dark-skinned people, the melanocytes produce greater quantities of melanin and the melanin in the keratinocytes breaks down more slowly. Thus, melanized cells may be seen throughout the epidermis, from stratum basale to stratum corneum. In light-skinned people, the melanin breaks down more rapidly and little of it is seen beyond the stratum basale, if even there.

The amount of melanin in the skin also varies with exposure to the ultraviolet (UV) rays of sunlight, which stimulate melanin synthesis and darken the skin. A suntan fades as melanin is degraded in older keratinocytes and as the keratinocytes migrate to the surface and exfoliate. The amount of melanin also varies substantially from place to place on the body. It is relatively concentrated in freckles and moles, on the dorsal surfaces of the hands and feet as compared with the palms and soles, on the nipple and surrounding area (areola) of the breast, around the anus, on the scrotum and penis, and on the lateral surfaces of the female genital folds (labia majora). The contrast between heavily melanized and lightly melanized regions of the skin is more pronounced in some races than others, but it exists to some extent in nearly everyone.

Other factors in skin color are hemoglobin and carotene. **Hemoglobin,** the red pigment of blood, imparts reddish to pinkish hues to the skin. Its color is lightened

---

[11]*hypo* = below + *derm* = skin
[12]*fasc* = band

[13]*eu* = true + *melan* = black + *in* = substance
[14]*pheo* = dusky + *melan* = black + *in* = substance

| TABLE 6.1 | Stratification of the Skin and Hypodermis |
|---|---|
| **Layer** | **Description** |
| *Epidermis* | Keratinized stratified squamous epithelium |
| Stratum corneum | Dead, keratinized cells of the skin surface |
| Stratum lucidum | Clear, featureless, narrow zone seen only in thick skin |
| Stratum granulosum | Two to five layers of cells with dark-staining keratohyalin granules; scanty in thin skin |
| Stratum spinosum | Many layers of keratinocytes, typically shrunken in fixed tissues but attached to each other by desmosomes, which give them a spiny look; progressively flattened the farther they are from the dermis. Dendritic cells occur here but are not visible in routinely stained preparations. |
| Stratum basale | Single layer of cuboidal to columnar cells resting on basement membrane; site of most mitosis; consists of keratinocytes, melanocytes, and tactile cells, but these are not distinguishable with routine stains. Melanin is conspicuous in keratinocytes of this layer in black to brown skin. |
| *Dermis* | Fibrous connective tissue, richly endowed with blood vessels and nerve endings. Sweat glands and hair follicles originate here and in hypodermis. |
| Papillary layer | Superficial one-fifth of dermis; composed of areolar tissue; often extends upward as dermal papillae |
| Reticular layer | Deeper four-fifths of dermis; dense irregular connective tissue |
| *Hypodermis* | Areolar or adipose tissue between skin and muscle |

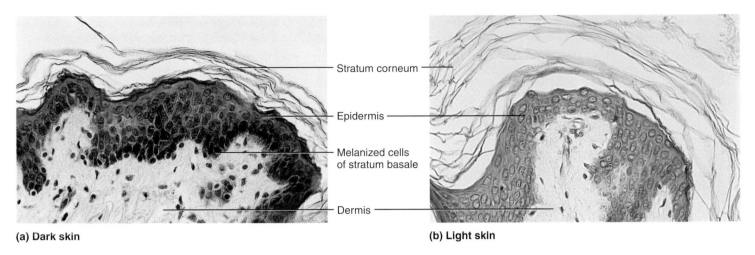

**(a) Dark skin**    **(b) Light skin**

**FIGURE 6.6** **Variations in Skin Pigmentation.** (a) Keratinocytes in and near the stratum basale have heavy deposits of melanin in dark skin. (b) Light skin shows little to no visible melanin in the basal keratinocytes.

▶ *Which of the five types of epidermal cells are the melanized cells in part (a)?*

by the white of the dermal collagen. The skin is redder in places such as the lips, where blood capillaries come closer to the surface and the hemoglobin shows through more vividly. **Carotene**[15] is a yellow pigment acquired from egg yolks and yellow and orange vegetables. Depending on the diet, it can become concentrated to various degrees in the stratum corneum and subcutaneous fat. It is often most conspicuous in skin of the heel and in "corns" or calluses of the feet, because this is where the stratum corneum is thickest.

The skin may also exhibit abnormal colors of diagnostic value:

- **Cyanosis**[16] is blueness of the skin resulting from a deficiency of oxygen in the circulating blood. Oxygen deficiency turns the hemoglobin a reddish violet color. It can result from conditions that prevent the blood from picking up a normal load of oxygen in the lungs, such as airway obstructions in drowning and choking, lung diseases such as emphysema, or respiratory arrest. Cyanosis also occurs in situations such as cold weather and cardiac arrest, when blood flows so slowly through the skin that most of its oxygen is extracted faster than freshly oxygenated blood arrives.

- **Erythema**[17] (ERR-ih-THEE-muh) is abnormal redness of the skin. It occurs in such situations as exercise, hot weather, sunburns, anger, and embarrassment. Erythema is caused by increased blood flow in dilated cutaneous blood vessels or by dermal pooling of red blood cells that have escaped from abnormally permeable capillaries, as in sunburn.

- **Pallor** is a pale or ashen color that occurs when there is so little blood flow through the skin that the white color of the dermal collagen shows through. It can result from emotional stress, low blood pressure, circulatory shock, cold temperatures, or severe anemia.

- **Albinism**[18] is a genetic lack of melanin that results in white hair, pale skin, and pink eyes. Melanin is synthesized from the amino acid tyrosine by the enzyme tyrosinase. People with albinism have inherited a recessive, nonfunctional tyrosinase allele from both parents.

- **Jaundice**[19] is a yellowing of the skin and whites of the eyes resulting from high levels of bilirubin in the blood. Bilirubin is a hemoglobin breakdown product. When erythrocytes get old, they disintegrate and release their hemoglobin. The liver converts hemoglobin to bilirubin and other pigments, which are excreted in the bile. Bilirubin can accumulate enough to discolor the skin, however, in such situations as a rapid rate of erythrocyte destruction; when diseases such as cancer, hepatitis, and cirrhosis interfere with liver function; and in premature infants, where the liver is not well enough developed to dispose of bilirubin efficiently.

- **Bronzing** is a golden-brown skin color that results from Addison disease, a deficiency of glucocorticoid hormones from the adrenal cortex. President John F. Kennedy had Addison disease and bronzing of the skin, which many people mistook for a suntan.

- A **hematoma**,[20] or bruise, is a mass of clotted blood showing through the skin. It is usually due to

---

[15]*carot* = carrot
[16]*cyan* = blue + *osis* = condition
[17]*eryth* = red + *em* = blood

[18]*alb* = white + *ism* = state, condition
[19]*jaun* = yellow
[20]*hemat* = blood + *oma* = mass

accidental trauma (blows to the skin), but it may indicate hemophilia, other metabolic or nutritional disorders, or physical abuse.

> **Think About It**
>
> *An infant brought to a clinic shows abnormally yellow skin. What sign could you look for to help decide whether this was due to jaundice or to a large amount of carotene from strained vegetables in the diet?*

## SKIN MARKINGS

The skin is marked by many lines, creases, ridges, and patches of accentuated pigmentation. **Friction ridges** are the markings on the fingertips that leave distinctive oily fingerprints on surfaces we touch (see fig. 6.2b). They are characteristic of most primates. They help prevent monkeys, for example, from slipping off a branch as they walk across it, and they enable us to manipulate small objects more easily. Friction ridges form during fetal development and remain essentially unchanged for life. Everyone has a unique pattern of friction ridges; not even identical twins have identical fingerprints.

**Flexion lines** (flexion creases) are the lines on the flexor surfaces of the digits, palms, wrists, elbows, and other places (see fig. B.8, p. 397). They mark sites where the skin folds during flexion of the joints. The skin is tightly bound to the deep fascia along these lines.

Freckles and moles are tan to black aggregations of melanocytes. **Freckles** are flat melanized patches that vary with heredity and exposure to the sun. A **mole** (nevus) is an elevated patch of melanized skin, often with hair. Moles are harmless and sometimes even regarded as "beauty marks," but they should be watched for changes in color, diameter, or contour that may suggest malignancy (skin cancer).

**Hemangiomas**[21] (he-MAN-jee-OH-mas), or birthmarks, are patches of discolored skin caused by benign tumors of the dermal blood capillaries. *Capillary hemangiomas* (strawberry birthmarks) are bright red to deep purple and are slightly swollen; they usually disappear in childhood. *Cavernous hemangiomas* (port wine stains) are flat and duller in color, and last for life.

## Before You Go On

*Answer the following questions to test your understanding of the preceding section:*

1. *What is the major histological difference between thick and thin skin? Where on the body is each type of skin found?*

2. *How does the skin help control body temperature?*

3. *List the five cell types of the epidermis. Describe their locations and functions.*

4. *List the five layers of epiderms from deep to superficial. What are the distinctive features of each layer?*

5. *What are the two layers of the dermis? What type of tissue composes each layer?*

6. *Name the pigments responsible for normal skin colors and explain how certain conditions can produce discolorations of the skin.*

# HAIR AND NAILS

### Objecives

When you have completed this section, you should be able to

- distinguish between three types of hair;
- describe the histology of a hair and its follicle;
- discuss some theories of the purposes served by various kinds of hair; and
- describe the strucure and function of nails.

The hair, nails, and cutaneous glands are the **accessory organs** (appendages) of the skin. Hair and nails are composed mostly of dead, keratinized cells. The stratum corneum of the skin is made of pliable **soft keratin,** but the hair and nails are composed mostly of **hard keratin.** Hard keratin is more compact than soft keratin and is toughened by numerous cross-linkages between the keratin molecules.

## HAIR

A hair is also known as a **pilus** (PY-lus); in the plural, *pili* (PY-lye). It has a slender filament of keratinized cells that grows from an oblique tube in the skin called a **hair follicle** (fig. 6.7).

### Distribution and Types

Hair is found almost everywhere on the body except the palms and soles, ventral and lateral surfaces of the fingers and toes, distal segment of the fingers, lips, nipples, and parts of the genitals. Hairless skin is sometimes called *glabrous*[22] *skin*. The limbs and trunk have about 55 to 70 hairs per square centimeter, and the face has about 10 times as many. There are about 30,000 hairs in a man's beard and about 100,000 hairs on the average person's scalp. The number of hairs in a given area does not differ much from one person to another or even between the sexes. Differences in apparent hairiness are due mainly to differences in the texture and pigmentation of the hair.

---

[21]*hem* = blood + *angi* = vessels + *oma* = mass, tumor

[22]*glab* = smooth

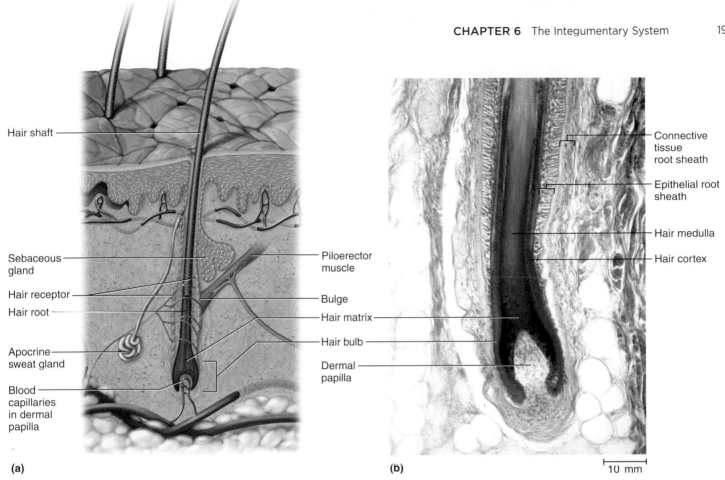

Hair shaft

Sebaceous
gland

Hair receptor

Hair root

Apocrine
sweat gland

Blood
capillaries
in dermal
papilla

Piloerector
muscle

Bulge

Hair matrix

Hair bulb

Dermal
papilla

**(a)**

Connective
tissue
root sheath

Epithelial root
sheath

Hair medulla

Hair cortex

**(b)**                                    10 mm

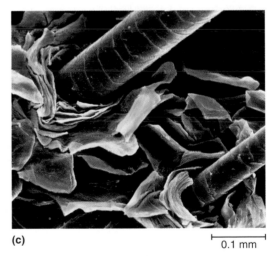

**(c)**                        0.1 mm

**FIGURE 6.7** **Structure of a Hair and Its Follicle.** (a) Anatomy of the follicle and associated structures. (b) Light micrograph of the base of a hair follicle. (c) Electron micrograph of two hairs emerging from their follicles. Note the exfoliating epidermal cells encircling the follicles like rose petals.

Not all hair is alike, even on one person. Over the course of our lives, we grow three kinds of hair: lanugo, vellus, and terminal hair. **Lanugo**[23] is fine, downy, unpigmented hair that appears on the fetus in the last 3 months of development. By the time of birth, most of it is replaced by **vellus,**[24] similarly fine, unpigmented hair. Vellus constitutes about two-thirds of the hair of women, one-tenth of the hair of men, and all of the hair of children except for the eyebrows, eyelashes, and hair of the scalp. **Terminal hair** is longer, coarser, and pigmented. It forms the eyebrows and eyelashes, covers the scalp, and after puberty, it forms the axillary and pubic hair, the male facial hair, and some of the hair on the trunk and limbs.

## Structure of the Hair and Follicle

A hair is divisible into three zones along its length: (1) the **bulb,** a swelling at the base where the hair originates in the dermis; (2) the **root,** which is the remainder of the hair within the follicle; and (3) the **shaft,** which is the portion

[23]*lan* = down, wool
[24]*vellus* = fleece

above the skin surface. Except near the bulb, all the tissue is dead. The hair bulb grows around a bud of vascular connective tissue called the **dermal papilla,** which provides the hair with its sole source of nutrition. Immediately above the papilla is a region of mitotically active cells, the **hair matrix,** which is the hair's growth center. All cells higher up are dead.

In cross section, a hair reveals three layers: (1) the **medulla,** a core of loosely arranged cells and air spaces found in thick hairs, but absent from thin ones; (2) the **cortex,** a layer of keratinized cuboidal cells; and (3) the **cuticle,** a surface layer of scaly cells that overlap each other like roof shingles, with their free edges directed upward (fig. 6.7c). Cells lining the follicle are like shingles facing in the opposite direction. They interlock with the scales of the hair cuticle and resist pulling on the hair. When a hair is pulled out, this layer of follicle cells comes with it.

The follicle is a diagonal tube that dips deeply into the dermis and sometimes extends as far as the hypodermis. It has two principal layers: an **epithelial root sheath** and a **connective tissue root sheath.** The epithelial root sheath, which is an extension of the epidermis, lies immediately adjacent to the hair root. The connective tissue root sheath, derived from the dermis, surrounds the epithelial sheath and is somewhat denser than the adjacent dermal connective tissue.

Associated with the follicle are nerve and muscle fibers. Nerve fibers called **hair receptors** entwine each follicle and respond to hair movements. You can feel their effect by carefully moving a single hair with a pin or by lightly running your finger over the hairs of your forearm without touching the skin. Also associated with each hair is a **piloerector muscle** (arrector pili[25]), a bundle of smooth muscle cells extending from dermal collagen fibers to the connective tissue root sheath of the follicle (see figs. 6.1 and 6.8). In response to cold, fear, or other stimuli, the sympathetic nervous system stimulates these muscles to contract and thereby makes the hair stand on end. In other mammals, this traps an insulating layer of warm air next to the skin or makes the animal appear larger and less vulnerable to a potential enemy. In humans, it pulls the follicles into a vertical position and causes "goose bumps" but serves no useful purpose.

## Hair Texture and Color

The texture of hair is related to differences in cross-sectional shape (fig. 6.8)—straight hair is round, wavy hair is oval, and tightly curly hair is relatively flat. Hair color is due to pigment granules in the cells of the cortex. Brown and black hair are rich in eumelanin. Red hair has a slight amount of eumelanin but a high concentration of pheomelanin. Blond hair has an intermediate amount of pheomelanin but very little eumelanin. Gray and white hair result from a scarcity or absence of melanins in the cortex and the presence of air in the medulla.

## Hair Growth and Loss

A given hair grows in a **hair cycle** consisting of three stages: anagen, catagen, and telogen. In the **anagen**[26] stage, stem cells from the bulge in the follicle multiply and travel downward, pushing the dermal papilla deeper into the skin and forming the epithelial root sheath. Root sheath cells directly above the dermal papilla form the

---

---

**INSIGHT 6.3**    Clinical Application

### The Science and Pseudoscience of Hair Analysis

Laboratory analysis of hair can provide important clues to metabolic diseases and poisoning. A deficiency of zinc in the hair may indicate malnutrition. A deficiency of both magnesium and calcium is a sign of phenylketonuria (PKU), a hereditary defect in metabolism. A deficiency of calcium along with an excess of sodium may indicate cystic fibrosis.

Several poisons accumulate in the hair. Arsenic, cadmium, lead, and mercury, for example, can be 10 times as concentrated in hair as in blood or urine, and hair analysis can therefore aid in establishing a cause of death. DNA extracted from a single hair can be enough to implicate a suspect in a crime through the technique of *DNA fingerprinting.*

Although hair analysis has valuable medical and legal applications, it is also exploited by health-food charlatans who charge high fees for hair analysis and special diets based on supposed deficiencies that the analysis "reveals." In reality, however, hair analysis has proven to be unreliable as a basis for dietary recommendations.

---

**hair matrix.** Here, sheath cells transform into hair cells, which synthesize keratin and then die as they are pushed upward away from the papilla.

In the **catagen**[27] phase, epithelial root sheath cells below the bulge undergo apoptosis. The follicle shrinks, the dermal papilla is drawn up toward the bulge, and the hair loses its anchorage. When the papilla reaches the bulge, the hair goes into a resting stage called the **telogen**[28] phase. The hair may fall out during catagen or telogen. About 50 to 100 scalp hairs are lost daily. Eventually, anagen begins anew and the cycle repeats itself.

In a young adult, scalp follicles typically spend 6 to 8 years in anagen, 2 to 3 weeks in catagen, and 1 to 3 months in telogen. About 90% of the scalp follicles at any given time are in anagen. Scalp hairs grow at a rate of about 1 mm per 3 days (10–18 cm/yr) during this phase.

Hair grows fastest from adolescence until the 40s. After that, an increasing percentage of follicles are in the catagen and telogen phases rather than the growing anagen phase. Follicles also shrink and begin producing wispy vellus hairs instead of thicker terminal hairs. Thinning of the hair, or baldness, is called **alopecia**[29] (AL-oh-PEE-she-uh). It occurs to some degree in both sexes and may be worsened by disease, poor nutrition, fever, emotional trauma, radiation, or chemotherapy. In the great majority of cases, however, it is simply a matter of aging.

---

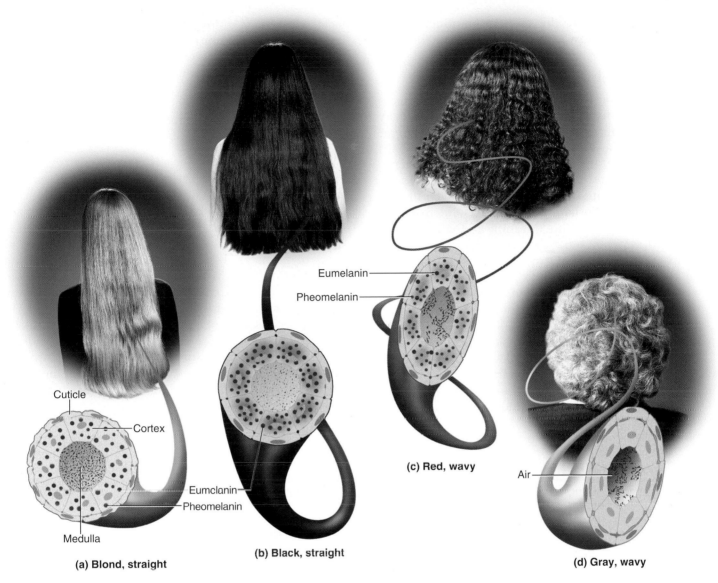

Cuticle

Cortex

Eumclanin

Pheomelanin

Medulla

**(a) Blond, straight**

Eumelanin

Pheomelanin

**(c) Red, wavy**

Air

**(b) Black, straight**

**(d) Gray, wavy**

**FIGURE 6.8 Basis of Hair Color and Texture.** Straight hair (a and b) is round in cross section, and curly hair (c and d) is more oval. Blond hair (a) has scanty eumelanin and a moderate amount of pheomelanin. Eumelanin predominates in black and brown hair (b). Red hair (c) derives its color predominantly from pheomelanin. Gray and white hair (d) lack pigment and have air in the medulla.

**Pattern baldness** is the condition in which hair is lost from specific regions of the scalp rather than thinning uniformly across the entire scalp. It results from a combination of genetic and hormonal influences. The relevant gene has two alleles: one for uniform hair growth and a baldness allele for patchy hair growth. The baldness allele is dominant in males and is expressed only in the presence of the high level of testosterone that is characteristic of men. In men who are either heterozygous or homozygous for the baldness allele, testosterone causes the terminal hair of the scalp to be replaced by thinner vellus, beginning on top of the head and later the sides. In women, the baldness allele is recessive. Homozygous dominant and heterozygous women show normal hair distribution; only homozygous recessive women are at risk of pattern baldness. Even then, they exhibit the trait only if their testosterone levels are abnormally high for a woman (for example, because of a tumor of the adrenal gland, a woman's principal source of testosterone). Such characteristics in which an allele is dominant in one sex and recessive in the other are called *sex-influenced traits.*

Excessive or undesirable hairiness in areas that are not usually hairy, especially in women and children, is called **hirsutism.**[30] It tends to run in families and usually results from either masculinizing ovarian tumors or hypersecretion of testosterone by the adrenal cortex. It is often associated with menopause.

Contrary to popular misconceptions, hair and nails do not continue to grow after a person dies, cutting hair does

---

[30]*hirsut* = shaggy

not make it grow faster or thicker, and emotional stress cannot make the hair turn white overnight.

## Functions of Hair

Compared with other mammals, the relative hairlessness of humans is so unusual that it raises the question, Why do we have any hair at all? What purpose does it serve? There are different answers for the different types of hair; furthermore, some of the answers would make little sense if we limited our frame of reference to industrialized societies, where barbers and hairdressers are engaged to alter the natural state of the hair. It is more useful to take a comparative approach to this question and consider the purposes hair serves in other species of mammals.

Most hair of the human trunk and limbs is probably best interpreted as vestigial, with little present function. Body hair undoubtedly served to keep our ancestors warm, but in modern humans it is too scanty for this purpose. Stimulation of the hair receptors, however, alerts us to parasites crawling on the skin, such as lice and fleas.

The scalp is normally the only place where the hair is thick enough to retain heat. Heat loss from a bald scalp can be substantial and quite uncomfortable. The brain receives a rich supply of warm blood, and most of the scalp lacks an insulating fat layer. Heat is easily conducted through the bone of the skull and lost to the surrounding air. In addition, without hair there is nothing to break the wind and stop it from carrying away heat. Hair also protects the scalp from sunburn, since the scalp is otherwise most directly exposed to the sun's rays. These may be the reasons humans have retained thick hair on their heads while losing most of it from the rest of the body.

Tufts and patches of hair, sometimes with contrasting colors, are important among mammals for advertising species, age, sex, and individual identity. For the less groomed members of the human species, scalp hair may play a similar role. The indefinitely growing hair of a man's scalp and beard, for example, could provide a striking contrast to a face that is otherwise almost hairless. It creates a badge of recognition instantly visible at a distance.

The beard and pubic and axillary hair signify sexual maturity and aid in the transmission of sexual scents. We will further reflect on this in a later discussion of apocrine sweat glands, whose distribution and function add significant evidence to support this theory.

Stout protective **guard hairs,** or **vibrissae** (vy-BRISS-ee), guard the nostrils and ear canals and prevent foreign particles from entering easily. The eyelashes can shield the eye from windblown debris with a quick blink. In windy or rainy conditions, we can squint so that the eyelashes protect the eyes without completely obstructing our vision.

The eyebrows are often presumed to keep sweat or debris out of the eyes, but this seems a negligible role. It is more plausible that they function mainly to enhance facial expression. Movements of the eyebrows are an important means of nonverbal communication in humans of all cultures, and we even have special *frontalis* muscles for this purpose. Eyebrow expressiveness is not unique to humans; many species of monkeys and apes use quick flashes of the eyebrows to greet each other, assert their dominance, and break up quarrels.

## NAILS

Fingernails and toenails are clear, hard derivatives of the stratum corneum. They are composed of very thin, dead, scaly cells, densely packed together and filled with parallel fibers of hard keratin. Most mammals have claws, whereas flat nails are one of the distinguishing characteristics of primates. Flat nails allow for more fleshy and sensitive fingertips, while they also serve as strong keratinized "tools" that can be used for digging, grooming, picking apart food, and other manipulations.

Fingernails grow at a rate of about 1 mm per week and toenails somewhat more slowly. New cells are added to the nail plate by mitosis in the *nail matrix* at its proximal end. Contrary to some advertising claims, adding gelatin to the diet has no effect on the growth or hardness of the nails.

The anatomical features of a nail are shown in figure 6.9. The most important of these are the aforementioned **nail matrix,** a growth zone concealed beneath the skin at the proximal edge of the nail, and the **nail plate,** which is the visible portion covering the fingertip. The nail groove and the space beneath the free edge accumulate dirt and bacteria and require special attention when scrubbing for duty in an operating room or nursery.

The appearance of the nails can be valuable to medical diagnosis. An iron deficiency, for example, may cause the nails to become flat or concave (spoonlike) rather than convex. The nails and fingertips become clubbed in conditions of long-term hypoxemia—deficiency of oxygen in the blood—resulting from congenital heart defects and other causes.

### Before You Go On

*Answer the following questions to test your understanding of the preceding section:*

8. *What is the difference between vellus and terminal hair?*

9. *State the functions of the hair papilla, hair receptors, and piloerector.*

10 *Describe what happens in the anagen, catagen, and telogen phases of the hair cycle.*

11. *State some reasonable theories for the different functions of hair of the eyebrows, eyelashes, scalp, nostrils, and axilla.*

12. *Define or describe the nail plate, nail fold, eponychium, hyponychium, and nail matrix.*

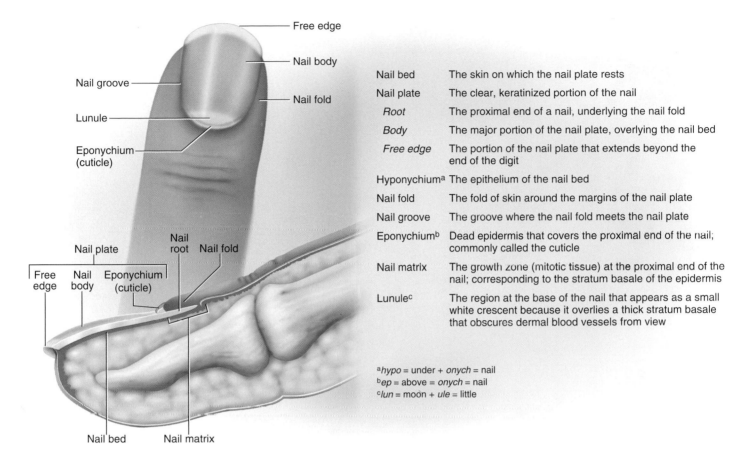

| | |
|---|---|
| Nail bed | The skin on which the nail plate rests |
| Nail plate | The clear, keratinized portion of the nail |
| *Root* | The proximal end of a nail, underlying the nail fold |
| *Body* | The major portion of the nail plate, overlying the nail bed |
| *Free edge* | The portion of the nail plate that extends beyond the end of the digit |
| Hyponychium[a] | The epithelium of the nail bed |
| Nail fold | The fold of skin around the margins of the nail plate |
| Nail groove | The groove where the nail fold meets the nail plate |
| Eponychium[b] | Dead epidermis that covers the proximal end of the nail; commonly called the cuticle |
| Nail matrix | The growth zone (mitotic tissue) at the proximal end of the nail; corresponding to the stratum basale of the epidermis |
| Lunule[c] | The region at the base of the nail that appears as a small white crescent because it overlies a thick stratum basale that obscures dermal blood vessels from view |

[a]*hypo* = under + *onych* = nail
[b]*ep* = above = *onych* = nail
[c]*lun* = moon + *ule* = little

**FIGURE 6.9**  Anatomy of a Fingernail.

# Cutaneous Glands

### Objectives

When you have completed this section, you should be able to

- name two types of sweat glands, and describe the structure and function of each;
- describe the location, structure, and function of sebaceous and ceruminous glands; and
- discuss the distinction between breasts and mammary glands, and explain their respective functions.

The skin has five types of glands: *merocrine sweat glands, apocrine sweat glands, sebaceous glands, ceruminous glands,* and *mammary glands.*

## SWEAT GLANDS

Sweat glands, or **sudoriferous**[31] (soo-dor-IF-er-us) **glands,** are of two kinds, described in chapter 5: merocrine and apocrine. **Merocrine (eccrine) sweat glands,** the most numerous glands of the skin, produce watery perspiration that serves primarily to cool the body (fig. 6.10a). There are 3 to 4 million of these in the adult skin, with a total weight about equal to that of a kidney. They are especially abundant on the palms, soles, and forehead, but they are widely distributed over the rest of the body as well. Each is a simple tubular gland with a twisted coil in the dermis or hypodermis and an undulating or coiled duct leading to a sweat pore on the skin surface. This duct is lined by a stratified cuboidal epithelium in the dermis and by keratinocytes in the epidermis. Amid the secretory cells at the deep end of the gland, there are specialized **myoepithelial**[32] **cells** with properties similar to smooth muscle. They contract in response to stimuli from the sympathetic nervous system and squeeze perspiration up the duct.

Sweat begins as a protein-free filtrate of the blood plasma produced by the deep secretory portion of the gland. Most sodium chloride is reabsorbed from this filtrate as the secretion passes through the duct. Potassium ions, urea, lactic acid, ammonia, and some sodium chloride remain in the sweat. Some drugs are also excreted in the perspiration. On average, sweat is 99% water and has a pH

---

[31]*sudor* = sweat + *fer* = carry, bear

[32]*myo* = muscle

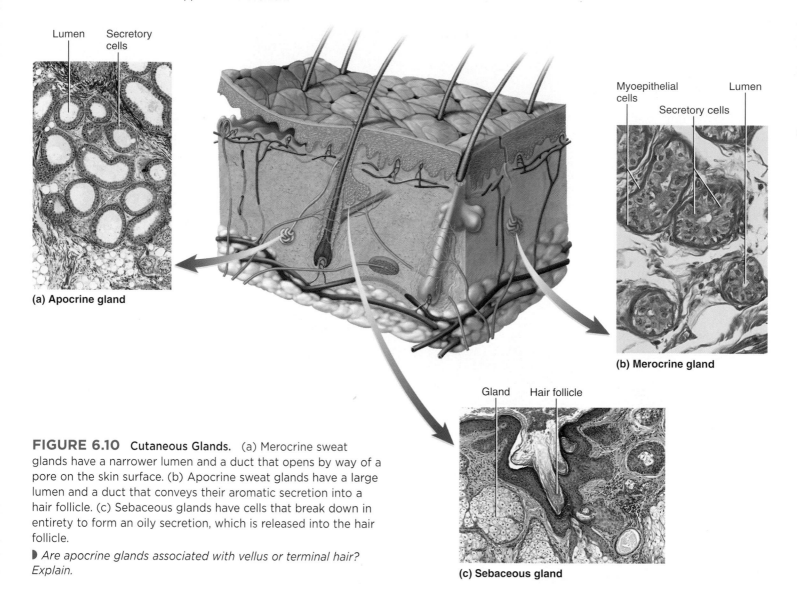

Lumen    Secretory cells

**(a) Apocrine gland**

Myoepithelial cells    Lumen

Secretory cells

**(b) Merocrine gland**

Gland    Hair follicle

**(c) Sebaceous gland**

**FIGURE 6.10  Cutaneous Glands.**  (a) Merocrine sweat glands have a narrower lumen and a duct that opens by way of a pore on the skin surface. (b) Apocrine sweat glands have a large lumen and a duct that conveys their aromatic secretion into a hair follicle. (c) Sebaceous glands have cells that break down in entirety to form an oily secretion, which is released into the hair follicle.

▶ *Are apocrine glands associated with vellus or terminal hair? Explain.*

ranging from 4 to 6. Each day, the sweat glands secrete about 500 mL of **insensible perspiration,** which does not produce noticeable wetness of the skin. Sweating with wetness of the skin is called **diaphoresis**[33] (DY-uh-foe-REE-sis). Under conditions of exercise or heat, a person may lose as much as a liter of perspiration each hour. In fact, so much fluid can be lost from the bloodstream by sweating as to cause circulatory shock.

**Apocrine sweat glands** occur in the groin, anal region, axilla, and areola, and in mature males, the beard area. They are absent from the axillary region of Koreans, however, and are sparse in the Japanese. Their ducts lead into nearby hair follicles rather than opening directly onto the skin surface (fig. 6.10, center figure). They produce their secretion in the same way that merocrine glands do—by exocytosis. The secretory part of an apocrine gland, however, has a much larger lumen than that of a merocrine gland, so these glands have continued to be

referred to as apocrine to distinguish them functionally and histologically from the merocrine type. Apocrine sweat is thicker and more milky than merocrine sweat because it has more fatty acids in it.

Apocrine sweat glands are scent glands that respond especially to stress and sexual stimulation. They do not develop until puberty, and they apparently correspond to the scent glands that develop in other mammals on attainment of sexual maturity. In women, they enlarge and shrink in phase with the menstrual cycle. Apocrine sweat does not have a disagreeable odor, and indeed it is considered attractive or arousing in some cultures, where it is as much a part of courtship as artificial perfume is to other people. Clothing, however, traps stale sweat long enough for bacteria to degrade the secretion and release free fatty acids with a rancid odor. Disagreeable body odor is called *bromhidrosis*.[34] It occasionally indicates a metabolic disorder, but more often it reflects poor hygiene.

[33]*dia* = through + *phoresis* = carrying

[34]*brom* = stench + *hidros* = sweat

Many mammals have apocrine scent glands associated with specialized tufts of hair. In humans, apocrine glands are found mainly in the regions covered by the pubic hair, axillary hair, and beard. This suggests that like other mammalian scent glands, they serve to produce *pheromones,* chemicals that influence the physiology or behavior of other members of the species (see Insight 16.1, p. 597). The hair serves to retain the aromatic secretion and regulate its rate of evaporation from the skin. Thus, it seems no mere coincidence that women's faces lack both apocrine scent glands and a beard.

## SEBACEOUS GLANDS

**Sebaceous**[35] (see-BAY-shus) **glands** produce an oily secretion called **sebum** (SEE-bum). They are flask-shaped, with short ducts that usually open into a hair follicle (fig. 6.10c), although some of them open directly onto the skin surface. These are holocrine glands with little visible lumen. Their secretion consists of broken-down cells that are replaced by mitosis at the base of the gland. Sebum keeps the skin and hair from becoming dry, brittle, and cracked. The sheen of well-brushed hair is due to sebum distributed by the hairbrush. Ironically, we go to great lengths to wash sebum from the skin, only to replace it with various skin creams and hand lotions made of little more than lanolin, which is sheep sebum.

## CERUMINOUS GLANDS

**Ceruminous** (seh-ROO-mih-nus) **glands** are found only in the external ear canal, where their secretion combines with sebum and dead epidermal cells to form earwax, or **cerumen.**[36] They are simple, coiled, tubular glands with ducts leading to the skin surface. Cerumen keeps the eardrum pliable, waterproofs the auditory canal, and has a bactericidal effect.

## MAMMARY GLANDS

The **mammary glands** and breasts (mammae) are often mistakenly regarded as one and the same. Breasts, however, are present in both sexes, and even in females they rarely contain more than small traces of mammary gland. Some authorities regard the female breast as one of the *secondary sex characteristics*—anatomical features whose function lies primarily in their appeal to the opposite sex. The mammary glands, by contrast, are the milk-producing glands that develop within the female breast only during pregnancy and lactation. Mammary glands are modified apocrine sweat glands that produce a richer secretion and channel it through ducts to a nipple for more efficient conveyance to the offspring. The anatomy

| TABLE 6.2 | Cutaneous Glands |
|---|---|
| **Gland Type** | **Definition** |
| Sudoriferous glands | Sweat glands |
|   Merocrine | Sweat glands that function in evaporative cooling; widely distributed over the body surface; open by ducts onto the skin surface |
|   Apocrine | Sweat glands that function as scent glands; found in the regions covered by the pubic, axillary, and male facial hair; open by ducts into hair follicles |
| Sebaceous glands | Oil glands associated with hair follicles |
| Ceruminous glands | Glands of the ear canal that produce cerumen (earwax) |
| Mammary glands | Milk-producing glands located in the breasts |

and physiology of the mammary gland are discussed in more detail in chapter 28.

In most mammals, two rows of mammary glands form along lines called the *mammary ridges,* or *milk lines.* Primates have dispensed with all but two of these glands. A few people, however, develop additional nipples or mammae along the milk line inferior to the primary mammae. In the Middle Ages and colonial America, this condition, called *polythelia,*[37] was used to incriminate women as supposed witches.

The glands of the skin are summarized in table 6.2.

### Before You Go On

*Answer the following questions to test your understanding of the preceding section:*

13. How do merocrine and apocrine sweat glands differ in structure and function?

14. What other type of gland is associated with hair follicles? How does its mode of secretion differ from that of sweat glands?

15. What is the difference between the breast and mammary gland?

# Skin Disorders

### Objectives

When you have completed this section, you should be able to

- describe the three most common forms of skin cancer; and
- describe the three classes of burns and the priorities in burn treatment.

---

[35]*seb* = fat, tallow + *aceous* = possessing
[36]*cer* = wax

[37]*poly* = many + *theli* = nipples

Because it is the most exposed of all our organs, skin is not only the most vulnerable to injury and disease but is also the one place where we are most likely to notice anything out of the ordinary. Skin diseases become increasingly common in old age, and the great majority of people over age 70 have complaints about their integumentary system. Aging of the skin is discussed more fully on page 1128. The healing of cuts and other injuries to the skin occurs by the process described at the end of chapter 5. We focus here on two particularly common and serious disorders: skin cancer and burns. Other skin diseases are briefly summarized in table 6.3.

## SKIN CANCER

Skin cancer is induced by the ultraviolet rays of the sun. It occurs most often on the head and neck, where exposure to the sun is greatest. It is most common in fair-skinned people and the elderly, who have had the longest lifetime UV exposure (Insight 6.4). The ill-advised popularity of suntanning, however, has caused an alarming increase in skin cancer among younger people. Skin cancer is one of the most common cancers, but it is also one of the easiest to treat and has one of the highest survival rates when it is detected and treated early.

### Think About It

*Skin cancer is relatively rare in people with dark skin. Other than possible differences in behavior, such as lack of intentional suntanning, why do you think this is so?*

## INSIGHT 6.4    Clinical Application

### UVA, UVB, and Sunscreens

Some people distinguish between two forms of ultraviolet radiation and argue, fallaciously, that one is less harmful than the other. UVA has wavelengths ranging from 320 to 400 nm and UVB has wavelengths from 290 to 320 nm. (Visible light starts at about 400 nm, the deepest violet we can see.) UVA and UVB are sometimes called "tanning rays" and "burning rays," respectively. Tanning salons often advertise that the UVA rays they use are safe, but public health authorities know better. UVA can burn as well as tan, and it inhibits the immune system, while both UVA and UVB are now thought to initiate skin cancer. As dermatologists say, there is no such thing as a healthy suntan.

Whether or not sunscreens help to protect against skin cancer remains unproven. As the sale of sunscreen has risen in recent decades, so has the incidence of skin cancer. Indeed, recent studies have shown that people who use sunscreens have a higher incidence of basal cell carcinoma than people who do not, while data relating malignant melanoma to sunscreen use are still contradictory and inconclusive. Some of the chemicals used in sunscreens damage DNA and generate harmful free radicals when exposed to UV—chemicals such as zinc oxide, titanium dioxide, and para-aminobenzoic acid (PABA) (now discontinued). The jury is still out, however, on how sunscreens react on and with the skin; authorities cannot say with certainty, yet, whether sunscreens provide any protection or do any harm. Not enough data are available yet on older people who have used sunscreen their entire lives. Given the uncertainty about their effectiveness, it is best not to assume that sunscreens protect you from skin cancer and, above all, not to assume that using a sunscreen means that you can safely stay out longer in the sun.

| TABLE 6.3 | Some Disorders of the Integumentary System |
|---|---|
| Acne | Inflammation of the sebaceous glands, especially beginning at puberty; follicle becomes blocked with keratinocytes and sebum and develops into a whitehead *(comedo)* composed of these and bacteria; continued inflammation of follicle results in pus production and appearance of pimples, and oxidation of sebum turns a whitehead into a blackhead. |
| Dermatitis | Any inflammation of the skin, typically marked by itching and redness; often *contact dermatitis,* caused by exposure to toxins such as poison ivy. |
| Eczema (ECK-zeh-mah) | Itchy, red, "weeping" skin lesions caused by an allergy, usually beginning before age 5; may progress to thickened, leathery, darkly pigmented patches of skin. |
| Psoriasis (so-RY-ah-sis) | Recurring, reddened plaques covered with silvery scale; sometimes disfiguring; possibly caused by an autoimmune response; runs in families. |
| Rosacea (ro-ZAY-she-ah) | A red rashlike area, often in the area of the nose and cheeks, marked by fine networks of dilated blood vessels; worsened by hot drinks, alcohol, and spicy food. |
| Seborrheic (seb-oh-REE-ik) dermatitis | Recurring patches of scaly white or yellowish inflammation often on the head, face, chest, and back; called *cradle cap* (yellow, crusty scalp lesion) in infants. Cause unknown, but correlated with genetic and climatic factors. |
| Tinea | Any fungal infection of the skin; common in moist areas such as the axilla, groin, and foot *(athlete's foot).* Misnamed *ringworm* because of the circular, wormlike growth pattern sometimes exhibited. |

***Disorders described elsewhere***

Abnormal skin coloration p. 195

Baldness p. 198–199

Birthmarks p. 196

Burns p. 206

Genital warts p. 1060

Hirsutism p. 199

Pemphigus vulgaris p. 175

Polythelia p. 203

Skin cancer p. 204

There are three types of skin cancer, named for the epidermal cells in which they originate: *basal cell carcinoma, squamous cell carcinoma,* and *malignant melanoma.* The three types are also distinguished from each other by the appearance of their **lesions**[38] (zones of tissue injury).

**Basal cell carcinoma**[39] is the most common type, but it is also the least dangerous because it seldom metastasizes. It arises from cells of the stratum basale and eventually invades the dermis. On the surface, the lesion first appears as a small, shiny bump. As the bump enlarges, it often develops a central depression and a beaded "pearly" edge (fig. 6.11a).

**Squamous cell carcinoma** arises from keratinocytes of the stratum spinosum. Lesions usually appear on the scalp, ears, lower lip, or back of the hand. They have a raised, reddened, scaly appearance, later forming a concave ulcer with raised edges (fig. 6.11b). The chance of recovery is good with early detection and surgical removal, but if it goes unnoticed or is neglected, this cancer tends to metastasize to the lymph nodes and can be lethal.

**Malignant melanoma** is the most deadly skin cancer but accounts for only 5% of all cases. It often arises from the melanocytes of a preexisting mole. It metastasizes quickly and is often fatal if not treated immediately. The risk for malignant melanoma is greatest in people who experienced severe sunburns as children, especially redheads. Men have a significantly higher incidence of malignant melanoma than women do.

About two-thirds of cases of malignant melanoma in men result from an oncogene called *BRAF.* In women, *BRAF* does not appear to trigger malignant melanoma, but it has been linked to some breast and ovarian cancers. *BRAF* mutations are commonly found in moles. *BRAF* was recently discovered in the course of the new Cancer Genome Project, a multinational effort to identify cancer genes.

It is important to distinguish a mole from malignant melanoma. A mole usually has a uniform color and even contour, and it is no larger in diameter than the end of a pencil eraser (about 6 mm). If it becomes malignant, however, it forms a large, flat, spreading lesion with a scalloped border (fig. 6.11c). The American Cancer Society suggests an "ABCD rule" for recognizing malignant melanoma: *A* for asymmetry (one side of the lesion looks different from the other); *B* for border irregularity (the contour is not uniform but wavy or scalloped); *C* for color (often a mixture of brown, black, tan, and sometimes red and blue); and *D* for diameter (greater than 6 mm).

Skin cancer is treated by surgical excision, radiation therapy, or destruction of the lesion by heat (electrodesiccation) or cold (cryosurgery).

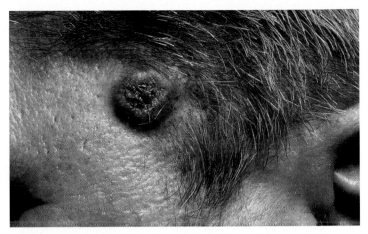

**(a) Basal cell carcinoma**

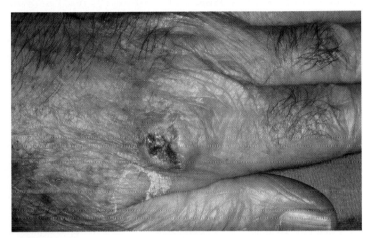

**(b) Squamous cell carcinoma**

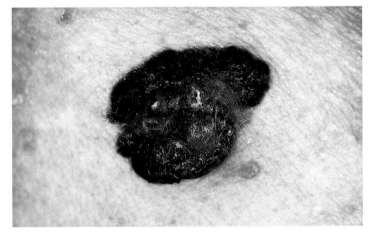

**(c) Malignant melanoma**

**FIGURE 6.11** Skin Cancer.

▶ *Which of the ABCD rules can you identify in part (c)?*

---

[38]*lesio* = injure
[39]*carcin* = cancer + *oma* = tumor

# BURNS

**Burns** are the leading cause of accidental death. They are usually caused by fires, kitchen spills, or excessively hot bath water, but they also can be caused by sunlight, ionizing radiation, strong acids and bases, or electrical shock. Burn deaths result primarily from fluid loss, infection, and the toxic effects of **eschar** (ESS-car)—the burned, dead tissue.

Burns are classified according to the depth of tissue involvement (fig. 6.12). **First-degree burns** involve only the epidermis and are marked by redness, slight edema, and pain. They heal in a few days and seldom leave scars. Most sunburns are first-degree burns.

**Second-degree burns** involve the epidermis and part of the dermis but leave at least some of the dermis intact. First- and second-degree burns are therefore also known as **partial-thickness burns.** A second-degree burn may be red, tan, or white and is blistered and very painful. It may take from 2 weeks to several months to heal and may leave scars. The epidermis regenerates by division of epithelial cells in the hair follicles and sweat glands and around the edges of the lesion. Some sunburns and many scalds are second-degree burns.

**Third-degree burns** are also called **full-thickness burns** because the epidermis, dermis, and often some deeper tissue are completely destroyed. Since no dermis remains, the skin can regenerate only from the edges of the wound. Third-degree burns often require skin grafts (Insight 6.5). If a third-degree burn is left to itself to heal, contracture (abnormal connective tissue fibrosis) and severe disfigurement may result.

> **Think About It**
>
> *A third-degree burn may be surrounded by painful areas of first- and second-degree burns, but the region of the third-degree burn is painless. Explain the reason for this lack of pain.*

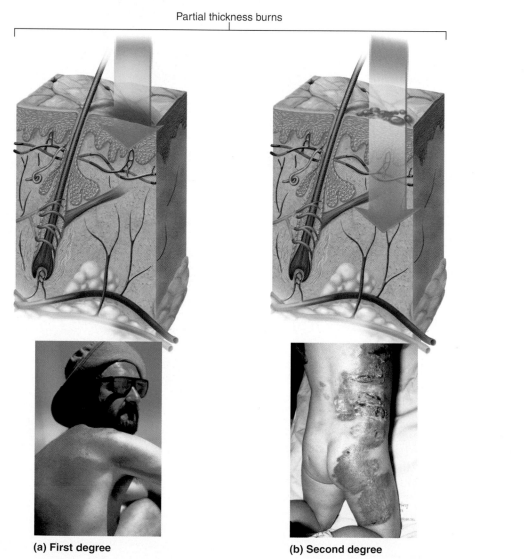
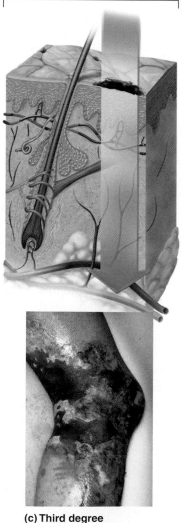

**(a) First degree**    **(b) Second degree**    **(c) Third degree**

**FIGURE 6.12** **Burns.** (a) First-degree burn, involving only the epidermis. (b) Second-degree burn, involving the epidermis and part of the dermis. (c) Third-degree burn, extending through the entire dermis and often involving even deeper tissue.

## INSIGHT 6.5    Clinical Application

### Skin Grafts and Artificial Skin

Third-degree burns leave no dermal tissue to regenerate what was lost, and therefore they generally require skin grafts. The ideal graft is an *autograft*[40]—tissue taken from another location on the same person's body—because it is not rejected by the immune system. An autograft is performed by taking epidermis and part of the dermis from an undamaged area such as the thigh or buttock and grafting it to a burned area. This method is called a *split-skin graft* because part of the dermis is left behind to proliferate and replace the epidermis that was removed—the same way a second-degree burn heals.

An autograft may not be possible, however, if the burns are too extensive. The best treatment option in this case is an *isograft*,[41] which uses skin from an identical twin. Because the donor and recipient are genetically identical, the recipient's immune system is unlikely to reject the graft. Since identical twins are rare, however, the best one can hope for in most cases is donor skin from another close relative.

An *allograft*,[42] or *homograft*,[43] is a graft from any other person. Skin banks provide skin from deceased persons for this purpose. The immune system attempts to reject allografts, but they suffice as temporary coverings for the burned area. They can be replaced by autografts when the patient is well enough for healthy skin to be removed from an undamaged area of the body.

Pig skin is sometimes used on burn patients but presents the same problem of immune rejection. A graft of tissue from a different species is called a *heterograft*,[44] or *xenograft*.[45] This is a special case of a *heterotransplant*, which also includes transplantation of organs such as baboon hearts or livers into humans. Heterografts and heterotransplants are short-term methods of maintaining a patient until a better, long-term solution is possible. The immune reaction can be suppressed by drugs called *immunosuppressants*. This procedure is risky, however, because it lowers a person's resistance to infection, which is already compromised in a burn patient.

Some alternatives to skin grafts are also being used. Burns are sometimes temporarily covered with amnion (the membrane that surrounds a developing fetus) obtained from afterbirths. In addition, tiny keratinocyte patches cultured with growth stimulants have produced sheets of epidermal tissue as large as the entire body surface. These can replace large areas of burned tissue. Dermal fibroblasts also have been successfully cultured and used for autografts. A drawback to these approaches is that the culture process requires 3 or 4 weeks, which is too long a wait for some patients with severe burns.

Various kinds of artificial skin have also been developed as a temporary burn covering. One concept is a sheet with an upper layer of silicone and a lower layer of collagen and chondroitin sulfate. It stimulates growth of connective tissue and blood vessels from the patient's underlying tissue. The artificial skin can be removed after about 3 weeks and replaced with a thin layer of cultured or grafted epidermis. At least two bioengineering companies have developed artificial skins approved in 1997-98 by the U.S. Food and Drug Administration for patient use. The manufacture of one such product begins by culturing fibroblasts on a collagen gel to produce a dermis, then culturing keratinocytes on this dermal substrate to produce an epidermis. Such products are now being used to treat burn patients as well as leg and foot ulcers that result from diabetes mellitus. This is one aspect of the larger field of *tissue engineering*, which biotechnology companies hope will lead, within a few decades, even to engineering replacement livers and other organs (see Insight 5.3).

---

The two most urgent considerations in treating a burn patient are fluid replacement and infection control. A patient can lose several liters of water, electrolytes, and protein each day from the burned area. As fluid is lost from the tissues, more is transferred from the bloodstream to replace it, and the volume of circulating blood declines. A patient may lose up to 75% of the blood plasma within a few hours, potentially leading to circulatory shock and cardiac arrest—the principal cause of death in burn patients. Intravenous fluid must be given to make up for this loss. A severely burned patient may also require thousands of extra calories daily to compensate for protein loss and the demands of tissue repair. Supplementary nutrients are given intravenously or through a gastric tube.

Infection is controlled by keeping the patient in an aseptic (germ-free) environment and administering antibiotics. The eschar is sterile for the first 24 hours, but then it quickly becomes infected and may have toxic effects on the digestive, respiratory, and other systems. Its removal, called **debridement**[46] (deh-BREED-ment), is essential to infection control.

### Before You Go On

*Answer the following questions to test your understanding of the preceding section:*

16. *What types of cells are involved in each type of skin cancer?*

17. *Which type of skin cancer is most dangerous? What are its early warning signs?*

18. *What are the differences between a first-, second-, and third-degree burn?*

19. *What are the two most urgent priorities in treating a burn victim? How are these needs dealt with?*

---

[40]*auto* = self
[41]*iso* = same
[42]*allo* = different, other
[43]*homo* = same
[44]*hetero* = different
[45]*xeno* = strange, alien

[46]*de* = un + *bride* = bridle

# CONNECTIVE ISSUES

## Interactions Between the
## INTEGUMENTARY SYSTEM
## and Other Organ Systems

■ indicates ways in which this system affects other systems

■ indicates ways in which other systems affect this system

## SKELETAL SYSTEM

Role of skin in vitamin D synthesis promotes calcium absorption needed for bone growth and maintenance

Supports skin at scalp and other places where bone lies close to surface

## MUSCULAR SYSTEM

Vitamin D synthesis promotes absorption of calcium needed for muscle contraction; skin dissipates heat generated by muscles

Active muscles generate heat and warm the skin; contractions of skeletal muscles pull on skin and produce facial expressions

## NERVOUS SYSTEM

Sensory impulses from skin transmitted to nervous system

Regulates diameter of cutaneous blood vessels; stimulates perspiration and contraction of piloerector muscles

## ENDOCRINE SYSTEM

Vitamin $D_3$ acts as a hormone

Sex hormones cause changes in integumentary features at puberty; some hormone imbalances have pathological effects on skin

## CIRCULATORY SYSTEM

Dermal vasoconstriction diverts blood to other organs; skin prevents loss of fluid from cardiovascular system; dermal mast cells cause vasodilation and increased blood flow

Delivers $O_2$, nutrients, and hormones to skin and carries away wastes; hemoglobin colors skin

### ALL SYSTEMS

The integumentary system serves all other systems by providing a physical barrier to environmental hazards

## LYMPHATIC/IMMUNE SYSTEMS

Dendritic cells detect foreign substances

Lymphatic system controls fluid balance and prevents edema; immune cells protect skin from infection and promote tissue repair

## RESPIRATORY SYSTEM

Nasal hairs filter particles that might otherwise be inhaled

Provides $O_2$ and removes $CO_2$

## URINARY SYSTEM

Skin complements urinary system by excreting salts and some nitrogenous wastes in sweat

Disposes of wastes and maintains electrolyte and pH balance

## DIGESTIVE SYSTEM

Vitamin D synthesis promotes intestinal absorption of calcium

Provides nutrients needed for integumentary development and function; nutritional disorders often reflected in appearance of skin, hair, and nails

## REPRODUCTIVE SYSTEM

Cutaneous receptors respond to erotic stimuli; mammary glands produce milk to nourish infants; apocrine glands produce scents with subtle sexual functions

Gonadal sex hormones promote growth, maturation, and maintenance of skin

## CHAPTER REVIEW

# Review of Key Concepts

### The Skin and Subcutaneous Tissue (p. 188)

1. *Dermatology* is the study of the *integumentary system,* a system that includes the skin *(integument),* hair, nails, and cutaneous glands.

2. The skin is composed of a superficial *epidermis* of keratinized stratified squamous epithelium, and a deeper *dermis* of fibrous connective tissue. Beneath the skin is a connective tissue *hypodermis.*

3. Skin ranges from less than 0.5 mm to as much as 6 mm thick. *Thick skin* is named for a thick, heavily keratinized epidermis, not necessarily for total thickness; it is found on the palms, soles, and corresponding surfaces of the digits, and it is hairless. The rest of the body is covered with *thin skin,* in which the epidermis is more lightly keratinized.

4. Functions of the skin include resistance to trauma and infection, water retention, vitamin D synthesis, sensation, thermoregulation, and nonverbal communication.

5. The epidermis has five types of cells: *stem cells, keratinocytes, melanocytes, tactile cells,* and *dendritic cells.*

6. Layers of the epidermis from deep to superficial are the *stratum basale, stratum spinosum, stratum granulosum, stratum lucidum* (in thick skin only), and *stratum corneum.*

7. Keratinocytes are the majority of epidermal cells. They originate by mitosis of stem cells in the stratum basale and push the older keratinocytes upward. Keratinocytes flatten and produce *membrane-coating vesicles* and cytoskeletal filaments as they migrate upward. In the stratum granulosum, the cytoskeletal filaments are transformed to keratin, the membrane-coating vesicles release lipids that help to render the cells water-resistant, and the cells undergo apoptosis. Above the stratum granulosum, dead keratinocytes become compacted into the stratum corneum.

Thirty to 40 days after its mitotic birth, the average keratinocyte flakes off the epidermal surface. This loss of dead cells is called *exfoliation.*

8. The dermis is 0.2 to 4 mm thick. It is composed mainly of collagen but also includes elastic and reticular fibers, fibroblasts, and other cell types. It contains blood vessels, sweat glands, sebaceous glands, nerve endings, hair follicles, nail roots, smooth muscle, and in the face, skeletal muscle.

9. In most places, upward projections of the dermis called *dermal papillae* interdigitate with downward *epidermal ridges* to form a wavy boundary. The papillae form the friction ridges of the fingertips and irregular ridges, separated by furrows, elsewhere.

10. The dermis is composed of a superficial *papillary layer,* which is composed of areolar tissue and forms the dermal papillae; and a thicker, deeper *reticular layer* composed of dense irregular connective tissue. The reticular layer provides toughness to the dermis, while the papillary layer forms an arena for the mobilization of defenses against pathogens that breach the epidermis.

11. The *hypodermis* (subcutaneous tissue) is composed of more areolar and adipose tissue than the reticular layer of dermis. It pads the body and binds the skin to underlying muscle or other tissues. In areas composed mainly of adipocytes, it is called *subcutaneous fat.*

12. Normal skin colors result from various proportions of *eumelanin, pheomelanin,* the hemoglobin the blood, the white collagen of the dermis, and dietary *carotene.* Pathological conditions with abnormal skin coloration include *cyanosis, erythema, pallor, albinism, jaundice, bronzing, hematomas,* and *hemangiomas* (birthmarks).

13. Skin markings include *friction ridges* of the fingertips (the source of oily fingerprints); *flexion lines* of the palm, wrist, and other places; *freckles* and *moles;* and *hemangiomas* (birthmarks).

### Hair and Nails (p. 196)

1. Hair and nails are composed of compact, highly cross-linked *hard keratin.*

2. A hair *(pilus)* is a slender filament of keratinized cells growing from an oblique *hair follicle.*

3. The three types of hair are *lanugo,* present only prenatally; *vellus,* a fine unpigmented body hair; and the coarser, pigmented *terminal hair* of the eyebrows, scalp, beard, and other areas.

4. Deep in the follicle, a hair begins with a dilated *bulb,* continues as a narrower *root* below the skin surface, and extends above the skin as the *shaft.* The bulb contains a *dermal papilla* of vascular connective tissue. The *hair matrix* just above the papilla is the site of hair growth by mitosis of the matrix cells. In cross section, a hair exhibits a thin outer *cuticle,* a thicker layer of keratinized cells forming the hair *cortex,* and a core called the *medulla.*

5. A hair follicle consists of an inner *epithelial root sheath* (an extension of the epidermis) and an outer *connective tissue root sheath.* It is supplied by nerve endings called *hair receptors* that detect hair movements, and a bundle of smooth muscle called the *piloerector muscle,* which erects the hair.

6. Differences in hair texture are attributable to differences in cross-sectional shape—straight hair is round, wavy hair is oval, and tightly curly hair is relatively flat.

7. Variations in hair color arise from the relative amounts of eumelanin and pheomelanin.

8. A hair has a life cycle consisting of a growing *anagen* stage, a shrinking *catagen* stage, and a resting *telogen* stage. Hairs usually fall out during the catagen or telogen stage. A scalp hair typically lives 6 to 8 years and grows about 10 to 18 cm/yr.

9. Generalized thinning of hair is called *alopecia.* Loss of hair from specific regions of scalp is called *pattern baldness,* and is a hereditary sex-influenced condition. Excessive hairiness is called *hirsutism.*

10. The functions of hair include thermal insulation, protection from the sun and foreign objects, sensation, facial expression, signaling sexual maturity, and regulating the dispersal of pheromones.

11. Fingernails and toenails are hard plates of densely packed, dead, keratinized cells. They arise from a growth zone called the *nail matrix.*

## Cutaneous Glands (p. 201)

1. The most abundant and widespread sweat glands are *merocrine sweat glands,* which produce a watery secretion that cools the body.

Merocrine glands release their product by exocytosis.

2. *Apocrine sweat glands* are associated with hair follicles in the pubic and anal regions, axilla, areola, and beard. They develop at puberty along with the appearance of hair in these regions, and apparently function to secrete sex pheromones. Apocrine sweat glands also release their secretion by exocytosis.

3. *Sebaceous glands,* also usually associated with hair follicles, produce an oily secretion called *sebum,* which keeps the skin and hair pliable. These are holocrine glands; their cells break down in entirety to form the secretion.

4. *Ceruminous glands* are found in the auditory canal. *Cerumen,* or earwax, is a mixture of ceruminous gland secretion, sebum, and dead epidermal cells. It keeps the eardrum pliable, waterproofs the auditory canal, and kills bacteria.

5. *Mammary glands* are modified apocrine sweat glands that develop in the breast during pregnancy and lactation and produce milk.

## Skin Disorders (p. 203)

1. The three forms of skin cancer are defined by the types of cells in which they originate: *basal cell carcinoma* (the most common but least dangerous form), *squamous cell carcinoma,* and *malignant melanoma* (the least common but most dangerous form).

2. Burns can be first-degree (epidermal damage only, as in a mild sunburn), second-degree (some dermal damage, as in a scald), or third-degree (complete penetration of the dermis, as in burns from fire).

3. People with serious burns must be treated first to restore fluid balance and second to control infection.

# Testing Your Recall

1. Cells of the _____ are keratinized and dead.
   a. papillary layer
   b. stratum spinosum
   c. stratum basale
   d. stratum corneum
   e. stratum granulosum

2. Which of the following terms is *least* related to the rest?
   a. subcutaneous fat
   b. superficial fascia
   c. reticular layer
   d. hypodermis
   e. subcutaneous tissue

3. Which of the following skin conditions or appearances would most likely result from liver failure?
   a. pallor
   b. erythema
   c. pemphigus vulgaris
   d. jaundice
   e. melanization

4. All of the following interfere with microbial invasion of the skin *except*
   a. the acid mantle.
   b. melanization.
   c. inflammation.
   d. keratinization.
   e. sebum.

5. The hair on a 6-year-old's arms is
   a. vellus.
   b. lanugo.

   c. alopecia.
   d. terminal hair.
   e. rosacea.

6. Which of the following terms is *least* related to the rest?
   a. lunule
   b. nail plate
   c. hyponychium
   d. free edge
   e. cortex

7. Which of the following is a scent gland?
   a. eccrine gland
   b. sebaceous gland
   c. apocrine gland
   d. ceruminous gland
   e. merocrine gland

8. _____ are skin cells with a sensory role.
   a. Tactile cells
   b. Dendritic cells
   c. Prickle cells
   d. Melanocytes
   e. Keratinocytes

9. Which of the following glands produce the acid mantle?
   a. merocrine sweat glands
   b. apocrine sweat glands
   c. mammary glands
   d. ceruminous glands
   e. sebaceous glands

10. Which of the following skin cells alert the immune system to pathogens?
    a. fibroblasts
    b. melanocytes
    c. keratinocytes
    d. dendritic cells
    e. tactile cells

11. _____ is sweating without noticeable wetness of the skin.

12. A muscle that causes a hair to stand on end is called a/an _____.

13. The process of removing burned skin from a patient is called _____.

14. Blueness of the skin due to low oxygen concentration in the blood is called _____.

15. Projections of the dermis toward the epidermis are called _____.

16. Cerumen is more commonly known as _____.

17. The holocrine glands that secrete into a hair follicle are called _____.

18. Hairs grow only during the _____ phase of the hair cycle.

19. A hair is nourished by blood vessels in a connective tissue projection called the _____.

20. A _____ burn destroys the entire dermis.

*Answers in Appendix B*

# True or False

*Determine which five of the following statements are false, and briefly explain why.*

1. Dander consists of dead keratinocytes.

2. The term *integument* means only the skin, but *integumentary system* includes the skin, hair, nails, and cutaneous glands.

3. The dermis is composed mainly of keratin.

4. Vitamin D synthesis begins in certain cutaneous glands.

5. Cells of the stratum granulosum cannot undergo mitosis.

6. Dermal papillae are better developed in skin subjected to a lot of mechanical stress than in skin subjected to less stress.

7. The three layers of the skin are the epidermis, dermis, and hypodermis.

8. People of African descent have a much higher density of epidermal melanocytes than do people of northern European descent.

9. Pallor indicates a genetic lack of melanin.

10. Apocrine sweat glands develop at the same time in life as the pubic and axillary hair.

*Answers in Appendix B*

# Testing Your Comprehension

1. Many organs of the body contain numerous smaller organs, perhaps even thousands. Describe an example of this in the integumentary system.

2. Certain aspects of human form and function are easier to understand when viewed from the perspective of comparative anatomy and evolution. Discuss examples of this in the integumentary system.

3. Explain how the complementarity of form and function is reflected in the fact that the dermis has two histological layers and not just one.

4. Cold weather does not normally interfere with oxygen uptake by the blood, but it can cause cyanosis anyway. Why?

5. Why is it important for the epidermis to be effective, but not *too* effective, in screening out UV radiation?

*Answers at www.mhhe.com/saladin4*

# www.mhhe.com/saladin4

*The textbook website provides a wealth of interactive study materials fully organized and integrated by chapter. You will find practice quizzes, labeling exercises, and much more that will complement your learning and understanding of anatomy and physiology. The website also includes tools designed to enhance your* **Anatomy & Physiology | REVEALED** *experience.*

*Spongy bone of the human femur*

# BONE TISSUE

## CHAPTER OUTLINE

## INSIGHTS

## Brushing Up

To understand this chapter, it is important that you understand or brush up on the following concepts:

Anatomy & Physiology | REVEALED®

In art and history, nothing has symbolized death more than a skull or skeleton.[1] The dry bones presented for laboratory study suggest that the skeleton is a nonliving scaffold for the body, like the steel girders of a building. Seeing it in such a sanitized form makes it easy to forget that the living skeleton is made of dynamic tissues, full of cells—that it continually remodels itself and interacts physiologically with all of the other organ systems of the body. The skeleton is permeated with nerves and blood vessels, which attests to its sensitivity and metabolic activity.

Osteology,[2] the study of bone, is the subject of these next three chapters. In this chapter, we study bone as a tissue—its composition, its functions, how it develops and grows, how its metabolism is regulated, and some of its disorders. This will provide a basis for understanding the skeleton, joints, and muscles in the chapters that follow.

# Tissues and Organs of the Skeletal System

### Objectives

When you have completed this section, you should be able to

- name the tissues and organs that compose the skeletal system;
- state several functions of the skeletal system;
- distinguish between bone as a tissue and as an organ;
- describe four types of bones classified by shape; and
- describe the general features of a long bone and a flat bone.

The **skeletal system** is composed of bones, cartilages, and ligaments joined tightly to form a strong, flexible framework for the body. Cartilage, the forerunner of most bones in embryonic and childhood development, covers many joint surfaces in the mature skeleton. Ligaments hold bones together at the joints and are discussed in chapter 9. Tendons are structurally similar to ligaments but attach muscle to bone; they are discussed with the muscular system in chapter 10.

[1]*skelet* = dried up
[2]*osteo* = bone + *logy* = study of

## FUNCTIONS OF THE SKELETON

The skeleton plays at least six roles:

1. **Support.** Bones of the legs, pelvis, and vertebral column hold up the body; nearly all bones provide support for the muscles; the mandible and maxilla support the teeth.
2. **Protection.** Bones enclose and protect the brain, spinal cord, heart, lungs, pelvic viscera, and bone marrow.
3. **Movement.** Limb movements, breathing, and other movements are produced by the action of muscles on the bones.
4. **Electrolyte balance.** The skeleton stores calcium and phosphate ions and releases them into the tissue fluid and blood according to the body's physiological needs.
5. **Acid–base balance.** Bone tissue buffers the blood against excessive pH changes by absorbing or releasing alkaline salts.
6. **Blood formation.** Red bone marrow is the major producer of blood cells, including blood cells of the immune system.

## BONES AND OSSEOUS TISSUE

Bone, or **osseous**[3] **tissue,** is a connective tissue in which the matrix is hardened by the deposition of calcium phosphate and other minerals. The hardening process is called **mineralization,** or **calcification.** (Bone is not the hardest substance in the body; that distinction goes to tooth enamel.) Osseous tissue is only one of the tissues that make up a bone. Also present are blood, bone marrow, cartilage, adipose tissue, nervous tissue, and fibrous connective tissue. The word *bone* can denote an organ composed of all these tissues, or it can denote just the osseous tissue.

## THE SHAPES OF BONES

Bones are classified into four groups according to their shapes and corresponding functions (fig. 7.1):

1. **Long bones** are conspicuously longer than wide. Like crowbars, they serve as rigid levers that are acted upon by the skeletal muscles to produce body movements. Long bones include the humerus of the arm, the radius and ulna of the forearm, the metacarpals and phalanges of the hand, the femur of the thigh, the tibia and fibula of the leg, and the metatarsals and phalanges of the feet.
2. **Short bones** are nearly equal in length and width. They include the carpal (wrist) and tarsal (ankle) bones. They have limited motion and merely glide across one another, enabling the ankles and wrists to bend in multiple directions.

[3]*os, osse, oste* = bone

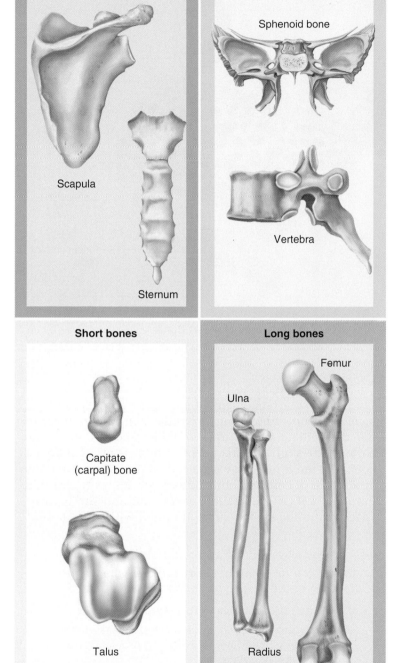

**FIGURE 7.1  Classification of Bones by Shape.**  The light blue areas are articular cartilages.

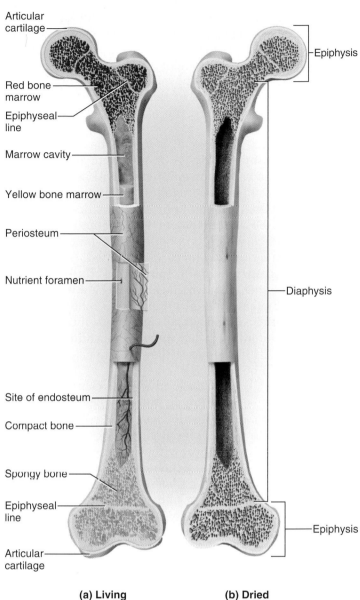

**(a) Living**          **(b) Dried**

**FIGURE 7.2   Anatomy of a Long Bone.**  (a) The femur (thighbone) with its soft tissues including bone marrow, articular cartilage, blood vessels, and periosteum. (b) A dried femur in longitudinal section.

❱ *What is the functional significance of a long bone being wider at the epiphyses than at the diaphysis?*

4. **Irregular bones** have elaborate shapes that do not fit into any of the preceding categories. They include the vertebrae and some skull bones, such as the sphenoid and ethmoid.

## GENERAL FEATURES OF BONES

Figure 7.2 shows a typical long bone. Much of it is composed of a cylinder of dense white osseous tissue called **compact (dense) bone.** The cylinder encloses a space called the **medullary** (MED-you-lerr-ee) **cavity,** or **marrow**

3. **Flat bones** enclose and protect soft organs and provide broad surfaces for muscle attachment. They include most cranial bones and the ribs, sternum (breastbone), scapula (shoulder blade), and os coxae (hipbone).

**cavity,** which contains bone marrow. At the ends of the bone, the central space is occupied by a more loosely organized form of osseous tissue called **spongy (cancellous) bone.** The skeleton is about three-quarters compact bone and one-quarter spongy bone by weight. Spongy bone is found at the ends of the long bones and in the middle of nearly all others. It is always enclosed by more durable compact bone.

The principal features of a long bone are its shaft, called the **diaphysis**[4] (dy-AF-ih-sis), and an expanded head at each end called the **epiphysis**[5] (eh-PIF-ih-sis). The diaphysis provides leverage, and the epiphysis is enlarged to strengthen the joint and provide added surface area for the attachment of tendons and ligaments. The joint surface where one bone meets another is covered with a layer of hyaline cartilage called the **articular cartilage.** Together with a lubricating fluid secreted between the bones, this cartilage enables a joint to move far more easily than it would if one bone rubbed directly against the other. Blood vessels penetrate into the bone through minute holes called **nutrient foramina** (for-AM-ih-nuh); we will trace where they go when we consider the histology of bone.

Externally, a bone is covered with a sheath called the **periosteum.**[6] This has a tough, outer *fibrous layer* of collagen and an inner *osteogenic layer* of bone-forming cells described later in the chapter. Some collagen fibers of the outer layer are continuous with the tendons that bind muscle to bone, and some penetrate into the bone matrix as **perforating (Sharpey**[7]**) fibers.** The periosteum thus provides strong attachment and continuity from muscle to tendon to bone. The osteogenic layer is important to the growth of bone and healing of fractures. There is no periosteum over the articular cartilage. The internal surface of a bone is lined with **endosteum,**[8] a thin layer of reticular connective tissue and *osteogenic cells* that give rise to other types of bone cells.

In children and adolescents, an **epiphyseal** (EP-ih-FIZZ-ee-ul) **plate** of hyaline cartilage separates the marrow spaces of the epiphysis and diaphysis (fig 7.2a). On X rays, it appears as a transparent line at the end of a long bone (see fig. 7.12). The epiphyseal plate is a zone where the bones elongate by a growth process detailed later in the chapter. In adults, the epiphyseal plate is depleted and the bones no longer grow in length, but an *epiphyseal line* marks where the plate used to be.

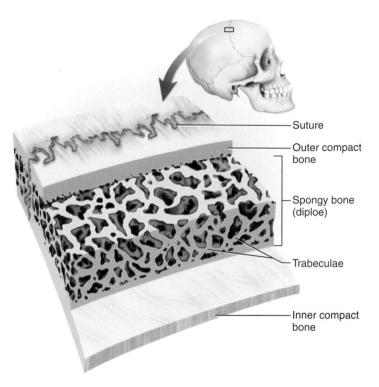

**FIGURE 7.3**   **Anatomy of a Flat Bone.**

Figure 7.3 shows a flat bone of the cranium. It has a sandwichlike arrangement of spongy bone covered by a layer of compact bone on each side. The spongy layer in the cranium is called **diploe**[9] (DIP-lo-ee). A moderate blow to the skull can fracture the outer layer of compact bone, but the diploe may absorb the impact and leave the inner layer of compact bone unharmed. Both surfaces of a flat bone are covered with periosteum, and the marrow spaces amid the spongy bone are lined with endosteum.

## Before You Go On

*Answer the following questions to test your understanding of the preceding section:*

1. *Name at least five tissues found in a bone.*

2. *List three or more functions of the skeletal system other than supporting the body and protecting some of the internal organs.*

3. *Name the four bone shapes and give an example of each.*

4. *Explain the difference between compact and spongy bone, and describe their spatial relationship to each other.*

5. *State the anatomical terms for the shaft, head, growth zone, and fibrous covering of a long bone.*

---

[4]*dia* = across + *physis* = growth; originally named for a ridge on the shaft of the tibia
[5]*epi* = upon, above + *physis* = growth
[6]*peri* = around + *oste* = bone
[7]William Sharpey (1802–80), Scottish histologist
[8]*endo* = within + *oste* = bone

[9]*diplo* = double

# Histology of Osseous Tissue

### Objectives

When you have completed this section, you should be able to

- list and describe the cells, fibers, and ground substance of bone tissue;
- state the importance of each constituent of bone tissue;
- compare the histology of the two types of bone tissue; and
- distinguish between the three types of bone marrow.

## BONE CELLS

Like any other connective tissue, bone consists of cells, fibers, and ground substance. There are four principal types of bone cells (fig. 7.4):

1. **Osteogenic**[10] **cells** are stem cells that develop from fibroblasts and then give rise to most other bone cells. They are found in the endosteum, the inner layer of the periosteum, and in the central canals. They multiply continually, and some of them go on to become the *osteoblasts* described next. Osteoblasts are nonmitotic, so the only source of new ones is mitosis and differentiation of the osteogenic cells.

2. **Osteoblasts**[11] are bone-forming cells. They are roughly cuboidal or angular, and line up in a single layer on the bone surface under the endosteum and periosteum. They synthesize the soft organic matter of the bone matrix, which then hardens by mineral deposition. Stress and fractures stimulate osteogenic cells to multiply more rapidly and quickly generate increased numbers of osteoblasts, which reinforce or rebuild the bone.

3. **Osteocytes** are former osteoblasts that have become trapped in the matrix they deposited. They reside in tiny cavities called **lacunae**,[12] which are interconnected by slender channels called **canaliculi**[13] (CAN-uh-LIC-you-lye). Each osteocyte has delicate

---

[10]*osteo* = bone + *genic* = producing

[11]*osteo* = bone + *blast* = form, produce
[12]*lac* = lake, hollow + *una* = little
[13]*canal* = canal, channel + *icul* = little

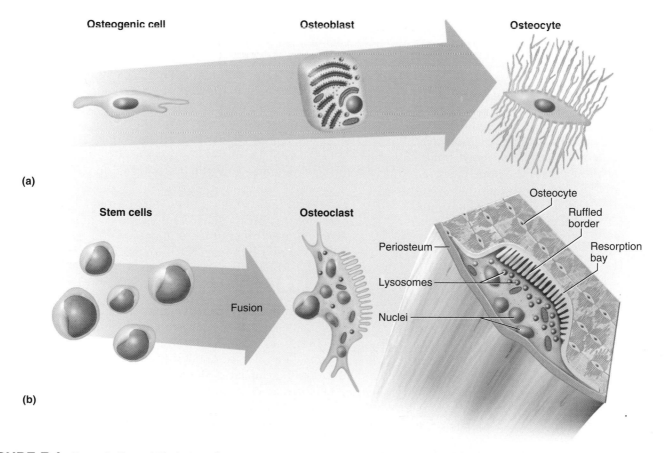

(a)

Osteogenic cell        Osteoblast        Osteocyte

(b)

Stem cells        Osteoclast

Fusion

Osteocyte

Ruffled border

Periosteum

Resorption bay

Lysosomes

Nuclei

**FIGURE 7.4  Bone Cells and Their Development.**  (a) Osteogenic cells give rise to osteoblasts, which deposit matrix around themselves and transform into osteocytes. (b) Bone marrow stem cells fuse to form osteoclasts.

fingerlike cytoplasmic processes that reach into the canaliculi to contact the processes from neighboring osteocytes. Some of them also contact osteoblasts on the bone surface. Neighboring osteocytes are connected by gap junctions where their processes meet, so they can pass nutrients and chemical signals to one another and pass their metabolic wastes to the nearest blood vessel for disposal.

Osteocytes have multiple functions. Some resorb bone matrix and others deposit it, so they contribute to the homeostatic maintenance of both bone density and blood concentrations of calcium and phosphate ions. Perhaps even more importantly, they are strain sensors. When a load is applied to a bone, it produces a flow in the extracellular fluid of the lacunae and canaliculi. This stimulates the osteocytes to secrete biochemical signals that may regulate bone remodeling—adjustments in shape and bone density to adapt to stress.

4. **Osteoclasts**[14] are bone-dissolving cells found on the bone surface. They develop from the bone marrow stem cells that also give rise to monocytes and some other white blood cells. Thus, osteogenic cells, osteoblasts, and osteocytes all belong to one cell lineage (fig. 7.4a), but osteoclasts have an independent origin (fig. 7.4b). Each osteoclast is formed by the fusion of several stem cells, so osteoclasts are unusually large cells (up to 150 μm in diameter, visible to the naked eye). They typically have 3 or 4 nuclei, but sometimes up to 50, each contributed by one stem cell. One side of the osteoclast, facing the bone surface, has a *ruffled border* with many deep infoldings of the plasma membrane. These increase the cell surface area and thus enhance the efficiency of bone resorption. Osteoclasts often reside in pits called *resorption bays (Howship*[15] *lacunae)* that they have etched into the bone surface. Bone remodeling results from the combined action of these bone-dissolving osteoclasts and bone-depositing osteoblasts.

## THE MATRIX

The matrix of osseous tissue is, by dry weight, about one-third organic and two-thirds inorganic matter. The organic matter, synthesized by the osteoblasts, includes collagen and various protein–carbohydrate complexes such as glycosaminoglycans, proteoglycans, and glycoproteins. The inorganic matter is about 85% **hydroxyapatite,** a crystallized calcium phosphate salt $[Ca_{10}(PO_4)_6(OH)_2]$, 10% calcium carbonate ($CaCO_3$), and lesser amounts of magnesium, sodium, potassium, fluoride, sulfate, carbonate, and hydroxide ions. Several foreign elements behave chemically like bone minerals and become incorporated into osseous tissue as contaminants, sometimes with deadly results (Insight 7.1).

> **Think About It**
>
> *What two organelles do you think are especially prominent in osteoblasts? (Hint: Consider the major substances that osteoblasts synthesize.)*

Bone is in a class of materials that engineers call a **composite**—a combination of two basic structural materials, in this case a ceramic and a polymer. A composite can combine the optimal mechanical properties of each component. Consider a fiberglass fishing rod, for example, made of a ceramic (glass fibers) and a polymer (resin). The resin alone would be too brittle and the fibers alone too limp to serve the purpose of a fishing rod, but together they produce a material of great strength and flexibility.

In bone, the polymer is the collagen and the ceramic is the hydroxyapatite and other minerals. The ceramic component enables a bone to support the weight of the body without sagging. If the minerals are dissolved out of a bone with acid, the remaining bone becomes rubbery. When the bones are deficient in calcium salts, they are soft and bend easily. This is the central problem in the childhood disease *rickets,* in which the soft bones of the lower limbs bend under the body's weight and become permanently deformed.

The protein component gives bone a degree of flexibility. Without protein, a bone is excessively brittle, as in *osteogenesis imperfecta,* or *brittle bone disease* (see table 7.3, p. 235). Without collagen, a jogger's bones would shatter under the impact of running. But normally, when a bone bends slightly toward one side, the tensile strength of the collagen fibers on the opposite side holds the bone together and prevents it from snapping like a stick of chalk.

Unlike fiberglass, bone varies from place to place in its ratio of minerals to collagen. Osseous tissue is thus adapted to different amounts of tension and compression exerted on different parts of the skeleton.

## COMPACT BONE

The histological study of compact bone usually uses slices that have been dried, cut with a saw, and ground to translucent thinness. This procedure destroys the cells but reveals fine details of the matrix (fig. 7.5). Such sections show onionlike **concentric lamellae**—layers of matrix concentrically arranged around a **central (haversian**[16] **or osteonic) canal** and connected with each other by canaliculi. A central canal and its lamellae constitute an **osteon** (haversian

---

[14]*osteo* = bone + *clast* = destroy, break down
[15]J. Howship (1781–1841), English surgeon

[16]Clopton Havers (1650–1702), English anatomist

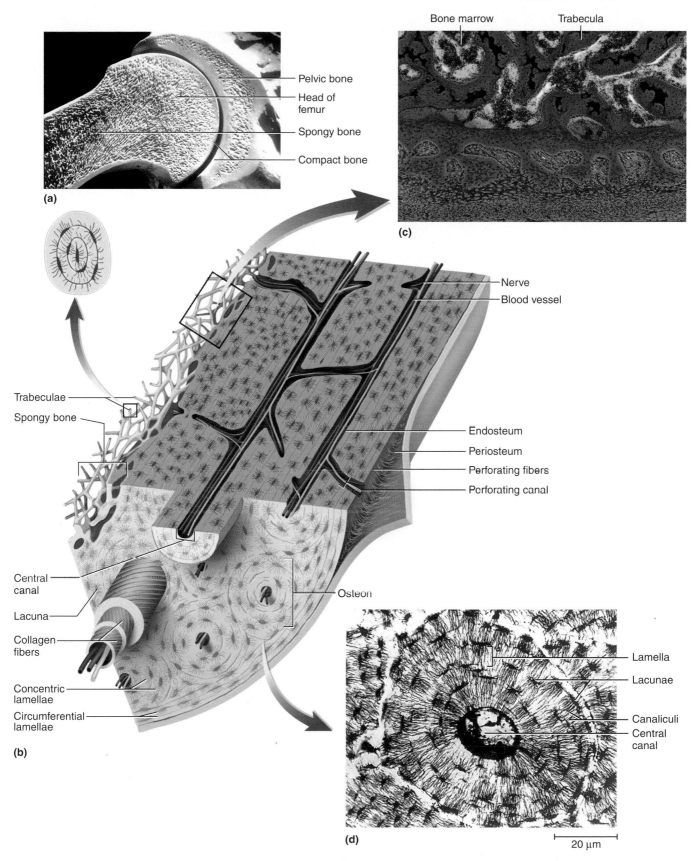

**(a)**

Pelvic bone
Head of femur
Spongy bone
Compact bone

Bone marrow    Trabecula

**(c)**

Trabeculae
Spongy bone

Nerve
Blood vessel

Endosteum
Periosteum
Perforating fibers
Perforating canal

Central canal
Lacuna
Collagen fibers
Concentric lamellae
Circumferential lamellae

Osteon

**(b)**

Lamella
Lacunae
Canaliculi
Central canal

**(d)**    20 μm

**FIGURE 7.5**  **The Histology of Osseous Tissue.**   (a) Compact and spongy bone in a frontal section of the hip joint. (b) The three-dimensional structure of compact bone. Lamellae of one osteon are telescoped to show their alternating arrangement of collagen fibers. (c) Microscopic appearance of spongy bone. (d) Microscopic appearance of a cross section of compact bone.

▶ *Which type of bone, spongy or compact, has more surface area exposed to osteoclast action?*

## Bone Contamination

When Marie and Pierre Curie and Henri Becquerel received their 1903 Nobel Prize for the discovery of radioactivity (see Insight 2.1, p. 57), radiation captured the public imagination. Not for several decades did anyone realize its dangers. For example, factories employed women to paint glow-in-the-dark numbers on watch and clock dials with radium paint. The women moistened their paint brushes with their tongues to keep them finely pointed and ingested radium in the process. The radium accumulated in their bones and caused many of them to develop a form of bone cancer called osteosarcoma.

Even more horrific, in the wisdom of our hindsight, was a deadly health fad in which people drank "tonics" made of radium-enriched water. One famous enthusiast was the millionaire playboy and championship golfer Eben Byers (1880–1932), who drank several bottles of radium tonic each day and praised its virtues as a wonder drug and aphrodisiac. Like the factory women, Byers contracted osteosarcoma. By the time of his death, holes had formed in his skull and doctors had removed his entire upper jaw and most of his mandible in an effort to halt the spreading cancer. Byer's bones and teeth were so radioactive they could expose photographic film in the dark. Brain damage left him unable to speak, but he remained mentally alert to the bitter end. His tragic decline and death shocked the world and put an end to the radium tonic fad.

system)—the basic structural unit of compact bone. In longitudinal views and three-dimensional reconstructions, we can see that an osteon is actually a cylinder of tissue surrounding a central canal. Along their length, central canals are joined by transverse or diagonal passages.

Collagen fibers "corkscrew" down the matrix of a given lamella in a helical arrangement like the threads of a screw. In the adjacent lamella, they angle in the opposite direction—alternating between right- and left-handed helices from lamella to lamella (fig. 7.5b). This enhances the strength of bone on the same principle as plywood, made of thin layers of wood with the grain running in different directions from one layer to the next. The helices tend to be more stretched out along the longitudinal axis of bones that must resist tension (bending), but are tighter and run more nearly across the bone in those that must resist compression.

The skeleton receives about half a liter of blood per minute. Blood vessels, along with nerves, enter the bone tissue through nutrient foramina on the surface. These open into narrow **perforating (Volkmann[17]) canals** that cross the matrix and feed into the central canals. The innermost osteocytes around each central canal receive nutrients from these blood vessels and pass them along through their gap junctions to neighboring osteocytes. They also receive wastes from their neighbors and convey them to the central canal for removal by the bloodstream. Thus, the cytoplasmic processes of the osteocytes maintain a two-way flow of nutrients and wastes between the central canal and the outermost cells of the osteon.

Not all of the matrix is organized into osteons. The inner and outer boundaries of dense bone are arranged in *circumferential lamellae* that run parallel to the bone surface. Between osteons, we can find irregular regions called *interstitial lamellae,* the remains of old osteons that broke down as the bone grew and remodeled itself.

## SPONGY BONE

Spongy bone (fig. 7.5c) consists of a lattice of slender rods, plates, and spines called **trabeculae.**[18] Although calcified and hard, spongy bone is named for its spongelike appearance (see p. 213). It is permeated by spaces filled with bone marrow. The matrix is arranged in lamellae like those of compact bone, but there are few osteons. Central canals are not needed here because no osteocyte is very far from the marrow. Spongy bone is well designed to impart strength to a bone while adding a minimum of weight. Its trabeculae are not randomly arranged as they might seem at a glance, but develop along the bone's lines of stress (fig. 7.6).

## BONE MARROW

**Bone marrow** is a general term for soft tissue that occupies the marrow cavity of a long bone, the spaces amid the trabeculae of spongy bone, and the larger central canals. There are two kinds of marrow—red and yellow. We can best appreciate their differences by considering how marrow changes over a person's lifetime.

In a child, the marrow cavity of nearly every bone is filled with **red bone marrow** (myeloid tissue). This is a *hemopoietic*[19] (HE-mo-poy-ET-ic) tissue—that is, it produces blood cells. Red bone marrow looks like blood but with a thicker consistency. It consists of a delicate mesh of reticular tissue saturated with immature blood cells and scattered adipocytes. The structure of red bone marrow is described in more detail in chapter 21.

In adults, most of this red marrow turns to fatty **yellow bone marrow,** like the fat at the center of a ham bone. Yellow bone marrow no longer produces blood, although in the event of severe or chronic anemia, it can transform back into red marrow. In adults, red marrow is limited to the skull, vertebrae, ribs, sternum, part of the pelvic (hip)

---

[18]*trabe* = plate + *cul* = little
[19]*hemo* = blood + *poietic* = forming

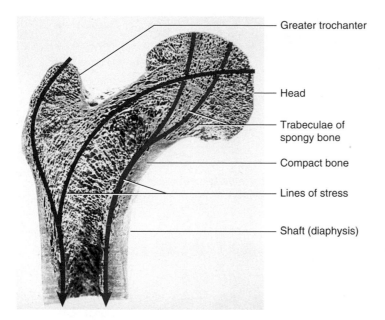

**FIGURE 7.6** **Spongy Bone Structure in Relation to Mechanical Stress.** In this frontal section of the femur (thighbone), the trabeculae of spongy bone can be seen oriented along lines of mechanical stress applied by the weight of the body.

girdle, and the proximal heads of the humerus and femur (fig. 7.7).

## Before You Go On

*Answer the following questions to test your understanding of the preceding section:*

6. *Suppose you had unlabeled electron micrographs of the four kinds of bone cells and their neighboring tissues. Name each of the four cells and explain how you could visually distinguish it from the other three.*

7. *Name three organic components of the bone matrix.*

8. *What are the mineral crystals of bone called, and what are they made of?*

9. *Sketch a cross section of an osteon and label its major parts.*

10. *What are the two kinds of bone marrow? What does hemopoietic tissue mean? Which type of bone marrow fits this description?*

# Bone Development

### Objectives
When you have completed this section, you should be able to

- describe two mechanisms of bone formation; and
- explain how mature bone continues to grow and remodel itself.

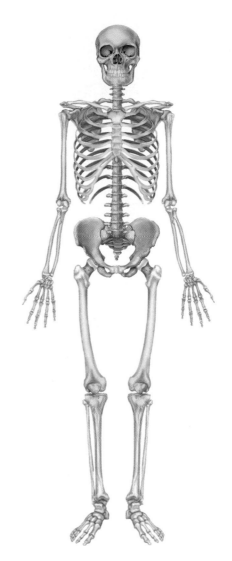

**FIGURE 7.7** **Distribution of Red and Yellow Bone Marrow.** In an adult, red bone marrow occupies the marrow cavities of the axial skeleton and proximal heads of the humerus and femur. Yellow bone marrow occurs in the long bones of the limbs.
▶ *What would be the most accessible places to draw red bone marrow from an adult?*

The formation of bone is called **ossification** (OSS-ih-fih-CAY-shun), or **osteogenesis.** In the human fetus and infant, bone develops by two methods called *intramembranous* and *endochondral* ossification, which we examine in the following sections.

## INTRAMEMBRANOUS OSSIFICATION

**Intramembranous**[20] (IN-tra-MEM-bra-nus) **ossification** produces the flat bones of the skull and most of the clavicle (collarbone). Such bones develop within a fibrous sheet similar to the dermis of the skin, so they are sometimes

---

[20]*intra* = within + *membran* = membrane

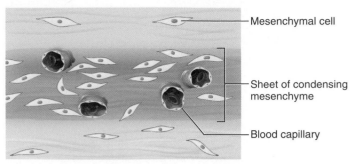

Mesenchymal cell

Sheet of condensing mesenchyme

Blood capillary

① Condensation of mesenchyme into soft sheet permeated with blood capillaries

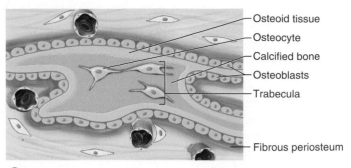

Osteoid tissue

Osteocyte

Calcified bone

Osteoblasts

Trabecula

Fibrous periosteum

② Deposition of osteoid tissue by osteoblasts on mesenchymal surface; entrapment of first osteocytes; formation of periosteum

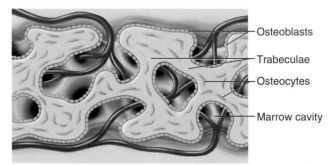

Osteoblasts

Trabeculae

Osteocytes

Marrow cavity

③ Honeycomb of bony trabeculae formed by continued mineral deposition; creation of spongy bone

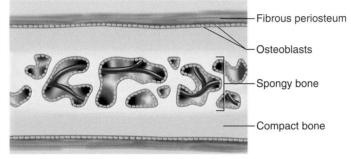

Fibrous periosteum

Osteoblasts

Spongy bone

Compact bone

④ Surface bone filled in by bone deposition, converting spongy bone to compact bone. Persistence of spongy bone in the middle layer.

**FIGURE 7.8**   Stages of Intramembranous Ossification.

called *dermal bones.* Figure 7.8 shows the stages of the process.

① Some of the embryonic connective tissue (mesenchyme) condenses into a layer of soft tissue with a dense supply of blood capillaries. The mesenchymal cells enlarge and differentiate into osteogenic cells, and regions of mesenchyme become a network of soft sheets called trabeculae.

② Osteogenic cells gather on these trabeculae and differentiate into osteoblasts. These cells deposit an organic matrix called **osteoid**[21] **tissue**—soft collagenous tissue similar to bone except for a lack of minerals (fig. 7.9). As the trabeculae grow thicker, calcium phosphate is deposited in the matrix. Some osteoblasts become trapped in the matrix and are now osteocytes. Mesenchyme close to the surface of a trabecula remains uncalcified, but becomes denser and more fibrous, forming a periosteum.

③ Osteoblasts continue to deposit minerals, producing a honeycomb of bony trabeculae. Some trabeculae persist as permanent spongy bone, while osteoclasts

resorb and remodel others to form a marrow cavity in the middle of the bone.

④ Trabeculae at the surface continue to calcify until the spaces between them are filled in, converting the spongy bone to compact bone. This process gives rise to the sandwichlike arrangement typical of mature flat bones.

## ENDOCHONDRAL OSSIFICATION

**Endochondral**[22] (EN-doe-CON-drul) **ossification** is a process in which a bone develops from a preexisting model composed of hyaline cartilage. It begins around the sixth week of fetal development and continues into a person's 20s. Most bones of the body, including the vertebrae, ribs, sternum, scapula, pelvis, and bones of the limbs, develop in this way. Figure 7.10 shows the following steps in endochondral ossification, using the example of a *metacarpal bone* in the palmar region of the hand.

① Mesenchyme develops into a body of hyaline cartilage, covered with a fibrous perichondrium, in the location of a future bone. For a time, the perichon-

---

[21]*oste* = bone + *oid* = like, resembling

[22]*endo* = within + *chondr* = cartilage

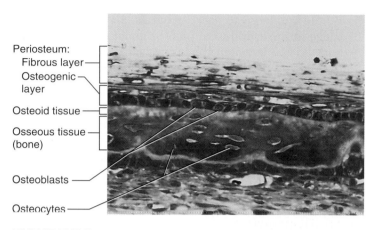

**FIGURE 7.9** **Intramembranous Ossification of a Cranial Bone of the Human Fetus.** Note the layers of osteoid tissue, osteoblasts, and fibrous periosteum on both sides of the bone.

drium produces chondrocytes and the cartilage model grows in thickness.

② Eventually, the perichondrium stops producing chondrocytes and begins producing osteoblasts. These deposit a thin collar of bone around the middle of the cartilage model, encircling it like a napkin ring and providing physical reinforcement. The former peri

chondrium is now considered to be a periosteum. Meanwhile, chondrocytes in the middle of the model enlarge and the matrix between their lacunae is reduced to thin walls. This region of chondrocyte enlargement is called the **primary ossification center.** The walls of matrix between the lacunae calcify and block nutrients from reaching the chondrocytes. The cells die and their lacunae merge into a single cavity in the middle of the model.

③ Blood vessels penetrate the bony collar and invade the primary ossification center. As the center of the model is hollowed out and filled with blood and stem cells, it becomes the **primary marrow cavity.** Various stem cells introduced with the blood give rise to osteoblasts and osteoclasts. Osteoblasts line the cavity, begin depositing osteoid tissue, and calcify it to form a temporary network of bony trabeculae. As the bony collar under the periosteum thickens and elongates, a wave of cartilage death progresses toward the ends of the bone. Osteoclasts in the marrow cavity follow this wave, dissolving calcified cartilage remnants and enlarging the marrow cavity of the diaphysis. The region of transition from cartilage to bone at each end of the primary marrow cavity is called a **metaphysis.**

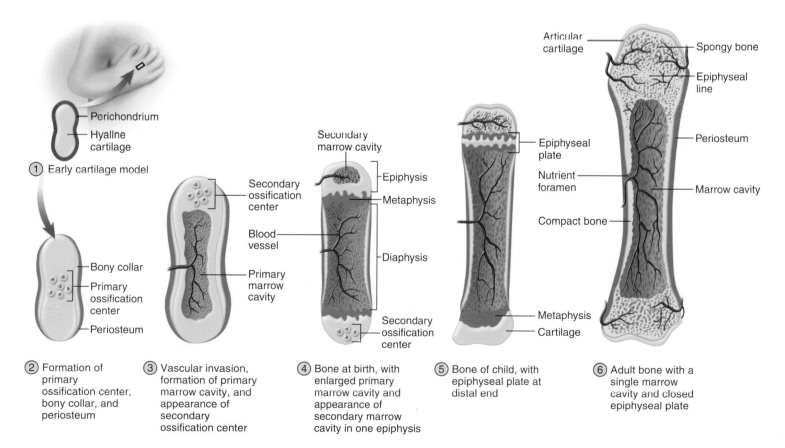

**FIGURE 7.10** **Stages of Endochondral Ossification.**

Soon, chondrocyte enlargement and death occur in the epiphysis of the model as well, creating a **secondary ossification center.** In the metacarpal bones, as illustrated in figure 7.10, this occurs in only one epiphysis. In longer bones of the arms, forearms, legs, and thighs, it occurs at both ends.

④ The secondary ossification center becomes hollowed out by the same process as the diaphysis, generating a **secondary marrow cavity** in the epiphysis. This cavity expands outward from the center, in all directions. At the time of birth, the bone typically looks like step 4 in figure 7.10. In bones with two secondary ossification centers, one center lags behind the other in development, so at birth there is a secondary marrow cavity at one end while chondrocyte growth has just begun at the other. The joints of the limbs are still cartilaginous at birth (fig. 7.11).

⑤ During infancy and childhood, the epiphyses fill with spongy bone. Cartilage is then limited to the articular cartilage covering each joint surface, and to an **epiphyseal** (EP-ih-FIZ-ee-ul) **plate,** a thin wall of cartilage separating the primary and secondary marrow cavities at one or both ends of the bone. The epiphyseal plate persists through childhood and adolescence and serves as a growth zone for bone elongation. This growth process is described in the next section.

⑥ By the late teens to early twenties, all remaining cartilage in the epiphyseal plate is generally consumed and the gap between the epiphysis and diaphysis closes. The primary and secondary marrow cavities then unite into a single cavity, and the bone can no longer grow in length.

# BONE GROWTH

Ossification does not end at birth, but continues throughout life with the growth and remodeling of bones. Bones grow in two directions: length and width.

## Bone Elongation

To understand growth in length, we must return to the epiphyseal plates mentioned earlier (see fig. 7.10, step 5). From infancy through adolescence, an epiphyseal plate is present at one or both ends of a long bone, at the junction between the diaphysis and epiphysis. On X rays, it appears as a translucent line across the end of a bone, since it is not yet ossified (fig. 7.12; compare the X ray of an adult hand in fig. 8.34). The epiphyseal plate is a region of transition from cartilage to bone, and functions as a growth zone where the bones elongate. Growth here is responsible for a person's increase in height.

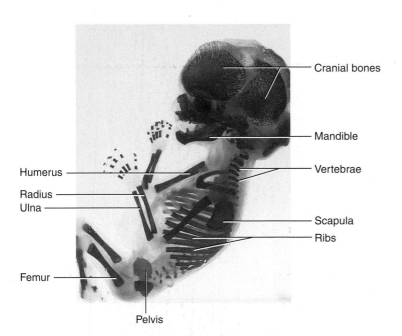

Cranial bones

Mandible

Vertebrae

Humerus

Radius

Ulna

Scapula

Ribs

Femur

Pelvis

**FIGURE 7.11**  **The Fetal Skeleton at 12 Weeks.**  The red-stained regions are calcified at this age, whereas the elbow, wrist, knee, and ankle joints appear translucent because they are still cartilaginous.

▶ *Why are the joints of an infant weaker than those of an older child?*

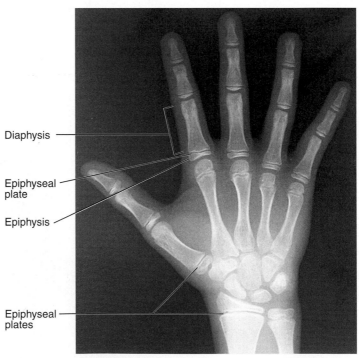

Diaphysis

Epiphyseal plate

Epiphysis

Epiphyseal plates

**FIGURE 7.12**  **X Ray of a Child's Hand.**  The cartilaginous epiphyseal plates are evident at the ends of the long bones. These will disappear, and the epiphyses will fuse with the diaphyses, by adulthood. Long bones of the hand and fingers develop only one epiphyseal plate.

The epiphyseal plate consists of typical hyaline cartilage in the middle, with a transitional zone on each side where cartilage is transforming into bone. The transitional zone, facing the marrow cavity, is called the **metaphysis** (meh-TAF-ih-sis). In figure 7.10, step 4, the cartilage is the blue region and each metaphysis is violet. Figure 7.13 shows the histological structure of the metaphysis and the following steps in the conversion of cartilage to bone.

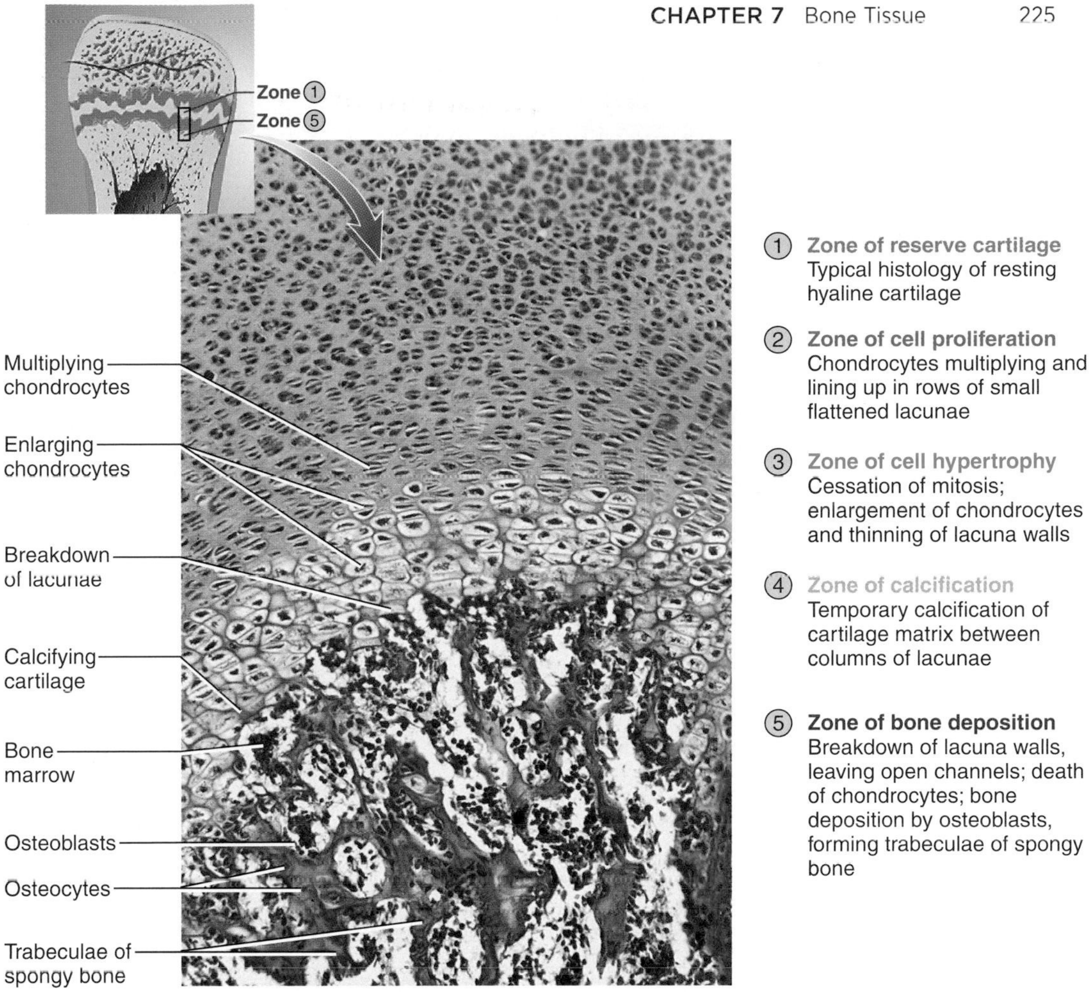
Zone ①
Zone ⑤
Multiplying chondrocytes
Enlarging chondrocytes
Breakdown of lacunae
Calcifying cartilage
Bone marrow
Osteoblasts
Osteocytes
Trabeculae of spongy bone
① Zone of reserve cartilage
Typical histology of resting hyaline cartilage
② Zone of cell proliferation
Chondrocytes multiplying and lining up in rows of small flattened lacunae
③ Zone of cell hypertrophy
Cessation of mitosis; enlargement of chondrocytes and thinning of lacuna walls
④ Zone of calcification
Temporary calcification of cartilage matrix between columns of lacunae
⑤ Zone of bone deposition
Breakdown of lacuna walls, leaving open channels; death of chondrocytes; bone deposition by osteoblasts, forming trabeculae of spongy bone

**FIGURE 7.13 Zones of the Metaphysis.** This micrograph shows the transition from cartilage to bone in the growth zone of a long bone.

① **Zone of reserve cartilage.** This region, farthest from the marrow cavity, consists of typical hyaline cartilage that as yet shows no sign of transforming into bone.

② **Zone of cell proliferation.** A little closer to the marrow cavity, chondrocytes multiply and arrange themselves into longitudinal columns of flattened lacunae.

③ **Zone of cell hypertrophy.** Next, the chondrocytes cease to divide and begin to hypertrophy (enlarge), much like they do in the primary ossification center of the fetus. The walls of matrix between lacunae become very thin.

④ **Zone of calcification.** Minerals are deposited in the matrix between the columns of lacunae and calcify the cartilage. These are not the permanent mineral deposits of bone, but only a temporary support for the cartilage that would otherwise soon be weakened by the breakdown of the enlarged lacunae.

⑤ **Zone of bone deposition.** Within each column, the walls between the lacunae break down and the chondrocytes die. This converts each column into a longitudinal channel (white spaces in the figure), which is immediately invaded by blood vessels and marrow from the marrow cavity. Osteoblasts line up along the walls of these channels and begin depositing concentric lamellae of matrix, while osteoclasts dissolve the temporarily calcified cartilage.

The process of bone deposition in zone 5 creates a region of spongy bone at the end of the marrow cavity facing the metaphysis. This spongy bone remains for life, although with extensive lifelong remodeling. But around the perimeter of the marrow cavity, continuing ossification converts this spongy bone to compact bone. Osteoblasts lining the aforementioned channels deposit layer after layer of bone matrix, so the channel grows narrower and narrower. These layers become the concentric lamellae of an osteon. Finally only a slender channel persists, the central canal of a new osteon. As usual, osteoblasts trapped in the matrix become osteocytes.

### Think About It

*In a given osteon, which lamellae are the oldest—those immediately adjacent to the central canal or those around the perimeter of the osteon? Explain your answer.*

How does a child or adolescent grow in height? Chondrocyte multiplication in zone 2 and hypertrophy in zone 3 continually push the zone of reserve cartilage (1) toward the ends of the bone, so the bone elongates. In the lower limbs, this process causes a person to grow in height, while bones of the upper limbs grow proportionately.

Thus, bone elongation is really a result of cartilage growth. Cartilage growth from within, by the multiplication of chondrocytes and deposition of new matrix in the interior, is called **interstitial growth.**[23] The most common form of dwarfism results from a failure of cartilage growth in the long bones (Insight 7.2).

In the late teens to early twenties, all the cartilage of the epiphyseal plate is depleted. The primary and secondary marrow cavities now unite into one cavity (fig. 7.10, step 6). The junctional region where they meet is filled with spongy bone, and the site of the original epiphyseal plate is marked with a line of slightly denser spongy bone called the **epiphyseal line** (see figs. 7.2, 7.6, and 7.10, step 6). Often a delicate ridge on the bone surface marks the location of this line. When the epiphyseal plate is depleted, we say that the epiphyses have "closed," and a person can grow no taller. The epiphyseal plates close at different ages in different bones and in different regions of the same bone. The state of closure in various bones is often used in forensic science to estimate the age at death of a subadult skeleton.

## Bone Widening and Thickening

Bones also continually grow throughout life in diameter and thickness. This involves a process called **appositional growth,**[24] the deposition of new tissue at the surface. Cartilages can enlarge by both interstitial and appositional growth, but bone is limited to the appositional method. Embedded in a calcified matrix, the osteocytes have little room to spare for the deposition of more matrix internally.

Appositional growth is similar to intramembranous ossification. Osteoblasts in the inner layer of periosteum deposit osteoid tissue on the bone surface, calcify it, and become trapped in it as osteocytes—much like the process in figure 7.9. They lay down matrix in layers parallel to the surface, not in cylindrical osteons like those deeper in the bone. This process produces the surface layers of bone called *circumferential lamellae,* described earlier. As a bone increases in diameter, its marrow cavity also widens. This is achieved by osteoclasts of the endosteum dissolving tissue on the inner bone surface.

## BONE REMODELING

In addition to their growth, bones are continually remodeled throughout life by the absorption of old bone and deposition of new. This process replaces about 10% of the skeletal tissue per year. It repairs microfractures, releases minerals into the blood, and reshapes bones in response

---

## INSIGHT 7.2    Clinical Application

### Achondroplastic Dwarfism

*Achondroplastic*[25] (a-con-dro-PLAS-tic) *dwarfism* is a condition in which the long bones of the limbs stop growing in childhood, while the growth of other bones is unaffected. As a result, a person has a short stature but a normal-sized head and trunk (fig. 7.14). As its name implies, achondroplastic dwarfism results from a failure of cartilage growth—specifically, failure of the chondrocytes in zones 2 and 3 of the metaphysis to multiply and enlarge. This is different from *pituitary dwarfism,* in which a deficiency of growth hormone stunts the growth of all of the bones, and a person has short stature but normal proportions throughout the skeletal system.

Achondroplastic dwarfism results from a spontaneous mutation that can arise any time DNA is replicated. Two people of normal height with no family history of dwarfism can therefore have a child with achondroplastic dwarfism. The mutant allele is dominant, so the children of a heterozygous achondroplastic dwarf have at least a 50% chance of exhibiting dwarfism, depending on the genotype of the other parent.

**FIGURE 7.14  Achondroplastic Dwarfism.**  The student on the right, pictured with her roommate of normal height, is an achondroplastic dwarf with a height of about 122 cm (48 in.). Her parents were of normal height. Note the normal proportion of head to trunk but shortening of the limbs.

---

to use and disuse. **Wolff's**[26] **law of bone** states that the architecture of a bone is determined by the mechanical stresses placed upon it, and the bone thereby adapts to withstand those stresses. Wolff's law is a fine example of the complementarity of form and function, showing that the form of a bone is shaped by its functional experience. It is admirably demonstrated by figure 7.6, in which we see that the trabeculae of spongy bone have developed along the lines of stress placed on the femur. Wolff observed that these stress lines were very similar to the ones that engineers saw in mechanical cranes.

Bone remodeling comes about through the collaborative action of osteoblasts and osteoclasts. If a bone is little used, osteoclasts remove matrix and get rid of unnecessary mass. If a bone is heavily used, or a stress is consistently applied to a particular region of a bone, osteoblasts deposit new osseous tissue and thicken the bone. Consequently, the comparatively smooth bones of an infant or toddler develop a variety of surface bumps, ridges, and spines (described in chapter 8) as the child begins to walk. The greater trochanter of the femur, for example (see figs. 7.6 and 8.38), is a massive outgrowth of bone stimulated by the pull of tendons from several powerful hip muscles employed in walking.

On average, bones have greater density and mass in athletes and people engaged in heavy manual labor than they do in sedentary people. Anthropologists who study ancient skeletal remains use evidence of this sort to help distinguish between members of different social classes, such as distinguishing royalty from laborers. Even in studying modern skeletal remains, as in investigating a suspicious death, Wolff's law comes into play as the bones give evidence of a person's sex, race, height, weight, work or exercise habits, nutritional status, and medical history.

The orderly remodeling of bone depends on a precise balance between deposition and resorption, between osteoblasts and osteoclasts. If one process outpaces the other, or both processes occur too rapidly, various bone deformities, developmental abnormalities, and other disorders occur, such as *osteitis deformans* (Paget disease), *osteogenesis imperfecta* (brittle bone disease) and *osteoporosis* (see table 7.3 and Insight 7.4).

## Before You Go On

*Answer the following questions to test your understanding of the preceding section:*

11. *Describe the stages of intramembranous ossification. Name a bone that is formed in this way.*

12. *Describe the five zones of a metaphysis and the major distinctions between them.*

13. *How does Wolff's law explain some of the structural differences between the bones of a young child and the bones of a young adult?*

# Physiology of Osseous Tissue

### Objectives
When you have completed this section, you should be able to

- describe the processes by which minerals are added to and removed from bone tissue;
- discuss the role of the bones in regulating blood calcium and phosphate levels; and
- name several hormones that regulate bone physiology and describe their effects.

Even after a bone is fully formed, it remains a metabolically active organ with many roles to play. Not only is it involved in its own maintenance, growth, and remodeling, it also exerts a profound influence on the rest of the body by exchanging minerals with the tissue fluid. Disturbances of calcium homeostasis in the skeleton can disrupt the functioning of other organ systems, especially the nervous and muscular systems. For reasons explained later, such disturbances can even cause death by suffocation. At this point, we turn our attention to the physiology of mature osseous tissue.

## MINERAL DEPOSITION

**Mineral deposition** (mineralization) is a crystallization process in which calcium, phosphate, and other ions are taken from the blood plasma and deposited in bone tissue. It begins in fetal ossification and continues throughout life.

Osteoblasts begin the process by laying down collagen fibers in a helical pattern along the length of the osteon. These fibers then become encrusted with minerals—especially calcium phosphate—that harden the matrix. Calcium phosphate crystals do not form unless the product of calcium and phosphate concentration in the tissue fluids, represented $[Ca^{2+}] \cdot [PO_4^{3-}]$, reaches a critical value called the **solubility product.** Most tissues have inhibitors to prevent this, so they do not become calcified. Osteoblasts, however, apparently neutralize these inhibitors and thus allow the salts to precipitate in the bone matrix. The first few hydroxyapatite crystals to form act as "seed crystals" that attract more calcium and phosphate from solution. The more hydroxyapatite that forms, the more it attracts additional minerals from the tissue fluid, until the matrix is thoroughly calcified.

Osseous tissue sometimes forms in the lungs, brain, eyes, muscles, tendons, arteries, and other organs. Such abnormal calcification of tissues is called **ectopic**[27] **ossification.** One example of this is arteriosclerosis, or "hardening of the arteries," which results from calcification of the arterial walls. A calcified mass in an otherwise soft organ such as the lungs is called a **calculus**.[28]

> ⌐ **Think About It**
>
> *What positive feedback process can you recognize in bone deposition?*

## MINERAL RESORPTION

**Mineral resorption** is the process of dissolving bone. It releases minerals into the blood and makes them available for other uses. Resorption is carried out by osteoclasts. Hydrogen pumps in the ruffled border of the osteoclast secrete hydrogen ions into the extracellular fluid, and chloride ions follow by electrical attraction. The space between the osteoclast and the bone thus becomes filled with concentrated hydrochloric acid with a pH of about 4. The acid dissolves the bone minerals. The osteoclast also secretes an enzyme called **acid phosphatase** that digests the collagen of the bone matrix. This enzyme is named for its ability to function in a highly acidic environment.

When orthodontic appliances (braces) are used to reposition teeth, a tooth moves because osteoclasts dissolve bone ahead of the tooth (where the appliance creates greater pressure of the tooth against the bone) and osteoblasts deposit bone in the low-pressure zone behind it.

## CALCIUM AND PHOSPHATE HOMEOSTASIS

Calcium and phosphate are used for much more than bone structure. Phosphate groups are a component of DNA, RNA, ATP, phospholipids, and many other compounds. Phosphate ions also help to correct acid–base imbalances in the body fluids (Insight 7.3). Calcium plays

roles in communication among neurons, and in muscle contraction, blood clotting, and exocytosis. It is also a second messenger in many cell-signaling processes and a cofactor for some enzymes.

The skeleton is a reservoir for these minerals. Minerals are deposited in the skeleton when the supply is ample and withdrawn when they are needed for other purposes.

The adult body contains about 1,100 g of calcium, with 99% of it in the bones. Bone has two calcium reserves: (1) a stable pool of calcium, which is incorporated into hydroxyapatite and is not easily exchanged with the blood, and (2) exchangeable calcium, which is 1% or less of the total but is easily released to the tissue fluid. The adult skeleton exchanges about 18% of its calcium with the blood each year.

The calcium concentration in the blood plasma is normally 9.2 to 10.4 mg/dL. This is a rather narrow margin of safety, as we shall soon see. About 45% of it is in the ionized form ($Ca^{2+}$), which can diffuse through capillary walls and affect neighboring cells. The rest of it is bound to plasma proteins and other solutes. It is not physiologically active, but it serves as a reserve from which free $Ca^{2+}$ can be obtained as needed.

The average adult has 500 to 800 g of phosphorus, of which 85% to 90% is in the bones. The phosphorus concentration in the plasma ranges between 3.5 and 4.0 mg/dL. It occurs in two principal forms, $HPO_4^{2-}$ and $H_2PO_4^{-}$ (monohydrogen and dihydrogen phosphate ions, respectively).

Changes in phosphate concentration have little immediate effect on the body, but changes in calcium can be serious. A blood calcium deficiency is called **hypocalcemia**[29] (HY-po-cal-SEE-me-uh). It causes excessive excitability of the nervous system and leads to muscle tremors, spasms, or **tetany**—inability of the muscle to relax. Tetany begins to occur as the plasma $Ca^{2+}$ concentration falls to 6 mg/dL. One sign of hypocalcemia is a tetany of the hands and feet called *carpopedal spasm* (fig. 7.15). At 4 mg/dL, muscles of the larynx contract tightly,

---

[29]*hypo* = below normal + *calc* = calcium + *emia* = blood condition

## INSIGHT 7.3    Clinical Application

### Osseous Tissue and pH Balance

The urinary, respiratory, and skeletal systems cooperate to maintain the body's acid–base balance. If the pH of the blood drops below 7.35, a state of acidosis exists, triggering corrective mechanisms in these three organ systems. The role of the skeleton is to release calcium phosphate. As a base, calcium phosphate helps to prevent the blood pH from dropping lower. Patients with chronic kidney disease may have impaired hydrogen ion excretion in the urine. Their pH stabilizes at a level below normal but is kept from dropping indefinitely by the buffering action of the skeleton. This can have adverse effects on the skeleton, however, and lead to rickets or osteomalacia. Treatment of the acidosis with intravenous bicarbonate restores the pH to normal and prevents damage to the bones.

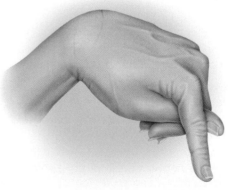

**FIGURE 7.15   Carpopedal Spasm.** Such muscle tetany occurring in the hands and feet can be a sign of hypocalcemia.

---

[28]*calc* = stone + *ulus* = little

a condition called *laryngospasm,* which can shut off air flow and cause suffocation.

The reason for hypocalcemic excitability is this: Calcium ions normally bind and neutralize negatively charged groups on the cell surface, contributing to the difference between the positively charged outer surface of the membrane and the negatively charged inner surface. In hypocalcemia, fewer calcium ions are present, so there is less charge difference between the two sides of the membrane. Sodium channels in the plasma membrane are sensitive to this charge difference, and when the difference is diminished, they open more easily and stay open longer. This allows sodium ions to enter the cell too freely. As you will see in chapters 11 and 12, an inflow of sodium is the normal process that excites nerve and muscle cells. In hypocalcemia, this excitation is excessive and results in the aforementioned tetany.

A blood calcium excess is called **hypercalcemia.**[30] In this condition, excessive amounts of calcium bind to the cell surface, increasing the charge difference across the membrane and making sodium channels less responsive. Thus, nerve and muscle cells are less excitable than normal. At 12 mg/dL and higher, hypercalcemia causes depression of the nervous system, emotional disturbances, muscle weakness, sluggish reflexes, and sometimes cardiac arrest.

You can see how critical blood calcium level is, but what causes it to deviate from the norm, and how does the body correct such imbalances? Hypercalcemia is rare, but hypocalcemia can result from a wide variety of causes including vitamin D deficiency, diarrhea, thyroid tumors,

[30]*hyper* = above normal + *calc* = calcium + *emia* = blood condition

or underactive parathyroid glands. Pregnancy and lactation put women at risk of hypocalcemia because of the calcium demanded by ossification of the fetal skeleton and synthesis of milk. The leading cause of hypocalcemic tetany is accidental removal of the parathyroid glands during thyroid surgery. Without hormone replacement therapy, the lack of parathyroid glands can lead to fatal tetany within 4 days.

Calcium phosphate homeostasis depends on a balance between dietary intake, urinary and fecal losses, and exchanges with the osseous tissue. It is regulated by three hormones: *calcitriol, calcitonin,* and *parathyroid hormone* (fig. 7.16).

## Calcitriol

**Calcitriol** (CAL-sih-TRY-ol) is a form of vitamin D produced by the sequential action of the skin, liver, and kidneys (fig. 7.17):

1. Epidermal keratinocytes use ultraviolet radiation from sunlight to convert a steroid, 7-dehydrocholesterol, to previtamin $D_3$. Over another 3 days, the warmth of sunlight on the skin further converts this to vitamin $D_3$, and a transport protein carries this to the bloodstream.

2. The liver adds a hydroxyl group to the molecule, converting it to *calcidiol.*

3. The kidney then adds another hydroxyl group, converting calcidiol to calcitriol, the most active form of vitamin D.

Calcitriol behaves as a hormone—a blood-borne chemical messenger from one organ to another. It is called a vitamin

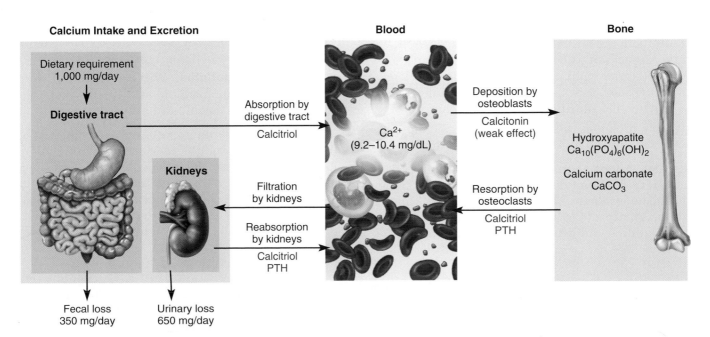

**FIGURE 7.16  Hormonal Control of Calcium Balance.**   Calcitriol, parathyroid hormone (PTH), and, to some extent, calcitonin maintain the blood calcium concentration at 9.2 to 10.4 mg/dL.

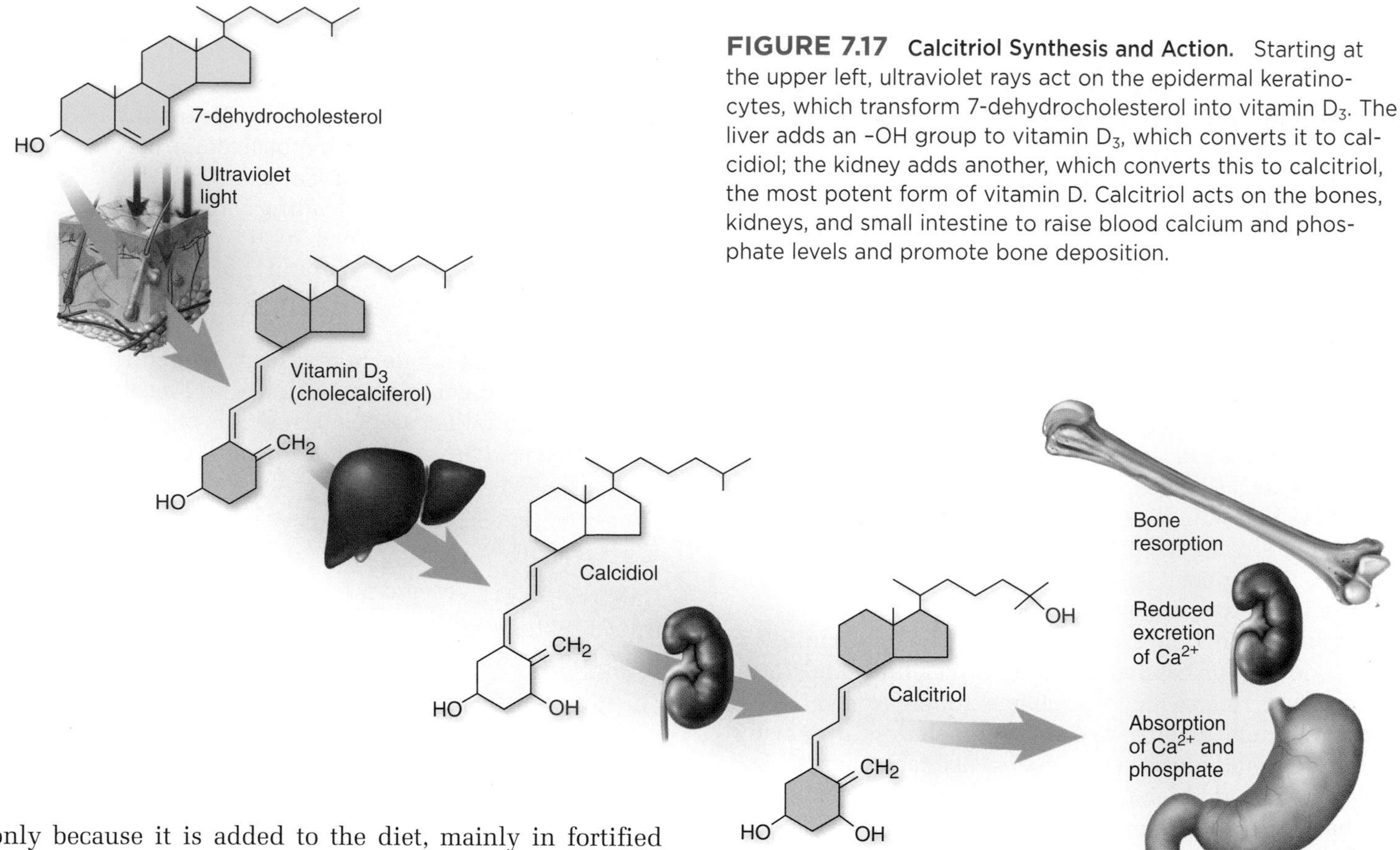

**FIGURE 7.17 Calcitriol Synthesis and Action.** Starting at the upper left, ultraviolet rays act on the epidermal keratinocytes, which transform 7-dehydrocholesterol into vitamin $D_3$. The liver adds an –OH group to vitamin $D_3$, which converts it to calcidiol; the kidney adds another, which converts this to calcitriol, the most potent form of vitamin D. Calcitriol acts on the bones, kidneys, and small intestine to raise blood calcium and phosphate levels and promote bone deposition.

only because it is added to the diet, mainly in fortified milk, as a safeguard for people who do not get enough sunlight to initiate its synthesis in the skin.

The principal function of calcitriol is to raise the blood calcium concentration. It does this in three ways:

1. It increases calcium absorption by the small intestine. (It increases the absorption of phosphate and magnesium ions as well.)

2. It increases calcium (and phosphate) resorption from the skeleton. Calcitriol binds to osteoblasts, which release another chemical messenger called RANKL. This is the ligand (L) for a receptor named RANK[31] on the surfaces of osteoclast-producing stem cells. This messenger stimulates the stem cells to differentiate into osteoclasts. The new osteoclasts then liberate calcium and phosphate ions from bone.

3. It weakly promotes the reabsorption of calcium ions by the kidneys, so less calcium is lost in the urine.

Although calcitriol promotes bone resorption, it is also necessary for bone deposition. Without it, calcium and phosphate levels in the blood are too low for normal deposition. The result is a softness of the bones called **rickets** in children and **osteomalacia**[32] in adults.

## Calcitonin

**Calcitonin** is secreted by *C cells* (*C* for "calcitonin") of the thyroid gland, a large endocrine gland in the neck (see fig.

17.9). It is secreted when the blood calcium concentration rises too high, and it lowers the concentration by two principal mechanisms (fig. 7.18a):

1. **Osteoclast inhibition.** Within 15 minutes after it is secreted, calcitonin reduces osteoclast activity by as much as 70%, so osteoclasts liberate less calcium from the skeleton.

2. **Osteoblast stimulation.** Within an hour, calcitonin increases the number and activity of osteoblasts, which deposit calcium into the skeleton.

Calcitonin plays an important role in children but has little effect in most adults. The osteoclasts of children are highly active in skeletal remodeling and release 5 g or more of calcium into the blood each day. By inhibiting this activity, calcitonin can significantly lower the blood calcium level in children. In adults, however, the osteoclasts release only about 0.8 g of calcium per day. Calcitonin cannot change adult blood calcium very much by suppressing this lesser contribution. Calcitonin deficiency is not known to cause any adult disease. Calcitonin may, however, prevent bone loss in pregnant and lactating women.

## Parathyroid Hormone

**Parathyroid hormone (PTH)** is secreted by the parathyroid glands, which adhere to the posterior surface of the

---

[31]RANK = receptor activator of nuclear factor kappa B
[32]*osteo* = bone + *malacia* = softening

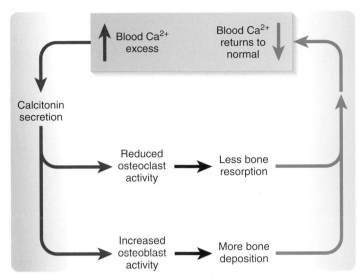

**(a) Correction for hypercalcemia**

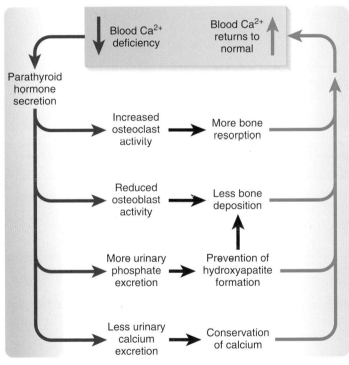

**(b) Correction for hypocalcemia**

**FIGURE 7.18** **Negative Feedback Loops in Calcium Homeostasis.** (a) The correction of hypercalcemia by calcitonin. (b) The correction of hypocalcemia by parathyroid hormone.

thyroid gland (see fig. 17.10). These glands release PTH when blood calcium is low. A mere 1% drop in the blood calcium level doubles the secretion of PTH. PTH raises the blood calcium level by four mechanisms (fig. 7.18b):

1. PTH binds to receptors on the osteoblasts, stimulating them to secrete RANKL, which in turn raises the osteoclast population and promotes bone resorption.

2. PTH promotes calcium reabsorption by the kidneys, so less calcium is lost in the urine.

3. PTH promotes the final step of calcitriol synthesis in the kidneys, thus enhancing the calcium-raising effect of calcitriol.

4. PTH inhibits collagen synthesis by osteoblasts, thus inhibiting bone deposition.

Notwithstanding these normal effects of PTH, the intermittent secretion (or injection) of PTH at low levels can stimulate bone deposition. PTH can therefore increase or decrease bone mass, depending on other factors such as exercise, stress on the bones, calcium and phosphate availability, and the action of vitamin D and other hormones.

> **Think About It**
>
> *While raising blood calcium levels, PTH lowers blood phosphate levels by promoting urinary excretion of phosphate. Explain why this is important for achieving the purpose of PTH.*

## OTHER FACTORS AFFECTING BONE

At least 20 more hormones, growth factors, and vitamins affect osseous tissue in complex ways that are still not well understood (table 7.1). Bone growth is especially rapid in puberty and adolescence, when surges of growth hormone, estrogen, and testosterone promote ossification. These hormones stimulate rapid multiplication of osteogenic cells, matrix deposition by osteoblasts, and multiplication and hypertrophy of the chondrocytes in the metaphyses. Adolescent girls grow faster than boys and attain their full height earlier, not only because they begin puberty earlier but also because estrogen has a stronger effect than testosterone. Since males grow for a longer time, however, they usually grow taller. A deficiency or excess of these steroids can therefore cause abnormalities ranging from stunted growth to very tall stature (see chapter 17). The use of anabolic steroids by adolescent athletes can cause premature closure of the epiphyseal plates and result in abnormally short adult stature (see p. 85).

### Before You Go On

*Answer the following questions to test your understanding of the preceding section:*

14. Describe the role of collagen and seed crystals in bone mineralization.

15. Why is it important to regulate blood calcium concentration within such a narrow range?

16. What effect does calcitonin have on blood calcium concentration, and how does it produce this effect? Answer the same questions for parathyroid hormone.

17. How is vitamin D synthesized, and what effect does it have on blood calcium concentration?

| TABLE 7.1 | Agents Affecting Calcium and Bone Metabolism |
|---|---|
| **Name** | **Effect** |
| *Hormones* | |
| Calcitonin | Little effect in adults; promotes mineralization and lowers blood $Ca^{2+}$ concentration in children; may prevent bone loss in pregnant and lactating women |
| Calcitriol (vitamin D) | Promotes intestinal absorption of $Ca^{2+}$ and phosphate; reduces urinary excretion of both; promotes both resorption and mineralization; stimulates osteoclast activity |
| Cortisol | Inhibits osteoclast activity, but if secreted in excess (Cushing disease), can cause osteoporosis by reducing bone deposition (inhibiting cell division and protein synthesis) |
| Estrogen | Stimulates osteoblasts and adolescent growth; prevents osteoporosis |
| Growth hormone | Stimulates bone elongation and cartilage proliferation at epiphyseal plate; increases urinary excretion of $Ca^{2+}$ but also increases intestinal $Ca^{2+}$ absorption, which compensates for the loss |
| Insulin | Stimulates bone formation; significant bone loss occurs in untreated diabetes mellitus |
| Parathyroid hormone | Indirectly activates osteoclasts, which resorb bone and raise blood $Ca^{2+}$ concentration; inhibits urinary $Ca^{2+}$ excretion; promotes calcitriol synthesis |
| Testosterone | Stimulates osteoblasts and promotes protein synthesis, thus promoting adolescent growth and epiphyseal closure |
| Thyroid hormone | Essential to bone growth; enhances effects of growth hormone, but excesses can cause hypercalcemia, increased $Ca^{2+}$ excretion in urine, and osteoporosis |
| *Growth Factors* | At least 12 hormonelike substances produced in bone itself that stimulate neighboring bone cells, promote collagen synthesis, stimulate epiphyseal growth, and produce many other effects |
| *Vitamins* | |
| Vitamin A | Promotes glycosaminoglycan (chondroitin sulfate) synthesis |
| Vitamin C (ascorbic acid) | Promotes collagen cross-linking, bone growth, and fracture repair |
| Vitamin D | Normally functions as a hormone (see calcitriol) |

| TABLE 7.2 | Classification of Fractures |
|---|---|
| **Type** | **Description** |
| Closed | Skin is not broken (formerly called a *simple* fracture) |
| Open | Skin is broken; bone protrudes through skin, or wound extends to fractured bone (formerly called a *compound* fracture) |
| Complete | Bone is broken into two or more pieces |
| Incomplete | Partial fracture that extends only partway across bone; pieces remain joined |
| Greenstick | Bone is bent on one side and has incomplete fracture on opposite side |
| Hairline | Fine crack in which sections of bone remain aligned; common in skull |
| Comminuted | Bone is broken into three or more pieces |
| Displaced | The portions of a fractured bone are out of anatomical alignment (see fig. 7.21a) |
| Nondisplaced | The portions of bone are still in correct anatomical alignment |
| Impacted | One bone fragment is driven into the marrow cavity or spongy bone of the other |
| Depressed | Broken portion of bone forms a concavity, as in skull fractures |
| Linear | Fracture parallel to long axis of bone |
| Transverse | Fracture perpendicular to long axis of bone |
| Oblique | Diagonal fracture, between linear and transverse |
| Spiral | Fracture spirals around axis of long bone, the result of a twisting stress, often produced when an abusive adult roughly picks up a child by the arm |
| Epiphyseal | Epiphysis is separated from diaphysis along the epiphyseal plate; seen in juveniles |
| Avulsion | Body part (such as a finger) is completely severed |
| Colles[33] | Fracture of the distal end of the radius and ulna; common in osteoporosis |
| Pott[34] | Fracture at the distal end of the tibia, fibula, or both; a common sports injury |

## Bone Disorders

### Objectives

When you have completed this section, you should be able to

- name and describe several bone diseases;
- name and describe the types of fractures;
- explain how a fracture is repaired; and
- discuss some clinical treatments for fractures and other skeletal disorders.

Most people probably give little thought to their skeletal systems unless they break a bone. This section describes bone fractures, their healing, and their treatment, followed by a summary of other bone diseases and a clinical insight on osteoporosis.

## FRACTURES AND THEIR REPAIR

There are multiple ways of classifying bone fractures. A **stress fracture** is a break caused by abnormal trauma to a bone, such as fractures incurred in falls, athletics, and military combat. A **pathologic fracture** is a break in a bone weakened by some other disease, such as bone cancer or osteoporosis, usually caused by a stress that would not normally fracture a bone. Fractures are also classified according to the direction of the fracture line, whether the skin is broken, and whether a bone is merely cracked or broken into separate pieces (table 7.2; fig. 7.19).

---

[33]Abraham Colles (1773–1843), Irish surgeon
[34]Sir Percivall Pott (1714–88), British surgeon

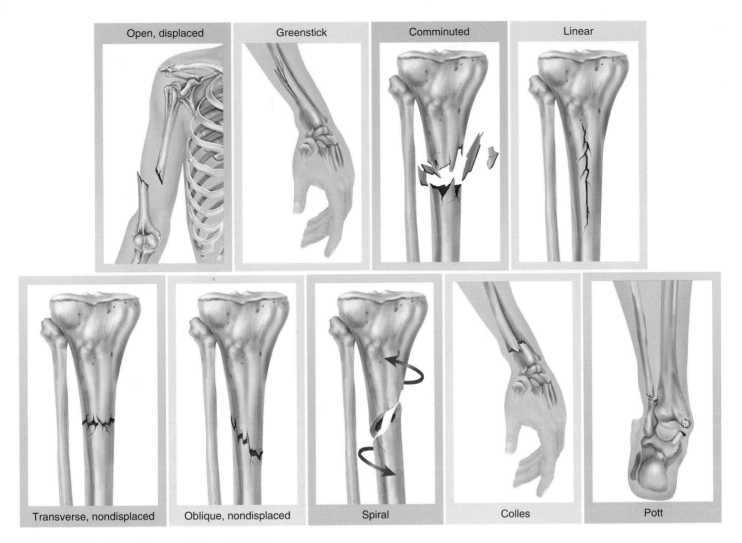

**FIGURE 7.19** Some Types of Bone Fractures.

## The Healing of Fractures

An uncomplicated fracture heals in about 8 to 12 weeks, but complex fractures take longer and all fractures heal more slowly in older people. The healing process occurs in the following stages (fig. 7.20):

① **Formation of hematoma and granulation tissue.** A bone fracture severs blood vessels of the bone and periosteum, causing bleeding and the formation of a blood clot *(fracture hematoma)*. Blood capillaries soon grow into the clot, while fibroblasts, macrophages, osteoclasts, and osteogenic cells invade the tissue from both the periosteal and medullary sides of the fracture. Osteogenic cells become very abundant within 48 hours of the injury. All of this capillary and cellular invasion converts the blood clot to a soft fibrous mass called **granulation tissue.**

② **Formation of a soft callus.**[35] Fibroblasts deposit collagen in the granulation tissue, while some osteo-genic cells become chondroblasts and produce patches of fibrocartilage called the **soft callus.**

③ **Conversion to hard callus.** Other osteogenic cells differentiate into osteoblasts, which produce a bony collar called the **hard callus** around the fracture. The hard callus is cemented to the dead bone around the injury site and acts as a temporary splint to join the broken ends or bone fragments together. It takes about 4 to 6 weeks for a hard callus to form. During this period, it is important that a broken bone be immobilized by traction or a cast to prevent reinjury.

④ **Remodeling.** The hard callus persists for 3 to 4 months. Meanwhile, osteoclasts dissolve small fragments of broken bone, and osteoblasts deposit spongy bone to bridge the gap between the broken ends. This spongy bone gradually fills in to become compact bone, in a manner similar to intramembranous ossification. Usually the fracture leaves a slight thickening of the bone visible by X ray, but in some

---

[35]*call* = hard, tough

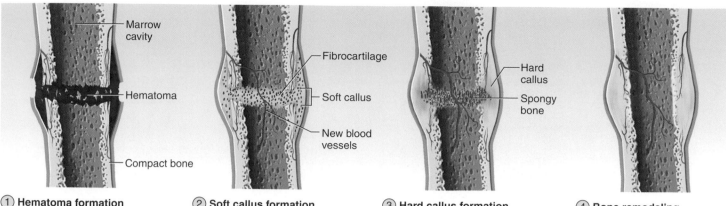

① **Hematoma formation**
The hematoma is converted to granulation tissue by invasion of cells and blood capillaries.

② **Soft callus formation**
Deposition of collagen and fibrocartilage converts granulation tissue to a soft callus

③ **Hard callus formation**
Osteoblasts deposit a temporary bony collar around the fracture to unite the broken pieces while ossification occurs

④ **Bone remodeling**
Small bone fragments are removed by osteoclasts, while osteoblasts deposit spongy bone and then convert it to compact bone

**FIGURE 7.20**   The Healing of a Bone Fracture.

cases healing is so complete that no trace of the fracture can be found.

## The Treatment of Fractures

Most fractures are set by **closed reduction,** a procedure in which the bone fragments are manipulated into their normal positions without surgery. **Open reduction** involves the surgical exposure of the bone and the use of plates, screws, or pins to realign the fragments (fig. 7.21b). To stabilize the bone during healing, fractures are set in casts. Traction is

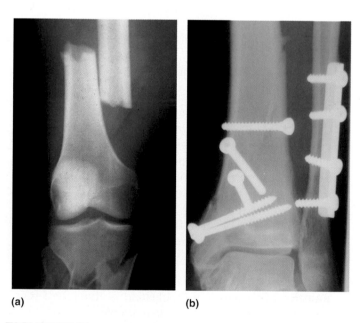

**(a)**              **(b)**

**FIGURE 7.21**   X Rays of Bone Fractures.   (a) A displaced fracture of the femur. (b) An ankle fracture involving both the tibia and fibula. This fracture has been set by open reduction, a process of surgically exposing the bone and realigning the fragments with plates and screws.

used to treat fractures of the femur in children. It aids in the alignment of the bone fragments by overriding the force of the strong thigh muscles. Traction is rarely used for elderly patients, however, because the risks from long-term confinement to bed outweigh the benefits. Hip fractures are usually pinned, and early ambulation (walking) is encouraged because it promotes blood circulation and healing. Fractures that take longer than 2 months to heal may be treated with electrical stimulation, which accelerates repair by suppressing the effects of parathyroid hormone.

**Orthopedics**[36] is the branch of medicine that deals with the prevention and correction of injuries and disorders of the bones, joints, and muscles. As the word suggests, this field originated as the treatment of skeletal deformities in children, but it is now much more extensive. It includes the design of artificial joints and limbs and the treatment of athletic injuries.

## OTHER BONE DISORDERS

Several additional bone disorders are summarized in table 7.3. The most common bone disease, **osteoporosis,**[37] receives special consideration in Insight 7.4. The effects of aging on the skeletal system are described on pages 1128–1129.

### Before You Go On

*Answer the following questions to test your understanding of the preceding section:*

18. *Name and describe any five types of bone fractures.*

19. *Why would osteomyelitis be more likely to occur in an open fracture than in a closed fracture?*

20. *What is a callus? How does it contribute to fracture repair?*

---

[36]*ortho* = straight + *ped* = child, foot
[37]*osteo* = bone + *por* = porous + *osis* = condition

| TABLE 7.3 | Bone Diseases |
|---|---|
| Rickets | Defective mineralization of bone in children, usually as a result of insufficient sunlight or vitamin D, sometimes due to a dietary deficiency of calcium or phosphate, or to liver or kidney diseases that interfere with calcitriol synthesis. Causes bone softening and deformity, especially in the weight-bearing bones of the lower limbs. |
| Osteomalacia | Adult form of rickets, most common in poorly nourished women who have had multiple pregnancies. Bones become softened, deformed, and more susceptible to fractures. |
| Osteoporosis | Loss of bone mass, especially spongy bone, usually as a result of lack of exercise or a deficiency of estrogen after menopause. It results in increasing brittleness and susceptibility to fractures (see Insight 7.4 for details). |
| Osteitis deformans (Paget[38] disease) | Excessive proliferation of osteoclasts and resorption of excess bone, with osteoblasts attempting to compensate by depositing extra bone. This results in rapid, disorderly bone remodeling and weak, deformed bones. Osteitis deformans usually passes unnoticed, but in some cases it causes pain, disfigurement, and fractures. It is most common in males over the age of 50. |
| Osteomyelitis[39] | Inflammation of osseous tissue and bone marrow as a result of bacterial infection. This disease was often fatal before the discovery of antibiotics and is still very difficult to treat. |
| Osteogenesis imperfecta (brittle bone disease) | A defect in collagen deposition that renders bones exceptionally brittle, resulting in fractures present at birth or occurring with extraordinary frequency during childhood; also causing tooth deformity, and hearing loss due to deformity of middle-ear bones. |
| Osteoma[40] | A benign bone tumor, especially in the flat bones of the skull; may grow into the orbits or sinuses. |
| Osteochondroma[41] | A benign tumor of bone and cartilage; often forms spurs at the ends of long bones. |
| Osteosarcoma[42] (osteogenic sarcoma) | The most common and deadly form of bone cancer. It occurs most often in the tibia, femur, and humerus of males between the ages of 10 and 25. In 10% of cases, it metastasizes to the lungs or other organs; if untreated, death occurs within 1 year. |
| Chondrosarcoma | A slow-growing cancer of hyaline cartilage, most common in middle age. It requires surgical removal; chemotherapy is ineffective. |

*Disorders described elsewhere*

## INSIGHT 7.4    Clinical Application

### Osteoporosis

The most common bone disease is osteoporosis (OSS-tee-oh-pore-OH-sis), literally, "porous bones." This is a condition in which the bones lose mass to such an extent that they become abnormally brittle and subject to pathologic fractures in response to relatively mild stresses. It involves loss of both organic matrix and minerals, and it affects spongy bone in particular, since this is the most metabolically active type and has the greatest surface area subject to osteoclast action (fig. 7.22a).

Fractures are the most serious consequence of osteoporosis. They occur especially in the hip, wrist, and vertebral column and under stresses as slight as sitting down too quickly. Hip fractures usually occur at the neck of the femur, while wrist fractures occur at the distal end of the radius and ulna (Colles fracture). About 275,000 elderly Americans fracture their hips each year. About 1 in 5 of these people die within a year from complications of immobility such as infections and thrombosis (blood clotting), and survivors often face a long, costly recovery. As the weight-bearing bodies of the vertebrae lose spongy bone, they become compressed like marshmallows (fig. 7.22b). Consequently, many people lose height after middle age, and in some women especially, the spine becomes deformed into a "widow's hump," a condition called kyphosis (fig. 7.22c).

Postmenopausal white women are at greatest risk for osteoporosis because women have less bone mass than men to begin with, begin losing it earlier (starting around age 40), and lose it faster than men do. By age 70, the average white woman loses 30% of her bone tissue, and some as much as 50%. Young black women develop denser bones on average and rarely suffer osteoporosis. Although they, too, lose bone after menopause, the loss usually does not reach the threshold for osteoporosis and pathologic fractures. In men, bone loss begins around age 60 and seldom exceeds 25%. Aside from age, race, and sex,

[38]Sir James Paget (1814–99), English surgeon
[39]osteo = bone + myel = marrow + itis = inflammation
[40]oste = bone + oma = tumor
[41]osteo = bone + chondr = cartilage + oma = tumor
[42]osteo = bone + sarc = flesh + oma = tumor

some risk factors for osteoporosis include smoking, diabetes mellitus, and diets poor in calcium, protein, vitamin C, and vitamin D. Once osteoporosis has set in, milk and other calcium sources and moderate exercise can slow its progression, but only slightly.

Estrogen has an osteoclast-inhibiting effect and is important in maintaining bone density in both sexes. It does not directly stimulate bone deposition, but inhibits its resorption. However, when the blood estrogen level drops below 30 ng/mL, osteoclast activity outpaces the osteoblasts and there is a net loss of bone. In women, this becomes a problem especially after menopause, when the ovaries stop producing estrogen and women lose this important brake on bone resorption. Men produce estrogen in the testes and adrenal glands. By age 70, 50% of men have estrogen levels below the threshold needed to maintain bone density. About 20% of osteoporosis patients are men.

Although osteoporosis has become an increasing public health problem because of the rising age of the population, it is not limited to the elderly. Ironically, it also occurs among young female runners, dancers, and gymnasts in spite of their vigorous exercise. Their percentage of body fat is so low that they stop ovulating and the ovaries secrete unusually low levels of estrogen. *Disuse osteoporosis* can occur at any age as a result of immobilization or inadequate weight-bearing exercise. In early long-term space flights, astronauts developed disuse osteoporosis because their bones were subjected to so little stress in that microgravity environment. This is one reason that exercise equipment is now standard on space shuttles and stations.

As widespread and disabling as osteoporosis is, it is a field of intensive medical research, resulting in new diagnostic and treatment strategies. Osteoporosis is now diagnosed with *dual-energy X-ray absorptiometry* (DEXA), which uses low-dose X rays to measure bone density. DEXA allows for early diagnosis and more effective drug treatment. However, the severity of osteoporosis and risk of fractures depend not on bone density alone, but also on the degree of connectivity between the trabeculae of spongy bone, which is lost as trabeculae deteriorate.

Neither DEXA nor any other diagnostic method yet available can detect this.

Treatments for osteoporosis are aimed at slowing the rate of bone resorption. Estrogen-replacement therapy was once the treatment of choice for postmenopausal women, but it fell out of favor in 2002 when a women's health study found that it increased the risk of breast cancer, stroke, and coronary artery disease. One of the current preferred treatments is a family of drugs called *bis-phosphonates* (trade names Fosamax, Actonel), which destroy osteoclasts. They have been shown to increase bone mass by 5% to 10% over 3 years, and to reduce the incidence of fractures by 50%. Their long-term safety is still being evaluated.

Oddly enough, parathyroid hormone (PTH) is also given to slow the rate of bone loss. This seems unexpected, since we know that PTH promotes osteoclast activity. That effect occurs, however, when PTH secretion is continuous. If it is given in a pulsed or intermittent fashion by daily injection, it seems to stimulate bursts of osteoblast development and inhibit the death of established osteoblasts. PTH and derivatives such as *teriparatide* (trade name Forteo) have been shown to increase bone density by 10% within 1 year, and to reduce the incidence of fractures by as much as 75%. However, teriparatide is used for no longer than 2 years because of concerns, based on animal studies, that longer use may promote bone cancer (osteogenic sarcoma). Thus there is still a need for safer drugs. Some researchers are working on antibodies or other ways of blocking the action of RANKL, the chemical messenger that promotes osteoclast development.

As is so often true, an ounce of prevention is worth a pound of cure. Drug treatments for osteoporosis are very expensive (around $7,000 per year in the United States as of 2004); exercise and a good bone-building diet are far less costly. The optimal time to prevent osteoporosis is between the ages of 25 and 40, when the skeleton is building to its maximum mass. The more bone a person has going into middle age, the less he or she will be affected by osteoporosis later. Ample exercise and calcium intake are the best preventive measures.

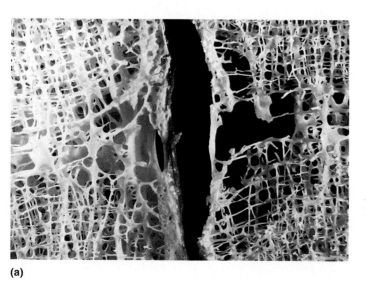
(a)

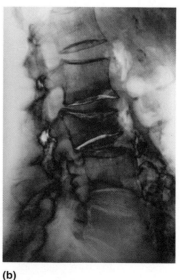
(b)

(c)

**FIGURE 7.22** **Spinal Osteoporosis.** (a) Spongy bone in the body of a vertebra in good health (left) and with osteoporosis (right). (b) Colorized X ray of lumbar vertebrae severely damaged by osteoporosis. (c) Abnormal thoracic spinal curvature (kyphosis) due to compression of thoracic vertebrae with osteoporosis.

# CONNECTIVE ISSUES

## Interactions Between the
## SKELETAL SYSTEM
## and Other Organ Systems

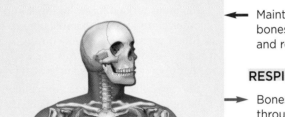

■ indicates ways in which this system affects other systems

■ indicates ways in which other systems affect this system

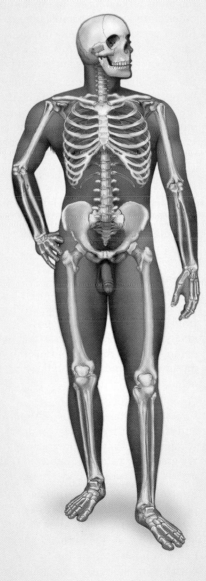

### INTEGUMENTARY SYSTEM

Bones lying close to body surfaces shape the skin

Initiates synthesis of vitamin D needed for bone deposition

### MUSCULAR SYSTEM

Bones provide leverage and sites of attachment for muscles; provide calcium needed for muscle contraction

Muscles move bones; stress produced by muscles affects patterns of ossification and remodeling, as well as shape of mature bones

### NERVOUS SYSTEM

Cranium and vertebral column protect brain and spinal cord; bones provide calcium needed for neural function

Sensory receptors provide sensations of body position and pain from bones and joints

### ENDOCRINE SYSTEM

Bones protect endocrine organs in head, chest, and pelvis

Hormones regulate mineral deposition and resorption, bone growth, and skeletal mass and density

### CIRCULATORY SYSTEM

Myeloid tissue forms blood cells; bone matrix stores calcium needed for cardiac muscle activity

Delivers $O_2$, nutrients, and hormones to bone tissue and carries away wastes; delivers blood cells to marrow

### LYMPHATIC/IMMUNE SYSTEMS

Most types of blood cells produced in myeloid tissue function as part of immune system

Maintains balance of interstitial fluid in bones; lymphocytes assist in defense and repair of bones

### RESPIRATORY SYSTEM

Bones form respiratory passageway through nasal cavity; protect lungs and aid in ventilation

Provides $O_2$ and removes $CO_2$

### URINARY SYSTEM

Skeleton physically supports and protects organs of urinary system

Kidneys activate vitamin D and regulate calcium and phosphate excretion

### DIGESTIVE SYSTEM

Skeleton provides bony protection for digestive organs

Provides nutrients needed for bone growth and maintenance

### REPRODUCTIVE SYSTEM

Skeleton protects some reproductive organs

Gonads produce hormones that affect bone growth and closure of epiphyseal plates

# CHAPTER REVIEW

# Review of Key Concepts

**Tissues and Organs of the Skeletal System (p. 214)**

1. *Osteology* is the study of bone. The *skeletal system* consists of bones, cartilages, and ligaments.

2. The functions of the skeletal system include bodily support, protection of internal organs, movement, blood formation, and electrolyte and acid–base balance.

3. *Bone (osseous tissue)* is a connective tissue with a mineralized matrix.

4. Bones are classified by shape as long, short, flat, and irregular bones.

5. A long bone is a cylinder of *compact bone* enclosing more porous *spongy bone* and a central *marrow cavity.*

6. The shaft of a long bone is the *diaphysis,* and each expanded end is an *epiphysis.* The epiphysis is covered with *articular cartilage* where it meets an adjacent bone. An *epiphyseal plate* separates the marrow cavities of the diaphysis and epiphysis in children and adolescents.

7. A bone is externally covered with a fibrous *periosteum* and internally lined with a thin *endosteum.*

8. A flat bone consists of a layer of spongy bone sandwiched between two layers of compact bone.

**Histology of Osseous Tissue (p. 217)**

1. Osseous tissue has four types of cells: stem cells called *osteogenic cells,* matrix-depositing cells called *osteoblasts,* strain detectors called *osteocytes* enclosed in lacunae of the matrix, and matrix-dissolving cells called *osteoclasts.*

2. The bone matrix is a composite of collagen fibers and other protein–carbohydrate complexes embedded in a ground substance of *hydroxyapatite,* calcium carbonate, and other minerals. The minerals enable bones to resist compression and support weight, and the proteins enable bones to bend slightly without breaking.

3. *Compact bone* consists mostly of *osteons.* An osteon is a cylindrical group of concentric *lamellae* of matrix surrounding a *central canal* occupied by a nerve and blood vessels. *Interstitial lamellae,* the remains of old osteons, occupy some of the spaces between osteons, and *circumferential lamellae* parallel to the bone surface form the inner and outer boundaries of compact bone. Blood vessels reach the central canals by way of *perforating canals* that open on the bone surface.

4. *Spongy bone* is a porous lattice of bony *trabeculae* with lamellae but few osteons. The trabeculae are oriented along lines of stress, giving spongy bone strength without excessive weight.

5. The marrow cavity and the spaces between spongy bone trabeculae are occupied by bone marrow. *Red bone marrow* is a *hemopoietic* (blood-forming) tissue and *yellow bone marrow* is adipose tissue.

**Bone Development (p. 221)**

1. *Intramembranous ossification* is a process in which flat bones develop from sheets of embryonic mesenchyme. Osteoblasts deposit an organic matrix that transforms the mesenchyme into a soft collagenous *osteoid tissue.*

2. Osteoblasts then deposit calcium salts that harden the osteoid tissue and transform it to spongy bone. The surface zones of this spongy bone are gradually filled in with calcified matrix to become compact bone.

3. *Endochondral ossification* is a process in which bones develop from hyaline cartilage. Most of the fetal skeleton forms in this manner.

4. Endochondral ossification begins with enlargement of the lacunae and death of chondrocytes in the *primary ossification center,* which is soon hollowed out to form a *primary marrow space.* Osteoblasts populate this space and create a temporary scaffold of spongy bone.

5. Between the epiphysis and the primary marrow space is a *metaphysis,* which exhibits five zones of histological transformation from cartilage to bone.

6. *Secondary ossification centers* appear later in one or both epiphyses of a long bone. The cartilaginous epiphyseal plate persists through adolescence as a growth zone for the long bone.

7. Mature bones continue to be remodeled throughout life, employing *appositional growth* to grow in diameter and thickness and to reshape bone surfaces in response to stress.

**Physiology of Osseous Tissue (p. 227)**

1. The addition of inorganic salts to osseous tissue is called *mineral deposition.* Osteoblasts produce collagen fibers on which hydroxyapatite and other minerals crystallize. Mineralization requires a critical ratio of calcium to inorganic phosphate called the *solubility product.*

2. The removal of salts from osseous tissue is called *mineral resorption.* Osteoclasts resorb bone by secreting hydrochloric acid, which dissolves the inorganic salts of the matrix, and acid phosphatase, an enzyme that digests the organic matrix.

3. The skeleton is the body's major reservoir of calcium, which is also needed for many other physiological processes. The $Ca^{2+}$ concentration of the blood plasma is maintained within narrow limits by resorbing $Ca^{2+}$ from the skeleton or depositing excess $Ca^{2+}$ into it.

4. *Hypocalcemia,* a $Ca^{2+}$ deficiency, can cause potentially fatal muscle tetany, whereas *hypercalcemia,* a $Ca^{2+}$ excess, can depress neuromuscular function. These imbalances are normally prevented by the actions of calcitriol, calcitonin, and parathyroid hormone.

5. *Calcitriol* is a form of vitamin D synthesized by the sequential action of the skin, liver, and kidneys. It raises blood $Ca^{2+}$ concentration by promoting absorption of $Ca^{2+}$ by the small intestine, conservation of $Ca^{2+}$ by the kidneys, and resorption of $Ca^{2+}$ from bone by osteoclasts.

6. *Calcitonin* is secreted by the thyroid gland in response to hypercalcemia. It lowers blood $Ca^{2+}$ concentration by inhibiting osteoclasts and stimulating the depositional activity of osteoblasts.

7. *Parathyroid hormone* is secreted by the parathyroid glands in response to hypocalcemia. It raises blood $Ca^{2+}$ levels by indirectly stimulating osteoclasts, inhibiting osteoblasts,

promoting calcitriol synthesis, and promoting $Ca^{2+}$ conservation by the kidneys.

8. Many other hormones, growth factors, and vitamins influence bone physiology, such as growth hormone, testosterone, thyroid hormone, estrogen, insulin, and vitamins A and C.

### Bone Disorders (p. 232)

1. The repair of a fractured bone begins when fibroblasts, osteogenic cells, and other cells invade the blood clot (fracture hematoma) and transform it into *granulation tissue.* Collagen and cartilage deposition convert this to a *soft callus.* This is followed by the formation of a calcified *hard callus* that encircles the fracture and

reunites the broken bone pieces. The hard callus is remodeled by osteoclasts and osteoblasts over several months.

2. *Closed reduction* is the realignment of the parts of a broken bone without surgery. *Open reduction* involves surgery and fragment realignment with the aid of plates and screws.

3. The most common bone disease is *osteoporosis,* a loss of bone mass resulting in brittleness and abnormal vulnerability to fractures. It is most common in postmenopausal white women. The risk of osteoporosis can be minimized with diet (adequate calcium intake) and exercise.

## Testing Your Recall

1. Which cells have a ruffled border and secrete hydrochloric acid?
   a. C cells
   b. osteocytes
   c. osteogenic cells
   d. osteoblasts
   e. osteoclasts

2. The marrow cavity of an adult bone may contain
   a. myeloid tissue.
   b. hyaline cartilage.
   c. periosteum.
   d. osteocytes.
   e. articular cartilages.

3. The spurt of growth in puberty results from cell proliferation and hypertrophy in
   a. the epiphysis.
   b. the epiphyseal line.
   c. the dense bone.
   d. the epiphyseal plate.
   e. the spongy bone.

4. Osteoclasts are most closely related, by common descent, to
   a. osteoprogenitor cells.
   b. osteogenic cells.
   c. monocytes.
   d. fibroblasts.
   e. osteoblasts.

5. The walls between cartilage lacunae break down in the zone of
   a. cell proliferation.
   b. calcification.
   c. reserve cartilage.
   d. bone deposition.
   e. cell hypertrophy.

6. Which of these is *not* an effect of PTH?
   a. rise in blood phosphate level
   b. reduction of calcium excretion
   c. increased intestinal calcium absorption
   d. increased number of osteoclasts
   e. increased calcitriol synthesis

7. A child jumps to the ground from the top of a playground "jungle gym." His leg bones do not shatter mainly because they contain
   a. an abundance of glycosaminoglycans.
   b. young, resilient osteocytes.
   c. an abundance of calcium phosphate.
   d. collagen fibers.
   e. hydroxyapatite crystals.

8. One long bone meets another at its
   a. diaphysis.
   b. epiphyseal plate.
   c. periosteum.
   d. metaphysis.
   e. epiphysis.

9. Calcitriol is made from
   a. calcitonin.
   b. 7-dehydrocholesterol.
   c. hydroxyapatite.
   d. estrogen.
   e. PTH.

10. One sign of osteoporosis is
    a. osteosarcoma.
    b. osteomalacia.
    c. osteomyelitis.

    d. a Colles fracture.
    e. hypocalcemia

11. Calcium phosphate crystallizes in bone as a mineral called _____.

12. Osteocytes contact each other through channels called _____ in the bone matrix.

13. A bone increases in diameter only by _____ growth, the addition of new surface lamellae.

14. Seed crystals of hydroxyapatite form only when the levels of calcium and phosphate in the tissue fluid exceed the _____.

15. A calcium deficiency called _____ can cause death by suffocation.

16. _____ are cells that secrete collagen and stimulate calcium phosphate deposition.

17. The most active form of vitamin D, produced mainly by the kidneys, is _____.

18. The most common bone disease is _____.

19. The transitional region between epiphyseal cartilage and the primary marrow cavity of a young bone is called the _____.

20. A pregnant, poorly nourished woman may suffer a softening of the bones called _____.

*Answers in Appendix B*

# True or False

*Determine which five of the following statements are false, and briefly explain why.*

1. Spongy bone is always covered by compact bone.

2. Most bones develop from hyaline cartilage.

3. Fractures are the most common bone disorder.

4. The growth zone of the long bones of adolescents is the articular cartilage.

5. Osteoclasts develop from osteoblasts.

6. Osteocytes develop from osteoblasts.

7. The protein of the bone matrix is called hydroxyapatite.

8. Blood vessels travel through the central canals of compact bone.

9. Vitamin D promotes bone deposition, not resorption.

10. Parathyroid hormone promotes bone resorption and raises blood calcium concentration.

*Answers in Appendix B*

# Testing Your Comprehension

1. Most osteocytes of an osteon are far removed from blood vessels, but still receive blood-borne nutrients. Explain how this is possible.

2. A 50-year-old business executive decides he has not been getting enough exercise for the last several years. He takes up hiking and finds that he really loves it. Within 2 years, he is spending many of his weekends hiking with a heavy backpack and camping in the mountains. Explain what changes in his anatomy could be predicted from Wolff's law of bone.

3. How does the regulation of blood calcium concentration exemplify negative feedback and homeostasis?

4. Describe how the arrangement of trabeculae in spongy bone demonstrates the unity of form and function.

5. Identify two bone diseases you would expect to see if the epidermis were a completely effective barrier to UV radiation and a person took no dietary supplements to compensate for this. Explain your answer.

*Answers at www.mhhe.com/saladin4*

# www.mhhe.com/saladin4

*The textbook website provides a wealth of interactive study materials fully organized and integrated by chapter. You will find practice quizzes, labeling exercises, and much more that will complement your learning and understanding of anatomy and physiology. The website also includes tools designed to enhance your* **Anatomy & Physiology | REVEALED** *experience.*

CHAPTER

# 8

*Vertebral column and ribs*

# THE SKELETAL SYSTEM

## Brushing Up

To understand this chapter, it is important that you understand or brush up on the following concepts:

- Directional terminology (p. 31)
- Body regions and cavities (pp. 32–37)

Knowledge of skeletal anatomy will be useful as you study later chapters. It provides a foundation for studying the gross anatomy of other organ systems because many organs are named for their relationships to nearby bones. The subclavian artery and vein, for example, are located beneath the clavicles; the temporalis muscle is attached to the temporal bone; the ulnar nerve and radial artery travel beside the ulna and radius, respectively, of the forearm; and the frontal, parietal, temporal, and occipital lobes of the brain are named for bones of the cranium. An understanding of how the muscles produce body movements also depends on knowledge of skeletal anatomy. Additionally, the positions, shapes, and processes of bones can serve as landmarks for a clinician in determining where to give an injection or record a pulse, what to look for in an X ray, or how to perform physical therapy and other medical procedures.

# Overview of the Skeleton

### Objectives

When you have completed this section, you should be able to

- state the approximate number of bones in the adult body;

- explain why this number varies with age and from one person to another; and

- define several terms that denote surface features of bones.

Students typically begin by examining an *articulated*[1] skeleton (dried bones held together by wires and rods to show their spatial relationships to one another) or a *disarticulated* one (one that is taken apart so that the anatomy of individual bones can be studied in more detail). The skeleton is shown in figure 8.1. Note that it is divided into two regions: the **axial skeleton** and the **appendicular skeleton.** The axial skeleton, which forms the central supporting axis of the body, includes the skull, auditory ossicles, hyoid bone, vertebral column, and thoracic cage (ribs and sternum). The appendicular skeleton includes the bones of the upper limb and pectoral girdle, and the bones of the lower limb and pelvic girdle.

## BONES OF THE SKELETAL SYSTEM

It is often stated that there are 206 bones in the skeleton, but this is only a typical adult count. At birth there are

[1]*artic* = joint

about 270, and even more bones form during childhood. With age, however, the number decreases as separate bones fuse. For example, each half of the adult pelvis is a single bone called the *os coxae,* which results from the fusion of three childhood bones: the ilium, ischium, and pubis. The fusion of several bones, completed by late adolescence to the mid-20s, brings about the average adult number of 206. These bones are listed in table 8.1.

This number varies even among adults. One reason is the development of **sesamoid**[2] **bones**—bones that form within some tendons in response to stress. The patella (kneecap) is the largest of these; most of the others are

[2]*sesam* = sesame seed + *oid* = resembling

| **TABLE 8.1** | Bones of the Adult Skeletal System |
|---|---|
| **Axial Skeleton** | |
| ***Skull (22 bones)*** | ***Auditory Ossicles (6 bones)*** |
| Cranial bones | Malleus (2) |
|   Frontal bone (1) | Incus (2) |
|   Parietal bone (2) | Stapes (2) |
|   Occipital bone (1) | |
|   Temporal bone (2) | ***Hyoid Bone (1 bone)*** |
|   Sphenoid bone (1) | |
|   Ethmoid bone (1) | ***Vertebral Column (26 bones)*** |
| Facial bones | Cervical vertebrae (7) |
|   Maxilla (2) | Thoracic vertebrae (12) |
|   Palatine bone (2) | Lumbar vertebrae (5) |
|   Zygomatic bone (2) | Sacrum (1) |
|   Lacrimal bone (2) | Coccyx (1) |
|   Nasal bone (2) | |
|   Vomer (1) | ***Thoracic Cage (25 bones)*** |
|   Inferior nasal concha (2) | Ribs (24) |
|   Mandible (1) | Sternum (1) |
| **Appendicular Skeleton** | |
| ***Pectoral Girdle (4 bones)*** | ***Pelvic Girdle (2 bones)*** |
| Scapula (2) | Os coxae (2) |
| Clavicle (2) | |
| ***Upper Limb (60 bones)*** | ***Lower Limb (60 bones)*** |
| Humerus (2) | Femur (2) |
| Radius (2) | Patella (2) |
| Ulna (2) | Tibia (2) |
| Carpals (16) | Fibula (2) |
| Metacarpals (10) | Tarsals (14) |
| Phalanges (28) | Metatarsals (10) |
| | Phalanges (28) |
| **Grand Total: 206 bones** | |

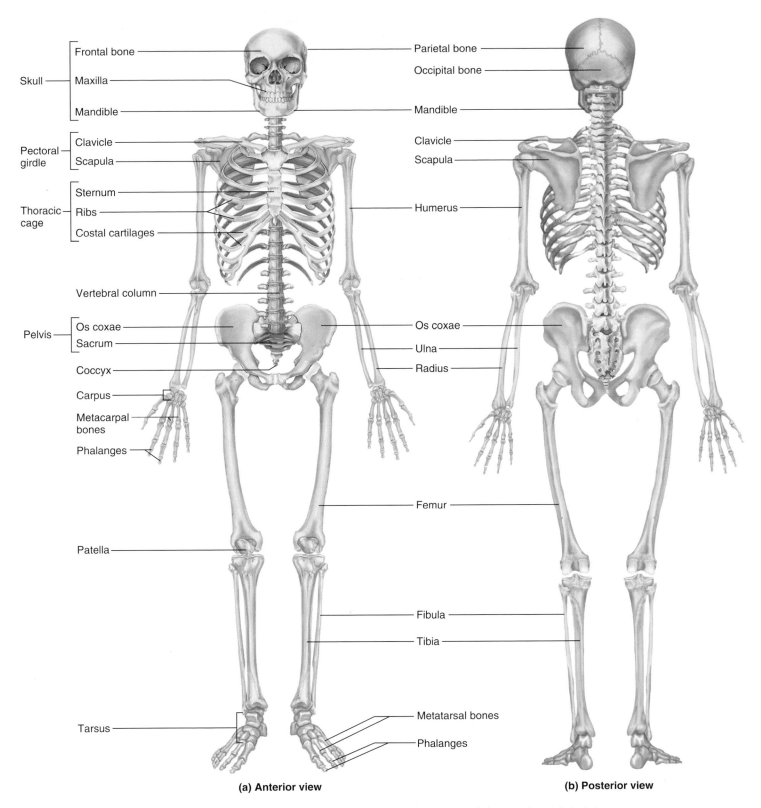

Skull
- Frontal bone
- Maxilla
- Mandible

Pectoral girdle
- Clavicle
- Scapula

Thoracic cage
- Sternum
- Ribs
- Costal cartilages

Vertebral column

Pelvis
- Os coxae
- Sacrum

Coccyx

Carpus

Metacarpal bones

Phalanges

Patella

Tarsus

Parietal bone
Occipital bone
Mandible
Clavicle
Scapula
Humerus

Os coxae
Ulna
Radius

Femur

Fibula
Tibia

Metatarsal bones
Phalanges

**(a) Anterior view**              **(b) Posterior view**

**FIGURE 8.1**  **The Adult Skeleton.**  The appendicular skeleton is colored green, and the rest is axial skeleton.

small, rounded bones in such locations as the knuckles. Another reason for adult variation is that some people have extra bones in the skull called **sutural** (SOO-chure-ul), or **wormian**,[3] **bones** (see fig. 8.6).

## SURFACE FEATURES OF BONES

Bone surfaces exhibit a variety of ridges, spines, bumps, depressions, canals, pores, slits, and articular surfaces. It is important to know the names of these *surface markings* because later descriptions of joints, muscle attachments, and the routes traveled by nerves and blood vessels are based on this terminology. The terms for the most common of these features are listed in table 8.2, and several of them are illustrated in figure 8.2.

The rest of this chapter is divided into four sections: (1) the skull, (2) the vertebral column and thoracic cage, (3) the pectoral girdle and upper limb, and (4) the pelvic girdle and lower limb. At the end of each section, you will find a review and checklist of skeletal features you should know (see tables 8.4, 8.6, 8.7, and 8.9). As you study this chapter, use yourself as a model. You can easily palpate (feel) many of the bones and some of their details through the skin. Rotate your forearm, cross your legs, palpate your skull, and think about what is happening beneath the surface or what you can feel through the skin. You will gain the most from this chapter (and indeed, the entire book) if you are conscious of your own body in relation to what you are studying.

### Before You Go On

*Answer the following questions to test your understanding of the preceding section:*

1. *Name the major components of the axial skeleton. Name those of the appendicular skeleton.*

2. *Explain why an adult does not have as many bones as a child does. Explain why one adult may have more bones than another adult of the same age has.*

3. *Briefly describe each of the following bone features: a condyle, epicondyle, process, tubercle, fossa, sulcus, and foramen.*

# The Skull

### Objectives

When you have completed this section, you should be able to

- name the bones of the skull and their anatomical features; and
- describe the development of the skull from infancy through childhood.

[3]Ole Worm (1588–1654), Danish physician

| TABLE 8.2 | Surface Features (Markings) of Bones |
|---|---|
| **Term** | **Description and Example** |
| *Articulations* | |
| Condyle | A rounded knob that articulates with another bone (occipital condyles of the skull) |
| Facet | A smooth, flat, slightly concave or convex articular surface (articular facets of the vertebrae) |
| Head | The prominent expanded end of a bone, sometimes rounded (head of the femur) |
| *Extensions and Projections* | |
| Crest | A narrow ridge (iliac crest of the pelvis) |
| Epicondyle | A projection superior to a condyle (medial epicondyle of the femur) |
| Line | A slightly raised, elongated ridge (nuchal lines of the skull) |
| Process | Any bony prominence (mastoid process of the skull) |
| Protuberance | A bony outgrowth or protruding part (mental protuberance of the chin) |
| Spine | A sharp, slender, or narrow process (spine of the scapula) |
| Trochanter | Two massive processes unique to the femur |
| Tubercle | A small, rounded process (greater tubercle of the humerus) |
| Tuberosity | A rough elevated surface (tibial tuberosity) |
| *Depressions* | |
| Alveolus | A pit or socket (tooth socket) |
| Fossa | A shallow, broad, or elongated basin (mandibular fossa) |
| Fovea | A small pit (fovea capitis of the femur) |
| Sulcus | A groove for a tendon, nerve, or blood vessel (intertubercular sulcus of the humerus) |
| *Passages* | |
| Canal | A tubular passage or tunnel in a bone (condylar canal of the skull) |
| Fissure | A slit through a bone (orbital fissures behind the eye) |
| Foramen | A hole through a bone, usually round (foramen magnum of the skull) |
| Meatus | An opening into a canal (acoustic meatus of the ear) |

The skull is the most complex part of the skeleton. Figures 8.3 to 8.6 present an overview of its general anatomy. Although it may seem to consist only of the mandible (lower jaw) and "the rest," in reality the skull is composed of 22 bones (and sometimes more). Most of these are connected by immovable joints called **sutures** (SOO-chures), which are visible as seams on the surface (fig. 8.4). These are important landmarks in the descriptions that follow.

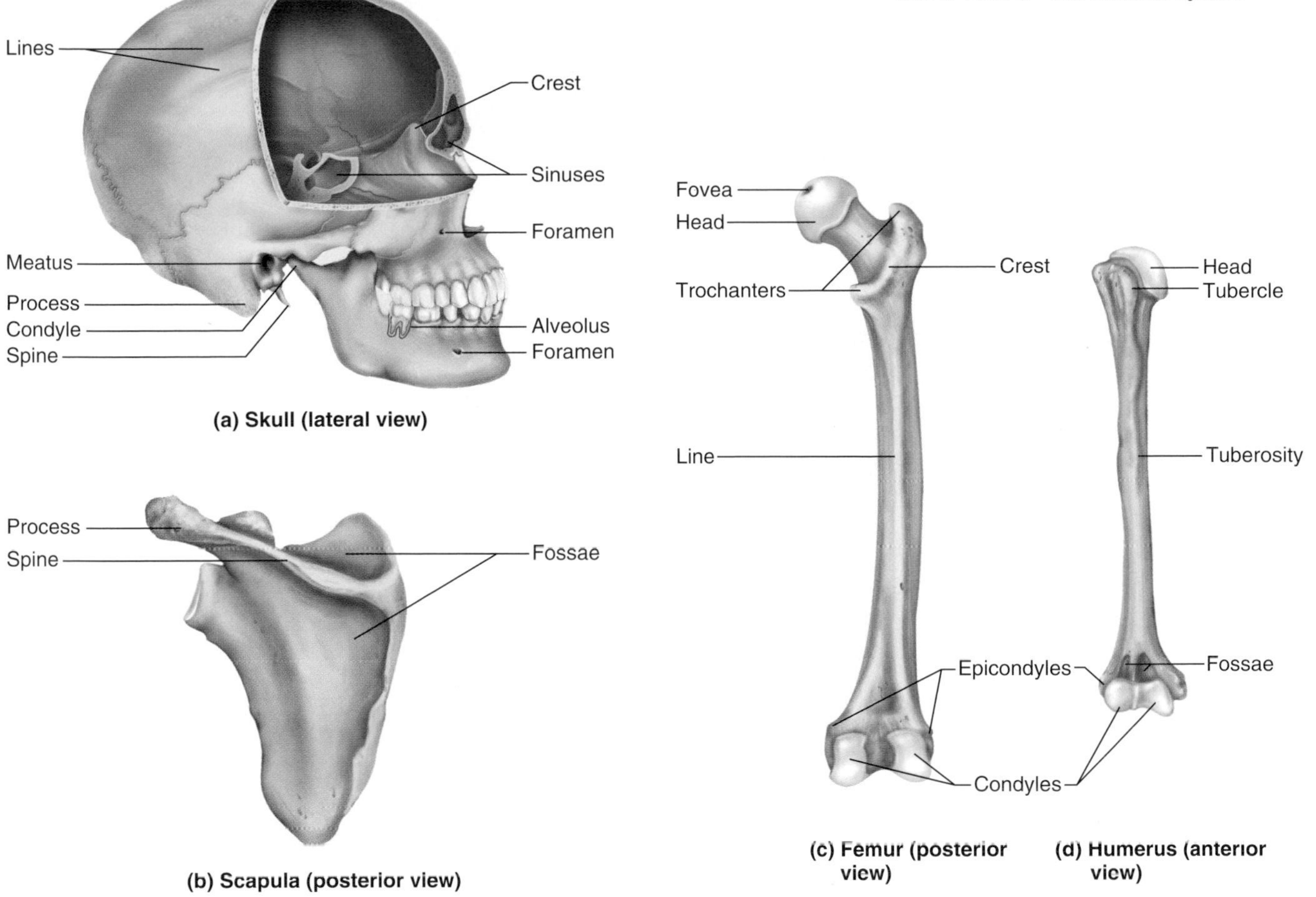

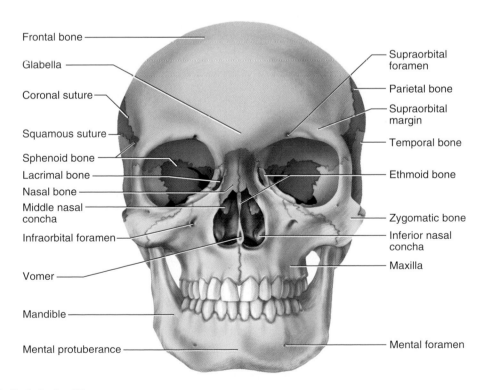

Lines
Crest
Sinuses
Foramen
Meatus
Process
Condyle
Spine
Alveolus
Foramen
(a) Skull (lateral view)
Process
Spine
Fossae
(b) Scapula (posterior view)
Fovea
Head
Trochanters
Crest
Line
Epicondyles
Condyles
(c) Femur (posterior view)
Head
Tubercle
Tuberosity
Fossae
Epicondyles
Condyles
(d) Humerus (anterior view)

**FIGURE 8.2** **Surface Features of Some Representative Bones.** Most of these features also occur on many other bones of the body.

Frontal bone
Glabella
Coronal suture
Squamous suture
Sphenoid bone
Lacrimal bone
Nasal bone
Middle nasal concha
Infraorbital foramen
Vomer
Mandible
Mental protuberance
Supraorbital foramen
Parietal bone
Supraorbital margin
Temporal bone
Ethmoid bone
Zygomatic bone
Inferior nasal concha
Maxilla
Mental foramen

**FIGURE 8.3** The Skull, Anterior View.

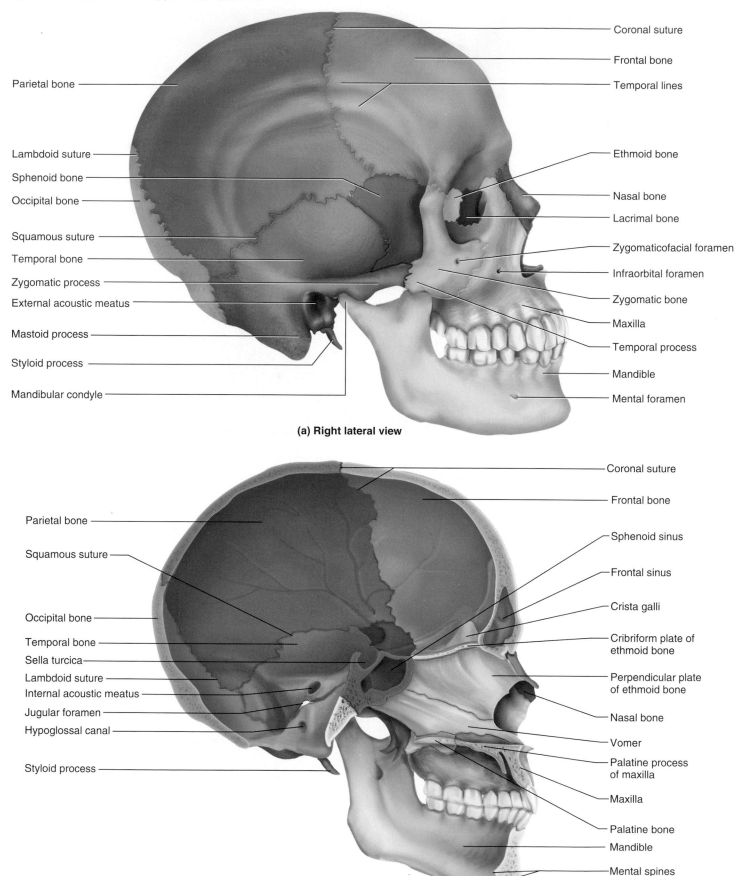

Parietal bone

Coronal suture

Frontal bone

Temporal lines

Lambdoid suture

Sphenoid bone

Occipital bone

Ethmoid bone

Nasal bone

Lacrimal bone

Squamous suture

Temporal bone

Zygomatic process

External acoustic meatus

Zygomaticofacial foramen

Infraorbital foramen

Zygomatic bone

Maxilla

Temporal process

Mastoid process

Styloid process

Mandibular condyle

Mandible

Mental foramen

**(a) Right lateral view**

Parietal bone

Coronal suture

Frontal bone

Squamous suture

Sphenoid sinus

Frontal sinus

Crista galli

Occipital bone

Temporal bone

Sella turcica

Lambdoid suture

Internal acoustic meatus

Jugular foramen

Hypoglossal canal

Cribriform plate of
ethmoid bone

Perpendicular plate
of ethmoid bone

Nasal bone

Vomer

Styloid process

Palatine process
of maxilla

Maxilla

Palatine bone

Mandible

Mental spines

**FIGURE 8.4** The Skull, Lateral Views
(External and Internal).

**(b) Median section**

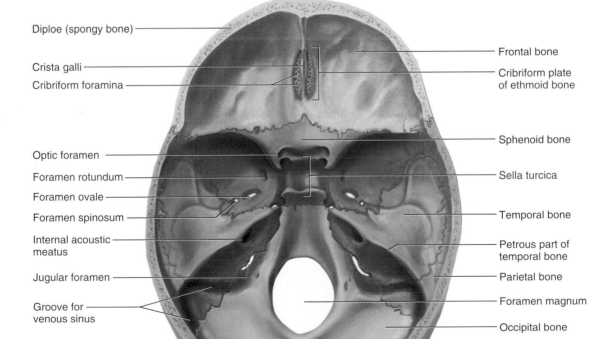

**(a) Inferior view**

Incisive foramen

Zygomatic bone

Zygomatic arch

Nasal choana

Vomer

Sphenoid bone

Mandibular fossa

Styloid process

External acoustic meatus

Occipital condyle

Mastoid process

Mastoid notch

Temporal bone

Condylar canal

Parietal bone

Inferior nuchal line

Superior nuchal line

Occipital bone

Palatine process of maxilla

Intermaxillary suture

Palatine bone

Greater palatine foramen

Medial pterygoid plate

Lateral pterygoid plate

Foramen ovale

Foramen spinosum

Foramen lacerum

Basilar part of occipital bone

Carotid canal

Stylomastoid foramen

Jugular foramen

Foramen magnum

Mastoid foramen

Lambdoid suture

External occipital protuberance

Diploe (spongy bone)

Crista galli

Cribriform foramina

Optic foramen

Foramen rotundum

Foramen ovale

Foramen spinosum

Internal acoustic meatus

Jugular foramen

Groove for venous sinus

Frontal bone

Cribriform plate of ethmoid bone

Sphenoid bone

Sella turcica

Temporal bone

Petrous part of temporal bone

Parietal bone

Foramen magnum

Occipital bone

**(b) Superior view of cranial floor**

**FIGURE 8.5** The Base of the Skull.

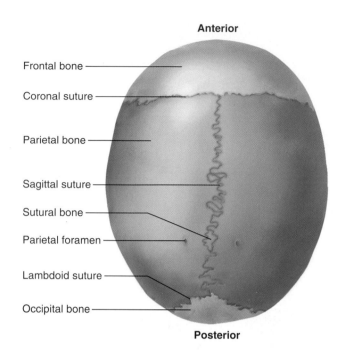

**FIGURE 8.6**   The Calvaria (Skullcap), Superior View.

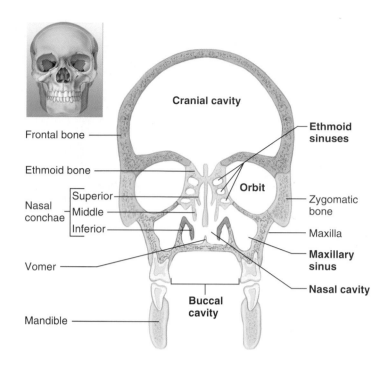

**FIGURE 8.7**   Major Cavities of the Skull, Frontal Section.

The skull contains several prominent cavities (fig. 8.7). The largest, with an adult volume of about 1,350 mL, is the **cranial cavity,** which encloses the brain. Other cavities include the **orbits** (eye sockets), **nasal cavity, buccal (BUCK-ul) cavity** (mouth), **middle-** and **inner-ear cavities,** and **paranasal sinuses.** The paranasal sinuses are named for the bones in which they occur (fig. 8.8)—the **frontal, sphenoid, ethmoid,** and **maxillary sinuses.** These cavities are connected with the nasal cavity, lined by a mucous membrane, and filled with air. They lighten the anterior portion of the skull and act as chambers that add resonance to the voice. The latter effect can be sensed in the way your voice changes when you have a cold and mucus obstructs the travel of sound into the sinuses and back.

Bones of the skull have especially conspicuous **foramina**—singular, *foramen* (fo-RAY-men)—holes that allow passage for nerves and blood vessels. The major foramina are summarized in table 8.3. The details of this table will mean more to you when you study cranial nerves and blood vessels in later chapters.

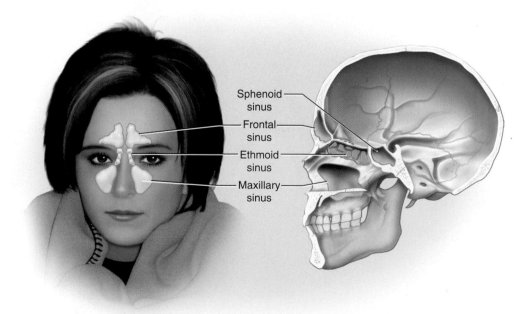

**FIGURE 8.8**   The Paranasal Sinuses.

| TABLE 8.3 | Foramina of the Skull and the Nerves and Blood Vessels Transmitted Through Them |
|---|---|
| **Bones and Their Foramina**[*] | **Structures Transmitted** |
| *Frontal Bone* | |
| Supraorbital foramen or notch | Supraorbital nerve, artery, and vein; ophthalmic nerve |
| *Parietal Bone* | |
| Parietal foramen | Emissary vein of superior longitudinal sinus |
| *Temporal Bone* | |
| Carotid canal | Internal carotid artery |
| External acoustic meatus | Sound waves to eardrum |
| Internal acoustic meatus | Vestibulocochlear nerve; internal auditory vessels |
| Stylomastoid foramen | Facial nerve |
| Mastoid foramen | Meningeal artery; vein from sigmoid sinus |
| *Temporal-Occipital Region* | |
| Jugular foramen | Internal jugular vein; glossopharyngeal, vagus, and accessory nerves |
| *Temporal-Occipital-Sphenoid Region* | |
| Foramen lacerum | No major nerves or vessels; closed by cartilage |
| *Occipital Bone* | |
| Foramen magnum | Spinal cord; accessory nerve; vertebral arteries |
| Hypoglossal canal | Hypoglossal nerve to muscles of tongue |
| Condylar canal | Vein from transverse sinus |
| *Sphenoid Bone* | |
| Foramen ovale | Mandibular division of trigeminal nerve; accessory meningeal artery |
| Foramen rotundum | Maxillary division of trigeminal nerve |
| Foramen spinosum | Middle meningeal artery; spinosal nerve; part of trigeminal nerve |
| Optic foramen | Optic nerve; ophthalmic artery |
| Superior orbital fissure | Oculomotor, trochlear, and abducens nerves; ophthalmic division of trigeminal nerve; ophthalmic veins |
| *Ethmoid Bone* | |
| Cribriform foramina | Olfactory nerves |
| *Maxilla* | |
| Infraorbital foramen | Infraorbital nerve and vessels; maxillary division of trigeminal nerve |
| Incisive foramen | Nasopalatine nerves |
| *Maxilla-Sphenoid Region* | |
| Inferior orbital fissure | Infraorbital nerve; zygomatic nerve; infraorbital vessels |
| *Lacrimal Bone* | |
| Lacrimal foramen | Tear duct leading to nasal cavity |
| *Palatine Bone* | |
| Greater palatine foramen | Palatine nerves |
| *Zygomatic Bone* | |
| Zygomaticofacial foramen | Zygomaticofacial nerve |
| Zygomaticotemporal foramen | Zygomaticotemporal nerve |
| *Mandible* | |
| Mental foramen | Mental nerve and vessels |
| Mandibular foramen | Inferior alveolar nerves and vessels to the lower teeth |

*When two or more bones are listed together (for example, temporal–occipital), it indicates that the foramen passes between them.

# CRANIAL BONES

The cranial cavity is enclosed by the **cranium**[4] (braincase), which protects the brain and associated sensory organs. The cranium is composed of eight **cranial bones:**

| | |
|---|---|
| 1 frontal bone | 1 occipital bone |
| 2 parietal bones | 1 sphenoid bone |
| 2 temporal bones | 1 ethmoid bone |

The cranium is a rigid structure with an opening, the **foramen magnum** (literally "large hole"), where the spinal cord enters. An important consideration in treatment of head injuries is swelling of the brain. Since the cranium cannot enlarge, swelling puts pressure on the brain and results in even more tissue damage. Severe swelling may force the brainstem out through the foramen magnum, usually with fatal consequences.

The delicate brain tissue does not directly contact the cranial bones but is separated from them by three membranes called the *meninges* (meh-NIN-jeez) (see chapter 14). The thickest and toughest of these, the *dura mater*[5] (DUE-rah MAH-tur), lies loosely against the inside of the cranium in most places but is firmly attached to it at a few points.

The cranium consists of two major parts: the calvaria and the base. The **calvaria**[6] (skullcap) forms the roof and

---

[4]*crani* = helmet
[5]*dura* = tough, strong + *mater* = mother
[6]*calvar* = bald, skull

walls (see fig. 8.6). In study skulls it is often sawed so that part of it can be lifted off for examination of the interior. This reveals the **base** (floor) of the cranial cavity (see fig. 8.5b), which is divided into three basins corresponding to the contour of the inferior surface of the brain (fig. 8.9). The relatively shallow **anterior cranial fossa** is crescent-shaped and accommodates the frontal lobes of the brain. The **middle cranial fossa,** which drops abruptly deeper, is shaped like a pair of outstretched bird's wings and accommodates the temporal lobes. The **posterior cranial fossa** is deepest and houses a large posterior division of the brain called the cerebellum.

We now consider the eight cranial bones and their distinguishing features.

## Frontal Bone

The **frontal bone** extends from the forehead back to a prominent *coronal suture,* which crosses the crown of the head from right to left and joins the frontal bone to the parietal bones (see figs. 8.3 and 8.4). It forms the anterior wall and about one-third of the roof of the cranial cavity, and it turns inward to form nearly all of the anterior cranial fossa and the roof of the orbit. Deep to the eyebrows it has a ridge called the **supraorbital margin.** The center of each margin is perforated by a single **supraorbital foramen** (see figs. 8.3 and 8.14), which provides passage for a nerve, artery, and veins. In some people, the edge of this foramen breaks through the margin of the orbit and forms a *supraorbital notch.* A person may have a foramen on one supraorbital margin and a notch on the other. The

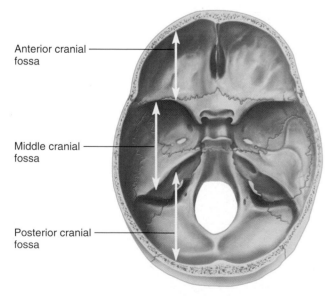

**(a) Superior view**

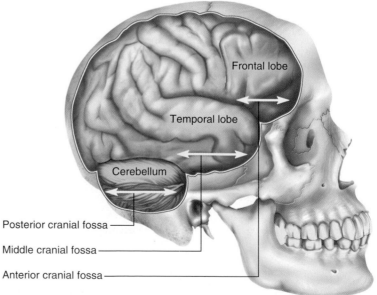

**(b) Lateral view**

**FIGURE 8.9** Cranial Fossae. The three fossae conform to the contours of the base of the brain.

smooth area of the frontal bone just above the root of the nose is called the **glabella.**[7] The frontal bone also contains the frontal sinuses. You may not see these on some study skulls. They are absent from some people, and on some skulls the calvaria is cut too high to show them. Along the cut edge of the calvaria, you can see the diploe—the layer of spongy bone in the middle of the cranial bones (see fig. 8.5b).

## Parietal Bones

The right and left **parietal** (pa-RY-eh-tul) **bones** form most of the cranial roof and part of its walls (see figs. 8.4 and 8.6). Each is bordered by four sutures that join it to the neighboring bones: (1) the **sagittal suture** between the parietal bones; (2) the **coronal**[8] **suture** at the anterior margin; the **lambdoid**[9] (LAM-doyd) **suture** at the posterior margin; and (4) the **squamous suture** laterally. Small sutural (wormian) bones are often seen along the sagittal and lambdoid sutures, like little islands of bone with the suture lines passing around them. Internally, the parietal and frontal bones have markings that look a bit like aerial photographs of river tributaries (see fig. 8.4b). These represent places where the bone has been molded around blood vessels of the meninges.

Externally, the parietal bones have few features. A **parietal foramen** sometimes occurs near the corner of the lambdoid and sagittal sutures (see fig. 8.6). A pair of slight thickenings, the superior and inferior **temporal lines,** form an arc across the parietal and frontal bones (see fig. 8.4a). They mark the attachment of the large, fan-shaped *temporalis* muscle, a chewing muscle that passes between the zygomatic arch and temporal bone and inserts on the mandible.

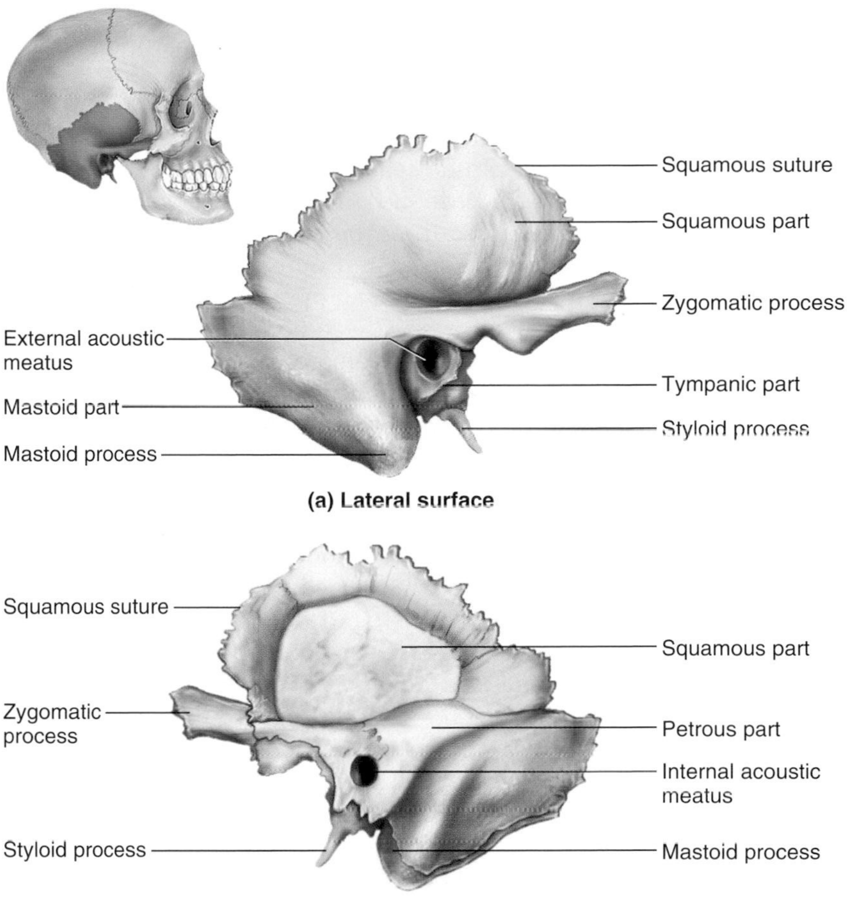

## Temporal Bones

If you palpate your skull just above and anterior to the ear—that is, the temporal region—you can feel the **temporal bone,** which forms much of the lower wall and part of the floor of the cranial cavity (fig. 8.10). The temporal bone derives its name from the fact that people often develop their first gray hairs on the temples with the passage of time.[10] The relatively complex shape of the temporal bone is best understood by dividing it into four parts:

1. The **squamous**[11] **part** (which you just palpated) is relatively flat and vertical. It is encircled by the squamous suture. It bears two prominent features: (1) the **zygomatic process,** which extends anteriorly

**(a) Lateral surface**

Squamous suture
Squamous part
Zygomatic process
External acoustic meatus
Tympanic part
Mastoid part
Styloid process
Mastoid process

**(b) Medial surface**

Squamous suture
Squamous part
Zygomatic process
Petrous part
Internal acoustic meatus
Styloid process
Mastoid process

**FIGURE 8.10** **The Right Temporal Bone.** The lateral surface faces the scalp and external ear; the medial surface faces the brain.

▶ *List all the bones that articulate with the temporal bone.*

to form part of the **zygomatic arch** (cheekbone), and (2) the **mandibular fossa,** a depression where the mandible articulates with the cranium.

2. The **tympanic**[12] **part** is a small ring of bone that borders the **external acoustic meatus** (me-AY-tus), the opening into the ear canal. It has a pointed spine on its inferior surface, the **styloid process,** named for its resemblance to the stylus used by ancient Greeks and Romans to write on wax tablets. The styloid process provides attachment for muscles of the tongue, pharynx, and hyoid bone.

3. The **mastoid**[13] **part** lies posterior to the tympanic part. It bears a heavy **mastoid process,** which you can palpate as a prominent lump behind the earlobe. It is filled with small air sinuses that communicate with the middle-ear cavity. These sinuses are subject to infection and inflammation *(mastoiditis),* which can erode the bone and spread to the brain. A groove called the **mastoid notch** lies medial to the mastoid

---

[7]*glab* = smooth
[8]*corona* = crown
[9]Shaped like the Greek letter lambda (λ)
[10]*tempor* = time
[11]*squam* = flat + *ous* = characterized by

[12]*tympan* = drum (eardrum) + *ic* = pertaining to
[13]*mast* = breast + *oid* = resembling

process (see fig. 8.5a). It is the origin of the *digastric muscle,* which opens the mouth. The notch is perforated by the **stylomastoid foramen** at its anterior end and the **mastoid foramen** at its posterior end.

4. The **petrous**[14] **part** can be seen in the cranial floor, where it resembles a little mountain range separating the middle cranial fossa from the posterior fossa (fig. 8.10b). It houses the middle- and inner-ear cavities. The **internal acoustic meatus,** an opening on its posteromedial surface, allows passage of the vestibulocochlear (vess-TIB-you-lo-COC-lee-ur) nerve, which carries sensations of hearing and balance from the inner ear to the brain. On the inferior surface of the petrous part are two prominent foramina named for the major blood vessels that pass through them (see fig. 8.5a): (1) The **carotid canal** is a passage for the internal carotid artery, a major blood supply to the brain. This artery is so close to the inner ear that you can sometimes hear the pulsing of its blood when your ear is resting on a pillow or your heart is beating hard. (2) The **jugular foramen** is a large, irregular opening just medial to the styloid process, between the temporal and occipital bones. Blood from the brain drains through this foramen into the internal jugular vein of the neck. Three cranial nerves also pass through this foramen.

## Occipital Bone

The **occipital** (oc-SIP-ih-tul) **bone** forms the rear of the skull *(occiput)* and much of its base (see fig. 8.5). Its most conspicuous feature, the foramen magnum, admits the spinal cord to the cranial cavity and provides a point of attachment for the dura mater. The bone continues anterior to the foramen magnum as a thick medial plate, the **basilar part.** On either side of the foramen magnum is a smooth knob called the **occipital condyle** (CON-dile), where the skull rests on the vertebral column. At the anterolateral edge of each condyle is a **hypoglossal**[15] **canal,** named for the hypoglossal nerve that passes through it to supply the muscles of the tongue. A **condylar** (CON-dih-lur) **canal** sometimes occurs posterior to each occipital condyle.

Internally, the occipital bone displays impressions left by large venous sinuses that drain blood from the brain (see fig. 8.5b). One of these grooves travels along the midsagittal line. Just before reaching the foramen magnum, it branches into right and left grooves that wrap around the occipital bone like outstretched arms before terminating at the jugular foramina.

Other features of the occipital bone can be palpated on the back of your head. One is a prominent medial bump called the **external occipital protuberance**—the attach-

ment for the **nuchal**[16] (NEW-kul) **ligament,** which binds the skull to the vertebral column. A ridge, the **superior nuchal line,** can be traced horizontally from the external occipital protuberance toward the mastoid process (see fig. 8.5a). It defines the superior limit of the neck and provides attachment to the skull for several neck and back muscles. By pulling down on the occipital bone, some of these muscles help to keep the head erect. The **inferior nuchal line** provides attachment for some of the deep neck muscles. This inconspicuous ridge cannot be palpated on the living body but is visible on an isolated skull.

## Sphenoid Bone

The **sphenoid**[17] (SFEE-noyd) **bone** has a complex shape with a thick medial **body** and outstretched **greater** and **lesser wings,** which give the bone as a whole a somewhat ragged mothlike shape. Most of it is best seen from the superior perspective (fig. 8.11a). In this view, the lesser wings form the posterior margin of the anterior cranial fossa and end at a sharp bony crest, where the sphenoid drops abruptly to the greater wings. These form about half of the middle cranial fossa (the temporal bone forming the rest) and are perforated by several foramina to be discussed shortly.

The greater wing forms part of the lateral surface of the cranium just anterior to the temporal bone (see fig. 8.4a). The lesser wing forms the posterior wall of the orbit and contains the **optic foramen,** which permits passage of the optic nerve and ophthalmic artery (see fig. 8.14). Superiorly, a pair of bony spines of the lesser wing called the **anterior clinoid processes** appear to guard the optic foramina. A gash in the posterior wall of the orbit, the **superior orbital fissure,** angles upward lateral to the optic foramen. It serves as a passage for nerves that supply the muscles that move the eyes.

The body of the sphenoid has a saddlelike prominence named the **sella turcica**[18] (SEL-la TUR-sih-ca). It consists of a deep pit called the *hypophyseal fossa,* which houses the pituitary gland (hypophysis); a raised anterior margin called the *tuberculum sellae* (too-BUR-cu-lum SEL-lee); and a posterior margin called the *dorsum sellae.* In life, a fibrous membrane is stretched over the sella turcica. A stalk penetrates the membrane to connect the pituitary gland to the base of the brain.

Lateral to the sella turcica, the sphenoid is perforated by several foramina (see fig. 8.5a). The **foramen rotundum** and **foramen ovale** (oh-VAY-lee) are passages for two branches of the trigeminal nerve. The **foramen spinosum,** about the diameter of a pencil lead, provides passage for an artery of the meninges. An irregular gash called the **foramen lacerum**[19] (LASS-eh-rum) occurs at the junction of the

---

[14]*petr* = stone, rock + *ous* = like
[15]*hypo* = below + *gloss* = tongue

[16]*nucha* = back of the neck
[17]*sphen* = wedge + *oid* = resembling
[18]*sella* = saddle + *turcica* = Turkish
[19]*lacerum* = torn, lacerated

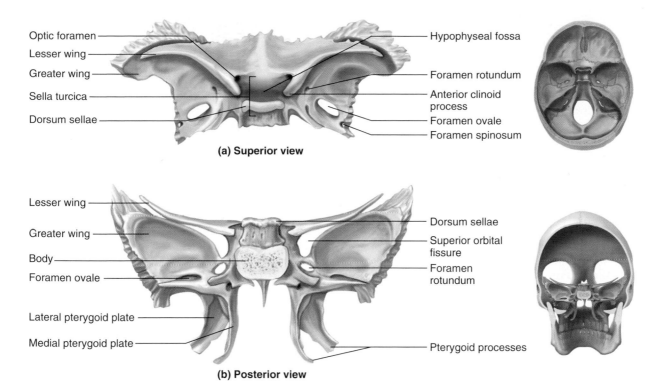

Optic foramen
Lesser wing
Greater wing
Sella turcica
Dorsum sellae

Hypophyseal fossa
Foramen rotundum
Anterior clinoid process
Foramen ovale
Foramen spinosum

**(a) Superior view**

Lesser wing
Greater wing
Body
Foramen ovale
Lateral pterygoid plate
Medial pterygoid plate

Dorsum sellae
Superior orbital fissure
Foramen rotundum
Pterygoid processes

**(b) Posterior view**

**FIGURE 8.11**   The Sphenoid Bone.

sphenoid, temporal, and occipital bones. It is filled with cartilage in life and transmits no major vessels or nerves.

In an inferior view, the sphenoid can be seen just anterior to the basilar part of the occipital bone. The internal openings of the nasal cavity seen here are called the **nasal choanae**[20] (co-AH-nee), or **internal nares.** Lateral to each choana, the sphenoid bone exhibits a pair of parallel plates: the **medial** and **lateral pterygoid**[21] (TERR-ih-goyd) **plates** (see fig. 8.5a). Each plate has a narrower inferior extension called the **pterygoid process.** These plates and processes provide attachment for some of the jaw muscles. The sphenoid sinus occurs within the body of the sphenoid bone.

## Ethmoid Bone

The **ethmoid**[22] (ETH-moyd) **bone** is an anterior cranial bone located between the eyes (figs. 8.7 and 8.12). It contributes to the medial wall of the orbit, the roof and walls of the nasal cavity, and the nasal septum. It is a very porous and delicate bone, with three major portions:

1. The vertical **perpendicular plate,** a thin median plate of bone that forms the superior two-thirds of the nasal septum (see fig. 8.4b). (The lower part is formed by the *vomer,* discussed later.) The septum divides the

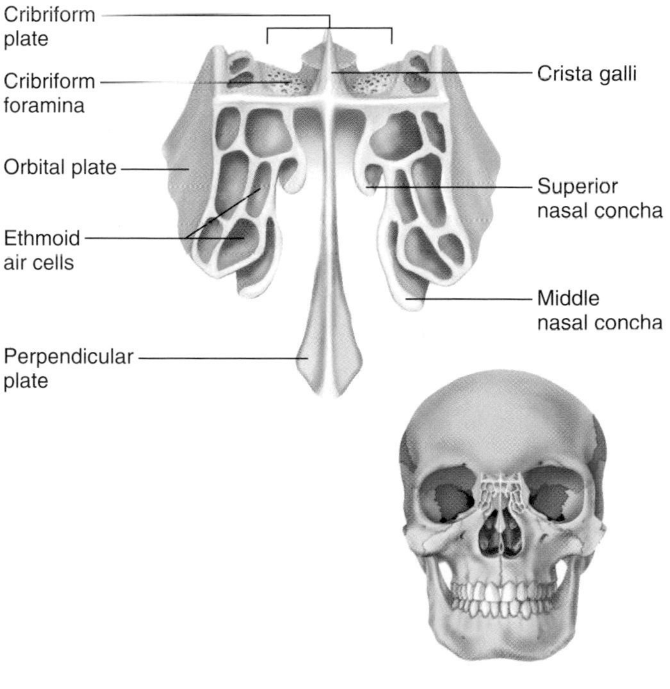

Cribriform plate
Cribriform foramina
Orbital plate
Ethmoid air cells
Perpendicular plate

Crista galli
Superior nasal concha
Middle nasal concha

**FIGURE 8.12**   The Ethmoid Bone, Anterior View.
❯ *List all the bones that articulate with the ethmoid bone.*

[20]*choana* = funnel
[21]*pterygo* = wing
[22]*ethmo* = sieve, strainer + *oid* = resembling

nasal cavity into right and left air spaces called the **nasal fossae** (FOSS-ee). The septum is often curved, or deviated, toward one nasal fossa or the other.

2.  A horizontal **cribriform**[23] (CRIB-rih-form) **plate,** which forms the roof of the nasal cavity. This plate has a median crest called the **crista galli**[24] (GAL-eye), an attachment point for membranes *(meninges)* that enclose the brain. On each side of the crista is an elongated depressed area perforated with numerous holes, the **cribriform foramina.** A pair of *olfactory bulbs* of the brain, concerned with the sense of smell, rest in these depressions, and the foramina allow passage for olfactory nerves from the nasal cavity to the bulbs (Insight 8.1).

3.  The **labyrinth,** a large mass on each side of the perpendicular plate. The labyrinth is named for the fact that internally, it has a maze of air spaces called the **ethmoidal cells.** Collectively, these constitute the *ethmoid sinus* discussed earlier. The lateral surface of the labyrinth is a smooth, slightly concave **orbital plate** seen on the medial wall of the orbit (see fig. 8.14). The medial surface of the labyrinth gives rise to two curled, scroll-like plates of bone called the **superior** and **middle nasal conchae**[25] (CON-kee). These project into the nasal fossa from its lateral wall

---

[23]*cribri* = sieve + *form* = in the shape of
[24]*crista* = crest + *galli* = of a rooster
[25]*concha* = conch (large marine snail)

## INSIGHT 8.1   Clinical Application

### Injury to the Ethmoid Bone

The ethmoid bone is very delicate and is easily injured by a sharp upward blow to the nose, such as a person might suffer by striking an automobile dashboard in a collision. The force of a blow can drive bone fragments through the cribriform plate into the meninges or brain tissue. Such injuries are often evidenced by leakage of cerebrospinal fluid into the nasal cavity, and may be followed by the spread of infection from the nasal cavity to the brain. Blows to the head can also shear off the olfactory nerves that pass through the ethmoid bone and cause *anosmia,* an irreversible loss of the sense of smell and a great reduction in the sense of taste (most of which depends on smell). This not only deprives life of some of its pleasures, but can also be dangerous, as when a person fails to smell smoke, gas, or spoiled food.

toward the septum (see figs. 8.7 and 8.13). There is also a separate bone, the *inferior nasal concha,* discussed later. The three of them together occupy most of the space in the nasal cavity. By filling space and creating turbulence in the flow of inhaled air, they ensure that the air contacts the mucous membranes that cover these bones, which cleanse, humidify, and warm the inhaled air before it reaches the lungs. The superior concha and adjacent part of the nasal septum also bear the sensory cells of smell.

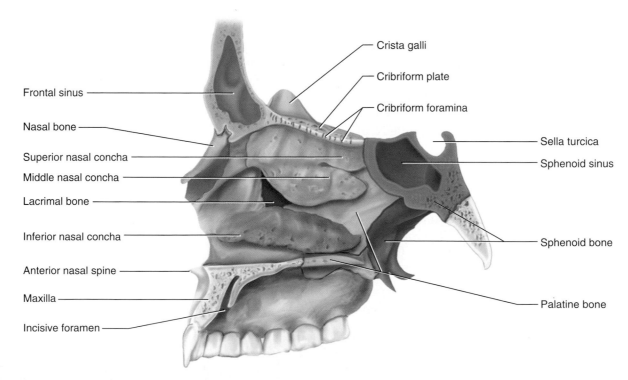

**FIGURE 8.13**   The Right Nasal Cavity, Sagittal Section.

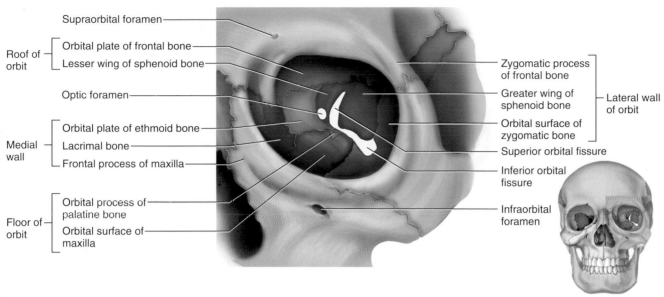

FIGURE 8.14  The Left Orbit, Anterior View.

In studying an intact skull, all that can be seen of the ethmoid is the perpendicular plate, by looking into the nasal cavity (see fig. 8.3); the orbital plate, by looking at the medial wall of the orbit (see fig. 8.14); and the crista galli and cribriform plate, viewed from within the cranial cavity (see fig. 8.5b).

## FACIAL BONES

The **facial bones** are those that have no direct contact with the brain or meninges. They support the teeth, give shape and individuality to the face, form part of the orbital and nasal cavities, and provide attachment for the muscles of facial expression and mastication. There are 14 facial bones:

| | |
|---|---|
| 2 maxillae | 2 nasal bones |
| 2 palatine bones | 2 inferior nasal conchae |
| 2 zygomatic bones | 1 vomer |
| 2 lacrimal bones | 1 mandible |

### Maxillae

The **maxillae** (mac-SILL-ee) form the upper jaw and meet each other at a median *intermaxillary suture* (see figs. 8.3, 8.4a, and 8.5a). Small points of maxillary bone called **alveolar processes** grow into the spaces between the bases of the teeth. The root of each tooth is inserted into a deep socket, or **alveolus.** If a tooth is lost or extracted so that chewing no longer puts stress on the maxilla, the alveolar processes are resorbed and the alveolus fills in with new bone, leaving a smooth area on the maxilla. The teeth are discussed in detail in chapter 25.

> **Think About It**
>
> *Suppose you were studying a skull with some teeth missing. How could you tell whether the teeth had been lost after the person's death or years before it?*

Each maxilla extends from the teeth to the inferomedial wall of the orbit. Just below the orbit, it exhibits an **infraorbital foramen,** which provides passage for a blood vessel to the face and a nerve that receives sensations from the nasal region and cheek. This nerve emerges through the foramen rotundum into the cranial cavity. The maxilla forms part of the floor of the orbit, where it exhibits a gash called the **inferior orbital fissure** that angles downward and medially (fig. 8.14). The inferior and superior orbital fissures form a sideways V whose apex lies near the optic foramen. The inferior orbital fissure is a passage for blood vessels and sensory nerves from the face.

The **palate** forms the roof of the mouth and floor of the nasal cavity. It consists of a bony **hard palate** anteriorly and a fleshy **soft palate** posteriorly. Most of the hard palate is formed by horizontal extensions of the maxilla called **palatine** (PAL-uh-tine) **processes** (see fig. 8.5a). Near the anterior margin of each palatine process, just behind the incisors, is an **incisive foramen.** The palatine processes normally meet at the intermaxillary suture at about 12 weeks of gestation. Failure to join results in a *cleft palate,* often accompanied by a *cleft lip* lateral to the midline. Cleft palate and lip can be surgically corrected with good cosmetic results, but a cleft palate makes it difficult for an infant to generate the suction needed for nursing.

### Evolutionary Significance of the Palate

In most vertebrates, the nasal passages open into the oral cavity. Mammals, by contrast, have a palate that separates the nasal cavity from the oral cavity. In order to maintain our high metabolic rate, we must digest our food rapidly; in order to do this, we chew it thoroughly to break it up into small, easily digested particles before swallowing it. The palate allows us to continue breathing during this prolonged chewing.

## Palatine Bones

The **palatine bones** form the rest of the hard palate, part of the wall of the nasal cavity, and part of the floor of the orbit (see figs. 8.5a and 8.13). At the posterolateral corners of the hard palate are two large **greater palatine foramina.**

## Zygomatic Bones

The **zygomatic**[26] **bones** form the angles of the cheeks at the inferolateral margins of the orbits and part of the lateral wall of each orbit; they extend about halfway to the ear (see figs. 8.4a and 8.5a). Each zygomatic bone has an inverted T shape and usually a small **zygomaticofacial** (ZY-go-MAT-ih-co-FAY-shul) **foramen** near the intersection of the stem and crossbar of the T. The prominent zygomatic arch that flares from each side of the skull is formed mainly by the union of the zygomatic process of the temporal bone and the *temporal process* of the zygomatic bone (see fig. 8.4a).

## Lacrimal Bones

The **lacrimal**[27] (LACK-rih-mul) **bones** form part of the medial wall of each orbit (fig. 8.14). A depression called the **lacrimal fossa** houses a membranous *lacrimal sac* in life. Tears from the eye collect in this sac and drain into the nasal cavity.

## Nasal Bones

Two small rectangular **nasal bones** form the bridge of the nose (see fig. 8.3) and support cartilages that give shape to the lower portion of the nose. If you palpate the bridge, you can easily feel where the nasal bones end and the cartilages begin. The nasal bones are often fractured by blows to the nose.

## Inferior Nasal Conchae

There are three conchae in the nasal cavity. The superior and middle conchae, as discussed earlier, are parts of the ethmoid bone. The **inferior nasal concha**—the largest of the three—is a separate bone (see fig. 8.13).

## Vomer

The **vomer** forms the inferior half of the nasal septum (see figs. 8.3 and 8.4b). Its name literally means "plowshare," which refers to its resemblance to the blade of a plow. The superior half of the nasal septum is formed by the perpendicular plate of the ethmoid bone, as mentioned earlier. The vomer and perpendicular plate support a wall of *septal cartilage* that forms most of the anterior part of the nasal septum.

## Mandible

The **mandible** (fig. 8.15) is the strongest bone of the skull and the only one that can move. It supports the lower teeth and provides attachment for muscles of mastication and facial expression. The horizontal portion is called the **body;** the vertical-to-oblique posterior portion is the **ramus** (RAY-mus)—plural, *rami* (RAY-my); and these two portions meet at a corner called the **angle.** The mandible develops as separate right and left bones in the fetus, joined by a midsagittal cartilaginous joint called the **mental symphysis** (SIM-fih-sis). This joint ossifies in early childhood, uniting the two halves into a single bone. The inner (posterior) surface of the mandible in this region has a pair of small points, the **mental spines,** which serve for attachment of certain chin muscles (see fig. 8.4b). The point of the chin is the **mental protuberance.** Like the maxilla, the mandible has pointed alveolar processes between the teeth. On the anterolateral surface of the body, the **mental foramen** permits the passage of nerves and blood vessels of the chin. The inner surface of the body has a number of shallow depressions and ridges to accommodate muscles and salivary glands. The angle of the mandible has a rough lateral surface for insertion of the *masseter,* a muscle of mastication.

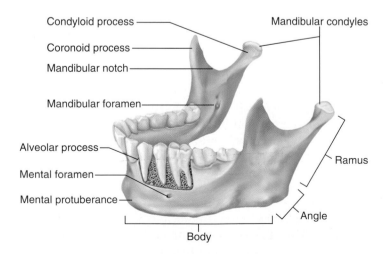

Condyloid process — Mandibular condyles
Coronoid process
Mandibular notch
Mandibular foramen
Ramus
Alveolar process
Mental foramen
Mental protuberance
Angle
Body

**FIGURE 8.15** The Mandible.

---

[26]*zygo* = to join, unite
[27]*lacrim* = tear, to cry

The ramus is somewhat Y-shaped. Its posterior branch, called the **condyloid (CON-dih-loyd) process,** bears the **mandibular condyle**—an oval knob that articulates with the mandibular fossa of the temporal bone. The hinge of the mandible is the **temporomandibular joint (TMJ).** The anterior branch of the ramus, called the **coronoid process,** is the point of insertion for the temporalis muscle, which pulls the mandible upward when you bite. The U-shaped arch between the two processes is called the **mandibular notch.** Just below the notch, on the medial surface of the ramus, is the **mandibular foramen.** The nerve and blood vessels that supply the lower teeth enter this foramen. Dentists inject anesthetic near here to deaden sensation from the lower teeth.

## BONES ASSOCIATED WITH THE SKULL

Seven bones are closely associated with the skull but not considered part of it. These are the three auditory ossicles in each middle-ear cavity and the hyoid bone beneath the chin. The **auditory ossicles**[28]—named the **malleus** (hammer), **incus** (anvil), and **stapes** (STAY-peez) (stirrup)—are discussed in connection with hearing in chapter 16. The **hyoid**[29] **bone** is a slender bone between the chin and larynx (fig. 8.16). It is one of the few bones that does not articulate with any other. It is suspended from the styloid processes of the skull, somewhat like a hammock, by the small *stylohyoid muscles* and *stylohyoid ligaments.* The median **body** of the hyoid is flanked on either side by hornlike projections called the **greater** and **lesser cornua**[30] (CORN-you-uh)—singular, *cornu* (COR-new). The hyoid bone serves for attachment of several muscles that control the mandible, tongue, and larynx. Forensic pathologists look for a fractured hyoid as evidence of strangulation.

[28] *os* – bone + *icle* – little
[29] *hy* = the letter *U* + *oid* = resembling
[30] *cornu* = horn

## THE SKULL IN INFANCY AND CHILDHOOD

The head of an infant could not fit through the mother's pelvic outlet at birth were it not for the fact that the bones of its skull are not yet fused. The shifting of the skull bones during birth may cause the infant to appear deformed, but the head soon assumes a more normal shape. Spaces between the unfused cranial bones are called **fontanels,**[31] after the fact that pulsation of the infant's blood can be felt there. The bones are joined at these points only by fibrous membranes, in which intramembranous ossification is completed later. Four of these sites are especially prominent and regular in location: the **anterior, posterior, sphenoid (anterolateral),** and **mastoid (posterolateral) fontanels** (fig. 8.17). Most

[31] *fontan* = fountain + *el* = little

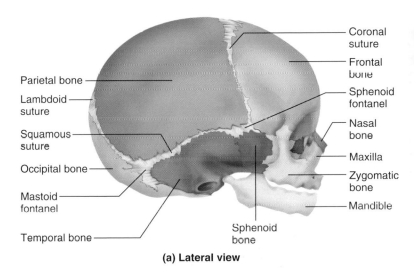

Parietal bone — Lambdoid suture — Squamous suture — Occipital bone — Mastoid fontanel — Temporal bone — Coronal suture — Frontal bone — Sphenoid fontanel — Nasal bone — Maxilla — Zygomatic bone — Mandible — Sphenoid bone

**(a) Lateral view**

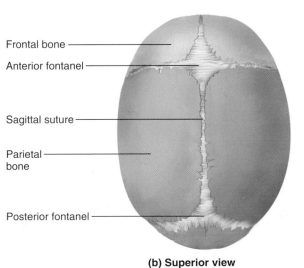

Frontal bone — Anterior fontanel — Sagittal suture — Parietal bone — Posterior fontanel

**(b) Superior view**

**FIGURE 8.17** The Fetal Skull Near the Time of Birth.

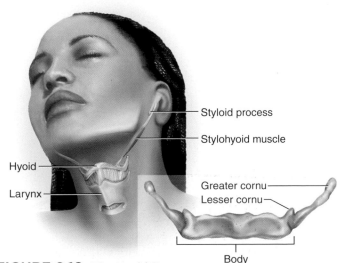

Styloid process — Stylohyoid muscle — Hyoid — Larynx — Greater cornu — Lesser cornu — Body

**FIGURE 8.16** The Hyoid Bone.

**INSIGHT 8.3**    Clinical Application

## Cranial Assessment of the Newborn

Obstetric nurses routinely assess the fontanels of newborns by palpation. In a difficult delivery, one cranial bone may override another along a suture line, which calls for close monitoring of the infant. Abnormally wide sutures may indicate hydrocephalus, the accumulation of excessive amounts of cerebrospinal fluid, which causes the cranium to swell. Bulging fontanels suggest abnormally high intracranial pressure, while depressed fontanels indicate dehydration.

7.  *State which bone has each of these features: a squamous part, hypoglossal foramen, greater cornu, greater wing, condyloid process, and cribriform plate.*

8.  *For each of the following bones, name all the other bones with which it articulates: parietal, temporal, zygomatic, and ethmoid.*

9.  *Determine which of the following structures cannot normally be palpated on a living person: the mastoid process, crista galli, superior orbital fissure, palatine processes, zygomatic bone, mental protuberance, and stapes. You may find it useful to palpate some of these on your own skull as you try to answer.*

fontanels ossify by the time the infant is a year old, but the largest one—the anterior fontanel—can still be palpated 18 to 24 months after birth.

The frontal bone and mandible are separate right and left bones at birth, but fuse medially in early childhood. The frontal bones usually fuse by age 5 or 6, but in some children a *metopic*[32] *suture* persists between them. Traces of this suture are evident in some adult skulls.

The face of a newborn is flat and the cranium relatively large. To accommodate the growing brain, the skull grows more rapidly than the rest of the skeleton during childhood. It reaches about half its adult size by 9 months of age, three-quarters by age 2, and nearly final size by 8 or 9 years. The heads of babies and children are therefore much larger in proportion to the trunk than the heads of adults—an attribute thoroughly exploited by cartoonists and advertisers who draw big-headed characters to give them a more endearing or immature appearance. In humans and other animals, the large rounded heads of the young are thought to promote survival by stimulating parental caregiving instincts.

Table 8.4 summarizes the bones of the skull.

## Before You Go On

*Answer the following questions to test your understanding of the preceding section:*

4.  *Name the paranasal sinuses and state their locations. Name any four other cavities in the skull.*

5.  *Explain the difference between a cranial bone and a facial bone. Give four examples of each.*

6.  *Draw an oval representing a superior view of the calvaria. Draw lines representing the coronal, lambdoid, and sagittal sutures. Label the four bones separated by these sutures.*

---

[32]*met* = beyond + *op* = the eyes

# The Vertebral Column and Thoracic Cage

### Objectives

When you have completed this section, you should be able to

*   describe the general features of the vertebral column and those of a typical vertebra;

*   describe the special features of vertebrae in different regions of the vertebral column, and discuss the functional significance of the regional differences; and

*   describe the anatomy of the sternum and ribs and how the ribs articulate with the thoracic vertebrae.

## GENERAL FEATURES OF THE VERTEBRAL COLUMN

The **vertebral** (VUR-teh-brul) **column** physically supports the skull and trunk, allows for their movement, protects the spinal cord, and absorbs stresses produced by walking, running, and lifting. It also provides attachment for the limbs, thoracic cage, and postural muscles. Although commonly called the backbone, it does not consist of a single bone but a chain of 33 **vertebrae** with **intervertebral discs** of fibrocartilage between most of them. The adult vertebral column averages about 71 cm (28 in.) long, with the 23 intervertebral discs accounting for about one-quarter of the length.

Most people are about 1% shorter when they go to bed at night than when first rising in the morning. This is because during the day, the weight of the body compresses the intervertebral discs and squeezes water out of them. When one is sleeping, with the weight off the spine, the discs reabsorb water and swell.

## TABLE 8.4     Anatomical Checklist for the Skull and Associated Bones

### Cranial Bones

**Frontal Bone (figs. 8.3 to 8.7)**
Supraorbital margin
Supraorbital foramen or notch
Glabella
Frontal sinus

**Parietal Bones (figs. 8.4 to 8.6)**
Temporal lines
Parietal foramen

**Temporal Bones (figs. 8.4, 8.5, and 8.10)**
Squamous part
  Zygomatic process
  Mandibular fossa
Tympanic part
  External acoustic meatus
  Styloid process
Mastoid part
  Mastoid process
  Mastoid notch
  Mastoid foramen
  Stylomastoid foramen
Petrous part
  Internal acoustic meatus
  Carotid canal
  Jugular foramen

**Occipital Bone (figs. 8.4 and 8.5)**
Foramen magnum
Basilar part
Occipital condyles
Hypoglossal canal

**Occipital Bone—(Cont.)**
Condylar canal
External occipital protuberance
Superior nuchal line
Inferior nuchal line

**Sphenoid Bone (figs. 8.4, 8.5, and 8.11)**
Body
Lesser wing
  Optic foramen
  Anterior clinoid process
  Superior orbital fissure
Greater wing
  Foramen ovale
  Foramen rotundum
  Foramen spinosum
  Foramen lacerum
Medial and lateral pterygoid plates
Pterygoid processes
Nasal choanae
Sphenoid sinus
Sella turcica
Dorsum sellae

**Ethmoid Bone (figs. 8.4b, 8.7, and 8.12)**
Perpendicular plate
Superior nasal concha
Middle nasal concha
Ethmoid sinus (ethmoidal cells)
Crista galli
Cribriform plate

### Facial Bones

**Maxilla (figs. 8.3, 8.4, and 8.5a)**
Alveoli
Alveolar processes
Infraorbital foramen
Inferior orbital fissure
Palatine processes
Incisive foramen
Maxillary sinus

**Palatine Bones (figs. 8.5a and 8.13)**
Greater palatine foramen

**Zygomatic Bones (figs. 8.4a and 8.5a)**
Zygomaticofacial foramen
Temporal process

**Lacrimal Bones (figs. 8.4a and 8.14)**
Lacrimal fossa

**Nasal Bones (figs. 8.3 and 8.13)**

**Inferior Nasal Concha (fig. 8.13)**

**Vomer (figs. 8.3 and 8.4b)**

**Mandible (figs. 8.3, 8.4, and 8.15)**
Body
  Mental symphysis
  Mental protuberance
  Mental spines
  Mental foramen
Angle
Ramus
  Condyloid process
  Mandibular condyle
  Coronoid process
  Mandibular notch
  Mandibular foramen

### Bones Associated with the Skull

**Auditory Ossicles**
Malleus (hammer)
Incus (anvil)
Stapes (stirrup)

**Hyoid Bone (fig. 8.16)**
Body
Greater cornu
Lesser cornu

**Anterior view**    **Posterior view**

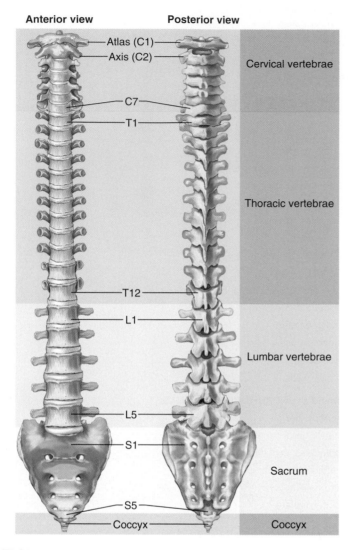

FIGURE 8.18   The Vertebral Column.

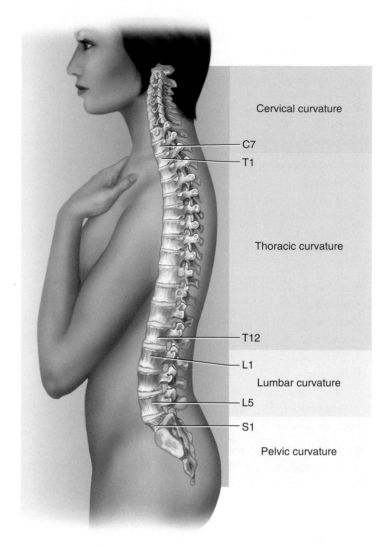

FIGURE 8.19   Curvatures of the Adult Vertebral Column.

As shown in figure 8.18, the vertebrae are divided into five groups, usually numbering 7 *cervical* (SUR-vih-cul) *vertebrae* in the neck, 12 *thoracic vertebrae* in the chest, 5 *lumbar vertebrae* in the lower back, 5 *sacral vertebrae* at the base of the spine, and 4 tiny *coccygeal* (coc-SIDJ-ee-ul) *vertebrae.* To help remember the numbers of cervical, thoracic, and lumbar vertebrae—7, 12, and 5—you might think of a typical work day: go to work at 7, have lunch at 12, and go home at 5.

Variations in this arrangement occur in about 1 person in 20. For example, the last lumbar vertebra is sometimes incorporated into the sacrum, producing four lumbar and six sacral vertebrae. In other cases, the first sacral vertebra fails to fuse with the second, producing six lumbar and four sacral vertebrae. The coccyx usually has four but sometimes five vertebrae. The cervical and thoracic vertebrae are more constant in number.

Beyond the age of 3 years, the vertebral column is slightly S-shaped, with four bends called the **cervical, thoracic, lumbar,** and **pelvic curvatures** (fig. 8.19). These are not present in the newborn, whose spine exhibits one continuous C-shaped curve (fig. 8.20) as it does in monkeys, apes, and most other four-legged animals. As an infant begins to crawl and lift its head, the cervical region becomes curved toward the posterior side, enabling an infant on its belly to look forward. As a toddler begins walking, another curve develops in the same direction in the lumbar region. The resulting S shape makes sustained bipedal walking possible (see Insight 8.5, p. 284). The thoracic and pelvic curvatures are called *primary curvatures* because they are remnants of the original infantile curvature. The cervical and lumbar curvatures are called *secondary curvatures* because they develop later, in the child's first few years of crawling and walking.

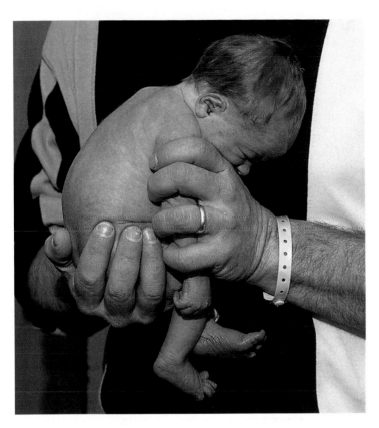

**FIGURE 8.20** Spinal Curvature of the Newborn Infant. At this age, the spine forms a single C-shaped curve.

# GENERAL STRUCTURE OF A VERTEBRA

A representative vertebra and intervertebral disc are shown in figure 8.22. The most obvious feature of a vertebra is the **body** (centrum)—a mass of spongy bone and red bone marrow covered with a thin layer of compact bone. This is the weight-bearing portion of the vertebra. Its rough superior and inferior surfaces provide firm attachment to the intervertebral discs.

> **Think About It**
> *The vertebral bodies and intervertebral discs get progressively larger as we look lower and lower on the vertebral column. What is the functional significance of this trend?*

Posterior (dorsal) to the body of each vertebra is a triangular canal called the **vertebral foramen.** The vertebral foramina collectively form the **vertebral canal,** a passage for the spinal cord. The foramen is bordered by a bony **vertebral arch** composed of two parts on each side: a pillarlike **pedicle**[33] and platelike **lamina.**[34] Extending from

---

[33]*ped* = foot + *icle* = little
[34]*lamina* = plate

---

## Abnormal Spinal Curvatures

Abnormal spinal curvatures (fig. 8.21) can result from disease; weakness or paralysis of the trunk muscles; poor posture; pregnancy; or congenital defects in vertebral anatomy. The most common deformity is an abnormal lateral curvature called *scoliosis.* It occurs most often in the thoracic region, particularly among adolescent girls. It sometimes results from a developmental abnormality in which the body and arch fail to develop on one side of a vertebra. If the person's skeletal growth is not yet complete, scoliosis can be corrected with a back brace.

An exaggerated thoracic curvature is called *kyphosis* (hunchback, in lay language). It is usually a result of osteoporosis, but it also occurs in people with osteomalacia or spinal tuberculosis and in adolescent boys who engage heavily in such spine-loading sports as wrestling and weightlifting. An exaggerated lumbar curvature is called *lordosis* (swayback, in lay language). It may have the same causes as kyphosis, or it may result from added abdominal weight in pregnancy or obesity.

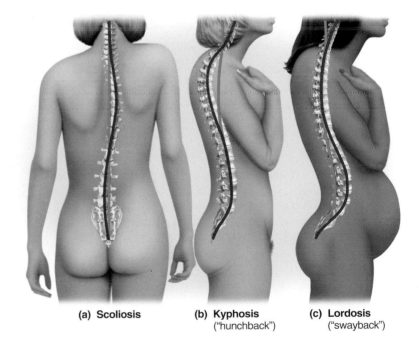

(a) **Scoliosis**    (b) **Kyphosis** ("hunchback")    (c) **Lordosis** ("swayback")

**FIGURE 8.21** Abnormal Spinal Curvatures.
(a) Scoliosis, an abnormal lateral deviation.
(b) Kyphosis, an exaggerated thoracic curvature common in old age. (c) Lordosis, an exaggerated lumbar curvature common in pregnancy and obesity.

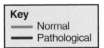

| Key | |
|---|---|
| —— | Normal |
| —— | Pathological |

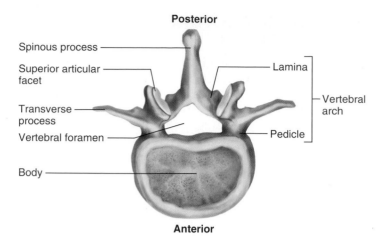

**Posterior**

Spinous process

Superior articular facet

Transverse process

Vertebral foramen

Body

Lamina

Vertebral arch

Pedicle

**Anterior**

**(a) 2nd lumbar vertebra (L2)**

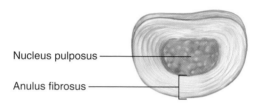

Nucleus pulposus

Anulus fibrosus

**(b) Intervertebral disc**

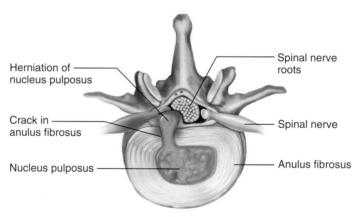

Herniation of nucleus pulposus

Crack in anulus fibrosus

Nucleus pulposus

Spinal nerve roots

Spinal nerve

Anulus fibrosus

**(c) Herniated disc**

**FIGURE 8.22 A Representative Vertebra and Intervertebral Disc, Superior Views.** (a) A typical vertebra. (b) An intervertebral disc, oriented the same way as the vertebral body in part (a) for comparison. (c) A herniated disc, showing compression of the spinal nerve roots by the nucleus pulposus oozing from the disc.

the apex of the arch, a projection called the **spinous process** is directed dorsally and downward. You can see and feel the spinous processes as a row of bumps along the spine. A **transverse process** extends laterally from the point where the pedicle and lamina meet. The spinous and transverse processes provide points of attachment for ligaments and spinal muscles.

A pair of **superior articular processes** project upward from one vertebra and meet a similar pair of inferior artic-

ular processes that project downward from the vertebra just above (fig. 8.23a). Each process has a flat articular surface (facet) facing that of the adjacent vertebra. These processes restrict twisting of the vertebral column, which could otherwise severely damage the spinal cord.

When two vertebrae are joined, they exhibit an opening between their pedicles called the **intervertebral foramen.** This allows passage for spinal nerves that connect with the spinal cord at regular intervals. Each foramen is formed by an **inferior vertebral notch** in the pedicle of the superior vertebra and a **superior vertebral notch** in the pedicle of the one just below it (fig. 8.23b).

## INTERVERTEBRAL DISCS

An **intervertebral disc** is a pad consisting of an inner gelatinous **nucleus pulposus** surrounded by a ring of fibrocartilage, the **anulus fibrosus** (see fig. 8.22b). The discs help to bind adjacent vertebrae together, support the weight of the body, and absorb shock. Under stress—for example, when you lift a heavy weight—the discs bulge laterally. Excessive stress can crack the anulus and cause the nucleus to ooze out. This is called a *herniated disc* ("ruptured" or "slipped" disc in lay terms) and may put painful pressure on the spinal cord or a spinal nerve. To relieve the pressure, a procedure called a *laminectomy* may be performed—each lamina is cut and the laminae and spinous processes are removed. This procedure is also used to expose the spinal cord for anatomical study or surgery.

## REGIONAL CHARACTERISTICS OF VERTEBRAE

We are now prepared to consider how vertebrae differ from one region of the vertebral column to another and from the generalized anatomy just described. Knowing these variations will enable you to identify the region of the spine from which an isolated vertebra was taken. More importantly, these modifications in form reflect functional differences among the vertebrae.

### Cervical Vertebrae

The cervical vertebrae (C1–C7) are the smallest and lightest ones other than the coccygeals. The first two (C1 and C2) have unique structures that allow for head movements (fig. 8.24). Vertebra C1 is called the **atlas** because it supports the head in a manner reminiscent of the Titan of Greek mythology who was condemned by Zeus to carry the world on his shoulders. It scarcely resembles the typical vertebra; it has no body, and is little more than a delicate ring surrounding a large vertebral foramen. On each side is a **lateral mass** with a deeply concave **superior articular facet** that articulates with the occipital condyle of the skull. A nodding motion of the skull, as in gesturing "yes," causes the occipital condyles to rock back and forth

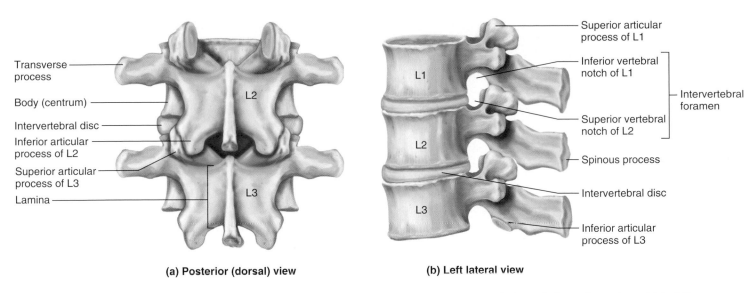

Transverse process

Body (centrum)

Intervertebral disc

Inferior articular process of L2

Superior articular process of L3

Lamina

L2

L3

**(a) Posterior (dorsal) view**

Superior articular process of L1

Inferior vertebral notch of L1

Intervertebral foramen

Superior vertebral notch of L2

Spinous process

Intervertebral disc

Inferior articular process of L3

L1

L2

L3

**(b) Left lateral view**

**FIGURE 8.23  Articulated Vertebrae.** (a) Posterior view of vertebrae L2 and L3. (b) Left lateral view of vertebrae L1 to L3.

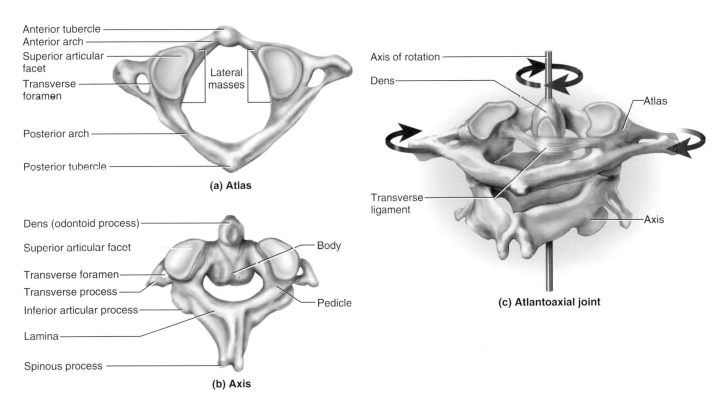

Anterior tubercle

Anterior arch

Superior articular facet

Transverse foramen

Posterior arch

Posterior tubercle

Lateral masses

**(a) Atlas**

Dens (odontoid process)

Superior articular facet

Transverse foramen

Transverse process

Inferior articular process

Lamina

Spinous process

Body

Pedicle

**(b) Axis**

Axis of rotation

Dens

Transverse ligament

Atlas

Axis

**(c) Atlantoaxial joint**

**FIGURE 8.24  The Atlas and Axis, Cervical Vertebrae C1 and C2.** (a) The atlas, superior view. (b) The axis, posterosuperior view. (c) Articulation of the atlas and axis and rotation of the atlas. This movement turns the head from side to side, as in gesturing "no." Note the transverse ligament holding the dens of the axis in place.

on these facets. The **inferior articular facets,** which are comparatively flat or only slightly concave, articulate with C2. The lateral masses are connected by an **anterior arch** and a **posterior arch,** which bear slight protuberances called the **anterior** and **posterior tubercle,** respectively.

Vertebra C2, the **axis,** allows rotation of the head as in gesturing "no." Its most distinctive feature is a prominent

knob called the **dens** (denz), or **odontoid**[35] **process,** on its anterosuperior side. No other vertebra has a dens. It begins to form as an independent ossification center during the first year of life and fuses with the axis by the age

[35]*dens = odont* = tooth + *oid* = resembling

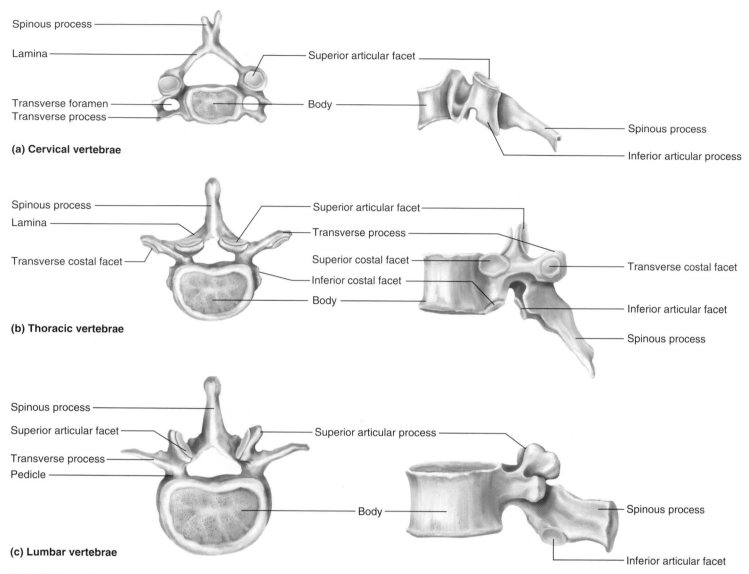

**(a) Cervical vertebrae**

Spinous process
Lamina
Transverse foramen
Transverse process
Superior articular facet
Body
Spinous process
Inferior articular process

**(b) Thoracic vertebrae**

Spinous process
Lamina
Transverse costal facet
Superior articular facet
Transverse process
Superior costal facet
Inferior costal facet
Body
Transverse costal facet
Inferior articular facet
Spinous process

**(c) Lumbar vertebrae**

Spinous process
Superior articular facet
Transverse process
Pedicle
Superior articular process
Body
Spinous process
Inferior articular facet

**FIGURE 8.25** **Typical Cervical, Thoracic, and Lumbar Vertebrae.** The left-hand figures are superior views, and the right-hand figures are left lateral views.

▶ *Construct a three-column table headed C4, T4, and L4. In each column, list all anatomical features that would distinguish that vertebra from the other two.*

of 3 to 6 years. It projects into the vertebral foramen of the atlas, where it is nestled in a facet and held in place by a **transverse ligament** (fig. 8.24c). A heavy blow to the top of the head can cause a fatal injury in which the dens is driven through the foramen magnum into the brainstem. The articulation between the atlas and the cranium is called the **atlanto-occipital joint;** the one between the atlas and axis is called the **atlantoaxial joint.**

The axis is the first vertebra that exhibits a spinous process. In vertebrae C2 to C6, the process is forked, or *bifid*,[36] at its tip (fig. 8.25a). This fork provides attachment for the *nuchal ligament* of the back of the neck. All seven cervical vertebrae have a prominent round **transverse foramen** in each transverse process. These foramina pro-

vide passage and protection for the *vertebral arteries,* which supply blood to the brain. Transverse foramina occur in no other vertebrae and thus provide an easy means of recognizing a cervical vertebra.

**Think About It**

*How would head movements be affected if vertebrae C1 and C2 had the same structure as C3? What is the functional advantage of the lack of a spinous process in C1?*

Cervical vertebrae C3 to C6 are similar to the typical vertebra described earlier, with the addition of the transverse foramina and bifid spinous processes. Vertebra C7 is a little different—its spinous process is not bifid, but it is especially long and forms a prominent bump on the lower back of the neck. This feature is a convenient landmark

---

[36]*bifid* = cleft into two parts

for counting vertebrae. Because it is so conspicuous, C7 is sometimes called the *vertebra prominens.*

## Thoracic Vertebrae

There are 12 **thoracic vertebrae** (T1–T12), corresponding to the 12 pairs of ribs attached to them. They lack the transverse foramina and bifid processes that distinguish the cervicals, but possess the following distinctive features of their own (fig. 8.25b):

- The spinous processes are relatively pointed and angle sharply downward.
- The body is somewhat heart-shaped, more massive than in the cervical vertebrae but less than in the lumbar vertebrae.
- The body has small, smooth, slightly concave spots called *costal facets* (to be described shortly) for attachment of the ribs.
- Vertebrae T1 to T10 have a shallow, cuplike **transverse costal**[37] **facet** at the end of each transverse process. These provide a second point of articulation for ribs 1 to 10. There are no transverse costal facets on T11 and T12 because ribs 11 and 12 attach only to the bodies of the vertebrae.

No other vertebrae have ribs articulating with them. Thoracic vertebrae vary among themselves mainly because of variations in the way the ribs articulate. In most cases, a rib inserts between two vertebrae, so each vertebra contributes one-half of the articular surface. A rib articulates with the **inferior costal facet** (FASS-et) of the upper vertebra and the **superior costal facet** of the vertebra below that. This terminology may be a little confusing, but note that the superior and inferior facets are named for their position on the vertebral body, not for which part of the rib's articulation they provide. Vertebrae T1 and T10 to T12, however, have complete costal facets on the bodies for ribs 1 and 10 to 12, which articulate on the vertebral body instead of between vertebrae. Vertebrae T11 and T12, as noted, have no transverse costal facets. These variations will be more functionally understandable after you have studied the anatomy of the ribs, so we will return then to the details of these articular surfaces.

Each thoracic vertebra has a pair of **superior articular facets** that face posteriorly and a pair of **inferior articular facets** that face anteriorly (except in vertebra T12). Thus the superior facets of one vertebra articulate with the inferior facets of the next one below it. In vertebra T12, however, the inferior articular facets face somewhat laterally instead of anteriorly. This positions them to articulate with the medially facing superior articular facets of the first lumbar vertebra. T12 thus show an anatomical transition between the thoracic and lumbar pattern, described next.

## Lumbar Vertebrae

There are five **lumbar vertebrae** (L1–L5). Their most distinctive features are a thick, stout body and a blunt, squarish spinous process (fig. 8.25c). In addition, their articular processes are oriented differently than on other vertebrae. The superior processes face medially (like the palms of your hands about to clap), and the inferior processes face laterally, toward the superior processes of the next vertebra. This arrangement makes the lumbar region of the spine especially resistant to twisting. These differences are best observed on an articulated skeleton.

## Sacrum

The **sacrum** (SACK-rum, SAY-krum) is a bony plate that forms the posterior wall of the pelvic cavity (fig. 8.26). It is named for the fact that it was once considered the seat of the soul.[38] In children, there are five separate **sacral vertebrae** (S1–S5). They begin to fuse around age 16 and are fully fused by age 26.

The anterior surface of the sacrum is relatively smooth and concave and has four transverse lines that indicate where the five vertebrae have fused. This surface exhibits four pairs of large **anterior sacral (pelvic) foramina,** which allow for passage of nerves and arteries to the pelvic organs. The posterior surface is very rough. The spinous processes of the vertebrae fuse into a ridge called the **median sacral crest.** The transverse processes fuse into a less prominent **lateral sacral crest** on each side of the median crest. Again on the posterior side of the sacrum, there are four pairs of openings for spinal nerves, the **posterior sacral foramina.** The nerves that emerge here supply the gluteal region and lower limb.

A **sacral canal** runs through the sacrum and ends in an inferior opening called the sacral hiatus (hy-AY-tus). This canal contains spinal nerve roots in life. On each side of the sacrum is an ear-shaped region called the **auricular**[39] (aw-RIC-you-lur) **surface.** This articulates with a similarly shaped surface on the os coxae (see fig. 8.36b) and forms the strong, nearly immovable **sacroiliac** (SAY-cro-ILL-ee-ac) **joint.** At the superior end of the sacrum, lateral to the median crest, are a pair of **superior articular processes** that articulate with vertebra L5. Lateral to these are a pair of large, rough, winglike extensions called the **alae**[40] (AIL-ee).

## Coccyx

The **coccyx**[41] (fig. 8.26) usually consists of four (sometimes five) small vertebrae, Co1 to Co4, which fuse by the age of 20 to 30 into a single triangular bone. Vertebra Co1

---

[37]*costa* = rib + *al* = pertaining to

[38]*sacr* = sacred
[39]*auri* = ear + *cul* = little + *ar* = pertaining to
[40]*alae* = wings
[41]*coccyx* = cuckoo (named for resemblance to a cuckoo's beak)

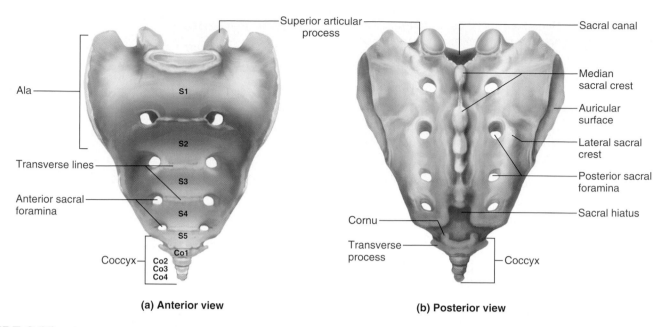

**(a) Anterior view**   **(b) Posterior view**

**FIGURE 8.26   The Sacrum and Coccyx.**   (a) The anterior surface, which faces the viscera of the pelvic cavity. (b) The posterior surface, whose processes can be palpated in the sacral region.

has a pair of hornlike projections, the **cornua,** which serve as attachment points for ligaments that bind the coccyx to the sacrum. The coccyx can be fractured by a difficult childbirth or a hard fall to the buttocks. Although it is the vestige of a tail, it is not entirely useless; it provides attachment for muscles of the pelvic floor.

## THE THORACIC CAGE

The **thoracic cage** (fig. 8.27) consists of the thoracic vertebrae, sternum, and ribs. It forms a more or less conical enclosure for the lungs and heart and provides attachment for the pectoral girdle and upper limb. It has a broad base and a somewhat narrower superior apex; it is rhythmically expanded by the respiratory muscles to create a vacuum that draws air into the lungs. The inferior border of the thoracic cage is formed by a downward arc of the ribs called the **costal margin.** The ribs protect not only the thoracic organs but also the spleen, most of the liver, and to some extent the kidneys.

### Sternum

The **sternum** (breastbone) is a bony plate anterior to the heart. It is subdivided into three regions: the manubrium, body, and xiphoid process. The **manubrium**[42] (ma-NOO-bree-um) is the broad superior portion. It has a median **suprasternal notch** (jugular notch), which you can easily palpate between your clavicles (collarbones), and right and left **clavicular notches,** where it articulates with the

clavicles. The **body,** or **gladiolus,**[43] is the longest part of the sternum. It joins the manubrium at the **sternal angle,** which can be palpated as a transverse ridge at the point where the sternum projects farthest forward. In some people, however, the angle is rounded or concave. The second rib attaches here, making the sternal angle a useful landmark for counting ribs in a physical examination. The manubrium and body have scalloped lateral margins where cartilages of the ribs are attached. At the inferior end of the sternum is a small, pointed **xiphoid**[44] (ZIF-oyd) **process** that provides attachment for some of the abdominal muscles.

### Ribs

There are 12 pairs of **ribs,** with no difference between the sexes. Each is attached at its posterior (proximal) end to the vertebral column. A strip of hyaline cartilage called the **costal cartilage** extends from the anterior (distal) ends of ribs 1 to 7 to the sternum. Ribs 1 to 7 are thus called **true ribs.** Ribs 8, 9, and 10 attach to the costal cartilage of rib 7, and ribs 11 and 12 do not attach to anything at the distal end but are embedded in muscle. Ribs 8 to 12 are therefore called **false ribs,** and ribs 11 and 12 are also called **floating ribs** for lack of any connection to the sternum.

Ribs 1 to 10 each have a proximal **head** and **tubercle,** connected by a narrow **neck;** ribs 11 and 12 have a head

---

[42]*manubrium* = handle

[43]*gladiolus* = sword
[44]*xipho* = sword + *oid* = resembling

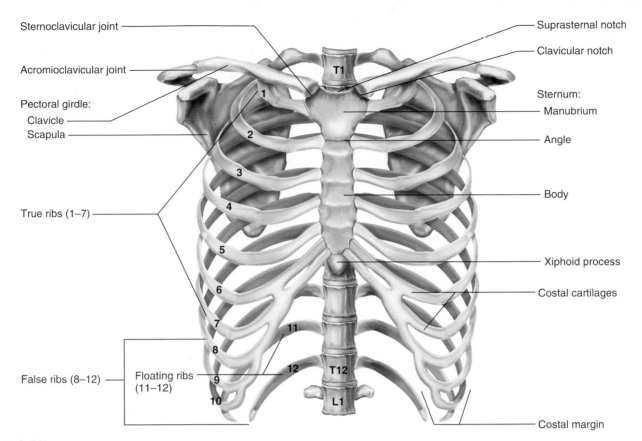

**FIGURE 8.27** The Thoracic Cage and Pectoral Girdle, Anterior View.

Sternoclavicular joint
Acromioclavicular joint
Pectoral girdle:
Clavicle
Scapula
True ribs (1–7)
False ribs (8–12)
Floating ribs (11–12)

Suprasternal notch
Clavicular notch
Sternum:
Manubrium
Angle
Body
Xiphoid process
Costal cartilages
Costal margin

T1
T12
L1

only (fig. 8.28). Ribs 2 to 9 have beveled heads that come to a point between a **superior articular facet** above and an **inferior articular facet** below. Rib 1, unlike the others, is a flat horizontal plate. Ribs 2 to 10 have a sharp turn called the **angle,** distal to the tubercle, and the remainder consists of a flat blade called the **shaft.** Along the inferior margin of the shaft is a **costal groove** that marks the path of the intercostal blood vessels and nerve.

Variations in rib anatomy relate to the way different ribs articulate with the vertebrae. Once you observe these articulations on an intact skeleton, you will be better able to understand the anatomy of isolated ribs and vertebrae. Vertebra T1 has a complete superior costal facet on the body that articulates with rib 1, as well as a small inferior costal facet that provides half of the articulation with rib 2. Ribs 2 through 9 all articulate between two vertebrae, so these vertebrae have both superior and inferior costal facets on the respective margins of the body. The inferior costal facet of each vertebra articulates with the superior articular facet of the rib, and the superior costal facet of the next vertebra articulates with the inferior articular facet of the same rib (fig. 8.29a). Ribs 10 through 12 each articulate with a single costal facet on the bodies of the respective vertebrae.

Ribs 1 to 10 each have a second point of attachment to the vertebrae: the tubercle of the rib articulates with the costal facet of the same-numbered vertebra (fig. 8.29b). Ribs 11 and 12 articulate only with the vertebral bodies; they do not have tubercles and vertebrae T11 and T12 do not have costal facets.

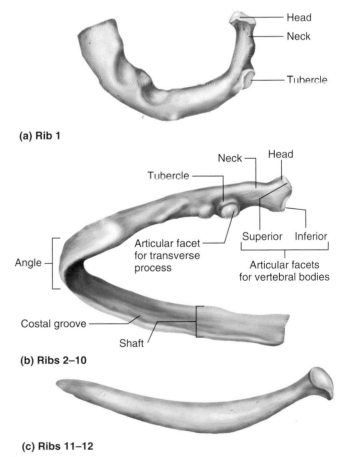

**(a) Rib 1**

Head
Neck
Tubercle

**(b) Ribs 2–10**

Neck
Head
Tubercle
Articular facet for transverse process
Angle
Costal groove
Shaft
Superior
Inferior
Articular facets for vertebral bodies

**(c) Ribs 11–12**

**FIGURE 8.28** **Anatomy of the Ribs.** (a) Rib 1 is an atypical flat plate. (b) Typical features of ribs 2 to 10. (c) Appearance of the floating ribs, 11 and 12.

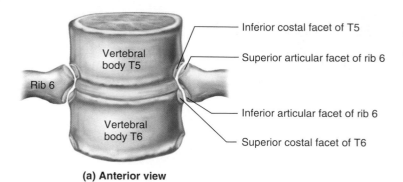

**(a) Anterior view**

Vertebral body T5

Rib 6

Vertebral body T6

Inferior costal facet of T5

Superior articular facet of rib 6

Inferior articular facet of rib 6

Superior costal facet of T6

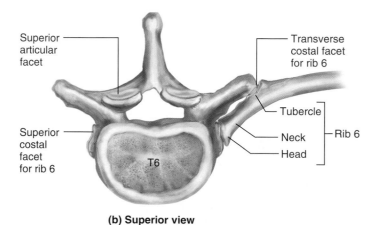

**(b) Superior view**

Superior articular facet

Superior costal facet for rib 6

T6

Transverse costal facet for rib 6

Tubercle

Neck — Rib 6

Head

**FIGURE 8.29**   Articulation of Rib 6 with Vertebrae T5 and T6. (a) Anterior view. Note the relationships of the articular facets of the rib with the costal facets of the two vertebrae. (b) Superior view. Note that the rib articulates with a vertebra at two points: the costal facet on the vertebral body and the transverse costal facet on the transverse process.

Table 8.5 summarizes these variations. Table 8.6 provides a checklist that you can use to review your knowledge of the vertebral column and thoracic cage.

## Before You Go On

*Answer the following questions to test your understanding of the preceding section:*

10. Make a table with three columns headed "cervical," "thoracic," and "lumbar." In each column, list the identifying characteristics of each type of vertebra.

11. Describe how rib 5 articulates with the spine. How do ribs 1 and 12 differ from this and from each other in their modes of articulation?

12. Distinguish between true, false, and floating ribs. State which ribs fall into each category.

13. Name the three divisions of the sternum and list the sternal features that can be palpated on a living person.

# The Pectoral Girdle and Upper Limb

### Objective

When you have completed this section, you should be able to

- identify and describe the features of the clavicle, scapula, humerus, radius, ulna, and bones of the wrist and hand.

| TABLE 8.5 | | Articulations of the Ribs | | | |
|---|---|---|---|---|---|
| **Rib** | **Type** | **Costal Cartilage** | **Articulating Vertebral Bodies** | **Articulating with a Transverse Costal Facet?** | **Rib Tubercle** |
| 1 | True | Individual | T1 | Yes | Present |
| 2 | True | Individual | T1 and T2 | Yes | Present |
| 3 | True | Individual | T2 and T3 | Yes | Present |
| 4 | True | Individual | T3 and T4 | Yes | Present |
| 5 | True | Individual | T4 and T5 | Yes | Present |
| 6 | True | Individual | T5 and T6 | Yes | Present |
| 7 | True | Individual | T6 and T7 | Yes | Present |
| 8 | False | Shared with rib 7 | T7 and T8 | Yes | Present |
| 9 | False | Shared with rib 7 | T8 and T9 | Yes | Present |
| 10 | False | Shared with rib 7 | T10 | Yes | Present |
| 11 | False, floating | None | T11 | No | Absent |
| 12 | False, floating | None | T12 | No | Absent |

| **TABLE 8.6** | Anatomical Checklist for the Vertebral Column and Thoracic Cage |
|---|---|

## Vertebral Column

### Spinal Curvatures (fig. 8.19)
Cervical curvature

Thoracic curvature

Lumbar curvature

Pelvic curvature

### General Vertebral Structure (figs. 8.22 and 8.23)
Body (centrum)

Vertebral foramen

Vertebral canal

Vertebral arch

   Pedicle

   Lamina

Spinous process

Transverse process

Superior articular process

Inferior articular process

Intervertebral foramen

   Inferior vertebral notch

   Superior vertebral notch

### Intervertebral Discs (fig. 8.22)
Anulus fibrosus

Nucleus pulposus

### Cervical Vertebrae (figs. 8.24 and 8.25a)
Transverse foramina

Bifid spinous process

Atlas

   Anterior arch

   Anterior tubercle

   Posterior arch

### Cervical Vertebrae—(Cont.)
Atlas (cont.)

   Posterior tubercle

   Lateral mass

      Superior articular facet

      Inferior articular facet

Axis

   Dens (odontoid process)

### Thoracic Vertebrae (fig. 8.25b)
Superior costal facet

Inferior costal facet

Transverse costal facet

### Lumbar Vertebrae (figs. 8.23 and 8.25c)

### Sacral Vertebrae (fig. 8.26)
Sacrum

Anterior sacral foramina

Posterior sacral foramina

Median sacral crest

Lateral sacral crest

Sacral canal

Sacral hiatus

Auricular surface

Superior articular process

Alae

### Coccygeal Vertebrae (fig. 8.26)
Coccyx

Cornu

## Thoracic Cage

### Sternum (fig. 8.27)
Manubrium

   Suprasternal notch

   Clavicular notch

   Sternal angle

Body (gladiolus)

Xiphoid process

### Rib Types (fig. 8.27)
True ribs (ribs 1–7)

False ribs (ribs 8–12)

Floating ribs (ribs 11 and 12)

### Rib Features (fig. 8.28)
Head

Superior articular facet

Inferior articular facet

Neck

Tubercle

Angle

Shaft

Costal groove

Costal cartilage

# PECTORAL GIRDLE

The **pectoral girdle** (shoulder girdle) supports the arm. It consists of two bones on each side of the body: the *clavicle* (collarbone) and the *scapula* (shoulder blade). The medial end of the clavicle articulates with the sternum at the **sternoclavicular joint,** and its lateral end articulates with the scapula at the **acromioclavicular joint** (see fig. 8.27). The scapula also articulates with the humerus at the **humeroscapular joint.** These are loose attachments that result in a shoulder far more flexible than that of most other mammals, but they also make the shoulder joint easy to dislocate.

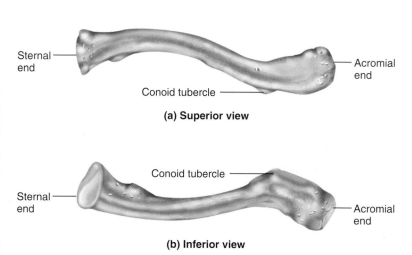

Sternal end

Conoid tubercle

Acromial end

**(a) Superior view**

Conoid tubercle

Sternal end

Acromial end

**(b) Inferior view**

**FIGURE 8.30**   The Right Clavicle (Collarbone).

┌ **Think About It**

*How is the unusual flexibility of the human shoulder joint related to the habitat of our primate ancestors?*

## Clavicle

The **clavicle**[45] (fig. 8.30) is a slightly S-shaped bone, somewhat flattened vertically, and easily seen and palpated on the upper thorax (see fig. B.1b, p. 391). The superior surface is relatively smooth, whereas the inferior surface is marked by grooves and ridges for muscle attachment. The medial **sternal end** has a rounded, hammerlike head, and the lateral **acromial end** is markedly flattened. Near the acromial end is a rough tuberosity called the **conoid tubercle**—a ligament attachment that faces toward the rear and slightly downward. The clavicle braces the shoulder and is thickened in people who do heavy manual labor. Without it, the pectoralis major muscles would pull the shoulders forward and medially, as occurs when a clavicle is fractured. Indeed, the clavicle is the most commonly fractured bone in the body because it is so close to the surface and because people often reach out with their arms to break a fall.

## Scapula

The **scapula**[46] (fig. 8.31), named for its resemblance to a spade or shovel, is a triangular plate that posteriorly

---

[45]*clav* = hammer, club, key + *icle* = little
[46]*scap* = spade, shovel + *ula* = little

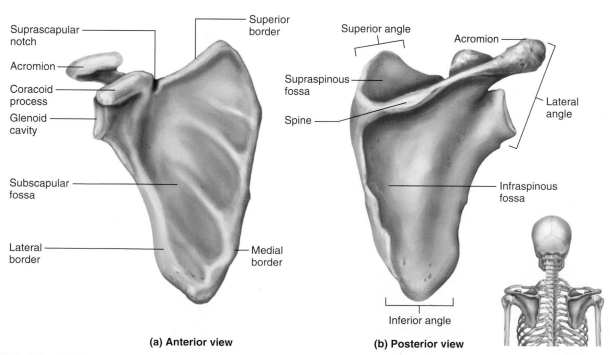

Suprascapular notch

Acromion

Coracoid process

Glenoid cavity

Subscapular fossa

Lateral border

Superior border

Medial border

Superior angle

Acromion

Supraspinous fossa

Spine

Lateral angle

Infraspinous fossa

Inferior angle

**(a) Anterior view**

**(b) Posterior view**

**FIGURE 8.31**  The Right Scapula.

overlies ribs 2 to 7. The three sides of the triangle are called the **superior, medial (vertebral),** and **lateral (axillary) borders,** and its three angles are the **superior, inferior,** and **lateral angles.** A conspicuous **suprascapular notch** in the superior border provides passage for a nerve. The broad anterior surface of the scapula, called the subscapular fossa, is slightly concave and relatively featureless. The posterior surface has a transverse ridge called the **spine,** a deep indentation superior to the spine called the **supraspinous fossa,** and a broad surface inferior to it called the **infraspinous fossa.**[47]

The most complex region of the scapula is its lateral angle, which has three main features:

1. The **acromion**[48] (ah-CRO-me-on) is a platelike extension of the scapular spine that forms the apex of the shoulder. It articulates with the clavicle—the sole point of attachment of the arm and scapula to the rest of the skeleton.

2. The **coracoid**[49] (COR-uh-coyd) **process** is shaped like a finger but named for a vague resemblance to a crow's beak; it provides attachment for the biceps brachii and other muscles of the arm.

3. The **glenoid**[50] (GLEN-oyd) **cavity** is a shallow socket that articulates with the head of the humerus.

## UPPER LIMB

The upper limb is divided into four regions containing a total of 30 bones per limb:

1. The **brachium**[51] (BRAY-kee-um), or arm proper, extends from shoulder to elbow. It contains only one bone, the *humerus.*

2. The **antebrachium,**[52] or forearm, extends from elbow to wrist and contains two bones: the *radius* and *ulna.* In anatomical position, these bones are parallel and the radius is lateral to the ulna.

3. The **carpus,**[53] or wrist, contains eight small bones arranged in two rows.

4. The **manus,**[54] or hand, contains 19 bones in two groups: 5 *metacarpals* in the palm and 14 *phalanges* in the fingers.

### Humerus

The **humerus** has a hemispherical **head** that articulates with the glenoid cavity of the scapula (fig. 8.32). The

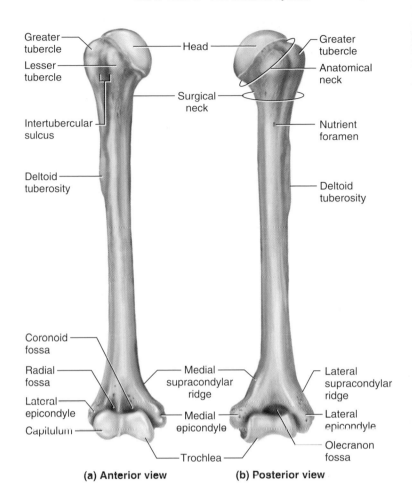

**(a) Anterior view**    **(b) Posterior view**

**FIGURE 8.32** The Right Humerus.

smooth surface of the head (covered with articular cartilage in life) is bordered by a groove called the **anatomical neck.** Other prominent features of the proximal end are muscle attachments called the **greater** and **lesser tubercles** and an **intertubercular sulcus** between them that accommodates a tendon of the biceps muscle. The **surgical neck,** a common fracture site, is a narrowing of the bone just distal to the tubercles, at the transition from the head to the shaft.

The shaft has a rough area called the **deltoid tuberosity** on its lateral surface. This is an insertion for the deltoid muscle of the shoulder. The distal end of the humerus has two smooth condyles. The lateral one, called the **capitulum**[55] (ca-PIT-you-lum), is shaped somewhat like a flat tire and articulates with the radius. The medial one, called the **trochlea**[56] (TROCK-lee-uh), is pulleylike and articulates with the ulna. Immediately proximal to these condyles, the humerus flares out to form two bony processes, the **lateral** and **medial epicondyles.** The medial epicondyle protects the ulnar nerve, which passes

---

[47] *supra* = above; *infra* = below
[48] *acr* = extremity, point + *omi* = shoulder
[49] *corac* = crow + *oid* = resembling
[50] *glen* = pit, socket + *oid* = resembling
[51] *brachi* = arm
[52] *ante* = before
[53] *carp* = wrist
[54] *man* = hand

[55] *capit* = head + *ulum* = little
[56] *troch* = wheel, pulley

close to the surface across the back of the elbow. This epicondyle is popularly known as the "funny bone" because striking the elbow on the edge of a table stimulates the ulnar nerve and produces a sharp tingling sensation. Just above these epicondyles, the margins of the humerus are called the **lateral** and **medial supracondylar ridges.** These are attachments for certain forearm muscles.

The distal end of the humerus also shows three deep pits: two anterior and one posterior. The posterior pit, called the **olecranon** (oh-LEC-ruh-non) **fossa,** accommodates the olecranon of the ulna when the arm is extended. On the anterior surface, a medial pit called the **coronoid fossa** accommodates the coronoid process of the ulna when the arm is flexed. The lateral pit is the **radial fossa,** named for the nearby head of the radius.

## Radius

The proximal head of the **radius** (fig. 8.33) is a distinctive disc that rotates freely on the humerus when the palm is turned forward and back. It articulates with the capitulum of the humerus and radial notch of the ulna. On the shaft, immediately distal to the head, is a medial rough **tuberosity,** which is the insertion of the biceps muscle. The distal end of the radius has the following features, from lateral to medial:

1. a bony point, the **styloid process,** which can be palpated proximal to the thumb;

2. two shallow depressions (articular facets) that articulate with the scaphoid and lunate bones of the wrist; and

3. the **ulnar notch,** which articulates with the end of the ulna.

## Ulna

At the proximal end of the **ulna** (fig. 8.33) is a deep, C-shaped **trochlear notch** that wraps around the trochlea of the humerus. The posterior side of this notch is formed by a prominent **olecranon**—the bony point where you rest your elbow on a table. The anterior side is formed by a less prominent **coronoid process.** Laterally, the head of the ulna has a less conspicuous **radial notch,** which accommodates the head of the radius. At the distal end of the ulna is a medial **styloid process.** The bony lumps you can palpate on each side of your wrist are the styloid processes of the radius and ulna.

The radius and ulna are attached along their shafts by a ligament called the **interosseous** (IN-tur-OSS-ee-us) **membrane,** which is attached to an angular ridge called the **interosseous margin** on each bone. Most fibers of the interosseous membrane (IM) are oriented obliquely, slanting upward from the ulna to the radius. If you lean forward on a table supporting your weight on your hands,

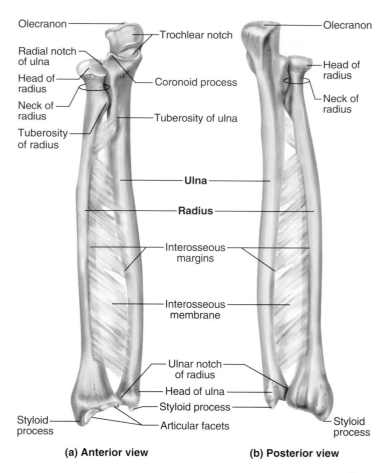

Olecranon — Trochlear notch
Radial notch of ulna
Head of radius — Coronoid process
Neck of radius
Tuberosity of radius — Tuberosity of ulna

Olecranon
Head of radius
Neck of radius

**Ulna**

**Radius**

Interosseous margins

Interosseous membrane

Ulnar notch of radius
Head of ulna
Styloid process
Styloid process — Articular facets — Styloid process

**(a) Anterior view**    **(b) Posterior view**

**FIGURE 8.33** The Right Radius and Ulna.

about 80% of the force is borne by the radius. This tenses the IM, which pulls the ulna upward and transfers some of this force through the ulna to the humerus. The IM thereby enables the two elbow joints (between humerus and radius, and humerus and ulna) to share the load and reduces the wear and tear that one joint would otherwise have to bear alone. The IM also serves as an attachment for several forearm muscles.

## Carpal Bones

The **carpal bones,** which form the wrist, are arranged in two rows of four bones each (fig. 8.34). These short bones allow movements of the wrist from side to side and anterior to posterior. The carpal bones of the proximal row, starting at the lateral (thumb) side, are the **scaphoid** (navicular), **lunate, triquetrum** (tri-QUEE-trum), and **pisiform**—Latin for boat-, moon-, triangle-, and pea-shaped, respectively. Unlike the other carpal bones, the pisiform is a sesamoid bone; it develops within the tendon of the *flexor carpi ulnaris muscle.*

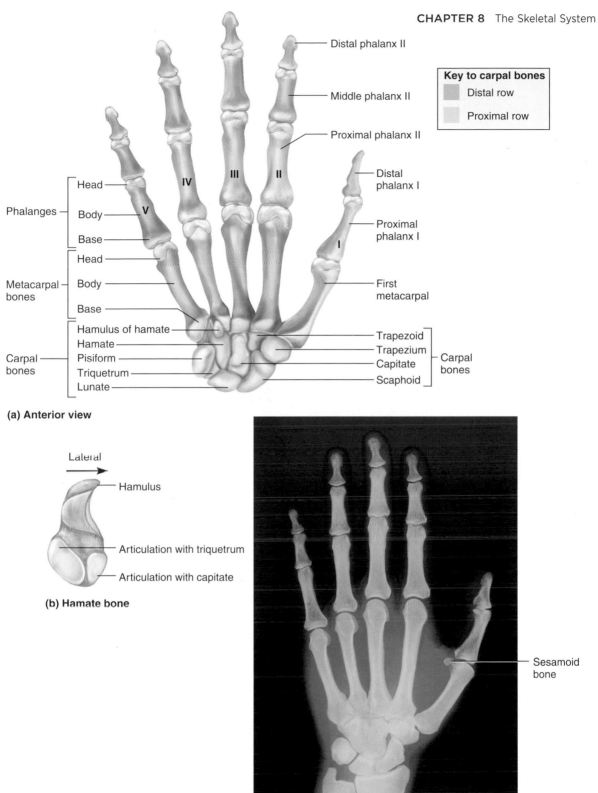

**(a) Anterior view**

Distal phalanx II

Middle phalanx II

Proximal phalanx II

Distal phalanx I

Proximal phalanx I

**Key to carpal bones**
Distal row
Proximal row

Head
Phalanges — Body
Base
Head
Metacarpal bones — Body
Base
Hamulus of hamate
Hamate
Carpal bones — Pisiform
Triquetrum
Lunate

First metacarpal

Trapezoid
Trapezium
Capitate — Carpal bones
Scaphoid

**(b) Hamate bone**

Lateral

Hamulus

Articulation with triquetrum

Articulation with capitate

**(c) X ray of adult hand**

Sesamoid bone

**FIGURE 8.34 The Right Wrist and Hand, Anterior (Palmar) View.** (a) Some people remember the names of the carpal bones with the mnemonic, "Sally left the party to take Charlie home." The first letters of these words correspond to the first letters of the carpal bones, from lateral to medial, proximal row first. (b) The right hamate bone, viewed from the proximal side of the wrist to show its distinctive hook. This unique bone is a useful landmark for locating the others when studying the skeleton. (c) Color-enhanced X ray of an adult hand. Identify the unlabeled bones in the X ray by comparing it to the drawing in part (a).

▶ *How does part (c) differ from the X ray of a child's hand, figure 7.12?*

The bones of the distal row, again starting on the lateral side, are the **trapezium,**[57] **trapezoid, capitate,**[58] and **hamate.**[59] The hamate can be recognized by a prominent hook called the **hamulus** on the palmar side (fig. 8.34b).

## Metacarpal Bones

Bones of the palm are called **metacarpals.**[60] Metacarpal I is located at the base of the thumb and metacarpal V at the base of the little finger. On a skeleton, the metacarpals look like extensions of the fingers, so that the fingers seem much longer than they really are. The proximal end of a metacarpal bone is called the **base,** the shaft is called the **body,** and the distal end is called the **head.** The heads of the metacarpals form knuckles when you clench your fist.

## Phalanges

The bones of the fingers are called **phalanges** (fah-LAN-jeez); in the singular, *phalanx* (FAY-lanks). There are two phalanges in the **pollex** (thumb) and three in each of the other digits. Phalanges are identified by Roman numerals preceded by *proximal, middle,* and *distal.* For example, proximal phalanx I is in the basal segment of the thumb (the first segment beyond the web between the thumb and palm); the left proximal phalanx IV is where people usually wear wedding rings; and distal phalanx V forms the tip of the little finger. The three parts of a phalanx are the same as in a metacarpal: base, body, and head. The ventral surface of a phalanx is slightly concave from end to end and flattened from side to side; the dorsal surface is rounder and slightly convex.

Table 8.7 summarizes the bones of the pectoral girdle and upper limb.

## Before You Go On

*Answer the following questions to test your understanding of the preceding section:*

14. *Describe how to distinguish the medial and lateral ends of the clavicle from each other, and how to distinguish its superior and inferior surfaces.*

15. *Name the three fossae of the scapula and describe the location of each.*

16. *What three bones meet at the elbow? Identify the fossae, articular surfaces, and processes of this joint and state to which bone each of these features belongs.*

17. *Name the four bones of the proximal row of the carpus from lateral to medial, and then the four bones of the distal row in the same order.*

18. *Name the four bones from the tip of the little finger to the base of the hand on that side.*

---

[57]*trapez* = table, grinding surface
[58]*capit* = head + *ate* = possessing
[59]*ham* = hook + *ate* = possessing
[60]*meta* = beyond + *carp* = wrist

| TABLE 8.7 | Anatomical Checklist for the Pectoral Girdle and Upper Limb |
|---|---|
| **Pectoral Girdle** | |
| *Clavicle (fig. 8.30)* | *Scapula—(Cont.)* |
| Sternal end | Suprascapular notch |
| Acromial end | Spine |
| Conoid tubercle | Fossae |
| | Subscapular fossa |
| *Scapula (fig. 8.31)* | Supraspinous fossa |
| Borders | Infraspinous fossa |
| Superior border | Acromion |
| Medial (vertebral) border | Coracoid process |
| Lateral (axillary) border | Glenoid cavity |
| Angles | Olecranon |
| Superior angle | |
| Inferior angle | |
| Lateral angle | |
| **Upper Limb** | |
| *Humerus (fig. 8.32)* | *Ulna—(Cont.)* |
| Proximal end | Radial notch |
| Head | Styloid process |
| Anatomical neck | Interosseous border |
| Surgical neck | Interosseous membrane |
| Greater tubercle | |
| Lesser tubercle | *Carpal Bones (fig. 8.34)* |
| Intertubercular sulcus | Proximal group |
| Shaft | Scaphoid |
| Deltoid tuberosity | Lunate |
| Distal end | Triquetrum |
| Capitulum | Pisiform |
| Trochlea | Distal group |
| Lateral epicondyle | Trapezium |
| Medial epicondyle | Trapezoid |
| Lateral supracondylar ridge | Capitate |
| Medial supracondylar ridge | Hamate |
| Olecranon fossa | |
| Coronoid fossa | *Bones of the Hand (fig. 8.34)* |
| Radial fossa | Metacarpal bones I–V |
| | Base |
| *Radius (fig. 8.33)* | Body |
| Head | Head |
| Tuberosity | Phalanges I–V |
| Styloid process | Proximal phalanx |
| Articular facets | Middle phalanx |
| Ulnar notch | Distal phalanx |
| *Ulna (fig. 8.33)* | |
| Trochlear notch | |
| Coronoid process | |

# The Pelvic Girdle and Lower Limb

### Objectives

When you have completed this section, you should be able to

- identify and describe the features of the pelvic girdle, femur, patella, tibia, fibula, and bones of the foot; and

- compare the anatomy of the male and female pelvis and explain the functional significance of the differences.

## PELVIC GIRDLE

Each adult hip bone is called an **os coxae**[61] (oss COC-see). Together, the two *ossa coxae* (plural) are called the **pelvic girdle,** and combined with the sacrum, they form the **pelvis**[62] (fig. 8.35). Another term for the os coxae—

arguably the most self-contradictory term in anatomy—is the *innominate*[63] (ih-NOM-ih-nate) *bone,* "the bone with no name." The pelvis supports the trunk on the legs and encloses and protects viscera of the pelvic cavity—mainly the lower colon, urinary bladder, and reproductive organs.

Each os coxae is joined to the vertebral column at the sacroiliac joint, where its **auricular surface** matches the one on the sacrum (fig. 8.36b). On the anterior side of the pelvis is the **pubic symphysis,**[64] the point where the right and left ossa coxae are joined by a pad of fibrocartilage (the *interpubic disc*). The symphysis can be palpated immediately above the genitalia.

The pelvis has a bowl-like shape with the broad **greater (false) pelvis** between the flare of the hips, and the narrower **lesser (true) pelvis** below. The two are separated by a somewhat round margin called the **pelvic brim.** The opening circumscribed by the brim is called the **pelvic inlet**—an entry into the lesser pelvis through which an infant's head passes during birth. The lower margin of the lesser pelvis is called the **pelvic outlet.**

---

[61]*os* = bone + *coxae* = of the hip
[62]*pelv* = basin, bowl

[63]*in* = without + *nomin* = name + *ate* = having
[64]*sym* = together + *physis* = growth

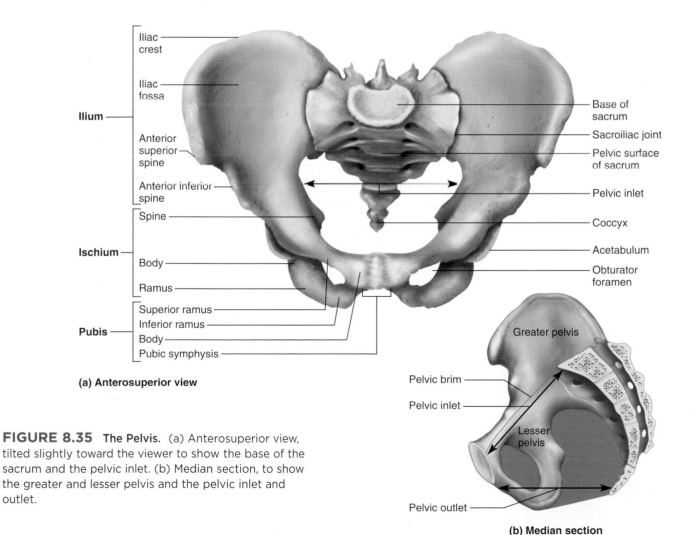

**(a) Anterosuperior view**

**FIGURE 8.35 The Pelvis.** (a) Anterosuperior view, tilted slightly toward the viewer to show the base of the sacrum and the pelvic inlet. (b) Median section, to show the greater and lesser pelvis and the pelvic inlet and outlet.

**(b) Median section**

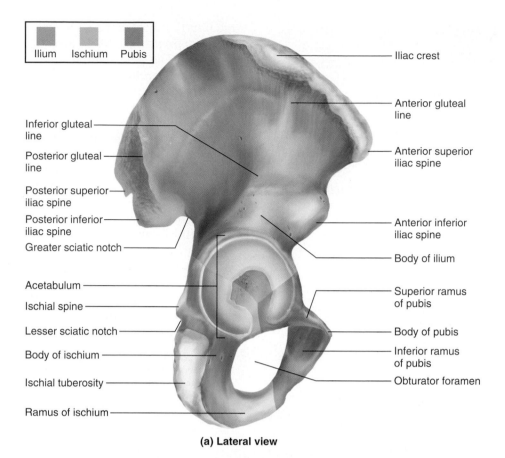

Ilium　Ischium　Pubis

Iliac crest

Anterior gluteal line

Inferior gluteal line

Posterior gluteal line

Posterior superior iliac spine

Posterior inferior iliac spine

Greater sciatic notch

Anterior superior iliac spine

Anterior inferior iliac spine

Body of ilium

Acetabulum

Ischial spine

Lesser sciatic notch

Body of ischium

Ischial tuberosity

Ramus of ischium

Superior ramus of pubis

Body of pubis

Inferior ramus of pubis

Obturator foramen

**(a) Lateral view**

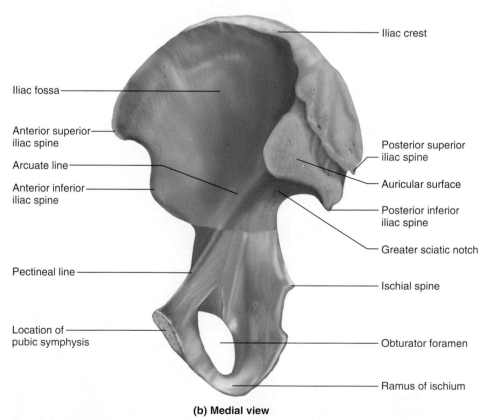

Iliac crest

Iliac fossa

Anterior superior iliac spine

Arcuate line

Anterior inferior iliac spine

Pectineal line

Location of pubic symphysis

Posterior superior iliac spine

Auricular surface

Posterior inferior iliac spine

Greater sciatic notch

Ischial spine

Obturator foramen

Ramus of ischium

**(b) Medial view**

**FIGURE 8.36**　The Right Os Coxae.

The os coxae has three distinctive features that will serve as landmarks for further description. These are the **iliac**[65] crest (superior crest of the hip); **acetabulum**[66] (ASS-eh-TAB-you-lum) (the hip socket—named for its resemblance to vinegar cups used on ancient Roman dining tables); and **obturator**[67] **foramen** (a large round-to-triangular hole below the acetabulum, closed by a ligament called the *obturator membrane* in life).

The adult os coxae forms by the fusion of three childhood bones called the *ilium* (ILL-ee-um), *ischium* (ISS-kee-um), and *pubis* (PEW-biss), identified by color in figure 8.36. The largest of these is the **ilium,** which extends from the iliac crest to the superior wall of the acetabulum. The iliac crest extends from a point or angle on the anterior side, called the **anterior superior spine,** to a sharp posterior angle, called the **posterior superior spine.** In a lean person, the anterior superior spines form visible anterior protrusions, and the posterior superior spines are sometimes marked by dimples above the buttocks where connective tissue attached to the spines pulls inward on the skin.

Below the superior spines are the **anterior** and **posterior inferior spines.** Below the posterior inferior spine is a deep **greater sciatic** (sy-AT-ic) **notch,** named for the sciatic nerve that passes through it and continues down the posterior side of the thigh.

The posterolateral surface of the ilium is relatively rough-textured because it serves for attachment of several muscles of the buttocks and thighs. The anteromedial surface, by contrast, is the smooth, slightly concave **iliac fossa,** covered in life by the broad *iliacus* muscle. Medially, the ilium exhibits an *auricular surface* that matches the one on the sacrum, so that the two bones form the sacroiliac joint.

The **ischium** is the inferoposterior portion of the os coxae. Its heavy **body** is marked with a prominent **spine.** Inferior to the spine is a slight indentation, the **lesser sciatic notch,** and then the thick, rough-surfaced **ischial tuberosity,** which supports your body when you are sitting. The tuberosity can be palpated by sitting on your fingers. The **ramus** of the ischium joins the inferior ramus of the pubis anteriorly.

The **pubis** (pubic bone) is the most anterior portion of the os coxae. It has a **superior** and **inferior ramus** and a triangular **body.** The body of one pubis meets the body of the other at the pubic symphysis. The pubis and ischium encircle the obturator foramen.

The female pelvis is adapted to the needs of pregnancy and childbirth. Some of the differences between the male and female pelves are described in table 8.8 and illustrated in figure 8.37.

# LOWER LIMB

The number and arrangement of bones in the lower limb are similar to those of the upper limb. In the lower limb, however, they are adapted for weight-bearing and locomotion and are therefore shaped and articulated differently. The lower limb is divided into four regions containing a total of 30 bones per limb:

1. The **femoral region,** or thigh, extends from hip to knee and contains the *femur.* The *patella* (kneecap) is a sesamoid bone at the junction of the femoral and crural regions.
2. The **crural** (CROO-rul) **region,** or leg proper, extends from knee to ankle and contains two bones, the medial *tibia* and lateral *fibula.*
3. The **tarsal region** (tarsus), or ankle, is the union of the crural region with the foot. The tarsal bones are treated as part of the foot.
4. The **pedal region** (pes), or foot, is composed of 7 *tarsal bones,* 5 *metatarsals,* and 14 *phalanges* in the toes.

## Femur

The **femur** (FEE-mur) is the longest and strongest bone in the body (fig. 8.38). It has a hemispherical head that articulates with the acetabulum of the pelvis, forming a quintessential *ball-and-socket joint.* A ligament extends from the acetabulum to a pit, the **fovea capitis**[68] (FOE-vee-uh CAP-ih-tiss), in the head of the femur. Distal to the head is a constricted **neck** and then two massive, rough processes called the **greater** and **lesser trochanters** (tro-CAN-turs), which are insertions for the powerful muscles of the hip. The trochanters are connected on the posterior side by a thick oblique ridge of bone, the **intertrochanteric crest,** and on the anterior side by a more delicate **intertrochanteric line.**

The primary feature of the shaft is a posterior ridge called the **linea aspera**[69] (LIN-ee-uh ASS-peh-ruh) at its midpoint. At its upper end, the linea aspera forks into a medial **spiral (pectineal) line** and a lateral **gluteal tuberosity.** The gluteal tuberosity is a rough ridge (sometimes a depression) that serves for attachment of the powerful *gluteus maximus* muscle of the buttock. At its lower end, the linea aspera forks into **medial** and **lateral supracondylar lines,** which continue down to the respective epicondyles.

The **medial** and **lateral epicondyles** are the widest points of the femur at the knee. These and the supracondylar lines are attachments for certain thigh and leg muscles and knee ligaments. At the distal end of the femur are two smooth round surfaces of the knee joint, the **medial** and **lateral condyles,** separated by a groove

---

[65]*ili* = flank, loin + *ac* = pertaining to
[66]*acetabulum* = vinegar cup
[67]*obtur* = to close, stop up + *ator* = that which

[68]*fovea* = pit + *capitis* = of the head
[69]*linea* + line + *asper* = rough

| TABLE 8.8 | Comparison of the Male and Female Pelves | |
| --- | --- | --- |
| Feature | Male | Female |
| General Appearance | More massive; rougher; heavier processes | Less massive; smoother; more delicate processes |
| Tilt | Upper end of pelvis relatively vertical | Upper end of pelvis tilted forward |
| Ilium, Greater Pelvis | Deeper; projects farther above sacroiliac joint | Shallower; does not project as far above sacroiliac joint |
| Lesser Pelvis | Narrower and deeper | Wider and shallower |
| Sacrum | Narrower and longer | Shorter and wider |
| Coccyx | Less movable; more vertical | More movable; tilted posteriorly |
| Width of Greater Pelvis | Anterior superior spines closer together; hips less flared | Anterior superior spines farther apart; hips more flared |
| Pelvic Inlet | Heart-shaped | Round or oval |
| Pelvic Outlet | Smaller | Larger |
| Greater Sciatic Notch | Narrower | Wider |
| Obturator Foramen | Round | Triangular to oval |
| Acetabulum | Faces more laterally, larger | Faces slightly anteriorly, smaller |
| Pubic arch | Usually 90° or less | Usually greater than 100° |

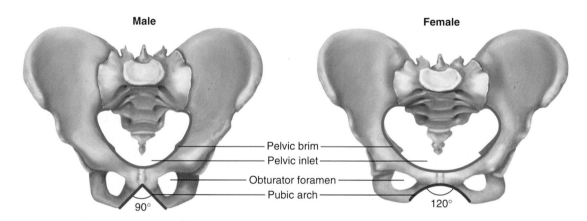

**Male**    **Female**

Pelvic brim
Pelvic inlet
Obturator foramen
Pubic arch

90°    120°

**FIGURE 8.37    Comparison of the Male and Female Pelvic Girdles.**  The pelvic brim is the margin highlighted in blue. Compare with table 8.8.

called the **intercondylar** (IN-tur-CON-dih-lur) **fossa.** On the anterior side of the femur, a smooth medial depression called the **patellar surface** articulates with the patella. On the posterior side is a flat or slightly depressed area called the **popliteal surface.**

## Patella

The **patella,**[70] or kneecap (fig. 8.38), is a roughly triangular sesamoid bone embedded in the tendon of the knee. It is cartilaginous at birth and ossifies at 3 to 6 years of age. It has a broad superior **base,** a pointed inferior **apex,** and a pair of shallow **articular facets** on its posterior surface where it articulates with the femur. The lateral facet is usually larger than the medial. The *quadriceps femoris*

*tendon* extends from the anterior muscle of the thigh (the *quadriceps femoris*) to the patella, and it continues as the *patellar ligament* from the patella to the tibia.

## Tibia

The leg has two bones: a thick, strong tibia (TIB-ee-uh) and a slender, lateral fibula (FIB-you-luh) (fig. 8.39). The **tibia,** on the medial side, is the only weight-bearing bone of the crural region. Its broad superior head has two fairly flat articular surfaces, the **medial** and **lateral condyles,** separated by a ridge called the **intercondylar eminence.** The condyles of the tibia articulate with those of the femur. The rough anterior surface of the tibia, the **tibial tuberosity,** can be palpated just below the patella. This is an attachment for the powerful thigh muscles that extend (straighten) the knee. Distal to this, the shaft has a sharply

[70]*pat* = pan + *ella* = little

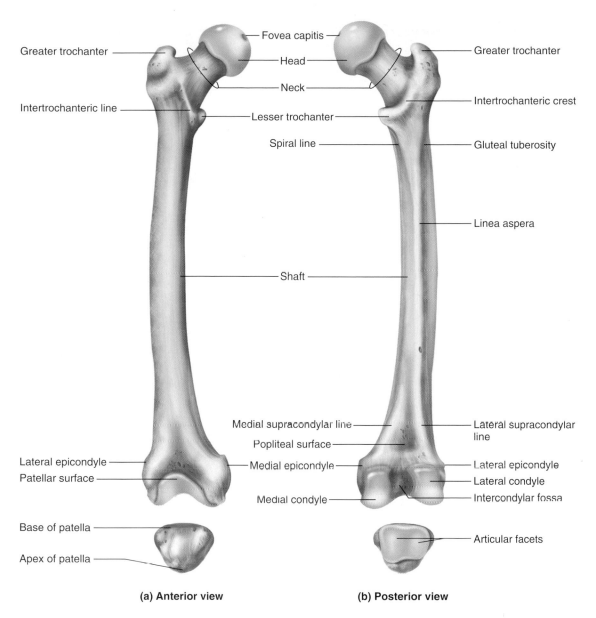

Greater trochanter
Fovea capitis
Head
Neck
Intertrochanteric line
Lesser trochanter
Spiral line

Greater trochanter
Intertrochanteric crest
Gluteal tuberosity

Linea aspera

Shaft

Medial supracondylar line
Popliteal surface
Lateral epicondyle
Medial epicondyle
Patellar surface
Medial condyle

Lateral supracondylar line
Lateral epicondyle
Lateral condyle
Intercondylar fossa

Base of patella
Apex of patella

Articular facets

**(a) Anterior view**

**(b) Posterior view**

**FIGURE 8.38   The Right Femur and Patella.**

angular **anterior crest,** which can be palpated in the shin. At the ankle, just above the rim of a standard dress shoe, you can palpate a prominent bony knob on each side. These are the **medial** and **lateral malleoli**[71] (MAL-ee-OH-lie). The medial malleolus is part of the tibia, and the lateral malleolus is the part of the fibula.

## Fibula

The **fibula** (fig. 8.39) is a slender lateral bone that helps to stabilize the ankle. It does not bear any of the body's weight; indeed, orthopedic surgeons sometimes remove part of the fibula and use it to replace damaged or missing bone elsewhere in the body. The fibula is somewhat thicker and broader at its proximal end, the **head,** than at

the distal end. The point of the head is called the **apex.** The distal expansion is the lateral malleolus.

Like the radius and ulna, the tibia and fibula are joined by an interosseous membrane along their shafts.

## The Ankle and Foot

The **tarsal bones** of the ankle are arranged in proximal and distal groups somewhat like the carpal bones of the wrist (fig. 8.40). Because of the load-bearing role of the ankle, however, their shapes and arrangement are conspicuously different from those of the carpal bones, and they are thoroughly integrated into the structure of the foot. The largest tarsal bone is the **calcaneus**[72] (cal-CAY-nee-us), which forms the heel. Its posterior end is the

[71]*malle* = hammer + *olus* = little

[72]*calc* = stone, chalk

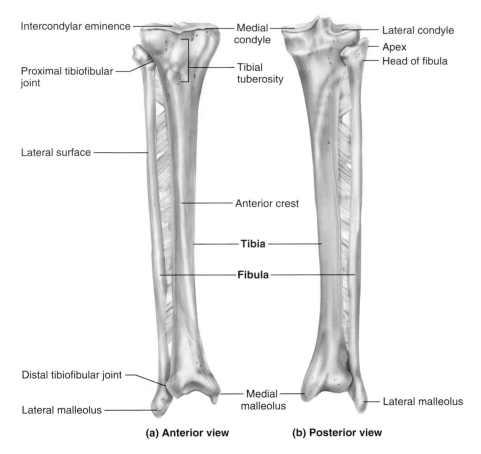

Intercondylar eminence

Proximal tibiofibular joint

Lateral surface

Anterior crest

**Tibia**

**Fibula**

Distal tibiofibular joint

Lateral malleolus

Medial condyle

Tibial tuberosity

Lateral condyle

Apex

Head of fibula

Medial malleolus

Lateral malleolus

**(a) Anterior view**    **(b) Posterior view**

**FIGURE 8.39** The Right Tibia and Fibula.

▶ *Why is the distal end of the tibia broader than that of the fibula?*

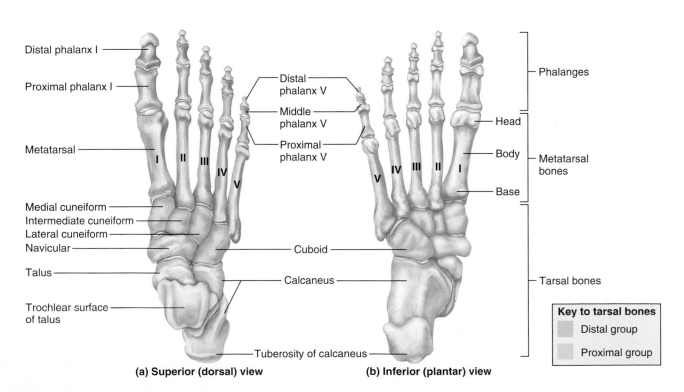

Distal phalanx I

Proximal phalanx I

Metatarsal

Medial cuneiform

Intermediate cuneiform

Lateral cuneiform

Navicular

Talus

Trochlear surface of talus

Distal phalanx V

Middle phalanx V

Proximal phalanx V

Cuboid

Calcaneus

Tuberosity of calcaneus

Phalanges

Head

Body

Base

Tarsal bones

Metatarsal bones

**Key to tarsal bones**

Distal group

Proximal group

**(a) Superior (dorsal) view**    **(b) Inferior (plantar) view**

**FIGURE 8.40** The Right Foot.

point of attachment for the **calcaneal (Achilles) tendon** from the calf muscles. The second-largest tarsal bone, and the most superior, is the **talus.** It has three articular surfaces: an inferoposterior one that articulates with the calcaneus, a superior **trochlear surface** that articulates with the tibia, and an anterior surface that articulates with a short, wide tarsal bone called the **navicular.** The talus, calcaneus, and navicular are considered the proximal row of tarsal bones. (*Navicular* is also used as a synonym for the scaphoid bone of the wrist.)

The distal group forms a row of four bones. Proceeding from medial to lateral, these are the **medial, intermediate,** and **lateral cuneiforms**[73] (cue-NEE-ih-forms) and the **cuboid.** The cuboid is the largest.

The remaining bones of the foot are similar in arrangement and name to those of the hand. The proximal **metatarsals**[74] are similar to the metacarpals. They are **metatarsals I** to **V** from medial to lateral, metatarsal I being proximal to the great toe. Metatarsals I to III articulate with the first through third cuneiforms; metatarsals IV and V both articulate with the cuboid.

Bones of the toes, like those of the fingers, are called phalanges. The great toe is the **hallux** and contains only two bones, the proximal and distal phalanx I. The other toes each contain a proximal, middle, and distal phalanx,

---

[73]*cunei* = wedge + *form* = in the shape of
[74]*meta* = beyond + *tars* = ankle

and are numbered II through V from medial to lateral. Thus, middle phalanx V, for example, would be the middle bone of the smallest toe. The metatarsal and phalangeal bones each have a base, body, and head, like the bones of the hand. All of them, especially the phalanges, are slightly concave on the ventral side.

Note that Roman numeral I represents the *medial* group of bones in the foot but the *lateral* group in the hand. In both cases, however, Roman numeral I refers to the largest digit of the limb. The reason for the difference between the hand and foot lies in a rotation of the limbs that occurs in the seventh week of embryonic development. Early in the seventh week, the limbs extend anteriorly from the body, the foot is a paddlelike *foot plate,* and the hand is also more or less paddlelike with the finger buds showing early separation (fig. 8.41a). The future thumb and great toe are both directed superiorly, and the future palms and soles face each other, medially. But then each limb rotates about 90° in opposite directions. The upper limb rotates laterally. To visualize this, hold your arms straight out in front of you with the palms facing each other as if you were about to clap hands. Then rotate your forearms so the thumbs face away from each other (laterally) and the palms face upward. The lower limbs rotate in the opposite direction, medially, so that the soles face downward and the great toes become medial. So even though the thumb and great toe (digit I of the hand and foot) start out facing in the same direction, these opposite

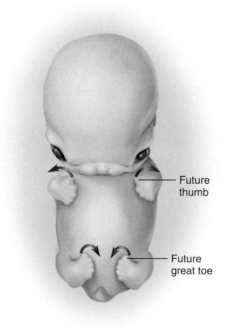

**(a) Seven weeks**

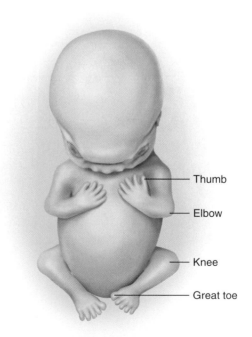

**(b) Eight weeks**

**FIGURE 8.41 Embryonic Limb Rotation.** In the seventh week of development, the forelimbs and hindlimbs of the embryo rotate about 90° in opposite directions. This explains why the largest digits (digit I) are on opposite sides of the hand and foot, and why the elbow and knee flex in opposite directions.

rotations result in their being on opposite sides of the hand and foot (fig. 8.41b). This rotation also explains why the elbow flexes posteriorly and the knee flexes anteriorly, and why (as you will see in chapter 10) the muscles that flex the elbow are on the anterior side of the arm, whereas those that flex the knee are on the posterior side of the thigh.

The sole of the foot normally does not rest flat on the ground; rather, it has three springy arches that absorb the stress of walking (fig. 8.42). The **medial longitudinal arch,** which essentially extends from heel to hallux, is formed from the calcaneus, talus, navicular, cuneiforms, and metatarsals I to III. The **lateral longitudinal arch** extends from heel to little toe and includes the calcaneus, cuboid, and metatarsals IV and V. The **transverse arch** includes the cuboid, cuneiforms, and proximal heads of the metatarsals. These arches are held together by short, strong ligaments. Excessive weight, repetitive stress, or congenital weakness of these ligaments can stretch them, resulting in *pes planus* (commonly called flat feet or

fallen arches). This condition makes a person less tolerant of prolonged standing and walking. A comparison of the flat-footed apes with humans underscores the significance of the human foot arches (Insight 8.5).

Table 8.9 summarizes the pelvic girdle and lower limb.

## Before You Go On

*Answer the following questions to test your understanding of the preceding section:*

19. Name the bones of the adult pelvic girdle. What three bones of a child fuse to form the os coxae of an adult?

20. Name any four structures of the pelvis that you can palpate and describe where to palpate them.

21. What parts of the femur are involved in the hip joint? What parts are involved in the knee joint?

22. Name the prominent knobs on each side of your ankle. What bones contribute to these structures?

23. Name all the bones that articulate with the talus and describe the location of each.

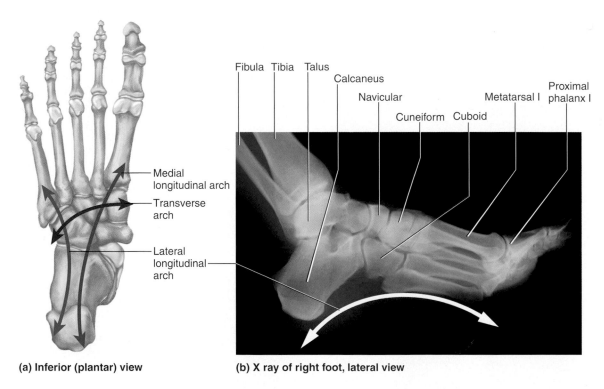

**(a) Inferior (plantar) view**    **(b) X ray of right foot, lateral view**

**FIGURE 8.42** Arches of the Foot.

## TABLE 8.9   Anatomical Checklist for the Pelvic Girdle and Lower Limb

### Pelvic Girdle

**Os Coxae (figs. 8.35 and 8.36)**
Pubic symphysis
Greater (false) pelvis
Lesser (true) pelvis
Pelvic brim
Pelvic inlet
Pelvic outlet
Acetabulum
Obturator foramen
Ilium
  Iliac crest
  Anterior superior spine
  Anterior inferior spine
  Posterior superior spine
  Posterior inferior spine

**Ilium—(Cont.)**
  Greater sciatic notch
  Iliac pillar
  Iliac fossa
  Auricular surface
Ischium
  Body
  Ischial spine
  Lesser sciatic notch
  Ischial tuberosity
  Ramus
Pubis
  Superior ramus
  Inferior ramus
  Body

### Lower Limb

**Femur (fig. 8.38)**
Proximal end
  Head
  Fovea capitis
  Neck
  Greater trochanter
  Lesser trochanter
  Intertrochanteric crest
  Intertrochanteric line
Shaft
  Spiral line
  Gluteal tuberosity
  Linea aspera
  Medial supracondylar line
  Lateral supracondylar line
Distal end
  Medial condyle
  Lateral condyle
  Intercondylar fossa
  Medial epicondyle
  Lateral epicondyle
  Patellar surface
  Popliteal surface

**Patella (fig. 8.38)**
Base
Apex
Articular facets

**Tibia (fig. 8.39)**
Medial condyle
Lateral condyle
Intercondylar eminence
Tibial tuberosity

**Tibia—(Cont.)**
Anterior crest
Medial malleolus

**Fibula (fig. 8.39)**
Head
Apex (styloid process)
Lateral malleolus

**Tarsal Bones (fig. 8.40)**
Proximal group
  Calcaneus
  Talus
  Navicular
Distal group
  Medial cuneiform
  Intermediate cuneiform
  Lateral cuneiform
  Cuboid

**Bones of the Foot (figs. 8.40 and 8.42)**
Metatarsal bones I–V
Phalanges
  Proximal phalanx
  Middle phalanx
  Distal phalanx
Arches of the foot
  Medial longitudinal arch
  Lateral longitudinal arch
  Transverse arch

## Skeletal Adaptations for Bipedalism

Some mammals can stand, hop, or walk briefly on their hind legs, but humans are the only mammals that are habitually bipedal. Footprints preserved in a layer of volcanic ash in Tanzania indicate that hominids walked upright as early as 3.6 million years ago. This bipedal locomotion is possible only because of several adaptations of the human feet, legs, spine, and skull (fig. 8.43). These features are so distinctive that paleoanthropologists (those who study human fossil remains) can tell with considerable certainty whether a fossil species was able to walk upright.

As important as the hand has been to human evolution, the foot may be an even more significant adaptation. Unlike other mammals, humans support their entire body weight on two feet. While apes are flat-footed, humans have strong, springy foot arches that absorb shock as the body jostles up and down during walking and running. The tarsal bones are tightly articulated with one another, and the calcaneus is strongly developed. The hallux (great toe) is not opposable as it is in most Old World monkeys and apes, but it is highly developed so that it

provides the "toe-off" that pushes the body forward in the last phase of the stride (fig. 8.43a). For this reason, loss of the hallux has a more crippling effect than the loss of any other toe.

While the femurs of apes are nearly vertical, in humans they angle medially from the hip to the knee (fig. 8.43b). This places our knees closer together, beneath the body's center of gravity. We lock our knees when standing, allowing us to maintain an erect posture with little muscular effort. Apes cannot do this, and they cannot stand on two legs for very long without tiring—much as you would if you tried to maintain an erect posture with your knees slightly bent.

In apes and other quadrupedal (four-legged) mammals, the abdominal viscera are supported by the muscular wall of the abdomen. In humans, the viscera bear down on the floor of the pelvic cavity, and a bowl-shaped pelvis is necessary to support their weight. This has resulted in a narrower pelvic outlet—a condition quite incompatible with the fact that we, including our infants, are such a large-brained species. The pain of childbirth seems unique to humans and, one might say, is a price we must pay for having both a large brain and a bipedal stance.

The largest muscle of the buttock, the *gluteus maximus,* serves in apes primarily as an abductor of the thigh—that is, it moves the leg laterally. In humans, however, the ilium has

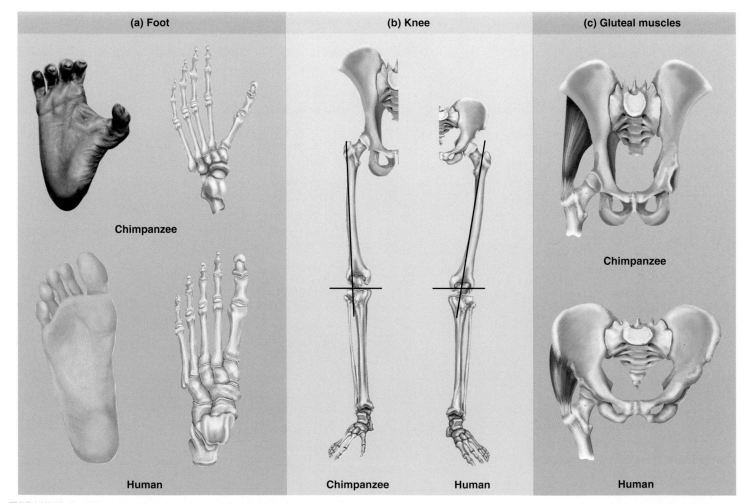

| (a) Foot | (b) Knee | (c) Gluteal muscles |

Chimpanzee

Human

Chimpanzee    Human

Chimpanzee

Human

**FIGURE 8.43  Skeletal Adaptations for Bipedalism.**  Human adaptations for bipedalism are best understood by comparison to our closest living relative, the chimpanzee, which is not adapted for a comfortable or sustained erect stance. See the text for the relevance of each comparison.

expanded posteriorly, so the gluteus maximus originates behind the hip joint. This changes the function of the muscle—instead of abducting the thigh, it pulls the thigh back in the second half of a stride (pulling back on your right thigh, for example, when your left foot is off the ground and swinging forward). Two other buttock muscles, the *gluteus medius* and *gluteus minimus,* extend laterally in humans from the surface of the ilium to the greater trochanter of the femur (fig. 8.43c). In walking, when one foot is lifted from the ground, these muscles shift the body weight over the other foot so we do not fall over. The actions of all three gluteal muscles, and the corresponding evolutionary remodeling of the pelvis, account for the smooth, efficient stride of a human as compared with the awkward, shuffling gait of a chimpanzee or gorilla when it is walking upright. The posterior growth of the ilium (fig. 8.43d) is the reason the greater sciatic notch is so deeply concave.

The lumbar curvature of the human spine allows for efficient bipedalism by shifting the body's center of gravity to the rear, above and slightly behind the hip joint (fig. 8.43e). Because of their C-shaped spines, chimpanzees cannot stand as easily. Their center of gravity is anterior to the hip joint when they stand; they must exert a continual muscular effort to keep from falling forward, and fatigue sets in relatively quickly. Humans, by contrast, require little muscular effort to keep their balance. Our australopithecine ancestors probably could travel all day with relatively little fatigue.

The human head is balanced on the vertebral column with the gaze directed forward. The cervical curvature of the spine and remodeling of the skull have made this possible. The foramen magnum has moved to a more inferior and anterior location, and the face is much flatter than in an ape (fig. 8.43f), so there is less weight anterior to the occipital condyles. Being balanced on the spine, the head does not require strong muscular attachments to hold it erect. Apes have prominent supraorbital ridges for the attachment of muscles that pull back on the skull. In humans these ridges are much lighter and the muscles of the forehead serve only for facial expression, not to hold the head up.

The forelimbs of apes are longer than the hindlimbs; indeed, some species such as the orangutan and gibbons hold their long forelimbs over their heads when they walk on their hind legs. By contrast, our arms are shorter than our legs and far less muscular than the forelimbs of apes. No longer needed for locomotion, our forelimbs have become better adapted for carrying objects, holding things closer to the eyes, and manipulating them more precisely.

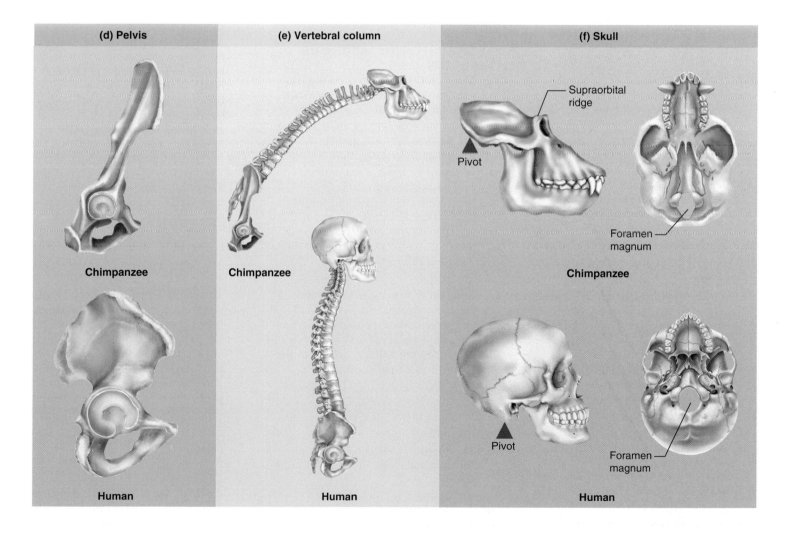

**(d) Pelvis**

**Chimpanzee**

**Human**

**(e) Vertebral column**

**Chimpanzee**

**Human**

**(f) Skull**

Supraorbital ridge

Pivot

Foramen magnum

**Chimpanzee**

Pivot

Foramen magnum

**Human**

## CHAPTER REVIEW

# Review of Key Concepts

### Overview of the Skeleton (p. 242)

1. The skeletal system is divisible into the central *axial skeleton* (skull, vertebral column, and thoracic cage) and the *appendicular skeleton* (bones of the upper and lower limbs and their supporting girdles).

2. There are typically 206 named bones in the adult (table 8.1), but the number varies from person to person, it is higher in newborns, and it increases in childhood before bone fusion leads to the adult number of bones.

3. Before studying individual bones, one must be familiar with the terminology of bone surface features (table 8.2).

### The Skull (p. 244)

1. The skull consists of eight *cranial bones,* which contact the meninges around the brain, and 14 *facial bones,* which do not.

2. The skull encloses several spaces: the cranial, nasal, buccal, middle-ear, and inner-ear cavities, the orbits, and the paranasal sinuses (frontal, sphenoid, ethmoid, and maxillary).

3. Bones of the skull are perforated by numerous *foramina,* which allow passage for cranial nerves and blood vessels.

4. Some prominent features of the skull in general are the *foramen magnum* where the spinal cord joins the brainstem; the *calvaria,* which forms a roof over the cranial cavity; the *orbits,* which house the eyes; the three *cranial fossae,* which form the floor of the cranial cavity; the *hard palate,* forming the roof of the mouth; and the zygomatic arches, or "cheekbones."

5. The cranial bones are the frontal, parietal, temporal, occipital, sphenoid, and ethmoid bones. The parietal and temporal bones are paired, and the others single.

6. The facial bones are the maxillae; the palatine, zygomatic, lacrimal, and nasal bones; the inferior nasal conchae; and the vomer and mandible. All but the last two are

paired. The mandible is the only movable bone of the skull.

7. Features of the individual bones are summarized in table 8.4.

8. Associated with the skull are the *hyoid bone* in the neck and the three *auditory ossicles (malleus, incus, and stapes)* in each middle ear.

9. The skull of the fetus and infant is marked by six gaps, or *fontanels,* where the cranial bones have not fully fused: one anterior, one posterior, two sphenoid, and two mastoid fontanels. A child's skull attains nearly adult size by the age of 8 or 9 years.

### The Vertebral Column and Thoracic Cage (p. 258)

1. The *vertebral column* normally consists of 33 vertebrae and 23 cartilaginous intervertebral discs. It is slightly S-shaped, with four curvatures: *cervical, thoracic, lumbar,* and *pelvic.*

2. A typical vertebra exhibits a *body,* a *vertebral foramen,* a *spinous process,* and two *transverse processes.* The shapes and proportions of these features, and some additional features, distinguish vertebrae from different regions of the vertebral column (table 8.6).

3. There are five classes of vertebrae, numbering 7 cervical, 12 thoracic, 5 lumbar, 5 sacral, and 4 coccygeal vertebrae in most people. In adults, the sacral vertebrae are fused into a single *sacrum* and the coccygeal vertebrae into a single *coccyx.*

4. An intervertebral disc is composed of a gelatinous *nucleus pulposus* enclosed in a fibrous ring, the *anulus fibrosus.*

5. The *thoracic cage* consists of the thoracic vertebrae, the sternum, and the ribs.

6. The sternum has three parts: *manubrium, body,* and *xiphoid process.*

7. There are 12 pairs of ribs. Ribs 1 through 7 are called *true ribs* because each has its own *costal car-*

*tilage* connecting it to the sternum; 8 through 12 are called *false ribs,* and 11 and 12, the only ones with no costal cartilages, are also called *floating ribs.*

### The Pectoral Girdle and Upper Limb (p. 268)

1. The *pectoral girdle* attaches the upper limb to the axial skeleton. It consists of a *scapula* (shoulder blade) and *clavicle* (collar bone) on each side. The clavicle articulates with the sternum and the scapula articulates with the humerus.

2. The upper limb bones are the *humerus* in the brachium; the lateral *radius* and medial *ulna* in the antebrachium (forearm); eight *carpal bones* in the wrist; five *metacarpal bones* in the hand; two *phalanges* in the thumb; and three phalanges in each of the other four digits.

### The Pelvic Girdle and Lower Limb (p. 275)

1. The *pelvic girdle* attaches the lower limb to the axial skeleton. It consists of two *ossa coxae.* Each adult os coxae results from the fusion of three bones of the child: the *ilium, ischium,* and *pubis.* The ossa coxae and sacrum are collectively called the *pelvis.*

2. The pelvis forms two basinlike structures: a superior, wide *false (greater) pelvis* and an inferior, narrower *true (lesser) pelvis.* The passage from the false to the true pelvis is called the *pelvic inlet* and its margin is the *pelvic brim;* the exit from the true pelvis is called the *pelvic outlet.*

3. Two other major features of the os coxae are the *iliac crest,* which forms the flare of the hip, and the *acetabulum,* the cuplike socket for the femur.

4. The lower limb bones are the *femur* in the thigh; the lateral *fibula* and larger, medial *tibia* in the leg proper; seven *tarsal* (ankle) *bones* forming the posterior half of the foot; five *metatarsal bones* in its anterior half; two phalanges in the great toe; and three phalanges in each of the other digits.

# Testing Your Recall

1. Which of these is *not* a paranasal sinus?
   a. frontal
   b. temporal
   c. sphenoid
   d. ethmoid
   e. maxillary

2. Which of these is a facial bone?
   a. frontal
   b. ethmoid
   c. occipital
   d. temporal
   e. lacrimal

3. Which of these *cannot* be palpated on a living person?
   a. the crista galli
   b. the mastoid process
   c. the zygomatic arch
   d. the superior nuchal line
   e. the hyoid bone

4. All of the following are groups of vertebrae *except* for ____, which is a spinal curvature.
   a. thoracic
   b. cervical
   c. lumbar
   d. pelvic
   e. sacral

5. Thoracic vertebrae do *not* have
   a. transverse foramina.
   b. costal facets.
   c. spinous processes.
   d. transverse processes.
   e. pedicles.

6. The tubercle of a rib articulates with
   a. the sternal notch.
   b. the margin of the gladiolus.
   c. the costal facets of two vertebrae.
   d. the body of a vertebra.
   e. the transverse process of a vertebra.

7. The disc-shaped head of the radius articulates with the ____ of the humerus.
   a. radial tuberosity
   b. trochlea
   c. capitulum
   d. olecranon
   e. glenoid cavity

8. All of the following are carpal bones, *except* the ____ , which is a tarsal bone.
   a. trapezium
   b. cuboid
   c. trapezoid
   d. triquetrum
   e. pisiform

9. The bone that supports your body weight when you are sitting down is
   a. the acetabulum.
   b. the pubis.
   c. the ilium.
   d. the coccyx.
   e. the ischium.

10. Which of these is the bone of the heel?
    a. cuboid
    b. calcaneus
    c. navicular
    d. trochlear
    e. talus

11. Gaps between the cranial bones of an infant are called _____.

12. The external auditory canal is a passage in the _____ bone.

13. Bones of the skull are joined along lines called _____.

14. The _____ bone has greater and lesser wings and protects the pituitary gland.

15. A herniated disc occurs when a ring called the _____ cracks.

16. The transverse ligament of the atlas holds the _____ of the axis in place.

17. The sacroiliac joint is formed where the _____ surface of the sacrum articulates with that of the ilium.

18. The _____ processes of the radius and ulna form bony protuberances on each side of the wrist.

19. The thumb is also known as the _____ and the great toe is also known as the _____.

20. The _____ arch of the foot extends from the heel to the great toe.

*Answers in Appendix B*

# True or False

*Determine which five of the following statements are false, and briefly explain why.*

1. Not everyone has frontal sinuses.

2. The hands have more phalanges than the feet.

3. As an adaptation to pregnancy, the female's pelvis is deeper than the male's.

4. There are more carpal bones than tarsal bones.

5. On a living person, it would be possible to palpate the muscles in the infraspinous fossa but not those of the subscapular fossa.

6. If you rest your chin on your hands with your elbows on a table, the olecranon of the ulna rests on the table.

7. The lumbar vertebrae do not articulate with any ribs and therefore do not have transverse processes.

8. The most frequently broken bone is the humerus.

9. In strict anatomical terminology, the words *arm* and *leg* both refer to regions with only one bone.

10. The pisiform bone and patella are both sesamoid bones.

*Answers in Appendix B*

# Testing Your Comprehension

1. A child was involved in an automobile collision. She was not wearing a safety restraint, and her chin struck the dashboard hard. When the physician looked into her auditory canal, he could see into her throat. What do you infer from this about the nature of her injury?

2. By palpating the hind leg of a cat or dog or by examining a laboratory skeleton, you can see that cats and dogs stand on the heads of their metatarsal bones; the calcaneus does not touch the ground. How is this similar to the stance of a woman wearing high-heeled shoes? How is it different?

3. Contrast the tarsal bones with the carpal bones. Which ones are similar in name, location, or both? Which ones are different?

4. In adolescents, trauma sometimes separates the head of the femur from the neck. Why do you think this is more common in adolescents than in adults?

5. Andy, a 55-year-old, 75 kg (165 lb) roofer, is shingling the steeply pitched roof of a new house when he loses his footing and slides down the roof and over the edge, feet first. He braces himself for the fall, and when he hits ground he cries out and doubles up in excruciating pain. Emergency medical technicians called to the scene tell him he has broken his hips. Describe, more specifically, where his fractures most likely occurred. On the way to the hospital, Andy says, "You know it's funny, when I was a kid, I used to jump off roofs that high, and I never got hurt." Why do you think Andy was more at risk of a fracture as an adult than he was as a boy?

*Answers at www.mhhe.com/saladin4*

# www.mhhe.com/saladin4

*The textbook website provides a wealth of interactive study materials fully organized and integrated by chapter. You will find practice quizzes, labeling exercises, and much more that will complement your learning and understanding of anatomy and physiology. The website also includes tools designed to enhance your* **Anatomy & Physiology** | **REVEALED** *experience.*

*Lateral view of the knee (colorized MRI)*

# JOINTS

## CHAPTER OUTLINE

## INSIGHTS

## Brushing Up

To understand this chapter, it is important that you understand or brush up on the following concepts:

- Anatomical planes (p. 31)
- Names of all major bones (table 8.1, p. 242; fig. 8.1, p. 243)
- Surface features of bones, especially of their articular surfaces (table 8.2, p. 244)

In order for the skeleton to serve the purposes of protection and movement, the bones must be joined together. A **joint,** or **articulation,** is any point at which two bones meet, regardless of whether it is movable. Your knee, for example, is a very movable joint, whereas the skull sutures described in chapter 8 are immovable joints.

The science of joint structure, function, and dysfunction is called **arthrology.** The study of musculoskeletal movement is **kinesiology** (kih-NEE-see-OL-oh-jee). Kinesiology is a branch of **biomechanics,** which deals with a broad range of motions and mechanical processes in the body, including the physics of blood circulation, respiration, and hearing.

This chapter describes the joints of the skeleton and the types of joint movements relevant to the actions of skeletal muscles described in chapter 10.

# Joints and Their Classification

### Objectives

When you have completed this section, you should be able to

- explain what joints are, how they are named, and what functions they serve;

- name and describe the four major classes of joints;

- describe the three types of fibrous joints and give an example of each;

- distinguish between the three types of sutures;

- describe the two types of cartilaginous joints and give an example of each; and

- name some joints that become synostoses as they age.

Joints such as the shoulder, elbow, and knee are remarkable specimens of biological design—self-lubricating, almost frictionless, and able to bear heavy loads and withstand compression while executing smooth and precise movements (fig. 9.1). Yet, it is equally important that other joints be less movable or even immovable. Such joints are better able to support the body and provide protection for delicate organs. The vertebral column, for example, must provide a combination of support and flexibility; thus its joints are only moderately movable. The immovable joints between the cranial bones afford the best possible protection for the brain and sense organs.

**FIGURE 9.1  Joint Flexibility.**  This gymnast demonstrates the flexibility, precision, and weight-bearing capacity of the body's joints.

The name of a joint is typically derived from the names of the bones involved. For example, the *atlanto-occipital joint* is where the occipital condyles meet the atlas, the *humeroscapular joint* is where the humerus meets the scapula, and the *coxal joint* is where the femur meets the os coxae.

Joints are classified according to the manner in which the adjacent bones are bound to each other, with corresponding differences in how freely the bones can move. Authorities differ in their classification schemes, but one common view places the joints in four major categories: *bony, fibrous, cartilaginous,* and *synovial joints.* This section will describe the first three of these and the subclasses of each. The remainder of the chapter will then be concerned primarily with synovial joints.

## BONY JOINTS

A **bony joint,** or **synostosis**[1] (SIN-oss-TOE-sis), is an immovable joint formed when the gap between two bones ossifies and they become, in effect, a single bone. Bony joints can form by ossification of either fibrous or cartilaginous joints. An infant is born with right and left frontal and mandibular bones, for example, but these soon fuse seamlessly into a single frontal bone and mandible. In old age, some cranial sutures become obliterated by ossification and the adjacent cranial bones, such as the parietal bones, fuse. The epiphyses and diaphyses of the long bones are joined by cartilaginous joints in childhood and adolescence, and these become synostoses in early adulthood. The attachment of the first rib to the sternum also becomes a synostosis with age.

## FIBROUS JOINTS

A **fibrous joint** is also called a **synarthrosis**[2] (SIN-ar-THRO-sis) or **synarthrodial joint.** It is a point at which adjacent bones are bound by collagen fibers that emerge from one bone, cross the space between them, and penetrate into the other (fig. 9.2). There are three kinds of fibrous joints: *sutures, gomphoses,* and *syndesmoses.* In sutures and gomphoses, the fibers are very short and allow for little or no movement. In syndesmoses, the fibers are longer and the attached bones are more movable.

### Sutures

**Sutures** are immovable fibrous joints that closely bind the bones of the skull to each other; they occur nowhere else.

[1] *syn* = together + *ost* = bone + *osis* = condition

[2] *syn* = together + *arthr* = joined + *osis* = condition

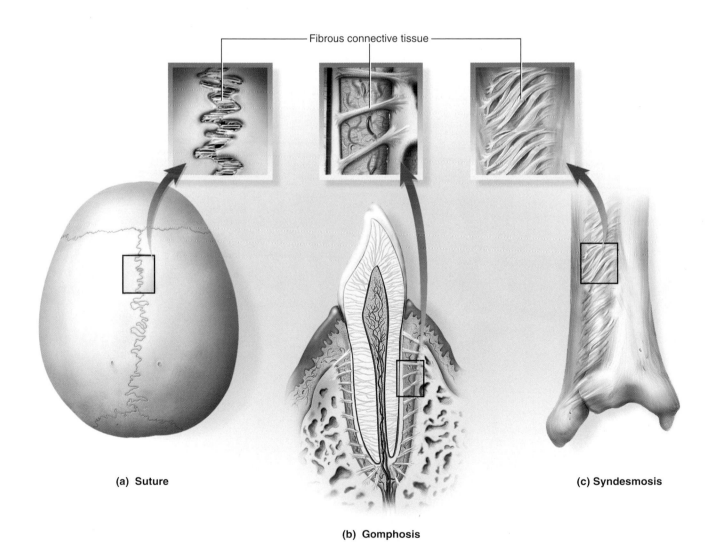

**FIGURE 9.2  Fibrous Joints.**  (a) A suture between the parietal bones. (b) A gomphosis between a tooth and the jaw. (c) A syndesmosis between the tibia and fibula.

In chapter 8 we did not take much notice of the differences between one suture and another, but some differences may have caught your attention as you studied the diagrams in that chapter or examined laboratory specimens. Sutures can be classified as *serrate, lap,* and *plane sutures.* Readers with some knowledge of woodworking may recognize that the structures and functional properties of these sutures have something in common with basic types of carpentry joints (fig. 9.3).

**Serrate sutures** appear as wavy lines along which the adjoining bones firmly interlock with each other by their serrated margins, like pieces of a jigsaw puzzle. Serrate sutures are analogous to a dovetail wood joint. Examples include the coronal, sagittal, and lambdoid sutures that border the parietal bones.

**Lap (squamous) sutures** occur where two bones have overlapping beveled edges, like a miter joint in carpentry. On the surface, a lap suture appears as a relatively smooth (nonserrated) line. An example is the squamous suture between the temporal and parietal bones.

**Plane (butt) sutures** occur where two bones have straight, nonoverlapping edges. The two bones merely border on each other, like two boards glued together in a butt joint. This type of suture is seen between the palatine processes of the maxillae in the roof of the mouth.

## Gomphoses

Even though the teeth are not bones, the attachment of a tooth to its socket is classified as a joint called a **gomphosis** (gom-FOE-sis). The term refers to its similarity to a nail hammered into wood.[3] The tooth is held firmly in place by a fibrous **periodontal ligament,** which consists of collagen fibers that extend from the bone matrix of the jaw into the dental tissue (see fig. 9.2b). The periodontal ligament allows the tooth to move or give a little under the stress of chewing. This allows us to sense how hard we are biting or to sense a particle of food stuck between the teeth.

---

[3]*gomph* = nail, bolt + *osis* = condition

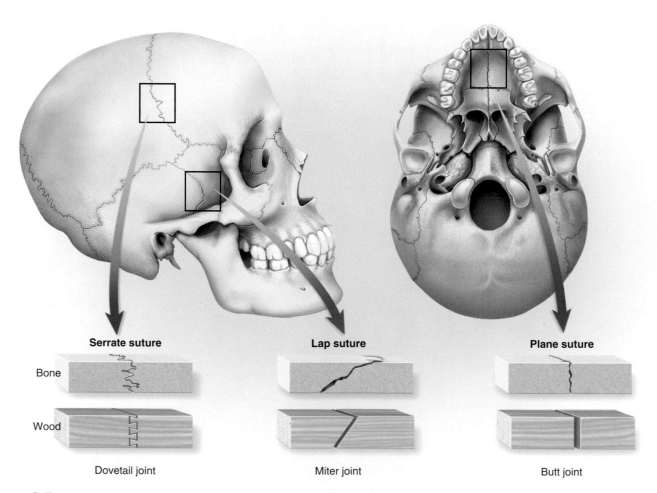

**Figure 9.3  Sutures.**   Serrate, lap, and plane sutures compared to some common wood joints.

## Syndesmoses

A **syndesmosis**[4] (SIN-dez-MO-sis) is a fibrous joint at which two bones are bound by longer collagenous fibers than in a suture or gomphosis, giving the bones more mobility. While the range of motion differs greatly among syndesmoses, all of them are more mobile than sutures or gomphoses. One of the less movable syndesmoses is the joint that binds the distal ends of the tibia and fibula together, side by side. A more movable one exists between the shafts of the radius and ulna, which are joined by a broad fibrous sheet called an **interosseous membrane** that allows for movement such as pronation and supination of the forearm (see fig. 9.2c).

[4]*syn* = together + *desm* = band + *osis* = condition

## CARTILAGINOUS JOINTS

A **cartilaginous joint** is also called an **amphiarthrosis**[5] (AM-fee-ar-THRO-sis) or **amphiarthrodial joint.** In these joints, two bones are linked by cartilage (fig. 9.4). The two types of cartilaginous joints are *synchondroses* and *symphyses*.

### Synchondroses

A **synchondrosis**[6] (SIN-con-DRO-sis) is a joint in which the bones are bound by hyaline cartilage. An example is the temporary joint between the epiphysis and diaphysis of a long bone in a child, formed by the cartilage of the epiphyseal plate. Another is the attachment of the first rib to the sternum by a hyaline costal cartilage (fig. 9.4a).

[5]*amphi* = on all sides + *arthr* = joined + *osis* = condition
[6]*syn* = together + *chondr* = cartilage + *osis* = condition

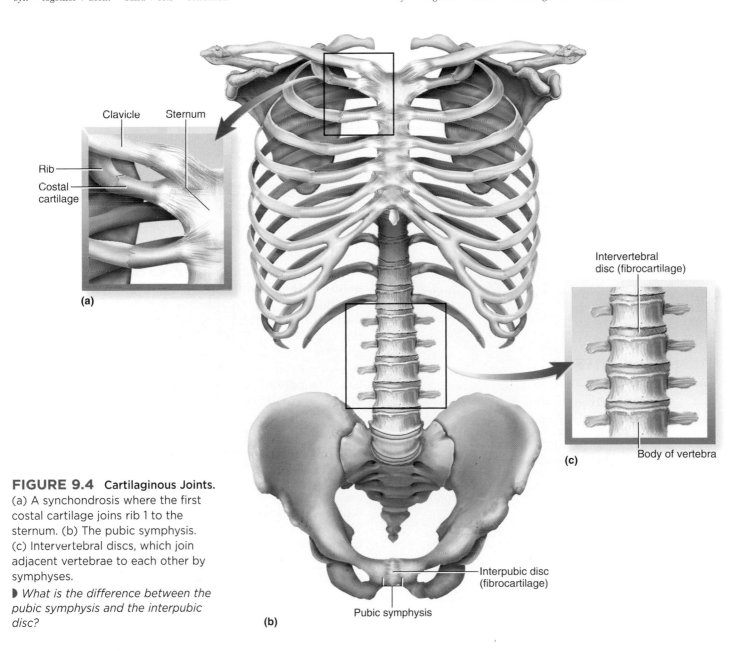

**FIGURE 9.4  Cartilaginous Joints.** (a) A synchondrosis where the first costal cartilage joins rib 1 to the sternum. (b) The pubic symphysis. (c) Intervertebral discs, which join adjacent vertebrae to each other by symphyses.

▶ *What is the difference between the pubic symphysis and the interpubic disc?*

## Symphyses

In a **symphysis**[7] (SIM-fih-sis), two bones are joined by fibro-cartilage (fig. 9.4b, c). One example is the pubic symphysis, in which the right and left pubic bones are joined by the cartilaginous interpubic disc. Another is the joint between the bodies of two vertebrae, united by an intervertebral disc. The surface of each vertebral body is covered with hyaline cartilage. Between the vertebrae, this cartilage becomes infiltrated with collagen bundles to form fibrocartilage. Each intervertebral disc permits only slight movement between adjacent vertebrae, but the collective effect of all 23 discs gives the spine considerable flexibility.

> **Think About It**
>
> *The intervertebral joints are symphyses only in the cervical through the lumbar region. How would you classify the intervertebral joints of the sacrum and coccyx in a middle-aged adult?*

### Before You Go On

*Answer the following questions to test your understanding of the preceding section:*

1. *What is the difference between arthrology and kinesiology?*
2. *Distinguish between a synostosis, synarthrosis, and amphiarthrosis.*
3. *Define suture, gomphosis, and syndesmosis, and explain what these three joints have in common.*
4. *Name the three types of sutures and describe how they differ.*
5. *Name two synchondroses and two symphyses.*
6. *Give some examples of joints that become synostoses with age.*

# Synovial Joints

### Objectives

When you have completed this section, you should be able to

- identify the anatomical components of a typical synovial joint;
- classify any given joint action as a first-, second-, or third-class lever;
- explain how mechanical advantage relates to the power and speed of joint movement;
- discuss the factors that determine a joint's range of motion;

- describe the primary axes of rotation that a bone can have and relate this to a joint's degrees of freedom;
- name and describe six classes of synovial joints; and
- use the correct standard terminology for various joint movements.

A **synovial** (sih-NO-vee-ul) **joint** can be defined as one in which two bones are separated by a film of slippery *synovial fluid*. They are the body's most familiar and freely movable joints—for example, the jaw, elbow, hip, and knee joints. They are the most important joints for such professionals as physical and occupational therapists, athletic coaches, nurses, and fitness trainers to understand well. Their mobility makes the synovial joints especially important to the quality of life. Reflect, for example, on the performance extremes of a young athlete, the decline in flexibility that comes with age, and the crippling effect of rheumatoid arthritis. The rest of this chapter is concerned with synovial joints.

## GENERAL ANATOMY

The study of dry bones and models in the laboratory can easily give the impression that the knee, elbow, or hip is a point where one bone rubs against another. In the living body, however, the bones do not touch each other; rather, fluid and soft tissues separate them and hold them in proper alignment.

Synovial joints are the most structurally complex of all joints. Figure 9.5 shows a relatively simple example, the interphalangeal joint between two bones of a finger. The bones are separated by a space called the **joint (articular) cavity,** containing the synovial fluid. **Synovial fluid** is a lubricant rich in albumin and hyaluronic acid. It is named for the similarity of its viscous, slippery texture to raw egg white.[8] It nourishes the joint cartilages and removes their wastes, and it contains phagocytes that clean up tissue debris resulting from cartilage wear and tear. The adjoining bone surfaces are each covered with **articular cartilage,** a layer of hyaline cartilage about 2 mm thick in young, healthy joints. The cartilages and synovial fluid make joint movements almost friction-free.

A **joint (articular) capsule** encloses the cavity and retains the fluid. It consists of an outer *fibrous capsule* continuous with the periosteum of the adjoining bones, and an inner *synovial membrane* of areolar tissue, which secretes the synovial fluid.

In a few joints, cartilage grows inward from the joint capsule and forms a pad called the **articular disc,** which separates the two bones and divides the joint cavity into two spaces. Articular discs occur in the temporomandibu-

---

[7]*sym* = together + *physis* = growth

[8]*ovi* = egg

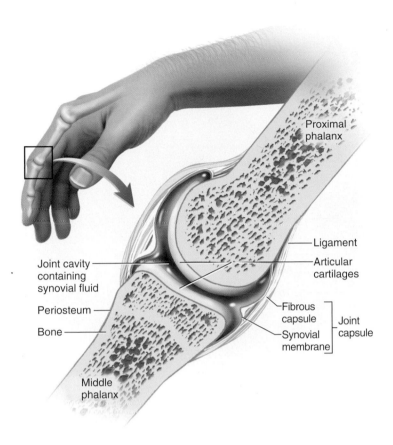

**FIGURE 9.5** Structure of a Simple Synovial Joint.

▶ *Why is a meniscus unnecessary in an interphalangeal joint?*

lar, sternoclavicular, acromioclavicular, and distal radioulnar joints. In the knee, two C-shaped cartilages called **menisci**[9] face each other across the joint. Unlike an articular disc, a meniscus extends only partway across a joint; there is a gap between them (see fig. 9.28d). Functionally, however, menisci are the same as articular discs elsewhere. All of these cartilage pads absorb shock and pressure, guide the moving bones across each other, reduce the chance of dislocation, and distribute force across the entire joint instead of at just a few points of contact.

> **Think About It!**
>
> *Given that joints are typically named for the two articulating bones or bone parts, describe exactly where you would find the temporomandibular, sternoclavicular, acromioclavicular, and distal radioulnar joints mentioned in the preceding paragraph.*

The accessory structures of a synovial joint include tendons, ligaments, and bursae. A **tendon** is a strip or sheet of tough collagenous connective tissue that attaches a muscle to a bone. A **ligament** is a similar tissue that attaches one bone to another. Several ligaments are named and illustrated later in this chapter, and tendons are more fully considered in chapter 10 along with the gross anatomy of the muscles.

A **bursa**[10] is a fibrous sac filled with synovial fluid, located between adjacent muscles or where a tendon passes over a bone (see fig. 9.23). Bursae cushion muscles, help tendons slide more easily over the joints, and sometimes enhance the mechanical effect of a muscle by modifying the direction in which its tendon pulls. **Tendon sheaths** are elongated cylindrical bursae wrapped around a tendon, seen especially in the hand and foot (fig. 9.6).

## JOINTS AND LEVER SYSTEMS

Long bones act as levers to enhance the speed or power of limb movements. A lever is any elongated, rigid object that rotates around a fixed point called the fulcrum (fig. 9.7). Rotation occurs when an effort applied to one point on the lever overcomes a resistance (load) at some other point. The portion of a lever from the fulcrum to the point of effort is called the **effort arm,** and the part from the fulcrum to the point of resistance is called the **resistance arm.** In skeletal anatomy, the fulcrum is a joint; the effort is applied by a muscle; and the resistance can be an object against which the body is working (as in weight lifting), the weight of the limb itself, or the tension in an opposing muscle.

---

**INSIGHT 9.1**    Clinical Application

### Exercise and Articular Cartilage

When synovial fluid is warmed by exercise, it becomes thinner (less viscous) and more easily absorbed by the articular cartilage. The cartilage then swells and provides a more effective cushion against compression. For this reason, a warm-up period before vigorous exercise helps protect the articular cartilage from undue wear and tear.

Because cartilage is nonvascular, repetitive compression during exercise is important to its nutrition and waste removal. Each time a cartilage is compressed, fluid and metabolic wastes are squeezed out of it. When weight is taken off the joint, the cartilage absorbs synovial fluid like a sponge, and the fluid carries oxygen and nutrients to the chondrocytes. Without exercise, articular cartilages deteriorate more rapidly from lack of nutrition, oxygenation, and waste removal.

Weight-bearing exercise builds bone mass and strengthens the muscles that stabilize many of the joints, thus reducing the risk of joint dislocations. Excessive joint stress, however, can hasten the progression of osteoarthritis by damaging the articular cartilage (see Insight 9.4, p. 315). Swimming is a good way of exercising the joints with minimal damage.

---

[9]*men* = moon, crescent + *isc* = little

[10]*bursa* = purse

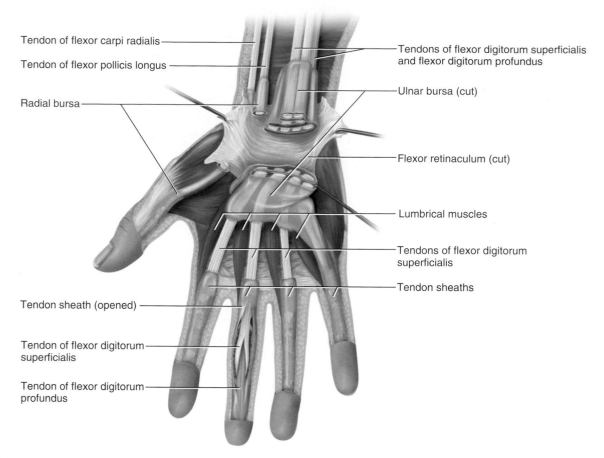

**FIGURE 9.6**   Tendon Sheaths and Other Bursae in the Hand and Wrist.

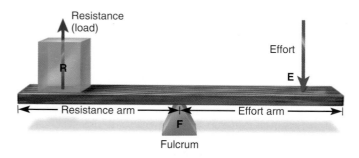

**FIGURE 9.7**   **The Basic Components of a Lever.**   This example is a first-class lever.

▶ *What would be the mechanical advantage of the lever shown here? Where would you put the fulcrum to increase the mechanical advantage (MA) without changing the lever class?*

## Mechanical Advantage

The advantage conferred by a lever can be either of two kinds: (1) to exert more force against a resisting object than the force applied to the lever—for example, when a crowbar is used to move a heavy object; or (2) to move the resisting object farther or faster than the effort arm is moved, as in rowing a boat. A single lever cannot confer both advantages. There is a trade-off between force on one

hand and speed or distance on the other—as one increases, the other decreases.

The **mechanical advantage (MA)** of a lever is the ratio of its output force to its input force. If $L_E$ is the length of the effort arm and $L_R$ is the length of the resistance arm, $MA = L_E/L_R$. If MA is greater than 1.0, the lever produces more force, but less speed or distance, than the force exerted on it. If MA is less than 1.0, the lever produces more speed or distance, but less force, than the input.

Consider the forearm, for example (fig. 9.8a). Its resistance arm is longer than its effort arm, so we know from the preceding formula that MA must be less than 1.0. The figure shows some representative values for $L_E$ and $L_R$ that yield MA = 0.15. The biceps brachii muscle puts more power into the lever than we get out of it, but the hand moves farther and faster than the point where the biceps tendon inserts on the radius. Most musculoskeletal levers operate with an MA much less than 1, but figure 9.8b shows a case with MA greater than 1.

In chapter 10, you will find that two or more muscles often act on the same joint, seemingly producing the same effect. This may seem redundant, but it makes sense if the tendinous insertions of the muscles are at different points on a bone and produce different mechanical advantages. A sprinter taking off from the starting line, for example,

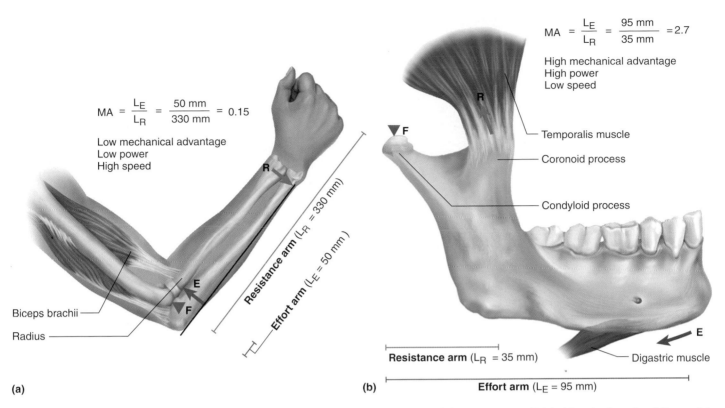

**FIGURE 9.8  Mechanical Advantage (MA).** MA is calculated from the length of the effort arm divided by the length of the resistance arm. (a) The forearm acts as a third-class lever during flexion of the elbow. (b) The mandible acts as a second-class lever when the jaw is forcibly opened. The digastric muscle and others provide the effort, while tension in the temporalis muscle and others provide resistance.

uses "low-gear" (high-MA) muscles that do not generate maximum speed, but have the power to overcome the inertia of the body. The runner then "shifts into high gear" by using muscles with different insertions that have a lower mechanical advantage but produce more speed. This is analogous to the way an automobile transmission uses one gear to get a car moving and other gears to cruise at higher speeds.

## Types of Levers

Physicists recognize three classes of levers that differ with respect to the relative locations of the fulcrum (F), effort (E), and resistance (R) (fig. 9.9).

1. A **first-class lever** is one with the fulcrum in the middle (RFE). A seesaw is a common example. An anatomical example is the atlanto-occipital joint of the neck, where the muscles of the back of the neck pull down on the occipital bone of the skull and oppose the tendency of the head to tip forward. Loss of muscle tone here can be embarrassing if you nod off in class.

2. A **second-class lever** has the resistance in the middle (FRE). Lifting the handles of a wheelbarrow, for example, causes it to pivot on the axle of the wheel at the opposite end and lift a load in the middle. The

mandible acts as a second-class lever when the *digastric muscle* pulls down on the chin to open the mouth. The fulcrum is the temporomandibular (jaw) joint, the effort is applied at the chin by the digastric muscle, and the resistance is the tension in muscles such as the *temporalis,* which tends to hold the mouth shut. This example is upside down relative to a wheelbarrow, but the mechanics remain the same.

3. In a **third-class lever,** the effort is applied between the fulcrum and resistance (REF). A pair of forceps, for example, consists of two third-class levers joined at the fulcrum. Most musculoskeletal levers are third-class. The forearm acts as a third-class lever when you flex your elbow. The fulcrum is the joint between the ulna and humerus, the effort is applied in part by the biceps brachii muscle, and the resistance can be any weight in the hand or the weight of the forearm itself.

The classification of a lever changes as it makes different actions. We use the forearm as a third-class lever when we flex the elbow, as in weight lifting; but we use it as a first-class lever when we extend it, as in hammering nails. The mandible is a second-class lever when we open the mouth and a third-class lever when we close it to bite off a piece of food.

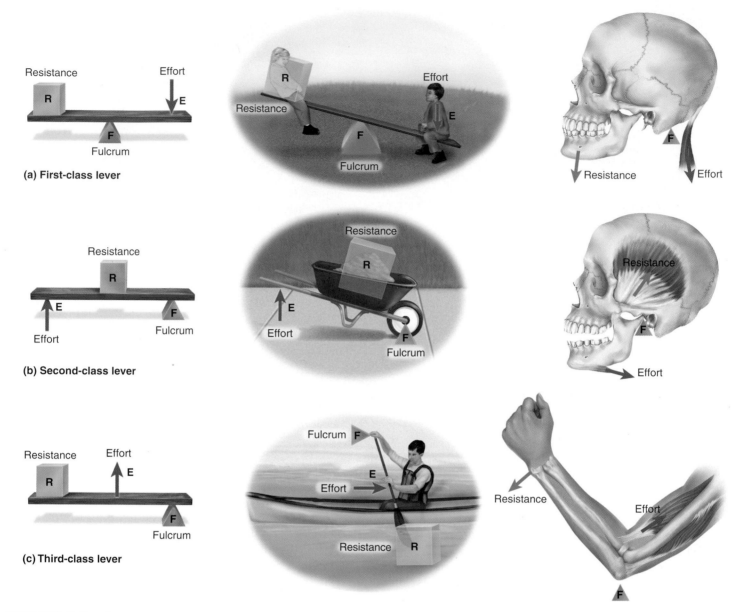

**FIGURE 9.9   The Three Classes of Levers.** *Left:* The lever classes defined by the relative positions of the resistance (load), fulcrum, and effort. *Center:* Mechanical examples. *Right:* Anatomical examples. (a) Muscles of the back of the neck pull the skull downward to oppose the tendency of the head to drop forward. The fulcrum is the occipital condyles. (b) To open the mouth, the digastric muscle pulls down on the chin. It is resisted by the temporalis muscle on the side of the head. The fulcrum is the temporomandibular (jaw) joint. (c) In flexing the elbow, the biceps brachii muscle exerts an effort on the radius. Resistance is provided by the weight of the forearm or anything held in the hand. The fulcrum is the elbow joint.

## Range of Motion

One aspect of joint performance and physical assessment of a patient is a joint's flexibility, or **range of motion (ROM)**—the degrees through which a joint can move. The knee, for example, can flex through an arc of 130° to 140°, the metacarpophalangeal joint of the index finger about 90°, and the ankle about 74°. The ROM of a joint is normally determined by the following factors:

- **Structure of the articular surfaces of the bones.** In many cases, joint movement is limited by the shapes

of the bone surfaces. For example, you cannot straighten your elbow beyond 180° or so because, as it straightens, the olecranon of the ulna swings into the olecranon fossa of the humerus and the fossa prevents it from moving any farther. In the jaw, the condyle of the mandible is elongated and fits into a complementary elongated depression, the mandibular fossa, in the temporal bone. This *temporomandibular joint* (TMJ) hinges quite freely in the sagittal plane, but prevents the mandible from swiveling very far from side to side.

- **Strength and tautness of ligaments and joint capsules.** Some bone surfaces impose little if any limitation on joint movement. The articulations of the phalanges are an example; as one can see by examining a dry skeleton, an interphalangeal joint can bend through a broad arc. In life, however, these bones are joined by ligaments which limit their movement. As you flex one of your knuckles, ligaments on the anterior (palmar) side of the joint go slack, but ligaments on the posterior (dorsal) side tighten and prevent the joint from flexing beyond 90° or so. The knee is another case in point. In kicking a football, the knee rapidly extends to about 180°, but it can go no farther. Its motion is limited in part by a *cruciate ligament* and other knee ligaments described later. Gymnasts, dancers, and acrobats increase the ROM of their synovial joints by gradually stretching their ligaments during training. "Double-jointed" people have unusually large ROMs at some joints, not because the joint is actually double or fundamentally different from normal in its anatomy, but because the ligaments are unusually long or slack.

- **Action of the muscles and tendons.** Extension of the knee is also limited by the *hamstring muscles* on the posterior side of the thigh. In many other joints, too, pairs of muscles oppose each other and moderate the speed and range of joint motion. Even a resting muscle maintains a state of tension called *muscle tone,* which serves in many cases to stabilize a joint. One of the major factors preventing dislocation of the shoulder joint, for example, is tension in the *biceps brachii* muscle, whose tendons cross the joint, insert on the scapula, and hold the head of the humerus against the glenoid cavity. The nervous system continually monitors and adjusts joint angles and muscle tone to maintain joint stability and limit unwanted movements.

## Axes of Rotation

In solid geometry, we recognize three mutually perpendicular axes, *x, y,* and *z.* In anatomy, these correspond to the transverse, frontal, and sagittal planes of the body. Just as we can describe any point in space by its *x, y,* and *z* coordinates, we can describe any joint movement by reference to these three anatomical planes.

A moving bone has a relatively stationary **axis of rotation** that passes through the bone in a direction perpendicular to the plane of movement. Think of a door for comparison; it moves horizontally as it opens and closes and it rotates on hinges that are oriented on the vertical axis. Now consider the shoulder joint, where the convex head of the humerus inserts into the concave glenoid cavity of the scapula. If you raise your arm to one side of your body (*abduction,* a motion defined later), the humerus rotates on an axis that passes from anterior to posterior;

the arm rises in the frontal plane whereas its axis of rotation is in the sagittal plane (fig. 9.10a). If you lift your arm to point straight in front of you, it moves through the sagittal plane whereas its axis of rotation is on the frontal plane, passing through the shoulder from lateral to medial (fig. 9.10b). And if you swing your arm in a horizontal arc, for example to fold it across your chest, the arm rotates in the transverse plane and its axis of rotation passes vertically through the joint (fig. 9.10c).

Because the arm can move in all three anatomical planes, the shoulder joint is said to have three **degrees of freedom,** or to be a **multiaxial** joint. Other joints move through only one or two planes; they have one or two degrees of freedom and are called **monaxial** and **biaxial** joints, respectively. Degrees of freedom are a factor used in classifying the synovial joints.

## Classes of Synovial Joints

There are six fundamental types of synovial joints, distinguished by the shapes of their articular surfaces and their degrees of freedom. We will begin by looking at these six types in simple terms, but then see that this is an imperfect classification for reasons discussed at the end. All six types can be found in the upper limb (fig. 9.11). They are listed here in descending order of mobility: one multiaxial type (ball-and-socket), three biaxial types (condyloid, saddle, and gliding), and two monaxial types (hinge and pivot).

1. **Ball-and-socket joints.** These are the shoulder and hip joints: the only multiaxial joints in the body. In both cases, one bone (the humerus and femur) has a smooth hemispherical head that fits into a cuplike socket on the other (the glenoid cavity of the scapula and the acetabulum of the os coxae).

2. **Condyloid (ellipsoid) joints.** These joints exhibit an oval convex surface on one bone that fits into a similarly shaped depression on the other. The radiocarpal joint of the wrist and metacarpophalangeal (MET-uh-CAR-po-fah-LAN-jee-ul) joints at the bases of the fingers are examples. They are biaxial joints, capable of movement in two planes. To demonstrate this, hold your hand with the palm facing you. Make a fist, and these joints flex in the sagittal plane. Fan your fingers apart, and they move in the frontal plane.

3. **Saddle joints.** Here, both bones have a saddle-shaped surface—concave in one direction (like the front to rear curvature of a horse's saddle) and convex in the other (like the left to right curvature of a saddle). The clearest example of this is the trapeziometacarpal joint at the base of the thumb. Saddle joints are biaxial. The thumb, for example, moves in a frontal plane when you spread the fingers apart, and in a sagittal plane when you move it as if to

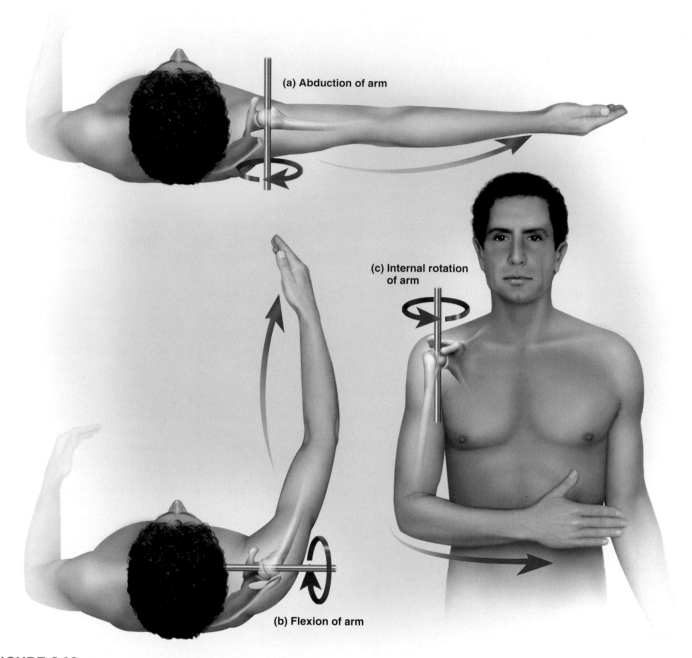

**FIGURE 9.10    Axes of Joint Rotation.**    All three axes are represented in movements of the multiaxial ball-and-socket joint of the shoulder.

grasp a tool such as a hammer. This range of motion gives us and other primates that anatomical hallmark, the opposable thumb. Another saddle joint is the sternoclavicular joint, where the clavicle articulates with the sternum. The clavicle moves vertically in the frontal plane at this joint when you lift a suitcase, and moves horizontally in the transverse plane when you reach forward to push open a door.

4. **Gliding joints.** Here the bone surfaces are flat or only slightly concave and convex. The adjacent bones slide over each other and have relatively limited

movement; they are amphiarthroses in contrast to the other types listed here, which are diarthroses. Gliding joints are found between the carpal bones of the wrist, the tarsal bones of the ankle, and the articular processes of the vertebrae. Their movements, although slight, are complex. They are usually biaxial. For example, when the head is tilted forward and back, the articular facets of the vertebrae slide anteriorly and posteriorly; when the head is tilted from side to side, the facets slide laterally. Although any one joint moves only slightly, the combined

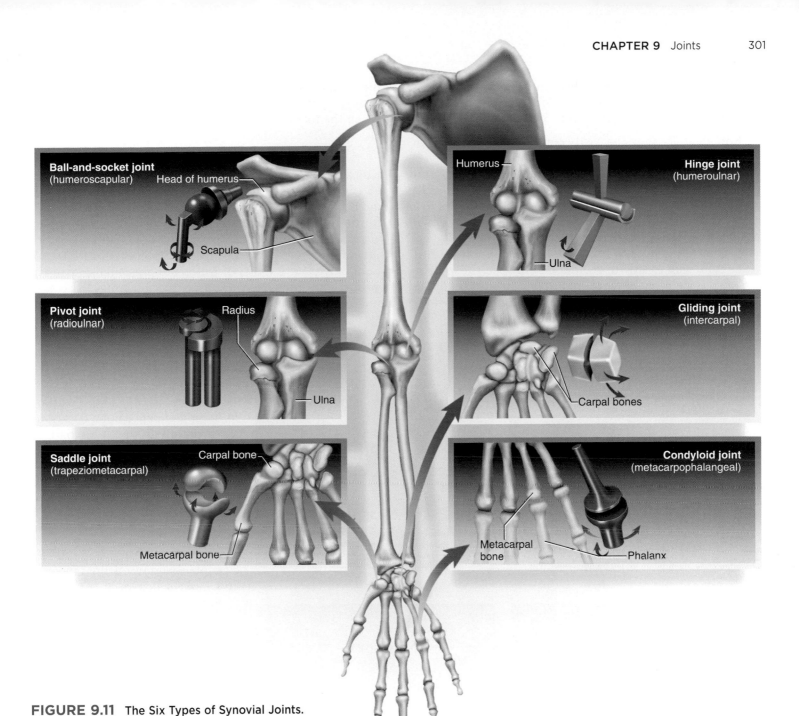

**FIGURE 9.11  The Six Types of Synovial Joints.**
All six have representatives in the forelimb. Mechanical models show the types of motion possible at each joint.

action of the many joints in the wrist, ankle, and vertebral column allows for a significant amount of overall movement.

5. **Hinge joints.** These are essentially monaxial joints, moving freely in one plane with very little movement in any other. Examples include the elbow, knee, and interphalangeal (finger and toe) joints. In these cases, one bone has a convex (but not hemispherical) surface, such as the trochlea of the humerus and the condyles of the femur. This fits into a concave depression on the other bone, such as the trochlear notch of the ulna and the condyles of the tibia.

6. **Pivot joints.** These are monaxial joints in which a bone spins on its longitudinal axis. There are two principal examples: the atlantoaxial joint between the first two vertebrae, and the radioulnar joint at the elbow. At the atlantoaxial joint, the dens of the axis projects into the vertebral foramen of the atlas and is held against the anterior arch of the atlas by the transverse ligament (see fig. 8.24). As the head rotates left and right, the skull and atlas pivot around the dens. At the radioulnar joint, the anular ligament of the ulna wraps around the neck of the radius. During pronation and supination of the forearm, the disclike

radial head pivots like a wheel turning on its axle. The edge of the wheel spins against the radial notch of the ulna like a car tire spinning in snow.

Some joints cannot be easily classified into any one of these six categories. The temporomandibular joint, for example, has some aspects of condyloid, hinge, and gliding joints. It clearly has an elongated condyle where it meets the temporal bone of the cranium, but it moves in a hingelike fashion when the mandible moves up and down in speaking, biting, and chewing; it glides slightly forward when the jaw juts (protracts) to take a bite; and it glides from side to side to grind food between the molars. The knee is a classic hinge joint, but has an element of the pivot type; when we lock our knees to stand more effortlessly, the femur pivots slightly on the tibia. The humeroradial joint (between humerus and radius) acts as a hinge joint when the elbow flexes and a pivot joint when the forearm pronates.

## MOVEMENTS OF SYNOVIAL JOINTS

Kinesiology, physical therapy, and other medical and scientific fields have a specific vocabulary for the movements of synovial joints. The following terms form a basis for describing the muscle actions in chapter 10 and may also be indispensible to your advanced coursework or intended career. This section introduces the terms for joint movements, many of which are presented in pairs or groups with opposite or contrasting meanings. This section relies on familiarity with the three cardinal anatomical planes and the directional terms in atlas A. All directional terms used here refer to a person in standard anatomical position. When one is standing in anatomical position, each joint is said to be in its **zero position.** Joint movements can be described as deviation from the zero position or returning to it.

### Flexion and Extension

**Flexion** (fig. 9.12) is a movement that decreases a joint angle, usually in the sagittal plane. This is particularly common at hinge joints—for example, bending of the elbow or knee—but it occurs in other types of joints as well. For example, if you hold out your hands with the palms up, flexion of the wrist tips your palms toward you. The meaning of the *flexion* is perhaps least obvious in the ball-and-socket joints of the shoulder and hip. At the shoulder, it means to raise your arm as if pointing at something directly in front of you or to continue in that arc and point toward the sky. At the hip, it means to raise the thigh, for example to place your foot on the next higher step when ascending a flight of stairs.

**Extension** is a movement that straightens a joint and generally returns a body part to the zero position—for example, straightening the elbow, wrist, or knee, or returning the arm or thigh back to zero position. In stair climbing, both the hip and knee extend when lifting the body to the next higher step.

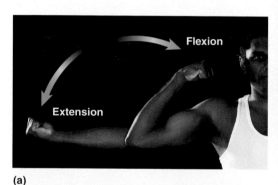
(a)

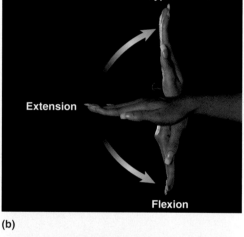
(b)

(c)

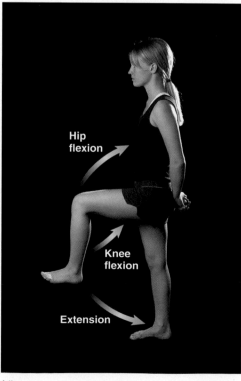
(d)

**FIGURE 9.12**   Flexion and Extension.
(a) Flexion and extension of the elbow.
(b) Flexion, extension, and hyperextension of the wrist. (c) Flexion and hyperextension of the shoulder. (d) Flexion and extension of the hip and knee.

Extreme extension of a joint, beyond the zero position, is called **hyperextension.**[11] For example, if you hold your hand in front of you with the palm down, then raise the back of your hand as if you were admiring a new ring, you hyperextend the wrist. Hyperextension of the upper or lower limb means to move the limb to a position behind the frontal plane of the trunk, as if reaching around with your arm to scratch your back. Each backswing of the lower limb when you walk hyperextends the hip.

Flexion and extension occur at nearly all diarthroses, but hyperextension is limited to only a few. At most diarthroses, ligaments or bone structure prevents hyperextension.

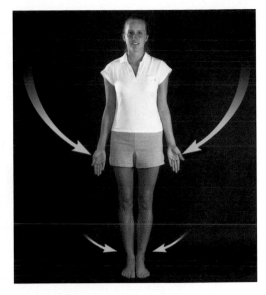

**(a) Abduction**                                    **(b) Adduction**

**FIGURE 9.13**   Abduction and Adduction.

## Abduction and Adduction

**Abduction**[12] (ab-DUC-shun) (fig. 9.13a) is the movement of a body part in the frontal plane away from the midline of the body—for example, moving the feet apart to stand spread-legged, or raising an arm to one side of the body. **Adduction**[13] (fig. 9.13b) is movement in the frontal plane back toward the midline. Some joints can be **hyperadducted,** as when you stand with your ankles crossed, cross your fingers, or hyperadduct the shoulder to stand with your elbows straight and your hands clasped below your waist. You **hyperabduct** the arm if you raise it high enough to cross slightly over the front or back of your head.

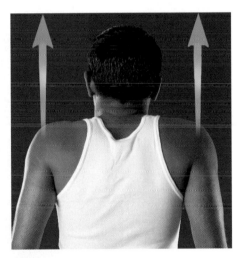
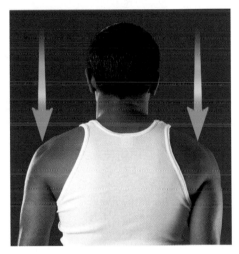

**(a) Elevation**                                    **(b) Depression**

**FIGURE 9.14**   Elevation and Depression.

## Elevation and Depression

**Elevation** (fig. 9.14a) is a movement that raises a body part vertically in the frontal plane. **Depression** (fig. 9.14b) lowers a body part in the same direction. For example, to lift a heavy suitcase from the floor, you elevate your scapula; in setting it down again, you depress the scapula.

## Protraction and Retraction

**Protraction**[14] (fig. 9.15) is the anterior movement of a body part in the transverse (horizontal) plane, and **retraction**[15] is posterior movement. Your shoulder protracts, for example, when you reach in front of you to push a door open. It retracts when you return it to the resting (zero)

position or pull the shoulders back to stand at military attention. Such exercises as rowing a boat, bench presses, and push-ups involve repeated protraction and retraction of the shoulders.

## Circumduction

In **circumduction,**[16] (fig. 9.16) one end of an appendage remains fairly stationary while the other end makes a circular motion. If an artist standing at an easel reaches

---

[11]*hyper* = excessive, beyond normal
[12]*ab* = away + *duc* = to lead or carry
[13]*ad* = toward + *duc* = to lead or carry
[14]*pro* = forward + *trac* = to pull or draw
[15]*re* = back + *trac* = to pull or draw

[16]*circum* = around + *duc* = to carry, lead

forward and draws a circle on a canvas, she circumducts the upper limb; the shoulder remains stationary while the hand moves in a circle. A baseball player winding up for the pitch circumducts the upper limb in a more extreme "windmill" fashion. One can also circumduct an individual finger, the hand, the thigh, the foot, the trunk, and the head.

**(a) Protraction**

**(b) Retraction**

**FIGURE 9.15**  Protraction and Retraction.

**FIGURE 9.16**  Circumduction.

**Think About It**

*Choose any example of circumduction and explain why this motion is actually a sequence of flexion, abduction, extension, and adduction.*

## Rotation

In one sense, the term *rotation* applies to any bone turning around a fixed axis, as described earlier. But in the terminology of specific joint movements, **rotation** (fig. 9.17) is a movement in which a bone spins on its longitudinal axis. For example, if you stand with bent elbow and reach out with your right arm to grasp something on the right side of the body, your humerus spins in a motion called **lateral (external) rotation.** If you make the opposite motion, for example to place your palm against your chest, your humerus exhibits **medial (internal) rotation.** If you turn your right foot so your toes are pointing away from your left foot, and then turn it so your toes are pointing toward your left foot, your femur undergoes lateral and medial rotation, respectively. Other examples are given in the ensuing discussions of forearm and head movements.

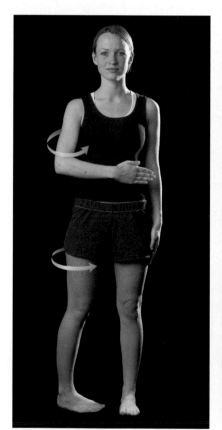

**(a) Medial (internal) rotation**            **(b) Lateral (external) rotation**

**FIGURE 9.17**  Medial (Internal) and Lateral (External) Rotation.   (a) Medial rotation of the humerus and femur. (b) Lateral rotation of the humerus and femur.

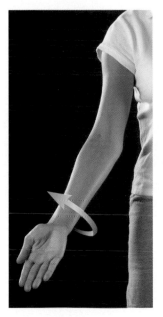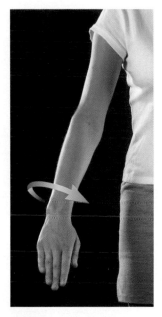

**(a) Supination**          **(b) Pronation**

**FIGURE 9.18**    Supination and Pronation.

## Supination and Pronation

Supination and pronation (fig. 9.18) are known primarily as forearm movements, but see also the later discussion of foot movements. **Supination**[17] (SOO-pih-NAY-shun) of the forearm is a movement that turns the palm to face anteriorly or upward; in anatomical position, the arm is supine and the radius is parallel to the ulna. **Pronation**[18] is the opposite movement, causing the palm to face posteriorly or downward, and the radius to cross the ulna like an X. During these movements, the concave end of the disc-shaped head of the radius spins on the capitulum of the humerus, and the edge of the disc spins in the radial notch of the ulna. The ulna remains relatively stationary.

As an aid to remembering these terms, think of it this way: You are *prone* to stand in the most comfortable position, which is with the forearm *pronated.* But if you were holding a bowl of *soup* in your hand, you would need to *supinate* the forearm to keep from spilling it.

Chapter 10 describes the muscles that perform these actions. Of these, the *supinator* is the most powerful. Supination is the type of movement you would usually make with your right hand to turn a doorknob clockwise or to drive a screw into a piece of wood. The threads of screws and bolts are designed with the relative strength of the supinator in mind, so the greatest power can be applied when driving them with a screwdriver.

---

[17]*supin* = to lay back
[18]*pron* = to bend forward

We will now consider a few body regions that combine the foregoing motions, or that have unique movements and terminology.

## Movements of the Head and Trunk

*Flexion* of the vertebral column produces forward-bending movements, as in tilting the head forward or bending at the waist in a toe-touching exercise (fig. 9.19). *Extension* of the vertebral column straightens the trunk or the neck, as in standing up or returning the head to a forward-looking (zero) position. *Hyperextension* is employed in looking up toward the sky or bending over backward.

**Lateral flexion** is tilting the head or the trunk to the right or left of the midline. Twisting at the waist or turning of the head is called **right rotation** or **left rotation** when the chest or the face turns to the right or left of the forward-facing zero position. Powerful right and left rotation at the waist is important in baseball pitching, discus throwing, and other sports.

## Movements of the Mandible

Movements of the mandible are concerned especially with biting and chewing (fig. 9.20). Imagine taking a bite of raw carrot. Most people have some degree of overbite; at rest, the upper incisors (front teeth) overhang the lower ones. For effective biting, however, the chisel-like edges of the incisors must meet. In preparation to bite, we therefore protract the mandible to bring the lower incisors forward. After the bite is taken, we retract it. To actually take the bite, we must depress the mandible to open the mouth, then elevate it so the incisors can cut off the piece of food.

Next, to chew the food, we do not simply raise and lower the mandible as if hammering away at the food between the teeth; rather, we exercise a grinding action that shreds the food between the broad, bumpy surfaces of the premolars and molars. This entails a side-to-side movement of the mandible called **lateral excursion** (movement to the left or right of the zero position) and **medial excursion** (movement back to the median, zero position).

## Movements of the Hand and Digits

The hand moves anteriorly and posteriorly by flexion and extension of the wrist (fig. 9.21). It can also move in the frontal plane. **Ulnar flexion** tilts the hand toward the little finger, and **radial flexion** tilts it toward the thumb. We often use such motions when waving hello to someone with a side-to-side wave of the hand, or when washing windows or polishing furniture.

Movements of the digits are more varied, especially those of the thumb. *Flexion* of the fingers is curling them; *extension* is straightening them. Most people cannot hyperextend their fingers. Spreading the fingers apart is *abduction,* and bringing them together again so they touch along their surfaces is *adduction.*

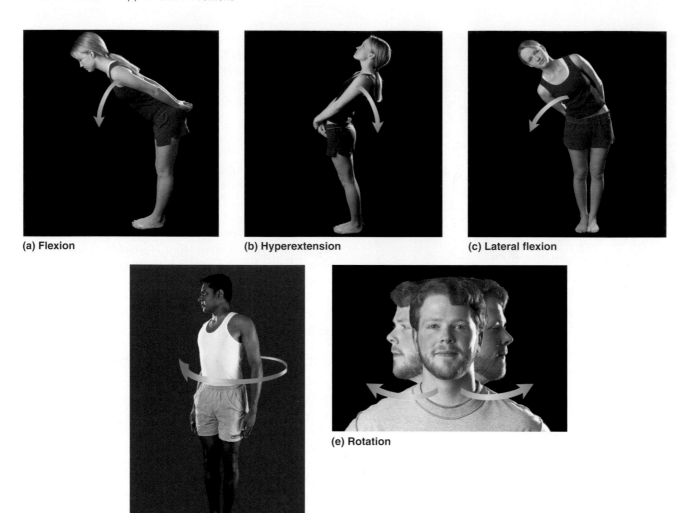

(a) Flexion

(b) Hyperextension

(c) Lateral flexion

(d) Right rotation

(e) Rotation

FIGURE 9.19    Movements of the Head and Trunk.

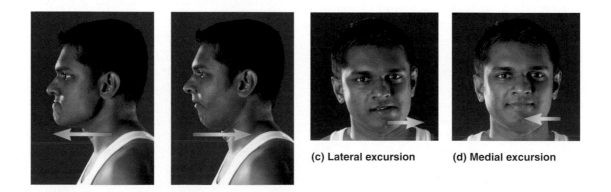

(a) Protraction

(b) Retraction

(c) Lateral excursion

(d) Medial excursion

FIGURE 9.20    Movements of the Mandible.

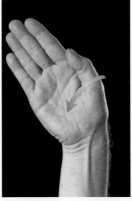
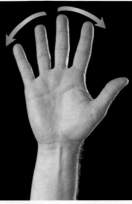

**(a) Radial flexion**    **(b) Ulnar flexion**    **(c) Abduction of fingers**

**(d) Abduction of thumb**    **(e) Opposition of thumb**

**FIGURE 9.21** **Movements of the Hand and Digits.** (a) Radial flexion of the wrist. (b) Ulnar flexion of the wrist. (c) Abduction of fingers II–V (a and b show adduction of the fingers). (d) Abduction of the thumb. (e) Opposition of the thumb (reposition is shown in a and b).

The thumb is different, however, because in embryonic development it rotates nearly 90° from the rest of the hand. If you hold your hand in a completely relaxed position, you will probably see that the plane that contains your thumb and index finger is about 90° to the plane that contains the index through little finger. Much of the terminology of thumb movement therefore differs from that of the other four fingers. *Flexion* of the thumb is bending the

joints so the tip of the thumb is directed toward the palm, and *extension* is straightening the thumb. If your place the palm of your hand on a table top with all five digits parallel and touching, the thumb is extended. Keeping your hand there, if you move your thumb away from the index finger so they form a 90° angle (but both are in the plane of the table top), the thumb is *hyperextended* at the trapeziometacarpal joint. It may seem as if that would be considered abduction, like spreading apart the other four digits. However, because of the anatomical rotation of the thumb, thumb *abduction* means, in anatomical position, to move it away from the plane of the hand so it points anteriorly (fig. 9.21d), and adduction means to bring it back to touch the base of the index finger.

Two terms are unique to the thumb: **Opposition**[19] means to move the thumb to touch the tip of any of the other four fingers. **Reposition**[20] is the return to zero position.

## Movements of the Foot

A few movement terms are also unique to the foot. **Dorsiflexion** (DOR-sih-FLEC-shun) is a movement in which the toes are elevated, as one might do in applying toenail polish (fig. 9.22). In each step you take, the foot dorsiflexes as it comes forward. This prevents you from scraping your toes on the ground and results in the characteristic *heel strike* of human locomotion when the foot touches down in front of you. **Plantar flexion** is movement of the foot so the toes point downward, as in pressing the gas pedal of a car or standing on tiptoes. This motion also produces the *toe-off* in each step you take, as the heel of the foot behind you lifts off the ground. Plantar flexion can be a very powerful motion, epitomized by high jumpers and the jump shots of basketball players.

[19]*op* = against + *posit* = to place
[20]*re* = back + *posit* = to place

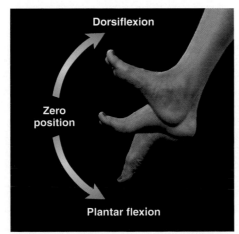
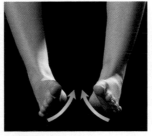
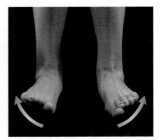

**(a) Flexion of ankle**    **FIGURE 9.22** Movements of the Foot.

**Inversion**[21] is a foot movement that tips the soles medially, somewhat facing each other, and **eversion**[22] is a movement that tips the soles laterally, away from each other. These movements are common in fast sports such as tennis and football, and sometimes cause ankle sprains. These terms also refer to congenital deformities of the feet, which are often corrected by orthopedic shoes or braces.

*Pronation* and *supination,* while used mainly for forearm movements, also apply to the feet but refer here to a more complex combination of movements. Pronation of the foot is a combination of dorsiflexion, eversion, and abduction—that is, the toes are elevated and turned away from the other foot and the sole is tilted away from the other foot. Supination of the foot is a combination of plantar flexion, inversion, and adduction—the toes are lowered and turned toward the other foot and the sole is tilted toward it. These may seem a little difficult to visualize and perform, but they are ordinary motions during walking, running, and crossing uneven surfaces such as stepping stones.

You can perhaps understand why these terms apply to the feet if you place the palms of your hands on a table and pretend they are your soles. Tilt your hands so the inner edge (thumb side) of each is raised from the table. This is like raising the medial edge of your foot from the ground, and as you can see, it involves a slight supination of your forearms. Resting your hands palms down on a table, your forearms are already pronated; but if you raise the outer edges of your hands (the little finger side), like pronating the feet, you will see that it involves a continuation of the pronation movement of the forearm.

## Before You Go On

*Answer the following questions to test your understanding of the preceding section:*

7. *Describe the roles of articular cartilage and synovial fluid in joint mobility.*

8. *Give an anatomical example of each class of levers and explain why each example belongs in that class.*

9. *Give an example of each of the six classes of synovial joints and state how many axes of rotation each example has.*

10. *Suppose you reach overhead and screw a lightbulb into a ceiling fixture. Name each joint that would be involved and the joint actions that would occur.*

11. *Where are the effort, fulcrum, and resistance in the act of dorsiflexion? What class of lever does the foot act as during dorsiflexion? Would you expect it to have a mechanical advantage greater or less than 1.0? Why?*

# Anatomy of Selected Diarthroses

### Objectives

When you have completed this section, you should be able to

- identify the major anatomical features of the shoulder, elbow, hip, and knee joints; and
- explain how the anatomical differences between these joints are related to differences in function.

We now examine the gross anatomy of certain diarthroses. It is beyond the scope of this book to discuss all of them, but the ones selected here most often require medical attention and many of them have a strong bearing on athletic performance and everyday mobility.

## THE HUMEROSCAPULAR JOINT

The **humeroscapular (glenohumeral) joint,** or shoulder joint, is where the hemispherical head of the humerus articulates with the glenoid cavity of the scapula (fig. 9.23). Together, the shoulder and elbow joints serve to position the hand for the performance of a task; without a hand, shoulder and elbow movements are almost useless. The relatively loose shoulder joint capsule and shallow glenoid cavity sacrifice stability to maximize mobility. The cavity, however, has a ring of fibrocartilage called the **glenoid labrum**[23] around its margin, making it somewhat deeper than it looks on a dried skeleton.

---

**INSIGHT 9.2**    Clinical Application

### Shoulder Dislocation

Shoulder dislocations are very painful and sometimes cause permanent damage. The most common dislocation is downward displacement of the humerus, because (1) the rotator cuff protects the joint in all directions except inferiorly, and (2) the joint is protected from above by the coracoid process, acromion, and clavicle. Dislocations most often occur when the arm is abducted and then receives a blow from above—for example, when the outstretched arm is struck by heavy objects falling off a shelf. They also occur in children who are jerked off the ground by one arm or forced to follow by a hard tug on the arm. Children are especially prone to such injury not only because of the inherent stress caused by such abuse, but also because a child's shoulder is not fully ossified and the rotator cuff is not strong enough to withstand very much stress. Because this joint is so easily dislocated, one should never attempt to move an immobilized person by pulling on his or her arm.

---

[21] *in* = inward + *version* = turning
[22] *e* = outward + *version* = turning

[23] *labrum* = lip

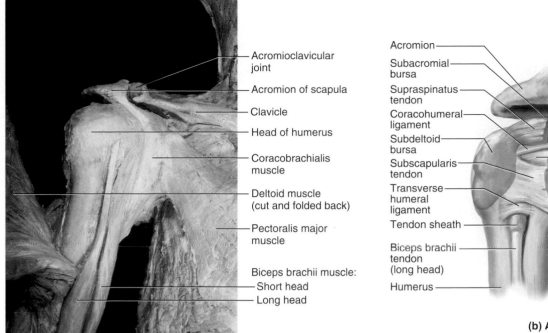

**(a) Anterior dissection**

- Acromioclavicular joint
- Acromion of scapula
- Clavicle
- Head of humerus
- Coracobrachialis muscle
- Deltoid muscle (cut and folded back)
- Pectoralis major muscle
- Biceps brachii muscle:
  - Short head
  - Long head

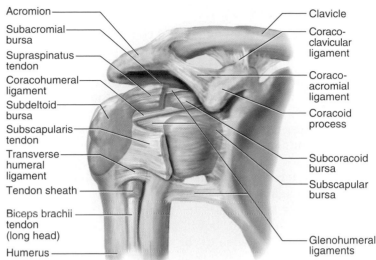

**(b) Anterior view**

- Acromion
- Subacromial bursa
- Supraspinatus tendon
- Coracohumeral ligament
- Subdeltoid bursa
- Subscapularis tendon
- Transverse humeral ligament
- Tendon sheath
- Biceps brachii tendon (long head)
- Humerus
- Clavicle
- Coraco-clavicular ligament
- Coraco-acromial ligament
- Coracoid process
- Subcoracoid bursa
- Subscapular bursa
- Glenohumeral ligaments

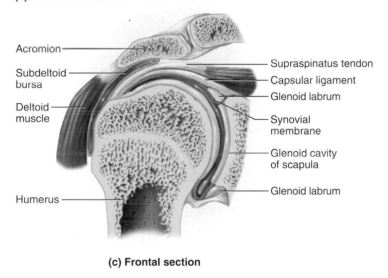

**(c) Frontal section**

- Acromion
- Subdeltoid bursa
- Deltoid muscle
- Humerus
- Supraspinatus tendon
- Capsular ligament
- Glenoid labrum
- Synovial membrane
- Glenoid cavity of scapula
- Glenoid labrum

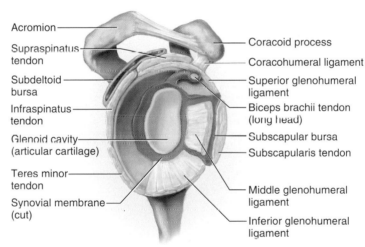

**(d) Lateral view, humerus removed**

- Acromion
- Supraspinatus tendon
- Subdeltoid bursa
- Infraspinatus tendon
- Glenoid cavity (articular cartilage)
- Teres minor tendon
- Synovial membrane (cut)
- Coracoid process
- Coracohumeral ligament
- Superior glenohumeral ligament
- Biceps brachii tendon (long head)
- Subscapular bursa
- Subscapularis tendon
- Middle glenohumeral ligament
- Inferior glenohumeral ligament

**FIGURE 9.23** The Humeroscapular (Shoulder) Joint.

The shoulder is stabilized mainly by the biceps brachii muscle on the anterior side of the arm. One of its tendons arises from the *long head* of the muscle (see chapter 10), passes through the intertubercular groove of the humerus, and inserts on the superior margin of the glenoid cavity. It acts as a taut strap that presses the humeral head against the glenoid cavity. Four additional muscles help to stabilize this joint: the *supraspinatus, infraspinatus, teres minor,* and *subscapularis.* Their tendons form the **rotator cuff,** which is fused to the joint capsule on all sides except the inferior (see fig. 10.24). Chapter 10 further describes the rotator cuff and its injuries.

Five principal ligaments also support this joint. Three of them, called the **glenohumeral ligaments,** are relatively weak and sometimes absent. The other two are the **coracohumeral ligament,** which extends from the coracoid process of the scapula to the greater tubercle of the humerus, and the **transverse humeral ligament,** which extends from the greater to the lesser tubercle of the humerus and forms a tunnel housing the tendon from the long head of the biceps.

Four bursae occur at the shoulder. Their names describe their locations: the **subdeltoid, subacromial, subcoracoid,** and **subscapular bursae.** The *deltoid* is the large muscle that caps the shoulder, and the other bursae are named for parts of the scapula described in chapter 8.

# THE ELBOW JOINT

The elbow contains a hinge joint composed of two articulations: the **humeroulnar joint,** where the trochlea of the humerus joins the trochlear notch of the ulna, and the **humeroradial joint,** where the capitulum of the humerus meets the head of the radius (fig. 9.24). Both are enclosed in a single joint capsule. On the posterior side of the elbow, there is a prominent **olecranon bursa** to ease the movement of tendons over the joint. Side-to-side motions of the elbow joint are restricted by a pair of ligaments: the **radial (lateral) collateral ligament** and **ulnar (medial) collateral ligament.**

Another joint occurs in the elbow region, the **proximal radioulnar joint,** but it is not involved in the hinge. At this joint, the edge of the disclike head of the radius fits into the radial notch of the ulna. It is held in place by the **anular ligament,** which encircles the radial head and is attached at each end to the ulna. The radial head rotates like a wheel against the ulna as the forearm is pronated or supinated.

# THE COXAL JOINT

The **coxal (hip) joint** is the point where the head of the femur inserts into the acetabulum of the os coxae (fig. 9.25). Because the coxal joints bear much of the body's weight, they have deep sockets and are much more stable than the shoulder joint. The depth of the socket is somewhat greater than you see on dried bones because of a horseshoe-shaped ring of fibrocartilage, the **acetabular labrum,** attached to its rim. Dislocations of the hip are rare, but some infants suffer

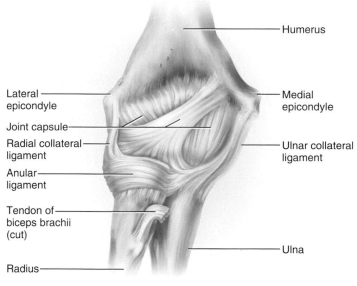

**(a) Anterior view**

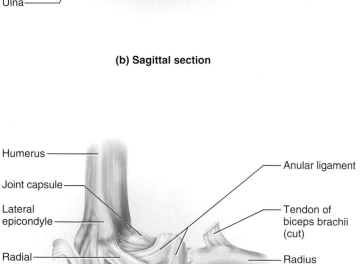

**(b) Sagittal section**

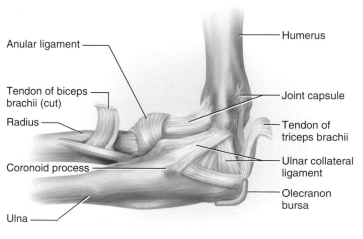

**(c) Medial view**

**(d) Lateral view**

**FIGURE 9.24**   The Elbow Joint.

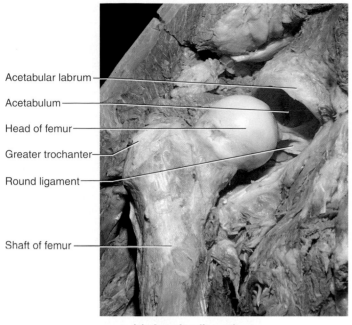

Acetabular labrum
Acetabulum
Head of femur
Greater trochanter
Round ligament
Shaft of femur

(a)  **Anterior dissection**

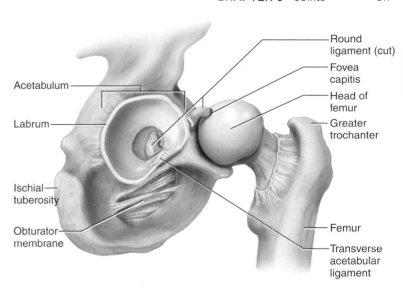

Acetabulum
Labrum
Ischial tuberosity
Obturator membrane

Round ligament (cut)
Fovea capitis
Head of femur
Greater trochanter
Femur
Transverse acetabular ligament

(b)  **Lateral view, femur retracted**

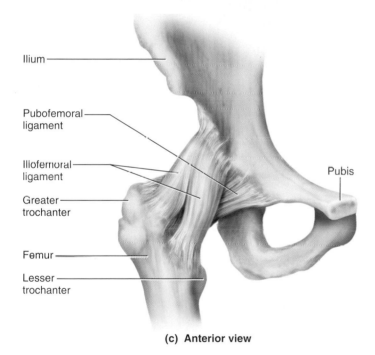

Ilium
Pubofemoral ligament
Iliofemoral ligament
Greater trochanter
Femur
Lesser trochanter
Pubis

(c)  **Anterior view**

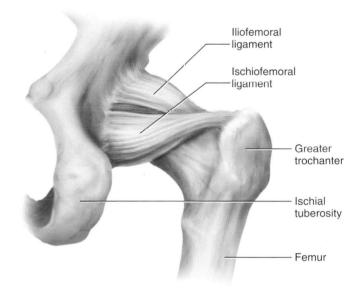

Iliofemoral ligament
Ischiofemoral ligament
Greater trochanter
Ischial tuberosity
Femur

(d)  **Posterior view**

**FIGURE 9.25**  The Coxal (Hip) Joint.

congenital dislocations because the acetabulum is not deep enough to hold the head of the femur in place. This condition can be treated by placing the infant in traction until the acetabulum develops enough strength to support the body's weight (fig. 9.26).

Ligaments that support the coxal joint include the **iliofemoral** (ILL-ee-oh-FEM-oh-rul) and **pubofemoral** (PYU-bo-FEM-or-ul) **ligaments** on the anterior side and the **ischiofemoral** (ISS-kee-oh-FEM-or-ul) **ligament** on the posterior side. The name of each ligament refers to the bones to which it attaches—the femur and the ilium, pubis, or ischium. When you stand up, these ligaments

become twisted and pull the head of the femur tightly into the acetabulum. The head of the femur has a conspicuous pit called the **fovea capitis.** The **round ligament,** or **ligamentum teres**[24] (TERR-eez), arises here and attaches to the lower margin of the acetabulum. This is a relatively slack ligament, so it is doubtful that it plays a significant role in holding the femur in its socket. It does, however, contain an artery that supplies blood to the head of the femur. A **transverse acetabular ligament** bridges a gap in the inferior margin of the acetabular labrum.

[24]*teres* = round

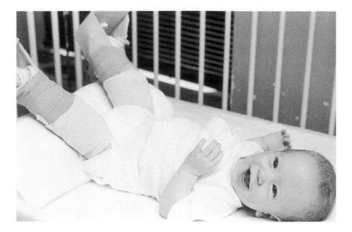

**FIGURE 9.26** Treatment of Congenital Hip Dislocation. Infants are sometimes placed in traction to treat this condition.

## THE KNEE JOINT

The **tibiofemoral (knee) joint** is the largest and most complex diarthrosis of the body (figs. 9.27 and 9.28). It is primarily a hinge joint, but when the knee is flexed it is also capable of slight rotation and lateral gliding. The patella and patellar ligament also articulate with the femur to form a gliding **patellofemoral joint.**

The joint capsule encloses only the lateral and posterior aspects of the knee joint, not the anterior. The anterior aspect is covered by the patellar ligament and the *lateral* and *medial patellar retinacula* (not illustrated). These are extensions of the tendon of the *quadriceps femoris* muscle, the large anterior muscle of the thigh. The knee is stabilized mainly by the quadriceps tendon in front and the tendon of the *semimembranosus* muscle on the rear of the thigh. Developing strength in these muscles therefore reduces the risk of knee injury.

The joint cavity contains two cartilages called the **lateral meniscus** and **medial meniscus,** joined by a **transverse ligament.** These menisci absorb the shock of the body weight jostling up and down on the knee and prevent the femur from rocking from side to side on the tibia.

The posterior **popliteal** (pop-LIT-ee-ul) **region** of the knee is supported by a complex array of *extracapsular ligaments* external to the joint capsule and two *intracapsular ligaments* within it. The extracapsular ligaments include two collateral ligaments that prevent the knee from rotating when the joint is extended—the **fibular (lateral) collateral ligament** and the **tibial (medial) collateral ligament**—and other ligaments not illustrated.

The two intracapsular ligaments lie deep within the joint. The synovial membrane folds around them, however, so that they are excluded from the fluid-filled synovial cavity. These ligaments cross each other in the form of an X; hence, they are called the **anterior cruciate**[25]

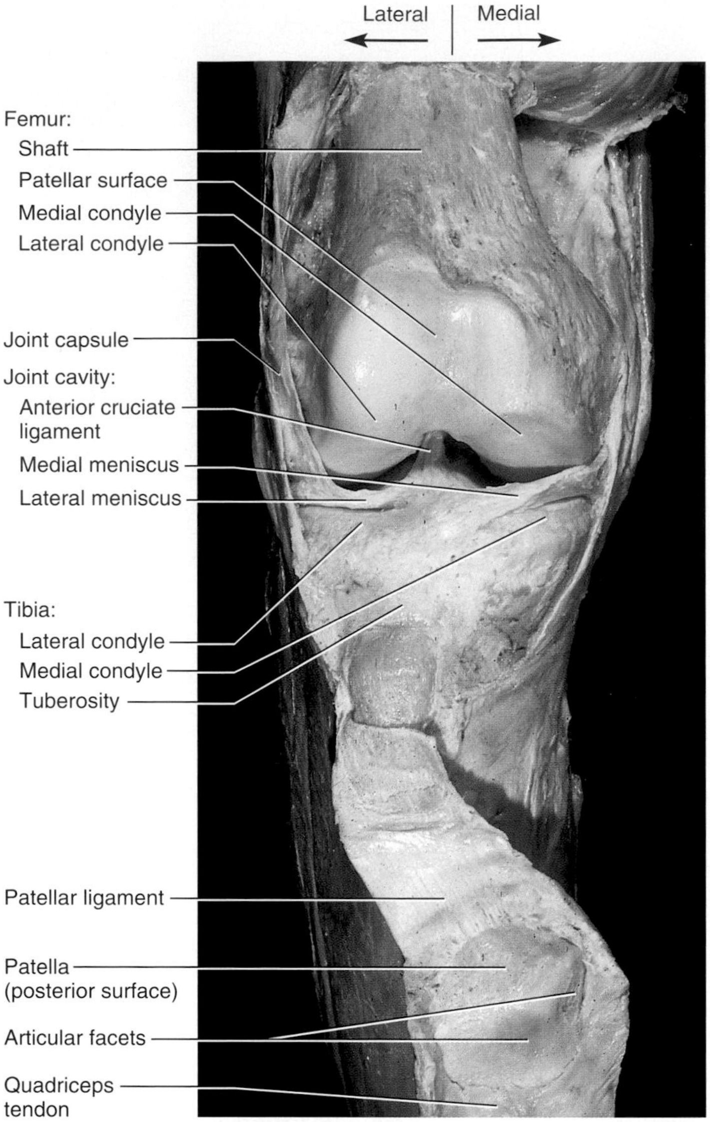

25*cruci* = cross + *ate* = characterized by

Lateral | Medial

**Femur:**
  Shaft
  Patellar surface
  Medial condyle
  Lateral condyle

Joint capsule

Joint cavity:
  Anterior cruciate ligament
  Medial meniscus
  Lateral meniscus

**Tibia:**
  Lateral condyle
  Medial condyle
  Tuberosity

Patellar ligament

Patella (posterior surface)

Articular facets

Quadriceps tendon

**FIGURE 9.27** The Right Knee, Anterior Dissection. The quadriceps tendon has been cut and folded (reflected) downward to expose the joint cavity and the posterior surface of the patella.

▶ *Identify the tibial collateral ligament.*

(CROO-she-ate) **ligament (ACL)** and **posterior cruciate ligament (PCL).** These are named according to whether they attach to the anterior or posterior side of the tibia, not for their attachments to the femur. When the knee is extended, the ACL is pulled tight and prevents hyperextension. The PCL prevents the femur from sliding off the front of the tibia and prevents the tibia from being displaced backward.

An important aspect of human bipedalism is the ability to "lock" the knees and stand erect without tiring the extensor muscles of the leg. When the knee is extended to the fullest degree allowed by the ACL, the femur rotates medially on the tibia. This action locks the knee, and in this state all the major knee ligaments are twisted and taut. To unlock the knee, the *popliteus* muscle rotates the femur laterally and untwists the ligaments.

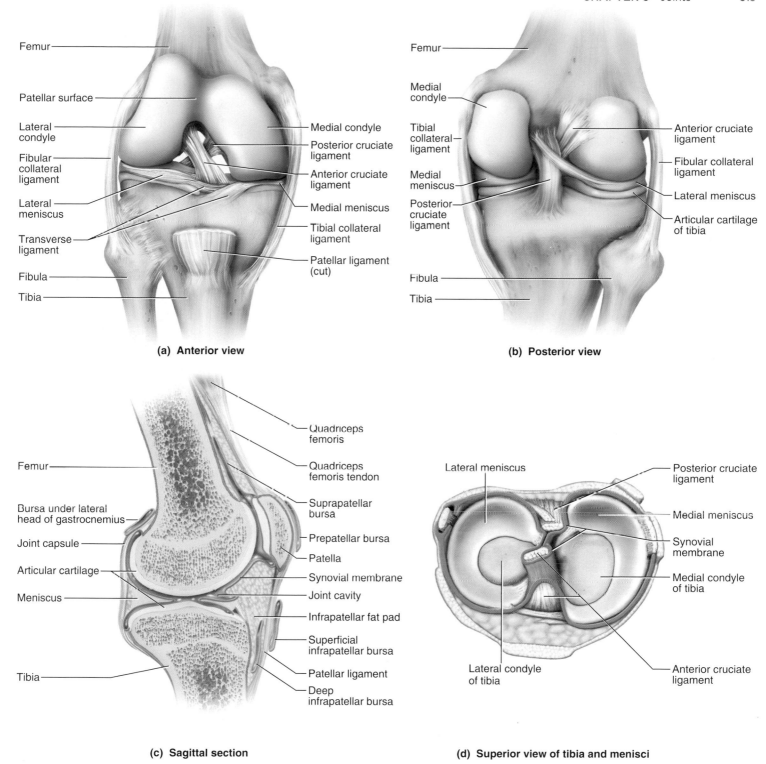

**(a) Anterior view**

Femur
Patellar surface
Lateral condyle
Fibular collateral ligament
Lateral meniscus
Transverse ligament
Fibula
Tibia
Medial condyle
Posterior cruciate ligament
Anterior cruciate ligament
Medial meniscus
Tibial collateral ligament
Patellar ligament (cut)

**(b) Posterior view**

Femur
Medial condyle
Tibial collateral ligament
Medial meniscus
Posterior cruciate ligament
Fibula
Tibia
Anterior cruciate ligament
Fibular collateral ligament
Lateral meniscus
Articular cartilage of tibia

**(c) Sagittal section**

Femur
Bursa under lateral head of gastrocnemius
Joint capsule
Articular cartilage
Meniscus
Tibia
Quadriceps femoris
Quadriceps femoris tendon
Suprapatellar bursa
Prepatellar bursa
Patella
Synovial membrane
Joint cavity
Infrapatellar fat pad
Superficial infrapatellar bursa
Patellar ligament
Deep infrapatellar bursa

**(d) Superior view of tibia and menisci**

Lateral meniscus
Lateral condyle of tibia
Posterior cruciate ligament
Medial meniscus
Synovial membrane
Medial condyle of tibia
Anterior cruciate ligament

**FIGURE 9.28** The Right Tibiofemoral (Knee) Joint.

The knee joint has at least 13 bursae. Four of these are anterior: the **superficial infrapatellar, suprapatellar, prepatellar,** and **deep infrapatellar.** Located in the popliteal region are the *popliteal bursa* and *semimembranosus bursa* (not illustrated). At least seven more bursae are found on the lateral and medial sides of the knee joint. From figure 9.28a, your knowledge of the relevant word elements (*infra-, supra-, pre-*), and the terms *superficial* and *deep,* you should be able to work out the reasoning behind most of these names and develop a system for remembering the locations of these bursae.

Knee injuries are common and often require surgery (Insight 9.3). Other joint disorders are discussed in Insight 9.4 (arthritis) and table 9.1

**INSIGHT 9.3**    Clinical Application

## Knee Injuries and Arthroscopic Surgery

Although the knee can bear a lot of weight, it is highly vulnerable to rotational and horizontal stress, especially when the knee is flexed (as in skiing or running) and receives a blow from behind or from the lateral side. The most common injuries are to a meniscus or the anterior cruciate ligament (ACL) (fig. 9.29). Knee injuries heal slowly because ligaments and tendons have a very scant blood supply and cartilage has no blood vessels at all.

The diagnosis and surgical treatment of knee injuries has been greatly improved by *arthroscopy,* a procedure in which the interior of a joint is viewed with a pencil-thin instrument, the *arthroscope,* inserted through a small incision. The arthroscope has a light source, a lens, and fiber optics that allow a viewer to see into the cavity, take photographs or videotapes of the joint, and withdraw samples of synovial fluid. Saline is often introduced through one incision to expand the joint and provide a clearer view of its structures. If surgery is required, additional small incisions can be made for the surgical instruments and the procedures can be observed through the arthroscope or on a monitor. Arthroscopic surgery produces much less tissue damage than conventional surgery and enables patients to recover more quickly.

Orthopedic surgeons now often replace a damaged ACL with a graft from the patellar ligament or a hamstring tendon. The surgeon "harvests" a strip from the middle of the patient's ligament (or tendon), drills a hole into the femur and tibia within the joint cavity, threads the ligament through the holes, and fastens it with screws. The grafted ligament is more taut and "competent" than the damaged ACL. It becomes ingrown with blood vessels and serves as a substrate for the deposition of more collagen, which further strengthens it in time. Following arthroscopic ACL reconstruction, a patient typically must use crutches for 7 to 10 days and undergo supervised physical therapy for 6 to 10 weeks, followed by self-directed exercise therapy. Healing is completed in about 9 months.

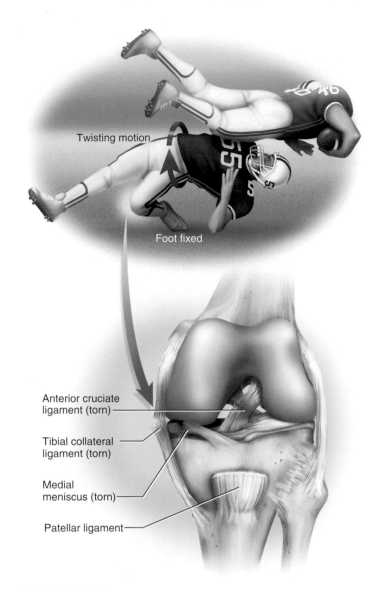

Twisting motion

Foot fixed

Anterior cruciate ligament (torn)

Tibial collateral ligament (torn)

Medial meniscus (torn)

Patellar ligament

**FIGURE 9.29**    **Some Common Knee Injuries.**

| TABLE 9.1 | Some Common Joint Disorders |
|---|---|
| Arthritis | Broad term embracing more than 100 types of joint rheumatism. |
| Bursitis | Inflammation of a bursa, usually due to overexertion of a joint. |
| Dislocation | Displacement of a bone from its normal position at a joint, usually accompanied by a sprain of the adjoining connective tissues. Most common at the fingers, thumb, shoulder, and knee. |
| Gout | A hereditary disease, most common in men, in which uric acid crystals accumulate in the joints and irritate the articular cartilage and synovial membrane. Causes gouty arthritis, with swelling, pain, tissue degeneration, and sometimes fusion of the joint. Most commonly affects the great toe. |
| Rheumatism | Broad term for any pain in the supportive and locomotory organs of the body, including bones, ligaments, tendons, and muscles. |
| Sprain | Torn ligament or tendon, sometimes with damage to a meniscus or other cartilage. |
| Strain | Painful overstretching of a tendon or muscle without serious tissue damage. Often results from inadequate warm-up before exercise. |
| Synovitis | Inflammation of a joint capsule, often as a complication of a sprain. |
| Tendinitis | A form of bursitis in which a tendon sheath is inflamed. |

***Disorders described elsewhere***

Hip dislocation p. 310          Osteoarthritis p. 315          Rotator cuff injury p. 382

Knee injuries p. 314          Rheumatoid arthritis p. 315          Shoulder dislocation p. 308

## INSIGHT 9.4   Clinical Application

### Arthritis and Artificial Joints

**Arthritis**[26] is a broad term for pain and inflammation of a joint and embraces more than a hundred different diseases of largely obscure or unknown causes. In all of its forms, it is the most common crippling disease in the United States; nearly everyone past middle age develops arthritis to some degree. Physicians who treat arthritis and other joint disorders are called *rheumatologists*.

The most common form of arthritis is *osteoarthritis (OA)*, also called "wear-and-tear arthritis" because it is apparently a normal consequence of years of wear on the joints. As joints age, the articular cartilage softens and degenerates. As the cartilage becomes roughened by wear, joint movement may be accompanied by crunching or crackling sounds called *crepitus*. OA affects especially the fingers, intervertebral joints, hips, and knees. As the articular cartilage wears away, exposed bone tissue often develops spurs that grow into the joint cavity, restrict movement, and cause pain. OA rarely occurs before age 40, but it affects about 85% of people older than 70, especially those who are overweight. It usually does not cripple, but in severe cases it can immobilize the hip.

*Rheumatoid arthritis (RA)*, which is far more severe than osteoarthritis, results from an autoimmune attack against the joint tissues. It begins when the body produces antibodies to fight an infection. Failing to recognize the body's own tissues, a misguided antibody known as *rheumatoid factor* also attacks the synovial membranes. Inflammatory cells accumulate in the synovial fluid and produce enzymes that degrade the articular cartilage. The synovial membrane thickens and adheres to the articular cartilage, fluid accumulates in the joint capsule, and

the capsule is invaded by fibrous connective tissue. As articular cartilage degenerates, the joint begins to ossify, and sometimes the bones become solidly fused and immobilized, a condition called *ankylosis*[27] (fig. 9.30). The disease tends to develop symmetrically—if the right wrist or hip develops RA, so does the left.

Rheumatoid arthritis is named for the fact that symptoms tend to flare up and subside (go into remission) periodically.[28] It affects women far more often than men, and because RA typically begins as early as age 30 to 40, it can cause decades of pain and disability. There is no cure, but joint damage can be slowed with hydrocortisone or other steroids. Because long-term use of steroids weakens the bone, however, aspirin is the treatment of first choice to control the inflammation. Physical therapy is also used to preserve the joint's range of motion and the patient's functional ability.

*Arthroplasty*,[29] a treatment of last resort, is the replacement of a diseased joint with an artificial device called a *prosthesis*.[30] Joint prostheses were first developed to treat injuries in World War II and the Korean War. Total hip replacement (THR), first performed in 1963 by English orthopedic surgeon Sir John Charnley, is now the most common orthopedic procedure for the elderly. The first knee replacements were performed in the 1970s. Joint prostheses are now available for finger, shoulder, and elbow joints, as well as the hip and knee. Arthroplasty is performed on over 250,000 patients per year in the United States, primarily to relieve pain and restore function in elderly people with OA or RA.

Arthroplasty presents ongoing challenges for biomedical engineering. An effective prosthesis must be strong, nontoxic, and corrosion-resistant. In addition, it must bond firmly to the patient's bones and enable a normal range of motion with a minimum of friction. The heads of long bones are usually replaced with prostheses made of a metal alloy such as

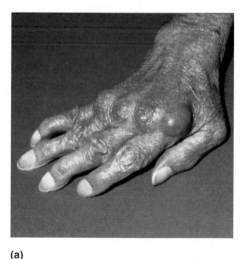
(a)

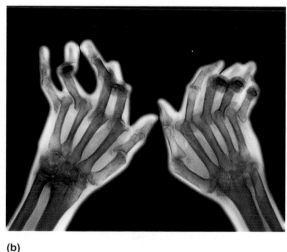
(b)

**FIGURE 9.30   Rheumatoid Arthritis (RA).**   (a) A severe case with ankylosis of the joints. (b) X ray of RA of the hands.

---

[26]*arthr* = joint + *it is* = inflammation

[27]*ankyl* = bent, crooked + *osis* = condition
[28]*rheumat* = tending to change
[29]*arthro* = joint + *plasty* = surgical repair
[30]*prosthe* = something added

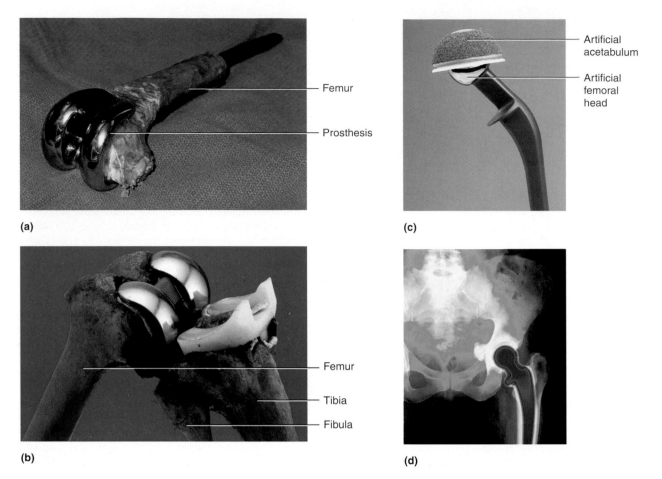

**FIGURE 9.31** **Joint Prostheses.** (a) An artificial femoral head inserted into the femur. (b) An artificial knee joint bonded to a natural femur and tibia. (c) A porous-coated hip prosthesis. The caplike portion replaces the acetabulum of the os coxae, and the ball and shaft shown below are bonded to the proximal end of the femur. (d) Colorized X ray of a patient with a total hip replacement.

cobalt–chrome, titanium alloy, or stainless steel. Joint sockets are made of polyethylene (fig. 9.31). Prostheses are bonded to the patient's bone with screws or bone cement.

Improvements in technology have resulted in long-lasting prostheses. Over 75% of artificial knees last 20 years, nearly 85% last 15 years, and over 90% last 10 years. The most common form of failure is detachment of the prosthesis from the bone. This problem has been reduced by using *porous-coated prostheses,* which become infiltrated by the patient's own bone and create a firmer bond. A prosthesis is not as strong as a natural joint, however, and is not an option for many young, active patients.

Arthroplasty has been greatly improved by *computer-assisted design and manufacture (CAD/CAM).* A computer scans X rays from the patient and presents several design possibilities for review. Once a design is selected, the computer generates a program to operate the machinery that produces the prosthesis. CAD/CAM has reduced the waiting period for a prosthesis from about 12 weeks to about 2 weeks and has lowered the cost dramatically.

## Before You Go On

*Answer the following questions to test your understanding of the preceding section:*

12. *Explain how the biceps brachii tendon braces the shoulder joint.*

13. *Identify the three joints found at the elbow and name the movements in which each joint is involved.*

14. *What structure elsewhere in the skeletal system has a structure and function similar to the acetabular labrum of the os coxae?*

15. *What keeps the femur from slipping backward off the tibia?*

## CHAPTER REVIEW

# Review of Key Concepts

**Joints and Their Classification (p. 290)**

1. A *joint (articulation)* is any point at which two bones meet. Not all joints are movable.

2. The sciences dealing with joints include arthrology, kinesiology, and biomechanics.

3. Joints are typically named for the bones involved, such as the humero-scapular joint.

4. Joints are classified according to the manner in which the bones are joined and corresponding differences in how freely the bones can move.

5. *Bony joints (synostoses)* are joints at which the original gap between two bones becomes ossified and adjacent bones become, in effect, a single bone—for example, union of the two frontal bones into a single bone.

6. *Fibrous joints (synarthroses)* are joints at which two bones are united by collagenous fibers. The three types of fibrous joints are *sutures* (which are of *serrate, lap,* and *plane* types), *gomphoses* (teeth in their sockets), and *syndesmoses* (exemplified by long bones joined along their shafts by interosseous membranes).

7. *Cartilaginous joints (amphiarthroses)* are joints at which two bones are united by cartilage. In *synchondroses,* the cartilage is hyaline (as in the epiphyseal plates of juvenile bones), and in *symphyses,* it is fibrocartilage (as in the intervertebral discs and pubic symphysis).

**Synovial Joints (p. 294)**

1. In a *synovial joint,* two bones are separated by a *joint cavity* containing lubricating *synovial fluid.* The articulating surfaces of the adjacent bones are covered by a hyaline *articular cartilage* and are sometimes held apart by a cartilaginous *articular disc* or a pair of *menisci.*

2. Synovial joints commonly exhibit *tendons* (which join muscle to bone), *ligaments* (which join one bone to another), and *bursae* (sacs of synovial fluid outside the joint cavity).

3. A long bone acts as a lever, with a stationary *fulcrum,* an *effort arm,* and a *resistance arm.*

4. The *mechanical advantage (MA)* of a lever is equal to the ratio of the length of the effort arm to the length of the resistance arm. A lever with an MA > 1 puts out more force than is put into it, and one with an MA < 1 puts out less force, but produces movements of greater speed or distance than the input.

5. Levers of the body are classified as *first-, second-,* or *third-class* depending on the relative positions of the fulcrum, effort, and resistance.

6. Joints vary in their *range of motion,* determined by the structure of the articulating bones, the tautness of ligaments and joint capsules, and the action of muscles and tendons.

7. A bone can rotate on as many as three mutually perpendicular *axes of rotation* (*x, y,* and *z*) corresponding to the three anatomical planes of the body. A joint has one to three *degrees of freedom* depending on how many of these axes of rotation it has. A joint is classified as *monaxial, biaxial,* or *multiaxial* depending on its degrees of freedom.

8. *Ball-and-socket joints* at the shoulder and hip are the only multiaxial joints.

9. *Condyloid joints,* for example at the articulation of the proximal phalanges with the metacarpals, are biaxial.

10. *Saddle joints* are also biaxial, occurring at the articulations of metacarpal I with the trapezium, and the clavicle with the sternum.

11. *Gliding joints* are biaxial and have limited gliding movement, for example between the carpal bones and the articular facets of the vertebrae.

12. *Hinge joints* are monaxial and represented by the elbow and knee.

13. *Pivot joints* are monaxial and occur where a bone spins on its longitudinal axis, such as the head of the radius articulating with the ulna.

14. Joint movements are described by a specific terminology that is widely used in kinesiology, physical therapy, and other fields. Many are described in reference to the *zero position* that a joint has when a person stands in anatomical position.

15. *Flexion* is a movement that decreases the angle between two bones, as in flexing the elbow; *extension* straightens a joint, usually returning a body part to zero position; and *hyperextension* extends a joint beyond the zero position. The vertebral column exhibits flexion and extension in bending forward and returning to zero position, respectively; hyperextension when bending backward; and *lateral flexion* when bending to one side.

16. *Abduction* moves a bone away from the median plane of the body, as in spreading the legs, and *adduction* moves it back. Abduction beyond a 180° angle is *hyperabduction,* as in lifting an arm to cross behind the head, and adduction beyond the zero position is *hyperadduction,* as in standing with the ankles crossed.

17. *Elevation* raises a bone such as the scapula and *depression* lowers it.

18. *Protraction* thrusts a bone such as the clavicle or mandible forward and *retraction* draws it back.

19. *Circumduction* moves the distal end of a bone in a circle while the proximal end remains relatively stationary.

20. *Rotation* turns a bone on its longitudinal axis. For example, folding the forearm across the abdomen and then rotating it in the opposite direction, with the elbow bent, cause *medial* and *lateral rotation* of the humerus. Turning the head or twisting at the waist is called *right rotation* and *left rotation.*

21. *Supination* rotates the forearm so the palm faces forward or upward, and *pronation* turns it so the palm faces toward the rear or downward.

22. The mandible exhibits some of the foregoing movements when biting and chewing, such as protraction, retraction, elevation, and depression; but also *lateral* and *medial excursion* when grinding food between the molars.

23. The wrist exhibits unique movements called *ulnar* and *radial flexion* when it bends medially and laterally, and the thumb exhibits unique movements called *opposition* and *reposition* when it crosses the palm to approach the fingertips and then returns to zero position. Thumb abduction and adduction occur in a plane 90° to the other digits because of embryonic rotation of the thumb to the sagittal plane.

24. The foot exhibits *dorsiflexion* and *plantar flexion* in the sagittal plane; *inversion* and *eversion* in the frontal plane; and a complex combination of movements collectively called *pronation* and *supination,* but with only slight resemblance to the same-named movements of the forearm.

**Anatomy of Selected Diarthroses (p. 308)**

1. The *humeroscapular joint* is a ball-and-socket joint at which the head of the humerus articulates with the glenoid cavity of the scapula. It is supported by a cartilaginous *glenoid labrum,* two major ligaments and three minor (weak) ones, the biceps brachii tendon, and tendons of the four *rotator cuff* muscles. It has four bursae.

2. The elbow has three joints of the hinge and pivot types: the *humero-ulnar joint* where the trochlear notch of the ulna wraps around the trochlea of the humerus; the *humeroradial joint* where the head of the radius pivots on the capitulum of the humerus; and the *proximal radioulnar joint* where the edge of the radial head spins against the radial notch of the ulna. The first and second joints are involved in the hinge action of the elbow, and the second and third in the pivot action. An *anular ligament* arises from the ulna and encircles the neck of the radius. The joint is also stabilized by *radial* and *ulnar collateral ligaments* and has a posterior *olecranon bursa.*

3. The *coxal joint* is a ball-and-socket joint where the head of the femur articulates with the acetabulum of the os coxae. The acetabulum is deepened by a cartilaginous *acetabular labrum.* Three major ligaments attach the femur to the pubis, ischium, and ilium.

4. The knee has two principal joints, the *tibiofemoral* hinge joint and the *patellofemoral* gliding joint. It is enclosed laterally and posteriorly by the joint capsule and anteriorly by a pair of *patellar retinacula.* The joint is stabilized mainly by the *quadriceps femoris* and *semimembranosus* muscles of the thigh. The most important structures within the joint cavity are the two *menisci* and the *anterior* and *posterior cruciate ligaments,* but the knee has several other stabilizing ligaments and at least 13 bursae.

# Testing Your Recall

1. Internal and external rotation of the humerus are made possible by a _____ joint.
   a. pivot
   b. condyloid
   c. ball-and-socket
   d. saddle
   e. hinge

2. Which of the following is the least movable?
   a. a diarthrosis
   b. a synarthrosis
   c. a symphysis
   d. a synovial joint
   e. a condyloid joint

3. Which of the following movements are unique to the foot?
   a. dorsiflexion and inversion
   b. elevation and depression
   c. circumduction and rotation
   d. abduction and adduction
   e. opposition and reposition

4. Which of the following joints cannot be circumducted?
   a. carpometacarpal
   b. metacarpophalangeal
   c. humeroscapular
   d. coxal
   e. interphalangeal

5. Which of the following terms denotes a general condition that includes the other four?
   a. gout
   b. arthritis
   c. rheumatism
   d. osteoarthritis
   e. rheumatoid arthritis

6. In the adult, the ischium and pubis are united by
   a. a synchondrosis.
   b. a diarthrosis.
   c. a synostosis.
   d. an amphiarthrosis.
   e. a symphysis.

7. In a second-class lever, the effort
   a. is applied to the end opposite the fulcrum.
   b. is applied to the fulcrum itself.
   c. is applied between the fulcrum and resistance.
   d. always produces an MA less than 1.0.
   e. is applied on one side of the fulcrum to move a resistance on the other side.

8. Which of the following joints has anterior and posterior cruciate ligaments?
   a. the shoulder
   b. the elbow
   c. the hip
   d. the knee
   e. the ankle

9. To bend backward at the waist involves _____ of the vertebral column.
   a. rotation
   b. hyperextension
   c. dorsiflexion
   d. abduction
   e. flexion

10. The rotator cuff includes the tendons of all of the following muscles *except*
    a. the subscapularis.
    b. the supraspinatus.
    c. the infraspinatus.
    d. the biceps brachii.
    e. the teres minor.

11. The lubricant of a diarthrosis is _____.

12. A fluid-filled sac that eases the movement of a tendon over a bone is called a/an _____.

13. A _____ joint allows one bone to swivel on another.

14. _____ is the science of movement.

15. The joint between a tooth and the mandible is called a/an _____.

16. In a _____ suture, the articulating bones have interlocking wavy margins, somewhat like a dovetail joint in carpentry.

17. In kicking a football, what type of action does the knee joint exhibit?

18. The angle through which a joint can move is called its _____.

19. Both the shoulder and hip joints are somewhat deepened and supported by a ring of fibrocartilage called a/an _____.

20. The femur is prevented from slipping sideways off the tibia in part by a pair of cartilages called the lateral and medial _____.

*Answers in Appendix B*

## True or False

*Determine which five of the following statements are false, and briefly explain why.*

1. More people get rheumatoid arthritis than osteoarthritis.

2. A doctor who treats arthritis is called a kinesiologist.

3. Synovial joints are also known as synarthroses.

4. There is no meniscus in the elbow joint.

5. Reaching behind you to take something out of your hip pocket involves hyperextension of the shoulder.

6. The anterior cruciate ligament normally prevents hyperextension of the knee.

7. The femur is held tightly in the acetabulum mainly by the round ligament.

8. The knuckles are diarthroses.

9. Synovial fluid is secreted by the bursae.

10. Unlike most ligaments, the periodontal ligaments do not attach one bone to another.

*Answers in Appendix B*

## Testing Your Comprehension

1. All second-class levers produce a mechanical advantage greater than 1.0 and all third-class levers produce a mechanical advantage less than 1.0. Explain why.

2. For each of the following joint movements, state what bone the axis of rotation passes through and which of the three anatomical planes contains the axis of rotation. You might find it helpful to produce some of these actions on an articulated laboratory skeleton so you can more easily visualize the axis of rotation. (a) Plantar flexion; (b) flexion of the hip; (c) abduction of the thigh; (d) flexion of the knee; (e) flexion of the first interphalangeal joint of the index finger. (Do not bend the fingers of a wired laboratory skeletal hand, because they can break off.)

3. In order of occurrence, list the joint actions (flexion, pronation, etc.) and the joints where they would occur as you (a) sit down at a table, (b) reach out and pick up an apple, (c) take a bite, and (d) chew it. Assume that you start in anatomical position.

4. The deltoid muscle inserts on the deltoid tuberosity of the humerus and abducts the arm. Imagine a person holding a weight in the the hand and abducting the arm. On a laboratory skeleton, identify the fulcrum; measure the effort arm and resistance arm; determine the mechanical advantage of this movement; and determine which of the three lever types the upper limb acts as when performing this movement.

5. List the six types of synovial joints, and for each one, if possible, identify a joint in the upper limb and a joint in the lower limb that falls into each category. Which of these six joints have no examples in the lower limb?

*Answers at www.mhhe.com/saladin4*

# www.mhhe.com/saladin4

*The textbook website provides a wealth of interactive study materials fully organized and integrated by chapter. You will find practice quizzes, labeling exercises, and much more that will complement your learning and understanding of anatomy and physiology. The website also includes tools designed to enhance your* **Anatomy & Physiology | REVEALED** *experience.*

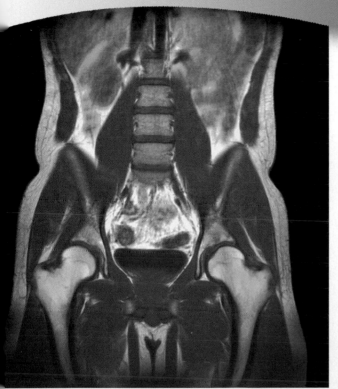

*Colorized MRI scan showing muscles of the lumbar, pelvic, and upper femoral regions*

# THE MUSCULAR SYSTEM

## CHAPTER OUTLINE

## Brushing Up

To understand this chapter, it is important that you understand or brush up on the following concepts:

- Gross anatomy of the skeleton (chapter 8)

- Movements of synovial joints (pp. 302–308)

## INSIGHTS

Anatomy & Physiology | REVEALED

The **muscular system** consists of about 600 skeletal muscles—striated muscles, most of which are attached to bones. (The term does not include smooth or cardiac muscle.) The form and function of the muscular system occupy a place of central importance in several fields of health care and fitness. Physical and occupational therapists must be well acquainted with the muscular system to design and carry out rehabilitation programs. Nurses and other health-care providers often move patients who are physically incapacitated, and to do this safely and effectively requires an understanding of joints and muscles. Even to give intramuscular injections safely requires a knowledge of the muscles and the nerves and blood vessels associated with them. Coaching, movement science, sports medicine, and dance benefit from a knowledge of skeletomuscular anatomy and mechanics.

**Myology,**[1] the study of muscles, is closely related to what we have covered in the preceding chapters. It relates muscle attachments to the bone structures described in chapter 8, and muscle function to the joint movements described in chapter 9. In this chapter, we consider the gross anatomy of the muscular system and how it relates to joint movements. In chapter 11, we examine the mechanisms of muscle contraction at the cellular and molecular levels.

# The Structural and Functional Organization of Muscles

### Objectives
When you have completed this section, you should be able to

- list several functions of muscles;
- describe the connective tissues associated with a skeletal muscle;
- explain what is meant by the origin, insertion, belly, action, and innervation of a muscle;
- describe the various shapes of skeletal muscles and relate this to their functions;
- describe the ways that muscles work in groups to aid, oppose, or moderate each other's actions;
- distinguish between intrinsic and extrinsic muscles; and

[1]myo = muscle + logy = study of

- translate several Latin words commonly used in the naming of muscles.

## THE FUNCTIONS OF MUSCLES

A muscle is an organ specialized to move a body part. It converts the chemical energy of ATP into the mechanical energy of motion and exerts a useful pull on another tissue. More specifically, muscle contraction serves the following overlapping functions:

- **Movement.** Most obviously, muscles enable us to move from place to place and to move individual body parts. Muscular contractions also move body contents in the course of respiration, circulation, digestion, defecation, urination, and childbirth.
- **Stability.** Muscles maintain posture by resisting the pull of gravity and preventing unwanted movements. They hold some articulating bones in place by maintaining tension on the tendons.
- **Communication.** Muscles are used for facial expression, other body language, writing, and speech.
- **Control of body openings and passages.** Ringlike *sphincter muscles* around the eyelids, pupils, and mouth control the admission of light, food, and drink into the body; others that encircle the urethral and anal orifices control elimination of waste; and other sphincters control the movement of food, bile, and other materials through the body.
- **Heat production.** The skeletal muscles produce as much as 85% of our body heat, which is vital to the functioning of enzymes and therefore to all of our metabolism.

Some of these functions are shared by skeletal, cardiac, and smooth muscle. The remainder of this chapter, however, is concerned only with skeletal muscles.

## CONNECTIVE TISSUES OF A MUSCLE

A skeletal muscle is composed of both muscular tissue and connective tissue (fig. 10.1). A skeletal muscle cell *(muscle fiber)* is about 10 to 100 μm in diameter and up to 30 cm long. It is surrounded by a sparse layer of areolar connective tissue called the **endomysium**[2] (EN-doe-MIZ-ee-um), which allows room for blood capillaries and nerve fibers to reach each muscle fiber. Muscle fibers are grouped in bundles called **fascicles**[3] (FASS-ih-culs), which are visible to the naked eye as parallel strands. These are the "grain" in a cut of meat; tender meat is easily pulled apart along its fascicles. Each fascicle is separated from neighboring ones by a thicker connective tissue sheath called the **perimysium.**[4] The muscle as a

[2]endo = within + mys = muscle
[3]fasc = bundle + icle = little
[4]peri = around + mys = muscle

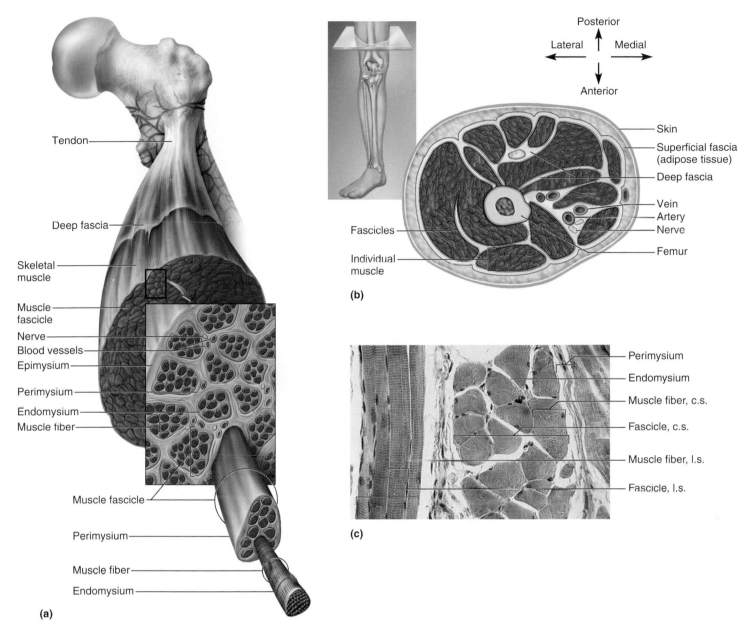

Posterior

Lateral ← → Medial

Anterior

Tendon

Deep fascia

Skeletal muscle

Muscle fascicle

Nerve
Blood vessels
Epimysium

Perimysium

Endomysium
Muscle fiber

Muscle fascicle

Perimysium

Muscle fiber

Endomysium

**(a)**

Skin
Superficial fascia (adipose tissue)
Deep fascia
Vein
Artery
Nerve
Femur

Fascicles

Individual muscle

**(b)**

Perimysium
Endomysium
Muscle fiber, c.s.
Fascicle, c.s.
Muscle fiber, l.s.
Fascicle, l.s.

**(c)**

**FIGURE 10.1  Connective Tissues of a Muscle.**   (a) The muscle–bone attachment. There is a continuity of connective tissues from the endomysium around the muscle fibers, to the perimysium, epimysium, deep fascia, and tendon, grading into the periosteum and finally the matrix of the bone. (b) A cross section of the thigh showing the relationship of neighboring muscles to fascia and bone. (c) Muscle fascicles in the tongue. Vertical fascicles passing between the dorsal and ventral surfaces of the tongue are seen alternating with cross-sectioned horizontal fascicles that pass from the tip to the rear of the tongue. A fibrous perimysium can be seen between the fascicles, and endomysium between the muscle fibers within each fascicle (c.s. = cross section; l.s. = longitudinal section).

whole is surrounded by still another connective tissue layer, the **epimysium.**[5] The epimysium grades imperceptibly into connective tissue sheets called fasciae (FASH-ee-ee)—**deep fasciae** between adjacent muscles and a **superficial fascia** (hypodermis) between the muscles and skin. The superficial fascia is very adipose in areas such as the buttocks and abdomen, but the deep fasciae are devoid of fat.

Connective tissue also attaches a muscle to a bone. There are two forms of such attachment. In a **direct (fleshy) attachment,** collagen fibers of the epimysium are continuous with the periosteum, the fibrous sheath around a bone. The red muscle tissue appears to emerge directly from the bone. The *intercostal muscles* between the ribs show this type of attachment. In an **indirect attachment,** the collagen fibers of the epimysium continue as a strong fibrous **tendon** that merges into the periosteum of a nearby bone (fig. 10.1a). The tendon creates a conspicuous gap between the

[5]*epi* = upon, above + *mys* = muscle

muscular tissue and bone. The attachment of the *biceps brachii muscle* to the scapula is one of many examples. Some collagen fibers of the periosteum continue into the bone matrix as *perforating fibers* (see chapter 7), so there is a strong structural continuity from endomysium to perimysium to epimysium to tendon to periosteum to bone matrix. Excessive stress is more likely to tear a tendon than to pull it loose from the muscle or bone.

In some cases, the epimysium of one muscle attaches to the fascia or tendon of another, or to collagen fibers of the dermis. The ability of a muscle to produce facial expressions depends on the latter type of attachment. Some muscles are connected to a broad sheetlike tendon called an **aponeurosis**[6] (AP-oh-new-RO-sis). This term originally referred to the tendon located beneath the scalp, but now it also refers to similar tendons associated with certain abdominal, lumbar, hand, and foot muscles (see figs. 10.15a and 10.16).

In some places, groups of tendons from separate muscles pass under a band of connective tissue called a **retinaculum.**[7] One of these covers each surface of the wrist like a bracelet, for example. The tendons of several forearm muscles pass under them on their way to the hand.

## GENERAL ANATOMY OF SKELETAL MUSCLES

Most skeletal muscles are attached to a different bone at each end, so the muscle or its tendon spans at least one joint. When a muscle contracts, the bone at one end usually remains still while the bone at the other end moves. Figure 10.2 shows a traditional terminology that describes many muscles: a relatively stationary attachment called its **origin;** a thick midregion called the **belly;** and a relatively mobile attachment called the **insertion.**

The terminology of origins and insertions, however, is imperfect and sometimes misleading. One end of a muscle might function as its stationary origin during one action, but function as its moving insertion during a different action. For example, consider the *quadriceps* muscle on the anterior side of the thigh. It is a powerful extensor of the knee, connected at its proximal end mainly to the femur and at its distal end to the tibia, just below the knee. If you kick a soccer ball, the tibia moves more than the femur, so the tibia would be considered the insertion of the quadriceps muscle and the femur would be considered its origin. But when you sit down in a chair, the tibia remains stationary and the femur moves, with the quadriceps acting as a brake so you don't sit down too abruptly and hard. By the foregoing definitions, the tibia would now be considered the origin of the quadriceps, and the femur would be its insertion.

In many such cases, the moving and nonmoving ends of the muscle are reversed when different actions are performed. Consider the difference, for example, in the relative movements of the humerus and ulna when flexing the elbow to lift dumbbells as compared with flexing the elbow to perform chin-ups or scale a climbing wall. For such reasons, some anatomists are abandoning origin and insertion terminology and speaking instead of a muscle's proximal and distal or superior and inferior attachments, especially in the limbs. Nevertheless, the tables in this chapter use the traditional, if flawed, descriptions.

The strength of a muscle and the direction in which it pulls are determined partly by the orientation of its fascicles, illustrating the complementarity of form and function. Differences in fascicle orientation are the basis for classifying muscles into five types (fig. 10.3):

1. **Fusiform**[8] muscles are thick in the middle and tapered at each end. Their contractions are moderately strong. The *biceps brachii* of the arm and *gastrocnemius* of the calf are examples of this type.

2. **Parallel muscles** are long, straplike muscles of uniform width and parallel fascicles. They can span a great distance and shorten more than other muscle types, but they are weaker than fusiform muscles. Examples include the *rectus abdominis* of the abdomen, *sartorius* of the thigh, and *zygomaticus major* of the face.

3. **Convergent muscles** are fan-shaped—broad at the origin and converging toward a narrower insertion.

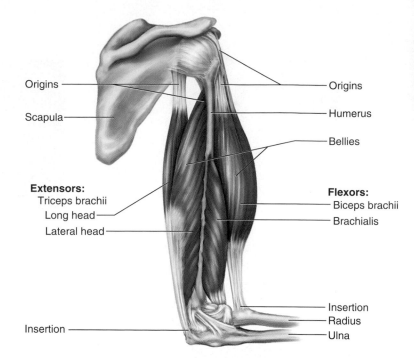

**FIGURE 10.2** **Synergistic and Antagonistic Muscle Pairs.** The biceps brachii and brachialis muscles are synergists in elbow flexion. The triceps brachii is an antagonist of those two muscles and is the prime mover in elbow extension.

Labels: Origins; Scapula; Origins; Humerus; Bellies; **Extensors:** Triceps brachii, Long head, Lateral head; **Flexors:** Biceps brachii, Brachialis; Insertion; Radius; Ulna; Insertion

---

[6]*apo* = upon, above + *neuro* = nerve
[7]*retinac* = retainer, bracelet + *cul* = little

[8]*fusi* = spindle + *form* = shape

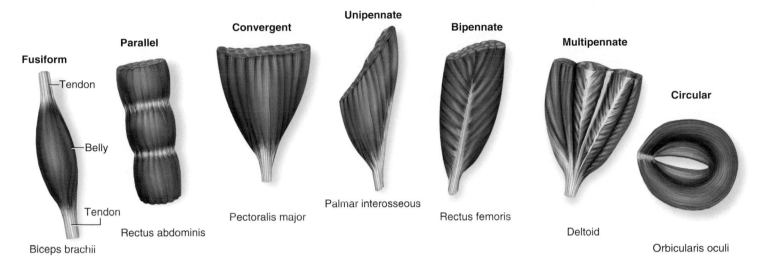

**FIGURE 10.3** **Classification of Muscles According to Fascicle Orientation.** The fascicles are the "grain" visible in each illustration.

These muscles are relatively strong because all of their fascicles exert their tension on a relatively small insertion. The *pectoralis major* in the chest is a muscle of this type.

4. **Pennate**[9] **muscles** are feather-shaped. Their fascicles insert obliquely on a tendon that runs the length of the muscle, like the shaft of a feather. There are three types of pennate muscles: *unipennate,* in which all fascicles approach the tendon from one side (for example, the *palmar interosseous muscles* of the hand and *semimembranosus* of the thigh); *bipennate,* in which fascicles approach the tendon from both sides (for example, the *rectus femoris* of the thigh); and *multipennate,* shaped like a bunch of feathers with their quills converging on a single point (for example, the *deltoid* of the shoulder).

5. **Circular muscles (sphincters)** form rings around body openings. These include the *orbicularis oculi* of the eyelids and the *urethral* and *anal sphincters.*

## COORDINATED ACTION OF MUSCLE GROUPS

The movement produced by a muscle is called its **action.** Skeletal muscles seldom act independently; instead, they function in groups whose combined actions produce the coordinated motion of a joint. Muscles can be classified into four categories according to their actions, but it must be stressed that a particular muscle can act in a certain way during one joint action and in a different way during other actions of the same joint:

1. The **prime mover (agonist)** is the muscle that produces the most force during a particular joint action.

In flexing the elbow, for example, the prime mover is the brachialis.

2. A **synergist**[10] (SIN-ur-jist) is a muscle that aids the prime mover. Several synergists acting on a joint can produce more power than a single larger muscle. The biceps brachii, for example, overlies the brachialis and works with it as a synergist to flex the elbow. The actions of a prime mover and its synergist are not necessarily identical and redundant. If the prime mover worked alone at a joint, it might cause rotation or other undesirable movements of a bone. A synergist may stabilize a joint and restrict these movements, or modify the direction of a movement, so that the action of the prime mover is more coordinated and specific.

3. An **antagonist**[11] is a muscle that opposes the prime mover. In some cases, it relaxes to give the prime mover almost complete control over an action. More often, however, the antagonist moderates the speed or range of the agonist, thus preventing excessive movement, joint injury, or inappropriate actions. If you extend your arm to reach out and pick up a cup of tea, for example, your *triceps brachii* serves as the prime mover of elbow extension and your brachialis acts as an antagonist to slow the extension and stop it at the appropriate point. If you extend your arm rapidly to throw a dart, the brachialis must be quite relaxed. The brachialis and triceps brachii represent an **antagonistic pair** of muscles that act on opposite sides of a joint (see fig. 10.2). We need antagonistic pairs at a joint because a muscle can only pull, not push—a single muscle cannot flex *and* extend the

---

[9]*penna* = feather

[10]*syn* = together + *erg* = work
[11]*ant* = against + *agonist* = competitor

elbow, for example. Which member of the pair acts as the agonist depends on the motion under consideration. In flexion of the elbow, the brachialis is the agonist and the triceps is the antagonist; when the elbow is extended, their roles are reversed.

4. A **fixator** is a muscle that prevents a bone from moving. To *fix* a bone means to hold it steady, allowing another muscle attached to it to pull on something else. For example, consider again the flexion of the elbow by the biceps brachii. The biceps originates on the scapula, crosses both the shoulder and elbow joints, and inserts on the radius. The scapula is loosely attached to the axial skeleton, so when the biceps contracts, it seems that it would pull the scapula laterally. There are fixator muscles (the *rhomboids*) attached to the scapula, however, that contract at the same time. By holding the scapula firmly in place, they ensure that the force generated by the biceps moves the radius rather than the scapula.

## INTRINSIC AND EXTRINSIC MUSCLES

In places such as the tongue, larynx, back, hand, and foot, anatomists distinguish between intrinsic and extrinsic muscles. An **intrinsic muscle** is entirely contained within a particular region, having both its origin and insertion there. An **extrinsic muscle** acts upon a designated region but has its origin elsewhere. For example, some movements of the fingers are produced by extrinsic muscles in the forearm, whose long tendons reach to the phalanges; other finger movements are produced by the intrinsic muscles of the hand, located between the metacarpal bones.

## MUSCLE INNERVATION

**Innervation** means the nerve supply to an organ. Knowing the innervation to each muscle enables clinicians to diagnose nerve and spinal cord injuries from their effects on muscle function, and to set realistic goals for rehabilitation. The muscle tables of this chapter identify the innervation of each muscle. This information will be more meaningful after you have studied the peripheral nervous system in chapters 13 and 14, but a brief orientation will be helpful here. Muscles of the head and neck are supplied by *cranial nerves* that arise from the base of the brain and emerge through the skull foramina. Cranial nerves are identified by numerals I to XII, although not all 12 of them innervate skeletal muscles. Muscles elsewhere are supplied by *spinal nerves,* which originate in the spinal cord, emerge through the intervertebral foramina, and branch into a *dorsal* and *ventral ramus.*[12] The spinal

nerves are identified by letters and numbers that refer to the vertebrae—for example, T6 for the sixth thoracic nerve and S2 for the second sacral nerve. You will note references to nerve numbers and rami in many of the muscle tables. The term *plexus* in some of the tables refers to weblike networks of spinal nerves adjacent to the vertebral column. All of the nerves named here are illustrated, and most are also discussed, in chapters 13 and 14 (see tables 13.3–13.6 and 14.2).

## HOW MUSCLES ARE NAMED

Most of this chapter is a descriptive inventory of muscles, including their location, action, origin, insertion, and innervation. Learning the names of the muscles is much easier when you have some appreciation of the meanings behind the words. The Latin and English muscle names in this chapter are from the *Terminologia Anatomica* (T.A.) (see chapter 1). Although this book gives most terms in English rather than Latin, the customary English names for skeletal muscles are, at most, only slight modifications of the Latin—for example, *anterior scalene muscle* is a derivative of the T.A. term, *musculus scalenus anterior.*

Some muscle names are several words long—for example, the flexor digiti minimi brevis, a "short muscle that flexes the little finger." Such names may seem intimidating at first, but they are really more of a help than an obstacle to understanding if you gain a little insight into the most commonly used Latin words. Several of these are interpreted in table 10.1, and others are explained in footnotes throughout the chapter. Familiarity with these terms will help you translate muscle names and remember the location, appearance, and action of the muscles.

## A LEARNING STRATEGY

In the remainder of this chapter, we consider about 160 muscles. Many of the relatively superficial ones are shown in figure 10.4. The following suggestions may help you develop a rational strategy for learning the muscular system:

- Examine models, cadavers, dissected animals, or a photographic atlas as you read about these muscles. Visual images are often easier to remember than words, and direct observation of a muscle may stick in your memory better than descriptive text or two-dimensional drawings.

- When studying a particular muscle, palpate it on yourself if possible. Contract the muscle to feel it bulge and sense its action. This makes muscle locations and actions less abstract. Atlas B following this chapter shows where you can see and palpate several of these muscles on the living body.

---

[12]*ramus* = branch

| TABLE 10.1 | Words Commonly Used to Name Muscles | |
|---|---|---|
| **Criterion** | **Term and Meaning** | **Examples of Usage** |
| Size | Major (large) | Pectoralis major |
| | Maximus (largest) | Gluteus maximus |
| | Minor (small) | Pectoralis minor |
| | Minimus (smallest) | Gluteus minimus |
| | Longus (long) | Abductor pollicis longus |
| | Brevis (short) | Extensor pollicis brevis |
| Shape | Rhomboideus (rhomboidal) | Rhomboideus major |
| | Trapezius (trapezoidal) | Trapezius |
| | Teres (round, cylindrical) | Pronator teres |
| | Deltoid (triangular) | Deltoid |
| Location | Capitis (of the head) | Splenius capitis |
| | Cervicis (of the neck) | Semispinalis cervicis |
| | Pectoralis (of the chest) | Pectoralis major |
| | Thoracis (of the thorax) | Spinalis thoracis |
| | Intercostal (between the ribs) | External intercostals |
| | Abdominis (of the abdomen) | Rectus abdominis |
| | Lumborum (of the lower back) | Quadratus lumborum |
| | Femoris (of the femur, or thigh) | Quadriceps femoris |
| | Peroneus (of the fibula) | Peroneus longus |
| | Brachii (of the arm) | Biceps brachii |
| | Carpi (of the wrist) | Flexor carpi ulnaris |
| | Digiti (of a finger or toe, singular) | Extensor digiti minimi |
| | Digitorum (of the fingers or toes, plural) | Flexor digitorum profundus |
| | Pollicis (of the thumb) | Opponens pollicis |
| | Indicis (of the index finger) | Extensor indicis |
| | Hallucis (of the great toe) | Abductor hallucis |
| | Superficialis (superficial) | Flexor digitorum superficialis |
| | Profundus (deep) | Flexor digitorum profundus |
| Number of Heads | Biceps (two heads) | Biceps femoris |
| | Triceps (three heads) | Triceps brachii |
| | Quadriceps (four heads) | Quadriceps femoris |
| Orientation | Rectus (straight) | Rectus abdominis |
| | Transversus (transverse) | Transversus abdominis |
| | Oblique (slanted) | External abdominal oblique |
| Action | Adductor (adducts a body part) | Adductor pollicis |
| | Abductor (abducts a body part) | Abductor digiti minimi |
| | Flexor (flexes a joint) | Flexor carpi radialis |
| | Extensor (extends a joint) | Extensor carpi radialis |
| | Pronator (pronates forearm) | Pronator teres |
| | Supinator (supinates forearm) | Supinator |
| | Levator (elevates a body part) | Levator scapulae |
| | Depressor (depresses a body part) | Depressor anguli oris |

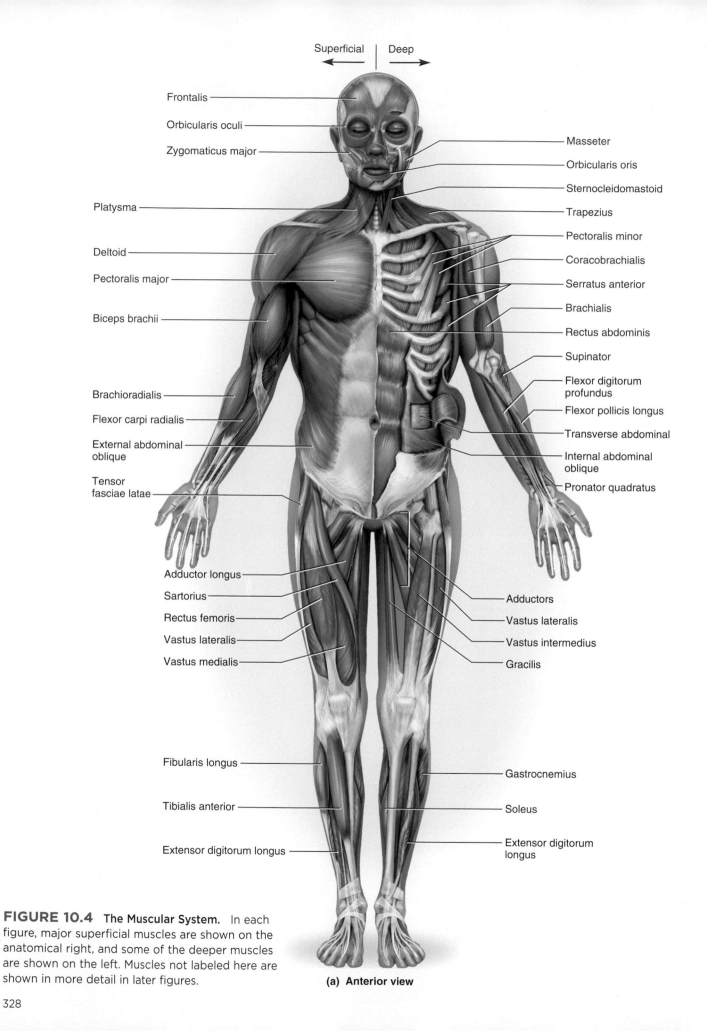

Superficial | Deep

Frontalis
Orbicularis oculi
Zygomaticus major
Platysma
Deltoid
Pectoralis major
Biceps brachii
Brachioradialis
Flexor carpi radialis
External abdominal oblique
Tensor fasciae latae
Adductor longus
Sartorius
Rectus femoris
Vastus lateralis
Vastus medialis
Fibularis longus
Tibialis anterior
Extensor digitorum longus

Masseter
Orbicularis oris
Sternocleidomastoid
Trapezius
Pectoralis minor
Coracobrachialis
Serratus anterior
Brachialis
Rectus abdominis
Supinator
Flexor digitorum profundus
Flexor pollicis longus
Transverse abdominal
Internal abdominal oblique
Pronator quadratus
Adductors
Vastus lateralis
Vastus intermedius
Gracilis
Gastrocnemius
Soleus
Extensor digitorum longus

**FIGURE 10.4  The Muscular System.**  In each figure, major superficial muscles are shown on the anatomical right, and some of the deeper muscles are shown on the left. Muscles not labeled here are shown in more detail in later figures.

**(a) Anterior view**

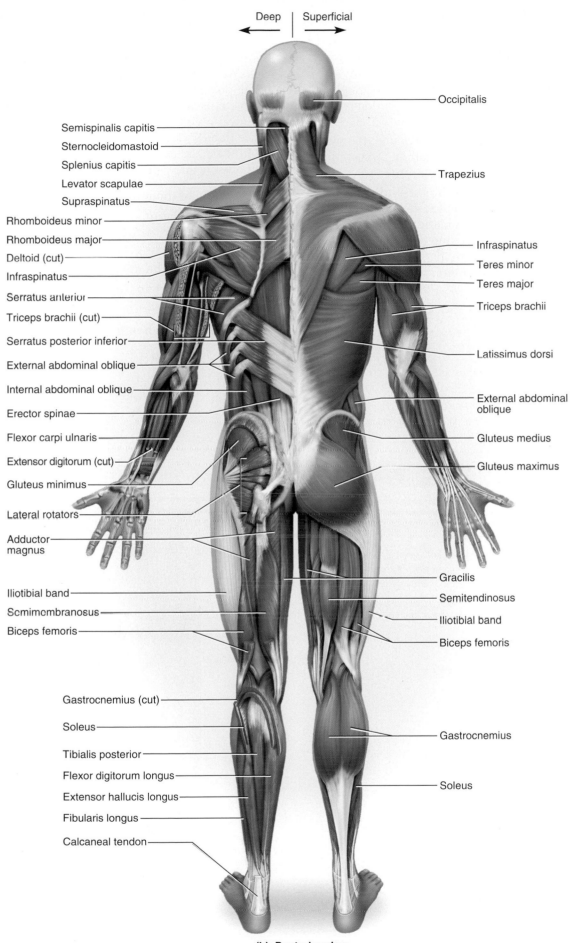

Deep | Superficial

Occipitalis

Semispinalis capitis

Sternocleidomastoid

Splenius capitis

Levator scapulae

Supraspinatus

Rhomboideus minor

Rhomboideus major

Deltoid (cut)

Infraspinatus

Serratus anterior

Triceps brachii (cut)

Serratus posterior inferior

External abdominal oblique

Internal abdominal oblique

Erector spinae

Flexor carpi ulnaris

Extensor digitorum (cut)

Gluteus minimus

Lateral rotators

Adductor magnus

Iliotibial band

Semimembranosus

Biceps femoris

Gastrocnemius (cut)

Soleus

Tibialis posterior

Flexor digitorum longus

Extensor hallucis longus

Fibularis longus

Calcaneal tendon

Trapezius

Infraspinatus

Teres minor

Teres major

Triceps brachii

Latissimus dorsi

External abdominal oblique

Gluteus medius

Gluteus maximus

Gracilis

Semitendinosus

Iliotibial band

Biceps femoris

Gastrocnemius

Soleus

**(b) Posterior view**

- Locate the origins and insertions of muscles on an articulated skeleton. Some study skeletons are painted and labeled to show these. This helps you visualize the locations of muscles and understand how they produce particular joint actions.

- Study the derivation of each muscle name; the name usually describes the muscle's location, appearance, origin, insertion, or action.

- Say the names aloud to yourself or a study partner. It is harder to remember and spell terms you cannot pronounce, and silent pronunciation is not nearly as effective as speaking and hearing the names. Pronunciation guides are provided in the muscle tables for all but the most obvious cases.

### Before You Go On

*Answer the following questions to test your understanding of the preceding section:*

1. *List some functions of the muscular system other than movement of the body.*

2. *Describe the relationship of endomysium, perimysium, and epimysium to each other. Which of these separates one fascicle from another? Which separates one muscle from another?*

3. *Distinguish between direct and indirect muscle attachments to bones.*

4. *Define* origin, insertion, belly, action, *and* innervation.

5. *Describe the five basic muscle shapes (fascicle arrangements).*

6. *Distinguish between a synergist, antagonist, and fixator. Explain how each of these may affect the action of an agonist.*

7. *In muscle names, what do the words* brevis, teres, digitorum, pectoralis, triceps, *and* profundus *mean?*

# Muscles of the Head and Neck

### Objectives

When you have completed this section, you should be able to

- name and locate the muscles that produce facial expressions;

- name and locate the muscles used for chewing and swallowing;

- name and locate the neck muscles that move the head; and

- identify the origin, insertion, action, and innervation of any of these muscles.

Figure 10.5 shows some of the muscles of the head and neck. We will treat the muscles of this region from a functional perspective, thus placing them in the following groups: muscles of facial expression, muscles of chewing and swallowing, and muscles that move the head as a whole (tables 10.2–10.4).

In these tables and throughout the rest of the chapter, each muscle entry provides the following information:

- the name of the muscle;

- the pronunciation of the name, unless it is self-evident or uses words whose pronunciations have been provided in a recent entry;

- the actions of the muscle;

- the muscle's origin, indicated by the letter O;

- the muscle's insertion, indicated by the letter I; and

- the muscles innervation, indicated by the letter N.

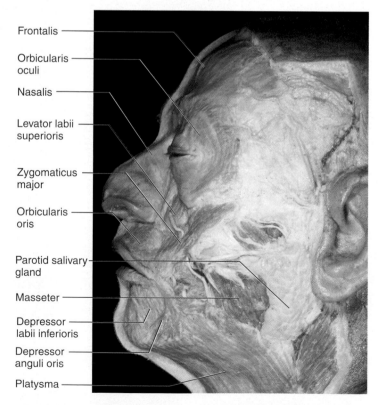

Frontalis

Orbicularis oculi

Nasalis

Levator labii superioris

Zygomaticus major

Orbicularis oris

Parotid salivary gland

Masseter

Depressor labii inferioris

Depressor anguli oris

Platysma

**FIGURE 10.5**  **Some Muscles of Facial Expression in the Cadaver.**

| TABLE 10.2 | Muscles of Facial Expression |
|---|---|

Humans have much more expressive faces than other mammals because of a complex array of muscles that insert in the dermis and subcutaneous tissues. These muscles tense the skin and produce such expressions as a pleasant smile, a threatening scowl, a puzzled frown, or a flirtatious wink (fig. 10.6). They add subtle shades of meaning to our spoken words. Facial muscles also contribute directly to speech, chewing, and other oral functions. All but one of these muscles are innervated by the facial nerve (cranial nerve VII). This nerve is especially vulnerable to injury from lacerations and skull fractures, which can paralyze the muscles and cause parts of the face to sag.

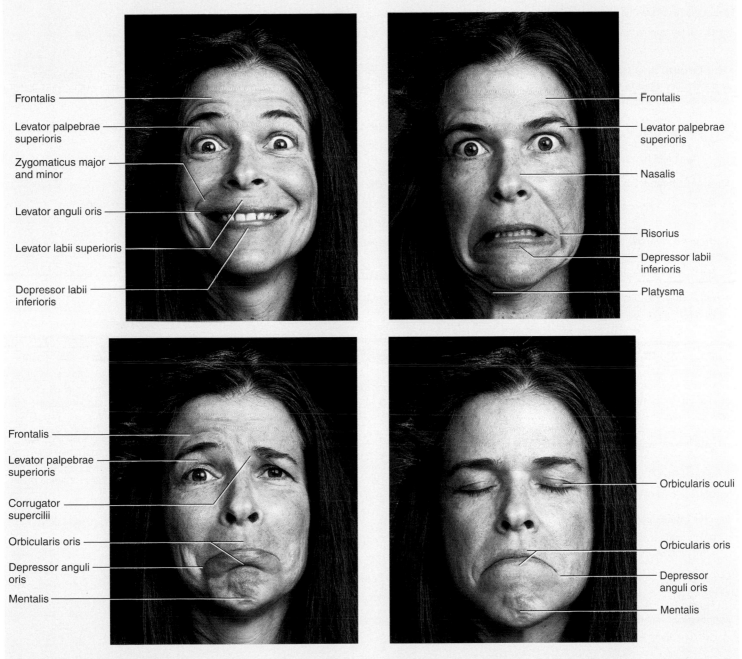

**FIGURE 10.6 Expressions Produced by Several of the Facial Muscles.** The ordinary actions of these muscles are usually more subtle than these demonstrations.

| **TABLE 10.2** | Muscles of Facial Expression *(cont.)* |
|---|---|

**The Scalp.**    The *occipitofrontalis* overlies the dome of the cranium. It is divided into the *frontalis* of the forehead and *occipitalis* at the rear of the head, named for the frontal and occipital bones underlying them. They are connected to each other by a broad aponeurosis, the **galea aponeurotica**[13] (gay-LEE-UH APO-oh-new-ROT-ih-cuh) (fig. 10.7).

**Occipitalis (oc-SIP-ih-TAY-lis)**    Retracts scalp

| **O:** Superior nuchal line and temporal bone | **I:** Galea aponeurotica | **N:** Facial n. |
|---|---|---|

**Frontalis (frun-TAY-lis)**    Elevates eyebrows in glancing upward and expressions of surprise or fright; draws scalp forward

| **O:** Galea aponeurotica | **I:** Subcutaneous tissue of eyebrows | **N:** Facial n. |
|---|---|---|

**The Orbital and Nasal Regions.**    The *orbicularis oculi* is a sphincter of the eyelid that encircles and closes the eye. The *levator palpebrae superioris* lies deep to the orbicularis oculi, in the eyelid and orbital cavity, and opens the eye. Other muscles in this group move the eyelids and skin of the forehead and dilate the nostrils. Muscles within the orbit that move the eyeball itself are discussed in chapter 16.

**Orbicularis Oculi[14] (or-BIC-you-LERR-is OC-you-lye)**    Sphincter of the eye; closes eye in blinking, squinting, and sleep; aids in flow of tears across eye

| **O:** Lacrimal bone, adjacent regions of frontal bone and maxilla, medial angle of eyelids | **I:** Upper and lower eyelids, skin around margin of orbit | **N:** Facial n. |
|---|---|---|

**Levator Palpebrae Superioris[15] (leh-VAY-tur pal-PEE-bree soo-PEER-ee-OR-is)**    Elevates upper eyelid, opens eye

| **O:** Lesser wing of sphenoid in posterior wall of orbit | **I:** Upper eyelid | **N:** Oculomotor n. |
|---|---|---|

**Corrugator Supercilii[16] (COR-oo-GAY-tur SOO-per-SIL-ee-eye)**    Draws eyebrows medially and downward in frowning and concentration; reduces glare of bright sunlight

| **O:** Medial end of supraorbital margin | **I:** Skin of eyebrow | **N:** Facial n. |
|---|---|---|

**Nasalis[17] (nay-ZAIL-is)**    Widens nostrils; narrows internal air passage between vestibule and nasal cavity

| **O:** Maxilla just lateral to nose | **I:** Bridge and alar cartilages of nose | **N:** Facial n. |
|---|---|---|

**The Oral Region.**    The mouth is the most expressive part of the face, and movement of the lips is necessary for intelligible speech; thus it is not surprising that the muscles here are especially diverse. The *orbicularis oris* is a complex of muscles in the lips that encircle the mouth; until recently it was misinterpreted as a sphincter or circular muscle, but it is actually composed of four independent quadrants that interlace and give only an appearance of circularity. Other muscles in this region approach the lips from all directions and thus draw the lips or angles (corners) of the mouth upward, laterally, and downward. Some of these have origins or insertions in a complex cord called the **modiolus**[18] just lateral to each angle of the lips (see fig. 10.6). Named for the hub of a cartwheel, the modiolus is a point of convergence and divergence of several muscles of the lower face. You can palpate it by inserting one finger just inside the corner of your lips and pinching the corner between the finger and thumb.

**Orbicularis Oris[19] (or-BIC-you-LERR-is OR-is)**    Encircles mouth, closes lips, protrudes lips as in kissing; uniquely developed in humans for speech

| **O:** Modiolus of mouth | **I:** Submucosa and dermis of lips | **N:** Facial n. |
|---|---|---|

**Levator Labii Superioris[20] (leh-VAY-tur LAY-bee-eye soo-PEER-ee-OR-is)**    Elevates and everts upper lip in sad, sneering, or serious expressions

| **O:** Zygomatic bone and maxilla near inferior margin of orbit | **I:** Muscles of upper lip | **N:** Facial n. |
|---|---|---|

**Levator Anguli Oris[21] (leh-VAY-tur ANG-you-lye OR-is)**    Elevates angle of mouth as in smiling

| **O:** Maxilla just below infraorbital foramen | **I:** Muscles at angle of mouth | **N:** Facial n. |
|---|---|---|

**Zygomaticus[22] Major (ZY-go-MAT-ih-cus)**    Draws angle of mouth upward and laterally in laughing

| **O:** Zygomatic bone | **I:** Superolateral angle of mouth | **N:** Facial n. |
|---|---|---|

**Zygomaticus Minor**    Elevates upper lip, exposes upper teeth in smiling or sneering

| **O:** Zygomatic bone | **I:** Muscles of upper lip | **N:** Facial n. |
|---|---|---|

[13]*galea* = helmet + *apo* = above + *neuro* = nerves, the brain
[14]*orb* = circle + *ocul* = eye
[15]*levator* = that which raises + *palpebr* = eyelid + *superior* = upper
[16]*corrug* = wrinkle + *supercilii* = of the eyebrow
[17]*nas* = of the nose

[18]*modiolus* = hub
[19]*orb* = circle + *or* = mouth
[20]*levat* = to raise + *labi* = lip + *superior* = upper
[21]*angul* = angle, corner + *or* = mouth
[22]*zygo* = join, unite (refers to zygomatic bone)

**FIGURE 10.7** Muscles of Facial Expression.

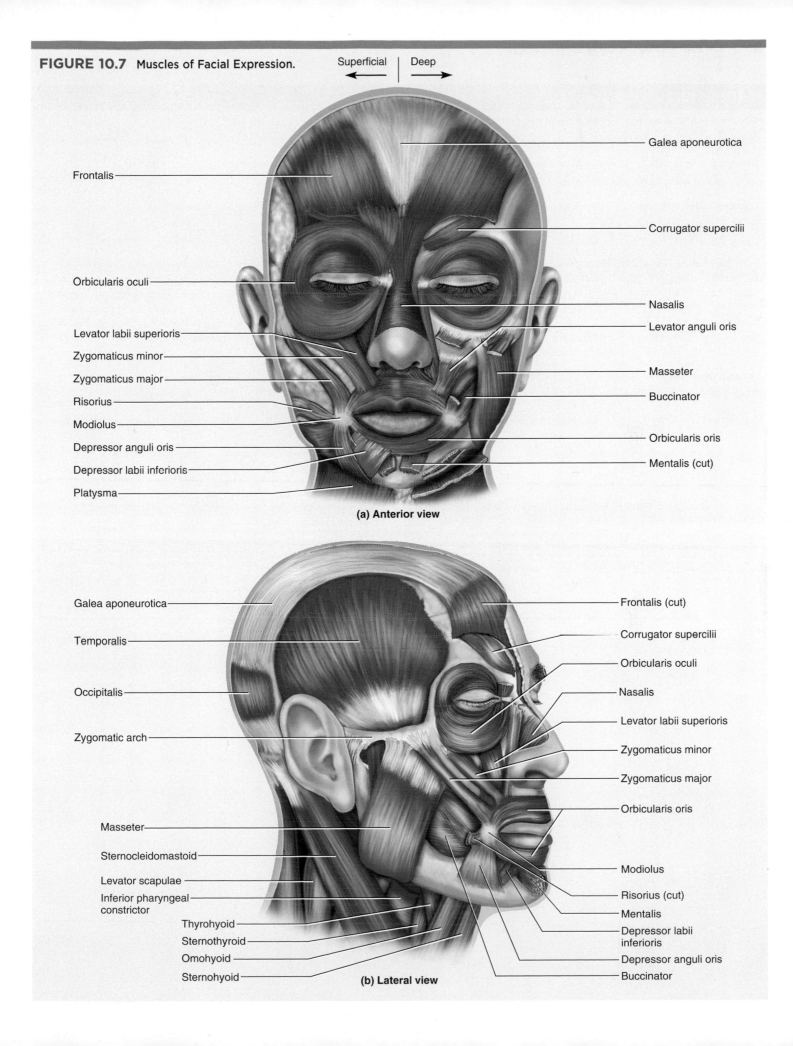

Superficial ← | → Deep

Galea aponeurotica

Frontalis

Corrugator supercilii

Orbicularis oculi

Nasalis

Levator anguli oris

Levator labii superioris

Zygomaticus minor

Zygomaticus major

Masseter

Buccinator

Risorius

Modiolus

Orbicularis oris

Depressor anguli oris

Mentalis (cut)

Depressor labii inferioris

Platysma

**(a) Anterior view**

Galea aponeurotica

Frontalis (cut)

Corrugator supercilii

Temporalis

Orbicularis oculi

Nasalis

Occipitalis

Levator labii superioris

Zygomaticus minor

Zygomatic arch

Zygomaticus major

Orbicularis oris

Masseter

Sternocleidomastoid

Modiolus

Levator scapulae

Risorius (cut)

Inferior pharyngeal constrictor

Mentalis

Thyrohyoid

Depressor labii inferioris

Sternothyroid

Omohyoid

Depressor anguli oris

Sternohyoid

Buccinator

**(b) Lateral view**

| **TABLE 10.2** | Muscles of Facial Expression *(cont.)* |
|---|---|

**Risorius**[23] **(rih-SOR-ee-us)**    Draws angle of mouth laterally in expressions of laughing, horror, or disdain

| **O:** Zygomatic arch, fascia near ear | **I:** Modiolus of mouth | **N:** Facial n. |
|---|---|---|

**Depressor Anguli Oris**[24]    Draws angle of mouth laterally and downward in opening mouth or sad expressions

| **O:** Inferior margin of mandibular body | **I:** Modiolus of mouth | **N:** Facial n. |
|---|---|---|

**Depressor Labii Inferioris**[25]    Draws lower lip downward and laterally in chewing and expressions of melancholy or doubt

| **O:** Near mental protuberance | **I:** Skin and mucosa of lower lip | **N:** Facial n. |
|---|---|---|

**The Mental and Buccal Regions.**    Adjacent to the oral orifice are the mental region (chin) and buccal region (cheeks). In addition to muscles already discussed that act directly on the lower lip, the mental region has a pair of small *mentalis muscles* extending from the upper margin of the mandible to the skin of the chin. In some people, these muscles are especially thick and have a visible dimple between them called the *mental cleft*. The *buccinator* is the muscle in the cheek. It has multiple functions in chewing, sucking, and blowing. If the cheek is inflated with air, compression of the buccinator blows it out. Sucking is achieved by contracting the buccinators to draw the cheeks inward, and then relaxing them. This action is especially important to nursing infants. To feel this action, hold your fingertips lightly on your cheeks as you make a kissing noise. You will notice the relaxation of the buccinators at the moment air is sharply drawn in through the pursed lips.

**Mentalis (men-TAY-lis)**    Elevates and protrudes lower lip in drinking, pouting, and expressions of doubt or disdain; elevates and wrinkles skin of chin

| **O:** Mandible near inferior incisors | **I:** Skin of chin at mental protuberance | **N:** Facial n. |
|---|---|---|

**Buccinator**[26] **(BUC-sin-AY-tur)**    Compresses cheek against teeth and gums; directs food between molars; retracts cheek from teeth when mouth is closing to prevent biting cheek; expels air and liquid

| **O:** Alveolar processes on lateral surfaces of maxilla and mandible | **I:** Orbicularis oris; submucosa of cheek and lips | **N:** Facial n. |
|---|---|---|

**The Cervical and Mental Region.**    The *platysma* is a thin superficial muscle of the upper chest and lower face. It is relatively unimportant, but when men shave they tend to tense the platysma to make the concavity between the jaw and neck shallower and the skin tauter.

**Platysma**[27] **(plah-TIZ-muh)**    Draws lower lip and angle of mouth downward in expressions of horror or surprise; may aid in opening mouth widely

| **O:** Fascia of deltoid and pectoralis major | **I:** Mandible; skin and subcutaneous tissue of lower face | **N:** Facial n. |
|---|---|---|

---

[23]*risor* = laughter
[24]*depress* = to lower + *angul* = angle, corner + *or* = mouth
[25]*labi* = lip + *inferior* = lower

[26]*buccinator* = trumpeter
[27]*platy* = flat

| TABLE 10.3 | Muscles of Chewing and Swallowing |
|---|---|

The following muscles contribute to facial expression and speech but are primarily concerned with the manipulation of food, including tongue movements, chewing, and swallowing.

**Extrinsic Muscles of the Tongue.**    The tongue is a very agile organ. It pushes food between the molars for chewing (mastication) and later forces the food into the pharynx for swallowing (deglutition); it is also, of course, of crucial importance to speech. Both intrinsic and extrinsic muscles are responsible for its complex movements. The intrinsic muscles consist of a variable number of vertical fascicles that extend from the superior to the inferior sides of the tongue, transverse fascicles that extend from right to left, and longitudinal fascicles that extend from root to tip. The extrinsic muscles listed below connect the tongue to other structures in the head (fig. 10.8).

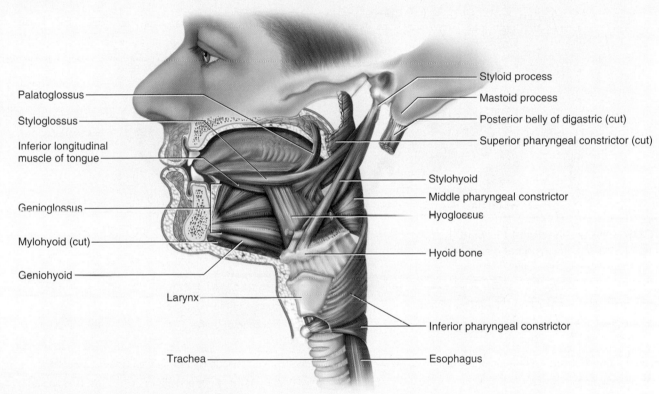

**FIGURE 10.8**  Muscles of the Tongue and Pharynx.

**Genioglossus**[28] **(JEE-nee-oh-GLOSS-us)**    Unilateral action draws tongue to one side; bilateral action depresses midline of tongue or protrudes tongue

**O:** Superior mental spine on posterior surface of mental protuberance    **I:** Ventral surface of tongue from root to apex    **N:** Hypoglossal n.

**Hyoglossus**[29] **(HI-oh-GLOSS-us)**    Depresses tongue

**O:** Body and greater horn of hyoid    **I:** Lateral and ventral surfaces of tongue    **N:** Hypoglossal n.

**Styloglossus**[30] **(STY-lo-GLOSS-us)**    Draws tongue upward and posteriorly

**O:** Styloid process of temporal bone and ligament from styloid to mandible    **I:** Dorsolateral surface of tongue    **N:** Hypoglossal n.

**Palatoglossus**[31] **(PAL-a-toe-GLOSS-us)**    Elevates root of tongue and closes oral cavity off from pharynx; forms palatoglossal arch at rear of oral cavity

**O:** Soft palate    **I:** Lateral surface of tongue    **N:** Accessory and vagus nn.

---

[28]*genio* = chin + *gloss* = tongue
[29]*hyo* = hyoid bone + *gloss* = tongue

[30]*stylo* = styloid process + *gloss* = tongue
[31]*palato* = palate + *gloss* = tongue

## TABLE 10.3    Muscles of Chewing and Swallowing *(cont.)*

**Muscles of Chewing.**    Four pairs of muscles (fig. 10.9) produce the biting and chewing movements of the mandible: the *temporalis, masseter,* and two pairs of *pterygoid* muscles. Their actions include *depression* to open the mouth for receiving food; *elevation* for biting off a piece of food or crushing it between the teeth; *protraction* so that the incisors meet in cutting off a piece of food; *retraction* to draw the lower incisors behind the upper incisors and make the rear teeth meet; and *lateral* and *medial excursion,* the side-to-side movements that grind food between the rear teeth.

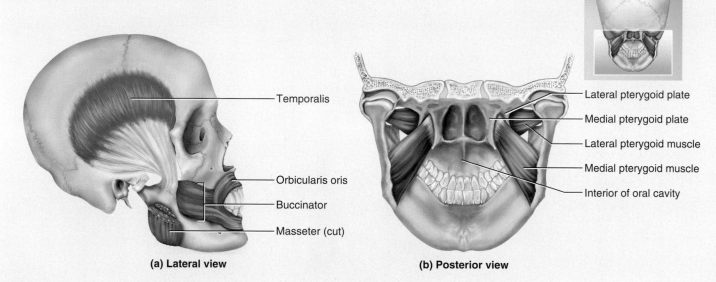

**(a) Lateral view**

Temporalis
Orbicularis oris
Buccinator
Masseter (cut)

**(b) Posterior view**

Lateral pterygoid plate
Medial pterygoid plate
Lateral pterygoid muscle
Medial pterygoid muscle
Interior of oral cavity

**FIGURE 10.9  Muscles of Chewing.** (a) Right lateral view. In order to expose the insertion of the temporalis muscle on the mandible, part of the zygomatic arch and masseter muscle are removed. (b) View of the pterygoid muscles looking into the oral cavity from behind the skull.

**Temporalis[32] (TEM-po-RAY-liss)**    Elevation, retraction, and lateral and medial excursion of the mandible

| **O:** Temporal lines and temporal fossa of cranium | **I:** Coronoid process and anterior border of mandibular ramus | **N:** Mandibular n. |

**Masseter[33] (ma-SEE-tur)**    Elevation of the mandible, with smaller roles in protraction, retraction, and lateral and medial excursion

| **O:** Zygomatic arch | **I:** Lateral surface of mandibular ramus and angle | **N:** Mandibular n. |

**Medial Pterygoid[34] (TERR-ih-goyd)**    Elevation, protraction, and lateral and medial excursion of the mandible

| **O:** Medial surface of lateral pterygoid plate, palatine bone, lateral surface of maxilla near molar teeth | **I:** Medial surface of mandibular ramus and angle | **N:** Mandibular n. |

**Lateral Pterygoid**    Depression (in wide opening of the mouth), protraction, and lateral and medial excursion of the mandible

| **O:** Lateral surfaces of lateral pterygoid plate and greater wing of sphenoid | **I:** Neck of mandible (just below condyle); articular disc and capsule of temporomandibular joint | **N:** Mandibular n. |

**Hyoid Muscles—Suprahyoid Group.**    Several aspects of chewing and swallowing are aided by eight pairs of *hyoid muscles* associated with the hyoid bone (fig. 10.10). The *suprahyoid group* is composed of the four muscles superior to the hyoid—the *digastric, geniohyoid, mylohyoid,* and *stylohyoid.* The digastric is an unusual muscle, named for its two bellies. Its *posterior belly* arises from the mastoid notch of the cranium and slopes downward and forward. The *anterior belly* arises from a trench called the *digastric fossa* on the inner surface of the mandibular body. It slopes downward and backward. The two bellies meet at a constriction, the *intermediate tendon.* This tendon passes through a connective tissue loop, the *fascial sling,* attached to the hyoid bone. Thus, when the two bellies of the digastric contract, they pull upward on the hyoid; but if the hyoid is fixed from

---

[32]*temporalis* = of the temporal region of the head
[33]*masset* = chew

[34]*pteryg* = wing + *oid* = resembling; pterygoid plate of sphenoid

below, the digastric aids in wide opening of the mouth. The lateral pterygoids are more important in wide mouth opening, with the digastrics coming into play only in extreme opening, as in yawning or taking a large bite of an apple.

**Digastric**[35]  Depresses mandible when hyoid is fixed; opens mouth widely, as when ingesting food or yawning; elevates hyoid when mandible is fixed

**O:** Mastoid notch of temporal bone, digastric fossa of mandible
**I:** Hyoid bone, via fascial sling
**N:** Posterior belly: facial n. Anterior belly: mylohyoid n. (a branch of the trigeminal n.)

**Geniohyoid**[36] **(JEE-nee-oh-HY-oyd)**  Depresses mandible when hyoid is fixed; elevates and protracts hyoid when mandible is fixed

**O:** Inferior mental spine of mandible
**I:** Hyoid bone
**N:** Spinal nerve C1 via hypoglossal n.

**Mylohyoid**[37]  Spans mandible from side to side and forms floor of mouth; elevates floor of mouth in initial stage of swallowing

**O:** Mylohyoid line near inferior margin of mandible
**I:** Hyoid bone
**N:** Mylohyoid n.

**Stylohyoid**  Elevates and retracts hyoid, elongating floor of mouth; roles in speech, chewing, and swallowing are not yet clearly understood

**O:** Styloid process of temporal bone
**I:** Hyoid bone
**N:** Facial n.

### Hyoid Muscles—Infrahyoid Group.

The infrahyoid muscles are inferior to the hyoid bone. By fixing the hyoid from below, they enable the suprahyoid muscles to open the mouth. The *omohyoid* is unusual in that it arises from the shoulder, passes under the sternocleidomastoid, and then ascends to the hyoid bone. Like the digastric, it has two bellies. The *thyrohyoid,* named for the hyoid bone and the large shield-shaped *thyroid cartilage* of the larynx, helps prevent choking. It elevates the thyroid cartilage during swallowing so that its opening is sealed by a flap of tissue, the *epiglottis.* You can feel this effect by placing your finger on the "Adam's apple" (the anterior prominence of the thyroid cartilage) and feeling it bob up as you swallow. The *sternothyroid* muscle then pulls the larynx down again so you can resume breathing; it is the only infrahyoid muscle with no connection to the hyoid bone. The infrahyoid muscles that act on the larynx are regarded as the *extrinsic muscles* of the larynx. The *intrinsic muscles,* considered in chapter 22, are concerned with control of the vocal cords and laryngeal opening. The *ansa cervicalis,* which innervates three of these muscles, is a loop of nerve on the side of the neck formed by certain fibers from cervical nerves 1 to 3.

**Omohyoid**[38]  Depresses hyoid after it has been elevated

**O:** Superior border of scapula
**I:** Hyoid bone
**N:** Ansa cevicalis

**Sternohyoid**[39]  Depresses hyoid after it has been elevated

**O:** Manubrium of sternum, medial end of clavicle
**I:** Hyoid bone
**N:** Ansa cevicalis

**Thyrohyoid**[40]  Depresses hyoid; with hyoid fixed, elevates larynx as in singing high notes

**O:** Thyroid cartilage of larynx
**I:** Hyoid bone
**N:** Spinal nerve C1 via hypoglossal n.

**Sternothyroid**  Protracts larynx after it has been elevated in swallowing and speech; aids in singing low notes

**O:** Manubrium of sternum, costal cartilage 1
**I:** Thyroid cartilage of larynx
**N:** Ansa cevicalis

### Muscles of the Pharynx.

The three pairs of *pharyngeal constrictors* encircle the pharynx on its posterior and lateral sides, forming a muscular funnel (see fig. 10.8).

**Pharyngeal Constrictors (three muscles)**  During swallowing, contract in order from *superior* to *middle* to *inferior constrictor* to drive food into esophagus

**O:** Medial pterygoid plate, mandible, hyoid, stylohyoid ligament, cricoid and thyroid cartilages of larynx
**I:** Median pharyngeal raphe (seam on posterior side of pharynx); basilar part of occipital bone
**N:** Glossopharyngeal and vagus nn.

---

[35]*di* = two + *gastr* = bellies
[36]*genio* = chin
[37]*mylo* = mill, molar tooth
[38]*omo* = shoulder
[39]*sterno* = chest, sternum
[40]*thyro* = shield, thyroid cartilage

**TABLE 10.3**    Muscles of Chewing and Swallowing *(cont.)*

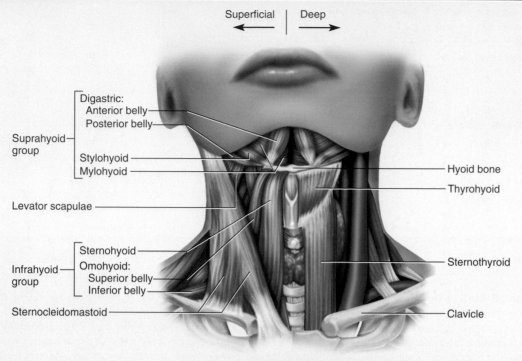

Superficial ← | → Deep

Digastric:
  Anterior belly
  Posterior belly

Suprahyoid group

  Stylohyoid
  Mylohyoid

Levator scapulae

Infrahyoid group
  Sternohyoid
  Omohyoid:
    Superior belly
    Inferior belly

Sternocleidomastoid

Hyoid bone

Thyrohyoid

Sternothyroid

Clavicle

**(a) Anterior view**

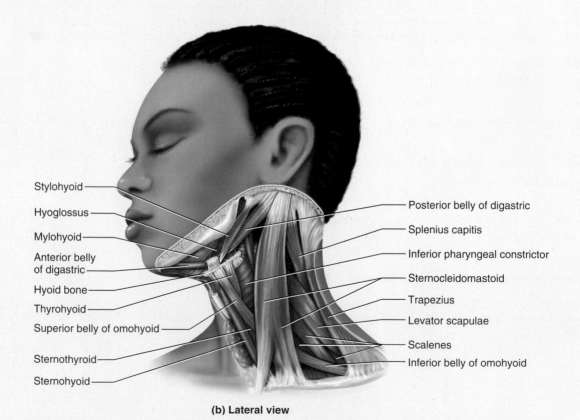

Stylohyoid

Hyoglossus

Mylohyoid

Anterior belly of digastric

Hyoid bone

Thyrohyoid

Superior belly of omohyoid

Sternothyroid

Sternohyoid

Posterior belly of digastric

Splenius capitis

Inferior pharyngeal constrictor

Sternocleidomastoid

Trapezius

Levator scapulae

Scalenes

Inferior belly of omohyoid

**(b) Lateral view**

**FIGURE 10.10    Muscles of the Neck.**    The geniohyoid is deep to the mylohyoid and can be seen in figure 10.8.

## TABLE 10.4        Muscles Acting on the Head

Muscles that move the head originate on the vertebral column, thoracic cage, and pectoral girdle and insert on the cranial bones. Their actions include flexion (tipping the head forward), extension (holding the head erect), hyperextension (as in looking upward), lateral flexion (tilting the head to one side), and rotation (turning the head to look left or right). Flexion, extension, and hyperextension involve simultaneous action of the right and left muscles of a pair; the other actions require the muscle on one side to contract more strongly than its mate. Many head actions result from a combination of these movements—for example, looking up over the shoulder involves a combination of rotation and hyperextension.

Depending on the relations of the muscle origins and insertions, a muscle may cause a *contralateral* movement of the head (toward the opposite side, as when contraction of a muscle on the left turns the face toward the right) or an *ipsilateral* movement (toward the same side as the muscle, as when contraction of a muscle on the left tilts the head to the left).

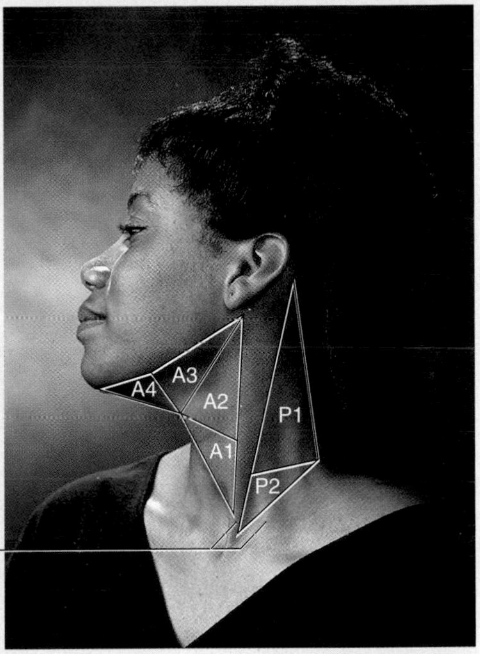

**Flexors of the Neck.**    The prime mover of neck flexion is the *sternocleidomastoid,* a thick muscular cord that extends from the upper chest (sternum and clavicle) to the mastoid process behind the ear (see fig. 10.10). This is most easily seen when the head is rotated to one side and slightly extended. To visualize the action of a single sternocleidomastoid, place the index finger of your left hand on your left mastoid process and the index finger of your right hand on your sternal notch. Now contract your left sternocleidomastoid to bring your two fingertips as close together as possible. You will see that this tilts your face downward and to the right.

As the sternocleidomastoid passes obliquely across the neck, it divides it into *anterior* and *posterior triangles* (fig. 10.11). Other muscles and landmarks subdivide each of these into smaller triangles of surgical importance.

The three *scalenes*[41] (anterior, middle, and posterior), located on the side of the neck, are named for being arranged somewhat like a staircase. Their actions are similar so they are considered collectively below.

**Anterior triangles**
A1. Muscular
A2. Carotid
A3. Submandibular
A4. Suprahyoid

**Posterior triangles**
P1. Occipital
P2. Omoclavicular

Sternocleidomastoid

**FIGURE 10.11   Triangles of the Neck.**   The sternocleidomastoid muscle separates the anterior triangles from the posterior triangles.

**Sternocleidomastoid[42] (STIR-no-CLY-do-MAST-oyd)**    Unilateral action tilts head forward and toward the opposite side, as in looking down toward one foot. Rotates head directly to the side opposite the muscle when other muscles prevent the forward tilt. Bilateral action draws the head straight forward and down, as when eating or looking between the feet. Aids in deep breathing when head is fixed.

| **O:** Manubrium of sternum, medial one-third of clavicle | **I:** Mastoid process and lateral half of superior nuchal line | **N:** Accessory n., spinal nerves C2–C3 |

**Scalenes (SCAY-leens) (three muscles)**    Unilateral contraction causes ipsilateral flexion or contralateral rotation (tilts head toward same shoulder, or rotates face away), depending on action of other muscles. Bilateral contraction flexes head. If spine is fixed, scalenes elevate ribs 1–2 and aid in breathing.

| **O:** Transverse processes of all cervical vertebrae (C1–C7) | **I:** Ribs 1–2 | **N:** Ventral rami of C3–C8 |

**Extensors of the Neck.**    The extensors are located mainly in the nuchal region (back of the neck; fig. 10.12) and therefore tend to hold the head erect or draw it back. The *trapezius* is the most superficial of these. It extends from the nuchal region over the shoulders and halfway down the back. It is named for the fact that the right and left trapezii together form a diamond or trapezoidal shape (see fig. 10.4b). The *splenius* is a deeper, elongated muscle with *splenius capitis* and *splenius cervicis* regions in the head and neck, respectively. It is nicknamed the "bandage muscle" because of the way it wraps around still deeper neck muscles. One of those deeper muscles is the semispinalis, another elongated muscle with head, neck, and thoracic regions. Only the *semispinalis capitis* and *cervicis* are tabulated here; the semispinalis thoracis does not act on the neck, but is included in table 10.7.

**Trapezius[43] (tra-PEE-zee-us)**    Extends and laterally flexes head. See also roles in scapular movement in table 10.9

| **O:** External occipital protuberance, medial one-third of superior nuchal line, nuchal ligament, spinous processes of vertebrae C7–T3 or T4 | **I:** Acromion and spine of scapula, lateral one-third of clavicle | **N:** Accessory n., ventral rami of C3–C4 |

---

[41]*scal* = staircase
[42]*sterno* = chest, sternum + *cleido* = hammer, clavicle + *masto* = breastlike, mastoid process

[43]*trapez* = table, trapezoid

| TABLE 10.4 | Muscles Acting on the Head *(cont.)* |
|---|---|

**Splenius Capitis[44] (SPLEE-nee-us CAP-ih-tiss) and Splenius Cervicis[45] (SIR-vih-sis)**    Acting unilaterally, produce ipsilateral flexion and slight rotation of head; extend head when acting bilaterally

|  |  |  |
|---|---|---|
| **O:** Inferior half of nuchal ligament, spinous processes of vertebrae C7–T6 | **I:** Mastoid process and occipital bone just inferior to superior nuchal line; cervical vertebrae C1 to C2 or C3 | **N:** Dorsal rami of middle cervical nn. |

**Semispinalis Capitis (SEM-ee-spy-NAY-lis) and Semispinalis Cervicis**    Extend and contralaterally rotate head

|  |  |  |
|---|---|---|
| **O:** Articular processes of vertebrae C4–C7, transverse processes of T1–T6 | **I:** Occipital bone between nuchal lines, spinous processes of vertebrae C2–C5 | **N:** Dorsal rami of cervical and thoracic nn. |

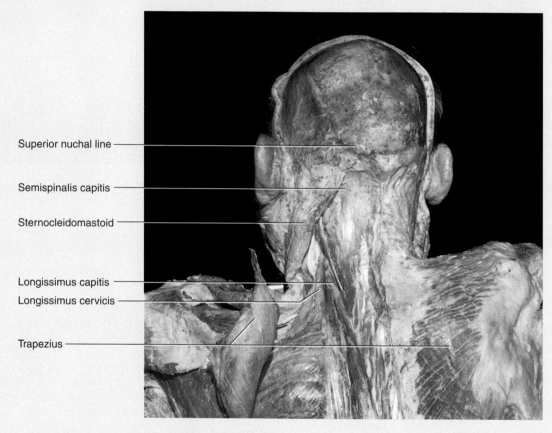

Superior nuchal line

Semispinalis capitis

Sternocleidomastoid

Longissimus capitis
Longissimus cervicis

Trapezius

**FIGURE 10.12**   Muscles of the Shoulder and Nuchal Regions.

## Think About It

*Of the muscles you have studied so far, name three that you would consider intrinsic muscles of the head and three that you would classify as extrinsic. Explain your reason for each.*

## Before You Go On

*Answer the following questions to test your understanding of the preceding section:*

8. *Name two muscles that elevate the upper lip and two that depress the lower lip.*

9. *Name the four paired muscles of mastication and state where they insert on the mandible.*

10. *Distinguish between the functions of the suprahyoid and infrahyoid muscles.*

[44]*splenius* = bandage + *capitis* = of the head
[45]*cervicis* = of the neck

# Muscles of the Trunk

## Objectives

When you have completed this section, you should be able to

- name and locate the muscles of respiration and explain how they affect abdominal pressure;
- name and locate the muscles of the abdominal wall, back, and pelvic floor; and

- identify the origin, insertion, action, and innervation of any of these muscles.

In this section, we will examine muscles of the trunk of the body in three functional groups concerned with respiration, support of the abdominal wall and pelvic floor, and movement of the vertebral column (tables 10.5–10.8). In the illustrations, you will note some major muscles that are not discussed in the associated tables—for example, the pectoralis major and serratus anterior. Although they are *located in* the trunk, they *act upon* the limbs and limb girdles, and are further discussed in table 10.9.

| TABLE 10.5 | Muscles of Respiration |
|---|---|

We breathe primarily by means of muscles that enclose the thoracic cavity—the diaphragm, external intercostal muscles, and internal intercostal muscles (fig. 10.13).

The *diaphragm* is a muscular dome between the thoracic and abdominal cavities, bulging upward against the base of the lungs. It has openings for passage of the esophagus, major blood vessels, and nerves between the two cavities. Its fibers converge from the margins toward a fibrous **central tendon.** When the diaphragm contracts, it flattens slightly and enlarges the thoracic cavity, causing air intake *(inspiration);* when it relaxes, it rises and shrinks the thoracic cavity, expelling air *(expiration).*

Eleven pairs of *external intercostal muscles* lie superficially between the ribs, extending from the rib tubercle posteriorly almost to the beginning of the costal cartilage anteriorly. Each one slopes downward and anteriorly from one rib to the next inferior one. The 11 pairs of *internal intercostal muscles* lie deep to the external intercostals and extend from the margin of the sternum to the angles of the rib. They are thickest in the region between the costal cartilages and grow thinner in the region where they overlap the internal intercostals. Their fibers slope downward and posteriorly from each rib to the one below, at nearly right angles to the external intercostals. The primary function of the intercostal muscles is to stiffen the thoracic cage during respiration so that it does not cave inward when the diaphragm descends. However, they also contribute to enlargement and contraction of the thoracic cage and thus add to the air volume that ventilates the lungs.

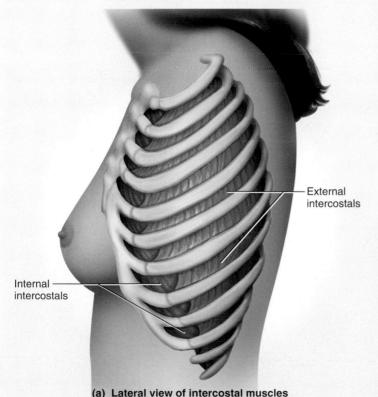

**(a) Lateral view of intercostal muscles**

External intercostals

Internal intercostals

**FIGURE 10.13   Muscles of Respiration.**   (a) The intercostal muscles. (b) The diaphragm.

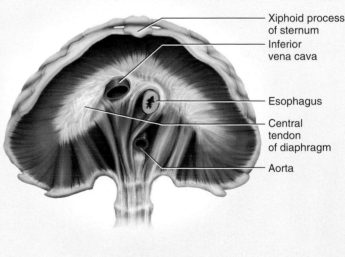

Xiphoid process of sternum

Inferior vena cava

Esophagus

Central tendon of diaphragm

Aorta

**(b) Inferior view of diaphragm**

| **TABLE 10.5** | Muscles of Respiration *(cont.)* |
|---|---|

Many other muscles of the chest and abdomen contribute significantly to breathing: the sternocleidomastoid and scalenes of the neck; pectoralis major and serratus anterior of the chest; latissimus dorsi of the lower back; internal and external abdominal obliques and transverse abdominal muscle; and even some of the anal muscles. The respiratory actions of all these muscles are described in chapter 22.

**Diaphragm[46] (DY-ah-fram)**   Prime mover of inspiration (responsible for about two-thirds of air intake); contracts in preparation for sneezing, coughing, crying, laughing, and weight lifting; contraction compresses abdominal viscera and aids in childbirth and expulsion of urine and feces

| **O:** Xiphoid process of sternum; ribs and costal cartilages 7–12; lumbar vertebrae | **I:** Central tendon of diaphragm | **N:** Phrenic nn. |
|---|---|---|

**External Intercostals[47] (IN-tur-COSS-tulz)**   When scalenes fix rib 1, external intercostals elevate and protract ribs 2–12; this expands the thoracic cavity and creates a partial vacuum causing inflow of air. Exercise a braking action during expiration so that expiration is not overly abrupt.

| **O:** Inferior margins of ribs 1–11 | **I:** Each inserts on superior margin of next lower rib | **N:** Intercostal nn. |
|---|---|---|

**Internal Intercostals**   When quadratus lumborum and other muscles fix rib 12, internal intercostals depress and retract ribs; this compresses the thoracic cavity and expels air. Used only in forceful expiration, not in relaxed breathing.

| **O:** Superior margins and costal cartilages of ribs 2–12; margin of sternum | **I:** Each inserts on inferior margin of next higher rib | **N:** Intercostal nn. |
|---|---|---|

---

[46]*dia* = across + *phragm* = partition
[47]*inter* = between + *costa* = rib

## Think About It

*What muscles are eaten as "spare ribs"? What is the tough fibrous membrane between the meat and the bone?*

## TABLE 10.6    Muscles of the Anterior Abdominal Wall

Unlike the thoracic cavity, the abdominal cavity has little skeletal support. It is enclosed, however, in layers of broad flat muscles whose fibers run in different directions, strengthening the abdominal wall on the same principle as the alternating layers of plywood. Three layers of muscle enclose the lateral abdominal region and extend about halfway across the anterior abdomen (fig. 10.14). The most superficial layer is the *external abdominal oblique.* Its fibers pass downward and anteriorly. The next deeper layer is the *internal abdominal oblique,* whose fibers pass upward and anteriorly, roughly perpendicular to those of the external oblique. The deepest layer is the *transverse abdominal (transversus abdominis),* with horizontal fibers. Anteriorly, a pair of vertical *rectus abdominis* muscles extend from sternum to pubis. These are divided into segments by three transverse **tendinous intersections,** giving them an appearance that body builders nickname the "six pack."

The tendons of the oblique and transverse muscles are *aponeuroses*—broad fibrous sheets which continue medially and inferiorly (figs. 10.15 and 10.16). At the rectus abdominis, they diverge and pass around its anterior and posterior sides, enclosing the muscle in a vertical sleeve called the **rectus sheath.** They meet again at a median line called the **linea alba** between the rectus muscles. Another line, the **linea semilunaris,** marks the lateral boundary where the rectus sheath meets the aponeurosis. The aponeurosis of the external oblique also forms a cordlike **inguinal ligament** at its inferior margin. This extends obliquely from the anterior superior spine of the ilium to the pubis. The linea alba, linea semilunaris, and inguinal ligament are externally visible on a person with good muscle definition (see fig. B.2 in atlas B following this chapter).

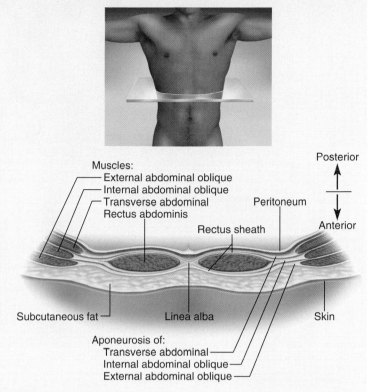

**FIGURE 10.14   Cross Section of the Anterior Abdominal Wall.**

**External Abdominal Oblique**    Supports abdominal viscera against pull of gravity; stabilizes vertebral column during heavy lifting; maintains posture; compresses abdominal organs, thus aiding in forceful expiration; aids in childbirth, urination, defecation, and vomiting. Unilateral contraction causes contralateral rotation of waist

| | | |
|---|---|---|
| **O:** Ribs 5–12 | **I:** Anterior half of iliac crest, pubic symphysis, and superior margin of pubis | **N:** Ventral rami of spinal nerves T7–T12 |

**Internal Abdominal Oblique**    Same as external oblique except that unilateral contraction causes ipsilateral rotation of waist

| | | |
|---|---|---|
| **O:** Inguinal ligament, iliac crest, and thoracolumbar fascia | **I:** Ribs 10–12, costal cartilages 7–10, pubis | **N:** Ventral rami of spinal nerves T7–L1 |

**Transverse Abdominal**    Compresses abdominal contents, with same effects as external oblique, but does not contribute to movements of vertebral column

| | | |
|---|---|---|
| **O:** Inguinal ligament, iliac crest, thoracolumbar fascia, costal cartilages 7–12 | **I:** Linea alba, pubis, aponeurosis of internal oblique | **N:** Ventral rami of spinal nerves T7–L1 |

**Rectus Abdominis (REC-tus ab-DOM-ih-nis)**    Flexes lumbar region of vertebral column, producing forward bending at the waist

| | | |
|---|---|---|
| **O:** Pubic symphysis and superior margin of pubis | **I:** Xiphoid process, costal cartilages 5–7 | **N:** Ventral rami of spinal nerves T6–T12 |

**TABLE 10.6** | Muscles of the Anterior Abdominal Wall *(cont.)*

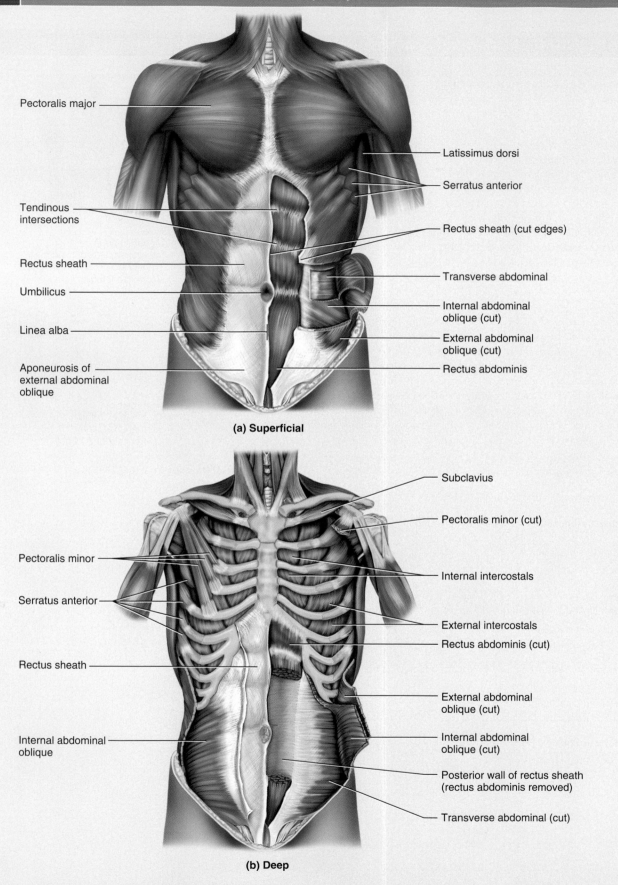

Pectoralis major

Tendinous
intersections

Rectus sheath

Umbilicus

Linea alba

Aponeurosis of
external abdominal
oblique

Latissimus dorsi

Serratus anterior

Rectus sheath (cut edges)

Transverse abdominal

Internal abdominal
oblique (cut)

External abdominal
oblique (cut)

Rectus abdominis

**(a) Superficial**

Pectoralis minor

Serratus anterior

Rectus sheath

Internal abdominal
oblique

Subclavius

Pectoralis minor (cut)

Internal intercostals

External intercostals

Rectus abdominis (cut)

External abdominal
oblique (cut)

Internal abdominal
oblique (cut)

Posterior wall of rectus sheath
(rectus abdominis removed)

Transverse abdominal (cut)

**(b) Deep**

**FIGURE 10.15  Thoracic and Abdominal Muscles.**   (a) Superficial muscles. The left rectus sheath is cut away to expose the rectus abdominis muscle. (b) Deep muscles. On the anatomical right, the external oblique has been removed to expose the internal oblique and the pectoralis major removed to expose the pectoralis minor. On the anatomical left, the internal oblique has been cut to expose the transverse abdominal, and the middle of the rectus abdominis has been cut out to expose the posterior rectus sheath.

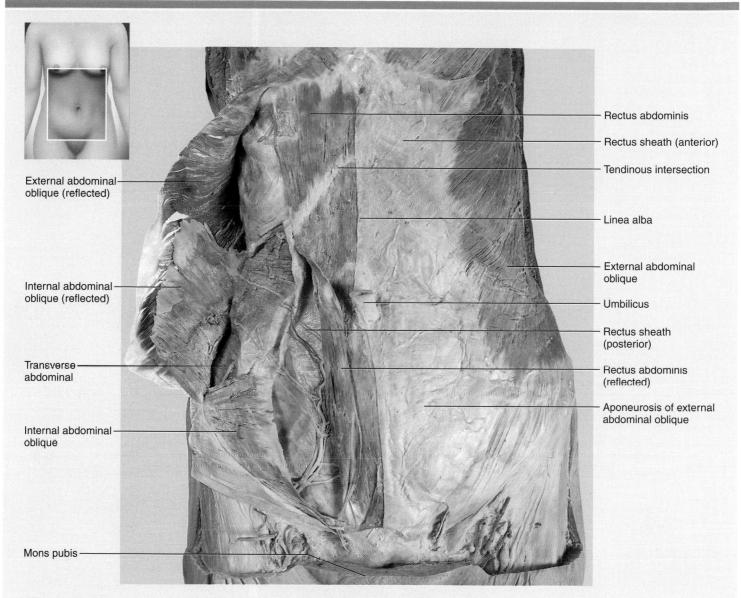

External abdominal oblique (reflected)

Internal abdominal oblique (reflected)

Transverse abdominal

Internal abdominal oblique

Mons pubis

Rectus abdominis

Rectus sheath (anterior)

Tendinous intersection

Linea alba

External abdominal oblique

Umbilicus

Rectus sheath (posterior)

Rectus abdominis (reflected)

Aponeurosis of external abdominal oblique

**FIGURE 10.16** Thoracic and Abdominal Muscles of the Cadaver.

## TABLE 10.7    Muscles of the Back

Muscles of the back primarily extend, rotate, and laterally flex the vertebral column. The most prominent superficial back muscles are the latissimus dorsi and trapezius (fig. 10.17), but they are concerned with upper limb movements and covered in tables 10.9 and 10.10. Deep to these are the *serratus posterior superior* and *inferior* (fig. 10.18). They extend from the vertebrae to the ribs. Their function and significance remain unknown, so we will not consider them further.

Deep to these is a prominent muscle, the **erector spinae,** which runs vertically for the entire length of the back from the cranium to the sacrum. It is a thick muscle, easily palpated on each side of the vertebral column in the lumbar region. (Pork chops and T-bone steaks are erector spinae muscles.) As it ascends, it divides in the upper lumbar region into three parallel columns (figs. 10.18 and 10.19). The most lateral of these is the **iliocostalis,** which from inferior to superior is divided into the *iliocostalis lumborum, iliocostalis thoracis,* and *iliocostalis cervicis* (lumbar, thoracic, and cervical regions). The next medial column is the **longissimus,** divided from inferior to superior into the *longissimus thoracis, longissimus cervicis, and longissimus capitis* (thoracic, cervical, and cephalic regions). The most medial column is the **spinalis,** divided into *spinalis thoracis, spinalis cervicis,* and *spinalis capitis.* The functions of all three columns are sufficiently similar that we will treat them collectively as the erector spinae.

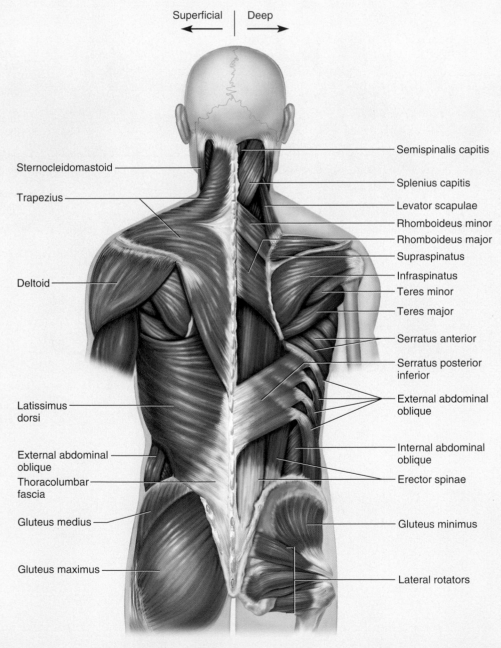

**FIGURE 10.17    Neck, Back, and Gluteal Muscles.**    The most superficial muscles are shown on the left, and the next deeper layer on the right.

The major deep muscles are the *semispinalis thoracis* in the thoracic region and *quadratus lumborum* in the lumbar region. The erector spinae and quadratus lumborum are enclosed in a fibrous sheath called the **thoracolumbar fascia,** which is the origin of some of the abdominal and lumbar muscles. The *multifidus* is actually a collective name for a series of tiny muscles that connect adjacent vertebrae to each other from the cervical to lumbar region.

**Erector Spinae (eh-REC-tur SPY-nee)**    Extension and lateral flexion of vertebral column; the longissimus capitis also produces ipsilateral rotation of the head

**O:** Nuchal ligament, ribs 3–12, thoracic and lumbar vertebrae, median and lateral sacral crests, thoracolumbar fascia

**I:** Mastoid process, cervical and thoracic vertebrae, and all ribs

**N:** Dorsal rami of cervical to lumbar spinal nerves

**Semispinalis Thoracis (SEM-ee-spy-NAY-liss tho-RA-sis)**    Extension and contralateral rotation of vertebral column

**O:** Vertebrae T6–T10

**I:** Vertebrae C6–T4

**N:** Dorsal rami of cervical and thoracic spinal nerves

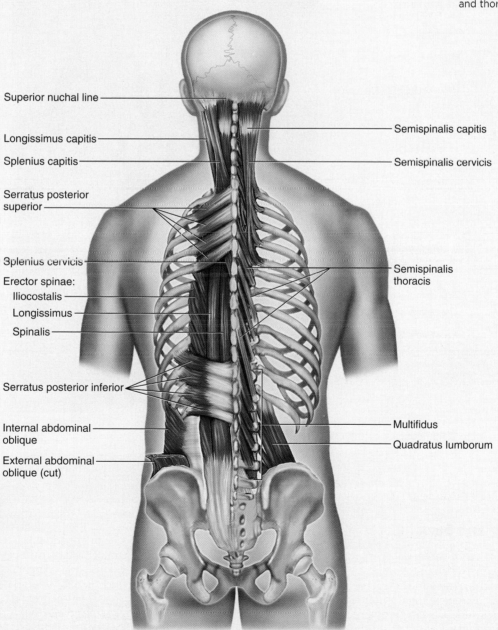

Superior nuchal line

Longissimus capitis

Splenius capitis

Serratus posterior superior

Splenius cervicis

Erector spinae:
  Iliocostalis
  Longissimus
  Spinalis

Serratus posterior inferior

Internal abdominal oblique

External abdominal oblique (cut)

Semispinalis capitis

Semispinalis cervicis

Semispinalis thoracis

Multifidus

Quadratus lumborum

**FIGURE 10.18  Muscles Acting on the Vertebral Column.**    These are deeper than the muscles in figure 10.17, and those on the right are deeper than those on the left.

| TABLE 10.7 | Muscles of the Back *(cont.)* |

**Quadratus Lumborum**[48] **(quad-RAY-tus lum-BORE-um)**    Aids respiration by fixing rib 12 and stabilizing inferior attachments of diaphragm. Unilateral contraction causes ipsilateral flexion of lumbar vertebral column; bilateral contraction extends lumbar vertebral column.

**O:** Iliac crest, iliolumbar ligament        **I:** Rib 12 and vertebrae L1–L4        **N:** Ventral rami of spinal nerves T12–L4

**Multifidus**[49] **(mul-TIFF-ih-dus)**    Stabilization of adjacent vertebrae, maintenance of posture, control of vertebral movement when erector spinae acts on vertebral column

**O:** Vertebrae C4–L5, posterior superior iliac spine, sacrum, aponeurosis of erector spinae        **I:** Laminae and spinous processes of vetrebrae superior to origins        **N:** Dorsal rami of cervical to lumbar spinal nerves

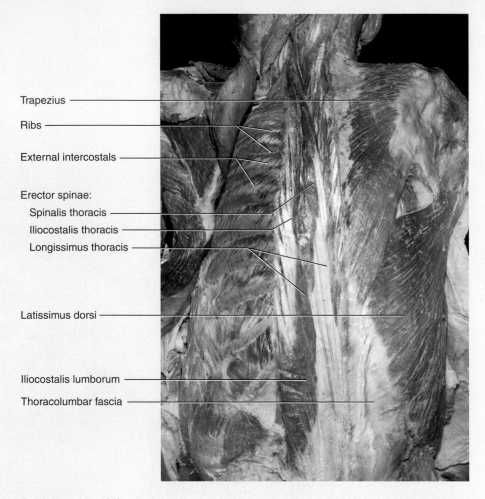

Trapezius
Ribs
External intercostals
Erector spinae:
  Spinalis thoracis
  Iliocostalis thoracis
  Longissimus thoracis
Latissimus dorsi
Iliocostalis lumborum
Thoracolumbar fascia

**FIGURE 10.19  Deep Back Muscles of the Cadaver.**

---

### INSIGHT 10.1    Clinical Application

#### Heavy Lifting and Back Injuries

When you are fully bent over forward, as in touching your toes, the erector spinae is fully stretched. Because of the *length-tension relationship* explained in chapter 11, muscles that are stretched to such extremes cannot contract very effectively. Standing up from such a position is therefore initiated by the hamstring muscles on the back of the thigh and the gluteus maximus of the buttocks. The erector spinae joins in the action when it is partially contracted.

Standing too suddenly or improperly lifting a heavy weight, however, can strain the erector spinae, cause painful muscle spasms, tear tendons and ligaments of the lower back, and rupture intervertebral discs. The lumbar muscles are adapted for maintaining posture, not for lifting. This is why it is important, in heavy lifting, to kneel and use the powerful extensor muscles of the thighs and buttocks to lift the load.

---

[48]*quadrat* = four-sided + *lumborum* = of the lumbar region        [49]*multi* = many + *fid* = branched, sectioned

| TABLE 10.8 | Muscles of the Pelvic Floor |
|---|---|

The floor of the pelvic cavity is formed by three layers of muscles and fasciae that span the pelvic outlet and support the viscera. It is penetrated by the anal canal, urethra, and vagina, which open into a diamond-shaped region between the thighs called the **perineum** (PERR-ih-NEE-um). The perineum is bordered by four bony landmarks: the pubic symphysis anteriorly, the coccyx posteriorly, and the ischial tuberosities laterally. The anterior half of the perineum is the **urogenital triangle** and the posterior half is the **anal triangle** (fig. 10.20). These are especially important landmarks in obstetrics.

**Superficial Perineal Space.**    The pelvic floor is divided into three layers or *compartments.* The one just deep to the skin, called the **superficial perineal space** (fig. 10.20a), contains three muscles: the ischiocavernosus, bulbospongiosus, and superficial transverse perineal. The *ischiocavernosus* muscles converge like a V from the ischial tuberosities toward the penis or clitoris. In males, the *bulbospongiosus (bulbocavernosus)* forms a sheath around the base (bulb) of the penis, and in females it encloses the vagina like a pair of parentheses. *Cavernosus* in these names refers to the spongy, cavernous structure of the tissues in the penis and clitoris. The *superficial transverse perineal muscle* extends from the ischial tuberosities to a strong median fibromuscular anchorage, the **perineal body.** It is a weakly developed muscle and not always present, so it is not tabulated below. The other two muscles of this layer primarily serve sexual functions.

**Ischiocavernosus (ISS-kee-oh-CAV-er-NO-sus)**    Maintains erection of the penis or clitoris by compressing deep structures of the organ and forcing blood forward into its body

| **O:** Ramus and tuberosity of ischium | **I:** Ensheaths deep structures of penis or clitoris | **N:** Pudendal n. |
|---|---|---|

**Bulbospongiosus (BUL-bo-SPUN-jee-OH-sus)**    Expels remaining urine from urethra after bladder has emptied. Aids in erection of penis or clitoris. In male, spasmodic contractions expel semen during ejaculation. In female, contractions constrict vaginal orifice and expel secretions of greater vestibular glands

| **O:** Perineal body and median raphe | **I:** Ensheaths root of penis or clitoris | **N:** Pudendal n. |
|---|---|---|

**The Middle Compartment.**    In the middle compartment, the urogenital triangle is spanned by a thin triangular sheet called the **urogenital diaphragm.** This is composed of a fibrous membrane and two muscles: the *deep transverse perineal muscle* and the *external urethral sphincter* (fig. 10.20b). The anal triangle has one muscle at this level, the *external anal sphincter.* The deep transverse perineal is another weakly developed muscle of little importance, not tabulated below.

**External Urethral Sphincter**    Retains urine in bladder until voluntarily voided

| **O:** Right and left ischiopubic rami | **I:** Encircles urethral orifice | **N:** Pudendal n., pelvic splanchnic n. |
|---|---|---|

**External Anal Sphincter**    Retains feces in rectum until voluntarily voided

| **O:** Coccyx, perineal body | **I:** Encircles anal canal and orifice | **N:** Pudendal n., S4 |
|---|---|---|

**The Pelvic Diaphragm.**    The deepest compartment, the **pelvic diaphragm,** consists of two muscle pairs shown in figure 10.20c: the *levator ani* and *coccygeus.* The levator ani is a broad muscle with attachments encircling the pelvic outlet; it forms most of the pelvic floor.

**Levator Ani[50] (leh-VAY-tur AY-nye)**    Compresses anal canal and reinforces external anal and urethral sphincters; supports uterus and other pelvic viscera; aids in the falling away of the feces; vertical movements affect pressure differences between abdominal and thoracic cavities and thus aid in breathing

| **O:** Inner surface of lesser pelvis from pubis to coccyx | **I:** Perineal body, walls of urethra, vagina, and anal canal | **N:** Pudendal n., S2–S3 |
|---|---|---|

**Coccygeus (coc-SIDJ-ee-us)**    Aids levator ani

| **O:** Ischium | **I:** Coccyx and vertebra S5 | **N:** S3–S4 |
|---|---|---|

---

[50]*levat* = to elevate + *ani* = of the anus

**TABLE 10.8**      Muscles of the Pelvic Floor *(cont.)*

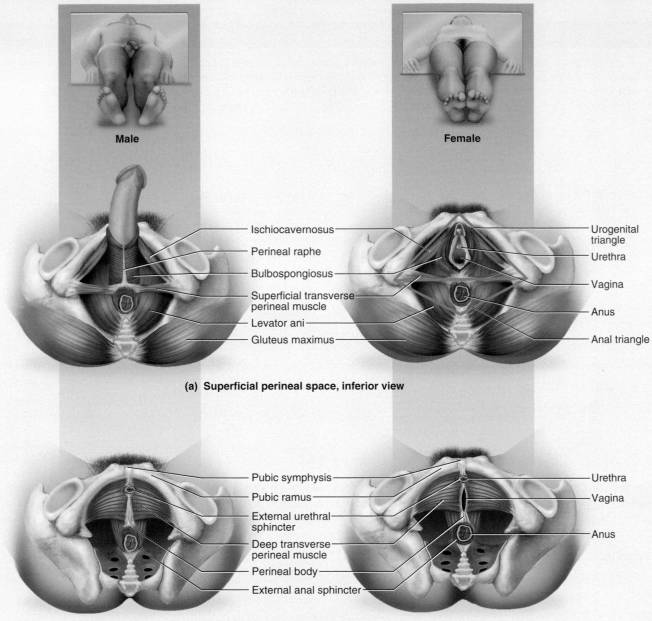

**Male**

**Female**

Ischiocavernosus

Perineal raphe

Bulbospongiosus

Superficial transverse perineal muscle

Levator ani

Gluteus maximus

Urogenital triangle

Urethra

Vagina

Anus

Anal triangle

**(a) Superficial perineal space, inferior view**

Pubic symphysis

Pubic ramus

External urethral sphincter

Deep transverse perineal muscle

Perineal body

External anal sphincter

Urethra

Vagina

Anus

**(b) Urogenital diaphragm, inferior view**

Urogenital diaphragm

Pelvic diaphragm:

Levator ani

Coccygeus

Coccyx

Piriformis

Urethra

Vagina

Anus

**Female**

**(c) Pelvic diaphragm, superior view**

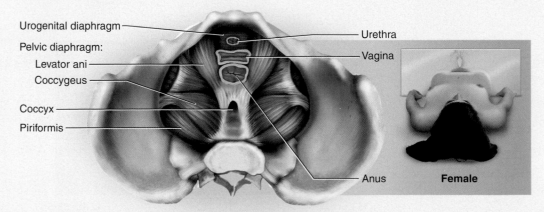

**FIGURE 10.20**   **Muscles of the Pelvic Floor.**   (a) The superficial perineal space, with triangles of the perineum marked on the right. (b) The urogenital diaphragm; this is the next deeper layer after the muscles in (a). (c) Superior view of the female pelvic diaphragm, the deepest layer, seen from within the pelvic cavity.

## Hernias

A hernia is any condition in which the viscera protrude through a weak point in the muscular wall of the abdominopelvic cavity. The most common type to require treatment is an *inguinal hernia*. In the male fetus, each testis descends from the pelvic cavity into the scrotum by way of a passage called the *inguinal canal* through the muscles of the groin. This canal remains a weak point in the pelvic floor, especially in infants and children. When pressure rises in the abdominal cavity, it can force part of the intestine or bladder into this canal or even into the scrotum. This also sometimes occurs in men who hold their breath while lifting heavy weights. When the diaphragm and abdominal muscles contract, pressure in the abdominal cavity can soar to 1,500 pounds per square inch—more than 100 times the normal pressure and quite sufficient to produce an inguinal hernia, or "rupture." Inguinal hernias rarely occur in women.

Two other sites of hernia are the diaphragm and navel. A *hiatal hernia* is a condition in which part of the stomach protrudes through the diaphragm into the thoracic cavity. This is most common in overweight people over 40. It may cause heartburn due to the regurgitation of stomach acid into the esophagus, but most cases go undetected. In an *umbilical hernia,* abdominal viscera protrude through the navel.

## Before You Go On

*Answer the following questions to test your understanding of the preceding section:*

11. Which muscles are used more often, the external intercostals or internal intercostals? Explain.
12. Explain how pulmonary ventilation affects abdominal pressure and vice versa.
13. Name a major superficial muscle and two major deep muscles of the back.
14. Define perineum, urogenital triangle, *and* anal triangle.
15. Name one muscle in the superficial perineal space, one in the urogenital diaphragm, and one in the pelvic diaphragm. State the function of each.

# Muscles Acting on the Shoulder and Upper Limb

### Objectives
When you have completed this section, you should be able to

- name and locate the muscles that act on the pectoral girdle, shoulder, elbow, wrist, and hand;
- relate the actions of these muscles to the joint movements described in chapter 9; and
- describe the origin, insertion, and innervation of each muscle.

The upper limb is used for a broad range of both powerful and subtle actions, ranging from climbing, grasping, and throwing to writing, playing musical instruments, and manipulating small objects. It therefore has an especially complex array of muscles, but the muscles fall into logical groups that make their functional relationships and names easier to understand. Tables 10.9 through 10.16 group these into muscles that act on the scapula, those that act on the humerus and shoulder joint, those that act on the forearm and elbow joint, extrinsic (forearm) muscles that act on the wrist and hand, and intrinsic (hand) muscles that act on the fingers.

| **TABLE 10.9** | Muscles Acting on the Scapula |
|---|---|

Muscles that act on the pectoral girdle originate on the axial skeleton and insert on the clavicle and scapula. The scapula is only loosely attached to the thoracic cage and is capable of considerable movement (fig. 10.21)—rotation (as in raising and lowering the apex of the shoulder), elevation and depression (as in shrugging and lowering the shoulders), and protraction and retraction (pulling the shoulders forward and back). The clavicle braces the shoulder and moderates these movements.

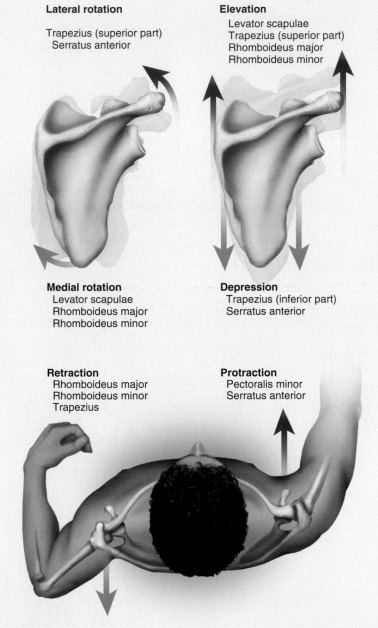

**Lateral rotation**

Trapezius (superior part)
Serratus anterior

**Elevation**

Levator scapulae
Trapezius (superior part)
Rhomboideus major
Rhomboideus minor

**Medial rotation**
Levator scapulae
Rhomboideus major
Rhomboideus minor

**Depression**
Trapezius (inferior part)
Serratus anterior

**Retraction**
Rhomboideus major
Rhomboideus minor
Trapezius

**Protraction**
Pectoralis minor
Serratus anterior

**FIGURE 10.21  Actions of Some Thoracic Muscles on the Scapula.**   Note than an individual muscle can contribute to multiple actions, depending on which fibers contract and what synergists act with it.

▶ *In the two upper figures, mark the insertion of each of the named muscles.*

**Anterior Group.** Muscles of the pectoral girdle fall into anterior and posterior groups (figs. 10.22 and 10.23 in table 10.10). The major muscles of the anterior group are the *pectoralis minor* and *serratus anterior* (see fig. 10.15b). The pectoralis minor arises by three heads from ribs 3 to 5 and converges on the coracoid process of the scapula. The serratus anterior arises from separate heads on all or nearly all of the ribs, wraps laterally around the chest and passes across the back between the rib cage and scapula, and inserts on the medial (vertebral) border of the scapula. Thus, when it contracts, the scapula glides laterally and slightly forward around the ribs.

**Pectoralis Minor (PECK-toe-RAY-liss)** With serratus anterior, draws scapula laterally and forward around chest wall; with other muscles, rotates scapula and depresses apex of shoulder, as in reaching down to pick up a suitcase

    **O:** Ribs 3–5 and overlying fascia     **I:** Coracoid process     **N:** Medial and lateral pectoral nn.

**Serratus[51] Anterior (serr-AY-tus)** With pectoralis minor, draws scapula laterally and forward around chest wall; protracts scapula, and is the prime mover in all forward-reaching and pushing actions; aids in rotating scapula to elevate apex of shoulder; fixes scapula during abduction of arm

    **O:** All or nearly all ribs     **I:** Medial border of scapula     **N:** Long thoracic n.

**Posterior Group.** The posterior muscles that act on the scapula include the large, superficial trapezius, already discussed (table 10.4), and three deep muscles: the *levator scapulae, rhomboideus major,* and *rhomboideus minor.* The action of the trapezius depends on whether its superior, middle, or inferior fibers contract and whether it acts alone or with other muscles. The levator scapulae and superior fibers of the trapezius rotate the scapula in opposite directions if either of them acts alone. If both act together, their opposite rotational effects balance each other and they elevate the scapula and shoulder, as when you lift a suitcase from the floor. Depression of the scapula occurs mainly by gravitational pull, but the trapezius and serratus anterior can depress it more rapidly and forcefully, as in swimming, hammering, and rowing

**Trapezius (tra-PEE-zee-us)** Stabilizes scapula and shoulder during arm movements; elevates and depresses apex of shoulder; acts with other muscles to rotate and retract scapula. See also roles in head and neck movements in table 10.4.

    **O:** External occipital protuberance, medial one-third of superior nuchal line, nuchal ligament, spinous processes of vertebrae C7–T3 or T4     **I:** Acromion and spine of scapula, lateral one-third of clavicle     **N:** Accessory n., ventral rami of C3–C4

**Levator Scapulae (leh-VAY-tur SCAP-you-lee)** Elevates scapula if cervical vertebrae are fixed; flexes neck laterally if scapula is fixed; retracts scapula and braces shoulder; rotates scapula and depresses apex of shoulder

    **O:** Transverse processes of vertebrae C1–C4     **I:** Superior angle to medial border of scapula     **N:** C3–C4, and C5 via dorsal scapular n.

**Rhomboideus Minor (rom-BOY-dee-us)** Retracts scapula and braces shoulder; fixes scapula during arm movements

    **O:** Spinous processes of vertebrae C7–T1, nuchal ligament     **I:** Medial border of scapula     **N:** Dorsal scapular n.

**Rhomboideus Major**

Same as rhomboideus minor

    **O:** Spinous processes of vertebrae T2–T5     **I:** Medial border of scapula     **N:** Dorsal scapular n.

---

[51]*serrate* = scalloped, zigzag

## TABLE 10.10    Muscles Acting on the Humerus

### Axial Muscles.
Nine muscles cross the shoulder joint and insert on the humerus. Two are considered **axial muscles** because they originate primarily on the axial skeleton—the *pectoralis major* and *latissimus dorsi* (figs. 10.15, 10.22, and 10.23). The pectoralis major is the thick, fleshy muscle of the mammary region, and the latissimus dorsi is a broad muscle of the back that extends from the waist to the axilla. These muscles bear the primary responsibility for attaching the arm to the trunk and are the prime movers of the shoulder joint.

**Pectoralis Major (PECK-toe-RAY-liss)**    Flexes, adducts, and medially rotates humerus, as in climbing or hugging. Aids in deep inspiration.

**O:** Medial half of clavicle, costal cartilages 1–7, aponeurosis of external oblique    **I:** Lateral lip of intertubercular sulcus of humerus    **N:** Medial and lateral pectoral nn.

**Latissimus Dorsi**[52] **(la-TISS-ih-mus DOR-sye)**    Adducts and medially rotates humerus; extends the shoulder joint as in pulling on the oars of a rowboat; produces backward swing of arm in such actions as walking and bowling. With hands grasping overhead objects, pulls body forward and upward, as in climbing. Aids in deep inspiration, sudden expiration such as sneezing and coughing; and prolonged forceful expiration as in singing or blowing a sustained note on a wind instrument.

**O:** Vertebrae T7–L5, lower three or four ribs, iliac crest, thoracolumbar fascia    **I:** Floor of intertubercular sulcus of humerus    **N:** Thoracodorsal n.

### Scapular Muscles.
The other seven muscles of the shoulder are considered **scapular muscles** because they originate on the scapula. Four of them form the rotator cuff and are treated in the next section. The most conspicuous scapular muscle is the *deltoid,* the thick triangular muscle that caps the shoulder. This is a commonly used site of drug injections. Its anterior, lateral, and posterior fibers act like three different muscles.

**Deltoid**    Anterior fibers flex and medially rotate arm; lateral fibers abduct arm; posterior fibers extend and laterally rotate arm. Involved in arm swinging during such actions as walking or bowling, and in adjustment of hand height for various manual tasks.

**O:** Acromion and spine of scapula; clavicle    **I:** Deltoid tuberosity of humerus    **N:** Axillary n.

**Teres Major (TERR-eez)**    Extends and medially rotates humerus; contributes to arm swinging

**O:** Inferior angle of scapula    **I:** Medial lip of intertubercular sulcus of humerus    **N:** Lower subscapular n.

**Coracobrachialis (COR-uh-co-BRAY-kee-AL-iss)**    Flexes and medially rotates arm; resists deviation of arm from frontal plane during abduction

**O:** Coracoid process    **I:** Medial aspect of humeral shaft    **N:** Musculocutaneous n.

---

[52]*latissimus* = broadest + *dorsi* = of the back

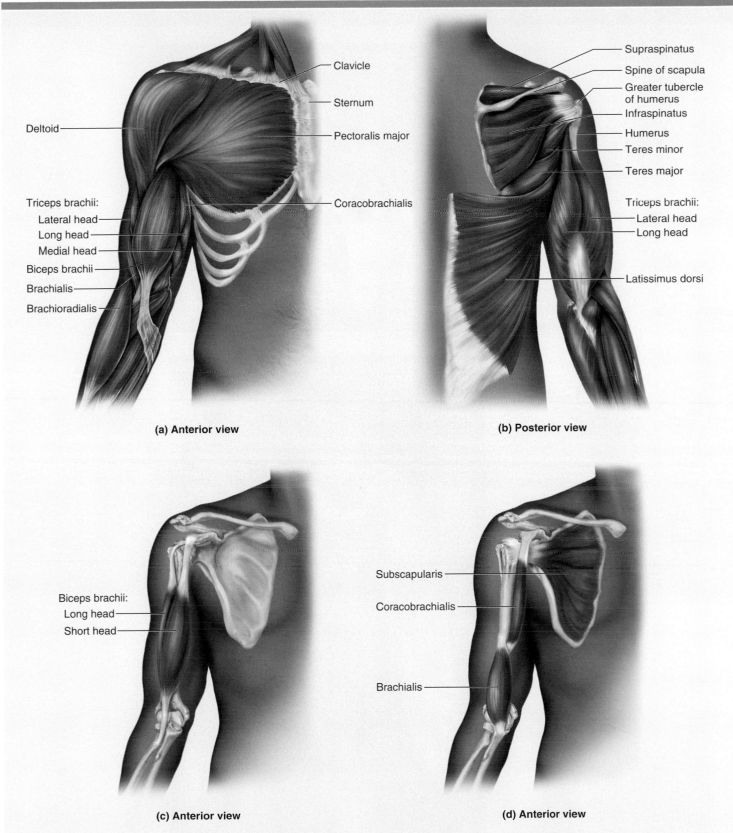

**(a) Anterior view**

Clavicle
Sternum
Pectoralis major
Deltoid
Coracobrachialis
Triceps brachii:
  Lateral head
  Long head
  Medial head
Biceps brachii
Brachialis
Brachioradialis

**(b) Posterior view**

Supraspinatus
Spine of scapula
Greater tubercle of humerus
Infraspinatus
Humerus
Teres minor
Teres major
Triceps brachii:
  Lateral head
  Long head
Latissimus dorsi

**(c) Anterior view**

Biceps brachii:
  Long head
  Short head

**(d) Anterior view**

Subscapularis
Coracobrachialis
Brachialis

**FIGURE 10.22  Pectoral and Brachial Muscles.**   (a) Superficial muscles, anterior view. (b) Superficial muscles, posterior view. (c) The biceps brachii, the superficial flexor of the elbow. (d) The brachialis, the deep flexor of the elbow, and the coracobrachialis and subscapularis, which act on the humerus.

**TABLE 10.10**    Muscles Acting on the Humerus *(cont.)*

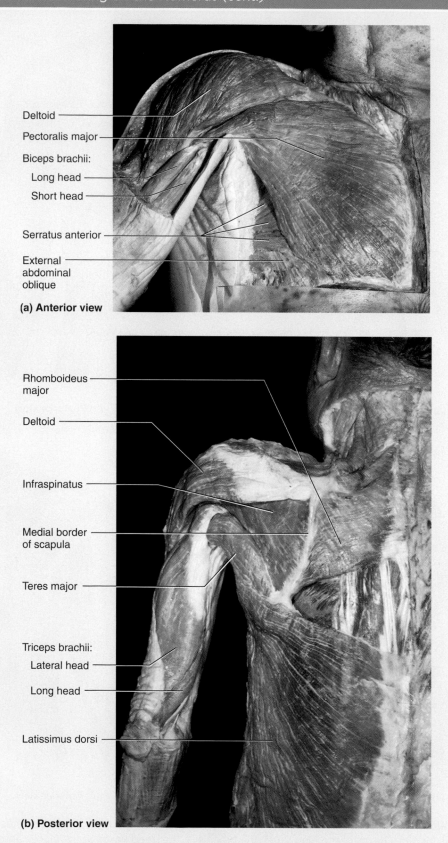

Deltoid

Pectoralis major

Biceps brachii:

Long head

Short head

Serratus anterior

External
abdominal
oblique

**(a) Anterior view**

Rhomboideus
major

Deltoid

Infraspinatus

Medial border
of scapula

Teres major

Triceps brachii:

Lateral head

Long head

Latissimus dorsi

**(b) Posterior view**

**FIGURE 10.23**    Muscles of the Chest and Arm of the Cadaver.

**The Rotator Cuff.**    Tendons of the remaining four scapular muscles form the **rotator cuff** (fig. 10.24). These muscles are nicknamed the "SITS muscles" for the first letters of their names—*supraspinatus, infraspinatus, teres minor,* and *subscapularis.* The first three muscles lie on the posterior side of the scapula (see fig. 10.22b). The supraspinatus and infraspinatus occupy the supraspinous and infraspinous fossae, above and below the scapular spine. The teres minor lies inferior to the infraspinatus. The subscapularis occupies the subscapular fossa on the anterior surface of the scapula, between the scapula and ribs. The tendons of these muscles merge with the joint capsule of the shoulder as they cross it en route to the humerus. They insert on the proximal end of the humerus, forming a partial sleeve around it. The rotator cuff reinforces the joint capsule and holds the head of the humerus in the glenoid cavity. The rotator cuff, especially the supraspinatus tendon, is easily damaged by strenuous circumduction or hard blows to the shoulder (see Insight 10.4).

Since the humeroscapular joint is capable of such a wide range of movements and is acted upon by so many muscles, its actions are summarized in table 10.11.

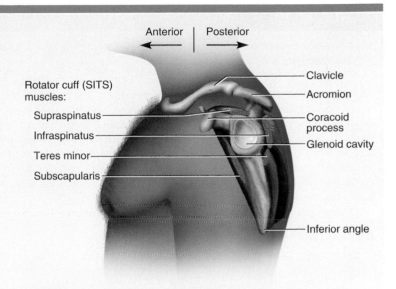

**FIGURE 10.24  Rotator Cuff Muscles in Relation to the Scapula.**   Lateral view. For posterior and anterior views of these muscles see figures 10.22b and d.

**Supraspinatus[53] (SOO-pra-spy-NAY-tus)**    Aids deltoid in abduction of arm; resists downward slippage of humeral head when arm is relaxed or when carrying weight

   **O:** Supraspinous fossa of scapula        **I:** Greater tubercle of humerus        **N:** Suprascapular n.

**Infraspinatus[54] (IN-fra-spy-NAY-tus)**    Modulates action of deltoid, preventing humeral head from sliding upward; rotates humerus laterally

   **O:** Infraspinous fossa of scapula        **I:** Greater tubercle of humerus        **N:** Suprascapular n.

**Teres Minor (TERR-eez)**    Modulates action of deltoid, preventing humeral head from sliding upward as arm is abducted; rotates humerus laterally

   **O:** Lateral border and adjacent posterior        **I:** Greater tubercle of humerus; posterior        **N:** Axillary n.
       surface of scapula                 surface of joint capsule

**Subscapularis[55] (SUB-SCAP-you-LERR-iss)**    Modulates action of deltoid, preventing humeral head from sliding upward as arm is abducted; rotates humerus medially

   **O:** Subscapular fossa of scapula        **I:** Lesser tubercle of humerus, anterior surface        **N:** Upper and lower
                of joint capsule                 subscapular nn.

[53]*supra* = above + *spin* = spine of scapula
[54]*infra* = below, under + *spin* = spine of scapula
[55]*sub* = below, under

## TABLE 10.11      Actions of the Shoulder (Humeroscapular) Joint

Italics indicate prime movers; others are synergists. Parentheses indicate only a slight effect.

| Flexion | Extension | Abduction | Adduction | Medial Rotation | Lateral Rotation |
|---|---|---|---|---|---|
| *Anterior deltoid* | *Posterior deltoid* | *Lateral deltoid* | *Pectoralis major* | *Subscapularis* | *Infraspinatus* |
| *Pectoralis major* | *Latissimus dorsi* | Supraspinatus | *Latissimus dorsi* | Teres major | *Teres minor* |
| Coracobrachialis | Teres major | | Coracobrachialis | Latissimus dorsi | Deltoid |
| Biceps brachii | | | Triceps brachii | Deltoid | |
| | | | Teres major | Pectoralis major | |
| | | | (Teres minor) | | |

| **TABLE 10.12** | Muscles Acting on the Forearm |
|---|---|

The elbow and forearm are capable of four motions—flexion, extension, pronation, and supination—carried out by muscles in both the brachium and antebrachium (arm and forearm) (table 10.13).

**Muscles with Bellies in the Arm (Brachium).**    The principal elbow flexors are on the anterior side of the humerus—the brachialis and biceps brachii (see fig. 10.22c–d). The *biceps brachii* appears as a large anterior bulge on the arm and commands considerable interest among body builders, but the *brachialis* underlying it generates about 50% more power and is thus the prime mover of elbow flexion. The biceps is not only a flexor but also a powerful forearm supinator. It is named for its two heads: a *short head* whose tendon arises from the coracoid process of the scapula, and a *long head* whose tendon originates on the superior margin of the glenoid cavity, loops over the shoulder, and braces the humerus against the glenoid cavity (see chapter 9). The two heads converge close to the elbow on a single distal tendon that inserts on the radius and on the fascia of the medial side of the upper forearm. Note that *biceps* is the singular term; there is no such word as *bicep*. To refer to the biceps muscles of both arms, the plural is *bicipites* (by-SIP-ih-teez).

The triceps brachii is a three-headed muscle on the posterior side of the humerus, and is the prime mover of elbow extension (see fig. 10.22b).

**Brachialis (BRAY-kee-AL-iss)**    Prime mover of elbow flexion

**O:** Anterior surface of distal half of humerus    **I:** Coronoid process and tuberosity of ulna    **N:** Musculocutaneous n., radial n.

**Biceps Brachii (BY-seps BRAY-kee-eye)**    Rapid or forceful supination of forearm; synergist in elbow flexion; slight shoulder flexion; tendon of long head stabilizes shoulder by holding humeral head against glenoid cavity

**O:** Long head: superior margin of glenoid cavity    **I:** Tuberosity of radius, fascia of forearm    **N:** Musculocutaneous n.
Short head: coracoid process

**Triceps Brachii (TRI-seps BRAY-kee-eye)**    Extends elbow; assists in adduction of humerus

**O:** Long head: inferior margin of glenoid cavity and joint capsule    **I:** Olecranon, fascia of forearm    **N:** Radial n.

Lateral head: posterior surface of proximal end of humerus

Medial head: posterior surface of entire humeral shaft

**Muscles with Bellies in the Forearm (Antebrachium).**    Most forearm muscles act on the wrist and hand, but two of them are synergists in elbow flexion and extension and three of them function in pronation and supination. The *brachioradialis* is the large fleshy mass of the lateral (radial) side of the forearm just distal to the elbow (see figs. 10.22a and 10.28a). Its origin is on the distal end of the humerus and its insertion on the distal end of the radius. With the insertion so far from the fulcrum of the elbow, it does not generate as much force as the brachialis and biceps; it is effective mainly when those muscles have already partially flexed the elbow. The *anconeus* is a weak synergist of elbow extension on the posterior side of the elbow (see fig. 10.29). Pronation is achieved by the *pronator teres* near the elbow and *pronator quadratus* (the prime mover) near the wrist. Supination is usually achieved by the *supinator* of the upper forearm, with the biceps brachii aiding when additional speed or power is required (fig. 10.25).

**Brachioradialis (BRAY-kee-oh-RAY-dee-AL-iss)**    Flexes elbow

**O:** Lateral supracondylar ridge of humerus    **I:** Lateral surface of radius near styloid process    **N:** Radial n.

**Anconeus[56] (an-CO-nee-us)**    Extends elbow; may help to control ulnar movement during pronation

**O:** Lateral epicondyle of humerus    **I:** Olecranon and posterior surface of ulna    **N:** Radial n.

**Pronator Quadratus (PRO-nay-tur quad-RAY-tus)**    Prime mover of forearm pronation; also resists separation of radius and ulna when force is applied to forearm through wrist, as in doing push-ups

**O:** Anterior surface of distal ulna    **I:** Anterior surface of distal radius    **N:** Median n.

**Pronator Teres (PRO-nay-tur TERR-eez)**    Assists pronator quadratus in pronation, but only in rapid or forceful action; weakly flexes elbow

**O:** Humeral shaft near medial epicondyle; coronoid process of ulna    **I:** Lateral surface of radial shaft    **N:** Median n.

**Supinator (SOO-pih-NAY-tur )**    Supinates forearm

**O:** Lateral epicondyle of humerus; supinator crest and fossa of ulna just distal to radial notch; anular and radial collateral ligaments of elbow    **I:** Proximal one-third of radius    **N:** Posterior interosseous n.

---

[56]*anconeus* = elbow

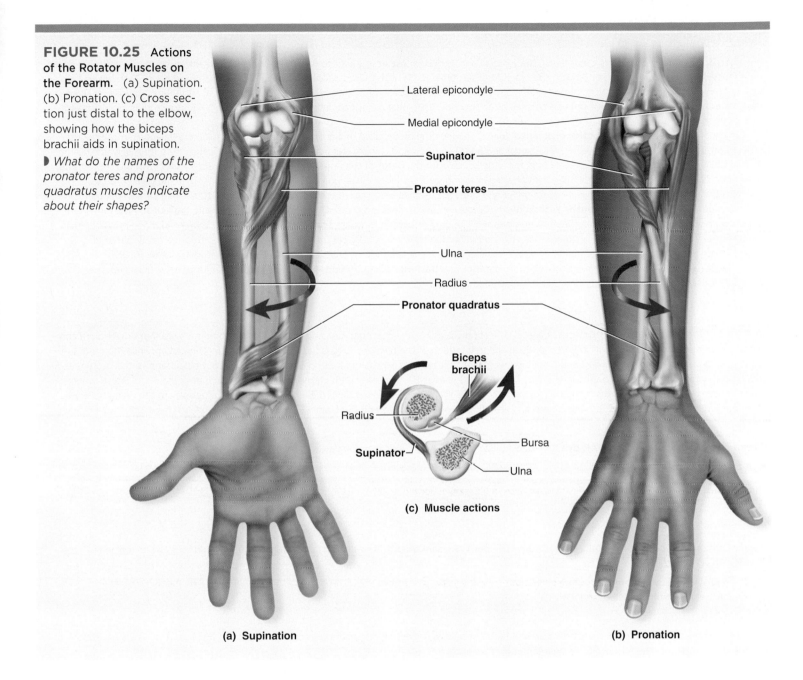

**FIGURE 10.25** Actions of the Rotator Muscles on the Forearm. (a) Supination. (b) Pronation. (c) Cross section just distal to the elbow, showing how the biceps brachii aids in supination.

▶ *What do the names of the pronator teres and pronator quadratus muscles indicate about their shapes?*

Lateral epicondyle
Medial epicondyle
**Supinator**
**Pronator teres**
Ulna
Radius
**Pronator quadratus**

**Biceps brachii**
Radius
**Supinator**
Bursa
Ulna

(c) Muscle actions

(a) Supination

(b) Pronation

| TABLE 10.13 | Actions of the Forearm |
|---|---|

Italics indicate prime movers; others are synergists. Parentheses indicate only a slight effect.

| Flexion | Extension | Pronation | Supination |
|---|---|---|---|
| *Brachialis* | *Triceps brachii* | *Pronator teres* | *Supinator* |
| Biceps brachii | Anconeus | Pronator quadratus | Biceps brachii |
| Brachioradialis | | | |
| Flexor carpi radialis | | | |
| (Pronator teres) | | | |

## TABLE 10.14    Muscles Acting on the Wrist and Hand

The hand is acted upon by extrinsic muscles in the forearm and intrinsic muscles in the hand itself. The bellies of the extrinsic muscles form the fleshy roundness of the upper forearm (along with the brachioradialis, table 10.13), with their tendons extending into the wrist and hand. Their actions are mainly flexion and extension of the wrist and digits, but also include abduction, adduction, radial and ulnar deviation, and thumb opposition. These muscles are numerous and complex, but their names often describe their location, appearance, and function. Table 10.15 groups these muscles by their actions.

Many of them act on the **metacarpophalangeal joints,** between the metacarpal bones of the hand and the proximal phalanges of the fingers, and the **interphalangeal joints,** between the proximal and middle or the middle and distal phalanges (or proximal-distal in the thumb, which has no middle phalanx). The metacarpophalangeal joints form the knuckles at the bases of the fingers, and the interphalangeal joints form the second and third knuckles. Some tendons cross multiple joints before inserting on a middle or distal phalanx, and can flex or extend all the joints it crosses.

Most tendons of the extrinsic muscles pass under a fibrous, braceletlike sheet called the **flexor retinaculum** (transverse carpal ligament) on the anterior side of the wrist or the **extensor retinaculum** (dorsal carpal ligament) on the posterior side. These ligaments prevent the tendons from standing up like taut bowstrings when the muscles contract. The **carpal tunnel** is a tight space between the flexor retinaculum and carpal bones. The flexor tendons passing through the tunnel are enclosed in tendon sheaths that enable them to slide back and forth quite easily, although this region is very subject to painful inflammation—*carpal tunnel syndrome*—resulting from repetitive motion (see Insight 10.3 and fig. 10.29).

The deep fasciae divide the forearm muscles into **anterior** and **posterior compartments** and each compartment into superficial and deep layers (fig. 10.26). The muscles will be described below in these four groups.

### Anterior Compartment, Superficial Layer.
Most muscles of the anterior compartment are wrist and finger flexors that arise from a common tendon on the humerus (fig. 10.27). At the distal end, the tendon of the *palmaris longus* passes over the flexor retinaculum while the other tendons pass beneath it, through the carpal tunnel. The two prominent tendons you can palpate at the wrist belong to the palmaris longus on the medial side and the *flexor carpi radialis* on the lateral side (see fig. B.8 in atlas B). The latter is an important landmark for finding the radial artery, where the pulse is usually taken. The palmaris longus is absent on one or both sides (most commonly the left) in about 14% of people. To see if you have one, flex your wrist and touch the tips of your thumb and little finger together. If present, the palmaris longus tendon will stand up prominently on the wrist.

**Flexor Carpi Radialis**[57] **(FLEX-ur CAR-pye RAY-dee-AL-iss)**    Flexes wrist; aids in radial deviation of hand

| | | |
|---|---|---|
| **O:** Medial epicondyle of humerus | **I:** Base of metacarpals II–III | **N:** Median n. |

**Flexor Carpi Ulnaris**[58] **(ul-NAY-ris)**    Flexes wrist; aids in ulnar deviation of hand

| | | |
|---|---|---|
| **O:** Medial epicondyle of humerus; medial margin of olecranon; posterior surface of ulna | **I:** Pisiform, hamate, metacarpal V | **N:** Ulnar n. |

**Flexor Digitorum Superficialis**[59] **(DIDJ-ih-TOE-rum SOO-per-FISH-ee-AY-lis)**    Flexes wrist, metacarpophalangeal, and interphalangeal joints depending on action of other muscles

| | | |
|---|---|---|
| **O:** Medial epicondyle of humerus; ulnar collateral ligament; coronoid process; superior half of radius | **I:** Middle phalanges II–V | **N:** Median n. |

**Palmaris Longus (pal-MERR-iss)**    Anchors skin and fascia of palmar region; resists shearing forces when stress is applied to skin by such actions as climbing and tool use. Weakly developed and sometimes absent.

| | | |
|---|---|---|
| **O:** Medial epicondyle of humerus | **I:** Flexor retinaculum, palmar aponeurosis | **N:** Median n. |

### Anterior Compartment, Deep Layer.
The following two flexors constitute the deep layer (fig. 10.27c). The *flexor digitorum profundus* flexes fingers II–V while the thumb (pollex) has a flexor of its own—one of several muscles serving exclusively for thumb movements.

**Flexor Digitorum Profundus**    Flexes wrist, metacarpophalangeal, and interphalangeal joints; sole flexor of the distal interphalangeal joints

| | | |
|---|---|---|
| **O:** Proximal three-quarters of ulna; coronoid process; interosseous membrane | **I:** Distal phalanges II–V | **N:** Median n., ulnar n. |

**Flexor Pollicis Longus (PAHL-ih-sis)**    Flexes phalanges of thumb

| | | |
|---|---|---|
| **O:** Radius, interosseous | **I:** Distal phalanx I | **N:** Median n. |

---

[57]*carpi* = of the wrist + *radialis* = of the radius
[58]*ulnaris* = of the ulna

[59]*digitorum* = of the digits + *superficialis* = shallow, near the surface

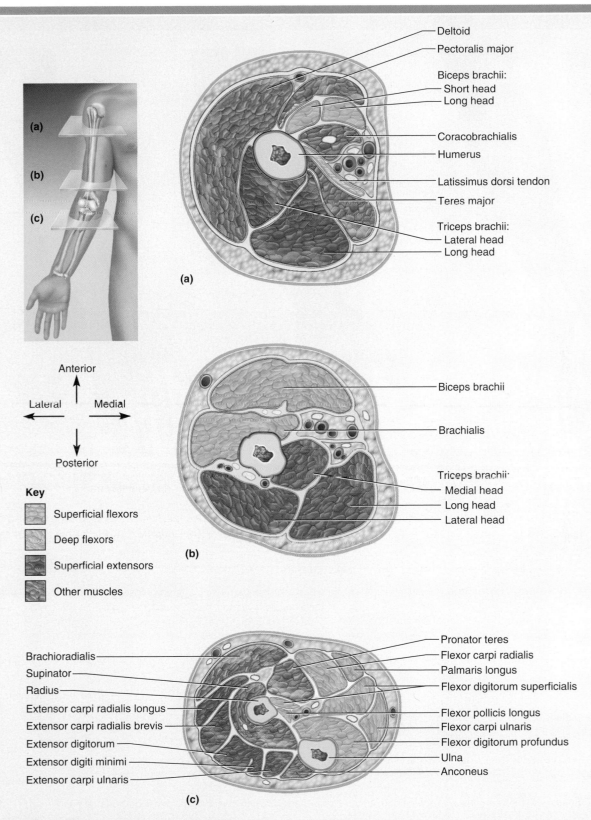

**FIGURE 10.26 Serial Cross Sections Through the Upper Limb.** Each section is taken at the correspondingly lettered level in the figure at the left and is pictured with the posterior muscle compartment facing the bottom of the page, as if viewing a person's right limb extended toward you with the palm up.

▶ *Why are the extensor pollicis longus and extensor indicis not seen in part (c)?*

**TABLE 10.14**    Muscles Acting on the Wrist and Hand *(cont.)*

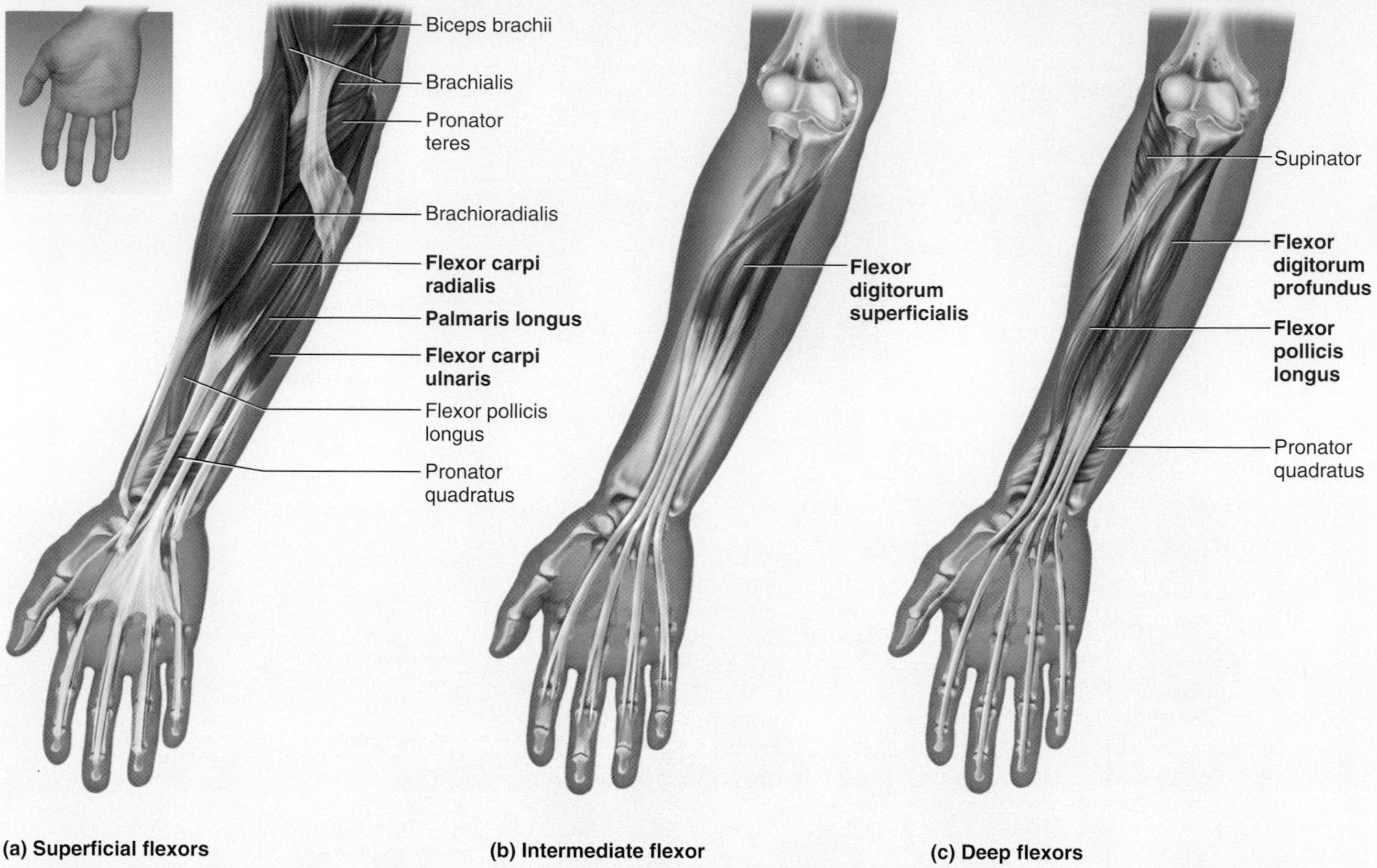

(a) **Superficial flexors**    (b) **Intermediate flexor**    (c) **Deep flexors**

**FIGURE 10.27**    **Flexors of the Wrist and Hand.** Anterior views of the forearm. (a) Superficial flexors. (b) The flexor digitorum superficialis, deep to the muscles in part (a) but also classified as a superficial flexor. (c) Deep flexors. Flexor muscles of each compartment are labeled in boldface.

**Posterior Compartment, Superficial Layer.**    Muscles of the posterior compartment are mostly wrist and finger extensors, and share a single tendon arising from the humerus (fig. 10.28). The first of these, the *extensor digitorum,* has four distal tendons that can easily be seen and palpated on the back of the hand when the fingers are strongly extended (see fig. B.8b in atlas B). It serves digits II through V, and the other muscles in this group each serve a single digit.

**Extensor Digitorum**    Extends wrist, metacarpophalangeal, and interphalangeal joints; tends to spread digits apart when extending metacarpophalangeal joints

| **O:** Lateral epicondyle of humerus | **I:** Dorsal surfaces of phalanges II–V | **N:** Posterior interosseous n. |

**Extensor Carpi Radialis Longus**    Extends wrist; aids in radial deviation of hand

| **O:** Lateral supracondylar ridge of humerus | **I:** Base of metacarpal II | **N:** Radial n. |

**Extensor Carpi Radialis Brevis (BREV-iss)**    Extends wrist; aids in radial deviation of hand

| **O:** Lateral epicondyle of humerus | **I:** Base of metacarpal III | **N:** Posterior interosseous n. |

**Extensor Carpi Ulnaris**    Extends and fixes wrist when fist is clenched or hand grips an object; aids in ulnar deviation of hand

| **O:** Lateral epicondyle of humerus; posterior surface of ulnar shaft | **I:** Base of metacarpal V | **N:** Posterior interosseous n. |

**Extensor Digiti Minimi**[60] **(DIDJ-ih-ty MIN-ih-my)**    Extends wrist and all joints of little finger

| **O:** Lateral epicondyle of humerus | **I:** Proximal phalanx V | **N:** Posterior interosseous n. |

**Posterior Compartment, Deep Layer.** The deep muscles below serve only the thumb and index finger. By strongly abducting and extending the thumb into a hitchhiker's position, you may see a deep dorsolateral pit at the base of the thumb, with a taut tendon on each side of it (see fig. B.8 in atlas B). This depression is called the *anatomical snuffbox* because it was once fashionable to place a pinch of snuff here and inhale it. It is bordered laterally by the tendons of the *abductor pollicis longus* and *extensor pollicis brevis,* and medially by the tendon of the *extensor pollicis longus.*

**Abductor Pollicis Longus**    Abducts thumb in frontal (palmar) plane; extends thumb at carpometacarpal joint

**O:** Posterior surfaces of radius and ulna; interosseous membrane     **I:** Trapezium, base of metacarpal I     **N:** Posterior interosseous n.

**Extensor Pollicis Brevis**    Extends metacarpal and proximal phalanx of thumb

**O:** Shaft of radius, interosseous membrane     **I:** Proximal phalanx I     **N:** Posterior interosseous n.

**Extensor Pollicis Longus**    Extends distal phalanx I; aids in extending proximal phalanx I and metacarpal I; adducts and laterally rotates thumb

**O:** Posterior surface of ulna, interosseous membrane     **I:** Distal phalanx I     **N:** Posterior interosseous n.

**Extensor Indicis (IN-dih-sis)**    Extends wrist and index finger

**O:** Posterior surface of ulna, interosseous membrane     **I:** Middle and distal phalanges of index finger     **N:** Posterior interosseous n.

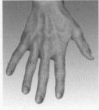
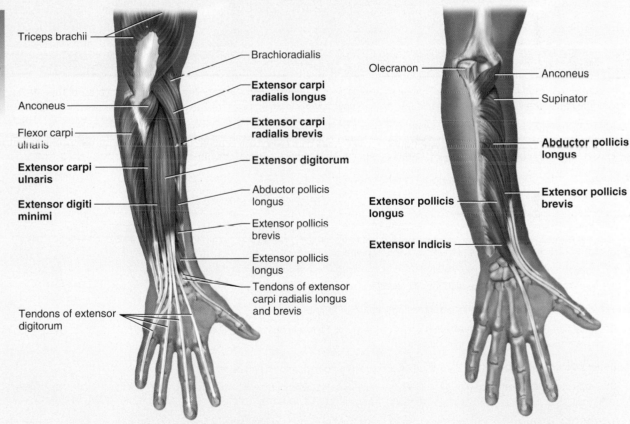

(a) Superficial extensors        (b) Deep extensors

**FIGURE 10.28 Extensors of the Wrist and Hand.** Posterior views of the forearm. Extensor muscles of each compartment are labeled in boldface.

---

[60]*digit* = finger + *minim* = smallest

## TABLE 10.15    Actions of the Wrist and Hand

Italics indicate prime movers; others are synergists. Parentheses indicate only a slight effect.

| Wrist Flexion | Wrist Extension | Wrist Abduction | Wrist Adduction |
|---|---|---|---|
| *Flexor carpi radialis* | *Extensor digitorum* | Flexor carpi radialis | Flexor carpi ulnaris |
| *Flexor carpi ulnaris* | Extensor carpi radialis longus | Extensor carpi radialis longus | Extensor carpi ulnaris |
| *Flexor digitorum superficialis* | Extensor carpi radialis brevis | Extensor carpi radialis brevis | |
| (Palmaris longus) | Extensor carpi ulnaris | Abductor pollicis longus | |
| (Flexor pollicis longus) | | | |

| Finger Flexion | Finger Extension | Thumb Opposition |
|---|---|---|
| Flexor digitorum superficialis | Extensor pollicis longus | Opponens pollicis |
| Flexor digitorum profundus | Extensor pollicis brevis | Opponens digiti minimi |
| Flexor pollicis longus | Extensor digitorum | |
| | Extensor indicis | |

## TABLE 10.16    Intrinsic Muscles of the Hand

The intrinsic muscles of the hand assist the flexors and extensors in the forearm and make finger movements more precise. They are divided into three groups: the *thenar group* at the base of the thumb, the *hypothenar group* at the base of the little finger, and the *midpalmar group* between these (fig. 10.30).

**Thenar Group.**    The thenar group of muscles form the thick fleshy mass *(thenar eminence)* at the base of the thumb, except for the *adductor pollicis,* which forms the web between the thumb and palm. All are concerned with thumb movements.

**Adductor Pollicis**    Draws thumb toward palm as in gripping a tool

**O:** Capitate; bases of metacarpals II–III; anterior ligaments of wrist; tendon sheath of flexor carpi radialis
**I:** Medial surface of proximal phalanx I
**N:** Ulnar n.

**Abductor Pollicis Brevis**    Abducts thumb in sagittal plane

**O:** Mainly flexor retinaculum; also scaphoid, trapezium, and abductor pollicis longus tendon
**I:** Lateral surface of proximal phalanx I
**N:** Median n.

**Flexor Pollicis Brevis**    Flexes metacarpophalangeal joint of thumb

**O:** Trapezium, trapezoid, capitate, anterior ligaments of wrist, flexor retinaculum
**I:** Proximal phalanx I
**N:** Median n., ulnar n.

**Opponens Pollicis (op-PO-nenz)**    Flexes metacarpal I to oppose thumb to fingertips

**O:** Trapezium, flexor retinaculum
**I:** Metacarpal I
**N:** Median n.

**Hypothenar Group.**    The hypothenar group forms the fleshy mass *(hypothenar eminence)* at the base of the little finger. All of these muscles are concerned with movement of that digit.

**Abductor Digiti Minimi**    Abducts little finger, as in spreading fingers apart

**O:** Pisiform, tendon of flexor carpi ulnaris
**I:** Medial surface of proximal phalanx V
**N:** Ulnar n.

**Flexor Digiti Minimi Brevis**    Flexes little finger at metacarpophalangeal joint

**O:** Hamulus of hamate bone, flexor retinaculum
**I:** Medial surface of proximal phalanx V
**N:** Ulnar n.

**Opponens Digiti Minimi**    Flexes metacarpal V at carpometacarpal joint when little finger is moved into opposition with tip of thumb; deepens palm of hand

**O:** Hamulus of hamate bone, flexor retinaculum
**I:** Medial surface of metacarpal V
**N:** Ulnar n.

**Midpalmar Group.**     The midpalmar group occupies the hollow of the palm. It has 11 small muscles divided into three groups.

**Dorsal Interosseous[61] Muscles (IN-tur-OSS-ee-us) (four muscles)**     Abduct fingers; strongly flex metacarpophalangeal joints but extend interphalangeal joints, depending on action of other muscles; important for grip strength

   **O:** Each with two heads arising from facing          **I:** Proximal phalanges II–IV                **N:** Ulnar n.
        surfaces of adjacent metacarpals

**Palmar Interosseous Muscles (three muscles)**     Adduct fingers; other actions same as for dorsal interosseous muscles

   **O:** Metacarpals I, II, IV, V          **I:** Proximal phalanges II, IV, V          **N:** Ulnar n.

**Lumbricals[62] (LUM-brick-ulz) (four muscles)**     Extend interphalangeal joints; contribute to ability to pinch objects between fleshy pulp of thumb and finger instead of these digits meeting by the edges of their nails

   **O:** Tendons of flexor digitorum profundus          **I:** Proximal phalanges II–V          **N:** Median n., ulnar n.

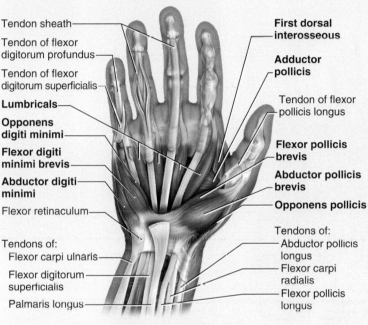

**(a) Palmar aspect, superficial**

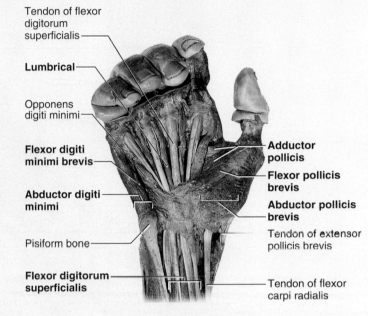

**(b) Palmar dissection, superficial**

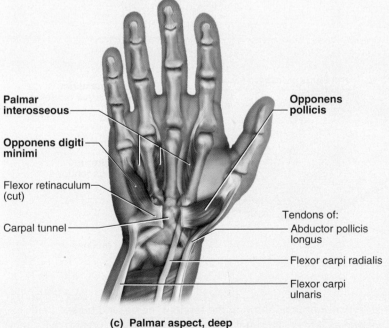

**(c) Palmar aspect, deep**

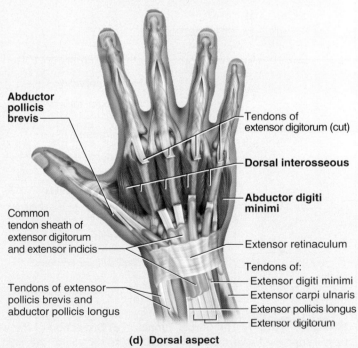

**(d) Dorsal aspect**

**FIGURE 10.29    Intrinsic Muscles of the Hand.**    Boldface labels in parts (a), (c), and (d) indicate the muscles of each layer.

---

[61]*inter* = between + *osse* = bones          [62]*lumbrical* = resembling an earthworm

## INSIGHT 10.3    Clinical Application

### Carpal Tunnel Syndrome

Prolonged, repetitive motions of the wrist and fingers can cause tissues in the carpal tunnel to become inflamed, swollen, or fibrotic. Since the carpal tunnel cannot expand, swelling puts pressure on the median nerve of the wrist, which passes through the carpal tunnel with the flexor tendons (fig. 10.30).

This pressure causes tingling and muscular weakness in the palm and medial side of the hand and pain that may radiate to the arm and shoulder. This condition, called *carpal tunnel syndrome,* is common among keyboard operators, pianists, meat cutters, and others who spend long hours making repetitive wrist motions. Carpal tunnel syndrome is treated with aspirin and other anti-inflammatory drugs, immobilization of the wrist, and sometimes surgical removal of part or all of the flexor retinaculum to relieve pressure on the nerve.

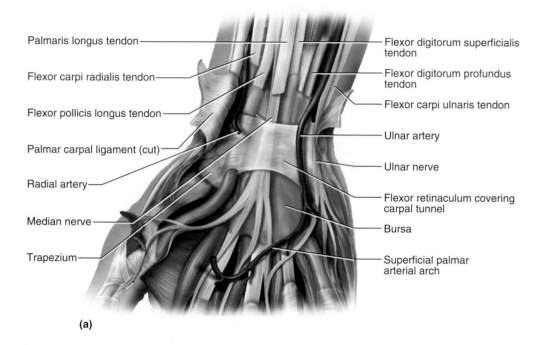

(a)

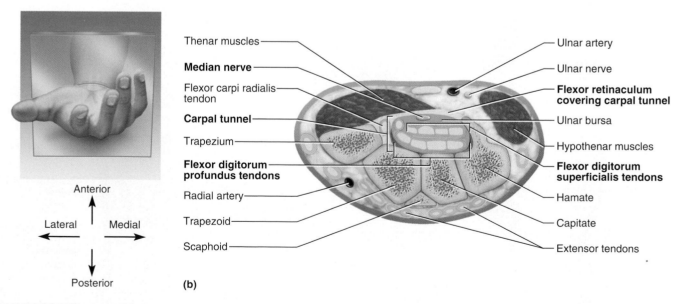

(b)

**FIGURE 10.30    The Carpal Tunnel.**    (a) Dissection of the wrist (anterior aspect) showing the tendons, nerve, and bursae that pass under the flexor retinaculum. (b) Cross section of the wrist, viewed as if from the distal end of a person's right forearm extended toward you with the palm up. Note how the flexor tendons and median nerve are confined in the tight space between the carpal bones and flexor retinaculum.

## Before You Go On

*Answer the following questions to test your understanding of the preceding section:*

16. *Name a muscle that inserts on the scapula and plays a significant role in each of the following actions: (a) pushing a stalled car, (b) paddling a canoe, (c) squaring the shoulders in military attention, (d) lifting the shoulder to carry a heavy box on it, and (e) lowering the shoulder to lift a suitcase.*

17. *Describe three contrasting actions of the deltoid muscle.*

18. *Name the four rotator cuff muscles and identify the scapular surfaces against which they lie.*

19. *Name the prime movers of elbow flexion and extension.*

20. *Identify three functions of the biceps brachii.*

21. *Name three extrinsic muscles and two intrinsic muscles that flex the phalanges.*

# Muscles Acting on the Hip and Lower Limb

### Objectives

When you have completed this section, you should be able to

- name and locate the muscles that act on the hip, knee, ankle, and toe joints;
- relate the actions of these muscles to the joint movements described in chapter 9; and
- describe the origin, insertion, and innervation of each muscle.

The largest muscles are found in the lower limb. Unlike those of the upper limb, they are adapted less for precision than for the strength needed to stand, maintain balance, walk, and run. Several of them cross and act upon two or more joints, such as the hip and knee. To avoid confusion in this discussion, remember that in the anatomical sense the word *leg* refers only to that part of the limb between the knee and ankle. The term *foot* includes the tarsal region (ankle), metatarsal region, and toes. Tables 10.17 through 10.20 group the muscles of the lower limb into those that act on the femur and hip joint, those that act on the leg and knee joint, extrinsic (leg) muscles that act on the foot and ankle joint, and intrinsic (foot) muscles that act on the arches and toes.

| TABLE 10.17 | Muscles Acting on the Hip and Femur |
|---|---|

**Anterior Muscles of the Hip.**    Most muscles that act on the femur originate on the os coxae. The two principal anterior muscles are the *iliacus,* which fills most of the broad iliac fossa of the pelvis, and the *psoas major,* a thick rounded muscle that arises mainly from the lumbar vertebrae (fig. 10.31). Collectively, they are called the **iliopsoas** and share a common tendon to the femur.

**Iliacus[63] (ih-LY-uh-cus)**    Flexes thigh at hip when trunk is fixed; flexes trunk at hip when thigh is fixed, as in bending forward in a chair or sitting up in bed; balances trunk during sitting

| | | |
|---|---|---|
| **O:** Iliac crest and fossa, superolateral region of sacrum, ventral sacroiliac and iliolumbar ligaments | **I:** Lesser trochanter and nearby shaft of femur | **N:** Femoral n. |

**Psoas[64] Major (SO-ass)**    Same as iliacus

| | | |
|---|---|---|
| **O:** Bodies and intervertebral discs of vertebrae T12–L5, transverse processes of lumbar vertebrae | **I:** Lesser trochanter and nearby shaft of femur | **N:** Ventral rami of lumbar spinal nn. |

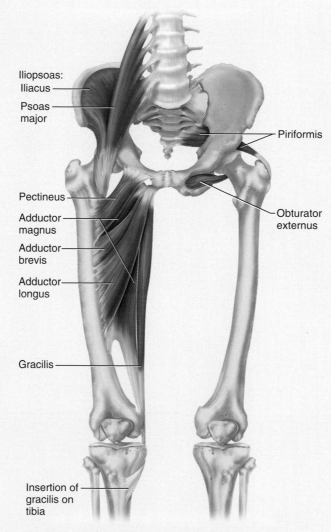

Iliopsoas:
Iliacus
Psoas major
Piriformis
Pectineus
Adductor magnus
Adductor brevis
Adductor longus
Obturator externus
Gracilis
Insertion of gracilis on tibia

**FIGURE 10.31    Muscles Acting on the Hip and Femur.**    Anterior view.

[63]*ili* = loin, flank

[64]*psoa* = loin

**Lateral and Posterior Muscles of the Hip.**     On the lateral and posterior sides of the hip are the *tensor fasciae latae* and three gluteal muscles. The **fascia lata** is a fibrous sheath that encircles the thigh like a subcutaneous stocking and tightly binds its muscles. On the lateral surface, it combines with the tendons of the gluteus maximus and tensor fasciae latae to form the **iliotibial band,** which extends from the iliac crest to the lateral condyle of the tibia (see fig. 10.34, table 10.18). The tensor fasciae latae tautens the iliotibial band and braces the knee, especially when the opposite foot is lifted.

The gluteal muscles are the *gluteus maximus, gluteus medius,* and *gluteus minimus* (figs. 10.32 and 10.33). The gluteus maximus is the largest of these and forms most of the lean mass of the buttock. It is an extensor of the hip joint that produces the backswing of the leg in walking and provides most of the lift when you climb stairs. It generates its maximum force when the thigh is flexed at a 45° angle to the trunk. This is the advantage in starting a foot race from a crouched position. The gluteus medius is deep and lateral to the gluteus maximus. Its name refers to its size, not its position. The gluteus minimus is the smallest and deepest of the three.

**Tensor Fasciae Latae**[65] **(TEN-sur FASH-ee-ee LAY-tee)**    Extends knee, laterally rotates tibia, aids in abduction and medial rotation of femur; during standing, steadies pelvis on femoral head and steadies femoral condyles on tibia

**O:** Iliac crest, anterior superior spine, deep surface of fascia lata

**I:** Lateral condyle of tibia via iliotibial band

**N:** Superior gluteal n.

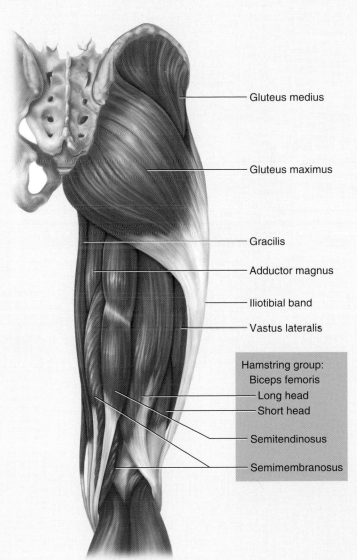

**FIGURE 10.32  Superficial Gluteal and Thigh Muscles.**   Posterior view.

---

[65]*fasc* = band + *lat* = broad

| **TABLE 10.17** | Muscles Acting on the Hip and Femur *(cont.)* |
|---|---|

**Gluteus Maximus**[66]    Extends thigh at hip as in stair climbing (rising to next step) or running and walking (backswing of limb); abducts thigh; elevates trunk after stooping; prevents trunk from pitching forward during walking and running; helps stabilize femur on tibia

| **O:** Posterolateral region of ilium from iliac crest posterior superior spine; coccyx; dorsal surface of lower sacrum; aponeurosis of erector spinae | **I:** Gluteal tuberosity of femur; lateral condyle of tibia via iliotibial band | **N:** Inferior gluteal n. |
|---|---|---|

**Gluteus Medius and Gluteus Minimus**    Abduct and medially rotate thigh; during walking, shift weight of trunk toward limb with foot on the ground as other foot is lifted

| **O:** Most of lateral surface of ilium between crest and acetabulum | **I:** Greater trochanter of femur | **N:** Superior gluteal n. |
|---|---|---|

**Lateral Rotators.**    Inferior to the gluteus minimus and deep to the other two gluteal muscles are six muscles called the **lateral rotators,** named for their action on the femur (fig. 10.33). Their action is most clearly visualized when you cross your legs to rest an ankle on your knee, causing your femur to rotate and the knee to point laterally. Thus, they oppose medial rotation by the gluteus medius and minimus. Most of them also abduct or adduct the femur. The abductors are important in walking because when we lift one foot from the ground, they shift the body weight to other leg and prevent us from falling.

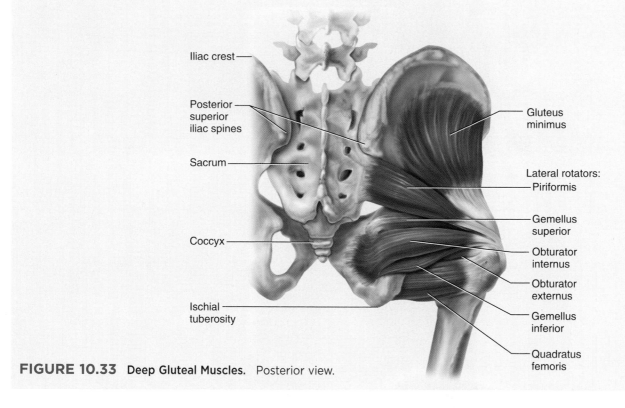

Iliac crest

Posterior superior iliac spines

Sacrum

Coccyx

Ischial tuberosity

Gluteus minimus

Lateral rotators:
Piriformis

Gemellus superior

Obturator internus

Obturator externus

Gemellus inferior

Quadratus femoris

**FIGURE 10.33  Deep Gluteal Muscles.**   Posterior view.

---

[66]*glut* = buttock + *maxim* = largest

**Gemellus[67] Superior (jeh-MEL-us)**    Laterally rotates extended thigh; abducts flexed thigh. Sometimes absent.

**O:** Ischial spine     **I:** Greater trochanter of femur     **N:** Nerve to obturator internus

**Gemellus Inferior**    Same actions as gemellus superior

**O:** Ischial tuberosity     **I:** Greater trochanter of femur     **N:** Nerve to quadratus femoris

**Obturator[68] Externus (OB-too-RAY-tur)**    Not well understood; thought to laterally rotate thigh in climbing

**O:** External surface of obturator membrane; pubic and ischial rami     **I:** Femur between head and greater trochanter     **N:** Obturator n.

**Obturator Internus**    Not well understood; thought to laterally rotate extended thigh and abduct flexed thigh

**O:** Ramus of ischium; inferior ramus of pubis; anteromedial surface of lesser pelvis     **I:** Greater trochanter of femur     **N:** Nerve to obturator internus

**Piriformis[69] (PIR-ih-FOR-mis)**    Laterally rotates extended thigh; abducts flexed thigh

**O:** Anterior surface of sacrum; gluteal surface of ilium; capsule of sacroiliac joint     **I:** Greater trochanter of femur     **N:** Spinal nn. L5–S2

**Quadratus Femoris[70] (quad-RAY-tus FEM-oh-ris)**    Laterally rotates thigh

**O:** Ischial tuberosity of femur     **I:** Intertrochanteric crest of femur     **N:** Nerve to quadratus femoris

### Medial (Adductor) Compartment of the Thigh.
Deep fasciae divide the thigh into three compartments, each with its own nerve and blood supply: the *anterior (extensor) compartment, posterior (flexor) compartment,* and *medial (adductor) compartment.* Muscles of the anterior and posterior compartments function mainly as extensors and flexors of the knee, respectively, and are treated in table 10.18. The five muscles of the medial compartment act primarily as adductors of the thigh (see fig. 10.31), but some of them cross both the hip and knee joints and have additional actions as follows.

**Adductor Brevis**    Adducts thigh

**O:** Body and inferior ramus of pubis     **I:** Linea aspera and spiral line of femur     **N:** Obturator n.

**Adductor Longus**    Adducts and medially rotates thigh; flexes thigh at hip

**O:** Body and inferior ramus of pubis     **I:** Linea aspera of femur     **N:** Obturator n.

**Adductor Magnus**    Adducts and medially rotates thigh; extends thigh at hip

**O:** Inferior ramus of pubis; ramus and tuberosity of ischium     **I:** Linea aspera, gluteal tuberosity, and medial supracondylar line of femur     **N:** Obturator n., tibial n.

**Gracilis[71] (GRASS-ih-lis)**    Flexes and medially rotates tibia at knee

**O:** Body and inferior ramus of pubis; ramus of ischium     **I:** Medial surface of tibia just below condyle     **N:** Obturator n.

**Pectineus[72] (pec-TIN-ee-us)**    Flexes and adducts thigh

**O:** Superior ramus of pubis     **I:** Spiral line of femur     **N:** Femoral n.

---

[67]*gemellus* = twin
[68]*obtur* = to close, stop up
[69]*piri* = pear + *form* = shaped

[70]*quadrat* = four-sided + *femoris* = of the thigh or femur
[71]*gracil* = slender
[72]*pectin* = comb

## TABLE 10.18     Muscles Acting on the Knee and Leg

The following muscles form most of the mass of the thigh and produce their most obvious actions on the knee joint. Some of them, however, cross both the hip and knee joints and produce actions at both, moving the femur, tibia, and fibula.

### Anterior (Extensor) Compartment of the Thigh.

The anterior compartment of the thigh contains the large quadriceps femoris muscle, the prime mover of knee extension and the most powerful muscle of the body (figs. 10.34 and 10.35). As the name implies, it has four heads: the *rectus femoris, vastus lateralis, vastus medialis,* and *vastus intermedius.* All four converge on a single **quadriceps (patellar) tendon,** which extends to the patella, then continues as the **patellar ligament** and inserts on the tibial tuberosity. (Remember that a tendon usually extends from muscle to bone, and a ligament from bone to bone.) The patellar ligament is struck with a rubber reflex hammer to test the knee-jerk reflex. The quadriceps extends the knee when you stand up, take a step, or kick a ball. It is very important in running because, together with the iliopsoas, it flexes the hip in each airborne phase of the leg's cycle of motion. The rectus femoris also flexes the hip in such actions as high kicks, stair climbing, or simply in drawing the leg forward during a stride.

Crossing the quadriceps from the lateral side of the hip to the medial side of the knee is the narrow, straplike *sartorius,* the longest muscle of the body. It flexes the hip and knee joints and laterally rotates the thigh, as in crossing the legs. It is colloquially called the "tailor's muscle" after the cross-legged stance of a tailor supporting his work on the raised knee.

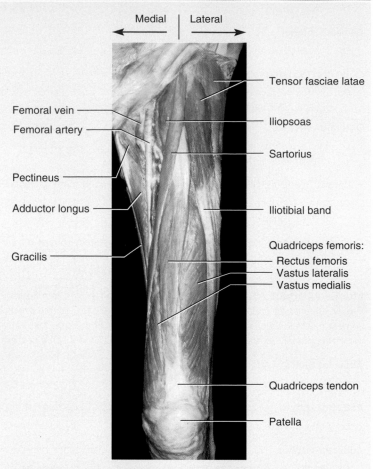

**FIGURE 10.34   Superficial Anterior Muscles of the Thigh of the Cadaver.**

### Quadriceps Femoris (QUAD-rih-seps FEM-oh-ris)     All heads insert on tibia through a common tendon and extend the knee, in addition to the actions of individual heads below.

**Rectus femoris**    Extends knee; flexes thigh at hip; flexes trunk on hip if thigh is fixed

| | | |
|---|---|---|
| **O:** Ilium at anterior inferior spine and superior margin of acetabulum; capsule of hip joint | **I:** Patella, tibial tuberosity, lateral and medial condyles of tibia | **N:** Femoral n. |

**Vastus[73] lateralis**    Extends knee; retains patella in groove on femur during knee movements

| | | |
|---|---|---|
| **O:** Femur at greater trochanter and intertrochanteric line, gluteal tuberosity, and linea aspera | **I:** Same as rectus femoris | **N:** Same as rectus femoris |

**Vastus medialis**    Same as vastus lateralis

| | | |
|---|---|---|
| **O:** Femur at intertrochanteric line, spiral line, linea aspera, and medial supracondylar line | **I:** Same as rectus femoris | **N:** Same as rectus femoris |

**Vastus intermedius**    Extends knee

| | | |
|---|---|---|
| **O:** Anterior and lateral surfaces of femoral shaft | **I:** Same as rectus femoris | **N:** Same as rectus femoris |

**Sartorius[74]**    Aids in knee and hip flexion, as in sitting or climbing; abducts and laterally rotates thigh

| | | |
|---|---|---|
| **O:** On and near anterior superior spine of ilium | **I:** Medial surface of proximal end of tibia | **N:** Femoral n. |

---

[73]*vastus* = large, extensive          [74]*sartor* = tailor

**Posterior (Flexor) Compartment of the Thigh.**     The posterior compartment contains the *biceps femoris, semimembranosus,* and *semitendinosus* (see fig. 10.32). These are colloquially known as the **hamstring muscles.** The pit at the back of the knee, known anatomically as the popliteal fossa, is colloquially called the *ham.* The tendons of these muscles can be felt as prominent cords on both sides of the pit—the biceps tendon on the lateral side and the semimembranosus and semitendinosus tendons on the medial side. When wolves attack large prey, they often attempt to sever the hamstring tendons, because this renders the prey helpless. The hamstrings flex the knee, and aided by the gluteus maximus, they extend the hip during walking and running. The semitendinosus is named for its unusually long tendon. This muscle also is usually bisected by a transverse tendinous band. The semimembranosus is named for the flat shape of its superior attachment.

**FIGURE 10.35  Muscles of the Thigh.**   Anterior view.
(a) Superficial muscles. (b) Rectus femoris and other muscles removed to expose the other three heads of the quadriceps femoris.

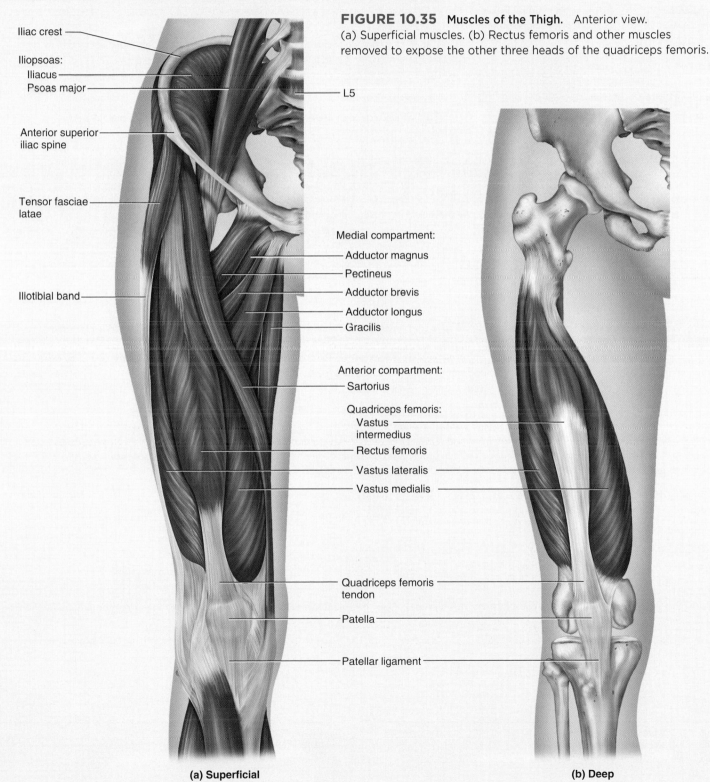

Iliac crest

Iliopsoas:
  Iliacus
  Psoas major

L5

Anterior superior iliac spine

Tensor fasciae latae

Iliotibial band

Medial compartment:
  Adductor magnus
  Pectineus
  Adductor brevis
  Adductor longus
  Gracilis

Anterior compartment:
  Sartorius

Quadriceps femoris:
  Vastus intermedius
  Rectus femoris
  Vastus lateralis
  Vastus medialis

Quadriceps femoris tendon

Patella

Patellar ligament

**(a) Superficial**                    **(b) Deep**

| TABLE 10.18 | Muscles Acting on the Knee and Leg (cont.) |
|---|---|

**Biceps Femoris**    Flexes knee; extends hip; elevates trunk from stooping posture; laterally rotates tibia on femur when knee is flexed; laterally rotates femur when hip is extended; counteracts forward bending at hips

| **O:** Long head: ischial tuberosity Short head: linea aspera and lateral supracondylar line of femur | **I:** Head of fibula | **N:** Tibial n., common fibular n. |
|---|---|---|

**Semimembranosus**[75] **(SEM-ee-MEM-bran-OH-sus)**    Flexes knee; medially rotates tibia on femur when knee is flexed; medially rotates femur when hip is extended; counteracts forward bending at hips

| **O:** Ischial tuberosity | **I:** Medial condyle and nearby margin of tibia; intercondylar line and lateral condyle of femur; ligament of popliteal region | **N:** Tibial n. |
|---|---|---|

**Semitendinosus**[76] **(SEM-ee-TEN-din-OH-sus)**    Same as semimembranosus

| **O:** Ischial tuberosity | **I:** Medial surface of upper tibia | **N:** Tibial n. |
|---|---|---|

**Posterior Compartment of the Leg.**    Most muscles in the posterior compartment of the leg act on the ankle and foot and are reviewed in table 10.19, but the *popliteus* acts on the knee (see fig. 10.39a).

**Popliteus**[77] **(pop-LIT-ee-us)**    Rotates tibia medially on femur if femur is fixed (as in sitting down), or rotates femur laterally on tibia if tibia is fixed (as in standing up); unlocks knee to allow flexion; may prevent forward dislocation of femur during crouching

| **O:** Lateral condyle of femur; lateral meniscus and joint capsule | **I:** Posterior surface of upper tibia | **N:** Tibial n. |
|---|---|---|

| TABLE 10.19 | Muscles Acting on the Foot |
|---|---|

The fleshy mass of the leg is formed by a group of **crural muscles,** which act on the foot (fig. 10.36). These muscles are tightly bound by deep fasciae, which compress them and aid in the return of blood from the legs. The fasciae separate the crural muscles into anterior, lateral, and posterior compartments, each with its own nerve and blood supply (see fig. 10.40b).

**Anterior Compartment of the Leg.**    Muscles of the anterior compartment dorsiflex the ankle and prevent the toes from scuffing the ground during walking. These are the *extensor digitorum longus* (extensor of toes II–V), *extensor hallucis longus* (extensor of the great toe), *fibularis (peroneus) tertius,* and *tibialis anterior.* Their tendons are held tightly against the ankle and kept from bowing by two **extensor retinacula** similar to the one at the wrist (fig. 10.37).

**Extensor Digitorum Longus (DIDJ-ih-TOE-rum)**    Extends toes, dorsiflexes foot, tautens plantar aponeurosis

| **O:** Lateral condyle of tibia, shaft of fibula, interosseous membrane | **I:** Middle and distal phalanges II–V | **N:** Deep fibular n. |
|---|---|---|

**Extensor Hallucis Longus (ha-LOO-sis)**    Extends great toe, dorsiflexes foot

| **O:** Anterior surface of middle of fibula, interosseous membrane | **I:** Distal phalanx I | **N:** Deep fibular n. |
|---|---|---|

**Fibularis (Peroneus**[78]**) Tertius**[79] **(FIB-you-LERR-iss TUR-she-us)**    Dorsiflexes and everts foot during walking, helps toes clear the ground during forward swing

| **O:** Medial surface of lower one-third of fibula, interosseous membrane | **I:** Metatarsal V | **N:** Deep fibular n. |
|---|---|---|

---

[75]*semi* = half + *membranosus* = membranous
[76]*semi* = half + *tendinosus* = tendinous
[77]*poplit* = ham (pit) of the knee

[78]*perone* = pinlike (fibula)
[79]*fibularis* = of the fibula + *tert* = third

**Tibialis[80] Anterior (TIB-ee-AY-lis)**   Dorsiflexes and inverts foot; resists backward tipping of body (as when standing on a moving boat deck); helps support medial longitudinal arch of foot

**O:** Lateral condyle and lateral margin of proximal half of tibia; interosseous membrane

**I:** Medial cuneiform, metatarsal I

**N:** Deep fibular n.

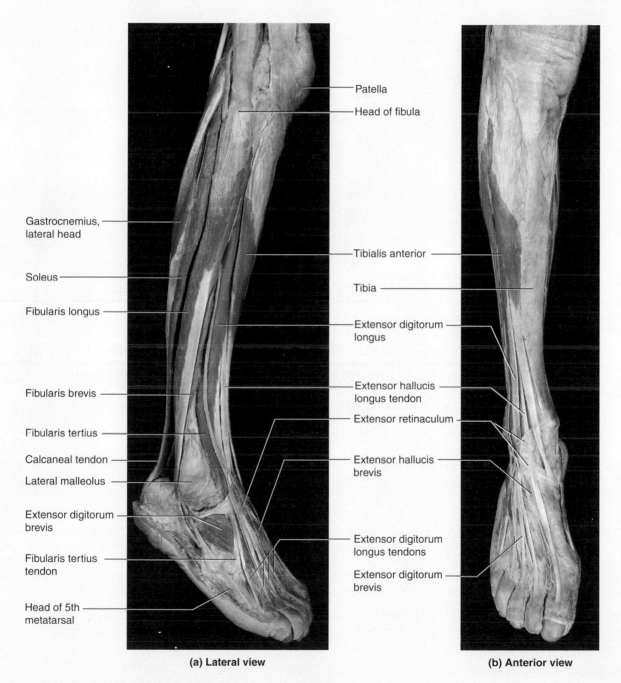

Patella
Head of fibula

Gastrocnemius, lateral head

Soleus

Fibularis longus

Fibularis brevis

Fibularis tertius

Calcaneal tendon

Lateral malleolus

Extensor digitorum brevis

Fibularis tertius tendon

Head of 5th metatarsal

Tibialis anterior

Tibia

Extensor digitorum longus

Extensor hallucis longus tendon

Extensor retinaculum

Extensor hallucis brevis

Extensor digitorum longus tendons

Extensor digitorum brevis

**(a) Lateral view**

**(b) Anterior view**

**FIGURE 10.36** Superficial Muscles of the Right Leg of the Cadaver.

---

[80]*tibialis* = of the tibia

## TABLE 10.19    Muscles Acting on the Foot (cont.)

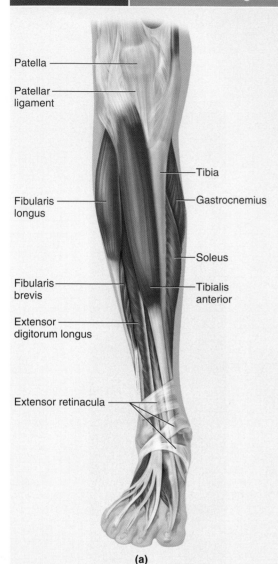

- Patella
- Patellar ligament
- Fibularis longus
- Fibularis brevis
- Extensor digitorum longus
- Extensor retinacula
- Tibia
- Gastrocnemius
- Soleus
- Tibialis anterior

(a)

**FIGURE 10.37    Muscles of the Leg, Anterior Compartment.** (a) Superficial anterior view of the leg. Some muscles of the posterior and lateral compartments are also partially visble. (b–d) Individual muscles of the anterior compartment of the leg and dorsal aspect of the foot.

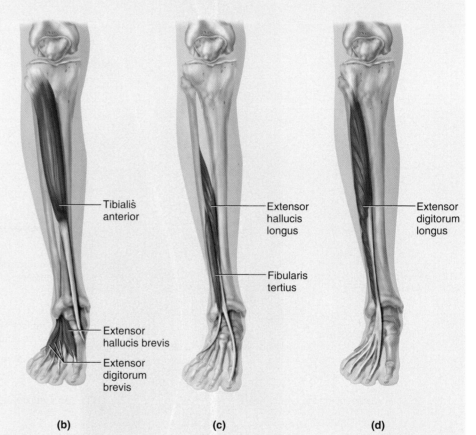

- Tibialis anterior
- Extensor hallucis brevis
- Extensor digitorum brevis

(b)

- Extensor hallucis longus
- Fibularis tertius

(c)

- Extensor digitorum longus

(d)

### Posterior Compartment of the Leg, Superficial Group.

The posterior compartment has superficial and deep muscle groups. The three muscles of the superficial group are plantar flexors: the *gastrocnemius, soleus,* and *plantaris* (fig. 10.38). The first two of these, collectively known as the *triceps surae,*[81] insert on the calcaneus by way of the **calcaneal (Achilles) tendon.** This is the strongest tendon of the body but is nevertheless a common site of sports injuries resulting from sudden stress. The plantaris, a weak synergist of the triceps surae, is a relatively unimportant muscle and is absent from many people; it is not tabulated here. Surgeons often use the plantaris tendon for tendon grafts needed in other parts of the body.

**Gastrocnemius[82] (GAS-trock-NEE-me-us)**    Plantar flexes foot, flexes knee; active in walking, running, and jumping

| **O:** Condyles and popliteal surface of femur, lateral supracondylar line, capsule of knee joint | **I:** Calcaneus | **N:** Tibial n. |
|---|---|---|

**Soleus[83] (SO-lee-us)**    Plantar flexes foot; steadies leg on ankle during standing

| **O:** Posterior surface of head and proximal one-fourth of fibula; middle one-third of tibia; interosseous membrane | **I:** Calcaneus | **N:** Tibial n. |
|---|---|---|

[81]*sura* = calf of leg
[82]*gastro* = belly + *cnem* = leg

[83]Named for its resemblance to a flatfish (sole)

**FIGURE 10.38 Superficial Muscles of the Leg, Posterior Compartment.** (a) The gastrocnemius. (b) The soleus, deep to the gastrocnemius and sharing the calcaneal tendon with it.

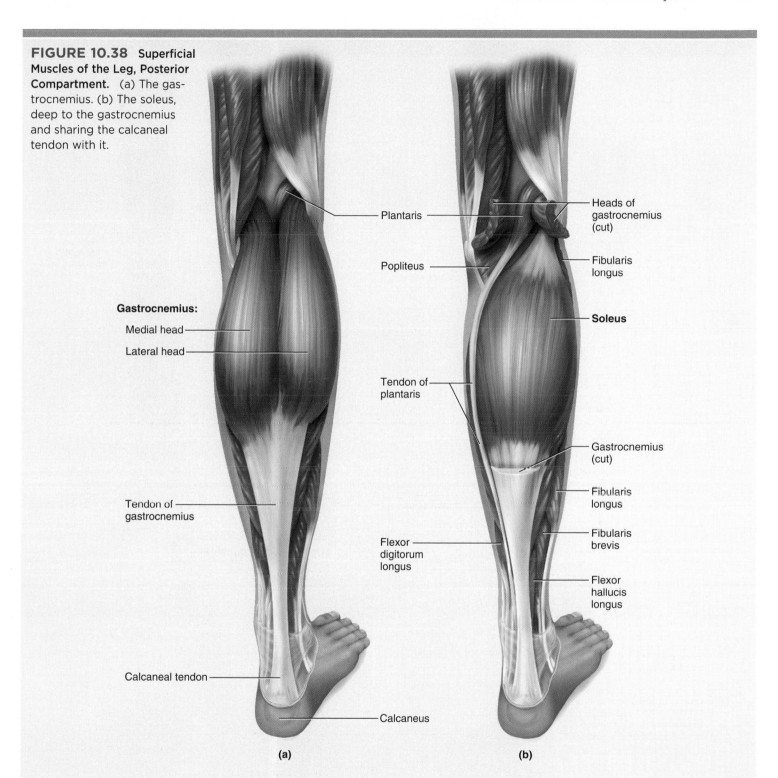

Plantaris

Popliteus

**Gastrocnemius:**

Medial head

Lateral head

Tendon of gastrocnemius

Calcaneal tendon

Calcaneus

Heads of gastrocnemius (cut)

Fibularis longus

**Soleus**

Tendon of plantaris

Gastrocnemius (cut)

Fibularis longus

Fibularis brevis

Flexor digitorum longus

Flexor hallucis longus

(a)                                    (b)

**Posterior Compartment of the Leg, Deep Group.** There are four muscles in the deep group (fig. 10.39). The *flexor digitorum longus, flexor hallucis longus,* and *tibialis posterior* are plantar flexors. The fourth muscle, the *popliteus,* is described in table 10.18 because it acts on the knee rather than on the foot.

**Flexor Digitorum Longus** Flexes phalanges of digits II–V as foot is raised from ground; stabilizes metatarsal heads and keeps distal pads of toes in contact with ground in toe-off and tiptoe movements

**O:** Posterior surface of tibial shaft            **I:** Distal phalanges II–V            **N:** Tibial n.

**TABLE 10.19**   Muscles Acting on the Foot *(cont.)*

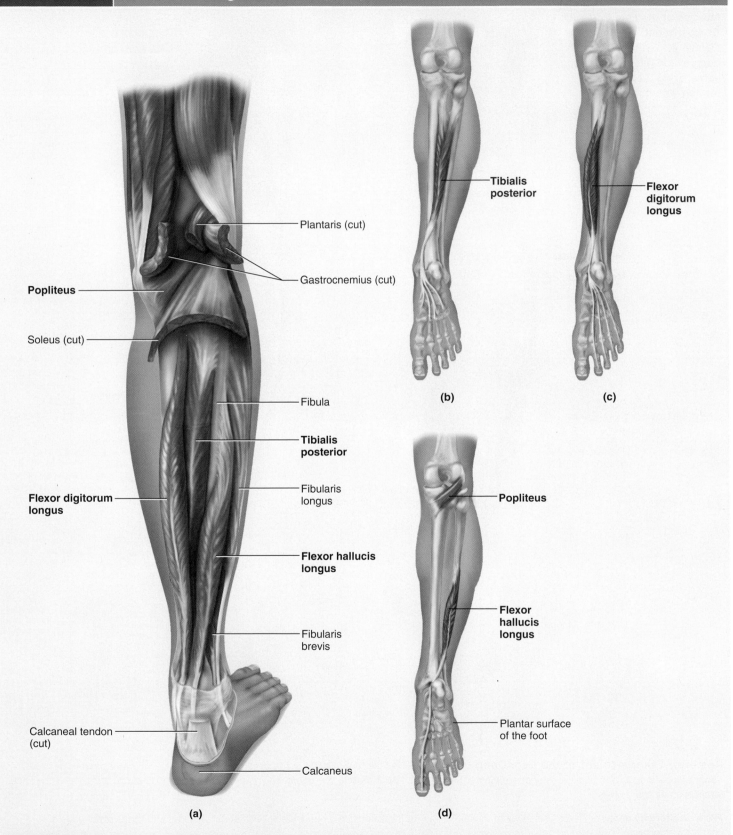

**FIGURE 10.39   Deep Muscles of the Leg, Posterior and Lateral Compartments.**   (a) Muscles deep to the soleus. (b–d) Exposure of some individual deep muscles with foot plantar flexed.

**Flexor Hallucis Longus**    Same actions as flexor digitorum longus, but for great toe (digit I)

**O:** Inferior two-thirds of fibula and interosseous membrane        **I:** Distal phalanx I        **N:** Tibial n.

**Tibialis Posterior**    Inverts foot; may assist in strong plantar flexion or control pronation of foot during walking

**O:** Posterior surface of proximal half of tibia, fibula, and interosseous membrane        **I:** Navicular, medial cuneiform, metatarsals II–IV        **N:** Tibial n.

### Lateral (Fibular) Compartment of the Leg.

The lateral compartment includes the *fibularis brevis* and *fibularis longus* (figs. 10.36, 10.37a, 10.40b). They plantar flex and evert the foot. Plantar flexion is important not only in standing on tiptoes but in providing lift and forward thrust each time you take a step.

**Fibularis (Peroneus) Brevis**    Maintains concavity of sole during toe-off and tiptoeing; may evert foot and limit inversion and help steady leg on foot

**O:** Lateral surface of distal two-thirds of fibula        **I:** Base of metatarsal V        **N:** Superficial fibular n.

**Fibularis (Peroneus) Longus**    Maintains concavity of sole during toe-off and tiptoeing; everts and plantar flexes foot

**O:** Head and lateral surface of proximal two-thirds of fibula        **I:** Medial cuneiform, metatarsal I        **N:** Superficial fibular n.

**Key a**

 Anterior compartment

 Medial compartment

 Posterior compartment (hamstrings)

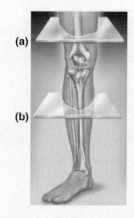

(a)

(b)

**Key b**

 Anterior compartment

 Lateral (fibular) compartment

 Posterior superficial compartment

 Posterior deep compartment

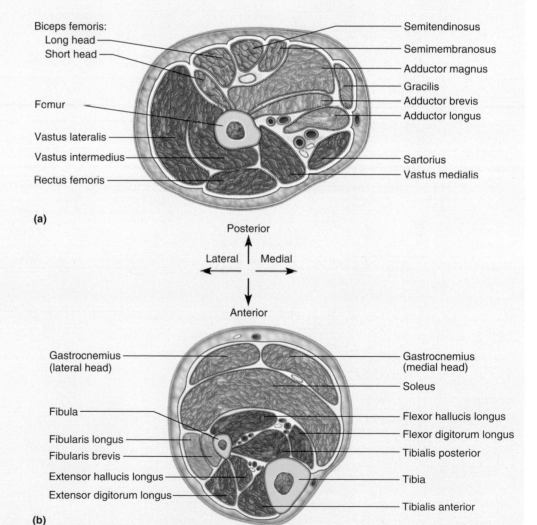

**FIGURE 10.40  Serial Cross Sections Through the Lower Limb.** Each section is taken at the correspondingly lettered level in the figure at the left.

▶ *Which of these muscles are named for the adjacent bones?*

| **TABLE 10.20** | Intrinsic Muscles of the Foot |

The intrinsic muscles of the foot help to support the arches and act on the toes in ways that aid locomotion. Several of them are similar in name and location to the intrinsic muscles of the hand.

### Dorsal Aspect of Foot.
Only one of the intrinsic muscles, the *extensor digitorum brevis,* is on the dorsal side of the foot. The medial slip of this muscle, serving the great toe, is sometimes called the *extensor hallucis brevis.*

**Extensor Digitorum Brevis**   Extends proximal phalanx I and all phalanges of digits II–IV

| | | |
|---|---|---|
| **O:** Calcaneus, inferior extensor retinaculum of ankle | **I:** Proximal phalanx I, tendons of extensor digitorum longus to II–IV | **N:** Deep fibular n. |

### Ventral Layer 1 (Most Superficial).
All remaining intrinsic muscles are on the ventral aspect of the foot or between the metatarsal bones. They are grouped in four layers (fig. 10.41). Dissecting into the foot from the plantar surface, one first encounters a tough fibrous sheet, the **plantar aponeurosis,** between the skin and muscles. It diverges like a fan from the calcaneus to the bases of all the toes, and serves as an origin for several ventral muscles. The ventral muscles include the stout *flexor digitorum brevis* on the midline of the foot, with four tendons that supply all digits except the hallux. It is flanked by the *abductor digiti minimi* laterally and the *abductor hallucis* medially.

**Flexor Digitorum Brevis**   Flexes digits II–IV; supports arches of foot

| | | |
|---|---|---|
| **O:** Calcaneus, plantar aponeurosis | **I:** Middle phalanges II–V | **N:** Medial plantar n. |

**Abductor Digiti Minimi**[84]   Abducts and flexes little toe; supports arches of foot

| | | |
|---|---|---|
| **O:** Calcaneus, plantar aponeurosis | **I:** Proximal phalanx V | **N:** Lateral plantar n. |

**Abductor Hallucis**   Abducts great toe; supports arches of foot

| | | |
|---|---|---|
| **O:** Calcaneus, plantar aponeurosis, flexor retinaculum | **I:** Proximal phalanx I | **N:** Medial plantar n. |

### Ventral Layer 2.
The next deeper layer consists of the thick *quadratus plantae* in the middle of the foot and the four *lumbrical* muscles located between the metatarsals.

**Quadratus Plantae**[85] **(quad-RAY-tus PLAN-tee)**   Same as flexor digitorum longus (table 10.19); flexion of digits II–V and associated locomotor functions

| | | |
|---|---|---|
| **O:** Two heads on the medial and lateral sides of calcaneus | **I:** Distal phalanges II–V via flexor digitorum longus tendons | **N:** Lateral plantar n. |

**Lumbricals (LUM-brick-ulz)**   Flex toes II–V

| | | |
|---|---|---|
| **O:** Tendon of flexor digitorum longus | **I:** Proximal phalanges II–V | **N:** Lateral and medial plantar nn. |

### Ventral Layer 3.
The muscles of this layer serve only the great and little toes. They are the *flexor digiti minimi brevis, flexor hallucis brevis,* and *adductor hallucis.* The adductor hallucis has an *oblique head* that extends diagonally from the midplantar region to the base of the great toe, and a transverse head that passes across the bases of digits II–IV and meets the long head at the base of the great toe.

**Adductor Hallucis**   Adducts great toe

| | | |
|---|---|---|
| **O:** Metatarsals II–IV, fibularis longus tendon, ligaments at bases of digits III–V | **I:** Proximal phalanx I | **N:** Lateral plantar n. |

**Flexor Digiti Minimi Brevis**   Flexes little toe

| | | |
|---|---|---|
| **O:** Metatarsal V, sheath of fibularis longus | **I:** Proximal phalanx V | **N:** Lateral plantar n. |

**Flexor Hallucis Brevis**   Flexes great toe

| | | |
|---|---|---|
| **O:** Cuboid, lateral cuneiform, tibialis posterior tendon | **I:** Proximal phalanx I | **N:** Medial plantar n. |

### Ventral Layer 4 (Deepest).
This layer consists only of the small interosseous muscles located between the metatarsal bones—four dorsal and three plantar. Each dorsal interosseous muscle is bipennate and originates on two adjacent metatarsals. The plantar interosseous muscles are unipennate and originate on only one metatarsal each.

**Dorsal Interosseous Muscles (four muscles)**   Abduct toes II–IV

| | | |
|---|---|---|
| **O:** Each with two heads arising from facing surfaces of two adjacent metatarsals | **I:** Proximal phalanges II–IV | **N:** Lateral plantar n. |

**Plantar Interosseous Muscles (three muscles)**   Adduct toes III–V

| | | |
|---|---|---|
| **O:** Medial aspect of metatarsals III–V | **I:** Proximal phalanges III–V | **N:** Lateral plantar n. |

---

[84]*digit* = toe + *minim* = smallest          [85]*quadrat* = four-sided + *plantae* = of the plantar region

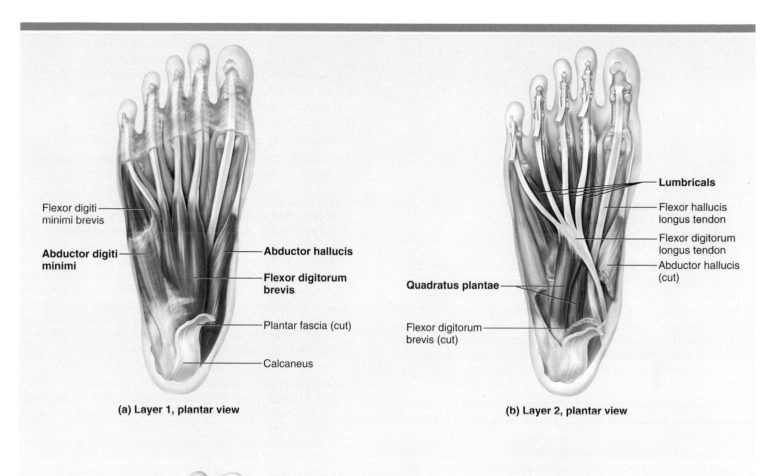

Flexor digiti minimi brevis

**Abductor digiti minimi**

**Abductor hallucis**

**Flexor digitorum brevis**

Plantar fascia (cut)

Calcaneus

**(a) Layer 1, plantar view**

**Lumbricals**

Flexor hallucis longus tendon

Flexor digitorum longus tendon

Abductor hallucis (cut)

**Quadratus plantae**

Flexor digitorum brevis (cut)

**(b) Layer 2, plantar view**

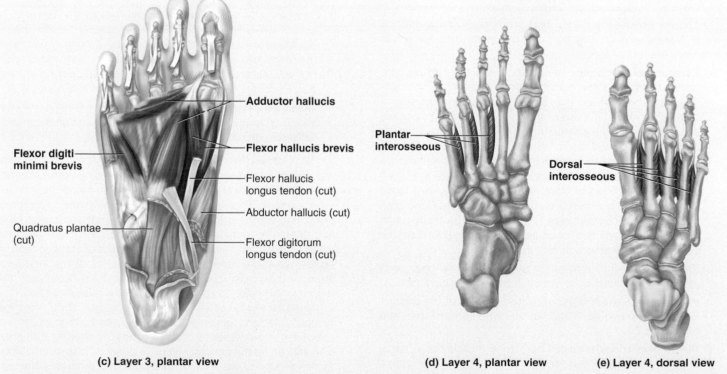

**Adductor hallucis**

**Flexor hallucis brevis**

Flexor hallucis longus tendon (cut)

Abductor hallucis (cut)

Flexor digitorum longus tendon (cut)

**Flexor digiti minimi brevis**

Quadratus plantae (cut)

**(c) Layer 3, plantar view**

**Plantar interosseous**

**Dorsal interosseous**

**(d) Layer 4, plantar view**

**(e) Layer 4, dorsal view**

**FIGURE 10.41 Intrinsic Muscles of the Foot.** (a–d) First through fourth layers, respectively, in ventral (plantar) views. (e) Fourth layer, dorsal view. The muscles belonging to each layer are shown in color and with boldface labels.

Think About It

*Not everyone has the same muscles. From the information provided in this chapter, identify two muscles that are lacking in some people.*

Before You Go On

*Answer the following questions to test your understanding of the preceding section:*

22. *In the middle of a stride, you have one foot on the ground and you are about to swing the other leg forward. What muscles produce the movements of that leg?*

23. *Name the muscles that cross both the hip and knee joints and produce actions at both.*

24. *List the major actions of the muscles of the anterior, medial, and posterior compartments of the thigh.*

25. *Describe the role of plantar flexion and dorsiflexion in walking. What muscles produce these actions?*

## INSIGHT 10.4    Clinical Application

### Athletic Injuries

Although the muscular system is subject to fewer diseases than most organ systems, it is particularly vulnerable to injuries resulting from sudden and intense stress placed on muscles and tendons. Each year, thousands of athletes from the high school to professional level sustain some type of injury to their muscles, as do the increasing numbers of people who have taken up running and other forms of physical conditioning. Overzealous exertion without proper conditioning and warm-up is frequently the cause. Some of the more common athletic injuries are:

**Baseball finger**—tears in the extensor tendons of the fingers resulting from the impact of a baseball with the extended fingertip.

**Blocker's arm**—abnormal calcification in the lateral margin of the forearm as a result of repeated impact with opposing players.

**Charley horse**—any painful tear, stiffness, or blood clotting in a muscle. A charley horse of the quadriceps femoris is often caused by football tackles.

**Compartment syndrome**—a condition in which overuse, contusion, or muscle strain damages blood vessels in a compartment of the arm or leg. Since the fasciae enclosing a compartment are tight and cannot stretch, blood or tissue fluid accumulating in the compartment can put pressure on the muscles, nerves, and blood vessels. The lack of blood flow in the compartment can cause destruction of nerves if untreated within 2 to 4 hours and death of muscle tissue if it goes untreated for 6 hours or more. Nerves can regenerate if blood flow is restored, but muscle damage is irreversible. Depending on its severity, compartment syndrome may be treated with immobilization and rest, or an incision to drain fluid from the compartment or otherwise relieve the pressure.

**Pitcher's arm**—inflammation at the origin of the flexor carpi resulting from hard wrist flexion in releasing a baseball.

**Pulled groin**—strain in the adductor muscles of the thigh. It is common in gymnasts and dancers who perform splits and high kicks.

**Pulled hamstrings**—strained hamstring muscles or a partial tear in their tendinous origins, often with a hematoma (blood clot) in the fascia lata. This condition is frequently caused by repetitive kicking (as in football and soccer) or long, hard running.

**Rider's bones**—abnormal calcification in the tendons of the adductor muscles of the medial thigh. It results from prolonged abduction of the thighs when riding horses.

**Rotator cuff injury**—a tear in the tendon of any of the SITS (rotator cuff) muscles, most often the tendon of the supraspinatus. Such injuries are caused by strenuous circumduction of the arm, shoulder dislocation, hard falls or blows to the shoulder, or repetitive use of the arm in a position above horizontal. They are common among baseball pitchers and third basemen, bowlers, swimmers, weight lifters, and in racquet sports. Recurrent inflammation of a SITS tendon can cause a tendon to degenerate and then to rupture in response to moderate stress. Injury causes pain and makes the shoulder joint unstable and subject to dislocation.

**Shinsplints**—a general term embracing several kinds of injury with pain in the crural region: tendinitis of the tibialis posterior muscle, inflammation of the tibial periosteum, and anterior compartment syndrome. Shinsplints may result from unaccustomed jogging, walking on snowshoes, or any vigorous activity of the legs after a period of relative inactivity.

**Tennis elbow**—inflammation at the origin of the extensor carpi muscles on the lateral epicondyle of the humerus. It occurs when these muscles are repeatedly tensed during backhand strokes and then strained by sudden impact with the tennis ball. Any activity that requires rotary movements of the forearm and a firm grip of the hand (for example, using a screwdriver) can cause the symptoms of tennis elbow.

**Tennis leg**—a partial tear in the lateral origin of the gastrocnemius muscle. It results from repeated strains put on the muscle while supporting the body weight on the toes.

Most athletic injuries can be prevented by proper conditioning. A person who suddenly takes up vigorous exercise may not have sufficient muscle and bone mass to withstand the stresses such exercise entails. These must be developed gradually. Stretching exercises keep ligaments and joint capsules supple and therefore reduce injuries. Warm-up exercises promote more efficient and less injurious musculoskeletal function in several ways, discussed in chapter 11. Most of all, moderation is important, as most injuries simply result from overuse of the muscles. "No pain, no gain" is a dangerous misconception.

Muscular injuries can be treated initially with "RICE": **r**est, **i**ce, **c**ompression, and **e**levation. Rest prevents further injury and allows repair processes to occur; ice reduces swelling; compression with an elastic bandage helps to prevent fluid accumulation and swelling; and elevation of an injured limb promotes drainage of blood from the affected area and limits further swelling. If these measures are not enough, anti-inflammatory drugs may be employed, including corticosteroids as well as aspirin and other nonsteroidal agents. Serious injuries, such as compartment syndrome, require emergency attention by a physician.

# Interactions Between the
## MUSCULAR SYSTEM
### and Other Organ Systems

■ indicates ways in which this system affects other systems     ■ indicates ways in which other systems affect this system

## INTEGUMENTARY SYSTEM

Facial muscles pull on skin to provide facial expressions

Covers and protects superficial muscles; initiates synthesis of calcitriol, which promotes absorption of $Ca^{2+}$ needed for muscle contraction; dissipates heat generated by muscles

## SKELETAL SYSTEM

Muscles move and stabilize joints and produce stress that affects ossification, bone remodeling, and shapes of bones

Provides levers on which muscles act; stores $Ca^{2+}$ needed for muscle contraction

## NERVOUS SYSTEM

Muscles give expression to thoughts, emotions, and motor commands that arise in the central nervous system

Stimulates muscle contraction; monitors and adjusts muscle tension; adjusts cardiopulmonary functions to meet needs of muscles during exercise

## ENDOCRINE SYSTEM

Skeletal muscles protect some endocrine organs

Hormones stimulate growth and development of muscles and regulate levels of glucose and electrolytes important for muscle contraction

## CIRCULATORY SYSTEM

Muscle contractions help to move blood through veins; exercise stimulates growth of new blood vessels

Delivers $O_2$ and nutrients; removes muscle wastes and heat cardiovascular efficiency, RBC count, hemoglobin level, and density of blood capillaries in muscle greatly affect muscular endurance

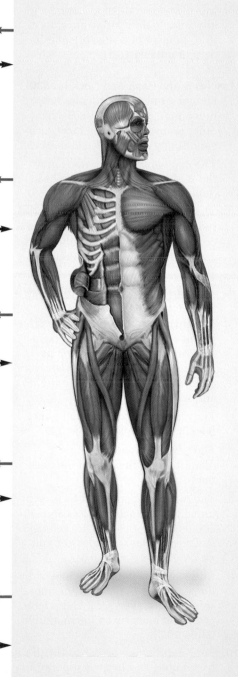

## LYMPHATIC AND IMMUNE SYSTEMS

Muscle contractions promote lymph flow; exercise elevates level of immune cells and antibodies; excess exercise inhibits immune responses

Lymphatic system drains fluid from muscles; immune system protects muscles from pathogens and promotes tissue repair

## RESPIRATORY SYSTEM

Muscle contractions ventilate lungs; muscles of larynx and pharynx regulate air flow; $CO_2$ generated by exercise stimulates respiratory rate and depth

Provides $O_2$ and removes $CO_2$; respiratory efficiency greatly affects muscular endurance

## URINARY SYSTEM

Muscles control voluntary urination; muscles of pelvic floor support bladder

Eliminates wastes generated by muscles; regulates levels of electrolytes important for muscle contraction

## DIGESTIVE SYSTEM

Muscles enable chewing and swallowing; control voluntary defecation; abdominal and lumbar muscles protect lower digestive organs

Absorbs nutrients needed by muscles; liver regulates blood glucose levels and metabolizes lactic acid generated by anaerobic muscle metabolism

## REPRODUCTIVE SYSTEM

Muscles contribute to erection and ejaculation; abdominal and pelvic muscles aid childbirth

Gonadal steroids affect muscular growth and development

## CHAPTER REVIEW

# Review of Key Concepts

**The Structural and Functional Organization of Muscles (p. 322)**

1. The muscular system consists of the skeletal muscles. The study of this system is *myology.*

2. The muscular system serves for body movements, stability, communication, control of body openings and passages, and heat production.

3. Each skeletal muscle fiber (cell) is enclosed in a fibrous *endomysium.* A *fascicle* is a bundle of muscle fibers enclosed in a fibrous *perimysium.* The muscle as a whole is enclosed in a fibrous *epimysium.* Connective tissue *deep fasciae* separate neighboring muscles from each other, and the *superficial fascia* separates the muscles from the skin.

4. A muscle may have a *direct attachment* to a bone in which the muscle fibers reach all the way to the bone, or an *indirect attachment* in which a tendon intervenes between the muscle and bone.

5. A muscle typically has a stationary *origin (head)* on one bone, a mobile *insertion* on another bone, and a thicker *belly* between the origin and insertion.

6. Muscles are classified according to the orientation of their fascicles as *fusiform, parallel, convergent, pennate,* and *circular* (fig. 10.3). Circular muscles are also called *sphincters.*

7. The motion produced by a muscle is called its *action.* Muscles work in groups across a joint. In a given joint action, the *prime mover* produces most of the force, a *synergist* aids the prime mover or modifies its action, an *antagonist* opposes that action, and a *fixator* prevents a bone from moving. These muscles may change roles in different actions at the same joint.

8. *Intrinsic muscles* are entirely contained in a region of study such as the head or hand; *extrinsic muscles* have their origin in a region other than the site of their action, such as forearm muscles that move the fingers.

9. Muscles are innervated by cranial nerves and spinal nerves.

10. Muscle names, often in Latin, usually refer to their size, shape, location, number of heads, orientation, or action (table 10.1).

**Muscles of the Head and Neck (p. 330)**

1. Humans and other primates have much more expressive faces than other animals, and they have correspondingly complex facial muscles.

2. The *occipitalis* and *frontalis* move the scalp, eyebrows, and forehead (table 10.2).

3. The eyelid and other tissues around the eye are moved by the *orbicularis oculi, levator palpebrae superioris,* and *corrugator supercilii* (table 10.2).

4. The *nasalis* muscle flares and compresses the nostrils (table 10.2).

5. The lips are acted upon by the *orbicularis oris, levator labii superioris, levator anguli oris, zygomaticus major* and *minor, risorius, depressor anguli oris, depressor labii inferioris,* and *mentalis* (table 10.2).

6. The cheeks are acted upon by the *buccinator* muscles (table 10.2).

7. The *platysma* acts upon the mandible and the skin of the neck (table 10.2).

8. The tongue is controlled by a set of unnamed *intrinsic muscles* and several *extrinsic muscles:* the *genioglossus, hyoglossus, styloglossus,* and *palatoglossus* (table 10.3).

9. Biting and chewing are achieved by the actions of the *temporalis, masseter, medial pterygoid,* and *lateral pterygoid* muscles on the mandible (table 10.3).

10. Four muscles are associated with the hyoid bone and located superior to it, and are thus called the *suprahyoid group:* the *digastric, geniohyoid, mylohyoid,* and *stylohyoid* (table 10.3). These muscles act on the mandible and hyoid bone to forcibly open the mouth and to aid in swallowing.

11. Another four muscles associated with the hyoid bone are inferior to it and therefore called the *infrahyoid group:* the *omohyoid, sternohyoid, thyrohyoid,* and *sternothyroid* (table 10.3). These muscles depress or fix the hyoid and elevate or depress the larynx, especially in association with swallowing.

12. The *superior, middle,* and *inferior pharyngeal constrictors* contract in sequence to force food down and into the esophagus (table 10.3).

13. The *sternocleidomastoid* and three *scalene* muscles flex the neck. The *trapezius, splenius capitis,* and *semispinalis capitis* are the major extensors of the neck. Some of these are also employed in rotation of the head (table 10.4).

**Muscles of the Trunk (p. 341)**

1. Breathing is achieved by the muscles of respiration, especially the *diaphragm, external intercostals,* and *internal intercostals* (table 10.5).

2. The abdominal wall is supported by the sheetlike *external abdominal oblique, internal abdominal oblique,* and *transverse abdominal,* and by the *rectus abdominis* muscles. These support the abdominal viscera, stabilize the vertebral column during lifting, and aid in respiration, urination, defecation, vomiting, and childbirth (table 10.6).

3. The back has numerous complex muscles that extend, rotate, and abduct the vertebral column and aid in breathing. The major superficial muscle is the *erector spinae* (which is subdivided into the *iliocostalis, longissimus,* and *spinalis* muscle columns). The deep back muscles include the *semispinalis thoracis, quadratus lumborum,* and *multifidus* (table 10.7).

4. The pelvic floor is spanned by three layers of muscles and fasciae (table 10.8). The anal canal, urethra, and vagina penetrate the pelvic floor muscles and open into the *perineum,* a diamond-shaped space between the thighs bordered by the pubic symphysis, coccyx, and ischial tuberosities. The anterior half of the perineum is the *urogenital triangle,* and the posterior half is the *anal triangle.*

5. The most superficial compartment of the pelvic floor is the *superficial perineal space.* It contains two major muscles: the *ischiocavernosus* and *bulbospongiosus.*

6. The middle compartment of the pelvic floor contains the *external urethral sphincter* and *external anal sphincter.*

7. The deepest compartment of the pelvic floor is the *pelvic diaphragm.* It consists of two muscles: the *levator ani* and *coccygeus.*

## Muscles Acting on the Shoulder and Upper Limb (p. 351)

1. Muscles that act on the pectoral girdle originate on the axial skeleton and insert on the clavicle and scapula. They fall into an anterior group that includes the *pectoralis minor* and *serratus anterior,* and a posterior group that includes the superficial *trapezius* and three deeper muscles, the *levator scapulae, rhomboideus major,* and *rhomboideus minor* (table 10.9).

2. Muscles that act on the arm (humerus) originate mainly on the pectoral girdle and axial skeleton and cross the shoulder joint. These include the *pectoralis major, latissimus dorsi, deltoid, teres major, coracobrachialis,* and four *rotator cuff (SITS)* muscles: the *supraspinatus, infraspinatus, teres minor,* and *subscapularis* (table 10.10).

3. Muscles that act on the forearm and elbow are located in both the arm (the *brachialis, biceps brachii,* and *triceps brachii*) and the forearm (the *brachioradialis, anconeus, pronator quadratus, pronator teres,* and *supinator*) (table 10.13).

4. Most muscles whose origins and bellies are in the forearm have insertions in, and act upon, the wrist and hand (table 10.14). Several of them, however, have origins on the humerus and also cross the elbow joint. Therefore, they also contribute slightly to actions of the elbow. Deep fasciae divide the forearm muscles into *anterior* and *posterior compartments* and separate those of each compartment into superficial and deep layers.

5. Anterior compartment muscles are mainly flexors of the wrist and hand. Superficial muscles of the anterior compartment include the *palmaris longus* and *flexor carpi radialis,* which form the two most prominent tendons of the anterior wrist, and the *flexor carpi ulnaris* and *flexor digitorum superficialis.* The deep muscles of the anterior compartment include the *flexor digitorum profundus* and *flexor pollicis longus.*

6. Posterior compartment muscles are mainly extensors of the wrist and hand. Superficial muscles of the posterior compartment include the *extensor digitorum, extensor carpi radialis longus, extensor carpi radialis brevis, extensor carpi ulnaris,* and *extensor digiti minimi.* Deep muscles of the posterior compartment include the *abductor pollicis longus, extensor pollicis brevis, extensor pollicis longus,* and *extensor indicis.*

7. Most tendons of the forearm muscles pass under a *flexor retinaculum* on the anterior side of the wrist or an *extensor retinaculum* on the posterior side of the wrist. The space between the flexor retinaculum and carpal bones is called the *carpal tunnel.*

8. Intrinsic muscles of the hand assist the forearm muscles and make movements of the digits more precise. They are divided into a thenar, hypothenar, and midpalmar group (table 10.16).

9. The *thenar group* of muscles form the thick fleshy mass at the base of the thumb and the web between the thumb and palm. They move the thumb. They include the *adductor pollicis, abductor pollicis brevis, flexor pollicis brevis,* and *opponens pollicis.*

10. The *hypothenar group* muscles form the fleshy *hypothenar eminence* at the base of the little finger, and are concerned with movements of that digit. They include the *abductor digiti minimi, flexor digiti minimi brevis,* and *opponens digiti minimi.*

11. The *midpalmar group* of muscles span the palm and include four *dorsal interosseous muscles,* three *palmar interosseous muscles,* and four *lumbrical muscles,* located between the metacarpal bones. They act on digits II through V.

## Muscles Acting on the Hip and Lower Limb (p. 367)

1. Most muscles that act on the femur originate on the os coxae (table 10.17). The two major anterior muscles of this group are the *iliacus* and the *psoas major,* collectively called the *iliopsoas.*

2. Superficial muscles on the lateral and posterior sides of the hip include the *tensor fasciae latae, gluteus maximus, gluteus medius,* and *gluteus minimus.* The tendons of the first two of these muscles join the *fascia lata* to form the fibrous *iliotibial band* on the lateral aspect of the thigh.

3. Deep muscles on the lateral aspect of the hip, known as the *lateral rotators,* include the *gemellus superior, gemellus inferior, obturator externus, obturator internus, piriformis,* and *quadratus femoris.* The principal actions of these muscles are abduction and lateral rotation of the femur.

4. Deep fasciae divide the thigh muscles into an *anterior (extensor) compartment, medial (adductor) compartment,* and *posterior (flexor) compartment.*

5. Muscles of the medial compartment act as adductors of the femur. These include the *adductor brevis, adductor longus, adductor magnus, gracilis,* and *pectineus* (table 10.17).

6. Muscles of the anterior compartment act mainly as extensors of the knee. These include the four heads of the *quadriceps femoris—rectus femoris, vastus lateralis, vastus medialis,* and *vastus intermedius—*and the *sartorius.*

7. Muscles of the posterior compartment act as extensors of the hip and flexors of the knee. These are the *biceps femoris, semimembranosus,* and *semitendinosus,* known colloquially as the *hamstring muscles* (table 10.18)

8. Muscles of the leg are divided into anterior, posterior, and lateral compartments (table 10.19). Most of them act on the foot.

9. Anterior compartment muscles of the leg include the *extensor digitorum longus, extensor hallucis longus, fibularis tertius,* and *tibialis anterior.*

10. Superficial posterior compartment muscles include the *popliteus,* which acts on the knee, and two muscles, the *gastrocnemius* and *soleus,* collectively also known as the *triceps surae* (these share the calcaneal tendon to the heel).

11. Deep posterior compartment muscles include the *flexor digitorum longus, flexor hallucis longus,* and *tibialis posterior.*

12. Lateral compartment muscles include the *fibularis brevis* and *fibularis longus.*

13. Intrinsic muscles of the foot support the arches and act on the toes, and resemble intrinsic muscles of the hand. The *extensor digitorum brevis* is located dorsally. The others are ventral and are arranged in layers (table 10.20).

14. The layer 1 (most superficial) intrinsic muscles of the foot are the *flexor digitorum brevis, abductor digiti minimi,* and *abductor hallucis;* layer 2 comprises the *quadratus plantae* and four *lumbrical muscles;* layer 3 includes the *adductor hallucis, flexor digiti minimi brevis,* and *flexor hallucis brevis;* and layer 4 (deepest) includes four *dorsal interosseous muscles* and three *plantar interosseous muscles.*

# Testing Your Recall

1. Which of the following muscles is the prime mover in spitting out a mouthful of liquid?
   a. platysma
   b. buccinator
   c. risorius
   d. masseter
   e. palatoglossus

2. Each muscle fiber has a sleeve of areolar connective tissue around it called
   a. the deep fascia.
   b. the superficial fascia.
   c. the perimysium.
   d. the epimysium.
   e. the endomysium.

3. Which of these is *not* a suprahyoid muscle?
   a. genioglossus
   b. geniohyoid
   c. stylohyoid
   d. mylohyoid
   e. digastric

4. Which of these muscles is an extensor of the neck?
   a. external oblique
   b. sternocleidomastoid
   c. splenius capitis
   d. iliocostalis
   e. latissimus dorsi

5. Which of these muscles of the pelvic floor is the deepest?
   a. superficial transverse perineal
   b. bulbospongiosus
   c. ischiocavernosus
   d. deep transverse perineal
   e. levator ani

6. Which of these actions is *not* performed by the trapezius?
   a. extension of the neck
   b. depression of the scapula
   c. elevation of the scapula
   d. rotation of the scapula
   e. adduction of the humerus

7. Both the hands and feet are acted upon by a muscle or muscles called
   a. the extensor digitorum.
   b. the abductor digiti minimi.
   c. the flexor digitorum profundus.
   d. the abductor hallucis.
   e. the flexor digitorum longus.

8. Which of the following muscles does *not* extend the hip joint?
   a. quadriceps femoris
   b. gluteus maximus
   c. biceps femoris
   d. semitendinosus
   e. semimembranosus

9. Both the gastrocnemius and _____ muscles insert on the heel by way of the calcaneal tendon.
   a. semimembranosus
   b. tibialis posterior
   c. tibialis anterior
   d. soleus
   e. plantaris

10. Which of the following muscles raises the upper lip?
    a. levator palpebrae superioris
    b. orbicularis oris
    c. zygomaticus minor
    d. masseter
    e. mentalis

11. The _____ of a muscle is the point where it attaches to a relatively stationary bone.

12. A bundle of muscle fibers surrounded by perimysium is called a/an _____.

13. The _____ is the muscle primarily responsible for a given movement at a joint.

14. The three large muscles on the posterior side of the thigh are commonly known as the _____ muscles.

15. Connective tissue bands called _____ prevent flexor tendons of the forearm and leg from rising like bowstrings.

16. The anterior half of the perineum is a region called the _____.

17. The abdominal aponeuroses converge on a median fibrous band on the abdomen called the _____.

18. A muscle that works with another to produce the same or similar movement is called a/an _____.

19. A muscle somewhat like a feather, with fibers obliquely approaching its tendon from both sides, is called a/an _____ muscle.

20. A circular muscle that closes a body opening is called a/an _____.

*Answers in Appendix B*

# True or False

*Determine which five of the following statements are false, and briefly explain why.*

1. Cutting the phrenic nerves would paralyze the prime mover of respiration.

2. The orbicularis oculi is a sphincter.

3. The origin of the sternocleidomastoid muscle is the mastoid process of the skull.

4. To push someone away from you, you would use the serratus anterior more than the trapezius.

5. Both the extensor digitorum and the extensor digiti minimi extend the little finger.

6. Curling the toes employs the quadratus plantae.

7. The scalenes are superficial to the trapezius.

8. Exhaling requires contraction of the internal intercostal muscles.

9. Hamstring injuries often result from rapid flexion of the knee.

10. The tibialis anterior and tibialis posterior are synergists.

*Answers in Appendix B*

# Testing Your Comprehension

1. Radical mastectomy, once a common treatment for breast cancer, involved removal of the pectoralis major along with the breast. What functional impairments would result from this? What synergists could a physical therapist train a patient to use to recover some lost function?

2. Removal of cancerous lymph nodes from the neck sometimes requires removal of the sternocleidomastoid on that side. How would this affect a patient's range of head movement?

3. In a disease called tick paralysis, the saliva from a tick bite paralyzes skeletal muscles beginning with the lower limbs and progressing superiorly. What would be the most urgent threat to the life of a tick paralysis patient?

4. Women who habitually wear high heels may suffer painful "high heel syndrome" when they go barefoot or wear flat shoes. What muscle(s) and tendon(s) are involved? Explain.

5. A student moving out of a dormitory kneels down, in correct fashion, to lift a heavy box of books. What prime movers are involved as he straightens his legs to lift the box?

*Answers at www.mhhe.com/saladin4*

# www.mhhe.com/saladin4

*The textbook website provides a wealth of interactive study materials fully organized and integrated by chapter. You will find practice quizzes, labeling exercises, and much more that will complement your learning and understanding of anatomy and physiology. The website also includes tools designed to enhance your* **Anatomy & Physiology | REVEALED** *experience.*

# Atlas B

# SURFACE ANATOMY

**Volume 1** Skeletal and Muscular Systems

# The Importance of External Anatomy

In the study of human anatomy, it is easy to become so preoccupied with internal structure that we forget the importance of what we can see and feel externally. Yet external anatomy and appearance are major concerns in giving a physical examination and in many aspects of patient care. A knowledge of the body's surface landmarks is essential to one's competence in physical therapy, cardiopulmonary resuscitation, surgery, making X rays and electrocardiograms, giving injections, drawing blood, listening to heart and respiratory sounds, measuring the pulse and blood pressure, and finding pressure points to stop arterial bleeding, among other procedures. A misguided attempt to perform some of these procedures while disregarding or misunderstanding external anatomy can be very harmful and even fatal to a patient.

Having just studied skeletal and muscular anatomy in the preceding chapters, this is an opportune time for you to study the body surface. Much of what we see there reflects the underlying structure of the superficial bones and muscles. A broad photographic overview of surface anatomy is given in atlas A (see fig. A.5). In the following pages, we examine the body literally from head (fig. B.1) to toe (fig. B.14), studying its regions in more detail. To make the most profitable use of this atlas, refer to the skeletal and muscular anatomy in chapters 8 to 10. Relate drawings of the clavicles in chapter 8 to the photograph in figure B.1, for example. Study the shape of the scapula in chapter 8 and see how much of it you can trace on the photographs in figure B.3. See if you can relate the ten-dons visible on the hand (fig. B.8) to the muscles of the forearm illustrated in chapter 10, and the external markings of the pelvis (fig. B.4) to bone structure in chapter 8.

For learning surface anatomy, there is a resource available to you that is far more valuable than any laboratory model or textbook illustration—your own body. For the best understanding of human structure, compare the art and photographs in this book with your body or with structures visible on a study partner. In addition to bones and muscles, you can palpate a number of superficial arteries, veins, tendons, ligaments, and cartilages, among other structures. By palpating regions such as the shoulder, elbow, or ankle, you can develop a mental image of the subsurface structures better than you can obtain by looking at two-dimensional textbook images. And the more you can study with other people, the more you will appreciate the variations in human structure and be able to apply your knowledge to your future patients or clients, who will not look quite like any textbook diagram or photograph you have ever seen. Through comparisons of art, photography, and the living body, you will get a much deeper understanding of the body than if you were to study this atlas in isolation from the earlier chapters.

At the end of this atlas, you can test your knowledge of externally visible muscle anatomy. The two photographs in figure B.15 have 30 numbered muscles and a list of 26 names, some of which are shown more than once in the photographs and some of which are not shown at all. Identify the muscles to your best ability without looking back at the previous illustrations, and then check your answers in appendix B at the back of the book.

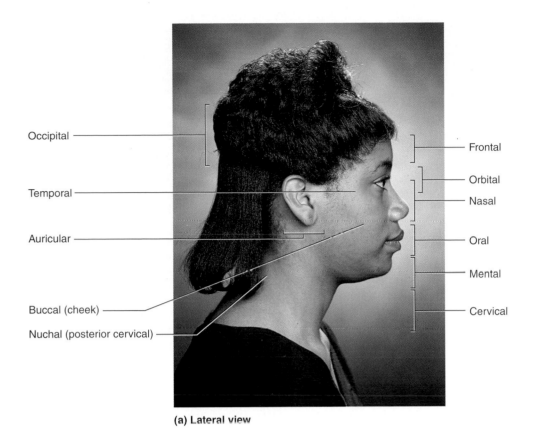

Occipital

Frontal

Temporal

Orbital

Nasal

Auricular

Oral

Mental

Buccal (cheek)

Cervical

Nuchal (posterior cervical)

**(a) Lateral view**

Frons (forehead)

Root of nose

Bridge of nose

Superciliary ridge

Lateral commissure

Superior palpebral sulcus

Medial commissure

Inferior palpebral sulcus

Dorsum nasi

Apex of nose

Auricle (pinna) of ear

Ala nasi

Philtrum

Mentolabial sulcus

Labia (lips)

Mentum (chin)

Trapezius muscle

Sternoclavicular joints

Supraclavicular fossa

Clavicle

Suprasternal notch

Sternum

**(b) Anterior view**

**FIGURE B.1**  **The Head and Neck.**  (a) Anatomical regions of the head. (b) Features of the facial region and upper thorax.

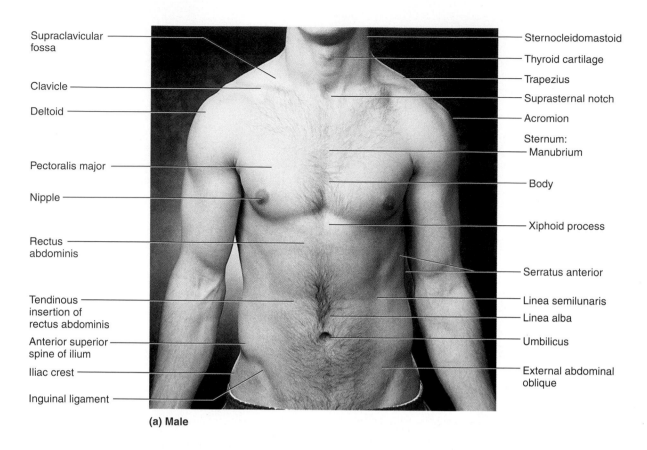

Supraclavicular fossa

Clavicle

Deltoid

Pectoralis major

Nipple

Rectus abdominis

Tendinous insertion of rectus abdominis

Anterior superior spine of ilium

Iliac crest

Inguinal ligament

Sternocleidomastoid

Thyroid cartilage

Trapezius

Suprasternal notch

Acromion

Sternum:
Manubrium

Body

Xiphoid process

Serratus anterior

Linea semilunaris

Linea alba

Umbilicus

External abdominal oblique

**(a) Male**

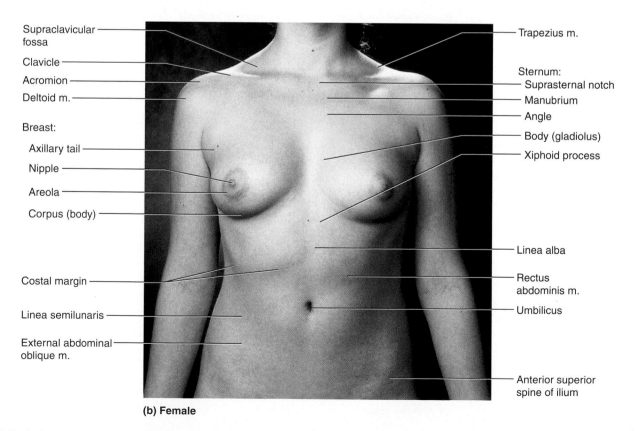

Supraclavicular fossa

Clavicle

Acromion

Deltoid m.

Breast:

Axillary tail

Nipple

Areola

Corpus (body)

Costal margin

Linea semilunaris

External abdominal oblique m.

Trapezius m.

Sternum:
Suprasternal notch

Manubrium

Angle

Body (gladiolus)

Xiphoid process

Linea alba

Rectus abdominis m.

Umbilicus

Anterior superior spine of ilium

**(b) Female**

**FIGURE B.2** **The Thorax and Abdomen.** Anterior view. All of the features labeled are common to both sexes, though some are labeled only on the photograph that shows them best.

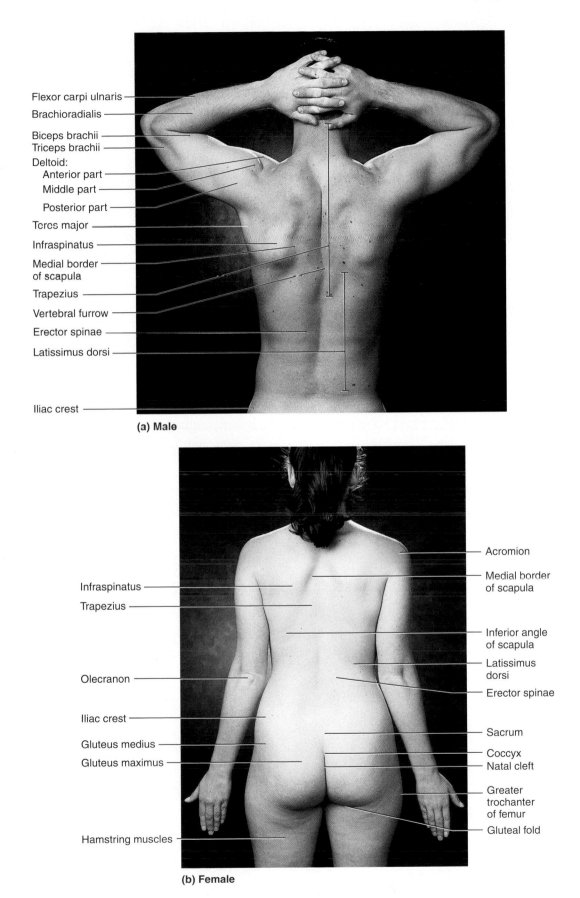

Flexor carpi ulnaris

Brachioradialis

Biceps brachii
Triceps brachii
Deltoid:
   Anterior part
   Middle part
   Posterior part
Teres major
Infraspinatus
Medial border
of scapula
Trapezius
Vertebral furrow
Erector spinae
Latissimus dorsi

Iliac crest

**(a) Male**

Infraspinatus
Trapezius

Olecranon

Iliac crest
Gluteus medius
Gluteus maximus

Hamstring muscles

Acromion
Medial border
of scapula

Inferior angle
of scapula
Latissimus
dorsi
Erector spinae

Sacrum
Coccyx
Natal cleft
Greater
trochanter
of femur
Gluteal fold

**(b) Female**

**FIGURE B.3  The Back and Gluteal Region.**  All of the features labeled are common to both sexes, though some are labeled only on the photograph that shows them best.

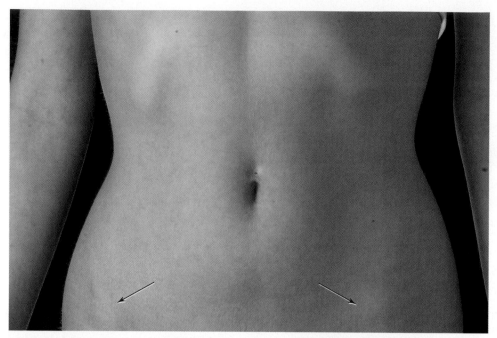

**(a) Anterior view**

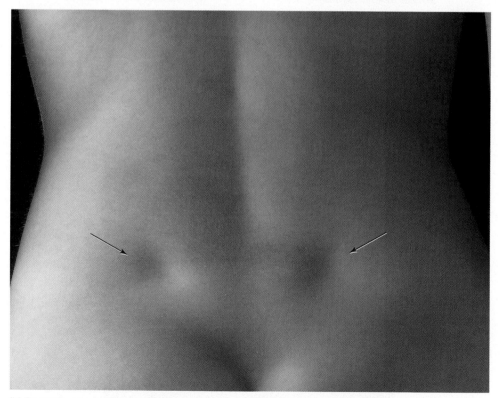

**(b) Posterior view**

**FIGURE B.4    The Pelvic Region.**  (a) The anterior superior spines of the ilium are marked by anterolateral protuberances (arrows). (b) The posterior superior spines are marked in some people by dimples in the sacral region (arrows).

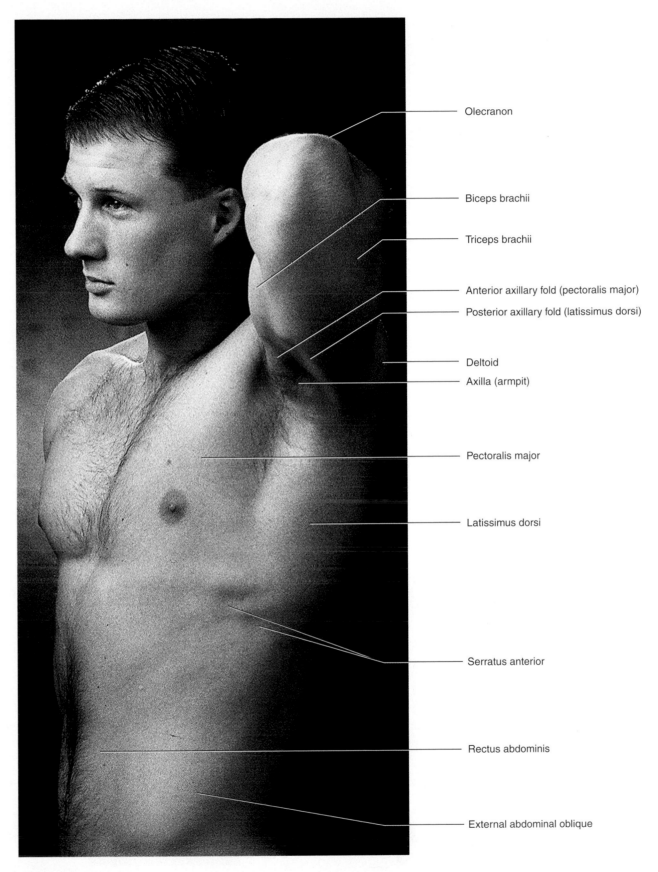

Olecranon

Biceps brachii

Triceps brachii

Anterior axillary fold (pectoralis major)

Posterior axillary fold (latissimus dorsi)

Deltoid

Axilla (armpit)

Pectoralis major

Latissimus dorsi

Serratus anterior

Rectus abdominis

External abdominal oblique

**FIGURE B.5** The Axillary Region.

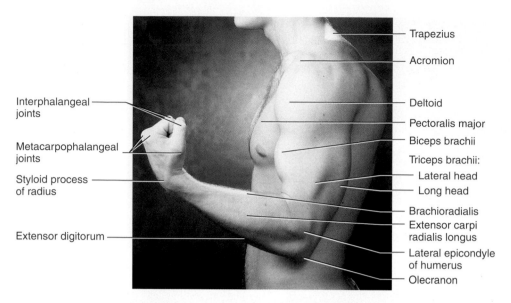

Trapezius
Acromion
Deltoid
Pectoralis major
Biceps brachii
Triceps brachii:
    Lateral head
    Long head
Brachioradialis
Extensor carpi radialis longus
Lateral epicondyle of humerus
Olecranon

Interphalangeal joints
Metacarpophalangeal joints
Styloid process of radius
Extensor digitorum

**FIGURE B.6**  The Upper Limb, Lateral View.

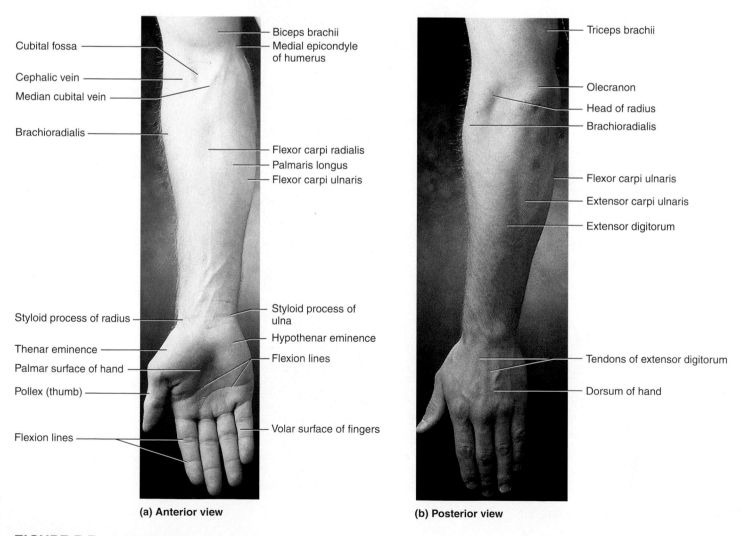

Cubital fossa
Cephalic vein
Median cubital vein
Brachioradialis

Biceps brachii
Medial epicondyle of humerus

Flexor carpi radialis
Palmaris longus
Flexor carpi ulnaris

Styloid process of radius
Thenar eminence
Palmar surface of hand
Pollex (thumb)
Flexion lines

Styloid process of ulna
Hypothenar eminence
Flexion lines

Volar surface of fingers

**(a) Anterior view**

Triceps brachii
Olecranon
Head of radius
Brachioradialis

Flexor carpi ulnaris
Extensor carpi ulnaris
Extensor digitorum

Tendons of extensor digitorum
Dorsum of hand

**(b) Posterior view**

**FIGURE B.7**  The Right Antebrachium (Forearm).

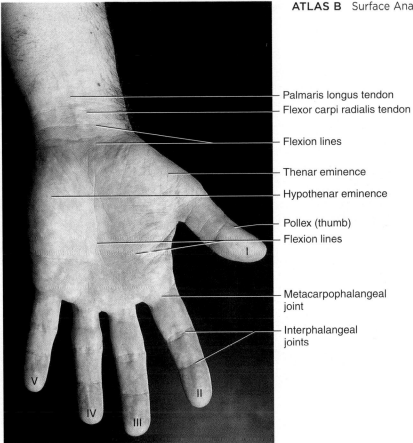

Palmaris longus tendon
Flexor carpi radialis tendon
Flexion lines
Thenar eminence
Hypothenar eminence
Pollex (thumb)
Flexion lines
Metacarpophalangeal joint
Interphalangeal joints

**(a) Anterior (Palmar) view**

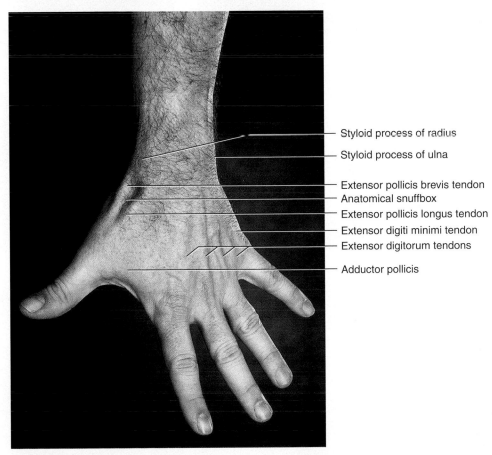

Styloid process of radius
Styloid process of ulna
Extensor pollicis brevis tendon
Anatomical snuffbox
Extensor pollicis longus tendon
Extensor digiti minimi tendon
Extensor digitorum tendons
Adductor pollicis

**(b) Posterior (Dorsal) view**

**FIGURE B.8**
**The Left Wrist and Hand.**

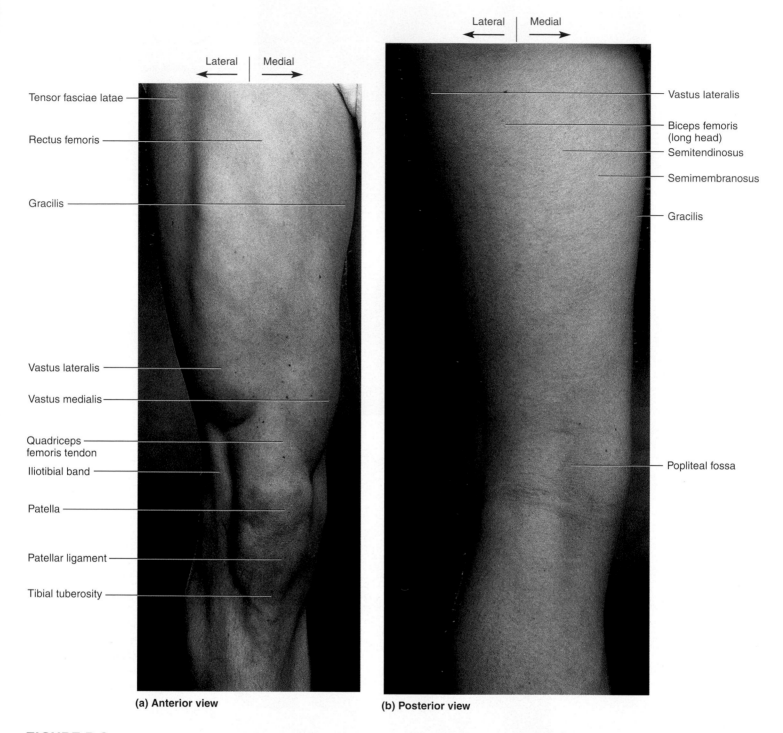

Lateral | Medial

Tensor fasciae latae

Rectus femoris

Gracilis

Vastus lateralis

Vastus medialis

Quadriceps femoris tendon

Iliotibial band

Patella

Patellar ligament

Tibial tuberosity

**(a) Anterior view**

Lateral | Medial

Vastus lateralis

Biceps femoris (long head)

Semitendinosus

Semimembranosus

Gracilis

Popliteal fossa

**(b) Posterior view**

**FIGURE B.9** **The Thigh and Knee.** Locations of posterior thigh muscles are indicated, but the boundaries of the individual muscles are rarely visible on a living person.

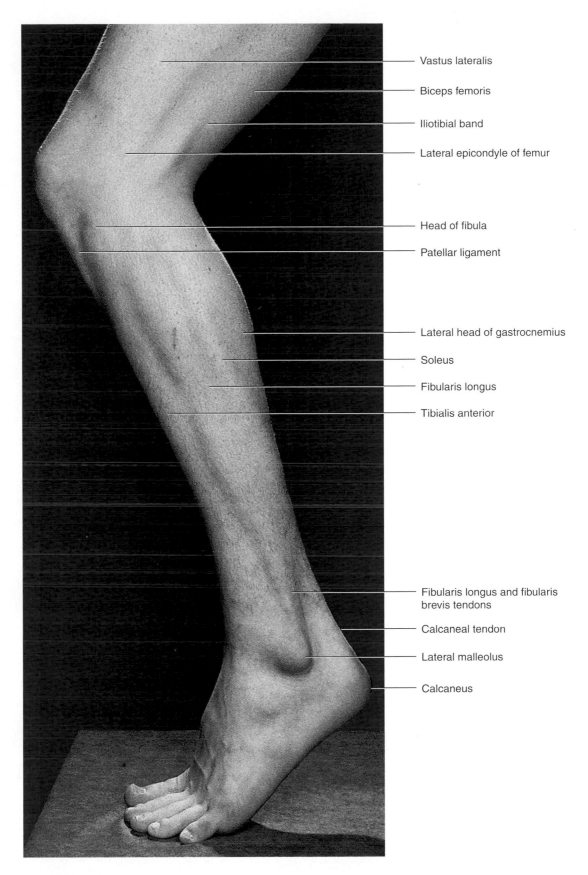

Vastus lateralis

Biceps femoris

Iliotibial band

Lateral epicondyle of femur

Head of fibula

Patellar ligament

Lateral head of gastrocnemius

Soleus

Fibularis longus

Tibialis anterior

Fibularis longus and fibularis brevis tendons

Calcaneal tendon

Lateral malleolus

Calcaneus

**FIGURE B.10  The Left Leg and Foot.**  Lateral view.

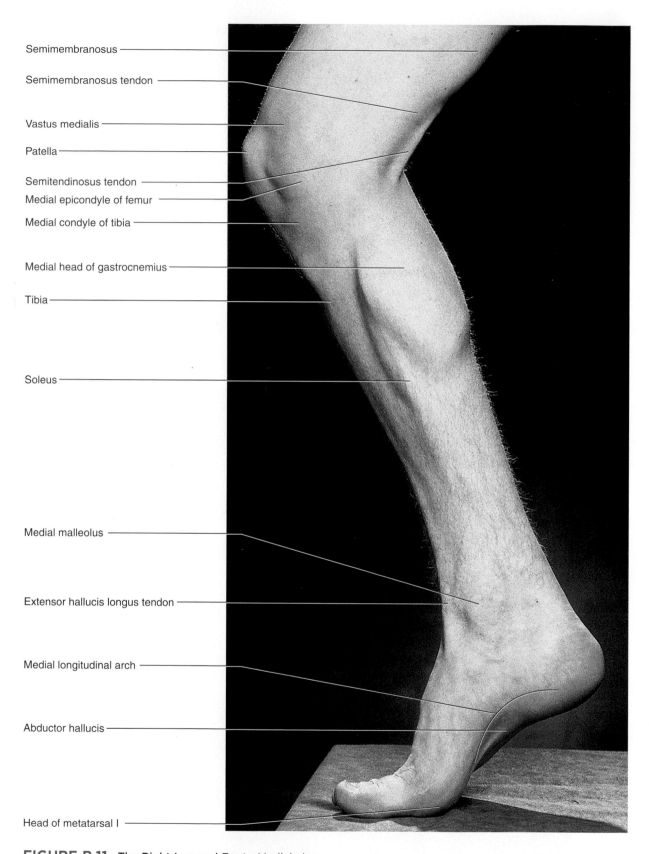

Semimembranosus

Semimembranosus tendon

Vastus medialis

Patella

Semitendinosus tendon

Medial epicondyle of femur

Medial condyle of tibia

Medial head of gastrocnemius

Tibia

Soleus

Medial malleolus

Extensor hallucis longus tendon

Medial longitudinal arch

Abductor hallucis

Head of metatarsal I

**FIGURE B.11   The Right Leg and Foot.** Medial view.

Lateral | Medial
←         →

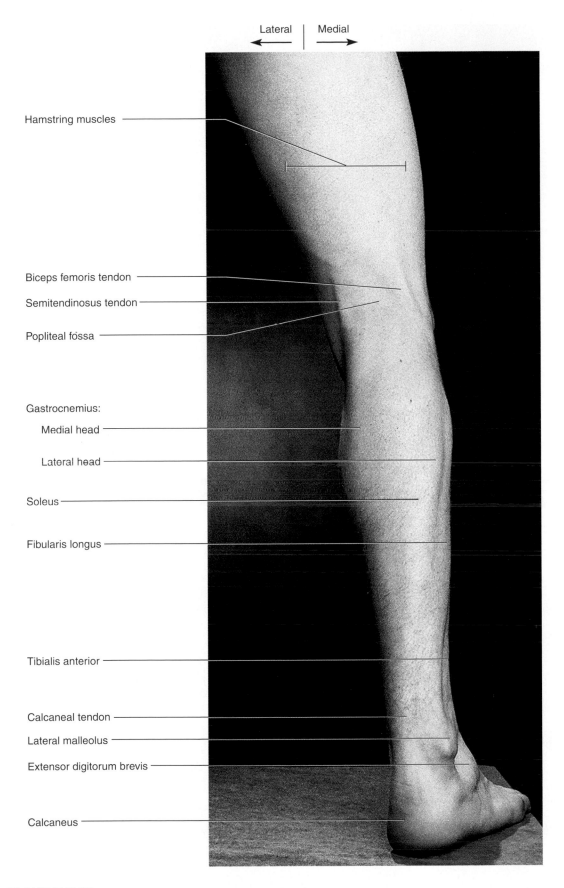

Hamstring muscles

Biceps femoris tendon

Semitendinosus tendon

Popliteal fossa

Gastrocnemius:

Medial head

Lateral head

Soleus

Fibularis longus

Tibialis anterior

Calcaneal tendon

Lateral malleolus

Extensor digitorum brevis

Calcaneus

**FIGURE B.12** **The Right Leg and Foot.** Posterior view.

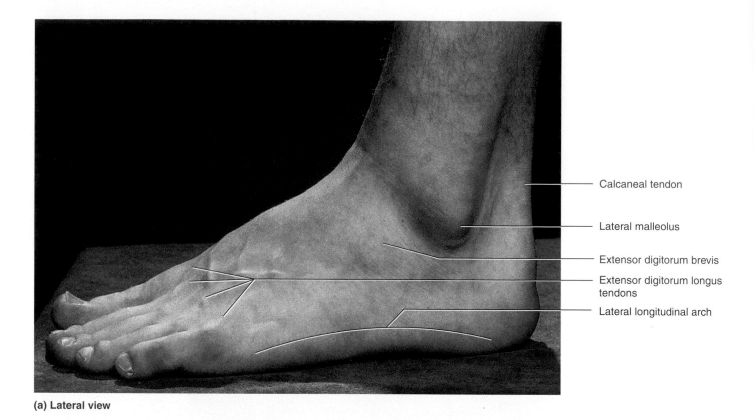

Calcaneal tendon

Lateral malleolus

Extensor digitorum brevis

Extensor digitorum longus tendons

Lateral longitudinal arch

**(a) Lateral view**

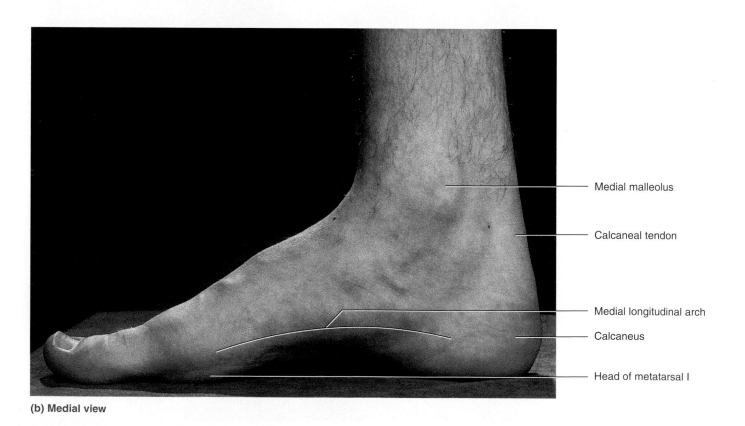

Medial malleolus

Calcaneal tendon

Medial longitudinal arch

Calcaneus

Head of metatarsal I

**(b) Medial view**

**FIGURE B.13**  The Foot.

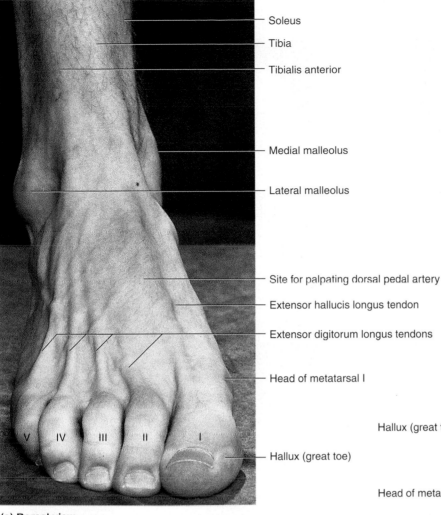

Soleus

Tibia

Tibialis anterior

Medial malleolus

Lateral malleolus

Site for palpating dorsal pedal artery

Extensor hallucis longus tendon

Extensor digitorum longus tendons

Head of metatarsal I

Hallux (great toe)

V    IV    III    II    I

**(a) Dorsal view**

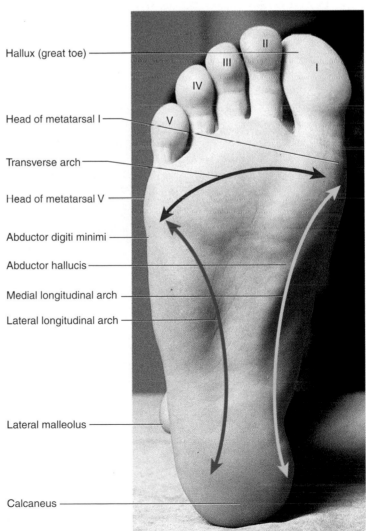

Hallux (great toe)

Head of metatarsal I

Transverse arch

Head of metatarsal V

Abductor digiti minimi

Abductor hallucis

Medial longitudinal arch

Lateral longitudinal arch

Lateral malleolus

Calcaneus

II

III

IV

I

V

**(b) Plantar view**

**FIGURE B.14** **The Foot.** Compare the arches in part (b) to the skeletal anatomy in figure 8.42.

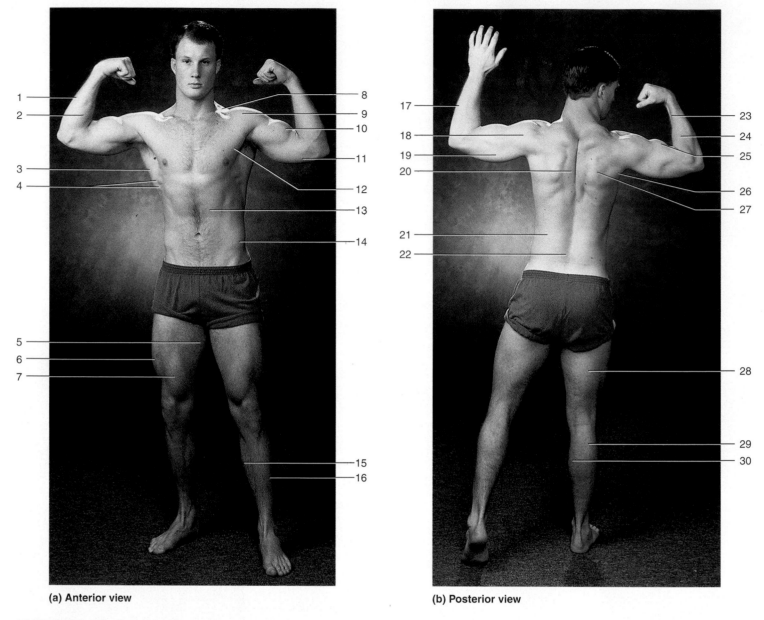

**(a) Anterior view**

**(b) Posterior view**

**FIGURE B.15**   **Muscle Test.**   To test your knowledge of muscle anatomy, match the 30 labeled muscles on these photographs to the alphabetical list of muscles below. Answer as many as possible without referring to the previous illustrations. Some of these names will be used more than once, since the same muscle may be shown from different perspectives, and some of these names will not be used at all. The answers are in appendix B.

a.  biceps brachii

b.  brachioradialis

c.  deltoid

d.  erector spinae

e.  external abdominal oblique

f.  flexor carpi ulnaris

g.  gastrocnemius

h.  gracilis

i.  hamstrings

j.  infraspinatus

k.  latissimus dorsi

l.  pectineus

m.  pectoralis major

n.  rectus abdominis

o.  rectus femoris

p.  serratus anterior

q.  soleus

r.  splenius capitis

s.  sternocleidomastoid

t.  subscapularis

u.  teres major

v.  tibialis anterior

w.  transverse abdominal

x.  trapezius

y.  triceps brachii

z.  vastus lateralis

*Neuromuscular junctions (SEM)*

# MUSCULAR TISSUE

## INSIGHTS

## Brushing Up

To understand this chapter, it is important that you understand or brush up on the following concepts:

- Aerobic and anaerobic metabolism (p. 83)
- The functions of membrane proteins, especially receptors and ion gates (p. 96)
- Structure of a neuron (p. 171)
- General histology of the three types of muscle (p. 173)
- Desmosomes and gap junctions (p. 175)
- Connective tissues of a muscle (p. 322)

Movement is a fundamental characteristic of all living organisms, from bacteria to humans. Even plants and other seemingly immobile organisms move cellular components from place to place. Across the entire spectrum of life, the molecular mechanisms of movement are very similar, involving motor proteins such as myosin and dynein. But in animals, movement has developed to the highest degree, with the evolution of **muscular tissue** specialized for this function. A muscle is essentially a device for converting the chemical energy of ATP into the mechanical energy of movement.

The three types of muscle tissue—skeletal, cardiac, and smooth—were described and compared in chapter 5. Cardiac and smooth muscle are further described in this chapter, and cardiac muscle is discussed most extensively in chapter 19. Most of the present chapter, however, concerns **skeletal muscle,** the type which holds the body erect against the pull of gravity and produces its outwardly visible movements.

This chapter treats the structure, contraction, and metabolism of skeletal muscle at the molecular, cellular, and tissue levels of organization. Understanding muscle at these levels provides an indispensable basis for understanding such aspects of motor performance as warm-up, quickness, strength, endurance, and fatigue. Such factors have obvious relevance to athletic performance, and they become very important when a lack of physical conditioning, old age, or injury interferes with a person's ability to carry out everyday tasks or meet the extra demands for speed or strength that we all occasionally encounter.

# Types and Characteristics of Muscular Tissue

### Objectives

When you have completed this section, you should be able to

- describe the physiological properties that all muscle types have in common;
- list the defining characteristics of skeletal muscle; and
- describe the elastic functions of the connective tissue components of a muscle.

## UNIVERSAL CHARACTERISTICS OF MUSCLE

The functions of the muscular system were detailed in the preceding chapter: movement, stability, communication, control of body openings and passages, and heat production. To carry out those functions, all muscular tissue has the following characteristics:

- **Responsiveness** (excitability). Responsiveness is a property of all living cells, but muscle and nerve cells have developed this property to the highest degree. When stimulated by chemical signals (neurotransmitters), stretch, and other stimuli, muscle cells respond with electrical changes across the plasma membrane.

- **Conductivity.** Stimulation of a muscle fiber produces more than a local effect. The local electrical change triggers a wave of excitation that travels rapidly along the fiber and initiates processes leading to muscle contraction.

- **Contractility.** Muscle fibers are unique in their ability to shorten substantially when stimulated. This enables them to pull on bones and other tissues and create movements of the body and its parts.

- **Extensibility.** In order to contract, a muscle cell must also be extensible—able to stretch again between contractions. Most cells rupture if they are stretched even a little, but skeletal muscle fibers can stretch to as much as three times their contracted length.

- **Elasticity.** When a muscle cell is stretched and then the tension is released, it recoils to its original resting length. Elasticity, commonly misunderstood as the ability to stretch, refers to this tendency of a muscle cell (or other structures) to return to the original length when tension is released.

## SKELETAL MUSCLE

**Skeletal muscle** may be defined as voluntary striated muscle that is usually attached to one or more bones. A skeletal muscle cell exhibits alternating light and dark transverse bands, or **striations,** that result from an overlapping arrangement of their internal contractile proteins (fig. 11.1). Skeletal muscle is called **voluntary** because it is usually subject to conscious control. The other types of muscle are **involuntary** (not usually under conscious control), and they are never attached to bones.

A typical skeletal muscle cell is about 100 μm in diameter and 3 cm long; some are as thick as 500 μm and as long as 30 cm. Because of their extraordinary length, skeletal muscle cells are usually called *muscle fibers* or *myofibers.*

Recall from chapter 10 that a skeletal muscle is composed not only of muscular tissue, but also of fibrous con-

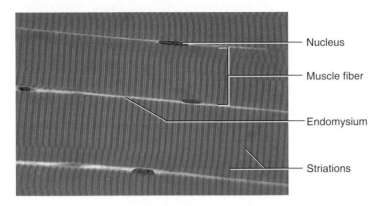

Nucleus

Muscle fiber

Endomysium

Striations

**FIGURE 11.1** **Skeletal Muscle Fibers.** Note the striations.

nective tissue: the *endomysium* that surrounds each muscle fiber, the *perimysium* that bundles muscle fibers together into fascicles, and the *epimysium* that encloses the entire muscle. These connective tissues are continuous with the collagen fibers of tendons and those, in turn, with the collagen of the bone matrix. Thus, when a muscle fiber contracts, it pulls on these collagen fibers and usually moves a bone.

Collagen is not excitable or contractile, but it is somewhat extensible and elastic. The collagenous components of a muscle fall into two groups. One is the *parallel elastic component,* which surrounds and parallels the contractile parts of the muscle. This includes the endomysium, perimysium, and epimysium. The other is the *series elastic component,* which is joined end to end with the contractile part and constitutes the tendons or other connections that bind the muscles to the bones. Stretching of a muscle, for example by extending a joint, lengthens these components and generates resistance to further stretch, thus protecting the muscle from injury. Upon relaxation of the muscle, elastic recoil of these components helps to return the muscle to its resting length and keeps it from becoming too flaccid.

Elastic recoil also adds significantly to the power output and efficiency of a muscle. Some of the pull exerted on a bone results from this recoil rather than from the expenditure of ATP to drive the muscle's contractile mechanism. When you are running, for example, recoil of the calcaneal (Achilles) tendon helps to lift the heel from the ground and produces some of the thrust as you push off with the toes. Each time the foot strikes the ground, the tendon is stretched again and ready once more for this elastic recoil on the next toe-off.

## Before You Go On

*Answer the following questions to test your understanding of the preceding section:*

1. *Define* responsiveness, conductivity, contractility, extensibility, *and* elasticity. *State why each of these properties is necessary for muscle function.*

2. *How is skeletal muscle different from the other types of muscle?*

3. *Why would the skeletal muscles perform poorly without their series elastic components?*

# Microscopic Anatomy of Skeletal Muscle

### Objectives

When you have completed this section, you should be able to

- describe the structural components of a muscle fiber;

- relate the striations of a muscle fiber to the overlapping arrangement of its protein filaments; and

- name the major proteins of a muscle fiber and state the function of each.

## THE MUSCLE FIBER

In order to understand muscle function, you must know how the organelles and macromolecules of a muscle fiber are arranged. Perhaps more than any other cell, a muscle fiber exemplifies the adage: Form follows function. It has a complex, tightly organized internal structure in which even the spatial arrangement of protein molecules is closely tied to its contractile function.

Muscle fibers have multiple flattened or sausage-shaped nuclei pressed against the inside of the plasma membrane. This unusual condition results from their embryonic development—several stem cells called **myoblasts**[1] fuse to produce each muscle fiber, with each myoblast contributing a nucleus to the mature fiber. Some myoblasts remain as unspecialized **satellite cells** between the muscle fiber and endomysium. When a muscle is injured, satellite cells can multiply and produce new muscle fibers to some degree. Most muscle repair, however, is by fibrosis rather than regeneration of functional muscle.

The plasma membrane of a muscle fiber is called the **sarcolemma,**[2] and its cytoplasm is called the **sarcoplasm.** The sarcoplasm is occupied mainly by long protein bundles called **myofibrils** about 1 μm in diameter (fig. 11.2). It also contains an abundance of **glycogen,** a carbohydrate which provides energy for the cell during heightened levels of exercise, and the red pigment **myoglobin,** which stores oxygen until needed for muscular activity.

Most other organelles of the cell, such as mitochondria and smooth endoplasmic reticulum (ER), are packed

---

[1]*myo* = muscle + *blast* = precursor
[2]*sarco* = flesh, muscle + *lemma* = husk

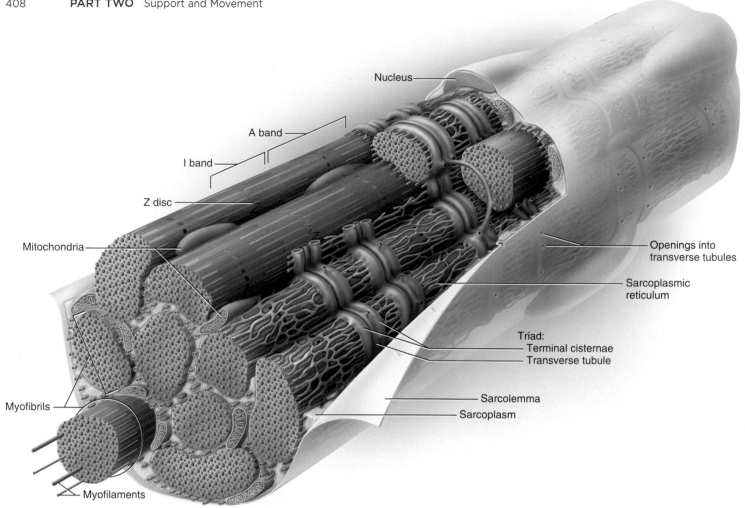

**FIGURE 11.2   Structure of a Skeletal Muscle Fiber.** This is a single cell containing 11 myofibrils (9 shown at the left end and 2 cut off at midfiber). A few myofilaments are shown projecting from the myofibril at the left. Their finer structure is shown in figure 11.3.

into the spaces between the myofibrils. The smooth ER of a muscle fiber is called **sarcoplasmic reticulum (SR).** It forms a network around each myofibril and periodically exhibits dilated end-sacs called **terminal cisternae,** which traverse the muscle fiber. The sarcolemma has tubular infoldings called **transverse (T) tubules,** which penetrate the interior of the cell and emerge on the other side. Each T tubule is closely associated with two terminal cisternae, running alongside it on each side. A T tubule and the adjacent terminal cisternae constitute a **triad.** The SR is a reservoir of calcium ions; it has gated channels in its membrane that open at the right times to release a flood of $Ca^{2+}$ into the cytosol, where the calcium activates the muscle contraction process. The T tubule signals the SR when to release these calcium bursts.

## MYOFILAMENTS

Let's return to the myofibrils just mentioned—the long protein cords that fill most of the muscle cell—and look at their structure at a finer, molecular level. It is here that the key to muscle contraction lies. Each myofibril is a bundle of parallel protein microfilaments called **myofilaments** (see the left end of figure 11.2). There are three kinds of myofilaments:

1. **Thick filaments** (fig. 11.3a, b) are about 15 nm in diameter. Each is made of several hundred molecules of a protein called **myosin.** A myosin molecule is shaped like a golf club, with two polypeptides intertwined to form a shaftlike *tail* and a double globular *head* projecting from it at an angle. A thick filament may be likened to a bundle of 200 to 500 such "golf clubs," with their heads directed outward in a spiral array around the bundle. The heads on one half of the thick filament angle to the left, and the heads on the other half angle to the right; in the middle is a *bare zone* with no heads.

2. **Thin filaments** (fig. 11.3c, d), 7 nm in diameter, are composed primarily of two intertwined strands of a protein called **fibrous (F) actin.** Each F actin is like a bead necklace—a string of subunits called **globular (G) actin.** Each G actin has an **active site** that can bind to the head of a myosin molecule. A thin fila-

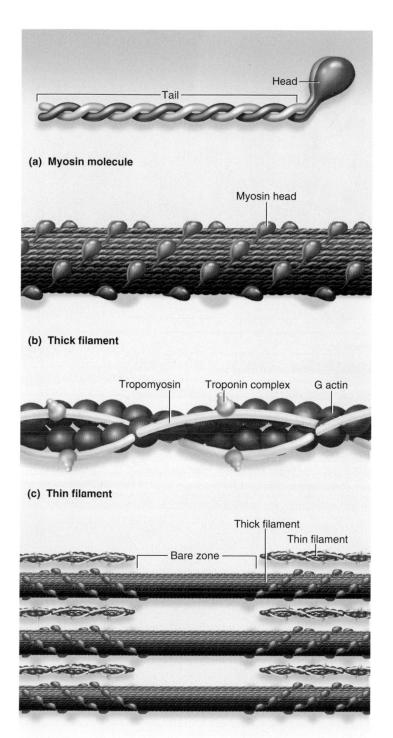

**(a) Myosin molecule**

Head

Tail

**(b) Thick filament**

Myosin head

**(c) Thin filament**

Tropomyosin    Troponin complex    G actin

**(d) Portion of a sarcomere showing the overlap of thick and thin filaments**

Thick filament

Thin filament

Bare zone

**FIGURE 11.3  Molecular Structure of Thick and Thin Filaments.** (a) A single myosin molecule consists of two intertwined polypeptides forming a twisted filamentous tail and a double globular head. (b) A thick filament consists of 200 to 500 myosin molecules bundled together with the heads projecting outward in a spiral array. (c) A thin filament consists of two intertwined chains of G actin molecules, smaller filamentous tropomyosin molecules, and a calcium-binding protein called troponin associated with the tropomyosin. (d) A region of overlap between the thick and thin myofilaments.

ment also has 40 to 60 molecules of yet another protein called **tropomyosin.** When a muscle fiber is relaxed, tropomyosin blocks the active sites of six or seven G actins and prevents myosin from binding to them. Each tropomyosin molecule, in turn, has a smaller calcium-binding protein called **troponin** bound to it.

3. **Elastic filaments** (fig. 11.4b), 1 nm in diameter, are made of a huge springy protein called **titin**[3] (connectin). They flank each thick filament and anchor it to a structure called the *Z disc.* This helps to stabilize the thick filament, center it between the thin filaments, and prevent overstretching.

At least seven other accessory proteins occur in or are associated with the thick and thin filaments. They anchor the myofilaments, regulate their length, and keep them aligned with each other for optimal contractile effectiveness. The most clinically important of these is **dystrophin,** an enormous protein located just under the sarcolemma in the vicinity of each I band. It links the actin filaments to proteins immediately external to the muscle fiber, and ultimately to the endomysium. Thus it is dystrophin that links the shortening of components within the muscle fiber to a mechanical pull on the parallel elastic components external to the fiber. Genetic defects in dystrophin are responsible for the disabling disease, **muscular dystrophy** (see Insight 11.4).

Myosin and actin are called the **contractile proteins** of muscle because they do the work of shortening the muscle fiber. Tropomyosin and troponin are called the **regulatory proteins** because they act like a switch to determine when it can contract and when it cannot. Several clues as to how they do this may be apparent from what has already been said—calcium ions are released into the sarcoplasm to activate contraction; calcium binds to troponin; troponin is also bound to tropomyosin; and tropomyosin blocks the active sites of actin, so that myosin cannot bind to it when the muscle is not stimulated. Perhaps you are already forming some idea of the contraction mechanism to be explained shortly.

## STRIATIONS

Myosin and actin are not unique to muscle; these proteins occur in all cells, where they function in cellular motility, mitosis, and transport of intracellular materials. In skeletal and cardiac muscle they are especially abundant, however, and are organized in a precise array that accounts for the striations of these two muscle types (fig. 11.4).

Striated muscle has dark **A bands** alternating with lighter **I bands.** (*A* stands for *anisotropic* and *I* for *isotropic,* which refers to the way these bands affect polarized light. To help remember which band is which,

[3]*tit* = giant + *in* = protein

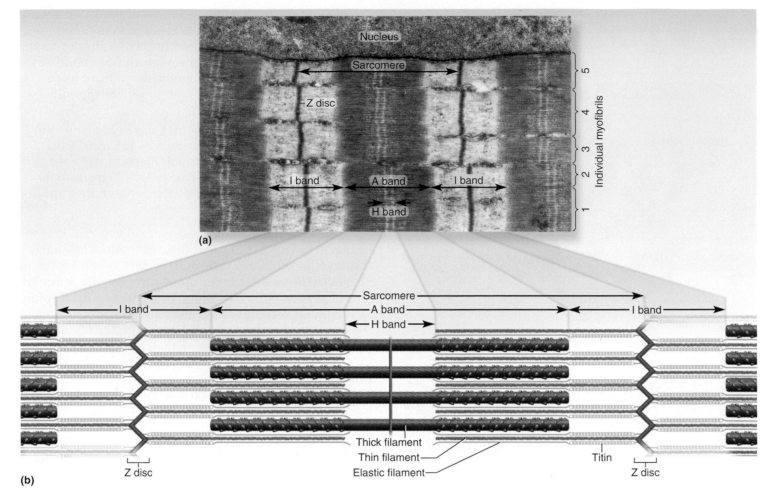

**FIGURE 11.4** **Muscle Striations and Their Molecular Basis.** (a) Five myofibrils of a single muscle fiber, showing the striations in the relaxed state. (b) The overlapping pattern of thick and thin myofilaments that accounts for the striations seen in part (a).

think "d**A**rk" and "l**I**ght.") Each A band consists of thick filaments lying side by side. Part of the A band, where thick and thin filaments overlap, is especially dark. In this region, each thick filament is surrounded by thin filaments. In the middle of the A band, there is a lighter region called the **H band,**[4] into which the thin filaments do not reach.

Each light I band is bisected by a dark narrow **Z disc**[5] (Z line) composed of titin. The Z disc provides anchorage for the thin filaments and elastic filaments. Each segment of a myofibril from one Z disc to the next is called a **sarcomere**[6] (SAR-co-meer), the functional contractile unit of the muscle fiber. A muscle shortens because its individual sarcomeres shorten and pull the Z discs closer to each other, and the Z discs are connected to the sarcolemma by

way of the cytoskeleton. As the Z discs are pulled closer together during contraction, they pull on the sarcolemma to achieve overall shortening of the cell.

The terminology of muscle fiber structure is reviewed in table 11.1; this table may be a useful reference as you study the mechanism of contraction.

### Before You Go On

*Answer the following questions to test your understanding of the preceding section:*

4. *What special terms are given to the plasma membrane, cytoplasm, and smooth ER of a muscle cell?*

5. *What is the difference between a myofilament and a myofibril?*

6. *List five proteins of the myofilaments and describe their physical arrangement.*

7. *Sketch the overlapping pattern of myofilaments to explain how they account for the A bands, I bands, H bands, and Z discs.*

---

[4]H = *helle* = bright
[5]Z = *Zwichenscheibe* = "between disc"
[6]*sarco* = muscle + *mere* = part, segment

| TABLE 11.1 | Structural Components of a Muscle Fiber |
|---|---|
| **Term** | **Definition** |
| *General Structure and Contents of the Muscle Fiber* | |
| Sarcolemma | The plasma membrane of a muscle fiber |
| Sarcoplasm | The cytoplasm of a muscle fiber |
| Glycogen | An energy-storage polysaccharide abundant in muscle |
| Myoglobin | An oxygen-storing red pigment of muscle |
| T tubule | A tunnel-like extension of the sarcolemma extending from one side of the muscle fiber to the other; conveys electrical signals from the cell surface to its interior |
| Sarcoplasmic reticulum | The smooth ER of a muscle fiber; a $Ca^{2+}$ reservoir |
| Terminal cisternae | The dilated ends of sarcoplasmic reticulum adjacent to a T tubule |
| *Myofibrils* | |
| Myofibril | A bundle of protein microfilaments (myofilaments) |
| Myofilament | A threadlike complex of several hundred contractile protein molecules |
| Thick filament | A myofilament about 11 nm in diameter composed of bundled myosin molecules |
| Elastic filament | A myofilament about 1 nm in diameter composed of a giant protein, titin, that flanks a thick filament and anchors it to a Z disc |
| Thin filament | A myofilament about 5 to 6 nm in diameter composed of actin, troponin, and tropomyosin |
| Myosin | A protein with a long shaftlike tail and a globular head; constitutes the thick myofilament |
| F actin | A fibrous protein made of a long chain of G actin molecules twisted into a helix; main protein of the thin myofilament |
| G actin | A globular subunit of F actin with an active site for binding a myosin head |
| Regulatory proteins | Troponin and tropomyosin, proteins that do not directly engage in the sliding filament process of muscle contraction but regulate myosin-actin binding |
| Tropomyosin | A regulatory protein that lies in the groove of F actin and, in relaxed muscle, blocks the myosin-binding active sites |
| Troponin | A regulatory protein associated with tropomyosin that acts as a calcium receptor |
| Titin | A springy protein that forms the elastic filaments and Z discs |
| Dystrophin | A large protein that links thin filaments and Z discs to extracellular proteins; transfers the force of sarcomere contraction to the parallel elastic component of muscle |
| *Striations and Sarcomeres* | |
| Striations | Alternating light and dark transverse bands across a myofibril |
| A band | Dark band formed by parallel thick filaments that partly overlap the thin filaments |
| H band | A lighter region in the middle of an A band that contains thick filaments only; thin filaments do not reach this far into the A band in relaxed muscle |
| I band | A light band composed of thin filaments only |
| Z disc | A protein disc to which thin filaments and elastic filaments are anchored at each end of a sarcomere; appears as a narrow dark line in the middle of the I band |
| Sarcomere | The distance from one Z disc to the next; the contractile unit of a muscle fiber |

# The Nerve–Muscle Relationship

### Objectives

When you have completed this section, you should be able to

- explain what a motor unit is and how it relates to muscle contraction;
- describe the structure of a junction where a nerve fiber meets a muscle fiber; and
- explain why a cell has an electrical charge difference across its plasma membrane and, in general terms, how this relates to muscle contraction.

Skeletal muscle never contracts unless it is stimulated by a nerve (or artificially with electrodes). If its nerve connections are severed or poisoned, a muscle is paralyzed. If innervation is not restored, the paralyzed muscle undergoes a shrinkage called *denervation atrophy*. Thus, muscle contraction cannot be understood without first understanding the relationship between nerve and muscle cells.

## MOTOR NEURONS

Skeletal muscles are innervated by *somatic motor neurons.* The cell bodies of these neurons are in the brainstem and spinal cord. Their axons, called **somatic motor fibers,** lead to the skeletal muscles. At its distal end, each somatic motor fiber branches about 200 times on average, with each branch leading to a different muscle fiber. Each muscle fiber is innervated by only one motor neuron.

## THE MOTOR UNIT

When a nerve signal approaches the end of an axon, it spreads out over all of its terminal branches and stimulates all the muscle fibers supplied by them. Thus, these muscle fibers contract in unison. Since they behave as a single functional unit, one nerve fiber and all the muscle fibers innervated by it are called a **motor unit.** The muscle fibers of a single motor unit are not all clustered together but are dispersed throughout a muscle (fig. 11.5). Thus, when they are stimulated, they cause a weak contraction over a wide area—not just a localized twitch in one small region.

Earlier it was stated that a motor nerve fiber supplies about 200 muscle fibers, but this is just a representative number. Where fine control is needed, we have *small motor units.* In the muscles of eye movement, for example, there are only 3 to 6 muscle fibers per nerve fiber. Small motor units are not very strong, but they provide the fine degree of control needed for subtle movements. They also have small neurons that are easily stimulated. Where strength is more important than fine control, we

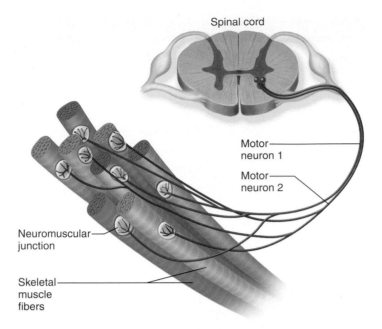

**FIGURE 11.5   Motor Units.** A motor unit consists of one motor neuron and all skeletal muscle fibers that it innervates. Two motor units are here represented by the red and blue nerve and muscle fibers. Note that the muscle fibers of a motor unit are not clustered together, but distributed through the muscle and commingled with the fibers of other motor units.

have large motor units. The gastrocnemius muscle of the calf, for example, has about 1,000 muscle fibers per nerve fiber. Large motor units are much stronger, but have larger neurons that are harder to stimulate, and they do not produce such fine control.

One advantage of having multiple motor units in a muscle is that they are able to work in shifts. Muscle fibers fatigue when subjected to continual stimulation. If all of the fibers in one of your postural muscles fatigued at once, for example, you might collapse. To prevent this, other motor units take over while the fatigued ones rest, and the muscle as a whole can sustain long-term contraction. The role of motor units in muscular strength is discussed later in the chapter.

## THE NEUROMUSCULAR JUNCTION

The functional connection between a nerve fiber and its target cell is called a **synapse** (SIN-aps). When the second cell is a muscle fiber, the synapse is called a **neuromuscular junction** (fig. 11.6). Each branch of a motor nerve fiber ends in a bulbous swelling called a **synaptic** (sih-NAP-tic) **knob,** which is nestled in a depression on the sarcolemma called the **motor end plate.** The two cells do not actually touch each other but are separated by a tiny gap, the **synaptic cleft,** about 60 to 100 nm wide. A third

**FIGURE 11.6 Innervation of a Skeletal Muscle.** (a) Neuromuscular junctions, with muscle fibers slightly teased apart (LM). (b) Structure of a single neuromuscular junction.

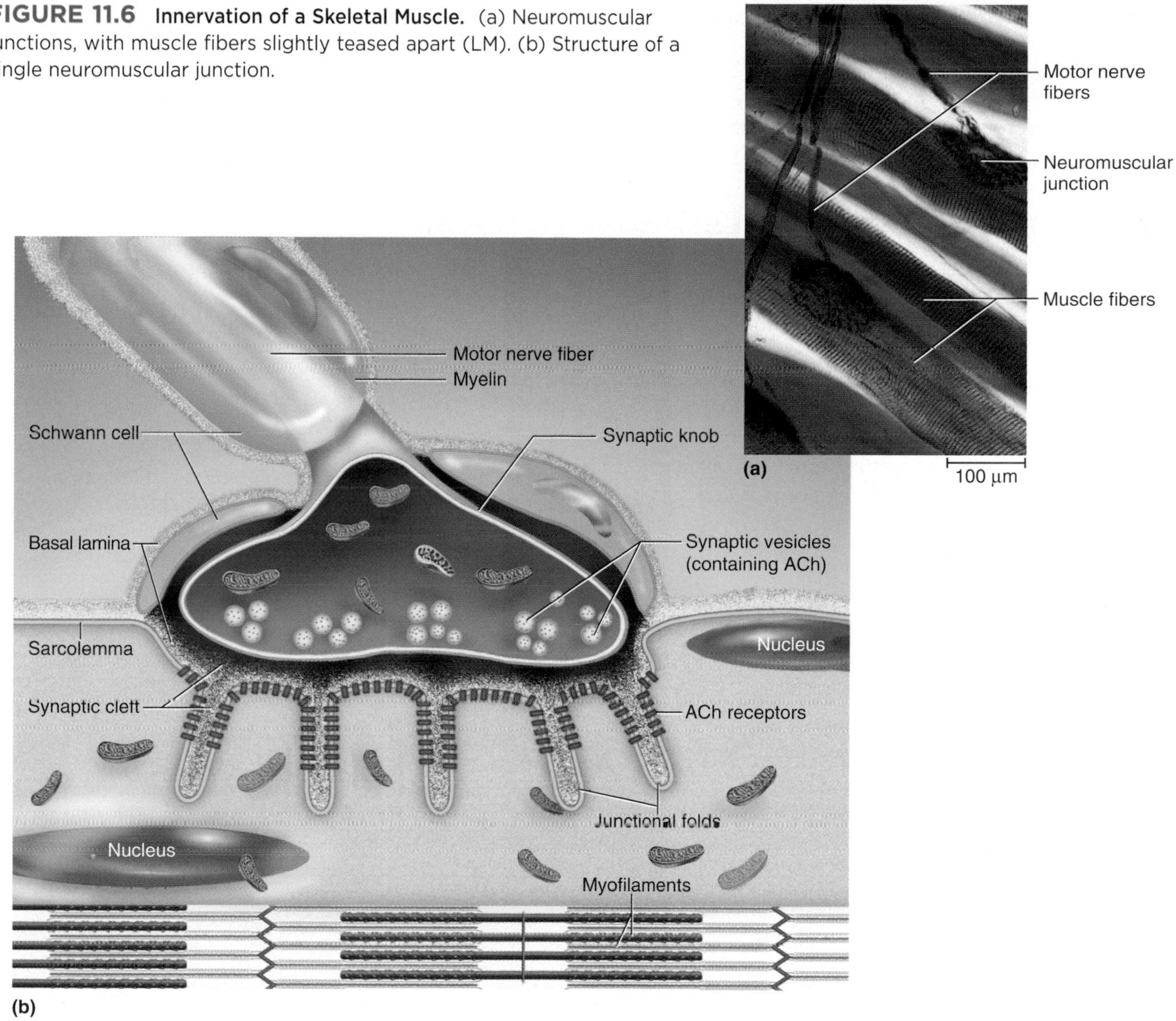

cell, called a *Schwann cell,* envelops the entire neuromuscular junction and isolates it from the surrounding tissue fluid.

The electrical signal (nerve impulse) traveling down a nerve fiber cannot cross the synaptic cleft like a spark jumping between two electrodes—rather, it causes the nerve fiber to release a neurotransmitter that stimulates the next cell. Although many chemicals function as neurotransmitters, the one released at the neuromuscular junction is **acetylcholine (ACh)** (ASS-eh-till-CO-leen or ah-SEE-tul-CO-leen). ACh is stored in spherical organelles called **synaptic vesicles.**

Directly across from the synaptic vesicles, the sarcolemma of the muscle cell exhibits infoldings called *junctional folds,* about 1 μm deep. The muscle fiber has about 50 million membrane proteins called **ACh receptors,** which bind the acetylcholine released by the nerve fiber. Most ACh receptors are concentrated in and near these junctional folds; very few are found anywhere else on a muscle fiber. Junctional folds increase the surface area for receptor sites and ensure a more effective response to ACh. The muscle nuclei beneath the junctional folds are specifically dedicated to the synthesis of ACh receptors and other proteins of the motor end plate. A deficiency of ACh receptors leads to muscle paralysis in the disease *myasthenia gravis* (see Insight 11.4, p. 436).

The entire muscle fiber is surrounded by a *basal lamina,* a thin layer composed partially of collagen and glycoproteins, which passes through the synaptic cleft and virtually fills it. Both the sarcolemma and that part of the basal lamina in the cleft contain an enzyme called **acetylcholinesterase (AChE)** (ASS-eh-till-CO-lin-ESS-ter-ase or ah-SEE-tul-CO-lin-ESS-ter-ase). AChE breaks down ACh, shuts down the stimulation of muscle fibers, and allows a muscle to relax (see Insight 11.1).

You must be very familiar with the foregoing terms to understand how a nerve stimulates a muscle fiber and how the fiber contracts. They are summarized in table 11.2 for your later reference.

## ELECTRICALLY EXCITABLE CELLS

Muscle fibers and neurons are regarded as *electrically excitable cells* because their plasma membranes exhibit voltage changes in response to stimulation. The study of the electrical activity of cells, called **electrophysiology,** is a key to understanding nervous activity, muscle contraction, the heartbeat, and other physiological phenomena. The details of electrophysiology are presented in chapter 12, but a few fundamental principles must be introduced here so you can understand muscle excitation.

In an unstimulated (resting) cell, there are more anions (negative ions) on the inside of the plasma membrane than on the outside. Thus, the plasma membrane is electrically **polarized,** or charged, like a little battery. In a resting muscle cell, there is an excess of sodium ions ($Na^+$) in the extracellular fluid (ECF) outside the cell and an excess of potassium ions ($K^+$) in the intracellular fluid (ICF) within the cell. Also in the ICF, and unable to penetrate the plasma membrane, are anions such as proteins, nucleic acids, and phosphates. These anions make the inside of the plasma membrane negatively charged by comparison to its outer surface.

A difference in electrical charge from one point to another is called an electrical potential, or voltage. The

---

**INSIGHT 11.1**    Clinical Application

### Neuromuscular Toxins and Paralysis

Toxins that interfere with synaptic function can paralyze the muscles. Some pesticides, for example, contain *cholinesterase inhibitors* that bind to acetylcholinesterase and prevent it from degrading ACh. This causes *spastic paralysis,* a state of continual contraction of the muscle that poses the danger of suffocation if the laryngeal and respiratory muscles are affected. A person poisoned by a cholinesterase inhibitor must be kept lying down and calm, and sudden noises or other disturbances must be avoided. A minor startle response can escalate to dangerous muscle spasms in a poisoned individual.

*Tetanus* ("lockjaw") is a form of spastic paralysis caused by a toxin from the bacterium *Clostridium tetani.* In the spinal cord, an inhibitory neurotransmitter called glycine stops motor neurons from producing unwanted muscle contractions. The tetanus toxin blocks glycine release and thus allows overstimulation of the muscles. (At the cost of some confusion, the word *tetanus* also refers to a completely different and normal muscle phenomenon discussed later in this chapter.)

*Flaccid paralysis* is a state in which the muscles are limp and cannot contract. It can cause respiratory arrest when it affects the thoracic muscles. Flaccid paralysis can be caused by poisons such as curare (cue-RAH-ree) that compete with ACh for receptor sites but do not stimulate the muscle. Curare is extracted from certain plants and used by some South American natives to poison blowgun darts. It has been used to treat muscle spasms in some neurological disorders and to relax abdominal muscles for surgery, but other muscle relaxants have now replaced curare for most purposes.

---

| TABLE 11.2 | Components of the Neuromuscular Junction |
|---|---|
| **Term** | **Definition** |
| Neuromuscular junction | A functional connection between the distal end of a nerve fiber and the middle of a muscle fiber; consists of a synaptic knob and motor end plate |
| Synaptic knob | The dilated tip of a nerve fiber; contains synaptic vesicles |
| Motor end plate | A depression in the sarcolemma, near the middle of the muscle fiber, that receives the synaptic knob; contains acetylcholine receptors |
| Synaptic cleft | A gap of about 60 to 100 nm between the synaptic knob and motor end plate |
| Synaptic vesicle | A secretory vesicle in the synaptic knob; contains acetylcholine |
| Junctional folds | Invaginations of the membrane of the motor end plate where ACh receptors are especially concentrated |
| Acetylcholine (ACh) | The neurotransmitter released by a somatic motor fiber that stimulates a skeletal muscle fiber (also used elsewhere in the nervous system) |
| ACh receptor | A transmembrane protein in the sarcolemma of the motor end plate that binds to ACh |
| Acetylcholinesterase (AChE) | An enzyme in the sarcolemma and basal lamina of the muscle fiber in the synaptic region; responsible for degrading ACh and stopping the stimulation of the muscle fiber |

difference is typically 12 volts (V) for a car battery and 1.5 V for a flashlight battery, for example. On the sarcolemma of a muscle cell, the voltage is much smaller, about −90 millivolts (mV), but critically important to life. (The negative sign refers to the relatively negative charge on the intracellular side of the membrane.) This voltage is called the **resting membrane potential (RMP).** It is maintained by the sodium–potassium pump, as explained in chapter 3.

When a nerve or muscle cell is stimulated, dramatic things happen electrically, as we shall soon see in our study of the excitation of muscle. Ion gates in the plasma membrane open and Na⁺ instantly diffuses down its concentration gradient into the cell. These cations override the negative charges in the ICF, so the inside of the plasma membrane briefly becomes positive. This change is called **depolarization** of the membrane. Immediately, Na⁺ gates close and K⁺ gates open. K⁺ rushes out of the cell, partly because it is repelled by the positive sodium charge and partly because it is more concentrated in the ICF than in the ECF, so it diffuses down its concentration gradient when it has the opportunity. The loss of positive potassium ions from the cell turns the inside of the membrane negative again **(repolarization).** This quick up-and-down voltage shift, from the negative RMP to a positive value and then back to a negative value again, is called an **action potential.** The RMP is a stable voltage seen in a "waiting" cell, whereas the action potential is a quickly fluctuating voltage seen in an active, stimulated cell. Chapter 12 explains the mechanism of action potentials more fully.

Action potentials have a way of perpetuating themselves—an action potential at one point on a plasma membrane causes another one to happen immediately in front of it, which triggers another one a little farther along, and so forth. A wave of action potentials spreading along a nerve fiber like this is called a *nerve impulse* or *nerve signal.* Such signals also travel along the sarcolemma of a muscle fiber. We will see shortly how this leads to muscle contraction.

## Before You Go On

*Answer the following questions to test your understanding of the preceding section:*

8. What differences would you expect to see between a motor unit where muscular strength is more important than fine control and another motor unit where fine control is more important?

9. Distinguish between acetylcholine, an acetylcholine receptor, and acetylcholinesterase. State where each is found and describe the function it serves.

10. What accounts for the resting membrane potential seen in unstimulated nerve and muscle cells?

11. What is the difference between a resting membrane potential and an action potential?

# Behavior of Skeletal Muscle Fibers

### Objectives

When you have completed this section, you should be able to

- explain how a nerve fiber stimulates a skeletal muscle fiber;
- explain how stimulation of a muscle fiber activates its contractile mechanism;
- explain the mechanism of muscle contraction;
- explain how a muscle fiber relaxes; and
- explain why the force of a muscle contraction depends on its length prior to stimulation.

The process of muscle contraction and relaxation can be viewed as having four major phases: (1) excitation, (2) excitation–contraction coupling, (3) contraction, and (4) relaxation. Each phase occurs in several smaller steps, which we now examine in detail. The steps are numbered in the following descriptions to correspond to those in figures 11.7 to 11.10.

## EXCITATION

**Excitation** is the process in which action potentials in the nerve fiber lead to action potentials in the muscle fiber. The steps in excitation are shown in figure 11.7.

1. A nerve signal arrives at the synaptic knob and stimulates voltage-gated calcium channels to open. Calcium ions enter the synaptic knob.

2. Calcium stimulates exocytosis of the synaptic vesicles, which release acetylcholine (ACh) into the synaptic cleft. One action potential causes exocytosis of about 60 synaptic vesicles, and each vesicle releases about 10,000 molecules of ACh.

3. ACh diffuses across the synaptic cleft and binds to receptor proteins on the sarcolemma.

4. These receptors are *ligand-gated ion channels.* When ACh (the ligand) binds to them, they change shape and open an ion channel through the middle of the receptor protein. Each channel allows Na⁺ to diffuse quickly into the cell and K⁺ to diffuse outward. As a result of these ion movements, the sarcolemma reverses polarity—its voltage quickly jumps from the RMP of −90 mV to a peak of +75 mV as Na⁺ enters, and then falls back to a level close to the RMP as K⁺ diffuses out. This rapid fluctuation in membrane voltage at the motor end plate is called the **end-plate potential (EPP).**

5. Areas of sarcolemma next to the end plate have *voltage-gated ion channels* that open in response to the EPP. Some of these are specific for Na⁺ and admit it to the cell, while others are specific for K⁺ and allow it to leave. These ion movements create an *action potential.* The muscle fiber is now excited.

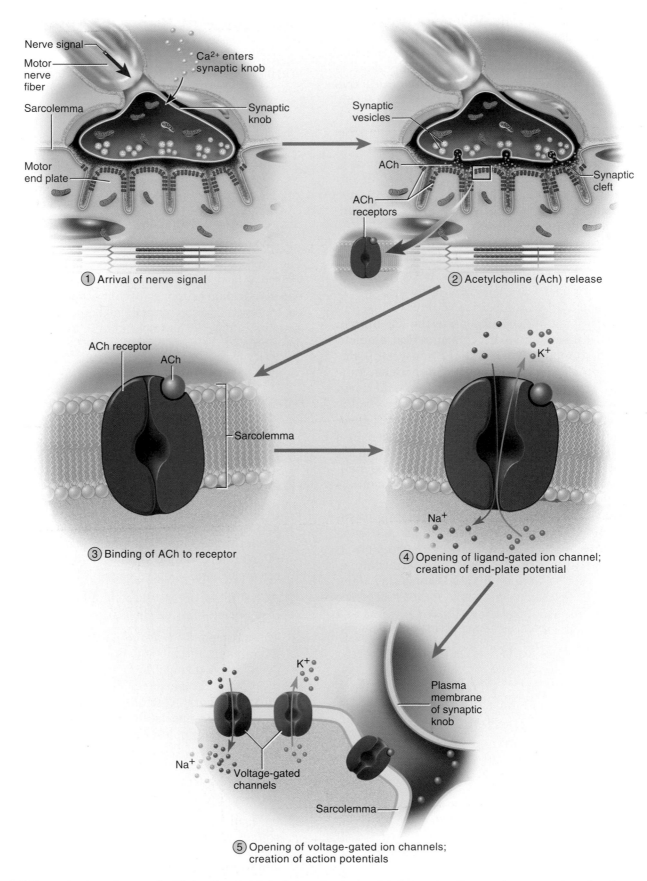

**FIGURE 11.7** **Excitation of a Muscle Fiber.** These events link action potentials in a nerve fiber to the generation of action potentials in the muscle fiber. See the correspondingly numbered steps in the text for explanation.

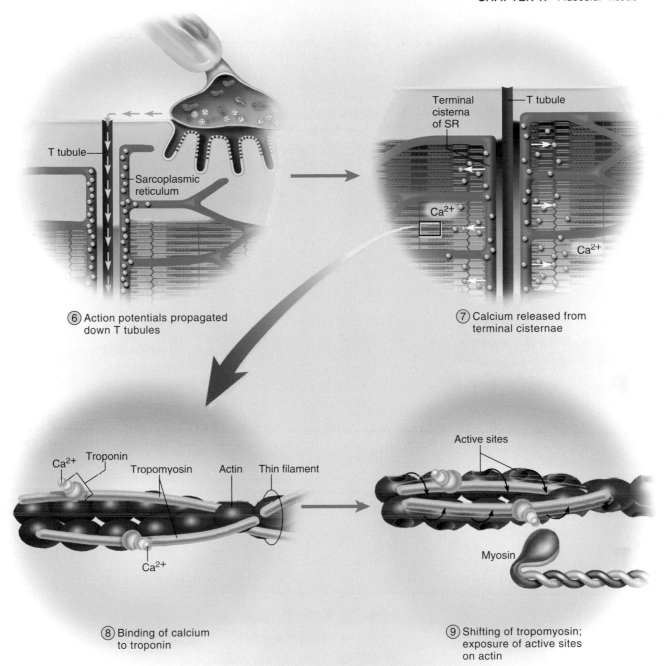

⑥ Action potentials propagated down T tubules

⑦ Calcium released from terminal cisternae

⑧ Binding of calcium to troponin

⑨ Shifting of tropomyosin; exposure of active sites on actin

**FIGURE 11.8 Excitation-Contraction Coupling.** These events link action potentials in the muscle fiber to the release and binding of calcium ions. See the correspondingly numbered steps in the text for explanation. The numbers in this figure begin where figure 11.7 ended.

# EXCITATION–CONTRACTION COUPLING

**Excitation–contraction coupling** refers to the events that link the action potentials on the sarcolemma to activation of the myofilaments, thereby preparing them to contract. The steps in the coupling process are shown in figure 11.8.

⑥ A wave of action potentials spreads from the end plate in all directions, like ripples on a pond. When this wave of excitation reaches the T tubules, it continues down them into the sarcoplasm.

⑦ Action potentials open voltage-gated ion channels in the T tubules. These are physically linked to calcium channels in the terminal cisternae of the sarcoplasmic reticulum (SR). Thus, gates in the SR open as well and calcium diffuses out of the SR, down its concentration gradient and into the cytosol.

⑧ Calcium binds to the troponin of the thin filaments.

⑨ The troponin–tropomyosin complex changes shape and sinks deeper into the groove of the thin filament. This exposes the active sites on the actin filaments and makes them available for binding to myosin heads.

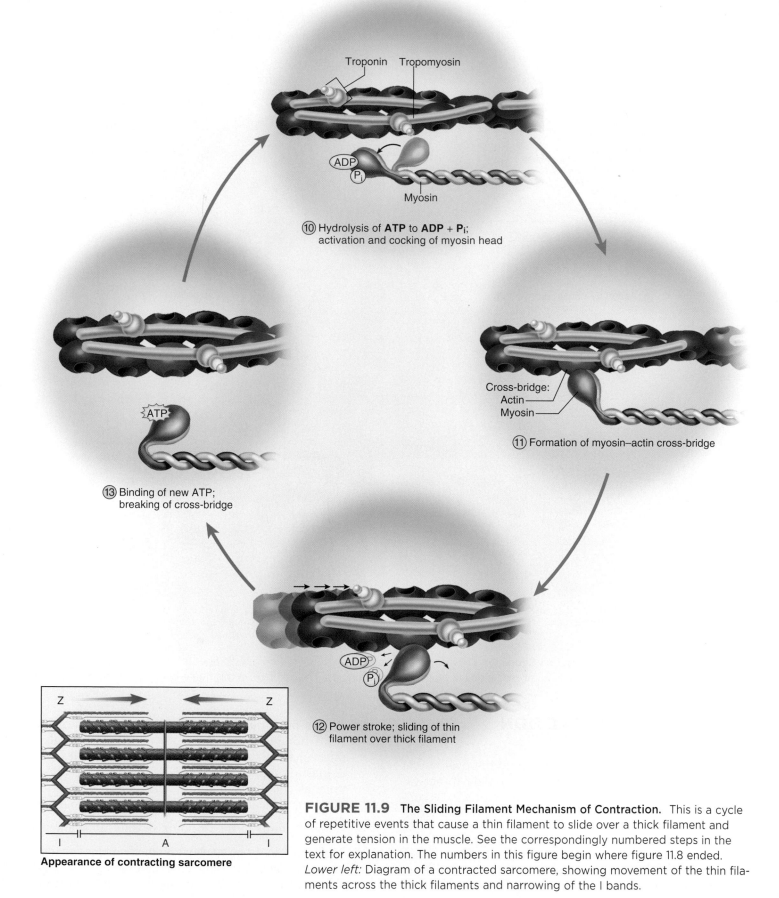

**FIGURE 11.9   The Sliding Filament Mechanism of Contraction.** This is a cycle of repetitive events that cause a thin filament to slide over a thick filament and generate tension in the muscle. See the correspondingly numbered steps in the text for explanation. The numbers in this figure begin where figure 11.8 ended. *Lower left:* Diagram of a contracted sarcomere, showing movement of the thin filaments across the thick filaments and narrowing of the I bands.

# CONTRACTION

Contraction is the step in which the muscle fiber develops tension and may shorten. (Muscles often "contract," or develop tension, without shortening, as we see later.) How a fiber shortens remained a mystery until sophisticated techniques in electron microscopy enabled cytologists to see the molecular organization of muscle fibers. In 1954, two researchers at the Massachusetts Institute of Technology—Jean Hanson and Hugh Huxley—reported convincing evidence for a model now called the **sliding filament theory.** This theory holds that the myofilaments do not become any shorter during contraction; rather, the thin filaments slide over the thick ones and pull the Z discs behind them, causing each sarcomere as a whole to shorten. The individual steps in this mechanism are shown in figure 11.9.

⑩ The myosin head must have an ATP molecule bound to it to initiate the contraction process. **Myosin ATPase,** an enzyme in the head, hydrolyzes this ATP. The energy released by this process activates the head, which "cocks" into an extended, high-energy position. The head temporarily keeps the ADP and phosphate group bound to it.

⑪ The cocked myosin binds to an exposed active site on the thin filament, forming a **cross-bridge** between the myosin and actin.

⑫ Myosin releases the ADP and phosphate and flexes into a bent, low-energy position, tugging the thin filament along with it. This is called the **power stroke.** The head remains bound to actin until it binds a new ATP.

⑬ Upon binding more ATP, myosin releases the actin. It is now prepared to repeat the whole process—it will hydrolyze the ATP, recock (the **recovery stroke**), attach to a new active site farther down the thin filament, and produce another power stroke.

It might seem as if releasing the thin filament at step 13 would simply allow it to slide back to its previous position, so that nothing would have been accomplished. Think of the sliding filament mechanism, however, as being similar to the way you would pull in a boat anchor hand over hand. When the myosin head cocks, it is like your hand reaching out to grasp the anchor rope. When it flexes back into the low-energy position, it is like your elbow flexing to pull on the rope and draw the anchor up a little bit. When you let go of the rope with one hand, you hold onto it with the other, alternating hands until the anchor is pulled in. Similarly, when one myosin head releases the actin in preparation for the recovery stroke, there are many other heads on the same thick filament holding onto the thin filament so that it doesn't slide back. At any given moment during contraction, about half

of the heads are bound to the thin filament and the other half are extending forward to grasp the filament farther down. That is, the myosin heads of a thick filament do not all stroke at once but contract sequentially.

Each myosin head acts in a jerky manner, but hundreds of them working together produce a smooth, steady pull on the thin filament. This is similar to the locomotion of a millipede—a wormlike animal with a few hundred tiny legs. Each leg takes individual jerky steps, but all the legs working together produce a smooth gliding movement. Note that even though the muscle fiber contracts, the *myofilaments do not become shorter* any more than a rope becomes shorter as you pull in an anchor. The thin filaments slide over the thick ones, as the name of the theory implies.

A single cycle of power and recovery strokes by all the myosin heads in a muscle fiber would shorten the fiber by about 1%. A fiber, however, may shorten by as much as 40% of its resting length, so obviously the cycle of power and recovery must be repeated many times by each myosin head. Each head carries out about five strokes per second, and each stroke consumes one molecule of ATP.

# RELAXATION

When its work is done, a muscle fiber relaxes and returns to its resting length. This is achieved by the steps shown in figure 11.10.

⑭ Nerve signals stop arriving at the neuromuscular junction, so the synaptic knob stops releasing ACh.

⑮ As ACh dissociates (separates) from its receptor, AChE breaks it down into fragments that cannot stimulate the muscle. The synaptic knob reabsorbs these fragments for recycling. All of this happens continually while the muscle is being stimulated, too, but when nerve signals stop, no new ACh is released to replace that which is broken down. Therefore, stimulation of the muscle fiber by ACh ceases.

⑯ Active transport pumps in the SR begin to pump $Ca^{2+}$ from the cytosol back into the cisternae. Here, the calcium binds to a protein called **calsequestrin** (CAL-see-QUES-trin) and is stored until the fiber is stimulated again. Since active transport requires ATP, you can see that *ATP is needed for muscle relaxation as well as for muscle contraction* (see Insight 11.2).

⑰ As calcium ions dissociate from troponin, they are pumped into the SR and are not replaced.

⑱ Tropomyosin moves back into the position where it blocks the active sites of the actin filament. Myosin can no longer bind to actin, and the muscle fiber ceases to produce or maintain tension.

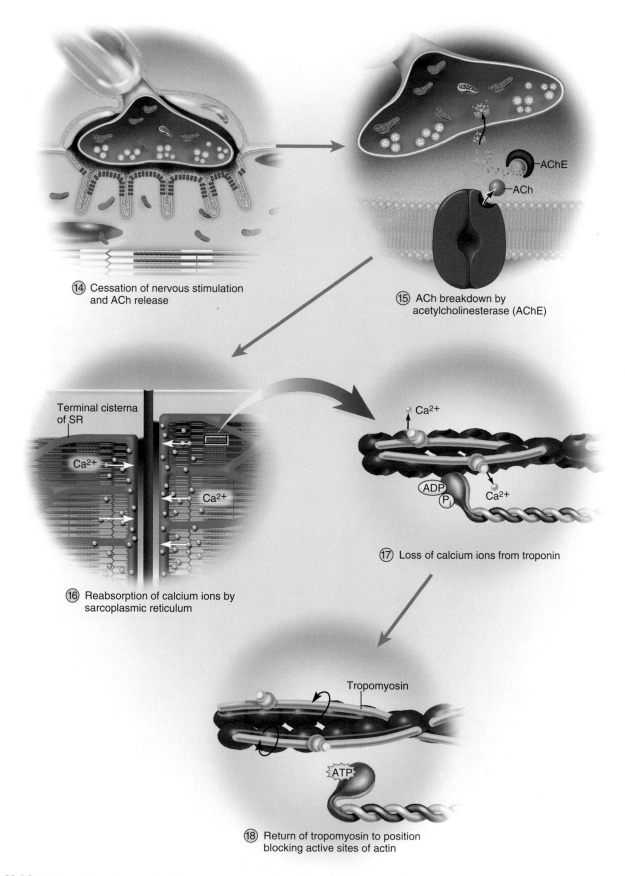

⑭ Cessation of nervous stimulation and ACh release

⑮ ACh breakdown by acetylcholinesterase (AChE)

AChE

ACh

Terminal cisterna of SR

Ca²⁺

Ca²⁺

⑯ Reabsorption of calcium ions by sarcoplasmic reticulum

Ca²⁺

ADP
Pi

Ca²⁺

⑰ Loss of calcium ions from troponin

Tropomyosin

ATP

⑱ Return of tropomyosin to position blocking active sites of actin

**FIGURE 11.10  Relaxation of a Muscle Fiber.**  These events lead from the cessation of a nerve signal to the release of thin filaments by myosin. See the correspondingly numbered steps in the text for explanation. The numbers in this figure begin where figure 11.9 ended.

Clinical Application

## Rigor Mortis

*Rigor mortis*[7] is the hardening of the muscles and stiffening of the body that begins 3 to 4 hours after death. It occurs partly because the deteriorating sarcoplasmic reticulum releases calcium into the cytosol, and the deteriorating sarcolemma admits more calcium from the extracellular fluid. The calcium activates myosin–actin cross-bridging. Once bound to actin, myosin cannot release it without first binding an ATP molecule, and of course no ATP is produced in a dead body. Thus, the thick and thin filaments remain rigidly cross-linked until the myofilaments begin to decay. Rigor mortis peaks about 12 hours after death and then diminishes over the next 48 to 60 hours.

A muscle returns to its resting length with the aid of two forces: (1) like a recoiling rubber band, the parallel and series elastic components stretch it; and (2) since muscles often occur in antagonistic pairs, the contraction of an antagonist lengthens the relaxed muscle. Contraction of the triceps brachii, for example, extends the elbow and lengthens the biceps brachii.

### Think About It

*Chapter 2 noted that one of the most important properties of proteins is their ability to change shape repeatedly. Identify at least two muscle proteins that must change shape in order for a muscle to contract and relax.*

## THE LENGTH–TENSION RELATIONSHIP AND MUSCLE TONE

The amount of tension generated by a muscle, and therefore the force of its contraction, depends on how stretched or contracted it was before it was stimulated, among other factors. This principle is called the **length–tension relationship.** The reasons for it can be seen in figure 11.11. If a fiber is overly contracted at rest, its thick filaments are rather close to the Z discs. The stimulated muscle may contract a little, but then the thick filaments butt against the Z discs and can go no farther. The contraction is therefore a weak one. On the other hand, if a muscle fiber is too stretched before it is stimulated, there is relatively little overlap between its thick and thin filaments. When the muscle is stimulated, the myosin heads cannot "get a good grip" on the thin filaments, and again the contraction is weak. (As mentioned in chapter 10, this is one reason you should not bend at the waist to pick up a heavy object. Muscles of the back become overly stretched and cannot contract effectively to straighten your spine against a heavy resistance.)

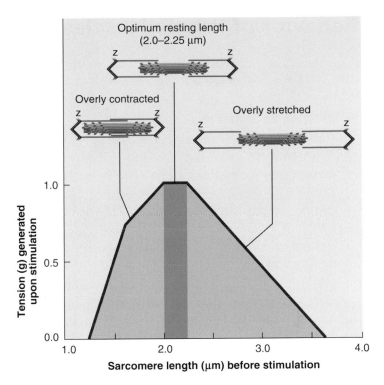

**FIGURE 11.11** **The Length-Tension Relationship.** *Center:* In a resting muscle fiber, the sarcomeres are usually 2.0 to 2.25 μm long, the optimum length for producing maximum tension when the muscle is stimulated to contract. Note how this relates to the degree of overlap between the thick and thin filaments. *Left:* If the muscle is overly contracted, the thick filaments butt against the Z discs and the fiber cannot contract very much more when it is stimulated. *Right:* If the muscle is overly stretched, there is so little overlap between the thick and thin myofilaments that few cross-bridges can form between myosin and actin.

Between these extremes, there is an optimum resting length at which a muscle responds with the greatest force. The central nervous system continually monitors and adjusts the length of the resting muscles, maintaining a state of partial contraction called **muscle tone.** This maintains optimum length and makes the muscles ideally ready for action. The elastic filaments of the sarcomere also help to maintain enough myofilament overlap to ensure an effective contraction when the muscle is called into action.

### Before You Go On

*Answer the following questions to test your understanding of the preceding section:*

12. *What change does ACh cause in an ACh receptor? How does this electrically affect the muscle fiber?*

13. *How do troponin and tropomyosin regulate the interaction between myosin and actin?*

14. *Describe the roles played by ATP in the power and recovery strokes of myosin.*

15. *What steps are necessary for a contracted muscle to return to its resting length?*

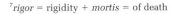

[7]*rigor* = rigidity + *mortis* = of death

# Behavior of Whole Muscles

### Objectives

When you have completed this section, you should be able to

- describe the stages of a muscle twitch;
- explain why muscle does not contract in an all-or-none manner;
- explain how successive muscle twitches add up to produce stronger muscle contractions;
- distinguish between isometric and isotonic contraction; and
- distinguish between concentric and eccentric contraction.

Now you know how an individual muscle cell shortens. Our next objective is to move up to the organ grade of construction and consider how this relates to the action of the muscle as a whole.

## THRESHOLD, LATENT PERIOD, AND TWITCH

Muscle contraction has often been studied and demonstrated using the gastrocnemius (calf) muscle of a frog, which can easily be isolated from the leg along with its connected sciatic nerve (see Insight 11.3). This nerve–muscle preparation can be attached to stimulating electrodes and to a recording device that produces a *myogram,* a chart of the timing and strength of the muscle's contraction.

A sufficiently weak electrical stimulus to a muscle causes no contraction. By gradually increasing the voltage and stimulating the muscle again, we can determine the **threshold,** or minimum voltage necessary to generate an action potential in the muscle fiber and produce a contrac-

---

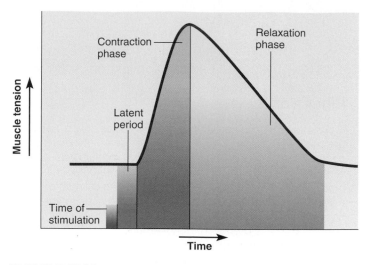

**FIGURE 11.12**   A Muscle Twitch.   What role does ATP play during the relaxation phase?

---

tion. The action potential triggers the release of a pulse of $Ca^{2+}$ into the cytosol and activates the sliding filament mechanism. At threshold or higher, a stimulus thus causes a quick cycle of contraction and relaxation called a **twitch** (fig. 11.12).

There is a delay, or **latent period,** of about 2 milliseconds (msec) between the onset of the stimulus and the onset of the twitch. This is the time required for excitation, excitation–contraction coupling, and tensing of the elastic components of the muscle. The force generated during this time is called *internal tension.* It is not visible on the myogram because it causes no shortening of the muscle.

Once the elastic components are taut, the muscle begins to produce *external tension* and move a resisting object, or load. This is called the **contraction phase** of the twitch. In the frog gastrocnemius preparation, the load is the sensor of the recording apparatus; in the body, it is usually a bone. By analogy, imagine lifting a weight from a table with a rubber band. At first, internal tension would stretch the rubber band. Then as the rubber band became taut, external tension would lift the weight.

The contraction phase is short-lived, because the SR quickly reabsorbs $Ca^{2+}$ before the muscle develops maximal force. As the $Ca^{2+}$ level in the cytoplasm falls, myosin releases the thin filaments and muscle tension declines. This is seen in the myogram as the **relaxation phase.** As shown by the asymmetry of the myogram, the muscle is quicker to contract than it is to relax. The entire twitch lasts from about 7 to 100 msec.

## CONTRACTION STRENGTH OF TWITCHES

We have seen that a weak stimulus induces no muscle contraction at all, but at threshold intensity, a twitch is produced. Increasing the stimulus voltage still more, however, produces twitches no stronger than those at threshold. Superficially, the muscle fiber seems to be giving its maximum response once the stimulus intensity is at threshold or higher. For this reason, it has been com-

---

mon to say a muscle fiber obeys an *all-or-none law,* either contracting to its maximum possible extent or not at all. It is true that the electrical *excitation* of a muscle or nerve fiber follows an all-or-none law, and this is further discussed in chapter 12. But it is not true that muscle fibers exhibit all-or-none twitches in response to that excitation. On the contrary, even for a constant stimulus voltage, twitches vary in strength. This is so for a variety of reasons, some of which are causally linked to each other:

- Twitch strength varies with stimulation frequency; stimuli arriving close together produce stronger twitches than stimuli arriving at longer time intervals. See the phenomena of *treppe* and *tetanus* described shortly.

- Twitches vary with the concentration of $Ca^{2+}$ in the sarcoplasm, which in turn can vary with stimulus frequency.

- Twitch strength depends on how stretched the muscle was just before it was stimulated, as we have already seen in the length–tension relationship.

- Twitches vary with the temperature of the muscle; a warmed-up muscle contracts more strongly because enzymes such as the myosin heads work more quickly.

- Twitches are weaker when the pH of the sarcoplasm falls below normal. This and other factors produce a weakening of muscle contraction called *fatigue,* discussed later in this chapter.

- Twitches vary with the state of hydration of a muscle, which affects the overlap between thick and thin filaments and the ability of myosin to form cross-bridges with actin.

It should not be surprising that muscle twitches vary in strength. Indeed, an individual twitch is not strong enough to do any useful work. Muscles must be able to contract with variable strength for different tasks, such as lifting a glass of champagne compared with lifting barbells at the gym.

Let us examine more closely the contrasting effects of stimulus *intensity* versus stimulus *frequency* on contraction strength. Suppose we apply a stimulating electrode to a motor nerve that supplies a muscle, such as the frog sciatic nerve–gastrocnemius preparation. Stimulus voltages below threshold produce no response (fig. 11.13). At threshold, we see a weak twitch (at *3* in the top row of the figure), and if we continue to raise the voltage, we see stronger twitches. The reason for this is that higher voltages excite more and more nerve fibers in the motor nerve (middle row of the figure), and thus stimulate more and more motor units to contract. The process of bringing more motor units into play is called **recruitment,** or **multiple motor unit (MMU) summation.** This is seen not just in artificial stimulation, but is part of the way the nervous system behaves naturally to produce varying muscle contractions.

But even when stimulus intensity (voltage) remains constant, twitch strength can vary with stimulus fre-

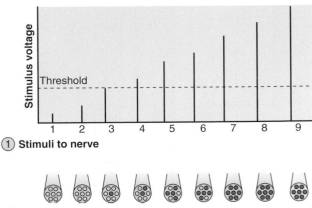

**① Stimuli to nerve**

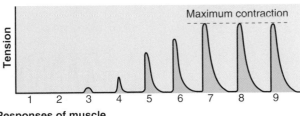

**② Proportion of nerve fibers excited**

**③ Responses of muscle**

**FIGURE 11.13** **The Relationship Between Stimulus Intensity (voltage) and Muscle Tension.** *Top row:* Nine stimuli of increasing strength. *Middle row:* Cross section of a motor nerve with seven nerve fibers. The colored nerve fibers are the excited ones. *Bottom row:* Graph of muscle tension. Weak stimuli (1–2) fail to excite any nerve fibers and therefore produce no muscle contraction. When stimuli reach or exceed threshold (3–7), they excite more and more nerve fibers and motor units, thus they produce stronger and stronger contractions. This is multiple motor unit summation (recruitment). Once all of the nerve fibers are stimulated (7–9), further increases in stimulus strength produce no further increase in muscle tension.

quency. High-frequency stimulation produces stronger twitches than low-frequency stimulation. In figure 11.14a, we see that when a muscle is stimulated at a low frequency (up to 10 stimuli/sec in this example), it produces an identical twitch for each stimulus and fully recovers between twitches.

Between 10 and 20 stimuli per second, the muscle still recovers fully between twitches, but each twitch develops more tension than the one before. This pattern of increasing tension with repetitive stimulation is called **treppe**[8] (TREP-eh), or the *staircase phenomenon,* after the appearance of the myogram (fig. 11.14b). One cause of treppe is that when stimuli arrive so rapidly, the SR does not have time between stimuli to completely reabsorb all the $Ca^{2+}$ it released. Thus, the calcium concentration in the cytosol rises higher and higher with each stimulus and causes subsequent twitches to be stronger. Another factor is that

---

[8]*treppe* = staircase

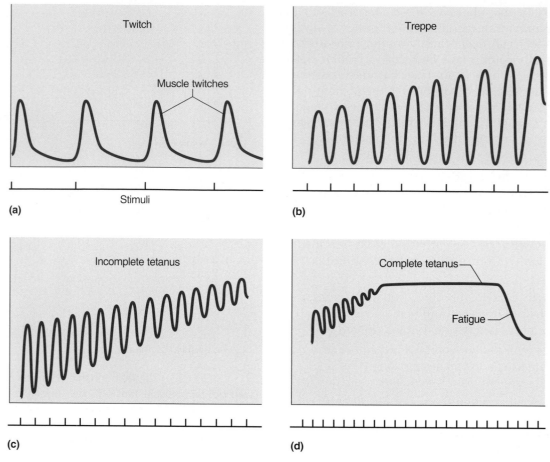

**FIGURE 11.14**  **The Relationship Between Stimulus Frequency and Muscle Tension.**  (a) Twitch: At low frequency, the muscle relaxes completely between stimuli and shows twitches of uniform strength. (b) Treppe: At a moderate frequency of stimulation, the muscle relaxes fully between contractions, but successive twitches are stronger. (c) Wave summation and incomplete tetanus: At still higher stimulus frequency, the muscle does not have time to relax completely between twitches, and the force of each twitch builds on the previous one. (d) Complete tetanus: At high stimulus frequency, the muscle does not have time to relax at all between stimuli and exhibits a state of continual contraction with about four times as much tension as a single twitch. Tension declines as the muscle fatigues. Only conditions (b) and (c) occur in the human body; (a) and (d) are produced only by artificial stimulation below or above the range of nerve firing frequencies.

the heat released by each twitch causes muscle enzymes such as myosin ATPase to work more efficiently and produce stronger twitches as the muscle warms up.

> ⌐ **Think About It**
>
> *Explain why a rising concentration of $Ca^{2+}$ in the sarcoplasm would enable more myosin–actin cross-bridges to form, and thus result in stronger twitches.*

At a still higher stimulus frequency (20–40 stimuli/ sec in fig. 11.14c), each new stimulus arrives before the previous twitch is over. Each new twitch "rides piggyback" on the previous one and generates higher tension. This phenomenon goes by two names: **temporal[9] summation,** because it results from two stimuli arriving close

together, or **wave summation,** because it results from one wave of contraction added to another. Wave is added upon wave, so each twitch reaches a higher level of tension than the one before, and the muscle relaxes only partially between stimuli. This effect produces a state of sustained fluttering contraction called **incomplete tetanus.**

At a still higher frequency, such as 40 to 50 stimuli per second, the muscle has no time to relax at all between stimuli, and the twitches fuse into a smooth, prolonged contraction called **complete tetanus** (fig. 11.14d). A muscle in complete tetanus produces about four times as much tension as a single twitch. This type of tetanus should not be confused with the disease of the same name caused by the tetanus toxin, explained in Insight 11.1.

Complete tetanus is a phenomenon seen in artificial stimulation of a muscle, however, and rarely if ever occurs in the body. Even during the most intense muscle contractions, the frequency of stimulation by a motor neuron rarely

---

[9]*tempor* = time

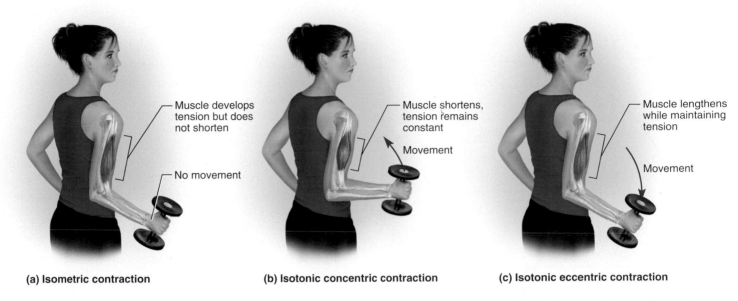

**(a) Isometric contraction**

Muscle develops tension but does not shorten

No movement

**(b) Isotonic concentric contraction**

Muscle shortens, tension remains constant

Movement

**(c) Isotonic eccentric contraction**

Muscle lengthens while maintaining tension

Movement

**FIGURE 11.15** **Isometric and Isotonic Contraction.** (a) Isometric contraction, in which a muscle develops tension but does not shorten. This occurs at the beginning of any muscle contraction but is prolonged in actions such as lifting heavy weights. (b) Isotonic concentric contraction, in which the muscle shortens while maintaining a constant degree of tension. In this phase, the muscle moves a load. (c) Isotonic eccentric contraction, in which the muscle maintains tension while it lengthens, allowing a muscle to relax without going suddenly limp.

▶ *Name a muscle that undergoes eccentric contraction as you sit down in a chair.*

exceeds 25 per second, which is far from sufficient to produce complete tetanus. The reason for the smoothness of muscle contractions is that motor units function asynchronously; when one motor unit relaxes, another contracts and takes over so that the muscle does not lose tension.

## ISOMETRIC AND ISOTONIC CONTRACTION

In muscle physiology, "contraction" does not always mean the shortening of a muscle—it may mean only that the muscle is producing internal tension while an external resistance causes it to stay the same length or even become longer. Thus, physiologists speak of different kinds of muscle contraction as *isometric* versus *isotonic* and *concentric* versus *eccentric.*

Suppose you lift a heavy dumbbell. When you first contract the muscles of your arms, you can feel the tension building in them even though the dumbbell is not yet moving. At this point, your muscles are contracting at a cellular level, but their tension is being absorbed by the series-elastic components and is resisted by the weight of the load; the muscle as a whole is not producing any external movement. This phase is called **isometric**[10] **contraction**—contraction without a change in length (fig. 11.15a). Isometric contraction is not merely a prelude to movement. The isometric contraction of antagonistic

muscles at a single joint is important in maintaining joint stability at rest, and the isometric contraction of postural muscles is what keeps us from sinking in a heap to the floor. **Isotonic**[11] **contraction**—contraction with a change in length but no change in tension—begins when internal tension builds to the point that it overcomes the resistance. The muscle now shortens, moves the load, and maintains essentially the same tension from then on (fig. 11.15b). Isometric and isotonic contraction are both phases of normal muscular action (fig. 11.16).

There are two forms of isotonic contraction: concentric and eccentric. In **concentric contraction,** a muscle shortens as it maintains tension—for example, when the biceps brachii contracts and flexes the elbow. In **eccentric contraction,** a muscle lengthens as it maintains tension. If you set that dumbbell down again (fig. 11.15c), your biceps brachii lengthens as you extend your elbow, but it maintains tension to act as a brake and keep you from simply dropping the weight. A weight lifter uses concentric contraction when lifting a dumbbell and eccentric contraction when lowering it.

In summary, during isometric contraction, a muscle develops tension without changing length, and in isotonic contraction, it changes length while maintaining constant tension. In concentric contraction, a muscle maintains tension as it shortens, and in eccentric contraction, it maintains tension while it is lengthening.

---

[10]*iso* = same, uniform + *metr* = length

[11]*iso* = same, uniform + *ton* = tension

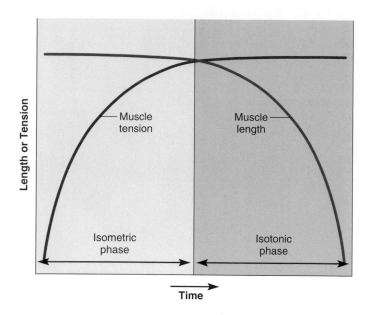

**FIGURE 11.16    Isometric and Isotonic Phases of Contraction.**
At the beginning of a contraction (isometric phase), muscle tension rises but the length remains constant (the muscle does not shorten). When tension overcomes the resistance of the load, the tension levels off and the muscle begins to shorten and move the load (isotonic phase).

▶ *How would you extend this graph in order to show eccentric contraction?*

## Before You Go On

*Answer the following questions to test your understanding of the preceding section:*

16. *State three or more reasons why muscle twitch strength can vary even when stimulus intensity remains constant.*

17. *Explain the role of tetanus in normal muscle action.*

18. *Describe an everyday activity* not *involving the arms in which your muscles would switch from isometric to isotonic contraction.*

19. *Describe an everyday activity* not *involving the arms that would involve concentric contraction and one that would involve eccentric contraction.*

# Muscle Metabolism

### Objectives
When you have completed this section, you should be able to

• explain how skeletal muscle meets its energy demands during rest and exercise;

• explain the basis of muscle fatigue and soreness;

• define *oxygen debt* and explain why extra oxygen is needed even after an exercise has ended;

• distinguish between two physiological types of muscle fibers, and explain their functional roles;

• discuss the factors that affect muscular strength; and

• discuss the effects of resistance and endurance exercises on muscle.

## ATP SOURCES

All muscle contraction depends on ATP; no other energy source can serve in its place. The supply of ATP depends, in turn, on the availability of oxygen and organic energy sources such as glucose and fatty acids. To understand how muscle manages its ATP budget, you must be familiar with the two main pathways of ATP synthesis: *anaerobic fermentation* and *aerobic respiration* (see fig. 2.31, p. 84). Each of these has advantages and disadvantages. Anaerobic fermentation enables a cell to produce ATP in the absence of oxygen, but the ATP yield is very limited and the process produces a toxic end product, lactic acid, which is a major factor in muscle fatigue. By contrast, aerobic respiration produces far more ATP and less toxic end products (carbon dioxide and water), but it requires a continual supply of oxygen. Although aerobic respiration is best known as a pathway for glucose oxidation, it is also used to extract energy from other organic compounds. In a resting muscle, most ATP is generated by the aerobic respiration of fatty acids.

During the course of exercise, different mechanisms of ATP synthesis are used depending on the exercise duration. We will view these mechanisms from the standpoint of immediate, short-term, and long-term energy, but it must be stressed that muscle does not make sudden shifts from one mechanism to another like an automobile transmission shifting gears. Rather, these mechanisms blend and overlap as the exercise continues (fig. 11.17).

### Immediate Energy

In a short, intense exercise such as a 100-m dash, the respiratory and cardiovascular systems cannot deliver extra oxygen to the muscles quickly enough for aerobic respiration to meet the increased ATP demand. The myoglobin in a muscle fiber supplies oxygen for a limited amount of aerobic respiration, but in brief exercises a muscle meets most of its ATP demand by borrowing phosphate groups ($P_i$) from other molecules and transferring them to ADP. Two enzyme systems control these phosphate transfers (fig. 11.18):

1. **Myokinase** (MY-oh-KY-nase) transfers $P_i$ from one ADP to another, converting the latter to ATP.

2. **Creatine kinase** (CREE-uh-tin KY-nase) obtains $P_i$ from a phosphate-storage molecule, **creatine phosphate (CP),** and donates them to ADP to make ATP. This is a fast-acting system that helps to maintain the ATP level while other ATP-generating mechanisms are being activated.

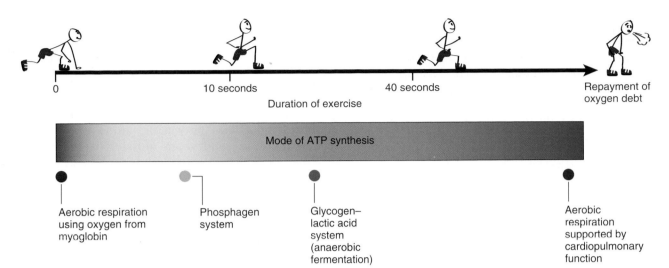

**FIGURE 11.17**  Modes of ATP Synthesis During Exercise.

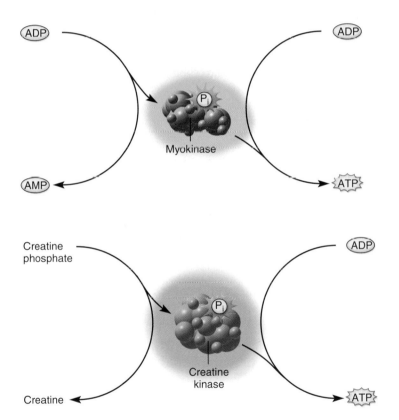

**FIGURE 11.18**  **The Phosphagen System.**  Two enzymes, myokinase and creatine kinase, generate ATP in the absence of oxygen. Myokinase borrows phosphate groups from ADP, and creatine kinase borrows them from creatine phosphate, to convert an ADP to ATP.

ATP and CP, collectively called the **phosphagen system,** provide nearly all the energy used for short bursts of intense activity. Muscle contains about 5 millimoles of ATP and 15 millimoles of CP per kilogram of tissue, which is enough to power about 1 minute of brisk walking or 6 seconds of sprinting or fast swimming. The phosphagen system is especially important in activities requiring brief but maximal effort, such as football, baseball, and weight lifting.

## Short-Term Energy

As the phosphagen system is exhausted, the muscles shift to anaerobic fermentation to "buy time" until cardiopulmonary function can catch up with the muscles' oxygen demand. During this period, the muscles obtain glucose from the blood and their own stored glycogen. You may recall from chapter 2 that in the absence of oxygen, the pathway of glycolysis can generate a net gain of 2 ATP for every glucose molecule consumed, as it converts glucose to lactic acid. The pathway from glycogen to lactic acid, called the **glycogen–lactic acid system,** produces enough ATP for 30 to 40 seconds of maximum activity. To play basketball or to run completely around a baseball diamond, for example, depends heavily on this energy-transfer system.

## Long-Term Energy

After 40 seconds or so, the respiratory and cardiovascular systems "catch up" and deliver oxygen to the muscles fast enough for aerobic respiration to meet most of the ATP demand. Aerobic respiration produces much more ATP than glycolysis does—typically another 36 ATPs per glucose. Thus it is a very efficient means of meeting the ATP demands of prolonged exercise. One's rate of oxygen consumption rises for 3 to 4 minutes and then levels off at a *steady state* in which aerobic ATP production keeps pace with the demand. In exercises lasting more than 10 minutes, more than 90% of the ATP is produced aerobically.

Little lactic acid accumulates under steady state conditions, but this does not mean that aerobic exercise can continue indefinitely or that it is limited only by a person's willpower. The depletion of glycogen and blood glucose, together with the loss of fluid and electrolytes through sweating, set limits to endurance and performance even when lactic acid does not.

# FATIGUE AND ENDURANCE

Muscle **fatigue** is the progressive weakness and loss of contractility that results from prolonged use of the muscles. For example, if you hold this book at arm's length for a minute, you will feel your muscles growing weaker and eventually you will be unable to hold it up. Repeatedly squeezing a rubber ball, pushing a video game button, or trying to take lecture notes from a fast-talking professor produces fatigue in the hand muscles. Fatigue has multiple causes:

- ATP synthesis declines as glycogen is consumed.

- The ATP shortage slows down the sodium–potassium pumps, compromising their ability to maintain the resting membrane potential and excitability of the muscle fibers.

- Lactic acid lowers the pH of the sarcoplasm, which inhibits the enzymes involved in contraction, ATP synthesis, and other aspects of muscle function.

- Each action potential releases $K^+$ from the sarcoplasm to the extracellular fluid. The accumulation of extracellular $K^+$ lowers the membrane potential (hyperpolarizes the cell) and makes the muscle fiber less excitable.

- Motor nerve fibers use up their ACh, which leaves them less capable of stimulating muscle fibers. This is called *junctional fatigue*.

- The central nervous system, where all motor commands originate, fatigues by processes not yet understood, so there is less signal output to the skeletal muscles.

## Think About It

*Suppose you repeatedly stimulated the sciatic nerve in a frog nerve-muscle preparation until the muscle stopped contracting. What simple test could you do to determine whether this was due to junctional fatigue or to one of the other fatigue mechanisms?*

The ability to maintain high-intensity exercise for more than 4 to 5 minutes is determined in large part by one's **maximum oxygen uptake ($V_{O_2}$max)**—the point at which the rate of oxygen consumption reaches a plateau and does not increase further with an added workload. $V_{O_2}$max is proportional to body size; it peaks at around age 20; it is usually greater in males than in females; and it can be twice as great in a trained endurance athlete as in an untrained person (see the later discussion on effects of conditioning). A typical sedentary adult has a $V_{O_2}$max of about 35 mL/min/kg. Such a person weighing 73 kg (160 pounds) and exercising at maximum intensity might therefore "burn" about 2.6 L of oxygen per minute, which sets a limit to his rate of ATP production. Elite endurance athletes can have a $V_{O_2}$max of about 70 mL/min/kg. The world's top-rated endurance athlete of the early twenty-first century has been Lance Armstrong,

history's only seven-time winner of the grueling Tour de France bicycle race. Armstrong's $V_{O_2}$max has been measured as 83.8 mL/min/kg. Relating this to maximum ATP production, and ATP supply to muscle performance, it is easy to see why he has such outstanding endurance.

Physical endurance also depends on the supply of organic nutrients—fatty acids, amino acids, and especially glucose. Many endurance athletes use a dietary strategy called *carbohydrate loading* to pack as much as 5 g of glycogen into every 100 g of muscle. This can significantly increase endurance, but an extra 2.7 g of water is also stored with each added gram of glycogen. Some athletes feel that the resulting sense of heaviness and other side effects outweigh the benefits of carbohydrate loading.

## OXYGEN DEBT

You have probably noticed that you breathe heavily not only during a strenuous exercise but also for several minutes afterwards. This is because your body accrues an oxygen debt that must be "repaid." **Oxygen debt** is the difference between the resting rate of oxygen consumption and the elevated rate following an exercise; it is also known as *excess postexercise oxygen consumption (EPOC)*. The total amount of extra oxygen consumed after a strenuous exercise is typically about 11 L. It is used for the following purposes:

- *Replacing the body's oxygen reserves* that were depleted in the first minute of exercise. These include the oxygen bound to muscle myoglobin and blood hemoglobin, oxygen dissolved in the blood plasma and other extracellular fluids, and oxygen in the air in the lungs.

- *Replenishing the phosphagen system.* This involves synthesizing ATP and using some of it to donate phosphate groups back to creatine until the resting levels of ATP and CP are restored.

- *Oxidizing lactic acid.* About 80% of the lactic acid produced by muscle enters the bloodstream and is reconverted to pyruvic acid in the kidneys, the cardiac muscle, and especially the liver. Some of this pyruvic acid enters the aerobic (mitochondrial) pathway to make ATP, but the liver converts most of it back to glucose. Glucose is then available to replenish the glycogen stores of the muscle.

- *Serving the elevated metabolic rate.* As long as the body temperature remains elevated by exercise, the total metabolic rate remains high, and rapid metabolism consumes extra oxygen.

## PHYSIOLOGICAL CLASSES OF MUSCLE FIBERS

Not all muscle fibers are metabolically alike or adapted to perform the same task. Some respond slowly but are relatively resistant to fatigue, while others respond more

| TABLE 11.3 | Classification of Skeletal Muscle Fibers | |
|---|---|---|
| | **Fiber Type** | |
| **Properties** | **Slow Oxidative** | **Fast Glycolytic** |
| Relative diameter | Smaller | Larger |
| ATP synthesis | Aerobic | Anaerobic |
| Fatigue resistance | Good | Poor |
| ATP hydrolysis | Slow | Fast |
| Glycolysis | Moderate | Fast |
| Myoglobin content | Abundant | Low |
| Glycogen content | Low | Abundant |
| Mitochondria | Abundant and large | Fewer and smaller |
| Capillaries | Abundant | Fewer |
| Color | Red | White, pale |
| *Representative Muscles in Which Fiber Type Is Predominant* | Soleus | Gastrocnemius |
| | Erector spinae | Biceps brachii |
| | Quadratus lumborum | Muscles of eye movement |

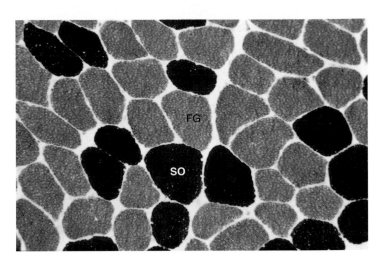

**FIGURE 11.19  Muscle Stained to Distinguish Fast Glycolytic (FG) from Slow Oxidative (SO) Fibers.**  Cross section.

quickly but also fatigue quickly (table 11.3). Each primary type of fiber goes by several names:

- **Slow oxidative (SO), slow-twitch, red,** or **type I fibers.** These fibers have relatively abundant mitochondria, myoglobin, and blood capillaries, and therefore a relatively deep red color. They are well adapted to aerobic respiration, which does not generate lactic acid. Thus, these fibers do not fatigue easily. However, in response to a single stimulus, they exhibit a relatively long twitch, lasting about 100 msec. The soleus muscle of the calf and the postural muscles of the back are composed mainly of these SO, high-endurance fibers.

- **Fast glycolytic (FG), fast-twitch, white,** or **type II fibers.** These fibers are well adapted for quick responses but not for fatigue resistance. They are rich in enzymes of the phosphagen and glycogen–lactic acid systems. Their sarcoplasmic reticulum releases and reabsorbs $Ca^{2+}$ quickly, which partially accounts for their quick, forceful contractions. They are poorer than SO fibers in mitochondria, myoglobin, and blood capillaries, so they are relatively pale (hence the expression *white* fibers). These fibers produce twitches as short as 7.5 msec, but because of the lactic acid they generate, they fatigue more easily than SO fibers. Thus, they are especially important in sports such as basketball that require stop-and-go activity and frequent changes of pace. The gastrocnemius muscle of the calf, biceps brachii of the arm, and the muscles of eye movement consist mainly of FG fibers.

Some authorities recognize two subtypes of FG fibers called types IIA and IIB. Type IIB is the common type just described, while IIA, or **intermediate fibers,** combine fast-twitch responses with aerobic fatigue-resistant metabolism. Type IIA fibers, however, are relatively rare except in some endurance-trained athletes. The fiber types can be differentiated histologically by using stains for certain mitochondrial enzymes and other cellular components (fig. 11.19). All muscle fibers of one motor unit belong to the same physiological type.

Nearly all muscles are composed of both SO and FG fibers, but the proportions of these fiber types differ from one muscle to another. Muscles composed mainly of SO fibers are called *red muscles* and those composed mainly of FG fibers are called *white muscles.* People with different types and levels of physical activity differ in the proportion of one fiber type to another even in the same muscle, such as the *quadriceps femoris* of the anterior thigh (table 11.4). It is thought that people are born with a genetic predisposition for a certain ratio of fiber types. Those who go into competitive sports discover the sports at which they can excel and gravitate toward those for

| TABLE 11.4 | Proportion of Slow Oxidative (SO) and Fast Glycolytic (FG) Fibers in the Quadriceps Femoris Muscle of Male Athletes | |
|---|---|---|
| **Sample Population** | **SO** | **FG** |
| Marathon runners | 82% | 18% |
| Swimmers | 74 | 26 |
| Average males | 45 | 55 |
| Sprinters and jumpers | 37 | 63 |

which heredity has best equipped them. One person might be a "born sprinter" and another a "born marathoner."

We noted earlier that sometimes two or more muscles act across the same joint and superficially seem to have the same function. We have already seen some reasons why such muscles are not as redundant as they seem. Another reason is that they may differ in the proportion of SO to FG fibers. For example, the gastrocnemius and soleus muscles of the calf both insert on the calcaneus through the same tendon, the calcaneal tendon, so they exert the same pull on the heel. The gastrocnemius, however, is a white, predominantly FG muscle adapted for quick, powerful movements such as jumping, whereas the soleus is a red, predominantly SO muscle that does most of the work in endurance exercises such as jogging and skiing.

# MUSCULAR STRENGTH AND CONDITIONING

We have far more muscular strength than we normally use. The gluteus maximus can generate 1,200 kg of tension, and all the muscles of the body can produce a total tension of 22,000 kg (nearly 25 tons). Indeed, the muscles can generate more tension than the bones and tendons can withstand—a fact that accounts for many injuries to the patellar and calcaneal tendons. Muscular strength depends on a variety of anatomical and physiological factors:

- **Muscle size.** The strength of a muscle depends primarily on its size; this is why weight lifting increases the size and strength of a muscle simultaneously. A muscle can exert a tension of about 3 to 4 kg per square cm of cross-sectional area.

- **Fascicle arrangement.** Pennate muscles such as the quadriceps femoris are stronger than parallel muscles such as the sartorius, which in turn are stronger than circular muscles such as the orbicularis oculi.

- **Size of active motor units.** Large motor units produce stronger contractions than small ones.

- **Multiple motor unit summation.** When a stronger muscle contraction is desired, the nervous system activates more motor units. This process is the *recruitment,* or *MMU summation,* described earlier. It can produce extraordinary feats of strength under desperate conditions—rescuing a loved one pinned under an automobile, for example. Getting "psyched up" for athletic competition is also partly a matter of MMU summation.

- **Temporal summation.** Nerve impulses usually arrive at a muscle in a series of closely spaced action potentials. Because of the *temporal summation* described earlier, the greater the frequency of stimulation, the more strongly a muscle contracts.

- **The length–tension relationship.** As noted earlier, a muscle resting at optimum length is prepared to contract more forcefully than a muscle that is excessively contracted or stretched.

- **Fatigue.** Fatigued muscles contract more weakly than rested ones.

**Resistance exercise,** such as weight lifting, is the contraction of muscles against a load that resists movement. A few minutes of resistance exercise at a time, a few times each week, is enough to stimulate muscle growth. Growth results primarily from cellular enlargement, not cellular division. The muscle fibers synthesize more myofilaments and the myofibrils grow thicker. Myofibrils split longitudinally when they reach a certain size, so a well-conditioned muscle has more myofibrils than a poorly conditioned one. Muscle fibers themselves are incapable of mitosis, but there is some evidence that as they enlarge, they too may split longitudinally. A small part of muscle growth may therefore result from an increase in the number of fibers, but most results from the enlargement of fibers that have existed since childhood.

> **Think About It**
>
> *Is muscle growth mainly the result of hypertrophy or hyperplasia?*

**Endurance (aerobic) exercise,** such as jogging and swimming, improves the fatigue resistance of the muscles. Slow-twitch fibers, especially, produce more mitochondria and glycogen and acquire a greater density of blood capillaries as a result of conditioning. Endurance exercise also improves skeletal strength, increases the red blood cell count and the oxygen transport capacity of the blood, and enhances the function of the cardiovascular, respiratory, and nervous systems.

Endurance training does not significantly increase muscular strength, and resistance training does not improve endurance. Optimal performance and skeletomuscular health require **cross-training,** which incorporates elements of both types. If muscles are not kept sufficiently active, they become *deconditioned*—weaker and more easily fatigued.

## Before You Go On

*Answer the following questions to test your understanding of the preceding section:*

20. From which two molecules can ADP borrow a phosphate group to become ATP? What is the enzyme that catalyzes each transfer?

21. In a long period of intense exercise, why does muscle generate ATP anaerobically at first and then switch to aerobic respiration?

22. List four causes of muscle fatigue.

23. List three causes of oxygen debt.

24. What properties of fast glycolytic and slow oxidative fibers adapt them for different physiological purposes?

# Cardiac and Smooth Muscle

### Objectives

When you have completed this section, you should be able to

- describe the structural and physiological differences between cardiac muscle and skeletal muscle;
- explain why these differences are important to cardiac function;
- describe the structural and physiological differences between smooth muscle and skeletal muscle; and
- relate the unique properties of smooth muscle to its locations and functions.

In this section, we compare cardiac muscle and smooth muscle to skeletal muscle. As you will find, cardiac and smooth muscle have special structural and physiological properties related to their distinctive functions. They also have certain properties in common with each other. The muscle cells of both cardiac and smooth muscle are called **myocytes.** By comparison to the long multinucleate fibers of skeletal muscle, these are relatively short cells with only one nucleus. Cardiac and smooth muscle are *involuntary* muscle tissues, not usually subject to our conscious control.

## CARDIAC MUSCLE

**Cardiac muscle** constitutes most of the heart. Its structure and function are discussed extensively in chapter 19 so that you will be able to relate these to the actions of the heart. Here, we only briefly compare it with skeletal and smooth muscle (table 11.5).

Cardiac muscle is striated like skeletal muscle, but its myocytes *(cardiocytes)* are shorter and thicker, shaped like a log with uneven, notched ends (see fig. 19.13). Each myocyte is joined to several others at its ends through linkages called **intercalated** (in-TUR-kuh-LAY-ted) **discs.** These appear as thick dark lines in stained tissue sections. An intercalated disc has electrical *gap junctions*

| TABLE 11.5 | Comparison of Skeletal, Cardiac, and Smooth Muscle | | |
|---|---|---|---|
| **Feature** | **Skeletal Muscle** | **Cardiac Muscle** | **Smooth Muscle** |
| Location | Associated with skeletal system | Heart | Walls of viscera and blood vessels, iris of eye, piloerector of hair follicles |
| Cell shape | Long threadlike fibers | Short, slightly branched cells | Short fusiform cells |
| Cell length | 100 μm–30 cm | 50–100 μm | 50–200 μm |
| Cell width | 10–100 μm | 10–20 μm | 2–10 μm |
| Striations | Present | Present | Absent |
| Nuclei | Multiple nuclei, adjacent to sarcolemma | Usually one nucleus, near middle of cell | One nucleus, near middle of cell |
| Connective tissues | Endomysium, perimysium, epimysium | Endomysium only | Endomysium only |
| Sarcoplasmic reticulum | Abundant | Present | Scanty |
| T tubules | Present, narrow | Present, wide | Absent |
| Gap junctions | Absent | Present in intercalated discs | Present in single-unit smooth muscle |
| Autorhythmicity | Absent | Present | Present in single-unit smooth muscle |
| Thin filament attachment | Z discs | Z discs | Dense bodies |
| Regulatory proteins | Tropomyosin, troponin | Tropomyosin, troponin | Calmodulin, light-chain myokinase |
| $Ca^{2+}$ source | Sarcoplasmic reticulum | Sarcoplasmic reticulum and extracellular fluid | Mainly extracellular fluid |
| $Ca^{2+}$ receptor | Troponin of thin filament | Troponin of thin filament | Calmodulin of thick filament |
| Innervation and control | Somatic motor fibers (voluntary) | Autonomic fibers (involuntary) | Autonomic fibers (involuntary) |
| Nervous stimulation required? | Yes | No | No |
| Effect of nervous stimulation | Excitatory only | Excitatory or inhibitory | Excitatory or inhibitory |
| Mode of tissue repair | Limited regeneration, mostly fibrosis | Limited regeneration, mostly fibrosis | Relatively good capacity for regeneration |

that allow each myocyte to directly stimulate its neighbors, and mechanical junctions that keep the myocytes from pulling apart when the heart contracts. The sarcoplasmic reticulum is less developed than in skeletal muscle, but the T tubules are larger and admit supplemental Ca²⁺ from the extracellular fluid. Damaged cardiac muscle is repaired by fibrosis. Cardiac muscle has no satellite cells, and even though mitosis has recently been detected in cardiac myocytes following heart attacks, it does not produce a significant amount of regenerated functional muscle.

Unlike skeletal muscle, cardiac muscle can contract without the need of nervous stimulation. It contains a built-in **pacemaker** that rhythmically sets off a wave of electrical excitation. This wave travels through the muscle and triggers the contraction of the heart chambers. Cardiac muscle is said to be **autorhythmic**[12] because of this ability to contract rhythmically and independently. The heart does, however, receive fibers from the *autonomic nervous system* that can either increase or decrease the heart rate and contraction strength. Cardiac muscle does not exhibit quick twitches like skeletal muscle. Rather, it maintains tension for about 200 to 250 msec, giving the heart time to expel blood.

Cardiac muscle uses aerobic respiration almost exclusively. It is very rich in myoglobin and glycogen, and it has especially large mitochondria that fill about 25% of the cell, compared with smaller mitochondria occupying about 2% of a skeletal muscle fiber. Cardiac muscle is very adaptable with respect to the fuel used, but very vulnerable to interruptions in oxygen supply. Because it makes little use of anaerobic fermentation, cardiac muscle is highly resistant to fatigue.

## SMOOTH MUSCLE

**Smooth muscle** is composed of myocytes with a fusiform shape, about 30 to 200 μm long, 5 to 10 μm wide at the middle, and tapering to a point at each end. There is only one nucleus, located near the middle of the cell. Although thick and thin filaments are both present, they are not aligned with each other and produce no visible striations or sarcomeres; this is the reason for the name *smooth* muscle. Z discs are absent. In their place are protein plaques on the inner face of the plasma membrane, and a well-ordered array of protein masses called **dense bodies** in the cytoplasm. The cytoplasm also contains an extensive cytoskeleton of intermediate filaments. These intermediate filaments as well as the thin myofilaments attach to the membrane plaques and dense bodies, thus providing mechanical linkages between the thin myofilaments and the plasma membrane (see fig. 11.23). The role of this linkage in smooth muscle contraction is explained later.

[12]*auto* = self

The sarcoplasmic reticulum is scanty, and there are no T tubules. The Ca²⁺ needed to activate smooth muscle contraction comes mainly from the extracellular fluid (ECF) by way of Ca²⁺ channels in the sarcolemma. During relaxation, Ca²⁺ is pumped back out of the cell. Some smooth muscle has no nerve supply, but when nerve fibers are present, they are autonomic (like those of cardiac muscle) and not somatic motor fibers.

Unlike skeletal and cardiac muscle, smooth muscle is capable of mitosis and hyperplasia. Thus, an organ such as the pregnant uterus can grow by adding more myocytes, and injured smooth muscle regenerates well.

### Types of Smooth Muscle

There are two functional categories of smooth muscle called *multiunit* and *single-unit* types (fig. 11.20). **Multiunit smooth muscle** occurs in some of the largest arteries and pulmonary air passages, in the piloerector muscles of the hair follicles, and in the iris of the eye. Its innervation, although autonomic, is otherwise similar to that of skeletal muscle—the terminal branches of a nerve fiber synapse with individual myocytes and form a motor

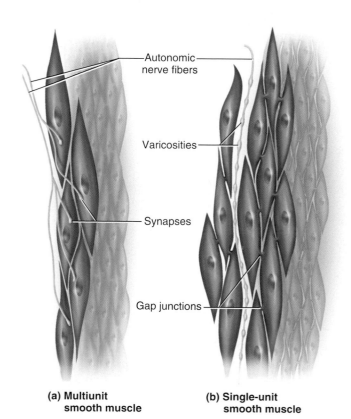

**(a) Multiunit smooth muscle**

**(b) Single-unit smooth muscle**

**FIGURE 11.20   Multiunit and Single-Unit Smooth Muscle.** (a) Multiunit smooth muscle, in which each muscle cell receives its own nerve supply and contracts independently. (b) Single-unit smooth muscle, in which a nerve fiber passes through the tissue without synapsing with any specific muscle cell, and muscle cells are coupled by electrical gap junctions.

unit. Each motor unit contracts independently of the others, hence the name of this muscle type.

**Single-unit smooth muscle** is more widespread. It occurs in most blood vessels and in the digestive, respiratory, urinary, and reproductive tracts—thus, it is also called **visceral muscle.** In many of the hollow viscera, it forms two or more layers—typically an *inner circular layer,* in which the myocytes encircle the organ, and an outer *longitudinal layer,* in which the myocytes run lengthwise along the organ (fig. 11.21). The name *single-unit* refers to the fact that the myocytes of this type of muscle are electrically coupled to each other by gap junctions. Thus, they directly stimulate each other and a large number of cells contract as a unit, almost as if they were a single cell.

## Stimulation of Smooth Muscle

Like cardiac muscle, smooth muscle is involuntary and capable of contracting without nervous stimulation. Some smooth muscle contracts in response to chemical stimuli such as hormones, carbon dioxide, low pH, and oxygen deficiency and in response to stretch (as in a full stomach or bladder). Some single-unit smooth muscle, especially in the stomach and intestines, has pacemaker cells that spontaneously depolarize and set off waves of contraction throughout an entire layer of muscle. Such smooth muscle is autorhythmic, like cardiac muscle, although with a much slower rhythm.

But like cardiac muscle, most smooth muscle is innervated by autonomic nerve fibers that can trigger or modify its contractions. Autonomic nerve fibers stimulate smooth muscle with either acetylcholine or norepinephrine. The nerve fibers have contrasting effects on smooth muscle in different locations. They relax the smooth muscle of arteries while contracting the smooth muscle in the bronchioles of the lungs, for example.

In single-unit smooth muscle, each autonomic nerve fiber has up to 20,000 beadlike swellings called **varicosities** along its length (figs. 11.20 and 11.22). Each varicosity contains synaptic vesicles and a few mitochondria. Instead of closely approaching any one myocyte, the nerve fiber passes amid several myocytes and stimulates all of them at once when it releases its neurotransmitter. The muscle cells do not have motor end plates or any other specialized area of sarcolemma to bind the neurotransmitter; rather, they have receptor sites scattered throughout the surface. Such nerve–muscle relationships are called **diffuse junctions** because there is no one-to-one relationship between a nerve fiber and a myocyte.

## Contraction and Relaxation

Smooth muscle resembles the other muscle types in that contraction is triggered by $Ca^{2+}$, energized by ATP, and achieved by the sliding of thin filaments over the thick filaments. The mechanism of excitation–contraction coupling, however, is very different. Little of the $Ca^{2+}$ comes from the sarcoplasmic reticulum; most comes from the

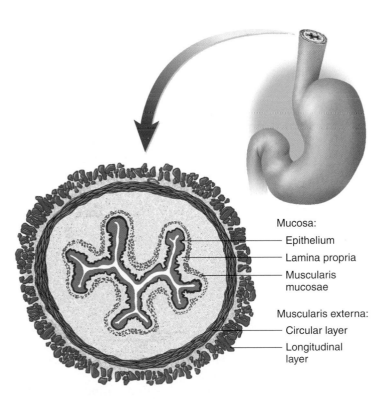

Mucosa:
- Epithelium
- Lamina propria
- Muscularis mucosae

Muscularis externa:
- Circular layer
- Longitudinal layer

**FIGURE 11.21** **Layers of Visceral Muscle in a Cross Section of the Esophagus.** Many hollow organs have alternating circular and longitudinal layers of smooth muscle.

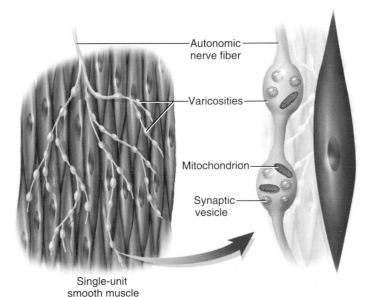

- Autonomic nerve fiber
- Varicosities
- Mitochondrion
- Synaptic vesicle

Single-unit smooth muscle

**FIGURE 11.22** **Varicosities of an Autonomic Nerve Fiber in Single-Unit Smooth Muscle.**

extracellular fluid and enters the cell through calcium channels in the sarcolemma. Some of these channels are voltage-gated and open in response to changes in membrane voltage; some are ligand-gated and open in response to hormones and neurotransmitters; and some are mechanically gated and open in response to stretching of the cell.

### Think About It

*How is smooth muscle contraction affected by the drugs called calcium channel blockers? (see p. 97)*

(see p. 97)

Smooth muscle has no troponin. Calcium binds instead to a similar protein called **calmodulin**[13] (cal-MOD-you-lin), associated with the thick filaments. Calmodulin then activates an enzyme called **light-chain myokinase,** which transfers a phosphate group from ATP to the head of the myosin. This activates the myosin ATPase and enables it to bind to actin, but in order to execute a power stroke, the myosin must bind and hydrolyze yet another ATP. It then produces power and recovery strokes like those of skeletal muscle.

As thick filaments pull on the thin ones, the thin filaments pull on the dense bodies and membrane plaques. Through the dense bodies and cytoskeleton, force is transferred to the plasma membrane and the entire cell shortens. When a smooth muscle cell contracts, it puckers and twists somewhat like wringing out a wet towel (fig. 11.23).

In skeletal muscle, there is typically a 2-msec latent period between stimulation and the onset of contraction. In smooth muscle, by contrast, the latent period is 50 to 100 msec long. Tension peaks about 500 msec (0.5 sec) after the stimulus and then declines over a period of 1 to 2 seconds. The effect of all this is that compared with skeletal muscle, smooth muscle is very slow to contract and relax. It is slow to contract because its myosin ATPase is a slow enzyme. It is slow to relax because the pumps that remove $Ca^{2+}$ from the cell are also slow. As the $Ca^{2+}$ level falls, myosin is dephosphorylated and is no longer able to hydrolyze ATP and execute power strokes. However, it does not necessarily detach from actin immediately. It has a *latch-bridge mechanism* that enables it to remain attached to actin for a prolonged time without consuming more ATP.

Smooth muscle often exhibits tetanus and is very resistant to fatigue. It makes most of its ATP aerobically, but its ATP requirement is small and it has relatively few mitochondria. Skeletal muscle requires 10 to 300 times as much ATP as smooth muscle to maintain the same amount of tension. The fatigue-resistance and latch-bridge mechanism of smooth muscle are important in

---

[13]acronym for *cal*cium *modu*lating pro*tein*

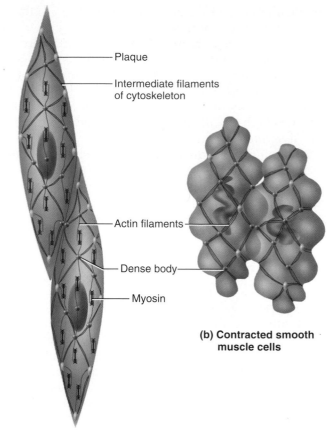

**(a) Relaxed smooth muscle cells**

**(b) Contracted smooth muscle cells**

**FIGURE 11.23    Smooth Muscle Contraction.** (a) Relaxed cells. Actin myofilaments are anchored to dense bodies in the sarcoplasm and on the plasma membrane, rather than to Z discs. (b) Contracted cells.

enabling it to maintain a state of continual **smooth muscle tone** (tonic contraction). Tonic contraction keeps the arteries in a state of partial constriction called *vasomotor tone.* A loss of muscle tone in the arteries can cause a dangerous drop in blood pressure. Smooth muscle tone also keeps the intestines partially contracted. The intestines are much longer in a cadaver than they are in a living person because of the loss of muscle tone at death.

### Response to Stretch

Stretch alone sometimes causes smooth muscle to contract by opening mechanically gated calcium channels in the sarcolemma. Distension of the esophagus with food or the colon with feces, for example, evokes a wave of contraction called **peristalsis** (PERR-ih-STAL-sis) that propels the contents along the organ.

Smooth muscle exhibits a reaction called the **stress-relaxation** (or **receptive relaxation**) **response.** When stretched, it briefly contracts and resists, but then relaxes. The significance of this response is apparent in the urinary bladder, whose wall consists of three layers of smooth muscle. If the stretched bladder contracted and did not soon relax, it would expel urine almost as soon as it began to fill, thus failing to store the urine until an opportune time.

Remember that skeletal muscle cannot contract very forcefully if it is overstretched. Smooth muscle, however, is not subject to the limitations of the length–tension relationship. It must be able to contract forcefully even when greatly stretched, so that hollow organs such as the stomach and bladder can fill and then expel their contents efficiently. Skeletal muscle must be within 30% of optimum length in order to contract strongly when stimulated. Smooth muscle, by contrast, can be anywhere from half to twice its resting length and still contract powerfully. There are three reasons for this: (1) there are no Z discs, so thick filaments cannot butt against them and stop the contraction; (2) since the thick and thin filaments are not arranged in orderly sarcomeres, stretching of the muscle does not cause a situation where there is too little overlap for cross-bridges to form; and (3) the thick filaments of smooth muscle have myosin heads along their entire length (there is no bare zone), so cross-bridges can form anywhere, not just at the ends. Smooth muscle also

exhibits **plasticity**—the ability to adjust its tension to the degree of stretch. Thus, a hollow organ such as the bladder can be greatly stretched yet not become flabby when it is empty.

The muscular system suffers fewer diseases than any other organ system, but several of its more common dysfunctions are listed in table 11.6. The effects of aging on the muscular system are described on page 1129.

## Before You Go On

*Answer the following questions to test your understanding of the preceding section:*

25. Explain why intercalated discs are important to cardiac muscle function.
26. Explain why it is important for cardiac muscle to have longer-lasting contractions than skeletal muscle.
27. How do single-unit and multiunit smooth muscle differ in innervation and contractile behavior?
28. How does smooth muscle differ from skeletal muscle with respect to its source of calcium and its calcium receptor?
29. Explain why the stress-relaxation response is an important factor in smooth muscle function.

| TABLE 11.6 | Some Disorders of the Muscular System |
|---|---|
| Contracture | Abnormal muscle shortening not caused by nervous stimulation. Can result from failure of the calcium pump to remove $Ca^{2+}$ from the sarcoplasm or from contraction of scar tissue, as in burn patients. |
| Cramps | Painful muscle spasms triggered by heavy exercise, extreme cold, dehydration, electrolyte loss, low blood glucose, or lack of blood flow. |
| Crush syndrome | A shocklike state following the massive crushing of muscles; associated with high and potentially fatal fever, cardiac irregularities resulting from $K^+$ released from the muscle, and kidney failure resulting from blockage of the renal tubules with myoglobin released by the traumatized muscle. Myoglobinuria (myoglobin in the urine) is a common sign. |
| Delayed-onset muscle soreness | Pain, stiffness, and tenderness felt from several hours to a day after strenuous exercise. Associated with microtrauma to the muscles, with disrupted Z discs, myofibrils, and plasma membranes, and with elevated levels of myoglobin, creatine kinase, and lactate dehydrogenase in the blood. |
| Disuse atrophy | Reduction in the size of muscle fibers as a result of nerve damage or muscular inactivity, for example in limbs in a cast and in patients confined to a bed or wheelchair. Muscle strength can be lost at a rate of 3% per day of bed rest. |
| Fibromyalgia | Diffuse, chronic muscular pain and tenderness, often associated with sleep disturbances and fatigue; often misdiagnosed as chronic fatigue syndrome. Can be caused by various infectious diseases, physical or emotional trauma, or medications. Most common in women 30 to 50 years old. |
| Myositis | Muscle inflammation and weakness resulting from infection or autoimmune disease. |

*Disorders described elsewhere*

| | | |
|---|---|---|
| Back injuries p. 348 | Compartment syndrome p. 382 | Pitcher's arm p. 382 |
| Baseball finger p. 382 | Hernia p. 351 | Pulled hamstrings p. 382 |
| Carpal tunnel syndrome p. 366 | Muscular dystrophy p. 436 | Rotator cuff injury p. 382 |
| | Myasthenia gravis p. 436 | Tennis elbow p. 382 |
| Charley horse p. 382 | Paralysis p. 414 | Tennis leg p. 382 |

## Muscular Dystrophy and Myasthenia Gravis

*Muscular dystrophy*[14] is a collective term for several hereditary diseases in which the muscles degenerate, weaken, and are gradually replaced by fat and fibrous scar tissue. The most common form of the disease is *Duchenne*[15] *muscular dystrophy (DMD)*, a sex-linked recessive trait affecting about 1 out of every 3,500 live-born boys.

DMD is not evident at birth, but begins to exhibit its effects as a child shows difficulty keeping up with other children, falls frequently, and finds it hard to stand again. It is typically diagnosed between the ages of 2 and 10 years. It affects the muscles of the hips first, then the legs, and then progresses to the abdominal, spinal, and respiratory muscles as well as cardiac muscle. Muscles shorten as they atrophy, causing postural abnormalities such as scoliosis. Persons with DMD are usually wheelchair-dependent by the age of 10 or 12, and seldom live past the age of 20. For obscure reasons, they also frequently suffer a progressive decline in mental ability. Death usually results from respiratory insufficiency, pulmonary infection, or heart failure. DMD is incurable, but is treated with exercise to slow the atrophy of the muscles and with braces to reinforce the weakened hips and maintain posture.

The underlying cause of DMD is a mutation in the gene for the muscle protein dystrophin (see p. 409)—a large gene highly vulnerable to mutation. Without dystrophin, there is no coupling between the thin myofilaments and the sarcolemma. The sacromeres move independently of the sarcolemma, creating tears in the membrane. The torn membrane admits excess $Ca^{2+}$ into the cell, which activates intracellular proteases (protein-digesting enzymes). These enzymes degrade the contractile proteins of the muscle, leading to weakness and cellular necrosis. Dying muscle fibers are replaced with scar tissue, which blocks blood circulation in the muscle and thereby contributes to still further necrosis. Muscle degeneration accelerates in a fatal spiral of positive feedback.

Genetic screening can identify heterozygous carriers of DMD, allowing for counseling of prospective parents on the risk of having a child with the disease. However, about one out of three cases arises by a new spontaneous mutation and therefore cannot be predicted by genetic testing.

A less severe form of muscular dystrophy is *facioscapulohumeral (Landouzy-Dejerine*[16]*) MD*, an autosomal dominant trait that begins in adolescence and affects both sexes equally. It involves the facial and shoulder muscles more than the pelvic muscles and cripples some individuals while it barely affects others. A third form, *limb-girdle dystrophy*, is a combination of several diseases of intermediate severity that affect the shoulder, arm, and pelvic muscles.

*Myasthenia gravis*[17] (MY-ass-THEE-nee-uh GRAV-is) *(MG)* usually occurs in women between the ages of 20 and 40. It is an autoimmune disease in which antibodies attack the neuromuscular junctions and bind ACh receptors together in clusters. The muscle fiber then removes the clusters from the sarcolemma by endocytosis. As a result, the muscle fibers become less and less sensitive to ACh. The effects often appear first in the facial muscles and commonly include drooping eyelids (*ptosis,* fig. 11.24) and double vision (due to *strabismus,* inability

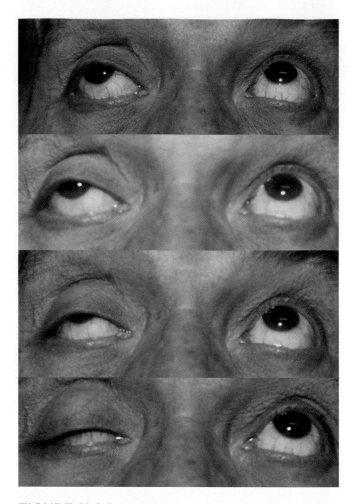

**FIGURE 11.24  Myasthenia Gravis.** These photographs were taken when the patient was first told to gaze upward (top photo), and then after 30, 60, and 90 seconds. Note the inability to keep the right eyelid open. This shows the ptosis that is diagnostic of myasthenia gravis.

to fixate on the same point with both eyes). The initial symptoms are often followed by difficulty in swallowing, weakness of the limbs, and poor physical endurance. Some people with MG die quickly as a result of respiratory failure, but others have normal life spans. One method of assessing the progress of the disease is to use *bungarotoxin,* a protein from cobra venom that binds to ACh receptors. The amount that binds is proportional to the number of receptors that are still functional. The muscle of an MG patient sometimes binds less than one-third as much bungarotoxin as normal muscle does.

Myasthenia gravis is often treated with cholinesterase inhibitors. These drugs retard the breakdown of ACh in the neuromuscular junction and enable it to stimulate the muscle longer. Immunosuppressive agents such as prednisone and azathioprine (imuram) may be used to suppress the production of the antibodies that destroy ACh receptors. Since certain immune cells are stimulated by hormones from the thymus, removal of the thymus (*thymectomy*) helps to dampen the overactive immune response that causes myasthenia gravis. Also, a technique called *plasmapheresis* may be used to remove harmful antibodies from the blood plasma.

[14]*dys* = bad, abnormal + *trophy* = growth
[15]Guillaume B. A. Duchenne (1806–75), French physician
[16]Louis T. J. Landouzy (1845–1917) and Joseph J. Dejerine (1849–1917), French neurologists
[17]*my* = muscle + *asthen* = weakness + *grav* = severe

# CHAPTER REVIEW

# Review of Key Concepts

## Types and Characteristics of Muscular Tissue (p. 406)

1. Muscular tissue has the properties of responsiveness, conductivity, contractility, extensibility, and elasticity.

2. Skeletal muscle is voluntary striated muscle that is usually attached to one or more bones.

3. A skeletal muscle cell, or muscle fiber, is a threadlike cell typically 100 mm in diameter and 3 cm long.

4. A muscle has *parallel* and *series elastic components* of fibrous, noncontractile tissue including the endomysium, perimysium, epimysium, and its tendons or other attachments to the skeleton. These components resist excessive stretching and injury to the muscle, and their elastic recoil enhances the power output of a muscle.

## Microscopic Anatomy of Skeletal Muscle (p. 407)

1. A muscle fiber forms by the fusion of many stem cells called *myoblasts,* and is thus multinucleate.

2. The *sarcolemma* (plasma membrane) exhibits tunnel-like infoldings called *transverse (T) tubules* that cross from one side of the cell to the other.

3. The *sarcoplasm* (cytoplasm) is occupied mainly by protein bundles called *myofibrils.* Mitochondria, glycogen, and myoglobin are packed between the myofibrils.

4. The fiber has an extensive *sarcoplasmic reticulum (SR)* that serves as a $Ca^{2+}$ reservoir. On each side of a T tubule, the SR expands into a *terminal cisterna.*

5. A myofibril is a bundle of two kinds of protein *myofilaments* called thick and thin filaments.

6. *Thick filaments* are composed of bundles of *myosin* molecules, each of which has a filamentous tail and a globular head.

7. Thin filaments are composed mainly of a double strand of *actin,* with a myosin-binding *active site* on each of its globular subunits. In the groove between the two actin strands are two regulatory proteins, *tropomyosin* and *troponin.*

8. Elastic filaments composed of *titin* flank each thick filament and attach to Z discs.

9. Several other accessory proteins are associated with the thick and thin filaments. One of these, *dystrophin,* links the thin filaments to the extracellular parallel elastic components and thus links the movement of the contractile proteins to the pull exerted on a bone.

10. Skeletal and cardiac muscle exhibit alternating light and dark bands, or *striations,* that result from the pattern of overlap between thick and thin filaments. The principal striations are a dark *A band* with a light *H zone* in the middle, and a light *I band* with a dark line, the *Z disc,* in the middle.

11. The functional unit of a muscle fiber is the *sarcomere,* which is a segment from one Z disc to the next.

## The Nerve–Muscle Relationship (p. 412)

1. Skeletal muscle contracts only when it is stimulated by a *somatic motor nerve fiber.*

2. One somatic motor fiber branches at the end and innervates from 3 to 1,000 muscle fibers. The nerve fiber and its muscle fibers are called a *motor unit.* Small motor units (few muscle fibers per nerve fiber) are found in muscles where fine control of movement is important, and large motor units in muscles where strength is more important than precision.

3. The point where a nerve fiber meets a muscle fiber is a type of synapse called the *neuromuscular junction.* It consists of the *synaptic knob* (a dilated tip of the nerve fiber) and a *motor end plate* (a folded depression in the sarcolemma). The gap between the knob and end plate is the *synaptic cleft.*

4. *Synaptic vesicles* in the knob release a neurotransmitter called *acetylcholine (ACh),* which diffuses across the cleft and binds to *ACh receptors* on the end plate.

5. An unstimulated nerve, muscle, or other cell has a difference in positive and negative charges on the two sides of its plasma membrane; it is *polarized.* The charge difference, called the *resting membrane potential,* is typically about −90 mV on a muscle fiber.

6. When a nerve or muscle fiber is stimulated, a quick, self-propagating voltage shift called an *action potential* occurs. Action potentials form nerve signals and activate muscle contraction.

## Behavior of Skeletal Muscle Fibers (p. 415)

1. The first stage of muscle action is *excitation.* An arriving nerve signal triggers ACh release, ACh binds to receptors on the motor end plate and triggers a voltage change called an *end-plate potential (EPP),* and the EPP triggers action potentials in adjacent regions of the sarcolemma.

2. The second stage is *excitation–contraction coupling.* Action potentials spread along the sarcolemma and down the T tubules, and trigger $Ca^{2+}$ release from the terminal cisternae of the SR. $Ca^{2+}$ binds to troponin of the thin filaments, and tropomyosin shifts position to expose the active sites on the actin.

3. The third stage is *contraction.* A myosin head binds to an active site on actin, flexes, tugs the thin filament closer to the A band, then releases the actin and repeats the process. Each cycle of binding and release consumes one ATP.

4. The fourth and final stage is *relaxation*. When nerve signals cease, ACh release ceases. The enzyme acetylcholinesterase degrades the ACh already present, halting stimulation of the muscle fiber. The SR pumps $Ca^{2+}$ back into it for storage. In the absence of $Ca^{2+}$, tropomyosin blocks the active sites of actin so myosin can no longer bind to them, and the muscle relaxes.

5. Overly contracted and overly stretched muscle fibers respond poorly to stimulation. A muscle responds best when it is slightly contracted before it is stimulated, so that there is optimal overlap between the resting thick and thin filaments. This is the *length–tension relationship. Muscle tone* maintains an optimal resting length and readiness to respond.

## Behavior of Whole Muscles (p. 422)

1. A stimulus must be of at least *threshold* strength to make a muscle contract. After a short *latent period,* the muscle responds to a single stimulus with a brief contraction called a *twitch.*

2. Muscle twitches vary in strength even when stimulus intensity is constant. Reasons for such variation include stimulus frequency, varying $Ca^{2+}$ concentration in the cytosol, the length–tension relationship, and the temperature, pH, fatigue, and hydration of the muscle.

3. A single twitch does no useful work for the body. In *recruitment,* however, multiple motor units are activated at once to produce a stronger muscle contraction. In high-frequency stimulation, successive twitches become progressively stronger; this is called *treppe* when the muscle completely relaxes between twitches and *incomplete tetanus* when it relaxes only partially and each twitch "piggybacks" on the previous ones to achieve greater tension.

4. In *isometric contraction,* a muscle develops tension without changing length; in *isotonic contraction,* it changes length while maintaining constant tension. In *concentric contraction,* a muscle maintains tension as it shortens; in *eccentric contraction,* it maintains tension as it lengthens.

## Muscle Metabolism (p. 426)

1. A muscle must have ATP in order to contract. It generates ATP by different mechanisms over the duration of a period of exercise.

2. At the outset, muscle uses oxygen from its myoglobin to generate ATP by aerobic respiration.

3. As the stored oxygen is depleted, muscle regenerates ATP from ADP by adding a phosphate ($P_i$) to it. It gets this $P_i$ either from another ADP, using the enzyme myokinase to transfer the phosphate, or from creatine phosphate, using the enzyme creatine kinase to do so. This is the *phosphagen system* for regenerating ATP.

4. Further into an exercise, as the phosphagen system is depleted, a muscle shifts to anaerobic fermentation (the *glycogen–lactic acid system*) to generate ATP.

5. Still later, the respiratory and circulatory systems may catch up with the demands of a muscle and deliver enough oxygen for aerobic respiration to meet the muscle's ATP demand.

6. Muscle fatigue results from several factors: ATP and ACh depletion, loss of membrane excitability, lactic acid accumulation, and central nervous system mechanisms.

7. The ability to maintain high-intensity exercise depends partly on one's *maximum oxygen uptake* ($VO_{2max}$), which varies with body size, age, sex, and physical condition.

8. Prolonged exercise produces an *oxygen debt* that is "repaid" by continued heavy breathing after the exercise is over. The extra $O_2$ breathed during this time goes mainly to restore oxygen reserves in the myoglobin and blood, replenish the phosphagen system, oxidize lactic acid, and meet the needs of a metabolic rate elevated by the high postexercise body temperature.

9. *Slow oxidative* muscle fibers are adapted for aerobic respiration and relatively resistant to fatigue, but produce relatively slow responses. *Fast glycolytic* muscle fibers respond more quickly but fatigue sooner. *Intermediate fibers* are relatively rare but combine fast responses with fatigue resistance.

10. The strength of a muscle depends on its size, fascicle arrangement, size of its motor units, multiple motor unit summation, temporal summation of twitches, prestimulation length, and fatigue.

11. Resistance exercise stimulates muscle growth and increases strength; endurance exercise increases fatigue resistance.

## Cardiac and Smooth Muscle (p. 431)

1. Cardiac muscle consists of relatively short, slightly branched, striated cells joined physically and electrically by *intercalated discs.*

2. Cardiac muscle is *autorhythmic* and thus contracts even without nervous stimulation.

3. Cardiac muscle is rich in myoglobin, glycogen, and large mitochondria; uses aerobic respiration almost exclusively; and is very fatigue-resistant.

4. Smooth muscle consists of short, fusiform, nonstriated cells.

5. Smooth muscle has no T tubules and little sarcoplasmic reticulum; it gets $Ca^{2+}$ from the extracellular fluid.

6. In multiunit smooth muscle, each cell is separately innervated by an autonomic nerve fiber and contracts independently. In single-unit smooth muscle, the muscle cells are connected by gap junctions and respond as a unit. Nerve fibers do not synapse with any specific muscle cells in the latter type.

7. In smooth muscle, $Ca^{2+}$ binds to calmodulin rather than troponin. This activates a kinase, which phosphorylates myosin and triggers contraction.

8. Smooth muscle lacks Z discs. Its myofilaments indirectly pull on *dense bodies* and cause the cell to contract in a twisting fashion.

9. Smooth muscle has a latch-bridge mechanism that enables it to maintain tonic contraction with little ATP expenditure.

10. Smooth muscle is not subject to the length–tension relationship. Its unusual ability to stretch and maintain responsiveness allows such organs as the stomach and urinary bladder to expand greatly without losing contractility.

# Testing Your Recall

1. To make a muscle contract more strongly, the nervous system can activate more motor units. This process is called
   a. recruitment.
   b. summation.
   c. incomplete tetanus.
   d. twitch.
   e. treppe.

2. The _____ is a depression in the sarcolemma that receives a motor nerve ending.
   a. T tubule
   b. terminal cisterna
   c. sarcomere
   d. motor end plate
   e. synapse

3. Before a muscle fiber can contract, ATP must bind to
   a. a Z disc.
   b. the myosin head.
   c. tropomyosin.
   d. troponin.
   e. actin.

4. Before a muscle fiber can contract, $Ca^{2+}$ must bind to
   a. calsequestrin.
   b. the myosin head.
   c. tropomyosin.
   d. troponin.
   e. actin.

5. Skeletal muscle fibers have _____, whereas smooth muscle cells do not.
   a. T tubules
   b. ACh receptors
   c. thick myofilaments
   d. thin myofilaments
   e. dense bodies

6. Smooth muscle cells have_____, whereas skeletal muscle fibers do not.
   a. sarcoplasmic reticulum
   b. tropomyosin
   c. calmodulin
   d. Z discs
   e. myosin ATPase

7. ACh receptors are found mainly in
   a. synaptic vesicles.
   b. terminal cisternae.
   c. thick filaments.
   d. thin filaments.
   e. junctional folds.

8. Single-unit smooth muscle cells can stimulate each other because they have
   a. a latch-bridge.
   b. diffuse junctions.
   c. gap junctions.
   d. tight junctions.
   e. cross-bridges.

9. A person with a high $V_{O_2max}$
   a. needs less oxygen than someone with a low $V_{O_2max}$.
   b. has stronger muscles than someone with a low $V_{O_2max}$.
   c. is less likely to show muscle tetanus than someone with a low $V_{O_2max}$.
   d. has fewer muscle mitochondria than someone with a low $V_{O_2max}$.
   e. experiences less muscle fatigue during exercise than someone with a low $V_{O_2max}$.

10. Slow oxidative fibers have all of the following except
    a. an abundance of myoglobin.
    b. an abundance of glycogen.
    c. high fatigue resistance.
    d. a red color.
    e. a high capacity to synthesize ATP aerobically.

11. The minimum stimulus intensity that will make a muscle contract is called _____.

12. A state of prolonged maximum contraction is called _____.

13. Parts of the sarcoplasmic reticulum called _____ lie on each side of a T tubule.

14. Thick myofilaments consist mainly of the protein _____.

15. The neurotransmitter that stimulates skeletal muscle is _____.

16. Muscle contains an oxygen-binding pigment called _____.

17. The _____ of skeletal muscle play the same role as dense bodies in smooth muscle.

18. In autonomic nerve fibers that stimulate single-unit smooth muscle, the neurotransmitter is contained in swellings called _____.

19. A state of continual partial muscle contraction is called _____.

20. _____ is an end product of anaerobic fermentation that causes muscle fatigue.

*Answers in Appendix B*

# True or False

*Determine which five of the following statements are false, and briefly explain why.*

1. Each motor neuron supplies just one muscle fiber.

2. To initiate muscle contraction, calcium ions must bind to the myosin heads.

3. Slow oxidative fibers are relatively resistant to fatigue.

4. Thin filaments are found in both the A bands and I bands of striated muscle.

5. Thin filaments do not change length when a muscle contracts.

6. Smooth muscle lacks striations because it does not have thick and thin myofilaments.

7. A muscle must contract to the point of complete tetanus if it is to move a load.

8. If no ATP were available to a muscle fiber, the excitation stage of muscle action could not occur.

9. For the first 30 seconds of an intense exercise, muscle gets most of its energy from lactic acid.

10. Cardiac and some smooth muscle are autorhythmic, but skeletal muscle is not.

*Answers in Appendix B*

# Testing Your Comprehension

1. Without ATP, relaxed muscle cannot contract and a contracted muscle cannot relax. Explain why.

2. Slight pH variations can cause enzymes to change conformation and can reduce enzyme activity. Explain how this relates to muscle fatigue.

3. Why would skeletal muscle be unsuitable for the wall of the urinary bladder? Explain how this illustrates the complementarity of form and function at a cellular and molecular level.

4. As skeletal muscle contracts, one or more bands of the sarcomere become narrower and disappear, and one or more of them remain the same width. Which bands will change—A, H, or I—and why?

5. Botulism occurs when a bacterium, *Clostridium botulinum*, releases a neurotoxin that prevents motor neurons from releasing ACh. In view of this, what early signs of botulism would you predict? Explain why a person with botulism could die of suffocation.

*Answers at www.mhhe.com/saladin4*

# www.mhhe.com/saladin4

*The textbook website provides a wealth of interactive study materials fully organized and integrated by chapter. You will find practice quizzes, labeling exercises, and much more that will complement your learning and understanding of anatomy and physiology. The website also includes tools designed to enhance your* **Anatomy & Physiology | REVEALED** *experience.*

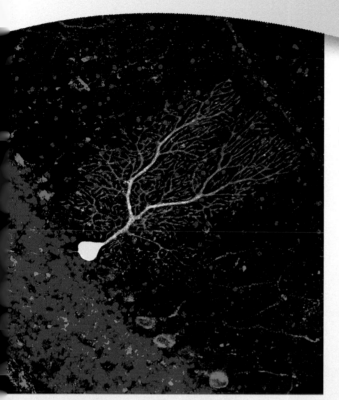

*A Purkinje cell, a neuron from the cerebellum of the brain*

# NERVOUS TISSUE

## CHAPTER OUTLINE

## INSIGHTS

## Brushing Up

To understand this chapter, it is important that you understand or brush up on the following concepts:

- Cations and anions (p. 58)
- Ligand- and voltage-regulated gates (p. 97)
- Cyclic AMP as a second messenger (p. 97)
- Simple diffusion (p. 102)
- Active transport and the sodium–potassium pump (p. 106)
- Basic structure of nerve cells (p. 171)

The next five chapters are concerned with the science of **neurobiology**—the study of the nervous system. This is a system of great complexity and mystery. It is the foundation of all our conscious experience, personality, and behavior. It profoundly intrigues biologists, physicians, psychologists, and even philosophers. Understanding this fascinating system is regarded by many as the ultimate challenge facing the behavioral and life sciences.

We begin our study at the simplest organizational level—the nerve cells *(neurons)* and cells called *neuroglia* that support their function in various ways. We then progress to the organ level to examine the spinal cord (chapter 13), brain (chapter 14), autonomic nervous system (chapter 15), and sense organs (chapter 16).

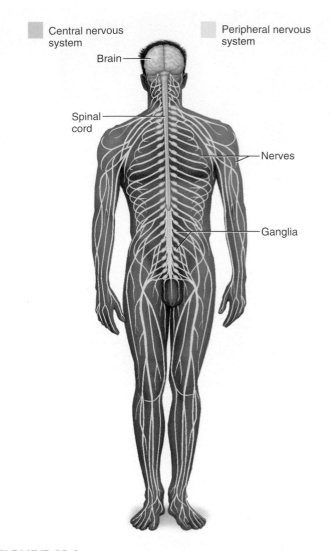

FIGURE 12.1 The Nervous System.

# Overview of the Nervous System

### Objectives
When you have completed this section, you should be able to

- describe the function of the nervous system;
- describe the major anatomical subdivisions of the nervous system;
- state the general functions of the nervous system and how these relate to the general classes of nerve cells; and
- describe the basic physiological properties of nerve cells that enable them to carry out their functions.

If the body is to maintain homeostasis and function effectively, its trillions of cells must work together in a coordinated fashion. If each cell behaved without regard to what others are doing, the result would be physiological chaos and death. We have two organ systems dedicated to maintaining internal coordination—the **endocrine system,** which communicates by means of chemical messengers (hormones) secreted into the blood, and the **nervous system** (fig. 12.1), which employs electrical and chemical means to send messages very quickly from cell to cell.

The nervous system carries out its coordinating task in three basic steps: (1) Through sense organs and simple sensory nerve endings, it receives information about changes in the body and the external environment and transmits coded messages to the spinal cord and brain. (2) The spinal cord and brain process this information, relate it to past experience, and determine what response, if any, is appropriate to the circumstances. (3) The spinal cord and brain issue commands primarily to muscle and gland cells to carry out such responses.

The nervous system has two major anatomical subdivisions (fig. 12.2):

- The **central nervous system (CNS)** consists of the brain and spinal cord, which are enclosed and protected by the cranium and vertebral column.
- The **peripheral nervous system (PNS)** consists of all the nervous system except the brain and spinal cord. It is composed of nerves and ganglia. A **nerve** is a bundle of nerve fibers (axons) wrapped in fibrous connective tissue. Nerves emerge from the CNS through foramina of the skull and vertebral column and carry signals to and from other organs of the body. A **ganglion**[1] (plural, *ganglia*) is a knotlike swelling in a nerve where the cell bodies of neurons are concentrated.

---

[1]*gangli* = knot

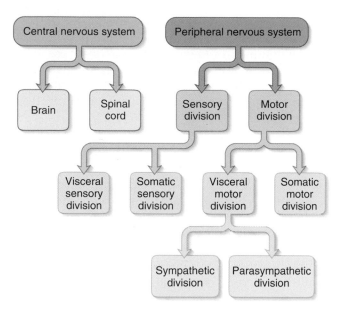

**FIGURE 12.2** Subdivisions of the Nervous System.

The peripheral nervous system is functionally divided into *sensory* and *motor* divisions, and each of these is further divided into *somatic* and *visceral* subdivisions.

- The **sensory (afferent[2]) division** carries sensory signals from various **receptors** (sense organs and simple sensory nerve endings) to the CNS. This is the pathway that informs the CNS of stimuli within and around the body.

    - The **somatic[3] sensory division** carries signals from receptors in the skin, muscles, bones, and joints.

    - The **visceral sensory division** carries signals mainly from the viscera of the thoracic and abdominal cavities, such as the heart, lungs, stomach, and urinary bladder.

- The **motor (efferent[4]) division** carries signals from the CNS to gland and muscle cells that carry out the body's responses. Cells and organs that respond to commands from the nervous system are called **effectors.**

    - The **somatic motor division** carries signals to the skeletal muscles. This output produces muscular contractions that are under voluntary control, as well as involuntary muscle contractions called *somatic reflexes.*

    - The **visceral motor division (autonomic[5]) nervous system** carries signals to glands, cardiac muscle, and smooth muscle. We usually have no voluntary control over these effectors, and this

system operates at an unconscious level. The responses of this system and its effectors are *visceral reflexes.* The autonomic nervous system has two further divisions:

- The **sympathetic division** tends to arouse the body for action, for example by accelerating the heartbeat and increasing respiratory airflow, but it inhibits digestion.

- The **parasympathetic division** tends to have a calming effect, slowing down the heartbeat, for example, but stimulating digestion.

The foregoing terminology may give the impression that the body has several nervous systems—central, peripheral, sensory, motor, somatic, and visceral. These are just terms of convenience, however. There is only one nervous system, and these subsystems are interconnected parts of the whole.

## Before You Go On

*Answer the following questions to test your understanding of the preceding section:*

1. *What is a receptor? Give two examples of effectors.*

2. *Distinguish between the central and peripheral nervous systems, and between visceral and somatic divisions of the sensory and motor systems.*

3. *What is another name for the visceral motor nervous system? What are the two subdivisions of this system?*

# Nerve Cells (Neurons)

### Objectives

When you have completed this section, you should be able to

- describe three functional properties found in all neurons;
- define the three most basic functional categories of neurons;
- identify the parts of a neuron; and
- explain how neurons transport materials between the cell body and tips of the axon.

## UNIVERSAL PROPERTIES

The communicative role of the nervous system is carried out by nerve cells, or **neurons.** These cells have three fundamental physiological properties that enable them to communicate with other cells:

1. **Excitability** (irritability). All cells are excitable—that is, they respond to environmental changes called **stimuli.** Neurons have developed this property to the highest degree.

---

[2]*af = ad* = toward + *fer* = to carry
[3]*somat* = body + *ic* = pertaining to
[4]*ef = ex* = out, away + *fer* = to carry
[5]*auto* = self + *nom* = law, governance

2. **Conductivity.** Neurons respond to stimuli by producing electrical signals that are quickly conducted to other cells at distant locations.

3. **Secretion.** When the electrical signal reaches the end of a nerve fiber, the neuron secretes a chemical *neurotransmitter* that crosses the gap and stimulates the next cell.

> **Think About It**
>
> *What basic physiological properties do a nerve cell and a skeletal muscle fiber have in common? Name a physiological property of each that the other one lacks.*

## FUNCTIONAL CLASSES

There are three general classes of neurons (fig. 12.3) corresponding to the three major aspects of nervous system function listed earlier:

1. **Sensory (afferent) neurons** are specialized to detect stimuli such as light, heat, pressure, and chemicals, and transmit information about them to the CNS.

Such neurons begin in almost every organ of the body and end in the CNS; the word *afferent* refers to signal conduction *toward* the CNS. Some receptors, such as those for pain and smell, are themselves neurons. In other cases, such as taste and hearing, the receptor is a separate cell that communicates directly with a sensory neuron.

2. **Interneurons**[6] (association neurons) lie entirely within the CNS. They receive signals from many other neurons and carry out the integrative function of the nervous system—that is, they process, store, and retrieve information and "make decisions" that determine how the body responds to stimuli. About 90% of our neurons are interneurons. The word *interneuron* refers to the fact that they lie *between,* and interconnect, the incoming sensory pathways and the outgoing motor pathways of the CNS.

3. **Motor (efferent) neurons** send signals predominantly to muscle and gland cells, the effectors that carry out the body's responses to stimuli. These neurons are called *motor* neurons because most of them lead to muscle cells, and *efferent* neurons to signify the signal conduction *away from* the CNS.

## STRUCTURE OF A NEURON

There are several varieties of neurons, as we shall see, but a good starting point for discussing neuron structure is a motor neuron of the spinal cord (fig. 12.4). The control center of the neuron is its **soma**,[7] also called the **cell body** or **perikaryon**[8] (PERR-ih-CARE-ee-on). It has a single, centrally located nucleus with a large nucleolus. The cytoplasm contains mitochondria, lysosomes, a Golgi complex, numerous inclusions, and an extensive rough endoplasmic reticulum and cytoskeleton. The cytoskeleton consists of a dense mesh of microtubules and **neurofibrils** (bundles of actin filaments) which compartmentalize the rough ER into dark-staining regions called **Nissl**[9] **bodies** (fig. 12.4c, d). Nissl bodies are unique to neurons and a helpful clue to identifying them in tissue sections with mixed cell types. Mature neurons have no centrioles and apparently undergo no further mitosis after adolescence; however, they are unusually long-lived cells, capable of functioning for over a hundred years. Even into old age there are unspecialized stem cells in the CNS that can divide and develop into new neurons (see Insight 4.2, p. 139).

The major cytoplasmic inclusions in a neuron are glycogen granules, lipid droplets, melanin, and a golden brown pigment called *lipofuscin*[10] (LIP-oh-FEW-sin), produced when lysosomes digest worn-out organelles and

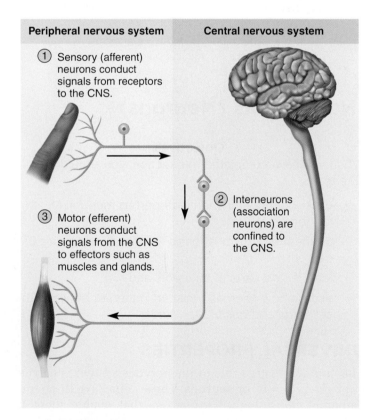

**Peripheral nervous system**    **Central nervous system**

① Sensory (afferent) neurons conduct signals from receptors to the CNS.

② Interneurons (association neurons) are confined to the CNS.

③ Motor (efferent) neurons conduct signals from the CNS to effectors such as muscles and glands.

**FIGURE 12.3   Functional Classes of Neurons.** Sensory (afferent) neurons carry signals to the central nervous system (CNS); interneurons are contained entirely within the CNS and carry signals from one neuron to another; and motor (efferent) neurons carry signals from the CNS to muscles and glands.

---

[6]*inter* = between
[7]*soma* = body
[8]*peri* = around + *karyo* = nucleus
[9]Franz Nissl (1860–1919), German neuropathologist
[10]*lipo* = fat, lipid + *fusc* = dusky, brown

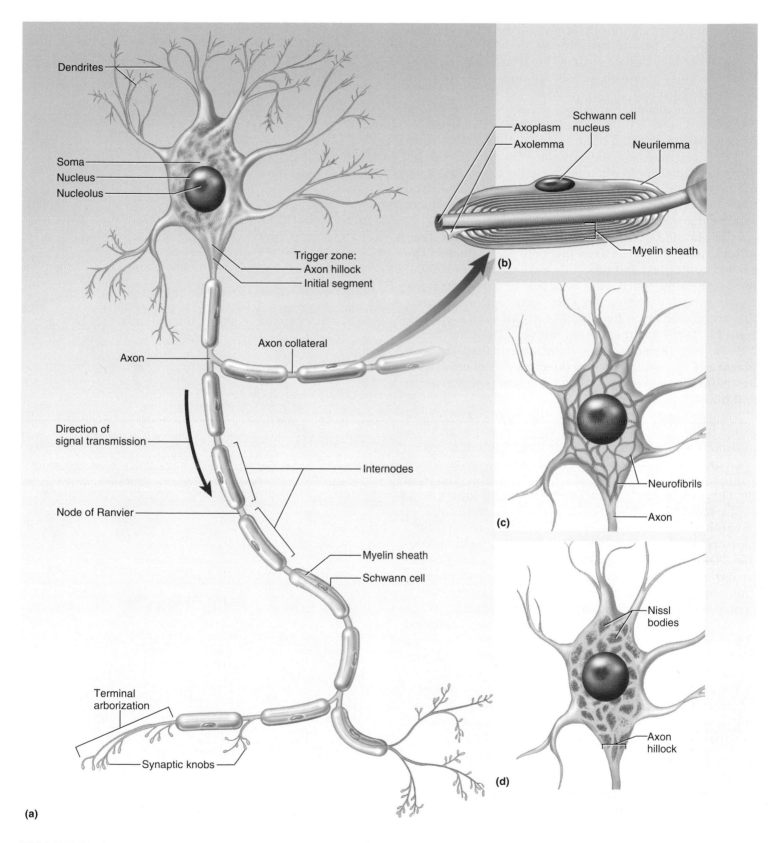

**FIGURE 12.4  A Representative Neuron.**   The Schwann cells and myelin sheath are explained later in this chapter. (a) A multipolar neuron such as a spinal motor neuron. (b) Detail of the myelin sheath. (c) Neurofibrils of the soma. (d) Nissl bodies, stained masses of rough ER separated by the bundles of neurofibrils shown in part (c).

other products. Lipofuscin accumulates with age and pushes the nucleus to one side of the cell. Lipofuscin granules are also called "wear-and-tear granules" because they are most abundant in old neurons, but they are apparently harmless.

The soma of a neuron usually gives rise to a few thick processes that branch into a vast number of **dendrites**[11]— named for their striking resemblance to the bare branches of a tree in winter. The dendrites are the primary site for receiving signals from other neurons. Some neurons have only one dendrite and some have thousands. The more dendrites a neuron has, the more information it can receive and incorporate into its decision making.

On one side of the soma is a mound called the **axon hillock,** from which the **axon** (nerve fiber) originates. The axon is cylindrical and relatively unbranched for most of its length, although it may give rise to a few branches called *axon collaterals* along the way, and most axons branch extensively at their distal end. An axon is specialized for rapid conduction of nerve signals to points remote from the soma. Its cytoplasm is called the **axoplasm** and its membrane the **axolemma.**[12] A neuron never has more than one axon, and some neurons in the retina and brain have none.

Somas range from 5 to 135 μm in diameter, while axons range from 1 to 20 μm in diameter and from a few millimeters to more than a meter long. Such dimensions are more impressive when we scale them up to the size of familiar objects. If the soma of a spinal motor neuron were the size of a tennis ball, its dendrites would form a huge bushy mass that could fill a 30-seat classroom from floor to ceiling. Its axon would be up to a mile long but a little narrower than a garden hose. This is quite a point to ponder. The neuron must assemble molecules and organelles in its "tennis ball" soma and deliver them through its "mile-long garden hose" to the end of the axon. How it achieves this remarkable feat is explained shortly.

At the distal end, an axon usually has a **terminal arborization**[13]—an extensive complex of fine branches. Each branch ends in a **synaptic knob** (terminal button), a little swelling that forms a junction **(synapse**[14]**)** with the next cell. It contains **synaptic vesicles** full of neurotransmitter.

Not all neurons fit the preceding description. Neurons are classified structurally according to the number of processes extending from the soma (fig. 12.5):

- **Multipolar neurons** are those, like the preceding, that have one axon and multiple dendrites. This is the most common type and includes most neurons of the brain and spinal cord.

---

[11]*dendr* = tree, branch + *ite* = little
[12]*axo* = axis, axon + *lemma* = husk, peel, sheath
[13]*arbor* = tree
[14]*syn* = together + *aps* = to touch, join

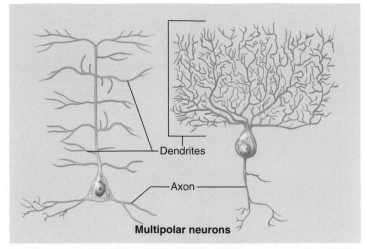

**Multipolar neurons**

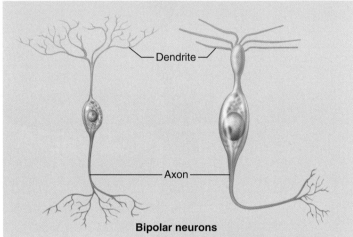

**Bipolar neurons**

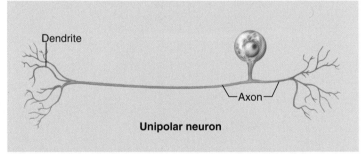

**Unipolar neuron**

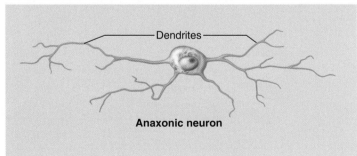

**Anaxonic neuron**

**FIGURE 12.5** **Variation in Neuron Structure.** *Top row, left to right:* Two multipolar neurons of the brain: a pyramidal cell and Purkinje cell. *Second row, left to right:* Two bipolar neurons: a bipolar cell of the retina and an olfactory neuron. *Third row:* A unipolar neuron of the type involved in the senses of touch and pain. *Bottom row:* An anaxonic neuron (amacrine cell) of the retina.

- **Bipolar neurons** have one axon and one dendrite. Examples include olfactory cells of the nasal cavity, some neurons of the retina, and sensory neurons of the inner ear.

- **Unipolar neurons** have only a single process leading away from the soma. They are represented by the neurons that carry sensory signals to the spinal cord. These neurons are also called *pseudounipolar* because they start out as bipolar neurons in the embryo, but their two processes fuse into one as the neuron matures. A short distance away from the soma, the process branches like a T, with a *peripheral fiber* carrying signals from the source of sensation and a *central fiber* continuing into the spinal cord. In most other neurons, a dendrite carries signals toward the soma and an axon carries them away. In unipolar neurons, however, there is one long fiber that bypasses the soma and carries nerve signals directly to the spinal cord. The dendrites are the branching receptive endings in the skin or other place of origin, while the rest of the fiber is considered to be the axon (defined in these neurons by the presence of myelin and the ability to generate action potentials—two concepts explained later in this chapter).

- **Anaxonic neurons** have multiple dendrites but no axon. They communicate through their dendrites and do not produce action potentials. Some anaxonic neurons are found in the brain, retina, and adrenal medulla. In the retina, they help in visual processes such as the perception of contrast.

## AXONAL TRANSPORT

All of the proteins needed by a neuron must be made in the soma, where the protein-synthesizing organelles such as the nucleus, ribosomes, and rough endoplasmic reticulum are located. Yet many of these proteins are needed in the axon, for example to repair and maintain the axolemma, to serve as ion gates in the membrane, or to act in the synaptic knob as enzymes and signaling molecules. Other substances are transported from the axon terminals back to the soma for disposal or recycling. The two-way passage of proteins, organelles, and other materials along an axon is called **axonal transport.** Movement away from the soma down the axon is called **anterograde**[15] **transport** and movement up the axon toward the soma is called **retrograde**[16] **transport.**

Materials travel along microtubules of the cytoskeleton, which act like railroad tracks to guide them to their destination. But what is the "motor" that drives them along the tracks? Anterograde transport employs a motor protein called *kinesin,*[17] while retrograde transport uses one called *dynein*[18] (the same protein we encountered earlier in cilia and flagella; see chapter 3). These proteins carry materials "on their backs" while they reach out, like the myosin heads of muscle (see chapter 11), to bind repeatedly to the microtubules and crawl along them.

There are two types of axonal transport: fast and slow.

1. **Fast axonal transport** occurs at a rate of 20 to 400 mm/day and may be either anterograde or retrograde:
   - *Fast anterograde transport* moves mitochondria, synaptic vesicles, other organelles, components of the axolemma, calcium ions, enzymes such as acetylcholinesterase, and small molecules such as glucose, amino acids, and nucleotides.
   - *Fast retrograde transport* returns used synaptic vesicles and other materials to the soma and informs the soma of conditions at the axon terminals. Some pathogens exploit this process to invade the nervous system. They enter the distal tips of an axon and travel to the soma by retrograde transport. Examples include the herpes simplex, rabies, and polio viruses and tetanus toxin. In such infections, the delay between infection and the onset of symptoms corresponds to the time needed for the pathogens to reach the somas.

2. **Slow axonal transport,** also called *axoplasmic flow,* occurs at a rate of 0.5 to 10 mm/day and is always anterograde. It moves enzymes and cytoskeletal components down the axon, renews worn-out axoplasmic components in mature neurons, and supplies new axoplasm for developing or regenerating neurons. Damaged nerve fibers regenerate at a speed governed by slow axonal transport.

### Before You Go On

*Answer the following questions to test your understanding of the preceding section:*

4. *Sketch a multipolar neuron and label its soma, dendrites, axon, terminal arborization, synaptic knobs, myelin sheath, and nodes of Ranvier.*

5. *Explain the difference between a sensory neuron, motor neuron, and interneuron.*

6. *What is the functional difference between a dendrite and an axon?*

7. *How do proteins and other chemicals synthesized in the soma get to the synaptic knobs? By what process can a virus that invades a peripheral nerve fiber get to the soma of that neuron?*

---

[15]*antero* = forward + *grad* = to walk, to step
[16]*retro* = back + *grad* = to walk, to step

[17]*kines* = motion + *in* = protein
[18]*dyne* = force + *in* = protein

# Supportive Cells (Neuroglia)

### Objectives

When you have completed this section, you should be able to

- name the cells that aid neurons and state their functions;
- describe the myelin sheath that is found around certain nerve fibers and explain its importance;
- describe the relationship of unmyelinated nerve fibers to their supportive cells; and
- explain how damaged nerve fibers regenerate.

There are about a trillion ($10^{12}$) neurons in the nervous system—10 times as many neurons in your body as there are stars in our galaxy! Yet they are outnumbered as much as 50 to 1 by supportive cells called **neuroglia** (noo-ROG-lee-uh), or **glial** (GLEE-ul) **cells.** Glial cells protect the neurons and help them function. The word *glia,* which means "glue," implies one of their roles—to bind neurons together and provide a supportive framework for the nervous tissue. In the fetus, glial cells form a scaffold that guides young migrating neurons to their destinations. Wherever a mature neuron is not in synaptic contact with another cell, it is covered with glial cells. This prevents neurons from contacting each other except at points specialized for signal transmission, and thus gives precision to their conduction pathways.

## TYPES OF NEUROGLIA

There are six kinds of neuroglia, each with a unique function (table 12.1). Four types occur in the central nervous system (fig. 12.6):

1. **Oligodendrocytes**[19] (OL-ih-go-DEN-dro-sites) somewhat resemble an octopus; they have a bulbous body with as many as 15 armlike processes. Each process reaches out to a nerve fiber and spirals around it like electrical tape wrapped repeatedly around a wire. This spiral wrapping, called the *myelin sheath,* insulates the nerve fiber from the extracellular fluid. For reasons explained later, it speeds up signal conduction in the nerve fiber.

2. **Ependymal**[20] (ep-EN-dih-mul) **cells** resemble a cuboidal epithelium lining the internal cavities of the brain and spinal cord. Unlike epithelial cells, however, they have no basement membrane and they exhibit rootlike processes that penetrate into the underlying nervous tissue. Ependymal cells produce *cerebrospinal fluid (CSF),* a clear liquid that bathes

| TABLE 12.1 | Types of Glial Cells |
|---|---|
| **Types** | **Functions** |
| ***Neuroglia of CNS*** | |
| Oligodendrocytes | Form myelin in brain and spinal cord |
| Ependymal cells | Line cavities of brain and spinal cord; secrete and circulate cerebrospinal fluid |
| Microglia | Phagocytize and destroy microorganisms, foreign matter, and dead nervous tissue |
| Astrocytes | Cover brain surface and nonsynaptic regions of neurons; form supportive framework in CNS; induce formation of blood–brain barrier; nourish neurons; produce growth factors that stimulate neurons; communicate electrically with neurons and may influence synaptic signalling; remove K$^+$ and some neurotransmitters from ECF of brain and spinal cord; help to regulate composition of ECF; form scar tissue to replace damaged nervous tissue |
| ***Neuroglia of PNS*** | |
| Schwann cells | Form neurilemma around all PNS nerve fibers and myelin around most of them; aid in regeneration of damaged nerve fibers |
| Satellite cells | Surround somas of neurons in the ganglia; function uncertain |

the CNS and fills its internal cavities. They have patches of cilia on their apical surfaces that help to circulate the CSF. Ependymal cells and CSF are considered in more detail in chapter 14.

3. **Microglia** are small macrophages that develop from white blood cells called monocytes. They wander through the CNS and phagocytize dead nervous tissue, microorganisms, and other foreign matter. They become concentrated in areas damaged by infection, trauma, or stroke. Pathologists look for clusters of microglia in brain tissue as a clue to sites of injury.

4. **Astrocytes**[21] are the most abundant glial cells in the CNS and constitute over 90% of the tissue in some areas of the brain. They cover the entire brain surface and most nonsynaptic regions of the neurons in the gray matter of the CNS. They are named for their many-branched, somewhat starlike shape. They have the most diverse functions of any glia:

- They form a supportive framework for the nervous tissue.
- They have extensions called *perivascular feet,* which contact the blood capillaries and stimulate

---

[19]*oligo* = few + *dendro* = branches + *cyte* = cell
[20]*ependyma* = upper garment

[21]*astro* = star + *cyte* = cell

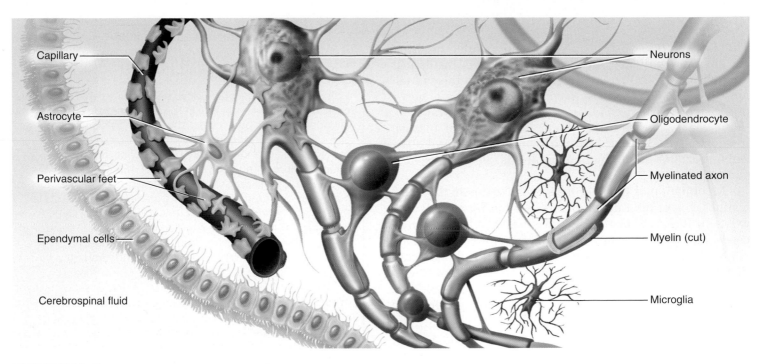

**FIGURE 12.6** Neuroglia of the Central Nervous System.

them to form a tight seal called the *blood–brain barrier.* This barrier isolates the blood from the brain tissue and limits what substances are able to get to the brain cells, thus protecting the neurons (see chapter 14).

- They convert blood glucose to lactate and supply this to the neurons for nourishment.

- They secrete proteins called *nerve growth factors* (see Insight 12.3) that promote neuron growth and synapse formation.

- They communicate electrically with neurons and may influence future synaptic signaling between neurons.

- They regulate the chemical composition of the tissue fluid. When neurons transmit signals, they release neurotransmitters and potassium ions. Astrocytes absorb these substances and prevent them from reaching excessive levels in the tissue fluid.

- When neurons are damaged, astrocytes form hardened scar tissue and fill space formerly occupied by the neurons. This process is called *astrocytosis* or *sclerosis.*

The other two types of glial cells occur in the peripheral nervous system:

5. **Schwann**[22] (shwon) **cells** envelop nerve fibers of the PNS. In most cases, a Schwann cell winds repeat-edly around a nerve fiber and produces a myelin sheath similar to the one produced by oligodendrocytes in the CNS. There are some important differences between the CNS and PNS in the way myelin is produced, which we consider shortly. In addition to myelinating peripheral nerve fibers, Schwann cells assist in the regeneration of damaged fibers, which also is discussed later.

6. **Satellite cells** surround the neuron cell bodies in ganglia of the PNS. Little is known of their function.

[22]Theodor Schwann (1810–82), German histologist

[23]*glia* = glial cells + *oma* = tumor

# MYELIN

The **myelin** (MY-eh-lin) **sheath** is an insulating layer around a nerve fiber, somewhat like the rubber insulation on a wire. It is formed by oligodendrocytes in the central nervous system and Schwann cells in the peripheral nervous system. Since it consists of the plasma membranes of these glial cells, its composition is like that of plasma membranes in general. It is about 20% protein and 80% lipid, the latter including phospholipids, glycolipids, and cholesterol.

Myelination of the nervous system begins in the fourteenth week of fetal development, yet hardly any myelin exists in the brain at the time of birth. Myelination proceeds rapidly in infancy and isn't completed until late adolescence. Since myelin has such a high lipid content, dietary fat is important to early nervous system development. It is best not to give children under 2 years old the sort of low-fat diets (skimmed milk, etc.) that may be beneficial to an adult.

In the PNS, a Schwann cell spirals repeatedly around a single nerve fiber, laying down as many as a hundred compact layers of its own membrane with almost no cytoplasm between the membranes (fig. 12.7a). These layers constitute the myelin sheath. The Schwann cell spirals outward as it wraps the nerve fiber, finally ending with a thick outermost coil called the **neurilemma**[24] (noor-ih-LEM-ah). Here, the bulging body of the Schwann cell contains its nucleus and most of its cytoplasm. External to the neurilemma is a basal lamina and then a thin sleeve of fibrous connective tissue called the *endoneurium.* To visualize this myelination process, imagine that you wrap an almost-empty tube of toothpaste tightly around a pencil. The pencil represents the axon, and the spiral layers of toothpaste tube (with the toothpaste squeezed out) represent the myelin. The toothpaste would be forced to one end of the tube, which would form a bulge on the external surface of the wrapping, like the body of the Schwann cell.

In the CNS, each oligodendrocyte reaches out to myelinate several nerve fibers in its immediate vicinity (fig. 12.7b). Since it is anchored to multiple nerve fibers, it cannot migrate around any one of them like a Schwann cell does. It must push newer layers of myelin under the older ones, so myelination spirals inward toward the nerve fiber. Nerve fibers of the CNS have no neurilemma or endoneurium.

In both the PNS and CNS, a nerve fiber is much longer than the reach of a single glial cell, so it requires many Schwann cells or oligodendrocytes to cover one nerve fiber. Consequently, the myelin sheath is segmented. The gaps between the segments are called **nodes of Ranvier**[25]

---

## INSIGHT 12.2    Clinical Application

### Diseases of the Myelin Sheath

Multiple sclerosis and Tay–Sachs disease are degenerative disorders of the myelin sheath. In *multiple sclerosis*[26] *(MS),* the oligodendrocytes and myelin sheaths of the CNS deteriorate and are replaced by hardened scar tissue, especially between the ages of 20 and 40. Nerve conduction is disrupted, with effects that depend on what part of the CNS is involved—double vision, blindness, speech defects, neurosis, tremors, or numbness, for example. Patients experience variable cycles of milder and worse symptoms until they eventually become bedridden. Most die from 7 to 32 years after the onset of the disease. The cause of MS remains uncertain; most theories suggest that it is an autoimmune disorder triggered by a virus in genetically susceptible individuals. There is no cure.

*Tay–Sachs*[27] disease is a hereditary disorder seen mainly in infants of Eastern European Jewish ancestry. It results from the abnormal accumulation of a glycolipid called $GM_2$ (ganglioside) in the myelin sheath. $GM_2$ is normally decomposed by a lysosomal enzyme, but this enzyme is lacking from people who are homozygous recessive for the Tay–Sachs allele. As $GM_2$ accumulates, it disrupts the conduction of nerve signals and the victim typically suffers blindness, loss of coordination, and dementia. Signs begin to appear before the child is a year old, and most victims die by the age of 3 or 4. Asymptomatic adult carriers can be identified by a blood test and advised by genetic counselors on the risk of their children having the disease.

---

(RON-vee-AY), and the myelin-covered segments from one gap to the next are called **internodes** (see fig. 12.4). The internodes are about 0.2 to 1.0 mm long. The short section of nerve fiber between the axon hillock and the first glial cell is called the **initial segment.** Since the axon hillock and initial segment play an important role in initiating a nerve signal, they are collectively called the **trigger zone.**

## UNMYELINATED NERVE FIBERS

Many nerve fibers in the CNS and PNS are unmyelinated. In the PNS, however, even the unmyelinated fibers are enveloped in Schwann cells. In this case, one Schwann cell harbors from 1 to 12 small nerve fibers in grooves in its surface (fig. 12.8). The Schwann cell's plasma membrane does not spiral repeatedly around the fiber as it does in a myelin sheath, but folds once around each fiber and somewhat overlaps itself along the edges. This wrapping is the neurilemma. A basal lamina surrounds the entire Schwann cell along with its nerve fibers.

---

[24]*neuri* = nerve + *lemma* = husk, peel, sheath
[25]L. A. Ranvier (1835–1922), French histologist and pathologist

[26]*scler* = hard, tough + *osis* = condition
[27]Warren Tay (1843–1927), English physician; Bernard Sachs (1858–1944), American neurologist

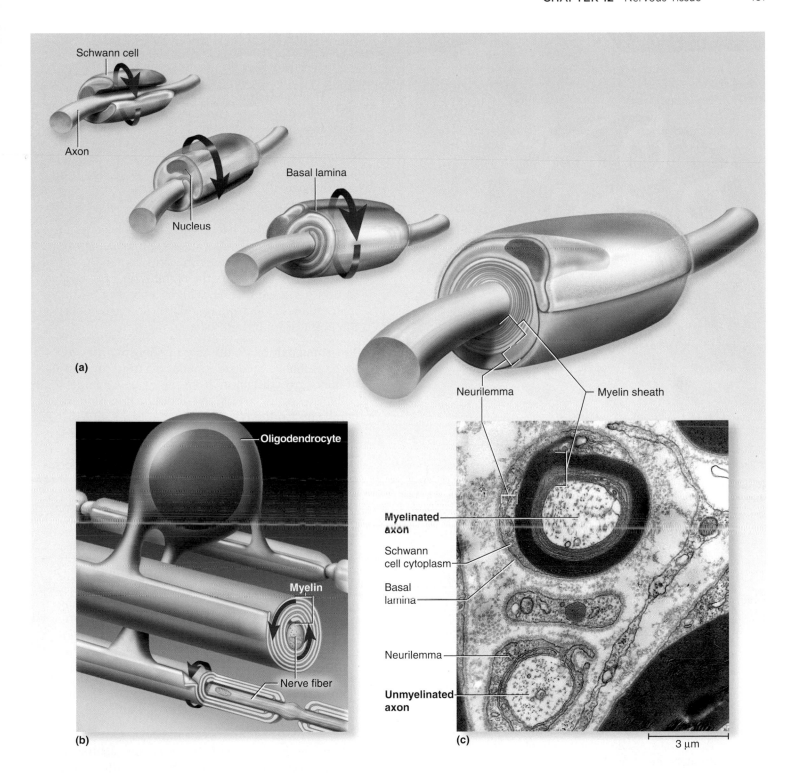

**FIGURE 12.7 Myelination.** (a) A Schwann cell of the PNS, wrapping repeatedly around an axon to form the multilayered myelin sheath. The myelin spirals outward away from the axon as it is laid down. The outermost coil of the Schwann cell constitutes the neurilemma. (b) An oligodendrocyte of the CNS wrapping around the axons of multiple neurons. Here, the myelin spirals inward toward the axon as it is laid down. (c) A myelinated axon (top) and unmyelinated axon (bottom) (TEM).

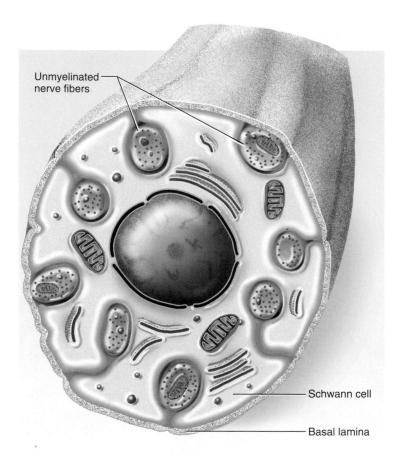

Unmyelinated nerve fibers

Schwann cell

Basal lamina

**FIGURE 12.8**  **Unmyelinated Nerve Fibers.**  Multiple unmyelinated fibers are enclosed in channels in the surface of a single Schwann cell.

## CONDUCTION SPEED OF NERVE FIBERS

The speed at which a nerve signal travels along a nerve fiber depends on two factors: the diameter of the fiber and the presence or absence of myelin. Signal conduction occurs along the surface of a fiber, not deep within its axoplasm. Large fibers have more surface area and conduct signals more rapidly than small fibers. Myelin further speeds signal conduction for reasons discussed later. Nerve signals travel about 0.5 to 2.0 m/sec in small unmyelinated fibers (2–4 µm in diameter) and 3 to 15 m/sec in myelinated fibers of the same size. In large myelinated fibers (up to 20 µm in diameter) they travel as fast as 120 m/sec. One might wonder why all of our nerve fibers are not large, myelinated, and fast, but if this were so, our nervous system would be impossibly bulky or limited to far fewer fibers. Slow unmyelinated fibers are quite sufficient for processes in which quick responses are not particularly important, such as secreting stomach acid or dilating the pupil. Fast myelinated fibers are employed where speed is more important, as in motor commands to the skeletal muscles and sensory signals for vision and balance.

## REGENERATION OF NERVE FIBERS

Nerve fibers of the PNS are vulnerable to cuts, crushing injuries, and other trauma. A damaged peripheral nerve fiber may regenerate, however, if its soma is intact and at least some neurilemma remains. Figure 12.9 shows the process of regeneration, taking as its example a somatic motor neuron:

① In the normal nerve fiber, note the size of the soma and the size of the muscle fibers for comparison to later stages.

② When a nerve fiber is cut, the fiber distal to the injury cannot survive because it is incapable of protein synthesis. Protein-synthesizing organelles are limited to the soma. As the distal fiber degenerates, so do its Schwann cells, which depend on it for their maintenance. Macrophages clean up tissue debris at the point of injury and beyond.

③ The soma exhibits a number of abnormalities of its own, probably because it is cut off from the supply of nerve growth factors from the neuron's target cells. The soma swells, the endoplasmic reticulum breaks up (so the Nissl bodies disperse), and the nucleus moves off center. Not all damaged neurons survive; some die at this stage. But often, the axon stump sprouts multiple growth processes as the severed distal end shows continued degeneration of the axon and Schwann cells. Muscle fibers deprived of their nerve supply exhibit a shrinkage called *denervation atrophy*.

④ Near the injury, Schwann cells, the basal lamina, and the neurilemma form a **regeneration tube.** The Schwann cells produce cell-adhesion molecules and *nerve growth factors* (see Insight 12.3) that enable a neuron to regrow to its original destination. When one growth process finds its way into the tube, it grows rapidly (3–5 mm/day), and the other growth processes are retracted.

⑤ The regeneration tube guides the growing sprout back to the original target cells, reestablishing synaptic contact.

⑥ When contact is established, the soma shrinks and returns to its original appearance, and the reinnervated muscle fibers regrow.

Regeneration is not perfect. Some nerve fibers connect to the wrong muscle fibers or never find a muscle fiber at all, and some damaged motor neurons simply die. Nerve injury is therefore often followed by some degree of loss of fine motor control. Damaged nerve fibers in the CNS cannot regenerate at all, but since the CNS is enclosed in bone, it suffers less trauma than the PNS.

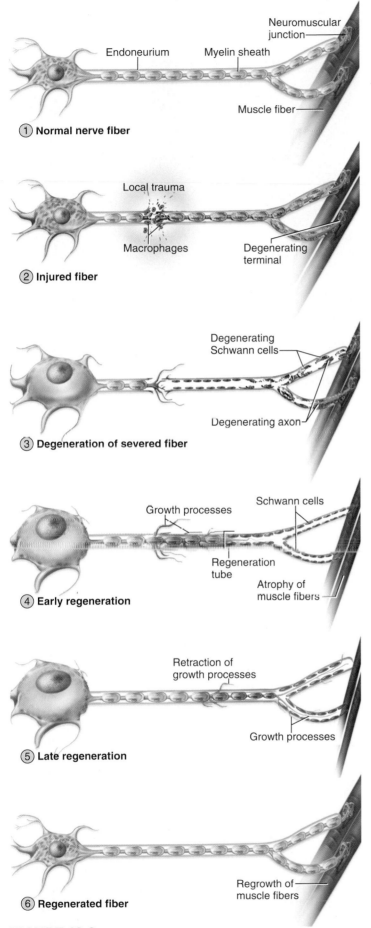

① Normal nerve fiber

Endoneurium  Myelin sheath  Neuromuscular junction

Muscle fiber

② Injured fiber

Local trauma

Macrophages  Degenerating terminal

③ Degeneration of severed fiber

Degenerating Schwann cells

Degenerating axon

④ Early regeneration

Growth processes  Schwann cells

Regeneration tube

Atrophy of muscle fibers

⑤ Late regeneration

Retraction of growth processes

Growth processes

⑥ Regenerated fiber

Regrowth of muscle fibers

**FIGURE 12.9  Regeneration of a Damaged Nerve Fiber.**  See text for explanation.

┌ **Think About It**

*What features essential to regeneration, present in the PNS, are lacking from the CNS?*

## Before You Go On

*Answer the following questions to test your understanding of the preceding section:*

8. How is a glial cell different from a neuron? List the six types of glial cells and discuss their functions.

9. How is myelin produced? How does myelin production in the CNS differ from that in the PNS?

10. How can a severed peripheral nerve fiber find its way back to the cells it originally innervated?

# Electrophysiology of Neurons

### Objectives

When you have completed this section, you should be able to

- explain why a cell has an electrical charge difference (voltage) across its membrane;
- explain how stimulation of a neuron causes a local electrical response in its membrane;
- explain how local responses generate a nerve signal; and
- explain how the nerve signal is transmitted down an axon.

The nervous system has intrigued scientists and philosophers since ancient times. The Roman physician Galen thought that the brain pumped a vapor called *psychic pneuma* through hollow nerves and squirted it into the muscles to make them contract. The French philosopher René Descartes still argued for this theory in the seventeenth century. It finally fell out of favor in the eighteenth century, when Luigi Galvani discovered the role of electricity in muscle contraction. Further progress had to await improvements in microscope technology and histological staining methods. Italian histologist Camillo Golgi (1843–1926) developed an important method for the staining of neurons with silver. This enabled Spanish histologist Santiago Ramón y Cajal (1852–1934), with tremendous skill and patience, to trace the course of nerve fibers through tissue sections. He demonstrated that the nervous pathway was not a continuous "wire" or tube, but a series of cells separated by the gaps we now call synapses. Golgi and Ramón y Cajal shared the 1906 Nobel Prize for Physiology or Medicine for this important discovery.

Ramón y Cajal's theory, now called the *neuron doctrine,* suggested another direction for research—how do neurons communicate? Two key issues in neurophysiology are (1) How does a neuron generate an electrical signal and (2) How does it transmit a meaningful message

## Nerve Growth Factor—From Bedroom Science to Nobel Prize

It is remarkable what odds can be overcome by intelligent self-confidence. Such a case was neurobiologist Rita Levi-Montalcini (1909–    ) (fig. 12.10). Born of a cultured and accomplished Italian Jewish family, Rita and her twin sister Paola aspired to attend a college-preparatory high school. Their traditional father, however, believed the only fitting roles for women were wife and mother; he sent them instead to a high school with no mathematics or science classes for girls. The girls decided to attend university anyway, and hired their own tutor to prepare them for the admissions exam. Rita entered medical school at Turin in 1930, graduated *summa cum laude* in medicine and surgery, and embarked on advanced study in psychiatry and neurology. Paola became a renowned artist.

But then arose the sinister specter of antisemitism, as fascist dictator Mussolini issued a 1936 decree barring Jews from academic or professional careers. Levi-Montalcini returned to live with her family, despairing of being able to pursue medical practice or research. A visiting college friend, however, reminded her of how much the neuroanatomist Ramón y Cajal had done under very primitive conditions. That inspired her to set up a little laboratory in her bedroom, where she studied nervous system development in chick embryos. She had read of work by Viktor Hamburger at Washington University in St. Louis, Missouri, showing that early removal of the limbs from chick embryos stopped the growth of motor nerves from the spinal cord. Hamburger believed the limb tissues must secrete a chemical that attracts nerves to grow into them. Levi-Montalcini, however, believed that nerves start to grow into the limb stumps, but soon die for lack of a substance needed to sustain them.

Fleeing first from Allied bombing and then from Hitler's invasion of Italy, Levi-Montalcini had to abandon her work and go into hiding with her family until the end of the war. At war's end, Hamburger invited her to join him at Washington University, where they found her hypothesis to be correct. Levi-Montalcini and Stanley Cohen isolated the nerve-sustaining substance in the 1950s and named it *nerve growth factor* (NGF).

**FIGURE 12.10**   Rita Levi-Montalcini (1909–    ).

NGF is a protein secreted by gland and muscle cells and picked up by the axon terminals of neurons. It prevents apoptosis (programmed cell death) in growing neurons and thus enables them to establish connections with their target cells.

Levi-Montalcini divided the rest of her career between the United States and Italy. In 1986, the work first begun under wartime duress in a humble bedroom laboratory, leading to the identification of NGF at St. Louis, earned her and Cohen the Nobel Prize for Physiology or Medicine.

NGF was the first of many cell growth factors discovered, and launched what is today a vibrant field of biomedical research in the use of growth factors to stimulate tissue development and repair (see Insight 12.4 on the use of NGF in Alzheimer disease, for example). Levi-Montalcini has followed this discovery with other pioneering work in apoptosis. She and her sister Paola also created the Levi-Montalcini Foundation to support the career development of young people and especially the scientific education of women in Africa.

to the next cell? These are the questions to which this section and the next are addressed.

## ELECTRICAL POTENTIALS AND CURRENTS

Neural communication, like muscle excitation, is based on electrophysiology—cellular mechanisms for producing electrical potentials and currents. An **electrical potential** is a difference in the concentration of charged particles between one point and another. It is a form of potential energy that, under the right circumstances, can produce a current. An electrical **current** is a flow of charged particles from the one point to another. A new flashlight battery, for example, typically has a potential, or charge, of 1.5 volts (V). If the two poles of the battery are connected by a wire, electrons flow through the wire from one pole of the battery to the other, creating a current that lights the bulb. As long as the battery has a potential, we say it is **polarized.**

Living cells are also polarized. As we saw in chapter 11, the charge difference across the plasma membrane is called the **resting membrane potential (RMP).** It is much less than the potential of a flashlight battery—typically about −70 millivolts (mV) in an unstimulated, "resting" neuron. The negative value means there are more negatively charged particles on the inside of the membrane than on the outside.

We do not have free electrons in the body as we do in an electrical circuit. Electrical currents in the body are created, instead, by the flow of ions such as Na⁺ and K⁺ through gated channels in the plasma membrane. As we saw in chapters 3 and 11, gated channels can be opened and closed by various stimuli. This enables cells to turn electrical currents on and off.

## THE RESTING MEMBRANE POTENTIAL

The reason that a cell has a resting membrane potential is that electrolytes are unequally distributed between the extracellular fluid (ECF) on the outside of the plasma membrane and the intracellular fluid (ICF) on the inside. The RMP results from the combined effect of three factors: (1) the diffusion of ions down their concentration gradients through the plasma membrane; (2) selective permeability of the plasma membrane, allowing some ions to pass more easily than others; and (3) the electrical attraction of cations and anions to each other.

Potassium ions (K⁺) have the greatest influence on the RMP, because the plasma membrane is more permeable to K⁺ than to any other ion. Imagine a hypothetical cell in which all the K⁺ starts out in the ICF, with none in the ECF. Also in the ICF are a number of cytoplasmic anions that cannot escape from the cell because of their size or charge—phosphates, sulfates, small organic acids, proteins, ATP, and RNA. Assuming K⁺ can diffuse freely through channels in the plasma membrane, it diffuses down its concentration gradient and out of the cell, leaving these cytoplasmic anions behind (fig. 12.11). As a result, the ICF grows more and more negatively charged. But as the ICF becomes more negative, it exerts a stronger attraction for the positive potassium ions and attracts

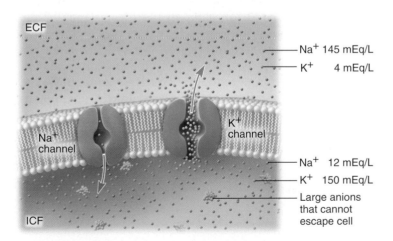

**FIGURE 12.11 Ionic Basis of the Resting Membrane Potential.** Note that sodium ions are much more concentrated in the extracellular fluid (ECF) than in the intracellular fluid (ICF), while potassium ions are more concentrated in the ICF. Large anions unable to penetrate the plasma membrane give the cytoplasm a negative charge relative to the ECF.

▶ If we suddenly increased the concentration of Cl⁻ ions in the ICF, would the membrane potential become higher or lower?

some of them back into the cell. Eventually an *equilibrium* (balance) is reached in which K⁺ is moving out of the cell (diffusion down its concentration gradient) and into the cell (by electrical attraction) at equal rates. The *net* diffusion of K⁺ then stops. At the point of equilibrium, K⁺ is about 40 times as concentrated in the ICF as in the ECF.

If K⁺ were the only ion affecting the RMP, it would give the membrane a potential of about −90 mV. However, sodium ions (Na⁺) also enter the picture. Sodium is about 12 times as concentrated in the ECF as in the ICF. The resting plasma membrane is much less permeable to Na⁺ than to K⁺, but Na⁺ does diffuse down its concentration gradient into the cell, attracted by the negative charge in the ICF. This sodium leak is only a trickle (fig. 12.11), but it is enough to neutralize some of the negative charge and reduce the voltage across the membrane.

Sodium leaks into the cell and potassium leaks out, but the sodium–potassium (Na⁺–K⁺) pump described in chapter 3 continually compensates for this leakage. It pumps 3 Na⁺ out of the cell for every 2 K⁺ it brings in, consuming 1 ATP for each exchange cycle. By removing more cations from the cell than it brings in, it contributes about −3 mV to the resting membrane potential. The net effect of all this—K⁺ diffusion out of the cell, Na⁺ diffusion inward, and the Na⁺–K⁺ pump—is the resting membrane potential of −70 mV.

The Na⁺–K⁺ pump accounts for about 70% of the energy (ATP) requirement of the nervous system. Every signal generated by a neuron slightly upsets the distribution of Na⁺ and K⁺, so the pump must work continually to restore equilibrium. This is why nervous tissue has one of the highest rates of ATP consumption of any tissue in the body, and why it demands so much glucose and oxygen. Although a neuron is said to be resting when it is not producing signals, it is highly active maintaining its RMP and "waiting," as it were, for something to happen.

## LOCAL POTENTIALS

We now consider the disturbances in membrane potential that occur when a neuron is stimulated. Typically (but with exceptions), the response of a neuron begins at a dendrite, spreads through the soma, travels down the axon, and ends at the synaptic knobs. We consider the process in that order.

Neurons can be stimulated by chemicals, light, heat, or mechanical distortion of the plasma membrane. We'll take as our example a neuron being chemically stimulated on its dendrite (fig. 12.12). The chemical—perhaps a pain signal from a damaged tissue or odor molecule in a breath of air—binds to receptors on the neuron. These receptors are ligand-regulated sodium gates that open and allow Na⁺ to rush into the cell. The inflow of Na⁺ neutralizes some of the internal negative charge, so the voltage across the membrane drifts toward zero. Any such case in which membrane voltage shifts to a less negative value is called **depolarization.** The incoming Na⁺ diffuses for short distances along the inside of

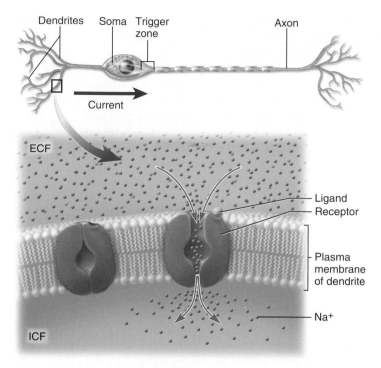

**FIGURE 12.12   Excitation of a Neuron by a Chemical Stimulus.**   When the chemical (ligand) binds to a receptor on the neuron, the receptor acts as a ligand-regulated gate which opens and allows $Na^+$ to diffuse into the cell. This depolarizes the plasma membrane.

the plasma membrane and produces a current that travels from the point of stimulation toward the cell's trigger zone. This short-range change in voltage is called a **local potential.**

There are four characteristics that distinguish local potentials from the action potentials we will study shortly (table 12.2). You will appreciate these distinctions more fully after you have studied action potentials.

1. Local potentials are **graded,** meaning that they vary in magnitude (voltage) according to the strength of the stimulus. A more intense or prolonged stimulus opens more ion gates than a weaker stimulus. Thus, more $Na^+$ enters the cell and the voltage changes more than it does with a weaker stimulus.

2. Local potentials are **decremental,** meaning they get weaker as they spread from the point of stimulation. The decline in strength occurs because as $Na^+$ spreads out under the plasma membrane and depolarizes it, $K^+$ flows out and reverses the effect of the $Na^+$ inflow. Therefore, the voltage shift caused by $Na^+$ diminishes rapidly with distance. This prevents local potentials from having any long-distance effects.

3. Local potentials are **reversible,** meaning that if stimulation ceases, $K^+$ diffusion out of the cell quickly returns the membrane voltage to its resting potential.

4. Local potentials can be either **excitatory** or **inhibitory.** So far, we have considered only excitatory local potentials, which depolarize a cell and make a neuron more likely to produce an action potential. Acetylcholine usually has this effect. Other neurotransmitters, such as glycine, cause an opposite effect—they **hyperpolarize** a cell, or make the membrane more negative. The neuron is then less sensitive and less likely to produce an action potential. A balance between excitatory and inhibitory potentials is very important to information processing in the nervous system, and we explore this more fully later in the chapter.

## ACTION POTENTIALS

An **action potential** is a more dramatic change produced by voltage-regulated ion gates in the plasma membrane. Action potentials occur only where there is a high enough density of voltage-regulated gates. Most of the soma has only 50 to 75 gates per square micrometer ($\mu m^2$) and cannot generate action potentials. The trigger zone, however, has 350 to 500 gates per $\mu m^2$. If an excitatory local poten-

| TABLE 12.2 | Comparison of Local Potentials and Action Potentials |
|---|---|
| **Local Potential** | **Action Potential** |
| Produced by ligand-regulated or mechanically regulated gates on the dendrites and soma | Produced by voltage-regulated gates on the trigger zone and axon |
| May be a positive (depolarizing) or negative (hyperpolarizing) voltage change | Always begins with depolarization |
| Graded; proportional to stimulus strength | All-or-none; either does not occur at all or exhibits the same peak voltage regardless of stimulus strength |
| Reversible; returns to RMP if stimulation ceases before threshold is reached | Irreversible; goes to completion once it begins |
| Local; has effects for only a short distance from point of origin | Self-propagating; has effects a great distance from point of origin |
| Decremental; signal grows weaker with distance | Nondecremental; signal maintains same strength regardless of distance |

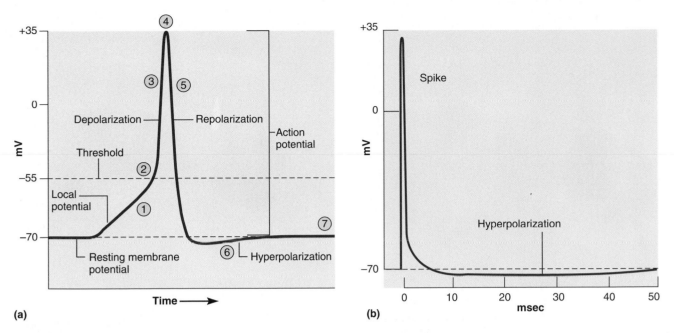

**FIGURE 12.13  An Action Potential.** (a) Diagrammed with a distorted timescale to make details of the action potential visible. Numbers correspond to stages discussed in the text. (b) On an accurate timescale, the local potential is so brief it is imperceptible, the action potential appears as a spike, and the hyperpolarization is very prolonged.

tial spreads all the way to the trigger zone and is still strong enough when it arrives, it can open these gates and generate an action potential.

The action potential is a rapid up-and-down shift in membrane voltage. Figure 12.13a shows an action potential numbered to correspond to the following description.

① When sodium ions arrive at the axon hillock, they depolarize the membrane at that point. This appears as a steadily rising local potential.

② For anything more to happen, this local potential must rise to a critical voltage called the **threshold** (typically about −55 mV). This is the minimum needed to open voltage-regulated gates.

③ The neuron now "fires," or produces an action potential. At threshold, voltage-regulated Na⁺ gates open quickly, while K⁺ gates open more slowly. The initial effect on membrane potential is therefore due to Na⁺. Initially, only a few Na⁺ gates open but as Na⁺ enters the cell, it further depolarizes the membrane. This stimulates still more voltage-regulated Na⁺ gates to open and admit even more Na⁺. Thus, a positive feedback cycle is created that makes the membrane voltage rise rapidly.

④ As the rising membrane potential passes 0 mV, Na⁺ gates are *inactivated* and begin closing. By the time they all close and Na⁺ inflow ceases, the voltage peaks at approximately +35 mV. (The peak is as low as 0 mV in some neurons and as high as 50 mV in others.) The membrane is now positive on the inside

and negative on the outside—its polarity is reversed compared to the RMP.

⑤ By the time the voltage peaks, the slow K⁺ gates are fully open. Potassium ions, repelled by the positive intracellular fluid, exit the cell. Their outflow **repolarizes** the membrane—that is, it shifts the voltage back into the negative numbers. The action potential consists of the up-and-down voltage shifts that occur from the time the threshold is reached to the time the voltage returns to the RMP.

⑥ Potassium gates stay open longer than Na⁺ gates, so the amount of K⁺ that leaves the cell is greater than the amount of Na⁺ that entered. Therefore, the membrane voltage drops to 1 or 2 mV more negative than the original RMP, producing a negative overshoot called *hyperpolarization.*

⑦ As you can see, Na⁺ and K⁺ switch places across the membrane during an action potential. During hyperpolarization, ion diffusion through the membrane and (in the CNS) the removal of extracellular K⁺ by the astrocytes gradually restores the original resting membrane potential.

Figure 12.14 correlates these voltage changes with events in the plasma membrane. At the risk of being misleading, this figure is drawn as if most of the Na⁺ and K⁺ had traded places. In reality, only about one ion in a million crosses the membrane to produce an action potential, and an action potential affects ion distribution only in a thin layer close to the membrane. If the illustration tried

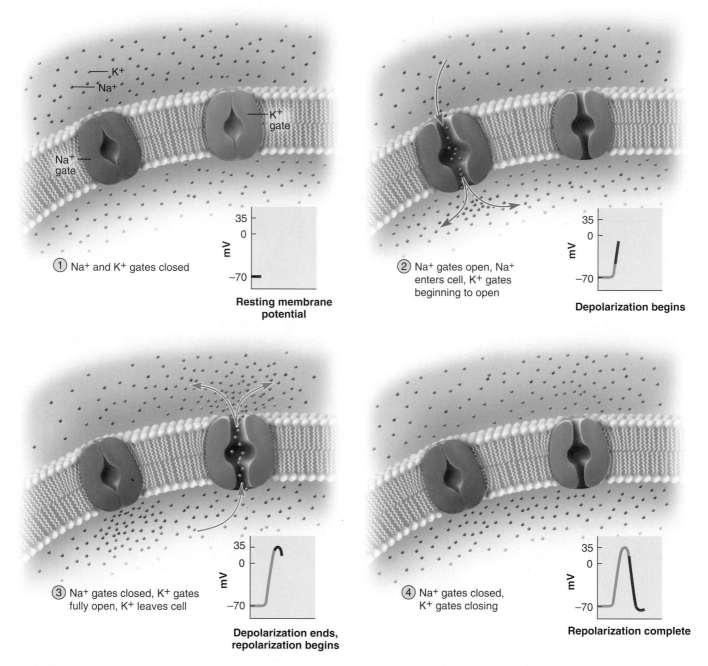

**FIGURE 12.14** **Actions of the Sodium and Potassium Gates During an Action Potential.** The red part of each graph shows the point in the action potential where the events of steps 1 through 4 are occurring.

to represent these points accurately, the difference would be so slight you could not see it, indeed the changes in ion concentrations inside and outside the cell are so slight they cannot be measured in the laboratory unless a neuron has been stimulated to produce numerous action potentials in quick succession. Even after thousands of action potentials, the cytosol still has a higher concentration of $K^+$ and a lower concentration of $Na^+$ than the ECF does.

Figure 12.13a also is deliberately distorted. In order to demonstrate the different phases of the local potential and action potential, the magnitudes of the local potential and hyperpolarization are exaggerated, the local potential is stretched out to make it seem longer, and the duration of hyperpolarization is shrunken so the graph will fit the page. When these events are plotted on a more realistic timescale, they look like figure 12.13b. The local potential is so brief it is unnoticeable, and hyperpolarization is very long but only slightly more negative than the RMP. An action potential is often called a *spike;* it is easy to see why from this figure.

Earlier we saw that local potentials are graded, decremental, and reversible. We can now examine how action potentials compare on these points.

- Action potentials follow an **all-or-none law.** If a stimulus depolarizes the neuron to threshold, the neuron fires at its maximum voltage (such as +35 mV); if threshold is not reached, the neuron does not fire at all. Above threshold, stronger stimuli do not produce stronger action potentials. Thus, action potentials are not graded (proportional to stimulus strength).

- Action potentials are **nondecremental.** For reasons to be examined shortly, they do not get weaker with distance. The last action potential at the end of a nerve fiber is just as strong as the first action potential in the trigger zone up to a meter away.

- Action potentials are **irreversible.** If a neuron reaches threshold, the action potential goes to completion; it cannot be stopped once it begins.

In some respects, we can compare the firing of a neuron to the firing of a gun. As the trigger is squeezed, a gun either fires with maximum force or does not fire at all (analogous to the all-or-none law). You cannot fire a fast bullet by squeezing the trigger hard or a slow bullet by squeezing it gently—once the trigger is pulled to its "threshold," the bullet always leaves the muzzle at the same velocity. And, like an action potential, the firing of a gun is irreversible once the threshold is reached. Table 12.2 further contrasts a local potential with an action potential, including some characteristics of action potentials explained in the following sections.

## THE REFRACTORY PERIOD

During an action potential and for a few milliseconds after, it is difficult or impossible to stimulate that region of a neuron to fire again. This period of resistance to re-stimulation is called the **refractory period.** It is divided into two phases: an *absolute refractory period* in which no stimulus of any strength will trigger a new action potential, followed by a *relative refractory period* in which it is possible to trigger a new action potential, but only with an unusually strong stimulus (fig. 12.15).

The absolute refractory period lasts from the start of the action potential until the membrane returns to the resting potential—that is, for as long as the Na$^+$ gates are open and subsequently inactivated. The relative refractory period lasts until hyperpolarization ends. During this period, K$^+$ gates are still open. A new stimulus tends to admit Na$^+$ and depolarize the membrane, but K$^+$ diffuses out through the open gates as Na$^+$ comes in, and thus opposes the effect of the stimulus. It requires an especially strong stimulus to override the K$^+$ outflow and depolarize the cell enough to set off a new action potential. By the end of hyperpolarization, K$^+$ gates are closed and the cell is as responsive as ever.

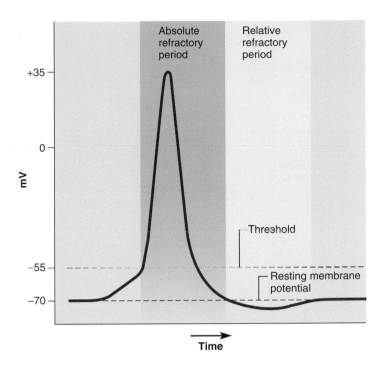

**FIGURE 12.15** The Absolute and Relative Refractory Periods in Relation to the Action Potential.

The refractory period refers only to a small patch of membrane where an action potential has already begun, not to the entire neuron. Other parts of the neuron can still be stimulated while a small area of it is refractory, and even this area quickly recovers once the nerve signal has passed on.

## SIGNAL CONDUCTION IN NERVE FIBERS

If a neuron is to communicate with another cell, a signal has to travel to the end of the axon. We can now examine how this is achieved.

### Unmyelinated Fibers

Unmyelinated fibers present a relatively simple case of signal conduction, easy to understand based on what we have already covered (fig. 12.16). An unmyelinated fiber has voltage-regulated ion gates along its entire length. When an action potential occurs at the trigger zone, Na$^+$ enters the axon and diffuses to adjacent regions just beneath the plasma membrane. The resulting depolarization excites voltage-regulated gates immediately distal to the action potential. Sodium and potassium gates open and close just as they did at the trigger zone, and a new action potential is produced. By repetition, this excites the membrane immediately distal to that. This chain reaction continues until the traveling signal reaches the end of the axon.

Note that an action potential itself does not travel along an axon; rather, it stimulates the production of a new action potential in the membrane just ahead of it.

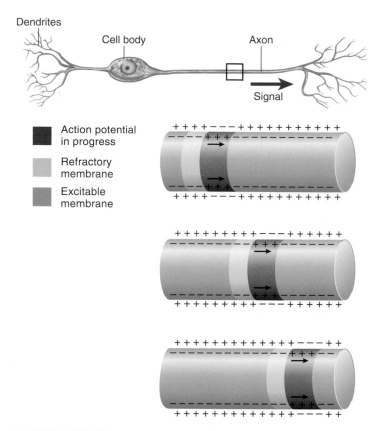

**FIGURE 12.16   Conduction of a Nerve Signal in an Unmyelinated Fiber.** Note that the membrane polarity is reversed in the region of the action potential (red). A region of membrane in its refractory period (yellow) trails the action potential and prevents the nerve signal from going backward toward the soma. The other membrane areas (green) are fully polarized and ready to respond.

Thus, we can distinguish an *action potential* from a *nerve signal.* The nerve signal is a traveling wave of excitation produced by self-propagating action potentials. It is a little like a line of falling dominoes. No one domino travels to the end of the line, but each domino pushes over the next one and there is a transmission of energy from the first domino to the last. Similarly, no one action potential travels to the end of an axon; a nerve signal is a chain reaction of action potentials.

If one action potential stimulates the production of a new one next to it, you might think that the signal could also start traveling backward and return to the soma. This does not occur, however, because the membrane behind the nerve signal is still in its refractory period and cannot be restimulated. Only the membrane ahead is sensitive to stimulation. The refractory period thus ensures that nerve signals are conducted in the proper direction, from the soma to the synaptic knobs.

A traveling nerve signal is an electrical current, but it is not the same as a current traveling through a wire. A current in a wire travels millions of meters per second and is decremental—it gets weaker with distance. A nerve signal is much slower (not more than 2 m/sec in unmyelinated fibers), but it is *nondecremental.* Even in the longest axons, the last action potential generated at a synaptic knob has the same voltage as the first one generated at the trigger zone. To clarify this concept, we can compare the nerve signal to a burning fuse. When a fuse is lit, the heat ignites powder immediately in front of this point, and this event repeats itself in a self-propagating fashion until the end of the fuse is reached. At the end, the fuse burns just as hotly as it did at the beginning. In a fuse, the combustible powder is the source of potential energy that keeps the process going in a nondecremental fashion. In an axon, the potential energy comes from the ion gradient across the plasma membrane. Thus, the signal does not grow weaker with distance; it is self-propagating, like the burning of a fuse.

## Myelinated Fibers

Matters are somewhat different in myelinated fibers. Voltage-regulated ion gates are scarce in the myelin-covered internodes—fewer than $25/\mu m^2$ in these regions compared with 2,000 to $12,000/\mu m^2$ at the nodes of Ranvier. There would be little point in having ion gates in the internodes—myelin insulates the fiber from the ECF at these points, and $Na^+$ from the ECF could not flow into the cell even if more gates were present.

The only way a nerve signal can travel along an internode is for $Na^+$ that enters at the previous node to diffuse down the fiber under the axolemma (fig. 12.17a). This is a very fast process, but the nerve fiber resists their flow (just as a wire resists a current) and the signal becomes weaker the farther it goes. Therefore, this aspect of conduction is decremental. The signal cannot travel much farther than 1 mm before it becomes too weak to open any voltage-regulated gates. But fortunately, there is another node of Ranvier every millimeter or less along the axon, where the axolemma is exposed to ECF and there is an abundance of voltage-regulated gates. When the diffusing ions reach this point, the signal is just strong enough to open these gates and create a new action potential. This action potential has the same strength as the one at the previous node, so each node of Ranvier boosts the signal back to its original strength (+35 mV). This mode of signal conduction is called **saltatory**[28] **conduction**—the propagation of a nerve signal that seems to jump from node to node (fig. 12.17b).

In the internodes, saltatory conduction is therefore based on a process that is very fast (diffusion of ions along the fiber) but decremental. In the nodes, conduction is slower but nondecremental. Since most of the axon is covered with myelin, conduction occurs mainly by the fast longitudinal diffusion process. This is why myelinated fibers transmit signals much faster (up to 120 m/sec) than unmyelinated ones (up to 2 m/sec).

---

[28]from *saltare* = to leap, to dance

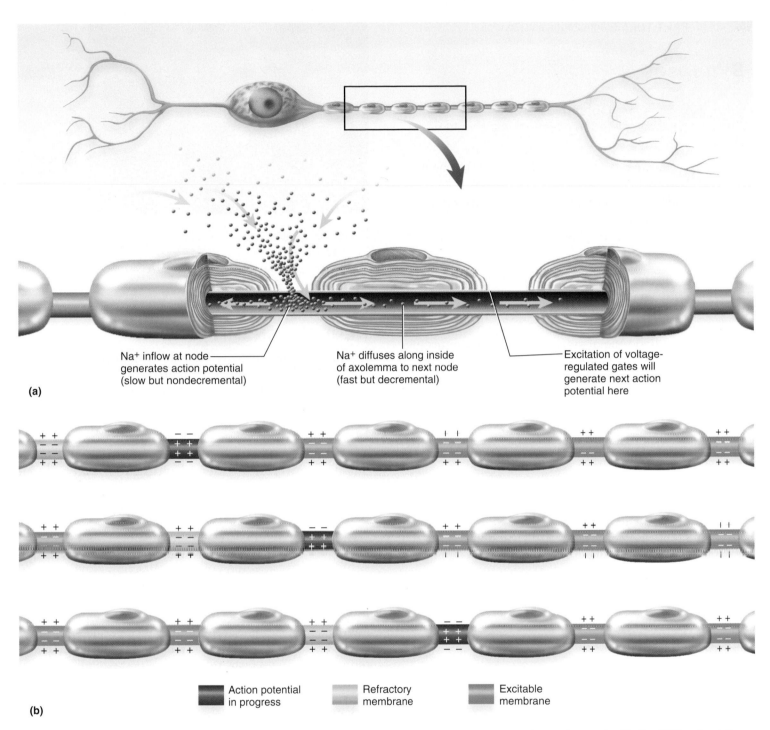

**(a)**

Na⁺ inflow at node generates action potential (slow but nondecremental)

Na⁺ diffuses along inside of axolemma to next node (fast but decremental)

Excitation of voltage-regulated gates will generate next action potential here

**(b)**

Action potential in progress

Refractory membrane

Excitable membrane

**FIGURE 12.17 Saltatory Conduction of a Nerve Signal in a Myelinated Fiber.** (a) Ions can be exchanged with the ECF only at the nodes of Ranvier. In the internodes, the nerve signal travels by the rapid diffusion of ions along the inside of the plasma membrane. (b) Action potentials occur only at the nodes, and the nerve signal (red) therefore appears to jump from node to node.

## Before You Go On

*Answer the following questions to test your understanding of the preceding section:*

11. What causes K⁺ to diffuse out of a resting cell? What attracts it into the cell?

12. What happens to Na⁺ when a neuron is stimulated on its dendrite? Why does the movement of Na⁺ raise the voltage on the plasma membrane?

13. What does it mean to say a local potential is graded, decremental, and reversible?

14. How does the plasma membrane at the trigger zone differ from that on the soma? How does it resemble the membrane at a node of Ranvier?

15. What makes an action potential rise to +35 mV? What makes it drop again after this peak?

16. List four ways in which an action potential is different from a local potential.

17. Explain why myelinated fibers transmit signals much faster than unmyelinated fibers.

# Synapses

## Objectives

When you have completed this section, you should be able to

- explain how messages are transmitted from one neuron to another;
- give examples of neurotransmitters and describe their actions; and
- explain how stimulation of a postsynaptic cell is stopped.

A nerve signal soon reaches the end of an axon and can go no farther. In most cases, it triggers the release of a neurotransmitter that stimulates a new wave of electrical activity in the next cell across the synapse. The most thoroughly studied type of synapse is the neuromuscular junction described in chapter 11, but here we consider synapses between two neurons. The first neuron in the signal path is the **presynaptic neuron,** which releases the neurotransmitter. The second is the **postsynaptic neuron,** which responds to it (fig. 12.18a).

The presynaptic neuron may synapse with a dendrite, the soma, or the axon of a postsynaptic neuron and form an *axodendritic, axosomatic,* or *axoaxonic synapse,* respectively (fig. 12.18b). A neuron can have an enormous number of synapses (fig. 12.19). For example, a spinal motor neuron is covered with about 10,000 synaptic knobs from other neurons—8,000 ending on its dendrites and another 2,000 on the soma. In part of the brain called the cerebellum, one neuron can have as many as 100,000 synapses.

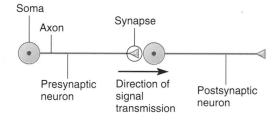

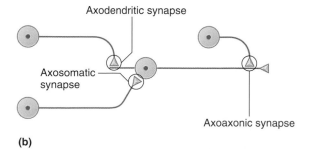

**FIGURE 12.18   Synaptic Relationships Between Neurons.**
(a) Pre- and postsynaptic neurons. (b) Types of synapses defined by the site of contact on the postsynaptic neuron.

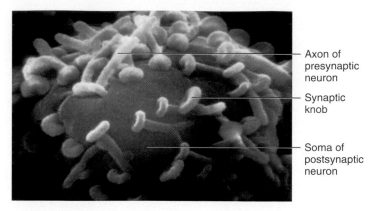

**FIGURE 12.19**   Synaptic Knobs on the Soma of a Neuron in a Marine Slug, *Aplysia* (SEM).
❱ *Are these synapses axodendritic, axosomatic, or axoaxonic?*

## THE DISCOVERY OF NEUROTRANSMITTERS

As the neuron doctrine became generally accepted, it raised the question of how neurons communicate with each other. In the early twentieth century, biologists assumed that synaptic communication was electrical—a logical hypothesis given that neurons seemed to touch each other and signals were transmitted so quickly from one to the next. Ramón y Cajal's careful histological examinations, however, revealed a 20- to 40-nm gap between neurons—the **synaptic cleft**—casting doubt on the possibility of electrical conduction.

In 1921, the German pharmacologist Otto Loewi (1873–1961) conclusively demonstrated that neurons communicate by releasing chemicals. The *vagus nerves* supply the heart, among other organs, and slow it down. Loewi opened two frogs and flooded the hearts with saline to keep them moist. He stimulated the vagus nerve of one frog, and its heart rate dropped as expected. He then removed some of the saline from that heart and squirted it onto the heart of the second frog. The solution alone reduced that frog's heart rate. Evidently it contained something released by the vagus nerve of the first frog. Loewi called it *Vagusstoffe* ("vagus substance") and it was later renamed acetylcholine—the first known neurotransmitter.

**Think About It**

*As described, does the previous experiment conclusively prove that the second frog's heart slowed as a result of something released by the vagus nerves? If you were Loewi, what control experiment would you do to rule out alternative explanations?*

Following Loewi's work, the idea of electrical communication between cells fell into disrepute. Now, however, we realize that some neurons, neuroglia, and cardiac and single-unit smooth muscle (see chapter 11) do indeed have **electrical synapses,** where adjacent cells are joined by gap junctions and ions can diffuse directly from one cell into the next. These junctions have the advantage of quick transmission because there is no delay for the release and binding of neurotransmitter. Their disadvantage, however, is that they cannot integrate information and make decisions. The ability to do that is a property of **chemical synapses,** in which neurons communicate by neurotransmitters.

## STRUCTURE OF A CHEMICAL SYNAPSE

The synaptic knob (fig. 12.20) was described in chapter 11. It contains synaptic vesicles, many of which are "docked" at release sites on the plasma membrane, ready to release their neurotransmitter on demand. A reserve pool of synaptic vesicles is located a little farther away from the membrane, clustered near the release sites and tethered to the cytoskeleton by protein microfilaments.

The postsynaptic neuron does not show such conspicuous specializations. At this end, the neuron has no synaptic vesicles and cannot release neurotransmitters. Its membrane does, however, contain proteins that function as receptors and ligand-regulated ion gates.

## NEUROTRANSMITTERS AND RELATED MESSENGERS

More than 100 confirmed or suspected neurotransmitters have been identified since Loewi discovered acetyl-

choline. Neurotransmitters fall into four major categories according to chemical composition (fig. 12.21). Some of the best-known ones are listed in table 12.3. Parts of the brain referred to in this table will become familiar to you as you study chapter 14, and you may wish to refer back to this table then to enhance your understanding of brain function.

1. **Acetylcholine** is in a class by itself. It is formed from acetic acid (acetate) and choline.
2. **Amino acid** neurotransmitters include glycine, glutamate, aspartate, and γ-aminobutyric acid (GABA).
3. **Monoamines** (biogenic amines) are synthesized from amino acids by removal of the –COOH group. They retain the –NH₂ (amino group), hence their name. The major monoamines are epinephrine, norepinephrine, dopamine, histamine, and serotonin (5-hydroxytryptamine, or 5-HT). The first three of these are in a subclass of monoamines called **catecholamines** (CAT-eh-COAL-uh-meens).
4. **Neuropeptides** are chains of 2 to 40 amino acids. Some examples are β-endorphin and substance P. Neuropeptides typically act at lower concentrations and have longer lasting effects than other neurotransmitters, and they are stored in *secretory granules*

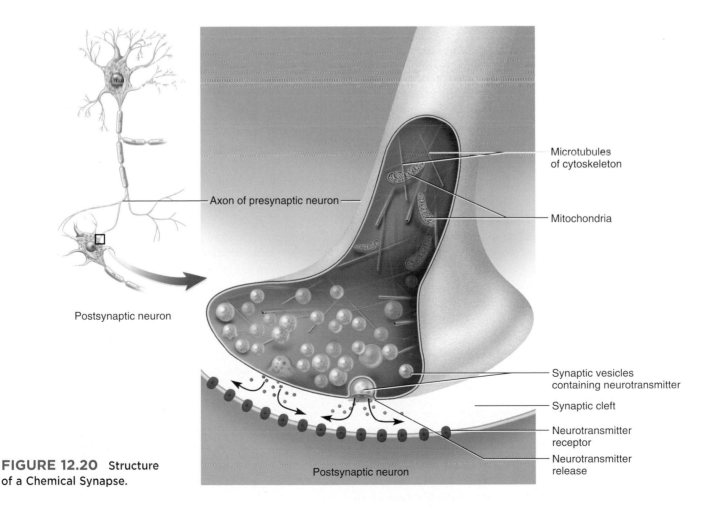

**FIGURE 12.20** Structure of a Chemical Synapse.

Microtubules of cytoskeleton
Mitochondria
Synaptic vesicles containing neurotransmitter
Synaptic cleft
Neurotransmitter receptor
Neurotransmitter release
Axon of presynaptic neuron
Postsynaptic neuron
Postsynaptic neuron

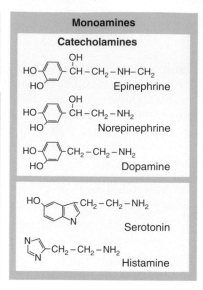

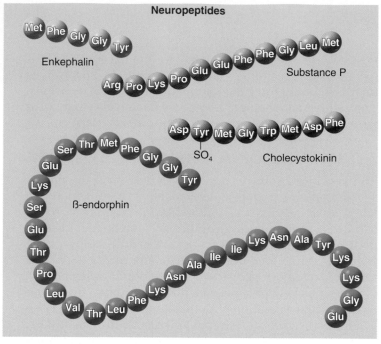

**FIGURE 12.21** Classification of Some Neurotransmitters.
The neuropeptides are chains of amino acids, each identified by its three-letter code. Table 2.8 explains the codes.

| TABLE 12.3 | Neurotransmitters and Neuropeptides |
|---|---|
| **Name** | **Locations and Actions** |
| *Acetylcholine (Ach)* | Neuromuscular junctions, most synapses of autonomic nervous system, retina, and many parts of the brain; excites skeletal muscles, inhibits cardiac muscle, and has excitatory or inhibitory effects on smooth muscle and glands depending on location |
| *Amino Acids* | |
| Excitatory Amino Acids | |
|   Glutamate (glutamic acid) | Cerebral cortex and brainstem; retina; accounts for about 75% of all excitatory synaptic transmission in the brain; involved in learning and memory |
|   Aspartate (aspartic acid) | Spinal cord; effects similar to those of glutamate |
| Inhibitory Amino Acids | |
|   Glycine | Inhibitory neurons of the brain, spinal cord, and retina; most common inhibitory neurotransmitter in the spinal cord |
|   GABA (γ-aminobutyric acid) | Thalamus, hypothalamus, cerebellum, occipital lobes of cerebrum, and retina; most common inhibitory neurotransmitter in the brain |
| *Monoamines (biogenic amines)* | |
| Catecholamines | |
|   Norepinephrine | Sympathetic nervous system, cerebral cortex, hypothalamus, brainstem, cerebellum, and spinal cord; involved in dreaming, waking, and mood; excites cardiac muscle; can excite or inhibit smooth muscle and glands depending on location |
|   Epinephrine | Hypothalamus, thalamus, spinal cord, and adrenal medulla; effects similar to those of norepinephrine |
|   Dopamine | Hypothalamus, limbic system, cerebral cortex, and retina; highly concentrated in substantia nigra of midbrain; involved in elevation of mood and control of skeletal muscles |
| Other monoamines | |
|   Serotonin | Hypothalamus, limbic system, cerebellum, retina, and spinal cord; also secreted by blood platelets and intestinal cells; involved in sleepiness, alertness, thermoregulation, and mood |
|   Histamine | Hypothalamus; also a potent vasodilator released by mast cells of connective tissue and basophils of the blood |
| *Neuropeptides* | |
|   Substance P | Basal nuclei, midbrain, hypothalamus, cerebral cortex, small intestine, and pain-receptor neurons; mediates pain transmission |
|   Enkephalins | Hypothalamus, limbic system, pituitary, pain pathways of spinal cord, and nerve endings of digestive tract; act as analgesics (pain-relievers) by inhibiting substance P; inhibit intestinal motility; secretion increases sharply in women in labor |
|   β-endorphin | Digestive tract, spinal cord, and many parts of the brain; also secreted as a hormone by the pituitary; suppresses pain; reduces perception of fatigue and may produce "runner's high" in athletes |
|   Cholecystokinin (CCK) | Cerebral cortex and small intestine; suppresses appetite |

*(dense-core vesicles)* that are about 100 nm in diameter, twice as large as typical synaptic vesicles. Some neuropeptides also function as hormones or as **neuromodulators,** whose action is discussed later in this chapter. Some neuropeptides are produced not only by neurons but also by the digestive tract; thus they are known as *gut-brain peptides.* Some of these cause cravings for specific nutrients such as fat or sugar and may be associated with certain eating disorders (see p. 1005).

> **Think About It**
>
> *Neuropeptides can be synthesized only in the soma and must be transported to the synaptic knobs. Why is their synthesis limited to the soma?*

The more we learn about neurotransmitters, hormones, and other chemical messengers, the harder it is to distinguish them from each other or even to rigorously define "neurotransmitter." Traditionally, neurotransmitters have been conceived as small organic compounds that function at the synapse as follows: (1) they are synthesized by the presynaptic neuron, (2) they are released in response to stimulation, (3) they bind to specific receptors on the postsynaptic cell, and (4) they alter the physiology of that cell. Neuropeptides are an exception to the small size of neurotransmitters, and neurons have additional means of communication, such as the gas nitric oxide, that fall outside the scope of this traditional concept.

We will see, especially in chapter 15, that a given neurotransmitter does not have the same effect everywhere in the body. There are multiple receptor types in the body for a particular neurotransmitter—over 14 receptor types for serotonin, for example—and it is the receptor that governs what effect a neurotransmitter has on its target cell.

## SYNAPTIC TRANSMISSION

Neurotransmitters are quite diverse in their action. Some are excitatory, some are inhibitory, and for some the effect depends on what kind of receptor the postsynaptic cell has. Some neurotransmitters open ligand-regulated ion gates while others act through second-messenger systems. Bearing this diversity in mind, we will here examine three kinds of synapses with different modes of action: an *excitatory cholinergic synapse,* an *inhibitory GABA-ergic synapse,* and an *excitatory adrenergic synapse.*

### An Excitatory Cholinergic Synapse

A **cholinergic**[29] (CO-lin-UR-jic) synapse employs acetylcholine (ACh) as its neurotransmitter. ACh excites some postsynaptic cells (such as skeletal muscle; chapter 11)

---

[29]*cholin* = acetylcholine + *erg* = work, action

and inhibits others, but this discussion will describe an excitatory action. The steps in transmission at such a synapse are as follows (fig. 12.22):

① The arrival of a nerve signal at the synaptic knob opens voltage-regulated calcium gates.

② $Ca^{2+}$ enters the knob and triggers exocytosis of the synaptic vesicles, releasing ACh.

③ Empty vesicles drop back into the cytoplasm to be refilled with ACh, while synaptic vesicles in the reserve pool move to the active sites and release their ACh—a bit like a line of Revolutionary War soldiers firing their muskets and falling back to reload as another line moves to the fore.

④ Meanwhile, ACh diffuses across the synaptic cleft and binds to ligand-regulated gates on the postsynaptic neuron. These gates open, allowing $Na^+$ to

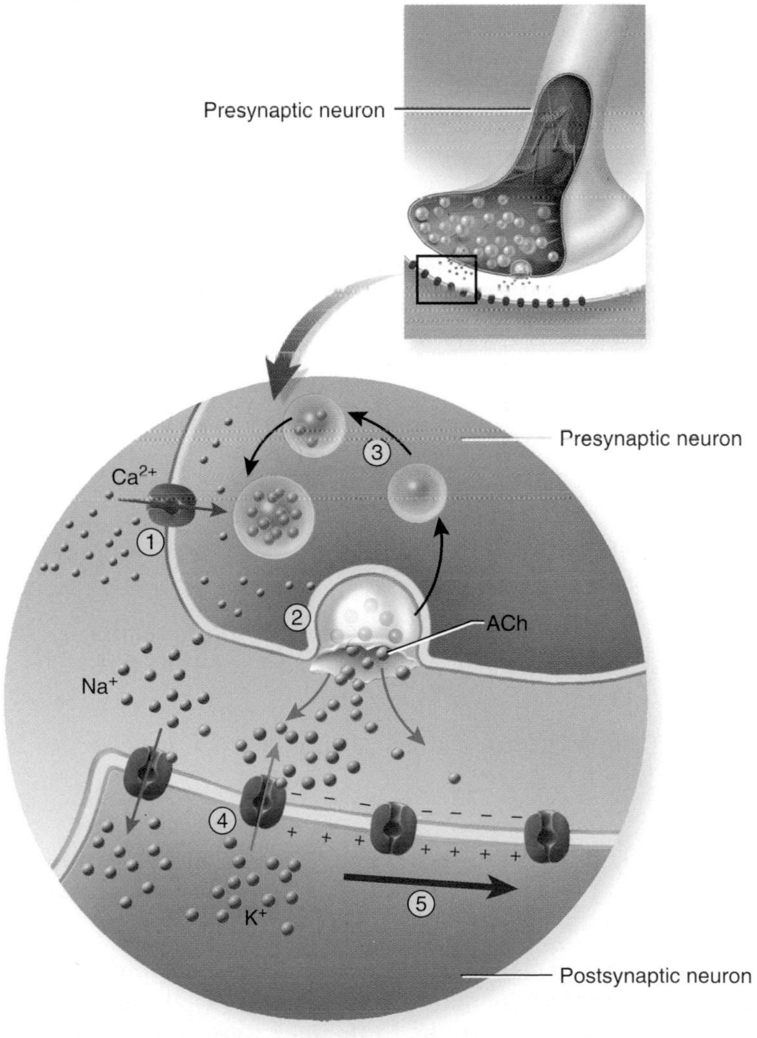

**FIGURE 12.22 Transmission at a Cholinergic Synapse.** Acetylcholine directly opens ion gates in the plasma membrane of the postsynaptic neuron. Numbered steps correspond to the description in the text.

enter the cell and $K^+$ to leave. Although illustrated separately, $Na^+$ and $K^+$ pass in opposite directions through the same gates.

⑤ As $Na^+$ enters the cell, it spreads out along the inside of the plasma membrane and depolarizes it, producing a local potential called the **postsynaptic potential.** Like other local potentials, if this is strong and persistent enough (that is, if enough $Na^+$ makes it to the axon hillock), it opens voltage-regulated ion gates in the trigger zone and causes the postsynaptic neuron to fire.

## An Inhibitory GABA-ergic Synapse

A **GABA-ergic** synapse employs γ-aminobutyric acid (GABA) as its neurotransmitter. Amino acid neurotransmitters work by the same mechanism as ACh—binding to ion gates and causing immediate changes in membrane potential. The release of GABA and binding to its receptor are similar to the preceding case. The GABA receptor, however, is a chloride channel. When this channel opens, $Cl^-$ enters the cell and makes the inside even more negative than the resting membrane potential. The neuron is therefore inhibited, or less likely to fire.

## An Excitatory Adrenergic Synapse

An **adrenergic synapse** employs the neurotransmitter norepinephrine (NE), also called noradrenaline. NE, other monoamines, and neuropeptides act through second-messenger systems such as cyclic AMP (cAMP). The receptor is not an ion gate but a transmembrane protein associated with a G protein. Figure 12.23 shows some ways in which an adrenergic synapse can function, numbered to correspond to the following:

① The unstimulated NE receptor is bound to a G protein.

② Binding of NE to the receptor causes the G protein to dissociate from it.

③ The G protein binds to adenylate cyclase, which activates this enzyme and induces it to convert ATP to cyclic AMP.

④ Cyclic AMP can induce several alternative effects in the cell.

⑤ One effect is to produce an internal chemical that binds to a ligand-regulated ion gate from the inside of the membrane, opening the gate and depolarizing the cell.

⑥ Another is to activate preexisting cytoplasmic enzymes, which can lead to diverse metabolic changes (for example, inducing a liver cell to break down glycogen and release glucose into the blood).

⑦ Yet another is for cAMP to induce genetic transcription, so that the cell produces new cytoplasmic

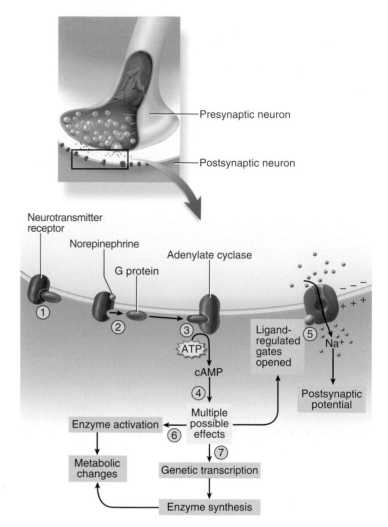

**FIGURE 12.23   Transmission at an Adrenergic Synapse.** The norepinephrine receptor is not an ion channel. It activates a second-messenger system with a variety of possible effects in the postsynaptic cell. Numbered steps correspond to the description in the text.

enzymes, again leading to diverse possible metabolic effects.

Although slower to respond than cholinergic and GABA-ergic synapses, adrenergic synapses do have an advantage—**enzyme amplification.** A single NE molecule binding to a receptor can induce the formation of many cAMPs, each of those can activate many enzyme molecules or induce the transcription of a gene into numerous mRNA molecules, and each of those can result in the production of a vast number of metabolic products such as glucose molecules.

As complex as synaptic events may seem, they typically require only 0.5 msec or so—an interval called **synaptic delay.** This is the time from the arrival of a signal at the axon terminal of a presynaptic cell to the beginning of an action potential in the postsynaptic cell.

## CESSATION OF THE SIGNAL

It is important not only to stimulate a postsynaptic cell but also to turn off the stimulus in due time. Otherwise the postsynaptic neuron could continue firing indefinitely, causing a breakdown in physiological coordination. Here we examine some ways this is done.

A neurotransmitter molecule binds to its receptor for only 1 msec or so, then dissociates from it. If the presynaptic cell continues to release neurotransmitter, one molecule is quickly replaced by another and the postsynaptic cell is restimulated. This immediately suggests a way of stopping synaptic transmission—stop adding new neurotransmitter and get rid of that which is already there. The first step is achieved simply by the cessation of signals in the presynaptic nerve fiber. The second can be achieved in the following ways:

- **Diffusion.** Neurotransmitter escapes from the synapse into the nearby ECF, where astrocytes absorb it and return it to the neurons.

- **Reuptake.** The synaptic knob reabsorbs amino acids and monoamines by endocytosis and breaks them down with an enzyme called **monoamine oxidase (MAO).** Some antidepressant drugs work by inhibiting MAO (see Insight 15.2, p. 579).

- **Degradation in the synaptic cleft.** The enzyme acetylcholinesterase (AChE), located in the synaptic cleft and on the postsynaptic membrane, breaks ACh down into acetate and choline. These breakdown products have no stimulatory effect on the postsynaptic cell. The synaptic knob reabsorbs the choline and uses it to synthesize more ACh.

## NEUROMODULATORS

**Neuromodulators** are hormones, neuropeptides, and other messengers that modify synaptic transmission. They may stimulate a neuron to increase the number of receptors in the postsynaptic membrane, thus adjusting its sensitivity to a neurotransmitter, or they may alter the rate of neurotransmitter synthesis, release, reuptake, or breakdown. One example—a rather recent and surprising discovery— is nitric oxide (NO), a lightweight gas that is released by postsynaptic neurons in some areas of the brain concerned with learning and memory. NO diffuses into the presynaptic neuron and stimulates it to release more neurotransmitter—like one neuron's way of telling the other, "Give me more." Thus, there is at least some chemical communication that goes backward across a synapse.

### Before You Go On

*Answer the following questions to test your understanding of the preceding section:*

18. *Concisely describe five steps that occur between the arrival of an action potential at the synaptic knob and the beginning of a new action potential in the postsynaptic neuron.*

19. *Contrast the actions of acetylcholine, GABA, and norepinephrine at their respective synapses.*

20. *Describe three mechanisms that stop synaptic transmission.*

21. *What is the function of neuromodulators?*

# Neural Integration

### Objectives

When you have completed this section, you should be able to

- explain how a neuron "decides" whether or not to produce action potentials;

- explain how the nervous system translates complex information into a simple code;

- explain how neurons work together in groups to process information and produce effective output; and

- describe how memory works at a cellular and molecular level.

Synaptic delay slows the transmission of nerve signals; the more synapses there are in a neural pathway, the longer it takes information to get from its origin to its destination. You might wonder, therefore, why we have synapses—why a nervous pathway is not, indeed, a continuous "wire" as biologists believed before the neuron doctrine was accepted. The presence of synapses is not due to limitations on the length of a neuron—after all, one nerve fiber can reach from your toes to your brainstem; imagine how long some nerve fibers may be in a giraffe or a whale! We also have seen that cells can communicate through gap junctions much more quickly than they can through chemical synapses. So why have chemical synapses at all?

What we value most about our nervous system is its ability to process information, store it, and make decisions—and chemical synapses are the decision-making devices of the system. The more synapses a neuron has, the greater its information-processing capability. At this moment, you are using certain *pyramidal cells* of the cerebral cortex (see fig. 12.5) to read and comprehend this passage. Each pyramidal cell has about 40,000 synaptic contacts with other neurons. The cerebral cortex alone (the main information-processing tissue of your brain) is estimated to have 100 trillion ($10^{14}$) synapses. To get some impression of this number, imagine trying to count them. Even if you could count two synapses per second, day and night without stopping, and you were immortal, it would take you 1.6 million years. The ability of your neurons to process information, store and recall it, and make decisions is called **neural integration.**

# POSTSYNAPTIC POTENTIALS

Neural integration is based on the postsynaptic potentials produced by neurotransmitters. Remember that a typical neuron has a resting membrane potential (RMP) of about −70 mV and a threshold of about −55 mV. A neuron has to be depolarized to this threshold in order to produce action potentials. Any voltage change in that direction makes a neuron more likely to fire and is therefore called an **excitatory postsynaptic potential (EPSP)** (fig. 12.24a). EPSPs usually result from Na$^+$ flowing into the cell and canceling some of the negative charge on the inside of the membrane.

In other cases, a neurotransmitter hyperpolarizes the postsynaptic cell and makes it more negative than the RMP. Since this makes the postsynaptic cell less likely to fire, it is called an **inhibitory postsynaptic potential (IPSP)** (fig. 12.24b). Some IPSPs are produced by a neurotransmitter opening ligand-regulated chloride gates, causing Cl$^-$ to flow into the cell and make the cytosol more negative. A less common way is to open selective K$^+$ gates, increasing the diffusion of K$^+$ out of the cell.

We must recognize that because of ion leakage through their membranes, all neurons fire at a certain background rate even when they are not being stimulated. EPSPs and IPSPs do not determine whether or not a neuron fires, but only change the rate of firing by stimulating or inhibiting the production of more action potentials.

Glutamate and aspartate are excitatory brain neurotransmitters that produce EPSPs. Glycine and GABA produce IPSPs and are therefore inhibitory. Acetylcholine (ACh) and norepinephrine are excitatory to some cells and inhibitory to others, depending on the type of receptors present on the target cells. For example, ACh excites skeletal muscle but inhibits cardiac muscle because the two types of muscle have different types of ACh receptors. This is discussed more fully in chapter 15.

# SUMMATION, FACILITATION, AND INHIBITION

One neuron may receive input from thousands of other neurons simultaneously. Some incoming nerve fibers may produce EPSPs while others produce IPSPs. Whether or not the neuron fires depends on whether the *net* input is excitatory or inhibitory. If the EPSPs override the IPSPs, threshold may be reached and the neuron will fire; if the IPSPs prevail, the neuron will not fire. **Summation** is the process of adding up postsynaptic potentials and responding to their net effect. It occurs in the trigger zone.

Suppose, for example, you are working in the kitchen and accidentally touch a hot cooking pot. EPSPs in your motor neurons might cause you to jerk your hand back quickly and avoid being burned. Yet a moment later, you might nonchalantly pick up a cup of tea that is even hotter than the pot. Since you are expecting the teacup to be hot, you do not jerk your hand away. You have learned that it will not injure you, so at some level of the nervous system, IPSPs prevail and inhibit the motor response.

It is fundamentally a balance between EPSPs and IPSPs that enables the nervous system to make decisions. A postsynaptic neuron is like a little cellular democracy acting on the "majority vote" of hundreds or thousands of presynaptic cells. In the teacup example, some presynaptic neurons are sending messages that signify "hot! danger!" in the form of EPSPs that may activate a hand-withdrawal reflex, while at the same time, others are producing IPSPs that signify "safe" and suppress the withdrawal reflex. Whether the postsynaptic neurons cause you to jerk your hand back depends on whether the EPSPs override the IPSPs or vice versa.

A single action potential in a synaptic knob does not produce enough activity to make a postsynaptic cell fire. An EPSP may be produced, but it decays before reaching threshold. A typical EPSP is a voltage change of only

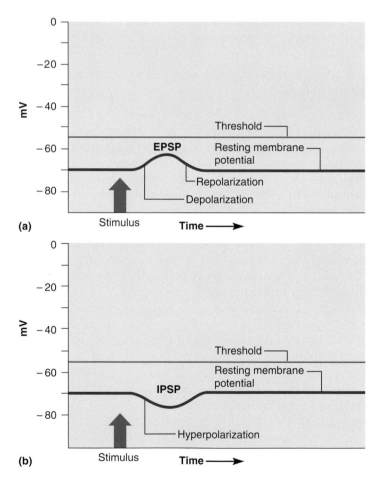

**FIGURE 12.24 Postsynaptic Potentials.** (a) An excitatory postsynaptic potential (EPSP), which shifts the membrane voltage closer to threshold and makes the cell more likely to fire. (b) An inhibitory postsynaptic potential (IPSP), which shifts the membrane voltage farther away from threshold and makes the cell less likely to fire. The sizes of these postsynaptic potentials are greatly exaggerated here for clarity; compare figure 12.26.

▶ *Why is a single EPSP insufficient to make a neuron fire?*

0.5 mV and lasts only 15 to 20 msec. If a neuron has an RMP of −70 mV and a threshold of −55 mV, it needs at least 30 EPSPs to reach threshold and fire. There are two ways in which EPSPs can add up to do this, and both may occur simultaneously.

1. **Temporal summation** (fig. 12.25a). This occurs when a single synapse generates EPSPs so quickly that each is generated before the previous one decays. This allows the EPSPs to add up over time to a threshold voltage that triggers an action potential (fig. 12.26). Temporal summation can occur if even one presynaptic neuron stimulates the postsynaptic neuron intensely enough.

2. **Spatial summation** (fig. 12.25b). This occurs when EPSPs from several different synapses add up to threshold at the axon hillock. Any one synapse may admit only a moderate amount of Na⁺ into the cell, but several synapses acting together admit enough Na⁺ to reach a threshold. The presynaptic neurons cooperate to induce the postsynaptic neuron to fire.

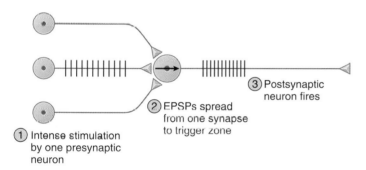

**(a) Temporal summation**

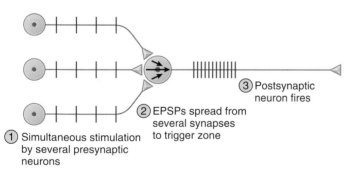

**(b) Spatial summation**

**FIGURE 12.25 Temporal and Spatial Summation.** Vertical lines on the nerve fibers indicate relative firing frequency. Arrows within the postsynaptic neuron indicate Na⁺ diffusion. (a) In temporal summation, a single presynaptic neuron stimulates the postsynaptic neuron so intensely that its EPSPs add up to threshold and make it fire. (b) In spatial summation, multiple inputs to the postsynaptic cell each produce a moderate amount of stimulation, but collectively they produce enough EPSPs to add up to threshold at the trigger zone and make the cell fire.

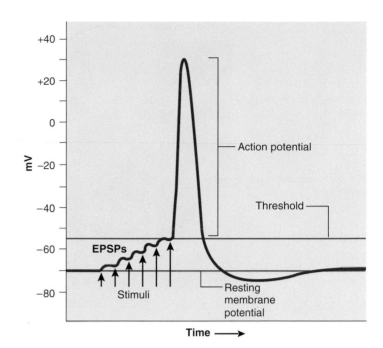

**FIGURE 12.26 Summation of EPSPs.** Each stimulus (arrows) produces one EPSP. If enough EPSPs arrive at the trigger zone faster than they decay, they can build on each other to bring the neuron to threshold and trigger an action potential.

Neurons routinely work in groups to modify each other's actions. **Facilitation** is a process in which one neuron enhances the effect of another one. In spatial summation, for example, one neuron acting alone may be unable to induce a postsynaptic neuron to fire. But when they cooperate, their combined "effort" does induce firing in the postsynaptic cell. They each enhance one another's effect, or *facilitate* each other.

**Presynaptic inhibition** is the opposite of facilitation, a mechanism in which one presynaptic neuron suppresses another one. This mechanism is used to reduce or halt unwanted synaptic transmission. In figure 12.27, we see three neurons which we will call neuron *S* for the stimulator, neuron *I* for the inhibitor, and neuron *R* for the responder. Neuron I forms an axoaxonic synapse with S (synapses with the axon of S). When presynaptic inhibition is not occurring, neuron S releases its neurotransmitter and triggers a response in R. But when there is a need to block transmission across this pathway, neuron I releases the inhibitory neurotransmitter GABA. GABA prevents the voltage-regulated calcium gates of neuron S from opening. Consequently, neuron S releases less neurotransmitter or none, and fails to stimulate neuron R.

## NEURAL CODING

The nervous system must interpret and pass along both quantitative and qualitative information about its environment—whether a light is dim or bright, red or

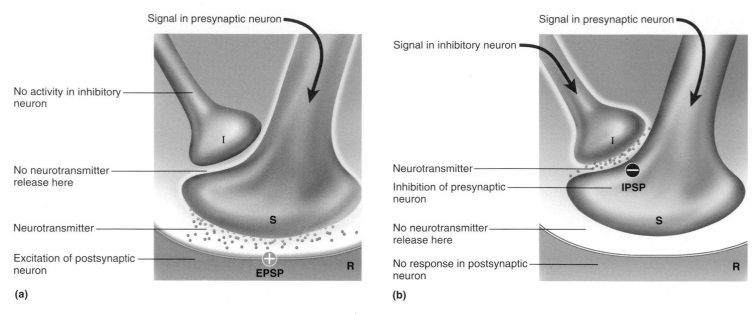

**FIGURE 12.27  Presynaptic Inhibition.**    S, stimulating neuron. I, inhibitory neuron. R, responding postsynaptic neuron. +, excitation (EPSP). –, inhibition (IPSP). (a) In the absence of inhibition, neuron S releases neurotransmitter and stimulates neuron R. (b) In presynaptic inhibition, neuron I suppresses the release of neurotransmitter by S, and S cannot stimulate R.

green; whether a taste is mild or intense, salty or sour; whether a sound is loud or soft, high-pitched or low. Considering the complexity of information to be communicated about conditions in and around the body, it is a marvel that it can be done in the form of something as simple as action potentials—particularly since all the action potentials of a given neuron are identical. Yet when we considered the genetic code in chapter 4, we saw that complex messages can indeed be expressed in simple codes. The way in which the nervous system converts information to a meaningful pattern of action potentials is called **neural coding** (or *sensory coding* when it occurs in the sense organs).

The most important mechanism for transmitting qualitative information is the **labeled line code.** This code is based on the fact that each nerve fiber to the brain leads from a receptor that specifically recognizes a particular stimulus type. Nerve fibers in the optic nerve, for example, carry signals only from light receptors in the eye; these fibers will never carry information about taste, sound, or touch. The brain therefore interprets any signals in those fibers in terms of light—even if the signals result from artificial stimulation of the nerve, or from a blow to the eye that excites the optic nerve and makes one see flashes of light. Electrical stimulation of the auditory nerve can enable deaf people to hear sounds of different frequencies, even though receptors in the inner ear are nonfunctional. Thus, each nerve fiber to the brain is a line of communication "labeled," or recognized by the brain,

as representing a particular stimulus quality—the color of a light, the pitch of a sound, or the salty or sour quality of a taste, for example.

Quantitative information—information about the intensity of a stimulus—is encoded in two ways. One depends on the fact that different neurons have different thresholds of excitation. A weak stimulus excites neurons with the lowest thresholds, while a strong stimulus excites less sensitive high-threshold neurons. Bringing additional neurons into play as the stimulus becomes stronger is called **recruitment.** It enables the nervous system to judge stimulus strength by which neurons, and how many of them, are firing.

Another way of encoding stimulus strength depends on the fact that the more strongly a neuron is stimulated, the more frequently it fires. A weak stimulus may cause a neuron to generate 6 action potentials per second, and a strong stimulus, 600 per second. Thus, the central nervous system can judge stimulus strength from the firing frequency of afferent neurons (fig. 12.28).

There is a limit to how often a neuron can fire, set by its absolute refractory period. Think of an electronic camera flash by analogy. If you take a photograph and your flash unit takes 15 seconds to recharge, then you cannot take more than four photographs per minute. Similarly, if a nerve fiber takes 1 msec to repolarize after it has fired, then it cannot fire more than 1,000 times per second. Refractory periods may be as short as 0.5 msec, which sets a theoretical limit to firing frequency of 2,000 action

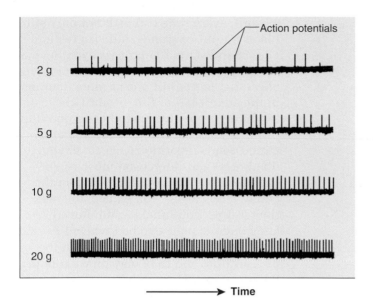

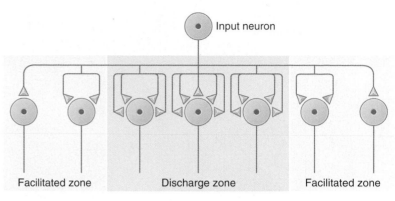

**FIGURE 12.29 Facilitated and Discharge Zones in a Neural Pool.** In the discharge zone, the presynaptic input neuron has so many synaptic contacts with each postsynaptic neuron that it alone can induce the postsynaptic cell to fire, employing spatial summation as in figure 12.25b. In a facilitated zone, the presynaptic neuron lacks enough synaptic contacts with a postsynaptic neuron to induce firing by itself. However, it can collaborate with other presynaptic neurons, facilitating each other in making the postsynaptic cell fire.

**FIGURE 12.28 An Example of Neural Coding.** This figure is based on recordings made from a sensory fiber of the frog sciatic nerve as the gastrocnemius muscle was stretched by suspending weights from it. As the stimulus strength (weight) and stretch increase, the firing frequency of the neuron increases. Firing frequency is a coded message that informs the CNS of stimulus intensity.

▶ *In what other way is the CNS informed of stimulus intensity?*

potentials per second. The highest frequencies actually observed, however, are between 500 and 1,000 per second.

In summary, a mild stimulus excites sensitive, low-threshold nerve fibers. As the stimulus intensity rises, these fibers fire at a higher and higher frequency, up to a certain maximum. If the stimulus intensity exceeds the capacity of these low-threshold fibers, it may recruit less sensitive, high-threshold fibers to begin firing. Still further increases in intensity cause these high-threshold fibers to fire at a higher and higher frequency.

> **Think About it**
>
> *How is neural recruitment related to the process of multiple motor unit summation described in chapter 11?*

## NEURAL POOLS AND CIRCUITS

So far, we have dealt with interactions involving only two or three neurons at a time. Actually, neurons function in larger ensembles called **neural pools,** each of which consists of thousands to millions of interneurons concerned with a particular body function—one to control the rhythm of your breathing, one to move your limbs rhythmically as you walk, one to regulate your sense of hunger,

and another to interpret smells, for example. At this point, we explore a few ways in which neural pools collectively process information.

Information arrives at a neural pool through one or more input neurons, which branch repeatedly and synapse with numerous interneurons in the pool. Some input neurons form multiple synapses with a single postsynaptic cell. They can produce EPSPs at all points of contact with that cell and, through spatial summation, make it fire more easily than if they synapsed with it at only one point. Within the **discharge zone** of an input neuron, an input neuron acting alone can make the postsynaptic cells fire (fig. 12.29). But in a broader **facilitated zone,** it synapses with still other neurons in the pool, with fewer synapses on each of them. It can stimulate those neurons to fire only with the assistance of other input neurons; that is, it facilitates the others. It "has a vote" on what the postsynaptic cells in the facilitated zone will do, but it cannot determine the outcome alone. Such arrangements, repeated thousands of times throughout the central nervous system, give neural pools great flexibility in integrating input from several sources and "deciding" on an appropriate output.

The functioning of a radio can be understood from a circuit diagram showing its components and their connections. Similarly, the functions of a neural pool are partly determined by its **neural circuit**—the pathways among its neurons. Just as a wide variety of electronic devices are constructed from a relatively limited number of circuit types, a wide variety of neural functions result

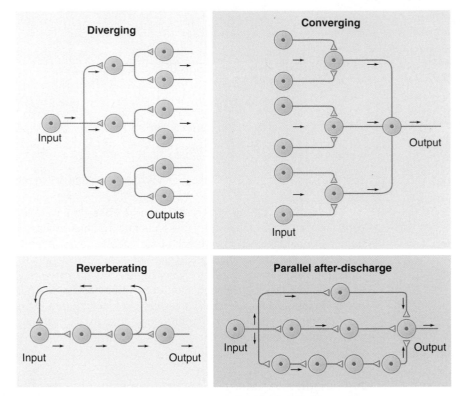

**FIGURE 12.30**  **Four Types of Neural Circuits.**  Arrows indicate the direction of the nerve signal.

▶ *Which of these four circuits is likely to fire the longest after a stimulus ceases? Why?*

from the operation of four principal kinds of neural circuits (fig. 12.30):

1. In a **diverging circuit,** one nerve fiber branches and synapses with several postsynaptic cells. Each of those may synapse with several more, so input from just one neuron may produce output through hundreds of neurons. Such a circuit allows one motor neuron of the brain, for example, to ultimately stimulate thousands of muscle fibers.

2. A **converging circuit** is the opposite of a diverging circuit—input from many different nerve fibers is funneled to one neuron or neural pool. Such an arrangement allows input from your eyes, inner ears, and stretch receptors in your neck to be channeled to an area of the brain concerned with the sense of balance. Also through neural convergence, a respiratory center in your brainstem receives input from other parts of your brain, from receptors for blood chemistry in your arteries, and from stretch receptors in your lungs. The respiratory center can then produce an output that takes all of these factors into account and sets an appropriate pattern of breathing.

3. In a **reverberating circuit,** neurons stimulate each other in a linear sequence such as A → B → C → D, but neuron C sends an axon collateral back to A. As

a result, every time C fires it not only stimulates output neuron D, but also restimulates A and starts the process over. Such a circuit produces a prolonged or repetitive effect that lasts until one or more neurons in the circuit fail to fire, or until an inhibitory signal from another source stops one of them from firing. A reverberating circuit sends repetitious signals to your diaphragm and intercostal muscles, for example, to make you inhale. When the circuit stops firing, you exhale; the next time it fires, you inhale again. Reverberating circuits may also be involved in short-term memory, as discussed in the next section, and they may play a role in the uncontrolled "storms" of neural activity that occur in epilepsy.

4. In a **parallel after-discharge circuit,** an input neuron diverges to stimulate several chains of neurons. Each chain has a different number of synapses, but eventually they all reconverge on a single output neuron. Since each pathway differs in total synaptic delay, their signals arrive at the output neuron at different times, and the output neuron may go on firing for some time after input has ceased. Unlike a reverberating circuit, this type has no feedback loop. Once all the neurons in the circuit have fired, the output ceases. Continued firing after the stimulus stops is called *after-discharge.* It explains why you can stare at a lamp, then close your eyes and continue to see an image of it for a while. Such a circuit is also important to withdrawal reflexes, in which a brief pain produces a longer-lasting output to the limb muscles and causes you to draw back your hand or foot from danger.

## MEMORY AND SYNAPTIC PLASTICITY

You may have wondered as you studied this chapter, How am I going to remember all of this? It seems fitting that we end with the subject of how memory works, for you now have the information necessary to understand its cellular and chemical basis.

The things we learn and remember are not stored in individual "memory cells" in the brain. We do not have a neuron assigned to remember our phone number and another assigned to remember our mother's birthday, for example. Instead, the physical basis of memory is a *pathway* through the brain called a **memory trace (engram[30]),** in which new synapses have formed or existing synapses have been modified to make transmission easier. In other

[30]*en* = inner + *gram* = mark, trace, record

words, synapses are not fixed for life; in response to experience, they can be added, taken away, or modified to make transmission easier or harder. Indeed, synapses can be formed or deleted in as little as 1 to 2 hours. This ability of synapses to change is called **synaptic plasticity.**

Think about when you learned as a child to tie your shoes. The procedure was very slow, confusing, and laborious at first, but eventually it became so easy you could do it with little thought—like a motor program playing out in your brain without requiring your conscious attention. It became easier to do because the synapses in a certain pathway were modified to allow signals to travel more easily across them than across "untrained" synapses. The process of making transmission easier is called **synaptic potentiation** (one form of synaptic plasticity).

Neuroscientists still argue about how to classify the various forms of memory, but three kinds often recognized are *immediate memory, short-term memory,* and *long-term memory.* We also know of different modes of synaptic potentiation that last from just a few seconds to a lifetime, and we can correlate these at least tentatively with different forms of memory.

## Immediate Memory

**Immediate memory** is the ability to hold something in mind for just a few seconds. By remembering what just happened, we get a feeling for the flow of events and a sense of the present. Immediate memory is indispensible to the ability to read; you must remember the earliest words of a sentence until you get to its end in order to extract any meaning from the sentence. You could not make any sense of what you read if you forgot each word as soon as you moved on to the next one. Immediate memory might be based on reverberating circuits. Our impression of what just happened can thus reecho in our minds for a few seconds as we experience the present moment and anticipate the next one.

## Short-Term Memory

**Short-term memory (STM)** lasts from a few seconds to a few hours. Information stored in STM may be quickly forgotten if we stop mentally reciting it, we are distracted, or we have to remember something new. **Working memory** is a form of STM that allows us to hold an idea in mind long enough to carry out an action such as calling a telephone number we just looked up, working out the steps of a mathematics problem, or searching for a lost set of keys while remembering where we have already looked. It is limited to a few bits of information such as the digits of a telephone number. These short-term memory tasks may be carried out by reverberating circuits of neurons.

Somewhat longer-lasting memories, however, probably involve a synaptic effect called **facilitation** (different from the facilitation of one neuron by another that we studied earlier in the chapter). This form of facilitation is induced by *tetanic stimulation,* the rapid arrival of repetitive signals at a synapse. Each signal causes a certain amount of $Ca^{2+}$ to enter the synaptic knob. If signals arrive very rapidly; the neuron cannot pump out all the $Ca^{2+}$ admitted by one action potential before the next action potential occurs. More and more $Ca^{2+}$ accumulates in the knob. Since $Ca^{2+}$ is what triggers the release of neurotransmitter, each signal releases more neurotransmitter than the one before. With more neurotransmitter, the EPSPs in the postsynaptic cell become stronger and stronger, and that cell is more likely to fire. Thus, tetanic stimulation facilitates the synapse and makes it easier for the postsynaptic cell to fire.

Memories lasting for a few hours, such as remembering what someone said to you earlier in the day or remembering an upcoming appointment, may involve **posttetanic potentiation.** In this process, the $Ca^{2+}$ level in the synaptic knob stays elevated for so long that another signal, coming well after the tetanic stimulation has ceased, releases an exceptionally large burst of neurotransmitter. That is, if a synapse has been heavily used in the recent past, a new stimulus can excite the postsynaptic cell more easily. Thus your memory may need only a slight jog to recall something from several hours earlier.

## Long-Term Memory

**Long-term memory (LTM)** lasts up to a lifetime and is less limited than STM in the amount of information it can store. LTM allows you to memorize the lines of a play, the words of a favorite song, or (one hopes!) textbook information for an exam. On a still longer timescale, it enables you to remember your name, the route to your home, and your childhood experiences.

There are two forms of long-term memory: declarative and procedural. **Declarative memory** is the retention of events and facts that you can put into words—numbers, names, dates, and so forth. **Procedural memory** is the retention of motor skills—how to tie your shoes, play a musical instrument, or type on a keyboard. These forms of memory involve different regions of the brain but are probably similar at the cellular level.

Some LTM involves the physical remodeling of synapses or the formation of new ones through the growth and branching of axon terminals and dendrites. In the pyramidal cells of the brain, the dendrites are studded with knoblike *dendritic spines* that increase the area of synaptic contact. Studies on fish and other experimental animals have shown that social and sensory deprivation causes these spines to decline in number, while a richly stimulatory environment causes them to proliferate—an intriguing clue to the importance of a stimulating environment to infant and child development. In some cases of LTM, a new synapse grows beside the original one, giving the presynaptic cell twice as much input into the postsynaptic cell.

LTM can also be grounded in molecular changes called **long-term potentiation.** This involves *NMDA*[31] *receptors,* which occur on the synaptic knobs of the pyramidal cells and bind the neurotransmitter glutamate. NMDA receptors are usually blocked by magnesium ions ($Mg^{2+}$), but when they bind glutamate *and* are simultaneously subjected to tetanic stimulation, they expel the $Mg^{2+}$ and open to admit $Ca^{2+}$ into the dendrite. When $Ca^{2+}$ enters, it acts as a second messenger that leads to a variety of effects:

- The neuron produces an increased number of NMDA receptors, which makes it more sensitive to glutamate in the future.
- It synthesizes proteins concerned with physically remodeling a synapse.
- It releases nitric oxide, which diffuses back to the presynaptic neuron and triggers events that ultimately increase glutamate release.

You can see that in all of these ways, long-term potentiation can increase transmission across "experienced"

---

## INSIGHT 12.4    Clinical Application

### Alzheimer and Parkinson Diseases

Alzheimer and Parkinson diseases are degenerative disorders of the brain associated with neurotransmitter deficiencies.

*Alzheimer*[32] *disease (AD)* may begin before the age of 50 with symptoms so slight and ambiguous that early diagnosis is difficult. One of its first symptoms is memory loss, especially for recent events. A person with AD may ask the same questions repeatedly, show a reduced attention span, and become disoriented and lost in previously familiar places. Family members often feel helpless and confused as they watch their loved one's personality gradually deteriorate beyond recognition. The AD patient may become moody, confused, paranoid, combative, or hallucinatory—he or she may ask irrational questions such as Why is the room full of snakes? The patient may eventually lose even the ability to read, write, talk, walk, and eat. Death ensues from pneumonia or other complications of confinement and immobility.

AD affects about 11% of the U.S. population over the age of 65; the incidence rises to 47% by age 85. It accounts for nearly half of all nursing home admissions and is a leading cause of death among the elderly. AD claims about 100,000 lives per year in the United States.

Diagnosis of AD is confirmed on autopsy. There is atrophy of some of the gyri (folds) of the cerebral cortex and the hippocampus, an important center of memory. Nerve cells exhibit *neurofibrillary tangles*—dense masses of broken and twisted cytoskeleton (fig. 12.31). Alois Alzheimer first observed these in 1907 in the brain of a patient who had died of senile dementia. The more severe the signs of disease, the more neurofibrillary tangles are seen at autopsy. In the intercellular spaces, there are

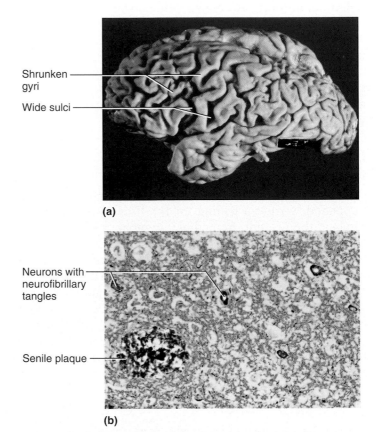

(a)

(b)

**FIGURE 12.31   Alzheimer Disease.**   (a) Brain of a person who died of AD. Note the shrunken folds of cerebral tissue (gyri) and wide gaps (sulci) between them. (b) Cerebral tissue from a person with AD. Neurofibrillary tangles are present within the neurons, and a senile plaque is evident in the extracellular matrix.

---

[31] *N*-methyl-D-aspartate, a chemical similar to glutamate
[32] Alois Alzheimer (1864–1915), German neurologist

synapses. Remodeling a synapse or increasing the number of neurotransmitter receptors has longer-lasting effects than facilitation or posttetanic potentiation.

## Before You Go On

*Answer the following questions to test your understanding of the preceding section:*

22. Contrast the two types of summation at a synapse.

23. *Describe how the nervous system communicates quantitative and qualitative information about stimuli.*

24. *List the four types of neural circuits and describe their similarities and differences. Discuss the unity of form and function in these four types—that is, explain why each type would not perform as it does if its neurons were connected differently.*

25. *How does long-term potentiation enhance the transmission of nerve signals along certain pathways?*

*senile plaques* consisting of aggregations of cells, altered nerve fibers, and a core of *β-amyloid protein*—the breakdown product of a glycoprotein of plasma membranes. Amyloid protein is rarely seen in elderly people without AD.

AD is marked by the degeneration of cholinergic neurons and a deficiency of ACh. Treatment with ACh precursors is ineffective, but therapy with cholinesterase inhibitors to slow down the degradation of existing ACh has been of some value. AD patients show a deficiency of nerve growth factor (NGF; see Insight 12.3) in some regions of the brain. NGF stimulates ACh synthesis; it helps to retard brain degeneration in humans and other animals and improves memory in some AD patients. Intense biomedical research efforts are currently geared toward identifying the cause of AD and developing treatment strategies. Researchers have identified three genes on chromosomes 1, 14, and 21 for various forms of early- and late-onset AD.

Parkinson[33] disease (PD), also called *paralysis agitans* or *parkinsonism,* is a progressive loss of motor function beginning in a person's 50s or 60s. It is due to degeneration of dopamine-releasing neurons in a portion of the brain called the *substantia nigra.* A gene has recently been identified for a hereditary form of PD, but most cases are nonhereditary and of little-known cause; some authorities suspect environmental neurotoxins.

Dopamine (DA) is an inhibitory neurotransmitter that normally prevents excessive activity in motor centers of the brain called the *basal nuclei.* Degeneration of the dopamine-releasing neurons leads to an excessive ratio of ACh to DA, leading to hyperactivity of the basal nuclei. As a result, a person with PD suffers involuntary muscle contractions. These take such forms as shaking of the hands (tremor) and compulsive "pill-rolling" motions of the thumb and fingers. In addition, the facial muscles may become rigid and produce a staring, expressionless face with a slightly open mouth. The patient's range of motion diminishes. He or she takes smaller steps and develops a slow, shuffling gait with a forward-bent posture and a tendency to fall forward. Speech becomes slurred and handwriting becomes cramped and eventually illegible. Tasks such as buttoning clothes and preparing food become increasingly laborious.

Patients cannot be expected to recover from PD, but its effects can be alleviated with drugs and physical therapy. Treatment with dopamine is ineffective because it cannot cross the blood–brain barrier, but its precursor, levodopa (L-dopa), does cross the barrier and has been used to treat PD since the 1960s. L-dopa affords some relief from symptoms, but it does not slow progression of the disease and it has undesirable side effects on the liver and heart. It is effective for only 5 to 10 years of treatment. A newer drug, deprenyl, is a monoamine oxidase (MAO) inhibitor that retards neural degeneration and delays the development of symptoms. Modest improvement has been obtained by implanting other dopamine-producing tissues into the brains of PD patients—namely, adrenal medulla and fetal brain tissue. Even though the latter tissue has not come from elective abortions, this approach has triggered ethical controversy.

A surgical technique called *pallidotomy* has been used since the 1940s to alleviate severe tremors. It involves the destruction of a small portion of cerebral tissue in an area called the *globus pallidus.* Pallidotomy fell out of favor in the late 1960s when L-dopa came into common use. By the early 1990s, however, the limitations of L-dopa had become apparent, while MRI- and CT-guided methods had improved surgical precision and reduced the risks of brain surgery. Pallidotomy has thus made a comeback. Other surgical treatments for parkinsonism target brain areas called the *subthalamic nucleus* and the *ventral intermediate nucleus* of the thalamus, and involve either the destruction of tiny areas of tissue or the implantation of a stimulating electrode.

---

[33]James Parkinson (1755–1824), British physician

# CONNECTIVE ISSUES

## Interactions Between the NERVOUS SYSTEM and Other Organ Systems

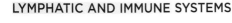

■ indicates ways in which this system affects other systems

■ indicates ways in which other systems affect this system

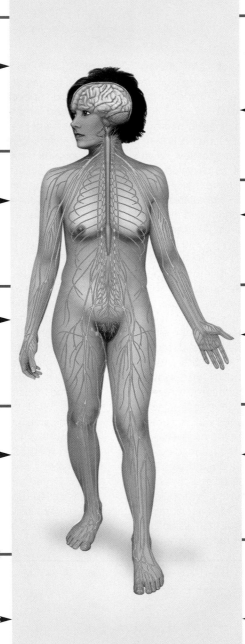

## INTEGUMENTARY SYSTEM

Nervous system regulates piloerection and sweating; controls cutaneous blood flow to regulate heat loss

Provides sensations of heat, cold, pressure, pain, and vibration; protects peripheral nerves

## SKELETAL SYSTEM

Nervous stimulation generates muscle tension essential for bone development and remodeling

Serves as reservoir of $Ca^{2+}$ needed for neural function; protects CNS and some peripheral nerves

## MUSCULAR SYSTEM

Somatic nervous system activates skeletal muscles and maintains muscle tone

Gives expression to thoughts, emotions, and motor commands that arise in the CNS

## ENDOCRINE SYSTEM

Hypothalamus controls pituitary gland; sympathetic nervous system stimulates adrenal medulla

Many hormones affect neuron growth and metabolism; hormones control electrolyte balance essential for neural function

## CIRCULATORY SYSTEM

Nervous system regulates heartbeat, blood vessel diameters, blood pressure, and routing of blood; influences blood clotting

Delivers $O_2$ and carries away wastes; transports hormones to and from CNS; CSF is produced from and returned to blood

## LYMPHATIC AND IMMUNE SYSTEMS

Nerves innervate lymphoid organs and influence development and activity of immune cells; nervous system plays a role in regulating immune response; emotional states influence susceptibility to infection

Immune cells provide protection and promote tissue repair

## RESPIRATORY SYSTEM

Nervous system regulates rate and depth of respiration

Provides $O_2$, removes $CO_2$, and helps to maintain proper pH for neural function

## URINARY SYSTEM

Nervous system regulates renal blood flow, thus affecting rate of urine formation; controls emptying of bladder

Disposes of wastes and maintains electrolyte and pH balance

## DIGESTIVE SYSTEM

Nervous system regulates appetite, feeding behavior, digestive secretion and motility, and defecation

Provides nutrients; liver provides stable level of blood glucose for neural function during periods of fasting

## REPRODUCTIVE SYSTEM

Nervous system regulates sex drive, arousal, and orgasm; secretes or stimulates pituitary release of many hormones involved in menstrual cycle, sperm production, pregnancy, and lactation

Sex hormones influence CNS development and sexual behavior; hormones of the menstrual cycle stimulate or inhibit hypothalamus

## CHAPTER REVIEW

# Review of Key Concepts

**Overview of the Nervous System (p. 442)**

1. The nervous and endocrine systems are the body's two main systems of internal communication and physiological coordination.

2. The nervous system receives information from *receptors, integrates* information, and issues commands to *effectors.*

3. The nervous system is divided into the *central nervous system (CNS)* and *peripheral nervous system (PNS).* The PNS has *sensory* and *motor* divisions, and each of these has *somatic* and *visceral* subdivisions.

4. The visceral motor division is also called the *autonomic nervous system,* which has *sympathetic* and *parasympathetic* divisions.

**Nerve Cells (Neurons) (p. 443)**

1. Neurons have the properties of *excitability, conductivity,* and *secretion.*

2. Neurons are classified by their location and function into *sensory (afferent) neurons, interneurons* (association neurons), and *motor (efferent) neurons.*

3. A neuron has a *soma* where its nucleus and most other organelles are located; usually multiple *dendrites* that receive signals and conduct them to the soma; and one *axon* (nerve fiber) that carries nerve signals away from the soma.

4. The axon branches at the distal end into a *terminal arborization,* and each branch ends in a *synaptic knob.* The synaptic knob contains *synaptic vesicles,* which contain neurotransmitters.

5. Neurons are described as *multipolar, bipolar,* or *unipolar* depending on the number of dendrites present, or *anaxonic* if they have no axon.

6. Neurons move material along the axon by *axonal transport,* which can be *fast* or *slow, anterograde* (away from the soma) or *retrograde* (toward the soma).

**Supportive Cells (Neuroglia) (p. 448)**

1. Supportive cells called *neuroglia* greatly outnumber neurons. There are six kinds of neuroglia: *oligodendrocytes, ependymal cells, microglia,* and *astrocytes* in the CNS, and *Schwann cells* and *satellite cells* in the PNS.

2. *Oligodendrocytes* produce the myelin sheath around CNS nerve fibers.

3. *Ependymal cells* line the inner cavities of the CNS and secrete and circulate cerebrospinal fluid.

4. *Microglia* are macrophages that destroy microorganisms, foreign matter, and dead tissue in the CNS.

5. *Astrocytes* play a wide variety of protective, nutritional, homeostatic, and communicative roles for the neurons, and form scar tissue when CNS tissue is damaged.

6. *Schwann cells* cover nerve fibers in the PNS and produce myelin around many of them.

7. *Satellite cells* surround somas of the PNS neurons and have an uncertain function.

8. *Myelin* is a multilayered coating of oligodendrocyte or Schwann cell membrane around a nerve fiber, with periodic gaps called *nodes of Ranvier* between the glial cells.

9. Signal transmission is relatively slow in small nerve fibers, unmyelinated fibers, and at nodes of Ranvier. It is much faster in large nerve fibers and myelinated segments *(internodes)* of a fiber.

10. Damaged nerve fibers in the PNS sometimes regenerate if the soma is unharmed. Repair requires a *regeneration tube* composed of Schwann cells, *basal lamina,* and *neurilemma,* which are present only in the PNS. Damaged fibers in the CNS cannot regenerate.

**Electrophysiology of Neurons (p. 453)**

1. An *electrical potential* is a difference in electrical charge between two points. When a cell has a charge difference between the two sides of the plasma membrane, it is *polarized.* The charge difference is called the *resting membrane potential (RMP).* For a resting neuron, it is typically $-70$ mV (negative on the intracellular side).

2. A *current* is a flow of charge particles—especially, in living cells, $Na^+$ and $K^+$. Resting cells have more $K^+$ inside than outside the cell, and more $Na^+$ outside than inside. A current occurs when gates in the plasma membrane open and allow these ions to diffuse across the membrane, down their concentration gradients.

3. When a neuron is stimulated on the dendrites or soma, $Na^+$ gates open and allow $Na^+$ to enter the cell. This slightly depolarizes the membrane, creating a *local potential.* Short-distance diffusion of $Na^+$ inside the cell allows local potentials to spread to nearby areas of membrane.

4. Local potentials are *graded, decremental, reversible,* and can be *excitatory* or *inhibitory.*

5. The trigger zone and unmyelinated regions of a nerve fiber have voltage-regulated $Na^+$ and $K^+$ gates that open in response to changes in membrane potential and allow these ions through.

6. If a local potential reaches *threshold,* voltage-regulated gates open. The inward movement of $Na^+$ followed by the outward movement of $K^+$ creates a quick voltage change called an *action potential.* The cell *depolarizes* as the membrane potential becomes less negative, and *repolarizes* as it returns toward the RMP.

7. Unlike local potentials, action potentials follow an *all-or-none law* and are *nondecremental* and *irreversible.* Following an action potential, a patch of cell membrane has a

*refractory period* in which it cannot respond to another stimulus.

8. One action potential triggers another in the plasma membrane just distal to it. By repetition of this process, a chain of action potentials, or *nerve signal,* travels the entire length of an unmyelinated axon. The refractory period of the recently active membrane prevents this signal from traveling backward toward the soma.

9. In a myelinated fiber, only the initial segment and nodes of Ranvier have voltage-regulated gates. In the internodes, the signal travels rapidly by Na$^+$ diffusing along the intracellular side of the membrane. At each node, new action potentials occur, slowing the signal somewhat, but restoring signal strength. Myelinated nerve fibers are said to show *saltatory conduction* because the signal seems to jump from node to node.

### Synapses (p. 462)

1. At the distal end of a nerve fiber is a *synapse* where it meets the next cell (usually another neuron or a muscle or gland cell).

2. The *presynaptic* neuron must release chemical signals called *neurotransmitters* to cross the synaptic cleft and stimulate the next *(postsynaptic)* cell.

3. Neurotransmitters include acetylcholine (ACh), monoamines such as norepinephrine (NE) and serotonin, amino acids such as glutamate and GABA, and neuropeptides such as β-endorphin and substance P. A single neurotransmitter can affect different cells differently, because of the variety of receptors for it that various cells possess.

4. Some synapses are excitatory, as when ACh triggers the opening of Na$^+$–K$^+$ gates and depolarizes the postsynaptic cell, or when NE triggers the synthesis of the second messenger cAMP.

5. Some synapses are inhibitory, as when GABA opens a Cl$^-$ gate and the inflow of Cl$^-$ hyperpolarizes the postsynaptic cell.

6. In some synapses, a neurotransmitter such as NE does not directly open an ion gate, but acts through second-messenger systems such as cAMP.

7. Synaptic transmission ceases when the neurotransmitter diffuses away from the synaptic cleft, is reabsorbed by the presynaptic cell, or is degraded by an enzyme in the cleft such as acetylcholinesterase (AChE).

8. Hormones, neuropeptides, nitric oxide (NO), and other chemicals can act as *neuromodulators,* which alter synaptic function by altering neurotransmitter synthesis, release, reuptake, or breakdown.

### Neural Integration (p. 467)

1. Synapses slow down communication in the nervous system, but their role in *neural integration* (information processing and decision making) overrides this drawback.

2. Neural integration is based on the relative effects of small depolarizations called *excitatory postsynaptic potentials (EPSPs)* and small hyperpolarizations called *inhibitory postsynaptic potentials (IPSPs)* in the postsynaptic membrane. EPSPs make it easier for the postsynaptic neuron to fire, and IPSPs make it harder.

3. Some combinations of neurotransmitter and receptor produce EPSPs and some produce IPSPs. The postsynaptic neuron can fire only if EPSPs override IPSPs enough for the membrane voltage to reach threshold.

4. One neuron receives input from thousands of others, some producing EPSPs and some producing IPSPs. *Summation,* the adding up of these potentials, occurs in the trigger zone. Two types of summation are *temporal* (based on how frequently a presynaptic neuron is stimulating the postsynaptic one) and *spatial* (based on how many presynaptic neurons are simultaneously stimulating the postsynaptic one).

5. One presynaptic neuron can *facilitate* another, making it easier for the second to stimulate a postsynaptic cell, or it can produce *presynaptic inhibition,* making it harder for the second one to stimulate the postsynaptic cell.

6. Neurons encode qualitative and quantitative information by means of *neural coding.* Stimulus type (qualitative information) is represented by which nerve cells are firing (the *labeled line code).* Stimulus intensity (quantitative information) is represented both by which nerve cells are firing and by their firing frequency.

7. The refractory period sets an upper limit on how frequently a neuron can fire.

8. Neurons work in groups called *neural pools.*

9. A presynaptic neuron can, by itself, cause postsynaptic neurons in its *discharge zone* to fire. In its *facilitated zone,* it can only get a postsynaptic cell to fire by collaborating with other presynaptic neurons (facilitating each other).

10. Signals can travel *diverging, converging, reverberating,* or *parallel after-discharge circuits* of neurons.

11. Memories are formed by neural pathways of modified synapses. The ability of synapses to change with experience is called *synaptic plasticity,* and changes that make synaptic transmission easier are called *synaptic potentiation.*

12. Immediate memory may be based on reverberating circuits. *Short-term memory (STM)* may employ these circuits as well as *synaptic facilitation,* which is thought to involve an accumulation of Ca$^{21}$ in the synaptic knob.

13. *Long-term memory (LTM)* involves the remodeling of synapses, or modification of existing synapses so that they release more neurotransmitter or have more receptors for a neurotransmitter. The two forms of LTM are *declarative* and *procedural memory.*

# Testing Your Recall

1. The integrative functions of the nervous system are performed mainly by
   a. afferent neurons.
   b. efferent neurons.
   c. neuroglia.
   d. sensory neurons.
   e. interneurons.

2. The highest density of voltage-regulated ion gates is found on the _____ of a neuron.
   a. dendrites
   b. soma
   c. nodes of Ranvier
   d. internodes
   e. synaptic knobs

3. The soma of a mature neuron lacks
   a. a nucleus.
   b. endoplasmic reticulum.
   c. lipofuscin.
   d. centrioles.
   e. ribosomes.

4. The glial cells that fight infections in the CNS are
   a. microglia.
   b. satellite cells.
   c. ependymal cells.
   d. oligodendrocytes.
   e. astrocytes.

5. Posttetanic potentiation of a synapse increases the amount of _____ in the synaptic knob.
   a. neurotransmitter
   b. neurotransmitter receptors
   c. calcium
   d. sodium
   e. NMDA

6. An IPSP is _____ of the postsynaptic neuron.
   a. a refractory period
   b. an action potential
   c. a depolarization
   d. a repolarization
   e. a hyperpolarization

7. Saltatory conduction occurs only
   a. at chemical synapses.
   b. in the initial segment of an axon.
   c. in both the initial segment and axon hillock.
   d. in myelinated nerve fibers.
   e. in unmyelinated nerve fibers.

8. Some neurotransmitters can have either excitatory or inhibitory effects depending on the type of
   a. receptors on the postsynaptic neuron.
   b. synaptic vesicles in the axon.
   c. synaptic potentiation that occurs.
   d. postsynaptic potentials on the synaptic knob.
   e. neuromodulator involved.

9. Differences in the volume of a sound are likely to be encoded by differences in _____ in nerve fibers from the inner ear.
   a. neurotransmitters
   b. signal conduction velocity
   c. types of postsynaptic potentials
   d. firing frequency
   e. voltage of the action potentials

10. Motor effects that depend on repetitive output from a neural pool are most likely to use
    a. parallel after-discharge circuits.
    b. reverberating circuits.
    c. facilitated circuits.
    d. diverging circuits.
    e. converging circuits.

11. Neurons that convey information to the CNS are called sensory, or _____, neurons.

12. To perform their role, neurons must have the properties of excitability, secretion, and _____.

13. The _____ is a period of time in which a neuron is producing an action potential and cannot respond to another stimulus of any strength.

14. Neurons receive incoming signals by way of specialized processes called _____.

15. In the CNS, myelin is produced by glial cells called _____.

16. A myelinated nerve fiber can produce action potentials only in specialized regions called _____.

17. The trigger zone of a neuron consists of its _____ and _____.

18. The neurotransmitter secreted at an adrenergic synapse is _____.

19. A presynaptic nerve fiber cannot cause other neurons in its _____ to fire, but it can make them more sensitive to stimulation from other presynaptic fibers.

20. _____ are substances released along with a neurotransmitter that modify the neurotransmitter's effect.

*Answers in Appendix B*

# True or False

*Determine which five of the following statements are false, and briefly explain why.*

1. A neuron never has more than one axon.

2. Oligodendrocytes perform the same function in the brain as Schwann cells do in the peripheral nerves.

3. A resting neuron has a higher concentration of $K^+$ in its cytoplasm than in the extracellular fluid surrounding it.

4. During an action potential, a neuron is repolarized by the outflow of $Na^+$.

5. Excitatory postsynaptic potentials lower the threshold of a neuron and thus make it easier to stimulate.

6. The absolute refractory period sets an upper limit on how often a neuron can fire.

7. A given neurotransmitter has the same effect no matter where in the body it is secreted.

8. Nerve signals travel more rapidly through the nodes of Ranvier than through the internodes.

9. The synaptic contacts in the nervous system are fixed by the time of birth and cannot be changed thereafter.

10. Mature neurons are incapable of mitosis.

*Answers in Appendix B*

# Testing Your Comprehension

1. Schizophrenia is sometimes treated with drugs such as chlorpromazine that inhibit dopamine receptors. A side effect is that patients begin to develop muscle tremors, speech impairment, and other disorders similar to Parkinson disease. Explain.

2. Hyperkalemia is an excess of potassium in the extracellular fluid. What effect would this have on the resting membrane potentials of the nervous system and on neural excitability?

3. Suppose the $Na^+$–$K^+$ pumps of nerve cells were to slow down because of some metabolic disorder. How would this affect the resting membrane potentials of neurons? Would it make neurons more excitable than normal, or make them more difficult to stimulate? Explain.

4. The unity of form and function is an important concept in understanding synapses. Give two structural reasons why nerve signals cannot travel backward across a chemical synapse. What might be the consequences if signals did travel freely in both directions?

5. The local anesthetics tetracaine and procaine (Novocain) prevent voltage-regulated $Na^+$ gates from opening. Explain why this would block the conduction of pain signals in a sensory nerve.

*Answers at www.mhhe.com/saladin4*

# www.mhhe.com/saladin4

*The textbook website provides a wealth of interactive study materials fully organized and integrated by chapter. You will find practice quizzes, labeling exercises, and much more that will complement your learning and understanding of anatomy and physiology. The website also includes tools designed to enhance your **Anatomy & Physiology | REVEALED** experience.*

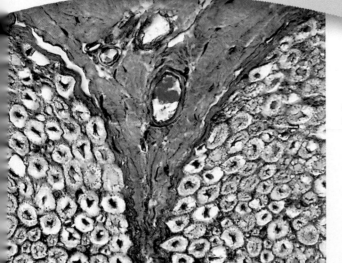

*Cross section through two fascicles (bundles) of nerve fibers in a nerve*

# THE SPINAL CORD, SPINAL NERVES, AND SOMATIC REFLEXES

## CHAPTER OUTLINE

## INSIGHTS

## Brushing Up

To understand this chapter, it is important that you understand or brush up on the following concepts:

- Function of antagonistic muscles (p. 325)
- Basic neuron structure (p. 444)
- EPSPs and IPSPs (p. 468)
- Parallel after-discharge circuits (p. 472)

We studied the nervous system at a cellular level in chapter 12. In these next two chapters, we move up the structural hierarchy to study the nervous system at the organ and system levels of organization.

The *spinal cord* is an "information highway" between the brain and the trunk and limbs. It is about as thick as a finger, and extends through the vertebral canal as far as the first lumbar vertebra. At regular intervals, it gives off a pair of *spinal nerves* that receive sensory input from the skin, muscles, bones, joints, and viscera, and that issue motor commands back to muscle and gland cells. The spinal cord is a component of the central nervous system and the spinal nerves a component of the peripheral nervous system, but these central and peripheral components are so closely linked structurally and functionally that it is appropriate that we consider them together in this chapter. The brain and cranial nerves will be discussed in chapter 14.

# The Spinal Cord

### Objectives
When you have completed this section, you should be able to

- state the three principal functions of the spinal cord;
- describe the gross and microscopic structure of the spinal cord; and
- trace the pathways (tracts) followed by nervous signals traveling up and down the spinal cord.

## FUNCTIONS

The spinal cord serves three principal functions:

1. **Conduction.** It contains bundles of nerve fibers that conduct information up and down the cord, connecting different levels of the trunk with each other and with the brain. This enables sensory information to reach the brain, motor commands to reach the effectors, and input received at one level of the cord to affect output from another level.

2. **Locomotion.** Walking involves repetitive, coordinated contractions of several muscle groups in the limbs. Motor neurons in the brain initiate walking and determine its speed, distance, and direction, but the simple repetitive muscle contractions that put one foot in front of another, over and over, are coordinated by groups of neurons called **central pattern generators** in the cord. These neural circuits produce the sequence of outputs to the extensor and flexor muscles that cause alternating movements of the legs.

3. **Reflexes.** Reflexes are involuntary stereotyped responses to stimuli, such as the withdrawal of a hand from pain. They involve the brain, spinal cord, and peripheral nerves.

## GROSS ANATOMY

The **spinal cord** (fig. 13.1) is a cylinder of nervous tissue that begins at the foramen magnum of the skull and passes through the vertebral canal as far as the inferior margin of the first lumbar vertebra (L1) or slightly beyond. In adults, it averages about 1.8 cm thick and 45 cm long. Early in fetal development, the spinal cord extends for the full length of the vertebral column. However, the vertebral column grows faster than the spinal cord, so the cord extends only to L3 by the time of birth and to L1 in an adult. Thus, it occupies only the upper two-thirds of the vertebral canal; the lower one-third is described shortly. The cord gives rise to 31 pairs of spinal nerves that pass through the intervertebral foramina. Although the spinal cord is not visibly segmented, the part supplied by each pair of spinal nerves is called a *segment.* The cord exhibits longitudinal grooves on its ventral and dorsal sides—the *ventral median fissure* and *dorsal median sulcus,* respectively.

The spinal cord is divided into **cervical, thoracic, lumbar,** and **sacral regions.** It may seem odd that it has a sacral region when the cord itself ends well above the sacrum. These regions, however, are named for the level of the vertebral column from which the spinal nerves emerge, not for the vertebrae that contain the cord itself.

In two areas, the cord is a little thicker than elsewhere. In the inferior cervical region, a **cervical enlargement** of the cord gives rise to nerves of the upper limbs. In the lumbosacral region, there is a similar **lumbar enlargement** where nerves to the pelvic region and lower limbs arise. Inferior to the lumbar enlargement, the cord tapers to a point called the **medullary cone** (conus medullaris). The lumbar enlargement and medullary cone give rise to a bundle of nerve roots that occupy the vertebral canal from L2 to S5. This bundle, named the **cauda equina**[1] (CAW-duh ee-KWY-nah) for its resemblance to a horse's tail, innervates the pelvic organs and lower limbs.

> **Think About It**
>
> *Spinal cord injuries commonly result from fractures of vertebrae C5 to C6, but never from fractures of L3 to L5. Explain both observations.*

---

[1]*cauda* = tail + *equin* = horse

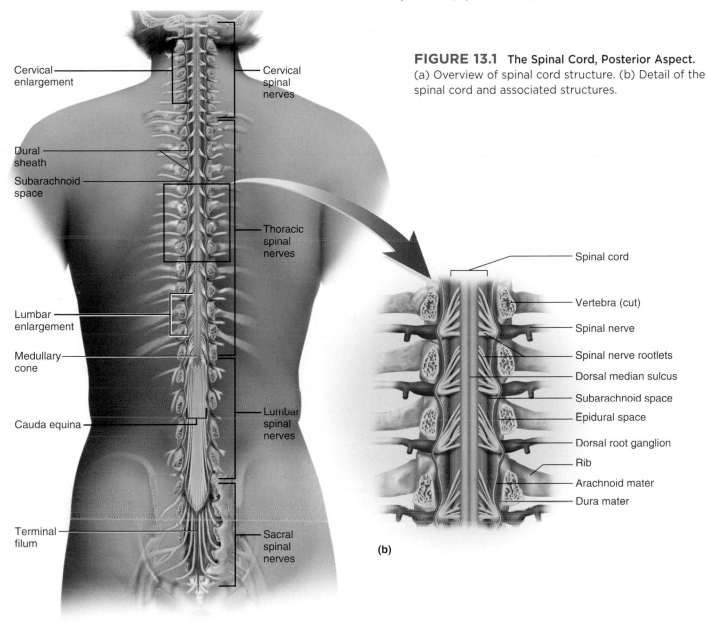

**FIGURE 13.1  The Spinal Cord, Posterior Aspect.**
(a) Overview of spinal cord structure. (b) Detail of the spinal cord and associated structures.

Cervical enlargement

Cervical spinal nerves

Dural sheath

Subarachnoid space

Thoracic spinal nerves

Lumbar enlargement

Medullary cone

Cauda equina

Lumbar spinal nerves

Terminal filum

Sacral spinal nerves

(a)

Spinal cord

Vertebra (cut)

Spinal nerve

Spinal nerve rootlets

Dorsal median sulcus

Subarachnoid space

Epidural space

Dorsal root ganglion

Rib

Arachnoid mater

Dura mater

(b)

## MENINGES OF THE SPINAL CORD

The spinal cord and brain are enclosed in three fibrous connective tissue membranes called **meninges**[2] (meh-NIN-jeez)—singular, *meninx* (MEN-inks). These membranes separate the soft tissue of the central nervous system from the bones of the vertebrae and skull. From superficial to deep, they are the dura mater, arachnoid mater, and pia mater.

The **dura mater**[3] (DOO-ruh MAH-tur) forms a loose-fitting sleeve called the **dural sheath** around the spinal cord. It is a tough collagenous membrane about as thick as a rubber kitchen glove. The space between the sheath and

vertebral bones, called the **epidural space,** is occupied by blood vessels, adipose tissue, and loose connective tissue (fig. 13.2a). Anesthetics are sometimes introduced to this space to block pain signals during childbirth or surgery; this procedure is called *epidural anesthesia.*

The **arachnoid**[4] (ah-RACK-noyd) **mater** adheres to the dural sheath. It consists of a simple squamous epithelium, the *arachnoid membrane,* adhering to the inside of the dura, and a loose mesh of collagenous and elastic fibers spanning the gap between the arachnoid membrane and the pia mater. This gap, called the **subarachnoid space,** is filled with cerebrospinal fluid (CSF), a clear liquid discussed in chapter 14.

[2]*menin* = membrane
[3]*dura* = tough + *mater* = mother, womb

[4]*arachn* = spider, spider web + *oid* = resembling

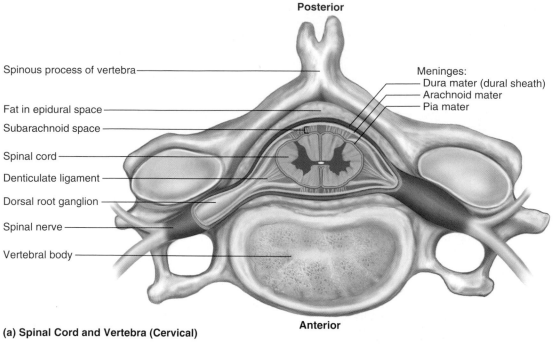

**Posterior**

Spinous process of vertebra

Fat in epidural space
Subarachnoid space

Spinal cord

Denticulate ligament

Dorsal root ganglion

Spinal nerve

Vertebral body

Meninges:
Dura mater (dural sheath)
Arachnoid mater
Pia mater

**Anterior**

**(a) Spinal Cord and Vertebra (Cervical)**

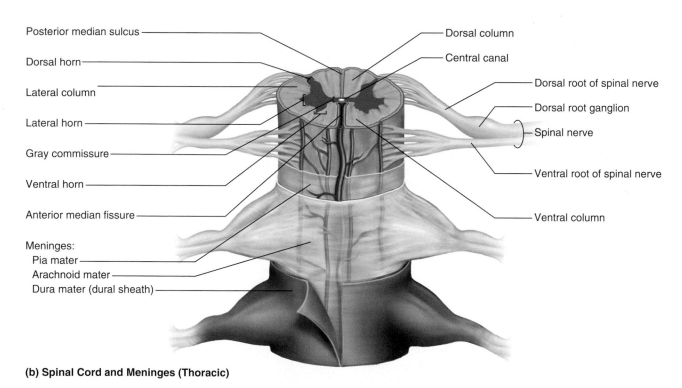

Posterior median sulcus

Dorsal horn

Lateral column

Lateral horn

Gray commissure

Ventral horn

Anterior median fissure

Meninges:
Pia mater
Arachnoid mater
Dura mater (dural sheath)

Dorsal column

Central canal

Dorsal root of spinal nerve

Dorsal root ganglion

Spinal nerve

Ventral root of spinal nerve

Ventral column

**(b) Spinal Cord and Meninges (Thoracic)**

**FIGURE 13.2**  **Cross Section of Spinal Cord at the Level of the Second Lumbar Vertebra.**  (a) Relationship to the vertebra, meninges, and spinal nerve. (b) Detail of the spinal cord, meninges, and spinal nerves.

The **pia**[5] (PEE-uh) **mater** is a delicate, translucent membrane that closely follows the contours of the spinal cord. It continues beyond the medullary cone as a

fibrous strand, the **terminal filum,** forming part of the *coccygeal ligament* that anchors the cord to vertebra Co1. At regular intervals along the cord, extensions of the pia called **denticulate ligaments** extend through the arachnoid to the dura, anchoring the cord and limiting side-to-side movements.

[5]*pia* = tender, soft

## INSIGHT 13.1    Clinical Application

### Spina Bifida

About one baby in 1,000 is born with *spina bifida* (SPY-nuh BIF-ih-duh), a congenital defect in which one or more vertebrae fail to form a complete vertebral arch for enclosure of the spinal cord. This is especially common in the lumbosacral region. One form, *spina bifida occulta,*[6] involves only one to a few vertebrae and causes no functional problems. Its only external sign is a dimple or hairy pigmented spot. *Spina bifida cystica*[7] is more serious. A sac protrudes from the spine and may contain meninges, cerebrospinal fluid, and parts of the spinal cord and nerve roots (fig. 13.3). In extreme cases, inferior spinal cord function is absent, causing lack of bowel control and paralysis of the lower limbs and urinary bladder. The last of these conditions can lead to chronic urinary infections and renal failure. An ample dietary intake of folic acid (a B vitamin) reduces the risk of bearing a child with spina bifida. Unfortunately, it is now thought that by the time a woman knows she is pregnant, it is already too late for a folic acid supplement to have this preventive effect. Thus, folic acid should be part of a healthy diet for all women of childbearing age. Good sources include green leafy vegetables, black beans, lentils, and enriched bread and pasta.

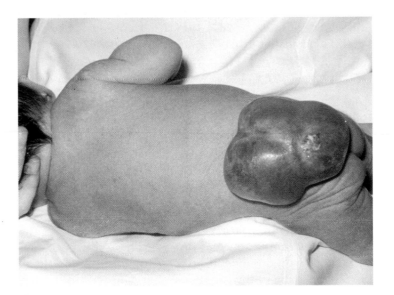

**FIGURE 13.3    Spina Bifida Cystica.** The sac in the lumbar region is called a myelomeningocele.

## CROSS-SECTIONAL ANATOMY

Figure 13.2a shows the relationship of the spinal cord to a vertebra and spinal nerve, and figure 13.2b shows the cord itself in more detail. The spinal cord, like the brain, consists of two kinds of nervous tissue called gray and white matter. **Gray matter** has a relatively dull color because it contains little myelin. It contains the somas, dendrites, and proximal parts of the axons of neurons. It is the site of synaptic contact between neurons, and therefore the site of all synaptic integration (information processing) in the central nervous system. **White matter,** by contrast, has a bright, pearly white appearance, which it gets from an abundance of myelin. It is composed of bundles of axons, called **tracts,** that carry signals from one part of the CNS to another. Both gray and white matter also have an abundance of glial cells. In silver-stained nervous tissue, gray matter tends to have a brown or golden color and white matter a lighter tan to yellow color.

### Gray Matter

The spinal cord has a central core of gray matter that looks somewhat butterfly- or H-shaped in cross sections. The core consists mainly of two **dorsal (posterior) horns,** which extend toward the dorsolateral surfaces of the cord, and two thicker **ventral (anterior) horns,** which extend toward the ventrolateral surfaces. The right and left sides are connected by a **gray commissure.** In the middle of the

commissure is the **central canal,** which is collapsed in most areas of the adult spinal cord, but in some places (and in young children) remains open, lined with ependymal cells, and filled with CSF.

As a spinal nerve approaches the cord, it branches into a *dorsal root* and *ventral root.* The dorsal root carries sensory nerve fibers, which enter the dorsal horn of the cord and sometimes synapse with an interneuron there. Such interneurons are especially numerous in the cervical and lumbar enlargements and are quite evident in histological sections at these levels. The ventral horns contain the large somas of the somatic motor neurons. Axons from these neurons exit by way of the ventral root of the spinal nerve and lead to the skeletal muscles. The spinal nerve roots are described more fully later in this chapter.

An additional **lateral horn** is visible on each side of the gray matter from the second thoracic through first lumbar segments of the cord. It contains neurons of the sympathetic nervous system, which send their axons out of the cord by way of the ventral root along with the somatic efferent fibers.

### White Matter

The white matter of the spinal cord surrounds the gray matter. It consists of bundles of axons which course up and down the cord and provide avenues of communication between different levels of the CNS. These bundles are arranged in three pairs called **columns,** or **funiculi**[8]

---

[6]*bifid* = divided, forked + *occult* = hidden
[7]*cyst* = sac, bladder

[8]*funicul* = little rope, cord

(few-NIC-you-lie)—a **dorsal (posterior)**, **lateral**, and **ventral (anterior) column** on each side. Each column consists of subdivisions called **tracts,** or **fasciculi**[9] (fah-SIC-you-lye).

## SPINAL TRACTS

Knowledge of the locations and functions of the spinal tracts is essential in diagnosing and managing spinal cord injuries. **Ascending tracts** carry sensory information up the cord, and **descending tracts** conduct motor impulses down. All nerve fibers in a given tract have a similar origin, destination, and function. Many of these fibers, as you will see, have their origin or destination in a region called the *brainstem.* Described more fully in chapter 14 (see fig. 14.2), this is a vertical stalk that supports the large *cerebellum* at the rear of the head and, even larger, two globes called the *cerebral hemispheres* that dominate the brain. In the following discussion, you will find references to brainstem and other regions where spinal cord tracts begin and end. Spinal cord anatomy will grow in meaning as you study the brain.

Several of these tracts undergo **decussation**[10] (DEE-cuh-SAY-shun) as they pass up or down the brainstem and spinal cord—meaning that they cross over from the left side of the body to the right, or vice versa. As a result, the left side of the brain receives sensory information from the right side of the body and sends motor commands to that side, while the right side of the brain senses and controls the left side of the body. A stroke that damages motor centers of the right side of the brain can thus cause paralysis of the left limbs, and vice versa.

When the origin and destination of a tract are on opposite sides of the body, we say they are **contralateral**[11] to each other. When a tract does not decussate, so the origin and destination of its fibers are on the same side of the body, we say they are **ipsilateral.**[12]

The major spinal cord tracts are summarized in table 13.1 and figure 13.4. Bear in mind that each tract is repeated on the right and left sides of the spinal cord.

## Ascending Tracts

Ascending tracts carry sensory signals up the spinal cord. Sensory signals typically travel across three neurons from their origin in the receptors to their destination in the sensory areas of the brain: a **first-order neuron** that detects a stimulus and transmits a signal to the spinal cord or brainstem; a **second-order neuron** that continues as far as a "gateway" called the *thalamus* at the upper end of the brainstem; and a **third-order neuron** that carries the signal

| TABLE 13.1 | Major Spinal Tracts | | |
|---|---|---|---|
| **Tract** | **Column** | **Decussation** | **Functions** |
| *Ascending (Sensory) Tracts* | | | |
| Cuneate fasciculus | Dorsal | In medulla | Sensations of limb and trunk position and movement, deep touch, visceral pain, and vibration, from level T6 up |
| Gracile fasciculus | Dorsal | In medulla | Same as cuneate fasciculus, below level T6 |
| Spinothalamic | Lateral and ventral | In spinal cord | Sensations of light touch, tickle, itch, temperature, pain, and pressure |
| Spinoreticular | Lateral and ventral | In spinal cord (some fibers) | Sensation of pain from tissue injury |
| Dorsal spinocerebellar | Lateral | None | Feedback from muscles (proprioception) |
| Ventral spinocerebellar | Lateral | In spinal cord | Same as dorsal spinocerebellar |
| *Descending (Motor) Tracts* | | | |
| Lateral corticospinal | Lateral | In medulla | Fine control of limbs |
| Ventral corticospinal | Ventral | None | Fine control of limbs |
| Tectospinal | Ventral | In midbrain | Reflexive head-turning in response to visual and auditory stimuli |
| Lateral reticulospinal | Lateral | None | Balance and posture; regulation of awareness of pain |
| Medial reticulospinal | Ventral | None | Same as lateral reticulospinal |
| Lateral vestibulospinal | Ventral | None | Balance and posture |
| Medial vestibulospinal | Ventral | In medulla (some fibers) | Control of head position |

---

[9]*fascicul* = little bundle
[10]*decuss* = to cross, form an X

[11]*contra* = opposite + *later* = side
[12]*ipsi* = the same + *later* = side

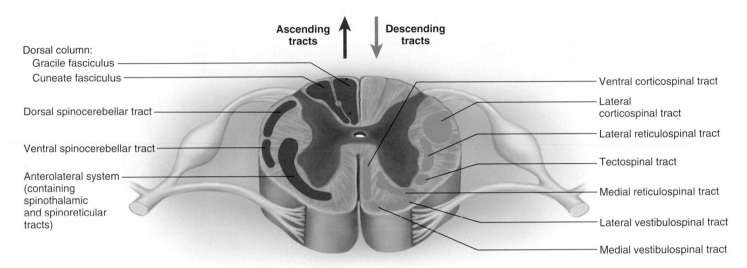

**FIGURE 13.4  Tracts of the Spinal Cord.**  All of the illustrated tracts occur on both sides of the cord, but only the ascending sensory tracts are shown on the left (red), and only the descending motor tracts on the right (green).

▶ *If you were told that this cross section is either at level T4 or T10, how could you determine which is correct?*

the rest of the way to the sensory region of the cerebral cortex. The axons of these neurons are called the *first-through third-order nerve fibers*. Deviations from the pathway described here will be noted for some of the sensory systems to follow.

The major ascending tracts are as follows. The names of most of them consist of the prefix *spino-* followed by a root denoting the destination of its fibers in the brain.

- The **gracile**[13] **fasciculus** (GRAS-el fah-SIC-you-lus) carries signals from the midthoracic and lower parts of the body. Below vertebra T6, it composes the entire dorsal column. At T6, it is joined by the cuneate fasciculus, discussed next. It consists of first-order nerve fibers that travel up the ipsilateral side of the spinal cord and terminate at the *gracile nucleus* in the medulla oblongata of the brainstem. These fibers carry signals for vibration, visceral pain, deep and discriminative touch (touch whose location one can precisely identify), and especially *proprioception*[14] from the lower limbs and lower trunk. (Proprioception is the nonvisual sense of the position and movements of the body.)

- The **cuneate**[15] **fasciculus** (fig. 13.5a) joins the gracile fasciculus at the T6 level. It occupies the lateral portion of the dorsal column and forces the gracile fasciculus medially. It carries the same type of sensory signals, originating from level T6 and up (from the upper limb and chest). Its fibers end in the *cuneate nucleus* on the ipsilateral side of the medulla oblongata. In the medulla, second-order fibers of the gracile and cuneate systems decussate

and form the **medial lemniscus**[16] (lem-NIS-cus), a tract of nerve fibers that leads the rest of the way up the brainstem to the thalamus. Third-order fibers go from the thalamus to the cerebral cortex. Because of decussation, the signals carried by the gracile and cuneate fasciculi ultimately go to the contralateral cerebral hemisphere.

- The **spinothalamic** (SPY-no-tha-LAM-ic) **tract** fig. 13.5b) and some smaller tracts form the *anterolateral system*, which passes up the anterior and lateral columns of the spinal cord. The spinothalamic tract carries signals for pain, temperature, pressure, tickle, itch, and light or crude touch. Light touch is the sensation produced by stroking hairless skin with a feather or cotton wisp, without indenting the skin; crude touch is touch whose location one can only vaguely identify. In this pathway, first-order neurons end in the dorsal horn of the spinal cord near the point of entry. Here they synapse with second-order neurons, which decussate to the opposite side of the spinal cord and form the ascending spinothalamic tract. These fibers lead all the way to the thalamus. Third-order neurons continue from there to the cerebral cortex. Because of its decussation, the spinothalamic tract ultimately sends its signals to the contralateral cerebral hemisphere.

- The **spinoreticular tract** also travels up the anterolateral system. It carries pain signals resulting from tissue injury. The first-order sensory neurons enter the dorsal horn and immediately synapse with second-order neurons. These decussate to the opposite anterolateral system, ascend the cord, and end in a loosely organized core of gray matter called the

[13]*gracil* = thin, slender
[14]*proprio* = one's own + *ception* = sensation
[15]*cune* = wedge

[16]*lemniscus* = ribbon

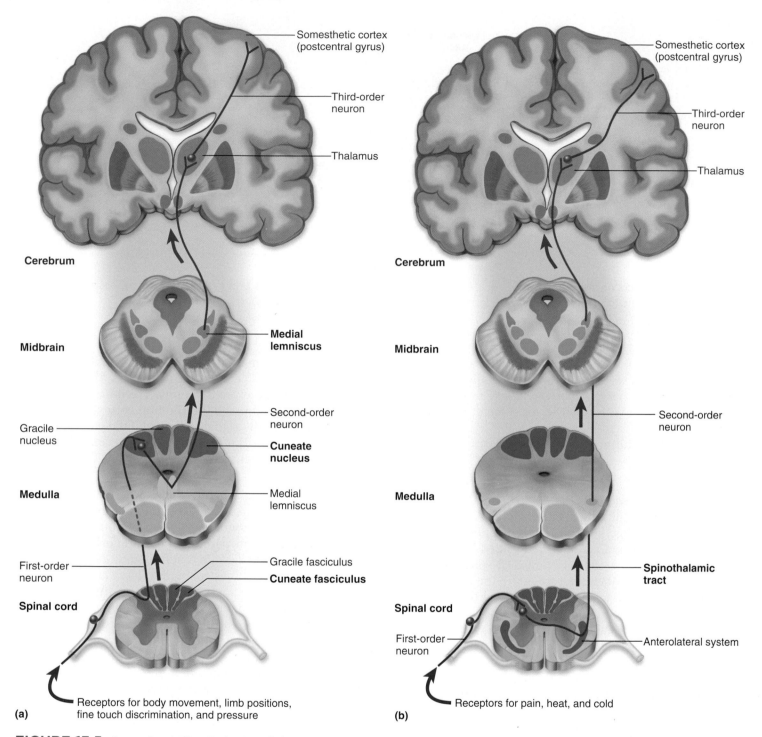

**FIGURE 13.5** **Some Ascending Pathways of the CNS.**   The spinal cord, medulla, and midbrain are shown in cross section and the cerebrum and thalamus (top) in frontal section. Nerve signals enter the spinal cord at the bottom of the figure and carry somatosensory information up to the cerebral cortex. (a) The cuneate fasciculus. (b) The spinothalamic tract.

*reticular formation* in the medulla and pons. Third-order neurons continue from the pons to the thalamus, and fourth-order neurons complete the path from there to the cerebral cortex. The reticular formation is further described in chapter 14, and the role of the spinoreticular tract in pain sensation is further discussed in chapter 16.

• The **dorsal** and **ventral spinocerebellar** (SPY-no-SERR-eh-BEL-ur) **tracts** travel through the lateral column and carry proprioceptive signals from the limbs and trunk to the cerebellum, a large motor control area at the rear of the brain. The first-order neurons of this system originate in the muscles and tendons and end in the dorsal horn of the spinal cord.

Second-order neurons send their fibers up the spinocerebellar tracts and end in the cerebellum. Fibers of the dorsal tract travel up the ipsilateral side of the spinal cord. Those of the ventral tract cross over and travel up the contralateral side but then cross back in the brainstem to enter the ipsilateral side of the cerebellum.

## Descending Tracts

Descending tracts carry motor signals down the brainstem and spinal cord. A descending motor pathway typically involves two neurons called the upper and lower motor neurons. The **upper motor neuron** begins with a soma in the cerebral cortex or brainstem and has an axon that terminates on a **lower motor neuron** in the brainstem or spinal cord. The axon of the lower motor neuron then leads the rest of the way to the muscle or other target organ. The names of most descending tracts consist of a word root denoting the point of origin in the brain, followed by the suffix -*spinal*. The major descending tracts are described here.

- The **corticospinal** (COR-tih-co-SPY-nul) **tracts** carry motor signals from the cerebral cortex for precise, finely coordinated limb movements. The fibers of this system form ridges called *pyramids* on the ventral surface of the medulla oblongata, so these tracts were once called *pyramidal tracts.* Most corticospinal fibers decussate in the lower medulla and form the **lateral corticospinal tract** on the contralateral side of the spinal cord. A few fibers remain uncrossed and form the **ventral (anterior) corticospinal tract** on the ipsilateral side (fig. 13.6). Fibers of the ventral tract decussate lower in the spinal cord, however, so even they control contralateral muscles.

- The **tectospinal** (TEC-toe-SPY-nul) **tract** begins in a midbrain region called the *tectum* ("roof") and crosses to the contralateral side of the midbrain. It descends through the brainstem to the upper spinal cord on that side, going only as far as the neck. It is involved in reflex turning of the head, especially in response to sights and sounds.

- The **lateral** and **medial reticulospinal** (reh-TIC-you-lo-SPY-nul) **tracts** originate in the *reticular formation* of the brainstem. They control muscles of the upper and lower limbs, especially to maintain posture and balance. They also contain *descending analgesic pathways* that reduce the transmission of pain signals to the brain (see chapter 16).

- The **lateral** and **medial vestibulospinal** (vess-TIB-you-lo-SPY-nul) **tracts** begin in the brainstem *vestibular nuclei,* which receive impulses for balance from the inner ear. The lateral vestibulospinal tract passes down the ventral column of the spinal cord and facilitates neurons that control the extensor

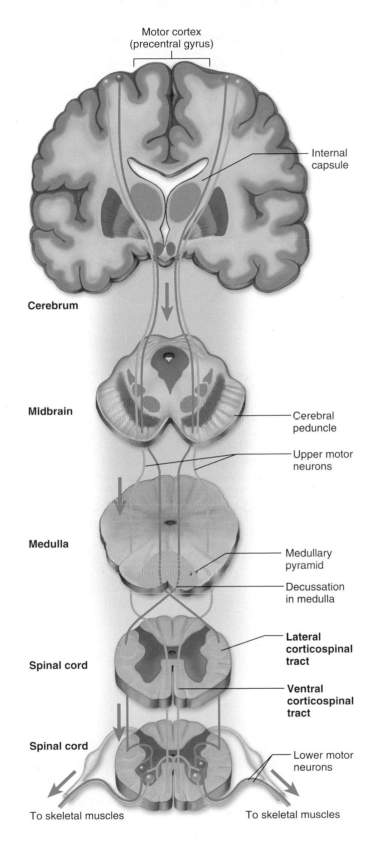

**FIGURE 13.6  Two Descending Pathways of the CNS.** The lateral and ventral corticospinal tracts, which carry signals for voluntary muscle contraction. Nerve signals originate in the cerebral cortex at the top of the figure and carry motor commands down the spinal cord.

## Poliomyelitis and Amyotrophic Lateral Sclerosis

*Poliomyelitis*[17] and *amyotrophic lateral sclerosis*[18] (ALS) are two diseases that involve destruction of motor neurons. In both diseases, the skeletal muscles atrophy from lack of innervation.

Poliomyelitis is caused by the poliovirus, which destroys motor neurons in the brainstem and ventral horn of the spinal cord. Signs of polio include muscle pain, weakness, and loss of some reflexes, followed by paralysis, muscular atrophy, and sometimes respiratory arrest. The virus spreads by fecal contamination of water. Historically, polio afflicted many children who contracted the virus from swimming in contaminated pools. The polio vaccine has nearly eliminated new cases.

ALS is also known as Lou Gehrig[19] disease after the baseball player who had to retire from the sport because of it. It is marked not only by the degeneration of motor neurons and atrophy of the muscles, but also sclerosis of the lateral regions of the spinal cord—hence its name. Most cases occur when astrocytes fail to reabsorb the neurotransmitter glutamate from the tissue fluid, allowing it to accumulate to a neurotoxic level. The early signs of ALS include muscular weakness and difficulty in speaking, swallowing, and using the hands. Sensory and intellectual functions remain unaffected, as evidenced by the accomplishments of astrophysicist and best-selling author Stephen Hawking (figure 13.7), who was stricken with ALS while he was in college. Despite near-total paralysis, he remains highly productive and communicates with the aid of a speech synthesizer and computer. Tragically, many people are quick to assume that those who have lost most of their ability to communicate their ideas and feelings have no ideas and feelings to communicate. To a victim, this may be more unbearable than the loss of motor function itself.

**FIGURE 13.7  Stephen Hawking (1942—    ), Lucasian Professor of Mathematics at Cambridge University.**

muscles of the limbs, thus inducing the limbs to stiffen and straighten. This is an important reflex in responding to body tilt and keeping one's balance. The medial vestibulospinal tract splits into ipsilateral and contralateral fibers that descend through the ventral column on both sides of the spinal cord and terminate in the neck. It plays a role in the control of head position.

*Rubrospinal tracts* are prominent in other mammals, where they aid in muscle coordination. Although often pictured in illustrations of human anatomy, they are almost nonexistent in humans and have little functional importance.

> **Think About It**
>
> *You are blindfolded and either a tennis ball or an iron ball is placed in your right hand. What spinal tract(s) would carry the signals that enable you to discriminate between these two objects?*

## Before You Go On

*Answer the following questions to test your understanding of the preceding section:*

1. *Name the four major regions and two enlargements of the spinal cord.*

2. *Describe the distal (inferior) end of the spinal cord and the contents of the vertebral canal from level L2 to S5.*

3. *Sketch a cross section of the spinal cord showing the dorsal and ventral horns. Where are the gray and white matter? Where are the columns and tracts?*

4. *Give an anatomical explanation of why a stroke in the right cerebral hemisphere can paralyze the limbs on the left side of the body.*

# The Spinal Nerves

### Objectives

When you have completed this section, you should be able to

- describe the attachment of a spinal nerve to the spinal cord;

- trace the branches of a spinal nerve distal to its attachment;

- name the five plexuses of spinal nerves and describe their general anatomy;

- name some major nerves that arise from each plexus; and

- explain the relationship of dermatomes to the spinal nerves.

---

[17]*polio* = gray matter + *myel* = spinal cord + *itis* = inflammation
[18]*a* = without + *myo* = muscle + *troph* = nourishment; *sclerosis* = hardening
[19]Lou Gehrig (1903–41), New York Yankees baseball player

# GENERAL ANATOMY OF NERVES AND GANGLIA

The spinal cord communicates with the rest of the body by way of the spinal nerves. Before we discuss those specific nerves, however, it is necessary to be familiar with the structure of nerves and ganglia in general.

A **nerve** is a cord composed of numerous nerve fibers (axons) bound together by connective tissue (fig. 13.8). If we compare a *nerve fiber* to a wire carrying an electrical current in one direction, a *nerve* would be comparable to an electrical cable composed of thousands of wires carrying currents in opposite directions. A nerve contains anywhere from a few nerve fibers to more than a million. Nerves usually have a pearly white color and resemble frayed string as they divide into smaller and smaller branches.

Nerve fibers of the peripheral nervous system are ensheathed in Schwann cells, which form a neurilemma and often a myelin sheath around the axon (see chapter 12). External to the neurilemma, each fiber is surrounded by a basal lamina and then a thin sleeve of loose connective tissue called the **endoneurium.** In most nerves, the nerve fibers are gathered in bundles called **fascicles,** each

wrapped in a sheath called the **perineurium.** The perineurium is composed of several layers of overlapping, squamous, epithelium-like cells. Several fascicles are then bundled together and wrapped in an outer **epineurium** to compose the nerve as a whole. The epineurium is composed of dense irregular connective tissue and protects the nerve from stretching and injury. Nerves have a high metabolic rate and need a plentiful blood supply. Blood vessels penetrate as far as the perineurium, and oxygen and nutrients diffuse through the extracellular fluid from there to the nerve fibers.

## Think About It

*How does the structure of a nerve compare to that of a skeletal muscle? Which of the descriptive terms for nerves have similar counterparts in muscle histology?*

As we have seen in chapter 12, peripheral nerve fibers are of two kinds: sensory (afferent) fibers carry signals from sensory receptors to the CNS, and motor (efferent) fibers carry signals from the CNS to muscles and glands. Both sensory and motor fibers can also be described as

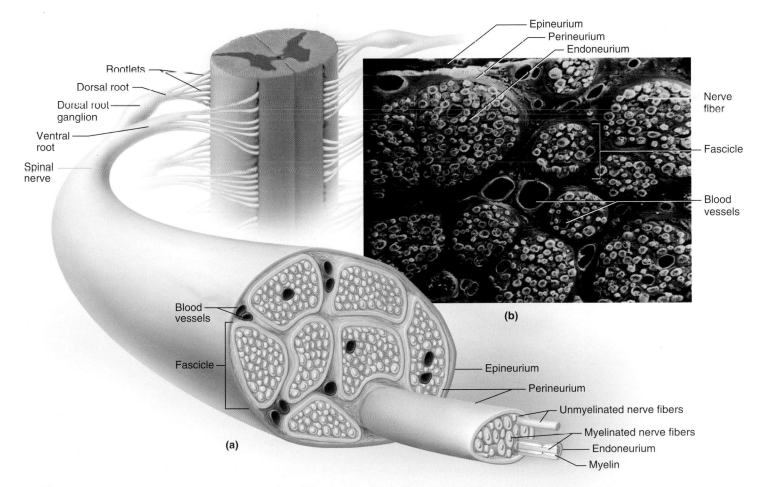

**FIGURE 13.8 Anatomy of a Nerve.** (a) A spinal nerve and its association with the spinal cord. (b) Cross section of a nerve (SEM). Myelinated nerve fibers appear in the photograph as white rings and unmyelinated fibers as solid gray. [(b) From Richard E. Kessel and Randy H. Kardon, *Tissues and Organs: A Text-Atlas of Scanning Electron Microscopy,* 1979, W. H. Freeman and Company.]

*somatic* or *visceral* and as *general* or *special* depending on the organs they innervate (table 13.2).

A **mixed nerve** consists of both sensory and motor fibers and thus transmits signals in two directions, although any one nerve fiber within the nerve transmits signals one way only. Most nerves are mixed. **Sensory nerves,** composed entirely of sensory axons, are less common; they include the olfactory and optic nerves discussed in chapter 14. **Motor nerves** carry only motor fibers. Many nerves often described as motor are actually mixed because they carry sensory signals of proprioception from the muscle back to the CNS.

If a nerve resembles a thread, a **ganglion**[20] resembles a knot in the thread. A ganglion is a cluster of cell bodies (somas) outside the CNS. It is enveloped in an epineurium continuous with that of the nerve. Among the somas are bundles of nerve fibers leading into and out of the ganglion. Figure 13.9 shows a type of ganglion called the *dorsal root ganglion* associated with the spinal nerves.

| TABLE 13.2 | The Classification of Nerve Fibers |
|---|---|
| **Class** | **Description** |
| Afferent fibers | Carry sensory signals from receptors to the CNS |
| Efferent fibers | Carry motor signals from the CNS to effectors |
| Somatic fibers | Innervate skin, skeletal muscles, bones, and joints |
| Visceral fibers | Innervate blood vessels, glands, and viscera |
| General fibers | Innervate widespread organs such as muscles, skin, glands, viscera, and blood vessels |
| Special fibers | Innervate more localized organs in the head, including the eyes, ears, olfactory and taste receptors, and muscles of chewing, swallowing, and facial expression |

[20]*gangli* = knot

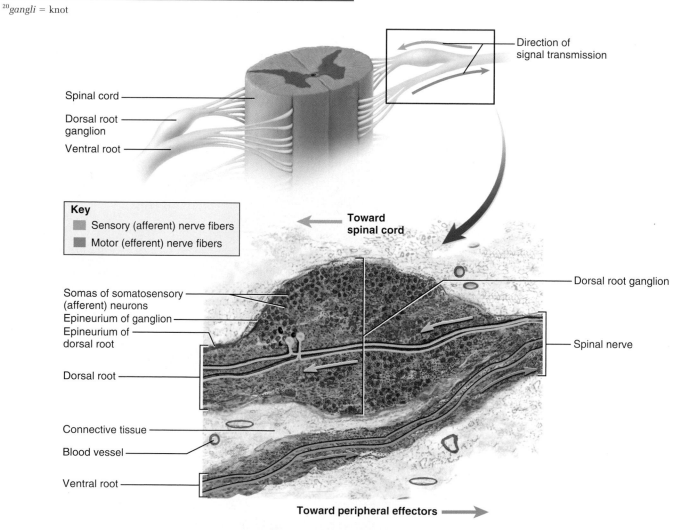

**FIGURE 13.9   Anatomy of a Ganglion (longitudinal section).**   The dorsal root ganglion contains the somas of unipolar sensory neurons conducting signals from peripheral sense organs toward the spinal cord. Below this is the ventral root of the spinal nerve, which conducts motor signals away from the spinal cord, toward peripheral effectors. (The ventral root is not part of the ganglion.)

▶ *Where are the somas of the motor neurons located?*

# SPINAL NERVES

There are 31 pairs of **spinal nerves:** 8 cervical (C1–C8), 12 thoracic (T1–T12), 5 lumbar (L1–L5), 5 sacral (S1–S5), and 1 coccygeal (Co) (fig. 13.10). The first cervical nerve emerges between the skull and atlas, and the others emerge through intervertebral foramina, including the anterior and posterior foramina of the sacrum.

## Proximal Branches

Each spinal nerve has two points of attachment to the spinal cord (fig. 13.11). Dorsally, a branch of the spinal nerve called the **dorsal root** divides into six to eight *nerve rootlets* that enter the spinal cord (fig. 13.12). A little distal to the rootlets is a swelling called the **dorsal root ganglion,** which contains the somas of unipolar afferent neurons.

Ventrally, another row of six to eight rootlets leave the spinal cord and converge to form the **ventral root.**

The dorsal and ventral roots merge, penetrate the dural sac, enter the intervertebral foramen, and there form the spinal nerve proper.

Spinal nerves are mixed nerves, with a two-way traffic of afferent (sensory) and efferent (motor) signals. Afferent signals approach the cord by way of the dorsal root and enter the dorsal horn of the gray matter. Efferent signals begin at the somas of motor neurons in the ventral horn and leave the spinal cord via the ventral root. Some viruses invade the central nervous system by way of these roots (see Insight 13.3).

The dorsal and ventral roots are shortest in the cervical region and become longer inferiorly. The roots that arise from segments L2 to Co of the cord form the cauda equina.

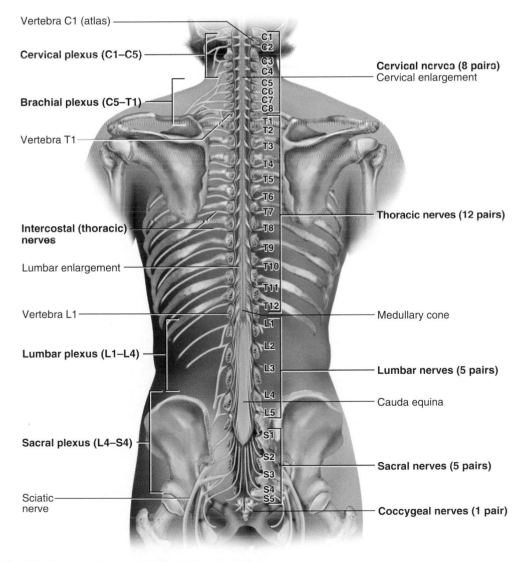

**FIGURE 13.10   The Spinal Nerve Roots and Plexuses.** Posterior aspect.

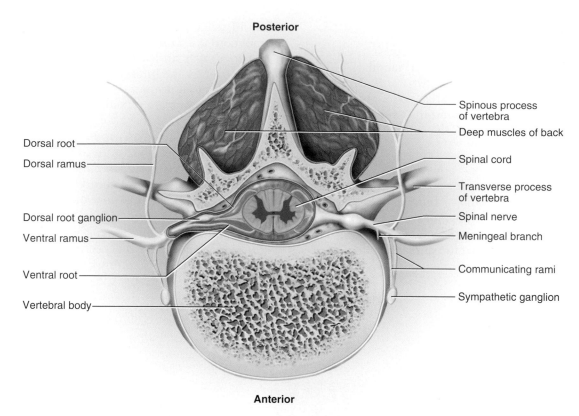

**Posterior**

Dorsal root

Dorsal ramus

Dorsal root ganglion

Ventral ramus

Ventral root

Vertebral body

Spinous process of vertebra

Deep muscles of back

Spinal cord

Transverse process of vertebra

Spinal nerve

Meningeal branch

Communicating rami

Sympathetic ganglion

**Anterior**

**FIGURE 13.11**  **Branches of a Spinal Nerve in Relation to the Spinal Cord and Vertebra.**  Cross section.

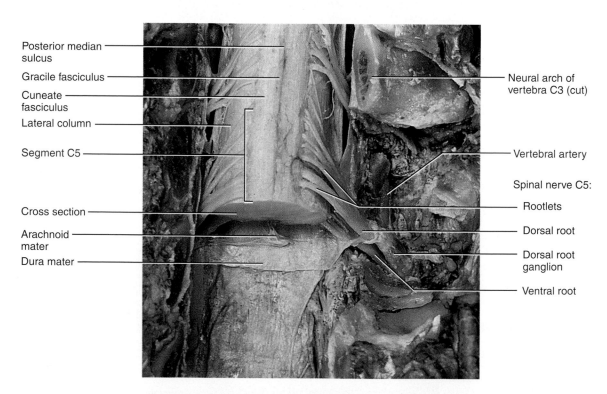

Posterior median sulcus

Gracile fasciculus

Cuneate fasciculus

Lateral column

Segment C5

Cross section

Arachnoid mater

Dura mater

Neural arch of vertebra C3 (cut)

Vertebral artery

Spinal nerve C5:

Rootlets

Dorsal root

Dorsal root ganglion

Ventral root

**FIGURE 13.12**  **The Point of Entry of Two Spinal Nerves into the Spinal Cord.**  Posterior (dorsal) view with vertebrae cut away. Note that each dorsal root divides into several rootlets that enter the spinal cord. A segment of the spinal cord is the portion receiving all the rootlets of one spinal nerve.

▶ *In the labeled rootlets of spinal nerve C5, are the nerve fibers afferent or efferent? How do you know?*

Clinical Application

## Shingles

Chickenpox *(varicella),* a common disease of early childhood, is caused by the *varicella-zoster* virus. It produces an itchy rash that usually clears up without complications. The virus, however, remains for life in the dorsal root ganglia, kept in check by the immune system. If the immune system is compromised, however, the virus can travel down the sensory nerves by fast axonal transport and cause *shingles (herpes zoster).* This is particularly common after the age of 50. Shingles is characterized by a painful trail of skin discoloration and fluid-filled vesicles along the path of the nerve. These signs usually appear in the chest and waist, often on just one side of the body. There is no cure, and the vesicles generally heal spontaneously within 1 to 3 weeks. In the meantime, aspirin and steroidal ointments can help to relieve the pain and inflammation of the lesions. Antiviral drugs such as acyclovir can shorten the course of an episode of shingles, but only if taken within the first 2 to 3 days of outbreak. Even after the lesions disappear, however, some people suffer intense pain along the course of the nerve *(postherpetic neuralgia, PHN),* lasting for months or even years. PHN has proven very difficult to treat, but pain relievers and antidepressants are of some help. Childhood vaccination against varicella reduces the risk of shingles later in life.

## Distal Branches

Distal to the vertebrae, the branches of a spinal nerve are more complex (fig. 13.13). Immediately after emerging from the intervertebral foramen, the nerve divides into a **dorsal ramus,**[21] a **ventral ramus,** and a small **meningeal branch.** The meningeal branch (see fig. 13.11) reenters the vertebral canal and innervates the meninges, vertebrae, and spinal ligaments. The dorsal ramus innervates the muscles and joints in that region of the spine and the skin of the back. The ventral ramus innervates the ventral and lateral skin and muscles of the trunk, and gives rise to nerves of the limbs.

> **Think About It**
>
> *Do you think the meningeal branch is sensory, motor, or mixed? Explain your reasoning.*

The ventral ramus differs from one region of the trunk to another. In the thoracic region, it forms an **intercostal nerve,** which travels along the inferior margin of a rib and innervates the skin and intercostal muscles (thus contributing to breathing). It also innervates the internal oblique, external oblique, and transverse abdominal muscles. All other ventral rami form the *nerve plexuses* described next.

[21]*ramus* = branch

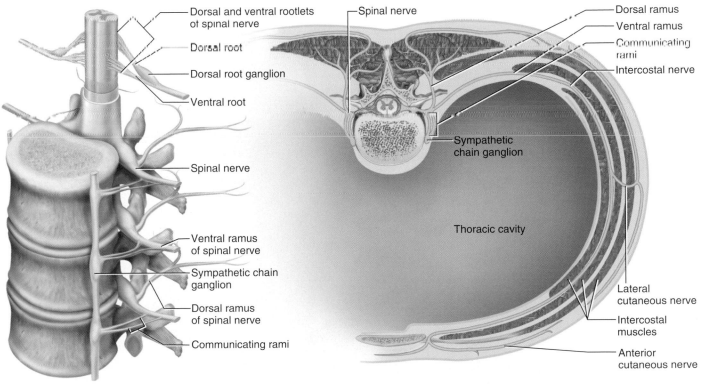

**FIGURE 13.13   Rami of the Spinal Nerves.**   (a) Anterolateral view of the spinal nerves and their subdivisions in relation to the spinal cord and vertebrae. (b) Cross section of the thorax showing innervation of muscles and skin of the chest and back. This section is cut through the intercostal muscles between two ribs.

(a)                                                                          (b)

## NERVE PLEXUSES

Except in the thoracic region, the ventral rami branch and anastomose (merge) repeatedly to form five weblike nerve plexuses: the small **cervical plexus** deep in the neck, the **brachial plexus** near the shoulder, the **lumbar plexus** of the lower back, the **sacral plexus** immediately inferior to this, and finally the tiny **coccygeal plexus** adjacent to the lower sacrum and coccyx. A general view of these plexuses is shown in figure 13.10; they are illustrated and described in tables 13.3 through 13.6. Since many of the nerves tabulated here innervate a large number of muscles, some innervations are described only in general terms. The individual muscles can be found by referring back to innervations listed in the muscle tables of chapter 10. Those tables also describe the muscle actions controlled by these spinal nerves.

| TABLE 13.3 | The Cervical Plexus |
|---|---|

The cervical plexus (fig. 13.14) receives fibers from the ventral rami of nerves C1 to C5 and gives rise to the nerves listed below, in order from superior to inferior. The most important of these are the *phrenic*[22] (FREN-ic) *nerves,* which travel down each side of the mediastinum, innervate the diaphragm, and play an essential role in breathing (see fig. 15.3). In addition to the major nerves listed here, there are several motor branches that innervate the geniohyoid, thyrohyoid, scalene, levator scapulae, trapezius, and sternocleidomastoid muscles.

| Nerve | Composition | Innervation |
|---|---|---|
| Lesser occipital nerve | Somatosensory | Skin of lateral scalp and dorsal part of external ear |
| Great auricular nerve | Somatosensory | Skin of and around external ear |
| Transverse cervical nerve | Somatosensory | Skin of anterior and lateral neck |
| Ansa cervicalis | Motor | Omohyoid, sternohyoid, and sternothyroid |
| Supraclavicular nerves | Somatosensory | Skin of lower ventral and lateral neck, shoulder, and anterior chest |
| Phrenic nerve | Motor | Diaphragm |

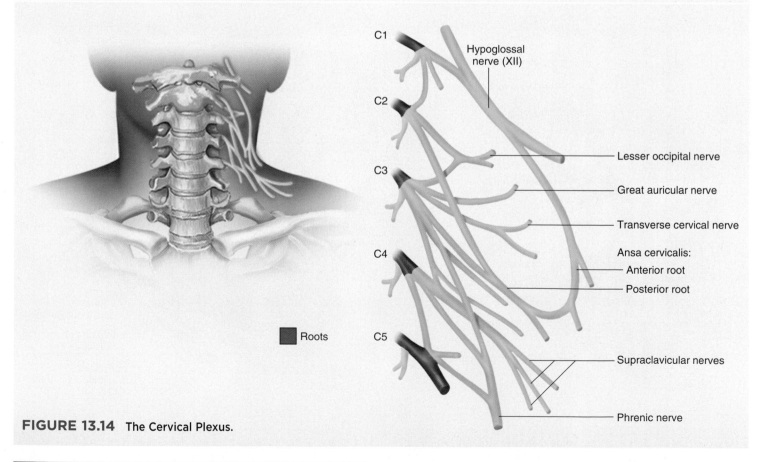

**FIGURE 13.14** The Cervical Plexus.

[22]*phren* = diaphragm

**TABLE 13.4**  The Brachial Plexus

The brachial plexus (figs. 13.15 and 13.16) is formed by the ventral rami of nerves C4 to T2. It passes over the first rib into the axilla and innervates the upper limb and some muscles of the neck and shoulder.

The subdivisions of this plexus are called *roots, trunks, divisions,* and *cords* (color-coded in figure 13.15). The five **roots** are the ventral rami of nerves C5 through T1, which provide most of the fibers to this plexus (C4 and T2 contribute partially). The five roots unite to form the **upper, middle,** and **lower trunks.** Each trunk divides into an **anterior** and **posterior division,** and finally the six divisions merge to form three large fiber bundles: the **posterior, medial,** and **lateral cords.** From these cords arise the following major nerves, which serve for cutaneous sensation, muscle contraction, and proprioception from the joints and muscles.

| Nerve | Composition | Cord of origin | Sensory innervation | Motor innervation |
|---|---|---|---|---|
| Axillary nerve | Motor and somatosensory | Posterior cord of brachial plexus | Skin of lateral shoulder and arm; shoulder joint | Deltoid and teres minor |
| Radial nerve | Motor and somatosensory | Posterior cord of brachial plexus | Skin of posterior arm, forearm, and wrist; joints of elbow, wrist, and hand | Mainly extensor muscles of posterior arm and forearm (see tables 10.13 and 10.14) |
| Musculocutaneous nerve | Motor and somatosensory | Lateral cord of brachial plexus | Skin of lateral forearm | Brachialis, biceps brachii, and coracobrachiallis |
| Median nerve | Motor and somatosensory | Medial cord of brachial plexus | Skin of lateral two-thirds of hand, joints of hand | Mainly forearm flexors; thenar muscles and lumbricals I–II of hand (see tables 10.13, 10.14, and 10.16) |
| Ulnar nerve | Motor and somatosensory | Medial cord of brachial plexus | Skin of medial hand; joints of hand | Some forearm flexors; adductor pollicis, hypothenar, and interosseous muscles; lumbricals III–IV (see tables 10.14 and 10.16) |

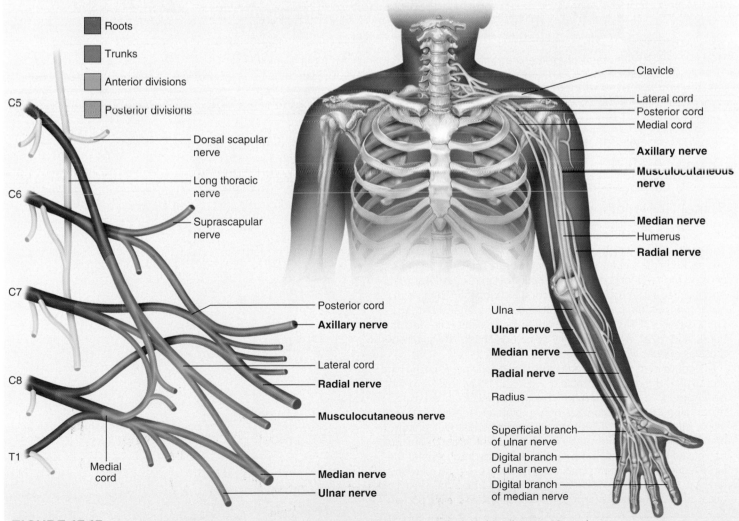

**FIGURE 13.15**  **The Brachial Plexus.** The labeled nerves innervate muscles tabulated in chapter 10, and those in boldface are further detailed here.

| **TABLE 13.4** | The Brachial Plexus *(cont.)* |
|---|---|

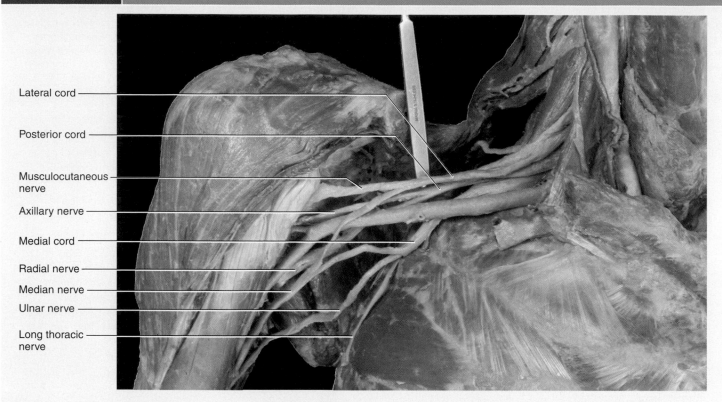

Lateral cord

Posterior cord

Musculocutaneous nerve

Axillary nerve

Medial cord

Radial nerve

Median nerve

Ulnar nerve

Long thoracic nerve

**FIGURE 13.16**   **Photograph of the Brachial Plexus.**   Anterior view of the right shoulder.

---

**INSIGHT 13.4**   Clinical Application

## Spinal Nerve Injuries

The radial and sciatic nerves are especially vulnerable to injury. The radial nerve, which passes through the axilla, may be compressed against the humerus by improperly adjusted crutches, causing *crutch paralysis.* A similar injury often resulted from the now-discredited practice of correcting a dislocated shoulder by putting a foot in a person's armpit and pulling on the arm. One consequence of radial nerve injury is *wrist drop*—the fingers, hand, and wrist are chronically flexed because the extensor muscles supplied by the radial nerve are paralyzed.

Because of its position and length, the sciatic nerve of the hip and thigh is the most vulnerable nerve in the body. Trauma to this nerve produces *sciatica,* a sharp pain that travels from the gluteal region along the posterior side of the thigh and leg as far as the ankle. Ninety percent of cases result from a herniated intervertebral disc or osteoarthritis of the lower spine, but sciatica can also be caused by pressure from a pregnant uterus, dislocation of the hip, injections in the wrong area of the buttock, or sitting for a long time on the edge of a hard chair. Men sometimes suffer sciatica because of the habit of sitting on a wallet carried in the hip pocket.

TABLE 13.5 | The Lumbar Plexus

The lumbar plexus (fig. 13.17) is formed from the ventral rami of nerves L1 through L4 and some fibers from T12. With only five roots and two divisions, it is less complex than the brachial plexus. It gives rise to the following nerves.

| Nerve | Composition | Sensory Innervation | Motor innervation |
|---|---|---|---|
| Iliohypogastric nerve | Motor and somatosensory | Skin of anterior abdominal wall | Internal and external obliques and transverse abdominal |
| Ilioinguinal nerve | Motor and somatosensory | Skin of upper medial thigh; male scrotum and root of penis; female labia majora | Joins iliohypogastric nerve and innervates the same muscles |
| Genitofemoral nerve | Somatosensory | Skin of middle anterior thigh; male scrotum and cremaster muscle; female labia majora | (None) |
| Lateral femoral cutaneous nerve | Somatosensory | Skin of lateral thigh | (None) |
| Femoral nerve | Motor and somatosensory | Skin of anterior and lateral thigh; medial leg and foot | Iliacus, pectineus, quadriceps femoris, and sartorius |
| Saphenous nerve | Somatosensory | Skin of medial leg and foot; knee joint | (None) |
| Obturator nerve | Motor and somatosensory | Skin of upper medial thigh; hip and knee joint | Medial thigh muscles (see table 10.17) |

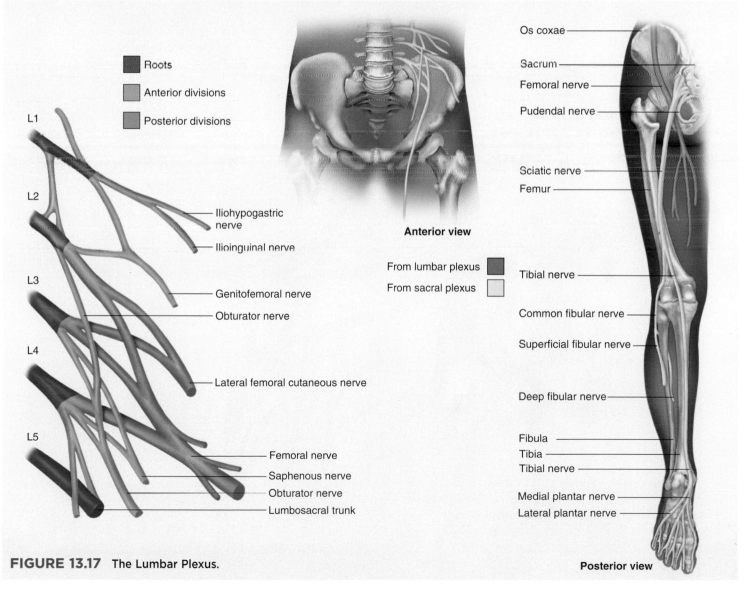

Roots

Anterior divisions

Posterior divisions

L1

L2

L3

L4

L5

Iliohypogastric nerve

Ilioinguinal nerve

Genitofemoral nerve

Obturator nerve

Lateral femoral cutaneous nerve

Femoral nerve

Saphenous nerve

Obturator nerve

Lumbosacral trunk

**Anterior view**

From lumbar plexus

From sacral plexus

Os coxae

Sacrum

Femoral nerve

Pudendal nerve

Sciatic nerve

Femur

Tibial nerve

Common fibular nerve

Superficial fibular nerve

Deep fibular nerve

Fibula

Tibia

Tibial nerve

Medial plantar nerve

Lateral plantar nerve

**Posterior view**

**FIGURE 13.17** The Lumbar Plexus.

| TABLE 13.6 | The Sacral and Coccygeal Plexuses |
|---|---|

The sacral plexus is formed from the ventral rami of nerves L4, L5, and S1 through S4. It has six roots and anterior and posterior divisions. Since it is connected to the lumbar plexus by fibers that run through the lumbosacral trunk, the *lumbar* and *sacral plexuses* are sometimes referred to collectively as the *lumbosacral plexus*. The coccygeal plexus is a tiny plexus formed from the ventral rami of S4, S5, and Co (fig. 13.18).

The *tibial* and *common fibular* nerves listed in this table travel together through a connective tissue sheath; they are referred to collectively as the **sciatic** (sy-AT-ic) **nerve.** The sciatic nerve passes through the greater sciatic notch of the pelvis, extends for the length of the thigh, and ends at the popliteal fossa. Here, the tibial and common fibular nerves diverge and follow their separate paths into the leg. The tibial nerve descends through the leg and then gives rise to medial and plantar nerves in the foot. The common fibular nerve divides into deep and superficial fibular nerves.

| Nerve | Composition | Sensory innervation | Motor innervation |
|---|---|---|---|
| Superior gluteal nerve | Motor | (None) | Gluteus minimus, gluteus medius, and tensor fasciae latae |
| Inferior gluteal nerve | Motor | (None) | Gluteus maximus |
| Posterior cutaneous nerve | Somatosensory | Skin of inferior lateral buttock, anal region, superior posterior thigh, superior calf, scrotum, and labia majora | (None) |
| Tibial nerve | Motor and somatosensory | Skin of posterior leg and sole of foot; knee and foot joints | Hamstrings; posterior muscles of leg (see tables 10.18 and 10.19); most intrinsic muscles of foot (via plantar nerves) (see table 10.20) |
| Fibular (peroneal) nerve (common, deep, and superficial) | Motor and somatosensory | Skin of anterior distal one-third of leg, dorsum of foot, and toes I and II; knee joint | Biceps femoris; anterior and lateral muscles of leg; extensor digitorum brevis of foot (see tables 10.19 and 10.20) |
| Pudendal nerve | Motor and somatosensory | Skin of penis and scrotum of male; clitoris, labia majora and minora, and lower vagina of female | Muscles of perineum (see table 10.8) |

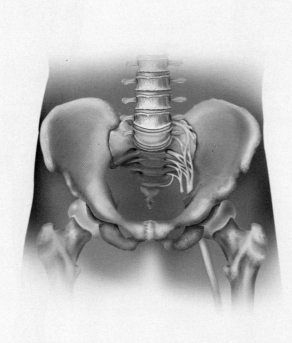
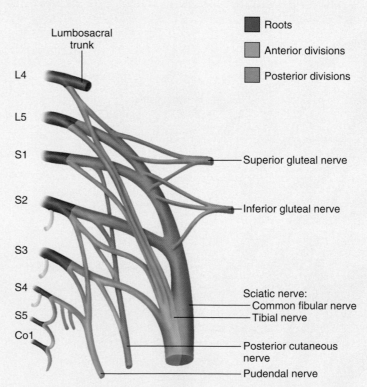

**FIGURE 13.18**  The Sacral and Coccygeal Plexuses.

# CUTANEOUS INNERVATION AND DERMATOMES

Each spinal nerve except C1 receives sensory input from a specific area of skin called a **dermatome**.[23] A *dermatome map* (fig. 13.19) is a diagram of the cutaneous regions innervated by each spinal nerve. Such a map is oversimplified, however, because the dermatomes overlap at their edges by as much as 50%. Therefore, severance of one sensory nerve root does not entirely deaden sensation from a dermatome. It is necessary to sever or anesthetize three successive spinal nerves to produce a total loss of sensation from one dermatome. Spinal nerve damage is assessed by testing the dermatomes with pinpricks and noting areas in which the patient has no sensation.

## Before You Go On

*Answer the following questions to test your understanding of the preceding section:*

5. What is meant by the dorsal and ventral roots of a spinal nerve? Which of these is sensory and which is motor?

6. Where are the somas of the dorsal root located? Where are the somas of the ventral root?

7. List the five plexuses of spinal nerves and state where each one is located.

8. State which plexus gives rise to each of the following nerves: axillary, ilioinguinal, obturator, phrenic, pudendal, radial, and sciatic.

# Somatic Reflexes

### Objectives

When you have completed this section, you should be able to

• define *reflex* and explain how reflexes differ from other motor actions;

• describe the general components of a typical reflex arc; and

• explain how the basic types of somatic reflexes function.

Most of us have had our reflexes tested with a little rubber hammer; a tap near the knee produces an uncontrollable jerk of the leg, for example. In this section, we discuss what reflexes are and how they are produced by an assembly of receptors, neurons, and effectors. We also survey the different types of neuromuscular reflexes and how they are important to motor coordination.

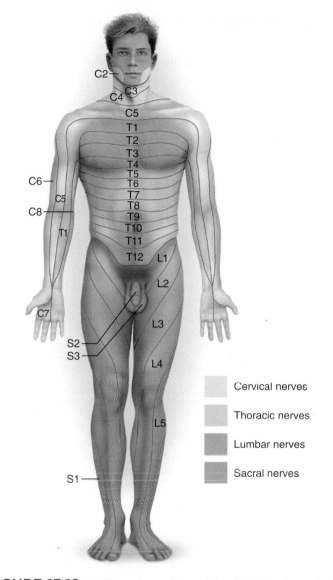

**FIGURE 13.19** **A Dermatome Map of the Anterior Aspect of the Body.** Each zone of the skin is innervated by sensory branches of the spinal nerves indicated by the labels. Nerve C1 does not innervate the skin.

---

[23]*derma* = skin + *tome* = segment, part

# THE NATURE OF REFLEXES

**Reflexes** are quick, involuntary, stereotyped reactions of glands or muscles to stimulation. This definition sums up four important properties of a reflex:

1. Reflexes *require stimulation*—they are not spontaneous actions but responses to sensory input.

2. Reflexes are *quick*—they generally involve few if any interneurons and minimum synaptic delay.

3. Reflexes are *involuntary*—they occur without intent, often without our awareness, and they are difficult to suppress. Given an adequate stimulus, the response is essentially automatic. You may become conscious of the stimulus that evoked a reflex, and this awareness may enable you to correct or avoid a potentially dangerous situation, but awareness is not a part of the reflex itself. It may come after the reflex action has been completed, and somatic reflexes can occur even if the spinal cord has been severed so that no stimuli reach the brain.

4. Reflexes are *stereotyped*—they occur in essentially the same way every time; the response is very predictable.

Reflexes include glandular secretion and contractions of all three types of muscle. They also include some learned responses, such as the salivation of dogs in response to a sound they have come to associate with feeding time, first studied by Ivan Pavlov and named *conditioned reflexes.* In this section, however, we are concerned with unlearned skeletal muscle reflexes that are mediated by the brainstem and spinal cord. They result in the involuntary contraction of a muscle—for example, the quick withdrawal of your hand from a hot stove or the lifting of your foot when you step on something sharp. These are **somatic reflexes,** since they involve the somatic nervous system. Chapter 15 concerns *visceral reflexes.* The somatic reflexes have traditionally been called *spinal reflexes,* although this is a misleading expression for two reasons: (1) Spinal reflexes are not exclusively somatic; the autonomic (visceral) reflexes also involve the spinal cord. (2) Some somatic reflexes are mediated more by the brain than by the spinal cord.

A somatic reflex employs a **reflex arc,** in which signals travel along the following pathway:

1. *somatic receptors* in the skin, a muscle, or a tendon;

2. *afferent nerve fibers,* which carry information from these receptors into the dorsal horn of the spinal cord or to the brainstem;

3. an *integrating center,* a point of synaptic contact between neurons in the gray matter of the spinal cord or brainstem. In most reflex arcs, there are one or more interneurons in the integrating center. Synaptic events in the integrating center determine whether the efferent (output) neuron issues a signal to the muscle.

4. *efferent nerve fibers,* which carry motor impulses to the skeletal muscles; and

5. *skeletal muscles,* the somatic effectors that carry out the response.

# THE MUSCLE SPINDLE

Many somatic reflexes involve stretch receptors called **muscle spindles** embedded in the skeletal muscles. These are among the body's **proprioceptors,** sense organs specialized to monitor the position and movement of body parts. The function of muscle spindles is to inform the brain of muscle length and body movements. This enables the brain to send motor commands back to the muscles that control coordinated movement, corrective reflexes (for example, to keep one's balance), muscle tone, and posture. Spindles are especially abundant in muscles that require fine control. Hand and foot muscles have 100 or more spindles per gram of muscle, whereas there are relatively few in large muscles with coarse movements, and none at all in the middle-ear muscles.

Muscle spindles are named for their fusiform shape (fig. 13.20). They are about 4 to 10 mm long, thick in the middle and tapered at the ends, and scattered through the fleshy part of a muscle with their long axes parallel to the muscle fibers. Spindles are especially concentrated at the ends of a muscle, near the tendons. A spindle contains 3 to 12 modified muscle fibers and a few nerve fibers, all wrapped in a fibrous capsule that blends at each end into the endomysium of the muscle. The muscle fibers within the spindle are called **intrafusal**[24] **fibers,** while those that make up the rest of the muscle and do its work are called **extrafusal fibers.**

Intrafusal fibers are modified muscle cells that have sarcomeres and contractile ability only at the two ends; the middle of the fiber lacks sarcomeres and cannot contract. There are two classes of intrafusal fibers:

1. Typically above five **nuclear chain fibers,** which have a single file of nuclei in the noncontractile region.

2. Typically two or three **nuclear bag fibers,** which are fatter, about twice as long, and have nuclei clustered in the baglike middle region.

---

[24]*intra* = within + *fus* = spindle

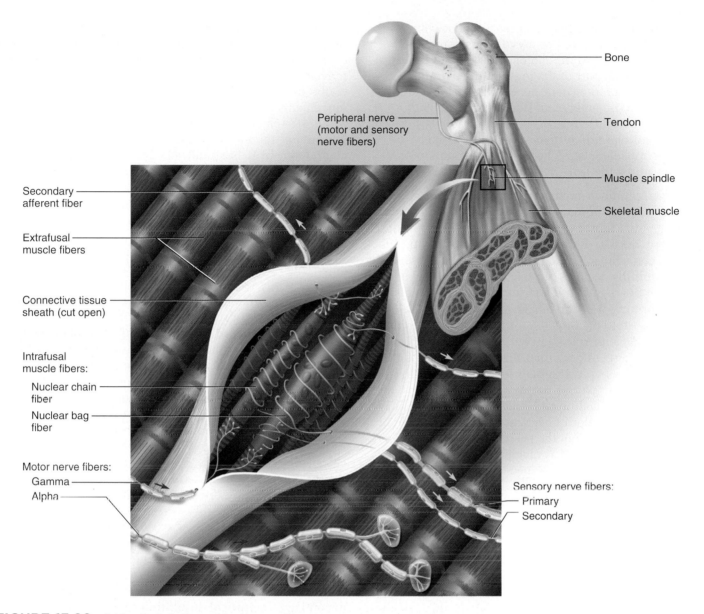

**FIGURE 13.20** A Muscle Spindle and Its Innervation.

Muscle spindles have three principal classes of nerve fibers; the first two are sensory fibers that detect changes in muscle length, and the third is a motor fiber that adjusts the length of the muscle spindle itself:

1. One **primary afferent (group Ia) fiber,** which arises from springlike *annulospiral endings* that coil around the middle of both nuclear chain and nuclear bag fibers. This is a large, fast nerve fiber, very sensitive to small changes in muscle length and to sudden body movements.

2. Up to eight **secondary afferent (group II) fibers,** which wind primarily around the nuclear chain fibers adjacent to the region of the annulospiral endings. They are intermediate-sized fibers with slower conduction speed than the primary afferents. These fibers inform the brain of muscle length but are less responsive to the *rate* of muscle shortening or lengthening.

3. **Gamma motor neurons,** which originate in the ventral horn of the spinal cord and lead to the contractile

ends of the intrafusal fibers. These are relatively small-diameter, slow motor nerve fibers compared with the large, fast **alpha motor neurons** that innervate the working part of the muscle (the extrafusal fibers). Gamma motor neurons stimulate shortening of the contractile ends of an intrafusal fiber. This stretches the middle part of the fiber and makes the sensory neurons fire, or makes them more likely to fire if the muscle is stretched. Thus, gamma motor neurons adjust the sensitivity of the muscle spindle.

A muscle spindle is a remarkably complex and sophisticated device, with functional subtypes of nuclear bag fibers and gamma motor neurons, but we will not delve into all these details. Fundamentally, a muscle spindle does this: When a muscle is at rest, relaxed, and stretched, the muscle spindle itself is stretched, and both types of sensory nerve fibers transmit a steady stream of signals to the brain informing it of the length of the muscle. This is called a *steady-state* or *tonic* response (further explained in chapter 16). As a muscle contracts, the spindles shorten along with it. The secondary fiber fires at lower frequency and the primary fiber completely ceases firing; this is called a *dynamic* or *phasic* response. The primary fiber resumes firing, but at a slow rate, when the muscle is fully contracted and stable. Both fiber types inform the brain of the length of the muscle, but the primary fiber also informs it of how fast muscle length is changing. These responses quickly inform the brain of the speed of body movements and allow it to initiate quick corrective reflexes—for example when your body tilts a little bit and you adjust muscle tension to keep your balance.

## THE STRETCH REFLEX

When a muscle is stretched, it "fights back"—it contracts, maintains increased tonus, and feels stiffer than an unstretched muscle. This response, called the **stretch (myotatic[25]) reflex,** helps to maintain equilibrium and posture. For example, if your head starts to tip forward, it stretches muscles such as the semispinalis and splenius capitis of the nuchal region (back of your neck). This stimulates their muscle spindles, which send afferent signals to the cerebellum by way of the brainstem. The cerebellum integrates this information and relays it to the cerebral cortex, and the cortex sends signals back to the nuchal muscles. The muscles contract and raise your head.

Stretch reflexes often feed back not to a single muscle but to a set of synergists and antagonists. Since the contraction of a muscle on one side of a joint stretches the antagonist on the other side, the flexion of a joint creates a stretch reflex in the extensors, and extension creates a stretch reflex in the flexors. Consequently, stretch reflexes are valuable in stabilizing joints by balancing the tension of the extensors and flexors. They also dampen (smooth out) muscle action. Without stretch reflexes, a person's movements tend to be jerky. Stretch reflexes are especially important in coordinating vigorous and precise movements such as dance.

A stretch reflex is mediated primarily by the brain and is not, therefore, strictly a spinal reflex, but a weak component of it is spinal and occurs even if the spinal cord is severed from the brain. The spinal component can be more pronounced if a muscle is stretched very suddenly. This occurs in a **tendon reflex**—the reflexive contraction of a muscle when its tendon is tapped, as in the familiar knee-jerk (patellar) reflex. Tapping the patellar ligament with a reflex hammer suddenly stretches the quadriceps femoris muscle of the thigh (fig. 13.21). This stimulates numerous muscle spindles in the quadriceps and sends an intense volley of signals to the spinal cord, mainly by way of primary afferent fibers.

In the spinal cord, the primary afferent fibers synapse directly with the alpha motor neurons that return to the muscle, thus forming **monosynaptic reflex arcs.** That is, there is only one synapse between the afferent and efferent neuron, so there is little synaptic delay and a very prompt response. The alpha motor neurons excite the quadriceps muscle, making it contract and creating the knee jerk.

There are many other tendon reflexes. A tap on the calcaneal tendon causes plantar flexion of the foot, a tap on the triceps brachii tendon causes extension of the elbow, and a tap on the masseter causes clenching of the jaw. Testing somatic reflexes is valuable in diagnosing many diseases that cause exaggeration, inhibition, or absence of reflexes—for example, neurosyphilis, diabetes mellitus, multiple sclerosis, alcoholism, electrolyte imbalances, and lesions of the nervous system.

Stretch reflexes and other muscle contractions often depend on **reciprocal inhibition,** a reflex phenomenon that prevents muscles from working against each other by inhibiting antagonists. In the knee jerk, for example, the quadriceps femoris would not produce much joint movement if its antagonists, the hamstring muscles, contracted at the same time. But reciprocal inhibition prevents that from happening. Some branches of the sensory fibers from the muscle spindles in the quadriceps stimulate spinal cord interneurons which, in turn, *inhibit* the alpha motor neurons of the hamstrings (fig. 13.21). The hamstring muscles remain relaxed and allow the quadriceps to extend the knee.

## THE FLEXOR (WITHDRAWAL) REFLEX

A **flexor reflex** is the quick contraction of flexor muscles resulting in the withdrawal of a limb from an injurious stimulus. For example, suppose you are wading in a lake and step on a broken bottle with your right foot

---

[25]*myo* = muscle + *tat* (from *tasis*) = stretch

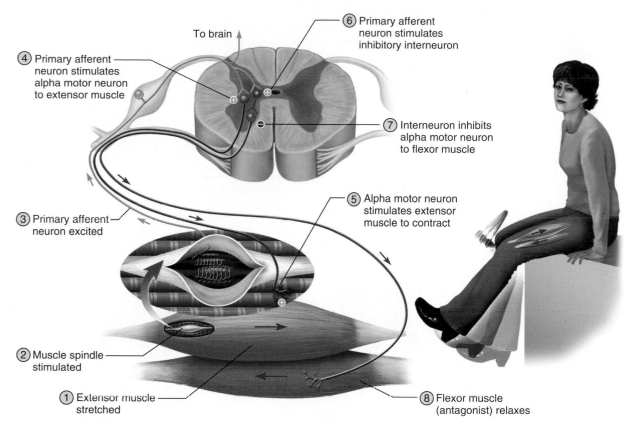

**FIGURE 13.21** **The Patellar Tendon Reflex Arc and Reciprocal Inhibition of the Antagonistic Muscle.** Plus signs indicate excitation of a postsynaptic cell (EPSPs), and minus signs indicate inhibition (IPSPs). The tendon reflex occurs in the quadriceps femoris muscle (large red arrow), while the hamstring muscles exhibit reciprocal inhibition (large blue arrow) so they do not contract and oppose the quadriceps.

▶ *Why is no IPSP shown at point 8 if the contraction of this muscle is being inhibited?*

(fig. 13.22). Even before you are consciously aware of the pain, you quickly pull your foot away before the glass penetrates any deeper. This action involves contraction of the flexors and relaxation of the extensors in that limb; the latter is another case of reciprocal inhibition.

The protective function of this reflex requires more than a quick jerk like a tendon reflex, so it involves more complex neural pathways. Sustained contraction of the flexors is produced by a parallel after-discharge circuit in the spinal cord (see fig. 12.30, p. 472). This circuit is part of a **polysynaptic reflex arc**—a pathway in which signals travel over many synapses on their way back to the muscle. Some signals follow routes with only a few synapses and return to the flexor muscles quickly. Others follow routes with more synapses, and therefore more delay, so they reach the flexor muscles a little later. Consequently, the flexor muscles receive prolonged output from the spinal cord and not just one sudden stimulus as in a stretch reflex. By the time these efferent signals begin to die out, you will probably be consciously aware of the pain and begin taking voluntary action to prevent further harm.

## THE CROSSED EXTENSION REFLEX

In the preceding situation, if *all* you did was to quickly lift the injured leg from the lake bottom, you would fall over. To prevent this and maintain your balance, other reflexes shift your center of gravity over the leg that is still on the ground. The **crossed extension reflex** is the contraction of extensor muscles in the limb opposite from the one that is withdrawn (fig. 13.22). It extends that limb and enables you to keep your balance. To produce this reflex, branches of the afferent nerve fibers cross from the stimulated side of the body to the contralateral side of the spinal cord. There, they synapse with interneurons, which, in turn, excite or inhibit alpha motor neurons to the muscles of the contralateral limb.

In the ipsilateral leg (the side that was hurt), you would contract your flexors and relax your extensors to lift the leg from the ground. On the contralateral side, you would relax your flexors and contract the extensors to stiffen that leg, since it must suddenly support your entire body. At the same time, signals travel up the spinal cord

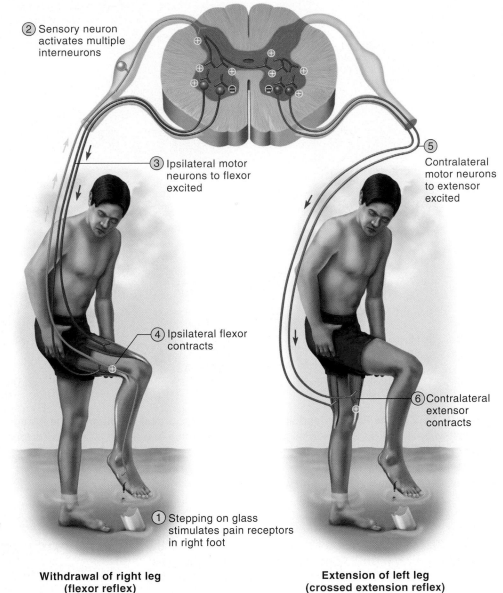

② Sensory neuron activates multiple interneurons

⑤ Contralateral motor neurons to extensor excited

③ Ipsilateral motor neurons to flexor excited

④ Ipsilateral flexor contracts

⑥ Contralateral extensor contracts

① Stepping on glass stimulates pain receptors in right foot

**Withdrawal of right leg (flexor reflex)**    **Extension of left leg (crossed extension reflex)**

**FIGURE 13.22   The Flexor and Crossed Extension Reflexes.** A pain stimulus triggers a withdrawal reflex, which results in contraction of flexor muscles of the injured limb. At the same time, a crossed extension reflex results in contraction of extensor muscles of the opposite limb. The latter reflex aids in balance when the injured limb is raised. Note that for each limb, while the agonist contracts, the alpha motor neuron to its antagonist is inhibited, as indicated by the red minus signs in the spinal cord.

▶ *Would you expect this reflex arc to show more synaptic delay, or less, than the ones in figure 13.21? Why?*

and cause contraction of contralateral muscles of the hip and abdomen to shift your center of gravity over the extended leg. To a large extent, the coordination of all these muscles and maintenance of equilibrium is mediated by the cerebellum and cerebral cortex.

The flexor reflex employs an **ipsilateral reflex arc**—one in which the sensory input and motor output are on the same sides of the spinal cord. The crossed extension reflex employs a **contralateral reflex arc,** in which the input and output are on opposite sides. An **intersegmental reflex arc** is one in which the input and output occur at different levels (segments) of the spinal cord—for example, when pain to the foot causes contractions of abdominal and hip muscles higher up the body. Note that all of these reflex arcs can function simultaneously to produce a coordinated protective response to pain.

## THE GOLGI TENDON REFLEX

**Golgi tendon organs** are proprioceptors located in a tendon near its junction with a muscle (fig. 13.23). A tendon organ is about 1 mm long and consists of a tangle of knobby nerve endings entwined in the collagen fibers of the tendon. As long as the tendon is slack, its collagen fibers are slightly spread and they put little pressure on the nerve endings woven among them. When muscle contraction pulls on the tendon, the collagen fibers come together like the two sides of a stretched rubber band and squeeze the nerve endings between them. The nerve fiber sends signals to the spinal cord that provide the CNS with feedback on the degree of muscle tension at the joint.

The **Golgi tendon reflex** is a response to excessive tension on the tendon. It inhibits alpha motor neurons to the muscle so the muscle does not contract as strongly. This serves to moderate muscle contraction before it tears a tendon or pulls it loose from the muscle or bone. Nevertheless, strong muscles and quick movements sometimes damage a tendon before the reflex can occur, causing such athletic injuries as a ruptured calcaneal tendon.

The Golgi tendon reflex also functions when some parts of a muscle contract more than others. It inhibits the fibers connected with overstimulated tendon organs so that their contraction is more comparable to the contraction of the rest of the muscle. This reflex spreads the workload more evenly over the entire muscle, which is beneficial in such actions as maintaining a steady grip on a tool.

Table 13.7 and Insight 13.5 describe some injuries and other disorders of the spinal cord and spinal nerves.

## Before You Go On

*Answer the following questions to test your understanding of the preceding section:*

9. Name five structural components of a typical somatic reflex arc. Which of these is absent from a monosynaptic arc?

10. State the function of each of the following in a muscle spindle: intrafusal fiber, annulospiral ending, and gamma motor neuron.

11. Explain how nerve fibers in a tendon sense the degree of tension in a muscle.

12. Why must the withdrawal reflex, but not the stretch reflex, involve a polysynaptic reflex arc?

13. Explain why the crossed extension reflex must accompany a withdrawal reflex of the leg.

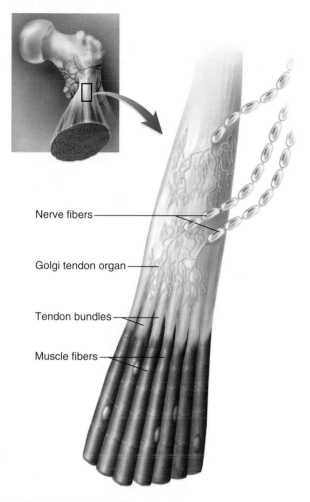

**FIGURE 13.23**  A Golgi Tendon Organ.

Nerve fibers

Golgi tendon organ

Tendon bundles

Muscle fibers

| TABLE 13.7 | Some Disorders of the Spinal Cord and Spinal Nerves |
|---|---|
| Guillain–Barré syndrome | An acute demyelinating nerve disorder often triggered by viral infection, resulting in muscle weakness, elevated heart rate, unstable blood pressure, shortness of breath, and sometimes death from respiratory paralysis |
| Neuralgia | General term for nerve pain, often caused by pressure on spinal nerves from herniated intervertebral discs or other causes |
| Paresthesia | Abnormal sensations of prickling, burning, numbness, or tingling; a symptom of nerve trauma or other peripheral nerve disorders |
| Peripheral neuropathy | Any loss of sensory or motor function due to nerve injury; also called *nerve palsy* |
| Rabies (hydrophobia) | A disease usually contracted from animal bites, involving viral infection that spreads via somatic motor nerve fibers to the CNS and then autonomic nerve fibers; leads to seizures, coma, and death; invariably fatal if not treated before CNS symptoms appear |
| Spinal meningitis | Inflammation of the spinal meninges due to viral, bacterial, or other infection |

**Disorders described elsewhere**

## Spinal Cord Trauma

Each year in the United States, 10,000 to 12,000 people become paralyzed by spinal cord trauma, usually as a result of vertebral fractures. The group at greatest risk is males from 16 to 30 years old, because of their high-risk behaviors. Fifty-five percent of their injuries are from automobile and motorcycle accidents, 18% from sports, and 15% from gunshot and stab wounds. Elderly people are also at above-average risk because of falls, and in times of war, battlefield injuries account for many cases.

### Effects of Injury

Complete *transection* (severance) of the spinal cord causes immediate loss of motor control at and below the level of the injury. Transection superior to segment C4 presents a threat of respiratory failure. Victims also lose all sensation from the level of injury and below, although some patients temporarily feel burning pain within one or two dermatomes of the level of the lesion.

In the early stage, victims exhibit a syndrome (a suite of signs and symptoms) called *spinal shock.* The muscles below the level of injury exhibit flaccid paralysis and an absence of reflexes because of the lack of stimulation from higher levels of the CNS. For 8 days to 8 weeks after the accident, the patient typically lacks bladder and bowel reflexes and thus retains urine and feces. Lacking sympathetic stimulation to the blood vessels, a patient may exhibit *neurogenic shock* in which the vessels dilate and blood pressure drops dangerously low. Fever may occur because the hypothalamus cannot induce sweating to cool the body. Spinal shock can last from a few days to several weeks (usually 7 to 20 days).

As spinal shock subsides, somatic reflexes begin to reappear, at first in the toes and progressing to the feet and legs. Autonomic reflexes also reappear. Contrary to the earlier urinary and fecal retention, a patient now has the opposite problem, incontinence, as the rectum and bladder empty reflexively in response to stretch. Both the somatic and autonomic nervous systems typically exhibit exaggerated reflexes, a state called *hyperreflexia* or the *mass reflex reaction.* Stimuli such as a full bladder or cutaneous touch can trigger an extreme cardiovascular reaction. The systolic blood pressure, normally about 120 mmHg, jumps to as high as 300 mmHg. This causes intense headaches and sometimes a stroke. Pressure receptors in the major arteries sense this rise in blood pressure and activate a reflex that slows the heart, sometimes to a rate as low as 30 or 40 beats/minute *(bradycardia),* compared with a normal rate of 70 to 80. The patient may also experience profuse sweating and blurred vision. Men at first lose the capacity for erection and ejaculation. They may recover these functions later and become capable of ejaculating and fathering children, but without sexual sensation. In females, menstruation may become irregular or cease.

The most serious permanent effect of spinal cord trauma is paralysis. The flaccid paralysis of spinal shock later changes to spastic paralysis as spinal reflexes are regained, but lack inhibitory control from the brain. Spastic paralysis typically starts with chronic flexion of the hips and knees (flexor spasms) and progresses to a state in which the limbs become straight and rigid (extensor spasms). Three forms of muscle paralysis are *paraplegia,* a paralysis of both lower limbs resulting from spinal cord lesions at levels T1 to L1; *quadriplegia,* the paralysis of all four limbs resulting from lesions above level C5; and *hemi-*

*plegia,* paralysis of one side of the body, usually resulting not from spinal cord injuries but from a stroke or other brain lesion. Spinal cord lesions from C5 to C7 can produce a state of partial quadriplegia—total paralysis of the lower limbs and partial paralysis (*paresis,* or weakness) of the upper limbs.

### Pathogenesis

Spinal cord trauma produces two stages of tissue destruction. The first is instantaneous—the destruction of cells by the traumatic event itself. The second wave of destruction, involving tissue death by necrosis and apoptosis, begins in minutes and lasts for days. It is far more destructive than the initial injury, typically converting a lesion in one spinal cord segment to a lesion that spans four or five segments, two above and two below the original site.

Microscopic hemorrhages appear in the gray matter and pia mater within minutes and grow larger over the next 2 hours. The white matter becomes edematous (swollen). This hemorrhaging and edema spread to adjacent segments of the cord and can fatally affect respiration or brainstem function when it occurs in the cervical region. *Ischemia* (iss-KEE-me-uh), the lack of blood, quickly leads to tissue necrosis. The white matter regains circulation in about 24 hours, but the gray matter remains ischemic. Inflammatory cells (leukocytes and macrophages) infiltrate the lesion as the circulation recovers, and while they clean up necrotic tissue, they also contribute to the damage by releasing destructive free radicals and other toxic chemicals. The necrosis worsens, and is accompanied by another form of cell death, apoptosis (see chapter 5). Apoptosis of the spinal oligodendrocytes, the myelinating glial cells of the CNS, results in demyelination of spinal nerve fibers, followed by death of the neurons.

In as little as 4 hours, this second wave of destruction, called *posttraumatic infarction,* consumes about 40% of the cross-sectional area of the spinal cord; within 24 hours, it destroys 70%. As many as five segments of the cord become transformed into a fluid-filled cavity, which is replaced with collagenous scar tissue over the next 3 to 4 weeks. This scar is one of the obstacles to the regeneration of lost nerve fibers.

### Treatment

The first priority in treating a spinal injury patient is to immobilize the spine to prevent further injury to the cord. Respiratory or other life support may also be required. Methylprednisolone, a steroid, dramatically improves recovery. Given within 3 hours of the trauma, it reduces injury to cell membranes and inhibits inflammation and apoptosis.

After these immediate requirements are met, reduction (repair) of the fracture is important. If a CT or MRI scan indicates spinal cord compression by the vertebral canal, a *decompression laminectomy* may be performed, in which the vertebral arch is removed from the affected region. CT and MRI have helped a great deal in recent decades for assessing vertebral and spinal cord damage, guiding surgical treatment, and improving recovery. Physical therapy is important for maintaining muscle and joint function as well as promoting the patient's psychological recovery.

Treatment strategies for spinal cord injuries are a lively area of medical research today. Some current interests are the use of antioxidants to reduce free radical damage, and the implantation of embryonic stem cells, which has produced significant (but not perfect) recovery from spinal cord lesions in rats. Public hopes have often been raised by promising studies reported in the scientific literature and news media, only to be dashed by the inability of other laboratories to repeat and confirm the results.

# Review of Key Concepts

## The Spinal Cord (p. 482)

1. The spinal cord conducts signals up and down the body, contains *central pattern generators* that control locomotion, and mediates many reflexes.

2. The spinal cord occupies the vertebral canal from vertebra C1 to L1. A bundle of nerve roots called the *cauda equina* occupies the vertebral canal from L2 to S5.

3. The cord is divided into cervical, thoracic, lumbar, and sacral regions, named for the levels of the vertebral column through which the spinal nerves emerge. The portion served by each spinal nerve is called a *segment* of the cord.

4. *Cervical* and *lumbar enlargements* are wide points in the cord marking the emergence of nerves that control the limbs.

5. The spinal cord is enclosed in three fibrous *meninges*. From superficial to deep, these are the dura mater, arachnoid mater, and pia mater. An *epidural space* exists between the dura mater and vertebral bone, and a *subarachnoid space* between the arachnoid and pia mater.

6. The pia mater issues periodic *denticulate ligaments* that anchor it to the dura, and continues inferiorly as a *coccygeal ligament* that anchors the cord to vertebra L2.

7. In cross section, the spinal cord exhibits a central H-shaped core of *gray matter* surrounded by *white matter*. The gray matter contains the somas, dendrites, and synapses while the white matter consists of nerve fibers (axons).

8. The *dorsal horn* of the gray matter receives afferent (sensory) nerve fibers from the dorsal root of the spinal nerve. The *ventral horn* contains the somas that give rise to the efferent (motor) nerve fibers of the ventral root of the nerve. A *lateral horn* in the thoracic and lumbar regions contains somas of the sympathetic neurons.

9. The white matter is divided into *dorsal, lateral,* and *ventral columns* on each side of the cord. Each column consists of multiple *tracts,* or bundles of nerve fibers. The nerve fibers in a given tract are similar in origin, destination, and function.

10. *Ascending tracts* carry sensory information up the cord to the brain. Their names and functions are listed in table 13.1.

11. From receptor to cerebral cortex, sensory signals typically travel through three neurons (first- through third-order) and cross over *(decussate)* from one side of the body to the other in the spinal cord or brainstem. Thus, the right cerebral cortex receives sensory input from the left side of the body (from the neck down) and vice versa.

12. *Descending tracts* carry motor commands from the brain downward. Their names and functions are also listed in table 13.1.

13. Motor signals typically begin in an *upper motor neuron* in the cerebral cortex and travel to a *lower motor neuron* in the brainstem or spinal cord. The latter neuron's axon leaves the CNS in a cranial or spinal nerve leading to a muscle.

## The Spinal Nerves (p. 490)

1. A nerve is a cord composed of nerve fibers (axons) and connective tissue.

2. Each nerve fiber is enclosed in its own fibrous sleeve called an *endoneurium*. Nerve fibers are bundled in groups called *fascicles* separated from each other by a *perineurium*. A fibrous *epineurium* covers the entire nerve.

3. Nerve fibers are classified as *afferent* or *efferent* depending on the direction of signal conduction; *somatic* or *visceral* depending on the types of organs they innervate; and *special* or *general* depending on the locations of the organs they innervate (table 13.2).

4. A *sensory nerve* is composed of afferent fibers only, a *motor nerve* of efferent fibers only, and a *mixed nerve* is composed of both. Most nerves are mixed.

5. A *ganglion* is a swelling along the course of a nerve containing the cell bodies of the peripheral neurons.

6. There are 31 pairs of *spinal nerves,* which enter and leave the spinal cord and emerge mainly through the intervertebral foramina. Within the vertebral canal, each branches into a *dorsal root* which carries sensory signals to the dorsal horn of the spinal cord, and a *ventral root* which receives motor signals from the ventral horn. The dorsal root has a swelling, the *dorsal root ganglion,* containing unipolar neurons of somatic sensory neurons.

7. Distal to the intervertebral foramen, each spinal nerve branches into a *dorsal ramus, ventral ramus,* and *meningeal branch.*

8. The ventral ramus gives rise to *intercostal nerves* in the thoracic region and *nerve plexuses* in all other regions. The nerve plexuses are weblike networks adjacent to the vertebral column: the *cervical, brachial, lumbar, sacral,* and *coccygeal plexus.* The nerves arising from each are described in tables 13.3 through 13.6.

## Somatic Reflexes (p. 501)

1. A reflex is a quick, involuntary, stereotyped reaction of a gland or muscle to a stimulus.

2. *Somatic (spinal) reflexes* are responses of skeletal muscles. The nerve signals in a somatic reflex travel by way of a *reflex arc* from a receptor via an afferent neuron to the spinal cord or brainstem, sometimes through interneurons in the CNS, then via an efferent neuron to a skeletal muscle.

3. Many somatic reflexes are initiated by *proprioceptors,* organs that

monitor the position and movements of body parts.

4. *Muscle spindles* are proprioceptors embedded in the skeletal muscles that respond to stretching of the muscle. They are composed of modified *intrafusal muscle fibers, primary* and *secondary afferent nerve fibers*, and *gamma motor neurons*, all enclosed in a fibrous sheath.

5. The *stretch reflex* is the tendency of a muscle to contract when it is stretched, as in the patellar tendon (knee-jerk) reflex. Stretch reflexes smooth joint actions and maintain equilibrium and posture. Many stretch reflexes travel via *monosynaptic* pathways so there is minimal synaptic delay and a very quick response.

6. A stretch reflex is often accompanied by *reciprocal inhibition,* a reflex that prevents an antagonistic muscle from contracting and interfering with the reflex action.

7. The *flexor reflex* is the withdrawal of a limb from an injurious stimulus, as in pulling back from a hot stove. It employs a polysynaptic reflex arc that produces a sustained response in the muscle.

8. The *crossed extension reflex* is contraction of the extensors on one side of the body when the flexors are contracted on the other side. It shifts the body weight so that one does not fall over.

9. The *Golgi tendon reflex* is the inhibition of a muscle contraction that occurs when its tendon is excessively stretched. Stretching stimulates a receptor in the tendon called a *Golgi tendon organ.* The reflex prevents tendon injuries and helps to distribute workload across a muscle.

# Testing Your Recall

1. Below L2, the vertebral canal is occupied by a bundle of spinal nerve roots called
   a. the terminal filum.
   b. the descending tracts.
   c. the gracile fasciculus.
   d. the medullary cone.
   e. the cauda equina.

2. The brachial plexus gives rise to all of the following nerves *except*
   a. the axillary nerve.
   b. the radial nerve.
   c. the saphenous nerve.
   d. the median nerve.
   e. the ulnar nerve.

3. Nerve fibers that adjust the tension in a muscle spindle are called
   a. intrafusal fibers.
   b. extrafusal fibers.
   c. alpha motor neurons.
   d. gamma motor neurons.
   e. annulospiral fibers.

4. A stretch reflex requires the action of _____ to prevent an antagonistic muscle from interfering with the agonist.
   a. gamma motor neurons
   b. a withdrawal reflex
   c. a crossed extension reflex
   d. reciprocal inhibition
   e. a contralateral reflex

5. A patient has a gunshot wound that caused a bone fragment to nick the spinal cord. The patient now feels no pain or temperature sensations from that level of the body down. Most likely, the _____ was damaged.
   a. gracile fasciculus
   b. medial lemniscus
   c. tectospinal tract
   d. lateral corticospinal tract
   e. spinothalamic tract

6. Which of these is *not* a region of the spinal cord?
   a. cervical
   b. thoracic
   c. pelvic
   d. lumbar
   e. sacral

7. In the spinal cord, the somas of the lower motor neurons are found in
   a. the cauda equina.
   b. the dorsal horns.
   c. the ventral horns.
   d. the dorsal root ganglia.
   e. the fasciculi.

8. The outermost connective tissue wrapping of a nerve is called the
   a. epineurium.
   b. perineurium.
   c. endoneurium.
   d. arachnoid mater.
   e. dura mater.

9. The intercostal nerves between the ribs arise from which spinal nerve plexus?
   a. cervical
   b. brachial
   c. lumbar
   d. sacral
   e. none of them

10. All somatic reflexes share all of the following properties *except*
    a. they are quick.
    b. they are monosynaptic.
    c. they require stimulation.
    d. they are involuntary.
    e. they are stereotyped.

11. Outside the CNS, the somas of neurons are clustered in swellings called _____.

12. Distal to the intervertebral foramen, a spinal nerve branches into a dorsal and ventral _____.

13. The cerebellum receives feedback from the muscles and joints by way of the _____ tracts of the spinal cord.

14. In the _____ reflex, contraction of flexor muscles in one limb is accompanied by the contraction of extensor muscles in the contralateral limb.

15. Modified muscle fibers serving primarily to detect stretch are called _____.

16. The _____ nerves arise from the cervical plexus and innervate the diaphragm.

17. The crossing of a nerve fiber or tract from the right side of the CNS to the left, or vice versa, is called _____.

18. The nonvisual awareness of the body's position and movements is called _____.

19. The _____ ganglion contains the somas of neurons that carry sensory signals to the spinal cord.

20. The sciatic nerve is a composite of two nerves, the _____ and _____.

*Answers in Appendix B*

# True or False

*Determine which five of the following statements are false, and briefly explain why.*

1. The gracile fasciculus is a descending spinal tract.

2. At the inferior end, the adult spinal cord ends before the vertebral column does.

3. Each spinal cord segment has only one pair of spinal nerves.

4. Some spinal nerves are sensory and others are motor.

5. The dura mater adheres tightly to the bone of the vertebral canal.

6. The dorsal and ventral horns of the spinal cord are composed of gray matter.

7. The corticospinal tracts carry motor signals down the spinal cord.

8. The dermatomes are nonoverlapping regions of skin innervated by different spinal nerves.

9. Somatic reflexes are those that do not involve the brain.

10. The Golgi tendon reflex acts to inhibit muscle contraction.

*Answers in Appendix B*

# Testing Your Comprehension

1. Jillian is thrown from a horse. She strikes the ground with her chin, causing severe hyperextension of the neck. Emergency medical technicians properly immobilize her neck and transport her to a hospital, but she dies 5 minutes after arrival. An autopsy shows multiple fractures of vertebrae C1, C6, and C7 and extensive damage to the spinal cord. Explain why she died rather than being left quadriplegic.

2. Wallace is the victim of a hunting accident. A bullet grazed his vertebral column and bone fragments severed the left half of his spinal cord at segments T8 through T10. Since the accident, Wallace has had a condition called *dissociated sensory loss*, in which he feels no sensations of deep touch or limb position on the *left* side of his body below the injury, and no sensations of pain or heat from the *right* side. Explain what spinal tract(s) the injury has affected and why these sensory losses are on opposite sides of the body.

3. Anthony gets into a fight between rival gangs. As an attacker comes at him with a knife, he turns to flee, but stumbles. The attacker stabs him on the medial side of the right gluteal fold and Anthony collapses. He loses all use of his right limb, being unable to extend his hip, flex his knee, or move his foot. He never fully recovers these lost functions. Explain what nerve injury Anthony has most likely suffered.

4. Stand with your right shoulder, hip, and foot firmly against a wall. Raise your left foot from the floor without losing contact with the wall at any point. What happens? Why? What principle of this chapter does this demonstrate?

5. When a patient needs a tendon graft, surgeons sometimes use the tendon of the palmaris longus, a relatively dispensable muscle of the forearm. The median nerve lies nearby and looks very similar to this tendon. There have been cases where a surgeon mistakenly removed a section of this nerve instead of the tendon. What effects do you think such a mistake would have on the patient?

*Answers at www.mhhe.com/saladin4*

# www.mhhe.com/saladin4

*The textbook website provides a wealth of interactive study materials fully organized and integrated by chapter. You will find practice quizzes, labeling exercises, and much more that will complement your learning and understanding of anatomy and physiology. The website also includes tools designed to enhance your* **Anatomy & Physiology | REVEALED** *experience.*

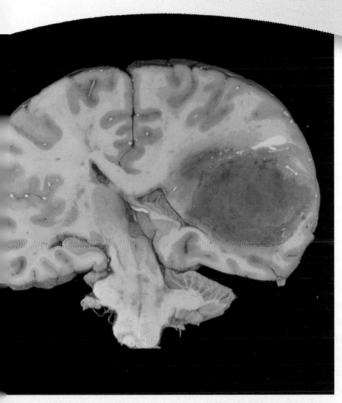

*Frontal section of the brain with a large tumor (glioblastoma) in the left cerebral hemisphere.*

# THE BRAIN AND CRANIAL NERVES

## CHAPTER OUTLINE

## Brushing Up

To understand this chapter, it is important that you understand or brush up on the following concepts:

The mystique of the brain continues to intrigue modern biologists and psychologists even as it did the philosophers of antiquity. Aristotle thought that the brain was a radiator for cooling the blood, but generations earlier, Hippocrates had expressed a more accurate view. "Men ought to know," he said, "that from the brain, and from the brain only, arise our pleasures, joy, laughter and jests, as well as our sorrows, pains, griefs and tears. Through it, in particular, we think, see, hear, and distinguish the ugly from the beautiful, the bad from the good, the pleasant from the unpleasant."

Brain function is so strongly associated with what it means to be alive and human that the cessation of brain activity is taken as a clinical criterion of death even when other organs of the body are still functioning. With its hundreds of neural pools and trillions of synapses, the brain performs sophisticated tasks beyond our present understanding. Still, all of our mental functions, no matter how complex, are ultimately based on the cellular activities described in chapter 12. The relationship of the mind or personality to the cellular function of the brain is a question that will provide fertile ground for scientific study and philosophical debate long into the future.

This chapter is a study of the brain and the cranial nerves directly connected to it. Here we will plumb some of the mysteries of motor control, sensation, emotion, thought, language, personality, memory, dreams, and plans. Your study of this chapter is one brain's attempt to understand itself.

# Overview of the Brain

### Objectives

When you have completed this section, you should be able to

- describe the major subdivisions and anatomical landmarks of the brain;
- describe the locations of the gray and white matter of the brain; and
- describe the embryonic development of the CNS and relate this to adult brain anatomy.

## MAJOR LANDMARKS

Before we consider the form and function of specific regions of the brain, it will help to get a general overview of its major landmarks (figs. 14.1 and 14.2). These will provide important points of reference as we progress through a more detailed study.

Two directional terms often used to describe brain anatomy are *rostral* and *caudal*. **Rostral**[1] means "toward the nose" and **caudal**[2] means "toward the tail." These are apt descriptions for an animal such as a laboratory rat, on which so much neuroscience research has been done. The terms are retained for human neuroanatomy as well, but in reference to the human brain, *rostral* means toward the forehead and *caudal* means toward the spinal cord. In the spinal cord and brainstem, which are vertically oriented, *rostral* means higher and *caudal* means lower.

The average adult brain weighs about 1,600 g (3.5 lb) in men and 1,450 g in women. Its size is proportional to body size, not intelligence—Neanderthals had larger brains than modern humans.

The brain is divided into three major portions: the *cerebrum, cerebellum,* and *brainstem.* The **cerebrum** (seh-REE-brum or SER-eh-brum) constitutes about 83% of its volume and consists of a pair of half-globes called the **cerebral hemispheres.** Each hemisphere is marked by thick folds called **gyri**[3] (JY-rye; singular, *gyrus*) separated by shallow grooves called **sulci**[4] (SUL-sye; singular, *sulcus*). A very deep groove, the **longitudinal fissure,** separates the right and left hemispheres from each other. At the bottom of this fissure, the hemispheres are connected by a thick bundle of nerve fibers called the **corpus callosum**[5]—a prominent landmark for anatomical description (fig. 14.2).

The **cerebellum**[6] (SER-eh-BEL-um) lies inferior to the cerebrum and occupies the posterior cranial fossa. It is also marked by gyri, sulci, and fissures. The cerebellum is the second-largest region of the brain; it constitutes about 10% of its volume but contains over 50% of its neurons.

Authorities differ on how they define the **brainstem.** The definition adopted here is that it is what remains of the brain if the cerebrum and cerebellum are removed. Its major components, from rostral to caudal, are the *diencephalon, midbrain, pons,* and *medulla oblongata.* The most common alternative definition includes only the last three of these.

In a living person, the brainstem is oriented like a vertical stalk with the cerebrum perched on top of it like a mushroom cap. Postmortem changes give it a more oblique angle in the cadaver and consequently in many medical illustrations. Caudally, the brainstem ends at the foramen magnum of the skull, and the central nervous system (CNS) continues below this as the spinal cord.

---

[1]*rostr* = nose
[2]*caud* = tail
[3]*gy* = turn, twist
[4]*sulc* = furrow, groove
[5]*corpus* = body + *call* = thick
[6]*cereb* = brain + *ellum* = little

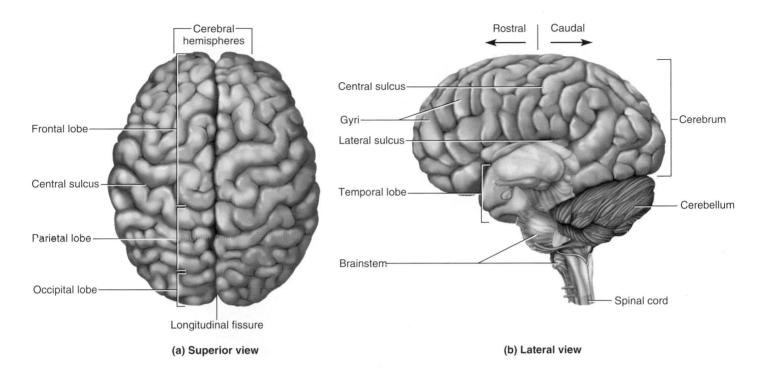

Cerebral hemispheres

Frontal lobe

Central sulcus

Parietal lobe

Occipital lobe

Longitudinal fissure

**(a) Superior view**

Rostral | Caudal

Central sulcus

Gyri

Lateral sulcus

Temporal lobe

Brainstem

Cerebrum

Cerebellum

Spinal cord

**(b) Lateral view**

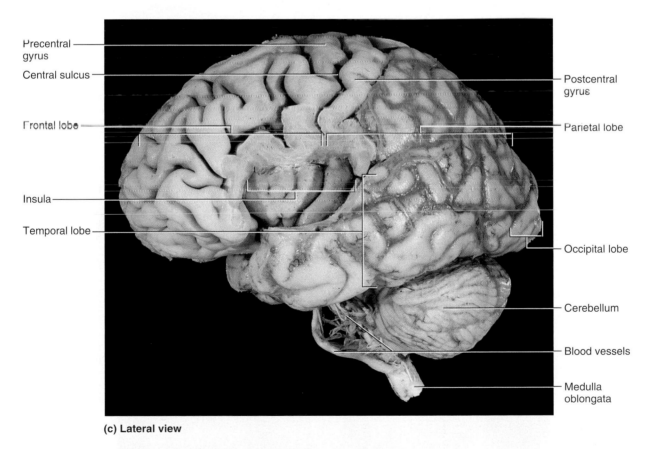

Precentral gyrus

Central sulcus

Frontal lobe

Insula

Temporal lobe

Postcentral gyrus

Parietal lobe

Occipital lobe

Cerebellum

Blood vessels

Medulla oblongata

**(c) Lateral view**

**FIGURE 14.1  Surface Anatomy of the Brain.**  (a) Superior view of the cerebral hemispheres. (b) Left lateral view, with the brainstem in yellow. The portion of the brainstem above the cerebellum is represented as showing through the cerebrum to convey its location. (c) The partially dissected brain of a cadaver. Part of the left hemisphere is cut away to expose the insula. The arachnoid mater is removed from the anterior (rostral) half of the brain to expose the gyri and sulci. The arachnoid with its blood vessels is seen in the posterior (caudal) half. Blood vessels of the brainstem are left in place.

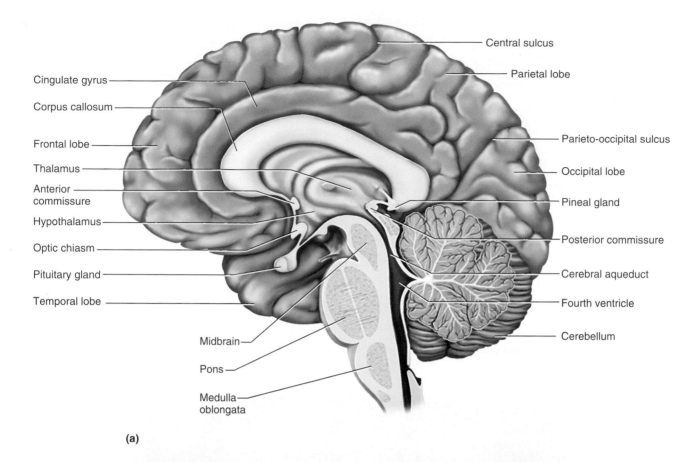

Central sulcus

Cingulate gyrus

Corpus callosum

Frontal lobe

Thalamus

Anterior commissure

Hypothalamus

Optic chiasm

Pituitary gland

Temporal lobe

Parietal lobe

Parieto-occipital sulcus

Occipital lobe

Pineal gland

Posterior commissure

Cerebral aqueduct

Fourth ventricle

Cerebellum

Midbrain

Pons

Medulla oblongata

(a)

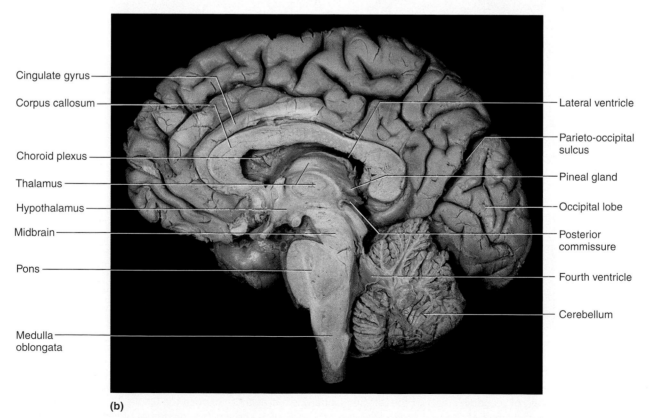

Cingulate gyrus

Corpus callosum

Choroid plexus

Thalamus

Hypothalamus

Midbrain

Pons

Medulla oblongata

Lateral ventricle

Parieto-occipital sulcus

Pineal gland

Occipital lobe

Posterior commissure

Fourth ventricle

Cerebellum

(b)

**FIGURE 14.2** **Medial Aspect of the Brain.** (a) Major anatomical landmarks of the medial surface. (b) Median section of the cadaver brain.

## GRAY AND WHITE MATTER

The brain, like the spinal cord, is composed of gray and white matter. Gray matter—the seat of the neuron cell bodies, dendrites, and synapses—forms a surface layer called the **cortex** over the cerebrum and cerebellum, and deeper masses called **nuclei** surrounded by white matter. The white matter lies deep to the cortical gray matter of the brain, opposite from the relationship of gray and white matter in the spinal cord. As in the spinal cord, the white matter is composed of **tracts,** or bundles of axons, which here connect one part of the brain to another. These are described later.

## EMBRYONIC DEVELOPMENT

To understand the terminology of mature brain anatomy, it is necessary to know the embryonic development of the CNS. The nervous system develops from ectoderm, the outermost germ layer of an embryo. By the third week of development, a dorsal streak called the *neuroectoderm* appears along the length of the embryo and thickens to form a **neural plate** (fig. 14.3). This is destined to give rise to all neurons and glial cells except microglia, which come from mesoderm. As development progresses, the neural plate sinks and forms a **neural groove** with a raised **neural**

fold along each side. The neural folds then fuse along the midline, somewhat like a closing zipper. By 4 weeks, this process creates a hollow channel called the **neural tube.** The neural tube now separates from the overlying ectoderm, sinks a little deeper, and grows lateral processes that later form motor nerve fibers. The lumen of the neural tube develops into fluid-filled spaces called the *central canal* of the spinal cord and *ventricles* of the brain.

As the neural tube develops, some ectodermal cells that originally lay along the margin of the groove separate from the rest and form a longitudinal column on each side called the **neural crest.** Some neural crest cells become sensory neurons, while others migrate to other locations and become sympathetic neurons, Schwann cells, and other cell types.

By the fourth week, the neural tube exhibits three anterior dilations, or *primary vesicles,* called the **forebrain, midbrain,** and **hindbrain,** also known, respectively, as the prosencephalon[7] (PROSS-en-SEF-uh-lon), mesencephalon[8] (MEZ-en-SEF-uh-lon), and rhombencephalon[9]

---

[7] *pros* = before, in front + *encephal* = brain
[8] *mes* = middle + *encephal* = brain
[9] *rhomb* = rhombus + *encephal* = brain

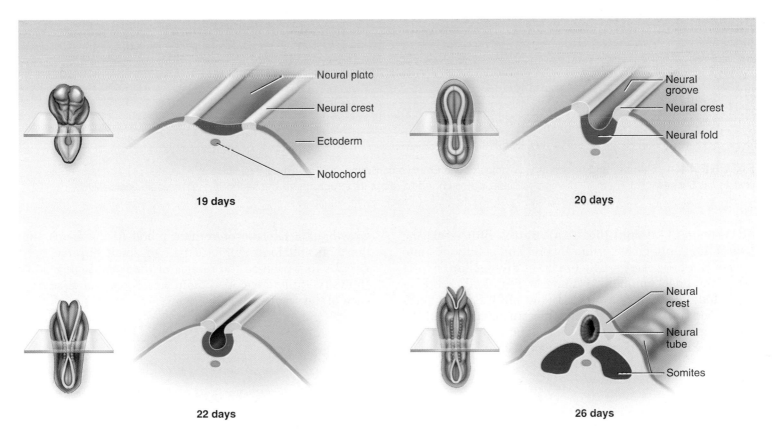

**FIGURE 14.3** **Formation of the Embryonic Neural Tube.** The left-hand figure in each case is a dorsal view of the embryo, and the right-hand figure is a three-dimensional representation of the tissues at the indicated level of the respective embryo.

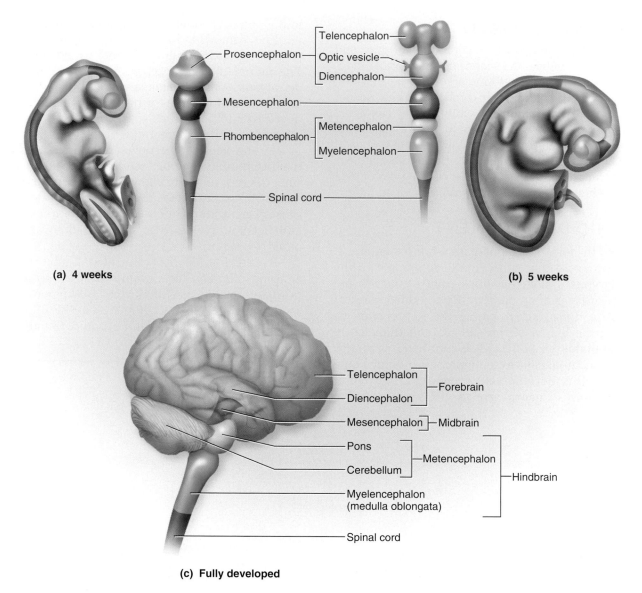

**FIGURE 14.4** **Primary and Secondary Vesicles of the Embryonic Brain.** (a) The primary vesicles at 4 weeks. (b) The secondary vesicles at 5 weeks. (c) The fully developed brain, color-coded to relate its structures to the secondary embryonic vesicles.

(ROM-ben-SEF-uh-lon) (fig. 14.4). By the fifth week, the neural tube undergoes further flexion and subdivides into five *secondary vesicles.* The forebrain divides into two of them, the **telencephalon**[10] (TEL-en-SEF-uh-lon) and **diencephalon**[11] (DY-en-SEF-uh-lon); the midbrain remains undivided and retains the name **mesencephalon;** and the hindbrain divides into two vesicles, the **metencephalon**[12] (MET-en-SEF-uh-lon) and **myelencephalon**[13] (MY-el-en-SEF-uh-lon). The telencephalon has a pair of lateral out-

growths that later become the cerebral hemispheres, and the diencephalon exhibits a pair of small cuplike *optic vesicles* that become the retinas of the eyes. Figure 14.4c shows the fully developed brain structures that arise from each of the secondary vesicles.

## Before You Go On

*Answer the following questions to test your under-standing of the preceding section:*

1. List the three major parts of the brain and describe their locations.

2. Define gyrus *and* sulcus.

3. Explain how the five secondary brain vesicles arise from the neural tube.

---

[10]*tele* = end, remote + *encephal* = brain
[11]*di* = through, between + *encephal* = brain
[12]*met* = behind, beyond, distal to + *encephal* = brain
[13]*myel* = spinal cord + *encephal* = brain

# Meninges, Ventricles, Cerebrospinal Fluid, and Blood Supply

### Objectives

When you have completed this section, you should be able to

- describe the meninges of the brain;
- describe the fluid-filled chambers within the brain;
- discuss the production, circulation, and function of the cerebrospinal fluid that fills these chambers; and
- explain the significance of the brain barrier system.

## MENINGES

Like the spinal cord, the brain is enveloped in connective tissue membranes, the meninges, which lie between the nervous tissue and bone. The meninges of the brain are basically the same as those of the spinal cord—dura mater, arachnoid mater, and pia mater—although there are some differences in the dura (fig. 14.5). In the cranial cavity, it consists of two layers: an outer *periosteal layer,* equivalent to the periosteum of the cranial bone, and an inner *meningeal layer.* Only the meningeal layer continues into the vertebral canal. The cranial dura mater lies closely against the cranial bone, with no intervening epidural space like that of the spinal cord. In some places, the two layers are separated by **dural sinuses,** spaces that collect blood that has circulated through the brain. Two major dural sinuses are the **superior sagittal sinus,** found just under the cranium along the midsagittal line, and the **transverse sinus,** which runs horizontally from the rear of the head toward each ear. These sinuses meet like an inverted T at the back of the brain and ultimately empty into the internal jugular veins of the neck.

In certain places, the meningeal layer of the dura folds inward to separate major parts of the brain from each other: the *falx*[14] *cerebri* (falks SER-eh-bry) extends into the longitudinal fissure between the right and left cerebral hemispheres; the *tentorium*[15] (ten-TOE-ree-um) *cerebelli* stretches like a roof over the posterior cranial fossa and separates the cerebellum from the overlying cerebrum; and the *falx cerebelli* partially separates the right and left halves of the cerebellum on the inferior side.

---

[14]*falx* = sickle
[15]*tentorium* = tent

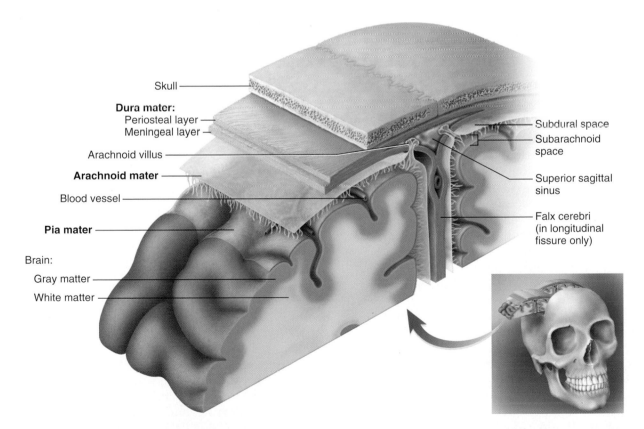

**FIGURE 14.5** **The Meninges of the Brain.** Frontal section of the head.

INSIGHT 14.1    Clinical Application

## Meningitis

*Meningitis*—inflammation of the meninges—is one of the most serious diseases of infancy and childhood. It occurs especially between 3 months and 2 years of age. Meningitis is caused by a variety of bacteria and viruses that invade the CNS by way of the nose and throat, often following respiratory, throat, or ear infections. The pia mater and arachnoid are most likely to be affected, and from here the infection can spread to the adjacent nervous tissue. In bacterial meningitis, the brain swells, the ventricles enlarge, and the brainstem may hemorrhage. Signs include a high fever, stiff neck, drowsiness, and intense headache and may progress to vomiting, loss of sensory and motor functions, and coma. Death can occur within hours of the onset. Infants and toddlers with a high fever should therefore receive immediate medical attention.

Meningitis is diagnosed partly by examining the cerebrospinal fluid (CSF) for bacteria and white blood cells. The CSF is obtained by making a *lumbar puncture (spinal tap)* between two lumbar vertebrae and drawing fluid from the subarachnoid space. This site is chosen because it has an abundance of CSF and there is no risk of injury to the spinal cord, which does not extend into the lower lumbar vertebrae.

The arachnoid mater and pia mater are similar to those of the spinal cord. A *subarachnoid space* separates the arachnoid from the pia, and in some places, a *subdural space* separates the dura from the arachnoid.

## VENTRICLES AND CEREBROSPINAL FLUID

The brain has four internal chambers called **ventricles** (fig. 14.6). The most rostral ones are the **lateral ventricles,** which form a large arc in each cerebral hemisphere. Through a tiny passage called the **interventricular foramen,** each lateral ventricle is connected to the **third ventricle,** a narrow medial space inferior to the corpus callosum. From here, a canal called the **cerebral aqueduct** passes down the core of the midbrain and leads to the **fourth ventricle,** a small chamber between the pons and cerebellum. Caudally, this space narrows and forms a **central canal** that extends through the medulla oblongata into the spinal cord.

These ventricles and canals are lined with ependymal cells, a type of neuroglia that resembles a simple cuboidal epithelium. Each ventricle contains a **choroid** (CO-royd) **plexus,** named for its histological resemblance to a fetal membrane called the chorion. The choroid plexus is a network of blood capillaries anchored to the floor or wall of the ventricle and covered by ependymal cells.

A clear, colorless liquid called **cerebrospinal fluid (CSF)** fills the ventricles and canals of the CNS and bathes its external surface. The brain produces about 500 mL of CSF per day, but the fluid is constantly reabsorbed at the same rate and only 100 to 160 mL is normally present at one time (but see Insight 14.2). About 40% of it is formed in the subarachnoid space external to the brain, 30% by the general ependymal lining of the brain ventricles, and 30% by the choroid plexuses. CSF forms partly by the filtration of blood plasma through the choroid plexuses and other capillaries of the brain. The ependymal cells modify this filtrate, however, so the CSF has more sodium and chloride than the blood plasma, but less potassium, calcium, and glucose and very little protein.

Cerebrospinal fluid serves three purposes:

1. **Buoyancy.** Because the brain and CSF are similar in density, the brain neither sinks nor floats in the CSF but remains suspended in it—that is, the brain has *neutral buoyancy.* A human brain removed from the body weighs about 1,500 g, but when suspended in CSF its effective weight is only about 50 g. By analogy, consider how much easier it is to lift another person when you are standing in a lake than it is on land. Neutral buoyancy allows the brain to attain considerable size without being impaired by its own weight. If the brain rested heavily on the floor of the cranium, the pressure would kill the nervous tissue.

2. **Protection.** CSF also protects the brain from striking the cranium when the head is jolted. If the jolt is severe, however, the brain still may strike the inside of the cranium or suffer shearing injury from contact with the angular surfaces of the cranial floor. This is one of the common findings in child abuse (shaken child syndrome) and in head injuries (concussions) from auto accidents, boxing, and the like.

3. **Chemical stability.** CSF is secreted into each ventricle of the brain and is ultimately absorbed into the bloodstream. It provides a means of rinsing metabolic wastes from the CNS and homeostatically regulating its chemical environment. Slight changes in its composition can cause malfunctions of the nervous system. For example, a high glycine concentration disrupts temperature and blood pressure control, and a high pH causes dizziness and fainting.

The CSF is not a stationary fluid but continually flows through and around the CNS, driven partly by its own pressure, partly by the beating of ependymal cilia, and partly by rhythmic pulsations of the brain produced by each heartbeat. The CSF secreted in the lateral ventricles flows through the interventricular foramina into the third ventricle (fig. 14.7) and then down the cerebral aqueduct to the fourth ventricle. The third and fourth ventricles and their choroid plexuses add more CSF along the way. A small amount of CSF fills the central canal of the spinal cord, but ultimately, all of it escapes through three pores in the walls of the fourth ventricle—a *median aperture* and two *lateral apertures*. These lead into the subarachnoid space on the brain and spinal cord surface. From this space, the CSF is reabsorbed by **arachnoid villi,** cauli-flower-like extensions of the arachnoid meninx of the brain. The villi protrude through the dura mater into the superior sagittal sinus. CSF penetrates the walls of the villi and mixes with the blood in the sinus.

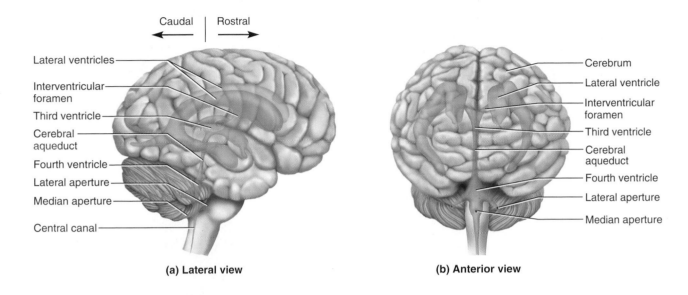

**(a) Lateral view**

**(b) Anterior view**

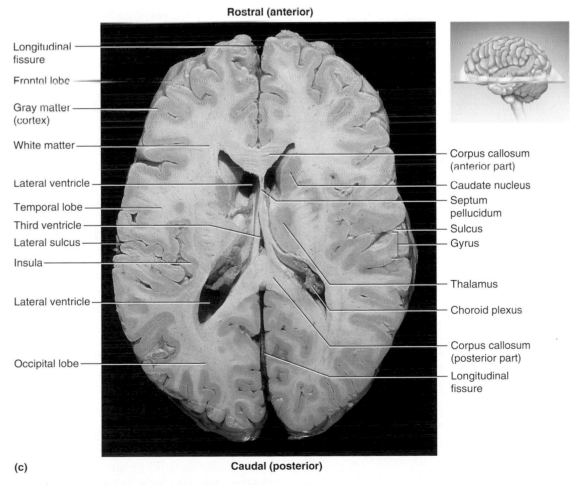

**(c)**

**FIGURE 14.6   Ventricles of the Brain.** (a) Right lateral view. (b) Anterior view. (c) Superior view of a horizontal section of the cadaver brain, showing the lateral ventricles and some other features of the cerebrum.

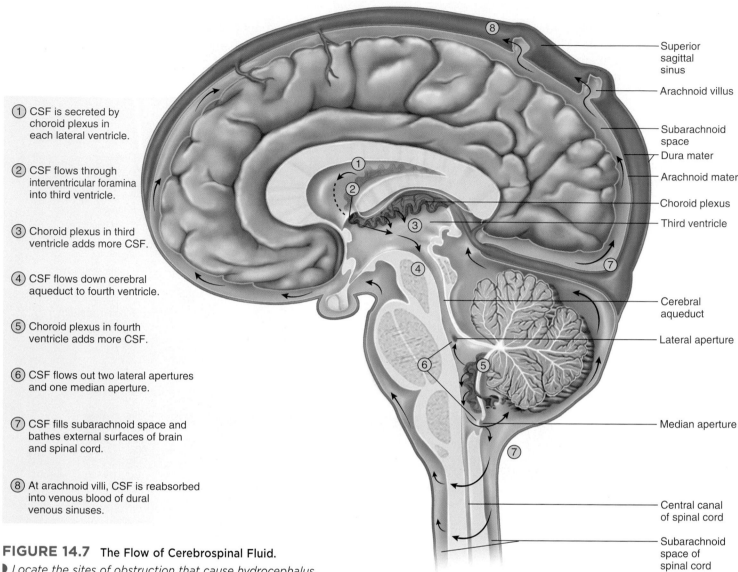

① CSF is secreted by choroid plexus in each lateral ventricle.

② CSF flows through interventricular foramina into third ventricle.

③ Choroid plexus in third ventricle adds more CSF.

④ CSF flows down cerebral aqueduct to fourth ventricle.

⑤ Choroid plexus in fourth ventricle adds more CSF.

⑥ CSF flows out two lateral apertures and one median aperture.

⑦ CSF fills subarachnoid space and bathes external surfaces of brain and spinal cord.

⑧ At arachnoid villi, CSF is reabsorbed into venous blood of dural venous sinuses.

Superior sagittal sinus
Arachnoid villus
Subarachnoid space
Dura mater
Arachnoid mater
Choroid plexus
Third ventricle
Cerebral aqueduct
Lateral aperture
Median aperture
Central canal of spinal cord
Subarachnoid space of spinal cord

**FIGURE 14.7** The Flow of Cerebrospinal Fluid.

▶ *Locate the sites of obstruction that cause hydrocephalus.*

---

**INSIGHT 14.2** Clinical Application

### Hydrocephalus

*Hydrocephalus*[16] is the abnormal accumulation of CSF in the brain, usually resulting from a blockage in its route of flow and reabsorption. Such obstructions usually occur in the interventricular foramen, cerebral aqueduct, and the apertures of the fourth ventricle. The accumulated CSF expands the ventricles and compresses the nervous tissue, with potentially fatal consequences. In a fetus or infant, it can cause the entire head to enlarge because the cranial bones are not yet fused. Good recovery can be achieved if a tube (shunt) is inserted to drain fluid from the ventricles into a vein of the neck.

---

[16]*hydro* = water + *cephal* = head

## BLOOD SUPPLY AND THE BRAIN BARRIER SYSTEM

Although the brain is only 2% of the adult body weight, it receives 15% of the blood (about 750 mL/min) and consumes 20% of the body's oxygen and glucose. Because neurons have such a high demand for ATP, and therefore glucose and oxygen, the constancy of blood supply is especially critical to the nervous system. A mere 10-second interruption in blood flow can cause loss of consciousness; an interruption of 1 to 2 minutes can significantly impair neural function; and 4 minutes without blood causes irreversible brain damage.

Despite its critical importance to the brain, the blood is also a source of antibodies, macrophages, and other potentially harmful agents. Consequently, there is a **brain barrier system** that strictly regulates what substances can get from the bloodstream into the tissue fluid of the brain. There are two potential points of entry that must be guarded: the blood capillaries throughout the brain tissue and the capillaries of the choroid plexuses.

At the former site, the brain is well protected by the **blood–brain barrier (BBB),** which consists mainly of the tightly joined endothelial cells that form the capillaries and partly of the basement membrane surrounding them. In the developing brain, astrocytes reach out and contact the capillaries with their perivascular feet. They stimulate the endothelial cells to form tight junctions, which seal off the capillaries and ensure that anything leaving the blood must pass through the cells and not between them. At the choroid plexuses, the brain is protected by a similar **blood–CSF barrier,** composed of ependymal cells joined by tight junctions. Tight junctions are absent from ependymal cells elsewhere, because it is important to allow exchanges between the brain tissue and CSF. That is, there is no brain–CSF barrier.

The brain barrier system (BBS) is highly permeable to water, glucose, and lipid-soluble substances such as oxygen, carbon dioxide, alcohol, caffeine, nicotine, and anesthetics. It is slightly permeable to sodium, potassium, chloride, and the waste products urea and creatinine. While the BBS is an important protective device, it is an obstacle to the delivery of drugs for brain diseases. Trauma and inflammation sometimes damage the BBS and allow pathogens to enter the brain tissue. Furthermore, there are places called **circumventricular organs (CVOs)** in the third and fourth ventricles where the barrier system is absent, and the blood does have direct access to the brain. These enable the brain to monitor and respond to fluctuations in blood glucose, pH, osmolarity, and other variables. Unfortunately, the CVOs also afford a route of invasion by the human immunodeficiency virus (HIV, the AIDS virus).

### Before You Go On

*Answer the following questions to test your understanding of the preceding section:*

4. *Name the three meninges from superficial to deep.*

5. *Describe three functions of the cerebrospinal fluid.*

6. *Where does the CSF originate and what route does it take through and around the CNS?*

7. *Name the two components of the brain barrier system and explain the importance of this system.*

# The Hindbrain and Midbrain

### Objectives

When you have completed this section, you should be able to

- list the components of the hindbrain and midbrain and their functions; and

- describe the location and functions of the reticular formation.

Our study of the brain in the following pages will be organized around the five secondary vesicles of the embryonic brain and their mature derivatives. We will proceed in a caudal to rostral direction, beginning with the hindbrain and its relatively simple functions, then progressing to the forebrain, the seat of such complex functions as thought, memory, and emotion.

## THE MEDULLA OBLONGATA

As noted earlier, the embryonic hindbrain differentiates into two subdivisions: the myelencephalon and metencephalon (see fig. 14.4). The myelencephalon becomes just one adult structure, the **medulla oblongata** (meh-DULL-uh OB-long-GAH-ta). The medulla is about 3 cm long and superficially looks like an extension of the spinal cord, but slightly wider. Significant differences are seen, however, on closer inspection of its gross and microscopic anatomy.

The anterior surface bears a pair of clublike ridges, the **pyramids.** Resembling two side-by-side baseball bats, the pyramids are wider at the rostral end, taper caudally, and are separated by an *anterior median fissure* continuous with that of the spinal cord (fig. 14.8). The pyramids contain *corticospinal tracts* of nerve fibers that carry motor signals from the cerebrum to the spinal cord, ultimately to stimulate the skeletal muscles. Most of these motor nerve fibers decussate at a visible point near the caudal end of the pyramids. As a result, each side of the brain controls muscles on the contralateral side of the body. (This pertains only to muscles below the head.)

Lateral to each pyramid is a mound called the **olive.** It contains a wavy layer of gray matter, the **inferior olivary nucleus** (fig. 14.9c). This is a center that receives information from many levels of the brain and spinal cord and relays it mainly to the cerebellum. Dorsally, the medulla exhibits ridges called the *gracile fasciculus* and *cuneate fasciculus.* These are continuations of the spinal cord tracts of the same names that carry sensory signals to the brain.

In addition to ascending and descending nerve tracts, the medulla contains sensory nuclei that receive input from the taste buds, pharynx, and viscera of the thoracic and abdominal cavities, and motor nuclei that control several primitive visceral and somatic functions. Some motor nuclei of the medulla discussed later in the book are:

- the **cardiac center,** which regulates the rate and force of the heartbeat;

- the **vasomotor center,** which adjusts blood vessel diameter; this response regulates blood pressure and reroutes blood from one part of the body to another; and

- two **respiratory centers,** which regulate the rate and depth of breathing.

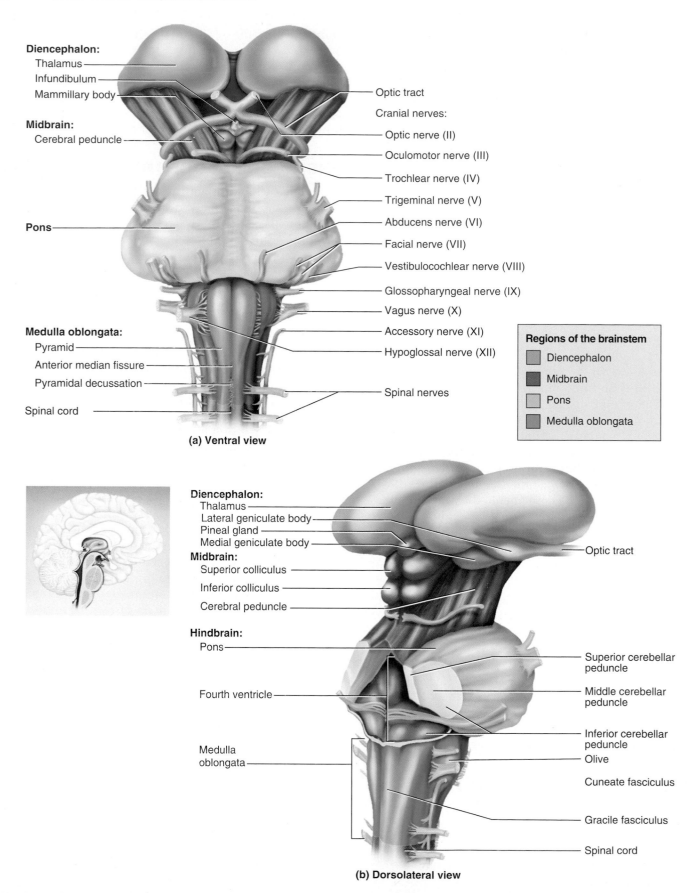

**Diencephalon:**
Thalamus
Infundibulum
Mammillary body

**Midbrain:**
Cerebral peduncle

**Pons**

**Medulla oblongata:**
Pyramid
Anterior median fissure
Pyramidal decussation

Spinal cord

Optic tract

Cranial nerves:

Optic nerve (II)
Oculomotor nerve (III)
Trochlear nerve (IV)
Trigeminal nerve (V)
Abducens nerve (VI)
Facial nerve (VII)
Vestibulocochlear nerve (VIII)
Glossopharyngeal nerve (IX)
Vagus nerve (X)
Accessory nerve (XI)
Hypoglossal nerve (XII)

Spinal nerves

**Regions of the brainstem**
Diencephalon
Midbrain
Pons
Medulla oblongata

**(a) Ventral view**

**Diencephalon:**
Thalamus
Lateral geniculate body
Pineal gland
Medial geniculate body
**Midbrain:**
Superior colliculus
Inferior colliculus
Cerebral peduncle

**Hindbrain:**
Pons

Fourth ventricle

Medulla
oblongata

Optic tract

Superior cerebellar peduncle
Middle cerebellar peduncle
Inferior cerebellar peduncle
Olive
Cuneate fasciculus
Gracile fasciculus
Spinal cord

**(b) Dorsolateral view**

**FIGURE 14.8   The Brainstem.** These illustrations are color-coded to match the embryonic origins in figure 14.4. The boundary between the middle and inferior cerebellar peduncles is indistinct. Authorities vary as to whether to include the diencephalon in the brainstem.

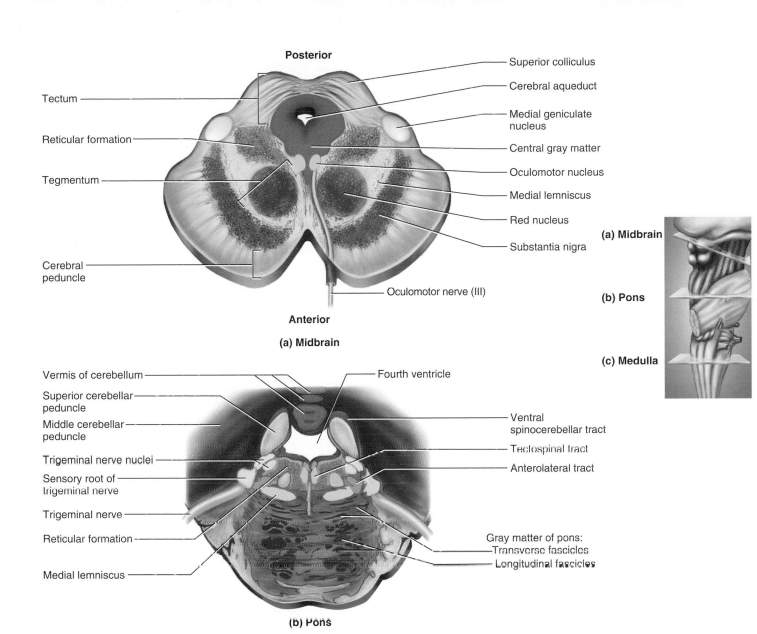

**Posterior**

Tectum

Reticular formation

Tegmentum

Cerebral peduncle

**Anterior**

**(a) Midbrain**

Superior colliculus

Cerebral aqueduct

Medial geniculate nucleus

Central gray matter

Oculomotor nucleus

Medial lemniscus

Red nucleus

Substantia nigra

Oculomotor nerve (III)

**(a) Midbrain**

**(b) Pons**

**(c) Medulla**

Vermis of cerebellum

Superior cerebellar peduncle

Middle cerebellar peduncle

Trigeminal nerve nuclei

Sensory root of trigeminal nerve

Trigeminal nerve

Reticular formation

Medial lemniscus

Fourth ventricle

Ventral spinocerebellar tract

Tectospinal tract

Anterolateral tract

Gray matter of pons:
Transverse fascicles
Longitudinal fascicles

**(b) Pons**

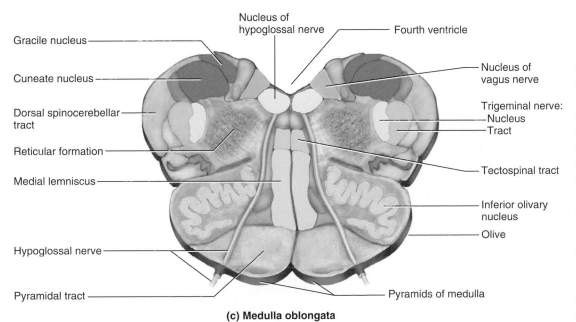

Gracile nucleus

Cuneate nucleus

Dorsal spinocerebellar tract

Reticular formation

Medial lemniscus

Hypoglossal nerve

Pyramidal tract

Nucleus of hypoglossal nerve

Fourth ventricle

Nucleus of vagus nerve

Trigeminal nerve:
Nucleus
Tract

Tectospinal tract

Inferior olivary nucleus

Olive

Pyramids of medulla

**(c) Medulla oblongata**

**FIGURE 14.9 Cross Sections of the Brainstem.** The level of each section is shown in the figure on the right. (a) The midbrain, cut obliquely to pass through the superior colliculi. (b) The pons. The straight edges indicate cut edges of the peduncles where the cerebellum was removed. (c) The medulla oblongata.

▶ *Trace the route taken through all three of these figures by fibers from the gracile and cuneate fasciculi described in chapter 13.*

Other nuclei of the medulla are concerned with coughing, sneezing, salivation, swallowing, gagging, vomiting, gastrointestinal secretion, sweating, speech, and movements of the tongue and head. Many of the medulla's sensory and motor functions are mediated through the last four cranial nerves, which begin or end here: cranial nerves IX (glossopharyngeal), X (vagus), XI (accessory), and XII (hypoglossal), discussed later in this chapter.

## THE PONS

The embryonic metencephalon develops into two structures, the pons and cerebellum. The **pons**[17] appears as an anterior bulge in the brainstem rostral to the medulla (figs. 14.8 and 14.9b). Its white matter includes tracts that conduct signals from the cerebrum down to the cerebellum and medulla, and tracts that carry sensory signals up to the

thalamus. Cranial nerve V (trigeminal) arises from the pons, and cranial nerves VI (abducens), VII (facial), and VIII (vestibulocochlear) arise from the junction of the pons and medulla. The pons contains nuclei that relay signals from the cerebrum to the cerebellum, and nuclei concerned with sleep, hearing, equilibrium, taste, eye movements, facial expressions, facial sensation, respiration, swallowing, bladder control, and posture.

## THE CEREBELLUM

The **cerebellum** is the largest part of the hindbrain and second-largest part of the brain as a whole (fig. 14.10). It consists of right and left **cerebellar hemispheres** connected by a narrow bridge called the **vermis**.[18] Each hemisphere exhibits slender, transverse, parallel folds called **folia**[19]

---

[17]*pons* = bridge

[18]*verm* = worm
[19]*foli* = leaf

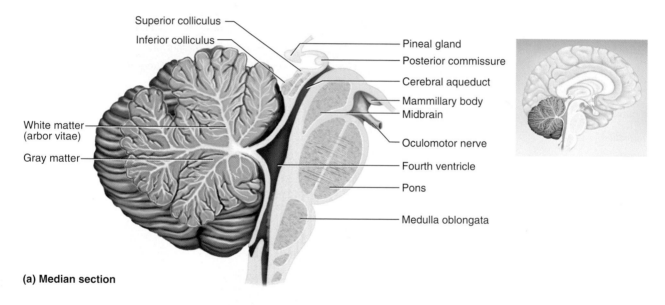

Superior colliculus
Inferior colliculus
Pineal gland
Posterior commissure
Cerebral aqueduct
Mammillary body
Midbrain
White matter (arbor vitae)
Gray matter
Oculomotor nerve
Fourth ventricle
Pons
Medulla oblongata

**(a) Median section**

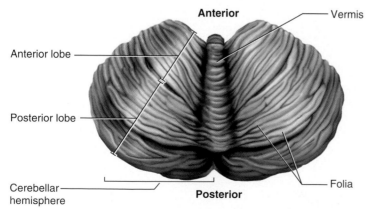

Anterior — Vermis
Anterior lobe
Posterior lobe
Cerebellar hemisphere
Folia
**Posterior**

**(b) Superior view**

**FIGURE 14.10  The Cerebellum.**
(a) Median section, showing relationship to the brainstem. (b) Superior view.

(gyri) separated by shallow sulci. The cerebellum has a surface cortex of gray matter and a deeper layer of white matter. In a sagittal section, the white matter, called the **arbor vitae,**[20] exhibits a branching, fernlike pattern. Each hemisphere has four **deep nuclei,** masses of gray matter embedded in the white matter. All input to the cerebellum goes to the cortex and all of its output comes from the deep nuclei.

Although the cerebellum is only about 10% of the mass of the brain, it has about 60% as much surface area as the cerebral cortex and it contains more than half of all brain neurons—about 100 billion of them. Its tiny, densely spaced *granule cells* are the most abundant type of neuron in the entire brain. Its most distinctly shaped neurons, however, are the unusually large, globose *Purkinje*[21] (pur-KIN-jee) *cells.* These have a tremendous profusion of dendrites, compressed into a single plane like a flat tree (see fig. 12.5, p. 446 and the photograph on p. 441). The Purkinje cells are arranged in a single file, with these thick dendritic planes parallel to each other, like books on a shelf. Their axons travel to the deep nuclei, where they synapse on output neurons that issue fibers to the brainstem.

The cerebellum is connected to the brainstem by three pairs of stalks called **cerebellar peduncles**[22] (peh-DUN-culs): two *inferior peduncles* connecting it to the medulla oblongata, two *middle peduncles* to the pons, and two *superior peduncles* to the midbrain. These are composed of thick bundles of nerve fibers that carry signals into and out of the cerebellum. Connections between the cerebellum and brainstem regions are very complex, but bearing in mind that there are some exceptions to each of these, we can draw a few generalizations. Most input to the cerebellum from the spinal cord enters by way of the inferior peduncles; most input from the rest of the brain enters by way of the middle peduncles; and the superior peduncles carry most of the cerebellum's output.

Through the inferior peduncles, the cerebellum receives signals from the spinocerebellar tracts of the spinal cord. These tracts convey proprioceptive information about muscle performance so that the cerebellum can monitor body movements and mediate some somatic reflexes. The middle peduncles carry signals from the cerebrum about what the muscles were commanded to do, so the cerebellum can compare the command with the performance. These peduncles also carry signals from the inner ear concerned with some aspects of hearing and with equilibrium, a sense of the body's movements and orientation of the head relative to gravity. Output through the superior peduncles leads to various points in the midbrain and thalamus. The thalamus relays cerebellar out-

put to the cerebral cortex so that the cerebrum can make fine adjustments in muscle performance.

The function of the cerebellum was unknown in the 1950s. By the 1970s, it had come to be regarded as a central control point of motor coordination. Now, however, positron emission tomography (PET) and functional magnetic resonance imaging (fMRI), as well as behavioral studies of people with cerebellar lesions, have created a much more expansive view of cerebellar function. It appears that the main role of the cerebellum is the evaluation of certain kinds of sensory input, and monitoring the movements of muscles is only a part of that general role. People with cerebellar lesions exhibit serious deficits in coordination and locomotor ability; more will be said later in this chapter about the role of the cerebellum in movement. But lesions also affect several sensory, linguistic, emotional, and other nonmotor functions.

The cerebellum is highly active when a person is asked to explore objects with the fingertips, for example to compare the textures of two objects without looking at them. (Tactile nerve fibers from a rat's snout and a cat's forepaws also project strongly to the cerebellum.) Some spatial perception also resides here. The cerebellum is much more active when a person is required to solve a pegboard puzzle than when moving pegs randomly around the same puzzle board. People with cerebellar lesions also have difficulty identifying different views of the same three-dimensional object as belonging to the same object.

The cerebellum is also a timekeeping center of the brain. PET scans show increased cerebellar activity during tasks in which a person must judge the elapsed time between two stimuli. People with cerebellar lesions have difficulty with rhythmic finger-tapping tasks and other tests of temporal judgment. An important aspect of cerebellar timekeeping is the ability to predict where a moving object will be in the next second or so. You can imagine the importance of this to a predator chasing its prey, to a tennis player, or to driving a car in heavy traffic. The cerebellum also helps to predict how much the eyes must move in order to compensate for head movements and remain visually fixated on an object.

Even hearing has some newly discovered and surprising cerebellar components. Cerebellar lesions impair a person's ability to judge differences in pitch between two tones and to distinguish between similar-sounding words such as *rabbit* and *rapid.* Language output also involves the cerebellum. If a person is given a noun such as *apple* and told to think of a related verb such as *eat,* the cerebellum shows higher PET activity than if the person is told merely to repeat the word *apple.*

People with cerebellar lesions also have difficulty planning and scheduling tasks. They tend to overreact emotionally and have difficulty with impulse control. Many children with attention deficit–hyperactivity disorder (ADHD) have abnormally small cerebellums.

---

[20]*arbor* = tree + *vitae* = of life
[21]Johannes E. von Purkinje (1787–1869), Bohemian anatomist
[22]*ped* = foot + *uncle* = little

# THE MIDBRAIN

The embryonic mesencephalon becomes just one mature brain structure, the **midbrain**—a short segment of the brainstem that connects the hindbrain and forebrain (see figs. 14.2 and 14.8). It contains the cerebral aqueduct and gives rise to two cranial nerves that control eye movements: cranial nerve III (oculomotor) and IV (trochlear). Some major regions of the midbrain are (fig. 14.9a):

- The **cerebral peduncles,** which anchor the cerebrum to the brainstem. The corticospinal tracts pass through the peduncles on their way to the medulla.

- The **tegmentum,**[23] the main mass of the midbrain, located between the cerebral peduncles and cerebral aqueduct. It contains the **red nucleus,** which has a pink color in life because of its high density of blood vessels. Fibers from the red nucleus form the *rubrospinal tract* in most mammals, but in humans its connections go mainly to and from the cerebellum, with which it collaborates in fine motor control.

- The **substantia nigra**[24] (sub-STAN-she-uh NY-gruh), a dark gray to black nucleus pigmented with melanin, located between the peduncles and the tegmentum. This is a motor center that relays inhibitory signals to the thalamus and basal nuclei (both of which are discussed later). Degeneration of the neurons in the substantia nigra leads to the muscle tremors of Parkinson disease (see Insight 12.4, p. 474).

- The **central (periaqueductal) gray matter,** a large arrowhead-shaped region of gray matter surrounding the cerebral aqueduct. It is involved with the *reticulospinal tracts* in controlling our awareness of pain (see chapter 16).

- The **tectum,**[25] a rooflike region dorsal to the aqueduct. It consists of four nuclei, the **corpora quadrigemina,**[26] which bulge from the midbrain roof. The two superior nuclei, called the **superior colliculi**[27] (col-LIC-you-lye), function in visual attention, visually tracking moving objects, and such reflexes as blinking, focusing, pupillary dilation and constriction, and turning the eyes and head in response to a visual stimulus (for example, to look at something that you catch sight of in your peripheral vision). The two **inferior colliculi** receive signals from the inner ear and relay them to other parts of the brain, especially the thalamus. Among other functions, they mediate the reflexive turning of the head in response to a sound and jumping when startled by a sudden noise.

- The **medial lemniscus,** a continuation of the gracile and cuneate tracts of the spinal cord and brainstem.

---

[23]*tegmen* = cover
[24]*substantia* = substance + *nigra* = black
[25]*tectum* = roof, cover
[26]*corpora* = bodies + *quadrigemina* = quadruplets
[27]*coli* = hill + *cul* = little

The *reticular formation,* discussed next, is a prominent feature of the midbrain but is not limited to this region.

> **Think About It**
>
> *Why are the inferior colliculi shown in figure 14.10a but not in figure 14.9a? How are these two figures related?*

# THE RETICULAR FORMATION

The **reticular**[28] **formation** is a loosely organized web of gray matter that runs vertically through all levels of the brainstem and projects to many areas of the cerebrum (fig. 14.11). It occupies much of the space between the white fiber tracts and the more anatomically distinct brainstem nuclei. It consists of more than 100 small neural networks without well

---

[28]*ret* = network + *icul* = little

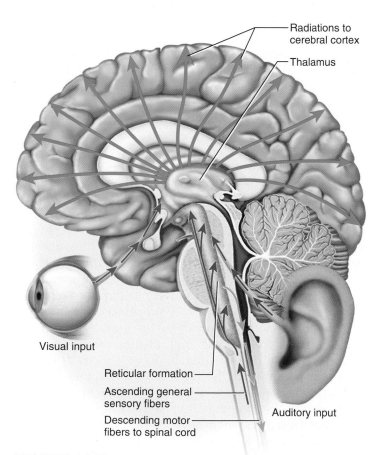

**FIGURE 14.11 The Reticular Formation.** The formation consists of over 100 nuclei scattered through the brainstem. Red arrows indicate routes of input to the reticular formation; blue arrows indicate the radiating relay of signals from the thalamus to the cerebral cortex; and green arrows indicate output from the reticular formation to the spinal cord.

▶ *Locate components of the reticular formation in all three parts of figure 14.9.*

defined boundaries. The functions of these networks include the following:

- **Somatic motor control.** Some motor neurons of the cerebral cortex send their axons to reticular formation nuclei, which then give rise to the *reticulospinal tracts* of the spinal cord. These tracts modulate (adjust) muscle contraction to maintain tone, balance, and posture. The reticular formation also relays signals from the eyes and ears to the cerebellum so the cerebellum can integrate visual, auditory, and vestibular (balance and motion) stimuli into its role in motor coordination. Other reticular formation nuclei include *gaze centers,* which enable the eyes to track and fixate on objects, and **central pattern generators**—neural pools that produce rhythmic signals to the muscles of breathing and swallowing.

- **Cardiovascular control.** The reticular formation includes the cardiac and vasomotor centers of the medulla oblongata.

- **Pain modulation.** The reticular formation is one route by which pain signals from the lower body reach the cerebral cortex. It is also the origin of the *descending analgesic pathways* mentioned in the description of the reticulospinal tracts on page 489. The nerve fibers in these pathways act in the spinal cord to block the transmission of pain signals to the brain (see chapter 16).

- **Sleep and consciousness.** The reticular formation has projections to the thalamus and cerebral cortex that allow it some control over what sensory signals reach the cerebrum and come to our conscious attention. It plays a central role in states of consciousness such as alertness and sleep. Injury to the reticular formation can result in irreversible coma.

    The reticular formation also is involved in **habituation**—a process in which the brain learns to ignore repetitive, inconsequential stimuli while remaining sensitive to others. In a noisy city, for example, a person can sleep through traffic sounds but wake promptly to the sound of an alarm clock or a crying baby. Reticular formation nuclei that modulate activity of the cerebral cortex are called the *reticular activating system* or *extrathalamic cortical modulatory system.*

## Before You Go On

*Answer the following questions to test your understanding of the preceding section:*

8. *Name the visceral functions controlled by nuclei of the medulla.*

9. *Describe the general functions of the cerebellum.*

10. *What are some functions of the midbrain nuclei?*

11. *Describe the reticular formation and list several of its functions.*

# The Forebrain

### Objectives

When you have completed this section, you should be able to

- name the three major components of the diencephalon and describe their locations and functions;
- identify the five lobes of the cerebrum;
- describe the three types of tracts in the cerebral white matter;
- describe the distinctive cell types and histological arrangement of the cerebral cortex; and
- describe the location and functions of the basal nuclei and limbic system.

The forebrain consists of the diencephalon and telencephalon. The diencephalon encloses the third ventricle and is the most rostral part of the brainstem. The telencephalon develops chiefly into the cerebrum.

## THE DIENCEPHALON

The embryonic diencephalon has three major derivatives: the *thalamus, hypothalamus,* and *epithalamus.*

### The Thalamus

Each side of the brain has a **thalamus,**[29] an ovoid mass perched at the superior end of the brainstem beneath the cerebral hemisphere (see figs. 14.6c, 14.8, and 14.16). The thalami constitute about four-fifths of the diencephalon. Each one protrudes medially into the third ventricle and laterally into the lateral ventricle. In about 70% of people, the two thalami are joined medially by a narrow *intermediate mass.*

The thalamus is composed of at least 23 nuclei, but we will consider only five major functional groups into which most of these are classified: the anterior, posterior, medial, lateral, and ventral groups. These regions and their functions are shown in figure 14.12a.

Generally speaking, the thalamus is the "gateway to the cerebral cortex." Nearly all sensory input and other information going to the cerebrum passes by way of synapses in the thalami nuclei, including signals for taste, smell, hearing, equilibrium, vision, and such somesthetic senses as touch, pain, pressure, heat, and cold. The thalamic nuclei filter this information and relay only a small portion of it to the cerebral cortex.

The thalamus also plays a key role in motor control by relaying signals from the cerebellum to the cerebrum and by providing feedback loops between the cerebral cortex and the *basal nuclei* (deep cerebral motor centers described later). Finally, the thalamus is involved in the

---

[29]*thalamus* = chamber, inner room

memory and emotional functions of the *limbic system,* a complex of structures (also described later) that includes cerebral cortex of the temporal and frontal lobes and some of the anterior thalamic nuclei. The role of the thalamus in motor and sensory circuits is further discussed later in this chapter and in chapter 16.

## The Hypothalamus

The **hypothalamus** (fig. 14.12b) forms part of the walls and floor of the third ventricle. It extends anteriorly to the *optic chiasm* (ky-AZ-um), where the optic nerves meet, and posteriorly to a pair of humps called the *mam-*

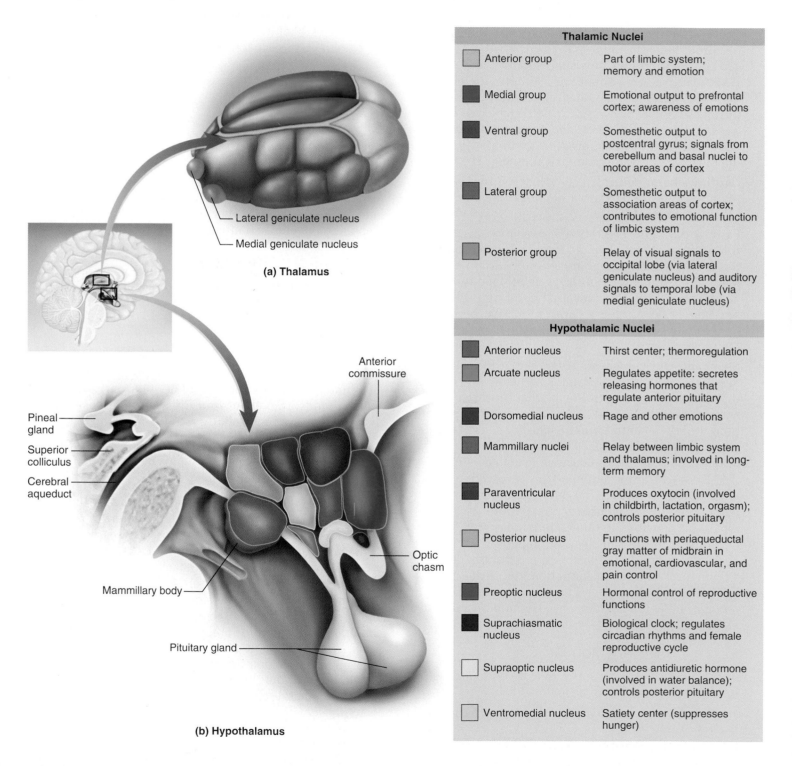

| Thalamic Nuclei | |
|---|---|
| Anterior group | Part of limbic system; memory and emotion |
| Medial group | Emotional output to prefrontal cortex; awareness of emotions |
| Ventral group | Somesthetic output to postcentral gyrus; signals from cerebellum and basal nuclei to motor areas of cortex |
| Lateral group | Somesthetic output to association areas of cortex; contributes to emotional function of limbic system |
| Posterior group | Relay of visual signals to occipital lobe (via lateral geniculate nucleus) and auditory signals to temporal lobe (via medial geniculate nucleus) |

| Hypothalamic Nuclei | |
|---|---|
| Anterior nucleus | Thirst center; thermoregulation |
| Arcuate nucleus | Regulates appetite: secretes releasing hormones that regulate anterior pituitary |
| Dorsomedial nucleus | Rage and other emotions |
| Mammillary nuclei | Relay between limbic system and thalamus; involved in long-term memory |
| Paraventricular nucleus | Produces oxytocin (involved in childbirth, lactation, orgasm); controls posterior pituitary |
| Posterior nucleus | Functions with periaqueductal gray matter of midbrain in emotional, cardiovascular, and pain control |
| Preoptic nucleus | Hormonal control of reproductive functions |
| Suprachiasmatic nucleus | Biological clock; regulates circadian rhythms and female reproductive cycle |
| Supraoptic nucleus | Produces antidiuretic hormone (involved in water balance); controls posterior pituitary |
| Ventromedial nucleus | Satiety center (suppresses hunger) |

(a) Thalamus

Lateral geniculate nucleus
Medial geniculate nucleus

Anterior commissure

Pineal gland
Superior colliculus
Cerebral aqueduct

Mammillary body

Optic chasm

Pituitary gland

(b) Hypothalamus

**FIGURE 14.12    The Diencephalon.** Only some of the nuclei of the thalamus and hypothalamus are shown, and some of their functions listed. These lists are by no means complete.

*millary*[30] *bodies.* The mammillary bodies each contain three to four *mammillary nuclei.* Their primary function is to relay signals from the limbic system to the thalamus. The pituitary gland is attached to the hypothalamus by a stalk *(infundibulum)* between the optic chiasm and mammillary bodies.

The hypothalamus is the major control center of the autonomic nervous system and endocrine system. It plays an essential role in the homeostatic regulation of nearly all organs of the body. Its nuclei include centers concerned with a wide variety of visceral functions:

- **Hormone secretion.** The hypothalamus secretes hormones that control the anterior pituitary gland. Acting through the pituitary, it regulates growth, metabolism, reproduction, and stress responses. The hypothalamus also produces two hormones that are stored in the posterior pituitary gland, concerned with labor contractions, lactation, and water conservation. These relationships are explored especially in chapter 17.

- **Autonomic effects.** The hypothalamus is a major integrating center for the autonomic nervous system. It sends descending fibers to nuclei lower in the brainstem that influence heart rate, blood pressure, gastrointestinal secretion and motility, and pupillary diameter, among other functions.

- **Thermoregulation.** The *hypothalamic thermostat* is a nucleus that monitors blood temperature. When the temperature becomes too high or too low, the thermostat signals other hypothalamic nuclei—the *heat-losing center* or *heat-producing center,* respectively, which control cutaneous vasodilation and vasoconstriction, sweating, shivering, and piloerection.

- **Food and water intake.** Neurons of the *hunger* and *satiety centers* monitor blood glucose and amino acid levels and produce sensations of hunger and satiety (satisfaction of the appetite). Hypothalamic neurons called *osmoreceptors* monitor the osmolarity of the blood and stimulate the hypothalamic *thirst center* when the body is dehydrated. Dehydration also stimulates the hypothalamus to produce *antidiuretic hormone,* which conserves water by reducing urine output.

- **Sleep and circadian rhythms.** The caudal part of the hypothalamus is part of the reticular formation. It contains nuclei that regulate falling asleep and waking. Superior to the optic chiasm, the hypothalamus contains a *suprachiasmatic nucleus* that controls our 24-hour (circadian) rhythm of activity.

- **Memory.** The mammillary nuclei lie in the pathway of signals traveling from the hippocampus, an important memory center of the brain, to the thalamus. Thus they are important in memory, and lesions to the mammillary nuclei interrupt the memory pathways and cause memory deficits. (Memory is discussed more fully later in this chapter.)

- **Emotional behavior.** Hypothalamic centers are involved in a variety of emotional responses including anger, aggression, fear, pleasure, and contentment; and in sexual drive, copulation, and orgasm.

## The Epithalamus

The **epithalamus** is a very small mass of tissue composed mainly of the **pineal gland** (an endocrine gland discussed in chapter 17), the **habenula** (a relay from the limbic system to the midbrain), and a thin roof over the third ventricle.

# THE CEREBRUM

The embryonic telencephalon becomes the cerebrum, the largest and most conspicuous part of the human brain. Your cerebrum enables you to turn these pages, read and comprehend the words, remember ideas, talk about them with your peers, and take an examination. It is the seat of your sensory perception, memory, thought, judgment, and voluntary motor actions. It is the most challenging frontier of neurobiology.

## Gross Anatomy

The surface of the cerebrum, including its gray matter (cerebral cortex) and part of the white matter, is folded into gyri that allow a greater amount of cortex to fit in the cranial cavity. These folds give the cerebrum a surface area of about 2,500 cm², comparable to 4½ pages of this book. If the cerebrum were smooth-surfaced, it would have only one-third as much area and proportionately less information-processing capability. This extensive folding is one of the greatest differences between the human brain and the relatively smooth-surfaced brains of most other mammals.

Some gyri have consistent and predictable anatomy, while others vary from brain to brain and from the right hemisphere to the left. Certain unusually prominent sulci divide each hemisphere into five anatomically and functionally distinct lobes, listed next. The first four lobes are visible superficially and are named for the cranial bones overlying them (fig. 14.13); the fifth lobe is not visible from the surface.

1. The **frontal lobe** lies immediately behind the frontal bone, superior to the eyes. Its posterior boundary is the **central sulcus.** It is chiefly concerned with voluntary motor functions, motivation, foresight, planning, memory, mood, emotion, social judgment, and aggression.

2. The **parietal lobe** forms the uppermost part of the brain and underlies the parietal bone. Its rostral

---

[30]*mammill* = nipple

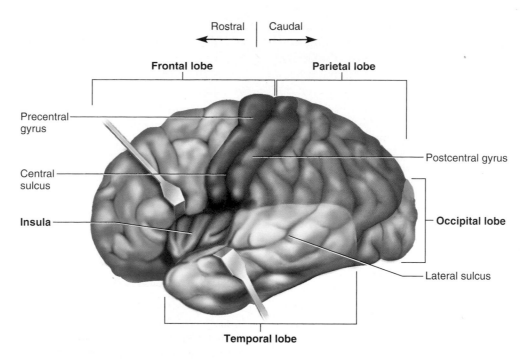

Rostral | Caudal

**Frontal lobe**    **Parietal lobe**

Precentral gyrus

Postcentral gyrus

Central sulcus

**Insula**    **Occipital lobe**

Lateral sulcus

**Temporal lobe**

**FIGURE 14.13   Lobes of the Cerebrum.** The frontal and temporal lobes are retracted slightly to reveal the insula.

boundary is the central sulcus and its caudal boundary is the **parieto-occipital sulcus,** visible on the medial surface of each hemisphere (see fig. 14.2). This lobe is concerned with the sensory reception and integration of somesthetic, taste, and some visual information.

3. The **occipital lobe** is at the rear of the head underlying the occipital bone. It is the principal visual center of the brain.

4. The **temporal lobe** is a lateral, horizontal lobe deep to the temporal bone, separated from the parietal lobe above it by a deep **lateral sulcus.** It is concerned with hearing, smell, learning, memory, visual recognition, and emotional behavior.

5. The **insula**[31] is a small mass of cortex deep to the lateral sulcus, made visible only by retracting or cutting away some of the overlying cerebrum (see figs. 14.1c, 14.6c, and 14.13). It is not as accessible to study in living people as other parts of the cortex and is still little-known territory. It apparently plays roles in understanding spoken language, in the sense of taste, and in integrating sensory information from visceral receptors.

## The Cerebral White Matter

Most of the volume of the cerebrum is white matter. This is composed of glia and myelinated nerve fibers that transmit signals from one region of the cerebrum to another and between the cerebrum and lower brain centers. These fibers form bundles, or *tracts,* of three kinds (fig. 14.14):

1. **Projection tracts** extend vertically between higher and lower brain and spinal cord centers, and carry information between the cerebrum and the rest of the body. The corticospinal tracts, for example, carry motor signals from the cerebrum to the brainstem and spinal cord. Other projection tracts carry signals upward to the cerebral cortex. Superior to the brainstem, such tracts form a broad, dense sheet called the *internal capsule* between the thalamus and basal nuclei (described shortly), then radiate in a diverging, fanlike array (the *corona radiata*[32]) to specific areas of the cortex.

2. **Commissural tracts** cross from one cerebral hemisphere to the other through bridges called **commissures** (COM-ih-shurs). The great majority of them pass through the large C-shaped corpus callosum (see fig. 14.1d), which forms the floor of the longitudinal fissure. A few tracts pass through the much smaller **anterior** and **posterior commissures.** Commissural tracts enable the two sides of the cerebrum to communicate with each other.

3. **Association tracts** connect different regions within the same cerebral hemisphere. *Long association fibers* connect different lobes of a hemisphere to each other, whereas *short association fibers* connect different gyri within a single lobe. Among their

---

[31]*insula* = island

[32]*corona* = crown + *radiata* = radiating

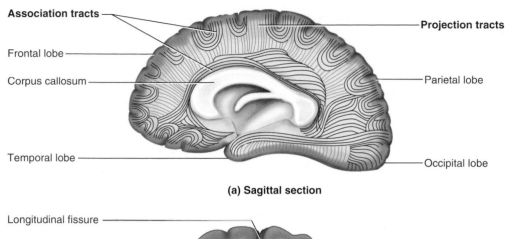

**(a) Sagittal section**

Association tracts
Frontal lobe
Corpus callosum
Temporal lobe
Projection tracts
Parietal lobe
Occipital lobe

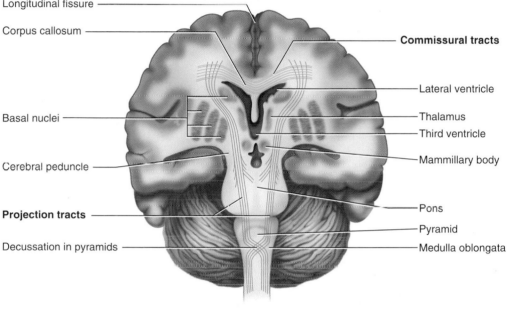

**(b) Frontal section**

Longitudinal fissure
Corpus callosum
Basal nuclei
Cerebral peduncle
Projection tracts
Decussation in pyramids
Commissural tracts
Lateral ventricle
Thalamus
Third ventricle
Mammillary body
Pons
Pyramid
Medulla oblongata

**FIGURE 14.14   Tracts of Cerebral White Matter.** (a) Sagittal section, showing association and projection tracts. (b) Frontal section, showing commissural and projection tracts.

▶ *What route can commissural tracts take other than the one shown here?*

roles, association tracts link perceptual and memory centers of the brain; for example, they enable you to smell a rose, name it, and picture what it looks like.

## The Cerebral Cortex

Neural integration is carried out in the gray matter of the cerebrum, which is found in three places: the cerebral cortex, basal nuclei, and limbic system. We begin with the **cerebral cortex,**[33] a layer covering the surface of the hemispheres. Even though it is only 2 to 3 mm thick, the cortex constitutes about 40% of the mass of the brain and contains 14 to 16 billion neurons. It is composed of two

principal types of neurons (fig. 14.15): (1) **Stellate cells** have spheroidal somas with dendrites projecting for short distances in all directions. They are concerned largely with receiving sensory input and processing information on a local level. (2) **Pyramidal cells** are tall and conical (triangular in tissue sections). Their apex points toward the brain surface and has a thick dendrite with many branches and small, knobby *dendritic spines.* The base gives rise to horizontally oriented dendrites and an axon that passes into the white matter. Pyramidal cells are the output neurons of the cerebrum—they transmit signals to other parts of the CNS. Their axons have collaterals that synapse with other neurons in the cortex or in deeper regions of the brain.

About 90% of the human cerebral cortex is a six-layered tissue called **neocortex**[34] because of its relatively recent evolutionary origin. Although vertebrate animals have existed for about 600 million years, the neocortex did not develop significantly until about 60 million years ago, when there was a sharp increase in the diversity of mammals. It attained its highest development by far in the primates. The six layers of neocortex, numbered in figure 14.15, vary from one part of the cerebrum to another in relative thickness, cellular composition, synaptic connections, size of the neurons, and destination of their axons. Layer IV is thickest in sensory regions and layer V in motor regions, for example. All axons that leave the cortex and enter the white matter arise from layers III, V, and VI.

Some regions of cerebral cortex have fewer than six layers. The earliest type of cortex to appear in vertebrate evolution was a one- to five-layered tissue called *paleocortex* (PALE-ee-oh-cor-tex), limited in humans to part of the insula and certain areas of the temporal lobe concerned with smell. The next to evolve was a three-layered *archicortex* (AR-kee-cor-tex), found in the human hippocampus, a memory-forming center in the temporal lobe. The neocortex was the last to evolve.

---

[33]*cortex* = bark, rind

[34]*neo* = new

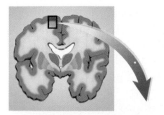

**FIGURE 14.15**  Histology of the Neocortex.  Neurons are arranged in six layers.

▶ *Are the long processes leading upward from each pyramidal cell body dendrites or axons?*

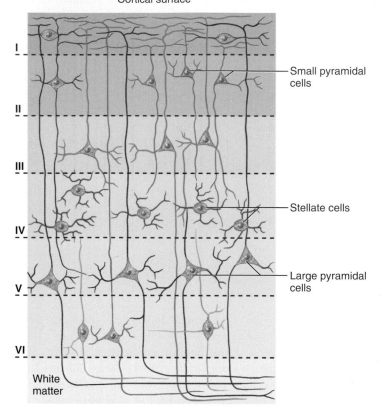

Cortical surface

I

II

III

IV

V

VI

White matter

Small pyramidal cells

Stellate cells

Large pyramidal cells

## The Basal Nuclei

The basal nuclei are masses of cerebral gray matter buried deep in the white matter, lateral to the thalamus (fig. 14.16). They are often called *basal ganglia,* but the word *ganglion* is best restricted to clusters of neurons outside the CNS. Neuroanatomists disagree on how many brain centers to classify as basal nuclei, but agree on at least three: the **caudate**[35] **nucleus, putamen,**[36] and **globus pallidus.**[37] The putamen and globus pallidus are also collectively called the *lentiform*[38] *nucleus,* because together they form a lens-shaped body. The three nuclei are also collectively called the *corpus striatum*[39] after their striped appearance. The basal nuclei receive input from the substantia nigra of the midbrain and motor areas of the cerebral cortex, and send signals back to both of these locations. They are involved in motor control and are further discussed in a later section on that topic.

## The Limbic System

The **limbic**[40] **system,** is one of the brain's most important centers of emotion and learning. It was originally described in the 1850s as a ring of structures on the medial side of the cerebral hemisphere, encircling the corpus callosum and thalamus. Its most anatomically

[35]*caudate* = tailed, tail-like
[36]*putam* = pod, husk
[37]*glob* = globe, ball + *pall* = pale
[38]*lenti* = lens + *form* = shape
[39]*corpus* = body + *striat* = stripe
[40]*limbus* = border

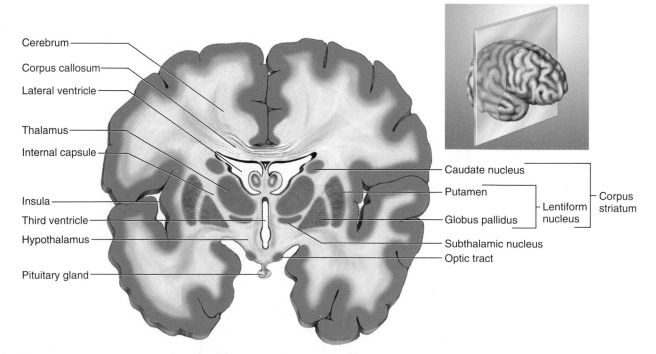

Cerebrum

Corpus callosum

Lateral ventricle

Thalamus

Internal capsule

Insula

Third ventricle

Hypothalamus

Pituitary gland

Caudate nucleus

Putamen

Globus pallidus

Subthalamic nucleus

Optic tract

Lentiform nucleus

Corpus striatum

**FIGURE 14.16**  The Basal Nuclei.  Frontal section of the brain.

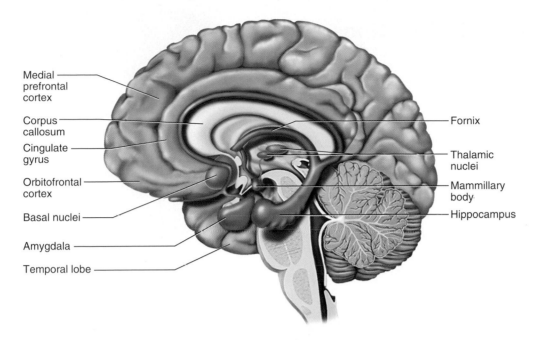

Medial prefrontal cortex

Corpus callosum

Cingulate gyrus

Orbitofrontal cortex

Basal nuclei

Amygdala

Temporal lobe

Fornix

Thalamic nuclei

Mammillary body

Hippocampus

**FIGURE 14.17** **The Limbic System.** This ring of structures includes important centers of learning and emotion. In the frontal lobe, there is no sharp rostral boundary to limbic system components.

prominent components are the **cingulate**[41] (SING-you-let) **gyrus** that arches over the top of the corpus callosum in the frontal and parietal lobes, the **hippocampus**[42] in the medial temporal lobe (fig. 14.17), and the **amygdala**[43] (ah-MIG-da-luh) immediately rostral to the hippocampus, also in the temporal lobe. There are still differences of opinion on what structures to consider as parts of the limbic system, but these three are agreed upon; other components include the mammillary nuclei and other hypothalamic nuclei, some thalamic nuclei, parts of the basal nuclei, and parts of the prefrontal cortex. Limbic system components are interconnected through a complex loop of fiber tracts allowing for somewhat circular patterns of feedback among its nuclei and cortical neurons. All of these structures are bilaterally paired; there is a limbic system in each cerebral hemisphere.

The limbic system was long thought to be associated with smell because of its close association with olfactory pathways, but beginning in the early 1900s and continuing even now, experiments have abundantly demonstrated more significant roles in emotion and memory. Most limbic system structures have centers for both gratification and aversion. Stimulation of a gratification center produces a sense of pleasure or reward; stimulation of an aversion center produces unpleasant sensations such as fear or sorrow. Gratification centers dominate some limbic structures, such as the *nucleus accumbens* (not illustrated), while aversion centers dominate others such as the amygdala. The roles of the amygdala in emotion and the hippocampus in memory are described in the next section, on higher forebrain functions.

## Before You Go On

*Answer the following questions to test your understanding of the preceding section:*

12. What are the three major components of the diencephalon? Which ventricle does the diencephalon enclose?

13. What is the role of the thalamus in sensory function?

14. List at least six functions of the hypothalamus.

15. Name the five lobes of the cerebrum and describe their locations and boundaries.

16. Distinguish between commissural, association, and projection tracts of the cerebrum.

17. Where are the basal nuclei located? What is their general function?

18. Where is the limbic system located? What component of it is involved in emotion? What component is involved in memory?

---

[41]*cingul* = girdle
[42]*hippocampus* = sea horse, named for its shape
[43]*amygdala* = almond

# Higher Forebrain Functions

### Objectives

When you have completed this section, you should be able to

- list the types of brain waves and discuss their relationship to sleep and other mental states;
- explain how the brain controls the skeletal muscles;
- identify the parts of the cerebrum that receive and interpret somatic sensory signals;
- identify the parts of the cerebrum that receive and interpret signals from the special senses;
- describe the locations and functions of the language centers;
- discuss the brain regions involved in memory; and
- discuss the functional relationship between the right and left cerebral hemispheres.

This section concerns such brain functions as sleep, memory, cognition, emotion, sensation, motor control, and language. These are associated especially with the cerebral cortex, but not exclusively; they involve interactions between the cerebral cortex and such areas as the cerebellum, basal nuclei, limbic system, hypothalamus, and reticular formation. Most of these functions are not associated with or limited to only one brain region, so they cannot be discussed in a simple one-to-one relationship with the anatomical parts described up to this point. Some of them present the most difficult challenges for neurobiology, but they are the most intriguing functions of the brain and involve its largest areas.

## THE ELECTROENCEPHALOGRAM

**Brain waves** are rhythmic voltage changes resulting predominantly from synchronized postsynaptic potentials in the superficial layers of the cerebral cortex. They are recorded from electrodes on the scalp. The recording, called an **electroencephalogram**[44] **(EEG),** is useful in studying normal brain functions such as sleep and consciousness, and in diagnosing degenerative brain diseases, metabolic abnormalities, brain tumors, sites of trauma, and so forth. States of consciousness ranging from high alert to deep sleep are correlated with changes in the EEG. The complete and persistent absence of brain waves is often used as a clinical and legal criterion of brain death.

There are four types of brain waves, distinguished by differences in amplitude (mV) and frequency. Frequency is expressed in hertz (Hz), or cycles per second (fig. 14.18):

1. **Alpha (α) waves** have a frequency of 8 to 13 Hz and are recorded especially in the parieto-occipital area. They occur when a person is awake and resting, with the eyes closed and the mind wandering. They

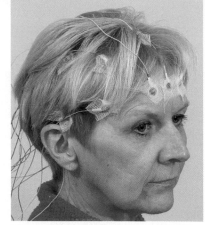

**(a)**

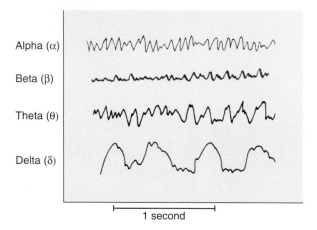

1 second

**(b)**

**FIGURE 14.18   The Electroencephalogram (EEG).** (a) An EEG is recorded from electrodes on the forehead and scalp. (b) Four classes of brain waves are seen in EEGs.

are suppressed when a person opens the eyes, receives specific sensory stimulation, or engages in a mental task such as performing mathematical calculations. They are absent during deep sleep.

2. **Beta (β) waves** have a frequency of 14 to 30 Hz and occur in the frontal to parietal region. They become accentuated during mental activity and sensory stimulation.

3. **Theta (θ) waves** have a frequency of 4 to 7 Hz. They are normal in children and in drowsy or sleeping adults, but a predominance of theta waves in awake adults suggests emotional stress or brain disorders.

4. **Delta (δ) waves** are high-amplitude "slow waves" with a frequency of less than 3.5 Hz. Infants exhibit delta waves when awake, and adults exhibit them in deep sleep. A predominance of delta waves in awake adults indicates serious brain damage.

## SLEEP

Multiple functions in the human body, even in individual cells, occur in cycles called **circadian**[45] (sur-CAY-dee-an)

---

[44]*electro* = electricity + *encephalo* = brain + *gram* = record

[45]*circa* = approximately + *dia* = a day, 24 hours

**rhythms,** so-named because they are marked by events that reoccur at intervals of about 24 hours. Circadian rhythms are regulated in part by a "biological clock" located in a neural network called the **suprachiasmatic** (SOO-pra-KY-az-MAT-ic) **nucleus (SCN).** The SCN lies in the anterior hypothalamus just above the optic chiasm (see fig. 14.12b). It receives some input from the eyes which it uses to synchronize body rhythms with the external rhythm of night and day. The SCN regulates not only sleep but also circadian rhythms of hormone secretion, urine production, body temperature, and other functions.

The most obvious circadian rhythm in humans and many other animals is the cycle of sleep and waking. **Sleep** can be defined as a temporary state of unconsciousness from which a person can awaken when stimulated. It is characterized by a stereotyped posture (usually lying down with the eyes closed) and inhibition of muscular activity *(sleep paralysis).* It superficially resembles other states of prolonged unconsciousness such as coma and animal hibernation, except that individuals cannot be aroused from those states by sensory stimulation.

Sleep occurs in distinct stages recognizable from changes in the EEG. In the first 30 to 45 minutes of sleep, the EEG waves drop in frequency but increase in amplitude as one passes through four sleep stages (fig. 14.19a):

- **Stage 1.** We feel drowsy, close our eyes and begin to relax. Thoughts come and go, and we often feel a drifting sensation. We awaken easily if stimulated. The EEG is dominated by alpha waves.

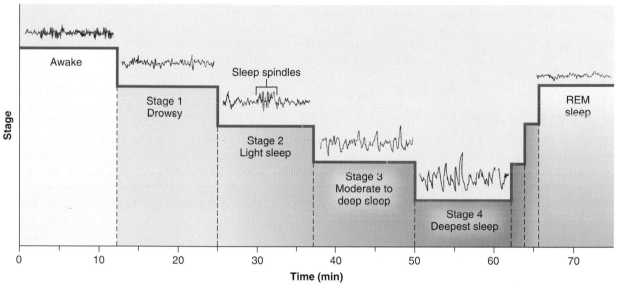

**(a) One sleep cycle**

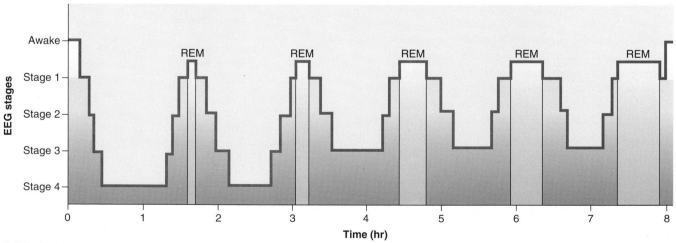

**(b) Typical 8-hour sleep period**

**FIGURE 14.19  Sleep Stages and Brain Activity.**  (a) A single sleep cycle, from waking to deep sleep, followed by 10 minutes of REM sleep. (b) Stages of sleep over an 8-hour night in a typical young adult. Stage 4 is attained only in the first two cycles. Periods of REM sleep increase from about 10 minutes in the first cycle to as long as 50 minutes in the last hour of sleep. Most dreaming occurs during REM sleep.

- **Stage 2.** We pass into light sleep. The EEG declines in frequency but increases in amplitude. Occasionally it exhibits 1 or 2 seconds of *sleep spindles,* high spikes resulting from interactions between neurons of the thalamus and cerebral cortex.

- **Stage 3.** This is moderate to deep sleep, typically beginning about 20 minutes after stage 1. Sleep spindles occur less often, and theta and delta waves appear in the EEG. The muscles relax, and the *vital signs* (body temperature, blood pressure, and heart and respiratory rates) fall.

- **Stage 4.** This is also called *slow-wave sleep (SWS),* because the EEG is dominated by low-frequency, high-amplitude delta waves. The muscles are now very relaxed, vital signs are at their lowest levels, and it is difficult to awaken the sleeper.

Young adults typically spend about 5% of their sleep time in stage 1, 50% to 60% in stage 2, 15% to 20% in stages 3 and 4, and 20% to 25% in the REM sleep described next. These proportions are significantly different in children and elderly people.

About five times a night, a sleeper backtracks from stage 3 or 4 to stage 2 and exhibits bouts of **rapid eye movement (REM) sleep** (fig. 14.19b). This is so-named because the eyes oscillate back and forth as if watching a movie. REM sleep is also called *paradoxical sleep* because the EEG resembles that of the waking state, and yet the sleeper is harder to arouse than in any other stage. The vital signs increase during REM sleep, and the brain consumes even more oxygen than when a person is awake. Sleep paralysis, other than in the muscles of eye movement, is especially strong during REM sleep. Paralysis may serve to prevent the sleeper from acting out his dreams and may have prevented our tree-dwelling ancestors from falling during their sleep.

Dreams occur in both REM and non-REM sleep, but REM dreams tend to be longer and more vivid and emotional than non-REM dreams. The parasympathetic nervous system is very active during REM sleep, leading to constriction of the pupils and erection of the penis or clitoris. In men, penile erection accompanies about 80% to 95% of REM sleep, but it is seldom associated with sexual dream content; only about 12% of men's dreams are sexual.

Neuroscientists have had more success in describing the rhythm of sleep than in determining its neurological mechanism or its ultimate purpose. The cycle of sleep and waking is controlled by a complex interaction among nuclei in the hypothalamus, reticular formation, thalamus, and cerebral cortex. Nuclei in the upper reticular formation, near the junction of the pons and midbrain, induce arousal, whereas nuclei below the pons induce sleep. Sleep is also induced by a nucleus in the hypothalamus called the *ventrolateral preoptic nucleus,* which inhibits the arousal neurons of the upper reticular formation. These centers exert their effects apparently by modulating communication between the thalamus and cerebral cortex. The suprachiasmatic nucleus of the hypothalamus does not in itself induce either sleep or waking, but does determine the time of day that a person or animal sleeps. Destruction of this nucleus produces an animal that sleeps the same number of hours per day, but at random times with no relationship to night or day.

We still know little about the function of sleep and dreaming. Non-REM sleep seems to have a restorative effect on the body, and prolonged sleep deprivation is fatal to experimental animals. Yet it is unclear why quiet bed rest alone cannot serve the purpose—that is, why we must lose consciousness. One hypothesis is that sleep serves as a time to replenish such energy sources as glycogen and ATP. In animals, brain glycogen levels decline during the waking hours and rebound during sleep. ATP consumption during waking may result in an accumulation of adenosine, which is known to induce sleepiness (see p. 580). Some researchers have suggested that REM sleep is a period in which the brain either "consolidates" and strengthens memories by reinforcing synaptic connections, or purges superfluous information from memory by eliminating other synapses. When people are given a new motor task to learn and their brain activity is monitored during subsequent sleep, stage 4 sleep is seen in the regions involved in learning the task, and not uniformly throughout the brain. From comparative zoology and evolutionary theory, there also is evidence that sleep may have evolved to motivate animals to find a safe place and remain inactive during the most dangerous times of day, when such risks as predation may outweigh such benefits as food-finding. But while intriguing hypotheses for the purposes of sleep and dreaming abound, the evidence for any of them is still scanty.

## COGNITION

**Cognition**[46] refers to mental processes such as awareness, perception, thinking, knowledge, and memory. It is an integration of information between the points of sensory input and motor output. Seventy-five percent of our brain tissue consists of cognitive **association areas** of the cerebrum. We know the functions of these areas largely through studies of people with brain lesions—local injuries resulting from such causes as infection, trauma, cancer, and stroke. A few examples of the effects of cerebral lesions reveal some functions of the association areas:

- Parietal lobe lesions can cause people to become unaware of objects, or even their own limbs, on the other side of the body—a condition called *contralateral neglect syndrome.* In typical cases, men shave only one half of the face, women apply makeup to only one side, patients dress only half of the body, and some people deny that one arm or leg belongs to them.

---

[46]*cognit* = to know

## INSIGHT 14.3 Medical History

### The Accidental Lobotomy of Phineas Gage

Accidental but nonfatal destruction of parts of the brain has afforded many clues to the function of various regions. One of the most famous incidents occurred in 1848 to Phineas Gage, a laborer on a railroad construction project in Vermont. Gage was packing blasting powder into a hole with a $3\frac{1}{2}$-ft tamping iron when the powder prematurely exploded. The tamping rod was blown out of the hole and passed through Gage's maxilla, orbit, and the frontal lobe of his brain before emerging from his skull near the hairline and landing 50 feet away (fig. 14.20). Gage went into convulsions, but later sat up and conversed with his crewmates as they drove him to a physician in an oxcart. On arrival, he stepped out on his own and told the physician, "Doctor, here is business enough for you." His doctor, John Harlow, reported that he could insert his index finger all the way into Gage's wound. Yet 2 months later, Gage was walking around town carrying on his normal business.

He was not, however, the Phineas Gage people had known. Before the accident, he had been a competent, responsible, financially prudent man, well liked by his associates. In an 1868 publication on the incident, Harlow said that following the accident, Gage was "fitful, irreverent, indulging at times in the grossest profanity." He became irresponsible, lost his job, worked for a while as a circus sideshow attraction, and died a vagrant 12 years later.

A 1994 computer analysis of Gage's skull indicated that the brain injury was primarily to the ventromedial region of both frontal lobes. In Gage's time, scientists were reluctant to attribute social behavior and moral judgment to any region of the brain. These functions were strongly tied to issues of religion and ethics and were considered inaccessible to scientific analysis. Based partly on Phineas Gage and other brain-injury patients like him, neuroscientists today recognize that planning, moral judgment, and emotional control are among the functions of the prefrontal cortex.

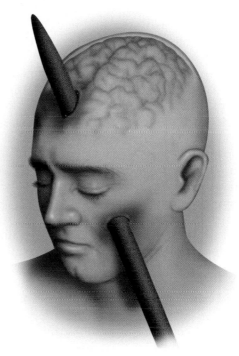

**FIGURE 14.20** **Phineas Gage's 1848 Accident.** The tamping bar flew upward, entered Gage's maxilla, passed through the orbit, destroyed brain tissue in the medial regions of each frontal lobe, and passed out of his skull near the hairline. The accident resulted in permanent personality changes that helped define some functions of the frontal lobes.

- Temporal lobe lesions often result in *agnosia*[47] (ag-NO-zee-ah), the inability to recognize, identify, and name familiar objects. In *prosopagnosia*,[48] a person cannot remember familiar faces, even his or her own reflection in a mirror.

- Frontal lobe lesions are especially devastating to the qualities we think of as personality. The *frontal association area* (prefrontal cortex) is well developed only in primates, especially humans. It integrates information from sensory and motor regions of the cortex and from other association areas. It gives us a sense of our relationship to the rest of the world, enabling us to think about it and to plan and execute appropriate behavior. It is responsible for giving appropriate expression to our emotions. Lesions here may produce profound personality disorders and socially inappropriate behaviors (see Insight 14.3).

As a broad generalization, we can conclude that the parietal association cortex is responsible for perceiving and attending to stimuli, the temporal association cortex for identifying them, and the frontal association cortex for planning our responses.

## MEMORY

We studied the neuronal and molecular mechanisms of memory in chapter 12. Now that you have been introduced to the gross anatomy of the brain, we can consider where those processes occur anatomically.

Our subject is really a little broader than memory per se. Information management by the brain entails learning (acquiring new information), memory proper (information storage and retrieval), and forgetting (eliminating trivial information). Forgetting is as important as remembering. People with a pathological inability to forget trivial information have great difficulty in reading comprehension and other functions that require us to separate what is important from what is not. More often, though, brain-injured

---

[47]*a* = without + *gnos* = knowledge
[48]*prosopo* = face, person

people are either unable to store new information **(anterograde amnesia)** or to recall things they knew before the injury **(retrograde amnesia).** *Amnesia* refers to defects in *declarative* memory (such as the ability to describe past events), not *procedural* memory (such as the ability to tie your shoes) (see definitions on p. 473).

The **hippocampus** of the limbic system is an important memory-forming center (see fig. 14.17). It does not store memories, but organizes sensory and cognitive experiences into a unified long-term memory. The hippocampus learns from sensory input while an experience is happening, but it has a short memory. Later, perhaps when one is sleeping, it plays this memory repeatedly to the cerebral cortex, which is a "slow learner" but forms longer-lasting memories through the processes described in chapter 12. This process of "teaching the cerebral cortex" until a long-term memory is established is called **memory consolidation.** Long-term memories are stored in various areas of cortex. Our vocabulary and memory of faces and familiar objects, for example, are stored in the superior temporal lobe, and memories of our plans and social roles are stored in the prefrontal cortex.

Lesions of the hippocampus can cause profound anterograde amnesia. For example, in 1953, a famous patient known as H. M. underwent surgical removal of a large portion of both temporal lobes, including both hippocampi, in an effort to treat his severe epilepsy. The operation had no adverse effect on his intelligence, procedural memory, or declarative memory for things that had happened early in his life, but it left him with an inability to establish new memories. He could hold a conversation with his psychologist, but a few minutes later deny that it had taken place. He worked with the same psychologist for more than 40 years after his operation, yet had no idea who she was from day to day.

Other parts of the brain involved in memory include the cerebellum, with a role in learning motor skills, and the amygdala, with a role in emotional memory—both of which are examined shortly.

## EMOTION

The prefrontal cortex is the seat of judgment, intent, and control over the expression of our emotions. However we may feel, it is here that we decide the appropriate way to show those feelings. But the feelings themselves, and emotional memories, form in deeper regions of the brain—the hypothalamus and amygdala. Here are the nuclei that stimulate us to recoil in fear from a rattlesnake or yearn for a lost love. The amygdala also seems to be involved in a broad range of functions including food intake, sexual activity, and drawing our attention to novel stimuli.

Emotional control centers of the brain have been identified not only by studying people with brain lesions, but also by such techniques as surgical removal, ablation (destruction) of small regions with electrodes, and stimulation with electrodes and chemical implants, especially in experimental animals. Changes in behavior following removal, ablation, or stimulation give clues to the functions that a region performs. However, interpretation of the results is difficult and controversial because of the complex connections between the emotional brain and other regions.

Many important aspects of personality depend on an intact, functional amygdala and hypothalamus. When specific regions are destroyed or artificially stimulated, humans and other animals exhibit blunted or exaggerated expressions of anger, fear, aggression, self-defense, pleasure, pain, love, sexuality, and parental affection, as well as abnormalities in learning, memory, and motivation. Lesions of the amygdala, for example, can completely abolish the sense of fear.

Much of our behavior is shaped by learned associations between stimuli, our responses to them, and the rewards or punishments that result. Nuclei involved in the senses of reward and punishment have been identified in the hypothalamus of cats, rats, monkeys, and other animals. In a representative experiment, an electrode is implanted in an area of an animal's hypothalamus called the *median forebrain bundle (MFB)*. The animal is placed in a chamber with a foot pedal wired to that electrode. When the animal steps on the pedal, it receives a mild electrical stimulus to the MFB. Apparently the sensation is strongly rewarding, because the animal soon learns to press the pedal over and over and may spend most of its time doing so—even to the point of neglecting food and water. Rats have been known to bar-press 5,000 to 12,000 times an hour, and monkeys up to 17,000 times an hour, to stimulate their MFBs.

These animals cannot tell us what they are feeling, but electrode implants have also been used to treat people who suffer otherwise incurable schizophrenia, pain, or epilepsy. These patients also repeatedly press a button to stimulate the MFB, but they do not report feelings of joy or ecstasy. Some are unable to explain why they enjoy the stimulus, and others report "relief from tension" or "a quiet, relaxed feeling." With electrodes misplaced in other areas of the hypothalamus, subjects report feelings of fear or terror when stimulated.

## SENSATION

Sensory functions fall into two categories:

1. **somesthetic**[49] (somatosensory, somatic) sensation, which comes from receptors widely distributed over the body for such stimuli as touch, pressure, stretch, movement, heat, cold, and pain; and

2. **special senses,** which originate in sense organs of the head and include vision, hearing, equilibrium, taste, and smell.

---

[49]*som* = body + *esthet* = feeling

## Somesthetic Sensation

Somesthetic nerve signals travel up the spinal cord and brainstem to the thalamus, which routes them to the **postcentral gyrus.** This is the most anterior gyrus of the parietal lobe, immediately posterior to the central sulcus. The cortex of this gyrus is the **primary somesthetic cortex (somatosensory area)** (fig. 14.21). Somesthetic fibers decussate on their way to the thalamus, so the right postcentral gyrus receives signals from the left side of the body and the left gyrus receives signals from the right.

Each gyrus is like an upside-down sensory map of the contralateral side of the body, traditionally diagrammed as a *sensory homunculus*[50] (fig. 14.21b). As the diagram shows, receptors in the lower limb project to the superior and medial parts of the gyrus and receptors in the face project to the inferior and lateral parts of the gyrus. Such a point-for-point correspondence between an area of the body and an area of the CNS is called **somatotopy.**[51] The reason for the bizarre, distorted appearance of the homunculus is that the amount of cerebral tissue devoted to a given body region is proportional to how richly innervated and sensitive that part of the body is. Thus, the very sensitive hands and face are represented by a much larger area of cortex than the less sensitive trunk.

[50]*homunculus* = little man
[51]*somato* = body + *topy* = place

## The Special Senses

The receptors of taste also project to the postcentral gyrus, but the organs of smell, vision, hearing, and equilibrium project to other specialized regions of the brain. Their pathways are described in chapter 16. For now, we will simply identify the **primary sensory areas** where the signals are ultimately received (fig. 14.22)—that is, what parts of the brain initially process the input:

- **taste** (gustation)—near the inferior lateral end of the postcentral gyrus and part of the insula;
- **smell** (olfaction)—medial surface of the temporal lobe and inferior surface of the frontal lobe;
- **vision**—posterior occipital lobe;
- **hearing**—superior temporal lobe and nearby insula; and
- **equilibrium**—mainly the cerebellum, but also via the thalamus to the roof of the lateral sulcus and lower end of the central sulcus.

## Sensory Association Areas

Each primary sensory area of the cerebral cortex lies adjacent to an **association area** (secondary sensory cortex) that interprets sensory information and makes it identifiable

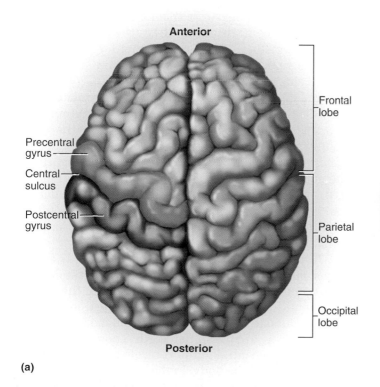

(a)

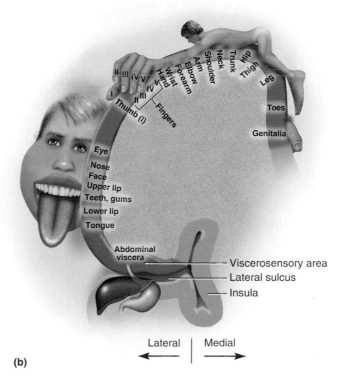

(b)

**FIGURE 14.21 The Primary Somesthetic Cortex (postcentral gyrus).** (a) Superior view of the brain showing the location of the postcentral gyrus (violet). (b) The sensory homunculus, drawn so that body parts are in proportion to the amount of cortex dedicated to their sensation. This gyrus also includes centers for visceral sensation from the intra-abdominal organs (the viscerosensory area).

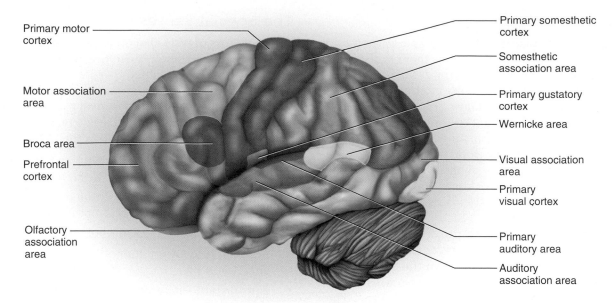

**FIGURE 14.22** **Some Functional Regions of the Cerebral Cortex.** The Broca and Wernicke areas for language abilities are found in only one hemisphere, usually the left. The other regions shown here are mirrored in both hemispheres.

and useful (fig. 14.22). The largest and best-known association areas are:

- The **somesthetic association area** is located in the parietal lobe immediately posterior to the postcentral gyrus. It makes us aware of the positions of our limbs, the location of a touch or pain, and the shape, weight, and texture of an object in our hand, for example.

- The **visual association area** occupies all of the occipital lobe anterior to the primary visual cortex, some of the posterior parietal lobe (concerned with spatial perception), and much of the inferior temporal lobe (where we recognize faces and other familiar objects).

- The **auditory association area** occupies areas of the temporal lobe inferior to the primary auditory cortex and deep within the lateral sulcus. It enables us to remember the name of a piece of music and to identify a person by his or her voice.

- The **orbitofrontal cortex** (see fig. 14.17) is an association area in the frontal lobe that integrates gustatory, olfactory, and visual information to create a sense of the overall flavor and desirability (or rejection) of food.

## MOTOR CONTROL

The intention to contract a skeletal muscle begins in the **motor association (premotor) area** of the frontal lobes (fig. 14.22). This is where we plan our behavior—where neurons compile a program for the degree and sequence of muscle contractions required for an action such as dancing,

typing, or speaking. The program is then transmitted to neurons of the **precentral gyrus** (primary motor area), which is the most posterior gyrus of the frontal lobe, immediately anterior to the central sulcus (fig. 14.23a). Neurons here send signals to the brainstem and spinal cord that ultimately result in muscle contractions.

The precentral gyrus, like the postcentral one, exhibits somatotopy. The neurons for toe movements, for example, are deep in the longitudinal fissure on the medial side of the gyrus. The summit of the gyrus controls the trunk, shoulder, and arm, and the inferolateral region controls the facial muscles. This map is diagrammed as a *motor homunculus* (fig. 14.23b). Like the sensory homunculus, it has a distorted look because the amount of cortex devoted to a given body region is proportional to the number of muscles and motor units in that region, not to the size of the region. Areas of fine control, such as the hands, have more muscles than such areas as the trunk and thigh, more motor units per muscle, and larger areas of motor cortex to control them.

The pyramidal cells of the precentral gyrus are called **upper motor neurons.** Their fibers project caudally, with about 19 million fibers ending in nuclei of the brainstem and 1 million forming the corticospinal tracts. These tracts decussate in the pyramids of the medulla oblongata on their way to the spinal cord. Therefore, below the neck, each precentral gyrus controls muscles on the contralateral side of the body. In the brainstem or spinal cord, the fibers from the upper motor neurons synapse with **lower motor neurons** whose axons innervate the skeletal muscles.

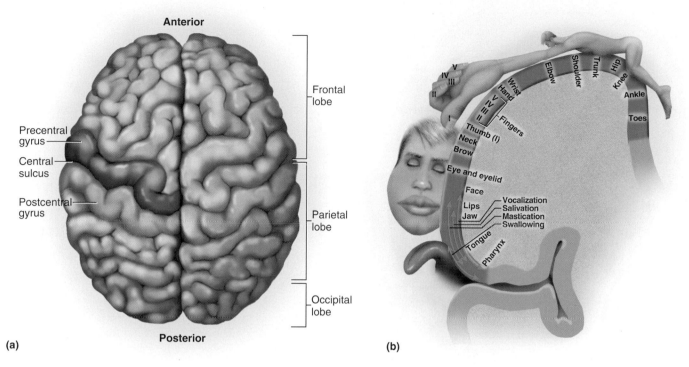

**FIGURE 14.23  The Primary Motor Cortex (precentral gyrus).**  (a) Superior view of the brain showing the location of the precentral gyrus (blue). (b) Motor homunculus, drawn so that body parts are in proportion to the amount of primary motor cortex dedicated to their control.
▶ *Which body regions are controlled by the largest areas of motor cortex—regions with a few large muscles, or regions with numerous small muscles?*

Other areas of the brain important in muscle control are the basal nuclei and cerebellum. The basal nuclei lie along a feedback pathway that receives signals from the cerebral cortex and directs their output back to the cortex. Nearly all areas of cerebral cortex, except for the primary visual and auditory areas, send signals to the basal nuclei. The basal nuclei process these and issue their output to the thalamus, which relays these signals back to the cerebral cortex—especially to the prefrontal cortex, motor association area, and precentral gyrus. The basal nuclei thus lie in a feedback circuit involved in the planning and execution of movement.

Among other functions, the basal nuclei assume control of highly practiced behaviors that one carries out with little thought—writing, typing, driving a car, using scissors, or tying one's shoes, for example. They also control the onset and cessation of planned movements, and the repetitive movements at the shoulder and hip that occur during walking.

Lesions of the basal nuclei cause movement disorders called **dyskinesias.**[52] Some dyskinesias are characterized by abnormally inhibited movements—for example, difficulty rising from a chair or beginning to walk, and a slow shuffling walk, as seen in Parkinson disease (see p. 475).

Smooth, easy movements require the excitation of agonistic muscles and inhibition of their antagonists. In Parkinson disease, the antagonists are not inhibited. Therefore, opposing muscles at a joint fight each other, making it a struggle to move as one wishes. Other dyskinesias are characterized by exaggerated or unwanted movements, such as flailing of the limbs *(ballismus)* in Huntington disease.

The cerebellum is highly important in motor coordination. It aids in learning motor skills, maintains muscle tone and posture, smooths muscle contractions, coordinates eye and body movements, and coordinates the motions of different joints with each other (such as the shoulder and elbow in pitching a baseball). The cerebellum acts as a comparator in motor control. It receives information from the upper motor neurons of the cerebrum about what movements are intended, and information from proprioceptors in the muscles and joints about the actual performance of the movement (fig. 14.24). The Purkinje cells of the cerebellum compare the two. If there is a discrepancy between the intent and the performance, they signal the deep cerebellar nuclei, which in turn relay signals to the thalamus and cerebral cortex. Motor neurons in the cortex then correct the muscle performance to match the intent. Lesions of the cerebellum can result in a clumsy, awkward gait *(ataxia)* and make some tasks such as climbing stairs virtually impossible.

---

[52]*dys* = bad, abnormal, difficult + *kines* = movement

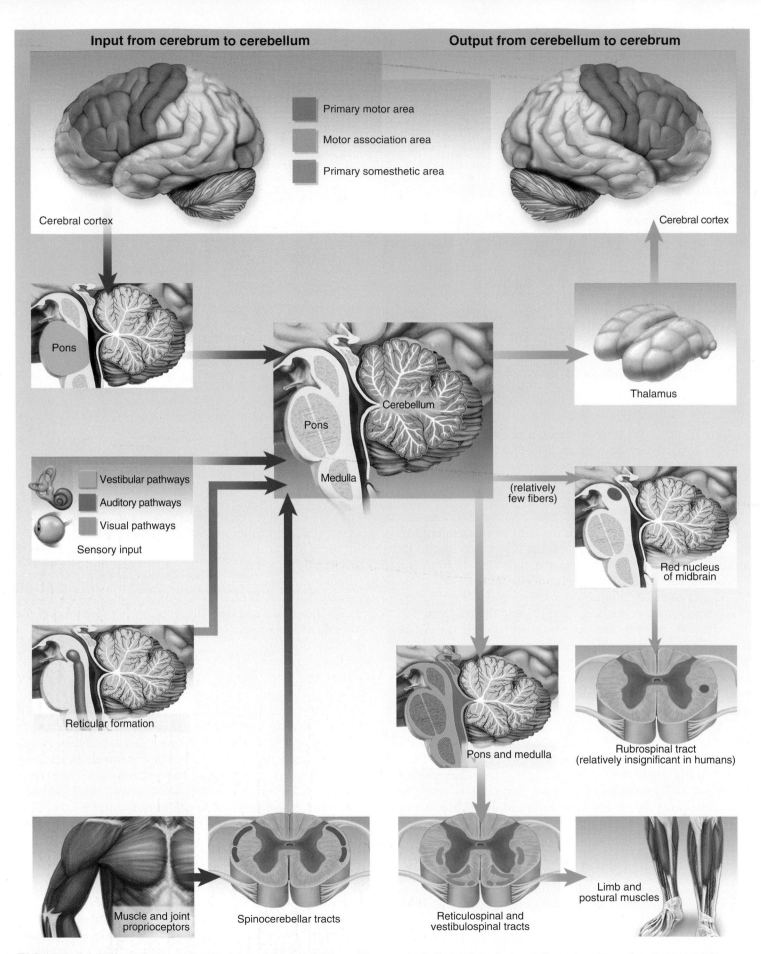

Primary motor area

Motor association area

Primary somesthetic area

Cerebral cortex

Cerebral cortex

Pons

Thalamus

Cerebellum

Pons

Medulla

(relatively few fibers)

Vestibular pathways

Auditory pathways

Visual pathways

Sensory input

Red nucleus of midbrain

Reticular formation

Pons and medulla

Rubrospinal tract (relatively insignificant in humans)

Muscle and joint proprioceptors

Spinocerebellar tracts

Reticulospinal and vestibulospinal tracts

Limb and postural muscles

**FIGURE 14.24 Motor Pathways Involving the Cerebellum.** The cerebellum receives its input from the afferent pathways (red) on the left and sends its output through the efferent pathways (green) on the right.

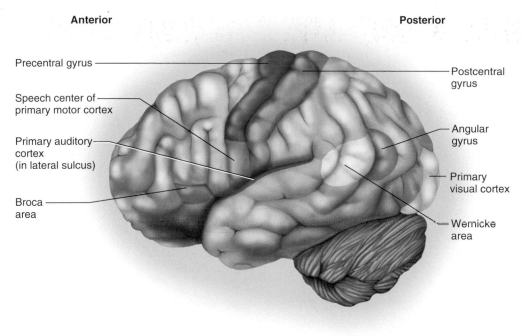

**Anterior**

Precentral gyrus

Speech center of primary motor cortex

Primary auditory cortex (in lateral sulcus)

Broca area

**Posterior**

Postcentral gyrus

Angular gyrus

Primary visual cortex

Wernicke area

**FIGURE 14.25** Language Centers of the Left Hemisphere.

# LANGUAGE

Language includes several abilities—reading, writing, speaking, and understanding words—assigned to different regions of cerebral cortex (fig. 14.25). The **Wernicke**[53] (WUR-ni-keh) **area** is responsible for the recognition of spoken and written language. It lies just posterior to the lateral sulcus, usually in the left hemisphere. It is a sensory association area that receives visual and auditory input from the respective regions of primary sensory cortex. The *angular gyrus* just posterior to the Wernicke area processes the words we read into a form that we can speak.

The Wernicke area formulates phrases according to learned rules of grammar and transmits a plan of speech to the **Broca**[54] **area,** located in the inferior prefrontal cortex in the same hemisphere. The Broca area generates a motor program for the muscles of the larynx, tongue, cheeks, and lips to produce speech. It transmits this program to the primary motor cortex, which executes it—that is, it issues commands to the lower motor neurons that supply the relevant muscles. PET scans show a rise in the metabolic activity of the Broca area as we prepare to speak.

The emotional aspect of language is controlled by regions in the opposite hemisphere that mirror the Wernicke and Broca areas. Opposite the Broca area is the *affective language area.* Lesions to this area result in *aprosody*—flat, emotionless speech. The cortex opposite the Wernicke area is concerned with recognizing the emotional content of another person's speech. Lesions here can result in such problems as the inability to understand a joke.

**Aphasia**[55] (ah-FAY-zee-uh) is any language deficit resulting from lesions in the hemisphere (usually the left) containing the Wernicke and Broca areas. The many forms of aphasia are difficult to classify. *Nonfluent (Broca) aphasia,* due to a lesion to the Broca area, results in slow speech, difficulty in choosing words, or use of words that only approximate the correct word. For example, a person may say "tssair" when asked to identify a picture of a chair. In extreme cases, the person's entire vocabulary consists of two or three words, sometimes those that were being spoken when a stroke occurred. Such patients feel very frustrated with themselves and often maintain a tight-lipped reluctance to talk. A lesion to the Wernicke area may cause *fluent (Wernicke) aphasia,* in which a person speaks normally and sometimes excessively, but uses jargon and invented words that make little sense (for example, "choss" for chair). Such a person also cannot comprehend written and spoken words. In *anomic aphasia,* a person can speak normally and understand speech, but cannot identify written words or pictures. Shown a picture of a chair, the person may say, "I know what it is. . . . I have a lot of them," but be unable to name the object.

This represents only a small sample of the complex and puzzling linguistic effects of brain lesions. Other lesions to small areas of cortex can cause impaired mathematical ability, a tendency to write only consonants, or difficulty understanding the second half of each word a person reads.

# CEREBRAL LATERALIZATION

The two cerebral hemispheres look identical at a glance, but close examination reveals a number of differences. For example, in women the left temporal lobe is longer than the right. In left-handed people, the left frontal, parietal, and occipital lobes are usually wider than those on the right. The two hemispheres also differ in some of their functions (fig. 14.26). Neither hemisphere is "dominant," but each is specialized for certain tasks. This difference in function is called **cerebral lateralization.**

One hemisphere, usually the left, is called the *categorical hemisphere.* It is specialized for spoken and written language and for the sequential and analytical reasoning employed in such fields as science and mathematics. This hemisphere seems to break information into fragments and

[53]Karl Wernicke (1848–1905), German neurologist
[54]Pierre Paul Broca (1824–80), French surgeon and anthropologist
[55]a = without + phas = speech

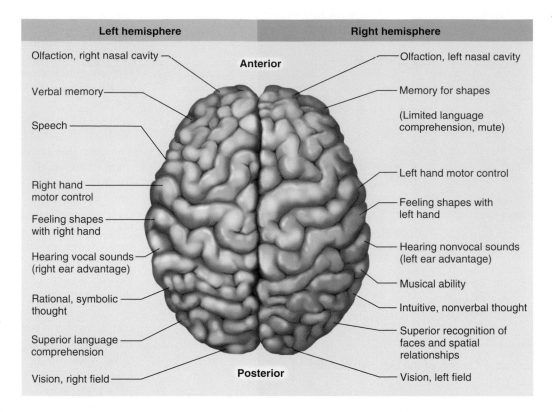

**FIGURE 14.26** **Lateralization of Cerebral Functions.** The two cerebral hemispheres are not functionally identical.

analyze it in a linear way. The other hemisphere, usually the right, is called the *representational hemisphere.* It perceives information in a more integrated, holistic way. It is the seat of imagination and insight, musical and artistic skill, perception of patterns and spatial relationships, and comparison of sights, sounds, smells, and tastes.

Cerebral lateralization is highly correlated with handedness. The left hemisphere is the categorical one in 96% of right-handed people, and the right hemisphere in 4%. Among left-handed people, the right hemisphere is categorical in 15%, the left in 70%, and in the remaining 15% neither hemisphere is distinctly specialized.

Lateralization develops with age. In young children, if one cerebral hemisphere is damaged or removed (for example, because of brain cancer), the other hemisphere can often take over its functions. Adult males exhibit more lateralization than females and suffer more functional loss when one hemisphere is damaged. When the left hemisphere is damaged, men are three times as likely as women to become aphasic. The reason for this difference is not yet clear, but it may be related to the corpus callosum. In men, the corpus callosum has a fairly uniform thickness, but in

women its caudal portion is thickened by additional commissural fibers, suggesting that women have a more extensive communication between their hemispheres.

Not surprisingly, the brain is the most structurally complex organ of the body. We have barely scratched the surface in this chapter. Table 14.1 summarizes the gross anatomy of the brain and will help you put its structures into context.

## Before You Go On

*Answer the following questions to test your understanding of the preceding section:*

19. *Suppose you are reading a novel and gradually fall asleep and begin to dream. How would your brain waves change during this sequence of events?*

20. *Describe the locations and functions of the somesthetic, visual, auditory, and frontal association areas.*

21. *Describe the somatotopy of the primary motor area and primary sensory area.*

22. *What are the roles of the Wernicke area, Broca area, and precentral gyrus in language?*

| TABLE 14.1 | Anatomical Checklist for the Brain |
|---|---|
| *Meninges* | *Reticular Formation* |
| Dura mater | *Prosencephalon (forebrain)* |
|   Falx cerebri | *Diencephalon* |
|   Falx cerebelli |   Thalamus |
|   Tentorium cerebelli |   Hypothalamus |
| Arachnoid mater |   Epithalamus |
|   Arachnoid villi |     Pineal gland |
| Pia mater |     Habenula |
| *Ventricle System* | *Telencephalon (cerebrum)* |
| Lateral ventricles |   Cerebral hemispheres |
| Interventricular foramen |   Major fissure and sulci |
| Third ventricle |     Longitudinal fissure |
| Cerebral aqueduct |     Central sulcus |
| Fourth ventricle |     Parieto-occipital sulcus |
| Median and lateral apertures |     Lateral sulcus |
| Central canal |   Lobes |
| Choroid plexuses |     Frontal lobe |
| *Rhombencephalon (hindbrain)* |     Parietal lobe |
| *Myelencephalon* |     Occipital lobe |
|   Medulla oblongata |     Temporal lobe |
|     Pyramids |     Insula |
|     Olive |   Gray matter (cerebral cortex) |
|       Inferior olivary nucleus |   White matter |
| *Metencephalon* |     Projection tracts |
|   Pons |     Commissural tracts |
|   Cerebellum |       Corpus callosum |
|     Cerebellar hemispheres |       Anterior commissure |
|     Vermis |       Posterior commissure |
|     Cerebellar peduncles |     Association tracts |
|     Folia |   Major gyri |
|     Arbor vitae |     Precentral gyrus |
|     Deep nuclei |     Postcentral gyrus |
| *Mesencephalon (midbrain)* |     Cingulate gyrus |
| Cerebral peduncles |   Basal nuclei |
|   Tegmentum |     Caudate nucleus |
|     Red nucleus |     Putamen |
|   Substantia nigra |     Globus pallidus |
|   Central gray matter |   Limbic system |
|   Tectum |     Hippocampus |
|     Corpora quadrigemina |     Amygdala |
|       Superior colliculi |     Fornix |
|       Inferior colliculi | |

# The Cranial Nerves

### Objectives

When you have completed this section, you should be able to

- list the 12 cranial nerves by name and number;
- identify where each cranial nerve originates and terminates; and
- state the functions of each cranial nerve.

To be functional, the brain must communicate with the rest of the body. Most of its input and output travels by way of the spinal cord, but it also communicates by way of 12 pairs of **cranial nerves.** These arise from the base of the brain, exit the cranium through its foramina, and lead to muscles and sense organs located mainly in the head and neck. The cranial nerves are numbered I to XII starting with the most rostral pair (fig. 14.27). Each nerve also has a descriptive name such as *optic nerve* and *vagus nerve.* Page 557 suggests some aids to remembering their names and numbers.

## CRANIAL NERVE PATHWAYS

Most motor fibers of the cranial nerves begin in nuclei of the brainstem and lead to glands and muscles. The sensory fibers begin in receptors located mainly in the head and neck and lead mainly to the brainstem. These include both the special senses such as vision and hearing, and general senses such as touch and proprioception. Pathways for the special senses are described in chapter 16. Sensory fibers for proprioception begin in the muscles innervated by the motor fibers of the cranial nerves, but they often travel to the brain in a different nerve than the one which supplies the motor innervation.

Most cranial nerves carry fibers between the brainstem and ipsilateral receptors and effectors. Thus, a lesion in one side of the brainstem causes a sensory or motor deficit on the same side of the head. This contrasts with lesions to the motor and somesthetic cortex of the cerebrum, which, as we saw earlier, cause sensory and motor deficits on the *contralateral* side of the body. The exceptions are the optic nerve (II), where half the fibers decussate to the opposite side of the brain (see chapter 16), and the trochlear nerve (IV), in which all efferent fibers lead to a muscle of the contralateral eye.

## CRANIAL NERVE CLASSIFICATION

Cranial nerves are traditionally classified as sensory (I, II, and VIII), motor (III, IV, VI, XI, and XII), or mixed (V, VII, IX, and X). In reality, only cranial nerves I and II (for

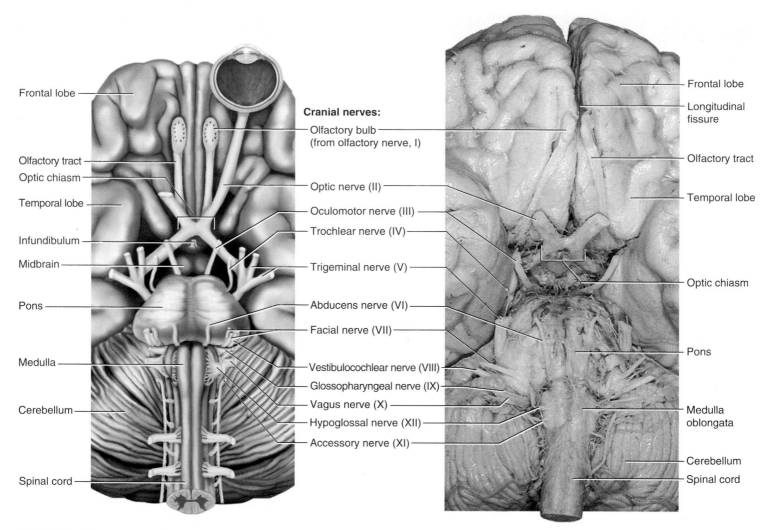

**FIGURE 14.27   The Cranial Nerves.**   (a) Base of the brain, showing the 12 cranial nerves. (b) Photograph of the cranial nerves.

smell and vision) are purely sensory, whereas all of the rest contain both afferent and efferent fibers and are therefore mixed nerves. Those traditionally classified as motor not only stimulate muscle contractions but also contain afferent fibers of proprioception, which provide the brain with feedback for controlling muscle action and make one consciously aware of such things as the position of the tongue and orientation of the head. Cranial nerve VIII, concerned with hearing and equilibrium, is traditionally classified as sensory, but it also has motor fibers that return signals to the inner ear and "tune" it to sharpen the sense of hearing. The nerves traditionally classified as mixed have sensory functions quite unrelated to their motor functions. For example, the facial

nerve (VII) has a sensory role in taste and a motor role in controlling facial expressions.

## CRANIAL NERVE TABLE

Table 14.2 describes and illustrates the 12 pairs of cranial nerves. For each nerve, it describes the composition (sensory, motor, or mixed); its functions; its course from origin to termination and its path through the cranium; signs and symptoms of nerve damage; and some clinical tests used to evaluate its function. In order to teach the traditional classification (which is relevant for such purposes as board examinations and comparison to other books), yet remind you that all but two of these nerves are mixed, the table describes many of the nerves as *predominantly* sensory or motor.

## TABLE 14.2 | The Cranial Nerves

Origins of proprioceptive fibers are not tabulated. Nerves listed as mixed or sensory are agreed by all authorities to be either mixed or purely sensory nerves. Nerves classified as *predominantly* motor or sensory are traditionally classified that way but contain some fibers of the other type.

### I. Olfactory Nerve

This is the nerve for the sense of smell. It consists of several separate fascicles that pass independently through the cribriform plate in the roof of the nasal cavity. It is not visible on brains removed from the skull because these fascicles are severed by removal of the brain.

| Composition | Function | Origin | Termination | Cranial Passage | Effect of Damage | Clinical Test |
|---|---|---|---|---|---|---|
| Sensory | Smell | Olfactory mucosa in nasal cavity | Olfactory bulbs | Cribriform foramina of ethmoid bone | Impaired sense of smell | Determine whether subject can smell (not necessarily identify) aromatic substances such as coffee, vanilla, clove oil, or soap |

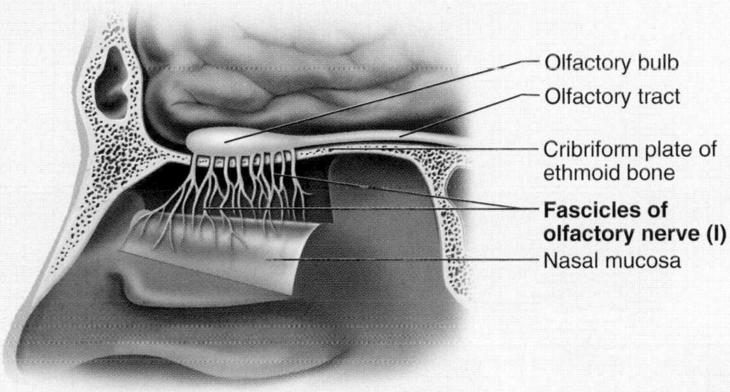

- Olfactory bulb
- Olfactory tract
- Cribriform plate of ethmoid bone
- **Fascicles of olfactory nerve (I)**
- Nasal mucosa

**FIGURE 14.28**  The Olfactory Nerve (I).

### II. Optic Nerve

This is the nerve for vision.

| Composition | Function | Origin | Termination | Cranial Passage | Effect of Damage | Clinical Test |
|---|---|---|---|---|---|---|
| Sensory | Vision | Retina | Thalamus and midbrain | Optic foramen | Blindness in part or all of visual field | Inspect retina with ophthalmoscope; test peripheral vision and visual acuity |

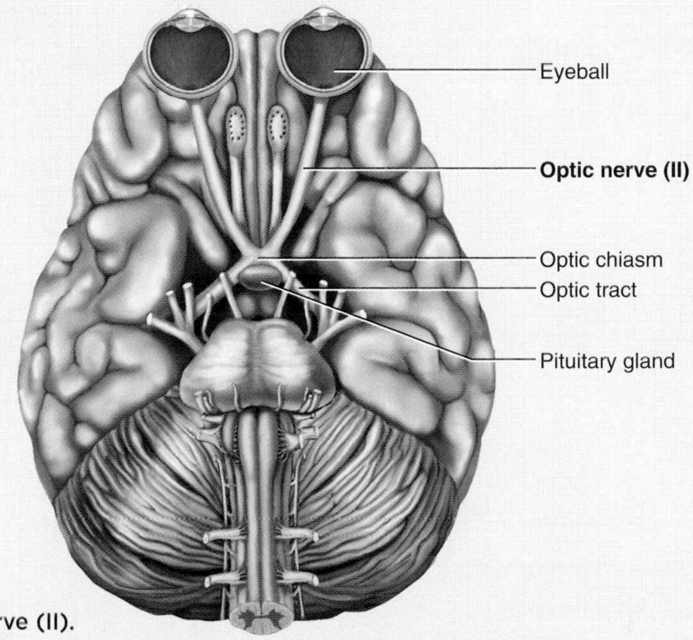

- Eyeball
- **Optic nerve (II)**
- Optic chiasm
- Optic tract
- Pituitary gland

**FIGURE 14.29**  The Optic Nerve (II).

## TABLE 14.2 | The Cranial Nerves (cont.)

### III. Oculomotor Nerve (OC-you-lo-MO-tur)

This nerve controls muscles that turn the eyeball up, down, and medially, as well as controlling the iris, lens, and upper eyelid.

| Composition | Function | Origin | Termination | Cranial Passage | Effect of Damage | Clinical Test |
|---|---|---|---|---|---|---|
| Predominantly motor | Eye movements, opening of eyelid, pupillary constriction, focusing | Midbrain | Somatic fibers to levator palpebrae superioris, superior, medial, and inferior rectus muscles and inferior oblique muscles of eye. Autonomic fibers enter eyeball and lead to constrictor of iris and ciliary muscle of lens. | Superior orbital fissure | Drooping eyelid; dilated pupil; inability to move eye in some directions; tendency of eye to rotate laterally at rest; double vision; difficulty focusing | Look for differences in size and shape of right and left pupils; test pupillary response to light; test ability to track moving objects |

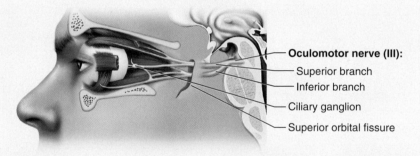

**FIGURE 14.30**  The Oculomotor Nerve (III).

### IV. Trochlear Nerve (TROCK-lee-ur)

This nerve controls a muscle that rotates the eyeball medially and slightly depresses the eyeball when the head turns.

| Composition | Function | Origin | Termination | Cranial Passage | Effect of Damage | Clinical Test |
|---|---|---|---|---|---|---|
| Predominantly motor | Eye movements | Midbrain | Superior oblique muscle of eye | Superior orbital fissure | Double vision and inability to rotate eye inferolaterally; eye points superolaterally and subject tends to tilt head toward affected side | Test ability of eye to rotate inferolaterally |

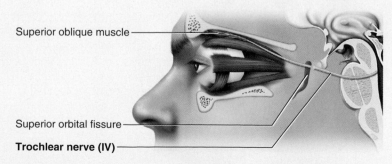

**FIGURE 14.31**  The Trochlear Nerve (IV).

### V. Trigeminal Nerve (tri-JEM-ih-nul)

This is the largest of the cranial nerves, and the most important sensory nerve of the face. It forks into three divisions: *ophthalmic* ($V_1$), *maxillary* ($V_2$), and *mandibular* ($V_3$).

| Composition | Function | Origin | Termination | Cranial Passage | Effect of Damage | Clinical Test |
|---|---|---|---|---|---|---|
| **$V_1$, Ophthalmic Division** | | | | | | |
| Sensory | Touch, temperature, and pain sensations from upper face | Superior region of face as illustrated; surface of eyeball; lacrimal (tear gland); superior nasal mucosa; frontal and ethmoid sinuses | Pons | Superior orbital fissure | Loss of sensation from upper face | Test corneal reflex (blinking in response to light touch to eyeball) |
| **$V_2$, Maxillary Division** | | | | | | |
| Sensory | Same as $V_1$, lower on face | Middle region of face as illustrated; nasal mucosa; maxillary sinus; palate; upper teeth and gums | Pons | Foramen rotundum and infraorbital foramen | Loss of sensation from middle face | Test sense of touch, pain, and temperature with light touch, pinpricks, and hot and cold objects |
| **$V_3$, Mandibular Division** | | | | | | |
| Mixed | *Sensory:* Same as $V_1$ and $V_2$, lower on face  *Motor:* mastication | *Sensory:* Inferior region of face as illustrated; anterior two-thirds of tongue (but not taste buds); lower teeth and gums; floor of mouth; dura mater  *Motor:* Pons | *Sensory:* Pons  *Motor:* Anterior belly of digastric; masseter, temporalis, mylohyoid, and pterygoid muscles; tensor tympani muscle of middle ear | Foramen ovale | Loss of sensation; impaired chewing | Assess motor functions by palpating masseter and temporalis while subject clenches teeth; test ability to move mandible from side to side and to open mouth against resistance |

---

## INSIGHT 14.4 Clinical Application

### Some Cranial Nerve Disorders

*Trigeminal neuralgia*[56] (tic douloureux[57]) is a syndrome characterized by recurring episodes of intense stabbing pain in the trigeminal nerve. The cause is unknown; there is no visible change in the nerve. It usually occurs after the age of 50 and mostly in women. The pain lasts only a few seconds to a minute or two, but it strikes at unpredictable intervals and sometimes up to a hundred times a day. The pain usually occurs in a specific zone of the face, such as around the mouth and nose. It may be triggered by touch, drinking, tooth brushing, or washing the face. Analgesics (pain relievers) give only limited relief. Severe cases are treated by cutting the nerve, but this also deadens most other sensation in that side of the face.

*Bell*[58] *palsy* is a degenerative disorder of the facial nerve, probably due to a virus. It is characterized by paralysis of the facial muscles on one side with resulting distortion of the facial features, such as sagging of the mouth or lower eyelid. The paralysis may interfere with speech, prevent closure of the eye, and cause excessive tear secretion. There may also be a partial loss of the sense of taste. Bell palsy may appear abruptly, sometimes overnight, and often disappears spontaneously within 3 to 5 weeks.

---

[56]*neur* = nerve + *algia* = pain
[57]*douloureux* = painful

[58]Sir Charles Bell (1774–1842), Scottish physician

## TABLE 14.2    The Cranial Nerves (*cont.*)

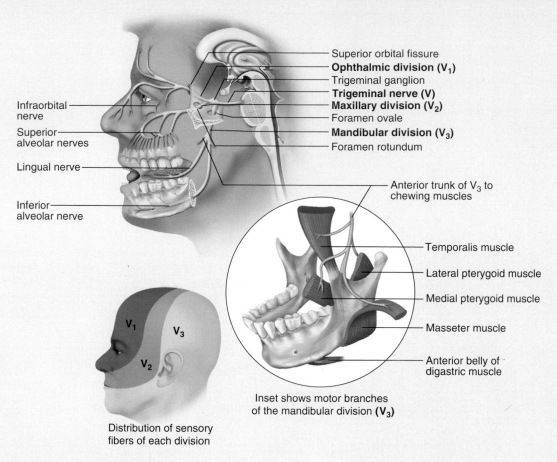

**FIGURE 14.32   The Trigeminal Nerve (V).**

### VI. Abducens Nerve (ab-DOO-senz)

This nerve controls a muscle that turns the eyeball laterally.

| Composition | Function | Origin | Termination | Cranial Passage | Effect of Damage | Clinical Test |
|---|---|---|---|---|---|---|
| Predominantly motor | Lateral eye movement | Inferior pons | Lateral rectus muscle of eye | Superior orbital fissure | Inability to turn eye laterally; at rest, eye turns medially because of action of antagonistic muscles | Test lateral eye movement |

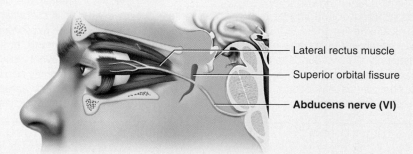

**FIGURE 14.33   The Abducens Nerve (VI).**

## VII. Facial Nerve

This is the major motor nerve of the facial muscles. It divides into five prominent branches: *temporal, zygomatic, buccal, mandibular,* and *cervical.*

| Composition | Function | Origin | Termination | Cranial Passage | Effect of Damage | Clinical Test |
|---|---|---|---|---|---|---|
| Mixed | *Sensory:* Taste<br><br>*Motor:* Facial expression; secretion of tears, saliva, nasal and oral mucus | *Sensory:* Taste buds of anterior two-thirds of tongue<br><br>*Motor:* Pons | *Sensory:* Thalamus<br><br>*Motor:* Somatic fibers to digastric muscle, stapedius muscle of middle ear, muscles of facial expression. Autonomic fibers to sub-mandibular and sublingual salivary glands, tear glands, nasal and palatine glands | Stylomastoid foramen | Inability to control facial muscles; sagging due to loss of muscle tone; distorted sense of taste, especially for sweets | Test anterior two-thirds of tongue with substances such as sugar, salt, vinegar, and quinine; test response of tear glands to ammonia fumes; test subject's ability to smile, frown, whistle, raise eyebrows, close eyes, etc. |

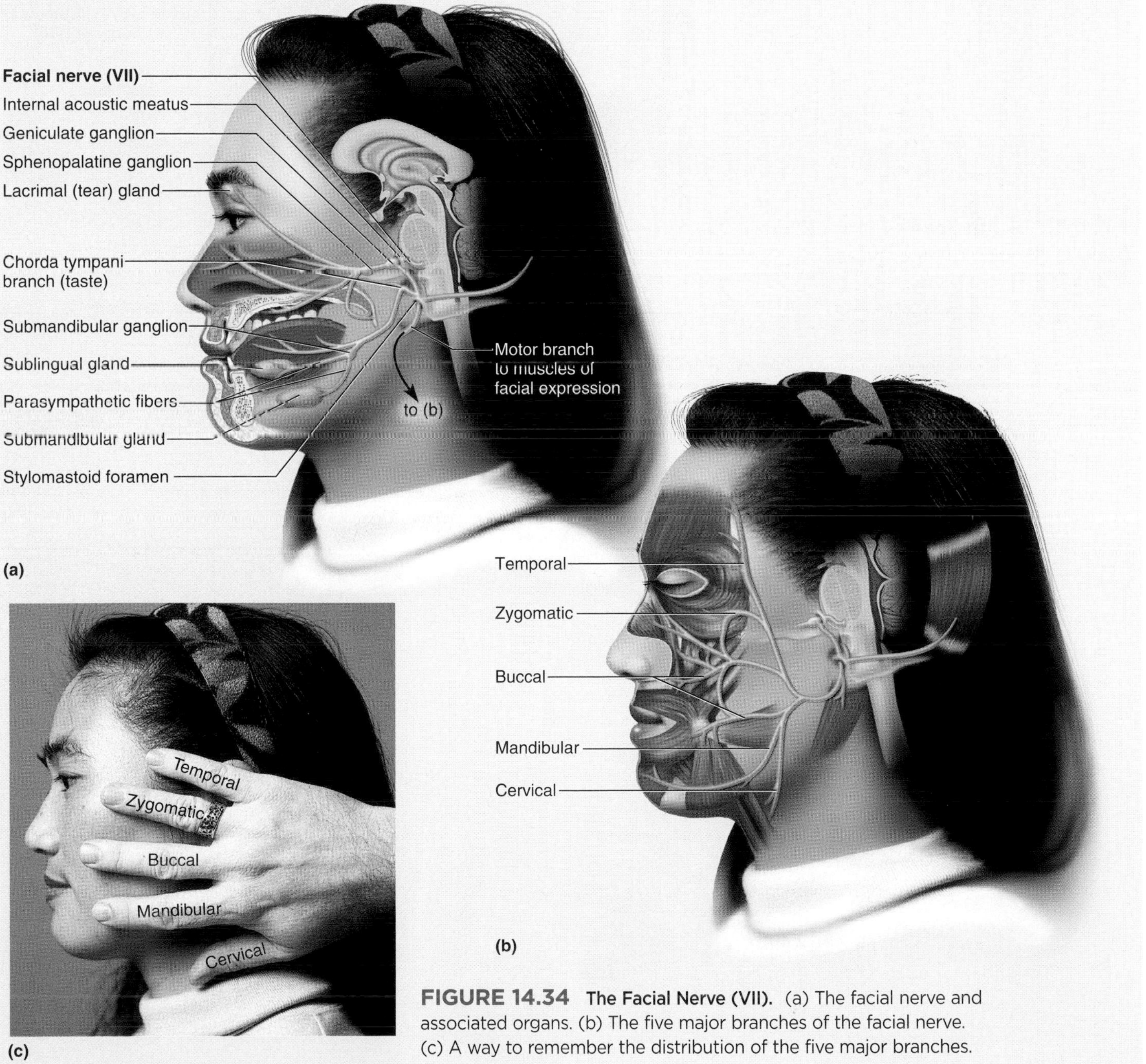

**(a)**

Facial nerve (VII)
Internal acoustic meatus
Geniculate ganglion
Sphenopalatine ganglion
Lacrimal (tear) gland
Chorda tympani branch (taste)
Submandibular ganglion
Sublingual gland
Parasympathetic fibers
Submandibular gland
Stylomastoid foramen

Motor branch to muscles of facial expression

to (b)

**(b)**

Temporal
Zygomatic
Buccal
Mandibular
Cervical

**(c)**

Temporal
Zygomatic
Buccal
Mandibular
Cervical

**FIGURE 14.34   The Facial Nerve (VII).** (a) The facial nerve and associated organs. (b) The five major branches of the facial nerve. (c) A way to remember the distribution of the five major branches.

**TABLE 14.2** | The Cranial Nerves (*cont.*)

## VIII. Vestibulocochlear Nerve (vess-TIB-you-lo-COC-lee-ur)

This is the nerve of hearing and equilibrium, but it also has motor fibers that lead to cells of the cochlea that tune the sense of hearing (see chapter 16).

| Composition | Function | Origin | Termination | Cranial Passage | Effect of Damage | Clinical Test |
|---|---|---|---|---|---|---|
| Predominantly sensory | Hearing and equilibrium | *Sensory:* Cochlea, vestibule, and semicircular ducts of inner ear<br><br>*Motor:* Pons | *Sensory:* Fibers for hearing end in medulla; fibers for equilibrium end at junction of medulla and pons<br><br>*Motor:* Outer hair cells of cochlea of inner ear | Internal acoustic meatus | Nerve deafness, dizziness, nausea, loss of balance, and nystagmus (involuntary oscillation of eyes from side to side) | Look for nystagmus; test hearing, balance, and ability to walk a straight line |

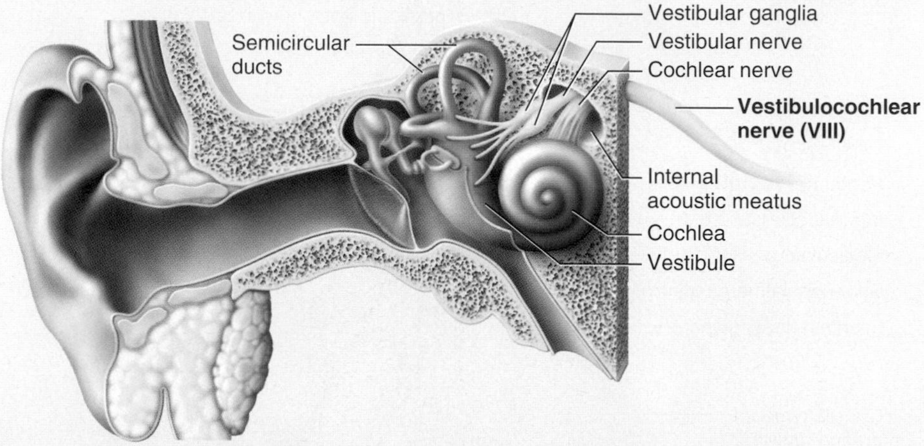

**FIGURE 14.35** The Vestibulocochlear Nerve (VIII).

## IX. Glossopharyngeal Nerve (GLOSS-oh-fah-RIN-jee-ul)

This is the complex, mixed nerve with numerous sensory and motor functions in the head, neck, and thoracic regions including sensation from the tongue, throat, and outer ear; control of food ingestion; and some aspects of cardiovascular and respiratory function.

| Composition | Function | Origin | Termination | Cranial Passage | Effect of Damage | Clinical Test |
|---|---|---|---|---|---|---|
| Mixed | *Sensory:* Taste; touch, pressure, pain and temperature sensations from tongue and outer ear; regulation of blood pressure and respiration<br><br>*Motor:* Salivation, swallowing, gagging | *Sensory:* Pharynx; middle and outer ear; posterior one-third of tongue (including taste buds); internal carotid artery<br><br>*Motor:* Medulla oblongata | *Sensory:* Medulla oblongata<br><br>*Motor:* Parotid salivary gland; glands of posterior tongue; stylopharyngeal muscle (which dilates pharynx during swallowing) | Jugular foramen | Loss of bitter and sour taste; impaired swallowing | Test gag reflex, swallowing, and coughing; note any speech impediments; test posterior one-third of tongue with bitter and sour substances |

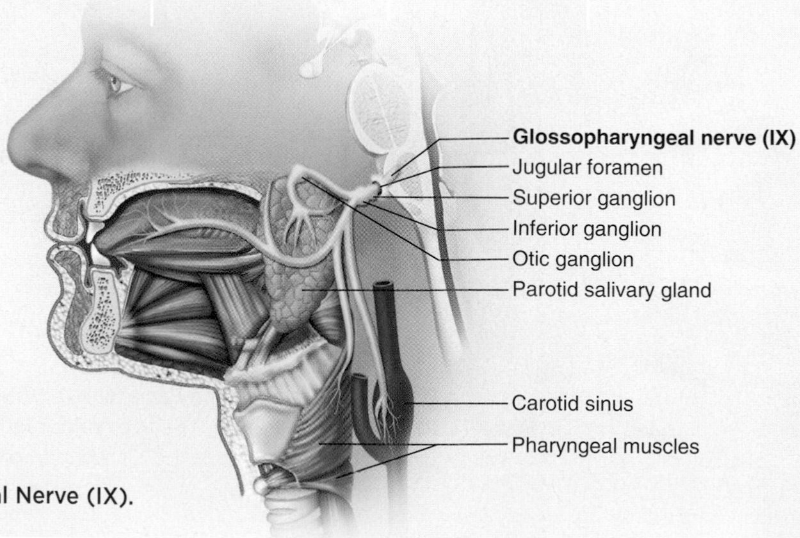

**FIGURE 14.36** The Glossopharyngeal Nerve (IX).

## X. Vagus Nerve (VAY-gus)

The vagus has the most extensive distribution of any cranial nerve, supplying not only organs in the head and neck but also supplying most viscera of the thoracic and abdominal body cavities. It plays major roles in the control of cardiac, pulmonary, digestive, and urinary functions.

| Composition | Function | Origin | Termination | Cranial Passage | Effect of Damage | Clinical Test |
|---|---|---|---|---|---|---|
| Mixed | *Sensory:* Taste; sensations of hunger, fullness, and gastrointestinal discomfort<br><br>*Motor:* Swallowing, speech, deceleration of heart, bronchoconstriction, gastrointestinal secretion and motility | *Sensory:* Thoracic and abdominal viscera, root of tongue, pharynx, larynx, epiglottis, outer ear, dura mater<br><br>*Motor:* Medulla oblongata | *Sensory:* Medulla oblongata<br><br>*Motor:* Tongue, palate, pharynx, larynx, lungs, heart, liver, spleen, digestive tract, kidney, ureter | Jugular foramen | Hoarseness or loss of voice; impaired swallowing and gastrointestinal motility; fatal if both vagus nerves are damaged | Same tests as for cranial nerve IX |

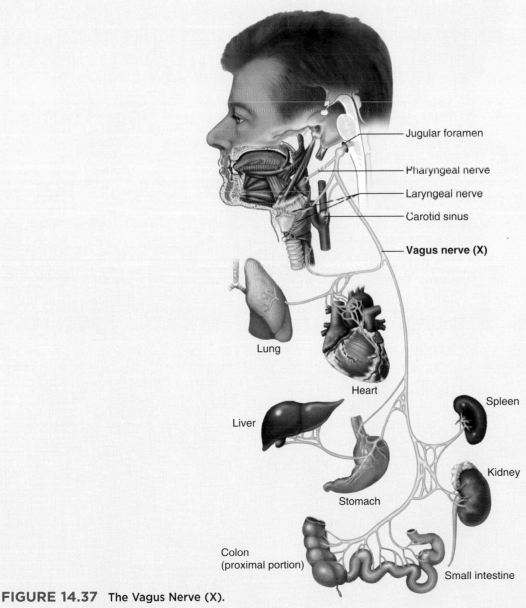

- Jugular foramen
- Pharyngeal nerve
- Laryngeal nerve
- Carotid sinus
- **Vagus nerve (X)**
- Lung
- Heart
- Liver
- Spleen
- Kidney
- Stomach
- Colon (proximal portion)
- Small intestine

**FIGURE 14.37   The Vagus Nerve (X).**

**TABLE 14.2** | The Cranial Nerves (*cont.*)

## XI. Accessory Nerve

This nerve takes an unusual path. Unlike any other cranial nerve, it does not arise entirely from the brain. A small root does arise from the medulla oblongata, but a larger root arises from the cervical spinal cord. The latter root ascends alongside the spinal cord and enters the cranial cavity through the foramen magnum. It then joins the smaller root for a short distance and they exit the cranium together through the jugular foramen, bundled with the vagus and glossopharyngeal nerves. The accessory nerve controls mainly swallowing and neck and shoulder muscles.

| Composition | Function | Origin | Termination | Cranial Passage | Effect of Damage | Clinical Test |
|---|---|---|---|---|---|---|
| Predominantly motor | Swallowing; head, neck, and shoulder movements | Medulla oblongata and spinal cord segments C1 to C6 | Palate, pharynx, trapezius and sterno-cleidomastoid muscles | Jugular foramen | Impaired movement of head, neck, and shoulders; difficulty shrugging shoulder on damaged side; paralysis of sternocleido-mastoid causing head to turn toward injured side | Test ability to rotate head and shrug shoulders against resistance |

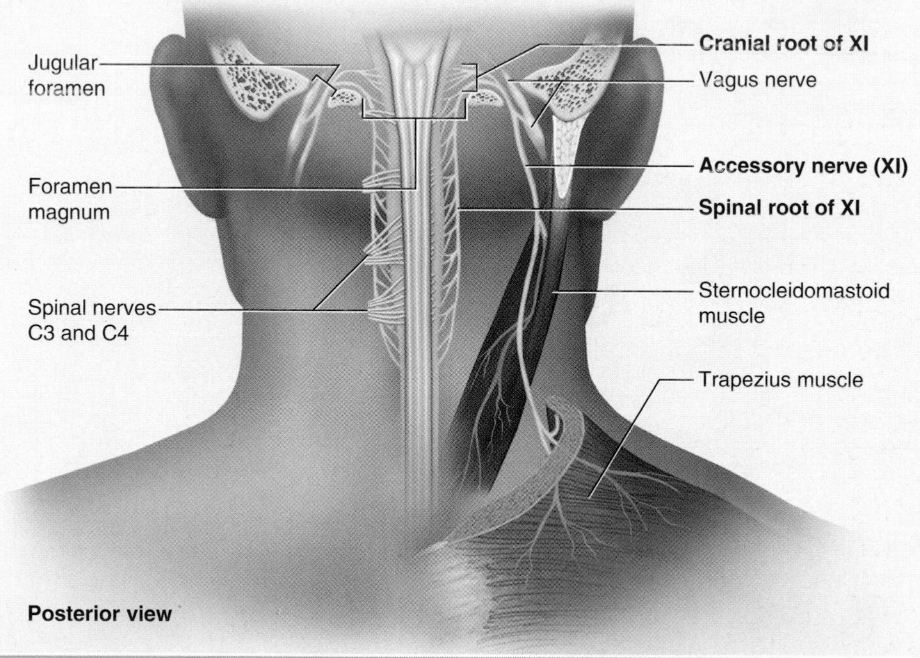

**FIGURE 14.38**
The Accessory Nerve (XI).

**Posterior view**

## XII. Hypoglossal Nerve (HY-po-GLOSS-ul)

This nerve controls tongue movements.

| Composition | Function | Origin | Termination | Cranial Passage | Effect of Damage | Clinical Test |
|---|---|---|---|---|---|---|
| Predominantly motor | Tongue movements of speech, food manipulation, and swallowing | Medulla oblongata | Intrinsic and extrinsic muscles of tongue; thyrohyoid and geniohyoid muscles | Hypoglossal canal | Impaired speech and swallowing; inability to protrude tongue if both right and left nerves are damaged; deviation of tongue toward injured side, and atrophy on that side, if only one nerve damaged | Note deviations of tongue as subject protrudes and retracts it; test ability to protrude tongue against resistance |

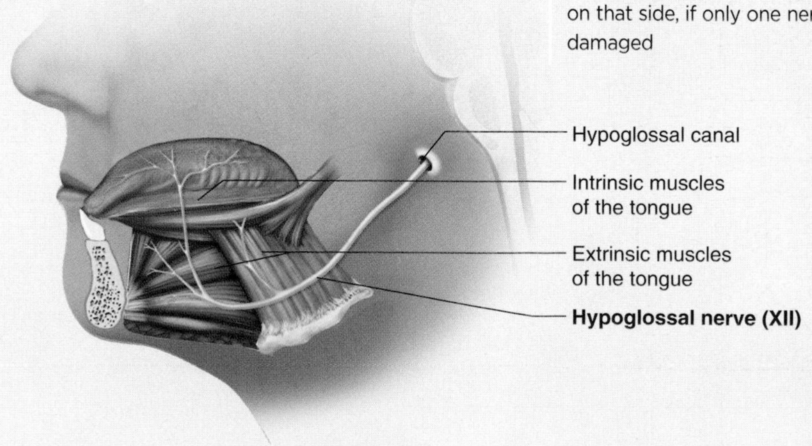

**FIGURE 14.39**
The Hypoglossal Nerve (XII).

| TABLE 14.3 | Some Disorders Associated with the Brain and Cranial Nerves |
|---|---|
| Cerebral palsy | Muscular incoordination resulting from damage to the motor areas of the brain during fetal development, birth, or infancy; causes include prenatal rubella infection, drugs, or radiation exposure; oxygen deficiency during birth; and hydrocephalus |
| Concussion | Damage to the brain typically resulting from a blow, often with loss of consciousness, disturbances of vision or equilibrium, and short-term amnesia |
| Encephalitis | Inflammation of the brain, accompanied by fever, usually caused by mosquito-borne viruses or herpes simplex virus; causes neuronal degeneration and necrosis; can lead to delirium, seizures, and death |
| Epilepsy | Disorder causing sudden, massive discharge of neurons (seizures) resulting in motor convulsions, sensory and psychic disturbances, and often impaired consciousness; may result from birth trauma, tumors, infections, drug or alcohol abuse, or congenital brain malformation |
| Migraine headache | Recurring headaches often accompanied by nausea, vomiting, dizziness, and aversion to light, often triggered by such factors as weather changes, stress, hunger, red wine, or noise; more common in women and sometimes running in families |
| Schizophrenia | A thought disorder involving delusions, hallucinations, inappropriate emotional responses to situations, incoherent speech, and withdrawal from society, resulting from hereditary or developmental abnormalities in neural networks |

**Disorders described elsewhere**

| | | |
|---|---|---|
| Alzheimer disease p. 474 | Brain tumors p. 449 | Multiple sclerosis p. 450 |
| Amnesia p. 540 | Cerebellar ataxia p. 543 | Parkinson disease pp. 475, 543 |
| Aphasia p. 545 | Cranial nerve injuries p. 551 | Poliomyelitis p. 490 |
| Aprosody p. 545 | Hydrocephalus pp. 258, 522 | Tay-Sachs disease p. 450 |
| Bell palsy p. 551 | Meningitis p. 520 | Trigeminal neuralgia p. 551 |

## AN AID TO MEMORY

Generations of biology and medical students have relied on mnemonic (memory-aiding) phrases and ditties, ranging from the sublimely silly to the unprintably ribald, to help them remember the cranial nerves and other anatomy. An old classic began, "On old Olympus' towering tops . . . ," with the first letter of each word matching the first letter of each cranial nerve (olfactory, optic, oculomotor, etc.). Some cranial nerves have changed names, however, since that passage was devised. One of the author's former students[†] devised a better mnemonic that can remind you of the first two to four letters of most cranial nerves:

| | |
|---|---|
| **Ol**d | **ol**factory (I) |
| **Op**ie | **op**tic (II) |
| **oc**casionally | **oc**ulomotor (III) |
| **tr**ies | **tr**ochlear (IV) |
| **trig**onometry | **trig**eminal (V) |
| **a**nd | **ab**ducens (VI) |
| **f**eels | **f**acial (VII) |
| **ve**ry | **ve**stibulocochlear (VIII) |
| **glo**omy, | **glo**ssopharyngeal (IX) |
| **vag**ue, | **vag**us (X) |
| **a**nd | **a**ccessory (XI) |
| **hypo**active | **hypo**glossal (XII) |

Another student's mnemonic, using only the first letter of each nerve's name, is "Oh, once one takes the anatomy final, very good vacation ahead."[‡] The first two letters of *ahead* represent nerves XI and XII.

Like a machine with a great number of moving parts, the nervous system is highly subject to malfunctions. Table 14.3 lists a few well-known brain and cranial nerve dysfunctions. The effects of aging on the CNS are described on page 1129.

### Before You Go On

*Answer the following questions to test your understanding of the preceding section:*

23. List the purely sensory cranial nerves and state the function of each.
24. What is the only cranial nerve to extend beyond the head-neck region?
25. If the oculomotor, trochlear, or abducens nerve were damaged, the effect would be similar in all three cases. What would that effect be?
26. Which cranial nerve carries sensory signals from the greatest area of the face?
27. Name two cranial nerves involved in the sense of taste and describe where their sensory fibers originate.

## INSIGHT 14.5    Clinical Application

### Images of the Mind

Enclosed as it is in the cranium, there is no easy way to observe a living brain directly. This has long frustrated neurobiologists, who once had to content themselves with glimpses of brain function afforded by electroencephalograms, patients with brain lesions, and patients who remained awake and conversant during brain surgery and consented to experimentation while the brain was exposed. New imaging methods, however, are yielding dramatic perspectives on brain function. Two of these—positron emission tomography (PET) and magnetic resonance imaging (MRI)—were explained in Insight 1.5 at the end of chapter 1. Both techniques rely on transient increases in blood flow to parts of the brain called into action to perform specific tasks. By monitoring these changes, neuroscientists can identify which parts of the brain are involved in specific tasks.

To produce a PET scan of the brain, the subject is given an injection of radioactively labeled glucose and a scan is made in a *control state* before any specific mental task is begun. Then the subject is given a task. For example, the subject may be instructed to read the word *car* and speak a verb related to it, such as *drive*. New PET scans are made in the *task state* while the subject performs this task. Neither control- nor task-state images are very revealing by themselves, but the computer subtracts the control-state data from the task-state data and presents a color-coded image of the difference. To compensate for chance events and individual variation, the computer also produces an image that is either averaged from several trials with one person or from trials with several different people.

In such averaged images, the busiest areas of the brain seem to "light up" from moment to moment as the task is performed (fig. 14.40). This identifies the regions used for various stages of the task, such as reading the word, thinking of a verb to go with it, planning to say *drive*, and actually saying it. Among other things, such experiments demonstrate that the Broca and Wernicke areas are not involved in simply repeating

words; they are active, however, when a subject must evaluate a word and choose an appropriate response—that is, they function in formulating the new word the subject is going to say. PET scans also show that different neuronal pools take over a task as we practice and become more proficient at it.

*Functional magnetic resonance imaging (fMRI)* depends on the role of astrocytes in brain metabolism. The main excitatory neurotransmitter secreted by cerebral neurons is glutamate. After a neuron releases glutamate and glutamate stimulates the next neuron, astrocytes quickly remove it from the synapse and convert it to glutamine. Astrocytes acquire the energy for this from the anaerobic fermentation of glucose. High activity in an area of cortex thus requires an increased blood flow to supply this glucose, but it does not elevate oxygen consumption from that blood. Thus, the oxygen supply exceeds demand in that part of the brain, and blood leaving the region contains more oxygen than the blood leaving less active regions. Since the magnetic properties of hemoglobin depend on how much oxygen is bound to it, fMRI can detect changes in brain circulation.

fMRI is more precise than PET and pinpoints regions of brain activity with a precision of 1 to 2 mm. It also has the advantages of requiring no injected substances and no exposure to radioisotopes. While it takes about 1 minute to produce a PET scan, fMRI produces images much more quickly, which makes it more useful for determining how the brain responds immediately to sensory input or mental tasks.

PET and fMRI scanning have enhanced our knowledge of neurobiology by identifying shifting patterns of brain activity associated with attention and consciousness, sensory perception, memory, emotion, motor control, reading, speaking, musical judgment, planning a chess strategy, and so forth. In addition to their contribution to basic neuroscience, these techniques have proven very valuable to neurosurgery and psychopharmacology. They also are enhancing our understanding of brain dysfunctions such as depression, schizophrenia, and attention deficit–hyperactivity disorder (ADHD). We have entered an exciting era in the safe visualization of normal brain function, producing pictures of the mind at work.

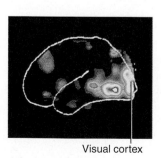

Primary auditory cortex

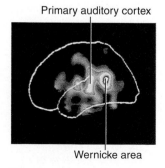

Premotor area

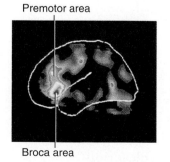

Primary motor cortex

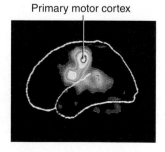

Visual cortex

Wernicke area

Broca area

① The word *car* is seen in the visual cortex.

② Wernicke area conceives of the verb *drive* to go with it.

③ Broca area compiles a motor program to speak the word *drive*.

④ The primary motor cortex executes the program and the word is spoken.

**FIGURE 14.40    PET Scans of the Brain Made During the Performance of a Language Task.**    These images show the cortical regions that are active when a person reads words and then speaks them. The most active areas are shown in red and the least active areas are shown in blue.

# CHAPTER REVIEW

# Review of Key Concepts

## Overview of the Brain (p. 514)

1. The adult brain weighs 1,450 to 1,600 g. It is divided into the *cerebrum, cerebellum,* and *brainstem.*

2. The cerebrum and cerebellum exhibit folds called *gyri* separated by grooves called *sulci.* The groove between the cerebral hemispheres is the *longitudinal fissure.*

3. The cerebrum and cerebellum have gray matter in their surface *cortex* and deeper *nuclei,* and white matter deep to the cortex.

4. Embryonic development of the brain progresses through *neural plate* and *neural tube* stages in the first 4 weeks. The anterior neural tube then begins to bulge and differentiate into forebrain, midbrain, and hindbrain. By the fifth week, the forebrain and hindbrain show further subdivision into two secondary vesicles each.

5. The five embryonic vesicles, in order from rostral to caudal, then differentiate into the mature cerebrum; diencephalon; midbrain; pons and cerebellum; and medulla oblongata.

## Meninges, Ventricles, Cerebrospinal Fluid, and Blood Supply (p. 519)

1. Like the spinal cord, the brain is surrounded by a dura mater, arachnoid mater, and pia mater. The dura mater is divided into two layers, *periosteal* and *meningeal,* which in some places are separated by a blood-filled *dural sinus.* In some places, a *subdural space* also separates the dura from the arachnoid.

2. The brain has four internal, interconnected cavities: two *lateral ventricles* in the cerebral hemispheres, a *third ventricle* between the hemispheres, and a *fourth ventricle* between the pons and cerebellum.

3. The ventricles and canals of the CNS are lined with ependymal cells, and each ventricle contains a *choroid plexus* of blood capillaries.

4. These spaces are filled with cerebrospinal fluid (CSF), which is produced by the ependyma and choroid plexuses and in the subarachnoid space around the brain. The CSF of the ventricles flows from the lateral to the third and then fourth ventricle, out through foramina in the fourth, into the subarachnoid space around the brain and spinal cord, and finally returns to the blood by way of arachnoid villi.

5. CSF provides buoyancy, physical protection, and chemical stability for the CNS.

6. The brain has a high demand for glucose and oxygen and thus receives a copious blood supply.

7. The *blood–brain barrier* and *blood–CSF barrier* tightly regulate what substances can escape the blood and reach the nervous tissue.

## The Hindbrain and Midbrain (p. 523)

1. The *medulla oblongata* is the most caudal part of the brain, just inside the foramen magnum. It conducts signals up and down the brainstem and between the brainstem and cerebellum, and contains nuclei involved in vasomotion, respiration, coughing, sneezing, salivation, swallowing, gagging, vomiting, gastrointestinal secretion, sweating, and control of tongue and head movements. Cranial nerves IX through XII arise from the medulla.

2. The *pons* is immediately rostral to the medulla. It conducts signals up and down the brainstem and between the brainstem and cerebellum, and contains nuclei involved in sleep, hearing, equilibrium, taste, eye movements, facial expression and sensation, respiration, swallowing, bladder control, and posture. Cranial nerve V arises from the pons, and nerves VI through VIII arise between the pons and medulla.

3. The *cerebellum* is the largest part of the hindbrain. It is composed of two hemispheres joined by a vermis, and has three pairs of *cerebellar peduncles* that attach it to the medulla, pons, and midbrain and carry signals between the brainstem and cerebellum.

4. Histologically, the cerebellum exhibits a fernlike pattern of white matter called the *arbor vitae, deep nuclei* of gray matter embedded in the white matter, and unusually large neurons called *Purkinje cells.*

5. The cerebellum is concerned largely with evaluation of sensory information. It performs roles in equilibrium, motor coordination, tactile sensation, spatial perception, timekeeping, language abilities, planning, and impulse control.

6. The *midbrain* is rostral to the pons. It conducts signals up and down the brainstem and between the brainstem and cerebellum, and contains nuclei involved in motor control, pain, visual attention, and auditory reflexes. It gives rise to cranial nerves III and IV.

7. The *reticular formation* is an elongated cluster of nuclei extending throughout the brainstem, including some of the nuclei already mentioned. It is involved in the control of skeletal muscles, the visual gaze, breathing, swallowing, cardiac and vasomotor control, pain, sleep, consciousness, and sensory awareness.

## The Forebrain (p. 529)

1. The forebrain consists of the diencephalon and cerebrum.

2. The *diencephalon* is composed of the thalamus, hypothalamus, and epithalamus.

3. The *thalamus* constitutes four-fifths of the diencephalon. It processes sensory information of all kinds and relays signals to specific regions of the cerebral cortex. It also lies within the feedback loops between the cerebrum and cerebellum, and between the cerebral cortex and basal nuclei, and participates in the memory and emotional functions of the limbic system.

4. The *hypothalamus* is inferior to the thalamus and forms the walls and floor of the third ventricle. It is a major homeostatic control center. It synthesizes some pituitary hormones and controls the timing of pituitary secretion, and it has nuclei concerned with heart rate, blood pressure, gastrointestinal secretion and motility, pupillary diameter, thermoregulation, hunger and thirst, sleep and circadian rhythms, memory, sexual function, and emotion.

5. The *epithalamus* lies above the thalamus and includes the pineal gland (an endocrine gland) and habenula (a relay from limbic system to midbrain).

6. The cerebrum is the largest part of the brain. It is divided into two hemispheres, and each hemisphere into five lobes: *frontal, parietal, occipital,* and *temporal lobes* and the *insula.*

7. Nerve fibers of the cerebral white matter are bundled in tracts of three kinds: *projection tracts* that extend between higher and lower brain centers; *commissural tracts* that cross between the right and left cerebral hemispheres through the *corpus callosum* and the *anterior* and *posterior commissures;* and *association tracts* that connect different lobes and gyri within a single hemisphere.

8. The cerebral cortex is gray matter with two types of neurons: stellate cells and pyramidal cells. All output from the cortex travels by way of axons of the pyramidal cells. Most of the cortex is *neocortex,* in which there are six layers of nervous tissue. Evolutionarily older parts of the cerebrum have one- to five-layered *paleocortex* and *archicortex.*

9. The *basal nuclei* are masses of cerebral gray matter lateral to the thalamus, concerned with motor control. They include the *caudate nucleus, putamen,* and *globus pallidus.*

10. The *limbic system* is a loop of specialized cerebral cortex on the medial border of the temporal lobe. Some of its parts are the *hippocampus, amygdala, fornix,* and *cingulate gyrus.* It is important in smell, emotion, and memory.

## Higher Forebrain Functions (p. 536)

1. The cerebral cortex generates *brain waves* that can be recorded as an *electroencephalogram (EEG).* Different types of brain waves (alpha, beta, theta, delta) predominate in various states of consciousness and certain brain disorders.

2. The cycle of sleep and waking is controlled by the *suprachiasmatic nucleus* of the hypothalamus and the reticular formation of the lower brainstem. Sleep progresses from stage 1 to stage 4 with characteristic changes in the EEG and other physiological values. Most dreaming occurs during a fifth type of sleep called *rapid eye movement (REM) sleep.*

3. *Cognition* (consciousness, thought, etc.) involves several *association areas* of the cerebral cortex, especially in the parietal, temporal, and frontal lobes.

4. The hippocampus of the limbic system processes information and organizes it into long-term memories *(memory consolidation).* These memories are then stored in other regions of the cerebral cortex, including the prefrontal cortex and the temporal lobe. The cerebellum is also involved in procedural memory (learning motor skills) and the amygdala in emotional memory.

5. The amygdala, hippocampus, and hypothalamus are important emotional centers of the brain, involved in such feelings as love, fear, anger, pleasure, and pain, and in learning to associate behaviors with reward and punishment.

6. *Somesthetic sensation* is controlled by the postcentral gyrus, where there is a point-for-point correspondence *(somatotopy)* with specific regions on the contralateral side of the body.

7. Special senses other than equilibrium are controlled by other areas of *primary sensory cortex:* taste in the insula and parietal lobe, smell in the temporal and frontal lobes, vision in the occipital lobe, and hearing in the temporal lobe and insula. Taste signals go with somesthetic senses to the postcentral gyrus and equilibrium signals to the cerebellum.

8. The primary sensory areas are surrounded with *sensory association areas* that process sensory input, relate it to memory, and identify the stimuli.

9. Motor control resides in the *motor association area* and *precentral gyrus* of the frontal lobe. The precentral gyrus shows a somatotopic correspondence with muscles on the contralateral side of the body.

10. The basal nuclei and cerebellum play important roles in motor coordination and the conduct of learned motor skills.

11. Language is coordinated largely by the Wernicke and Broca areas. Recognizing language and formulating what one will say or write occur in the Wernicke area; compiling the motor program of speech resides in the Broca area; and commands to the muscles of speech originate in the precentral gyrus.

12. The brain exhibits *cerebral lateralization;* some functions are coordinated mainly by the left hemisphere and others by the right. The *categorical hemisphere* (in most people, the left) is responsible for verbal and mathematical skills and logical, linear thinking. The *representational hemisphere* (usually the right) is a seat of imagination, insight, spatial perception, musical skill, and other "holistic" functions.

## The Cranial Nerves (p. 547)

1. Twelve pairs of *cranial nerves* arise from the floor of the brain, pass through foramina of the skull, and lead primarily to structures in the head and neck.

2. Cranial nerves I and II are purely sensory. All the rest are mixed, although the sensory components of some are only proprioceptive and aid in motor control, so they are often classified as motor nerves (III, IV, VI, XI, and XII).

3. The olfactory nerve (I) carries the sensory signals for smell.

4. The optic nerve (II) carries the sensory signals for vision.

5. The oculomotor nerve (III) carries motor signals that control eye movements, opening of the eyelids, pupillary constriction, and focusing of the lens.

6. The trochlear nerve (IV) carries motor signals for eye movement.

7. The trigeminal nerve (V) is a large nerve with three branches (ophthalmic, maxillary, and mandibular). It is the main sensory nerve of the face and it carries motor signals for mastication.

8. The abducens nerve (VI) carries motor signals for eye movement. Cranial nerves III, IV, and VI control different muscles of the eye.

9. The facial nerve (VII) carries sensory signals for taste and motor signals that control facial expression and the secretion of saliva, tears, and nasal and oral mucus.

10. The vestibulocochlear nerve (VIII) carries sensory signals of hearing and equilibrium and carries motor signals to the inner ear for tuning the sense of hearing.

11. The glossopharyngeal nerve (IX) carries sensory signals for taste, touch, pressure, pain, and temperature sensations from the tongue; for touch, pain, and temperature sensations from the outer ear; and for regulation of blood pressure and respiration. It also carries motor signals for salivation, swallowing, and gagging.

12. The vagus nerve (X) has the most extensive distribution of all cranial nerves, with branches leading not only to organs in the head and neck but also to most viscera of the thoracic and abdominal cavities. It carries sensory signals for taste, hunger, fullness, and gastrointestinal discomfort, and motor signals for swallowing, speech, and regulation of pulmonary, cardiovascular, and gastrointestinal functions.

13. The accessory nerve (XI) carries motor signals for swallowing and for head, neck, and shoulder movements.

14. The hypoglossal nerve (XII) carries motor signals for the tongue movements of speech, food manipulation, and swallowing.

# Testing Your Recall

1. Which of these is caudal to the hypothalamus?
   a. the thalamus
   b. the optic chiasm
   c. the cerebral aqueduct
   d. the pituitary gland
   e. the corpus callosum

2. If the telencephalon were removed from a 5-week-old embryo, which of the following structures would fail to develop in the fetus?
   a. cerebral hemispheres
   b. the thalamus
   c. the midbrain
   d. the medulla oblongata
   e. the spinal cord

3. The blood–CSF barrier is formed by
   a. blood capillaries.
   b. endothelial cells.
   c. protoplasmic astrocytes.
   d. oligodendrocytes.
   e. ependymal cells.

4. The pyramids of the medulla oblongata contain
   a. descending corticospinal fibers.
   b. commissural fibers.
   c. ascending spinocerebellar fibers.
   d. fibers going to and from the cerebellum.
   e. ascending spinothalamic fibers.

5. Which of the following does *not* receive any input from the eyes?
   a. the hypothalamus
   b. the frontal lobe
   c. the thalamus
   d. the occipital lobe
   e. the midbrain

6. While studying in a noisy cafeteria, you get sleepy and doze off for a few minutes. You awaken with a start and realize that all the cafeteria sounds have just "come back." While you were dozing, this auditory input was blocked from reaching your auditory cortex by
   a. the temporal lobe.
   b. the thalamus.
   c. the reticular activating system.
   d. the medulla oblongata.
   e. the vestibulocochlear nerve.

7. Because of a brain lesion, a certain patient never feels full, but eats so excessively that she now weighs nearly 270 kg (600 lb). The lesion is most likely in her
   a. hypothalamus.
   b. amygdala.
   c. hippocampus.
   d. basal nuclei.
   e. pons.

8. The _____ is most closely associated with the cerebellum in embryonic development and remains its primary source of input fibers throughout life.
   a. telencephalon
   b. thalamus
   c. midbrain
   d. pons
   e. medulla

9. Damage to the _____ nerve could result in defects of eye movement.
   a. optic
   b. vagus
   c. trigeminal
   d. facial
   e. abducens

10. All of the following *except* the _____ nerve begin or end in the orbit.
    a. optic
    b. oculomotor
    c. trochlear
    d. abducens
    e. accessory

11. The right and left cerebral hemispheres are connected to each other by a thick C-shaped bundle of fibers called the _____.

12. The brain has four chambers called _____ filled with _____ fluid.

13. On a sagittal plane, the cerebellar white matter exhibits a branching pattern called the _____.

14. Abnormal accumulation of cerebrospinal fluid in the ventricles can cause a condition called _____.

15. Cerebrospinal fluid is secreted partly by a mass of blood capillaries called the _____ in each ventricle.

16. The primary motor area of the cerebrum is the _____ gyrus of the frontal lobe.

17. Your personality is determined mainly by which lobe of the cerebrum?

18. Areas of cerebral cortex that identify or interpret sensory information are called _____.

19. Linear, analytical, and verbal thinking occurs in the _____ hemisphere of the cerebrum, which is on the left in most people.

20. The motor pattern for speech is generated in an area of cortex called _____ and then transmitted to the primary motor cortex to be carried out.

*Answers in Appendix B*

## True or False

*Determine which five of the following statements are false, and briefly explain why.*

1. The two hemispheres of the cerebellum are separated by the longitudinal fissure.

2. The cerebral hemispheres would fail to develop if the neural crests of the embryo were destroyed.

3. The midbrain is caudal to the thalamus.

4. The Broca area is ipsilateral to the Wernicke area.

5. Most of the cerebrospinal fluid is produced by the choroid plexuses.

6. Hearing is a function of the occipital lobe.

7. Respiration is controlled by nuclei in both the pons and medulla oblongata.

8. The trigeminal nerve carries sensory signals from a larger area of the face than the facial nerve does.

9. Unlike other cranial nerves, the vagus nerve extends far beyond the head-neck region.

10. The optic nerve controls movements of the eye.

*Answers in Appendix B*

## Testing Your Comprehension

1. Which cranial nerve conveys pain signals to the brain in each of the following situations: (*a*) sand blows into your eye; (*b*) you bite the rear of your tongue; and (*c*) your stomach hurts from eating too much?

2. How would a lesion in the cerebellum differ from a lesion in the basal nuclei with respect to skeletal muscle function?

3. Suppose that a neuroanatomist performed two experiments on an animal with the same basic spinal brainstem structure as a human's: In experiment 1, he selectively transected (cut across) the pyramids on the ventral side of the medulla oblongata, and in experiment 2, he selectively transected the gracile and cuneate fasciculi on the dorsal side. How would the outcomes of the two experiments differ?

4. A person can survive destruction of an entire cerebral hemisphere but cannot survive destruction of the hypothalamus, which is a much smaller mass of brain tissue. Explain this difference and describe some ways that destruction of a cerebral hemisphere would affect one's quality of life.

5. What would be the most obvious effects of lesions that destroyed each of the following: (*a*) the hippocampus, (*b*) the amygdala, (*c*) the Broca area, (*d*) the occipital lobe, and (*e*) the hypoglossal nerve?

*Answers at www.mhhe.com/saladin4*

# www.mhhe.com/saladin4

*The textbook website provides a wealth of interactive study materials fully organized and integrated by chapter. You will find practice quizzes, labeling exercises, and much more that will complement your learning and understanding of anatomy and physiology. The website also includes tools designed to enhance your* **Anatomy & Physiology | REVEALED** *experience.*

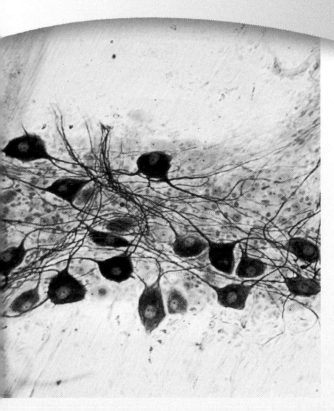

*Autonomic neurons in the myenteric plexus of the digestive tract*

# THE AUTONOMIC NERVOUS SYSTEM AND VISCERAL REFLEXES

## CHAPTER OUTLINE

## Brushing Up

To understand this chapter, it is important that you understand or brush up on the following concepts:

- Innervation of smooth muscle (p. 433)
- Neurotransmitters and synaptic transmission (pp. 462–466)
- Spinal nerves (p. 493)
- The hypothalamus and limbic system (pp. 530, 534)
- Cranial nerves (p. 547)

We have studied the somatic nervous system and somatic reflexes, and we now turn to the *autonomic nervous system (ANS)* and visceral reflexes—reflexes that regulate such primitive functions as blood pressure, heart rate, body temperature, digestion, energy metabolism, respiratory airflow, pupillary diameter, defecation, and urination. In short, the ANS quietly manages a myriad of unconscious processes responsible for the body's homeostasis. Not surprisingly, many drug therapies are based on alteration of autonomic function; some examples are discussed at the end of this chapter.

Harvard Medical School physiologist Walter Cannon, who coined such expressions as *homeostasis* and the *fight or flight* reaction, dedicated his career to the physiology of the autonomic nervous system. Cannon found that an animal can live without a functional sympathetic nervous system, but it must be kept warm and free of stress; it cannot survive on its own or tolerate any strenuous exertion. The autonomic nervous system is more necessary to survival than many functions of the somatic nervous system; an absence of autonomic function is fatal because the body cannot maintain homeostasis. We are seldom aware of what our autonomic nervous system is doing, much less able to control it; indeed, it is difficult to consciously alter or suppress autonomic responses, and for this reason they are the basis for polygraph ("lie detector") tests. Nevertheless, for an understanding of bodily function and health care, we must be well aware of how this system works.

# General Properties of the Autonomic Nervous System

### Objectives

When you have completed this section, you should be able to

- explain how the autonomic and somatic nervous systems differ in form and function; and
- explain how the two divisions of the autonomic nervous system differ in general function.

The **autonomic nervous system (ANS)** can be defined as a motor nervous system that controls glands, cardiac muscle, and smooth muscle. It is also called the **visceral motor system** to distinguish it from the somatic motor system that controls the skeletal muscles. The primary target organs of the ANS are the viscera of the thoracic and abdominal cavities and some structures of the body wall, including cutaneous blood vessels, sweat glands, and piloerector muscles.

*Autonomic* literally means "self-governed."[1] The ANS usually carries out its actions involuntarily, without our conscious intent or awareness, in contrast to the voluntary nature of the somatic motor system. This voluntary-involuntary distinction is not, however, as clear-cut as it once seemed. Some skeletal muscle responses are quite involuntary, such as the somatic reflexes, and some skeletal muscles are difficult or impossible to control, such as the middle-ear muscles. On the other hand, therapeutic uses of biofeedback show that some people can learn to voluntarily control such visceral functions as blood pressure (see Insight 15.1).

Visceral effectors do not depend on the autonomic nervous system to function, but only to adjust their activity to the body's changing needs. The heart, for example, goes on beating even if all autonomic nerves to it are severed, but the ANS modulates the heart rate in conditions of rest or exercise. If the somatic nerves to a skeletal muscle are severed, the muscle exhibits flaccid paralysis—it no longer functions. But if the autonomic nerves to cardiac or smooth muscle are severed, the muscle exhibits exaggerated responses *(denervation hypersensitivity)*.

---

**INSIGHT 15.1**    Clinical Application

### Biofeedback

*Biofeedback* is a technique in which an instrument produces auditory or visual signals in response to changes in a subject's blood pressure, heart rate, muscle tone, skin temperature, brain waves, or other physiological variables. It gives the subject awareness of changes that he or she would not ordinarily notice. Some people can be trained to control these variables in order to produce a certain acoustic tone or color of light from the apparatus. Eventually they can control blood pressure or other autonomic functions without the aid of the monitor. Biofeedback is not a quick, easy, infallible, or inexpensive cure for all ills, but it has been used successfully to treat hypertension, stress, and migraine headaches.

---

## VISCERAL REFLEXES

The ANS is responsible for the body's **visceral reflexes**—unconscious, automatic, stereotyped responses to stimulation, much like the somatic reflexes discussed in chapter 13, but involving visceral receptors and effectors and somewhat slower responses. Some authorities regard

---

[1]*auto* = self + *nom* = rule

the visceral afferent (sensory) pathways as part of the ANS, while most prefer to limit the term *ANS* to the efferent (motor) pathways. Regardless of this preference, however, autonomic activity involves a visceral reflex arc that includes receptors (nerve endings that detect stretch, tissue damage, blood chemicals, body temperature, and other internal stimuli), afferent neurons leading to the CNS, interneurons in the CNS, efferent neurons carrying motor signals away from the CNS, and finally effectors.

For example, high blood pressure activates a visceral *baroreflex.*[2] It stimulates stretch receptors called *baroreceptors* in the carotid arteries and aorta, and they transmit signals via the glossopharyngeal nerves to the medulla oblongata (fig. 15.1). The medulla integrates this input with other information and transmits efferent signals back to the heart by way of the vagus nerves. The vagus nerves slow down the heart and reduce blood pressure, thus completing a homeostatic negative feedback loop. A separate autonomic reflex arc accelerates the heart when blood pressure drops below normal.

## DIVISIONS OF THE AUTONOMIC NERVOUS SYSTEM

The ANS has two subsystems: the sympathetic and parasympathetic divisions. These divisions differ in anatomy and function, but they often innervate the same target organs and may have cooperative or contrasting effects on them. The **sympathetic division** adapts the body in many ways for physical activity—it increases alertness, heart rate, blood pressure, pulmonary airflow, blood glucose concentration, and blood flow to cardiac and skeletal muscle, but at the same time, it reduces blood flow to the skin and digestive tract. Cannon referred to extreme sympathetic responses as the "fight or flight" reaction because they come into play when an animal must attack, defend itself, or flee from danger. In our own lives, this reaction occurs in many situations involving arousal, competition, stress, danger, anger, or fear. Ordinarily, however, the sympathetic division has more subtle effects that we notice barely, if at all. The **parasympathetic division,** by comparison, has a calming effect on many body functions. It is associated with reduced energy expenditure and normal bodily maintenance, including such functions as digestion and waste elimination. This can be thought of as the "resting and digesting" state.

This does not mean that the body alternates between states where one system or the other is active. Normally both systems are active simultaneously. They exhibit a background rate of activity called **autonomic tone,** and the balance between *sympathetic tone* and *parasympathetic tone* shifts in accordance with the body's changing needs. Parasympathetic tone, for example, maintains

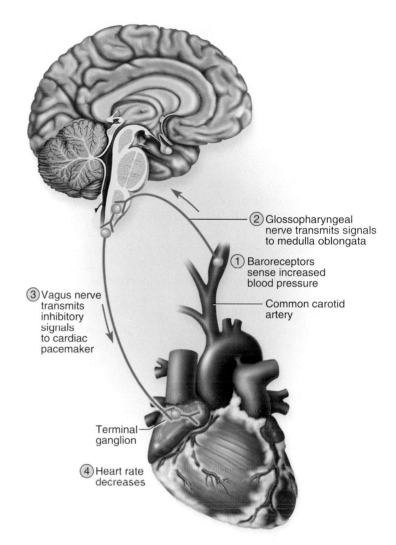

**FIGURE 15.1 An Autonomic Reflex Arc in the Regulation of Blood Pressure.** In this example, a rise in blood pressure is detected by baroreceptors in the carotid artery. The glossopharyngeal nerve transmits signals to the medulla oblongata, resulting in parasympathetic output from the vagus nerve that reduces the heart rate and lowers blood pressure.

smooth muscle tone in the intestines and holds the resting heart rate down to about 70 to 80 beats/minute. If the parasympathetic vagus nerves to the heart are cut, the heart beats at its own intrinsic rate of about 100 beats/min. Sympathetic tone keeps most blood vessels partially constricted and thus maintains blood pressure. A loss of sympathetic tone can cause such a rapid drop in blood pressure that a person goes into shock.

Neither division has universally excitatory or calming effects. The sympathetic division, for example, excites the heart but inhibits digestive and urinary functions, while the parasympathetic division has the opposite effects. We will later examine how differences in neurotransmitters and their receptors account for these differences of effect.

---

[2]*baro* = pressure

# NEURAL PATHWAYS

The ANS has components in both the central and peripheral nervous systems. It includes control nuclei in the hypothalamus and other regions of the brainstem, motor neurons in the spinal cord and peripheral ganglia, and nerve fibers that travel through the cranial and spinal nerves you have already studied.

The autonomic motor pathway to a target organ differs significantly from somatic motor pathways. In somatic pathways, a motor neuron in the brainstem or spinal cord issues a myelinated axon that reaches all the way to a skeletal muscle. In autonomic pathways, the signal must travel across two neurons to get to the target organ, and it must cross a synapse where these two neurons meet in an autonomic ganglion (fig. 15.2). The first neuron, called the **preganglionic neuron,** has a soma in the brainstem or spinal cord; its axon terminates in the ganglion. It synapses there with a **postganglionic neuron** whose axon extends the rest of the way to the target cells. (Some call this cell the *ganglionic neuron* since its soma is in the ganglion and only its axon is truly postganglionic.) The axons of these neurons are called the *pre-* and *postganglionic fibers.*

In summary, the autonomic nervous system is a division of the nervous system responsible for homeostasis, acting through the mostly unconscious and involuntary control of glands, smooth muscle, and cardiac muscle. Its target organs are mostly the thoracic and abdominal viscera, but also include some cutaneous and other effectors. It acts through motor pathways that involve two neurons, preganglionic and postganglionic, reaching from CNS to effector. The ANS has two divisions, sympathetic and parasympathetic, that often have cooperative or contrasting effects on the same target organ. Both divisions have excitatory effects on some target cells and inhibitory effects on others. These and other differences between the somatic and autonomic nervous systems are summarized in table 15.1.

## Before You Go On

*Answer the following questions to test your understanding of the preceding section:*

1. *How does the autonomic nervous system differ from the somatic motor system?*

2. *How do the general effects of the sympathetic division differ from those of the parasympathetic division?*

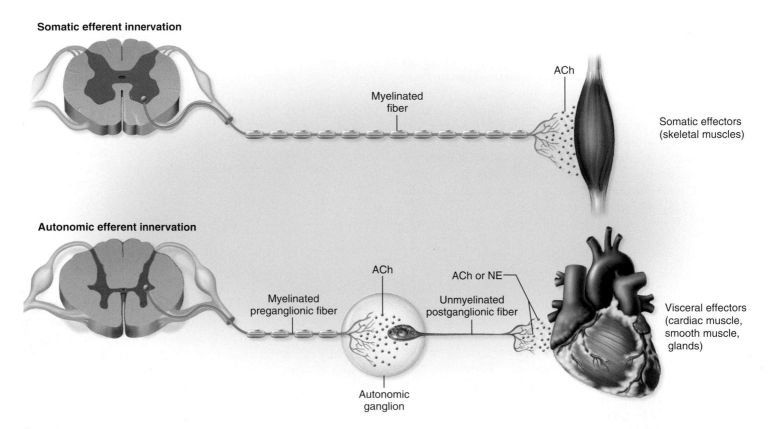

**Somatic efferent innervation**

Myelinated fiber

ACh

Somatic effectors (skeletal muscles)

**Autonomic efferent innervation**

Myelinated preganglionic fiber

ACh

ACh or NE

Unmyelinated postganglionic fiber

Visceral effectors (cardiac muscle, smooth muscle, glands)

Autonomic ganglion

**FIGURE 15.2 Comparison of Somatic and Autonomic Efferent Pathways.** The entire distance from CNS to effector is spanned by one neuron in the somatic system and two neurons in the autonomic system. Only acetylcholine (ACh) is employed as a neurotransmitter by the somatic neuron and the autonomic preganglionic neuron, but autonomic postganglionic neurons can employ either ACh or norepinephrine (NE).

| TABLE 15.1 | Comparison of the Somatic and Autonomic Nervous Systems | |
|---|---|---|
| Feature | Somatic | Autonomic |
| Effectors | Skeletal muscle | Glands, smooth muscle, cardiac muscle |
| Efferent pathways | One nerve fiber from CNS to effector; no ganglia | Two nerve fibers from CNS to effector; synapse at a ganglion |
| Neurotransmitters | Acetylcholine (ACh) | ACh and norepinephrine (NE) |
| Effect on target cells | Always excitatory | Excitatory or inhibitory |
| Effect of denervation | Flaccid paralysis | Denervation hypersensitivity |
| Control | Usually voluntary | Usually involuntary |

# Anatomy of the Autonomic Nervous System

### Objectives

When you have completed this section, you should be able to

- identify the anatomical components and nerve pathways of the sympathetic and parasympathetic divisions; and
- discuss the relationship of the adrenal glands to the sympathetic nervous system.

## THE SYMPATHETIC DIVISION

The sympathetic division is also called the *thoracolumbar division* because it arises from the thoracic and lumbar regions of the spinal cord. It has relatively short preganglionic and long postganglionic fibers. The preganglionic somas are in the lateral horns and nearby regions of the gray matter of the spinal cord. Their fibers exit by way of spinal nerves T1 to L2 and lead to the nearby **sympathetic chain** of ganglia (paravertebral[3] ganglia) along each side of the vertebral column (figs. 15.3 and 15.4). Although these chains receive input from only the thoracolumbar region of the cord, they extend into the cervical and sacral to coccygeal regions as well. Some nerve fibers entering the chain at levels T1 to L2 travel up or down the chain to reach these cervical and sacral ganglia. The number of ganglia varies from person to person, but usually there are 3 cervical (*superior, middle,* and *inferior*), 11 thoracic, 4 lumbar, 4 sacral, and 1 coccygeal ganglion in each chain.

In the thoracolumbar region, each paravertebral ganglion is connected to a spinal nerve by two branches called

[3]*para* = next to + *vertebr* = vertebral column

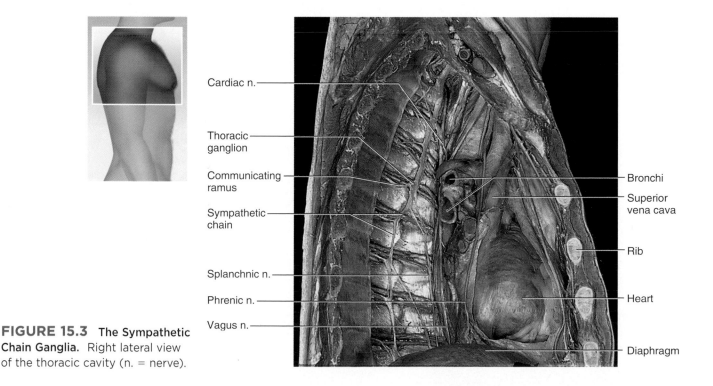

**FIGURE 15.3  The Sympathetic Chain Ganglia.** Right lateral view of the thoracic cavity (n. = nerve).

Cardiac n. — Thoracic ganglion — Communicating ramus — Sympathetic chain — Splanchnic n. — Phrenic n. — Vagus n. — Bronchi — Superior vena cava — Rib — Heart — Diaphragm

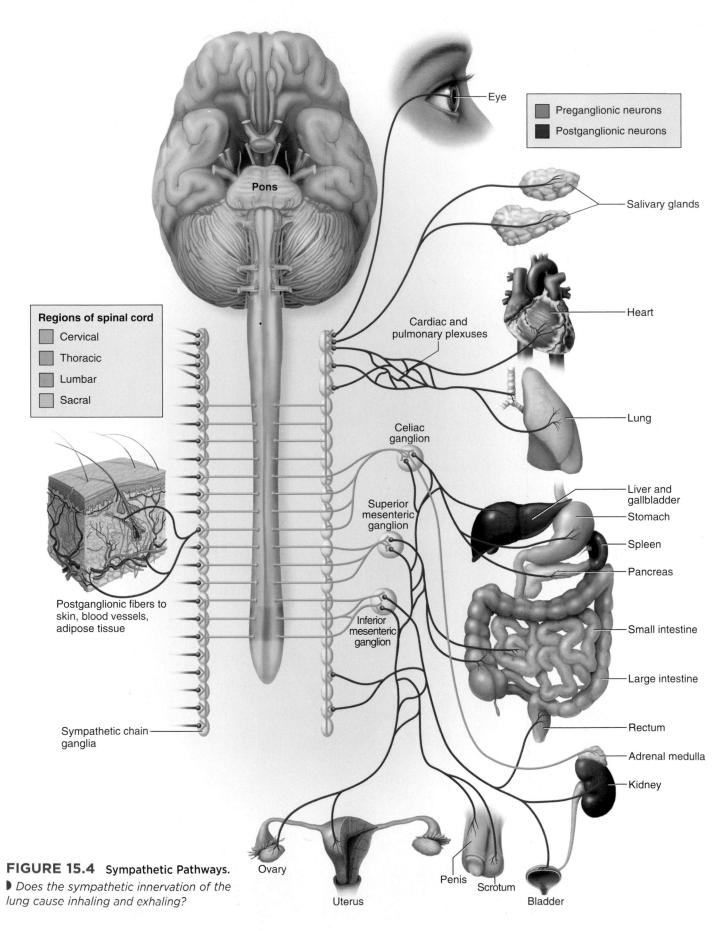

**Regions of spinal cord**
- Cervical
- Thoracic
- Lumbar
- Sacral

Pons

Eye

Preganglionic neurons
Postganglionic neurons

Salivary glands

Cardiac and pulmonary plexuses

Heart

Lung

Celiac ganglion

Liver and gallbladder

Stomach

Superior mesenteric ganglion

Spleen

Pancreas

Postganglionic fibers to skin, blood vessels, adipose tissue

Small intestine

Inferior mesenteric ganglion

Large intestine

Rectum

Adrenal medulla

Kidney

Sympathetic chain ganglia

Ovary

Uterus

Penis

Scrotum

Bladder

**FIGURE 15.4** **Sympathetic Pathways.**

▶ *Does the sympathetic innervation of the lung cause inhaling and exhaling?*

*communicating rami* (fig. 15.5). The preganglionic fibers are small myelinated fibers that travel from the spinal nerve to the ganglion by way of the **white communicating ramus,**[4] which gets its color and name from the myelin. Unmyelinated postganglionic fibers leave the ganglion by way of the **gray communicating ramus,** named for its lack of myelin and duller color, and by other routes. These long fibers extend the rest of the way to the target organ.

> **Think About It**
>
> *Would autonomic postganglionic fibers have faster or slower conduction speeds than somatic motor fibers? Why? (See hints in chapter 12.)*

[4]*ramus* = branch

After entering the sympathetic chain, preganglionic fibers may follow any of three courses:

- Some end in the ganglion that they enter and synapse immediately with a postganglionic neuron.

- Some travel up or down the chain and synapse in ganglia at other levels. It is these fibers that link the paravertebral ganglia into a chain. They are the only route by which ganglia at the cervical, sacral, and coccygeal levels receive input.

- Some pass through the chain without synapsing and continue as *splanchnic* (SPLANK-nic) *nerves,* to be considered shortly.

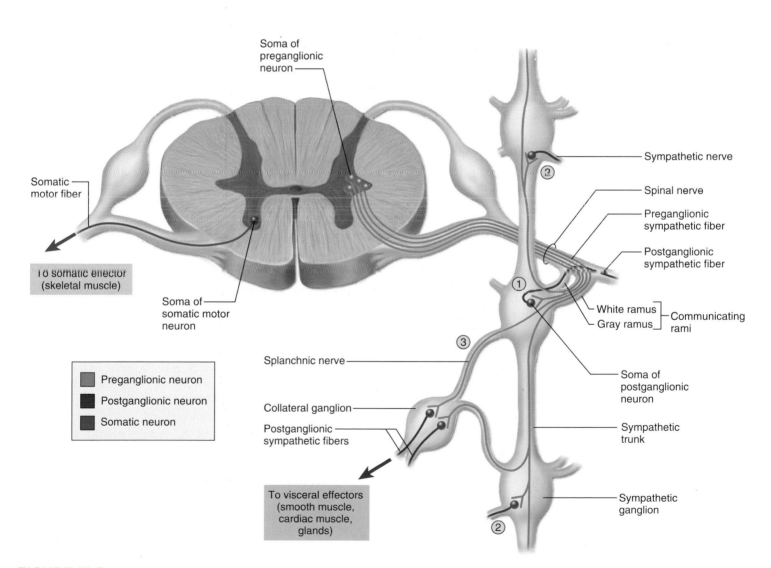

**FIGURE 15.5** **Sympathetic Pathways (right) Compared with the Somatic Efferent Pathway (left).** Sympathetic fibers can follow any of the three numbered routes: (1) the spinal nerve route, (2) the sympathetic nerve route, or (3) the splanchnic nerve route.

▶ *Name the parts of the spinal cord where the somas of the sympathetic and somatic efferent neurons are located.*

There is no simple one-to-one relationship between preganglionic and postganglionic neurons in the sympathetic division. For one thing, each postganglionic cell may receive synapses from multiple preganglionic cells, thus exhibiting the principle of *neuronal convergence* discussed in chapter 12. Furthermore, each preganglionic fiber branches and synapses with multiple postganglionic fibers, thus showing *neuronal divergence.* There are about 17 postganglionic neurons for every preganglionic neuron in the sympathetic division. This means that when one preganglionic neuron fires, it can excite multiple postganglionic fibers leading to different target organs. The sympathetic division thus tends to have relatively widespread effects—as suggested by the name *sympathetic.*[5]

Nerve fibers leave the sympathetic chain by three routes: spinal, sympathetic, and splanchnic nerves. These are numbered in figure 15.5 to correspond to the following descriptions:

① **The spinal nerve route.** Some postganglionic fibers exit a ganglion by way of the gray ramus, return to the spinal nerve or its subdivisions, and travel the rest of the way to the target organ. This is the route to most sweat glands, piloerector muscles, and blood vessels of the skin and skeletal muscles.

② **The sympathetic nerve route.** Other postganglionic fibers leave by way of **sympathetic nerves** that extend to the heart, lungs, esophagus, and thoracic blood vessels. These nerves form a plexus around each carotid artery of the neck and issue fibers from there to effectors in the head—including sweat, salivary, and nasal glands; piloerector muscles; blood vessels; and dilators of the iris. Some fibers from the superior and middle cervical ganglia form the *cardiac nerves* to the heart. (The cardiac nerves also contain parasympathetic fibers.)

③ **The splanchnic[6] nerve route.** Some of the fibers that arise from spinal nerves T5 to T12 pass through the sympathetic ganglia without synapsing. Beyond the ganglia, they continue as **splanchnic nerves,** which lead to a second set of ganglia called **collateral (prevertebral) ganglia.** Here the preganglionic fibers synapse with the postganglionics.

The collateral ganglia contribute to a network called the **abdominal aortic plexus** wrapped around the aorta (fig. 15.6). There are three major collateral ganglia in this plexus—the **celiac, superior mesenteric,** and **inferior mesenteric ganglion**—located at points where arteries of the same names branch off the aorta. The postganglionic fibers accompany these arteries and their branches to the target organs. Table 15.2 summarizes the innervation to and from the three major collateral ganglia.

The term *solar plexus* is used by some authorities as a collective name for the celiac and superior mesenteric ganglia, and by others as a synonym for the celiac ganglion only. The term comes from the nerves radiating from the ganglion like rays of the sun.

In summary, effectors in the muscles and body wall are innervated mainly by sympathetic fibers in the spinal nerves; effectors in the head and thoracic cavity by sympathetic nerves; and effectors in the abdominal cavity by splanchnic nerves.

## THE ADRENAL GLANDS

The paired **adrenal[7] glands** rest like hats on the superior pole of each kidney (fig. 15.6). Each adrenal is actually two glands with different functions and embryonic origins. The outer rind, the **adrenal cortex,** secretes steroid hormones discussed in chapter 17. The inner core, the **adrenal medulla,** is essentially a sympathetic ganglion. It consists of modified postganglionic neurons without dendrites or axons. Sympathetic preganglionic fibers penetrate through the cortex and terminate on these cells. The sympathetic nervous system and adrenal medulla are so closely related in development and function that they are referred to collectively as the *sympathoadrenal system.*

When stimulated, the adrenal medulla secretes a mixture of hormones into the bloodstream—about 85% epinephrine (adrenaline), 15% norepinephrine (noradrenaline), and a trace of dopamine. These hormones, the *catecholamines,* were briefly considered in chapter 12 because they also function as neurotransmitters.

## THE PARASYMPATHETIC DIVISION

The parasympathetic division is also called the *craniosacral division* because it arises from the brain and sacral region of the spinal cord; its fibers travel in certain cranial and sacral nerves. Somas of the preganglionic neurons are located in the midbrain, pons, medulla oblongata, and segments S2 to S4 of the spinal cord (fig. 15.7). They issue long preganglionic fibers which end in **terminal ganglia** in or near the target organ (see fig. 15.1). (If a terminal ganglion is embedded within the wall of a target organ, it is also called an *intramural[8] ganglion.*) Thus, the parasympathetic division has long preganglionic fibers, reaching almost all the way to the target cells, and short postganglionic fibers that cover the rest of the distance.

There is some neuronal divergence in the parasympathetic division, but much less than in the sympathetic. The parasympathetic division has a ratio of about two

---

[5]*sym* = together + *path* = feeling
[6]*splanchn* = viscera

[7]*ad* = near + *ren* = kidney
[8]*intra* = within + *mur* = wall

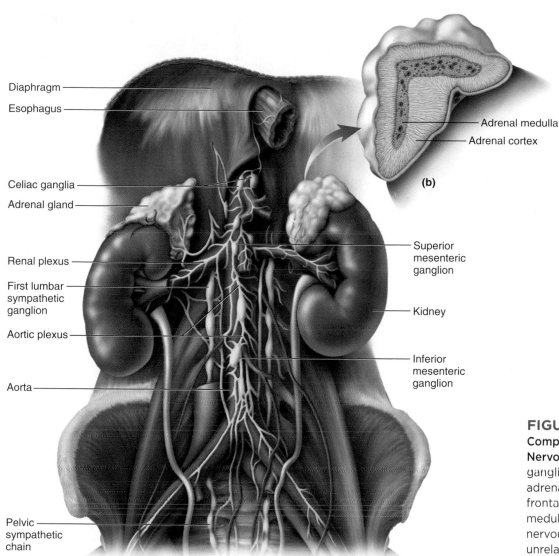

Diaphragm

Esophagus

Adrenal medulla

Adrenal cortex

**(b)**

Celiac ganglia

Adrenal gland

Renal plexus

First lumbar sympathetic ganglion

Aortic plexus

Aorta

Pelvic sympathetic chain

Superior mesenteric ganglion

Kidney

Inferior mesenteric ganglion

**FIGURE 15.6  Abdominal Components of the Sympathetic Nervous System.**  (a) The collateral ganglia, abdominal aortic plexus, and adrenal glands. (b) An adrenal gland, frontal section. Only the adrenal medulla plays a role in the sympathetic nervous system; the adrenal cortex has unrelated roles described in chapter 17.

**(a)**

| TABLE 15.2 | Innervation to and from the Collateral Ganglia | | | | |
|---|---|---|---|---|---|
| **Sympathetic Ganglia** | $\longrightarrow$ | **Collateral Ganglion** | $\longrightarrow$ | **Postganglionic Target Organs** | |
| Thoracic ganglia 5 to 9 or 10 | $\longrightarrow$ | Celiac ganglion | $\longrightarrow$ | Stomach, spleen, liver, small intestine, and kidneys | |
| Thoracic ganglia 9 and 10 | $\longrightarrow$ | Celiac and superior mesenteric ganglia | $\longrightarrow$ | Small intestine and colon | |
| Lumbar ganglia | $\longrightarrow$ | Inferior mesenteric ganglia | $\longrightarrow$ | Rectum, urinary bladder, and reproductive organs | |

postganglionic fibers to every preganglionic. Furthermore, the preganglionic fiber reaches the target organ before even this slight divergence occurs. The parasympathetic division is therefore relatively selective in its stimulation of target organs.

Parasympathetic fibers leave the brainstem by way of the following four cranial nerves. The first three supply all parasympathetic innervation to the head, and the last one supplies viscera of the thoracic and abdominal cavities.

1. **Oculomotor nerve (III).** The oculomotor nerve carries parasympathetic fibers that control the lens and pupil of the eye. The preganglionic fibers enter the orbit and terminate in the *ciliary ganglion* behind the eyeball. Postganglionic fibers enter the eyeball and innervate the *ciliary muscle,* which thickens the lens, and the *pupillary constrictor,* which narrows the pupil.

2. **Facial nerve (VII).** The facial nerve carries parasympathetic fibers that regulate the tear glands, salivary

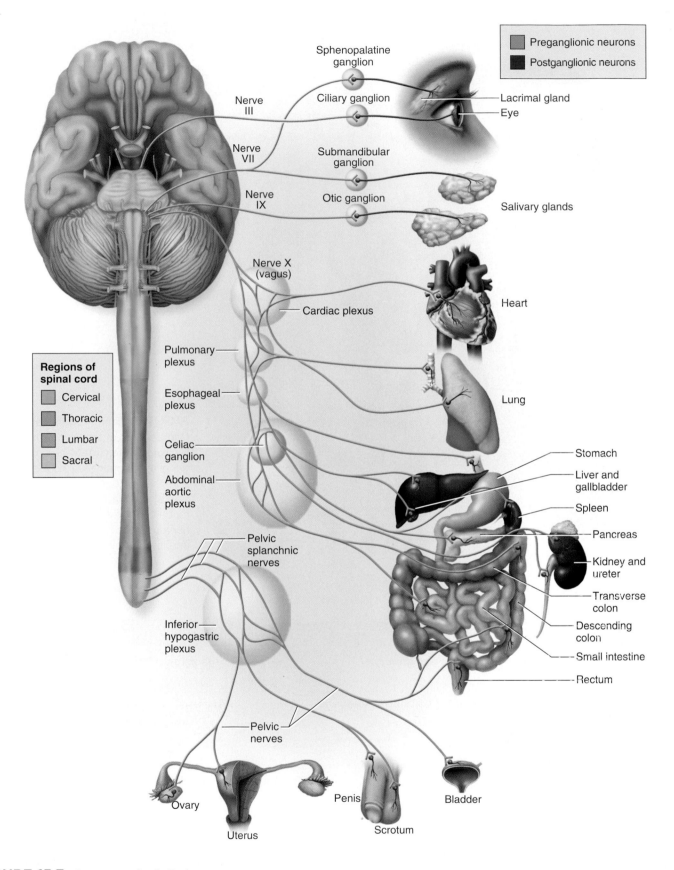

Sphenopalatine ganglion

Ciliary ganglion

Nerve III

Nerve VII

Submandibular ganglion

Nerve IX

Otic ganglion

Nerve X (vagus)

Cardiac plexus

Pulmonary plexus

Esophageal plexus

Celiac ganglion

Abdominal aortic plexus

Pelvic splanchnic nerves

Inferior hypogastric plexus

Pelvic nerves

Ovary

Uterus

Penis

Scrotum

Bladder

Lacrimal gland

Eye

Salivary glands

Heart

Lung

Stomach

Liver and gallbladder

Spleen

Pancreas

Kidney and ureter

Transverse colon

Descending colon

Small intestine

Rectum

**Preganglionic neurons**

**Postganglionic neurons**

**Regions of spinal cord**

Cervical

Thoracic

Lumbar

Sacral

**FIGURE 15.7    Parasympathetic Pathways.**

▶ *Which nerve carries the most parasympathetic nerve fibers?*

glands, and nasal glands. Soon after the facial nerve emerges from the pons, its parasympathetic fibers split away and form two smaller branches. The upper branch ends at the *sphenopalatine ganglion* near the junction of the maxilla and palatine bone. Postganglionic fibers then continue to the tear glands and glands of the nasal cavity, palate, and other areas of the oral cavity. The lower branch crosses the middle-ear cavity and ends at the *submandibular ganglion* near the angle of the mandible. Postganglionic fibers from here supply salivary glands in the floor of the mouth.

3. **Glossopharyngeal nerve (IX).** The glossopharyngeal nerve carries parasympathetic fibers concerned with salivation. The preganglionic fibers leave this nerve soon after its origin and form the *tympanic nerve.* A continuation of this nerve crosses the middle-ear cavity and ends in the *otic*[9] *ganglion* near the foramen ovale. The postganglionic fibers then follow the trigeminal nerve to the *parotid salivary gland* just in front of the earlobe.

4 **Vagus nerve (X).** The vagus nerve carries about 90% of all parasympathetic preganglionic fibers. It travels down the neck and forms three networks in the mediastinum of the chest—the **cardiac plexus,** which supplies fibers to the heart; the **pulmonary plexus,** whose fibers accompany the bronchi and blood vessels into the lungs; and the **esophageal plexus,** whose fibers regulate swallowing.

At the lower end of the esophagus, these plexuses give off anterior and posterior **vagal trunks,** each of which contains fibers from both the right and left vagus. These penetrate the diaphragm, enter the abdominal cavity, and contribute to the extensive *abdominal aortic plexus* mentioned earlier. As we have seen, sympathetic fibers synapse here. The parasympathetic fibers, however, pass through the plexus without synapsing and lead to the liver,

pancreas, stomach, small intestine, kidney, ureter, and proximal half of the colon.

The remaining parasympathetic fibers arise from levels S2 to S4 of the spinal cord. They travel a short distance in the ventral rami of the spinal nerves and then form **pelvic splanchnic nerves** that lead to the **inferior hypogastric (pelvic) plexus.** Some parasympathetic fibers synapse here, but most pass through this plexus and travel by way of **pelvic nerves** to the terminal ganglia in their target organs: the distal half of the colon, the rectum, urinary bladder, and reproductive organs. The parasympathetic system does not innervate body wall structures (sweat glands, piloerector muscles, or cutaneous blood vessels).

The sympathetic and parasympathetic divisions of the ANS are compared in table 15.3.

> **Think About It**
>
> *Would autonomic functions be affected if the ventral roots of the cervical spinal nerves were damaged? Why or why not?*

## THE ENTERIC NERVOUS SYSTEM

The digestive tract has a nervous system of its own called the **enteric**[10] **nervous system.** Unlike the ANS proper, it does not arise from the brainstem or spinal cord, but like the ANS, it innervates smooth muscle and glands. Thus, opinions differ on whether it should be considered part of the ANS. It consists of about 100 million neurons embedded in the wall of the digestive tract—perhaps more neurons than there are in the spinal cord—and it has its own reflex arcs. The enteric nervous system regulates the motility of the esophagus, stomach, and intestines and the secretion of digestive enzymes and acid. To function normally, however, these digestive activities also require regulation by the sympathetic and parasympathetic systems. The enteric nervous system is discussed in more detail in chapter 25.

---

[9]*ot* = ear + *ic* = pertaining to

[10]*enter* = intestines + *ic* = pertaining to

| TABLE 15.3 | Comparison of the Sympathetic and Parasympathetic Divisions | |
|---|---|---|
| **Feature** | **Sympathetic** | **Parasympathetic** |
| Origin in CNS | Thoracolumbar | Craniosacral |
| Location of ganglia | Paravertebral ganglia adjacent to spinal column and prevertebral ganglia anterior to it | Terminal ganglia near or within target organs |
| Fiber lengths | Short preganglionic | Long preganglionic |
| | Long postganglionic | Short postganglionic |
| Neuronal divergence | Extensive (about 1:17) | Minimal (about 1:2) |
| Effects of system | Often widespread and general | More specific and local |

3. *Explain why the sympathetic division is also called the thoracolumbar division even though its paravertebral ganglia extend all the way from the cervical to the sacral region.*

4. *Describe or diagram the structural relationships among the following: preganglionic fiber, postganglionic fiber, gray ramus, white ramus, and sympathetic ganglion.*

5. *Explain in anatomical terms why the parasympathetic division affects target organs more selectively than the sympathetic division does.*

6. *Trace the pathway of a parasympathetic fiber of the vagus nerve from the medulla oblongata to the small intestine.*

# Autonomic Effects on Target Organs

### Objectives
When you have completed this section, you should be able to

- name the neurotransmitters employed at different synapses of the ANS;
- name the receptors for these neurotransmitters and explain how they relate to autonomic effects;
- explain how the ANS controls many target organs through dual innervation; and
- explain how control is exerted in the absence of dual innervation.

## NEUROTRANSMITTERS

As noted earlier, the divisions of the ANS often have contrasting effects on an organ. The sympathetic division accelerates the heartbeat and the parasympathetic division slows it down, for example. But each division of the ANS also can have contrasting effects on different organs. For example, while the parasympathetic division inhibits the contraction of cardiac muscle, it stimulates the contraction of intestinal smooth muscle.

The key to understanding such contradictory effects lies in knowing which neurotransmitters the autonomic neurons release and what kind of receptors occur on the target cells. The ANS has both **cholinergic fibers,** which secrete acetylcholine (ACh), and **adrenergic fibers,** which secrete norepinephrine (NE). Cholinergic fibers include the preganglionic fibers of both divisions, the postganglionic fibers of the parasympathetic division, and a few sympathetic postganglionic fibers (those that innervate

| TABLE 15.4 | Locations of Cholinergic and Adrenergic Fibers in the ANS | |
|---|---|---|
| **Division** | **Preganglionic Fibers** | **Postganglionic Fibers** |
| Sympathetic | Always cholinergic | Mostly adrenergic; a few cholinergic |
| Parasympathetic | Always cholinergic | Always cholinergic |

sweat glands and some blood vessels). Most sympathetic postganglionic fibers are adrenergic (table 15.4).

The sympathetic nervous system tends to have longer-lasting effects than the parasympathetic. After ACh is secreted by the parasympathetic fibers, it is quickly broken down by acetylcholinesterase (AChE) in the synapse and its effect lasts only a few seconds. The NE released by sympathetic nerve fibers, however, has various fates: (1) Some is reabsorbed by the nerve fiber where it is either reused or broken down by *monoamine oxidase (MAO)*. (2) Some diffuses into the surrounding tissues, where it is degraded by another enzyme, *catechol-O-methyltransferase (COMT)*. (3) Much of it is picked up by the bloodstream, where MAO and COMT are absent. This NE, along with epinephrine from the adrenal gland, circulates throughout the body and exerts a prolonged effect.

ACh and NE are not the only neurotransmitters employed by the ANS. Although all autonomic fibers secrete one of these, many of them also secrete neuropeptides that modulate ACh or NE function. Sympathetic fibers may also secrete enkephalin, substance P, neuropeptide Y, somatostatin, neurotensin, or gonadotropin-releasing hormone. Some parasympathetic fibers relax blood vessels by stimulating their endothelial cells to release the gas nitric oxide (NO). NO inhibits smooth muscle tone in the vessel wall, thus allowing the vessel to dilate. This increases blood flow through the vessel. Among other functions, this mechanism is crucial to penile erection (see Insight 27.4, p. 1058).

## RECEPTORS

Both the sympathetic and parasympathetic divisions have excitatory effects on some target cells and inhibitory effects on others. For example, the parasympathetic division contracts the wall of the urinary bladder but relaxes the internal urinary sphincter—both of which are necessary for the expulsion of urine. It employs ACh for both purposes. Similarly, the sympathetic division constricts most blood vessels but dilates the coronary arteries, and it achieves both effects with NE. Clearly, the difference is not due to the neurotransmitter. Rather, it is due to the fact that different effector cells have *different kinds of receptors* for it. The receptors for ACh and NE are called **cholinergic** and **adren-**

ergic receptors, respectively. Knowledge of these receptor types is essential to the field of neuropharmacology (see Insight 15.2, p. 579).

## Cholinergic Receptors

Acetylcholine binds to two classes of cholinergic receptors—nicotinic (NIC-oh-TIN-ic) and muscarinic (MUSS-cuh-RIN-ic) receptors—named for toxins that were used to identify and distinguish them. Nicotine binds only to the former type, while muscarine, a mushroom poison, binds only to the latter. Other drugs also selectively bind to one type or the other—atropine binds only to muscarinic receptors and curare only to nicotinic receptors, for example.

Nicotinic receptors occur on the postsynaptic cells in all ganglia of the ANS, in the adrenal medulla, and in neuromuscular junctions. Muscarinic receptors occur on all gland, smooth muscle, and cardiac muscle cells that receive cholinergic innervation. All cells with nicotinic receptors are excited by ACh, but some cells with muscarinic receptors are excited while others are inhibited by it. For example, by acting on different subclasses of muscarinic receptors, ACh excites intestinal smooth muscle but inhibits cardiac muscle. Nicotinic receptors work by opening ligand-gated ion channels and changing the postsynaptic potential of the target cell. Muscarinic receptors work through a variety of second-messenger systems.

## Adrenergic Receptors

The contrasting effects of norepinephrine on different target cells also is accounted for by different classes of receptors. NE receptors fall into two broad classes called alpha-($\alpha$-)adrenergic and beta-($\beta$-)adrenergic receptors. The binding of NE to $\alpha$-adrenergic receptors is usually excitatory, and its binding to $\beta$-adrenergic receptors is usually inhibitory, but there are exceptions to both. For example, NE binds to $\beta$ receptors in cardiac muscle but has an excitatory effect.

The exceptions result from the existence of subclasses of each receptor type, called $\alpha_1$, $\alpha_2$, $\beta_1$, and $\beta_2$ receptors. All four types function by means of second messengers. Both $\beta$ receptors activate the production of cyclic AMP (cAMP), $\alpha_2$ receptors suppress cAMP production, and $\alpha_1$ receptors employ calcium ions as the second messenger. Some target cells have both $\alpha$ and $\beta$ receptors.

The binding of NE to $\alpha$-adrenergic receptors in blood vessels causes vasoconstriction. The arteries that supply the heart and skeletal muscles, however, have $\beta$-adrenergic receptors, which dilate the arteries, thereby increasing blood flow to these organs. NE also relaxes the bronchioles when it binds to $\beta$-adrenergic receptors, thus enhancing respiratory airflow. These effects are obviously appropriate to the exercise state, in which the sympathetic nervous system is most active. The autonomic effects on glandular secretion are often achieved through the adjustment of blood flow to the gland rather than direct stimulation of the gland cells. Many glandular secretions begin as a filtrate of the blood.

Figure 15.8 summarizes the locations of these receptor types. Table 15.5 compares the effects of sympathetic and parasympathetic stimulation on various target organs.

### Think About It

*Table 15.5 notes that the sympathetic nervous system has an $\alpha$-adrenergic effect on blood platelets and promotes clotting. How can the sympathetic nervous system stimulate platelets, considering that platelets are drifting cell fragments in the bloodstream with no nerve fibers leading to them?*

## DUAL INNERVATION

Most of the viscera receive nerve fibers from both the sympathetic and parasympathetic divisions and thus are said to have **dual innervation.** In such cases, the two divisions may have either *antagonistic* or *cooperative* effects on the same organ.

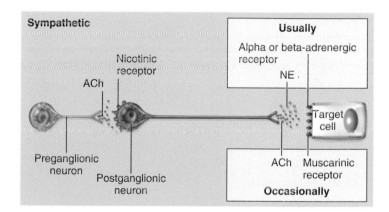

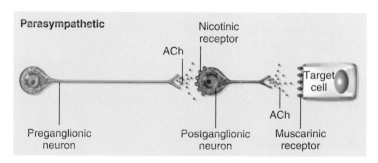

**FIGURE 15.8 Neurotransmitters and Receptors of the Autonomic Nervous System.** A given postganglionic fiber releases either ACh or NE, but not both. Both are shown in the top illustration only to emphasize that some sympathetic fibers are adrenergic and some are cholinergic. In the parasympathetic division (bottom), all fibers are cholinergic.

**TABLE 15.5**   Effects of the Sympathetic and Parasympathetic Nervous Systems

| Target | Sympathetic Effect and Receptor Type | Parasympathetic Effect (all muscarinic) |
|---|---|---|
| **Eye** | | |
| Iris | Pupillary dilation ($\alpha_1$) | Pupillary constriction |
| Ciliary muscle and lens | Relaxation for far vision ($\beta_2$) | Contraction for near vision |
| Lacrimal (tear) gland | None | Secretion |
| **Integumentary System** | | |
| Merocrine sweat glands (cooling) | Secretion (muscarinic) | No effect |
| Apocrine sweat glands (scent) | Secretion ($\alpha_1$) | No effect |
| Piloerector muscles | Hair erection ($\alpha_1$) | No effect |
| **Adipose Tissue** | Decreased fat breakdown ($\alpha_2$) | No effect |
| | Increased fat breakdown ($\alpha_1$, $\beta_1$) | |
| **Adrenal Medulla** | Hormone secretion (nicotinic) | No effect |
| **Circulatory System** | | |
| Heart rate and force | Increased ($\beta_1$, $\beta_2$) | Decreased |
| Deep coronary arteries | Vasodilation ($\beta_2$) | Slight vasodilation |
| | Vasoconstriction ($\alpha_1$, $\alpha_2$) | |
| Blood vessels of most viscera | Vasoconstriction ($\alpha_1$) | Vasodilation |
| Blood vessels of skeletal muscles | Vasodilation ($\beta_2$) | No effect |
| Blood vessels of skin | Vasoconstriction ($\alpha_1$, $\alpha_2$) | Vasodilation, blushing |
| Platelets (blood clotting) | Increased clotting ($\alpha_2$) | No effect |
| **Respiratory System** | | |
| Bronchi and bronchioles | Bronchodilation ($\beta_2$) | Bronchoconstriction |
| Mucous glands | Decreased secretion ($\alpha_1$) | No effect |
| | Increased secretion ($\beta_2$) | |
| **Urinary System** | | |
| Kidneys | Reduced urine output ($\alpha_1$, $\alpha_2$) | No effect |
| Bladder wall | No effect | Contraction |
| Internal urethral sphincter | Contraction, urine retention ($\alpha_1$) | Relaxation, urine release |
| **Digestive System** | | |
| Salivary glands | Thick mucous secretion ($\alpha_1$) | Thin serous secretion |
| Gastrointestinal motility | Decreased ($\alpha_1$, $\alpha_2$, $\beta_1$, $\beta_2$) | Increased |
| Gastrointestinal secretion | Decreased ($\alpha_2$) | Increased |
| Liver | Glycogen breakdown ($\alpha_1$, $\beta_2$) | Glycogen synthesis |
| Pancreatic enzyme secretion | Decreased ($\alpha_1$) | Increased |
| Pancreatic insulin secretion | Decreased ($\alpha_2$) | No effect |
| | Increased ($\beta_2$) | |
| **Reproductive System** | | |
| Penile or clitoral erection | No effect | Stimulation |
| Glandular secretion | No effect | Stimulation |
| Orgasm, smooth muscle roles | Stimulation ($\alpha_1$) | No effect |
| Uterus | Relaxation ($\beta_2$) | No effect |
| | Labor contractions ($\alpha_1$) | |

**Antagonistic effects** oppose each other. For example, the sympathetic division speeds up the heart and the parasympathetic division slows it down; the sympathetic division inhibits digestion and the parasympathetic division stimulates it; the sympathetic division dilates the pupil and the parasympathetic division constricts it. In some cases, these effects are exerted through dual innervation of the same effector cells, as in the heart, where nerve fibers of both divisions terminate on the same muscle cells. In other cases, antagonistic effects arise because each division innervates different effector cells with opposite effects on organ function. In the iris of the eye, for example, sympathetic fibers innervate the pupillary

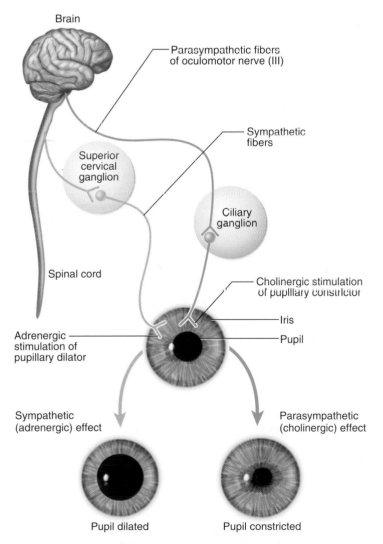

**FIGURE 15.9  Dual Innervation of the Iris.** Antagonistic effects of the sympathetic (yellow) and parasympathetic (blue) divisions on the iris.

dilator and parasympathetic fibers innervate the constrictor (fig. 15.9).

**Cooperative effects** are seen when the two divisions act on different effectors to produce a unified overall effect. Salivation is a good example. The parasympathetic division stimulates serous cells of the salivary glands to secrete a watery, enzyme-rich secretion, while the sympathetic division stimulates mucous cells of the same glands to secrete mucus. The enzymes and mucus are both necessary components of the saliva.

Even when both divisions innervate a single organ, they do not always innervate it equally or exert equal influence. For example, the parasympathetic division forms an extensive plexus in the wall of the digestive tract and exerts much more influence over it than the sympathetic division does. In the ventricles of the heart, by contrast, there is much less parasympathetic than sympathetic innervation.

## CONTROL WITHOUT DUAL INNERVATION

Dual innervation is not always necessary for the ANS to produce opposite effects on an organ. The adrenal medulla, piloerector muscles, sweat glands, and many blood vessels receive only sympathetic fibers. The most significant example of control without dual innervation is regulation of blood pressure and routes of blood flow. The sympathetic fibers to a blood vessel have a baseline sympathetic tone which keeps the vessels in a state of partial constriction called **vasomotor tone** (fig. 15.10). An increase in firing rate constricts a vessel by increasing smooth muscle contraction. A drop in firing frequency dilates a vessel by allowing the smooth muscle to relax. The blood pressure in the vessel, pushing outward on its wall, then dilates the vessel. Thus, the sympathetic division alone exerts opposite effects on the vessels.

Sympathetic control of vasomotor tone can shift blood flow from one organ to another according to the changing needs of the body. In times of emergency, stress, or exercise, the skeletal muscles and heart receive a high priority and the sympathetic division dilates the arteries that supply them. Such processes as digestion, nutrient absorption, and urine formation can wait; thus the sympathetic division constricts arteries to the gastrointestinal tract and kidneys. It also reduces blood flow through the skin, which may help to minimize bleeding in the event that the stress-producing situation leads to injury. Furthermore, since there is not enough blood in the body to supply all the organ systems equally, it is necessary to temporarily divert blood away from some organs in order to supply an adequate amount to the muscular system.

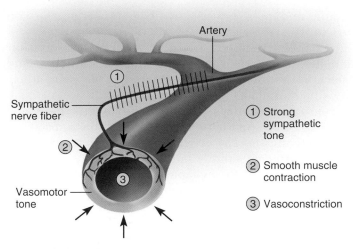

Artery

Sympathetic
nerve fiber

① Strong
   sympathetic
   tone

② Smooth muscle
   contraction

Vasomotor
tone

③ Vasoconstriction

**(a) Vasoconstriction**

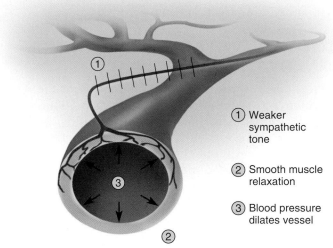

① Weaker
   sympathetic
   tone

② Smooth muscle
   relaxation

③ Blood pressure
   dilates vessel

**(b) Vasodilation**

**FIGURE 15.10   Sympathetic and Vasomotor Tone.**
(a) Vasoconstriction in response to a high rate of sympathetic
nerve firing. (b) Vasodilation in response to a low rate of sympathetic nerve firing. Black lines crossing each nerve fiber represent
firing frequency.

## Before You Go On

*Answer the following questions to test your under-
standing of the preceding section:*

7. *What neurotransmitters are secreted by adrenergic
   and cholinergic fibers?*

8. *Why do sympathetic effects last longer than parasympathetic effects?*

9. *How can the sympathetic division cause smooth muscle to relax in some organs but contract in others?*

10. *What are the two ways in which the sympathetic and
    parasympathetic systems can affect each other when they
    both innervate the same target organ? Give examples.*

11. *How can the sympathetic nervous system have contrast-
    ing effects in a target organ without dual innervation?*

# Central Control of Autonomic Function

### Objective
When you have completed this section, you should be able to

- describe how the autonomic nervous system is regulated by the brain and somatic nervous system.

In spite of its name, the ANS is not an independent
nervous system. All of its output originates in the CNS,
and it receives input from the cerebral cortex, hypothalamus, medulla oblongata, and somatic branch of the
peripheral nervous system. In this section we briefly consider how the ANS is influenced by these other levels of
the nervous system.

## THE CEREBRAL CORTEX

Even if we usually cannot consciously control the ANS, it
is clear that the mind does influence it. Anger raises the
blood pressure, fear makes the heart race, thoughts of
good food make the stomach rumble, sexual thoughts or
images increase blood flow to the genitals, and anxiety
inhibits sexual function. The limbic system (p. 534), an
ancient part of the cerebral cortex, is involved in many
emotional responses and has extensive connections with
the hypothalamus, a site of several nuclei of autonomic
control. Thus, the limbic system provides a pathway connecting sensory and mental experiences with the autonomic nervous system.

## THE HYPOTHALAMUS

While the major site of CNS control over the somatic
motor system is the primary motor cortex, the major
control center of the visceral motor system is the hypothalamus. This small but vital region in the floor of the
brain contains many nuclei for primitive functions,
including hunger, thirst, thermoregulation, emotions,
and sexuality. Artificial stimulation of different regions
of the hypothalamus can activate the fight or flight
response typical of the sympathetic nervous system or
have the calming effects typical of the parasympathetic.
Output from the hypothalamus travels largely to nuclei
in more caudal regions of the brainstem, and from there
to the cranial nerves and the sympathetic neurons in
the spinal cord.

## THE MIDBRAIN, PONS, AND MEDULLA OBLONGATA

These regions of the brainstem house numerous autonomic
nuclei described in chapter 14: centers for cardiac and vasomotor control, salivation, swallowing, sweating, gastrointestinal secretion, bladder control, pupillary constriction

| TABLE 15.6 | Some Disorders of the Autonomic Nervous System |
|---|---|
| Horner syndrome | Chronic unilateral pupillary constriction, sagging of the eyelid, withdrawal of the eye into the orbit, flushing of the skin, and lack of facial perspiration, resulting from lesions in the cervical ganglia, upper thoracic spinal cord, or brainstem that interrupt sympathetic innervation of the head. |
| Raynaud disease | Intermittent attacks of paleness, cyanosis, and pain in the fingers and toes, caused when cold or emotional stress triggers excessive vasoconstriction in the digits; most common in young women. In extreme cases, causes gangrene and may require amputation. Sometimes treated by severing sympathetic nerves to the affected regions. |

*Disorders described elsewhere*

Autonomic effects of cranial nerve injuries pp. 549–556

Mass reflex reaction p. 508

Orthostatic hypotension p. 800

and dilation, and other primitive functions. Many of these nuclei belong to the reticular formation, which extends from the medulla to the hypothalamus. Autonomic output from these nuclei travels by way of the spinal cord and the oculomotor, facial, glossopharyngeal, and vagus nerves.

## THE SPINAL CORD

Such autonomic reflexes as defecation, micturition (urination), erection, and ejaculation are integrated in the spinal cord (see details in chapters 23, 25, and 27). Fortunately, the brain is able to inhibit defecation and urination consciously, but when injuries sever the spinal cord from the brain, the autonomic spinal reflexes alone control the elimination of urine and feces.

Table 15.6 describes some dysfunctions of the autonomic nervous system.

### Before You Go On

*Answer the following questions to test your understanding of the preceding section:*

12. What system in the brain connects our conscious thoughts and feelings with the autonomic control centers of the hypothalamus?

13. List some autonomic responses that are controlled by nuclei in the hypothalamus.

14. What is the role of the midbrain, pons, and medulla in autonomic control?

15. Name some visceral reflexes controlled by the spinal cord.

---

**INSIGHT 15.2**   Clinical Application

### Drugs and the Nervous System

*Neuropharmacology* is a branch of medicine that deals with the effects of drugs on the nervous system, especially drugs that mimic, enhance, or inhibit the action of neurotransmitters. A few examples will illustrate the clinical relevance of neurotransmitter and receptor functions.

A number of drugs work by stimulating adrenergic and cholinergic neurons or receptors. *Sympathomimetics*[11] are drugs that enhance sympathetic action. They stimulate adrenergic receptors or promote norepinephrine release. For example phenylephrine, found in such cold medicines as Chlor-Trimeton and Dimetapp, aids breathing by stimulating $\alpha_1$ receptors, dilating the bronchioles, and constricting nasal blood vessels, thus reducing swelling in the nasal mucosa. *Sympatholytics*[12] are drugs that suppress sympathetic action by inhibiting norepi-

nephrine release, or by binding to adrenergic receptors without stimulating them. Propranolol, for example, is a *beta-blocker*. It reduces hypertension partly by blocking β-adrenergic receptors. This interferes with the effects of epinephrine and norepinephrine on the heart and blood vessels. (It also reduces the production of *angiotensin II,* a hormone that stimulates vasoconstriction and raises blood pressure.)

*Parasympathomimetics* enhance parasympathetic effects. Pilocarpine, for example, relieves glaucoma (excessive pressure in the eyeball) by dilating a vessel that drains fluid from the eye. *Parasympatholytics* inhibit ACh release or block its receptors. Atropine, for example, blocks muscarinic receptors. It is sometimes used to dilate the pupils for eye examinations and to dry the mucous membranes of the respiratory tract before inhalation anesthesia. It is an extract of the deadly nightshade plant, *Atropa belladonna.*[13] Women of the Middle Ages used nightshade to dilate their pupils, which was regarded as a beauty enhancement.

---

[11]*mimet* = imitate, mimic
[12]*lyt* = break down, destroy

[13]*bella* = beautiful, fine + *donna* = woman

The drugs we have mentioned so far act on the peripheral nervous system and its effectors. Many others act on the central nervous system. Strychnine, for example, blocks the inhibitory action of glycine on spinal motor neurons. The neurons then overstimulate the muscles, causing spastic paralysis and sometimes death by suffocation.

Sigmund Freud predicted that psychiatry would eventually draw upon biology and chemistry to deal with emotional problems once treated only by counseling and psychoanalysis. A branch of neuropharmacology called *psychopharmacology* has fulfilled his prediction. This field dates to the 1950s when chlorpromazine, an antihistamine, was accidentally found to relieve the symptoms of schizophrenia.

The management of clinical depression is one example of how contemporary psychopharmacology has supplemented counseling approaches. Some forms of depression result from deficiencies of the monoamine neurotransmitters. Thus, they yield to drugs that prolong the effects of the monoamines already present at the synapses. One of the earliest discovered antidepressants was imipramine, which blocks the synaptic reuptake of serotonin and norepinephrine. However, it produces undesirable side effects such as dry mouth and irregular cardiac rhythms. It has been largely replaced by Prozac (fluoxetine), which blocks serotonin reuptake and prolongs its mood-elevating effect; thus Prozac is called a *selective serotonin reuptake inhibitor (SSRI)*. It is also used to treat fear of rejection, excess sensitivity to criticism, lack of self-esteem, and inability to experience pleasure, all of which were long handled only through counseling, group therapy, or psychoanalysis. After monoamines are taken up from the synapse, they are degraded by monoamine oxidase (MAO). Drugs called *MAO inhibitors* interfere with the breakdown of monoamine neurotransmitters and provide another pharmacological approach to depression.

Our growing understanding of neurochemistry also gives us deeper insight into the action of addictive drugs of abuse such as amphetamines and cocaine. Amphetamines ("speed") chemically resemble norepinephrine and dopamine, two neurotransmitters associated with elevated mood. Dopamine is especially important in sensations of pleasure. Cocaine blocks dopamine reuptake and thus produces a brief rush of good feelings. But when dopamine is not reabsorbed by the neurons, it diffuses out of the synaptic cleft and is degraded elsewhere. Cocaine thus depletes the neurons of dopamine faster than they can synthesize it, so that finally there is no longer an adequate supply to maintain normal mood. The postsynaptic neurons make new dopamine receptors as if "searching" for the neurotransmitter—all of which leads ultimately to anxiety, depression, and the inability to experience pleasure without the drug.

Caffeine exerts its stimulatory effect by competing with adenosine. Adenosine, which you know as a component of DNA, RNA, and ATP, also functions as an inhibitory neuromodulator in the brain. One theory of sleepiness is that it results when prolonged metabolic activity breaks down so much ATP that the accumulated adenosine has a noticeably inhibitory effect in the brain. Caffeine has enough structural similarity to adenosine (fig. 15.11) to bind to its receptors, but it does not produce the inhibitory effect. Thus, it prevents adenosine from exerting its effect and a person feels more alert.

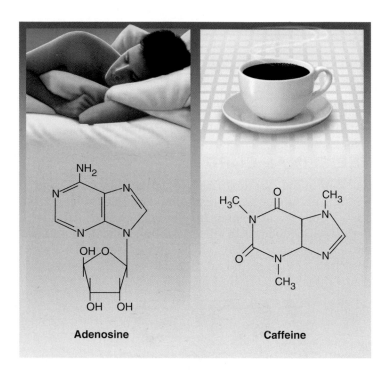

**FIGURE 15.11 Adenosine and Caffeine.** Adenosine is an inhibitory neuromodulator that inhibits ACh release and produces a sense of sleepiness. Caffeine is similar enough in structure to bind to adenosine receptors and block the action of adenosine. This results in increased ACh release and heightened arousal.

## CHAPTER REVIEW

# Review of Key Concepts

**General Properties of the Autonomic Nervous System (p. 564)**

1. The autonomic nervous system (ANS) carries out many visceral reflexes that are crucial to homeostasis. It is a visceral motor system that acts on cardiac muscle, smooth muscle, and glands.

2. Functions of the ANS are largely, but not entirely, unconscious and involuntary.

3. The *sympathetic division* of the ANS prepares the body for physical activity and is especially active in stressful "fight or flight" situations.

4. The *parasympathetic division* has a calming effect on many body functions, but stimulates digestion; it is especially active in "resting and digesting" states.

5. Although the balance of activity may shift from one division to the other, both divisions are normally active simultaneously. Each maintains a background level of activity called *autonomic tone.*

6. The ANS is composed of nuclei in the brainstem, motor neurons in the spinal cord and ganglia, and nerve fibers in the cranial and spinal nerves.

7. Most autonomic efferent pathways, unlike somatic motor pathways, involve two neurons: a *preganglionic neuron* whose axon travels to a peripheral ganglion and synapses with a *postganglionic neuron,* whose axon leads the rest of the way to the target cells.

**Anatomy of the Autonomic Nervous System (p. 567)**

1. Sympathetic preganglionic neurons arise from thoracic and lumbar segments of the spinal cord and travel through spinal nerves T1 through L2 to a *sympathetic chain* of ganglia adjacent to the vertebral column.

2. Most preganglionic fibers synapse with postganglionic neurons in one of the ganglia of this chain, sometimes at a higher or lower level than the ganglion at which they enter. Some fibers pass through the chain without synapsing.

3. Sympathetic pathways show substantial neuronal divergence, with the average preganglionic neuron synapsing with 17 postganglionic neurons. Sympathetic stimulation therefore tends to have widespread effects on multiple target organs.

4. Postganglionic fibers leave the sympathetic chain by way of either the spinal nerve route or the sympathetic nerve route. Preganglionic fibers that pass through the chain without synapsing travel by way of *splanchnic nerves* to various *collateral ganglia,* and synapse there with postganglionic neurons.

5. The *adrenal medulla* is a modified sympathetic ganglion composed of anaxonic cells. These cells secrete mainly epinephrine and norepinephrine into the blood when stimulated.

6. The parasympathetic division issues relatively long preganglionic fibers through cranial nerves III, VII, IX, and X, and spinal nerves S2 through S4, to their target organs. The vagus nerve carries about 90% of all parasympathetic preganglionic fibers.

7. Parasympathetic preganglionic fibers end in *terminal ganglia* in or near the target organ. Relatively short postganglionic fibers complete the route to specific target cells.

8. The wall of the digestive tract contains an *enteric nervous system,* sometimes considered part of the ANS because it innervates smooth muscle and glands of the tract.

**Autonomic Effects on Target Organs (p. 574)**

1. The autonomic effect on a target cell depends on the neurotransmitter released and the type of receptors that the target cell has.

2. *Cholinergic* fibers secrete acetylcholine (ACh) and include all preganglionic fibers, all parasympathetic postganglionic fibers, and some sympathetic postganglionic fibers. Most sympathetic postganglionic fibers are *adrenergic* and secrete norepinephrine (NE).

3. ACh breaks down quickly, and parasympathetic effects are therefore usually short-lived. NE persists longer, and sympathetic effects tend to be longer-lasting.

4. Autonomic neurons also employ a broad range of other neurotransmitters and neuromodulators ranging from the peptide enkephalin to the inorganic gas nitric oxide.

5. ACh binds to two classes of receptors called *nicotinic* and *muscarinic* receptors. The binding of ACh to a nicotinic receptor always excites a target cell, but binding to a muscarinic receptor can have excitatory effects on some cells and inhibitory effects on others, owing to different subclasses of muscarinic receptors.

6. NE binds to two major classes of receptors called alpha and beta receptors. Binding to an $\alpha$-adrenergic receptor is usually excitatory, and binding to a $\beta$-adrenergic receptor is usually inhibitory, but there are exceptions to both owing to subclasses of each receptor type.

7. Many organs receive *dual innervation* by both sympathetic and parasympathetic fibers. In such cases, the two divisions may have either antagonistic or cooperative effects on the organ.

8. The sympathetic division can have contrasting effects on an organ even without dual innervation, by increasing or decreasing the firing rate of the sympathetic neuron.

**Central Control of Autonomic Function (p. 578)**

1. All autonomic output originates in the CNS and is subject to control by multiple levels of the CNS.

2. The hypothalamus is an especially important center of autonomic control, but the cerebral cortex, midbrain, pons, and medulla oblongata are also involved.

3. Some autonomic reflexes such as defecation and micturition are regulated by the spinal cord.

# Testing Your Recall

1. The autonomic nervous system innervates all of these *except*
   a. cardiac muscle.
   b. skeletal muscle.
   c. smooth muscle.
   d. salivary glands.
   e. blood vessels.

2. Muscarinic receptors bind
   a. epinephrine.
   b. norepinephrine.
   c. acetylcholine.
   d. cholinesterase.
   e. neuropeptides.

3. All of the following cranial nerves except the _____ carry parasympathetic fibers.
   a. vagus
   b. facial
   c. oculomotor
   d. glossopharyngeal
   e. hypoglossal

4. Which of the following cranial nerves carries sympathetic fibers?
   a. oculomotor
   b. facial
   c. trigeminal
   d. vagus
   e. none of them

5. Which of these ganglia is *not* involved in the sympathetic division?
   a. intramural
   b. superior cervical
   c. paravertebral
   d. inferior mesenteric
   e. celiac

6. Epinephrine is secreted by
   a. sympathetic preganglionic fibers.
   b. sympathetic postganglionic fibers.
   c. parasympathetic preganglionic fibers.
   d. parasympathetic postganglionic fibers.
   e. the adrenal medulla.

7. The major autonomic control center within the CNS is
   a. the cerebral cortex.
   b. the limbic system.
   c. the midbrain.
   d. the hypothalamus.
   e. the sympathetic chain ganglia.

8. The gray communicating ramus contains
   a. visceral sensory fibers.
   b. parasympathetic motor fibers.
   c. sympathetic preganglionic fibers.
   d. sympathetic postganglionic fibers.
   e. somatic motor fibers.

9. Throughout the autonomic nervous system, the neurotransmitter released by the preganglionic neuron binds to _____ receptors on the postganglionic neuron.
   a. nicotinic
   b. muscarinic
   c. adrenergic
   d. $\alpha_1$
   e. $\beta_2$

10. Which of these does *not* result from sympathetic stimulation?
   a. dilation of the pupil
   b. acceleration of the heart

   c. digestive secretion
   d. enhanced blood clotting
   e. piloerection

11. Nerve fibers that secrete norepinephrine are called _____ fibers.

12. _____ is a state in which a target organ receives both sympathetic and parasympathetic fibers.

13. _____ is a state of continual background activity of the sympathetic and parasympathetic divisions.

14. Most parasympathetic preganglionic fibers are found in the _____ nerve.

15. The digestive tract has a semi-independent nervous system called the _____ nervous system.

16. MAO and COMT are enzymes that break down _____ at certain ANS synapses.

17. The adrenal medulla consists of modified postganglionic neurons of the _____ nervous system.

18. The sympathetic nervous system has short _____ and long _____ nerve fibers.

19. Adrenergic receptors classified as $\alpha_2$, $\beta_1$, and $\beta_2$ act by changing the level of _____ in the target cell.

20. Sympathetic fibers to blood vessels maintain a state of partial vasoconstriction called _____.

*Answers in Appendix B*

# True or False

*Determine which five of the following statements are false, and briefly explain why.*

1. The parasympathetic nervous system shuts down when the sympathetic nervous system is active, and vice versa.

2. Blood vessels of the skin receive no parasympathetic innervation.

3. Voluntary control of the ANS is not possible.

4. The sympathetic nervous system stimulates digestion.

5. Some sympathetic postganglionic fibers are cholinergic.

6. Urination and defecation cannot occur without signals from the brain to the bladder and rectum.

7. Some parasympathetic nerve fibers are adrenergic.

8. Parasympathetic effects are more localized and specific than sympathetic effects.

9. The parasympathetic division shows less neuronal divergence than the sympathetic division does.

10. The two divisions of the ANS have antagonistic effects on the iris.

*Answers in Appendix B*

# Testing Your Comprehension

1. You are dicing raw onions while preparing dinner, and the vapor makes your eyes water. Describe the afferent and efferent pathways involved in this response.

2. Suppose you are walking alone at night when you hear a dog growling close behind you. Describe the ways your sympathetic nervous system would prepare you to deal with this situation.

3. Suppose that the cardiac nerves were destroyed. How would this affect the heart and the body's ability to react to a stressful situation?

4. What would be the advantage to a wolf in having its sympathetic nervous system stimulate the piloerector muscles? What happens in a human when the sympathetic system stimulates these muscles?

5. Pediatric literature has reported many cases of poisoning in children with Lomotil, an antidiarrheic medicine. Lomotil works primarily by means of the morphine-like effects of its chief ingredient, diphenoxylate, but it also contains atropine. Considering the mode of action described for atropine in Insight 5.2, why might atropine contribute to the antidiarrheic effect of Lomotil? In atropine poisoning, would you expect the pupils to be dilated or constricted? The skin to be moist or dry? The heart rate to be elevated or depressed? The bladder to retain urine or void uncontrollably? Explain each answer. Atropine poisoning is treated with physostigmine, a cholinesterase inhibitor. Explain the rationale of this treatment.

*Answers at www.mhhe.com/saladin4*

# www.mhhe.com/saladin4

*The textbook website provides a wealth of interactive study materials fully organized and integrated by chapter. You will find practice quizzes, labeling exercises, and much more that will complement your learning and understanding of anatomy and physiology. The website also includes tools designed to enhance your* **Anatomy & Physiology | REVEALED** *experience.*

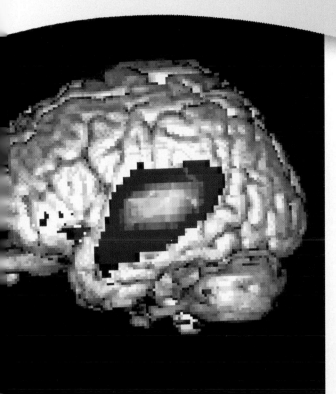

PET scan of the brain responding to
sound with activity in the temporal lobe

# SENSE ORGANS

## CHAPTER OUTLINE

## INSIGHTS

## Brushing Up

To understand this chapter, it is
important that you understand or
brush up on the following concepts.

- Excitatory and inhibitory postsynaptic potentials (EPSPs and
  IPSPs) (p. 468)

- Spatial summation (p. 469)

- Neural coding (p. 469)

- Converging circuits of neurons
  (p. 472)

- Spinal cord tracts (p. 486)

- Decussation (p. 486)

Anyone who enjoys music, art, fine food, or a good conversation appreciates the human senses. Yet their importance extends beyond deriving pleasure from the environment. In the 1950s, behavioral scientists at Princeton University studied the methods used by Soviet Communists to extract confessions from political prisoners, including solitary confinement and sensory deprivation. Student volunteers were immobilized in dark soundproof rooms or suspended in dark chambers of water. In a short time, they experienced visual, auditory, and tactile hallucinations, incoherent thought patterns, deterioration of intellectual performance, and sometimes morbid fear or panic. Similar effects have been seen in burn patients who are immobilized and extensively bandaged (including the eyes) and thus suffer prolonged lack of sensory input. Patients connected to life-support equipment and confined under oxygen tents sometimes become delirious. Sensory input is vital to the integrity of personality and intellectual function.

Furthermore, much of the information communicated by the sense organs never comes to our conscious attention—blood pressure, body temperature, and muscle tension, for example. By monitoring such conditions, however, the sense organs initiate somatic and visceral reflexes that are indispensable to homeostasis and to our very survival in a ceaselessly changing and challenging environment.

# Properties and Types of Sensory Receptors

### Objectives
When you have completed this section, you should be able to

- define *receptor* and *sense organ;*
- list the four kinds of information obtained from sensory receptors, and describe how the nervous system encodes each type; and
- outline three ways of classifying receptors.

A sensory **receptor** is any structure specialized to detect a stimulus. Some receptors are simple nerve endings, whereas others are **sense organs**—nerve endings combined with connective, epithelial, or muscular tissues that enhance or moderate the response to a stimulus. Our eyes and ears are obvious examples of sense organs, but there are also innumerable microscopic sense organs in our skin, muscles, joints, and viscera.

## GENERAL PROPERTIES OF RECEPTORS

The general purpose of a sensory receptor is to convert the energy of a stimulus into the energy of a nerve signal, so that the central nervous system (CNS) can detect and interpret stimuli acting on and within the body. Any device that converts one form of energy into another is called a *transducer.* Transducers include not only sensory receptors but also such everyday devices as a microphone, lightbulb, or gasoline engine. In the nervous system, the process of converting stimulus energy to nerve energy is called **sensory transduction.**

The initial effect of a stimulus on the nervous system is a small local electrical change called a **receptor potential,** occurring in a receptor cell. In many cases, such as the senses of touch and smell, the receptor cell is a neuron. If the receptor potential is strong enough, the sensory neuron fires with a volley of action potentials, thus generating a nerve signal to the CNS. In other cases, such as taste and hearing, the receptor cell is not a neuron but an epithelial cell. Nevertheless, it has synaptic vesicles at its base, it releases a neurotransmitter in response to a stimulus, and it stimulates an adjacent neuron. That neuron then generates signals to the CNS.

When nerve signals arrive at the CNS, we may experience a sensation—a conscious awareness of the stimulus. However, most sensory signals delivered to the CNS produce no conscious sensation at all. This is partly because most of them are filtered out at the thalamus or other levels of the brainstem, thus keeping us from being consciously attentive to a vast multitude of unimportant stimuli, and partly because many of the body's responses do not require conscious awareness. We respond physiologically to changes in blood pH, muscle tension, and other stimuli without being aware of them.

Sensory receptors transmit four kinds of information—*modality, location, intensity,* and *duration:*

1. **Modality** refers to the type of stimulus or the sensation it produces. Vision, hearing, and taste are examples of sensory modalities. The nervous system distinguishes modalities from each other partly by means of the *labeled line code* described on page 470. Each sensory nerve fiber to the CNS is a "line" leading ultimately from a specific kind of receptor. Even though the action potentials in any nerve fiber are identical to action potentials anywhere else, the brain interprets signals from a given fiber in terms of the receptor type symbolized by that fiber. Thus, the action potentials for vision and smell are identical, but the brain interprets them differently becuse they arrive via different lines.

2. **Location** is also encoded by which nerve fibers are firing. A sensory neuron receives input from an area called its **receptive field.** The brain's ability to deter-

mine the location of a stimulus depends on the size of this field. In tactile (touch) neurons, for example, a receptive field on one's back may be as big as 7 cm in diameter. Any touch within that area stimulates one neuron, so it is difficult to tell precisely where the touch occurs. Being touched at two points 5 cm apart within the same field would feel like a single touch. On the fingertips, by contrast, receptive fields may be less than 1 mm in diameter. Two points of contact just 2 mm apart would thus be felt separately (fig. 16.1). Thus, we say the fingertips have finer *two-point discrimination* than the skin on the back. This is crucial to such functions as feeling textures and reading Braille. The receptive field is relevant not only to touch, but also to other senses such as vision.

Sensory projection is the ability of the brain to identify the site of stimulation, including very small and specific areas within a receptor such as the retina. The pathways followed by sensory signals to their ultimate destinations in the CNS are called **projection pathways.**

3. **Intensity** is encoded in three ways: (a) as stimulus intensity rises, the firing frequencies of sensory nerve fibers rise (see fig. 12.28, p. 471); (b) intense stimuli recruit greater numbers of nerve fibers to fire; and (c) weak stimuli activate only the most sensitive nerve fibers, whereas strong stimuli can activate a less sensitive group of fibers with higher thresholds. Thus, the brain can distinguish intensities based on the number and kind of fibers that are firing and the time intervals between action potentials. These concepts were discussed under *neural coding* in chapter 12.

4. **Duration** is encoded in the way nerve fibers change their firing frequencies over time. **Phasic receptors** generate a burst of action potentials when first stimulated, then quickly adapt and sharply reduce or halt signal transmission even if the stimulus continues. Some of them fire again when the stimulus ceases. Lamellated corpuscles, tactile receptors, hair receptors, and smell receptors are rapidly adapting phasic receptors. **Tonic receptors** adapt slowly and generate nerve impulses more steadily. Proprioceptors are among the most slowly adapting tonic receptors because the brain must always be aware of body position, muscle tension, and joint motions. All receptors, however, exhibit sensory **adaptation**—if the stimulus is prolonged, firing frequency and conscious sensation decline. Adapting to hot bath water is an example.

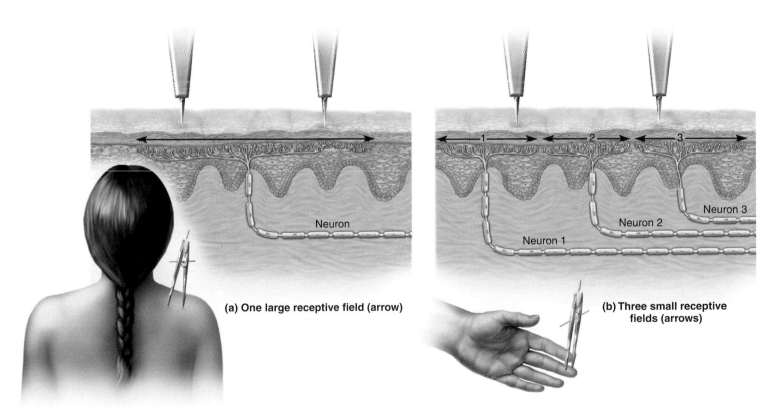

(a) **One large receptive field (arrow)**

(b) **Three small receptive fields (arrows)**

**FIGURE 16.1** **Receptive Fields.** (a) A neuron with a large receptive field, as found in the skin of the back. If the skin is touched in two close-together places within this receptive field, the brain will sense it as only one point of contact. (b) Neurons with small receptive fields, as found in the fingertips. Two close-together points of contact here are likely to stimulate two different neurons and to be felt as separate touches.

▶ *If the receptive field in part (a) is 7 cm in diameter, is it possible for two touches 1 cm apart to be felt separately?*

## CLASSIFICATION OF RECEPTORS

Receptors can be classified by several overlapping systems, some of which have been introduced in earlier chapters:

1. By stimulus modality:
   - **Chemoreceptors** respond to chemicals, including odors, tastes, and body fluid composition.
   - **Thermoreceptors** respond to heat and cold.
   - **Nociceptors**[1] (NO-sih-SEP-turs) are pain receptors; they respond to tissue damage resulting from trauma (blows, cuts), ischemia (poor blood flow), or agents such as heat and chemicals.
   - **Mechanoreceptors** respond to physical deformation caused by vibration, touch, pressure, stretch, or tension. They include the organs of hearing and balance and many receptors of the skin, viscera, and joints.
   - **Photoreceptors,** the eyes, respond to light.

2. By the origins of the stimuli:
   - **Interoceptors** detect stimuli in the internal organs and produce feelings of visceral pain, nausea, stretch, and pressure.
   - **Proprioceptors** sense the position and movements of the body or its parts. They occur in muscles, tendons, and joint capsules.
   - **Exteroceptors** sense stimuli external to the body; they include the receptors for vision, hearing, taste, smell, touch, and cutaneous pain.

3. By the distribution of receptors in the body. There are two broad classes of senses:
   - **General (somesthetic) senses,** with receptors that are widely distributed in the skin, muscles, tendons, joint capsules, and viscera. These include the sense of touch, pressure, stretch, heat, cold, and pain, as well as many stimuli that we do not perceive consciously, such as blood pressure and composition.
   - **Special senses,** which are limited to the head and innervated by the cranial nerves. The special senses are vision, hearing, equilibrium, taste, and smell.

# The General Senses

Receptors for the general senses are relatively simple in structure and physiology. They consist of one or a few sensory nerve fibers and, usually, a sparse amount of connective tissue. These receptors are shown in figure 16.2 and described in table 16.1.

## UNENCAPSULATED NERVE ENDINGS

**Unencapsulated nerve endings** are sensory dendrites that are not wrapped in connective tissue. They include free nerve endings, tactile discs, and hair receptors:

- **Free nerve endings** include *warm receptors,* which respond to rising temperature; *cold receptors,* which respond to falling temperature; and *nociceptors* (pain receptors). They are bare dendrites that have no special association with specific accessory cells or tissues. They are most abundant in epithelia and connective tissue.
- **Tactile (Merkel)**[2] **discs** are tonic receptors for light touch, thought to sense textures, edges, and shapes. They are flattened nerve endings associated with specialized *tactile (Merkel) cells* at the base of the epidermis.

---

[1]*noci* = pain

[2]Friedrich S. Merkel (1845–1919), German anatomist and physiologist

- **Hair receptors** (peritrichial[3] endings) monitor the movements of hairs. They consist of a few dendrites entwined around the base of a hair follicle. They respond to any light touch that bends a hair. Because they adapt quickly, we are not constantly annoyed by our clothing bending the body hairs. However, when an ant crawls across our skin, bending one hair after another, we are very aware of it.

## ENCAPSULATED NERVE ENDINGS

**Encapsulated nerve endings** are nerve fibers wrapped in glial cells or connective tissue. Most of them are mechanoreceptors for touch, pressure, and stretch. The connective tissues around a sensory dendrite enhance the sensitivity or specificity of the receptor. We have already considered some encapsulated nerve endings in chapter 13—muscle spindles and Golgi tendon organs. Others are as follows:

- **Tactile (Meissner[4]) corpuscles** are phasic receptors for light touch and texture. They occur in the dermal papillae of the skin and are limited to sensitive hairless areas such as the fingertips, palms, eyelids, lips, nipples, and parts of the genitals. They are tall, ovoid- to pear-shaped, and consist of two or three nerve fibers meandering upward through a mass of flattened Schwann cells. Tactile corpuscles enable you to tell the difference between silk and sandpaper, for example, by light strokes of your fingertips.
- **Krause[5] end bulbs** are similar to tactile corpuscles but occur in mucous membranes rather than in the skin.
- **Lamellated (pacinian[6]) corpuscles** are phasic receptors for deep pressure, stretch, tickle, and vibration. They are large receptors, up to 1 or 2 mm long, and look like a sliced onion in section. A single sensory

---

[3]*peri* = around + *trich* = hair

[4]Georg Meissner (1829–1905), German histologist
[5]Wilhelm J. F. Krause (1833–1910), German anatomist
[6]Filippo Pacini (1812–83), Italian anatomist

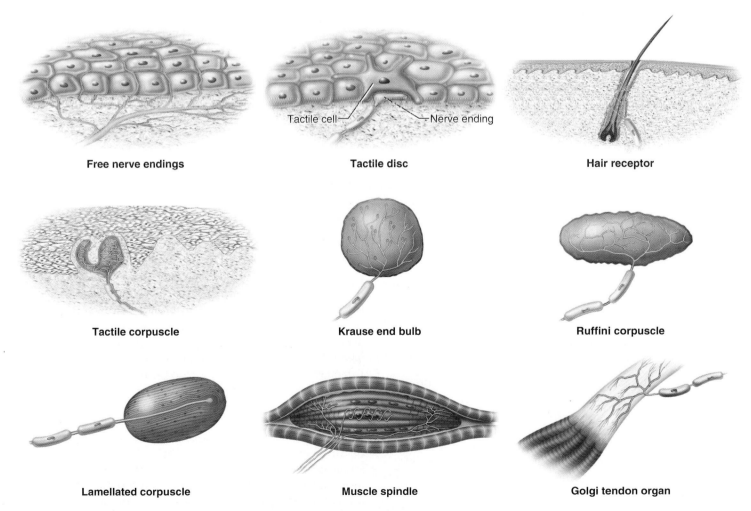

**Free nerve endings**    **Tactile disc** (Tactile cell — Nerve ending)    **Hair receptor**

**Tactile corpuscle**    **Krause end bulb**    **Ruffini corpuscle**

**Lamellated corpuscle**    **Muscle spindle**    **Golgi tendon organ**

**FIGURE 16.2    Receptors of the General (somesthetic) Senses.**   See table 16.1 for functions.

| TABLE 16.1 | Receptors of the General Senses | |
|---|---|---|
| **Receptor Type** | **Locations** | **Modality** |
| *Unencapsulated Endings* | | |
| Free nerve endings | Widespread, especially in epithelia and connective tissues | Pain, heat, cold |
| Tactile discs | Stratum basale of epidermis | Light touch, pressure |
| Hair receptors | Around hair follicle | Light touch, movement of hairs |
| *Encapsulated Nerve Endings* | | |
| Tactile (Meissner) corpuscles | Dermal papillae of fingertips, palms, eyelids, lips, tongue, nipples, and genitals | Light touch, texture, low-frequency vibration |
| Krause end bulbs | Mucous membranes | Similar to tactile corpuscles |
| Ruffini corpuscles | Dermis, subcutaneous tissue, and joint capsules | Heavy continuous touch or pressure; joint movements |
| Lamellated (pacinian) corpuscles | Dermis, joint capsules, periosteum, breasts, genitals, and some viscera | Deep pressure, stretch, high-frequency vibration |
| Muscle spindles | Skeletal muscles near tendon | Muscle stretch (proprioception) |
| Golgi tendon organs | Tendons | Tension on tendons (proprioception) |

dendrite travels through the center of the corpuscle. The innermost onionlike layers around it are flattened Schwann cells, but the greater bulk of the corpuscle consists of concentric layers of fibroblasts with narrow fluid-filled spaces between them. These receptors occur in the periosteum of bone; in joint capsules; in the pancreas and some other viscera; and deep in the dermis, especially on the hands, feet, breasts, and genitals.

- **Ruffini**[7] **corpuscles** are tonic receptors for heavy touch, pressure, stretching of the skin, and joint movements. They are flattened, elongated capsules containing a few nerve fibers and are located in the dermis, subcutaneous tissue, ligaments, tendons, and joint capsules.

## SOMESTHETIC PROJECTION PATHWAYS

From the receptor to the final destination in the brain, most somesthetic signals travel by way of three neurons called the **first-, second-,** and **third-order neurons.** Their axons are called first- through third-order nerve fibers. The first-order fibers for touch, pressure, and proprioception are large, myelinated, and fast; those for heat and

cold are small, unmyelinated or lightly myelinated, and slower.

Somesthetic signals from the head, such as facial sensations, travel by way of several cranial nerves (especially V, the trigeminal nerve) to the pons and medulla oblongata. In the brainstem, the first-order fibers of these neurons synapse with second-order neurons that decussate and end in the contralateral thalamus. Third-order neurons then complete the route to the cerebrum. Proprioceptive signals from the head are an exception, as the second-order fibers carry these signals to the cerebellum.

Below the head, the first-order fibers enter the dorsal horn of the spinal cord. Signals ascend the spinal cord in the spinothalamic and other pathways as detailed in chapter 13 (see table 13.1 and fig. 13.5). These pathways decussate either at or near the point of entry into the spinal cord, or in the brainstem, so the primary somesthetic cortex in each cerebral hemisphere receives signals from the contralateral side of the body.

Signals for proprioception below the head travel up the spinocerebellar tracts to the cerebellum. Signals from the thoracic and abdominal viscera travel to the medulla oblongata by way of sensory fibers in the vagus nerve (X). Recent research has shown that visceral pain signals also ascend the spinal cord in the gracile fasciculus.

## PAIN

Pain is a discomfort caused by tissue injury or noxious stimulation, and typically leading to evasive action. As undesirable as pain may seem, we would be far worse off without it. We see evidence of its value in such diseases as leprosy and diabetes mellitus, where the sense of pain is lost because of nerve damage *(neuropathy)*. The absence of pain makes people unaware of minor injuries. They do not take care of them, so the injuries can become infected and grow worse, to the point that the victim may lose fingers, toes, or entire limbs. In short, pain is an adaptive and necessary sensation. It is mediated by its own specialized nerve fibers, the nociceptors. These are especially dense in the skin and mucous membranes and occur in virtually all organs, although not in the brain. In some brain surgery, the patient must remain conscious and able to talk with the surgeon; such patients need only a local anesthetic. Nociceptors do occur in the meninges of the brain and play an important role in headaches.

---

[7]Angelo Ruffini (1864–1929), Italian anatomist

There are two types of nociceptors corresponding to different pain sensations. Myelinated pain fibers conduct at speeds of 12 to 30 m/sec and produce the sensation of **fast (first) pain**—a feeling of sharp, localized, stabbing pain perceived at the time of injury. Unmyelinated pain fibers conduct at speeds of 0.5 to 2.0 m/sec and produce the **slow (second) pain** that follows—a longer-lasting, dull, diffuse feeling. Pain from the skin, muscles, and joints is called **somatic pain,** and pain from the viscera is called **visceral pain.** The latter often results from stretch, chemical irritants, or ischemia, and is often accompanied by nausea.

Injured tissues release several chemicals that stimulate the nociceptors and trigger pain. **Bradykinin** is the most potent pain stimulus known; it hurts intensely when injected under the skin. It not only makes us aware of injuries but also activates a cascade of reactions that promote healing. Serotonin, prostaglandins, and histamine also stimulate nociceptors, as do potassium ions and ATP released from ruptured cells.

## Projection Pathways for Pain

It is notoriously difficult for clinicians to locate the origin of a patient's pain because pain travels by such diverse and complex routes, and the sensation can originate anywhere along any of the routes. Pain signals reach the brain by two main pathways, but there are multiple subroutes within each of them:

1. Pain signals from the head travel to the brainstem by way of four cranial nerves: mainly the trigeminal (V), but also the facial (VII), glossopharyngeal (IX), and vagus (X) nerves. Trigeminal fibers enter the pons and descend to synapses in the medulla. Pain fibers of the other three cranial nerves also end here. Second-order neurons arise in the medulla and ascend to the thalamus, which relays the message to the cerebral cortex. We will consider the relay from thalamus to cortex shortly.

2. Pain signals from the neck down travel by way of three ascending spinal cord tracts: the spinothalamic tract, spinoreticular tract, and gracile fasciculus. These pathways are described in chapter 13 and need not be repeated here (see table 13.1 and fig. 13.5). The spinothalamic tract is the most significant pain pathway and carries most of the somatic pain signals that ultimately reach the cerebral cortex, making us conscious of pain. The spinoreticular tract carries pain signals to the reticular formation of the brainstem, and these are ultimately relayed to the hypothalamus and limbic system. These pain signals activate visceral, emotional, and behavioral reactions to pain, such as nausea, fear, and some reflex responses. The gracile fasciculus has not been recognized as a pain pathway until recently. It car-

ries signals to the thalamus for visceral pain, such as the pain of a stomachache or from passing a kidney stone. Figure 16.3 shows the spinothalamic and spinoreticular pain pathways.

When the thalamus receives pain signals from the foregoing sources, it relays most of them through third-order neurons to their final destination in the postcentral gyrus of the cerebrum. Exactly what part of this gyrus receives the signals depends on where the pain originated; recall the concept of somatotopy and the sensory homunculus in chapter 14 (p. 541). Most of this gyrus is somesthetic—that is, it receives signals for somatic pain and other senses. A region of the gyrus deep within the lateral sulcus of the brain, however, is a viscerosensory area, which receives the signals of visceral pain conveyed by the gracile fasciculus (see fig. 14.21).

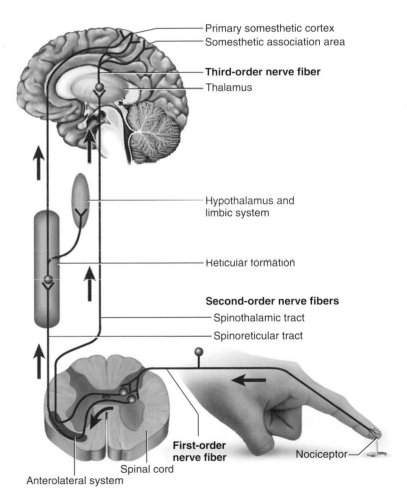

Primary somesthetic cortex
Somesthetic association area

**Third-order nerve fiber**
Thalamus

Hypothalamus and limbic system

Reticular formation

**Second-order nerve fibers**
Spinothalamic tract
Spinoreticular tract

**First-order nerve fiber**

Nociceptor

Spinal cord
Anterolateral system

**FIGURE 16.3  Projection Pathways for Pain.** A first-order neuron conducts a pain signal to the dorsal horn of the spinal cord, a second-order neuron conducts it to the thalamus, and a third-order neuron conducts it to the cerebral cortex. Signals from the spinothalamic tract pass through the thalamus. Signals from the spinoreticular tract bypass the thalamus on the way to the sensory cortex.

Pain in the viscera is often mistakenly thought to come from the skin or other superficial sites—for example when the pain of a heart attack is felt "radiating" along the left shoulder and medial side of the arm. This phenomenon is called **referred pain** (fig. 16.4). It results from the convergence of neural pathways in the CNS. In the case of cardiac pain, for example, spinal cord segments T1 to T5 receive input from the heart as well as the chest and arm. Pain fibers from the heart and skin in this region converge on the same spinal interneurons, then follow the same pathway from there to the thalamus and cerebral cortex. The brain cannot distinguish which source the arriving signals are coming from. It acts as if it assumes that signals arriving by this path are most likely coming from the skin, since skin has more pain receptors than the heart and suffers injury more often. Knowledge of the origins of referred pain is essential to the skillful diagnosis of organ dysfunctions.

## CNS Modulation of Pain

A person's physical and mental state can greatly affect his or her perception of pain. Many mortally wounded soldiers, for example, report little or no pain. The central nervous system (CNS) has **analgesic**[8] (pain-relieving) mechanisms that are just beginning to be understood. The discovery of these mechanisms is tied to the long-known analgesic effects of opium, morphine, and heroin. In 1974, neurophysiologists discovered receptor sites in the brain for these drugs. Since these opiates do not occur naturally in the body, physiologists wondered what normally binds to these receptors. They soon found two analgesic oligopeptides with 200 times the potency of morphine and named them **enkephalins.**[9] Larger analgesic neuropeptides, the **endorphins**[10] and **dynorphins,**[11] were discovered later. All three are known as **endogenous opioids** (which means "internally produced opium-like substances").

These opioids are secreted by the CNS, pituitary gland, digestive tract, and other organs in states of stress or exercise. In the CNS, they are found especially in the central (periaqueductal) gray matter of the midbrain (see fig. 14.9a) and the dorsal horn of the spinal cord. They are *neuromodulators* (see p. 467) that can block the transmission of pain signals and produce feelings of pleasure and euphoria. They may be responsible for the "second wind" ("runner's high") experienced by athletes and for the aforementioned battlefield reports. Their secretion rises sharply in women giving birth. Efforts to employ them in pain therapy have been disappointing, but exercise is an effective part of therapy for chronic pain and may help because it stimulates opioid secretion.

How do these opioids block pain? For pain to be perceived, signals from the nociceptors must get beyond the dorsal horn of the spinal cord and travel to the brain. Through mechanisms called **spinal gating**, pain signals can be stopped at the dorsal horn. Two of these mechanisms are described here.

One mechanism involves **descending analgesic fibers**—nerve fibers that arise in the brainstem, travel down the spinal cord in the reticulospinal tract, and block pain signals from traveling up the cord to the brain. Figure 16.5 depicts one relatively simple mechanism of pain modulation in the dorsal horn of the spinal cord. The normal route of pain transmission is indicated by the red arrows and steps 1 through 3:

① A nociceptor stimulates a second-order nerve fiber. The neurotransmitter at this synapse is called **substance P** (think *P* for "pain"[12]).

② The second-order nerve fiber transmits signals up the spinothalamic tract to the thalamus.

③ The thalamus relays the signals through a third-order neuron to the cerebral cortex, where one becomes conscious of pain.

The pathway for pain blocking, or modulation, is indicated by descending green arrows and steps 4 through 8:

④ Signals from the hypothalamus and cerebral cortex feed into the central gray matter of midbrain, allowing both autonomic and conscious influences on pain perception.

---

[8]*an* = without + *alges* = pain

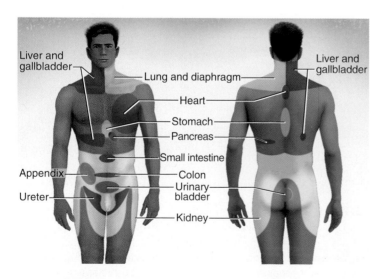

**FIGURE 16.4    Referred Pain.**   Pain from the viscera is often felt in specific areas of the skin.

Liver and gallbladder
Lung and diaphragm
Heart
Stomach
Pancreas
Small intestine
Appendix
Colon
Ureter
Urinary bladder
Kidney
Liver and gallbladder

---

[9]*en* = within + *kephal* = head
[10]acronym from *end*ogenous *m*orphinelike substance
[11]*dyn* = pain
[12]Named *substance P* because it was first discovered in a *powdered* extract of brain and intestine

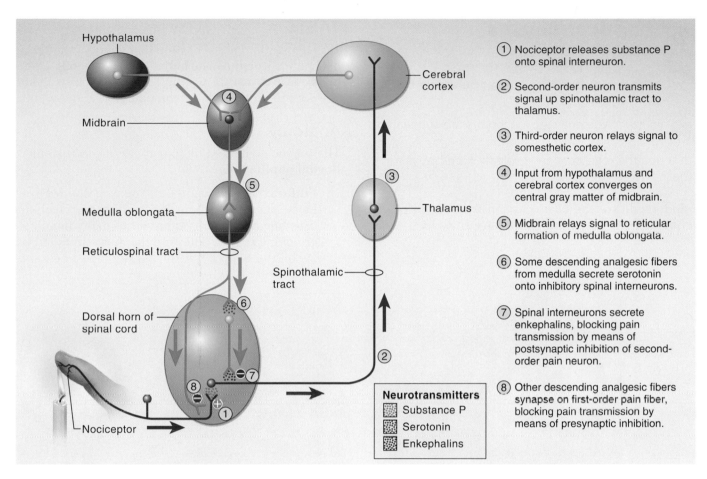

**FIGURE 16.5** Spinal Gating of Pain Signals.

The following labels appear on the figure:

Hypothalamus

Midbrain

Medulla oblongata

Reticulospinal tract

Dorsal horn of spinal cord

Nociceptor

Cerebral cortex

Thalamus

Spinothalamic tract

Neurotransmitters
- Substance P
- Serotonin
- Enkephalins

① Nociceptor releases substance P onto spinal interneuron.

② Second-order neuron transmits signal up spinothalamic tract to thalamus.

③ Third-order neuron relays signal to somesthetic cortex.

④ Input from hypothalamus and cerebral cortex converges on central gray matter of midbrain.

⑤ Midbrain relays signal to reticular formation of medulla oblongata.

⑥ Some descending analgesic fibers from medulla secrete serotonin onto inhibitory spinal interneurons.

⑦ Spinal interneurons secrete enkephalins, blocking pain transmission by means of postsynaptic inhibition of second-order pain neuron.

⑧ Other descending analgesic fibers synapse on first-order pain fiber, blocking pain transmission by means of presynaptic inhibition.

⑤ The midbrain relays signals to certain nuclei in the reticular formation of the medulla oblongata.

⑥ The medulla issues descending, serotonin-secreting analgesic fibers to the spinal cord. These fibers travel the reticulospinal tract and terminate in the dorsal horn at all levels of the cord.

⑦ In the dorsal horn, some of the descending analgesic fibers synapse on short spinal interneurons, which in turn synapse on the second-order pain fiber. These interneurons secrete enkephalins to inhibit the second-order neuron. This is an example of postsynaptic inhibition, working on the downstream side of the synapse between the first- and second-order pain neurons.

⑧ Some fibers from the medulla also exert presynaptic inhibition, synapsing on the axons of the nociceptors and blocking the release of substance P.

Another mechanism of spinal gating is one you may often have consciously employed without knowing why it worked. Have you ever banged your elbow on the edge of a table or pinched your finger in a door, and found that you could ease the pain by rubbing or massaging the injured area? This works because pain-inhibiting interneurons of the dorsal horn, like the one at step 7 in figure 16.5, also receive input from mechanoreceptors in the skin and deeper tissues. When you rub an injured area, you stimulate those mechanoreceptors; they stimulate the spinal interneurons; the interneurons secrete enkephalins; and enkephalins inhibit the second-order pain neurons.

The clinical control of pain has had a particularly interesting history, some of which is retold in Insight 16.5 (p. 630).

## Before You Go On

*Answer the following questions to test your understanding of the preceding section:*

5. What stimulus modalities are detected by free nerve endings?

6. Name any four encapsulated nerve endings and identify their stimulus modalities.

7. Where do most second-order somesthetic neurons synapse with third-order neurons?

8. Explain the phenomenon of referred pain in terms of the neural pathways involved.

9. Explain the roles of bradykinin, substance P, and enkephalins in the perception of pain.

# The Chemical Senses

### Objectives

When you have completed this section, you should be able to

- explain how taste and smell receptors are stimulated; and
- describe the receptors and projection pathways for these two senses.

Taste and smell are the chemical senses. In both cases, environmental chemicals stimulate sensory cells.

## TASTE

Taste (**gustation**) is a sensation that results from the action of chemicals on the **taste buds.** There are about 4,000 taste buds located mainly on the tongue, but also inside the cheeks and on the soft palate, pharynx, and epiglottis.

### Anatomy

The tongue is marked by four types of protrusions called **lingual papillae** (fig. 16.6a):

1. **Filiform**[13] **papillae** are tiny spikes without taste buds. They are responsible for the rough feel of a cat's tongue and are important to many mammals for grooming the fur. They are the most abundant papillae on the human tongue, but they are small and

---

[13]*fili* = thread + *form* = shaped

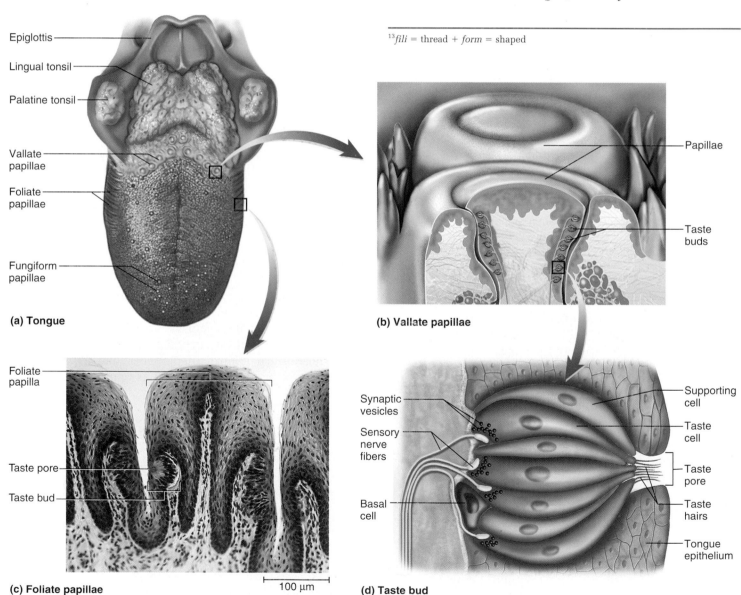

**FIGURE 16.6** Taste (gustatory) Receptors. (a) Dorsal view of the tongue and locations of its papillae. (b) Detail of the vallate papillae. (c) Taste buds on the walls of two adjacent foliate papillae. (d) Structure of a taste bud.

play no gustatory role. They are, however, important to a sense of the texture of food.

2. **Foliate[14] papillae** are also weakly developed in humans. They form parallel ridges on the sides of the tongue about two-thirds of the way back from the tip. Most of their taste buds degenerate by the age of 2 or 3 years.

3. **Fungiform[15] (FUN-jih-form) papillae** are shaped somewhat like mushrooms. Each has about three taste buds, located mainly on the apex. These papillae are widely distributed but especially concentrated at the tip and sides of the tongue.

4. **Vallate[16] (circumvallate) papillae** are large papillae arranged in a V at the rear of the tongue. Each is surrounded by a deep circular trench. There are only 7 to 12 of them, but they contain up to half of all taste buds—around 250 each, located on the wall of the papilla facing the trench (fig. 16.6b).

Regardless of location and sensory specialization, all taste buds look alike (fig. 16.6c, d). They are lemon-shaped groups of 40 to 60 cells of three kinds: *taste cells, supporting cells,* and *basal cells.* **Taste (gustatory) cells** are more or less banana-shaped and have a tuft of apical microvilli called **taste hairs,** which serve as receptor surfaces for taste molecules. The hairs project into a pit called a **taste pore** on the epithelial surface of the tongue. Taste cells are epithelial cells, not neurons, but they synapse with sensory nerve fibers at their base. A taste cell lives 7 to 10 days and is then replaced by mitosis and differentiation of basal stem cells. Supporting cells have a similar shape but no taste hairs. They lie between the taste cells.

## Physiology

To be tasted, molecules must dissolve in the saliva and flood the taste pore. On a dry tongue, sugar or salt has as little taste as a sprinkle of sand. Physiologists currently recognize five primary taste sensations:

1. **Salty,** produced by metal ions such as sodium and potassium. Since these are vital electrolytes, there is obvious value in our ability to taste them and in having an appetite for salt. Electrolyte deficiencies can cause a craving for salt; many animals such as deer, elephants, and parrots thus seek salt deposits when necessary. Pregnancy can lower a woman's electrolyte concentrations and create a craving for salty food.

2. **Sweet,** produced by many organic compounds, especially sugars. Sweetness is associated with carbo-

hydrates and foods of high caloric value. Many flowering plants have evolved sweet nectar and fruits that entice animals to eat them and disperse their pollen and seeds. Thus, our fondness for fruit has coevolved with plant reproductive strategies.

3. **Sour,** usually associated with acids in such foods as citrus fruits.

4. **Bitter,** associated with spoiled foods and alkaloids such as nicotine, caffeine, quinine, and morphine. Alkaloids are often poisonous, and the bitter taste sensation usually induces a human or animal to reject a food. While flowering plants make their fruits temptingly sweet, they often load their leaves with bitter, toxic alkaloids to deter animals from eating them.

5. **Umami** is a "meaty" taste produced by amino acids such as aspartic and glutamic acids. The taste is best known from the salt of glutamic acid, monosodium glutamate (MSG). Pronounced "ooh-mommy," the word is Japanese slang for "delicious" or "yummy."

The many flavors we perceive are not simply a mixture of these five primary tastes, but are also influenced by food texture, aroma, temperature, appearance, and one's state of mind, among other things. Many flavors depend on smell; without its aroma, cinnamon merely has a faintly sweet taste, and coffee and peppermint are bitter. Some flavors such as pepper are due to stimulation of free endings of the trigeminal nerve rather than taste buds. Food scientists refer to the texture of food as *mouthfeel.* Filiform and fungiform papillae of the tongue are innervated by the *lingual nerve* (a branch of the trigeminal) and are sensitive to texture.

All of the primary tastes can be detected throughout the tongue, but certain regions are more sensitive to one category than to others. The tip of the tongue is most sensitive to sweet tastes, which trigger such responses as licking, salivation, and swallowing. The lateral margins of the tongue are the most sensitive areas for salty and sour tastes. Taste buds in the vallate papillae at the rear of the tongue are especially sensitive to bitter compounds, which tend to trigger rejection responses such as gagging to protect against the ingestion of toxins. The threshold for the bitter taste is the lowest of all—that is, we can taste lower concentrations of alkaloids than of acids, salts, and sugars. The senses of sweet and salty are the least sensitive. It is not yet known whether umami stimulates any particular region of the tongue more than other regions.

Sugars, alkaloids, and glutamate stimulate taste cells by binding to receptors on the membrane surface, which then activate G proteins and second-messenger systems within the cell. Sodium and acids penetrate into the cell and depolarize it directly. By either mechanism, taste cells then release neurotransmitters that stimulate the sensory dendrites at their base.

---

[14]*foli* = leaf + *ate* = like
[15]*fungi* = mushroom + *form* = shaped
[16]*vall* = wall + *ate* = like, possessing

## Projection Pathways

Taste buds stimulate the facial nerve (VII) in the anterior two-thirds of the tongue, the glossopharyngeal nerve (IX) in the posterior one-third, and the vagus nerve (X) in the palate, pharynx, and epiglottis. All taste fibers project to the *solitary nucleus* in the medulla oblongata. Second-order neurons from this nucleus relay the signals to two destinations: (1) nuclei in the hypothalamus and amygdala that activate autonomic reflexes such as salivation, gagging, and vomiting, and (2) the thalamus, which relays signals to the insula and postcentral gyrus of the cerebrum. That is where we become conscious of the taste.

## SMELL

The receptor cells for **olfaction** (smell) form a patch of epithelium, the **olfactory mucosa,** in the roof of the nasal cavity (fig. 16.7). This location places the olfactory cells close to the brain, but it is poorly ventilated; forcible sniffing is often needed to identify an odor or locate its source. Nevertheless, the sense of smell is highly sensitive. We can detect odor concentrations as low as a few parts per trillion. Most people can distinguish 2,000 to 4,000 odors, and some can distinguish up to 10,000. On average, women are more sensitive to odors than men are, and they are measurably more sensitive to some odors

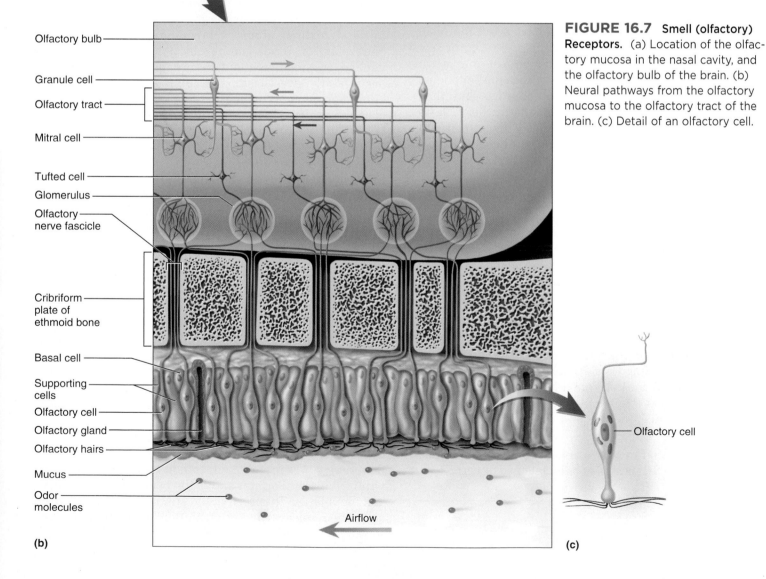

**FIGURE 16.7   Smell (olfactory) Receptors.** (a) Location of the olfactory mucosa in the nasal cavity, and the olfactory bulb of the brain. (b) Neural pathways from the olfactory mucosa to the olfactory tract of the brain. (c) Detail of an olfactory cell.

near the time of ovulation as opposed to other phases of the menstrual cycle. Olfaction is highly important in the social interactions of other animals and, in more subtle ways, to humans (see Insight 16.1).

## Anatomy

The olfactory mucosa covers about 5 cm$^2$ of the superior concha, cribriform plate, and nasal septum of each nasal fossa. It consists of 10 to 20 million **olfactory cells** as well as epithelial supporting cells and basal stem cells. The rest of the nasal cavity is lined by a nonsensory *respiratory mucosa.*

Unlike taste cells, which are epithelial, olfactory cells are neurons. They are shaped a little like bowling pins (fig. 16.7c). The widest part, the soma, contains the nucleus. The neck and head of the cell are a modified dendrite with a swollen tip. The head bears 10 to 20 cilia called **olfactory hairs.** These cilia are immobile, but they have binding sites for odor molecules. They lie in a tangled mass embedded in a thin layer of mucus. The basal end of each cell tapers to become an axon. These axons collect into small fascicles that leave the nasal cavity through pores *(cribriform foramina)* in the ethmoid bone. Collectively, the fascicles are regarded as cranial nerve I (the olfactory nerve).

Olfactory cells are the only neurons in the body directly exposed to the external environment. Apparently this is hard on them, because they have a life span of only 60 days. Unlike most neurons, however, they are replaceable. The basal cells continually divide and differentiate into new olfactory cells.

## Physiology

Humans have a poor sense of smell compared with most other mammals; it seems to have declined as the primate visual system increased in significance. A 3-kg cat, for example, has about 20 cm$^2$ of olfactory mucosa, whereas a 70-kg human has only 10 cm$^2$. Nevertheless, we can smell a great multitude of odorant molecules, some of them at very low concentrations. Attempts to group these into odor classes have been inconclusive and controversial; suggested classes have included pungent, floral, musky, and earthy. Olfaction researchers Linda Buck and Richard Axel received the 2004 Nobel Prize for Physiology or Medicine for applying methods of molecular genetics to identify olfactory receptor proteins and their genes. Rats and mice have up to 1,200 functional olfactory genes, but in humans, two-thirds of them have mutated to the point of being inoperative; we have only about 350 kinds of olfactory receptors left. Each olfactory cell has only one receptor type and therefore binds only one odorant.

The first step in smell is that an odorant molecule must bind to a receptor on one of the olfactory hairs. Hydrophilic odorants diffuse freely through the mucus of the olfactory epithelium and bind directly to a receptor. Hydrophobic odorants are transported to the receptor by an *odorant-binding protein* in the mucus. When the receptor binds an odorant, it activates a G protein and through it, the cyclic adenosine monophosphate (cAMP) second-messenger system. The cAMP system ultimately opens ion channels in the plasma membrane, admitting cations (Na$^+$ or Ca$^{2+}$) into the cell and depolarizing it, creating a receptor potential. This triggers action potentials in the axon of the olfactory cell, and a signal is transmitted to the brain.

Some odorants, however, act on nociceptors of the trigeminal nerve rather than on olfactory cells. These odorants include ammonia, menthol, chlorine, and the capsaicin of hot peppers. "Smelling salts" revive unconscious persons by strongly stimulating the trigeminal nerve with ammonia fumes.

The sense of smell adapts quickly. We may therefore be unaware of our own body odors or have difficulty locating a gas leak in a room because the smell quickly "goes away." Adaptation does not occur in the sensory cells, but is due to synaptic inhibition in the *olfactory bulbs,* which we are about to study.

## Projection Pathways

When olfactory fibers pass through the roof of the nose, they enter a pair of **olfactory bulbs** beneath the frontal lobes of the brain (see fig. 14.27, p. 548). Here they synapse with the dendrites of two types of neurons higher in the bulb, called *mitral cells* and *tufted cells.* Olfactory axons reach up, and mitral and dendritic cell dendrites reach down, to meet each other in spherical clusters called *glomeruli* (fig. 16.7b). All olfactory fibers leading to any one glomerulus come from cells with the same receptor type; thus each glomerulus is dedicated to a particular

---

## INSIGHT 16.1  Evolutionary Medicine

### Human Pheromones

There is an abundance of anecdote, but no clear experimental evidence, that human body odors affect sexual behavior. There is more adequate evidence, however, that a person's sweat and vaginal secretions affect other people's sexual physiology, even when the odors cannot be consciously smelled. Experimental data show that a woman's apocrine sweat can influence the timing of other women's menstrual cycles. This can produce a so-called *dormitory effect,* in which women who live together tend to have synchronous menstrual cycles. The presence of men seems to influence female ovulation. Conversely, when a woman is ovulating or close to it, and therefore fertile, her vaginal secretions contain pheromones called *copulines,* which have been shown to raise men's testosterone levels.

odor. Higher brain centers interpret complex odors such as chocolate, perfume, or coffee by decoding signals from a combination of odor-specific glomeruli. This is similar to the way our visual system decodes all the colors of the rainbow using input from just three color-specific receptor cells of the eye.

The tufted and mitral cells receive output from the glomeruli. Their axons form bundles called **olfactory tracts,** which course posteriorly along the underside of the frontal lobes. Most fibers of the olfactory tracts end in various neighboring regions of the inferior surface of the temporal lobe (fig. 16.8); collectively we can regard all these regions as the **primary olfactory cortex.** It is noteworthy that olfactory signals reach the cerebral cortex directly, without first passing through the thalamus; this is not true of any of our other senses. Even in olfaction, however, some signals from the primary olfactory cortex continue to a relay in the thalamus on their way to olfactory association areas elsewhere.

From the primary olfactory cortex, signals travel to several other secondary destinations in the cerebrum and brainstem. Two important cerebral destinations are the insula and orbitofrontal cortex. The orbitofrontal cortex, which lies on the floor of the frontal lobe just above the eyes (see fig. 14.17, p. 535), seems to be the site where we identify and discriminate among odors. It receives inputs for both taste and smell and integrates these into our overall perception of flavor. Other secondary destinations for olfactory signals include the hippocampus, amygdala, and hypothalamus. Considering the roles of these brain areas, it is not surprising that the odor of certain foods, a perfume, a hospital, or decaying flesh can evoke strong memories, emotional responses, and visceral reactions such as sneezing or coughing, the secretion of saliva and stomach acid, or vomiting.

Most areas of olfactory cortex also send fibers back to the olfactory bulbs, where they synapse on *granule cells* and in the glomeruli. Granule cells, in turn, inhibit the mitral and tufted cells. An effect of this feedback is that odors can change in quality and significance under different conditions. Food may smell more appetizing when you are hungry, for example, than when you have just eaten or when you are ill.

**Think About It**

*Which taste sensations could be lost after damage to (1) the facial nerve or (2) the glossopharyngeal nerve? A fracture of which cranial bone would most likely eliminate the sense of smell?*

### Before You Go On

*Answer the following questions to test your understanding of the preceding section:*

10. *What is the difference between a lingual papilla and a taste bud? Which is visible to the naked eye?*

11. *List the primary taste sensations and discuss their adaptive significance (survival value).*

12. *Which cranial nerves carry gustatory impulses to the brain?*

13. *What part of an olfactory cell binds odor molecules?*

# Hearing and Equilibrium

### Objectives

When you have completed this section, you should be able to

- identify the properties of sound waves that account for pitch and loudness;
- describe the gross and microscopic anatomy of the ear;
- explain how the ear converts vibrations to nerve signals and discriminates between sounds of different intensity and pitch;
- explain how the vestibular apparatus enables the brain to interpret the body's position and movements; and
- describe the pathways taken by auditory and vestibular signals to the brain.

*Hearing* is a response to vibrating air molecules and *equilibrium* is the sense of motion and balance. These senses

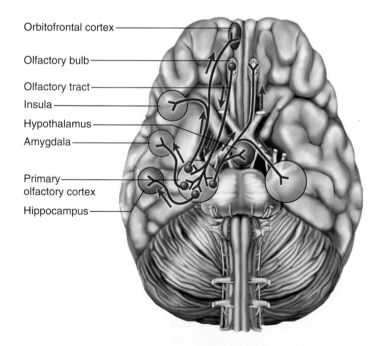

Orbitofrontal cortex
Olfactory bulb
Olfactory tract
Insula
Hypothalamus
Amygdala
Primary olfactory cortex
Hippocampus

**FIGURE 16.8**  Olfactory Projection Pathways in the Brain.

reside in the inner ear, a maze of fluid-filled passages and sensory cells. This section explains how the fluid is set in motion and how the sensory cells convert this motion into an informative pattern of action potentials.

## THE NATURE OF SOUND

To understand the physiology of hearing, it is necessary to know some basic properties of sound. **Sound** is any audible vibration of molecules. It can be transmitted through water, solids, or air, but not through a vacuum. Our discussion is limited to airborne sound.

Sound is produced by a vibrating object such as a tuning fork, a loudspeaker, or the vocal cords. Consider a loudspeaker producing a pure tone. When the speaker cone moves forward, it pushes air molecules ahead of it. They collide with other molecules just ahead of them, and energy is thus transferred from molecule to molecule until it reaches the eardrum. No one molecule moves very far; they simply collide with each other like a series of billiard balls until finally, some molecules collide with the eardrum and make it vibrate. The sensations we perceive as the pitch and loudness of the sound are related to the physical properties of these vibrations.

### Pitch

**Pitch** is our sense of whether a sound is "high" (treble) or "low" (bass). It is determined by the frequency at which the sound source, eardrum, and other parts of the ear vibrate. One movement of a vibrating object back and forth is called a *cycle,* and the number of cycles per second (cps or hertz, Hz) is called **frequency.** The lowest note on a piano, for example, is 27.5 Hz, middle C is 261 Hz, and the highest note is 4,176 Hz. The most sensitive human ears can hear frequencies from 20 to 20,000 Hz. The *infrasonic* frequencies below 20 Hz are not detected by the ear, but we sense them through vibrations of the skull and skin, and they play a significant role in our appreciation of music. The inaudible vibrations above 20,000 Hz are *ultrasonic.* Human ears are most sensitive to frequencies ranging from 1,500 to 5,000 Hz. In this range, we can hear sounds of relatively low energy (volume), whereas sounds above or below this range must be louder to be audible (fig. 16.9). Normal speech falls within this frequency range. Most of the hearing loss suffered with age is in the range of 250 to 2,050 Hz.

### Loudness

**Loudness** is the perception of sound energy, intensity, or **amplitude** of vibration. In the speaker example, amplitude is a measure of how far forward and back the cone vibrates on each cycle and how much it compresses the air molecules in front of it. Loudness is expressed in

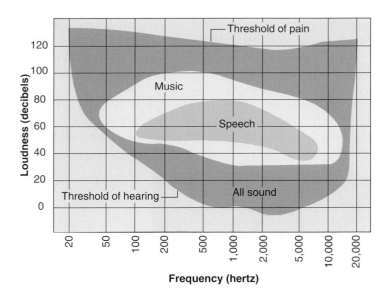

**FIGURE 16.9  The Range of Human Hearing.** People with very sensitive ears can hear sounds from 20 to 20,000 hertz (Hz), but to be heard, sounds at these extremes must be louder than those in the midrange. Our ears are most sensitive to frequencies of 1,500 to 5,000 Hz, where we can hear relatively soft sounds. Thus, the threshold of hearing varies with the frequency of the sound. Most sounds above 120 decibels (dB) are painful to the ear.

❱ *How would the shape of this graph change in a case of moderate hearing loss between 200 and 5,000 Hz?*

decibels (dB), with 0 dB being the threshold of human hearing. Every 10 dB step up the scale represents a sound with 10 times greater intensity. Thus, 10 dB is 10 times threshold, 20 dB is 100 times threshold, 30 dB is 1,000 times threshold, and so forth. Normal conversation has a loudness of about 60 dB. At most frequencies, the threshold of pain is 120 to 140 dB, approximately the intensity of a loud thunderclap. Prolonged exposure to sounds greater than 90 dB can cause permanent loss of hearing.

## ANATOMY OF THE EAR

The ear has three sections called the *outer, middle,* and *inner ear.* The first two are concerned only with transmitting sound to the inner ear, which houses the transducer that converts fluid motion to action potentials.

### Outer Ear

The **outer (external) ear** is essentially a funnel for conducting air vibrations to the eardrum. It begins with the fleshy **auricle,** or **pinna,** on the side of the head, shaped and supported by elastic cartilage except for the earlobe. It has a predictable arrangement of named whorls and recesses that direct sound into the auditory canal (fig. 16.10).

The **auditory canal** is the passage leading through the temporal bone to the eardrum. Beginning at the external opening, the **external acoustic meatus,** it follows a slightly S-shaped course for about 3 cm (fig. 16.11). It is lined with skin and supported by fibrocartilage at its opening and by the temporal bone for the rest of its length. Ceruminous and sebaceous glands in the canal produce secretions that mix with dead skin cells and form *cerumen* (earwax). Cerumen waterproofs the auditory canal, keeps the eardrum pliable, and inhibits microbial growth. Cerumen normally dries and falls from the canal, but sometimes it becomes impacted and interferes with hearing.

## Middle Ear

The **middle ear** is located in the **tympanic cavity** of the temporal bone. It begins with the eardrum, or **tympanic[17] membrane,** which closes the inner end of the auditory canal and separates it from the middle ear. The membrane is about 1 cm in diameter and slightly concave on its outer surface. It is suspended in a ring-shaped groove in the temporal bone and vibrates freely in response to sound. It is innervated by sensory branches of the vagus and trigeminal nerves and is highly sensitive to pain.

Posteriorly, the tympanic cavity is continuous with the mastoidal air cells in the mastoid process. It is filled with air that enters by way of the **auditory (eustachian[18]) tube,** a passageway to the nasopharynx. (Be careful not to confuse *auditory tube* with *auditory canal.*) The auditory tube is normally flattened and closed, but swallowing or yawning opens it and allows air to enter or leave the tympanic cavity. This equalizes

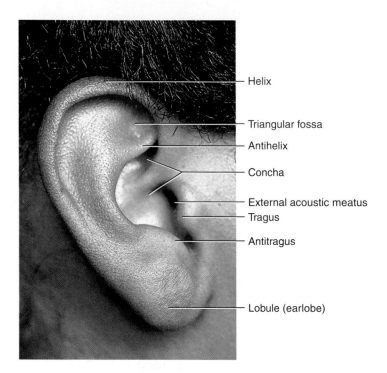

**FIGURE 16.10   Anatomy of the Auricle (pinna) of the Ear.**

Helix

Triangular fossa

Antihelix

Concha

External acoustic meatus

Tragus

Antitragus

Lobule (earlobe)

---

[17]*tympan* = drum
[18]Bartholomeo Eustachio (1524–74), Italian anatomist

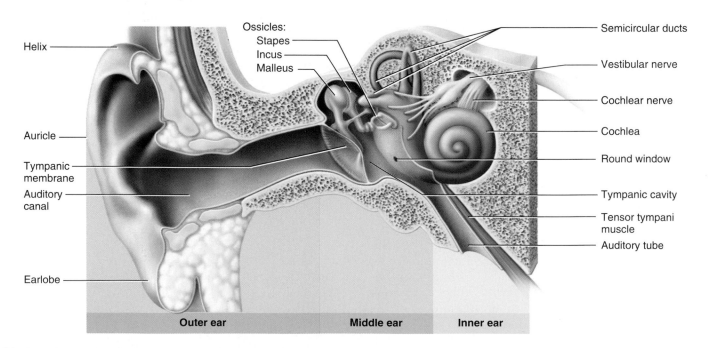

Helix

Auricle

Tympanic membrane

Auditory canal

Earlobe

Ossicles:
Stapes
Incus
Malleus

Semicircular ducts

Vestibular nerve

Cochlear nerve

Cochlea

Round window

Tympanic cavity

Tensor tympani muscle

Auditory tube

**Outer ear**          **Middle ear**          **Inner ear**

**FIGURE 16.11   Internal Anatomy of the Ear.**

air pressure on both sides of the eardrum and allows it to vibrate freely. Excessive pressure on one side or the other dampens the sense of hearing. Unfortunately, the auditory tube also allows throat infections to spread to the middle ear (see Insight 16.2).

The tympanic cavity, a space only 2 to 3 mm wide between the outer and inner ear, contains the three smallest bones and two smallest skeletal muscles of the body. The bones, called the **auditory ossicles,**[19] connect the eardrum to the inner ear. Progressing inward, the first is the **malleus,**[20] which has an elongated *handle* attached to the inner surface of the eardrum; a *head,* which is suspended by a ligament from the wall of the tympanic cavity; and a *short process,* which articulates with the next ossicle. The second bone, the **incus,**[21] has a somewhat cuboidal body that articulates with the malleus, and an elongated process that articulates with the stapes. The **stapes**[22] (STAY-peez) has an arch and *footplate* that give it a shape like a stirrup. The footplate, shaped like the sole of a steam iron, is held by a ringlike ligament in an opening called the **oval window,** where the inner ear begins.

The muscles of the middle ear are the stapedius and tensor tympani. The **stapedius** (stay-PEE-dee-us) arises from the posterior wall of the cavity and inserts on the stapes. The **tensor tympani** (TEN-sor TIM-pan-eye) arises from the wall of the auditory tube, travels alongside it, and inserts on the malleus. The function of these muscles is discussed under the physiology of hearing.

## INSIGHT 16.2 Clinical Application

### Middle-Ear Infection

Otitis[23] media (middle-ear infection) is common in children because their auditory tubes are relatively short and horizontal. It allows upper respiratory infections to spread easily from the throat to the tympanic cavity and mastoidal air cells. Fluid accumulates in the cavity and produces pressure, pain, and impaired hearing. If otitis media goes untreated, it may spread from the mastoidal air cells and cause meningitis, a potentially deadly infection (see Insight 14.1). Otitis media can also cause fusion of the middle-ear bones and result in hearing loss. It is sometimes necessary to drain fluid from the tympanic cavity by lancing the eardrum and inserting a tiny drainage tube—a procedure called *myringotomy.*[24] The tube, which is eventually sloughed out of the ear, relieves the pressure and permits the infection to heal.

## Inner Ear

The **inner ear** is housed in a maze of temporal bone passageways called the **bony labyrinth,** which is lined by a system of fleshy tubes called the **membranous labyrinth** (fig. 16.12). Between the bony and membranous labyrinths is a cushion of fluid called **perilymph** (PER-ih-limf), similar to cerebrospinal fluid. Within the membranous labyrinth is a fluid called **endolymph,** similar to intracellular fluid.

The labyrinths begin with a chamber called the **vestibule,** which contains organs of equilibrium to be discussed later. The organ of hearing is the **cochlea**[25] (COC-lee-uh), a coiled tube that arises from the anterior side of the vestibule. In other vertebrates, the cochlea is straight or slightly curved. In most mammals, however, it assumes the form of a snaillike spiral, which allows a longer cochlea to fit in a compact space. In humans, the spiral is about 9 mm wide at the base and 5 mm high. Its apex points anterolaterally. The cochlea winds for about 2.5 coils around an axis of spongy bone called the **modiolus**[26] (mo-DY-oh-lus). The modiolus is shaped like a screw; its threads form a spiral platform that supports the fleshy tube of the cochlea.

A vertical section cuts through the cochlea about five times (fig. 16.13a). A single cross section looks like figure 16.13b. It is important to realize that the structures seen in cross section actually have the form of spiral strips winding around the modiolus from base to apex.

The cochlea has three fluid-filled chambers separated by membranes. The superior chamber is called the **scala**[27] **vestibuli** (SCAY-la vess-TIB-you-lye) and the inferior one is the **scala tympani.** These are filled with perilymph and communicate with each other through a narrow channel at the apex of the cochlea. The scala vestibuli begins near the oval window and spirals to the apex; from there, the scala tympani spirals back down to the base and ends at the **round window** (see fig. 16.12). The round window is covered by a membrane called the *secondary tympanic membrane.*

The middle chamber is a triangular space, the **cochlear duct** (scala media). It is separated from the scala vestibuli above by a thin **vestibular membrane** and from the scala tympani below by a much thicker **basilar membrane.** Unlike those chambers, it is filled with endolymph rather than perilymph. The vestibular membrane separates the endolymph from the perilymph and helps to maintain a chemical difference between them. Within the cochlear duct, supported on the basilar membrane, is the **spiral organ,** also known as the *acoustic organ* or *organ of Corti*[28] (COR-tee). The acoustic organ is a thick epithelium of sensory and supporting cells and associated membranes (fig. 16.13c). It is the transducer that converts vibrations

[19]*oss* = bone + *icle* = little
[20]*malleus* = hammer
[21]*incus* = anvil
[22]*stapes* = stirrup
[23]*ot* = ear + *itis* = inflammation
[24]*myringo* = eardrum + *tomy* = cutting

[25]*cochlea* = snail
[26]*modiolus* = hub
[27]*scala* = staircase
[28]Alfonso Corti (1822–88), Italian anatomist

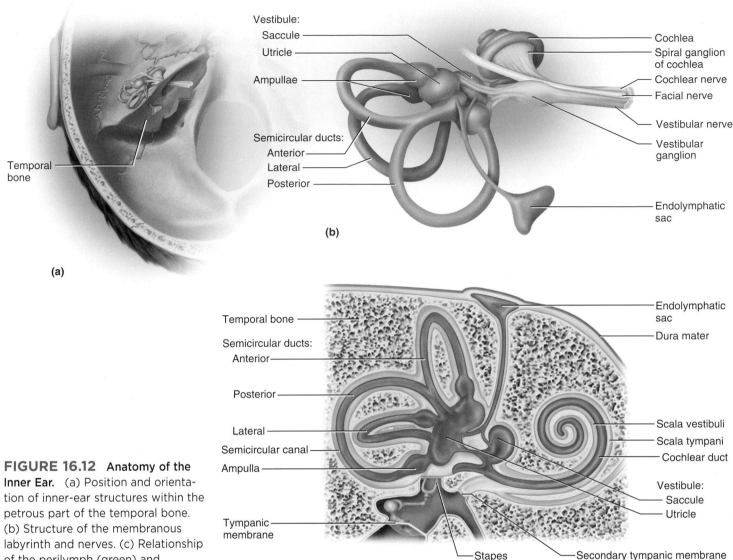

**(a)**

**(b)**

Vestibule:
Saccule
Utricle
Ampullae

Semicircular ducts:
Anterior
Lateral
Posterior

Temporal
bone

Cochlea
Spiral ganglion
of cochlea
Cochlear nerve
Facial nerve
Vestibular nerve
Vestibular
ganglion

Endolymphatic
sac

Temporal bone
Semicircular ducts:
Anterior
Posterior
Lateral
Semicircular canal
Ampulla
Tympanic
membrane

Endolymphatic
sac
Dura mater

Scala vestibuli
Scala tympani
Cochlear duct

Vestibule:
Saccule
Utricle

Stapes
in oval window
Secondary tympanic membrane
in round window

**(c)**

**FIGURE 16.12   Anatomy of the Inner Ear.** (a) Position and orientation of inner-ear structures within the petrous part of the temporal bone. (b) Structure of the membranous labyrinth and nerves. (c) Relationship of the perilymph (green) and endolymph (blue) to the labyrinth.

into nerve impulses, so we must pay particular attention to its structural details.

The spiral organ has an epithelium composed of **hair cells** and **supporting cells.** Hair cells are named for the long, stiff microvilli called **stereocilia**[29] on their apical surfaces. (Stereocilia should not be confused with true cilia. They do not have an axoneme of microtubules as seen in cilia, and they do not move by themselves.) Resting on top of the stereocilia is a gelatinous **tectorial**[30] **membrane.**

The spiral organ has four rows of hair cells spiraling along its length (fig. 16.14). About 3,500 of these, called **inner hair cells (IHCs),** are arranged in a row on the medial side of the basilar membrane (facing the modiolus). Each of these has a cluster of 50 to 60 stereocilia, graded from short to tall. Another 20,000 **outer hair cells (OHCs)** are neatly

arranged in three rows across from the inner hair cells. Each outer hair cell has about 100 stereocilia arranged in the form of a V, with their tips embedded in the tectorial membrane. All that we hear comes from the IHCs, which supply 90% to 95% of the sensory fibers of the cochlear nerve. The function of the OHCs is to adjust the response of the cochlea to different frequencies and enable the IHCs to work with greater precision. We will see shortly how this is done. Hair cells are not neurons, but synapse with nerve fibers at their base—the OHCs with both sensory and motor neurons and the IHCs with sensory neurons only.

## THE PHYSIOLOGY OF HEARING

We can now examine the way in which sound affects the ear and produces action potentials. Sound waves enter the auditory canal on one side, and nerve signals exit the inner ear on the other. Connecting these is the middle ear, so we begin with an analysis of its contribution.

[29]*stereo* = solid
[30]*tect* = roof

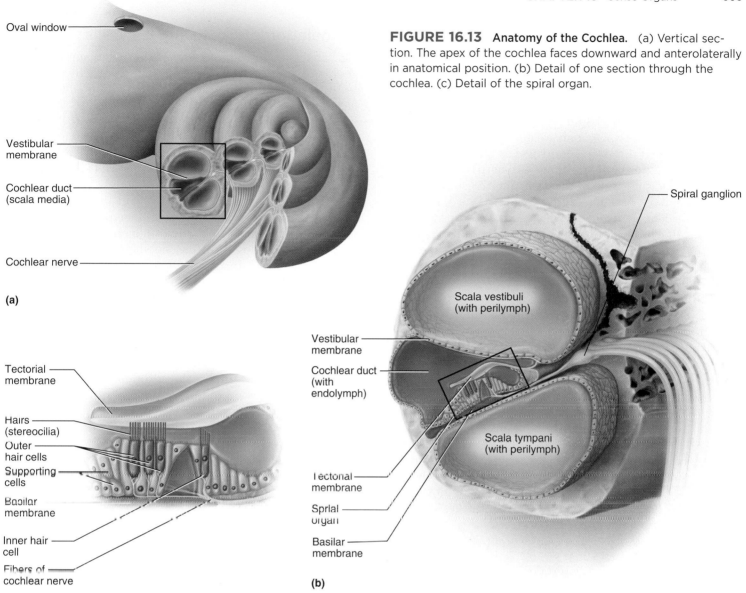

**FIGURE 16.13** **Anatomy of the Cochlea.** (a) Vertical section. The apex of the cochlea faces downward and anterolaterally in anatomical position. (b) Detail of one section through the cochlea. (c) Detail of the spiral organ.

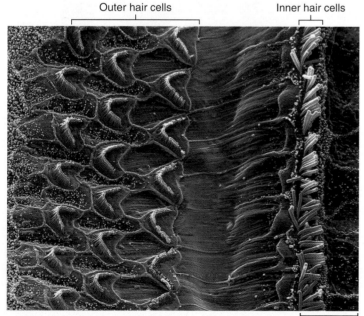

**FIGURE 16.14** **Apical Surfaces of the Cochlear Hair Cells.** All signals that we hear come from the inner hair cells on the right.

▶ *What is the function of the outer hair cells?*

## The Middle Ear

One might wonder why we have a middle ear at all—why the eardrum doesn't simply vibrate against the fluid-filled labyrinth of the inner ear. The reason is that the eardrum, which moves in air, vibrates quite easily, whereas the stapes footplate must vibrate against the fluid of the inner ear. This fluid puts up a much greater resistance to motion than air does. If airborne sound waves struck the stapes footplate directly, or if the inner-ear perilymph directly contacted the inner side of the eardrum, the sound waves would not have enough energy to move the perilymph adequately. The eardrum, however, has 18 times the area of the oval window. By concentrating the energy of the vibrating eardrum on an area $^{1}/_{18}$ that size, the ossicles create a greater force per unit area at the oval window and overcome the inertia of the endolymph.

The auditory ossicles do not, however, provide any mechanical advantage, any amplification of sound. Vibrations of the stapes against the inner ear normally have the same amplitude as vibrations of the eardrum against the malleus. Why, then, have a lever system composed of three auditory ossicles? Why not simply have one ossicle concentrating the mechanical energy of the eardrum directly on the inner ear?

The answer is that the chain of ossicles serves at times to *lessen* the transfer of energy to the inner ear. They and their muscles also have a protective function. In response to a loud noise, the tensor tympani pulls the eardrum inward and tenses it, while the stapedius reduces the motion of the stapes. This **tympanic reflex** muffles the transfer of vibrations from the eardrum to the oval window. The reflex probably evolved in part for protection from loud but slowly building natural sounds such as thunder. The reflex has a latency of about 40 msec, which is not quick enough to protect the inner ear from sudden artificial noises such as gunshots. The tympanic reflex also does not adequately protect the ears from sustained loud noises such as factory noise or loud music. Such noises can irreversibly damage the hair cells of the inner ear by fracturing their stereocilia. It is therefore imperative to wear ear protection when using firearms or working in noisy environments.

The middle-ear muscles also help to coordinate speech with hearing. Without them, the sound of your own speech would be so loud it could damage your inner ear, and it would drown out soft or high-pitched sounds from other sources. Just as you are about to speak, however, the brain signals these muscles to contract. This dampens the sense of hearing in phase with the inflections of your own voice and makes it easier to hear other people while you are speaking.

### Think About It

*What type of muscle fibers—slow oxidative or fast glycolytic (see p. 429)—do you think constitute the stapedius and tensor tympani? That is, which type would best suit the purpose of these muscles?*

## Stimulation of Cochlear Hair Cells

The next step in hearing is based on movement of the cochlear hair cells relative to stationary structures nearby. In this section, we will see how movements of the inner-ear fluids and basilar membrane move the hair cells and, just as importantly, why it is important that the tectorial membrane near the hair cells remain relatively still.

A simple mechanical model of the ear can help in visualizing how this happens (fig. 16.15). (The vestibular membrane is omitted from the model for simplicity, because it has no significant effect on the mechanics of the cochlea.) As you listen to your favorite music, each inward movement of the tympanic membrane pushes the middle-ear ossicles inward. The stapes, in turn, pushes on the perilymph in the scala vestibuli. Perilymph, like other liquids, cannot be compressed, so it flows away from the stapes footplate. The resulting pressure in the scala vestibuli pushes the vestibular membrane downward; this pushes on the endolymph in the cochlear duct; the endolymph pushes down on the basilar membrane; the basilar membrane pushes on the perilymph in the scala tympani; and finally, the secondary tympanic membrane bulges outward to relieve the pressure. As the cycle of vibration continues, the stapes pulls back from the oval window and all of this happens in reverse.

In short, as the stapes goes in-out-in, the secondary tympanic membrane goes out-in-out, and the basilar membrane goes down-up-down. It is not difficult to see how this happens—the only thing hard to imagine is that

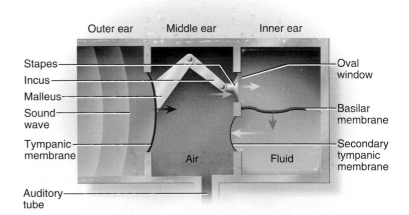

**FIGURE 16.15  Mechanical Model of the Ear.** Each inward movement of the tympanic membrane pushes inward on the auditory ossicles of the middle ear and fluid of the inner ear. This pushes down on the basilar membrane, and pressure is relieved by an outward bulge of the secondary tympanic membrane. Thus the basilar membrane vibrates up and down in synchrony with the vibrations of the tympanic membrane.

▶ *Why would high air pressure in the middle ear reduce the movements of the basilar membrane of the inner ear?*

it can happen as often as 20,000 times per second! The important thing about all this is that the hair cells, affixed to the basilar membrane, are going along for the ride, bobbing up and down as the basilar membrane moves.

To understand how all of this leads to electrical excitation of the hair cells, we must seemingly digress for a moment to examine the tips of the hair cells. These are bathed in a high-potassium fluid, the endolymph. Why is this fluid so rich in potassium, and why is that important? Potassium is secreted into the endolymph by cells around the circumference of the cochlear duct (on the wall opposite from the modiolus). The vestibular membrane retains this fluid in the duct. Relative to the perilymph, the endolymph has an electrical potential of about +80 mV, and the interior of the hair cell about −40 mV. Thus there is an exceptionally strong electrochemical gradient from the endolymph to the hair cell cytoplasm. This gradient provides the potential energy that ultimately enables the hair cell to work.

On the inner hair cells—the ones that generate all the signals we hear—each stereocilium has a single transmembrane protein at its tip that functions as a mechanically gated ion channel. A fine, stretchy protein filament called a **tip link** extends like a spring from the ion channel of one stereocilium to the sidewall of the stereocilium next to it (fig. 16.16). The stereocilia increase in height progressively, so that every stereocilium but the tallest one has a tip link leading to the next taller stereocilium beside it.

Now what of the tectorial membrane? This is a conspicuous structure of the spiral organ which is anchored to the core of the cochlea and remains relatively still as the hair cells dance up and down to the beat of the music. Each time the basilar membrane rises upward toward the tectorial membrane, the hair cell stereocilia are pushed against that membrane and tilt toward the tallest one. As each taller stereocilium bends over, it pulls on the tip link. The tip link, connected to the ion channel of the next shorter stereocilium, pulls the channel open and allows ions to flood into the cell. The channel is nonselective, but since the predominant ion of the endolymph is $K^+$, the primary effect of this gating is to allow a quick burst of $K^+$ to flow into the hair cell. This depolarizes the hair cell while the channel is open, and when the basilar membrane drops and the stereocilium bends the other way, its channel closes and the cell becomes briefly hyperpolarized. During each moment of depolarization, a hair cell releases a burst of neurotransmitter from its base, exciting the sensory dendrite with which the hair cell synapses. Each depolarization thus generates action potentials in the cochlear nerve.

To summarize this process: Each sound wave pushes the eardrum and ossicles inward, creating pressure on the perilymph in the upper chamber (scala vestibuli) of the cochlea. The pressure wave in the inner-ear fluids pushes down on the basilar membrane and is then relieved by the outward bulge of the secondary tympanic membrane. As the eardrum vibrates outward, all of this happens in

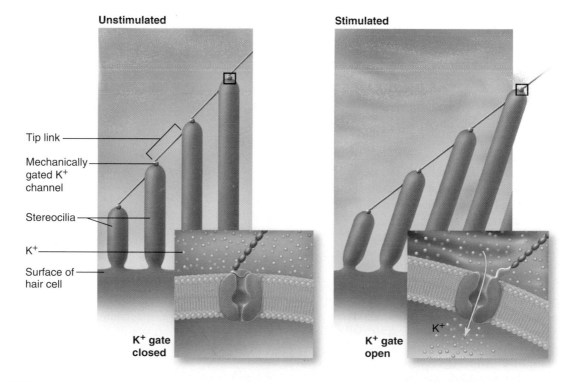

**Unstimulated**     **Stimulated**

Tip link

Mechanically gated $K^+$ channel

Stereocilia

$K^+$

Surface of hair cell

$K^+$ gate closed      $K^+$ gate open

$K^+$

**FIGURE 16.16 Potassium Gates of the Cochlear Hair Cells.** Each stereocilium has a gated $K^+$ channel at its tip. Vibrations of the cochlea cause each stereocilium to bend and, with its tip link, pull open the $K^+$ channel of the adjacent stereocilium. The inflow of $K^+$ depolarizes the hair cell.

reverse—the basilar membrane moves up and the secondary tympanic membrane vibrates inward.

Each upward movement of the basilar membrane pushes the inner hair cells closer to the stationary tectorial membrane. This forces the stereocilia to bend in the direction of the tallest one. Each stereocilium has a tip link connecting it to an ion channel at the top of the next shorter stereocilium. When the taller one bends over, it pulls the ion channel open. Potassium ions flow into the hair cell and depolarize it. The hair cell releases a burst of neurotransmitter, exciting the sensory processes of the cochlear nerve cells below it. Thus a signal is generated in the cochlear nerve and transmitted to the brain.

## Sensory Coding

For sounds to carry any meaning, we must distinguish differences in loudness and pitch. Our ability to do so stems from the fact that the cochlea responds differently to sounds of different amplitude and frequency. Variations in loudness (amplitude) cause variations in the intensity of cochlear vibration. A soft sound produces relatively slight up-and-down movements of the basilar membrane. Hair cells are stimulated only moderately, and a given sound frequency stimulates hair cells in a relatively limited, or focused, region of the spiral organ. A louder sound makes the basilar membrane vibrate more vigorously. Hair cells respond more intensely, generating a higher firing frequency in the cochlear nerve, and a given sound frequency excites a greater number of hair cells over a broader expanse of the spiral organ. If the brain detects moderate firing rates associated with hair cells in relatively narrow bands of the spiral organ, it interprets this as a soft sound. If it detects a high firing frequency in nerve fibers associated with broad bands of the spiral organ, it interprets this as a loud sound.

Frequency discrimination is more sophisticated. At its proximal end (the base of the cochlea), the basilar membrane is attached, narrow, and stiff. At its distal end (the apex of the cochlea), it is unattached, five times wider than at the base, and more flexible. Think of the basilar membrane as analogous to a rope stretched tightly between two posts. If you pluck the rope at one end, a wave of vibration travels down its length and back. This produces a standing wave, with some regions of the rope vertically displaced more than others. Similarly, a sound causes a standing wave in the basilar membrane. The peak amplitude of this wave is near the distal end in the case of low-frequency sounds and nearer the proximal (attached) end with sounds of higher frequencies. When the brain receives signals mainly from inner hair cells at the distal end, it interprets the sound as low-pitched; when signals come mainly from the proximal end, it interprets the sound as high-pitched (fig. 16.17). Speech, music, and other everyday sounds, of course, are not pure tones—they create complex patterns of vibration in the basilar membrane that must be decoded by the brain.

## Cochlear Tuning

Just as we tune a radio to receive a certain frequency, we also tune our cochlea to receive some frequencies better than others. The outer hair cells (OHCs) are supplied with a few sensory fibers (5%–10% of those in the cochlear nerve), but more importantly, they receive motor fibers from the brain.

In response to sound, the OHCs send nerve signals to the medulla by way of the sensory neurons, and the pons sends signals immediately back to the OHCs by way of the motor neurons. In response, the hair cells contract by about 10% to 15% of their height. Remember that an OHC is anchored to the basilar membrane below and its stereocilia are embedded in the tectorial membrane above. Therefore, contraction of an OHC reduces the basilar membrane's freedom to vibrate. This results in some regions of the spiral organ sending fewer signals to the brain than neighboring regions, so the brain can better distinguish between the more active and less active hair cells and sound frequencies. When OHCs are experimentally incapacitated, the inner hair cells (IHCs) respond much less precisely to differences in pitch.

There is another mechanism of cochlear tuning involving the inner hair cells. The pons sends efferent fibers to the cochlea that synapse with the sensory nerve fibers near the base of the IHCs. The efferent fibers can inhibit the sensory fibers from firing in some areas of the cochlea, and thus enhance the contrast between signals from the more responsive and less responsive regions. Combined with the previously described role of the OHCs, this sharpens the tuning of the cochlea and our ability to discriminate sounds of different pitch.

---

**INSIGHT 16.3**    Clinical Application

## Deafness

*Deafness* means any hearing loss, from mild and temporary to complete and irreversible. *Conductive deafness* results from any condition that interferes with the transmission of vibrations to the inner ear. Such conditions include a damaged eardrum, otitis media, blockage of the auditory canal, and otosclerosis. *Otosclerosis*[31] is fusion of the auditory ossicles to each other, or fusion of the stapes to the oval window. Either way, it prevents the bones from vibrating freely. *Sensorineural (nerve) deafness* results from the death of hair cells or any of the nervous elements concerned with hearing. It is a common occupational disease of factory and construction workers, musicians, and other people exposed to frequent or sustained loud sounds. Deafness leads some people to develop delusions of being talked about, disparaged, or cheated. Beethoven said his deafness drove him nearly to suicide.

---

[31]*oto* = ear + *scler* = hardening + *osis* = process, condition

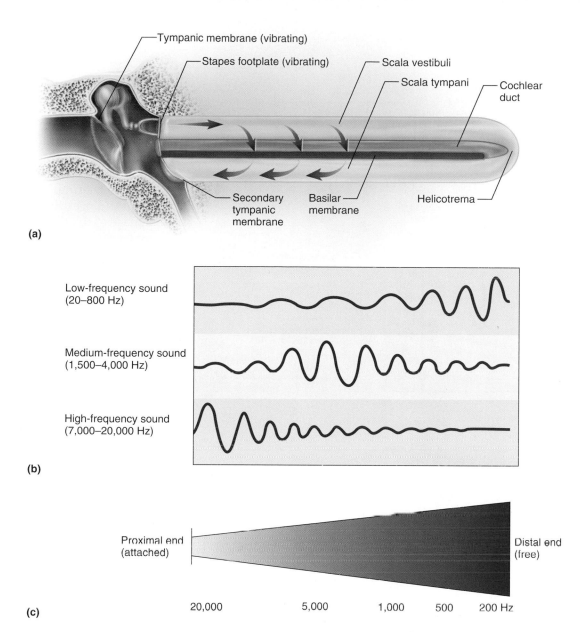

**FIGURE 16.17** **Frequency Response of the Basilar Membrane of the Cochlea.** (a) The cochlea, uncoiled and laid out straight. (b) Sounds produce a standing wave of vibration along the basilar membrane. The peak amplitude of the wave varies with the frequency of the sound, as shown here. The amount of vibration is greatly exaggerated in this diagram to clarify the standing wave. (c) The taper of the basilar membrane and its correlation with sound frequencies. High frequencies (7,000–20,000 Hz) are best detected by hair cells near the narrow proximal end at the left, and low frequencies (20–800 Hz) by hair cells near the wider distal end at the right.

## The Auditory Projection Pathway

The sensory nerve fibers beginning at the bases of the hair cells belong to bipolar sensory neurons whose somas form a coil, the **spiral ganglion,** around the modiolus (see fig. 16.12). The axons of these cells form the **cochlear nerve,** leading away from the cochlea. This nerve joins the *vestibular nerve,* discussed later, and the two together become the *vestibulocochlear nerve* (cranial nerve VIII).

The cochlear nerve fibers from each ear lead to *cochlear nuclei* on both sides of the pons. There, they synapse with second-order neurons that ascend to the nearby *superior olivary nucleus* of the pons (fig. 16.18). By way of cranial nerve VIII, the superior olivary nucleus issues the efferent fibers back to the cochlea that are involved in cochlear tuning. By way of cranial nerves $V_3$ and VII, it also issues motor fibers to the tensor tympani and stapedius muscles, respectively. The superior olivary nucleus also functions in **binaural**[32] **hearing**—comparing signals from the right and left ears to identify the direction from which a sound is coming.

Other fibers from the cochlear nuclei ascend to the inferior colliculi of the midbrain. The inferior colliculi help to locate the origin of a sound in space, process fluctuations in pitch that are important for such purposes as understanding another person's speech, and mediate the startle response

---

[32]*bin* = two + *aur* = ears

607

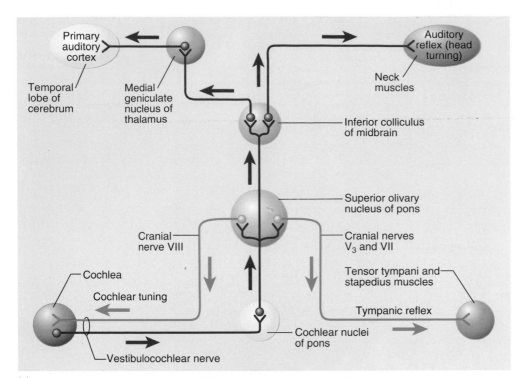

**(a)**

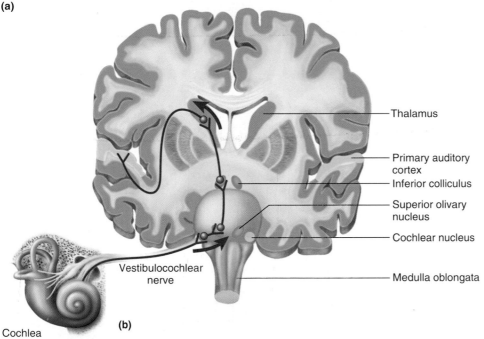

**(b)**

Cochlea

**FIGURE 16.18** **Auditory Pathways in the Brain.** (a) Schematic. (b) Brainstem and frontal section of the cerebrum, showing the locations of auditory processing centers.

and rapid head turning that occur in reaction to loud or sudden noises.

Third-order neurons begin in the inferior colliculi and lead to the thalamus. Fourth-order neurons begin here and complete the pathway to the primary auditory cortex; thus the auditory pathway, unlike most other sensory pathways, involves not three but four neurons from receptor to cere-

bral cortex. The primary auditory cortex lies in the superior margin of the temporal lobe deep within the lateral sulcus (see photo on p. 585). The temporal lobe is the site of conscious perception of sound, and it completes the information processing essential to binaural hearing. Because of extensive decussation in the auditory pathway, damage to the right or left cortex does not cause a unilateral loss of hearing.

## EQUILIBRIUM

The original function of the ear in vertebrate evolution was not hearing, but **equilibrium**—coordination, balance, and orientation in three-dimensional space. Only later did vertebrates evolve the cochlea, middle-ear structures, and auditory function of the ear. In humans, the receptors for equilibrium constitute the **vestibular apparatus,** which consists of three **semicircular ducts** and two chambers—an anterior **saccule** (SAC-yule) and a posterior **utricle**[33] (YOU-trih-cul) (see fig. 16.12b, c).

The sense of equilibrium is divided into **static equilibrium,** the perception of the orientation of the head when the body is stationary, and **dynamic equilibrium,** the perception of motion or acceleration. Acceleration is divided into *linear acceleration,* a change in velocity in a straight line, as when riding in a car or elevator, and *angular acceleration,* a change in the rate of rotation, as when your car turns a corner or you swivel in a rotating chair. The saccule and utricle are responsible for static equilibrium and the sense of linear acceleration; the semicircular ducts detect only angular acceleration.

## The Saccule and Utricle

Each of these chambers has a 2-by-3 mm patch of hair cells and supporting cells called a **macula.**[34] The **macula sacculi** lies nearly vertically on the wall of the saccule, and the **macula utriculi** lies nearly horizontally on the floor of the utricle (fig. 16.19a).

---

[33]*saccule* = little sac; *utricle* = little bag
[34]*macula* = spot

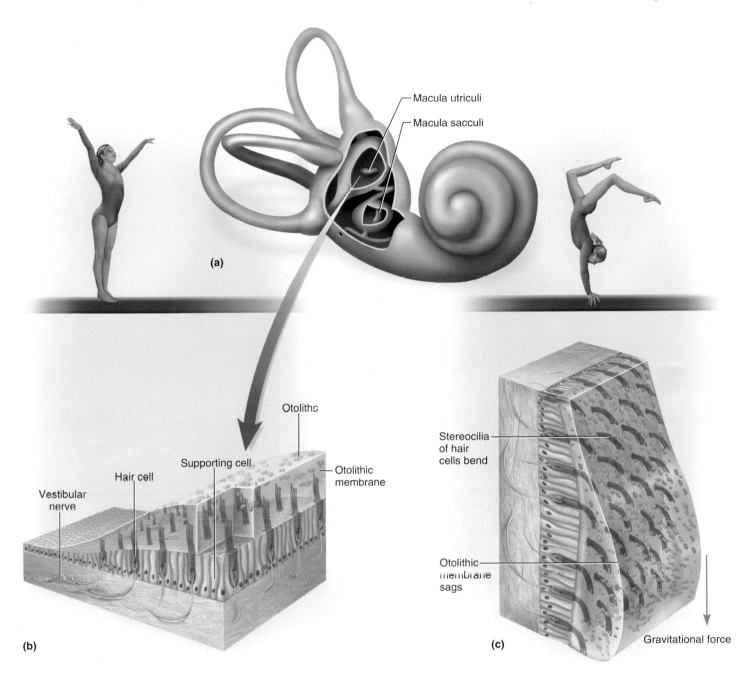

**FIGURE 16.19  The Saccule and Utricle.** (a) Locations of the macula sacculi and macula utriculi. (b) Structure of a macula. (c) Action of the otolithic membrane on the hair cells when the head is tilted.

Each hair cell of a macula has 40 to 70 stereocilia and one true cilium called a **kinocilium.**[35] The tips of the stereocilia and kinocilium are embedded in a gelatinous **otolithic membrane.** This membrane is weighted with calcium carbonate–protein granules called **otoliths**[36] (fig. 16.19b), which add to the density and inertia of the membrane and enhance the sense of gravity and motion.

Figure 16.19c shows how the macula utriculi detects tilt of the head. With the head erect, the otolithic membrane bears directly down on the hair cells, and stimulation is minimal. When the head is tilted, however, the weight of the membrane bends the stereocilia and stimulates the hair cells. Any orientation of the head causes a combination of stimulation to the utricles and saccules of the two ears. The brain interprets head orientation by comparing these inputs to each other and to other input from the eyes and stretch receptors in the neck, to know if only the head is tilted or if the entire body is tipping.

The inertia of the otolithic membranes is especially important in detecting linear acceleration. Suppose you are sitting in a car at a stoplight and then begin to move.

[35]*kino* = moving
[36]*oto* = ear + *lith* = stone

The heavy otolithic membrane of the macula utriculi briefly lags behind the rest of the tissues, bends the stereocilia backward, and stimulates the cells. When you stop at the next light, the macula stops but the otolithic membrane keeps going for a moment, bending the stereocilia forward. The hair cells convert this pattern of stimulation to nerve signals, and the brain is thus advised of changes in your linear velocity.

If you are standing in an elevator and it begins to move up, the otolithic membrane of the vertical macula sacculi lags behind briefly and pulls down on the hairs. When the elevator stops, the otolithic membrane keeps going for a moment and bends the hairs upward. The macula sacculi thus detects vertical acceleration. These sensations are important in such ordinary actions as sitting down, falling, and walking (when the head bobs up and down with each step).

## The Semicircular Ducts

Rotational acceleration is detected by the three *semicircular ducts* (fig. 16.20), each housed in an osseous *semicircular canal* of the temporal bone. The **anterior** and **posterior semicircular ducts** are positioned vertically, at right angles to each other. The **lateral semicircular duct** is about 30° from the horizontal plane. The orientation of

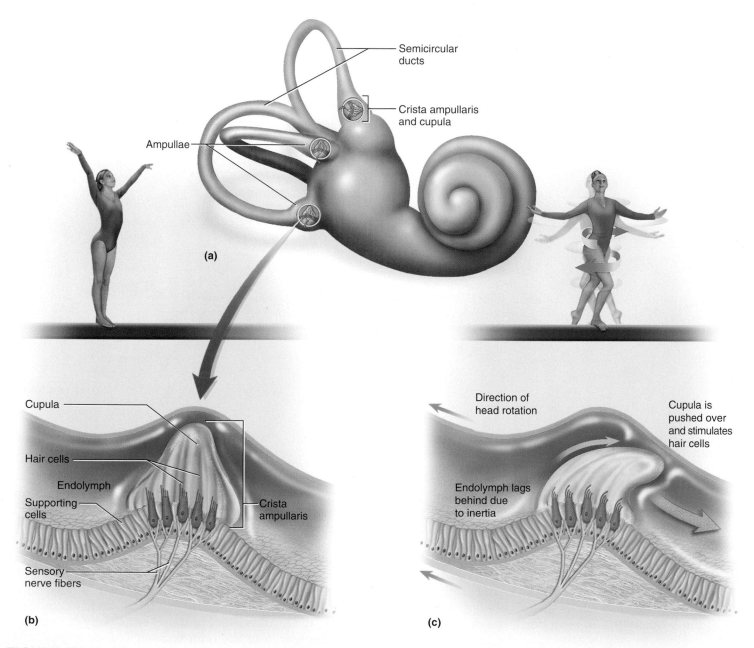

**FIGURE 16.20  The Semicircular Ducts.**   (a) Structure of the semicircular ducts, with each ampulla opened to show the crista ampullaris and cupula. (b) Detail of the crista ampullaris. (c) Action of the endolymph on the cupula and hair cells when the head is rotated.

the ducts causes a different duct to be stimulated by rotation of the head in different planes—turning it from side to side as in gesturing "no," nodding up and down as in gesturing "yes," or tilting it from side to side as in touching your ears to your shoulders.

The semicircular ducts are filled with endolymph. Each duct opens into the utricle and has a dilated sac at one end called an **ampulla.**[37] Within the ampulla is a mound of hair cells and supporting cells called the **crista**[38] **ampullaris.** The hair cells have stereocilia and a kinocilium embedded in a gelatinous cap called the **cupula,**[39] which extends from the crista to the roof of the ampulla. When the head turns the duct rotates, but the endolymph lags behind. It pushes the cupula, bends the stereocilia, and stimulates the hair cells. After 25 to 30 seconds of continual rotation, however, the endolymph catches up with the movement of the duct and stimulation of the hair cells ceases.

### Think About It

*The semicircular ducts do not detect motion itself, but only acceleration—a change in the rate of motion. Explain.*

## Projection Pathways

Hair cells of the macula sacculi, macula utriculi, and semicircular ducts synapse at their bases with sensory fibers of the **vestibular nerve.** This and the cochlear nerve merge to form the vestibulocochlear nerve (cranial nerve VIII). Fibers of the vestibular apparatus lead to a complex of four **vestibular nuclei** on each side of the pons and medulla. Nuclei on the right and left sides of the brainstem communicate extensively with each other, so each receives input from both the right and left ears. They process signals about the position and movement of the body and relay information to five targets (fig. 16.21):

1. The cerebellum, which integrates vestibular information into its control of head movements, eye movements, muscle tone, and posture.

2. Nuclei of the oculomotor, trochlear, and abducens nerves (cranial nerves III, IV, and VI). These nerves produce eye movements that compensate for movements of the head (the *vestibulo-ocular reflex*). To

observe this effect, hold this book in front of you at a comfortable reading distance and fix your gaze on the middle of the page. Move the book left and right about once per second, and you will be unable to read it. Now hold the book still and shake your head from side to side at the same rate. This time you will be able to read the page because the vestibulo-ocular reflex compensates for your head movements and keeps your eyes fixed on the target. This reflex enables you to keep your vision fixed on a distant object as you walk or run toward it.

3. The reticular formation, which is thought to adjust breathing and blood circulation to changes in posture.

4. The spinal cord, where fibers descend the two vestibulospinal tracts on each side (see fig. 13.4, p. 487) and synapse on motor neurons that innervate the extensor (antigravity) muscles. This pathway allows you to make quick movements of the trunk and limbs to keep your balance.

5. The thalamus, which relays signals to two areas of the cerebral cortex. One is at the inferior end of the postcentral gyrus adjacent to sensory regions for the face. It is here that we become consciously aware of body position and movement. The other is slightly rostral to this, at the inferior end of the central sulcus in the transitional zone for primary sensory to motor cortex. This area is thought to be involved in motor control of the head and body.

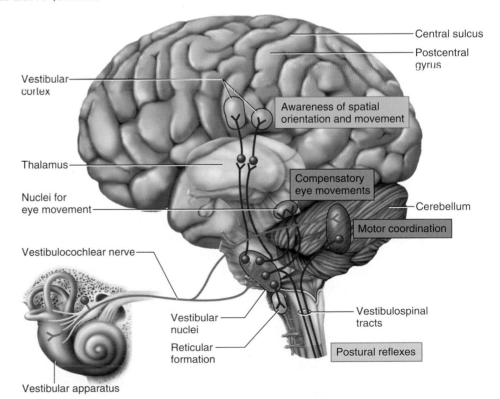

**FIGURE 16.21  Vestibular Projection Pathways in the Brain.**

---

[37]*ampulla* = little jar
[38]*crista* = crest, ridge
[39]*cupula* = little tub

## Before You Go On

*Answer the following questions to test your understanding of the preceding section:*

14. *What physical properties of sound waves correspond to the sensations of loudness and pitch?*

15. *What is the benefit of having auditory ossicles and muscles in the middle ear?*

16. *Explain how vibration of the tympanic membrane ultimately produces fluctuations of membrane voltage in a cochlear hair cell.*

17. *How does the brain recognize the difference between the musical notes high C and middle C? Between a loud sound and a soft one?*

18. *How does the function of the semicircular ducts differ from the function of the saccule and utricle?*

19. *How is sensory transduction in the semicircular ducts similar to that in the saccule and utricle?*

# Vision

### Objectives

When you have completed this section, you should be able to

- describe the anatomy of the eye and its accessory structures;

- discuss the structure of the retina and its receptor cells;

- explain how the optical system of the eye creates an image on the retina;

- discuss how the retina converts this image to nerve impulses;

- explain why different types of receptor cells and neural circuits are required for day and night vision;

- describe the mechanism of color vision; and

- trace the visual projection pathways in the brain.

## LIGHT AND VISION

*Vision* (sight) is the perception of objects in the environment by means of the light that they emit or reflect. *Light* is visible electromagnetic radiation. Human vision is limited to wavelengths ranging from about 400 to 700 nm. The *ultraviolet (UV)* radiation just below 400 nm and the *infrared (IR)* radiation just above 700 nm are invisible to us, although some animals can see a little farther into those ranges than we can. Most solar radiation that reaches the surface of the earth falls within this range; most radiation of shorter and longer wavelengths is filtered out by ozone, carbon dioxide, and water vapor in the atmosphere. Vision is thus adapted to take advantage of the radiation that is most available to us.

Yet there is further reason for vision to be limited to this range of wavelengths. To produce a physiological response, light must cause a *photochemical reaction*—a change in chemical structure caused by light energy. When an electron absorbs a photon of light, it is boosted to a higher energy level (orbit) around its nucleus and may transfer to another atom. The transfer of an electron from one atom to another is the essence of a chemical reaction. Ultraviolet radiation has so much energy that it ionizes organic molecules and kills cells. It is useful for sterilizing food and instruments, but it has too much energy for the biochemical processes of vision. Infrared radiation has too little energy to activate the visual process. It warms the tissues (heat lamps are based on this principle) but does not usually cause chemical reactions.

## ACCESSORY STRUCTURES OF THE ORBIT

Before considering the eye itself, let's survey the accessory structures located in and around the orbit (figs. 16.22 and 16.23). These include the *eyebrows, eyelids, conjunctiva, lacrimal apparatus,* and *extrinsic eye muscles:*

- The **eyebrows** probably serve mainly to enhance facial expressions and nonverbal communication (see p. 200), but they may also protect the eyes from glare and help to keep perspiration from running into the eye.

- The **eyelids,** or **palpebrae** (pal-PEE-bree), block foreign objects from the eye, prevent visual stimuli from disturbing our sleep, and blink periodically to moisten the eye with tears and sweep debris from the surface. The eyelids are separated from each other by the **palpebral fissure** and meet each other at

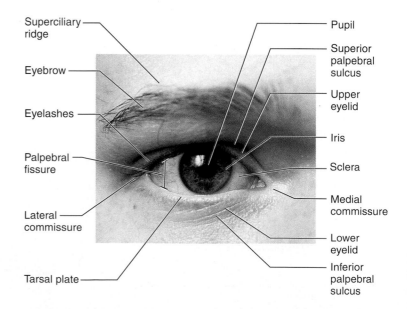

**FIGURE 16.22**  External Anatomy of the Orbital Region.

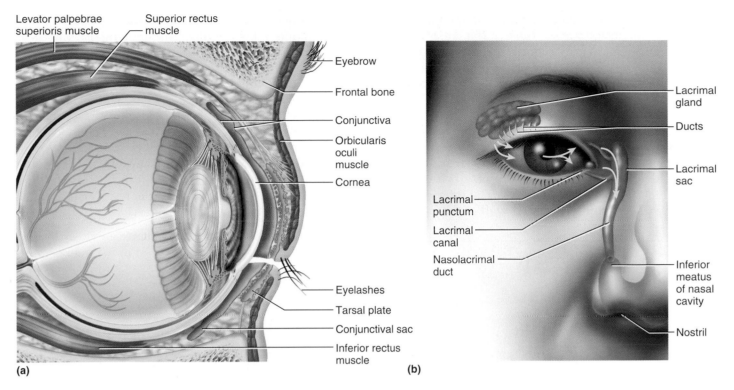

**FIGURE 16.23** Accessory Structures of the Orbit. (a) Sagittal section of the eye and orbit. (b) The lacrimal apparatus. The arrows indicate the flow of tears from the lacrimal gland, across the front of the eye, into the lacrimal sac, and down the nasolacrimal duct to the nose.

the corners called the **medial** and **lateral commissures** (canthi). The eyelid consists largely of the orbicularis oculi muscle covered with skin (fig. 16.23a). It also contains a supportive fibrous **tarsal plate**, which is thickened along the margin of the eyelid. Within the plate are 20 to 25 **tarsal glands** that open along the margin. They secrete an oil that coats the eye and reduces tear evaporation. The **eyelashes** are guard hairs that help to keep debris from the eye. Touching the eyelashes stimulates hair receptors and triggers the blink reflex.

- The **conjunctiva** (CON-junk-TY-vuh) is a transparent mucous membrane that covers the inner surface of the eyelid and anterior surface of the eyeball, except for the cornea. Its primary purpose is to secrete a thin mucous film that prevents the eyeball from drying. It is richly innervated and highly sensitive to pain. It is also very vascular, which is especially evident when the vessels are dilated and the eyes are "bloodshot." Because it is vascular and the cornea is not, the conjunctiva heals more readily than the cornea when injured.

- The **lacrimal**[40] **apparatus** (fig. 16.23b) consists of the lacrimal (tear) gland and a series of ducts that drain the tears into the nasal cavity. The **lacrimal gland,**

about the size and shape of an almond, is nestled in a shallow fossa of the frontal bone in the superolateral corner of the orbit. About 12 short ducts lead from the lacrimal gland to the surface of the conjunctiva. Tears function to cleanse and lubricate the eye surface, deliver oxygen and nutrients to the conjunctiva, and prevent infection by means of a bactericidal enzyme, *lysozyme.* Periodic blinking spreads the tears across the eye surface. On the margin of each eyelid near the medial commissure is a tiny pore, the **lacrimal punctum.**[41] This is the opening to a short **lacrimal canal,** which leads to the **lacrimal sac** in the medial wall of the orbit. From this sac, a **nasolacrimal duct** carries the tears to the inferior meatus of the nasal cavity—thus an abundance of tears from crying or watery eyes can result in a runny nose. Once the tears enter the nasal cavity, they normally flow back to the throat and we swallow them. When we have a cold, the nasolacrimal ducts become swollen and obstructed, the tears cannot drain, and they may overflow from the brim of the eye.

- The **extrinsic eye muscles** are the six muscles attached to the walls of the orbit and to the external surface of the eyeball. *Extrinsic* means arising externally; it distinguishes these from the *intrinsic*

[40]*lacrim* = tear

[41]*punct* = point

muscles inside the eyeball, to be considered later. The extrinsic muscles move the eye (fig. 16.24). They include four *rectus* ("straight") muscles and two *oblique* muscles. The **superior, inferior, medial, and lateral rectus** originate on the posterior wall of the orbit and insert on the anterior region of the eyeball, just beyond the visible "white of the eye." They move the eye up, down, medially, and laterally. The **superior oblique** travels along the medial wall of the orbit. Its tendon passes through a fibrocartilage ring, the **trochlea**[42] (TROCK-lee-uh), and inserts on the superolateral aspect of the eyeball. The **inferior oblique** extends from the medial wall of the orbit to

the inferolateral aspect of the eye. To visualize the function of the oblique muscles, suppose you turn your eyes to the right. The superior oblique muscle will slightly depress your right eye, while the inferior oblique slightly elevates the left eye. The opposite occurs when you look to the left. This is the primary function of the oblique muscles, but they also slightly rotate the eyes, turning the "twelve o'clock pole" of each eye slightly toward or away from the nose. Most of the extrinsic muscles are supplied by the oculomotor nerve (cranial nerve III), but the superior oblique is innervated by the trochlear nerve (IV) and the lateral rectus by the abducens nerve (VI).

The eye is surrounded on the sides and back by **orbital fat.** It cushions the eye, allows it to move freely, and pro-

[42] *trochlea* = pulley

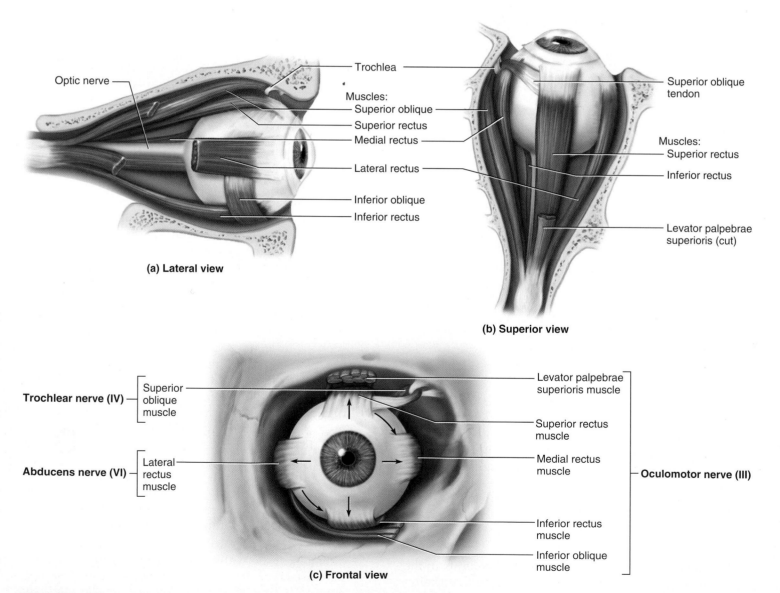

**FIGURE 16.24  Extrinsic Muscles of the Eye.**  (a) Lateral view of the right eye. The lateral rectus muscle is cut to show a portion of the optic nerve. (b) Superior view of the right eye. (c) Innervation of the extrinsic muscles; arrows indicate the eye movement produced by each muscle.

tects blood vessels and nerves as they pass through the rear of the orbit.

## ANATOMY OF THE EYE

The eyeball itself is a sphere about 24 mm in diameter (fig. 16.25) with three principal components: (1) three layers (tunics) that form the wall of the eyeball; (2) optical components that admit and focus light; and (3) neural components, the retina and optic nerve. The retina is not only a neural component but also part of the inner tunic. The cornea is part of the outer tunic as well as one of the optical components.

### The Tunics

The three tunics of the eyeball are as follows:

- The outer **fibrous layer** (tunica fibrosa). This is divided into two regions: the sclera and cornea. The **sclera**[43] (white of the eye) covers most of the eye surface and consists of dense collagenous connective tissue perforated by blood vessels and nerves. The **cornea** is the anterior transparent region of modified sclera that admits light into the eye.

- The middle **vascular layer** (tunica vasculosa), also called the **uvea**[44] (YOU-vee-uh) because it resembles a peeled grape in fresh dissection. It consists of three regions—the choroid, ciliary body, and iris. The **choroid** (CO-royd) is a highly vascular, deeply pigmented layer of tissue behind the retina. It gets its name from a histological resemblance to the chorion of the pregnant uterus. The **ciliary body,** a thickened extension of the choroid, forms a muscular ring around the lens. It supports the iris and lens and secretes a fluid called aqueous humor. The **iris** is an adjustable diaphragm that controls the diameter of the **pupil,** its central opening. The iris has two pigmented layers. One is a posterior *pigment epithelium* that blocks stray light from reaching the retina. The other is the *anterior border layer,* which contains pigmented cells called **chromatophores.**[45] High concentrations of melanin in the chromatophores give the iris a black, brown, or hazel color. If the melanin is scanty, light reflects from the posterior pigment epithelium and gives the iris a blue, green, or gray color.

- The **inner layer** (tunica interna), which consists of the retina and beginning of the optic nerve.

[43]*scler* = hard, tough

[44]*uvea* = grape
[45]*chromato* = color + *phore* = bearer

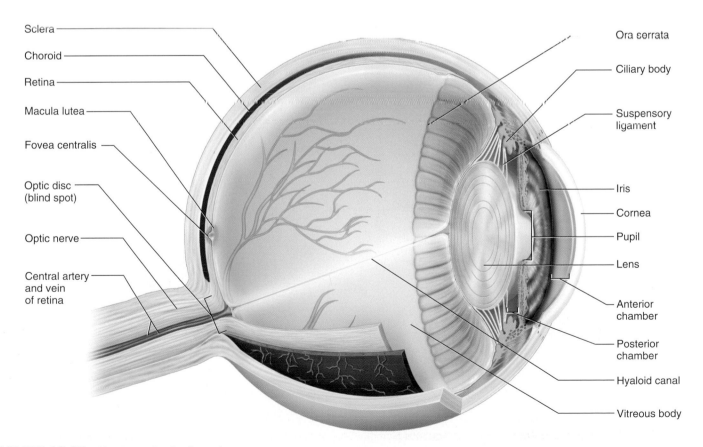

**FIGURE 16.25** **The Eye.** Sagittal section.

## The Optical Components

The optical components of the eye are transparent elements that admit light rays, bend (refract) them, and focus images on the retina. They include the *cornea, aqueous humor, lens,* and *vitreous body.* The cornea has been described already.

- The **aqueous humor** is a serous fluid secreted by the ciliary body into the **posterior chamber,** a space between the iris and lens (fig. 16.26). It flows through the pupil into the **anterior chamber** between the cornea and iris. From here, it is reabsorbed by a ringlike blood vessel called the **scleral venous sinus** (canal of Schlemm[46]). Normally the rate of reabsorption balances the rate of secretion (see Insight 16.4 for an important exception).

- The **lens** is suspended behind the pupil by a ring of fibers called the **suspensory ligament** (figs. 16.25 and 16.27), which attaches it to the ciliary body. Tension on the ligament somewhat flattens the lens so it is about 9.0 mm in diameter and 3.6 mm thick at the middle. When the lens is removed from the eye and not under tension, it relaxes into a more spheroid shape and resembles a plastic bead.

- The **vitreous**[47] **body** (vitreous humor) is a transparent jelly that fills the large space behind the lens. An oblique channel through this body called the *hyaloid canal* is the remnant of a *hyaloid artery* present in the embryo (see fig. 16.25).

[46]Friedrich S. Schlemm (1795–1858), German anatomist
[47]*vitre* = glassy

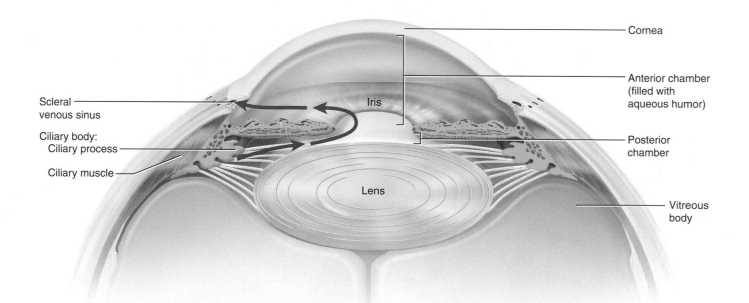

**FIGURE 16.26  Production and Reabsorption of Aqueous Humor.**    Blue arrows indicate the flow of aqueous humor from the ciliary body into the posterior chamber, through the pupil into the anterior chamber, and finally into the scleral venous sinus, the vein that reabsorbs the fluid.

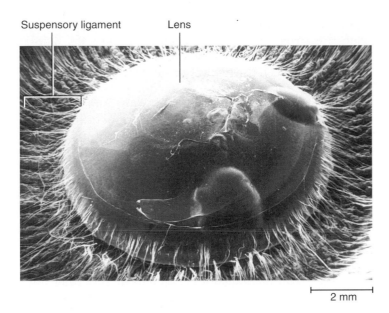

Suspensory ligament    Lens

2 mm

**FIGURE 16.27** The Lens of the Eye, Posterior View (SEM).

## The Neural Components

The neural components are the retina and optic nerve. The **retina** forms from a cup-shaped outgrowth of the diencephalon (see chapter 14); it is actually a part of the brain—the only part that can be viewed without dissection. It is a thin transparent membrane attached to the rest of the eye at only two points: the **optic disc,** where the optic nerve leaves the rear *(fundus)* of the eye, and its scalloped anterior margin, the **ora serrata.**[50] The vitreous body presses the retina smoothly against the rear of the eyeball. The retina can separate from the wall of the eyeball because of blows to the head or insufficient pressure from the vitreous body. Such a *detached retina* may cause blurry areas in the field of vision. It leads to blindness if the retina remains separated for too long from the choroid, on which it depends for oxygen, nutrition, and waste removal.

The retina is examined with an illuminating and magnifying instrument called an *ophthalmoscope.* Directly posterior to the center of the lens, on the visual axis of the eye, is a patch of cells called the **macula lutea**[51] about 3 mm in diameter (fig. 16.28). In the center of the macula is a tiny pit, the **fovea**[52] **centralis,** which produces the most finely detailed images for reasons that will be apparent later. About 3 mm medial to the macula lutea is the optic disc. Nerve fibers from all regions of the retina converge on this point and exit the eye to form the optic nerve. Blood vessels enter and leave the eye by way of the optic disc. Eye examinations serve for more than evaluating the visual system; they allow for a direct, noninvasive exami-

**(a)**

Arteriole
Venule
Fovea centralis
Macula lutea
Optic disc

**(b)**

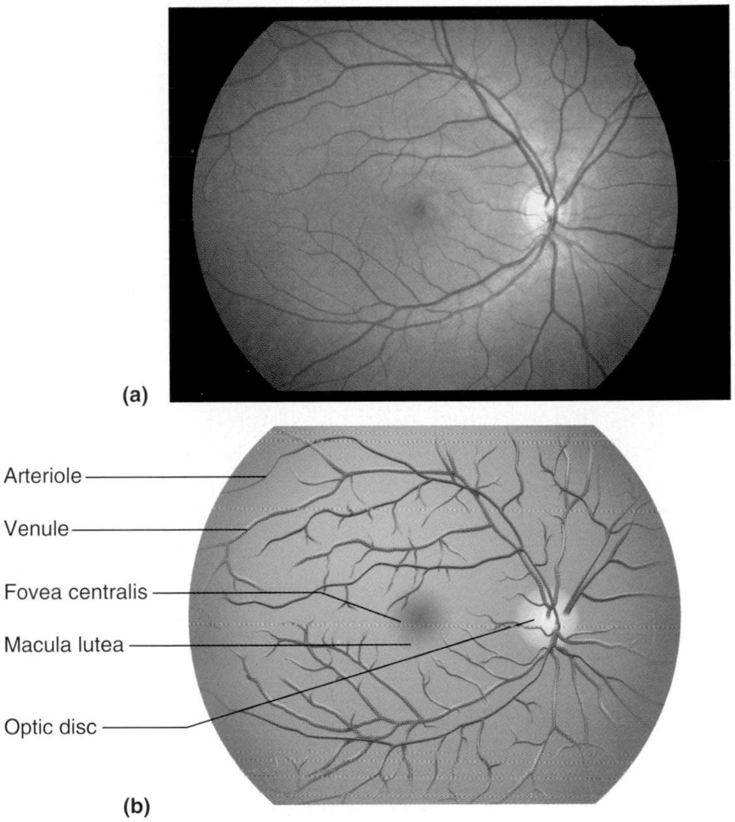

**FIGURE 16.28** The Fundus (rear) of the Eye. (a) As seen with an ophthalmoscope. (b) Anatomical features of the fundus. Note the blood vessels diverging from the optic disc, where they enter the eye with the optic nerve. An eye examination also serves as a partial check on cardiovascular health.

▶ *Is this the subject's right or left eye? How can you tell?*

nation of blood vessels for signs of hypertension, diabetes mellitus, atherosclerosis, and other vascular diseases.

The optic disc contains no receptor cells, so it produces a **blind spot** in the visual field of each eye. You can detect your blind spot and observe an interesting visual phenomenon with the help of figure 16.29. Close or cover your right eye and hold the page about 30 cm (1 ft) from your face. Fixate on the X with your left eye. Without taking your gaze off the X, move the page slightly forward and back, or right and left, until the red dot disappears. This occurs because the image of the dot is falling on the blind spot of your left eye.

You should notice something else happen at the same time as the dot disappears—a phenomenon called **visual**

X

**FIGURE 16.29** Demonstration of the Blind Spot and Visual Filling. See text for explanation of how to conduct this demonstration.

[50]*ora* = border, margin + *serrata* = notched, serrated
[51]*macula* = spot + *lutea* = yellow
[52]*fovea* = pit, depression

**filling.** The green bar seems to fill in the space where the dot used to be. This occurs because the brain uses the image surrounding the blind spot to fill in the area with similar, but imaginary, information. The brain acts as if it is better to assume that the unseen area probably looks like its surroundings than to allow a dark blotch to disturb your vision.

## FORMATION OF AN IMAGE

The visual process begins when light rays enter the eye, focus on the retina, and produce a tiny inverted image. When fully dilated, the pupil admits five times as much light as it does when fully constricted. Its diameter is controlled by two sets of contractile elements in the iris: (1) The **pupillary constrictor** consists of smooth muscle cells that encircle the pupil. When stimulated by the parasympathetic nervous system, it narrows the pupil and admits less light to the eye. (2) The **pupillary dilator** consists of a spokelike arrangement of modified contractile epithelial cells called *myoepithelial cells.* When stimulated by the sympathetic nervous system, these cells contract, widen the pupil, and admit more light to the eye (see fig. 15.9, p. 577). Pupillary constriction and dilation occur in two situations: when light intensity changes and when we shift our gaze between distant and nearby objects. Pupillary constriction in response to light is called the **photopupillary reflex.** It is also described as a *consensual light reflex* because both pupils constrict even if only one eye is illuminated. Constriction in response to a shift in gaze is part of the *near response* described later.

The photopupillary reflex is mediated by a parasympathetic reflex arc. When light intensity rises, signals are transmitted from the eye to the *pretectal region* just rostral to the tectum of the midbrain. Preganglionic parasympathetic fibers originate in the midbrain and travel by way of the oculomotor nerve to the ciliary ganglion in the orbit. From here, postganglionic parasympathetic fibers continue into the eye, where they stimulate the pupillary constrictor. Sympathetic innervation to the pupil originates, like all other sympathetic efferents, in the spinal cord. Preganglionic fibers lead from the thoracic cord to the superior cervical ganglion. From there, postganglionic fibers follow the carotid arteries into the head and lead ultimately to the pupillary dilator.

## Refraction

Image formation depends on **refraction,** the bending of light rays. Light travels at a speed of 300,000 km/sec (186,000 mi/sec) in a vacuum, but it slows down slightly in air, water, glass, and other media. The *refractive index* of a medium *(n)* is a measure of how much it retards light rays relative to air. The refractive index of air is arbitrarily set at $n = 1.00$. If light traveling through air strikes a medium of higher refractive index at a 90° *angle of incidence,* it slows

down but does not change course—the light rays are not bent. If it strikes at any other angle, however, the light is refracted (fig. 16.30a). The greater the difference in refractive index between the two media, and the greater the angle of incidence, the stronger the refraction is.

As light enters the eye, it passes from a medium with $n = 1.00$ (air) to one with $n = 1.38$ (the cornea). Light rays striking the very center of the cornea pass straight through, but because of the curvature of the cornea, rays striking off-center are bent toward the center (fig. 16.30b). The aqueous humor has a refractive index of 1.33 and does not greatly alter the path of the light. The lens has a refractive index of 1.40. As light passes from air to cornea, the refractive index changes by 0.38; but as it passes from aqueous humor to lens, the refractive index changes by only 0.07. Thus, the cornea refracts light more than the lens does. The lens merely fine-tunes the image, especially as you shift your focus between near and distant objects.

## The Near Response

**Emmetropia**[53] (EM-eh-TRO-pee-uh) is a state in which the eye is relaxed and focused on an object more than 6 m (20

[53]*em* = in + *metr* = measure + *opia* = vision

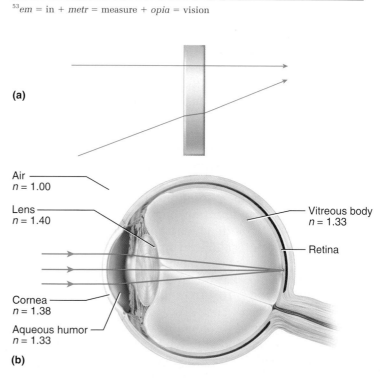

(a)

Air $n = 1.00$

Lens $n = 1.40$

Cornea $n = 1.38$

Aqueous humor $n = 1.33$

Vitreous body $n = 1.33$

Retina

(b)

**FIGURE 16.30 Principles of Refraction.** (a) A refractive medium does not bend light rays that strike it at a 90° angle, but does bend rays that enter or leave it at any other angle. (b) Refractive indices of the media from air to retina. The greater the difference between the refractive indices of two media, the more strongly light rays are refracted when passing from one to the next. In vision, most refraction (focusing) occurs as light passes from air to cornea. The lens makes only fine adjustments in the image.

ft) away, the light rays coming from that object are essentially parallel, and the rays are focused on the retina without effort. (An *emmetropic eye* does not need a corrective lens to focus the image.) If the gaze shifts to something closer, light rays from the source are too divergent to be focused without effort. In other words, the eye is automatically focused on things in the distance unless you make an effort to focus elsewhere. For a wild animal or our prehistoric ancestors, this arrangement would be adaptive because it allows for alertness to predators or prey at a distance.

The **near response** (fig. 16.31), or adjustment to close-range vision, involves three processes to focus an image on the retina:

1. **Convergence of the eyes.** Move your finger gradually closer to a baby's nose and the baby will go cross-eyed. This **convergence** of the eyes orients the visual axis of each eye toward the object in order to focus its image on each fovea. If the eyes cannot converge accurately—for example, when the extrinsic muscles are weaker in one eye than in the other—double vision, or *diplopia,*[54] results. The images fall on different parts of the two retinas and the brain sees two images. You can simulate this effect by pressing gently on one eyelid as you look at this page; the image of the print will fall on non-corresponding regions of the two eyes and cause you to see double.

[54]*dipl* = double + *opia* = vision

2. **Constriction of the pupil.** Lenses cannot refract light rays at their edges as well as they can closer to the center. The image produced by any lens is therefore somewhat blurry around the edges; this *spherical aberration* is quite evident in an inexpensive microscope. It can be minimized by screening out these peripheral light rays and looking only at the better-focused center. For this purpose, the pupil constricts as you focus on nearby objects. Like the diaphragm setting (f-stop) of a camera, the pupil thus has a dual purpose: to adjust the eye to variations in brightness and to reduce spherical aberration.

3. **Accommodation of the lens. Accommodation** is a change in the curvature of the lens that enables you to focus on a nearby object. When you look at something nearby, the ciliary muscle surrounding the lens contracts. This narrows the diameter of the ciliary body, relaxes the fibers of the suspensory ligament, and allows the lens to relax into a more convex shape (fig. 16.32). In emmetropia, the lens is about 3.6 mm thick at the center; in accommodation, it thickens to about 4.5 mm. A more convex lens refracts light more strongly and focuses the divergent light rays onto the retina. The closest an object can be and still come into focus is called the **near point of vision.** It depends on the flexibility of the lens. The lens stiffens with age, so the near point averages about 9 cm at the age of 10 and 83 cm by the age of 60.

Some common defects in image formation are listed in table 16.2.

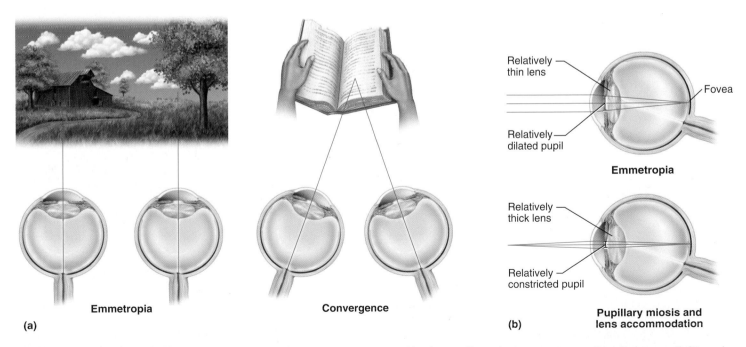

**FIGURE 16.31 Emmetropia and the Near Response.** (a) Superior view of both eyes fixated on a scene more than 6 m away (left), and both eyes fixated on an object closer than 6 m (right). (b) Lateral view of the eye fixated on a distant object (top) and nearby object (bottom).

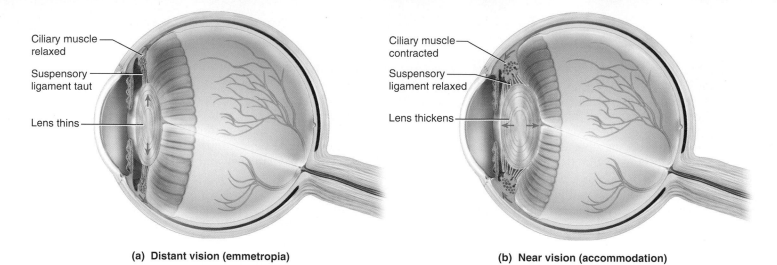

**(a) Distant vision (emmetropia)**

**(b) Near vision (accommodation)**

**FIGURE 16.32 Accommodation of the Lens.** (a) In the emmetropic eye, the ciliary muscle is relaxed and dilated. It puts tension on the suspensory ligament and flattens the lens. (b) In accommodation, the ciliary muscle contracts and narrows in diameter. This reduces tension on the suspensory ligament and allows the lens to relax into a more convex shape.

| TABLE 16.2 | Common Defects of Image Formation |
|---|---|
| Astigmatism | Inability to simultaneously focus light rays that enter the eye on different planes. Focusing on vertical lines, such as the edge of a door, may cause horizontal lines, such as a tabletop, to go out of focus. Caused by a deviation in the shape of the cornea so that it is shaped like the back of a spoon rather than part of a sphere. Corrected with cylindrical lenses, which refract light more in one plane than another. |
| Hyperopia | Farsightedness—a condition in which the eyeball is too short. The retina lies in front of the focal point of the lens, and the light rays have not yet come into focus when they reach the retina (see top of fig. 16.33b). Causes the greatest difficulty when viewing nearby objects. Corrected with convex lenses, which cause light rays to converge slightly before entering the eye. |
| Myopia | Nearsightedness—a condition in which the eyeball is too long. Light rays come into focus before they reach the retina and begin to diverge again by the time they fall on it (see top of fig. 16.33c). Corrected with concave lenses, which cause light rays to diverge slightly before entering the eye. |
| Presbyopia | Reduced ability to accommodate for near vision with age. Caused by declining flexibility of the lens. Results in difficulty reading and doing close handwork. Corrected with bifocal lenses or reading glasses. |

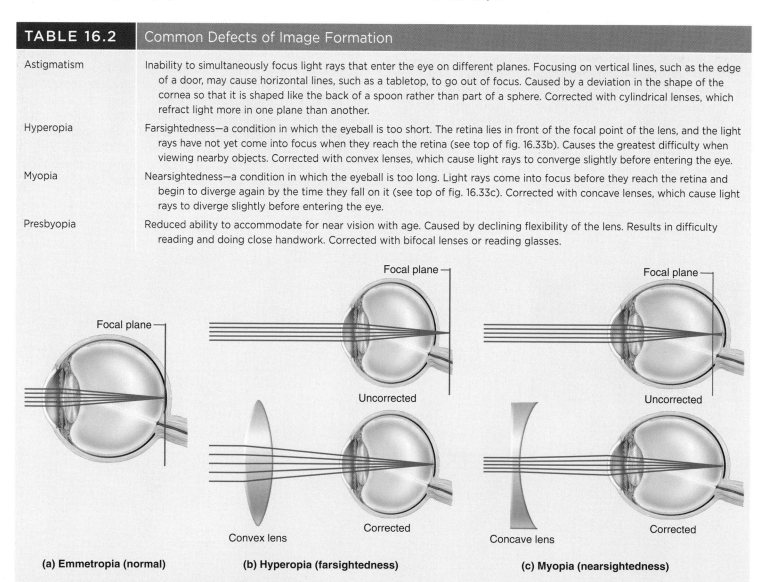

**(a) Emmetropia (normal)**

**(b) Hyperopia (farsightedness)**

**(c) Myopia (nearsightedness)**

**FIGURE 16.33 Two Common Visual Defects and the Effects of Corrective Lenses.** (a) The normal emmetropic eye, with light rays converging on the retina. (b) Hyperopia (farsightedness) and the corrective effect of a convex lens. (c) Myopia (nearsightedness) and the corrective effect of a concave lens.

## SENSORY TRANSDUCTION IN THE RETINA

The conversion of light energy into action potentials occurs in the retina. We begin our exploration of this process with the cellular layout of the retina (fig. 16.34). From there we go to the pigments that absorb light and then to what happens when light is absorbed.

The most posterior layer of the retina is the **pigment epithelium,** a layer of darkly pigmented cuboidal cells whose basal processes interdigitate with the photoreceptor cells. The pigment here is not involved in nerve signaling; rather, its purpose is to absorb light that is not absorbed first by the receptor cells. This prevents stray light from reflecting back into the eye and degrading the visual image. It acts like the blackened inside of a camera to reduce reflection.

The neural components of the retina consist of three principal cell layers. Progressing from the rear of the eye forward, these are composed of *photoreceptor cells, bipolar cells,* and *ganglion cells:*

1. **Photoreceptor cells.** The photoreceptors are cells that absorb light and generate a chemical or electrical signal. There are three kinds: rods, cones, and some of the ganglion cells. Only the rods and cones produce visual images; the ganglion cells are discussed shortly. **Rods** and **cones** are derived from the same stem cells that produce ependymal cells of the brain. Each rod or cone has an **outer segment** that points toward the wall of the eye and an **inner segment** facing the interior (fig. 16.35). The two segments are separated by a narrow constriction containing nine pairs of microtubules; the outer segment is actually a highly modified cilium specialized to absorb light.

**FIGURE 16.34 Histology of the Retina.** (a) Photomicrograph. (b) Schematic of the layers and synaptic relationships of the retinal cells.

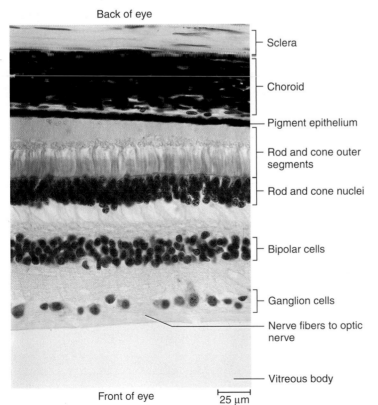

Back of eye

Sclera

Choroid

Pigment epithelium

Rod and cone outer segments

Rod and cone nuclei

Bipolar cells

Ganglion cells

Nerve fibers to optic nerve

Vitreous body

Front of eye

25 µm

(a)

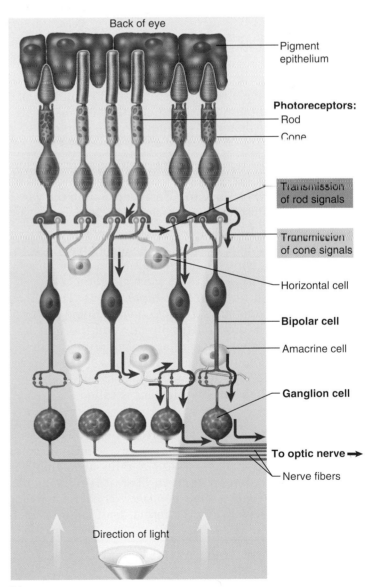

Back of eye

Pigment epithelium

**Photoreceptors:**
Rod
Cone

Transmission of rod signals

Transmission of cone signals

Horizontal cell

**Bipolar cell**

Amacrine cell

**Ganglion cell**

**To optic nerve →**

Nerve fibers

Direction of light

(b)

The inner segment contains mitochondria and other organelles. At its base, it gives rise to a cell body, which contains the nucleus, and to processes that synapse with retinal neurons in the next layer.

In a rod, the outer segment is cylindrical and resembles a stack of coins in a paper roll—there is a plasma membrane around the outside and a neatly arrayed stack of about 1,000 membranous discs inside. Each disc is densely studded with globular proteins—the visual pigment *rhodopsin,* to be discussed later. The membranes hold these pigment molecules in a position that results in the most efficient light absorption. Rod cells are responsible for **night (scotopic[55]) vision;** they cannot distinguish colors from each other.

A cone cell is similar except that the outer segment tapers to a point, and the discs are not detached from the plasma membrane but are parallel infoldings of it. Cones function in bright light; they are responsible for **day (photopic[56]) vision** as well as color vision.

2. **Bipolar cells.** Rods and cones synapse with the dendrites of **bipolar cells,** the first-order neurons of the visual pathway. They in turn synapse with the ganglion cells described next (see fig. 16.34b).

3. **Ganglion cells. Ganglion cells** are the largest neurons of the retina, arranged in a single layer close to the vitreous body. They are the second-order neurons of the visual pathway. Most ganglion cells receive input from multiple bipolar cells. The ganglion cell axons form the optic nerve. Some of the ganglion cells absorb light directly and transmit signals to brainstem nuclei that control pupillary diameter and the body's circadian rhythms. They do not contribute to visual images but detect only light intensity. Their sensory pigment is called **melanopsin.**

There are approximately 130 million rods and 6.5 million cones in one retina, but only 1.2 million nerve fibers in the optic nerve. With a ratio averaging 114 receptor cells to one optic nerve fiber, it is obvious that there must be substantial *neuronal convergence* and information processing in the retina itself before signals

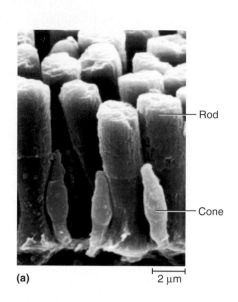

**(a)**    2 μm

**FIGURE 16.35   Rod and Cone Cells.**   (a) Rods and cones of a salamander retina (SEM). (b) Structure of human rods and cones.

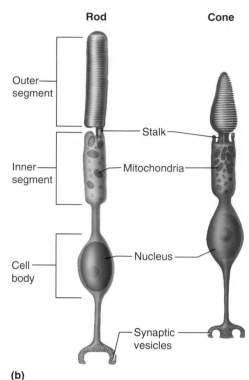

**(b)**

are transmitted to the brain proper. Convergence begins where multiple rod or cone cells synapse with one bipolar cell, and occurs again where multiple bipolar cells synapse with one ganglion cell. Later we will examine how convergence functions in visual resolution and night vision.

There are other retinal cells, but they do not form layers of their own. **Horizontal cells** and **amacrine[57] cells** form horizontal connections among rod, cone, and bipolar cells and intervene in the pathways from receptor cells to ganglion cells. They play diverse roles in enhancing the perception of contrast, the edges of objects, moving objects, and changes in light intensity. In addition, much of the mass of the retina is composed of astrocytes and other types of glial cells.

## Visual Pigments

The visual pigment of the rods is called **rhodopsin** (ro-DOP-sin), or *visual purple.* Each molecule consists of two major parts (moieties): a protein called **opsin** and a vitamin A derivative called **retinal** (rhymes with "pal"), also known as **retinene** (fig. 16.36). Opsin is embedded in the disc membranes of the rod's outer segment. All rod cells contain a single kind of rhodopsin with an absorption peak at a wavelength of 500 nm. The rods are less sensitive to light of other wavelengths.

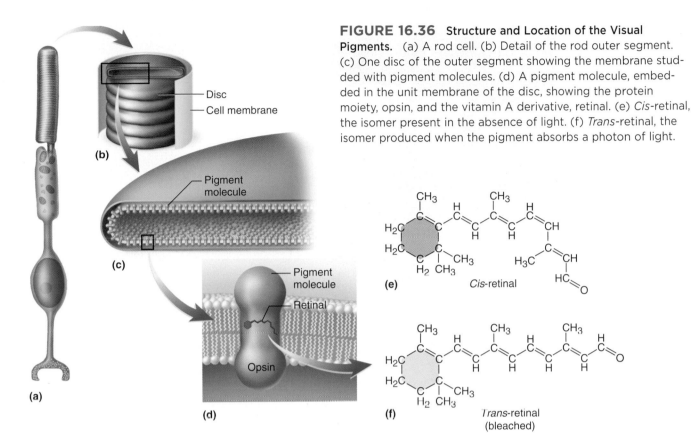

**FIGURE 16.36 Structure and Location of the Visual Pigments.** (a) A rod cell. (b) Detail of the rod outer segment. (c) One disc of the outer segment showing the membrane studded with pigment molecules. (d) A pigment molecule, embedded in the unit membrane of the disc, showing the protein moiety, opsin, and the vitamin A derivative, retinal. (e) *Cis*-retinal, the isomer present in the absence of light. (f) *Trans*-retinal, the isomer produced when the pigment absorbs a photon of light.

In cones, the pigment is called **photopsin** (iodopsin). Its retinal moiety is the same as that of rhodopsin, but the opsin moiety has a different amino acid sequence that determines which wavelengths of light the pigment absorbs. There are three kinds of cones, which are identical in appearance but optimally absorb different wavelengths of light. These differences, as you will see shortly, enable us to perceive different colors.

## The Photochemical Reaction

The events of sensory transduction are probably the same in rods and cones, but rods and rhodopsin have been better studied than cones and photopsin. In the dark, retinal has a bent shape called *cis*-**retinal.** When it absorbs light, it changes to a straight form called ***trans*-retinal,** and the retinal dissociates from the opsin (fig. 16.37). Purified rhodopsin changes from violet to colorless when this happens, so the process is called **bleaching.**

For a rod to continue functioning, it must regenerate rhodopsin at a rate that keeps pace with bleaching. When *trans*-retinal dissociates from opsin, it is transported to the pigment epithelium, converted back to *cis*-retinal, returned to the rod outer segment, and reunited with opsin. It takes about 5 minutes to regenerate 50% of the bleached rhodopsin. Cone cells are less dependent on the pigment epithelium and regenerate half of their pigment in about 90 seconds.

## Generating the Optic Nerve Signal

In the dark, rods do not sit quietly doing nothing. They produce a **dark current,** a steady flow of sodium ions into the outer segment, and as long as this is happening, they release a neurotransmitter, glutamate, from the basal end of the cell (fig. 16.38a). When a rod absorbs light, the dark current and glutamate secretion cease (fig. 16.38b). The on-and-off glutamate secretion influences the bipolar cells in ways we will examine shortly, but first we will explore why the dark current occurs and why it stops in the light.

The outer segment of the rod has ligand-regulated $Na^+$ gates that bind cyclic guanosine monophosphate (cGMP) on their intracellular side. In the dark, cGMP opens the gate and permits the inflow of $Na^+$. This $Na^+$ current reduces the membrane potential of the rod from the $-70$ mV typical of neurons to about $-40$ mV. The cell responds by secreting glutamate. The sodium that enters the outer segment is removed by $Na^+$–$K^+$ pumps in the inner segment.

Why does the dark current cease when a rod absorbs light? The intact rhodopsin molecule is essentially a dormant enzyme. When it bleaches, it becomes enzymatically active and triggers reactions that ultimately break down several hundred thousand molecules of cGMP. As cGMP is degraded, the $Na^+$ gates in the outer segment close, the dark current ceases, and the $Na^+$–$K^+$ pump shifts the membrane voltage toward 270 mV. This shift causes the rod to stop secreting glutamate. The sudden

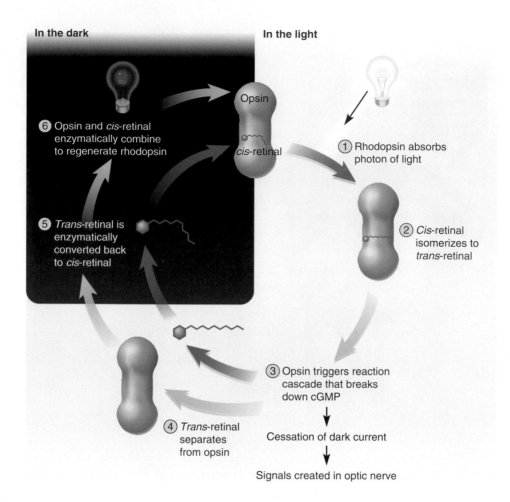

**In the dark**    **In the light**

Opsin

⑥ Opsin and *cis*-retinal enzymatically combine to regenerate rhodopsin

*cis*-retinal

① Rhodopsin absorbs photon of light

⑤ *Trans*-retinal is enzymatically converted back to *cis*-retinal

② *Cis*-retinal isomerizes to *trans*-retinal

③ Opsin triggers reaction cascade that breaks down cGMP

④ *Trans*-retinal separates from opsin

Cessation of dark current

Signals created in optic nerve

**FIGURE 16.37**  **The Bleaching and Regeneration of Rhodopsin.**  Numbers 1 through 4 indicate the bleaching events that occur in the light; numbers 5 and 6 indicate the regenerative events that are independent of light. The regenerative events occur in light and dark, but in the light, they are outpaced by bleaching.

drop in glutamate secretion informs the bipolar cell that the rod has absorbed light.

There are two kinds of bipolar cells. One type is inhibited (hyperpolarized) by glutamate and thus excited (depolarized) when its secretion drops. This type of cell is excited by *rising* light intensity. The other type is excited by glutamate and inhibited when its secretion drops, so it is excited by *falling* light intensity. As your eye scans a scene, it passes areas of greater and lesser brightness. Their images on the retina cause a rapidly changing pattern of bipolar cell responses as the light intensity on a patch of retina rises and falls.

When bipolar cells detect fluctuations in light intensity, they stimulate ganglion cells either directly (by synapsing with them) or indirectly (via pathways that go through amacrine cells). Ganglion cells are the only retinal cells that produce action potentials; all other retinal cells produce only graded local potentials. The ganglion cells respond with rising and falling firing frequencies which, via the optic nerve, provide the brain with a basis for interpreting the image on the retina.

## LIGHT AND DARK ADAPTATION

**Light adaptation** occurs when you go from a dark or dimly lit area into brighter light. If you wake up in the night and turn on a lamp, at first you see a harsh glare; you may experience discomfort from the overstimulated retinas. Your pupils quickly constrict to reduce the intensity of stimulation, but color vision and visual acuity (the ability to see fine detail) remain below normal for 5 to 10 minutes—the time needed for pigment bleaching to adjust retinal sensitivity to this light intensity. The rods bleach quickly in bright light, and cones take over. Even in typical indoor light, rod vision is nonfunctional.

On the other hand, suppose you are sitting in a bright room at night and there is a power failure. Your eyes must undergo **dark adaptation** before you can see well enough to find your way in the dark. There is not enough light for cone (photopic) vision, and it takes a little time for the rods (scotopic vision) to adjust to the dark. Your rod pigment was bleached by the lights in the room while the power was on, but now in the rela-

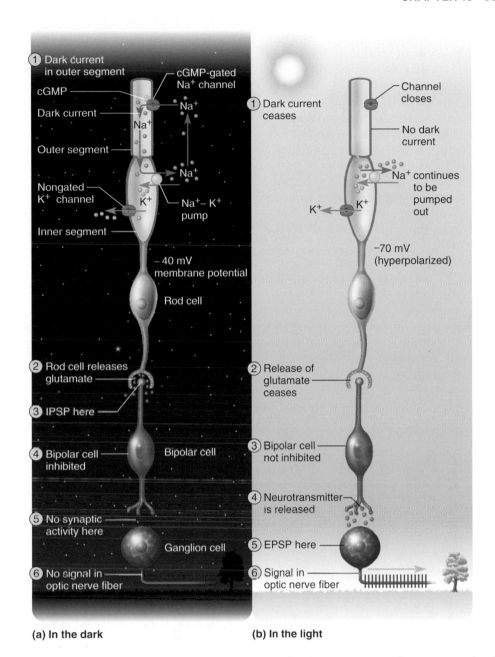

**(a) In the dark**
**(b) In the light**

**FIGURE 16.38 Mechanism of Generating Visual Signals.** (a) In the dark, cGMP opens a sodium gate, and a dark current in the rod cell stimulates glutamate release. (b) In the light, cGMP breaks down and its absence shuts off the dark current and glutamate secretion. The bipolar cell in this case is inhibited by glutamate and stimulates the ganglion cell when glutamate secretion decreases.

tive absence of light, rhodopsin regenerates faster than it bleaches. In a minute or two, scotopic vision begins to function, and after 20 to 30 minutes, the amount of regenerated rhodopsin is sufficient for your eyes to reach essentially maximum sensitivity. Dilation of the pupils also helps by admitting more light to the eye.

## THE DUPLICITY THEORY

You may wonder why we have both rods and cones. Why can't we simply have one type of receptor cell that produces detailed color vision, both day and night? The **duplicity theory** of vision holds that a single type of receptor cell cannot produce both high sensitivity and high resolution. It takes one type of cell and neural circuit to provide sensitive night vision and a different type to provide high-resolution daytime vision.

The high sensitivity of rods in dim light stems partly from the cascade of reactions leading to cGMP breakdown described earlier; a single photon leads to the breakdown of hundreds of thousands of cGMP molecules. But the sensitivity of scotopic (rod) vision is also due to the extensive neuronal convergence that occurs between the rods and ganglion cells. Up to 600 rods converge on each bipolar cell, and many bipolar cells converge on each ganglion cell. This allows for a high degree of *spatial summation* in the scotopic system (fig. 16.39a). Weak stimulation of numerous rod cells can produce an additive effect on one bipolar cell, and several bipolar cells can collaborate to excite one ganglion cell. Thus, a ganglion cell can respond in dim light that only weakly stimulates any individual rod. Scotopic vision is functional even at a light intensity less than starlight reflected from a sheet of white paper. A shortcoming of this system is that it cannot resolve finely detailed images. One ganglion cell receives input from all the rods in about 1 mm$^2$ of retina—its receptive field. What the brain perceives is therefore a coarse, grainy image similar to an overenlarged newspaper photograph.

Around the edges of the retina, receptor cells are especially large and widely spaced. If you fixate on the middle of this page, you will notice that you cannot read the words near the margins. Visual acuity decreases rapidly as the image falls away from the fovea centralis. Our peripheral vision is a low-resolution system that serves mainly to alert us to motion in the periphery and to stimulate us to look that way to identify what is there.

When you look directly at something, its image falls on the fovea, which is occupied by about 4,000 tiny cones and no rods. The other neurons of the fovea are displaced to one side so they won't interfere with light falling on the cones. The smallness of these cones is like the smallness of the dots in a fine-grained photograph; it is partially responsible for the high-resolution images formed at the fovea. In addition, the cones here show no neuronal convergence. Each cone synapses with only one bipolar cell and each bipolar cell with only one ganglion cell. This gives each foveal cone a "private line to the brain," and each ganglion cell of the fovea reports to the brain on a receptive field of just 2 μm$^2$ of retinal area (fig. 16.39b). Cones distant from the fovea exhibit some neuronal convergence but not nearly as much as rods do. The price of this lack of convergence at the fovea, however, is that cone cells have little spatial summation, and the cone system therefore has less sensitivity to light. The threshold of photopic (cone) vision lies between the intensity of starlight and moonlight reflected from white paper.

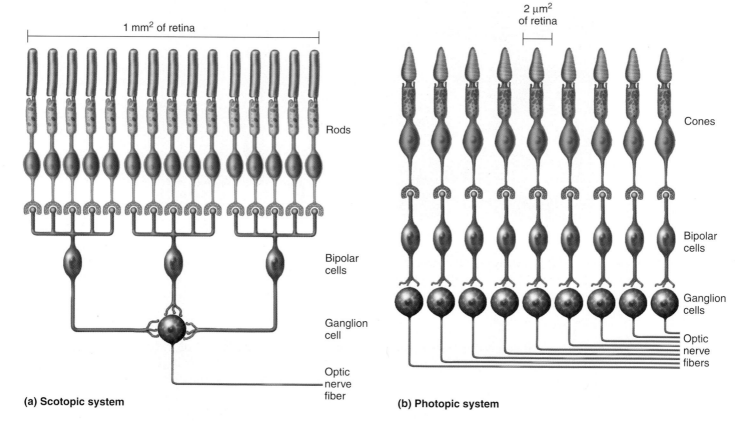

**(a) Scotopic system**

**(b) Photopic system**

**FIGURE 16.39   The Duplicity Theory of Vision.**   (a) In the scotopic (night vision) system, many rods converge on each bipolar cell and many bipolar cells converge on each ganglion cell (via amacrine cells, not shown). This allows extensive spatial summation—many rods add up their effects to stimulate a ganglion cell even in dim light. However, it means that each ganglion cell (and its optic nerve fiber) represents a relatively large area of retina and produces a grainy image. (b) In the photopic (day vision) system, there is little neuronal convergence. In the fovea, represented here, each cone has a "private line" to the brain, so each optic nerve fiber represents a tiny area of retina, and vision is relatively sharp. However, the lack of convergence prevents spatial summation. Photopic vision does not function well in dim light because weakly stimulated cones cannot collaborate to stimulate a ganglion cell.

## COLOR VISION

Most nocturnal vertebrates have only rod cells, but many diurnal animals are endowed with cones and color vision. Color vision is especially well developed in primates for evolutionary reasons discussed in chapter 1. It is based on three kinds of cones named for the absorption peaks of their photopsins: **blue cones,** with peak sensitivity at a wavelength of 420 nm; **green cones,** which peak at 531 nm; and **red cones,** which peak at 558 nm. Red cones do not peak in the red part of the spectrum (558 nm light is perceived as orange-yellow), but they are the only cones that respond at all to red light. Our perception of different colors is based on a mixture of nerve signals representing cones with different absorption peaks. In figure 16.40,

note that light at 400 nm excites only the blue cones. At 500 nm, all three types of cones are stimulated; the red cones respond at 60% of their maximum capacity, green cones at 82% of their maximum, and blue cones at 20%. The brain interprets this mixture of signals as blue-green. The table in figure 16.40 shows how some other color sensations are generated by other response ratios.

Some individuals have a hereditary lack of one photopsin or another and consequently exhibit **color blindness.** The most common form is *red-green color blindness,* which results from a lack of either red or green cones and renders a person incapable of distinguishing these and related shades from each other. For example, a person with normal *trichromatic* color vision sees figure 16.41 as showing the number 16, whereas a person with red-green color blindness sees no number. Red-green color blindness is a sex-linked recessive trait. It occurs in about 8% of males and 0.5% of females. (See p. 146 to review sex linkage and the reason such traits are more common in males.)

## STEREOSCOPIC VISION

**Stereoscopic vision** (stereopsis) is depth perception—the ability to judge how far away objects are. It depends on having two eyes with overlapping visual fields, which allows each eye to look at the same object from a different angle. Stereoscopic vision contrasts with the *panoramic vision* of mammals such as rodents and horses, in which

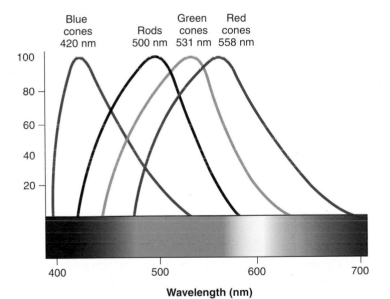

| Wavelength (nm) | Percent of maximum cone response (red:green:blue) | Perceived hue |
|---|---|---|
| 400 | 0 : 0 : 50 | Violet |
| 450 | 0 : 30 : 72 | Blue |
| 500 | 60 : 82 : 20 | Blue-green |
| 550 | 97 : 85 : 0 | Green |
| 625 | 35 : 3 : 0 | Orange |
| 675 | 5 : 0 : 0 | Red |

**FIGURE 16.40** Absorption Spectra of the Retinal Cells. In the middle column of the table, each number indicates how strongly the respective cone cells respond as a percentage of their maximum capability. At 550 nm, for example, red cones respond at 97% of their maximum, green cones at 85%, and blue cones not at all. The result is a perception of green light.

▶ *If you were to add another row to this table for 600 nm, what would you enter in the middle and right-hand columns?*

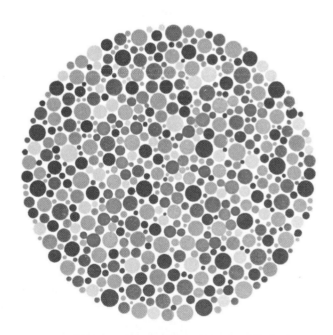

**FIGURE 16.41** A Test for Red–Green Color Blindness. Persons with normal vision see the number 16. Persons with red-green color blindness see no discernible number. [Reproduced from *Ishihara's Tests for Colour Deficiency,* published by Kanehara Trading, Inc., Tokyo, Japan. Tests for color deficiency cannot be conducted with this material. For accurate testing, the original plates should be used.]

the eyes are on opposite sides of the head. Although stereoscopic vision covers a smaller visual field than panoramic vision and provides less alertness to sneaky predators, it has the advantage of depth perception. The evolutionary basis of depth perception in primates was considered in chapter 1 (p. 10).

When you fixate on something within 30 m (100 ft) away, each eye views it from a slightly different angle and focuses its image on the fovea centralis. The point on which the eyes are focused is called the *fixation point.* Objects farther away than the fixation point cast an image somewhat medial to the foveas, and closer objects cast their images more laterally (fig. 16.42). The distance of an image from the two foveas provides the brain with information used to judge the position of other points relative to the fixation point.

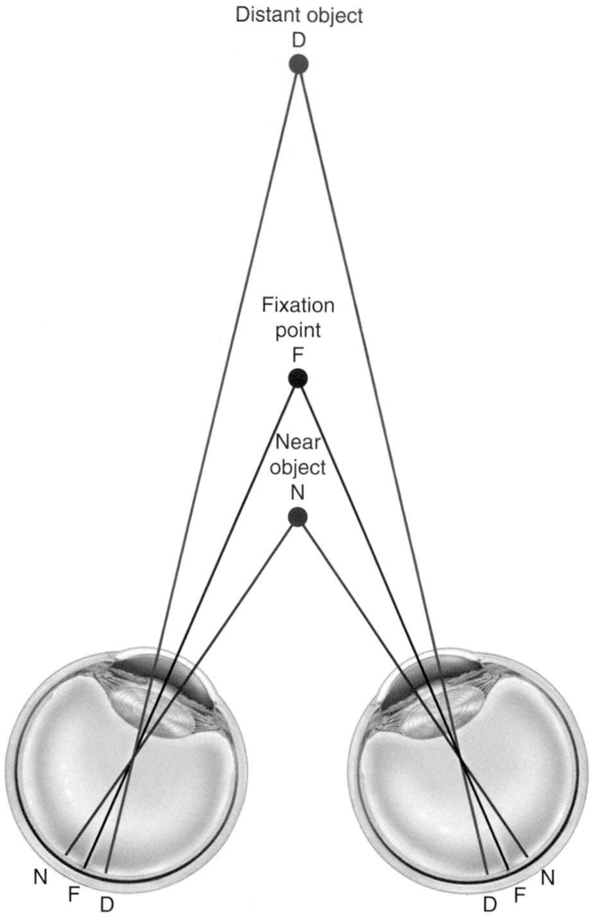

## THE VISUAL PROJECTION PATHWAY

The first-order neurons in the visual pathway are the bipolar cells of the retina. The second-order neurons are the retinal ganglion cells, whose axons are the fibers of the optic nerve. The optic nerves leave each orbit through the optic foramen and then converge on each other to form an X, the **optic chiasm**[58] (ky-AZ-um), immediately inferior to the hypothalamus and anterior to the pituitary. Beyond this, the fibers continue as a pair of **optic tracts.**

Within the chiasm, half the fibers of each optic nerve cross over to the opposite side of the brain (fig. 16.43). This is called **hemidecussation,**[59] since only half of the fibers decussate. As a result, objects in the left visual field, whose images fall on the right half of each retina (the medial half of the left eye and lateral half of the right eye), are perceived by the right cerebral hemisphere. Objects in the right visual field are perceived by the left hemisphere. Since the right brain controls motor responses on the left side of the body and vice versa, each side of the brain needs to see what is on the side of the body where it exerts motor control. In animals with panoramic vision, nearly 100% of the optic nerve fibers of the right eye decussate to the left brain and vice versa.

The optic tracts pass laterally around the hypothalamus, and most of their axons end in the **lateral geniculate**[60] (jeh-NIC-you-late) **nucleus** of the thalamus. Third-order neurons arise here and form the **optic radiation** of fibers in the white matter of the cerebrum. These project to the primary visual cortex of the occipital lobe, where the conscious visual sensation occurs. A lesion in the occipital lobe can cause blindness even if the eyes are fully functional.

A few optic nerve fibers take a different route in which they project to the midbrain and terminate in the

**FIGURE 16.42  The Retinal Basis of Stereoscopic Vision (depth perception).**  When the eyes are fixated on the fixation point (F), more distant objects (D) are focused on the retinas medial to the fovea and the brain interprets them as being farther away than the fixation point. Nearby objects (N) are focused lateral to the fovea and interpreted as being closer.

superior colliculi and pretectal nuclei. The superior colliculi control the visual reflexes of the extrinsic eye muscles, and the pretectal nuclei are involved in the photopupillary and accommodation reflexes.

Space does not allow us to consider much about the very complex processes of visual information processing in the brain. Some processing, such as contrast, brightness, motion, and stereopsis, begins in the retina. The primary visual cortex in the occipital lobe is connected by association tracts to nearby visual association areas in the posterior part of the parietal lobe and inferior part of the temporal lobe. These association areas process retinal data in ways beyond our present consideration to extract information about the location, motion, color, shape, boundaries, and other qualities of the objects we look at. They also store visual memories and enable the brain to identify what we are seeing—for example, to recognize printed words or name the objects we see. What is yet to be learned about visual processing promises to have important implications for biology, medicine, psychology, and even philosophy.

[58]*chiasm* = cross, X
[59]*hemi* = half + *decuss* = to cross, form an X
[60]*geniculate* = bent like a knee

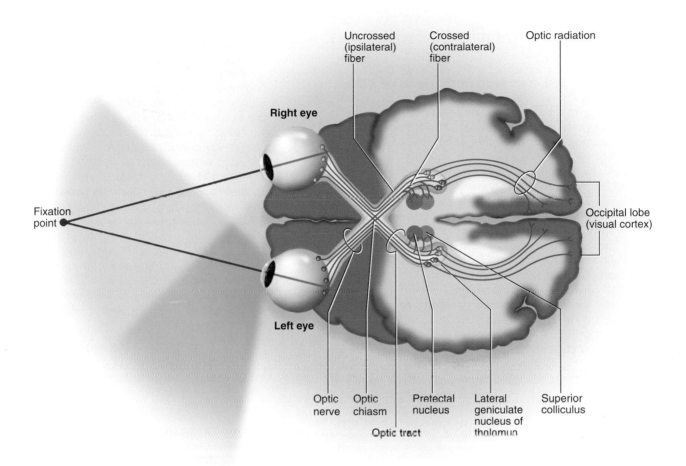

**FIGURE 16.43** **The Visual Projection Pathway.** Diagram of hemidecussation and projection to the primary visual cortex. Blue and yellow indicate the receptive fields of the left and right eyes; green indicates the area of overlap and stereoscopic vision. Nerve fibers from the medial side of the right eye descussate to the left side of the brain, while fibers from the lateral side remain on the right side of the brain. The converse is true of the left eye. The right occipital lobe thus monitors the left side of the visual field and the left occipital lobe monitors the right side.

▶ If a stroke destroyed the optic radiation of the right cerebral hemisphere, how would it affect a person's vision? Would it affect the person's visual reflexes?

## Before You Go On

Answer the following questions to test your understanding of the preceding section:

20. Why can't we see wavelengths below 350 nm or above 750 nm?

21. Why are light rays bent (refracted) more by the cornea than by the lens?

22. List as many structural and functional differences between rods and cones as you can.

23. Explain how the absorption of a photon of light leads to depolarization of a bipolar retinal cell.

24. Discuss the duplicity theory of vision, summarizing the advantage of having separate types of retinal photoreceptor cells for photopic and scotopic vision.

## Anesthesia—From Ether Frolics to Modern Surgery

Surgery is as old as civilization. People from the Stone Age to the pre-Columbian civilizations of the Americas practiced *trephination*—cutting a hole in the skull to let out "evil spirits" that were thought to cause headaches. The ancient Hindus were expert surgeons for their time, and the Greeks and Romans pioneered military surgery. But until the nineteenth century, surgery was a miserable and dangerous business, done only as a last resort and with little hope of the patient's survival. Surgeons rarely attempted anything more complex than amputations or kidney stone removal. A surgeon had to be somewhat indifferent to the struggles and screams of his patient. Most operations had to be completed in 3 minutes or less, and a strong arm and stomach were more important qualifications for a surgeon than extensive anatomical knowledge.

At least three things were needed for surgery to be more effective: better knowledge of anatomy, *asepsis*[61] for the control of infection, and *anesthesia*[62] for the control of pain. Early efforts to control surgical pain were crude and usually ineffective, such as choking a patient into unconsciousness and trying to complete the surgery before he or she awoke. Alcohol and opium were often used as anesthetics, but the dosage was poorly controlled; some patients were underanesthetized and suffered great pain anyway, and others died of overdoses. Often there was no alternative but for a few strong men to hold the struggling patient down as the surgeon worked. Charles Darwin originally intended to become a physician, but left medical school because he was sickened by observing "two very bad operations, one on a child," in the days before anesthesia.

In 1799, Sir Humphry Davy suggested using nitrous oxide to relieve pain. His student, Michael Faraday, suggested ether. Neither of these ideas caught on for several decades, however. Nitrous oxide ("laughing gas") was a popular amusement in the 1800s, when traveling showmen went from town to town demonstrating its effects on volunteers from the audience. In 1841, at a medicine show in Georgia, some students were impressed with the volunteers' euphoric giggles and antics and asked a young local physician, Crawford W. Long, if he could make some nitrous oxide for them. Long lacked the equipment to synthesize it, but he recommended they try ether. Ether was commonly used in small oral doses for toothaches and "nervous ailments," but its main claim to popularity was its use as a party drug for so-called ether frolics. Long himself was a bit of a bon vivant who put on demonstrations for some of the young ladies, with the disclaimer that he could not be held responsible for whatever he might do under the influence of ether (such as stealing a kiss).

At these parties, Long noted that people sometimes suffered considerable injuries without feeling pain. In 1842, he had a patient who was terrified of pain but needed a tumor removed from his neck. Long excised the tumor without difficulty as his patient sniffed ether from a towel. The operation created a sensation in town, but other physicians ridiculed Long and pronounced anesthesia dangerous. His medical practice declined as people grew afraid of him, but over the next 4 years he performed eight more minor surgeries on patients under ether. Struggling to overcome criticisms that the effects he saw were due merely to hypnotic suggestion or individual variation in sensitivity to pain, Long even compared surgeries done on the same person with and without ether.

Long failed to publish his results quickly enough, and in 1844 he was scooped by a Connecticut dentist, Horace Wells, who had tried nitrous oxide as a dental anesthetic. Another dentist, William Morton of Boston, had tried everything from champagne to opium to kill pain in his patients. He too became interested in ether and gave a public demonstration at Massachusetts General Hospital, where he etherized a patient and removed a tumor. Within a month of this successful and sensational demonstration, ether was being used in other cities of the United States and England. Morton patented a "secret formula" he called Morton's Letheon,[63] which smelled suspiciously of ether, but eventually he went broke trying to monopolize ether anesthesia and he died a pauper. His grave near Boston bears the epitaph:

WILLIAM T. G. MORTON
*Inventor and Revealer of Anaesthetic Inhalation*
*Before Whom, in All Time, Surgery was Agony.*
*By Whom Pain in Surgery Was Averted and Annulled.*
*Since Whom Science Has Control of Pain.*

Wells, who had engaged in a bitter feud to establish himself as the inventor of ether anesthesia, committed suicide at the age of 33. Crawford Long went on to a successful career as an Atlanta pharmacist, but to his death he remained disappointed that he had not received credit as the first to perform surgery on etherized patients.

Ether and chloroform became obsolete when safer anesthetics such as cyclopropane, ethylene, and nitrous oxide were developed. These are *general anesthetics* that render a patient unconscious by crossing the blood–brain barrier and blocking nervous transmission through the brainstem. Many general anesthetics deaden pain by activating GABA receptors and causing an inflow of $Cl^-$, which hyperpolarizes neurons and makes them less likely to fire. Diazepam (Valium) also employs this mechanism. Other general anesthetics work on nicotinic ACh receptors and glutamate receptors. *Local anesthetics* such as procaine (Novocain) and tetracaine selectively deaden specific nerves. They decrease the permeability of membranes to $Na^+$, thereby reducing their ability to produce action potentials.

A sound knowledge of anatomy, control of infection and pain, and development of better tools converged to allow surgeons time to operate more carefully. As a result, surgery became more intellectually challenging and interesting. It attracted a more educated class of practitioner, which put it on the road to becoming the remarkable lifesaving approach that it is today.

---

[61]*a* = without + *sepsis* = infection
[62]*an* = without + *esthesia* = feeling, sensation

[63]*lethe* = oblivion, forgetfulness

## CHAPTER REVIEW

# Review of Key Concepts

**Properties and Types of Sensory Receptors (p. 586)**

1. Sensory *receptors* range from simple nerve endings to complex sense organs.

2. *Sensory transduction* is the conversion of stimulus energy into a pattern of action potentials.

3. Transduction begins with a *receptor potential* which, if it reaches threshold, triggers the production of action potentials.

4. Receptors transmit four kinds of information about stimuli: *modality, location, intensity,* and *duration.*

5. Receptors can be classified by modality as *chemoreceptors, thermoreceptors, nociceptors, mechanoreceptors,* and *photoreceptors.*

6. Receptors can also be classified by the origins of their stimuli as *interoceptors, proprioceptors,* and *exteroceptors.*

7. General *(somesthetic) senses* have receptors widely distributed over the body and include the senses of touch, pressure, stretch, temperature, and pain. *Special senses* have receptors in the head only and include vision, hearing, equilibrium, taste, and smell.

**The General Senses (p. 588)**

1. Unencapsulated nerve endings are simple sensory nerve fibers not enclosed in specialized connective tissue; they include *free nerve endings, tactile discs,* and *hair receptors.*

2. Encapsulated nerve endings are nerve fibers enclosed in glial cells or connective tissues that modify their sensitivity. They include *muscle spindles, Golgi tendon organs, tactile corpuscles, Krause end bulbs, lamellated corpuscles,* and *Ruffini corpuscles.*

3. Somesthetic signals from the head travel the trigeminal and other cranial nerves to the brainstem, and those below the head travel up the spinothalamic tract and other pathways. Most signals reach the con-tralateral primary somesthetic cortex, but proprioceptive signals travel to the cerebellum.

4. Pain is a sensation that occurs when nociceptors detect tissue damage or potentially injurious situations.

5. *Fast pain* is a relatively quick, localized response mediated by myelinated nerve fibers; it may be followed by a less localized *slow pain* mediated by unmyelinated fibers.

6. *Somatic pain* arises from the skin, muscles, and joints, and may be *superficial* or *deep pain. Visceral pain* arises from the viscera; it is less localized and is often associated with nausea.

7. Injured tissues release bradykinin, serotonin, prostaglandins, and other chemicals that stimulate nociceptors.

8. Pain signals travel from the receptor to the cerebral cortex by way of *first-* through *third-order neurons.* Signals from the head travel through the trigeminal, facial, glossopharyngeal, and vagus nerves to the medulla oblongata, and are then relayed through the thalamus to the cerebrum. Signals from below the head ascend the spinothalamic and spinoreticular tracts and gracile fasciculus of the spinal cord. The thalamus relays pain signals to the postcentral gyrus of the cerebrum.

9. Pain signals in the spinoreticular tract travel to the reticular formation and from there to the hypothalamus and limbic system, producing visceral and emotional responses to pain.

10. *Referred pain* is the brain's misidentification of the location of pain resulting from convergence in sensory pathways.

11. Pain awareness can be reduced by *spinal gating* mechanisms that block the transmission of pain signals up the spinal cord. Analgesic neuropeptides *(endogenous opioids)* called *enkephalins, endorphins,* and *dynor-phins* exert pain-blocking effects at multiple levels of the spinal cord and brainstem.

**The Chemical Senses (p. 594)**

1. Taste *(gustation)* results from the action of chemicals on the *taste buds,* which are groups of sensory cells located on some of the *lingual papillae* and in the palate, pharynx, and epiglottis.

2. *Foliate, fungiform,* and *vallate papillae* have taste buds; *filiform papillae* lack taste buds but sense the texture of food.

3. The primary taste sensations are salty, sweet, sour, bitter, and umami. Flavor is a combined effect of these tastes and the texture, aroma, temperature, and appearance of food. Some flavors result from the stimulation of free nerve endings.

4. Some taste chemicals (sugars, alkaloids, and glutamate) bind to surface receptors on the taste cells and activate second messengers in the cell; sodium and acids penetrate into the taste cell and depolarize it.

5. Taste signals travel from the tongue through the facial and glossopharyngeal nerves, and from the palate, pharynx, and epiglottis through the vagus nerve. They travel to the medulla oblongata and then by one route to the hypothalamus and amygdala, and by another route to the thalamus and cerebral cortex.

6. Smell *(olfaction)* results from the action of chemicals on *olfactory cells* in the roof of the nasal cavity.

7. Odor molecules bind to surface receptors on the *olfactory hairs* of the olfactory cells and activate second messengers in the cell.

8. Nerve fibers from the olfactory cells assemble into fascicles that collectively constitute cranial nerve I, pass through foramina of the cribriform plate, and end in the olfactory bulbs beneath the frontal lobes of the cerebrum.

9. Olfactory signals travel the *olfactory tracts* from the bulbs to the temporal lobes, and are relayed to the insula, orbitofrontal cortex, hippocampus, amygdala, and hypothalamus. The cerebral cortex also sends signals back to the bulbs that moderate one's perception of smell.

### Hearing and Equilibrium (p. 598)

1. Sound is generated by vibrating objects. The *amplitude* of the vibration determines the *loudness* of a sound, measured in *decibels (db)*, and the *frequency* of vibration determines the *pitch*, measured in *hertz (Hz)*.

2. Humans hear best at frequencies of 1,500 to 5,000 Hz, but sensitive ears can hear sounds from 20 Hz to 20,000 Hz. The threshold of hearing is 0 db and the threshold of pain is about 140 db; most conversation is about 60 db.

3. The *outer ear* consists of the *auricle* and *auditory canal*. The *middle ear* consists of the tympanic membrane and an air-filled tympanic cavity containing three bones *(malleus, incus,* and *stapes)* and two muscles *(tensor tympani* and *stapedius)*. The inner ear consists of fluid-filled chambers and tubes (the *membranous labyrinth*) including the *vestibule, semicircular ducts,* and *cochlea.*

4. The most important part of the cochlea, the organ of hearing, is the *spiral organ,* which includes sensory *hair cells.* A row of 3,500 *inner hair cells* generates the signals we hear, and three rows of *outer hair cells* tune the cochlea to enhance its pitch discrimination.

5. Vibrations in the ear move the *basilar membrane* of the cochlea up and down. As the hair cells move up and down, their stereocilia bend against the relatively stationary tectorial membrane above them. This opens $K^+$ channels at the tip of each stereocilium, and the inflow of $K^+$ depolarizes the cell. This triggers neurotransmitter release, which initiates a nerve signal.

6. *Loudness* determines the amplitude of basilar membrane vibration and the firing frequency of the associated auditory neurons. *Pitch* determines which regions of the basilar membrane vibrate more than others, and

which auditory nerve fibers respond most strongly.

7. The cochlear nerve joins the vestibular nerve to become cranial nerve VIII. Cochlear nerve fibers project to the pons and from there to the inferior colliculi of the midbrain, then the thalamus, and finally the primary auditory cortex of the temporal lobes.

8. The pons transmits excitatory nerve signals back to the outer hair cells of the cochlea and inhibitory nerve signals back to the sensory dendrites associated with the inner hair cells, acting in both cases to sharpen the contrast between responses in adjacent regions of the spinal organ. This "tunes" the cochlea to enhance one's ability to discriminate between sounds of similar pitch.

9. *Static equilibrium* is the sense of the orientation of the head; *dynamic equilibrium* is the sense of linear or angular acceleration of the head.

10. The *saccule* and *utricle* are chambers in the vestibule of the inner ear, each with a *macula* containing sensory hair cells. The *macula sacculi* is nearly vertical and the *macula utriculi* is nearly horizontal.

11. The hair cell stereocilia are capped by a weighted gelatinous *otolithic membrane.* When pulled by gravity or linear acceleration of the body, these membranes stimulate the hair cells.

12. Any orientation of the head causes a combination of stimulation to the four maculae, sending signals to the brain that enable it to sense the orientation. Vertical acceleration also stimulates each macula sacculi, and horizontal acceleration stimulates each macula utriculi.

13. Each inner ear also has three *semicircular ducts* with a sensory patch of hair cells, the *crista ampullaris,* in each duct. The stereocilia of these hair cells are embedded in a gelatinous *cupula.*

14. Tilting or rotation of the head moves the ducts relative to the fluid (endolymph) within, causing the fluid to push the cupula and stimulate the hair cells. The brain detects angular acceleration of the head from the combined input from the six ducts.

15. Signals from the utricle, saccule, and semicircular ducts travel the *vestibular nerve,* which joins the cochlear nerve in cranial nerve VIII. Vestibular nerve fibers lead to four pairs of *vestibular nuclei* in the pons and medulla. These nuclei relay signals to the cerebellum; nuclei that control eye movements; the reticular formation; the spinal cord; and via the thalamus to the cerebrum.

### Vision (p. 612)

1. Vision is a response to electromagnetic radiation with wavelengths from about 400 to 750 nm.

2. Accessory structures of the orbit include the eyebrows, eyelids, conjunctiva, lacrimal apparatus, and extrinsic eye muscles.

3. The wall of the eyeball is composed of an outer *fibrous layer* composed of *sclera* and *cornea;* a middle *vascular layer* composed of *choroid, ciliary body,* and *iris;* and an *inner layer* composed of the *retina* and beginning of the *optic nerve.*

4. The optical components of the eye admit and bend (refract) light rays and bring images to a focus on the retina. They include the *cornea, aqueous humor, lens,* and *vitreous body.* Most refraction occurs at the air–cornea interface, but the lens adjusts the focus.

5. The neural components of the eye absorb light and encode the stimulus in action potentials transmitted to the brain. They include the *retina* and *optic nerve.* The sharpest vision occurs in a region of retina called the *fovea centralis,* while the *optic disc,* where the optic nerve originates, is a blind spot with no receptor cells.

6. The relaxed *(emmetropic)* eye focuses on objects 6 m or more away. A *near response* is needed to focus on closer objects. This includes convergence of the eyes, constriction of the pupil, and *accommodation* (thickening) of the lens.

7. Light falling on the retina is absorbed by visual pigments in the *outer segments* of the *rod* and *cone* cells. Rods function at low light intensities (producing night, or *scotopic,* vision) but produce monochromatic images with poor resolution. Cones require higher light intensities (producing day, or

*photopic,* vision) and produce color images with finer resolution.

8. Light absorption bleaches the *rhodopsin* of rods or the *photopsins* of the cones. In rods (and probably cones), this stops the *dark current* of $Na^+$ flow into the cell and the release of glutamate from the inner end of the cell.

9. Rods and cones synapse with *bipolar cells,* which respond to changes in glutamate secretion. Bipolar cells, in turn, stimulate *ganglion cells.* Ganglion cells are the first cells in the pathway that generate action potentials; their axons form the optic nerve.

10. The eyes respond to changes in light intensity by *light adaptation* (pupillary constriction and pigment bleaching) and *dark adaptation* (pupillary dilation and pigment regeneration).

11. The *duplicity theory* explains that a single type of receptor cell cannot produce both high light sensitivity (like the rods) and high resolution (like the cones). The neuronal convergence responsible for the sensitivity of rod pathways limits resolution, while the lack of convergence responsible for the high resolution of cones limits light sensitivity.

12. Three types of cones—blue, green, and red—have slight differences in their photopsins that result in peak absorption in different regions of the spectrum. This results in the ability to distinguish colors.

13. *Stereoscopic vision* (depth perception) results from each eye viewing an object from a slightly different angle, so its image falls on different areas of the two retinas.

14. Fibers of the optic nerves *hemidecussate* at the *optic chiasm,* so images in the left visual field project from both eyes to the right cerebral hemisphere, and images on the right project to the left hemisphere.

15. Beyond the optic chiasm, most nerve fibers end in the *lateral geniculate nucleus* of the thalamus. Here they synapse with third-order neurons whose fibers form the *optic radiation* leading to the primary visual cortex of the occipital lobe.

16. Some fibers of the optic nerve lead to the superior colliculi and pretectal nuclei of the midbrain. These midbrain nuclei control visual reflexes of the extrinsic eye muscles, pupillary reflexes, and accommodation of the lens in near vision.

## Testing Your Recall

1. Hot and cold stimuli are detected by
   a. free nerve endings.
   b. proprioceptors.
   c. Krause end bulbs.
   d. lamellated corpuscles.
   e. tactile corpuscles.

2. _____ is a neurotransmitter that transmits pain sensations to second-order spinal neurons.
   a. Endorphin
   b. Enkephalin
   c. Substance P
   d. Acetylcholine
   e. Norepinephrine

3. _____ is a neuromodulator that blocks the transmission of pain sensations to second-order spinal neurons.
   a. Endorphin
   b. Enkephalin
   c. Substance P
   d. Acetylcholine
   e. Norepinephrine

4. Taste buds of the vallate papillae are most sensitive to
   a. bitter.
   b. sour.
   c. sweet.
   d. umami.
   e. salty.

5. The higher the frequency of a sound,
   a. the louder it sounds.
   b. the harder it is to hear.
   c. the more it stimulates the distal end of the spiral organ.
   d. the faster it travels through air.
   e. the higher its pitch.

6. Cochlear hair cells rest on
   a. the tympanic membrane.
   b. the secondary tympanic membrane.
   c. the tectorial membrane.
   d. the vestibular membrane.
   e. the basilar membrane.

7. The acceleration you feel when an elevator begins to rise is sensed by
   a. the anterior semicircular duct.
   b. the spiral organ.
   c. the crista ampullaris.
   d. the macula sacculi.
   e. the macula utriculi.

8. The color of light is determined by
   a. its velocity.
   b. its amplitude.
   c. its wavelength.
   d. refraction.
   e. how strongly it stimulates the rods.

9. The retina receives its oxygen supply from
   a. the hyaloid artery.
   b. the vitreous body.
   c. the choroid.
   d. the pigment epithelium.
   e. the scleral venous sinus.

10. Which of the following statements about photopic vision is false?
    a. It is mediated by the cones.
    b. It has a low threshold.
    c. It produces fine resolution.
    d. It does not function in starlight.
    e. It does not employ rhodopsin.

11. The most finely detailed vision occurs when an image falls on a pit in the retina called the _____ .

12. The only cells of the retina that generate action potentials are the _____ cells.

13. The retinal dark current results from the flow of _____ into the receptor cells.

14. The gelatinous membranes of the macula sacculi and macula utriculi are weighted by calcium carbonate and protein granules called _____.

15. Three rows of _____ in the cochlea have V-shaped arrays of stereocilia and tune the frequency sensitivity of the cochlea.

16. The _____ is a tiny bone that vibrates in the oval window and thereby transfers sound vibrations to the inner ear.

17. The _____ of the midbrain receive auditory input and trigger the head-turning auditory reflex.

18. The apical stereocilia of a gustatory cell are called _____ .

19. Olfactory neurons synapse with mitral cells and tufted cells in the

_____ , which lies inferior to the frontal lobe.

20. In the phenomenon of _____ , pain from the viscera is perceived as coming from an area of the skin.

*Answers in Appendix B*

## True or False

*Determine which five of the following statements are false, and briefly explain why.*

1. The sensory (afferent) nerve fibers for touch end in the thalamus.

2. Things we touch with the left hand are perceived only in the right cerebral hemisphere.

3. Things we see with the left eye are perceived only in the right cerebral hemisphere.

4. Some chemoreceptors are interoceptors and some are exteroceptors.

5. The vitreous body occupies the posterior chamber of the eye.

6. Descending analgesic fibers prevent pain signals from reaching the spinal cord.

7. Cranial nerve VIII carries signals for both hearing and balance.

8. The tympanic cavity is filled with air, but the membranous labyrinth is filled with liquid.

9. Rods and cones release their neurotransmitter in the dark, not in the light.

10. All of the extrinsic muscles of the eye are controlled by the oculomotor nerve.

*Answers in Appendix B*

## Testing Your Comprehension

1. The principle of neural convergence is explained on page 472. Discuss its relevance to referred pain and scotopic vision.

2. What type of cutaneous receptor enables you to feel an insect crawling through your hair? What type enables you to palpate a patient's pulse? What type enables a blind person to read braille?

3. Contraction of a muscle usually puts more tension on a structure, but contraction of the ciliary muscle puts less tension on the lens. Explain how.

4. Janet has terminal ovarian cancer and is in severe pelvic pain that has not yielded to any other treatment. A neurosurgeon performs an *anterolateral cordotomy*, cutting across the

anterolateral region of her lumbar spinal cord. Explain the rationale of this treatment and its possible side effects.

5. What would be the benefit of a drug that blocks the receptors for substance P?

*Answers at www.mhhe.com/saladin4*

# www.mhhe.com/saladin4

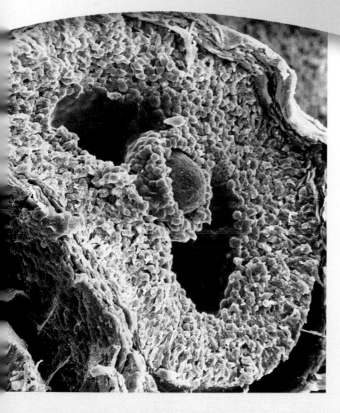

An ovarian follicle (SEM). Layered cells of the follicle wall secrete steroid hormones. The egg is the large pink cell.

# THE ENDOCRINE SYSTEM

## Brushing Up

To understand this chapter, it is important that you understand or brush up on the following concepts:

- Structure and function of the plasma membrane (p. 94)
- G proteins, cAMP, and other second messengers (p. 98)
- Active transport and the transport maximum (p. 106)
- Protein synthesis and secretion (pp. 131–137)
- Monoamines, especially catecholamines (p. 463)
- The hypothalamus (p. 530)

Anatomy & Physiology **REVEALED**

**Volume 4** Endocrine System

For the body to maintain homeostasis, cells must be able to communicate and integrate their activities with each other. For the last five chapters, we have examined how this is achieved through the nervous system. We now turn to two modes of chemical communication called *endocrine* and *paracrine* signaling, with an emphasis on the former. This chapter is primarily about **endocrinology,** the study of the endocrine system and the diagnosis and treatment of its dysfunctions.

You probably have at least some prior acquaintance with this system. Perhaps you have heard of the pituitary gland and thyroid gland, secretions such as growth hormone, estrogen, and insulin, and endocrine disorders such as dwarfism, goiter, and diabetes mellitus. Fewer readers, perhaps, are familiar with what hormones are or exactly how they work. Therefore, this chapter starts with the relatively familiar—a survey of the endocrine glands, their hormones, and the principal effects of these hormones. We will then work our way down to the finer and less familiar details—the chemical identity of hormones, how they are made and transported, and how they produce their effects on their target cells. Shorter sections at the end of the chapter discuss the role of the endocrine system in adapting to stress, some hormonelike paracrine secretions, and the pathologies that result from endocrine dysfunction.

# Overview of the Endocrine System

## Objectives

When you have completed this section, you should be able to

- define *hormone* and *endocrine system;*
- list the major organs of the endocrine system;
- recognize the standard abbreviations for many hormones; and
- compare and contrast the nervous and endocrine systems.

Cells communicate with each other in four ways:

1. **Gap junctions** join single-unit smooth muscle, cardiac muscle, epithelial, and other cells to each other. They enable cells to pass nutrients, electrolytes, and signaling molecules directly from the cytoplasm of one cell to the cytoplasm of the next through adjacent pores in their plasma membranes (fig. 5.28, p. 174).

2. **Neurotransmitters** are released by neurons, diffuse across a narrow synaptic cleft, and bind to receptors on the surface of the next cell.

3. **Paracrines**[1] are secreted into the tissue fluid by a cell, diffuse to nearby cells in the same tissue, and stimulate their physiology. They are sometimes called *local hormones.*

4. **Hormones**[2] are chemical messengers that are secreted into the bloodstream and stimulate the physiology of cells in another tissue or organ, often a considerable distance away. Hormones produced by the pituitary gland in the head, for example, can act on organs in the abdominal and pelvic cavities. (Some authorities define *hormone* so broadly as to include paracrines and neurotransmitters. This book adopts the stricter definition of hormones as blood-borne messengers secreted by endocrine cells.)

Our focus in this chapter will be primarily on hormones and the **endocrine**[3] **glands** that secrete them (fig. 17.1). The **endocrine system** is composed of these glands as well as hormone-secreting cells in many organs not usually thought of as glands, such as the brain, heart, and small intestine. Hormones travel anywhere the blood goes, but they affect only those cells that have receptors for them. These are called the **target cells** for a particular hormone.

In chapter 5, we saw that glands can be classified as exocrine or endocrine. One way in which these differ is that exocrine glands have ducts to carry their secretion to the body surface (as in sweat) or to the cavity of another organ (as in digestive enzymes). Endocrine glands have no ducts but do have dense blood capillary networks. Endocrine cells release their hormones into the surrounding tissue fluid, and then the bloodstream quickly picks up and distributes the hormones. Exocrine secretions have extracellular effects such as the digestion of food, whereas endocrine secretions have intracellular effects—they alter the metabolism of their target cells.

## COMPARISON OF THE NERVOUS AND ENDOCRINE SYSTEMS

Although the nervous and endocrine systems both serve for internal communication, they are not redundant; they complement rather than duplicate each other's function (table 17.1). The systems differ in their means of communication—both electrical and chemical in the nervous system and solely chemical in the endocrine system (fig. 17.2)—yet as we shall see, they have many similarities on this point as well. They differ also in how quickly they start and stop responding to stimuli.

---

[1] *para* = next to + *crin* = secrete
[2] *hormone* = to excite, set in motion
[3] *endo* = into + *crin* = secrete

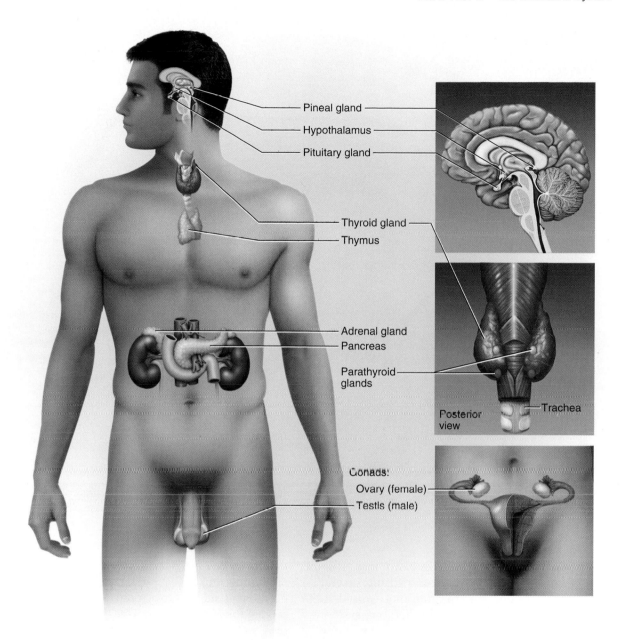

**FIGURE 17.1  Major Organs of the Endocrine System.**  This system also includes gland cells in many other organs not shown here.

| TABLE 17.1 | Comparison of the Nervous and Endocrine Systems |
|---|---|
| **Nervous System** | **Endocrine System** |
| Communicates by means of electrical impulses and neurotransmitters | Communicates by means of hormones |
| Releases neurotransmitters at synapses at specific target cells | Releases hormones into bloodstream for general distribution throughout body |
| Usually has relatively local, specific effects | Sometimes has very general, widespread effects |
| Reacts quickly to stimuli, usually within 1–10 msec | Reacts more slowly to stimuli, often taking seconds to days |
| Stops quickly when stimulus stops | May continue responding long after stimulus stops |
| Adapts relatively quickly to continual stimulation | Adapts relatively slowly; may continue responding for days to weeks of stimulation |

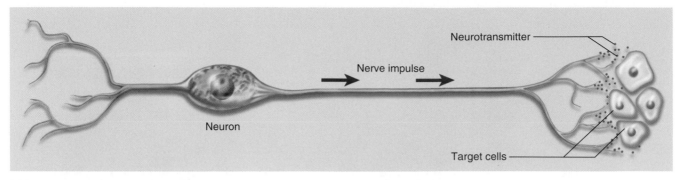

**(a) Nervous system**

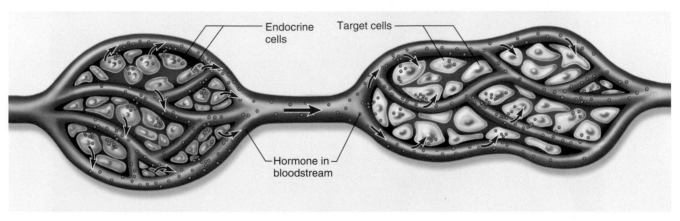

**(b) Endocrine system**

**FIGURE 17.2  Communication by the Nervous and Endocrine Systems.**  (a) A neuron has a long fiber that delivers its neurotransmitter to the immediate vicinity of its target cells. (b) Endocrine cells secrete a hormone into the bloodstream. The hormone binds to target cells at places often remote from the gland cells.

The nervous system typically responds in just a few milliseconds, whereas hormone release may follow from several seconds to several days after the stimulus that caused it. Furthermore, when a stimulus ceases, the nervous system stops responding almost immediately, whereas some endocrine effects persist for several days or even weeks. On the other hand, under long-term stimulation, neurons soon adapt and their response declines. The endocrine system shows more persistent responses. For example, thyroid hormone secretion rises in cold weather and remains elevated as long as it remains cold. Another difference between the two systems is that an efferent nerve fiber innervates only one organ and a limited number of cells within that organ; its effects, therefore, are precisely targeted and relatively specific. Hormones, by contrast, circulate throughout the body and some of them, such as growth hormone, epinephrine, and thyroid hormone, have more widespread effects than does the output of any one nerve fiber.

But these differences should not blind us to the similarities between the two systems. Several chemicals function as both neurotransmitters and hormones, including norepi-nephrine, cholecystokinin, thyrotropin-releasing hormone, dopamine, and antidiuretic hormone (= vasopressin). Some hormones, such as oxytocin and the catecholamines, are secreted by **neuroendocrine cells**—neurons that release their secretions into the extracellular fluid. Some hormones and neurotransmitters produce overlapping effects on the same target cells. For example, norepinephrine and glucagon cause glycogen hydrolysis in the liver. The nervous and endocrine systems continually regulate each other as they coordinate the activities of other organ systems. Neurons often trigger hormone secretion, and hormones often stimulate or inhibit neurons.

## HORMONE NOMENCLATURE

Many hormones are denoted by standard abbreviations which are used repeatedly in this chapter. These abbreviations are listed alphabetically in table 17.2 so that you can use this as a convenient reference while you work through the chapter. This is by no means a complete list. It does not include hormones that have no abbreviation, such as estrogen and insulin, and it omits hormones that are not discussed much in this chapter. Synonyms used by many authors are indicated in parentheses, but the first name listed is the one that is used in this book.

## Before You Go On

*Answer the following questions to test your understanding of the preceding section:*

1. *Define the word* hormone *and distinguish a hormone from a neurotransmitter. Why is this an imperfect distinction?*

2. *Describe some ways in which endocrine glands differ from exocrine glands.*

3. *Name some sources of hormones other than purely endocrine glands.*

4. *List some similarities and differences between the endocrine and nervous systems.*

# The Hypothalamus and Pituitary Gland

## Objectives
When you have completed this section, you should be able to

- describe the anatomical relationship of the pituitary gland to the hypothalamus;
- name the hormones produced by the hypothalamus and pituitary gland and state the effects of each;
- discuss the actions of growth hormone; and
- explain how the pituitary gland is controlled by the hypothalamus and feedback from its target organs.

There is no "master control center" that regulates the entire endocrine system, but the pituitary gland and a nearby region of the brain, the hypothalamus, have a more wide-ranging influence than any other part of the system. This is an appropriate place to begin a survey of the endocrine system.

## ANATOMY

The hypothalamus forms the floor and walls of the third ventricle of the brain (see fig. 14.12b, p. 530). It regulates primitive functions of the body ranging from water balance to childbirth. Many of its functions are carried out by way of the pituitary gland, which is closely associated with it.

The **pituitary gland** (hypophysis[4]) is suspended from the hypothalamus by a stalk (*infundibulum*[5]) and housed in the sella turcica of the sphenoid bone. It is usually about 1.3 cm in diameter, but grows about 50% larger in pregnancy. It is actually composed of two structures—the adenohypophysis and neurohypophysis—that arise independently in the embryo and have entirely separate functions. The adenohypophysis arises from a *hypophyseal pouch* that grows upward from the pharynx, while the neurohypophysis arises as a downgrowth of the brain, the *neurohypophyseal bud* (fig. 17.3). They come to lie side by side and are so closely joined that they look like a single gland.

| TABLE 17.2 | Names and Abbreviations for Hormones | |
|---|---|---|
| **Abbreviation** | **Name** | **Source** |
| ACTH | Adrenocorticotropic hormone (corticotropin) | Anterior pituitary |
| ADH | Antidiuretic hormone (vasopressin) | Posterior pituitary |
| ANP | Atrial natriuretic peptide | Heart |
| CRH | Corticotropin-releasing hormone | Hypothalamus |
| DHEA | Dehydroepiandrosterone | Adrenal cortex |
| EPO | Erythropoietin | Kidney, liver |
| FSH | Follicle-stimulating hormone | Anterior pituitary |
| GH | Growth hormone (somatotropin) | Anterior pituitary |
| GHRH | Growth hormone–releasing hormone | Hypothalamus |
| GnRH | Gonadotropin-releasing hormone | Hypothalamus |
| IGFs | Insulin-like growth factors (somatomedins) | Liver, other tissues |
| LH | Luteinizing hormone | Anterior pituitary |
| NE | Norepinephrine | Adrenal medulla |
| OT | Oxytocin | Posterior pituitary |
| PIH | Prolactin-inhibiting hormone (dopamine) | Hypothalamus |
| PRH | Prolactin-releasing hormone | Hypothalamus |
| PRL | Prolactin | Anterior pituitary |
| PTH | Parathyroid hormone (parathormone) | Parathyroids |
| $T_3$ | Triiodothyronine | Thyroid |
| $T_4$ | Thyroxine (tetraiodothyronine) | Thyroid |
| TH | Thyroid hormone ($T_3$ and $T_4$) | Thyroid |
| TRH | Thyrotropin-releasing hormone | Hypothalamus |
| TSH | Thyroid-stimulating hormone | Anterior pituitary |

---

[4]*hypo* = below + *physis* = growth
[5]*infundibulum* = funnel

The **adenohypophysis**[6] (AD-eh-no-hy-POFF-ih-sis) constitutes the anterior three-quarters of the pituitary (figs. 17.4a, 17.5a). It has two parts: a large **anterior lobe,** also called the *pars distalis* ("distal part") because it is

most distal to the pituitary stalk, and the *pars tuberalis,* a small mass of cells adhering to the anterior side of the stalk. In the fetus there is also a *pars intermedia,* a strip of tissue between the anterior lobe and neurohypophysis. During subsequent development, its cells mingle with those of the anterior lobe; in adults, there is no longer a separate pars intermedia.

---

[6]*adeno* = gland

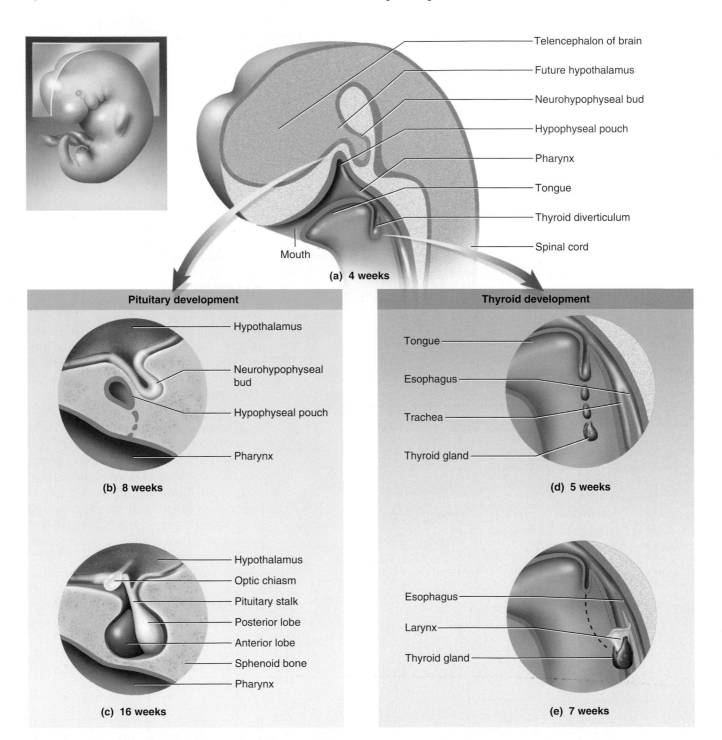

**FIGURE 17.3    Embryonic Development of the Pituitary and Thyroid Glands.** (a) Sagittal section of the head showing the neural and pharyngeal origins of the glands. (b–c) Development of the two lobes of the pituitary from their neural and pharyngeal origins. (d–e) Descent of the thyroid as it loses its connection with the root of the tongue and associates with the larynx.

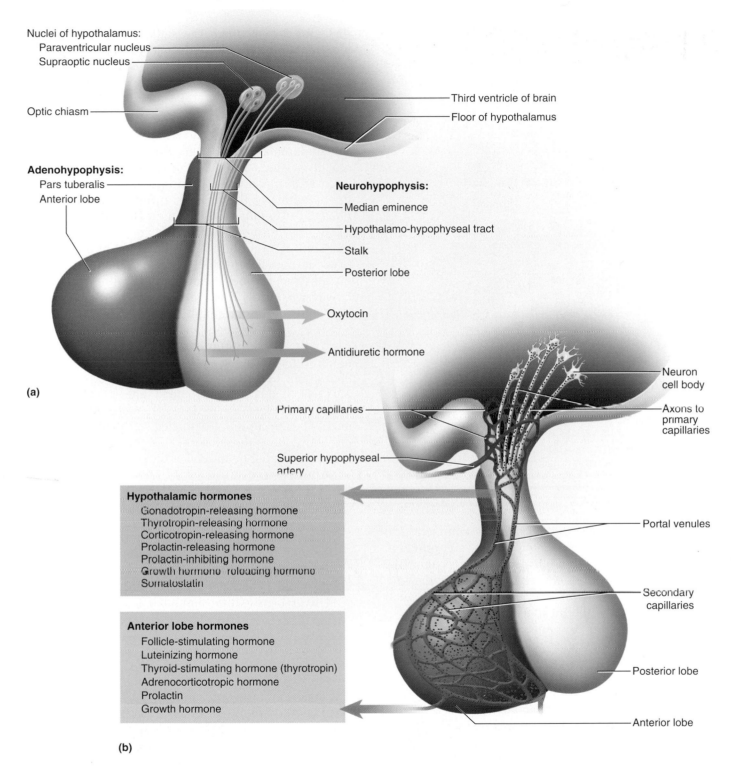

Nuclei of hypothalamus:
   Paraventricular nucleus
   Supraoptic nucleus

Optic chiasm

Third ventricle of brain
Floor of hypothalamus

**Adenohypophysis:**
   Pars tuberalis
   Anterior lobe

**Neurohypophysis:**
   Median eminence
   Hypothalamo-hypophyseal tract
   Stalk
   Posterior lobe

Oxytocin

Antidiuretic hormone

**(a)**

Neuron cell body

Axons to primary capillaries

Primary capillaries

Superior hypophyseal artery

**Hypothalamic hormones**
   Gonadotropin-releasing hormone
   Thyrotropin-releasing hormone
   Corticotropin-releasing hormone
   Prolactin-releasing hormone
   Prolactin-inhibiting hormone
   Growth hormone–releasing hormone
   Somatostatin

Portal venules

Secondary capillaries

**Anterior lobe hormones**
   Follicle-stimulating hormone
   Luteinizing hormone
   Thyroid-stimulating hormone (thyrotropin)
   Adrenocorticotropic hormone
   Prolactin
   Growth hormone

Posterior lobe

Anterior lobe

**(b)**

**FIGURE 17.4  Anatomy of the Pituitary Gland.** (a) Major structures of the pituitary and hormones of the neurohypophysis. Note that these hormones are produced by two nuclei in the hypothalamus and later released from the posterior lobe of the pituitary. (b) The hypophyseal portal system, which regulates the anterior lobe of the pituitary. The hormones in the violet box are secreted by the hypothalamus and travel in the portal system to the anterior pituitary. The hormones in the pink box are secreted by the anterior pituitary under the control of the hypothalamic releasers and inhibitors.

⏵ *Which lobe of the pituitary is essentially composed of brain tissue?*

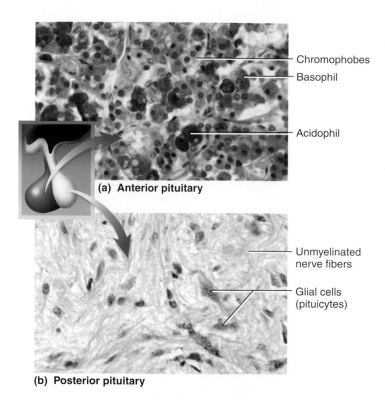

**(a) Anterior pituitary**

- Chromophobes
- Basophil
- Acidophil

- Unmyelinated nerve fibers
- Glial cells (pituicytes)

**(b) Posterior pituitary**

**FIGURE 17.5**  **Histology of the Pituitary Gland.**  (a) The anterior lobe. Chromophobes are inactive cells. Basophils include gonadotropes, thyrotropes, and corticotropes. Acidophils include somatotropes and lactotropes. These subtypes are not distinguishable with this histological stain. (b) The posterior lobe.

The anterior pituitary has no nervous connection to the hypothalamus but is connected to it by a complex of blood vessels called the **hypophyseal portal system** (fig. 17.4b). This begins with a network of *primary capillaries* in the hypothalamus, leading to *portal venules* (small veins) that travel down the pituitary stalk to a complex of *secondary capillaries* in the anterior pituitary. The primary capillaries pick up hormones from the hypothalamus, the venules deliver them to the anterior pituitary, and the hormones leave the circulation at the secondary capillaries. The pituitary and other endocrine glands have an especially porous type of blood capillaries called *fenestrated capillaries,* described in chapter 20 (see fig. 20.6). In essence, they have numerous "windows" in their walls that allow for very rapid uptake and release of hormones and other substances.

The **neurohypophysis** constitutes the posterior one-quarter of the pituitary. It has three parts: an extension of the hypothalamus called the *median eminence;* the *stalk;* and the largest part, the **posterior lobe** (pars nervosa). The neurohypophysis is not a true gland but a mass of neuroglia and nerve fibers (fig. 17.5b). The nerve fibers arise from cell bodies in the hypo-

thalamus, travel down the stalk as a bundle called the **hypothalamo-hypophyseal tract,** and end in the posterior lobe. The hypothalamic neurons synthesize hormones, transport them down the stalk, and store them in the posterior pituitary until a nerve signal triggers their release.

## HYPOTHALAMIC HORMONES

The hypothalamus produces nine hormones important to our discussion. Seven of them, listed in figure 17.4 and table 17.3, travel through the portal system and regulate the activities of the anterior pituitary. Five of these are *releasing hormones* that stimulate the anterior pituitary to secrete its hormones, and two are *inhibiting hormones* that suppress pituitary secretion. Most of these hypothalamic hormones control the release of just one anterior pituitary hormone. Gonadotropin-releasing hormone, however, controls the release of both follicle-stimulating hormone and luteinizing hormone.

The other two hypothalamic hormones are secreted by way of the posterior pituitary. These are **oxytocin (OT)** and **antidiuretic hormone (ADH).** OT is produced mainly by neurons in the **paraventricular**[7] **nuclei** of the hypothalamus, so-called because they lie in the walls of the third ventricle (the nuclei are paired right and left). ADH is produced mainly by the **supraoptic**[8] **nuclei,** so-called because they lie just above the optic chiasm on each side. Each nucleus also produces smaller quantities of the other hormone.

## PITUITARY HORMONES

The secretions of the pituitary gland are as follows:

- The *anterior lobe* synthesizes and secretes six principal hormones: follicle-stimulating hormone (FSH), luteinizing hormone (LH), thyroid-stimulating hormone (TSH), adrenocorticotropic hormone (ACTH), growth hormone (GH), and prolactin (PRL)

---

[7]*para* = next to + *ventricular* = pertaining to the ventricle
[8]*supra* = above

| TABLE 17.3 | Hypothalamic Releasing and Inhibiting Hormones That Regulate the Anterior Pituitary |
|---|---|
| **Hormone** | **Principal Effects** |
| TRH: Thyrotropin-releasing hormone | Promotes TSH and PRL secretion |
| CRH: Corticotropin-releasing hormone | Promotes ACTH secretion |
| GnRH: Gonadotropin-releasing hormone | Promotes FSH and LH secretion |
| PRH: Prolactin-releasing hormone | Promotes PRL secretion |
| PIH: Prolactin-inhibiting hormone | Inhibits PRL secretion |
| GHRH: Growth hormone–releasing hormone | Promotes GH secretion |
| Somatostatin | Inhibits GH and TSH secretion |

(table 17.4). The first five of these are **tropic,** or **trophic,**[9] **hormones**—pituitary hormones that stimulate endocrine cells elsewhere to release their own hormones. More specifically, the first two are called **gonadotropins** because their target organs are the gonads.

The hormonal relationship between the hypothalamus, pituitary, and a more remote endocrine gland is called an *axis*. There are three such axes: the **hypothalamo–pituitary–gonadal axis** involving GnRH, FSH, and LH; the **hypothalamo–pituitary–thyroid axis** involving TRH and TSH; and the **hypothalamo–pituitary–adrenal axis** involving CRH and ACTH (fig. 17.6).

- The *pars intermedia* is absent from the adult human pituitary, but is present in other animals and the human fetus. In other species, it secretes *melanocyte-stimulating hormone (MSH),* which influences pigmentation of the skin, hair, or feathers. Humans, however, apparently produce no circulating MSH. Some anterior pituitary cells derived from the pars intermedia produce a large polypeptide called *pro-opiomelanocortin (POMC).* POMC is

---

[9]*trop* = to turn, change; *troph* = to feed, nourish

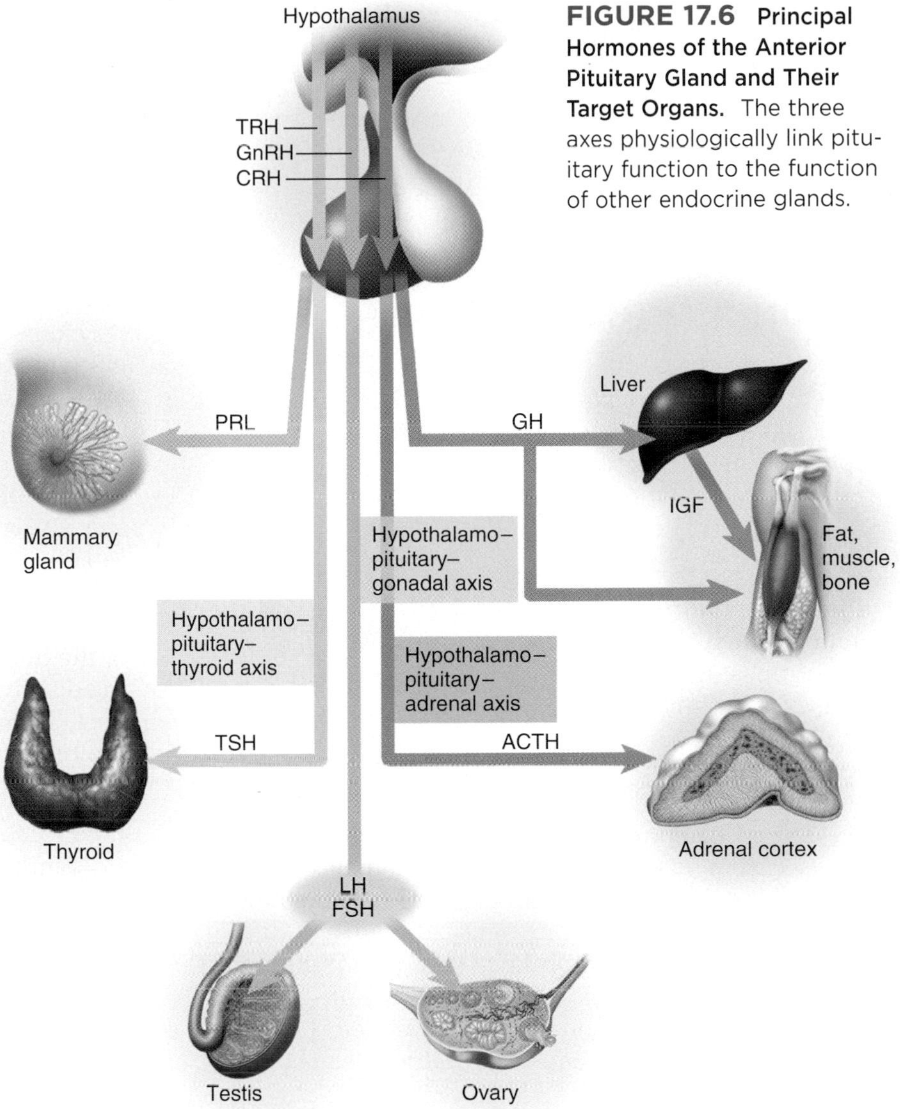

**FIGURE 17.6 Principal Hormones of the Anterior Pituitary Gland and Their Target Organs.** The three axes physiologically link pituitary function to the function of other endocrine glands.

| TABLE 17.4 | Pituitary Hormones | |
|---|---|---|
| **Hormone** | **Target Organ or Tissue** | **Principal Effects** |
| *Anterior Pituitary* | | |
| FSH: Follicle-stimulating hormone | Ovaries, testes | Female: growth of ovarian follicles and secretion of estrogen<br>Male: sperm production |
| LH: Luteinizing hormone | Ovaries, testes | Female: ovulation, maintenance of corpus luteum<br>Male: testosterone secretion |
| TSH: Thyroid-stimulating hormone | Thyroid gland | Growth of thyroid, secretion of thyroid hormone |
| ACTH: Adrenocorticotropic hormone | Adrenal cortex | Growth of adrenal cortex, secretion of corticosteroids |
| PRL: Prolactin | Mammary glands, testes | Female: milk synthesis<br>Male: increased LH sensitivity and testosterone secretion |
| GH: Growth hormone (somatotropin) | Liver, bone, muscle, fat | Somatomedin secretion, widespread tissue growth |
| *Posterior Pituitary* | | |
| ADH: Antidiuretic hormone | Kidneys | Water retention |
| OT: Oxytocin | Uterus, mammary glands | Labor contractions, milk release; possibly involved in ejaculation, sperm transport, and sexual affection |

not secreted but is processed within the pituitary to yield smaller fragments such as ACTH and endorphins.

- The *posterior lobe* produces no hormones of its own but only stores and releases OT and ADH. Since they are released into the blood by the posterior pituitary, however, these are treated as pituitary hormones for convenience.

## ACTIONS OF THE PITUITARY HORMONES

Now for a closer look at what all of these pituitary hormones do. Most of them receive their fullest treatment in later chapters on such topics as the urinary and reproductive systems, but growth hormone gets its fullest treatment here.

### Anterior Lobe Hormones

*Follicle-Stimulating Hormone (FSH).* FSH, one of the gonadotropins, is secreted by pituitary cells called gonadotropes. Its target organs are the ovaries and testes. In the ovaries, it stimulates the development of eggs and the follicles that contain them. In the testes, it stimulates sperm production.

*Luteinizing Hormone (LH).* LH, the other gonadotropin, is also secreted by the gonadotropes. In females, it stimulates ovulation (the release of an egg). LH is named for the fact that after ovulation, the remainder of a follicle is called the corpus luteum ("yellow body"). LH stimulates the corpus luteum to secrete estrogen and progesterone, hormones important to pregnancy. In males, LH stimulates interstitial cells of the testes to secrete testosterone.

*Thyroid-Stimulating Hormone (TSH).* Thyrotropin or TSH is secreted by pituitary cells called *thyrotropes*. It stimulates growth of the thyroid gland and the secretion of thyroid hormone, which has widespread effects on the body's metabolism considered later in this chapter.

*Adrenocorticotropic Hormone (ACTH).* Corticotropin or ACTH is secreted by pituitary cells called *corticotropes*. ACTH stimulates the adrenal cortex to secrete its hormones *(corticosteroids),* especially cortisol, which regulates glucose, fat, and protein metabolism. ACTH plays a central role in the body's response to stress, which we will examine more fully later in this chapter.

*Prolactin[10] (PRL).* PRL is secreted by *lactotropes* (mammotropes), which increase greatly in size and number during pregnancy. PRL level rises during pregnancy, but it has no effect until after a woman gives birth. Then, it stimulates the mammary glands to synthesize milk. In males, PRL has a gonadotropic effect that makes the testes more sensitive to LH. Thus, it indirectly enhances their secretion of testosterone.

*Growth Hormone (GH), or Somatotropin.* GH is secreted by *somatotropes,* the most numerous cells in the anterior pituitary. The pituitary produces at least a thousand times as much GH as any other hormone. The general effect of GH is to promote mitosis and cellular differentiation and thus to promote widespread tissue growth. Unlike the foregoing hormones, GH is not targeted to any one or few organs, but has widespread effects on the body, especially on cartilage, bone, muscle, and fat. It exerts these effects both directly and indirectly. GH itself directly stimulates these tissues, but it also induces the liver and other tissues to produce growth stimulants called **insulin-like growth factors** (IGF-I and II), or **somatomedins,**[11] which then stimulate target cells in diverse tissues. Most of these effects are caused by IGF-I, but IGF-II is important in fetal growth.

Hormones have a **half-life,** the time required for half of the hormone to be cleared from the blood. GH is short-lived; it has a half-life of 6 to 20 minutes. IGFs, by contrast, have half-lives of about 20 hours, so they greatly prolong the effect of GH. The mechanisms of GH–IGF action include:

- **Protein synthesis.** Tissue growth requires protein synthesis, and protein synthesis needs two things: amino acids for building material, and messenger RNA (mRNA) for instructions. Within minutes of GH secretion, preexisting mRNA is translated and proteins synthesized; within a few hours, DNA is transcribed and more mRNA is produced. GH enhances amino acid transport into cells, and to ensure that protein synthesis outpaces breakdown, it suppresses protein catabolism.

- **Lipid metabolism.** To provide energy for growing tissues, GH stimulates adipocytes to catabolize fat and release free fatty acids (FFAs) and glycerol into the blood. GH has a **protein-sparing effect**—by liberating FFAs and glycerol for energy, it makes it unnecessary for cells to consume their proteins.

- **Carbohydrate metabolism.** GH also has a **glucose-sparing effect.** Its role in mobilizing FFAs reduces the body's dependence on glucose, which is used instead for glycogen synthesis and storage.

- **Electrolyte balance.** GH promotes $Na^+$, $K^+$, and $Cl^-$ retention by the kidneys, enhances $Ca^{2+}$ absorption by the small intestine, and makes these electrolytes available to the growing tissues.

The most conspicuous effects of GH are on bone, cartilage, and muscle growth, especially during childhood and adolescence. IGF-I stimulates bone growth at the epiphyseal

---

[10]*pro* = favoring + *lact* = milk

[11]Acronym for *somato*tropin *med*iating prote*in*

plates. It promotes the multiplication of chondrocytes and osteogenic cells and stimulates protein deposition in the cartilage and bone matrix. In adulthood, it stimulates osteoblast activity and the appositional growth of bone; thus, it continues to influence bone thickening and remodeling.

The blood GH concentration declines gradually with age—averaging about 6 ng/mL (ng = nanograms) in adolescence and one-quarter of that in very old age. The resulting decline in protein synthesis may contribute to aging of the tissues, including wrinkling of the skin and decreasing muscular mass and strength. At age 30, the average adult body is 10% bone, 30% muscle, and 20% fat; at age 75, it averages 8% bone, 15% muscle, and 40% fat.

GH concentration fluctuates greatly over the course of a day. It rises to 20 ng/mL or higher during the first 2 hours of deep sleep and may reach 30 ng/mL in response to vigorous exercise. Smaller peaks occur after high-protein meals, but high-carbohydrate meals tend to suppress GH secretion. Trauma, hypoglycemia (low blood sugar), and other conditions also stimulate GH secretion.

## Posterior Lobe Hormones

*Antidiuretic*[12] *Hormone (ADH).* ADH acts on the kidneys to increase water retention, reduce urine volume, and help prevent dehydration. We will study this hormone more extensively when we deal with the urinary system. ADH is also called *vasopressin* because it causes vasoconstriction at high concentrations. These concentrations are so unnatural for the human body, however, that this effect is of doubtful significance except in pathological states. ADH also functions as a brain neurotransmitter and is usually called vasopressin, or arginine vasopressin (AVP), in the neurobiology literature.

*Oxytocin*[13] *(OT).* OT has various reproductive roles. In childbirth, it stimulates smooth muscle of the uterus to contract, thus contributing to the labor contractions that expel the infant. In lactating mothers, it stimulates musclelike cells of the mammary glands to squeeze on the glandular acini and force milk to flow down the ducts to the nipple. In both sexes, OT secretion surges during sexual arousal and orgasm. It may play a role in the propulsion of semen through the male reproductive tract, in uterine contractions that help transport sperm up the female reproductive tract, and in feelings of sexual satisfaction and emotional bonding.

Hormones of the pituitary gland are summarized in table 17.4.

## CONTROL OF PITUITARY SECRETION

Pituitary hormones are not secreted at a steady rate. GH is secreted mainly at night, LH peaks at the middle of the

menstrual cycle, and OT surges during labor and nursing, for example. The timing and amount of pituitary secretion are regulated by the hypothalamus, other brain centers, and feedback from the target organs.

## Hypothalamic and Cerebral Control

Both lobes of the pituitary gland are strongly subject to control by the brain. As we have seen, the anterior lobe is regulated by releasing and inhibiting hormones from the hypothalamus. Thus, the brain can monitor conditions within and outside of the body and stimulate or inhibit the release of anterior lobe hormones appropriately. For example, in cold weather, the hypothalamus stimulates the pituitary to secrete TSH, which indirectly helps generate body heat; in times of stress, it triggers ACTH secretion, which indirectly mobilizes materials needed for tissue repair; during pregnancy, it induces prolactin secretion so a woman will be prepared to lactate; after a high-protein meal, it triggers the release of growth hormone so we can best use the amino acids for tissue growth.

The posterior lobe of the pituitary is controlled by **neuroendocrine reflexes**—the release of hormones in response to signals from the nervous system. For example, the suckling of an infant stimulates nerve endings in the nipple. Sensory signals are transmitted through the spinal cord and brainstem to the hypothalamus and from there to the posterior pituitary. This causes the release of oxytocin, which results in milk ejection.

Antidiuretic hormone (ADH) is also controlled by a neuroendocrine reflex. Dehydration raises the osmolarity of the blood, which is detected by hypothalamic neurons called *osmoreceptors*. The osmoreceptors trigger ADH release, and ADH promotes water conservation. Excessive blood pressure, by contrast, stimulates stretch receptors in the heart and certain arteries. By another neuroendocrine reflex, this inhibits ADH release, increases urine output, and brings blood volume and pressure back to normal.

Neuroendocrine reflexes can also involve higher brain centers. For example, the milk-ejection reflex can be triggered when a lactating mother simply hears a baby cry. Emotional stress can affect the secretion of gonadotropins, thus disrupting ovulation, the menstrual rhythm, and fertility.

> ### Think About It
> *Which of the unifying themes at the end of chapter 1 (p. 22) is best exemplified by the neuroendocrine reflexes that govern ADH secretion?*

## Feedback from Target Organs

The regulation of other endocrine glands by the pituitary is not simply a system of "command from the top down." Those target organs also regulate the pituitary and hypothalamus through various feedback loops.

---

[12]*anti* = against + *diuret* = to pass through, urinate
[13]*oxy* = quick + *toc* = childbirth

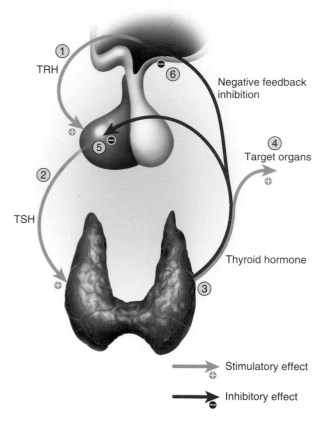

TRH

TSH

Negative feedback inhibition

Target organs

Thyroid hormone

⑤

⑥

①

②

③

④

⟶ Stimulatory effect
+

⟹ Inhibitory effect
−

**FIGURE 17.7  Negative Feedback Inhibition in the Pituitary–Thyroid Axis.**  See text for explanation of numbered steps.

Most often, this takes the form of **negative feedback inhibition**—the pituitary stimulates another endocrine gland to secrete its hormone, and that hormone feeds back to the pituitary and inhibits further secretion of the tropic hormone. All of the pituitary axes are controlled this way. Figure 17.7 shows negative feedback inhibition in the pituitary–thyroid axis as an example. The figure is numbered to correspond to the following description:

① The hypothalamus secretes thyrotropin-releasing hormone (TRH).

② TRH stimulates the anterior pituitary to secrete thyroid-stimulating hormone (TSH).

③ TSH stimulates the thyroid gland to secrete the two thyroid hormones, $T_3$ and $T_4$.

④ $T_3$ and $T_4$ stimulate the metabolism of most cells throughout the body.

⑤ $T_3$ and $T_4$ also *inhibit* the release of TSH by the pituitary.

⑥ To a lesser extent, $T_3$ and $T_4$ also *inhibit* the release of TRH by the hypothalamus.

Steps 5 and 6 are negative feedback inhibition of the pituitary and hypothalamus. These steps ensure that when thyroid hormone (TH) levels are high, TSH secre-

tion remains low. If TH secretion drops, TSH secretion rises and stimulates the thyroid to secrete more TH. This negative feedback keeps TH levels oscillating around a set point in typical homeostatic fashion.

⌐ **Think About It**

*If the thyroid gland were removed from a cancer patient, would you expect the level of TSH to rise or fall? Why?*

Feedback from a target organ is not always inhibitory. During labor, oxytocin triggers a positive feedback cycle. Uterine stretching sends a nerve signal to the brain that stimulates OT release. OT stimulates uterine contractions, which push the infant downward. This stretches the lower end of the uterus some more, which results in a nerve signal that stimulates still more OT release. This positive feedback cycle continues until the infant is born (see fig. 1.12, p. 19).

## Before You Go On

*Answer the following questions to test your understanding of the preceding section:*

5. *What are two good reasons for considering the pituitary to be two separate glands?*

6. *Name three anterior lobe hormones that have reproductive functions and three that have nonreproductive roles. What target organs are stimulated by each of them?*

7. *Briefly contrast hypothalamic control of the anterior pituitary with its control of the posterior pituitary.*

8. *In what sense does the pituitary "take orders" from the target organs under its command?*

# Other Endocrine Glands

### Objectives

When you have completed this section, you should be able to

• describe the structure and location of the remaining organs of the endocrine system; and

• name the hormones these endocrine organs produce and state their functions.

## THE PINEAL GLAND

The **pineal**[14] (PIN-ee-ul) **gland** is a pine cone–shaped growth on the roof of the third ventricle of the brain, beneath the posterior end of the corpus callosum (see fig. 17.1). The philosopher René Descartes (1596–1650) thought it was the seat of the human soul. If so, children must have more

―――――――――――
[14]*pineal* = pine cone

soul than adults—a child's pineal gland is about 8 mm long and 5 mm wide, but after age 7 it regresses rapidly and is no more than a tiny shrunken mass of fibrous tissue in the adult. Such shrinkage of an organ is called **involution.**[15] Pineal secretion peaks between the ages of 1 and 5 years and declines 75% by the end of puberty.

We no longer look for the human soul in the pineal gland, but this little organ remains an intriguing mystery. It produces **serotonin** by day and **melatonin** at night. Melatonin has been implicated in some human mood disorders, although the evidence remains inconclusive (Insight 17.1). In animals with seasonal breeding, it regulates the gonads and the annual breeding cycle. Melatonin may suppress gonadotropin secretion; removal of the pineal from animals causes premature sexual maturation. Some physiologists think that the pineal gland may similarly regulate the timing of puberty in humans, but a clear demonstration of its role has remained elusive. Pineal tumors cause premature onset of puberty in boys, but such tumors also damage the hypothalamus, so we cannot be sure the effect is due specifically to pineal damage.

## THE THYMUS

The **thymus** is located in the mediastinum superior to the heart (fig. 17.8). Like the pineal, it is large in infants and children but involutes after puberty. In elderly people, it is a shriveled vestige of its former self, with most of its parenchyma replaced by fibrous and adipose tissue. The

---

[15]*in* = inward + *volution* = rolling or turning

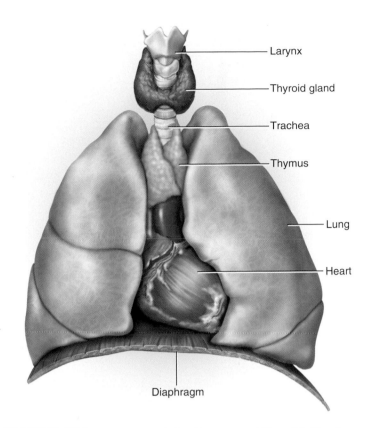

**FIGURE 17.8** Locations of the Thymus and Thyroid Gland.
▶ *Which of these glands will be markedly smaller than the other in a 50-year-old?*

thymus secretes *thymopoietin, thymosin,* and other hormones that regulate the development and later activation of disease-fighting blood cells called T lymphocytes (*T* for *thymus*). This is discussed in detail in chapter 21.

## THE THYROID GLAND

The **thyroid gland** is the largest endocrine gland; it weighs 20 to 25 g and receives one of the body's highest rates of blood flow per gram of tissue. It arises from the root of the embryonic tongue and descends into the neck over the next 3 weeks (see fig. 17.3d–e). After birth, it is wrapped around the anterior and lateral aspects of the trachea, immediately below the larynx. It consists of two large lobes, one on each side of the trachea, connected by a narrow anterior *isthmus* (fig. 17.9a).

Histologically, the thyroid is composed mostly of sacs called **thyroid follicles** (fig. 17.9b). Each is filled with a protein-rich colloid and lined by a simple cuboidal epithelium of **follicular cells.** These cells secrete two main thyroid hormones: $T_3$, or **triiodothyronine** (try-EYE-oh-doe-THY-ro-neen), and **thyroxine,** also known as $T_4$ or **tetraiodothyronine** (TET-ra-EYE-oh-doe-THY-ro-neen). These names refer to the fact that the two hormones contain three ($T_3$) and four ($T_4$) iodine atoms. The expression *thyroid hormone* refers to $T_3$ and $T_4$ collectively.

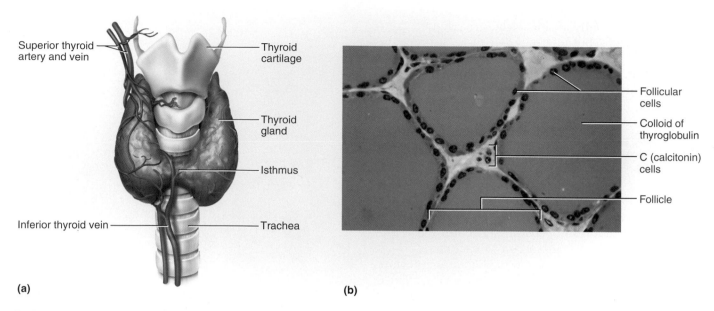

(a)                              (b)

**FIGURE 17.9**  **Anatomy of the Thyroid Gland.**  (a) Anterior view, gross anatomy. (b) Histology, showing the saccular thyroid follicles (source of thyroid hormone) and a nest of C cells (source of calcitonin).

Thyroid hormone is secreted in response to TSH from the pituitary. The primary effect of TH is to increase the body's metabolic rate. As a result, it raises oxygen consumption and has a **calorigenic**[16] **effect**—it increases heat production. TH secretion rises in cold weather and thus helps to compensate for increased heat loss. To ensure an adequate blood and oxygen supply to meet this increased metabolic demand, thyroid hormone also raises the heart rate and contraction strength and raises the respiratory rate. It stimulates the appetite and accelerates the breakdown of carbohydrates, fats, and protein for fuel. Thyroid hormone promotes alertness; bone growth and remodeling; development of the skin, hair, nails, and teeth; and fetal nervous system and skeletal development. It also stimulates the pituitary gland to secrete growth hormone.

**Calcitonin** is another hormone produced by the thyroid gland. It comes from **C (calcitonin) cells,** also called *parafollicular cells,* found in clusters between the thyroid follicles. Calcitonin is secreted when blood calcium level rises. It antagonizes the action of parathyroid hormone (described shortly) and promotes calcium deposition and bone formation by stimulating osteoblast activity. Calcitonin is important mainly to children. It has relatively little effect in adults for reasons explained earlier (p. 230).

## THE PARATHYROID GLANDS

The **parathyroid glands** are partially embedded in the posterior surface of the thyroid (fig. 17.10). There are usually four, each about 3 to 8 mm long and 2 to 5 mm wide.

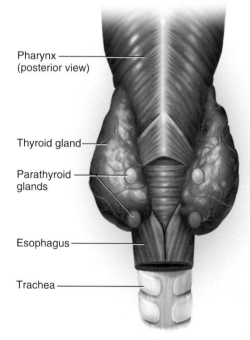

**FIGURE 17.10**  **Location of the Parathyroid Glands.** Posterior view.

They secrete **parathyroid hormone (PTH)** in response to hypocalcemia. PTH raises blood calcium levels by promoting the synthesis of calcitriol, which in turn promotes intestinal calcium absorption; by inhibiting urinary calcium excretion; by promoting phosphate excretion (so the phosphate does not combine with calcium and deposit into the bones); and by indirectly stimulating osteoclasts to resorb bone. PTH and calcium metabolism are discussed in more detail in chapter 7.

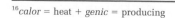

[16]*color* = heat + *genic* = producing

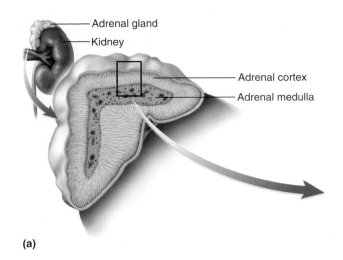

**(a)**

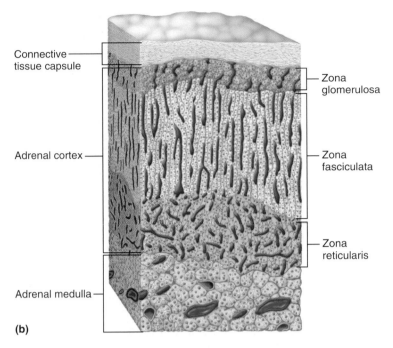

**(b)**

**FIGURE 17.11** **Anatomy of the Adrenal Gland.** (a) Gross anatomy. (b) Histology. The adrenal cortex consists of three layers that each produce their own family of steroid hormones. The adrenal medulla is composed of modified neurons of the sympathetic nervous system.

# THE ADRENAL GLANDS

An **adrenal (suprarenal) gland** sits like a cap on the superior pole of each kidney (fig. 17.11). In adults, the adrenal is about 5 cm (2 in.) long, 3 cm (1.2 in.) wide, and weighs about 4 g, half as much as it weighs at birth. Like the pituitary gland, the adrenal gland is formed by the merger of two fetal glands with different origins and functions. Its inner core, the *adrenal medulla,* is a small portion of the total gland. Surrounding it is a much thicker *adrenal cortex.*

## The Adrenal Medulla

The **adrenal medulla** was discussed as part of the sympathetic nervous system in chapter 15. It arises from the neural crest and is not fully formed until the age of 3. It is actually a sympathetic ganglion consisting of modified neurons, called *chromaffin cells,* that lack dendrites and axons. These cells are richly innervated by sympathetic preganglionic fibers and respond to stimulation by secreting catecholamines, especially epinephrine and norepinephrine. About 85% of the output is epinephrine.

These hormones supplement the effects of the sympathetic nervous system. They increase alertness, anxiety, or even fear, and prepare the body in several ways for physical activity. They mobilize high-energy fuels such as lactate, fatty acids, and glucose. Glucose levels are boosted by **glycogenolysis** (hydrolysis of glycogen to glucose) and **gluconeogenesis** (conversion of fatty acids, amino acids, and other noncarbohydrates to glucose). In order to further ensure an adequate supply of glucose to the brain, epinephrine inhibits insulin secretion. Without insulin, the muscles and other insulin-dependent organs cannot absorb and consume glucose. Thus, epinephrine is said to have a *glucose-sparing effect,* sparing it from needless consumption by organs that can use alternative fuels or that have their own intracellular reserves of glycogen (as muscle cells do).

The adrenal catecholamines also raise the heart rate and blood pressure, stimulate circulation to the muscles, increase pulmonary airflow, and raise the metabolic rate. At the same time, they *inhibit* such temporarily inessential functions as digestion and urine production so that they do not compete for blood flow and energy.

The medulla and cortex are not as functionally independent as once thought. The boundary between them is indistinct, and some cells of the medulla extend into the cortex. When stress activates the sympathetic nervous system, these medullary cells secrete catecholamines that stimulate the cortex to secrete corticosterone.

## The Adrenal Cortex

The **adrenal cortex** has three layers of glandular tissue (fig. 17.11b): an outer **zona glomerulosa**[17] (glo-MER-you-LO-suh) composed of globular cell clusters; a thick middle **zona fasciculata**[18] (fah-SIC-you-LAH-ta) composed of cell columns separated by blood sinuses; and an inner **zona reticularis**[19] (reh-TIC-you-LAR-iss), where the cells form a network. The cortex synthesizes more than 25 steroid hormones known collectively as the **corticosteroids** (corticoids). The three tissue layers, in the same order, secrete the following corticosteroids:

1. **Mineralocorticoids** (zona glomerulosa only), which act on the kidneys to control electrolyte balance. The principal mineralocorticoid is **aldosterone,** which

[17]*zona* = zone + *glomerul* = little balls + *osa* = full of
[18]*fascicul* = little cords + *ata* = possessing
[19]*reticul* = little network + *aris* = like

promotes Na⁺ retention and K⁺ excretion by the kidneys. Aldosterone is discussed more fully in chapter 24.

2. **Glucocorticoids** (mainly zona fasciculata), especially **cortisol** (hydrocortisone); **corticosterone** is a less potent relative. Glucocorticoids stimulate fat and protein catabolism, gluconeogenesis, and the release of fatty acids and glucose into the blood. This helps the body adapt to stress and repair damaged tissues. Glucocorticoids also have an anti-inflammatory effect and are widely used in ointments to relieve swelling and other signs of inflammation. Long-term secretion, however, suppresses the immune system for reasons we will see later in the discussion of stress.

3. **Sex steroids** (mainly zona reticularis), including weak **androgens** and smaller amounts of **estrogens.** Androgens control many aspects of male development and reproductive physiology. The principal adrenal androgen is **dehydroepiandrosterone (DHEA)** (de-HY-dro-EP-ee-an-DROSS-tur-own). DHEA has weak hormonal effects in itself, but more importantly, other tissues convert it to the more potent androgen, **testosterone.** This source is relatively unimportant in men because the testes produce so much more testosterone than the adrenals do. In women, however, the adrenal glands meet about 50% of the total androgen requirement. In both sexes, androgens stimulate the development of pubic and axillary hair and apocrine

scent glands at puberty, and they sustain the libido (sex drive) throughout adult life.

Adrenal estrogen (estradiol) is of minor importance to women of reproductive age because its quantity is small compared with estrogen from the ovaries. After menopause, however, the ovaries no longer function and the adrenals are the only remaining estrogen source. Both androgens and estrogens promote adolescent skeletal growth and help to sustain adult bone mass.

### Think About It

*Which could a person more easily live without—the adrenal medulla or adrenal cortex? Why?*

## THE PANCREAS

The elongated spongy **pancreas** is located inferior and dorsal to the stomach (fig. 17.12); most of the gland is retroperitoneal. It is approximately 15 cm long and 2.5 cm thick. Most of it is an exocrine digestive gland, but scattered through the exocrine tissue are endocrine cell clusters called **pancreatic islets** (islets of Langerhans[20]). There are 1 to 2 million islets, but they constitute only about 2% of the pancreatic tissue. They secrete at least five hormones and paracrine products,

[20]Paul Langerhans (1847–88), German anatomist

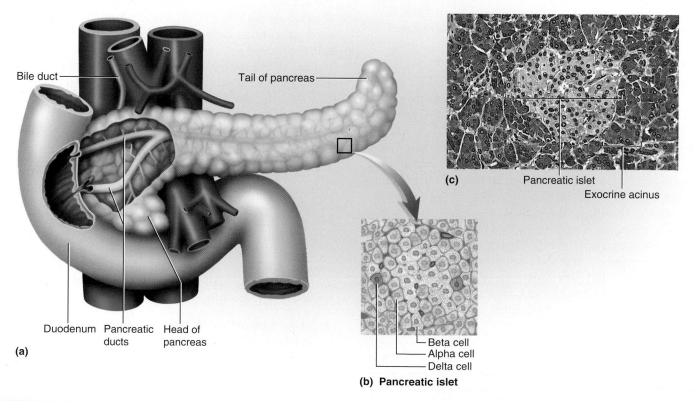

**(a)**
Bile duct

Tail of pancreas

Duodenum   Pancreatic ducts   Head of pancreas

Beta cell
Alpha cell
Delta cell

**(b) Pancreatic islet**

**(c)**   Pancreatic islet
Exocrine acinus

**FIGURE 17.12   Anatomy of the Pancreas.** (a) Gross anatomy and relationship to the duodenum and other nearby organs. (b) The alpha, beta, and delta cells of a pancreatic islet. (c) Light micrograph of a pancreatic islet. The exocrine acini around it produce digestive enzymes.

the most important of which are insulin, glucagon, and somatostatin.

- **Insulin** is secreted by the **beta (β) cells** of the islets when we digest a meal and the level of glucose and amino acids in the blood rises. In such times of plenty, insulin stimulates cells to absorb these nutrients from the blood and stimulates muscle and adipose tissue to store glycogen and fat. While stimulating cells to store excess nutrients for later use, it suppresses the use of already-stored fuels. The stored nutrients are then available for use between meals and overnight. By promoting glycogen, fat, and protein synthesis, insulin enhances cell growth and differentiation. It also antagonizes the effects of glucagon. Some cells and organs do not depend on insulin for glucose uptake: the kidneys, brain, liver, and red blood cells. Insulin does, however, promote the liver's synthesis of glycogen from the absorbed glucose.

- **Glucagon** is secreted by **alpha (α) cells** when blood glucose concentration falls between meals. In the liver, it stimulates gluconeogenesis, glycogenolysis, and the release of glucose into circulation. In adipose tissue, it stimulates fat catabolism and the release of free fatty acids. Glucagon is also secreted in response to rising amino acid levels in the blood after a high-protein meal. By promoting amino acid absorption, it provides cells with raw material for gluconeogenesis.

- **Somatostatin** is secreted by the **delta (δ) cells** when blood glucose, fatty acids, and amino acids rise after a meal. It travels briefly in the blood and inhibits various digestive functions, but also acts locally in the pancreas as a *paracrine* secretion—a chemical messenger that diffuses through the tissue fluid to target cells a short distance away. Somatostatin inhibits the secretion of glucagon and insulin by the neighboring alpha and beta cells. The reason for this is still somewhat a matter of speculation. It has been hypothesized that it serves to prolong the absorption of nutrients by the tissues and thus to prevent excessively quick depletion of blood-borne nutrients.

Any hormone that raises blood glucose concentration is called a *hyperglycemic hormone.* You may have noticed that glucagon is not the only hormone that does so; so do growth hormone, epinephrine, norepinephrine, cortisol, and corticosterone. Insulin is called a *hypoglycemic hormone* because it lowers blood glucose levels.

## THE GONADS

Like the pancreas, the **gonads** are both endocrine and exocrine. Their exocrine products are eggs and sperm, and their endocrine products are the gonadal hormones, most of which are steroids.

Each follicle of the ovary contains an egg cell surrounded by a wall of **granulosa cells** (fig. 17.13a). The

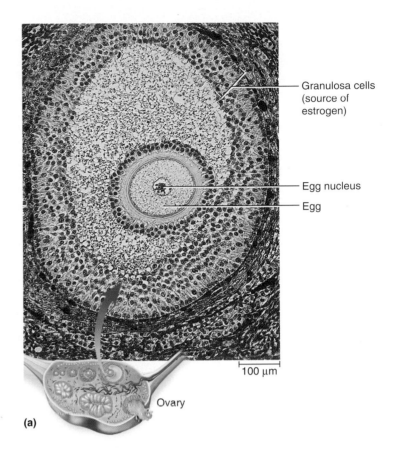

Granulosa cells (source of estrogen)

Egg nucleus

Egg

100 μm

Ovary

**(a)**

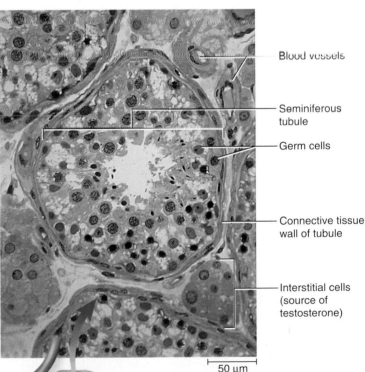

Blood vessels

Seminiferous tubule

Germ cells

Connective tissue wall of tubule

Interstitial cells (source of testosterone)

50 μm

Testis

**(b)**

**FIGURE 17.13 The Gonads.** (a) A follicle of the ovary. Compare with the SEM micrograph on page 635. (b) Histology of the testis. The granulosa cells of the ovary and interstitial cells of the testis are endocrine cells.

granulosa cells produce an estrogen called **estradiol** in the first half of the menstrual cycle. After ovulation, the corpus luteum secretes estradiol and **progesterone** for 12 days or so, or for 8 to 12 weeks in the event of pregnancy. The functions of estradiol and progesterone are discussed in chapter 28. In brief, they contribute to the development of the reproductive system and feminine physique, they promote adolescent bone growth, they regulate the menstrual cycle, they sustain pregnancy, and they prepare the mammary glands for lactation. The follicle and corpus luteum also secrete **inhibin,** which suppresses FSH secretion by means of negative feedback inhibition of the anterior pituitary.

The testis consists mainly of microscopic *seminiferous tubules* that produce sperm. Nestled between them are clusters of **interstitial cells** (fig. 17.13b), which produce testosterone and lesser amounts of weaker androgens and estrogen. Testosterone stimulates development of the male reproductive system in the fetus and adolescent, the development of the masculine physique in adolescence, and the sex drive. It sustains sperm production and the sexual instinct throughout adult life. **Sustentacular (Sertoli[21]) cells** of the testis secrete inhibin, which suppresses FSH secretion and thus homeostatically stabilizes the rate of sperm production.

# ENDOCRINE FUNCTIONS OF OTHER ORGANS

Several other organs have hormone-secreting cells:

- **The heart.** Rising blood pressure stretches the heart wall and stimulates muscle cells in the atria to secrete **atrial natriuretic[22] peptide (ANP).** ANP increases sodium excretion and urine output and opposes the action of angiotensin II, described shortly. Together, these effects lower the blood pressure.

- **The skin.** Keratinocytes of the epidermis produce vitamin $D_3$, the first step in the synthesis of **calcitriol.** Its synthesis is completed by the liver and kidneys, as described in the following paragraphs.

- **The liver.** The liver is involved in the production of at least five hormones: (1) **Erythropoietin (EPO)** (eh-RITH-ro-POY-eh-tin) is secreted by the liver (15%) and kidneys (85%). It stimulates the bone marrow to produce red blood cells (see chapter 18). (2) **Angiotensin II** is a hormone that raises blood pressure. Its production begins with a polypeptide called *angiotensinogen,* secreted by the liver. This is converted in two steps to angiotensin II by enzymes in the kidneys and lungs (see figure 23.14 for details). (3) **Calcitriol,** a hormone that raises blood calcium concentration, was discussed in chapter 7. It is synthesized in three stages involving the skin, liver, and

kidneys (see fig. 7.17). The role of the liver is to convert an intermediate, vitamin $D_3$, to *calcidiol.* (4) **Insulin-like growth factor I (IGF-I),** a hormone that mediates the action of growth hormone, is produced by the liver and other sources. (5) **Hepcidin,** discovered in 2003, promotes the intestinal absorption and mobilization of iron for hemoglobin production and other uses. It is regarded as the principal hormonal mechanism of iron homeostasis.

- **The kidneys.** The kidneys, as noted in the preceding paragraph, play endocrine roles in the production of EPO, angiotensin, and calcitriol. They secrete about 85% of the body's EPO. They convert antiogensinogen to angiotensin I (which the lungs then convert to angiotensin II). They convert the calcidiol produced by the liver into calcitriol, the most active form of vitamin D. Calcitriol promotes calcium absorption by the small intestine, somewhat inhibits its urinary excretion, and thus makes more calcium available for bone deposition and other metabolic needs.

- **The stomach and small intestine.** These have various *enteroendocrine cells,*[23] which secrete at least 10 **enteric hormones.** In general, they coordinate the different regions and glands of the digestive system with each other (see chapter 25).

- **The placenta.** This organ performs many functions in pregnancy, including fetal nutrition and waste removal. But it also secretes estrogens (estriol and estradiol), progesterone, and other hormones that regulate pregnancy and stimulate development of the fetus and the mother's mammary glands (see chapter 28).

You can see that the endocrine system is extensive. It includes numerous discrete glands as well as individual cells in the tissues of other organs. The endocrine organs and tissues other than the hypothalamus and pituitary are reviewed in table 17.5.

## Before You Go On

*Answer the following questions to test your understanding of the preceding section:*

9. Name three endocrine glands that are larger in children than in adults.

10. What is the value of the calorigenic effect of thyroid hormone?

11. Name a glucocorticoid, a mineralocorticoid, and a catecholamine secreted by the adrenal gland.

12. Does the action of glucocorticoids more closely resemble that of glucagon or insulin? Explain.

13. What is the difference between a gonadal hormone and a gonadotropin?

---

[21]Enrico Sertoli (1842–1910), Italian histologist
[22]*natri* = sodium + *uretic* = pertaining to urine

[23]*entero* = intestine

TABLE 17.5

| Hormone | Target | Principal Effects |
|---------|--------|-------------------|
| **_Pineal Gland_** | | |
| Melatonin and serotonin | Brain | Influence mood; may regulate the timing of puberty |
| **_Thymus_** | | |
| Thymopoietin and thymosins | T lymphocytes | Stimulate T lymphocytes |
| **_Thyroid_** | | |
| Triiodothyronine ($T_3$) and thyroxine ($T_4$) | Most tissues | Elevate metabolic rate, $O_2$ consumption, and heat production; stimulate circulation and respiration; promote nervous system and skeletal development |
| Calcitonin | Bone | Promotes $Ca^{2+}$ deposition and ossification; reduces blood $Ca^{2+}$ level |
| **_Parathyroids_** | | |
| Parathyroid hormone (PTH) | Bone, kidneys | Increases blood $Ca^{2+}$ level by stimulating bone resorption and calcitriol synthesis and reducing urinary $Ca^{2+}$ excretion |
| **_Adrenal Medulla_** | | |
| Epinephrine, norepinephrine, dopamine | Most tissues | Complement effects of sympathetic nervous system |
| **_Adrenal Cortex_** | | |
| Aldosterone | Kidney | Promotes $Na^+$ retention and $K^+$ excretion, maintains blood pressure and volume |
| Cortisol and corticosterone | Most tissues | Stimulate fat and protein catabolism, gluconeogenesis, stress resistance, and tissue repair; inhibit immune system |
| Androgen (DHEA) and estrogen | Bone, muscle, integument, many other tissues | Growth of pubic and axillary hair, bone growth, sex drive, male prenatal development |
| **_Pancreatic Islets_** | | |
| Insulin | Most tissues | Stimulates glucose and amino acid uptake; lowers blood glucose level; promotes glycogen, fat, and protein synthesis |
| Glucagon | Primarily liver | Stimulates gluconeogenesis, glycogen and fat breakdown, release of glucose and fatty acids into circulation |
| **_Ovaries_** | | |
| Estradiol | Many tissues | Stimulates female reproductive development, regulates menstrual cycle and pregnancy, prepares mammary glands for lactation |
| Progesterone | Uterus, mammary glands | Regulates menstrual cycle and pregnancy, prepares mammary glands for lactation |
| Inhibin | Anterior pituitary | Inhibits FSH secretion |
| **_Testes_** | | |
| Testosterone | Many tissues | Stimulates reproductive development, skeletomuscular growth, sperm production, and libido |
| Inhibin | Anterior pituitary | Inhibits FSH secretion |
| **_Heart_** | | |
| Atrial natriuretic peptide | Kidney | Lowers blood volume and pressure by promoting $Na^+$ and water loss |
| **_Skin_** | | |
| Vitamin $D_3$ | — | First step in calcitriol synthesis (see kidneys) |
| **_Liver_** | | |
| Erythropoietin | Red bone marrow | Promotes red blood cell production |
| Angiotensinogen (a prohormone) | Blood vessels | Precursor of angiotensin II, a vasoconstrictor |
| Calcidiol | — | Second step in calcitriol synthesis (see kidneys) |
| IGF-I | Many tissues | Mediates action of growth hormone |
| Hepcidin | Small intestine | Iron absorption and mobilization |

| TABLE 17.5 | Hormones from Sources Other Than the Hypothalamus and Pituitary—*(cont.)* | |
|---|---|---|
| **Hormone** | **Target** | **Principal Effects** |
| *Kidneys* | | |
| Calcitriol | Small intestine, kidneys | Promotes bone deposition by increasing calcium and phosphate absorption in small intestine and reducing their urinary loss |
| Erythropoietin | Red bone marrow | Promotes red blood cell production |
| *Stomach and Small Intestine* | | |
| Enteric hormones | Stomach and intestines | Coordinate digestive motility and secretion |
| *Placenta* | | |
| Estrogen, progesterone, and others | Many tissues of mother and fetus | Enhance effects of ovarian hormones on fetal development, maternal reproductive system, and preparation for lactation |

# Hormones and Their Actions

## Objectives

When you have completed this section, you should be able to

- identify the chemical classes to which various hormones belong;
- describe how hormones are synthesized and transported to their target organs;
- describe how hormones stimulate their target cells;
- explain how target cells regulate their sensitivity to circulating hormones;
- discuss how hormones are removed from circulation after they have performed their roles; and
- describe how hormones affect each other when two or more of them stimulate the same target cells.

Having surveyed the body's major hormones and their effects, we are left with some deeper questions: Exactly what is a hormone? How are hormones synthesized and transported to their destinations? How does a hormone produce its effects on a target organ? Thus, we now address endocrinology at the molecular and cellular levels.

## HORMONE CHEMISTRY

Most hormones fall into three chemical classes: *steroids, peptides,* and *monoamines* (table 17.6, fig. 17.14).

1. **Steroid hormones** are derived from cholesterol. They include sex steroids produced by the testes and

| TABLE 17.6 | Chemical Classification of Hormones |
|---|---|
| **Steroids and Steroid Derivatives** | |
| Aldosterone | Estrogens |
| Calcitriol | Progesterone |
| Corticosterone | Testosterone |
| Cortisol | |
| **Oligopeptides (3–10 amino acids)** | |
| Angiotensin II | Oxytocin |
| Antidiuretic hormone | Thyrotropin-releasing hormone |
| Gonadotropin-releasing hormone | |
| **Polypeptides (14–199 amino acids)** | |
| Adrenocorticotropic hormone | Growth hormone–releasing hormone |
| Atrial natriuretic peptide | Hepcidin |
| Calcitonin | Insulin |
| Corticotropin-releasing hormone | Parathyroid hormone |
| Glucagon | Prolactin |
| Growth hormone | Somatostatin |
| **Glycoproteins** | |
| (92 amino acids in the α chain, 112–118 amino acids in the β chain) | |
| Follicle-stimulating hormone | Luteinizing hormone |
| Human chorionic gonadotropin | Thyroid-stimulating hormone |
| Inhibin | |
| **Monoamines** | |
| Dopamine | Serotonin |
| Epinephrine | Thyroxine ($T_4$) |
| Melatonin | Triiodothyronine ($T_3$) |
| Norepinephrine | |

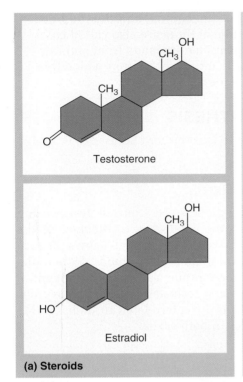

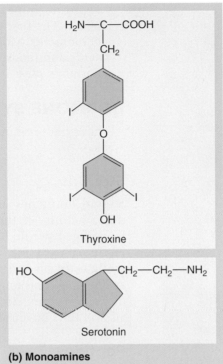

**(a) Steroids**

**(b) Monoamines**

**FIGURE 17.14 The Chemical Classes of Hormones.** (a) Two steroid hormones, defined by their four-membered ring derived from cholesterol. (b) Two monoamines, derived from amino acids and defined by their one -NH$_2$ (amino) group. (c) Two peptides. The three-letter labels are standard symbols for the various amino acids (see table 2.8).

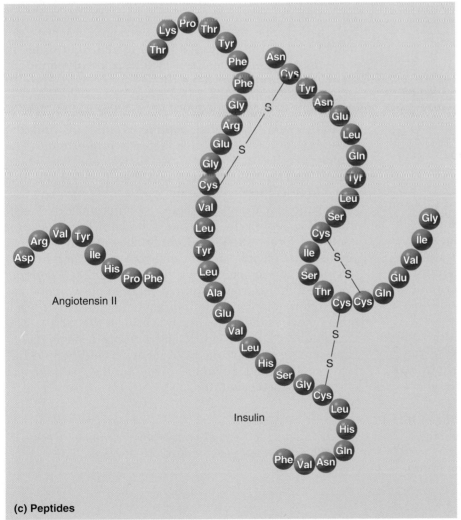

**(c) Peptides**

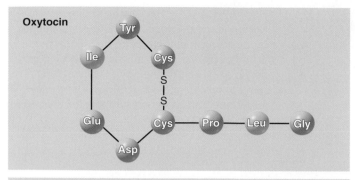

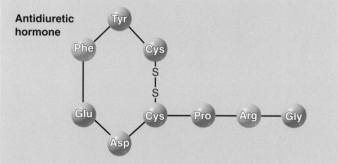

**FIGURE 17.15** **Oxytocin and Antidiuretic Hormone.** Note the structural similarity of these two hormones of the posterior pituitary.

ovaries (such as estrogens, progesterone, and testosterone) and corticosteroids produced by the adrenal gland (such as cortisol, corticosterone, aldosterone, and DHEA). Calcitriol, the calcium-regulating hormone, is not a steroid but is derived from one and has the same hydrophobic character and mode of action as the steroids.

2. **Peptide hormones** are chains of 3 to 200 or more amino acids. The two posterior pituitary hormones, oxytocin and antidiuretic hormone, are very similar oligopeptides; they differ in only two of their nine amino acids (fig. 17.15). Except for dopamine, the releasing and inhibiting hormones produced by the hypothalamus are polypeptides. Most hormones of the anterior pituitary are polypeptides or glycoproteins—polypeptides conjugated with short carbohydrate chains. All glycoprotein hormones have an identical α chain of 92 amino acids and a variable β chain that distinguishes them from each other.

3. **Monoamines** (biogenic amines) were introduced in chapter 12, since this class also includes several neurotransmitters (see fig. 12.21, p. 464). The monoamine hormones include epinephrine, norepinephrine, dopamine, melatonin, and thyroid hormone. The first three of these are also called *catecholamines*. Monoamines are made from amino acids and retain an amino group, from which this hormone class gets its name.

## HORMONE SYNTHESIS

All hormones are made from either cholesterol or amino acids.

### Steroids

Steroid hormones are synthesized from cholesterol and differ mainly in the functional groups attached to the four-ringed steroid backbone. Figure 17.16 shows the synthetic pathway for several steroid hormones. Notice that while estrogen and progesterone are typically thought of as "female" hormones and testosterone as a "male" hormone, these sex steroids are interrelated in their synthesis and thus have roles in both sexes.

### Peptides

Peptide hormones are synthesized the same way as any other protein. The gene for the hormone is transcribed to form a molecule of mRNA, and ribosomes translate the mRNA and assemble amino acids in the right order to make the hormone. The newly synthesized polypeptide is an inactive **preprohormone.** It has a signal peptide of hydrophobic amino acids that guide it into the cisterna of the rough endoplasmic reticulum (as explained on p. 135). Here, the signal peptide is split off and the remainder of the polypeptide is now a **prohormone.** The prohormone is transferred to the Golgi complex, which may further cut and splice it and then package the hormone for secretion.

Insulin, for example, begins as *preproinsulin.* When the signal peptide is removed, the chain folds back on itself and forms three disulfide bridges. It is now called *proinsulin.* Enzymes in the Golgi complex then remove a large middle segment called the *connecting (C) peptide.* The remainder is now insulin, composed of two polypeptide chains totaling 51 amino acids, connected to each other by two of the three disulfide bridges (fig. 17.17). The C peptide is not wasted; it has recently been found to have some hormonal effects of its own, binding to cellular receptors and reducing some of the pathological effects of diabetes mellitus.

**Think About It**

*During the synthesis of glycoprotein hormones, where in the cell would the carbohydrate be added? (See chapter 4.)*

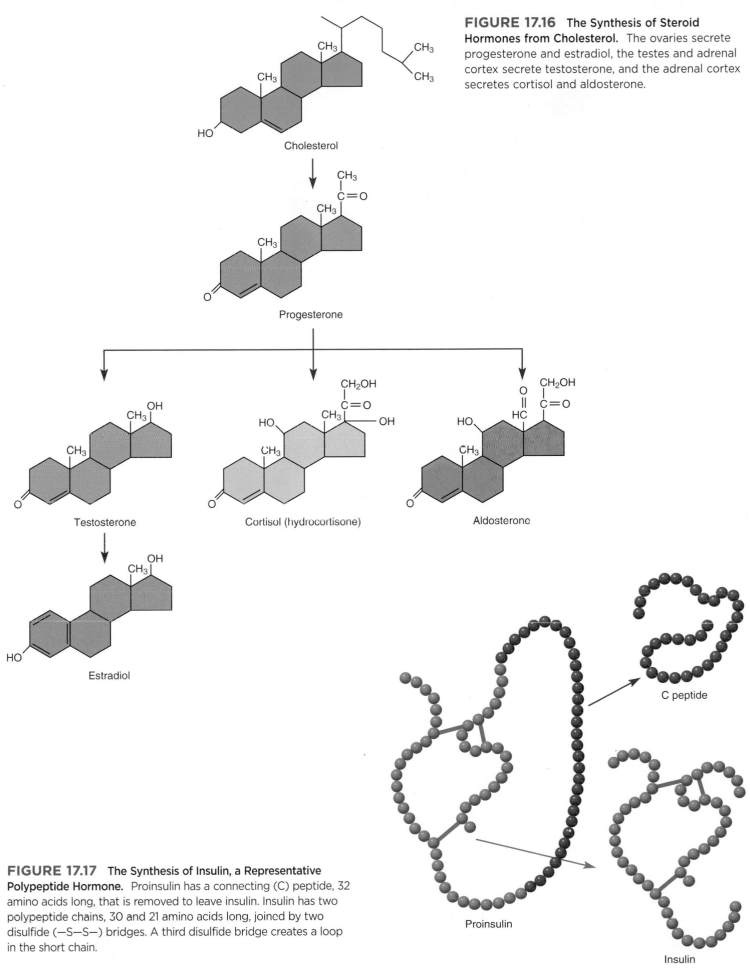

**FIGURE 17.16** **The Synthesis of Steroid Hormones from Cholesterol.** The ovaries secrete progesterone and estradiol, the testes and adrenal cortex secrete testosterone, and the adrenal cortex secretes cortisol and aldosterone.

Cholesterol

Progesterone

Testosterone

Cortisol (hydrocortisone)

Aldosterone

Estradiol

C peptide

Proinsulin

Insulin

**FIGURE 17.17** **The Synthesis of Insulin, a Representative Polypeptide Hormone.** Proinsulin has a connecting (C) peptide, 32 amino acids long, that is removed to leave insulin. Insulin has two polypeptide chains, 30 and 21 amino acids long, joined by two disulfide (—S—S—) bridges. A third disulfide bridge creates a loop in the short chain.

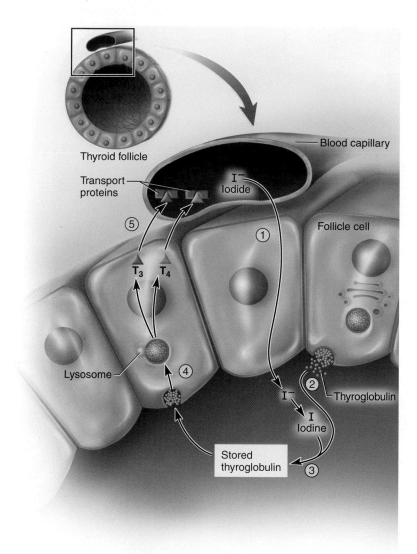

Thyroid follicle

Transport proteins

Blood capillary

I⁻ Iodide

Follicle cell

⑤

T₃    T₄

①

Lysosome

④

I⁻
I Iodine

②
Thyroglobulin

Stored thyroglobulin

③

① Iodide absorption and oxidation

② Thyroglobulin synthesis and secretion

③ Iodine added to tyrosines of thyroglobulin

④ Thyroglobulin uptake and hydrolysis

⑤ Release of $T_3$ and $T_4$ into the blood

**FIGURE 17.18    Thyroid Hormone Synthesis, Storage, and Secretion.** See figure 17.19 for details of steps 3 and 4.

## Monoamines

Melatonin is synthesized from the amino acid tryptophan and the other monoamines from the amino acid tyrosine. Thyroid hormone is unusual in that each molecule is composed of *two* tyrosines. Figure 17.18 shows its synthesis, storage, and secretion at a cellular level, numbered to correspond to the following description:

① Follicular cells absorb iodide (I⁻) ions from the blood plasma and secrete them into the lumen of the follicle. Here, iodide is oxidized to neutral iodine (I) atoms.

② Meanwhle, the follicular cells synthesize a large protein called **thyroglobulin** through the usual mechanism involving the rough endoplasmic reticulum and Golgi complex. They release this into the follicular lumen by exocytosis; in the lumen, it forms the colloid with the pink-staining material in figure 17.9b. Each thyroglobulin molecule has

123 tyrosines in its amino acid chain, but only 4 to 8 of them are destined to become thyroid hormone (TH).

③ In the lumen, tyrosine and iodine combine to form the two types of thyroid hormone, $T_3$ and $T_4$, through a process to be described shortly. However, the TH remains bound to thyroglobulin and stored in the follicle until the cell receives a signal (thyroid-stimulating hormone) inducing it to secrete TH.

④ When stimulated by TSH, the follicular cells absorb droplets of thyroglobulin by pinocytosis. A lysosome fuses with the pinocytotic vesicle and contributes an enzyme that hydrolyzes the thyroglobulin chain, liberating $T_3$ and $T_4$ from it.

⑤ The two hormones are released into the blood as a mixture of about 10% $T_3$ and 90% $T_4$. They bind to various transport proteins in the blood plasma, which carry them to their target cells.

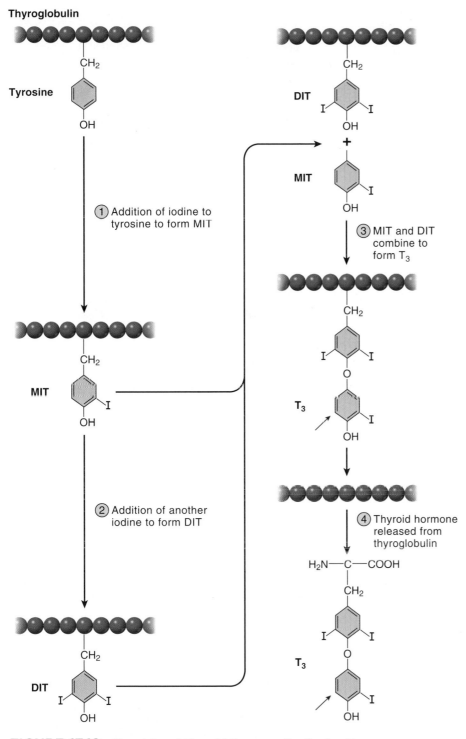

**Thyroglobulin**

**Tyrosine**

① Addition of iodine to tyrosine to form MIT

**MIT**

② Addition of another iodine to form DIT

**DIT**

**DIT**

**MIT**

③ MIT and DIT combine to form $T_3$

$T_3$

④ Thyroid hormone released from thyroglobulin

$H_2N$—C—COOH

$T_3$

**FIGURE 17.19  Chemistry of Thyroid Hormone Synthesis.** These processes occur at steps 3 and 4 of figure 17.18.

Figure 17.19 takes a closer look at the processes that occur in the follicular lumen to combine iodine and tyrosine into $T_3$ and $T_4$.

① When one iodine atom binds to the tyrosine ring, it converts it to *monoiodotyrosine* (MON-oh-eye-OH-do-TY-ro-seen), or *MIT.*

② Some (not all) of the MIT goes on to bind a second iodine atom, converting it to *diiodotyrosine (DIT).*

③ Either two DITs, or an MIT and a DIT, become linked to each other through an oxygen atom of one of their rings. The rest of the peptide chain splits away from one of the tyrosines. The combination of a DIT and MIT forms triiodothyronine, with three iodine atoms, as shown in the figure. The combination of two DITs forms $T_4$ (tetraiodothyronine), with a fourth iodine atom at the point indicated by the red arrow in the figure.

④ When the cell is signaled to release thyroid hormone, the lysosomal enzyme mentioned earlier degrades the peptide chain of thyroglobulin and releases $T_3$ and $T_4$ from it.

# HORMONE TRANSPORT

To get from an endocrine cell to a target cell, a hormone must travel in the blood, which is mostly water. Most of the monoamines and peptides are hydrophilic, so mixing with the blood plasma presents no problem for them. Steroids and thyroid hormone, however, are hydrophobic and must bind to hydrophilic **transport proteins** to get to their destination. The transport proteins are albumins and globulins synthesized by the liver. A hormone attached to a transport protein is called a **bound hormone,** and one that is not attached is an **unbound (free) hormone.** Only the unbound hormone can leave a blood capillary and get to a target cell (fig. 17.20).

Transport proteins not only enable hydrophobic hormones to travel in the blood, they also prolong their half-lives. They protect circulating hormones from being broken down by enzymes in the blood plasma and liver and from being filtered out of the blood by the kidneys. Free hormone may be broken down or removed from the blood in a few minutes, whereas bound hormone may circulate for hours to weeks.

Thyroid hormone binds to three transport proteins in the blood plasma: *albumin,* an albumin-like protein called *thyretin,* and an α-globulin named *thyroxine-binding globulin (TBG).* TBG binds the greatest amount. About 99.8% of $T_3$ and 99.98% of $T_4$ are protein-bound. Bound TH serves as a long-lasting blood reservoir, so even if the thyroid is surgically removed (as for cancer surgery), no signs of TH deficiency appear for about 2 weeks.

Steroid hormones bind to globulins such as *transcortin,* the transport protein for cortisol. Aldosterone is unusual. It has no specific transport protein, but binds weakly to albumin and others. However, 85% of it remains unbound, and correspondingly, it has a half-life of only 20 minutes.

# HORMONE RECEPTORS AND MODE OF ACTION

Hormones stimulate only those cells that have receptors for them. The receptors are protein or glycoprotein molecules located on the plasma membrane, on mitochondria and other organelles in the cytoplasm, or in the nucleus. They act like switches to turn certain metabolic pathways on or off when the hormone binds to them. A target cell usually has a few thousand receptors for a given hormone. Receptor defects lie at the heart of several endocrine diseases (see Insight 17.2).

Receptor–hormone interactions are similar to the enzyme–substrate interactions described in chapter 2.

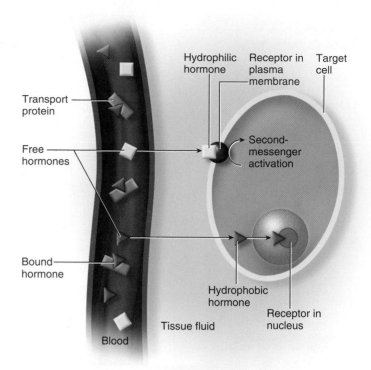

**FIGURE 17.20**   **Transport and Action of Hormones.**
Hydrophilic hormones (yellow squares) usually travel in the blood as free (unbound) hormones. Upon leaving the blood, they bind to receptors on the cell surface and activate second messengers. Hydrophobic hormones (blue triangles) are carried in the blood by transport proteins. Upon dissociating from a protein, they leave the blood, diffuse into the target cell, and bind to receptors in the nucleus or cytoplasm.

Unlike enzymes, receptors do not chemically change their ligands, but they do exhibit enzymelike specificity and saturation. *Specificity* means that the receptor for one hormone will not bind other hormones. *Saturation* is the condition in which all the receptor molecules are occupied by hormone molecules. Adding more hormone cannot produce any greater effect.

## Steroids and Thyroid Hormone

The hydrophobic steroid and thyroid hormones easily penetrate the phospholipid plasma membrane of a target cell and enter the cytoplasm. Steroids enter the nucleus and bind to a receptor associated with the DNA. The receptor has three functional regions that explain its action on the DNA: one that binds the hormone, one that binds to an *acceptor site* on the chromatin, and one that activates DNA transcription at that site. Transcription produces new mRNA that leads to the synthesis of

## INSIGHT 17.2 Clinical Application

### Hormone Receptors and Therapy

In treating endocrine disorders, it is essential to understand the role of hormone receptors. For example, type II diabetes mellitus has long been thought to result from an insulin receptor defect or deficiency (among other possible causes). No amount of insulin replacement can correct this. And while growth hormone is now abundantly available thanks to genetic engineering, it is useless to children with *Laron dwarfism*, who have a hereditary defect in their GH receptors. *Androgen insensitivity syndrome* is due to an androgen receptor defect or deficiency; it causes genetic males to develop feminine genitalia and other features (see Insight 27.1, p. 1037). Estrogen stimulates the growth of some malignant tumors with estrogen receptors. For this reason, estrogen replacement therapy should not be used for women with estrogen-dependent cancer.

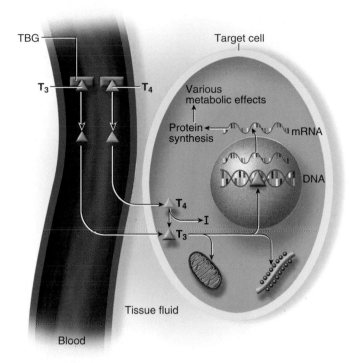

**FIGURE 17.21 The Action of Thyroid Hormone on a Target Cell.** $T_3$ and $T_4$ dissociate from thyroxine-binding globulin (TBG), leave the bloodstream, and enter the target cell cytoplasm. Here, $T_4$ is converted to $T_3$ by the removal of one iodine atom. $T_3$ may bind to receptors on the mitochondria or ribosomes, or enter the nucleus and activate genetic transcription.

proteins, which then alter the metabolism of the target cell. Estrogen, for example, stimulates cells of the uterine lining to synthesize proteins that act as progesterone receptors. Progesterone, which comes later in the menstrual cycle, then binds to these receptors and stimulates the cells to produce enzymes that synthesize glycogen. Glycogen prepares the uterus to nourish an embryo in the event of pregnancy.

Even though $T_4$ constitutes 90% of the secreted thyroid hormone, it has little direct metabolic effect on the target cells. Unbound $T_3$ and $T_4$ enter the target cell cytoplasm, where an enzyme converts the $T_4$ to $T_3$ by removing one iodine. $T_3$ binds to receptors in three sites: on mitochondria, where it increases the rate of aerobic respiration; on ribosomes, where it stimulates the translation of mRNA and thus increases the rate of protein synthesis; and in the nucleus, where it binds to receptors in the chromatin and stimulates DNA transcription (mRNA synthesis) (fig. 17.21). One of the proteins produced under the influence of $T_3$ is $Na^+$–$K^+$ ATPase—the sodium–potassium pump. As we saw in chapter 3 (p. 108), one of the functions of $Na^+$–$K^+$ ATPase is to generate heat, thus accounting for the calorigenic effect of thyroid hormone.

### Peptides and Catecholamines

Peptides and catecholamines (hydrophilic hormones) cannot penetrate into a target cell, so they must stimulate its physiology indirectly. They bind to cell-surface receptors, which are linked to second-messenger systems on the other side of the plasma membrane (see fig. 17.20). The best-known second messenger is cyclic adenosine monophosphate (cAMP). When glucagon binds to liver cell receptors, for example, the receptor activates a G protein, which in turn activates adenylate cyclase, the membrane enzyme that produces cAMP. cAMP leads ultimately to the activation of enzymes that hydrolyze glycogen stored in the liver cell (fig. 17.22). Somatostatin, however, works by *inhibiting* cAMP synthesis. Atrial natriuretic peptide (ANP) works through a similar second messenger, cyclic guanosine monophosphate (cGMP). Second messengers do not linger in the cell for long. cAMP, for example, is broken down very quickly by an enzyme called **phosphodiesterase.**

Other second messengers include **diacylglycerol** (diglyceride) and **inositol triphosphate.** These act on cell

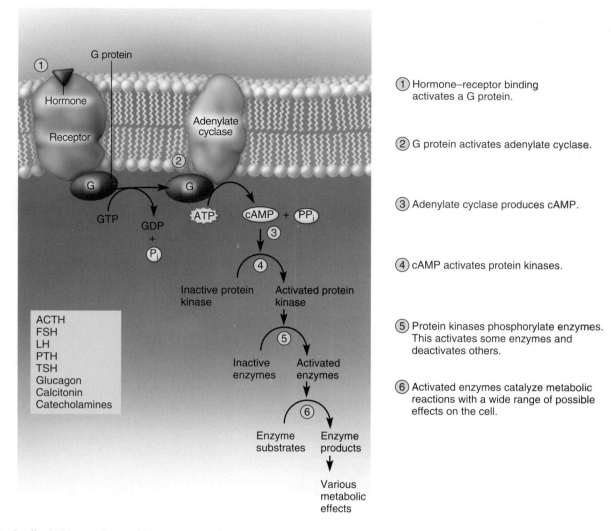

ACTH
FSH
LH
PTH
TSH
Glucagon
Calcitonin
Catecholamines

① Hormone–receptor binding activates a G protein.

② G protein activates adenylate cyclase.

③ Adenylate cyclase produces cAMP.

④ cAMP activates protein kinases.

⑤ Protein kinases phosphorylate enzymes. This activates some enzymes and deactivates others.

⑥ Activated enzymes catalyze metabolic reactions with a wide range of possible effects on the cell.

**FIGURE 17.22   Cyclic AMP as a Second Messenger.**   The green box lists some hormones that act in this manner.
▶ *Why are no steroid hormones listed in this figure?*

metabolism in a variety of ways. The numbered steps in fig. 17.23 correspond to the following description. The diacylgycerol (DAG) pathway is depicted on the left side of the figure:

① A hormone binds to its receptor, which activates a G protein.

② The G protein migrates to a phospholipase molecule and activates it.

③ Phospholipase removes the phosphate-containing group from the head of a membrane phospholipid, leaving DAG, which remains embedded in the plasma membrane.

④ DAG activates protein kinase (PK), an enzyme that phosphorylates other enzymes. (See fig. 3.9 for more detail on this.) By doing so, PK can turn metabolic pathways on or off and have a wide variety of effects on cell metabolism.

The inositol triphosphate (IP$_3$) pathway is depicted on the right side of the figure. Steps 1 and 2 are the same as in the DAG pathway. New aspects of the IP$_3$ pathway are:

⑤ The phosphate-containing group removed at step 3 is IP$_3$. IP$_3$ raises calcium concentration in the cytosol in two ways (steps 6–7).

⑥ IP$_3$ opens gated channels in the plasma membrane and admits Ca$^{2+}$ to the cell from the ECF.

⑦ IP$_3$ opens gated channels in the sarcoplasmic reticulum and releases Ca$^{2+}$ from storage. Calcium, a third messenger, can have three effects (steps 8–10).

⑧ Ca$^{2+}$ may bind to other gated membrane channels and alter the membrane potential of the cell or its permeability to various solutes.

⑨ Ca$^{2+}$ may activate cytoplasmic enzymes that alter cell metabolism.

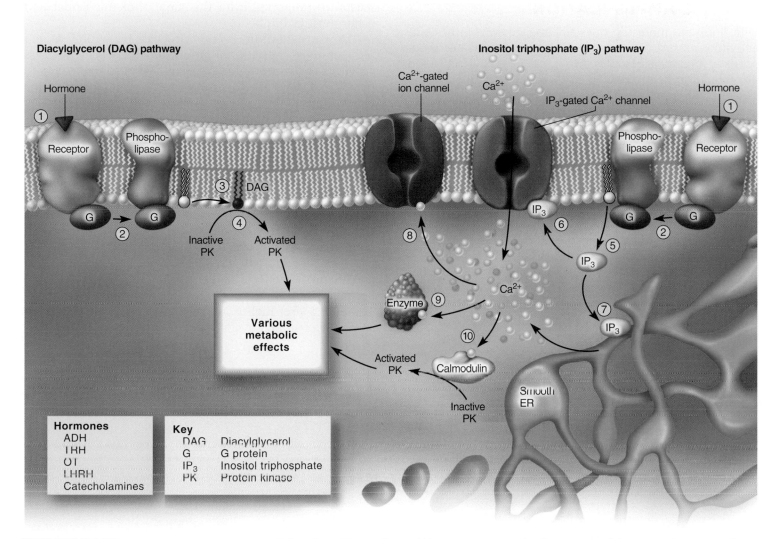

**FIGURE 17.23  Diacylglycerol and Inositol Triphosphate (IP₃) as Second Messengers.** The blue box at the left lists some hormones that act in this manner. See the text for explanation of the numbered steps.

⑩ $Ca^{2+}$ may bind to the cytoplasmic receptor protein **calmodulin,** which in turn activates a protein kinase with effects the same as noted in step 4.

The general point of all this is that hydrophilic hormones such as those listed in the box at the left side of the figure cannot enter the target cell. Yet by merely "knocking on the door" (binding to a surface receptor), they can initiate a flurry of metabolic activity within the cell. The initial steps in this process are activation of a G protein and phospholipase. From there, divergent pathways are taken that involve DAG, $IP_3$, and $Ca^{2+}$ as second and third messengers. Ultimately these pathways lead to metabolic pathways being switched on or off within the cell.

As an example, oxytocin, the labor-inducing hormone, binds to a surface receptor on smooth muscle cells that activates the $IP_3$ pathway. Through the action of $IP_3$, the intracellular calcium concentration rises

sharply. As you saw in chapter 11, calcium triggers muscle contraction—in this case, the smooth muscle of the uterus.

A given hormone doesn't always employ the same second messenger. Antidiuretic hormone (ADH) employs the $IP_3$–calcium system in smooth muscle but the cAMP system in kidney tubules. Insulin is unusual in comparison with other peptide hormones. Rather than using a second-messenger system, it binds to a plasma membrane enzyme, tyrosine kinase, that directly phosphorylates cytoplasmic proteins.

**Think About It**

*From the moment either hormone enters the bloodstream, insulin works much more quickly than estrogen. In view of how each hormone acts on its target cells, explain why.*

# ENZYME AMPLIFICATION

One hormone molecule does not trigger the synthesis or activation of just one enzyme molecule. It activates thousands of enzyme molecules through a cascade effect called **enzyme amplification** (fig. 17.24). To put it in a simplistic but illustrative way, suppose one glucagon molecule triggered the formation of 1,000 molecules of cAMP; cAMP activated a protein kinase; each protein kinase activated 1,000 other enzyme molecules; and each of those enzymes produced 1,000 molecules of a reaction product. These are modest numbers as chemical reactions go, and yet even at this low estimate, that one glucagon molecule would have triggered the production of 1 billion molecules of reaction product. Whatever the actual numbers may be, you can see how enzyme amplification enables a very small stimulus to produce a very large effect. Hormones are therefore powerfully effective in minute quantities. Their circulating concentrations are very low compared with other blood substances: on the order of nanograms per deciliter. Blood glucose, for example, is about 100 million times this concentrated. Because of enzyme amplification, target cells do not need a great number of hormone receptors.

# MODULATION OF TARGET CELL SENSITIVITY

Target cells can adjust their sensitivity to a hormone by changing the number of receptors for it. In **up-regulation,** a cell increases the number of hormone receptors and becomes more sensitive to the hormone (fig. 17.25a). In late pregnancy, for example, the uterus produces oxytocin

receptors, preparing it for the surge of oxytocin that will occur during childbirth.

**Down-regulation** is the process in which a cell reduces its receptor population and thus becomes less sensitive to a hormone (fig. 17.25b). This sometimes happens in response to long-term exposure to a high hormone concentration. For example, adipocytes down-regulate when exposed to high concentrations of insulin, and cells of the testis down-regulate in response to high concentrations of luteinizing hormone.

Hormone therapy often involves long-term use of abnormally high *pharmacological doses* of hormone, which may have undesirable side effects. Long-term treatment of inflammation with hydrocortisone, for example, has undesirable effects on bone metabolism. Two ways in which these abnormal effects can be produced are: (1) excess hormone may bind to receptor sites for other related hormones and mimic their effects, and (2) a target cell may convert one hormone into another, such as testosterone to estrogen. Thus, long-term high doses of testosterone can, paradoxically, have feminizing effects.

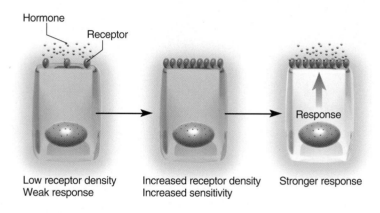

**(a) Up-regulation**

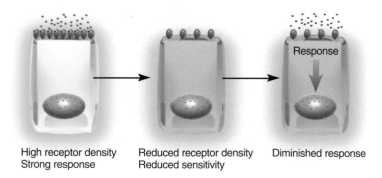

**(b) Down-regulation**

**FIGURE 17.25**   Modulation of Target Cell Sensitivity. (a) Up-regulation, in which a cell produces more receptors and increases its own sensitivity to a hormone. (b) Down-regulation, in which a cell reduces the density of its receptors and lessens its sensitivity to a hormone.

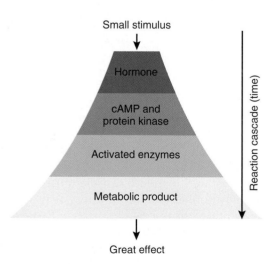

**FIGURE 17.24**   **Enzyme Amplification.**  A single hormone molecule can trigger the production of many cAMP molecules and activation of many molecules of protein kinase. Every protein kinase molecule can phosphorylate and activate many other enzymes. Each of those enzyme molecules can produce many molecules of a metabolic product. Amplification of the process at each step allows for a very small hormonal stimulus to cause a very large metabolic effect.

# HORMONE INTERACTIONS

No hormone travels in the bloodstream alone, and no cell is exposed to only one hormone. Rather, there are many hormones in the blood and tissue fluid at once. Cells ignore the majority of them because they have no receptors for them, but most cells are sensitive to more than one. In these cases, the hormones may have three kinds of interactive effects:

1. **Synergistic effects,** in which two or more hormones act together to produce an effect that is greater than the sum of their separate effects. Neither FSH nor testosterone alone, for example, can stimulate significant sperm production. When they act together, however, the testes produce some 300,000 sperm per minute.

2. **Permissive effects,** in which one hormone enhances the target organ's response to a second hormone that is secreted later. Estrogen stimulates the up-regulation of progesterone receptors in the uterus. The uterus would respond poorly to progesterone, if at all, had it not been primed by the first hormone. Estrogen thus has a permissive effect on progesterone action.

3. **Antagonistic effects,** in which one hormone opposes the action of another. For example, insulin lowers blood glucose level and glucagon raises it. During pregnancy, estrogen from the placenta inhibits the mammary glands from responding to prolactin; thus milk is not secreted until the placenta is shed at birth.

## HORMONE CLEARANCE

Hormonal signals, like nervous signals, must be turned off when they have served their purpose. Most hormones are taken up and degraded by the liver and kidneys and then excreted in the bile or urine. Some are degraded by their target cells. As noted earlier, hormones that bind to transport proteins are removed from the blood much more slowly than hormones that do not employ transport proteins. The rate of hormone removal is called the *metabolic clearance rate (MCR),* and the length of time required to clear 50% of the hormone from the blood is the half-life. The faster the MCR, the shorter is the half-life.

## Before You Go On

*Answer the following questions to test your understanding of the preceding section:*

14. *What are the three chemical classes of hormones? Name at least one hormone in each class.*

15. *Why do corticosteroids and thyroid hormones require transport proteins to travel in the bloodstream?*

16. *Explain how MIT, DIT, $T_3$, and $T_4$ relate to each other structurally.*

17. *Where are hormone receptors located in target cells? Name one hormone that employs each receptor location.*

18. *Explain how one hormone molecule can activate millions of enzyme molecules.*

# Stress and Adaptation

## Objectives

When you have completed this section, you should be able to

- give a physiological definition of stress; and
- discuss how the body adapts to stress through its endocrine and sympathetic nervous systems.

Stress affects us all from time to time, and we react to it in ways that are mediated mainly by the endocrine and sympathetic nervous systems. **Stress** is defined as any situation that upsets homeostasis and threatens one's physical or emotional well-being. Physical causes of stress *(stressors)* include injury, surgery, hemorrhage, infection, intense exercise, temperature extremes, pain, and malnutrition. Emotional causes include anger, grief, depression, anxiety, and guilt.

Whatever the cause, the body reacts to stress in a fairly consistent way called the **stress response** or **general adaptation syndrome (GAS).** The response typically involves elevated levels of epinephrine and glucocorticoids, especially cortisol; some physiologists now define stress as any situation that raises the cortisol level. A pioneering researcher on stress physiology, Canadian biochemist Hans Selye, showed in 1936 that the GAS typically occurs in three stages, which he called the *alarm reaction,* the *stage of resistance,* and the *stage of exhaustion.*

## THE ALARM REACTION

The initial response to stress is an **alarm reaction** mediated mainly by norepinephrine from the sympathetic nervous system and epinephrine from the adrenal medulla. These catecholamines prepare the body to take action such as fighting or escaping danger. One of their effects, the consumption of stored glycogen, is particularly important to the transition to the next stage of the stress response. Aldosterone and angiotensin levels also rise during the alarm reaction. Angiotensin helps to raise the blood pressure, and aldosterone promotes sodium and water conservation, which helps to offset possible losses by sweating and bleeding.

## THE STAGE OF RESISTANCE

After a few hours, the body's glycogen reserves are exhausted, and yet the nervous system continues to demand glucose. If a stressful situation is not resolved before the glycogen is gone, the body enters the **stage of resistance,** in which the first priority is to provide alternative fuels for metabolism. This stage is dominated by cortisol. The hypothalamus secretes corticotropin-releasing hormone (CRH), the pituitary responds by secreting adrenocorticotropic hormone (ACTH), and this, in turn, stimulates the adrenal cortex to secrete cortisol and other glucocorticoids. Cortisol promotes the breakdown of fat and protein into glycerol, fatty acids, and amino acids, providing the liver with raw material for gluconeogenesis (glucose synthesis). Like epinephrine, cortisol inhibits glucose uptake by most organs and thus has a glucose-sparing effect. It also inhibits protein synthesis, leaving the free amino acids available for gluconeogenesis.

Unfortunately, this has adverse effects on the immune system, which depends heavily on the synthesis of antibodies and other proteins. Immunity is depressed by long-term cortisol exposure. Lymphoid tissues atrophy, antibody levels drop, the number of circulating leukocytes declines, and inflammatory cells such as *mast cells* (see chapter 21) release less histamine and other inflammatory chemicals. Wounds heal poorly, and a person under chronic stress becomes more susceptible to infections and some forms of cancer.

Cortisol stimulates gastric secretion, which may aggravate the ulcers that occur in chronic stress, but it suppresses the secretion of sex hormones such as estrogen, testosterone, and luteinizing hormone, causing disturbances of fertility and sexual function.

## THE STAGE OF EXHAUSTION

The body's fat reserves can carry it through months of stress, but when fat is depleted, stress overwhelms homeostasis. The **stage of exhaustion** sets in, often marked by rapid decline and death. With its fat stores gone, the body now relies primarily on protein breakdown to meet its energy needs. Thus, there is a progressive wasting away of the muscles and weakening of the body. After prolonged stimulation, the adrenal cortex may stop producing glucocorticoids, making it all the more difficult to maintain glucose homeostasis. Aldosterone sometimes promotes so much water retention that it creates a state of hypertension, and while it conserves sodium, it hastens the elimination of potassium and hydrogen ions. This creates a state of hypokalemia (potassium deficiency in the blood) and alkalosis (excessively high blood pH), resulting in nervous and muscular system dysfunctions. Death frequently results from heart failure, kidney failure, or overwhelming infection.

## Before You Go On

*Answer the following questions to test your understanding of the preceding section:*

19. Define stress *from the standpoint of endocrinology.*

20. Describe the stages of the general adaptation syndrome.

21. List six hormones that show increased secretion in the stress response. Describe how each one contributes to recovery from stress.

# Eicosanoids and Paracrine Signaling

### Objectives

When you have completed this section, you should be able to

- explain what eicosanoids are and how they are produced;
- identify some classes and functions of eicosanoids; and
- describe several physiological roles of prostaglandins.

Neurotransmitters and hormones are not the only chemical messengers in the body. Here we briefly consider the *paracrine* messengers. These are chemical signals released by cells into the tissue fluid; they do not travel to their target cells by way of the blood, but diffuse from their source to nearby cells in the same tissue. Histamine, for example, is released by mast cells that lie alongside the blood vessels in a connective tissue. It diffuses to the smooth muscle of the blood vessel, relaxing it and allowing vasodilation. Nitric oxide, another paracrine vasodilator, is released by the endothelial cells of the blood vessel itself. In the pancreas, somatostatin acts as a paracrine signal when it is released by delta cells and diffuses to the alpha and beta cells in the same islet, inhibiting their secretion of glucagon and insulin. Catecholamines diffuse from the adrenal medulla to the cortex to stimulate corticosterone secretion. A single chemical can act as a hormone, a paracrine, or even a neurotransmitter in different locations and circumstances.

The **eicosanoids**[24] (eye-CO-sah-noyds) are an important family of paracrine secretions. They have 20-carbon backbones derived from a polyunsaturated fatty acid called **arachidonic** (ah-RACK-ih-DON-ic) **acid.** Some peptide hormones and other stimuli liberate arachidonic acid from one of the phospholipids of the plasma membrane, and the following two enzymes then convert it to various eicosanoids (fig. 17.26).

**Lipoxygenase** helps to convert arachidonic acid to **leukotrienes,** eicosanoids that mediate allergic and inflammatory reactions (see chapter 21). **Cyclooxygenase** converts arachidonic acid to three other types of eicosanoids:

1. **Prostacyclin** is produced by the walls of the blood vessels, where it inhibits blood clotting and vasoconstriction.

2. **Thromboxanes** are produced by blood platelets. In the event of injury, they override prostacyclin and stimulate vasoconstriction and clotting.

---

[24]*eicosa* (variation of *icosa*) = 20

Prostacyclin and thromboxanes are further discussed in chapter 18.

3. **Prostaglandins (PGs)** are the most diverse eicosanoids. They have a five-sided carbon ring in their backbone. They are named PG for *prostaglandin,* plus a third letter that indicates the type of ring structure (PGE, PGF, etc.) and a subscript that indicates the number of C=C double bonds in the side chain. They were first found in bull semen and the prostate gland, hence their name, but they are now thought to be produced in most

organs of the body. The PGEs are usually antagonized by PGFs. For example, the PGE family relaxes smooth muscle in the bladder, intestines, bronchioles, and uterus and stimulates contraction of the smooth muscle of blood vessels. $PGF_{2\alpha}$ has precisely the opposite effects. Some other roles of prostaglandins are described in table 17.7.

Understanding the pathways of eicosanoid synthesis makes it possible to understand the action of several familiar drugs (see Insight 17.3). The roles of prostaglandins and other eicosanoids are further explored in later chapters on blood, immunity, and reproduction.

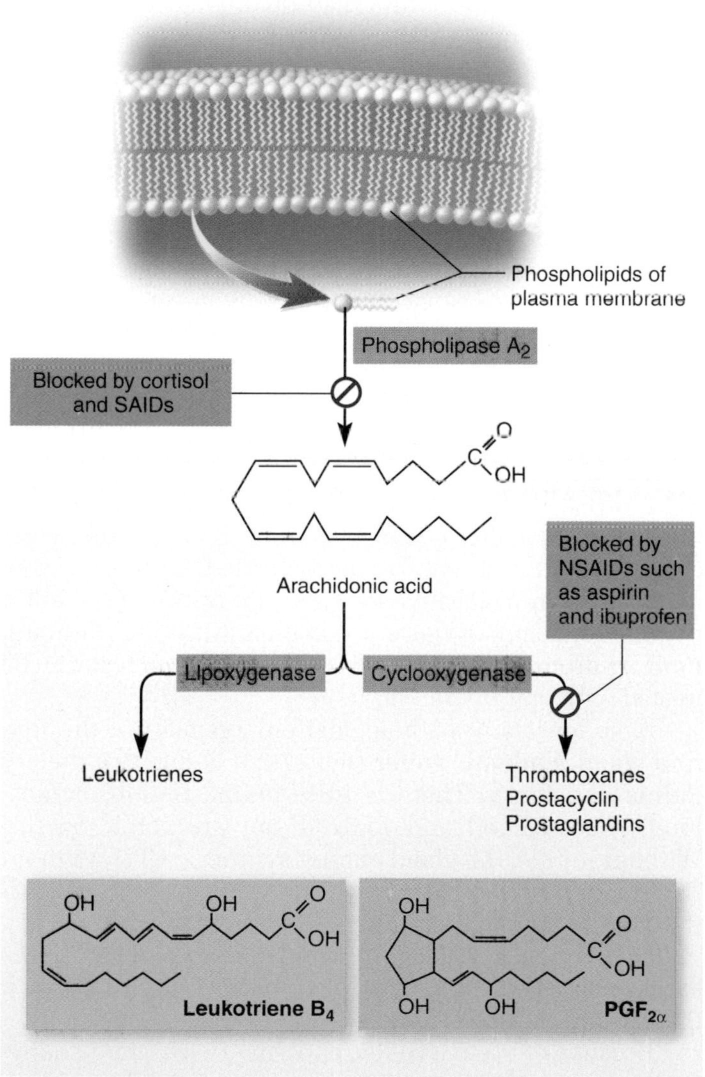

**FIGURE 17.26  Eicosanoid Synthesis and Related Drug Actions.** SAIDs are steroidal anti-inflammatory drugs such as hydrocortisone; NSAIDs are nonsteroidal anti-inflammatory drugs such as aspirin and ibuprofen.

▶ *How would the body be affected by a drug that inhibited lipoxygenase?*

| TABLE 17.7 | Some of the Roles of Prostaglandins |
|---|---|

*Inflammatory:* Promote fever and pain, two cardinal signs of inflammation

*Endocrine:* Mimic effects of TSH, ACTH, and other hormones; alter sensitivity of anterior pituitary to hypothalamic hormones; work with glucagon, catecholamines, and other hormones in regulation of fat mobilization

*Nervous:* Function as neuromodulators, altering the release or effects of neurotransmitters in the brain

*Reproductive:* Promote ovulation and formation of corpus luteum; induce labor contractions

*Gastrointestinal:* Inhibit gastric secretion

*Vascular:* Act as vasodilators and vasoconstrictors

*Respiratory:* Constrict or dilate bronchioles

*Renal:* Promote blood circulation through the kidney, increase water and electrolyte excretion

---

**INSIGHT 17.3**    Clinical Application

## Anti-Inflammatory Drugs

Cortisol and corticosterone are *steroidal anti-inflammatory drugs (SAIDs).* They inhibit inflammation by blocking the release of arachidonic acid from the plasma membrane, thus inhibiting the synthesis of all eicosanoids. Their main disadvantage is that prolonged use causes side effects that mimic Cushing syndrome (see p. 669). Aspirin and ibuprofen (Motrin) are *nonsteroidal anti-inflammatory drugs (NSAIDs)* with more selective effects. They stop the action of cyclooxygenase, thus blocking prostaglandin synthesis without affecting lipoxygenase or the leukotrienes. For similar reasons, aspirin inhibits blood clotting (see chapter 18). One theory of fever is that it results from the action of prostaglandins on the hypothalamus. Most antipyretic (fever-reducing) drugs work by inhibiting cyclooxygenase.

# Endocrine Disorders

### Objectives

When you have completed this section, you should be able to

- explain some general causes and examples of hormone hyposecretion and hypersecretion;
- briefly describe some common disorders of pituitary, thyroid, parathyroid, and adrenal function; and
- in more detail, describe the causes and pathology of diabetes mellitus.

Hormones are very potent chemicals; as we saw in the discussion of enzyme amplification, a little hormone goes a long way. It is therefore necessary to tightly regulate their secretion and blood concentration. Variations in hormone concentration and target cell sensitivity often have very noticeable effects on the body. This section deals with some of the better-known dysfunctions of the endocrine system. The effects of aging on this system are described on page 1130.

## HYPOSECRETION AND HYPERSECRETION

Inadequate hormone release is called **hyposecretion.** It can result from tumors or lesions that destroy an endocrine gland or interfere with its ability to receive signals from another gland. For example, a fractured sphenoid bone can sever the hypothalamo-hypophyseal tract and thus prevent the transport of oxytocin and antidiuretic hormone (ADH) to the posterior pituitary. The resulting ADH hyposecretion disables the water-conserving capability of the kidneys and leads to **diabetes insipidus,** a condition of chronic polyuria without glucose in the urine. (*Insipidus* means "without taste" and refers to the lack of sweetness of the glucose-free urine, in contrast to the sugary urine of diabetes mellitus.) Autoimmune diseases can also lead to hormone hyposecretion when misguided antibodies (autoantibodies) attack endocrine cells. This is thought to be one of the causes of diabetes mellitus, as explained shortly.

Excessive hormone release, called **hypersecretion,** has multiple causes. Some tumors result in the overgrowth of functional endocrine tissue. A **pheochromocytoma** (FEE-o-CRO-mo-sy-TOE-muh), for example, is a tumor of the adrenal medulla that secretes excessive amounts of epinephrine and norepinephrine (table 17.8, p. 672). Some tumors in nonendocrine organs produce hormones. For example, some lung tumors secrete ACTH and thus overstimulate cortisol secretion by the adrenal gland. Whereas certain autoimmune disorders can cause endocrine hyposecretion, others cause hypersecretion. An example of this is **toxic goiter** (Graves[26] disease), in which autoantibodies mimic the effect of TSH on the thyroid, causing thyroid hypersecretion (table 17.8). Endocrine hypersecretion disorders can also be mimicked by excess or long-term clinical administration of hormones such as cortisol.

Following are brief descriptions of some of the better-known disorders of the major endocrine glands. Table 17.8 provides further details on some of these and lists some additional endocrine disorders. Diabetes mellitus, by far the most prevalent endocrine disease, receives a more extended discussion.

## PITUITARY DISORDERS

The hypersecretion of growth hormone (GH) in adults causes **acromegaly**—thickening of the bones and soft tissues with especially noticeable effects on the hands, feet, and face (fig. 17.27). When it begins in childhood or adolescence, GH hypersecretion causes **gigantism** and hyposecretion causes **pituitary dwarfism** (table 17.8). Now that growth hormone is plentiful, made by genetically engineered bacteria containing the human GH gene, pituitary dwarfism has become rare.

## THYROID AND PARATHYROID DISORDERS

**Congenital hypothyroidism** is thyroid hyposecretion present from birth; it was formerly called *cretinism,* now regarded as an insensitive term. Severe or prolonged adult hypothyroidism can cause **myxedema** (MIX-eh-DEE-muh). Both syndromes are described in table 17.8, and both can be treated with oral thyroid hormone.

A *goiter* is any pathological enlargement of the thyroid gland. **Endemic goiter** (fig. 17.28) is due to a dietary iodine deficiency. There is little iodine in soil or most foods, but seafood and iodized salt are good sources. Without iodine, the gland cannot synthesize TH. Without TH, the pituitary gland receives no feedback and acts as if the thyroid were understimulated. It produces extra TSH, which stimulates hypertrophy of the thyroid gland. In the earlier-described toxic goiter, by contrast, the overgrown thyroid produces functional TH.

Because of their location and small size, the parathyroids are sometimes accidentally removed in thyroid surgery. Without hormone replacement therapy, the resulting **hypoparathyroidism** causes a rapid decline in blood calcium level and leads to fatal tetany within 3 or 4 days. **Hyperparathyroidism,** excess PTH secretion, is usually caused by a parathyroid tumor. It causes the bones to become

---

[26]Robert James Graves (1796–1853), Irish physician

| Age 9 | Age 16 | Age 33 | Age 52 |

**FIGURE 17.27  Acromegaly, a Condition Caused by Growth Hormone Hypersecretion in Adulthood.**  These are four photographs of the same person taken at different ages. Note the characteristic thickening of the face and hands.
▶ *How would she have been affected if GH hypersecretion began at age 9?*

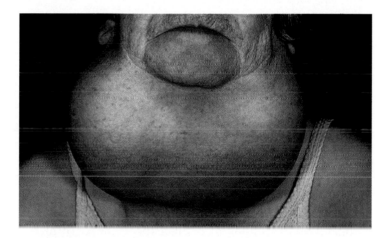

**FIGURE 17.28  Endemic Goiter.**  An iodine deficiency in this person's diet resulted in a lack of thyroid hormone. For lack of negative feedback inhibition, the pituitary secreted elevated levels of thyroid-stimulating hormone (TSH). This resulted in hypertrophy of the thyroid gland.

soft, deformed, and fragile; it raises the blood levels of calcium and phosphate ions; and it promotes the formation of *renal calculi* (kidney stones) composed of calcium phosphate. Chapter 7 further describes the relationship between parathyroid function, blood calcium, and bone tissue.

## ADRENAL DISORDERS

**Cushing[27] syndrome** is excess cortisol secretion owing to any of several causes: ACTH hypersecretion by the pituitary, ACTH-secreting tumors, or hyperactivity of the adrenal cortex independently of ACTH. Cushing syndrome disrupts carbohydrate and protein metabolism,

leading to hyperglycemia, hypertension, muscular weakness, and edema. Muscle and bone mass are lost rapidly as protein is catabolized. Some patients exhibit abnormal fat deposition between the shoulders ("buffalo hump") or in the face ("moon face") (fig. 17.29). Long-term hydrocortisone therapy can have similar effects.

**Adrenogenital syndrome (AGS),** the hypersecretion of adrenal androgens, commonly accompanies Cushing syndrome. In children, AGS often causes enlargement of the penis or clitoris and the premature onset of puberty. Prenatal AGS can result in newborn girls exhibiting masculinized genitalia and being misidentified as boys (fig. 17.30). In women, AGS produces such masculinizing effects as increased body hair, deepening of the voice, and beard growth.

## DIABETES MELLITUS

**Diabetes mellitus[28] (DM)** warrants special consideration. This is the world's most prevalent metabolic disease, and is the leading cause of adult blindness, renal failure, gangrene, and the necessity for limb amputations.

Diabetes mellitus can be defined as a disruption of carbohydrate, fat, and protein metabolism resulting from the hyposecretion or inaction of insulin. Its classic signs and symptoms are "the three polys": **polyuria[29]** (excessive urine output), **polydipsia[30]** (intense thirst), and **polyphagia[31]** (ravenous hunger). We can add to this list three clinical signs revealed by blood and urine tests: **hyperglycemia[32]** (elevated blood glucose), **glycosuria[33]** (glucose in the urine), and **ketonuria** (ketones in the urine).

[27]Harvey Cushing (1869–1939), American physician

[28]*diabet* = to flow through + *melli* = honey
[29]*poly* = much, excessive + *uri* = urine
[30]*dipsia* = drinking
[31]*phagia* = eating
[32]*hyper* = excess + *glyc* = sugar, glucose + *emia* = blood condition
[33]*glyco* = glucose, sugar + *uria* = urine condition

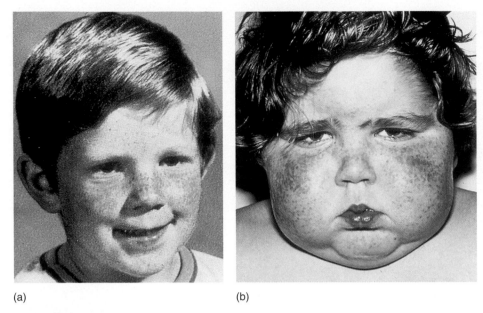

(a)                                        (b)

**FIGURE 17.29** **Cushing Syndrome.** (a) Patient before the onset of the syndrome. (b) The same boy, only 4 months later, showing the "moon face" characteristic of Cushing syndrome.

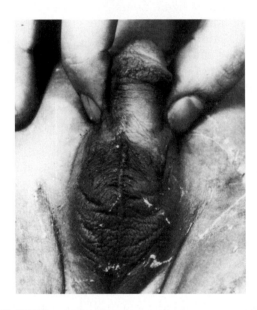

**FIGURE 17.30** **Adrenogenital Syndrome (AGS).** These are the genitals of a baby girl with AGS, masculinized by prenatal hypersecretion of adrenal androgens. Note the fusion of the labia majora to resemble a scrotum and enlargement of the clitoris to resemble a penis. Such infants are easily mistaken for boys and raised as such.

A little knowledge of kidney physiology is necessary to understand why glycosuria and polyuria occur. The kidneys filter blood plasma and convert the filtrate to urine. Normally, the kidney tubules remove all glucose from the filtrate and return it to the blood, so there is lit-tle or no glucose in the urine of a healthy person. Water follows the glucose and other solutes by osmosis, so the tubules also reclaim most of the water in the filtrate.

But like any other carrier-mediated transport system, there is a limit to how fast the glucose transporters of the kidney can reabsorb glucose. The maximum rate of reabsorption is called the transport maximum, $T_m$ (see p. 106). In diabetes mellitus, glucose enters the tubules so rapidly that it exceeds the $T_m$ and the tubules cannot reabsorb it all. The excess passes through into the urine. Glucose and ketones in the tubules also raise the osmolarity of the tubular fluid and cause **osmotic diuresis**—water remains in the tubules with these solutes, so large amounts of water are passed in the urine. This accounts for the polyuria, dehydration, and thirst of diabetes. Diabetics often pass 10 to 15 L of urine per day, compared with 1 or 2 L in a healthy person.

## Types and Treatment

There are two forms of diabetes mellitus: type I and type II. **Type I,** or **insulin-dependent diabetes mellitus (IDDM),** accounts for about 10% of cases. The cause is still little understood, but at least some cases appear to result from the destruction of beta cells by autoantibodies. Several genes are implicated in IDDM, but heredity alone is not the cause; IDDM results from the interaction of hereditary and environmental factors. Up to a point, the body can tolerate the loss of beta cells. Signs of diabetes begin to appear when 80% to 90% of them are destroyed and insulin falls to critically low levels. However, the signs are not caused by a deficiency of insulin alone. The relative level of glucagon is elevated in IDDM, and it is the abnormally low *ratio* of insulin to glucagon that causes the signs. IDDM is most often diagnosed around age 12 and used to be called *juvenile diabetes,* but it can also appear later in life. IDDM must be treated through meal planning, exercise, the self-monitoring of blood glucose level by the patient, and periodic insulin injections or the continual subcutaneous delivery of insulin by a pump worn on the body.

Most diabetics (90%) have **type II,** or **non-insulin-dependent diabetes mellitus (NIDDM).** Nearly 7% of U.S. residents are diagnosed with NIDDM. Type II diabetics may have normal or even elevated insulin levels; the problem is *insulin resistance*—a failure of target cells to respond to insulin. There may be multiple reasons for resistance. One hypothesis is that target cells have a shortage of insulin receptors, although research has failed to find any consistent relationship between NIDDM and defective receptors, so this hypothesis is being increasingly questioned. The three major risk factors for NIDDM are heredity, age, and obesity. It tends to run in families; if an identical twin contracts NIDDM it is virtually certain that the other twin will too. NIDDM has a gradual onset, with signs usually not appearing until age 40 or beyond; it used to be called *adult-onset DM.* Type II diabetics are usually overweight and the incidence of NIDDM is increasing as people eat too much, exercise too little, and obesity becomes more prevalent. In obesity, the adipocytes produce a hormonelike secretion that indirectly interferes with glucose transport into most kinds of cells. NIDDM can often be managed through a weight-loss program of diet and exercise. Some patients are helped with oral medications that improve insulin secretion or target cell sensitivity or help in other ways to lower blood glucose.

## Pathogenesis

When cells cannot absorb glucose, they rely on fat and protein for energy. Fat and protein breakdown results in emaciation, muscular atrophy, and weakness. Before insulin therapy was introduced in 1922, the victims of type I diabetes wasted away to an astonishing extent (see Insight 17.4). Diabetes was described in the first century as "a melting down of the flesh and limbs into urine." Adult patients weighed as little as 27 to 34 kg (60–75 lb) and looked like victims of a concentration camp or severe famine. Their breath had a sickeningly sweet ketone smell, like rotten apples. One typical patient was described by medical historian Michael Bliss as "barely able to lift his head from his pillow, crying most of the time from pain, hunger, and despair." In the terminal stage of the disease, patients became increasingly drowsy,

gasped for air, became comatose, and then died within a few hours. Most diabetic children lived less than 1 year after diagnosis—a year of utmost misery at that.

Rapid fat catabolism elevates blood concentrations of free fatty acids and their breakdown products, the ketone bodies (acetoacetic acid, acetone, and β-hydroxybutyric acid). Ketonuria promotes osmotic diuresis and flushes $Na^+$ and $K^+$ from the body, thus creating electrolyte deficiencies that can lead to abdominal pain, vomiting, irregular heartbeat, and neurological dysfunction. As acids, ketones lower the pH of the blood and produce a condition called **ketoacidosis.** Ketoacidosis causes a deep, gasping breathing called *Kussmaul*[34] *respiration,* typical of terminal diabetes. It also depresses the nervous system and produces diabetic coma.

Diabetes mellitus also leads to long-term degenerative cardiovascular and neurological diseases—signs that were seldom seen before insulin therapy, when patients died too quickly to show the chronic effects. Despite years of research and debate, it remains unclear exactly how diabetes mellitus causes cardiovascular disease. It appears that chronic hyperglycemia activates a metabolic reaction cascade that leads to cellular damage in small to medium blood vessels and peripheral nerves. Nerve damage *(diabetic neuropathy),* the most common complication of diabetes mellitus, can lead to impotence, incontinence, and loss of sensation from affected areas. The last effect makes a patient dangerously unaware of minor injuries, which can thus fester from neglect and contribute (along with lowered resistance to infection) to gangrene and the necessity of amputation. Many diabetics lose their toes, feet, or legs to the disease. Diabetes also promotes *atherosclerosis,* the blockage of blood vessels with fatty deposits, causing poor circulation. The effects include degeneration of the small arteries of the retina and the kidneys, leading to blindness and kidney failure as common complications. Atherosclerosis also contributes to gangrene. People with type I diabetes are much more likely to die of kidney failure than those with type II. In type II diabetes, the most common cause of death is heart failure stemming from atherosclerosis of the coronary arteries.

## Before You Go On

*Answer the following questions to test your understanding of the preceding section:*

25. *Explain some causes of hormone hyposecretion, and give examples. Do the same for hypersecretion.*

26. *In diabetes mellitus, explain the chain of events that lead to (a) osmotic diuresis, (b) ketoacidosis and coma, and (c) gangrene of the lower limbs.*

---

[34]Adolph Kussmaul (1822–1902), German physician

| TABLE 17.8 | Some Disorders of the Endocrine System |
|---|---|
| Addison[25] disease | Hyposecretion of adrenal glucocorticoids and mineralocorticoids, causing hypoglycemia, hypotension, weight loss, weakness, loss of stress resistance, darkening or bronzing (metallic discoloration) of the skin, and potentially fatal dehydration and electrolyte imbalances |
| Congenital hypothyroidism | Thyroid hormone hyposecretion present from birth, resulting in stunted physical development, thickened facial features, low body temperature, lethargy, and irreversible brain damage in infancy |
| Diabetes insipidus | Chronic polyuria due to ADH hyposecretion. Can result from tumors, skull fractures, or infections that destroy hypothalamic tissue or the hypothalamo-hypophyseal tract |
| Hyperinsulinism | Insulin excess caused by islet hypersecretion or injection of excess insulin, causing hypoglycemia, weakness, hunger, and sometimes *insulin shock,* which is characterized by disorientation, convulsions, or unconsciousness |
| Myxedema | A syndrome occurring in severe or prolonged adult hypothyroidism, characterized by low metabolic rate, sluggishness and sleepiness, weight gain, constipation, dry skin and hair, abnormal sensitivity to cold, hypertension, and tissue swelling |
| Pheochromocytoma | A tumor of the adrenal medulla that secretes excess epinephrine and norepinephrine. Causes hypertension, elevated metabolic rate, nervousness, indigestion, hyperglycemia, and glycosuria |
| Toxic goiter (Graves disease) | Thyroid hypertrophy and hypersecretion, occurring when autoantibodies mimic the effect of TSH and overstimulate the thyroid. Results in elevated metabolic rate and heart rate, nervousness, sleeplessness, weight loss, abnormal heat sensitivity and sweating, and bulging of the eyes (exophthalmos) resulting from eyelid retraction and edema of the orbital tissues |

*Disorders described elsewhere*

| | | |
|---|---|---|
| Acromegaly p. 668 | Cushing syndrome p. 669 | Hyperparathyroidism p. 668 |
| Adrenogenital syndrome p. 669 | Endemic goiter p. 668 | Hypoparathyroidism p. 668 |
| Androgen-insensitivity syndrome p. 1037 | Gigantism p. 668 | Pituitary dwarfism p. 668 |

[25]Thomas Addison (1793–1860), English physician

---

**INSIGHT 17.4**   Medical History

## The Discovery of Insulin

At the start of the twentieth century, physicians felt nearly help-less in the face of diabetes mellitus. They put patients on use-less diets—the oatmeal cure, the potato cure, and others—or on starvation diets as low as 750 Cal per day so as not to "stress the system." They were resigned to the fact that their patients were doomed to die, and simple starvation seemed to produce the least suffering.

After the cause of diabetes was traced to the pancreatic islets in 1901, European researchers tried treating patients and experimental animals with extracts of pancreas, but became discouraged by the severe side effects of impurities in the extracts. They lacked the resources to pursue the problem to completion, and by 1913, the scientific community showed signs of giving up on diabetes.

But in 1920, Frederick Banting (1891–1941), a young Canadian physician with a failing medical practice, became intrigued with a possible method for isolating the islets from the pancreas and testing extracts of the islets alone. He returned to his alma mater, the University of Toronto, to pres-ent his idea to Professor J. J. R. Macleod (1876–1935), a leading authority on carbohydrate metabolism. Macleod was unim-pressed with Banting, finding his knowledge of the diabetes lit-erature and scientific method superficial. Nevertheless, he felt Banting's idea was worth pursuing and thought that with his military surgical training, Banting might be able to make some progress where others had failed. He offered Banting labora-tory space for the summer, giving him a marginal chance to test his idea. Banting was uncertain whether to accept, but when his fiancée broke off their engagement and an alterna-tive job offer fell through, he closed his medical office, moved to Toronto, and began work. Little did either man realize that in two years' time, they would share a Nobel Prize and would so thoroughly detest each other they would scarcely be on speaking terms.

### A Modest Beginning

Macleod advised Banting on an experimental plan of attack and gave him an assistant, Charles Best (1899–1978). Best had just received his B.A. in physiology and looked forward to an inter-esting summer job with Banting before starting graduate school. He received no pay for his work, and Banting himself was des-perately poor, living in a 7- by 9-foot room and supporting him-self on $2 a week earned by performing tonsillectomies. Over the summer of 1921, Banting and Best removed the pancreases from some dogs to render them diabetic, and tied off the pancreatic ducts in other dogs to make most of the pancreas degenerate while leaving the islets intact. Their plan was to treat the diabetic dogs with extracts made from the degenerated pancreases of the others.

It was a difficult undertaking. Their laboratory was tiny, filthy, unbearably hot, and reeked of dog excrement. The pancreatic ducts were very small and difficult to tie, and it was hard to tell if all the pancreas had been removed from the dogs intended to become diabetic. Several dogs died of overanesthesia, infection, and bleeding from Banting's clumsy surgical technique. Banting was also careless in reading his data and interpreting the results, and he had little interest in reading the literature to see what other diabetes researchers were doing. In Banting and Best's first publication, in early 1922, the data in their discussion disagreed with the data in the tables, and both disagreed with the data in their laboratory notebooks. These were not the signs of promising researchers.

In spite of themselves, Banting and Best achieved modest positive results over the summer. Crude pancreatic extracts brought one dog back from a diabetic coma and reduced the hyperglycemia and glycosuria of others. Buoyed by these results, Banting demanded a salary, a better laboratory, and another assistant. Macleod grudgingly obtained salaries for the pair, but Banting began to loathe him for their disagreement over his demands, and he and Macleod grew in mutual contempt as the project progressed.

## Success and Conflict

Macleod brought biochemist J. B. Collip (1892–1965) into the project that fall in hopes that he could produce purer extracts. More competent in experimental science, Collip was the first to show that pancreatic extracts could eliminate ketosis and restore the liver's ability to store glycogen. He obtained better and better results in diabetic rabbits until, by January 1922, the group felt ready for human trials. Banting was happy to have Collip on the team initially, but grew intensely jealous of him as Collip not only achieved better results than he had, but also developed a closer relationship with Macleod. Banting, who had no qualifications to perform human experiments, feared he would be pushed aside as the project moved to its clinical phase. At one point, the tension between Banting and Collip nearly erupted into a fistfight in the laboratory.

Banting insisted that the first human trial be done with an extract he and Best prepared, not with Collip's. The patient was a 14-year-old boy who weighed only 29 kg (65 lb) and was on the verge of death. He was injected on January 11 with the Banting and Best extract, described by one observer as "a thick brown muck." The trial was an embarrassing failure, with only a slight lowering of his blood glucose and a severe reaction to the impurities in the extract. On January 23, the same boy was treated again, but with Collip's extract. This time, his ketonuria and glycosuria were almost completely eliminated and his blood sugar dropped 77%. This was the first successful clinical trial of insulin. Six more patients were treated in February 1922 and quickly became stronger, more alert, and in better spirits. In April, the Toronto group began calling the product insulin, and

at a medical conference in May, they gave the first significant public report of their success.

But Banting felt increasingly excluded from the project. He quit coming to the laboratory, stayed drunk much of the time, and day-dreamed of leaving diabetes research to work on cancer. He remained only because Best pleaded with him to stay. Banting briefly operated a private diabetes clinic, but fearful of embarrassment over alienating the discoverer of insulin, the university soon lured him back with a salaried appointment and hospital privileges.

Banting had a number of high-profile, successful cases in 1922, such as 14-year-old Elizabeth Hughes, who weighed only 20 kg (45 lb) before treatment. She began treatment in August and showed immediate, dramatic improvement. She was a spirited, optimistic, and articulate girl who kept enthusiastic diaries of being allowed to eat bread, potatoes, and macaroni and cheese for the first time since the onset of her illness. "Oh it is simply too wonderful for words this stuff," she exuberantly wrote to her mother—even though the still-impure extracts caused her considerable pain and swelling. The world quickly beat a path to Toronto begging for insulin. The pharmaceutical firm of Eli Lilly and Company entered into an agreement with the University of Toronto for the mass production of insulin, and by the fall of 1923, over 25,000 patients were being treated at more than 60 Canadian and U.S. clinics.

## The Bitter Fruits of Success

Banting's self-confidence was restored. He had become a public hero, and the Canadian parliament awarded him an endowment generous enough to ensure him a life of comfort. Several distinguished physiologists nominated Banting and Macleod for the 1923 Nobel Prize, and they won. When the award was announced, Banting was furious about having to share it with Macleod. At first, he threatened to refuse it, but when he cooled down, he announced that he would split his share of the prize money with Best. Macleod quickly announced that half of his share would go to Collip.

Banting subsequently made life at the university so unbearable that Macleod left in 1928 to accept a university post in Scotland. Banting stayed on at Toronto. Although now wealthy and surrounded by admiring students, he achieved nothing significant in science for the rest of his career. He was killed in a plane crash in 1941. Best replaced Macleod on the Toronto faculty, led a distinguished career, and developed the anticoagulant heparin. Collip went on to play a lead role in the isolation of PTH, ACTH, and other hormones.

Insulin made an industry giant of Eli Lilly and Company. It became the first protein whose amino acid sequence was determined, for which Frederick Sanger received a Nobel Prize in 1958. Diabetics today no longer depend on a limited supply of insulin extracted from beef and pork pancreas. Human insulin is now in plentiful supply, made by genetically engineered bacteria. Paradoxically, while insulin has dramatically reduced the suffering caused by diabetes mellitus, it has increased the number of people who have the disease—because, thanks to insulin, diabetics are now able to live long enough to raise families and pass on the diabetes genes.

# CONNECTIVE ISSUES

## Interactions Between the ENDOCRINE SYSTEM and Other Organ Systems

 indicates ways in which this system affects other systems

 indicates ways in which other systems affect this system

### ALL SYSTEMS

Growth hormone, insulin-like growth factors, insulin, thyroid hormone, and glucocorticoids affect the development and metabolism of most tissues

### INTEGUMENTARY SYSTEM

Sex hormones affect skin pigmentation, development of body hair and apocrine glands, and subcutaneous fat deposition

Skin synthesizes calcitriol precursor

### SKELETAL SYSTEM

Many hormones affect bone development

Protects some endocrine glands; stores calcium needed for endocrine function

### MUSCULAR SYSTEM

Growth hormone and testosterone stimulate muscular growth; insulin regulates carbohydrate metabolism in muscle; other hormones affect electrolyte balance, which is critical to muscle function

Skeletal muscles protect some endocrine glands

### NERVOUS SYSTEM

Exerts negative feedback inhibition on hypothalamus; several hormones affect nervous system development, mood, and behavior; hormones regulate electrolyte balance, which is critical to neuron function

Hypothalamus regulates secretion of pituitary hormones and synthesizes the posterior pituitary hormones; sympathetic nervous system triggers secretion by adrenal medulla

### CIRCULATORY SYSTEM

Angiotensin II, aldosterone, ADH, and ANP regulate blood volume and pressure; epinephrine, TH, and other hormones affect heart rate and contraction force

Blood transports hormones to their target organs; blood pressure and osmolarity variations trigger secretion of some hormones

### LYMPHATIC/IMMUNE SYSTEMS

Thymosin and other hormones activate immune cells; glucocorticoids suppress immunity and inflammation

Lymphatic system maintains fluid balance in endocrine glands; lymphocytes protect endocrine glands from infection

### RESPIRATORY SYSTEM

Epinephrine and norepinephrine increase pulmonary airflow

Provides $O_2$ and removes $CO_2$; lungs convert angiotensin I to angiotensin II

### URINARY SYSTEM

ADH regulates water excretion; PTH, calcitriol, and aldosterone regulate electrolyte excretion

Degrades and excretes hormones

### DIGESTIVE SYSTEM

PTH affects intestinal $Ca^{2+}$ absorption; insulin and glucagon modulate nutrient storage and metabolism; enteric hormones regulate gastrointestinal secretion and motility

Provides nutrients; nutrient absorption triggers insulin secretion; gut–brain peptides act on hypothalamus and stimulate specific hungers; liver degrades and excretes hormones

### REPRODUCTIVE SYSTEM

Gonadotropins and sex steroids regulate sexual development, sperm and egg production, sex drive, menstrual cycle, pregnancy, fetal development, and lactation

Sex hormones inhibit secretion by hypothalamus and pituitary gland

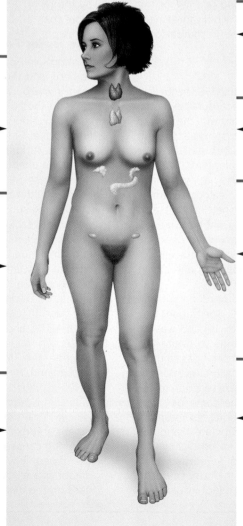

## CHAPTER REVIEW

# Review of Key Concepts

**Overview of the Endocrine System (p. 636)**

1. Intercellular communication is necessary for homeostasis. Cells communicate through gap junctions, neurotransmitters, paracrines, and hormones.

2. *Endocrinology* is the study of the endocrine system, which is composed of hormone-secreting cells and *endocrine glands.*

3. Hormones circulate throughout the body but stimulate only *target cells,* which have receptors for them.

4. Endocrine glands have no ducts, but secrete their products into the bloodstream.

5. The nervous and endocrine systems collaborate and have several overlapping functions and properties, but are not redundant. Differences are listed in table 17.1.

**The Hypothalamus and Pituitary Gland (p. 639)**

1. The hypothalamus and pituitary gland have a more wide-ranging influence on the body than any other endocrine gland.

2. The pituitary gland is connected to the hypothalamus by a stalk. It is divided into an anterior *adenohypophysis,* whose main part is the *anterior lobe,* and a posterior *neurohypophysis,* whose main part is the *posterior lobe* of the pituitary.

3. The adenohypophysis is connected to the hypothalamus by a system of blood vessels called the hypophyseal portal system. The neurohypophysis is connected to the hypothalamus by a bundle of nerve fibers, the hypothalamo-hypophyseal tract.

4. The hypothalamus secretes seven hormones that travel through the portal system and regulate the anterior lobe. Five of these trigger hormone release and two of them inhibit hormone release by the anterior lobe (see table 17.3).

5. The hypothalamus synthesizes two hormones that are stored in the posterior pituitary and released in response to nerve signals: oxytocin (OT) and antidiuretic hormone (ADH).

6. The anterior lobe secretes prolactin (PRL), follicle-stimulating hormone (FSH), luteinizing hormone (LH), thyroid-stimulating hormone (TSH), adrenocorticotropic hormone (ACTH), and growth hormone (GH). Their functions are summarized in table 17.4.

7. GH has an especially broad effect on the body, notably to stimulate the growth of cartilage, bone, muscle, and fat through its influence on protein, lipid, and carbohydrate metabolism. Many of its effects are mediated through insulin-like growth factors (IGFs) secreted by the liver.

8. The posterior lobe stores OT and ADH until the brain stimulates their release. OT promotes labor contractions and milk release and possibly contributes to semen propulsion, uterine contractions, and sexual emotion. ADH promotes water conservation.

9. The hypothalamus and pituitary are regulated by positive and negative feedback from their target organs.

**Other Endocrine Glands (p. 646)**

1. The *pineal gland* secretes *serotonin* and *melatonin* and may contribute to the timing of puberty.

2. The *thymus* secretes *thymopoietin,* *thymosin,* and other hormones that regulate immunity.

3. The *thyroid gland* secretes *triiodothyronine* ($T_3$) and *thyroxine* ($T_4$), which stimulate catabolism and heat production, tissue growth, nervous system development, and alertness. It also secretes calcitonin, which promotes bone deposition and lowers blood $Ca^{2+}$ levels.

4. The four *parathyroid glands* secrete *parathyroid hormone,* which acts in several ways to increase blood $Ca^{2+}$ levels.

5. The *adrenal medulla* secretes mainly *epinephrine* and *norepinephrine,* which raise blood glucose levels and help the body adapt to physical activity and stress.

6. The *adrenal cortex* secretes *aldosterone,* which promotes $Na^+$ retention and $K^+$ excretion; *cortisol* and *corticosterone,* which raise blood glucose and fatty acids levels and aid in stress adaptation and tissue repair; and *androgens* and *estrogens,* which contribute to reproductive development and physiology.

7. The *pancreatic islets* secrete *insulin,* which promotes glucose uptake by the tissues; *glucagon,* which stimulates the release of glucose from storage; and *somatostatin,* which regulates the insulin- and glucagon-secreting cells and inhibits some digestive functions.

8. The ovaries secrete estradiol and progesterone, which regulate reproductive development and physiology, and inhibin, which suppresses FSH secretion by the pituitary. The testes secrete testosterone and inhibin, with corresponding roles in the male.

9. Many other hormones are produced by cells that populate other organs but do not form discrete glands; *atrial natriuretic peptide* by the heart; *calcitriol* by sequential action of the skin, liver, and kidneys; *erythropoietin* by the liver and kidneys; *angiotensinogen,* IGF-I, and hepcidin by the liver; several *enteric hormones* by the stomach and small intestine; and *estrogens* and *progesterone* by the placenta.

**Hormones and Their Actions (p. 654)**

1. There are three chemical classes of hormones: steroids, peptides, and monoamines (table 17.6).

2. The steroids are synthesized from cholesterol.

3. Peptide hormones are synthesized by ribosomes and pass through inactive stages called the *preprohormone* and *prohormone* before being converted to active hormone.

4. Monoamines are synthesized from the amino acids tryptophan and tyrosine. Thyroid hormone is an unusual case in that its synthesis begins with a large protein, thyroglobulin. Iodine is added to its tyrosine residues, the modified tyrosines are linked in pairs, and then the tyrosines are cleaved from the protein to yield the two thyroid hormones, $T_3$ and $T_4$.

5. Peptides and most monoamines travel easily in the watery blood plasma, but the hydrophobic steroids and thyroid hormone are carried by *transport proteins,* which also prolong the half-life of the hormones.

6. Hormone receptors are found in the plasma membranes, on the mitochondria and other organelles, and in the nuclei of the target cells. Steroids bind to nuclear receptors; $T_3$ to receptors in the nucleus, mitochondria, and ribosomes; and peptides and catecholamines to plasma membrane receptors, which activate formation of second messengers in the cytosol.

7. Second messengers for hormone action include cAMP, cGMP, diacylglycerol, and inositol triphosphate.

8. Because of *enzyme amplification,* small amounts of hormone can have great physiological effects.

9. Target cells can adjust their sensitivity to a hormone by *up-regulation* (increasing the number of hormone receptors) and *down-regulation* (decreasing the number of receptors).

10. Hormones do not act in isolation, but influence each other. One hormone can have *synergistic, permissive,* or *antagonistic effects* on the action of another.

11. Hormones are cleared from the blood by metabolism in the liver and kidneys and excretion in the bile and urine.

## Stress and Adaptation (p. 665)

1. *Stress* is any situation that upsets homeostasis and threatens well-being.

2. Regardless of the cause of stress, the body reacts in a fairly consistent way called the *stress response* (general adaptation syndrome).

3. The first stage of the stress response, the *alarm reaction,* is characterized by elevated levels of norepinephrine, epinephrine, aldosterone, and angiotensin, resulting in glycogen breakdown, fluid and electrolyte retention, and elevated blood pressure.

4. The second stage, the *stage of resistance,* sets in when stored glycogen is exhausted. It is characterized by elevated levels of ACTH and cortisol, and the breakdown of fat and protein for fuel.

5. The third stage, the *stage of exhaustion,* sets in if fat reserves are depleted. It is characterized by increasing protein breakdown, weakening of the body, and often death.

## Eicosanoids and Paracrine Signaling (p. 666)

1. *Paracrines* are hormonelike messengers that travel only short distances and stimulate nearby cells.

2. Many paracrines are *eicosanoids,* which have 20-carbon backbones and are derived from arachidonic acid. Eicosanoids include leukotrienes, prostacyclin, thromboxanes, and prostaglandins.

## Endocrine Disorders (p. 668)

1. Endocrine dysfunctions often stem from hormone hyposecretion or hypersecretion.

2. Some pituitary disorders include *acromegaly* and *gigantism* from GH hypersecretion, *pituitary dwarfism* from GH hyposecretion, and *diabetes insipidus* from ADH hyposecretion.

3. *Myxedema* and *endemic goiter* are thyroid hyposecretion disorders and *toxic goiter* is a thyroid hypersecretion disorder.

4. *Hypoparathyroidism* leads to low blood $Ca^{2+}$ level and sometimes fatal tetany; *hyperparathyroidism* leads to excessive blood $Ca^{2+}$ and bone fragility.

5. Hypersecretion by the adrenal cortex can result in *Cushing syndrome* (excess cortisol) or *adrenogenital syndrome* (excess androgen).

6. Diabetes mellitus (DM) is the most common endocrine dysfunction, resulting from the hyposecretion or inaction of insulin—type I or insulin-dependent DM (IDDM) and type II or non-insulin-dependent DM (NIDDM), respectively.

7. DM is characterized by *polyuria, polydipsia,* and *polyphagia* and clinically confirmed by findings of *hyperglycemia, glycosuria,* and *ketonuria.* Glycosuria and osmotic diuresis result in the extreme polyuria, dehydration, and thirst typical of DM.

8. IDDM results from destruction of insulin-secreting beta cells of the pancreas and is treated with insulin. NIDDM results from insensitivity of target cells to insulin and is treated with exercise, weight-loss diets, and medications to enhance insulin sensitivity.

9. Untreated DM results in severe wasting away of the body, degenerative cardiovascular and neurological disease, gangrene, blindness, kidney failure, ketoacidosis, coma, and death.

# Testing Your Recall

1. CRH secretion would *not* raise the blood concentration of
   a. ACTH.
   b. thyroxine.
   c. cortisol.
   d. corticosterone.
   e. glucose.

2. Which of the following hormones has the least in common with the others?
   a. adrenocorticotropic hormone
   b. follicle-stimulating hormone
   c. thyrotropin
   d. thyroxine
   e. prolactin

3. Which hormone would no longer be secreted if the hypothalamo-hypophyseal tract were destroyed?
   a. oxytocin
   b. follicle-stimulating hormone
   c. growth hormone
   d. adrenocorticotropic hormone
   e. corticosterone

4. Which of the following is *not* a hormone?
   a. prolactin
   b. prolactin-inhibiting factor
   c. thyroxine-binding globulin
   d. atrial natriuretic factor
   e. cortisol

5. Where are the receptors for insulin located?
   a. in the pancreatic beta cells
   b. in the blood plasma
   c. on the target cell membrane
   d. in the target cell cytoplasm
   e. in the target cell nucleus

6. What would be the consequence of defective ADH receptors?
   a. diabetes mellitus
   b. adrenogenital syndrome
   c. dehydration
   d. seasonal affective disorder
   e. none of these

7. Which of these has more exocrine than endocrine tissue?
   a. the pineal gland
   b. the adenohypophysis
   c. the thyroid gland
   d. the pancreas
   e. the adrenal gland

8. Which of these cells stimulate bone deposition?
   a. alpha cells
   b. beta cells
   c. C cells
   d. G cells
   e. T cells

9. Which of these hormones relies on cAMP as a second messenger?
   a. ACTH
   b. progesterone
   c. thyroxine
   d. testosterone
   e. epinephrine

10. Prostaglandins are derived from
    a. proopiomelanocortin.
    b. cyclooxygenase.
    c. leukotriene.
    d. lipoxygenase.
    e. arachidonic acid.

11. The _____ develops from the hypophyseal pouch of the embryo.

12. Thyroxine ($T_4$) is synthesized by combining two molecules of the amino acid _____.

13. Growth hormone hypersecretion in adulthood causes a disease called _____.

14. The dominant hormone in the stage of resistance of the stress response is _____.

15. Adrenal steroids that regulate glucose metabolism are collectively called _____.

16. Sex steroids are secreted by the _____ cells of the ovary and _____ cells of the testis.

17. Target cells can reduce pituitary secretion by a process called _____.

18. Hypothalamic releasing factors are delivered to the anterior pituitary by way of a network of blood vessels called the _____.

19. A hormone is said to have a/an _____ effect when it stimulates the target cell to develop receptors for other hormones to follow.

20. _____ is a process in which a cell increases its number of receptors for a hormone.

*Answers in Appendix B*

# True or False

*Determine which five of the following statements are false, and briefly explain why.*

1. Castration would raise a man's blood gonadotropin concentration.

2. Hormones in the glycoprotein class cannot have cytoplasmic or nuclear receptors in their target cells.

3. Epinephrine and thyroid hormone have the same effects on metabolic rate and blood pressure.

4. Tumors can lead to either hyposecretion or hypersecretion of various hormones.

5. All hormones are secreted by endocrine glands.

6. An atherosclerotic deposit that blocked blood flow in the hypophyseal portal system would cause the testes and ovaries to malfunction.

7. The pineal gland and thymus become larger as one gets older.

8. A deficiency of dietary iodine would lead to negative feedback inhibition of the hypothalamo–pituitary–thyroid axis.

9. The tissue at the center of the adrenal gland is called the zona reticularis.

10. Of the endocrine organs covered in this chapter, only the adrenal glands are paired; the rest are single.

*Answers in Appendix B*

# Testing Your Comprehension

1. Propose a model of enzyme amplification for the effect of a steroid hormone and construct a diagram similar to figure 17.24 for your model. A review of protein synthesis in chapter 4 may help.

2. Suppose you were browsing in a health-food store and saw a product advertised: "Put an end to heart disease. This herbal medicine will totally rid your body of cholesterol!" Would you buy it? Why or why not? If the product were as effective as claimed, what are some other effects it would produce?

3. A person with toxic goiter tends to sweat profusely. Explain this in terms of homeostasis.

4. How is the action of a peptide hormone similar to the action of the neurotransmitter norepinephrine?

5. Review the effects of anabolic steroid abuse (see Insight 2.5, p. 85), and explain how some of these effects may relate to the concept of down-regulation explained in this chapter.

*Answers at www.mhhe.com/saladin4*

# www.mhhe.com/saladin4

*The textbook website provides a wealth of interactive study materials fully organized and integrated by chapter. You will find practice quizzes, labeling exercises, and much more that will complement your learning and understanding of anatomy and physiology. The website also includes tools designed to enhance your* **Anatomy & Physiology | REVEALED** *experience.*

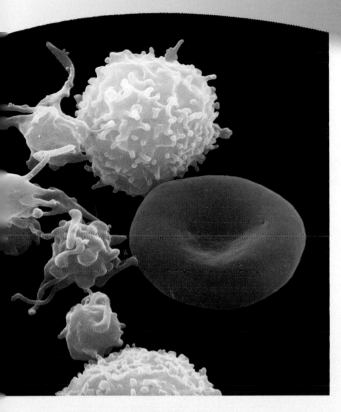

A red blood cell, white blood cells, and four platelets (SEM)

# THE CIRCULATORY SYSTEM: BLOOD

## CHAPTER OUTLINE

## INSIGHTS

## Brushing Up

To understand this chapter, it is important that you understand or brush up on the following concepts:

- Polypeptides and conjugated proteins (p. 78)
- Filtration (p. 102)
- Osmosis and osmolarity (pp. 103–105)
- Dominant and recessive alleles (p. 144)
- Sex linkage (p. 146)

Blood has always had a special mystique. From time immemorial, people have seen blood flow from the body and with it, the life of the individual. People thus presumed that blood carried a mysterious "vital force," and Roman gladiators drank it to fortify themselves for battle. Even today, we become especially alarmed when we find ourselves bleeding, and the emotional impact of blood is enough to make many people faint at the sight of it. From ancient Egypt to nineteenth-century America, physicians drained "bad blood" from their patients to treat everything from gout to headaches, from menstrual cramps to mental illness. It was long thought that hereditary traits were transmitted through the blood, and people still use such unfounded expressions as "I have one-quarter Cherokee blood."

Scarcely anything meaningful was known about blood until blood cells were seen with the first microscopes. Even though blood is a uniquely accessible tissue, most of what we know about it dates only to the last 50 years. Recent developments in **hematology**[1]—the study of blood—have empowered us to save and improve the lives of countless people who would otherwise suffer or die.

# Introduction

### Objectives

When you have completed this section, you should be able to

- describe the functions and major components of the circulatory system;
- describe the components and physical properties of blood;
- describe the composition of blood plasma;
- explain the significance of blood viscosity and osmolarity; and
- describe in general terms how blood is produced.

## FUNCTIONS OF THE CIRCULATORY SYSTEM

The **circulatory system** consists of the heart, blood vessels, and blood. The term **cardiovascular system**[2] refers only to the heart and blood vessels, which are the subject of chapters 19 and 20.

The fundamental purpose of the circulatory system is to transport substances from place to place in the blood. Blood is the liquid medium in which these materials travel, blood vessels ensure the proper routing of blood to its destinations, and the heart is the pump that keeps the blood flowing.

More specifically, the functions of the circulatory system are as follows:

### TRANSPORT
- The blood carries oxygen from the lungs to all of the body's tissues, while it picks up carbon dioxide from those tissues and carries it to the lungs to be removed from the body.
- It picks up nutrients from the digestive tract and delivers them to all of the body's tissues.
- It carries other metabolic wastes to the kidneys for removal.
- It carries hormones from endocrine cells to their target cells.
- It transports a variety of stem cells from the bone marrow and other origins to the tissues where they lodge and mature.
- It helps to regulate body temperature by carrying heat to the body surface for removal.

### PROTECTION
- The blood plays several roles in inflammation, a mechanism for limiting the spread of infection.
- White blood cells destroy microorganisms and cancer cells.
- Antibodies and other blood proteins neutralize toxins and help to destroy pathogens.
- Platelets secrete factors that initiate blood clotting and other processes for minimizing blood loss.

### REGULATION
- By absorbing or giving off fluid under different conditions, the blood capillaries help to stabilize fluid distribution in the body.
- By buffering acids and bases, blood proteins help to stabilize the pH of the extracellular fluids.

Considering the importance of efficiently transporting nutrients, wastes, hormones, and especially oxygen from place to place, it is easy to understand why an excessive loss of blood is quickly fatal, and why the circulatory system needs mechanisms for minimizing such losses.

## COMPONENTS AND GENERAL PROPERTIES OF BLOOD

All of the foregoing functions depend, of course, on the characteristics of the blood. Most adults have 4 to 6 L of blood. It is a liquid connective tissue with two main

---

[1]*hem, hemato* = blood + *logy* = study of
[2]*cardio* = heart + *vas* = vessel

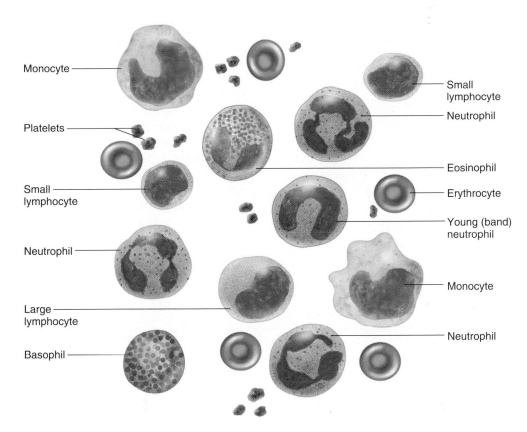

**FIGURE 18.1** **The Formed Elements of Blood.**

▶ *What do erythrocytes and platelets lack that the other formed elements have?*

components: the plasma and formed elements. **Plasma** is a clear extracellular matrix. It is no longer present on prepared slides of blood and would not be visible even if it were. **Formed elements,** by contrast, have a visible structure. They are cells and cell fragments: red blood cells, white blood cells, and platelets (fig. 18.1).

The formed elements are classified as follows:

Erythrocytes[3] (red blood cells, RBCs)

Platelets

Leukocytes[4] (white blood cells, WBCs)

  Granulocytes

    Neutrophils

    Eosinophils

    Basophils

  Agranulocytes

    Lymphocytes

    Monocytes

Thus, there are seven kinds of formed elements: the erythrocytes, platelets, and five kinds of leukocytes. The

five leukocyte types are divided into two categories, the *granulocytes* and *agranulocytes,* on grounds explained later.

The ratio of formed elements to plasma can be seen by taking a sample of blood in a tube and spinning it for a few minutes in a centrifuge (fig. 18.2). Erythrocytes are the densest elements, settle to the bottom of the tube, and typically constitute about 45% of the total volume. This value is called the *hematocrit* or *packed cell volume.* WBCs and platelets make up a narrow cream- or buff-colored zone called the *buffy coat* just above the RBCs; they total 1% or less of the blood volume. At the top of the tube is the plasma, which has a pale yellow color and accounts for nearly 55% of the total volume. Table 18.1 lists several properties of the blood.

**Think About It**

*Based on your body weight, estimate the volume (in liters) and weight (in kilograms) of your own blood, using the data in table 18.1.*

## BLOOD PLASMA

Even though blood plasma has no anatomy that we can study visually, we cannot ignore its importance as the matrix of this liquid connective tissue we call blood. Plasma is a complex mixture of water, proteins, nutrients,

[3]*erythro* = red + *cyte* = cell
[4]*leuko* = white + *cyte* = cell

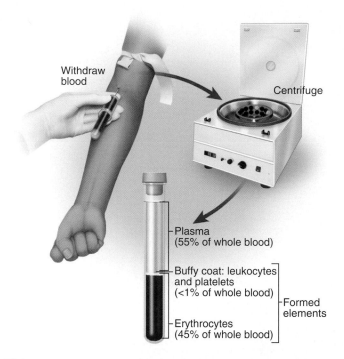

**FIGURE 18.2   The Hematocrit.** A sample of blood is spun in a centrifuge to separate the cells from the plasma. The percent volume of red cells (hematocrit) is then measured. In this example, the hematocrit is 45%.

| TABLE 18.1 | General Properties of Blood* |
|---|---|
| **Characteristic** | **Average Value for Healthy Adult** |
| Mean fraction of body weight | 8% |
| Volume in adult body | Female: 4-5 L; male: 5-6 L |
| Volume/body weight | 80-85 mL/kg |
| Mean temperature | 38°C (100.4°F) |
| pH | 7.35-7.45 |
| Viscosity (relative to water) | Whole blood: 4.5-5.5; plasma: 2.0 |
| Osmolarity | 280-296 mOsm/L |
| Mean salinity (mainly NaCl) | 0.9% |
| Hematocrit (packed cell volume) | Female: 37-48%<br>    Male: 45-52% |
| Hemoglobin | Female: 12-16 g/dL<br>    Male: 13-18 g/dL |
| Mean RBC count | Female: 4.2-5.4 million/μL<br>    Male: 4.6-6.2 million/μL |
| Platelet count | 130,000-360,000/μL |
| Total WBC count | 5,000-10,000/μL |

*Values vary slightly depending on the testing methods used.

electrolytes, nitrogenous wastes, hormones, and gases (table 18.2). When the blood clots and the solids are removed, the remaining fluid is the blood **serum.** Serum is essentially identical to plasma except for the absence of the clotting protein fibrinogen.

Protein is the most abundant plasma solute by weight, totaling 6 to 9 g/dL. Plasma proteins play a variety of roles including clotting, defense, and transport of other solutes such as iron, copper, lipids, and hydrophobic hormones. There are three major categories of proteins: the albumins, globulins, and fibrinogen (table 18.3). Many other plasma proteins are indispensable to survival, but they account for less than 1% of the total.

**Albumin** is the smallest and most abundant plasma protein. It serves to transport various plasma solutes and buffer the pH of the blood plasma. It also makes a major contribution to two physical properties of blood: its *viscosity* and *osmolarity,* discussed shortly. Through its effects on these two variables, changes in albumin concentration can significantly affect blood volume, pressure, and flow. **Globulins** are divided into three subclasses; from smallest to largest in molecular weight, they are the alpha (α), beta (β), and gamma (γ) globulins. Globulins play various roles in solute transport, clotting, and immunity. **Fibrinogen** is a soluble precursor of *fibrin,* a sticky protein that forms the framework of a blood clot. Some of the other plasma proteins are enzymes involved in the clotting process.

The liver produces as much as 4 g of plasma protein per hour, contributing all of the major proteins except gamma globulins. The gamma globulins come from *plasma cells*—connective tissue cells that are descended from white blood cells called *B lymphocytes.*

**Think About It**

*How could a disease such as liver cancer or hepatitis result in impaired blood clotting?*

In addition to protein, the blood plasma contains such nitrogen-containing compounds as free amino acids and nitrogenous wastes. **Nitrogenous wastes** are toxic end products of catabolism. The most abundant is *urea,* a product of amino acid catabolism. These wastes are normally excreted by the kidneys at a rate that balances their production.

The plasma also transports nutrients absorbed by the digestive tract, including glucose (blood sugar), amino acids, fats, cholesterol, phospholipids, vitamins, and minerals. It transports some of the oxygen and carbon dioxide carried by the blood and carries a substantial amount of dissolved nitrogen. Free nitrogen normally has no physiological role in the body, but it becomes important under circumstances such as scuba diving and aviation.

Electrolytes are another important component of the blood plasma. Sodium ions constitute about 90% of the plasma cations. Sodium is more important than any other solute for the osmolarity of the blood. As such, it has a major influence on blood volume and pressure; people with high blood pressure are often advised to limit their sodium intake. Electrolyte concentrations are carefully regulated by the body and have rather stable concentrations in the plasma.

| TABLE 18.2 | Composition of Blood Plasma* |
|---|---|
| **Blood Component** | **Average Value for Healthy Adults** |
| *Water* | 92% by weight |
| *Proteins* | Total 6-9 g/dL |
| Albumins | 60% of total protein, 3.2-5.5 g/dL |
| Globulins | 36% of total protein, 2.3-3.5 g/dLFibrinogen |
| | 4% of total protein, 0.2-0.3 g/dL |
| *Nutrients* | |
| Glucose (dextrose) | 70-110 mg/dL |
| Amino acids | 33-51 mg/dL |
| Lactic acid | 6-16 mg/dL |
| Total lipid | 450-850 mg/dL |
| Cholesterol | 120-220 mg/dL |
| Fatty acids | 190-420 mg/dL |
| High-density lipoprotein (HDL) | 30-80 mg/dL |
| Low-density lipoprotein (LDL) | 62-185 mg/dL |
| Triglycerides (neutral fats) | 40-150 mg/dL |
| Phospholipids | 6-12 mg/dL |
| Iron | 50-150 μg/dl |
| Trace elements | Traces |
| Vitamins | Traces |
| *Electrolytes* | |
| Sodium ($Na^+$) | 135-145 mEq/L |
| Calcium ($Ca^{2+}$) | 9.2-10.4 mEq/L |
| Potassium ($K^+$) | 3.5-5.0 mEq/L |
| Magnesium ($Mg^{2+}$) | 1.3-2.1 mEq/L |
| Chloride ($Cl^-$) | 100-106 mEq/L |
| Bicarbonate ($HCO_3^-$) | 23.1-26.7 mEq/L |
| Phosphate ($HPO_4^{2-}$) | 1.4-2.7 mEq/L |
| Sulfate ($SO_4^{2-}$) | 0.6-1.2 mEq/L |
| *Nitrogenous Wastes* | |
| Urea | 10-20 mg/dL |
| Uric acid | 1.5-8.0 mg/dL |
| Creatinine | 0.6-1.5 mg/dL |
| Creatine | 0.2-0.8 mg/dL |
| Ammonia | 0.02-0.09 mg/dL |
| Bilirubin | 0-1.0 mg/dL |
| *Other Components* | |
| Respiratory gases ($O_2$, $CO_2$, $N_2$) | — |
| Enzymes of diagnostic value | — |
| Hormones | — |

*This table is limited to substances of greatest relevance to this and later chapters. Concentrations refer to plasma only, not to whole blood.

| TABLE 18.3 | Major Proteins of the Blood Plasma |
|---|---|
| **Proteins** | **Functions** |
| *Albumins (60%)** | Responsible for colloid osmotic pressure; major contributor to blood viscosity; transport lipids, hormones, calcium, and other solutes; buffer blood pH |
| *Globulins (36%)** | |
| *Alpha (α) Globulins* | |
| Haptoglobulin | Transports hemoglobin released by dead erythrocytes |
| Ceruloplasmin | Transports copper |
| Prothrombin | Promotes blood clotting |
| Others | Transport lipids, fat-soluble vitamins, and hormones |
| *Beta (β) Globulins* | |
| Transferrin | Transports iron |
| Complement proteins | Aid in destruction of toxins and microorganisms |
| Others | Transport lipids |
| *Gamma (γ) Globulins* | Antibodies; combat pathogens |
| *Fibrinogen (4%)** | Becomes fibrin, the major component of blood clots |

*Mean percentage of the total plasma protein by weight.

## BLOOD VISCOSITY AND OSMOLARITY

Two important properties of blood—viscosity and osmolarity—arise from the formed elements and plasma composition. **Viscosity** is the resistance of a fluid to flow, resulting from the cohesion of its particles. Loosely speaking, it is the thickness or stickiness of a fluid. At a given temperature, mineral oil is more viscous than water, for example, and honey is more viscous than mineral oil. Whole blood is 4.5 to 5.5 times as viscous as water. This is due mainly to the RBCs; plasma alone is 2.0 times as viscous as water, mainly because of its protein. Viscosity is important in circulatory function because it partially governs the flow of blood through the vessels. An RBC or protein deficiency reduces viscosity and causes blood to flow too easily, whereas an excess causes blood to flow too sluggishly. Either of these conditions puts a strain on the heart that may lead to serious cardiovascular problems if not corrected.

The **osmolarity** of blood (total molarity of those dissolved particles that cannot pass through the blood vessel wall) is another important factor in cardiovascular function. In order to nourish surrounding cells and remove their wastes, substances must pass between the bloodstream and tissue fluid through the capillary walls. This transfer of fluids depends on a balance between the filtration of fluid from the capillary and its reabsorption by osmosis (see fig. 3.15). The rate of reabsorption is governed by the relative osmolarity of the blood versus the tissue fluid. If the osmolarity of the blood is too high, the bloodstream absorbs too much water. This raises the blood volume, resulting in elevated blood pressure and a potentially dangerous strain on the heart and arteries. If its osmolarity drops too low, too much water remains in the tissues. They become edematous (swollen) and the blood pressure may drop to dangerously low levels because of the amount of water lost from the bloodstream.

It is therefore important that the blood maintain an optimal osmolarity. The osmolarity of the blood is a product mainly of its sodium ions, protein, and erythrocytes. The contribution of protein to blood osmotic pressure—called the **colloid osmotic pressure (COP)**—is especially important, as we see from the effects of extremely low-protein diets (see Insight 18.1).

## HOW BLOOD IS PRODUCED

We lose blood continually, not only from bleeding but also as blood cells grow old and die, and plasma components are consumed or excreted from the body. Therefore, we must continually replace it. An adult typically produces 400 billion platelets, 200 billion RBCs, and 10 billion WBCs every day. The production of blood, especially its formed elements, is called **hemopoiesis**[5] (HE-mo-poy-EE-sis). A knowledge of this process provides an indispensible foundation for understanding leukemia, anemia, and other blood disorders.

The tissues that produce blood cells are called **hemopoietic tissues.** The first hemopoietic tissues of the human embryo form in the *yolk sac,* a membrane associated with all vertebrate embryos. In most vertebrates (fish, amphibians, reptiles, and birds), this sac encloses the egg yolk, transfers its nutrients to the growing embryo, and produces the forerunners of the first blood cells. Even animals that don't lay eggs, however, have a yolk sac that retains its hemopoietic function. (It is also the source of cells that later produce eggs and sperm.) Cell clusters called *blood islands* form here by the third week of human development. They produce primitive *stem cells* that migrate into the embryo proper and colonize the bone marrow, liver, spleen, and thymus. Here, the stem cells multiply and give rise to blood cells throughout fetal development. The liver stops producing blood cells around the time of birth. The spleen stops producing RBCs soon after, but it continues to produce lymphocytes for life.

From infancy onward, the red bone marrow produces all seven kinds of formed elements, while lymphocytes are produced not only there but also in the lymphatic tissues and organs—especially the thymus, tonsils, lymph nodes, spleen, and patches of lymphatic tissue in the mucous membranes. Blood formation in the bone marrow and lymphatic organs is called, respectively, **myeloid**[6] and **lymphoid hemopoiesis.**

[5]*hemo* = blood + *poiesis* = formation
[6]*myel* = bone marrow

## INSIGHT 18.1    Clinical Application

### Starvation and Plasma Protein Deficiency

Several conditions can lead to *hypoproteinemia,* a deficiency of plasma protein: extreme starvation or dietary protein deficiency, liver diseases that interfere with protein synthesis, kidney diseases that result in protein loss through the urine, and severe burns that result in protein loss through the body surface. As the protein content of the blood plasma drops, so does its osmolarity. The bloodstream loses more fluid to the tissues than it reabsorbs by osmosis. Thus, the tissues become edematous and a pool of fluid may accumulate in the abdominal cavity—a condition called *ascites* (ah-SY-teez).

Children who suffer severe dietary protein deficiencies often exhibit a condition called *kwashiorkor* (KWASH-ee-OR-cor) (fig. 18.3). The arms and legs are emaciated for lack of muscle, the skin is shiny and tight with edema, and the abdomen is swollen by ascites. *Kwashiorkor* is an African word for a "deposed" or "displaced" child who is no longer breast-fed. Symptoms appear when a child is weaned and placed on a diet consisting mainly of rice or other cereals. Children with kwashiorkor often die of diarrhea and dehydration.

**FIGURE 18.3   Children of Angola with Kwashiorkor.**   Note the thin limbs and fluid-distended abdomens.

All formed elements trace their origins to a common type of bone marrow stem cell, the **pluripotent**[7] **stem cell (PPSC)** (formerly called a hemocytoblast[8]). PPSCs are so-named because they have the potential to develop into multiple mature cell types. They multiply at a relatively slow rate and thus maintain a small population in the bone marrow. Some of them go on to differentiate into a variety of more specialized cells called **colony-forming units (CFUs),** each type destined to produce one or another class of formed elements. The specific processes leading from a PPSC to RBCs, WBCs, and platelets are described at later points in this chapter.

[7]*pluri* = many, multiple + *potent* = able
[8]*hemo* = blood + *cyto* = cell + *blast* = precursor

Blood plasma also requires continual replacement. It is composed mainly of water, which it obtains primarily by absorption from the digestive tract. Its electrolytes and organic nutrients are also acquired there, and its gamma globulins come from connective tissue plasma cells and its other proteins from the liver.

## Before You Go On

*Answer the following questions to test your understanding of the preceding section:*

1. *Identify at least two each of the transport, protective, and regulatory functions of the circulatory system.*

2. *What are the two principal components of the blood?*

3. *List the three major classes of plasma proteins. Which one is absent from blood serum?*

4. *Define the viscosity and osmolarity of blood. Explain why each of these is important for human survival.*

5. *What does hemopoiesis mean? After birth, what one cell type is the starting point for all hemopoiesis?*

# Erythrocytes

### Objectives

When you have completed this section, you should be able to

- discuss the structure and function of erythrocytes (RBCs);

- describe the structure and function of hemoglobin;

- state and define some clinical measurements of RBC and hemoglobin quantities;

- describe the life cycle of erythrocytes; and

- name and describe the types, causes, and effects of RBC excesses and deficiencies.

**Erythrocytes,** or **red blood cells (RBCs),** have two principal functions: (1) to pick up oxygen from the lungs and deliver it to tissues elsewhere, and (2) to pick up carbon dioxide from the tissues and unload it in the lungs. RBCs are the most abundant formed elements of the blood and therefore the most obvious things one sees upon its microscopic examination. They are also the most critical to survival; although a severe deficiency of leukocytes or platelets can be fatal within a few days, a severe deficiency of erythrocytes can be fatal within a few minutes. It is the deficiency of life-giving oxygen, carried by erythrocytes, that leads rapidly to death in cases of major trauma or hemorrhage.

## FORM AND FUNCTION

An erythrocyte is a discoid cell with a thick rim and a thin sunken center. It is about 7.5 μm in diameter and 2.0 μm thick at the rim (fig. 18.4). Although most cells,

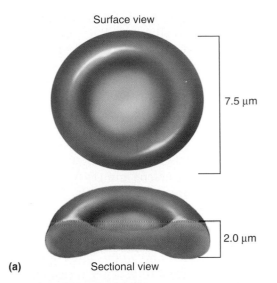

Surface view

7.5 μm

2.0 μm

**(a)** Sectional view

**(b)**

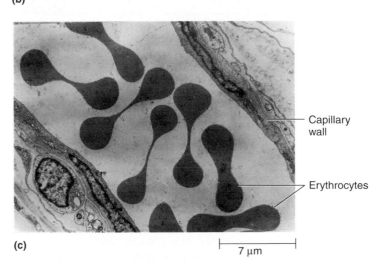

Capillary wall

Erythrocytes

**(c)** 7 μm

**FIGURE 18.4 The Structure of Erythrocytes.**
(a) Dimensions and shape of an erythrocyte. (b) Erythrocytes on the tip of a hypodermic needle (SEM). (c) Erythrocytes in a blood capillary (TEM). Note the absence of organelles or other internal features in the cells.

including white blood cells, have an abundance of organelles, RBCs lose nearly all organelles during their development and are thus remarkably devoid of internal structure. When viewed with the transmission electron microscope, the interior of an RBC appears uniformly gray. Lacking mitochondria, RBCs rely exclusively on anaerobic fermentation to produce ATP. The lack of aerobic respiration prevents them from consuming the oxygen that they must transport to other tissues. Erythrocytes are the only cells in the body that carry on anaerobic fermentation indefinitely. Lacking a nucleus and DNA, RBCs also are incapable of protein synthesis and mitosis.

The plasma membrane of a mature RBC has glycoproteins and glycolipids on the outer surface that determine a person's blood type. On its inner surface are two cytoskeletal proteins, *spectrin* and *actin,* that give the membrane resilience and durability. This is especially important when RBCs pass through small blood capillaries and sinusoids. Many of these passages are narrower than the diameter of an RBC, forcing the RBCs to stretch, bend, and fold as they squeeze through. When they enter larger vessels, RBCs spring back to their discoid shape.

The cytoplasm of an RBC consists mainly of a 33% solution of hemoglobin (about 280 million molecules per cell). This is the red pigment that gives an RBC its color and name. It is known especially for its oxygen-transport function, but it also aids in the transport of carbon dioxide and the buffering of blood pH. Although the lack of a nucleus makes an RBC unable to repair itself, it has an overriding advantage: The biconcave shape gives the cell a much greater ratio of surface area to volume, which enables $O_2$ and $CO_2$ to diffuse quickly to and from the hemoglobin.

The cytoplasm also contains an enzyme, *carbonic anhydrase (CAH)*, that catalyzes the reaction $CO_2 + H_2O \leftrightarrows H_2CO_3$. The role of CAH in gas transport and pH balance is discussed in chapters 22 and 24.

## HEMOGLOBIN

Hemoglobin consists of four protein chains called **globins** (fig. 18.5). Two of these, the *alpha* (α) *chains,* are 141 amino acids long, and the other two, the *beta* (β) *chains,* are 146 amino acids long. Each chain is conjugated with a nonprotein moiety called the **heme** group, which binds oxygen to a ferrous ion ($Fe^{2+}$) at its center. Each heme can carry one molecule of $O_2$; thus, the hemoglobin molecule as a whole can transport up to 4 $O_2$. About 5% of the $CO_2$ in the bloodstream is also transported by hemoglobin but is bound to the globin moiety rather than to the heme. Gas transport by hemoglobin is discussed in detail in chapter 22.

Hemoglobin exists in several forms with slight differences in the globin chains. The form we have just described is called *adult hemoglobin (HbA)*. About 2.5% of an adult's hemoglobin, however, is of a form called HbA$_2$, which has two *delta* (δ) *chains* in place of the beta chains. The fetus produces a form called *fetal hemoglobin (HbF),*

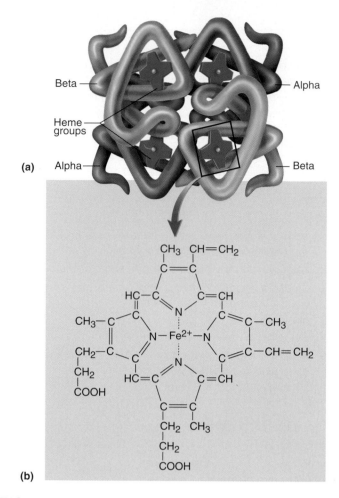

**(a)**

**(b)**

**FIGURE 18.5    The Structure of Adult Hemoglobin (HbA).**
(a) The hemoglobin molecule consists of two alpha proteins and two beta proteins, each conjugated to a nonprotein heme group. (b) Structure of the heme group. Oxygen binds to $Fe^{2+}$ at the center of the heme.

▶ *In what way does this exemplify a quaternary protein structure? What is the prosthetic group of hemoglobin?*

which has two *gamma* (γ) *chains* in place of the beta chains. The delta and gamma chains are the same length as the beta chains but differ in amino acid sequence. HbF binds oxygen more tightly than HbA does; thus it enables the fetus to extract oxygen from the mother's bloodstream.

## QUANTITIES OF ERYTHROCYTES AND HEMOGLOBIN

The RBC count and hemoglobin concentration are important clinical data because they determine the amount of oxygen the blood can carry. Three of the most common measurements are hematocrit, hemoglobin concentration, and RBC count. The **hematocrit**[9] (packed cell volume, PCV) is the percentage of whole blood volume composed of RBCs (see fig. 18.2). In men, it normally ranges between 42% and 52%; in women, between 37% and 48%. The **hemoglobin concentration** of whole blood is normally 13

---

[9]*hemato* = blood + *crit* = to separate

## INSIGHT 18.2   Evolutionary Medicine

### The Packaging of Hemoglobin

The gas-transport pigments of earthworms, snails, and many other animals are dissolved in the plasma rather than contained in blood cells. You might wonder why human hemoglobin must be contained in RBCs. The main reason is osmotic. Remember that the osmolarity of blood depends on the number of particles in solution. A "particle," for this purpose, can be a sodium ion, an albumin molecule, or a whole cell. If all the hemoglobin contained in the RBCs were free in the plasma, it would drastically increase blood osmolarity, since each RBC contains about 280 million molecules of hemoglobin. The circulatory system would become enormously congested with fluid, and circulation would be severely impaired. The blood simply could not contain that much free hemoglobin and support life. On the other hand, if it contained a safe level of free hemoglobin, it could not transport enough oxygen to support the high metabolic demand of the human body. By having our hemoglobin packaged in RBCs, we are able to have much more of it and hence to have more efficient gas transport and more active metabolism.

Another reason for packaging the hemoglobin in RBCs is that some of the body's capillaries (*fenestrated capillaries* found in the kidneys and endocrine glands, for example) are permeable to proteins. Hemoglobin would leak out into the tissues if it were not contained in cells too big to pass through the capillary wall. Hemoglobin liberated by the rapid breakdown of RBCs can plug the kidney tubules and cause renal failure.

to humans. From the evolutionary standpoint, the adaptive value of these differences may lie in the fact that male animals fight more than females and suffer more injuries. The traits described here may serve to minimize or compensate for their blood loss.

**Think About It**

*Explain why the hemoglobin concentration could appear deceptively high in a patient who is dehydrated.*

## THE ERYTHROCYTE LIFE CYCLE

An erythrocyte lives for an average of 120 days from the time it is produced in the red bone marrow until it dies, breaks up, and its fragments are phagocytized by cells in the spleen and elsewhere. In a state of balance and stable RBC count, the birth and death of RBCs amounts to about 2.5 million cells per second, or a packed cell volume of 20 mL of RBCs per day.

### Erythrocyte Production

Erythrocyte production is called **erythropoiesis** (eh-RITH-ro-poy-EE-sis). The process normally takes 3 to 5 days and involves four major developments: a reduction in cell size, an increase in cell number, the synthesis of hemoglobin, and the loss of the nucleus and most other organelles. It begins when a pluripotent stem cell (PPSC) becomes an *erythrocyte colony-forming unit (ECFU)* (fig. 18.6), which has receptors for the hormone **erythropoietin (EPO)**. EPO stimulates the ECFU to transform into an *erythroblast.* Erythroblasts multiply and synthesize hemoglobin. When this task is completed, the nucleus shrivels and is discharged from the cell. The cell is now called a *reticulocyte,* named for a temporary network (reticulum) composed of ribosome clusters (polyribosomes).

Reticulocytes leave the bone marrow and enter the circulating blood. In a day or two, the last of the polyribosomes disintegrate and disappear, and the cell is a mature erythrocyte. Normally, about 0.5% to 1.5% of the circulating RBCs are reticulocytes, but this percentage rises under certain

to 18 g/dL in men and 12 to 16 g/dL in women. The RBC count is normally 4.6 to 6.2 million RBCs/μL in men and 4.2 to 5.4 million/μL in women. This is often expressed as cells per cubic millimeter (mm³); 1 μL = 1 mm³.

Notice that these values tend to be lower in women than in men. There are three physiological reasons for this: (1) androgens stimulate RBC production, and men have higher androgen levels than women; (2) women of reproductive age have periodic menstrual losses; and (3) the hematocrit is inversely proportional to percent body fat, which is higher in women than in men. In men, the blood also clots faster and the skin has fewer blood vessels than in women. Such differences are not limited

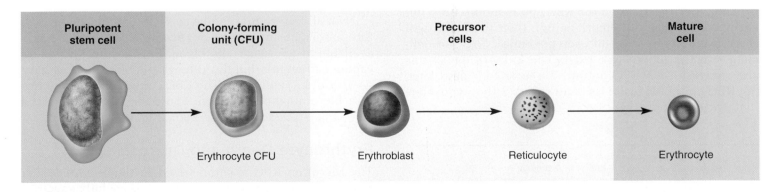

| Pluripotent stem cell | Colony-forming unit (CFU) | Precursor cells | | Mature cell |
|---|---|---|---|---|
| | Erythrocyte CFU | Erythroblast | Reticulocyte | Erythrocyte |

**FIGURE 18.6**   **Erythropoiesis.**   Stages in the development of a red blood cell.

circumstances. Blood loss, for example, stimulates accelerated erythropoiesis and leads to an increasing number of reticulocytes in circulation—as if the bone marrow were in such a hurry to replenish the lost RBCs that it releases many developing RBCs into circulation a little early.

## Iron Metabolism

Iron is a critical part of the hemoglobin molecule and therefore one of the key nutritional requirements for erythropoiesis. Men lose about 0.9 mg of iron per day through the urine, feces, and bleeding, and women of reproductive age lose an average of 1.7 mg/day because of the added factor of menstruation. Since we absorb only a fraction of the iron in our food, we must consume 5 to 20 mg/day to replace our losses. Pregnant women need 20 to 48 mg/day, especially in the last 3 months, to meet not only their own need but also that of the fetus.

Dietary iron exists in two forms: ferric ($Fe^{3+}$) and ferrous ($Fe^{2+}$) ions. Stomach acid converts most $Fe^{3+}$ to $Fe^{2+}$, the only form that can be absorbed by the small intestine (fig. 18.7). A protein called **gastroferritin**,[10] produced by the stomach, then binds $Fe^{2+}$ and transports it to the small intestine. Here, it is absorbed into the blood, binds to a plasma protein called **transferrin,** and travels to the bone marrow, liver, and other tissues. Bone marrow uses $Fe^{2+}$ for hemoglobin synthesis; muscle uses it to make the oxygen-storage protein myoglobin; and nearly all cells use iron to make electron-transport molecules called cytochromes in their mitochondria. The liver binds surplus iron to a protein called **apoferritin**,[11] forming an iron-storage complex called **ferritin.** It releases $Fe^{2+}$ into circulation when needed.

Some other nutritional requirements for erythropoiesis are vitamin $B_{12}$ and folic acid, required for the rapid cell division and DNA synthesis that occurs in erythropoiesis, and vitamin C and copper, which are cofactors for some of the enzymes that synthesize hemoglobin. Copper is transported in the blood by an alpha globulin called *ceruloplasmin*.[12]

## Erythrocyte Homeostasis

The RBC count is maintained in a classic negative feedback manner (fig. 18.8). If the RBC count should drop (for example, because of hemorrhaging), then the blood will carry less oxygen—a state of **hypoxemia**[13] (oxygen deficiency in the blood) will exist. The kidneys detect this and increase their EPO output. Three or four days later, the RBC count begins to rise and reverses the hypoxemia that started the process.

---

[10]*gastro* = stomach + *ferrit* = iron + *in* = protein
[11]*apo* = separated from + *ferrit* = iron + *in* = protein
[12]*cerulo* = blue-green, the color of oxidized copper + *plasm* = blood plasma + *in* = protein
[13]*hyp* = below normal + *ox* = oxygen + *emia* = blood condition

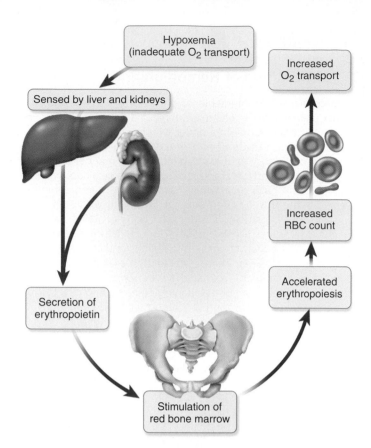

**FIGURE 18.7**   The Negative Feedback Correction of Hypoxemia by Erythropoiesis.

Hypoxemia has many causes other than blood loss. Another cause is a low level of oxygen in the atmosphere. If you were to move from Miami to Denver, for example, the lower $O_2$ level at the high altitude of Denver would produce temporary hypoxemia and stimulate EPO secretion and erythropoiesis. The blood of an average adult has about 5 million RBCs/$\mu$L, but people who live at high altitudes may have counts of 7 to 8 million RBCs/$\mu$L. Another cause of hypoxemia is an abrupt increase in the body's oxygen consumption. If a lethargic person suddenly takes up tennis or aerobics, for example, the muscles consume oxygen more rapidly and create a state of hypoxemia that stimulates erythropoiesis. Endurance-trained athletes commonly have RBC counts as high as 6.5 million RBCs/$\mu$L.

Not all hypoxemia can be corrected by increasing erythropoiesis. In emphysema, for example, less lung tissue is available to oxygenate the blood. Raising the RBC count cannot correct this, but the kidneys and bone marrow have no way of knowing this. The RBC count continues to rise in a futile attempt to restore homeostasis, resulting in a dangerous excess called *polycythemia,* discussed shortly.

## Erythrocyte Death and Disposal

The life of an RBC is summarized in figure 18.9. As an RBC ages and its membrane proteins (especially spectrin) deteriorate, the membrane grows increasingly fragile.

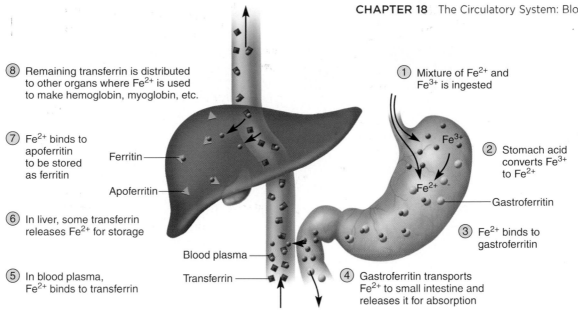

⑧ Remaining transferrin is distributed to other organs where $Fe^{2+}$ is used to make hemoglobin, myoglobin, etc.

⑦ $Fe^{2+}$ binds to apoferritin to be stored as ferritin

Ferritin

Apoferritin

⑥ In liver, some transferrin releases $Fe^{2+}$ for storage

Blood plasma

⑤ In blood plasma, $Fe^{2+}$ binds to transferrin

Transferrin

① Mixture of $Fe^{2+}$ and $Fe^{3+}$ is ingested

$Fe^{3+}$

② Stomach acid converts $Fe^{3+}$ to $Fe^{2+}$

$Fe^{2+}$

Gastroferritin

③ $Fe^{2+}$ binds to gastroferritin

④ Gastroferritin transports $Fe^{2+}$ to small intestine and releases it for absorption

**FIGURE 18.8** Iron Metabolism.

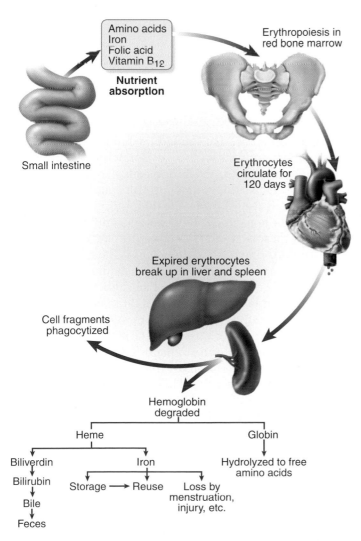

Amino acids
Iron
Folic acid
Vitamin $B_{12}$

**Nutrient absorption**

Erythropoiesis in red bone marrow

Small intestine

Erythrocytes circulate for 120 days

Expired erythrocytes break up in liver and spleen

Cell fragments phagocytized

Hemoglobin degraded

Heme — Globin

Biliverdin — Iron — Hydrolyzed to free amino acids

Bilirubin — Storage → Reuse — Loss by menstruation, injury, etc.

Bile

Feces

**FIGURE 18.9** **The Life and Death of Erythrocytes.** Note especially the stages of hemoglobin breakdown and disposal.

Without a nucleus or ribosomes, an RBC cannot synthesize new spectrin. Many RBCs die in the spleen, which has been called the "erythrocyte graveyard." The spleen has channels as narrow as 3 μm that severely test the ability of old, fragile RBCs to squeeze through the organ. Old cells become trapped, broken up, and destroyed. An enlarged and tender spleen may indicate diseases in which RBCs are rapidly breaking down.

Table 18.4 outlines the process of disposing of old erythrocytes and hemoglobin. **Hemolysis**[14] (he-MOLL-ih-sis), the rupture of RBCs, releases hemoglobin and leaves empty plasma membranes. The membrane fragments are easily digested by macrophages in the liver and spleen, but hemoglobin disposal is a bit more complicated. It must be disposed of efficiently, however, or it can block kidney tubules and cause renal failure. Macrophages begin the disposal process by separating the heme from the globin. They hydrolyze the globin into free amino acids, which become part of the body's general pool of amino acids available for protein synthesis or energy-releasing catabolism.

Disposing of the heme is another matter. First, the macrophage removes the iron and releases it into the blood, where it combines with transferrin and is used or stored in the same way as dietary iron. The macrophage converts the rest of the heme into a greenish pigment called **biliverdin**[15] (BIL-ih-VUR-din), then further converts most of this to a yellow-green pigment called **bilirubin.**[16] Bilirubin is released by the macrophages and binds to albumin in the blood plasma. The liver removes bilirubin from the albumin and secretes it into the bile, to

---

[14]*hemo* = blood + *lysis* = splitting, breakdown
[15]*bili* = bile + *verd* = green + *in* = substance
[16]*bili* = bile + *rub* = red + *in* = substance

| TABLE 18.4 | The Fate of Expired Erythrocytes and Hemoglobin |
|---|---|

1. RBCs lose elasticity with age
2. RBCs break down while squeezing through blood capillaries and sinusoids
3. Cell fragments are phagocytized by macrophages in the spleen and liver
4. Hemoglobin decomposes into:

   Globin portion—hydrolyzed to amino acids, which can be reused

   Heme portion—further decomposed into:

   Iron

   1. Transported by transferrin to bone marrow and liver
   2. Some used in bone marrow to make new hemoglobin
   3. Excess stored in liver as ferritin

   Biliverdin

   1. Converted to bilirubin and bound to albumin
   2. Removed by liver and secreted in bile
   3. Stored and concentrated in gallbladder
   4. Discharged into small intestine
   5. Converted by intestinal bacteria to urobilinogen
   6. Excreted in feces

which it imparts a dark green color as the bile becomes concentrated in the gallbladder. Biliverdin and bilirubin are collectively known as **bile pigments.** The gallbladder discharges the bile into the small intestine, where bacteria convert bilirubin to *urobilinogen,* responsible for the brown color of the feces. Another hemoglobin breakdown pigment, *urochrome,* produces the yellow color of urine. A high level of bilirubin in the blood causes *jaundice,* a yellowish cast in light-colored skin and the whites of eyes. Jaundice may be a sign of rapid hemolysis or a liver disease or bile duct obstruction that interferes with bilirubin disposal.

## ERYTHROCYTE DISORDERS

Any imbalance between the rates of erythropoiesis and RBC destruction may produce an excess or deficiency of red cells. An RBC excess is called *polycythemia*[17] (POL-ee-sy-THEE-me-uh), and a deficiency of either RBCs or hemoglobin is called *anemia.*[18]

### Polycythemia

**Primary polycythemia** (polycythemia vera) is due to cancer of the erythropoietic line of the red bone marrow. It can result in an RBC count as high as 11 million RBCs/μL

and a hematocrit as high as 80%. Polycythemia from all other causes, called **secondary polycythemia,** is characterized by RBC counts as high as 6 to 8 million RBCs/μL. It can result from dehydration because water is lost from the bloodstream while erythrocytes remain and become abnormally concentrated. More often, it is caused by smoking, air pollution, emphysema, high altitude, strenuous physical conditioning, or other factors that create a state of hypoxemia and stimulate erythropoietin secretion.

The principal dangers of polycythemia are increased blood volume, pressure, and viscosity. Blood volume can double in primary polycythemia and cause the circulatory system to become tremendously engorged. Blood viscosity may rise to three times normal. Circulation is poor, the capillaries are clogged with viscous blood, and the heart is dangerously strained. Chronic (long-term) polycythemia can lead to embolism, stroke, or heart failure. The deadly consequences of emphysema and some other lung diseases are due in part to polycythemia.

### Anemia

The causes of **anemia** fall into three categories: (1) inadequate erythropoiesis or hemoglobin synthesis, (2) **hemorrhagic anemia** from bleeding, and (3) **hemolytic anemia** from RBC destruction. Table 18.5 gives specific examples and causes for each category. We give special attention to the deficiencies of erythropoiesis and some forms of hemolytic anemia.

Anemia often results from kidney failure, because RBC production depends on erythropoietin, which is produced mainly by the kidneys. Erythropoiesis also declines with age, simply because the kidneys atrophy and produce less and less EPO as we get older. Compounding this problem, elderly people tend to get less exercise and eat less well, and both of these factors reduce erythropoiesis.

*Nutritional anemia* results from a dietary deficiency of any of the requirements for erythropoiesis discussed earlier. Its most common form is **iron-deficiency anemia. Pernicious anemia** can result from a deficiency of vitamin $B_{12}$, but this vitamin is so abundant in meat that a $B_{12}$ deficiency is rare except in strict vegetarians. More often, it occurs when glands of the stomach fail to produce a substance called **intrinsic factor** that the small intestine needs to absorb vitamin $B_{12}$. This becomes more common in old age because of atrophy of the stomach. Pernicious anemia can also be hereditary. It is treatable with vitamin $B_{12}$ injections; oral $B_{12}$ would be useless because the digestive tract cannot absorb it without intrinsic factor.

*Hypoplastic*[19] *anemia* is caused by a decline in erythropoiesis, whereas the complete failure or destruction of the myeloid tissue produces *aplastic anemia,* a complete cessation of erythropoiesis. Aplastic anemia leads to

[17]*poly* = many + *cyt* = cell + *hem* = blood + *ia* = condition
[18]*an* = without + *em* = blood + *ia* = condition

[19]*hypo* = below normal + *plas* = formation + *tic* = pertaining to

grotesque tissue necrosis and blackening of the skin. Most victims die within a year. About half of all cases are of unknown or hereditary cause, especially in adolescents and young adults. Other causes are given in table 18.5.

Anemia has three potential consequences:

1. The tissues suffer **hypoxia** (oxygen deprivation). The individual is lethargic and becomes short of breath upon physical exertion. The skin is pallid because of the deficiency of hemoglobin. Severe anemic hypoxia can cause life-threatening necrosis of brain, heart, and kidney tissues.

2. Blood osmolarity is reduced. More fluid is thus transferred from the bloodstream to the intercellular spaces, resulting in edema.

3. Blood viscosity is reduced. Because the blood puts up so little resistance to flow, the heart beats faster than normal and cardiac failure may ensue. Blood pressure also drops because of the reduced volume and viscosity.

## Sickle Cell Disease

Sickle cell disease and thalassemia are hereditary hemoglobin defects that occur mostly among people of African and Mediterranean descent, respectively. About 1.3% of African Americans have **sickle cell disease.** This disorder is caused by a recessive allele that modifies the structure of hemoglobin. Sickle cell hemoglobin (HbS) differs from normal HbA only in the sixth amino acid of the beta chain, where HbA has glutamic acid and HbS has valine. People who are homozygous for HbS exhibit sickle cell disease. People who are heterozygous for it—about 8.3% of African Americans—have *sickle cell trait* but rarely have severe symptoms. However, if two carriers reproduce, their children each have a 25% chance of being homozygous and having the disease.

Without treatment, a child with sickle cell disease has little chance of living to age 2, but even with the best available treatment, few victims live to the age of 50. HbS does not bind oxygen very well. At low oxygen concentrations, it becomes deoxygenated, polymerizes, and forms a gel that causes the erythrocytes to become elongated and pointed at the ends (fig. 18.10), hence the name of the disease. Sickled erythrocytes are sticky; they **agglutinate**[20] (clump together) and block small blood vessels, causing

---

[20]*ag* = together + *glutin* = glue

| TABLE 18.5 | Causes of Anemia |
|---|---|
| **Categories of Anemia** | **Causes or Examples** |
| *Inadequate Erythropoiesis* | |
| Iron-deficiency anemia | Dietary iron deficiency |
| Other nutritional anemias | Dietary folic acid, vitamin $B_{12}$, or vitamin C deficiency |
| Anemia due to renal insufficiency | Deficiency of EPO secretion |
| Pernicious anemia | Deficiency of intrinsic factor |
| Hypoplastic and aplastic anemia | Destruction of myeloid tissue by radiation, viruses, some drugs and poisons (arsenic, benzene, mustard gas), or autoimmune disease |
| Anemia of old age | Declining erythropoiesis due to nutritional deficiencies, reduced physical activity, gastric atrophy (reduced intrinsic factor secretion), or renal atrophy (depressed EPO secretion) |
| *Blood Loss (Hemorrhagic Anemia)* | Trauma, hemophilia, menstruation, ulcer, ruptured aneurysm, etc. |
| *RBC Destruction (Hemolytic Anemia)* | |
| Drug reactions | Penicillin allergy |
| Poisoning | Mushroom toxins, snake and spider venoms |
| Parasitic infection | RBC destruction by malaria parasites |
| Hereditary hemoglobin defects | Sickle cell disease, thalassemia |
| Blood type incompatabilities | Hemolytic disease of the newborn |

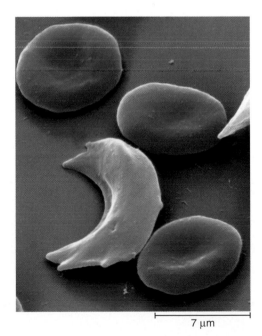

7 μm

**FIGURE 18.10 Blood of a Person with Sickle Cell Disease.** Shows one deformed, pointed erythrocyte and three normal erythrocytes.

intense pain in oxygen-starved tissues. Blockage of the circulation can also lead to kidney or heart failure, stroke, rheumatism, or paralysis. Hemolysis of the fragile cells causes anemia and hypoxemia, which triggers further sickling in a deadly positive feedback loop. Chronic hypoxemia also causes fatigue, weakness, mental deficiency, and deterioration of the heart and other organs. In a futile effort to counteract the hypoxemia, the hemopoietic tissues become so active that bones of the cranium and elsewhere become enlarged and misshapen. The spleen reverts to a hemopoietic role, while also disposing of dead RBCs, and becomes enlarged and fibrous. Sickle cell disease is a prime example of *pleiotropy*—the occurrence of multiple phenotypic effects from a change in a single gene (see p. 146).

Why does sickle cell disease exist? In Africa, where it originated, vast numbers of people die of malaria. Malaria is caused by a parasite that invades the RBCs and feeds on hemoglobin. Sickle cell hemoglobin, HbS, is indigestible to the parasites, and people heterozygous for sickle cell disease are resistant to malaria. The lives saved by this gene far outnumber the deaths of homozygous individuals, so the gene persists in the population.

## Before You Go On

*Answer the following questions to test your understanding of the preceding section:*

6. Describe the size, shape, and contents of an erythrocyte, and explain how it acquires its unusual shape.
7. What is the function of hemoglobin? What are its protein and nonprotein moieties called?
8. Define hematocrit, hemoglobin concentration, and RBC count, and give the units of measurement in which each is expressed.
9. List the stages in the production of an RBC and describe how each stage differs from the previous one.
10. What is the role of erythropoietin in the regulation of RBC count? What is the role of gastroferritin?
11. What happens to each component of an RBC and its hemoglobin when it dies and disintegrates?
12. What are the three primary causes or categories of anemia? What are its three primary consequences?

# Blood Types

## Objectives

When you have completed this section, you should be able to

- explain what determines a person's ABO and Rh blood types and how this relates to transfusion compatibility;

- describe the effect of an incompatibility between mother and fetus in Rh blood type; and
- list some blood groups other than ABO and Rh and explain how they may be useful.

Blood types and transfusion compatibility are a matter of interactions between plasma proteins and erythrocytes. Ancient Greek physicians attempted to transfuse blood from one person to another by squeezing it from a pig's bladder through a porcupine quill into the recipient's vein. Although some patients benefited from the procedure, it was fatal to others. The reason some people have compatible blood and some do not remained obscure until 1900, when Karl Landsteiner discovered blood types A, B, and O—a discovery that won him a Nobel Prize in 1930; type AB was discovered later. World War II stimulated great improvements in transfusions, blood banking, and blood substitutes (see Insight 18.3).

Blood types are based on interactions between large molecules called *antigens* and *antibodies*. Explained more fully in chapter 21, these will be introduced only briefly here. **Antigens** are complex molecules such as proteins, glycoproteins, and glycolipids that are genetically unique to each individual (except identical twins). They occur on the surfaces of all cells and enable the body to distinguish its own cells from foreign matter. When the body detects an antigen of foreign origin, it activates an immune response. This response consists partly of the *plasma cells*, mentioned earlier, secreting proteins (gamma globulins) called **antibodies.**

Antibodies bind to antigens and mark them, or the cells bearing them, for destruction. One method of antibody action is **agglutination,** in which each antibody molecule binds to two or more antigen molecules and sticks them together. Repetition of this process produces large **antigen–antibody complexes** that immobilize the antigens until certain immune cells can break them down. Whole cells may become immobilized and clumped together when antibodies bind to their surface antigens.

Blood types are based on antigens called **agglutinogens** (ah-glue-TIN-oh-jens) on the surfaces of the RBCs, and antibodies called **agglutinins** (ah-GLUE-tih-nins) in the blood plasma. These names reflect the fact that they interact to agglutinate RBCs in the event of a mismatched transfusion.

## THE ABO GROUP

Blood types A, B, AB, and O form the **ABO blood group** (table 18.6). Your ABO blood type is determined by the hereditary presence or absence of antigens A and B on your RBCs. The genetic determination of blood types is explained on page 145. Figure 18.12 shows how the carbohydrate moieties of RBC surface antigens determine the ABO blood types.

## INSIGHT 18.3 Medical History

### Charles Drew—Blood Banking Pioneer

Charles Drew (fig. 18.11) was a scientist who lived and died in the grip of irony. After receiving his M.D. from McGill University of Montreal in 1933, Drew became the first black person to pursue the advanced degree of Doctor of Science in Medicine, for which he studied transfusion and blood-banking at Columbia University. He became the director of a new blood bank at Columbia Presbyterian Hospital in 1939 and organized numerous blood banks during World War II.

Drew saved countless lives by convincing physicians to use plasma rather than whole blood for battlefield and other emergency transfusions. Whole blood could be stored for only a week and given only to recipients with compatible blood types. Plasma could be stored longer and was less likely to cause transfusion reactions.

When the U.S. War Department issued a directive forbidding the mixing of Caucasian and Negro blood in military blood banks, Drew denounced the order and resigned his position. He became a professor of surgery at Howard University in Washington, D.C., and later chief of staff at Freedmen's Hospital. He was a mentor for numerous young black physicians and campaigned to get them accepted into the medical community. The American Medical Association, however, firmly refused to admit black members, even Drew himself.

Late one night in 1950, Drew and three colleagues set out to volunteer their medical services to an annual free clinic in Tuskegee, Alabama. Drew fell asleep at the wheel and was critically injured in the resulting accident. Doctors at the nearest hospital administered blood and attempted unsuccessfully to revive him. For all the lives he saved through his pioneering work in blood transfusion, Drew himself bled to death at the age of 45.

**FIGURE 18.11** Charles Drew (1904–50).

| **TABLE 18.6** | The ABO Blood Group | | | |
|---|---|---|---|---|
| | **ABO Blood Type** | | | |
| **Characteristics** | **Type O** | **Type A** | **Type B** | **Type AB** |
| Possible genotypes* | $ii$ | $I^A I^A$ or $I^A i$ | $I^B I^B$ or $I^B i$ | $I^A I^B$ |
| RBC antigen | None | A | B | A,B |
| Plasma antibody | Anti-A, anti-B | Anti-B | Anti-A | None |
| Compatible donor RBCs | O | O, A | O, B | O, A, B, AB |
| Incompatible donor RBCs | A, B, AB | B, AB | A, AB | None |
| Frequency in U.S. population | | | | |
| White | 45% | 40% | 11% | 4% |
| Black | 49% | 27% | 20% | 4% |
| Hispanic | 63% | 14% | 20% | 3% |
| Japanese | 31% | 38% | 22% | 9% |
| Native American | 79% | 16% | 4% | <1% |

*$I^A$ is the dominant allele for agglutinogen A; $I^B$ is the dominant allele for agglutinogen B; and allele $i$ is recessive to both of these.

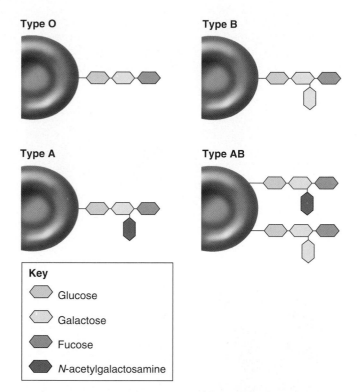

**Key**

⬡ Glucose

⬡ Galactose

⬡ Fucose

⬡ *N*-acetylgalactosamine

**FIGURE 18.12   Chemical Basis of the ABO Blood Types.**
The terminal carbohydrates of the antigenic glycolipids are shown. All of them end with galactose and fucose (not to be confused with fructose). In type A, the galactose also has an *N*-acetylgalactosamine added to it; in type B, it has another galactose; and in type AB, both of these chain types are present.

⌐ **Think About It**

*Suppose you could develop an enzyme that selectively split N-acetylgalactosamine off the glycolipid of type A blood cells (fig. 18.12). What would be the potential benefit of this product to blood banking and transfusion?*

Antibodies of the ABO group begin to appear in the plasma 2 to 8 months after birth. They reach their maximum concentrations between 8 and 10 years of age and then slowly decline for the rest of one's life. They are produced mainly in response to the bacteria that inhabit our intestines, but they cross-react with RBC antigens and are therefore best known for their significance in transfusions.

Antibodies of the ABO group react against any A or B antigen except those on one's own RBCs. The antibody that reacts against antigen A is called alpha *agglutinin,* or *anti-A;* it is present in the plasma of people with type O or type B blood—that is, anyone who does *not* possess antigen A. The antibody that reacts against antigen B is beta *agglutinin,* or *anti-B,* and is present in type O and type A individuals—those who do not possess antigen B. Each antibody molecule has 10 binding sites where it can attach to either an A or B antigen. An antibody can therefore attach to several RBCs at once and agglutinate them (fig. 18.13).

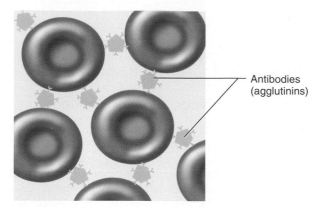

Antibodies (agglutinins)

**FIGURE 18.13   Agglutination of RBCs by an Antibody.**
Anti-A and anti-B have 10 binding sites, located at the 2 tips of each of the 5 Ys, and can therefore bind multiple RBCs to each other.

A person's ABO blood type can be determined by placing one drop of blood in a pool of anti-A serum and another drop in a pool of anti-B serum. Blood type AB exhibits conspicuous agglutination in both antisera; type A or B agglutinates only in the corresponding antiserum; and type O does not agglutinate in either one (fig. 18.14).

Type O blood is the most common and AB is the rarest in the United States. Percentages differ from one region of the world to another and among ethnic groups because people tend to marry within their locality and ethnic group and perpetuate statistical variations particular to that group (see table 18.6).

In giving transfusions, it is imperative that the donor's RBCs not agglutinate as they enter the recipient's bloodstream. For example, if type B blood were transfused into a type A recipient, the recipient's anti-B antibodies would immediately agglutinate the donor's RBCs (fig. 18.15). A mismatched transfusion causes a **transfusion reaction**—the agglutinated RBCs block small blood vessels, hemolyze, and release their hemoglobin over the next few hours to days. Free hemoglobin can block the kidney tubules and cause death from acute renal failure within a week or so. For this reason, a person with type A (anti-B) blood must never be given a transfusion of type B or AB blood. A person with type B (anti-A) must never receive type A or AB blood. Type O (anti-A and anti-B) individuals cannot safely receive type A, B, or AB blood.

Type AB is sometimes called the *universal recipient* because this blood type lacks both anti-A and anti-B antibodies; thus, it will not agglutinate donor RBCs of any ABO type. However, this overlooks the fact that the *donor's* plasma can agglutinate the *recipient's* RBCs if it contains anti-A, anti-B, or both. For similar reasons, type O is sometimes called the *universal donor.* The plasma of a type O donor, however, can agglutinate the RBCs of a type A, B, or AB recipient. There are procedures for reducing the risk of a transfusion reaction in certain mismatches, however, such as giving packed RBCs with a minimum of plasma.

Contrary to some people's belief, blood type is not changed by transfusion. It is fixed at conception and remains the same for life.

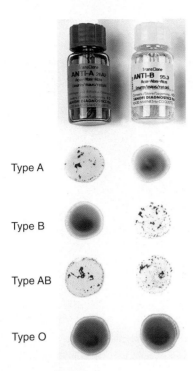

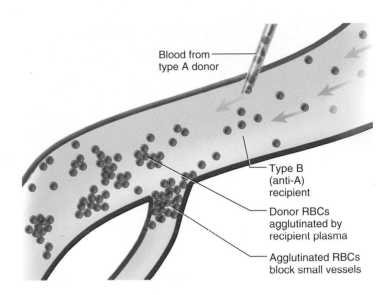

Blood from type A donor

Type B (anti-A) recipient

Donor RBCs agglutinated by recipient plasma

Agglutinated RBCs block small vessels

**FIGURE 18.15** **Effects of a Mismatched Transfusion.** Donor RBCs become agglutinated in the recipient's blood plasma. The agglutinated RBCs lodge in smaller blood vessels downstream from this point and cut off the blood flow to vital tissues.

**FIGURE 18.14** **ABO Blood Typing.** Each row shows the appearance of a drop of blood mixed with anti-A and anti-B antisera. Blood cells become clumped if they possess the antigens for the antiserum (top row left, second row right, third row both) but otherwise remain uniformly mixed. Thus type A agglutinates only in anti-A; type B agglutinates only in anti-B; type AB agglutinates in both; and type O agglutinates in neither of them.

## THE Rh GROUP

The **Rh blood group** is named for the rhesus monkey, in which the Rh antigens were discovered in 1940. This group is determined by three genes called C, D, and E, each of which has two alleles: *C, c, D, d, E, e.* Whatever other alleles a person may have, anyone with genotype *DD* or *Dd* has D antigens on his or her RBCs and is classified as *Rh-positive (Rh⁺).* In *Rh-negative (Rh⁻)* people, the D antigen is lacking. The Rh blood type is tested by using an anti-D reagent. The Rh type is usually combined with the ABO type in a single expression such as O⁺ for type O, Rh-positive, or AB⁻ for type AB, Rh-negative. About 85% of white Americans are Rh⁺ and 15% are Rh⁻. ABO blood type has no influence on Rh type, or vice versa. If the frequency of type O whites in the United States is 45%, and 85% of these are also Rh⁻, then the frequency of O⁺ individuals is the product of these separate frequencies: $0.45 \times 0.85 = 0.38$, or 38%. Rh frequencies vary among ethnic groups just as ABO frequencies do. About 99% of Asians are Rh⁺, for example.

> **Think About It**
>
> *Predict what percentage of Japanese Americans have type B⁻ blood.*

In contrast to the ABO group, anti-D antibodies are not normally present in the blood. They form only in Rh⁻ individuals who are exposed to Rh⁺ blood. If an Rh⁻ person receives an Rh⁺ transfusion, the recipient produces anti-D. Since anti-D does not appear instantaneously, this presents little danger in the first mismatched transfusion. But if that person should later receive another Rh⁺ transfusion, his or her anti-D could agglutinate the donor's RBCs.

A related condition sometimes occurs when an Rh⁻ woman carries an Rh⁺ fetus. The first pregnancy is likely to be uneventful because the placenta normally prevents maternal and fetal blood from mixing. However, at the time of birth, or if a miscarriage occurs, placental tearing exposes the mother to Rh⁺ fetal blood. She then begins to produce anti-D antibodies (fig. 18.16). If she becomes pregnant again with an Rh⁺ fetus, her anti-D antibodies may pass through the placenta and agglutinate the fetal erythrocytes. Agglutinated RBCs hemolyze, and the baby is born with a severe anemia called **hemolytic disease of the newborn (HDN),** or **erythroblastosis fetalis.** Not all HDN is due to Rh incompatibility, however. About 2% of cases result from incompatibility of ABO and other blood types. About 1 out of 10 cases of ABO incompatibility between mother and fetus results in HDN.

HDN, like so many other disorders, is easier to prevent than to treat. If an Rh⁻ woman gives birth to (or miscarries) an Rh⁺ child, she can be given an *Rh immune globulin* (sold under trade names such as RhoGAM and Gamulin). The immune globulin binds fetal RBC antigens so they cannot stimulate her immune system to produce anti-D. It is now common to give immune globulin at 28 to 32 weeks' gestation and at birth in any pregnancy in which the mother is Rh⁻.

If an Rh⁻ woman has had one or more previous Rh⁺ pregnancies, her subsequent Rh⁺ children have about a

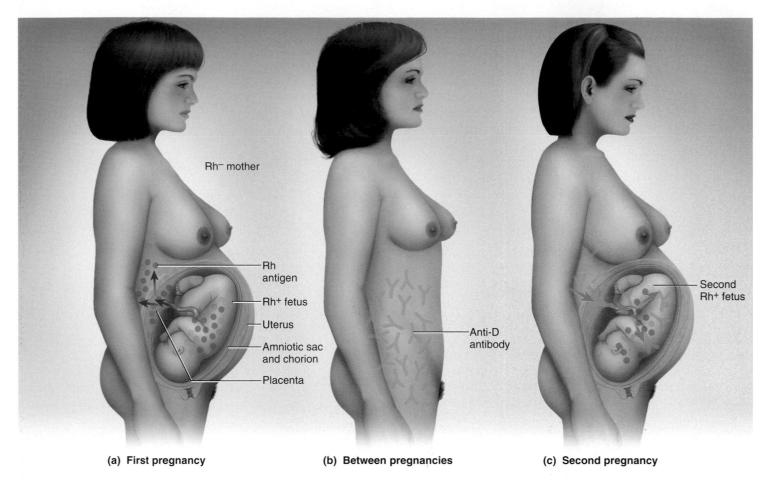

**(a) First pregnancy**    **(b) Between pregnancies**    **(c) Second pregnancy**

**FIGURE 18.16    Hemolytic Disease of the Newborn (HDN).** (a) When an Rh⁻ woman is pregnant with an Rh⁺ fetus, she is exposed to D (Rh) antigens, especially during childbirth. (b) Following that pregnancy, her immune system produces anti-D antibodies. (c) If she later becomes pregnant with another Rh⁺ fetus, her anti-D antibodies can cross the placenta and agglutinate the blood of that fetus, causing that child to be born with HDN.

17% probability of being born with HDN. Infants with HDN are usually severely anemic. As the fetal hemopoietic tissues respond to the need for more RBCs, erythroblasts (immature RBCs) enter the circulation prematurely—hence the name *erythroblastosis fetalis.* Hemolyzed RBCs release hemoglobin, which is converted to bilirubin. High bilirubin levels can cause *kernicterus,* a syndrome of toxic brain damage that may kill the infant or leave it with motor, sensory, and mental deficiencies. HDN can be treated with *phototherapy*—exposing the infant to ultraviolet light, which degrades bilirubin as blood passes through the capillaries of the skin. In more severe cases, an *exchange transfusion* may be given to completely replace the infant's Rh⁺ blood with Rh⁻. In time, the infant's hemopoietic tissues will replace the donor's RBCs with Rh⁺ cells, and by then the mother's antibody will have disappeared from the infant's blood.

## Think About It

*A baby with HDN typically has jaundice and an enlarged spleen. Explain these effects.*

## OTHER BLOOD GROUPS

In addition to the ABO and Rh groups, there are at least 100 other known blood groups with a total of more than 500 antigens, including the MN, Duffy, Kell, Kidd, and Lewis groups. These rarely cause transfusion reactions, but they are useful for such legal purposes as paternity and criminal cases and for research in anthropology and population genetics. Reactions in the Kell, Kidd, and Duffy groups occasionally cause HDN.

### Before You Go On

*Answer the following questions to test your understanding of the preceding section:*

13. *What are antibodies and antigens? How do they interact to cause a transfusion reaction?*

14. *What antibodies and antigens are present in people with each of the four ABO blood types?*

15. *Describe the cause, prevention, and treatment of HDN.*

16. *Why might someone be interested in determining a person's blood type other than ABO/Rh?*

# Leukocytes

## Objectives

When you have completed this section, you should be able to

- explain the function of leukocytes in general and the individual role of each leukocyte type;
- describe the appearance and relative abundance of each type of leukocyte;
- describe the formation and life history of leukocytes; and
- discuss the types, causes, and effects of leukocyte excesses and deficiencies.

## FORM AND FUNCTION

**Leukocytes,** or **white blood cells (WBCs),** are the least abundant formed elements, totaling only 5,000 to 10,000 WBCs/μL. Yet we cannot live long without them, because they afford protection against infectious microorganisms and other pathogens. WBCs are easily recognized in stained blood films because they have conspicuous nuclei that stain from light violet to dark purple with the most common blood stains. They are much more abundant in the body than their low number in blood films would suggest, because they spend only a few hours in the bloodstream, then migrate through the walls of the capillaries and venules and spend the rest of their lives in the connective tissues. It's as if the bloodstream were merely the subway that the WBCs take to work; In blood films, we see only the ones on their way to work, not the WBCs already at work in the tissues.

Leukocytes differ from erythrocytes in that they retain their organelles throughout life; thus, when viewed with the transmission electron microscope, they show a complex internal structure (fig. 18.17). Among these organelles are the usual instruments of protein synthesis—the nucleus, rough endoplasmic reticulum, ribosomes, and Golgi complex—for leukocytes must synthesize proteins in order to carry out their functions. Some of these proteins are packaged into lysosomes and other organelles, which appear as conspicuous cytoplasmic granules that distinguish one WBC type from another.

## TYPES OF LEUKOCYTES

As outlined at the beginning of this chapter, there are five kinds of leukocytes (table 18.7). They are distinguished from each other by their relative size and abundance, the size and shape of their nuclei, the presence or absence of cytoplasmic granules, the coarseness and staining properties of those granules, and most importantly by their functions. Three of the WBC types—the neutrophils, eosinophils, and basophils—are called **granulocytes** because their cytoplasm contains lysosomes and other membrane-bounded organelles that appear as conspicuous

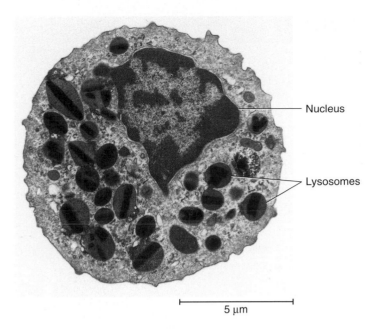

**FIGURE 18.17   The Structure of a Leukocyte (TEM).**   This example is an eosinophil. The lysosomes seen here are the coarse pink granules seen in the cytoplasm of the second cell in table 18.7.

colored granules in stained blood films. The other two WBC types—lymphocytes and monocytes—are called **agranulocytes** because cytoplasmic granules are scarce or absent.

## Granulocytes

- **Neutrophils** are the most abundant WBCs—generally about 4,150 cells/μL and constituting 60% to 70% of the circulating leukocytes. The nucleus is clearly visible and in a mature neutrophil, typically consists of three to five lobes connected by slender nuclear strands. These strands are sometimes so delicate that they are scarcely visible, and the neutrophil may seem as if it had multiple nuclei. Young neutrophils have an undivided nucleus shaped like a band or a knife puncture; thus they are called *band cells* or *stab cells.* Neutrophils are also called *polymorphonuclear leukocytes (PMNs)* because of their varied nuclear shapes.

  The cytoplasm contains fine reddish to violet granules, which contain lysozyme, peroxidase, and other antibiotic agents. Neutrophils are named for the way these granules take up histological stains at pH 7—some stain with acidic dyes and others with basic dyes, and the combined effect gives the cytoplasm a pale lilac color.

- **Eosinophils** (EE-oh-SIN-oh-fills) are harder to find in a blood film, because they are only 2% to 4% of the WBC total, typically numbering about 170 cells/μL. The eosinophil count fluctuates greatly, however, from day to night, seasonally, and with the phase of the menstrual cycle. It rises (*eosinophilia*) in allergies,

| TABLE 18.7 | The White Blood Cells (Leukocytes) |
| --- | --- |

### Neutrophils

| Percent of WBCs | 60-70% |
| --- | --- |
| Mean count | 4,150 cells/μL |
| Diameter | 9-12 μm |

*Appearance**

• Nucleus usually with 3-5 lobes in S- or C-shaped array

• Fine reddish to violet granules in cytoplasm

*Differential Count*

• Increases in bacterial infections

*Functions*

• Phagocytize bacteria

• Release antimicrobial chemicals

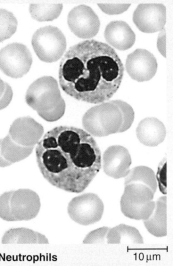

**Neutrophils**    10 μm

### Eosinophils

| Percent of WBCs | 2-4% |
| --- | --- |
| Mean count | 165 cells/μL |
| Diameter | 10-14 μm |

*Appearance**

• Nucleus usually has two large lobes connected by thin strand

• Large orange-pink granules in cytoplasm

*Differential Count*

• Fluctuates greatly from day to night, seasonally, and with phase of menstrual cycle

• Increases in parasitic infections, allergies, collagen diseases, and diseases of spleen and central nervous system

*Functions*

• Phagocytize antigen–antibody complexes, allergens, and inflammatory chemicals

• Release enzymes that weaken or destroy parasites such as worms

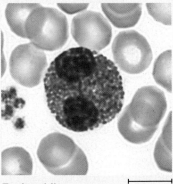

**Eosinophil**    10 μm

### Basophils

| Percent of WBCs | < 0.5-1% |
| --- | --- |
| Mean count | 44 cells/μL |
| Diameter | 8-10 μm |

*Appearance**

• Nucleus large and U- to S-shaped, but typically pale and obscured from view

• Coarse, abundant, dark violet granules in cytoplasm

*Differential Count*

• Relatively stable

• Increases in chicken pox, sinusitis, diabetes mellitus, myxedema, and polycythemia

*Functions*

• Secrete histamine (a vasodilator), which increases blood flow to a tissue

• Secrete heparin (an anticoagulant), which promotes mobility of other WBCs by preventing clotting

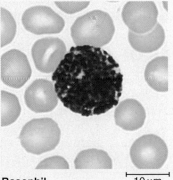

**Basophil**    10 μm

## Lymphocytes

| | |
|---|---|
| Percent of WBCs | 25–33% |
| Mean count | 2,185 cells/μL |
| Diameter | |
|   Small class | 5–8 μm |
|   Medium class | 10–12 μm |
|   Large class | 14–17 μm |

*Appearance\**

- Nucleus round, ovoid, or slightly dimpled on one side, of uniform dark violet color
- In small lymphocytes, nucleus fills nearly all of the cell and leaves only a scanty rim of clear, light blue cytoplasm
- In larger lymphocytes, cytoplasm is more abundant; large lymphocytes may be hard to differentiate from monocytes

*Differential Count*

- Increases in diverse infections and immune responses

*Functions*

- Several functional classes usually indistinguishable by light microscopy
- Destroy cancer cells, cells infected with viruses, and foreign cells
- "Present" antigens to activate other cells of immune system
- Coordinate actions of other immune cells
- Secrete antibodies
- Serve in immune memory

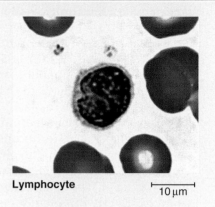

**Lymphocyte**  ⊢ 10 μm ⊣

## Monocytes

| | |
|---|---|
| Percent of WBCs | 3–8% |
| Mean count | 456 cells/μL |
| Diameter | 12–15 μm |

*Appearance\**

- Nucleus ovoid, kidney-shaped, or horseshoe-shaped; light violet
- Abundant cytoplasm with sparse, fine granules
- Sometimes very large with stellate or polygonal shapes

*Differential Count*

- Increases in viral infections and inflammation

*Functions*

- Differentiate into macrophages (large phagocytic cells of the tissues)
- Phagocytize pathogens, dead neutrophils, and debris of dead cells
- "Present" antigens to activate other cells of immune system

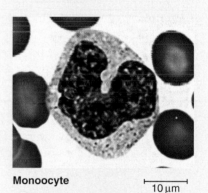

**Monoocyte**  ⊢ 10 μm ⊣

\*Appearance pertains to blood films dyed with Wright's stain.

parasitic infections, collagen diseases, and diseases of the spleen and central nervous system. Although relatively scanty in the blood, eosinophils are abundant in the mucous membranes of the respiratory, digestive, and lower urinary tracts. The eosinophil nucleus usually has two large lobes connected by a thin strand, and the cytoplasm has an abundance of coarse rosy to orange-colored granules.

- **Basophils** are the rarest of the WBCs and, indeed, of all formed elements. They number about 40 cells/μL and constitute from less than 0.5% to about 1% of the WBC count. They can be recognized mainly by an abundance of very coarse, dark violet cytoplasmic granules. The nucleus is largely hidden from view by these granules, but is large, pale, and typically S- or U-shaped.

## Agranulocytes

- **Lymphocytes** are second to neutrophils in abundance and are thus quickly spotted when you examine a blood film. They number about 2,200 cells/μL and are 25% to 33% of the WBC count. They include the smallest WBCs; at 5 to 17 μm in diameter, they range from smaller than RBCs to two and a half times as large. They are sometimes classified into three size classes (table 18.7), but there are gradations between them. Medium and large lymphocytes are usually seen in fibrous connective tissues and only occasionally in the circulating blood. The lymphocytes seen in blood films are mostly in the small size class. The lymphocyte nucleus is round, ovoid, or slightly dimpled on one side, and usually stains dark violet. In small lymphocytes, it fills nearly the entire cell and leaves only a narrow rim of light blue cytoplasm, often barely detectable, around the cell perimeter. The cytoplasm is more abundant in medium and large lymphocytes.

  Small lymphocytes are sometimes difficult to distinguish from basophils, but most basophils are conspicuously grainy, whereas the lymphocyte nucleus is uniform or merely mottled. Basophils also lack the rim of clear cytoplasm seen in most lymphocytes. Large lymphocytes are sometimes difficult to distinguish from monocytes.

  There are several subclasses of lymphocytes with different immune functions (see chapter 21), but they look alike through the light microscope.

- **Monocytes** are the largest WBCs, often two or three times the diameter of an RBC. They number about 460 cells/μL and about 3% to 8% of the WBC count. The nucleus is large and clearly visible, often a relatively light violet, and typically ovoid, kidney-shaped, or horseshoe-shaped. The cytoplasm is abundant and contains sparse, fine granules. In prepared blood films, monocytes often assume sharply angular to spiky shapes (see fig. 18.1).

## THE LEUKOCYTE LIFE CYCLE

**Leukopoiesis** (LOO-co-poy-EE-sis), production of white blood cells, begins with the same pluripotent stem cells (PPSCs) as erythropoiesis. Some PPSCs differentiate into the distinct types of colony-forming units (CFUs), which then go on to produce the following cell lines (fig. 18.18), each of them now irreversibly committed to a certain outcome.

1. *Myeloblasts,* which ultimately differentiate into the three types of granulocytes (neutrophils, eosinophils, and basophils).

2. *Monoblasts, w*hich look identical to myeloblasts but lead ultimately to monocytes.

---

3. *Lymphoblasts,* which give rise to three types of lymphocytes (B lymphocytes, T lymphocytes, and natural killer cells).

Committed cells have receptors for colony-stimulating factors (CSFs). Mature lymphocytes and macrophages secrete several types of CSFs in response to infections and other immune challenges. Each CSF stimulates a different WBC type to develop in response to specific needs. Thus, a bacterial infection may trigger the production of neutrophils, whereas an allergy triggers the production of eosinophils, each process working through its own CSF.

The red bone marrow stores granulocytes and monocytes until they are needed and contains 10 to 20 times more of these cells than the circulating blood does. Lymphocytes begin developing in the bone marrow but do not stay there. Some types mature there and others migrate to the thymus to complete their development. Mature lymphocytes from both locations then colonize the spleen, lymph nodes, and other lymphoid organs and tissues.

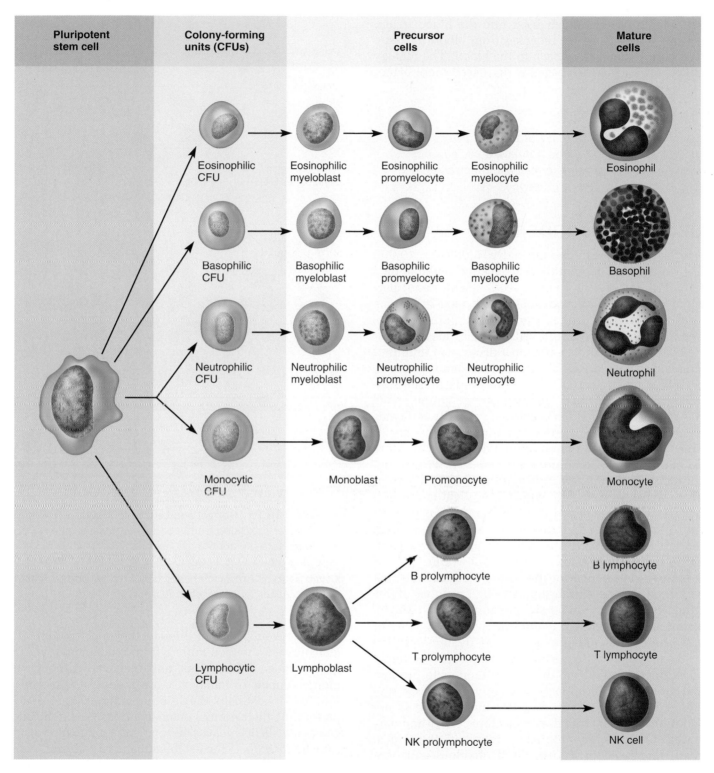

| Pluripotent stem cell | Colony-forming units (CFUs) | Precursor cells | | | Mature cells |
|---|---|---|---|---|---|
| | Eosinophilic CFU | Eosinophilic myeloblast | Eosinophilic promyelocyte | Eosinophilic myelocyte | Eosinophil |
| | Basophilic CFU | Basophilic myeloblast | Basophilic promyelocyte | Basophilic myelocyte | Basophil |
| | Neutrophilic CFU | Neutrophilic myeloblast | Neutrophilic promyelocyte | Neutrophilic myelocyte | Neutrophil |
| | Monocytic CFU | Monoblast | Promonocyte | | Monocyte |
| | Lymphocytic CFU | Lymphoblast | B prolymphocyte | | B lymphocyte |
| | | | T prolymphocyte | | T lymphocyte |
| | | | NK prolymphocyte | | NK cell |

**FIGURE 18.18   Leukopoiesis.**   Stages in the development of white blood cells. The pluripotent stem cell at the left also is the ultimate source of red blood cells (see fig. 18.6) and platelet-producing cells.

Circulating leukocytes do not stay in the blood for very long. Granulocytes circulate for 4 to 8 hours and then migrate into the tissues, where they live another 4 or 5 days. Monocytes travel in the blood for 10 to 20 hours, then migrate into the tissues and transform into a variety of **macrophages** (MAC-ro-fay-jes). Macrophages can live as long as a few years.

Lymphocytes, responsible for long-term immunity, survive from a few weeks to decades; they leave the bloodstream for the tissues and eventually enter the lymphatic system, which empties them back into the bloodstream. Thus, they are continually recycled from blood to tissue fluid to lymph and finally back to the blood. The biology of leukocytes and macrophages is discussed more extensively in chapter 21.

> ┌ **Think About It**
>
> *Some authorities state that RBCs do not live as long as WBCs because RBCs do not have a nucleus and therefore cannot repair and maintain themselves. Explain the flaw in this argument.*

## LEUKOCYTE DISORDERS

The total WBC count is normally 5,000 to 10,000 WBCs/μL. A count below this range, called **leukopenia**[21] (LOO-co-PEE-nee-uh), is seen in lead, arsenic, and mercury poisoning; radiation sickness; and such infectious diseases as measles, mumps, chicken pox, poliomyelitis, influenza, typhoid fever, and AIDS. It can also be produced by glucocorticoids, anticancer drugs, and immunosuppressant drugs given to organ transplant patients. Since WBCs are protective cells, leukopenia presents an elevated risk of infection and cancer. A count above 10,000 WBCs/μL, called **leukocytosis,**[22] usually indicates infection, allergy, or other diseases but can also occur in response to dehydration or emotional disturbances. More useful than a total WBC count is a *differential WBC count,* which identifies what percentage of the total WBC count consists of each type of leukocyte. A high neutrophil count is a sign of bacterial infection; neutrophils become sharply elevated in appendicitis, for example. A high eosinophil count usually indicates an allergy or a parasitic infection such as hookworms or tapeworms.

**Leukemia** is a cancer of the hemopoietic tissues that usually produces an extraordinarily high number of circulating leukocytes and their precursors (fig. 18.19). Leukemia is classified as myeloid or lymphoid, acute or chronic. **Myeloid leukemia** is marked by uncontrolled granulocyte production, whereas **lymphoid leukemia** involves uncontrolled lymphocyte or monocyte production. **Acute leukemia** appears suddenly, progresses rapidly, and causes death within a few months if it is not treated. **Chronic leukemia** develops more slowly and may go undetected for many months; if untreated, the typical survival time is about 3 years. Both myeloid and lymphoid leukemia occur in acute and chronic forms. The greatest success in treatment and cure has been with acute lymphoblastic leukemia, the most common type of childhood cancer. Treatment employs chemotherapy and marrow transplants along with the control of side effects such as anemia, hemorrhaging, and infection.

---

[21]*leuko* = white + *penia* = deficiency
[22]*leuko* = white + *cyt* = cell + *osis* = condition

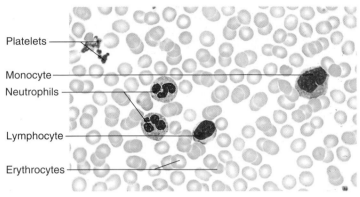

**(a)**

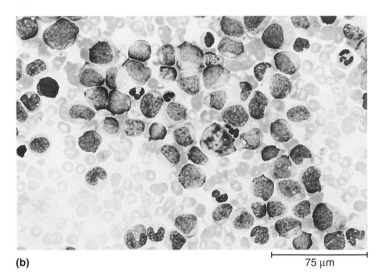

**(b)**                                                          75 μm

**FIGURE 18.19   Normal and Leukemic Blood.**   (a) A normal blood smear. (b) Blood from a patient with acute monocytic leukemia. Note the abnormally high number of white blood cells, especially monocytes, in part (b).

❱ *With all these extra white cells, why isn't the body's infection-fighting capability increased in leukemia?*

As leukemic cells proliferate, they replace normal bone marrow and a person suffers from a deficiency of normal granulocytes, erythrocytes, and platelets. Although enormous numbers of leukocytes are produced and spill over into the bloodstream, they are immature cells incapable of performing their normal defensive roles. The deficiency of competent WBCs leaves the patient vulnerable to *opportunistic infection*—the establishment of pathogenic organisms that usually cannot get a foothold in people with healthy immune systems. The RBC deficiency renders the patient anemic and fatigued, and the platelet deficiency results in hemorrhaging and impaired blood clotting. The immediate cause of death is usually hemorrhage or opportunistic infection. Cancerous hemopoietic tissue often metastasizes from the bone marrow or lymph nodes to other organs of the body, where the cells displace or compete with normal cells. Metastasis to the bone tissue itself is common and leads to bone and joint pain.

### Before You Go On

*Answer the following questions to test your understanding of the preceding section:*

17. *What is the overall function of leukocytes?*

18. *List the five kinds of leukocytes in order of abundance, identify whether each is a granulocyte or agranulocyte, and describe how to identify each one.*

19. *What does leukopoiesis have in common with erythropoiesis? How does it differ?*

20. *What can cause an abnormally high or low WBC count?*

21. *WBCs fight infection, and leukemia is an abnormally high WBC count. Yet leukemia patients are very vulnerable to infection. Explain this seeming paradox.*

# Platelets and Hemostasis— The Control of Bleeding

### Objectives

When you have completed this section, you should be able to

- describe the body's mechanisms for controlling bleeding;
- list the functions of platelets;
- describe two reaction pathways that produce blood clots;
- explain what happens to blood clots when they are no longer needed;
- explain what keeps blood from clotting in the absence of injury; and
- describe some disorders of blood clotting.

Circulatory systems developed very early in animal evolution, and with them evolved mechanisms for stopping leaks, which are potentially fatal. **Hemostasis**[23] is the cessation of bleeding. Although hemostatic mechanisms may not stop a hemorrhage from a large blood vessel, they are quite effective at closing breaks in small ones. Platelets play multiple roles in hemostasis, so we begin with a consideration of their form and function.

## PLATELET FORM AND FUNCTION

**Platelets** are not cells but small fragments of marrow cells called *megakaryocytes.* Platelets are the second most abundant formed elements, after erythrocytes; a normal platelet count in blood from a fingerstick ranges from 130,000 to 400,000 platelets/µL (averaging about 250,000). The platelet count can vary greatly, however, under different physiological conditions and in blood

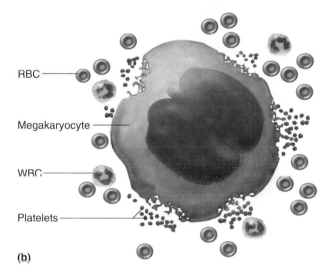

**(a)**   2 µm

**(b)**

**FIGURE 18.20   Platelets.** (a) Structure of blood platelets (TEM). (b) Platlets being produced by a megakaryocyte in the bone marrow. Note the enormous size of the megakaryocyte compared with RBCs and WBCs.

samples taken from various places in the body. In spite of their numbers, platelets are so small (2 to 4 µm in diameter) that they contribute even less than the WBCs to the blood volume.

Platelets have a complex internal structure that includes lysosomes, mitochondria, microtubules and microfilaments, **granules** filled with platelet secretions, and a system of channels called the **open canalicular system,** which opens onto the platelet surface (fig. 18.20a). They have no nucleus. When activated, they form pseudopods and are capable of ameboid movement.

Despite their small size, platelets have a greater variety of functions than any of the true blood cells:

- They secrete *vasoconstrictors,* chemicals that cause spasmodic constriction of broken vessels and thus help reduce blood loss.

- They stick together to form temporary *platelet plugs* to seal small breaks in injured blood vessels.

[23]*hemo* = blood + *stasis* = stability

- They secrete *procoagulants,* or clotting factors, which promote blood clotting.
- They initiate the formation of a clot-dissolving enzyme that dissolves blood clots that have outlasted their usefulness.
- They secrete chemicals that attract neutrophils and monocytes to sites of inflammation.
- They phagocytize and destroy bacteria.
- They secrete *growth factors* that stimulate mitosis in fibroblasts and smooth muscle and thus help to maintain and repair blood vessels.

## PLATELET PRODUCTION

The production of platelets is a division of hemopoiesis called **thrombopoiesis.** (Platelets are occasionally called *thrombocytes,*[24] but this term is now usually reserved for nucleated true cells in other animals such as birds and reptiles.) Some pluripotent stem cells produce receptors for the hormone *thrombopoietin,* thus becoming megakaryoblasts, cells committed to the platelet-producing line. The megakaryoblast duplicates its DNA repeatedly without undergoing nuclear or cytoplasmic division. The result is a **megakaryocyte**[25] (meg-ah-CAR-ee-oh-site), a gigantic cell up to 150 μm in diameter, visible to the naked eye, with a huge multilobed nucleus and multiple sets of chromosomes (fig. 18.20b). Most megakaryocytes

live in the bone marrow, but some of them colonize the lungs and produce platelets there.

A megakaryocyte develops infoldings of the plasma membrane that divide its marginal cytoplasm into little compartments. The cytoplasm breaks up along these lines of weakness into tiny fragments that enter the bloodstream. Some of these fragments are already functional platelets, while others are larger particles that break up into platelets as they pass through the lungs. About 25% to 40% of the platelets are stored in the spleen and released as needed. The remainder circulate freely in the blood and live for about 10 days.

## HEMOSTASIS

There are three hemostatic mechanisms—*vascular spasm, platelet plug formation,* and *blood clotting* (coagulation) (fig. 18.21). Platelets play an important role in all three.

### Vascular Spasm

The most immediate protection against blood loss is **vascular spasm,** a prompt constriction of the broken vessel. Several things trigger this reaction. An injury stimulates pain receptors, some of which directly innervate nearby blood vessels and cause them to constrict. This effect lasts only a few minutes, but other mechanisms take over by the time it subsides. Injury to the smooth muscle of the blood vessel itself causes a longer-lasting vasoconstriction, and platelets release serotonin, a chemical vasoconstrictor. Thus, the vascular spasm is maintained long enough for the other two hemostatic mechanisms to come into play.

---

[24]*thrombo* = clotting + *cyte* = cell
[25]*mega* = giant + *karyo* = nucleus + *cyte* = cell

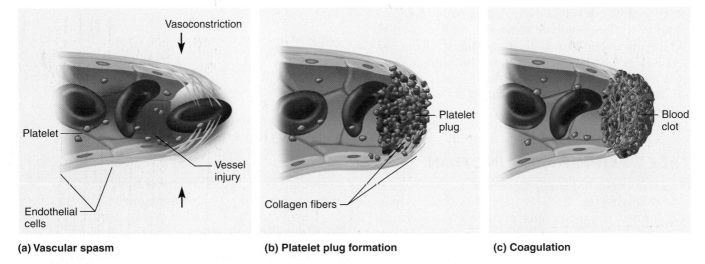

(a) **Vascular spasm**    (b) **Platelet plug formation**    (c) **Coagulation**

**FIGURE 18.21   Hemostasis.**  (a) Vasoconstriction of a broken vessel reduces bleeding. (b) A platelet plug forms as platelets adhere to exposed collagen fibers of the vessel wall. The platelet plug temporarily seals the break. (c) A blood clot forms as platelets become enmeshed in fibrin threads. This forms a longer-lasting seal and gives the vessel a chance to repair itself.

▶ *How does a blood clot differ from a platelet plug?*

## Platelet Plug Formation

Platelets will not adhere to the endothelium (inner lining) of undamaged blood vessels. The endothelium is normally very smooth and coated with **prostacyclin,** a platelet repellent. When a vessel is broken, however, collagen fibers of its wall are exposed to the blood. Upon contact with collagen or other rough surfaces, platelets put out long spiny pseudopods that adhere to the vessel and to other platelets; the pseudopods then contract and draw the walls of the vessel together. The mass of platelets thus formed, called a **platelet plug,** may reduce or stop minor bleeding.

As platelets aggregate, they undergo **degranulation**—the exocytosis of their cytoplasmic granules and release of factors that promote hemostasis. Among these are serotonin, a vasoconstrictor; adenosine diphosphate (ADP), which attracts more platelets to the area and stimulates their degranulation; and **thromboxane $A_2$,** an eicosanoid that promotes platelet aggregation, degranulation, and vasoconstriction. Thus, a positive feedback cycle is activated that can quickly seal a small break in a blood vessel.

## Coagulation

**Coagulation** (clotting) of the blood is the last but most effective defense against bleeding. It is important for the blood to clot quickly when a vessel has broken, but equally important for it not to clot in the absence of vessel damage. Because of this delicate balance, coagulation is one of the most complex processes in the body, involving over 30 chemical reactions. It is presented here in a very simplified form.

Perhaps clotting is best understood if we first consider its goal. The objective is to convert the plasma protein fibrinogen into **fibrin,** a sticky protein that adheres to the walls of a vessel. As blood cells and platelets arrive, they stick to the fibrin like insects sticking to a spider web (fig. 18.22). The resulting mass of fibrin, blood cells, and platelets ideally seals the break in the blood vessel. The complexity of clotting lies in how the fibrin is formed.

There are two reaction pathways to coagulation (fig. 18.23). One of them, the **extrinsic mechanism,** is initiated by clotting factors released by the damaged blood vessel and perivascular[26] tissues. The word *extrinsic* refers to the fact that these factors come from sources other than the blood itself. Blood may also clot, however, without these tissue factors—for example, when platelets adhere to a fatty plaque of atherosclerosis or to a test tube. The reaction pathway in this case is called the **intrinsic mechanism** because it uses only clotting factors found in the blood itself. In most cases of bleeding, both the extrinsic and intrinsic mechanisms work simultaneously to contribute to hemostasis.

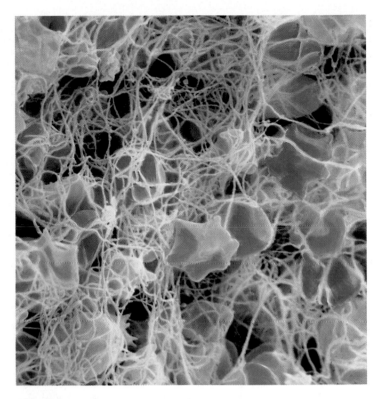

**FIGURE 18.22   A Blood Clot (SEM).**   The gray filaments are fibrin. Platelets are shown in orange.

Clotting factors (table 18.8) are called **procoagulants,** in contrast to the **anticoagulants** discussed later (see Insight 18.6, p. 710). Most procoagulants are proteins produced by the liver. They are always present in the plasma in inactive form, but when one factor is activated, it functions as an enzyme that activates the next one in the pathway. That factor activates the next, and so on, in a sequence called a **reaction cascade**—a series of reactions, each of which depends on the product of the preceding one. Many of the clotting factors are identified by Roman numerals, which indicate the order in which they were discovered, not the order of the reactions. Factors IV and VI are not included in table 18.8. These terms were abandoned when it was found that factor IV was calcium and factor VI was activated factor V. The last four procoagulants in the table are called *platelet factors* ($PF_1$ through $PF_4$) because they are produced by the platelets.

***Initiation of Coagulation***   The extrinsic mechanism is diagrammed on the top left side of figure 18.23. The damaged blood vessel and perivascular tissues release a lipoprotein mixture called **tissue thromboplastin**[27] (factor III). Factor III combines with factor VII to form a complex which, in the presence of $Ca^{2+}$, then activates factor X. The extrinsic and intrinsic pathways differ only in how

---

[26]*peri* = around + *vas* = vessel + *cular* = pertaining to

[27]*thrombo* = clot + *plast* = forming + *in* = substance

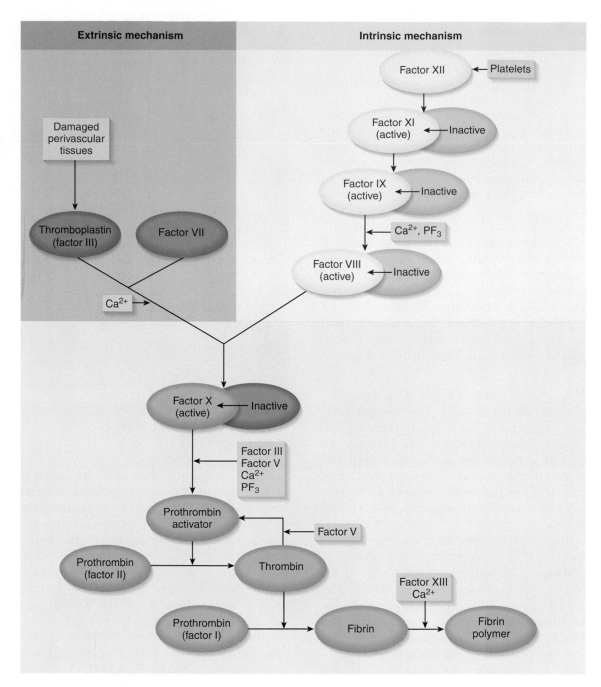

**FIGURE 18.23   The Pathways of Coagulation.**   Most clotting factors act as enzymes that convert the next factor from an inactive form (shaded ellipse) to an active form (lighter ellipse).

▶ *Would hemophilia C (see p. 709) affect the extrinsic mechanism, the intrinsic mechanism, or both?*

they arrive at active factor X. Therefore, before examining their common pathway from factor X to the end, let's consider how the intrinsic pathway reaches this step.

The intrinsic mechanism is diagrammed on the top right side of figure 18.23. Everything needed to initiate it is present in the plasma or platelets. When platelets degranulate, they release factor XII (Hageman factor, named for the patient in whom it was discovered). Through a cascade of reactions, this leads to activated fac-

tors XI, IX, and VIII, in that order—each serving as an enzyme that catalyzes the next step—and finally to factor X. This pathway also requires $Ca^{2+}$ and $PF_3$.

***Completion of Coagulation***   Once factor X is activated, the remaining events are identical in the intrinsic and extrinsic mechanisms. Factor X combines with factors III and V in the presence of $Ca^{2+}$ and $PF_3$ to produce *prothrombin activator.* This enzyme acts on a globulin called

| TABLE 18.8 | Clotting Factors (Procoagulants) | | |
|---|---|---|---|
| Number | Name | Origin | Function |
| I | Fibrinogen | Liver | Precursor of fibrin |
| II | Prothrombin | Liver | Precursor of thrombin |
| III | Tissue thromboplastin | Perivascular tissue | Activates factor VII |
| V | Proaccelerin | Liver | Activates factor VII; combines with factor X to form prothrombin activator |
| VII | Proconvertin | Liver | Activates factor X in extrinsic pathway |
| VIII | Antihemophiliac factor A | Liver | Activates factor X in intrinsic pathway |
| IX | Antihemophiliac factor B | Liver | Activates factor VIII |
| X | Thrombokinase | Liver | Combines with factor V to form prothrombin activator |
| XI | Antihemophiliac factor C | Liver | Activates factor IX |
| XII | Hageman factor | Liver, platelets | Activates factor XI and plasmin; converts prekallikrein to kallikrein |
| XIII | Fibrin-stabilizing factor | Platelets, plasma | Cross-links fibrin filaments to make fibrin polymer and stabilize clot |
| $PF_1$ | Platelet factor 1 | Platelets | Same role as factor V; also accelerates platelet activation |
| $PF_2$ | Platelet factor 2 | Platelets | Accelerates thrombin formation |
| $PF_3$ | Platelet factor 3 | Platelets | Aids in activation of factor VIII and prothrombin activator |
| $PF_4$ | Platelet factor 4 | Platelets | Binds heparin during clotting to inhibit its anticoagulant effect |

**prothrombin** (factor II) and converts it to the enzyme **thrombin.** Thrombin then chops up fibrinogen into shorter strands of fibrin. Factor XIII cross-links these fibrin strands to create a dense aggregation called *fibrin polymer,* which forms the structural framework of the blood clot.

Once a clot begins to form, it launches a self-accelerating positive feedback process that seals off the damaged vessel more quickly. Thrombin works with factor V to accelerate the production of prothrombin activator, which in turn produces more thrombin.

The cascade of enzymatic reactions acts as an amplifying mechanism to ensure the rapid clotting of blood (fig. 18.24). Each activated enzyme in the pathway produces a larger number of enzyme molecules at the following step. One activated molecule of factor XII at the start of the intrinsic pathway, for example, causes thousands of fibrin molecules to be produced very quickly. Note the similarity of this process to the *enzyme amplification* that occurs in hormone action (see fig. 17.24).

Notice that the extrinsic mechanism requires fewer steps to activate factor X than the intrinsic mechanism does; it is a "shortcut" to coagulation. It takes 3 to 6 minutes for a clot to form by the intrinsic pathway but only 15 seconds or so by the extrinsic pathway. For this reason, when a small wound bleeds, you can stop the bleeding sooner by massaging the site. This releases thromboplastin from the perivascular tissues and activates or speeds up the extrinsic pathway.

A number of laboratory tests are used to evaluate the efficiency of coagulation. Normally, the bleeding of a fin-

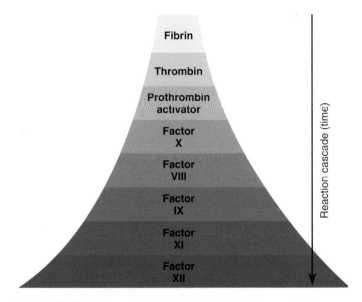

**FIGURE 18.24   Enzyme Amplification in Blood Clotting.** Each clotting factor produces many molecules of the next one, so the number of active clotting factors increases rapidly and a large amount of fibrin is quickly formed. The example shown here is for the intrinsic mechanism.

▶ *How does this compare with hormone action (chapter 17)?*

gerstick should stop within 2 to 3 minutes, and a sample of blood in a clean test tube should clot within 15 minutes. Other techniques are available that can separately assess the effectiveness of the intrinsic and extrinsic mechanisms.

# THE FATE OF BLOOD CLOTS

After a clot has formed, spinous pseudopods of the platelets adhere to strands of fibrin and contract. This pulls on the fibrin threads and draws the edges of the broken vessel together, like a drawstring closing a purse. Through this process of **clot retraction,** the clot becomes more compact within about 30 minutes.

Platelets and endothelial cells secrete a mitotic stimulant named *platelet-derived growth factor (PDGF).* PDGF stimulates fibroblasts and smooth muscle cells to multiply and repair the damaged blood vessel. Fibroblasts also invade the clot and produce fibrous connective tissue, which helps to strengthen and seal the vessel while the repairs take place.

Eventually, tissue repair is completed and the clot must be disposed of. **Fibrinolysis,** the dissolution of a clot, is achieved by a small cascade of reactions with a positive feedback component (fig. 18.25). In addition to promoting clotting, factor XII catalyzes the formation of a plasma enzyme called **kallikrein** (KAL-ih-KREE-in). Kallikrein, in turn, converts the inactive protein *plasminogen* into **plasmin,** a fibrin-dissolving enzyme that breaks up the clot. Thrombin also activates plasmin, and plasmin indirectly promotes the formation of more kallikrein, thus completing a positive feedback loop.

# PREVENTION OF INAPPROPRIATE CLOTTING

Precise controls are required to prevent coagulation when it is not needed. These include the following:

- **Platelet repulsion.** As noted earlier, platelets do not adhere to the smooth prostacyclin-coated endothelium of undamaged blood vessels.

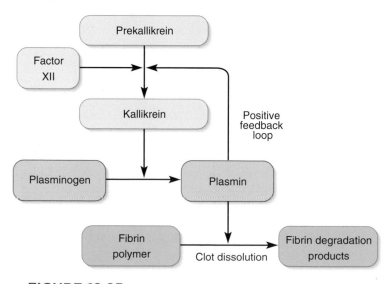

**FIGURE 18.25** **The Mechanism for Dissolving Blood Clots.** Prekallikrein is converted to kallikrein. Kallikrein is an enzyme that catalyzes the formation of plasmin. Plasmin is an enzyme that dissolves the blood clot.

- **Dilution.** Small amounts of thrombin form spontaneously in the plasma, but at normal rates of blood flow the thrombin is diluted so quickly that a clot has little chance to form. If flow decreases, however, enough thrombin can accumulate to cause clotting. This can happen in circulatory shock, for example, when output from the heart is diminished and circulation slows down.

- **Anticoagulants.** Thrombin formation is suppressed by anticoagulants that are present in the plasma. **Antithrombin,** secreted by the liver, deactivates thrombin before it can act on fibrinogen. **Heparin,** secreted by basophils and mast cells, interferes with the formation of prothrombin activator, blocks the action of thrombin on fibrinogen, and promotes the action of antithrombin. Heparin is given by injection to patients with abnormal clotting tendencies.

# CLOTTING DISORDERS

In a process as complex as coagulation, it is not surprising that things can go wrong. Clotting deficiencies can result from causes as diverse as malnutrition, leukemia, and gallstones (see Insight 18.5).

A deficiency of any clotting factor can shut down the coagulation cascade. This happens in **hemophilia,** a family of hereditary diseases characterized by deficiencies of one factor or another. Because of its sex-linked recessive mechanism of heredity, most hemophilia occurs predominantly in males. They can inherit it only from their mothers, however, as happened with the descendants of Queen Victoria. The lack of factor VIII causes *classical hemophilia (hemophilia A),* which accounts for about 83% of cases and afflicts 1 in 5,000 males worldwide. Lack of factor IX causes *hemophilia B,* which accounts for 15% of cases and occurs in about 1 out of 30,000 males. Factors VIII and IX are therefore known as *antihemophilic factors*

---

**INSIGHT 18.5**    Clinical Application

## Liver Disease and Blood Clotting

Proper blood clotting depends on normal liver function for two reasons. First, the liver synthesizes most of the clotting factors. Therefore, diseases such as hepatitis, cirrhosis, and cancer that degrade liver function result in a deficiency of clotting factors. Second, the synthesis of clotting factors II, VII, IX, and X require vitamin K. The absorption of vitamin K from the diet requires bile, a liver secretion. Gallstones can lead to a clotting deficiency by obstructing the bile duct and thus interfering with bile secretion and vitamin K absorption. Efficient blood clotting is especially important in childbirth, since both the mother and infant bleed from the trauma of birth. Therefore, pregnant women should take vitamin K supplements to ensure fast clotting, and newborn infants may be given vitamin K injections.

| TABLE 18.9 | Some Disorders of the Blood |
|---|---|
| Disseminated intravascular coagulation (DIC) | Widespread clotting within unbroken vessels, limited to one organ or occurring throughout the body. Usually triggered by septicemia but also occurs when blood circulation slows markedly (as in cardiac arrest). Marked by widespread hemorrhaging, congestion of the vessels with clotted blood, and tissue necrosis in blood-deprived organs. |
| Infectious mononucleosis | Infection of B lymphocytes with Epstein-Barr virus, most commonly in adolescents and young adults. Usually transmitted by exchange of saliva, as in kissing. Causes fever, fatigue, sore throat, inflamed lymph nodes, and leukocytosis. Usually self-limiting and resolves within a few weeks. |
| Septicemia | Bacteremia (bacteria in the bloodstream) accompanying infection elsewhere in the body. Often causes fever, chills, and nausea, and may cause DIC or septic shock (see p. 744). |
| Thalassemia | A group of hereditary anemias most common in Greeks, Italians, and others of Mediterranean descent; shows a deficiency or absence of alpha or beta hemoglobin and RBC counts that may be less than 2 million/$\mu$L. |
| Thrombocytopenia | A platelet count below 100,000/mL. Causes include bone marrow destruction by radiation, drugs, poisons, or leukemia. Signs include small hemorrhagic spots in the skin or hematomas in response to minor trauma. |

*Disorders described elsewhere*

| | | |
|---|---|---|
| Anemia p. 690 | Hypoproteinemia p. 684 | Polycythemia p. 690 |
| Embolism p. 709 | Hypoxemia p. 688 | Sickle cell disease p. 691 |
| Hematoma pp. 195, 709 | Leukemia p. 702 | Thrombosis p. 709 |
| Hemolytic disease of the newborn p. 695 | Leukocytosis p. 702 | Transfusion reaction p. 695 |
| Hemophilia p. 708 | Leukopenia p. 702 | |

*A* and *B*. A rarer form called *hemophilia C* (factor XI deficiency) is autosomal, not sex-linked, so it occurs equally in both sexes.

Before purified factor VIII became available in the 1960s, more than half of those with hemophilia died before age 5 and only 10% lived to age 21. Physical exertion causes bleeding into the muscles and joints. Excruciating pain and eventual joint immobility can result from intramuscular and joint **hematomas**[28] (masses of clotted blood in the tissues). Hemophilia varies in severity, however. Half of the normal level of clotting factor is enough to prevent the symptoms, and the symptoms are mild even in individuals with as little as 30% of the normal amount. Such cases may go undetected even into adulthood. Bleeding can be relieved for a few days by transfusion of plasma or purified clotting factors.

⌐ **Think About It**

*Why is it important for people with hemophilia not to use aspirin? (Hint: See p. 667.)*

Far more people die from unwanted blood clotting than from clotting failure. Most strokes and heart attacks are due to **thrombosis**—the abnormal clotting of blood in an unbroken vessel. A **thrombus** (clot) may grow large enough to obstruct a small vessel, or a piece of it may break loose and begin to travel in the bloodstream as an **embolus**.[29] An embolus may lodge in a small artery and block blood flow from that point on. If that vessel supplies a vital organ such as the heart, brain, lung, or kidney, *infarction* (tissue death) may result. About 650,000 Americans die annually of *thromboembolism* (traveling blood clots) in the cerebral, coronary, and pulmonary arteries.

Thrombosis is more likely to occur in veins than in arteries because blood flows more slowly in the veins and does not dilute thrombin and fibrin as rapidly. It is especially common in the leg veins of inactive people and patients immobilized in a wheelchair or bed. Most venous blood flows directly to the heart and then to the lungs. Therefore, blood clots arising in the legs or arms commonly lodge in the lungs and cause *pulmonary embolism*. When blood cannot circulate freely through the lungs, it cannot receive oxygen and a person may die of hypoxia.

Table 18.9 describes some additional disorders of the blood. The effects of aging on the blood are described on page 1130.

## *Before You Go On*

*Answer the following questions to test your understanding of the preceding section:*

25. *What are the three basic mechanisms of hemostasis?*

26. *How do the extrinsic and intrinsic mechanisms of coagulation differ? What do they have in common?*

27. *In what respect does blood clotting represent a negative feedback loop? What part of it is a positive feedback loop?*

28. *Describe some of the mechanisms that prevent clotting in undamaged vessels.*

29. *Describe a common source and effect of pulmonary embolism.*

---

[28]*hemato* = blood + *oma* = mass
[29]*em* = in, within + *bolus* = ball, mass

## INSIGHT 18.6    Clinical Application

# Clinical Management of Blood Clotting

For many cardiovascular patients, the goal of treatment is to prevent clotting or to dissolve clots that have already formed. Several strategies employ inorganic salts and products of bacteria, plants, and animals with anticoagulant and clot-dissolving effects.

## Preventing Clots from Forming

Since calcium is an essential requirement for blood clotting, blood samples can be kept from clotting by adding a few crystals of sodium oxalate, sodium citrate, or EDTA[30]—salts that bind calcium ions and prevent them from participating in the coagulation reactions. Blood-collection equipment such as hematocrit tubes may also be coated with heparin, a natural anticoagulant whose action was explained earlier.

Since vitamin K is required for the synthesis of clotting factors, anything that antagonizes vitamin K usage makes the blood clot less readily. One vitamin K antagonist is *coumarin*[31] (COO-muh-rin), a sweet-smelling extract of tonka beans, sweet clover, and other plants, used in perfume. Taken orally by patients at risk for thrombosis, coumarin takes up to 2 days to act, but it has longer-lasting effects than heparin. A similar vitamin K antagonist is the pharmaceutical preparation *warfarin*[32] *(Coumadin)*, which was originally developed as a pesticide—it makes rats bleed to death. Obviously, such anticoagulants must be used in humans with great care.

As explained in chapter 17, aspirin suppresses the formation of the eicosanoid thromboxane $A_2$, a factor in platelet aggregation. Low daily doses of aspirin can therefore suppress thrombosis and prevent heart attacks.

Many parasites feed on the blood of vertebrates and secrete anticoagulants to keep the blood flowing. Among these are segmented worms known as leeches. Leeches secrete a local anesthetic that makes their bites painless; therefore, as early as 1567 BCE, physicians used them for bloodletting. This method was less painful and repugnant to their patients than *phlebotomy*[33]—cutting a vein—and indeed, leeching became very popular. In seventeenth-century France it was quite the rage; tremendous numbers of leeches were used to treat headaches, insomnia, whooping cough, obesity, tumors, menstrual cramps, mental illness, and almost anything else doctors or their patients imagined to be caused by "bad blood."

The first known anticoagulant was discovered in the saliva of the medicinal leech, *Hirudo medicinalis,* in 1884. Named *hirudin,* it is a polypeptide that prevents clotting by inhibiting thrombin. It causes the blood to flow freely while the leech feeds and for as long as an hour thereafter. While the doctrine of bad blood is now discredited, leeches have lately reentered medical usage (fig. 18.26). A major problem in reattaching a

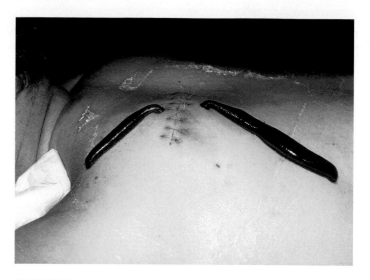

**FIGURE 18.26**    **A Modern Use of Leeching.**    Two medicinal leeches are being used to remove clotted blood from a postsurgical hematoma. Despite their formidable size, the leeches secrete a natural anesthetic and produce a painless bite.

severed body part such as a finger or ear is that the tiny veins draining these organs are too small to reattach surgically. Since arterial blood flows into the reattached organ and cannot flow out, it pools and clots there. This inhibits the regrowth of veins and the flow of fresh blood through the organ, and thus often leads to necrosis. Some vascular surgeons now place leeches on the reattached part. Their anticoagulant keeps the blood flowing freely and allows new veins to grow. After 5 to 7 days, venous drainage is restored and leeching can be stopped.

Anticoagulants also occur in the venom of some snakes. *Arvin,* for example, is obtained from the venom of the Malayan viper. It rapidly breaks down fibrinogen and may have potential as a clinical anticoagulant.

## Dissolving Clots That Have Already Formed

When a clot has already formed, it can be treated with clot-dissolving drugs such as *streptokinase,* an enzyme made by certain bacteria (streptococci). Intravenous streptokinase is used to dissolve blood clots in coronary vessels, for example. It is nonspecific, however, and digests almost any protein. *Tissue plasminogen activator (TPA)* works faster, is more specific, and is now made by transgenic bacteria. TPA converts plasminogen into the clot-dissolving enzyme plasmin. Some anticoagulants of animal origin also work by dissolving fibrin. A giant Amazon leech, *Haementeria,* produces one such anticoagulant named *hementin.* This, too, has been successfully produced by genetically engineered bacteria and used to dissolve blood clots in cardiac patients.

---

[30]ethylenediaminetetraacetic acid
[31]*coumaru*, tonka bean tree
[32]acronym from Wisconsin Alumni Research Foundation
[33]*phlebo* = vein + *tomy* = cutting

# CHAPTER REVIEW

# Review of Key Concepts

**Introduction (p. 680)**

1. The circulatory system consists of the heart, blood vessels, and blood.

2. Blood serves to transport $O_2$, $CO_2$, nutrients, wastes, hormones, stem cells, and heat; to protect the body by means of antibodies, leukocytes, platelets, and its roles in inflammation; and it helps to stabilize the body's fluid balance and pH.

3. The blood is about 55% *plasma* and 45% *formed elements* by volume. The formed elements include erythrocytes, platelets, and five kinds of leukocytes.

4. Blood plasma is the clear fluid in which the formed elements are suspended. It is a mixture of water, nutrients, electrolytes, wastes, hormones, gases, and proteins. Its major proteins are albumin, globulins, and fibrinogen.

5. The viscosity of blood, stemming mainly from its RBCs and proteins, is an important factor in blood flow.

6. The osmolarity of blood, stemming mainly from its RBCs, proteins, and $Na^+$, governs its water content and is thus a major factor in blood volume and pressure. The protein contribution to osmolarity is the *colloid osmotic pressure*.

7. *Hemopoiesis* is production of the formed elements of blood. It begins in the embryonic yolk sac and continues in the fetal bone marrow, liver, spleen, and thymus. From infancy onward, it occurs in the bone marrow *(myeloid hemopoiesis)* and lymphoid tissues *(lymphoid hemopoiesis)*.

8. All formed elements derive ultimately from a *pluripotent stem cell* in the red bone marrow.

**Erythrocytes (p. 685)**

1. RBCs transport $O_2$ and $CO_2$. They are discoid cells with a sunken center and no organelles, but they do have a cytoskeleton of *spectrin* and *actin* that reinforces the plasma membrane.

2. The most important components of the cytoplasm are *hemoglobin (Hb)* and *carbonic anhydrase (CAH)*. Hb transports nearly all of the $O_2$ and some of the $CO_2$ in the blood, and CAH catalyzes the reversible reaction $CO_2 + H_2O \leftrightarrows H_2CO_3$.

3. Hemoglobin consists of four peptide chains—two alpha and two beta—each with a *heme* moiety.

4. Oxygen binds to the $Fe^{2+}$ at the center of each heme. About 5% of the $CO_2$ in the blood binds to the globin moiety for transport.

5. The quantities of RBCs and Hb are clinically important. They are measured in terms of hematocrit (percent of the blood volume composed of RBCs), hemoglobin concentration (g/dL), and RBC count (RMC μL of blood). Normal averages are lower in women than in men.

6. *Erythropoiesis*, or RBC production, proceeds through the following stages: *pluripotent stem cell → erythocyte colony-forming unit → erthroblast → reticulocyte → erythrocyte*. These transformations entail a reduction in cells size, increase in cell number, synthesis of hemoglobin, and loss of organelles.

7. Iron, in the form of ferrous ions $(Fe^{2+})$, is essential for hemoglobin synthesis. Dietary $Fe^{3+}$ is converted to $Fe^{2+}$ by stomach acid, then binds to *gastroferritin*, is absorbed into the blood, and binds with the plasma protein *transferrin*. Transferrin transports $Fe^{2+}$ to the myeloid tissue and liver. The liver stores excess iron in *ferritin*.

8. Erythropoiesis is stimulated by the hormone *erythropoietin*. It is regulated by a negative feedback loop that responds to *hypoxemia* with increased EPO secretion, and thus increased erythropoiesis.

9. An RBC lives for about 120 days, grows increasingly fragile, and then breaks apart, especially in the spleen. *Hemolysis*, the rupture of RBCs, releases cell fragments and free hemoglobin.

10. Hemoglobin is broken down into its globin and heme moieties. The globin is hydrolyzed into its free amino acids, which are reused. The heme is broken down into its $Fe^{2+}$ and organic components. The $Fe^{2+}$ is recycled or stored, and the organic component eventually becomes *biliverdin* and *bilirubin* (bile pigments), which are excreted as waste.

11. An excessive RBC count is *polycythemia*. *Primary polycythemia* results from cancer of the bone marrow, and *secondary polycythemia* from many other causes, such as dehydration, smoking, high altitude, and habitual strenuous exercise. Polycythemia increases blood volume, pressure, and viscosity to sometimes dangerous levels.

12. A deficiency of RBCs is *anemia*. Causes of anemia, classified and described in table 18.5, essentially fall into three categories: inadequate erythropoiesis, hemorrhage, or hemolysis.

13. The effects of anemia include tissue hypoxia and necrosis, reduced blood osmolarity, and reduced blood viscosity.

14. Sickle cell disease and thalassemia are hereditary hemoglobin defects that result in severe anemia and multiple other effects.

**Blood Types (p. 692)**

1. Blood types are determined by antigenic glycoproteins and glycolipids on the RBC surface. Incompatibility of one person's blood with another results from the action of plasma antibodies against these RBC antigens.

2. Blood types A, B, AB, and O form the ABO blood group. The first two have antigen A or B on the RBC surface, the third has both A and B, and type O has neither.

3. A few months after birth, a person develops anti-A and anti-B antibodies

against intestinal bacteria. These antibodies cross-react with foreign ABO antigens and thus limit transfusion compatibility.

4. When anti-A reacts with type A or AB red cells, or anti-B reacts with type B or AB red cells, the red cells agglutinate and hemolyze, causing a severe *transfusion reaction* that can lead to renal failure and death.

5. The Rh blood group is inherited through genes called C, D, and E. Anyone with genotype *DD* or *Dd* is Rh-positive (Rh⁺).

6. An Rh-negative (Rh⁻) person who is exposed to Rh⁺ RBCs through transfusion or childbirth develops an anti-D antibody. Later exposures to Rh⁺ red cells can cause a transfusion reaction.

7. Rh incompatibility between a sensitized Rh⁻ woman and an Rh⁺ fetus can cause *hemolytic disease of the newborn,* a severe neonatal anemia that must be treated by phototherapy or transfusion.

8. Many other blood groups besides ABO and Rh exist. They rarely cause transfusion reactions but are useful in paternity and criminal cases and for studies of population genetics.

### Leukocytes (p. 697)

1. WBCs play various roles in defending the body from pathogens. Neutrophils, eosinophils, and basophils are classified as *granulocytes,* and lymphocytes and monocytes are classified as *agranulocytes.* The appearance and function of each type are detailed in table 18.7.

2. *Leukopoiesis,* the production of WBCs, follows three principal lines stemming from pluripotent stem cells (PPSCs). PPSCs develop into myeloblasts (precursors of the granulocytes), monoblasts (precursors of monocytes), and lymphoblasts (precursors of lymphocytes).

3. Circulating WBCS remain in the bloodstream for only a matter of hours and spend most of their lives in other tissues. Lymphocytes cycle repeatedly from blood to tissue fluids to lymph and back to the blood.

4. A WBC deficiency, called *leukopenia,* may result from chemical or radiation

poisoning, various infections, and certain drugs. It reduces a person's resistance to infection and cancer.

5. A WBC excess, called *leukocytosis,* may result from infection, allergy, dehydration, or emotional disorders, or from *leukemia* (cancer of the hemopoietic tissues).

6. Leukemia is classified by site of origin as *myeloid* or *lymphoid,* and by speed of progression as *acute* or *chronic.* Leukemia increases the risk of *opportunistic infection* and is typically accompanied by RBC and platelet deficiencies.

### Platelets and Hemostasis—The Control of Bleeding (p. 703)

1. Platelets are not cells, but small, mobile, phagocytic fragments of megakaryocyte cytoplasm, second only to RBCs in abundance. They are filled with diverse organelles but have no nuclei.

2. Platelets have a wide variety of functions: secretion of vasoconstrictors, formation of platelet plugs to plug small broken blood vessels, secretion of clotting factors, dissolving old blood clots, attracting neutropils and monocytes to inflamed tissue, phagocytizing bacteria, and secretion growth factors that stimulate tissue repair and maintenance.

3. *Thrombopoiesis,* the production of platelets, also begins with PPSCs of the bone marrow. The hormone *thrombopoietin* induces the differentiation of large cells called *megakaryocytes,* which pinch off bits of peripheral cytoplasm that break up into platelets.

4. Breakage of a blood vessel leads first to *vascular spasm,* then formation of a *platelet plug,* and third but most effectively, *coagulation* (formation of a blood clot).

5. The objective of coagulation is to form a mesh of sticky protein called *fibrin.* There are two biochemical pathways to fibrin production, called the *extrinsic* and *intrinsic mechanisms.* Both pathways involve a self-amplifying chain reaction, or *reaction cascade,* of chemicals called *procoagulants.*

6. The extrinsic mechanism depends on chemicals released by damaged cells outside the bloodstream. It begins with release of a lipoprotein called *tissue thromboplastin* and leads to activation of a procoagulant called *factor X.*

7. The intrinsic mechanism employs only factors found in the blood plasma or platelets. It begins with *factor XII* and likewise ends with the activation of factor X.

8. Beyond the activation of factor X, events are identical regardless of intrinsic or extrinsic beginnings. The remaining steps include activation of the enzyme *thrombin,* which cuts plasma fibrinogen into fibrin. Fibrin then polymerizes to form the weblike matrix of the blood clot.

9. Positive feedback and enzyme amplification ensure rapid clotting and the production of a large amount of fibrin in spite of small amounts of the other procoagulants that drive the process.

10. Once formed, a blood clot undergoes *clot retraction,* a process that helps to seal the wound. *Plateletderived growth factor* promotes repair of the damaged blood vessel and surrounding connective tissues. Tissue repair is followed by *fibrinolysis,* in which the blood clot, no longer needed, is dissolved by the enzyme *plasmin.*

11. Inappropriate coagulation is normally prevented by the repulsion of platelets by prostacyclin on the blood vessel endothelium, dilution of the small amounts of thrombin that form spontaneously, and anticoagulants such as *heparin.*

12. Clotting deficiency can result from *thrombocytopenia* (low platelet count) or *hemophilia* (hereditary deficiency in procoagulant structure and function, especially in factor VIII).

13. Unwanted clotting in unbroken blood vessels is called *thrombosis.* A *thrombus* (clot) can break loose and become a traveling *embolus,* which can cause a sometimes fatal obstruction of small blood vessels.

# Testing Your Recall

1. Antibodies belong to a class of plasma proteins called
   a. albumins.
   b. gamma globulins.
   c. alpha globulins.
   d. procoagulants.
   e. agglutinins.

2. Serum is blood plasma minus its
   a. sodium ions.
   b. calcium ions.
   c. clotting proteins.
   d. globulins.
   e. albumins.

3. Which of the following conditions is most likely to cause hemolytic anemia?
   a. folic acid deficiency
   b. iron deficiency
   c. mushroom poisoning
   d. alcoholism
   e. hypoxemia

4. It is impossible for a type $O^+$ baby to have a type _____ mother.
   a. $AB^-$
   b. $O^-$
   c. $O^+$
   d. $A^!$
   e. $B^+$

5. Which of the following is *not* a component of hemostasis?
   a. platelet plug formation
   b. agglutination
   c. clot retraction
   d. a vascular spasm
   e. degranulation

6. Which of the following contributes most to the viscosity of blood?
   a. albumin
   b. sodium
   c. globulins
   d. erythrocytes
   e. fibrin

7. Which of these is a granulocyte?
   a. a monocyte
   b. a lymphocyte
   c. a macrophage
   d. an eosinophil
   e. an erythrocyte

8. Excess iron is stored in the liver as a complex called
   a. gastroferritin.
   b. transferrin.
   c. ferritin.
   d. hepatoferritin.
   e. erythropoietin.

9. Pernicious anemia is a result of
   a. hypoxemia.
   b. iron deficiency.
   c. malaria.
   d. lack of intrinsic factor.
   e. Rh incompatibility.

10. The first clotting factor that the intrinsic and extrinsic pathways have in common is
    a. thromboplastin.
    b. Hageman factor.
    c. factor X.
    d. prothrombin activator.
    e. factor VIII.

11. Production of all the formed elements of blood is called _____.

12. The percentage of blood volume composed of RBCs is called the _____.

13. The extrinsic pathway of coagulation is activated by _____ from damaged perivascular tissues.

14. The RBC antigens that determine transfusion compatibility are called _____.

15. The hereditary lack of factor VIII causes a disease called _____.

16. The overall cessation of bleeding, involving several mechanisms, is called _____.

17. _____ results from a mutation that changes one amino acid in the hemoglobin molecule.

18. An excessively high RBC count is called _____.

19. Intrinsic factor enables the small intestine to absorb _____.

20. The kidney hormone _____ stimulates RBC production.

*Answers in Appendix B*

# True or False

*Determine which five of the following statements are false, and briefly explain why.*

1. By volume, the blood usually contains more plasma than blood cells.

2. An increase in the albumin concentration of the blood would tend to increase blood pressure.

3. Anemia is caused by a low oxygen concentration in the blood.

4. Hemostasis, coagulation, and clotting are three terms for the same process.

5. A man with blood type $A^+$ and a woman with blood type $B^+$ could have a baby with type $O^-$.

6. Lymphocytes are the most abundant WBCs in the blood.

7. Calcium ions are required for blood clotting.

8. All formed elements of the blood come ultimately from pluripotent stem cells.

9. When RBCs die and break down, the globin moiety of hemoglobin is excreted and the heme is recycled to new RBCs.

10. Leukemia is a severe deficiency of white blood cells.

*Answers in Appendix B*

# Testing Your Comprehension

1. Why would erythropoiesis not correct the hypoxemia resulting from lung cancer?

2. People with chronic kidney disease often have hematocrits of less than half the normal value. Explain why.

3. An elderly white woman is hit by a bus and severely injured. Accident investigators are informed that she lives in an abandoned warehouse where her few personal effects include several empty wine bottles and an expired driver's license indicating she is 72 years old. At the hospital, she is found to be severely anemic. List all the factors you can think of that may contribute to her anemia.

4. How is coagulation different from agglutination?

5. Although fibrinogen and prothrombin are equally necessary for blood clotting, fibrinogen is about 4% of the plasma protein whereas prothrombin is present only in small traces. In light of the roles of these clotting factors and your knowledge of enzymes, explain this difference in their abundance.

*Answers at www.mhhe.com/saladin4*

# www.mhhe.com/saladin4

The textbook website provides a wealth of interactive study materials fully organized and integrated by chapter. You will find practice quizzes, labeling exercises, and much more that will complement your learning and understanding of anatomy and physiology. The website also includes tools designed to enhance your **Anatomy & Physiology | REVEALED** experience.

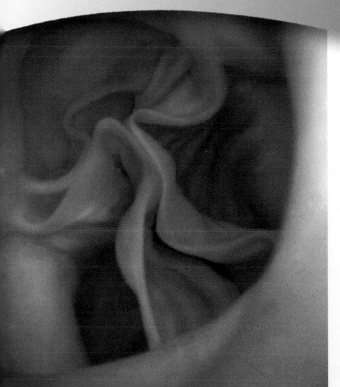

*A semilunar valve of the heart (endoscopic photo)*

# THE CIRCULATORY SYSTEM: THE HEART

## CHAPTER OUTLINE

## INSIGHTS

## Brushing Up

To understand this chapter, it is important that you understand or brush up on the following concepts.

- Desmosomes and gap junctions (p. 175)
- Ultrastructure of striated muscle (pp. 407–410)
- Excitation-contraction coupling in muscle (p. 417)
- Length-tension relationship in muscle fibers (p. 421)
- Action potentials (p. 456)

We are more conscious of our heart than we are of most organs, and more wary of its failure. Speculation about the heart is at least as old as written history. Some ancient Chinese, Egyptian, Greek, and Roman scholars correctly surmised that the heart is a pump for filling the vessels with blood. Aristotle's views, however, were a step backward. Perhaps because the heart quickens its pace when we are emotionally aroused, and because grief causes "heartache," he regarded it primarily as the seat of emotion, as well as a source of heat to aid digestion. During the Middle Ages, Western medical schools clung dogmatically to the ideas of Aristotle. Perhaps the only significant advance came from Muslim medicine, when thirteenth-century physician Ibn an-Nafis described the role of the coronary blood vessels in nourishing the heart. The sixteenth-century dissections and anatomical charts of Vesalius, however, greatly improved knowledge of cardiovascular anatomy and set the stage for a more scientific study of the heart and treatment of its disorders—the science we now call **cardiology.**[1]

In the early decades of the twentieth century, little could be recommended for heart disease other than bed rest. Then nitroglycerin was found to improve coronary circulation and relieve the pain resulting from physical exertion, digitalis proved effective for treating abnormal heart rhythms, and diuretics were first used to reduce hypertension. Coronary bypass surgery, replacement of diseased valves, clot-dissolving enzymes, heart transplants, artificial pacemakers, and artificial hearts have made cardiology one of the most dramatic and attention-getting fields of medicine in the last quarter-century.

# Overview of the Cardiovascular System

### Objectives

When you have completed this section, you should be able to

- define and distinguish between the pulmonary and systemic circuits;
- describe the general location, size, and shape of the heart; and
- describe the pericardial sac that encloses the heart.

[1]*cardio* = heart + *logy* = study

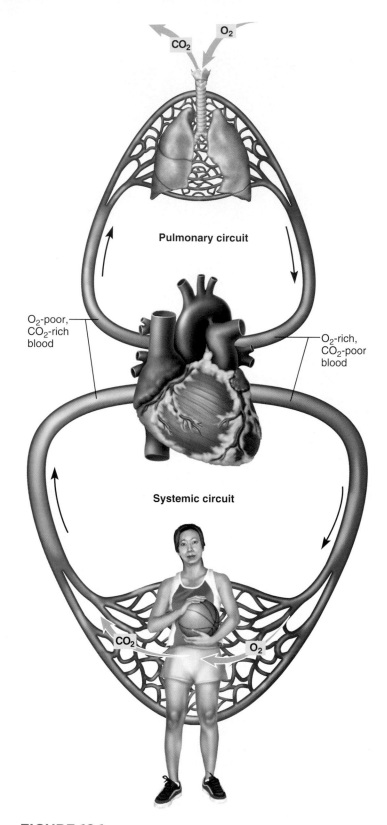

**FIGURE 19.1** General Schematic of the Cardiovascular System.

## THE PULMONARY AND SYSTEMIC CIRCUITS

The **cardiovascular system** consists of the heart and the blood vessels that carry the blood to and from the body's organs (fig. 19.1). The system has two major divisions: a

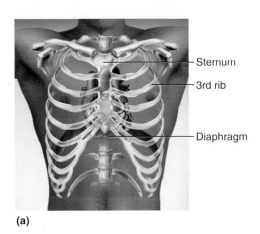

(a)

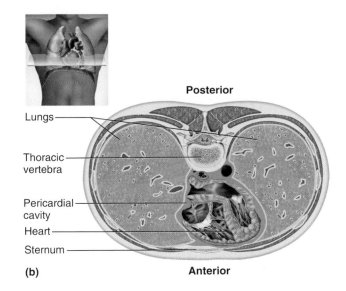

(b)

**FIGURE 19.2  Position of the Heart in the Thoracic Cavity.**  (a) Relationship to the thoracic cage. (b) Cross section of the thorax at the level of the heart. (c) Frontal view with the lungs slightly retracted and the pericardial sac opened.

▶ *Does most of the heart lie to the right or left of the median plane?*

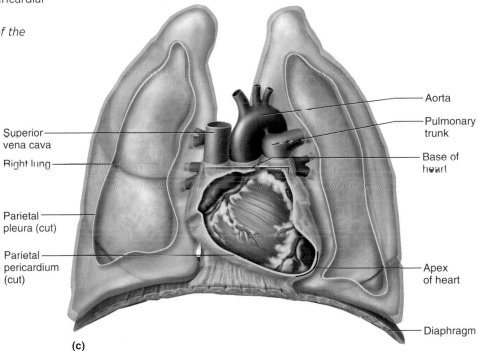

(c)

**pulmonary circuit,** which carries blood to the lungs for gas exchange and returns it to the heart, and a **systemic circuit,** which supplies blood to every organ of the body, including other parts of the lungs and the wall of the heart itself. The right side of the heart serves the pulmonary circuit. It receives blood that has circulated through the body, unloaded its oxygen and nutrients, and picked up a load of carbon dioxide and other wastes. It pumps this oxygen-poor blood into a large artery, the *pulmonary trunk,* which immediately divides into right and left *pulmonary arteries.* These transport blood to the air sacs *(alveoli)* of the lungs, where carbon dioxide is unloaded and oxygen is picked up. The oxygen-rich blood then flows by way of the *pulmonary veins* to the left side of the heart.

The left side serves the systemic circuit. Blood leaves it by way of another large artery, the *aorta.* The aorta takes a sharp inverted U-turn, the *aortic arch,* and passes downward, dorsal to the heart. The aortic arch gives off arteries that supply the head, neck, and upper limbs. The aorta then travels through the thoracic and abdominal cavities and issues smaller arteries to the other organs. After circulating through the body, the now-deoxygenated systemic blood returns to the right side of the heart mainly by way of two large veins: the *superior vena cava* (draining the head, neck, upper limbs, and thoracic organs) and *inferior vena cava* (draining the organs below the diaphragm). The

major arteries and veins entering and leaving the heart are called the *great vessels* (great arteries and veins) because of their relatively large diameters.

## POSITION, SIZE, AND SHAPE OF THE HEART

The heart is located in the thoracic cavity in the mediastinum, between the lungs and deep to the sternum. From its superior to inferior midpoints, it is tilted toward the left, so about two-thirds of the heart lies to the left of the median plane (figs A.18–A.19 and 19.2). The broad superior portion of the heart, called the **base,** is the point of attachment for the great vessels described previously.

The inferior end tapers to a blunt point, the **apex** of the heart, immediately above the diaphragm.

The adult heart is about 9 cm (3.5 in.) wide at the base, 13 cm (5 in.) from base to apex, and 6 cm (2.5 in.) from anterior to posterior at its thickest point—roughly the size of one's fist. It weighs about 300 g (10 oz).

## THE PERICARDIUM

The heart is enclosed in a double-walled sac called the **pericardium.**[2] The outer wall, called the **parietal pericardium** (pericardial sac), has a tough, superficial *fibrous layer* of dense irregular connective tissue and a deep, thin *serous layer*. The serous layer turns inward at the base of the heart and forms the **visceral pericardium** (epicardium) covering the heart surface (fig. 19.3). The pericardial sac is anchored by ligaments to the diaphragm below and the sternum anterior to it, and more loosely anchored by fibrous connective tissue to mediastinal tissue dorsal to the heart.

Between the parietal and visceral membranes is a space called the **pericardial cavity** (see figs. 19.2b and

19.3). It contains 5 to 30 mL of **pericardial fluid,** exuded by the serous pericardium. The pericardial fluid lubricates the membranes and allows the heart to beat almost without friction. In *pericarditis*—inflammation of the pericardium—the membranes may become dry and produce a painful *friction rub* with each heartbeat. In addition to reducing friction, the pericardium isolates the heart from other thoracic organs, allows the heart room to expand, yet resists excessive expansion. (See *cardiac tamponade* in table 19.3, p. 745).

### Before You Go On

*Answer the following questions to test your understanding of the preceding section:*

1. Distinguish between the pulmonary and systemic circuits and state which part of the heart serves each one.

2. Make a two-color sketch of the pericardium; use one color for the fibrous pericardium and another for the serous pericardium and show their relationship to the heart wall and pericardial cavity.

# Gross Anatomy of the Heart

### Objectives

When you have completed this section, you should be able to

- describe the three layers of the heart wall;
- identify the four chambers of the heart;
- identify the surface features of the heart and correlate them with its internal four-chambered anatomy;
- identify the four valves of the heart;
- trace the flow of blood through the four chambers of the heart and adjacent blood vessels; and
- describe the arteries that nourish the myocardium and the veins that drain it.

## THE HEART WALL

The heart wall consists of three layers: a thin *epicardium* covering its external surface, a thick muscular *myocardium* in the middle, and a thin *endocardium* lining the interior of the chambers.

The **epicardium**[3] (visceral pericardium) is a serous membrane on the heart surface. It consists mainly of a simple squamous epithelium overlying a thin layer of areolar tissue. In some places, it also includes a thick layer of adipose tissue, whereas in other areas it is fat-free and

---

[2]*peri* = around + *cardi* = heart

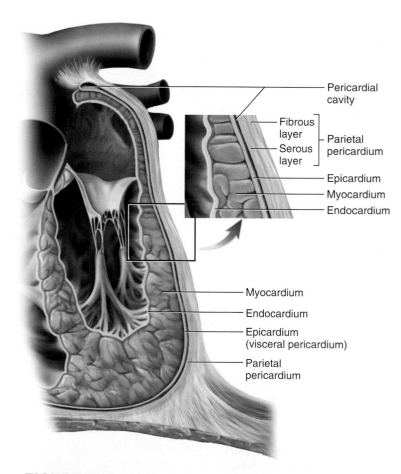

**FIGURE 19.3**  **The Pericardium and Heart Wall.**  The inset shows the layers of the heart wall in relation to the pericardium.

PeUcardial cavity — Fibrous layer / Serous layer — Parietal pericardium — Epicardium — Myocardium — Endocardium

Myocardium — Endocardium — Epicardium (visceral pericardium) — Parietal pericardium

---

[3]*epi* = upon + *cardi* = heart

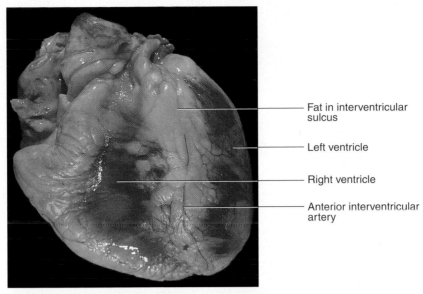

Fat in interventricular sulcus

Left ventricle

Right ventricle

Anterior interventricular artery

**(a) Anterior view, external anatomy**

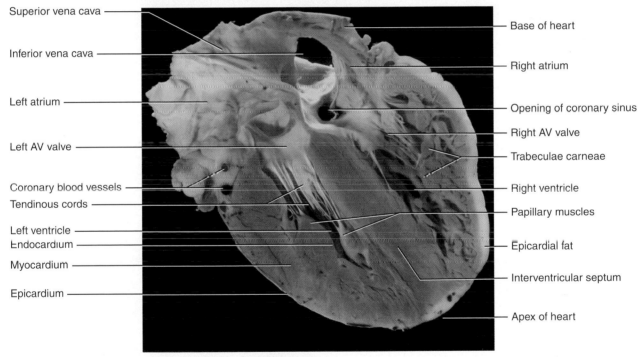

Superior vena cava

Inferior vena cava

Left atrium

Left AV valve

Coronary blood vessels

Tendinous cords

Left ventricle

Endocardium

Myocardium

Epicardium

Base of heart

Right atrium

Opening of coronary sinus

Right AV valve

Trabeculae carneae

Right ventricle

Papillary muscles

Epicardial fat

Interventricular septum

Apex of heart

**(b) Posterior view, internal anatomy**

**FIGURE 19.4** The Human Heart.

translucent, so the muscle of the underlying myocardium shows through (figs. 19.4a and 19.5). The largest branches of the coronary blood vessels travel through the epicardium. A similar layer, the **endocardium**,[4] lines the interior of the heart chambers (see fig. 19.4b). It is a simple squamous endothelium overlying a thin areolar tissue layer; it has no adipose tissue. The endocardium covers the valve surfaces and is continuous with the endothelium of the blood vessels.

The layer between these two, the **myocardium**,[5] is composed of cardiac muscle. This is by far the thickest layer and performs the work of the heart. Its thickness varies greatly from one heart chamber to another and is proportional to the workload on the individual chambers. Its muscle spirals around the heart (fig. 19.6), so when the ventricles contract, they exhibit a twisting or wringing motion. The microscopic structure of the cardiac muscle cells (*cardiac myocytes* or *cardiocytes*) is detailed later.

[4]*endo* = internal + *cardi* = heart

[5]*myo* = muscle + *cardi* = heart

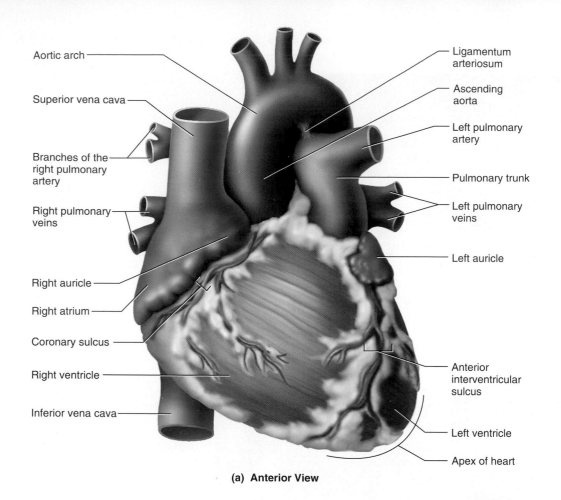

Aortic arch

Superior vena cava

Branches of the right pulmonary artery

Right pulmonary veins

Right auricle

Right atrium

Coronary sulcus

Right ventricle

Inferior vena cava

Ligamentum arteriosum

Ascending aorta

Left pulmonary artery

Pulmonary trunk

Left pulmonary veins

Left auricle

Anterior interventricular sulcus

Left ventricle

Apex of heart

**(a) Anterior View**

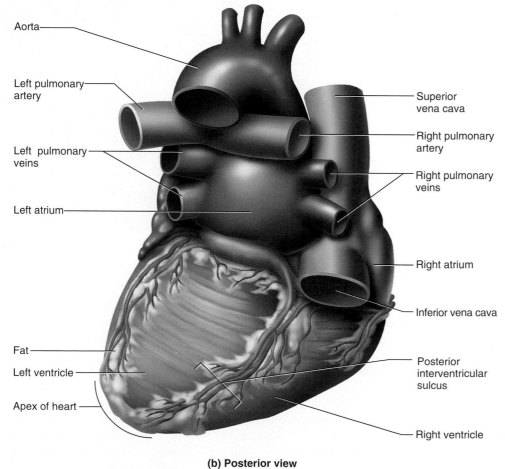

Aorta

Left pulmonary artery

Left pulmonary veins

Left atrium

Fat

Left ventricle

Apex of heart

Superior vena cava

Right pulmonary artery

Right pulmonary veins

Right atrium

Inferior vena cava

Posterior interventricular sulcus

Right ventricle

**(b) Posterior view**

**FIGURE 19.5** **Surface Anatomy of the Heart.** The coronary blood vessels on the heart surface are identified in figure 19.11.

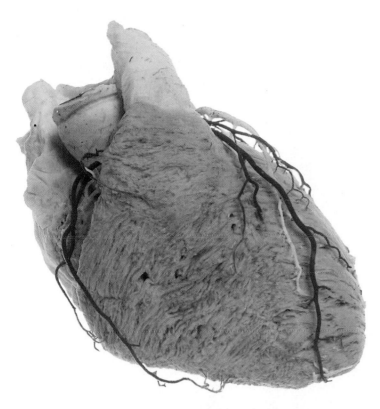

**FIGURE 19.6** Spiral Orientation of Myocardial Muscle.
A heart with the epicardium stripped off to expose the bundles of myocardial muscle.

The heart also has a meshwork of collagenous and elastic fibers that make up the **fibrous skeleton.** This tissue is especially concentrated in the walls (septa) between the heart chambers, in *fibrous rings* (anuli fibrosi) around the openings of the heart valves, and in sheets of tissue that interconnect these rings. The fibrous skeleton has multiple functions: (1) It provides structural support for the heart, especially around the valves and the openings of the great vessels; it holds the valve orifices open and prevents them from excessively stretching when blood surges through them. (2) It anchors the myocytes and gives them something to pull against. (3) As a nonconductor of electricity, it serves as electrical insulation between the atria and the ventricles, so the atria cannot stimulate the ventricles directly. This insulation is important to the timing and coordination of electrical and contractile activity. (4) Some authorities think (while others disagree) that elastic recoil of the fibrous skeleton may aid in refilling the heart with blood after each beat, like a hollow rubber ball that expands when you relax your grip.

⌐ **Think About It**

*Parts of the fibrous skeleton sometimes become calcified in old age. How would you expect this to affect cardiac function?*

## THE CHAMBERS

The heart has four chambers, best seen in a frontal section (figs. 19.4b and 19.7). The two at the superior pole (base) of the heart are the **right** and **left atria** (AY-tree-uh; singular *atrium*[6]). They are thin-walled receiving chambers for blood returning to the heart by way of the great veins. Most of the mass of each atrium is on the posterior side of the heart, so only a small portion is visible from the anterior view. Here, each atrium has a small earlike extension called an *auricle*[7] that slightly increases its volume.

The two inferior heart chambers, the **right** and **left ventricles,**[8] are the pumps that eject blood into the arteries and keep it flowing around the body. The right ventricle constitutes most of the anterior aspect of the heart, whereas the left ventricle forms the apex and inferoposterior aspect.

On the surface, the boundaries of the four chambers are marked by three sulci (grooves). The sulci are occupied largely by fat and coronary blood vessels (see fig. 19.5a). The **coronary**[9] **(atrioventricular) sulcus** encircles the heart near the base and separates the atria above from the ventricles below. It can be exposed by lifting the margins of the atria. The other two sulci extend obliquely down the heart from the coronary sulcus toward the apex, one on the front of the heart called the **anterior interventricular sulcus** and one on the back called the **posterior interventricular sulcus.** These sulci overlie an internal *interventricular septum* that divides the right ventricle from the left. The coronary sulcus and two interventricular sulci harbor the largest of the coronary blood vessels.

The atria exhibit thin flaccid walls corresponding to their light workload—all they do is pump blood into the ventricles immediately below (see fig. 19.7). They are separated from each other by a wall called the **interatrial septum.** The right atrium and both auricles exhibit internal ridges of myocardium called **pectinate**[10] **muscles.** The **interventricular septum** is a much more muscular, vertical wall between the ventricles. The right ventricle pumps blood only to the lungs and back, so its wall is only moderately muscular. The wall of the left ventricle is two to four times as thick because it bears the greatest workload of all four chambers, pumping blood through the entire body. Both ventricles exhibit internal ridges called **trabeculae carneae**[11] (trah-BEC-you-lee CAR-nee-ee).

## THE VALVES

To pump blood effectively, the heart needs valves that ensure a predominantly one-way flow. There is a valve between each atrium and its ventricle and another at

---

[6]*atrium* = entryway
[7]*auricle* = little ear
[8]*ventr* = belly, lower part + *icle* = little
[9]*coron* = crown + *ary* = pertaining to
[10]*pectin* = comb + *ate* = like
[11]*trabecula* = little beam + *carne* = flesh, meat

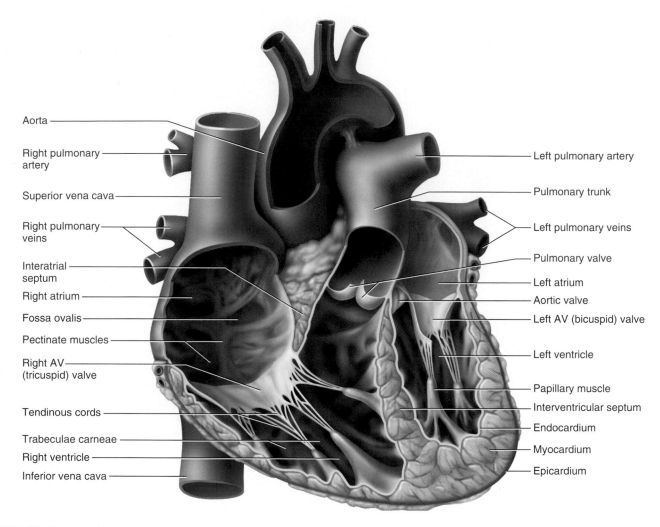

**FIGURE 19.7**  **Internal Anatomy of the Heart.**  Anterior view.

▶ *Do the atrial pectinate muscles more nearly resemble the ventricular papillary muscles or the trabeculae carneae?*

the exit from each ventricle into its great artery (see fig. 19.7), but there are no valves where the great veins empty into the atria. Each valve consists of two or three fibrous flaps of tissue called **cusps** or **leaflets,** covered with endothelium.

The **atrioventricular (AV) valves** regulate the openings between the atria and ventricles. The **right AV (tricuspid) valve** has three cusps and the **left AV (bicuspid) valve** has two (fig. 19.8). The left AV valve is also known as the **mitral** (MY-trul) **valve** after its resemblance to a miter, the head-dress of a church bishop. Stringlike **tendinous cords** (chordae tendineae), reminiscent of the shroud lines of a parachute, connect the valve cusps to conical **papillary**[12] **muscles** on the floor of the ventricle.

The **semilunar**[13] **valves** (pulmonary and aortic valves) regulate the flow of blood from the ventricles into the great arteries. The **pulmonary valve** controls the opening from the right ventricle into the pulmonary trunk, and the **aortic valve** controls the opening from the left ventricle into the aorta. Each has three cusps shaped somewhat like shirt pockets (see photograph on p. 715). There are no tendinous cords on the semilunar valves.

The opening and closing of heart valves is the result of pressure gradients between the upstream and downstream sides of the valve (fig. 19.9). When the ventricles are relaxed and their pressure is low, the AV valve cusps hang down limply and both AV valves are open. Blood flows freely from the atria into the ventricles even before

---

[12]*papill* = nipple + *ary* = like, shaped

[13]*semi* = half + *lunar* = like the moon

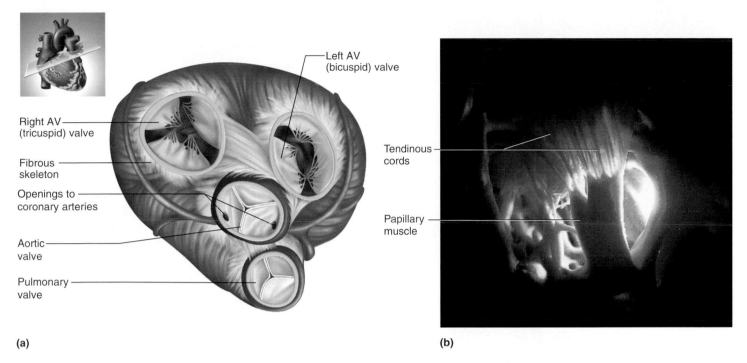

**FIGURE 19.8  The Heart Valves.** (a) Superior view of the heart with the atria removed. (b) Papillary muscle and tendinous cords seen from within the right ventricle. The upper ends of the cords are attached to the cusps of the right AV valve.

the atria contract. When the ventricles have filled with blood and begin to contract, their internal pressure rises and blood surges against the AV valves. This pushes their cusps together, seals the openings, and prevents blood from flowing back into the atria. The papillary muscles contract with the rest of the ventricular myocardium and tug on the tendinous cords, which prevents the valves from bulging excessively (prolapsing) into the atria or turning inside out like windblown umbrellas (see *mitral valve prolapse* in Insight 19.1).

As the pressure rises in the contracting ventricles, it soon exceeds the pressure downstream from it in the pulmonary trunk and aorta. The ventricular blood then forces the pulmonary and aortic valves open, and blood is ejected from the heart. Then as the ventricles relax again and their pressure falls below that in the arteries, arterial blood briefly flows backward and fills the pocketlike cusps of the semilunar valves. The three cusps meet in the middle of the orifice and seal it, thereby preventing blood from reentering the heart.

**Think About It**

*How would valvular stenosis (see Insight 19.1) affect the amount of blood pumped into the aorta? How might this affect a person's physical stamina? Explain your reasoning.*

**INSIGHT 19.1    Clinical Application**

### Valvular Insufficiency

*Valvular insufficiency* (incompetence) refers to any failure of a valve to prevent *reflux* (regurgitation)—the backward flow of blood. *Valvular stenosis*[14] is a form of insufficiency in which the cusps are stiffened and the opening is constricted by scar tissue. It frequently results from rheumatic fever, an autoimmune disease in which antibodies produced to fight a bacterial infection also attack the mitral and aortic valves. As the valves become scarred and constricted, the heart is overworked by the effort to force blood through the openings and may become enlarged. Regurgitation of blood through the incompetent valves creates turbulence that can be heard as a *heart murmur.*

*Mitral valve prolapse (MVP)* is an insufficiency in which one or both mitral valve cusps bulge into the atrium during ventricular contraction. It is often hereditary and affects about 1 out of 40 people, especially young women. In many cases, it causes no serious dysfunction, but in some people it causes chest pain, fatigue, and shortness of breath. An incompetent valve can eventually lead to heart failure. A defective valve can be replaced with an artificial valve or a valve transplanted from a pig heart.

[14]*sten* = narrow + *osis* = condition

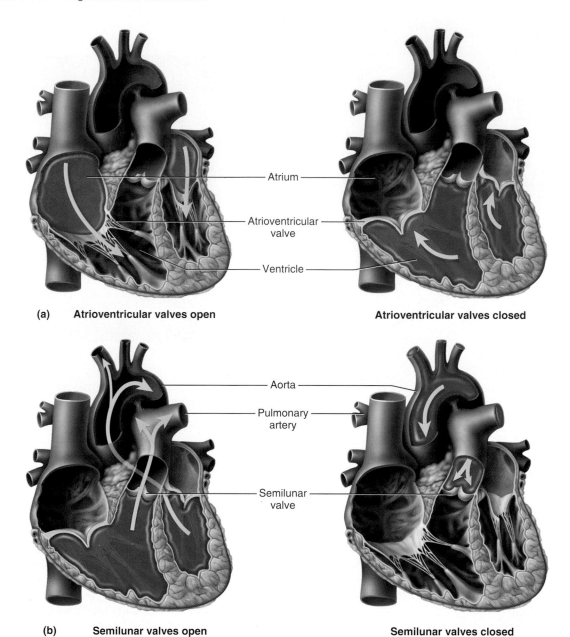

(a)    **Atrioventricular valves open**            **Atrioventricular valves closed**

Atrium

Atrioventricular
valve

Ventricle

Aorta

Pulmonary
artery

Semilunar
valve

(b)    **Semilunar valves open**            **Semilunar valves closed**

**FIGURE 19.9  Operation of the Heart Valves.**  (a) The atrioventricular valves. When atrial pressure is greater than ventricular pressure, the valve opens and blood flows through (green arrows). When ventricular pressure rises above atrial pressure, the blood in the ventricle pushes the valve cusps closed. (b) The semilunar valves. When the pressure in the ventricle is greater than the pressure in the artery, the valve is forced open and blood is ejected. When ventricular pressure is lower than arterial pressure, arterial blood holds the valve closed.

## BLOOD FLOW THROUGH THE HEART CHAMBERS

Until the sixteenth century, blood was thought to flow directly from the right ventricle into the left through invisible pores in the septum. This of course is not true. Blood is kept entirely separate on the right and left sides of the heart. Figure 19.10 shows the pathway of the blood as it travels from the right atrium through the body and back to the starting point.

## THE CORONARY CIRCULATION

If your heart lasts for 80 years and beats an average of 75 times a minute, it will beat more than 3 billion times and pump more than 200 million liters of blood. It is, in short, a remarkably hard-working organ, and understandably, it needs an abundant supply of oxygen and nutrients. These needs are not met to any appreciable extent by the blood in the heart chambers, because the diffusion of nutrients from there to the myocardium would be too

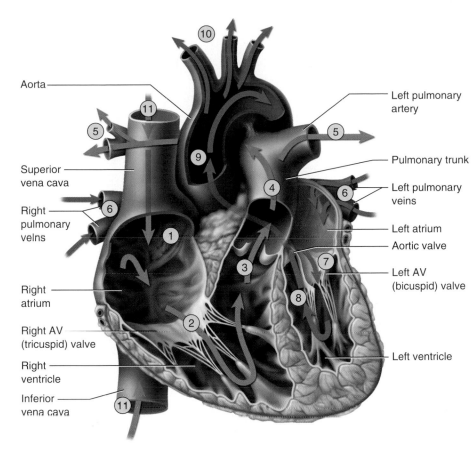

1. Blood enters right atrium from superior and inferior venae cavae.

2. Blood in right atrium flows through right AV valve into right ventricle.

3. Contraction of right ventricle forces pulmonary valve open.

4. Blood flows through pulmonary valve into pulmonary trunk.

5. Blood is distributed by right and left pulmonary arteries to the lungs, where it unloads $CO_2$ and loads $O_2$.

6. Blood returns from lungs via pulmonary arteries to left atrium.

7. Blood in left atrium flows through left AV valve into left ventricle.

8. Contraction of left ventricle (simultaneous with step 3) forces aortic valve open.

9. Blood flows through aortic valve into ascending aorta.

10. Blood in aorta is distributed to every organ in the body, where it unloads $O_2$ and loads $CO_2$.

11. Blood returns to heart via venae cavae.

**FIGURE 19.10  The Pathway of Blood Flow Through the Heart.** The pathway from 4 through 6 is the pulmonary circuit, and the pathway from 9 through 11 is the systemic circuit. Violet arrows indicate oxygen-poor blood; orange arrows indicate oxygen-rich blood.

slow. Instead, the myocardium has its own supply of arteries and capillaries that deliver blood to every muscle cell. The blood vessels of the heart wall constitute the **coronary circulation.**

At rest, the coronary blood vessels supply the myocardium with about 250 mL of blood per minute. Approximately 5% of the circulating blood goes to meet the metabolic needs of the heart, even though the heart is only 0.5% of the body's weight. It receives 10 times its "fair share" to sustain its strenuous workload.

## Arterial Supply

The coronary circulation is the most variable aspect of cardiac anatomy. The following description covers only the largest coronary blood vessels and describes only the pattern seen in about 70% to 85% of persons.

Immediately after the aorta leaves the left ventricle, it gives off a right and left coronary artery. The orifices of these two arteries lie deep in the pockets formed by the aortic valve cusps (see fig. 19.8a). The **left coronary artery (LCA)** travels through the coronary sulcus under the left auricle and divides into two branches (fig. 19.11):

1. The **anterior interventricular branch** travels down the anterior interventricular sulcus to the apex,

rounds the bend, and travels a short distance up the posterior side of the heart. There it anastomoses with (joins) the posterior interventricular branch described shortly. Clinically, it is also called the *left anterior descending (LAD) branch.* This artery supplies blood to both ventricles and the anterior two-thirds of the interventricular septum.

2. The **circumflex branch** continues around the left side of the heart in the coronary sulcus. It gives off a **left marginal branch** that passes down the left margin of the heart and furnishes blood to the left ventricle. The circumflex branch then ends on the posterior side of the heart. It supplies blood to the left atrium and posterior wall of the left ventricle.

The **right coronary artery (RCA)** supplies the right atrium and sinoatrial node (pacemaker), continues along the coronary sulcus under the right auricle, and gives off two branches of its own:

1. The **right marginal branch** runs toward the apex of the heart and supplies the lateral aspect of the right atrium and ventricle.

2. The RCA continues around the right margin of the heart to the posterior side, sends a small branch to the atrioventricular node, then gives off a large

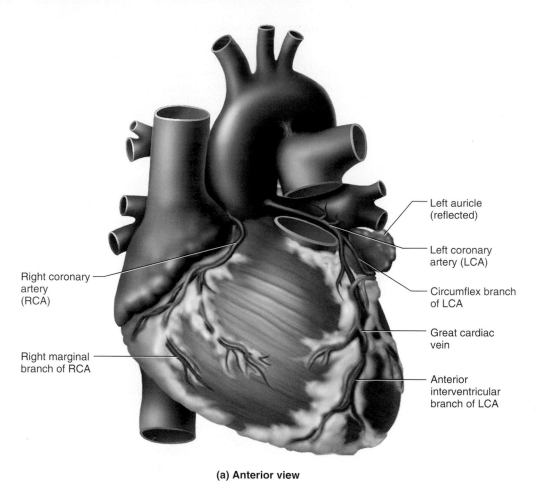

**(a) Anterior view**

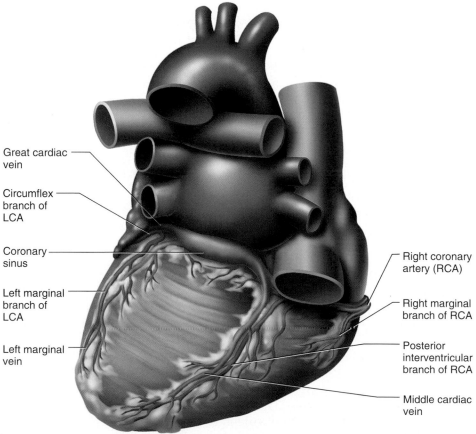

**FIGURE 19.11** The Coronary Blood Vessels.
(a) Anterior view. (b) Posterior view.

**(b) Posterior view**

**posterior interventricular branch.** This branch travels down the corresponding sulcus and supplies the posterior walls of both ventricles as well as the posterior portion of the interventricular septum. It ends by anastomosing with the circumflex and anterior interventricular branches of the LCA.

The energy demand of the cardiac muscle is so critical that an interruption of the blood supply to any part of the myocardium can cause necrosis within minutes. A fatty deposit or blood clot in a coronary artery can cause a **myocardial infarction**[15] **(MI),** or heart attack (see Insight 19.2). Some protection from MI is provided by the aforementioned *anastomoses* (ah-NASS-tih-MO-seez), points where two arteries come together and combine their blood flow to points farther downstream. Anastomoses provide an alternative route, called **collateral circulation,** that can supply the heart tissue with blood if the primary route becomes obstructed.

Most organs receive more arterial blood when the ventricles contract than when they relax, but the opposite is true in the coronary arteries. There are three reasons for this. (1) Contraction of the myocardium compresses the

arteries and obstructs blood flow. (2) During *ventricular systole* (contraction of the ventricles), the aortic valve is forced open and the valve cusps cover the openings to the coronary arteries, blocking blood from flowing into them. (3) During *ventricular diastole* (relaxation), blood in the aorta briefly surges back toward the heart. It fills the aortic valve cusps and some of it flows into the coronary arteries, like sand filling a shirt pocket and flowing out through a hole in the bottom. In the coronary blood vessels, therefore, diastolic blood flow is greater than systolic blood flow.

## Venous Drainage

**Venous drainage** refers to the route by which blood leaves an organ. After flowing through capillaries of the heart wall, about 20% of the coronary blood empties directly from multiple small *thebesian*[18] veins into the right atrium and ventricle. The other 80% returns to the right atrium by the following route (fig. 19.11):

- The **great cardiac vein** collects blood from the anterior aspect of the heart and travels alongside the anterior interventricular artery. It carries blood from the apex toward the coronary sulcus, then arcs around the left side of the heart and empties into the coronary sinus.

- The **posterior interventricular (middle cardiac) vein,** found in the posterior sulcus, collects blood from the posterior aspect of the heart. It, too, carries blood from the apex upward and drains into the same sinus.

- The **left marginal vein** travels from a point near the apex up the left margin, and also empties into the coronary sinus.

- The **coronary sinus,** a large transverse vein in the coronary sulcus on the posterior side of the heart, collects blood from all three of the aforementioned veins as well as some smaller ones. It empties blood into the right atrium.

---

**INSIGHT 19.2** Clinical Application

### Angina and Heart Attack

An obstruction of the coronary blood flow can cause a chest pain known as *angina pectoris*[16] (an-JY-na PEC-toe-riss) or, more seriously, *myocardial infarction* (heart attack). Angina is a sense of heaviness or pain in the chest resulting from temporary and reversible *ischemia*[17] (iss-KEE-me-ah), or deficiency of blood flow to the cardiac muscle. It typically occurs when a partially blocked coronary artery constricts. The oxygen-deprived myocardium shifts to anaerobic fermentation, producing lactic acid, which stimulates pain receptors in the heart. The pain abates when the artery relaxes and normal blood flow resumes.

Myocardial infarction (MI), on the other hand, is the sudden death of a patch of mycardium resulting from long-term obstruction of the coronary circulation. Coronary arteries often become obstructed by a blood clot or a fatty deposit called an *atheroma* (see Insight 19.5). As cardiac muscle downstream from the obstruction dies, the individual commonly feels a sense of heavy pressure or squeezing pain in the chest, often "radiating" to the shoulder and arm. Infarctions weaken the heart wall and disrupt electrical conduction pathways, potentially leading to fibrillation and cardiac arrest (discussed later in this chapter). MI is responsible for about half of all deaths in the United States.

### Before You Go On

*Answer the following questions to test your understanding of the preceding section:*

3. Name the three layers of the heart and describe their structural differences.

4. What are the functions of the fibrous skeleton?

5. Trace the flow of blood through the heart, naming each chamber and valve in order.

6. What are the three principal branches of the left coronary artery? Where are they located on the heart

---

[15]*infarct* = to stuff
[16]*angina* = to choke, strangle + *pectoris* = of the chest
[17]*isch* = holding back + *em* = blood + *ia* = condition

[18]Adam Christian Thebesius (1686–1732), German physician

*surface? What are the branches of the right coronary artery, and where are they located?*

7. *What is the medical significance of anastomoses in the coronary arterial system?*

8. *Why do the coronary arteries carry a greater blood flow during ventricular diastole than they do during ventricular systole?*

9. *What are the three major veins that empty into the coronary sinus?*

# The Cardiac Conduction System and Cardiac Muscle

### Objectives
When you have completed this section, you should be able to

- describe the nerve supply to the heart;
- describe the internal electrical system of the heart;
- describe the unique structural and metabolic characteristics of cardiac muscle; and
- explain the nature and functional significance of the intercellular junctions between cardiac muscle cells.

The most obvious physiological fact about the heart is its rhythmicity. It contracts at regular intervals, typically about 75 beats per minute (bpm) in a resting adult. Among invertebrates such as clams, crabs, and insects, each heartbeat is triggered by a pacemaker in the nervous system. The vertebrate heartbeat, however, is said to be *myogenic*[19] because the signal originates within the heart itself. Indeed, we can sever the nerves to the heart, or even remove the heart from the body and keep it in aerated saline, and it will beat for hours. Cut the heart into little pieces, and each piece continues its own rhythmic pulsations. Thus it is obviously not dependent on the nervous system for its rhythm. The heart has its own pacemaker and electrical system. We now turn our attention to the heart's nerve supply, its pacemaker, its internal electrical conduction system, and the structure and metabolism of cardiac muscle—the anatomical foundations for its electrical activity and rhythmic beat.

## NERVE SUPPLY TO THE HEART

Even though the heart has its own pacemaker, it does receive both sympathetic and parasympathetic nerves, which modify the heart rate and contraction strength. Sympathetic stimulation is able to raise the heart rate to as high as 230 bpm, and parasympathetic stimulation can slow the heart rate to as low as 20 bpm or even stop the heart for a few seconds.

The sympathetic pathway to the heart originates with neurons in the lower cervical to upper thoracic spinal cord. Efferent fibers from these neurons pass from the spinal cord to the sympathetic chain and travel up the chain to the three cervical ganglia. **Cardiac nerves** arise from the cervical ganglia (see fig. 15.4) and lead mainly to the ventricular myocardium, where they increase the force of contraction. Some fibers, however, innervate the atria. Sympathetic fibers to the coronary arteries dilate them and increase coronary blood flow during exercise.

The parasympathetic pathway to the heart is through the vagus nerves. The right vagus nerve innervates mainly an electrical center of the heart called the *SA node,* and the left vagus nerve mainly innervates another center called the *AV node,* although there is some cross-innervation from each nerve to both nodes. (These nodes are described more fully in the next section.) The ventricles receive little or no vagal stimulation. The vagus nerves slow the heartbeat. Without this influence, the average resting heart rate would be about 100 bpm, but steady background firing of the vagus nerves, called *vagal tone,* normally holds the resting rate down to about 70 to 80 bpm.

In summary, the sympathetic and parasympathetic nerves to the heart are not what makes it beat. The heartbeat is set off by the heart's own internal pacemaker, but these nerves can modify it.

## THE CONDUCTION SYSTEM

Cardiac myocytes (muscle cells) are said to be **autorhythmic**[20] because they depolarize spontaneously at regular time intervals. Some of them lose the ability to contract and become specialized, instead, for generating action potentials. These cells constitute the **cardiac conduction system,** which controls the route and timing of electrical conduction to ensure that the four chambers are coordinated with each other. Electrical signals arise and travel through the cardiac conduction system in the following order (fig. 19.12):

1. The **sinoatrial (SA) node,** a patch of modified myocytes in the right atrium, just under the epicardium near the superior vena cava. This is the **pacemaker** that initiates each heartbeat and determines the heart rate. Signals from the SA node spread throughout the atria, as shown by the red arrows in the figure.

2. The **atrioventricular (AV) node,** located near the right AV valve at the lower end of the interatrial septum. This node acts as an electrical gateway to the ventricles; the fibrous skeleton acts as an insulator to prevent currents from getting to the ventricles by any other route.

---

[19]*myo* = muscle + *genic* = arising from

[20]*auto* = self

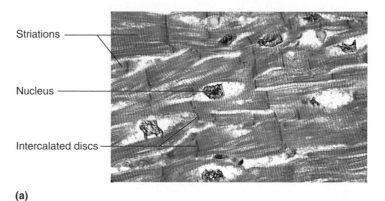

Striations

Nucleus

Intercalated discs

(a)

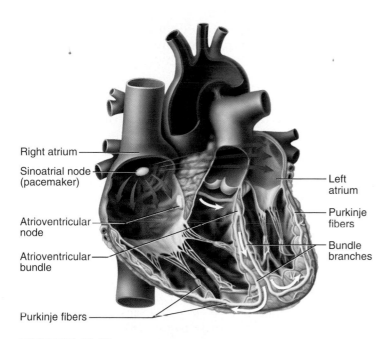

Right atrium

Sinoatrial node (pacemaker)

Atrioventricular node

Atrioventricular bundle

Purkinje fibers

Left atrium

Purkinje fibers

Bundle branches

**FIGURE 19.12 The Cardiac Conduction System.** Electrical signals travel along the pathways indicated by the arrows.

▶ *Which atrium is first to receive the signal that induces it to contract?*

3. The **atrioventricular (AV) bundle** *(bundle of His[21])*, a pathway by which signals leave the AV node. The AV bundle soon forks into **right** and **left bundle branches**, which enter the interventricular septum and descend toward the apex.

4. **Purkinje[22]** (pur-KIN-jee) **fibers,** nervelike processes that arise from the lower end of the bundle branches and turn upward to spread throughout the ventricular myocardium. Purkinje fibers distribute the electrical excitation to the myocytes of the ventricles. They form a more elaborate network in the left ventricle than in the right.

## STRUCTURE OF CARDIAC MUSCLE

The traveling electrical signal does not end with the Purkinje fibers, and Purkinje fibers do not reach every myocyte. Rather, the myocytes pass the signal from cell to cell. This is something that skeletal muscle cannot do, so to understand how the heartbeat is coordinated, one must understand microscopic anatomy of cardiac myocytes and how they differ from skeletal muscle.

Cardiac muscle is striated like skeletal muscle but otherwise differs from it in many structural and physiological ways. Cardiac myocytes, or *cardiocytes,* are relatively short, thick, branched cells, typically 50 to 100 μm long and 10 to 20 μm wide (fig. 19.13). They usually have only

[21]Wilhelm His, Jr. (1863–1934), German physiologist
[22]Johannes E. Purkinje (1787–1869), Bohemian physiologist

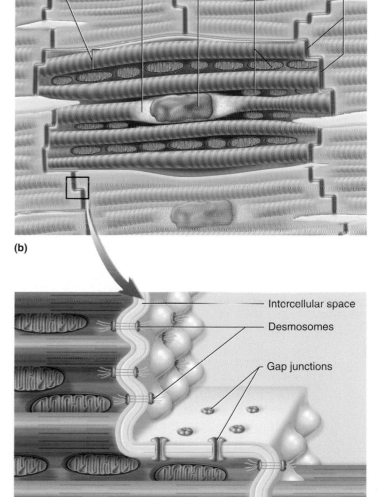

Striated myofibril  Glycogen  Nucleus  Mitochondria  Intercalated discs

(b)

Intercellular space

Desmosomes

Gap junctions

(c)

**FIGURE 19.13 Cardiac Muscle.** (a) Light micrograph. (b) Structure of a cardiac myocyte and its relationship to adjacent myocytes. Note that the myocyte is notched at the ends and typically linked to two or more neighboring myocytes by the mechanical and electrical junctions of the intercalated discs. (c) Structure of an intercalated disc.

one, centrally placed nucleus. The sarcoplasmic reticulum (SR) is less developed than in skeletal muscle; it lacks terminal cisternae, although it does have footlike sacs associated with the T tubules. The T tubules are much larger than in skeletal muscle. During excitation of the cell, they admit supplemental calcium ions from the extracellular fluid to activate muscle contraction. Cardiocytes have especially large mitochondria for reasons noted shortly.

The myocytes are joined end to end by thick connections called **intercalated** (in-TUR-ka-LAY-ted) **discs,** which appear as dark lines (thicker than the striations) in properly stained tissue sections. An intercalated disc is a complex steplike structure with three distinctive features not found in skeletal muscle:

1. **Interdigitating folds.** The plasma membrane at the end of the cell is folded somewhat like the bottom of an egg carton. The folds of adjoining cells interlock with each other and increase the surface area of intercellular contact.

2. **Mechanical junctions.** The cells are tightly joined by two types of mechanical junctions: the fascia adherens and desmosomes. The *fascia adherens*[23] (FASH-ee-ah ad-HEER-enz) is the most extensive. It is a broad band in which the actin of the thin myofilaments is anchored to the plasma membrane, and via transmembrane proteins, one cell is linked to the next. The fascia adherens is interrupted here and there by *desmosomes.* Described in more detail on p. 175, desmosomes are weldlike mechanical junctions between cells. In the contracting heart, they enable the myocytes to pull on each other without pulling apart.

3. **Electrical junctions.** The intercalated discs also contain *gap junctions,* which form channels that allow ions to flow from the cytoplasm of one cell directly into the next (see p. 175 for details). These junctions enable each myocyte to electrically stimulate its neighbors. Thus the entire myocardium of the two atria contracts in unison, almost as if it were a single cell. The entire myocardium of the two ventricles does likewise, but separately from the atria. This unified action is essential for the effective pumping of a heart chamber.

Skeletal muscle contains satellite cells that can divide and replace dead muscle fibers to some extent. Cardiac muscle lacks satellite cells, however, so the repair of dam-

aged cardiac muscle is almost entirely by fibrosis (scarring). A limited capacity for myocardial mitosis and regeneration was discovered in 2001.

## METABOLISM OF CARDIAC MUSCLE

Cardiac muscle depends almost exclusively on aerobic respiration to make ATP. It is very rich in myoglobin (a short-term source of stored oxygen for aerobic respiration) and glycogen (a stored energy source). It also has especially large mitochondria, which fill about 25% of the myocyte; skeletal muscle fibers, by comparison, have much smaller mitochondria that occupy only 2% of the fiber. Cardiac muscle is relatively adaptable with respect to the organic fuels used. At rest, the heart gets about 60% of its energy from fatty acids, 35% from glucose, and 5% from other fuels such as ketones, lactic acid, and amino acids. Cardiac muscle is more vulnerable to an oxygen deficiency than it is to the lack of any specific fuel. Because it makes little use of anaerobic fermentation or the oxygen debt mechanism, it is not prone to fatigue. You can easily appreciate this fact by clenching and opening your fist once every second for a minute or two. You will soon feel weakness and fatigue in your skeletal muscles and perhaps feel all the more grateful that cardiac muscle can maintain its rhythm, without fatigue, for a lifetime.

**Think About It**

*Why should mitochondria be larger and more abundant in cardiac muscle than in skeletal muscle?*

## Before You Go On

*Answer the following questions to test your understanding of the preceding section:*

10. *Why does the heart have a nerve supply, since it continues to beat even without one?*

11. *What is the timing mechanism that normally sets off each heartbeat and regulates its rhythm?*

12. *What organelle(s) are less developed in cardiac muscle than in skeletal muscle? What one(s) are more developed? What is the functional significance of these differences between muscle types?*

13. *What component of an intercalated disc enables one cardiac myocyte to directly stimulate another? What component keeps the myocytes from pulling apart when the muscle contracts?*

14. *Cardiac muscle rarely uses anaerobic fermentation to generate ATP. What benefit do we gain from this fact?*

---

[23]*fascia* = band + *adherens* = adhering

# Electrical and Contractile Activity of the Heart

### Objectives

When you have completed this section, you should be able to

- explain why the SA node fires spontaneously and rhythmically;
- explain how the SA node excites the myocardium;
- describe the unusual action potentials of cardiac muscle and relate them to the contractile behavior of the heart; and
- interpret a normal electrocardiogram.

In this section, we examine how the electrical events in the heart produce its cycle of contraction and relaxation. Contraction is called **systole** (SIS-toe-lee) and relaxation is **diastole** (dy-ASS-toe-lee). These terms can refer to a specific part of the heart (for example, atrial systole), but if no particular chamber is specified, they usually refer to the more conspicuous and important ventricular action, which ejects blood from the heart.

## THE CARDIAC RHYTHM

The normal heartbeat, triggered by the SA node, is called the **sinus rhythm.** At rest, the adult heart typically beats about 70 to 80 times per minute, although heart rates from 60 to 100 bpm are not unusual.

Stimuli such as hypoxia, electrolyte imbalances, caffeine, nicotine, and other drugs can cause other parts of the conduction system to fire before the SA node does, setting off an extra heartbeat called a *premature ventricular contraction (PVC),* or *extrasystole.* Any region of spontaneous firing other than the SA node is called an **ectopic**[24] **focus.** If the SA node is damaged, an ectopic focus may take over the governance of the heart rhythm. The most common ectopic focus is the AV node, which produces a slower heartbeat of 40 to 50 bpm called a **nodal rhythm.** If neither the SA nor AV node is functioning, other ectopic foci fire at rates of 20 to 40 bpm. The nodal rhythm is sufficient to sustain life, but a rate of 20 to 40 bpm provides too little flow to the brain to be survivable. This condition calls for an artificial pacemaker.

Any abnormal cardiac rhythm is called **arrhythmia**[25] (see Insight 19.3). One cause of arrhythmia is a **heart**

---

[24]*ec* = out of + *top* = place
[25]*a* = without + *ia* = condition

**block**—the failure of any part of the cardiac conduction system to transmit signals, usually as a result of disease and degeneration of conduction system fibers. A *bundle branch block,* for example, is due to damage to one or both bundle branches. Damage to the AV node causes *total heart block,* in which signals from the atria fail to reach the ventricles and the ventricles beat at their own intrinsic rhythm of 20 to 40 bpm.

## PACEMAKER PHYSIOLOGY

Why does the SA node spontaneously fire at regular intervals? Unlike skeletal muscle or neurons, cells of the SA node do not have a stable resting membrane potential. Their membrane potential starts at about −60 mV and drifts upward, showing a gradual depolarization called the **pacemaker potential** (fig. 19.14). This is thought to result from a slow inflow of $Na^+$ without a compensating outflow of $K^+$.

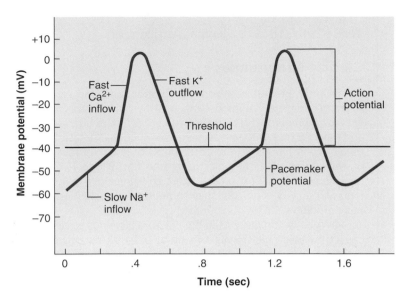

**FIGURE 19.14** Pacemaker Potentials and Action Potentials of the SA Node.

When the pacemaker potential reaches a threshold of $-40$ mV, voltage-regulated **fast calcium channels** open and $Ca^{2+}$ flows in from the extracellular fluid (ECF). This produces the rising (depolarizing) phase of the action potential, which peaks slightly above 0 mV. At that point, $K^+$ channels open and $K^+$ leaves the cell. This makes the cytosol increasingly negative and creates the falling (repolarizing) phase of the action potential. When repolarization is complete, the $K^+$ channels close and the pacemaker potential starts over, on its way to producing the next heartbeat. Each depolarization of the SA node sets off one heartbeat. When the SA node fires, it excites the other components in the conduction system; thus, the SA node serves as the system's pacemaker. At rest, it typically fires every 0.8 second or so, creating a heart rate of about 75 bpm.

## IMPULSE CONDUCTION TO THE MYOCARDIUM

Firing of the SA node excites atrial myocytes and stimulates the two atria to contract almost simultaneously. The signal travels at a speed of about 1 m/sec through the atrial myocardium and reaches the AV node in about 50 msec. In the AV node, the signal slows down to about 0.05 m/sec, partly because the myocytes here are thinner, but more importantly because they have fewer gap junctions over which the signal can be transmitted. This delays the signal at the AV node for about 100 msec—like highway traffic slowing down at a small town. This delay is essential because it gives the ventricles time to fill with blood before they begin to contract.

The ventricular myocardium has a conduction speed of only 0.3 to 0.5 m/sec. If this were the only route of travel for the excitatory signal, some myocytes would be stimulated much sooner than others. Ventricular contraction would not be synchronized and the pumping effectiveness of the ventricles would be severely compromised. But signals travel through the AV bundle and Purkinje fibers at a speed of 4 m/sec, the fastest in the conduction system. Consequently, the entire ventricular myocardium depolarizes within 200 msec after the SA node fires, causing the ventricles to contract in near unison.

Signals reach the papillary muscles before the rest of the myocardium. Thus, these muscles contract and begin taking up slack in the tendinous cords an instant before ventricular contraction causes blood to surge against the AV valves. Ventricular systole begins at the apex of the heart, which is first to be stimulated, and progresses upward—pushing the blood upward toward the semilunar valves. Because of the spiral arrangement of ventricular myocytes, the ventricles twist slightly as they contract, like someone wringing out a towel.

> **Think About It**
>
> *Some people have abnormal cords or bridges of myocardium that extend from atrium to ventricle, bypassing the AV node and other parts of the conduction system. How would you expect this to affect the cardiac rhythm?*

## ELECTRICAL BEHAVIOR OF THE MYOCARDIUM

The action potentials of cardiac myocytes are significantly different from those of neurons and skeletal muscle fibers (fig. 19.15). Cardiac myocytes have a stable resting potential of $-90$ mV and normally depolarize only when stimulated, unlike the cells of the SA node. A stimulus opens voltage-regulated sodium gates, causing an $Na^+$ inflow and depolarizing the cell to its threshold. The threshold voltage rapidly opens additional $Na^+$ gates and triggers a positive feedback cycle like the one seen in the firing of a neuron (see p. 457). The action potential peaks at nearly $+30$ mV. The $Na^+$ gates close quickly, and the rising phase of the action potential is very brief.

As action potentials spread over the plasma membrane, they open voltage-gated slow calcium channels, which admit a small amount of $Ca^{2+}$ from the extracellular fluid into the cell. This $Ca^{2+}$ binds to ligand-gated $Ca^{2+}$ channels on the sarcoplasmic reticulum (SR), opening them and releasing a greater quantity of $Ca^{2+}$ from the SR into the cytosol. This second wave of $Ca^{2+}$ binds to troponin and triggers contraction in the same way as it does in skeletal muscle (see chapter 11). The SR provides 90% to 98% of the $Ca^{2+}$ needed for myocardial contraction.

In skeletal muscle and neurons, an action potential falls back to the resting potential within 2 msec. In car-

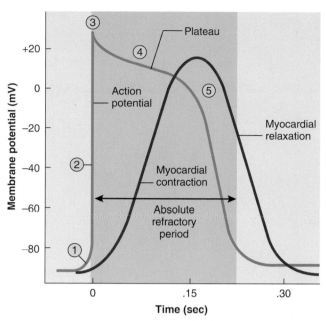

① Voltage-gated Na⁺ channels open.

② Na⁺ inflow depolarizes the membrane and triggers the opening of still more Na⁺ channels, creating a positive feedback cycle and a rapidly rising membrane voltage.

③ Na⁺ channels close when the cell depolarizes, and the voltage peaks at nearly +30 mV.

④ $Ca^{2+}$ entering through slow $Ca^{2+}$ channels prolongs depolarization of membrane, creating a plateau. Plateau falls slightly because of some K⁺ leakage, but most K⁺ channels remain closed until end of plateau.

⑤ $Ca^{2+}$ channels close and $Ca^{2+}$ is transported out of cell. K⁺ channels open, and rapid K⁺ outflow returns membrane to its resting potential.

**FIGURE 19.15** **Action Potential of a Ventricular Myocyte.** The red curve represents rising and falling muscle tension as the myocardium contracts and relaxes.

▶ *What is the advantage of having such a long absolute refractory period in cardiac muscle?*

diac muscle, however, the depolarization is prolonged for 200 to 250 msec (at a heart rate of 70–80 bpm) producing a long plateau in the action potential—perhaps because the $Ca^{2+}$ channels of the SR are slow to close or because the SR is slow to remove $Ca^{2+}$ from the cytosol.

As long as the action potential is in its plateau, the myocytes contract. Thus, in figure 19.15, you can see the development of muscle tension (myocardial contraction) following closely behind the depolarization and plateau. Rather than showing a brief twitch like skeletal muscle, cardiac muscle has a more sustained contraction necessary for expulsion of blood from the heart chambers. Both atrial and ventricular myocytes exhibit these plateaus, but they are more pronounced in the ventricles.

At the end of the plateau, $Ca^{2+}$ channels close and K⁺ channels open. Potassium diffuses rapidly out of the cell and $Ca^{2+}$ is transported back into the extracellular fluid and SR. Membrane voltage drops rapidly, and muscle tension declines soon afterward.

Cardiac muscle has an *absolute refractory period* of 250 msec, compared with 1 to 2 msec in skeletal muscle. This long refractory period prevents wave summation and tetanus, which would stop the pumping action of the heart.

⌐ **Think About It**

*With regard to the ions involved, how does the falling (repolarization) phase of a myocardial action potential differ from that of a neuron's action potential? (See p. 457.)*

## THE ELECTROCARDIOGRAM

We can detect electrical currents in the heart by means of electrodes (leads) applied to the skin. An instrument called the *electrocardiograph* amplifies these signals and produces a record, usually on a moving paper chart, called an **electrocardiogram**[26] (**ECG** or **EKG**[27]). To record an ECG, electrodes are typically attached to the wrists, ankles, and six locations on the chest. Several simultaneous recordings can be made from electrodes at different distances from the heart; collectively, they provide a comprehensive image of the heart's electrical activity. An ECG is a composite recording of all the action potentials produced by the nodal and myocardial cells—it should not be misconstrued as a tracing of a single action potential.

Figure 19.16 shows a typical ECG. It shows three principal deflections above and below the baseline: the *P wave, QRS complex,* and *T wave.* Figure 19.17 shows how these correspond to regions of the heart undergoing depolarization and repolarization.

The **P wave** is produced when a signal from the SA node spreads through the atria and depolarizes them. Atrial systole begins about 100 msec after the P wave begins, during the *PQ segment.* This segment is about 160 msec long and represents the time required for impulses to travel from the SA node to the AV node.

---

[26]*graph* = recording instrument; *graphy* = recording procedure; *gram* = record of
[27]EKG is from the German spelling, Elektrokardiogramm

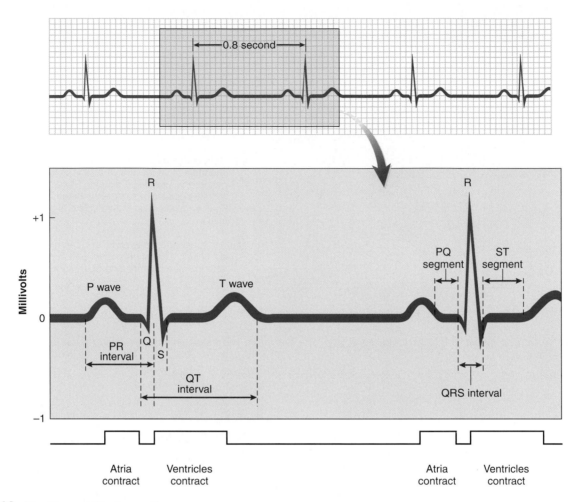

**FIGURE 19.16**   The Normal Electrocardiogram.

The **QRS complex** consists of a small downward deflection (Q), a tall sharp peak (R), and a final downward deflection (S). It is produced when the signal from the AV node spreads through the ventricular myodardium. Its complex shape is due to the different sizes of the two ventricles and the different times required for them to depolarize. Ventricular systole begins shortly after the QRS complex in the *ST segment*. Atrial repolarization and diastole also occur during the QRS interval, but atrial repolarization sends a relatively weak signal that is obscured by the electrical activity of the more muscular ventricles. The ST segment corresponds to the plateau in the myocardial action potential and thus represents the time during which the ventricles contract and eject blood.

The **T wave** is generated by ventricular repolarization immediately before diastole. The ventricles take longer to repolarize than to depolarize; the T wave is therefore smaller and more spread out than the QRS complex, and it has a rounder peak. Even in cases where the T wave is taller than the QRS complex, it can be recognized by its relatively rounded peak.

The ECG affords a wealth of information about the normal electrical activity of the heart. Deviations from normal are invaluable for diagnosing abnormalities in the conduction pathways, myocardial infarction, enlargement of the heart, and electrolyte and hormone imbalances. A few examples of abnormal ECGs are given in table 19.1 and figure 19.18.

## Before You Go On

*Answer the following questions to test your understanding of the preceding section:*

15. *Define* systole *and* diastole.

16. *How does the pacemaker potential of the SA node differ from the resting membrane potential of a neuron? Why is this important in creating the heart rhythm?*

17. *How does excitation-contraction coupling in cardiac muscle resemble that of skeletal muscle? How is it different?*

18. *What produces the plateau in the action potentials of cardiac myocytes? Why is this important to the pumping ability of the heart?*

19. *Name the waves of the ECG and explain what myocardial events produce each wave.*

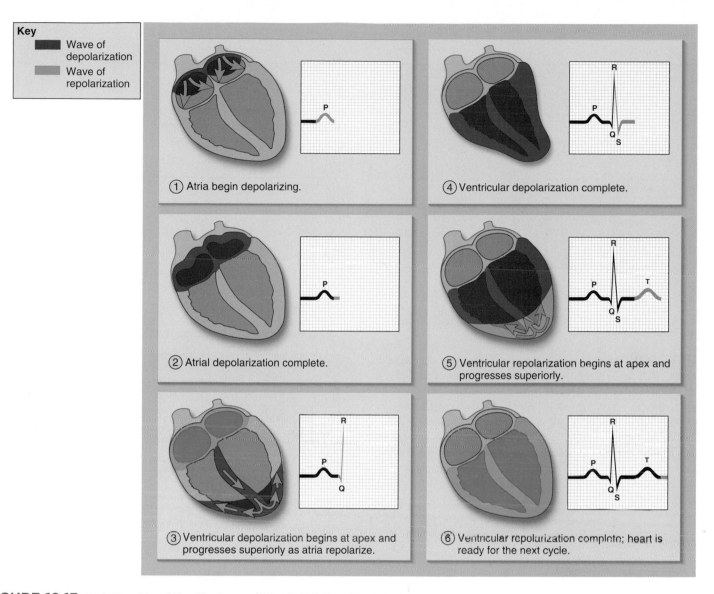

**FIGURE 19.17** **Relationship of the Electrocardiogram (ECG) to Electrical Activity and Contraction of the Myocardium.** Each heart diagram indicates the events occurring at the time of the colored segment of the ECG. Red indicates depolarizing or depolarized myocardium, and green indicates repolarizing or repolarized myocardium. Arrows indicate the direction in which a wave of depolarization or repolarization is traveling.

| TABLE 19.1 | Examples of the Diagnostic Interpretation of Abnormal Electrocardiograms |
|---|---|
| **Appearance** | **Suggested Meaning** |
| Enlarged P wave | Atrial hypertrophy, often a result of mitral valve stenosis |
| Missing or inverted P wave | SA node damage; AV node has taken over pacemaker role |
| Two or more P waves per cycle | Extrasystole; heart block |
| Extra, misshapen, sometimes inverted QRS not preceded by P wave | Premature ventricular contraction (PVC) (extrasystole) |
| Enlarged Q wave | Myocardial infarction |
| Enlarged R wave | Ventricular hypertrophy |
| Abnormal T waves | Flattened in hypoxia; elevated in hyperkalemia ($K^+$ excess) |
| Abnormally long PQ segment | Scarring of atrial myocardium, forcing impulses to bypass normal conduction pathways and take slower alternative routes to AV node |
| Abnormal ST segment | Elevated above baseline in myocardial infarction; depressed in myocardial hypoxia |

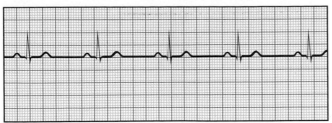

**(a) Sinus rhythm (normal)**

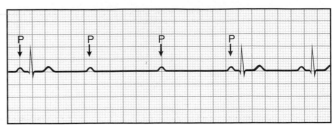

**(c) Heart block**

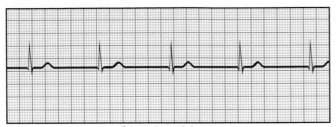

**(b) Nodal rhythm – no SA node activity**

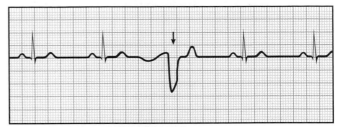

**(d) Premature ventricular contraction**

**FIGURE 19.18** Normal and Pathological Electrocardiograms.
(a) Normal sinus rhythm. (b) Nodal rhythm generated by the AV
node in the absence of SA node activity; note the lack of P waves.
(c) Heart block, in which some P waves are not transmitted through
the AV node and do not generate QRS complexes. (d) Premature
ventricular contraction (PVC), or extrasystole; note the inverted QRS
complex, misshapen QRS and T, and absence of a P wave preceding
this contraction. (e) Ventricular fibrillation, with grossly irregular
waves of depolarization. This is typically seen in a myocardial infarction (heart attack).

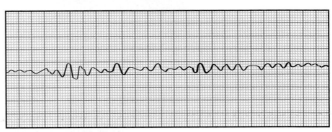

**(e) Ventricular fibrillation**

# Blood Flow, Heart Sounds, and the Cardiac Cycle

### Objectives

When you have completed this section, you should be
able to

- explain how pressure and resistance determine the
  flow of a fluid;
- explain what causes the sounds of the heartbeat;
- describe in detail one complete cycle of heart contraction and relaxation; and
- relate the events of the cardiac cycle to the volume
  of blood entering and leaving the heart.

A **cardiac cycle** consists of one complete contraction and
relaxation of all four heart chambers. We will examine
these events in detail to see how they relate to the entry
and expulsion of blood, but first we consider two related
issues: (1) some general principles of pressure changes
and how they affect the flow of blood, and (2) the heart
sounds produced during the cardiac cycle, which we can
then relate to the stages of the cycle.

## PRINCIPLES OF PRESSURE AND FLOW

A fluid is any liquid or gas—a state of matter that can flow
in bulk from one place to another. In this and some forth-
coming chapters, we are concerned with factors that govern the flow of fluids such as blood, lymph, air, and urine.
Some basic principles of fluid movement *(fluid dynamics)* are therefore important to understand at this time.
Fluid dynamics are governed by two main variables: pressure, which can cause a fluid to flow, and resistance,
which opposes flow.

### Measurement of Pressure

Pressure is often measured by observing how high it can
push a column of mercury (Hg) up an evacuated tube
called a *manometer.* Mercury is used because it is very
dense and enables us to measure pressure with shorter
columns than we would need with a less dense liquid
such as water. Because pressures are compared to the
force generated by a column of mercury, they are
expressed in terms of millimeters of mercury (mm Hg).
Blood pressure is usually measured with a **sphygmo-**

**manometer**[28] (SFIG-mo-ma-NOM-eh-tur)—a calibrated tube filled with mercury and attached to an inflatable pressure cuff wrapped around the arm. Blood pressure and the method of measuring it are discussed in greater detail in chapter 20.

## Pressure Gradients and Flow

For any fixed quantity (mass) of fluid, its pressure depends on the volume of space it occupies. The greater the volume, the lower the pressure, and vice versa. If we start with equal pressures inside and outside a container and then change the container volume, we create a pressure difference between the inside and outside. The heart chambers are containers whose volumes and pressures change repeatedly with each heartbeat.

In blood circulation, we are of course most concerned with flow—and it is a pressure difference between two points, or **pressure gradient,** that makes a fluid flow. Consider two hypothetical points in space, A and B, where the pressure at point A is higher than the pressure at point B. Assuming nothing blocks its way, a fluid will flow from A to B, down the pressure gradient. As it does, the pressure at point A will fall and the pressure at point B will rise, until the two are equal. At that time, there will be no pressure gradient between A and B, and flow will cease. Flow will also cease, of course, if something blocks its way—a point of obvious relevance where the opening and closing of heart valves are concerned.

By analogy, suppose you pull back the plunger of a syringe. The volume in the syringe barrel increases and its pressure falls (fig. 19.19a). Since pressure outside the syringe is greater than the pressure inside, air flows into it until the pressures inside and outside are equal. If you then push the plunger in (fig. 19.19b), pressure inside rises above the pressure outside, and air flows out—again following a gradient from high pressure to low.

The syringe barrel is analogous to a heart chamber such as the left ventricle. When the ventricle is expanding, its internal pressure falls. If the AV valve is open, blood flows into the ventricle from the atrium above. When the ventricle contracts, its internal pressure rises. When the aortic valve opens, blood is ejected from the ventricle into the aorta.

A pressure difference does not guarantee that a fluid will flow. There is always a positive blood pressure in the aorta, and if it is greater than the pressure in the ventricle, it holds the aortic valve closed and prevents the expulsion of blood. When continuing contraction causes ventricular pressure to rise above aortic pressure, however, the valve is forced open and blood is ejected into the aorta.

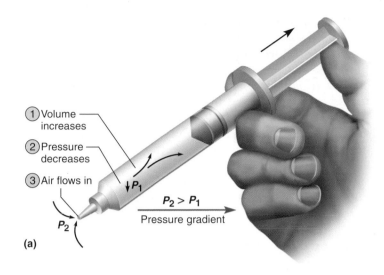

(a)

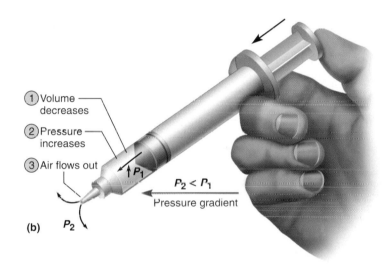

(b)

**FIGURE 19.19  Principles of Volume, Pressure, and Flow Illustrated with a Syringe.** (a) As the plunger is pulled back, the volume of the enclosed space increases, its pressure falls, and pressure inside the syringe ($P_1$) is lower than the pressure outside ($P_2$). The pressure gradient causes air to flow inward until the pressures are equal. This is analogous to the filling of an expanding heart chamber. (b) As the plunger is depressed, the volume of the enclosed space decreases, $P_1$ rises above $P_2$, and air flows out until the pressures are equal. This is analogous to the ejection of blood from a contracting heart chamber. In both cases, fluids flow down their pressure gradients.

## HEART SOUNDS

As we follow events through the cardiac cycle, we will note the occurrence of *heart sounds.* Listening to sounds made by the body is called **auscultation** (AWS-cul-TAY-shun). Each cardiac cycle generates two or three sounds that are audible with a stethoscope. The **first** and **second heart sounds,** symbolized $S_1$ and $S_2$, are often described as a

---

[28]*sphygmo* = pulse + *mano* = rare, sparse, roomy

"lubb-dupp"—$S_1$ is louder and longer and $S_2$ a little softer and sharper. In children and adolescents, it is normal to hear a **third heart sound** ($S_3$). This is rarely audible in people older than 30, but when it is, the heartbeat is said to show a *triple rhythm* or *gallop.* If the normal sounds are roughly simulated by drumming two fingers on a table, a triple rhythm sounds a little like drumming with three fingers. The heart valves themselves operate silently, but $S_1$ and $S_2$ occur in conjunction with the closing of the valves as a result of turbulence in the bloodstream and movements of the heart wall. The cause of each sound is not known with certainty, but the probable factors are discussed in the respective phases of the cardiac cycle.

## PHASES OF THE CARDIAC CYCLE

We now examine the phases of the cardiac cycle, the pressure changes that occur, and how the pressure changes and valves govern the flow of blood. A substantial amount of information about these events is summarized in figure 19.20, which is divided into colored bars numbered to correspond to the phases described here. Closely follow the figure as you study the following text. Where to begin when describing a circular chain of events is somewhat arbitrary. However, in this presentation we begin with the filling of the ventricles. Remember that all these events are completed in less than 1 second.

1. **Ventricular filling.** During diastole, the ventricles expand and their pressure drops below that of the atria. As a result, the AV valves open and blood flows into the ventricles, causing ventricular pressure to rise and atrial pressure to fall. Ventricular filling occurs in three phases: **(1a)** The first one-third is *rapid ventricular filling,* when blood enters especially quickly. **(1b)** The second one-third, called *diastasis* (di-ASS-tuh-sis), is marked by slower filling. The P wave of the electrocardiogram occurs at the end of diastasis, marking the depolarization of the atria. **(1c)** In the last one-third, *atrial systole* completes the filling process. The right atrium contracts slightly before the left because it is the first to receive the signal from the SA node. As the ventricles fill, the flaccid cusps of the AV valves float up toward the closed position. At the end of ventricular filling, each ventricle contains an **end-diastolic volume (EDV)** of about 130 mL of blood. Only 40 mL (31%) of this is contributed by atrial systole.

2. **Isovolumetric contraction.** The atria repolarize, relax, and remain in diastole for the rest of the cardiac cycle. The ventricles depolarize, generate the QRS complex, and begin to contract. Pressure in the ventricles rises sharply and reverses the pressure gradient between atria and ventricles. The AV valves close as ventricular blood surges back against the cusps. Heart sound $S_1$ occurs at the beginning of this phase and is produced mainly by the left ventricle; the right ventricle is thought to make little contribution. Causes

of the sound are thought to include the tensing of ventricular tissues, acceleration of the ventricular wall, turbulence in the blood as it surges against the closed AV valves, and impact of the heart against the chest wall.

This phase is called *isovolumetric*[29] because even though the ventricles contract, they do not eject blood yet, and there is no change in their volume. This is because pressures in the aorta (80 mm Hg) and pulmonary trunk (10 mm Hg) are still greater than the pressures in the respective ventricles and thus oppose the opening of the semilunar valves. The myocytes exert force, but with all four valves closed, the blood cannot go anywhere.

3. **Ventricular ejection.** The ejection of blood begins when ventricular pressure exceeds arterial pressure and forces the semilunar valves open. The pressure typically peaks at 120 mm Hg in the left ventricle and 25 mm Hg in the right. Blood spurts out of each ventricle rapidly at first *(rapid ejection)* and then flows out more slowly under less pressure *(reduced ejection).* By analogy, suppose you were to shake up a bottle of soda pop and remove the cap. The soda would spurt out rapidly at high pressure and then more would dribble out at lower pressure, much like the blood leaving the ventricles. Ventricular ejection lasts about 200 to 250 msec, which corresponds to the plateau of the myocardial action potential but lags somewhat behind it (review the tension curve in fig. 19.15). The T wave occurs late in this phase, beginning at the moment of peak ventricular pressure.

    The ventricles do not expel all their blood. In an average resting heart, each ventricle contains an EDV of 130 mL. The amount ejected, about 70 mL, is called the **stroke volume (SV).** The percentage of the EDV ejected, about 54%, is the **ejection fraction.** The blood remaining behind, about 60 mL in this case, is called the **end-systolic volume (ESV).** Note that EDV − SV = ESV. In vigorous exercise, the ejection fraction may be as high as 90%. Ejection fraction is an important measure of cardiac health. A diseased heart may eject much less than 50% of the blood it contains.

4. **Isovolumetric relaxation.** This is early ventricular diastole, when the T wave ends and the ventricles begin to expand. There are competing hypotheses as to how they expand. One is that the blood flowing into the ventricles "inflates" them. Another is that contraction of the ventricles deforms the fibrous skeleton, which subsequently springs back like a rubber ball that has been squeezed and released. This elastic recoil and expansion would cause pressure to drop rapidly and suck blood into the ventricles.

    At the beginning of ventricular diastole, blood from the aorta and pulmonary trunk briefly flows backward

---

[29]*iso* = same

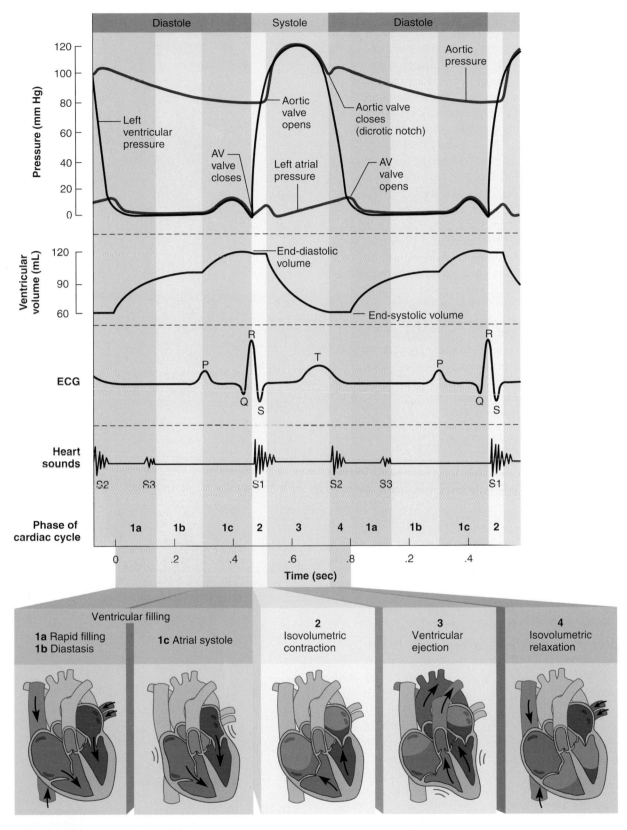

**FIGURE 19.20  Events of the Cardiac Cycle.** Two complete cycles are shown. The phases are numbered across the bottom to correspond to the text description.

▶ *Explain why the aortic pressure curve begins to rise abruptly at about 0.5 second.*

through the semilunar valves. The backflow, however, quickly fills the cusps and closes them, creating a slight pressure rebound that appears as the *dicrotic notch* of the aortic pressure curve (fig. 19.20). Heart sound $S_2$ occurs as blood rebounds from the closed semilunar valves and the ventricles expand. This phase is called *isovolumetric* because the semilunar valves are closed, the AV valves have not yet opened, and the ventricles are therefore taking in no blood. When the AV valves open, ventricular filling (phase 1) begins again. Heart sound $S_3$, if it occurs, is thought to result from the transition from expansion of the empty ventricles to their sudden filling with blood.

In a resting person, atrial systole lasts about 0.1 second; ventricular systole, 0.3 second; and the *quiescent period* (when all four chambers are in diastole), 0.4 second. Total duration of the cardiac cycle is therefore 0.8 second (800 msec) in a heart beating at 75 bpm.

## OVERVIEW OF VOLUME CHANGES

An additional perspective on the cardiac cycle can be gained if we review the volume changes that occur. This "balance sheet" is from the standpoint of one ventricle; both ventricles have equal volumes. The volumes vary somewhat from one person to another and depend on a person's state of activity.

| | |
|---|---|
| End-systolic volume (ESV, left from previous heartbeat) | 60 mL |
| Passively added to the ventricle during atrial diastole | +30 mL |
| Added by atrial systole | +40 mL |
| *Total:* End-diastolic volume (EDV) | 130 mL |
| Stroke volume (SV) ejected by ventricular systole | −70 mL |
| *Leaves:* End-systolic volume (ESV) | 60 mL |

Notice that the ventricle pumps as much blood as it received during diastole: 70 mL in this example.

Both ventricles eject the same amount of blood even though pressure in the right ventricle is only about one-fifth the pressure in the left. Blood pressure in the pulmonary trunk is relatively low, so the right ventricle does not need to generate very much pressure to overcome it. It is essential that both ventricles have the same output. If the right ventricle pumped more blood into the lungs than the left side of the heart could handle on return, blood would accumulate in the lungs and cause pulmonary hypertension and edema (fig. 19.21a). This would put a person at risk of suffocation as fluid filled the lungs and interfered with gas exchange. Conversely, if the left ventricle pumped out more blood than the right heart could handle on return, blood would accumulate in the systemic circuit and cause hypertension and edema there (fig. 19.21b). Over the long

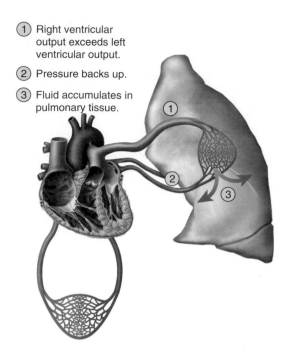

① Right ventricular output exceeds left ventricular output.

② Pressure backs up.

③ Fluid accumulates in pulmonary tissue.

**(a) Pulmonary edema**

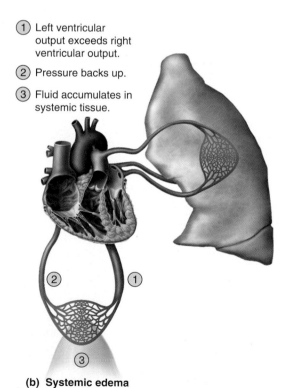

① Left ventricular output exceeds right ventricular output.

② Pressure backs up.

③ Fluid accumulates in systemic tissue.

**(b) Systemic edema**

**FIGURE 19.21** **The Necessity of Balanced Ventricular Output.** (a) If the left ventricle pumps less blood than the right, blood pressure backs up into the lungs and causes pulmonary edema. (b) If the right ventricle pumps less blood than the left, pressure backs up in the systemic circulation and causes systemic edema. To maintain homeostasis, both ventricles must pump the same average amount of blood.

Clinical Application

## Congestive Heart Failure

*Congestive heart failure (CHF)* results from the failure of either ventricle to eject blood effectively. It is usually due to a heart weakened by myocardial infarction, chronic hypertension, valvular insufficiency, or congenital defects in cardiac structure. If the left ventricle fails, blood backs up into the lungs and causes pulmonary edema (fluid in the lungs), shortness of breath, and a sense of suffocation. If the right ventricle fails, blood backs up into the venae cavae and causes systemic, or generalized, edema (formerly called *dropsy*). Systemic edema is marked by enlargement of the liver, ascites (the pooling of fluid in the abdominal cavity), distension of the jugular veins, and swelling of the fingers, ankles, and feet. Failure of one ventricle eventually increases the workload on the other ventricle, which stresses it and leads to its eventual failure as well.

term, this could lead to aneurysms (weakened, bulging arteries), stroke, kidney failure, or heart failure (see Insight 19.4). To maintain homeostasis, the two ventricles must have equal output.

### Before You Go On

*Answer the following questions to test your understanding of the preceding section:*

20. Explain how a pressure gradient across a heart valve determines whether a ventricle ejects blood.

21. What factors are thought to cause the first and second heart sounds? When do these sounds occur?

22. What phases of the cardiac cycle are isovolumetric? Explain what this means.

# Cardiac Output

### Objectives

When you have completed this section, you should be able to

- define *cardiac output* and explain its importance;
- identify the factors that govern cardiac output;
- discuss some of the nervous and chemical factors that alter heart rate, stroke volume, and cardiac output;
- explain how the right and left ventricles achieve balanced output; and
- describe some effects of exercise on cardiac output.

The entire point of all the cardiac physiology we have considered thus far is to eject blood from the heart. The amount ejected by each ventricle in 1 minute is called the **cardiac output (CO).** If HR is heart rate (beats/min) and SV is stroke volume (mL/beat), $CO = HR \times SV$. At typical resting values, $CO = 75$ beats/min $\times 70$ mL/beat $= 5{,}250$ mL/min. Thus, the body's total volume of blood (4–6 L) passes through the heart every minute; or to look at it another way, an RBC leaving the left ventricle will, on average, arrive back at the left ventricle in about 1 minute.

Cardiac output is not constant but varies with the body's state of activity. Vigorous exercise increases CO to as much as 21 L/min in a person in good condition, and up to 35 L/min in world-class athletes. The difference between the maximum and resting cardiac output is called **cardiac reserve.** People with severe heart disease may have little or no cardiac reserve and little tolerance of physical exertion.

Given that cardiac output equals HR × SV, you can see that there are only two ways to change it: change the heart rate or change the stroke volume. We will consider factors that influence each of these variables, but bear in mind that heart rate and stroke volume are somewhat interdependent. They usually change together and in opposite directions. As heart rate goes up, stroke volume goes down, and vice versa.

## HEART RATE

Heart rate is most easily measured by taking a person's **pulse** at some point where an artery runs close to the body surface, such as the *radial artery* in the wrist or *common carotid artery* in the neck. Each beat of the heart produces a surge of pressure that can be felt by palpating a superficial artery with the fingertips. Heart rate can be obtained by counting the number of pulses in 15 seconds and multiplying by 4 to get the beats per minute. In newborn infants, the resting heart rate is commonly 120 bpm or greater. It declines steadily with age, averaging 72 to 80 bpm in young adult females and 64 to 72 bpm in young adult males. It rises again in the elderly.

**Tachycardia**[30] is a persistent, resting adult heart rate above 100 bpm. It can be caused by stress, anxiety, drugs, heart disease, or fever. Heart rate also rises to compensate to some extent for a drop in stroke volume. Thus, the heart races when the body has lost a significant quantity of blood or when there is damage to the myocardium.

**Bradycardia**[31] is a persistent, resting adult heart rate below 60 bpm. It is common during sleep and in endurance-trained athletes. Endurance training enlarges the heart and increases its stroke volume. Thus, it can maintain the same cardiac output with fewer beats. Hypothermia (low body temperature) also slows the heart rate and may be deliberately induced in preparation for cardiac surgery. Diving mammals such as whales and seals exhibit bradycardia during the dive, as do humans to some extent when the face is immersed in cool water.

Factors that raise the heart rate are called *positive chronotropic*[32] *agents,* and factors that lower it are *negative*

---

[30]*tachy* = speed, fast + *card* = heart + *ia* = condition
[31]*brady* = slow + *card* = heart + *ia* = condition
[32]*chrono* = time + *trop* = turn, change, influence

*chronotropic agents.* We next consider some chronotropic effects of the autonomic nervous system, hormones, electrolytes, and blood gases.

## Chronotropic Effects of the Autonomic Nervous System

Although the nervous system does not initiate the heartbeat, it does modulate its rhythm and force. The medulla oblongata contains a **cardiac center** composed of two neural pools: a cardioacceleratory center and cardioinhibitory center. The **cardioacceleratory center** sends signals by way of sympathetic **cardiac nerves** to the SA node, AV node, and myocardium. These nerves secrete norepinephrine, which binds to β-adrenergic receptors in the heart and increases the heart rate. Cardiac output peaks when the heart rate is 160 to 180 bpm, although the sympathetic nervous system can get the heart rate up to as much as 230 bpm. This limit is set mainly by the refractory period of the SA node; it cannot fire any more frequently. At such a high rate, however, the ventricles beat so rapidly that they have little time to fill between beats; therefore, the stroke volume and cardiac output are less than they are at rest. At a heart rate of 65 bpm, ventricular diastole lasts about 0.62 seconds, but at 200 bpm, it lasts only 0.14 seconds. At that high rate, there is less time available for refilling between beats.

The **cardioinhibitory center** sends signals by way of parasympathetic fibers in the vagus nerves to the SA and AV nodes. The vagus nerves secrete acetylcholine, which binds to muscarinic receptors and opens $K^+$ channels in the nodal cells. As $K^+$ leaves the cells, the cells become hyperpolarized and fire less frequently, so the heart slows down.

The vagus nerves maintain a background firing rate called **vagal tone** that inhibits the nodes. If the vagus nerves to the heart are severed, the SA node fires at its own intrinsic frequency of about 100 times per minute. With the vagus nerve intact, however, vagal tone holds the heart rate down to the usual 70 to 80 bpm. Maximum vagal stimulation can reduce the heart rate to as low as 20 bpm.

The cardiac center receives and integrates input from multiple sources. Sensory and emotional stimuli can act on it by way of the cerebral cortex, limbic system, and hypothalamus; therefore, heart rate can climb even as you anticipate taking the first plunge on a roller coaster, and it is influenced by emotions such as love and anger. The cardiac center also receives input from receptors in the muscles, joints, arteries, and brainstem:

- **Proprioceptors** in the muscles and joints quickly inform the cardiac center of changes in physical activity. Thus, the heart can increase its output even before the metabolic demands of the muscles rise.

- **Baroreceptors** (pressoreceptors) are pressure sensors in the aorta and internal carotid arteries (see fig. 15.1, p. 565). They send a continual stream of signals to the cardiac center. When the heart rate rises,

cardiac output increases and raises the blood pressure at the baroreceptors. The baroreceptors increase their signaling to the cardiac center, and depending on circumstances, the cardioinhibitory center may lower the heart rate. Conversely, when blood pressure at the baroreceptors drops, their signaling rate falls. The cardioacceleratory center may respond to this by increasing the heart rate, bringing cardiac output and blood pressure back up to normal. Either way, a negative feedback loop prevents the blood pressure from deviating too far from normal.

- **Chemoreceptors** sensitive to blood pH, carbon dioxide, and oxygen are found in the aortic arch, carotid arteries, and medulla oblongata. They are more important in respiratory control than in cardiovascular control, but they do influence the heart rate. If circulation to the tissues is too slow to remove $CO_2$ as fast as the tissues produce it, then $CO_2$ accumulates in the blood and cerebrospinal fluid (CSF) and produces a state of *hypercapnia* ($CO_2$ excess). Furthermore, $CO_2$ generates hydrogen ions by reacting with water: $CO_2$ 1 $H_2O \rightarrow HCO_3^- + H^+$. The hydrogen ions lower the pH of the blood and CSF and may create a state of acidosis (pH < 7.35). Hypercapnia and acidosis stimulate the cardiac center to increase the heart rate, thus improving perfusion of the tissues and restoring homeostasis. The chemoreceptors also respond to extreme *hypoxemia* (oxygen deficiency), such as in suffocation, but the effect is usually to slow down the heart, perhaps so the heart does not compete with the brain for the limited oxygen supply.

Such responses to fluctuations in blood chemistry and blood pressure, called **chemoreflexes** and **baroreflexes,** are good examples of negative feedback loops. They are discussed more fully in chapter 20.

## Chronotropic Effects of Chemicals

The heart rate is affected by numerous chemicals, including the epinephrine and norepinephrine secreted by the cardiac nerves and adrenal medulla. These catecholamines bind to adrenergic receptors on the myocytes and trigger the intracellular production of the second messenger, cyclic adenosine monophosphate (cAMP). cAMP then activates an enzyme that phosphorylates a calcium channel in the plasma membrane. This speeds up the flow of $Ca^{2+}$ into the myocyte and accelerates contraction of the cell. Through other mechanisms, cAMP also accelerates relaxation of the cell: It speeds the uptake of $Ca^{2+}$ by the sarcoplasmic reticulum and inhibits the binding of $Ca^{2+}$ to the troponin of the thin myofilaments (concepts that can be reviewed at pp. 417 to 420). We can easily understand how, by accelerating *both* contraction and relaxation, cAMP increases the heart rate.

The chronotropic action of some other chemicals can be understood from their relationships to this catecholamine–cAMP mechanism. Caffeine and the related stimulants in tea and chocolate accelerate the heart by inhibiting cAMP breakdown, thus prolonging its effect. Nicotine accelerates the heart by stimulating catecholamine secretion. Thyroid hormone increases the number of adrenergic receptors, thus making the heart more responsive to sympathetic stimulation and increasing the heart rate. An excess of thyroid hormone (hyperthyroidism) is marked by tachycardia, which in the long run, can weaken the heart and cause heart failure.

The electrolyte with the greatest chronotropic effect is potassium ($K^+$). In a potassium excess, called *hyperkalemia*,[33] $K^+$ diffuses rapidly into the myocytes, making the membrane potential less negative and interfering with myocyte repolarization. The myocardium becomes less excitable, the heart rate becomes slow and irregular, and the heart may arrest in diastole. In $K^+$ deficiency, called *hypokalemia*, $K^+$ diffuses out of the myocytes. The myocytes then become hyperpolarized—their membrane potential is more negative than normal—and they are harder to stimulate. These potassium imbalances are very dangerous and require emergency medical treatment.

Calcium also affects heart rate. The heart beats abnormally slowly in a state of calcium excess *(hypercalcemia)* and rapidly in a state of calcium deficiency *(hypocalcemia)*. These calcium imbalances are relatively rare, however, and when they do occur, their primary effect is on contraction strength, which is considered in the coming section on contractility. Chapter 24, further explores the causes and effects of imbalances in potassium, calcium, and other electrolytes.

## STROKE VOLUME

The other factor in cardiac output is stroke volume. Stroke volume, in turn, is governed by three factors called *preload, contractility,* and *afterload.* Increased preload or contractility increases stroke volume, whereas increased afterload opposes the emptying of the ventricles and reduces stroke volume.

### Preload

The amount of tension in the ventricular myocardium immediately before it begins to contract is called the **preload.** To understand how this influences stroke volume, imagine yourself engaged in heavy exercise. As active muscles massage your veins, they drive more blood back to the heart, increasing *venous return.* As more blood enters the heart, it stretches the myocardium. Because of the length–tension relationship of striated muscle explained in chapter 11, moderate stretch enables the myocytes to generate more tension when they begin to contract—that is, it increases the preload. If the ventricles contract more forcefully, they expel more blood, thus adjusting your cardiac output to the increase in venous return.

This principle is summarized by the **Frank–Starling law of the heart.**[34] In a concise, symbolic way, it states that stroke volume is proportional to the end-diastolic volume: $SV \propto EDV$. In other words, the ventricles tend to eject as much blood as they receive. Within limits, the more they are stretched, the harder they contract when stimulated.

While relaxed skeletal muscle is normally at an optimum length for the most forceful contraction, relaxed cardiac muscle is at less than optimum length. Additional stretch therefore produces a significant increase in contraction force on the next beat. This helps balance the output of the two ventricles. For example, if the right ventricle begins to pump an increased amount of blood, this soon arrives at the left ventricle, stretches it more than before, and causes it to increase its stroke volume to match that of the right.

## Contractility

**Contractility** refers to how hard the myocardium contracts *for a given preload.* It does not describe an increase in tension produced by stretching the muscle, but rather an increase caused by factors that make the myocytes more responsive to stimulation. Factors that increase contractility are called *positive inotropic*[35] *agents,* and those that reduce it are *negative inotropic agents.*

Calcium has a strong, positive inotropic effect—it increases the strength of each contraction of the heart. This is not surprising, since $Ca^{2+}$ not only is essential to the excitation–contraction coupling of muscle, but also prolongs the plateau of the myocardial action potential. Calcium imbalances therefore affect not only heart rate, as we have already seen, but also contraction strength. In a state of calcium excess *(hypercalcemia),* extra $Ca^{2+}$ diffuses into the myocytes and produces strong, prolonged contractions. In extreme cases, it can cause cardiac arrest in systole. In a calcium deficiency *(hypocalcemia),* the myocytes lose $Ca^{2+}$ to the extracellular fluid, leading to a weak, irregular heartbeat and potentially to cardiac arrest in diastole. However, as explained in chapter 7, severe hypocalcemia is likely to kill through skeletal muscle paralysis and suffocation before the cardiac effects are felt.

Agents that affect calcium availability not only have the chronotropic effects already examined, but also inotropic effects. Thus, epinephrine and norepinephrine increase not only heart rate but also contraction strength. The pancreatic hormone glucagon exerts an inotropic effect by stimulating cAMP production; a solution of glucagon and calcium chloride is sometimes used for the emergency treatment of heart attacks. Digitalis, a cardiac stimulant from the foxglove plant, also raises the intracellular calcium level and contraction strength; it is used to treat congestive heart failure.

---

[33]*kal* = potassium (Latin, *kalium*)

[34]Otto Frank (1865–1944), German physiologist; Ernest Henry Starling (1866–1927), English physiologist
[35]*ino* = fiber + *trop* = to change, affect

A potassium excess, hyperkalemia, has a negative inotropic effect because it reduces the strength of myocardial action potentials and thus reduces the release of $Ca^{2+}$ into the sarcoplasm. The heart becomes dilated and flaccid. Hypokalemia, however, has little effect on contractility.

The vagus nerves have a negative inotropic effect on the atria, but they provide so little innervation to the ventricular myocytes that they have little effect on the ventricles. There are other chronotropic and inotropic agents too numerous to mention here. The ones we have discussed are summarized in table 19.2.

> ⌐ **Think About It**
>
> *Suppose a person has a heart rate of 70 bpm and a stroke volume of 70 mL. A negative inotropic agent then reduces the stroke volume to 50 mL. What would the new heart rate have to be to maintain the same cardiac output?*

## Afterload

The blood pressure in the arteries just outside the semilunar valves, called the **afterload,** opposes the opening of these valves. An increased afterload therefore reduces stroke volume. Thus, hypertension increases afterload and opposes ventricular ejection. Anything that impedes arterial circulation can also increase the afterload. For example, in some lung diseases, scar tissue forms in the lungs and restricts pulmonary circulation. This increases the afterload in the pulmonary trunk. As the right ventricle works harder to overcome this resistance, it gets larger like any other muscle. Stress and hypertrophy of a ventricle can eventually cause it to weaken and fail. Right ventricular failure due to obstructed pulmonary circulation is called *cor pulmonale*[36] (CORE PUL-mo-NAY-lee). It is a common complication of emphysema, chronic bronchitis, and black lung disease (see chapter 22).

## EXERCISE AND CARDIAC OUTPUT

It is no secret that exercise makes the heart work harder, and it should come as no surprise that this increases cardiac output. The main reason the heart rate increases at the beginning of exercise is that proprioceptors in the muscles and joints transmit signals to the cardiac center, signifying that the muscles are active and will quickly need an increased blood flow. As the exercise progresses, muscular activity increases venous return. This increases the preload on the right ventricle and is soon reflected in the left ventricle as more blood flows through the pulmonary circuit and reaches the left heart. As the heart rate and stroke volume rise, cardiac output rises, which compensates for the increased venous return.

A sustained program of exercise causes hypertrophy of the ventricles, which increases their stroke volume. As explained earlier, this allows the heart to beat more slowly and still maintain a normal resting cardiac output. Endurance athletes commonly have resting heart rates as low as 40 to 60 bpm, but because of the higher stroke volume, their resting cardiac output is about the same as that of an untrained person. The champion cyclist Lance Armstrong has an astonishingly low resting heart rate of 32 to 34 bpm. Such athletes have greater cardiac reserve, so they can tolerate more exertion than a sedentary person can.

The effects of aging on the heart are discussed on pp. 1130–1131, and some common heart diseases are listed in table 19.3. Disorders of the blood and blood vessels are described in chapters 18 and 20.

| **TABLE 19.2** | Some Chronotropic and Inotropic Agents |
|---|---|
| **Chronotropic Agents (influence heart rate)** | |
| **Positive** | **Negative** |
| Sympathetic stimulation | Parasympathetic stimulation |
| Epinephrine and norepinephrine | Acetylcholine |
| Thyroid hormone | Hyperkalemia |
| Hypocalcemia | Hypokalemia |
| Hypercapnia and acidosis | Hypercalcemia |
| Digitalis | Hypoxia |
| **Inotropic Agents (influence contraction strength)** | |
| **Positive** | **Negative** |
| Sympathetic stimulation | (Parasympathetic effect negligible) |
| Epinephrine and norepinephrine | Hyperkalemia |
| Hypercalcemia | Hypocalcemia |
| Digitalis | Myocardial hypoxia |
| Glucagon | Myocardial hypercapnia |

### Before You Go On

*Answer the following questions to test your understanding of the preceding section:*

23. Define cardiac output *in words and with a simple formula.*

24. Describe the cardiac center and innervation of the heart.

25. Explain what is meant by positive and negative chronotropic and inotropic agents. Give two examples of each.

26. How do preload, contractility, and afterload influence stroke volume and cardiac output?

27. Explain the principle behind the Frank–Starling law of the heart. How does this mechanism normally prevent pulmonary or systemic congestion?

---

[36]*cor* = heart + *pulmo* = lung

| TABLE 19.3 | Some Disorders of the Heart |
|---|---|
| Acute pericarditis | Inflammation of the pericardium, sometimes due to infection, radiation therapy, or connective tissue disease, causing pain and friction rub |
| Cardiac tamponade | Compression of the heart by an abnormal accumulation of fluid in the pericardial cavity, interfering with ventricular filling; may result from pericarditis |
| Cardiomyopathy | Any disease of the myocardium not resulting from coronary artery disease, valvular dysfunction, or other cardiovascular disorders; can cause dilation and failure of the heart, thinning of the heart wall, or thickening of the interventricular septum |
| Infective endocarditis | Inflammation of the endocardium, usually due to infection, especially streptococcus and staphylococcus bacterial infections |
| Myocardial ischemia | Inadequate blood flow to the myocardium, usually because of coronary atherosclerosis; can lead to myocardial infarction |
| Pericardial effusion | Seepage of fluid from the pericardium into the pericardial cavity, often resulting from pericarditis and sometimes causing cardiac tamponade |
| Septal defects | Abnormal openings in the interatrial or interventricular septum, resulting in blood from the right atrium flowing directly into the left atrium, or blood from the left ventricle returning to the right ventricle; results in pulmonary hypertension, difficulty breathing, and fatigue. Often fatal in childhood if uncorrected |

*Disorders described elswhere*

| | | |
|---|---|---|
| Angina pectoris p. 727 | Cor pulmonale p. 744 | Myocardial infarction p. 727 |
| Atrial flutter p. 731 | Coronary artery disease pp. 746–747 | Premature ventricular contraction p. 731 |
| Bradycardia p. 741 | Familial hypercholesterolemia p. 746 | Tachycardia p. 741 |
| Bundle branch block p. 731 | Friction rub p. 718 | Total heart block p. 731 |
| Cardiac arrest p. 731 | Heart murmur p. 723 | Valvular stenosis p. 723 |
| Congestive heart failure p. 741 | Mitral valve prolapse p. 723 | Ventricular fibrillation p. 731 |

## INSIGHT 19.5    Clinical Application

# Coronary Atherosclerosis

*Atherosclerosis*[37] is a disorder in which fatty deposits form in an arterial wall, obstruct the lumen, and cause deterioration of the wall. It is especially critical when it occurs in the coronary arteries and threatens to cut off the blood supply to the myocardium. Atherosclerosis is also a leading contributor to stroke and kidney failure.

## Cause and Pathogenesis

According to one theory, the stage is set for atherosclerosis when the endothelium of a blood vessel is damaged by hypertension, viral infection, diabetes mellitus, or other causes. Monocytes adhere to the damaged endothelium, penetrate beneath it, and transform into macrophages. Macrophages and smooth muscle cells absorb cholesterol and neutral fats from the arterial tissue and acquire a frothy appearance; they are then called *foam cells* and are visible as a *fatty streak* on the vessel wall.

Platelets also adhere to areas of endothelial damage, degranulate, and release platelet-derived growth factor (PDGF); some PDGF also comes from macrophages and endothelial cells. PDGF stimulates mitosis of smooth muscle, leading eventually to a mass of lipid, smooth muscle, and macrophages called an *atheroma* (atherosclerotic plaque). The muscular and elastic tissue of the artery become increasingly replaced with scar tissue. When atheromas become calcified, they are called *complicated plaques.* Such plaques cause a state of arterial rigidity called *arteriosclerosis.*

Atherosclerosis is caused in large part by a combination of too much *low-density lipoprotein (LDL)* in the blood plasma and defective LDL receptors in the arteries. LDLs are small protein-coated droplets of cholesterol, neutral fat, free fatty acids, and phospholipids (see chapter 26). Most cells have LDL receptors, take up these droplets from the blood by receptor-mediated endocytosis, and stop when they have enough cholesterol. In atherosclerosis, arterial cells have dysfunctional receptors that continue taking up plasma lipids and cause the cells to accumulate excess cholesterol.

As an atheroma grows, more and more of the arterial lumen becomes obstructed (fig. 19.22). Angina pectoris and other symptoms begin to occur when the lumen of a major coronary artery is reduced by at least 75%. When platelets adhere to lesions of the arterial wall, they release clotting factors, so an atheroma can become a focus for thrombosis. A clot can block what remains of the lumen, or it can break free and become an embolus that travels downstream until it lodges in a smaller artery. Part of an atheroma itself can also break loose and travel as a *fatty embolus.*

Atheromas also contribute to coronary artery spasms. Healthy endothelial cells secrete nitric oxide (NO), which causes the arteries to dilate. Vessels damaged by atherosclerosis release less NO, and the coronary arteries exhibit spasms. With much of the lumen already obstructed by the atheroma and perhaps a thrombus, an arterial spasm can temporarily shut off the remaining flow and precipitate an attack of angina.

## Risk and Prevention

*Risk factors* are personal characteristics or elements of the environment that predispose an individual to a particular disease. Some risk factors for atherosclerosis cannot be avoided—for example, aging, heredity, and being male. One form of hereditary atherosclerosis is *familial hypercholesterolemia* ("elevated blood cholesterol levels running in the family"). Most people have two recessive alleles *(hh)* of the gene for LDL receptors, which leads to the synthesis of normal receptors. One person in 500 is heterozygous *(Hh)* and makes only half the normal number of LDL receptors; one in a million is homozygous dominant *(HH)* and makes no LDL receptors. When the body's cells have few or no LDL receptors, they fail to absorb LDLs from the blood. LDL levels therefore remain high—six times normal in *HH* individuals. Foam cells, however, absorb LDLs even without these receptors, and with excess LDL in the blood plasma, atheromas grow rapidly. Heterozygous individuals usually suffer heart attacks by age 35, and homozygous dominant individuals usually have heart attacks in childhood, sometimes before age 2.

Most risk factors for atherosclerosis, however, are preventable. A sedentary lifestyle promotes LDL formation, whereas exercise promotes the formation of *high-density lipoproteins (HDLs),* which not only don't contribute to coronary disease but also help to lower blood cholesterol. Obesity is a risk factor that can be reduced by exercise. Aggressiveness, anxiety, and emotional stress promote hypertension and atherosclerosis. Smoking is another avoidable risk factor. The incidence of coronary heart disease is proportional to the number of cigarettes smoked per day and the number of years a person has been a smoker. This is reversible; people who quit smoking drop to normal risk levels within 5 years.

Diet, of course, is an overwhelmingly important factor. Eating animal fat reduces the number of LDL receptors and raises plasma LDL levels. Foods high in soluble fiber (such as beans, apples, and oat bran) lower blood cholesterol by an interesting mechanism. The liver normally converts cholesterol to bile acids, which it secretes into the small intestine to aid fat digestion. The bile acids are reabsorbed farther down the intestine and recycled to the liver for reuse. Soluble fiber, however, binds bile acids and carries them out in the feces. To replace them, the liver must synthesize more, thus using more cholesterol and lowering the blood cholesterol.

In the 1970s, scientists found that the Inuit people of Greenland had unusually low rates of coronary atherosclerosis despite the fact that their diet consisted entirely of meat—averaging a pound of whale meat and a pound of fish per day. Japanese and other groups with large amounts of fish in their diets also show low blood cholesterol levels. It is suspected that this is due to *omega-3 polyunsaturated fatty acids (PUFAs)* in

---

[37]*athero* = fat, fatty + *sclerosis* = hardening

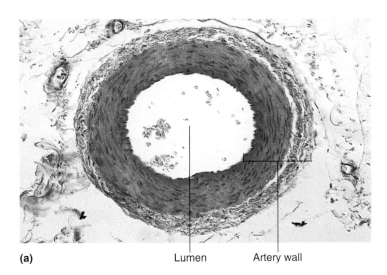

**(a)**  Lumen        Artery wall

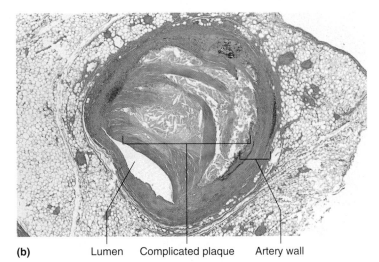

**(b)**  Lumen    Complicated plaque    Artery wall

**FIGURE 19.22  Atherosclerosis.** (a) Cross section of a healthy artery. (b) Cross section of an artery with advanced atherosclerosis. The lumen is reduced to a small space that can easily be blocked by thrombosis, embolism, or vasoconstriction. Most of the original lumen is obstructed by a complicated plaque composed of calcified scar tissue. (c) Coronary angiogram showing 60% obstruction of the anterior interventricular artery (arrow).

fish oil. PUFAs increase the fluidity of plasma membranes and enable cells to remove more lipid from the blood. However, a daily capsule of fish oil does not hold much promise for controlling cholesterol. Doses of PUFAs high enough to reduce blood cholesterol would be prohibitively expensive and have undesirable side effects, including suppression of the immune system. Studies on the effectiveness of PUFAs remain inconclusive.

### Treatment Options

The first pioneering approach to treating atherosclerosis, and still a common standby, is *coronary artery bypass surgery.* Sections of the great saphenous vein of the leg or small arteries from the thoracic cavity are used to construct a detour from the aorta to a point on a coronary artery beyond the obstruction.

*Balloon angioplasty*[38] is a technique in which a thin, flexible catheter is threaded into a coronary artery to the point of obstruction, and then a balloon at its tip is inflated to press the atheroma against the arterial wall, opening up the lumen. Its usefulness is limited to well-localized atheromas. In another method, *laser angioplasty,* an illuminated catheter enables the surgeon to see inside a diseased artery on a monitor and to use a laser to vaporize atheromas and reopen the artery. These methods are cheaper and less risky than bypass surgery. However, there is some concern that these procedures may cause new injuries to the arterial walls, which may be foci for the development of new atheromas. Also, angioplasty is often followed by *restenosis*—atheromas grow back and reobstruct

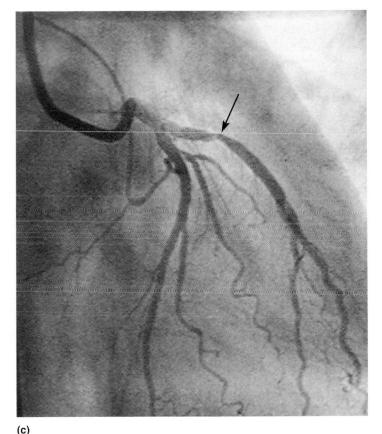

**(c)**

the artery months later. Insertion of a tube called a *stent* into the artery can prevent restenosis, ensuring that the vessel remains open.

Clearly, prevention is the least expensive, least risky, and most effective approach to the threat of coronary artery disease.

---

[38]*angio* = vessel + *plasty* = surgical repair

# CHAPTER REVIEW

# Review of Key Concepts

## Overview of the Cardiovascular System (p. 716)

1. The cardiovascular system is divided into a pulmonary circuit served by the right heart and a systemic circuit served by the left.

2. The heart is located in the mediastinum between the lungs, with about two-thirds of it to the left of the median plane.

3. The heart is enclosed in a fibrous, two-layered *pericardium.* The space between the parietal and visceral pericardium is the pericardial cavity, and contains lubricating pericardial fluid.

## Gross Anatomy of the Heart (p. 718)

1. The heart wall is composed of a thin outer *epicardium,* a thick muscular *myocardium,* and a thin inner *endocardium.*

2. The heart has a connective tissue *fibrous skeleton,* which supports the myocardium and valves, anchors the myocytes, electrically insulates the ventricles from the atria, and may aid in ventricular filling by means of elastic recoil.

3. The two upper chambers of the heart are the *atria,* which serve to receive blood from the venae cavae and pulmonary veins. The two lower chambers are the *ventricles,* which eject blood into the pulmonary trunk and aorta. The ventricles are much more muscular than the atria. The atrioventricular and interventricular sulci on the heart surface mark the boundaries of these chambers.

4. The chambers are internally separated by an *interatrial septum* between the atria and *interventricular septum* between the ventricles.

5. The passages between the atria and ventricles are regulated by the atrioventricular valves (*tricuspid valve* on the right and *bicuspid,* or *mitral, valve* on the left). The cusps of these valves are connected by *tendinous cords* to *papillary muscles* on the floor of the ventricles.

6. The openings into the pulmonary trunk and aorta are regulated by the *semilunar* (*pulmonary* and *aortic*) *valves.* The opening and closing of the heart valves is caused by changes in the pressure difference on the two sides of a valve.

7. Systemic blood is received by the right atrium and flows into the right ventricle. The right ventricle pumps it into the pulmonary trunk, from which it flows to the lungs. Pulmonary blood returning from the lungs is received by the left atrium and flows into the left ventricle. The left ventricle pumps it into the aorta, the beginning of the systemic circulation.

8. The myocardium has a high workload and metabolic rate and needs an abundant oxygen and nutrient supply. It gets this not from the blood in its chambers but from a system of blood vessels called the *coronary circulation.*

9. The *left coronary artery* arises behind an aortic valve cusp near the beginning of the aorta, and gives rise mainly to the *anterior interventricular* and *circumflex branches.* The circumflex gives off a *left marginal branch.*

10. The *right coronary artery* gives off mainly a *right marginal branch* and *posterior interventricular branch.*

11. Obstruction of a coronary artery deprives the downstream myocardium of a blood supply and may cause *myocardial infarction* (heart attack). The risk of this is reduced to some extent by several *anastomoses* in the coronary circulation.

12. The myocardium is drained mainly by the *great cardiac, posterior interventricular,* and *left marginal veins,* all of which empty into the *coronary sinus.* The coronary sinus and several small *thebesian veins* empty into the right atrium and ventricle.

## The Cardiac Conduction System and Cardiac Muscle (p. 728)

1. Although the heart can beat independently of the nervous system, it is innervated by the autonomic nervous system, which modifies the heart rate and contraction force. Sympathetic nerves supply mainly the ventricular myocardium and parasympathetic (vagus) nerves supply the SA and AV nodes. Sympathetic stimulation increases heart rate and contraction strength, and parasympathetic stimulation reduces the heart rate.

2. The cardiac rhythm is set by its own internal pacemaker, the *sinoatrial (SA) node.* Electrical signals originating here spread through the atrial myocardium and then travel via the *atrioventricular (AV) node, AV bundle, bundle branches,* and *Purkinje fibers* to reach the ventricular myocytes.

3. Cardiac myocytes are striated muscle cells with a single nucleus, a poorly developed sarcoplasmic reticulum, very large T tubules, and large abundant mitochondria. They are joined end to end by *interdigitating folds* in their *intercalated discs.* The discs also include intercellular mechanical junctions (*fascia adherens* and *desmosomes*) and electrical (gap) junctions. The latter enable cardiocytes to electrically communicate directly with each other.

4. Cardiac muscle uses almost exclusively aerobic respiration, and has large, abundant mitochondria and abundant glycogen and myoglobin to meet this demand. It employs fatty acids, glucose, and other organic fuels.

## Electrical and Contractile Activity of the Heart (p. 731)

1. *Systole* is the contraction of any heart chamber, and *diastole* is relaxation.

2. A cardiac rhythm activated by the SA node is the *sinus rhythm.* Irritation of the heart or damage to the SA node can cause other areas to take over control. When the AV node takes over, the heart beats with a *nodal rhythm* that is slower than the sinus rhythm. Any abnormal cardiac rhythm is called *arrhythmia.*

3. Cells of the SA node exhibit a *pacemaker potential* in which the membrane voltage starts at −60 mV and drifts spontaneously toward a threshold of −40 mV. At this point, *fast calcium channels* open, and the inflow of $Ca^{2+}$ sets off an action potential. In a resting sinus rhythm of 70 to 80 beats/min, this process repeats itself about every 0.8 second.

4. Firing of the SA node excites the atria and causes atrial systole. The spreading wave of excitation slows down at the AV node, then quickly spreads to the ventricular myocytes and triggers ventricular systole.

5. Papillary muscles contract and pull on the tendinous cords just before the rest of the ventricle contracts; the cords prevent the AV valves from prolapsing when pressure rises in the ventricles.

6. Ordinary cardiac myocytes have a resting potential of −90 mV. Upon excitation, $Na^+$ enters the cells and sets off an action potential that peaks around +30 mV. $Ca^{2+}$ channels then open, admitting $Ca^{2+}$ into the cytosol from the ECF and SR and triggering muscle contraction.

7. The action potential of a cardiac myocyte has a sustained plateau of 200 to 250 msec, causing prolonged contraction rather than a muscle twitch. The plateau ensures that contraction is sustained long enough to expel blood from the ventricles.

8. At the end of the plateau, $Ca^{2+}$ channels close and $K^+$ channels open. The membrane potential drops rapidly as $K^+$ leaves the cell. Contraction is followed by a long refractory period that prevents wave summation and tetanus in the heart.

9. The electrical events of the myocardium as a whole generate the *electrocardiogram (ECG)*. The P wave of the ECG indicates atrial depolarization; atrial systole occurs during the PQ segment. The QRS complex indicates ventricular depolarization, but atrial repolarization occurs at the same time. Ventricular systole begins in the ST segment. The T wave indicates ventricular repolarization and is followed by ventricular diastole.

## Blood Flow, Heart Sounds, and the Cardiac Cycle (p. 736)

1. Blood pressure is measured with a sphygmomanometer and expressed in millimeters of mercury (mm Hg).

2. Fluids flow from a point of higher pressure to a point of lower pressure. The pressure of any fluid such as blood is inversely proportional to the volume it occupies. Therefore, when a heart chamber expands, its pressure falls, and blood tends to flow into the chamber from the veins. When a chamber contracts, its pressure rises, and it tends to eject blood into the arteries.

3. Adults normally have two heart sounds: $S_1$ and $S_2$. Listening to these sounds is called *auscultation.*

4. A *cardiac cycle* is one complete cycle of contraction and relaxation of all four chambers. It can be divided into four phases.

5. In phase 1, ventricular filling, the ventricles expand, the AV valves open, and blood flows into the ventricles, rapidly at first and then more slowly. The P wave occurs, and the atria contract and contribute the last one-third of the blood to the ventricles. At the end of phase 1, each ventricle contains an *end-diastolic volume (EDV)* of about 130 mL.

6. In phase 2, isovolumetric contraction, the QRS wave occurs, the atria repolarize and relax, and the ventricles begin contracting. The AV valves close and heart sound $S_1$ occurs. The semilunar valves remain closed and no blood is expelled yet.

7. In phase 3, ventricular ejection, the semilunar valves open and blood is ejected, rapidly at first and then more slowly. Each ventricle ejects a *stroke volume* of about 70 mL, which is an *ejection fraction* of about 54% of the EDV. The blood remaining behind, about 60 mL, is the *end-systolic volume (ESV)*. The T wave begins around the middle of this phase.

8. In phase 4, isovolumetric relaxation, the ventricles repolarize and relax. The semilunar valves close and heart sound $S_2$ occurs. The AV valves remain closed and no blood enters the ventricles until the next phase 1.

9. Each ventricle ejects the same amount of blood. If they ejected unequal amounts, fluid would accumulate in the tissues of either the pulmonary or systemic circuit.

## Cardiac Output (p. 741)

1. *Cardiac output (CO)* is the volume of blood pumped by each ventricle in 1 minute. It is a product of heart rate × stroke volume, and averages about 5.25 L/min.

2. Heart rate is typically about 70 to 80 bpm in young adults, but higher in children and the elderly. A persistent high resting heart rate is *tachycardia* and a persistent low resting rate is *bradycardia.*

3. Heart rate is raised or lowered, respectively, by *positive* and *negative chronotropic agents.*

4. The *cardioacceleratory center* raises the heart rate through sympathetic cardiac nerves. The *cardioinhibitory center* reduces heart rate through parasympathetic fibers in the vagus nerve. Both centers constitute the *cardiac center* of the medulla oblongata.

5. The cardiac center receives input from proprioceptors, baroreceptors, and chemoreceptors. It adjusts the heart rate to maintain blood pressure, pH, and blood $O_2$ and $CO_2$ levels within homeostatic limits.

6. Several chemicals exert positive chronotropic effects by acting through cAMP, which influences the admission of $Ca^{2+}$ into the cardiac myocytes. These include epinephrine, norepinephrine, caffeine, nicotine, and thyroid hormone. Potassium imbalances affect heart rate by altering the membrane potential of the myocytes.

7. Stroke volume is determined by the relationship of preload, contractility, and afterload. *Preload* is the amount of tension in the myocardium just before contraction; *contractility* is the amount of force that the contracting myocardium generates for a given preload; and *afterload* is resistance from blood pressure in the major arteries attached to the heart.

8. Factors that increase or decrease contractility are called *positive* and *negative inotropic agents,* respectively.

9. Calcium exerts a strong, positive inotropic effect by activating the excitation–contraction coupling process of cardiac muscle and by prolonging the plateau of the myocardial action potential. Agents that increase $Ca^{2+}$ availability—such as epinephrine, norepinephrine, glucagon, and digitalis—therefore also have positive inotropic effects. Hyperkalemia has a negative inotropic effect.

10. Exercise influences cardiac output through its effects on the proprioceptors and on the venous return of blood to the heart. Habitual endurance exercise increases ventricular size and stroke volume, and reduces resting heart rate.

# Testing Your Recall

1. The cardiac conduction system includes all of the following *except*
   a. the SA node.
   b. the AV node.
   c. the bundle branches.
   d. the tendinous cords.
   e. the Purkinje fibers.

2. To get from the right atrium to the right ventricle, blood flows through
   a. the pulmonary valve.
   b. the tricuspid valve.
   c. the bicuspid valve.
   d. the aortic valve.
   e. the mitral valve.

3. Assume that one ventricle of a child's heart has an EDV of 90 mL, an ESV of 60 mL, and a cardiac output of 2.55 L/min. What are the child's stroke volume (SV), ejection fraction (EF), and heart rate (HR)?
   a. SV = 60 mL; EF = 33%; HR = 85 bpm
   b. SV = 30 mL; EF = 60%; HR = 75 bpm
   c. SV = 150 mL; EF = 67%; HR = 42 bpm
   d. SV = 30 mL; EF = 33%; HR = 85 bpm
   e. There is not enough information to calculate these.

4. A heart rate of 45 bpm and an absence of P waves suggest
   a. damage to the SA node.
   b. ventricular fibrillation.
   c. cor pulmonale.
   d. extrasystole.
   e. heart block.

5. There is/are _____ pulmonary vein(s) emptying into the right atrium of the heart.
   a. no
   b. one
   c. two
   d. four
   e. more than four

6. The coronary blood vessels are part of the _____ circuit of the circulatory system.
   a. cardiac
   b. pulmonary
   c. systematic
   d. systemic
   e. cardiovascular

7. The atria contract during
   a. the first heart sound.
   b. the second heart sound.
   c. the QRS complex.
   d. the PQ segment.
   e. the ST segment.

8. Cardiac muscle does not exhibit tetanus because it has
   a. fast $Ca^{2+}$ channels.
   b. scanty sarcoplasmic reticulum.
   c. a long absolute refractory period.
   d. electrical synapses.
   e. exclusively aerobic respiration.

9. The blood contained in a ventricle during isovolumetric relaxation is
   a. the end-systolic volume.
   b. the end-diastolic volume.
   c. the stroke volume.
   d. the ejection fraction.
   e. none of these; the ventricle is empty then.

10. Drugs that increase the heart rate have a _____ effect.
    a. myogenic
    b. negative inotropic
    c. positive inotropic
    d. negative chronotropic
    e. positive chronotropic

11. The contraction of any heart chamber is called _____ and its relaxation is called _____.

12. The circulatory route from aorta to the venae cavae is the _____ circuit.

13. The circumflex artery travels in a groove called the _____.

14. The pacemaker potential of the SA node cells results from the slow inflow of _____.

15. Electrical signals pass quickly from one cardiac myocyte to another through the _____ of the intercalated discs.

16. Repolarization of the ventricles produces the _____ of the electrocardiogram.

17. The _____ nerves innervate the heart and tend to reduce the heart rate.

18. The death of cardiac tissue from lack of blood flow is commonly known as a heart attack, but clinically called _____.

19. Blood in the heart chambers is separated from the myocardium by a thin membrane called the _____.

20. The Frank–Starling law of the heart explains why the _____ of the left ventricle is the same as that of the right ventricle.

*Answers in Appendix B*

# True or False

*Determine which five of the following statements are false, and briefly explain why.*

1. The blood supply to the myocardium is the coronary circulation; everything else is called the systemic circuit.

2. There are no valves at the point where venous blood flows into the atria.

3. No blood can enter the ventricles until the atria contract.

4. The vagus nerves reduce the heart rate but have little effect on the strength of ventricular contraction.

5. A high blood $CO_2$ level and low blood pH stimulate an increase in heart rate.

6. The first heart sound occurs at the time of the P wave of the electrocardiogram.

7. If all nerves to the heart were severed, the heart would instantly stop beating.

8. If the two pulmonary arteries were clamped shut, systemic edema would follow.

9. Ventricular myocytes have a stable resting membrane potential but myocytes of the SA node do not.

10. An electrocardiogram is a tracing of the action potential of a cardiac myocyte.

*Answers in Appendix B*

# Testing Your Comprehension

1. Verapimil is a calcium channel blocker used to treat hypertension. It selectively blocks slow calcium channels. Would you expect it to have a positive or negative inotropic effect? Explain. (See p. 97 to review calcium channel blockers.)

2. To temporarily treat tachycardia and restore the normal resting sinus rhythm, a physician may massage a patient's carotid artery near the angle of the mandible. Propose a mechanism by which this treatment would have the desired effect.

3. Becky, age 2, was born with a hole in her interventricular septum (*ventricular septal defect,* or *VSD*). Considering that the blood pressure in the left ventricle is significantly higher than blood pressure in the right ventricle, predict the effect of the VSD on Becky's pulmonary blood pressure, systemic blood pressure, and long-term changes in the ventricular walls.

4. In ventricular systole, the left ventricle is the first to begin contracting, but the right ventricle is the first to expel blood. Aside from the obvious fact that the pulmonary valve opens before the aortic valve, how can you explain this difference?

5. In dilated cardiomyopathy of the left ventricle, the ventricle can become enormously enlarged. Explain why this might lead to regurgitation of blood through the mitral valve (blood flowing from the ventricle back into the left atrium) during ventricular systole.

*Answers at www.mhhe.com/saladin4*

# www.mhhe.com/saladin4

*The textbook website provides a wealth of interactive study materials fully organized and integrated by chapter. You will find practice quizzes, labeling exercises, and much more that will complement your learning and understanding of anatomy and physiology. The website also includes tools designed to enhance your* **Anatomy & Physiology | REVEALED** *experience.*

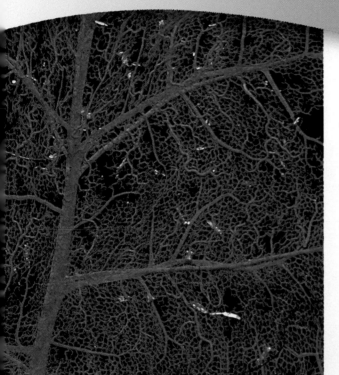

*Beds of blood capillaries*

# THE CIRCULATORY SYSTEM: BLOOD VESSELS AND CIRCULATION

## CHAPTER OUTLINE

## INSIGHTS

## Brushing Up

To understand this chapter, it is important that you understand or brush up on the following concepts:

- Set point and dynamic equilibrium in homeostasis (p. 17)
- Diffusion (p. 102)
- Equilibrium between filtration and osmosis (p. 104)
- Transcytosis (p. 110)
- Viscosity and osmolarity of blood (pp. 683–684)
- Principles of pressure and flow (p. 736)
- Autonomic effects on the heart (p. 742)

The route taken by the blood after it leaves the heart was a point of much confusion for many centuries. Chinese emperor Huang Ti (2697–2597 B.C.E.) correctly believed that it flowed in a complete circuit around the body and back to the heart. But in the second century, Roman physician Claudius Galen (129–c. 199) argued that it flowed back and forth in the veins, like air in the bronchial tubes. He believed that the liver received food from the small intestine and converted it to blood, the heart pumped the blood through the veins to all other organs, and those organs consumed it.

Huang Ti was right, but the first experimental demonstration of this did not come until the seventeenth century. English physician William Harvey (see p. 000) studied the filling and emptying of the heart in snakes, tied off the vessels above and below the heart to observe the effects on cardiac filling and output, and measured cardiac output in a variety of living animals. He concluded that (1) the heart pumps more blood in half an hour than there is in the entire body, (2) not enough food is consumed to account for the continual production of so much blood, and (3) since the planets orbit the sun and (as he believed) the human body is modeled after the solar system, it follows that the blood orbits the body. So for a peculiar combination of experimental and superstitious reasons, Harvey argued that the blood returns to the heart rather than being consumed by the peripheral organs. He could not explain how, since the microscope had yet to be developed and he did not know of capillaries— later discovered by Antony van Leeuwenhoek and Marcello Malpighi.

Harvey's work was the first experimental study of animal physiology and a landmark in the history of biology and medicine. But so entrenched were the ideas of Aristotle and Galen in the medical community, and so strange was the idea of doing experiments on living animals, that Harvey's contemporaries rejected his ideas. Indeed, some of them regarded him as a crackpot because his conclusion flew in the face of common sense—if the blood was continually recirculated and not consumed by the tissues, they reasoned, then what purpose could it serve? We now know, of course, that he was right. Harvey's case is one of the most interesting in biomedical history, for it shows how empirical science overthrows old theories and spawns better ones, and how common sense and blind allegiance to authority can interfere with the acceptance of truth.

# General Anatomy of the Blood Vessels

## Objectives

When you have completed this section, you should be able to

- describe the structure of a blood vessel;
- describe the different types of arteries, capillaries, and veins;
- trace the general route usually taken by the blood from the heart and back again; and
- describe some variations on this route.

There are three principal categories of blood vessels: arteries, veins, and capillaries (fig. 20.1). **Arteries** are the

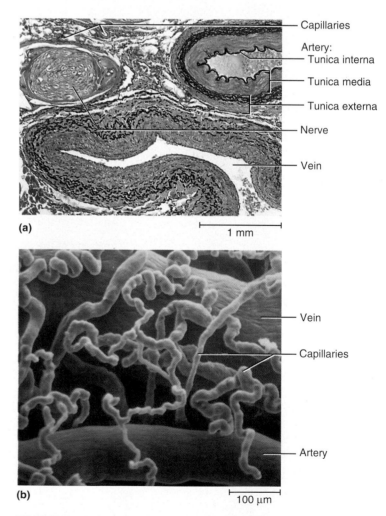

**FIGURE 20.1  Micrographs of Blood Vessels.**  (a) A neurovascular bundle, composed of a small artery, vein, and nerve in a common sheath of connective tissue (LM). (b) A cast of blood vessels of the eye (SEM). This was prepared by injecting the vessels with a polymer, digesting away all tissue to leave a replica of the vessels, and photographing the cast with the electron microscope.

efferent vessels of the cardiovascular system—that is, vessels that carry blood away from the heart. **Veins** are the afferent vessels—vessels that carry blood back to the heart. **Capillaries** are microscopic, thin-walled vessels that connect the smallest arteries to the smallest veins. Aside from their general location and direction of blood flow, these three categories of vessels also differ in the histological structure of their walls.

## THE VESSEL WALL

The walls of arteries and veins are composed of three layers called *tunics* (fig. 20.2):

1. The **tunica interna** (tunica intima) lines the inside of the vessel and is exposed to the blood. It consists of a simple squamous epithelium called the **endothelium,** overlying a basement membrane and a sparse layer of

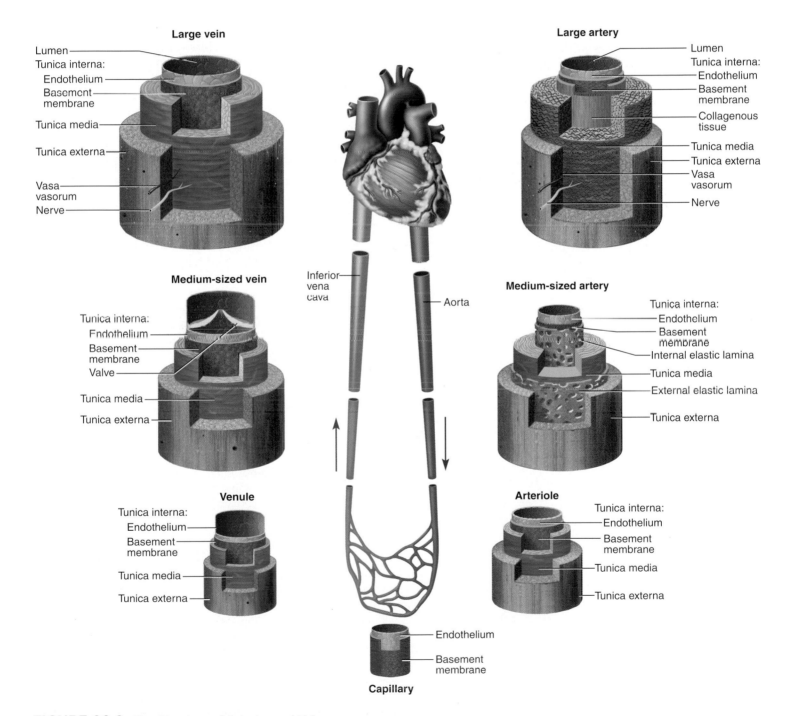

**FIGURE 20.2** The Structure of Arteries and Veins.

▶ *Why are elastic laminae found in arteries but not in veins?*

loose connective tissue. The endothelium acts as a selectively permeable barrier to materials entering or leaving the bloodstream; it secretes chemicals that stimulate dilation or constriction of the vessel; and it normally repels blood cells and platelets so that they flow freely without sticking to the vessel wall. When the endothelium is damaged, however, platelets may adhere to it and form a blood clot; and when the tissue around a vessel is inflamed, the endothelial cells produce *cell-adhesion molecules* that induce leukocytes to adhere to the surface. This causes leukocytes to congregate in tissues where their defensive actions are needed.

2. The **tunica media,** the middle layer, is usually the thickest. It consists of smooth muscle, collagen, and in some cases, elastic tissue. The relative amounts of smooth muscle and elastic tissue vary greatly from one vessel to another and form a basis for classifying vessels as described in the next section. The principal functions of the tunica media are to strengthen the vessels and prevent the blood pressure from rupturing them, and to provide for **vasomotion,** changes in the diameter of a blood vessel.

3. The **tunica externa** (tunica adventitia[1]) is the outermost layer. It consists of loose connective tissue that often merges with that of neighboring blood vessels, nerves, or other organs (see fig. 20.1a). It anchors the vessel and provides passage for small nerves, lymphatic vessels, and smaller blood vessels. Small vessels called the **vasa vasorum**[2] (VAY-za vay-SO-rum) supply blood to at least the outer half of the wall of a larger vessel. Tissues of the inner half of the wall are thought to be nourished by diffusion from blood in the lumen.

# ARTERIES

Arteries are sometimes called the *resistance vessels* of the cardiovascular system because they have a relatively strong, resilient tissue structure that resists high blood pressure. Each beat of the ventricles creates a surge of pressure in the arteries as blood is ejected into them. Arteries are constructed to withstand these surges. They are more muscular than veins, so they retain their round shape even when empty, and they appear relatively circular in tissue sections. They are divided into three categories by size, but there is a smooth transition from one category to the next.

1. **Conducting (elastic** or **large) arteries** are the biggest. The aorta, common carotid and subclavian arteries, pulmonary trunk, and common iliac arteries are examples of conducting arteries. There is a very thin layer of elastic fibers called the *internal elastic lamina* at the border between the intima and media, but it is sparse and not very visible microscopically. The tunica media consists of 40 to 70 layers of elastic sheets, perforated like slices of Swiss cheese, alternating with thin layers of smooth muscle, collagen, and elastic fibers. In histological sections, the view is dominated by this elastic tissue. There is a thin *external elastic lamina* at the border between the media and externa. The tunica externa is relatively thick and well supplied with vasa vasorum.

Conducting arteries expand during ventricular systole to receive blood, and recoil during diastole. Their expansion takes some of the pressure off the blood so that smaller arteries downstream are subjected to less systolic stress. Their recoil between heartbeats prevents the blood pressure from dropping too low while the heart is relaxing and refilling. These effects lessen the fluctuations in blood pressure that would otherwise occur. Arteries stiffened by atherosclerosis cannot expand and recoil as freely. Consequently, the downstream vessels are subjected to greater stress and are more likely to develop aneurysms and rupture (see Insight 20.1).

2. **Distributing (muscular** or **medium) arteries** are smaller branches that distribute blood to specific organs. You could compare a conducting artery to an interstate highway and distributing arteries to the exit ramps and state highways that serve specific towns. Most arteries that have specific anatomical names are in these first two size classes. The brachial, femoral, renal, and splenic arteries are examples of distributing arteries. Distributing arteries typically have up to 40 layers of smooth muscle constituting about three-quarters of the wall thickness. This smooth muscle is more conspicuous than

---

## INSIGHT 20.1    Clinical Application

### Aneurysm

An *aneurysm*[3] is a weak point in an artery or in the heart wall. It forms a thin-walled, bulging sac that pulsates with each beat of the heart and may eventually rupture. In a *dissecting aneurysm,* blood pools between the tunics of an artery and separates them, usually because of degeneration of the tunica media. The most common sites of aneurysms are the abdominal aorta, renal arteries, and the arterial circle at the base of the brain. Even without hemorrhaging, aneurysms can cause pain or death by putting pressure on brain tissue, nerves, adjacent veins, pulmonary air passages, or the esophagus. Other consequences include neurological disorders, difficulty in breathing or swallowing, chronic cough, or congestion of the tissues with blood. Aneurysms sometimes result from congenital weakness of the blood vessels and sometimes from trauma or bacterial infections such as syphilis. The most common cause, however, is the combination of atherosclerosis and hypertension.

---

[1]*advent* = added to
[2]*vasa* = vessels + *vasorum* = of the vessels

[3]*aneurysm* = widening

the elastic tissue in histological specimens of distributing arteries. Both the internal and external elastic laminae, however, are thick and often histologically conspicuous in distributing arteries.

3. **Resistance (small) arteries** are usually too variable in number and location to be given individual names. They exhibit up to 25 layers of smooth muscle and relatively little elastic tissue. Their tunica media is thicker in proportion to the lumen than that of larger arteries. The smallest of these arteries, about 40 to 200 μm in diameter and with only one to three layers of smooth muscle, are called **arterioles.** Arterioles have very little tunica externa.

**Metarterioles**[4] are short vessels that link arterioles and capillaries. Instead of a continuous tunica media, they have individual muscle cells spaced a short distance apart, each forming a **precapillary sphincter** that encircles the entrance to one capillary (fig. 20.3). Constriction of these sphincters reduces or shuts off blood flow through their respective capillaries and diverts blood to tissues or organs elsewhere.

## Arterial Sense Organs

Certain major arteries above the heart have sensory structures in their walls that monitor blood pressure and chemistry (fig. 20.4). They transmit information to the

[4]*meta* = beyond, next in a series

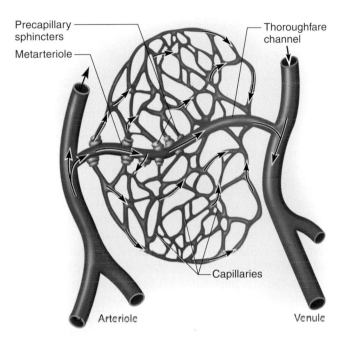

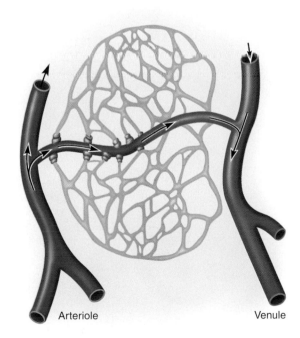

**(a) Sphincters open**

**(b) Sphincters closed**

**FIGURE 20.3 Perfusion of a Capillary Bed.** (a) Precapillary sphincters dilated and capillaries well perfused. (b) Precapillary sphincters closed, with most blood bypassing the capillaries.

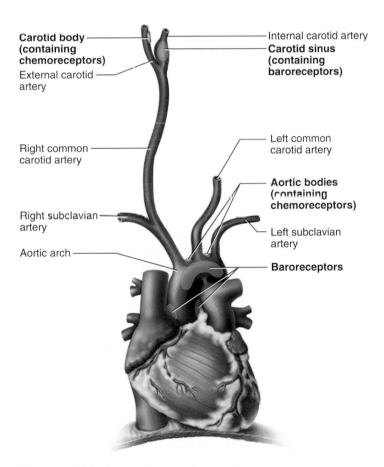

**FIGURE 20.4 Locations of Arterial Baroreceptors and Chemoreceptors.** Chemoreceptors are located in the carotid bodies and aortic bodies. Baroreceptors are located in the colored regions of the ascending aorta, aortic arch, and carotid sinus. The structures shown here in the right carotid arteries are repeated in the left carotids.

brainstem that is used to regulate the heartbeat, vasomotion, and respiration. The sensory receptors are of three kinds:

1. **Carotid sinuses.** These are *baroreceptors* (pressure sensors) that respond to changes in blood pressure. Ascending the neck on each side is a *common carotid artery* which branches near the angle of the mandible, forming the *internal carotid artery* to the brain and *external carotid artery* to the face. The carotid sinuses are located in the wall of the internal carotid artery just above the branch point. The carotid sinus has a relatively thin tunica media and an abundance of glossopharyngeal nerve fibers in the tunica externa. A rise in blood pressure easily stretches the thin media and stimulates these nerve fibers. The glossopharyngeal nerve then transmits signals to the vasomotor and cardiac centers of the brainstem, and the brainstem responds by lowering the heart rate and dilating the blood vessels, thereby lowering the blood pressure.

2. **Carotid bodies.** Also located near the branch of the common carotid arteries, these are oval receptors about $3 \times 5$ mm in size, innervated by sensory fibers of the vagus and glossopharyngeal nerves. They are *chemoreceptors* that monitor changes in blood composition. They primarily transmit signals to the brainstem respiratory centers, which adjust breathing to stabilize the blood pH and its $CO_2$ and $O_2$ levels.

3. **Aortic bodies.** These are one to three chemoreceptors located in the aortic arch near the arteries to the head and arms. They are structurally similar to the carotid bodies and have the same function.

# CAPILLARIES

For the blood to serve any purpose, materials such as nutrients, wastes, and hormones must be able to pass between the blood and the tissue fluids, through the walls of the blood vessels. There are only two places in the circulation where this can occur: the capillaries and some venules. Since capillaries greatly exceed venules in number and permeability, they are the more important of the two. Capillaries are sometimes called the *exchange vessels* of the cardiovascular system. We can think of them as the "business end" of the system, because all the rest of the circulatory system exists to serve the exchange processes that occur in the capillaries and smallest of the venules.

Blood capillaries (see figs. 20.1–20.3) are composed of only an endothelium and basement membrane. Their walls are as thin as 0.2 to 0.4 µm. They average about 5 µm in diameter at the proximal end (where they receive arterial blood), they widen to about 9 µm at the distal end (where they empty into a small vein), and they often branch along the way. Since an erythrocyte is about 7 µm

in diameter, erythrocytes often have to stretch into elongated shapes to squeeze through the smallest capillaries.

The number of capillaries has been estimated at a billion and their total surface area at 6,300 m². But a more important point is that scarcely any cell in the body is more than 60 to 80 µm (about four to six cell widths) away from the nearest capillary (see page 753). There are a few exceptions: Capillaries are scarce in tendons and ligaments and absent from epithelia, the cornea and lens of the eye, and most cartilage.

## Types of Capillaries

There are three types of capillaries, distinguished by the ease with which they allow substances to pass through their walls and by structural differences that account for their greater or lesser permeability.

1. **Continuous capillaries** (fig. 20.5) occur in most tissues, such as skeletal muscle. Their endothelial cells, held together by tight junctions, form an uninterrupted tube. A thin protein-carbohydrate layer, the **basal lamina,** surrounds the endothelium and separates it from the adjacent connective tissues. The endothelial cells are separated by narrow **intercellular clefts** about 4 nm wide. Small solutes such as glucose can pass through these clefts, but most plasma protein, other large molecules, and formed elements of the blood are held back. The continuous capillaries of the brain lack intercellular clefts and

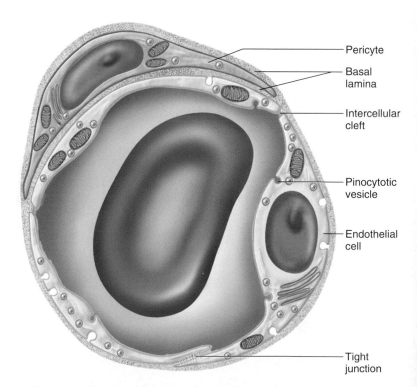

**FIGURE 20.5**    **Structure of a Continuous Capillary.** Cross section.

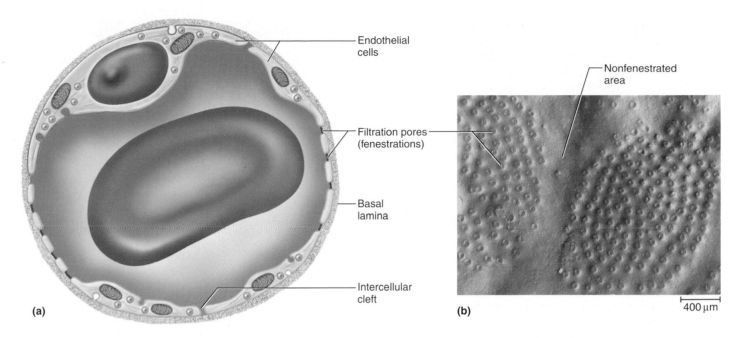

(a) — Endothelial cells / Filtration pores (fenestrations) / Basal lamina / Intercellular cleft

(b) — Nonfenestrated area / 400 μm

**FIGURE 20.6** **Structure of a Fenestrated Capillary.** (a) Cross section of the capillary wall. (b) Surface view of a fenestrated endothelial cell. The cell has patches of fenestrations separated by nonfenestrated areas.

have more complete tight junctions that form the blood–brain barrier discussed in chapter 14.

Some continuous capillaries exhibit cells called **pericytes** that lie external to the endothelium. Pericytes have elongated tendrils that wrap around the capillary. They contain the same contractile proteins as muscle, and it is thought that they contract and regulate blood flow through the capillaries. They also can differentiate into endothelial and smooth muscle cells and thus contribute to vessel growth and repair.

2. **Fenestrated capillaries** have endothelial cells riddled with holes called **filtration pores** (fenestrations[5]) (fig. 20.6). Filtration pores are about 20 to 100 nm in diameter and are often spanned by a glycoprotein membrane that is much thinner than the cell's plasma membrane. These pores allow for the rapid passage of molecules, even proteins, through the capillary wall, but still retain blood cells and platelets in the bloodstream. Fenestrated capillaries are important in organs that engage in rapid absorption or filtration—the kidneys, endocrine glands, small intestine, and choroid plexuses of the brain, for example.

3. **Sinusoids** (discontinuous capillaries) are irregular blood-filled spaces in the liver, bone marrow, spleen, and some other organs (fig. 20.7). They are twisted, tortuous passageways, typically 30 to 40 μm wide, that conform to the shape of the surrounding tissue.

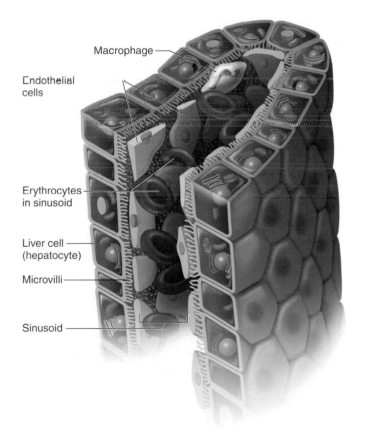

Macrophage / Endothelial cells / Erythrocytes in sinusoid / Liver cell (hepatocyte) / Microvilli / Sinusoid

**FIGURE 20.7** **A Sinusoid of the Liver.** Large gaps between the endothelial cells allow blood plasma to directly contact the liver cells, but retain blood cells in the lumen of the sinusoid.

[5]*fenestra* = window

The endothelial cells are separated by wide gaps with no basal lamina, and the cells also frequently have especially large fenestrations through them. Even proteins and blood cells can pass through these pores; this is how albumin, clotting factors, and other proteins synthesized by the liver enter the blood, and how newly formed blood cells enter the circulation from the bone marrow and lymphatic organs. Sinusoids may contain macrophages or other specialized cells.

## Capillary Beds

Capillaries are organized into networks called **capillary beds**—usually 10 to 100 capillaries supplied by a single metarteriole (see p. 753 and fig. 20.3). Beyond the origins of the capillaries, the metarteriole continues as a **thoroughfare channel** leading directly to a venule. Capillaries empty into the distal end of the thoroughfare channel or directly into the venule.

When the precapillary sphincters are open, the capillaries are well perfused with blood and engage in exchanges with the tissue fluid. When the sphincters are closed, blood bypasses the capillaries, flows through the thoroughfare channel to a venule, and engages in relatively little fluid exchange. There is not enough blood in the body to fill the entire vascular system at once; consequently, about three-quarters of the body's capillaries are shut down at any given time. In the skeletal muscles, for example, about 90% of the capillaries have little to no blood flow during periods of rest. During exercise, they receive an abundant blood flow while capillary beds elsewhere—for example, in the skin and intestines—shut down to compensate for this.

## VEINS

Veins are sometimes called the *capacitance vessels* of the cardiovascular system because they are relatively thin-walled and flaccid, and expand easily to accommodate an increased volume of blood; that is, they have a greater capacity for blood containment than arteries do. At rest, about 54% of the blood is found in the systemic veins as compared with only 11% in the systemic arteries (fig. 20.8). The reason that veins are so thin-walled and accommodating is that, being distant from the ventricles of the heart, they are subjected to relatively low blood pressure. In large arteries, blood pressure averages 90 to 100 mm Hg (millimeters of mercury) and surges to 120 mm Hg during systole, whereas in veins it averages about 10 mm Hg. Furthermore, the blood flow in the veins is steady, rather than pulsating with the heartbeat like the flow in the arteries. Veins therefore do not require thick, pressure-resistant walls. Veins collapse when empty and thus have relatively flattened, irregular shapes in histological sections (see fig. 20.1a).

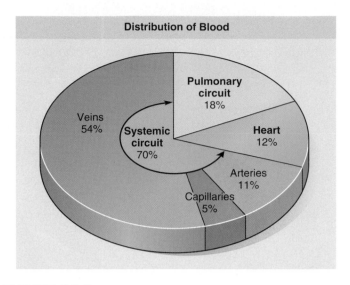

**FIGURE 20.8**   Typical Blood Distribution in a Resting Adult.

As we trace blood flow in the arteries, we find it splitting off repeatedly into smaller and smaller *branches* of the arterial system. In the venous system, conversely, we find small veins merging to form larger and larger ones as they approach the heart. We refer to the smaller veins as *tributaries,* by analogy to the streams that converge and act as tributaries to rivers. In examining the types of veins, we will follow the direction of blood flow, working up from the smallest to the largest vessels.

1. **Postcapillary venules** are the smallest of the veins, beginning with diameters of about 15 to 20 μm. They receive blood from capillaries directly or by way of the distal ends of the thoroughfare channels. They have a tunica interna with only a few fibroblasts around it, and like capillaries, they are often surrounded by pericytes. Postcapillary venules are even more porous than capillaries; therefore venules also exchange fluid with the surrounding tissues. Most leukocytes emigrate from the bloodstream through the venule walls.

2. **Muscular venules** receive blood from the postcapillary venules. They are up to 1 mm in diameter. They have a tunica media of one or two layers of smooth muscle, and a thin tunica externa.

3. **Medium veins** range up to 10 mm in diameter. Most veins with individual names are in this category, such as the radial and ulnar veins of the forearm and the small and great saphenous veins of the lower limb. Medium veins have a tunica interna with an endothelium, basement membrane, loose connective tissue, and sometimes a thin internal elastic lamina. The tunica media is much thinner than it is in medium arteries; it exhibits bundles of smooth muscle, but not a continuous muscular layer as seen in

arteries. The muscle is interrupted by regions of collagenous, reticular, and elastic tissue. The tunica externa is relatively thick.

Many medium veins, especially in the limbs, exhibit infoldings of the tunica interna that meet in the middle of the lumen, forming **venous valves** directed toward the heart (see fig. 20.19). The pressure in the veins is not high enough to push blood upward against the pull of gravity in a standing or sitting person. The upward flow of blood in these vessels depends partly on the massaging action of skeletal muscles and the ability of these valves to keep the blood from dropping down again when the muscles relax. When the muscles surrounding a vein contract, they force blood through these valves. The propulsion of venous blood by muscular massaging, aided by the venous valves, is a mechanism of blood flow called the *skeletal muscle pump.* Varicose veins result in part from the failure of the venous valves (see Insight 20.2). Such valves are absent from very small and large veins, veins of the abdominal and thoracic cavities, and veins of the brain.

4. **Venous sinuses** are veins with especially thin walls, large lumens, and no smooth muscle. Examples include the coronary sinus of the heart and the dural sinuses of the brain. Unlike other veins, they are not capable of vasomotion.

5. **Large veins** have diameters greater than 10 mm. They have a relatively thin tunica media with only a moderate amount of smooth muscle; the tunica externa is the thickest layer and contains longitudinal bundles of smooth muscle. Some smooth muscle is found in all three layers of these veins. Large veins include the venae cavae, pulmonary veins, internal jugular veins, and renal veins.

## INSIGHT 20.2 Clinical Application

### Varicose Veins

In people who stand for long periods, such as barbers and cashiers, blood tends to pool in the lower limbs and stretch the veins. This is especially true of superficial veins, which are not surrounded by supportive tissue. Stretching pulls the cusps of the venous valves farther apart until the valves become incompetent to prevent the backflow of blood. As the veins become further distended, their walls grow weak and they develop into *varicose veins* with irregular dilations and twisted pathways. Obesity and pregnancy also promote development of varicose veins by putting pressure on large veins of the pelvic region and obstructing drainage from the legs. Varicose veins sometimes develop because of hereditary weakness of the valves. With less drainage of blood, tissues of the leg and foot may become edematous and painful. *Hemorrhoids* are varicose veins of the anal canal.

## CIRCULATORY ROUTES

The simplest and most common route of blood flow is heart → arteries → capillaries → veins → heart. Blood usually passes through only one network of capillaries from the time it leaves the heart until the time it returns (fig. 20.9a) but there are exceptions, notably portal systems and anastomoses.

In a **portal system** (fig. 20.9b), blood flows through two consecutive capillary networks before returning to the heart. Portal systems occur in the kidneys (chapter 23); connecting the hypothalamus and anterior pituitary (chapter 17); and connecting the intestines to the liver (table 20.13).

An **anastomosis** is a point where two blood vessels merge. In an **arteriovenous anastomosis** (shunt), blood flows from an artery directly into a vein and bypasses the capillaries (fig. 20.9c). Shunts occur in the fingers, palms, toes, and ears, where they reduce heat loss in cold weather by allowing warm blood to bypass these exposed surfaces. Unfortunately, this makes these poorly perfused areas more susceptible to frostbite. The most common

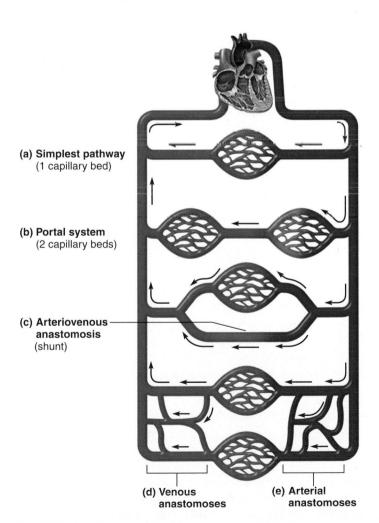

(a) **Simplest pathway** (1 capillary bed)

(b) **Portal system** (2 capillary beds)

(c) **Arteriovenous anastomosis** (shunt)

(d) **Venous anastomoses**

(e) **Arterial anastomoses**

**FIGURE 20.9** Variations in Circulatory Pathways.

anastomoses are **venous anastomoses,** in which one vein empties directly into another (fig. 20.9d). These provide several alternative routes of drainage from an organ. **Arterial anastomoses,** in which two arteries merge (fig. 20.9e), provide *collateral* (alternative) routes of blood supply to a tissue. Those of the coronary circulation were mentioned in chapter 19. They are also common around joints where movement may temporarily obstruct one pathway. Several arterial and venous anastomoses are described later in this chapter.

## Before You Go On

*Answer the following questions to test your understanding of the preceding section:*

1. *Name the three tunics of a typical blood vessel and explain how they differ from each other.*

2. *Contrast the tunica media of a conducting artery, arteriole, and venule and explain how the histological differences are related to the functional differences between these vessels.*

3. *Describe the differences between a continuous capillary, a fenestrated capillary, and a sinusoid.*

4. *Describe the differences between a medium vein and a medium (muscular) artery. State the functional reasons for these differences.*

5. *Describe three routes by which substances can escape the bloodstream and pass through a capillary wall into the tissue fluid.*

# Blood Pressure, Resistance, and Flow

### Objectives

When you have completed this section, you should be able to

- explain the relationship between blood pressure, resistance, and flow;
- describe how blood pressure is expressed and how pulse pressure and mean arterial pressure are calculated;
- describe three factors that determine resistance to blood flow;
- explain how vasomotion influences blood flow; and
- describe some local, neural, and hormonal influences on vasomotion.

To sustain life, the circulatory system must deliver oxygen and nutrients to the tissues, and remove their wastes, at a rate that keeps pace with tissue metabolism. A lack of oxygen or nutrients, or the accumulation of unremoved wastes, can lead within minutes to tissue necrosis and possibly death of the individual. Thus, it is crucial that the cardiovascular system respond promptly to the needs of the tissues and ensure that they have an adequate blood supply at all times. This section of the chapter explores the mechanisms for achieving this.

The blood supply to a tissue can be expressed in terms of *flow* and *perfusion.* **Flow** is the amount of blood flowing through an organ, tissue, or blood vessel in a given time (such as mL/min). **Perfusion** is the flow per given volume or mass of tissue (such as mL/min/g). Thus a large organ such as the femur could have a *greater flow* but *less perfusion* than a small organ such as the ovary, because the ovary receives much more blood per gram of tissue.

In a resting individual, *total* flow is quite constant and is equal to cardiac output (typically 5.25 L/min). Flow through individual organs, however, varies from minute to minute as blood is redirected from one organ to another. Digestion, for example, requires abundant flow to the intestines, and the cardiovascular system reduces flow through other organs such as the kidneys in order to make it available to the digestive system. When digestion and nutrient absorption are over, blood flow to the intestines declines and a higher priority is given to the kidneys and other organs. Great variations in regional flow can occur with little or no change in total flow.

*Hemodynamics,* the physical principles of blood flow, are based mainly on pressure and resistance. These relationships can be concisely summarized by the proportionality $F \propto \Delta P/R$. In other words, the greater the pressure difference ($\Delta P$) between two points, the greater the flow; the greater the resistance $(R)$, the less the flow. Therefore, to understand the flow of blood, we must consider the factors that affect pressure and resistance.

## BLOOD PRESSURE

**Blood pressure (BP)** is the force that the blood exerts against a vessel wall. It can be measured within a blood vessel or the heart by inserting a catheter or a needle connected to an external manometer (pressure-measuring device). For routine clinical purposes, however, the measurement of greatest interest is the systemic arterial BP at a point close to the heart. We customarily measure it with a sphygmomanometer connected to an inflatable cuff wrapped around the arm. The *brachial artery* passing through this region is sufficiently close to the heart that the BP recorded here reflects the maximum arterial BP found anywhere in the body.

Two pressures are recorded: **systolic pressure** is the peak arterial BP attained during ventricular contraction, and **diastolic pressure** is the minimum arterial BP occurring during the ventricular relaxation between heartbeats. For a healthy person age 20 to 30, these pressures are typically about 120 and 75 mm Hg, respectively. Arterial BP is written as a ratio of systolic over diastolic pressure: 120/75.

The difference between systolic and diastolic pressure is called **pulse pressure** (not to be confused with pulse *rate*). For the preceding BP, pulse pressure would be 120 − 75 = 45 mm Hg. This is an important measure of the stress exerted on small arteries by the pressure surges generated by the heart. Another measure of stress on the blood vessels is the **mean arterial pressure (MAP)**—the mean pressure you would obtain if you took measurements at several intervals (say every 0.1 sec) throughout the cardiac cycle. MAP is not simply an average of systolic and diastolic pressures, because the low-pressure diastole lasts longer than the high-pressure systole. The best estimate of MAP is diastolic pressure plus one-third of the pulse pressure. For a blood pressure of 120/75, MAP ≈ 75 + 45/3 = 90 mm Hg. This is typical for vessels at the level of the heart, but MAP varies with the influence of gravity. In a standing adult, it is about 62 mm Hg in the major arteries of the head and 180 mm Hg in major arteries of the ankle.

**Hypertension** (high BP) is commonly considered to be a chronic resting blood pressure higher than 140/90. (*Transient* high BP resulting from emotion or exercise is not hypertension.) Among other effects, hypertension can weaken the small arteries and cause aneurysms, and it promotes the development of atherosclerosis (see Insight 20.4, p. 801). **Hypotension** is chronic low resting BP. It may be a consequence of blood loss, dehydration, anemia, or other factors and is normal in people approaching death.

The importance of preventing excessive blood pressure is clear. One of the body's chief mechanisms for doing so is the ability of the arteries to distend and recoil during the cardiac cycle. If the arteries were rigid tubes, pressure would rise much higher in systole and drop to nearly zero in diastole. Blood throughout the circulatory system would flow and stop, flow and stop, and put great stress on the small vessels. But healthy conducting arteries expand with each systole and absorb some of the force of the ejected blood. Then, when the heart is in diastole, their elastic recoil exerts pressure on the blood and prevents the BP from dropping to zero. This combination of expansion and recoil maintains a steady flow of blood downstream, in the capillaries, throughout the cardiac cycle. Thus, the elastic arteries "smooth out" the pressure fluctuations and reduce stress on the smaller arteries.

Nevertheless, blood flow in the arteries is *pulsatile*. In the aorta, blood rushes forward at 120 cm/sec during systole and has an average speed of 40 cm/sec over the cardiac cycle. When measured farther away from the heart, systolic and diastolic pressures are lower and there is less difference between them (fig. 20.10). In capillaries and veins, the blood flows at a steady speed without pulsation because the pressure surges have been damped out by the distance traveled and the elasticity of the arteries. This is why an injured vein exhibits relatively slow, steady bleeding, whereas blood jets intermittently from a severed artery. In the inferior vena cava near the heart, however, venous flow fluctuates with the respiratory cycle for rea-

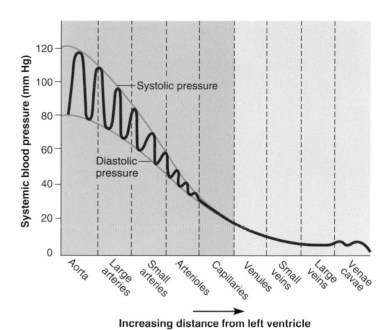

**FIGURE 20.10   Changes in Blood Pressure Related to Distance from the Heart.** Because of arterial elasticity and the effect of friction against the vessel wall, all measures of blood pressure decline with distance—systolic pressure, diastolic pressure, pulse pressure, and mean arterial pressure. There is no pulse pressure beyond the arterioles, but there are slight pressure oscillations in the venae cavae caused by the respiratory pump described later in this chapter.

sons explained later, and there is some fluctuation in the jugular veins of the neck.

**Think About It**

*Explain how the histological structure of large arteries relates to their ability to stretch during systole and recoil during diastole.*

As we get older, our arteries become less distensible and absorb less systolic force. Blood pressure therefore rises with age (table 20.1). Atherosclerosis also stiffens the arteries and raises the blood pressure.

Blood pressure is determined by three principal variables:

1. Cardiac output, which was discussed in chapter 19.
2. Blood volume, which is regulated mainly by the kidneys. Perhaps surprisingly, the kidneys have a greater influence than any other organ on blood pressure (assuming the heart is beating). Their role is discussed in chapters 23 and 24.
3. Resistance to flow, which results from the friction of the blood against the walls of the vessels. Resistance, in turn, hinges on three variables which we will now consider: blood viscosity, vessel length, and vessel radius.

| TABLE 20.1 | Normal Arterial Blood Pressure at Various Ages (mm Hg)* | |
|---|---|---|
| **Age (years)** | **Male** | **Female** |
| 1 | 96/66 | 95/65 |
| 5 | 92/62 | 92/62 |
| 10 | 103/69 | 103/70 |
| 15 | 112/75 | 112/76 |
| 20 | 123/76 | 116/72 |
| 25 | 125/78 | 117/74 |
| 30 | 126/79 | 120/75 |
| 40 | 129/81 | 127/80 |
| 50 | 135/83 | 137/84 |
| 60 | 142/85 | 144/85 |
| 70 | 145/82 | 159/85 |
| 80 | 145/82 | 157/83 |

*Average for healthy individuals

## PERIPHERAL RESISTANCE

**Peripheral resistance** is the opposition to flow that the blood encounters in vessels away from the heart. A moving fluid has no pressure unless it encounters at least some resistance. Thus, pressure and resistance are not independent factors in blood flow—rather, pressure is affected by resistance, and flow is affected by both.

### Blood Viscosity

Chapter 18 discusses the factors that affect the viscosity ("thickness") of the blood (p. 683). The most significant of these are the erythrocyte count and albumin concentration. A deficiency of erythrocytes (anemia) or albumin (hypoproteinemia) reduces viscosity and speeds up blood flow. On the other hand, viscosity increases and flow declines in such conditions as polycythemia and dehydration.

### Vessel Length

The farther a liquid travels through a tube, the more cumulative friction it encounters; thus, pressure and flow decline with distance. Partly for this reason, if you were to measure mean arterial pressure in a reclining person, you would obtain a higher value in the arm, for example, than in the ankle. (This would not be true in a standing person because of the influence of gravity, explained earlier.) A strong pulse in the *dorsal pedal artery* of the foot is a good sign of adequate cardiac output. If perfusion is good at that distance from the heart, it is likely to be good elsewhere in the systemic circulation.

## Vessel Radius

In a healthy individual, blood viscosity is quite stable, and of course vessel lengths do not change in the short term. Therefore, the only significant way of controlling peripheral resistance from moment to moment is by vasomotion—adjusting the radius of the blood vessels. This includes **vasoconstriction,** the narrowing of a vessel, and **vasodilation,** the widening of a vessel. Vasoconstriction occurs when the smooth muscle of the tunica media contracts. Vasodilation occurs when this muscle relaxes and allows the blood pressure within the vessel to push its walls outward.

The effect of vessel radius on blood flow stems from the friction of the moving blood against the walls of the vessel. Blood normally exhibits smooth, silent **laminar[6] flow.** That is, it flows in "layers"—faster near the center of a vessel, where it encounters less friction, and slower near the walls, where it drags against the vessel. You can observe a similar effect from the vantage point of a riverbank. The current may be very swift in the middle of a river but quite sluggish near shore, where the water encounters more friction against the riverbank and bottom. When a blood vessel dilates, a greater portion of the blood is in the middle of the stream and the average flow may be quite swift. When the vessel constricts, more of the blood is close to the wall and the average flow is slower (fig. 20.11).

Thus the radius of a vessel markedly affects blood velocity. Indeed, blood flow is proportional not merely to

---

[6]*lamina* = layer

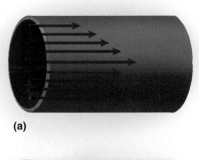

(a)

(b)

**FIGURE 20.11  Laminar Flow and the Effect of Vessel Radius.**  Blood flows more slowly near the vessel wall, as indicated by shorter arrows, than it does near the center of the vessel. (a) When vessel radius is large, the average velocity of flow is high. (b) When the radius is less, the average velocity is lower because a larger portion of the blood is slowed down by friction against the vessel wall.

vessel radius but to the *fourth power* of radius—that is, $F \propto r^4$. This makes vessel radius a very potent factor in the control of flow. For the sake of simplicity, consider a hypothetical blood vessel with a 1-mm radius when maximally constricted and a 3-mm radius when completely relaxed. At a 1-mm radius, suppose the blood travels 1 mm/sec. By the formula $F \propto r^4$, consider how the velocity would change as radius changed:

| | | |
|---|---|---|
| $r = 1$ mm | $r^4 = 1^4 = 1$ | $F = 1$ mm/sec (given) |
| $r = 2$ mm | $r^4 = 2^4 = 16$ | $F = 16$ mm/sec |
| $r = 3$ mm | $r^4 = 3^4 = 81$ | $F = 81$ mm/sec |

These actual numbers do not matter; what matters is that a mere 3-fold increase in radius has produced an 81-fold increase in velocity—a demonstration that vessel radius exerts a very powerful influence over flow. Blood vessels are, indeed, capable of substantial changes in radius. The arteriole in figure 20.12, for example, has constricted to one-third of its relaxed diameter under the influence of a drop of epinephrine. Since blood viscosity and vessel length do not change from moment to moment,

vessel radius is the most adjustable of all variables that govern peripheral resistance.

> **Think About It**
>
> *Suppose a vessel with a radius of 1 mm had a flow of 5 mm/sec, and then the vessel dilated to a radius of 5 mm. What would be the new flow rate?*

To integrate this information, consider how the velocity of blood flow differs from one part of the systemic circuit to another (table 20.2). Flow is fastest in the aorta because it is a large vessel close to the pressure source, the left ventricle. From aorta to capillaries, velocity diminishes for three reasons: (1) The blood has traveled a greater distance, so friction has slowed it down. (2) The arterioles and capillaries have smaller radii and therefore put up more resistance. (3) Even though the radii of individual vessels become smaller as we progress farther from the heart, the number of vessels and their *total* cross-sectional area becomes greater and greater. The aorta has a cross-sectional area of 3 to 5 cm$^2$, whereas the total cross-sectional area of all the capillaries is about 4,500 to 6,000 cm$^2$. Thus, a given volume of aortic blood is distributed over a greater total area in the capillaries, which *collectively* form a wider path in the bloodstream. Just as water slows down when a narrow mountain stream flows into a lake, blood slows down as it enters pathways with a greater total width.

From capillaries to vena cava, velocity rises again. One reason for this is that the veins have larger diameters than the capillaries, so they create less resistance. Furthermore, since many capillaries converge on one venule, and many venules on a larger vein, a large amount of blood is being forced into a progressively smaller channel—like water flowing from a lake into an outlet stream and thus flowing faster again. Note, however, that blood in the veins never regains the velocity it had in the large arteries. This is because the veins are farther from the heart and the pressure is much lower.

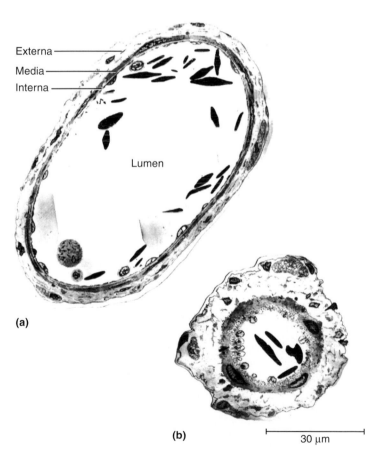

Externa
Media
Interna

Lumen

**(a)**

**(b)**            30 μm

**FIGURE 20.12  The Capacity for Vasoconstriction in an Arteriole.**  (a) A dilated arteriole. (b) The same arteriole, at a point just 1 mm from the area photographed in part (a), stimulated by a single drop of epinephrine. The diameter of the dilated region is about three times that of the constricted region.

| TABLE 20.2 | Blood Velocity in the Systemic Circuit | |
|---|---|---|
| **Vessel** | **Typical Lumen Diameter** | **Velocity*** |
| Aorta | 2.5 cm | 1,200 mm/sec |
| Arterioles | 20–50 μm | 15 mm/sec |
| Capillaries | 5–9 μm | 0.4 mm/sec |
| Venules | 20 μm | 5 mm/sec |
| Inferior vena cava | 3 cm | 80 mm/sec |

*Peak systolic velocity in the aorta; mean or steady velocity in other vessels

# REGULATION OF BLOOD PRESSURE AND FLOW

Vasomotion, we have seen, is a quick and powerful way of altering blood pressure and flow. There are three ways of controlling vasomotion: local, neural, and hormonal mechanisms. We now consider each of these influences in turn.

## Local Control

**Autoregulation** is the ability of tissues to regulate their own blood supply. According to the *metabolic theory of autoregulation,* if a tissue is inadequately perfused, it becomes hypoxic and its metabolites (waste products) accumulate—$CO_2$, $H^+$, $K^+$, lactic acid, and adenosine, for example. These factors stimulate vasodilation, which increases perfusion. As the bloodstream delivers oxygen and carries away the metabolites, the vessels constrict. Thus, a homeostatic dynamic equilibrium is established that adjusts perfusion to the tissue's metabolic needs.

In addition, blood platelets, endothelial cells, and the perivascular tissues secrete a variety of **vasoactive chemicals**—substances that stimulate vasomotion. Histamine, bradykinin, and prostaglandins stimulate vasodilation under such conditions as trauma, inflammation, and exercise. Endothelial cells secrete prostacyclin and nitric oxide, which are vasodilators, and polypeptides called *endothelins,* which are vasoconstrictors.

If a tissue's blood supply is cut off for a time and then restored, it often exhibits **reactive hyperemia**—an increase above the normal level of flow. This may be due to the accumulation of metabolites during the period of ischemia. Reactive hyperemia is seen when the skin flushes after a person comes in from the cold. It also occurs in the forearm if a blood pressure cuff is inflated for too long and then loosened.

Over a longer time, a hypoxic tissue can increase its own perfusion by **angiogenesis**[7]—the growth of new blood vessels. (This term also refers to embryonic development of blood vessels.) Three situations in which this is important are the regrowth of the uterine lining after each menstrual period, the development of a higher density of blood capillaries in the muscles of well-conditioned athletes, and the growth of arterial bypasses around obstructions in the coronary circulation. Several growth factors and inhibitors control angiogenesis, but physiologists are not yet sure how it is regulated. There is great clinical importance in finding out. Malignant tumors secrete growth factors that stimulate a dense network of blood vessels to grow into them and provide nourishment to the cancer cells. Oncologists are interested in finding a way to block tumor angiogenesis, which would choke off a tumor's blood supply and perhaps shrink or kill it.

[7]*angio* = vessels + *genesis* = production of

## Neural Control

In addition to local control, the blood vessels are under remote control by the autonomic nervous system. The **vasomotor center** of the medulla oblongata exerts sympathetic control over blood vessels throughout the body. (Precapillary sphincters have no innervation, however, and respond only to local and hormonal stimuli.) Sympathetic nerve fibers stimulate most blood vessels to constrict, but they dilate the vessels of skeletal and cardiac muscle in order to meet the metabolic demands of exercise. The role of sympathetic tone and vasomotor tone in controlling vessel diameter is explained in chapter 15.

The vasomotor center is an integrating center for three autonomic reflexes—*baroreflexes, chemoreflexes,* and the *medullary ischemic reflex.* A **baroreflex**[8] is an autonomic, negative feedback response to changes in blood pressure. The changes are detected by the carotid sinuses described earlier (p. 758). Glossopharyngeal nerve fibers from these sinuses transmit signals continually to the brainstem. When the blood pressure rises, their signaling rate rises. This input *inhibits* the sympathetic cardiac and vasomotor neurons and reduces sympathetic tone, and it *excites* the vagal fibers to the heart. Thus, it reduces the heart rate and cardiac output, dilates the arteries and veins, and reduces the blood pressure (fig. 20.13). When blood

[8]*baro* = pressure

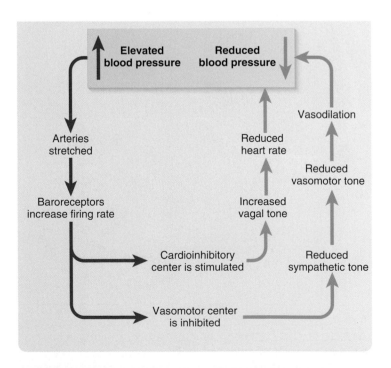

**FIGURE 20.13   Negative Feedback Control of Blood Pressure.** High blood pressure activates this cycle of reactions that return blood pressure to normal.

pressure drops below normal, on the other hand, the opposite reactions occur and BP rises back to normal.

Baroreflexes are important chiefly in short-term regulation of BP, for example in adapting to changes in posture. Perhaps you have jumped quickly out of bed and felt a little dizzy for a moment. This occurs because gravity draws the blood into the large veins of the abdomen and lower limbs when you stand, which reduces venous return to the heart and cardiac output to the brain. Normally, the baroreceptors respond quickly to this drop in pressure and restore cerebral perfusion. Baroreflexes are not effective in correcting chronic hypertension, however. Apparently they adjust their set point to the higher BP and maintain dynamic equilibrium at this new level.

A **chemoreflex** is an autonomic response to changes in blood chemistry, especially its pH and concentrations of $O_2$ and $CO_2$. It is initiated by the chemoreceptors called aortic bodies and carotid bodies (see p. 758). The primary role of chemoreflexes is to adjust respiration to changes in blood chemistry, but they have a secondary role in stimulating vasomotion. Hypoxemia (blood $O_2$ deficiency), hypercapnia ($CO_2$ excess), and acidosis (low blood pH) stimulate the chemoreceptors and act through the vasomotor center to induce widespread vasoconstriction. This increases overall BP, thus increasing perfusion of the lungs and the rate of gas exchange. Chemoreceptors also stimulate breathing, so increased ventilation of the lungs matches their increased perfusion. Increasing one without the other would be of little use.

The **medullary ischemic** (iss-KEE-mic) **reflex** is an autonomic response to a drop in perfusion of the brain. Within seconds, the cardiac and vasomotor centers of the medulla oblongata send sympathetic signals to the heart and blood vessels that induce (1) an increase in heart rate and contraction force and (2) widespread vasoconstriction. These actions raise the blood pressure and, ideally, restore normal perfusion of the brain. The cardiac and vasomotor centers also receive input from other brain centers. Thus stress, anger, and arousal can also raise the blood pressure. The hypothalamus acts through the vasomotor center to redirect blood flow in response to exercise or changes in body temperature.

## Hormonal Control

All of the following hormones influence blood pressure:

- **Angiotensin II.** This is a potent vasoconstrictor that raises the blood pressure. Its synthesis and action are detailed in chapter 23 (see fig. 23.14). One of the enzymes required for its synthesis is *angiotensin-converting enzyme (ACE).* Hypertension is often treated with drugs called *ACE inhibitors,* which block the action of this enzyme, thus lowering angiotensin II levels and blood pressure.

- **Aldosterone.** This "salt-retaining hormone" primarily promotes $Na^+$ retention by the kidneys. Since water follows sodium osmotically, $Na^+$ retention promotes water retention, thus promoting a higher blood volume and pressure.

- **Atrial natriuretic peptide.** ANP, secreted by the heart, antagonizes aldosterone. It increases $Na^+$ excretion by the kidneys, thus reducing blood volume and pressure. It also has a generalized vasodilator effect that contributes to lowering the blood pressure.

- **Antidiuretic hormone.** ADH primarily promotes water retention, but at pathologically high concentrations it is also a vasoconstrictor—hence its alternate name, *vasopressin.* Both of these effects raise blood pressure.

- **Epinephrine and norepinephrine.** These adrenal and sympathetic catecholamines bind to α-adrenergic receptors on the smooth muscle of most blood vessels. This stimulates the muscle to contract, thus producing vasoconstriction and raising the blood pressure. In the coronary blood vessels and blood vessels of the skeletal muscles, however, these chemicals bind to β-adrenergic receptors and cause vasodilation, thus increasing blood flow to the myocardium and muscular system during exercise.

## VASOMOTION AND ROUTING OF BLOOD FLOW

If a chemical such as epinephrine causes widespread vasoconstriction, or if it causes vasoconstriction in a large system such as the integumentary or digestive system, it can produce an overall rise in blood pressure. Localized vasoconstriction, however, has a very different effect. If a particular artery constricts, pressure downstream from the constriction drops and pressure upstream from it rises. If blood can travel by either of two routes and one route puts up more resistance than the other, most blood follows the path of least resistance. This mechanism enables the body to redirect blood from one organ to another.

For example, if you are dozing in an armchair after a big meal, vasoconstriction shuts down blood flow to 90% or more of the capillaries in the muscles of your lower limbs. This raises the BP above the limbs, where the aorta gives off a branch, the superior mesenteric artery, supplying the small intestine. High resistance in the circulation of the legs and low resistance in the superior mesenteric artery routes blood to the small intestine, where it is needed to absorb digested nutrients (fig. 20.14a).

On the other hand, during vigorous exercise, the arteries in your lungs, coronary circulation, and muscles dilate. To increase the circulation in these routes, flow must be reduced elsewhere, such as the kidneys and digestive tract (figs. 20.14b, 20.15). Thus, changes in peripheral resistance can shift blood flow from one organ system to another to meet the changing metabolic priorities of the body.

The arterioles are the most significant point of control over peripheral resistance and blood flow because (1) they

are on the proximal sides of the capillary beds, so they are best positioned to regulate flow into the capillaries; (2) they greatly outnumber any other class of arteries and thus provide the most numerous control points; and (3) they are more muscular in proportion to their diameters than any other class of blood vessels and are highly capable of

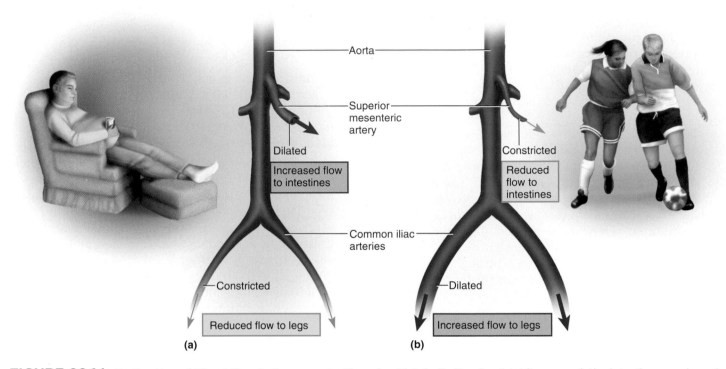

**FIGURE 20.14** **Redirection of Blood Flow in Response to Changing Metabolic Needs.** (a) After a meal, the intestines receive priority and the skeletal muscles receive relatively little flow. (b) During exercise, the muscles receive higher priority. Although vasodilation and vasoconstriction are shown here in major arteries for illustration purposes, most control occurs at a microscopic level in the arterioles.

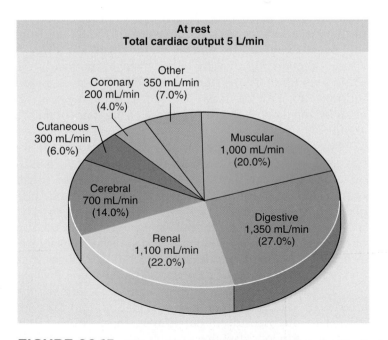

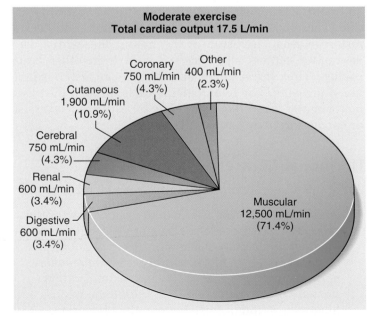

**FIGURE 20.15** Differences in Systemic Blood Flow During Rest and Exercise.

vasomotion. Arterioles alone account for about half of the total peripheral resistance of the circulatory system. However, larger arteries and veins are also capable of considerable vasomotion and control of peripheral resistance.

## Before You Go On

*Answer the following questions to test your understanding of the preceding section:*

6. *For a healthy 15-year-old girl at rest, what would be typical readings for systolic pressure, diastolic pressure, pulse pressure, and mean arterial pressure?*

7. *Explain why arterial blood flow is pulsatile and venous flow is not.*

8. *What three variables affect peripheral resistance to blood flow? Which of these is most able to change from one minute to the next?*

9. *What are the three primary mechanisms for controlling vessel radius? Briefly explain each.*

10. *Explain how the baroreflex serves as an example of homeostasis and negative feedback.*

11. *Explain how the body can shift the flow of blood from one organ system to another.*

# Capillary Exchange

### Objectives

When you have completed this section, you should be able to

- describe how materials get from the blood to the surrounding tissues;

- describe and calculate the forces that enable capillaries to give off and reabsorb fluid; and

- describe the causes and effects of edema.

Only 250 to 300 mL of blood is in the capillaries at any given time. This is the most important blood in the body, however, for it is mainly across capillary walls that exchanges occur between the blood and surrounding tissues. **Capillary exchange** refers to this two-way movement of fluid.

Chemicals given off by the capillaries to serve the perivascular tissues include oxygen; glucose, amino acids, lipids, and other organic nutrients; minerals; antibodies; and hormones. Chemicals taken up by the capillaries include carbon dioxide, ammonia, and other wastes, and many of the same substances as they give off: glucose and fatty acids released from storage in the liver and adipose tissue; calcium and other minerals released from bone; antibodies secreted by connective tissue immune cells; and hormones secreted by the endocrine glands. Thus many of these chemicals have a two-way traffic between the blood and connective tissue, leaving the capillaries at one point and entering at another. Along with all these solutes, there is substantial movement of water in and out of the bloodstream across the capillary walls.

The mechanisms of capillary exchange are difficult to study quantitatively because it is hard to measure pressure and flow in such small vessels. For this reason, theories of capillary exchange remain in dispute. Few capillaries of the human body are accessible to direct, noninvasive observation, but those of the fingernail bed and eponychium (cuticle) at the base of the nails can be observed with a stereomicroscope and have been the basis for a number of studies. Their BP has been measured at 32 mm Hg at the arterial end and 15 mm Hg at the venous end, 1 mm away. Capillary BP drops rapidly because of the substantial friction the blood encounters in such narrow vessels. It takes 1 to 2 seconds for an RBC to pass through a nail bed capillary, traveling about 0.7 mm/sec.

Chemicals pass through the capillary wall by three routes (fig. 20.16).

1. Through the endothelial cell cytoplasm.
2. Intercellular clefts between the endothelial cells.
3. Filtration pores (fenestrations) of the fenestrated capillaries.

The mechanisms of movement through the capillary wall are *diffusion, transcytosis, filtration*, and *reabsorption*, which we will examine in that order.

## DIFFUSION

The most important mechanism of exchange is diffusion. Glucose and oxygen, being more concentrated in the systemic blood than in the tissue fluid, diffuse out of the

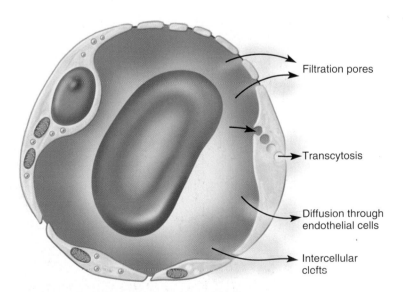

**FIGURE 20.16 Routes of Capillary Fluid Exchange.**
Materials move through the capillary wall through filtration pores (in fenestrated capillaries only), by transcytosis, by diffusion through the endothelial cells, and through intercellular clefts.

Filtration pores

Transcytosis

Diffusion through endothelial cells

Intercellular clefts

blood. Carbon dioxide and other wastes, being more concentrated in the tissue fluid, diffuse into the blood. (Oxygen and carbon dioxide diffuse in the opposite directions in the pulmonary circuit.) Such diffusion is only possible if the solute can either permeate the plasma membranes of the endothelial cells or find passages large enough to pass through—namely, the filtration pores and intercellular clefts. Such lipid-soluble substances as steroid hormones, $O_2$, and $CO_2$ diffuse easily through the plasma membranes. Substances insoluble in lipids, such as glucose and electrolytes, must pass through membrane channels, pores, or clefts. Large molecules such as proteins are usually held back.

## TRANSCYTOSIS

Transcytosis is a process in which endothelial cells pick up droplets of fluid on one side of the plasma membrane by pinocytosis, transport the vesicles across the cell, and discharge the fluid on the other side by exocytosis (see fig. 3.23, p. 111). This probably accounts for only a small fraction of solute exchange across the capillary wall, but fatty acids, albumin, and some hormones such as insulin move across the endothelium by this mechanism. Cytologists think that the filtration pores of fenestrated capillaries

might be only a chain of pinocytotic vesicles that have temporarily fused to form a continuous channel through the cell, as the suggestive appearance of fig. 20.16 conveys.

## FILTRATION AND REABSORPTION

The equilibrium between filtration and osmosis discussed in chapter 3 becomes particularly relevant when we consider capillary fluid exchange. Typically, fluid filters out of the arterial end of a capillary and osmotically reenters it at the venous end (fig. 20.17). This fluid delivers materials to the cells and removes their metabolic wastes. It may seem odd that a capillary could give off fluid at one point and reabsorb it at another. This comes about as the result of a shifting balance between hydrostatic and osmotic forces. **Hydrostatic pressure** is the physical force exerted against a surface, such as a capillary wall, by a liquid. Blood pressure is one example of hydrostatic pressure.

A typical capillary has a blood hydrostatic pressure of about 30 mm Hg at the arterial end. The hydrostatic pressure of the interstitial space has been difficult to measure and remains a point of controversy, but a typical value accepted by many authorities is −3 mm Hg. The negative value indicates that this is a slight suction, which helps

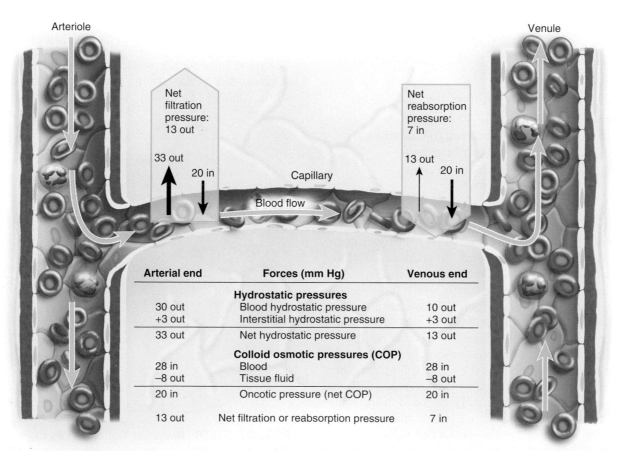

| Arterial end | Forces (mm Hg) | Venous end |
|---|---|---|
| | **Hydrostatic pressures** | |
| 30 out | Blood hydrostatic pressure | 10 out |
| +3 out | Interstitial hydrostatic pressure | +3 out |
| 33 out | Net hydrostatic pressure | 13 out |
| | **Colloid osmotic pressures (COP)** | |
| 28 in | Blood | 28 in |
| −8 out | Tissue fluid | −8 out |
| 20 in | Oncotic pressure (net COP) | 20 in |
| 13 out | Net filtration or reabsorption pressure | 7 in |

**FIGURE 20.17**  **The Forces of Capillary Filtration and Reabsorption.**  Note the shift from net filtration at the arterial end (left) to net reabsorption at the venous end (right).

draw fluid out of the capillary. (This force will be represented hereafter as $3_{out}$.) In this case, the positive hydrostatic pressure within the capillary and the negative interstitial pressure work in the same direction, creating a total outward force of about 33 mm Hg.

These forces are opposed by **colloid osmotic pressure (COP),** the portion of the osmotic pressure due to protein. The blood has a COP of about 28 mm Hg, due mainly to albumin. Tissue fluid has less than one-third the protein concentration of blood plasma and has a COP of about 8 mm Hg. The difference between the COP of blood and COP of tissue fluid is called **oncotic pressure:** $28_{in} - 8_{out} = 20_{in}$. Oncotic pressure tends to draw water into the capillary by osmosis, opposing hydrostatic pressure. These opposing forces produce a **net filtration pressure (NFP)** of 13 mm Hg out, as follows:

**Hydrostatic pressure**

| Blood pressure | | $30_{out}$ |
|---|---|---|
| Interstitial pressure | + | $3_{out}$ |
| Net hydrostatic pressure | | $33_{out}$ |

**Colloid osmotic pressure**

| Blood COP | | $28_{in}$ |
|---|---|---|
| Tissue fluid COP | − | $8_{out}$ |
| Oncotic pressure | | $20_{in}$ |

**Net filtration pressure**

| Net hydrostatic pressure | | $33_{out}$ |
|---|---|---|
| Oncotic pressure | − | $20_{in}$ |
| Net filtration pressure | | $13_{out}$ |

The NFP of 13 mm Hg causes about 0.5% of the blood plasma to leave the capillaries at the arterial end.

At the venous end, however, capillary blood pressure is lower—about 10 mm Hg. All the other pressures are unchanged. Thus, we get:

**Hydrostatic pressure**

| Blood pressure | | $10_{out}$ |
|---|---|---|
| Interstitial pressure | + | $3_{out}$ |
| Net hydrostatic pressure | | $13_{out}$ |

**Net reabsorption pressure**

| Oncotic pressure | | $20_{in}$ |
|---|---|---|
| Net hydrostatic pressure | − | $13_{out}$ |
| Net reabsorption pressure | | $7_{in}$ |

The prevailing force is inward at the venous end because osmotic pressure overrides filtration pressure. The **net reabsorption pressure** of 7 mm Hg inward causes the capillary to reabsorb fluid at this end.

Now you can see why a capillary gives off fluid at one end and reabsorbs it at the other. The only pressure that changes from the arterial end to the venous end is the capillary blood pressure, and this change is responsible for the shift from filtration to reabsorption. With a reabsorption pressure of 7 mm Hg and a net filtration pressure of 13 mm Hg, it might appear that far more fluid would leave the capillaries than reenter them. However, since capillaries branch along their length, there are more of them at the venous end than at the arterial end, which partially compensates for the difference between filtration and reabsorption pressures. They also typically have nearly twice the diameter at the venous end that they have at the arterial end, so there is more capillary surface area available to reabsorb fluid than to give it off. Consequently, capillaries reabsorb about 85% of the fluid they filter. The other 15% is absorbed and returned to the blood by way of the lymphatic system, as described in chapter 21.

Of course, water is not the only substance that crosses the capillary wall by filtration and reabsorption. Chemicals dissolved in the water are "dragged" along with it and pass through the capillary wall if they are not too large. This process, called **solvent drag,** will be important to our discussions of kidney and intestinal function in later chapters.

## Variations in Capillary Filtration and Reabsorption

The figures used in the preceding discussion serve only as examples; circumstances differ from place to place in the body and from time to time in the same capillaries. Capillaries usually reabsorb most of the fluid they filter, but this is not always the case. The kidneys have capillary networks called *glomeruli* in which there is little or no reabsorption; they are entirely devoted to filtration. Alveolar capillaries of the lungs, by contrast, are almost entirely dedicated to absorption so that fluid does not fill the air spaces.

Capillary activity also varies from moment to moment. In a resting tissue, most precapillary sphincters are constricted and the capillaries are collapsed. Capillary BP is very low (if there is any flow at all), and reabsorption predominates. When a tissue becomes metabolically active, its capillary flow increases. In active muscles, capillary pressure rises to the point that it overrides reabsorption along the entire length of the capillary. Fluid accumulates in the muscle, and exercising muscles increase in size by as much as 25%. Capillary permeability is also subject to chemical influences. Traumatized tissue releases such chemicals as substance P, bradykinin, and histamine, which increase permeability and filtration.

## EDEMA

**Edema** is the accumulation of excess fluid in a tissue. It often shows as swelling of the face, fingers, abdomen, or ankles, but also occurs in internal organs where its effects are hidden from view. Edema occurs when fluid filters into a tissue faster than it is reabsorbed. It has three fundamental causes:

1. **Increased capillary filtration.** Numerous conditions can increase the rate of capillary filtration and accumulation of fluid in the tissues. Kidney failure, for example, leads to water retention and hypertension, thus raising capillary blood pressure and filtration rate. Histamine dilates arterioles and raises capillary pressure and also makes the capillary wall more permeable. Capillaries generally become more permeable in old age as well, putting elderly people at increased risk for edema. Capillary blood pressure also rises in cases of poor venous return—the flow of blood from the capillaries back to the heart. As we will see in the next section, good venous return depends on muscular activity. Therefore, edema is a common problem among people confined to bed or a wheelchair. Failure of the right ventricle of the heart tends to cause pressure to back up in the systemic veins and capillaries, thus resulting in systemic edema. Failure of the left ventricle causes pressure to back up in the lungs, causing pulmonary edema.

2. **Reduced capillary reabsorption.** Capillary reabsorption depends on oncotic pressure, which is proportional to the concentration of blood albumin. Therefore a deficiency of albumin (hypoproteinemia) produces edema by reducing the reabsorption of tissue fluid. Since albumin is produced by the liver, liver diseases such as cirrhosis tend to lead to hypoproteinemia and edema. Edema is commonly seen in regions of famine due to dietary protein deficiency (see kwashiorkor, p. 684). Hypoproteinemia and edema also commonly result from severe burns, owing to the loss of protein from body surfaces no longer covered with skin, and from kidney diseases that allow protein to escape in the urine.

3. **Obstructed lymphatic drainage.** The lymphatic system, described in detail in chapter 21, is a system of one-way vessels that collect fluid from the tissues and return it to the bloodstream. Obstruction of these vessels or the surgical removal of lymph nodes can interfere with fluid drainage and lead to the accumulation of tissue fluid distal to the obstruction (see elephantiasis, fig. 21.2).

Edema has multiple pathological consequences. As the tissues become congested with fluid, oxygen delivery and waste removal are impaired and the tissues may begin to die. Pulmonary edema presents a threat of suffocation as fluid replaces air in the lungs, and cerebral edema can produce headaches, nausea, and sometimes seizure and coma. In severe edema, so much fluid may transfer from the blood vessels to the tissue spaces that blood volume and pressure drop low enough to cause circulatory shock (described later in this chapter).

## Before You Go On

*Answer the following questions to test your understanding of the preceding section:*

12. List the three mechanisms of capillary exchange and relate each one to the structure of capillary walls.

13. What forces favor capillary filtration? What forces favor reabsorption?

14. How can a capillary shift from a predominantly filtering role at one time to a predominantly reabsorbing role at another?

15. State the three fundamental causes of edema and explain why edema can be dangerous.

# Venous Return and Circulatory Shock

## Objectives

When you have completed this section, you should be able to

- explain how blood in the veins is returned to the heart;
- discuss the importance of physical activity in venous return;
- discuss several causes of circulatory shock; and
- name and describe the stages of shock.

Hieronymus Fabricius (1537–1619) discovered the valves of the veins and argued that they would allow blood to flow in only one direction, not back and forth as Galen had thought. One of his medical students was William Harvey, who performed simple experiments on the valves that you can easily reproduce. In figure 20.18, by

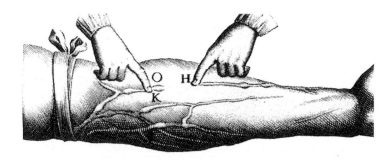

**FIGURE 20.18   An Illustration from William Harvey's** *De Motu Cordis* **(1628).** These experiments demonstrate the existence of one-way valves in veins of the arms. See text for explanation.

▶ *In the space between* ***O*** *and* ***H***, *what (if anything) would happen if the experimenter lifted his finger from point* ***O***? *What if he lifted his finger from point* ***H***? *Why?*

Harvey, the experimenter has pressed on a vein at point H to block flow from the wrist toward the elbow. With another finger, he has milked the blood out of it up to point O, the first valve proximal to H. When he tries to force blood downward, it stops at that valve. It can go no farther, and it causes the vein to swell at that point. Blood can flow from right to left through that valve but not from left to right.

You can easily demonstrate the action of these valves in your own hand. Hold your hand still, below waist level, until veins stand up on the back of it. (Do not apply a tourniquet!) Press on a vein close to your knuckles, and while holding it down, use another finger to milk that vein toward the wrist. It collapses as you force the blood out of it, and if you remove the second finger, it will not refill. The valves prevent blood from flowing back into it from above. When you remove the first finger, however, the vein fills from below.

## MECHANISMS OF VENOUS RETURN

The flow of blood back to the heart, called **venous return**, is achieved by five mechanisms:

1. **The pressure gradient.** Pressure generated by the heart is the most important force in venous flow, even though it is substantially weaker in the veins than in the arteries. Pressure in the venules ranges from 12 to 18 mm Hg, and pressure at the point where the venae cavae enter the heart, called **central venous pressure**, averages 4.6 mm Hg. Thus, there is a venous pressure gradient ($\Delta P$) of about 7 to 13 mm Hg favoring the flow of blood toward the heart. The pressure gradient and venous return increase when blood volume increases. Venous return decreases when the veins constrict *(venoconstriction)* and oppose flow, and increases when they dilate and offer less resistance. However, it increases if *all* the body's blood vessels constrict, because this reduces the "storage capacity" of the circulatory system and raises blood pressure and flow.

2. **Gravity.** When you are sitting or standing, blood from your head and neck returns to the heart simply by "flowing downhill" through the large veins above the heart. Thus the large veins of the neck are normally collapsed or nearly so, and their venous pressure is close to zero. The dural sinuses of the brain, however, have more rigid walls and cannot collapse. Their pressure is as low as $-10$ mm Hg, creating a risk of *air embolism* if they are punctured (see Insight 20.3).

3. **The skeletal muscle pump.** In the limbs, the veins are surrounded and massaged by the muscles. They squeeze the blood out of the compressed part of a vein, and the valves ensure that this blood can go in only one direction—toward the heart (fig. 20.19).

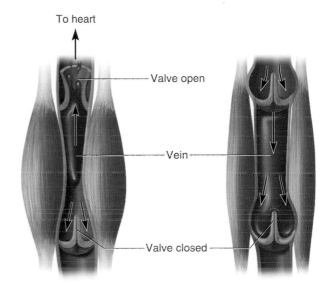

**(a) Contracted skeletal muscles**      **(b) Relaxed skeletal muscles**

**FIGURE 20.19**  The Skeletal Muscle Pump.  (a) When the muscles contract and compress a vein, blood is squeezed out of it and flows upward toward the heart; valves below the point of compression prevent backflow of the blood. (b) When the muscles relax, blood flows back downward under the pull of gravity but can only flow as far as the nearest valve.

4. **The thoracic (respiratory) pump.** This mechanism aids the flow of venous blood from the abdominal to the thoracic cavity. When you inhale, your thoracic cavity expands and its internal pressure drops, while downward movement of the diaphragm raises the pressure in your abdominal cavity. The *inferior vena cava (IVC),* your largest vein, is a flexible tube passing through both of these cavities. If abdominal pressure on the IVC rises while thoracic pressure on it drops, then blood is squeezed upward toward the

heart. It is not forced back into the lower limbs because the venous valves there prevent this. Because of the thoracic pump, central venous pressure fluctuates from 2 mm Hg when you inhale to 6 mm Hg when you exhale, and blood flows faster when you inhale.

5. **Cardiac suction.** During ventricular systole, the tendinous cords pull the AV valve cusps downward, slightly expanding the atrial space. This creates a slight suction that draws blood into the atria from the venae cavae and pulmonary veins.

## VENOUS RETURN AND PHYSICAL ACTIVITY

Exercise increases venous return for many reasons. The heart beats faster and harder, increasing cardiac output and blood pressure. Blood vessels of the skeletal muscles, lungs, and coronary circulation dilate, increasing flow. The increase in respiratory rate and depth enhances the action of the thoracic pump. Muscle contractions increase venous return by means of the skeletal muscle pump. Increased venous return increases cardiac output, which is important in perfusion of the muscles just when they need it most.

Conversely, when a person is still, blood accumulates in the limbs because venous pressure is not high enough to override the weight of the blood and drive it upward. Such accumulation of blood is called **venous pooling.** To demonstrate this effect, hold one hand below your waist for about a minute and (just for comparison) hold the other hand over your head. Then, quickly bring your two hands together and compare the palms. The hand held overhead usually appears pale because its blood has drained out of it; the hand held below the waist appears redder than normal because of venous pooling in its veins and capillaries. Venous pooling is troublesome to people who must stand for prolonged periods. If enough blood accumulates in the limbs, cardiac output may become so low that the brain is inadequately perfused and a person may experience dizziness or **syncope** (SIN-co-pee) (fainting). This can usually be prevented by periodically tensing the calf and other muscles to keep the skeletal muscle pump active. Military jet pilots often perform maneuvers that could cause the blood to pool in the abdomen and lower limbs, causing partial loss of vision or loss of consciousness. To prevent this, they wear pressure suits that inflate and tighten on the lower limbs during these maneuvers; in addition, they sometimes must tense their abdominal muscles to prevent venous pooling and blackout.

┌─ **Think About It**

*Why is venous pooling not a problem when you are sleeping and the skeletal muscle pump is inactive?*

## CIRCULATORY SHOCK

**Circulatory shock** (not to be confused with electrical or spinal shock) is any state in which cardiac output is insufficient to meet the body's metabolic needs. All forms of circulatory shock fall into two categories: (1) **cardiogenic shock,** caused by inadequate pumping by the heart, usually as a result of myocardial infarction, and (2) **low venous return (LVR) shock,** in which cardiac output is low because too little blood is returning to the heart.

There are three principal forms of LVR shock:

1. **Hypovolemic shock,** the most common form, is produced by a loss of blood volume as a result of hemorrhage, trauma, bleeding ulcers, burns, or dehydration. Dehydration is a major cause of death from heat exposure. In hot weather, the body produces as much as 1.5 L of sweat per hour. Water transfers from the bloodstream to replace lost tissue fluid, and blood volume may drop too low to maintain adequate circulation.

2. **Obstructed venous return shock** occurs when a growing tumor or aneurysm, for example, compresses a vein and impedes its blood flow.

3. **Venous pooling (vascular) shock** occurs when the body has a normal total blood volume, but too much of it accumulates in the limbs. This can result from long periods of standing or sitting or from widespread vasodilation. **Neurogenic shock** is a form of venous pooling shock that results from a sudden loss of vasomotor tone, allowing the vessels to dilate. This can result from causes as severe as brainstem trauma or as slight as an emotional shock.

Elements of both venous pooling and hypovolemic shock are present in certain cases, such as septic shock and anaphylactic shock, which involve both vasodilation and a loss of fluid through abnormally permeable capillaries. **Septic shock** occurs when bacterial toxins trigger vasodilation and increased capillary permeability. **Anaphylactic shock,** discussed more fully in chapter 21, results from exposure to an antigen to which a person is allergic, such as bee venom. Antigen–antibody complexes trigger the release of histamine, which causes generalized vasodilation and increased capillary permeability.

### Responses to Circulatory Shock

In **compensated shock,** several homeostatic mechanisms act to bring about spontaneous recovery. The hypotension resulting from low cardiac output triggers the baroreflex and the production of angiotensin II, both of which counteract shock by stimulating vasoconstriction. Furthermore, if a person faints and falls to a horizontal position, gravity restores blood flow to the brain. Even quicker recovery is achieved if the person's feet are elevated to promote drainage of blood from the legs.

If these mechanisms prove inadequate, **decompensated shock** ensues and several life-threatening positive feedback loops occur. Poor cardiac output results in myocardial ischemia and infarction, which further weakens the heart and reduces output. Slow circulation of the blood can lead to disseminated intravascular coagulation (DIC) (see chapter 18). As the vessels become congested with clotted blood, venous return grows even worse. Ischemia and acidosis of the brainstem depress the vasomotor and cardiac centers, causing loss of vasomotor tone, further vasodilation, and further drop in BP and cardiac output. Before long, damage to the cardiac and brain tissues may be too great to survive. About half of those who go into circulatory shock die from it.

### Before You Go On

*Answer the following questions to test your understanding of the preceding section:*

16. *Explain how respiration aids venous return.*

17. *Explain how muscular activity and venous valves aid venous return.*

18. *Define circulatory shock. What are some of the causes of low venous return shock?*

# Special Circulatory Routes

### Objectives

When you have completed this section, you should be able to

- explain how the brain maintains stable perfusion;
- discuss the causes and effects of strokes and transient ischemic attacks;
- explain the mechanisms that increase muscular perfusion during exercise; and
- contrast the blood pressure of the pulmonary circuit with that of the systemic circuit, and explain why the difference is important in pulmonary function.

Certain circulatory pathways have special physiological properties adapted to the functions of their organs. Two of these are described in other chapters: the coronary circulation in chapter 19 and fetal and placental circulation in chapter 29. Here we take a closer look at the circulation to the brain, skeletal muscles, and lungs.

## BRAIN

Total blood flow to the brain fluctuates less than that of any other organ (about 700 mL/min at rest). Such constancy is important because even a few seconds of oxygen deprivation causes loss of consciousness, and 4 or 5 minutes of anoxia is time enough to cause irreversible dam-

age. Although total cerebral perfusion is fairly stable, blood flow can be shifted from one part of the brain to another in a matter of seconds as different parts engage in motor, sensory, or cognitive functions.

The brain regulates its own blood flow in response to changes in BP and chemistry. The cerebral arteries dilate when the systemic BP drops and constrict when it rises, thus minimizing fluctuations in cerebral BP. Cerebral blood flow thus remains quite stable even when mean arterial pressure (MAP) fluctuates from 60 to 140 mm Hg. An MAP below 60 mm Hg produces syncope and an MAP above 160 mm Hg causes cerebral edema.

The main chemical stimulus for cerebral autoregulation is pH. Poor perfusion allows $CO_2$ to accumulate in the brain tissue. This lowers the pH of the tissue fluid and triggers local vasodilation, which improves perfusion. Extreme hypercapnia, however, depresses neural activity. The opposite condition, hypocapnia, raises the pH and stimulates vasoconstriction, thus reducing perfusion and giving $CO_2$ a chance to rise to a normal level. Hyperventilation (exhaling $CO_2$ faster than the body produces it) induces hypocapnia, which leads to cerebral vasoconstriction, ischemia, dizziness, and sometimes syncope.

Brief episodes of cerebral ischemia produce **transient ischemic attacks (TIAs)**, characterized by temporary dizziness, loss of vision or other senses, weakness, paralysis, headache, or aphasia. A TIA may result from spasms of diseased cerebral arteries. It lasts from just a moment to a few hours and is often an early warning of an impending stroke. People with TIAs should receive prompt medical attention to identify the cause using brain imaging and other diagnostic means. Immediate treatment should be initiated to prevent a stroke.

A **stroke,** or **cerebrovascular accident (CVA),** is the sudden death (infarction) of brain tissue caused by ischemia. Cerebral ischemia can be produced by atherosclerosis, thrombosis, or a ruptured aneurysm. The effects of a CVA range from unnoticeable to fatal, depending on the extent of tissue damage and the function of the affected tissue. Blindness, paralysis, loss of sensation, and loss of speech are common. Recovery depends on the ability of neighboring neurons to take over the lost functions and on the extent of collateral circulation to regions surrounding the cerebral infarction.

## SKELETAL MUSCLES

In contrast to the brain, the skeletal muscles receive a highly variable blood flow depending on their state of exertion. At rest, the arterioles are constricted, most of the capillary beds are shut down, and total flow through the muscular system is about 1 L/min. During exercise, the arterioles dilate in response to epinephrine and norepinephrine from the adrenal medulla and sympathetic nerves. Precapillary sphincters, which lack innervation, dilate in response to muscle metabolites such as lactic acid, $CO_2$, and adenosine. Blood flow can increase more than 20-fold during strenuous exercise, which requires that blood be diverted from other organs such as the digestive tract and kidneys to meet the needs of the working muscles.

Muscular contraction compresses the blood vessels and impedes flow. For this reason, isometric contraction causes fatigue more quickly than intermittent isotonic contraction. If you squeeze a rubber ball as hard as you can without relaxing your grip, you feel the muscles fatigue more quickly than if you intermittently squeeze and relax.

## LUNGS

After birth, the pulmonary circuit is the only route in which the arterial blood contains less oxygen than the venous blood. The pulmonary arteries have thin distensible walls with less elastic tissue than the systemic arteries. Thus, they have a BP of only 25/10. Capillary hydrostatic pressure is about 10 mm Hg in the pulmonary circuit as compared with an average of 17 mm Hg in systemic capillaries. This lower pressure has two implications for pulmonary circulation: (1) blood flows more slowly through the pulmonary capillaries, and therefore it has more time for gas exchange; and (2) oncotic pressure overrides hydrostatic pressure, so these capillaries are engaged almost entirely in absorption. This prevents fluid accumulation in the alveolar walls and lumens, which would interfere with gas exchange. In a condition such as mitral valve stenosis, however, BP may back up into the pulmonary circuit, raising the capillary hydrostatic pressure and causing pulmonary edema, congestion, and hypoxemia.

⌐ **Think About It**

*What abnormal skin coloration would result from pulmonary edema?*

Another unique characteristic of the pulmonary arteries is their response to hypoxia. Systemic arteries dilate in response to local hypoxia and improve tissue perfusion. By contrast, pulmonary arteries constrict. Pulmonary hypoxia indicates that part of the lung is not being ventilated well, perhaps because of mucous congestion of the airway or a degenerative lung disease. Vasoconstriction in poorly ventilated regions of the lung redirects blood flow to better ventilated regions.

### Before You Go On

*Answer the following questions to test your understanding of the preceding section:*

19. *In what conspicuous way does perfusion of the brain differ from perfusion of the skeletal muscles?*

20. *How does a stroke differ from a transient ischemic attack? Which of these bears closer resemblance to a myocardial infarction?*

21. *How does the low hydrostatic blood pressure in the pulmonary circuit affect the fluid dynamics of the capillaries there?*

22. *Contrast the vasomotor responses of the lungs versus skeletal muscles to hypoxia.*

# Anatomy of the Pulmonary Circuit

**Objective**

When you have completed this section, you should be able to

• trace the route of blood through the pulmonary circuit.

The next three sections of this chapter center on the names and pathways of the principal arteries and veins. The pulmonary circuit is described here, and the systemic arteries and veins are described in the two sections that follow.

The pulmonary circuit (fig. 20.20) begins with the **pulmonary trunk,** a large vessel that ascends diagonally from the right ventricle and branches into the right and left **pulmonary arteries.** Each pulmonary artery enters a medial indentation of the lung called the *hilum* and branches into one **lobar artery** for each lobe of the lung: three on the right and two on the left. These arteries lead ultimately to small basketlike capillary beds that surround the pulmonary alveoli. This is where the blood unloads $CO_2$ and loads $O_2$. After leaving the alveolar capillaries, the pulmonary blood flows into venules and veins, ultimately leading to the **pulmonary veins,** which exit the lung at the hilum. The left atrium of the heart receives two pulmonary veins on each side.

The purpose of the pulmonary circuit is to exchange $CO_2$ for $O_2$. It does not serve the metabolic needs of the lung tissue itself; there is a separate systemic supply to the lungs for that purpose, the *bronchial arteries,* discussed later.

### Before You Go On

*Answer the following questions to test your understanding of the preceding section:*

23. *Trace the flow of an RBC from right ventricle to left atrium and name the vessels along the way.*

24. *The lungs have two separate arterial supplies. Explain their functions.*

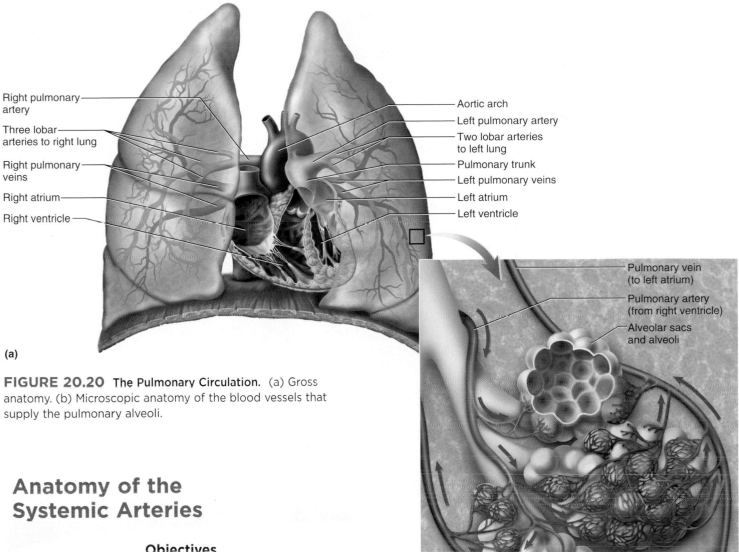

Right pulmonary artery

Three lobar arteries to right lung

Right pulmonary veins

Right atrium

Right ventricle

Aortic arch

Left pulmonary artery

Two lobar arteries to left lung

Pulmonary trunk

Left pulmonary veins

Left atrium

Left ventricle

Pulmonary vein (to left atrium)

Pulmonary artery (from right ventricle)

Alveolar sacs and alveoli

(a)

(b)

**FIGURE 20.20 The Pulmonary Circulation.** (a) Gross anatomy. (b) Microscopic anatomy of the blood vessels that supply the pulmonary alveoli.

# Anatomy of the Systemic Arteries

### Objectives
When you have completed this section, you should be able to

- identify the principal arteries of the systemic circuit; and

- trace the flow of blood from the heart to any major organ.

The systemic circuit (fig. 20.21) supplies oxygen and nutrients to all the organs and removes their metabolic wastes. Part of it, the coronary circulation, was described in chapter 19. The other systemic arteries are described in tables 20.3 through 20.8 (figs. 20.22–20.31). There is a great deal of anatomical variation in the circulatory system from one person to another. These tables and figures show only the most common patterns and remark on some of the exceptions. Many of the tables include not only a figure showing realistic anatomical relations of the vessels to each other, but also stylized flowcharts that make no attempt to resemble the vessels, but clarify the routes taken by the arterial blood.

The names of the blood vessels often describe their location by indicating the body region traversed (as in the *axillary* or *femoral* artery); an adjacent bone (as in *radial* or *temporal* artery); or the organ supplied or drained by the vessel (as in *hepatic* or *renal* vein).

In some places, major arteries come close enough to the body surface to be palpated. These places can be used to take a pulse, and they can serve as emergency **pressure points** where firm pressure can be applied to temporarily reduce arterial bleeding. One of these points is the **femoral triangle** of the upper medial thigh (fig. 20.32, p. 790). This is an important landmark for arterial supply, venous drainage, and innervation of the lower limbs. Its boundaries are the sartorius muscle laterally, the inguinal ligament superiorly, and the adductor longus muscle medially. The femoral artery, vein, and nerve run close to the surface at this point.

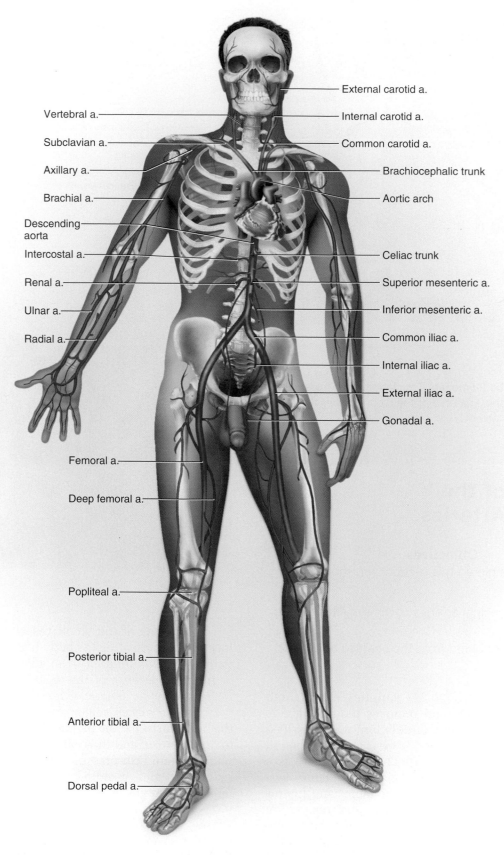

Vertebral a.

Subclavian a.

Axillary a.

Brachial a.

Descending
aorta

Intercostal a.

Renal a.

Ulnar a.

Radial a.

External carotid a.

Internal carotid a.

Common carotid a.

Brachiocephalic trunk

Aortic arch

Celiac trunk

Superior mesenteric a.

Inferior mesenteric a.

Common iliac a.

Internal iliac a.

External iliac a.

Gonadal a.

Femoral a.

Deep femoral a.

Popliteal a.

Posterior tibial a.

Anterior tibial a.

Dorsal pedal a.

**FIGURE 20.21**   **The Major Systemic Arteries.**   (a. = artery; aa. = arteries)

| **TABLE 20.3** | The Aorta and Its Major Branches |
|---|---|

All systemic arteries arise from the aorta, which has three principal regions (fig. 20.22):

1. The **ascending aorta** rises about 5 cm above the left ventricle. Its only branches are the coronary arteries, which arise behind two cusps of the aortic valve. Opposite each semilunar valve cusp is an **aortic sinus** containing baroreceptors.

2. The **aortic arch** curves to the left like an inverted U superior to the heart. It gives off three major arteries in this order: the **brachiocephalic**[9] (BRAY-kee-oh-seh-FAL-ic) **trunk, left common carotid** (cah-ROT-id) **artery,** and **left subclavian**[10] (sub-CLAY-vee-un) **artery,** which are further traced in tables 20.4 and 20.5.

3. The **descending aorta** passes downward dorsal to the heart, at first to the left of the vertebral column and then anterior to it, through the thoracic and abdominal cavities. It is called the **thoracic aorta** above the diaphragm and the **abdominal aorta** below. It ends in the lower abdominal cavity by forking into the *right* and *left common iliac arteries,* which are further traced in table 20.8.

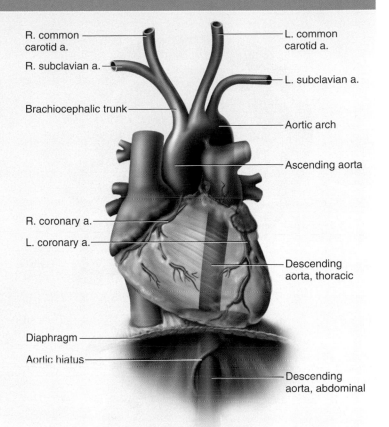

**FIGURE 20.22 Beginning of the Aorta.** (R. = right; L. = left; a. = artery)

| **TABLE 20.4** | Arterial Supply to the Head and Neck |
|---|---|

**Origins of the Head–Neck Arteries**

The head and neck receive blood from four pairs of arteries (fig. 20.23):

1. The **common carotid arteries.** The brachiocephalic trunk divides shortly after leaving the aortic arch and gives rise to the *right subclavian* and *right common carotid arteries.* The *left common carotid* artery arises directly from the aortic arch. The common carotids pass up the anterolateral aspect of the neck, alongside the trachea.

2. The vertebral arteries arise from the right and left subclavian arteries. Each travels up the neck through the transverse foramina of the cervical vertebrae and enters the cranial cavity through the foramen magnum.

3. The **thyrocervical**[11] trunks are tiny arteries that arise from the subclavian arteries lateral to the vertebral arteries; they supply the thyroid gland and some scapular muscles.

4. The **costocervical**[12] trunks (also illustrated in table 20.6) arise from the subclavian arteries a little farther laterally. They perfuse the deep neck muscles and some of the intercostal muscles of the superior rib cage.

---

[9]*brachio* = arm + *cephal* = head
[10]*sub* = below + *clavi* = clavicle, collarbone
[11]*thyro* = thyroid gland + *cerv* = neck
[12]*costo* = rib

| **TABLE 20.4** | Arterial Supply to the Head and Neck *(cont.)* |
|---|---|

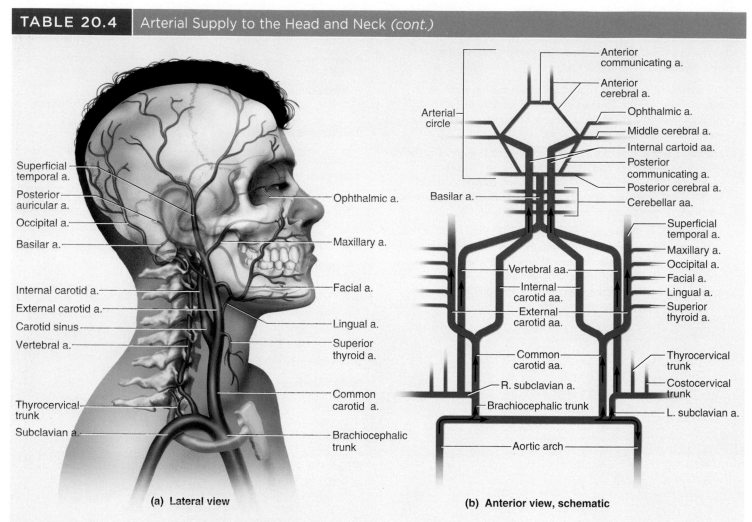

(a) Lateral view

(b) Anterior view, schematic

**FIGURE 20.23**   Arteries Supplying the Head and Neck.

▶ *List the arteries, in order, that an erythrocyte must travel to get from the left ventricle to the skin of the left side of the forehead.*

**Continuation of the Common Carotid Arteries**

The common carotid arteries have the most extensive distribution of all the head-neck arteries. Near the laryngeal prominence (Adam's apple), each common carotid branches into an *external carotid artery* and an *internal carotid artery:*

1. **The external carotid artery** ascends the side of the head external to the cranium and supplies most external head structures except the orbits. The external carotid gives rise to the following arteries, in ascending order:

   a. the **superior thyroid artery** to the thyroid gland and larynx,

   b. the **lingual artery** to the tongue,

   c. the **facial artery** to the skin and muscles of the face,

   d. the **occipital artery** to the posterior scalp,

   e. the **maxillary artery** to the teeth, maxilla, buccal cavity, and external ear, and

   f. the **superficial temporal artery** to the chewing muscles, nasal cavity, lateral aspect of the face, most of the scalp, and the dura mater surrounding the brain.

2. The **internal carotid artery** passes medial to the angle of the mandible and enters the cranial cavity through the carotid canal of the temporal bone. It supplies the orbits and about 80% of the cerebrum. Compressing the internal carotids near the mandible can therefore cause loss of consciousness.[13] The carotid sinus is located in the internal carotid just above the branch point; the carotid body is nearby. After entering the cranial cavity, each internal carotid artery gives rise to the following branches:

   a. the **ophthalmic artery,** with extensive branches to the eyeball, tear gland, eyelids, and other structures of the orbit; the nose and perinasal sinuses; the meninges; and the forehead;

   b. the **anterior cerebral artery** to the medial aspect of the cerebral hemisphere (see *arterial circle*); and

   c. the **middle cerebral artery,** which travels in the lateral sulcus of the cerebrum and supplies the lateral aspect of the temporal and parietal lobes.

### Continuation of the Vertebral Arteries

The vertebral arteries give rise to small branches in the neck that supply the spinal cord and other neck structures, then enter the foramen magnum and merge to form a single **basilar artery** along the anterior aspect of the brainstem. Branches of the basilar artery supply the cerebellum, pons, and inner ear. At the pons–midbrain junction, the basilar artery divides and gives rise to the *arterial circle.*

### The Arterial Circle

Blood supply to the brain is so critical that it is furnished by several arterial anastomoses, especially an array of arteries called the **arterial circle** (circle of Willis[14]), which surrounds the pituitary gland and optic chiasm. The arterial circle receives blood from the internal carotid and basilar arteries (fig. 20.24). Only 20% of people have a complete arterial circle. It consists of

1. two **posterior cerebral arteries,**

2. two **posterior communicating arteries,**

3. two **anterior cerebral arteries,** and

4. a single anterior **communicating artery.**

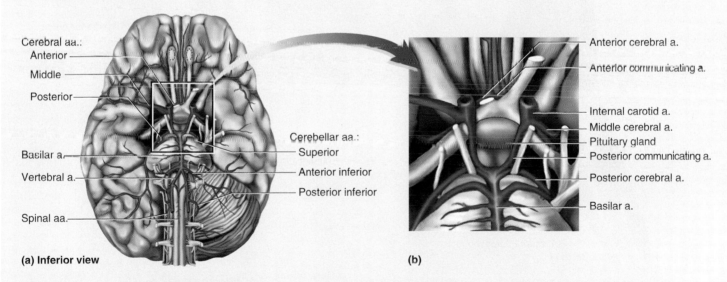

**(a) Inferior view**

**(b)**

**FIGURE 20.24  The Arterial Circle of the Brain.**  (a) The base of the brain showing the blood supply to the arterial circle. (b) Detail of the arterial circle

---

[13]*carot* = stupor
[14]Thomas Willis (1621–75), English anatomist

## TABLE 20.5 | Arterial Supply to the Upper Limb

**The Shoulder and Arm (Brachium)**

The origins of the subclavian arteries were described and illustrated in table 20.3. We now trace these further to examine the blood supply to the upper limb (fig. 20.25). This begins with a large artery that changes name from *subclavian* to *axillary* to *brachial* along its course:

1. The **subclavian artery** travels between the clavicle and first rib. It gives off several small branches to the thoracic wall and viscera, considered later.

2. The **axillary artery** is the continuation of the subclavian artery through the axillary region. It also gives off small thoracic branches, discussed later, and then ends at the neck of the humerus. Here, it gives off the **circumflex humeral artery,** which encircles the humerus. This loop supplies blood to the shoulder joint and deltoid muscle.

3. The **brachial** (BRAY-kee-ul) **artery** is the continuation of the axillary artery beyond the circumflex. It travels down the medial side of the humerus and ends just distal to the elbow, supplying the anterior flexor muscles of the brachium along the way. It exhibits several anastomoses near the elbow, two of which are noted next. This is the most commonly used artery for routine BP measurements.

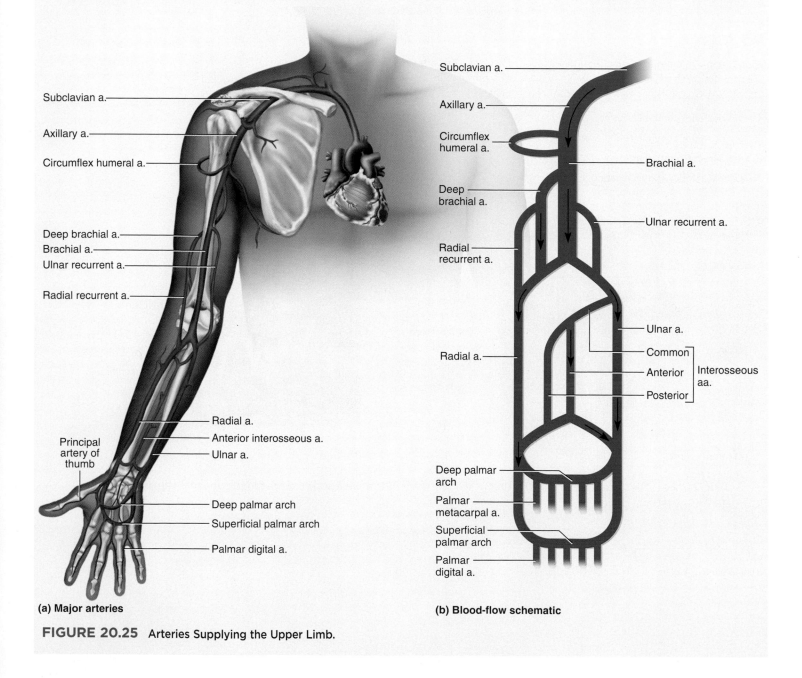

**(a) Major arteries**

**(b) Blood-flow schematic**

**FIGURE 20.25** Arteries Supplying the Upper Limb.

4. The **deep brachial artery** arises from the proximal end of the brachial artery and supplies the triceps brachii muscle.

5. The **ulnar recurrent artery** arises about midway along the brachial artery and anastomoses distally with the ulnar artery. It supplies the elbow joint and the triceps brachii.

6. The **radial recurrent artery** leads from the deep brachial artery to the radial artery and supplies the elbow joint and forearm muscles.

### The Forearm (Antebrachium)

Just distal to the elbow, the brachial artery divides into the **radial artery** and **ulnar artery,** which travel alongside the radius and ulna, respectively. The most common place to take a pulse is at the radial artery, just proximal to the thumb. Near its origin, the radial artery receives the deep brachial artery. The ulnar artery gives rise, near its origin, to the **anterior** and **posterior interosseous**[15] **arteries,** which travel between the radius and ulna. Structures supplied by these arteries are as follows:

1. Radial artery: lateral forearm muscles, wrist, thumb, and index finger.

2. Ulnar artery: medial forearm muscles, digits 3 to 5, and medial aspect of index finger.

3. Interosseous arteries: deep flexors and extensors.

### The Hand

At the wrist, the radial and ulnar arteries anastomose to form two *palmar arches:*

1. The **deep palmar arch** gives rise to the **palmar metacarpal arteries** of the hand.

2. The **superficial palmar arch** gives rise to the **palmar digital arteries** of the fingers.

| TABLE 20.6 | Arterial Supply to the Thorax |
|---|---|

The thoracic aorta begins distal to the aortic arch and ends at the **aortic hiatus** (hy-AY-tus), a passage through the diaphragm. Along the way, it sends off numerous small branches to viscera and structures of the body wall (fig. 20.26).

**Visceral Branches**

These supply the viscera of the thoracic cavity:

1. **Bronchial arteries.** Two of these on the left and one on the right supply the visceral pleura, esophagus, and bronchi of the lungs. They are the systemic blood supply to the lungs mentioned earlier.

2. **Esophageal arteries.** Four or five of these supply the esophagus.

3. **Mediastinal arteries.** Many small mediastinal arteries (not illustrated) supply structures of the posterior mediastinum.

**Parietal Branches**

The following branches supply chiefly the muscles, bones, and skin of the chest wall; only the first is illustrated:

1. **Posterior intercostal arteries.** Nine pairs of these course around the posterior aspect of the rib cage between the ribs and then anastomose with the anterior intercostal arteries (see following). They supply the skin and subcutaneous tissue, breasts, spinal cord and meninges, and the pectoralis, intercostal, and some abdominal muscles.

2. **Subcostal arteries.** A pair of these (not illustrated) arise from the aorta, inferior to the twelfth rib, and supply the posterior intercostal tissues, vertebrae, spinal cord, and deep muscles of the back.

3. **Superior phrenic**[16] (FREN-ic) **arteries.** These arteries (not illustrated) supply the posterior and superior aspects of the diaphragm.

---

[15]*inter* = between + *osse* = bones
[16]*phren* = diaphragm

| **TABLE 20.6** | Arterial Supply to the Thorax *(cont.)* |
| --- | --- |

The thoracic wall is also supplied by the following arteries. The first of these arises from the subclavian artery and the other three from the axillary artery:

1. The **internal thoracic (mammary) artery** supplies the breast and anterior thoracic wall and issues finer branches to the diaphragm and abdominal wall. Near its origin, it gives rise to the **pericardiophrenic artery,** which supplies the pericardium and diaphragm. As the internal thoracic artery descends alongside the sternum, it gives rise to anterior intercostal arteries that travel between the ribs and supply the ribs and intercostal muscles.

2. The **thoracoacromial**[17] (THOR-uh-co-uh-CRO-me-ul) **trunk** supplies the superior shoulder and pectoral regions.

3. The **lateral thoracic artery** supplies the lateral thoracic wall.

4. The **subscapular artery** supplies the scapula, latissimus dorsi, and posterior wall of the thorax.

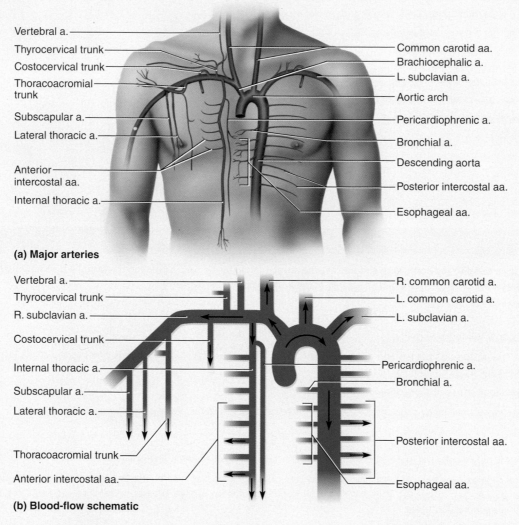

(a) **Major arteries**

(b) **Blood-flow schematic**

**FIGURE 20.26**  Arteries Supplying the Thorax.

▶ *Which artery in this figure supplies systemic blood to the lungs?*

---

[17]*thoraco* = chest + *acr* = tip + *om* = shoulder

| TABLE 20.7 | Arterial Supply to the Abdomen |
|---|---|

**Major Branches of Abdominal Aorta**

After passing through the aortic hiatus, the aorta descends through the abdominal cavity. The abdominal aorta is retroperitoneal. It gives off arteries in the order listed here (fig. 20.27). Those indicated in the plural are paired right and left, and those indicated in the singular are single median arteries:

1. The **inferior phrenic arteries** supply the inferior surface of diaphragm and issue a small **superior suprarenal artery** to each adrenal (suprarenal) gland.

2. The **celiac**[18] (SEE-lee-ac) **trunk** issues several branches to the upper abdominal viscera, further traced later in this table.

3. The **superior mesenteric artery** supplies the intestines (see mesenteric circulation later in this table).

4. The **middle suprarenal arteries** arise on either side of the superior mesenteric artery and supply the adrenal glands.

5. The **renal arteries** supply the kidneys and issue a small **inferior suprarenal artery** to each adrenal gland.

6. The **gonadal arteries** are long, narrow, winding arteries that descend from the midabdominal region to the female pelvic cavity or male scrotum. They are called the **ovarian arteries** in females and **testicular arteries** in males. The gonads begin their embryonic development near the kidneys. These arteries acquire their peculiar length and course by growing to follow the gonads as they descend to the pelvic cavity during fetal development.

7. The **inferior mesenteric artery** supplies the distal end of the large intestine (see mesenteric circulation).

8. The **lumbar arteries** arise from the lower aorta in four pairs and supply the posterior abdominal wall.

9. The **median sacral artery,** a tiny medial artery at the inferior end of the aorta, supplies the sacrum and coccyx.

10. The **common iliac arteries** arise as the aorta forks at its inferior end. They supply the lower abdominal wall, pelvic viscera (chiefly the urinary and reproductive organs), and lower limbs. They are further traced in table 20.8.

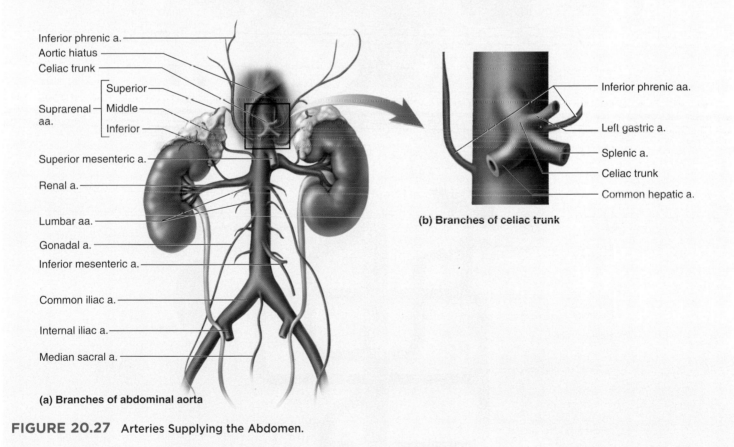

(a) Branches of abdominal aorta

(b) Branches of celiac trunk

**FIGURE 20.27**   Arteries Supplying the Abdomen.

---

[18]*celi* = belly, abdomen

**TABLE 20.7**   Arterial Supply to the Abdomen *(cont.)*

**Branches of the Celiac Trunk**

The celiac circulation to the upper abdominal viscera is perhaps the most complex route off the abdominal aorta. Because it has numerous anastomoses, the bloodstream does not follow a simple linear path but divides and rejoins itself at several points (fig. 20.28). As you study the following description, locate these branches in the figure and identify the points of anastomosis. The short, stubby celiac trunk is a median branch of the aorta. It immediately gives rise to three principal subdivisions—the *common hepatic, left gastric,* and *splenic arteries:*

1. The **common hepatic artery** issues two main branches:

   a. the **gastroduodenal artery,** which supplies the stomach, anastomoses with the right gastroepiploic artery (see following), and then continues as the **inferior pancreaticoduodenal** (PAN-cree-AT-ih-co-dew-ODD-eh-nul) **artery,** which supplies the duodenum and pancreas before anastomosing with the superior mesenteric artery; and

   b. the **proper hepatic artery,** which is the continuation of the common hepatic artery after it gives off the gastroduodenal artery. It enters the inferior surface of the liver and supplies the liver and gallbladder.

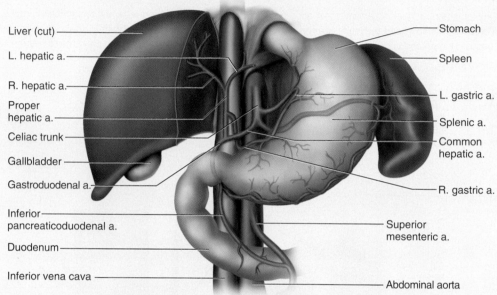

**(a) Branches of the celiac trunk**

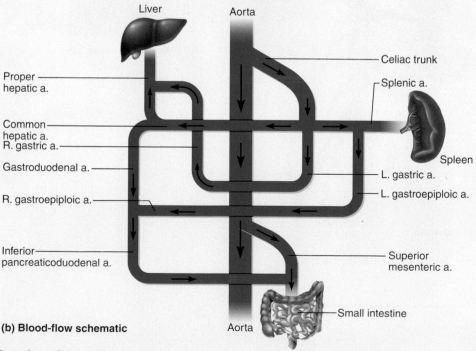

**(b) Blood-flow schematic**

**FIGURE 20.28**   Branches of the Celiac Trunk.

2. The **left gastric artery** supplies the stomach and lower esophagus, arcs around the *lesser curvature* of the stomach, becomes the **right gastric artery** (which supplies the stomach and duodenum), and then anastomoses with the proper hepatic artery.

3. The **splenic artery** supplies blood to the spleen, but gives off the following branches on its way there:

   a. the **pancreatic arteries** (not illustrated), which supply the pancreas; and

   b. the **left gastroepiploic**[19] (GAS-tro-EP-ih-PLO-ic) artery, which arcs around the *greater curvature* of the stomach, becomes the **right gastroepiploic artery,** and then anastomoses with the gastroduodenal artery. Along the way, it supplies blood to the stomach and *greater omentum* (a fatty membrane suspended from the greater curvature).

### Mesenteric Circulation

The mesentery (see atlas A, p. 37) contains numerous mesenteric arteries, veins, and lymphatic vessels that perfuse and drain the intestines. The arterial supply issues from the *superior* and *inferior mesenteric arteries* (fig. 20.29); numerous anastomoses between these ensure collateral circulation and adequate perfusion of the intestinal tract even if one route becomes obstructed. The following branches of the **superior mesenteric artery** serve the small intestine and most of the large intestine, among other organs:

1. The **inferior pancreaticoduodenal artery,** already mentioned, is an anastomosis from the gastroduodenal to the superior mesenteric artery; it supplies the pancreas and duodenum.

2. The **intestinal arteries** supply nearly all of the small intestine (jejunum and ileum).

3. The **ileocolic** (ILL-ee-oh-CO-lic) **artery** supplies the ileum of the small intestine and the appendix, cecum, and ascending colon.

4. The **right colic artery** supplies the ascending colon.

5. The **middle colic artery** supplies the transverse colon.

Branches of the *inferior mesenteric artery* serve the distal part of the large intestine:

1. The **left colic artery** supplies the transverse and descending colon.

2. The **sigmoid arteries** supply the descending and sigmoid colon.

3. The **superior rectal artery** supplies the rectum.

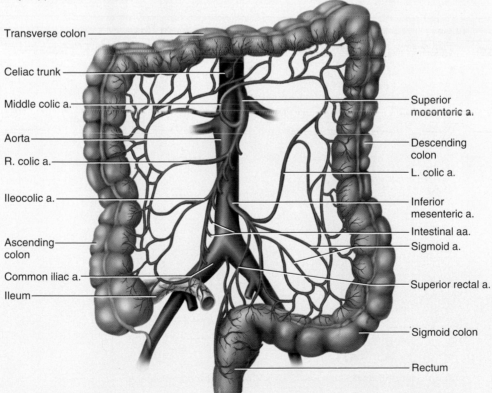

**FIGURE 20.29   The Mesenteric Arteries.**

---

[19] *gastro* = stomach + *epi* = upon, above + *ploic* = pertaining to the greater omentum

**TABLE 20.8**    Arterial Supply to the Pelvic Region and Lower Limb

The common iliac arteries arise from the aorta at the level of vertebra L4 and continue for about 5 cm. At the level of the sacroiliac joint, each divides into an internal and external iliac artery. The **internal iliac artery** supplies mainly the pelvic wall and viscera, and the **external iliac artery** supplies mainly the lower limb (figs. 20.30 and 20.31).

**Branches of the Internal Iliac Artery**

1. The **iliolumbar** and **lateral sacral arteries** supply the wall of the pelvic region.

2. The **middle rectal artery** supplies the rectum.

3. The **superior** and **inferior vesical**[20] **arteries** supply the urinary bladder.

4. The **uterine** and **vaginal arteries** supply the uterus and vagina.

5. The **superior** and **inferior gluteal arteries** supply the gluteal muscles.

6. The **obturator artery** supplies the adductor muscles of the medial thigh.

7. The **internal pudendal**[21] (pyu-DEN-dul) **artery** serves the perineum and external genitals; it supplies the blood for vascular engorgement during sexual arousal.

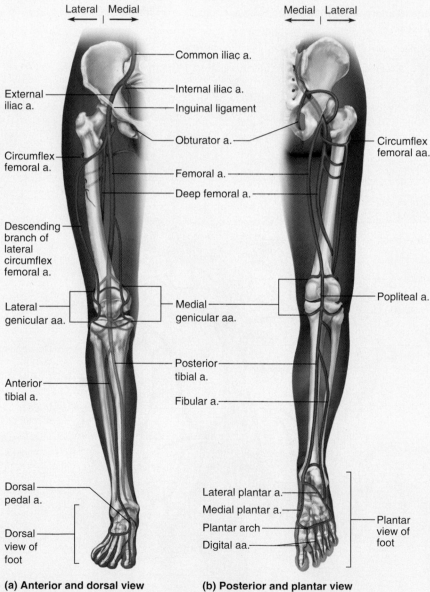

(a) Anterior and dorsal view    (b) Posterior and plantar view

**FIGURE 20.30    Arteries Supplying the Lower Limb.** (a) Anterior view of the right limb and dorsal aspect of the foot. (b) Posterior view of the right limb and plantar aspect of the foot.

---

[20]*vesic* = bladder
[21]*pudend* = literally "shameful parts"; the external genitals

### Branches of the External Iliac Artery

The external iliac artery sends branches to the skin and muscles of the abdominal wall and pelvic girdle. It then passes deep to the inguinal ligament and gives rise to branches that serve mainly the lower limbs:

1. The **femoral artery** passes through the femoral triangle of the upper medial thigh, where its pulse can be palpated. It gives off the following branches to supply the thigh region:

   a. The **deep femoral artery,** which supplies the hamstring muscles; and

   b. The **circumflex femoral arteries,** which encircle the neck of the femur and supply the femur and hamstring muscles.

2. The **popliteal artery** is a continuation of the femoral artery in the popliteal fossa at the rear of the knee. It produces anastomoses **(genicular arteries)** that supply the knee and then divides into the anterior and posterior tibial arteries.

3. The **anterior tibial artery** travels lateral to the tibia in the anterior compartment of the leg, where it supplies the extensor muscles. It gives rise to

   a. the **dorsal pedal artery,** which traverses the ankle and dorsum of the foot; and

   b. the **arcuate artery,** a continuation of the dorsal pedal artery that gives off the metatarsal arteries of the foot.

4. The **posterior tibial artery** travels through the posteromedial part of the leg and supplies the flexor muscles. It gives rise to

   a. the **fibular (peroneal) artery,** which arises from the proximal end of the posterior tibial artery and supplies the lateral peroneal muscles;

   b. the **lateral** and **medial plantar arteries,** which arise by bifurcation of the posterior tibial artery at the ankle and supply the plantar surface of the foot; and

   c. the **plantar arch,** an anastomosis from the lateral plantar artery to the dorsal pedal artery that gives rise to the digital arteries of the toes.

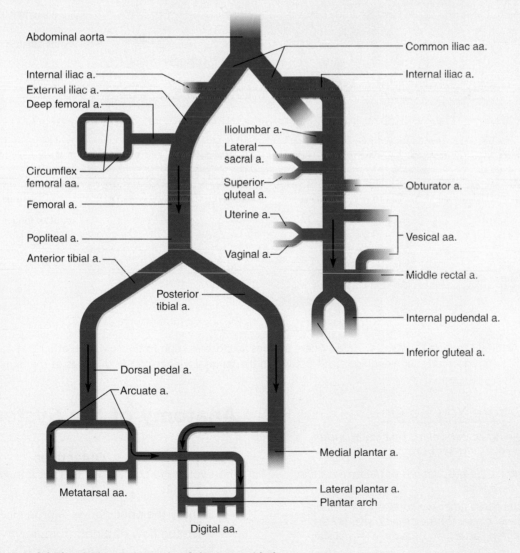

**FIGURE 20.31** Arterial Blood-Flow Schematic of the Lower Limb.

▶ *What arteries of the wrist and hand are most comparable to the arcuate artery and plantar arch of the foot?*

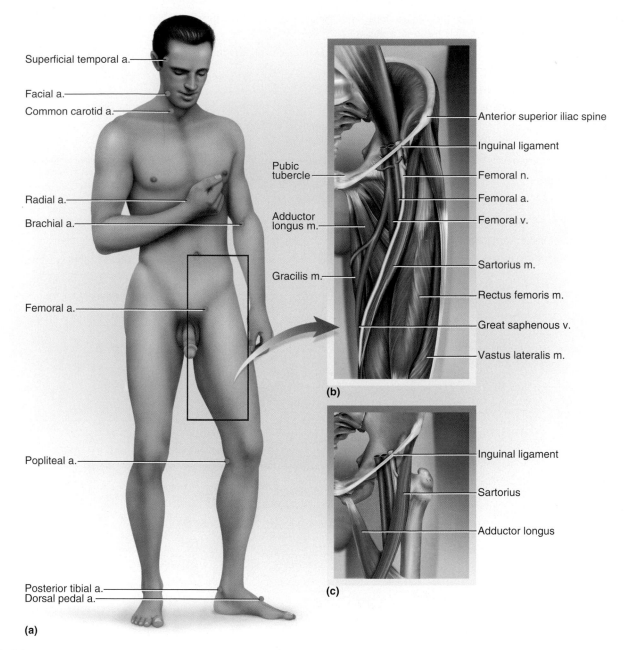

**FIGURE 20.32  Arterial Pressure Points.** (a) Areas where arteries lie close enough to the surface that a pulse can be palpated or pressure can be applied to reduce arterial bleeding. (b) Structures in the femoral triangle. (c) Boundaries of the femoral triangle.

## Before You Go On

*Answer the following questions to test your understanding of the preceding section:*

25. *Concisely contrast the destinations of the external and internal carotid arteries.*

26. *Briefly state the tissues that are supplied with blood by (a) the arterial circle, (b) the celiac trunk, (c) the superior mesenteric artery, and (d) the external iliac artery.*

27. *Trace the path of an RBC from the left ventricle to the metatarsal arteries. State two places along this path where you can palpate the arterial pulse.*

# Anatomy of the Systemic Veins

### Objectives

When you have completed this section, you should be able to

- identify the principal veins of the systemic circuit; and
- trace the flow of blood from any major organ to the heart.

The principal veins of the systemic circuit (fig. 20.33) are detailed in tables 20.9 through 20.14 (figs. 20.34–20.40).

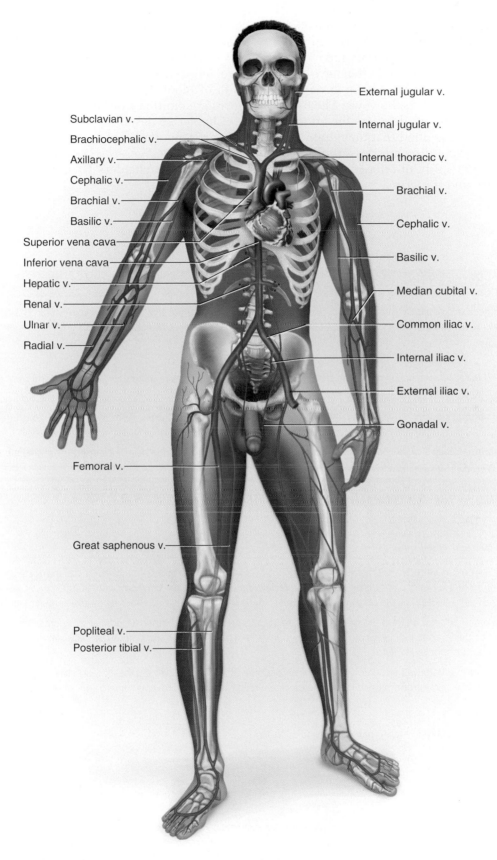

Subclavian v.

Brachiocephalic v.

Axillary v.

Cephalic v.

Brachial v.

Basilic v.

Superior vena cava

Inferior vena cava

Hepatic v.

Renal v.

Ulnar v.

Radial v.

Femoral v.

Great saphenous v.

Popliteal v.

Posterior tibial v.

External jugular v.

Internal jugular v.

Internal thoracic v.

Brachial v.

Cephalic v.

Basilic v.

Median cubital v.

Common iliac v.

Internal iliac v.

External iliac v.

Gonadal v.

**FIGURE 20.33  The Major Systemic Veins.**  (v. = vein; vv. = veins)

Although arteries are usually deep and well protected, veins occur in both deep and superficial groups; you may be able to see quite a few of them in your arms and hands. Deep veins run parallel to the arteries and often have similar names (*femoral artery* and *femoral vein,* for example); this is not true of the superficial veins, however. The deep veins are not described in as much detail as the arteries were, since it can usually be assumed that they drain the same structures as the corresponding arteries supply.

In general, we began the study of arteries with those lying close to the heart and progressed away. In the venous system, by contrast, we begin with those that are remote from the heart and follow the flow of blood as they join each other and approach the heart. Venous pathways have more anastomoses than arterial pathways, so the route of blood flow is often not as clear. Many anastomoses are omitted from the following figures for clarity.

Chapter 29 describes the effects of aging on the circulatory system, and table 20.15 (p. 800) lists some disorders of the blood vessels. Disorders of the blood and heart are tabulated in chapters 18 and 19.

| TABLE 20.9 | Venous Drainage of the Head and Neck |
|---|---|

Most blood of the head and neck is drained by three pairs of veins: the *internal jugular, external jugular,* and *vertebral veins.* This table traces their origins and drainage and follows them to the formation of the *brachiocephalic veins* and *superior vena cava* (fig. 20.34).

**Dural Sinuses**

Large thin-walled veins called **dural sinuses** occur within the cranial cavity between layers of dura mater. They receive blood from the brain and face and empty into the internal jugular veins:

1. The **superior** and **inferior sagittal sinuses** are found in the falx cerebri between the cerebral hemispheres; they receive blood that has circulated through the brain.

2. The **cavernous sinuses** occur on each side of the body of the sphenoid bone; they receive blood from the **superior ophthalmic vein** draining the orbit and the facial vein draining the nose and upper lip.

3. The **transverse (lateral) sinuses** encircle the inside of the occipital bone and lead to the jugular foramen on each side. They receive blood from the previously mentioned sinuses and empty into the internal jugular veins.

**Major Veins of the Neck**

Blood flows down the neck mainly through three veins on each side, all of which empty into the subclavian vein:

1. The **internal jugular**[22] (JUG-you-lur) **vein** courses down the neck, alongside the internal carotid artery, deep to the sternocleidomastoid muscle. It receives most of the blood from the brain, picks up blood from the **facial vein** and **superficial temporal vein** along the way, passes deep to the clavicle, and joins the subclavian vein. (Note that the facial vein empties into both the cavernous sinus and the internal jugular vein.)

2. The **external jugular vein** drains tributaries from the parotid gland, facial muscles, scalp, and other superficial structures. Some of this blood also follows venous anastomoses to the internal jugular vein. The external jugular vein courses down the side of the neck superficial to the sternocleidomastoid muscle and empties into the subclavian vein.

3. The **vertebral vein** travels with the vertebral artery in the transverse foramina of the cervical vertebrae. Although the companion artery leads to the brain, the vertebral vein does not come from there. It drains the cervical vertebrae, spinal cord, and some of the small deep muscles of the neck.

**Drainage from Shoulder to Heart**

From the shoulder region, blood takes the following path to the heart:

1. The **subclavian vein** drains the arm and travels inferior to the clavicle; receives the external jugular, vertebral, and internal jugular veins in that order; and ends where it receives the internal jugular.

2. The **brachiocephalic vein** is formed by union of the subclavian and internal jugular veins. It continues medially and receives tributaries draining the upper thoracic wall and breast.

3. The **superior vena cava** is formed by the union of the right and left brachiocephalic veins. It travels inferiorly for about 7.5 cm and empties into the right atrium. It drains all structures superior to the diaphragm except the pulmonary circuit and coronary circulation. It also receives considerable drainage from the abdominal cavity by way of the azygos system (see table 20.11).

[22]*jugul* = neck

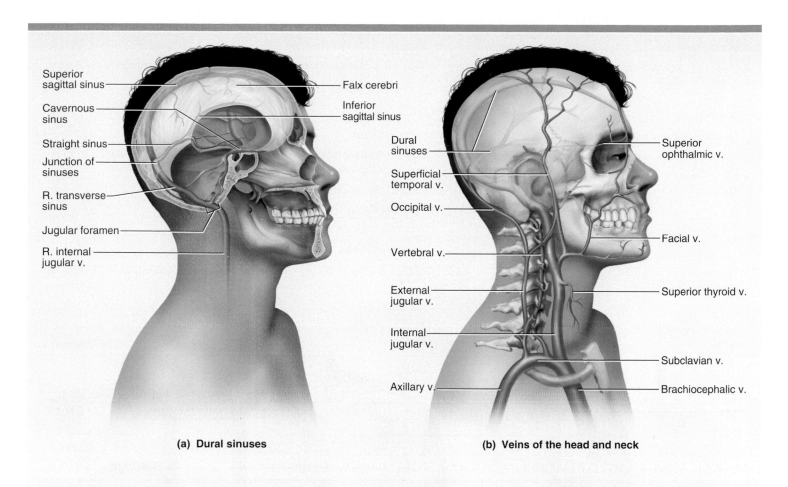

Superior sagittal sinus
Cavernous sinus
Straight sinus
Junction of sinuses
R. transverse sinus
Jugular foramen
R. internal jugular v.
Falx cerebri
Inferior sagittal sinus

**(a) Dural sinuses**

Dural sinuses
Superficial temporal v.
Occipital v.
Vertebral v.
External jugular v.
Internal jugular v.
Axillary v.
Superior ophthalmic v.
Facial v.
Superior thyroid v.
Subclavian v.
Brachiocephalic v.

**(b) Veins of the head and neck**

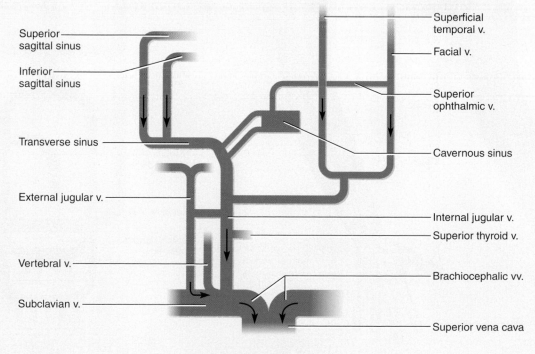

Superior sagittal sinus
Inferior sagittal sinus
Transverse sinus
External jugular v.
Vertebral v.
Subclavian v.
Superficial temporal v.
Facial v.
Superior ophthalmic v.
Cavernous sinus
Internal jugular v.
Superior thyroid v.
Brachiocephalic vv.
Superior vena cava

**(c) Venous flow schematic**

**FIGURE 20.34** Veins Draining the Head and Neck.

## TABLE 20.10    Venous Drainage of the Upper Limb

Table 20.9 briefly noted the subclavian veins that drain each arm. This table begins distally in the forearm and traces venous drainage to the subclavian vein (fig. 20.35).

**Deep Veins**

1. The **palmar digital veins** drain each finger into the **superficial venous palmar arch.**

2. The **metacarpal veins** parallel the metacarpal bones and drain blood from the hand into the **deep venous palmar arch.** Both the superficial and deep venous palmar arches are anastomoses between the next two veins, which are the major deep veins of the forearm.

3. The **radial vein** receives blood from the lateral side of both palmar arches and courses up the forearm alongside the radius.

4. The **ulnar vein** receives blood from the medial side of both palmar arches and courses up the forearm alongside the ulna.

5. The **brachial vein** is formed by the union of the radial and ulnar veins at the elbow; it courses up the brachium.

6. The **axillary vein** is formed at the axilla by the union of the brachial and basilic veins (the basilic vein is described in the next section).

7. The **subclavian vein** is a continuation of the axillary vein into the shoulder inferior to the clavicle. The further course of the subclavian is explained in the previous table.

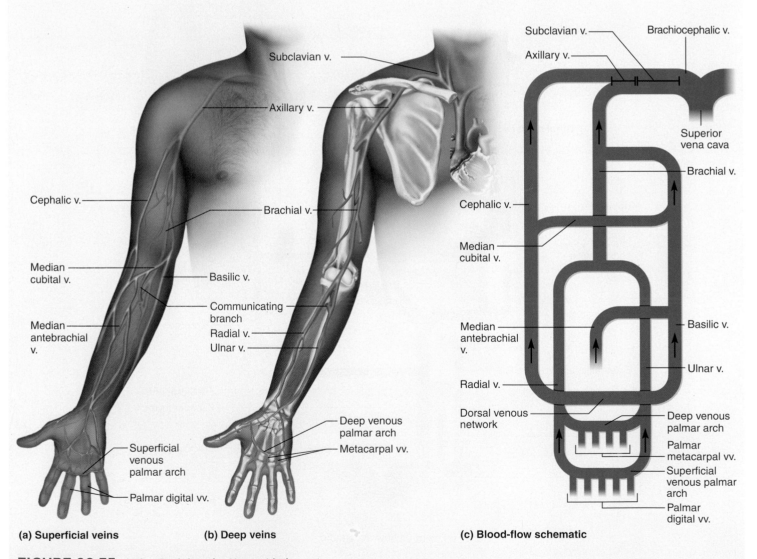

(a) **Superficial veins**            (b) **Deep veins**            (c) **Blood-flow schematic**

**FIGURE 20.35    Veins Draining the Upper Limb.**

▶ *Name three veins that are often visible through the skin of the upper limb.*

**Superficial Veins**

These are easily seen through the skin of most people and are larger in diameter than the deep veins:

1. The **dorsal venous network** is a plexus of veins visible on the back of the hand; it empties into the major superficial veins of the forearm, the cephalic and basilic.

2. The **cephalic vein** arises from the lateral side of the dorsal venous network, winds around the radius as it travels up the forearm, continues up the lateral aspect of the brachium to the shoulder, and joins the axillary vein there. Intravenous fluids are often administered through the distal end of this vein.

3. The **basilic**[23] (bah-SIL-ic) **vein** arises from the medial side of the dorsal venous network, travels up the posterior aspect of the forearm, and continues into the brachium. About midway up the brachium it turns deeper and runs beside the brachial artery. At the axilla it joins the brachial vein, and the union of these two gives rise to the axillary vein.

4. The **median cubital vein** is a short anastomosis between the cephalic and basilic veins that obliquely crosses the cubital fossa (anterior bend of the elbow). It is clearly visible through the skin and is the most common site for drawing blood.

5. The **median antebrachial vein** originates near the base of the thumb, travels up the forearm between the radial and ulnar veins, and terminates at the elbow; it empties into the cephalic vein in some people and into the basilic vein in others.

---

| TABLE 20.11 | The Azygos System |
|---|---|

The superior vena cava receives extensive drainage from the thoracic and abdominal walls by way of the **azygos** (AZ-ih-goss) **system** (fig. 20.36).

**Drainage of the Abdominal Wall**

A pair of **ascending lumbar veins** receive blood from the common iliac veins below and a series of short horizontal **lumbar veins** that drain the abdominal wall. The ascending lumbar veins anastomose with the inferior vena cava beside them and ascend through the diaphragm into the thoracic cavity.

**Drainage of the Thorax**

*Right side.* After penetrating the diaphragm, the right ascending lumbar vein becomes the **azygos**[24] vein of the thorax. The azygos receives blood from the **right posterior intercostal veins,** which drain the chest muscles, and from the **esophageal, mediastinal, pericardial,** and **right bronchial veins.** It then empties into the superior vena cava at the level of vertebra T4.

*Left side.* The left ascending lumbar vein continues into the thorax as the **hemiazygos**[25] vein. The hemiazygos drains the ninth through eleventh posterior intercostal veins and some esophageal and mediastinal veins on the left. At midthorax, it crosses over to the right side and empties into the azygos vein.

The **accessory hemiazygos vein** is a superior extension of the hemiazygos. It drains the fourth through eighth posterior intercostal veins and the left bronchial vein. It also crosses to the right side and empties into the azygos vein.

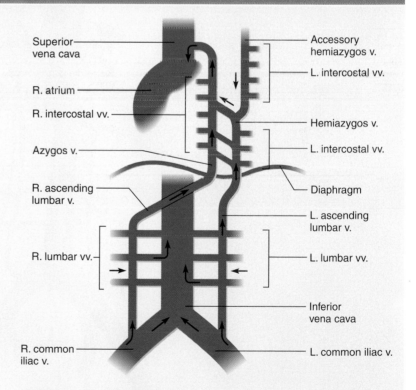

**FIGURE 20.36** Veins of the Azygos System.

---

[23]*basilic* = royal, prominent, important
[24]unpaired; from *a* = without + *zygo* = union, mate
[25]*hemi* = half

## TABLE 20.12 | Major Tributaries of the Inferior Vena Cava

The **inferior vena cava (IVC)** is formed by the union of the right and left common iliac veins at the level of vertebra L5. It is retroperitoneal and lies immediately to the right of the aorta. Its diameter of 3.5 cm is the largest of any vessel in the body. As it ascends the abdominal cavity, the IVC picks up blood from numerous tributaries in the order listed here (fig. 20.37):

1.  Some **lumbar veins** empty into the IVC as well as into the ascending lumbar veins described in table 20.11.

2.  The **gonadal veins** (**ovarian veins** in the female and **testicular veins** in the male) drain the gonads. The right gonadal vein empties directly into the IVC, whereas the left gonadal vein empties into the left renal vein.

3.  The **renal veins** drain the kidneys into the IVC. The left renal vein also receives blood from the left gonadal and left suprarenal veins.

4.  The **suprarenal veins** drain the adrenal (suprarenal) glands. The right suprarenal empties directly into the IVC, and the left suprarenal empties into the renal vein.

5.  The **hepatic veins** drain the liver; they extend a short distance from its superior surface to the IVC.

6.  The **inferior phrenic veins** drain the inferior aspect of the diaphragm.

After receiving these inputs, the IVC penetrates the diaphragm and enters the right atrium from below. It does not receive any thoracic drainage.

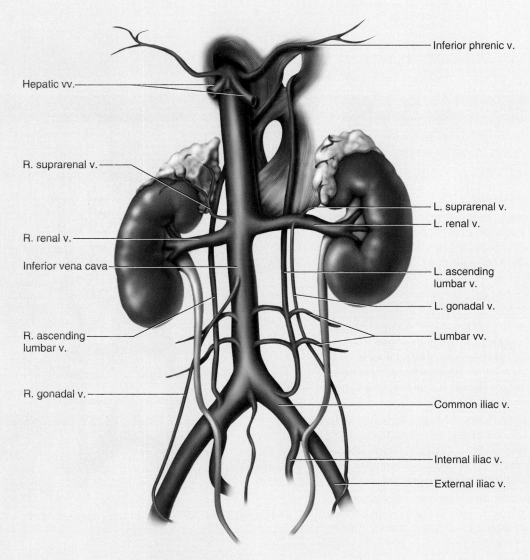

**FIGURE 20.37**  The Inferior Vena Cava and Its Tributaries.

## TABLE 20.13 | The Hepatic Portal System

The **hepatic portal system** connects capillaries of the intestines and other digestive organs to the **hepatic sinusoids** of the liver. The intestinal blood is richly laden with nutrients for a few hours following a meal. The hepatic portal system gives the liver "first claim" to these nutrients before the blood is distributed to the rest of the body. It also allows the blood to be cleansed of bacteria and toxins picked up from the intestines, an important function of the liver. The route from the intestines to the inferior vena cava follows (fig. 20.38):

1. The **inferior mesenteric vein** receives blood from the rectum and distal part of the large intestine. It converges in a fan-like array in the mesentery and empties into the splenic vein.

2. The **superior mesenteric vein** receives blood from the entire small intestine, ascending colon, transverse colon, and stomach. It, too, exhibits a fanlike arrangement in the mesentery and then joins the splenic vein to form the hepatic portal vein.

3. The **splenic vein** drains the spleen and travels across the abdominal cavity toward the liver. Along the way, it picks up the **pancreatic veins** from the pancreas, then the inferior mesenteric vein. It changes name when it joins the superior mesenteric vein, as explained next.

4. The **hepatic portal vein** is formed by convergence of the splenic and superior mesenteric veins. It travels about 8 cm up and to the right and then enters the inferior surface of the liver. Near this point it receives the **cystic vein** from the gallbladder. In the liver, the hepatic portal vein ultimately leads to the innumerable microscopic hepatic sinusoids. Blood from the sinusoids empties into the hepatic veins described earlier. Circulation within the liver is described in more detail in chapter 25.

5. The left and right **gastric veins** form an arch along the lesser curvature of the stomach and empty into the hepatic portal vein.

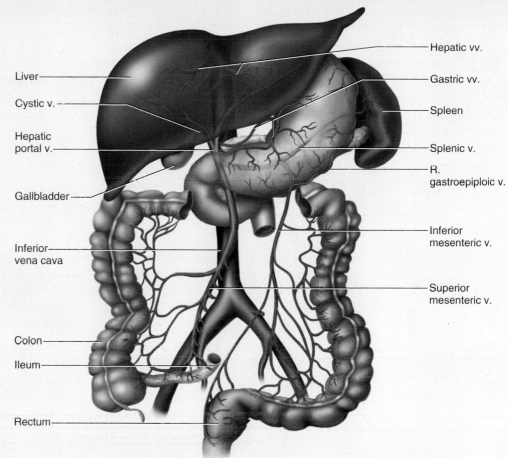

(a) Tributaries of the hepatic portal system

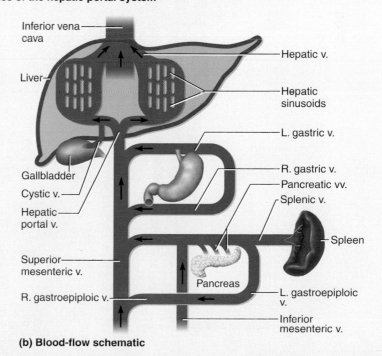

(b) Blood-flow schematic

**FIGURE 20.38** The Hepatic Portal System and Its Tributaries.

| **TABLE 20.14** | Venous Drainage of the Lower Limb and Pelvic Organs |
|---|---|

Drainage of the lower limb is described starting at the toes and following the flow of blood to the inferior vena cava (figs. 20.39 and 20.40). As in the upper limb, there are deep and superficial veins with anastomoses between them.

**Deep Veins**

1. The **plantar venous arch** drains the plantar aspect of the foot, receives blood from the **plantar digital veins** of the toes, and gives rise to the next vein.

2. The **posterior tibial vein** drains the plantar arch and passes up the leg embedded deep in the calf muscles; it receives drainage along the way from the **fibular (peroneal) vein.**

3. The **dorsal pedal vein** drains the dorsum of the foot.

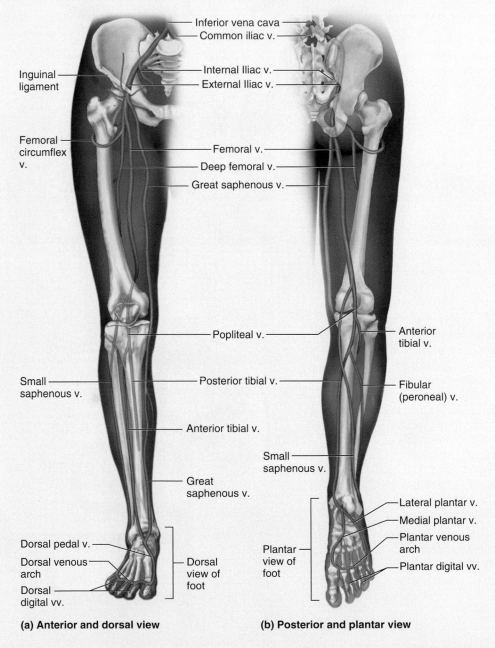

**(a) Anterior and dorsal view**        **(b) Posterior and plantar view**

**FIGURE 20.39   Veins Draining the Lower Limb.** (a) Anterior view of the right limb and dorsal aspect of the foot. (b) Posterior view of the right limb and plantar aspect of the foot.

4. The **anterior tibial vein** is a continuation of the dorsal pedal vein. It travels up the anterior compartment of the leg between the tibia and fibula.

5. The **popliteal vein** is formed at the back of the knee by the union of the anterior and posterior tibial veins.

6. The **femoral vein** is a continuation of the popliteal vein into the thigh. It receives drainage from the deep thigh muscles and femur.

7. The **external iliac vein,** superior to the inguinal ligament, is formed by the union of the femoral vein and great saphenous vein (one of the superficial veins described next).

8. The **internal iliac vein** follows the course of the internal iliac artery and its distribution. Its tributaries drain the gluteal muscles; the medial aspect of the thigh; the urinary bladder, rectum, prostate, and ductus deferens in the male; and the uterus and vagina in the female.

9. The **common iliac vein** is formed by the union of the external and internal iliac veins; it also receives blood from the ascending lumbar vein. The right and left common iliacs then unite to form the inferior vena cava.

### Superficial Veins

1. The **dorsal venous arch** is visible through the skin on the dorsum of the foot. It has numerous anastomoses similar to the dorsal venous network of the hand.

2. The **great saphenous**[26] (sah-FEE-nus) **vein,** the longest vein in the body, arises from the medial side of the dorsal venous arch. It traverses the medial aspect of the leg and thigh and terminates by emptying into the femoral vein, slightly inferior to the inguinal ligament. It is commonly used as a site for the long-term administration of intravenous fluids; it is a relatively accessible vein in infants and in patients in shock whose veins have collapsed. Portions of this vein are commonly excised and used as grafts in coronary bypass surgery.

3. The **small saphenous vein** arises from the lateral side of the dorsal venous arch, courses up the lateral aspect of the foot and through the calf muscles, and terminates at the knee by emptying into the popliteal vein. It has numerous anastomoses with the great saphenous vein. The great and small saphenous veins are among the most common sites of varicose veins.

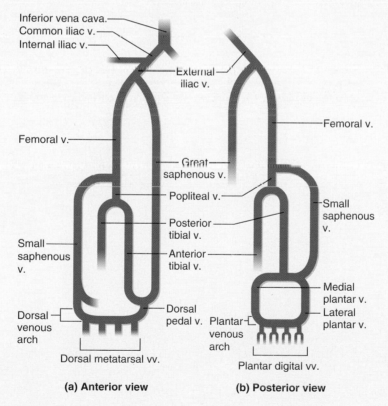

(a) Anterior view          (b) Posterior view

**FIGURE 20.40   Venous Blood-Flow Schematic of the Lower Limb.**

[26]*saphen* = standing

## Before You Go On

*Answer the following questions to test your under-standing of the preceding section:*

28. *If you were dissecting a cadaver, where would you look for the internal and external jugular veins? What muscle would help you distinguish one from the other?*

29. *How do the vertebral veins differ from the vertebral arteries in their superior terminations?*

30. *By what route does blood from the abdominal wall reach the superior vena cava?*

31. *Trace one possible path of an RBC from the fingertips to the right atrium and name the veins along the way.*

32. *State two ways in which the great saphenous vein has special clinical significance. Where is this vein located?*

| TABLE 20.15 | Some Disorders of the Arteries and Veins |
|---|---|
| Dissecting aneurysm | Splitting of the layers of an arterial wall from each other because of the accumulation of blood between layers. Results from either a tear in the tunica intima or rupture of the vasa vasorum. |
| Fat embolism | The presence of fat globules traveling in the bloodstream. Globules originate from bone fractures, fatty degeneration of the liver, and other causes and may block cerebral or pulmonary blood vessels. |
| Orthostatic hypotension | A decrease in blood pressure that occurs when one stands, often resulting in blurring of vision, dizziness, and syncope (fainting). Results from sluggish or inactive baroreflexes. |

***Disorders described elsewhere***

| | | |
|---|---|---|
| Aneurysm p. 756 | Embolism p. 773 | Stroke 773 |
| Atherosclerosis p. 746 | Hypertension pp. 763, 801 | Transient ischemic attack p. 775 |
| Circulatory shock p. 774 | Hypotension p. 763 | Varicose veins p. 761 |
| Edema p. 771 | | |

## INSIGHT 20.4  Clinical Application

## Hypertension—The "Silent Killer"

*Hypertension,* the most common cardiovascular disease, affects about 30% of Americans over age 50 and 50% by age 74. It is a "silent killer" that can wreak its destructive effects for 10 to 20 years before they are first noticed. Hypertension is the major cause of heart failure, stroke, and kidney failure. It damages the heart because it increases the afterload, which makes the ventricles work harder to expel blood. The myocardium enlarges up to a point (the *hypertrophic response*), but eventually it becomes excessively stretched and less efficient. Hypertension strains the blood vessels and tears the endothelium, thereby creating lesions that become focal points of atherosclerosis. Atherosclerosis then worsens the hypertension and establishes an insidious positive feedback cycle.

Another positive feedback cycle involves the kidneys. Their arterioles thicken in response to the stress, their lumens become narrower, and renal blood flow declines. In response to the resulting drop in blood pressure, the kidneys release renin, which leads to the formation of the vasoconstrictor angiotensin II and the release of aldosterone, a hormone that promotes salt retention (described in detail in chapter 24). These effects worsen the hypertension that already existed. If diastolic pressure exceeds 120 mm Hg, blood vessels of the eye hemorrhage, blindness ensues, the kidneys and heart deteriorate rapidly, and death usually follows within 2 years.

*Primary hypertension,* which accounts for 90% of cases, results from such a complex web of behavioral, hereditary, and other factors that it is difficult to sort out any specific underlying cause. It was once considered such a normal part of the "essence" of aging that it continues to be called by another name, *essential hypertension.* That term suggests a fatalistic resignation to hypertension as a fact of life, but this need not be. Many risk factors have been identified, and most of them are controllable.

One of the chief culprits is obesity. Each pound of extra fat requires miles of additional blood vessels to serve it, and all of this added vessel length increases peripheral resistance and blood pressure. Just carrying around extra weight, of course, also increases the workload on the heart. Even a small weight loss can significantly reduce blood pressure. Sedentary behav-

ior is another risk factor. Aerobic exercise helps to reduce hypertension by controlling weight, reducing emotional tension, and stimulating vasodilation.

Dietary factors are also significant contributors to hypertension. Diets high in cholesterol and saturated fat contribute to atherosclerosis. Potassium and magnesium reduce blood pressure; thus, diets deficient in these minerals promote hypertension. The relationship of salt intake to hypertension has been a very controversial subject. The kidneys compensate so effectively for excess salt intake that dietary salt has little effect on the blood pressure of most people. Reduced salt intake may, however, help to control hypertension in older people and in people with reduced renal function.

Nicotine makes a particularly devastating contribution to hypertension because it stimulates the myocardium to beat faster and harder, while it stimulates vasoconstriction and increases the afterload against which the myocardium must work. Just when the heart needs extra oxygen, nicotine causes coronary vasoconstriction and promotes myocardial ischemia.

Some risk factors cannot be changed at will—race, heredity, and sex. Hypertension runs in some families. A person whose parents or siblings have hypertension is more likely than average to develop it. The incidence of hypertension is about 30% higher, and the incidence of strokes about twice as high, among blacks as among whites. From ages 18 to 54, hypertension is more common in men, but above age 65, it is more common in women. Even people at risk from these factors, however, can minimize their chances of hypertension by changing risky behaviors.

Treatments for primary hypertension include weight loss, diet, and certain drugs. Diuretics lower blood volume and pressure by promoting urination. ACE inhibitors block the formation of the vasoconstrictor angiotensin II. Beta-blockers such as propranolol block the vasoconstrictive action of the sympathetic nervous system. Calcium channel blockers such as verapamil and nifedipine inhibit the inflow of calcium into cardiac and smooth muscle, thus inhibiting their contraction, promoting vasodilation, and reducing cardiac workload.

*Secondary hypertension,* which accounts for about 10% of cases, is high blood pressure that is secondary to (results from) other identifiable disorders. These include kidney disease (which may cause renin hypersecretion), atherosclerosis, hyperthyroidism, Cushing syndrome, and polycythemia. Secondary hypertension is corrected by treating the underlying disease.

# CONNECTIVE ISSUES

## Interactions Between the
## CIRCULATORY SYSTEM
### and Other Organ Systems

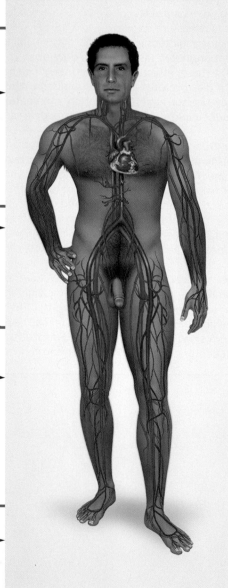

### ALL SYSTEMS

Circulatory system delivers $O_2$ and nutrients to all other systems and carries away wastes; carries heat from deeper organs to skin for elimination

### INTEGUMENTARY SYSTEM

Dermal blood flow affects sweat production

Serves as blood reservoir; helps to regulate blood temperature

### SKELETAL SYSTEM

Provides minerals for bone deposition; delivers erythropoietin to bone marrow and delivers hormones that regulate skeletal growth

Skeleton provides protective enclosure for heart and thoracic vessels; serves as reservoir of calcium needed for heart contractions; bone marrow carries out hemopoiesis

### MUSCULAR SYSTEM

Removes heat generated by exercise

Helps to regulate blood temperature; respiratory and limb muscles aid venous return; aerobic exercise enhances circulatory efficiency

### NERVOUS SYSTEM

Endothelial cells of blood vessels maintain blood-brain barrier and help to produce CSF

Modulates heart rate, strength of contraction, and vasomotion; governs routing of blood flow; monitors blood pressure and composition and activates homeostatic mechanisms to regulate these

### ENDOCRINE SYSTEM

Transports hormones to their target cells

Regulates blood volume and pressure; stimulates hemopoiesis

### LYMPHATIC/IMMUNE SYSTEM

Produces tissue fluid, which becomes lymph; provides the WBCs and plasma proteins employed in immunity

Lymphatic and circulatory systems jointly regulate fluid balance; lymphatic system returns fluid to bloodstream; spleen acts as RBC and platelet reservoir; lymphatic tissues produce lymphocytes; immune cells protect circulatory system from pathogens

### RESPIRATORY SYSTEM

Delivers and carries away respiratory gases; low capillary blood pressure keeps alveoli dry

Site of exchange for blood gases; helps to regulate blood pH; thoracic pump aids venous return

### URINARY SYSTEM

Blood pressure maintains kidney function

Controls blood volume, pressure, and composition; initiates renin-angiotensin-aldosterone mechanism; regulates RBC count by producing erythropoietin

### DIGESTIVE SYSTEM

Carries away absorbed nutrients; helps to reabsorb and recycle bile salts and minerals from intestines

Provides nutrients for hemopoiesis; affects blood composition

### REPRODUCTIVE SYSTEM

Distributes sex hormones; vasodilation causes erection

Estrogens may slow development of atherosclerosis in women; testosterone stimulates erythropoiesis

## CHAPTER REVIEW

# Review of Key Concepts

**General Anatomy of the Blood Vessels (p. 754)**

1. Blood flows away from the heart in *arteries* and back to the heart in *veins*. *Capillaries* connect the smallest arteries to the smallest veins.

2. The wall of a blood vessel has three layers: *tunica interna, tunica media,* and *tunica externa.* The tunica interna is lined with a simple squamous *endothelium.*

3. Arteries are classified as *conducting, distributing,* and *resistance arteries* from largest to smallest. Conducting arteries are subject to the highest blood pressure and have the most elastic tissue; distributing and resistance arteries contain more smooth muscle relative to their size.

4. The smallest of the resistance arteries are *arterioles. Metarterioles* link arterioles with capillaries.

5. Certain large arteries above the heart contain baroreceptors called *carotid sinuses* for monitoring blood pressure, and chemoreceptors called *carotid bodies* and *aortic bodies* for monitoring blood $CO_2$, $O_2$, and pH.

6. Capillaries are the primary point of fluid exchange with the tissues. Their wall is composed of endothelium and basement membrane only.

7. The three types of capillaries are *continuous capillaries,* which form an uninterrupted tube; *fenestrated capillaries* perforated by patches of filtration pores; and *sinusoids,* which are irregular, highly porous blood spaces in such tissues as liver, spleen, and bone marrow.

8. Capillaries are arranged in networks called *capillary beds,* supplied by a single metarteriole. *Precapillary sphincters* regulate blood flow through a capillary bed.

9. The smallest veins, or *venules,* also exchange fluid with the tissues. They converge to form medium veins, and medium veins converge to form large veins.

10. Venous sinuses are blood spaces with large lumens, thin walls, and no muscle. They occur in such places as the heart and brain.

11. Veins have relatively low blood pressure and therefore have thinner walls and less muscular and elastic tissue. Medium veins of the limbs have valves to prevent the backflow of blood.

12. Between the arteries and veins, blood normally flows through one capillary bed. Portal systems and anastomoses are exceptions to this rule.

**Blood Pressure, Resistance, and Flow (p. 762)**

1. Blood *flow* (mL/min) and *perfusion* (flow/g of tissue) vary with the metabolic needs of a tissue.

2. Flow *(F)* is directly proportional to the pressure difference between two points $(\Delta P)$ and inversely proportional to resistance *(R)*: $F \sim \Delta P/R$.

3. Blood pressure (BP) is usually measured with a sphygmomanometer. Arterial pressures are expressed as systolic over diastolic pressure—for example, 120/80 mm Hg.

4. *Pulse pressure* is systolic minus diastolic pressure. *Mean arterial pressure* is the average pressure in a vessel over the course of a cardiac cycle, estimated as diastolic pressure + 1/3 of pulse pressure.

5. Chronic, abnormally high BP is *hypertension* and low BP is *hypotension.*

6. The expansion and contraction of arteries during the cardiac cycle reduces the pulse pressure and eases the strain on smaller arteries, but arterial blood flow is nevertheless pulsatile. In capillaries and veins, flow is steady (without pulsation).

7. *Peripheral resistance* is opposition to blood flow in the blood vessels. Resistance is directly proportional to blood viscosity and vessel length,

and inversely proportional to vessel radius to the fourth power $(r^4)$. Changes in vessel radius *(vasomotion)* thus have the greatest influence on flow from moment to moment.

8. Blood flow is fastest in the aorta, slowest in the capillaries, and speeds up somewhat in the veins.

9. Blood pressure is controlled mainly by local, neural, and hormonal control of vasomotion.

10. *Autoregulation* is the ability of a tissue to regulate its own blood supply. Over the short term, local vasomotion is stimulated by *vasoactive chemicals* (histamine, nitric oxide, and others). Over the long term, autoregulation can be achieved by *angiogenesis,* the growth of new vessels.

11. Neural control of blood vessels is based in the *vasomotor center* of the medulla oblongata. This center integrates *baroreflexes, chemoreflexes,* and the *medullary ischemic reflex,* and issues signals to the blood vessels by way of sympathetic nerve fibers.

12. Blood pressure is regulated in various ways by the hormones angiotensin II, aldosterone, atrial natriuretic peptide, antidiuretic hormone, epinephrine, and norepinephrine.

13. Vasomotion often shifts blood flow from organs with less need of perfusion at a given time, to organs with greater need—for example, away from the intestines and to the skeletal muscles during exercise.

**Capillary Exchange (p. 769)**

1. *Capillary exchange* is a two-way movement of water and solutes between the blood and tissue fluids across the walls of the capillaries and venules.

2. Materials pass through the vessel wall by diffusion, transcytosis, filtration, and reabsorption, passing

through intercellular clefts, fenestrations, and the endothelial cell cytoplasm.

3. Fluid is forced out of the vessels by blood pressure and the negative hydrostatic pressure of the interstitial space. The force drawing fluid back into the capillaries is colloid osmotic pressure. The difference between the outward and inward forces is an outward *net filtration pressure* or an inward *net reabsorption pressure.*

4. Capillaries typically give off fluid at the arterial end, where the relatively high blood pressure overrides reabsorption; they reabsorb about 85% as much fluid at the venous end, where colloid osmotic pressure overrides the lower blood pressure.

5. About 15% of the tissue fluid is reabsorbed by the lymphatic system.

6. Fluid exchange dynamics vary from place to place in the body (some capillaries engage solely in filtration and some solely in reabsorption) and from moment to moment (as vasomotion shifts the balance between filtration and reabsorption).

7. Accumulation of excess tissue fluid is *edema.* It results from increased capillary filtration, reduced reabsorption, or obstructed lymphatic drainage.

## Venous Return and Circulatory Shock (p. 772)

1. *Venous return,* the flow of blood back to the heart, is driven by the venous blood pressure gradient, gravity, the skeletal muscle pump (aided by valves in the veins of the limbs), the thoracic pump, and cardiac suction.

2. Exercise increases venous return because the vessels dilate, the thoracic pump and skeletal muscle pump work more energetically, and cardiac output is elevated.

3. Inactivity allows blood to accumulate in low points in the body by gravity; this is called *venous pooling.* It can result in *syncope* (fainting) if too much blood drains away from the brain.

4. *Circulatory shock* is any state of inadequate cardiac output. Its two basic categories are *cardiogenic*

*shock* and *low venous return (LVR) shock.*

5. The main forms of LVR shock are *hypovolemic, obstructed venous return,* and *venous pooling shock.*

6. *Septic shock* and *anaphylactic shock* combine elements of hypovolemia and venous pooling.

7. *Compensated shock* is corrected by the body's homeostatic mechanisms. *Decompensated shock* is life threatening, incapable of self-correction, and requires clinical intervention.

## Special Circulatory Routes (p. 775)

1. The brain receives a relatively stable total blood flow of about 700 mL/min, but flow shifts rapidly from one part of the brain to another during varying cerebral activities.

2. The brain regulates its own blood flow in response to changes in BP and pH.

3. *Transient ischemic attacks* result from brief periods of cerebral ischemia (poor blood flow). A *cerebral vascular accident* (stroke) results from a permanent loss of perfusion due to arterial blockage or rupture.

4. Skeletal muscles receive highly variable flow depending on their state of activity. Most muscle capillary beds are shut down at rest. During exercise, flow increases in response to muscle metabolites and sympathetic vasodilation.

5. The pulmonary circuit is the only route in which arteries carry less oxygen than veins do.

6. Pulmonary arteries have relatively low BP and slow flow, which allows ample time for gas exchange and promotes capillary reabsorption. The latter prevents fluid from accumulating in the lungs.

7. Pulmonary arteries, unlike systemic arteries, constrict in response to hypoxia, so less blood is sent to poorly ventilated areas of the lung.

## Anatomy of the Pulmonary Circuit (p. 776)

1. The route of blood flow in the pulmonary circuit is right ventricle of the heart → pulmonary trunk → pulmonary arteries → lobar arteries → alveolar capillary beds → venules →

pulmonary veins → left atrium of the heart.

2. The pulmonary circuit serves only to exchange $CO_2$ for $O_2$ in the blood. The metabolic needs of the lung tissue are met by a separate systemic blood supply to the lungs, via the bronchial arteries.

## Anatomy of the Systemic Arteries (p. 777)

1. The systemic circulation begins with the ascending aorta. Table 20.3 describes the major branches of the aorta.

2. The head and neck receive blood from the common carotid and vertebral arteries. Table 20.4 describes the branches of these arteries.

3. The upper limbs receive blood from the subclavian arteries. Table 20.5 describes the branches of these arteries in the limb.

4. The thoracic organs receive blood from several small branches of the thoracic aorta and the subclavian and axillary arteries. Table 20.6 describes these branches.

5. After passing through the diaphragm, the descending aorta gives off a series of branches to the abdominal viscera. Table 20.7 describes these.

6. At its inferior end, the abdominal aorta forks into two common iliac arteries, whose distal branches supply the pelvic region and lower limb. Table 20.8 describes these.

7. Arteries tend to be deeper than veins, but there are several places where they come close enough to the surface to be palpated. These sites serve for taking a pulse and as emergency *pressure points* where compression can stop arterial bleeding.

## Anatomy of the Systemic Veins (p. 790)

1. In venous circulation, blood flows through smaller veins that join to form progressively larger ones. Veins that merge to create a larger one are called *tributaries.*

2. The head and neck are drained by the jugular and vertebral veins, which ultimately converge to form the *superior vena cava* leading to the right atrium of the heart. Table

20.9 describes the tributaries that drain the head and neck.

3. Table 20.10 describes tributaries in the upper limb that converge to drain the limb via the axillary and subclavian veins.

4. The thoracic viscera are drained by the *azygos system*, described in table 20.11.

5. The abdominal viscera are drained by tributaries of the *inferior vena cava (IVC)*, described in table 20.12.

6. The digestive system is drained by the *hepatic portal system* of veins, described in table 20.13.

7. Table 20.14 describes tributaries of the lower limbs, which ultimately converge on the *common iliac veins*. The two common iliac veins join to form the IVC.

# Testing Your Recall

1. Blood normally flows into a capillary bed from
   a. the distributing arteries.
   b. the conducting arteries.
   c. a metarteriole.
   d. a thoroughfare channel.
   e. the venules.

2. Plasma solutes enter the tissue fluid most easily from
   a. continuous capillaries.
   b. fenestrated capillaries.
   c. arteriovenous anastomoses.
   d. collateral vessels.
   e. venous anastomoses.

3. A blood vessel adapted to withstand a high pulse pressure would be expected to have
   a. an elastic tunica media.
   b. a thick tunica intima.
   c. one-way valves.
   d. a flexible endothelium.
   e. a rigid tunica media.

4. The substance most likely to cause a rapid drop in blood pressure is
   a. epinephrine.
   b. norepinephrine.
   c. angiotensin II.
   d. serotonin.
   e. histamine.

5. A person with a systolic blood pressure of 130 mm Hg and a diastolic pressure of 85 mm Hg would have a mean arterial pressure of about
   a. 85 mm Hg.
   b. 100 mm Hg.
   c. 108 mm Hg.
   d. 115 mm Hg.
   e. 130 mm Hg.

6. The velocity of blood flow decreases if
   a. vessel radius increases.
   b. blood pressure increases.
   c. viscosity increases.
   d. afterload increases.
   e. vasomotion decreases.

7. Blood flows faster in a venule than in a capillary because venules
   a. have one-way valves.
   b. exhibit vasomotion.
   c. are closer to the heart.
   d. have higher blood pressures.
   e. have larger diameters.

8. In a case where interstitial hydrostatic pressure is negative, the only force causing capillaries to reabsorb fluid is
   a. colloid osmotic pressure of the blood.
   b. colloid osmotic pressure of the tissue fluid.
   c. capillary hydrostatic pressure.
   d. interstitial hydrostatic pressure.
   e. net filtration pressure.

9. Intestinal blood flows to the liver by way of
   a. the superior mesenteric artery.
   b. the celiac trunk.
   c. the inferior vena cava.
   d. the azygos system.
   e. the hepatic portal system.

10. The brain receives blood from all of the following vessels *except* the _____ artery or vein.
    a. basilar
    b. vertebral
    c. internal carotid
    d. internal jugular
    e. anterior communicating

11. The highest arterial blood pressure attained during ventricular contraction is called _____ pressure. The lowest attained during ventricular relaxation is called _____ pressure.

12. The capillaries of skeletal muscles are of the structural type called _____.

13. _____ shock occurs as a result of exposure to an antigen to which one is hypersensitive.

14. The role of breathing in venous return is called the _____.

15. The difference between the colloid osmotic pressure of blood and that of the tissue fluid is called _____.

16. Movement across the capillary endothelium by the uptake and release of fluid droplets is called _____.

17. All efferent fibers of the vasomotor center belong to the _____ division of the autonomic nervous system.

18. The pressure sensors in the major arteries near the head are called _____.

19. Most of the blood supply to the brain comes from a ring of arterial anastomoses called _____.

20. The major superficial veins of the arm are the _____ on the medial side and _____ on the lateral side.

*Answers in Appendix B*

# True or False

*Determine which five of the following statements are false, and briefly explain why.*

1. In some circulatory pathways, blood can get from an artery to a vein without going through capillaries.

2. In some cases, a blood cell may pass through two capillary beds in a single trip from left ventricle to right atrium.

3. The body's longest blood vessel is the great saphenous vein.

4. Arteries have a series of valves that ensure a one-way flow of blood.

5. If the radius of a blood vessel doubles and all other factors remain the same, blood flow through that vessel also doubles.

6. The femoral triangle is bordered by the inguinal ligament, sartorius muscle, and adductor longus muscle.

7. The lungs receive both pulmonary and systemic blood.

8. Blood capillaries must reabsorb all the fluid they emit, or else edema will occur.

9. An aneurysm is a ruptured blood vessel.

10. Anaphylactic shock is a form of hypovolemic shock.

*Answers in Appendix B*

# Testing Your Comprehension

1. It is a common lay perception that systolic blood pressure should be 100 plus a person's age. Evaluate the validity of this statement.

2. Calculate the net filtration or reabsorption pressure at a point in a hypothetical capillary assuming a hydrostatic blood pressure of 28 mm Hg, an interstitial hydrostatic pressure of −2 mm Hg, a blood COP of 25 mm Hg, and an interstitial COP of 4 mm Hg. Give the magnitude (in mm Hg) and direction (in or out) of the net pressure.

3. Aldosterone secreted by the adrenal gland must be delivered to the kidney immediately below. Trace the route that an aldosterone molecule must take from the adrenal gland to the kidney, naming all major blood vessels in the order traveled.

4. People in shock commonly exhibit paleness, cool skin, tachycardia, and a weak pulse. Explain the physiological basis for each of these signs.

5. Discuss why it is advantageous to have baroreceptors in the aortic arch and carotid sinus rather than in some other location such as the common iliac arteries.

*Answers at www.mhhe.com/saladin4*

# www.mhhe.com/saladin4

*The textbook website provides a wealth of interactive study materials fully organized and integrated by chapter. You will find practice quizzes, labeling exercises, and much more that will complement your learning and understanding of anatomy and physiology. The website also includes tools designed to enhance your* **Anatomy & Physiology | REVEALED** *experience.*

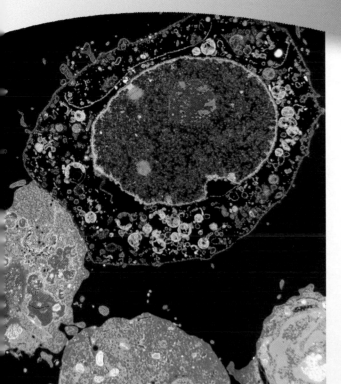

*T lymphocytes (green) attacking a cancer cell (with blue nucleus) (TEM)*

# THE LYMPHATIC AND IMMUNE SYSTEMS

## CHAPTER OUTLINE

## INSIGHTS

## Brushing Up

To understand this chapter, it is important that you understand or brush up on the following concepts:

- Endocytosis and exocytosis (p. 108)
- The general structure of glandular capsule, septa, stroma, and parenchyma (p. 176)
- The general nature of antigens and antibodies (p. 692)
- Leukocyte types (pp. 698–699)
- Mechanisms of venous blood flow (p. 773)

It may come as a surprise to know that the human body harbors about 10,000 times as many bacterial cells as human cells. But it shouldn't be surprising. After all, human homeostasis works wonderfully not only to sustain our lives, but also to provide a predictable, warm, wet, nutritious habitat for our internal guests. Indeed, it is a wonder that the body is not overrun and consumed by our microbial guests—which indeed quickly happens when one dies and homeostasis ceases.

Many of these guest microbes are beneficial or even necessary to human health, but some have the potential to cause disease if they get out of hand. Furthermore, we are constantly exposed to new invaders through the food and water we consume, the air we breathe, and even the surfaces we touch. We must have a means of keeping such would-be colonists in check.

One of these defenses was discovered in 1882 by a moody, intense, Russian zoologist, Elie Metchnikoff (1845–1916). When studying the tiny transparent larvae of starfish, he observed mobile cells wandering throughout their bodies. He thought at first that they must be digestive cells, but when he saw similar cells in sea anemones ingest nonnutritive dye particles, he thought they must play a defensive role. Metchnikoff knew that mobile cells exist in human blood and pus and quickly surround a splinter introduced through the skin, so he decided to experiment to see if the starfish cells would do the same. He impaled a starfish larva on a rose thorn, and the next morning he found the thorn crawling with cells that seemed to be trying to devour it. He later saw similar cells devouring and digesting infectious yeast in tiny transparent crustaceans called water fleas. He coined the word *phagocytosis* for this reaction and termed the wandering cells *phagocytes*—terms we still use today.

Metchnikoff showed that animals from simple sea anemones and starfish to humans actively defend themselves against disease agents. His observations marked the founding of cellular and comparative immunology, and won him the scientific respect he had so long coveted. Indeed, he shared the 1908 Nobel Prize for Physiology or Medicine with Paul Ehrlich, who had developed the theory of humoral immunity, a process also discussed in this chapter.

This chapter focuses largely on the *immune system,* which is not an organ system but rather a population of cells that inhabit all of our organs and defend the body from agents of disease. But immune cells are especially concentrated in a true organ system, the *lymphatic system.* This is a network of organs and veinlike vessels that recover tissue fluid, inspect it for disease agents, activate immune responses, and return the fluid to the bloodstream. It is with the lymphatic system that we will begin this chapter's exploration.

# The Lymphatic System

## Objectives

When you have completed this section, you should be able to

- list the functions of the lymphatic system;
- explain how lymph is formed and returned to the bloodstream;
- name the major types of cells in the lymphatic system and state their functions;
- name and describe the types of lymphatic tissues; and
- describe the form and function of red bone marrow, thymus, lymph nodes, tonsils, and spleen.

The **lymphatic system** (fig. 21.1) is composed of a network of vessels that penetrate nearly every tissue of the body, and a collection of tissues and organs that produce immune cells. The lymphatic system has three functions:

1. **Fluid recovery.** Fluid continually filters from our blood capillaries into the tissue spaces. The blood capillaries reabsorb about 85% of it, but the 15% that they do not absorb would amount, over the course of a day, to 2 to 4 L of water and one-quarter to one-half of the plasma protein. One would die of circulatory failure within hours if this water and protein were not returned to the bloodstream. One task of the lymphatic system is to reabsorb this excess and return it to the blood. Even partial interference with lymphatic drainage can lead to severe edema and sometimes even more grotesque consequences (fig. 21.2).

2. **Immunity.** As the lymphatic system recovers excess tissue fluid, it also picks up foreign cells and chemicals from the tissues. On its way back to the bloodstream, the fluid passes through lymph nodes, where immune cells stand guard against foreign matter. When they detect it, they activate a protective immune response.

3. **Lipid absorption.** In the small intestine, special lymphatic vessels called *lacteals* absorb dietary lipids that are not absorbed by the blood capillaries (see chapter 25).

The components of the lymphatic system are (1) *lymph,* the recovered fluid; (2) *lymphatic vessels,* which transport the lymph; (3) *lymphatic tissue,* composed of aggregates of lymphocytes and macrophages that populate many organs of the body; and (4) *lymphatic organs,* in which these cells are especially concentrated and which are set off from surrounding organs by connective tissue capsules.

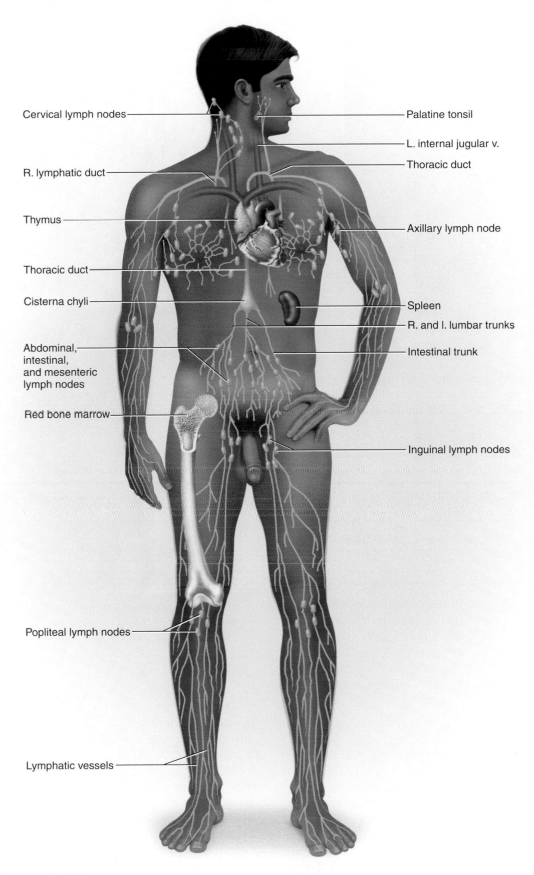

Cervical lymph nodes

Palatine tonsil

L. internal jugular v.

R. lymphatic duct

Thoracic duct

Thymus

Axillary lymph node

Thoracic duct

Cisterna chyli

Spleen

R. and l. lumbar trunks

Abdominal, intestinal, and mesenteric lymph nodes

Intestinal trunk

Red bone marrow

Inguinal lymph nodes

Popliteal lymph nodes

Lymphatic vessels

**FIGURE 21.1** The Lymphatic System.

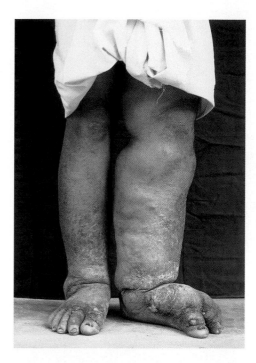

**FIGURE 21.2 Elephantiasis, a Tropical Disease Caused by Lymphatic Obstruction.** Mosquito-borne roundworms infect the lymph nodes and block the flow of lymph and recovery of tissue fluid. The resulting chronic edema leads to fibrosis and elephant-like thickening of the skin. The extremities are typically affected as shown here; the scrotum of men and breasts of women are often similarly affected.

## LYMPH AND THE LYMPHATIC VESSELS

**Lymph** is usually a clear, colorless fluid, similar to blood plasma but low in protein. It originates as tissue fluid that has been taken up by the lymphatic vessels. Its composition varies substantially from place to place. After a meal, for example, lymph draining from the small intestine has a milky appearance because of its lipid content. Lymph leaving the lymph nodes contains a large number of lymphocytes—indeed, this is the main supply of lymphocytes to the bloodstream. Lymph may also contain macrophages, hormones, bacteria, viruses, cellular debris, or even traveling cancer cells.

## Lymphatic Vessels

Lymph flows through a system of **lymphatic vessels** (lymphatics) similar to blood vessels. These begin with microscopic **lymphatic capillaries** (terminal lymphatics), which penetrate nearly every tissue of the body but are absent from the central nervous system, cartilage, cornea, bone, and bone marrow. They are closely associated with blood capillaries, but unlike them, they are closed at one end (fig. 21.3). A lymphatic capillary consists of a sac of thin endothelial cells that loosely overlap each other like the shingles of a roof. The cells are tethered to surrounding tissue by protein filaments that prevent the sac from collapsing.

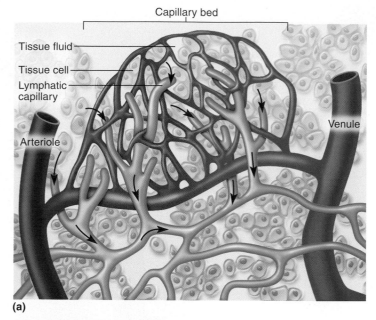

**(a)**

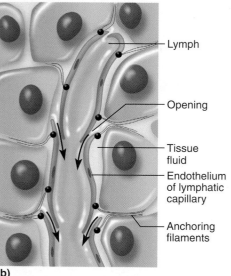

**(b)**

**FIGURE 21.3 Lymphatic Capillaries.** (a) Relationship of the lymphatic capillaries to a bed of blood capillaries. (b) Uptake of tissue fluid by a lymphatic capillary.

Unlike the endothelial cells of blood capillaries, lymphatic endothelial cells are not joined by tight junctions, nor do they have a continuous basement membrane; indeed, the gaps between them are so large that bacteria, lymphocytes, and other cells and particles can enter along with the tissue fluid. Thus, the composition of lymph arriving at a lymph node is like a report on the state of the upstream tissues.

The overlapping edges of the endothelial cells act as valvelike flaps that can open and close. When tissue fluid pressure is high, it pushes the flaps inward (open) and fluid flows into the capillary. When pressure is higher in the lymphatic capillary than in the tissue fluid, the flaps are pressed outward (closed).

## Think About It

*Contrast the structure of a lymphatic capillary with that of a continuous blood capillary. Explain why their structural difference is related to their functional difference.*

Lymphatic vessels form in the embryo by budding from the veins, so it is not surprising that the larger ones have a similar histology. They have a *tunica interna* with an endothelium and valves (fig. 21.4), a *tunica media* with elastic fibers and smooth muscle, and a thin outer *tunica externa*. Their walls are thinner and their valves are closer together than those of the veins.

As the lymphatic vessels converge along their path, they become larger and larger vessels with changing names. The route from the tissue fluid back to the bloodstream is: lymphatic capillaries → collecting vessels → six lymphatic trunks → two collecting vessels → subclavian veins. Thus, there is a continual recycling of fluid from blood to tissue fluid to lymph and back to the blood (fig. 21.5).

The lymphatic capillaries converge to form **collecting vessels.** These often travel alongside veins and arteries and share a common connective tissue sheath with them. At irregular intervals, the collecting vessels empty into lymph nodes. The lymph trickles slowly through the node, where bacteria are phagocytized and immune cells monitor the fluid for foreign antigens. It leaves the other side of the node through another collecting vessel, traveling on and often encountering additional lymph nodes before it finally returns to the bloodstream.

Eventually, the collecting vessels converge to form larger **lymphatic trunks,** each of which drains a major portion of the body. There are six principal lymphatic

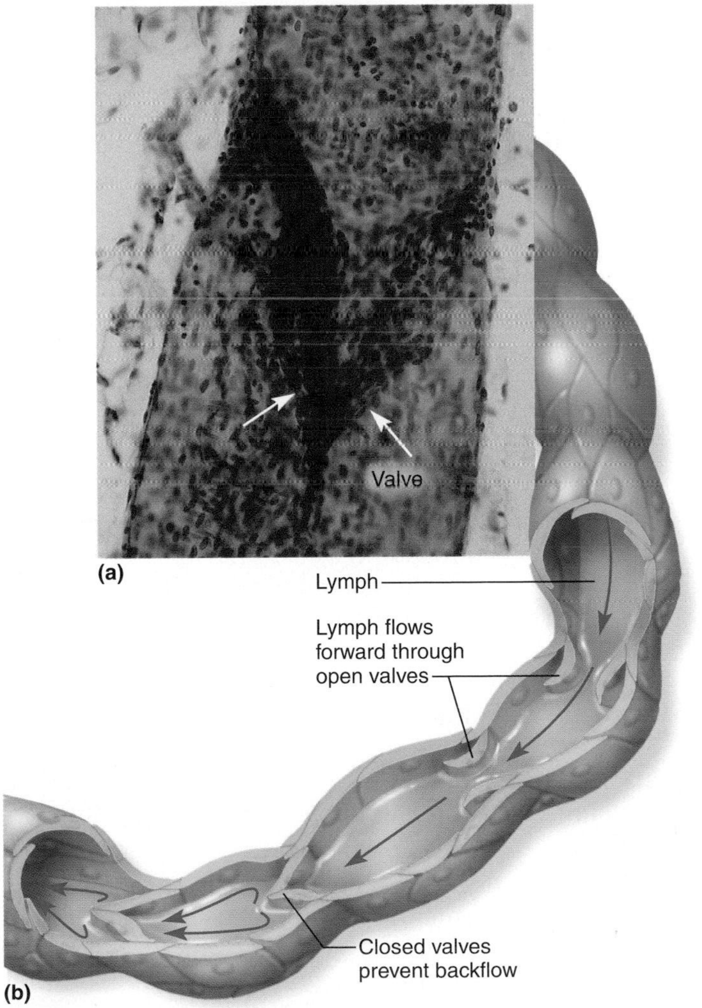

**FIGURE 21.4** **Valves in the Lymphatic Vessels.**
(a) Photograph of a lymphatic valve. (b) Operation of the valves to ensure a one-way flow of lymph.

▶ *What would be the consequence if these valves did not exist?*

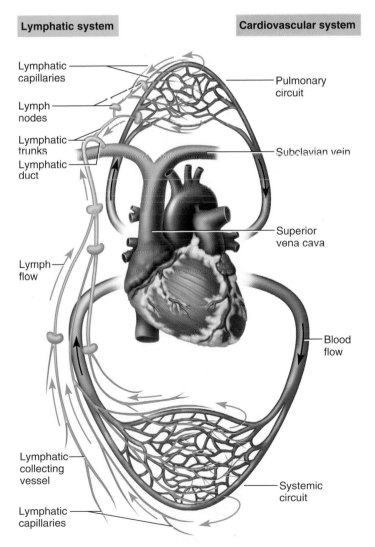

**FIGURE 21.5** **Fluid Exchange Between the Circulatory and Lymphatic Systems.** Blood capillaries lose fluid to the tissue spaces. The lymphatic system picks up excess tissue fluid and returns it to the bloodstream.

trunks whose names indicate their locations and parts of the body they drain: the *jugular, subclavian, bronchomediastinal, intercostal, intestinal,* and *lumbar trunks.* The lumbar trunk drains not only the lumbar region but also the lower limbs.

The lymphatic trunks converge to form two **collecting ducts,** the largest of the lymphatic vessels (fig. 21.6):

1. The **right lymphatic duct** is formed by the convergence of the right jugular, subclavian, and bronchomediastinal trunks in the right thoracic cavity. It receives lymphatic drainage from the right arm and right side of the thorax and head and empties into the right subclavian vein.

2. The **thoracic duct,** on the left, is larger and longer. It begins just below the diaphragm anterior to the vertebral column at the level of the second lumbar vertebra. Here, the two lumbar trunks and the intestinal trunk join and form a prominent sac called the **cisterna chyli** (sis-TUR-nuh KY-lye), named for the large amount of *chyle* (fatty intestinal lymph) that it collects after a meal. The thoracic duct then passes through the diaphragm with the aorta and ascends the mediastinum, adjacent to the vertebral column. As it passes through the thorax, it receives additional lymph from the left bronchomediastinal, left subclavian, and left jugular trunks, then empties into the left subclavian vein. Collectively, this duct therefore drains all of the body below the diaphragm, and the left upper limb and left side of the head, neck, and thorax.

## Flow of Lymph

Lymph flows under forces similar to those that govern venous return, except that the lymphatic system has no pump like the heart, and it flows at even lower pressure and speed than venous blood. The primary mechanism of flow is rhythmic contractions of the lymphatic vessels themselves, which contract when the flowing lymph stretches them. The valves of lymphatic vessels, like those of veins, prevent the fluid from flowing backward. Lymph flow is also produced by skeletal muscles squeezing the lymphatic vessels, like the skeletal muscle pump that moves venous blood. Since lymphatic vessels are often wrapped with an artery in a common connective tissue sheath, arterial pulsation may also rhythmically squeeze the lymphatic vessels and contribute to lymph flow. A thoracic (respiratory) pump promotes the flow of lymph from the abdominal to the thoracic cavity as one inhales, just as it does in venous return. Finally, at the point where the collecting ducts empty into the subclavian veins, the rapidly flowing bloodstream draws the lymph into it. Considering these mechanisms of lymph flow, it should be apparent that physical exercise significantly increases the rate of lymphatic return.

**Think About It**

*Why does it make more functional sense for the collecting ducts to connect to the subclavian veins than it would for them to connect to the subclavian arteries?*

## LYMPHATIC CELLS

Another component of the lymphatic system is the lymphatic tissues, which range from the loosely scattered cells in the mucous membranes of the respiratory, digestive, urinary, and reproductive tracts, to compact cell populations encapsulated in lymphatic organs. These tissues are composed of a variety of lymphocytes and other cells with various roles in defense and immunity:

1. **Natural killer (NK) cells** are large lymphocytes that attack and destroy bacteria, transplanted tissue cells, and *host cells* (cell's of one's own body) that have either become infected with viruses or turned cancerous. They are responsible for a mode of defense, discussed later, called *immune surveillance.*

2. **T lymphocytes** (T cells) are lymphocytes that mature in the thymus and later depend on thymic hormones; the *T* stands for *thymus-dependent.* There are several subclasses of T cells that will be introduced later.

3. **B lymphocytes** (B cells) are lymphocytes that differentiate into *plasma cells*—connective tissue cells that secrete the antibodies of the immune system. They are named for an organ in chickens (the *bursa of Fabricius*[1]) in which they were first discovered. However, you may find it more helpful to think of *B* for *bone marrow,* the site where these cells mature.

4. **Macrophages** are very large, avidly phagocytotic cells of the connective tissues. They develop from monocytes that have emigrated from the bloodstream. They phagocytize and destroy tissue debris, dead neutrophils, bacteria, and other foreign matter (fig. 21.7). They also process foreign matter and display antigenic fragments of it to certain T cells, thus alerting the immune system to the presence of an enemy. Macrophages and other cells that do this are collectively called **antigen-presenting cells (APCs).**

5. **Dendritic cells** are branched, mobile APCs found in the epidermis, mucous membranes, and lymphatic organs. (In the skin, they are often called *Langerhans*[2] *cells.*) They play an important role in alerting

---

[1]Hieronymus Fabricius (Girolamo Fabrizzi) (1537–1619), Italian anatomist
[2]Theodor Langerhans (1839–1915), German pathologist

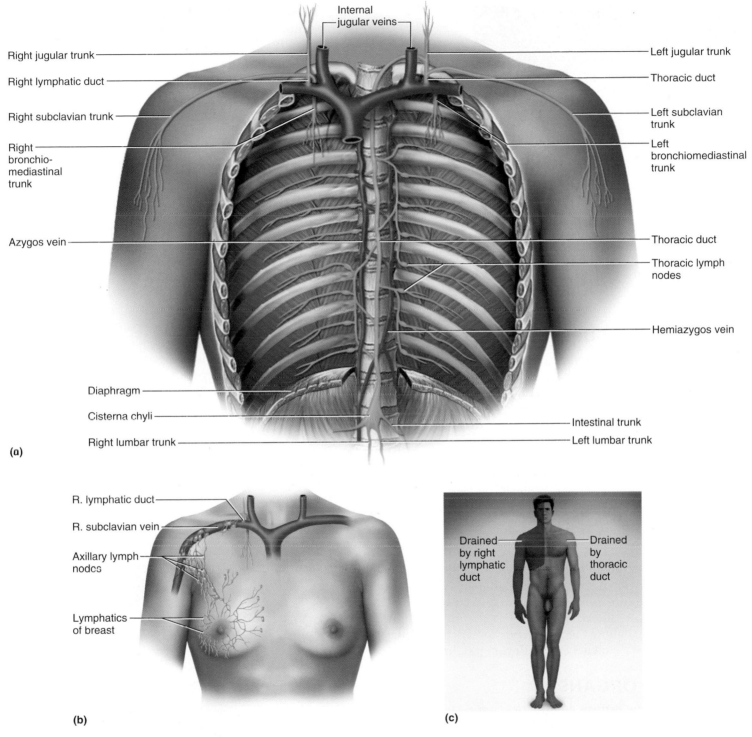

Internal jugular veins

Right jugular trunk

Right lymphatic duct

Right subclavian trunk

Right bronchio-mediastinal trunk

Azygos vein

Diaphragm

Cisterna chyli

Right lumbar trunk

Left jugular trunk

Thoracic duct

Left subclavian trunk

Left bronchiomediastinal trunk

Thoracic duct

Thoracic lymph nodes

Hemiazygos vein

Intestinal trunk

Left lumbar trunk

(a)

R. lymphatic duct

R. subclavian vein

Axillary lymph nodes

Lymphatics of breast

(b)

Drained by right lymphatic duct

Drained by thoracic duct

(c)

**FIGURE 21.6** **Lymphatic Drainage of the Thoracic Region.**   (a) Drainage of lymphatics into the subclavian veins. (b) Drainage of the right mammary and axillary regions. (c) Regions of the body drained by the right lymphatic duct and thoracic duct.

the immune system to pathogens that have breached the body surfaces. They engulf foreign matter by receptor-mediated endocytosis rather than phagocytosis, but otherwise function like macrophages and are included in the *macrophage system.*

6. **Reticular cells** are branched stationary cells that contribute to the stroma (connective tissue framework) of the lymphatic organs and act as APCs in the thymus (see fig. 21.10). (They should not be confused with reticular *fibers,* which are fine, branched collagen fibers common in lymphatic organs.)

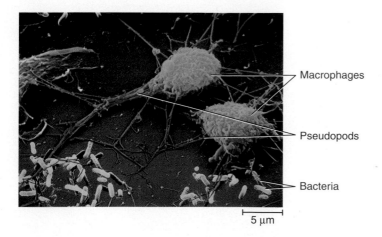

**FIGURE 21.7  Macrophages Phagocytizing Bacteria.**
Filamentous pseudopods of the macrophages snare the rod-shaped bacteria and draw them to the cell surface, where they are phagocytized.

## LYMPHATIC TISSUES

**Lymphatic (lymphoid) tissues** are aggregations of lymphocytes in the connective tissues of mucous membranes and various organs. The simplest form is **diffuse lymphatic tissue,** in which the lymphocytes are scattered rather than densely clustered. It is particularly prevalent in body passages that are open to the exterior—the respiratory, digestive, urinary, and reproductive tracts—where it is called **mucosa-associated lymphatic tissue (MALT).** (In the respiratory and digestive tracts, it is sometimes called bronchus-associated and gut-associated lymphatic tissue, BALT and GALT, respectively.)

In some places, lymphocytes and macrophages congregate in dense masses called **lymphatic nodules** (follicles), which come and go as pathogens invade the tissues and the immune system answers the challenge. Abundant lymphatic nodules are, however, a relatively constant feature of the lymph nodes (see fig. 21.12), tonsils, and appendix. In the ileum, the distal portion of the small intestine, they form clusters called **Peyer[3] patches.**

## LYMPHATIC ORGANS

In contrast to the diffuse lymphatic tissue, **lymphatic (lymphoid) organs** have well-defined anatomical sites and at least partial connective tissue capsules that separate the lymphatic tissue from neighboring tissues. These organs include the red bone marrow, thymus, lymph nodes, tonsils, and spleen. The red bone marrow and thymus are regarded as *primary lymphatic organs* because they are the sites where T and B lymphocytes, respectively, become *immunocompetent*—that is, able to recognize and respond to antigens. The lymph nodes, tonsils,

[3]Johann Conrad Peyer (1653–1712), Swiss anatomist

and spleen are called *secondary lymphatic organs* because they are populated with immunocompetent lymphocytes only after the cells have matured in the primary lymphatic organs.

## Red Bone Marrow

As discussed in chapter 7, there are two kinds of bone marrow: red and yellow. Red bone marrow is involved in hemopoiesis (blood formation) and immunity; yellow bone marrow can be disregarded for our present purposes. In children, red bone marrow occupies the medullary spaces of nearly the entire skeleton. In adults, it is limited to parts of the axial skeleton and the proximal heads of the humerus and femur. Red bone marrow is an important supplier of lymphocytes to the immune system. Its role in the life history of lymphocytes is described later.

Red bone marrow is a soft, loosely organized, highly vascular material, separated from osseous tissue by the endosteum of the bone. It produces all classes of formed elements of the blood; its red color comes from the abundance of erythrocytes. Numerous small arteries enter *nutrient foramina* on the bone surface, penetrate the bone, and empty into large *sinusoids* (45 to 80 μm wide) in the marrow (fig. 21.8). The sinusoids drain into a

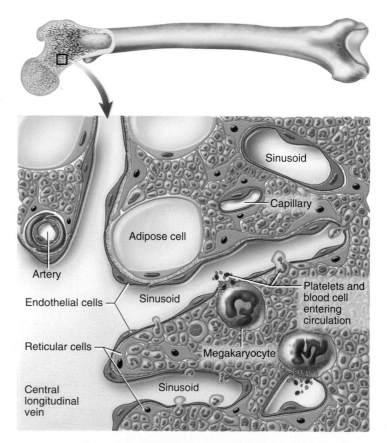

**FIGURE 21.8  Red Bone Marrow.** The formed elements of blood squeeze through the endothelial cells into the sinusoids, which converge on a central longitudinal vein at the lower left.

*central longitudinal vein* that exits the bone via the same route that the arteries entered it. The sinusoids are lined by endothelial cells, like other blood vessels, and are surrounded by reticular cells and reticular fibers. The reticular cells secrete colony-stimulating factors that induce the formation of various leukocyte types. In the long bones of the limbs, aging reticular cells accumulate fat and transform into adipose cells, eventually replacing red bone marrow with yellow bone marrow.

The spaces between the sinusoids are occupied by *islands* or *cords* of hemopoietic cells, composed of macrophages and blood cells in all stages of development. The macrophages destroy malformed blood cells and the nuclei discarded by developing erythrocytes. As blood cells mature, they push their way through the reticular and endothelial cells to enter the sinus and flow away in the bloodstream.

## Thymus

The **thymus** is a member of both the lymphatic and endocrine systems. It houses developing lymphocytes and secretes hormones that regulate their later activity. It is located between the sternum and aortic arch in the superior mediastinum. The thymus is very large in the fetus and grows slightly during childhood, when it is most active. After age 14, however, it begins to undergo involution (shrinkage) so that it is quite small in adults (fig. 21.9). In the elderly, the thymus is replaced almost entirely by fibrous and fatty tissue and is barely distinguishable from the surrounding tissues.

The fibrous capsule of the thymus gives off trabeculae that divide the gland into several angular lobules. Each lobule has a dense, dark-staining *cortex* and a lighter *medulla* populated by T lymphocytes (fig. 21.10). **Reticular epithelial cells** seal off the cortex from the medulla and surround blood vessels and lymphocyte clusters in the cortex. They thereby form a *blood–thymus barrier* that isolates developing lymphocytes from blood-borne antigens. After developing in the cortex, the T cells migrate to the medulla, where they spend another 3 weeks. There is no blood–thymus barrier in the medulla; mature T cells enter blood or lymphatic vessels here and leave the thymus. In the medulla, the reticular epithelial cells form whorls called *thymic (Hassall[4]) corpuscles,* useful for identifying the thymus histologically but of no known function.

Besides forming the blood–thymus barrier, reticular epithelial cells produce several signaling molecules that promote the development and action of T cells, including *thymosin, thymopoietin, thymulin, interleukins,* and an *interferon.* If the thymus is removed from newborn mammals, they waste away and never develop immunity. Other lymphatic organs also seem to depend on thymic hormones or T cells and develop poorly in thymectomized animals. The role of the thymus in T cell development is discussed later.

---

[4]Arthur H. Hassall (1817–94), British chemist and physician

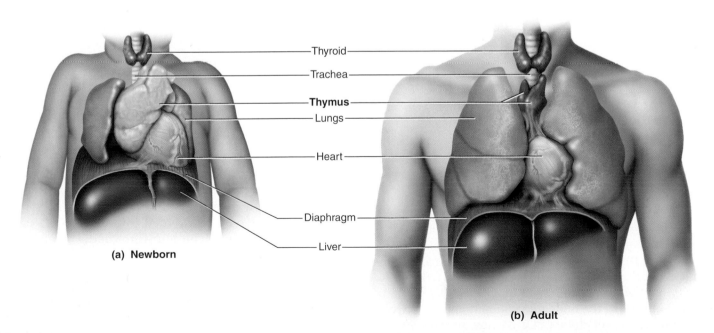

**FIGURE 21.9 Involution of the Thymus.** (a) The thymus is extremely large in a newborn infant. (b) The adult thymus is atrophied and often barely noticeable.

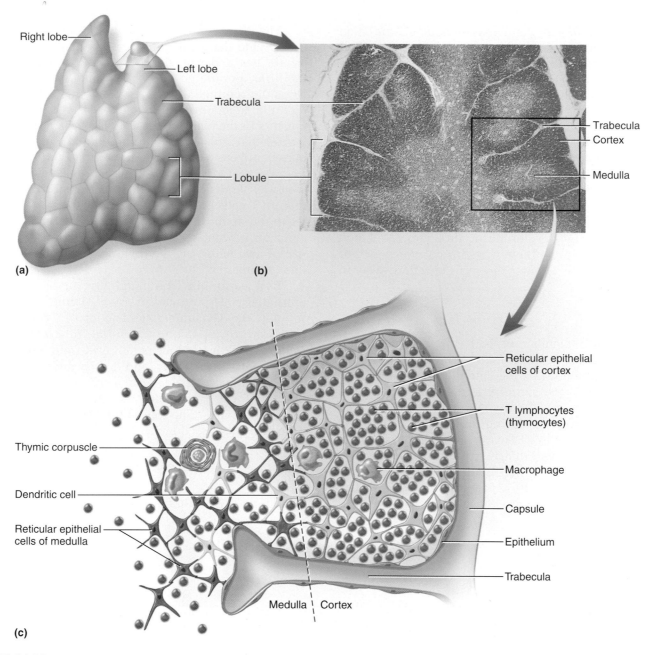

**FIGURE 21.10   Anatomy of the Thymus.** (a) Gross anatomy. (b) Tissue section. (c) Cellular architecture of a lobule. The reticular epithelial cells are interconnected to form a blood–thymus barrier.

## Lymph Nodes

**Lymph nodes** (fig. 21.11) are the most numerous lymphatic organs, numbering in the hundreds. They serve two functions: to cleanse the lymph and act as a site of T and B cell activation. A lymph node is an elongated or bean-shaped structure, usually less than 3 cm long, often with an indentation called the *hilum* on one side (fig. 21.12). It is enclosed in a fibrous capsule with trabeculae that partially divide the interior of the node into compartments. Between the capsule and parenchyma is a narrow, relatively clear space called the *subcapsular sinus,* which contains reticular fibers, macrophages, and dendritic cells. Deep to this,

the gland consists mainly of a stroma of reticular connective tissue (reticular fibers and reticular cells) and a parenchyma of lymphocytes and antigen-presenting cells.

The parenchyma is divided into an outer C-shaped **cortex** that encircles about four-fifths of the organ, and an inner **medulla** that extends to the surface at the hilum. The cortex consists mainly of ovoid to conical lymphatic nodules. When the lymph node is fighting a pathogen, these nodules acquire light-staining **germinal centers** where B cells multiply and differentiate into plasma cells. The medulla consists largely of a branching network of *medullary cords* composed of lymphocytes, plasma cells, macrophages, reticular cells, and reticular

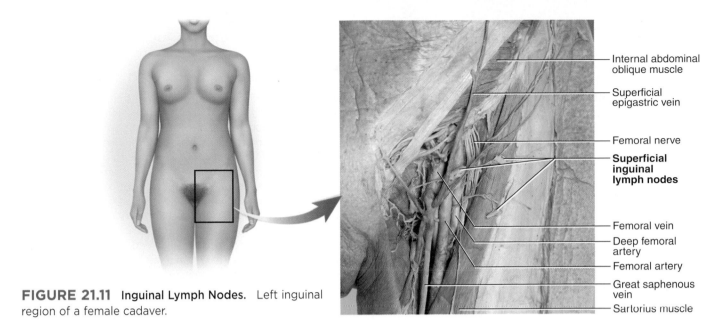

**FIGURE 21.11 Inguinal Lymph Nodes.** Left inguinal region of a female cadaver.

Internal abdominal oblique muscle
Superficial epigastric vein
Femoral nerve
**Superficial inguinal lymph nodes**
Femoral vein
Deep femoral artery
Femoral artery
Great saphenous vein
Sartorius muscle

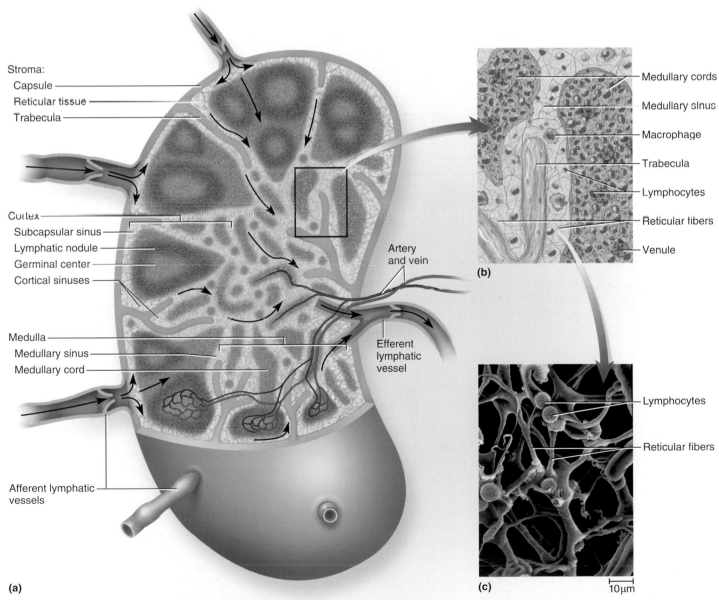

Stroma:
Capsule
Reticular tissue
Trabecula

Cortex
Subcapsular sinus
Lymphatic nodule
Germinal center
Cortical sinuses

Medulla
Medullary sinus
Medullary cord

Afferent lymphatic vessels

Artery and vein

Efferent lymphatic vessel

Medullary cords
Medullary sinus
Macrophage
Trabecula
Lymphocytes
Reticular fibers
Venule

**(b)**

Lymphocytes
Reticular fibers

**(a)**

**(c)**

10 μm

**FIGURE 21.12 Anatomy of a Lymph Node.** (a) Partially bisected lymph node showing pathway of lymph flow. (b) Detail of the boxed region in part (a). (c) Reticular fiber stroma and immune cells in a medullary sinus (SEM).

fibers. The cortex and medulla also contain lymph-filled sinuses continuous with the subcapsular sinus.

Several **afferent lymphatic vessels** lead into the node along its convex surface. Lymph flows from these vessels into the subcapsular sinus, percolates slowly through the sinuses of the cortex and medulla, and leaves the node through one to three **efferent lymphatic vessels** that emerge from the hilum. No other lymphatic organs have afferent lymphatic vessels; lymph nodes are the only organs that filter lymph as it flows along its course. The lymph node is a bottleneck that slows down lymph flow and allows time for cleansing it of foreign matter. The macrophages and reticular cells of the sinuses remove about 99% of the impurities before the lymph leaves the node. On its way to the bloodstream, lymph flows through one lymph node after another and thus becomes quite thoroughly cleansed of most impurities.

Blood vessels also penetrate the hilum of a lymph node. Arteries follow the medullary cords and give rise to capillary beds in the medulla and cortex. In the *deep cortex* near the junction with the medulla, lymphocytes can emigrate from the bloodstream into the parenchyma of the node. Most lymphocytes in the deep cortex are T cells.

Lymph nodes are widespread but especially concentrated in the following locations (see fig. 21.1):

- *Cervical lymph nodes* occur in deep and superficial groups in the neck, and monitor lymph coming from the head and neck.

- *Axillary lymph nodes* are concentrated in the armpit (axilla) and receive lymph from the upper limb and the female breast (fig. 21.6b).

- *Thoracic lymph nodes* occur in the thoracic cavity and receive lymph from the lungs, airway, and mediastinum.

- *Abdominal lymph nodes* monitor lymph from the urinary and reproductive systems.

- *Intestinal* and *mesenteric lymph nodes* monitor lymph from the digestive tract.

- *Inguinal lymph nodes* occur in the groin (see fig. 21.11) and receive lymph from the entire lower limb.

- *Popliteal lymph nodes* occur at the back of the knee and receive lymph from the leg proper.

When a lymph node is under challenge from a foreign antigen, it may become swollen and painful to the touch—a condition called **lymphadenitis**[5] (lim-FAD-en-EYE-tis). Physicians routinely palpate the accessible lymph nodes of the cervical, axillary, and inguinal regions for swelling. The collective term for all lymph node diseases is **lymphadenopathy**[6] (lim-FAD-eh-NOP-a-thee). Lymph nodes are common sites of metastatic cancer (see Insight 21.1).

[5]*lymph* = water + *adeno* = gland + *itis* = inflammation
[6]*lymph* = water + *adeno* = gland + *pathy* = disease

### Lymph Nodes and Metastatic Cancer

*Metastasis* is a phenomenon in which cancerous cells break free of the original *primary tumor,* travel to other sites in the body, and establish new tumors. Because of the high permeability of lymphatic capillaries, metastasizing cancer cells easily enter them and travel in the lymph. They tend to lodge in the first lymph node they encounter and grow there, eventually destroying the node. Cancerous lymph nodes are swollen but relatively firm and usually painless. Cancer of a lymph node is called *lymphoma.*

Once a tumor is well established in one node, cells may emigrate from there and travel to the next. However, if the metastasis is detected early enough, cancer can sometimes be eradicated by removing not only the primary tumor, but also the nearest lymph nodes downstream from that point. For example, breast cancer is often treated with a combination of lumpectomy or mastectomy along with removal of the nearby axillary lymph nodes.

## Tonsils

The **tonsils** are patches of lymphatic tissue located at the entrance to the pharynx, where they guard against ingested and inhaled pathogens. Each is covered by an epithelium and has deep pits called **tonsillar crypts** lined by lymphatic nodules (fig. 21.13). The crypts often contain food debris, dead leukocytes, bacteria, and antigenic chemicals. Below the crypts, the tonsils are partially separated from underlying connective tissue by an incomplete fibrous capsule.

There are three main sets of tonsils: (1) a single medial **pharyngeal tonsil** (adenoids) on the wall of the pharynx just behind the nasal cavity, (2) a pair of **palatine tonsils** at the posterior margin of the oral cavity, and (3) numerous **lingual tonsils,** each with a single crypt, concentrated in a patch on each side of the root of the tongue (see fig. 25.5a, p. 959).

The palatine tonsils are the largest and most often infected. *Tonsillitis* is an acute inflammation of the palatine tonsils, usually caused by a *Streptococcus* infection. Their surgical removal, called *tonsillectomy,* used to be one of the most common surgical procedures performed on children, but it is done less often today. Tonsillitis is now usually treated with antibiotics.

## Spleen

The **spleen** is the body's largest lymphatic organ. It is located in the left hypochondriac region, just inferior to the diaphragm and dorsolateral to the stomach (fig. 21.14; see also fig. A.15, p. 44). It is protected by ribs 10 through 12. The spleen fits snugly between the diaphragm, stomach, and kidney and has indentations called the *gastric area* and *renal area* where it presses against these adjacent viscera. It has a medial hilum penetrated by the splenic artery, splenic vein, and lymphatic vessels.

The parenchyma exhibits two types of tissue named for their appearance in fresh specimens (not in stained sections): **red pulp,** which consists of sinuses gorged

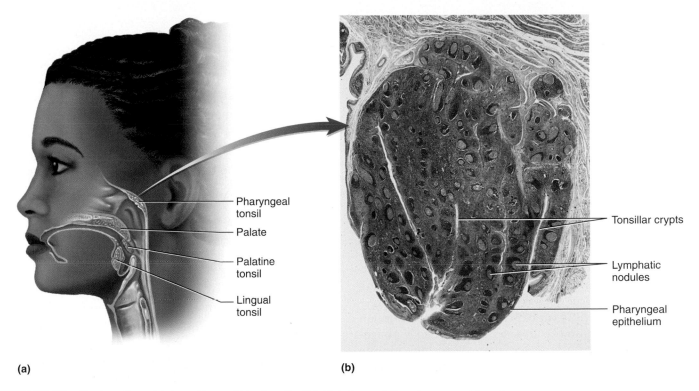

Pharyngeal tonsil
Palate
Palatine tonsil
Lingual tonsil

Tonsillar crypts
Lymphatic nodules
Pharyngeal epithelium

(a)

(b)

**FIGURE 21.13** **The Tonsils.** (a) Locations of the tonsils. (b) Histology of the pharyngeal tonsil.

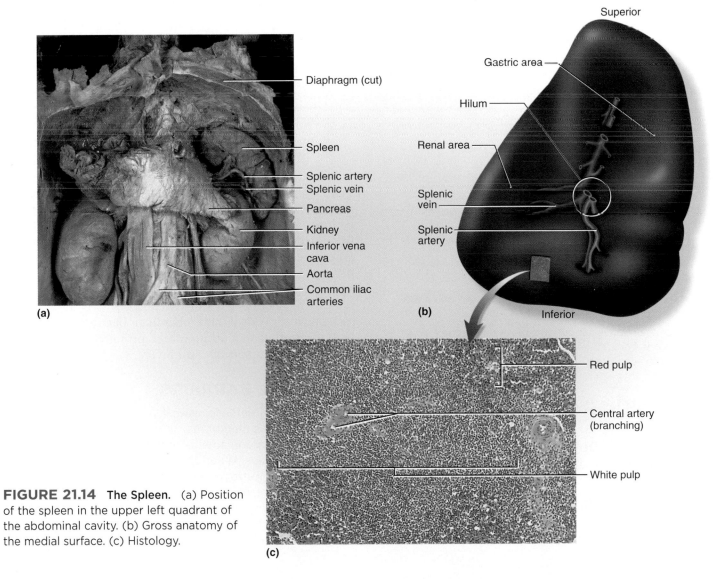

Diaphragm (cut)

Spleen

Splenic artery
Splenic vein

Pancreas

Kidney

Inferior vena cava

Aorta

Common iliac arteries

(a)

Superior

Gastric area

Hilum

Renal area

Splenic vein

Splenic artery

(b)

Inferior

Red pulp

Central artery (branching)

White pulp

(c)

**FIGURE 21.14** **The Spleen.** (a) Position of the spleen in the upper left quadrant of the abdominal cavity. (b) Gross anatomy of the medial surface. (c) Histology.

with concentrated erythrocytes, and **white pulp,** which consists of lymphocytes and macrophages aggregated like sleeves along small branches of the splenic artery. In tissue sections, white pulp appears as an ovoid mass of lymphocytes with an arteriole passing through it. However, it is important to bear in mind that the three-dimensional shape is not egglike but cylindrical.

These two tissue types reflect the multiple functions of the spleen. It produces blood cells in the fetus and may resume this role in adults in the event of extreme anemia. Lymphocytes and macrophages of the white pulp monitor the blood for foreign antigens, much like the lymph nodes do the lymph. The splenic blood capillaries are very permeable; they allow RBCs to leave the bloodstream, accumulate in the sinuses of the red pulp, and reenter the bloodstream later. The spleen is an "erythrocyte graveyard"—old, fragile RBCs rupture as they squeeze through the capillary walls into the sinuses. Macrophages phagocytize their remains, just as they dispose of blood-borne bacteria and other cellular debris. The spleen also helps to stabilize blood volume by transferring excess plasma from the bloodstream into the lymphatic system.

The spleen is highly vascular and vulnerable to trauma and infection. A ruptured spleen can hemorrhage fatally, but is difficult to repair surgically. Therefore a common procedure in such cases is its removal, *splenectomy.* A person can live without a spleen, but is somewhat more vulnerable to infections.

## Before You Go On

*Answer the following questions to test your understanding of the preceding section:*

1. *List the primary functions of the lymphatic system. What do you think would be the most noticeable effect of clamping the right lymphatic duct closed?*

2. *How does fluid get into the lymphatic system? What prevents it from draining back out?*

3. *What do NK, T, and B cells have in common? How do their functions differ?*

4. *List five major cell types of lymphatic tissues and state the function of each.*

5. *Predict the relative seriousness of removing the following organs from a 2-year-old child: (a) a lymph node, (b) the spleen, (c) the thymus, (d) the palatine tonsils.*

# Nonspecific Resistance

### Objectives
When you have completed this section, you should be able to

- identify the body's three lines of defense against pathogens;

- contrast nonspecific resistance with immunity;

- describe the defensive functions of each kind of leukocyte;

- describe the role of the complement system in resistance and immunity;

- describe the process of inflammation and explain what accounts for its cardinal signs; and

- describe the body's other nonspecific defenses.

For all living organisms, one of the greatest survival challenges is coping with **pathogens**[7]—environmental agents capable of producing disease. Pathogens include infectious organisms, toxic chemicals, and radiation. The human body has three lines of defense against pathogens:

1. The **first line of defense** consists of external barriers, notably the skin and mucous membranes, which are impenetrable to most of the pathogens that daily assault us.

2. The **second line of defense** consists of several nonspecific defense mechanisms against pathogens that break through the skin or mucous membranes. These include leukocytes and macrophages, antimicrobial proteins, immune surveillance, inflammation, and fever. These mechanisms are present from birth, are broadly effective against a wide range of pathogens, and work even against pathogens to which the body has never before been exposed.

3. The **third line of defense** is the immune system, which not only defeats a pathogen but leaves the body with a "memory" of it, enabling us to defeat it so quickly in future encounters that the pathogen causes no illness.

The first two mechanisms are called **nonspecific resistance** because they guard equally against a broad range of pathogens and their effectiveness does not depend on prior exposure to a pathogen. Immunity is called a **specific defense** because it results from prior exposure to a pathogen and usually provides future protection only against that particular pathogen.

In this section we study mechanisms of nonspecific resistance—physical and chemical barriers, leukocytes and macrophages, antimicrobial proteins, immune surveillance, inflammation, and fever.

## EXTERNAL BARRIERS

Our first line of defense against pathogens consists of the skin and mucous membranes, which make it mechanically difficult for microorganisms to enter the body and spread through its tissues. When the skin is broken by a scrape or animal bite or destroyed by a burn, one of the most urgent treatment concerns is the prevention of infection. This attests to the importance of intact skin as a barrier. The

---

[7]*patho* = disease, suffering + *gen* = producing

skin surface is composed mainly of keratin, a tough protein that few pathogens can penetrate. Furthermore, the surface is hostile to microbial reproduction. With exceptions such as the axillary and pubic areas, it is too dry and poor in nutrients to support much microbial growth. The skin is also coated with antimicrobial chemicals such as defensins and lactic acid. *Defensins* are peptides that kill microbes by creating holes in their membranes; they are produced by neutrophils and other cells and are found on the skin surface. The skin is also coated with a thin film of lactic acid (the *acid mantle*) from sweat, which also inhibits bacterial growth.

The digestive, respiratory, urinary, and reproductive tracts are open to the exterior, making them vulnerable to invasion, but they are protected by mucous membranes. Mucus physically ensnares microbes. Microbes trapped in the respiratory mucus are moved by cilia to the pharynx, swallowed, and destroyed by stomach acid. Microbes also are flushed from the upper digestive tract by saliva and from the lower urinary tract by urine. Mucus, tears, and saliva also contain **lysozyme,** an enzyme that destroys bacteria by dissolving their cell walls.

Beneath the epithelia of the skin and mucous membranes, there is a layer of areolar tissue. The ground substance of this tissue contains a giant glycosaminoglycan called **hyaluronic acid,** which gives it a viscous consistency. It is normally difficult for microbes to migrate through this sticky tissue gel. Some organisms overcome this obstacle, however, by producing an enzyme called *hyaluronidase,* which breaks it down to a thinner consistency that is more easily penetrated. Hyaluronidase occurs in some snake venoms and bacterial toxins and is produced by some parasitic protozoans to facilitate their invasion of the connective tissues.

## LEUKOCYTES AND MACROPHAGES

If microbes get past the physical barrier of the skin and mucous membranes, they are attacked by **phagocytes** (phagocytic cells) that have a voracious appetite for foreign matter. Leukocytes and macrophages play especially important roles in both nonspecific defense and specific immunity, and therefore, in both the second and third lines of defense.

### Leukocytes

The five types of leukocytes are described in table 18.7 (pp. 698–699). We will now examine their contributions to resistance and immunity in more detail.

1. **Neutrophils** spend most of their lives wandering in the connective tissues killing bacteria. One of their methods is simple phagocytosis and digestion. The other is a more complex process that produces a cloud of bactericidal chemicals. When a neutrophil detects bacteria in the immediate area, its lysosomes migrate to the cell surface and *degranulate,* or discharge their enzymes into the tissue fluid. Here, the enzymes catalyze a reaction called the **respiratory burst:** The neutrophil rapidly absorbs oxygen and reduces it to highly toxic *superoxide anions* ($O_2 \bullet^-$), and superoxide then reacts with $H^+$ to form hydrogen peroxide ($H_2O_2$). Another lysosomal enzyme produces hypochlorite (HClO), the active ingredient in chlorine bleach, from chloride ions in the tissue fluid. Superoxide, hydrogen peroxide, and hypochlorite are highly toxic; they form a chemical **killing zone** around the neutrophil that destroys far more bacteria than the neutrophil can destroy by phagocytosis alone. Unfortunately for the neutrophil, the killing zone is also deadly to it, and it dies in the course of the attack. These potent oxidizing agents can also damage connective tissues and sometimes contribute to rheumatoid arthritis.

2. **Eosinophils** are found especially in the mucous membranes, standing guard against parasites, allergens (allergy-causing antigens), and other pathogens. They also become especially concentrated at sites of allergy, inflammation, or parasitic infection. They help to kill parasites such as tapeworms and roundworms by producing superoxide, hydrogen peroxide, and various toxic proteins including a neurotoxin. They promote the action of basophils and mast cells (see next paragraph). They phagocytize and degrade antigen–antibody complexes. Finally, they secrete enzymes that degrade and limit the action of histamine and other inflammatory chemicals that, unchecked, could cause tissue damage.

3. **Basophils** aid the mobility and action of other leukocytes by secreting two chemicals: the vasodilator histamine, which increases blood flow and speeds the delivery of leukocytes to the area, and the anticoagulant heparin, which inhibits the formation of clots that would impede the mobility of other leukocytes. These substances are also produced by **mast cells,** a type of connective tissue cell very similar to basophils. Eosinophils promote basophil and mast cell action by stimulating them to release these secretions.

4. **Lymphocytes** all look more or less alike in blood films, but there are several functional types. Three basic categories have already been mentioned: natural killer (NK) cells, T cells, and B cells. In the circulating blood, about 80% of the lymphocytes are T cells, 15% B cells, and 5% NK and stem cells. The roles of these lymphocyte types are too diverse for easy generalizations here, but are described in later sections on immune surveillance and specific immunity.

5. **Monocytes** are leukocytes that emigrate from the blood into the connective tissues and transform into macrophages. All of the body's avidly phagocytotic cells except leukocytes are called the **macrophage**

**(lymphoid-macrophage) system.** Dendritic cells are included in this system even though they employ receptor-mediated endocytosis instead of phagocytosis to internalize foreign matter. Some of these phagocytes are wandering cells that actively seek pathogens, whereas reticular cells and others are fixed in place and phagocytize only those pathogens that come to them—although they are strategically positioned for this to occur. Macrophages are widely distributed in the loose connective tissues (where they are sometimes called *histiocytes*), but there are also specialized forms with more specific localities:

- **dendritic cells** of the epidermis, oral mucosa, esophagus, vagina, and lymphatic organs;
- **microglia** in the central nervous system (see chapter 14);
- **alveolar macrophages** in the lungs (see chapter 22); and
- **hepatic macrophages** in the liver (see chapter 25).

# ANTIMICROBIAL PROTEINS

Multiple types of proteins inhibit microbial reproduction and provide short-term, nonspecific resistance to pathogenic bacteria and viruses. Two families of antimicrobial proteins are described here—*interferons* and the *complement system*—and others are described in the ensuing discussion of immune surveillance.

## Interferons

When certain cells (especially leukocytes) are infected with viruses, they secrete proteins called **interferons.** These act to alert neighboring cells and prevent them from becoming infected. Interferons bind to surface receptors on a cell and activate second-messenger systems within. The alerted cell then synthesizes various proteins that defend it from infection by such means as breaking down viral genes or preventing their replication. Interferons also activate NK cells and macrophages, which destroy infected cells before they can liberate a swarm of newly replicated viruses. Interferons confer resistance not only to viruses but also to cancer, as the activated NK cells destroy malignant cells.

## Complement System

The **complement system** is a group of 30 or more globulins that make powerful contributions to both nonspecific resistance and specific immunity. Immunology pioneer Paul Ehrlich (1854–1915) named it *complement* because it "completes the action of antibody," and indeed this is the principal means of pathogen destruction in antibody-mediated immunity. Since Ehrlich's time, however, the complement system has also been found important in nonspecific defense.

Complement proteins are synthesized mainly by the liver. They circulate in the blood in inactive form and are activated in the presence of pathogens. In their inactive form, the proteins are named with the letter *C* and a number, such as C3. Activation splits them into fragments, which are further identified by lowercase letters (C3a and C3b, for example).

Activated complement brings about four methods of pathogen destruction: inflammation, immune clearance, phagocytosis, and cytolysis. We will examine the pathways of complement activation with a view to understanding how each of these goals is achieved. There are three such routes (fig. 21.15): the classical, alternative, and lectin pathways.

The **classical pathway** requires an antibody molecule to get it started; thus it is part of our specific immunity. The antibody binds to an antigen on the surface of a pathogenic organism, forming an antigen–antibody (Ag–Ab) complex. This changes the antibody's shape, exposing a pair of *complement-binding sites* (see fig. 21.27a). Binding of the first complement (C1) to these sites sets off a reaction cascade. Like the cascade of blood-clotting reactions, each step generates an enzyme that catalyzes the production of many more molecules at the next step; each step is an amplifying process, so many molecules of product result from a small beginning. In the classical pathway, the cascade is called **complement fixation,** since it results in the attachment of a chain of complement proteins to the antibody.

The alternative and lectin pathways do not require an antibody and thus belong to our nonspecific defenses. Complement C3 slowly and spontaneously breaks down in the blood into C3a and C3b. In the **alternative pathway,** C3b binds directly to targets such as human tumor cells, viruses, bacteria, and yeasts. This, too, triggers a reaction cascade—this time with an *autocatalytic effect* in which C3b eventually leads to the accelerated splitting of more C3 and production of even more C3b.

**Lectins** are plasma proteins that bind to carbohydrates. In the **lectin pathway,** a lectin binds to certain sugars of a microbial cell surface and sets off yet another reaction cascade leading to C3b production.

As we can see (fig. 21.15), the splitting of C3 into C3a and C3b is an intersection where all three pathways converge. These two C3 fragments then produce, directly or indirectly, the end results of the complement system:

1. **Inflammation.** C3a stimulates mast cells and basophils to secrete histamine and other inflammatory chemicals. It also activates and attracts neutrophils and macrophages, the two key cellular agents of pathogen destruction in inflammation. The exact roles of these chemicals and cells are explained in the section on inflammation to follow.

2. **Immune clearance.** C3b binds Ag–Ab complexes to red blood cells. As these RBCs circulate through the

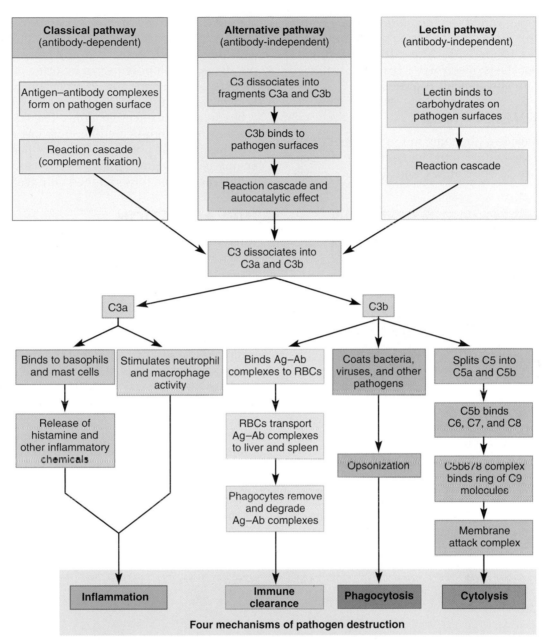

**FIGURE 21.15 Complement Activation.** The classical, alternative, and lectin pathways all lead to the cleavage of complement C3 into C3a and C3b. Those two fragments activate processes that lead to enhanced inflammation, immune clearance, phagocytosis, and cytolysis.

liver and spleen, the macrophages of those organs strip off and destroy the Ag–Ab complexes, leaving the RBCs unharmed. This is the principal means of clearing foreign antigens from the bloodstream.

3. **Phagocytosis.** Bacteria, viruses, and other pathogens are phagocytized and destroyed by neutrophils and macrophages. However, those phagocytes cannot easily internalize "naked" microorganisms. C3b assists them by the process of **opsonization:**[8] it coats microbial cells and serves as binding sites for phagocyte

attachment. The way Elie Metchnikoff described this, opsonization "butters up" the foreign cells to make them more appetizing.

4. **Cytolysis.**[9] C3b splits another complement protein, C5, into C5a and C5b. C5a joins C3a in its proinflammatory actions, but C5b plays a more important role in pathogen destruction. It binds to the enemy cell and then attracts complements C6, C7, and C8. This conglomeration of proteins (now called C5b678) goes on to bind up to 17 molecules of complement C9,

---

[8]*opson* = to prepare food

[9]*cyto* = cell + *lysis* = split apart, break down

which form a ring called the *membrane attack complex* (fig. 21.16). The complex forms a hole in the target cell up to 10 nm wide. The cell can no longer maintain homeostasis; electrolytes leak out, water flows rapidly in, and the cell ruptures.

## IMMUNE SURVEILLANCE

**Immune surveillance** is a phenomenon in which NK cells continually patrol the body "on the lookout" for pathogens or diseased host cells. They attack and destroy bacteria,

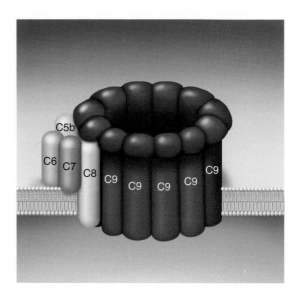

**FIGURE 21.16** **The Membrane Attack Complex.** Complement proteins C5b, C6, C7, and C8 form an organizing center around which many C9 molecules create a ring that opens a lethal hole in the enemy cell membrane.

▶ *In what way does the action of the membrane attack complex resemble the action of perforins?*

cells of transplanted organs and tissues, cells infected with viruses, and cancer cells. Upon recognition of an enemy cell, the NK cell binds to it and releases proteins called **perforins,** which polymerize in a ring and create a hole in its plasma membrane (fig. 21.17). This hole allows a rapid flow of water and salts into the enemy cell, which may kill it but is not very effective alone. The NK cell also secretes a group of protein-degrading enzymes called **granzymes,** which enter the pore made by the perforins. Inside the enemy cell, the granzymes destroy cellular enzymes and induce apoptosis (programmed cell death).

## INFLAMMATION

**Inflammation** is a local defensive response to tissue injury of any kind, including trauma and infection. Its general purposes are (1) to limit the spread of pathogens and ultimately destroy them, (2) to remove the debris of damaged tissue, and (3) to initiate tissue repair. Inflammation is characterized by four **cardinal signs:** redness, swelling, heat, and pain. Some authorities list impaired use as a fifth sign, but this may or may not occur and when it does, it is mostly because of the pain.

Words ending in the suffix *-itis* denote inflammation of specific organs and tissues: *arthritis, encephalitis, peritonitis, gingivitis,* and *dermatitis,* for example. Inflammation can occur anywhere in the body, but it is most common and observable in the skin, which is subject to more trauma than any other organ. Examples of cutaneous inflammation include an itchy mosquito bite, sunburn, a poison ivy rash, and the redness and blistering produced by manual labor, tight shoes, or a kitchen burn.

The following discussion of the process of inflammation will account for the four cardinal signs and explain how its three purposes are achieved. These processes are mediated by several types of cells and inflammatory

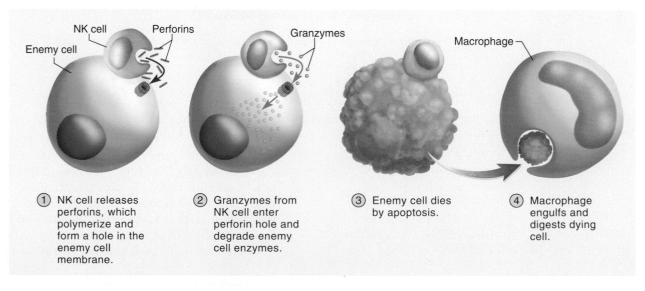

1. NK cell releases perforins, which polymerize and form a hole in the enemy cell membrane.

2. Granzymes from NK cell enter perforin hole and degrade enemy cell enzymes.

3. Enemy cell dies by apoptosis.

4. Macrophage engulfs and digests dying cell.

**FIGURE 21.17** **The Action of a Natural Killer (NK) Cell.**

| TABLE 21.1 | Agents of Inflammation |
|---|---|
| **Cellular agents** | |
| Basophils | Secrete histamine, heparin, kinins, and leukotrienes |
| Endothelial cells | Produce cell-adhesion molecules to recruit leukocytes; secrete platelet-derived growth factor |
| Eosinophils* | Produce antiparasitic oxidizing agents and toxic proteins; stimulate basophils and mast cells; limit action of histamine and other inflammatory chemicals |
| Fibroblasts | Rebuild damaged tissue by secreting collagen, ground substance, and other tissue components |
| Helper T cells* | Secrete chemotactic factors and colony-stimulating factors |
| Macrophages* | Clean up tissue damage; phagocytize bacteria, tissue debris, dead and dying leukocytes and pathogens |
| Mast cells | Same actions as basophils |
| Monocytes | Emigrate into inflamed tissue and become macrophages |
| Neutrophils | Phagocytize bacteria; secrete bactericidal oxidizing agents; secrete cytokines that activate more leukocytes |
| Platelets | Secrete clotting factors and platelet-derived growth factor |
| **Chemical agents** | |
| Bradykinin | A kinin that stimulates vasodilation and capillary permeability; a neutrophil chemotactic factor; stimulates pain receptors |
| Chemotactic factor | Any chemical that provides a trail that neutrophils and other leukocytes can follow to specific sites of infection and tissue injury; includes bradykinin, leukotrienes, and some complement proteins |
| Colony-stimulating factor | Any hormone that raises the WBC count by stimulating leukopoiesis |
| Complement* | Proteins that promote phagocytosis and cytolysis of pathogens, activate and attract neutrophils and macrophages, and stimulate basophils and mast cells to secrete inflammatory chemicals |
| Cytokine | Any small protein that serves for local intercellular communication |
| Fibrinogen | Filters into inflamed tissue and forms fibrin, thus inducing clotting of tissue fluid, walling off pathogens, and limiting their spread from the site of infection; forms temporary scaffold for tissue rebuilding |
| Heparin | An anticoagulant that inhibits clotting of tissue fluid within the area walled off by fibrin, thus promoting free mobility of leukocytes that attack infectious microorganisms |
| Histamine | Stimulates vasodilation and capillary permeability |
| Kinin | Any plasma protein (bradykinin and others) that is activated by tissue injury and then promotes vasodilation and capillary permeability |
| Leukotriene | Stimulates vasodilation and capillary permeability; a neutrophil chemotactic factor |
| Prostaglandin | Stimulates pain receptors; enhances histamine and bradykinin action; promotes neutrophil diapedesis |
| Selectin | A cell-adhesion molecule of endothelial cells that adheres to circulating leukocytes, promoting margination and diapedesis |

*These agents of inflammation have additional roles in specific immunity described in table 21.4.

chemicals that are summarized in table 21.1. Many of the chemicals that regulate inflammation and immunity are in a class called **cytokines**[10]—small proteins that are secreted by certain types of cells and alter the physiology or behavior of the receiving cell. Cytokines usually act only at short range, either on neighboring cells or on the same cell that secretes them. These are called *paracrine*[11] and *autocrine*[12] effects, respectively, to distinguish them from the long-distance *endocrine* effects of hormones. Cytokines include interferons, interleukins, tumor necrosis factor, chemotactic factors, and other chemicals you will soon encounter in this discussion.

Inflammation involves three major processes: mobilization of the body's defenses, containment and destruction of pathogens, and tissue cleanup and repair.

## Mobilization of Defenses

The most immediate requirement for dealing with tissue injury is to get defensive leukocytes to the site quickly. The way to do this is local **hyperemia**—increasing blood flow beyond its normal rate. This is achieved by local vasodilation.

To bring this about, certain cells secrete *vasoactive* chemicals that dilate the blood vessels. Among these are histamine, kinins, and leukotrienes. The cells that secrete them include basophils of the blood, mast cells of the connective tissue, and cells damaged by the trauma, toxins, or organisms that triggered the inflammation.

[10]*cyto* = cell + *kin* = to set in motion
[11]*para* = next to + *crin* = to secrete
[12]*auto* = self + *crin* = to secrete

Hyperemia not only results in the more rapid delivery of leukocytes, but also washes toxins and metabolic wastes from the tissue more rapidly.

In addition to dilating the local blood vessels, the vasoactive chemicals cause endothelial cells of the blood capillaries to separate a little, widening the intercellular clefts between them and increasing capillary permeability. This allows for the easier movement of fluid, leukocytes, and plasma proteins from the bloodstream into the surrounding tissue. Among the helpful proteins filtering from the blood are complement, antibodies, and clotting proteins, all of which aid in combating pathogens as described shortly.

Endothelial cells of the blood vessels actively aid in the recruitment of leukocytes. In the area of injury, they produce cell-adhesion molecules called **selectins,** which make their membranes sticky and snag leukocytes arriving in the bloodstream. Leukocytes adhere loosely to the selectins and slowly tumble along the endothelium, sometimes coating it so thickly that they obstruct blood flow. This adhesion to the vessel wall is called **margination.** The leukocytes then crawl through the gaps between the endothelial cells—an action called **diapedesis,**[13] or **emigration**—and thus enter the tissue fluid among the cells of the damaged tissue (fig. 21.18). Most diapedesis occurs across the walls of the postcapillary venules. Cells and chemicals that have left the bloodstream are said to be *extravasated.*[14]

In the events that have already transpired, we can see the basis for the four cardinal signs of inflammation: (1) the heat results from the hyperemia; (2) redness is also due to hyperemia and in some cases, such as sunburn, to extravasated erythrocytes in the tissue; (3) swelling (edema) is due to the increased fluid filtration from the capillaries; and (4) pain results from direct injury to the nerves, pressure on the nerves from the edema, and stimulation of pain receptors by prostaglandins, some bacterial toxins, and a kinin called **bradykinin.**

> ⌐ **Think About It**
>
> *Review eicosanoid synthesis (p. 666) and explain why aspirin eases the pain of inflammation.*

## Containment and Destruction of Pathogens

One priority in inflammation is to prevent pathogens from spreading through the body. The fibrinogen that filters into the tissue fluid clots in areas adjacent to the injury, forming a sticky mesh that sequesters (walls off) bacteria and other microbes. Heparin, the anticoagulant, prevents clotting in the area of the injury itself, so bacteria or other pathogens are essentially trapped in a fluid pocket surrounded by a gelatinous capsule of clotted fluid. They are attacked by antibodies, phagocytes, and

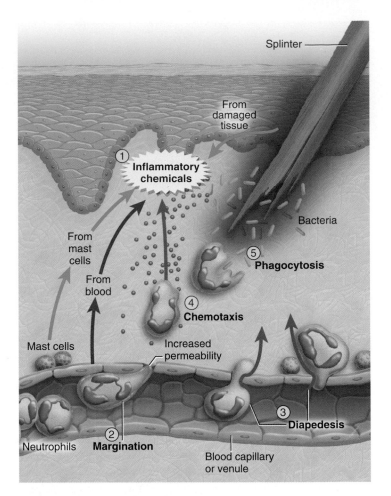

**FIGURE 21.18   Neutrophil Behavior in Inflammation.** Chemical messengers are released by basophils, mast cells, blood plasma, and damaged tissue. These inflammatory chemicals stimulate leukocyte margination (adhesion to the capillary wall), diapedesis (crawling through the capillary wall), chemotaxis (movement toward the source of the inflammatory chemicals), and phagocytosis (engulfing bacteria or other pathogens).

other defenses, while the surrounding areas of clotted tissue fluid prevent them from escaping this onslaught.

The chief enemies of bacteria are the neutrophils, which begin to accumulate in the inflamed tissue within an hour of injury. After emigrating from the bloodstream, they exhibit **chemotaxis**—attraction to chemicals such as bradykinin and leukotrienes that guide them to the site of injury or infection. As they encounter bacteria, neutrophils avidly phagocytize and digest them, and destroy many more by the respiratory burst described earlier. The four major stages of neutrophil action are summarized in figure 21.18.

Neutrophils also recruit macrophages and additional neutrophils by secreting cytokines, like shouting "Over here!" to bring in reinforcements. Activated macrophages and T cells in the inflamed tissue also secrete cytokines called *colony-stimulating factors,* which promote the production of more leukocytes (leukopoiesis) by the red bone marrow. Within a few hours of the onset of inflammation, the neutrophil count in the blood can rise from the normal 4,000 or 5,000 cells/μL to as high as 25,000 cells/μL, a condition

---

[13]*dia* = through + *pedesis* = stepping
[14]*extra* = outside + *vas* = vessel

called **neutrophilia.** In the case of an allergy or parasitic infection, an elevated eosinophil count, or **eosinophilia,** may also occur. The task of eosinophils was described earlier.

## Tissue Cleanup and Repair

Monocytes are major agents of tissue cleanup and repair. They arrive within 8 to 12 hours of an injury, emigrate from the bloodstream, and turn into macrophages. Macrophages engulf and destroy bacteria, damaged host cells, and dead and dying neutrophils. They also act as antigen-presenting cells, activating specific immune responses described later in the chapter.

Edema also contributes to tissue cleanup. The swelling compresses veins and reduces venous drainage, while it forces open the valves of lymphatic capillaries and promotes lymphatic drainage. The lymphatics can collect and remove bacteria, dead cells, proteins, and tissue debris better than blood capillaries can.

As the battle progresses, all of the neutrophils and most of the macrophages die. These dead cells, other tissue debris, and tissue fluid form a pool of yellowish fluid called **pus,** which accumulates in a tissue cavity called an **abscess.**[15] Pus is usually absorbed, but sometimes it forms a blister between the epidermis and dermis and may be released by its rupture.

Blood platelets and endothelial cells in an area of injury secrete **platelet-derived growth factor,** an agent that stimulates fibroblasts to multiply and synthesize collagen. Hyperemia, at the same time, delivers the oxygen, amino acids, and other necessities of protein synthesis, while the heat of inflamed tissue increases metabolic rate and the speed of mitosis and tissue repair. The fibrin clot in inflamed tissue may provide a scaffold for tissue recon-

struction. Pain also contributes importantly to recovery. It is an important alarm signal that calls our attention to the injury and makes us limit the use of a body part so it has a chance to rest and heal.

## FEVER

**Fever** is an abnormal elevation of body temperature. It is also known as **pyrexia,** and the term **febrile** means pertaining to fever (as in a "febrile attack"). Fever results from trauma, infections, drug reactions, brain tumors, and several other causes. Because of variations in human body temperature, there is no exact criterion for what constitutes a fever—a temperature that is febrile for one person may be normal for another.

Fever was long regarded as an undesirable side effect of illness, and efforts were (and still are) made to reduce it for the sake of comfort. It is now recognized, however, as an adaptive defense mechanism that, in moderation, does more good than harm. People with colds, for example, recover more quickly and are less infective to others when they allow a fever to run its course rather than using **antipyretic** (fever-reducing) medications such as aspirin. Fever is beneficial in that it (1) promotes interferon activity, (2) elevates metabolic rate and accelerates tissue repair, and (3) inhibits reproduction of bacteria and viruses.

When neutrophils and macrophages phagocytize bacteria, they secrete a **pyrogen**[16] (fever-producing agent) called *interleukin-1 (IL-1)*. IL-1 stimulates the anterior hypothalamus to secrete prostaglandin E (PGE). PGE, in turn, raises the hypothalamic set point for body temperature—say to 39°C (102°F) instead of the usual 37°C (fig. 21.19). Aspirin and ibuprofen reduce fever by inhibiting PGE synthesis.

---

[15]*ab* = away + *scess* (from *cedere*) = to go

[16]*pyro* = fire, heat + *gen* = producing

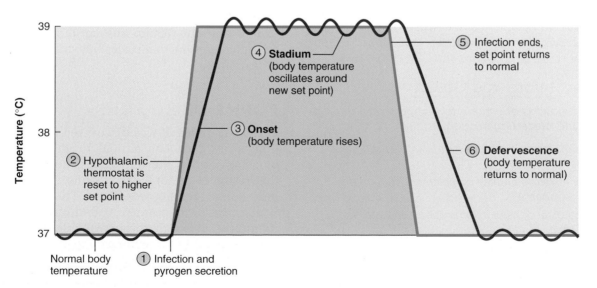

**FIGURE 21.19** The Course of a Fever.

### Reye Syndrome

In children younger than 15, an acute viral infection such as chickenpox or influenza is sometimes followed by a serious disorder called *Reye*[17] *syndrome.* First recognized in 1963, this disease is characterized by swelling of brain neurons and fatty infiltration of the liver and other viscera. Neurons die from hypoxia and the pressure of the swelling brain, which results in nausea, vomiting, disorientation, seizures, and coma. About 30% of victims die, and the survivors sometimes suffer mental retardation. Reye syndrome can be triggered by the use of aspirin to control fever; parents are now strictly advised never to give aspirin to children with chickenpox or flulike symptoms.

When the set point rises, a person shivers to generate heat and the cutaneous blood vessels constrict to reduce heat loss. In the stage of fever called *onset,* one has chills, feels cold and clammy to the touch, and has a rising temperature. In the next stage, *stadium,* the body temperature oscillates around the new set point for as long as the pathogen is present. The elevated temperature stimulates the liver and spleen to hoard zinc and iron, depriving bacteria of minerals they need to reproduce. When the infection is defeated, pyrogen secretion ceases and the hypothalamic thermostat is set back to normal. This activates heat-losing mechanisms, especially cutaneous vasodilation and sweating. The skin is warm and flushed during this phase. The phase of falling temperature is called *defervescence* in general, *crisis* (flush) if the temperature drops abruptly, or *lysis* if it falls slowly.

Even though most fevers are beneficial, excessively high temperature can be dangerous because it speeds up different enzymatic pathways to different degrees, thus causing metabolic discoordination and cellular dysfunction. Fevers above 40.5°C (105°F) can make a person delirious. Convulsions and coma ensue at higher temperatures, and death or irreversible brain damage commonly results from fevers that range from 44° to 46°C (111° to 115°F).

## Before You Go On

*Answer the following questions to test your understanding of the preceding section:*

6. *What are macrophages? Give four examples and state where they are found.*

7. *How do interferons and the complement system protect against disease?*

8. *List the cardinal signs of inflammation and state the cause of each.*

9. *Summarize the benefits of fever and the limits of these benefits.*

# General Aspects of Specific Immunity

### Objectives
When you have completed this section, you should be able to

- define *specific immunity;*
- contrast cellular and humoral immunity, active and passive immunity, and natural and artificial immunity;
- describe the chemical properties of antigens;
- describe and contrast the development of T and B lymphocytes; and
- describe the general roles played by lymphocytes, antigen-presenting cells, and interleukins in the immune response.

The remainder of this chapter is concerned with the immune system and specific immunity, the third line of defense. The **immune**[18] **system** is composed of a large population of widely distributed cells that recognize foreign substances and act to neutralize or destroy them. Two characteristics distinguish immunity from nonspecific resistance:

1. **Specificity.** Immunity is directed against a particular pathogen. Immunity to one pathogen usually does not confer immunity to others.

2. **Memory.** When reexposed to the same pathogen, the body reacts so quickly that there is no noticeable illness. The reaction time for inflammation and other nonspecific defenses, by contrast, is just as long for later exposures as for the initial one.

These properties of immunity were recognized even in the fifth century BCE, when Thucydides remarked that people who recover from a disease often become immune to that one but remain susceptible to others. A person might be immune to measles but still susceptible to polio, for example.

## FORMS OF IMMUNITY

In the late 1800s, it was discovered that immunity can be transferred from one animal to another by way of the blood serum. In the mid-1900s, however, it was found that serum does not always confer immunity; sometimes only donor lymphocytes do so. Thus, we now recognize two types of immunity, called cellular and humoral immunity, although these two interact extensively and often respond to the same pathogen.

---

[17]R. Douglas Reye (1912–77), Australian pathologist

[18]*immuno* = free

**Cellular (cell-mediated) immunity** employs lymphocytes that directly attack and destroy foreign cells or diseased host cells. It is a means of ridding the body of pathogens that reside inside human cells, where they are inaccessible to antibodies: for example, intracellular viruses, bacteria, yeast, and protozoans. Cellular immunity also acts against parasitic worms, cancer cells, and cells of transplanted tissues and organs.

**Humoral (antibody-mediated) immunity** is mediated by antibodies, which do not directly destroy a pathogen, but tag them for destruction by mechanisms described later. It is an indirect attack in which antibodies, not the immune cells, assault the pathogen. The expression *humoral* comes from the fact that many of the antibodies are dissolved in the body fluids, and body fluids were once called "humors." Humoral immunity is effective against extracellular viruses, bacteria, yeasts, and protozoans, and against molecular (noncellular) pathogens such as toxins, venoms, and allergens. In the unnatural event of a mismatched blood transfusion, it also destroys foreign erythrocytes.

Note that humoral immunity can work only against the *extracellular* stages of infectious microorganisms. When such microorganisms invade host cells, antibodies cannot get at them. However, the *intracellular* stages are still vulnerable to cellular immunity, which destroys them by killing the cells that harbor them. Thus, humoral and cellular immunity sometimes attack the same microorganism at different points in its life cycle.

You will find cellular and humoral immunity summarized and compared in table 21.5 (p. 840) following discussion of the details of the two processes.

Other ways of classifying immunity are active versus passive and natural versus artificial. In *active immunity* the body makes its own antibodies or T cells against a pathogen, whereas in *passive immunity* the body acquires them from another person or an animal. Either type of immunity can occur naturally or, for treatment and prevention purposes, it can be induced artificially. Thus we can recognize four classes of immunity under this scheme:

1. **Natural active immunity.** This is the production of one's own antibodies or T cells as a result of natural exposure to an antigen.
2. **Artificial active immunity.** This is the production of one's own antibodies or T cells as a result of **vaccination** against diseases such as smallpox, tetanus, or influenza. A **vaccine** consists of either dead or *attenuated* (weakened) pathogens which can stimulate an immune response but normally cause little or no discomfort or disease. In some cases, periodic *booster shots* are given to restimulate immune memory and maintain a high level of protection. Vaccination has eliminated smallpox worldwide and greatly reduced the incidence of life-threatening childhood diseases,

but many people continue to die from influenza and other diseases that could be prevented by vaccination.
3. **Natural passive immunity.** This is a temporary immunity that results from acquiring antibodies produced by another person. The only natural ways for this to happen are for a fetus to acquire antibodies from the mother through the placenta before birth, or for a baby to acquire them during breast-feeding.
4. **Artificial passive immunity.** This is a temporary immunity that results from the injection of an *immune serum* obtained from another person or from animals (such as horses) that produced antibodies against a certain pathogen. Immune serum is used for emergency treatment of snakebites, botulism, tetanus, rabies, and other diseases.

Only the two forms of active immunity involve immune memory and thus provide future protection. Passive immunity typically lasts for only 2 or 3 weeks, until the acquired antibody is degraded. The remaining discussion is based on natural active immunity.

## ANTIGENS

An **antigen**[19] **(Ag)** is any molecule that triggers an immune response. Some antigens are free molecules such as venoms and toxins; others are components of plasma membranes and bacterial cell walls. Small universal molecules such as glucose and amino acids are not antigenic; if they were, our immune systems would attack the nutrients and other molecules essential to our very survival. Most antigens have molecular weights over 10,000 amu and are generally complex molecules that are unique to each individual: proteins, polysaccharides, glycoproteins, and glycolipids. Their uniqueness enables the body to distinguish its own ("self") molecules from those of any other individual or organism ("nonself"). The immune system "learns" to distinguish self- from nonself-antigens prior to birth; thereafter, it normally attacks only nonself-antigens.

Only certain regions of an antigen molecule, called **epitopes** (antigenic determinants), stimulate immune responses. One antigen molecule typically has several different epitopes, however, that can stimulate the production of different antibodies.

Some molecules, called **haptens**,[20] are too small to be antigenic in themselves, but they can stimulate an immune response by binding to a host macromolecule and creating a unique complex that the body recognizes as foreign. After the first exposure, a hapten may stimulate an immune response without needing to bind to another molecule. Many people are allergic to haptens in

---

[19]acronym from *anti*body *gen*erating
[20]from *haptein* = to fasten

cosmetics, detergents, industrial chemicals, poison ivy, and animal dander. The most common drug allergy is to penicillin and its derivatives. Penicillin is a hapten that binds to host proteins in allergic individuals, creating modified proteins that bind to mast cells and trigger massive release of histamine and other inflammatory chemicals. This can cause death from *anaphylactic shock.*

## LYMPHOCYTES

The major cells of the immune system are lymphocytes and macrophages, which are especially concentrated at strategic places such as the lymphatic organs and mucous membranes. Lymphocytes fall into three classes: natural killer (NK) cells, T lymphocytes, and B lymphocytes. We have already discussed NK cells (see immune surveillance). Here, we must take a closer look at T and B lymphocytes.

### T Lymphocytes (T cells)

The life history of a T cell basically involves three stages and three anatomical stations in the body. We can loosely think of these as their "birth," their "training" or matura-

tion, and finally their "deployment" to locations where they will carry out their immune function.

T cells are "born" in the red bone marrow as descendents of the pluripotent stem cells (PPSCs) described in chapter 18. The bone marrow releases them into the blood as still-undifferentiated stem cells which colonize the thymus (fig. 21.20). The thymus is the "school" where they mature into fully functional T cells. In the thymic cortex, reticular epithelial (RE) cells secrete thymic hormones that stimulate the T cells to develop surface antigen receptors. With receptors in place, the T cells are now **immunocompetent,** capable of recognizing antigens presented to them by APCs. The RE cells then test these T cells by presenting self-antigens to them. There are two ways to fail the test: inability to recognize the RE cells at all (specifically, their MHC proteins, described later), or reacting to the self-antigens. Failure to recognize MHC antigens would mean the T cell would be incapable of recognizing a foreign attack on the body, but reacting to self-antigens would mean the T cells would attack one's own tissues. Either way, T cells that fail the test must be eliminated—a process called **negative selection.** There are two forms of negative selection: **clonal deletion,** in

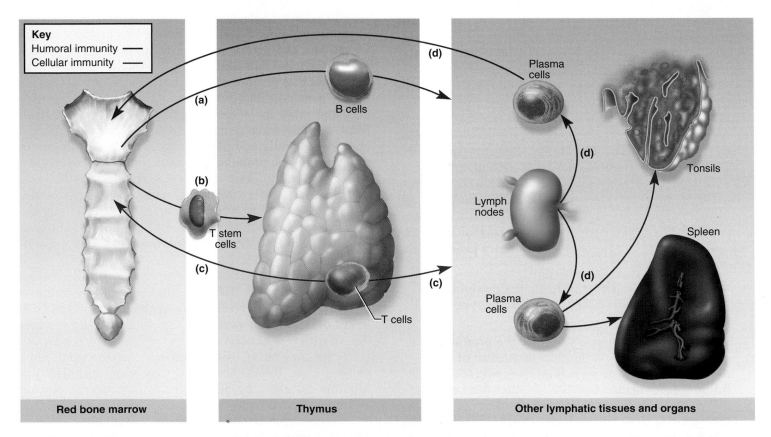

**FIGURE 21.20**   **The Life History and Migrations of B and T Cells.**   (a) B cells achieve immunocompetence in the red bone marrow, then many of them emigrate to other lymphatic tissues and organs. (b) T stem cells emigrate from the bone marrow and attain immunocompetence in the thymus. (c) Immunocompetent T cells emigrate from the thymus and disperse to the bone marrow and other organs. (d) Plasma cells develop in the lymph nodes (among other sites) and emigrate to the bone marrow and other lymphatic organs, where they spend a few days secreting antibodies.

which self-reactive T cells die and macrophages phagocytize them, and **anergy**,[21] in which they remain alive but unresponsive. Negative selection leaves the body in a state of **self-tolerance** in which the surviving, active T cells respond only to foreign antigens, not to one's own.

T cells that recognize the RE cells but do not react strongly to self-antigens have passed the test and now "graduate" to join the immune workforce. Only 2% of the T cells pass their graduation test. They move on to the medulla of the thymus, where they undergo **positive selection**—they multiply and form **clones** of identical T cells programmed to respond to a particular antigen. These cells, which are immunocompetent but have not yet encountered the "enemy" (foreign antigens), constitute the **naive lymphocytes pool**. Naive T cells leave the thymus and colonize lymphatic tissues and organs everywhere in the body (the bone marrow, lymph nodes, tonsils, and so forth); they are now ready to do battle.

> **Think About It**
>
> *Is clonal deletion a case of apoptosis or necrosis? Explain your answer. (Review these concepts in chapter 5 if necessary.)*

## B Lymphocytes (B cells)

Another group of fetal stem cells remain in the bone marrow to differentiate into B cells. Those that respond to self-antigens undergo either anergy or clonal deletion, much like self-reactive T cells. Self-tolerant B cells, on the other hand, go on to produce surface receptors for antigens, divide, and produce immunocompetent B cell clones. These cells disperse throughout the body, colonizing the same organs as T cells. They are abundant in the lymph nodes, spleen, bone marrow, and mucous membranes.

## ANTIGEN-PRESENTING CELLS

Although the function of T cells is to recognize and attack foreign antigens, they usually cannot recognize such antigens on their own. They require the help of **antigen-presenting cells (APCs)**. In addition to their other roles, B cells, macrophages, reticular cells, and dendritic cells function as APCs.

APC function hinges on a family of genes on chromosome 6 called the **major histocompatability complex (MHC)**. These genes code for **MHC proteins**—proteins on the APC surface that are shaped a little like hotdog buns, with an elongated groove for holding the "hotdog" of the foreign antigen. MHC proteins are structurally unique to every person except for identical twins. They act as "identification tags" that label every cell of your body as belonging to you.

When an APC encounters an antigen, it internalizes it by endocytosis, digests it into molecular fragments, and

---

[21]*an* = without + *erg* = action, work

displays the relevant fragments (its epitopes) in the grooves of the MHC proteins (fig. 21.21a). These steps are called **antigen processing.** Wandering T cells regularly inspect APCs for displayed antigens (fig. 21.21b). If an APC displays a self-antigen, the T cells disregard it. If it displays a

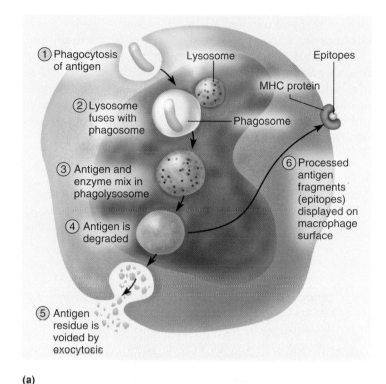

**(a)**

**(b)**  10μm

**FIGURE 21.21** **The Action of Antigen-Presenting Cells (APCs).** (a) Stages in the processing and presentation of an antigen by an APC such as a macrophage. (b) Macrophages and reticular cells presenting processed antigens to helper T cells. [From R. G. Kessel and R. H. Kardon, *Tissues and Organs: A Text-Atlas of Scanning Electron Microscopy,* W. H. Freeman & Co., 1979.]

nonself-antigen, however, the T cells initiate an immune attack. APCs thus alert the immune system to the presence of a foreign antigen. The key to a successful defense is then to quickly mobilize immune cells against the antigen.

With so many cell types involved in immunity, it is not surprising that they require chemical messengers to coordinate their activities. Lymphocytes and APCs talk to each other with cytokines called **interleukins**[22]—chemical signals from one leukocyte (or leukocyte derivative) to another.

With this introduction to the main actors in immunity, we can now look at the more specific features of cellular and humoral immunity. Since the terminology of immune cells and chemicals is quite complex, you may find it helpful to refer often to table 21.4 (p. 839) as you read the following discussions.

## Before You Go On

*Answer the following questions to test your understanding of the preceding section:*

10. *How does specific immunity differ from nonspecific defense?*

11. *How does humoral immunity differ from cellular immunity?*

12. *Contrast active and passive immunity. Give a natural and an artificial example of each.*

13. *What structural properties distinguish antigenic molecules from those that are not antigenic?*

14. *What is an immunocompetent lymphocyte? What does a lymphocyte have to produce in order to be immunocompetent?*

15. *What role does the thymus play in the life history of a T cell?*

16. *What role does an antigen-presenting cell play in the activation of a T cell?*

# Cellular Immunity

### Objectives

When you have completed this section, you should be able to

- list the types of lymphocytes involved in cellular immunity and describe the roles they play;
- describe the process of antigen presentation and T cell activation;
- describe how T cells destroy enemy cells; and
- explain the role of memory cells in cellular immunity.

Cellular (cell-mediated) immunity is a form of specific defense in which T lymphocytes directly attack and destroy diseased or foreign cells, and the immune system

then remembers the antigens of those invaders and prevents them from causing disease in the future. Cellular immunity employs three classes of T cells:

1. **Cytotoxic T ($T_C$) cells** are the "effectors" of cellular immunity that carry out the attack on enemy cells. They are also called *killer T cells,* but are not the same as *natural killer* cells.

2. **Helper T ($T_H$) cells** promote the action of $T_C$ cells as well as playing key roles in humoral immunity and nonspecific defense. All other T cells are involved in cellular immunity only.

3. **Memory T ($T_M$) cells** are descended from the cytotoxic T cells and are responsible for memory in cellular immunity.

$T_C$ cells are also known as T8, CD8, or CD8+ cells because they have a surface glycoprotein called CD8. $T_H$ cells are also known as T4, CD4, or CD4+ cells, after their glycoprotein, CD4. (CD stands for *cluster of differentiation,* a classification system for many cell-surface molecules.) These glycoproteins are cell-adhesion molecules that enable T cells to bind to other cells in the events to be described shortly.

Both cellular and humoral immunity occur in three stages that we can think of as recognition, attack, and memory (or "the three *R*s of immunity"—recognize, react, and remember). In cellular immunity, the events of each stage are as follows.

## RECOGNITION

The recognition phase has two aspects: antigen presentation and T cell activation.

## Antigen Presentation

When an antigen-presenting cell (APC) encounters and processes an antigen, it typically migrates to the nearest lymph node and displays it to T cells. Cytotoxic and helper T cells patrol the lymph nodes and other tissues as if looking for trouble. When they encounter a cell displaying an antigen on an MHC protein (MHCP), they initiate an immune response. T cells respond to two classes of MHCPs:

1. *MHC-I proteins* occur on every nucleated cell of the body (not erythrocytes). These proteins are constantly produced by the cell and transported to the plasma membrane. Along the way, they pick up small peptides in the cytoplasm and display these once they are installed in the membrane. If the peptides are normal self-antigens, they do not elicit a T cell response. If they are viral proteins or abnormal antigens made by cancer cells, however, they do. In this case, the Ag–MHCP complex is like a tag on the host cell that says, "I'm diseased; kill me." Infected or malignant cells are then destroyed before they can do further harm to the body.

2. *MHC-II proteins* (also called *human leukocyte antigens, HLAs*) occur only on APCs and display only foreign antigens. $T_C$ cells respond only to MHC-I proteins, and $T_H$ cells respond only to MHC-II (table 21.2).

## T Cell Activation

T cell activation is shown in figure 21.22. It begins when a $T_C$ or $T_H$ cell binds to an MHCP displaying an epitope that the T cell is programmed to recognize. Before the response can go any further, the T cell must bind to another APC protein, related to interleukins. In a sense, the T cell has to check twice to see if it really has bound to an APC displaying a foreign antigen. This signaling process, called **costimulation,** triggers a process called **clonal selection:** the activated T cell undergoes repeated mitosis, giving rise to a clone of identical T cells programmed against the same epitope. Some cells in the clone become effector cells that carry out an immune attack, and some become memory T cells.

## ATTACK

Helper and cytotoxic T cells play different roles in the attack phase.

## Helper T Cells

Most immune responses require the action of helper T cells, which play a central coordinating role in both humoral and cellular immunity (fig. 21.23). When a helper T cell recognizes an Ag–MHCP complex, it secretes interleukins that exert three effects: (1) they attract neutrophils and natural killer cells; (2) they attract macrophages, stimulate their phagocytic activity, and inhibit them from leaving the area; and (3) they stimulate T and B cell mitosis and maturation.

## Cytotoxic T Cells

Cytotoxic T ($T_C$) cells are the only T lymphocytes that directly attack and kill other cells (fig. 21.24). When a $T_C$ cell recognizes a complex of antigen and MHC-I protein on a diseased or foreign cell, it "docks" on that cell, delivers a **lethal hit** of cytotoxic chemicals that will destroy it, and goes off in search of other

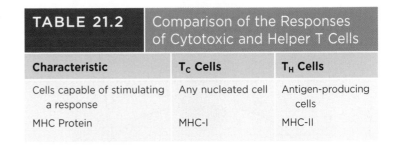

| TABLE 21.2 | Comparison of the Responses of Cytotoxic and Helper T Cells | |
|---|---|---|
| **Characteristic** | **$T_C$ Cells** | **$T_H$ Cells** |
| Cells capable of stimulating a response | Any nucleated cell | Antigen-producing cells |
| MHC Protein | MHC-I | MHC-II |

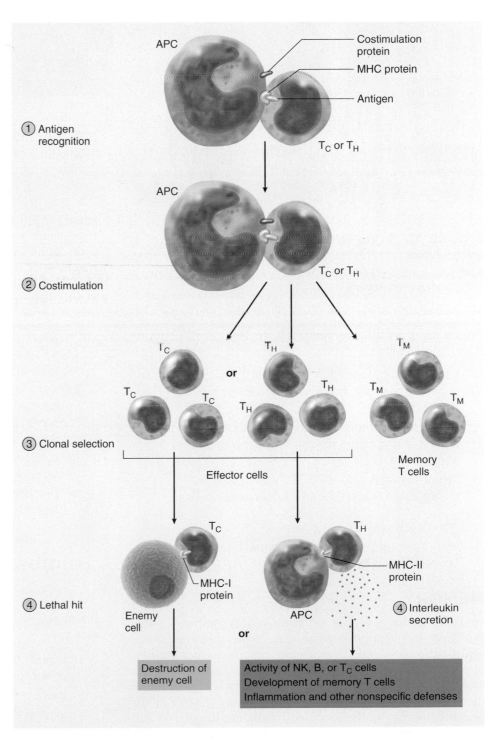

**FIGURE 21.22** T Cell Activation.

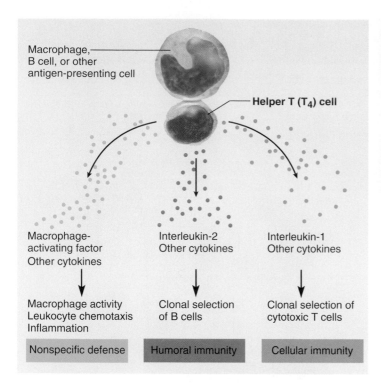

**FIGURE 21.23**   The Role of Helper T Cells in Defense and Immunity.

▶ *Why does AIDS reduce the effectiveness of all three defenses listed across the bottom?*

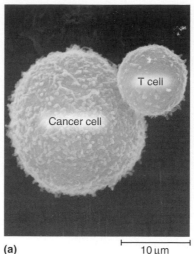 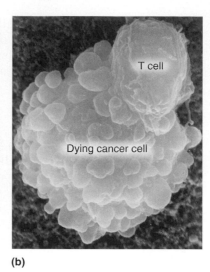

(a)                    ⊢—— 10 μm ——⊣     (b)

**FIGURE 21.24**   Destruction of a Cancer Cell by a Cytotoxic T Cell.   (a) T cell binding to cancer cell. (b) Death of the cancer cell due to the lethal hit by the T cell.

vated, and thus respond to antigens more rapidly. Upon reexposure to the same pathogen later in life, memory cells mount a quick attack called the **T cell recall response.** This time-saving response destroys a pathogen so quickly that no noticeable illness occurs—that is, the person is immune to the disease.

## Before You Go On

*Answer the following questions to test your understanding of the preceding section:*

17. *Name three types of lymphocytes that are involved in cellular immunity. Which of these is also essential to humoral immunity?*

18. *What are the three phases of an immune response?*

19. *Explain why cytotoxic T cells are activated by a broader range of host cells than are helper T cells.*

20. *Describe some ways in which cytotoxic T cells destroy target cells.*

enemy cells while the chemicals do their work. Among these chemicals are:

*   perforin and granzymes, which kill the target cell in the same manner as we saw earlier for NK cells (see fig. 21.17);

*   interferons, which inhibit viral replication and recruit and activate macrophages, among other effects; and

*   *tumor necrosis factor* (TNF), which aids in macrophage activation and kills cancer cells.

⌐ **Think About It**

*How is a cytotoxic T cell like a natural killer (NK) cell? How are they different?*

## MEMORY

As more and more cells are recruited by helper T cells, the immune response exerts an overwhelming force against the pathogen. The primary response, seen on first exposure to a particular pathogen, peaks in about a week and then gradually declines. It is followed by immune memory. Following clonal selection, some $T_C$ and $T_H$ cells become memory cells. Memory T cells are long-lived and much more numerous than naive T cells. Aside from their sheer numbers, they also require fewer steps to be acti-

# Humoral Immunity

### Objectives

When you have completed this section, you should be able to

*   explain how B cells recognize and respond to an antigen;

*   describe the structure, types, and actions of antibodies;

*   explain the mechanism of memory in humoral immunity; and

*   compare and contrast cellular and humoral immunity.

Humoral immunity is a more indirect method of defense than cellular immunity. Instead of directly attacking enemy cells, the B lymphocytes of humoral immunity produce antibodies that bind to antigens and tag them for destruction by other means. But like cellular immunity, humoral immunity works in three stages: recognition, attack, and memory.

## RECOGNITION

An immunocompetent B cell has thousands of surface receptors for one antigen. B cell activation begins when an antigen binds to several of these receptors, links them together, and is taken into the cell by receptor-mediated endocytosis. One reason small molecules are not antigenic is that they are too small to link multiple receptors together. After endocytosis, the B cell processes (digests) the antigen, links some of the epitopes to its MHC-II proteins, and displays these on the cell surface.

Usually, the B-cell response goes no further unless a helper T cell binds to this Ag–MHCP complex. (Some B cells are directly activated by antigens without the help of a $T_H$ cell.) When a $T_H$ cell does bind to the complex, it secretes interleukins that activate the B cell. This triggers clonal selection—B cell mitosis giving rise to a battalion of identical B cells programmed against the same antigen (fig. 21.25).

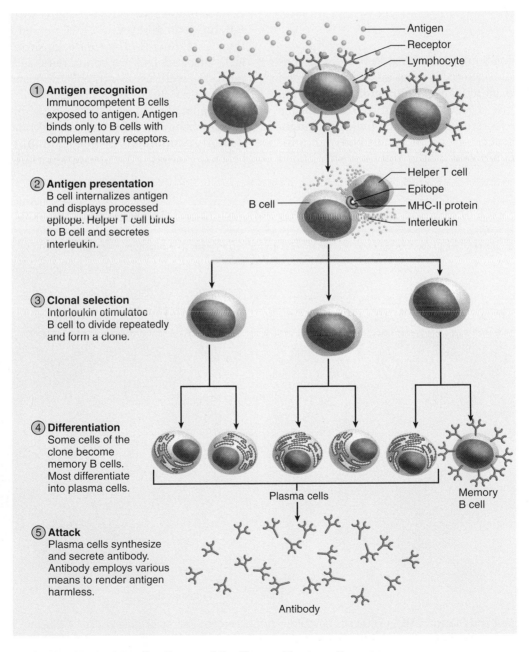

**FIGURE 21.25** Clonal Selection and Ensuing Events of the Humoral Immune Response.

Most cells of the clone differentiate into **plasma cells.** These are larger than B cells and contain an abundance of rough endoplasmic reticulum (fig. 21.26). Plasma cells develop mainly in the germinal centers of the lymphatic follicles of the lymph nodes. About 10% of them remain in the lymph node, but the rest leave the lymph nodes, take up residence in the bone marrow and elsewhere, and there produce antibodies until they die. A plasma cell secretes antibodies at the remarkable rate of 2,000 molecules per second over a life span of 4 to 5 days. These antibodies travel throughout the body in the blood and other body fluids. The first time you are exposed to a particular antigen, your plasma cells produce mainly an antibody class called IgM. In later exposures to the same antigen, they produce mainly IgG.

## ATTACK

We have said much about antibodies already, and it is now time to take a closer look at what an antibody is and how it works. Also called an **immunoglobulin (Ig),** an antibody is a defensive gamma globulin found in the blood plasma, body secretions, and some leukocyte membranes. The basic structural unit of an antibody, an **antibody monomer,** is composed of four polypeptides linked by disulfide (–S–S–) bonds (fig. 21.27). The two larger **heavy chains** are about 400 amino acids long, and the two **light chains** about half that long. Each heavy chain has a hinge region where the antibody is bent, giving the monomer a T or Y shape.

All four chains have a **variable (V) region,** which gives an antibody its uniqueness. The V regions of a heavy chain and light chain combine to form an **antigen-binding site** on each arm, which attaches to the epitope of an antigen molecule. The rest of each chain is a **constant (C) region,** which has the same amino acid sequence in all antibodies of a given class (within one person). The C region determines the mechanism of an antibody's action—for example, whether it can bind complement proteins.

There are five classes of antibodies named **IgA, IgD, IgE, IgG,** and **IgM** (table 21.3), named for the structures of their C regions (*alpha, delta, epsilon, gamma,* and *mu*). IgD, IgE, and IgG are monomers. IgA has a monomeric form as well as a dimer composed of two cojoined monomers. IgM is a pentamer composed of five monomers. The surface antigen receptors synthesized by a developing B cell are IgD and IgM molecules. IgG is particularly important in the immunity of the newborn because it crosses the placenta with relative ease. Thus, it

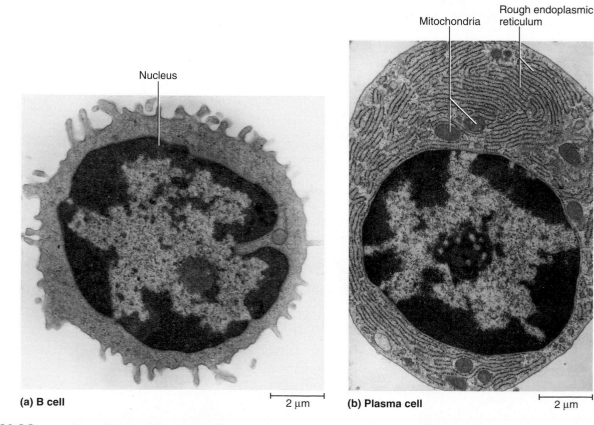

**(a) B cell**    Nucleus    2 μm

**(b) Plasma cell**    Mitochondria    Rough endoplasmic reticulum    2 μm

**FIGURE 21.26** **B Cell and Plasma Cell.** (a) B cells have little cytoplasm and scanty organelles. (b) A plasma cell, which differentiates from a B cell, has an abundance of rough endoplasmic reticulum.

▶ *What does this endoplasmic reticulum do in this plasma cell?*

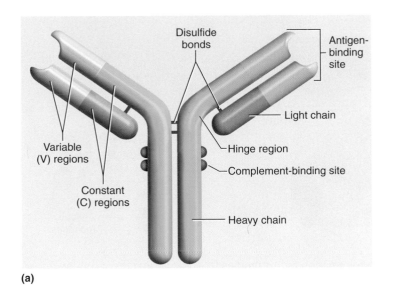

(a)

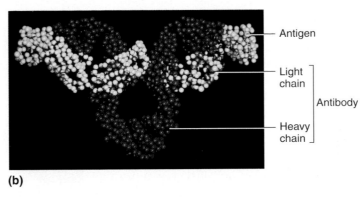

(b)

**FIGURE 21.27  Antibody Structure.** (a) A molecule of IgG, a monomer. (b) Computer-generated image of IgG bound to an antigen (lysozyme).

| Class | Structure | | Location and Function |
|---|---|---|---|
| **TABLE 21.3** | The Five Classes of Antibodies | | |
| IgA | Monomer | Dimer | Found as a monomer in blood plasma and mainly as a dimer in mucus, tears, milk, saliva, and intestinal secretions. Sometimes also forms trimers and tetramers. Prevents pathogens from adhering to epithelia and penetrating underlying tissues. Provides passive immunity to the newborn. |
| IgD | Monomer | | A transmembrane protein of B cells; thought to function in activation of B cells by antigens. |
| IgE | Monomer | | A transmembrane protein of basophils and mast cells. Stimulates them to release histamine and other chemical mediators of inflammation and allergy; important in immediate hypersensitivity reactions; attracts eosinophils to sites of parasitic infection. |
| IgG | Monomer | | Constitutes about 80% of circulating antibodies in plasma. The predominant antibody secreted in the secondary immune response. IgG and IgM are the only antibodies with significant complement fixation activity. Crosses placenta and confers temporary immunity on the fetus. Includes the anti-D antibodies of the Rh blood group. |
| IgM | Monomer | Pentamer | Constitutes about 10% of circulating antibodies in plasma. Monomer is a transmembrane protein of B cells, where it functions as part of the antigen receptor. Pentamer occurs in blood plasma and lymph. The predominant antibody secreted in the primary immune response; very strong agglutinating and complement-fixation abilities; includes the anti-A and anti-B agglutinins of the ABO blood group. |

transfers immunity from the mother to her fetus. In addition, an infant acquires some maternal IgA through breast milk and colostrum (the fluid secreted for the first 2 or 3 days of breast-feeding).

The human immune system is believed capable of producing at least 10 billion and perhaps up to 1 trillion different antibodies. Any one individual has a much smaller subset of these possibilities, but such an enormous potential helps to explain why we can deal with the tremendous diversity of antigens that must exist in our environment. Yet such huge numbers are puzzling, because we are accustomed to thinking of each protein in the body being encoded by one gene, and we have only about 35,000 genes, most of which have functions unrelated to immunity. How can so few genes generate so many antibodies? Obviously there cannot be a different gene for each one. One means of generating diversity is that the genome contains several hundred DNA segments that are shuffled and combined in various ways to produce antibody genes unique to each clone of B cells. This process is called **somatic recombination,** because it forms new combinations of DNA base sequences in somatic (nonreproductive) cells. Another mechanism of generating diversity is that B cells in the germinal centers of lymphatic nodules undergo exceptionally high rates of mutation, a process called **somatic hypermutation,** thereby not just recombining preexisting DNA but creating wholly new DNA sequences. These and other mechanisms explain how we can produce such a tremendous variety of antibodies with a limited number of genes.

Once released by a plasma cell, antibodies use four mechanisms to render antigens harmless:

1. **Neutralization.** Only certain regions of an antigen are pathogenic—for example, the parts of a toxin molecule or virus that enable these agents to bind to human cells. Antibodies can neutralize an antigen by binding to these active regions and masking them.

2. **Complement fixation.** Antibodies IgM and IgG bind to enemy cells and change shape, exposing their complement-binding sites (see fig. 21.27a). This initiates the binding of complement to the enemy cell surface and leads to inflammation, phagocytosis, immune clearance, and cytolysis, as described earlier. Complement fixation is the primary mechanism of defense against such foreign cells as bacteria and mismatched erythrocytes.

3. **Agglutination** was described in chapter 18 in the discussion of ABO and Rh blood types. It is effective not only in mismatched blood transfusions, but more importantly as a defense against bacteria. An antibody molecule has 2 to 10 binding sites; thus, it can bind to antigen molecules on two or more enemy cells at once and stick them together (fig. 21.28). This immobilizes microbes and antigen molecules and prevents them from spreading through the tissues.

4. **Precipitation** is a similar process in which antibodies link antigen molecules (not whole cells) together. This creates large Ag–Ab complexes that are too large to remain dissolved in solution. These complexes can be removed by immune clearance (see p. 822) or phagocytized by eosinophils in the connective tissue.

You will note that antibodies do not directly destroy an antigen in any of these mechanisms. They render it harmless by covering its pathogenic sites or agglutinating it, and they mark it for destruction by other agents such as complement, macrophages, or eosinophils.

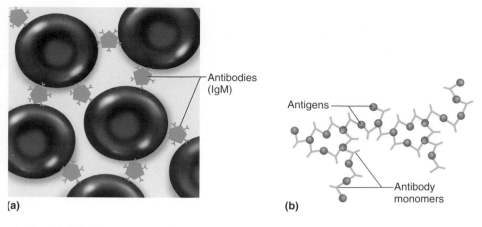

**FIGURE 21.28  Agglutination by Antibodies.**  (a) Agglutination of foreign erythrocytes by IgM, a pentamer. (b) An antigen–antibody complex involving a free molecular antigen and an antibody monomer such as IgG.

**Think About It**

*Explain why IgM has a stronger power of agglutination than antibodies of any other class.*

## MEMORY

When a person is exposed to a particular antigen for the first time, the immune reaction is called the **primary response.** The appearance of protective antibodies is delayed for 3 to 6 days while naive B cells multiply and differentiate into plasma cells. As the plasma cells begin secreting antibody, the **antibody titer** (level in the blood plasma) begins to rise (fig. 21.29). IgM appears first, peaks

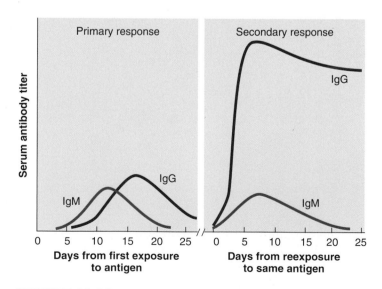

**FIGURE 21.29  The Primary and Secondary (Anamnestic) Responses in Humoral Immunity.**  The individual is exposed to antigen on day 0 in both cases. Note the differences in the speed of response, the height of the antibody titer, and the rate of decline in antibody titer.

in about 10 days, and soon declines. IgG levels rise as IgM declines, but even the IgG titer drops to a low level within a month.

The primary response, however, leaves one with an immune memory of the antigen. During clonal selection, some members of the clone become **memory B cells** rather than plasma cells (see fig. 21.25). Memory B cells, found mainly in the germinal centers of the lymph nodes, mount a very quick **secondary,** or **anamnestic**[23] (an-am-NESS-tic), **response** if reexposed to the same antigen. Plasma cells form within hours, so the IgG titer rises sharply and

peaks within a few days. The response is so rapid that the antigen has little chance to exert a noticeable effect on the body, and no illness results. A low level of IgM is also secreted and quickly declines, but IgG remains elevated for weeks to years, conferring lasting protection. Memory does not last as long in humoral immunity, however, as it does in cellular immunity.

Table 21.4 summarizes many of the cellular and chemical agents involved in humoral and cellular immunity. Table 21.5 compares the main features of humoral and cellular immunity. Remember that these two processes often occur simultaneously, and in conjunction with inflammation as a three-pronged attack on the same pathogen.

[23]*ana* = back | *mnes* = remember

| TABLE 21.4 | Agents of Specific Immunity |
|---|---|
| **Cellular agents** | |
| B Cells | Serve as antigen-presenting cells in humoral immunity; differentiate into antibody-secreting plasma cells |
| CD4 (T4) cells* | T lymphocytes with CD4 surface glycoproteins; helper T cells |
| CD8 (T8) cells | T lymphocytes with CD8 surface glycoproteins, cytotoxic T cells |
| Cytotoxic T (killer T, $T_C$, or CD8) cells | Effectors of cellular immunity; directly attack and destroy enemy cells, produce perforin, granzymes, interferon, tumor necrosis factor, and other cytokines |
| Eosinophils* | Phagocytize and degrade Ag–Ab complexes |
| Helper T ($T_H$ or CD4) cells* | Play a central regulatory role in nonspecific defense and humoral and cellular immunity; recognize antigen fragments displayed by APCs with MHC-II proteins; secrete interleukins that activate B, $T_C$, and NK cells, neutrophils, and macrophages |
| Macrophages* | Phagocytize pathogens and expended or damaged host cells; act as antigen presenting cells (APCs) |
| Memory B cells | Activated B cells that do not immediately differentiate into plasma cells; act as a pool of B cells that can execute a quick secondary response upon reexposure to the same antigen that initially activated them |
| Memory T cells | Activated T cells that do not immediately differentiate into effector T cells; act as a pool of T cells that can execute a quick T cell recall response upon reexposure to the same antigen that initially activated them |
| Naive lymphocytes | Immunocompetent lymphocytes that are capable of responding to an antigen but have not yet encountered one |
| Plasma cells | Develop from B cells that have been activated by helper T cells; synthesize and secrete antibodies |
| **Chemical agents** | |
| Antibody (Ab) | A gamma globulin produced by plasma cells in response to an antigen; counteracts antigen by means of complement fixation, neutralization of toxins, agglutination, or precipitation |
| Antigen (Ag) | Molecule capable of triggering an immune response; usually a protein, polysaccharide, glycolipid, or glycoprotein |
| Complement* | Group of plasma proteins that help to destroy pathogens by cytolysis, phagocytosis, immune clearance, or nonspecific defense (inflammation) |
| Granzyme | Proteolytic enzyme produced by NK and $T_C$ cells; enters the pore made by perforins, degrades enzymes of the enemy cell, and induces apoptosis |
| Hapten | Small molecule unable to trigger an immune response by itself but able to bind to host molecules and produce a complex that is antigenic |
| Interleukin | Cytokine produced by leukocytes and macrophages to stimulate other leukocytes |
| Perforin | A protein produced by NK and $T_C$ cells that binds to target cells, produces a hole, and admits granzymes into the cell |
| Tumor necrosis factor (TNF) | Cytokine secreted by $T_C$ cells that activates macrophages and kills cancer cells |

*These agents have additional roles in inflammation described in table 21.1.

| TABLE 21.5 | Some Comparisons Between Humoral and Cellular Immunity | |
|---|---|---|
| **Characteristic** | **Cellular Immunity** | **Humoral Immunity** |
| Pathogens | Intracellular viruses, bacteria, yeasts, and protozoans; parasitic worms; cancer cells; transplanted tissues and organs | Extracellular viruses, bacteria, yeasts and protozoans; toxins, venoms, and allergens; mismatched RBCs |
| Effector cells | Cytotoxic T cells | Plasma cells (develop from B cells) |
| Other cells involved in attack | Helper T cells | Helper T cells |
| Antigen-presenting cells | B cells, macrophages, nearly all cells | B cells |
| MHC proteins | MHC-I and MHC-II | MHC-II only |
| Chemical agents of attack | Perforins, granzymes, interferons, tumor necrosis factor | Antibodies, complement |
| Mechanisms of counteracting or destroying pathogens | Cytolysis, phagocytosis, apoptosis | Cytolysis, phagocytosis, immune clearance, inflammation, neutralization, agglutination, precipitation |
| Memory | T cell recall response | Secondary (anamnestic) response |

## Before You Go On

*Answer the following questions to test your understanding of the preceding section:*

21. What is the difference between a B cell and a plasma cell?

22. Describe four ways in which an antibody acts against an antigen.

23. Why does the secondary immune response prevent a pathogen from causing disease, while the primary immune response does not?

# Immune System Disorders

### Objectives

When you have completed this section, you should be able to

- distinguish between the four classes of immune hypersensitivity and give an example of each;

- explain the cause of anaphylaxis and distinguish local anaphylaxis from anaphylactic shock;

- state some reasons immune self-tolerance may fail, and give examples of the resulting diseases; and

- describe the pathology of immunodeficiency diseases, especially AIDS.

Because the immune system involves complex cellular interactions controlled by numerous chemical messengers, there are many points at which things can go wrong. The immune response may be too vigorous, too weak, or misdirected against the wrong targets. A few disorders are summarized here to illustrate the consequences.

# HYPERSENSITIVITY

**Hypersensitivity** is an excessive, harmful immune reaction to antigens that most people tolerate. It includes reactions to tissues transplanted from another person *(alloimmunity),* abnormal reactions to one's own tissues *(autoimmunity),* and **allergies,**[24] which are reactions to environmental antigens. Such antigens, called **allergens,** occur in mold, dust, pollen, vaccines, bee and wasp venoms, animal dander, toxins from poison ivy and other plants, and foods such as nuts, milk, eggs, and shellfish. Drugs such as penicillin, tetracycline, and insulin are allergenic to some people.

One classification system recognizes four kinds of hypersensitivity, distinguished by the types of immune agents (antibodies or T cells) involved and their methods of attack on the antigen. In this system, type I is also characterized as *acute (immediate) hypersensitivity* because the response is very rapid, while types II and III are characterized as *subacute* because they exhibit a slower onset (1–3 hours after exposure) and last longer (10–15 hours). Type IV is a delayed cell-mediated response whereas the other three are quicker antibody-mediated responses.

- **Type I (acute) hypersensitivity** includes the most common allergies. Some authorities use the word *allergy* for type I reactions only, and others use it for all four types. Type I is an IgE-mediated reaction that begins within seconds of exposure and usually subsides within 30 minutes, although it can be severe and even fatal. Allergens bind to IgE on the membranes of basophils and mast cells and stimulate them to secrete histamine and other inflammatory

---

[24]*allo* = altered + *erg* = action, reaction

and vasoactive chemicals. These chemicals trigger glandular secretion, vasodilation, increased capillary permeability, smooth muscle spasms, and other effects. The clinical signs include local edema, mucus hypersecretion and congestion, watery eyes, a runny nose, hives (red itchy skin), and sometimes cramps, diarrhea, and vomiting. Some examples of type I hypersensitivity are food allergies and **asthma,**[25] a local inflammatory reaction to inhaled allergens (see Insight 21.3).

**Anaphylaxis**[26] (AN-uh-fih-LAC-sis) is an immediate and severe type I reaction. Local anaphylaxis can be relieved with antihistamines. **Anaphylactic shock** is a severe, widespread acute hypersensitivity that occurs when an allergen such as bee venom or penicillin is introduced to the bloodstream of an allergic individual. It is characterized by bronchoconstriction, dyspnea (labored breathing), widespread vasodilation, circulatory shock, and sometimes sudden death. Antihistamines are inadequate to counter anaphylactic shock, but epinephrine relieves the symptoms by dilating the bronchioles, increasing cardiac output, and restoring blood pressure.

---

[25]*asthma* = panting
[26]*ana* = against + *phylax* = protection

- **Type II (antibody-dependent cytotoxic) hypersensitivity** occurs when IgG or IgM attacks antigens bound to cell surfaces. The reaction leads to complement activation and either lysis or opsonization of the target cell. Macrophages phagocytize and destroy opsonized platelets, erythrocytes, or other cells. Examples of cell destruction by type II reactions are blood transfusion reactions, pemphigus vulgaris (p. 175), and some drug reactions. In some other type II responses, an antibody binds to cell surface receptors and either interferes with their function (as in myasthenia gravis, p. 436) or overstimulates the cell (as in toxic goiter, p. 668).

- **Type III (immune complex) hypersensitivity** occurs when IgG or IgM forms antigen–antibody complexes that precipitate beneath the endothelium of the blood vessels or in other tissues. At the sites of deposition, these complexes activate complement and trigger intense inflammation, causing tissue destruction. Two examples of type III hypersensitivity are the autoimmune diseases acute glomerulonephritis (p. 924) and systemic lupus erythematosus, a widespread inflammation of the connective tissues (see table 21.6).

- **Type IV (delayed) hypersensitivity** is a cell-mediated reaction in which the signs appear about 12 to 72

---

## INSIGHT 21.3 Clinical Application

### Asthma

Asthma is the most common chronic illness of children, especially boys. It is the leading cause of school absenteeism and childhood hospitalization in the United States. About half of all cases develop before age 10 and only 15% after age 40. In the United States, it affects about 5% of adults and up to 10% of children, and takes about 5,000 lives per year. Moreover, asthma is on the rise; there are many more cases and deaths now than there were a few decades ago.

In *allergic (extrinsic) asthma,* the most common form, a respiratory crisis is triggered by allergies in pollen, mold, animal dander, food, dust mites, or cockroaches. The allergens stimulate plasma cells to secrete IgE, which binds to mast cells of the respiratory mucosa. Reexposure to the allergen causes the mast cells to release a complex mixture of histamine, interleukins, and several other inflammatory chemicals, which trigger intense airway inflammation. *Nonallergic (intrinsic) asthma* is not caused by allergens but can be triggered by infections, drugs, air pollutants, cold dry air, exercise, or emotions. This form is more common in adults over age 35 than in children. The effects, however, are much the same.

Within minutes, the bronchioles constrict spasmodically *(bronchospasm),* and a person exhibits severe coughing, wheezing, and sometimes fatal suffocation. A second respiratory crisis often occurs 6 to 8 hours later. Interleukins attract eosinophils to the bronchial tissue, where they secrete proteins that paralyze the respiratory cilia, severely damage the epithelium, and lead to scarring and extensive long-term damage to the lungs. The bronchioles also become edematous and plugged with thick, sticky mucus. People who die of asthmatic suffocation typically show airways so plugged with gelatinous mucus that they could not exhale. The lungs remain hyperinflated even at autopsy.

Asthma is treated with epinephrine and other β-adrenergic stimulants to dilate the airway and restore breathing, and with inhaled corticosteroids or nonsteroidal anti-inflammatory drugs to minimize airway inflammation and long-term damage. The treatment regimen can be very complicated, often requiring more than eight different medications daily, and compliance is therefore difficult for children and patients with low income or educational attainment.

Asthma runs in families and seems to result from a combination of hereditary factors and environmental irritants. In the United States, asthma is most common, paradoxically, in two groups: (1) inner-city children who are exposed to crowding, poor sanitation and ventilation, and who do not go outside very much or get enough exercise; and (2) children from extremely clean homes, perhaps because they have had too little opportunity to develop normal immunities. Asthma is also more common in countries where vaccines and antibiotics are widely used. It is less common in developing countries and in farm children of the United States.

hours after exposure. It begins when APCs in the lymph nodes display antigens to helper T cells, and these T cells secrete interferon and other cytokines that activate cytotoxic T cells and macrophages. The result is a mixture of nonspecific and immune responses. Type IV reactions include allergies to haptens in cosmetics and poison ivy, graft rejection, the tuberculosis skin test, and the beta cell destruction that causes insulin-dependent diabetes mellitus.

## AUTOIMMUNE DISEASES

**Autoimmune diseases** are failures of self-tolerance—the immune system fails to distinguish self-antigens from foreign ones and produces **autoantibodies** that attack the body's own tissues. There are at least three reasons why self-tolerance may fail:

1. **Cross-reactivity.** Some antibodies against foreign antigens react to similar self-antigens. In rheumatic fever, for example, a streptococcus infection stimulates production of antibodies that react not only against the bacteria but also against antigens of the heart tissue. It often results in scarring and stenosis (narrowing) of the mitral and aortic valves.

2. **Abnormal exposure of self-antigens to the blood.** Some of our native antigens are normally not exposed to the blood. For example, a blood–testis barrier (BTB) normally isolates sperm cells from the blood. Breakdown of the BTB can cause sterility when sperm first form in adolescence and activate the production of autoantibodies.

3. **Change in the structure of self-antigens.** Viruses and drugs may change the structure of self-antigens and cause the immune system to perceive them as foreign. One theory of type I diabetes mellitus is that a viral infection alters the antigens of the insulin-producing beta cells of the pancreatic islets, which leads to an autoimmune attack on the cells.

## IMMUNODEFICIENCY DISEASES

In the foregoing diseases, the immune system reacts too vigorously or directs its attack against the wrong targets. In immunodeficiency diseases, by contrast, the immune system fails to respond vigorously enough.

### Severe Combined Immunodeficiency Disease

**Severe combined immunodeficiency disease (SCID)** is a group of disorders caused by recessive alleles that result in a scarcity or absence of both T and B cells. Children with SCID are highly vulnerable to opportunistic infections and must live in protective enclosures. Perhaps the most publicized case was David Vetter, who spent his life

**FIGURE 21.30    Boy with Severe Combined Immunodeficiency Disease.**  David lived with SCID from 1971 to 1984. At the age of 6, he received a portable sterile suit designed by NASA that allowed him to leave the hospital for the first time.

in sterile plastic chambers and suits (fig. 21.30), finally succumbing at age 12 to cancer triggered by a viral infection. Children with SCID are sometimes helped by transplants of bone marrow or fetal thymus, but in some cases the transplanted cells fail to survive and multiply, or transplanted T cells attack the patient's tissues (the *graft-versus-host response*). David contracted a fatal virus from his sister through a bone marrow transplant.

### Acquired Immunodeficiency Syndrome

Acquired immunodeficiency diseases are nonhereditary diseases contracted after birth. The best-known example is **acquired immunodeficiency syndrome (AIDS),** a group of conditions that involve a severely depressed immune response caused by infection with the **human immunodeficiency virus (HIV).**

The structure of HIV is shown in figure 21.31a. Its inner core consists of a protein *capsid* enclosing two molecules of RNA, two molecules of an enzyme called *reverse transcriptase,* and a few other enzyme molecules. The capsid is enclosed in another layer of viral protein, the *matrix.* External to this is a *viral envelope* composed

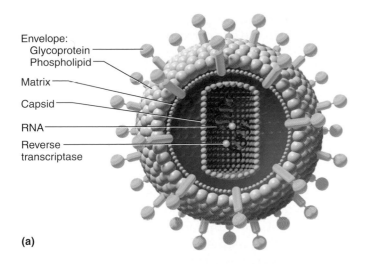

Envelope:
Glycoprotein
Phospholipid
Matrix
Capsid
RNA
Reverse
transcriptase

(a)

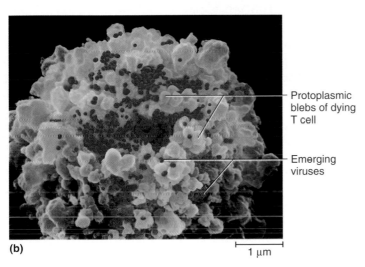

Protoplasmic
blebs of dying
T cell

Emerging
viruses

(b)

1 μm

**FIGURE 21.31** **The Human Immunodeficiency Virus (HIV).**
(a) Structure of the virus. (b) Viruses emerging from a dying
helper T cell. Each virus can now invade a new helper T cell and
produce a similar number of descendents.
▶ *Which of the molecules in part (a) is the target of the drug azidothymidine (AZT)? Why does AZT inhibit the spread of HIV?*

of phospholipids and glycoproteins derived from the
host cell. Like other viruses, HIV can be replicated only
by a living host cell. It invades helper T (CD4) cells, dendritic cells, and macrophages. HIV adheres to a target cell
by means of one of its envelope glycoproteins and
"tricks" the target cell into internalizing it by receptor-
mediated endocytosis. Within the host cell, reverse transcriptase uses the viral RNA as a template to synthesize
DNA—the opposite of the usual process of genetic transcription. Viruses that carry out this RNA → DNA reverse
transcription are called *retroviruses*.[31] The new DNA is

inserted into the host cell's DNA, where it may lie dormant for months to years. When activated, however, it
induces the host cell to produce new viral RNA, capsid
proteins, and matrix proteins. As the new viruses emerge
from the host cell (fig. 21.31b), they are coated with bits
of the cell's plasma membrane, forming the new viral
envelope. The new viruses then adhere to more host cells
and repeat the process.

By destroying $T_H$ cells, HIV strikes at a central coordinating agent of nonspecific defense, humoral immunity,
and cellular immunity (see fig. 21.23). The incubation
period—the time from infection to the occurrence of the
first symptoms—can range from a few months to 12 years.
Flulike episodes of chills and fever occur as HIV attacks $T_H$
cells. At first, antibodies against HIV are produced and the
$T_H$ count returns nearly to normal. As the virus destroys
more and more cells, however, the signs and symptoms
become more pronounced: night sweats, fatigue, headache,
extreme weight loss, and lymphadenitis.

A normal $T_H$ count is 600 to 1,200 cells/μL, but a criterion of AIDS is a $T_H$ count less than 200 cells/μL. With
such severe depletion of $T_H$ cells, a person succumbs to
opportunistic infections with such pathogens as
*Toxoplasma* (a protozoan previously known mainly for
causing birth defects), *Pneumocystis* (a group of respiratory fungi), herpes simplex virus, cytomegalovirus (which
can cause blindness), or tuberculosis bacteria. White
patches may appear in the mouth, caused by *Candida*
(thrush) or Epstein–Barr[32] virus (leukoplakia). A form of
cancer called Kaposi[33] sarcoma, common in AIDS
patients, originates in the endothelial cells of the blood
vessels and causes bruiselike purple lesions visible in the
skin (fig. 21.32).

Patients with full-blown AIDS show no response to
standard skin tests for delayed hypersensitivity. Slurred
speech, loss of motor and cognitive functions, and
dementia may occur as HIV invades the brain by way of
infected phagocytes (microglia) and induces them to
release toxins that destroy neurons and astrocytes. Death
from cancer or infection is inevitable, usually within a
few months but sometimes as long as 8 years after diagnosis. Some people, however, have been diagnosed as HIV-
positive and yet have survived for 10 years or longer
without developing AIDS.

HIV is transmitted through blood, semen, vaginal
secretions, and breast milk. It can be transmitted from
mother to fetus through the placenta or from mother to
infant during childbirth or nursing. HIV occurs in saliva
and tears, but is not believed to be transmitted by those
fluids. The most common means of transmission are sexual intercourse (vaginal, anal, or oral), contaminated
blood products, and drug injections with contaminated

[31]*retr* = an acronym from *reverse transcription*

[32]M. A. Epstein (1921– ), British physician; Y. M. Barr (1932– ), British virologist
[33]Moritz Kaposi (1837–1902), Austrian physician

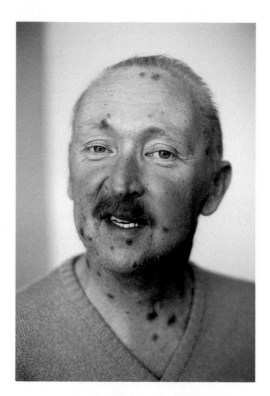

**FIGURE 21.32   Kaposi Sarcoma.** Typical lesions on the face and neck of a person with AIDS.

needles. Worldwide, about 75% of HIV infections are acquired through heterosexual, predominantly vaginal intercourse. In the United States, most cases occur in men who have sex with other men, but adolescents are the fastest rising group of AIDS patients because of the increasing exchange of unprotected sexual intercourse for drugs. The sharing of needles for drug use remains the chief means of transmission in urban ghettos. Many hemophiliacs became infected with HIV through blood transfusions before preventive measures were implemented in 1984, but all donated blood is now tested for HIV and the risk of infection is less than 1%. HIV cannot be contracted by donating blood, but irrational fear has resulted in an alarming drop in blood donors.

AIDS is not known to be transmitted through casual contact—for example, to family members, friends, coworkers, classmates, or medical personnel in charge of AIDS patients. It is not transmitted by kissing. Despite some speculation and fear, it has not been found to be transmitted by mosquitoes or other blood-sucking arthropods.

HIV survives poorly outside the human body. It is destroyed by laundering, dishwashing, exposure to heat (50°C [135°F] for at least 10 minutes), chlorination of swimming pools and hot tubs, and disinfectants such as bleach, Lysol, hydrogen peroxide, rubbing alcohol, and germicidal skin cleansers (Betadine and Hibiclens, for example). A properly used, undamaged latex condom is an effective barrier to HIV. Animal membrane condoms do not block HIV transmission because the viruses are smaller than the gaps in the membrane.

The AIDS epidemic has triggered an effort of unprecedented intensity to find a vaccine or cure. The strategies against HIV include efforts to prevent its binding to the CD4 proteins of $T_H$ cells, disrupting the action of reverse transcriptase, or inhibiting the assembly of new viruses or their release from host cells. HIV is a difficult pathogen to attack. Since it hides within host cells, it usually escapes recognition by the immune system. In the brain, it is protected by the blood–brain barrier.

About 1% of HIV's genes mutate every year. This rapid rate of mutation is a barrier to both natural immunity and development of a vaccine. Even when immune cells do become sensitized to HIV, the virus soon mutates and produces new surface antigens that escape recognition. The high mutation rate also would quickly make today's vaccine ineffective against tomorrow's strain of the virus. Another obstacle to treatment and prevention is the lack of animal models for vaccine and drug research and development. Most animals are not susceptible to HIV. The chimpanzee is an exception, but chimpanzees are difficult to maintain, and there are economic barriers and ethical controversies surrounding their use.

The first anti-HIV drug approved by the Food and Drug Administration (FDA) was azidothymidine (AZT, or Retrovir), which inhibited reverse transcriptase and prolonged the lives of some HIV-positive individuals. In 1996, a family of drugs called protease inhibitors became available, typically used in a "triple cocktail" combining these with two reverse transcriptase inhibitors. But by 1997, HIV had evolved resistance even to those drugs and they were failing in more than half of all patients. Today, more than 16 anti-HIV drugs are on the market, typically used in combinations of three or more. While such drug combinations have substantially reduced AIDS morbidity and mortality (disease and death) and reduced the hospital and hospice census, none of them can totally eliminate HIV from the body, and all of them have serious side effects that contraindicate their long-term use. The major unresolved questions in AIDS therapy today are when to start drug treatment, which drugs to use, when to switch drugs, how to improve patient compliance, and how to make therapy available in impoverished nations where AIDS is especially rampant.

There also remain a number of unanswered questions about the basic biology of HIV. It is still unknown, for example, why there are such strikingly different patterns of heterosexual versus homosexual transmission in different countries and why some people succumb so rapidly to infection, while others can be HIV-positive for years without developing AIDS. AIDS remains a stubborn problem sure to challenge virologists and epidemiologists for many years to come.

We have surveyed the major classes of immune system disorders and a few particularly notorious immune diseases. A few additional lymphatic and immune system disorders are described in table 21.6. The effects of aging on the lymphatic and immune systems are described on page 1131.

## Before You Go On

*Answer the following questions to test your understanding of the preceding section:*

24. How does subacute hypersensitivity differ from acute hypersensitivity? Give an example of each.

25. Aside from the time required for a reaction to appear, how does delayed hypersensitivity differ from the acute and subacute types?

26. State some reasons why antibodies may begin attacking self-antigens that they did not previously respond to. What are these self-reactive antibodies called?

27. What is the distinction between a person who has an HIV infection and a person who has AIDS?

28. How does a reverse transcriptase inhibitor such as AZT slow the progress of AIDS?

| TABLE 21.6 | Some Disorders of the Lymphatic and Immune Systems |
|---|---|
| Contact dermatitis | A form of delayed hypersensitivity that produces skin lesions limited to the site of contact with an allergen or hapten; includes responses to poison ivy, cosmetics, latex, detergents, industrial chemicals, and some topical medicines. |
| Hives (urticaria[27]) | An allergic skin reaction characterized by a "wheal and flare" reaction: white blisters (wheals) surrounded by reddened areas (flares), usually with itching. Caused by local histamine release in response to allergens. Can be triggered by food or drugs, but sometimes by nonimmunological factors such as cold, friction, or emotional stress. |
| Hodgkin[28] disease | A lymph node malignancy, with early symptoms including enlarged painful lymph nodes, especially in the neck, and fever of unknown origin; often progresses to neighboring lymph nodes. Radiation and chemotherapy cure about three out of four patients. |
| Splenomegaly[29] | Enlargement of the spleen, sometimes without underlying disease but often indicating infections, autoimmune diseases, heart failure, cirrhosis, Hodgkin disease, and other cancers. The enlarged spleen may "hoard" erythrocytes, causing anemia, and may become fragile and subject to rupture. |
| Systemic lupus erythematosus[30] | Formation of autoantibodies against DNA and other nuclear antigens, resulting in accumulation of antigen–antibody complexes in blood vessels and other organs, where they trigger widespread connective tissue inflammation. Named for skin lesions once likened to a wolf bite. Causes fever, fatigue, joint pain, weight loss, intolerance of bright light, and a "butterfly rash" across the nose and cheeks. Death may result from renal failure. |

*Disorders described elsewhere*

| | | |
|---|---|---|
| Acute glomerulonephritis p. 924 | Diabetes mellitus p. 669 | Rheumatic fever p. 842 |
| AIDS p. 842 | Elephantiasis p. 810 | Rheumatoid arthritis p. 315 |
| Allergy p. 840 | Lymphadenitis p. 818 | SCID p. 842 |
| Anaphylaxis p. 841 | Myasthenia gravis p. 436 | Toxic goiter p. 668 |
| Asthma p. 841 | Pemphigus vulgaris p. 175 | |

[27]*urtica* = nettle
[28]Thomas Hodgkin (1798–1866), British physician
[29]*megaly* = enlargement
[30]*lupus* = wolf + *erythema* = redness

**INSIGHT 21.4**    Clinical Application

## Neuroimmunology—The Mind–Body Connection

*Neuroimmunology* is a relatively new branch of medicine concerned with the relationship between mind and body in health and disease. It is attempting especially to understand how a person's state of mind influences health and illness through a three-way communication between the nervous, endocrine, and immune systems.

The sympathetic nervous system issues nerve fibers to the spleen, thymus, lymph nodes, and Peyer patches, where nerve fibers contact thymocytes, B cells, and macrophages. These immune cells have adrenergic receptors for norepinephrine and many other neurotransmitters such as neuropeptide Y, substance P, and vasoactive intestinal peptide (VIP). These neurotransmitters have been shown to influence immune cell activity in various ways. Epinephrine, for example, reduces the lymphocyte count and inhibits NK cell activity, thus suppressing immune surveillance and specific immunity. Cortisol, another stress hormone, inhibits T cell and macrophage activity, antibody production, and the secretion of inflammatory chemicals. It also promotes atrophy of the thymus, spleen, and lymph nodes and reduces the number of circulating lymphocytes, macrophages, and eosinophils. Thus, it is not surprising that prolonged stress increases susceptibility to illnesses such as infections and cancer.

The immune system also sends messages to the nervous and endocrine systems. Immune cells synthesize numerous hormones and neurotransmitters that we normally associate with endocrine and nerve cells. B lymphocytes produce adrenocorticotropic hormone (ACTH) and enkephalins; T lymphocytes produce growth hormone, thyroid-stimulating hormone, luteinizing hormone, and follicle-stimulating hormone. Monocytes secrete prolactin, VIP, and somatostatin. The interleukins and tumor-necrosis factor (TNF) produced by immune cells produce feelings of fatigue and lethargy when we are sick, and stimulate the hypothalamus to secrete corticotropin-releasing hormone, thus leading to ACTH and cortisol secretion. It remains uncertain and controversial whether the quantities of some of these substances produced by immune cells are enough to have far-reaching effects on the body, but it seems increasingly possible that immune cells may have wide-ranging effects on nervous and endocrine functions that affect recovery from illness.

Although neuroimmunology has met with some skepticism among physicians, there is less and less room for doubt about the importance of a person's state of mind to immune function. People under stress, such as medical students during examination periods and people caring for relatives with Alzheimer disease, show more respiratory infections than other people and respond less effectively to hepatitis and flu vaccines. The attitudes, coping abilities, and social support systems of patients significantly influence survival time even in such serious diseases as AIDS and breast cancer. Women with breast cancer die at markedly higher rates if their husbands cope poorly with stress. Attitudes such as optimism, cheer, depression, resignation, or despair in the face of disease significantly affect immune function. Religious beliefs can also influence the prospect of recovery. Indeed, ardent believers in voodoo sometimes die just from the belief that someone has cast a spell on them. The stress of hospitalization can counteract the treatment one gives to a patient, and neuroimmunology has obvious implications for treating patients in ways that minimize their stress and thereby promote recovery.

# Interactions Between the
## LYMPHATIC AND IMMUNE SYSTEMS
## and Other Organ Systems

| ■ | indicates ways in which this system affects other systems | ■ | indicates ways in which other systems affect this system |
|---|---|---|---|

## INTEGUMENTARY SYSTEM

Skin provides mechanical and chemical barriers to pathogens; has antigen-presenting cells in epidermis and dermis; and is a common site of inflammation

## SKELETAL SYSTEM

Lymphocytes and macrophages arise from bone marrow cells; skeleton protects thymus and spleen

## MUSCULAR SYSTEM

Skeletal muscle pump moves lymph through lymphatic vessels

## NERVOUS SYSTEM

Neuropeptides and emotional states affect immune function; blood–brain barrier prevents antibodies and immune cells from entering brain tissue

## ENDOCRINE SYSTEM

Lymph transports some hormones

Hormones from thymus stimulate development of lymphatic organs and T cells; stress hormones depress immunity and increase susceptibility to infection and cancer

## CIRCULATORY SYSTEM

Lymphatics return fluid to bloodstream; spleen disposes of expired RBCs; lymphatic organs prevent accumulation of debris in blood

Lymphatics develop from embryonic veins; WBCs serve in defense and immunity; blood transports immune cells and chemicals; endothelial cells mediate WBC margination and dispedesis; clotting restricts spread of pathogens

## NEARLY ALL SYSTEMS

Lymphatic system drains excess tissue fluid and removes cellular debris and pathogens. Immune system provides defense against pathogens and immune surveillance against cancer.

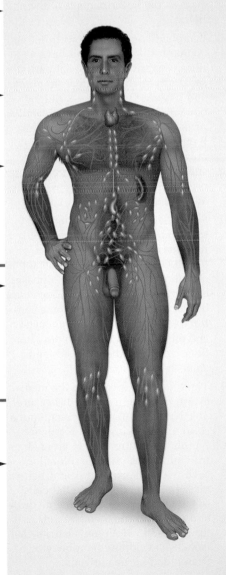

## RESPIRATORY SYSTEM

Alveolar macrophages remove debris from lungs

Provides immune system with $O_2$; disposes of $CO_2$; thoracic pump aids lymph flow; pharynx houses tonsils

## URINARY SYSTEM

Absorbs fluid and proteins in kidneys, which is essential to enabling kidneys to concentrate the urine and conserve water

Eliminates waste and maintains fluid and electrolyte balance important to lymphatic and immune function; urine flushes some pathogens from body; acidic pH of urine protects against urinary tract infection

## DIGESTIVE SYSTEM

Lymph absorbs and transports digested lipids

Nourishes lymphatic system and affects lymph composition; stomach acid destroys ingested pathogens

## REPRODUCTIVE SYSTEM

Immune system requires that the testes have a blood–testis barrier to prevent autoimmune destruction of sperm

Vaginal acidity inhibits growth of pathogens

# CHAPTER REVIEW

# Review of Key Concepts

**The Lymphatic System (p. 808)**

1. The lymphatic system serves to recover tissue fluid and maintain fluid balance; provide immune cells and monitor the body fluids for foreign matter; and transport dietary lipids from the small intestine to the blood.

2. The lymphatic system consists of lymph, lymphatic vessels, lymphatic tissues, and lymphatic organs.

3. Lymph is usually a colorless liquid similar to blood plasma, but is milky when absorbing digested lipids. It contains lymphocytes, macrophages, and hormones, and may contain metastasizing cancer cells, cellular debris, bacteria, and viruses.

4. Lymph originates in blind *lymphatic capillaries* that pick up tissue fluid throughout the body. The endothelial cells of lymphatic capillaries have large gaps between them that permit cells and other large particles to enter the lymph stream.

5. Lymphatic capillaries converge to form larger collecting vessels with a histology similar to that of veins. Lymph nodes lie at irregular intervals along the collecting vessels and filter the lymph on its way back to the blood.

6. Collecting vessels converge to form six lymphatic trunks that drain specific regions of the body. The lymphatic trunks then converge to form two collecting ducts—the *right lymphatic duct* and *thoracic duct*—which empty lymph into the subclavian veins.

7. There is no heartlike pump to move the lymph; lymph flows under forces similar to those that drive venous return, and like some veins, lymphatic vessels have valves to ensure a one-way flow.

8. The cells of lymphatic tissue are natural killer (NK) cells, T lympho-cytes, B lymphocytes, macrophages, dendritic cells, and reticular cells.

9. *Diffuse lymphatic tissue* is an aggregation of lymphatic cells in the walls of other organs, especially in the mucous membranes of the respiratory, digestive, urinary, and reproductive tracts—where it is called *mucosa-associated lymphatic tissue (MALT).* In some places, lymphocytes and macrophages form dense masses called *lymphatic nodules,* such as the *Peyer patches* of the ileum.

10. Lymphatic organs have well-defined anatomical locations and have a fibrous capsule that at least partially separates them from adjacent organs and tissues. Two of these, the red bone marrow and thymus, are called *primary lymphatic organs* because lymphocytes mature here before colonizing the other sites. The other sites, called *secondary lymphatic organs,* are the lymph nodes, tonsils, and spleen.

11. Red bone marrow is a hemopoietic tissue and the point of origin of all immune cells of the lymphatic system. It consists of *islands* of hemopoietic tissue composed of macrophages and developing blood cells, separated by *sinusoids* that converge on a *central longitudinal vein.* Lymphocytes and other formed elements pass from the islands into the sinusoids and enter the bloodstream.

12. The *thymus* is located in the mediastinum above the heart. It is a site of T lymphocyte development and a source of cytokines that regulate lymphocyte activity. It is divided into numerous polygonal lobules, each with a dense cortex and lighter medulla. Reticular epithelial cells separate the cortex from the medulla and surround the blood vessels, forming a blood–thymus barrier that isolates developing lymphocytes from blood-borne antigens.

13. Lymph nodes are numerous small, bean-shaped organs that receive lymph through *afferent lymphatic vessels,* filter it, and pass it along via *efferent lymphatic vessels* that exit the hilum. They monitor the lymph for foreign antigens, remove impurities before it returns to the bloodstream, contribute lymphocytes to the lymph and blood, and mount immune responses to foreign antigens.

14. The parenchyma of a lymph node exhibits an outer *cortex* composed mainly of lymphatic nodules, and a deeper *medulla* with a network of *medullary cords.* B cells multiply and differentiate into plasma cells in the germinal centers of the nodules. T cells are very concentrated in the *deep cortex* next to the medulla.

15. Lymph nodes are widespread but especially concentrated in cervical, axillary, thoracic, abdominal, intestinal, mesenteric, inguinal, and popliteal groups.

16. The *tonsils* encircle the pharynx and guard against inhaled and ingested pathogens. They include a medial *pharyngeal tonsil* in the nasopharynx, a pair of *palatine tonsils* at the rear of the oral cavity, and numerous *lingual tonsils* clustered in the root of the tongue. Their superficial surface is covered with epithelium and their deep surface with a fibrous partial capsule. They have deep pits called *tonsillar crypts* bordered by rows of lymphatic follicles.

17. The *spleen* lies in the left hypochondriac region between the diaphragm, stomach, and kidney. It monitors the blood for foreign antigens, activates immune responses to them, disposes of old RBCs, and helps to regulate blood volume. Its parenchyma is composed of *red pulp* containing concentrated RBCs and *white pulp* composed of lymphocytes and macrophages.

**Nonspecific Resistance (p. 820)**

1. The body has three lines of defense against pathogens: (1) external physical barriers, (2) nonspecific defenses, and (3) the immune system. The first two mechanisms are called *nonspecific resistance* because they guard equally against a broad range of pathogens and do not require prior exposure. Immunity is a *specific defense* limited to one pathogen or a few closely related ones.

2. The skin acts as a barrier to pathogens because of its tough keratinized surface, its relative dryness, and antimicrobial chemicals such as lactic acid and *defensins.*

3. Mucous membranes prevent most pathogens from entering the body because of the stickiness of the mucus, the antimicrobial action of *lysozyme,* and the viscosity of *hyaluronic acid.*

4. *Neutrophils,* the most abundant leukocytes, destroy bacteria by phagocytizing and digesting them and by a *respiratory burst* that produces a chemical *killing zone* of oxidizing agents.

5. *Eosinophils* produce antiparasitic agents, promote basophil and mast cell action, and control the action of inflammatory chemicals.

6. *Basophils* aid in defense by secreting *histamine* and *heparin,* as do their counterparts in the connective tissues, *mast cells.* These chemicals promote blood flow and leukocyte mobility.

7. *Lymphocytes* are of several kinds. Only one type, the natural killer (NK) cells, are involved in nonspecific defense. They are the agents of a defense mechanism called *immune surveillance.*

8. *Monocytes* develop into macrophages, which have voracious phagocytic activity and act as antigen-presenting cells. Some organs have specific types of macrophages called *dendritic cells, microglia,* and *alveolar* and *hepatic macrophages.*

9. *Interferons* are polypeptides secreted by cells in response to viral infection. They alert neighboring cells to synthesize antiviral proteins before

they become infected, and they activate NK cells and macrophages.

10. The *complement system* is a group of 30 or more globulins that are activated by pathogens and combat them by enhancing inflammation, *opsonizing* bacteria, and causing *immune clearance* of antigens and *cytolysis* of foreign cells.

11. *Immune surveillance* is a process in which NK cells nonspecifically detect and destroy foreign cells and diseased host cells. They employ *perforins* to create a hole in the enemy cell surface and *granzymes* to destroy its intracellular enzymes and induce cell death.

12. *Inflammation* is a defensive response to infection and trauma, characterized by redness, swelling, heat, and pain (the four *cardinal signs*).

13. Inflammation begins with a mobilization of defenses by vasoactive inflammatory chemicals such as histamine, kinins, and leukotrienes. These chemicals dilate blood vessels, increase blood flow, and make capillary walls more permeable, thus hastening the delivery of defensive cells and chemicals to the site of injury.

14. Leukocytes adhere to the vessel wall *(margination),* crawl between the endothelial cells into the connective tissues *(diapedesis),* and migrate toward sources of inflammatory chemicals *(chemotaxis).*

15. Inflammation continues with containment and destruction of the pathogens. This is achieved by clotting of the tissue fluid and attack by macrophages, leukocytes, and antibodies.

16. Inflammation concludes with tissue cleanup and repair, including phagocytosis of tissue debris and pathogens by macrophages, edema and lymphatic drainage of the inflamed tissue, and tissue repair stimulated by platelet-derived growth factor.

17. *Fever* (pyrexia) is induced by chemical *pyrogens* secreted by neutrophils and macrophages. The elevated body temperature inhibits the reproduc-

tion of pathogens and the spread of infection.

**General Aspects of Specific Immunity (p. 828)**

1. The *immune system* is a group of widely distributed cells that populate most body tissues and help to destroy pathogens.

2. Immunity is characterized by its *specificity* and *memory.*

3. The two basic forms of immunity are *cellular* (cell-mediated) and *humoral* (antibody-mediated).

4. Immunity can also be characterized as *active* (production of the body's own antibodies or immune cells) or *passive* (conferred by antibodies or lymphocytes donated by another individual), and as *natural* (caused by natural exposure to a pathogen) or *artificial* (induced by vaccination or injection of immune serum). Only active immunity results in immune memory and lasting protection.

5. *Antigens* are any molecules that induce immune responses. They are relatively large, complex, genetically unique molecules (proteins, polysaccharides, glycoproteins, and glycolipids).

6. The *antigenicity* of a molecule is due to specific regions of it called *epitopes.*

7. *Haptens* are small molecules that become antigenic by binding to larger host molecules.

8. *T cells* are lymphocytes that mature in the thymus, survive the process of *negative selection,* and go on to populate other lymphatic tissues and organs.

9. *B cells* are lymphocytes that mature in the bone marrow, survive negative selection, and then populate the same organs as T cells.

10. *Antigen-presenting cells (APCs)* are B cells, macrophages, reticular cells, and dendritic cells that process antigens, display the epitopes on their surface MHC proteins, and alert the immune system to the presence of a pathogen.

11. *Interleukins* are chemical signals by which immune cells communicate with each other.

## Cellular Immunity (p. 832)

1. Cellular immunity employs three classes of T lymphocytes: *cytotoxic (T_C), helper (T_H),* and *memory (T_M) T cells.*

2. Cellular immunity takes place in three stages: recognition, attack, and memory.

3. *Recognition:* APCs that detect foreign antigens typically migrate to the lymph nodes and display the epitopes there. $T_H$ and $T_C$ cells respond only to epitopes attached to MHC proteins (MHCPs).

4. MHC-I proteins occur on every nucleated cell of the body and display viral and cancer-related proteins from the host cell. $T_C$ cells respond only to antigens bound to MHC-I proteins.

5. MHC-II proteins occur only on APCs and display only foreign antigens. $T_H$ cells respond only to antigens bound to MHC-II proteins.

6. When a $T_C$ or $T_H$ cell recognizes an antigen-MHCP complex, it binds to a second site on the target cell. *Costimulation* by this site triggers *clonal selection,* multiplication of the T cell. Some daughter T cells carry out the attack on the invader and some become $T_M$ cells.

7. *Attack:* Activated $T_H$ cells secrete interleukins that attract neutrophils, NK cells, and macrophages, and stimulate T and B cell mitosis and maturation. Activated $T_C$ cells directly attack and destroy target cells, especially infected host cells, transplanted cells, and cancer cells. They employ a "lethal hit" of cytotoxic chemicals including *perforin, granzymes, interferons,* and *tumor necrosis factor.*

8. *Memory:* The primary response to the first exposure to a pathogen is followed by immune memory. Upon later reexposure, $T_M$ cells respond so quickly (the *T cell recall response*) that no noticeable illness occurs.

## Humoral Immunity (p. 834)

1. Humoral immunity is based on the production of antibodies rather than on lymphocytes directly contacting and attacking enemy cells. It also occurs in recognition, attack, and memory stages.

2. *Recognition:* An immunocompetent B cell binds and internalizes an antigen, processes it, and displays epitopes on its surface MHC-II proteins. A $T_H$ cell binds to the antigen–MHCP complex and secretes interleukins that activate the B cell.

3. The B cell divides repeatedly. Some daughter cells become memory B cells while others become antibody-synthesizing *plasma cells.*

4. *Attack:* Attack is carried out by antibodies (immunoglobulins). The basic *antibody monomer* is a T- or Y-shaped complex of four polypeptide chains (two heavy and two light chains). Each has a *constant (C) region* that is identical in all antibodies of a given class, and a *variable (V) region* that gives each antibody its uniqueness. Each has an *antigen-binding site* at the tip of each V region and can therefore bind two antigen molecules.

5. There are five classes of antibodies—IgA, IgD, IgE, IgG, and IgM—that differ in the number of antibody monomers (from one to five), structure of the C region, and immune function (table 21.3).

6. Antibodies inactivate antigens by *neutralization, complement fixation, agglutination,* and *precipitation.*

7. *Memory:* Upon reexposure to the same antigen, memory B cells mount a *secondary (anamnestic) response* so quickly that no illness results.

## Immune System Disorders (p. 840)

1. There are three principal dysfunctions of the immune system: too vigorous or too weak a response, or a response that is misdirected against the wrong target.

2. *Hypersensitivity* is an excessive reaction against antigens that most people tolerate. *Allergy* is the most common form of hypersensitivity.

3. *Type I (acute) hypersensitivity* is an IgE-mediated response that begins within seconds of exposure and subsides within about 30 minutes. Examples include asthma, anaphylaxis, and anaphylactic shock.

4. *Type II (antibody-dependent cytotoxic) hypersensitivity* occurs when IgG or IgM attacks antigens bound to a target cell membrane, as in a transfusion reaction.

5. *Type III (immune complex) hypersensitivity* results from widespread deposition of antigen–antibody complexes in various tissues, triggering intense inflammation, as in acute glomerulonephritis and systemic lupus erythematosus.

6. *Type IV (delayed) hypersensitivity* is a cell-mediated reaction (types I–III are antibody-mediated) that appears 12–72 hours after exposure, as in the reaction to poison ivy and the TB skin test.

7. *Autoimmune diseases* are disorders in which the immune system fails to distinguish self-antigens from foreign antigens and attacks the body's own tissues. They can occur because of cross-reactivity of antibodies, as in rheumatic fever; abnormal exposure of some self-antigens to the blood, as in one form of sterility resulting from sperm destruction; or changes in self-antigen structure, as in type I diabetes mellitus.

8. *Immunodeficiency diseases* are failures of the immune system to respond strongly enough to defend the body from pathogens. These include *severe combined immunodeficiency disease (SCID),* present at birth, and *acquired immunodeficiency disease (AIDS),* resulting from HIV infection.

9. HIV is a retrovirus that destroys $T_H$ cells. Since $T_H$ cells play a central coordinating role in cellular and humoral immunity and nonspecific defense, HIV knocks out the central control over multiple forms of defense and leaves a person vulnerable to *opportunistic infections* and certain forms of cancer.

# Testing Your Recall

1. The only lymphatic organ with both afferent and efferent lymphatic vessels is
   a. the spleen.
   b. a lymph node.
   c. a tonsil.
   d. a Peyer patch.
   e. the thymus.

2. Which of the following cells are involved in nonspecific resistance but not in specific defense?
   a. helper T cells
   b. cytotoxic T cells
   c. natural killer cells
   d. B cells
   e. plasma cells

3. The respiratory burst is used by _____ to kill bacteria.
   a. neutrophils
   b. basophils
   c. mast cells
   d. NK cells
   e. cytotoxic T cells

4. Which of these is a macrophage?
   a. a microglial cell
   b. a plasma cell
   c. a reticular cell
   d. a helper T cell
   e. a mast cell

5. The cytolytic action of the complement system is most similar to the action of
   a. interleukin-1.
   b. platelet-derived growth factor.
   c. granzymes.
   d. perforin.
   e. IgE.

6. _____ become antigenic by binding to larger host molecules.
   a. Epitopes
   b. Haptens
   c. Lymphokines
   d. Pyrogens
   e. Cell-adhesion molecules

7. Which of the following correctly states the order of events in humoral immunity? Let 1 = antigen display, 2 = antibody secretion, 3 = secretion of interleukin, 4 = clonal selection, and 5 = endocytosis of an antigen.
   a. 3–4–1–5–2
   b. 5–3–1–2–4
   c. 3–5–1–4–2
   d. 5–3–1–4–2
   e. 5–1–3–4–2

8. The cardinal signs of inflammation include all of the following *except*
   a. redness.
   b. swelling.
   c. heat.
   d. fever.
   e. pain.

9. A helper T cell can bind only to another cell that has
   a. MHC-II proteins.
   b. an epitope.
   c. an antigen-binding site.
   d. a complement-binding site.
   e. a CD4 protein.

10. Which of the following results from a lack of self-tolerance?
    a. SCID
    b. AIDS
    c. systemic lupus erythematosus
    d. anaphylaxis
    e. asthma

11. Any organism or substance capable of causing disease is called a/an _____.

12. Mucous membranes contain an antibacterial enzyme called _____.

13. _____ is a condition in which one or more lymph nodes are swollen and painful to the touch.

14. The movement of leukocytes through the capillary wall is called _____.

15. In the process of _____, complement proteins coat bacteria and serve as binding sites for phagocytes.

16. Any substance that triggers a fever is called a/an _____.

17. The chemical signals produced by leukocytes to stimulate other leukocytes are called _____.

18. Part of an antibody called the _____ binds to part of an antigen called the _____.

19. Self-tolerance results from a process called _____, in which lymphocytes programmed to react against self-antigens die.

20. Any disease in which antibodies attack one's own tissues is called a/an _____ disease.

*Answers in Appendix B*

# True or False

*Determine which five of the following statements are false, and briefly explain why.*

1. Some bacteria employ lysozyme to liquefy the tissue gel and make it easier for them to get around.

2. T lymphocytes undergo clonal deletion and anergy in the thymus.

3. Interferons help to reduce inflammation.

4. T lymphocytes are involved only in cell-mediated immunity.

5. The white pulp of the spleen gets its color mainly from lymphocytes and macrophages.

6. Perforins are employed in both nonspecific resistance and cellular immunity.

7. Histamine and heparin are secreted by basophils and mast cells.

8. A person who is HIV-positive and has a $T_H$ (CD4) count of 1,000 cells/$\mu$L does not have AIDS.

9. Anergy is often a cause of autoimmune diseases.

10. Interferons kill pathogenic bacteria by making holes in their cell walls.

*Answers in Appendix B*

# Testing Your Comprehension

1. Anti-D antibodies of an Rh$^-$ woman sometimes cross the placenta and hemolyze the RBCs of an Rh$^+$ fetus (see p. 695). Yet the anti-B antibodies of a type A mother seldom affect the RBCs of a type B fetus. Explain this difference based on your knowledge of the five immunoglobulin classes.

2. In treating a woman for malignancy in the right breast, the surgeon removes some of her axillary lymph nodes. Following surgery, the patient experiences edema of her right arm. Explain why.

3. A girl with a defective heart receives a new heart transplanted from another child who was killed in an accident. The patient is given an antilymphocyte serum containing antibodies against her lymphocytes. The transplanted heart is not rejected, but the patient dies of an overwhelming bacterial infection. Explain why the antilymphocyte serum was given and why the patient was so vulnerable to infection.

4. A burn research center uses mice for studies of skin grafting. To prevent graft rejection, the mice are thymec- tomized at birth. Even though B cells do not develop in the thymus, these mice show no humoral immune response and are very susceptible to infection. Explain why the removal of the thymus would improve the success of skin grafts but adversely affect humoral immunity.

5. Contrast the structure of a B cell with that of a plasma cell, and explain how their structural difference relates to their functional difference.

*Answers at www.mhhe.com/saladin4*

# www.mhhe.com/saladin4

*The textbook website provides a wealth of interactive study materials fully organized and integrated by chapter. You will find practice quizzes, labeling exercises, and much more that will complement your learning and understanding of anatomy and physiology. The website also includes tools designed to enhance your* **Anatomy & Physiology | REVEALED** *experience.*

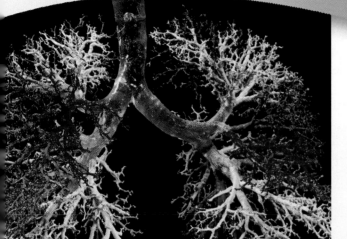

*The bronchial trees*

# THE RESPIRATORY SYSTEM

## CHAPTER OUTLINE

## INSIGHTS

## Brushing Up

To understand this chapter, it is important that you understand or brush up on the following concepts:

- Factors that affect simple diffusion (p. 103)
- Serous and mucous membranes (p. 178)
- Bones of the nasal region (p. 256)
- The muscles of respiration (p. 341)
- The structure of hemoglobin (p. 686)
- Principles of fluid pressure and flow (p. 736)
- Pulmonary blood circulation (p. 776)

Anatomy & Physiology | REVEALED®

Most metabolic processes of the body depend on ATP, and most ATP production requires oxygen and generates carbon dioxide as a waste product. The respiratory and cardiovascular systems collaborate to provide this oxygen and remove the carbon dioxide. Not only do these two systems have a close spatial relationship in the thoracic cavity, they also have such a close functional relationship that they are often considered jointly under the heading *cardiopulmonary*. A disorder that affects the lungs has direct and pronounced effects on the heart, and vice versa.

Furthermore, as discussed in the next two chapters, the respiratory system works closely with the urinary system to regulate the body's acid–base balance. Changes in the blood pH, in turn, trigger autonomic adjustments of the heart rate and blood pressure. Thus, the cardiovascular, respiratory, and urinary systems have an especially close physiological relationship. It is important that we now address the roles of the respiratory and urinary systems in the homeostatic control of blood gases, pH, blood pressure, and other variables related to the body fluids. This chapter deals with the respiratory system and chapter 23 with the urinary system.

# Anatomy of the Respiratory System

### Objectives

When you have completed this section, you should be able to

- state the functions of the respiratory system;
- name and describe the organs of this system;
- trace the flow of air from the nose to the pulmonary alveoli; and
- relate the function of any portion of the respiratory tract to its gross and microscopic anatomy.

The term **respiration** has three meanings: (1) ventilation of the lungs (breathing), (2) the exchange of gases between air and blood and between blood and tissue fluid, and (3) the use of oxygen in cellular metabolism. In this chapter, we are concerned with the first two processes. Cellular respiration was introduced in chapter 2 and is considered more fully in chapter 26.

The **respiratory system** is an organ system that rhythmically takes in air and expels it from the body, thereby supplying the body with oxygen and expelling the carbon

dioxide that it generates. However, it has a broader range of functions than are commonly supposed:

1. It provides for oxygen and carbon dioxide exchange between the blood and air.
2. It serves for speech and other vocalizations (laughing, crying).
3. It provides the sense of smell, which is important in social interactions, food selection, and avoiding danger (such as a gas leak or spoiled food).
4. By eliminating $CO_2$, it helps to control the pH of the body fluids. Excess $CO_2$ reacts with water and releases hydrogen ions ($CO_2 + H_2O \rightarrow H_2CO_3 \rightarrow HCO_3^- + H^+$); therefore, if the respiratory system does not keep pace with the rate of $CO_2$ production, $H^+$ accumulates and the body fluids have an abnormally low pH *(acidosis)*.
5. The lungs carry out a step in the synthesis of a vasoconstrictor called *angiotensin II*, which helps to regulate blood pressure.
6. Breathing creates pressure gradients between the thorax and abdomen that promote the flow of lymph and venous blood.
7. Breath-holding helps to expel abdominal contents during urination, defecation, and childbirth (the *Valsalva maneuver* described later).

The principal organs of the respiratory system are the nose, pharynx, larynx, trachea, bronchi, and lungs (fig. 22.1). Within the lungs, air flows along a dead-end pathway consisting essentially of bronchi → bronchioles → alveoli (with some refinements to be introduced later). Incoming air stops in the *alveoli* (millions of thin-walled, microscopic air sacs), exchanges gases with the bloodstream across the alveolar wall, and then flows back out.

The **conducting division** of the respiratory system consists of those passages that serve only for airflow, essentially from the nostrils through the bronchioles. The **respiratory division** consists of the alveoli and other distal gas-exchange regions. The airway from the nose through the larynx is often called the **upper respiratory tract** (that is, the respiratory organs in the head and neck), and the regions from the trachea through the lungs compose the **lower respiratory tract** (the respiratory organs of the thorax). However, these are inexact terms and various authorities place the dividing line between the upper and lower tracts at different points.

## THE NOSE

The **nose** has several functions: it warms, cleanses, and humidifies inhaled air; it detects odors in the airstream; and it serves as a resonating chamber that amplifies the voice. It extends from a pair of anterior openings called the **nostrils** or **anterior (external) nares** (NAIR-eze) to a pair of

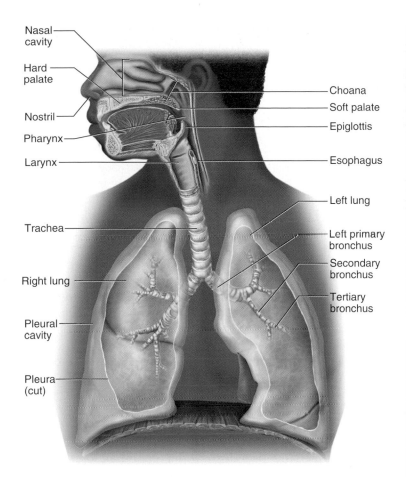

Labels for figure 22.1:
- Nasal cavity
- Hard palate
- Nostril
- Pharynx
- Larynx
- Trachea
- Right lung
- Pleural cavity
- Pleura (cut)
- Choana
- Soft palate
- Epiglottis
- Esophagus
- Left lung
- Left primary bronchus
- Secondary bronchus
- Tertiary bronchus

**FIGURE 22.1** The Respiratory System.

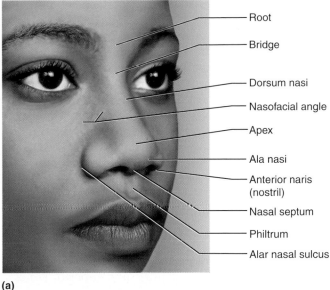

Labels for figure 22.2 (a):
- Root
- Bridge
- Dorsum nasi
- Nasofacial angle
- Apex
- Ala nasi
- Anterior naris (nostril)
- Nasal septum
- Philtrum
- Alar nasal sulcus

(a)

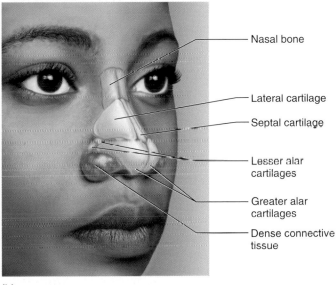

Labels for figure 22.2 (b):
- Nasal bone
- Lateral cartilage
- Septal cartilage
- Lesser alar cartilages
- Greater alar cartilages
- Dense connective tissue

(b)

**FIGURE 22.2** Anatomy of the Nasal Region. (a) External anatomy. (b) Connective tissues that shape the nose.

posterior openings called the **choanae**[1] (co-AH-nee), or **posterior (internal) nares**.

The facial part of the nose is shaped by bone and hyaline cartilage. Its superior half is supported by a pair of small nasal bones medially and the maxillae laterally. The inferior half is supported by the **lateral** and **alar cartilages** (fig. 22.2). By palpating your own nose, you can easily find the boundary between the bone and cartilage. The flared portion at the lower end of the nose, called the **ala nasi**[2] (AIL-ah NAZE-eye), is shaped by the alar cartilages and dense connective tissue.

The internal chamber of the nose, called the **nasal cavity,** is divided into right and left halves called **nasal fossae** (FAW-see). The dividing wall is a vertical plate, the **nasal septum,** composed of bone and hyaline cartilage. The vomer forms the inferior part of the septum, the perpendicular plate of the ethmoid bone forms its superior part, and the *septal cartilage* forms its anterior part. The

ethmoid and sphenoid bones compose the roof of the nasal cavity, and the palate forms its floor. The palate separates the nasal cavity from the oral cavity and allows you to breathe while there is food in your mouth. The paranasal sinuses (see chapter 8) and the nasolacrimal ducts of the orbits drain into the nasal cavity.

The nasal cavity begins with a small dilated chamber called the **vestibule** just inside the nostril, bordered by the ala nasi. This space is lined with stratified squamous epithelium like the facial skin, and has stiff **guard hairs,** or **vibrissae** (vy-BRISS-ee), that block insects and debris

---

[1]*choana* − funnel
[2]*ala* = wing + *nasi* = of the nose

from entering the nose. Posterior to the vestibule, the nasal cavity expands into a much larger chamber, but it does not have much open space. Most of it is occupied by three folds of tissue—the **superior, middle,** and **inferior nasal conchae**[3] (CON-kee)—that project from the lateral walls toward the septum (fig. 22.3). Beneath each concha is a narrow air passage called a **meatus** (me-AY-tus). The narrowness of these passages and the turbulence caused by the conchae ensure that most air contacts the mucous membrane on its way through. As it does, most dust in the air sticks to the mucus, and the air picks up moisture and heat from the mucosa. The conchae thus enable the nose to cleanse, warm, and humidify the air more effectively than if the air had an unobstructed flow through a cavernous space.

Odors are detected by sensory cells in the **olfactory mucosa,** a small patch of epithelium that covers the roof of the fossa and parts of the septum and superior concha. A ciliated pseudostratified **respiratory mucosa** not only lines the rest of the nasal cavity, but also extends deeply into the lungs. (In the lower reaches of the airway described later, it becomes a ciliated cuboidal epithelium.) It is a nonsensory epithelium with two principal types of cells: **goblet cells,** which secrete mucus, and **ciliated cells** which, in the nose, drive the mucus toward the posterior nares and into the pharynx so it can be swallowed and digested. The lamina propria also contains mucous glands, which supplement the mucus produced by the goblet cells. Pollen, dust, and other inhaled particles stick to the mucus, and lysozyme in the mucus destroys bacteria. The lamina propria is also well populated by lymphocytes that mount immune defenses against inhaled pathogens, and plasma cells that secrete antibodies into the tissue fluid.

The lamina propria contains large blood vessels that help to warm the air. The inferior concha has an especially extensive venous plexus called the **erectile tissue** (swell body). Every 30 to 60 minutes, the erectile tissue on one side becomes engorged with blood and restricts airflow through that fossa. Most air is then directed through the other nostril and fossa, allowing the engorged side time to recover from drying. Thus the preponderant flow of air shifts between the right and left nostrils once or twice each hour.

Nosebleed *(epistaxis)* usually results from trauma to the lower nasal septum, for example by nose-picking or blows to the face. In the absence of trauma, however, spontaneous nosebleeds often stem from the erectile tissue of the inferior nasal concha. This can be an early warning of hypertension.

## THE PHARYNX

The **pharynx** (FAIR-inks) is a muscular funnel extending about 13 cm (5 in.) from the choanae to the larynx. It has three regions: the *nasopharynx, oropharynx,* and *laryngopharynx* (fig. 22.3c).

The **nasopharynx,** which lies posterior to the choanae and dorsal to the soft palate, receives the auditory (eustachian) tubes from the middle ears and houses the pharyngeal tonsil. Inhaled air turns 90° downward as it passes through the nasopharynx. Relatively large particles (>10 μm) generally cannot make the turn because of their inertia. They collide with the posterior wall of the nasopharynx and stick to the mucosa near the tonsil, which is well positioned to respond to airborne pathogens.

The **oropharynx** is a space between the soft palate and root of the tongue that extends inferiorly as far as the hyoid bone. It contains the palatine and lingual tonsils. Its anterior border is formed by the base of the tongue and the *fauces* (FAW-seez), the opening of the oral cavity into the pharynx.

The **laryngopharynx** (la-RING-go-FAIR-inks) begins with the union of the nasopharynx and oropharynx at the level of the hyoid bone. It passes inferiorly and dorsal to the larynx and ends at the level of the *cricoid cartilage* at the inferior end of the larynx (described next). The esophagus begins at that point. The nasopharynx passes only air and is lined by pseudostratified columnar epithelium, whereas the oropharynx and laryngopharynx pass air, food, and drink and are lined by stratified squamous epithelium.

## THE LARYNX

The **larynx** (LAIR-inks), or "voicebox," is a cartilaginous chamber about 4 cm (1.5 in.) long (fig. 22.4). Its primary function is to keep food and drink out of the airway, but it evolved the additional role of sound production *(phonation)* in many animals and achieved its highest vocal sophistication in humans.

The superior opening of the larynx is guarded by a flap of tissue called the **epiglottis.**[4] During swallowing, *extrinsic muscles* of the larynx pull the larynx upward toward the epiglottis, the tongue pushes the epiglottis downward to meet it, and the epiglottis directs food and drink into the esophagus dorsal to the airway. The *vestibular folds* of the larynx, discussed shortly, play a greater role in keeping food and drink out of the airway, however. People who have had their epiglottis removed because of cancer do not choke any more than when it was present.

---

[3]*concha* = seashell

[4]*epi* = above, upon + *glottis* = back of the tongue

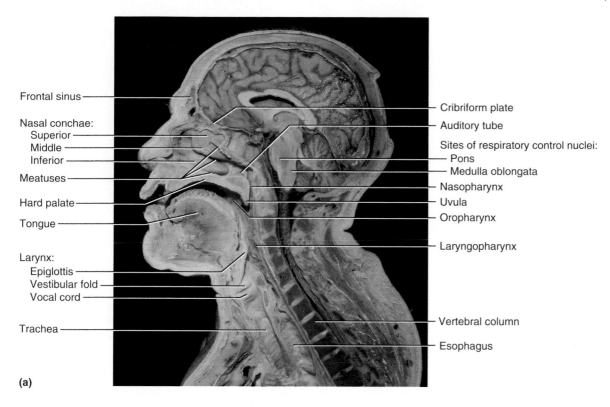

Frontal sinus

Nasal conchae:
  Superior
  Middle
  Inferior

Meatuses

Hard palate

Tongue

Larynx:
  Epiglottis
  Vestibular fold
  Vocal cord

Trachea

Cribriform plate

Auditory tube

Sites of respiratory control nuclei:
  Pons
  Medulla oblongata

Nasopharynx

Uvula

Oropharynx

Laryngopharynx

Vertebral column

Esophagus

(a)

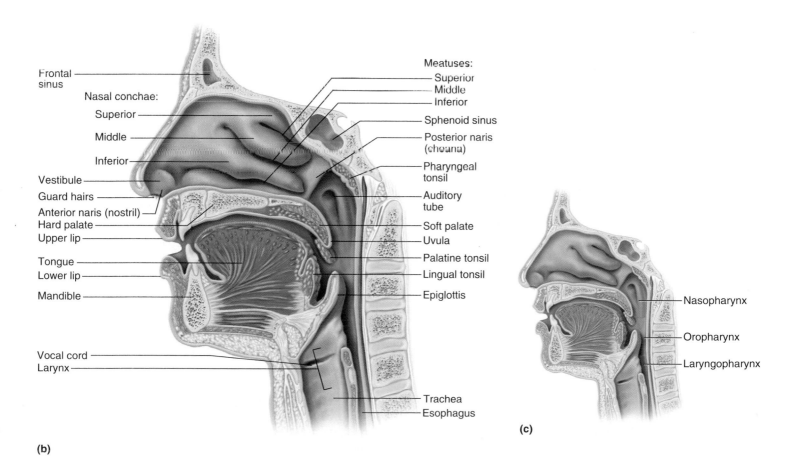

Frontal sinus

Nasal conchae:
  Superior
  Middle
  Inferior

Vestibule

Guard hairs

Anterior naris (nostril)

Hard palate

Upper lip

Tongue

Lower lip

Mandible

Vocal cord

Larynx

Meatuses:
  Superior
  Middle
  Inferior

Sphenoid sinus

Posterior naris (choana)

Pharyngeal tonsil

Auditory tube

Soft palate

Uvula

Palatine tonsil

Lingual tonsil

Epiglottis

Trachea

Esophagus

(b)

Nasopharynx

Oropharynx

Laryngopharynx

(c)

**FIGURE 22.3 Anatomy of the Upper Respiratory Tract.** (a) Median section of the head. (b) Internal anatomy. (c) Regions of the pharynx.
▶ *Why do throat infections so easily spread to the middle ear?*

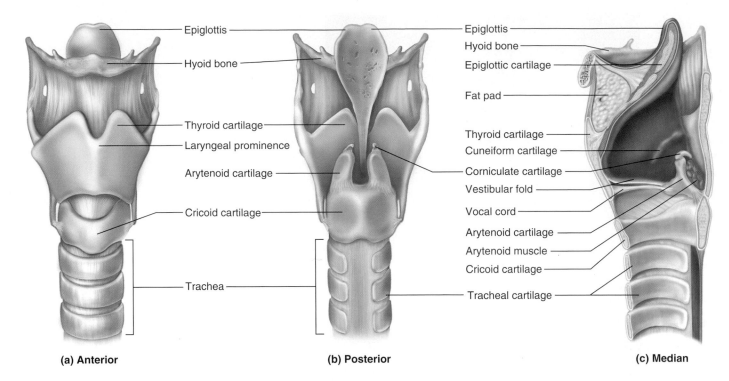

**(a) Anterior**    **(b) Posterior**    **(c) Median**

**FIGURE 22.4  Anatomy of the Larynx.**  Most muscles are removed in order to show the cartilages.

In infants, the larynx is relatively high in the throat and the epiglottis touches the soft palate. This creates a more or less continuous airway from the nasal cavity to the larynx and allows an infant to breathe continually while swallowing. The epiglottis deflects milk away from the airstream, like rain running off a tent while it remains dry inside. By age 2, the root of the tongue becomes more muscular and forces the larynx to descend to a lower position. It then becomes impossible to breathe and swallow at the same time without choking.

The framework of the larynx consists of nine cartilages. The first three are solitary and relatively large. The most superior one, the **epiglottic cartilage,** is a spoon-shaped supportive plate in the epiglottis. The largest, the **thyroid**[5] **cartilage,** is named for its shieldlike shape. It broadly covers the anterior and lateral aspects of the larynx. The "Adam's apple" is an anterior peak of the thyroid cartilage called the *laryngeal prominence.* Testosterone stimulates the growth of this prominence, which is therefore larger in males than in females. Inferior to the thyroid cartilage is a ringlike **cricoid**[6] (CRY-coyd) **cartilage,** which connects the larynx to the trachea. The thyroid and cricoid cartilages essentially constitute the "box" of the "voicebox."

The remaining cartilages are smaller and occur in three pairs. Posterior to the thyroid cartilage are the two **arytenoid**[7] (AR-ih-TEE-noyd) **cartilages,** and attached to

their upper ends are a pair of little horns, the **corniculate**[8] (cor-NICK-you-late) **cartilages.** The arytenoid and corniculate cartilages function in speech, as explained shortly. A pair of **cuneiform**[9] (cue-NEE-ih-form) **cartilages** support the soft tissues between the arytenoids and the epiglottis. The thyroid and cricoid cartilages and inferior part of the arytenoids are hyaline cartilage; the epiglottic, corniculate, and cuneiform cartilages and superior part of the arytenoids are elastic cartilage.

A group of fibrous ligaments bind the cartilages of the larynx together and to adjacent structures in the neck. Superiorly, a broad sheet called the **thyrohyoid ligament** joins the thyroid cartilage to the hyoid bone, and inferiorly, the **cricotracheal ligament** joins the cricoid cartilage to the trachea. These are collectively called the *extrinsic ligaments* because they link the larynx to other organs. The *intrinsic ligaments* are contained entirely within the larynx and link its nine cartilages to each other. Two pairs of intrinsic ligaments, the **vestibular** and **vocal ligaments,** extend from the thyroid cartilage anteriorly to the arytenoid cartilages posteriorly, and support the vestibular folds and vocal cords, described shortly.

The walls of the larynx are quite muscular. The deep *intrinsic muscles* operate the vocal cords, and the superficial *extrinsic muscles* connect the larynx to the hyoid bone and elevate the larynx during swallowing. The extrinsic muscles, also called the *infrahyoid group,* are named and described in chapter 10 (table 10.3).

---

[5]*thyr* = shield + *oid* = resembling
[6]*crico* = ring + *oid* = resembling
[7]*aryten* = ladle + *oid* = resembling

[8]*corni* = horn + *cul* = little + *ate* = possessing
[9]*cune* = wedge + *form* = shape

The interior wall of the larynx has two folds on each side that stretch from the thyroid cartilage in front to the arytenoid cartilages in back. The superior pair, called the **vestibular folds** (fig. 22.4c), play no role in speech but close the glottis during swallowing. They are supported by the aforementioned vestibular ligaments. The inferior pair, the **vocal cords** (vocal folds), produce sound when air passes between them. They contain the vocal ligaments and are covered with stratified squamous epithelium, best suited to endure vibration and contact between the cords. The vocal cords and the opening between them are collectively called the **glottis** (fig. 22.5).

The intrinsic muscles control the vocal cords by pulling on the corniculate and arytenoid cartilages, causing the cartilages to pivot. Depending on their direction of rotation, the arytenoid cartilages abduct or adduct the vocal cords (fig. 22.6). Air forced between the adducted vocal cords vibrates them, producing a high-pitched sound when the cords are relatively taut and a lower-pitched sound when they are more relaxed. In adult males, the vocal cords are usually longer and thicker, vibrate more slowly, and produce lower-pitched sounds than in females. Loudness is determined by the force of the air passing between the vocal cords. Although the vocal cords alone produce sound, they do not produce intelligible speech. The crude sounds coming from the larynx are formed into words by actions of the pharynx, oral cavity, tongue, and lips.

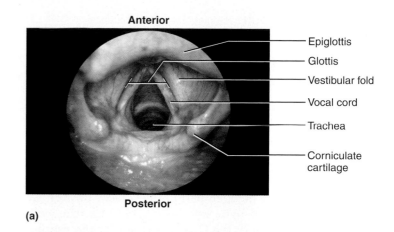

(a)

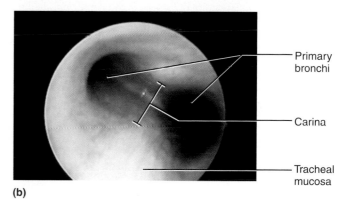

(b)

**FIGURE 22.5** Endoscopic Views of the Respiratory Tract.
(a) Superior view of the larynx, seen with a laryngoscope.
(b) Lower end of the trachea, where it forks into the two primary bronchi, seen with a bronchoscope.

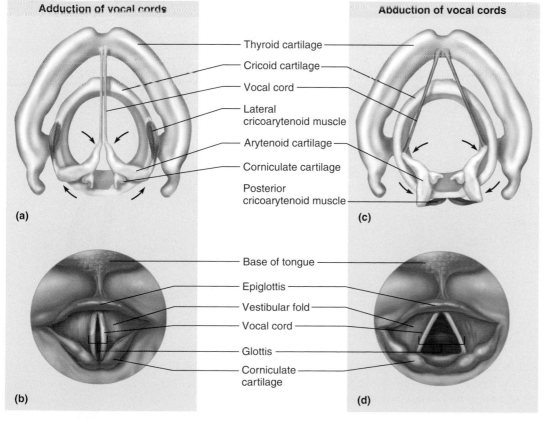

**FIGURE 22.6** Action of Some of the Intrinsic Laryngeal Muscles on the Vocal Cords.
(a) Adduction of the vocal cords by the lateral cricoarytenoid muscles. (b) Adducted vocal cords seen with the laryngoscope. (c) Abduction of the vocal cords by the posterior cricoarytenoid muscles. (d) Abducted vocal cords seen with the laryngoscope.

# THE TRACHEA

The **trachea** (TRAY-kee-uh), or "windpipe," is a rigid tube about 12 cm (4.5 in.) long and 2.5 cm (1 in.) in diameter, lying anterior to the esophagus (fig. 22.7a). It is supported by 16 to 20 C-shaped rings of hyaline cartilage, some of which you can palpate between your larynx and sternum. Like the wire spiral in a vacuum cleaner hose, the cartilage rings reinforce the trachea and keep it from collapsing when you inhale. The open part of the C faces posteriorly, where it is spanned by a smooth muscle, the **trachealis** (fig. 22.7b). The gap in the C allows room for the esophagus to expand as swallowed food passes by. The trachealis muscles contract or relax to adjust tracheal airflow.

The inner lining of the trachea is a pseudostratified columnar epithelium composed mainly of mucus-secreting goblet cells, ciliated cells, and short basal stem cells (figs. 22.7c and 22.8). The mucus traps inhaled particles, and the upward beating of the cilia drives the debris-laden mucus toward the pharynx, where it is swallowed. This mechanism of debris removal is called the **mucociliary escalator.**

## INSIGHT 22.1    Clinical Application

### TRACHEOSTOMY

The functional importance of the nasal cavity becomes especially obvious when it is bypassed. If the upper airway is obstructed, it may be necessary to make a temporary opening in the trachea inferior to the larynx and insert a tube to allow airflow—a procedure called *tracheostomy.* This prevents asphyxiation, but the inhaled air bypasses the nasal cavity and thus is not humidified. If the opening is left for long, the mucous membranes of the respiratory tract dry out and become encrusted, interfering with the clearance of mucus from the tract and promoting infection. When a patient is on a ventilator and air is introduced directly into the trachea, the air must be filtered and humidified by the apparatus to prevent respiratory tract damage.

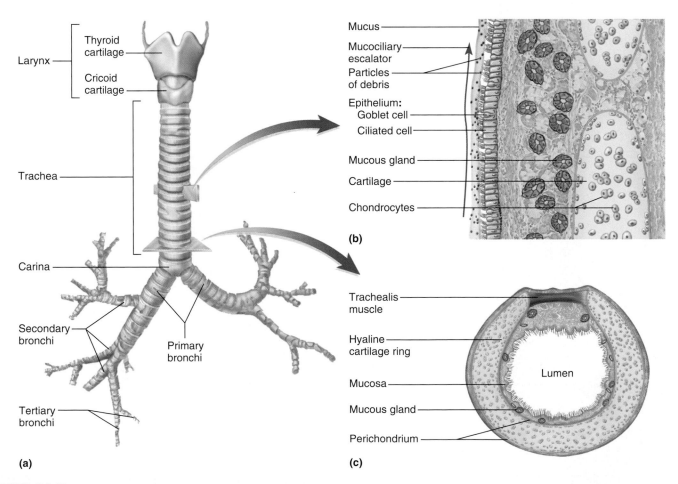

**FIGURE 22.7** **Anatomy of the Lower Respiratory Tract.** (a) Anterior view. (b) Longitudinal section of the trachea showing the action of the mucociliary escalator. (c) Cross section of the trachea showing the C-shaped tracheal cartilage.

▶ *Why do inhaled objects more often go into the right primary bronchus than into the left?*

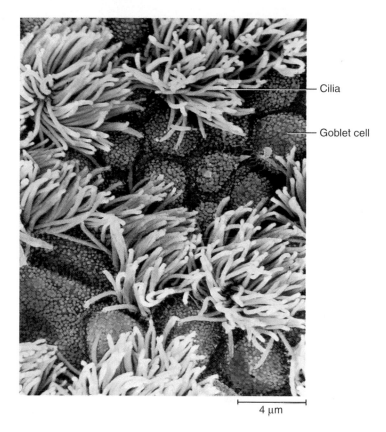

**FIGURE 22.8 The Tracheal Epithelium Showing Ciliated Cells and Nonciliated Goblet Cells.** The small bumps on the goblet cells are microvilli. (SEM)

Cilia

Goblet cell

4 μm

The connective tissue beneath the tracheal epithelium contains lymphatic nodules, mucous and serous glands, and the tracheal cartilages. The outermost layer of the trachea, called the **adventitia,** is fibrous connective tissue that blends into the adventitia of other organs of the mediastinum.

At its inferior end, the trachea forks into right and left *primary bronchi.* The lowermost tracheal cartilage has an internal median ridge called the **carina**[10] (ca-RY-na) that directs the airflow to the right and left (see fig. 22.5b). The bronchi are further traced in the discussion of the *bronchial tree* of the lungs.

## THE LUNGS AND BRONCHIAL TREE

Each **lung** is a somewhat conical organ with a broad, concave **base** resting on the diaphragm and a blunt peak called the **apex** projecting slightly above the clavicle (fig. 22.9). The broad **costal surface** is pressed against the rib cage, and the smaller concave **mediastinal surface** faces medially. The mediastinal surface exhibits a slit called the **hilum** through which the lung receives the primary bronchus, blood vessels, lymphatic vessels, and nerves. These structures constitute the **root** of the lung.

[10]*carina* = keel

The lungs are crowded by adjacent viscera and therefore neither fill the entire rib cage, nor are they symmetrical. Inferior to the lungs and diaphragm, much of the space within the rib cage is occupied by the liver, spleen, and stomach (see fig. A.14, p. 43). The right lung is shorter than the left because the liver rises higher on the right. The left lung, although taller, is narrower than the right because the heart tilts toward the left and occupies more space on this side of the mediastinum. On the medial surface, the left lung has an indentation called the **cardiac impression** where the heart presses against it. The right lung has three lobes—**superior, middle,** and **inferior**—separated by two fissures. The left lung has only a **superior** and **inferior lobe** and a single fissure.

### The Bronchial Tree

The lung has a spongy parenchyma containing the **bronchial tree,** a highly branched system of air tubes extending from the primary bronchus to about 65,000 *terminal bronchioles.* Two **primary bronchi** (BRONK-eye) arise from the trachea at the level of the angle of the sternum. Each continues for 2 to 3 cm and enters the hilum of its respective lung. The right bronchus is slightly wider and more vertical than the left; consequently, *aspirated* (inhaled) foreign objects lodge in the right bronchus more often than in the left. Like the trachea, the primary bronchi are supported by C-shaped hyaline cartilages. All divisions of the bronchial tree also have a substantial amount of elastic connective tissue, which is important in expelling air from the lungs.

Upon entering the hilum, the primary bronchus branches into one **secondary (lobar) bronchus** for each pulmonary lobe. Thus, there are two secondary bronchi in the left lung and three in the right. Each secondary bronchus divides into **tertiary (segmental) bronchi**—10 in the right lung and 8 in the left. The portion of the lung supplied by each tertiary bronchus is called a **bronchopulmonary segment.** Page 853 shows a cast of the bronchial trees made by injecting the airway with colored resins and then dissolving away the tissue. Each bronchopulmonary segment is identified by a different color of resin. Secondary and tertiary bronchi are supported by overlapping plates of cartilage, not rings.

Branches of the *pulmonary artery* closely follow the bronchial tree on their way to the alveoli. The bronchial tree itself is serviced by the *bronchial artery,* which arises from the aorta and carries systemic blood.

**Bronchioles** are continuations of the airway that lack supportive cartilage and are 1 mm or less in diameter. The portion of the lung ventilated by one bronchiole is called a **pulmonary lobule.** Bronchioles have a ciliated cuboidal epithelium and a well-developed layer of smooth muscle in their walls. Spasmodic contractions of this muscle at death cause the bronchioles to exhibit a wavy lumen in most histological sections.

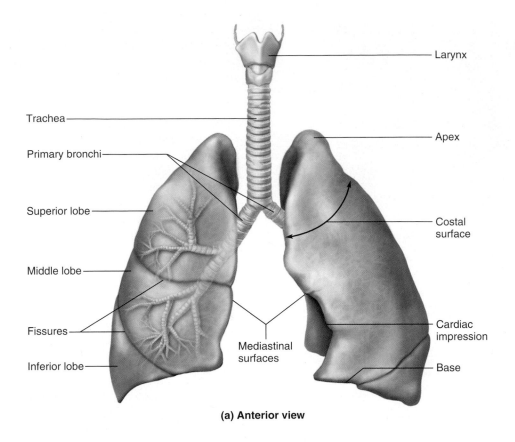

(a) Anterior view

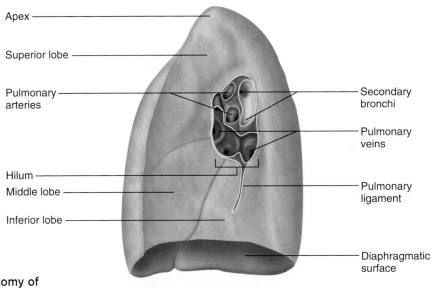

**FIGURE 22.9**   Gross Anatomy of
the Lungs.

(b) Mediastinal surface, right lung

Each bronchiole divides into 50 to 80 **terminal bronchioles,** the final branches of the conducting division. These measure 0.5 mm or less in diameter and have no mucous glands or goblet cells. They do have cilia, however, so that mucus draining into them from the higher passages can be driven back by the mucociliary escalator, thus preventing congestion of the terminal bronchioles and alveoli.

Each terminal bronchiole gives off two or more smaller **respiratory bronchioles,** which have alveoli budding from their walls. Respiratory bronchioles are the beginning of the respiratory division. Their walls have scanty smooth muscle, and the smallest of them are nonciliated. Each respiratory bronchiole divides into 2 to 10 elongated, thin-walled passages called **alveolar ducts,** which also have alveoli along their walls (fig. 22.10). The

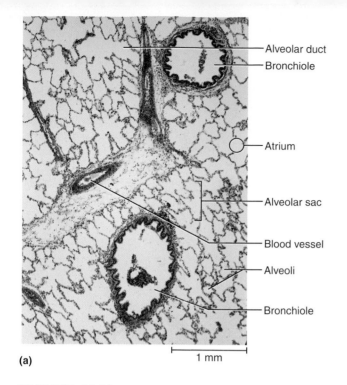

- Alveolar duct
- Bronchiole
- Atrium
- Alveolar sac
- Blood vessel
- Alveoli
- Bronchiole

(a)

|← 1 mm →|

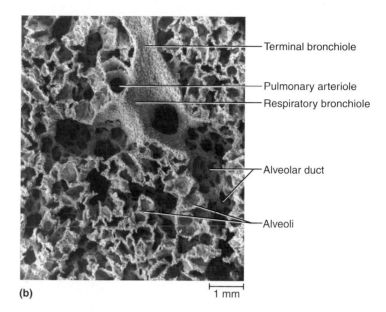

- Terminal bronchiole
- Pulmonary arteriole
- Respiratory bronchiole
- Alveolar duct
- Alveoli

(b)

|← 1 mm →|

**FIGURE 22.10   Lung Tissue.**   (a) Light micrograph. (b) SEM micrograph. Note the spongy texture of the lung.

alveolar ducts and smaller divisions have nonciliated simple squamous epithelia. The ducts end in **alveolar sacs,** which are grapelike clusters of alveoli arrayed around a central space called the *atrium.* The distinction between an alveolar duct and atrium is their shape—an elongated duct, or an atrium with about equal length and width. It is sometimes a subjective judgment whether to regard a space as an alveolar duct or atrium.

## Alveoli

The functional importance of human lung structure is best appreciated by comparison to the lungs of a few other animals. In frogs and other amphibians, the lung is a simple sac lined with blood vessels. This is sufficient to meet the oxygen needs of animals with relatively low metabolic rates. Mammals, with their high metabolic rates, could never have evolved with such a simple lung. Rather than consisting of one large sac, each human lung is a spongy mass composed of 150 million little sacs, the alveoli, which provide about 70 m² of surface for gas exchange.

An **alveolus** (AL-vee-OH-lus) is a pouch about 0.2 to 0.5 mm in diameter (fig. 22.11). Thin, broad cells called **squamous (type I) alveolar cells** cover about 95% of the alveolar surface area. Their thinness allows for rapid gas diffusion between the alveolus and bloodstream. The other 5% is covered by round to cuboidal **great (type II) alveolar cells.** Even though they cover less surface area, great alveolar cells considerably outnumber the squamous alveolar cells. Great alveolar cells have two functions: (1) they repair the alveolar epithelium when the squamous alveolar cells are damaged, and (2) they secrete *pulmonary surfactant,* a mixture of phospholipids and

protein that coats the alveoli and smallest bronchioles and prevents them from collapsing when one exhales. Without surfactant, the walls of a deflating alveolus would tend to cling together like sheets of wet paper, and it would be very difficult to reinflate them on the next inhalation.

The most numerous of all cells in the lung are **alveolar macrophages** (dust cells), which wander the lumens of the alveoli and the connective tissue between them. These cells keep the alveoli free of debris by phagocytizing dust particles that escape entrapment by mucus in the higher parts of the respiratory tract. In lungs that are infected or bleeding, the macrophages also phagocytize bacteria and loose blood cells. As many as 100 million alveolar macrophages perish each day as they ride up the mucociliary escalator to be swallowed and digested, thus ridding the lungs of their load of debris.

Each alveolus is surrounded by a basket of blood capillaries supplied by the pulmonary artery. The barrier between the alveolar air and blood, called the **respiratory membrane,** consists only of the squamous alveolar cell, the squamous endothelial cell of the capillary, and their shared basement membrane. These have a total thickness of only 0.5 μm, in contrast to the 7-μm diameter of the erythrocytes passing through the capillaries.

The pulmonary circulation has very low blood pressure. In alveolar capillaries, the mean blood pressure is 10 mm Hg and the oncotic pressure is 25 mm Hg. The osmotic uptake of water thus overrides filtration and keeps the alveoli free of fluid. The lungs also have a more extensive lymphatic drainage than any other organ in the body. The low capillary blood pressure also prevents the rupture of the delicate respiratory membrane.

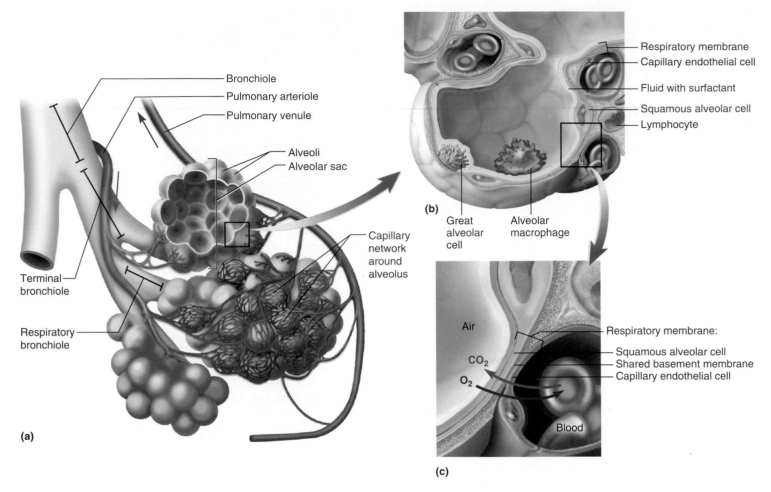

**FIGURE 22.11  Pulmonary Alveoli.**   (a) Clusters of alveoli and their blood supply. (b) Structure of an alveolus. (c) Structure of the respiratory membrane.

## THE PLEURAE

The surface of the lung is covered by a serous membrane, the **visceral pleura** (PLOOR-uh), which extends into the fissures. At the hilum, the visceral pleura turns back on itself and forms the **parietal pleura,** which adheres to the mediastinum, inner surface of the rib cage, and superior surface of the diaphragm (fig. 22.12). An extension of the parietal pleura, the *pulmonary ligament,* connects each lung to the diaphragm.

The space between the parietal and visceral pleurae is called the **pleural cavity.** The two membranes are normally separated only by a film of slippery **pleural fluid;** thus, the pleural cavity is only a *potential space,* meaning there is normally no room between the membranes. However, under pathological conditions this space can fill with air or liquid (see *pneumothorax,* p. 871).

The pleurae and pleural fluid have three functions:

1. **Reduction of friction.** Pleural fluid acts as a lubricant that enables the lungs to expand and contract with minimal friction. In some forms of *pleurisy,* the pleurae are dry and inflamed and each breath gives painful testimony to the function that the fluid should be serving.

2. **Creation of pressure gradient.** Pressure in the pleural cavity is lower than atmospheric pressure; as explained later, this assists in inflation of the lungs.

3. **Compartmentalization.** The pleurae, mediastinum, and pericardium compartmentalize the thoracic organs and prevent infections of one organ from spreading easily to neighboring organs.

⌐ **Think About It**

*In what ways do the structure and function of the pleurae resemble the structure and function of the pericardium?*

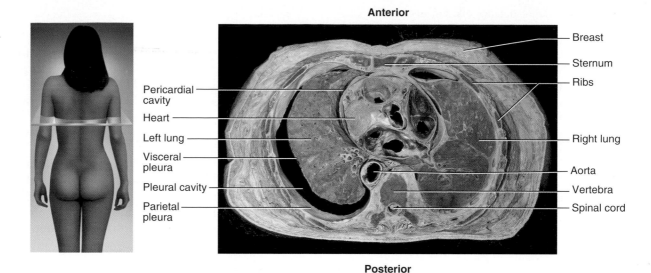

**Anterior**

Pericardial cavity

Heart

Left lung

Visceral pleura

Pleural cavity

Parietal pleura

Breast

Sternum

Ribs

Right lung

Aorta

Vertebra

Spinal cord

**Posterior**

**FIGURE 22.12  Cross Section Through the Thoracic Cavity.**  This photograph is oriented the same way as the reader's body. The pleural cavity is especially evident where the left lung has shrunken away from the thoracic wall, but in a living person the lung fully fills this space, the parietal and visceral pleurae are pressed together, and the pleural cavity is only a potential space between the membranes, as on the right side of this photograph.

## Before You Go On

*Answer the following questions to test your understanding of the preceding section:*

1. *A dust particle is inhaled and gets into an alveolus without being trapped along the way. Describe the path it takes, naming all air passages from external naris to alveolus. What would happen to it after arrival in the alveolus?*

2. *Describe the histology of the epithelium and lamina propria of the nasal cavity and the functions of the cell types present.*

3. *Describe the roles of the intrinsic muscles, corniculate cartilages, and arytenoid cartilages in speech.*

4. *Contrast the epithelium of the bronchioles with that of the alveoli and explain how the structural difference is related to their functional differences.*

# Pulmonary Ventilation

### Objectives

When you have completed this section, you should be able to

- name the muscles of respiration and describe their roles in breathing;

- describe the brainstem centers that control breathing and the inputs they receive from other levels of the nervous system;

- explain how pressure gradients account for the flow of air in and out of the lungs, and how those pressure gradients are produced;

- identify the sources of resistance to airflow and discuss their relevance to respiration;

- explain the significance of anatomical dead space to alveolar ventilation;

- define the clinical measurements of pulmonary volume and capacity; and

- define terms for various deviations from the normal pattern of breathing.

With the foregoing anatomical background, our next objective is to reach an understanding of how the lungs are ventilated. Breathing, or pulmonary ventilation, consists of a repetitive cycle of **inspiration** (inhaling) and **expiration** (exhaling). One complete inspiration and expiration is called a **respiratory cycle.** The following discussion will often distinguish between quiet and forced respiration. *Quiet respiration* refers to the way one breathes at rest, when not thinking about it. It is relaxed, unconscious, automatic breathing, the way one would breathe when reading a book or listening to a class lecture, and not thinking about breathing. *Forced respiration* is unusually deep or rapid breathing, as in a state of exercise or when blowing out a candle.

The most fundamental fact of pulmonary ventilation is that it requires a difference between the air pressure within the lungs and the air pressure external to the body; if there is no pressure difference, there can be no flow of air into and out of the lungs. Therefore, we begin our physiological investigation with the mechanism that produces these pressure differences—the respiratory muscles. Their anatomy is described in chapter 10 (see especially table 10.5, Muscles of Respiration, pp. 341–342). Here we will deal with their mode of action.

## THE RESPIRATORY MUSCLES

The lungs do not ventilate themselves. The only muscle they contain is smooth muscle in the walls of the bronchi and bronchioles. This muscle adjusts the diameter of those passages and affects the speed of airflow, but it does not expand or shrink the lungs or create the airflow. That job belongs to the skeletal muscles of the trunk, especially the diaphragm and intercostal muscles (fig. 22.13).

The prime mover of pulmonary ventilation is the diaphragm, the muscular dome that separates the thoracic cavity from the abdominal cavity. When relaxed, it bulges upward to its farthest extent, pressing against the base of the lungs. The lungs are at their minimum volume. When the diaphragm contracts, it tenses and flattens somewhat, dropping about 1.5 cm in relaxed inspiration and as much

as 7 cm in deep breathing. This enlarges the thoracic cavity and lungs. By the principles of fluid pressure and flow that we will examine later, this pulmonary expansion results in an inflow of air. When the diaphragm relaxes, it bulges upward again, compresses the lungs, and expels air. The diaphragm alone accounts for about two-thirds of the pulmonary airflow.

Several other muscles aid the diaphragm as synergists. Chief among these are the internal and external intercostal muscles between the ribs. Their primary action is to stiffen the thoracic cage during respiration and prevent it from caving inward when the diaphragm descends. However,

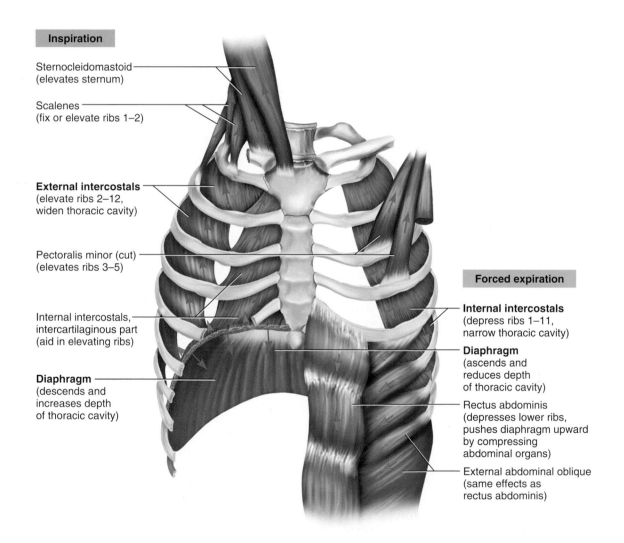

**FIGURE 22.13** **The Respiratory Muscles.** Boldface indicates the principal respiratory muscles; the others are accessory. Arrows indicate the direction of muscle action. Muscles listed on the left are active during inspiration and those on the right are active during forced expiration. Note that the diaphragm is active in both phases, and different parts of the internal intercostal muscles serve for inspiration and expiration. Some other accessory muscles not shown here are discussed in the text.

they also contribute to enlargement and contraction of the thoracic cage and add about one-third of the air that ventilates the lungs. During quiet breathing, the scalene muscles of the neck fix (hold stationary) ribs 1 and 2, while the external intercostal muscles pull the other ribs upward. Since most ribs are anchored at both ends—by their attachment to the vertebral column at the proximal end and their attachment through the costal cartilages to the sternum at the distal end—they swing upward like the handles on a bucket and thrust the sternum forward. These actions increase both the transverse (left to right) and anteroposterior (AP) diameters of the chest. In deep breathing, the AP diameter increases by as much as 20%.

Other muscles of the chest and abdomen also aid in breathing, especially during forced respiration; thus they are considered *accessory muscles* of respiration. Deep inspiration is aided by the erector spinae, which arches the back and increases AP chest diameter, and by several muscles that elevate the upper ribs: the sternocleidomastoids and scalenes of the neck, and the pectoralis minor, pectoralis major, and serratus anterior of the chest. (Although the scalenes merely fix the upper ribs during quiet respiration, they elevate them during forced inspiration.)

During forced expiration—for example, when singing or shouting, coughing or sneezing, or playing a wind instrument—the rectus abdominis pulls down on the sternum and lower ribs, while the internal intercostals pull the other ribs downward. These actions reduce the chest diameter and help to expel air more rapidly and thoroughly. Other muscles that contribute to forced expiration are the latissimus dorsi of the lower back, the transverse and oblique abdominal muscles, and even some of the pelvic muscles. They raise the pressure in the abdominal cavity and push some of the viscera, such as the stomach and liver, up against the diaphragm. This increases the pressure in the thoracic cavity and thus helps to expel air. Abdominal control of respiration is particularly important in singing and public speaking.

Not only does abdominal pressure affect thoracic pressure, but the opposite is also true. Depression of the diaphragm raises abdominal pressure and helps to expel the contents of certain abdominal organs, thus aiding in childbirth, urination, defecation, and vomiting. During such actions, we often consciously or unconsciously employ the **Valsalva**[11] **maneuver.** This consists of taking a deep breath, holding it by closing the glottis, and then contracting the abdominal muscles to raise abdominal pressure and push the organ contents out.

## NEURAL CONTROL OF BREATHING

The heartbeat and breathing are the two most conspicuously rhythmic processes in the body. The heart, we have seen, has an internal pacemaker and continues beating even if all nerves to it are severed. The lungs, by contrast, do not. Indeed, no autorhythmic pacemaker cells for respiration have been found that are analogous to those of the heart, and the exact mechanism for setting the rhythm of respiration remains unknown. But we do know that breathing depends on repetitive stimuli from the brain. It ceases if the nerve connections to the thoracic muscles are severed or if the spinal cord is severed high on the neck. There are two reasons for this dependence on the brain: (1) Skeletal muscles, unlike cardiac muscle, cannot contract without nervous stimulation. (2) Breathing involves the well-orchestrated action of multiple muscles and thus requires a central coordinating mechanism.

Breathing is controlled at two levels of the brain. One is cerebral and conscious, enabling us to inhale or exhale at will. Most of the time, however, we breathe without thinking about it—fortunately, for we otherwise could not go to sleep without fear of respiratory arrest (see Insight 22.3).

### INSIGHT 22.2   Clinical Application

#### Difficulty in Breathing

Asthma, emphysema, heart failure, and other conditions can cause *dyspnea,* difficulty catching one's breath. People with dyspnea make increased use of accessory muscles to aid the diaphragm and intercostals in breathing, and often lean on a table or chair back to breathe more deeply. This action fixes the clavicles and scapulae so that accessory muscles such as the pectoralis major and serratus anterior move the ribs instead of the bones of the pectoral girdle.

### INSIGHT 22.3   Clinical Application

#### Ondine's Curse

In German legend, there was a water nymph named Ondine who took a mortal lover. When her lover proved unfaithful, the king of the nymphs put a curse on him that took away his automatic physiological functions. Consequently, he had to remember to take each breath, and he could not go to sleep or he would die of suffocation—which, as exhaustion overtook him, was indeed his fate.

Some people suffer a disorder called *Ondine's curse,* in which the automatic respiratory functions are disabled—usually as a result of brainstem damage from poliomyelitis or as an accident of neurosurgery. Victims of Ondine's curse must remember to take each breath and cannot go to sleep without the aid of a mechanical ventilator.

[11]Antonio Maria Valsalva (1666–1723), Italian anatomist

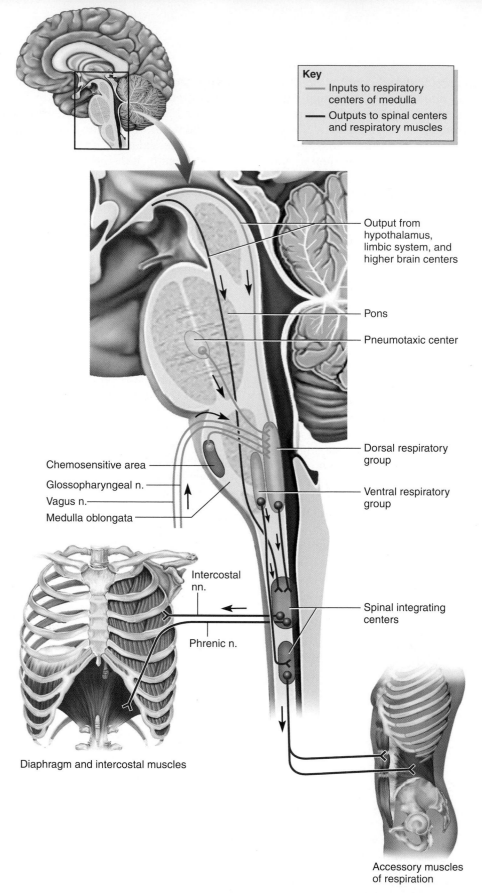

Key
— Inputs to respiratory centers of medulla
— Outputs to spinal centers and respiratory muscles

Output from hypothalamus, limbic system, and higher brain centers

Pons

Pneumotaxic center

Dorsal respiratory group

Ventral respiratory group

Chemosensitive area

Glossopharyngeal n.

Vagus n.

Medulla oblongata

Intercostal nn.

Spinal integrating centers

Phrenic n.

Diaphragm and intercostal muscles

Accessory muscles of respiration

**FIGURE 22.14 Respiratory Centers in the Central Nervous System.** The principal respiratory centers are the dorsal respiratory group (DRG) and ventral respiratory group (VRG), represented in blue in the medulla oblongata. The DRG, especially, receives input from the pneumotaxic center (green) in the pons, from the chemosensitive area (red) in the medulla, from higher brain centers involved in emotional and voluntary influences on respiration, and from peripheral chemoreceptors and stretch receptors by way of the vagus and glossopharyngeal nerves. Output from the DRG goes especially to integrating centers in the spinal cord (violet) that stimulate the intercostal muscles and diaphragm. Output from the VRG goes especially to spinal cord centers that stimulate abdominal and other accessory muscles of respiration.

## Brainstem Respiratory Centers

The automatic, unconscious cycle of breathing is controlled by three respiratory centers in the medulla oblongata and pons (fig. 22.14). Each of these is paired, left and right:

1. The **dorsal respiratory group (DRG)** is an elongated mass of neurons extending for much of the length of the medulla near the central canal. Neurons of this center are called **inspiratory (I) neurons** because their firing causes inhalation. Their axons decussate and descend the contralateral spinal cord, terminating in integrating centers of the cervical and thoracic regions. Lower motor neurons in these integrating centers issue axons to the diaphragm by way of the phrenic nerves and to the intercostal muscles by way of the intercostal nerves. DRG output begins weakly and builds in intensity over a period of about 2 sec. The inspiratory muscles contract with increasing force, and we inhale. DRG output then ceases abruptly, the inspiratory muscles relax, and elastic recoil of the thoracic cage causes us to exhale. About 3 sec later, the DRG begins firing again and the cycle repeats itself. Thus in quiet breathing, we typically take about 12 breaths per minute (5 sec/cycle).

2. The **ventral respiratory group (VRG)** is another elongated neural network just ventral to the DRG. It consists of both I neurons and **expiratory (E) neurons,** and is active during both inspiration and expiration. It shows little activity during quiet respiration, but comes into play in heavy breathing, such as during exercise. The VRG is especially important in stimulating the abdominal and other accessory muscles that produce deep breaths.

3. The **pneumotaxic center** is a nucleus in the pons that regulates the shift from inspiration to expiration. Its output inhibits the DRG and terminates inspiration. Therefore, strong output from the pneumotaxic center makes each breath shorter and the respiratory rate faster. Weak output results in long, slow breaths.

## Central and Peripheral Input to the Respiratory Centers

We are well aware that the respiratory rhythm is not constant; it varies with circumstances such as rest, exercise, and emotional state. Such adaptations are possible because the respiratory centers of the medulla and pons receive input from several other levels of the nervous system and therefore sense the body's varying physiological needs. Input from the hypothalamus and limbic system enables pain and emotions to affect breathing, for example in gasping, crying, and laughing. Anxiety can trigger a bout of uncontrollable *hyperventilation* in some people, a state in which breathing is so rapid that it expels $CO_2$ from the body faster than it is produced. As blood $CO_2$ levels drop, the pH rises and causes the cerebral arteries to constrict. This reduces cerebral perfusion and may cause dizziness or fainting. Hyperventilation can be brought under control by having a person rebreathe the expired $CO_2$ from a paper bag.

Multiple sensory receptors also provide information to the respiratory centers:

- **Central chemoreceptors** are brainstem neurons that respond especially to changes in the pH of the cerebrospinal fluid. They are concentrated on each side of the medulla oblongata at a point only 0.2 mm beneath its anterior surface. The pH of the CSF reflects the $CO_2$ level in the blood. By regulating respiration to maintain a stable pH, the respiratory centers also ensure a stable $CO_2$ level in the blood.

- **Peripheral chemoreceptors** are located in the carotid and aortic bodies of the large arteries above the heart (fig. 22.15). They respond to the $O_2$ and $CO_2$ content and the pH of the blood. The carotid bodies communicate with the brainstem by way of the glossopharyngeal nerves, and the aortic bodies by way of the vagus nerves. Sensory fibers in these nerves enter the medulla and synapse with neurons of the DRG.

- **Stretch receptors** are found in the smooth muscle of the bronchi and bronchioles, and in the visceral pleura. They respond to inflation of the lungs and signal the DRG by way of the vagus nerves. Excessive inflation triggers the **inflation (Hering–Breuer)**[12] **reflex,** a protective somatic reflex that strongly inhibits the I neurons and stops inspiration. In infants, this may be a normal mechanism of transition from inspiration to expiration, but after infancy it is activated only by extreme stretching of the lungs.

- **Irritant receptors** are nerve endings amid the epithelial cells of the airway. They respond to smoke, dust, pollen, chemical fumes, cold air, and excess mucus. They transmit signals by way of the vagus nerves to the DRG, and the DRG returns signals to the respiratory and bronchial muscles, resulting in such protective reflexes as bronchoconstriction, shallower breathing, breath-holding *(apnea),* or coughing.

## Voluntary Control of Breathing

Voluntary control of breathing is important in singing, speaking, breath-holding, and other circumstances. Such control originates in the motor cortex of the frontal lobe of

[12]Heinrich Ewald Hering (1866–1948), German physiologist; Josef Breuer (1842–1925), Austrian physician

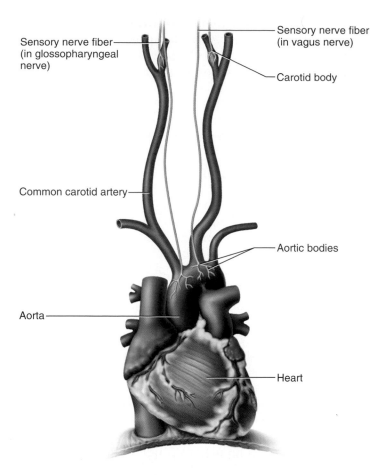

**FIGURE 22.15**  **The Peripheral Chemoreceptors of Respiration.**   Peripheral chemoreceptors in the aortic arch and carotid bodies monitor blood pH and gas concentrations. They send signals about blood chemistry to the respiratory centers of the medulla oblongata through sensory fibers in the vagus and glossopharyngeal nerves.

the cerebrum. The output neurons send impulses down the corticospinal tracts to the integrating centers in the spinal cord, bypassing the brainstem centers. There are limits to voluntary control. Temperamental children may threaten to hold their breath until they die, but it is impossible to do so. Holding one's breath raises the $CO_2$ level of the blood until a *breaking point* is reached where automatic controls override one's will. This forces a person to resume breathing even if he or she has lost consciousness.

## PRESSURE, RESISTANCE, AND AIRFLOW

Now that we know the neuromuscular aspects of breathing, let us see how the expansion and contraction of the thoracic cage produces airflow into and out of the lungs.

Understanding the ventilation of the lungs, the transport of gases in the blood, and the exchange of gases with the tissues draws in part on the gas laws of physics. These laws are named after their discoverers, therefore not intu-

| **TABLE 22.1** | The Gas Laws of Respiratory Physiology |
|---|---|
| Boyle's law[13] | The pressure of a given quantity of gas is inversely proportional to its volume (assuming a constant temperature). |
| Charles's law[14] | The volume of a given quantity of gas is directly porportional to its absolute temperature (assuming a constant pressure). |
| Dalton's law[15] | The total pressure of a gas mixture is equal to the sum of the partial pressures of its individual gases. |
| Henry's law[16] | At the air–water interface, the amount of gas that dissolves in water is determined by its solubility in water and its partial pressure in the air (assuming a constant temperature). |

itively easy to remember by name. Table 22.1 lists the gas laws used in this chapter and may be a helpful reference as you progress through respiratory physiology.

### Pressure and Airflow

Respiratory airflow is governed by the same principles of flow, pressure, and resistance as blood flow. As we saw in chapter 20, the flow ($F$) of a fluid is directly proportional to the pressure difference between two points ($\Delta P$) and inversely proportional to resistance ($R$): $F \propto \Delta P/R$. For the moment, we will focus especially on $\Delta P$, the pressure difference that produces airflow. We will deal with resistance later.

The pressure that drives respiration is **atmospheric (barometric) pressure**—the weight of the air above us. At sea level, a column of air as thick as the atmosphere (60 mi) and 1 in. square weighs 14.7 lb; it is thus said to exert a force of 14.7 pounds per square inch (psi). In standard international (SI) units, this is a column of air 100 km high exerting a force of $1.013 \times 10^6$ dynes/cm$^2$. This pressure, called *1 atmosphere* (1 atm), is enough to force a column of mercury (Hg) 760 mm up an evacuated tube; therefore, 1 atm = 760 mm Hg. This is the average atmospheric pressure at sea level; it fluctuates from day to day and is lower at higher altitudes.

One way to change the pressure of a gas is to change the volume of its container. This fact is summarized by **Boyle's law,** which states that at a constant temperature, *the pressure of a given quantity of gas is inversely proportional to its volume.* If the lungs contain a quantity of gas and lung volume increases, their internal pressure, called **intrapulmonary pressure,** falls. Conversely, if lung volume

---

[13]Robert Boyle (1627–91), English physicist
[14]Jacques A. C. Charles (1746–1823), French physicist
[15]John Dalton (1766–1844), British chemist
[16]William Henry (1774–1836), British chemist

decreases, intrapulmonary pressure rises. (Compare this to the syringe analogy on p. 737.)

We have seen that a fluid tends to flow from a point of higher pressure to a point of lower pressure. If the intrapulmonary pressure drops lower than the atmospheric pressure surrounding the body, then air tends to flow down its pressure gradient into the lungs. If the intrapulmonary pressure rises above the atmospheric pressure, air flows out of the lungs. Therefore, all we have to do to breathe is to cyclically raise and lower the intrapulmonary pressure, employing the neuromuscular mechanisms recently described.

The following discussion uses measurements of relative gas pressures within the body and in the atmosphere. Atmospheric pressure varies with location and weather, so we cannot assume one constant atmospheric pressure (760 mm Hg) pertaining to all situations. But what matters to airflow is the *difference* between atmospheric pressure and intrapulmonary pressure. Thus, we will think in terms of relative pressures. A relative pressure of −3 mm Hg, for example, means 3 mm Hg below atmospheric pressure; a relative pressure of +3 mm Hg is 3 mm Hg above atmospheric pressure. At an atmospheric pressure of 760 mm Hg, these would represent absolute pressures of 757 and 763 mm Hg, respectively.

## Inspiration

Now consider the flow of air into the lungs—inspiration. Figure 22.16 traces the events and pressure changes that occur throughout a respiratory cycle.

At the beginning, there is no movement of the thoracic cage, no difference between the air pressure within the lungs and external to the body, and no airflow. What happens when the thoracic cage expands? Why do the lungs not remain the same size and simply occupy less space in the chest? Consider the two layers of the pleura: the parietal pleura forms the inner lining of the rib cage, and the visceral pleura forms the outer layer of the lung. They are firmly connected to those structures. Although they are not anatomically attached to each other along their surfaces, they are wet and therefore tend to cling together like sheets of wet paper.

When the ribs swing upward and outward during inspiration, the parietal pleura follows them. The visceral pleura clings to it by the cohesion of water, and as it follows the parietal pleura it takes the lung surface along with it. Thus the entire lung expands along with the thoracic cage. As it increases in volume, its internal pressure drops, and air flows in.

The pressures involved are well known. Between the two layers of pleura, there is normally a slight vacuum called the **intrapleural pressure,** measuring about −4 mm Hg. This pressure drops to about −6 mm Hg during inspiration, as the parietal pleura pulls away. Some of this pressure change transfers to the interior of the lungs, where pressure within the alveoli, called **intrapulmonary pres-**

**sure,** drops to about −3 mm Hg. Thus, if the atmospheric pressure were 760 mm Hg, the intrapleural pressure would be 754 mm Hg and the intrapulmonary pressure 757 mm Hg. The difference between intrapleural and intrapulmonary pressure, 3 mm Hg, is the **transpulmonary pressure.** The pressure gradient of 760 → 757 mm Hg from atmosphere to alveoli makes air flow into the lungs.

Yet this is not the only force that expands the lungs. Another is warming of the inhaled air. As we see from **Charles's law** (see table 22.1), the volume of a given quantity of gas is directly proportional to its absolute temperature. On a day when the ambient temperature is 21°C (70°F), inhaled air is heated to 37°C (16°C warmer) by the time it reaches the alveoli. As the inhaled air expands, it helps to inflate the lungs.

When the respiratory muscles stop contracting, the inflowing air quickly achieves an intrapulmonary pressure equal to atmospheric pressure, and flow stops. In quiet breathing, the dimensions of the thoracic cage increase by only a few millimeters in each direction, but this is enough to increase its total volume by 500 mL. Thus, 500 mL of air flows into the respiratory tract.

> **Think About It**
>
> *When you inhale, does your chest expand because your lungs inflate, or do your lungs inflate because your chest expands? Explain.*

## Expiration

Relaxed expiration is a passive process; it involves no muscular effort except for a braking action described shortly. Many structures of the thorax are elastic: the costal cartilages, the ligaments that attach the ribs to the spine, the central tendon of the diaphragm, and elastic tissue in the bronchi and bronchioles. These structures spring back to their resting size when muscular tension subsides. The thoracic cage therefore returns to its resting volume and compresses the lungs. In quiet respiration, this raises the intrapulmonary pressure to about +3 mm Hg. Air thus flows down its pressure gradient, out of the lungs. In forced breathing, the accessory muscles raise intrapulmonary pressure to as high as +30 mm Hg.

If the respiratory muscles simply ceased contracting at the end of each inspiration, we would expect the thoracic structures to snap back abruptly and air to be exhaled explosively. In actuality, however, the phrenic nerves continue to stimulate the diaphragm at a reduced level. This produces a slight braking action that prevents the lungs from recoiling too suddenly, so it makes the transition from inspiration to expiration smoother.

The effect of pulmonary elasticity is evident in a pathological state of pneumothorax and atelectasis. **Pneumothorax** is the presence of air in the pleural cavity. If the thoracic wall is punctured, for example, inspiration sucks

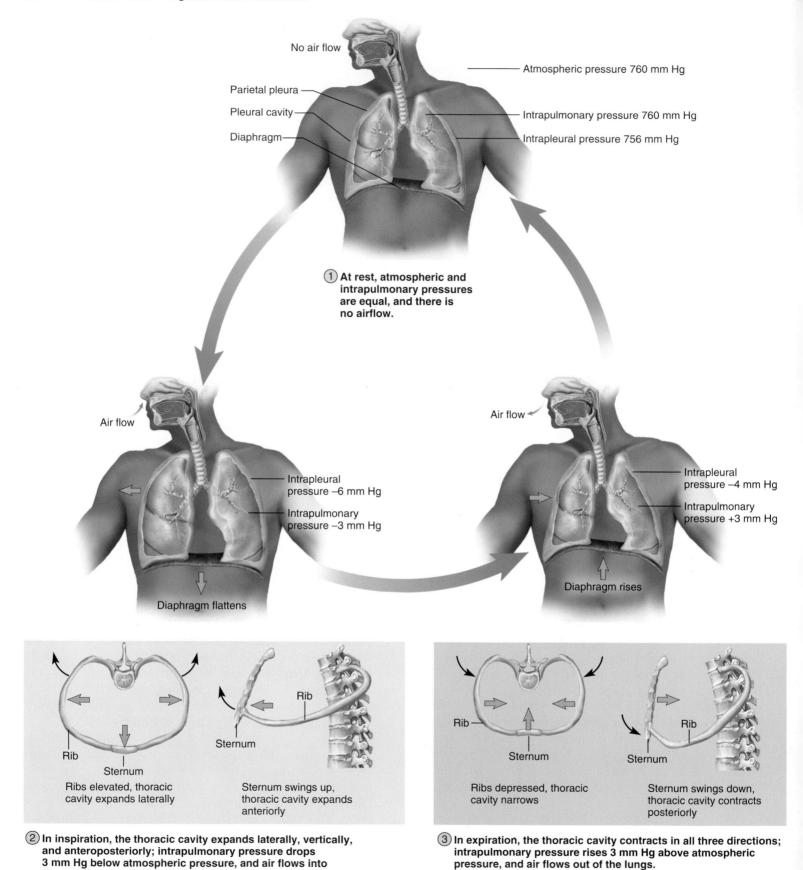

No air flow

Parietal pleura

Pleural cavity

Diaphragm

Atmospheric pressure 760 mm Hg

Intrapulmonary pressure 760 mm Hg

Intrapleural pressure 756 mm Hg

① **At rest, atmospheric and intrapulmonary pressures are equal, and there is no airflow.**

Air flow

Intrapleural pressure −6 mm Hg

Intrapulmonary pressure −3 mm Hg

Diaphragm flattens

Air flow

Intrapleural pressure −4 mm Hg

Intrapulmonary pressure +3 mm Hg

Diaphragm rises

Rib

Sternum

Ribs elevated, thoracic cavity expands laterally

Sternum

Rib

Sternum swings up, thoracic cavity expands anteriorly

Rib

Sternum

Ribs depressed, thoracic cavity narrows

Sternum

Rib

Sternum swings down, thoracic cavity contracts posteriorly

② **In inspiration, the thoracic cavity expands laterally, vertically, and anteroposteriorly; intrapulmonary pressure drops 3 mm Hg below atmospheric pressure, and air flows into the lungs.**

③ **In expiration, the thoracic cavity contracts in all three directions; intrapulmonary pressure rises 3 mm Hg above atmospheric pressure, and air flows out of the lungs.**

**FIGURE 22.16**  **The Respiratory Cycle.**  The pressures given here are based on an assumed atmospheric pressure of 760 mm Hg (1 atm). Values with plus and minus signs are relative to atmospheric pressure.

air through the wound into the pleural cavity, and the visceral and parietal pleurae separate; what was a *potential space* between them becomes an air-filled cavity. Without the negative intrapleural pressure to keep the lungs inflated, the lungs recoil and collapse. The collapse of part or all of a lung is called **atelectasis**[17] (AT-eh-LEC-ta-sis). Atelectasis can also result from airway obstruction—for example, by a lung tumor, aneurysm, swollen lymph node, or aspirated object. Blood absorbs gases from the alveoli distal to the obstruction, and that part of the lung collapses because it cannot be reventilated.

## Resistance to Airflow

Pressure is one determinant of airflow; the other is resistance. The greater the resistance, the slower the flow. But what governs resistance? Three factors are of particular importance:

1. **Diameter of the bronchioles.** Like arterioles, the large number of bronchioles, their small diameter, and their ability to change diameter make them the primary means of controlling resistance. The trachea and bronchi can also change diameter to a degree, but are more constrained by the supporting cartilages in their walls. An increase in the diameter of a bronchus or bronchiole is called **bronchodilation,** and a reduction in diameter is called **bronchoconstriction.** Epinephrine and the sympathetic nerves (norepinephrine) stimulate bronchodilation and increase airflow. Histamine, parasympathetic nerves (acetylcholine), cold air, and chemical irritants are among the factors that stimulate bronchoconstriction. Many people have suffocated from the extreme bronchoconstriction brought on by anaphylactic shock or asthma (see Insight 21.3).

2. **Pulmonary compliance.** This means the ease with which the lungs expand, or more exactly, the change in lung volume relative to a given pressure change. The thoracic cage may produce the same intrapleural pressure in two different people, but the lungs will expand less in a person with poorer pulmonary compliance (stiffer lungs). Compliance can be reduced by degenerative lung diseases such as tuberculosis and black lung disease, in which the lungs are stiffened by scar tissue. In such conditions, the thoracic cage expands normally and transpulmonary pressure falls, but the lungs expand relatively little.

3. **Surface tension of the alveoli and distal bronchioles.** The alveoli are relatively dry, but they have a thin film of water over the epithelium. This film is necessary for gas exchange, but creates a potential problem for pulmonary ventilation. Water molecules are attracted to each other by hydrogen bonds, creating surface tension, as we saw in chapter 2. You can appreciate the

strength of this attraction if you reflect on the difficulty of separating two sheets of wet paper compared with two sheets of dry paper. Such a force draws the walls of the alveoli inward toward the lumen. If it went unchecked, the alveoli would collapse with each expiration and would strongly resist reinflation.

The solution to this problem takes us back to the great alveolar cells and their surfactant. A *surfactant* is an agent that disrupts the hydrogen bonds of water and reduces surface tension; soaps and detergents are everyday examples. The pulmonary surfactant is composed of amphiphilic proteins and phospholipids. These molecules are partially hydrophobic, so they spread out over the surface of the water film, partially embedded in it like ice cubes floating in a bowl of water. As an alveolus begins to deflate, the surfactants are squeezed closer together, like the ice cubes being pushed together into a smaller area. If the alveolus were covered with a film of water only, it could continue collapsing, because water molecules can pile up into a thicker film of moisture. The physical structure of the surfactants resists compression, however. They cannot pile up into a thicker layer, because their hydrophilic regions resist separation from the water below. As they become crowded into a small area and resist layering, they retard and then halt the collapse of the alveolus.

The importance of this surfactant is especially apparent when it is lacking. Premature infants often have a deficiency of pulmonary surfactant and experience great difficulty breathing (see chapter 29). The resulting *respiratory distress syndrome* is often treated by administering artificial surfactant.

## ALVEOLAR VENTILATION

Air that actually enters the alveoli becomes available for gas exchange, but not all inhaled air gets that far. About 150 mL of it (typically 1 mL per pound of body weight) fills the conducting division of the airway. Since this air cannot exchange gases with the blood, it is called **dead air,** and the conducting division is called the **anatomical dead space.** In pulmonary diseases, some alveoli may be unable to exchange gases with the blood because they lack blood flow or because the pulmonary membrane is thickened by edema or fibrosis. **Physiological (total) dead space** is the sum of anatomical dead space and any pathological alveolar dead space that may exist. In healthy people, few alveoli are nonfunctional, and the anatomical and physiological dead spaces are identical.

The anatomical dead space varies with circumstances. In a state of relaxation, parasympathetic stimulation keeps the airway somewhat constricted. This minimizes the dead space so that more of the inhaled air ventilates the alveoli. In a state of arousal, by contrast, the sympathetic nervous system dilates the airway, which

---

[17]*atel* = imperfect, incomplete + *ectasis* = extension

increases airflow. The increased airflow outweighs the air that is wasted by filling the increased dead space.

If a person inhales 500 mL of air and 150 mL of it is dead air, then 350 mL of air ventilates the alveoli. Multiplying this by the respiratory rate gives the **alveolar ventilation rate (AVR)**—for example, 350 mL/breath × 12 breaths/min = 4,200 mL/min. Of all measures of pulmonary ventilation, this one is most directly relevant to the body's ability to get oxygen to the tissues and dispose of carbon dioxide.

The alveoli never completely empty during expiration. There is always some leftover air called the *residual volume,* typically about 1,300 mL even after maximum expiration. This air mixes with fresh air arriving on the next inspiration, so the same oxygen-depleted air does not remain in the lungs cycle after cycle.

## MEASUREMENTS OF VENTILATION

It is often important to measure a person's pulmonary ventilation in order to assess the severity of a respiratory disease or monitor a patient's improvement or deterioration. Measurements are made by having the subject breathe into a device called a **spirometer,**[18] which recaptures the expired breath and records such variables as the rate and depth of breathing, speed of expiration, and rate of oxygen consumption. Representative measurements for a healthy adult are given in table 22.2 and explained in figure 22.17. Four of them are called *respiratory volumes:* **tidal volume, inspiratory reserve volume, expiratory**

---

[18]*spiro* = breath + *meter* = measuring device

| TABLE 22.2 | Respiratory Volumes and Capacities for an Average Young Adult Male | |
|---|---|---|
| **Measurement** | **Typical Value** | **Definition** |
| ***Respiratory Volumes*** | | |
| Tidal volume (TV) | 500 mL | Amount of air inhaled or exhaled in one breath during quiet breathing |
| Inspiratory reserve volume (IRV) | 3,000 mL | Amount of air in excess of tidal volume that can be inhaled with maximum effort |
| Expiratory reserve volume (ERV) | 1,200 mL | Amount of air in excess of tidal volume that can be exhaled with maximum effort |
| Residual volume (RV) | 1,300 mL | Amount of air remaining in the lungs after maximum expiration; the amount that can never voluntarily be exhaled |
| ***Respiratory Capacities*** | | |
| Vital capacity (VC) | 4,700 mL | The amount of air that can be inhaled and then exhaled with maximum effort; the deepest possible breath (VC = ERV + TV + IRV) |
| Inspiratory capacity (IC) | 3,500 mL | Maximum amount of air that can be inhaled after a normal tidal expiration (IC = TV + IRV) |
| Functional residual capacity (FRC) | 2,500 mL | Amount of air remaining in the lungs after a normal tidal expiration (FRC = RV + ERV) |
| Total lung capacity (TLC) | 6,000 mL | Maximum amount of air the lungs can contain (TLC = RV + VC) |

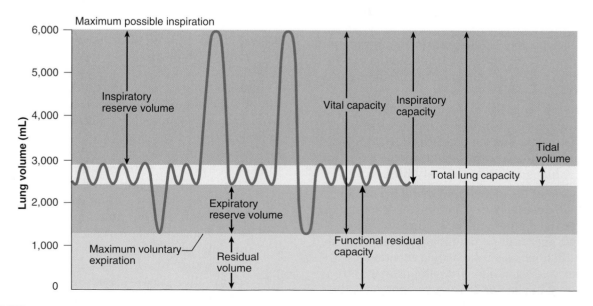

**FIGURE 22.17  Respiratory Volumes and Capacities.**  The wavy line indicates inspiration when it rises and expiration when it falls. Compare with table 22.2.

reserve volume, and residual volume. Four others, called *respiratory capacities,* are obtained by adding two or more of the respiratory volumes: **vital capacity, inspiratory capacity, functional residual capacity,** and **total lung capacity.** In general, respiratory volumes and capacities are proportional to body size; consequently, they are generally higher for men than for women.

**Spirometry,** the measurement of pulmonary function, is an aid to the diagnosis and assessment of *restrictive* and *obstructive* lung disorders. **Restrictive disorders** are those which reduce pulmonary compliance, thus limiting the amount to which the lungs can be inflated. They show in spirometry as a reduced vital capacity. Any disease that produces pulmonary fibrosis has a restrictive effect: black lung disease and tuberculosis, for example. **Obstructive disorders** are those that interfere with airflow by narrowing or blocking the airway. They make it harder to exhale a given amount of air. Asthma, emphysema, and chronic bronchitis are the most common examples. Obstructive disorders can be measured by having the subject exhale as rapidly as possible into a spirometer and measuring **forced expiratory volume (FEV)**—the percentage of the vital capacity that can be exhaled in a given time interval. A healthy adult should be able to expel 75% to 85% of the vital capacity in 1.0 second (a value called the $FEV_{1.0}$). At home, asthma patients and others can monitor their respiratory function by blowing into a handheld meter that measures **peak flow,** the maximum speed of expiration.

The amount of air inhaled per minute is the **minute respiratory volume (MRV).** MRV largely determines the alveolar ventilation rate. It can be measured directly with a spirometer or obtained by multiplying tidal volume by respiratory rate. For example, if a person has a tidal volume of 500 mL per breath and a rate of 12 breaths per minute, the MRV would be $500 \times 12 = 6,000$ mL/min. During heavy exercise, MRV may be as high as 125 to 170 L/min. This is called **maximum voluntary ventilation (MVV),** formerly called *maximum breathing capacity.*

## VARIATIONS IN THE RESPIRATORY RHYTHM

Relaxed, quiet breathing is called **eupnea**[19] (YOOP-nee-uh). It is typically characterized by a tidal volume of about 500 mL and a respiratory rate of 12 to 15 breaths per minute. Conditions ranging from exercise or anxiety to various disease states can cause deviations such as abnormally fast, slow, or labored breathing. Table 22.3 defines the clinical terms for several such variations. You should familiarize yourself with these before proceeding, because later discussions in this chapter assume a working knowledge of some of these terms.

Other variations in pulmonary ventilation serve the purposes of speaking, expressing emotion (laughing, cry-

ing), yawning, hiccuping, expelling noxious fumes, coughing, sneezing, and expelling abdominal contents. Coughing is induced by irritants in the lower respiratory tract. To cough, we close the glottis and contract the muscles of expiration, producing high pressure in the lower respiratory tract. We then suddenly open the glottis and release an explosive burst of air at speeds over 900 km/hr (600 mi/hr). This drives mucus and foreign matter toward the pharynx and mouth. Sneezing is triggered by irritants in the nasal cavity. Its mechanism is similar to coughing except that the glottis is continually open, the soft palate and tongue block the flow of air while thoracic pressure builds, and then the uvula (the conical projection of the posterior edge of the soft palate) is depressed to direct part of the airstream through the nose. These actions are coordinated by coughing and sneezing centers in the medulla oblongata.

| TABLE 22.3 | Variations in the Respiratory Rhythm |
|---|---|
| Apnea[20] (AP-nee-uh) | Temporary cessation of breathing (one or more skipped breaths) |
| Dyspnea[21] (DISP-nee-uh) | Labored, gasping breathing; shortness of breath |
| Hyperpnea[22] (HY-purp-NEE-uh) | Increased rate and depth of breathing in response to exercise, pain, or other conditions |
| Hyperventilation | Increased pulmonary ventilation in excess of metabolic demand, frequently associated with anxiety; expels $CO_2$ faster than it is produced, thus lowering the blood $CO_2$ concentration and raising the blood pH |
| Hypoventilation[23] | Reduced pulmonary ventilation; leads to an increase in blood $CO_2$ concentration if ventilation is insufficient to expel $CO_2$ as fast as it is produced |
| Kussmaul[24] respiration | Deep, rapid breathing often induced by acidosis; seen in diabetes mellitus |
| Orthopnea[25] (or-thop-NEE-uh) | Dyspnea that occurs when a person is lying down or in any position other than standing or sitting erect; seen in heart failure, asthma, emphysema, and other conditions |
| Respiratory arrest | Permanent cessation of breathing (unless there is medical intervention) |
| Tachypnea[26] (tack-ip-NEE-uh) | Accelerated respiration |

---

[19]*eu* = easy, normal + *pnea* = breathing

[20]*a* = without + *pnea* = breathing
[21]*dys* = difficult, abnormal, painful + *pnea* = breathing
[22]*hyper* = above normal
[23]*hypo* = below normal
[24]Adolph Kussmaul (1822–1902), German physician
[25]*ortho* = straight, erect + *pnea* = breathing
[26]*tachy* = fast + *pnea* = breathing

## Before You Go On

*Answer the following questions to test your understanding of the preceding section.*

5. Explain why contraction of the diaphragm causes inspiration *but contraction of the transverse abdominal muscle causes* expiration.

6. Which brainstem respiratory nucleus is indispensable to respiration? What do the other nuclei do?

7. Explain why Boyle's law is relevant to the action of the respiratory muscles.

8. Explain why eupnea requires little or no action by the muscles of expiration.

9. Identify a benefit and a disadvantage of normal (non-pathological) bronchoconstriction.

10. Suppose a healthy person has a tidal volume of 650 mL, an anatomical dead space of 160 mL, and a respiratory rate of 14 breaths per minute. Calculate her alveolar ventilation rate.

11. Suppose a person has a total lung capacity of 5,800 mL, a residual volume of 1,200 mL, an inspiratory reserve volume of 2,400 mL, and an expiratory reserve volume of 1,400 mL. Calculate his tidal volume.

# Gas Exchange and Transport

### Objectives

When you have completed this section, you should be able to

- define *partial pressure* and discuss its relationship to a gas mixture such as air;

- contrast the composition of inspired and alveolar air;

- discuss how partial pressure affects gas transport by the blood;

- describe the mechanisms of transporting $O_2$ and $CO_2$;

- describe the factors that govern gas exchange in the lungs and systemic capillaries;

- explain how gas exchange is adjusted to the metabolic needs of different tissues; and

- discuss the effect of blood gases and pH on the respiratory rhythm

Ultimately, respiration is about gases, especially oxygen and carbon dioxide. We will turn our attention now to the behavior of these gases in the human body: how oxygen is obtained from inspired air and delivered to the tissues, and how carbon dioxide is removed from the tissues and released into the expired air. First, however, it is necessary to understand the composition of the air we inhale and how gases behave in contact with the water film that lines the alveoli.

## COMPOSITION OF AIR

Air consists of about 78.6% nitrogen; 20.9% oxygen; 0.04% carbon dioxide; several quantitatively minor gases such as argon, neon, helium, methane, and ozone; and a variable amount of water vapor. Water vapor constitutes from 0% to 4%, depending on temperature and humidity; we will use a value of 0.5%, typical of a cool clear day.

The total atmospheric pressure is a sum of the contributions of these individual gases—a principle known as **Dalton's law** (see table 22.1). The separate contribution of each gas in a mixture is called its **partial pressure** and is symbolized with a *P* followed by the formula of the gas, such as $P_{N_2}$. If we assume the average sea-level pressure of 760 mm Hg and nitrogen is 78.6% of this, then $P_{N_2}$ is simply $0.786 \times 760$ mm Hg = 597 mm Hg. Applying Dalton's law to the mixture of gases in the preceding paragraph, $P_{N_2} + P_{O_2} + P_{H_2O} + P_{CO_2} \approx 597 + 159 + 3.7 + 0.3 = 760.0$ mm Hg.

This is the composition of the air we inhale, but it is not the composition of air in the alveoli. Alveolar air can be sampled with an apparatus that collects the last 10 mL of expired air. As we see in table 22.4, its composition differs from that of the atmosphere because of three influences: (1) It is humidified by contact with the mucous membranes during inspiration, so its $P_{H_2O}$ is more than 10 times higher than that of the inhaled air. (2) Freshly inspired air mixes with residual air left from the previous respiratory cycle, so its oxygen is diluted and it is enriched with $CO_2$ from the residual air. (3) Alveolar air exchanges $O_2$ and $CO_2$ with the blood. Thus, the $P_{O_2}$ of alveolar air is about 65% that of inhaled air, and its $P_{CO_2}$ is more than 130 times higher.

> ⌐ **Think About It**
>
> *Expired air, considered as a whole (not just the last 10 mL), contains about 116 mm Hg $O_2$ and 32 mm Hg $CO_2$. Why would these values differ from the ones for alveolar air?*

| TABLE 22.4 | Composition of Inspired (Atmospheric) and Alveolar Air | | | |
|---|---|---|---|---|
| **Gas** | **Inspired Air*** | | **Alveolar Air** | |
| $N_2$ | 78.6% | 597 mm Hg | 74.9% | 569 mm Hg |
| $O_2$ | 20.9% | 159 mm Hg | 13.7% | 104 mm Hg |
| $H_2O$ | 0.5% | 3.7 mm Hg | 6.2% | 47 mm Hg |
| $CO_2$ | 0.04% | 0.3 mm Hg | 5.3% | 40 mm Hg |
| Total | 100% | 760 mm Hg | 100% | 760 mm Hg |

*Typical values for a cool clear day; values vary with temperature and humidity. Other gases present in small amounts are disregarded.

# ALVEOLAR GAS EXCHANGE

Air in the alveolus is in contact with the film of water covering the alveolar epithelium. For oxygen to get into the blood, it must dissolve in this water and pass through the respiratory membrane separating the air from the bloodstream. For carbon dioxide to leave the blood, it must pass the other way and diffuse out of the water film into the alveolar air. This back-and-forth traffic of $O_2$ and $CO_2$ across the respiratory membrane is called **alveolar gas exchange.**

The reason that $O_2$ can diffuse in one direction and $CO_2$ in the other is that each gas diffuses down its own pressure gradient. Whenever air and water are in contact with each other, gases diffuse down their gradients until the partial pressure of each gas in the air is equal to its partial pressure in the water. If a gas is more abundant in the water than in the air, it diffuses into the air; the smell of chlorine near a swimming pool is evidence of this. If a gas is more abundant in the air, it diffuses into the water.

**Henry's law** states that *at the air–water interface, for a given temperature, the amount of gas that dissolves in the water is determined by its solubility in water and its partial pressure in the air* (fig. 22.18). Thus, the greater the $P_{O_2}$ in the alveolar air, the more $O_2$ the blood picks up. And since blood arriving at an alveolus has a higher $P_{CO_2}$ than air, it releases $CO_2$ into the air. At the alveolus, the blood is said to *unload* $CO_2$ and *load* $O_2$. Each gas in a mixture behaves independently; the diffusion of one gas does not influence the diffusion of another.

Both $O_2$ loading and $CO_2$ unloading involve erythrocytes (RBCs). The efficiency of these processes therefore depends on how long an RBC spends in an alveolar capillary compared with how long it takes for each gas to be fully loaded or unloaded—that is, for them to reach equilibrium concentrations in the capillary blood. It takes about 0.25 sec to reach equilibrium. At rest, when blood circulates at its slowest speed, an RBC passes through an alveolar capillary in about 0.75 sec—plenty of time. Even in vigorous exercise, when the blood flows faster, an erythrocyte is in the alveolar capillary for about 0.3 sec, which is still adequate.

Several variables affect the efficiency of alveolar gas exchange, and some of these can, under abnormal conditions, prevent the complete loading and unloading of these gases:

- **Pressure gradients of the gases.** The $P_{O_2}$ is about 104 mm Hg in the alveolar air and 40 mm Hg in the blood arriving at an alveolus. Oxygen therefore diffuses from the air into the blood, where it reaches a

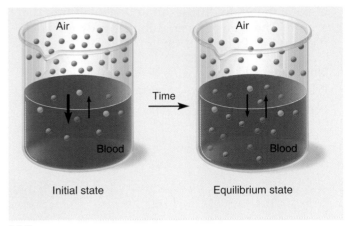

**(a) Oxygen**

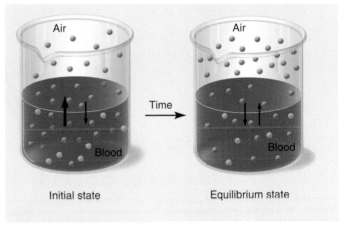

**(b) Carbon dioxide**

**FIGURE 22.18 Henry's Law and Its Relationship to Alveolar Gas Exchange.** (a) The $P_{O_2}$ of alveolar air is initially higher than the $P_{O_2}$ of the blood arriving at an alveolus. Oxygen diffuses into the blood until the two are in equilibrium. (b) The $P_{CO_2}$ of the arriving blood is initially higher than the $P_{CO_2}$ of alveolar air. Carbon dioxide diffuses into the alveolus until the two are in equilibrium. It takes about 0.25 sec for both gases to reach equilibrium.

$P_{O_2}$ of 104 mm Hg. Before the blood leaves the lung, however, this drops to about 95 mm Hg. This oxygen dilution occurs because the pulmonary veins anastomose with the bronchial veins in the lungs, so there is some mixing of the oxygen-rich pulmonary blood with the oxygen-poor systemic blood.

The $P_{CO_2}$ is about 46 mm Hg in blood arriving at the alveolus and 40 mm Hg in alveolar air. Carbon dioxide therefore diffuses from the blood into the alveoli. These changes are summarized here and at the middle of figure 22.19:

| Blood entering lungs | | Blood leaving lungs | |
|---|---|---|---|
| $P_{O_2}$ | 40 mm Hg | $P_{O_2}$ | 95 mm Hg |
| $P_{CO_2}$ | 46 mm Hg | $P_{CO_2}$ | 40 mm Hg |

These gradients differ under special circumstances such as high altitude and *hyperbaric oxygen therapy* (treatment with oxygen at greater than 1 atm of pressure) (fig. 22.20). At high altitudes, the partial pressures of all atmospheric gases are lower. Atmospheric $P_{O_2}$, for example, is 159 mm Hg at sea level and 110 mm Hg at 3,000 m (10,000 ft). The $O_2$

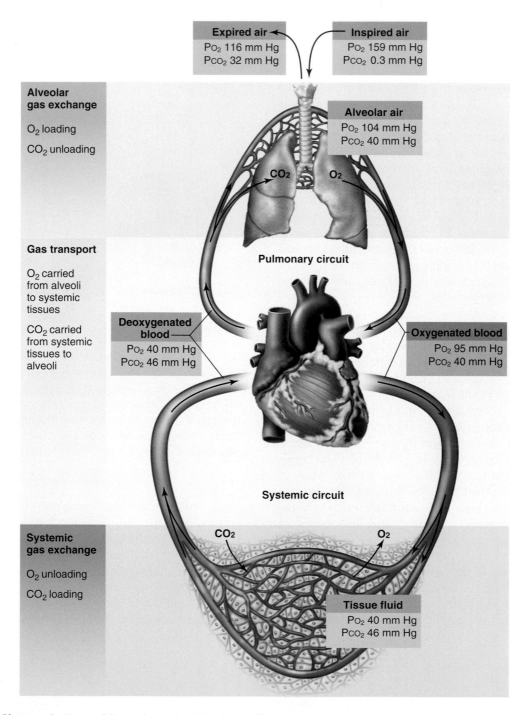

**FIGURE 22.19**   Changes in $P_{O_2}$ and $P_{CO_2}$ along the Circulatory Route.

▶ *Trace the partial pressure of oxygen from inspired air to expired air and explain each change in $P_{O_2}$ along the way. Do the same for $P_{CO_2}$.*

gradient from air to blood is proportionately less, and as we can predict from Henry's law, less $O_2$ diffuses into the blood. In a hyperbaric oxygen chamber, by contrast, a patient is exposed to 3 to 4 atm of oxygen to treat such conditions as gangrene (to kill anaerobic bacteria) and carbon monoxide poisoning (to displace the carbon monoxide from hemoglobin). The $P_{O_2}$ ranges between 2,300 and 3,000 mm Hg. Thus, there is a very steep gradient of $P_{O_2}$ from alveolus to blood and diffusion into the blood is accelerated.

- **Solubility of the gases.** Gases differ in their ability to dissolve in water. Carbon dioxide is about 20 times as soluble as oxygen, and oxygen is about twice as soluble as nitrogen. Even though the pressure gradient of $O_2$ is much greater than that of $CO_2$ across the

respiratory membrane, equal amounts of the two gases are exchanged because $CO_2$ is so much more soluble and diffuses more rapidly.

- **Membrane thickness.** The respiratory membrane between the blood and alveolar air is only 0.5 μm thick in most places—much less than the 7 to 8 mm diameter of a single RBC. Thus, it presents little obstacle to diffusion (fig. 22.21a). In such heart conditions as left ventricular failure, however, blood pressure backs up into the lungs and promotes capillary filtration into the connective tissues, causing the respiratory membranes to become edematous and thickened (similar to their condition in pneumonia; fig. 22.21b). The gases have farther to travel between blood and air and cannot equilibrate fast enough to

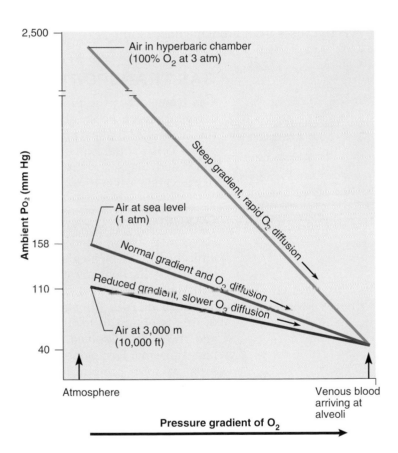

**FIGURE 22.20** Oxygen Loading in Relation to Concentration Gradient. The rate of loading depends on the steepness of the gradient from alveolar air to the venous blood arriving at the alveolar capillaries. Compared with the oxygen gradient at sea level (blue line), the gradient is less steep at high altitude (red line) because the $P_{O_2}$ of the atmosphere is lower. Thus oxygen loading of the pulmonary blood is slower. In a hyperbaric chamber with 100% oxygen, the gradient from air to blood is very steep (green line), and oxygen loading is correspondingly rapid. This is an illustration of Henry's law and has important effects in diving, aviation, mountain climbing, and oxygen therapy.

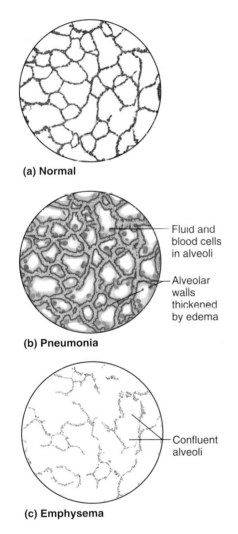

**FIGURE 22.21** Pulmonary Alveoli in Health and Disease. (a) In a healthy lung, the alveoli are small and have thin respiratory membranes. (b) In pneumonia, the respiratory membranes (alveolar walls) are thick with edema, and the alveoli contain fluid and blood cells. (c) In emphysema, alveolar membranes break down and neighboring alveoli join to form larger, fewer alveoli with less total surface area.

keep pace with blood flow. Under these circumstances, blood leaving the lungs has an unusually high $P_{CO_2}$ and low $P_{O_2}$.

- **Membrane area.** In good health, each lung has about 70 m² of respiratory membrane available for gas exchange. Since the alveolar capillaries contain a total of only 100 mL of blood at any one time, this blood is spread very thinly. Several pulmonary diseases, however, decrease the alveolar surface area and thus lead to low blood $P_{O_2}$—for example, emphysema (fig. 22.21c), lung cancer, and tuberculosis.

- **Ventilation–perfusion coupling.** Gas exchange requires not only good ventilation of the alveolus, but also good perfusion of its capillaries. As a whole, the lungs have a *ventilation–perfusion ratio* of about 0.8—a flow of 4.2 L of air and 5.5 L of blood per minute (at rest). The ratio is somewhat higher in the apex of the lung and lower in the base because more blood is drawn toward the base by gravity. **Ventilation–perfusion**

**coupling** is the ability to match ventilation and perfusion to each other (fig. 22.22). If part of a lung is poorly ventilated because of tissue destruction or airway obstruction, there is little point in directing much blood there. This blood would leave the lung carrying less oxygen than it should. But poor ventilation causes local constriction of the pulmonary arteries, reducing blood flow to that area and redirecting this blood to better-ventilated alveoli. Good ventilation, by contrast, dilates the arteries and increases perfusion so that most blood is directed to regions of the lung where it can pick up the most oxygen. This is opposite from the reactions of systemic arteries, where hypoxia ($O_2$ deficiency) causes vasodilation so that blood flow to a tissue will increase and reverse the hypoxia.

Ventilation is also adjustable. Poor ventilation causes local $CO_2$ accumulation, which stimulates local bronchodilation and improves airflow. Low $P_{CO_2}$ causes local bronchoconstriction.

# GAS TRANSPORT

**Gas transport** is the process of carrying gases from the alveoli to the systemic tissues and vice versa. This section explains how the blood loads and transports oxygen and carbon dioxide.

## Oxygen

Oxygen is transported almost entirely by the hemoglobin in the RBCs. Its concentration by volume in the arterial blood is about 20 mL/dL. About 98.5% of this is bound to hemoglobin and 1.5% is dissolved in the blood plasma. Hemoglobin consists of four protein (globin) chains, each with one heme group (see fig. 18.5, p. 686). Each heme group can bind 1 $O_2$ to the ferrous ion at its center; thus, one hemoglobin molecule can carry up to 4 $O_2$. If one or more molecules of $O_2$ are bound to hemoglobin, the compound is called **oxyhemoglobin (HbO$_2$),** whereas hemoglobin with no oxygen bound to it is **deoxyhemoglobin (HHb).** When hemoglobin is 100% saturated, every molecule of it carries 4 $O_2$; if it is 75% saturated, there is an average of 3 $O_2$ per hemoglobin

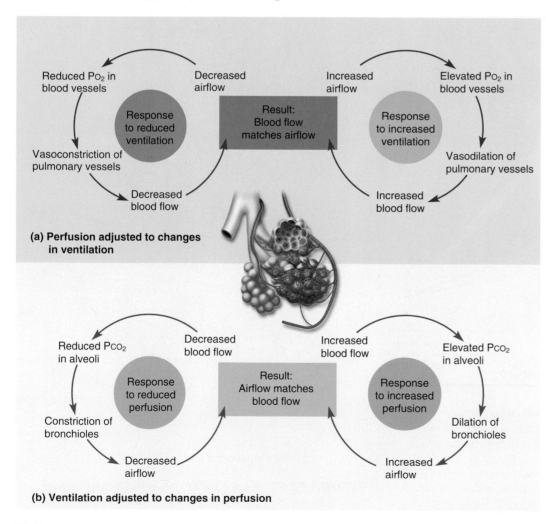

**FIGURE 22.22  Ventilation–Perfusion Coupling.**  Negative feedback loops adjust air flow and blood flow to each other. (a) Blood circulation can be adjusted to above- or below-normal ventilation of a part of the lung. (b) Ventilation can be adjusted to above- or below-normal blood circulation to a part of the lung.

## INSIGHT 22.4   Clinical Application

### Carbon Monoxide Poisoning

The lethal effect of carbon monoxide (CO) is well known. This colorless, odorless gas occurs in cigarette smoke, engine exhaust, and fumes from furnaces and space heaters. It binds to the ferrous ion of hemoglobin to form *carboxyhemoglobin (HbCO)*. Thus, it competes with oxygen for the same binding site. Not only that, but it binds 210 times as tightly as oxygen. Thus, CO tends to tie up hemoglobin for a long time. Less than 1.5% of the hemoglobin is occupied by carbon monoxide in most nonsmokers, but this figure rises to as much as 3% in residents of heavily polluted cities and 10% in heavy smokers. An atmospheric concentration of 0.1% CO, as in a closed garage, is enough to bind 50% of a person's hemoglobin, and an atmospheric concentration of 0.2% is quickly lethal.

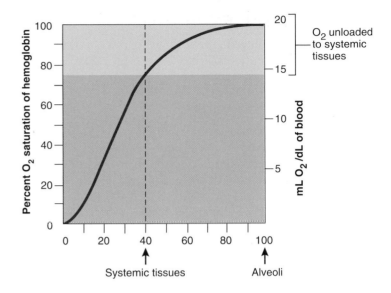

**Partial pressure of $O_2$ ($Po_2$) in mm Hg**

**FIGURE 22.23** **The Oxyhemoglobin Dissociation Curve.** This curve shows the relative amount of hemoglobin that is saturated with oxygen (*y*-axis) as a function of ambient (surrounding) oxygen concentration (*x*-axis). As it passes through the alveolar capillaries where the $Po_2$ is high, hemoglobin becomes saturated with oxygen. As it passes through the systemic capillaries where the $Po_2$ is low, it typically gives up about 22% of its oxygen (color bar at top of graph).

) *What would be the approximate utilization coefficient if the systemic tissues had a $Po_2$ of 20 mm Hg?*

molecule; if it is 50% saturated, there is an average of 2 $O_2$ per hemoglobin; and so forth. The poisonous effect of carbon monoxide stems from its competition for the $O_2$ binding site (see Insight 22.4).

The relationship between hemoglobin saturation and $Po_2$ is shown by an *oxyhemoglobin dissociation curve* (fig. 22.23). As you can see, it is not a simple linear relationship. At low $Po_2$, the curve rises slowly; then there is a rapid increase in oxygen loading as $Po_2$ rises further; finally, at high $Po_2$, the curve levels off as the hemoglobin approaches 100% saturation. This reflects the way hemoglobin loads oxygen. When the first heme group binds a molecule of $O_2$, hemoglobin changes shape in a way that facilitates uptake of the second $O_2$ by another heme group. This, in turn, promotes the uptake of the third and then the fourth $O_2$—hence the rapidly rising midportion of the curve.

> **Think About It**
>
> Is oxygen loading a positive or negative feedback process? Explain.

### Carbon Dioxide

Carbon dioxide is transported in three forms: carbonic acid, carbamino compounds, and dissolved gas.

1. About 90% of the $CO_2$ is hydrated (reacts with water) to form **carbonic acid,** which then dissociates into bicarbonate and hydrogen ions:

$$CO_2 + H_2O \rightarrow H_2CO_3 \rightarrow HCO_3^- + H^+$$

More will be said about this reaction shortly.

2. About 5% binds to the amino groups of plasma proteins and hemoglobin to form **carbamino compounds**—chiefly, **carbaminohemoglobin ($HbCO_2$).** The reaction with hemoglobin can be symbolized Hb + $CO_2 \rightarrow HbCO_2$. Carbon dioxide does not compete with oxygen because $CO_2$ and $O_2$ bind to different sites on the hemoglobin molecule—oxygen to the heme moiety and $CO_2$ to the polypeptide chains. Hemoglobin can therefore transport both $O_2$ and $CO_2$ simultaneously. As we will see, however, each gas somewhat inhibits transport of the other.

3. The remaining 5% of the $CO_2$ is carried in the blood as dissolved gas, like the $CO_2$ in soda pop.

The relative amounts of $CO_2$ exchanged between the blood and alveolar air differ from the percentages just given. About 70% of the *exchanged* $CO_2$ comes from carbonic acid, 23% from carbamino compounds, and 7% from the dissolved gas. That is, blood gives up the dissolved $CO_2$ gas and $CO_2$ from the carbamino compounds more easily than it gives up the $CO_2$ in bicarbonate.

## SYSTEMIC GAS EXCHANGE

Systemic gas exchange is the unloading of $O_2$ and loading of $CO_2$ at the systemic capillaries (see fig. 22.19, bottom, and fig. 22.24).

### Carbon Dioxide Loading

Aerobic respiration produces a molecule of $CO_2$ for every molecule of $O_2$ it consumes. The tissue fluid therefore contains a relatively high $P_{CO_2}$ and there is typically a $CO_2$ gradient of $46 \rightarrow 40$ mm Hg from tissue fluid to blood. Consequently, $CO_2$ diffuses into the bloodstream, where it is carried in the three forms noted. Most of it reacts with water to produce bicarbonate ($HCO_3^-$) and hydrogen ($H^+$) ions. This reaction occurs slowly in the blood plasma but much faster in the RBCs, where it is catalyzed by the enzyme *carbonic anhydrase.* An antiport called the *chloride–bicarbonate exchanger* then pumps most of the $HCO_3^-$ out of the RBC in exchange for $Cl^-$ from the blood plasma. This exchange is called the **chloride shift.** Most of the $H^+$ binds to hemoglobin or oxyhemoglobin, which thus buffers the intracellular pH.

### Oxygen Unloading

When $H^+$ binds to oxyhemoglobin ($HbO_2$), it reduces the affinity of hemoglobin for $O_2$ and tends to make hemoglobin release it. Oxygen consumption by respiring tissues keeps the $P_{O_2}$ of tissue fluid relatively low, and so there is typically a pressure gradient of $95 \rightarrow 40$ mm Hg of oxygen from the arterial blood to the tissue fluid. Thus, the liberated oxygen—along with some that was carried as dissolved gas in the plasma—diffuses from the blood into the tissue fluid.

As blood arrives at the systemic capillaries, its oxygen concentration is about 20 mL/dL and the hemoglobin is about 97% saturated. As it leaves the capillaries of a typical resting tissue, its oxygen concentration is about 15.6 mL/dL and the hemoglobin is about 75% saturated. Thus, it has given up 4.4 mL/dL—about 22% of its oxygen load. This fraction is called the **utilization coefficient.** The oxygen remaining in the blood after it passes through the capillary bed provides a **venous reserve** of oxygen, which can sustain life for 4 to 5 minutes even in the event of respiratory arrest. At rest, the circulatory system releases oxygen to the tissues at an overall rate of about 250 mL/min.

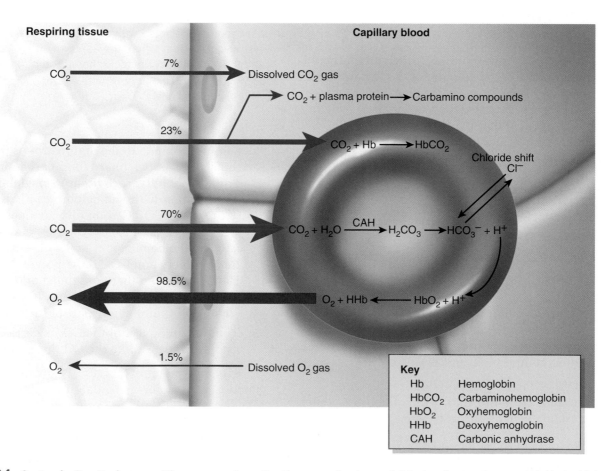

**FIGURE 22.24  Systemic Gas Exchange.**  Blue arrows show the three mechanisms of $CO_2$ loading and transport; their thickness represents the relative amounts of $CO_2$ transported in each of the three forms. Red arrows show the two mechanisms of $O_2$ unloading; their thickness indicates the relative amounts unloaded by each mechanism. Note that $CO_2$ loading releases hydrogen ions in the erythrocyte, and hydrogen ions promote $O_2$ unloading.

# ALVEOLAR GAS EXCHANGE REVISITED

The processes illustrated in figure 22.24 make it easier to understand alveolar exchange more fully. As shown in figure 22.25, the reactions that occur in the lungs are essentially the reverse of systemic gas exchange. As hemoglobin loads oxygen, its affinity for $H^+$ declines. Hydrogen ions dissociate from the hemoglobin and bind with bicarbonate ($HCO_3^-$) ions transported from the plasma into the RBCs. Chloride ions are transported back out of the RBC (a reverse chloride shift). The reaction of $H^+$ and $HCO_3^-$ reverses the hydration reaction and generates free $CO_2$. This diffuses into the alveolus to be exhaled—as does the $CO_2$ released from carbaminohemoglobin and $CO_2$ gas that was dissolved in the plasma.

# ADJUSTMENT TO THE METABOLIC NEEDS OF INDIVIDUAL TISSUES

Hemoglobin does not unload the same amount of oxygen to all tissues. Some tissues need more and some less, depending on their state of activity. Hemoglobin responds to such variations and unloads more oxygen to the tissues that need it most. In exercising skeletal muscles, for example, the utilization coefficient may be as high as 80%. Four factors adjust the rate of oxygen unloading to the metabolic rates of different tissues:

1. **Ambient $P_{O_2}$.** Since an active tissue consumes oxygen rapidly, the $P_{O_2}$ of its tissue fluid remains low. From the oxyhemoglobin dissociation curve (see fig. 22.23), you can see that at a low $P_{O_2}$, $HbO_2$ releases more oxygen.

2. **Temperature.** When temperature rises, the oxyhemoglobin dissociation curve shifts to the right (fig. 22.26a); in other words, elevated temperature promotes oxygen unloading. Active tissues are warmer than less active ones and thus extract more oxygen from the blood passing through them.

3. **The Bohr effect.** Active tissues also generate extra $CO_2$, which raises the $H^+$ concentration and lowers the pH of the blood. Like elevated temperatures, a drop in pH shifts the oxygen–hemoglobin dissociation curve to the right (fig. 22.26b) and promotes oxygen unloading. The increase in $HbO_2$ dissociation in

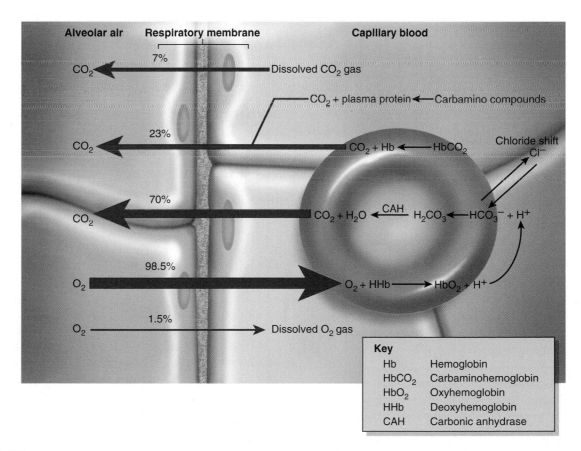

**FIGURE 22.25 Alveolar Gas Exchange.** Arrow colors and thicknesses represent the same variables as in figure 22.24. Note that $O_2$ loading promotes the decomposition of carbonic acid into $H_2O$ and $CO_2$, and most exhaled $CO_2$ comes from the erythrocytes.

▶ *In what fundamental way does this differ from the preceding figure? Following alveolar gas exchange, will the blood contain a higher or lower concentration of bicarbonate ions than it did before?*

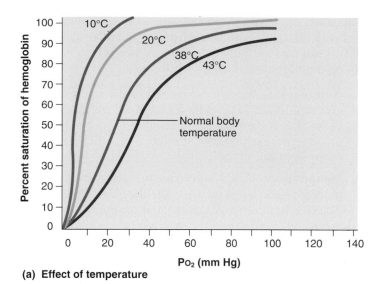

**(a)  Effect of temperature**

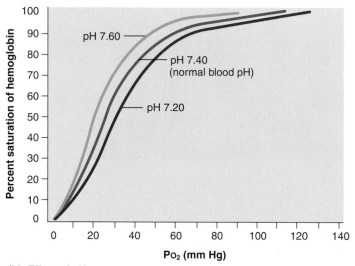

**(b)  Effect of pH**

**FIGURE 22.26  Effects of Temperature and pH on Oxyhemoglobin Dissociation.**  (a) For a given $P_{O_2}$, hemoglobin unloads more oxygen at higher temperatures. (b) For a given $P_{O_2}$, hemoglobin unloads more oxygen at lower pH (the Bohr effect). Both mechanisms cause hemoglobin to release more oxygen to tissues with higher metabolic rates.

▶ *Why is it physiologically beneficial to the body that the curves in part (a) shift to the right as temperature increases?*

response to low pH is called the **Bohr**[27] **effect.** It is less pronounced at the high $P_{O_2}$ present in the lungs, so pH has relatively little effect on pulmonary oxygen loading. In the systemic capillaries, however, $P_{O_2}$ is lower and the Bohr effect is more pronounced.

4. **BPG.** Erythrocytes have no mitochondria and meet their energy needs solely by anaerobic fermentation. One of their metabolic intermediates is **bisphospho-**

**glycerate** (BPG), which binds to hemoglobin and promotes oxygen unloading. An elevated body temperature (as in fever) stimulates BPG synthesis, as do thyroxine, growth hormone, testosterone, and epinephrine. All of these hormones thus promote oxygen unloading to the tissues.

The rate of $CO_2$ loading is also adjusted to varying needs of the tissues. A low level of oxyhemoglobin ($HbO_2$) enables the blood to transport more $CO_2$, a phenomenon known as the **Haldane effect.**[28] This occurs for two reasons: (1) $HbO_2$ does not bind $CO_2$ as well as deoxyhemoglobin (HHb) does. (2) HHb binds more hydrogen ions than $HbO_2$ does, and by removing $H^+$ from solution, HHb shifts the $H_2O + CO_2 \rightarrow HCO_3^- + H^+$ reaction to the right. A high metabolic rate keeps oxyhemoglobin levels relatively low and thus allows more $CO_2$ to be transported by these two mechanisms.

## BLOOD GASES AND THE RESPIRATORY RHYTHM

Normally the systemic arterial blood has a $P_{O_2}$ of 95 mm Hg, a $P_{CO_2}$ of 40 mm Hg, and a pH of $7.40 \pm 0.05$. The rate and depth of breathing are adjusted to maintain these values. This is possible only because the brainstem respiratory centers receive input from central and peripheral chemoreceptors that monitor the composition of the blood and CSF, as described on p. 869. Of these three chemical stimuli, the most potent stimulus for breathing is pH, followed by $CO_2$ and, perhaps surprisingly the least significant, $O_2$.

### Hydrogen Ions

Ultimately, pulmonary ventilation is adjusted to maintain the pH of the brain. The central chemoreceptors in the medulla oblongata produce about 75% of the change in respiration induced by pH shifts, and yet $H^+$ does not cross the blood–brain barrier very easily. However, $CO_2$ does, and once it is in the CSF, it reacts with water to produce carbonic acid, and the carbonic acid dissociates into bicarbonate and hydrogen ions. The CSF contains relatively little protein to buffer the hydrogen ions, so most $H^+$ remains free, and it strongly stimulates the central chemoreceptors. Hydrogen ions are also a potent stimulus to the peripheral chemoreceptors, which produce about 25% of the respiratory response to pH changes.

It is significant to ask how we know it is the $H^+$ that stimulates the central chemoreceptors and not primarily the $CO_2$ that diffuses into the CSF. Experimentally, it is possible to vary the pH or the $P_{CO_2}$ of the CSF while holding the other variable steady. When pH alone changes, there is a strong effect on respiration; when $P_{CO_2}$ alone changes, the effect is weaker. Therefore, even though these two variables usually change together, we can see that the chemoreceptors react primarily to the $H^+$.

[27]Christian Bohr (1855–1911), Danish physiologist

[28]John Scott Haldane (1860–1936), Scottish physiologist

A blood pH lower than 7.35 is called **acidosis,** and a pH greater than 7.45 is called **alkalosis.** The most common cause of acidosis is **hypercapnia,**[29] a $CO_2$ excess. The normal $P_{CO_2}$ of the arterial blood has a range of $40 \pm 3$ mm Hg. Hypercapnia is a $P_{CO_2}$ greater than 43 mm Hg. The opposite condition, **hypocapnia,** is a $P_{CO_2}$ less than 37 mm Hg. This is the usual cause of alkalosis. When these pH imbalances are due to a failure of pulmonary ventilation to match the body's rate of $CO_2$ production, they are called *respiratory acidosis* and *respiratory alkalosis* (further discussed in chapter 24).

The corrective homeostatic response to acidosis is hyperventilation, "blowing off" $CO_2$ faster than the body produces it. As $CO_2$ is eliminated from the body, the carbonic acid reaction shifts to the left:

$$CO_2 + H_2O \leftarrow H_2CO_3 \leftarrow HCO_3^- + H^+$$

Thus, the $H^+$ on the right is consumed, and as $H^+$ concentration declines, the pH rises and ideally returns the blood from the acidotic range to normal.

The corrective response to alkalosis is hypoventilation, which allows $CO_2$ to accumulate in the body fluids faster than we exhale it. Hypoventilation shifts the reaction to the right, raising the $H^+$ concentration and lowering the pH to normal:

$$CO_2 + H_2O \rightarrow H_2CO_3 \rightarrow HCO_3^- + H^+$$

Although pH changes usually result from $P_{CO_2}$ changes, they can have other causes. In diabetes mellitus, for example, rapid fat oxidation releases acidic ketone bodies, causing an abnormally low pH called *ketoacidosis* (see chapter 17). Ketoacidosis tends to induce a form of dyspnea called *Kussmaul respiration* (see table 22.3). Hyperventilation cannot reduce the level of ketone bodies in the blood, but by blowing off $CO_2$, it reduces the concentration of $CO_2$-generated $H^+$ and compensates to some degree for the $H^+$ released by the ketone bodies.

## Carbon Dioxide

Although the arterial $P_{CO_2}$ has a strong influence on respiration, we have seen that it is mostly an indirect one, mediated through its effects on the pH of the CSF. Yet the experimental evidence described earlier shows that $CO_2$ has some effect even when pH remains stable. At the beginning of exercise, the rising blood $CO_2$ level may directly stimulate the peripheral chemoreceptors and trigger an increase in ventilation more quickly than the central chemoreceptors do.

## Oxygen

The partial pressure of oxygen usually has little effect on respiration. Even in eupnea, the hemoglobin is 97% saturated with $O_2$, so little can be added by increasing pulmonary ventilation. Arterial $P_{O_2}$ significantly affects

respiration only if it drops below 60 mm Hg. At low altitudes, such a low $P_{O_2}$ seldom occurs even in prolonged holding of the breath. A moderate drop in $P_{O_2}$ does stimulate the peripheral chemoreceptors, but another effect overrides this: as the level of $HbO_2$ falls, hemoglobin binds more $H^+$ (see fig. 22.24). This raises the blood pH, which inhibits respiration and counteracts the effect of low $P_{O_2}$. At about 10,800 ft (3,300 m), $P_{O_2}$ falls to 60 mm Hg and the stimulatory effect of hypoxemia on the carotid bodies overrides the inhibitory effect of the pH increase. This produces immediate heavy breathing in people who are not acclimatized to high altitude. Long-term hypoxemia can lead to a condition called **hypoxic drive,** in which respiration is driven more by the low $P_{O_2}$ than by $CO_2$ or pH. This occurs in situations such as emphysema and pneumonia, which interfere with alveolar gas exchange, and in mountain climbing of at least 2 or 3 days' duration.

## Respiration and Exercise

It is common knowledge that we breathe more heavily during exercise, and it's tempting to think this occurs because exercise raises $CO_2$ levels, lowers the blood pH, and lowers blood $O_2$ levels. However, this is not true: all these values remain essentially the same in exercise as they do at rest. It appears that the increased respiration has other causes: (1) When the brain sends motor commands to the muscles (via the lower motor neurons of the spinal cord), it also sends this information to the respiratory centers, so they increase pulmonary ventilation in expectation of the needs of the exercising muscles. (2) Exercise stimulates proprioceptors of the muscles and joints, and they transmit excitatory signals to the brainstem respiratory centers. Thus, the respiratory centers increase breathing because they are informed that the muscles have been told to move or are actually moving. The increase in pulmonary ventilation keeps blood gas values at their normal levels in spite of the elevated $O_2$ consumption and $CO_2$ generation by the muscles.

In summary, the main chemical stimulus to pulmonary ventilation is the $H^+$ in the CSF and tissue fluid of the brain. These hydrogen ions arise mainly from $CO_2$ diffusing into the CSF and brain and generating $H^+$ through the carbonic acid reaction. Therefore the $P_{CO_2}$ of the arterial blood is an important driving force in respiration, even though its action on the chemoreceptors is indirect. Ventilation is adjusted to maintain arterial pH at about 7.40 and arterial $P_{CO_2}$ at about 40 mm Hg. This automatically ensures that the blood is at least 97% saturated with $O_2$ as well. Under ordinary circumstances, arterial $P_{O_2}$ has relatively little effect on respiration. When it drops below 60 mm Hg, however, it excites the peripheral chemoreceptors and stimulates an increase in ventilation. This can be significant at high altitudes and in certain lung diseases. The increase in respiration during exercise results from the expected or actual activity of the muscles, not from any change in blood gas pressures or pH.

[29]*capn* = smoke

## Before You Go On

*Answer the following questions to test your understanding of the preceding section:*

12. Why is the composition of alveolar air different from that of the atmosphere?

13. What four factors affect the efficiency of alveolar gas exchange?

14. Explain how perfusion of a pulmonary lobule changes if it is poorly ventilated.

15. How is oxygen transported in the blood, and why does carbon monoxide interfere with this?

16. What are the three ways in which blood transports $CO_2$?

17. Give two reasons why highly active tissues can extract more oxygen from the blood than less active tissues do.

18. Define hypocapnia *and* hypercapnia. *Name the pH imbalances that result from these conditions and explain the relationship between* $P_{CO_2}$ *and pH.*

19. What is the most potent chemical stimulus to respiration, and where are the most effective chemoreceptors for it located?

20. Explain how changes in pulmonary ventilation can correct pH imbalances.

# Respiratory Disorders

### Objectives

When you have completed this section, you should be able to

- describe the forms and effects of oxygen deficiency and oxygen excess;
- describe the chronic obstructive pulmonary diseases and their consequences; and
- explain how lung cancer begins, progresses, and exerts its lethal effects.

The delicate lungs are exposed to a wide variety of inhaled pathogens and debris; thus, it is not surprising that they are prone to a host of diseases. Several already have been mentioned in this chapter and some others are briefly described in table 22.5. The effects of aging on the respiratory system are discussed on page 1131.

## OXYGEN IMBALANCES

*Hypoxia* is a deficiency of oxygen in a tissue or the inability to use oxygen. It is not a respiratory disease in itself but is often a consequence of respiratory diseases. Hypoxia is classified according to cause:

- **Hypoxemic hypoxia,** a state of low arterial $P_{O_2}$, is usually due to inadequate pulmonary gas exchange. Some of its root causes include atmospheric deficiency of oxygen at high altitudes; impaired ventilation, as in drowning or aspiration of foreign matter; respiratory arrest; and the degenerative lung diseases discussed shortly. It also occurs in carbon monoxide poisoning, which prevents hemoglobin from transporting oxygen.

- **Ischemic hypoxia** results from inadequate circulation of the blood, as in congestive heart failure.

- **Anemic hypoxia** is due to anemia and the resulting inability of the blood to carry adequate oxygen.

- **Histotoxic hypoxia** occurs when a metabolic poison such as cyanide prevents the tissues from using the oxygen delivered to them.

Hypoxia is often marked by **cyanosis,** blueness of the skin. Whatever its cause, the primary effect of hypoxia is the necrosis of oxygen-starved tissues. This is especially critical in organs with the highest metabolic demands, such as the brain, heart, and kidneys.

An oxygen excess is also dangerous. You can safely breathe 100% oxygen at 1 atm for a few hours, but **oxygen toxicity** rapidly develops when pure oxygen is breathed at 2.5 atm or greater. Excess oxygen generates hydrogen peroxide and free radicals that destroy enzymes and damage nervous tissue; thus it can lead to seizures, coma, and death. This is why scuba divers breathe a mixture of oxygen and nitrogen rather than pure compressed oxygen (see Insight 22.5). Hyperbaric oxygen was formerly used to treat premature infants for respiratory distress syndrome, but it caused retinal deterioration and blinded many infants before the practice was discontinued.

## CHRONIC OBSTRUCTIVE PULMONARY DISEASES

**Chronic obstructive pulmonary disease (COPD)** refers to any disorder in which there is a long-term obstruction of airflow and a substantial reduction in pulmonary ventilation. The major COPDs are *asthma, chronic bronchitis,* and *emphysema.* In asthma, an allergen triggers the release of histamine and other inflammatory chemicals that cause intense bronchoconstriction and sometimes suffocation (see p. 841). The other COPDs are almost always caused by cigarette smoking but occasionally result from air pollution or occupational exposure to airborne irritants.

Beginning smokers exhibit inflammation and hyperplasia of the bronchial mucosa. In **chronic bronchitis,** the cilia are immobilized and reduced in number, while goblet cells enlarge and produce excess mucus. With extra mucus and fewer cilia to dislodge it, smokers develop a chronic cough that brings up **sputum** (SPEW-tum), a mixture of mucus and cellular debris. Thick, stagnant mucus in the respiratory tract provides a growth medium for bacteria, while cigarette smoke incapacitates the alveolar macrophages and reduces defense mechanisms against respiratory infections. Smokers therefore develop chronic infection and bronchial inflammation, with symptoms that include dyspnea, hypoxia, cyanosis, and attacks of coughing.

In **emphysema**[30] (EM-fih-SEE-muh), alveolar walls break down and the lung exhibits larger but fewer alveoli (see fig. 22.21c). Thus, there is much less respiratory membrane available for gas exchange. The lungs become fibrotic and less elastic. The air passages open adequately during inspiration, but they tend to collapse and obstruct the outflow of air. Air therefore becomes trapped in the lungs, and over a period of time a person becomes barrel-chested. The overly stretched thoracic muscles contract weakly, which further contributes to the difficulty of expiration. People with emphysema become exhausted because they expend three to four times the normal amount of energy just to breathe. Even slight physical exertion, such as walking across a room, can cause severe shortness of breath.

⌐ **Think About It**

*Explain how the length–tension relationship of skeletal muscle (see chapter 11) accounts for the weakness of the respiratory muscles in emphysema.*

---

[30]*emphys* = inflamed

All of the COPDs tend to reduce pulmonary compliance and vital capacity and cause hypoxemia, hypercapnia, and respiratory acidosis. Hypoxemia stimulates the kidneys to secrete erythropoietin, which leads to accelerated erythrocyte production and polycythemia, as discussed in chapter 18. COPD also leads to **cor pulmonale**—hypertrophy and potential failure of the right heart due to obstruction of the pulmonary circulation (see chapter 19).

## SMOKING AND LUNG CANCER

Lung cancer (fig. 22.27) accounts for more deaths than any other form of cancer. The most important cause of lung cancer is cigarette smoking, distantly followed by air pollution. Cigarette smoke contains at least 15 carcinogenic compounds. Lung cancer commonly follows or accompanies COPD.

There are three forms of lung cancer, the most common of which is **squamous-cell carcinoma.** In its early stage, basal cells of the bronchial epithelium multiply, and the ciliated pseudostratified columnar epithelium transforms into the stratified squamous type. As the dividing epithelial cells invade the underlying tissues of the bronchial wall, the bronchus develops bleeding lesions. Dense swirled masses of keratin appear in the lung parenchyma and replace functional respiratory tissue.

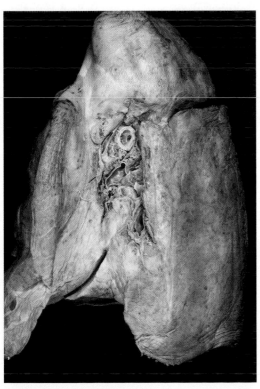

(a) Healthy lung, mediastinal surface

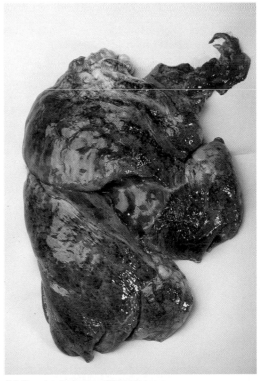

(b) Smoker's lung with carcinoma

**FIGURE 22.27**  Effect of Smoking.

A second form of lung cancer, nearly as common, is **adenocarcinoma,**[31] which originates in the mucous glands of the lamina propria. The least common (10%–20% of malignancies) but most dangerous form is **small-cell (oat-cell) carcinoma,** named for clusters of cells that resemble oat grains. This originates in the primary bronchi but invades the mediastinum and metastasizes quickly to other organs.

Over 90% of lung tumors originate in the mucous membranes of the large bronchi. As a tumor invades the bronchial wall and grows around it, it compresses the airway and may cause atelectasis (collapse) of more distal parts of the lung. Growth of the tumor produces a cough, but coughing is such an everyday occurrence among smokers it seldom causes much alarm. Often, the first sign of serious trouble is the coughing up of blood. Lung cancer metastasizes so rapidly that it has usually spread to other organs by the time it is diagnosed. Common sites of metastasis are the pericardium, heart, bones, liver, lymph nodes, and brain. The chance of recovery is poor, with only 7% of patients surviving for 5 years after diagnosis.

---

[31]*adeno* = gland + *carcino* = cancer + *oma* = tumor

## Before You Go On

*Answer the following questions to test your understanding of the preceding section:*

21. *Describe the four classes of hypoxia.*

22. *Name and compare two COPDs and describe some pathological effects that they have in common.*

23. *In what lung tissue does lung cancer originate? How does it kill?*

| TABLE 22.5 | Some Disorders of the Respiratory System |
|---|---|
| Acute rhinitis | The common cold. Caused by many types of viruses that infect the upper respiratory tract. Symptoms include congestion, increased nasal secretion, sneezing, and dry cough. Transmitted especially by contact of contaminated hands with mucous membranes; not transmitted orally. |
| Acute respiratory distress syndrome | Acute lung inflammation and alveolar injury stemming from trauma, infection, burns, aspiration of vomit, inhalation of noxious gases, drug overdoses, and other causes. Alveolar injury is accompanied by severe pulmonary edema and hemorrhage, followed by fibrosis that progressively destroys lung tissue. Fatal in about 40% of cases under age 60 and in 60% of cases over age 65. |
| Pneumonia | A lower respiratory infection caused by any of several viruses, fungi, or protozoans (most often the bacterium *Streptococcus pneumoniae*). Causes filling of alveoli with fluid and dead leukocytes and thickening of the respiratory membrane, which interferes with gas exchange and causes hypoxemia. Especially dangerous to infants, the elderly, and people with compromised immune systems, such as AIDS and leukemia patients. |
| Sleep apnea | Cessation of breathing for 10 sec or longer during sleep; sometimes occurs hundreds of times per night, often accompanied by restlessness and alternating with snoring. Can result from altered function of CNS respiratory centers, airway obstruction, or both. Over time, may lead to daytime drowsiness, hypoxemia, polycythemia, pulmonary hypertension, congestive heart failure, and cardiac arrhythmia. Most common in obese people and in men. |
| Tuberculosis (TB) | Pulmonary infection with the bacterium *Mycobacterium tuberculosis,* which invades the lungs by way of air, blood, or lymph. Stimulates the lung to form fibrous nodules called tubercles around the bacteria. Progressive fibrosis compromises the elastic recoil and ventilation of the lungs. Especially common among impoverished and homeless people and becoming increasingly common among people with AIDS. |

*Disorders described elsewhere*

Apnea p. 875

Asthma p. 841

Atelectasis p. 873

Carbon monoxide poisoning p. 881

Chronic bronchitis p. 886

Cor pulmonale p. 744

Decompression sickness p. 889

Dyspnea p. 875

Emphysema p. 887

Hypoxia p. 886

Lung cancer p. 887

Ondine's curse p. 867

Pleurisy p. 864

Pneumothorax p. 871

Pulmonary edema p. 879

Respiratory acidosis p. 946

Respiratory alkalosis p. 946

Respiratory distress syndrome p. 873

## Diving Physiology and Decompression Sickness

Because of the rise in popularity of scuba diving in recent years, many people now know something about the scientific aspects of breathing under high pressure. But diving is by no means a new fascination. As early as the fifth century B.C.E., Aristotle described divers using snorkels and taking containers of air underwater in order to stay down longer. Some Renaissance artists depicted divers many meters deep breathing from tubes to the water surface. In reality, this would be physically impossible. For one thing, such tubes would have so much dead space that fresh air from the surface would not reach the diver. The short snorkels used today are about the maximum length that will work for surface breathing. Another reason snorkels cannot be used at greater depths is that water pressure doubles for every 11 m of depth, and even at 1 m the pressure is so great that a diver cannot expand the chest muscles without help. This is one reason why scuba divers use pressurized air tanks. The tanks create a positive intrapulmonary pressure and enable the diver to inhale with only slight assistance from the thoracic muscles. Scuba tanks also have regulators that adjust the outflow pressure to the diver's depth and the opposing pressure of the surrounding water.

But breathing pressurized (hyperbaric) gas presents its own problems. Divers cannot use pure oxygen because of the problem of oxygen toxicity. Instead, they use compressed air—a mixture of 21% oxygen and 79% nitrogen. On land, nitrogen presents no physiological problems; it dissolves poorly in blood and it is physiologically inert. But under hyperbaric conditions, larger amounts of nitrogen dissolve in the blood. (Which of the gas laws applies here?) Even more dissolves in adipose tissue and the myelin of the brain, since nitrogen is more soluble in lipids. In the brain, it causes *nitrogen narcosis,* or what Jacques Cousteau termed "rapture of the deep." A diver can become dizzy, euphoric, and dangerously disoriented; for every 15 to 20 m of depth, the effect is said to be equivalent to one martini on an empty stomach.

Strong currents, equipment failure, and other hazards sometimes make scuba divers panic, hold their breath, and quickly swim to the surface (a *breath-hold ascent*). Ambient (surrounding) pressure falls rapidly as a diver ascends, and the air in the lungs expands just as rapidly. (Which gas law is demonstrated here?) It is imperative that an ascending diver keep his or her airway open to exhale the expanding gas; otherwise it is likely to cause *pulmonary barotrauma*—ruptured alveoli. Then, when the diver takes a breath of air at the surface, alveolar air goes directly into the bloodstream and causes air embolism. After passing through the heart, the emboli tend to enter the cerebral circulation because the diver is head-up and air bubbles rise in liquid. The resulting cerebral embolism can cause motor and sensory dysfunction, seizures, unconsciousness, and drowning.

Barotrauma can be fatal even at the depths of a backyard swimming pool. In one case, children trapped air in a bucket 1 m underwater and then swam under the bucket to breathe from the air space. Because the bucket was under water, the air in it was compressed. One child filled his lungs under the bucket, did a "mere" 1-m breath-hold ascent, and his alveoli ruptured. He died in the hospital, partly because the case was mistaken for drowning and not treated for what it really was. This would not have happened to a person who inhaled at the surface, did a breath-hold dive, and then resurfaced—nor is barotrauma a problem for those who do breath-hold dives to several meters. (Why? What is the difference?)

Even when not holding the breath, but letting the expanding air escape from the mouth, a diver must ascend slowly and carefully to allow for decompression of the nitrogen that has dissolved in the tissues. *Decompression tables* prescribe safe rates of ascent based on the depth and the length of time a diver has been down. When pressure drops, nitrogen dissolved in the tissues can go either of two places—it can diffuse into the alveoli and be exhaled, or it can form bubbles like the $CO_2$ in a bottle of soda when the cap is removed. The diver's objective is to ascend slowly, allowing for the former and preventing the latter. If a diver ascends too rapidly, nitrogen "boils" from the tissues—especially in the 3 m just below the surface, where the relative pressure change is greatest. A diver may double over in pain from bubbles in the joints, bones, and muscles—a disease called the "bends," or *decompression sickness (DCS)*. Nitrogen bubbles in the pulmonary capillaries cause "chokes"—substernal pain, coughing, and dyspnea. DCS is sometimes accompanied by mood changes, seizures, numbness, and itching. These symptoms usually occur within an hour of surfacing, but they are sometimes delayed for up to 36 hours. DCS is treated by putting the individual in a hyperbaric chamber to be recompressed and then *slowly* decompressed.

DCS is also called *caisson disease.* A caisson is a watertight underwater chamber filled with pressurized air. Caissons are used in underwater construction work on bridges, tunnels, ships' hulls, and so forth. Caisson disease was first reported in the late 1800s among workmen building the foundations of the Brooklyn Bridge.

# CONNECTIVE ISSUES

## Interactions Between the RESPIRATORY SYSTEM and Other Organ Systems

### INTEGUMENTARY SYSTEM

Nasal guard hairs reduce inhalation of dust and other foreign matter

### SKELETAL SYSTEM

Thoracic cage protects lungs; movement of ribs produces pressure changes that ventilate lungs

### MUSCULAR SYSTEM

Skeletal muscles ventilate lungs, control position of larynx during swallowing, control vocal cords during speech; exercise strongly stimulates respiration because of the $CO_2$ generated by active muscles

### NERVOUS SYSTEM

Produces the respiratory rhythm, monitors blood gases and pH, monitors stretching of lungs; phrenic, intercostal, and other nerves control respiratory muscles

### ENDOCRINE SYSTEM

Lungs produce angiotensin-converting enzyme (ACE), which converts angiotensin I to the hormone angiotensin II

Epinephrine and norepinephrine dilate bronchioles and stimulate ventilation

## ALL SYSTEMS

The respiratory system serves all other systems by supplying $O_2$, removing $CO_2$, and maintaining acid-base balance

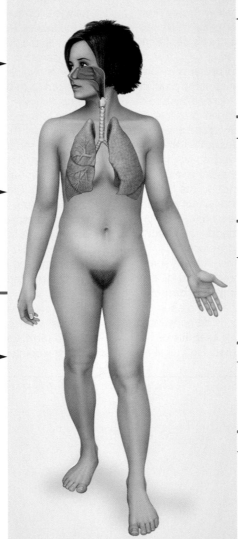

### CIRCULATORY SYSTEM

Regulates blood pH; thoracic pump aids in venous return; lungs produce blood platelets; production of angiotensin II by lungs is important in control of blood volume and pressure; obstruction of pulmonary circulation leads to right-sided heart failure

Blood transports $O_2$ and $CO_2$; mitral stenosis or left-sided heart failure can cause pulmonary edema; emboli from peripheral sites often lodge in lungs

### LYMPHATIC/IMMUNE SYSTEMS

Thoracic pump promotes lymph flow

Lymphatic drainage from lungs is important in keeping alveoli dry; immune cells protect lungs from infection

### URINARY SYSTEM

Valsalva maneuver aids in emptying bladder

Disposes of wastes from respiratory organs; collaborates with lungs in controlling blood pH

### DIGESTIVE SYSTEM

Valsalva maneuver aids in defecation

Provides nutrients for growth and maintenance of respiratory system

### REPRODUCTIVE SYSTEM

Valsalva maneuver aids in childbirth

Sexual arousal stimulates respiration

# Review of Key Concepts

## Anatomy of the Respiratory System (p. 854)

1. *Respiration* means ventilation of the lungs, gas exchange with the blood, and oxygen use by the tissues.

2. The respiratory system serves for gas exchange, vocalization, smell, pH balance, and angiotensin II synthesis, and aids in lymph and blood flow and expulsion of the contents of abdominal viscera.

3. The *conducting division* of the respiratory system consists of the nose, pharynx, larynx, trachea, bronchi, and most bronchioles; it serves only for airflow. The *respiratory division* consists of the alveoli and other distal gas-exchange regions of the lungs.

4. The upper respiratory tract consists of the organs in the head and neck (nose through larynx); the lower respiratory tract consists of the organs in the chest (trachea through alveoli).

5. The nose extends from the anterior nares to the choanae and is divided by the nasal septum into right and left *nasal fossae*.

6. Each fossa has three scroll-like *nasal conchae* covered with a ciliated mucous membrane. The conchae warm, humidify, and cleanse the air flowing over them.

7. The *pharynx* is a muscular passage divided into nasopharynx, oropharynx, and laryngopharynx.

8. The *larynx* is a cartilaginous chamber that contains the vocal cords and keeps food and drink out of the airway. Muscles of the larynx function in speech and swallowing.

9. The *trachea* is a 12-cm tube, supported by cartilage, ending where it branches into the two primary bronchi. Its ciliated mucosa acts as a *mucociliary escalator* to remove inhaled debris from the respiratory tract.

10. Each lung is a conical organ extending from the superior *apex* to the inferior, broad *base*. The left lung is divided into two lobes and the right lung into three. The bronchi, blood vessels, nerves, and lymphatics enter and leave by a slit called the *hilum* on the mediastinal surface.

11. One primary bronchus supplies each lung; it divides into one secondary bronchus for each lobe of the lung, and this divides into smaller tertiary bronchi.

12. *Bronchioles* are finer divisions of the airway lacking cartilage. The smallest members of the conducting division are the *terminal bronchioles.* Beyond this, thin-walled *respiratory bronchioles* begin the respiratory division. Respiratory bronchioles have alveoli along their walls and branch distally into alveolar ducts.

13. An alveolus is a thin-walled sac surrounded by a basket of blood capillaries. It is composed of squamous and great alveolar cells and contains alveolar macrophages, the last line of defense against inhaled debris. Alveoli are the site of gas exchange with the blood.

14. The surface of each lung is a serous membrane called the *visceral pleura.* It continues as the *parietal pleura,* which lines the inside of the rib cage. The space between the pleurae is the *pleural cavity,* and is lubricated with *pleural fluid.* The pleurae reduce friction during breathing, contribute to the pressure gradients that move air into and out of the lungs, and help compartmentalize the thoracic cavity.

## Pulmonary Ventilation (p. 865)

1. The lungs are ventilated in repetitive *respiratory cycles* of *inspiration* and *expiration.*

2. Airflow is driven by changes in lung volume produced by the respiratory muscles. Inspiration is achieved mainly by contraction of the diaphragm and external intercostal muscles. Expiration is achieved mainly by elastic recoil of the thoracic cage rather than muscular effort. Several thoracic and abdominal muscles act as *accessory muscles* of respiration in deep breathing.

3. Inspiration results from firing of *I neurons* in the *dorsal respiratory group (DRG)* in the medulla oblongata. Forced expiration results from firing of E neurons in the nearby *ventral respiratory group (VRG).* The *pneumotaxic center* in the pons regulates timing of the transition from inspiration to expiration, and thus regulates respiratory rate and depth.

4. The DRG and VRG receive input from *central chemoreceptors* in the medulla, *peripheral chemoreceptors* in the aortic arch and carotid bodies, stretch receptors in the airway, and irritant receptors in the respiratory mucosa. This enables adjustments of the respiratory rhythm to maintain blood chemistry and avoid irritants and overstretching of the lungs.

5. Voluntary control of respiration is achieved by cerebral output that bypasses the brainstem centers and goes directly to the lower motor neurons of the spinal cord, whose axons supply the respiratory muscles.

6. Airflow is directly proportional to the pressure difference between two points and inversely proportional to resistance.

7. When inspiratory muscles expand the chest, the lungs enlarge and the intrapulmonary pressure drops *(Boyle's law).* When intrapulmonary pressure is lower than atmospheric pressure, air flows down its pressure gradient into the lungs. The lungs are also inflated in part by the warming and expansion of inspired air *(Charles's law).*

8. Expiration occurs when elastic recoil of the thoracic cage compresses the lungs and raises the intrapulmonary pressure above atmospheric pressure. Air flows down its pressure gradient out of the lungs.

9. The speed of airflow depends on resistance of the airway, especially in the vast number of small, muscular bronchioles. Thus it is increased by *bronchodilation* and reduced by *bronchoconstriction.* Pulmonary compliance and alveolar surfactant help to minimize resistance.

10. In a typical 500-mL inspiration, about 150 mL fills the anatomical dead space. Only the other 350 mL ventilates the alveoli and participates in gas exchange with the blood. *Alveolar ventilation rate* is the product of the latter amount and the respiratory rate.

11. Pulmonary ventilation is measured with a device called a *spirometer.* Table 22.2 defines the volumes and capacities of principal interest.

12. *Restrictive disorders* of respiration reduce pulmonary compliance and vital capacity. *Obstructive disorders* reduce the speed of airflow, measured by such values as *forced expiratory volume.*

13. Normal quiet respiration is called *eupnea.* Variations such as *dyspnea, hyperventilation,* and others are defined in table 22.3.

### Gas Exchange and Transport (p. 876)

1. Air is composed of approximately 79% $N_2$, 21% $O_2$, and 0.04% $CO_2$. Water vapor is a significant but highly variable component (0–4%, typically 0.5% on a cool clear day).

2. Each of these gases contributes a *partial pressure* to the total gas pressure. Total pressure is the sum of the partial pressures of the gases in the mixture *(Dalton's law).*

3. Alveolar air is significantly higher in $CO_2$ and $H_2O$ than inspired air, and lower in $O_2$.

4. The $O_2$ in inspired air must dissolve in the water film on the alveolar wall before it can pass through the respiratory membrane into the blood. Henry's law states that the amount of gas that diffuses from air into water is proportional to its solubility and partial pressure.

5. The efficiency of $O_2$ loading and $CO_2$ unloading in the alveolus depends on the pressure gradients of the gases, their solubility in water,

thickness of the respiratory membrane, and alveolar surface area.

6. *Ventilation–perfusion coupling* matches airflow to blood flow in individual regions of the lung, thus ensuring optimal gas exchange between air and blood.

7. About 1.5% of the $O_2$ in the blood is dissolved in the plasma and 98.5% is bound to hemoglobin in the RBCs. Each hemoglobin can carry up to 4 $O_2$. It is called *oxyhemoglobin* ($HbO_2$) if it carries one or more $O_2$ molecules.

8. The relationship between oxygen partial pressure ($Po_2$) and percent $HbO_2$ is the *oxyhemoglobin dissociation curve* (see fig. 22.23). It shows that binding of the first oxygen to hemoglobin accelerates the binding of more $O_2$, until the Hb becomes saturated.

9. About 90% of the $CO_2$ in the blood is carried as bicarbonate ($HCO_3^-$) ions, 5% is bound to proteins as *carbamino* compounds, and 5% is dissolved in the blood plasma.

10. The loading of $CO_2$ from the tissue fluids is promoted by *carbonic anhydrase,* an enzyme in the RBCs that promotes the reaction of $CO_2$ and water to form carbonic acid. The carbonic acid breaks down to $HCO_3^-$ and $H^+$. Most of the $H^+$ binds to hemoglobin, while the $HCO_3^-$ is exchanged for $Cl^-$ from the blood plasma.

11. This binding of $H^+$ to hemoglobin promotes the unloading of $O_2$ to the systemic tissues. In one pass through the capillaries of a resting tissue, the blood gives up about 22% of its $O_2$ to the tissue (the *utilization coefficient*).

12. In the alveoli, Hb unloads $O_2$. This unloading causes $H^+$ to dissociate from the Hb and recombine with $HCO_3^-$ to produce carbonic acid. The carbonic acid is then broken down by carbonic anhydrase into water and $CO_2$. The $CO_2$ is exhaled.

13. Hemoglobin unloads varying amounts of $O_2$ to different tissues according to their needs. Hemoglobin adjusts $O_2$ unloading in response to variations in the tissue's $Po_2$, temperature, and pH (the *Bohr effect*), and the RBC's own temperature- and

hormone-sensitive concentration of bisphosphoglycerate (BPG).

14. $CO_2$ loading is enhanced by the unloading of $O_2$ (the *Haldane effect*), so hemoglobin picks up more $CO_2$ from highly active tissues than from less active ones.

15. The strongest chemical influence on breathing is $H^+$ sensed by the central chemoreceptors of the medulla. $CO_2$ and $H^+$ also stimulate breathing by acting on the peripheral chemoreceptors. $CO_2$ acts mainly by generating $H^+$ through the reaction $CO_2 + H_2O \leftrightarrows H_2CO_3 \leftrightarrows HCO_3^- + H^+$. A low $Po_2$ has little influence on breathing except at unusually low values (< 60 mm Hg).

16. Normally, the blood pH ranges from 7.35 to 7.45. A pH below 7.35 is called *acidosis* and a pH above 7.45 is *alkalosis.* These are usually caused, respectively, by a $CO_2$ excess *(hypercapnia)* or deficiency *(hypocapnia).* Thus, pH imbalances can be corrected by changing the respiratory rate.

17. Exercise increases respiration by means of input to the respiratory center from the cerebrum or from muscle and joint proprioceptors, not by changing blood chemistry.

### Respiratory Disorders (p. 886)

1. Hypoxia, a deficiency of $O_2$ in the tissues, can be of *hypoxemic, ischemic, anemic,* or *histotoxic* origin. It can cause cyanosis and, if severe and prolonged, tissue necrosis.

2. Oxygen excess can generate hydrogen peroxide and free radicals that cause oxygen toxicity.

3. The *chronic obstructive pulmonary diseases (COPDs)* are asthma, chronic bronchitis, and emphysema. Asthma is an allergic disease while the others are usually caused by tobacco smoke. Chronic bronchitis entails congestion of the airway with thick mucus, and susceptibility to respiratory infection. Emphysema entails destruction of pulmonary alveoli and air retention in expiration.

4. Lung cancer also is usually caused by tobacco smoke. Its variations are squamous-cell carcinoma, adenocarcinoma, and small-cell carcinoma. It tends to metastasize rapidly.

# Testing Your Recall

1. The nasal cavity is divided by the nasal septum into right and left
   a. nares.
   b. vestibules.
   c. fossae.
   d. choanae.
   e. conchae.

2. The intrinsic laryngeal muscles regulate speech by rotating
   a. the extrinsic laryngeal muscles.
   b. the corniculate cartilages.
   c. the arytenoid cartilages.
   d. the hyoid bone.
   e. the vocal cords.

3. The largest air passages that engage in gas exchange with the blood are
   a. the respiratory bronchioles.
   b. the terminal bronchioles.
   c. the primary bronchi.
   d. the alveolar ducts.
   e. the alveoli.

4. Respiratory arrest would most likely result from a tumor of the
   a. pons.
   b. midbrain.
   c. thalamus.
   d. cerebellum.
   e. medulla oblongata.

5. Which of these values would normally be the highest?
   a. tidal volume
   b. inspiratory reserve volume
   c. expiratory reserve volume
   d. residual volume
   e. vital capacity

6. The _____ protects the lungs from injury by excessive inspiration.
   a. pleura
   b. rib cage
   c. inflation reflex
   d. Haldane effect
   e. Bohr effect

7. According to _____, the warming of air as it is inhaled helps to inflate the lungs.
   a. Boyle's law
   b. Charles's law
   c. Dalton's law
   d. the Bohr effect
   e. the Haldane effect

8. Poor blood circulation causes _____ hypoxia.
   a. ischemic
   b. histotoxic
   c. hemolytic
   d. anemic
   e. hypoxemic

9. Most of the $CO_2$ that diffuses from the blood into an alveolus comes from
   a. dissolved gas.
   b. carbaminohemoglobin.
   c. carboxyhemoglobin.
   d. carbonic acid.
   e. expired air.

10. The duration of an inspiration is set by
    a. the pneumotaxic center.
    b. the phrenic nerves.
    c. the vagus nerves.
    d. the I neurons.
    e. the E neurons.

11. The superior opening into the larynx is guarded by a tissue flap called the _____.

12. Within each lung, the airway forms a branching complex called the _____.

13. The great alveolar cells secrete a phospholipid–protein mixture called _____.

14. Intrapulmonary pressure must be lower than _____ pressure for inspiration to occur.

15. _____ disorders reduce the flow of air through the airway.

16. Some inhaled air does not participate in gas exchange because it fills the _____ of the respiratory tract.

17. Inspiration depends on the ease of pulmonary inflation, called _____, whereas expiration depends on _____, which causes pulmonary recoil.

18. Inspiration is caused by the firing of I neurons in the _____ of the medulla oblongata.

19. The matching of airflow to blood flow in any region of the lung is called _____.

20. A blood pH > 7.45 is called _____ and can be caused by a $CO_2$ deficiency called _____.

*Answers in Appendix B*

# True or False

*Determine which five of the following statements are false, and briefly explain why.*

1. The phrenic nerves fire during inspiration only.

2. The lungs contain more respiratory bronchioles than terminal bronchioles.

3. In alveolar capillaries, oncotic pressure is greater than the mean blood pressure.

4. If you increase the volume of a given quantity of gas, its pressure increases.

5. Pneumothorax is the only cause of atelectasis.

6. Obstruction of the bronchial tree results in a reduced FEV.

7. At a given $Po_2$ and pH, hemoglobin carries less oxygen at warmer temperatures than it does at cooler temperatures.

8. Most of the air one inhales never makes it to the alveoli.

9. The greater the $Pco_2$ of the blood is, the lower its pH is.

10. Most of the $CO_2$ transported by the blood is in the form of dissolved gas.

*Answers in Appendix B*

## Testing Your Comprehension

1. Discuss how the different functions of the conducting division and respiratory division relate to differences in their histology.

2. State whether hyperventilation would raise or lower each of the following—the blood $P_{O_2}$, $P_{CO_2}$, and pH—and explain why. Do the same for emphysema.

3. Some competitive swimmers hyperventilate before a race, thinking they can "load up on extra oxygen" and hold their breaths longer underwater. While they can indeed hold their breaths longer, it is not for the reason they think. Furthermore, some have lost consciousness and drowned because of this practice. What is wrong with this thinking, and what accounts for the loss of consciousness?

4. Consider a man in good health with a 650 mL tidal volume and a respiratory rate of 11 breaths per minute. Report his minute respiratory volume in liters per minute. Assuming his anatomical dead space is 185 mL, calculate his alveolar ventilation rate in liters per minute.

5. An 83-year-old woman is admitted to the hospital, where a critical care nurse attempts to insert a nasoenteric tube ("stomach tube") for feeding. The patient begins to exhibit dyspnea, and a chest X ray reveals air in the right pleural cavity and a collapsed right lung. The patient dies 5 days later from respiratory complications. Name the conditions revealed by the X ray and explain how they could have resulted from the nurse's procedure.

*Answers at www.mhhe.com/saladin4*

# www.mhhe.com/saladin4

*The textbook website provides a wealth of interactive study materials fully organized and integrated by chapter. You will find practice quizzes, labeling exercises, and much more that will complement your learning and understanding of anatomy and physiology. The website also includes tools designed to enhance your* **Anatomy & Physiology** | **REVEALED** *experience.*

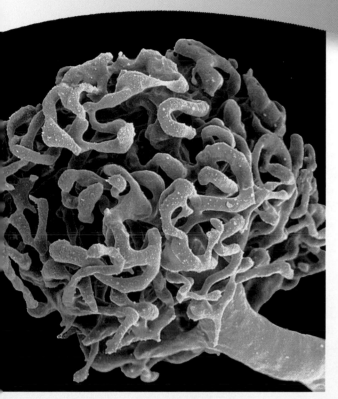

*The renal glomerulus, a mass of capillaries where the kidney filters the blood (SEM of a resin cast)*

# THE URINARY SYSTEM

## Brushing Up

To understand this chapter, it is important that you understand or brush up on the following concepts:

- Osmosis, tonicity, and osmolarity (pp. 103–105)
- Carrier-mediated transport mechanisms, especially symports and antiports (p. 106)
- Osmotic diuresis (p. 670)
- Blood pressure, resistance, and flow (p. 762)
- Capillary filtration and reabsorption (p. 770)

Anatomy & Physiology | REVEALED

The urinary system is well known for eliminating wastes from the body, but its role in homeostasis goes far beyond that. The kidneys also detoxify poisons, synthesize glucose, and play indispensable roles in controlling electrolyte and acid–base balance, blood pressure, erythrocyte count, and the $P_{O_2}$ and $P_{CO_2}$ of the blood. The urinary system thus has a very close physiological relationship with the endocrine, circulatory, and respiratory systems, covered in the preceding chapters.

Anatomically, the urinary system is closely associated with the reproductive system. In many animals the eggs and sperm are emitted through the urinary tract, and the two systems have a shared embryonic development and adult anatomical relationship. This is reflected in humans, where the systems develop together in the embryo and, in the male, the urethra continues to serve as a passage for both urine and sperm. Thus the urinary and reproductive systems are often collectively called the *urogenital (U–G) system,* and *urologists* treat both urinary and reproductive disorders. We examine the anatomical relationship between the urinary and reproductive systems in chapter 27, but the physiological link to the circulatory and respiratory systems is more important to consider at this time.

# Functions of the Urinary System

### Objectives
When you have completed this section, you should be able to

- name and locate the organs of the urinary system;
- list several functions of the kidneys in addition to urine formation;
- name the major nitrogenous wastes and identify their sources; and
- define *excretion* and identify the systems that excrete wastes.

The **urinary system** consists of six organs: two **kidneys,** two **ureters,** the **urinary bladder,** and the **urethra** (fig. 23.1). Most of our focus in this chapter is on the kidneys.

## FUNCTIONS OF THE KIDNEYS

Metabolism constantly produces a variety of waste products that can poison the body if not eliminated. The most

fundamental role of the kidneys is to eliminate these wastes and homeostatically regulate the volume and composition of the body fluids. All of the following processes are aspects of kidney function:

- They filter blood plasma, separate wastes from the useful chemicals, and eliminate the wastes while returning the rest to the bloodstream.

- They regulate blood volume and pressure by eliminating or conserving water as necessary.

- They regulate the osmolarity of the body fluids by controlling the relative amounts of water and solutes eliminated.

- They secrete the enzyme *renin,* which activates hormonal mechanisms that control blood pressure and electrolyte balance.

- They secrete the hormone *erythropoietin,* which controls the red blood cell count and oxygen-carrying capacity of the blood.

- They function with the lungs to regulate the $P_{CO_2}$ and acid–base balance of the body fluids.

- They contribute to calcium homeostasis through their role in synthesizing calcitriol (vitamin D) (see chapter 7).

- They detoxify free radicals and drugs with the use of peroxisomes.

- In times of starvation, they carry out *gluconeogenesis;* they *deaminate* amino acids (remove the $-NH_2$ group), excrete the amino group as ammonia ($NH_3$), and synthesize glucose from the rest of the molecule.

## NITROGENOUS WASTES

A **waste** is any substance that is useless to the body or present in excess of the body's needs. A **metabolic waste,** more specifically, is a waste substance produced by the body. Thus the food residue in feces, for example, is a waste but not a metabolic waste, since it was not produced by the body and, indeed, never entered the body's tissues.

Metabolism produces a great quantity of wastes that are lethal to cells if allowed to accumulate. Some of the most toxic examples are small nitrogen-containing compounds called **nitrogenous wastes** (fig. 23.2). About 50% of the nitrogenous waste is urea, a by-product of protein catabolism. Proteins are broken down to amino acids, and then the $-NH_2$ group is removed from each amino acid. The $-NH_2$ forms ammonia, which is exceedingly toxic but which the liver quickly converts to urea, $CO(NH_2)_2$, a less harmful waste:

$$2\,NH_3 + CO_2 \rightarrow H_2N\overset{\displaystyle O}{\overset{\displaystyle \|}{-C-}}NH_2 + H_2O$$

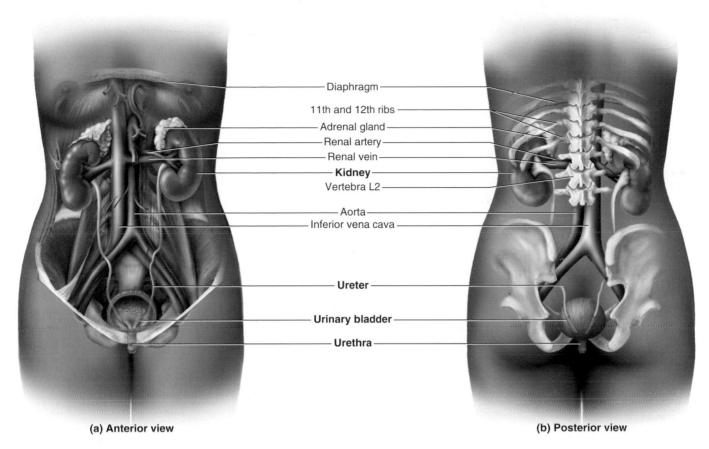

**(a) Anterior view**

**(b) Posterior view**

**FIGURE 23.1** **The Urinary System.** Organs of the urinary system are indicated in boldface.

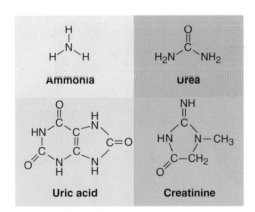

**FIGURE 23.2** **The Major Nitrogenous Wastes.**
▶ *How is each of these wastes produced in the body?*

Other nitrogenous wastes in the urine include **uric acid** and **creatinine** (cree-AT-ih-neen), produced by the catabolism of nucleic acids and creatine phosphate, respectively. Although less toxic than ammonia and less abundant than urea, these wastes are far from harmless.

The level of nitrogenous waste in the blood is typically expressed as **blood urea nitrogen (BUN)**. The urea concentration is normally 10 to 20 mg/dL. An abnormally elevated BUN is called **azotemia**[1] (AZ-oh-TEE-me-uh) and may indicate renal insufficiency. Azotemia may progress to **uremia** (you-REE-me-uh), a syndrome of diarrhea, vomiting, dyspnea, and cardiac arrhythmia stemming from the toxic effects of nitrogenous wastes. Convulsions, coma, and death can follow within a few days. Unless a kidney transplant is available, renal failure requires *hemodialysis* to remove nitrogenous wastes from the blood (see Insight 23.5, p. 925).

## EXCRETION

**Excretion** is the process of separating wastes from the body fluids and eliminating them. It is carried out by four organ systems:

1. The respiratory system excretes carbon dioxide, small amounts of other gases, and water.

2. The integumentary system excretes water, inorganic salts, lactic acid, and urea in the sweat.

3. The digestive system not only *eliminates* food residue (which is not a process of excretion) but also actively *excretes* water, salts, carbon dioxide, lipids, bile pigments, cholesterol, and other metabolic wastes.

[1]*azot* = nitrogen + *emia* = blood condition

4. The urinary system excretes a broad variety of metabolic wastes, toxins, drugs, hormones, salts, hydrogen ions, and water.

### Before You Go On

*Answer the following questions to test your understanding of the preceding section:*

1. *State four functions of the kidneys other than forming urine.*

2. *List four nitrogenous wastes and their metabolic sources.*

3. *Name some wastes eliminated by three systems other than the urinary system.*

# Anatomy of the Kidney

### Objectives
When you have completed this section, you should be able to

- describe the location and general appearance of the kidney;
- identify the major external and internal features of the kidney;
- trace the flow of fluid through the renal tubules;

- trace the flow of blood through the kidney;
- describe the nerve supply to the kidney;
- trace the flow of fluid through the renal tubules; and
- state the function of each segment of the renal tubule.

## POSITION AND ASSOCIATED STRUCTURES

The kidneys lie against the dorsal abdominal wall at the level of vertebrae T12 to L3. Rib 12 crosses the approximate middle of the kidney. The right kidney is slightly lower than the left because of the space occupied by the large right lobe of the liver above it. The kidneys are retroperitoneal, along with the ureters, urinary bladder, renal artery and vein, and the adrenal[2] glands (fig. 23.3). The left adrenal gland rests on the superior pole of that kidney, and the right adrenal gland lies against the superomedial surface of its kidney. Their functions (see chapter 17) are not as directly related to the kidneys as their spatial relationship might suggest, although the kidneys and adrenals do influence each other.

---

[2]*ad* = to, toward, near + *ren* = kidney + *al* = pertaining to

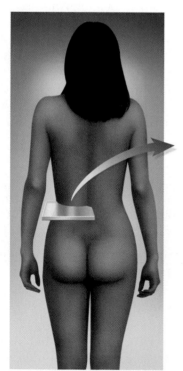
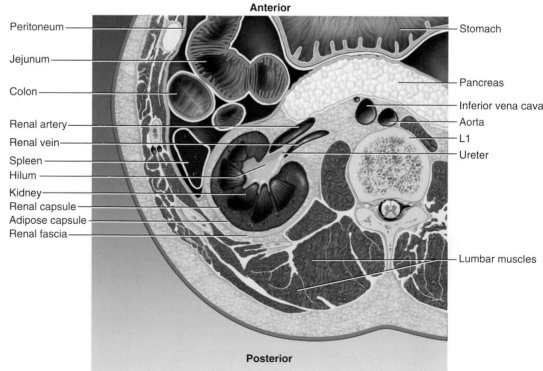

**FIGURE 23.3**  **Location of the Kidney.**  Cross section of the abdomen at the level of vertebra L1.
▶ *Why is the kidney described as retroperitoneal? Name another retroperitoneal organ in this figure.*

## GROSS ANATOMY

The kidney is a compound tubular gland containing about 1.2 million functional excretory units called **nephrons**[3] (NEF-rons). Each kidney weighs about 160 g and measures about 10 cm long, 5 cm wide, and 2.5 cm thick—about the size of a bar of bath soap. The lateral surface is convex, and the medial surface is concave and has a slit, the **hilum,** where it receives the renal nerves, blood vessels, lymphatics, and ureter (fig. 23.4).

The kidney is protected by three layers of connective tissue (see fig. 23.3): (1) A fibrous **renal fascia,** immediately deep to the parietal peritoneum, binds the kidney and associated organs to the abdominal wall; (2) the **adipose capsule,** a layer of fat, cushions the kidney and holds it in place; and (3) the **renal capsule,** a fibrous sac, encloses the kidney like a cellophane wrapper anchored at the hilum, and protects it from trauma and infection. Collagen fibers extend from the renal capsule, through the fat, to the renal fascia. The renal fascia is fused with the peritoneum ventrally and with the deep fascia of the lumbar muscles dorsally. Thus the kidneys are suspended in place. Nevertheless, they drop about 3 cm when one goes from a supine to a standing position, and under some circumstances they become detached and drift even lower, with pathological results (see nephroptosis, or "floating kidney," in table 23.3 at the end of this chapter).

The renal parenchyma—the glandular tissue that forms the urine—appears C-shaped in frontal section. It encircles a medial space, the **renal sinus,** occupied by blood and lymphatic vessels, nerves, and urine-collecting structures. Adipose tissue fills the remaining space in the sinus and holds these structures in place.

The parenchyma is divided into two zones: an outer **renal cortex** about 1 cm thick and an inner **renal medulla** facing the sinus. Extensions of the cortex called **renal columns** project toward the sinus and divide the medulla into 6 to 10 **renal pyramids.** Each pyramid is conical, with a broad base facing the cortex and a blunt point called the **renal papilla** facing the sinus. One pyramid and the overlying cortex constitute one *lobe* of the kidney.

The papilla of each renal pyramid is nestled in a cup called a **minor calyx**[4] (CAY-lix), which collects its urine. Two or three minor calyces (CAY-lih-seez) converge to form a **major calyx,** and two or three major calyces converge in the sinus to form the funnel-like **renal pelvis.**[5] The ureter is a tubular continuation of the renal pelvis that drains the urine down to the urinary bladder.

## CIRCULATION

Although the kidneys account for only 0.4% of the body weight, they receive about 21% of the cardiac output (the *renal fraction*). This is a hint of how important the kidneys are in regulating blood volume and composition.

[3]*nephro = kidney*

[4]*calyx = cup*
[5]*pelvis = basin*

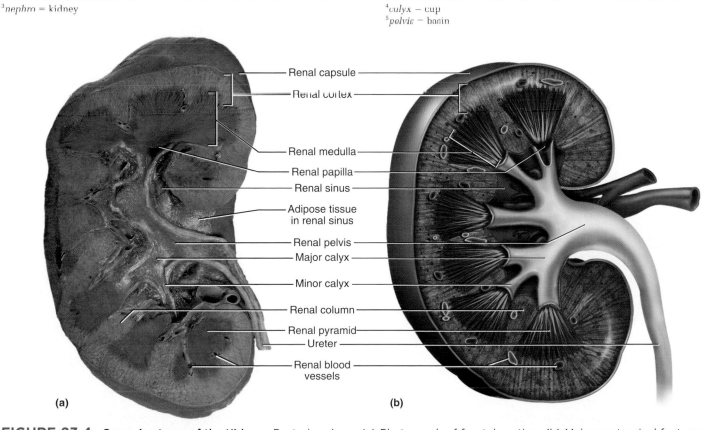

(a)                                    (b)

Renal capsule
Renal cortex
Renal medulla
Renal papilla
Renal sinus
Adipose tissue in renal sinus
Renal pelvis
Major calyx
Minor calyx
Renal column
Renal pyramid
Ureter
Renal blood vessels

**FIGURE 23.4  Gross Anatomy of the Kidney.**  Posterior views. (a) Photograph of frontal section. (b) Major anatomical features.

The larger divisions of the renal circulation are shown in figure 23.5. Each kidney is supplied by a **renal artery** arising from the aorta. Just before or after entering the hilum, the renal artery divides into a few **segmental arteries,** and each of these gives rise to a few **interlobar arteries.** An interlobar artery penetrates each renal column and travels between the pyramids toward the *corticomedullary junction,* the boundary between the cortex and medulla. Along the way, it branches again to form **arcuate arteries,** which make a sharp 90° bend and travel along the base of the pyramid. Each arcuate artery gives rise to several **interlobular arteries,** which pass upward into the cortex.

The finer branches of the renal circulation are shown in figure 23.6. As an interlobular artery ascends through the cortex, a series of **afferent arterioles** arise from it at nearly right angles like the limbs of a pine tree. Each afferent arteriole supplies one nephron. It leads to a spheroidal mass of capillaries called a **glomerulus**[6] (glo-MERR-you-lus), enclosed in a nephron structure called the *glomerular capsule,* to be discussed later. The glomerulus is drained by an **efferent arteriole.**

The afferent and efferent arterioles penetrate one side of the glomerular capsule together. Just outside the capsule, they contact the first part of the distal convoluted tubule and with it, form a **juxtaglomerular**[7] (JUX-tuh-glo-MER-you-lur) **apparatus.** This is a device that enables a nephron

[6]*glomer* = ball + *ulus* = little
[7]*juxta* = next to

to monitor and stabilize its own performance and compensate for fluctuations in blood pressure. It will be described in detail when we consider renal autoregulation.

The efferent arteriole usually leads to a plexus of **peritubular capillaries,** named for the fact that they form a network around the renal tubules. These capillaries pick up the water and solutes reabsorbed by the renal tubules. From the peritubular capillaries, blood flows to **interlobular veins, arcuate veins, interlobar veins,** and the **renal vein,** in that order. These veins travel parallel to the arteries of the same names. (There are, however, no segmental veins corresponding to the segmental arteries.) The renal vein leaves the hilum and drains into the inferior vena cava.

The renal medulla receives only 1% to 2% of the total renal blood flow, supplied by a network of vessels called the **vasa recta.**[8] These arise from the nephrons in the deep cortex, closest to the medulla *(juxtamedullary nephrons).* Here, the efferent arterioles descend immediately into the medulla and give rise to the vasa recta instead of peritubular capillaries. The capillaries of the vasa recta lead

[8]*vasa* = vessels + *recta* = straight

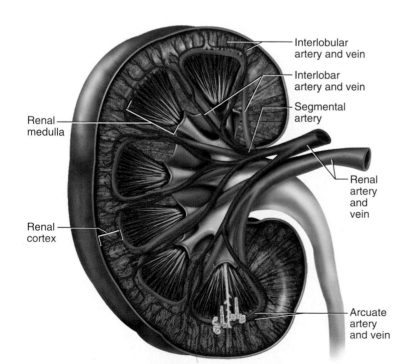

(a)

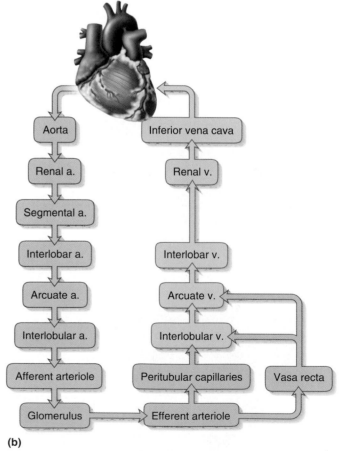

(b)

**FIGURE 23.5  Renal Circulation.**  (a) The larger blood vessels of the kidney. (b) Flow chart of renal circulation. The pathway through the vasa recta (instead of peritubular capillaries) applies only to the juxtamedullary nephrons.

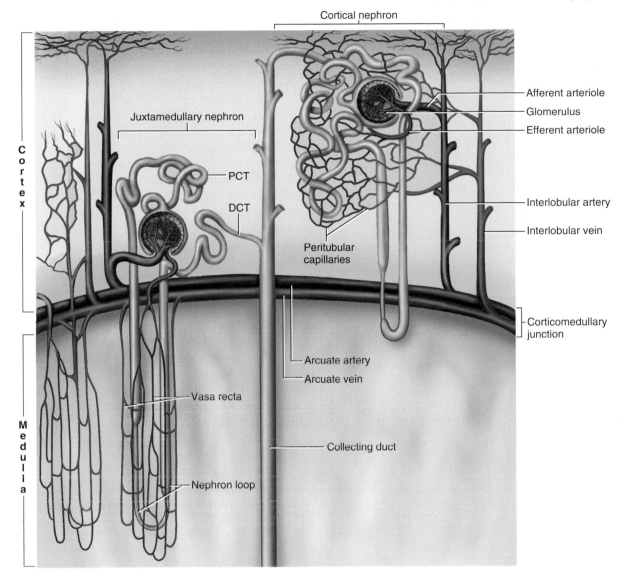

**FIGURE 23.6  Blood Circulation in the Nephron.**  For clarity, vasa recta are shown only on the left and peritubular capillaries only on the right. In the juxtamedullary nephron (left), the efferent arteriole gives rise to the vasa recta of the medulla. The renal tubule and blood vessels are separated for diagrammatic clarity. In the cortical nephron (right), the nephron loop barely dips into the renal medulla and the efferent arteriole gives rise to peritubular capillaries. DCT = distal proximal tubule; PCT = proximal convoluted tubule.

into venules that ascend and empty into the arcuate and interlobular veins. Capillaries of the vasa recta are wedged into the tight spaces between the medullary parts of the renal tubule, and carry away water and solutes reabsorbed by those sections of the tubule. Figure 23.5b summarizes the route of renal blood flow.

> **Think About It**
>
> *Can you identify a portal system in the renal circulation?*

## INNERVATION

**Renal nerves** arise from the superior mesenteric ganglion (see p. 568) and enter the hilum of each kidney. They follow branches of the renal artery and innervate the afferent and efferent arterioles. These nerves consist mostly of sympathetic fibers that regulate the blood flow into and out of each nephron, and thus control the rate of filtration and urine formation. If the blood pressure falls, they also stimulate the secretion of renin, an enzyme that activates hormonal mechanisms for restoring the blood pressure.

## THE NEPHRON

Each kidney contains about 1.2 million functional units called **nephrons** (NEF-rons) (fig. 23.6). If you can understand the function of one nephron, you will understand nearly everything about the function of the kidney. A nephron consists of two principal parts: a *renal corpuscle* where the blood plasma is filtered and a long *renal tubule* that processes this filtrate into urine.

## The Renal Corpuscle

The **renal corpuscle** (fig. 23.7) consists of the glomerulus described earlier and a two-layered **glomerular (Bowman's[9]) capsule** that encloses it. The parietal (outer) layer of the capsule is a simple squamous epithelium, and the visceral layer consists of elaborate cells called **podocytes[10]** wrapped around the capillaries of the glomerulus. The podocytes are described in detail later. The fluid that filters from the glomerular capillaries, called the **glomerular filtrate,** collects in the **capsular space** between the parietal and visceral layers and then flows into the renal tubule on one side of the capsule.

Opposite sides of the renal corpuscle are called the vascular and urinary poles. At the **vascular pole,** the afferent arteriole enters the capsule, bringing blood to the glomerulus, and the efferent arteriole leaves the capsule and carries blood away. The afferent arteriole is significantly larger than the efferent arteriole. Thus, the glomerulus has a large inlet and a small outlet—a point whose functional significance will be explained later. At the **urinary pole,** the parietal wall of the capsule turns away from the corpuscle and gives rise to the renal tubule described next. The simple squamous epithelium of the capsule becomes simple cuboidal in the tubule.

## The Renal Tubule

The **renal (uriniferous[11]) tubule** is a duct that leads away from the glomerular capsule and ends at the tip of a medullary pyramid. It is about 3 cm long and divided into four major regions: the *proximal convoluted tubule, nephron loop, distal convoluted tubule,* and *collecting duct* (fig. 23.8). Only the first three of these are parts of an individual nephron; the collecting duct receives fluid from many nephrons. Each region of the renal tubule has unique physiological properties and roles in the production of urine.

***The Proximal Convoluted Tubule***   The **proximal convoluted tubule (PCT)** arises from the glomerular capsule. It is the longest and most coiled of the four regions and thus dominates histological sections of renal cortex. The PCT has a simple cuboidal epithelium with prominent microvilli (a brush border), which attests to the great deal of absorption that occurs here. The microvilli give the epithelium a distinctively shaggy look in tissue sections.

***The Nephron Loop***   After coiling extensively near the renal corpuscle, the PCT straightens out and forms a long U-shaped **nephron loop** (loop of Henle[12]). The first portion of the loop, the **descending limb,** passes from the cortex into the medulla. At its deep end it turns 180° and forms an **ascending limb** that returns to the cortex. The nephron loop is divided into thick and thin segments. The **thick segments** have a simple cuboidal epithelium. They form the initial part of the descending limb and part or all of the ascending limb. The cells here are heavily engaged in active transport of salts, so they have very high metabolic activity and are loaded with mitochondria. The **thin segment** has a simple squamous epithelium. It forms

---

[9]Sir William Bowman (1816–92), British physician
[10]*podo* = foot + *cyte* = cell
[11]*urin* = urine + *fer* = to carry

[12]Friedrich G. J. Henle (1809–85), German anatomist

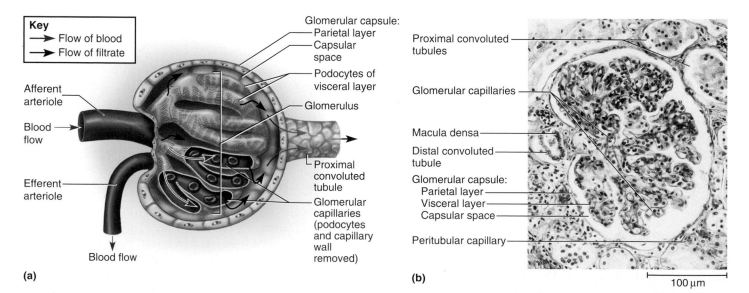

**FIGURE 23.7   The Renal Corpuscle.**   (a) Anatomy of the corpuscle. (b) Light micrograph.

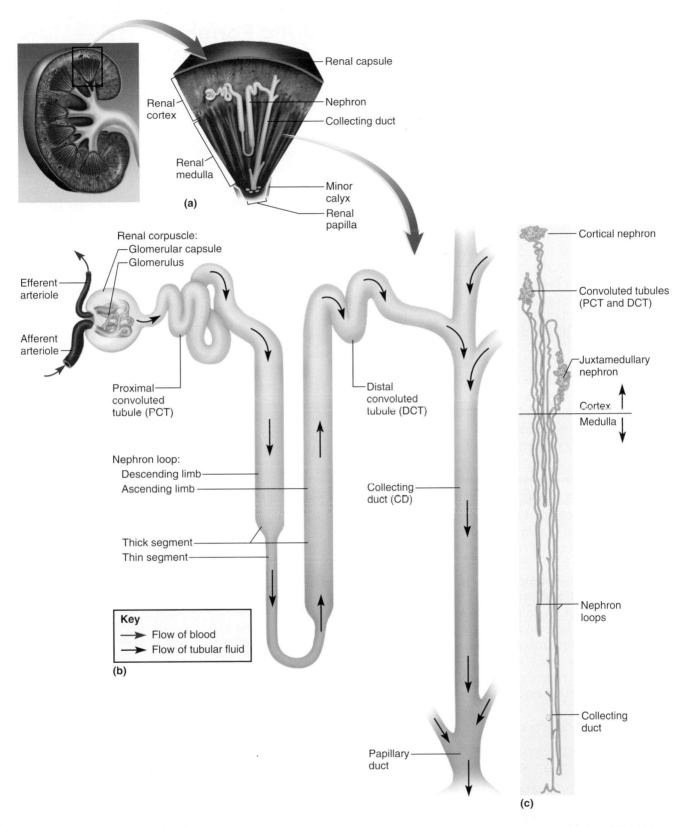

**FIGURE 23.8  Microscopic Anatomy of the Nephron.** (a) Location of the nephrons in one wedge-shaped lobe of the kidney. (b) Structure of a nephron. For clarity, the nephron is stretched out to separate the convoluted tubules. The nephron loop is greatly shortened for the purpose of illustration. (c) The true proportions of the nephron loops relative to the convoluted tubules. Three nephrons are shown. Their proximal and distal convoluted tubules are commingled in a single tangled mass in each nephron. Note the extreme lengths of the nephron loops.

the lower part of the descending limb, and in some nephrons, it rounds the bend and continues partway up the ascending limb. The cells here have low metabolic activity but are very permeable to water.

***The Distal Convoluted Tubule***    When the nephron loop returns to the cortex, it coils again and forms the **distal convoluted tubule (DCT).** This is shorter and less convoluted than the PCT, so fewer sections of it are seen in histological sections. It has a cuboidal epithelium with smooth-surfaced cells nearly devoid of microvilli. The DCT is the end of the nephron.

***The Collecting Duct***    The DCTs of several nephrons drain into a straight tubule called the **collecting duct,** which passes down into the medulla. Near the papilla, several collecting ducts merge to form a larger **papillary duct;** about 30 of these drain from each papilla into its minor calyx. The collecting and papillary ducts are lined with simple cuboidal epithelium.

The flow of fluid from the point where the glomerular filtrate is formed to the point where urine leaves the body is: glomerular capsule → proximal convoluted tubule → nephron loop → distal convoluted tubule → collecting duct → papillary duct → minor calyx → major calyx → renal pelvis → ureter → urinary bladder → urethra.

## Cortical and Juxtamedullary Nephrons

Nephrons just beneath the renal capsule, close to the kidney surface, are called **cortical nephrons.** They have relatively short nephron loops that dip only slightly into the outer medulla before turning back (see fig. 23.6) or turn back even before leaving the cortex. Some cortical nephrons have no nephron loops at all. Nephrons close to the medulla are called **juxtamedullary nephrons.** They have very long nephron loops that extend to the apex of the renal pyramid. As you will see later, nephron loops are responsible for maintaining a salinity gradient in the medulla that helps the body conserve water. Although only 15% of the nephrons are juxtamedullary, they are almost solely responsible for maintaining this gradient.

### Before You Go On

*Answer the following questions to test your understanding of the preceding section:*

4. *Arrange the following in order from the most numerous to the least numerous structures in a kidney: glomeruli, major calyces, minor calyces, interlobular arteries, interlobar arteries.*

5. *Trace the path taken by one red blood cell from the renal artery to the renal vein.*

6. *Consider one molecule of urea in the urine. Trace the route that it took from the point where it left the bloodstream to the point where it left the body.*

# Urine Formation I: Glomerular Filtration

### Objectives

When you have completed this section, you should be able to

- describe the glomerular filtration membrane and how it excludes blood cells and proteins from the urine;

- explain the forces that promote and oppose glomerular filtration, and calculate net filtration pressure if given the magnitude of these forces; and

- describe how the nervous system, hormones, and the kidney itself regulate glomerular filtration.

The kidney converts blood plasma to urine in three stages: glomerular filtration, tubular reabsorption and secretion, and water conservation (fig. 23.9). As we trace fluid through the nephron, we will refer to it by different names that reflect its changing composition: (1) The fluid in the capsular space, called **glomerular filtrate,** is similar to blood plasma except that it has almost no protein. (2) The

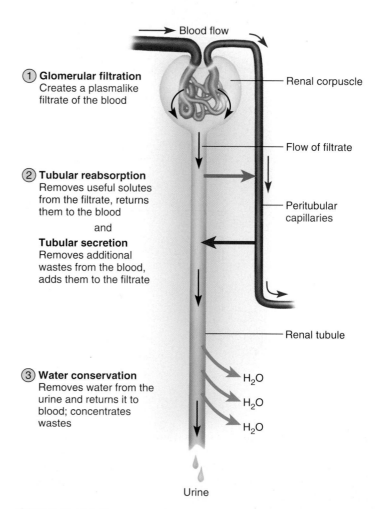

**FIGURE 23.9   Basic Steps in the Formation of Urine.**

fluid from the proximal convoluted tubule through the distal convoluted tubule will be called **tubular fluid.** It differs from the glomerular filtrate because of substances removed and added by the tubule cells. (3) The fluid will be called **urine** once it enters the collecting duct.

# THE FILTRATION MEMBRANE

**Glomerular filtration,** discussed in this section, is a special case of the capillary fluid exchange process described in chapter 20. It is a process in which water and some solutes in the blood plasma pass from the capillaries of the glomerulus into the capsular space of the nephron. To do so, fluid passes through three barriers that constitute the **filtration membrane** (fig. 23.10):

1. **The fenestrated endothelium of the capillary.**
   Endothelial cells of the glomerular capillaries are honeycombed with large filtration pores about 70 to 90 nm in diameter (see fig. 20.6, p. 759). Like fenestrated

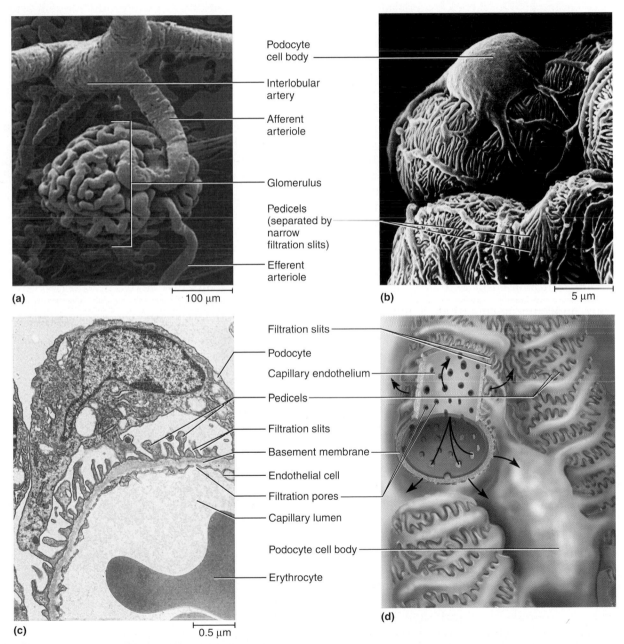

**FIGURE 23.10  Structure of the Glomerulus.**  (a) A resin cast of the glomerulus and nearby arteries (SEM). (b) Blood capillaries of the glomerulus closely wrapped in the spidery podocytes that form the visceral layer of the glomerular capsule (SEM). (c) A blood capillary and podocyte showing filtration pores and slits (TEM). (d) The production of glomerular filtrate by the passage of fluid through the endothelium and filtration slits. [Part (a) from R. G. Kessel and R. H. Kardon, *Tissues and Organs: A Text-Atlas of Scanning Electron Microscopy,* W. H. Freeman, 1979.]

▶ *Which is larger, the efferent arteriole or the afferent arteriole? How does the difference affect the function of the glomerulus?*

capillaries elsewhere, these are highly permeable, although their pores are small enough to exclude blood cells from the filtrate.

2. **The basement membrane.** This membrane consists of a proteoglycan gel. For large molecules to pass through it is like trying to pass sand through a kitchen sponge. A few particles may penetrate its small spaces, but most are held back. On the basis of size alone, the basement membrane would exclude any molecules larger than 8 nm. Some smaller molecules, however, are also held back by a negative electrical charge on the proteoglycans. Blood albumin is slightly less than 7 nm in diameter, but it is also negatively charged and thus repelled by the basement membrane. Although the blood plasma is 7% protein, the glomerular filtrate is only 0.03% protein. It has traces of albumin and smaller polypeptides, including some hormones.

3. **Filtration slits.** A podocyte of the glomerular capsule is shaped somewhat like an octopus, with a bulbous cell body and several thick arms. Each arm has numerous little extensions called **pedicels**[13] (foot processes) that wrap around the capillaries and interdigitate with each other, like wrapping your hands around a pipe and lacing your fingers together. The foot processes have negatively charged **filtration slits** about 30 nm wide between them, which are an additional obstacle to large anions.

Almost any molecule smaller than 3 nm can pass freely through the filtration membrane into the capsular space. This includes water, electrolytes, glucose, fatty acids, amino acids, nitrogenous wastes, and vitamins. Such substances have about the same concentration in the glomerular filtrate as in the blood plasma. Some substances of low molecular weight are retained in the bloodstream because they are bound to plasma proteins that cannot get through the membrane. For example, most calcium, iron, and thyroid hormone in the blood are bound to plasma proteins that retard their filtration by the kidneys. The small fraction that is unbound, however, passes freely through the filtration membrane and appears in the urine.

Kidney infections and trauma can damage the filtration membrane and allow albumin or blood cells to filter through. Kidney disease is sometimes marked by the presence of protein (especially albumin) or blood in the urine—conditions called **proteinuria** (albuminuria) and **hematuria,** respectively. Distance runners and swimmers often experience temporary proteinuria and hematuria. Strenuous exercise greatly reduces perfusion of the kidneys, and the glomerulus deteriorates under the prolonged hypoxia, thus leaking protein and sometimes blood into the filtrate.

---

[13]*pedi* = foot + *cel* = little

## FILTRATION PRESSURE

Glomerular filtration follows the same principles that govern filtration in other blood capillaries (see pp. 770–771), but there are significant differences in the magnitude of the forces involved:

- The blood hydrostatic pressure (BHP) is much higher here than elsewhere—about 60 mm Hg compared with 10 to 15 mm Hg in most other capillaries. This results from the fact that the afferent arteriole is substantially larger than the efferent arteriole, giving the glomerulus a large inlet and small outlet (fig. 23.10a).

- The hydrostatic pressure in the capsular space is about 18 mm Hg, compared with the slightly negative interstitial pressures elsewhere. This results from the high rate of filtration occurring here and the continual accumulation of fluid in the capsule.

- The colloid osmotic pressure (COP) of the blood is about the same here as anywhere else, 32 mm Hg.

- The glomerular filtrate is almost protein-free and has no significant COP. (This can change markedly in kidney diseases that allow protein to filter into the capsular space.)

On balance, then, we have a high outward pressure of 60 mm Hg, opposed by two inward pressures of 18 and 32 mm Hg (fig. 23.11), giving a net filtration pressure (NFP) of

$$60_{out} - 18_{in} - 32_{in} = 10 \text{ mm Hg}_{out}$$

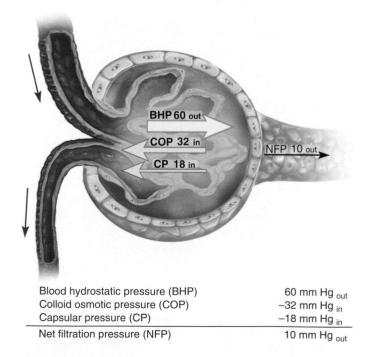

| | |
|---|---|
| Blood hydrostatic pressure (BHP) | 60 mm Hg $_{out}$ |
| Colloid osmotic pressure (COP) | −32 mm Hg $_{in}$ |
| Capsular pressure (CP) | −18 mm Hg $_{in}$ |
| Net filtration pressure (NFP) | 10 mm Hg $_{out}$ |

**FIGURE 23.11   The Forces Involved in Glomerular Filtration.**

In most blood capillaries, the BHP drops low enough at the venous end that osmosis overrides filtration and the capillaries reabsorb fluid. Although BHP also drops along the course of the glomerular capillaries, it remains high enough that these capillaries are engaged solely in filtration. They reabsorb little or no fluid.

The high blood pressure in the glomeruli makes the kidneys especially vulnerable to hypertension, which can have devastating effects on renal function. Hypertension ruptures glomerular capillaries and leads to scarring of the kidneys (nephrosclerosis). It promotes atherosclerosis of the renal blood vessels just as it does elsewhere in the body and thus diminishes renal blood supply. Over time, hypertension often leads to renal failure.

## GLOMERULAR FILTRATION RATE

**Glomerular filtration rate (GFR)** is the amount of filtrate formed per minute by the two kidneys combined. For every 1 mm Hg of net filtration pressure, the kidneys produce about 12.5 mL of filtrate per minute. This value, called the *filtration coefficient ($K_f$)*, depends on the permeability and surface area of the filtration barrier. $K_f$ is about 10% lower in women than in men. For the reference man (defined on p. 17),

$$GFR = NFP \times K_f = 10 \times 12.5 = 125 \text{ mL/min}$$

In the reference woman, the GFR is about 105 mL/min. This is a rate of about 180 L/day in males and 150 L/day in females—impressive numbers considering that this is about 60 times the amount of blood in the body and 50 to 60 times the amount of filtrate produced by all other capillaries combined. Obviously only a small portion of this is eliminated as urine. An average adult reabsorbs 99% of the filtrate and excretes 1 to 2 L of urine per day.

## REGULATION OF GLOMERULAR FILTRATION

GFR must be precisely controlled. If it is too high, fluid flows through the renal tubules too rapidly for them to reabsorb the usual amount of water and solutes. Urine output rises and creates a threat of dehydration and electrolyte depletion. If GFR is too low, fluid flows sluggishly through the tubules, they reabsorb wastes that should be eliminated in the urine, and azotemia may occur. The only way to adjust GFR from moment to moment is to change glomerular blood pressure. This is achieved by three homeostatic mechanisms: renal autoregulation, sympathetic control, and hormonal control.

### Renal Autoregulation

**Renal autoregulation** is the ability of the nephrons to adjust their own blood flow and GFR without external (nervous or hormonal) control. It enables them to main- tain a relatively stable GFR in spite of changes in arterial blood pressure. If the mean arterial pressure (MAP) rose from 100 to 125 mm Hg and there were no renal autoregulation, urine output would increase from the normal 1 to 2 L/day to more than 45 L/day. Because of renal autoregulation, however, urine output increases only a few percent even if MAP rises as high as 160 mm Hg. Renal autoregulation thus helps to ensure stable fluid and electrolyte balance in spite of the many circumstances that substantially alter one's blood pressure. There are two mechanisms of autoregulation: the myogenic mechanism and tubuloglomerular feedback.

***The Myogenic[14] Mechanism*** This mechanism of stabilizing the GFR is based on the tendency of smooth muscle to contract when stretched. When arterial blood pressure rises, it stretches the afferent arteriole. The arteriole contracts, and thus prevents blood flow into the glomerulus from changing very much. Conversely, when blood pressure falls, the afferent arteriole relaxes and allows blood to flow more easily into the glomerulus. Either way, glomerular blood flow and filtration remain fairly stable.

***Tubuloglomerular Feedback*** In this mechanism, the juxtaglomerular apparatus (JGA) monitors the fluid entering the distal convoluted tubule and adjusts the GFR to maintain homeostasis. An understanding of this mechanism requires a closer look at the components of the JGA (fig. 23.12).

1. The **juxtaglomerular (JG) cells** are enlarged smooth muscle cells found in the afferent arteriole and to some extent in the efferent arteriole. When stimulated by the macula densa (discussed next), they dilate or constrict the arterioles. They also contain granules of renin, which they secrete in response to a drop in blood pressure. This initiates negative feedback mechanisms, described later, that raise blood pressure.

2. The **macula densa[15]** is a patch of slender, closely spaced epithelial cells at the start of the distal convoluted tubule (DCT), directly across from the JG cells.

3. **Mesangial[16] (mez-AN-jee-ul) cells** are found in the cleft between the afferent and efferent arterioles and among capillaries of the glomerulus. Their role is not yet clearly understood, but they are connected to the macula densa and JG cells by gap junctions and perhaps mediate communication between those cells.

---

[14]*myo* = muscle + *genic* = produced by
[15]*macula* = spot, patch + *densa* = dense
[16]*mes* = in the middle + *angi* = vessel

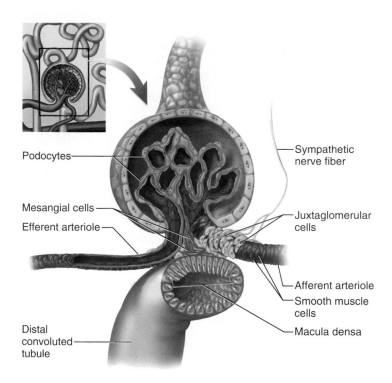

**FIGURE 23.12** The Juxtaglomerular Apparatus.

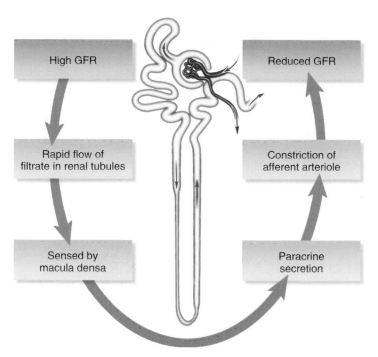

**FIGURE 23.13** Negative Feedback Control of Glomerular Filtration Rate.

The details of tubuloglomerular feedback are still obscure. If GFR rises, however, it increases the flow of tubular fluid and the rate of NaCl reabsorption. The macula densa apparently senses variations in flow or fluid composition and secretes a paracrine messenger that stimulates the JG cells. Contraction of the JG cells constricts the afferent arteriole, thus reducing GFR to normal (fig. 23.13). The mesangial cells amid the glomerular capillaries may also contract, constricting the capillaries and reducing filtration. Conversely, if GFR falls, the macula densa may secrete a different messenger, causing the afferent arteriole and mesangial cells to relax, blood flow to increase, and GFR to rise back to normal.

⌐ **Think About It**

*Describe or diagram a negative feedback loop similar to figure 23.13 to show how the macula densa could compensate for a drop in systemic blood pressure.*

Two important points must be noted about renal autoregulation. First, it does not completely prevent changes in the GFR. Like any other homeostatic mechanism, it maintains a *dynamic equilibrium;* the GFR fluctuates within narrow limits. Changes in blood pressure do affect the GFR and urine output. Second, renal autoregulation cannot compensate for extreme blood pressure variations. Over a MAP range of 90 to 180 mm Hg, the GFR remains quite stable. Below 70 mm Hg, however, glomerular filtration and urine output cease. This can happen in hypovolemic shock (p. 774).

## Sympathetic Control

Sympathetic nerve fibers richly innervate the renal blood vessels. In strenuous exercise or acute conditions such as circulatory shock, the sympathetic nervous system and adrenal epinephrine constrict the afferent arterioles. This reduces GFR and urine production, while redirecting blood from the kidneys to the heart, brain, and skeletal muscles, where it is more urgently needed. Under such conditions, GFR may be as low as a few milliliters per minute.

## The Renin–Angiotensin Mechanism

When blood pressure drops, the sympathetic nerves also stimulate the JG cells to secrete the enzyme **renin** (REE-nin). Renin acts on a plasma protein, *angiotensinogen,* to remove a fragment called angiotensin I, a chain of 10 amino acids. In the lungs and kidneys, **angiotensin-converting enzyme (ACE)** removes two more amino acids, converting it to **angiotensin II,** a hormone with multiple effects (fig. 23.14):

- It stimulates widespread vasoconstriction, which raises the MAP throughout the body.

- It constricts both the afferent and efferent arterioles. The net effect of this is to reduce GFR and water loss.

- It strongly stimulates NaCl and water reabsorption by the proximal convoluted tubule.

- It stimulates the adrenal cortex to secrete aldosterone, which in turn promotes sodium and water

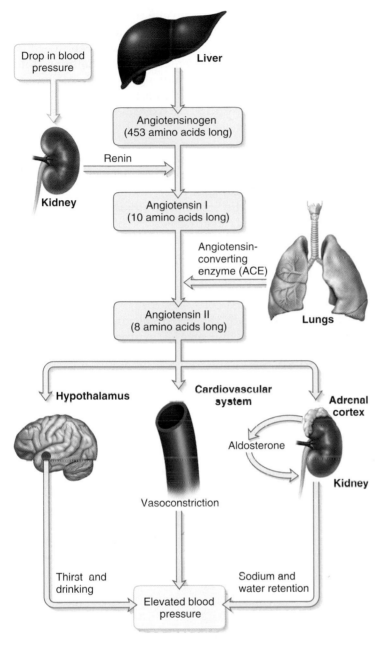

**FIGURE 23.14** **The Renin–Angiotensin–Aldosterone Mechanism.** This chain of events is activated by a drop in blood pressure and acts to raise it again.

retention by the distal convoluted tubule and collecting duct.

- It stimulates the secretion of antidiuretic hormone, which promotes water reabsorption.
- It stimulates the sense of thirst and encourages water intake.

Some of these effects are explained more fully later in this chapter and in chapter 24. Collectively, they act to raise blood pressure by reducing water loss, encouraging water intake, and constricting blood vessels.

---

**Think About It**

*What do you predict would be the effect of ACE inhibitors (see p. 767) on the tubular reabsorption of water by the kidneys?*

To summarize the events thus far: Glomerular filtration occurs because the high blood pressure of the glomerular capillaries overrides reabsorption. The filtration membrane allows most plasma solutes into the capsular space while retaining formed elements and protein in the bloodstream. Glomerular filtration is maintained at a fairly steady rate of about 125 mL/min in spite of variations in systemic blood pressure. This stability is achieved by renal autoregulation, sympathetic control, and hormonal control.

## Before You Go On

*Answer the following questions to test your understanding of the preceding section:*

7. Name the four major processes in urine production.

8. Trace the movement of a urea molecule from the blood to the capsular space, and name the barriers it passes through.

9. Calculate the net filtration pressure in a patient whose blood COP is only 10 mm Hg because of hypoproteinemia. Assume other relevant variables to be normal.

10. Assume a person is moderately dehydrated and has low blood pressure. Describe the homeostatic mechanisms that would help the kidneys maintain a normal GFR.

# Urine Formation II: Tubular Reabsorption and Secretion

### Objectives

When you have completed this section, you should be able to

- describe how the renal tubules reabsorb useful solutes from the glomerular filtrate and return them to the blood;
- describe how the tubules secrete solutes from the blood into the tubular fluid; and
- describe how the nephron regulates water excretion.

Conversion of the glomerular filtrate to urine involves the removal and addition of chemicals by tubular reabsorption and secretion, to be described in this section. Here we trace the course of the tubular fluid through the nephron, from proximal convoluted tubule through distal convoluted tubule, and see how the filtrate is modified at each point along the way. Refer to figure 23.9 to put these processes into perspective.

# THE PROXIMAL CONVOLUTED TUBULE

The proximal convoluted tubule (PCT) reabsorbs about 65% of the glomerular filtrate, while it also removes some substances from the blood and secretes them into the tubule for disposal in the urine. The importance of the PCT is reflected in its relatively great length and prominent microvilli, which increase its absorptive surface area. Its cells also contain abundant large mitochondria that provide ATP for active transport. Your PCTs alone account for about 6% of your resting ATP and calorie consumption.

**Tubular reabsorption** is the process of reclaiming water and solutes from the tubular fluid and returning them to the blood. The PCT reabsorbs a greater variety of chemicals than any other part of the nephron. There are two routes of reabsorption: (1) the **transcellular**[17] **route,** in which substances pass through the cytoplasm and out the base of the epithelial cells, and (2) the **paracellular**[18] **route,** in which substances pass between the epithelial

---

[17]*trans* = across
[18]*para* = next to

cells. The "tight" junctions between tubule epithelial cells are quite leaky and allow significant amounts of water, minerals, urea, and other matter to pass between the cells. Either way, such materials enter the extracellular fluid (ECF) at the base of the epithelium, and from there they are taken up by the peritubular capillaries. In the following discussion and figure 23.15, we examine mechanisms for the reabsorption of water and some individual solutes.

***Sodium*** Sodium reabsorption is the key to everything else, because it creates an osmotic and electrical gradient that drives the reabsorption of water and the other solutes. Sodium, the most abundant cation in the glomerular filtrate, is reabsorbed by both transcellular and paracellular routes. It has a concentration of 140 mEq/L in the fluid entering the PCT and only 12 mEq/L in the cytoplasm of the epithelial cells. Thus there is a steep concentration gradient favoring its facilitated diffusion into the epithelial cells.

In the first half of the proximal convoluted tubule, sodium is absorbed by several symport proteins that simultaneously bind glucose, amino acids, phosphate, or

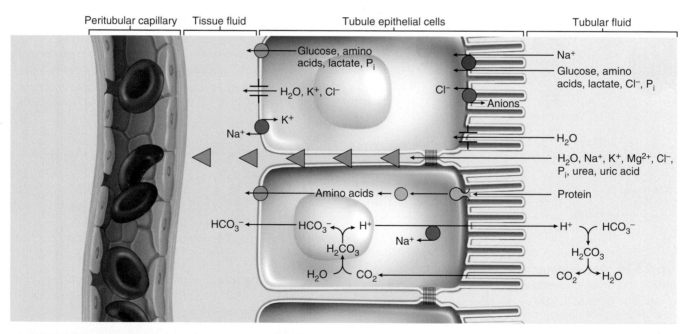

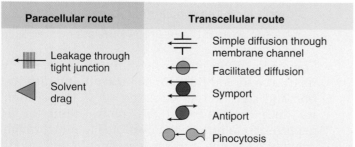

| Paracellular route | | Transcellular route | |
|---|---|---|---|
| ⊣ Leakage through tight junction | | ⊣⊢ | Simple diffusion through membrane channel |
| ◄ Solvent drag | | ●← | Facilitated diffusion |
| | | ●● | Symport |
| | | ●↘ | Antiport |
| | | ○–⤳ | Pinocytosis |

**FIGURE 23.15** Mechanisms of Reabsorption in the Proximal Convoluted Tubule.

▶ *How would increased Na⁺ reabsorption affect the pH of the urine? Why?*

lactate and transport them into the cell. In addition, $H^+$ ions are generated within the cell by the reaction $CO_2 + H_2O \rightarrow H_2CO_3 \rightarrow HCO_3^- + H^+$, and then an $Na^+-H^+$ antiport in the membrane transports $H^+$ out of the cell and $Na^+$ in. (The fate of the $HCO_3^-$ is explained shortly.) In the second half of the proximal tubule, the organic molecules in the tubular fluid have been largely depleted by reabsorption, but the chloride ion concentration is high. Thus, in this part of the tubule, $Na^+$ crosses the epithelium with $Cl^-$ through both transcellular and paracellular routes.

Sodium uptake is possible only because the $Na^+$ concentration in the tubule cells is much lower than in the tubular fluid. But with all this $Na^+$ entering the tubule cells, how does its cytoplasmic concentration remain so low? Why doesn't the inflow of $Na^+$ stop? The answer is that $Na^+-K^+$ pumps in the basal and lateral plasma membrane continually pump $Na^+$ out of the cell and into the extracellular fluid beneath the tubule epithelium. The transport of $Na^+$ and $Na^+$-linked solutes through the apical plasma membrane thus exemplifies *secondary active transport,* because even though the cotransport proteins here do not use ATP, they depend on the ATP-consuming $Na^+-K^+$ pumps in the basolateral part of the cell.

**Chloride**   Chloride is reabsorbed through both the paracellular and transcellular routes. Its reabsorption is favored by two factors: (1) negative chloride ions tend to follow the positive sodium ions by electrical attraction, and (2) water reabsorption raises the $Cl^-$ concentration in the tubular fluid, thereby creating a gradient favorable to $Cl^-$ reabsorption, especially in the second half of the tubule. In the transcellular route, $Cl^-$ is apically absorbed by various antiports that exchange $Cl^-$ for other anions. A $K^+-Cl^-$ symport transports the chloride ions out the basolateral cell surfaces.

**Bicarbonate**   Substantial amounts of bicarbonate ion ($HCO_3^-$) are filtered out of the blood by the glomerulus, and yet the urine is usually bicarbonate-free. Thus it would seem as if all the bicarbonate is reabsorbed by the nephron, but this is only an appearance. Bicarbonate ions do not actually cross the apical plasma membranes of the tubule cells. However, the tubule cells generate bicarbonate and hydrogen ions internally by the reaction of $CO_2$ and water. The hydrogen ions are pumped into the tubular fluid by the $Na^+-H^+$ antiport mentioned earlier, and neutralize the $HCO_3^-$ in the tubule. The bicarbonate ions are pumped out the base of the cell and enter the blood. Thus one $HCO_3^-$ disappears from the tubule fluid as one new $HCO_3^-$ appears in the blood, and the net effect is the same as if an $HCO_3^-$ ion had actually crossed the epithelium from tubular fluid to blood.

**Other Electrolytes**   Potassium, magnesium, and phosphate ($P_i$) ions diffuse through the paracellular route with water. Phosphate is also cotransported into the epithelial cells with $Na^+$ as noted earlier. Some calcium is reabsorbed through the paracellular route in the proximal tubule, but most $Ca^{2+}$ absorption occurs later in the nephron, as we will see. Sulfates and nitrates are not reabsorbed; thus they pass in the urine.

**Glucose**   Glucose is cotransported with $Na^+$ by carriers called **sodium–glucose transport proteins (SGLTs).** It is then removed from the basolateral surface of the cell by facilitated diffusion. Normally all glucose in the tubular fluid is reabsorbed and there is none in the urine.

**Nitrogenous Wastes**   Urea diffuses through the tubule epithelium with water. The nephron as a whole reabsorbs 40% to 60% of the urea in the tubular fluid, but since it reabsorbs 99% of the water, urine has a substantially higher urea concentration than blood or glomerular filtrate. When blood enters the kidney, its urea concentration is about 20 mg/dL; when it leaves the kidney, it is typically down to 10.4 mg/dL. Thus the kidney removes about half of the urea, keeping its concentration down to a safe level but not completely clearing the blood of it.

The PCT reabsorbs nearly all the uric acid entering it, but later parts of the nephron secrete it back into the tubular fluid. Creatinine is not reabsorbed at all. It is too large to diffuse through water channels in the plasma membrane, and there are no transport proteins for it. Therefore, all creatinine filtered by the glomerulus is excreted in the urine.

**Other Organic Solutes**   Some apical $Na^+$ carriers also bind and transport amino acids and lactate. Peptide hormones, other small peptides, and small amounts of larger proteins filter through the glomerulus. Although their rate of filtration is low, it would amount to a protein loss of 7.2 g/day if it were not reabsorbed. PCT cells partially degrade proteins to smaller peptides by means of enzymes on their brush border, then absorb the peptides and break them down the rest of the way to amino acids. Amino acids, lactate, and other small organics leave the basal side of the cell by facilitated diffusion.

**Water**   The kidneys reduce about 180 L of glomerular filtrate to 1 or 2 L of urine each day, so obviously water reabsorption is a significant function. About two-thirds of the water is reabsorbed by the PCT. The reabsorption of all the salt and organic solutes as just described makes the tubule cells and tissue fluid hypertonic to the tubular fluid. Water follows the solutes by osmosis through both the paracellular and transcellular routes. Transcellular absorption occurs by way of water channels called **aquaporins** in the plasma membrane.

Because the PCT reabsorbs proportionate amounts of solutes and water, the osmolarity of the tubular fluid remains unchanged here. Elsewhere in the nephron, the amount of water reabsorption is continually modulated by hormones according to the body's state of hydration. In the PCT, however, water is reabsorbed at a constant rate called **obligatory water reabsorption.**

## Uptake by the Peritubular Capillaries

After water and solutes leave the basal surface of the tubule epithelium, they are reabsorbed by the peritubular capillaries, thus returning to the bloodstream. The mechanisms of capillary absorption are osmosis and solvent drag. Three factors promote osmosis into these capillaries: (1) The accumulation of reabsorbed fluid around the basolateral sides of the epithelial cells creates a high interstitial fluid pressure that tends to drive water into the capillaries. (2) The narrowness of the efferent arteriole lowers the blood hydrostatic pressure (BHP) from 60 mm Hg in the glomerulus to only 8 mm Hg in the peritubular capillaries, so there is less capillary resistance to reabsorption here than in most systemic capillaries (fig. 23.16). (3) As blood passes through the glomerulus, a lot of water is filtered out but nearly all of the protein remains in the blood. Therefore, the blood has an elevated colloid osmotic pressure (COP) by the time it leaves the glomerulus. With a high COP and low BHP in

the capillaries and a high hydrostatic pressure in the tissue fluid, the balance of forces in the peritubular capillaries strongly favors reabsorption. Water on the basal side of the tubular epithelium therefore passes into the capillaries. The dissolved solutes enter the capillaries by **solvent drag**—the water "drags" them into the capillary with it.

## The Transport Maximum

There is a limit to the amount of solute that the renal tubule can reabsorb because there are a limited number of transport proteins in the plasma membranes. If all the transporters are occupied as solute molecules pass through, some solute will escape reabsorption and appear in the urine. The maximum rate of reabsorption is the *transport maximum ($T_m$)*, which is reached when the transporters are saturated (see p. 106). Each organic solute reabsorbed by the renal tubule has its own $T_m$. For glucose, for example, $T_m = 320$ mg/min. Glucose normally enters the renal tubule at a rate of 125 mg/min, well within the $T_m$; thus all of it is reabsorbed. But when the plasma concentration of glucose reaches a *threshold* of about 220 mg/dL, more glucose is filtered than the tubule can reabsorb, and we begin to see the excess glucose in the urine, a condition called **glycosuria**[19] (GLY-co-soo-ree-uh). In untreated diabetes mellitus, the plasma glucose concentration may exceed 400 mg/dL, so glycosuria is one of the classic signs of this disease.

## Tubular Secretion

**Tubular secretion** is a process in which the renal tubule extracts chemicals from the capillary blood and secretes them into the tubular fluid (see fig. 23.9). Tubular secretion in the distal convoluted tubule is discussed shortly. In the proximal convoluted tubule and nephron loop, it serves two purposes:

1. **Waste removal.** Urea, uric acid, bile acids, ammonia, catecholamines, prostaglandins, and a little creatinine are secreted into the tubule. Tubular secretion of uric acid compensates for its reabsorption earlier in the PCT and accounts for all of the uric acid in the urine. Tubular secretion also clears the blood of pollutants, morphine, penicillin, aspirin, and other drugs. One reason that so many drugs must be taken three or four times a day is to keep pace with this rate of clearance and maintain a therapeutically effective drug concentration in the blood.

2. **Acid–base balance.** Tubular secretion of hydrogen and bicarbonate ions serves to regulate the pH of the body fluids. The details are discussed in chapter 24.

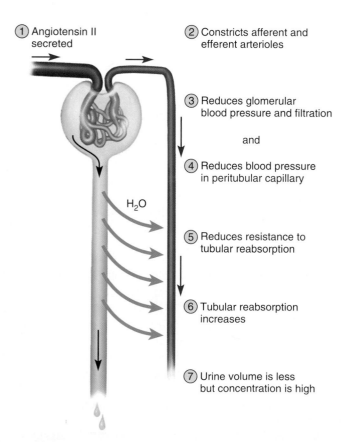

① Angiotensin II secreted

② Constricts afferent and efferent arterioles

③ Reduces glomerular blood pressure and filtration

and

④ Reduces blood pressure in peritubular capillary

$H_2O$

⑤ Reduces resistance to tubular reabsorption

⑥ Tubular reabsorption increases

⑦ Urine volume is less but concentration is high

**FIGURE 23.16** The Effect of Angiotensin II on Urine Volume and Concentration.

[19]*glycos* = sugar + *uria* = urine condition

# THE NEPHRON LOOP

The primary function of the nephron loop is to generate a salinity gradient that enables the collecting duct to concentrate the urine and conserve water, as discussed later. But in addition, the loop reabsorbs about 25% of the $Na^+$, $K^+$, and $Cl^-$ and 15% of the water in the glomerular filtrate. Cells in the thick segment of the loop have proteins in the apical membranes that simultaneously bind 1 $Na^+$, 1 $K^+$, and 2 $Cl^-$ from the tubular fluid and cotransport them into the cytoplasm. These ions leave the basolateral cell surfaces by active transport of $Na^+$ and diffusion of $K^+$ and $Cl^-$. Potassium reenters the cell by means of the $Na^+-K^+$ pump and then reenters the tubular fluid, but NaCl remains in the tissue fluid of the renal medulla. The thick segment of the loop is impermeable to water; thus water cannot follow the reabsorbed electrolytes, and tubular fluid becomes very dilute by the time it passes from the nephron loop into the distal convoluted tubule.

# THE DISTAL CONVOLUTED TUBULE AND COLLECTING DUCT

Fluid arriving in the DCT still contains about 20% of the water and 7% of the salts from the glomerular filtrate. If this were all passed as urine, it would amount to 36 L/day, so obviously a great deal of fluid reabsorption is still to come. The DCT and collecting duct reabsorb variable amounts of water and salts and are regulated by several hormones—particularly aldosterone, atrial natriuretic peptide, antidiuretic hormone, and parathyroid hormone. There are two kinds of cells in the DCT and collecting duct. The **principal cells** are the more abundant; they have receptors for these hormones and are involved chiefly in salt and water balance. The **intercalated cells** are fewer in number. They have a high density of mitochondria, reabsorb $K^+$, secrete $H^+$ into the tubule lumen, and are involved mainly in acid–base balance, as discussed in chapter 24.

The major hormonal influences on these parts of the nephron are as follows.

## Aldosterone

Aldosterone, the "salt-retaining hormone," is a steroid secreted by the adrenal cortex when the blood $Na^+$ concentration falls or its $K^+$ concentration rises. A drop in blood pressure also induces aldosterone secretion, but indirectly—it stimulates the kidney to secrete renin, this leads to the production of angiotensin II, and angiotensin II stimulates aldosterone secretion (see fig. 23.14).

Aldosterone acts on the thick segment of the ascending limb of the nephron loop, the DCT, and the cortical portion of the collecting duct. These regions of the nephron reabsorb more $Na^+$ and secrete more $K^+$. Water and $Cl^-$ follow the $Na^+$, so the net effect is that the body retains NaCl and water, the urine volume is reduced, and the urine has an elevated $K^+$ concentration. The retention of salt and water helps to maintain blood volume and pressure. Chapter 24 deals further with the action of aldosterone.

## Atrial Natriuretic Peptide

Atrial natriuretic peptide (ANP) is secreted by the atrial myocardium of the heart in response to high blood pressure. ANP has four actions that result in the excretion of more salt and water in the urine, thus reducing blood volume and pressure:

1. It dilates the afferent arteriole and constricts the efferent arteriole, thus increasing the glomerular filtration rate.
2. It antagonizes the angiotensin–aldosterone mechanism by inhibiting renin and aldosterone secretion.
3. It inhibits the secretion of antidiuretic hormone (ADH) and the action of ADH on the kidney.
4. It inhibits NaCl reabsorption by the collecting duct.

## Antidiuretic Hormone

ADH is secreted by the posterior lobe of the pituitary gland in response to dehydration and rising blood osmolarity. Its mechanism of action is explained later in more detail. Briefly, it makes the collecting duct more permeable to water, so water in the tubular fluid reenters the tissue fluid and bloodstream rather than being lost in the urine.

## Parathyroid Hormone

A calcium deficiency (hypocalcemia) stimulates the parathyroid glands to secrete parathyroid hormone (PTH). PTH acts in several ways to restore calcium homeostasis. Its effect on bone metabolism was described in chapter 7. In the kidney, PTH acts on the PCT to increase phosphate excretion, and acts on the thick segment of the ascending limb of the nephron loop and on the DCT to increase calcium reabsorption. Thus it increases the phosphate content and lowers the calcium content of the urine. This helps to minimize any further decline in the blood calcium level. Because phosphate is not retained along with the calcium, the calcium ions stay in circulation rather than precipitating into the bone tissue as calcium phosphate. Calcitriol and calcitonin have similar but weaker effects on the DCT. PTH also stimulates calcitriol synthesis by the epithelial cells of the PCT.

In summary, the PCT reabsorbs about 65% of the glomerular filtrate and returns it to the blood of the peritubular capillaries. Much of this reabsorption occurs by osmotic and cotransport mechanisms linked to the active transport of sodium ions. The nephron loop reabsorbs another 25% of the filtrate, although its primary role, detailed later, is to aid the function of the collecting duct. The DCT reabsorbs more sodium, chloride, and water, but its rates of reabsorption are subject to control by hormones, especially aldosterone and ANP. These tubules also extract drugs, wastes, and some other solutes from the blood and secrete them into the tubular fluid. The DCT essentially completes the process of determining the chemical composition of the urine. The principal function left to the collecting duct is to conserve body water.

### Before You Go On

*Answer the following questions to test your understanding of the preceding section:*

11. *The reabsorption of water, Cl⁻, and glucose by the PCT are all linked to the reabsorption of Na⁺, but in three very different ways. Contrast these three mechanisms.*

12. *Explain why a substance appears in the urine if its rate of glomerular filtration exceeds the $T_m$ of the renal tubule.*

13. *Contrast the effects of aldosterone and ANF on the renal tubule.*

# Urine Formation III: Water Conservation

### Objectives

When you have completed this section, you should be able to

- explain how the collecting duct and antidiuretic hormone regulate the volume and concentration of urine; and

- explain how the kidney maintains an osmotic gradient in the renal medulla that enables the collecting duct to function.

The kidney serves not just to eliminate metabolic waste from the body but to prevent excessive water loss in doing so, and thus to support the body's fluid balance. As the kidney returns water to the tissue fluid and bloodstream, the fluid remaining in the renal tubule becomes more and more concentrated. In this section, we examine the kidney's mechanisms for conserving water and concentrating the urine.

## THE COLLECTING DUCT

The collecting duct (CD) begins in the cortex, where it receives tubular fluid from numerous nephrons. As it passes through the medulla, it usually reabsorbs water and concentrates the urine. When urine enters the upper end of the CD, it is isotonic with blood plasma (300 mOsm/L), but by the time it leaves the lower end, it can be up to four times as concentrated—that is, highly hypertonic to the plasma. This ability to concentrate wastes and

**INSIGHT 23.1**   Evolutionary Medicine

### The Kidney and Life on Dry Land

Physiologists first suspected that the nephron loop plays a role in water conservation because of their studies of a variety of animal species. Animals that must conserve water have longer, more numerous nephron loops than animals with little need to conserve it. Fish and amphibians lack nephron loops and produce urine that is isotonic to their blood plasma. Aquatic mammals such as beavers have short nephron loops and only slightly hypertonic urine.

But the kangaroo rat, a desert rodent, provides an instructive contrast. It lives on seeds and other dry foods and need never drink water. Its kidneys are so efficient at conserving water that it can live entirely on the water produced by aerobic respiration. They have extremely long nephron loops and produce urine that is 10 to 14 times as concentrated as their blood plasma (compared with about 4 times, at most, in humans).

Comparative studies thus suggested a hypothesis for the function of the nephron loop and prompted many years of difficult research that led to the discovery of the countercurrent multiplier mechanism for water conservation. This shows how comparative anatomy provides suggestions and insights into function and why physiologists do not study human function in isolation from other species.

control water loss was crucial to the evolution of terrestrial animals such as ourselves (see Insight 23.1).

Two facts enable the collecting duct to produce such hypertonic urine: (1) the osmolarity of the extracellular fluid is four times as high in the lower medulla as it is in the cortex, and (2) the medullary portion of the CD is more permeable to water than to NaCl. Therefore, as urine passes down the CD through the increasingly salty medulla, water leaves the tubule by osmosis, most NaCl and other wastes remain behind, and the urine becomes more and more concentrated (fig. 23.17).

## CONTROL OF WATER LOSS

Just *how* concentrated the urine becomes depends on the body's state of hydration. For example, if you drink a large volume of water, you will soon produce a large volume of hypotonic urine. This response is called *water diuresis*[20] (DY-you-REE-sis). Under such conditions, the cortical portion of the CD reabsorbs NaCl but is impermeable to water. Thus salt is removed from the urine, water stays in the CD, and urine concentration may be as low as 50 mOsm/L.

---

[20]*diuresis* = passing urine

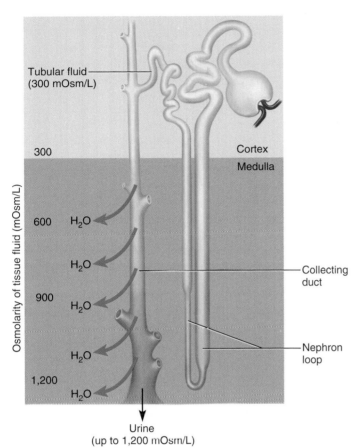

**FIGURE 23.17** **Water Reabsorption by the Collecting Duct.** Note that the osmolarity of the tissue fluid increases fourfold from 300 mOsm/L in the cortex to 1,200 mOsm/L deep in the medulla. When the collecting duct has open water channels, water leaves the duct by osmosis and urine concentration increases.

cantly reduce the glomerular filtration rate. When the GFR is low, fluid flows more slowly through the renal tubules and there is more time for tubular reabsorption. Less salt remains in the urine as it enters the collecting duct, so there is less opposition to the osmosis of water out of the duct and into the ECF. More water is reabsorbed and less urine is produced.

## THE COUNTERCURRENT MULTIPLIER

The ability of the CD to concentrate urine depends on the salinity gradient of the renal medulla. It may seem surprising that the ECF is four times as salty deep in the medulla as it is in the cortex. We would expect the salt to diffuse toward the cortex until it was evenly distributed through the kidney. However, there is a mechanism that overrides this—the nephron loop acts as a **countercurrent multiplier,** which continually recaptures salt and returns it to the deep medullary tissue. It is called a *multiplier* because it multiplies the salinity deep in the medulla, and a *countercurrent* mechanism because it is based on fluid flowing in opposite directions in two adjacent tubules—downward in the descending limb and upward in the ascending limb.

Figure 23.18 shows how the countercurrent multiplier works. Steps 2 through 5 form a positive feedback loop. As

Dehydration, on the other hand, causes your urine to be scanty and more concentrated. The high blood osmolarity of a dehydrated person stimulates the pituitary to release ADH. ADH induces the renal tubule cells to synthesize aquaporins (water-channel proteins) and install them in the plasma membrane, so more water can pass through the epithelial cells. The CD then reabsorbs more water, which is carried away by the vasa recta. Urine output is consequently reduced. By contrast, when you are well hydrated, ADH secretion falls and the tubule cells remove aquaporins from the plasma membrane. The duct is then less permeable to water, so more water remains in the duct and the kidney produces abundant, dilute urine.

In extreme cases, the blood pressure of a dehydrated person is low enough to signifi-

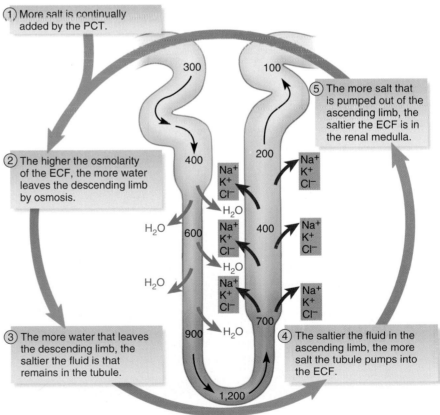

**FIGURE 23.18** **The Countercurrent Multiplier of the Nephron Loop.** The numbers in the tubule are in mOsm/L.

fluid flows down the descending limb of the nephron loop, it passes through an environment of increasing osmolarity. Most of the descending limb is very permeable to water but not to NaCl; therefore, water passes by osmosis from the tubule into the extracellular fluid (ECF), leaving the salts behind. The tubule contents increase in osmolarity, reaching about 1,200 mOsm/L by the time the fluid rounds the bend at the lower end of the loop.

Most or all of the ascending limb (its thick segment), by contrast, is impermeable to water, but has active transport pumps that cotransport $Na^+$, $K^+$, and $Cl^-$ into the ECF. This keeps the osmolarity of the renal medulla high. Since water remains in the tubule, the tubular fluid becomes more and more dilute as it approaches the cortex and is only about 100 mOsm/L at the top of the loop.

The collecting duct also helps to maintain the osmotic gradient (fig. 23.19). Its lower end is somewhat permeable to urea, which diffuses down its concentration gradient, out of the duct and into the ECF. Some of this urea enters the descending thin segment of the nephron loop and travels to the DCT. Neither the thick segment of the loop nor the DCT is permeable to urea, so urea remains in the tubules and returns to the collecting duct. Combined with new urea being added continually by the glomerular filtrate, urea remains concentrated in the fluid of the collecting duct, and some of it always diffuses out into the medulla. Thus there is a continual recycling of urea from the collecting duct to the medulla and back. Urea accounts for about 40% of the high osmolarity deep in the medulla.

## THE COUNTERCURRENT EXCHANGE SYSTEM

The renal medulla must have a blood supply to meet its metabolic needs, and this creates a potential problem—capillaries of the medulla could carry away the urea and salt that produce the high osmolarity. The vasa recta that supply the medulla, however, form a countercurrent system of their own that prevents this from happening. Blood flows in opposite directions in adjacent parallel capillaries. These capillaries form a **countercurrent exchange system.** Blood in the vasa recta exchanges water for salt as it flows downward into the deep medulla—water diffuses out of the cap-

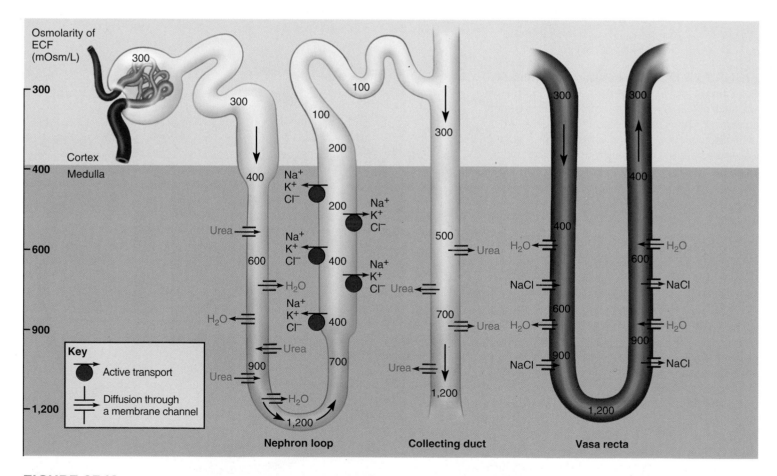

**FIGURE 23.19** Functional Relationship of the Nephron Loop, Vasa Recta, and Collecting Duct. These three structures work together to maintain a gradient of osmolarity in the renal medulla. The numbers in the tubule are in mOsm/L.

illaries and salt diffuses in. As the blood flows back toward the cortex, the opposite occurs; it exchanges salt for water. Thus, the vasa recta give the salt back and do not subtract from the osmolarity of the medulla. Indeed, they absorb more water on the way out than they unload on the way in; they carry away the water reabsorbed from the urine by the collecting duct and nephron loop.

To summarize what we have studied in this section, the collecting duct can adjust water reabsorption to produce urine as hypotonic as 50 mOsm/L or as hypertonic as 1,200 mOsm/L, depending on the body's need for water conservation or removal. In a state of hydration, ADH is not secreted and the cortical part of the CD reabsorbs salt without reabsorbing water; the water remains to be excreted in the dilute urine. In a state of dehydration, ADH is secreted, the medullary part of the CD reabsorbs water, and the urine is more concentrated. The CD is able to do this because it passes through a salinity gradient in the medulla from 300 mOsm/L near the cortex to 1,200 mOsm/L near the papilla. This gradient is produced by a countercurrent multiplier of the nephron loop, which concentrates NaCl in the lower medulla, and by the diffusion of urea from the collecting duct into the medulla. The vasa recta are arranged as a countercurrent exchange system that enables them to supply blood to the medulla without subtracting from its salinity gradient. Figure 23.20 summarizes the major solutes reabsorbed and secreted in each part of the renal tubule. Table 23.1 summarizes the hormones that affect renal function.

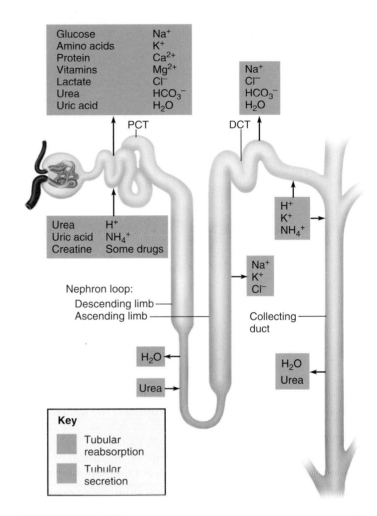

**FIGURE 23.20** Solutes Reabsorbed and Secreted in Each Portion of the Renal Tubule.

| TABLE 23.1 | Hormones Affecting Renal Function | |
|---|---|---|
| **Hormone** | **Renal targets** | **Effects** |
| Aldosterone | Nephron loop, DCT, CD | Promotes $Na^+$ reabsorption and $K^+$ secretion; indirectly promotes $Cl^-$ and $H_2O$ reabsorption; maintains blood volume and reduces urine volume |
| Angiotensin II | Afferent and efferent arterioles, PCT | Reduces water loss, encourages water intake, and constricts blood vessels, thus raising blood pressure. Acts as a generalized vasoconstrictor; reduces GFR; stimulates PCT to reabsorb NaCl and $H_2O$; stimulates aldosterone and ADH secretion; stimulates thirst |
| Antidiuretic hormone | Collecting duct | Promotes $H_2O$ reabsorption; reduces urine volume, increases concentration |
| Atrial natriuretic peptide | Afferent and efferent arterioles, collecting duct | Dilates afferent arteriole, constricts efferent arteriole, increases GFR; inhibits secretion of renin, ADH, and aldosterone; inhibits NaCl reabsorption by collecting duct; increases urine volume and lowers blood pressure |
| Calcitonin | DCT | Weak effects similar to those of PTH |
| Calcitriol | DCT | Weak effects similar to those of PTH |
| Epinephrine and norepinephrine | Juxtaglomerular apparatus, afferent arteriole | Induce renin secretion; constrict afferent arteriole; reduce GFR and urine volume |
| Parathyroid hormone | PCT, DCT, nephron loop | Promotes $Ca^{2+}$ reabsorption by loop and DCT; increases phosphate excretion by PCT; promotes calcitriol synthesis |

# Urine and Renal Function Tests

Medical diagnosis often rests on determining the current and recent physiological state of the tissues. No two fluids are as valuable for this purpose as blood and urine. **Urinalysis,** the examination of the physical and chemical properties of urine, is therefore one of the most routine procedures in medical examinations. The principal characteristics of urine and certain tests used to evaluate renal function are described here.

## COMPOSITION AND PROPERTIES OF URINE

The basic composition and properties of urine are as follows:

- **Appearance.** Urine varies from almost colorless to deep amber, depending on the body's state of hydration. The yellow color of urine is due to **urochrome,**[21] a pigment produced by the breakdown of hemoglobin from expired erythrocytes. Pink, green, brown, black, and other colors result from certain foods, vitamins, drugs, and metabolic diseases. Urine is normally clear but turns cloudy upon standing because of bacterial growth. Pus in the urine **(pyuria**[22]**)** makes it cloudy and suggests kidney infection. Blood in the urine (hematuria) may be due to a urinary tract infection, trauma, or kidney stones. Cloudiness or blood in a urine specimen sometimes, however, simply indicates contamination with semen or menstrual fluid.

- **Odor.** Fresh urine has a distinctive but not repellent odor. As it stands, however, bacteria multiply, degrade urea to ammonia, and produce the pungent odor typical of stale wet diapers. Asparagus and other foods can impart distinctive aromas to the urine. Diabetes mellitus gives it a sweet, fruity odor of acetone. A mousy odor suggests phenylketonuria (PKU), and a rotten odor may indicate urinary tract infection.

- **Specific gravity.** This is a ratio of the density (g/mL) of a substance to the density of distilled water. Distilled water has a specific gravity of 1.000, and urine ranges from 1.001 when it is very dilute to 1.028 when it is very concentrated. Multiplying the last two digits of the specific gravity by a proportionality constant of 2.6 gives an estimate of the grams of solid matter per liter of urine. For example, a specific gravity of 1.025 indicates a solute concentration of $25 \times 2.6 = 65$ g/L.

- **Osmolarity.** Urine can have an osmolarity as low as 50 mOsm/L in a very hydrated person or as high as 1,200 mOsm/L in a dehydrated person. Compared with the osmolarity of blood (300 mOsm/L), then, urine can be either hypotonic or hypertonic under different conditions.

- **pH.** The pH of urine ranges from 4.5 to 8.2 but is usually about 6.0 (mildly acidic). The regulation of urine pH is discussed extensively in chapter 24.

- **Chemical composition.** Urine averages 95% water and 5% solutes by volume (table 23.2). Normally, the most abundant solute is urea, followed by sodium chloride, potassium chloride, and lesser amounts of creatinine, uric acid, phosphates, sulfates, and traces of calcium, magnesium, and sometimes bicarbonate (table 23.2). Urine contains urochrome and a trace of bilirubin from the breakdown of hemoglobin and related products, and urobilin, a brown oxidized derivative of bilirubin. It is abnormal to find glucose, free hemoglobin, albumin, ketones, or bile pigments in the urine; their presence is an important indicator of disease.

## URINE VOLUME

An average adult produces 1 to 2 L of urine per day. An output in excess of 2 L/day is called diuresis or **polyuria**[23] (POL-ee-YOU-ree-uh). Fluid intake and some drugs can temporarily increase output to as much as 20 L/day. Chronic diseases such as diabetes (see next) can do so over a long term. **Oliguria**[24] (oll-ih-GUR-ee-uh) is an output of less than 500 mL/day, and **anuria**[25] is an output of

---

[21]*uro* = urine + *chrom* = color
[22]*py* = pus + *ur* = urine + *ia* = condition
[23]*poly* = many, much + *ur* = urine + *ia* = condition
[24]*oligo* = few, a little + *ur* = urine + *ia* = condition
[25]*an* = without + *ur* = urine + *ia* = condition

| TABLE 23.2 | Properties and Composition of Urine | |
|---|---|---|
| **Physical Properties** | | |
| Specific gravity | 1.001–1.028 | |
| Osmolarity | 50–1,200 mOsm/L | |
| pH | 6.0 (range 4.5–8.2) | |

| Solute | Concentration* | Output (g/day)** |
|---|---|---|
| *Inorganic ions* | | |
| Chloride | 533 mg/dL | 6.4 g/day |
| Sodium | 333 mg/dL | 4.0 g/day |
| Potassium | 166 mg/dL | 2.0 g/day |
| Phosphate | 83 mg/dL | 1 g/day |
| Ammonia | 60 mg/dL | 0.68 g/day |
| Calcium | 17 mg/dL | 0.2 g/day |
| Magnesium | 13 mg/dL | 0.16 g/day |
| *Nitrogenous Wastes* | | |
| Urea | 1.8 g/dL | 21 g/day |
| Creatinine | 150 mg/dL | 1.8 g/day |
| Uric acid | 40 mg/dL | 0.5 g/day |
| Urobilin | 125 µg/dL | 1.52 mg/day |
| Bilirubin | 20 µg/dL | 0.24 mg/day |
| *Other Organics* | | |
| Amino acids | 288 µg/dL | 3.5 mg/day |
| Ketones | 17 µg/dL | 0.21 mg/day |
| Carbohydrates | 9 µg/dL | 0.11 mg/day |
| Lipids | 1.6 µg/dL | 0.02 mg/day |

*Typical values for a reference man
**Assuming a urine output of 1.2 L/day

0 to 100 mL/day. Low output can result from kidney disease, dehydration, circulatory shock, prostate enlargement, and other causes. If urine output drops to less than 400 mL/day, the body cannot maintain a safe, low concentration of wastes in the blood plasma. The result is azotemia.

## Diabetes

**Diabetes**[26] is any metabolic disorder resulting in chronic polyuria. There are at least four forms of diabetes: *diabetes mellitus type I* and *type II, gestational diabetes,* and *diabetes insipidus.* In most cases, the polyuria results from a high concentration of glucose in the renal tubule. Glucose opposes the osmotic reabsorption of water, so more water is passed in the urine *(osmotic diuresis)* and a

person may become severely dehydrated. In diabetes mellitus and gestational diabetes, the high glucose concentration in the tubule is a result of hyperglycemia, a high concentration of glucose in the blood. About 1% to 3% of pregnant women experience gestational diabetes, in which pregnancy reduces the mother's insulin sensitivity, resulting in hyperglycemia and glycosuria. Diabetes insipidus results from ADH hyposecretion. Without ADH, the collecting duct does not reabsorb as much water as normal, so more water passes in the urine.

Diabetes mellitus and gestational diabetes are characterized by glycosuria. Before chemical tests for urine glucose were developed, physicians diagnosed diabetes mellitus[27] by tasting the patient's urine for sweetness. Tests for glycosuria are now as simple as dipping a chemical test strip into the urine specimen—an advance in medical technology for which urologists are no doubt grateful. In diabetes insipidus,[28] the urine contains no glucose and, by the old diagnostic method, does not taste sweet.

## Diuretics

**Diuretics** are chemicals that increase urine volume. They are used for treating hypertension and congestive heart failure because they reduce the body's fluid volume and blood pressure. Diuretics work by one of two mechanisms—increasing glomerular filtration or reducing tubular reabsorption. For example, caffeine, in the former category, dilates the afferent arteriole and increases GFR. Alcohol, in the latter category, inhibits ADH secretion. Also in the latter category are many osmotic diuretics, which reduce water reabsorption by increasing the osmolarity of the tubular fluid. Many diuretic drugs, such as furosemide (Lasix), produce osmotic diuresis by inhibiting sodium reabsorption.

## RENAL FUNCTION TESTS

There are several tests for diagnosing kidney diseases, evaluating their severity, and monitoring their progress. Here we examine two methods used to determine renal clearance and glomerular filtration rate.

### Renal Clearance

**Renal clearance** is the volume of blood plasma from which a particular waste is completely removed in 1 minute. It represents the net effect of three processes:

Glomerular filtration of the waste
+ Amount added by tubular secretion
− Amount removed by tubular reabsorption
———————————————————
Renal clearance

---

[26]*diabetes* = passing through

[27]*melli* = honey, sweet
[28]*insipid* = tasteless

In principle, we could determine renal clearance by sampling blood entering and leaving the kidney and comparing their waste concentrations. In practice, it is not practical to draw blood samples from the renal vessels, but clearance can be assessed indirectly by collecting samples of blood and urine, measuring the waste concentration in each, and measuring the rate of urine output.

Suppose the following values were obtained for urea:

U (urea concentration in urine)　= 6.0 mg/mL

V (rate of urine output)　　　　 = 2 mL/min

P (urea concentration in plasma) = 0.2 mg/mL

Renal clearance (C) is

$$C = UV/P$$

$$= (6.0 \text{ mg/mL})(2 \text{ mL/min})/0.2 \text{ mg/mL}$$

$$= 60 \text{ mL/min}$$

This means the equivalent of 60 mL of blood plasma is completely cleared of urea per minute. If this person has a normal GFR of 125 mL/min, then the kidneys have cleared urea from only 60/125 = 48% of the glomerular filtrate. This is a normal rate of urea clearance, however, and is sufficient to maintain safe levels of urea in the blood.

### Think About It

*What would you expect the value of renal clearance of glucose to be in a healthy individual? Why?*

## Glomerular Filtration Rate

Assessment of kidney disease often calls for a measurement of GFR. We cannot determine GFR from urea excretion for two reasons: (1) some of the urea in the urine is secreted by the renal tubule, not filtered by the glomerulus, and (2) much of the urea filtered by the glomerulus is reabsorbed by the tubule. To measure GFR ideally requires a substance that is not secreted or reabsorbed at all, so that all of it in the urine gets there by glomerular filtration.

There doesn't appear to be a single urine solute produced by the body that is not secreted or reabsorbed to some degree. However, several plants, including garlic and artichoke, produce a polysaccharide called inulin that is useful for GFR measurement. All inulin filtered by the glomerulus remains in the renal tubule and appears in the urine; none is reabsorbed, nor does the tubule secrete it. GFR can be measured by injecting inulin and subsequently measuring the rate of urine output and the concentrations of inulin in the blood and urine.

For inulin, GFR is equal to the renal clearance. Suppose, for example, that a patient's plasma concentration of inulin is P = 0.5 mg/mL, the urine concentration is U = 30 mg/mL, and urine output is V = 2 mL/min. This person has a normal GFR:

$$GFR = UV/P$$

$$= (30 \text{ mg/mL})(2 \text{ mL/min})/0.5 \text{ mg/mL}$$

$$= 120 \text{ mL/min}$$

In clinical practice, GFR is more often estimated from creatinine excretion. This has a small but acceptable error of measurement, and is an easier procedure than injecting inulin and drawing blood to measure its blood concentration.

A solute that is reabsorbed by the renal tubules will have a renal clearance *less* than the GFR (provided its tubular secretion is less than its rate of reabsorption). This is why the renal clearance of urea is about 60 mL/min. A solute that is secreted by the renal tubules will have a renal clearance *greater* than the GFR (provided its reabsorption does not exceed its secretion). Creatinine, for example, has a renal clearance of 140 mL/min.

### Before You Go On

*Answer the following questions to test your understanding of the preceding section:*

17. Define *oliguria* and *polyuria. Which of these is characteristic of diabetes?*

18. *Identify a cause of glycosuria other than diabetes mellitus.*

19. *How is the diuresis produced by furosemide like the diuresis produced by diabetes mellitus? How are they different?*

20. *Explain why GFR could not be determined by measuring the amount of NaCl in the urine.*

# Urine Storage and Elimination

### Objectives

When you have completed this section, you should be able to

- describe the functional anatomy of the ureters, urinary bladder, and male and female urethra; and
- explain how the nervous system and urethral sphincters control the voiding of urine.

Urine is produced continually, but fortunately it does not drain continually from the body. Urination is episodic—occurring when we allow it. This is made possible by an apparatus for storing urine and by neural controls for its timely release.

## THE URETERS

The renal pelvis funnels urine into the ureter, a retroperitoneal, muscular tube that extends to the urinary bladder. The ureter is about 25 cm long and reaches a maximum diameter of about 1.7 cm near the bladder. The ureters pass dorsal to the bladder and enter it from below, passing obliquely through its muscular wall and opening onto its floor. A small flap of mucosa acts as a valve at the opening of each ureter into the bladder.

The ureter has three layers: an adventitia, muscularis, and mucosa. The adventitia is a connective tissue layer that binds it to the surrounding tissues. The muscularis consists of two layers of smooth muscle over most of its length, but a third layer appears in the lower ureter. When urine enters the ureter and stretches it, the muscularis contracts and initiates a peristaltic wave that milks the urine from the renal pelvis down to the bladder. These contractions occur every few seconds to few minutes, proportional to the rate at which urine enters the ureter. The mucosa has a transitional epithelium that begins in the minor calyces of the kidney and extends from there through the bladder. The lumen of the ureter is very narrow and is easily obstructed or injured by kidney stones (see Insight 23.2).

## THE URINARY BLADDER

The urinary bladder (fig. 23.21) is a muscular sac on the floor of the pelvic cavity, inferior to the peritoneum and posterior to the pubic symphysis. It is covered by parietal peritoneum on its flattened superior surface and by a fibrous adventitia elsewhere. Its muscularis, called the **detrusor**[29] (deh-TROO-zur) **muscle**, consists of three layers of smooth muscle. The mucosa has a transitional epithelium, and in the relaxed bladder it has conspicuous

---

### INSIGHT 23.2   Clinical Application

#### Kidney Stones

A *renal calculus*[30] (kidney stone) is a hard granule of calcium, phosphate, uric acid, and protein. Renal calculi form in the renal pelvis and are usually small enough to pass unnoticed in the urine flow. Some, however, grow as large as several centimeters and block the renal pelvis or ureter, which can lead to the destruction of nephrons as pressure builds in the kidney. A large, jagged calculus passing down the ureter stimulates strong contractions that can be excruciatingly painful. It can also damage the ureter and cause hematuria. Causes of renal calculi include hypercalcemia, dehydration, pH imbalances, frequent urinary tract infections, or an enlarged prostate gland causing urine retention. Calculi are sometimes treated with stone-dissolving drugs, but often they require surgical removal. A nonsurgical technique called *lithotripsy*[31] uses ultrasound to pulverize the calculi into fine granules easily passed in the urine.

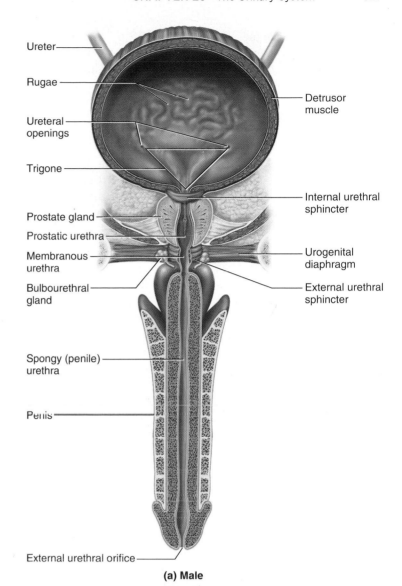

(a) Male

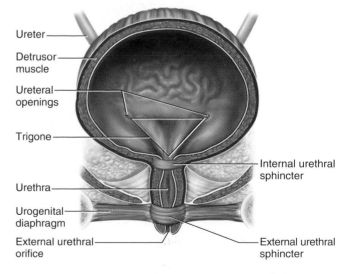

(b) Female

**FIGURE 23.21  The Urinary Bladder and Urethra.** Frontal sections.

▶ *Why are women more susceptible than men to bladder infections?*

---

[29]*de* = down + *trus* = push
[30]*calc* = calcium, stone + *ul* = little
[31]*litho* = stone + *tripsy* = crushing

wrinkles called **rugae**[32] (ROO-gee). The openings of the two ureters and the urethra mark a smooth-surfaced triangular area called the **trigone**[33] on the bladder floor. This is a common site of bladder infection (see Insight 23.3). For photographs of the relationship of the bladder and urethra to other pelvic organs in both sexes, see figure A.22 (p. 49).

The bladder is highly distensible. As it fills, it expands superiorly, the rugae flatten, and the epithelium thins from five or six cell layers to only two or three. A moderately full bladder contains about 500 mL of urine and extends about 12.5 cm from top to bottom. The maximum capacity is 700 to 800 mL.

## THE URETHRA

The urethra conveys urine out of the body. In the female, it is a tube 3 to 4 cm long bound to the anterior wall of the vagina by fibrous connective tissue. Its opening, the **external urethral orifice,** lies between the vaginal orifice and clitoris. The male urethra is about 18 cm long and has three regions: (1) The **prostatic urethra** begins at the urinary bladder and passes for about 2.5 cm through the prostate gland. During orgasm, it receives semen from the reproductive glands. (2) The **membranous urethra** is a short (0.5 cm), thin-walled portion where the urethra passes through the muscular floor of the pelvic cavity. (3) The **spongy (penile) urethra** is about 15 cm long and passes through the penis to the external urethral orifice. It is named for the *corpus spongiosum* of the penis, through which it passes. The male urethra assumes an **S**-shape: it passes downward from the bladder, turns anteriorly as it

enters the root of the penis, and then turns about 90° downward again as it enters the external, pendant part of the penis. The mucosa has a transitional epithelium near the bladder, a pseudostratified epithelium for most of its length, and finally a stratified squamous epithelium near the external urethral orifice. There are mucous **urethral glands** in its wall.

In both sexes, the detrusor muscle is thickened near the urethra to form an **internal urethral sphincter,** which compresses the urethra and retains urine in the bladder. Since this sphincter is composed of smooth muscle, it is under involuntary control. Where the urethra passes through the pelvic floor, it is encircled by an **external urethral sphincter** of skeletal muscle, which provides voluntary control over the voiding of urine.

## VOIDING URINE

Between acts of urination, when the bladder is filling, it is important that the detrusor muscle relax and the urethral sphincters remain tightly closed. This is ensured by sympathetic pathways that originate in the upper lumbar spinal cord. Postganglionic fibers travel through the hypogastric nerve to the detrusor muscle and internal urethral sphincter. Owing to different types of adrenergic receptors on these two muscles, these fibers *relax* the detrusor and *excite* the internal urethral sphincter. The external urethral sphincter is also held closed, but since this is skeletal muscle, it receives somatic innervation. Somatic motor fibers course from the upper sacral spinal cord via the pudendal nerve to the external sphincter.

The act of urinating, also called **micturition**[36] (MIC-too-RISH-un), is controlled partly by a spinal **micturition reflex.** The process is as follows (fig. 23.22):

Filling of the bladder to about 200 mL or more excites stretch receptors in the bladder wall. They transmit signals by way of the pelvic nerves to the sacral spinal cord (segments S2–S3 in some people, S3–S4 in others). These signals ascend the spinal cord to two destinations. Some terminate at inhibitory synapses on the sympathetic neurons that suppress urination; thus they incapacitate the sympathetic division from preventing urination. Others ascend further to a nucleus in the pons called the **micturition center.** The micturition center integrates information about the filling of the bladder with information from other brain centers such as the amygdala and cerebrum. Thus, urination can be prompted by fear or inhibited by knowledge that the circumstances are inappropriate for urination.

Fibers from the micturition center descend the spinal cord through the reticulospinal tracts. Some of these fibers end at inhibitory synapses on the sympathetic neurons,

---

### INSIGHT 23.3    Clinical Application

#### Urinary Tract Infections

Infection of the urinary bladder is called *cystitis.*[34] It is especially common in females because bacteria such as *Escherichia coli* can travel easily from the perineum up the short urethra. Because of this risk, young girls should be taught never to wipe the anus in a forward direction. If cystitis is untreated, bacteria can spread up the ureters and cause *pyelitis,*[35] infection of the renal pelvis. If it reaches the renal cortex and nephrons, it is called *pyelonephritis.* Kidney infections can also result from invasion by blood-borne bacteria. Urine stagnation due to renal calculi or prostate enlargement increases the risk of infection.

---

[32]*ruga* = fold, wrinkle
[33]*tri* = three + *gon* = angle
[34]*cyst* = bladder + *itis* = inflammation
[35]*pyel* = pelvis + *itis* = inflammation

[36]*mictur* = to urinate

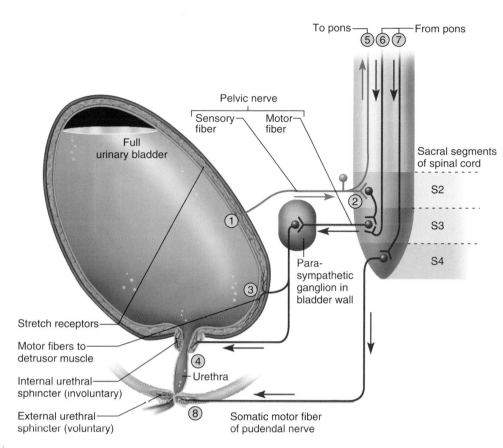

To pons — From pons

⑤ ⑥ ⑦

Pelvic nerve

Sensory fiber — Motor fiber

Full urinary bladder

Sacral segments of spinal cord

S2

②

S3

①

S4

Para-sympathetic ganglion in bladder wall

③

Stretch receptors

Motor fibers to detrusor muscle

④

Internal urethral sphincter (involuntary)

Urethra

External urethral sphincter (voluntary)

⑧

Somatic motor fiber of pudendal nerve

① Stretch receptors detect filling of bladder, transmit afferent signals to spinal cord.

② Signals return to bladder from spinal cord segments S2 and S3 via parasympathetic fibers in pelvic nerve.

③ Efferent signals excite detrusor muscle.

④ Efferent signals relax internal urethral sphincter. Urine is involuntarily voided if not inhibited by brain.

⑤ For voluntary control, micturition center in pons receives signals from stretch receptors.

⑥ If it is timely to urinate, pons returns signals to spinal interneurons that excite detrusor and relax internal urethral sphincter. Urine is voided.

⑦ If it is untimely to urinate, signals from pons excite spinal interneurons that keep external urethral sphincter contracted. Urine is retained in bladder.

⑧ If it is timely to urinate, signals from pons cease and external urethral sphincter relaxes. Urine is voided.

**FIGURE 23.22** Neural Control of Micturition.

thus reinforcing the effect noted at step 2. Others descend further to the lumbar spinal cord, where they excite parasympathetic neurons. Parasympathetic nerve fibers travel by way of the pelvic nerves to the detrusor (via a parasympathetic ganglion in or near the bladder). These are excitatory and stimulate the bladder to contract. The initial contraction further excites the stretch receptors that started this process, so a positive feedback loop is established that intensifies the bladder contraction as urination proceeds.

The final obstacle to urination, the external urethral sphincter, also must relax. Descending nerve fibers from the cerebral cortex travel the corticospinal tracts to the sacral spinal cord and inhibit the somatic motor neurons that supply that sphincter. It is this voluntary component of micturition that gives a person conscious control of when to urinate and the ability to stop urination in midstream. This voluntary component of micturition develops as the nervous system matures in early childhood. In infants, urination is controlled by the spinal micturition reflex alone. Males expel the last few milliliters of urine by voluntarily contracting the bulbocavernosus muscle that ensheaths the root of the penis. This helps to reduce the retention of urine in the longer male urethra.

If the urge to urinate arises and we must suppress it, the stretch receptors fatigue and stop firing. As bladder tension increases, however, the signals return with increasing frequency and persistence. Conversely, there are times when the bladder is not full enough to trigger the micturition reflex, but we wish to "go" anyway

## INSIGHT 23.4 Clinical Application

### Urination and Spinal Cord Injuries

Knowledge of the neural control of micturition is particularly important for understanding and treating persons with spinal cord injuries. Transection of the spinal cord, as in many cervical fractures, disconnects the *supraspinal* control centers (cerebrum and pons) from the spinal cord circuits that control urination. During the period of spinal shock (see p. 508), a person is generally incontinent—lacking any control over urination. Bladder control returns as the spinal cord recovers, but is limited to the involuntary micturition reflex. The bladder often cannot empty completely, and there is consequently an increased incidence of cystitis.

because of a long drive or lecture coming up. In this case, we use the Valsalva maneuver (p. 867) to compress the bladder and excite the stretch receptors early, thereby getting the reflex started. The Valsalva maneuver also aids in emptying the bladder.

The effects of aging on the urinary system are discussed on page 1131. Some disorders of this system are briefly described in table 23.3.

## Before You Go On

*Answer the following questions to test your understanding of the preceding section:*

21. *Describe the location and function of the detrusor muscle.*

22. *Compare and contrast the functions of the internal and external urethral sphincters.*

23. *In males, the sympathetic nervous system triggers ejaculation and at the same time, stimulates constriction of the internal urethral sphincter. What purpose is served by the latter action?*

| TABLE 23.3 | Some Disorders of the Urinary System |
|---|---|
| Acute glomerulonephritis | An autoimmune inflammation of the glomeruli, often following a streptococcus infection. Results in destruction of glomeruli leading to hematuria, proteinuria, edema, reduced glomerular filtration, and hypertension. Can progress to chronic glomerulonephritis and renal failure, but most individuals recover from acute glomerulonephritis without lasting effect. |
| Acute renal failure | An abrupt decline in renal function, often due to traumatic damage to the nephrons or a loss of blood flow stemming from hemorrhage or thrombosis. |
| Chronic renal failure | Long-term, progressive, irreversible loss of nephrons; see Insight 23.5 for a variety of causes. Requires a kidney transplant or hemodialysis. |
| Hydronephrosis[37] | Increase in fluid pressure in the renal pelvis and calyces owing to obstruction of the ureter by kidney stones, nephroptosis, or other causes. Can progress to complete cessation of glomerular filtration and atrophy of nephrons. |
| Nephroptosis[38] (NEFF-rop-TOE-sis) | Slippage of the kidney to an abnormally low position (floating kidney). Occurs in people with too little body fat to hold the kidney in place and in people who subject the kidneys to prolonged vibration, such as truck drivers, equestrians, and motorcyclists. Can twist or kink the ureter, which causes pain, obstructs urine flow, and potentially leads to hydronephrosis. |
| Nephrotic syndrome | Excretion of large amounts of protein in the urine ([greater than or equal to] 3.5 g/day) due to glomerular injury. Can result from trauma, drugs, infections, cancer, diabetes mellitus, lupus erythematosus, and other diseases. Loss of plasma protein leads to edema, ascites, hypotension, and susceptibility to infection (because of immunoglobulin loss). |
| Urinary incontinence | Inability to hold the urine; involuntary leakage from the bladder. Can result from incompetence of the urinary sphincters; bladder irritation; pressure on the bladder in pregnancy; an obstructed urinary outlet so that the bladder is constantly full and dribbles urine *(overflow incontinence);* uncontrollable urination due to brief surges in bladder pressure, as in laughing or coughing *(stress incontinence);* and neurological disorders such as spinal cord injuries. |

*Disorders described elsewhere*

| | | | |
|---|---|---|---|
| Azotemia p. 897 | Nephrosclerosis p. 907 | Proteinuria p. 906 | Uremia p. 897 |
| Hematuria p. 906 | Oliguria p. 918 | Pyuria p. 918 | Urinary tract infection p. 922 |
| Kidney stones p. 921 | | | |

---

[37]*hydro* = water + *nephr* = kidney + *osis* = medical condition
[38]*nephro* = kidney + *ptosis* = sagging, falling

## Renal Insufficiency and Hemodialysis

*Renal insufficiency* is a state in which the kidneys cannot maintain homeostasis due to extensive destruction of their nephrons. Some causes of nephron destruction include:

- Hypertension
- Chronic or repetitive kidney infections.
- Trauma from such causes as blows to the lower back or continual vibration from machinery.
- Prolonged ischemia and hypoxia, as in long-distance runners and swimmers.
- Poisoning by heavy metals such as mercury and lead and solvents such as carbon tetrachloride, acetone, and paint thinners. These are absorbed into the blood from inhaled fumes or by skin contact and then filtered by the glomeruli. They kill renal tubule cells.
- Blockage of renal tubules with proteins small enough to be filtered by the glomerulus—for example, myoglobin released by skeletal muscle damage and hemoglobin released by a transfusion reaction.
- Atherosclerosis, which reduces blood flow to the kidney.
- Glomerulonephritis, an autoimmune disease of the glomerular capillaries.

Nephrons can regenerate and restore kidney function after short-term injuries. Even when some of the nephrons are irreversibly destroyed, others hypertrophy and compensate for their lost function. Indeed, a person can survive on as little as one-third of one kidney. When 75% of the nephrons are lost, however, urine output may be as low as 30 mL/hr compared with the normal rate of 50 to 60 mL/hr. This is insufficient to maintain homeostasis and is accompanied by azotemia and acidosis. Uremia develops when there is 90% loss of renal function. Renal insufficiency also tends to cause anemia because the diseased kidneys produce too little erythropoietin (EPO), the hormone that stimulates red blood cell formation.

*Hemodialysis* is a procedure for artificially clearing wastes from the blood when the kidneys are not adequately doing so (fig. 23.23). Blood is pumped from the radial artery to a *dialysis machine* (artificial kidney) and returned to the patient by way of a vein. In the dialysis machine, the blood flows through a semipermeable cellophane tube surrounded by dialysis fluid. Urea, potassium, and other solutes that are more concentrated in the blood than in the dialysis fluid diffuse through the membrane into the fluid, which is discarded. Glucose, electrolytes, and drugs can be administered by adding them to the dialysis fluid so they will diffuse through the membrane into the blood. People with renal insufficiency also accumulate substantial amounts of excess body water between treatments, and dialysis serves also to remove it. Patients are typically given erythropoietin (EPO) to compensate for the lack of EPO from the failing kidneys.

Hemodialysis patients typically have three sessions per week for 4 to 8 hours per session. In addition to inconvenience, hemodialysis carries risks of infection and thrombosis. Blood tends to clot when exposed to foreign surfaces, so an anticoagulant such as heparin is added during dialysis. Unfortunately, this inhibits clotting in the patient's body as well, and dialysis patients sometimes suffer internal bleeding.

A procedure called *continuous ambulatory peritoneal dialysis (CAPD)* is more convenient. It can be carried out at home by the patient, who is provided with plastic bags of dialysis fluid. Fluid is introduced into the abdominal cavity through an indwelling catheter. Here, the peritoneum provides over 2 m² of blood-rich semipermeable membrane. The fluid is left in the body cavity for 15 to 60 minutes to allow the blood to equilibrate with it; then it is drained, discarded, and replaced with fresh dialysis fluid. The patient is not limited by a stationary dialysis machine and can go about most normal activities. CAPD is less expensive and promotes better morale than conventional hemodialysis, but it is less efficient in removing wastes and it is more often complicated by infection.

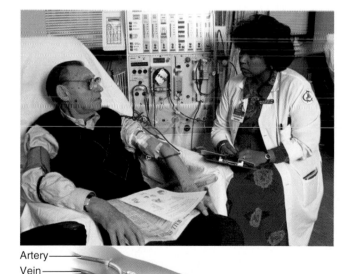

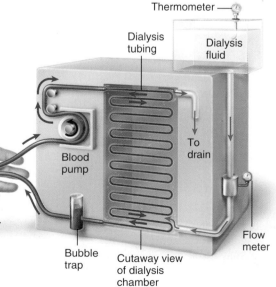

**FIGURE 23.23    Hemodialysis.**    Blood is pumped into a dialysis chamber, where it flows through a selectively permeable cellophane membrane surrounded by dialysis fluid. Blood leaving the chamber passes through a bubble trap to remove air before it is returned to the patient's body. The fluid picks up excess water and metabolic wastes from the patient's blood and may contain medications that diffuse into the blood.

# CONNECTIVE ISSUES

## Interactions Between the URINARY SYSTEM and Other Organ Systems

| ■ indicates ways in which this system affects other systems | ■ indicates ways in which other systems affect this system |
| --- | --- |

## INTEGUMENTARY SYSTEM

Renal control of fluid balance essential for sweat secretion

A barrier to fluid loss; profuse sweating can lead to oliguria; skin and kidneys help synthesize calcitriol

## SKELETAL SYSTEM

Renal control of calcium and phosphate balance and role in calcitriol synthesis are essential for bone deposition

Lower ribs and pelvis protect some urinary system organs

## MUSCULAR SYSTEM

Renal control of $Na^+$, $K^+$, and $Ca^{2+}$ balance important for muscle contraction

Some skeletal muscles aid or regulate micturition; muscles of pelvic floor support bladder

## NERVOUS SYSTEM

Sensitive to fluid, electrolyte, and acid–base imbalances that may result from renal dysfunction

Regulates GFR and micturition

## ENDOCRINE SYSTEM

Renin from kidneys promotes angiotensin and aldosterone secretion; kidneys produce erythropoietin

Regulates renal function through angiotensin II, aldosterone, atrial natriuretic factor, and antidiuretic hormone

## CIRCULATORY SYSTEM

Kidneys regulate blood composition, volume, pressure, and hematocrit; cardiac rhythm is sensitive to electrolyte imbalances that may result from renal dysfunction

### NEARLY ALL SYSTEMS

The urinary system serves all other systems by eliminating metabolic wastes and maintaining fluid, electrolyte, and acid–base balance

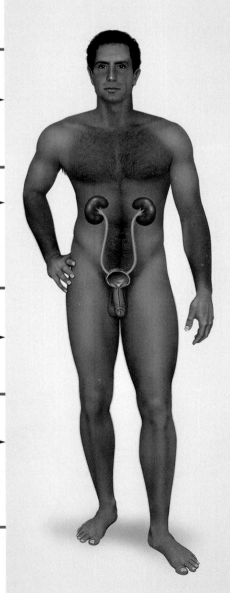

## CIRCULATORY SYSTEM (*cont.*)

Perfuses kidneys so wastes can be filtered; blood pressure influences GFR; blood reabsorbs water and solutes from renal tubules

## LYMPHATIC/IMMUNE SYSTEMS

Acidity of urine provides nonspecific defense against infection

Return of fluid to bloodstream maintains blood pressure and fluid balance essential for renal function; protect kidneys from infection

## RESPIRATORY SYSTEM

Acid secretion by kidneys affects blood pH and respiratory rhythm

Provides $O_2$ to meet metabolic demand of kidneys; pulmonary dysfunctions may require compensation by kidneys to maintain acid–base balance; inhaled fumes can damage kidneys

## DIGESTIVE SYSTEM

Kidneys excrete toxins absorbed by digestive tract; calcitriol from kidneys regulates $Ca^{2+}$ absorption by small intestine

Liver synthesizes urea, excreted by kidneys; urea contributes to osmotic gradient of renal medulla; liver and kidneys help to synthesize calcitriol

## REPRODUCTIVE SYSTEM

Male urethra serves as common passage for urine and sperm; urinary system of a pregnant woman eliminates fetal wastes

Enlarged prostate can cause urine retention and kidney damage in males; pregnant uterus compresses bladder and reduces its capacity in females

## CHAPTER REVIEW

# Review of Key Concepts

**Functions of the Urinary System (p. 896)**

1. The urinary system consists of two kidneys, two ureters, the urinary bladder, and the urethra.

2. The kidneys filter blood plasma, separate wastes from useful chemicals, regulate blood volume and pressure, secrete renin and erythropoietin, regulate blood pH, synthesize calcitriol, detoxify free radicals and drugs, and generate glucose in times of starvation.

3. Metabolic wastes are wastes produced by the body, such as $CO_2$ and *nitrogenous wastes*. The main human nitrogenous wastes are *urea, uric acid,* and *creatinine.*

4. The level of nitrogenous wastes in the blood is often expressed as *blood urea nitrogen (BUN).* An elevated BUN is called *azotemia,* and may progress to a serious syndrome called *uremia.*

5. *Excretion* is the process of separating wastes from the body fluids and eliminating them from the body. It is carried out by the respiratory, integumentary, digestive, and urinary systems.

**Anatomy of the Kidney (p. 898)**

1. The kidneys are located retroperitoneally against the superior dorsal abdominal wall. Each is provided with a renal artery and renal vein. The adrenal glands adhere to the superior to medial aspects of the kidneys.

2. The kidney has a slit called the *hilum* on its concave side, where it receives renal nerves, blood and lymphatic vessels, and the ureter.

3. From superficial to deep, the kidney is enclosed by the renal fascia, adipose capsule, and renal capsule.

4. The renal parenchyma is a C-shaped tissue enclosing a space called the renal sinus. The parenchyma is divided into an outer *renal cortex* and inner *renal medulla.* The medulla consists of 6 to 10 *renal pyramids.*

5. The apex, or papilla, of each pyramid projects into a receptacle called a minor calyx, which collects the urine from that pyramid. Minor calyces converge to form major calyces, and these converge on the renal pelvis, where the ureter arises.

6. Each kidney contains about 1.2 million functional units called *nephrons.*

7. The renal artery branches and gives rise to *segmental arteries, interlobar arteries, arcuate arteries,* and then *interlobular arteries,* which penetrate into the cortex. For each nephron, an *afferent arteriole* arises from the interlobular artery and supplies the capillaries of the *glomerulus.* An *efferent arteriole* leaves the glomerulus and usually gives rise to a bed of *peritubular capillaries* around the renal tubules. Blood then flows through a series of *interlobular, arcuate,* and *interlobar veins,* before leaving the kidney by way of the *renal vein.*

8. The renal medulla is supplied by vessels called the *vasa recta,* which arise from the efferent arterioles of juxtamedullary nephrons and empty into arcuate and interlobular veins.

9. *Renal nerves* follow the renal artery and innervate the afferent and efferent arterioles. They provide sympathetic control over blood flow to the glomerulus, and thus regulate the rate of glomerular filtration and urine formation. They also stimulate renin secretion.

10. A nephron begins with a double-walled *glomerular capsule* enclosing the glomerulus; the glomerulus and capsule constitute the *renal corpuscle.* The capsule consists of a parietal layer with a simple squamous epithelium and a visceral layer composed of *podocytes.* Podocytes have numerous *pedicels* that wrap around the glomerular capillaries.

11. Filtrate collects in the *capsular space* between the capsule layers and then flows into the *renal tubule* leading away from the capsule.

12. The renal tubule consists of a highly coiled *proximal convoluted tubule (PCT),* a U-shaped *nephron loop,* a coiled *distal convoluted tubule (DCT),* and a *collecting duct.* The first three of these belong to a single nephron; the collecting duct receives fluid from many nephrons.

**Urine Formation I: Glomerular Filtration (p. 904)**

1. A nephron produces urine in three stages: *glomerular filtration, tubular reabsorption* and *secretion,* and *water conservation.*

2. The first step in urine production is to filter the blood plasma, which occurs at the glomerulus.

3. In passing from the blood capillaries into the capsular space, fluid must pass through the filtraton pores of the capillary endothelium, the basement membrane, and filtration slits of the podocytes. These barriers hold back blood cells and most protein, but allow water and small solutes to pass.

4. Glomerular filtration is driven mainly by the high blood pressure in the glomerular capillaries.

5. Glomerular filtration rate (GFR), an important measure of renal health, is typically about 125 mL/min in men and 105 mL/min in women.

6. Renal autoregulation is the ability of the kidneys to maintain a stable GFR without nervous or hormonal control. There are a myogenic mechanism and a tubuloglomerular feedback mechanism of renal autoregulation.

7. The sympathetic nervous system also regulates GFR by controlling vasomotion of the afferent arterioles.

8. GFR is also controlled by hormones. A drop in blood pressure causes the kidneys to secrete renin. Renin and angiotensin-converting enzyme convert a plasma protein, angiotensinogen, into angiotensin II.

9. Angiotensin II helps to raise blood pressure by constricting the blood

vessels, reducing GFR, stimulating the PCT to reabsorb NaCl and water, promoting secretion of antidiuretic hormone (ADH) and aldosterone, and stimulating the sense of thirst.

10. ADH promotes water retention by the kidneys. Aldosterone promotes sodium retention, which in turn leads to water retention.

## Urine Formation II: Tubular Reabsorption and Secretion (p. 909)

1. The GFR is far in excess of the rate of urine output. Ninety-eight to 99% of the filtrate is reabsorbed by the renal tubules and only 1% to 2% is excreted as urine.

2. About 65% of the glomerular filtrate is reabsorbed by the PCT.

3. PCT cells absorb $Na^+$ from the tubular fluid through the apical cell surface and pump it out the basolateral cell surfaces by active transport. The reabsorption of other solutes—water, $Cl^-$, $HCO_3^-$, $K^+$, $Mg^{2+}$, phosphate, glucose, amino acids, lactate, urea, and uric acid—is linked in various ways to $Na^+$ reabsorption.

4. The peritubular capillaries pick up the reabsorbed water by osmosis, and other solutes follow by *solvent drag.*

5. The *transport maximum ($T_m$)* is the fastest rate at which the PCT can reabsorb a given solute. If a solute such as glucose is filtered by the glomerulus faster than the PCT can reabsorb it, the excess will pass in the urine (as in diabetes mellitus).

6. The PCT also carries out *tubular secretion,* removing solutes from the blood and secreting them into the tubular fluid. Secreted solutes include urea, uric acid, bile salts, ammonia, catecholamines, prostaglandins, creatinine, $H^+$, $HCO_3^-$, and drugs such as aspirin and penicillin.

7. The nephron loop serves mainly to generate an osmotic gradient in the renal medulla, which is necessary for collecting duct function; but it also reabsorbs a significant amount of water, $Na^+$, $K^+$, and $Cl^-$.

8. The DCT reabsorbs salt and water, and is regulated by several hormones.

9. Aldosterone stimulates the DCT to reabsorb $Na^+$ and secrete $K^+$. Water and $Cl^-$ are reabsorbed with $Na^+$.

10. Atrial natriuretic peptide increases salt and water excretion by increasing GFR, antagonizing aldosterone and ADH, and inhibiting NaCl reabsorption by the collecting duct.

11. Parathyroid hormone acts on the nephron loop and DCT to promote $Ca^{2+}$ reabsorption, and acts on the PCT to promote phosphate excretion. Calcitonin and calcitriol have similar but weaker effects on the DCT.

## Urine Formation III: Water Conservation (p. 914)

1. The collecting duct (CD) reabsorbs varying amounts of water to leave the urine as dilute as 50 mOsm/L or as concentrated as 1,200 mOsm/L.

2. The CD is permeable to water but not to NaCl. As it passes down the increasingly salty renal medulla, it loses water to the tissue fluid and the urine in the duct becomes more concentrated.

3. The rate of water loss from the CD is controlled by antidiuretic hormone (ADH). ADH stimulates the installation of aquaporins in the CD cells, increasing permeability of the CD to water. At high ADH concentrations, the urine is scanty and highly concentrated; at low ADH concentrations, the urine is dilute.

4. The salinity gradient of the renal medulla, which is essential to the ability of the CD to concentrate the urine, is maintained by the countercurrent multiplier mechanism of the nephron loop.

5. The vasa recta supply a blood flow to the renal medulla and employ a countercurrent exchange system to prevent them from removing salt from the medulla.

## Urine and Renal Function Tests (p. 918)

1. Urine normally has a yellow color due to *urochromes* derived from hemoglobin breakdown products.

2. Urine normally has a specific gravity from 1.001 to 1.028, an osmolarity from 50 to 1,200 mOsm/L, and a pH from 4.5 to 8.2.

3. A foul odor to the urine is abnormal and may result from bacterial degradation, some foods, urinary tract infection, or metabolic diseases such as diabetes mellitus or phenylketonuria.

4. The most abundant solutes in urine are urea, NaCl, and KCl. Urine normally contains no glucose, hemoglobin, albumin, ketones, or bile pigments, but may do so in some diseases.

5. Most adults produce 1 to 2 L of urine per day. Abnormally low urine output is *anuria* or *oliguria;* abnormally high output is *polyuria.*

6. *Diabetes* is any chronic polyuria of metabolic origin. Forms of diabetes include diabetes mellitus types I and II, gestational diabetes, and diabetes insipidus.

7. *Diuretics* are chemicals that increase urine output by increasing GFR or reducing tubular reabsorption. Caffeine and alcohol are diuretics, as are certain drugs used to reduce blood pressure.

8. Renal function can be assessed by making clinical measurements of GFR or *renal clearance.* The latter is the amount of blood completely freed of a given solute in 1 minute.

## Urine Storage and Elimination (p. 920)

1. Peristalsis of the ureters causes urine to flow from the kidneys to the urinary bladder.

2. The urinary bladder has a smooth muscle layer called the *detrusor muscle* with a thickened ring, the *internal urethral sphincter,* around the origin of the urethra.

3. The urethra is 3 to 4 cm long in the female, but in the male it is 18 cm long and divided into *prostatic, membranous,* and *spongy* (penile) segments. An *external urethral sphincter* of skeletal muscle encircles the urethra in both sexes where it passes through the pelvic floor.

4. Emptying of the bladder is controlled in part by a spinal *micturition reflex* initiated by stretch receptors in the bladder wall. Parasympathetic nerve fibers relax the internal urethral sphincter and stimulate the detrusor muscle to contract.

5. Supraspinal centers also regulate micturition. The *micturition center* in the pons inhibits the sympathetic neurons that suppress urination and stimulates the parasympathetic neurons that trigger it. The cerebrum issues signals that relax the external urethral sphincter.

# Testing Your Recall

1. Micturition occurs when the ___ contracts.
   a. detrusor muscle
   b. internal urethral sphincter
   c. external urethral sphincter
   d. muscularis of the ureter
   e. all of the above

2. The compact ball of capillaries in a nephron is called
   a. the nephron loop.
   b. the peritubular plexus.
   c. the renal corpuscle.
   d. the glomerulus.
   e. the vasa recta.

3. Which of these is the most abundant nitrogenous waste in the blood?
   a. uric acid
   b. urea
   c. ammonia
   d. creatinine
   e. albumin

4. Which of these lies closest to the renal cortex?
   a. the parietal peritoneum
   b. the renal fascia
   c. the renal capsule
   d. the adipose capsule
   e. the renal pelvis

5. Most sodium is reabsorbed from the glomerular filtrate by
   a. the vasa recta.
   b. the proximal convoluted tubule.
   c. the distal convoluted tubule.
   d. the nephron loop.
   e. the collecting duct.

6. A glomerulus and glomerular capsule make up one
   a. renal capsule.
   b. renal corpuscle.
   c. kidney lobule.
   d. kidney lobe.
   e. nephron.

7. The kidney has more ___ than any of the other structures listed.
   a. arcuate arteries
   b. minor calyces
   c. medullary pyramids
   d. afferent arterioles
   e. collecting ducts

8. The renal clearance of ___ is normally zero.
   a. sodium
   b. potassium
   c. uric acid
   d. urea
   e. amino acids

9. Beavers have relatively little need to conserve water and could therefore be expected to have ___ than humans do.
   a. fewer nephrons
   b. longer nephron loops
   c. shorter nephron loops
   d. longer collecting ducts
   e. longer convoluted tubules

10. Increased ADH secretion should cause the urine to have
    a. a higher specific gravity.
    b. a lighter color.
    c. a higher pH.
    d. a lower urea concentration.
    e. a lower potassium concentration.

11. The _____ reflex is an autonomic reflex activated by pressure in the urinary bladder.

12. _____ is the ability of a nephron to adjust its GFR independently of external nervous or hormonal influences.

13. The two ureters and the urethra form the boundaries of a smooth area called the _____ on the floor of the urinary bladder.

14. The _____ is a group of epithelial cells of the distal convoluted tubule that monitors the flow or composition of the tubular fluid.

15. To enter the capsular space, filtrate must pass between pedicels of the _____, cells that form the visceral layer of the glomerular capsule.

16. Glycosuria occurs if the rate of glomerular filtration of glucose exceeds the _____ of the proximal convoluted tubule.

17. _____ is a hormone that regulates the amount of water reabsorbed by the collecting duct.

18. The _____ sphincter is under involuntary control and relaxes during the micturition reflex.

19. Very little _____ is found in the glomerular filtrate because it is negatively charged and is repelled by the basement membrane of the glomerulus.

20. Blood flows through the _____ arteries just before entering the interlobular arteries.

*Answers in Appendix B*

# True or False

*Determine which five of the following statements are false, and briefly explain why.*

1. The proximal convoluted tubule is not subject to hormonal influence.

2. Sodium is the most abundant solute in the urine.

3. The kidney has more distal convoluted tubules than collecting ducts.

4. Tight junctions prevent material from leaking between the epithelial cells of the renal tubule.

5. All forms of diabetes are characterized by glucose in the urine.

6. If all other conditions remain the same, constriction of the afferent arteriole reduces the glomerular filtration rate.

7. Angiotensin II reduces urine output.

8. The minimum osmolarity of urine is 300 mOsm/L, equal to the osmolarity of the blood.

9. A sodium deficiency (hyponatremia) could cause glycosuria.

10. Micturition depends on contraction of the detrusor muscle.

*Answers in Appendix B*

# Testing Your Comprehension

1. How would glomerular filtration rate be affected by kwashiorkor (see p. 684)?

2. A patient produces 55 mL of urine per hour. Urea concentration is 0.25 mg/mL in her blood plasma and 8.6 mg/mL in her urine. (a) What is her rate of renal clearance for urea? (b) About 95% of adults excrete urea at a rate of 12.6 to 28.6 g/day. Is this patient above, within, or below this range? Show how you calculated your answers.

3. A patient with poor renal perfusion is treated with an ACE inhibitor and goes into renal failure. Explain the reason for the renal failure.

4. Drugs called *renin inhibitors* are used to treat hypertension. Explain how they would have this effect.

5. Discuss how the unity of form and function is exemplified by differences between the thin and thick segments of the nephron loop, between the proximal and distal convoluted tubules, and between the afferent and efferent arterioles.

*Answers at www.mhhe.com/saladin4*

# www.mhhe.com/saladin4

*The textbook website provides a wealth of interactive study materials fully organized and integrated by chapter. You will find practice quizzes, labeling exercises, and much more that will complement your learning and understanding of anatomy and physiology. The website also includes tools designed to enhance your* **Anatomy & Physiology** | **REVEALED** *experience.*

*Two glomeruli (green) and renal tubules (violet) of the kidney*

# WATER, ELECTROLYTE, AND ACID-BASE BALANCE

## CHAPTER OUTLINE

## Brushing Up

To understand this chapter, it is important that you understand or brush up on the following concepts:

• Electrolytes and milliequivalents/liter (p. 65)

• Acids, bases, and the pH scale (p. 66)

• Osmolarity (p. 105)

• Role of electrolytes in plasma membrane potentials (p. 455)

• Depolarization and hyperpolarization of plasma membranes (fig. 12.24, p. 468)

• The hypothalamus and posterior pituitary (p. 642)

• Influence of $CO_2$ and pH on pulmonary ventilation (pp. 884–885)

• Structure and physiology of the nephron (pp. 901–913)

Cellular function requires a fluid medium with a carefully controlled composition. If the quantity, osmolarity, electrolyte concentration, or pH of this medium is altered, life-threatening disorders of cellular function may result. Consequently, the body has several mechanisms for keeping these variables within narrow limits and maintaining three types of homeostatic balance:

1. *water balance,* in which average daily water intake and loss are equal;

2. *electrolyte balance,* in which the amount of electrolytes absorbed by the small intestine balance the amount lost from the body, chiefly through the urine; and

3. *acid–base balance,* in which the body rids itself of acid (hydrogen ions) at a rate that balances its metabolic production, thus maintaining a stable pH.

These balances are maintained by the collective action of the urinary, respiratory, digestive, integumentary, endocrine, nervous, cardiovascular, and lymphatic systems. This chapter describes the homeostatic regulation of water, electrolyte, and acid–base balance and shows the close relationship of these variables to each other.

# Water Balance

### Objectives

When you have completed this section, you should be able to

- name the major fluid compartments and explain how water moves from one to another;

- list the body's sources of water and routes of water loss;

- describe the mechanisms of regulating water intake and output; and

- describe some conditions in which the body has a deficiency or excess of water or an improper distribution of water among the fluid compartments.

We enter the world in a rather soggy condition, having swallowed, excreted, and floated in amniotic fluid for months. At birth, a baby's weight is as much as 75% water; infants normally lose a little weight in the first day or two as they excrete the excess. Young adult men average 55% to 60% water; women average slightly less because they have more adipose tissue, which is nearly free of water. Obese and elderly people are as little as 45%

water by weight. The **total body water (TBW)** content of a 70 kg (150 lb) young male is about 40 L.

## FLUID COMPARTMENTS

Body water is distributed among certain **fluid compartments,** areas separated by selectively permeable membranes and differing from each other in chemical composition. The major fluid compartments are:

65% *intracellular fluid (ICF)* and

35% *extracellular fluid (ECF),* subdivided into

25% *tissue (interstitial) fluid,*

8% *blood plasma* and *lymph,* and

2% *transcellular fluid,* a catch-all category for cerebrospinal, synovial, peritoneal, pleural, and pericardial fluids; vitreous and aqueous humors of the eye; bile; and fluid in the digestive, urinary, and respiratory tracts.

Fluid is continually exchanged between compartments by way of capillary walls and plasma membranes (fig. 24.1). Water moves by osmosis from the digestive tract to the bloodstream and by capillary filtration from the blood to the tissue fluid. From the tissue fluid, it may be reabsorbed by the capillaries, osmotically absorbed into cells, or taken up by the lymphatic system, which returns it to the bloodstream.

Because water moves so easily through plasma membranes, osmotic gradients between the ICF and ECF never last for very long. If a local imbalance arises, osmosis usually restores the balance within seconds so that intracellular and extracellular osmolarity are equal. If the osmolarity of the tissue fluid rises, water moves out of the cells; if it falls, water moves into the cells.

Osmosis from one fluid compartment to another is determined by the relative concentration of solutes in each compartment. The most abundant solute particles by far are the electrolytes—especially sodium salts in the ECF and potassium salts in the ICF. Electrolytes play the principal role in governing the body's water distribution and total water content; the subjects of water and electrolyte balance are therefore inseparable.

## WATER GAIN AND LOSS

A person is in a state of **water balance** when daily gains and losses are equal. We typically gain and lose about 2,500 mL/day (fig. 24.2). The gains come from two sources: **metabolic water** (about 200 mL/day), which is produced as a by-product of aerobic respiration and dehydration synthesis reactions, and **preformed water,** which is ingested in food (700 mL/day) and drink (1,600 mL/day).

The routes of water loss are more varied:

- 1,500 mL/day is excreted as urine.

- 200 mL/day is eliminated in the feces.

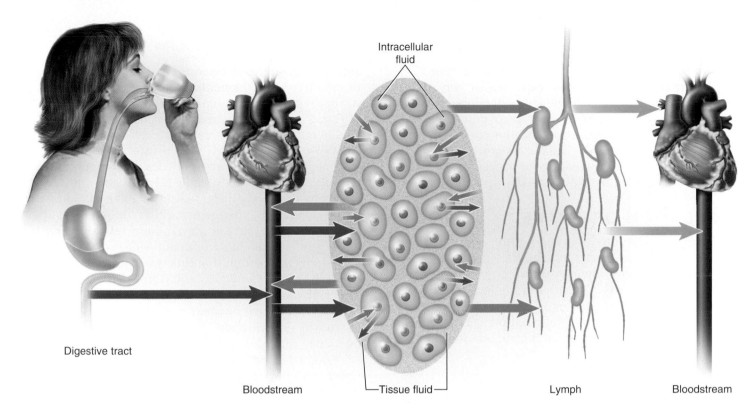

Intracellular fluid

Digestive tract

Bloodstream        Tissue fluid        Lymph        Bloodstream

**FIGURE 24.1** **The Movement of Water Between the Major Fluid Compartments.** Ingested water is absorbed by the bloodstream. There is a two-way exchange of water between the blood and tissue fluid and between the tissue and intracellular fluids. Excess tissue fluid is picked up by the lymphatic system, which returns it to the bloodstream.

▶ *In which of these places would fluid accumulate in edema?*

- 300 mL/day is lost in the expired breath. You can easily visualize this by breathing onto a cool surface such as a mirror.

- 100 mL/day of sweat is secreted by a resting adult at an ambient air temperature of 20°C (68°F).

- 400 mL/day is lost as **cutaneous transpiration,**[1] water that diffuses through the epidermis and evaporates. This is not the same as sweat; it is not a glandular secretion. A simple way to observe it is to cup the palm of your hand for a minute against a cool nonporous surface such as a laboratory benchtop or mirror. When you take your hand away, you will notice the water that transpired through the skin and condensed on that surface.

Water loss varies greatly with physical activity and environmental conditions. Respiratory loss increases in cold weather, for example, because cold air is drier and absorbs more body water from the respiratory tract. Hot, humid weather slightly reduces the respiratory loss but increases perspiration to as much as 1,200 mL/day. Prolonged, heavy work can raise the respiratory loss to 650 mL/day and perspiration to as much as 5 L/day, though it reduces urine output by nearly two-thirds.

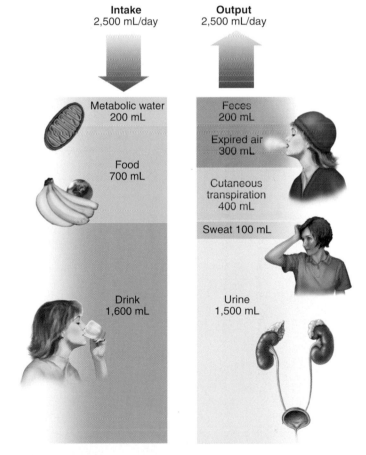

Intake
2,500 mL/day

Output
2,500 mL/day

Metabolic water
200 mL

Feces
200 mL

Expired air
300 mL

Food
700 mL

Cutaneous
transpiration
400 mL

Sweat 100 mL

Drink
1,600 mL

Urine
1,500 mL

**FIGURE 24.2** **Typical Water Intake and Output in a State of Fluid Balance.**

---

[1]*trans* = across, through + *spir* = to breathe

933

Output through the breath and cutaneous transpiration is called **insensible water loss** because we are not usually conscious of it. **Obligatory water loss** is output that is relatively unavoidable: expired air, cutaneous transpiration, sweat, fecal moisture, and the minimum urine output, about 400 mL/day, needed to prevent azotemia. Even dehydrated individuals cannot prevent such losses; thus they become further dehydrated.

## REGULATION OF INTAKE

Fluid intake is governed mainly by thirst, which is controlled by the mechanisms shown in figure 24.3. Dehydration reduces blood volume and pressure and raises blood osmolarity. The anterior nucleus of the hypothalamus has a neural pool called the **thirst center** that responds to multiple signs of dehydration: (1) angiotensin II, produced in response to falling blood pressure; (2)

antidiuretic hormone, released in response to rising blood osmolarity; and (3) signals from *osmoreceptors,* neurons in the hypothalamus that monitor the osmolarity of the ECF. A 2% to 3% increase in plasma osmolarity makes a person intensely thirsty, as does a 10% to 15% blood loss.

When we are thirsty, we salivate less. There are two reasons for this: (1) Sympathetic fibers from the thirst center inhibit the salivary glands. (2) Saliva is produced primarily by capillary filtration, but in a dehydrated person, this is opposed by the lower capillary blood pressure and higher osmolarity of the blood. Reduced salivation gives us a dry, sticky-feeling mouth and a desire to drink, but it is by no means certain that this is our primary motivation to drink. Some people do not secrete saliva, yet they do not drink more than normal individuals except when eating, when they need water to moisten the food. The same is true of experimental animals that have the salivary ducts tied off.

Long-term satiation of thirst depends on absorbing water from the small intestine and lowering the osmolarity of the blood. Reduced osmolarity stops the osmoreceptor response, promotes capillary filtration, and makes the saliva more abundant and watery. However, these changes require 30 minutes or longer to take effect, and it would be rather impractical if we had to drink that long while waiting to feel satisfied. Water intake would be grossly excessive. Fortunately, there are mechanisms that act more quickly to temporarily quench the thirst and allow time for the change in blood osmolarity to occur.

Experiments with rats and dogs have isolated the stimuli that quench the thirst. One of these is cooling and moistening the mouth; rats drink less if their water is cool than if it is warm, and simply moistening the mouth temporarily satisfies an animal even if the water is drained from its esophagus before it reaches the stomach. Distension of the stomach and small intestine is another inhibitor of thirst. If a dog is allowed to drink while the water is drained from its esophagus but its stomach is inflated with a balloon, its thirst is satisfied for a time. If the water is drained away but the stomach is not inflated, satiation does not last as long. Such fast-acting stimuli as coolness, moisture, and filling of the stomach stop an animal (and presumably a human) from drinking an excessive amount of liquid, but they are effective for only 30 to 45 minutes. If they are not soon followed by absorption of water into the bloodstream, the thirst soon returns. Only a drop in blood osmolarity produces a lasting effect.

## REGULATION OF OUTPUT

The only way to control water output significantly is through variations in urine volume. It must be realized, however, that the kidneys cannot completely prevent water loss, nor can they replace lost water or electrolytes. Therefore, they never restore fluid volume or osmolarity, but in dehydration they can support existing fluid levels

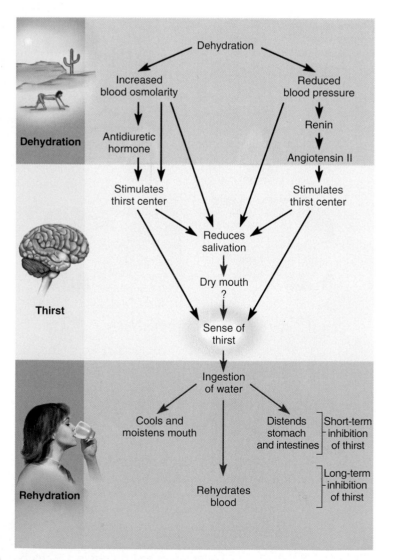

**FIGURE 24.3** Dehydration, Thirst, and Rehydration.

and slow down the rate of loss until water and electrolytes are ingested.

To understand the effect of the kidneys on water and electrolyte balance, it is also important to bear in mind that if a substance is reabsorbed by the kidneys, it is kept in the body and returned to the ECF, where it will affect fluid volume and composition. If a substance is filtered by the glomerulus or secreted by the renal tubules and not reabsorbed, then it is excreted in the urine and lost from the body fluids.

Changes in urine volume are usually linked to adjustments in sodium reabsorption. As sodium is reabsorbed or excreted, proportionate amounts of water accompany it. The total volume of fluid remaining in the body may change, but its osmolarity remains stable. Controlling water balance by controlling sodium excretion is best understood in the context of electrolyte balance, discussed later in the chapter.

Antidiuretic hormone (ADH), however, provides a means of controlling water output independently of sodium (fig. 24.4). In true dehydration (defined shortly), blood volume declines and sodium concentration rises. The increased osmolarity of the blood stimulates the hypothalamic osmoreceptors, which stimulate the posterior pituitary to release ADH. In response to ADH, cells of the collecting ducts of the kidneys synthesize the proteins called aquaporins. When installed in the plasma membrane, these serve as channels that allow water to diffuse out of the duct into the hypertonic tissue fluid of the renal medulla. Thus the kidneys reabsorb more water and produce less urine. Sodium continues to be excreted, so the *ratio* of sodium to water in the urine increases (the urine becomes more concentrated). By helping the kidneys retain water, ADH slows down the decline in blood volume and the rise in its osmolarity. Thus the ADH mechanism forms a negative feedback loop.

Conversely, if blood volume and pressure are too high or blood osmolarity is too low, ADH release is inhibited. The renal tubules reabsorb less water, urine output increases, and total body water declines. This is an effective way of compensating for hypertension. Since the lack of ADH increases the ratio of water to sodium in the urine, it raises the sodium concentration and osmolarity of the blood.

## DISORDERS OF WATER BALANCE

The body is in a state of fluid imbalance if there is an abnormality of total *volume, concentration,* or *distribution* of fluid among the compartments.

### Fluid Deficiency

Fluid deficiency arises when output exceeds intake over a long enough period of time. The two kinds of deficiency—volume depletion and dehydration—differ in the relative loss of water and electrolytes and the resulting osmolarity

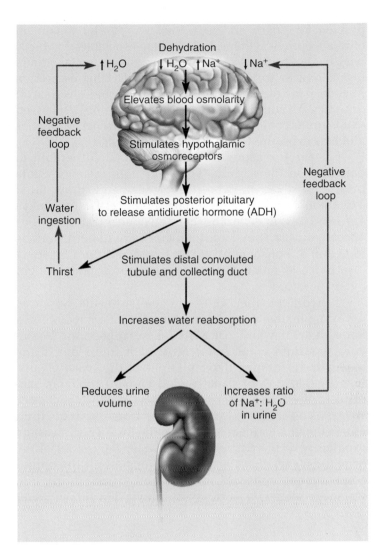

**FIGURE 24.4** The Action of Antidiuretic Hormone.
Pathways shown in red represent negative feedback.

of the ECF. This important distinction calls for different strategies of fluid replacement therapy (see Insight 24.2 at the end of the chapter).

**Volume depletion** (hypovolemia[2]) occurs when proportionate amounts of water *and* sodium are lost without replacement. Total body water declines but osmolarity remains normal. Volume depletion occurs in cases of hemorrhage, severe burns, and chronic vomiting or diarrhea. A less common cause is aldosterone hyposecretion (Addison disease), which results in inadequate sodium and water reabsorption by the kidneys.

**Dehydration** (negative water balance) occurs when the body eliminates significantly more water than sodium, so the ECF osmolarity rises. The simplest cause of dehydration is a lack of drinking water; for example, when stranded in a desert or at sea. It can be a serious

---

[2]*hypo* = below normal + *vol* = volume + *emia* = blood condition

problem for elderly and bedridden people who depend on others to provide them with water—especially for those who cannot express their need or whose caretakers are insensitive to it. Diabetes mellitus, ADH hyposecretion (diabetes insipidus), profuse sweating, and overuse of diuretics are additional causes of dehydration. Prolonged exposure to cold weather can dehydrate a person just as much as exposure to hot weather (see Insight 24.1).

For three reasons, infants are more vulnerable to dehydration than adults: (1) Their high metabolic rate produces toxic metabolites faster, and they must excrete more water to eliminate them. (2) Their kidneys are not fully mature and cannot concentrate urine as effectively. (3) They have a greater ratio of body surface to volume; consequently, compared with adults, they lose twice as much water per kilogram of body weight by evaporation.

Dehydration affects all fluid compartments. Suppose, for example, that you play a strenuous tennis match on a hot summer day and lose a liter of sweat per hour. Where does this fluid come from? Most of it filters out of the bloodstream through the capillaries of the sweat glands. In principle, 1 L of sweat would amount to about one-third of the blood plasma. However, as the blood loses water, its osmolarity rises and water from the tissue fluid enters the bloodstream to balance the loss. This raises the osmolarity of the tissue fluid, so water moves out of the cells to balance that (fig. 24.5). Ultimately, all three fluid compartments (the ICF, blood, and tissue fluid) lose water. To excrete 1 L of sweat, about 300 mL of water

would come from the tissue fluid and 700 mL from the ICF. Immoderate exercise without fluid replacement can lead to loss even greater than 1 L per hour.

The most serious effects of fluid deficiency are circulatory shock due to loss of blood volume and neurological dysfunction due to dehydration of brain cells. Volume depletion by diarrhea is a major cause of infant mortality, especially under unsanitary conditions that lead to intestinal infections such as cholera.

## Fluid Excess

Fluid excess is less common than fluid deficiency because the kidneys are highly effective at compensating for excessive intake by excreting more urine (fig. 24.6). Renal failure and other causes, however, can lead to excess fluid retention.

Fluid excesses are of two types called volume excess and hypotonic hydration. In **volume excess,** both sodium

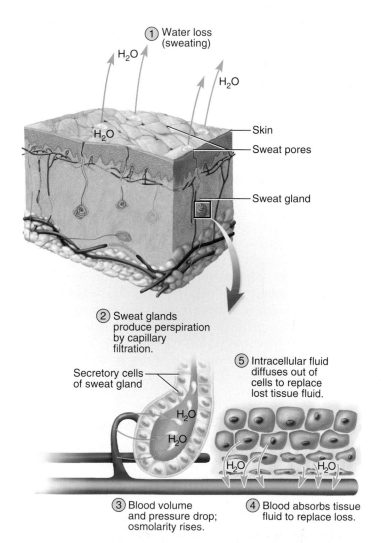

**FIGURE 24.5   Effects of Profuse Sweating on the Fluid Compartments.** In extreme dehydration, the loss of intracellular fluid can cause cellular shrinkage and dysfunction.

## INSIGHT 24.1   Clinical Application

### Fluid Balance in Cold Weather

Hot weather and profuse sweating are obvious threats to fluid balance, but so is cold weather. The body conserves heat by constricting the blood vessels of the skin and subcutaneous tissue, thus forcing blood into the deeper circulation. This raises the blood pressure, which inhibits the secretion of antidiuretic hormone and increases the secretion of atrial natriuretic peptide. These hormones increase urine output and reduce blood volume. In addition, cold air is relatively dry and increases respiratory water loss. This is why exercise causes the respiratory tract to "burn" more in cold weather than in warm.

These cold-weather respiratory and urinary losses can cause significant hypovolemia. Furthermore, the onset of exercise stimulates vasodilation in the skeletal muscles. In a hypovolemic state, there may not be enough blood to supply them, and a person may experience weakness, fatigue, or fainting (hypovolemic shock). In winter sports and other activities such as snow shoveling, it is important to maintain fluid balance. Even if you do not feel thirsty, it is beneficial to take ample amounts of warm liquids such as soup or cider. Coffee, tea, and alcohol, however, have diuretic effects that defeat the purpose of fluid intake.

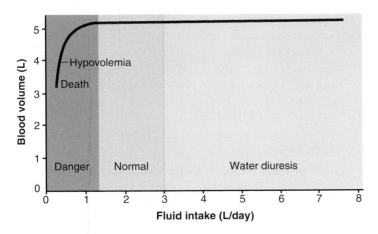

**FIGURE 24.6 The Relationship of Blood Volume to Fluid Intake.** The kidneys cannot compensate very well for inadequate fluid intake. Below an intake of about 1 L/day, blood volume drops significantly and there may be a threat of death from hypovolemic shock. The kidneys compensate very well, on the other hand, for abnormally high fluid intake; they eliminate the excess by water diuresis and maintain a stable blood volume.

and water are retained and the ECF remains isotonic. This can result from aldosterone hypersecretion or renal failure. In **hypotonic hydration** (also called water intoxication or positive water balance), more water than sodium is retained or ingested and the ECF becomes hypotonic. This can occur if you lose a large amount of water *and* salt through urine and sweat and you replace it by drinking plain water. Without a proportionate intake of electrolytes, water dilutes the ECF, makes it hypotonic, and causes cellular swelling. ADH hypersecretion can cause hypotonic hydration by stimulating excessive water retention as sodium continues to be excreted. Among the most serious effects of either type of fluid excess are pulmonary and cerebral edema.

## Fluid Sequestration

**Fluid sequestration**[3] (seh-ques-TRAY-shun) is a condition in which excess fluid accumulates in a particular location. Total body water may be normal, but the volume of circulating blood may drop to the point of causing circulatory shock. The most common form of sequestration is *edema,* the abnormal accumulation of fluid in the interstitial spaces, causing swelling of a tissue (discussed in detail in chapter 20). Hemorrhage can be another cause of fluid sequestration; blood that pools and clots in the tissues is lost to circulation. Yet another example is *pleural effusion,* caused by some lung infections, in which several liters of fluid accumulate in the pleural cavity.

The four principal forms of fluid imbalance are summarized and compared in table 24.1.

---

[3]*sequestr* = to isolate

| TABLE 24.1 | Forms of Fluid Imbalance | |
|---|---|---|
| **Form** | **Total Body Water** | **Osmolarity** |
| *Fluid Deficiency* | | |
| Volume depletion (hypovolemia) | Reduced | Isotonic (normal) |
| Dehydration (negative water balance) | Reduced | Hypertonic (elevated) |
| *Fluid Excess* | | |
| Volume excess | Elevated | Isotonic (normal) |
| Hypotonic hydration (positive water balance, water intoxication) | Elevated | Hypotonic (reduced) |

**Think About It**

*Some tumors of the brain, pancreas, and small intestine secrete ADH. What type of water imbalance would this produce? Explain why.*

## Before You Go On

*Answer the following questions to test your understanding of the preceding section:*

1. List five routes of water loss. Which one accounts for the greatest loss? Which one is most controllable?

2. Explain why even a severely dehydrated person inevitably experiences further fluid loss.

3. Suppose there were no mechanisms to stop the sense of thirst until the blood became sufficiently hydrated. Explain why we would routinely suffer hypotonic hydration.

4. Summarize the effect of ADH on total body water and blood osmolarity.

5. Name and define the four types of fluid imbalance, and give an example of a situation that could produce each type.

# Electrolyte Balance

### Objectives

When you have completed this section, you should be able to

- describe the physiological roles of sodium, potassium, calcium, chloride, and phosphate;

- describe the hormonal and renal mechanisms that regulate the concentrations of these electrolytes; and

- state the term for an excess or deficiency of each electrolyte and describe the consequences of these imbalances.

Electrolytes are physiologically important for multiple reasons: They are chemically reactive and participate in metabolism, they determine the electrical potential (charge difference) across cell membranes, and they strongly affect the osmolarity of the body fluids and the body's water content and distribution. Strictly speaking, electrolytes are salts such as sodium chloride, not just sodium or chloride ions. In common usage, however, the individual ions are often referred to as electrolytes. The major cations are sodium ($Na^+$), potassium ($K^+$), calcium ($Ca^{2+}$), and hydrogen ($H^+$), and the major anions are chloride ($Cl^-$), bicarbonate ($HCO_3^-$), and phosphates ($P_i$). Hydrogen and bicarbonate regulation are discussed later under acid–base balance. Here we focus on the other five.

The typical concentrations of these ions and the terms for electrolyte imbalances are listed in table 24.2. Blood plasma is the most accessible fluid for measurements of electrolyte concentration, so excesses and deficiencies are defined with reference to normal plasma concentrations. Concentrations in the tissue fluid differ only slightly from those in the plasma. The prefix *normo-* denotes a normal electrolyte concentration (for example, *normokalemia*), and *hyper-* and *hypo-* denote concentrations that are, respectively, sufficiently above or below normal to cause physiological disorders.

## SODIUM

### Functions

Sodium is one of the principal ions responsible for the resting membrane potentials of cells, and the inflow of sodium through gated membrane channels is an essential event in the depolarization that underlies nerve and muscle function. Sodium is the principal cation of the ECF; sodium salts account for 90% to 95% of its osmolarity. Sodium is therefore the most significant solute in determining total body water and the distribution of water among fluid compartments. Sodium gradients across the plasma membrane provide the potential energy that is tapped to cotransport other solutes such as glucose, potassium, and calcium. The $Na^+$–$K^+$ pump is an important mechanism for generating body heat. Sodium bicarbonate ($NaHCO_3$) plays a major role in buffering the pH of the ECF.

### Homeostasis

An adult needs about 0.5 g of sodium per day, whereas the typical American diet contains 3 to 7 g/day. Thus a dietary sodium deficiency is rare, and the primary concern is adequate excretion of the excess. This is one of the most important roles of the kidneys. There are multiple mechanisms for controlling sodium concentration, tied to its effects on blood pressure and osmolarity and coordinated by three hormones: aldosterone, antidiuretic hormone, and atrial natriuretic peptide.

Aldosterone, the "salt-retaining hormone," plays the primary role in adjustment of sodium excretion. Hyponatremia and hyperkalemia directly stimulate the adrenal cortex to secrete aldosterone, and hypotension stimulates its secretion by way of the renin–angiotensin mechanism (fig. 24.7).

Only cells in the ascending limb of the nephron loop, the distal convoluted tubule, and the cortical part of the collecting duct have aldosterone receptors. Aldosterone, a steroid, binds to nuclear receptors and activates transcription of a gene for the $Na^+$–$K^+$ pump. In 10 to 30 minutes, enough $Na^+$–$K^+$ pumps are synthesized and installed in the plasma membrane to produce a noticeable effect—sodium concentration in the urine begins to fall and potassium concentration rises as the tubules reabsorb more $Na^+$ and secrete more $H^+$ and $K^+$. Water and $Cl^-$ passively follow $Na^+$. Thus the primary effects of aldosterone are that the urine contains less NaCl and more $K^+$ and has a lower pH. An average adult male excretes 5 g of

| TABLE 24.2 | Electrolyte Concentrations and the Terminology of Electrolyte Imbalances | | | |
|---|---|---|---|---|
| | Mean Concentration (mEq/L)* | | Terms for Imbalances | |
| Electrolyte | Plasma | ICF | Deficiency | Excess |
| Sodium ($Na^+$) | 145 | 12 | Hyponatremia | Hypernatremia[4] |
| Potassium ($K^+$) | 4 | 150 | Hypokalemia | Hyperkalemia[5] |
| Calcium ($Ca^{2+}$) | 5 | <1 | Hypocalcemia | Hypercalcemia |
| Chloride ($Cl^-$) | 103 | 4 | Hypochloremia | Hyperchloremia |
| Phosphate ($PO_4^{3-}$) | 4 | 75 | Hypophosphatemia | Hyperphosphatemia |

*Concentrations in mmol/L are the same for $Na^+$, $K^+$, and $Cl^-$, one-half the above values for $Ca^{2+}$, and one-third the above values for $PO_4^{3-}$.

[4]*natr* = sodium + *emia* = blood condition

[5]*kal* = potassium + *emia* = blood condition

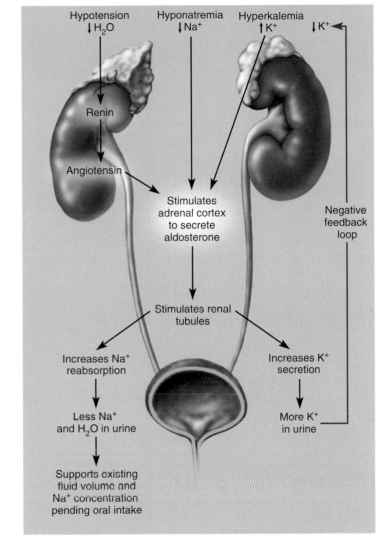

**FIGURE 24.7** **The Secretion and Effects of Aldosterone.** The pathway shown in red represents negative feedback.

▶ *What is required, in addition to aldosterone, to increase blood volume?*

sodium per day, but the urine can be virtually sodium-free when aldosterone level is high. Although aldosterone strongly influences sodium reabsorption, it has little effect on plasma sodium *concentration* because reabsorbed sodium is accompanied by a proportionate amount of water.

Hypertension inhibits the renin–angiotensin–aldosterone mechanism. The kidneys then reabsorb almost no sodium beyond the proximal convoluted tubule (PCT), and the urine contains up to 30 g of sodium per day.

Aldosterone has only slight effects on urine volume, blood volume, and blood pressure in spite of the tendency of water to follow sodium osmotically. Even in aldos-

terone hypersecretion, blood volume is rarely more than 5% to 10% above normal. An increase in blood volume increases blood pressure and glomerular filtration rate (GFR). Even though aldosterone increases the tubular reabsorption of sodium and water, this is offset by the rise in GFR and there is only a small drop in urine output.

Antidiuretic hormone modifies water excretion independently of sodium excretion. Thus, unlike aldosterone, it can change sodium *concentration*. A high concentration of sodium in the blood stimulates the posterior lobe of the pituitary gland to release ADH. Thus the kidneys reabsorb more water, which slows down any further increase in blood sodium concentration. ADH alone cannot lower the blood sodium concentration; this requires water ingestion, but remember that ADH also stimulates thirst. A drop in sodium concentration, by contrast, inhibits ADH release. More water is excreted, thereby raising the concentration of the sodium that remains in the blood.

Atrial natriuretic peptide (ANP) inhibits sodium and water reabsorption and the secretion of renin and ADH. The kidneys thus eliminate more sodium and water and lower the blood pressure.

Several other hormones also affect sodium homeostasis. Estrogens mimic the effect of aldosterone and cause women to retain water during pregnancy and part of the menstrual cycle. Progesterone reduces sodium reabsorption and has a diuretic effect. High levels of glucocorticoids promote sodium reabsorption and edema.

In some cases, sodium homeostasis is achieved by regulation of salt intake. A craving for salt occurs in people who are depleted of sodium; for example, by blood loss or Addison disease. Pregnant women sometimes develop a craving for salty foods. Salt craving is not limited to humans; many animals ranging from elephants to butterflies seek out wet salty soil where they can obtain this vital mineral.

## Imbalances

True imbalances in sodium concentration are relatively rare because sodium excess or depletion is almost always accompanied by proportionate changes in water volume. **Hypernatremia** is a plasma sodium concentration in excess of 145 mEq/L. It can result from the administration of intravenous saline (see Insight 24.2, p. 949). Its major consequences are water retention, hypertension, and edema. **Hyponatremia** (less than 130 mEq/L) is usually the result of excess body water rather than excess sodium excretion, as in the case mentioned earlier of a person who loses large volumes of sweat or urine and replaces it by drinking plain water. Usually, hyponatremia is quickly corrected by excretion of the excess water, but if uncorrected it produces the symptoms of hypotonic hydration described earlier.

# POTASSIUM

## Functions

Potassium is the most abundant cation of the ICF and is the greatest determinant of intracellular osmolarity and cell volume. Along with sodium, it produces the resting membrane potentials and action potentials of nerve and muscle cells (fig. 24.8a). Potassium is as important as sodium to the $Na^+$–$K^+$ pump and its functions of cotransport and thermogenesis (heat production). It is an essential cofactor for protein synthesis and some other metabolic processes.

## Homeostasis

Potassium homeostasis is closely linked to that of sodium. Regardless of the body's state of potassium balance, about 90% of the $K^+$ filtered by the glomerulus is reabsorbed by the PCT and the rest is excreted in the urine. Variations in potassium excretion are controlled later in the nephron by changing the amount of potassium returned to the tubular fluid by the distal convoluted tubule and cortical portion of the collecting duct (CD). When $K^+$ concentration is high, they secrete more $K^+$ into the filtrate and the urine may contain more $K^+$ than the glomerulus filters from the blood. When blood $K^+$ level is low, the CD secretes less. The distal convoluted tubule and collecting duct reabsorb $K^+$ through their intercalated cells.

Aldosterone regulates potassium balance along with sodium (see fig. 24.7). A rise in $K^+$ concentration stimulates the adrenal cortex to secrete aldosterone. Aldosterone stimulates renal secretion of $K^+$ at the same time that it stimulates reabsorption of sodium. The more sodium there is in the urine, the less potassium, and vice versa.

## Imbalances

Potassium imbalances are the most dangerous of all electrolyte imbalances. **Hyperkalemia** (>5.5 mEq/L) can have completely opposite effects depending on whether $K^+$ concentration rises quickly or slowly. It can rise quickly when, for example, a crush injury or hemolytic anemia releases large amounts of $K^+$ from ruptured cells. This can also result from a transfusion with outdated, stored blood because $K^+$ leaks from erythrocytes into the plasma during storage. A sudden increase in extracellular $K^+$ tends to

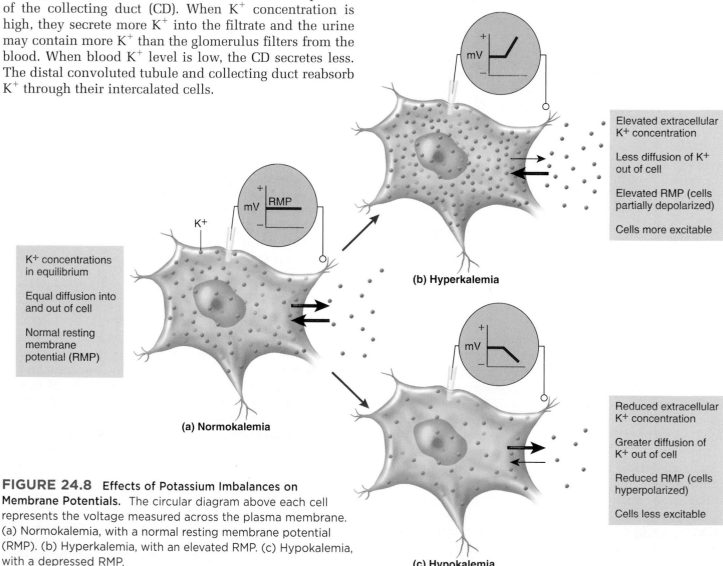

**(a) Normokalemia**

K⁺ concentrations in equilibrium

Equal diffusion into and out of cell

Normal resting membrane potential (RMP)

**(b) Hyperkalemia**

Elevated extracellular K⁺ concentration

Less diffusion of K⁺ out of cell

Elevated RMP (cells partially depolarized)

Cells more excitable

**(c) Hypokalemia**

Reduced extracellular K⁺ concentration

Greater diffusion of K⁺ out of cell

Reduced RMP (cells hyperpolarized)

Cells less excitable

**FIGURE 24.8   Effects of Potassium Imbalances on Membrane Potentials.** The circular diagram above each cell represents the voltage measured across the plasma membrane. (a) Normokalemia, with a normal resting membrane potential (RMP). (b) Hyperkalemia, with an elevated RMP. (c) Hypokalemia, with a depressed RMP.

make nerve and muscle cells abnormally excitable. Normally, $K^+$ continually passes in and out of cells at equal rates—leaving by diffusion and reentering by the $Na^+$–$K^+$ pump. But in hyperkalemia, there is less concentration difference between the ICF and ECF, so the outward diffusion of $K^+$ is reduced. More $K^+$ remains in the cell than normal, and the plasma membrane therefore has a less negative resting potential and is closer to the threshold at which it will set off action potentials (fig. 24.8b). This is a very dangerous condition that can quickly produce cardiac arrest. High-potassium solutions are sometimes used by veterinarians to euthanize animals and are used in some states as a lethal injection for capital punishment.

Hyperkalemia can also have a slower onset stemming from such causes as aldosterone hyposecretion, renal failure, or acidosis. (The relationship of acid–base imbalances to potassium imbalances is explained later.) Paradoxically, if the extracellular $K^+$ concentration rises slowly, nerve and muscle become *less* excitable. *Slow* depolarization of a cell inactivates voltage-gated $Na^+$ channels, and the channels do not become excitable again until the membrane repolarizes. Inactivated $Na^+$ channels cannot produce action potentials. For this reason, muscle cramps can be relieved by taking supplemental potassium.

**Hypokalemia** (<3.5 mEq/L) rarely results from a dietary deficiency, because most diets contain ample amounts of potassium; it can occur, however, in people with depressed appetites. Hypokalemia more often results from heavy sweating, chronic vomiting or diarrhea, excessive use of laxatives, aldosterone hypersecretion, or alkalosis. As ECF potassium concentration falls, more $K^+$ moves from the ICF to the ECF. With the loss of these cations from the cytoplasm, cells become hyperpolarized and nerve and muscle cells are less excitable (fig. 24.8c). This is reflected in muscle weakness, loss of muscle tone, depressed reflexes, and irregular electrical activity of the heart.

> **Think About It**
>
> *Some tumors of the adrenal cortex secrete excess aldosterone and may cause paralysis. Explain this effect and identify the electrolyte and fluid imbalances you would expect to observe in such a case.*

## CHLORIDE

### Functions

Chloride ions are the most abundant anions of the ECF and thus make a major contribution to its osmolarity. Chloride ions are required for the formation of stomach acid (HCl), and they are involved in the chloride shift that accompanies carbon dioxide loading and unloading by the erythrocytes (see chapter 22). By a similar mechanism explained later, $Cl^-$ plays a major role in the regulation of body pH.

### Homeostasis

$Cl^-$ is strongly attracted to $Na^+$, $K^+$, and $Ca^{2+}$. It would require great expenditure of energy to keep it separate from these cations, so $Cl^-$ homeostasis is achieved primarily as an effect of $Na^+$ homeostasis—as sodium is retained or excreted, $Cl^-$ passively follows.

### Imbalances

**Hyperchloremia** (>105 mEq/L) is usually the result of dietary excess or administration of intravenous saline. **Hypochloremia** (<95 mEq/L) is usually a side effect of hyponatremia but sometimes results from hypokalemia. In the latter case, the kidneys retain $K^+$ by excreting more $Na^+$, and $Na^+$ takes $Cl^-$ with it. The primary effects of chloride imbalances are disturbances in acid–base balance, but this works both ways—a pH imbalance arising from some other cause can also produce a chloride imbalance. Chloride balance is therefore discussed further in connection with acid–base balance.

## CALCIUM

### Functions

Calcium lends strength to the skeleton, activates the sliding filament mechanism of muscle contraction, serves as a second messenger for some hormones and neurotransmitters, activates exocytosis of neurotransmitters and other cellular secretions, and is an essential factor in blood clotting. Cells maintain a very low intracellular calcium concentration because they require a high concentration of phosphate ions (for reasons discussed shortly). If calcium and phosphate were both very concentrated in a cell, calcium phosphate crystals would precipitate in the cytoplasm. To maintain a high phosphate concentration but avoid crystallization of calcium phosphate, cells must pump out $Ca^{2+}$ and keep it at a low intracellular concentration, or else sequester $Ca^{2+}$ in the smooth ER and release it only when needed. Cells that store $Ca^{2+}$ often have a protein called *calsequestrin,* which binds the stored $Ca^{2+}$ and keeps it chemically unreactive.

### Homeostasis

Calcium concentration is regulated chiefly by parathyroid hormone, calcitriol, and in children, calcitonin. These hormones work through their effects on bone deposition and resorption, intestinal absorption of calcium, and urinary excretion. For flowcharts of calcium homeostasis, see figure 7.18 (p. 222); the associated discussion gives further details on the mode of action of these hormones.

### Imbalances

**Hypercalcemia** (>5.8 mEq/L) can result from alkalosis, hyperparathyroidism, or hypothyroidism. It reduces the $Na^+$ permeability of plasma membranes and inhibits the depolarization of nerve and muscle cells. At concentrations

$\geq$12 mEq/dL, hypercalcemia causes muscular weakness, depressed reflexes, and cardiac arrhythmia.

**Hypocalcemia** ($<$4.5 mEq/L) can result from vitamin D deficiency, diarrhea, pregnancy, lactation, acidosis, hypoparathyroidism, or hyperthyroidism. It increases the $Na^+$ permeability of plasma membranes, causing the nervous and muscular systems to be overly excitable. Tetany occurs when calcium concentration drops to 6 mg/dL and may be lethal at 4 mg/dL (2 mEq/L) due to laryngospasm and suffocation.

## PHOSPHATES
### Functions

The inorganic phosphates ($P_i$) of the body fluids are an equilibrium mixture of phosphate ($PO_4^{3-}$), monohydrogen phosphate ($HPO_4^{2-}$), and dihydrogen phosphate ($H_2PO_4^-$) ions. Phosphates are relatively concentrated in the ICF, where they are generated by the hydrolysis of ATP and other phosphate compounds. They are a component of nucleic acids, phospholipids, ATP, GTP, cAMP, creatine phosphate, and related compounds. Every process that depends on ATP depends on phosphate ions. Phosphates activate many metabolic pathways by phosphorylating enzymes and substrates such as glucose. They are also important as buffers that help stabilize the pH of the body fluids.

### Homeostasis

The average diet provides ample amounts of phosphate ions, which are readily absorbed by the small intestine. Plasma phosphate concentration is usually maintained at about 4 mEq/L, with continual loss of excess phosphate by glomerular filtration. If plasma phosphate concentration drops much below this level, however, the renal tubules reabsorb all filtered phosphate.

Parathyroid hormone increases the excretion of phosphate as part of the mechanism for increasing the concentration of free calcium ions in the ECF. Lowering the ECF phosphate concentration minimizes the formation of calcium phosphate and thus helps support plasma calcium concentration. Rates of phosphate excretion are also strongly affected by the pH of the urine, as discussed shortly.

### Imbalances

Phosphate homeostasis is not as critical as that of other electrolytes. The body can tolerate broad variations several times above or below the normal concentration with little immediate effect on physiology.

### Before You Go On

*Answer the following questions to test your understanding of the preceding section:*

6. *Which of these do you think would have the most serious effect, and why—a 5 mEq/L increase in the plasma concentration of sodium, potassium, chloride, or calcium?*

7. *Answer the same question for a 5 mEq/L decrease.*

8. *Explain why ADH is more likely than aldosterone to change the osmolarity of the blood plasma.*

9. *Explain why aldosterone hyposecretion could cause hypochloremia.*

10. *Why are more phosphate ions required in the ICF than in the ECF? How does this affect the distribution of calcium ions between these fluid compartments?*

# Acid–Base Balance

### Objectives
When you have completed this section, you should be able to

- define *buffer* and write chemical equations for the bicarbonate, phosphate, and protein buffer systems;
- discuss the relationship between pulmonary ventilation, pH of the extracellular fluids, and the bicarbonate buffer system;
- explain how the kidneys secrete hydrogen ions and how these ions are buffered in the tubular fluid;
- identify some types and causes of acidosis and alkalosis, and describe the effects of these pH imbalances; and
- explain how the respiratory and urinary systems correct acidosis and alkalosis, and compare their effectiveness and limitations.

As we saw in chapter 2, metabolism depends on the functioning of enzymes, and enzymes are very sensitive to pH. Slight deviations from the normal pH can shut down metabolic pathways as well as alter the structure and function of other macromolecules. Consequently, acid–base balance is one of the most important aspects of homeostasis.

The blood and tissue fluid normally have a pH of 7.35 to 7.45. Such a narrow range of variation is remarkable considering that our metabolism constantly produces acid: lactic acid from anaerobic fermentation, phosphoric acids from nucleic acid catabolism, fatty acids and ketones from fat catabolism, and carbonic acid from carbon dioxide. Here we examine mechanisms for resisting these challenges and maintaining acid–base balance.

**Think About It**

*In the systemic circulation, arterial blood has a pH of 7.40 and venous blood has a pH of 7.35. What do you think causes this difference?*

## ACIDS, BASES, AND BUFFERS

The pH of a solution is determined solely by its hydrogen ions ($H^+$). An acid is any chemical that releases $H^+$ in solution. A **strong acid** such as hydrochloric acid (HCl) ionizes freely, gives up most of its hydrogen ions, and can markedly lower the pH of a solution. A **weak acid** such as carbonic acid ($H_2CO_3$) ionizes only slightly and keeps most hydrogen in a chemically bound form that does not affect pH. A base is any chemical that accepts $H^+$. A **strong base** such as the hydroxide ion ($OH^-$) has a strong tendency to bind $H^+$ and raise the pH, whereas a **weak base** such as the bicarbonate ion ($HCO_3^-$) binds less of the available $H^+$ and has less effect on pH.

A **buffer,** broadly speaking, is any mechanism that resists changes in pH by converting a strong acid or base to a weak one. The body has both physiological and chemical buffers. A **physiological buffer** is a system—namely the respiratory or urinary system—that stabilizes pH by controlling the body's output of acids, bases, or $CO_2$. Of all buffer systems, the urinary system buffers the greatest quantity of acid or base, but it requires several hours to days to exert an effect. The respiratory system exerts an effect within a few minutes but cannot alter the pH as much as the urinary system can.

A **chemical buffer** is a substance that binds $H^+$ and removes it from solution as its concentration begins to rise, or releases $H^+$ into solution as its concentration falls. Chemical buffers can restore normal pH within a fraction of a second. They function as mixtures called **buffer systems** composed of a weak acid and a weak base. The three major chemical buffer systems of the body are the bicarbonate, phosphate, and protein systems.

The amount of acid or base that can be neutralized by a chemical buffer system depends on two factors: the concentration of the buffers and the pH of their working environment. Each system has an optimum pH at which it functions best; its effectiveness is greatly reduced if the pH of its environment deviates too far from this. The relevance of these factors will become apparent as you study the following buffer systems.

## The Bicarbonate Buffer System

The **bicarbonate buffer system** is a solution of carbonic acid and bicarbonate ions. Carbonic acid ($H_2CO_3$) forms by the hydration of carbon dioxide and then dissociates into bicarbonate ($HCO_3^-$) and $H^+$:

$$CO_2 + H_2O \leftrightarrows H_2CO_3 \leftrightarrows HCO_3^- + H^+$$

This is a reversible reaction. When it proceeds to the right, carbonic acid acts as a weak acid by releasing $H^+$ and lowering pH. When the reaction proceeds to the left, bicarbonate acts as a weak base by binding $H^+$, removing the ions from solution, and raising pH.

At a pH of 7.4, the bicarbonate system would not ordinarily have a particularly strong buffering capacity outside of the body. This is too far from its optimum pH of 6.1. If a strong acid were added to a beaker of carbonic acid–bicarbonate solution at pH 7.4, the preceding reaction would shift only slightly to the left. Much surplus $H^+$ would

remain and the pH would be substantially lower. In the body, by contrast, the bicarbonate system works quite well because the lungs and kidneys constantly remove $CO_2$ and prevent an equilibrium from being reached. This keeps the reaction moving to the left, and more $H^+$ is neutralized. Conversely, if there is a need to lower the pH, the kidneys excrete $HCO_3^-$, keep this reaction moving to the right, and elevate the $H^+$ concentration of the ECF. Thus you can see that the physiological and chemical buffers of the body function together in maintaining acid–base balance.

## The Phosphate Buffer System

The **phosphate buffer system** is a solution of $HPO_4^{2-}$ and $H_2PO_4^-$. It works in much the same way as the bicarbonate system. The following reaction can proceed to the right to liberate $H^+$ and lower pH, or it can proceed to the left to bind $H^+$ and raise pH:

$$H_2PO_4^- \leftrightarrows HPO_4^{2-} + H^+$$

The optimal pH for this system is 6.8, closer to the actual pH of the ECF. Thus the phosphate buffer system has a stronger buffering effect than an equal amount of bicarbonate buffer. However, phosphates are much less concentrated in the ECF than bicarbonate, so they are less important in buffering the ECF. They are more important in the renal tubules and ICF, where not only are they more concentrated, but the pH is lower and closer to their functional optimum. In the ICF, the constant production of metabolic acids creates pH values ranging from 4.5 to 7.4, probably averaging 7.0. The reason for the low pH in the renal tubules is discussed later.

## The Protein Buffer System

Proteins are more concentrated than either bicarbonate or phosphate buffers, especially in the ICF. The **protein buffer system** accounts for about three-quarters of all chemical buffering ability of the body fluids. The buffering ability of proteins is due to certain side groups of their amino acid residues. Some have carboxyl ($-COOH$) side groups, which release $H^+$ when pH begins to rise and thus lower pH:

$$-COOH \rightarrow -COO^- + H^+$$

Others have amino ($-NH_2$) side groups, which bind $H^+$ when pH falls too low, thus raising pH toward normal:

$$-NH_2 + H^+ \rightarrow -NH_3^+$$

**Think About It**

*What protein do you think is the most important buffer in blood plasma? In erythrocytes?*

## RESPIRATORY CONTROL OF pH

The equation for the bicarbonate buffer system shows that the addition of $CO_2$ to the body fluids raises $H^+$ concentration and lowers pH, while the removal of $CO_2$ has the opposite effects. This is the basis for the strong buffering capacity of the respiratory system. Indeed, this system can neutralize two or three times as much acid as the chemical buffers can.

Carbon dioxide is constantly produced by aerobic metabolism and is normally eliminated by the lungs at an equivalent rate. As explained in chapter 22, rising $CO_2$ concentration and falling pH stimulate peripheral and central chemoreceptors, which stimulate an increase in pulmonary ventilation. This expels excess $CO_2$ and thus reduces $H^+$ concentration. The free $H^+$ becomes part of the water molecules produced by this reaction:

$$HCO_3^- + H^+ \rightarrow H_2CO_3 \rightarrow CO_2 \text{ (expired)} + H_2O$$

Conversely, a drop in $H^+$ concentration raises pH and reduces pulmonary ventilation. This allows metabolic $CO_2$ to accumulate in the ECF faster than it is expelled, thus lowering pH to normal.

These are classic negative feedback mechanisms that result in acid–base homeostasis. Respiratory control of pH has some limitations, however, which are discussed later under acid–base imbalances.

## RENAL CONTROL OF pH

The kidneys can neutralize more acid or base than either the respiratory system or the chemical buffers. The essence of this mechanism is that the renal tubules secrete $H^+$ into the tubular fluid, where most of it binds to bicarbonate, ammonia, and phosphate buffers. Bound and free $H^+$ are then excreted in the urine. Thus the kidneys, in contrast to the lungs, actually expel $H^+$ from the body. The other buffer systems only reduce its concentration by binding it to another chemical.

Figure 24.9 shows how the renal tubule secretes and neutralizes $H^+$. The hydrogen ions are colored so you can trace them from the blood (step 1) to the tubular fluid (step 6). Notice that it is not a simple matter of transporting free $H^+$ across the tubule cells; rather, the $H^+$ travels in the form of carbonic acid and water molecules. The tubular secretion of $H^+$ takes place at step 6, where the ion is pumped out of the tubule cell into the tubular fluid. This can happen only if there is a steep enough concentration gradient between a high $H^+$ concentration within the tubule cell and a lower concentration in the tubular fluid.

1. $H^+$ in blood reacts with $HCO_3^-$ to form $H_2CO_3$.
2. $H_2CO_3$ decomposes into $H_2O$ and $CO_2$, which enter the tubule cell.
3. Tubule cells acquire $CO_2$ from blood, tubular fluid, and their own aerobic respiration.
4. Carbonic anhydrase (CAH) combines $H_2O$ and $CO_2$ to re-form $H_2CO_3$.
5. $H_2CO_3$ ionizes to form $HCO_3^-$ (which returns to the blood) and $H^+$.
6. $Na^+$– $H^+$ antiport exchanges $H^+$ for $Na^+$.
7. $NaHCO_3$ from glomerular filtrate decomposes into $Na^+$ and $HCO_3^-$. $Na^+$ is pumped into tubule cell.
8. $Na^+$ is removed by $Na^+$–$K^+$ pump at the base of the cell.
9. $HCO_3^-$ reacts with $H^+$ from tubule cell to form $H_2CO_3$.
10. CAH on brush border decomposes $H_2CO_3$ to $H_2O$ and $CO_2$ again.
11. $CO_2$ enters the tubular cell and $H_2O$ passes in the urine (carrying the $H^+$ that was originally in the blood).

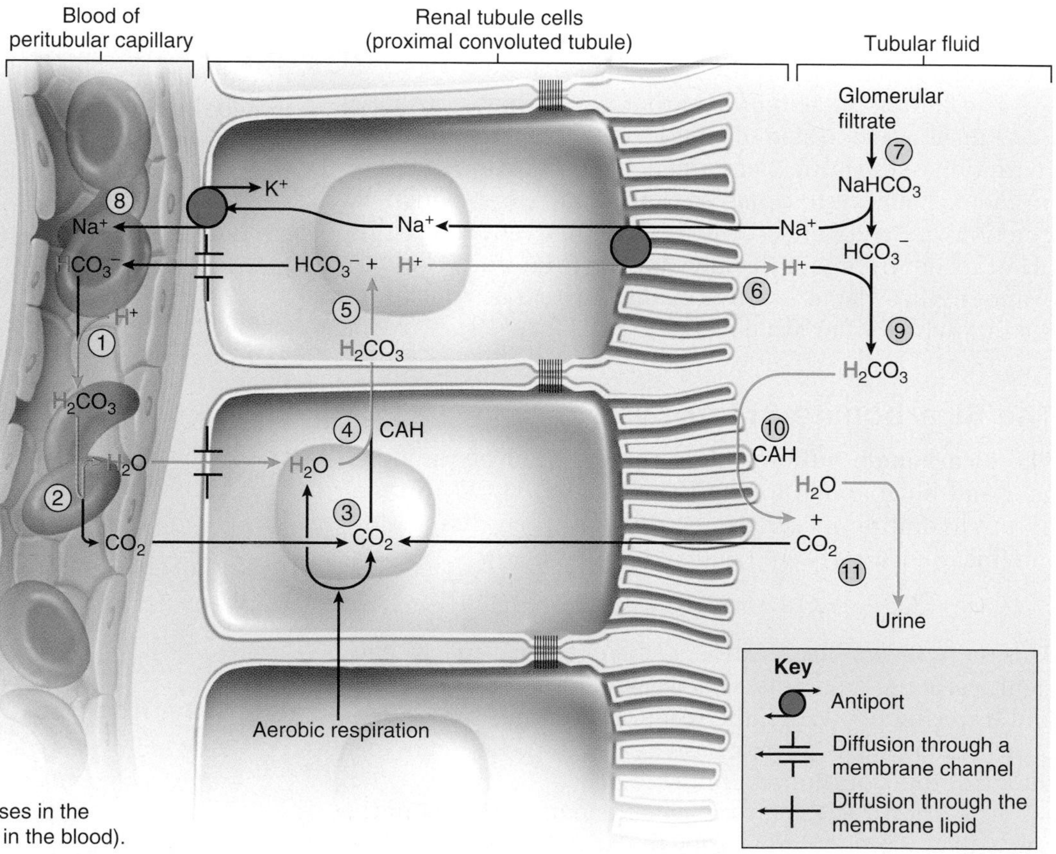


**FIGURE 24.9 Secretion and Neutralization of Hydrogen Ions in the Kidneys.** The colored hydrogen symbols allow you to trace hydrogen from $H^+$ in the blood to $H_2O$ in the urine.

*If the pH of the tubular fluid went down, how would its $Na^+$ concentration change?*

If the pH of the tubular fluid drops any lower than 4.5, $H^+$ concentration in the fluid is so high that tubular secretion ceases. Thus, pH 4.5 is the **limiting pH** for tubular secretion. This has added significance later in our discussion.

In a person with normal acid–base balance, all $HCO_3^-$ in the tubular fluid is consumed by neutralizing $H^+$; thus there is no $HCO_3^-$ in the urine. Bicarbonate ions are filtered by the glomerulus, gradually disappear from the tubular fluid, and appear in the peritubular capillary blood. It *appears* as if $HCO_3^-$ were reabsorbed by the renal tubules, but this is not the case; indeed, the renal tubules are incapable of reabsorbing $HCO_3^-$ directly. The cells of the proximal convoluted tubule, however, have carbonic anhydrase (CAH) on their brush borders facing the lumen. This breaks down the $H_2CO_3$ in the tubular fluid to $CO_2$ + $H_2O$ (step 10). It is the $CO_2$ that is reabsorbed, not the bicarbonate. For every $CO_2$ reabsorbed, however, a *new* bicarbonate ion is formed in the tubule cell and released into the blood (step 5). The effect is the same as if the tubule cells had reabsorbed bicarbonate itself.

Note that for every bicarbonate ion that enters the peritubular capillaries, a sodium ion does too. Thus the reabsorption of $Na^+$ by the renal tubules is part of the process of neutralizing acid. The more acid the kidneys excrete, the less sodium the urine contains.

The tubules secrete somewhat more $H^+$ than the available bicarbonate can neutralize. The urine therefore contains a slight excess of free $H^+$, which gives it a pH of about 5 to 6. Yet if all of the excess $H^+$ secreted by the tubules remained in this free ionic form, the pH of the tubular fluid would drop far below the limiting pH of 4.5, and $H^+$ secretion would stop. This must be prevented, and there are additional buffers in the tubular fluid to do so.

The glomerular filtrate contains $Na_2HPO_4$ (dibasic sodium phosphate), which reacts with some of the $H^+$ (fig. 24.10). A hydrogen ion replaces one of the sodium ions in the buffer, forming $NaH_2PO_4$ (monobasic sodium phosphate). This is passed in the urine and the displaced $Na^+$ is transported into the tubule cell and from there to the bloodstream.

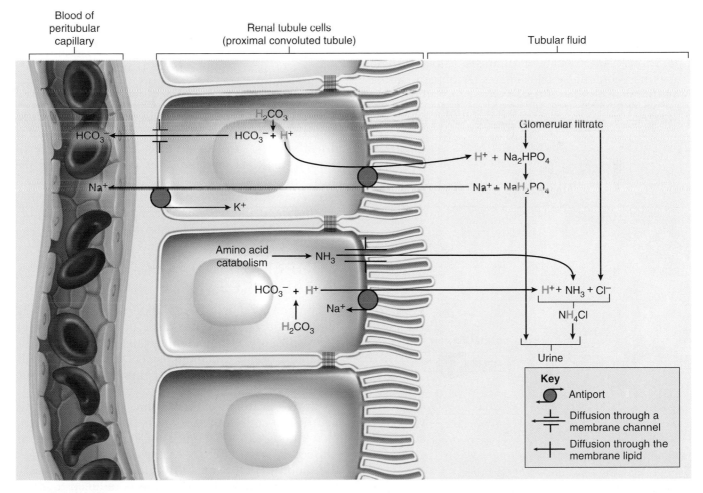

**FIGURE 24.10  Acid Buffering in the Urine.** Reactions in the tubule cells are the same as in figure 24.9 but are simplified in this diagram. The essential differences are the buffering mechanisms shown in the tubular fluid. Hydrogen symbols are colored to allow tracing them from carbonic acid to the urine.

In addition, tubule cells catabolize certain amino acids and release ammonia ($NH_3$) as a product (fig. 24.10). Ammonia diffuses into the tubular fluid, where it acts as a base to neutralize acid. It reacts with $H^+$ and $Cl^-$ (the most abundant anion in the glomerular filtrate) to form ammonium chloride ($NH_4Cl$), which is passed in the urine.

Since there is so much chloride in the tubular fluid, you might ask why $H^+$ is not simply excreted as hydrochloric acid (HCl). Why involve ammonia? The reason is that HCl is a strong acid—it dissociates almost completely, so most of its hydrogen would be in the form of free $H^+$. The pH of the tubular fluid would drop below the limiting pH and prevent excretion of more acid. Ammonium chloride, by contrast, is a weak acid—most of its hydrogen remains bound to it and does not lower the pH of the tubular fluid.

## DISORDERS OF ACID–BASE BALANCE

Figure 24.11 represents acid–base balance with an instructive metaphor to show its dependence on the bicarbonate buffer system. At a normal pH of 7.4, the ECF has a 20:1 ratio of $HCO_3^-$ to $H_2CO_3$. Excess hydrogen ions convert $HCO_3^-$ to $H_2CO_3$ and tip the balance to a lower pH. A pH below 7.35 is considered to be a state of **acidosis.** On the other hand, an $H^+$ deficiency causes $H_2CO_3$ to dissociate into $H^+$ and $HCO_3^-$, thus tipping the balance to a higher pH. A pH above 7.45 is a state of **alkalosis.** Either of these imbalances has potentially fatal effects. A person cannot live more than a few hours if the blood pH is below 7.0 or above 7.7; a pH below 6.8 or above 8.0 is quickly fatal.

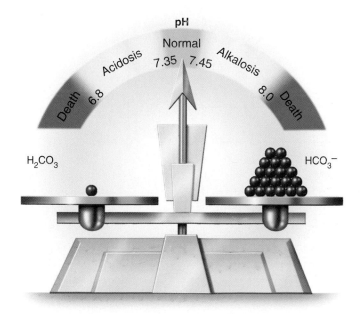

**FIGURE 24.11**  **The Relationship of Carbonic Acid-Bicarbonate Ratio to pH.**  At a normal pH of 7.40, there is a 20:1 ratio of bicarbonate ions ($HCO_3^-$) to carbonic acid ($H_2CO_3$) in the blood plasma. An excess of $HCO_3^-$ tips the balance toward alkalosis, whereas an excess of $H_2CO_3$ tips it toward acidosis.

In acidosis, $H^+$ diffuses down its concentration gradient into cells, and to maintain electrical balance, $K^+$ diffuses out (fig. 24.12a). The $H^+$ is buffered by intracellular proteins, so this exchange results in a net loss of cations from a cell. This makes the resting membrane potential more negative than usual (hyperpolarized) and makes nerve and muscle cells more difficult to stimulate. This is why acidosis depresses the central nervous system and causes such symptoms as confusion, disorientation, and coma.

In alkalosis, the extracellular $H^+$ concentration is low. Hydrogen ions diffuse out of the cells and $K^+$ diffuses in to replace them (fig. 24.12b). The net gain in positive intracellular charges shifts the membrane potential closer to firing level and makes the nervous system hyperexcitable. Neurons fire spontaneously and overstimulate skeletal muscles, causing muscle spasms, tetany, convulsions, or respiratory paralysis.

Acid–base imbalances fall into two categories, respiratory and metabolic (table 24.3). **Respiratory acidosis** occurs when the rate of alveolar ventilation fails to keep pace with the body's rate of $CO_2$ production. Carbon dioxide accumulates in the ECF and lowers its pH. This occurs in such conditions as emphysema, where there is a severe reduction in the number of functional alveoli. **Respiratory alkalosis** results from hyperventilation, in which $CO_2$ is eliminated faster than it is produced.

**Metabolic acidosis** can result from increased production of organic acids, such as lactic acid in anaerobic fermentation and ketone bodies in alcoholism and diabetes mellitus. It can also result from the ingestion of acidic drugs such as aspirin or from the loss of base due to chronic diarrhea or overuse of laxatives. Dying persons also typically exhibit acidosis. **Metabolic alkalosis** is rare but can result from overuse of bicarbonates (such as oral antacids and intravenous bicarbonate solutions) or from the loss of stomach acid by chronic vomiting.

## COMPENSATION FOR ACID–BASE IMBALANCES

In **compensated** acidosis or alkalosis, either the kidneys compensate for pH imbalances of respiratory origin, or the respiratory system compensates for pH imbalances of metabolic origin. Uncompensated acidosis or alkalosis is a pH imbalance that the body cannot correct without clinical intervention.

In **respiratory compensation,** changes in pulmonary ventilation correct the pH of the body fluids by expelling or retaining $CO_2$. If there is a $CO_2$ excess (hypercapnia), pulmonary ventilation increases to expel $CO_2$ and bring the blood pH back up to normal. If there is a $CO_2$ deficiency (hypocapnia), ventilation is reduced to allow $CO_2$ to accumulate in the blood and lower the pH to normal.

This is very effective in correcting pH imbalances due to abnormal $P_{CO_2}$ but not very effective in correcting other causes of acidosis and alkalosis. In diabetic acidosis, for example, the lungs cannot reduce the concentration of ketone bodies in the blood, although we can somewhat compensate for the $H^+$ that they release by increasing pulmonary ventilation and exhausting extra $CO_2$. The respiratory system can adjust a blood pH of 7.0 back to 7.2 or 7.3 but not all the way back to the normal 7.4. Although the respiratory system has a very powerful buffering effect, its ability to stabilize pH is therefore limited.

**Renal compensation** is an adjustment of pH by changing the rate of $H^+$ secretion by the renal tubules. The kidneys are slower to respond to pH imbalances but better at restoring a fully normal pH. Urine usually has a pH of 5 to 6, but in acidosis it may fall as low as 4.5 because of excess $H^+$, whereas in alkalosis it may rise as high as 8.2 because of excess $HCO_3^-$. The kidneys cannot act quickly enough to compensate for short-term pH imbalances, such as the acidosis that might result from an asthmatic attack lasting an hour or two, or the alkalosis resulting from a brief episode of emotional hyperventilation. They are effective, however, at compensating for pH imbalances that last for a few days or longer.

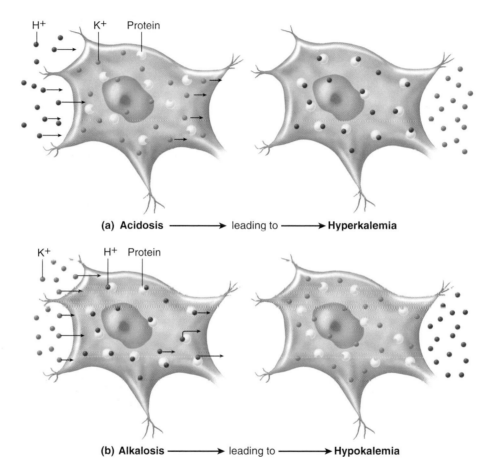

(a) **Acidosis** ⟶ leading to ⟶ **Hyperkalemia**

(b) **Alkalosis** ⟶ leading to ⟶ **Hypokalemia**

**FIGURE 24.12** **The Relationship Between Acid–Base Imbalances and Potassium Imbalances.** (a) In acidosis, $H^+$ diffuses into the cells and drives out $K^+$, elevating the $K^+$ concentration of the ECF. (b) In alkalosis, $H^+$ diffuses out of the cells and $K^+$ diffuses in to replace it, lowering the $K^+$ concentration of the ECF.

▶ *How would you change part (a) to show the effect of hyperkalemia on the pH of the ECF?*

| TABLE 24.3 | Some Causes of Acidosis and Alkalosis | |
|---|---|---|
| **Category** | **Acidosis** | **Alkalosis** |
| Respiratory | Hypoventilation, apnea, or respiratory arrest; asthma, emphysema, chronic bronchitis | Hyperventilation due to emotions or oxygen deficiency (as at high altitudes) |
| Metabolic | Excess production of organic acids, as in diabetes mellitus and long-term anaerobic fermentation; drugs such as aspirin and laxatives; chronic diarrhea | Rare but can result from chronic vomiting or overuse of bicarbonates (antacids) |

In acidosis, the renal tubules increase the rate of $H^+$ secretion. The extra $H^+$ in the tubular fluid must be buffered; otherwise, the fluid pH could exceed the limiting pH and $H^+$ secretion would stop. Therfore, in acidosis, the renal tubules secrete more ammonia to buffer the added $H^+$, and the amount of ammonium chloride in the urine may rise to 7 to 10 times normal.

## Think About It

*Suppose you measured the pH and ammonium chloride concentration of urine from a person with emphysema and urine from a healthy individual. How would you expect the two to differ, and why?*

In alkalosis, the bicarbonate concentration and pH of the urine are elevated. This is partly because there is more $HCO_3^-$ in the blood and glomerular filtrate and partly because there is not enough $H^+$ in the tubular fluid to neutralize all the $HCO_3^-$ in the filtrate.

## ACID–BASE IMBALANCES IN RELATION TO ELECTROLYTE AND WATER IMBALANCES

The foregoing discussion once again stresses a point made early in this chapter—we cannot understand or treat imbalances of water, electrolyte, or acid–base balance in isolation from each other, because each of these frequently affects the other two. Table 24.4 itemizes and explains a few of these interactions. This is by no means a complete list of how fluid, electrolytes, and pH affect each other, but it does demonstrate their interdependence. Note that many of these relationships are reciprocal—for example, acidosis can cause hyperkalemia, and conversely, hyperkalemia can cause acidosis.

## Before You Go On

*Answer the following questions to test your understanding of the preceding section.*

11. *Write two chemical equations that show how the bicarbonate buffer system compensates for acidosis and alkalosis and two equations that show how the phosphate buffer system compensates for these imbalances.*

12. *Why are phosphate buffers more effective in the cytoplasm than in the blood plasma?*

13. *Renal tubules cannot reabsorb $HCO_3^-$, and yet $HCO_3^-$ concentration in the tubular fluid falls while in the blood plasma it rises. Explain this apparent contradiction.*

14. *In acidosis, the renal tubules secrete more ammonia. Why?*

| TABLE 24.4 | | Some Relationships Among Fluid, Electrolyte, and Acid-Base Imbalances | |
|---|---|---|---|
| **Cause** | | **Potential Effect** | **Reason** |
| Acidosis | → | Hyperkalemia | $H^+$ diffuses into cells and displaces $K^+$ (see fig. 24.12a). As $K^+$ leaves the ICF, $K^+$ concentration in the ECF rises. |
| Hyperkalemia | → | Acidosis | Opposite from the above; high $K^+$ concentration in the ECF causes less $K^+$ to diffuse out of the cells than normally. $H^+$ diffuses out to compensate, and this lowers the extracellular pH. |
| Alkalosis | → | Hypokalemia | $H^+$ diffuses from ICF to ECF. More $K^+$ remains in the ICF to compensate for the $H^+$ loss, causing a drop in ECF $K^+$ concentration (see fig. 24.12b). |
| Hypokalemia | → | Alkalosis | Opposite from the above; low $K^+$ concentration in the ECF causes $K^+$ to diffuse out of cells. $H^+$ diffuses in to replace $K^+$, lowering the $H^+$ concentration of the ECF and raising its pH. |
| Acidosis | → | Hypochloremia | More $Cl^-$ is excreted as $NH_4Cl$ to buffer the excess acid in the renal tubules, leaving less $Cl^-$ in the ECF. |
| Alkalosis | → | Hyperchloremia | More $Cl^-$ is reabsorbed from the renal tubules, so ingested $Cl^-$ accumulates in the ECF rather than being excreted. |
| Hyperchloremia | → | Acidosis | More $H^+$ is retained in the blood to balance the excess $Cl^-$, causing hyperchloremic acidosis. |
| Hypovolemia | → | Alkalosis | More $Na^+$ is reabsorbed by the kidney. $Na^+$ reabsorption is coupled to $H^+$ secretion (see fig. 24.9), so more $H^+$ is secreted and pH of the ECF rises. |
| Hypervolemia | → | Acidosis | Less $Na^+$ is reabsorbed, so less $H^+$ is secreted into the renal tubules. $H^+$ retained in the ECF causes acidosis. |
| Acidosis | → | Hypocalcemia | Acidosis causes more $Ca^{2+}$ to bind to plasma protein and citrate ions, lowering the concentration of free, ionized $Ca^{2+}$ and causing symptoms of hypocalcemia. |
| Alkalosis | → | Hypercalcemia | Alkalosis causes more $Ca^{2+}$ to dissociate from plasma protein and citrate ions, raising the concentration of free $Ca^{2+}$. |

## INSIGHT 24.2   Clinical Application

### Fluid Replacement Therapy

One of the most significant problems in the treatment of seriously ill patients is the restoration and maintenance of proper fluid volume, composition, and distribution among the fluid compartments. Fluids may be administered to replenish total body water, restore blood volume and pressure, shift water from one fluid compartment to another, or restore and maintain electrolyte and acid–base balance.

Drinking water is the simplest method of fluid replacement, but it does not replace electrolytes. Heat exhaustion can occur when you lose water and salt in the sweat and replace the fluid by drinking plain water. Broths, juices, and sports drinks replace water, carbohydrates, and electrolytes.

If a patient cannot take fluids by mouth, they must be administered by alternative routes. Some can be given by enema and absorbed through the colon. All routes of fluid administration other than the digestive tract are called *parenteral*[6] routes. The most common of these is the intravenous (I.V.) route, but for various reasons, including inability to find a suitable vein, fluids are sometimes given by subcutaneous (sub-Q), intramuscular (I.M.), or other parenteral routes. Many kinds of sterile solutions are available to meet the fluid replacement needs of different patients.

In cases of extensive blood loss, there may not be time to type and cross-match blood for a transfusion. The more urgent need is to replenish blood volume and pressure. *Normal saline* (isotonic, 0.9% NaCl) is a relatively quick and simple way to raise blood volume while maintaining normal osmolarity, but it has significant shortcomings. It takes three to five times as much saline as whole blood to rebuild normal volume because much of the saline escapes the circulation into the interstitial fluid compartment or is excreted by the kidneys. In addition, normal saline can induce hypernatremia and hyperchloremia, because the body excretes the water but retains much of the NaCl. Hyperchloremia can, in turn, produce acidosis. Normal saline also lacks potassium, magnesium, and calcium. Indeed, it dilutes those electrolytes that are already present and creates a risk of cardiac arrest from hypocalcemia. Saline also dilutes plasma albumin and RBCs, creating still greater risks for patients who have suffered extensive blood loss. Nevertheless, the emergency maintenance of blood volume sometimes takes temporary precedence over these other considerations.

Fluid therapy is also used to correct pH imbalances. Acidosis may be treated with *Ringer's lactate solution,* which includes sodium to rebuild ECF volume, potassium to rebuild ICF volume, lactate to balance the cations, and enough glucose

to make the solution isotonic. Alkalosis can be treated with potassium chloride. This must be administered very carefully, because potassium ions can cause painful venous spasms, and even a small potassium excess can cause cardiac arrest. High-potassium solutions should never be given to patients in renal failure or whose renal status is unknown, because in the absence of renal excretion of potassium they can bring on lethal hyperkalemia. Ringer's lactate or potassium chloride also must be administered very cautiously, with close monitoring of blood pH, to avoid causing a pH imbalance opposite the one that was meant to be corrected. Too much Ringer's lactate causes alkalosis and too much KCl causes acidosis.

*Plasma volume expanders* are hypertonic solutions or colloids that are retained in the bloodstream and draw interstitial water into it by osmosis. They include albumin, sucrose, mannitol, and dextran. Plasma expanders are also used to combat hypotonic hydration by drawing water out of swollen cells, averting such problems as seizures and coma. A plasma expander can draw several liters of water out of the intracellular compartment within a few minutes.

Patients who cannot eat are often given isotonic 5% dextrose (glucose). A fasting patient loses as much as 70 to 85 g of protein per day from the tissues as protein is broken down to fuel the metabolism. Giving 100 to 150 g of I.V. glucose per day reduces this by half and is said to have a *protein-sparing effect*. More than glucose is needed in some cases—for example, if a patient has not eaten for several days and cannot be fed by nasogastric tube (due to lesions of the digestive tract, for example) or if large amounts of nutrients are needed for tissue repair following severe trauma, burns, or infections. In *total parenteral nutrition* (TPN), or *hyperalimentation*,[7] a patient is provided with complete I.V. nutritional support, including a protein hydrolysate (amino acid mixture), vitamins, electrolytes, 20% to 25% glucose, and on alternate days, a fat emulsion.

The water from parenteral solutions is normally excreted by the kidneys. If the patient has renal insufficiency, however, excretion may not keep pace with intake, and there is a risk of hypotonic hydration. Intravenous fluids are usually given slowly, by *I.V. drip,* to avoid abrupt changes or overcompensation for the patient's condition. In addition to pH, the patient's heart rate, blood pressure, hematocrit, and plasma electrolyte concentrations are monitored, and the patient is examined periodically for respiratory sounds indicating pulmonary edema.

The delicacy of fluid replacement therapy underscores the close relationships among fluids, electrolytes, and pH. It is dangerous to manipulate any one of these variables without close attention to the others. Parenteral fluid therapy is usually used for persons who are seriously ill. Their homeostatic mechanisms are already compromised and leave less room for error than in a healthy person.

[6]*para* = beside + *enter* = intestine

[7]*hyper* = above normal + *aliment* = nourishment

## CHAPTER REVIEW

# Review of Key Concepts

**Water Balance (p. 932)**

1. The young adult male body contains about 40 L of water. About 65% is in the intracellular fluid (ICF) and 35% in the extracellular fluid (ECF).

2. Water moves osmotically from one fluid compartment to another so that osmolarities of the ECF and ICF seldom differ.

3. In a state of water balance, average daily fluid gains and losses are equal (typically about 2,500 mL each). Water is gained from the metabolism and by ingestion of food and drink; water is lost in urine, feces, expired breath, sweat, and by cutaneous transpiration.

4. Fluid intake is governed mainly by the sense of thirst, controlled by the *thirst center* of the hypothalamus. This center responds to angiotensin II, ADH, and signals from *osmoreceptor* neurons that monitor blood osmolarity.

5. Long-term satiation of thirst depends on hydration of the blood, although the sense of thirst is briefly suppressed by wetting and cooling of the mouth and filling of the stomach.

6. Fluid loss is governed mainly by the factors that control urine output. ADH, for example, is secreted in response to dehydration and reduces urine output.

7. *Fluid deficiency* occurs when fluid output exceeds intake. In a form of fluid deficiency called *volume depletion* (hypovolemia), total body water is reduced but its osmolarity remains normal, because proportionate amounts of water and salt are lost. In the other form, true *dehydration*, volume is reduced and osmolarity is elevated because the body has lost more water than salt. Severe fluid deficiency can result in circulatory shock and death.

8. *Fluid excess* occurs in two forms called *volume excess* (retention of excess fluid with normal osmolarity) and *hypotonic hydration* (retention of more water than salt, so osmolarity is low).

9. *Fluid sequestration* is a state in which total body water may be normal, but the water is maldistributed in the body. *Edema* and *pleural effusion* are examples of fluid sequestration.

**Electrolyte Balance (p. 937)**

1. Sodium is the major cation of the ECF and is important in osmotic and fluid balance, nerve and muscle activity, cotransport, acid–base balance, and heat generation.

2. Aldosterone promotes $Na^+$ reabsorption. ADH reduces $Na^+$ concentration by promoting water reabsorption independently of $Na^+$. Atrial natriuretic peptide promotes $Na^+$ excretion.

3. A $Na^+$ excess *(hypernatremia)* tends to cause water retention, hypertension, and edema. A $Na^+$ deficiency *(hyponatremia)* is usually a result of hypotonic hydration.

4. Potassium is the major cation of the ICF. It is important for the same reasons as $Na^+$ and is a cofactor for some enzymes.

5. Aldosterone promotes $K^+$ excretion.

6. A $K^+$ excess *(hyperkalemia)* tends to cause nerve and muscle dysfunction, including cardiac arrest. A $K^+$ deficiency *(hypokalemia)* inhibits nerve and muscle function.

7. Chloride ions are the major anions of the ECF. They are important in osmotic balance, formation of stomach acid, and the chloride shift mechanism in respiratory and renal function.

8. $Cl^-$ follows $Na^+$ and other cations and is regulated as a side effect of $Na^+$ homeostasis. The primary effect of chloride imbalances *(hyper-* and *hypochloremia)* is a pH imbalance.

9. Calcium is necessary for muscle contraction, neurotransmission and other cases of exocytosis, blood clotting, some hormone actions, and bone and tooth formation.

10. Calcium homeostasis is regulated by parathyroid hormone, calcitonin, and calcitriol (see chapter 7).

11. Hypercalcemia causes muscular weakness, depressed reflexes, and cardiac arrhythmia. Hypocalcemia causes potentially fatal muscle tetany.

12. Inorganic phosphate ($P_i$) is a mixture of $PO_4^{3-}$, $HPO_4^{2-}$, and $H_2PO_4^-$ ions. $P_i$ is required for the synthesis of nucleic acids, phospholipids, ATP, GTP, and cAMP; it activates many metabolic pathways by phosphorylating such substances as enzymes and glucose; and it is an important acid–base buffer.

13. Phosphate levels are regulated by parathyroid hormone, but phosphate imbalances are not as critical as imbalances of other electrolytes.

**Acid–Base Balance (p. 942 )**

1. The pH of the ECF is normally maintained between 7.35 and 7.45.

2. pH is determined largely by the tendency of weak and strong acids to give up $H^+$ to solution, and weak and strong bases to absorb $H^+$.

3. A *buffer* is any system that resists changes in pH by converting a strong acid or base to a weak one. The *physiological buffers* are the urinary and respiratory systems; the *chemical buffers* are the bicarbonate, phosphate, and protein buffer systems.

4. The respiratory system buffers pH by adjusting pulmonary ventilation. Reduced ventilation allows $CO_2$ to accumulate in the blood and lower its pH by the reaction $CO_2 + H_2O \rightarrow H_2CO_3 \rightarrow HCO_3^- + H^+$ (generating the $H^+$ that lowers the pH). Increased ventilation expels $CO_2$, reversing this reaction, lowering $H^+$ concentration, and raising the pH.

5. The kidneys neutralize more acid or base than any other buffer system.

They secrete H⁺ into the tubular fluid, where it binds to chemical buffers and is voided from the body in the urine.

6. This $H^+$ normally neutralizes all the $HCO_3^-$ in the tubular fluid, making the urine bicarbonate-free. Excess $H^+$ in the tubular fluid can be buffered by phosphate and ammonia.

7. Acidosis is a pH <7.35. *Respiratory acidosis* occurs when pulmonary gas exchange is insufficient to expel $CO_2$ as fast as the body produces it. *Metabolic acidosis* is the result of lactic acid or ketone accumulation,

ingestion of acidic drugs such as aspirin, or loss of base in such cases as diarrhea.

8. Alkalosis is a pH >7.45. *Respiratory alkalosis* results from hyperventilation. *Metabolic alkalosis* is rare but can be caused by overuse of antacids or loss of stomach acid through vomiting.

9. *Uncompensated* acidosis or alkalosis is a pH imbalance that the body's homeostatic mechanisms cannot correct on their own; it requires clinical intervention.

10. *Compensated* acidosis or alkalosis is an imbalance that the body's homeostatic mechanisms can correct. *Respiratory compensation* is correction of the pH through changes in pulmonary ventilation. *Renal compensation* is correction of pH by changes in $H^+$ secretion by the kidneys.

11. Water, electrolyte, and acid–base imbalances are deeply interconnected; an imbalance in one area can cause or result from an imbalance in another (see table 24.4).

# Testing Your Recall

1. The greatest percentage of the body's water is in
   a. the blood plasma.
   b. the lymph.
   c. the intracellular fluid.
   d. the interstitial fluid.
   e. the extracellular fluid.

2. Hypertension is likely to increase the secretion of
   a. atrial natriuretic peptide.
   b. antidiuretic hormone.
   c. bicarbonate ions.
   d. aldosterone.
   e. ammonia.

3. _____ increases water reabsorption without increasing sodium reabsorption.
   a. Antidiuretic hormone
   b. Aldosterone
   c. Atrial natriuretic peptide
   d. Parathyroid hormone
   e. Calcitonin

4. Hypotonic hydration can result from
   a. ADH hypersecretion.
   b. ADH hyposecretion.
   c. aldosterone hypersecretion.
   d. aldosterone hyposecretion.
   e. *a* and *d* only.

5. Tetanus is most likely to result from
   a. hypernatremia.
   b. hypokalemia.
   c. hyperkalemia.
   d. hypocalcemia.
   e. *c* and *d* only.

6. The principal determinant of intracellular osmolarity and cellular volume is
   a. protein.
   b. phosphate.
   c. potassium.
   d. sodium.
   e. chloride.

7. Increased excretion of ammonium chloride in the urine most likely indicates
   a. hypercalcemia.
   b. hyponatremia.
   c. hypochloremia.
   d. alkalosis.
   e. acidosis.

8. The most effective buffer in the intracellular fluid is
   a. phosphate.
   b. protein.
   c. bicarbonate.
   d. carbonic acid.
   e. ammonia.

9. Tubular secretion of hydrogen is directly linked to
   a. tubular secretion of potassium.
   b. tubular secretion of sodium.
   c. tubular reabsorption of potassium.
   d. tubular reabsorption of sodium.
   e. tubular secretion of chloride.

10. Hyperchloremia is most likely to result in
    a. alkalosis.
    b. acidosis.
    c. hypernatremia.
    d. hyperkalemia.
    e. hypovolemia.

11. The most abundant cation in the ECF is _____.

12. The most abundant cation in the ICF is _____.

13. Water produced by the body's chemical reactions is called _____.

14. The skin loses water by two processes, sweating and _____.

15. Any abnormal accumulation of fluid in a particular place in the body is called _____.

16. An excessive concentration of potassium ions in the blood is called _____.

17. A deficiency of sodium ions in the blood is called _____.

18. A blood pH of 7.2 caused by inadequate pulmonary ventilation would be classified as _____.

19. Tubular secretion of hydrogen ions would cease if the acidity of the tubular fluid fell below a value called the _____.

20. Long-term satiation of thirst depends on a reduction of the _____ of the blood.

*Answers in Appendix B*

# True or False

*Determine which five of the following statements are false, and briefly explain why.*

1. Hypokalemia lowers the resting membrane potentials of nerve and muscle cells and makes them less excitable.

2. Aldosterone promotes sodium and water retention and can therefore greatly increase blood pressure.

3. Injuries that rupture a lot of cells tend to elevate the $K^+$ concentration of the ECF.

4. It is possible for a person to suffer circulatory shock even without losing a significant amount of fluid from the body.

5. Parathyroid hormone promotes calcium and phosphate reabsorption by the kidneys.

6. The bicarbonate system buffers more acid than any other chemical buffer.

7. The more sodium the renal tubules reabsorb, the more hydrogen ion they secrete into the tubular fluid.

8. The body does not compensate for respiratory acidosis by increasing the respiratory rate.

9. In true dehydration, the body fluids remain isotonic although total body water is reduced.

10. Aquaporins regulate the rate of water reabsorption in the proximal convoluted tubule.

*Answers in Appendix B*

# Testing Your Comprehension

1. A duck hunter is admitted to the hospital with a shotgun injury to the abdomen. He has suffered extensive blood loss but is conscious. He complains of being intensely thirsty. Explain the physiological mechanism connecting his injury to his thirst.

2. A woman living at poverty level finds bottled water at the grocery store next to the infant formula. The label on the water states that it is made especially for infants, and she construes this to mean that it can be used as a nutritional supplement. The water is much cheaper than formula, so she gives her baby several ounces of bottled water a day as a substitute for formula. After several days the baby has seizures and is taken to the hospital, where it is found to have edema, acidosis, and a plasma sodium concentration of 116 mEq/L. The baby is treated with anticonvulsants followed by normal saline and recovers. Explain each of the signs.

3. Explain why the respiratory and urinary systems are both necessary for the bicarbonate buffer system to work effectively in the blood plasma.

4. The left column indicates some increases or decreases in blood plasma values. In the right column, replace the question mark with an up or down arrow to indicate the expected effect. Explain each effect.

| Cause | Effect |
|---|---|
| a. $\uparrow H_2O$ | ? $Na^+$ |
| b. $\uparrow Na^+$ | ? $Cl^-$ |
| c. $\downarrow K^+$ | ? $H^+$ |
| d. $\uparrow H^+$ | ? $K^+$ |
| e. $\downarrow Ca^{2+}$ | ? $PO_4^{3-}$ |

5. A 4-year-old child is caught up in tribal warfare in Africa. In a refugee camp, the only drinking water is from a sewage-contaminated pond. The child soon develops severe diarrhea and dies 10 days later of cardiac arrest. Explain the possible physiological cause(s) of his death.

*Answers at www.mhhe.com/saladin4*

# www.mhhe.com/saladin4

*The textbook website provides a wealth of interactive study materials fully organized and integrated by chapter. You will find practice quizzes, labeling exercises, and much more that will complement your learning and understanding of anatomy and physiology. The website also includes tools designed to enhance your **Anatomy & Physiology | REVEALED** experience.*

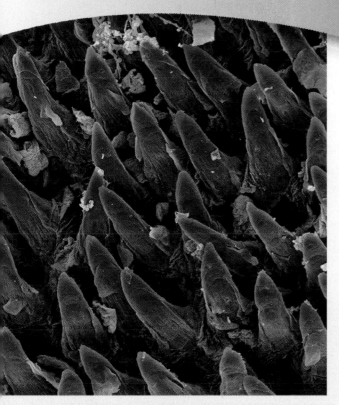

Filiform papillae of the human tongue
(SEM)

# THE DIGESTIVE SYSTEM

## CHAPTER OUTLINE

## INSIGHTS

## Brushing Up

To understand this chapter, it is
important that you understand or
brush up on the following concepts:

Anatomy & Physiology REVEALED®

**Volume 4** Digestive System

Most of the nutrients we eat cannot be used in their existing form. They must be broken down into smaller components, such as amino acids and monosaccharides, that are universal to all species. Consider what happens if you eat a piece of beef, for example. The myosin of beef differs very little from that of your own muscles, but the two are not identical, and even if they were, beef myosin could not be absorbed, transported in the blood, and incorporated into your muscles. Like any other dietary protein, it must be broken down into amino acids before it can be used. Since beef and human proteins are made of the same 20 amino acids, those of beef proteins might indeed become part of your own myosin but could equally well end up in your insulin, fibrinogen, collagen, or any other protein. The digestive system is essentially a disassembly line—its primary purpose is to break nutrients down into forms that can be used by the body and to absorb them so they can be distributed to the tissues. The study of the digestive tract and the diagnosis and treatment of its disorders is called **gastroenterology.**[1]

# General Anatomy and Digestive Processes

### Objectives
When you have completed this section, you should be able to

- list the functions and major physiological processes of the digestive system;
- distinguish between mechanical and chemical digestion;
- describe the basic chemical process underlying all chemical digestion, and name the major substrates and products of this process;
- list the regions of the digestive tract and the accessory organs of the digestive system;
- identify the layers of the digestive tract and describe its relationship to the peritoneum; and
- describe the general neural and chemical controls over digestive function.

## DIGESTIVE FUNCTION

The **digestive system** is the organ system that processes food, extracts nutrients from it, and eliminates the residue. It does this in four stages:

1. **ingestion,** the selective intake of food;
2. **digestion,** the mechanical and chemical breakdown of food into a form usable by the body;
3. **absorption,** the uptake of nutrient molecules into the epithelial cells of the digestive tract and then into the blood or lymph; and finally
4. **defecation,** the elimination of undigested residue.

The digestion stage itself has two facets, mechanical and chemical. **Mechanical digestion** is the physical breakdown of food into smaller particles. It is achieved by the cutting and grinding action of the teeth and the churning contractions of the stomach and small intestine. Mechanical digestion exposes more food surface to the action of digestive enzymes. **Chemical digestion** is a series of hydrolysis reactions that break dietary macromolecules into their monomers *(residues):* polysaccharides into monosaccharides, proteins into amino acids, fats into monoglycerides and fatty acids, and nucleic acids into nucleotides. It is carried out by digestive enzymes produced by the salivary glands, stomach, pancreas, and small intestine. Some nutrients are already present in usable form in the ingested food and are absorbed without being digested: vitamins, free amino acids, minerals, cholesterol, and water.

## GENERAL ANATOMY

The digestive system has two anatomical subdivisions, the digestive tract and the accessory organs (fig. 25.1). The **digestive tract** is a tube extending from mouth to anus, measuring about 9 m (30 ft) long in the cadaver. It is also known as the *alimentary*[2] *canal.* It includes the oral cavity, pharynx, esophagus, stomach, small intestine, and large intestine. Part of this, the stomach and intestines, constitute the *gastrointestinal (GI) tract.* The **accessory organs** are the teeth, tongue, salivary glands, liver, gallbladder, and pancreas.

The digestive tract is open to the environment at both ends. Most of the material in it has not entered any body tissues and is considered to be external to the body until it is absorbed by epithelial cells of the alimentary canal. In the strict sense, defecated food residue was never in the body.

Most of the digestive tract follows the basic structural plan shown in figure 25.2, with a wall composed of the

---

[1]*gastro* = stomach + *entero* = intestines + *logy* = study of

[2]*aliment* = food

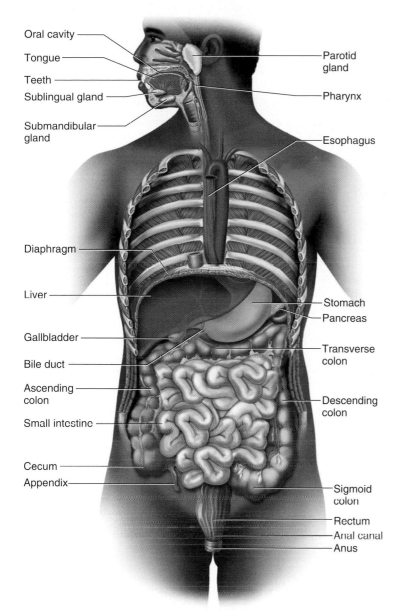

Oral cavity
Tongue
Teeth
Sublingual gland
Submandibular gland
Parotid gland
Pharynx
Esophagus
Diaphragm
Liver
Stomach
Pancreas
Gallbladder
Bile duct
Transverse colon
Ascending colon
Small intestine
Descending colon
Cecum
Appendix
Sigmoid colon
Rectum
Anal canal
Anus

**FIGURE 25.1**  The Digestive System.

following tissue layers, in order from the inner to the outer surface:

Mucosa
    Epithelium
    Lamina propria
    Muscularis mucosae
Submucosa
Muscularis externa
    Inner circular layer
    Outer longitudinal layer
Serosa
    Areolar tissue
    Mesothelium

Slight variations on this theme are found in different regions of the tract.

The **mucosa,** lining the lumen, consists of an inner epithelium, a loose connective tissue layer called the **lamina propria,** and a thin layer of smooth muscle called the **muscularis mucosae** (MUSS-cue-LERR-is mew-CO-see). The epithelium is simple columnar in most of the digestive tract, but stratified squamous from the oral cavity through the esophagus and in the lower anal canal, where the tract is subject to more abrasion. The muscularis mucosae tenses the mucosa, creating grooves and ridges that enhance its surface area and contact with food. This improves the efficiency of digestion and nutrient absorption.

The **submucosa** is a thicker layer of loose connective tissue containing blood vessels, lymphatic vessels, a nerve plexus, and in some places, glands that secrete lubricating mucus into the lumen.

The **muscularis externa** consists of usually two layers of smooth muscle near the outer surface. Cells of the inner layer encircle the tract while those of the outer layer run longitudinally. This layer is responsible for the motility that propels food and residue through the digestive tract.

The **serosa** is composed of a thin layer of areolar tissue topped by a simple squamous mesothelium. The serosa begins in the lower 3 to 4 cm of the esophagus and ends just before the rectum. The oral cavity, pharynx, most of the esophagus, and the rectum are surrounded by a fibrous connective tissue layer called the **adventitia.**

The esophagus, stomach, and intestines have a nervous network called the **enteric**[3] **nervous system,** which regulates digestive tract motility, secretion, and blood flow. This system is thought to have more neurons than the spinal cord. It can function completely independently of the central nervous system, although the CNS usually exerts a significant influence on its action. It is usually regarded as part of the autonomic nervous system, but opinions on this vary. The enteric nervous system is composed of two nerve networks: the **submucosal (Meissner**[4]**) plexus** in the submucosa and the **myenteric (Auerbach**[5]**) plexus** of parasympathetic ganglia and nerve fibers between the two layers of the muscularis externa. Parasympathetic preganglionic fibers terminate on the ganglia of the myenteric plexus. Postganglionic fibers arising in this plexus not only innervate the muscularis externa, but also pass through its inner circular layer and contribute to the submucosal plexus. The myenteric plexus controls peristalsis and other contractions of the

---

[3] *enter* = intestine
[4] Georg Meissner (1829–1905), German histologist
[5] Leopold Auerbach (1828–97), German anatomist

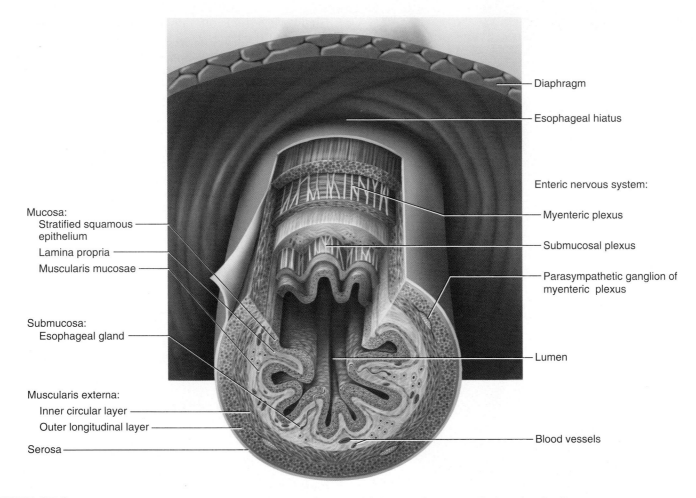

Mucosa:
- Stratified squamous epithelium
- Lamina propria
- Muscularis mucosae

Submucosa:
- Esophageal gland

Muscularis externa:
- Inner circular layer
- Outer longitudinal layer

Serosa

Diaphragm

Esophageal hiatus

Enteric nervous system:
- Myenteric plexus
- Submucosal plexus
- Parasympathetic ganglion of myenteric plexus

Lumen

Blood vessels

**FIGURE 25.2**  **Tissue Layers of the Digestive Tract.**   Cross section of the esophagus just below the diaphragm.

muscularis externa, and the submucosal plexus controls movements of the muscularis mucosae and glandular secretion of the mucosa.

## RELATIONSHIP TO THE PERITONEUM

In processing food, the stomach and intestines undergo such strenuous contractions that they need freedom to move in the abdominal cavity. Thus, they are not tightly bound to the abdominal wall, but over most of their length, they are loosely suspended from it by connective tissue sheets called **mesenteries** (see figs. A.9 and A.10, p. 37). The mesenteries also hold the abdominal viscera in their proper relationship to each other and prevent the small intestine, especially, from becoming twisted and tangled by changes in body position and by its own contractions. Furthermore, the mesenteries provide passage for the blood vessels and nerves that supply the digestive tract, and contain many lymph nodes and lymphatic vessels.

Along the dorsal midline of the abdominal cavity, the parietal peritoneum turns inward and forms a sheet of tissue, the **dorsal mesentery,** extending to the digestive tract. The membrane then folds around the digestive tract to form the serosa. In some places it continues beyond the digestive organs as a sheet of tissue called the **ventral mesentery,** which may hang freely in the abdominal cavity or attach to the ventral abdominal wall or other organs.

Along the right superior margin *(lesser curvature)* of the stomach, a ventral mesentery called the **lesser omentum** extends from the stomach to the liver (fig. 25.3). Another membrane, the **greater omentum,** hangs from the left inferior margin *(greater curvature)* of the stomach and loosely covers the small intestine like an apron. At its inferior margin, the greater omentum turns back on itself, passes upward, and forms serous membranes around the spleen and transverse colon. Beyond the transverse colon, it continues as a mesentery called the mesocolon, which anchors the colon to the posterior abdominal wall. The omenta have a loosely organized, lacy appearance due partly to many holes or gaps in the membranes and partly to an irregular distribution of adipose tissue. They also contain many lymph nodes, lymphatic vessels, blood vessels, and nerves. The omenta adhere to perforations or inflamed areas of the stomach or intestines, contribute immune cells to the site, and isolate infections that might otherwise give rise to peritonitis.

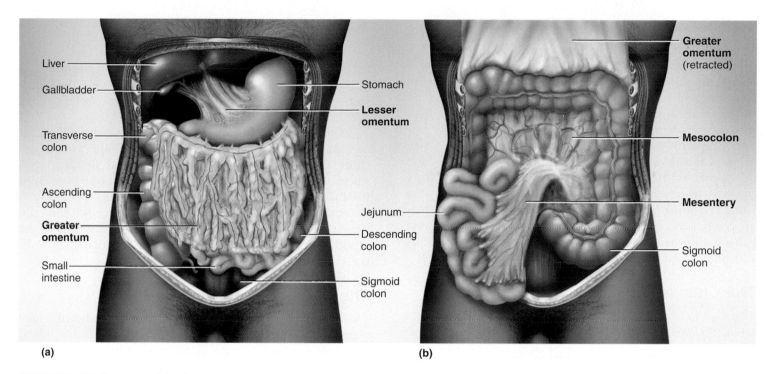

**FIGURE 25.3** **Serous Membranes Associated with the Digestive Tract.** (a) The greater and lesser omenta. (b) Greater omentum and small intestine retracted to show the mesocolon and mesentery. These membranes contain the mesenteric arteries and veins.

When an organ is enclosed by mesentery (serosa) on both sides, it is considered to be within the peritoneal cavity, or **intraperitoneal.** When an organ lies against the dorsal body wall and is covered by peritoneum on the ventral side only, it is said to be outside the peritoneal cavity, or **retroperitoneal.** The duodenum, most of the pancreas, and parts of the large intestine are retroperitoneal. The stomach, liver, and other parts of the small and large intestines are intraperitoneal.

## REGULATION OF THE DIGESTIVE TRACT

The motility and secretion of the digestive tract are controlled by neural, hormonal, and paracrine mechanisms. The neural controls include short and long autonomic reflexes. In **short (myenteric) reflexes,** stretching or chemical stimulation of the digestive tract acts through the myenteric nerve plexus to stimulate contractions in nearby regions of the muscularis externa, such as the *peristaltic* contractions of swallowing. **Long (vagovagal) reflexes** act through autonomic nerve fibers that carry sensory signals from the digestive tract to the central nervous system, and motor commands back to the digestive tract. Parasympathetic fibers of the vagus nerves are especially important in stimulating digestive motility and secretion by way of these long reflexes.

The digestive tract also produces numerous hormones such as *gastrin* and *secretin,* and paracrine secretions such as histamine and prostaglandins, that stimulate digestive function. The hormones are secreted into the blood and stimulate relatively distant parts of the digestive tract. The paracrine secretions diffuse through the tissue fluids and stimulate nearby target cells.

### Before You Go On

*Answer the following questions to test your understanding of the preceding section:*

1. *What is the term for the serous membrane that suspends the intestines from the abdominal wall?*

2. *Which physiological process of the digestive system truly moves a nutrient from the outside to the inside of the body?*

3. *What one type of reaction is the basis of all chemical digestion?*

4. *Name some nutrients that are absorbed without being digested.*

# The Mouth Through Esophagus

### Objectives

When you have completed this section, you should be able to

- describe the gross anatomy of the digestive tract from the mouth through the esophagus;

- describe the composition and functions of saliva; and

- describe the neural control of salivation and swallowing.

# THE MOUTH

The mouth is also known as the **oral,** or **buccal** (BUCK-ul), **cavity.** Its functions include ingestion (food intake), taste and other sensory responses to food, mastication (chewing), chemical digestion (starch is partially digested in the mouth), deglutition (swallowing), speech, and respiration. The mouth is enclosed by the cheeks, lips, palate, and tongue (fig. 25.4). Its anterior opening between the lips is the **oral orifice** and its posterior opening into the throat is the **fauces**[6] (FAW-seez). The oral cavity is lined with stratified squamous epithelium. This epithelium is keratinized in areas subject to the greatest food abrasion, such as the gums and hard palate, and nonkeratinized in other areas such as the floor of the mouth, the soft palate, and the inside of the cheeks and lips.

## The Cheeks and Lips

The cheeks and lips retain food and push it between the teeth for mastication. They are essential for articulate

---

[6]*fauces* = throat

speech and for sucking and blowing actions, including suckling by infants. Their fleshiness is due mainly to subcutaneous fat, the buccinator muscles of the cheeks, and the orbicularis oris muscle of the lips. Each lip is attached to the gum behind it by a midsagittal fold called the **labial frenulum.**[7] The **vestibule** is the space behind the cheeks and lips, external to the teeth.

Externally the lips are divided into a *cutaneous area* and *red area* (vermilion). Loosely speaking, you could think of these as the mustache area and lipstick area, respectively. The cutaneous area is colored like the rest of the face and has hair and sebaceous glands. The red area is more brightly colored and more sensitive because the skin here has unusually tall dermal papillae, which allow blood capillaries and nerves to come especially close to the surface. The red area has no hair or sebaceous glands.

## The Tongue

The tongue (fig. 25.5), although muscular and bulky, is a remarkably agile and sensitive organ. It manipulates food between the teeth while it avoids being bitten, it can extract food particles from the teeth after a meal, and it is sensitive enough to feel a stray hair in a bite of food. Its surface is covered with nonkeratinized stratified squamous epithelium and exhibits bumps and projections called **lingual papillae,** the site of the taste buds. The types of papillae and sense of taste are discussed in chapter 16, and the general anatomy of the tongue is shown in figure 25.5.

> **Think About It**
>
> *How does proprioception protect the tongue from being bitten?*

The anterior two-thirds of the tongue, called the **body,** occupies the oral cavity and the posterior one-third, the root, occupies the oropharynx. The boundary between them is marked by a V-shaped row of **vallate papillae** and, behind these, a groove called the **terminal sulcus.** The body is attached to the floor of the mouth by a midsagittal fold called the **lingual frenulum.**

The muscles of the tongue, which compose most of its mass, are described in chapter 10. The **intrinsic muscles,** contained entirely within the tongue, produce the relatively subtle tongue movements of speech. The **extrinsic muscles,** with origins elsewhere and insertions in the

---

[7]*labi* = lip + *frenulum* = little bridle

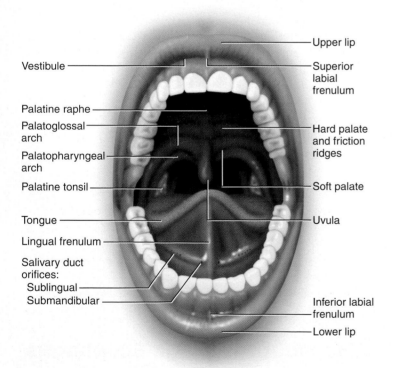

Vestibule

Palatine raphe

Palatoglossal arch

Palatopharyngeal arch

Palatine tonsil

Tongue

Lingual frenulum

Salivary duct orifices:
  Sublingual
  Submandibular

Upper lip

Superior labial frenulum

Hard palate and friction ridges

Soft palate

Uvula

Inferior labial frenulum

Lower lip

**FIGURE 25.4** **The Oral Cavity.** For a photographic medial view see figure A.17 (p. 46).

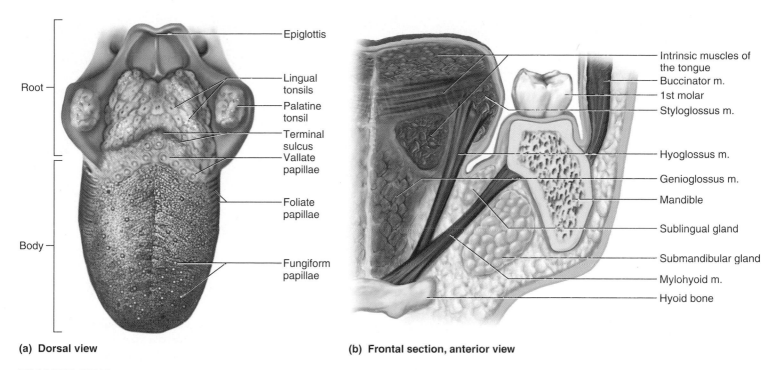

**(a) Dorsal view**

Root

Body

Epiglottis

Lingual tonsils

Palatine tonsil

Terminal sulcus

Vallate papillae

Foliate papillae

Fungiform papillae

**(b) Frontal section, anterior view**

Intrinsic muscles of the tongue

Buccinator m.

1st molar

Styloglossus m.

Hyoglossus m.

Genioglossus m.

Mandible

Sublingual gland

Submandibular gland

Mylohyoid m.

Hyoid bone

**FIGURE 25.5   The Tongue.**   (a) Dorsal surface. (b) Frontal section. For a sagittal section see figure 22.3.

tongue, produce the stronger movements of food manipulation. The extrinsic muscles include the *genioglossus, hyoglossus, palatoglossus,* and *styloglossus* (fig. 25.5b; see also fig. 10.8). Amid the muscles are serous and mucous **lingual glands,** which secrete a portion of the saliva. The lingual tonsils are contained in the root.

## The Palate

The palate, separating the oral cavity from the nasal cavity, makes it possible to breathe while chewing food. Its anterior portion, the **hard (bony) palate,** is supported by the palatine processes of the maxillae and by the smaller palatine bones. It has transverse *friction ridges (palatal rugae)* that aid the tongue in holding and manipulating food. Posterior to this is the **soft palate,** which has a more spongy texture and is composed mainly of skeletal muscle and glandular tissue, but no bone. It has a conical medial projection, the **uvula,**[8] visible at the rear of the oral cavity.

A pair of muscular arches on each side of the oral cavity begin dorsally near the uvula and follow the wall of the cavity to its floor. The anterior one is the **palatoglossal**

arch and the posterior one is the **palatopharyngeal arch.** The latter arch marks the beginning of the pharynx. The palatine tonsils are located on the wall between the arches.

## The Teeth

The teeth are collectively called the **dentition.** Adults normally have 16 teeth in the mandible and 16 in the maxilla. On each side of the midline, there are two incisors, a canine, two premolars, and three molars in each jaw (fig. 25.6b). The **incisors** are chisel-like cutting teeth used to bite off a piece of food. The **canines** are more pointed and act to puncture and shred it. They serve as weapons in many mammals but became reduced in the course of human evolution until they now project barely above the other teeth. The **premolars** and **molars** have relatively broad surfaces adapted to crushing and grinding.

The meeting of the teeth when the mouth closes is called **occlusion** (ah-CLUE-zhun), and the surfaces where they meet are called the **occlusal** (ah-CLUE-zul) **surfaces.** The occlusal surface of a premolar has two rounded bumps called **cusps;** thus the premolars are also known as **bicuspids.** The molars have four to five cusps. Cusps of the upper and lower premolars and molars mesh when the jaws are closed and slide over each other as the jaw makes

[8]*uvula* = little grape

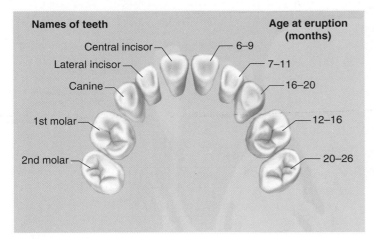

**(a) Deciduous (baby) teeth**

Names of teeth

Central incisor
Lateral incisor
Canine
1st molar
2nd molar

Age at eruption
(months)

6–9
7–11
16–20
12–16
20–26

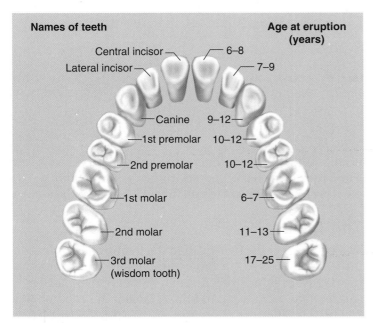

**(b) Permanent teeth**

Names of teeth

Central incisor
Lateral incisor
Canine
1st premolar
2nd premolar
1st molar
2nd molar
3rd molar
(wisdom tooth)

Age at eruption
(years)

6–8
7–9
9–12
10–12
10–12
6–7
11–13
17–25

**FIGURE 25.6    The Dentition.** Each figure shows only the upper teeth. The ages at eruption are composite ages for the corresponding upper and lower teeth. Generally, the lower (mandibular) teeth erupt somewhat earlier than their upper (maxillary) counterparts.

▶ *Which teeth are absent from a 3-year-old child?*

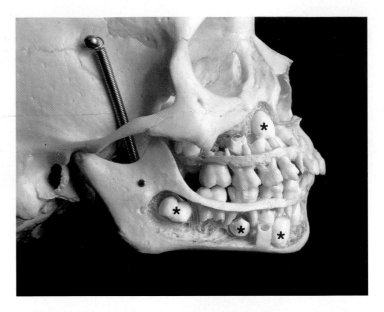

**FIGURE 25.7    Permanent and Deciduous Teeth in a Child's Skull.** This dissection shows erupted deciduous teeth and, below them and marked with asterisks, the permanent teeth waiting to erupt.

lateral chewing motions. This grinds and tears food more effectively than if the occlusal surfaces were flat.

Teeth develop beneath the gums and **erupt** (emerge) in predictable order. Twenty **deciduous teeth** (*milk teeth* or *baby teeth*) erupt from the ages of 6 to 30 months, beginning with the incisors (fig. 25.6a). Between 6 and 25 years of age, these are replaced by the 32 **permanent teeth.** As a permanent tooth grows below a deciduous tooth (fig. 25.7), the root of the deciduous tooth dissolves and leaves little more than the crown by the time it falls

out. The third molars (wisdom teeth) erupt around ages 17 to 25, if at all. Over the course of human evolution, the face became flatter and the jaws shorter, leaving little room for the third molars. Thus, they often remain below the gum and become *impacted*—so crowded against neighboring teeth and bone that they cannot erupt.

Each tooth is embedded in a socket called an **alveolus,** forming a joint called a *gomphosis* between the tooth and bone (fig. 25.8). The alveolus is lined by a **periodontal** (PERR-ee-oh-DON-tul) **ligament,** a modified periosteum whose collagen fibers penetrate into the bone on one side and into the tooth on the other. This anchors the tooth very firmly in the alveolus. The gum, or **gingiva** (JIN-jih-vuh), covers the alveolar bone. Regions of a tooth are defined by their relationship to the gingiva: the **crown** is the portion above the gum, the **root** is the portion below the gum, embedded in alveolar bone, and the **neck** is the point where the crown, root, and gum meet. The space between the tooth and gum is the **gingival sulcus.** The hygiene of this sulcus is especially important to dental health (see Insight 25.1).

Most of a tooth consists of hard yellowish tissue called **dentin,** covered with **enamel** in the crown and neck and **cementum** in the root. Dentin and cementum are living connective tissues with cells or cell processes embedded in a calcified matrix. Cells of the cementum *(cementocytes)* are scattered more or less randomly and occupy tiny cavities similar to the lacunae of bone. Cells of the dentin *(odontoblasts)* line the pulp cavity and have slender processes that travel through tiny parallel tunnels

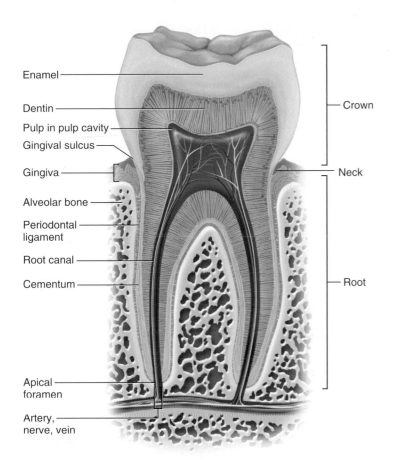

Enamel

Dentin

Pulp in pulp cavity

Gingival sulcus

Gingiva

Alveolar bone

Periodontal ligament

Root canal

Cementum

Apical foramen

Artery, nerve, vein

Crown

Neck

Root

**FIGURE 25.8** **Structure of a Tooth and Its Alveolus.** This particular example is a molar.

in the dentin. Enamel is not a tissue but a noncellular secretion produced before the tooth erupts. Damaged dentin and cementum can regenerate, but damaged enamel cannot—it must be artificially repaired.

Internally, a tooth has a dilated **pulp cavity** in the crown and a narrow **root canal** in the root. These spaces are occupied by **pulp**—a mass of loose connective tissue, blood and lymphatic vessels, and nerves. These nerves and vessels enter the tooth through a pore, the **apical foramen,** at the inferior end of each root canal.

## MASTICATION

**Mastication** (chewing) breaks food into pieces small enough to be swallowed and exposes more surface to the action of digestive enzymes. It is the first step in mechanical digestion. Mastication requires little thought because food stimulates receptors that trigger an involuntary chewing reflex. The tongue, buccinator, and orbicularis oris muscles manipulate food and push it between the teeth. The masseter and temporalis muscles produce the up-and-down crushing action of the teeth, and the lateral and medial pterygoid muscles and masseters produce side-to-side grinding action.

## SALIVA AND THE SALIVARY GLANDS

Saliva moistens the mouth, digests a little starch and fat, cleanses the teeth, inhibits bacterial growth, dissolves molecules so they can stimulate the taste buds, and moistens food and binds particles together to aid in swallowing. It is a hypotonic solution of 97.0% to 99.5% water and the following solutes:

- **salivary amylase,** an enzyme that begins starch digestion in the mouth;

- **lingual lipase,** an enzyme that is activated by stomach acid and digests fat after the food is swallowed;

- **mucus,** which binds and lubricates the food mass and aids in swallowing;

- **lysozyme,** an enzyme that kills bacteria;

- **immunoglobulin A** (IgA), an antibody that inhibits bacterial growth; and

- **electrolytes,** including sodium, potassium, chloride, phosphate, and bicarbonate salts.

Saliva has a pH of 6.8 to 7.0. There are striking differences in pH from one region of the digestive tract to another, with a powerful influence on the activity and deactivation of digestive enzymes. For example, salivary amylase works well at a neutral pH and is deactivated by the low pH of the stomach, whereas lingual lipase does

[9]*caries* = rottenness

not act in the mouth at all but is activated by the acidity of the stomach. Thus saliva begins to digest starch before the food is swallowed and fat after it is swallowed.

## The Salivary Glands

There are two kinds of salivary glands, intrinsic and extrinsic. The **intrinsic salivary glands** are an indefinite number of small glands dispersed amid the other oral tissues. They include *lingual glands* in the tongue, *labial glands* on the inside of the lips, and *buccal glands* on the inside of the cheeks. They secrete relatively small amounts of saliva at a fairly constant rate whether we are eating or not. This saliva contains lingual lipase and lysozyme and serves to moisten the mouth and inhibit bacterial growth.

The **extrinsic salivary glands** are three pairs of larger, more discrete organs located outside of the oral mucosa. They communicate with the oral cavity by way of ducts (fig. 25.9).

1. The **parotid**[10] **gland** is located just beneath the skin anterior to the earlobe. Its duct passes superficially over the masseter, pierces the buccinator, and opens

---

[10]*par* = next to + *ot* = ear

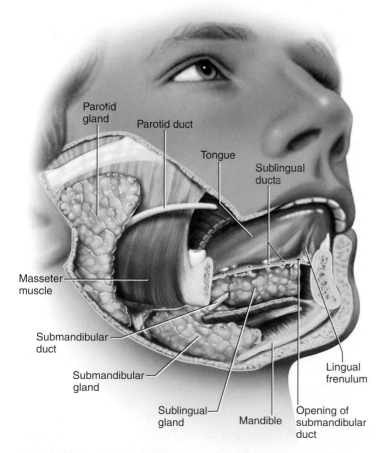

Parotid gland

Parotid duct

Tongue

Sublingual ducts

Masseter muscle

Submandibular duct

Submandibular gland

Sublingual gland

Mandible

Lingual frenulum

Opening of submandibular duct

**FIGURE 25.9  The Extrinsic Salivary Glands.**  Half of the mandible has been removed to expose the sublingual gland medial to it.

into the mouth opposite the second upper molar tooth. Mumps is inflammation and swelling of the parotid gland caused by a virus.

2. The **submandibular gland** is located halfway along the body of the mandible, medial to its margin, just deep to the mylohyoid muscle. Its duct empties into the mouth at a papilla on the side of the lingual frenulum, near the lower central incisors.

3. The **sublingual gland** is located in the floor of the mouth. It has multiple ducts that empty into the mouth posterior to the papilla of the submandibular duct.

These are all compound tubuloacinar glands with a treelike arrangement of branching ducts ending in acini (see chapter 5). Some acini have only mucous cells, some have only serous cells, and some have a mixture of both (fig. 25.10). Mucous cells secrete salivary mucus, and serous cells secrete a thinner fluid rich in amylase and electrolytes.

## Salivation

The extrinsic salivary glands secrete about 1.0 to 1.5 L of saliva per day. Cells of the acini filter water and electrolytes from the blood capillaries and add amylase, mucin, and lysozyme to it. The ducts slightly modify its electrolyte composition.

Food stimulates tactile, pressure, and taste receptors in the mouth, which transmit signals to a group of **salivatory nuclei** in the medulla oblongata and pons. These nuclei also receive input from higher brain centers, so even the odor, sight, or thought of food stimulates salivation. Irritation of the stomach and esophagus by spicy foods, stomach acid, or toxins also stimulates salivation, perhaps serving to dilute and rinse away the irritants.

The salivatory nuclei send signals to the glands by way of autonomic fibers in the facial and glossopharyngeal nerves. In response to such stimuli as the aroma or taste of food, the parasympathetic nervous system stimulates the glands to produce abundant, thin saliva rich in enzymes. Sympathetic stimulation, by contrast, causes the glands to produce less abundant, thicker saliva with more mucus. This is why the mouth may feel sticky or dry under conditions of stress. Dehydration also reduces salivation because it reduces capillary filtration.

Salivary amylase begins to digest starch as the food is chewed, while the mucus in the saliva binds food particles into a soft, slippery, easily swallowed mass called a **bolus.** Without mucus, one must drink a much larger volume of fluid to swallow food.

## THE PHARYNX

The pharynx, described in chapter 22, has a deep layer of longitudinally oriented skeletal muscle and a superficial layer of circular skeletal muscle. The circular muscle is divided into superior, middle, and inferior **pharyngeal constrictors,** which force food downward during swallowing.

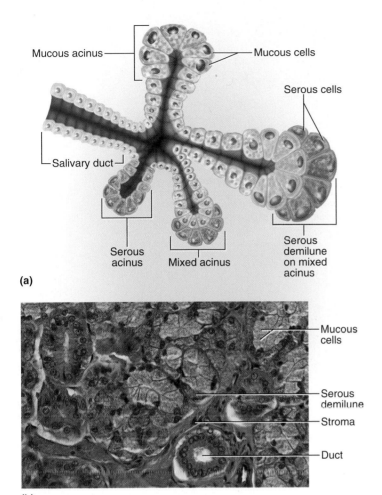

Mucous acinus — Mucous cells

Serous cells

Salivary duct

Serous acinus  Mixed acinus  Serous demilune on mixed acinus

**(a)**

Mucous cells

Serous demilune

Stroma

Duct

**(b)**

**FIGURE 25.10** Microscopic Anatomy of the Salivary Glands. (a) Duct and acini of a generalized salivary gland with a mixture of mucous and serous cells. Serous cells often form crescent-shaped caps called serous demilunes over the ends of mucous acini. (b) Histology of the sublingual salivary gland.

When food is not being swallowed, the inferior constrictor remains contracted to exclude air from the esophagus. This constriction is regarded as the **upper esophageal sphincter,** although it is not an anatomical feature of the esophagus. It disappears at the time of death when the muscle relaxes. Thus it is regarded as *physiological sphincter* rather than a constant anatomical structure.

## THE ESOPHAGUS

The **esophagus** is a straight muscular tube 25 to 30 cm long (see figs. 25.1 and 25.2). It begins at the level of the cricoid cartilage, inferior to the larynx and dorsal to the trachea. After passing downward through the mediastinum, the esophagus penetrates the diaphragm at an opening called the *esophageal hiatus,* continues another 3 to 4 cm, and meets the stomach at an opening called the **cardiac orifice** (named for its proximity to the heart). Food pauses briefly at this point before entering the stom-

ach because of a constriction called the **lower esophageal sphincter (LES).** The LES is also a physiological rather than an anatomical sphincter, and thus is not found in the cadaver. It is thought to be either a constriction of the diaphragm surrounding the esophageal hiatus, or muscle tone in the smooth muscle of the esophagus. The LES prevents stomach contents from regurgiating into the esophagus, thus protecting the esophageal mucosa from the corrosive effect of stomach acid. "Heartburn" has nothing to do with the heart, but is the burning sensation produced by acid reflux into the esophagus.

The wall of the esophagus is organized into the tissue layers described earlier, with some regional specializations. The mucosa has a nonkeratinized stratified squamous epithelium. The submucosa contains **esophageal glands** that secrete lubricating mucus into the lumen. When the esophagus is empty, the mucosa and submucosa are deeply folded into longitudinal ridges, giving the lumen a starlike shape in cross section.

The muscularis externa is composed of skeletal muscle in the upper one-third of the esophagus, a mixture of skeletal and smooth muscle in the middle one-third, and only smooth muscle in the lower one-third. This transition corresponds to a shift from voluntary to involuntary phases of swallowing as a food bolus passes down the esophagus.

Most of the esophagus is in the mediastinum. Here, it is covered with a connective tissue adventitia, which merges into the adventitias of the trachea and thoracic aorta. The short segment below the diaphragm is covered by a serosa.

## SWALLOWING

Swallowing, or **deglutition** (DEE-glu-TISH-un), is a complex action involving over 22 muscles in the mouth, pharynx, and esophagus, coordinated by the **swallowing center,** a nucleus in the medulla oblongata and pons. This center communicates with muscles of the pharynx and esophagus by way of the trigeminal, facial, glossopharyngeal, and hypoglossal nerves (cranial nerves V, VII, IX, and XII).

Swallowing occurs in two phases (fig. 25.11): (1) In the *buccal phase,* the tongue collects food, presses it against the palate to form a bolus, and pushes it back into the oropharynx. Here the bolus stimulates tactile receptors and activates the next phase. (2) In the *pharyngeal-esophageal phase,* three actions block food and drink from reentering the mouth or entering the nasal cavity or larynx: (a) the root of the tongue blocks the oral cavity, (b) the soft palate rises and blocks the nasopharynx, and (c) the infrahyoid muscles pull the larynx up, the epiglottis covers its opening, and the vestibular folds are adducted to close the airway. The food bolus is driven downward by constriction of the upper, then the middle, and finally the lower pharyngeal constrictors. As the bolus slides off the epiglottis into the esophagus, it stretches the esophagus and triggers **peristalsis,** a wave of muscular contraction that pushes the bolus ahead of it (fig. 25.11b).

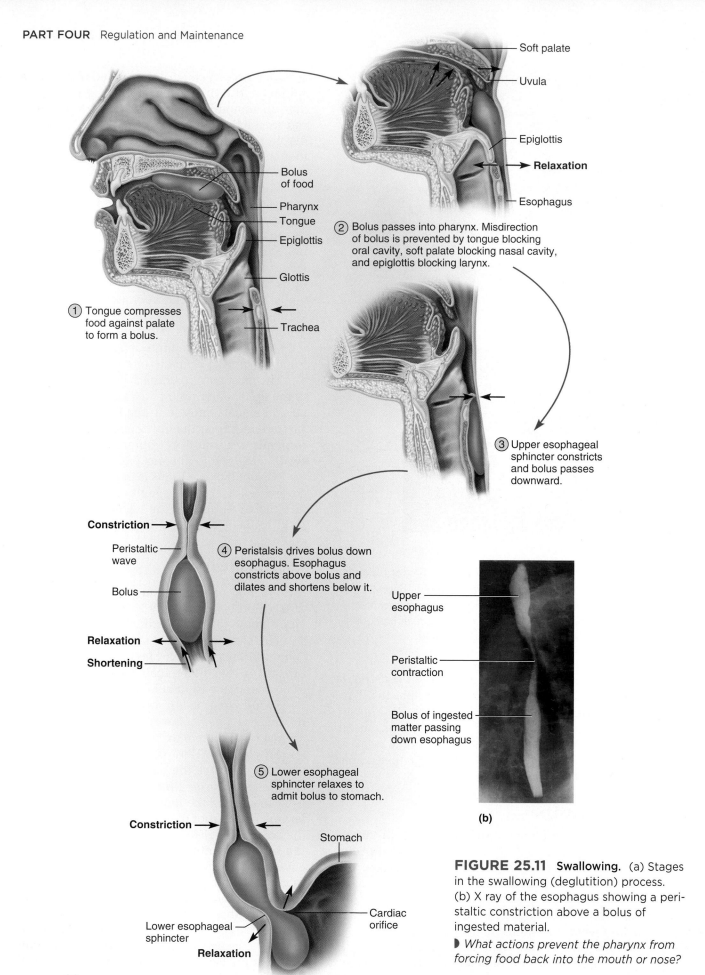

① Tongue compresses food against palate to form a bolus.

Bolus of food
Pharynx
Tongue
Epiglottis
Glottis
Trachea

Soft palate
Uvula
Epiglottis
**Relaxation**
Esophagus

② Bolus passes into pharynx. Misdirection of bolus is prevented by tongue blocking oral cavity, soft palate blocking nasal cavity, and epiglottis blocking larynx.

③ Upper esophageal sphincter constricts and bolus passes downward.

**Constriction**
Peristaltic wave
Bolus
**Relaxation**
**Shortening**

④ Peristalsis drives bolus down esophagus. Esophagus constricts above bolus and dilates and shortens below it.

Upper esophagus
Peristaltic contraction
Bolus of ingested matter passing down esophagus

**(b)**

⑤ Lower esophageal sphincter relaxes to admit bolus to stomach.

**Constriction**
Stomach
Lower esophageal sphincter
**Relaxation**
Cardiac orifice

**(a)**

**FIGURE 25.11  Swallowing.**  (a) Stages in the swallowing (deglutition) process. (b) X ray of the esophagus showing a peristaltic constriction above a bolus of ingested material.

▶ *What actions prevent the pharynx from forcing food back into the mouth or nose?*

Peristalsis is moderated partly by a short reflex through the myenteric nerve plexus. The bolus stimulates stretch receptors that feed into the nerve plexus, which transmits signals to the muscularis externa behind and ahead of the bolus. The circular muscle behind the bolus constricts and pushes it downward. Ahead of the bolus, the circular muscle relaxes while the longitudinal muscle contracts. The latter action pulls the wall of the esophagus slightly upward, which makes the esophagus a little shorter and wider and able to receive the descending food.

When we are standing or sitting upright, most food and liquid drop through the esophagus by gravity faster than the peristaltic wave can catch up to it. Peristalsis, however, propels more solid food pieces and ensures that you can swallow regardless of the body's position—even standing on your head! Liquid normally reaches the stomach in 1 to 2 seconds and a food bolus in 4 to 8 seconds. As a bolus reaches the lower end of the esophagus, the lower esophageal sphincter relaxes to let it pass into the stomach.

For a review of the anatomy up to this point, see table 25.1.

## Before You Go On

*Answer the following questions to test your understanding of the preceding section:*

5. List as many functions of the tongue as you can.

6. Imagine a line from the mandibular bone to the root canal of a tooth. Name the tissues, in order, through which this line would pass.

7. What is the difference in function and location between intrinsic and extrinsic salivary glands? Name the extrinsic salivary glands and describe their locations.

8. Describe the muscularis externa of the esophagus and its action in peristalsis.

9. Describe the mechanisms that prevent food from entering the nasal cavity and larynx during swallowing.

# The Stomach

## Objectives

When you have completed this section, you should be able to

- describe the gross and microscopic anatomy of the stomach;
- state the function of each type of epithelial cell in the gastric mucosa;
- identify the secretions of the stomach and state their functions;
- explain how the stomach produces hydrochloric acid and pepsin;
- describe the contractile responses of the stomach to food; and
- describe the three phases of gastric function and how gastric activity is activated and inhibited.

The stomach is a muscular sac in the upper left abdominal cavity immediately inferior to the diaphragm. It functions primarily as a food storage organ, with an internal

| TABLE 25.1 | Anatomical Checklist of the Digestive System from the Mouth Through the Esophagus |
|---|---|
| **Oral (buccal) Cavity** | |

| | |
|---|---|
| Oral orifice | Dentition (teeth) |
| Fauces | Developmental types |
| Cheeks | Deciduous teeth |
| Lips | Permanent teeth |
|   Cutaneous area | Functional types |
|   Red area (vermilion) | Incisors |
|   Labial frenulum | Canines |
| Vestibule | Premolars (bicuspids) |
| Tongue | Molars |
|   Body | Dental tissues |
|   Root | Dentin |
|   Terminal sulcus | Enamel |
|   Lingual frenulum | Cementum |
|   Lingual papillae | Pulp |
|   Taste buds | Anatomical features |
|   Lingual muscles | Crown |
|   Lingual tonsils |   Occlusal surface |
|   Lingual glands |   Cusps |
| Palate | Neck |
|   Hard palate | Root |
|   Soft palate | Pulp cavity |
|   Uvula | Root canal |
| Palatoglossal arch |   Apical foramen |
| Palatopharyngeal arch | Periodontal tissues |
| |   Alveolus |
| |   Periodontal ligament |
| |   Gingiva (gum) |
| |     Gingival sulcus |

| **Salivary Glands** |
|---|

| |
|---|
| Intrinsic |
| Extrinsic |
|   Parotid gland |
|   Submandibular gland |
|   Sublingual gland |

| **Pharynx** |
|---|

| |
|---|
| Pharyngeal constrictors |

| **Esophagus** |
|---|

| |
|---|
| Esophageal glands |
| Lower esophageal sphincter |
| Cardiac orifice |

volume of about 50 mL when empty and 1.0 to 1.5 L after a typical meal. When extremely full, it may hold up to 4 L and extend nearly as far as the pelvis.

Well into the nineteenth century, authorities regarded the stomach as essentially a grinding chamber, fermentation vat, or cooking pot. Some even attributed digestion to a supernatural spirit in the stomach. We now know that it mechanically breaks up food particles, liquefies the food, and begins the chemical digestion of proteins and a small amount of fat. This produces a soupy or pasty mixture of semidigested food called **chyme**[11] (kime). Most digestion occurs after the chyme passes on to the small intestine.

## GROSS ANATOMY

The stomach is J-shaped (fig. 25.12), relatively vertical in tall people, and more horizontal in short people. The **lesser curvature** of the stomach extends the short distance from esophagus to duodenum along the medial to superior aspect; the **greater curvature** extends the longer distance from esophagus to duodenum on the lateral to inferior aspect.

The stomach is divided into four regions: (1) The **cardiac region** (cardia) is a small area immediately inside the cardiac orifice. (2) The **fundic region** (fundus) is the dome-shaped portion superior to the esophageal attachment. (3) The **body** (corpus) makes up the greatest part of the stomach inferior to the cardiac orifice. (4) The **pyloric region** is a slightly narrower pouch at the inferior end; it is subdivided into a funnel-like **antrum**[12] and a narrower **pyloric canal.** The latter terminates at the **pylorus,**[13] a narrow passage into the duodenum. The pylorus is surrounded by a thick ring of smooth muscle, the **pyloric (gastroduodenal) sphincter,** which regulates the passage of chyme into the duodenum.

## INNERVATION AND CIRCULATION

The stomach receives parasympathetic nerve fibers from the vagus nerves and sympathetic fibers from the celiac ganglia (see p. 568). It is supplied with blood by branches of the celiac trunk (see p. 786). All blood drained from the stomach and intestines enters the hepatic portal circulation and filters through the liver before returning to the heart.

## THE STOMACH WALL

The stomach wall has tissue layers similar to those of the esophagus, with some variations. The mucosa is covered with a simple columnar glandular epithelium (fig. 25.13). The apical regions of its surface cells are filled with mucin, which swells with water and becomes mucus after it is secreted. The mucosa and submucosa are flat and smooth when the stomach is full, but as it empties, these

layers form conspicuous longitudinal wrinkles called **gastric rugae** (ROO-gee). The lamina propria is almost entirely occupied by tubular glands, to be described shortly. The muscularis externa has three layers, rather than two: an outer longitudinal, middle circular, and inner oblique layer (see fig. 25.12).

### Think About It

*Contrast the epithelium of the esophagus with that of the stomach. Why is each epithelial type best suited to the function of its respective organ?*

The gastric mucosa is pocked with depressions called **gastric pits** lined with the same columnar epithelium as the surface (fig. 25.13). Cells near the bottom of the gastric pits divide and produce new epithelial cells that continually migrate upward and replace old epithelial cells that are sloughed off into the chyme.

Two or three tubular glands open into the bottom of each gastric pit and span the rest of the lamina propria. In the cardiac and pyloric regions they are called **cardiac glands** and **pyloric glands,** respectively. In the rest of the stomach, they are called **gastric glands.** These three glands differ in cellular composition, as noted shortly. Collectively, they have the following cell types:

- **Mucous cells,** which secrete mucus, predominate in the cardiac and pyloric glands. In gastric glands, they are called *mucous neck cells* and are concentrated in the narrow *neck* of the gland, where it opens into the gastric pit.

- **Regenerative (stem) cells,** found in the base of the pit and neck of the gland, divide rapidly and produce a continual supply of new cells. Newly generated cells migrate upward to the gastric surface as well as downward into the glands to replace cells that die.

- **Parietal cells,** found mostly in the upper half of the gland, secrete *hydrochloric acid* and *intrinsic factor.* They are found mostly in the gastric glands, but a few occur in the pyloric glands.

- **Chief cells,** so-named because they are the most numerous, secrete *chymosin* and *lipase* in infancy and *pepsinogen* throughout life. They dominate the lower half of the gastric glands but are absent from cardiac and pyloric glands.

- **Enteroendocrine cells,** concentrated especially in the lower end of a gland, secrete hormones and paracrine messengers that regulate digestion. There are at least eight different kinds of enteroendocrine cells in the stomach, each of which produces a different chemical messenger.

In general, the cardiac and pyloric glands secrete mainly mucus; acid and enzyme secretion occur predominantly in the gastric glands; and hormones are secreted throughout the stomach.

[11]*chyme* = juice
[12]*antrum* = cavity
[13]*pylorus* = gatekeeper

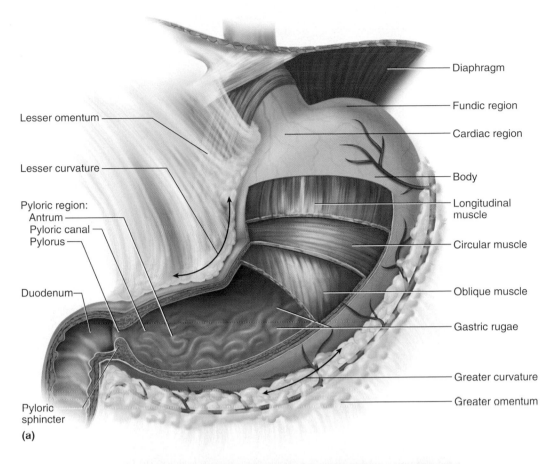

Diaphragm

Fundic region

Cardiac region

Body

Longitudinal muscle

Circular muscle

Oblique muscle

Gastric rugae

Greater curvature

Greater omentum

Lesser omentum

Lesser curvature

Pyloric region:
Antrum
Pyloric canal
Pylorus

Duodenum

Pyloric sphincter

**(a)**

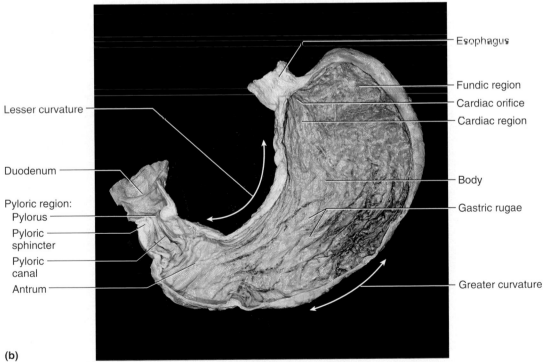

Esophagus

Fundic region

Cardiac orifice

Cardiac region

Body

Gastric rugae

Greater curvature

Lesser curvature

Duodenum

Pyloric region:
Pylorus

Pyloric sphincter

Pyloric canal

Antrum

**(b)**

**FIGURE 25.12    The Stomach.**    (a) Gross anatomy. (b) Photograph of the internal surface.

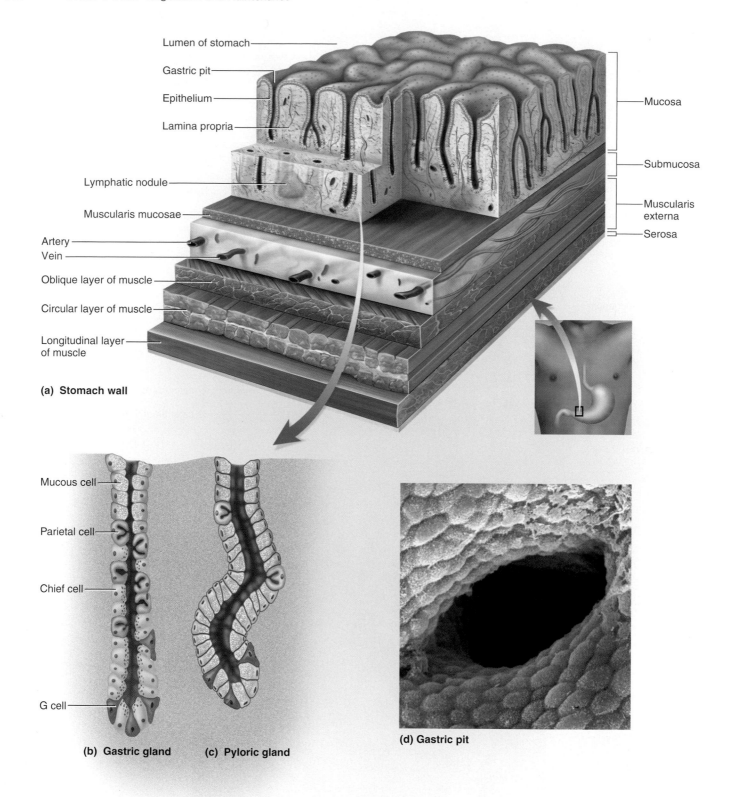

Lumen of stomach

Gastric pit

Epithelium

Lamina propria

Mucosa

Submucosa

Muscularis externa

Serosa

Lymphatic nodule

Muscularis mucosae

Artery

Vein

Oblique layer of muscle

Circular layer of muscle

Longitudinal layer of muscle

**(a) Stomach wall**

Mucous cell

Parietal cell

Chief cell

G cell

**(b) Gastric gland**     **(c) Pyloric gland**

**(d) Gastric pit**

**FIGURE 25.13** **Microscopic Anatomy of the Stomach Wall.**   (a) A block of tissue showing all layers from the mucosa (top) to the serosa (bottom). (b) A gastric gland, the most widespread type in the stomach. (c) A pyloric gland, from the inferior end of the stomach. Note the absence of chief cells and small number of parietal cells. (d) the opening of a gastric pit into the stomach, surrounded by the rounded apical surfaces of the columnar epithelial cells of the mucosa (SEM).

# GASTRIC SECRETIONS

The gastric glands produce 2 to 3 L of **gastric juice** per day, composed mainly of water, hydrochloric acid, and pepsin.

## Hydrochloric Acid

Gastric juice has a high concentration of hydrochloric acid (HCl) and a pH as low as 0.8. Such concentrated acid could cause a serious chemical burn to the skin. How, then, does the stomach produce and tolerate such acidity?

The reactions that produce HCl (fig. 25.14) may seem familiar by now because they have been discussed in previous chapters—most recently in connection with renal excretion of $H^+$ in chapter 24. Parietal cells contain carbonic anhydrase (CAH), which catalyzes the first step in the following reaction:

$$CO_2 + H_2O \xrightarrow{CAH} H_2CO_3 \rightarrow HCO_3^- + H^+$$

The $H^+$ produced by this reaction is pumped into the lumen of a gastric gland by an active transport protein similar to the $Na^+$–$K^+$ pump, called **$H^+$–$K^+$ ATPase**. This is an antiport that uses the energy of ATP to pump $H^+$ out and $K^+$ into the cell. HCl secretion does not affect the pH within the parietal cell because $H^+$ is pumped out as fast as it is generated. The bicarbonate ions ($HCO_3^-$) are exchanged for chloride ions ($Cl^-$) from the blood plasma—the same *chloride shift* process that occurs in the renal

tubules and red blood cells—and the $Cl^-$ is pumped into the lumen of the gastric gland to join the $H^+$.

Thus HCl accumulates in the stomach while bicarbonate ions accumulate in the blood. Because of the bicarbonate, blood leaving the stomach has a higher pH when digestion is occurring than when the stomach is empty. This high-pH blood is called the *alkaline tide*.

Stomach acid has several functions: (1) It activates the enzymes pepsin and lingual lipase, as discussed shortly. (2) It breaks up connective tissues and plant cell walls, helping to liquefy food and form chyme. (3) It converts ingested ferric ions ($Fe^{3+}$) to ferrous ions ($Fe^{2+}$), a form of iron that can be absorbed and used for hemoglobin synthesis. (4) It contributes to nonspecific disease resistance by destroying ingested bacteria and other pathogens.

## Pepsin

Several digestive enzymes are secreted as inactive proteins called **zymogens** and then converted to active enzymes by the removal of some of their amino acids. In the stomach, chief cells secrete a zymogen called **pepsinogen.** Hydrochloric acid removes some of its amino acids and converts it to **pepsin.** Since pepsin digests protein, and pepsinogen itself is a protein, pepsin has an *autocatalytic* effect—as some pepsin is formed, it converts pepsinogen into more pepsin (fig. 25.15). The ultimate function of pepsin, however, is to digest dietary proteins to shorter peptide chains, which then pass to the small intestine, where their digestion is completed.

## Other Enzymes

In infants, the chief cells also secrete **gastric lipase** and **chymosin** (rennin). Gastric lipase digests some of the butterfat of milk, and chymosin curdles milk by coagulating its proteins.

## Intrinsic Factor

Parietal cells also secrete a glycoprotein called **intrinsic factor** that is essential to the absorption of vitamin $B_{12}$ by the small intestine. Intrinsic factor binds vitamin $B_{12}$, and the intestinal cells then absorb this complex by receptor-mediated endocytosis. Without vitamin $B_{12}$, hemoglobin cannot be synthesized and pernicious anemia develops (see chapter 18). The secretion of intrinsic factor is the only indispensable function of the stomach. Digestion can continue following removal of the stomach *(gastrectomy),* but a person must then take vitamin $B_{12}$ by injection, or vitamin $B_{12}$ and intrinsic factor orally. As we age, the gastric mucosa atrophies, less intrinsic factor is secreted, and the risk of pernicious anemia rises.

## Chemical Messengers

The gastric and pyloric glands have various kinds of enteroendocrine cells that collectively produce as many

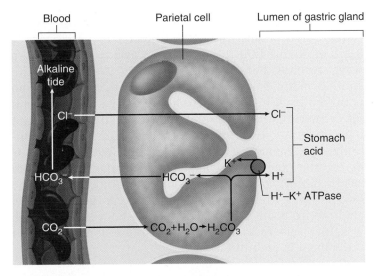

**FIGURE 25.14** **The Mode of Hydrochloric Acid Secretion.** The parietal cell combines water with $CO_2$ from the blood to form carbonic acid (bottom line of figure). Carbonic acid breaks down into bicarbonate ion ($HCO_3^-$) and hydrogen ion ($H^+$). $HCO_3^-$ returns to the blood. In exchange $Cl^-$ enters the lumen with $H^+$ and the two form hydrochloric acid.

▶ *What role does active transport play in this process?*

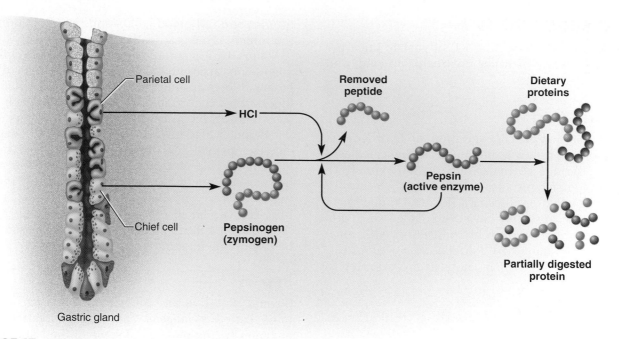

**FIGURE 25.15  The Production and Action of Pepsin.** The chief cells secrete pepsinogen and the parietal cells secrete HCl. HCl removes some of the amino acids from pepsinogen and converts it to pepsin. Pepsin catalyzes the production of more pepsin (an auto-catalytic effect), as well as partially digesting dietary protein.

as 20 chemical messengers. Most of these are hormones—they travel in the bloodstream and stimulate distant target cells. Some also behave as paracrine secretions, diffusing a short distance away and stimulating other cells in the gastric mucosa. Several of these are peptides produced in both the digestive tract and the central nervous system; thus they are called **gut-brain peptides.** These include substance P, vasoactive intestinal peptide (VIP), secretin, gastric inhibitory peptide (GIP), cholecystokinin, and neuropeptide Y (NPY). The functions of some of these peptides in digestion will be explained in the following sections, and their roles in appetite regulation are discussed in chapter 26.

Several of the gastric secretions are summarized in table 25.2. Some of the functions listed there are explained later in the chapter.

## GASTRIC MOTILITY

As you begin to swallow, the swallowing center of the medulla oblongata signals the stomach to relax, thus preparing it to receive food. The arriving food stretches the stomach and activates the *receptive-relaxation response* of smooth muscle: the stomach briefly resists stretching but then relaxes and is able to accommodate more food.

Soon, the stomach shows a rhythm of peristaltic contractions. These are governed by pacemaker cells in the longitudinal layer of the muscularis externa of the greater curvature. About every 20 seconds, a gentle ripple begins in the fundus and becomes stronger as it progresses toward the pyloric region, where the muscularis externa is thicker. After 30 minutes or so, these contractions become quite strong. They churn the food, mix it with gastric juice, and promote its physical breakup and chemical digestion.

The antrum holds about 30 mL of chyme. As a peristaltic wave passes down the antrum, it squirts about 3 mL of chyme into the duodenum at a time. When the wave reaches the pyloric sphincter, it squeezes the sphincter shut. Chyme that does not get through this time is turned back into the antrum and body of the stomach for further digestion. Allowing only small amounts into the duodenum at a time enables the duodenum to neutralize the stomach acid and digest nutrients little by little. If the duodenum becomes overfilled, it inhibits gastric motility and postpones receiving more chyme; the mechanism for this is discussed shortly. A typical meal is emptied from the stomach in about 4 hours, but it takes less time if the meal is more liquid and as long as 6 hours if the meal is high in fat.

## VOMITING

**Vomiting** is the forceful ejection of stomach and intestinal contents (chyme) from the mouth. It involves multiple muscular actions integrated by the **emetic**[14] **center** of the

---

[14]*emet* = vomiting

| TABLE 25.2 | Major Secretions of the Gastric Glands | |
|---|---|---|
| **Secretory Cells** | **Secretion** | **Function** |
| Mucous neck cells | Mucus | Protects mucosa from HCl and enzymes |
| Parietal cells | Hydrochloric acid | Activates pepsin and lingual lipase; helps liquefy food; reduces dietary iron to usable form ($Fe^{2+}$); destroys ingested pathogens |
| | Intrinsic factor | Enables small intestine to absorb vitamin $B_{12}$ |
| Chief cells | Pepsinogen | Converted to pepsin, which digests protein |
| | Chymosin | Coagulates milk proteins in infant stomach; not secreted in adults |
| | Gastric lipase | Digests fats in infant stomach; not secreted in adults |
| Enteroendocrine cells | Gastrin | Stimulates gastric glands to secrete HCl and enzymes; stimulates intestinal motility; relaxes ileocecal valve |
| | Serotonin | Stimulates gastric motility |
| | Histamine | Stimulates HCl secretion |
| | Somatostatin | Inhibits gastric secretion and motility; delays emptying of stomach; inhibits secretion by pancreas; inhibits gallbladder contraction and bile secretion; reduces blood circulation and nutrient absorption in small intestine |

medulla oblongata. Vomiting is commonly induced by overstretching of the stomach or duodenum; chemical irritants such as alcohol and bacterial toxins; visceral trauma (especially to the pelvic organs); intense pain; or psychological and sensory stimuli that activate the emetic center (thus vomiting can be induced by repugnant sights, smells, and thoughts).

Vomiting is usually preceded by nausea and retching. In **retching,** thoracic expansion and abdominal contraction create a pressure difference that dilates the esophagus. The lower esophageal sphincter relaxes while the stomach and duodenum contract spasmodically. Chyme enters the esophagus but then drops back into the stomach as the muscles relax; it does not get past the upper esophageal sphincter. Retching is often accompanied by tachycardia, profuse salivation, and sweating. Vomiting occurs when abdominal contraction and rising thoracic pressure force the upper esophageal sphincter open, the esophagus and body of the stomach relax, and chyme is driven out of the stomach and mouth by strong abdominal contraction combined with reserve peristalsis of the gastric antrum and the duodenum. **Projectile vomiting** is sudden vomiting with no prior nausea or retching. It may be caused by neurological lesions but is also common in infants after feeding.

Chronic vomiting can cause dangerous fluid, electrolyte, and acid–base imbalances. In cases of frequent vomiting, as in the eating disorder *bulimia,* the tooth enamel becomes severely eroded by the hydrochloric acid in the chyme. Aspiration (inhalation) of this acid is very destructive to the respiratory tract. Many have died from aspiration of vomit when they were unconscious or semiconscious. This is the reason that surgical anesthesia, which may induce nausea, must be preceded by fasting until the stomach and small intestine are empty.

## DIGESTION AND ABSORPTION

Salivary and gastric enzymes partially digest protein and small amounts of starch and fat in the stomach, but most digestion and nearly all nutrient absorption occur after the chyme passes into the small intestine. The stomach does not absorb any significant amount of nutrients but does absorb aspirin and some lipid-soluble drugs. Alcohol is absorbed mainly by the small intestine, so its intoxicating effect depends partly on how rapidly the stomach is emptied.

## PROTECTION OF THE STOMACH

You may wonder why the stomach does not digest itself. We can digest tripe (animal stomachs) as readily as any other meat. The living stomach, however, is protected in three ways from the harsh acidic and enzymatic environment it creates:

1. **Mucous coat.** The thick, highly alkaline mucus resists the action of acid and enzymes.

2. **Epithelial cell replacement.** The stomach's epithelial cells live only 3 to 6 days and are then sloughed off into the chyme and digested with the food. They are replaced just as rapidly, however, by cell division in the gastric pits.

3. **Tight junctions.** The epithelial cells are joined by tight junctions that prevent gastric juice from seeping between them and digesting the connective tissue of the lamina propria or beyond.

The breakdown of these protective mechanisms can result in inflammation and peptic ulcer (see Insight 25.2).

# REGULATION OF GASTRIC FUNCTION

The nervous and endocrine systems collaborate to increase gastric secretion and motility when food is eaten and suppress them as the stomach empties. Gastric activity is divided into three stages called the cephalic, gastric, and intestinal phases, based on whether the stomach is being controlled by the brain, by itself, or by the small intestine, respectively (fig. 25.17). These phases overlap and all three can occur simultaneously.

## The Cephalic Phase

The **cephalic phase** is the stage in which the stomach responds to the mere sight, smell, taste, or thought of food. These sensory and mental inputs converge on the hypothalamus, which relays signals to the medulla oblongata. Vagus nerve fibers from the medulla stimulate the enteric nervous system of the stomach which, in turn, stimulates gastric activity.

---

**INSIGHT 25.2**    Clinical Application

### Peptic Ulcer

Inflammation of the stomach, called *gastritis,* can lead to a *peptic ulcer* as pepsin and hydrochloric acid erode the stomach wall (fig. 25.16). Peptic ulcers occur even more commonly in the duodenum and occasionally in the esophagus. If untreated, they can perforate the organ and cause fatal hemorrhaging or peritonitis. Most such fatalities occur in people over age 65.

There is no evidence to support the popular belief that peptic ulcers result from psychological stress. Hypersecretion of acid and pepsin is sometimes involved, but even normal secretion can cause ulceration if the mucosal defense is compromised by other causes. Many or most ulcers involve an acid-resistant bacterium, *Helicobacter pylori,* that invades the mucosa of the stomach and duodenum and opens the way to chemical damage to the tissue. Other risk factors include smoking and the use of aspirin and other nonsteroidal anti-inflammatory drugs (NSAIDs). NSAIDs suppress the synthesis of prostaglandins, which normally stimulate the secretion of protective mucus and acid-neutralizing bicarbonate. Aspirin itself is an acid that directly irritates the gastric mucosa.

Until recently, the most widely prescribed drug in the United States was cimetidine (Tagamet), which was designed to treat peptic ulcers by reducing acid secretion. Histamine stimulates acid secretion by binding to sites on the parietal cells called $H_2$ *receptors;* cimetidine, an $H_2$ *blocker,* prevents this binding. Lately, however, ulcers have been treated more successfully with antibiotics against *Helicobacter* combined with bismuth suspensions such as Pepto-Bismol. This is a much shorter and less expensive course of treatment and permanently cures about 90% of peptic ulcers, as compared with a cure rate of only 20% to 30% for $H_2$ blockers.

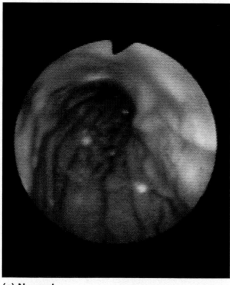
**(a) Normal**

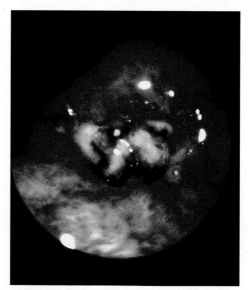
**(b) Peptic ulcer**

**FIGURE 25.16    Endoscopic Views of the Gastroesophageal Junction.**    The esophagus can be seen opening into the cardiac stomach. (a) A healthy gastric mucosa; the small white spots are reflections of light from the endoscope. (b) A bleeding peptic ulcer. A peptic ulcer typically has an oval shape and yellow-white color. Here the yellowish floor of the ulcer is partially obscured by black blood clots, and fresh blood is visible around the margin of the ulcer.

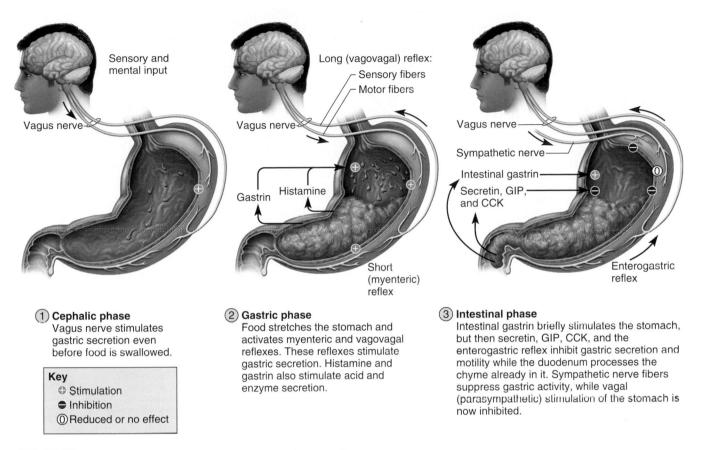

① **Cephalic phase**
Vagus nerve stimulates
gastric secretion even
before food is swallowed.

**Key**
⊕ Stimulation
⊖ Inhibition
Ⓞ Reduced or no effect

② **Gastric phase**
Food stretches the stomach and
activates myenteric and vagovagal
reflexes. These reflexes stimulate
gastric secretion. Histamine and
gastrin also stimulate acid and
enzyme secretion.

③ **Intestinal phase**
Intestinal gastrin briefly stimulates the stomach,
but then secretin, GIP, CCK, and the
enterogastric reflex inhibit gastric secretion and
motility while the duodenum processes the
chyme already in it. Sympathetic nerve fibers
suppress gastric activity, while vagal
(parasympathetic) stimulation of the stomach is
now inhibited.

**FIGURE 25.17** Neural and Hormonal Control of Gastric Secretion.

## The Gastric Phase

The **gastric phase** is a period in which swallowed food and semidigested protein (peptides and amino acids) activate gastric activity. About two-thirds of gastric secretion occurs during this phase. Ingested food stimulates gastric activity in two ways: by stretching the stomach and by raising the pH of its contents. Stretch activates two reflexes: a short reflex mediated through the myenteric nerve plexus, and a long reflex mediated through the vagus nerves and brainstem.

Gastric secretion is stimulated chiefly by three chemicals: acetylcholine (ACh), histamine, and gastrin. ACh is secreted by parasympathetic nerve fibers of both the short and long reflex pathways. Histamine is a paracrine secretion from enteroendocrine cells in the gastric glands. **Gastrin** is a hormone produced by enteroendocrine **G cells** in the pyloric glands.

All three of these stimulate parietal cells to secrete hydrochloric acid and intrinsic factor. The chief cells secrete pepsinogen in response to gastrin and especially ACh, and ACh also stimulates mucus secretion.

As dietary protein is digested, it breaks down into smaller peptides and amino acids, which directly stimulate the G cells to secrete even more gastrin—a positive feedback loop that accelerates protein digestion (fig. 25.18). Small peptides also buffer stomach acid so the pH does not fall excessively low. But as digestion continues and these peptides are emptied from the stomach, the pH drops lower and lower. Below pH 2, stomach acid inhibits the parietal cells and G cells—a negative feedback loop that winds down the gastric phase as the need for pepsin and HCl declines.

## The Intestinal Phase

The **intestinal phase** is a stage in which the duodenum responds to arriving chyme and moderates gastric activity through hormones and nervous reflexes. The duodenum initially enhances gastric secretion, but soon inhibits it. Stretching of the duodenum activates vagovagal reflexes that stimulate the stomach, and peptides and amino acids in the chyme stimulate G cells of the duodenum to secrete more gastrin, which further stimulates the stomach.

Soon, however, the acid and semidigested fats in the duodenum trigger the **enterogastric reflex**—the duodenum sends inhibitory signals to the stomach by way of the enteric nervous system, and sends signals to the medulla that (1) inhibit the vagal nuclei, thus reducing vagal stimulation of the stomach, and (2) stimulate sympathetic neurons, which send inhibitory signals to the stomach. Chyme also stimulates duodenal enteroendocrine cells to

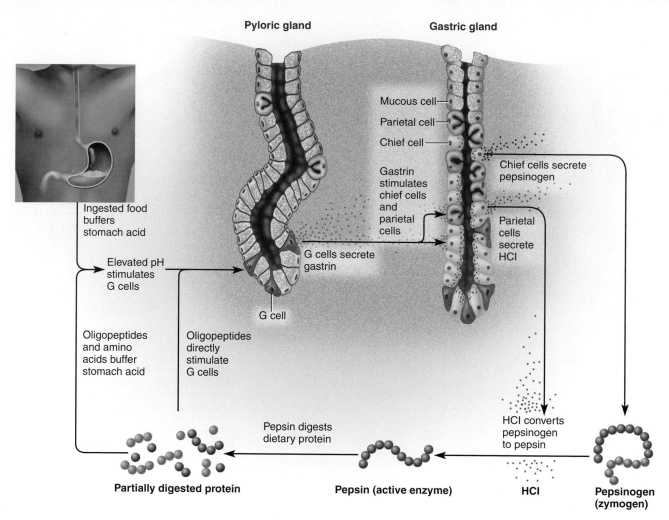

**Pyloric gland**

**Gastric gland**

Mucous cell
Parietal cell
Chief cell

Gastrin stimulates chief cells and parietal cells

Chief cells secrete pepsinogen

Parietal cells secrete HCl

G cells secrete gastrin

Ingested food buffers stomach acid

Elevated pH stimulates G cells

Oligopeptides and amino acids buffer stomach acid

Oligopeptides directly stimulate G cells

G cell

HCl converts pepsinogen to pepsin

Pepsin digests dietary protein

**Partially digested protein**    **Pepsin (active enzyme)**    **HCl**    **Pepsinogen (zymogen)**

**FIGURE 25.18** **Feedback Control of Gastric Secretion.** This positive feedback loop declines and stops as the stomach is emptied and the pH drops.

release **secretin, cholecystokinin** (CO-leh-SIS-toe-KY-nin) **(CCK),** and **gastric inhibitory peptide (GIP).** Secretin and CCK primarily stimulate the pancreas and gallbladder, as discussed later, but all three of these hormones suppress gastric secretion and motility. The effect of all this is that gastrin secretion declines and the pyloric sphincter contracts tightly to limit the admission of more chyme into the duodenum. This gives the duodenum time to work on the chyme it has already received before being loaded with more.

## Before You Go On

*Answer the following questions to test your understanding of the preceding section:*

10. *Name four types of epithelial cells of the gastric and pyloric glands and state what each one secretes.*

11. *Explain how the gastric glands produce hydrochloric acid and how this produces an alkaline tide.*

12. *What positive feedback cycle can you identify in the formation and action of pepsin?*

13. *How does food in the duodenum inhibit motility and secretion in the stomach?*

# The Liver, Gallbladder, and Pancreas

### Objectives

When you have completed this section, you should be able to

- describe the gross and microscopic anatomy of the liver, gallbladder, bile duct system, and pancreas;
- describe the digestive secretions and functions of the liver, gallbladder, and pancreas; and
- explain how hormones regulate secretions of the liver and pancreas.

The small intestine receives not only chyme from the stomach but also secretions from the liver and pancreas, which enter the digestive tract near the junction of the stomach and small intestine. These secretions are so important to the digestive processes of the small intestine that it is necessary to understand them before continuing with intestinal physiology.

# THE LIVER

The liver (fig. 25.19) is a reddish brown gland located immediately inferior to the diaphragm, filling most of the right hypochondriac and epigastric regions. It is the body's largest gland, weighing about 1.4 kg (3 lb). The liver has a tremendous variety of functions, but only one of them, the secretion of bile, contributes to digestion. Others are discussed in the following chapter, which provides a more thorough physiological basis for understanding liver function.

## Gross Anatomy

The liver has four lobes called the right, left, quadrate, and caudate lobes. From an anterior view, we see only a large **right lobe** and smaller **left lobe.** They are separated from each other by the **falciform**[15] **ligament,** a sheet of mesentery that suspends the liver from the diaphragm

---

[15]*falci* = sickle + *form* = shape

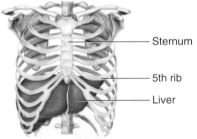

- Sternum
- 5th rib
- Liver

**(a) Location**

and anterior abdominal wall. The **round ligament** (ligamentum teres), also visible anteriorly, is a fibrous remnant of the umbilical vein, which carries blood from the umbilical cord to the liver of a fetus.

From the inferior view, we also see a squarish **quadrate lobe** next to the gallbladder and a tail-like **caudate**[16] **lobe** posterior to that. An irregular opening between these lobes, the **porta hepatis,**[17] is a point of entry for the hepatic portal vein and proper hepatic artery and a point of exit for the bile passages, all of which travel in the lesser omentum. The gallbladder adheres to a depression on the inferior surface of the liver between the right and quadrate lobes. The posterior aspect of the liver has a deep groove (sulcus) that accommodates the inferior vena cava. The superior surface has a *bare area* where it is attached to the diaphragm. The rest of the liver is covered by a serosa.

## Microscopic Anatomy

The interior of the liver is filled with innumerable tiny cylinders called **hepatic lobules,** about 2 mm long by 1 mm in diameter. A lobule consists of a **central vein** passing down its core, surrounded by radiating sheets of cuboidal cells called **hepatocytes** (fig. 25.20). Imagine spreading a book wide open until its front and back covers

---

[16]*caud* = tail
[17]*porta* = gateway, entrance + *hepatis* = of the liver

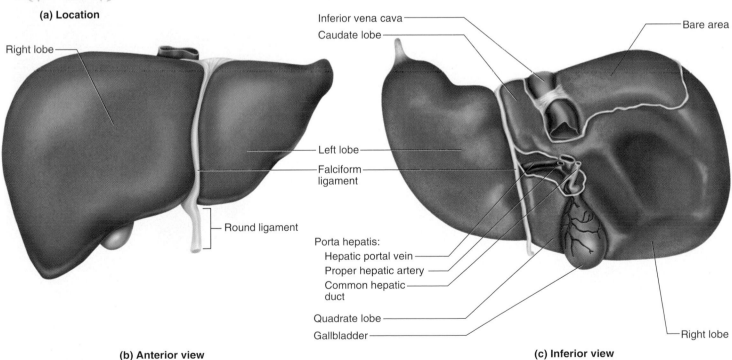

**(b) Anterior view**

- Right lobe
- Left lobe
- Falciform ligament
- Round ligament

**(c) Inferior view**

- Inferior vena cava
- Caudate lobe
- Bare area
- Porta hepatis:
  - Hepatic portal vein
  - Proper hepatic artery
  - Common hepatic duct
- Quadrate lobe
- Gallbladder
- Right lobe

**FIGURE 25.19  Gross Anatomy of the Liver.**

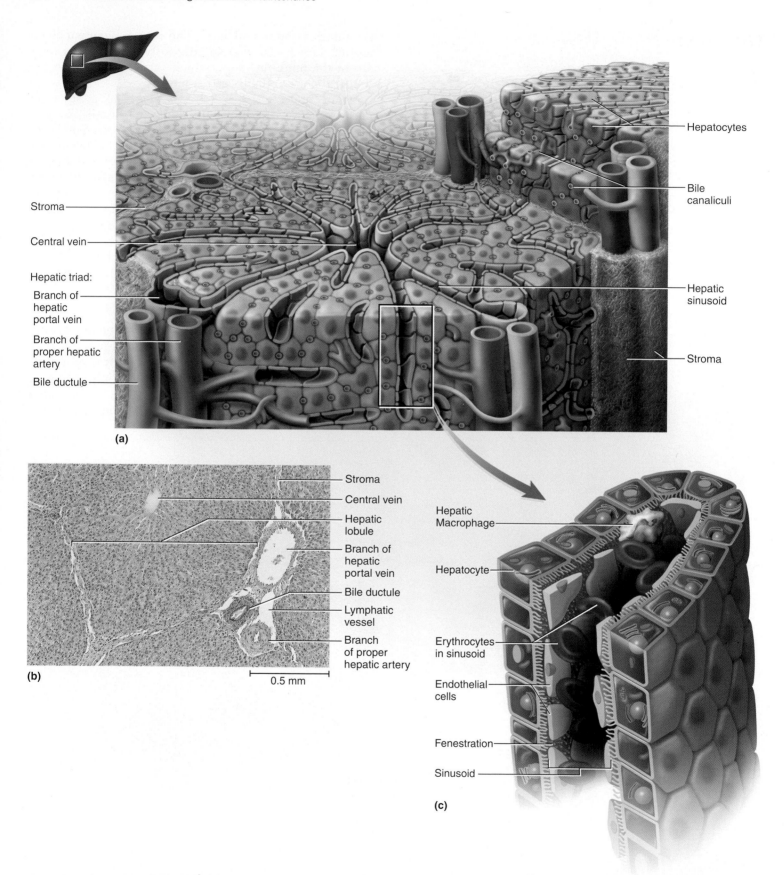

Stroma

Central vein

Hepatic triad:

Branch of hepatic portal vein

Branch of proper hepatic artery

Bile ductule

Hepatocytes

Bile canaliculi

Hepatic sinusoid

Stroma

(a)

Stroma

Central vein

Hepatic lobule

Branch of hepatic portal vein

Bile ductule

Lymphatic vessel

Branch of proper hepatic artery

0.5 mm

(b)

Hepatic Macrophage

Hepatocyte

Erythrocytes in sinusoid

Endothelial cells

Fenestration

Sinusoid

(c)

**FIGURE 25.20  Microscopic Anatomy of the Liver.** (a) The hepatic lobules and their relationship to the blood vessels and bile tributaries. (b) Histological section of the liver. (c) A hepatic sinusoid.

▶ *Identify two blood vessels in chapter 20 that supply blood to the hepatic sinusoids.*

touch. The pages of the book would fan out around the spine somewhat like the plates of hepatocytes fan out from the central vein of a liver lobule.

Each plate of hepatocytes is an epithelium one or two cells thick. The spaces between the plates are blood-filled channels called **hepatic sinusoids.** The sinusoids are lined by a fenestrated endothelium that separates the hepatocytes from the bloodstream, but allows blood plasma into the space between the hepatocytes and endothelium. The blood filtering through the sinusoids comes directly from the stomach and intestines. After a meal, the liver removes glucose, amino acids, iron, vitamins, and other nutrients from it for metabolism or storage. It also removes and degrades hormones, toxins, bile pigments, and drugs. At the same time, the liver secretes albumin, lipoproteins, clotting factors, angiotensinogen, and other products into the blood. Between meals, it breaks down stored glycogen and releases glucose into the circulation. The sinusoids also contain phagocytic cells called **hepatic macrophages** (Kupffer[18] cells), which remove bacteria and debris from the blood.

[18]Karl W. von Kupffer (1829–1902), German anatomist

The hepatic lobules are separated by a sparse connective tissue stroma. In cross sections, the stroma is especially visible in the triangular areas where three or more lobules meet. Here there is often a **hepatic triad** of two blood vessels and a bile ductule. The blood vessels are small branches of the proper hepatic artery and hepatic portal vein. Both of them supply blood to the sinusoids, which therefore receive a mixture of nutrient-laden venous blood from the intestines and freshly oxygenated arterial blood from the celiac trunk. After filtering through the sinusoids, this blood collects in the central vein. From here, it ultimately flows into the right and left hepatic veins, which leave the liver at its superior surface and drain immediately into the inferior vena cava.

The liver secretes bile into narrow channels, the **bile canaliculi,** between sheets of hepatocytes. Bile passes from there into the small **bile ductules** of the triads and ultimately into the **right** and **left hepatic ducts.** The two hepatic ducts converge on the inferior side of the liver to form the **common hepatic duct.** A short distance farther on, this is joined by the **cystic duct** coming from the gallbladder (fig. 25.21). Their union forms

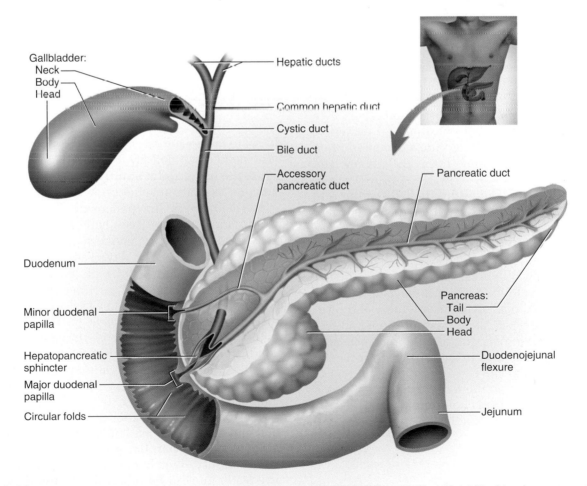

**FIGURE 25.21 Gross Anatomy of the Gallbladder, Pancreas, and Bile Passages.** The liver is omitted to show more clearly the gallbladder, which aheres to its inferior surface, and the hepatic ducts, which emerge from the liver tissue.

the **bile duct,** which descends through the lesser omentum toward the duodenum. Near the duodenum, the bile duct joins the duct of the pancreas and forms an expanded chamber called the **hepatopancreatic ampulla.** The ampulla terminates at a fold of tissue, the **major duodenal papilla,** on the duodenal wall. This papilla contains a muscular **hepatopancreatic sphincter** (sphincter of Oddi[19]), which regulates the passage of bile and pancreatic juice into the duodenum. Between meals, this sphincter is closed and prevents the release of bile into the intestine.

## THE GALLBLADDER AND BILE

The **gallbladder** is a sac on the underside of the liver that serves to store and concentrate bile. It is about 10 cm long and internally lined by a highly folded mucosa with a simple columnar epithelium. Its head *(fundus)* usually projects slightly beyond the inferior margin of the liver. Its neck *(cervix)* leads into the cystic duct, which leads in turn to the bile duct.

**Bile** is a yellow-green fluid containing minerals, cholesterol, neutral fats, phospholipids, bile pigments, and bile acids. The principal pigment is **bilirubin,** derived from the decomposition of hemoglobin. Bacteria of the large intestine metabolize bilirubin to **urobilinogen,** which is responsible for the brown color of feces. In the absence of bile secretion, the feces are grayish white and marked with streaks of undigested fat *(acholic feces).* **Bile acids** (bile salts) are steroids synthesized from cholesterol. Bile acids and lecithin, a phospholipid, aid in fat digestion and absorption, as discussed later. All other components of the bile are wastes destined for excretion in the feces. When these waste products become excessively concentrated, they may form gallstones (see Insight 25.3).

Bile gets into the gallbladder by first filling the bile duct, then overflowing into the gallbladder. Between meals, the gallbladder absorbs water and electrolytes from the bile and concentrates it by a factor of 5 to 20 times. The liver secretes about 500 to 1,000 mL of bile per day.

About 80% of the bile acids are reabsorbed in the ileum, the last portion of the small intestine, and returned to the liver, where the hepatocytes absorb them and resecrete them. This route of secretion, reabsorption, and resecretion, called the *enterohepatic circulation,* reabsorbs and reuses the bile acids two or more times during the digestion of an average meal. The 20% of the bile that is not reabsorbed is excreted in the feces. This is the body's only way of eliminating excess cholesterol. The liver synthesizes new bile acids from cholesterol to replace the quantity lost in the feces.

---

[19]Ruggero Oddi (1864–1913), Italian physician

### INSIGHT 25.3    Clinical Application

#### Gallstones

*Gallstones* (biliary calculi) are hard masses in the gallbladder or bile ducts, usually composed of cholesterol, calcium carbonate, and bilirubin. *Cholelithiasis,* the formation of gallstones, is most common in obese women over the age of 40 and usually results from excess cholesterol. The gallbladder may contain several gallstones, some over 1 cm in diameter. Gallstones cause excruciating pain when they obstruct the bile ducts or when the gallbladder or bile ducts contract. When they block the flow of bile into the duodenum, they cause jaundice (yellowing of the skin due to bile pigment accumulation), poor fat digestion, and impaired absorption of fat-soluble vitamins. Once treated only by surgical removal, gallstones are now often treated with stone-dissolving drugs or by *lithotripsy,* the use of ultrasonic vibration to pulverize them without surgery. Reobstruction can be prevented by inserting a stent (tube) into the bile duct, which keeps it distended and allows gallstones to pass while they are still small.

**Think About It**

*Certain drugs designed to reduce blood cholesterol work by blocking the reabsorption of bile acids in the ileum. Explain why they would have this cholesterol-lowering effect.*

## THE PANCREAS

The pancreas (fig. 25.21) is a spongy retroperitoneal gland dorsal to the greater curvature of the stomach. It has a globose *head* encircled by the duodenum, a midportion called the *body,* and a blunt, tapered *tail* on the left. The pancreas is both an endocrine and exocrine gland. Its endocrine part is the pancreatic islets, which secrete insulin and glucagon (see chapter 17). Most of the pancreas is exocrine tissue, which secretes 1,200 to 1,500 mL of **pancreatic juice** per day. The cells of the secretory acini exhibit a high density of rough ER and *zymogen granules,* which are vesicles filled with secretion (fig. 25.22). These acini open into a system of larger and larger ducts that eventually converge on the main **pancreatic duct.** This duct runs lengthwise through the middle of the gland and joins the bile duct at the hepatopancreatic ampulla. The hepatopancreatic sphincter thus controls the release of both bile and pancreatic juice into the duodenum. Usually, however, there is a smaller **accessory pancreatic duct** that branches from the main pancreatic duct and opens independently into the duodenum at the **minor duodenal papilla,** proximal to the major papilla. The accessory duct bypasses the sphincter and allows pancreatic juice to be released into the duodenum even when bile is not.

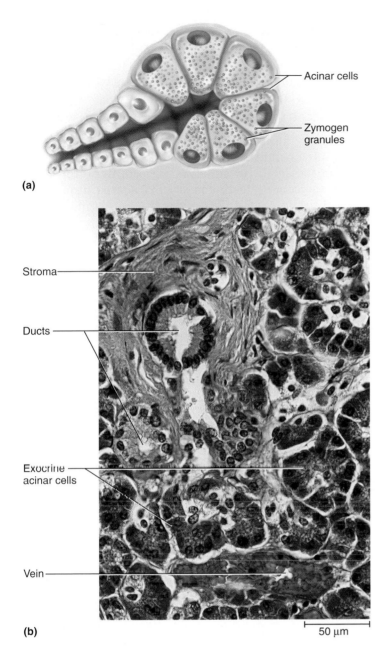

**(a)**

**(b)**

50 μm

**FIGURE 25.22   Histology of the Pancreas.** (a) An acinus. (b) Histological section of the exocrine tissue and some of the connective tissue stroma.

Pancreatic juice is an alkaline mixture of water, enzymes, zymogens, sodium bicarbonate, and other electrolytes. The acini secrete the enzymes and zymogens, whereas the ducts secrete the sodium bicarbonate. The bicarbonate buffers HCl arriving from the stomach.

The pancreatic zymogens are **trypsinogen** (trip-SIN-oh-jen), **chymotrypsinogen** (KY-mo-trip-SIN-o-jen), and **procarboxypeptidase** (PRO-car-BOC-see-PEP-tih-dase). When trypsinogen is secreted into the intestinal lumen, it is converted to **trypsin** by **enterokinase,** an enzyme on the surface of the intestinal epithelial cells (fig. 25.23).

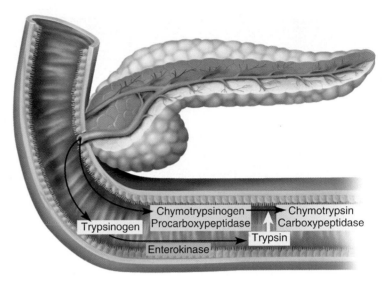

**FIGURE 25.23   The Activation of Pancreatic Enzymes in the Small Intestine.**   The pancreas secretes trypsinogen, and enterokinase in the intestinal brush border converts it to trypsin. Trypsin not only digests dietary protein but also activates two other pancreatic zymogens: chymotrypsinogen and procarboxypeptidase.

Trypsin then converts the other two zymogens into **chymotrypsin** and **carboxypeptidase,** in addition to its primary role of digesting dietary protein.

Other pancreatic enzymes include **pancreatic amylase,** which digests starch; **pancreatic lipase,** which digests fat; and **ribonuclease** and **deoxyribonuclease,** which digest RNA and DNA, respectively. Unlike the zymogens, these enzymes are not altered after secretion. They become active, however, only upon exposure to bile and ions in the intestinal lumen.

The exocrine secretions of the pancreas are summarized in table 25.3. Their specific digestive functions are explained later in more detail.

## REGULATION OF SECRETION

Bile and pancreatic juice are secreted in response to parasympathetic (vagal) stimulation and inhibited by sympathetic stimulation, and both are stimulated by the hormones cholecystokinin (CCK), gastrin, and secretin. The duodenum secretes CCK in response to acid and fat arriving from the stomach. CCK triggers three responses: (1) contraction of the gallbladder, which forces bile into the bile duct; (2) secretion of pancreatic enzymes; and (3) relaxation of the hepatopancreatic sphincter, which allows bile and pancreatic juice to be released into the duodenum. Gastrin from the stomach and duodenum stimulates gallbladder contraction and pancreatic enzyme secretion, but only half as strongly as CCK does. Acidic chyme also stimulates the duodenum to secrete **secretin,** the first hormone ever discovered (by William Bayliss and Ernest Starling in 1902). Secretin stimulates the hepatic

| **TABLE 25.3** | Exocrine Secretions of the Pancreas |
|---|---|
| **Secretion** | **Function** |
| *Sodium bicarbonate* | Neutralizes HCl |
| *Zymogens* | Converted to active digestive enzymes after secretion |
| Trypsinogen | Becomes trypsin, which digests protein |
| Chymotrypsinogen | Becomes chymotrypsin, which digests protein |
| Procarboxypeptidase | Becomes carboxypeptidase, which hydrolyzes the terminal amino acid from the carboxyl (—COOH) end of small peptides |
| *Enzymes* | |
| Pancreatic amylase | Digests starch |
| Pancreatic lipase | Digests fat |
| Ribonuclease | Digests RNA |
| Deoxyribonuclease | Digests DNA |

bile ducts and pancreatic ducts to secrete bicarbonate, so the bile and pancreatic juice both help to neutralize stomach acid in the duodenum.

**Think About It**

*Draw a negative feedback loop showing how secretin influences duodenal pH.*

### Before You Go On

*Answer the following questions to test your understanding of the preceding section:*

14. What does the liver contribute to digestion?

15. Trace the pathway taken by bile acids from the liver and back. What is this pathway called?

16. Name two hormones, four enzymes, and one buffer secreted by the pancreas, and state the function of each.

17. What stimulates cholecystokinin (CCK) secretion, and how does CCK affect other parts of the digestive system?

## The Small Intestine

### Objectives

When you have completed this section, you should be able to

- describe the gross and microscopic anatomy of the small intestine;

- state how the mucosa of the small intestine differs from that of the stomach, and explain the functional significance of the differences;
- define *contact digestion* and describe where it occurs; and
- describe the types of movement that occur in the small intestine.

Nearly all chemical digestion and nutrient absorption occur in the small intestine. To perform these roles efficiently, the small intestine must have a large surface area exposed to the chyme. Thus, it is the longest part of the digestive tract—about 6 to 7 m (19–23 ft) long in a cadaver, but because of muscle tone, only 2 m (6.6 ft) long in a living person. The term *small* intestine refers not to its length but to its diameter—about 2.5 cm (1 in.). Further enhancing its surface area, the mucosa of the small intestine is highly folded, with *circular folds* visible to the naked eye, barely visible projections called *villi*, and microscopic *microvilli* forming a brush border on its absorptive cells.

**Think About It**

*The small intestine exhibits some of the same structural adaptations as the proximal convoluted tubule of the kidney, and for the same reasons. Discuss what they have in common, the reasons for it, and how this relates to this book's theme of the unity of form and function.*

## GROSS ANATOMY

The small intestine is a coiled mass filling most of the abdominal cavity inferior to the stomach and liver. It is divided into three regions (fig. 25.24): the duodenum, jejunum, and ileum.

The **duodenum** (dew-ODD-eh-num, DEW-oh-DEE-num) constitutes the first 25 cm (10 in.). It begins at the pyloric valve, arcs around the head of the pancreas and passes to the left, and ends at a sharp bend called the **duodenojejunal flexure.** Its name refers to its length, about equal to the width of 12 fingers.[20] Along with the pancreas, most of it is retroperitoneal. The duodenum receives the stomach contents, pancreatic juice, and bile. Stomach acid is neutralized here, fats are physically broken up (emulsified) by the bile acids, pepsin is inactivated by the elevated pH, and pancreatic enzymes take over the job of chemical digestion.

The **jejunum** (jeh-JOO-num) is the next 2.5 m (8 ft), or by definition, the first 40% of the small intestine beyond the duodenum. Its name refers to the fact that early anatomists typically found it to be empty.[21] The jejunum

---

[20]*duoden* = 12
[21]*jejun* = empty, dry

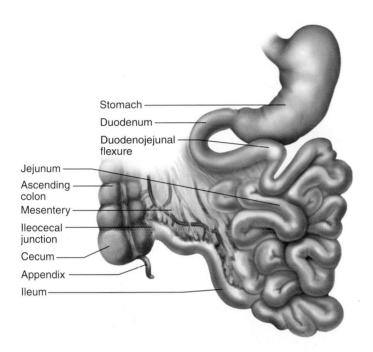

Stomach

Duodenum

Duodenojejunal
flexure

Jejunum

Ascending
colon

Mesentery

Ileocecal
junction

Cecum

Appendix

Ileum

**FIGURE 25.24  Gross Anatomy of the Small Intestine.**
The intestine is pulled aside to expose the mesentery and
ileocecal junction.

begins in the upper left quadrant of the abdomen but lies
mostly within the umbilical region (see fig. A.6). The
jejunum has large, tall, closely spaced circular folds. Most
digestion and nutrient absorption occur here. Its wall is
relatively thick and muscular, and it has an especially rich
blood supply which gives it a relatively red color.

The **ileum**[22] forms the last 3.6 m (12 ft), or 60% of the
post-duodenal small intestine. (The lengths given here are
for the cadaver.) The ileum occupies mainly the hypogastric
region and part of the pelvic cavity. Compared with the
jejunum, its wall is thinner, less muscular, and less vascular, and it has a paler pink color. On the side opposite from
its mesenteric attachment, the ileum has prominent lymphatic nodules in clusters called **Peyer**[23] **patches** (see chapter 21), which are readily visible to the naked eye and
become progressively larger approaching the large intestine.

The end of the small intestine is the **ileocecal** (ILL-ee-
oh-SEE-cul) **junction,** where the ileum joins the *cecum* of
the large intestine. The muscularis of the ileum is thickened at this point to form a sphincter, the **ileocecal** (ILL-
ee-oh-SEE-cul) **valve,** which protrudes into the cecum
and regulates the passage of food residue into the large
intestine. Both the jejunum and ileum are intraperitoneal
and thus covered with a serosa, which is continuous with
the complex, folded mesentery that suspends the small
intestine from the dorsal abdominal wall.

## MICROSCOPIC ANATOMY

The largest folds of the intestinal wall are transverse to
spiral ridges, up to 10 mm high, called **circular folds** (pli-
cae circulares) (see fig. 25.21). These involve only the
mucosa and submucosa; they are not visible on the external surface, which is smooth. They occur from the duodenum to the middle of the ileum, where they cause the
chyme to flow on a spiral path along the intestine. This
slows its progress, causes more contact with the mucosa,
and promotes more thorough mixing and nutrient absorption. Circular folds are relatively small and sparse in the
ileum and not found in the distal half, but most nutrient
absorption is completed by that point.

If the mucosa is examined closely it appears fuzzy,
like a terrycloth towel. This is due to projections called
**villi** (VIL-eye; singular, *villus*), about 0.5 to 1.0 mm high,
with tongue- to fingerlike shapes (fig. 25.25). The villi are
largest in the duodenum and become progressively
smaller in more distal regions of the small intestine. A villus is covered with two kinds of epithelial cells: columnar
**absorptive cells** (enterocytes) and mucus-secreting **goblet
cells.** Like epithelial cells of the stomach, those of the
small intestine are joined by tight junctions that prevent
digestive enzymes from seeping between them.

The core of a villus is filled with areolar tissue of the
lamina propria. Embedded in this tissue are an arteriole,
a capillary network, a venule, and a lymphatic capillary
called a **lacteal** (LAC-tee-ul). Most nutrients are absorbed
by the blood capillaries, but most fat is absorbed by the
lacteal and gives its contents the milky appearance for
which the lacteal is named.[24] The core of the villus also
has a few smooth muscle cells that contract periodically.
This enhances mixing of the chyme in the intestinal
lumen and milks lymph down the lacteal to the larger
lymphatic vessels of the submucosa.

Each absorptive cell of a villus has a fuzzy brush border of microvilli about 1 [mu]m high. The brush border
increases the absorptive surface area of the small intestine
and contains **brush border enzymes,** integral proteins of
the plasma membrane. One of these, enterokinase, activates pancreatic enzymes as explained earlier. Others
carry out some of the final stages of enzymatic digestion.
They are not released into the lumen; instead, the chyme
must contact the brush border for digestion to occur. This
process, called **contact digestion,** is one reason that thorough mixing of the chyme is so important.

On the floor of the small intestine, between the bases of
the villi, there are numerous pores that open into tubular
glands called **intestinal crypts** (crypts of Lieberkühn;[25]
LEE-ber-koohn). These crypts, similar to the gastric
glands, extend as far as the muscularis mucosae. In the

[22]from *eilos* = twisted
[23]Johann K. Peyer (1653–1712), Swiss anatomist
[24]*lact* = milk
[25]Johann N. Lieberkühn (1711–56), German anatomist

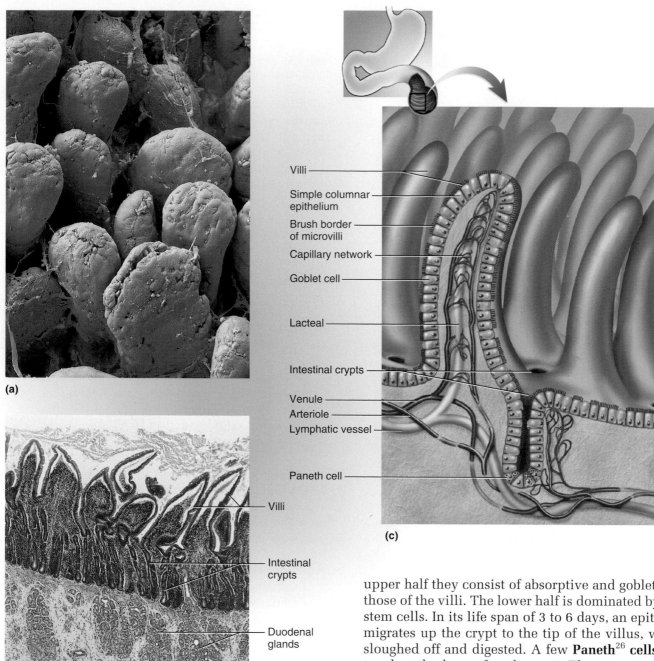

**(a)**

**(b)**

0.5 mm

**FIGURE 25.25   Intestinal Villi.** (a) Villi (SEM). Each villus is about 1 mm high. (b) Histological section of the duodenum showing villi, intestinal crypts, and duodenal glands. (c) Structure of a villus.

*Labels in (b):* Villi; Intestinal crypts; Duodenal glands; Muscularis externa; Serosa

*Labels in (c):* Villi; Simple columnar epithelium; Brush border of microvilli; Capillary network; Goblet cell; Lacteal; Intestinal crypts; Venule; Arteriole; Lymphatic vessel; Paneth cell

**(c)**

upper half they consist of absorptive and goblet cells like those of the villi. The lower half is dominated by dividing stem cells. In its life span of 3 to 6 days, an epithelial cell migrates up the crypt to the tip of the villus, where it is sloughed off and digested. A few **Paneth[26] cells** are clustered at the base of each crypt. They secrete lysozyme, phospholipase, and defensins, all of which protect against bacterial infection.

The duodenum has prominent **duodenal (Brunner[27]) glands** in the submucosa. They secrete an abundance of bicarbonate-rich mucus, which neutralizes stomach acid and shields the mucosa from its corrosive effects. Throughout the small intestine, the lamina propria and submucosa have a large population of lymphocytes that intercept pathogens before they can invade the bloodstream. In some places these are aggregated into conspicuous lymphatic nodules such as the Peyer patches of the ileum.

[26]Josef Paneth (1857–90), Austrian physician
[27]Johann C. Brunner (1653–1727), Swiss anatomist

The muscularis externa consists of a relatively thick inner circular layer and a thinner outer longitudinal layer. Ganglia of the myenteric nerve plexus occur between these layers.

## INTESTINAL SECRETION

The intestinal crypts secrete 1 to 2 L of **intestinal juice** per day, especially in response to acid, hypertonic chyme, and distension of the intestine. This fluid has a pH of 7.4 to 7.8. It contains water and mucus but relatively little enzyme. Most enzymes that function in the small intestine are found in the brush border and pancreatic juice.

## INTESTINAL MOTILITY

Contractions of the small intestine serve three functions: (1) to mix chyme with intestinal juice, bile, and pancreatic juice, allowing these fluids to neutralize acid and digest nutrients more effectively; (2) to churn chyme and bring it into contact with the mucosa for contact digestion and nutrient absorption; and (3) to move residue toward the large intestine.

**Segmentation** is a movement in which ringlike constrictions appear at several places along the intestine and then relax as new constrictions form elsewhere (fig. 25.26a). This is the most common movement of the small intestine. Its effect is to knead or churn the contents, which promotes the mixing of food and digestive secretions and enhances contact digestion. Pacemaker cells of the muscularis externa set the rhythm of segmentation, with contractions about 12 times per minute in the duodenum and 8 to 9 times per minute in the ileum. Since the contractions are less frequent distally, segmentation causes slow progression of the chyme toward the colon. The intensity (but not frequency) of contractions is modified by nervous and hormonal influences.

When most nutrients have been absorbed and little remains but undigested residue, segmentation declines and peristalsis begins. A peristaltic wave begins in the duodenum, travels 10 to 70 cm, and dies out, only to be followed by another wave that begins a little farther down the tract than the first one did (fig. 25.26b). These successive, overlapping waves of contraction are called a **migrating motor complex.** They milk the chyme toward the colon over a period of about 2 hours. A second complex then expels residue and bacteria from the small intestine, thereby helping to limit bacterial colonization. Refilling of the stomach at the next meal suppresses peristalsis and reactivates segmentation.

At the ileocecal junction, the muscularis of the ileum is thickened to form a sphincter, the **ileocecal** (ILL-ee-oh-SEE-cul) **valve,** which protrudes into the cecum like a doughnut. This valve is usually closed. Food in the stomach, however, triggers both the release of gastrin and the **gastroileal reflex,** both of which enhance segmentation in the ileum and relax the valve. As the cecum fills with residue, the pressure pinches the valve shut and prevents the reflux of cecal contents into the ileum.

Table 25.4 summarizes the anatomy of the stomach, accessory glands of the abdomen, and small intestine.

### Before You Go On

*Answer the following questions to test your understanding of the preceding section:*

18. What three structures increase the absorptive surface area of the small intestine?

19. Sketch a villus and label its epithelium, brush border, lamina propria, blood capillaries, and lacteal.

20. Distinguish between segmentation and the migrating motor complex of the small intestine. How do these differ in function?

**(a) Segmentation**

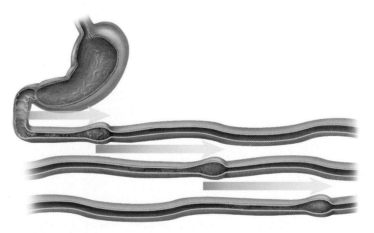

**(b) Peristalsis**

**FIGURE 25.26  Contractions of the Small Intestine.** (a) Segmentation, in which circular constrictions of the intestine cut into the contents, churning and mixing them. (b) The migrating motor complex of peristalsis, in which successive waves of peristalsis overlap each other. Each wave travels partway down the intestine and milks the contents toward the colon.

| **TABLE 25.4** | Anatomical Checklist of the Digestive System from the Stomach Through the Small Intestine |
| --- | --- |

**Stomach**

| | | |
| --- | --- | --- |
| *General features* | *Mucosa* | |
|   Lesser curvature |   Rugae | |
|   Greater curvature |   Gastric pits | |
|   Cardiac region |   Cardiac glands | |
|   Fundic region |   Pyloric glands | |
|   Body (corpus) |   Gastric glands | |
|   Pyloric region |     Mucous neck cells | |
|     Antrum |     Regenerative cells | |
|     Pyloric canal |     Parietal cells | |
|     Pylorus |     Chief cells | |
|     Pyloric sphincter |     Enteroendocrine cells | |
| | *Submucosa* | |
| | *Muscularis externa* | |
| | *Serosa* | |

**Liver**

| | |
| --- | --- |
| *Lobes* | *Bile tributaries* |
|   Right lobe |   Bile canaliculi |
|   Left lobe |   Bile ductules |
|   Quadrate lobe |   Hepatic ducts |
|   Caudate lobe |   Common hepatic duct |
| *Ligaments* | *Gallbladder and bile duct* |
|   Falciform ligament |   Head (fundus) |
|   Round ligament |   Neck (cervix) |
| *Porta hepatis* |   Cystic duct |
| *Microscopic anatomy* |   Bile duct |
|   Hepatocytes |   Hepatopancreatic ampulla |
|   Hepatic lobules |   Hepatopancreatic sphincter |
|   Central vein | |
|   Hepatic sinusoids | |
|   Hepatic macrophages | |
|   Hepatic triads | |

**Pancreas**

| | |
| --- | --- |
| *Head* | *Pancreatic islets* |
| *Body* | *Pancreatic duct* |
| *Tail* | *Accessory pancreatic duct* |
| *Exocrine acini* | |

**Small Intestine**

| | |
| --- | --- |
| *Duodenum* | *Mucosa* |
|   Major duodenal papilla |   Circular folds |
|   Minor duodenal papilla |   Villi |
|   Duodenal glands |     Goblet cells |
|   Duodenojejunal flexure |     Absorptive cells |
| *Jejunum* |     Microvilli |
| *Ileum* |     Lacteal |
|   Peyer patches |   Intestinal crypts |
|   Ileocecal valve |     Paneth cells |
|   Ileocecal junction | |

# Chemical Digestion and Absorption

### Objectives

When you have completed this section, you should be able to

- describe how each major class of nutrients is chemically digested, name the enzymes involved, and discuss the functional differences among these enzymes; and

- describe how each type of nutrient is absorbed by the small intestine.

Chemical digestion and nutrient absorption are essentially finished by the time food residue leaves the small intestine and enters the cecum. But before going on to the functions of the large intestine, we trace each major class of nutrients—especially carbohydrates, proteins, and fats—from the mouth through the small intestine to see how it is chemically degraded and absorbed.

# CARBOHYDRATES

Most digestible dietary carbohydrate is starch. Cellulose is indigestible and is not considered here, although its importance as dietary fiber is discussed in chapter 26. The amount of glycogen in the diet is negligible, but it is digested in the same manner as starch.

## Carbohydrate Digestion

Starch is digested first to oligosaccharides up to eight glucose residues long, then into the disaccharide maltose, and finally to glucose, which is absorbed by the small intestine. The process begins in the mouth, where salivary amylase hydrolyzes starch into oligosaccharides. Salivary amylase functions best at pH 6.8 to 7.0, typical of the oral cavity. It is quickly denatured upon contact with stomach acid, but it can digest starch for as long as 1 to 2 hours in the stomach as long as it is in the middle of a food mass and escapes contact with the acid. Amylase therefore works longer when the meal is larger, especially in the fundus, where gastric motility is weakest and a food bolus takes longer to break up. As acid, pepsin, and the churning contractions of the stomach break up the bolus, amylase is denatured; it does not function at a pH any lower than 4.5. Being a protein, amylase is then digested by pepsin along with the dietary proteins. About 50% of the dietary starch is digested before it reaches the small intestine.

Starch digestion resumes in the small intestine when the chyme mixes with pancreatic amylase (fig. 25.27). Starch is entirely converted to oligosaccharides and maltose within 10 minutes. Its digestion is completed as the chyme contacts the brush border of the absorptive cells. Two brush border enzymes, **dextrinase** and **glucoamylase,** hydrolyze oligosaccharides that are three or more residues long. The third, **maltase,** hydrolyzes maltose to glucose.

Maltose is also present in some foods, but the major dietary disaccharides are sucrose (cane sugar) and lactose (milk sugar). They are digested by the brush border enzymes **sucrase** and **lactase,** respectively, and the resulting monosaccharides (glucose and fructose from the former; glucose and galactose from the latter) are immediately absorbed. In most of the world population, however, lactase production ceases or declines to a low level after age 4 and lactose becomes indigestible (see Insight 25.4).

## Carbohydrate Absorption

The plasma membrane of the absorptive cells has transport proteins that absorb monosaccharides as soon as the brush border enzymes release them (fig. 25.28). About 80% of the absorbed sugar is glucose, which is taken up by a sodium–glucose transport protein (SGLT) like that of the kidney tubules (see p. 911). After a high-carbohydrate meal, however, solvent drag absorbs two to three times as much glucose as the SGLT. Sugars entering the extracellular fluid (ECF) at the base of the intestinal epithelium

---

### INSIGHT 25.4    Clinical Application

#### Lactose Intolerance

Humans are a strange species. Unique among mammals, we go on drinking milk in adulthood, and moreover, we drink the milk of other species! This odd habit is largely limited, however, to people of western and northern Europe, a few pastoral tribes of Africa, and their descendants in the Americas and elsewhere. They have an ancestral history of milking domestic animals, a practice that goes back about 10,000 years and has coincided with the continued production of lactase into adulthood.

People without lactase have *lactose intolerance.* If they consume milk, lactose passes undigested into the large intestine, increases the osmolarity of the intestinal contents, and causes colonic water retention and diarrhea. In addition, lactose fermentation by intestinal bacteria produces gas, resulting in painful cramps and flatulence.

---

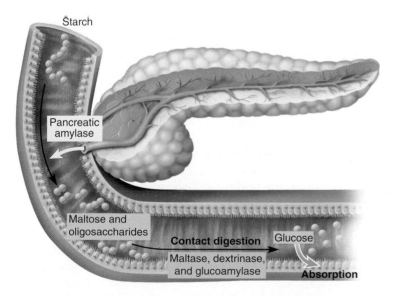

**FIGURE 25.27  Starch Digestion in the Small Intestine.** Pancreatic amylase digests starch into maltose and small oligosaccharides. Brush border enzymes (maltase, dextrinase, and glucoamylase) digest these to glucose, which is absorbed by the epithelial cells.

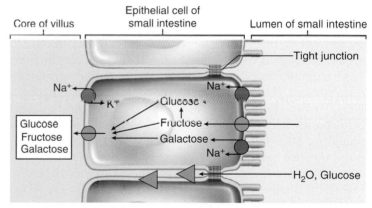
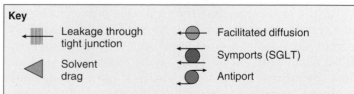

**FIGURE 25.28  Monosaccharide Absorption by the Small Intestine.** Glucose and galactose are absorbed by the SGLT cotransporter in the apical membrane of the absorptive cell *(right).* Glucose is also abosrbed along with water through the paracellular route *(between cells)* by solvent drag. Fructose is absorbed by a separate carrier using facilitated diffusion. Most fructose is converted to glucose within the epithelial cell. The monosaccharides pass through the basal membrane of the cell by facilitated diffusion *(left).*

increase its osmolarity. Water then passes osmotically from the lumen, through the now-leaky tight junctions between the epithelial cells, and into the ECF, carrying glucose and other nutrients with it.

The SGLT also absorbs galactose, whereas fructose is absorbed by facilitated diffusion using a separate carrier that does not depend on $Na^+$. Inside the epithelial cell, most fructose is converted to glucose. Glucose, galactose, and the small amount of remaining fructose are then transported out the base of the cell by facilitated diffusion and are absorbed by the blood capillaries of the villus. The hepatic portal system delivers them to the liver; chapter 26 follows the fate of these sugars from there.

Lactose intolerance occurs in about 15% of American whites; 90% of American blacks, who are predominantly descended from nonpastoral African tribes; 70% or more of Mediterraneans; and nearly all people of Asian descent, including those of us descended from the native migrants into North, Central, and South America. People with lactose intolerance can consume products such as yogurt and cheese, in which bacteria have broken down the lactose, and they can digest milk and ice cream with the aid of lactase drops or tablets.

## PROTEINS

The amino acids absorbed by the small intestine come from three sources: (1) dietary proteins, (2) digestive enzymes digested by each other, and (3) sloughed epithelial cells digested by these enzymes. The endogenous amino acids from the last two sources total about 30 g/day, compared with about 44 to 60 g/day from the diet.

Enzymes that digest proteins are called **proteases** (peptidases). They are absent from the saliva but begin work in the stomach. Here, pepsin hydrolyzes any peptide bond between tyrosine and phenylalanine, thus digesting 10% to 15% of the dietary protein into shorter polypeptides and a small amount of free amino acids (fig. 25.29). Pepsin has an optimal pH of 1.5 to 3.5; thus it is inactivated when it passes into the duodenum and mixes with the alkaline pancreatic juice (pH 8).

In the small intestine, the pancreatic enzymes trypsin and chymotrypsin take over protein digestion by hydrolyzing polypeptides into even shorter oligopeptides. Finally, these are taken apart one amino acid at a time by three more enzymes: (1) **carboxypeptidase** removes amino acids from the –COOH end of the chain; (2) **aminopeptidase** removes them from the $-NH_2$ end; and (3) **dipeptidase** splits dipeptides in the middle and releases the last two free amino acids. All three of these are brush border enzymes, while carboxypeptidase also occurs in the pancreatic juice.

Amino acid absorption is similar to that of monosaccharides. There are several sodium-dependent amino acid cotransporters for different classes of amino acids.

Dipeptides and tripeptides can also be absorbed, but they are hydrolyzed within the epithelial cells before their amino acids are released to the bloodstream. At the basal surfaces of the cells, amino acids behave like the monosaccharides discussed previously—they leave the cell by facilitated diffusion, enter the capillaries of the villus, and are thus carried away in the hepatic portal circulation.

The absorptive cells of infants can take up intact proteins by pinocytosis and release them to the blood by exocytosis. This allows IgA from breast milk to pass into an infant's bloodstream and confer passive immunity from mother to infant. It has the disadvantage, however, that intact proteins entering the infant's blood are detected as foreign antigens and sometimes trigger food allergies. As the intestine matures, its ability to pinocytose protein declines but never completely ceases.

## LIPIDS

The hydrophobic quality of lipids makes their digestion and absorption more complicated than that of carbohydrates and proteins (fig. 25.30). Fats are digested by enzymes called **lipases.** *Lingual lipase,* secreted by the intrinsic salivary glands of the tongue, is activated by acid in the stomach, where it digests as much as 10% of the ingested fat. In infants, the stomach also secretes *gastric lipase.* Most fat digestion, however, occurs in the small intestine through the action of *pancreatic lipase.*

As chyme enters the duodenum, its fat is in large globules exposed to lipase only at their surfaces. Fat digestion would be rather slow and inefficient if it remained this way. Instead, it is broken up into smaller **emulsification droplets** by certain components of the bile—lecithin (a phospholipid) and bile acids (steroids). These agents have hydrophobic regions attracted to the surface of a fat globule and hydrophilic regions attracted to the surrounding water. The agitation produced by intestinal segmentation breaks the fat up into droplets as small as 1 μm in diameter, and a coating of lecithin and bile acids keeps it broken up, exposing far more of its surface to enzymatic action.

There is enough pancreatic lipase in the small intestine after a meal to digest the average daily fat intake in as little as 1 or 2 minutes. When lipase acts on a triglyceride, it removes the first and third fatty acids from the glycerol backbone and usually leaves the middle one. The products of lipase action are therefore two free fatty acids (FFAs) and a monoglyceride.

The absorption of these products and other lipids depends on minute droplets in the bile called **micelles**[28] (my-SELLS). Micelles consist of 20 to 40 bile acid molecules aggregated with their hydrophilic side groups facing outward and their hydrophobic steroid rings facing

---

[28]*mic* = grain, crumb + *elle* = little

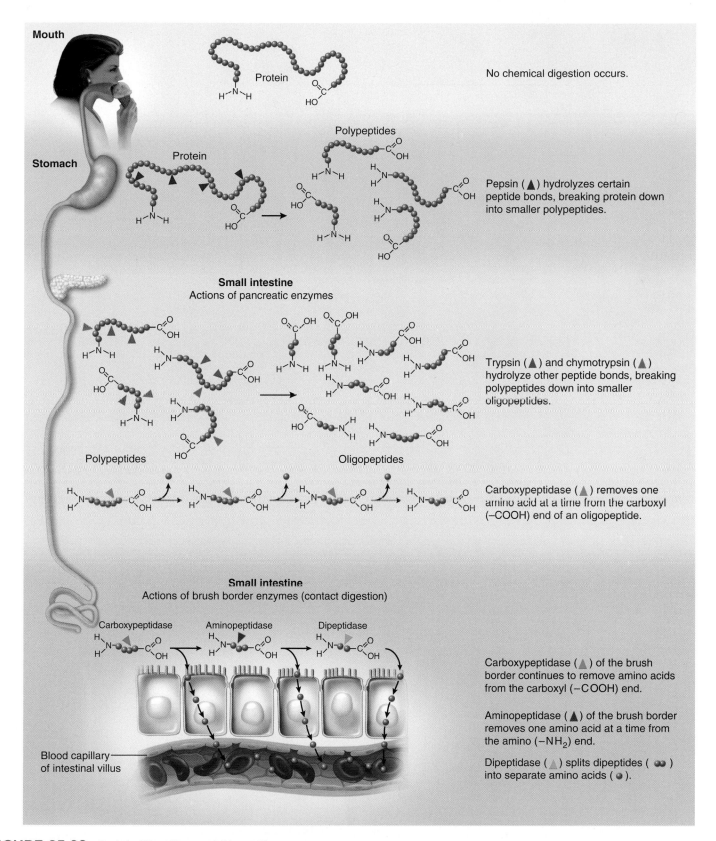

**Mouth**

Protein

No chemical digestion occurs.

**Stomach**

Protein

Polypeptides

Pepsin (▲) hydrolyzes certain peptide bonds, breaking protein down into smaller polypeptides.

**Small intestine**
Actions of pancreatic enzymes

Polypeptides

Oligopeptides

Trypsin (▲) and chymotrypsin (▲) hydrolyze other peptide bonds, breaking polypeptides down into smaller oligopeptides.

Carboxypeptidase (▲) removes one amino acid at a time from the carboxyl (–COOH) end of an oligopeptide.

**Small intestine**
Actions of brush border enzymes (contact digestion)

Carboxypeptidase        Aminopeptidase        Dipeptidase

Blood capillary of intestinal villus

Carboxypeptidase (▲) of the brush border continues to remove amino acids from the carboxyl (–COOH) end.

Aminopeptidase (▲) of the brush border removes one amino acid at a time from the amino (–NH$_2$) end.

Dipeptidase (▲) splits dipeptides ( ●● ) into separate amino acids ( ● ).

**FIGURE 25.29**  Protein Digestion and Absorption.

## Emulsification

Lecithin

Hydrophilic region
Hydrophobic region

Bile acid

Fat globule

Emulsification droplets

Fat globule is broken up and coated by lecithin and bile acids.

## Fat hydrolysis

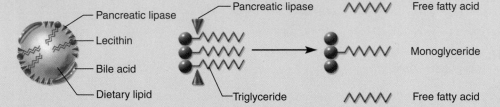

Pancreatic lipase

Lecithin

Bile acid

Dietary lipid

Pancreatic lipase

Triglyceride

Free fatty acid

Monoglyceride

Free fatty acid

Emulsification droplets are acted upon by pancreatic lipase, which hydrolyzes the first and third fatty acids from triglycerides, usually leaving the middle fatty acid.

## Lipid uptake by micelles

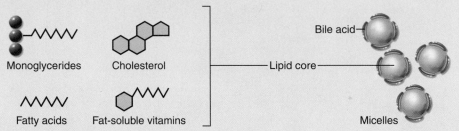

Monoglycerides

Cholesterol

Fatty acids

Fat-soluble vitamins

Bile acid

Lipid core

Micelles

Micelles in the bile pass to the small intestine and pick up several types of dietary and semidigested lipids.

## Chylomicron formation

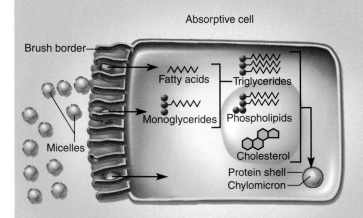

Absorptive cell

Brush border

Fatty acids

Triglycerides

Monoglycerides

Phospholipids

Cholesterol

Protein shell
Chylomicron

Micelles

Intestinal cells absorb lipids from micelles, resynthesize triglycerides, and package triglycerides, cholesterol, and phospholipids into protein-coated chylomicrons.

## Chylomicron exocytosis and lymphatic uptake

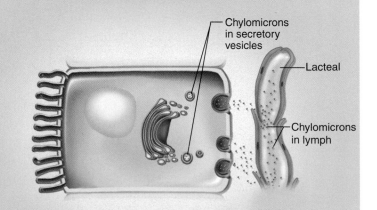

Chylomicrons in secretory vesicles

Lacteal

Chylomicrons in lymph

Golgi complex packages chylomicrons into secretory vesicles; chylomicrons are released from basal cell membrane by exocytosis and enter the lacteal (lymphatic capillary) of the villus.

**FIGURE 25.30** Fat Digestion and Absorption.

inward. Bile phospholipids and cholesterol diffuse into the center of the micelle to form its core. The micelles pass down the bile duct into the duodenum, where they absorb fat-soluble vitamins, more cholesterol, and the FFAs and monoglycerides produced by fat digestion. Because of their charged, hydrophilic surfaces, micelles remain dissolved in water more easily than free lipids do. They transport lipids to the surfaces of the intestinal absorptive cells, where the lipids leave the micelles and diffuse through the plasma membrane into the cells. The micelles are reused, picking up another cargo of lipids and ferrying them to the absorptive cells. Without micelles, the small intestine absorbs only about 40% to 50% of the dietary fat and almost no cholesterol.

Within the intestinal cell, the FFAs and monoglycerides are transported into the smooth endoplasmic reticulum and resynthesized into triglycerides. The Golgi complex combines these with a small amount of cholesterol and coats the complex with a film of phospholipids and protein, forming droplets about 60 to 750 nm in diameter called **chylomicrons**[29] (KY-lo-MY-crons). It packages chylomicrons into secretory vesicles that migrate to the basal surface of the cell and release their contents into the core of the villus. Although some FFAs enter the blood capillaries, chylomicrons are too large to penetrate the endothelium. They are taken up by the more porous lacteal into the lymph. The white, fatty intestinal lymph (chyle) flows through larger and larger lymphatic vessels of the mesenteries, eventually passing through the cisterna chyli (fig. 21.1) to the thoracic duct, then entering the bloodstream at the left subclavian vein. The further fate of dietary fat is described in chapter 26.

## Think About It

*Explain why the right lymphatic duct does not empty dietary fat into the bloodstream.*

## NUCLEIC ACIDS

The nucleic acids, DNA and RNA, are present in much smaller quantities than the polymers discussed previously. The **nucleases** (ribonuclease and deoxyribonuclease) of pancreatic juice hydrolyze these to their constituent nucleotides. **Nucleosidases** and **phosphatases** of the brush border then decompose the nucleotides into phosphate ions, ribose (from RNA) or deoxyribose (from DNA), and nitrogenous bases. These products are transported across the intestinal epithelium by membrane carriers and enter the capillary blood of the villus.

## VITAMINS

Vitamins are absorbed unchanged. The fat-soluble vitamins A, D, E, and K are absorbed with other lipids as just described. Therefore, if they are ingested without fat-containing food, they are not absorbed at all but are passed in the feces and wasted. Water-soluble vitamins (the B complex and vitamin C) are absorbed by simple diffusion, with the exception of vitamin $B_{12}$. This is an unusually large molecule that can only be absorbed if it binds to intrinsic factor from the stomach. The $B_{12}$–intrinsic factor complex then binds to receptors on absorptive cells of the distal ileum, where it is taken up by receptor-mediated endocytosis.

## MINERALS

Minerals (electrolytes) are absorbed along the entire length of the small intestine. Sodium ions are cotransported with sugars and amino acids. Chloride ions are actively transported in the distal ileum by a pump that exchanges them for bicarbonate ions, thus reversing the chloride–bicarbonate exchange that occurs in the stomach. Potassium ions are absorbed by simple diffusion. The $K^+$ concentration of the chyme rises as water is absorbed from it, creating a gradient favorable to $K^+$ absorption. In diarrhea, when water absorption is hindered, potassium ions remain in the intestine and pass with the feces; thus chronic diarrhea can lead to hypokalemia.

Iron and calcium are unusual in that they are absorbed in proportion to the body's need, whereas other minerals are absorbed at fairly constant rates regardless of need, leaving it to the kidneys to excrete any excess. Stimulated by hepcidin (see p. 652), the absorptive cells bind ferrous ions ($Fe^{2+}$) and internalize them by active transport; they are unable to absorb ferric ions ($Fe^{3+}$). $Fe^{2+}$ is transported to the basal surface of the cell and there taken up by the extracellular protein *transferrin*. The transferrin–iron complex diffuses into the blood and is carried to such places as the bone marrow for hemoglobin synthesis, muscular tissue for myoglobin synthesis, and the liver for storage (see fig. 18.8, p. 689). Excess dietary iron, if absorbed, binds irreversibly to ferritin in the epithelial cell and is held there until that cell sloughs off and passes in the feces.

## Think About It

*Young adult women have four times as many iron transport proteins in the intestinal mucosa as men have. Can you explain this?*

Calcium is absorbed especially by the duodenum and jejunum. Parathyroid hormone induces the kidneys to release active vitamin D (calcitriol), and vitamin D stimulates calcium absorption by the small intestine. $Ca^{2+}$ diffuses through calcium channels into the epithelial cells and binds to a cytoplasmic protein called *calbindin*. At the basal side of the epithelial cell, calcium is pumped out by a calcium-ATPase and a $Na^+$–$Ca^{2+}$ antiport. Vitamin D works by stimulating the synthesis of both calbindin and calcium-ATPase.

---

[29]*chyl* = juice + *micr* = small

## WATER

The digestive system is one of several systems involved in water balance. The digestive tract receives about 9 L of water per day—0.7 L in food, 1.6 L in drink, 6.7 L in the gastrointestinal secretions: saliva, gastric juice, bile, pancreatic juice, and intestinal juice. About 8 L of this is absorbed by the small intestine and 0.8 L by the large intestine, leaving 0.2 L voided in the daily fecal output. Water is absorbed by osmosis, following the absorption of salts and organic nutrients that create an osmotic gradient from the intestinal lumen to the ECF.

*Diarrhea* occurs when the large intestine absorbs too little water. This occurs when the intestine is irritated by bacteria and feces pass through too quickly for adequate reabsorption, or when the feces contain abnormally high concentrations of a solute such as lactose that opposes osmotic absorption of water. *Constipation* occurs when fecal movement is slow, too much water is reabsorbed, and the feces become hardened. This can result from lack of dietary fiber, lack of exercise, emotional upset, or long-term laxative abuse.

### Think About It

*Magnesium sulfate (epsom salt) is poorly absorbed by the intestines. In light of this, explain why it has a laxative effect.*

### Before You Go On

*Answer the following questions to test your understanding of the preceding section:*

21. What three polymers account for most of the dietary calories? What are the end products of enzymatic digestion of each?

22. What two nutrients are digested by saliva? Why is only one of them digested in the mouth?

23. Name as many enzymes of the intestinal brush border as you can, and identify the substrate or function of each.

24. Explain the distinctions between an emulsification droplet, a micelle, and a chylomicron.

25. What happens to digestive enzymes after they have done their job? What happens to dead epithelial cells that slough off the gastrointestinal mucosa? Explain.

# The Large Intestine

### Objectives

When you have completed this section, you should be able to

- describe the gross anatomy of the large intestine;
- contrast the mucosa of the colon with that of the small intestine;
- state the physiological significance of intestinal bacteria;
- discuss the types of contractions that occur in the colon; and
- explain the neurological control of defecation.

The large intestine (fig. 25.31) receives about 500 mL of indigestible food residue per day, reduces it to about 150 mL of feces by absorbing water and salts, and eliminates the feces by defecation.

## GROSS ANATOMY

The large intestine measures about 1.5 m (5 ft) long and 6.5 cm (2.5 in.) in diameter in the cadaver. It begins with the **cecum,**[30] a blind pouch in the lower right abdominal quadrant inferior to the ileocecal valve. Attached to the lower end of the cecum is the **appendix,** a blind tube 2 to 7 cm long. The appendix is densely populated with lymphocytes and is a significant source of immune cells.

The **colon** is that part of the large intestine between the ileocecal junction and anal canal. It is divided into the ascending, transverse, descending, and sigmoid regions. The **ascending colon** begins at the ileocecal valve and passes up the right side of the abdominal cavity. It makes a 90° turn at the **right colic (hepatic) flexure,** near the right lobe of the liver, and becomes the **transverse colon.** This passes horizontally across the upper abdominal cavity and turns 90° downward at the **left colic (splenic) flexure** near the spleen. Here it becomes the **descending colon,** which passes down the left side of the abdominal cavity. Ascending, transverse, and descending colons thus form a squarish, three-sided frame around the small intestine.

The pelvic cavity is narrower than the abdominal cavity, so at the pelvic inlet the colon turns medially and downward, forming a roughly S-shaped portion called the **sigmoid**[31] **colon.** (Visual examination of this region is performed with an instrument called a *sigmoidoscope.*) In the pelvic cavity, the large intestine straightens and forms the **rectum.**[32] The rectum has three internal transverse folds called **rectal valves** which enable it to retain feces while passing gas.

The final 3 cm of the large intestine is the **anal canal** (fig. 25.31b), which passes through the levator ani muscle of the pelvic floor and terminates at the anus. Here, the mucosa forms longitudinal ridges called **anal columns** with depressions between them called **anal sinuses.** As feces pass through the canal, they press against the sinuses and cause them to exude extra mucus and lubricate the canal during defecation. Large **hemorrhoidal veins** form superficial plexuses in the anal columns and around the orifice. Unlike veins in the extremities, they lack valves and are

---

[30]*cec* = blind
[31]*sigm* = sigma or S + *oid* = resembling
[32]*rect* = straight

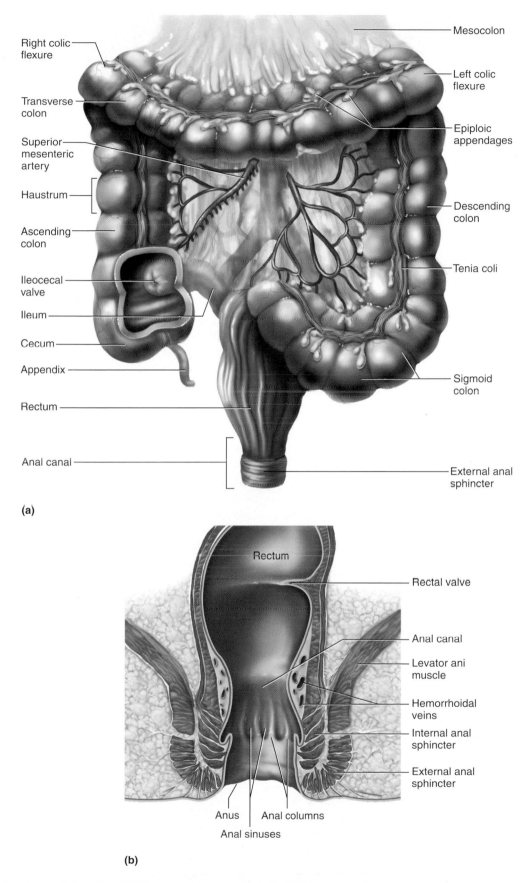

**(a)**

Right colic flexure

Transverse colon

Superior mesenteric artery

Haustrum

Ascending colon

Ileocecal valve

Ileum

Cecum

Appendix

Rectum

Anal canal

Mesocolon

Left colic flexure

Epiploic appendages

Descending colon

Tenia coli

Sigmoid colon

External anal sphincter

**(b)**

Rectum

Rectal valve

Anal canal

Levator ani muscle

Hemorrhoidal veins

Internal anal sphincter

External anal sphincter

Anus

Anal columns

Anal sinuses

**FIGURE 25.31   The Large Intestine.**   (a) Gross anatomy. (b) Detail of the anal canal.

▶ *Which anal sphincter is controlled by the autonomic nervous system? Which is controlled by the somatic nervous system?*

particularly subject to distension and venous pooling. *Hemorrhoids* are permanently distended veins that protrude into the anal canal or form bulges distal to the anus.

The muscularis externa of the colon is unusual. Although it completely encircles the colon just as it does the small intestine, its longitudinal fibers are especially concentrated in three thickened, ribbonlike strips. Each strip is called a **tenia coli** (TEE-nee-ah CO-lye) (plural, *teniae coli*). The muscle tone of the teniae coli contracts the colon lengthwise and causes its wall to form pouches called **haustra**[33] (HAW-stra; singular, *haustrum*). Haustra are not seen in the cadaver because they disappear when the muscle relaxes at death. In the rectum and anal canal, however, the longitudinal muscle forms a continuous sheet and haustra are absent. The anus, like the urethra, is regulated by two sphincters: an **internal anal sphincter** composed of smooth muscle of the muscularis externa, and an **external anal sphincter** composed of skeletal muscle of the pelvic diaphragm.

The ascending and descending colon are retroperitoneal, whereas the transverse and sigmoid colon are covered with serosa and anchored to the dorsal abdominal wall by the mesocolon. The serosa of these regions often has **epiploic**[34] **appendages,** clublike fatty pouches of peritoneum of unknown function.

## MICROSCOPIC ANATOMY

The mucosa of the large intestine has a simple columnar epithelium in all regions except the lower half of the anal canal, where it has a nonkeratinized stratified squamous epithelium. The latter provides more resistance to the abrasion caused by the passage of feces. There are no circular folds or villi in the large intestine, but there are intestinal crypts. They are deeper than in the small intestine and have a greater density of goblet cells; mucus is their only significant secretion.

The anatomical features of the digestive tract from the large intestine to the anus are summarized in table 25.5.

## BACTERIAL FLORA AND INTESTINAL GAS

The large intestine is densely populated with about 800 species of bacteria collectively called the **bacterial flora.**[35] We have a mutually beneficial relationship with many of these. We provide them with room and board, while they provide us with nutrients from our food that we are not equipped to extract on our own. For example, they digest cellulose, pectin, and other plant polysaccharides for which we have no digestive enzymes, and we absorb the resulting sugars. Thus we get more nutrition from our food because of these bacteria than we would get without them.

---

[33]*haustr* = to draw
[34]*epiploic* = pertaining to an omentum
[35]*flora* = flowers, plants

| **TABLE 25.5** | Anatomical Checklist of the Digestive System from the Large Intestine Through the Anus |
|---|---|
| Cecum | Anal canal |
| Appendix |   Internal anal sphincter |
| Ascending colon |   External anal sphincter |
| Right colic flexure |   Anal columns |
| Transverse colon |   Anal sinuses |
| Left colic flexure |   Hemorrhoidal veins |
| Descending colon | Taeniae coli |
| Sigmoid colon | Haustra |
| Rectum | Epiploic appendages |
|   Rectal valves | Mesocolon |

Indeed, one person may get more calories than another from the same food because of differences in their bacterial populations. Some bacteria also synthesize B vitamins and vitamin K, which are absorbed by the colon. This vitamin K is especially important because the diet alone usually does not provide enough to ensure adequate blood clotting.

One of the less desirable and sometimes embarrassing products of these bacteria is intestinal gas, or **flatus.** The average person expels about 500 mL of flatus per day. Most of this is swallowed air that has worked its way through the digestive tract, but the bacterial flora add to it. Painful cramping can result when undigested nutrients pass into the colon and furnish an abnormal substrate for bacterial action—for example, in lactose intolerance. Flatus is composed of nitrogen ($N_2$), carbon dioxide ($CO_2$), hydrogen ($H_2$), methane ($CH_4$), hydrogen sulfide ($H_2S$), and two amines: indole and skatole. Indole, skatole, and $H_2S$ produce the odor of flatus and feces, whereas the others are odorless. The hydrogen gas is combustible and has been known to explode in surgery that used electrical cauterization.

## ABSORPTION AND MOTILITY

The large intestine takes about 12 to 24 hours to reduce the residue of a meal to feces. It does not chemically change the residue but reabsorbs water and electrolytes (especially NaCl) from it. The feces consist of about 75% water and 25% solids. The solids are about 30% bacteria, 30% undigested dietary fiber, 10% to 20% fat, and smaller amounts of protein, sloughed epithelial cells, salts, mucus, and other digestive secretions. The fat is not from the diet but from broken-down epithelial cells and bacteria.

The most common type of colonic motility is a type of segmentation called **haustral contractions,** which occur about every 30 minutes. Distension of a haustrum with feces stimulates it to contract. This churns and mixes the residue, promotes water and salt absorption, and passes the residue distally to another haustrum. Stronger contractions called **mass movements** occur one to three times a day, last about 15 minutes, and move residue for several centimeters at a time. They are often triggered by the **gastrocolic** and **duodenocolic reflexes,** in which filling of the stomach and duodenum stimulates motility of the colon. Mass movements occur especially in the transverse to sigmoid colon, and often within an hour after breakfast.

## DEFECATION

Stretching of the rectum stimulates the defecation reflexes, which account for the urge to defecate that is often felt soon after a meal. The predictability of this response is useful in house-training pets and toilet training children. The process involves two reflexes:

1. The **intrinsic defecation reflex.** This reflex operates entirely within the myenteric nerve plexus. Stretch signals travel through the plexus to the muscularis of the descending and sigmoid colon and the rectum. This activates a peristaltic wave that drives feces downward, and it relaxes the internal anal sphincter. This reflex is relatively weak, however, and usually requires the cooperative action of the following reflex.

2. The **parasympathetic defecation reflex.** This is a spinal reflex. Its principal events (figure 25.32) are that stretch signals are transmitted to the spinal cord, and motor signals return by way of the pelvic nerves to intensify peristalsis in the descending and sigmoid colon and rectum and to relax the internal anal sphincter.

These reflexes are involuntary and are the sole means of controlling defecation in infants and some people with transecting spinal cord injuries. However, the external anal sphincter, like the external urethral sphincter controlling urination, is under voluntary control, enabling one to limit defecation to appropriate circumstances. Defecation normally occurs only when this sphincter is voluntarily relaxed. Defecation is also aided by the voluntary Valsalva maneuver, in which a breath-hold and contraction of the abdominal muscles increases abdominal pressure, compresses the rectum, and squeezes the feces from it. This maneuver can also initiate the defecation reflex by forcing feces from the descending colon into the rectum.

If the defecation urge is suppressed, contractions cease in a few minutes and the rectum relaxes. The defecation reflexes reoccur a few hours later or when another mass movement propels more feces into the rectum.

The effects of aging on the digestive system are discussed on page 1131. Table 25.6 lists and describes some common digestive disorders.

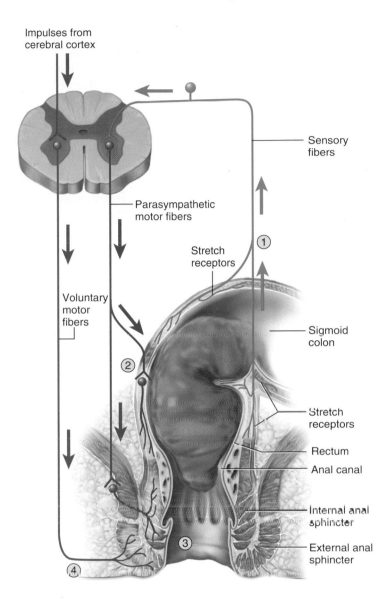

① Feces stretch the rectum and stimulate stretch receptors, which transmit signals to the spinal cord.

② A spinal reflex stimulates contraction of the rectum.

③ The spinal reflex also relaxes the internal anal sphincter.

④ Impulses from the brain prevent untimely defecation by keeping the external anal sphincter contracted. Defecation occurs only if this sphincter also relaxes.

**FIGURE 25.32 Neural Control of Defecation.**

### Before You Go On

*Answer the following questions to test your understanding of the preceding section:*

26. *How does the mucosa of the large intestine differ from that of the small intestine? How does the muscularis externa differ?*

27. *Name and briefly describe two types of contractions that occur in the colon and nowhere else in the alimentary canal.*

28. *Describe the reflexes that cause defecation in an infant. Describe the additional neural controls that function following toilet training.*

| TABLE 25.6 | Some Digestive System Diseases |
|---|---|
| Acute pancreatitis | Severe pancreatic inflammation perhaps caused by trauma leading to leakage of pancreatic enzymes into parenchyma, where they digest tissue and cause inflammation and hemorrhage. |
| Appendicitis | Inflammation of the appendix, with swelling, pain, and sometimes gangrene, perforation, and peritonitis. |
| Ascites | Accumulation of serous fluid in the peritoneal cavity, often causing extreme distension of the abdomen. Most often caused by cirrhosis of the liver (see Insight 26.3, p. 1021) and frequently associated with alcoholism. The diseased liver "weeps" fluid into the abdomen. About 25% of people who develop ascites as a consequence of cirrhosis die within one year. |
| Cancers | Digestive system is subject to cancer especially of the esophagus, stomach, colon, liver, and pancreas, with colon and pancreatic cancer being among the leading causes of cancer death in the United States. |
| Crohn disease | Inflammation of small and large intestines, similar to ulcerative colitis. Produces granular lesions and fibrosis of intestine, diarrhea, and lower abdominal pain. Often hereditary. |
| Diverticulitis | Presence of inflamed herniations (outpocketings, diverticula) of the colon, associated especially with low-fiber diets. Diverticula may rupture, leading to peritonitis. |
| Dysphagia | Difficulty swallowing. Can result from esophageal obstructions (tumors, constrictions) or impaired peristalsis (due to neuromuscular disorders). |
| Gluten-sensitive enteropathy | Formerly called *sprue* or *celiac disease.* Atrophy of intestinal villi triggered in genetically susceptible individuals by *gluten,* the protein component of cereal grains. Onset is usually in infancy or early childhood. Results in severe malabsorption of most nutrients, causing watery or fatty diarrhea, abdominal pain, diminished growth, and multiple problems tied to nutritional deficiencies. Treatable with intensive dietary management. |
| Hiatal hernia | Protrusion of part of the stomach into the thoracic cavity, where the negative thoracic pressure may cause it to balloon. Often causes gastroesophageal reflux (especially when a person is supine) and esophagitis (inflammation of the esophagus). |
| Ulcerative colitis | Chronic inflammation resulting in ulceration of the large intestine, especially the sigmoid colon and rectum. Tends to be hereditary but exact causes are not well known. |

**Disorders described elsewhere**

## The Man with a Hole in His Stomach

Perhaps the most famous episode in the history of digestive physiology began in 1822 on Mackinac Island in the strait between Lake Michigan and Lake Huron. Alexis St. Martin, a 19-year-old fur trapper, was standing outside a trading post when he was accidentally hit by a shotgun blast from 3 feet away. An Army doctor stationed at Fort Mackinac, William Beaumont (1785–1853), was summoned to examine St. Martin. As Beaumont later wrote, "a portion of the lung as large as a turkey's egg" protruded through St. Martin's lacerated and burnt flesh. Below that was a portion of the stomach with a puncture in it "large enough to receive my forefinger." Beaumont did his best to pick out bone fragments and dress the wound, though he did not expect St. Martin to survive.

Surprisingly, St. Martin lived. Over a period of months the wound extruded pieces of bone, cartilage, gunshot, and gun wadding. As the wound healed, a fistula (hole) remained in the stomach, so large that Beaumont had to cover it with a compress to prevent food from coming out. A fold of tissue later grew over the fistula, but it was easily opened. A year later, St. Martin was still feeble. Town authorities decided they could no longer support him on public funds and wanted to ship him 2,000 miles to his home in Canada. Beaumont, however, was imbued with a passionate sense of destiny. Very little was known about digestion, and he saw the accident as a unique opportunity to learn. He took St. Martin in at his personal expense and performed 238 experiments on him over several years. Beaumont had never attended medical school and had little idea how scientists work, yet he proved to be an astute experimenter. Under crude frontier conditions and with almost no equipment, he discovered many of the basic facts of gastric physiology discussed in this chapter.

"I can look directly into the cavity of the stomach, observe its motion, and almost see the process of digestion," Beaumont wrote. "I can pour in water with a funnel and put in food with a spoon, and draw them out again with a siphon." He put pieces of meat on a string into the stomach and removed them hourly for examination. He sent vials of gastric juice to the leading chemists of America and Europe, who could do little but report that it contained hydrochloric acid. He proved that digestion required HCl and could even occur outside the stomach, but he found that HCl alone did not digest meat; gastric juice must contain some other digestive ingredient. Theodor Schwann, one of the founders of the cell theory (see chapter 3), identified that ingredient as pepsin. Beaumont also demonstrated that gastric juice is secreted only in response to food; it did not accumulate between meals as previously thought. He disproved the idea that hunger is caused by the walls of the empty stomach rubbing against each other.

For his part, St. Martin felt helpless and humiliated by Beaumont's experiments. His fellow trappers taunted him as "the man with a hole in his stomach," and he longed to return to hunting and trapping in the wilderness. He had a wife and daughter in Canada whom he rarely got to see, and he ran away repeatedly to join them. He was once gone for 4 years before his poverty and physical disability made him yield to Beaumont's financial enticement to come back. Beaumont despised St. Martin's drunkenness and profanity and was quite insensitive to his embarrassment and discomfort over the experiments. Yet St. Martin's temper enabled Beaumont to make the first direct observations of the relationship between emotion and digestion. When St. Martin was particularly distressed, Beaumont noted little digestion occurring—as we now know, the sympathetic nervous system inhibits digestive activity.

Beaumont published a book in 1833 that laid the foundation for modern gastric physiology and dietetics. It was enthusiastically received by the medical community and had no equal until Russian physiologist Ivan Pavlov (1849–1936) performed his celebrated experiments on digestion in animals. Building on the methods pioneered by Beaumont, Pavlov received the 1904 Nobel Prize for Physiology or Medicine.

In 1853, Beaumont slipped on some ice, suffered a blow to the base of his skull, and died a few weeks later. St. Martin continued to tour medical schools and submit to experiments by other physiologists, whose conclusions were often less correct than Beaumont's. Some, for example, attributed chemical digestion to lactic acid instead of hydrochloric acid. St. Martin lived in wretched poverty in a tiny shack with his wife and several children, and died 28 years after Beaumont. By then he was senile and believed he had been to Paris, where Beaumont had often promised to take him.

# CONNECTIVE ISSUES

## Interactions Between the DIGESTIVE SYSTEM and Other Organ Systems

■ indicates ways in which this system affects other systems

■ indicates ways in which other systems affect this system

### ALL SYSTEMS

The digestive system provides all other systems with nutrients in a form usable for cellular metabolism and building of tissues

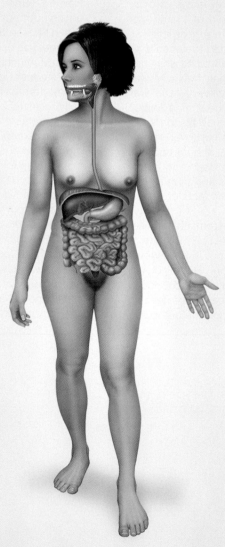

### INTEGUMENTARY SYSTEM

Skin helps synthesize calcitriol, needed for calcium and phosphorus absorption by small intestine

### SKELETAL SYSTEM

Small intestine adjusts calcium absorption in proportion to the needs of the skeletal system

Provides protective enclosure for some digestive organs, support for the teeth, and movements of mastication

### MUSCULAR SYSTEM

Liver disposes of lactic acid generated by muscles

Essential for chewing, swallowing, and defecation; muscles protect lower GI organs

### NERVOUS SYSTEM

Enteric and autonomic nervous systems regulate GI motility and secretion; somatic nervous system controls chewing, swallowing, and defecation; sense organs involved in food selection; hypothalamus contains centers for hunger, thirst, and satiety

### ENDOCRINE SYSTEM

Liver degrades hormones; enteroendocrine cells produce many hormones

Hormones regulate GI motility, secretion, and processing of nutrients

### CIRCULATORY SYSTEM

GI tract absorbs fluid needed to maintain blood volume; liver degrades heme from dead RBCs, secretes clotting factors, albumin, and other plasma proteins, and regulates blood glucose and iron levels

### CIRCULATORY SYSTEM (cont.)

Blood transports hormones that regulate GI activity; absorbs and distributes nutrients; vasomotion alters capillary filtration and salivation

### LYMPHATIC/IMMUNE SYSTEMS

GI mucosa is a site of lymphocyte production; acid, lysozyme, and other digestive enzymes provide nonspecific defense against pathogens; infant intestine absorbs maternal IgA to confer passive immunity on infant

Lymphatic capillaries (lacteals) absorb digested lipids; immune cells protect GI tract from infection

### RESPIRATORY SYSTEM

Pressure of digestive organs against diaphragm aids in expiration when abdominal muscles contract

Provides $O_2$, removes $CO_2$; Valsalva maneuver aids defecation

### URINARY SYSTEM

Intestines complement kidneys in water and electrolyte reabsorption; liver synthesizes urea and kidneys excrete it

Excretes bile pigments and other products of liver metabolism; completes the synthesis of calcitriol, needed for intestinal absorption of calcium and phosphorus

### REPRODUCTIVE SYSTEM

Provides nutrients for fetal growth

Developing fetus crowds digestive organs; may cause constipation and heartburn during pregnancy

## CHAPTER REVIEW

# Review of Key Concepts

### General Anatomy and Digestive Processes (p. 954)

1. The digestive system processes food, extracts nutrients, and eliminates the residue. It does this in four stages: ingestion, digestion (mechanical and chemical), absorption, and defecation.

2. The *digestive tract* is essentially a tube consisting of the oral cavity, pharynx, esophagus, stomach, and small and large intestines. The *accessory organs* are the teeth, tongue, salivary glands, liver, gallbladder, and pancreas.

3. In most areas, the wall of the digestive tract consists of an inner *mucosa,* a *submucosa,* a *muscularis externa,* and an outer *serosa.* In some areas a connective tissue *adventitia* replaces the serosa.

4. The *enteric nervous system* regulates much of digestive activity and consists of two nerve networks in the wall of the digestive tract: the *submucosal plexus* and the *myenteric plexus.*

5. In the abdominal cavity, the *dorsal mesentery* suspends the digestive tract from the body wall, wraps around it to form the serosa, and in some places continues as a *ventral mesentery.* The ventral mesentery includes the *greater* and *lesser omenta* and the *mesocolon.*

6. Digestive tract motility and secretion are regulated by hormones such as *gastrin* and *secretin,* paracrines such as *histamine* and *prostaglandins,* short *(myenteric) reflexes,* and *long (vagovagal) reflexes.*

### The Mouth Through Esophagus (p. 957)

1. The mouth (oral cavity) serves for ingestion, sensory responses to food, mastication, chemical digestion, deglutition, speech, and respiration.

2. The mouth extends from the oral orifice anteriorly to the fauces posteriorly. Its anatomical elements include the lips, cheeks, tongue, hard and soft palates, teeth, and intrinsic salivary glands (table 25.1). It also receives saliva from three pairs of extrinsic salivary glands: the *parotid, sublingual,* and *submandibular* glands.

3. Mastication breaks food into pieces small enough to be swallowed and exposes more food surface to the action of digestive enzymes, making digestion more efficient.

4. Saliva moistens the mouth, digests starch and fat, cleanses the teeth, inhibits bacterial growth, dissolves taste molecules, and binds food into a soft *bolus* to facilitate swallowing. It contains amylase, lipase, mucus, lysozyme, IgA, and electrolytes.

5. Salivation is controlled by *salivatory nuclei* in the medulla oblongata and pons and occurs in response to the thought, odor, sight, taste, or oral feel of food.

6. The pharynx is a muscular funnel in the throat where the respiratory and digestive tracts meet. Its wall contains three sets of *pharyngeal constrictor* muscles that aid in swallowing.

7. The esophagus extends from the pharynx to the *cardiac orifice* of the stomach. It is lined with a nonkeratinized stratified squamous epithelium, has a mixture of skeletal muscle (dominating the upper esophagus) and smooth muscle (dominating the lower), and is lubricated by mucous *esophageal glands* in the submucosa.

8. *Deglutition* (swallowing) requires the coordinated action of numerous muscles and is integrated by the *swallowing center* of the medulla oblongata and pons. It entails a *buccal phase* involving the tongue and a *pharyngeal–esophageal phase* involving the palate, infrahyoid muscles, and muscularis externa of the esophagus, among other structures.

9. Food and drink normally drop into the stomach by gravity alone, but esophageal peristalsis, mediated by the myenteric nerve plexus, can help drive food to the stomach.

### The Stomach (p. 965)

1. The stomach is primarily a food-storage organ, with a capacity of 4 L. It extends from the cardiac orifice above to the pylorus below (table 25.4).

2. The stomach wall is marked by *gastric pits,* with two or three tubular glands opening into the bottom of each pit. Most of the stomach has digestive *gastric glands,* whereas the cardiac and pyloric regions have mucous *cardiac glands* and *pyloric glands.* These various glands contain *mucous, regenerative, parietal, chief,* and *enteroendocrine cells.*

3. The stomach mechanically breaks up food, begins the chemical digestion of proteins, and converts ingested food to soupy *chyme.*

4. *Gastric juice* consists mainly of water, HCl, and pepsin.

5. The parietal cells secrete HCl, which activates pepsin and lingual lipase, breaks up ingested plant and animal tissues, converts dietary $Fe^{3+}$ into absorbable $Fe^{2+}$, and destroys ingested pathogens.

6. The parietal cells also secrete *intrinsic factor,* which is required for vitamin $B_{12}$ absorption. This is the only function of the stomach that is indispensable to life. Without intrinsic factor, a vitamin $B_{12}$ deficiency occurs and leads to pernicious anemia.

7. The chief cells secrete *pepsinogen,* a zymogen (inactive enzyme precursor) which HCl converts to *pepsin,* a protein-digesting enzyme. In infants, the chief cells also secrete the fat-digesting enzyme *gastric lipase* and the milk-curdling secretion *chymosin* (rennin).

8. Enteroendocrine cells of the stomach secrete several chemical messengers, including a variety of *gut–brain peptides,* which coordinate different regions of the digestive tract with each other.

9. Movements of the stomach begin with the *receptive-relaxation response* as it accommodates swallowed food, and progress to peristaltic waves that mix and break up the contents. Peristalsis drives the semidigested food, or *chyme,* a little at a time through the pylorus into the duodenum.

10. The stomach digests some protein and small amounts of fat and carbohydrate. It absorbs some drugs, but no significant nutrients.

11. The stomach is protected from its own acid and enzymes by its mucous coat, rapid epithelial cell replacement, and tight junctions between the epithelial cells.

12. In the *cephalic phase* of gastric activity, mental and sensory stimuli lead to stimulation of gastric secretion and motility through the vagus nerves.

13. In the *gastric phase,* swallowed food and semidigested protein stimulate gastric activity through short and long reflexes. The neurotransmitter acetylcholine, the paracrine histamine, and the hormone gastrin stimulate the secretion of HCl, intrinsic factor, and pepsin in this phase.

14. In the *intestinal phase,* chyme in the duodenum stimulates hormonal and nervous reflexes that initially stimulate the stomach, but soon activate an *enterogastric reflex* that inhibits it. Duodenal enteroendocrine cells secrete *secretin, cholecystokinin,* and *gastric inhibitory peptide,* all of which suppress gastric activity so that the stomach does not load chyme into the duodenum too rapidly.

## The Liver, Gallbladder, and Pancreas (p. 974)

1. The liver is the body's largest gland. It has four conspicuous lobes, each composed of innumerable microscopic, cylindrical *lobules* of liver cells *(hepatocytes).* The hepatocytes add some substances to the blood and remove others from it as the blood filters through the *hepatic sinusoids* between them.

2. Bile, secreted by the liver, fills the *bile duct* and *gallbladder.* The gallbladder concentrates the bile.

3. Bile is composed of water, minerals, cholesterol, fats, phospholipids, bile pigments, and bile acids. Most of these are body wastes. Bile pigments cause the brown color of the feces.

4. Bile acids and lecithin (a phospholipid) aid fat digestion; this is the liver's only digestive role.

5. The pancreas produces the hormones insulin and glucagon, and about 1.2 to 1.5 L of pancreatic juice per day. It secretes pancreatic juice through a duct that joins the bile duct before emptying into the duodenum.

6. Pancreatic juice contains sodium bicarbonate, which neutralizes stomach acid, and several digestive enzymes or enzyme precursors *(zymogens).* The pancreatic enzymes secreted in active form are *pancreatic amylase* (which digests starch), *pancreatic lipase* (which digests fat), *ribonuclease* (which digests RNA), and *deoxyribonuclease* (which digests DNA). Those secreted as zymogens and activated in the small intestine are *trypsin, chymotrypsin,* and *carboxypeptidase,* all of which digest proteins.

7. Bile and pancreatic juice are secreted in response to vagal stimulation, cholecystokinin, gastrin, and secretin.

## The Small Intestine (p. 980)

1. The small intestine begins at the pylorus (gateway from the stomach) and ends at the ileocecal junction (gateway to the large intestine). It consists of the duodenum (25 cm), jejunum (2.5 m), and ileum (3.6 m).

2. To carry out its digestive and nutrient-absorbing roles, the small intestine has a very large internal surface area stemming from its great length and three kinds of mucosal elaborations: the circular folds, villi, and microvilli.

3. Villi are covered with an epithelium of absorptive cells and goblet cells. Each villus contains a lipid-collecting lymphatic capillary called the *lacteal.* Glandular *intestinal crypts* open onto the floor of the intestine between the villi. The crypt epithelium is composed of absorptive and goblet cells, stem cells, and bacteria-fighting *Paneth cells.*

4. The enzymes that carry out intestinal digestion occur mostly in the brush borders of the absorptive cells and in the pancreatic juice.

5. Movements of the small intestine include chopping *segmentation* contractions and overlapping waves of peristalsis called the *migrating motor complex.*

## Chemical Digestion and Absorption (p. 984)

1. Salivary amylase begins digesting starch in the mouth and continues in the stomach until a food bolus breaks up and exposes the amylase to the hydrochloric acid. Starch digestion resumes in the small intestine by the action of pancreatic amylase, which breaks it down into oligosaccharides and maltose.

2. The brush border enzymes *dextrinase, glucoamylase,* and *maltase* finish digesting the breakdown products of starch, while *sucrase* and *lactase* break down the disaccharides sucrose and lactose.

3. The resulting monosaccharides (glucose, galactose, and fructose) are absorbed by the sodium–glucose transport protein, by a fructose carrier, and by solvent drag, then pass into the blood capillaries of the villus.

4. Protein digestion begins with the action of pepsin in the stomach and is continued by trypsin and chymotrypsin in the small intestine. The resulting small peptides are dissembled one amino acid at a time by brush border enzymes *carboxypeptidase, aminopeptidase,* and *dipeptidase.*

5. The amino acids are absorbed by sodium-dependent cotransport proteins, then pass into the blood capillaries of the villus.

6. Fat digestion is begun by lingual lipase in the saliva and continued, in infants, by gastric lipase in the stomach, but most fat is digested by pancreatic lipase in the small intestine.

7. Bile acids and lecithin help break fat into *emulsification droplets,* exposing more surface area to the action of lipase. Lipase hydrolyzes triglycerides into free fatty acids and monoglycerides.

8. These breakdown products, along with cholesterol and fat-soluble vitamins, then collect in small bile-salt-coated droplets called *micelles.* Micelles release the lipids into the intestinal absorptive cells.

9. The absorptive cells resynthesize triglycerides and package them, with cholesterol, into droplets called *chylomicrons,* coated with phospholipid and protein.

10. Chylomicrons are secreted from the basal surface of the absorptive cells and taken into the lacteal in the villus. They enter the bloodstream at the subclavian veins, where the lymph empties into the blood.

11. Dietary DNA and RNA are hydrolyzed into nucleotides by deoxyribonuclease and ribonuclease. Brush border *nucleosidases* and *phosphatases* then decompose these

into ribose, deoxyribose, phosphate, and nitrogenous bases, which are absorbed into the blood capillaries of the villus.

12. Vitamins are not digested, but are absorbed unchanged. Fat-soluble vitamins are absorbed with other dietary lipids; most water-soluble vitamins by simple diffusion; and vitamin $B_{12}$ binds to intrinsic factor and is then absorbed by receptor-mediated endocytosis.

13. Minerals are absorbed by cotransport with sugars and amino acids ($Na^+$), by active transport ($Cl^-$, $Fe^{2+}$), and by simple diffusion ($K^+$, $Ca^{2+}$).

14. Water is absorbed by osmosis, following an osmotic gradient created by the absorption of salts and organic nutrients.

**The Large Intestine (p. 990)**
1. The large intestine is about 1.5 m long and consists of the *cecum;* the *ascending, transverse, descending,* and *sigmoid colon; rectum;* and *anal canal.* The anal canal has an involuntary *internal anal sphincter* of

smooth muscle and a voluntary *external anal sphincter* of skeletal muscle.

2. The large intestine absorbs water and salts from the indigestible food residue and reduces the residue to feces.

3. Movements of the large intestine include frequent *haustral contractions* that move the feces only a short distance distally, and one to three daily *mass movements* that move the feces several centimeters.

4. Stretching of the rectum triggers the *intrinsic defecation reflex,* mediated by the myenteric nerve plexus. This reflex drives feces downward and relaxes the internal anal sphincter. A stronger *parasympathetic defecation reflex* involves a reflex arc through the spinal cord and parasympathetic fibers of the pelvic nerve.

5. These reflexes tend to cause defecation, although in individuals with bowel control, voluntary control of the external anal sphincter permits or prohibits defecation at will.

# Testing Your Recall

1. Which of the following enzymes acts in the stomach?
   a. chymotrypsin
   b. lingual lipase
   c. carboxypeptidase
   d. enterokinase
   e. dextrinase

2. Which of the following enzymes does *not* digest any nutrients?
   a. chymotrypsin
   b. lingual lipase
   c. carboxypeptidase
   d. enterokinase
   e. dextrinase

3. Which of the following is *not* an enzyme?
   a. chymotrypsin
   b. enterokinase
   c. secretin
   d. pepsin
   e. nucleosidase

4. The substance in question 3 that is *not* an enzyme is
   a. a zymogen.
   b. a nutrient.

   c. an emulsifier.
   d. a neurotransmitter.
   e. a hormone.

5. The lacteals absorb
   a. chylomicrons.
   b. micelles.
   c. emulsification droplets.
   d. amino acids.
   e. monosaccharides.

6. All of the following contribute to the absorptive surface area of the small intestine *except*
   a. its length.
   b. the brush border.
   c. haustra.
   d. circular folds.
   e. villi.

7. Which of the following is a periodontal tissue?
   a. the gingiva
   b. the enamel
   c. the cementum
   d. the pulp
   e. the dentin

8. The _____ of the stomach most closely resemble the _____ of the small intestine.
   a. gastric pits, intestinal crypts
   b. pyloric glands, intestinal crypts
   c. rugae, Peyer patches
   d. parietal cells, goblet cells
   e. gastric glands, duodenal glands

9. Which of the following cells secrete digestive enzymes?
   a. chief cells
   b. mucous neck cells
   c. parietal cells
   d. goblet cells
   e. enteroendocrine cells

10. What phase of gastric regulation includes inhibition by the enterogastric reflex?
    a. the intestinal phase
    b. the gastric phase
    c. the buccal phase
    d. the cephalic phase
    e. the pharyngo-esophageal phase

11. Cusps are a feature of the _____ surfaces of the molars and premolars.

12. The acidity of the stomach deactivates _____ but activates _____ of the saliva.

13. The _____ salivary gland is named for its proximity to the ear.

14. The submucosal and myenteric nerve plexuses collectively constitute the _____ nervous system.

15. Nervous stimulation of gastrointestinal activity is mediated mainly through the parasympathetic fibers of the _____ nerves.

16. Food in the stomach causes G cells to secrete _____, which in turn stimulates the secretion of HCl and pepsinogen.

17. Hepatic macrophages occur in blood-filled spaces of the liver called _____.

18. The brush border enzyme that finishes the job of starch digestion, producing glucose, is called _____. Its substrate is _____.

19. Fats are transported in the lymph and blood in the form of droplets called _____.

20. Within the absorptive cells of the small intestine, ferritin binds the nutrient _____.

*Answers in Appendix B*

## True or False

*Determine which five of the following statements are false, and briefly explain why.*

1. Fat is not digested until it reaches the duodenum.

2. A tooth is composed mostly of enamel.

3. Hepatocytes secrete bile into the hepatic sinusoids.

4. Cholecystokinin stimulates the release of bile into the duodenum.

5. Peristalsis is controlled by the myenteric nerve plexus.

6. Pepsinogen, trypsinogen, and pro-carboxypeptidase are enzymatically inactive zymogens.

7. The absorption of dietary iron depends on intrinsic factor.

8. Filling of the stomach stimulates contractions of the colon.

9. The duodenum secretes a hormone that inhibits contractions of the stomach.

10. Tight junctions of the small intestine prevent anything from leaking between the epithelial cells.

*Answers in Appendix B*

## Testing Your Comprehension

1. A physician plans to attend a cocktail party, but he is also on call that night and must avoid intoxication. Therefore he drinks a glass of cream just before leaving for the party. Explain.

2. Which of these do you think would have the most severe effect on digestion: surgical removal of the stomach, gallbladder, or pancreas? Explain.

3. What do carboxypeptidase and aminopeptidase have in common? Identify as many differences between them as you can.

4. What do micelles and chylomicrons have in common? Identify as many differences between them as you can.

5. Explain why most dietary lipids must be absorbed by the lacteals rather than by the blood capillaries of a villus.

*Answers at www.mhhe.com/saladin4*

# www.mhhe.com/saladin4

*The textbook website provides a wealth of interactive study materials fully organized and integrated by chapter. You will find practice quizzes, labeling exercises, and much more that will complement your learning and understanding of anatomy and physiology. The website also includes tools designed to enhance your* **Anatomy & Physiology | REVEALED** *experience.*

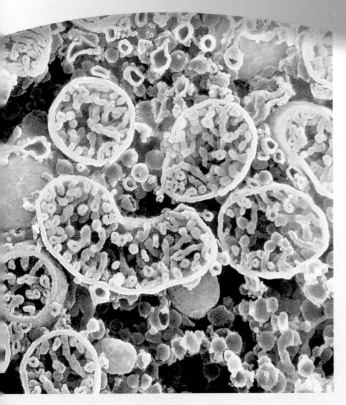

Mitochondria (green) and smooth endoplasmic reticulum in a cell of the ovary (SEM)

# NUTRITION AND METABOLISM

## CHAPTER OUTLINE

## INSIGHTS

## Brushing Up

To understand this chapter, it is important that you understand or brush up on the following concepts:

- Metabolism, catabolism, and anabolism (p. 69)
- Oxidation and reduction (p. 69)
- Saturated and unsaturated fats (p. 74)
- Metabolic pathways (p. 82)
- Structure and functions of ATP (p. 82)
- Receptor-mediated endocytosis (p. 109)
- Insulin and glucagon (p. 651)

Nutrition is the starting point and basis for all human form and function. From the time a single-celled, fertilized egg divides in two, nutrition provides the matter needed for cell division, growth, and development. It is the source of fuel that provides the energy for all biological work and of the raw materials for replacement of worn-out biomolecules and cells. The fact that it provides only the *raw* materials means, further, that chemical change—metabolism—lies at the foundation of form and function. In chapter 25, we saw how the digestive system breaks nutrients down into usable form and absorbs them into the blood and lymph. We now consider these nutrients in more depth, follow their fate after absorption, and explore related issues of metabolism and body heat.

# Nutrition

### Objectives

When you have completed this section, you should be able to

- describe some factors that regulate hunger and satiety;
- define *nutrient* and list the six major categories of nutrients;
- state the function of each class of macronutrients, the approximate amounts required in the diet, and some major dietary sources of each;
- name the blood lipoproteins, state their functions, and describe how they differ from each other; and
- name the major vitamins and minerals required by the body and the general functions they serve.

## BODY WEIGHT AND ENERGY BALANCE

The subject of nutrition quickly brings to mind the subject of body weight and the popular desire to control it. Weight is determined by the body's energy balance—if energy intake and output are equal, body weight is stable. We gain weight if intake exceeds output and lose weight if output exceeds intake. Although weight is determined by this ratio, it remains quite stable over many years' time and seems to have a homeostatic set point. This has been experimentally demonstrated in animals. If an animal is force-fed until it becomes obese and then allowed to feed

at will, it voluntarily reduces its intake and quickly returns to its former weight. Similarly, if an animal is undernourished until it loses much of its weight and then allowed to feed at will, it increases its intake and again returns quickly to its former weight.

In humans the set point varies greatly from person to person, and body weight results from a combination of hereditary and environmental influences. From studies of identical twins and other people, it appears that about 30% to 50% of the variation in human weight is due to heredity, and the rest to environmental factors such as eating and exercise habits.

## APPETITE

The struggle for weight control often seems to be a struggle against the appetite. Since the early 1990s, physiologists have discovered a still-growing list of peptide hormones and regulatory pathways that control short- and long-term appetite and body weight. Some of the hormones have been called *gut–brain peptides* because they act as chemical signals from the gastrointestinal tract to the brain. Only a few will be described here, but these suffice to give some idea of regulatory mechanisms known to date and where a great deal of research is currently focused.

### Short-Term Regulators of Appetite

The following three peptides work over periods of minutes to hours, making the individual feel hungry and begin eating, then to feel satiated and end a meal:

- **Ghrelin.**[1] This is secreted by parietal cells in the fundus of the stomach, especially when the stomach is empty. It produces the sensation of hunger and stimulates the hypothalamus to secrete growth-hormone–releasing hormone, thus priming the body to take best advantage of the nutrients about to be absorbed. Within an hour after eating, ghrelin secretion ceases.

- **Peptide YY (PYY).** This is a member of a family of hormones related to *neuropeptide Y* (NPY). It is secreted by enteroendocrine cells in the ileum and colon, but they sense that food has arrived even as it enters the stomach. They secrete PYY long before the chyme reaches the ileum, and in quantities proportional to the calories consumed. The primary effect of PYY is to signal satiety and terminate eating. Thus ghrelin is one of the signals that begins a meal, and PYY is one of the signals that ends it. PYY remains elevated well after a meal. It acts as an *ileal brake* that prevents the stomach from emptying too quickly, thus prolonging the sense of satiety.

---

[1]Named partly from *ghre* = growth, and partly as an acronym derived from **g**rowth-**h**ormone-**rele**asing hormone

- **Cholecystokinin (CCK).** As we saw in chapter 25, CCK is secreted by enteroendocrine cells in the duodenum and jejunum. It stimulates the secretion of bile and pancreatic enzymes, but also stimulates the brain and sensory fibers of the vagus nerves, producing an appetite-suppressing effect. Thus it joins PYY as a signal to stop eating.

## Long-Term Regulators of Appetite

Other peptides regulate appetite, metabolic rate, and body weight over the longer term, thus governing one's average rate of caloric intake and energy expenditure over periods of weeks to years. The following two members of this group work as "adiposity signals," informing the brain of how much adipose tissue the body has and activating mechanisms for adding or reducing fat.

- **Leptin.**[2] Leptin is secreted by adipocytes throughout the body. Its level is proportional to one's fat stores, so this is the brain's primary way of knowing how much body fat we have. Animals with a leptin deficiency or a defect in leptin receptors exhibit *hyperphagia* (overeating) and extreme obesity. With few exceptions, however, obese humans are not leptin-deficient or aided by leptin injections. More commonly, it seems that obesity is linked to unresponsiveness to leptin—a receptor defect rather than a hormone deficiency.

- **Insulin.** As we saw in chapter 17, insulin is secreted by the pancreatic beta cells. It stimulates glucose and amino acid uptake and promotes glycogen and fat synthesis. But it also has receptors in the brain and functions, like leptin, as an index of the body's fat stores. It has a weaker effect than leptin, however.

An important brain center for appetite regulation is the **arcuate nucleus** of the hypothalamus. All five of the aforementioned peptides have receptors in the arcuate nucleus, although they act on other target cells in the body as well. The arcuate nucleus has two groups of neurons involved in hunger. One group secretes **neuropeptide Y (NPY),** itself a potent appetite stimulant. The other secretes **melanocortin,** which inhibits eating. Ghrelin stimulates NPY secretion, whereas insulin, PYY, and leptin inhibit it. Leptin also stimulates melanocortin secretion (fig. 26.1) and inhibits the secretion of appetite stimulants called *endocannabinoids,* named for their resemblance to the tetrahydrocannabinol (THC) of marijuana.

> ⌐ **Think About It**
>
> *Suppose leptin or CCK could be economically produced and packaged as tablets to be taken orally. Would this be an effective diet drug? Why or why not?*

---

[2]*lept* = thin

---

┌─────────────────────────────────────────
**INSIGHT 26.1**    Clinical Application

## Obesity

*Obesity* is clinically defined as a weight more than 20% above the recommended norm for one's age, sex, and height. In the United States, about 30% of the population is obese and another 35% are overweight; there has lately been an alarming increase in the number of children who are morbidly obese by the age of 10. You can judge whether you are overweight or obese by calculating your *body mass index (BMI)*. If W is your weight in kilograms and H is your height in meters, $BMI = W/H^2$. (English–metric conversion factors can be found in Appendix C.) A BMI of 20 to 25 $kg/m^2$ is considered to be optimal for most people. A BMI over 27 $kg/m^2$ is considered overweight, and above 30 $kg/m^2$ is considered obese.

Excess weight shortens life expectancy and increases a person's risk of atherosclerosis, hypertension, diabetes mellitus, joint pain and degeneration, kidney stones, and gallstones; cancer of the breast, uterus, and liver in women; and cancer of the colon, rectum, and prostate gland in men. Excess thoracic fat impairs breathing and results in increased blood $P_{CO_2}$, sleepiness, and reduced vitality. Obesity is also a significant impediment to successful surgery.

Heredity plays as much role in obesity as in height, and even more than in many other disorders generally acknowledged to be hereditary. However, a predisposition to obesity is often greatly worsened by overfeeding in infancy and childhood. Consumption of excess calories in childhood causes adipocytes to increase in size and number. In adulthood, adipocytes do not multiply except in some extreme weight gains; their number remains constant while weight gains and losses result from changes in cell size (cellular hypertrophy).

As so many dieters learn, it is very difficult to substantially reduce one's adult weight. Most diets are unsuccessful over the long run as dieters lose and regain the same weight over and over. From an evolutionary standpoint, this is not surprising. The body's appetite- and weight-regulating mechanisms have evolved more to limit weight loss than weight gain, for a scarcity of food was surely a more common problem than a food surplus for our prehistoric ancestors. Were it not for the mechanisms that thwart weight loss, our ancestors might not have made it through the lean eons and we might not be here; but now that we are surrounded with a glut of tempting food, these survival mechanisms have become mechanisms of pathology.

Understandably, pharmaceutical companies are keenly interested in developing effective weight-control drugs. There could be an enormous profit, for example, in a drug that could inhibit ghrelin signaling or enhance or mimic leptin or melanocortin signaling. Such efforts have so far met with little success, but clearly a prerequisite to drug development is a better understanding of appetite-regulating peptides and their receptors. This subject is generating a prolific literature, and undoubtedly much more will be known about it before this chapter even gets to the printing press.

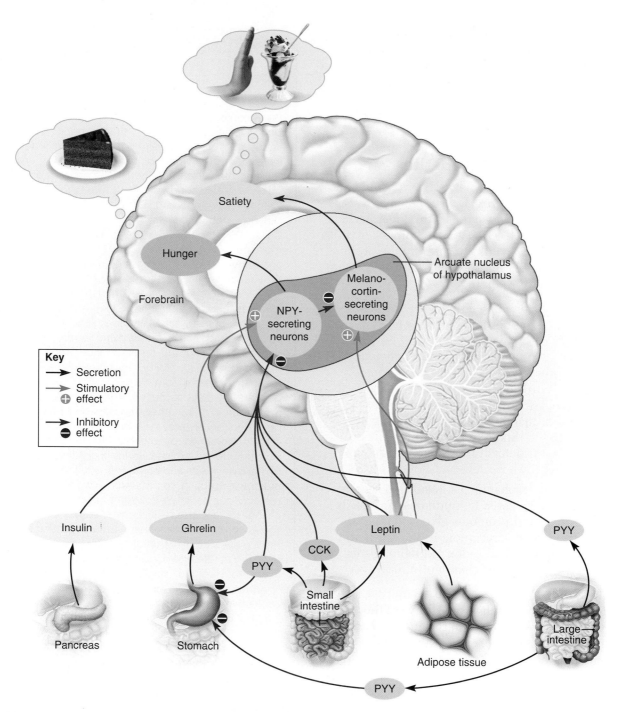

**FIGURE 26.1 Gut–Brain Peptides in Appetite Regulation.** Tissues and organs at the bottom of the figure are sources of peptides that stimulate or inhibit appetite-regulating neurons in the arcuate nucleus of the hypothalamus. Depending on the balance of stimulation and inhibition, those neurons secrete NPY or melanocortin to create a conscious sensation of hunger or satiety, respectively. CCK = cholecystokinin; NPY = neuropeptide Y; PYY = peptide YY.

Gut–brain peptides certainly are not the whole story behind appetite regulation. Hunger is also stimulated partly by gastric peristalsis. Mild **hunger contractions** begin soon after the stomach is emptied and increase in intensity over a period of hours. They can become quite a painful and powerful incentive to eat, yet they do not affect the amount of food consumed—this remains much the same even when nervous connections to the stomach and intestines are severed to cut off all conscious perception of hunger contractions. Food intake is terminated not only by PYY and CCK, but also in ways similar to the way that water intake slakes the thirst (see p. 934). Merely chewing and swallowing food briefly satisfies the appetite, even if the food is removed through an esophageal fistula (opening) before reaching the stomach. Inflating the stomach with a balloon inhibits hunger even in an animal that has not actually swallowed any food. Satiation produced by these mechanisms, however, is very short-lived compared with that produced by nutrient absorption.

Appetite is not merely a question of *how much* but also *what kind* of food is consumed. Even animals shift their diets from one kind of food to another, apparently because some foods provide nutrients that others do not. In humans, different neurotransmitters also seem to govern the appetite for different classes of nutrients. For example, *norepinephrine* stimulates the appetite for carbohydrates, *galanin* for fatty foods, and *endorphins* for protein.

## CALORIES

One calorie is the amount of heat that will raise the temperature of 1 g of water 1°C. One thousand calories is called a Calorie (capital *C*) in dietetics and a **kilocalorie (kcal)** in biochemistry. The relevance of calories to physiology is that they are a measure of the capacity to do biological work.

Nearly all dietary calories come from carbohydrates, proteins, and fats. Carbohydrates and proteins yield about 4 kcal/g when they are completely oxidized, and fats yield about 9 kcal/g. Alcohol (7.1 kcal/g) and sugary foods can promote malnutrition by providing "empty calories"—blunting the appetite for foods with greater nutrient diversity (see Insight 26.4, p. 1029). By suppressing the appetite but failing to provide other nutrients the body requires, they can contribute to malnutrition. In sound nutrition, the body's energy needs are met by more complex foods that simultaneously meet the need for proteins, lipids, vitamins, and other nutrients.

When a chemical is described as **fuel** in this chapter, we mean it is oxidized solely or primarily to extract energy from it. The extracted energy is usually used to make adenosine triphosphate (ATP), which then transfers the energy to other physiological processes (see fig. 2.30).

## NUTRIENTS

A **nutrient** is any ingested chemical that is used for growth, repair, or maintenance of the body. Nutrients fall into six major classes: water, carbohydrates, lipids, proteins, minerals, and vitamins (table 26.1). Water, carbohydrates, lipids, and proteins are considered **macronutrients** because they must be consumed in relatively large quantities. Minerals and vitamins are called **micronutrients** because only small quantities are required.

**Recommended daily allowances (RDAs)** of nutrients were first developed in 1943 by the National Research Council and National Academy of Sciences; they have been revised several times since. An RDA is a liberal but safe estimate of the daily intake that would meet the nutritional needs of most healthy people. Consuming less than the RDA of a nutrient does not necessarily mean you will be malnourished, but the probability of malnutrition increases in proportion to the amount of the deficit and how long it lasts.

Many nutrients can be synthesized by the body when they are unavailable from the diet. The body is incapable, however, of synthesizing minerals, most vitamins, eight of the amino acids, and one to three of the fatty acids. These are called **essential nutrients** because it is essential that they be included in the diet.

| TABLE 26.1 | Nutrient Classes and Their Principal Functions | |
|---|---|---|
| **Nutrient** | **Daily Requirement** | **Representative Functions** |
| Water | 2.5 L | Solvent; coolant; reactant or product in many metabolic reactions (especially hydrolysis and condensation); dilutes and eliminates metabolic wastes; supports blood volume and pressure |
| Carbohydrates | 125–175 g | Fuel; a component of nucleic acids, ATP and other nucleotides, glycoproteins, and glycolipids |
| Lipids | 80–100 g | Fuel; plasma membrane structure; myelin sheaths of nerve fibers; hormones; eicosanoids; bile salts; insulation; protective padding around organs; absorption of fat-soluble vitamins; vitamin D synthesis; some blood-clotting factors |
| Proteins | 44–60 g | Muscle contraction; ciliary and flagellar motility; structure of cellular membranes and extracellular material; enzymes; major component of connective tissues; transport of plasma lipids; some hormones; oxygen binding and transport pigments; blood-clotting factors; blood viscosity and osmolarity; antibodies; immune recognition; neuromodulators; buffers; emergency fuel |
| Minerals | 0.05–3,300 mg | Structure of bones and teeth; component of some structural proteins, hormones, ATP, phospholipids, and other chemicals; cofactors for many enzymes; electrolytes; oxygen transport by hemoglobin and myoglobin; buffers; stomach acid; osmolarity of body fluids |
| Vitamins | 0.002–60 mg | Coenzymes for many metabolic pathways; antioxidants; component of visual pigment; one hormone (vitamin D) |

# CARBOHYDRATES

A well-nourished adult has about 440 g of carbohydrate in the body, most of it in three places: about 325 g of muscle glycogen, 90 to 100 g of liver glycogen, and 15 to 20 g of blood glucose.

Sugars function as a structural component of other molecules including nucleic acids, glycoproteins, glycolipids, ATP, and related nucleotides (GTP, cAMP, etc.), and they can be converted to amino acids and fats. Most of the body's carbohydrate, however, serves as fuel—an easily oxidized source of chemical energy. Most cells meet their energy needs from a combination of carbohydrates and fats, but some cells, such as neurons and erythrocytes, depend almost exclusively on carbohydrates. Even a brief period of **hypoglycemia**[3] (deficiency of blood glucose) causes nervous system disturbances felt as weakness or dizziness.

Blood glucose concentration is therefore carefully regulated, mainly through the interplay of insulin and glucagon (see chapter 17 and later in this chapter). Among other effects, these hormones regulate the balance between glycogen and free blood glucose. If blood glucose concentration drops too low, the body draws on its stores of glycogen to meet its energy needs. If glycogen stores are depleted, physical endurance is greatly reduced. Thus it is important to consume enough carbohydrate to ensure that the body maintains adequate stores of glycogen for periods of exercise and fasting (including sleep).

Carbohydrate intake also influences the metabolism of other nutrients. Excess carbohydrate is converted to fat and conversely, fat is oxidized as fuel when glucose and glycogen levels are too low to meet our energy needs. This is why the consumption of starchy and sugary foods has a pronounced effect on body weight. It is unwise, however, to try to "burn off fat" by excessively reducing carbohydrate intake. As shown later in this chapter, the complete and efficient oxidation of fats depends on adequate carbohydrate intake and the presence of certain intermediates of carbohydrate metabolism. If these are lacking, fats are incompletely oxidized to ketone bodies, which may cause metabolic acidosis.

## Requirements

Because carbohydrates are rapidly oxidized, they are required in greater amounts than any other nutrient. The RDA is 125 to 175 g. The brain alone consumes about 120 g of glucose per day. Most Americans get about 40% to 50% of their calories from carbohydrates, but highly active people should get up to 60%.

Carbohydrate consumption in the United States has become excessive over the past century because of a combination of fondness for sweets, increased use of sugar in processed foods, and reduced physical activity (see Insight 26.2). A century ago, Americans consumed an

average of 1.8 kg (4 lb) of sugar per year. Now, with sucrose and high-fructose corn syrup so widely used in foods and beverages, the average American ingests 200 to 300 g of carbohydrate per day and the equivalent of 27 kg (60 lb) of table sugar and 21 kg (46 lb) of corn syrup per year. A single nondiet soft drink contains 38 to 43 g (about 8 teaspoons) of sugar per 355 mL (12 oz) serving.

Dietary carbohydrates come in three principal forms: monosaccharides, disaccharides, and polysaccharides (complex carbohydrates). The only nutritionally significant polysaccharide is starch. Although glycogen is a polysaccharide, only trivial amounts of it are present in cooked meats. Cellulose, another polysaccharide, is not considered a nutrient because it is not digested and never enters the human tissues. Its importance as dietary fiber, however, is discussed shortly.

The three major disaccharides are sucrose, lactose, and maltose. The monosaccharides—glucose, galactose, and fructose—arise mainly from the digestion of starch and disaccharides. The small intestine and liver convert fructose and galactose to glucose, so ultimately all carbohydrate digestion generates glucose. Outside of the hepatic portal system, glucose is the only monosaccharide present in the blood in significant quantity; thus it is known as *blood sugar*. Its concentration is normally maintained at 70 to 110 mg/dL in peripheral venous blood.

> **Think About It**
>
> *Glucose concentration is about 15 to 30 mg/dL higher in arterial blood than in most venous blood. Explain why.*

Ideally, most carbohydrate intake should be in the form of complex carbohydrates, primarily starch. This is partly because foods that provide starch also usually provide other nutrients. Simple sugars not only provide empty calories but also promote tooth decay. A typical American, however, now obtains only 50% of his or her carbohydrates from starch and the other 50% from sucrose and corn syrup.

---

**INSIGHT 26.2**    Evolutionary Medicine

### Evolution of the Sweet Tooth

Our craving for sugar doubtlessly originated in our prehistoric ancestors. Not only did they have to work much harder to survive than we do, but high-calorie foods were scarce in their environment and people were at constant risk of starvation. Those who were highly motivated to seek and consume sugary, high-calorie foods passed their "sweet tooth" on to us, their descendents—along with a similarly adaptive appetite for other rare but vital nutrients, namely fat and salt. The tastes that were essential to our ancestors' survival can now be a disadvantage in a culture where salty, fatty, and sugary foods are all too easy to obtain and the food industry is eager to capitalize on these tastes.

---

[3]*hypo* = below normal + *glyc* = sugar + *emia* = blood condition

## Dietary Sources

Nearly all dietary carbohydrates come from plants—particularly grains, legumes, fruits, and root vegetables. Sucrose is refined from sugarcane and sugar beets. Fructose is present in fruits and corn syrup. Maltose is present in some foods such as germinating cereal grains. Lactose is the most abundant solute in cow's milk (about 4.6% lactose by weight).

## Fiber

*Dietary fiber* refers to all fibrous materials of plant and animal origin that resist digestion. Most is plant matter—the carbohydrates cellulose and pectin and such noncarbohydrates as gums and lignin. Although it is not a nutrient, fiber is an essential component of the diet. The recommended daily allowance is about 30 g, but average intake varies greatly from country to country—from 40 to 150 g/day in India and Africa to only 12 g/day in the United States.

Fiber in the intestines absorbs water, swells, softens the stool, and increases its bulk by 40% to 100%. The last effect stretches the colon and stimulates peristalsis, thereby quickening the passage of feces from the colon.

Pectin is a **water-soluble fiber** found in oats, beans, peas, carrots, brown rice, and fruits. Soluble fiber reduces blood cholesterol and low-density lipoprotein (LDL) levels (see Insight 19.5, p. 746). Cellulose, hemicellulose, and lignin, called **water-insoluble fibers,** apparently have no effect on cholesterol or LDLs. Contrary to previous popular and scientific belief, dietary fiber is no longer thought to reduce the incidence of colorectal cancer. Excessive dietary fiber can actually have a deleterious effect by interfering with the absorption of iron, calcium, magnesium, phosphorus, and some trace elements.

## LIPIDS

The reference male and female are, respectively, about 15% and 25% fat by weight. Fat accounts for most of the body's stored energy. Lesser amounts of phospholipid, cholesterol, and other lipids also play vital structural and physiological roles.

A well-nourished adult meets 80% to 90% of his or her resting energy needs from fat. Fat is superior to carbohydrates for energy storage for two reasons: (1) carbohydrates are hydrophilic, absorb water, and thus expand and occupy more space in the tissues. Fat, however, is hydrophobic, contains almost no water, and is a more compact energy storage substance. (2) Fat is less oxidized than carbohydrate and contains over twice as much energy (9 kcal/g of fat compared with 4 kcal/g of carbohydrate). A man's typical fat reserves contain enough energy for 119 hours of running, whereas his carbohydrate stores would suffice for only 1.6 hours.

Fat has **glucose-sparing** and **protein-sparing effects**— as long as enough fat is available to meet the energy needs of the tissues, protein is not catabolized for fuel and glucose is spared for consumption by cells that cannot use fat, such as neurons.

Vitamins A, D, E, and K are fat-soluble vitamins, which depend on dietary fat for their absorption by the intestine. People who ingest less than 20 g of fat per day are at risk of vitamin deficiency because there is not enough fat in the intestine to transport these vitamins into the body tissues.

Phospholipids and cholesterol are major structural components of plasma membranes and myelin. Cholesterol is also important as a precursor of steroid hormones, bile acids, and vitamin D. Thromboplastin, an essential blood-clotting factor, is a lipoprotein. Two fatty acids—arachidonic acid and linoleic acid—are precursors of prostaglandins and other eicosanoids.

In addition to its metabolic and structural roles, fat has important protective and insulating functions described under adipose tissue in chapter 5.

## Requirements

Fat should account for no more than 30% of your daily caloric intake; no more than 10% of your fat intake should be saturated fat; and average cholesterol intake should not exceed 300 mg/day (one egg yolk contains about 240 mg). A typical American consumes 30 to 150 g of fat per day, obtains 40% to 50% of his or her calories from fat, and ingests twice as much cholesterol as the recommended limit.

Most fatty acids can be synthesized by the body. **Essential fatty acids** are those we cannot synthesize and therefore must obtain from the diet. These include linoleic acid and possibly linolenic and arachidonic acids; there are differences of opinion about the body's ability to synthesize the last two. As long as 1% to 2% of the total energy intake comes from linoleic acid, people do not develop signs of essential fatty acid deficiency. In the typical Western diet, linoleic acid provides about 6% of the energy.

## Sources

Saturated fats are predominantly of animal origin. They occur in meat, egg yolks, and dairy products but also in some plant products such as coconut and palm oil (common in nondairy coffee creamers and other products). Processed foods such as hydrogenated oils and vegetable shortening are also high in saturated fat, which is therefore abundant in many baked goods. Unsaturated fats predominate in nuts, seeds, and most vegetable oils. The essential fatty acids are amply provided by the vegetable oils in mayonnaise, salad dressings, and margarine and by whole grains and vegetables. Excessive consumption of saturated and unsaturated fats is a risk factor for diabetes mellitus, cardiovascular disease, and breast and colon cancer.

The richest source of cholesterol is egg yolks, but it is also prevalent in milk products; shellfish (especially shrimp); organ meats such as kidneys, liver, and brains; and other mammalian meat. Cholesterol does not occur in foods of plant origin. However, the serum cholesterol level is strongly influenced by the types and quantity of fatty acids in the diet. This relationship is explained in the next section.

## Cholesterol and Serum Lipoproteins

Lipids are an important part of the diet and must be transported to all cells of the body, yet they are hydrophobic and will not dissolve in the aqueous blood plasma. This problem is overcome by complexes called **lipoproteins**—tiny droplets with a core of cholesterol and triglycerides and a coating of proteins and phospholipids. The coating not only enables the lipids to remain suspended in the blood, but also serves as a recognition marker for cells that absorb them. The complexes are often referred to as *serum lipoproteins* because their concentrations are expressed in terms of a volume of blood serum, not whole blood.

Lipoproteins are classified into four major categories (and some lesser ones) by their density: **chylomicrons, high-density lipoproteins (HDLs), low-density lipoproteins (LDLs),** and **very low–density lipoproteins (VLDLs)** (fig. 26.2a). The higher the proportion of protein to lipid, the higher the density. These particles also differ considerably in size: chylomicrons range widely from 75 to 1,200 nm in diameter, but the others diminish in size from VLDLs (30–80 nm), to LDLs (18–25 mm), to HDLs (5–12 nm). Their most important differences, however, are in composition and function. Figure 26.2b shows the three primary pathways by which they are made and processed.

Chylomicrons form in the absorptive cells of the small intestine and then pass into the lymphatic system and ultimately the bloodstream (see chapter 25). Endothelial cells of the blood capillaries have a surface enzyme called **lipoprotein lipase** that hydrolyzes triglycerides into monoglycerides and free fatty acids (FFAs). These products can then pass through the capillary walls into adipocytes, where they are resynthesized into storage triglycerides. Some FFAs, however, remain in the blood plasma bound to albumin. The remainder of a chylomicron after the triglycerides have been extracted, called a *chylomicron remnant,* is removed and degraded by the liver.

VLDLs, produced by the liver, transport lipids to the adipose tissue for storage. When their triglycerides are removed in the adipose tissue, the VLDLs become LDLs and contain mostly cholesterol. Cells that need cholesterol (usually for membrane structure or steroid hormone synthesis) absorb LDLs by receptor-mediated endocytosis, digest them with lysosomal enzymes, and release the cholesterol for intracellular use.

HDL production begins in the liver, which produces an empty, collapsed protein shell. This shell travels in the blood and picks up cholesterol and phospholipids from other organs. The next time it circulates through the liver, the liver removes the cholesterol and eliminates it in the bile, either as cholesterol or as bile acids. HDLs are therefore a vehicle for removing excess cholesterol from the body.

It is desirable to maintain a total plasma cholesterol concentration of 200 mg/dL or less. From 200 to 239 mg/dL is considered borderline high, and levels over 240 mg/dL are undesirable.

Most of the body's cholesterol is endogenous (internally synthesized) rather than dietary, and the body compensates for variations in dietary intake. High intake steps down cholesterol synthesis by the liver, whereas a low dietary intake steps it up. Thus, lowering dietary cholesterol reduces the serum cholesterol level by no more than 5%. Much more important is the fact that certain saturated fatty acids (SFAs) raise serum cholesterol level. The 16-carbon SFA palmitic acid, for example, raises it by blocking cholesterol uptake by the tissues (yet stearic acid, an 18-carbon SFA, does not raise the cholesterol level). Some food advertising is very deceptive on this point. It may truthfully advertise a food as being cholesterol-free but neglect to mention that it contains SFAs that may raise the consumer's cholesterol level anyway. A moderate reduction of saturated fatty acid intake can lower blood cholesterol by 15% to 20%—considerably more effect than reducing dietary cholesterol per se.

Vigorous exercise also lowers blood cholesterol levels. The mechanism is somewhat roundabout: Exercise reduces the sensitivity of the right atrium of the heart to blood pressure, so the heart secretes less atrial natriuretic factor. Consequently, the kidneys excrete less sodium and water, and the blood volume rises. This dilutes the lipoproteins in the blood, and the adipocytes compensate by producing more lipoprotein lipase. Thus, the adipocytes consume more blood triglycerides. This shrinks the VLDL particles, which shed some of their cholesterol in the process, and HDLs pick up this free cholesterol for removal by the liver.

Blood cholesterol is not the only important measure of healthy lipid concentrations, however. A high LDL concentration is a warning sign because, as you can see from the function of LDLs described previously, it signifies a high rate of cholesterol deposition in the arteries. LDLs are elevated not only by saturated fats but also by cigarette smoking, coffee, and stress. A high proportion of HDL, on the other hand, is beneficial because it indicates that cholesterol is being removed from the arteries and transported to the liver for disposal. Thus it is desirable to increase your ratio of HDL to LDL. This is best done with a diet low in calories and saturated fats and is promoted by regular aerobic exercise.

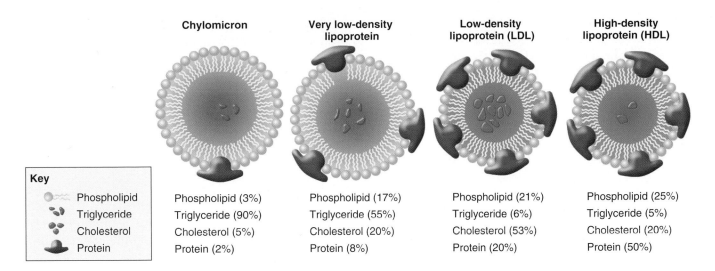

**Chylomicron**

Phospholipid (3%)
Triglyceride (90%)
Cholesterol (5%)
Protein (2%)

**Very low-density lipoprotein**

Phospholipid (17%)
Triglyceride (55%)
Cholesterol (20%)
Protein (8%)

**Low-density lipoprotein (LDL)**

Phospholipid (21%)
Triglyceride (6%)
Cholesterol (53%)
Protein (20%)

**High-density lipoprotein (HDL)**

Phospholipid (25%)
Triglyceride (5%)
Cholesterol (20%)
Protein (50%)

Key
- Phospholipid
- Triglyceride
- Cholesterol
- Protein

**(a) Lipoprotein types**

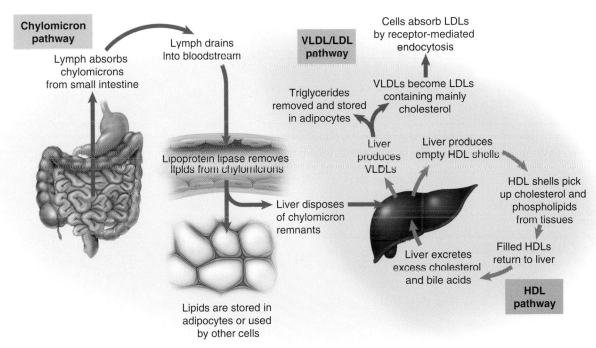

**(b) Lipoprotein processing pathways**

**FIGURE 26.2  Lipoprotein Processing.**  (a) The four types of serum lipoproteins. (b) The three pathways of lipoprotein processing.
▶ *Why is a high HDL:LDL ratio healthier than a high LDL:HDL ratio?*

## PROTEINS

Protein constitutes about 12% to 15% of the body's mass; 65% of it is in the skeletal muscles. Proteins are responsible for muscle contraction and the motility of cilia and flagella. They are a major structural component of all cellular membranes, with multiple important roles such as membrane receptors, pumps, ion channels, and cell-identity markers. Fibrous proteins such as collagen, elastin, and keratin make up much of the structure of bone, cartilage, tendons, ligaments, skin, hair, and nails. Globular proteins include antibodies, hormones, neuromodulators, hemoglobin, myoglobin, and about 2,000 enzymes that control nearly every aspect of cellular metabolism. They also include the albumin and other plasma proteins that maintain blood viscosity and osmolarity and transport lipids and some other plasma solutes. Proteins buffer the pH of body fluids and contribute to the resting membrane potentials of all cells. No other class of biomolecules has such a broad variety of functions.

## Requirements

For persons of average weight, the RDA of protein is 44 to 60 g, depending on age and sex. Multiplying your weight in pounds by 0.37 or your weight in kilograms by 0.8 gives an estimate of your RDA of protein in grams. A higher intake is recommended, however, under conditions of stress, infection, injury, and pregnancy. Infants and children require more protein than adults relative to body weight. Excessive protein intake, however, overloads the kidneys with nitrogenous waste and can cause renal damage. This is a risk in certain high-protein fad diets.

Total protein intake is not the only significant measure of dietary adequacy. The nutritional value of a protein depends on whether it supplies the right amino acids in the proportions needed for human proteins. Adults can synthesize 12 of the 20 amino acids from other organic compounds when they are not available from the diet, but there are 8 **essential amino acids** that we cannot synthesize: isoleucine, leucine, lysine, methionine, phenylalanine, threonine, tryptophan, and valine. (Infants also require histidine.) In addition, 2 amino acids can only be synthesized from essential amino acids: cysteine from methionine and tyrosine from phenylalanine. The other 10 (9 in infants) are called **inessential amino acids**—not because the body does not require them but because it can synthesize its own when the diet does not supply them.

Cells do not store surplus amino acids for later use. When a protein is to be synthesized, all of the amino acids necessary must be present at once, and if even one is missing, the protein cannot be made. High-quality **complete proteins** are those that provide all of the essential amino acids in the necessary proportions for human tissue growth, maintenance, and nitrogen balance. Lower-quality **incomplete proteins** lack one or more essential amino acids. For example, cereals are low in lysine, and legumes are low in methionine.

Protein quality is also determined by **net protein utilization**—the percent of the amino acids in a protein that the human body uses. We typically use 70% to 90% of animal protein but only 40% to 70% of plant protein. It therefore takes a larger serving of plant protein than animal protein to meet our needs—for example, we need 400 g (about 14 oz) of rice and beans to provide as much usable protein as 115 g (about 4 oz) of hamburger. However, reducing meat intake and increasing plant intake has advantages. Among other considerations, plant foods provide more vitamins, minerals, and fiber; less saturated fat; no cholesterol; and less pesticide. In an increasingly crowded world, it must also be borne in mind that it requires far more land to produce meat than to produce crops.

## Dietary Sources

The animal proteins of meat, eggs, and dairy products closely match human proteins in amino acid composition. Thus animal products provide high-quality complete protein, whereas plant proteins are incomplete. Nevertheless, this does not mean that your dietary protein *must* come from meat; indeed, about two-thirds of the world's population receives adequate protein nutrition from diets containing very little meat. We can combine plant foods so that one provides what another lacks—beans and rice, for example, are a complementary combination of legume and cereal. Beans provide the isoleucine and lysine lacking in grains, while rice provides the tryptophan and cysteine lacking in beans.

## Nitrogen Balance

Proteins are our chief dietary source of nitrogen. **Nitrogen balance** is a state in which the rate of nitrogen ingestion equals the rate of excretion (chiefly as nitrogenous wastes). Growing children exhibit a state of **positive nitrogen balance** because they ingest more than they excrete, thus retaining protein for tissue growth. Pregnant women and athletes in resistance training also show positive nitrogen balance. When excretion exceeds ingestion, a person is in a state of **negative nitrogen balance.** This indicates that body proteins are being broken down and used as fuel. Proteins of the muscles and liver are more easily broken down than others; thus negative nitrogen balance tends to be associated with muscle atrophy. Negative nitrogen balance may occur if carbohydrate and fat intake are insufficient to meet the need for energy. Carbohydrates and fats are said to have a protein-sparing effect because they prevent protein catabolism when present in sufficient amounts to meet energy needs.

Nitrogen balance is affected by some hormones. Growth hormone and sex steroids promote protein synthesis and positive nitrogen balance during childhood, adolescence, and pregnancy. Glucocorticoids, on the other hand, promote protein catabolism and negative nitrogen balance in states of stress.

> **Think About It**
>
> *Would you expect a person recovering from a long infectious disease to be in a state of positive or negative nitrogen balance? Why?*

## MINERALS AND VITAMINS

Minerals are inorganic elements that plants extract from soil or water and introduce into the food web. Vitamins are small dietary organic compounds that are necessary to metabolism. Neither is used as fuel, but both are essential to our ability to use other nutrients. With the exception of a few vitamins, these nutrients cannot be synthesized by the body and must be included in the diet. They are, however, required in relatively small quantities. Mineral RDAs range from 0.05 mg of chromium and selenium to 1,200 mg of calcium and phosphorus. Vitamin RDAs range from about 0.002 mg of vitamin $B_{12}$ to 60 mg of vitamin C. Despite the small quantities involved, minerals and vitamins have very potent effects on physiology. Indeed, excessive amounts are toxic and potentially lethal.

## Minerals

Minerals constitute about 4% of the body mass, with three-quarters of this being the calcium and phosphorus in the bones and teeth. Phosphorus is also a key structural component of phospholipids, ATP, cAMP, GTP, and creatine phosphate and is the basis of the phosphate buffer system (see chapter 24). Calcium, iron, magnesium, and manganese function as cofactors for enzymes. Iron is essential to the oxygen-carrying capacity of hemoglobin and myoglobin. Chlorine is a component of stomach acid (HCl). Many mineral salts function as electrolytes and thus govern the function of nerve and muscle cells, osmotically regulate the content and distribution of water in the body, and maintain blood volume.

Table 26.2 summarizes adult mineral requirements and dietary sources. Broadly speaking, the best sources of minerals are vegetables, legumes, milk, eggs, fish, shellfish, and some other meats. Cereal grains are a relatively poor source, but processed cereals may be mineral-fortified.

Sodium chloride has been both a prized commodity and a curse. Animal tissues contain relatively large amounts of salt, and carnivores rarely lack ample salt in their diets. Plants, however, are relatively poor in salt, so herbivores often must supplement their diet by ingesting salt from the soil. As humans developed agriculture and became more dependent on plants, they also became increasingly dependent on supplemental salt. Salt has often been used as a form of payment for goods and services—the word *salary* comes from *sal* (salt). Our fondness for salt and high sensitivity to it undoubtedly stem from its physiological importance and its scarcity in a largely vegetarian diet.

Now, however, this fondness has become a bane. The recommended sodium intake is 1.1 g/day, but a typical American diet contains about 4.5 g/day. This is due not just to the use of table salt but more significantly to the large amounts of salt in processed foods, much of it "disguised" in soy sauce, MSG (monosodium glutamate), baking soda, and baking powder. In some areas of Japan, salt intake averages 27 g/day and the great majority of people die before age 70 of stroke and other complications of hypertension.

| TABLE 26.2 | Mineral Requirements and Some Dietary Sources | |
|---|---|---|
| **Mineral** | **RDA (mg)** | **Some Dietary Sources*** |
| *Major Minerals* | | |
| Calcium | 1,200 | Milk, fish, shellfish, greens, tofu, orange juice |
| Phosphorus | 1,200 | Red meat, poultry, fish, eggs, milk, legumes, whole grains, nuts |
| Sodium | 1,500 | Table salt, processed foods; usually present in excess |
| Chloride | 2,300 | Table salt, some vegetables; usually present in excess |
| Magnesium | 280–350 | Milk, greens, whole grains, nuts, legumes, chocolate |
| Potassium | 4,700 | Red meat, poultry, fish, cereals, spinach, squash, bananas, apricots |
| Sulfur | Unknown | Meats, milk, eggs, legumes; almost any proteins |
| *Trace Minerals* | | |
| Zinc | 12–15 | Red meat, seafood, cereals, wheat germ, legumes, nuts, yeast |
| Iron | 10–15 | Red meat, liver, shellfish, eggs, dried fruits, legumes, nuts, molasses |
| Manganese | 2.5–5.0 | Greens, fruits, legumes, whole grains, nuts |
| Copper | 1.5–3.0 | Red meat, liver, shellfish, legumes, whole grains, nuts, cocoa |
| Fluoride | 1.5–4.0 | Fluoridated water and toothpaste, tea, seafood, seaweed |
| Iodine | 0.15 | Marine fish, fish oils, shellfish, iodized salt |
| Molybdenum | 0.07–0.25 | Beans, whole grains, nuts |
| Chromium | 0.05–0.25 | Meats, liver, cheese, eggs, whole grains, yeast, wine |
| Selenium | 0.05–0.07 | Red meats, organ meats, fish, shellfish, eggs, cereals |
| Cobalt | Unknown | Red meat, poultry, fish, liver, milk |

*"Red meat" refers to mammalian muscle such as beef and pork. "Organ meat" refers to brain, pancreas, heart, kidney, etc. Liver is specified separately and refers to beef, pork, and chicken livers, which are similar for most nutrients.

Hypertension is a leading cause of death among American blacks, who have twice the risk of hypertension and 10 times the risk of dying from it that American whites have. The reason for this is not excessive salt intake, but rather that people of West African descent have kidneys with an especially strong tendency to retain salt.

## Vitamins

Vitamins were originally named with letters in the order of their discovery, but they also have chemically descriptive names such as ascorbic acid (vitamin C) and riboflavin (vitamin $B_2$). Most vitamins must be obtained from the diet (table 26.3), but the body synthesizes some of them from precursors called *provitamins*—niacin from the amino acid tryptophan, vitamin D from cholesterol, and vitamin A from carotene, which is abundantly present in carrots, squash, and other yellow vegetables and fruits. Vitamin K, pantothenic acid, biotin, and folic acid are produced by the bacteria of the large intestine. The feces contain more biotin than food does.

Vitamins are classified as water-soluble or fat-soluble. **Water-soluble vitamins** are absorbed with water from the small intestine, dissolve freely in the body fluids, and are quickly excreted by the kidneys. They cannot be stored in the body and therefore seldom accumulate to excess. The water-soluble vitamins are ascorbic acid (vitamin C) and the B vitamins. Ascorbic acid promotes hemoglobin synthesis, collagen synthesis, and sound connective tissue structure, and it is an antioxidant that scavenges free radicals and possibly reduces the risk of cancer. The B vitamins function as coenzymes or parts of coenzyme molecules; they assist enzymes by transferring electrons from one metabolic reaction to another, making it possible for enzymes to catalyze these reactions. Some of their functions arise later in this chapter as we consider carbohydrate metabolism.

**Fat-soluble vitamins** are incorporated into lipid micelles in the small intestine and absorbed with dietary lipids. They are more varied in function than water-soluble vitamins. Vitamin A is a component of the visual pigments and promotes proteoglycan synthesis and epithelial maintenance. Vitamin D promotes calcium absorption and bone mineralization. Vitamin K is essential to prothrombin synthesis and blood clotting. Vitamins A and E are antioxidants, like ascorbic acid.

It is common knowledge that various diseases result from vitamin deficiencies, but it is less commonly known that **hypervitaminosis** (vitamin excesses) also causes disease. A *deficiency* of vitamin A, for example, can result in

| TABLE 26.3 | Vitamin Requirements and Some Dietary Sources | |
|---|---|---|
| **Mineral** | **RDA (mg)** | **Some Dietary Sources*** |
| *Water-Soluble Vitamins* | | |
| Ascorbic acid (C) | 60 | Citrus fruits, strawberries, tomatoes, greens, cabbage, cauliflower, broccoli, brussels sprouts |
| B complex | | |
|   Thiamine ($B_1$) | 1.5 | Red meat, organ meats, liver, eggs, greens, asparagus, legumes, whole grains, seeds, yeast |
|   Riboflavin ($B_2$) | 1.7 | Widely distributed, and deficiencies are rare; all types of meat, milk, eggs, greens, whole grains, apricots, legumes, mushrooms, yeast |
|   Pyridoxine ($B_6$) | 2.0 | Red meat, organ meats, fish, liver, greens, apricots, legumes, whole grains, seeds |
|   Cobalamin ($B_{12}$) | 0.002 | Red meat, organ meats, liver, shellfish, eggs, milk; absent from food plants |
|   Niacin (nicotinic acid) | 19 | Readily synthesized from tryptophan, which is present in any diet with adequate protein; red meat, organ meats, liver, poultry, fish, apricots, legumes, whole grains, mushrooms |
|   Pantothenic acid | 4–7 | Widely distributed, and deficiencies are rare; red meat, organ meats, liver, eggs, green and yellow vegetables, legumes, whole grains, mushrooms, yeast |
|   Folic acid (folacin) | 0.2 | Eggs, liver, greens, citrus fruits, legumes, whole grains, seeds |
|   Biotin | 0.03–0.10 | Red meat, organ meats, liver, eggs, cheese, cabbage, cauliflower, bananas, legumes, nuts |
| *Fat-Soluble Vitamins* | | |
| Retinol (A) | 1.0 | Fish oils, eggs, cheese, milk, greens, other green and yellow vegetables and fruits, margarine |
| Calcitriol (D) | 0.01 | Formed by exposure of skin to sunlight; fish, fish oils, milk |
| α-Tocopherol (E) | 10 | Fish oils, greens, seeds, wheat germ, vegetable oils, margarine, nuts |
| Phylloquinone (K) | 0.08 | Most of the RDA is met by synthesis by intestinal bacteria; liver, greens, cabbage, cauliflower |

*See footnote in table 26.2.

night blindness, dry skin and hair, a dry conjunctiva and cloudy cornea, and increased incidence of urinary, digestive, and respiratory infections. This is the world's most common vitamin deficiency. An *excess* of vitamin A, however, may cause anorexia, nausea and vomiting, headache, pain and fragility of the bones, hair loss, an enlarged liver and spleen, and birth defects. Vitamins $B_6$, C, D, and E have also been implicated in toxic hypervitaminosis.

Some people take *megavitamins*—doses 10 to 1,000 times the RDA—thinking that they will improve athletic performance. Since vitamins are not burned as fuel, and small amounts fully meet the body's metabolic needs, there is no evidence that vitamin supplements improve performance except when used to correct a dietary deficiency. Megadoses of fat-soluble vitamins can be especially harmful.

## Before You Go On

*Answer the following questions to test your understanding of the preceding section:*

1. Name two hormones that regulate short-term hunger and satiety. How does leptin differ from these in its effects?

2. Explain the following statement: Cellulose is an important part of a healthy diet but it is not a nutrient.

3. What class of nutrients provides most of the calories in the diet? What class of nutrients provides the body's major reserves of stored energy?

4. Contrast the functions of VLDLs, LDLs, and HDLs. Explain how this is related to the fact that a high blood HDL level is desirable, but a high VLDL–LDL level is undesirable.

5. Why do some proteins have more nutritional value than others?

# Carbohydrate Metabolism

### Objectives

When you have completed this section, you should be able to

- describe the principal reactants and products of each major step of glucose oxidation;
- contrast the functions and products of anaerobic fermentation and aerobic respiration;
- explain where and how cells produce ATP; and
- describe the production, function, and use of glycogen.

Most dietary carbohydrate is burned as fuel within a few hours of absorption. Although three monosaccharides are absorbed from digested food—glucose, galactose, and fructose—the last two are quickly converted to glucose, and all oxidative carbohydrate consumption is essentially a matter of glucose catabolism. The overall reaction for this is

$$C_6H_{12}O_6 + 6\ O_2 \rightarrow 6\ CO_2 + 6\ H_2O$$

The function of this reaction is not to produce carbon dioxide and water but to transfer energy from glucose to ATP.

Along the pathway of glucose oxidation are several links through which other nutrients—especially fats and amino acids—can also be oxidized as fuel. Carbohydrate catabolism therefore provides a central vantage point from which we can view the catabolism of all fuels and the generation of ATP.

## GLUCOSE CATABOLISM

If the preceding reaction were carried out in a single step, it would generate a short, intense burst of heat—like the burning of paper, which has the same chemical equation. Not only would this be useless to the body's metabolism, it would kill the cells. In the body, however, the process is carried out in a series of small steps, each controlled by a separate enzyme. Energy is released in small manageable amounts, and as much as possible is transferred to ATP. The rest is released as heat.

There are three major pathways of glucose catabolism:

1. **glycolysis,** which splits a glucose molecule into two molecules of pyruvic acid;

2. **anaerobic fermentation,** which occurs in the absence of oxygen and reduces pyruvic acid to lactic acid; and

3. **aerobic respiration,** which occurs in the presence of oxygen and oxidizes pyruvic acid to carbon dioxide and water.

You may find it helpful to review figure 2.31 (see p. 84) for a broad overview of these processes and their relationship to ATP production. Figures 26.3 to 26.6 examine these processes in closer detail. The first two figures are labeled with numbers that correspond to reaction steps described shortly.

Coenzymes are vitally important to these reactions. Enzymes remove electrons (as hydrogen atoms) from the intermediate compounds of these pathways, but they do not bind them. Instead, they transfer the hydrogen atoms to coenzymes, and the coenzymes donate them to other compounds later in one of the reaction pathways. Thus the enzymes of glucose catabolism cannot function without their coenzymes.

The two coenzymes of special importance to glucose catabolism are **NAD**$^+$ (nicotinamide adenine dinucleotide) and **FAD** (flavin adenine dinucleotide). Both are derived from B vitamins: NAD$^+$ from niacin and FAD from riboflavin. Hydrogen atoms are removed from metabolic intermediates in pairs—that is, two protons and two

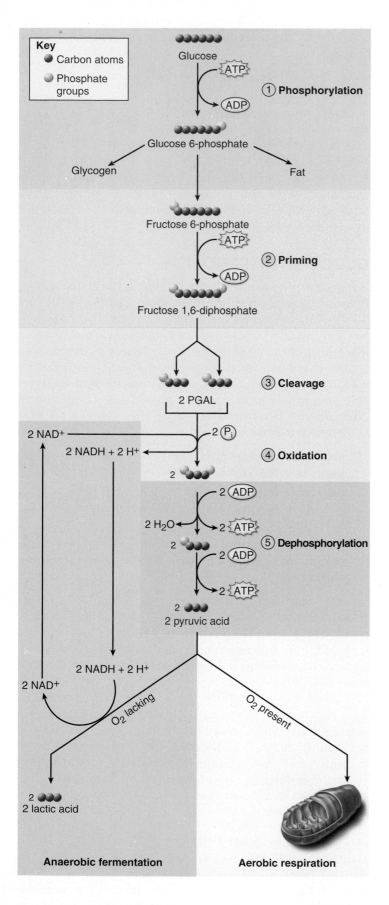

electrons (2 H$^+$ and 2 e$^-$) at a time—and transferred to a coenzyme. This produces a reduced coenzyme with a higher free energy content than it had before the reaction. Coenzymes thus become the temporary carriers of the energy extracted from glucose metabolites. The reactions for this are

$$FAD + 2 H \rightarrow FADH_2$$

and

$$NAD^+ + 2 H \rightarrow NADH + H^1$$

FAD binds two protons and two electrons to become FADH$_2$. NAD$^+$, however, binds the two electrons but only one of the protons to become NADH. The other proton remains a free hydrogen ion, H$^+$ (or H$_3$O$^+$, but it is represented in this chapter as H$^+$).

## GLYCOLYSIS

Upon entering a cell, glucose begins a series of conversions called glycolysis[4] (fig. 26.3):

① **Phosphorylation.** The enzyme *hexokinase* transfers an inorganic phosphate group (P$_i$) from ATP to glucose, producing glucose 6-phosphate (G6P). This has two effects:

- It keeps the intracellular concentration of glucose very low, thus maintaining a concentration gradient that favors the continued diffusion of more glucose into the cell.

- Phosphorylated compounds cannot pass through the plasma membrane, so this prevents the sugar from leaving the cell. In most cells, step 1 is irreversible because they lack the enzyme to convert G6P back to glucose. The few exceptions are cells that must be able to release free glucose to the blood: absorptive cells of the small intestine, proximal convoluted tubule cells in the kidney, and liver cells.

G6P is a versatile molecule that can be converted to fat or amino acids, polymerized to form glycogen for storage, or further oxidized to extract its energy. For

---

[4]*glyco* = sugar + *lysis* = splitting

**FIGURE 26.3   Glycolysis and Anaerobic Fermentation.**
Numbered reaction steps are explained in the text.

▶ *At what point would this reaction stop, and what reaction intermediate would accumulate, if NAD$^+$ were unavailable to the cell? What process replenishes the NAD$^+$ supply?*

now, we are mainly concerned with its further oxidation (glycolysis), the general effect of which is to split G6P (a six-carbon sugar, $C_6$) into two three-carbon ($C_3$) molecules of **pyruvic acid** (pyruvate). Continue tracing these steps in figure 26.3 as you read.

②  **Priming.** G6P is rearranged (isomerized) to form fructose 6-phosphate, which is phosphorylated again to form fructose 1,6-diphosphate. This "primes" the process by providing activation energy, somewhat like the heat of a match used to light a fireplace. Two molecules of ATP have already been consumed, but just as a fire gives back more heat than it takes to start it, aerobic respiration eventually gives back far more ATP than it takes to prime glycolysis.

③  **Cleavage.** The "lysis" part of glycolysis occurs when fructose 1,6-diphosphate splits into two 3-carbon ($C_3$) molecules. Through a slight rearrangement of one of them (not shown in the figure), this generates two molecules of **PGAL (phosphoglyceraldehyde,** or **glyceraldehyde 3-phosphate).**

④  **Oxidation.** Each PGAL molecule is then oxidized by removing a pair of hydrogen atoms. The electrons and one proton are picked up by $NAD^+$ and the other proton is released into the cytosol, yielding $NADH + H^+$. At this step, a phosphate ($P_i$) group is also added to each of the $C_3$ fragments. Unlike the earlier steps, this $P_i$ is not supplied by ATP but comes from the cell's pool of free phosphate ions.

⑤  **Dephosphorylation.** In the next two steps, phosphate groups are taken from the glycolysis intermediates and transferred to ADP, phosphorylating it to ATP. This converts the $C_3$ compound to pyruvic acid. The end products of glycolysis are therefore

2 pyruvic acid + 2 NADH + 2 $H^+$ + 2 ATP

Note that 4 ATP are actually produced (step 5), but 2 ATP were consumed to initiate glycolysis (steps 1 and 2), so the net gain is 2 ATP per glucose. Some of the energy originally in the glucose is contained in this ATP, some is in the NADH, and some is lost as heat. Most of the energy, however, remains in the pyruvic acid.

## ANAEROBIC FERMENTATION

The fate of pyruvic acid depends on whether or not oxygen is available. In an exercising muscle, the demand for ATP may exceed the supply of oxygen. The only ATP the cells can make under these circumstances is the 2 ATP produced by glycolysis. Cells without mitochondria, such as erythrocytes, are also restricted to making ATP by this method.

But glycolysis would quickly come to a halt if the reaction stopped at pyruvic acid. Why? Because it would use up the supply of $NAD^+$, which is needed to accept electrons at step 4 and keep glycolysis going. $NAD^+$ must be replenished.

In the absence of oxygen, a cell resorts to a one-step reaction called anaerobic fermentation. (This is often inaccurately called *anaerobic respiration,* but strictly speaking, human cells do not carry out anaerobic respiration; that is a process found only in certain bacteria.) In this pathway NADH donates a pair of electrons to pyruvic acid, thus reducing it to **lactic acid** and regenerating $NAD^+$.

**Think About It**
*Does lactic acid have more free energy than pyruvic acid or less? Explain.*

Lactic acid leaves the cells that generate it and travels by way of the bloodstream to the liver. When oxygen becomes available again, the liver oxidizes lactic acid back to pyruvic acid, which can then enter the aerobic pathway described shortly. The oxygen required to do this is part of the *oxygen debt* created by exercising skeletal muscles (see p. 428). The liver can also convert lactic acid back to G6P and can do either of two things with that: (1) polymerize it to form glycogen for storage or (2) remove the phosphate group and release free glucose into the blood.

Although anaerobic fermentation keeps glycolysis running a little longer, it has some drawbacks. One is that it is wasteful, because most of the energy of glucose is still in the lactic acid and has contributed no useful work. The other is that lactic acid is toxic and contributes to muscle fatigue.

Skeletal muscle is relatively tolerant of anaerobic fermentation, and cardiac muscle is less so. The brain employs almost no anaerobic fermentation. During birth, when the infant's blood supply is cut off, almost every organ of its body switches to anaerobic fermentation; thus they do not compete with the brain for the limited supply of oxygen.

## AEROBIC RESPIRATION

Most ATP is generated in the mitochondria, which require oxygen as the final electron acceptor. In the presence of oxygen, pyruvic acid enters the mitochondria and is oxidized by aerobic respiration. This occurs in two principal steps:

- a group of reactions we will call the **matrix reactions,** because their controlling enzymes are in the fluid of the mitochondrial matrix; and

- reactions we will call the **membrane reactions,** because their controlling enzymes are bound to the membranes of the mitochondrial cristae.

## The Matrix Reactions

The matrix reactions are shown in figure 26.4, where the reaction steps are numbered to resume where figure 26.3 ended. Most of the matrix reactions constitute a series called the **citric acid (Krebs[5]) cycle.** Preceding this, however, are three steps that prepare pyruvic acid to enter the cycle and thus link glycolysis to it.

⑥ Pyruvic acid is *decarboxylated;* that is, $CO_2$ is removed and pyruvic acid, a $C_3$ compound, becomes a $C_2$ compound.

⑦ $NAD^+$ removes hydrogen atoms from the $C_2$ compound (an oxidation reaction) and converts it to an **acetyl group** (acetic acid).

⑧ The acetyl group binds to coenzyme A, a derivative of pantothenic acid (a B vitamin). The result is **acetyl-coenzyme A** (acetyl-CoA). At this stage the $C_2$ remnant of the original glucose molecule is ready to enter the citric acid cycle.

⑨ At the beginning of the citric acid cycle, CoA hands off the acetyl ($C_2$) group to a $C_4$ compound, **oxaloacetic acid.** This produces the $C_6$ compound **citric acid,** for which the cycle is named.

⑩ Water is removed and the citric acid molecule is reorganized, but it still retains its six carbon atoms.

⑪ Hydrogen atoms are removed and accepted by $NAD^+$.

⑫ Another $CO_2$ is removed and the substrate becomes a five-carbon chain.

⑬–⑭ Steps 11 and 12 are essentially repeated, generating another free $CO_2$ molecule and leaving a four-carbon chain. No more carbon atoms are removed beyond this point; the substrate remains a series of $C_4$ compounds from here back to the start of the cycle. The three carbon atoms of pyruvic acid have all been removed as $CO_2$ at steps 6, 12, and 14. These *decarboxylation reactions* are the source of most of the $CO_2$ in your breath.

⑮ Some of the energy in the $C_4$ substrate goes to phosphorylate guanosine diphosphate (GDP) and convert it to guanosine triphosphate (GTP), a molecule similar to ATP. GTP quickly transfers the $P_i$ group to ADP to make ATP. Coenzyme A participates again in this step but is not shown in the figure.

⑯ Two hydrogen atoms are removed and accepted by the coenzyme FAD.

⑰ Water is added.

⑱ A final two hydrogen atoms are removed and transferred to $NAD^+$. This reaction generates oxaloacetic acid, which is available to start the cycle all over again.

[5]Sir Hans Krebs (1900–1981), German biochemist

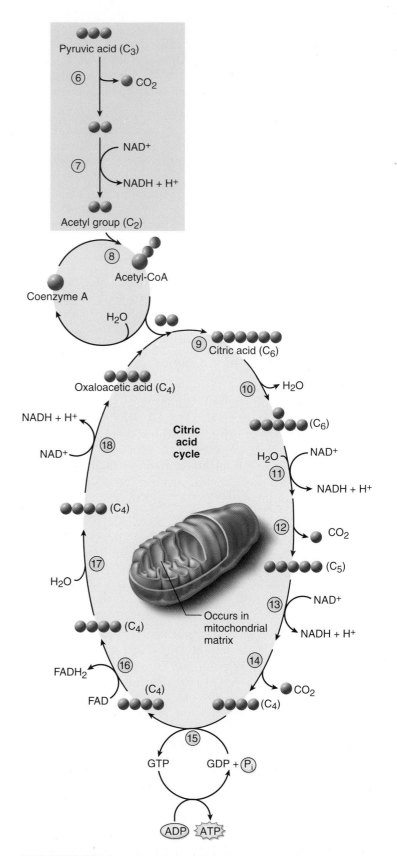

**FIGURE 26.4** **The Mitochondrial Matrix Reactions.** Numbered reaction steps are explained in the text.

It is important to remember that for every glucose molecule that entered glycolysis, all of these matrix reactions occur twice (once for each pyruvic acid). The matrix reactions can be summarized:

$$2 \text{ pyruvate} + 6 \text{ H}_2\text{O} \rightarrow 6 \text{ CO}_2$$

$$+ 2 \text{ ADP} + 2 \text{ P}_i \rightarrow 2 \text{ ATP}$$

$$+ 8 \text{ NAD}^+ + 8 \text{ H}_2 \rightarrow 8 \text{ NADH} + 8 \text{ H}^+$$

$$+ 2 \text{ FAD} + 2 \text{ H}_2 \rightarrow 2 \text{ FADH}_2$$

There is nothing left of the organic matter of the glucose; its carbon atoms have all been carried away as $CO_2$ and exhaled. Although still more of its energy is lost as heat along the way, some is stored in the additional 2 ATP, and most of it, by far, is in the reduced coenzymes—8 NADH and 2 $FADH_2$ molecules generated by the matrix reactions and 2 NADH generated by glycolysis. These must be oxidized to extract the energy from them.

The citric acid cycle not only oxidizes glucose metabolites but is also a pathway and a source of intermediates for the synthesis of fats and nonessential amino acids. The connections between the citric acid cycle and the metabolism of other nutrients are discussed later.

## The Membrane Reactions

The membrane reactions have two purposes: (1) to further oxidize NADH and $FADH_2$ and transfer their energy to ATP and (2) to regenerate $NAD^+$ and FAD and make them available again to earlier reaction steps. The membrane reactions are carried out by a series of compounds called the **mitochondrial electron-transport chain** (fig. 26.5). Most members of the chain are bound to the inner mitochondrial membrane. They are arranged in a precise order that enables each one to receive a pair of electrons from the member on one side of it (or, in two cases, from NADH and $FADH_2$) and pass these electrons along to the member on the other side—like a row of people passing along a hot potato. By the time the "potato" reaches the last member in the chain it is relatively "cool"—its energy has been used to make ATP.

The members of this transport chain are as follows:

- **Flavin mononucleotide (FMN),** a derivative of riboflavin similar to FAD, bound to a membrane protein. FMN accepts electrons from NADH.

- **Iron-sulfur (Fe-S) centers,** complexes of iron and sulfur atoms bound to membrane proteins.

- **Coenzyme Q (CoQ),** which accepts electrons from $FADH_2$. Unlike the other members, this is a relatively small, mobile molecule that moves about in the membrane.

- **Copper (Cu) ions** bound to two membrane proteins.

- **Cytochromes,**[6] five enzymes with iron cofactors, so-named because they are brightly colored in pure form. In order of participation in the chain, they are cytochromes b, $c_1$, c, a, and $a_3$.

*Electron Transport* Figure 26.5 shows the order in which electrons are passed along the chain. Hydrogen atoms are split apart as they transfer from coenzymes to the chain. The protons are pumped into the intermembrane space (fig. 26.6), and the electrons travel in pairs (2 $e^-$) along the transport chain. Each electron carrier in the chain becomes reduced when it receives an electron pair and oxidized again when it passes the electrons along to the next carrier. Energy is liberated at each transfer.

The final electron acceptor in the chain is oxygen. Each oxygen atom (half of an $O_2$ molecule) accepts two electrons (2 $e^-$) from cytochrome $a_3$ and two protons (2 $H^+$) from the mitochondrial matrix. The result is a molecule of water:

$$1/2 \text{ O}_2 + 2 \text{ e}^- + 2 \text{ H}^+ \rightarrow \text{H}_2\text{O}$$

This is the body's primary source of *metabolic water*—water synthesized in the body rather than ingested in food and

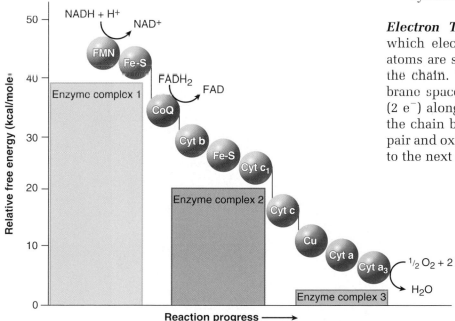

**FIGURE 26.5 The Mitochondrial Electron-Transport Chain.** Transport molecules are grouped into three enzyme complexes, each of which acts as a proton pump. Molecules at the upper left of the figure have a relatively high free energy content, and molecules at the lower right are relatively low in energy.

▶ *What two molecules import energy into this reaction chain, supplying the energy that becomes stored in ATP?*

[6]$cyto$ = cell + $chrom$ = color

drink. This reaction also explains why the body requires oxygen. Without it, this reaction stops and, like a traffic jam, stops all the other processes leading to it. As a result, a cell produces too little ATP to sustain life, and death ensues within a few minutes.

***The Chemiosmotic Mechanism***    Of primary importance is what happens to the energy liberated by the electrons as they pass along the chain. Some of it is unavoidably lost as heat, but some of it drives the **respiratory enzyme complexes.** The first complex includes FMN and five or more Fe–S centers; the second complex includes cytochromes b and $c_1$ and an Fe–S center; and the third complex includes two copper centers and cytochromes a and $a_3$. Each complex collectively acts as a **proton pump** that removes $H^+$ from the mitochondrial matrix and pumps it into the space between the inner and outer mitochondrial membranes (fig. 26.6). Coenzyme Q is a shuttle that transfers electrons from the first pump to the second, and cytochrome c shuttles electrons from the second pump to the third.

These pumps create a very high $H^+$ concentration (low pH) and positive charge between the membranes compared with a low $H^+$ concentration and negative charge in the mitochondrial matrix. That is, they create a steep electrochemical gradient across the inner mitochondrial membrane. If the inner membrane were freely permeable to $H^+$, these ions would have a strong tendency to diffuse down this gradient and back into the matrix.

The inner membrane, however, is permeable to $H^+$ only through specific channel proteins called **ATP synthase** (separate from the electron-transport system). As $H^+$ flows through these channels, it creates an electrical current (which, you may recall, is simply moving charged particles). ATP synthase harnesses the energy of this current to drive ATP synthesis. This process is called the **chemiosmotic[7] mechanism,** which suggests the "push" created by the electrochemical $H^+$ gradient.

---

[7]*chemi* = chemical + *osmo* = push

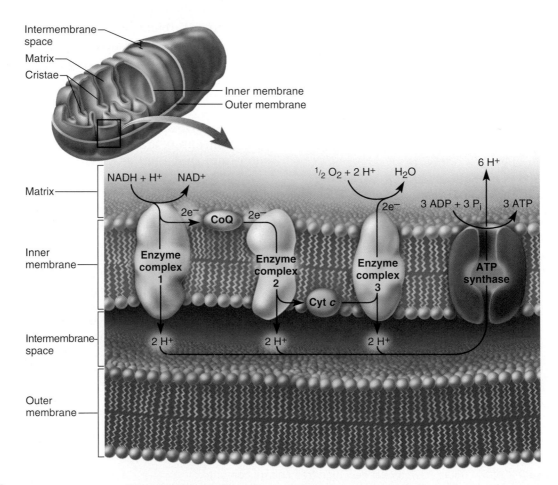

**FIGURE 26.6    The Chemiosmotic Mechanism of ATP Synthesis.**    Each enzyme complex pumps hydrogen ions into the space between the mitochondrial membranes. These hydrogen ions diffuse back into the matrix by way of ATP synthase, which taps their energy to synthesize ATP.

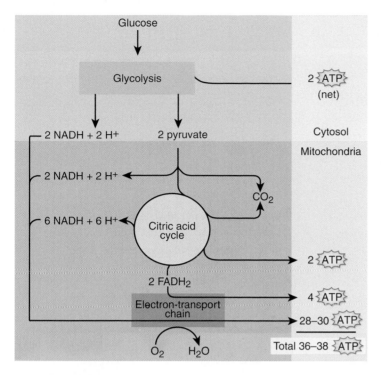

**FIGURE 26.7** Summary of the Sources of ATP Generated by the Complete Oxidation of Glucose.

## OVERVIEW OF ATP PRODUCTION

NADH releases its electron pairs (as hydrogen atoms) to FMN in the first proton pump of the electron-transport system. From there to the end of the chain, this generates enough energy to synthesize 3 ATP molecules per electron pair. FADH$_2$ releases its electron pairs to coenzyme Q, the shuttle between the first and second proton pumps. Therefore, it enters the chain at a point beyond the first pump and does not contribute energy to that pump. Each FADH$_2$ contributes enough energy to synthesize 2 ATP.

With that in mind, we can draw up an energy balance sheet to see how much ATP is produced by the complete aerobic oxidation of glucose to CO$_2$ and H$_2$O and where the ATP comes from; see also figure 26.7. This summary refers back to the reaction steps 1 to 18 in figures 26.3 and 26.4. For each glucose molecule, there are

  10 NADH produced at steps 4, 7, 11, 13, and 18
  × 3 ATP per NADH produced by the electron-
     transport chain

  **30 ATP** generated by NADH

Plus:  2 FADH$_2$ produced at step 16
    × 2 ATP per FADH$_2$ produced by the electron-
      transport chain

  = **4 ATP** generated by FADH$_2$

Plus:  **2 ATP** net amount generated by glycolysis
      (step 5 offset by step 2)
    **2 ATP** generated by the matrix reactions (step 15)

Total: **38 ATP** per glucose

This should be viewed as a theoretical maximum. There is some uncertainty about how much H$^+$ must be pumped between the mitochondrial membranes to generate 1 ATP, and some of the energy from the H$^+$ current is consumed by pumping ATP from the mitochondrial matrix into the cytosol and exchanging it for more raw materials (ADP and P$_i$) pumped from the cytosol into the mitochondria.

Furthermore, the NADH generated by glycolysis cannot enter the mitochondria and donate its electrons directly to the electron-transport chain. In liver, kidney, and myocardial cells, NADH passes its electrons to *malate,* a "shuttle" molecule that delivers the electrons to the beginning of the electron-transport chain. In this case, each NADH yields enough energy to generate 3 ATP. In skeletal muscle and brain cells, however, the glycolytic NADH transfers its electrons to *glycerol phosphate,* a different shuttle that donates the electrons farther down the electron-transport chain and results in the production of only 2 ATP. Therefore the amount of ATP produced per NADH differs from one cell type to another and is still unknown for others.

But if we assume the maximum ATP yield, every mole (180 g) of glucose releases enough energy to synthesize 38 moles of ATP. Glucose has an energy content of 686 kcal/mole and ATP has 7.3 kcal/mole (277.4 kcal in 38 moles). This means that aerobic respiration has an **efficiency** (a ratio of energy output to input) of up to 277.4/686 kcal = 40%. The other 60% (408.6 kcal) is body heat.

The pathways of glucose catabolism are summarized in table 26.4. The aerobic respiration of glucose can be represented in the summary equation:

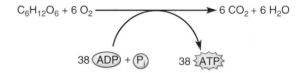

## GLYCOGEN METABOLISM

ATP is quickly used after it is synthesized—it is an *energy-transfer* molecule, not an *energy-storage* molecule. Therefore, if the body has an ample amount of ATP and there is still more glucose in the blood, it does not produce and store excess ATP but converts the glucose to other compounds better suited for energy storage—namely glycogen and fat. Fat synthesis is considered later. Here we consider the synthesis and use of glycogen. The average adult body contains about 400 to 450 g of glycogen: nearly one-quarter of it in the liver, three-quarters of it in the skeletal muscles, and small amounts in cardiac muscle and other tissues.

**Glycogenesis,** the synthesis of glycogen, is stimulated by insulin. Glucose 6-phosphate (G6P) is isomerized to glucose 1-phosphate (G1P). The enzyme *glycogen synthase* then cleaves off the phosphate group and attaches the glucose to a growing polysaccharide chain.

**Glycogenolysis,** the hydrolysis of glycogen, releases glucose between meals when new glucose is not being

| TABLE 26.4 | Pathways of Glucose Catabolism | | |
|---|---|---|---|
| **Stage** | **Principal Reactants** | **Principal Products** | **Purpose** |
| Glycolysis | Glucose, 2 ADP, 2 $P_i$, 2 $NAD^+$ | 2 pyruvic acid, 2 ATP, 2 NADH, 2 $H_2O$ | Reorganizes glucose and splits it in two in preparation for further oxidation by the mitochondria; sole source of ATP in anaerobic conditions |
| Anaerobic fermentation | 2 pyruvic acid, 2 NADH | 2 lactic acid, 2 $NAD^+$ | Regenerates $NAD^+$ so glycolysis can continue to function (and generate ATP) in the absence of oxygen |
| Aerobic respiration | | | |
| Matrix reactions | 2 pyruvic acid, 8 $NAD^+$, 2 FAD, 2 ADP, 2 $P_i$, 8 $H_2O$ | 6 $CO_2$, 8 NADH, 2 $FADH_2$, 2 ATP, 2 $H_2O$ | Remove electrons from pyruvic acid and transfer them to coenzymes $NAD^+$ and FAD; produce some ATP |
| Membrane reactions | 10 NADH, 2 $FADH_2$, 6 $O_2$ | 32–34 ATP, 12 $H_2O$ | Finish oxidation and produce most of the ATP of cellular respiration |

ingested. The process is stimulated by glucagon and epinephrine. The enzyme *glycogen phosphorylase* begins by phosphorylating a glucose residue and splitting it off the glycogen molecule as G1P. This is isomerized to G6P, which can then enter the pathway of glycolysis.

G6P usually cannot leave the cells that produce it. Liver cells, however, have an enzyme called *glucose 6-phosphatase,* which removes the phosphate group and produces free glucose. This can diffuse out of the cell into the blood, where it is available to any cells in the body. Although muscle cells cannot directly release glucose into the blood, they contribute indirectly to blood glucose concentration because they release pyruvic and lactic acids, which are converted to glucose by the liver.

**Gluconeogenesis**[8] is the synthesis of glucose from noncarbohydrates such as fats and amino acids. It occurs chiefly in the liver, but after several weeks of fasting, the kidneys also undertake this process and eventually produce just as much glucose as the liver does.

The processes described here are summarized in figure 26.8, and the distinctions among these similar terms are summarized in table 26.5.

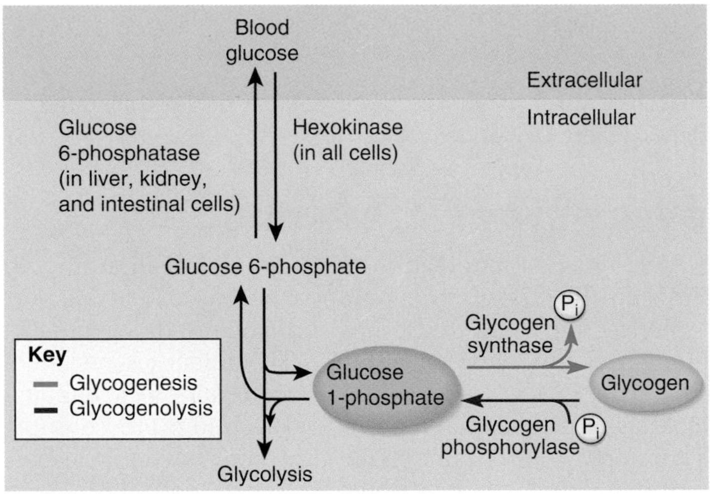

## FUNCTIONS OF THE LIVER

We have seen that the liver plays a central role in carbohydrate metabolism. Additional liver functions (table 26.6) were described in previous chapters and will be described later in this chapter. Except for phagocytosis, all of these are performed by the cuboidal hepatocytes described in chapter 25. Such functional diversity is remarkable in light of the uniform structure of these cells. Because of the wide range of functions performed by the liver, degenerative liver diseases such as hepatitis, cirrhosis, and liver cancer are especially life-threatening (see Insight 26.3).

**FIGURE 26.8**  Major Pathways of Glucose Storage and Use. In most cells, the glucose 1-phosphate generated by glycogenolysis can only undergo glycolysis. In liver, kidney, and intestinal cells, it can be converted back to free glucose and released into circulation.

| TABLE 26.5 | Some Terminology Related to Glucose and Glycogen Metabolism |
|---|---|
| ***Anabolic (Synthesis) Reactions*** | |
| Glycogenesis | The synthesis of glycogen by polymerizing glucose |
| Gluconeogenesis | The synthesis of glucose from noncarbohydrates such as fats and amino acids |
| ***Catabolic (Breakdown) Reactions*** | |
| Glycolysis | The splitting of glucose into two molecules of pyruvic acid in preparation for anaerobic fermentation or aerobic respiration |
| Glycogenolysis | The hydrolysis of glycogen to release free glucose or glucose 1-phosphate |

---

[8]*gluco* = sugar, glucose + *neo* = new + *genesis* = production of

| TABLE 26.6 | Functions of the Liver |
|---|---|

**Carbohydrate Metabolism**

Converts dietary fructose and galactose to glucose. Stabilizes blood glucose concentration by storing excess glucose as glycogen (glycogenesis), releasing glucose from glycogen when needed (glycogenolysis), and synthesizing glucose from fats and amino acids (gluconeogenesis) when glucose demand exceeds glycogen reserves. Receives lactic acid generated by anaerobic fermentation in skeletal muscle and other tissues and converts it back to pyruvic acid or glucose 6-phosphate.

**Lipid Metabolism**

Degrades chylomicron remnants. Carries out most of the body's lipogenesis (fat synthesis) and synthesizes cholesterol and phospholipids; produces VLDLs to transport lipids to adipose tissue and other tissues for storage or use; and stores fat in its own cells. Carries out most beta-oxidation of fatty acids; produces ketone bodies from excess acetyl-CoA. Produces HDL shells, which pick up excess cholesterol from other tissues and return it to the liver; excretes the excess cholesterol in bile.

**Protein and Amino Acid Metabolism**

Carries out most deamination and transamination of amino acids. Removes $-NH_2$ from glutamic acid and converts the resulting ammonia to urea by means of the ornithine cycle. Synthesizes nonessential amino acids by transamination reactions.

**Synthesis of Plasma Proteins**

Synthesizes nearly all the proteins of blood plasma, including albumin, alpha and beta globulins, fibrinogen, prothrombin, and several other clotting factors. (Does not synthesize plasma enzymes or gamma globulins.)

**Vitamin and Mineral Metabolism**

Converts vitamin $D_3$ to calcidiol, a step in the synthesis of calcitriol; stores a 3- to 4-month supply of vitamin D. Stores a 10-month supply of vitamin A and enough vitamin $B_{12}$ to last one to several years. Secretes hepcidin to regulate iron absorption; stores iron in ferritin and releases it as needed. Excretes excess calcium by way of the bile.

**Digestion**

Synthesizes bile acids and lecithin, which emulsify fat and promote its digestion.

**Disposal of Drugs, Toxins, and Hormones**

Detoxifies alcohol, antibiotics, and many other drugs. Metabolizes bilirubin from RBC breakdown and excretes it as bile pigments. Deactivates thyroxine and steroid hormones and excretes them or converts them to a form more easily excreted by the kidneys.

**Phagocytosis**

Macrophages cleanse blood of bacteria and other foreign matter.

---

## INSIGHT 26.3 Clinical Application

### Hepatitis and Cirrhosis

*Hepatitis,* inflammation of the liver, is usually caused by one of the five strains of hepatitis viruses. They differ in mode of transmission, severity of the resulting illness, affected age groups, and the best strategies for prevention. Hepatitis A is common and mild. Over 45% of people in urban areas of the United States have had it. It spreads rapidly in such settings as day-care centers and institutions for psychiatric patients, and it can be acquired by eating uncooked seafood such as oysters. Hepatitis B and C are far more serious. Both are transmitted sexually and through blood and other body fluids; the incidence of hepatitis C has surpassed AIDS as a sexually transmitted disease. Initial signs and symptoms of hepatitis include fatigue, malaise, nausea, vomiting, and weight loss. The liver becomes enlarged and tender. Jaundice, or yellowing of the skin, tends to follow as hepatocytes are destroyed, bile passages are blocked, and bile pigments accumulate in the blood. Hepatitis A causes up to six months of illness, but most people recover and then have permanent immunity to it. Hepatitis B and C, however, often lead to chronic hepatitis, which can progress to cirrhosis or liver cancer. More liver transplants are necessitated by hepatitis C than by any other cause.

*Cirrhosis* is an irreversible inflammatory liver disease. It develops slowly over a period of years, but has a high mortality rate and is one of the leading causes of death in the United States. It is characterized by a disorganized liver histology in which regions of scar tissue alternate with nodules of regenerating cells, giving the liver a lumpy or knobby appearance and hardened texture. As in hepatitis, blockage of the bile passages results in jaundice. Protein synthesis declines as the liver deteriorates, leading to ascites, impaired blood clotting, and other cardiovascular effects (see Insight 26.4). Obstruction of the hepatic circulation by scar tissue leads to *angiogenesis,* the growth of new blood vessels to bypass the liver. Deprived of blood, the condition of the liver worsens, with increasing necrosis and, often, liver failure. Most cases of cirrhosis result from alcohol abuse, but hepatitis, gallstones, pancreatic inflammation, and other conditions can also bring it about. The prognosis for recovery is often poor.

---

## Before You Go On

*Answer the following questions to test your understanding of the preceding section:*

6. Identify the reaction steps in figures 26.3 and 26.4 at which vitamins are essential for glucose catabolism.

7. In the laboratory, glucose can be oxidized in a single step to $CO_2$ and $H_2O$. Why is it done in so many little steps in cells?

8. Explain the origin of the word glycolysis *and why this is an appropriate name for the function of that reaction pathway.*

9. What are two advantages of aerobic respiration over anaerobic fermentation?

10. What important enzyme is found in the inner mitochondrial membrane other than those of the electron-transport chain? Explain how its function depends on the electron-transport chain.

11. Describe how the liver responds to (a) an excess and (b) a deficiency of blood glucose.

# Lipid and Protein Metabolism

### Objectives

When you have completed this section, you should be able to

- describe the processes of lipid catabolism and anabolism;
- describe the processes of protein catabolism and anabolism; and
- explain the metabolic source of ammonia and how the body disposes of it.

In the foregoing discussion, glycolysis and the mitochondrial reactions were treated from the standpoint of carbohydrate oxidation. These pathways also serve for the oxidation of proteins and lipids as fuel and as a source of metabolic intermediates that can be used for protein and lipid synthesis. Here we examine these related metabolic pathways.

## LIPIDS

Triglycerides are stored primarily in the body's adipocytes, where a given molecule remains for about 2 to 3 weeks. Although the total amount of stored triglyceride remains quite constant, there is a continual turnover as lipids are released, transported in the blood, and either oxidized for energy or redeposited in other adipocytes. Synthesizing fats from other types of molecules is called **lipogenesis,** and breaking down fat for fuel is called **lipolysis** (lih-POL-ih-sis).

## Lipogenesis

It is common knowledge that a diet high in sugars causes us to put on fat and gain weight. Lipogenesis employs compounds such as sugars and amino acids to synthesize glycerol and fatty acids, the triglyceride precursors. PGAL, one of the intermediates of glucose oxidation, can be converted to glycerol. As glucose and amino acids enter the citric acid cycle by way of acetyl-CoA, the acetyl-CoA can also be diverted to make fatty acids. The glycerol and fatty acids can then be condensed to form a triglyceride, which can be stored in the adipose tissue or converted to other lipids. These pathways are summarized in figure 26.9.

## Lipolysis

Lipolysis, also shown in figure 26.9, begins with the hydrolysis of a triglyceride into glycerol and fatty acids—a process stimulated by epinephrine, norepinephrine, glucocorticoids, thyroid hormone, and growth hormone. The glycerol and fatty acids are further oxidized by separate pathways. Glycerol is easily converted to PGAL and thus enters the pathway of glycolysis. It generates only half as much ATP as glucose, however, because it is a $C_3$

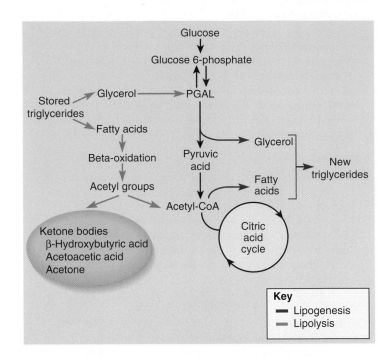

**FIGURE 26.9**  Pathways of Lipolysis and Lipogenesis in Relation to Glycolysis and the Citric Acid Cycle.

▶ *Name the acid–base imbalance that results from accumulation of the ketone bodies shown here.*

compound compared with glucose ($C_6$); thus it leads to the production of only half as much pyruvic acid.

The fatty acid component is catabolized in the mitochondrial matrix by a process called **beta-oxidation,** which removes two carbon atoms at a time. The resulting acetyl ($C_2$) groups are bonded to coenzyme A to make acetyl-CoA—the entry point into the citric acid cycle. A fatty acid of 16 carbon atoms can yield 129 molecules of ATP—obviously a much richer source of energy than a glucose molecule.

Excess acetyl groups can be metabolized by the liver in a process called **ketogenesis.** Two acetyl groups are condensed to form acetoacetic acid, and some of this is further converted to β-hydroxybutyric acid and acetone. These three products are the *ketone bodies.* Some cells convert acetoacetic acid back to acetyl-CoA and thus feed the $C_2$ fragments into the citric acid cycle to extract their energy. When the body is rapidly oxidizing fats, however, excess ketone bodies accumulate. This causes the ketoacidosis typical of insulin-dependent diabetes mellitus, in which cells must oxidize fats because they cannot absorb glucose.

Acetyl-CoA cannot go backward up the glycolytic pathway and produce glucose, because this pathway is irreversible past the point of pyruvic acid. Although glycerol can be used for gluconeogenesis, fatty acids cannot.

It was mentioned earlier that fats cannot be completely oxidized when there is not enough carbohydrate in the diet. This is because the mitochondrial reactions cannot proceed without oxaloacetic acid as a "pickup molecule" in the citric acid cycle. When carbohydrate is unavailable, oxaloacetic acid is converted to glucose and becomes unavailable to the citric acid cycle. Fat oxidation then produces excess ketones, leading to elevated blood ketones *(ketosis)* and potentially to a resulting pH imbalance *(ketoacidosis)*. Ketosis is a serious risk of extreme low-carbohydrate diets.

## PROTEINS

About 100 g of tissue protein breaks down each day into free amino acids. These combine with the amino acids from the diet to form an **amino acid pool** that cells can draw upon to make new proteins. The fastest rate of tissue protein turnover is in the intestinal mucosa, where epithelial cells are replaced at a very high rate. Dead cells are digested along with the food and thus contribute to the amino acid pool. Of all the amino acid absorbed by the small intestine, about 50% is from the diet, 25% from dead epithelial cells, and 25% from enzymes that have digested each other.

Some amino acids in the pool can be converted to others. Free amino acids also can be converted to glucose and fat or directly used as fuel. Such conversions involve three processes: (1) **deamination,** the removal of an amino group ($-NH_2$); (2) **amination,** the addition of $-NH_2$; or (3) **transamination,** the transfer of $-NH_2$ from one molecule to another. The following discussion shows how these processes are involved in amino acid metabolism.

### Use as Fuel

The first step in using amino acids as fuel is to deaminate them. After the $-NH_2$ group is removed, the remainder of the molecule is called a *keto acid.* Depending on which amino acid is involved, the resulting keto acid may be converted to pyruvic acid, acetyl-CoA, or one of the acids of the citric acid cycle (fig. 26.10). It is important to note that some of these reactions are reversible. When there is a deficiency of amino acids in the body, citric acid cycle intermediates can be aminated and converted to amino acids, which are then available for protein synthesis. In gluconeogenesis, keto acids are used to synthesize glucose, essentially through a reversal of the glycolysis reactions.

### Transamination, Ammonia, and Urea

When an amino acid is deaminated, its amino group is transferred to a citric acid cycle intermediate, α-ketoglutaric acid, converting it to glutamic acid. Such transamination reactions are the route by which several amino acids enter the citric acid cycle.

Glutamic acid can travel from any of the body's cells to the liver. Here its $-NH_2$ group is removed, converting it back to α-ketoglutaric acid. The $-NH_2$ becomes ammonia ($NH_3$), which is extremely toxic to cells and cannot be

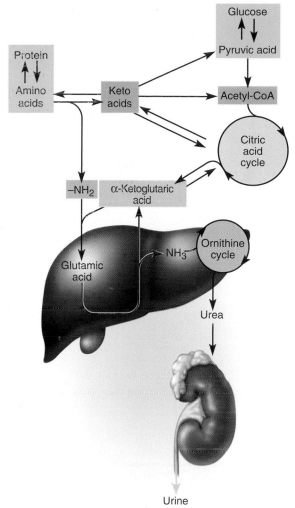

**FIGURE 26.10** Pathways of Amino Acid Metabolism in Relation to Glycolysis and the Citric Acid Cycle.
▶ *Find a pathway for gluconeogenesis in this diagram.*

allowed to accumulate. The liver quickly converts ammonia to a less toxic form, urea, by a pathway called the **ornithine cycle** (fig. 26.11). Urea is then excreted in the urine as one of the body's nitrogenous wastes. Other nitrogenous wastes and their sources are described in chapter 23 (see p. 896). When a diseased liver cannot carry out the ornithine cycle, $NH_3$ accumulates in the blood and death from *hepatic coma* may ensue within a few days.

### Protein Synthesis

Protein synthesis, described in detail in chapter 4, is a complex process involving DNA, mRNA, tRNA, ribosomes, and often the rough ER. It is stimulated by growth hormone, thyroid hormone, and insulin, and it requires a supply of all the amino acids necessary for a particular protein. The liver can make many of these amino acids from other amino acids or from citric acid cycle intermediates by transamination reactions. The essential amino acids, however, must be obtained from the diet.

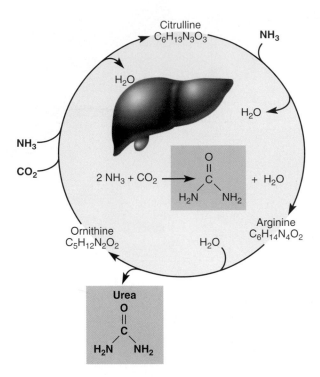

**FIGURE 26.11** Urea Synthesis in the Liver.    The ornithine cycle converts ammonia and carbon dioxide to urea, which is then excreted by the kidneys.

## Before You Go On

*Answer the following questions to test your understanding of the preceding section:*

12. Which of the processes in table 26.5 is most comparable to lipogenesis? Which is most comparable to lipolysis? Explain.

13. When fats are converted to glucose, only the glycerol component is used in this way, not the fatty acid. Explain why and what happens to the fatty acids.

14. What metabolic process produces ammonia? How does the body dispose of ammonia?

# Metabolic States and Metabolic Rate

## Objectives

When you have completed this section, you should be able to

- define the absorptive and postabsorptive states;
- explain what happens to carbohydrates, fats, and proteins in each of these states;
- describe the hormonal and nervous regulation of each state;

- define *metabolic rate* and *basal metabolic rate;* and
- describe some factors that alter the metabolic rate.

Your metabolism changes from hour to hour depending on how long it has been since your last meal. The **absorptive (fed) state** lasts about 4 hours during and after a meal. This is a time in which nutrients are being absorbed and may be used immediately to meet energy and other needs. The **postabsorptive (fasting) state** prevails in the late morning, late afternoon, and overnight. During this time the stomach and small intestine are empty and the body's energy needs are met from stored fuels. The two states are compared in table 26.7 and explained in the following discussion.

## THE ABSORPTIVE STATE

In the absorptive state, blood glucose is readily available for ATP synthesis. It serves as the primary fuel and spares the body from having to draw on stored fuels. The status of major nutrient classes during this phase is as follows:

- **Carbohydrates.** Absorbed sugars are transported by the hepatic portal system to the liver. Most glucose passes through the liver and becomes available to cells everywhere in the body. Glucose in excess of immediate need, however, is absorbed by the liver and may be converted to glycogen or fat. Most fat synthesized in the liver is released into the circulation; its further fate is comparable to that of dietary fats, discussed next.

- **Fats.** Fats enter the lymph as chylomicrons and initially bypass the liver. As described earlier, lipoprotein lipase removes fats from the chylomicrons for uptake by the tissues, especially adipose and muscular tissue. The liver disposes of the chylomicron remnants. Fats are the primary energy substrate for hepatocytes, adipocytes, and muscle cells.

- **Amino acids.** Amino acids, like sugars, circulate first to the liver. Most pass through and become available to other cells for protein synthesis. Some, however, are removed by the liver and have one of the following fates: (1) to be used for protein synthesis; (2) to be deaminated and used as fuel for ATP synthesis; or (3) to be deaminated and used for fatty acid synthesis.

## Regulation of the Absorptive State

The absorptive state is regulated largely by insulin, which is secreted in response to elevated blood glucose and amino acid levels and to the intestinal hormones gastrin, secretin, and cholecystokinin. Insulin regulates the rate of glucose uptake by nearly all cells except neurons, kidney cells, and erythrocytes, which have independent rates of

| TABLE 26.7 | Major Aspects of the Absorptive and Postabsorptive States | |
|---|---|---|
| | **Absorptive** | **Postabsorptive** |
| *Regulatory hormones* | Principally insulin | Principally glucagon |
| | Also gastrin, secretin, CCK | Also epinephrine, growth hormone |
| *Carbohydrate metabolism* | Blood glucose rising | Blood glucose falling |
| | Glucose stored by glycogenesis | Glucose released by glycogenolysis |
| | Gluconeogenesis suppressed | Gluconeogenesis stimulated |
| *Lipid metabolism* | Lipogenesis occurring | Lipolysis occurring |
| | Lipid uptake from chylomicrons | Fatty acids oxidized for fuel |
| | Lipid storage in fat and muscle | Glycerol used for gluconeogenesis |
| *Protein metabolism* | Amino acid uptake, protein synthesis | Amino acids oxidized if glycogen and fat |
| | Excess amino acids burned as fuel | stores are inadequate for energy needs |

uptake. On other target cells, insulin has the following effects:

- Within minutes, it increases the cellular uptake of glucose by as much as 20-fold. As cells absorb glucose, the blood glucose concentration falls.

- It stimulates glucose oxidation, glycogenesis, and lipogenesis.

- It inhibits gluconeogenesis, which makes sense since blood glucose concentration is already high and there is no immediate need for more.

- It stimulates the active transport of amino acids into cells and promotes protein synthesis.

- It acts on the brain as an adiposity signal, an index of the body's fat stores.

Following a high-protein, low-carbohydrate meal, it may seem that the amino acids would stimulate insulin secretion, insulin would accelerate both amino acid and glucose uptake, and since there was relatively little glucose in the ingested food, this would create a risk of hypoglycemia. In actuality, this is prevented by the fact that a high amino acid level stimulates the secretion of *both* insulin and glucagon. Glucagon, you may recall, is an insulin antagonist (see chapter 17). It supports an adequate level of blood glucose to meet the needs of the brain.

## THE POSTABSORPTIVE STATE

The essence of the postabsorptive state is to homeostatically regulate blood glucose concentration within about 90 to 100 mg/dL. This is especially critical to the brain, which cannot use alternative energy substrates except in cases of prolonged fasting. The postabsorptive status of major nutrients is as follows:

- **Carbohydrates.** Glucose is drawn from the body's glycogen reserves (glycogenolysis) or synthesized

from other compounds (gluconeogenesis). The liver usually stores enough glycogen after a meal to support 4 hours of postabsorptive metabolism before significant gluconeogenesis occurs.

- **Fats.** Adipocytes and hepatocytes hydrolyze fats and convert the glycerol to glucose. Free fatty acids (FFAs) cannot be converted to glucose, but they can favorably affect blood glucose concentration. As the liver oxidizes them to ketone bodies, other cells absorb and use these, or use FFAs directly, as their source of energy. By switching from glucose to fatty acid catabolism, they leave glucose for use by the brain (the glucose-sparing effect). After 4 to 5 days of fasting, the brain begins to use ketone bodies as supplemental fuel.

- **Proteins.** If glycogen and fat reserves are depleted, the body begins to use proteins as fuel. Some proteins are more resistant to catabolism than others. Collagen is almost never broken down for fuel, but muscle protein goes quickly. The extreme wasting away seen in cancer and some other chronic diseases, resulting from a loss of appetite (anorexia) as well as altered metabolism, is called **cachexia**[9] (ka-KEX-ee-ah).

## Regulation of the Postabsorptive State

Postabsorptive metabolism is more complex than the absorptive state. It is regulated mainly by the sympathetic nervous system and glucagon, but several other hormones are involved. As blood glucose level drops, insulin secretion declines and the pancreatic alpha cells secrete glucagon. Glucagon promotes glycogenolysis and gluconeogenesis, raising the blood glucose level, and it promotes lipolysis and a rise in FFA levels. Thus it makes both glucose and lipids available for fuel.

---

[9]*cac* = bad + *exia* = body condition

The sympathoadrenal system also promotes glycogenolysis and lipolysis, especially under conditions of injury, fear, anger, and other forms of stress. Adipose tissue is richly innervated by the sympathetic nervous system, while adipocytes, hepatocytes, and muscle cells also respond to epinephrine from the adrenal medulla. In circumstances where there is likely to be tissue injury and a need for repair, the sympathoadrenal system therefore mobilizes stored energy reserves and makes them available to meet the demands of tissue repair. Stress also stimulates the release of cortisol, which promotes fat and protein catabolism and gluconeogenesis (see chapter 17, p. 666).

Growth hormone is secreted in response to a rapid drop in blood glucose level and in conditions of prolonged fasting. It opposes insulin and raises blood glucose concentration.

## METABOLIC RATE

**Metabolic rate** means the amount of energy liberated in the body per unit of time, expressed in such terms as kcal/hr or kcal/day. Metabolic rate can be measured directly by putting a person in a **calorimeter,** a closed chamber with water-filled walls that absorb the heat given off by the body. The rate of energy release is measured from the temperature change of the water. Metabolic rate can also be measured indirectly with a spirometer, an apparatus described in chapter 22 that can be used to measure the amount of oxygen a person consumes. For every liter of oxygen, approximately 4.82 kcal of energy is released from organic nutrients. This is only an estimate, because the number of kilocalories per liter of oxygen varies slightly with the type of nutrients the person is oxidizing at the time of measurement.

Metabolic rate depends on physical activity, mental state, absorptive or postabsorptive status, thyroid hormone and other hormones, and other factors. The **basal metabolic rate (BMR)** is a baseline or standard of comparison that minimizes the effects of such variables. It is your metabolic rate when you are awake but relaxed, in a room at comfortable temperature, in a postabsorptive state 12 to 14 hours after your last meal. It is not the minimum metabolic rate needed to sustain life, however. When you are asleep, your metabolic rate is slightly lower than your BMR. **Total metabolic rate (TMR)** is the sum of BMR and energy expenditure for voluntary activities, especially muscular contractions.

The BMR of an average adult is about 2,000 kcal/day for a male and slightly less for a female. Roughly speaking, one must therefore consume at least 2,000 kcal/day to fuel essential metabolic tasks—active transport, muscle tone, brain activity, cardiac and respiratory rhythms, renal function, and other essential processes. Even a relatively sedentary lifestyle requires another 500 kcal/day to support a low level of physical activity, and someone who does hard physical labor may require as much as 5,000 kcal/day.

Aside from physical activity, some factors that raise the TMR and caloric requirements include pregnancy, anxiety (which stimulates epinephrine release and muscle tension), fever (TMR rises about 14% for each 1°C of body temperature), eating (TMR rises after a meal), and the catecholamine and thyroid hormones. TMR is relatively high in children and declines with age. Therefore, as we reach middle age we often find ourselves gaining weight with no apparent change in food intake.

Some factors that lower TMR include apathy, depression, and prolonged starvation. In weight-loss diets, loss is often rapid at first and then goes more slowly. This is partly because the initial loss is largely water and partly because the TMR drops over time, fewer dietary calories are "burned off," and there is more lipogenesis even with the same caloric intake.

### Before You Go On

*Answer the following questions to test your understanding of the preceding section:*

15. *Define the absorptive and postabsorptive states. In which state is the body storing excess fuel? In which state is it drawing from these stored fuel reserves?*

16. *What hormone primarily regulates the absorptive state, and what are the major effects of this hormone?*

17. *Explain why triglycerides have a glucose-sparing effect.*

18. *List a variety of factors and conditions that raise a person's total metabolic rate above basal metabolic rate.*

# Body Heat and Thermoregulation

### Objectives

When you have completed this section, you should be able to

- identify the principal sources of body heat;
- describe some factors that cause variations in body temperature;
- define and contrast the different forms of heat loss;
- describe how the hypothalamus monitors and controls body temperature; and
- describe conditions in which the body temperature is excessively high or low.

Heat generation must be matched by heat loss in order to maintain a stable internal body temperature. **Hypothermia,** an excessively low body temperature, can cause metabolism to slow down to the point that it cannot sustain life, whereas **hyperthermia,** an excessively high temperature, can disrupt the coordination of metabolic pathways and also lead to death. **Thermoregulation,** the balance between heat production and loss, is a critically important aspect of homeostasis. Thermoregulation was introduced in chapter 1 as an example of homeostasis.

# BODY TEMPERATURE

"Normal" body temperature depends on when, where, and in whom it is measured. Body temperature fluctuates about 1°C (1.8°F) in a 24-hour cycle. It tends to be lowest in the early morning and highest in the late afternoon. Temperature also varies from one place in the body to another.

The most important body temperature is the **core temperature**—the temperature of organs in the cranial, thoracic, and abdominal cavities. Rectal temperature is relatively easy to measure and gives an estimate of core temperature: usually 37.2° to 37.6°C (99.0°–99.7°F). It may be as high as 38.5°C (101°F) in active children and some adults.

**Shell temperature** is the temperature closer to the surface, especially skin and oral temperature. Here, heat is lost from the body and temperatures are slightly lower than rectal temperature. Adult oral temperature is typically 36.6° to 37.0°C (97.9°–98.6°F) but may be as high as 40°C (104°F) during hard exercise.

## HEAT PRODUCTION AND LOSS

Most body heat comes from exergonic (energy-releasing) chemical reactions such as nutrient oxidation and ATP use. A little heat is generated by joint friction, blood flow, and other movements. At rest, most heat is generated by the brain, heart, liver, and endocrine glands; the skeletal muscles contribute about 20% to 30% of the total resting heat. Increased muscle tone or exercise greatly increases heat generation in the muscles, however; in vigorous exercise, they produce 30 to 40 times as much heat as the rest of the body.

The body loses heat in three ways: radiation, conduction, and evaporation:

1. **Radiation.** In essence, heat means molecular motion. All molecular motion produces radiation in the infrared (IR) region of the electromagnetic spectrum. When IR radiation is absorbed by an object, it increases its molecular motion and raises its temperature. Therefore IR radiation removes heat from its source and adds heat to anything that absorbs it. The heat lamps used in bathrooms and restaurants work on this principle. Our bodies continually receive IR from the objects around us and give off IR to our surroundings. Since we are usually warmer than the objects around us, we usually lose more heat this way than we gain.

2. **Conduction.** As the molecules of our tissues vibrate with heat energy, they collide with other molecules and transfer kinetic energy to them. The warmth of your body therefore adds to the molecular motion and temperature of the clothes you wear, the chair you sit in, and the air around you. Conductive heat loss is aided by **convection,** the motion of a fluid due to uneven heating. Air is a fluid that becomes less dense and therefore rises as it is heated. Thus warm air rises from the body and is replaced by

cooler air from below. The same is true of water; for example, when you swim in a lake or take a cool bath. You can easily see convection when you heat water in a clear container. Convection of the air around the human body can be seen by a technique called schlieren photography (fig. 26.12). Convection

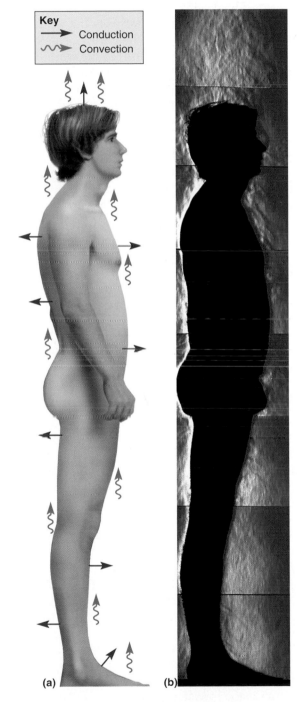

**FIGURE 26.12  Heat Loss from the Body.** (a) Heat transfers from the body to surrounding air molecules by conduction. Warm air then rises from the body by convection, carrying heat away. Cool air replaces this warm air from below. (b) Schlieren photograph of the column of warm air rising from the body.

is not a separate category of heat loss in itself, but increases the rate of conductive heat loss.

3. **Evaporation.** The cohesion of water molecules hampers their vibratory movement in response to heat input. If the temperature of water is raised sufficiently, however, its molecular motion becomes great enough for molecules to break free and evaporate. Evaporation of water thus carries a substantial amount of heat with it (0.5 kcal/g). This is the significance of perspiration. Sweat wets the skin surface and its evaporation carries heat away. In extreme conditions, the body can lose 2 L or more of sweat per hour, and dissipate up to 600 kcal of heat per hour by evaporative loss. Evaporative heat loss is increased by moving air, as you can readily feel when you are sweaty and stand in front of a fan. The rapid movement of heat-laden air away from the body **(forced convection)** accelerates heat removal, so a breeze or a fan enhances heat loss by conduction and evaporation. It has no effect on radiation.

The relative amounts of heat lost by different methods depend on prevailing conditions. A nude body at an air temperature of 21°C (70°F) loses about 60% of its heat by radiation, 18% by conduction, and 22% by evaporation. If air temperature is higher than skin temperature, evaporation becomes the only means of heat loss because radiation and conduction add more heat to the body than they remove. Hot, humid weather hinders even evaporative cooling because there is less of a humidity gradient from skin to air. Such conditions increase the risk of heatstroke (discussed shortly).

## THERMOREGULATION

Thermoregulation is achieved through several negative feedback loops. The preoptic area of the hypothalamus (anterior to the optic chiasm) functions as a **hypothalamic thermostat.** It monitors the temperature of the blood and receives signals also from **peripheral thermoreceptors** in the skin. In turn, it sends appropriate signals either to the **heat-losing center,** a nucleus still farther anterior in the hypothalamus, or to the **heat-promoting center,** a more posterior nucleus.

When the heat-losing center senses that the blood temperature is too high, it activates heat-losing mechanisms. The first and simplest of these is cutaneous vasodilation, which increases blood flow close to the body surface and thus promotes heat loss. If this fails to restore normal temperature, the heat-losing center triggers sweating. It also inhibits the heat-promoting center.

When the heat-promoting center senses that the blood temperature is too low, it activates mechanisms to conserve body heat or generate more. By way of the sympathetic nervous system, it causes cutaneous vasoconstriction. Warm blood is then retained deeper in the body and less heat is lost through the skin. The sympathetic sys-

tem in other mammals also stimulates the piloerector muscles, which make the hair stand on end. This traps an insulating blanket of still air near the skin. The human sympathetic nervous system attempts to do this as well, but since our body hair is so scanty, the only noticeable effect of this is goose bumps.

If dermal vasoconstriction cannot restore or maintain normal core temperature, the body resorts to **shivering thermogenesis.** If you leave a warm house on a cold day, you may notice that your muscles become tense, sometimes even painfully taut, and you begin to shiver. Shivering involves a spinal reflex that causes tiny alternating contractions in antagonistic muscle pairs. Every muscle contraction releases heat from ATP, and shivering can increase the body's heat production as much as fourfold.

**Nonshivering thermogenesis** is a more long-term mechanism for generating heat, used especially in the colder seasons of the year. The sympathetic nervous system and thyroid hormone stimulate an increase in metabolic rate, which can rise as much as 30% after several weeks of cold weather. More nutrients are burned as fuel, we consume more calories to "stoke the furnace," and consequently, we have greater appetites in the winter than in the summer. Infants can generate heat by breaking down *brown fat,* a tissue in which lipolysis is not linked to ATP synthesis, so all the energy released from the fat is in the form of heat.

In addition to these physiological mechanisms, and of even greater importance, humans and other animals practice **behavioral thermoregulation**—behaviors that raise or lower the body's heat gains and losses. Just getting out of the sun greatly cuts down heat gain by radiation, for example, while shedding heavy clothing or kicking off a blanket at night helps to cool the body.

In summary, you can see that thermoregulation is a function of multiple organs: the brain, autonomic nerves, thyroid gland, skin, blood vessels, and skeletal muscles.

## DISTURBANCES OF THERMOREGULATION

Chapter 21 described the mechanism of fever and its importance in combating infection. Fever is a normal protective mechanism that should be allowed to run its course if it is not excessively high. A body temperature above 42° to 43°C (108°–110°F), however, can be very dangerous. The high temperature elevates the metabolic rate, and the body generates heat faster than its heat-losing mechanisms can disperse it. Thus the metabolic rate increases the fever and the fever increases the metabolic rate in a dangerous positive feedback loop. At a core body temperature of 44° to 45°C (111°–113°F), metabolic dysfunction and neurological damage can be fatal.

Exposure to excessive heat causes heat cramps, heat exhaustion, and heatstroke. **Heat cramps** are painful muscle spasms that result from excessive electrolyte loss in the sweat. They occur especially when a person begins to relax after strenuous exertion and heavy sweating. **Heat exhaustion** results from more severe water and electrolyte loss and is characterized by hypotension, dizziness, vomiting, and sometimes fainting. Prolonged heat waves, especially if accompanied by high humidity, bring on many deaths from **heatstroke** (sunstroke). The body gains heat by radiation and conduction, but the humidity retards evaporative cooling. Heatstroke is clinically defined as a state in which the core body temperature is over 40°C (104°F); the skin is hot and dry; and the subject exhibits nervous system dysfunctions such as delirium, convulsions, or coma. It is also accompanied by tachycardia, hyperventilation, inflammation, and multiorgan dysfunction, and it is often fatal.

Hypothermia can result from exposure to cold weather or immersion in icy water. It, too, entails life-threatening positive feedback loops. If the core temperature falls below 33°C (91°F), the metabolic rate drops so low that heat production cannot keep pace with heat loss, and the temperature falls even more. Death from cardiac fibrillation may occur below 32°C (90°F), but some people survive body temperatures as low as 29°C (84°F) in a state of suspended animation. A body temperature below 24°C (75°F) is usually fatal. It is dangerous to give alcohol to someone in a state of hypothermia; it produces an illusion of warmth but actually accelerates heat loss by dilating cutaneous blood vessels.

## Before You Go On

*Answer the following questions to test your understanding of the preceding section:*

19. What is the primary source of body heat? What are some lesser sources?

20. What mechanisms of heat loss are aided by convection?

21. Describe the major heat-promoting and heat-losing mechanisms of the body.

22. Describe the positive feedback loops that can cause death from hyperthermia and hypothermia.

---

## INSIGHT 26.4　Clinical Application

### Alcohol and Alcoholism

Alcohol is not only a popular mind-altering drug but is also regarded in many cultures as a food staple. As a source of empty calories, an addictive drug, and a toxin, it can have a broad spectrum of adverse effects on the body.

#### Absorption and Metabolism

Alcohol is rapidly absorbed from the digestive tract—about 20% of it in the stomach and 80% in the proximal small intestine. Carbonation, as in beer and sparkling wines, increases its rate of absorption, whereas food reduces its absorption by delaying gastric emptying so the alcohol takes longer to reach its place of maximum absorption. Alcohol is soluble in both water and fat, so it is rapidly distributed to all body tissues and easily crosses the blood–brain barrier to exert its intoxicating effects on the brain.

Alcohol is detoxified by the hepatic enzyme *alcohol dehydrogenase,* which oxidizes it to acetaldehyde. This enters the citric acid cycle and is oxidized to $CO_2$ and $H_2O$. The average adult male can clear the blood of about 10 mL of 100% (200 proof) alcohol per hour—the amount in about 30 mL (1 oz) of whiskey or 355 mL (12 oz) of beer. Women have less alcohol dehydrogenase and clear alcohol from the bloodstream less quickly. They are also more vulnerable to alcohol-related illnesses such as cirrhosis of the liver (discussed shortly).

Tolerance to alcohol, the ability to "hold your liquor," results from two factors: behavioral modification, such as giving in less readily to lowered inhibitions, and increased levels of alcohol dehydrogenase in response to routine alcohol consumption. Alcohol dehydrogenase also deactivates other drugs, and drug dosages must be adjusted to compensate for this when treating alcoholics for other diseases.

#### Physiological Effects

***Nervous System***　Alcohol is a depressant that inhibits the release of norepinephrine and disrupts the function of GABA receptors. In low doses, it depresses inhibitory synapses and creates sensations of confidence, euphoria, and giddiness. As the dosage rises, however, the breakdown products of ethanol enhance the diffusion of $K^+$ out of neurons, hyperpolarizing them and making them less responsive to neurotransmitters. Thus, the timing and coordination of communication between neurons is impaired, resulting in such symptoms of intoxication as slurred speech, poor coordination, and slower reaction time. These symptoms begin to become significant at a blood alcohol level of 80 to 100 mg/dL—the legal criterion of intoxication in many states. Above 400 mg/dL, alcohol can so disrupt the electrophysiology of neurons as to induce coma and death.

***Liver***　The liver's role in metabolizing alcohol makes it especially susceptible to long-term toxic effects. Heavy drinking stresses the liver with a high load of acetaldehyde and acetate; this depletes its oxidizing agents and reduces its ability to catabolize these intermediates as well as fatty acids. Alcoholism often produces a greatly enlarged and fatty liver for multiple reasons: the calories provided by alcohol make it unnecessary to burn fat as fuel, fatty acids are poorly oxidized, and acetaldehyde is converted to new fatty acids. Acetaldehyde also causes inflammation of the liver and pancreas (*hepatitis* and *pancreatitis*), leading to disruption of digestive function. Acetaldehyde and other toxic intermediates destroy hepatocytes faster than they can be regenerated, leading to cirrhosis (see Insight 26.3). Many symptoms of alcoholism stem from deterioration of liver functions. Hepatic coma may occur as the liver becomes unable to produce urea, thus allowing ammonia to accumulate in the blood. Jaundice results from the liver's inability to excrete bilirubin.

*continued*

*Circulatory System*    Deteriorating liver functions exert several effects on the blood and cardiovascular system. Blood clotting is impaired because the liver cannot synthesize clotting factors adequately. Edema results from inadequate synthesis of blood albumin. Cirrhosis obstructs the hepatic portal blood circulation. Portal hypertension results, and combined with hypoproteinemia, this causes the liver and other organs to "weep" serous fluid into the peritoneal cavity. This leads to *ascites*[10] (ah-SY-teez)—swelling of the abdomen with as much as several liters of serous fluid. The combination of hypertension and impaired clotting often leads to hemorrhaging. *Hematemesis,*[11] the vomiting of blood, may occur as enlarged veins of the esophagus hemorrhage. Alcohol abuse also destroys myocardial tissue, reduces contractility of the heart, and causes cardiac arrhythmia.

*Digestive System and Nutrition*    Alcohol breaks down the protective mucous barrier of the stomach and the tight junctions between its epithelial cells. Thus it may cause gastritis and bleeding. Alcohol is commonly believed to be a factor in peptic ulcers, but there is little concrete evidence of this. Heavy drinking, especially in combination with smoking, increases the incidence of esophageal cancer. Malnutrition is a typical complication of alcoholism, partly because the empty calories of alcohol suppress the appetite for more nutritious foods. The average American gets about 4.5% of his or her calories from alcohol (more when nondrinkers are excluded), but heavy drinkers may obtain half or more of their calories from alcohol and have less appetite for foods that would meet their other nutritional requirements. In addition, acetaldehyde interferes with vitamin absorption and use. Thiamine deficiency is common in alcoholism, and thiamine is routinely given to alcoholics in treatment.

## Addiction

Alcohol is the most widely available addictive drug in America. In many respects it is almost identical to barbiturates in its toxic effects, its potential for tolerance and dependence, and the risk of overdose. The difference is that obtaining barbiturates usually requires a prescription, while obtaining alcohol requires, at most, proof of age.

*Alcoholism* is defined by a combination of criteria, including the pathological changes just described; physiological tolerance of high concentrations; impaired physiological, psychological, and social functionality; and withdrawal symptoms occurring when intake is reduced or stopped. Heavy drinking followed by a period of abstinence—for example, when a patient is admitted to the hospital and cannot get access to alcohol—may trigger *delirium tremens (DT),* characterized by restlessness, insomnia, confusion, irritability, tremors, incoherent speech, hallucinations, convulsions, and coma. DT has a 5% to 15% mortality rate.

Most alcoholism (type I) sets in after age 25 and is usually associated with stress or peer pressure. These influences lead to increased drinking, which can start a vicious cycle of illness, reduced job performance, family and social problems, arrest, and other stresses leading to still more drinking. A smaller number of alcoholics have type II alcoholism, which is at least partially hereditary. Most people with type II alcoholism are men who become addicted before age 25, especially the sons of other type II alcoholics. Type II alcoholics show abnormally rapid increases in blood acetaldehyde levels when they drink, and they have unusual brain waves (EEGs) even when not drinking. Children of alcoholics have a higher than average incidence of becoming alcoholic even when raised by nonalcoholic foster parents. It is by no means inevitable that such people will become alcoholics, but stress or peer pressure can trigger alcoholism more easily in those who are genetically predisposed to it.

Alcoholism is treated primarily through behavior modification—abstinence, peer support, avoidance or correction of the stresses that encourage drinking, and sometimes psychotherapy. The drug disulfiram (Antabuse) is used to support behavior modification programs. It inhibits the breakdown of acetaldehyde and thus heightens its short-term toxic effects. A person who takes Antabuse and drinks alcohol experiences headache, vomiting, tachycardia, chest pain, and hyperventilation and is less likely to look upon alcohol as a means of pleasure or escape.

---

[10]*asc* = bag + *ites* = like, resembling
[11]*hemat* = blood + *emesis* = vomiting

# CHAPTER REVIEW

## Review of Key Concepts

### Nutrition (p. 1002)

1. Body weight is stable when average daily energy intake and output are equal. Weight appears to have a homeostatic set point determined partly by heredity. About 30% to 50% of the difference in weight between people is hereditary and the rest is due to eating and exercise habits and other environmental variables.

2. Appetite is regulated in part by several *gut–brain peptides* that act as chemical signals from the gastrointestinal tract to the brain.

3. *Ghrelin* stimulates an immediate sensation of hunger and induces eating. *Peptide YY* and *cholecystokinin* create a feeling of satiety and terminate eating.

4. *Leptin* and *insulin* are adiposity signals that inform the brain of how much adipose tissue one has, and regulate long-term food intake, energy consumption, and body weight.

5. Many of the gut–brain peptides act through the *arcuate nucleus* of the hypothalamus, which responds to them by secreting the appetite stimulant neuropeptide Y or the appetite suppressant melanocortin.

6. Hunger contractions of the stomach also motivate eating, and chewing and swallowing alone produce a degree of very short-term reduction of hunger.

7. Some hormones stimulate cravings for specific categories of nutrients, such as *norepinephrine* for carbohydrates, *galanin* for fats, and *endorphins* for protein.

8. Dietary Calories (kilocalories) come predominantly from carbohydrates, fats, and proteins. "Empty calories" are calories gained from foods such as sugar and alcohol that provide little or no other nutrition.

9. *Nutrients* are ingested chemicals that provide material for growth, repair, and maintenance of the body. Dietary substances that never become part of the body's tissues (for example, fiber) are not considered nutrients but are nevertheless important components of a healthy diet. Some nutrients (water, minerals, vitamins) require no digestion and yield no calories.

10. Water, carbohydrates, lipids, and proteins are required in relatively large amounts and are thus called *macronutrients*. Minerals and vitamins are needed in small amounts and are thus called *micronutrients*.

11. *Essential nutrients* must be included in the diet because the body cannot synthesize them from other chemicals.

12. Carbohydrates are used as fuel and as structural components of many biological molecules.

13. In the body, the carbohydrate fuels are blood glucose and liver and muscle glycogen. The balance between glycogen and glucose is regulated by insulin and glucagon.

14. Starch is the most quantitatively significant digestible dietary carbohydrate, but significant quantities of lactose, sucrose, and fructose are ingested, especially in processed foods with added sweeteners.

15. Dietary fiber includes cellulose, pectin, gums, and lignin. Fiber promotes intestinal motility, and water-soluble fiber (pectin) lowers the levels of blood cholesterol and harmful low-density lipoproteins (LDLs).

16. Fat contains most of the body's stored energy. Being hydrophobic and less oxidized than carbohydrates, fats contain more than twice as many calories per gram as carbohydrates do.

17. The use of fats for fuel spares glucose and proteins for use by other tissues or for other purposes.

18. Other lipids important in human metabolism and structure include phospholipids, cholesterol, fat-soluble vitamins, prostaglandins, and eicosanoids.

19. Essential fatty acids—linoleic and possibly linolenic and arachidonic acids—must be included in the diet because the body cannot synthesize them.

20. Lipids are transported in the blood as *lipoproteins*—droplets of cholesterol and triglycerides coated with proteins and phospholipids. The types of lipoproteins are chylomicrons, which are formed in the small intestine and transport dietary lipids throughout the body; very low–density lipoproteins (VLDLs), which transport lipids from the liver to the adipose tissue; low-density lipoproteins (LDLs), which are the remainders of the VLDLs after triglycerides are removed, and which transport cholesterol to cells that need it; and high-density lipoproteins (HDLs), which transport excess cholesterol back to the liver for disposal.

21. A high LDL concentration indicates a heightened risk of cardiovascular disease, while a high HDL concentration is beneficial to cardiovascular health.

22. Protein typically constitutes 12% to 15% of the body mass and performs a wider variety of structural and physiological roles than any other class of biological molecules.

23. The nutritional value of a protein depends on whether it provides the right proportions of the various amino acids, especially the eight *essential amino acids. Complete proteins* supply all the essential amino acids in the proportions needed for the human body. The body makes more efficient use of animal proteins than of plant proteins.

24. *Nitrogen balance* is a state in which average daily nitrogen intake and output are equal. A greater intake than output *(positive nitrogen balance)* is typical of childhood, pregnancy, resistance training, and other states of tissue growth. Greater output

than intake (*negative nitrogen balance*) is typical in stress, muscle atrophy, and malnutrition.

25. *Minerals* are inorganic elements acquired from the soil by way of plants. Calcium and phosphorus are the body's most abundant minerals; sodium is a close third. Several others are present in relatively small quantities (table 26.2) but are vitally important.

26. *Vitamins* are small organic molecules that are not used for fuel (caloric content), but are necessary to metabolism. They act as coenzymes, antioxidants, components of visual pigments, and in other roles.

27. Vitamin C and the B vitamins are the *water-soluble vitamins;* vitamins A, D, E, and K are the *fat-soluble vitamins.*

28. Vitamin deficiencies cause a variety of illnesses, although vitamin excesses (*hypervitaminosis*) can also be quite harmful.

## Carbohydrate Metabolism (p. 1013)

1. The complete oxidation of glucose has the equation $C_6H_{12}O_6 + 6\ O_2 \rightarrow 6\ CO_2 + 6\ H_2O$.

2. The coenzymes $NAD^+$ and FAD are especially important in transferring electrons from one metabolic pathway to another in this process.

3. Glucose oxidation begins with a pathway called *glycolysis,* which splits glucose into two pyruvic acid molecules and has a net yield of 2 ATP per glucose.

4. In the absence of oxygen, pyruvic acid is reduced to lactic acid in a one-step reaction called *anaerobic fermentation.* The primary purpose of this is to regenerate $NAD^+$, which is needed to keep glycolysis running and producing ATP.

5. In the presence of oxygen, pyruvic acid enters a pathway called *aerobic respiration,* which produces much more ATP and has end products ($CO_2$ and $H_2O$) that are less toxic than lactic acid. Aerobic respiration occurs in the mitochondria.

6. The first principal group of reactions in aerobic respiration are the *matrix reactions,* mainly the *citric acid cycle.* This cycle breaks pyruvic acid down to $CO_2$, generates 2 ATP per glucose, and most importantly, generates 8 NADH and 2 $FADH_2$.

7. The final reactions of aerobic respiration are the *membrane reactions,* which occur on the inner mitochondrial membrane. Enzymes and other electron carriers here transport electrons from NADH and $FADH_2$ to oxygen, producing water as an end product. More importantly, the energy from these electron transfers drives *proton pumps,* which create a steep $H^+$ gradient between the mitochondrial membranes. This gradient drives a *chemiosmotic mechanism* by which *ATP synthase* generates ATP.

8. Glycolysis and aerobic respiration collectively produce up to 38 ATP per glucose, with the number varying slightly from one tissue type to another.

9. Glucose in excess of the body's immediate needs can be converted to fat or polymerized and stored as glycogen. *Glycogenesis* is the synthesis of glycogen. *Glycogenolysis* is the hydrolysis of glycogen to release glucose. *Gluconeogenesis* is the synthesis of glucose from glycerol (derived from fats) or amino acids.

10. The liver carries out these processes of carbohydrate metabolism among many other functions (table 26.6).

## Lipid and Protein Metabolism (p. 1022)

1. Adipocytes store and release most of the body's fat (triglycerides).

2. *Lipogenesis* is the synthesis of fats from precursors such as sugars and amino acids.

3. *Lipolysis* is the breakdown of fats, starting with hydrolysis and continuing with oxidation of the fatty acids and glycerol. Fatty acids are degraded by the process of *beta oxidation.* Oxidation of a typical fatty acid can yield 129 ATPs—much more than glucose oxidation.

4. Incomplete fatty acid oxidation produces acidic *ketone bodies,* a process called *ketogenesis.* Ketone bodies can be used as fuel but an excess can cause dangerous ketoacidosis, as it does in diabetes mellitus.

5. Proteins turn over at an average rate of about 100 g/day, with especially high turnover in the intestinal mucosa.

6. Free amino acids in the *amino acid pool* can be used to synthesize new proteins, converted to glucose or fat, or oxidized as fuel.

7. Amino acid catabolism entails *deamination,* the removal of the amino group. The amino group eventually becomes ammonia ($NH_3$). The liver combines ammonia and $CO_2$ to produce urea, which is a less toxic waste product than ammonia and the most abundant nitrogenous waste in the blood and urine.

## Metabolic States and Metabolic Rate (p. 1024)

1. The *absorptive state* lasts about 4 hours after a meal. During this time, nutrients are absorbed from the intestine and may be used immediately. Glucose level is high and excess glucose is stored as glycogen or converted to fat.

2. The absorptive state is regulated mainly by insulin, which promotes glucose uptake and oxidation, glycogenesis, and lipogenesis; promotes protein synthesis; and inhibits gluconeogenesis.

3. The *postabsorptive state* prevails between meals and overnight, when the stomach is empty and the body uses stored fuels. Glycogenolysis and gluconeogenesis maintain the blood glucose level during this state. Fatty acids derived from lipolysis are used as fuel by many cells.

4. The postabsorptive state is regulated by multiple hormones. Glucagon, epinephrine, and norepinephrine promote lipolysis and glycogenolysis; cortisol promotes fat and protein catabolism; cortisol and glucagon promote gluconeogenesis; and growth hormone raises blood glucose by antagonizing insulin.

5. *Metabolic rate* is the amount of energy released in the body in a given time, such as kcal/day. It varies according to metabolic state and physical, mental, and hormonal conditions. *Basal metabolic rate (BMR)* is a standard of reference based on a comfortable, resting, awake, postabsorptive state. *Total metabolic rate* is a higher nonresting rate that takes muscular activity into account.

6. BMR is about 2,000 kcal/day. A low level of physical activity increases daily energy needs to about 2,500

kcal/day, and hard physical labor can increase them to as much as 5,000 kcal/day. Metabolic rate also varies with age, sex, mental state, stress, and health or illness.

**Body Heat and Thermoregulation (p. 1026)**

1. *Thermoregulation* is the homeostatic control of body temperature. Excessively high or low body temperatures (*hyperthermia* and *hypothermia*) can be fatal.

2. *Core temperature* can be estimated from rectal temperature and is usually 37.2° to 37.6°C. Shell temperature, usually estimated from oral temperature, is usually 36.6° to 37.0°C.

3. Body heat is generated mainly by exergonic chemical reactions, especially in the brain, heart, liver, and endocrine glands at rest and in the skeletal muscles during activity. The body loses heat by radiation, conduction, and evaporation.

4. The *hypothalamic thermostat* monitors blood temperature and receives signals from peripheral thermoreceptors in the skin.

5. To rid the body of excess heat, the thermostat sends signals to a hypothalamic *heat-losing center,* which triggers cutaneous vasodilation and sweating.

6. To generate and retain heat, the thermostat sends signals to a hypothalamic *heat-promoting center,* which triggers shivering and cutaneous vasoconstriction.

7. Heat can also be produced by *nonshivering thermogenesis,* in which the metabolic rate is increased and releases more heat from organic fuels.

8. *Behavioral thermoregulation* includes behaviors that adjust body temperature, such as adding or removing clothes, or getting into the shade or sun.

9. *Heat cramps, heat exhaustion,* and *heatstroke* are three effects of hyperthermia. Heatstroke often progresses to fatal multiorgan dysfunction. Hypothermia may be fatal if the core temperature reaches 32°C or lower.

# Testing Your Recall

1. _____ are not used as fuel and are required in relatively small quantities.
   a. Micronutrients
   b. Macronutrients
   c. Essential nutrients
   d. Proteins
   e. Lipids

2. The only significant digestible polysaccharide in the diet is
   a. glycogen.
   b. cellulose.
   c. starch.
   d. maltose.
   e. fiber.

3. Which of the following stores the greatest amount of energy for the smallest amount of space in the body?
   a. glucose
   b. triglycerides
   c. glycogen
   d. proteins
   e. vitamins

4. The lipoproteins that remove cholesterol from the tissues are
   a. chylomicrons.
   b. lipoprotein lipases.
   c. VLDLs.
   d. LDLs.
   e. HDLs.

5. Which of the following is most likely to make you hungry?
   a. leptin
   b. ghrelin
   c. cholecystokinin
   d. peptide YY
   e. melanocortin

6. The primary function of B-complex vitamins is to act as
   a. structural components of cells.
   b. sources of energy.
   c. components of pigments.
   d. antioxidants.
   e. coenzymes.

7. FAD is reduced to $FADH_2$ in
   a. glycolysis.
   b. anaerobic fermentation.
   c. the citric acid cycle.
   d. the electron-transport chain.
   e. beta oxidation of lipids.

8. The primary, direct benefit of anaerobic fermentation is to
   a. regenerate $NAD^+$.
   b. produce $FADH_2$.
   c. produce lactic acid.
   d. dispose of pyruvic acid.
   e. produce more ATP than glycolysis does.

9. Which of these occurs in the mitochondrial matrix?
   a. glycolysis
   b. chemiosmosis
   c. the cytochrome reactions
   d. the citric acid cycle
   e. anaerobic fermentation

10. When the body emits more infrared energy than it absorbs, it is losing heat by
    a. convection.
    b. forced convection.
    c. conduction.
    d. radiation.
    e. evaporation.

11. A/an _____ protein lacks one or more essential amino acids.

12. In the postabsorptive state, glycogen is hydrolyzed to liberate glucose. This process is called _____.

13. Synthesis of glucose from amino acids or triglycerides is called _____.

14. The major nitrogenous waste resulting from protein catabolism is _____.

15. The organ that synthesizes the nitrogenous waste in question 14 is the _____.

16. The absorptive state is regulated mainly by the hormone _____.

17. The temperature of organs in the body cavities is called _____.

18. The appetite hormones ghrelin, leptin, CCK, and others act on part of the hypothalamus called the _____ nucleus.

19. The brightly colored, iron-containing, electron-transfer molecules of the inner mitochondrial membrane are called _____.

20. The flow of $H^+$ from the intermembrane space to the mitochondrial matrix creates an electrical current used by the enzyme _____ to make _____.

*Answers in Appendix B*

# True or False

*Determine which five of the following statements are false, and briefly explain why.*

1. Ghrelin and leptin are two hormones that stimulate the appetite.

2. Water is a nutrient, but oxygen and cellulose are not.

3. An extremely low-fat diet can cause vitamin-deficiency diseases.

4. Most of the body's cholesterol comes from the diet.

5. There is no harm in maximizing one's daily protein intake.

6. Aerobic respiration produces more ATP than anaerobic fermentation.

7. Reactions occurring on the mitochondrial inner membrane produce more ATP than glycolysis and the matrix reactions combined.

8. Gluconeogenesis occurs especially in the absorptive state during and shortly after a meal.

9. Brown fat generates more ATP than white fat and is therefore especially important for thermoregulation.

10. At a comfortable air temperature, the body loses more heat by infrared radiation than by any other means.

*Answers in Appendix B*

# Testing Your Comprehension

1. Cyanide blocks the transfer of electrons from cytochrome $\alpha_3$ to oxygen. In light of this, explain why it causes sudden death. Also explain whether cyanide poisoning could be treated by giving a patient supplemental oxygen, and justify your answer.

2. Chapter 17 defines and describes some hormone actions that are synergistic and antagonistic. Identify some synergistic and antagonistic hormone interactions in the postabsorptive state of metabolism.

3. Mrs. Jones, a 42-year-old, complains that, "Everything I eat goes to fat. But my husband and my son eat twice as much as I do, and they're both as skinny as can be." How would you explain this to her?

4. A television advertisement proclaims, "Feeling tired? Need more energy? Order your supply of Zippy Megavitamins and feel better fast!" Your friend Cathy is about to send in her order, and you try to talk her out of wasting her money. Summarize the argument you would use.

5. Explain why a patient whose liver has been extensively damaged by hepatitis could show elevated concentrations of thyroid hormone and bilirubin in the blood.

*Answers at www.mhhe.com/saladin4*

# www.mhhe.com/saladin4

*The textbook website provides a wealth of interactive study materials fully organized and integrated by chapter. You will find practice quizzes, labeling exercises, and much more that will complement your learning and understanding of anatomy and physiology. The website also includes tools designed to enhance your* **Anatomy & Physiology | REVEALED** *experience.*

*Seminiferous tubules, where sperm are produced. Sperm tails are seen as hair-like masses in the center of the tubules (SEM).*

# THE MALE REPRODUCTIVE SYSTEM

## Brushing Up

To understand this chapter, it is important that you understand or brush up on the following concepts:
- Chromosome structure (p. 126)
- Stages of mitosis (p. 141)
- The karyotype (p. 142)
- Muscles of the pelvic floor (p. 349)
- Hypothalamic releasing hormones (p. 642)
- Pituitary gonadotropins (pp. 643–644)
- Negative feedback inhibition of the pituitary (p. 646)
- Androgens (p. 650)

From all we have learned of the structure and function of the human body, it seems a wonder that it works at all! The fact is, however, that even with modern medicine we cannot keep it working forever. The body suffers various degenerative changes as we age, and eventually our time is up and we must say good-bye. Yet our genes live on in new containers—our offspring. The production of offspring is the subject of these last three chapters. In this chapter, we examine some general aspects of human reproductive biology and then focus on the role of the male in reproduction. Chapter 28 focuses on the female and chapter 29 on the embryonic development of humans and on changes at the other end of the life span—the changes of old age.

# Sexual Reproduction and Development

### Objectives

When you have completed this section, you should be able to

- identify the most fundamental biological distinction between male and female;

- define *primary sex organs, secondary sex organs,* and *secondary sex characteristics;*

- explain the role of the sex chromosomes in determining sex;

- explain how the Y chromosome determines the response of the fetal gonad to prenatal hormones;

- identify which of the male and female external genitalia are homologous to each other; and

- describe the descent of the testes and explain why it is important.

## THE ESSENCE OF SEX

Reproduction is one of the fundamental properties of all living things. Many organisms reproduce simply by fission (splitting in two) or by budding off miniature replicas of themselves. These are asexual processes that produce exact genetic copies of the parent. Most organisms, however, reproduce sexually. Sexual reproduction does not necessarily entail copulation or even physical contact between the parents—many species merely shed sex cells into the sea and the parents never meet. The essence of sexual reproduction is that each offspring has two parents and a combination of genes from both. Thus the offspring are not genetically identical to their parents and usually not even to each other. Genetic diversity provides the foundation for the survival and evolution of a species, and has been such a substantial advantage that it is employed by the great majority of organisms.

## THE TWO SEXES

To reproduce sexually means that the parents must produce **gametes**[1] (sex cells) that can meet and combine their genes in a **zygote**[2] (fertilized egg). The gametes must have two properties for reproduction to be successful: (1) motility, so they can achieve contact, and (2) nutrients for the developing embryo. A single cell cannot perform both of these roles very well, because to contain ample nutrients means to be relatively large and heavy, and this is inconsistent with the need for motility. Therefore, these tasks are usually apportioned to two kinds of gametes. The small motile one—little more than DNA with a propeller—is the **sperm** (spermatozoon), and the large nutrient-laden one is the **egg** (ovum).

It is usually easy to distinguish a human male and female from each other, but the difference is not so obvious in many other species. By definition, an individual that produces eggs is female and one that produces sperm is male. Even in humans, this criterion is not always that simple—for example, when considering some abnormalities in sexual development to be discussed later. Genetically, any human with a Y sex chromosome is classified as male and any individual lacking a Y chromosome (usually but not always having two X chromosomes) is classified as female.

In mammals, the female is also the parent that provides a sheltered internal environment for the development and prenatal nutrition of the embryo. For fertilization and development to occur in the female, the male must have a copulatory organ, the penis, for introducing his gametes into the female reproductive tract, and the female must have a copulatory organ, the vagina, for receiving the sperm. This is the most obvious difference between the sexes, but appearances can be deceiving (see fig. 17.30, p. 670 and Insight 27.1).

## OVERVIEW OF THE REPRODUCTIVE SYSTEM

The reproductive system consists of primary and secondary sex organs. The **primary sex organs,** or **gonads,**[3] are organs that produce the gametes—**testes** of the male and **ovaries** of the female. The **secondary sex organs** are organs other than the gonads that are necessary for reproduction. In the male, they constitute a system of ducts,

---

[1]*gam* = marriage, union
[2]*zygo* = yoke, union
[3]*gon* = seed

## Androgen-Insensitivity Syndrome

Occasionally, a girl shows all the usual changes of puberty except that she fails to menstruate. Physical examination shows the presence of testes in the abdomen and a karyotype reveals that she has the XY chromosomes of a male. The testes produce normal male levels of testosterone, but the target cells lack receptors for it. This is called *androgen-insensitivity syndrome* (AIS), or *testicular feminization.*

The external genitals develop female anatomy as if no testosterone were present. At puberty, breasts and other feminine secondary sex characteristics develop (fig. 27.1) because the testes secrete small amounts of estrogen and there is no overriding influence of testosterone. However, there is no uterus or menstruation. If the abdominal testes are not removed, the person has a high risk of testicular cancer.

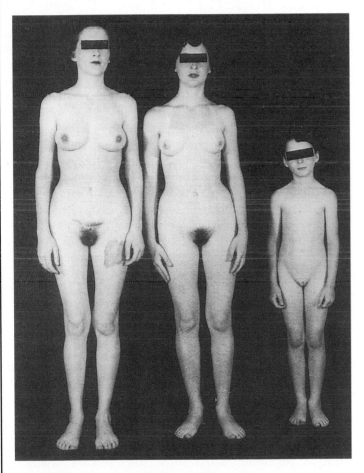

**FIGURE 27.1  Androgen-Insensitivity Syndrome.** These siblings are genetically male (XY). Testes are present and secrete testosterone, but the target cells lack receptors for it, so testosterone cannot exert its masculinizing effects. The external genitalia and secondary sex characteristics are feminine, but there are no ovaries, uterus, or vagina.

▶ *In what way is androgen-insensitivity syndrome similar to non-insulin-dependent diabetes mellitus?*

glands, and the penis, concerned with the storage, survival, and conveyance of sperm. In the female, they include the uterine tubes, uterus, and vagina, concerned with uniting the sperm and egg and harboring the developing fetus.

**Secondary sex characteristics** are features that develop at puberty, further distinguish the sexes, and play a role in mate attraction. They typically appear only as an animal approaches sexual maturity (during adolescence in humans). From the call of a bullfrog to the tail of a peacock, these are well known in the animal kingdom. In humans, the physical attributes that contribute to mate attraction are so culturally conditioned that it is more difficult to identify what secondary sex characteristics are biologically fundamental. Generally accepted as such in both sexes are the pubic and axillary hair and their associated scent glands, and the pitch of the voice. Other traits commonly regarded as male secondary sex characteristics are the facial hair, relatively coarse and visible hair on the torso and limbs, and the relatively muscular physique. In females, they include the distribution of body fat, enlargement of the breasts (independently of lactation), flare of the hips, and relatively hairless appearance of the skin.

## CHROMOSOMAL SEX DETERMINATION

What determines whether a zygote will develop into a male or female? The distinction begins with the combination of sex chromosomes bequeathed to the zygote. Most of our cells have 23 pairs of chromosomes: 22 pairs of *autosomes* and 1 pair of *sex chromosomes* (see fig. 4.16, p. 143). A sex chromosome can be either a large X chromosome or a small Y chromosome. Every egg contains an X chromosome, but half of the sperm carry an X and the other half carry a Y. If an egg is fertilized with an X-bearing sperm, it produces an XX zygote that is destined to become a female. If it is fertilized with a Y-bearing sperm, it produces an XY zygote destined to become a male. Thus the sex of a child is determined at conception (fertilization), and not by the mother's egg but by the sperm that fertilizes it (fig. 27.2).

## PRENATAL HORMONES AND SEXUAL DIFFERENTIATION

Sex determination does not end with fertilization, however. It requires an interaction between genetics and the hormones produced by the mother and fetus. Just as we have seen with other hormones (see chapter 17), those involved here require specific receptors on their target cells to exert an effect.

Up to a point, a fetus is sexually undifferentiated, or "noncommittal" as to which sex it will become. Its gonads begin to develop at 5 to 6 weeks as *gonadal ridges,* each lying alongside a primitive kidney, the *mesonephros,* which later degenerates. Adjacent to each gonadal ridge

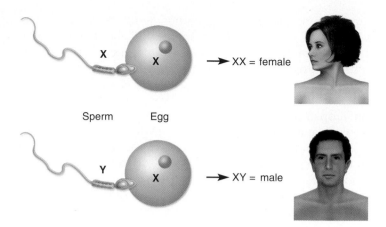

**FIGURE 27.2 Chromosomal Sex Determination.** All eggs carry the X chromosome. The sex of a child is determined by whether the egg is fertilized by an X-bearing sperm or a Y-bearing sperm.

are two ducts: the **mesonephric**[4] (MEZ-oh-NEF-ric) **(wolffian**[5]**) duct,** which originally serves the mesonephros, and the **paramesonephric**[6] **(müllerian**[7]**) duct.** In males, the mesonephric ducts develop into the reproductive tract and the paramesonephric ducts degenerate. In females, the opposite occurs (fig. 27.3).

But why? The Y chromosome has a gene called **SRY** (sex-determining region of the Y) that codes for a protein called **testis-determining factor (TDF).** TDF then interacts with genes on some of the other chromosomes, including a gene on the X chromosome for androgen receptors, and those genes initiate the development of male anatomy. By 8 to 9 weeks, the male has an identifiable testis that begins to secrete testosterone. Each testis stimulates the mesonephric duct on its own side to develop into the system of male reproductive ducts. By this time, the testis also secretes a hormone called **müllerian-inhibiting factor (MIF)** that causes atrophy of the paramesonephric (müllerian) duct on that side. Even an adult male, however, retains a tiny Y-shaped vestige of the paramesonephric ducts, like a vestigial uterus and uterine tubes, in the area of the prostatic urethra. It is named the *uterus masculinus.*

It may seem as if androgens should induce the formation of a male reproductive tract and estrogens induce a female reproductive tract. However, estrogen levels are always high during pregnancy, so if this mechanism were the case, they would feminize all fetuses. Thus the development of a female results from the absence of androgens, not the presence of estrogens.

## DEVELOPMENT OF THE EXTERNAL GENITALIA

You perhaps regard the external genitals as the most definitive characteristics of a male or female, yet there is more similarity between the sexes than most people realize. In the embryo, the genitals begin developing from identical structures in both sexes. By 6 weeks, the embryo has the following (fig. 27.4):

- a **genital tubercle,** an anterior bud destined to become the head (glans) of the penis or clitoris;
- **urogenital folds,** a pair of medial tissue folds slightly posterior to the genital tubercle; and
- **labioscrotal folds,** a larger pair of tissue folds lateral to the urogenital folds.

By the end of week 9, the fetus begins to show sexual differentiation, and either male or female genitalia are distinctly formed by the end of week 12. In the female, the three structures just listed become the clitoris, labia minora, and labia majora, respectively; all of these are more fully described in chapter 28. In the male, the genital tubercle elongates to form the *phallus;* the urogenital folds fuse to enclose the urethra, joining the phallus to form the penis; and the labioscrotal folds fuse to form the scrotum, a sac that will later contain the testes.

Male and female organs that develop from the same embryonic structure are said to be **homologous.** Thus the penis is homologous to the clitoris and the scrotum is homologous to the labia majora. This becomes strikingly evident in some abnormalities of sexual development. In the presence of excess androgen, the clitoris may become greatly enlarged and resemble a small penis. In other cases, the ovaries descend into the labia majora as if they were testes descending into a scrotum. Such abnormalities sometimes result in mistaken identification of the sex of an infant at birth.

## DESCENT OF THE TESTES

The testes begin development near the kidneys. How do they end up in the scrotum, and why? In the embryo, a connective tissue cord called the *gubernaculum*[8] (GOO-bur-NACK-you-lum) extends from the gonad to the floor of the abdominopelvic cavity. As it continues to grow, it passes between the internal and external abdominal oblique muscles and into the scrotal swelling. Independently of any migration of the testis, the peritoneum also develops a fold that extends into the scrotum as the *vaginal process.*[9] The gubernaculum and the vaginal process create a path of low resistance through the

---

[4]*meso* = middle + *nephr* = kidney; named for a temporary embryonic kidney, the mesonephros
[5]Kaspar F. Wolff (1733–94), German anatomist
[6]*para* = next to
[7]Johannes P. Müller (1801–58), German physician

[8]*gubern* = rudder, to steer, to guide
[9]*vagin* = sheath

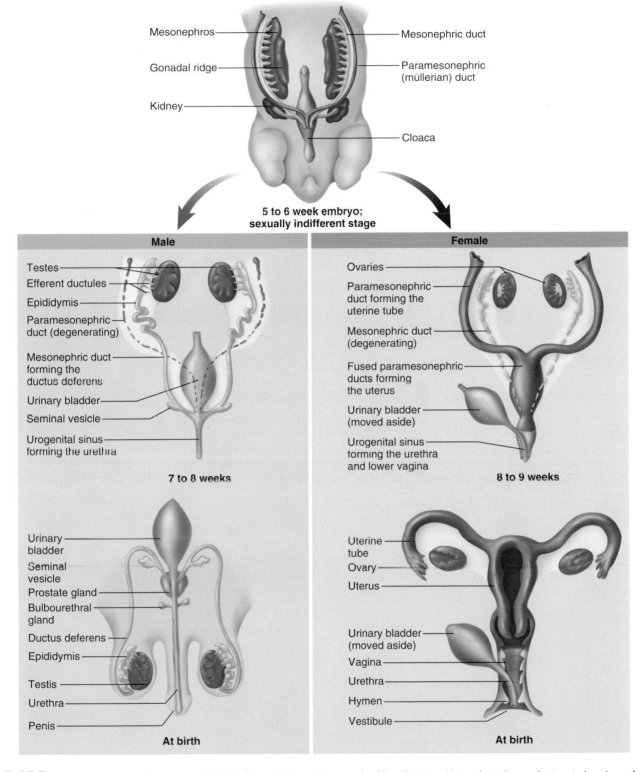

**FIGURE 27.3** **Embryonic Development of the Male and Female Reproductive Tracts.** Note that the male tract develops from the mesonephric duct and the female tract from the paramesonephric duct, while the other duct in each sex degenerates.

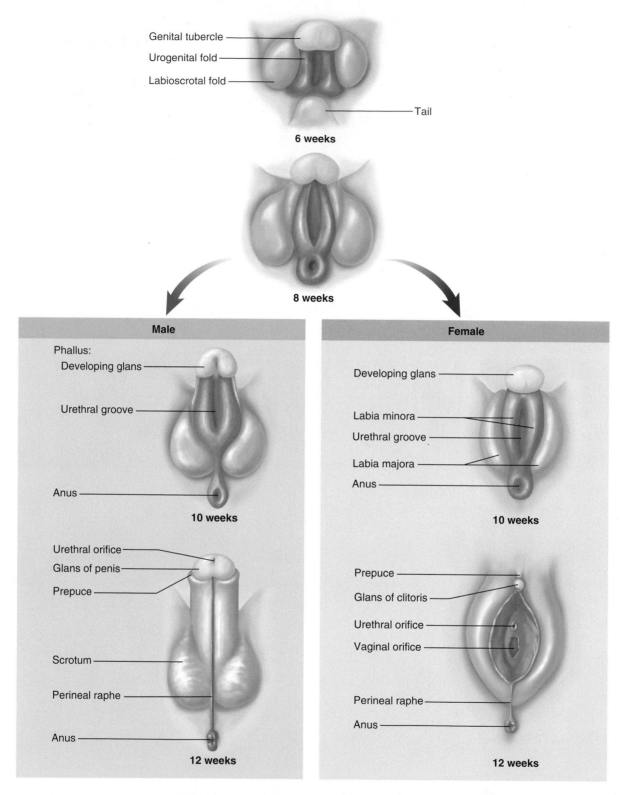

**FIGURE 27.4   Development of the External Genitalia.**   By 6 weeks, the embryo has three primordial structures—the genital tubercle, urogenital folds, and labioscrotal folds—which will become the male or female genitalia. At 8 weeks these structures have grown, but the sexes are still indistinguishable. Slight sexual differentiation is noticeable at 10 weeks, and the sexes are fully distinguishable by 12 weeks. Matching colors identify homologous structures of the male and female.

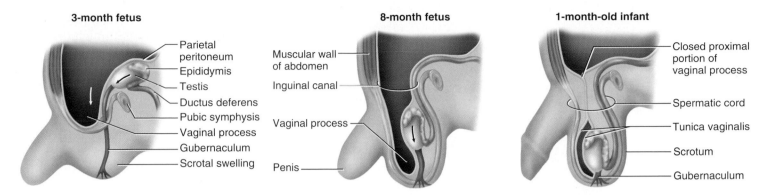

**FIGURE 27.5 Descent of the Testis.** Note that the testis and spermatic ducts are retroperitoneal. An extension of the peritoneum called the *vaginal process* follows the testis through the inguinal canal and becomes the tunica vaginalis.

▶ *Why is this structure of male anatomy called the tunica vaginalis?*

groin, anterior to the pubic symphysis, called the **inguinal canal**—the most common site of herniation in boys and men (*inguinal hernia; see* Insight 10.2, p. 351).

The **descent of the testes** (fig. 27.5) begins as early as week 6. The superior part of the embryonic gonad degenerates and its posterior part migrates inferiorly, guided by the gubernaculum. In the seventh month, the testes abruptly pass through the inguinal canals, anterior to the pubic symphysis, into the scrotum. As they descend, they are accompanied by ever-elongating testicular arteries and veins and by lymphatic vessels, nerves, spermatic ducts, and extensions of the internal abdominal oblique muscle. The vaginal process becomes separated from the peritoneal cavity and persists as a sac, the *tunica vaginalis,* enfolding the anterior side of the testis. Although multiple hypotheses have been offered, the actual mechanism of descent remains obscure. Testosterone stimulates it, but it is unknown how. The reason this descent is necessary, however, will be explained in the next section.

About 3% of boys are born with undescended testes, or **cryptorchidism.**[10] In most such cases, the testes descend within the first year of infancy, but if they do not, the condition can usually be corrected with a testosterone injection or a fairly simple surgery to dilate the inguinal canal and draw the testis into the scrotum. Uncorrected cryptorchidism, however, leads inevitably to sterility and sometimes to testicular cancer.

---

[10]*crypto* = hidden + *orchid* = testis + *ism* = condition

## Before You Go On

*Answer the following questions to test your understanding of the preceding section:*

1. *Define gonad and gamete. Explain the relationship between the terms.*

2. *Define male, female, sperm, and egg.*

3. *What are mesonephric and paramesonephric ducts? What factors determine which one develops and which one regresses in the fetus?*

4. *What male structure develops from the genital tubercle and urogenital folds? What develops from the labioscrotal folds?*

5. *Describe the pathway taken during descent of the male gonad.*

# Male Reproductive Anatomy

### Objectives

When you have completed this section, you should be able to

- describe the anatomy of the scrotum, testes, and penis;

- describe the pathway taken by a sperm cell from its formation to its ejaculation, naming all the passages it travels; and

- state the names, locations, and functions of the male accessory reproductive glands.

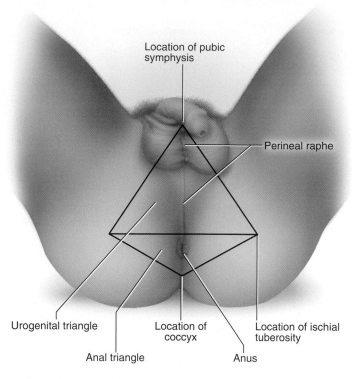

**FIGURE 27.6**  **The Male Perineum.**  Inferior view.

We will survey the male reproductive system beginning with the scrotum and testes, continuing through the spermatic ducts and accessory glands associated with them, and ending with the penis—thus taking anatomy in the order of sperm formation *(spermatogenesis)*, transport, and emission.

## THE SCROTUM

The scrotum and penis constitute the external genitalia of the male and occupy the **perineum** (PERR-ih-NEE-um). This is a diamond-shaped area between the thighs bordered by the pubic symphysis, ischial tuberosities, and coccyx (fig. 27.6). The **scrotum**[11] is the pendulous pouch containing the testes (fig. 27.7). The left testis is usually suspended lower than the right so the two are not compressed against each other between the thighs. The skin of the scrotum has sebaceous glands, sparse hair, rich sensory innervation, and somewhat darker pigmentation than skin elsewhere. The scrotum is divided into right and left compartments by an internal **median septum,**

---

[11]*scrotum = bag*

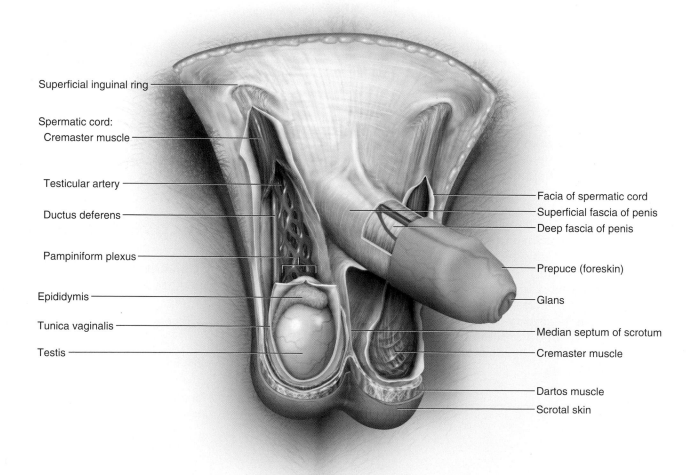

**FIGURE 27.7**  **The Scrotum and Spermatic Cord.**

which protects each testis from infections of the other one. The location of the septum is externally marked by a seam called the **perineal raphe**[12] (RAY-fee), which also extends anteriorly along the ventral side of the penis and posteriorly as far as the margin of the anus (see fig. 27.6).

The **spermatic cord** is a cord of connective tissue that passes through the scrotum superior to the testis, across the anterior side of the pubis, and into an opening called the *inguinal ring* in the muscles of the groin. From there, it travels about 4 cm through the inguinal canal and emerges into the pelvic cavity. It contains the *ductus (vas) deferens* (a sperm duct), blood and lymphatic vessels, and testicular nerves—structures that followed the testis as it descended through the canal. The cord is easily palpated through the skin of the scrotum.

Sperm cannot develop if the testes remain in the pelvic cavity; the core body temperature of 37°C is too warm. But the scrotum is about 2°C cooler. It is equipped with three mechanisms for maintaining this lower temperature:

1. The **cremaster**[13] **muscle** consists of strips of the internal abdominal oblique muscle that enmesh the spermatic cord. When it is cold, the cremaster contracts and draws the testes closer to the body to keep them warm. When it is warm, the cremaster relaxes and the testes are suspended farther from the body.

2. The **dartos**[14] **muscle** is a subcutaneous layer of smooth muscle. It, too, contracts when it is cold, and the scrotum becomes taut and wrinkled. The tautness of the scrotum helps to hold the testes snugly against the warm body and it reduces the surface area of the scrotum, thus reducing heat loss.

3. The **pampiniform**[15] **plexus** is an extensive network of veins from the testis that surround the testicular artery in the spermatic cord. As they pass through the inguinal canal, these veins converge to form the testicular vein, which emerges from the canal into the pelvic cavity. Without the pampiniform plexus, warm arterial blood would heat the testis and inhibit spermatogenesis. The pampiniform plexus, however, prevents this by acting as a *countercurrent heat exchanger* (fig. 27.8). Imagine that a house had uninsulated hot and cold water pipes running close to each other. Much of the heat from the hot water pipe would be absorbed by the cold water pipe next to it—especially if the water flowed in opposite directions so the cold water carried the heat away. In the spermatic cord, such a mechanism removes heat from the descending arterial blood, so by the time it reaches the testis this blood is 1.5° to 2.5°C cooler than the core body temperature.

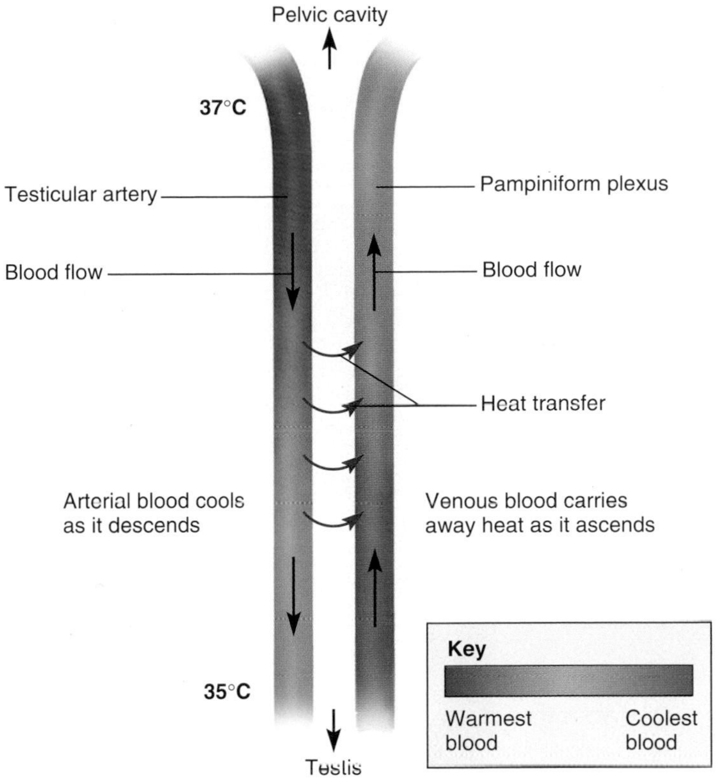

**FIGURE 27.8** **The Countercurrent Heat Exchanger.** Warm blood flowing down the testicular artery loses some of its heat to the cooler blood flowing in the opposite direction through the pampiniform plexus of veins (represented as a single vessel for simplicity). Arterial blood is about 2°C cooler by the time it reaches the testis.

## THE TESTES

Each testis is oval and slightly flattened, about 4 cm long, 3 cm from anterior to posterior, and 2.5 cm wide (fig. 27.9a). Its anterior and lateral surfaces are covered by the tunica vaginalis. The testis itself has a white fibrous capsule called the **tunica albuginea**[16] (TOO-nih-ca AL-byu-JIN-ee-uh). Connective tissue septa divide the organ into 250 to 300 wedge-shaped lobules. Each lobule contains one to three **seminiferous**[17] (SEM-ih-NIF-er-us) **tubules**—slender ducts up to 70 cm long in which the sperm are produced. Between the seminiferous tubules are clusters of **interstitial (Leydig**[18]**) cells,** the source of testosterone.

A seminiferous tubule has a narrow lumen lined by a thick **germinal epithelium** (fig. 27.10). The epithelium consists of several layers of **germ cells** in the process of becoming sperm and a much smaller number

---

[12]*raphe* = seam
[13]*cremaster* = suspender
[14]*dartos* = skinned
[15]*pampin* = tendril + *form* = shape

[16]*tunica* = coat + *alb* = white
[17]*semin* = seed, sperm + *fer* = to carry
[18]Franz von Leydig (1821–1908), German anatomist

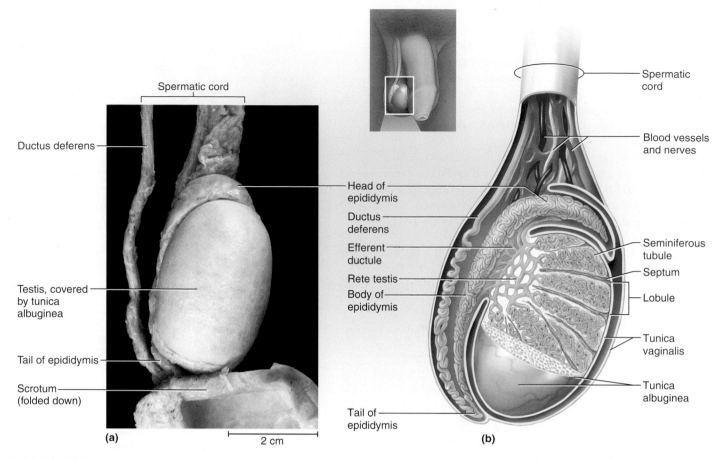

Spermatic cord

Ductus deferens

Testis, covered by tunica albuginea

Tail of epididymis

Scrotum (folded down)

2 cm

**(a)**

Spermatic cord

Blood vessels and nerves

Head of epididymis

Ductus deferens

Efferent ductule

Rete testis

Body of epididymis

Seminiferous tubule

Septum

Lobule

Tunica vaginalis

Tunica albuginea

Tail of epididymis

**(b)**

**FIGURE 27.9** **The Testis and Associated Structures.** (a) The scrotum is cut and folded downward to reveal the testis and associated organs. (b) Anatomy of the testis, epididymis, and spermatic cord.

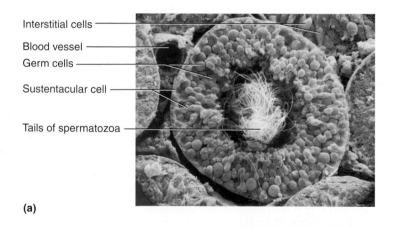

Interstitial cells

Blood vessel

Germ cells

Sustentacular cell

Tails of spermatozoa

**(a)**

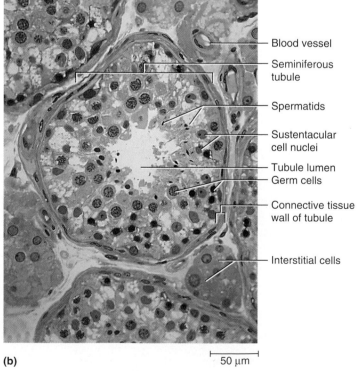

Blood vessel

Seminiferous tubule

Spermatids

Sustentacular cell nuclei

Tubule lumen

Germ cells

Connective tissue wall of tubule

Interstitial cells

50 μm

**(b)**

**FIGURE 27.10** **Histology of the Testis.** (a) Scanning electron micrograph. (b) Light micrograph. Part (b) is from a region of the tubule that did not have mature sperm at the time. [Part (a) from R. G. Kessel and R. H. Kardon, *Tissues and Organs: A Text-Atlas of Scanning Electron Microscopy,* W. H. Freeman, 1979.]

of tall **sustentacular**[19] (**Sertoli**[20]) **cells,** which protect the germ cells and promote their development. The germ cells depend on the sustentacular cells for nutrients, waste removal, growth factors, and other needs. The sustentacular cells also secrete a hormone, *inhibin,* that regulates the rate of sperm production, as we will see later.

A sustentacular cell is shaped a little like a tree trunk whose roots spread out over the basement membrane, forming the boundary of the tubule, and whose thick trunk reaches to the tubule lumen. Tight junctions between adjacent sustentacular cells form a **blood–testis barrier (BTB),** which prevents proteins and other large molecules in the blood and intercellular fluid from getting to the germ cells. This is important because the germ cells, being genetically different from other cells of the body, would otherwise be attacked by the immune system. Some cases of sterility occur when the BTB fails to form adequately in adolescence and the immune system produces autoantibodies against the germ cells.

⌐ **Think About It**

*Would you expect to find blood capillaries in the walls of the seminiferous tubules? Why or why not?*

The seminiferous tubules lead into a network called the **rete**[21] (REE-tee) **testis,** embedded in the capsule on the posterior side. Sperm partially mature in the rete. They are moved along by the flow of fluid secreted by the sustentacular cells and possibly by the cilia seen on some rete cells. Sperm do not swim while they are in the male.

Each testis is supplied by a **testicular artery** that arises from the abdominal aorta just below the renal artery. This is a very long, slender artery that winds its way down the posterior abdominal wall before passing through the inguinal canal into the scrotum (see fig. 27.7). Its blood pressure is very low, and indeed this is one of the few arteries to have no pulse. Consequently, blood flow to the testes is quite meager and the testes receive a poor oxygen supply. In response to this, the sperm develop unusually large mitochondria, which may precondition them for survival in the hypoxic environment of the female reproductive tract.

Blood leaves the testis by way of a **testicular vein.** The right testicular vein drains into the inferior vena cava and the left one drains into the left renal vein. Lymphatic vessels also drain each testis and lead to the inguinal lymph nodes. **Testicular nerves** lead to the gonads from spinal cord segment T10. They are mixed sensory and motor nerves containing predominantly sympathetic but also some parasympathetic fibers.

# THE SPERMATIC DUCTS

After leaving the testis, the sperm travel through a series of *spermatic ducts* to reach the urethra (fig. 27.11). These include the following:

- **Efferent ductules.** About 12 small efferent ductules arise from the posterior side of the testis and carry sperm to the epididymis. They have clusters of ciliated cells that help drive the sperm along.

- **Duct of the epididymis.** The **epididymis**[22] (EP-ih-DID-ih-miss; plural, *epididymides*) is a site of sperm maturation and storage. It adheres to the posterior side of the testis (see fig. 27.9), measures about 7.5 cm long, and consists of a clublike *head* at the superior pole of the testis, a long middle *body,* and a slender *tail* at its inferior end. It contains a single coiled duct, about 6 m (18 ft) long, embedded in connective tissue. This duct reabsorbs about 90% of the fluid secreted by the testis. Sperm are physiologically immature when they leave the testis but mature as they travel through the head and body of the epididymis. In 20 days or so, they reach the tail. They are stored here and in the adjacent portion of the ductus deferens. Stored sperm remain fertile for 40 to 60 days, but if they become too old without being ejaculated, they disintegrate and the epididymis reabsorbs them.

- **Ductus (vas) deferens.** The duct of the epididymis straightens out at the tail, turns 180°, and becomes the ductus deferens. This is a muscular tube about 45 cm long and 2.5 mm in diameter. It passes upward through the spermatic cord and inguinal canal and enters the pelvic cavity. There, it turns medially and approaches the urinary bladder. After passing between the bladder and ureter, the duct turns downward behind the bladder and widens into a terminal **ampulla.** The ductus deferens ends by uniting with the duct of the seminal vesicle, a gland considered later. The duct has a very narrow lumen and a thick wall of smooth muscle well innervated by sympathetic nerve fibers.

- **Ejaculatory duct.** Where the ductus deferens and duct of the seminal vesicle meet, they form a short (2 cm) ejaculatory duct, which passes through the prostate gland and empties into the urethra. The ejaculatory duct is the last of the spermatic ducts.

The male urethra is shared by the reproductive and urinary systems. It is about 20 cm long and consists of three regions: the *prostatic, membranous,* and *penile urethra.* Although it serves both urinary and reproductive roles, it cannot pass urine and semen simultaneously for reasons explained later.

---

[19]*sustentacul* = support
[20]Enrico Sertoli (1842–1910), Italian histologist
[21]*rete* = network
[22]*epi* = upon + *didym* = twins, testes

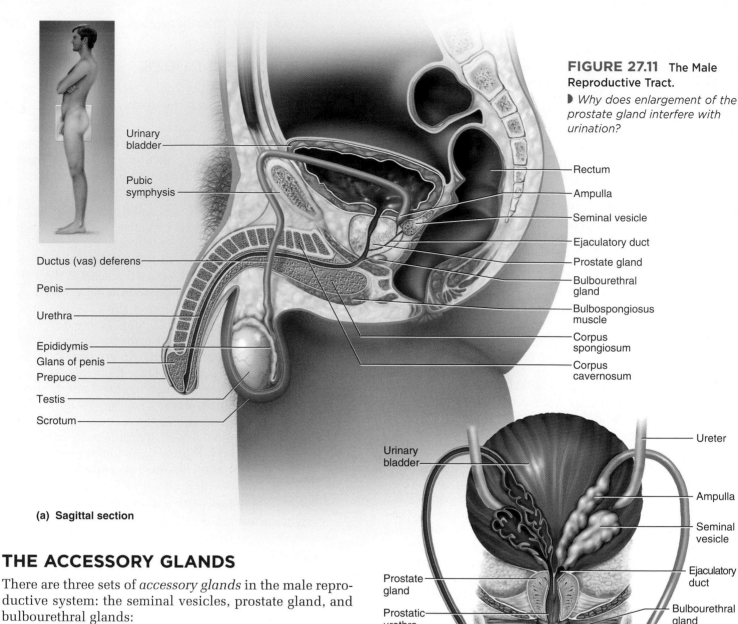

Urinary bladder

Pubic symphysis

Ductus (vas) deferens

Penis

Urethra

Epididymis

Glans of penis

Prepuce

Testis

Scrotum

Rectum

Ampulla

Seminal vesicle

Ejaculatory duct

Prostate gland

Bulbourethral gland

Bulbospongiosus muscle

Corpus spongiosum

Corpus cavernosum

**(a) Sagittal section**

# THE ACCESSORY GLANDS

There are three sets of *accessory glands* in the male reproductive system: the seminal vesicles, prostate gland, and bulbourethral glands:

1. The **seminal vesicles** are a pair of glands posterior to the urinary bladder; one is associated with each ductus deferens. A seminal vesicle is about 5 cm long, or approximately the dimensions of the little finger. It has a connective tissue capsule and underlying layer of smooth muscle. The secretory portion is a very convoluted duct with numerous branches that form a complex labyrinth. The duct empties into the ejaculatory duct. The yellowish secretion of the seminal vesicles constitutes about 60% of the semen; its composition and functions are discussed later.

2. The **prostate**[23] (PROSS-tate) **gland** surrounds the urethra and ejaculatory duct immediately inferior to the urinary bladder. It measures about 2 by 4 by 3 cm and is an aggregate of 30 to 50 compound tubuloacinar glands enclosed in a single fibrous capsule. These glands empty through about 20 pores in the urethral

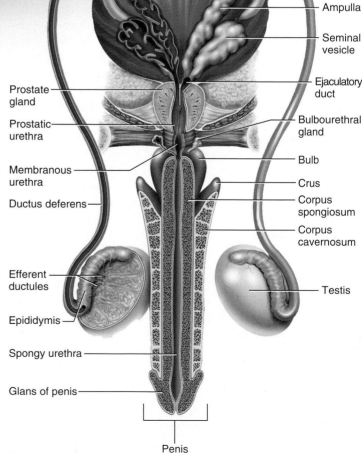

Urinary bladder

Ureter

Ampulla

Seminal vesicle

Ejaculatory duct

Prostate gland

Prostatic urethra

Bulbourethral gland

Membranous urethra

Bulb

Ductus deferens

Crus

Corpus spongiosum

Corpus cavernosum

Efferent ductules

Epididymis

Testis

Spongy urethra

Glans of penis

Penis

**(b) Posterior view**

---

[23]*pro* = before + *stat* = to stand; commonly misspelled and mispronounced "prostrate"

### Prostate Diseases

The prostate gland weighs about 20 g by age 20, remains at that weight until age 45 or so, and then begins to grow slowly again. By age 70, over 90% of men show some degree of *benign prostatic hyperplasia*—noncancerous enlargement of the gland. The major complication of this is that it compresses the urethra, obstructs the flow of urine, and may promote bladder and kidney infections.

Prostate cancer is the second most common cancer in men (after lung cancer), affecting about 9% of men over the age of 50. Prostate tumors tend to form near the periphery of the gland, where they do not obstruct urine flow; therefore, they often go unnoticed until they cause pain. Prostate cancer often metastasizes to nearby lymph nodes and then to the lungs and other organs. It is more common among American blacks than whites and very uncommon among Japanese. It is diagnosed by digital rectal examination and by detecting *prostate specific antigen (PSA)* and *acid phosphatase* (a prostatic enzyme) in the blood. Up to 80% of men with prostate cancer survive when it is detected and treated early, but only 10% to 50% survive if it spreads beyond the prostatic capsule.

wall. The stroma of the prostate consists of connective tissue and smooth muscle, like that of the seminal vesicles. The thin, milky secretion of the prostate constitutes about 30% of the semen. Its functions, too, are considered later. The position of the prostate immediately anterior to the rectum allows it to be palpated through the rectal wall to check for lumps suggestive of prostate cancer. This procedure is known as *digital rectal examination (DRE)* (Insight 27.2).

3. The **bulbourethral (Cowper[24]) glands** are named for their position near a dilated bulb at the inner end of the penis and their short (2.5 cm) ducts leading into the penile urethra. They are brownish, spherical glands about 1 cm in diameter. During sexual arousal, they produce a clear slippery fluid that lubricates the head of the penis in preparation for intercourse. Perhaps more importantly, though, it neutralizes the acidity of residual urine in the urethra, which would be harmful to the sperm.

## THE PENIS

The **penis**[25] serves to deposit semen in the vagina. Half of it is an internal **root** and half is the externally visible **shaft** and **glans**[26] (figs. 27.11 and 27.12). The external portion is

[24]William Cowper (1666–1709), British anatomist
[25]*penis* = tail
[26]*glans* = acorn

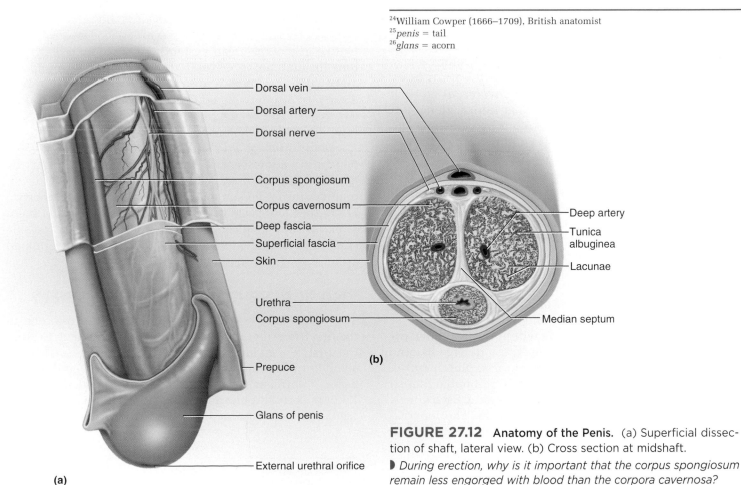

Dorsal vein
Dorsal artery
Dorsal nerve
Corpus spongiosum
Corpus cavernosum
Deep fascia
Superficial fascia
Skin
Urethra
Corpus spongiosum
Prepuce
Glans of penis
External urethral orifice

Deep artery
Tunica albuginea
Lacunae
Median septum

(a)

(b)

**FIGURE 27.12   Anatomy of the Penis.** (a) Superficial dissection of shaft, lateral view. (b) Cross section at midshaft.

▶ *During erection, why is it important that the corpus spongiosum remain less engorged with blood than the corpora cavernosa?*

about 8 to 10 cm (3–4 in.) long and 3 cm in diameter when flaccid (nonerect); the typical dimensions of an erect penis are 13 to 18 cm (5–7 in.) long and 4 cm in diameter. The glans is the expanded head at the distal end of the penis with the external urethral orifice at its tip.

Directional terminology may be a little confusing in the penis, because the *dorsal* side is the one that faces anteriorly, at least when the penis is flaccid, whereas the *ventral* side of the penis faces posteriorly. This is because in most mammals, the penis is horizontal, held against the abdomen by skin, and points anteriorly. The urethra passes through its lower, more obviously ventral, half. Directional terminology in the human penis follows the same convention as for other mammals, even though our bipedal posture and more pendulous penis change these anatomical relationships.

The skin is very loosely attached to the penile shaft, allowing for expansion during erection. It continues over the glans as the **prepuce,** or foreskin, which is often removed by circumcision. A ventral fold of tissue called the *frenulum* attaches the skin to the glans. The skin of the glans itself is thinner and firmly attached to the underlying erectile tissue. The glans and facing surface of the prepuce have sebaceous glands that produce a waxy secretion called **smegma.**[27]

The penis consists mainly of three cylindrical bodies called **erectile tissues,** which fill with blood during sexual arousal and account for its enlargement and erection. A single erectile body, the **corpus spongiosum,** passes along the ventral side of the penis and encloses the penile urethra. It expands at the distal end to fill the entire glans. Proximal to the glans, the dorsal side of the penis has a **corpus cavernosum** (plural, *corpora cavernosa*) on each side. Each is ensheathed in a fibrous **tunica albuginea,** and they are separated from each other by a **median septum.** (Note that the testes also have a tunica albuginea and the scrotum also has a median septum.)

All three cylinders of erectile tissue are spongy in appearance and contain numerous tiny blood sinuses called **lacunae.** The partitions between lacunae, called **trabeculae,** are composed of connective tissue and smooth **trabecular muscle.** In the flaccid penis, trabecular muscle tone collapses the lacunae, which appear as tiny slits in the tissue.

At the body surface, the penis turns 90° dorsally and continues inward as the root. The corpus spongiosum terminates internally as a dilated **bulb,** which is ensheathed in the bulbospongiosus muscle and attached to the lower surface of the perineal membrane within the urogenital triangle (see p. 350). The corpora cavernosa diverge like the arms of a Y. Each arm, called a **crus** (cruss; plural, *crura*), attaches the penis to the pubic arch (ischiopubic ramus) and perineal membrane on its respective side. Each crus is enveloped by an ischiocavernosus muscle. The innervation and blood supply to the penis are discussed later in connection with the mechanism of erection.

## Before You Go On

*Answer the following questions to test your understanding of the preceding section:*

6. State the names and locations of two muscles that help regulate the temperature of the testes.

7. Name three types of cells in the testis, and describe their locations and functions.

8. Name all the ducts that the sperm follow, in order, from the time they form in the testis to the time of ejaculation.

9. Describe the locations and functions of the seminal vesicles, prostate, and bulbourethral glands.

10. Name the erectile tissues of the penis, and describe their locations relative to each other.

# Puberty and Climacteric

**Objectives**
When you have completed this section, you should be able to

- describe the hormonal control of puberty;
- describe the resulting changes in the male body; and
- define and describe male climacteric and the effect of aging on male reproductive function.

Unlike any other organ system, the reproductive system remains dormant for several years after birth. Around age 10 to 12 in most boys and 8 to 10 in most girls, however, a surge of pituitary gonadotropins awakens the reproductive system and begins preparing it for adult reproductive function. This is the onset of puberty.

Definitions of *adolescence* and *puberty* vary. This book uses **adolescence**[28] to mean the period from the onset of gonadotropin secretion and reproductive development until a person attains full adult height. **Puberty**[29] is the first few years of adolescence, until the first menstrual period in girls or the first ejaculation of viable sperm in boys. In North America, this is typically attained around age 12 in girls and age 13 in boys.

## ENDOCRINE CONTROL OF PUBERTY

The testes secrete substantial amounts of testosterone in the first trimester (3 months) of fetal development. Even in the first few months of infancy, testosterone levels are about as high as they are in midpuberty, but then the testes become dormant for the rest of infancy and childhood. From puberty through adulthood, reproductive function is regulated by hormonal links between the hypothalamus,

---

[27]*smegma* = unguent, ointment, soap

[28]*adolesc* = to grow up
[29]*puber* = grown up

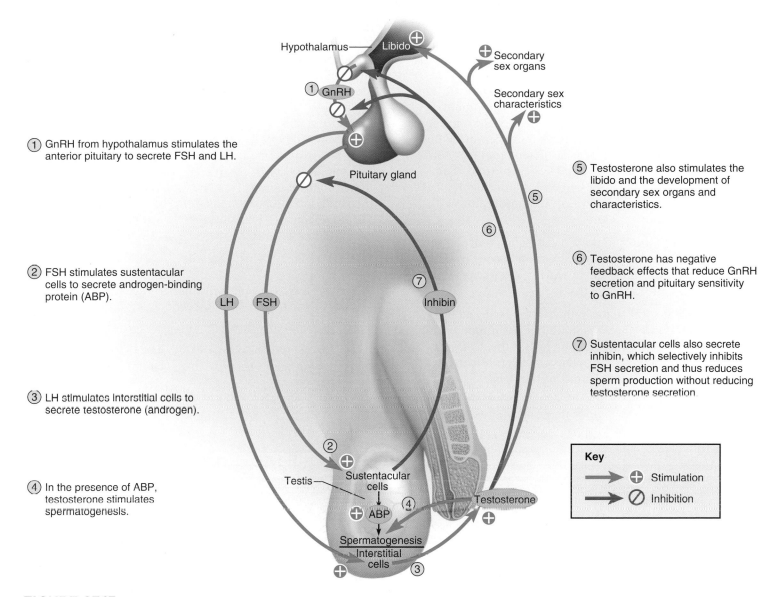

① GnRH from hypothalamus stimulates the anterior pituitary to secrete FSH and LH.

② FSH stimulates sustentacular cells to secrete androgen-binding protein (ABP).

③ LH stimulates interstitial cells to secrete testosterone (androgen).

④ In the presence of ABP, testosterone stimulates spermatogenesis.

⑤ Testosterone also stimulates the libido and the development of secondary sex organs and characteristics.

⑥ Testosterone has negative feedback effects that reduce GnRH secretion and pituitary sensitivity to GnRH.

⑦ Sustentacular cells also secrete inhibin, which selectively inhibits FSH secretion and thus reduces sperm production without reducing testosterone secretion.

**Key**

⊕ Stimulation

⊘ Inhibition

**FIGURE 27.13** Hormonal Relationships in the Brain–Testicular Axis.

pituitary gland, and gonads—the **brain–testicular axis** (fig. 27.13).

As the hypothalamus matures, it begins producing **gonadotropin-releasing hormone (GnRH),** which travels by way of the hypophyseal portal system to the anterior lobe of the pituitary. Here it stimulates cells called *gonadotropes* to secrete **follicle-stimulating hormone (FSH)** and **luteinizing hormone (LH),** two *gonadotropins* that stimulate different cells in the testis. Enlargement of the testes is the first sign of puberty.

LH stimulates the interstitial cells of the testis to secrete androgens, mainly testosterone. In the male, LH is sometimes called *interstitial cell–stimulating hormone (ICSH).* FSH stimulates the sustentacular cells to secrete **androgen-binding protein (ABP).** ABP is thought to raise testosterone levels in the seminiferous tubules and epididymis, but this remains unproven. Without FSH, however, testosterone has

no effect on the testis. Germ cells have no androgen receptors and do not respond to it. Nevertheless, testosterone normally has the following effects:

- It stimulates spermatogenesis in the presence of ABP. If testosterone secretion ceases, the sperm count and semen volume decline rapidly and a male becomes sterile.

- It inhibits GnRH secretion by the hypothalamus and reduces GnRH sensitivity of the pituitary. Consequently, FSH and LH secretion are held in check. Over the course of puberty, however, the pituitary becomes less sensitive to this negative feedback and secretes increasing amounts of gonadotropins.

- It stimulates development of the secondary sex characteristics and other somatic changes characteristic of puberty. Many of the physiological processes of puberty go unnoticed at first. The first visible sign is usually enlargement of the testes and scrotum around age 13. The penis continues to grow for about two more years. One of the androgens, dihydrotestosterone (DHT), stimulates development of the pubic, axillary, and facial hair. The skin becomes darker and thicker and secretes more sebum, which often leads to acne. The skin of acne patients contains 2 to 20 times as much DHT as normal. The apocrine sweat (scent) glands of the perineal, axillary, and beard areas develop in conjunction with the hair in those regions.

- It causes the ducts and accessory glands of the reproductive system to enlarge. Erections occur frequently and ejaculation may occur during sleep (nocturnal emissions, or "wet dreams").

- Testosterone stimulates a burst of growth, an increase in muscle mass, a higher basal metabolic rate, and a larger larynx. The last of these effects deepens the voice and makes the thyroid cartilage more prominent at the front of the neck.

- It stimulates erythropoiesis, which gives men a higher hematocrit and RBC count than those of women.

- Testosterone also stimulates the brain and awakens the **libido** (sex drive)—although, perhaps surprisingly, the neurons convert it to estrogen (a supposedly "female" sex hormone), which is what directly affects the behavior.

Throughout adulthood, testosterone sustains the male reproductive tract, spermatogenesis, and libido. At times, it is necessary to reduce FSH secretion and sperm production without reducing LH and testosterone secretion. To achieve this, the sustentacular cells secrete a hormone called **inhibin** that selectively suppresses FSH output from the pituitary. When the sperm count drops below 20 million sperm/mL, however, inhibin secretion drops and FSH secretion rises.

**Think About It**

*If a male animal is castrated, would you expect FSH and LH levels to rise, fall, or be unaffected? Why?*

## AGING AND SEXUAL FUNCTION

Testosterone secretion peaks at about 7 mg/day at age 20 and then declines steadily to as little as one-fifth of this level by age 80. There is a corresponding decline in the number and secretory activity of the interstitial cells (the source of testosterone) and sustentacular cells (the source of inhibin). As testosterone and inhibin levels decline, so does feedback inhibition of the pituitary. Consequently, FSH and LH levels rise significantly after age 50 and pro-

duce changes called **male climacteric.** Most men pass through climacteric with little or no effect, but in a few cases there are mood changes, hot flashes, and illusions of suffocation—symptoms similar to those of menopause in women. Despite references to "male menopause," the term *menopause* refers to the cessation of menstruation and therefore makes no sense in the context of male physiology.

About 20% of men in their 60s and 50% of men in their 80s experience *erectile dysfunction* (impotence), the frequent inability to produce or maintain an erection sufficient for intercourse (see table 27.1). Erectile dysfunction (ED) and declining sexual activity can have a major impact on older people's perception of the quality of life. Over 90% of men with ED, however, remain able to ejaculate.

### Before You Go On

*Answer the following questions to test your understanding of the preceding section:*

11. State the source, target organ, and effect of GnRH.

12. Identify the target cells and effects of FSH and LH.

13. Explain how testicular hormones affect the secretion of FSH and LH.

14. Describe the major effects of androgens on the body.

15. Define male climacteric.

## Sperm and Semen

### Objectives

When you have completed this section, you should be able to

- describe the stages of meiosis and contrast meiosis with mitosis;

- describe the sequence of cell types in spermatogenesis, and relate these to the stages of meiosis;

- describe the role of the sustentacular cell in spermatogenesis;

- draw or describe a sperm cell; and

- describe the composition of semen and functions of its components.

The most significant event of puberty is the onset of **spermatogenesis.** This is a process in which germ cells undergo two divisions called **meiosis** and the four daughter cells differentiate into spermatozoa (sperm). It occurs in the seminiferous tubules.

### MEIOSIS

In nearly all living organisms except bacteria, there are two forms of cell division: mitosis and meiosis. Mitosis, described in chapter 4, is the basis for division of the single-celled fertilized egg, growth of an embryo, and all postnatal growth and tissue repair. It is essentially the splitting of a cell with a distribution of chromosomes that results in two genetically identical daughter cells. It consists of four stages: prophase, metaphase, anaphase, and telophase.

You may find it beneficial to review figure 4.14 (p. 140) because of the important similarities and differences between mitosis and meiosis. There are three important differences:

1. In mitosis, each double-stranded chromosome divides into two single-stranded ones, but each daughter cell still has 46 chromosomes (23 pairs). Meiosis, by contrast, reduces the chromosome number by half. The parent cell is **diploid (2n)**, meaning it has 46 chromosomes in 23 homologous pairs (see fig. 4.16, p. 143), whereas the daughter cells are **haploid (n)**, with 23 unpaired chromosomes.

2. In mitosis, the chromosomes do not change their genetic makeup. In an early stage of meiosis, however, the chromosomes of each homologous pair join and exchange portions of their DNA. This creates new combinations of genes, so the chromosomes we pass to our offspring are not the same ones that we inherited from our parents.

3. In mitosis, each parent cell produces only two daughter cells. In meiosis, it produces four. In the male, four sperm therefore develop from each original germ cell. The situation is somewhat different in the female (see chapter 28).

Why use such a relatively complicated process for gametogenesis? Why not use mitosis, as we do for all other cell replication in the body? The answer is that sexual reproduction is, by definition, biparental. If we are going to combine gametes from two parents to make a child, there must be a mechanism for keeping the chromosome number constant from generation to generation. Mitosis would produce eggs and sperm with 46 chromosomes each. If these gametes combined, the zygote and the next generation would have 92 chromosomes per cell, the generation after that would have 184, and so forth. To prevent the chromosome number from doubling in every generation, the number is reduced by half during gametogenesis. Meiosis[30] is sometimes called *reduction division* for this reason.

The stages of meiosis are fundamentally the same in both sexes. Briefly, it consists of two cell divisions in succession and occurs in the following phases: prophase I, metaphase I, anaphase I, telophase I, interkinesis, prophase II, metaphase II, anaphase II, and telophase II. These events are detailed in figure 27.14, but let us note some of its unique and important aspects.

In prophase I, each pair of homologous chromosomes line up side by side and form a **tetrad** (*tetra* denoting the

four chromatids). One chromosome of each tetrad is from the individual's father (the paternal chromosome) and the other is from the mother (the maternal chromosome). The paternal and maternal chromosomes exchange segments of DNA in a process called **crossing-over.** This creates new combinations of genes and thus contributes to genetic variety in the offspring.

After crossing-over, the chromosomes line up at the midline of the cell in metaphase I, they separate at anaphase I, and the cell divides in two at telophase I. This looks superficially like mitosis, but there is an important difference: The centromeres do not divide and the chromatids do not separate from each other at anaphase I; rather, each homologous chromosome parts company with its twin. Therefore, at the conclusion of meiosis I, each chromosome is still double-stranded, but each daughter cell has only 23 chromosomes—it has become haploid.

Meiosis II is more like mitosis—the chromosomes line up on the cell equator again at metaphase II, the centromeres divide, and each chromosome separates into two chromatids. These chromatids are drawn to opposite poles of the cell at anaphase II. At the end of meiosis II, there are four haploid cells, each containing 23 single-stranded chromosomes. Fertilization combines 23 chromosomes from the father with 23 chromosomes from the mother and reestablishes the diploid number of 46 in the zygote.

## SPERMATOGENESIS

Now we will relate meiosis to sperm production (fig. 27.15). The first stem cells specifically destined to become sperm are **primordial germ cells.** Like the first blood cells, these form in the yolk sac, a membrane associated with the developing embryo. In the fifth to sixth week of development, they migrate into the embryo itself and colonize the gonadal ridges. Here they differentiate into **spermatogonia,** which lie along the periphery of the seminiferous tubule, outside the blood–testis barrier (BTB).

Spermatogonia multiply by mitosis, producing two types of daughter cells called type A and type B spermatogonia. Type A cells remain outside the BTB and continue to multiply from puberty until death. Thus men never exhaust their supply of gametes and normally remain fertile throughout old age.

Type B spermatogonia migrate closer to the tubule lumen and differentiate into slightly larger cells called **primary spermatocytes.** These cells must pass through the BTB and move toward the lumen of the tubule. Ahead of the primary spermatocyte, the tight junction between two sustentacular cells is dismantled, while a new tight junction forms on the other side, like closing the door behind the spermatocyte. The spermatocyte moves forward toward the lumen and is now separated from blood-borne agents, such as antibodies, held back by the tight junction.

Now safely isolated from the blood, the primary spermatocyte undergoes meiosis I, which gives rise to two

---

[30]*meio* = less, fewer

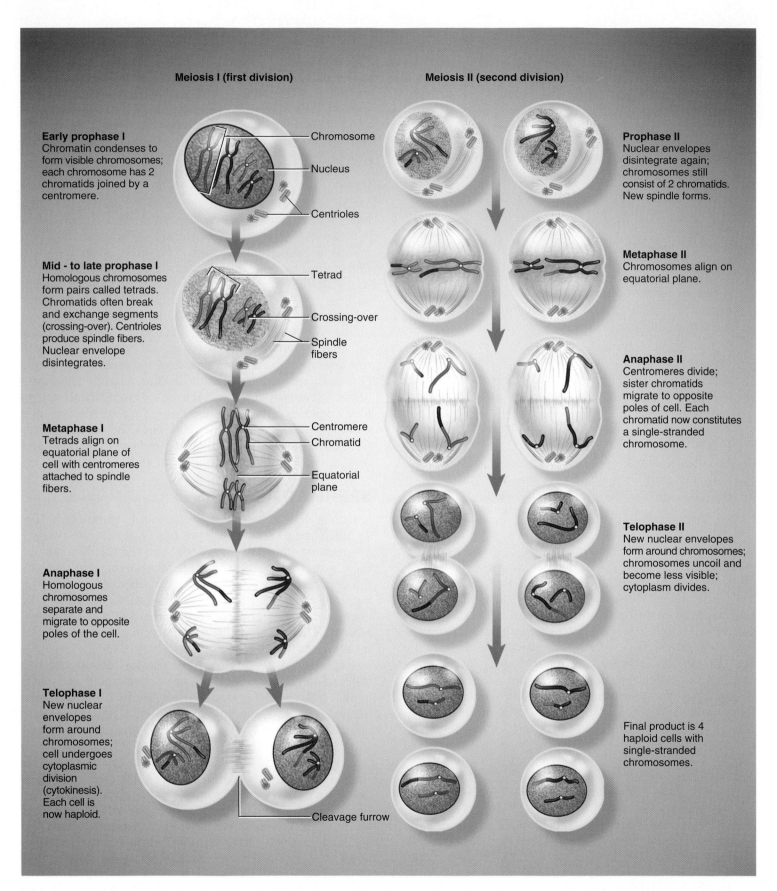

**Meiosis I (first division)**

**Meiosis II (second division)**

**Early prophase I**
Chromatin condenses to form visible chromosomes; each chromosome has 2 chromatids joined by a centromere.

- Chromosome
- Nucleus
- Centrioles

**Prophase II**
Nuclear envelopes disintegrate again; chromosomes still consist of 2 chromatids. New spindle forms.

**Mid - to late prophase I**
Homologous chromosomes form pairs called tetrads. Chromatids often break and exchange segments (crossing-over). Centrioles produce spindle fibers. Nuclear envelope disintegrates.

- Tetrad
- Crossing-over
- Spindle fibers

**Metaphase II**
Chromosomes align on equatorial plane.

**Metaphase I**
Tetrads align on equatorial plane of cell with centromeres attached to spindle fibers.

- Centromere
- Chromatid
- Equatorial plane

**Anaphase II**
Centromeres divide; sister chromatids migrate to opposite poles of cell. Each chromatid now constitutes a single-stranded chromosome.

**Anaphase I**
Homologous chromosomes separate and migrate to opposite poles of the cell.

**Telophase II**
New nuclear envelopes form around chromosomes; chromosomes uncoil and become less visible; cytoplasm divides.

**Telophase I**
New nuclear envelopes form around chromosomes; cell undergoes cytoplasmic division (cytokinesis). Each cell is now haploid.

- Cleavage furrow

Final product is 4 haploid cells with single-stranded chromosomes.

**FIGURE 27.14** **Meiosis.** For simplicity, the cell is shown with only two pairs of homologous chromosomes. Human cells begin meiosis with 23 pairs.

▶ *Although we pass the same genes to our offspring as we inherit from our parents, we do not pass on the same chromosomes. What process in this figure accounts for the latter fact?*

Cross section of
seminiferous tubules

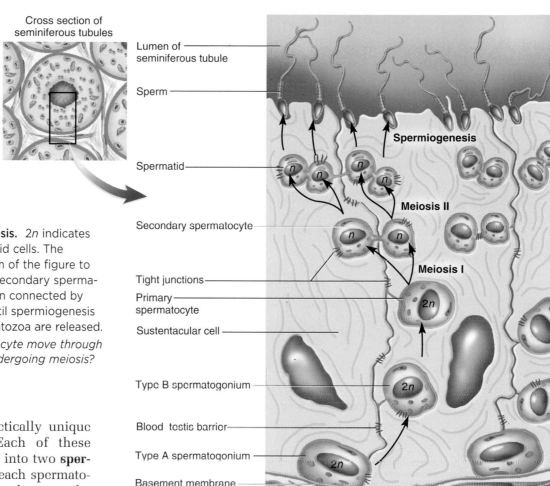

Lumen of
seminiferous tubule

Sperm

Spermiogenesis

Spermatid

n n n n

Meiosis II

Secondary spermatocyte

n n

Meiosis I

Tight junctions

Primary
spermatocyte

2n

Sustentacular cell

Type B spermatogonium

2n

Blood testis barrier

Type A spermatogonium

2n

Basement membrane
of seminiferous tubule

**FIGURE 27.15** **Spermatogenesis.** *2n* indicates diploid cells and *n* indicates haploid cells. The process proceeds from the bottom of the figure to the top. The daughter cells from secondary spermatocytes through spermatids remain connected by slender cytoplasmic processes until spermiogenesis is complete and individual spermatozoa are released.

▶ *Why must the primary spermatocyte move through the blood–testis barrier before undergoing meiosis?*

equal-sized, haploid and genetically unique **secondary spermatocytes.** Each of these undergoes meiosis II, dividing into two **spermatids**—or a total of four for each spermatogonium. Each stage is a little closer to the lumen of the tubule than the earlier stages. All stages on the lumenal side of the BTB are bound to the sustentacular cells by tight junctions and gap junctions, and are closely enveloped in tendrils of the sustentacular cells. Throughout these meiotic divisions, the daughter cells do not completely separate, but remain connected to each other by narrow cytoplasmic bridges.

The rest of spermatogenesis is called **spermiogenesis** (fig. 27.16). It involves no further cell division, but a gradual transformation of each spermatid into a spermatozoon. A spermatid sprouts a flagellum (tail) and discards most of its cytoplasm, becoming as small and lightweight as possible. Eventually, the sperm cell is released and is washed down the tubule by fluid from the sustentacular cells. It takes about 74 days for a spermatogonium to become a mature spermatozoon. A young man produces about 300,000 sperm per minute, or 400 million per day.

## THE SPERMATOZOON

The spermatozoon has two parts: a pear-shaped head and a long tail (fig. 27.17). The **head,** about 4 to 5 μm long and 3 μm wide at its broadest part, contains three structures: a nucleus, acrosome, and flagellar basal body. The most important of these is the nucleus, which fills most of the head and contains a haploid set of condensed, genetically

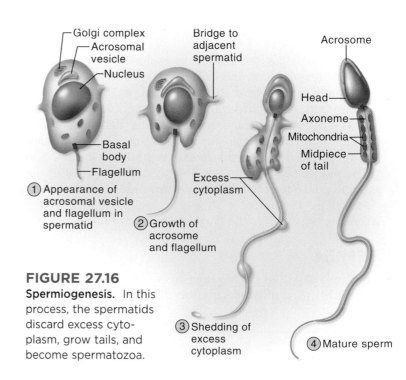

Golgi complex
Acrosomal
vesicle
Nucleus

Bridge to
adjacent
spermatid

Acrosome

Head

Axoneme

Mitochondria

Basal
body

Flagellum

Excess
cytoplasm

Midpiece
of tail

① Appearance of
acrosomal vesicle
and flagellum in
spermatid

② Growth of
acrosome
and flagellum

③ Shedding of
excess
cytoplasm

④ Mature sperm

**FIGURE 27.16**
**Spermiogenesis.** In this process, the spermatids discard excess cytoplasm, grow tails, and become spermatozoa.

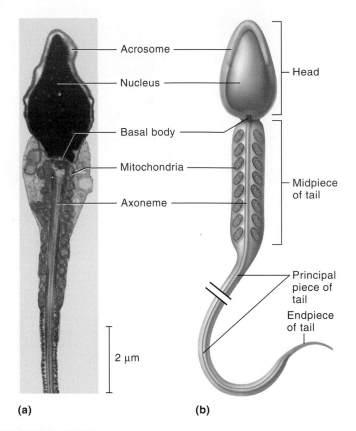

Acrosome
Nucleus
Basal body
Mitochondria
Axoneme

2 μm

**(a)**

Head
Midpiece of tail
Principal piece of tail
Endpiece of tail

**(b)**

**FIGURE 27.17    The Mature Spermatozoon.** (a) Head and part of the tail of a spermatozoon (TEM). (b) Sperm structure.

inactive chromosomes. The **acrosome**[31] is a lysosome in the form of a thin cap covering the apical half of the nucleus. It contains enzymes that are later used to penetrate the egg if the sperm is successful. The basal body of the tail flagellum is nestled in an indentation at the basal end of the nucleus.

The **tail** is divided into three regions called the midpiece, principal piece, and endpiece. The **midpiece,** a cylinder about 5 to 9 μm long and half as wide as the head, is the thickest part. It contains numerous large mitochondria that coil tightly around the axoneme of the flagellum. They produce the ATP needed for the beating of the tail when the sperm migrates up the female reproductive tract. The **principal piece,** 40 to 45 μm long, constitutes most of the tail and consists of the axoneme surrounded by a sheath of fibers. The **endpiece,** 4 to 5 μm long, consists of the axoneme only and is the narrowest part of the sperm.

## SEMEN

The fluid expelled during orgasm is called **semen**[32] (seminal fluid). A typical ejaculation discharges 2 to 5 mL of

semen, composed of about 10% sperm and spermatic duct secretions, 30% prostatic fluid, 60% seminal vesicle fluid, and a trace of bulbourethral fluid. Most of the sperm emerge in the first one or two jets of semen. The semen usually has a **sperm count** of 50 to 120 million sperm/mL. A sperm count any lower than 20 to 25 million sperm/mL is usually associated with **infertility** (sterility), the inability to fertilize an egg (see table 27.1). The prostate and seminal vesicles contribute the following constituents to the semen:

1. The prostate produces a thin, milky white fluid containing calcium, citrate, and phosphate ions, a clotting enzyme, and a protein-hydrolyzing enzyme called *serine protease* (also known as prostate-specific antigen, PSA; see Insight 27.2).

2. The seminal vesicles contribute a viscous yellowish fluid. This is the last component of the semen to emerge, and it flushes any remaining sperm from the urethra. This fluid contains fructose and other carbohydrates, citrate, prostaglandins (discovered in and named for the bovine prostate, but more abundant in the seminal vesicle fluid), and a protein called *proseminogelin.*

---

[31]*acro* = tip, peak + *some* = body
[32]*semen* = seed

A well-known property of semen is its stickiness, an adaptation that promotes fertilization. It arises when the clotting enzyme from the prostate activates proseminogelin, converting it to a sticky fibrinlike protein, **seminogelin.** Seminogelin entangles the sperm, sticks to the walls of the inner vagina and cervix, and ensures that the semen does not simply drain back out of the vagina. It may also promote the uptake of sperm-laden clots of semen into the uterus. Twenty to 30 minutes after ejaculation, the **serine protease** of the prostatic fluid breaks down seminogelin and liquifies the semen. The sperm, quiescent until then, now become very active, thrashing with their tails and crawling up the mucosa of the vagina and uterus. The prostaglandins of the semen may thin the mucus in the *cervical canal* of the female (see chapter 28), making it easier for sperm to migrate from the vagina into the uterus, and they may stimulate peristaltic waves in the uterus and uterine tubes that help to spread semen through the female reproductive tract.

Two requirements must be met for sperm motility: an elevated pH and an energy source. The pH of the vagina is about 3.5 to 4.0, and the male spermatic ducts are also quite acidic. The sperm remain quite still at such low pHs. But the prostatic fluid buffers the vaginal and seminal acidity, raising the pH to about 7.5 and activating the sperm. The sperm must synthesize a lot of ATP to power their movements. They get the energy for this from the fructose and other sugars contributed by the seminal vesicles.

### Before You Go On

*Answer the following questions to test your understanding of the preceding section:*

16. *State how many chromosomes a cell has, and how many chromatids each chromosome has, at the conclusion of meiosis I and meiosis II.*

17. *Name the stages of spermatogenesis from spermatogonium to spermatozoon.*

18. *Describe the three major parts of a spermatozoon and state what organelles or cytoskeletal components are contained in each.*

19. *List the major contributions of the seminal vesicles and prostate gland to the semen, and state the functions of these components.*

## Male Sexual Response

### Objectives

When you have completed this section, you should be able to

- describe the blood and nerve supply to the penis; and
- explain how these govern erection and ejaculation.

The physiology of sexual intercourse was unexplored territory before the 1950s because of repressive attitudes toward the subject. British psychologist Havelock Ellis (1859–1939) suffered severe professional sanctions merely for surveying people on their sexual behavior. In the 1950s, William Masters and Virginia Johnson daringly launched the first physiological studies of sexual response in the laboratory. In 1966, they published *Human Sexual Response,* which detailed measurements and observations on more than 10,000 sexual acts by nearly 700 volunteer men and women. Masters and Johnson then turned their attention to disorders of sexual function and pioneered modern therapy for sexual dysfunctions. They divided intercourse into four recognizable phases, which they called excitement, plateau, orgasm, and resolution. The following discussion is organized around this model, although other authorities have modified it or proposed alternatives. Sexual intercourse is also known as **coitus, coition,**[33] or **copulation.**[34]

## ANATOMICAL FOUNDATIONS

To understand male sexual function, we must give closer attention to the blood circulation and nerve supply to the penis.

Each internal iliac artery gives rise to an **internal pudendal (penile) artery,** which enters the root of the penis and divides in two. One branch, the **dorsal artery,** travels dorsally along the penis not far beneath the skin (see fig. 27.12). The other branch, the **deep artery,** travels through the core of the corpus cavernosum and gives off smaller **helicine**[35] **arteries,** which penetrate the trabeculae and empty into the lacunae. When the deep artery dilates, the lacunae fill with blood and the penis becomes erect. When the penis is flaccid, most of its blood supply comes from the dorsal arteries.

The penis is richly innervated by sensory and motor nerve fibers. The glans has an abundance of tactile, pressure, and temperature receptors, especially on its proximal margin and frenulum. Sensory fibers of the shaft, scrotum, perineum, and elsewhere are also highly important to erotic stimulation. They lead by way of a pair of prominent **dorsal nerves** of the penis to the **internal pudendal nerves,** then to the sacral plexus, and finally to the sacral region of the spinal cord.

Both autonomic and somatic motor fibers carry impulses from integrating centers in the spinal cord to the penis and other pelvic organs. Sympathetic fibers arise from levels T12 to L2, pass through the hypogastric and pelvic plexuses, and innervate the penile arteries, trabecular muscle, spermatic ducts, and accessory glands. They

---

[33]*coit* = to come together
[34]*copul* = link, bond
[35]*helic* = coil, helix

dilate the penile arteries and can induce erection even when the sacral region of the spinal cord is damaged. They also initiate erection in response to input to the special senses and to sexual thoughts.

Parasympathetic fibers extend from segments S2 to S4 of the spinal cord through the pudendal nerves to the arteries of the penis. They are involved in an autonomic reflex arc that causes erection in response to direct stimulation of the penis and other perineal organs.

## EXCITEMENT AND PLATEAU

The **excitement phase** is characterized by **vasocongestion** (swelling of the genitals with blood), **myotonia** (muscle tension), and increases in heart rate, blood pressure, and pulmonary ventilation (fig. 27.18). The bulbourethral glands secrete their fluid during this phase. The excitement phase can be initiated by a broad spectrum of erotic stimuli—sights, sounds, aromas, touch—and even by dreams or thoughts. Conversely, emotions can inhibit sexual response and make it difficult to function when a person is anxious, stressed, or preoccupied with other thoughts.

The most obvious manifestation of male sexual excitement is **erection** of the penis, which makes entry of the vagina possible. Erection is an autonomic reflex mediated predominantly by parasympathetic nerve fibers that travel alongside the deep and helicine arteries of the penis. These fibers trigger the secretion of nitric oxide (NO), which leads to the relaxation of the deep arteries and lacunae (Insight 27.4). Whether this is enough to cause erection, or whether it is also necessary to block the outflow of blood from the penis, is still debated. According to one hypothesis, as lacunae near the deep arteries fill with blood, they compress lacunae closer to the periphery of the erectile tissue. This is where blood leaves the erectile tissues, so the compression of the peripheral lacunae helps retain blood in the penis. Their compression is aided by the fact that each corpus cavernosum is wrapped in a tunica albuginea, which fits over the erectile tissue like a tight fibrous sleeve and contributes to its tension and firmness. In addition, the bulbospongiosus and ischiocavernosus muscles aid in erection by compressing the root of the penis and forcing blood forward into the shaft.

As the corpora cavernosa expand, the penis becomes enlarged, rigid, and elevated to an angle conducive to entry of the vagina. Once **intromission** (entry) is achieved, the tactile and pressure sensations produced by vaginal massaging of the penis further accentuate the erection reflex.

The corpus spongiosum has neither a central artery nor a tunica albuginea. It swells and becomes more visible as a cordlike ridge along the ventral surface of the penis, but it does not become nearly as engorged and hardened as the corpora cavernosa. Vasocongestion is not limited to the penis; the testes also become as much as 50% larger during excitement.

In the **plateau phase,** variables such as respiratory rate, heart rate, and blood pressure are sustained at a high level, or rise slightly, for a few seconds to a few minutes before orgasm. This phase may be marked by increased vasocongestion and myotonia.

> **Think About It**
>
> *Why is it important that the corpus spongiosum not become as engorged and rigid as the corpora cavernosa?*

## ORGASM AND EJACULATION

The **orgasm,**[36] or **climax,** is a short but intense reaction that lasts 3 to 15 seconds and usually is marked by the discharge of semen. The heart rate increases to as high as 180 beats/min, blood pressure rises proportionately, and the respiratory rate becomes as high as 40 breaths/min. From the standpoint of producing offspring, the most significant aspect of male orgasm is the **ejaculation**[37] of semen into the vagina.

Ejaculation occurs in two stages called emission and expulsion. In **emission,** the sympathetic nervous system stimulates peristalsis in the smooth muscle of the ductus deferens, which propels sperm from the tail of the epididymis, along the ductus, and into the ampulla. Contractions of the ampulla propel the sperm into the prostatic urethra, and contractions of smooth muscle in the prostate gland force prostatic fluid into the urethra. Secretions of the seminal vesicles join the semen soon after the prostatic secretion. The contractions and seminal flow of this phase create an urgent sensation that ejaculation is inevitable.

Semen in the urethra activates somatic and sympathetic reflexes that result in its expulsion. Sensory signals travel to the spinal cord via the internal pudendal nerve and reach an integrating center in the upper lumbar region. Sympathetic nerve fibers carry motor signals from here out to the prostate gland and seminal vesicles, causing the smooth muscle in their walls to express more fluid into the urethra. The sympathetic reflex also constricts the internal urethral sphincter so urine cannot enter the urethra and semen cannot enter the bladder.

Somatic motor signals leave the third and fourth sacral segments of the cord and travel to the bulbospongiosus, ischiocavernosus, and levator ani muscles. The bulbospongiosus, which envelops the root of the corpus spongiosum, undergoes five or six strong, spasmodic contractions that compress the urethra and forcibly expel the semen. Most sperm are ejected in the first milliliter of semen, mixed primarily with prostatic fluid. The seminal vesicle secretion

---

[36]*orgasm* = swelling
[37]*e* = ex = out + *jacul* = to throw

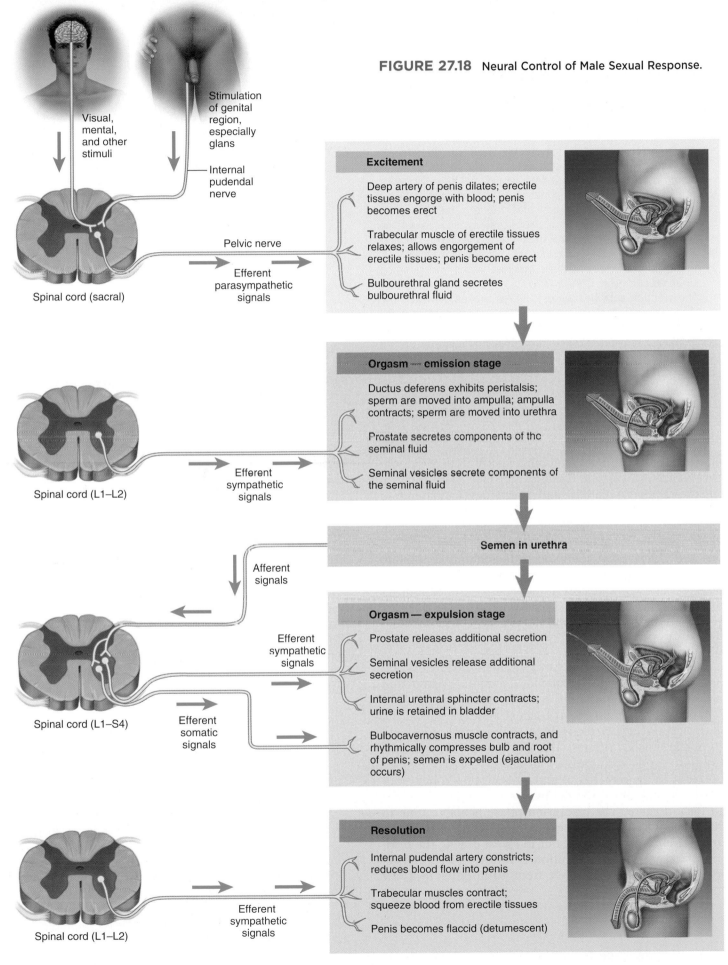

**FIGURE 27.18** Neural Control of Male Sexual Response.

Visual, mental, and other stimuli

Stimulation of genital region, especially glans

Internal pudendal nerve

Spinal cord (sacral)

Pelvic nerve

Efferent parasympathetic signals

**Excitement**

Deep artery of penis dilates; erectile tissues engorge with blood; penis becomes erect

Trabecular muscle of erectile tissues relaxes; allows engorgement of erectile tissues; penis become erect

Bulbourethral gland secretes bulbourethral fluid

Spinal cord (L1–L2)

Efferent sympathetic signals

**Orgasm — emission stage**

Ductus deferens exhibits peristalsis; sperm are moved into ampulla; ampulla contracts; sperm are moved into urethra

Prostate secretes components of the seminal fluid

Seminal vesicles secrete components of the seminal fluid

**Semen in urethra**

Afferent signals

Spinal cord (L1–S4)

Efferent sympathetic signals

Efferent somatic signals

**Orgasm — expulsion stage**

Prostate releases additional secretion

Seminal vesicles release additional secretion

Internal urethral sphincter contracts; urine is retained in bladder

Bulbocavernosus muscle contracts, and rhythmically compresses bulb and root of penis; semen is expelled (ejaculation occurs)

Spinal cord (L1–L2)

Efferent sympathetic signals

**Resolution**

Internal pudendal artery constricts; reduces blood flow into penis

Trabecular muscles contract; squeeze blood from erectile tissues

Penis becomes flaccid (detumescent)

## INSIGHT 27.4    Clinical Application

### Treating Erectile Dysfunction

Among the most lucrative drugs developed in the last decade are the popular treatments for erectile dysfunction: sildenafil (Viagra), vardenafil (Levitra), and tadalafil (Cialis). The basis for developing these drugs was a seemingly unrelated discovery: the role of nitric oxide in cell signaling. When sexual stimulation triggers NO secretion, NO activates the enzyme guanylate cyclase. Guanylate cyclase produces cyclic guanosine monophosphate (cGMP). cGMP then relaxes the smooth muscle of the deep arteries and lacunae of the corpora cavernosa, increasing blood flow into these erectile tissues and bringing about an erection (fig. 27.19).

The erection subsides when cGMP is broken down by the enzyme *phosphodiesterase type 5 (PDE5)*. The problem for many men and their partners is that it subsides too soon, and the solution is to prevent cGMP from breaking down so fast. The aforementioned drugs are in a family called *phosphodiesterase inhibitors*. By slowing down the action of PDE5, they prolong the life of cGMP and the duration of the erection.

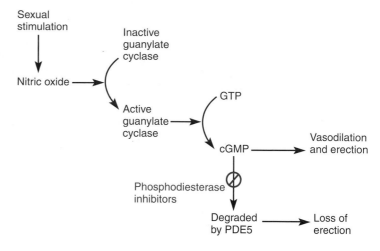

**FIGURE 27.19**    The Action of Phosphodiesterase Inhibitors.

---

follows and flushes most remaining sperm from the ejaculatory ducts and urethra. Some sperm may seep from the penis prior to ejaculation, and pregnancy can therefore result from genital contact even without orgasm.

Orgasm is accompanied by an intense feeling of release from tension. Ejaculation and orgasm are not the same. Although they usually occur together, it is possible to have all of the sensations of orgasm without ejaculating, and ejaculation occasionally occurs with little or no sensation of orgasm.

## RESOLUTION

Immediately following orgasm comes the **resolution** phase. Discharge of the sympathetic nervous system constricts the internal pudendal artery and reduces the flow of blood into the penis. It also causes contraction of the trabecular muscles, which squeeze blood from the lacunae of the erectile tissues. The penis may remain semi-erect long enough to continue intercourse, which may be important to the female's attainment of climax, but gradually the penis undergoes **detumescence**—it becomes soft and flaccid again. The resolution phase is also a time in which cardiovascular and respiratory functions return to normal. Many people break out in sweat during the resolution phase. In men, resolution is followed by a **refractory period,** lasting anywhere from 10 minutes to a few

hours, in which it is usually impossible to attain another erection and orgasm.

Men and women have many similarities and a few significant differences in sexual response. The response cycle of women is described in chapter 28.

Two of the most common concerns related to sex are sexually transmitted diseases (STDs) and contraception. Understanding most contraceptive methods requires a prior understanding of female anatomy and physiology, so contraceptive techniques for both sexes are discussed at the end of the next chapter, while STDs are discussed in this chapter (Insight 27.5). This is not to imply, of course, that STDs are only a male concern and contraception only a female concern. Reproductive disorders specific to males and females are briefly summarized in tables 27.1 and 28.5, respectively. The effects of aging on the reproductive system are described in chapter 29 (p. 1132).

### Before You Go On

*Answer the following questions to test your understanding of the preceding section:*

20. *Explain how penile blood circulation changes during sexual arousal and why the penis becomes enlarged and stiffened.*

21. *State the roles of the sympathetic, parasympathetic, and somatic nervous systems in male sexual response.*

| TABLE 27.1 | Some Male Reproductive Disorders |
|---|---|
| Breast cancer | Accounts for 0.2% of male cancers in the United States; usually seen after age 60 but sometimes in children and adolescents. About 175 females get breast cancer for every male who does so. Usually felt as a lump near the nipple, often with crusting and discharge from nipple. Often quite advanced by the time of diagnosis, with poor prospects for recovery, because of denial and delay in seeking treatment. |
| Erectile dysfunction (impotence) | Inability to maintain an erection adequate for vaginal entry in half or more of one's attempts. Can stem from aging and declining testosterone level as well as cardiovascular and neurological diseases, diabetes mellitus, medications, fear of failure, depression, and other causes. |
| Hypospadias[38] (HY-po-SPAY-dee-us) | A congenital defect in which the urethra opens on the ventral side or base of the penis rather than at the tip; usually corrected surgically at about 1 year of age. |
| Infertility | Inability to fertilize an egg because of a low sperm count (lower than 20 to 25 million/mL), poor sperm motility, or a high percentage of deformed sperm (two heads, defective tails, etc.). May result from malnutrition, gonorrhea and other infections, toxins, or testosterone deficiency. |
| Penile cancer | Accounts for 1% of male cancers in the United States; most common in black males aged 50 to 70 and of low income. Most often seen in men with nonretractable foreskins (phimosis) combined with poor penile hygiene; least common in men circumcised at birth. |
| Testicular cancer | The most common solid tumor in men 15 to 34 years old, especially white males of middle to upper economic classes. Typically begins as a painless lump or enlargement of the testis. Highly curable if detected early. Men should routinely palpate the testes for normal size and smooth texture. |
| Varicocele (VAIR-ih-co-seal) | Abnormal dilation of veins of the spermatic cord, so that they resemble a "bag of worms." Occurs in 10% of males in the United States. Caused by absence or incompetence of venous valves. Reduces testicular blood flow and often causes infertility. |

***Disorders described elsewhere***

Androgen-insensitivity syndrome  p. 1037

Benign prostatic hyperplasia  p. 1047

Chlamydia  p. 1059

Genital herpes  p. 1060

Genital warts  p. 1060

Gonorrhea  p. 1059

Prostate cancer  p. 1047

Syphilis  p. 1060

[38]*hypo* = below | *spad* = to draw off (the urine)

---

## INSIGHT 27.5   Clinical Application

### Sexually Transmitted Diseases

*Sexually transmitted diseases (STDs)* have been well known since the writings of Hippocrates and Galen. They have been called by a number of euphemisms aimed at downplaying their mode of transmission—for example, "social diseases" and "venereal diseases" (after Venus, the goddess of love).

Here we discuss three bacterial STDs—chlamydia, gonorrhea, and syphilis—and three viral STDs—genital herpes, genital warts, and hepatitis. AIDS, another important viral STD, is discussed in chapter 21. STDs often cause fetal deformity, stillbirth, and neonatal death. The focus here is on adults, whereas STDs of the newborn are discussed in chapter 29.

All of the STDs have certain points in common. They have an *incubation period* in which the pathogen multiplies in the body and begins to produce disease, but symptoms have not appeared yet, and they have a *communicable period* in which an infected person can transmit the disease to others. The communicable period can begin before symptoms are noticed, it can persist after symptoms have disappeared, or it can exist in *symptomless carriers*—people who show no evidence of the disease. STDs usually do not stimulate an immune response, because the pathogens live inside the host cells where the immune system cannot sense their presence. Thus, there are no vaccines to prevent them.

*Chlamydia* (cla-MID-ee-uh) is currently the most common bacterial STD in the United States, with 3 to 5 million cases per year. It is caused by *Chlamydia trachomatis*. After an incubation period of 1 to 3 weeks, there appears a scanty, watery discharge from the urethra and pain in the testes or in the rectal or abdominal region. *Nongonococcal urethritis (NGU)* is an STD caused by agents other than the gonorrhea bacterium (see next paragraph) including *Chlamydia, Mycoplasma hominis,* and *Ureaplasma urealyticum*. NGU is characterized by inflammation of the urethra, causing pain or discomfort on urination.

*Gonorrhea* (GON-oh-REE-uh), nicknamed the "clap" or "drip," is caused by the bacterium *Neisseria gonorrhoeae*. It acquired the name *gonorrhea* ("flow of seed") because Galen thought the pus discharged from the penis was semen. In addition to penile or vaginal discharge, gonorrhea causes abdominal discomfort, genital pain, painful urination (dysuria), and abnormal uterine bleeding. It may cause the uterine tubes to become scarred and obstructed, thus resulting in infertility. About 20% of infected women, however, are asymptomatic. Gonorrhea is treated with antibiotics. Gonorrhea and chlamydia frequently occur together and require treatment with one antibiotic for chlamydia and a different one for gonorrhea.

*Pelvic inflammatory disease (PID)* is a bacterial infection of the female pelvic organs, usually with *Chlamydia* or *Neisseria*. It often results in sterility and may require surgical removal of infected uterine tubes or other organs. The incidence of PID in the United States has increased from 17,800 cases in 1970 to about a million cases per year currently. PID has rendered tens of thousands of American women sterile.

*Syphilis* (SIFF-ih-liss), one of the most potentially devastating STDs, is named for a shepherd boy in a sixteenth-century poem by the physician Fracastoro. It is caused by a corkscrew-shaped bacterium named *Treponema pallidum*. After an incubation period of 2 to 6 weeks, an ulcer called a *chancre* (SHAN-kur) appears at the site of infection—usually on the penis of a male but sometimes out of sight in the vagina of a female. A chancre is a small, hard lesion with no discharge. It disappears in 4 to 6 weeks, ending the first stage of syphilis and often creating an illusion of recovery. In the second stage, however, the disease reappears, with a pink rash over the body, other skin eruptions, fever, joint pain, and hair loss. These symptoms disappear in 3 to 12 weeks. Symptoms then come and go for up to 5 years, but the infection is detectable by a blood test and the infected person is contagious even when symptoms are not occurring. The disease may progress to a third stage, *tertiary syphilis,* or *neurosyphilis,* in which there is damage to the blood vessels and heart valves, thickening of the meninges, and lesions of the brain that can cause paralysis and dementia. Syphilis is treated with antibiotics.

*Genital herpes* is the most common STD in the United States, with 20 to 40 million infected people and 500,000 new cases per year. It is usually caused by the herpes simplex virus type 2 (HSV-2). A close relative, HSV-1, causes cold sores (fever blisters) of the mouth and occasionally causes genital herpes, probably transmitted through oral–genital sex. After an incubation period of 4 to 10 days, the virus causes red blisters on the penis of the male; on the labia, vagina, or cervix of the female; and sometimes on the thighs and buttocks of either sex. Over a period of 2 to 10 days, these blisters rupture, seep fluid, and begin to form scabs. The initial infection may be painless or may cause intense pain, urethritis, and watery discharge from the penis or vagina. The lesions heal in 2 to 3 weeks and leave no scars.

During this time, however, the viruses travel by way of sensory nerve fibers to the dorsal root ganglia, where they become dormant. Later, they can migrate along the nerves and cause small epithelial lesions at various places on the body. The movement from place to place is the basis of the name *herpes.*[39] These secondary lesions are generally smaller and heal more quickly than the primary lesions. Most patients have five to seven recurrences, ranging from several times a year to several years apart. An infected person is contagious to a sexual partner when the lesions are present and sometimes even when they are not. Although mostly a painful nuisance, HSV has been implicated as a risk factor in cervical cancer. The drug acyclovir reduces the shedding of viruses and the spread of lesions but does not necessarily prevent recurrences.

*Genital warts* (condylomas) are one of the most rapidly increasing STDs today. There are about a million new cases per year, especially among young adults with multiple sex partners. Genital warts are caused by 60 or more viruses collectively called *human papillomaviruses (HPVs).* In the male, lesions usually appear on the penis, perineum, or anus, and in the female they are usually on the cervix, vaginal wall, perineum, or anus. Lesions are sometimes small and virtually invisible. A few of the 60 varieties of HPV have been implicated in cervical cancer and cancer of the penis, vagina, and anus. HPV is found in about 90% of all cervical cancers tested. About 90% of cases of genital warts, however, involve forms of HPV that have not been linked to cancer. There is still considerable difference of opinion on how to treat genital warts; they are sometimes treated with cryosurgery (freezing and excision), laser surgery, or interferon.

*Hepatitis B* and *C* are inflammatory liver diseases caused by the hepatitis B and C viruses (HBV, HCV), introduced in Insight 26.3 (p. 1021). Although they can be transmitted by means other than sex, they are becoming increasingly common as STDs. Hepatitis C threatens to become a major epidemic of the twenty-first century. It already far surpasses the prevalence of AIDS and is the leading reason for liver transplants in the United States. Although hepatitis B can be prevented by vaccination, there is no immediate prospect of a vaccine against hepatitis C, because HCV has a rapid mutation rate and quickly escapes the body's immune defenses. The potentially grave consequences of HBC and HCV infection are discussed in Insight 26.3.

---

[39]*herp* = to creep

# CHAPTER REVIEW

## Review of Key Concepts

### Sexual Reproduction and Development (p. 1036)

1. Sexual reproduction is the production of offspring that combine genes from two parents.

2. Sexual reproduction entails the union of two *gametes* to form a *zygote* (fertilized egg). The gametes are a small motile *sperm* produced by the male and a large, immobile, nutrient-laden *egg* produced by the female.

3. The *gonads* (testes and ovaries) are the *primary sex organs. Secondary sex organs* are other anatomical structures needed to produce offspring, such as the male glands, ducts, and penis and the female uterine tubes, uterus, and vagina. *Secondary sex characteristics* are features not essential to reproduction but which help to attract mates.

4. Human chromosomes include 22 pairs of *autosomes* and 1 pair of *sex chromosomes.* There are two types of sex chromosomes, a large X and a smaller Y chromosome. A person who inherits two X chromosomes is genetically female, and one who inherits an X and a Y is genetically male.

5. The Y chromosome bears a gene called *SRY,* which codes for a protein called *testis-determining factor (TDF).* TDF indirectly causes the fetal gonads to develop into testes. The testes secrete androgens, which stimulate the *mesonephric ducts* to develop into a male reproductive tract, and *müllerian-inhibiting factor,* which stimulates the *paramesonephric ducts* to degenerate.

6. In the absence of a Y chromosome, the gonads become ovaries, the mesonephric ducts degenerate, and the paramesonephric ducts develop into a female reproductive tract.

7. The external genitalia of both sexes begin as a *genital tubercle,* a pair of *urogenital folds,* and a pair of *labioscrotal folds.* The genital tubercle differentiates into the glans of the penis or clitoris; the urogenital folds enclose the urethra of the male or become the labia minora of a female; and the labioscrotal folds become the scrotum of a male or labia majora of a female.

8. During male development, the fetal testes follow a cord called the *gubernaculum* through the inguinal canal into the scrotum; this is called *descent of the testes.*

### Male Reproductive Anatomy (p. 1041)

1. The *scrotum* contains the testes and the *spermatic cord,* which is a bundle of connective tissue, testicular blood vessels, and a sperm duct, the *ductus deferens.* The spermatic cord passes up the back of the scrotum and through the inguinal ring into the pelvic cavity.

2. Sperm cannot develop at the core body temperature of 37°C. The testes are kept about 2°C cooler than this by three structures in the scrotum: the *cremaster muscle* of the spermatic cord, which relaxes when it is warm and contracts when it is cool, thus lowering or raising the scrotum and testes; the *dartos muscle* in the scrotal wall, which contracts and tautens the scrotum when it is cool; and the *pampiniform plexus* of blood vessels in the spermatic cord, which acts as a *countercurrent heat exchanger* to cool the blood on its way to the testis.

3. The testis has a fibrous capsule, the *tunica albuginea.* Fibrous septa divide the interior of the testis into 250 to 300 compartments called *lobules.* Each lobule contains 1 to 3 sperm-producing *seminiferous tubules.* Testosterone-secreting *interstitial cells* lie in clusters between the tubules.

4. The epithelium of a seminiferous tubule consists of *germ cells* and *sustentacular cells.* The germ cells develop into sperm, while the sustentacular cells support and nourish them, form a *blood–testis barrier* between the germ cells and nearest blood supply, and secrete *inhibin,* which regulates the rate of sperm production.

5. Each testis is supplied by a long, slender *testicular artery* and drained by a *testicular vein,* and is supplied with *testicular nerves* and lymphatic vessels.

6. *Spermatic ducts* carry sperm from the testis to the urethra. They include several *efferent ductules* leaving the testis; a single *duct of the epididymis,* a highly coiled structure adhering to the posterior side of the testis; a muscular *ductus deferens* that travels through the spermatic cord into the pelvic cavity; and a short *ejaculatory duct* that carries sperm and seminal vesicle secretions the last 2 cm to the urethra. The urethra completes the path of the sperm to the outside of the body.

7. The male has three sets of accessory glands: a pair of *seminal vesicles* posterior to the urinary bladder; a single *prostate gland* inferior to the bladder, enclosing the prostatic urethra; and a pair of small *bulbourethral glands* that secrete into the proximal end of the penile urethra. The seminal vesicles and prostate secrete most of the semen. The bulbourethral glands produce a small amount of clear slippery fluid that lubricates the urethra and neutralizes its pH.

8. The *penis* is divided into an internal *root* and an external *shaft* and *glans.* It is covered with loose skin that extends over the glans as the *prepuce,* or foreskin.

9. Internally, the penis consists mainly of three long *erectile tissues*—a pair of dorsal *corpora cavernosa,* which engorge with blood and produce most of the effect of erection, and a single ventral *corpus spongiosum,* which contains the urethra. All three tissues have blood sinuses called *lacunae* separated by *trabeculae* composed of connective tissue

and smooth muscle *(trabecular muscle).*

10. At the proximal end of the penis, the corpus spongiosum dilates into a *bulb* that receives the urethra and ducts of the bulbourethral glands, and the corpora cavernosa diverge into a pair of *crura* that anchor the penis to the pubic arch and perineal membrane.

## Puberty and Climacteric (p. 1048)

1. *Adolescence* is a period from the onset of gonadotropin secretion to the attainment of adulthood. The first few years of adolescence, until ejaculation begins, constitute *puberty.*

2. At puberty, the hypothalamus begins secreting gonadotropin-releasing hormone (GnRH), which induces the anterior pituitary to secrete follicle-stimulating hormone (FSH) and luteinizing hormone (LH). These hormones stimulate enlargement of the testes, an early sign of puberty.

3. LH stimulates the interstitial cells to produce androgens (especially testosterone), and FSH causes the sustentacular cells to produce *androgen-binding protein (ABP).*

4. Testosterone stimulates spermatogenesis; modulates the secretion of GnRH, FSH, and LH; stimulates the development of the secondary sex organs and sex characteristics; stimulates bodily growth and erythropoiesis; and activates sexual interest, or *libido.*

5. *Inhibin,* secreted by the sustentacular cells, selectively inhibits FSH secretion and sperm production without inhibiting LH and testosterone secretion.

6. After age 20, testosterone secretion steadily declines, as do the number and secretory activity of interstitial and sustentacular cells. Declining testosterone and inhibin levels reduce the negative feedback inhibition of the pituitary, which thus secretes elevated levels of FSH and LH. In some men, this results in physiological and mood changes called *male climacteric.*

7. In older age, *erectile dysfunction* becomes common, but the ability to ejaculate remains unaffected in most cases.

## Sperm and Semen (p. 1050)

1. Sexual reproduction requires a form of cell division called *meiosis* to reduce the chromosome number by half. This prevents a doubling of chromosome number in each generation when egg and sperm combine. Meiosis produces *haploid* gametes with 23 chromosomes each. The union of two such gametes restores the *diploid* number of 46 chromosomes in the zygote.

2. Meiosis consists of one cell division with stages called *prophase I, metaphase I, anaphase I,* and *telophase I;* then an interval called *interkinesis;* then a second cell division consisting of *prophase II, metaphase II, anaphase II,* and *telophase II.* The ultimate result is four haploid daughter cells.

3. Prophase I also involves *crossing-over,* an exchange of genes between an individual's homologous maternal and paternal chromosomes. As a consequence, the chromosomes passed to a child have new genetic combinations not inherited from one's parents.

4. The production of sperm is called *spermatogenesis.*

5. Spermatogenesis begins with *primordial germ cells,* which arise in the yolk sac, migrate to the gonads, and become *spermatogonia* by the time a boy is born.

6. From puberty to death, primary spermatogonia divide by mitosis into *type A* and *type B* spermatogonia. The latter develop into *primary spermatocytes,* then enter into meiosis and divide into *secondary spermatocytes* and finally *spermatids.*

7. In a process called *spermiogenesis,* spermatids shed excess cytoplasm and grow a tail (flagellum), thus becoming sperm.

8. The sperm consists of a head containing the nucleus and *acrosome,* and a tail composed of a mitochondria-stuffed *midpiece,* a long *principal piece,* and a short *endpiece.*

9. *Semen* is a mixture of sperm (10% of the volume) and fluids from the prostate (30%) and seminal vesicles (60%). It contains fructose, seminogelin, a clotting enzyme, prostaglandins, and serine protease, among other substances.

## Male Sexual Response (p. 1055)

1. The penis is supplied by a pair of internal pudendal arteries. One branch of this artery, the *dorsal artery,* travels dorsally under the skin of the penis and the other, the *deep artery,* travels through the corpus cavernosum and supplies blood to the lacunae. The dorsal arteries supply most of the blood when the penis is flaccid, and the deep arteries during erection.

2. Nerves of the penis converge on the *dorsal nerve,* which leads via the *internal pudendal nerve* to the sacral plexus and then the spinal cord. The penis receives sympathetic, parasympathetic, and somatic motor nerve fibers.

3. The *excitement phase* of male sexual response is marked by vasocongestion of the genitals; myotonia of the skeletal muscles; increases in heart rate, blood pressure, and pulmonary ventilation; secretion by the bulbourethral glands; and erection of the penis.

4. Erection occurs when parasympathetic nerve fibers trigger the secretion of nitric oxide, which dilates the deep arteries and lacunae. Blood flow into the erectile tissues engorges the lacunae, causing the penis to enlarge and stiffen.

5. The *plateau phase* is marked by elevated but stable or slightly rising heart rate, blood pressure, and respiration, and sometimes increased vasocongestion and myotonia.

6. *Orgasm* (climax) is marked by a sudden rise in heart rate, blood pressure, and respiration, and usually the ejaculation of semen. Ejaculation occurs in stages: *emission,* the peristaltic propulsion of sperm from the epididymis to the prostatic urethra, where it mixes with the prostatic and seminal vesicle secretions; and *expulsion,* in which five or six spasmodic contractions of the bulbospongiosus muscle compress the urethra and expel the semen from the penis.

7. Orgasm is followed by *resolution,* in which physiological values return to normal and the penis undergoes *detumescence,* or loss of engorgement and erection. This is typically followed by a *refractory period* in which it is impossible to attain another erection and orgasm.

# Testing Your Recall

1. The ductus deferens develops from the _____ of the embryo.
   a. mesonephric duct
   b. paramesonephric duct
   c. phallus
   d. labioscrotal folds
   e. urogenital folds

2. The protein that clots and causes the stickiness of semen is
   a. seminogelin.
   b. prostaglandin.
   c. fibrin.
   d. phosphodiesterase.
   e. serine protease.

3. The expulsion of semen occurs when the bulbospongiosus muscle is stimulated by
   a. somatic efferent neurons.
   b. somatic afferent neurons.
   c. sympathetic efferent neurons.
   d. parasympathetic efferent neurons.
   e. prostaglandins.

4. Prior to ejaculation, sperm are stored primarily in
   a. the seminiferous tubules.
   b. the rete testis.
   c. the epididymis.
   d. the seminal vesicles.
   e. the ejaculatory ducts.

5. The penis is attached to the pubic arch by crura of
   a. the corpora cavernosa.
   b. the corpus spongiosum.
   c. the perineal membrane.
   d. the bulbospongiosus.
   e. the ischiocavernosus.

6. The first hormone secreted at the onset of puberty is
   a. follicle–stimulating hormone.
   b. interstitial cell–stimulating hormone.
   c. human chorionic gonadotropin.
   d. gonadotropin-releasing hormone.
   e. testosterone.

7. When it is necessary to reduce sperm production without reducing testosterone secretion, the sustentacular cells secrete
   a. dihydrotestosterone.
   b. androgen-binding protein.
   c. LH.
   d. FSH.
   e. inhibin.

8. Four spermatozoa arise from each
   a. primordial germ cell.
   b. type A spermatogonium.
   c. type B spermatogonium.
   d. secondary spermatocyte.
   e. spermatid.

9. The point in meiosis at which sister chromatids separate from each other is
   a. prophase I.
   b. metaphase I.
   c. anaphase I.
   d. anaphase II.
   e. telophase II.

10. Blood is forced out of the penile lacunae by contraction of the _____ muscles.
    a. bulbospongiosus
    b. ischiocavernosus
    c. cremaster
    d. trabecular
    e. dartos

11. Under the influence of androgens, the embryonic _____ duct develops into the male reproductive tract.

12. Spermatozoa obtain energy for locomotion from _____ in the semen.

13. The _____, a network of veins in the spermatic cord, helps keep the testes cooler than the core body temperature.

14. All germ cells beginning with the _____ are genetically different from the rest of the body cells and therefore must be protected by the blood–testis barrier.

15. The corpora cavernosa as well as the testes have a fibrous capsule called the _____.

16. Over half of the semen consists of secretions from a pair of glands called the _____.

17. The blood-testis barrier is formed by tight junctions between the _____ cells.

18. The earliest haploid stage of spermatogenesis is the _____.

19. Erection of the penis occurs when nitric oxide causes the _____ arteries to dilate.

20. A sperm penetrates the egg by means of enzymes in its _____.

*Answers in Appendix B*

# True or False

*Determine which five of the following statements are false, and briefly explain why.*

1. The male scrotum is homologous to the female labia majora.

2. Sperm cannot develop at the core body temperature.

3. Luteinizing hormone stimulates the testes to secrete testosterone.

4. The testes and penis are the primary sex organs of the male.

5. A high testosterone level makes a fetus develop a male reproductive system, and a high estrogen level makes it develop a female reproductive system.

6. Most of the semen is produced by the seminal vesicles.

7. The pampiniform plexus serves to keep the testes warm.

8. Prior to ejaculation, sperm are stored mainly in the seminal vesicles.

9. Male menopause is the cessation of sperm production around the age of 55.

10. Erection is caused by parasympathetic stimulation of the penile arteries.

*Answers in Appendix B*

# Testing Your Comprehension

1. Explain why testosterone may be considered both an endocrine and a paracrine secretion of the testes. (Review paracrines in chapter 17 if necessary.)

2. A young man is in a motorcycle accident that severs his spinal cord at the neck and leaves him paralyzed from the neck down. When informed of the situation, his wife asks the physician if her husband will be able to have erections and father any children. What should the doctor tell her? Explain your answer.

3. Considering the temperature in the scrotum, would you expect hemoglobin to unload more oxygen to the testes, or less, than it unloads in the warmer internal organs? Why? (Hint: See figure 22.23.) How would you expect this fact to influence sperm development?

4. Why is it possible for spermatogonia to be outside the blood–testis barrier, yet necessary for primary spermatocytes and later stages to be within the barrier, isolated from the blood?

5. A 68-year-old man taking medication for hypertension complains to his physician that it has made him impotent. Explain why this could be an effect of antihypertension drugs.

*Answers at www.mhhe.com/saladin4*

# www.mhhe.com/saladin4

*The textbook website provides a wealth of interactive study materials fully organized and integrated by chapter. You will find practice quizzes, labeling exercises, and much more that will complement your learning and understanding of anatomy and physiology. The website also includes tools designed to enhance your* **Anatomy & Physiology | REVEALED** *experience.*

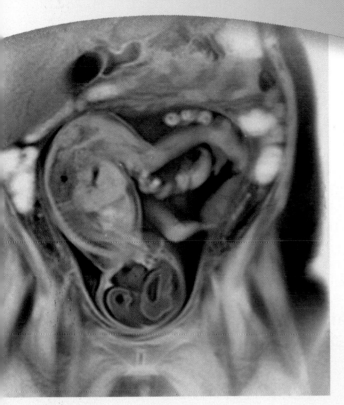

MRI scan of a 36-week-old fetus
in the uterus

# THE FEMALE
# REPRODUCTIVE SYSTEM

## CHAPTER OUTLINE

## INSIGHTS

## Brushing Up

To understand this chapter, it is
important that you understand or
brush up on the following concepts:

- Negative feedback inhibition of
  the pituitary gland (p. 646)

- Synergistic, permissive, and
  antagonistic hormone
  interactions (p. 665)

- Fetal development of the reproductive system (pp. 1037–1041)

- Meiosis (pp. 1050–1051)

The female reproductive system is more complex than the male's because it serves more purposes. Whereas the male needs only to produce and deliver gametes, the female must do this as well as provide nutrition and safe harbor for fetal development, then give birth and nourish the infant. Furthermore, female reproductive physiology is cyclic and female hormones are secreted in a more complex sequence compared with the relatively steady, simultaneous secretion of regulatory hormones in the male.

This chapter discusses the anatomy of the female reproductive system, the production of gametes and how it relates to the ovarian and menstrual cycles, the female sexual response, and the physiology of pregnancy, birth, and lactation. Embryonic and fetal development are treated in chapter 29.

# Reproductive Anatomy

### Objectives

When you have completed this section, you should be able to

- describe the structure of the ovary;
- trace the female reproductive tract and describe the gross anatomy and histology of each organ;
- identify the ligaments that support the female reproductive organs;
- describe the blood supply to the female reproductive tract;
- identify the external genitalia of the female; and
- describe the structure of the nonlactating breast.

## SEXUAL DIFFERENTIATION

The female reproductive system (fig. 28.1) is conspicuously different from that of the male, but as we saw earlier,

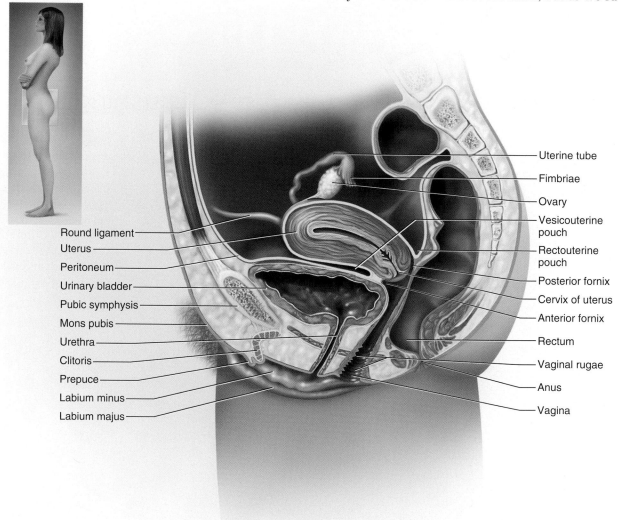

Round ligament
Uterus
Peritoneum
Urinary bladder
Pubic symphysis
Mons pubis
Urethra
Clitoris
Prepuce
Labium minus
Labium majus

Uterine tube
Fimbriae
Ovary
Vesicouterine pouch
Rectouterine pouch
Posterior fornix
Cervix of uterus
Anterior fornix
Rectum
Vaginal rugae
Anus
Vagina

**FIGURE 28.1**  The Female Reproductive System.

the two sexes are indistinguishable for the first 8 to 10 weeks of development (see fig. 27.4, p. 1040). The female reproductive tract develops from the paramesonephric duct not because of the positive action of any hormone, but because of the absence of testosterone and müllerian-inhibiting factor (MIF). Without testosterone, the mesonephric duct degenerates while the genital tubercle becomes a clitoris, the urogenital folds develop into labia minora, and the labioscrotal folds develop into labia majora. Without MIF, the paramesonephric duct develops into the uterine tubes, uterus, and vagina (see fig. 27.2, p. 1038). This developmental pattern can be disrupted, however, by abnormal hormonal exposure before birth, as happens in adrenogenital syndrome (see fig. 17.30, p. 670).

## THE OVARIES

The female gonads (primary sex organs) are the **ovaries,**[1] which produce egg cells (ova) and sex hormones. The ovary is an almond-shaped organ nestled in the *ovarian fossa,* a depression in the dorsal pelvic wall. It measures about 3 cm long, 1.5 cm wide, and 1 cm thick. Its capsule, like that of the testis, is called the **tunica albuginea.** The interior of the ovary is indistinctly divided into an outer **cortex,** where the germ cells develop, and a central **medulla** occupied by the major arteries and veins (fig. 28.2).

The ovary lacks ducts comparable to the seminiferous tubules of the testis. Instead, each egg develops in its own fluid-filled, bubblelike **follicle** and is released by *ovulation,* the bursting of the follicle. Figure 28.2 shows several types of follicles that coincide with different stages of egg maturation, as discussed later.

The ovary is held in place by several connective tissue ligaments (fig. 28.3). Its medial pole is attached to the uterus by the *ovarian ligament* and its lateral pole is attached to the pelvic wall by the *suspensory ligament.* The anterior margin of the ovary is anchored by a peritoneal fold called the *mesovarium.*[2] This ligament extends to a sheet of peritoneum called the *broad ligament,* which flanks the uterus and encloses the uterine tube in its superior margin.

---

[1]*ov* = egg + *ary* = place for

[2]*mes* = middle + *ovari* = ovary

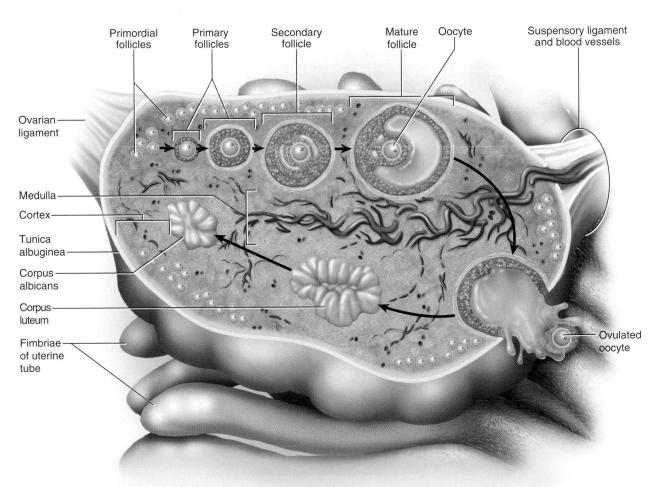

**FIGURE 28.2** Structure of the Ovary.

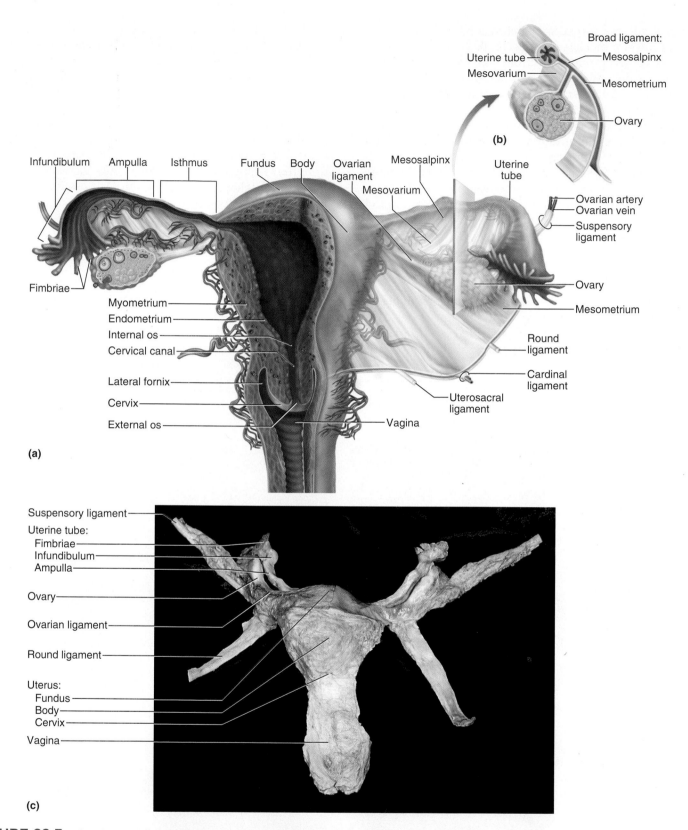

**FIGURE 28.3** **The Female Reproductive Tract and Supportive Ligaments.** (a) Dorsal view of the reproductive tract. (b) Relationship of the ligaments to the uterine tube and ovary. (c) Anterior view of the major female reproductive organs from a cadaver.

The ovary receives blood from two arteries: the **ovarian branch of the uterine artery,** which passes through the mesovarium and approaches the medial pole of the ovary, and the **ovarian artery,** which passes through the suspensory ligament and approaches the lateral pole. The ovarian artery is the female equivalent of the testicular artery described in chapter 27, arising high on the aorta and traveling down to the gonad along the dorsal body wall. The ovarian and uterine arteries anastomose along the margin of the ovary and give off multiple small arteries that enter the ovary on that side. Ovarian veins and nerves also travel through the suspensory ligament.

## THE GENITALIA

The internal genitalia are the uterine tubes, uterus, and vagina, which constitute a duct system from the vicinity of the ovary to the outside of the body. The external genitalia include principally the clitoris, labia minora, and labia majora. These occupy the perineum, which is defined by the same skeletal landmarks as in the male (see fig. 27.6, p. 1042). The internal and external genitalia constitute the secondary sex organs.

### The Uterine Tubes

The **uterine tube,** also called the **oviduct** or **fallopian**[3] **tube,** is a canal about 10 cm long from the ovary to the uterus. At the distal (ovarian) end, it flares into a trumpet-shaped **infundibulum**[4] with feathery projections called **fimbriae**[5] (FIM-bree-ee); the middle and longest part of the tube is the **ampulla;** and near the uterus it forms a narrower **isthmus.** The uterine tube is enclosed in the *mesosalpinx*[6] (MEZ-oh-SAL-pinks), which is the superior margin of the broad ligament.

The wall of the uterine tube is well endowed with smooth muscle. Its mucosa is extremely folded and convoluted and has an epithelium of ciliated cells and a smaller number of secretory cells (fig. 28.4). The cilia beat toward the uterus and, with the help of muscular contractions of the tube, convey the egg in that direction.

### The Uterus

The **uterus**[7] is a thick muscular chamber that opens into the roof of the vagina and usually tilts forward over the urinary bladder (see fig. 28.1). Its function is to harbor the fetus, provide a source of nutrition, and expel the fetus at the end of its development. It is somewhat pear-shaped, with a broad superior curvature called the **fundus,** a midportion called the **body** (corpus), and a cylindrical inferior end called the **cervix** (see fig. 28.3). The uterus

**FIGURE 28.4  Epithelial Lining of the Uterine Tube.** Secretory cells are shown in red and green, and cilia of the ciliated cells in yellow (SEM).

measures about 7 cm from cervix to fundus, 4 cm wide at its broadest point, and 2.5 cm thick, but it is somewhat larger in women who have been pregnant.

The lumen of the uterus is roughly triangular, with its two upper corners opening into the uterine tubes. In the nonpregnant uterus, the lumen is not a hollow cavity but rather a *potential space;* the mucous membranes of the opposite walls are pressed against each other with little room between them. The lumen communicates with the vagina by way of a narrow passage through the cervix called the **cervical canal.** The superior opening of this canal into the body of the uterus is the *internal os*[8] (oss) and its opening into the vagina is the *external os.* The canal contains **cervical glands** that secrete mucus, thought to prevent the spread of microorganisms from the vagina into the uterus. Near the time of ovulation, the mucus becomes thinner than usual and allows easier passage for sperm.

---

[3]Gabriele Fallopio (1523–62), Italian anatomist and physician
[4]*infundibulum* = funnel
[5]*fimbria* = fringe
[6]*meso* = mesentery + *salpin* = trumpet
[7]*uterus* = womb

[8]*os* = mouth

## INSIGHT 28.1    Clinical Application

### Pap Smears and Cervical Cancer

Cervical cancer is common among women between the ages of 30 and 50, especially those who smoke, who began sexual activity at an early age, and who have histories of frequent sexually transmitted diseases or cervical inflammation. It is often caused by the human papillomavirus (HPV), a sexually transmitted pathogen (see p. 1060). Cervical cancer usually begins in the epithelial cells of the lower cervix, develops slowly, and remains a local, easily removed lesion for several years. If the cancerous cells spread to the subepithelial connective tissue, however, the cancer is said to be *invasive* and is much more dangerous and potentially fatal.

The best protection against cervical cancer is early detection by means of a *Pap*[9] *smear*—a procedure in which loose cells are removed from the cervix and vagina with a small flat stick and cervical brush, then microscopically examined. The pathologist looks for cells with signs of *dysplasia* (abnormal development) or carcinoma (fig. 28.5). One system of grading Pap smears classifies abnormal results into three grades of *cervical intraepithelial neoplasia* (CIN). Findings are rated on the following scale, and further vigilance or treatment planned accordingly:

  ASCUS—atypical squamous cells of undetermined significance
  CIN I—mild dysplasia with cellular changes typically associated with HPV
  CIN II—moderate dysplasia with precancerous lesions
  CIN III—severe dysplasia, *carcinoma in situ* (preinvasive carcinoma of surface cells)

A rating of ASCUS or CIN I calls for a repeat Pap smear and visual examination of the cervix (*colposcopy*) in 3 to 6 months. CIN II calls for a biopsy, often done with an "electric scalpel" in a procedure called LEEP (loop electrosurgical excision procedure). A cone of tissue is removed to evaluate the depth of invasion by the malignant or premalignant cells. This in itself may be curative if all margins of the specimen are normal, indicating all abnormal cells were removed. CIN III may be cause for *hysterectomy*[10] or radiation therapy.

An average woman is typically advised to have annual Pap smears for 3 years and may then have them less often at the discretion of her physician. Women with any of the risk factors listed may be advised to have more frequent examinations.

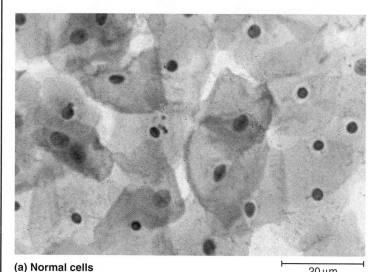

**(a) Normal cells**    20 μm

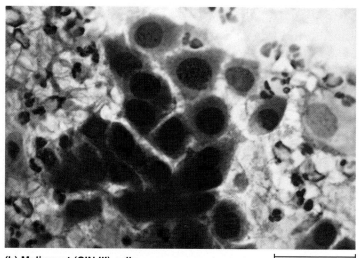

**(b) Malignant (CIN III) cells**    20 μm

**FIGURE 28.5  Pap smears.**  These are smears of squamous epithelial cells scraped from the cervix. In the cancerous cells, note the loss of cell volume and the greatly enlarged nuclei.

---

***Uterine Wall***  The uterine wall consists of an external serosa called the *perimetrium,* a middle muscular layer called the *myometrium,* and an inner mucosa called the *endometrium.* The **myometrium**[11] constitutes most of the wall; it is about 1.25 cm thick in the nonpregnant uterus.

It is composed of bundles of smooth muscle that sweep downward from the fundus and spiral around the body of the uterus. The myometrium is less muscular and more fibrous near the cervix; the cervix itself is almost entirely collagenous. The smooth muscle cells of the myometrium are about 40 μm long immediately after menstruation, but they are twice this long at the middle of the menstrual cycle and 10 times as long in pregnancy. The function of the myometrium is to produce the labor contractions that help to expel the fetus.

---

[9]George N. Papanicolaou (1883–1962), Greek-American physician and cytologist
[10]*hyster* = uterus + *ectomy* = cutting out
[11]*myo* = muscle + *metr* = uterus

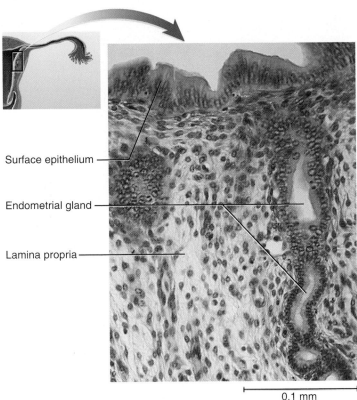

Surface epithelium

Endometrial gland

Lamina propria

0.1 mm

**FIGURE 28.6** Histology of the Endometrium.

The **endometrium**[12] is the mucosa. It has a simple columnar epithelium, compound tubular glands, and a stroma populated by leukocytes, macrophages, and other cells (fig. 28.6). The superficial half to two-thirds of it, called the **stratum functionalis,** is shed in each menstrual period. The deeper layer, called the **stratum basalis,** stays behind and regenerates a new functionalis in the next cycle. When pregnancy occurs, the endometrium is the site of attachment of the embryo and forms the maternal part of the *placenta* from which the fetus is nourished.

*Blood Supply*    The uterine blood supply is particularly important to the menstrual cycle and pregnancy. A **uterine artery** arises from each internal iliac artery and travels through the broad ligament to the uterus (fig. 28.7). It gives off several branches that penetrate into the myometrium and lead to **arcuate arteries.** Each arcuate artery travels in a circle around the uterus and anastomoses with the arcuate artery on the other side. Along its course, it gives rise to smaller arteries that penetrate the rest of the way through the myometrium, into the endometrium, and produce the **spiral arteries.** The spiral arteries wind tortuously between the endometrial glands toward the surface of the mucosa. They rhythmically constrict and dilate, making the mucosa alternately blanch and flush with blood.

---

[12]*endo* = inside + *metr* = uterus

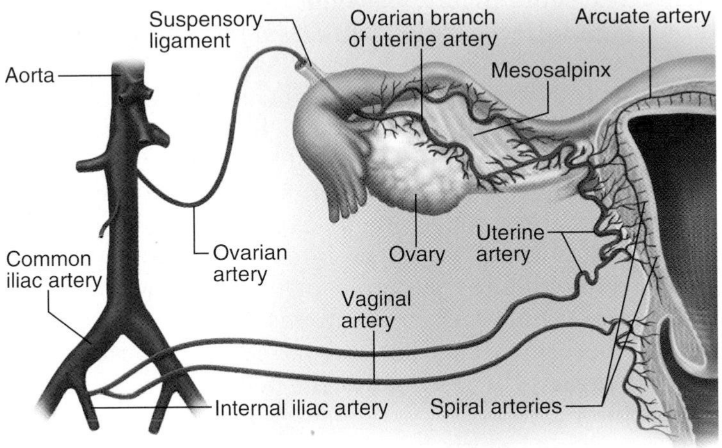

**FIGURE 28.7**    Blood Supply to the Female Reproductive Tract.

*Ligaments*    The uterus is supported by the muscular floor of the pelvic outlet and folds of peritoneum that form supportive ligaments around the organ, as they do for the ovary and uterine tube (see fig. 28.3a). The broad ligament has two parts: the mesosalpinx mentioned earlier and the *mesometrium* on each side of the uterus. The cervix and superior part of the vagina are supported by *cardinal (lateral cervical) ligaments* extending to the pelvic wall. A pair of *uterosacral ligaments* attach the dorsal side of the uterus to the sacrum, and a pair of *round ligaments* attach the ventral surface of the uterus to the abdominal wall. The round ligaments continue through the inguinal canals and terminate in the labia majora, much like the gubernaculum of the male terminating in the scrotum.

As the peritoneum folds around the various pelvic organs, it creates several dead-end recesses and pouches. Two major ones are the *vesicouterine*[13] *pouch,* which forms the space between the uterus and urinary bladder, and *rectouterine pouch* between the uterus and rectum (see fig. 28.1).

## The Vagina

The **vagina,**[14] or birth canal, is a tube about 8 to 10 cm long that allows for the discharge of menstrual fluid, receipt of the penis and semen, and birth of a baby. The vaginal wall is thin but very distensible. It consists of an outer adventitia, a middle muscularis, and an inner mucosa. The vagina tilts dorsally between the urethra and rectum; the urethra is bound to its anterior wall. The vagina has no glands, but it is lubricated by the *transudation* ("vaginal sweating") of serous fluid through its walls and by mucus from the cervical glands above it. The

---

[13]*vesico* = bladder
[14]*vagina* = sheath

vagina extends slightly beyond the cervix and forms blind-ended spaces called *fornices*[15] (FOR-nih-sees; singular, *fornix*) (see figs. 28.1 and 28.3a).

At its lower end, the vaginal mucosa folds inward and forms a membrane, the **hymen,** which stretches across the orifice. The hymen has one or more openings to allow menstrual fluid to pass through, but it usually must be ruptured to allow for intercourse. A little bleeding often accompanies the first act of intercourse; however, the hymen is commonly ruptured before then by tampons, medical examinations, or strenuous exercise. The lower end of the vagina also has transverse friction ridges, or **vaginal rugae,** which stimulate the penis and help induce ejaculation.

The vaginal epithelium is simple cuboidal in childhood, but the estrogens of puberty transform it into a stratified squamous epithelium. This is an example of *metaplasia,* the transformation of one tissue type to another. The epithelial cells are rich in glycogen. Bacteria ferment this to lactic acid, which produces a low vaginal pH (about 3.5–4.0) that inhibits the growth of pathogens. Recall from chapter 27 that this acidity is neutralized by the semen so it does not harm the sperm. The mucosa also has antigen-presenting cells called **dendritic cells,** which are a route by which HIV from infected semen invades the female body.

⌐ **Think About It**

*Why do you think the vaginal epithelium changes type at puberty? Of all types of epithelium it might become, why stratified squamous?*

## The External Genitalia

The external genitalia of the female are collectively called the **vulva**[16] (pudendum[17]); this includes the mons pubis, labia majora and minora, clitoris, vaginal orifice, and accessory glands and erectile tissues. It occupies most of the perineum (fig. 28.8a).

The **mons**[18] **pubis** consists mainly of an anterior mound of adipose tissue overlying the pubic symphysis, bearing most of the pubic hair (figs. A.5a and 28.8). The **labia majora**[19] (singular, *labium majus*) are a pair of thick folds of skin and adipose tissue inferior to the mons; the slit between them is the *pudendal cleft.* Pubic hair grows on the lateral surfaces of the labia majora at puberty, but the medial surfaces remain hairless. Medial to the labia majora are the much thinner, entirely hairless **labia minora**[20] (singular, *labium minus*). The area enclosed by them, called the **vestibule,** contains the urinary and vaginal orifices. At

---

[15]*fornix* = arch, vault
[16]*vulva* = covering
[17]*pudend* = shameful
[18]*mons* = mountain
[19]*labi* = lip + *major* = larger, greater
[20]*minor* = smaller, lesser

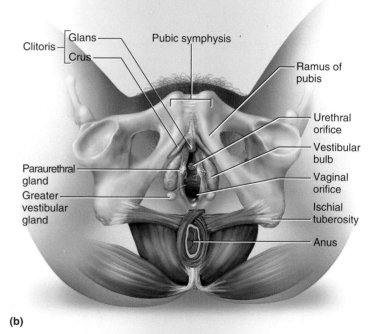

**(a)**

**(b)**

**FIGURE 28.8    The Female Perineum.** (a) Surface anatomy. (b) Subcutaneous structures.

▶ *Which of these glands is homologous to the male prostate gland?*

the anterior margin of the vestibule, the labia minora meet and form a hoodlike **prepuce** over the clitoris.

The **clitoris** is structured much like a miniature penis but has no urinary role. Its function is entirely sensory, serving as the primary center of erotic stimulation. Unlike the penis, it is almost entirely internal; it has no corpus spongiosum, and it does not enclose the urethra. Essentially, it is a pair of corpora cavernosa enclosed in connective tissue. Its head, the **glans,** protrudes slightly from the prepuce. The **body** (corpus) passes internally, inferior to the pubic symphysis (see fig. 28.1). At its internal end, the corpora cavernosa diverge like a Y as a pair of **crura,** which, like those of the penis, attach the clitoris to each side of the pubic arch. Like the penis, the clitoris is supplied by the internal pudendal arteries, also called the **clitoral arteries** in the female.

Just deep to the labia majora, a pair of subcutaneous erectile tissues called the **vestibular bulbs** bracket the vagina like parentheses. They become congested with blood during sexual excitement and cause the vagina to tighten somewhat around the penis, enhancing sexual stimulation.

On each side of the vagina is a pea-sized **greater vestibular (Bartholin[21]) gland** with a short duct opening into the vestibule or lower vagina (fig. 28.8b). These glands are homologous to the bulbourethral glands of the male. They keep the vulva moist, and during sexual excitement they provide most of the lubrication for intercourse. The vestibule is also lubricated by a number of **lesser vestibular glands.** A pair of mucous **paraurethral**

**(Skene[22]) glands,** homologous to the male prostate, open into the vestibule near the external urethral orifice.

## THE BREASTS AND MAMMARY GLANDS

The breast (fig. 28.9) is a mound of tissue overlying the pectoralis major. It develops at puberty and remains for life, but most of this time it contains very little mammary gland. The mammary gland develops within the breast during pregnancy, remains active in the lactating breast, and atrophies when a woman ceases to nurse.

The breast has two principal regions: the conical to pendulous **body,** with the nipple at its apex, and an extension toward the armpit called the **axillary tail.** Lymphatics of the axillary tail are especially important as a route of breast cancer metastasis.

The nipple is surrounded by a circular colored zone, the **areola.** Dermal blood capillaries and nerves come closer to the surface here than in the surrounding skin and make the areola more sensitive and more reddish in color. In pregnancy, the areola and nipple often darken and become more visible—a preparation for the indistinct vision of a nursing infant. Sensory nerve fibers of the areola are important in triggering a *milk ejection reflex* when an

---

[21]Caspar Bartholin (1655–1738), Danish anatomist

[22]Alexander J. C. Skene (1838–1900), American gynecologist

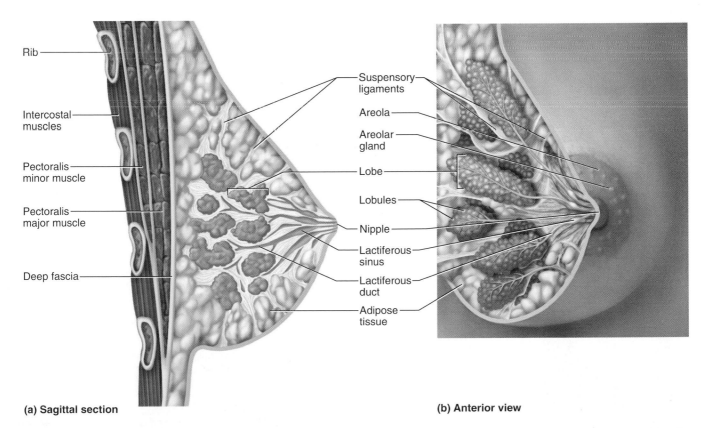

**(a) Sagittal section**

Rib
Intercostal muscles
Pectoralis minor muscle
Pectoralis major muscle
Deep fascia

Suspensory ligaments
Areola
Areolar gland
Lobe
Lobules
Nipple
Lactiferous sinus
Lactiferous duct
Adipose tissue

**(b) Anterior view**

**FIGURE 28.9 Anatomy of the Lactating Breast.** (a) Sagittal section. (b) Surface anatomy and cutaway view of the lobes of mammary gland; anterior view of right breast.

infant nurses. The areola has sparse hairs and **areolar glands,** visible as small bumps on the surface. These glands are intermediate between sweat glands and mammary glands in their degree of development. When a woman is nursing, the areola is protected from chapping and cracking by secretions of the areolar glands and sebaceous glands of the region. The dermis of the areola has smooth muscle fibers that contract in response to cold, touch, and sexual arousal, wrinkling the skin and erecting the nipple.

Internally, the nonlactating breast consists mostly of adipose and collagenous tissue (fig. 28.10). Breast size is determined by the amount of adipose tissue and has no relationship to the amount of milk the mammary gland can produce. **Suspensory ligaments** attach the breast to the dermis of the overlying skin and to the fascia of the pectoralis major. The nonlactating breast contains very little glandular tissue, but it does have a system of ducts branching through its connective tissue stroma and converging on the nipple. When the mammary gland develops during pregnancy, it exhibits 15 to 20 lobes arranged radially around the nipple, separated from each other by fibrous stroma. Each lobe is drained by a **lactiferous**[23] **duct,** which dilates to form a **lactiferous sinus** opening onto the nipple. Lactation and the associated changes in breast structure are described at the end of this chapter.

---

[23]*lact* = milk + *fer* = to carry

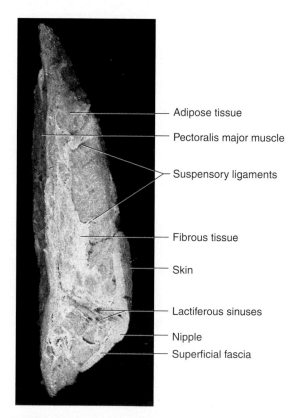

- Adipose tissue
- Pectoralis major muscle
- Suspensory ligaments
- Fibrous tissue
- Skin
- Lactiferous sinuses
- Nipple
- Superficial fascia

**FIGURE 28.10**   Sagittal Section of the Breast of a Cadaver.

## Breast Cancer

Breast cancer (fig. 28.11) occurs in one out of every eight or nine American women and is one of the leading causes of female mortality. Breast tumors begin with cells of the mammary ducts and may metastasize to other organs by way of the mammary and axillary lymphatics. Signs of breast cancer include a palpable lump (the tumor), puckering of the skin, changes in skin texture, and drainage from the nipple.

Two breast cancer genes were discovered in the 1990s, named *BRCA1* and *BRCA2,* but most breast cancer is nonhereditary. Some breast tumors are stimulated by estrogen. Consequently, breast cancer is more common among women who begin menstruating early in life and who reach menopause relatively late—that is, women who have a long period of fertility and estrogen exposure. Other risk factors include aging, exposure to ionizing radiation and carcinogenic chemicals, excessive alcohol and fat intake, and smoking. Over 70% of cases, however, lack any identifiable risk factors.

The majority of tumors are discovered during breast self-examination (BSE), which should be a monthly routine for all women. *Mammograms* (breast X rays), however, can detect tumors too small to be noticed by BSE. Although opinions vary, a schedule commonly recommended is to have a baseline mammogram in the late 30s and then have one every 2 years from ages 40 to 49 and every year beginning at age 50.

Treatment of breast cancer is usually by *lumpectomy* (removal of the tumor only) or *simple mastectomy* (removal of the breast tissue only or breast tissue and some axillary lymph nodes). *Radical mastectomy,* rarely done since the 1970s, involves the removal of not only the breast but also the underlying muscle, fascia, and lymph nodes. Although very disfiguring, it proved to be no more effective than simple mastectomy or lumpectomy. Surgery is generally followed by radiation or chemotherapy, and estrogen-sensitive tumors may also be treated with an estrogen blocker such as tamoxifen. A natural-looking breast can often be reconstructed from skin, fat, and muscle from other parts of the body.

## Before You Go On

*Answer the following questions to test your understanding of the preceding section:*

1. *How do the site of female gamete production and mode of release from the gonad differ from those in the male?*

2. *How is the structure of the uterine tube mucosa related to its function?*

3. *Contrast the function of the endometrium with that of the myometrium.*

4. *Describe the similarities and differences between the clitoris and penis.*

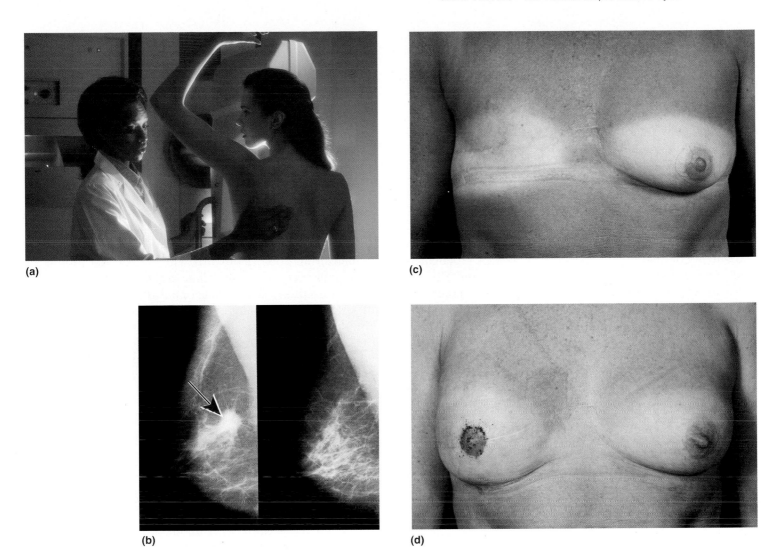

**FIGURE 28.11 Breast Cancer Screening and Treatment.** (a) Radiologic technologist assisting a patient in mammography. (b) Mammogram of a breast with a tumor visible at the arrow *(left)*, compared with the appearance of normal fibrous connective tissue of the breast *(right)*. (c) Patient following mastectomy of the right breast. (d) The same patient following surgical breast reconstruction.
▶ *What is the diagnostic benefit of compressing the breast for a mammogram?*

# Puberty and Menopause

### Objectives
When you have completed this section, you should be able to

- name the hormones that regulate female reproductive function, and state their roles;
- describe the principal signs of puberty;
- describe the hormonal changes of female climacteric and their effects; and
- define and describe menopause, and distinguish menopause from climacteric.

Puberty and menopause are physiological transitions at the beginning and end of a female's reproductive years.

## PUBERTY

Puberty begins at ages 9 to 10 for most girls in the United States and Europe, but significantly later in many countries. However, a 1997 study of more than 17,000 girls in the United States showed 3% of black girls and 1% of white girls beginning puberty by age 3, and 27% and 7%, respectively, by age 7. Puberty is triggered by the same hypothalamic and pituitary hormones in girls as it is in boys. Rising levels of gonadotropin-releasing hormone (GnRH) stimulate the anterior lobe of the pituitary to secrete follicle-stimulating hormone (FSH) and luteinizing hormone (LH). FSH, especially, stimulates development of the ovarian follicles, which, in turn, secrete estrogens, progesterone, inhibin, and a small amount of androgen. These hormone levels rise gradually from ages 8 to 12 and then more sharply in the early teens. The **estrogens**[24] are feminizing hormones with widespread effects on the body. They include *estradiol* (the most abundant), *estriol,* and *estrone.* Most of the visible changes at puberty result from estradiol and androgens.

The earliest noticeable sign of puberty is **thelarche**[25] (thee-LAR-kee), the onset of breast development. Estrogen, progesterone, and prolactin initially induce the formation of lobules and ducts in the breast. Duct development is completed under the influence of glucocorticoids and growth hormone, while adipose and fibrous tissue enlarge the breast. Breast development is complete around age 20, but minor changes occur in each menstrual cycle and major changes occur in pregnancy.

Thelarche is soon followed by **pubarche** (pyu-BAR-kee), the appearance of pubic and axillary hair, sebaceous glands, and axillary glands. Androgens from the ovaries and adrenal cortex stimulate pubarche as well as the libido. Women secrete about 0.5 mg of androgens per day, compared with 6 to 8 mg/day in men.

Next comes **menarche**[26] (men-AR-kee), the first menstrual period. In Europe and America, the average age at menarche declined from age 16.5 in 1860 to age 12 in 1997, mostly because of improved nutrition. Menarche cannot occur until a girl has attained at least 17% body fat, and adult menstruation generally ceases if a woman drops below 22% fat. This is about the minimum needed to sustain pregnancy and lactation; thus the body reacts as if to prevent a futile pregnancy when it is too lean.

As explained in chapter 26, the amount of body fat is detectable from the amount of leptin in the blood. In addition to its role in regulating appetite, leptin stimulates gonadotropin secretion. Therefore, if body fat and leptin levels drop too low, gonadotropin secretion declines and a girl's or woman's menstrual cycle may cease. Adolescent girls with very low body fat, such as many dancers and gymnasts, tend to begin menstruating at a later age than average.

Menarche does not necessarily signify fertility. A girl's first few menstrual cycles are typically *anovulatory* (no egg is ovulated). Most girls begin ovulating regularly about a year after they begin menstruating.

Estradiol stimulates many other changes of puberty. It causes the vaginal metaplasia described earlier. It stimulates growth of the ovaries and secondary sex organs. It stimulates growth hormone secretion and causes a rapid increase in height and widening of the pelvis. Estradiol is largely responsible for the feminine physique because it stimulates fat deposition in the mons pubis, labia majora, hips, thighs, buttocks, and breasts. It makes a girl's skin thicken, but the skin remains thinner, softer, and warmer than in males of corresponding age.

**Progesterone**[27] acts primarily on the uterus, preparing it for possible pregnancy in the second half of each menstrual cycle and playing roles in pregnancy discussed later. Estrogens and progesterone also suppress FSH and LH secretion through negative feedback inhibition of the anterior pituitary. **Inhibin** selectively suppresses FSH secretion.

Thus we see many hormonal similarities in males and females from puberty onward. The sexes differ less in the hormones that are present than in the relative amounts of those hormones—high levels of androgens and low levels of estrogens in males and the opposite in females. Another difference is that these hormones are secreted more or less continually and simultaneously in males, whereas in females secretion is distinctly cyclic and the hormones are secreted in sequence. This will be very apparent as you read about the ovarian and menstrual cycles.

## CLIMACTERIC AND MENOPAUSE

Women, like men, go through a midlife change in hormone secretion called the **climacteric.** In women, it is accompanied by **menopause,** the cessation of menstruation (see Insight 28.2).

A female is born with about 2 million eggs in her ovaries, each in its own follicle. The older she gets, the fewer follicles remain. Climacteric begins not at any specific age, but when she has only about 1,000 eggs left. Even the remaining follicles are less responsive to gonadotropins, so they secrete less estrogen and progesterone. Without these steroids, the uterus, vagina, and breasts atrophy. Intercourse may become uncomfortable, and vaginal infections more common, as the vagina becomes thinner, less distensible, and drier. The skin becomes thinner, cholesterol levels rise (increasing the risk of cardiovascular disease), and bone mass declines (increasing the risk of osteoporosis). Blood vessels constrict and dilate in response to shifting hormone balances, and the sudden dilation of cutaneous arteries may cause **hot flashes**—a spreading sense of heat from the abdomen to the thorax, neck, and face. Hot flashes may occur several times a day, sometimes accompanied by headaches resulting from the sudden vasodilation of arteries in the head. In some people, the changing hormonal profile also causes mood changes. Many physicians prescribe hormone replacement therapy (HRT)—low doses of estrogen

---

[24]*estro* = desire, frenzy + *gen* = to produce
[25]*thel* = breast, nipple + *arche* = beginning
[26]*men* = monthly

[27]*pro* = favoring + *gest* = pregnancy + *sterone* = steroid hormone

# Oogenesis and the Sexual Cycle

## INSIGHT 28.2 Evolutionary Medicine

### The Evolution of Menopause

There has been considerable speculation about why women do not remain fertile to the end of their lives, as men do. Some theorists argue that menopause served a biological purpose for our prehistoric foremothers. Human offspring take a long time to rear. Beyond a certain point, the frailties of age make it unlikely that a woman could rear another infant to maturity or even survive the stress of pregnancy. She might do better in the long run to become infertile and finish rearing her last child, or help to rear her grandchildren, instead of having more. In this view, menopause was biologically advantageous for our ancestors—in other words, an evolutionary adaptation.

Others argue against this hypothesis on the grounds that Ice Age skeletons indicate that early hominids rarely lived past age 40. If this is true, menopause setting in at 45 to 55 years of age could have served little purpose. In this view, Ice Age women may indeed have been fertile to the end of their lives; menopause now may be just an artifact of modern nutrition and medicine, which have made it possible for us to live much longer than our ancestors did.

and progesterone taken orally or by a skin patch—to relieve some of these symptoms. The risks and benefits of HRT are still being debated.

### Think About It

*FSH and LH secretion rise at climacteric and these hormones attain high concentrations in the blood. Explain this using the preceding information and what you know about the pituitary-gonadal axis.*

Menopause is the cessation of menstrual cycles, usually occurring between the ages of 45 and 55. The average age has increased steadily in the last century and is now about 52. It is difficult to precisely establish the time of menopause because the menstrual periods can stop for several months and then begin again. Menopause is generally considered to have occurred when there has been no menstruation for a year or more.

### Before You Go On

*Answer the following questions to test your understanding of the preceding section:*

5. Describe the similarities and differences between male and female puberty.

6. Describe the major changes that occur in female climacteric and the principal cause of these changes.

7. What is the difference between climacteric and menopause?

## Oogenesis and the Sexual Cycle

### Objectives

When you have completed this section, you should be able to

- describe the process of egg production (oogenesis);
- describe how the ovarian follicles change in relation to oogenesis;
- describe the hormonal events that regulate the ovarian cycle;
- describe how the uterus changes during the menstrual cycle; and
- construct a chart of the phases of the monthly sexual cycle showing the hormonal, ovarian, and uterine events of each phase.

The reproductive lives of women are conspicuously cyclic. They include the **reproductive cycle,** which encompasses the sequence of events from fertilization to giving birth, and the **sexual cycle,** which encompasses the events that recur every month when pregnancy does not intervene. The sexual cycle, in turn, consists of two interrelated cycles controlled by shifting patterns of hormone secretion: the **ovarian cycle,** consisting of events in the ovaries, and the **menstrual cycle,** consisting of parallel changes in the uterus. In this section, we examine the process of egg production, or oogenesis, and the monthly events of the ovarian and menstrual cycles.

### OOGENESIS

Egg production is called **oogenesis**[28] (OH-oh-JEN-eh-sis) (fig. 28.12). Like spermatogenesis, it produces a haploid gamete by means of meiosis. There are, however, numerous differences between oogenesis and spermatogenesis. The most obvious, perhaps, is that spermatogenesis goes on continually, while oogenesis is a distinctly cyclic event that normally produces only one egg per month. Oogenesis is accompanied by cyclic changes in hormone secretion and in the histological structure of the ovaries and uterus; the uterine changes result in the monthly menstrual flow.

The female germ cells arise, like those of the male, from the yolk sac of the embryo. They colonize the gonadal ridges in the first 5 to 6 weeks of development and then differentiate into **oogonia** (OH-oh-GO-nee-uh). Oogonia multiply until the fifth month, reach 6 to 7 million in number, and then go into a state of arrested development until shortly before birth. At that time, some of them transform into **primary oocytes** and go as far as

[28]*oo* = egg + *genesis* = producton

Development of the egg (oogenesis)                    Development of the follicle

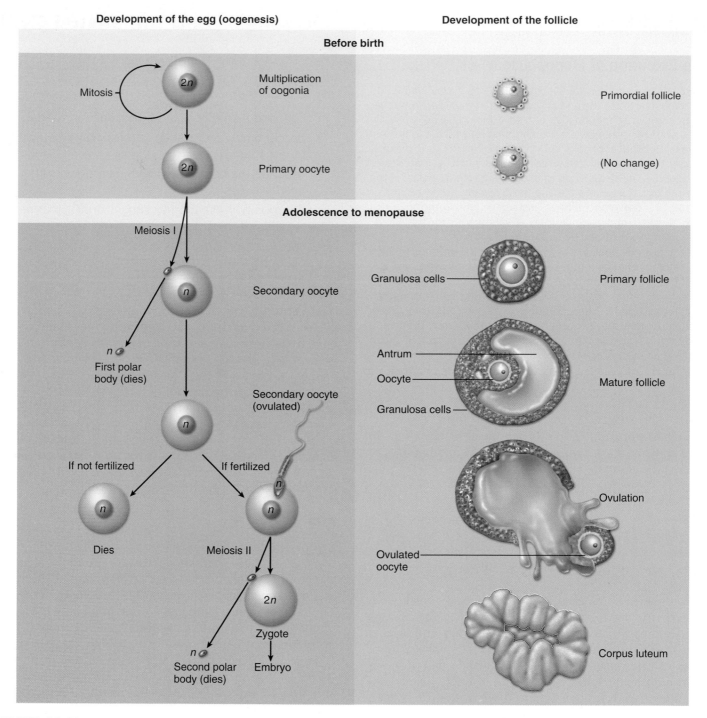

**FIGURE 28.12** Oogenesis (left) and Corresponding Development of the Follicle (right).

early meiosis I. Any stage from the primary oocyte to the time of fertilization can be called an egg, or **ovum.**

Most primary oocytes undergo a process of degeneration called **atresia** (ah-TREE-zhee-uh) before a girl is born. Only 2 million remain at the time of birth, and most of those undergo atresia during childhood. By puberty, only 400,000 oocytes remain. This is the female's lifetime supply of gametes, but it is more than ample; even if she ovu-

lated every 28 days from the ages of 14 to 50, she would ovulate only 480 times.

Beginning in adolescence, FSH stimulates primary oocytes to complete meiosis I, which yields two haploid daughter cells of unequal size and different destinies. In oogenesis it is important to produce an egg with as much cytoplasm as possible, because if it is fertilized it must divide repeatedly and produce numerous daughter cells.

Splitting each oocyte into four equal but small parts would run counter to this purpose. Therefore, meiosis I produces a large daughter cell called the **secondary oocyte** and a much smaller one called the **first polar body.** The polar body sometimes undergoes meiosis II but ultimately disintegrates. It is merely a means of discarding the extra haploid set of chromosomes.

The secondary oocyte proceeds as far as metaphase II and then arrests until after ovulation. If it is not fertilized, it dies and never finishes meiosis. If it is fertilized, it completes meiosis II and produces a **second polar body,** which disposes of one chromatid from each chromosome. The chromosomes of the large remaining egg unite with those of the sperm. Further development of the fertilized egg is discussed in chapter 29.

## THE SEXUAL CYCLE

The sexual cycle averages 28 days in length, which is the basis for the timetable described in the following pages. It commonly varies from 20 to 45 days, however, so be aware that the timetable given in this discussion may differ from person to person and from month to month. As you study this cycle, bear in mind that hormones of the hypothalamus and anterior pituitary gland regulate the ovaries; the ovaries, in turn, secrete hormones that regulate the uterus. That is, the basic hierarchy of control can be represented: hypothalamus → pituitary → ovaries → uterus. However, there is also feedback control from the ovaries to the hypothalamus and pituitary.

We begin with a brief preview of the sexual cycle as a whole. The cycle begins with a 2-week *follicular phase.* Menstruation occurs during the first 3 to 5 days, and then the uterus rebuilds the lost endometrial tissue. The ovarian follicles grow during this phase, and one of them ovulates around day 14. After ovulation, the remainder of the

follicle becomes a body called the corpus luteum. Over the next 2 weeks, called the *luteal phase,* the corpus luteum stimulates endometrial secretion, and the endometrium thickens still more. If pregnancy does not occur, it breaks down again in the last 2 days. As loose tissue and blood accumulate, menstruation begins and the cycle starts over.

## The Ovarian Cycle

We can now examine the ovarian cycle in more detail. We will see, step by step, what happens in the ovaries and in their relationship to the hypothalamus and pituitary gland.

The **follicular phase** of the cycle extends from the beginning of menstruation (day 1) to ovulation (day 14). This is the most variable part of the cycle and it is seldom possible to predict the date of ovulation reliably. We can divide the follicular phase into two periods, the *preantral* and *antral* phases, before and after the follicle develops a cavity called the *antrum.* The **preantral phase** in the life of any one follicle begins before birth, at 12 to 16 weeks of gestation. It includes two stages of follicular development:

1. **The primordial follicle.** This stage consists of an oocyte in early meiosis, surrounded by a single layer of thin follicular cells and a basement membrane that isolates the follicle from the surrounding stroma of the ovary (fig. 28.13a). The follicular cells are connected to the oocyte surface by fine cytoplasmic processes that allow for chemical signaling between cells. Primordial follicles are concentrated in the cortex of the ovary, close to the capsule.

2. **The primary follicle.** Some follicles advance to this stage as early as 21 weeks of gestation. The follicular cells become rounded or cuboidal, multiply, and

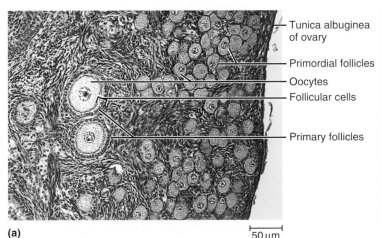

(a)  50 µm

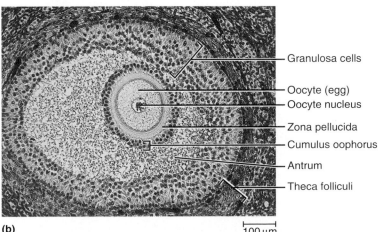

Tunica albuginea of ovary

Primordial follicles

Oocytes

Follicular cells

Primary follicles

Granulosa cells

Oocyte (egg)
Oocyte nucleus

Zona pellucida

Cumulus oophorus

Antrum

Theca folliculi

(b)  100 µm

**FIGURE 28.13** **Ovarian Follicles.** (a) Note the very thin layer of squamous cells around the oocyte in a primordial follicle, and the single layer of cuboidal cells in a primary follicle. (b) A mature (graafian) follicle. Just before ovulation, this follicle will grow to as much as 2.5 cm in diameter.

begin piling up in layers around the oocyte. The primary follicles in figure 28.13a are at an early stage, exhibiting cuboidal follicular cells that are not yet stratified. After stratification, they are called **granulosa cells.** These cells then secrete a glycoprotein gel around the oocyte called the **zona pellucida.**[29] The connective tissue around the granulosa cells condenses to form the **theca**[30] **folliculi** (THEE-ca fol-IC-you-lye). This is as far as any follicles develop prenatally; they resume development after menarche. Many primordial follicles persist into adultood and do not advance even to the primary follicle stage until then.

The preceding stages (1–2) occur autonomously, without stimulation by pituitary gonadotropins, but further development depends on FSH and LH. Soon after the midpoint (ovulation) of one ovarian cycle, a new group of follicles begin a race, the winner of which will ovulate in the cycle 2 months later. About 20 follicles descend from the cortex deeper into the ovary and progress to the **antral phase.** In the following stages (3–4), one of these emerges as the *dominant follicle* destined to ovulate.

3. **The secondary (antral) follicle.** In days 1 to 5 (concurrently with menstruation), granulosa cells begin to secrete **follicular fluid,** which accumulates in little pools amid the cells. These pools soon merge and become a single fluid-filled cavity, the **antrum;** the follicle is now considered to be a secondary (antral) follicle (fig. 28.13b). A mound of granulosa cells called the **cumulus oophorus**[31] covers the oocyte and secures it to the follicle wall. The innermost layer of cells in the cumulus, surrounding the zona pellucida, is the **corona radiata.**[32] These cells maintain communication with the oocyte by way of cytoplasmic processes that cross the zona pellucida. The theca folliculi also continues to differentiate, layering to form an outer fibrous capsule, the *theca externa,* and an inner, cellular, hormone-secreting layer, the *theca interna.* Theca interna cells provide the granulosa cells with androgens (androstenedione and testosterone), which the granulosa cells convert to estradiol.

4. **The dominant follicle.** By day 5, one of the secondary follicles emerges as the one destined to ovulate. Now 2 to 5 mm in diameter, this is called the **dominant follicle;** when fully developed and ready to ovulate, it is named the **mature (graafian**[33]**) follicle.**

Days 6 to 14 are considered the **preovulatory phase** of the cycle. Because of a complex exchange of hormones between the pituitary gland and follicle, the dominant follicle grows very rapidly during this phase. In brief, FSH stimulates the granulosa cells to secrete estradiol. That stimulates them to produce still more estradiol receptors as well as FSH and LH receptors. Estradiol, meanwhile, also stimulates the anterior pituitary to increase its LH output and reduce FSH output. One would think the declining FSH level would be detrimental to the ovarian follicles, and indeed it is—to all except the dominant follicle. That one has the richest blood supply and most FSH receptors of all, so it suffers least from the drop in FSH secretion. The lagging follicles of the group, however, suffer from the withdrawal of FSH and degenerate (undergo atresia).

**Ovulation,** the rupture of a follicle and release of the oocyte, typically occurs on day 14. Dramatic changes over the preceding day signify its imminence. In steps 5 through 7, the follicle swells rapidly and bursts.

5. Estradiol from the mature follicle stimulates a surge in pituitary LH secretion and a lesser spike in FSH secretion (fig. 28.14 and the midpoint of fig. 28.15). Several momentous things happen under the influence of

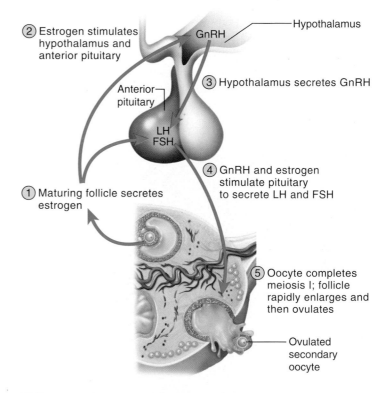

② Estrogen stimulates hypothalamus and anterior pituitary

Hypothalamus

GnRH

Anterior pituitary

③ Hypothalamus secretes GnRH

LH FSH

④ GnRH and estrogen stimulate pituitary to secrete LH and FSH

① Maturing follicle secretes estrogen

⑤ Oocyte completes meiosis I; follicle rapidly enlarges and then ovulates

Ovulated secondary oocyte

**FIGURE 28.14**  **Control of Ovulation by Hormones of the Pituitary–Ovarian Axis.**

[29]*zona* = zone + *pellucid* = clear, transparent
[30]*theca* = box, case
[31]*cumulus* = little mound + *oo* = egg + *phor* = to carry
[32]*corona* = crown + *radiata* = radiating
[33]Reijnier de Graaf (1641–73), Dutch physiologist and histologist

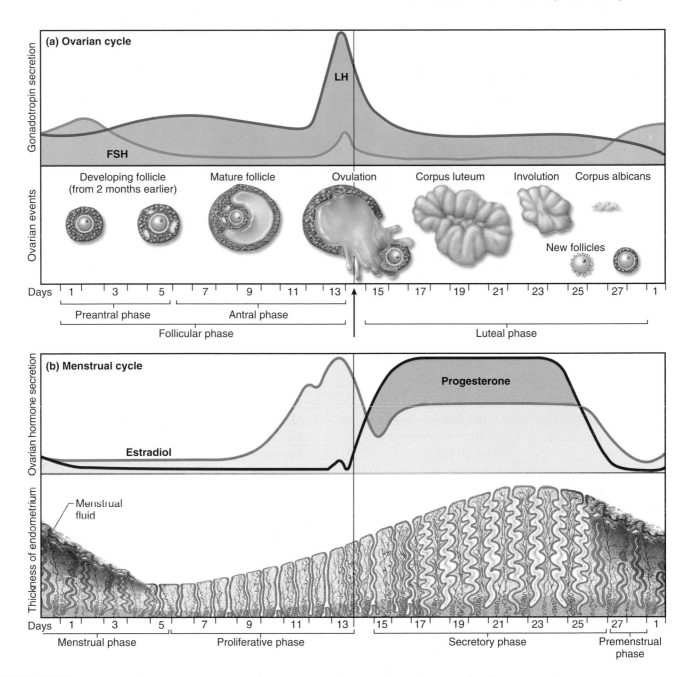

**FIGURE 28.15** **The Female Sexual Cycle.** (a) The ovarian cycle (events in the ovary). (b) The menstrual cycle (events in the uterus). The two hormone levels in part (a) are drawn to the same scale, but those in part (b) are not. The peak progesterone concentration is about 17 times as high as the peak estradiol concentration.

this stimulus. The oocyte resumes meiosis and completes meiosis I. It is now a secondary oocyte. Follicular fluid builds rapidly and the follicle increases in size, reaching a diameter of 20 mm in the last day or two before ovulation. The follicular wall and adjacent ovarian tissue are weakened by inflammation and a proteolytic enzyme, **plasmin** (the same enzyme that digests old blood clots). With mounting internal pressure and a weakening wall, the follicle approaches rupture.

6. Meanwhile, the uterine tube prepares to catch the oocyte when it emerges. It becomes edematous; its fimbriae envelop and caress the ovary in synchrony with the woman's heartbeat; and its cilia create a gentle current in the nearby peritoneal fluid.

7. Ovulation itself takes only 2 or 3 min. A nipplelike **stigma** appears on the ovarian surface over the follicle. It seeps follicular fluid for 1 or 2 min, and then the follicle bursts. The remaining fluid oozes out,

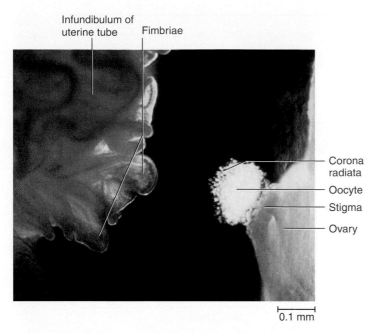

Infundibulum of uterine tube
Fimbriae
Corona radiata
Oocyte
Stigma
Ovary

0.1 mm

**FIGURE 28.16** Endoscopic View of Human Ovulation.

carrying the oocyte and corona radiata (fig. 28.16). These are normally swept up by the ciliary current and taken into the uterine tube, although many oocytes fall into the pelvic cavity and die.

## Think About It

*In chapter 17, review the concepts of up-regulation and the permissive effect in hormone interactions. Explain their relevance to the first 14 days of the ovarian cycle.*

## INSIGHT 28.3    Clinical Application

### Signs of Ovulation

If a couple is attempting to conceive a child or to avoid pregnancy, it is important to be able to tell when ovulation occurs. The signs are subtle but detectable. For one, the cervical mucus becomes thinner and more stretchy. Also, the resting body temperature *(basal temperature)* rises 0.2° to 0.3°C (0.4°–0.6°F). This is best measured first thing in the morning, before rising from bed; the change can be detected if basal temperatures are recorded for several days before ovulation in order to see the difference. The LH surge that occurs about 24 hours before ovulation can be detected with a home testing kit. Finally, some women experience twinges of ovarian pain known by the German name, *mittelschmerz,*[34] which last from a few hours to a day or so at the time of ovulation. The most likely time to become pregnant is within 24 hours after the cervical mucus changes consistency and the basal temperature rises.

The **luteal (postovulatory) phase** (steps 8–11) extends from day 15 to day 28, from just after ovulation to the onset of menstruation. Assuming pregnancy does not occur, the major events of this phase are as follows:

8. When the follicle ruptures, it collapses and bleeds into the antrum. As the clotted blood is slowly absorbed, granulosa and theca interna cells multiply and fill the antrum, and a dense bed of blood capillaries grows amid them. The ovulated follicle has now become a structure called the **corpus luteum,**[35] named for a yellow lipid that accumulates in the theca interna cells. These cells are now called **lutein cells.** This transformation from ruptured follicle to corpus luteum is regulated by LH; hence LH is also called *luteotropic hormone.*

9. LH stimulates the corpus luteum to continue growing and to secrete rising levels of estradiol and progesterone. The most important aspect of the luteal phase is a tenfold increase in progesterone level (see fig. 28.15). This hormone has a crucial role in preparing the uterus for the possibility of pregnancy.

10. Notwithstanding its luteinizing role, LH secretion declines steadily over the rest of the cycle, as does FSH. This is because the high levels of estradiol and progesterone, along with inhibin from the corpus luteum, have a negative feedback effect on the pituitary. (This fact is the basis for hormonal birth control; see Insight 28.4.)

11. If pregnancy does not occur, the corpus luteum begins a process of **involution,** or shrinkage, beginning around day 22 (8 days after ovulation). By day 26, involution is complete and the corpus luteum has become an inactive bit of scar tissue called the **corpus albicans.**[36] With the waning of ovarian steroid secretion, the pituitary is no longer inhibited and FSH levels begin to rise again, ripening a new cohort of follicles.

All of these events repeat themselves every month, but bear in mind that the follicles engaged in each monthly cycle began their development prenatally, as much as 50 years earlier, and the oocyte that ovulates each month began ripening 2 months earlier, not in the cycle when it ovulates or even the one immediately before. Table 28.1 summarizes the main events of the ovarian cycle and correlates them with events of the menstrual cycle, which we will now examine.

## The Menstrual Cycle

The menstrual cycle consists of a buildup of the endometrium through most of the sexual cycle, followed by its breakdown and vaginal discharge. The menstrual cycle is

---

[34]*mittel* = in the middle + *schmerz* = pain

[35]*corpus* = body + *lute* = yellow
[36]*corpus* = body + *alb* = white

| TABLE 28.1 | Phases of the Female Ovarian Cycle | |
|---|---|---|
| **Days** | **Phase** | **Major Features** |
| 1–14 | *Follicular phase* | Development of ovarian follicles and secretion primarily of estradiol. Coincides with menstrual and proliferative phases of the menstrual cycle. |
| | Preantral phase | Follicles lack an antrum |
| | Primordial follicles | Formed prenatally and many persist into adulthood. Consist of an oocyte surrounded by a single layer of squamous follicular cells. |
| | Primary follicles | Develop in cohorts from primordial follicles prenatally and throughout reproductive life. Consist of an oocyte surrounded by one or more layers of cuboidal follicular cells. Follicular cells stratify, become granulosa cells, and secrete a zona pellucida. Theca folliculi forms around follicle. One follicle becomes the dominant follicle during the menstrual phase of the menstrual cycle. |
| | Antral phase | Follicles possess an antrum |
| | Secondary follicles | Begin to develop from the cohort of primary follicles 2 months before one of them is scheduled to ovulate. Form an antrum filled with follicular fluid and exhibit a cumulus oophorus and corona radiata. |
| | Dominant follicle | The secondary follicle that is destined to ovulate. Present by the end of the menstrual phase. Hormonally dominates the rest of the cycle, while other follicles in the cohort undergo atresia. Secretes mainly estradiol. Coincides with the proliferative phase of the menstrual cycle, in which the uterine endometrium thickens by mitosis. |
| | Mature (graafian) follicle | The dominant follicle just prior to ovulation. Attains a diameter up to 20 mm and builds to high internal fluid pressure as adjacent ovarian wall weakens. |
| 14 | *Ovulation* | Rupture of mature follicle and release of oocyte |
| 15–28 | *Luteal (postovulatory) phase* | Dominated by corpus luteum. Coincides with secretory and premenstrual phases of the menstrual cycle. |
| | Corpus luteum | Develops from ovulated follicle by proliferation of granulosa and theca interna cells. Tenfold rise in progesterone level stimulates thickening of endometrium by secretion (secretory phase of the menstrual cycle). Begins to involute by day 22 in the absence of pregnancy; involution complete by day 26. |
| | Corpus albicans | Scar tissue left by involution of corpus luteum; not hormonally active. In the absence of progesterone, endometrium exhibits ischemia, necrosis, and sloughing of tissue. Necrotic endometrial tissue mixes with blood and forms menstrual fluid. |

divided into a *menstrual phase, proliferative phase, secretory phase,* and *premenstrual phase,* in that order (fig. 28.17). The menstrual phase averages 5 days long, and the first day of noticeable vaginal discharge is defined as day 1 of the sexual cycle. The reason for menstruation is best understood after you become acquainted with the buildup of endometrial tissue that precedes it, so we begin our study with the proliferative phase.

**Proliferative Phase** The **proliferative phase** is a time of rebuilding of endometrial tissue lost at the last menstruation. At the end of menstruation, around day 5, the endometrium is about 0.5 mm thick and consists of only the stratum basalis. The stratum functionalis is rebuilt by mitosis from day 6 to day 14. The principal processes in this phase are:

1. Estrogen from the ovaries stimulates mitosis in the stratum basalis as well as the prolific regrowth of

blood vessels. By day 14, the endometrium is about 2 to 3 mm thick.

2. Estrogen also stimulates the endometrium to produce progesterone receptors, thereby preparing it for the progesterone-dominated secretory phase.

**Secretory Phase** The **secretory phase** is a period of further endometrial thickening, but this results from secretion and fluid accumulation rather than mitosis. It extends from day 15 (after ovulation) to day 26 of a typical cycle. The principal processes in this phase are:

3. After ovulation, the corpus luteum secretes mainly progesterone. Progesterone stimulates the endometrial glands and cells of the stroma to accumulate glycogen. The glands grow wider, longer, and more coiled and secrete a glycogen-rich fluid into the lumen. The lamina propria swells with tissue fluid.

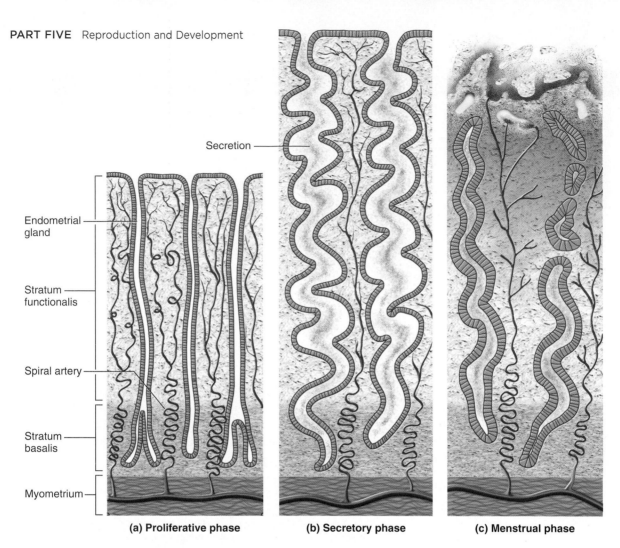

Secretion

Endometrial gland

Stratum functionalis

Spiral artery

Stratum basalis

Myometrium

**(a) Proliferative phase**    **(b) Secretory phase**    **(c) Menstrual phase**

**FIGURE 28.17** **Endometrial Changes Through the Menstrual Cycle.** (a) Late proliferative phase. The endometrium is 2 to 3 mm thick and has relatively straight, narrow endometrial glands. (b) Secretory phase. The endometrium thickens to 5 to 6 mm by accumulating glycogen and mucus. The endometrial glands are wider and more distinctly coiled, showing a zigzag or "sawtooth" appearance in tissue sections. (c) Menstrual phase. Ischemic tissue dies and falls away from the uterine wall, with bleeding from broken blood vessels and pooling of blood in the tissue and uterine lumen.

4. By the end of the secretory phase, the endometrium is about 5 to 6 mm thick—a soft, wet, nutritious bed available for embryonic development in the event of pregnancy.

*Premenstrual Phase* The **premenstrual phase** is a period of endometrial degeneration occurring in the last 2 days or so of the menstrual cycle.

5. As we have seen, in the absence of pregnancy, the corpus luteum atrophies and the progesterone level falls sharply. In the absence of progesterone, the spiral arteries of the endometrium exhibit spasmodic contractions that cause endometrial ischemia (interrupted blood flow). The premenstrual phase is therefore also called the **ischemic** (iss-KEE-mic) **phase.**

6. Ischemia causes tissue necrosis (and menstrual cramps). As the endometrial glands, stroma, and blood vessels degenerate, pools of blood accumulate in the stratum functionalis.

7. Necrotic endometrium falls away from the uterine wall, mixes with blood in the lumen, and forms the **menstrual fluid.**

*Menstrual Phase* The **menstrual phase** (menses) is the period in which blood, serous fluid, and degenerated endometrial tissue are discharged from the vagina. It commences when enough menstrual fluid accumulates in the uterus. The first day of external discharge marks day 1 of a new cycle. The average woman discharges about 40 mL of blood and 35 mL of serous fluid over a 5-day period. Menstrual fluid contains fibrinolysin, so it does not clot. The vaginal discharge of clotted blood may indicate uterine pathology rather than normal menstruation.

In summary, the ovaries go through a follicular phase characterized by growing follicles; then ovulation; and then a postovulatory (mostly luteal) phase dominated by the corpus luteum. The uterus, in the meantime, goes through a menstrual phase in which it discharges its stratum functionalis; then a proliferative phase in which it replaces that tissue by mitosis; then a secretory phase in which the endometrium thickens by the accumulation of secretions; and finally, a premenstrual (ischemic) phase in which the stratum functionalis breaks down again. The first half of the cycle is governed largely by follicle-stimulating hormone (FSH) from the pituitary gland and estrogen from the ovaries. Ovulation is triggered by luteinizing hormone (LH) from the pituitary, and the second half of the cycle is governed mainly by LH and progesterone, the latter secreted by the ovaries.

## Before You Go On

*Answer the following questions to test your understanding of the preceding section:*

8. *Name the sequence of cell types in oogenesis and identify the ways oogenesis differs from spermatogenesis.*

9. *Distinguish between a primordial, primary, and secondary follicle. Describe the major structures of a mature follicle.*

10. *Describe what happens in the ovary during the follicular and postovulatory phases.*

11. *Describe what happens in the uterus during the menstrual, proliferative, secretory, and premenstrual phases.*

12. *Describe the effects of FSH and LH on the ovary.*

13. *Describe the effects of estrogen and progesterone on the uterus, hypothalamus, and anterior pituitary.*

# Female Sexual Response

### Objectives

When you have completed this section, you should be able to

- describe the female sexual response at each phase of intercourse; and

- compare and contrast the female and male responses.

Female sexual response, the physiological changes that occur during intercourse, may be viewed in terms of the four phases identified by Masters and Johnson and discussed in chapter 27: excitement, plateau, orgasm, and resolution (fig. 28.18). The neurological and vascular controls of female sexual response are essentially the same as in the male (pp. 1056–1058) and need not be repeated here. The emphasis here is on ways the female response differs from that of the male.

## EXCITEMENT AND PLATEAU

Excitement is marked by myotonia, vasocongestion, and increased heart rate, blood pressure, and respiratory rate.

Although vasocongestion works by the same mechanism in both sexes, its effects are quite different in females. The labia minora become congested and often protrude beyond the labia majora. The labia majora become reddened and enlarged and then flatten and spread away from the vaginal orifice.

The vaginal wall becomes purple due to hyperemia, and serous fluid called the **vaginal transudate** seeps through the wall into the canal. Along with secretions of the greater vestibular glands, this moistens the vestibule and provides lubrication. The inner end of the vagina dilates and becomes cavernous, while the lower one-third of it constricts to form a narrow passage called the **orgasmic platform.** The narrower canal, combined with the vaginal rugae (friction ridges), enhances stimulation and helps induce orgasm in both partners.

The uterus, which normally tilts forward over the urinary bladder, stands more erect during excitement and the cervix withdraws from the vagina. In plateau, the uterus is nearly vertical and extends into the false pelvis. This is called the **tenting effect.**

Although the vagina is the female copulatory organ, the clitoris is more comparable to the penis in structure, physiology, and importance as the primary focus of erotic stimulation. It has a high concentration of sensory nerve endings, which, by contrast, are relatively scanty in the vagina. Recall that the penis and clitoris are homologous structures, both arising from the genital tubercle of the fetus. Both have a pair of corpora cavernosa with *deep arteries,* and both become engorged by the same mechanism. The glans and shaft of the clitoris swell to two or three times their unstimulated size, but since the clitoris cannot swing upward away from the body like the penis, it tends to withdraw beneath the prepuce. Thrusting of the penis in the vagina tugs on the labia minora and, by extension, pulls on the prepuce and stimulates the clitoris. The clitoris may also be stimulated by pressure between the pubic symphyses of the partners.

The breasts also become congested and swollen during the excitement phase, and the nipples become erect. Stimulation of the breasts also enhances sexual arousal.

## ORGASM

Late in plateau, many women experience involuntary pelvic thrusting, followed by 1 to 2 seconds of "suspension" or "stillness" preceding orgasm. Orgasm is commonly described as an intense sensation progressing from the clitoris through the pelvis, sometimes with pelvic throbbing and a spreading sense of warmth. The orgasmic platform gives three to five strong contractions about 0.8 seconds apart, while the cervix plunges spasmodically into the vagina and pool of semen, should this be present. The uterus exhibits peristaltic waves of contraction; it is still debated whether or not this helps to draw semen from the vagina. The anal and urethral sphincters constrict, and the paraurethral glands, which are homologous

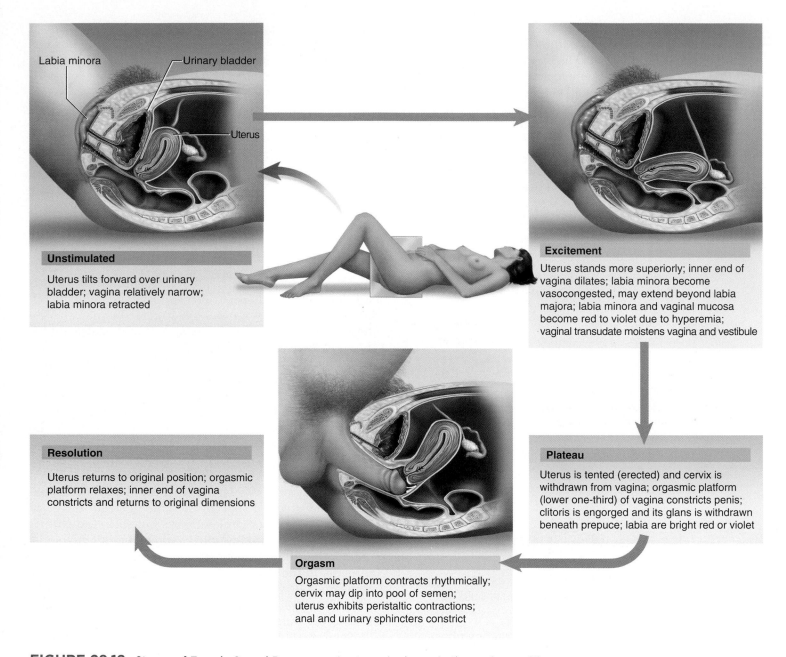

**FIGURE 28.18   Stages of Female Sexual Response.**  Anatomy is shown in the supine position.

to the prostate, sometimes expel fluid similar to prostatic fluid ("female ejaculation"). Tachycardia and hyperventilation occur; the breasts enlarge still more and the areolae often become engorged; and in many women a reddish, rashlike flush appears on the lower abdomen, chest, neck, and face.

## RESOLUTION

During resolution the uterus drops forward to its resting position. The orgasmic platform quickly relaxes, while the inner end of the vagina returns more slowly to its normal dimensions. The flush disappears quickly and the areolae and nipples undergo rapid detumescence, but it

may take 5 to 10 minutes for the breasts to return to their normal size. In many women (and men) there is a postorgasmic outbreak of perspiration. Unlike men, women do not have a refractory period and may quickly experience additional orgasms.

### Before You Go On

*Answer the following questions to test your understanding of the preceding section:*

14. *What are the female sources of lubrication in coitus?*

15. *What female tissues and organs become vasocongested?*

16. *Describe the actions of the uterus throughout the sexual response cycle.*

# Pregnancy and Childbirth

### Objectives

When you have completed this section, you should be able to

- list the major hormones that regulate pregnancy, and explain their roles;
- describe a woman's bodily adaptations to pregnancy;
- identify the physical and chemical stimuli that increase uterine contractility in late pregnancy;
- describe the mechanism of labor contractions;
- name and describe the three stages of labor; and
- describe the physiological changes that occur in the weeks following childbirth.

This section treats pregnancy from the maternal standpoint—that is, adjustments of the woman's body to pregnancy, and the mechanism of childbirth. Development of the fetus is described in chapter 29.

**Gestation** (pregnancy) lasts an average of 266 days from conception to childbirth, but the gestational calendar is usually measured from the first day of the woman's last menstrual period (LMP). Thus the birth is predicted to occur 280 days (about 40 weeks) from LMP. The duration of pregnancy, called its *term,* is commonly described in 3-month intervals called **trimesters.**

## PRENATAL DEVELOPMENT

A few fundamental facts of fetal development must be introduced as a foundation for understanding maternal physiology. All the products of conception—the embryo or fetus as well as the placenta and membranes associated with it—are collectively called the **conceptus.** The conceptus is a hollow ball called a *blastocyst* for much of the first 2 weeks, an *embryo* from 2 through 8 weeks, and a *fetus* from the beginning of week 9 until birth. The fetus is attached by way of an *umbilical cord* to a disc-shaped organ, the *placenta,* on the uterine wall. The placenta provides fetal nutrition and waste disposal, and secretes hormones that regulate pregnancy, mammary development, and fetal development. For the first 6 weeks after birth, the infant is called a *neonate.*[37]

## HORMONES OF PREGNANCY

The hormones with the strongest influences on pregnancy are estrogens, progesterone, human chorionic gonadotropin, and human chorionic somatomammotropin. These are secreted primarily by the placenta, but the corpus luteum is an important source of hormones in the

first 7 to 12 weeks. If the corpus luteum is removed before the seventh week, abortion almost always occurs. From weeks 7 to 17, the corpus luteum degenerates and the placenta takes over its endocrine functions.

## Human Chorionic Gonadotropin

**Human chorionic gonadotropin (HCG)** is secreted by the blastocyst and placenta. Its presence in the urine is the basis of pregnancy tests and can be detected with home testing kits as early as 8 or 9 days after conception. HCG secretion peaks around 10 to 12 weeks and then falls to a relatively low level for the rest of gestation (fig. 28.19). Like LH, it stimulates growth of the corpus luteum, which doubles in size and secretes increasing amounts of progesterone and estrogen. Without HCG, the corpus luteum would atrophy and the uterus would expel the conceptus.

## Estrogens

Estrogen secretion increases to about 30 times the usual amount by the end of gestation. The corpus luteum is an important source of estrogen for the first 12 weeks; after that, it comes mainly from the placenta. The adrenal glands of the mother and fetus secrete androgens, which the placenta converts to estrogens. The most abundant estrogen of pregnancy is estriol, but its effects are relatively weak; estradiol is less abundant but accounts for most of the estrogenic effects in pregnancy.

Estrogen stimulates tissue growth in the fetus and mother. It causes the mother's uterus and external genitalia to enlarge, the mammary ducts to grow, and the breasts to increase to nearly twice their former size. It

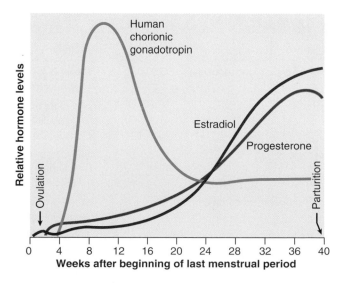

**FIGURE 28.19  Hormone Levels Over the Course of Pregnancy.**

▶ *How does the changing ratio of estradiol to progesterone relate to labor contractions?*

---

[37]*neo* = new + *nate* = born, birth

makes the pubic symphysis more elastic and the sacroiliac joints more limber, so the pelvis widens during pregnancy and the pelvic outlet expands during childbirth.

## Progesterone

The placenta secretes a great deal of progesterone, and early in the pregnancy, so does the corpus luteum. Progesterone and estrogen suppress pituitary secretion of FSH and LH, thereby preventing more follicles from developing during pregnancy. (This is the basis for contraceptive pills and implants; see Insight 28.4, p. 1096.) Progesterone also suppresses uterine contractions so the conceptus is not prematurely expelled. It prevents menstruation and promotes the proliferation of *decidual cells* of the endometrium, on which the blastocyst feeds. Once estrogen has stimulated growth of the mammary ducts, progesterone stimulates development of the secretory acini—another step toward lactation.

## Human Chorionic Somatomammotropin

The amount of **human chorionic somatomammotropin (HCS)** secreted in pregnancy is several times that of all the other hormones combined, yet its function is the least understood. The placenta begins secreting HCS around the fifth week and HCS output increases steadily from then until term, in direct proportion to the size of the placenta.

HCS is sometimes called *human placental lactogen* because, in other mammals, it causes mammary development and lactation; however, it does not induce lactation in humans. Its effects seem similar to those of growth hormone, but weaker. It also seems to reduce the mother's insulin sensitivity and glucose usage such that the mother consumes less glucose and leaves more of it for use by the fetus. HCS promotes the release of free fatty acids from the mother's adipose tissue, providing an alternative energy substrate for her cells to use in lieu of glucose.

## Other Hormones

Many other hormones induce additional bodily changes in pregnancy (table 28.2). A woman's pituitary gland grows about 50% larger during pregnancy and produces markedly elevated levels of thyrotropin, prolactin, and ACTH. The thyroid gland also becomes about 50% larger under the influence of HCG, pituitary thyrotropin, and *human chorionic thyrotropin* from the placenta. Elevated thyroid hormone secretion increases the metabolic rate of the mother and fetus. The parathyroid glands enlarge and stimulate osteoclast activity, liberating calcium from the mother's bones for fetal use. ACTH stimulates glucocorticoid secretion, which may serve primarily to mobilize amino acids for fetal protein synthesis. Aldosterone secretion rises and promotes fluid retention, contributing to the mother's increased blood volume. The corpus luteum and placenta secrete *relaxin,* which relaxes the pubic symphysis in other animals but does not seem to have this effect in humans. In humans, it synergizes with progesterone in stimulating the multiplication of decidual cells in early pregnancy and promotes the growth of blood vessels in the pregnant uterus.

## ADJUSTMENTS TO PREGNANCY

Pregnancy places considerable stress on a woman's body and requires adjustments in nearly all the organ systems. A few of the major adjustments and effects of pregnancy are described here.

| TABLE 28.2 | The Hormones of Pregnancy |
|---|---|
| **Hormone** | **Effects** |
| Human chorionic gonadotropin (HCG) | Prevents involution of corpus luteum and stimulates its growth and secretory activity; basis of pregnancy tests |
| Estrogens | Stimulate maternal and fetal tissue growth, including enlargement of uterus and maternal genitalia; stimulate development of mammary ducts; soften pubic symphysis and sacroiliac joints, facilitating pelvic expansion in pregnancy and childbirth; suppress FSH and LH secretion |
| Progesterone | Suppresses premature uterine contractions; prevents menstruation; stimulates proliferation of decidual cells, which nourish embryo; stimulates development of mammary acini; suppresses FSH and LH secretion |
| Human chorionic somatomammotropin (HCS) | Weak growth-stimulating effects similar to growth hormone; glucose-sparing effect on mother, making glucose more available to fetus; mobilization of fatty acids as maternal fuel |
| Pituitary thyrotropin | Stimulates thyroid activity and metabolic rate |
| Human chorionic thyrotropin | Same effect as pituitary thyrotropin |
| Parathyroid hormone | Stimulates osteoclasts and mobilizes maternal calcium |
| Adrenocorticotropic hormone | Stimulates glucocorticoid secretion; thought to mobilize amino acids for fetal protein synthesis |
| Aldosterone | Causes fluid retention, contributing to increased maternal blood volume |
| Relaxin | Promotes development of decidual cells and blood vessels in the pregnant uterus |

## Digestive System, Nutrition, and Metabolism

For many women, one of the first signs of pregnancy is morning sickness—nausea, especially after rising from bed—in the first few months of gestation. The cause of morning sickness is unknown. One hypothesis is that it stems from the reduced intestinal motility caused by the steroids of pregnancy. Another is that it is an evolutionary adaptation to protect the fetus from toxins. The fetus is most vulnerable to toxins at the same time that morning sickness peaks. Women with morning sickness tend to prefer bland foods and to avoid spicy and pungent foods, which are highest in toxic compounds. In some women, the nausea progresses to vomiting. Occasionally this is severe enough to require hospitalization (see *hyperemesis gravidarum* in table 28.5).

Constipation and heartburn are common in pregnancy. The former is another result of reduced intestinal motility. The latter is due to the enlarging uterus pressing upward on the stomach, causing the reflux of gastric contents into the esophagus.

The basal metabolic rate rises about 15% in the second half of gestation. Pregnant women often feel overheated because of this and the effort of carrying the extra weight. The appetite may be strongly stimulated, but a pregnant woman needs only 300 extra kcal/day even in the last trimester. With poor prenatal care and little self-control, however, some women greatly overeat and gain as much as 34 kg (75 lb) of weight compared with a healthy average of 11 kg (24 lb). Maternal nutrition should emphasize the quality of food eaten, not quantity.

During the last trimester, the fetus needs more nutrients than the mother's digestive tract can absorb. In preparation for this, the placenta stores nutrients early in gestation and releases them in the final trimester. The demand is especially high for protein, iron, calcium, and phosphates. A pregnant woman needs an extra 600 mg of iron for her own hemopoiesis and 375 mg for the fetus. She is likely to become anemic if she does not ingest enough iron during late pregnancy. Supplemental vitamin K is often given late in pregnancy to promote prothrombin synthesis in the fetus. In the United States, newborns are routinely given an injection of vitamin K to minimize the risk of neonatal hemorrhage, especially in the brain, caused by the stresses of birth. A vitamin D supplement helps to ensure adequate calcium absorption to meet fetal demands. Supplemental folic acid reduces the risk of neurological disorders in the fetus, such as spina bifida and anencephaly (failure of the cerebrum, cerebellum, and calvaria to develop), but it is effective only if taken habitually prior to conception (see Insight 13.1, p. 485).

## Circulatory System

By full term, the placenta requires about 625 mL of blood per minute from the mother. The mother's blood volume rises about 30% during pregnancy because of fluid retention and hemopoiesis; she eventually has about 1 to 2 L of extra blood. Cardiac output rises about 30% to 40% above normal by 27 weeks, but for unknown reasons, it falls almost to normal in the last 8 weeks. As the pregnant uterus puts pressure on the large pelvic blood vessels, it interferes with venous return from the legs and pelvic region. This can result in hemorrhoids, varicose veins, and edema of the feet.

## Respiratory System

Minute ventilation increases about 50% during pregnancy for two reasons: (1) Oxygen demands are about 20% higher by late pregnancy in order to supply the fetus and support the woman's increased metabolic rate. (2) Progesterone increases the sensitivity of the respiratory chemoreceptors to carbon dioxide, and ventilation is adjusted to keep the arterial $P_{CO_2}$ lower than normal. While there is a demand for increased ventilation, the expanding uterus pushes the abdominal viscera up against the diaphragm and interferes with breathing. Consequently, the respiratory rate increases to compensate for the lack of depth. Pressure on the diaphragm may be great enough to cause breathing difficulty (dyspnea) by late pregnancy. In the last month, however, the pelvis usually expands enough for the fetus to drop lower in the abdominopelvic cavity, taking some pressure off the diaphragm and allowing the woman to breathe more easily.

## Urinary System

Aldosterone and the steroids of pregnancy promote water and salt retention by the kidneys. Nevertheless, the glomerular filtration rate increases by 50% and urine output is slightly elevated. This enables a woman to dispose of both her own and the fetus's metabolic wastes. As the pregnant uterus compresses the bladder and reduces its capacity, urination becomes more frequent and some women experience uncontrollable leakage of urine (incontinence).

## Integumentary System

The skin must grow to accommodate expansion of the abdomen and breasts and added fat deposition in the hips and thighs. Stretching of the dermis often tears the connective tissue and causes *striae,* or *stretch marks.* These appear reddish at first but fade after pregnancy. Melanocyte activity increases in some areas and darkens the areolae and linea alba. The latter often becomes a dark line, the **linea nigra**[38] (LIN-ee-uh NY-gruh), from the umbilical to the pubic region. Some women also acquire a temporary blotchy darkening of the skin over the nose and cheeks called the "mask of pregnancy," or **chloasma**[39]

---

[38]*linea* = line + *nigra* = black
[39]*chloasma* = to be green

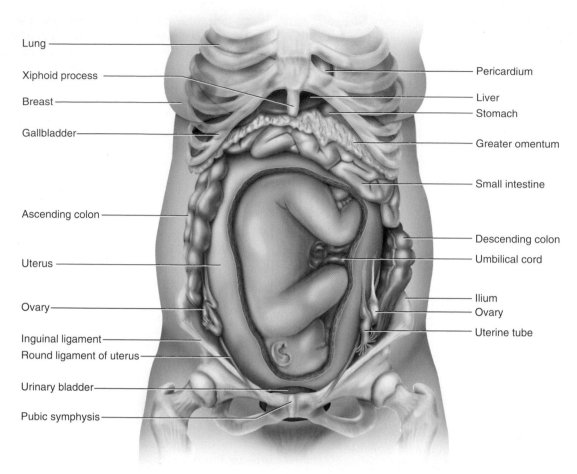

Lung
Xiphoid process
Breast
Gallbladder
Ascending colon
Uterus
Ovary
Inguinal ligament
Round ligament of uterus
Urinary bladder
Pubic symphysis

Pericardium
Liver
Stomach
Greater omentum
Small intestine
Descending colon
Umbilical cord
Ilium
Ovary
Uterine tube

**FIGURE 28.20**   **The Full-Term Fetus in Vertex Position.**   Note the displacement and compression of the abdominal viscera.

(clo-AZ-muh), which usually disappears when the pregnancy is over.

## Uterine Growth and Weight Gain

The uterus weighs about 50 g when a woman is not pregnant and about 900 g by the end of pregnancy. Its growth is monitored by palpating the fundus, which eventually reaches almost to the xiphoid process (fig. 28.20). Table 28.3 shows the sources of weight gain in pregnancy.

| TABLE 28.3 | Distribution of Weight Gain in Pregnancy |
|---|---|
| Fetus | 3 kg (7 lb) |
| Placenta, fetal membranes, and amniotic fluid | 1.8 kg (4 lb) |
| Blood and tissue fluid | 2.7 kg (6 lb) |
| Fat | 1.4 kg (3 lb) |
| Uterus | 0.9 kg (2 lb) |
| Breasts | 0.9 kg (2 lb) |
| **Total** | **11 kg (24 lb)** |

## CHILDBIRTH

In the seventh month of gestation, the fetus normally turns into a head-down *vertex position.* Consequently, most babies are born head first, the head acting as a wedge that widens the mother's cervix, vagina, and vulva during birth. The ancients thought that the fetus kicked against the uterus and pushed itself out head first. The fetus, however, is a rather passive player in its own birth; its expulsion is achieved only by the contractions of the mother's uterine and abdominal muscles. Yet there is evidence that the fetus may play some role in its birth by chemically stimulating labor contractions and perhaps even sending chemical messages that signify when it is developed enough to be born.

## Uterine Contractility

Over the course of gestation, the uterus exhibits relatively weak **Braxton Hicks**[40] **contractions.** These become stronger in late pregnancy and often send women rushing to the hospital with "false labor." At term, however, these

---

[40]John Braxton Hicks (1823–97), British gynecologist

contractions transform suddenly into the more powerful **labor contractions.** True labor contractions mark the onset of **parturition** (PAR-too-RISH-un), the process of giving birth.

Progesterone and estrogen (estradiol) balance may be one factor in this pattern of increasing contractility. Both hormone levels increase over the course of gestation. Progesterone inhibits uterine contractions, but its secretion levels off or declines slightly after 6 months, while estradiol secretion continues to rise (see fig. 28.19). Estradiol stimulates uterine contractions and may be a factor in the irritability of the uterus in late pregnancy.

Also, as the pregnancy nears full term, the posterior pituitary releases more oxytocin (OT) and the uterine muscle equips itself with more OT receptors. Oxytocin promotes labor in two ways: (1) It directly stimulates muscle of the myometrium, and (2) it stimulates the fetal membranes to secrete prostaglandins, which are synergists of OT in producing labor contractions. Labor is prolonged if OT or prostaglandins are lacking, and it may be induced or accelerated by giving a vaginal prostaglandin suppository or an intravenous OT "drip."

The conceptus itself may produce chemical stimuli promoting its own birth. Fetal cortisol secretion rises in late pregnancy and may enhance estrogen secretion by the placenta. The fetal pituitary gland also produces oxytocin, which does not enter the maternal circulation but may stimulate the fetal membranes to secrete prostaglandins.

Uterine stretching is also thought to play a role in initiating labor. Stretching any smooth muscle increases its contractility, and movements of the fetus produce the sort of intermittent stretch that is especially stimulatory to the myometrium. Twins are born an average of 19 days earlier than single infants, probably because of the greater stretching of the uterus. When the fetus is in the vertex position, its head pushes against the cervix, which is especially sensitive to stretch.

## Labor Contractions

Labor contractions begin about 30 minutes apart. As labor progresses, they become more intense and eventually occur every 1 to 3 minutes. It is important that they be intermittent rather than one long, continual contraction. Each contraction sharply reduces maternal blood flow to the placenta, so the uterus must periodically relax to restore flow and oxygen delivery to the fetus. Contractions are strongest in the fundus and body of the uterus and weaker near the cervix, thus pushing the fetus downward.

According to the **positive feedback theory of labor,** labor contractions are induced by stretching of the cervix. This triggers a reflex contraction of the uterine body that pushes the fetus downward and stretches the cervix still more. Thus there is a self-amplifying cycle of stretch and contraction. In addition, cervical stretching induces a neuroendocrine reflex through the spinal cord, hypothalamus, and posterior pituitary. The posterior pituitary releases oxytocin, which is carried in the blood and stimulates the uterine muscle both directly and through the action of prostaglandins. This, too, is a positive feedback cycle: cervical stretching → oxytocin secretion → uterine contraction → cervical stretching (see fig. 1.12, p. 19).

As labor progresses, a woman feels a growing urge to "bear down." A reflex arc extends from the uterus to the spinal cord and back to the skeletal muscles of the abdomen. Contraction of these muscles—partly reflexive and partly voluntary—aids in expelling the fetus, especially when combined with the Valsalva maneuver for increasing intra-abdominal pressure.

The pain of labor is due at first mainly to ischemia of the myometrium—muscle hurts when deprived of blood, and each labor contraction temporarily restricts uterine circulation. As the fetus enters the vaginal canal, the pain becomes stronger because of increased stretching of the cervix, vagina, and perineum and sometimes the tearing of vaginal tissue. At this stage, the obstetrician may perform an *episiotomy*—an incision in the vulva to widen the vaginal orifice and prevent random tearing. The pain of human childbirth, compared with the relative ease with which other mammals give birth, is an evolutionary product of two factors: the unusually large brain and head of the human infant, and the narrowing of the pelvic outlet, which helped to adapt hominids to bipedal locomotion (see p. 11).

## Stages of Labor

Labor occurs in three stages. The duration of each stage tends to be longer in a **primipara** (a woman giving birth for the first time) than in a **multipara** (a woman who has previously given birth).

1. **Dilation (First) Stage.** This is the longest stage, lasting 8 to 24 hours in a primipara but as little as a few minutes in a multipara. It is marked by the **dilation** (widening) of the cervical canal and **effacement** (thinning) of the cervix (fig. 28.21a, b). The cervix reaches a maximum diameter of about 10 cm (the diameter of the baby's head). During dilation, the fetal membranes usually rupture and the *amniotic fluid* is discharged (the "breaking of the waters").

2. **Expulsion (Second) Stage.** This stage typically lasts about 30 to 60 minutes in a primipara and as little as 1 minute in a multipara. It begins when the baby's head enters the vagina and lasts until the baby is entirely expelled (fig. 28.21c). The baby is said to be **crowning** when the top of its head is visible, stretching the vulva (fig. 28.22a). Delivery of the head is the most difficult part, with the rest of the body following much more easily. An episiotomy may be performed during this

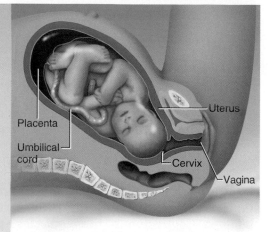

**(a) Early dilation stage**

Placenta

Umbilical cord

Uterus

Cervix

Vagina

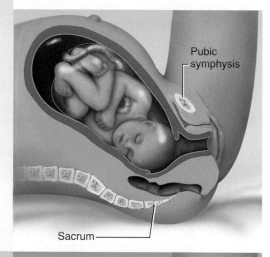

**(b) Late dilation stage**

Pubic symphysis

Sacrum

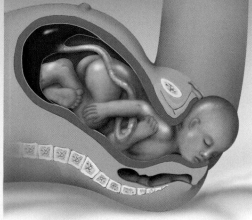

**(c) Expulsion stage**

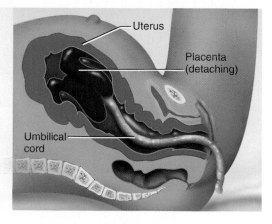

**(d) Placental stage**

Uterus

Placenta (detaching)

Umbilical cord

**FIGURE 28.21** The Stages of Childbirth.

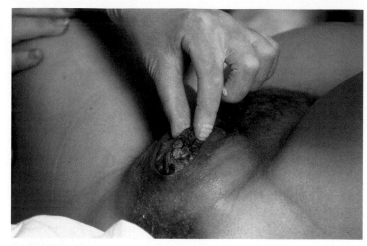

**(a) Crowning**

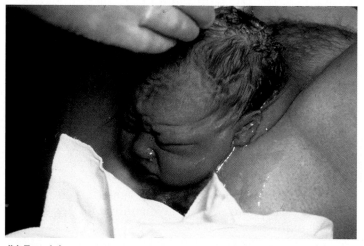

**(b) Expulsion stage**

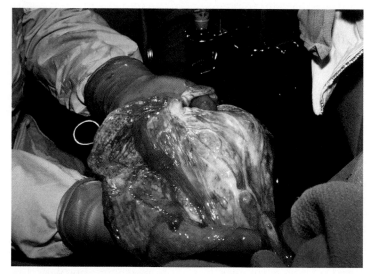

**(c) Afterbirth**

**FIGURE 28.22 Childbirth.** (a) Crowning. The baby's head, between the attendant's fingers, has begun to dilate the vulva. (b) Emergence of the head. (c) The placenta and fetal membranes (afterbirth).

stage. An attendant often uses a suction bulb to remove mucus from the baby's mouth and nose even before it is fully delivered. When the baby is fully expelled, an attendant drains the blood of the placental vein into the baby, clamps the umbilical cord in two places, and cuts the cord between the clamps.

3. **Placental (Third) Stage.** The uterus continues to contract after expulsion of the baby. The placenta, however, is a nonmuscular organ that cannot contract, so it buckles away from the uterine wall (see fig. 28.21d). About 350 mL of blood is typically lost at this stage, but contractions of the myometrium compress the blood vessels and prevent more extensive bleeding. The placenta, amnion, and other fetal membranes are expelled by uterine contractions, which may be aided by a gentle pull on the umbilical cord. The membranes *(afterbirth)* must be carefully inspected to be sure everything has been expelled (fig. 28.22c). If any of these structures remain in the uterus, they can cause postpartum hemorrhaging. The umbilical blood vessels are counted because an abnormal number in the cord may indicate cardiovascular abnormalities in the infant.

## PUERPERIUM

The first 6 weeks **postpartum** (after birth) are called the **puerperium**[41] (PYU-er-PEER-ee-um), a period in which the mother's anatomy and physiology stabilize and the reproductive organs return nearly to the pregravid state (their condition prior to pregnancy). The shrinkage of the uterus during this period is called **involution.** In a lactating woman, it loses about 50% of its weight in the first week and is nearly at its pregravid weight in 4 weeks. Involution is achieved by **autolysis** (self-digestion) of uterine cells by their own lysosomal enzymes. For about 10 days, this produces a vaginal discharge called **lochia,** which is bloody at first and then turns clear and serous. Breast-feeding promotes involution because (1) it suppresses estrogen secretion, which would otherwise cause the uterus to remain more flaccid; and (2) it stimulates oxytocin secretion, which causes the myometrium to contract and firm up the uterus sooner. It is important for the puerperium to be undisturbed, as emotional upset can inhibit lactation in some women.

### Before You Go On

*Answer the following questions to test your understanding of the preceding section:*

17. List the roles of HCG, estrogen, progesterone, and HCS in pregnancy.

18. What is the role of the corpus luteum in pregnancy? What eventually takes over this role?

19. List and briefly explain the special nutritional requirements of pregnancy.

20. How much weight does the average woman gain in pregnancy? What contributes to this weight gain other than the fetus?

21. Describe the positive feedback theory of labor.

22. What major events define the three stages of labor?

# Lactation

### Objectives

When you have completed this section, you should be able to

- describe development of the breasts in pregnancy;
- describe the shifting hormonal balance that regulates the onset and continuation of lactation;
- describe the mechanism of milk ejection;
- contrast colostrum with breast milk; and
- discuss the benefits of breast-feeding.

**Lactation** is the synthesis and ejection of milk from the mammary glands. It lasts for as little as a week postpartum in women who do not breast-feed their infants, but it can continue for many years as long as the breast is stimulated by a nursing infant or mechanical device (breast pump). Numerous studies conducted before the widespread marketing of artificial infant formulas suggest that worldwide, women traditionally nursed their infants until a median age of about 2.8 years.

## DEVELOPMENT OF THE MAMMARY GLANDS IN PREGNANCY

The high estrogen level in pregnancy causes the ducts of the mammary glands to grow and branch extensively. Growth hormone, insulin, glucocorticoids, and prolactin also contribute to this development. Once the ducts are complete, progesterone stimulates the budding and development of acini at the ends of the ducts. The fully developed mammary gland is of the compound tubuloacinar type. The acini are organized into grapelike clusters (lobules) within each lobe of the breast (see fig. 28.9).

## COLOSTRUM AND MILK SYNTHESIS

In late pregnancy, the mammary acini and ducts are distended with a secretion called **colostrum.** This is similar to breast milk in protein and lactose content but contains about one-third less fat. It is the infant's only natural source of nutrition for the first 1 to 3 days postpartum. Colostrum has a thin watery consistency and a cloudy yellowish color. The amount of colostrum secreted per day is at most 1% of the amount of milk secreted later, but

---

[41]*puer* = child + *per* (from *par*) = birth

since infants are born with excess body water and ample fat, they do not require high calorie and fluid intake at first. A major benefit of colostrum is that it contains immunoglobulins, especially IgA. IgA resists digestion and may protect the infant from gastroenteritis. It is also thought to be pinocytosed by the small intestine and to confer wider, systemic immunity to the neonate.

Milk synthesis is promoted by prolactin, a hormone of the anterior pituitary gland. In the nonpregnant state, dopamine (prolactin-inhibiting hormone) from the hypothalamus inhibits prolactin secretion. Prolactin secretion begins 5 weeks into the pregnancy, and by full term it is 10 to 20 times its normal level. Even so, prolactin has little effect on the mammary glands until after birth. While the steroids of pregnancy prepare the mammary glands for lactation, they antagonize prolactin and suppress milk synthesis. When the placenta is discharged at birth, the steroid levels abruptly drop and allow prolactin to have a stronger effect. Milk is synthesized in increasing quantity over the following week. Milk synthesis also requires the action of growth hormone, cortisol, insulin, and parathyroid hormone to mobilize the necessary amino acids, fatty acids, glucose, and calcium.

At the time of birth, baseline prolactin secretion drops to the nonpregnant level. Every time the infant nurses, however, it jumps to 10 to 20 times this level for the next hour and stimulates the synthesis of milk for the next feeding (fig. 28.23). These prolactin surges are accompanied by smaller increases in estrogen and progesterone secretion. If the mother does not nurse or these hormone surges are absent (due to pituitary damage, for example), the mammary glands stop producing milk in about a week. Even if she does nurse, milk production declines after 7 to 9 months.

Only 5% to 10% of women become pregnant again while breast-feeding. Apparently, either prolactin or nerve signals from the breast inhibit GnRH secretion, which, in turn, results in reduced gonadotropin secretion and ovarian cycling. This mechanism may have evolved as a natural means of spacing births, but breast-feeding is not a reliable means of contraception. Even in women who breast-feed, the ovarian cycle sometimes resumes several months postpartum. In those who do not breast-feed, the cycles resume in a few weeks, but for the first 6 months they are usually anovulatory.

## MILK EJECTION

Milk is continually secreted into the mammary acini, but it does not easily flow into the ducts. Its flow, called **milk ejection** (let-down), is controlled by a neuroendocrine reflex. The infant's suckling stimulates nerve endings of the nipple and areola, which in turn signal the hypothalamus and posterior pituitary to release oxytocin. Oxytocin stimulates **myoepithelial cells,** which form a basketlike mesh around each gland acinus (fig. 28.24). These cells are of epithelial origin, but are packed with actin and contract like smooth muscle to squeeze milk

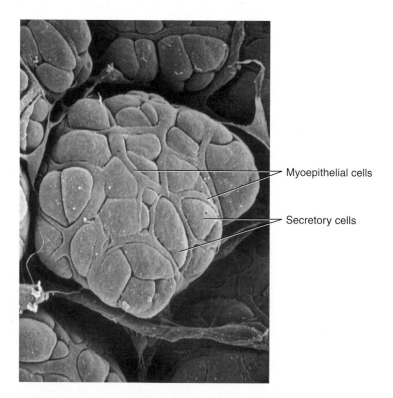

**FIGURE 28.24**  Acinus of a Mammary Gland.  Myoepithelial cells can be seen forming a mesh around the secretory cells. The myoepithelial cells contract and force milk from the acinus into the duct.

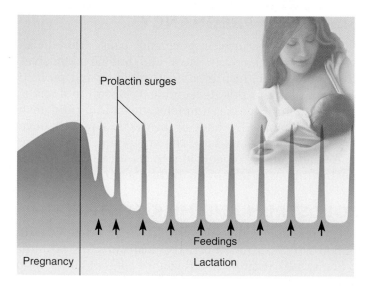

**FIGURE 28.23**  Prolactin Secretion in the Lactating Female. Each time the infant nurses, maternal prolactin secretion surges. This prolactin stimulates synthesis of the milk that will be available at the next feeding.

from the acinus into the duct. The infant does not get any milk for the first 30 to 60 seconds of suckling, but milk soon fills the ducts and lactiferous sinuses and is then easily sucked out.

⌐ **Think About It**

*When a woman is nursing her baby at one breast, would you expect only that breast, or both breasts, to eject milk? Explain why.*

## BREAST MILK

Table 28.4 compares the composition of colostrum, human milk, and cow's milk. Breast milk changes composition over the first 2 weeks, varies from one time of day to another, and changes even during the course of a single feeding. For example, at the end of a feeding there is less lactose and protein in the milk, but six times as much fat, as there is at the beginning.

Cow's milk is not a good substitute for human milk. It has one-third less lactose but three to five times as much protein and minerals. The excess protein forms a harder curd in the infant's stomach, so cow's milk is not digested and absorbed as efficiently as mother's milk. It also increases the infant's nitrogenous waste excretion, which increases the incidence and severity of diaper rash, par-

ticularly as bacteria in the diaper break urea down to ammonia, a skin irritant.

Colostrum and milk have a laxative effect that helps to clear the neonatal intestine of *meconium,* a greenish black, sticky fecal matter composed of bile, epithelial cells, and other wastes that accumulated during fetal development. By clearing bile and bilirubin from the body, breast-feeding also reduces the incidence and degree of jaundice in neonates. Breast milk promotes colonization of the neonatal intestine with beneficial bacteria and continues to supply antibodies that lend protection against infection by pathogenic bacteria. Breast-feeding also tends to promote a closer bond between mother and infant.

A woman nursing one baby eventually produces about 1.5 L of milk per day; women with twins produce more. Lactation places a great metabolic demand on the mother. It is equivalent to losing 50 g of fat, 100 g of lactose (made from her blood glucose), and 2 to 3 g of calcium phosphate per day. A woman is at greater risk of bone loss when breast-feeding than when she is pregnant, because much of the infant's skeleton is still cartilage at birth and becomes mineralized at her expense in the first year postpartum. If a nursing mother does not have enough calcium and vitamin D in her own diet, lactation stimulates parathyroid hormone secretion and osteoclast activity, taking calcium from her bones to supply her baby.

To conclude this chapter, table 28.5 briefly describes some of the common disorders of pregnancy. Other reproductive disorders are discussed elsewhere: cervical cancer in Insight 28.1, breast cancer on page 1074, and sexually transmitted diseases at the end of chapter 27.

| TABLE 28.4 | A Comparison of Colostrum, Human Milk, and Cow's Milk* | | |
|---|---|---|---|
| Nutrient | Human Colostrum | Human Milk | Cow's Milk |
| Total Protein (g/L) | 22.9 | 10.6 | 30.9 |
| Lactalbumin (g/L) | — | 3.7 | 25.0 |
| Casein (g/L) | — | 3.6 | 2.3 |
| Immunoglobulins (g/L) | 19.4 | 0.09 | 0.8 |
| Fat (g/L) | 29.5 | 45.4 | 38.0 |
| Lactose (g/L) | 57 | 71 | 47 |
| Calcium (mg/L) | 481 | 344 | 1370 |
| Phosphorus (mg/L) | 157 | 141 | 910 |

*Colostrum data are for the first day postpartum, and human milk data are for "mature milk" at about 15 days postpartum.

## Before You Go On

*Answer the following questions to test your understanding of the preceding section:*

23. *Why is little or no milk secreted while a woman is pregnant?*

24. *How does a lactating breast differ from a nonlactating breast in structure? What stimulates these differences to develop during pregnancy?*

25. *What is colostrum and what is its significance?*

26. *How does suckling stimulate milk ejection?*

27. *Why is breast milk superior to cow's milk for an infant?*

| **TABLE 28.5** | Some Disorders of Pregnancy |
|---|---|
| Abruptio placentae[42] | Premature separation of the placenta from the uterine wall, often associated with pre-eclampsia or cocaine use. May require birth by cesarian section. |
| Ectopic[43] pregnancy | Implantation of the conceptus anywhere other than the uterus; usually starts in the uterine tube (tubal pregnancy) and may progress to abdominal pregnancy if the tube ruptures. See Insight 29.2 for further details. |
| Gestational diabetes | A form of diabetes mellitus that develops in about 1–3% of pregnant women, characterized by insulin insensitivity, hyperglycemia, glycosuria, and a risk of excessive fetal size and birth trauma. Glucose metabolism often returns to normal after delivery of the infant, but 40–60% of women with gestational diabetes develop diabetes mellitus within 15 years after the pregnancy. |
| Hyperemesis gravidarum[44] | Prolonged vomiting, dehydration, alkalosis, and weight loss in early pregnancy, often requiring hospitalization to stabilize fluid, electrolyte, and acid–base balance; sometimes associated with liver damage. |
| Placenta previa[45] | Blockage of the cervical canal by the placenta, preventing birth of the infant before the placenta separates from the uterus. Requires birth by cesarian section. |
| Pre-eclampsia[46] | Gestational hypertension and proteinuria, often with edema of the face and hands, occurring especially in the third trimester in primiparas. Correlated with abnormal development of placental arteries, ultimately leading to widespread thrombosis and organ dysfunction in the mother. Occurs in 5–8% of pregnancies. Sometimes progresses to *eclampsia* (seizures), which may be fatal to the mother, fetus, or both. Eclampsia may occur postpartum. |
| Spontaneous abortion | Occurs in 10–15% of pregnancies, usually because of fetal deformities or chromosomal abnormalities incompatible with survival, but may also result from maternal abnormalities, infectious disease, and drug abuse. |

[42]*ab* = away + *rupt* = to tear + *placentae* = of the placenta
[43]*ec* = out of + *top* = place
[44]*hyper* = excessive + *emesis* = vomiting + *gravida* = pregnant woman
[45]*pre* = before + *via* = the way (obstructing the way)
[46]*ec* = forth + *lampsia* = shining

---

## INSIGHT 28.4   Clinical Application

## Methods of Contraception

The term *contraception* is used here to mean any procedure or device intended to prevent pregnancy (the presence of an implanted conceptus in the uterus). This essay describes the most common methods of contraception, some issues involved in choosing among them, and the relative reliability of the various methods. Several of those options are shown in figure 28.25.

### Behavioral Methods

*Abstinence* (refraining from intercourse) is, obviously, a completely reliable method if used consistently. The *rhythm method* (periodic abstinence) is based on avoiding intercourse near the time of expected ovulation. Among typical users, it has a 25% failure rate, partly due to lack of restraint and partly because it is difficult to predict the exact date of ovulation. Intercourse must be avoided for at least 7 days before ovulation so there will be no surviving sperm in the reproductive tract when the egg is ovulated, and for at least 2 days after ovulation so there will be no fertile egg present when sperm are introduced. The rhythm method is valuable, however, for couples who are trying to conceive a child by having intercourse at the time of apparent ovulation.

*Withdrawal* (coitus interruptus) requires the male to withdraw the penis before ejaculation. This often fails because of lack of willpower, because some sperm are present in the pre-ejaculatory fluid, and because sperm ejaculated anywhere in the female genital region can potentially get into the reproductive tract.

### Barrier and Spermicidal Methods

Barrier methods are designed to prevent sperm from getting into or beyond the vagina. They are most effective when used with chemical *spermicides*, available as over-the-counter foams, creams, and jellies.

The *male condom* is a sheath of latex, rubber, or animal membrane (lamb intestine) that is unrolled over the erect penis and collects the semen. It is inexpensive, convenient, and very reliable when used carefully. About 25% of American couples who use contraceptives use only condoms, which rank second to birth-control pills in popularity.

The *female condom* is less used. It is a polyurethane sheath with a flexible ring at each end. The inner ring fits over the cervix and the outer ring covers the external genitalia. Male and female condoms are the only contraceptives that also protect against disease transmission. Animal membrane condoms, however, are porous to HIV and hepatitis B viruses and do not afford dependable protection from disease.

The *diaphragm* is a latex or rubber dome that is placed over the cervix to block sperm transmission. It requires a physical examination and prescription to ensure a proper fit, but is otherwise comparable to the condom in convenience and reliabil-

Male condom

Female condom

Diaphragm with contraceptive jelly

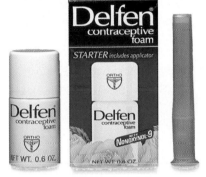

Contraceptive foam with vaginal applicator

Birth-control pills

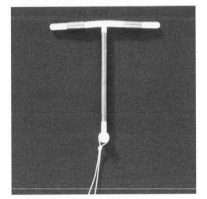

Intrauterine device (IUD)

**FIGURE 28.25 Contraceptive Devices.**

ity, provided it is used with a spermicide. Without a spermicide, it is not very effective.

The *sponge* is a foam disc inserted before intercourse to cover the cervix. It is impregnated with a spermicide and acts by trapping and killing the sperm. It requires no prescription or fitting. The sponge provides protection for up to 24 hours, and must be left in place for 6 hours after the last act of intercourse.

Contraceptive sponges and other barrier methods date to antiquity. The ancient Egyptians and Greeks used vaginal sponges soaked in lemon juice, which had a mild spermicidal effect. Some Egyptian women used vaginal suppositories made of crocodile dung and honey.

### Hormonal Methods

*Birth-control pills* are composed of estrogen and progesterone. They mimic the negative feedback effect of ovarian hormones on FSH secretion, thus preventing follicle development and ovulation. They are effective for most women with minimal complications, but they can increase the risk of heart attack or stroke in smokers and in women with a history of diabetes, hypertension, or clotting disorders. Birth-control pills are the most widely used contraceptives in the United States. Efforts to develop a birth-control pill for men have so far been unsuccessful, but are continuing.

*Medroxyprogesterone* (Depo-Provera) is a synthetic progesterone administered by injection two to four times per year. It

provides highly reliable, long-term contraception without the need of daily pills or implants, although in some women it causes headaches, nausea, or weight gain, and fertility may not return immediately when its use is discontinued.

A contraceptive skin patch has been available in the United States since 2001, but it has lately come under scrutiny because of an elevated rate of deaths from blood clots among users.

### Surgical Sterilization

People who are confident that they do not want more children (or any) often elect to be surgically sterilized. This entails the cutting and tying or clamping of the genital ducts, thus blocking the passage of sperm or eggs. Surgical sterilization has the advantage of convenience, since it requires no further attention. Its initial cost is higher, however, and for people who later change their minds, surgical reversal is much more expensive than the original procedure and is often unsuccessful. *Vasectomy* is the severing of the ductus (vas) deferens, done through a small incision in the back of the scrotum. In *tubal ligation,*[47] the uterine tubes are cut. This can be done through small abdominal incisions to admit a cutting instrument and laparoscope (viewing device).

### Preventing Implantation

The *intrauterine device (IUD)* is a springy plastic device inserted through the cervical canal into the uterus and left in place for 1

[47]*ligat* = to tie

to 4 years. It prevents the implantation of a blastocyst in the uterine wall. Some IUDs were removed from the market because they caused serious complications such as uterine perforation, and many other models were removed for fear of legal liability. Only two T-shaped models are currently available in the United States.

Some drugs can be taken orally after intercourse to prevent implantation of a conceptus. These are called emergency contraceptive pills (ECPs), or "morning after pills" (trade names Plan B, Levonelle). An ECP is a high dose of estrogen and progesterone or a synthetic progesterone alone. It can be taken within 72 hours after intercourse, and induces menstruation within 2 weeks. ECPs work on several fronts: inhibiting ovulation; inhibiting sperm or egg transport in the uterine tube; or preventing implantation of a blastocyst. They do not work if a blastocyst has already implanted. ECPs are available without a prescription in some states, but availability has been limited or delayed elsewhere by political controversy.

## Issues in Choosing a Contraceptive

Many issues enter into the appropriate choice of a contraceptive, including personal preference, pattern of sexual activity, medical history, religious views, convenience, initial and ongoing costs, and disease prevention. For most people, however, the two primary issues are safety and reliability.

The following table shows the expected rates of failure for several types of contraception as reported in the *Physician's Desk Reference*. Each column shows the number of sexually active women who typically become pregnant within 1 year while they or their partners are using the indicated contraceptives. The lowest rate (perfect use) is for those who use the method correctly and consistently, while the higher rate (typical use) is based on random surveys of users and takes human error (lapses and incorrect usage) into account.

We have not considered all the currently available methods of contraception or all the issues important to the choice of a contraceptive. No one contraceptive method can be recommended as best for all people. Further information necessary to a sound choice and proper use of contraceptives should be sought from a health department, college health service, physician, or other such sources.

## Failure Rates of Contraceptive Methods

| Method | Rate of Failure (pregnancies per 100 users) | |
| --- | --- | --- |
| | Perfect use | Typical use |
| No protection | 85 | 85 |
| Rhythm method | 1–9 | 25 |
| Withdrawal | 4 | 19 |
| Spermicide alone | 6 | 26 |
| Condom (male or female) | 3–5 | 14–21 |
| Diaphragm with spermicide | 6 | 20 |
| Birth-control pill | 0.1–0.5 | 5 |
| Medroxyprogesterone | 0.3 | 0.3 |
| Vasectomy | 0.10 | 0.15 |
| Tubal ligation | 0.5 | 0.5 |
| Intrauterine device | 0.1–2.0 | 0.1–1.5 |
| Vaginal sponge | 9 | 20 |

# CONNECTIVE ISSUES

## Interactions Between the
## REPRODUCTIVE SYSTEM
## and Other Organ Systems

■ indicates ways in which this system affects other systems

■ indicates ways in which other systems affect this system

### INTEGUMENTARY SYSTEM

At puberty, androgens stimulate development of body hair and apocrine glands, and increased sebaceous secretion; estrogens stimulate fat deposition and breast development in females; pregnancy may cause pigmentation changes and stretch marks

Sensory stimulation of skin important to sexual arousal; mammary glands nourish infant

### SKELETAL SYSTEM

Androgens and estrogens stimulate adolescent skeletal growth and maintain adult bone mass

Encloses and protects pelvic organs; provides minerals for fetal growth and lactation; narrow pelvic outlet makes childbirth difficult

### MUSCULAR SYSTEM

Gonadal steroids stimulate muscle growth

Pelvic muscles support reproductive organs, aid in erection and orgasm; abdominal muscles aid in childbirth; cremaster helps to maintain temperature of testes

### NERVOUS SYSTEM

Androgens stimulate libido; hormones from gonads and placenta exert negative feedback control on hypothalamus

Hypothalamus initiates gonadotropin function and lactation

### ENDOCRINE SYSTEM

Gonadal and placental hormones exert feedback control on anterior pituitary

Hormones regulate puberty, gametogenesis, libido, pregnancy, lactation, and climacteric

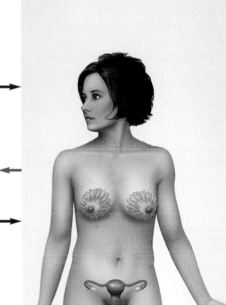

### CIRCULATORY SYSTEM

Androgens stimulate erythropoiesis; estrogens may inhibit development of atherosclerosis in females; pregnancy increases blood volume and may cause varicose veins

Changes in blood flow produce vasocongestion and erection in sexual arousal; blood distributes sex hormones, transports nutrients to fetus, and removes fetal wastes; pampiniform plexus prevents overheating of testes

### LYMPHATIC/IMMUNE SYSTEMS

Blood-testis barrier isolates sperm and protects them from immune system

Immune cells protect reproductive organs; IgA in colostrum and milk confers passive immunity on neonate

### RESPIRATORY SYSTEM

Sexual arousal increases pulmonary ventilation; pregnancy reduces depth of inspiration but increases respiratory rate

Provides $O_2$, removes $CO_2$; Valsalva maneuver aids childbirth

### URINARY SYSTEM

Sexual arousal constricts internal urinary sphincter; prostatic hyperplasia may impede urine flow; pregnancy crowds urinary bladder and often causes incontinence

Disposes of maternal and fetal wastes; urethra serves as passageway for semen

### DIGESTIVE SYSTEM

Fetus crowds digestive organs, contributing to heartburn and constipation

Provides nutrients for gametogenesis and fetal development

# CHAPTER REVIEW

# Review of Key Concepts

## Reproductive Anatomy (p. 1066)

1. In the female fetus, the absence of testosterone and müllerian inhibiting factor results in the paramesonephric duct developing into the uterine tubes, uterus, and vagina and the external genitalia developing into a clitoris, labia minora, and labia majora.

2. The ovary has a central *medulla,* a surface layer of parenchyma called the *cortex,* and an outer fibrous capsule, the *tunica albuginea.* The ovary is supported by three ligaments and supplied by an ovarian artery, ovarian veins, and ovarian nerves.

3. Each egg develops in its own bubblelike *follicle.* Follicles are located primarily in the cortex.

4. The *uterine (fallopian) tube* is a ciliated duct that extends from the ovary to the uterus.

5. The uterus is a thick muscular chamber superior to the urinary bladder. It consists of an upper *fundus,* middle *corpus* (body), and lower *cervix* (neck), where it meets the vagina.

6. The uterine wall consists of an outer serosa called the *perimetrium,* a thick muscular *myometrium,* and an inner mucosa called the *endometrium.* The endometrium contains numerous tubular glands and is divided into two layers: a thick superficial *stratum functionalis,* which is shed in each menstrual period, and a thinner basal *stratum basalis,* which is retained from cycle to cycle.

7. The uterus is anchored by four pairs of ligaments and supplied with blood by a *uterine artery* that arises from each internal iliac artery.

8. The *vagina* tilts dorsally between the urethra and rectum. It has no glands but is moistened by transudation of serous fluid through the vaginal wall and by mucus from glands in the cervical canal.

9. The *vulva* (pudendum or external genitalia) include the *mons pubis, labia majora* and *minora, clitoris,* vaginal orifice, accessory glands (*greater* and *lesser vestibular glands* and *paraurethral glands*), and erectile tissues *(vestibular bulbs).* The urethra also opens into the vulva.

10. The breast is internally divided into lobes, each with a *lactiferous duct* that conveys milk to the nipple. Outside of pregnancy or lactation, the breast contains only small traces of mammary gland.

11. Breast cancer strikes a high percentage of women. Two breast cancer genes are known, although most cases are nonhereditary and have no association with identifiable risk factors.

## Puberty and Menopause (p. 1075)

1. In the United States and Europe, female puberty typically begins around age 9 or 10. Rising GnRH levels trigger the secretion of FSH and LH. In response to FSH, the ovaries secrete estrogens, progesterone, inhibin, and androgens.

2. The earliest visible sign of puberty is breast development, or *thelarche,* which is stimulated by estrogen, progesterone, prolactin, glucocorticoids, and growth hormone.

3. *Pubarche* is the development of pubic and axillary hair, sebaceous glands, and axillary sweat glands. Androgens induce pubarche and activate the female libido.

4. *Menarche* is a girl's first menstrual period, occurring at an average age of 12 in the United States and Europe. The first few menstrual cycles are usually anovulatory; ovulation becomes regular about a year after menarche.

5. The estradiol of puberty induces development of the ovaries and secondary sex organs, has feminizing effects on the external anatomy, and stimulates bone growth.

Progesterone acts primarily on the uterus, and inhibin modulates FSH secretion.

6. With age, the number of ovarian follicles declines and with it, the source of estrogen and progesterone. The decline in levels of steroids brings on a transitional period of *climacteric,* lasting a few years and marked by *menopause,* the eventual cessation of ovulation and menstruation.

## Oogenesis and the Sexual Cycle (p. 1077)

1. The *sexual cycle* is the monthly cycle of events that occurs when a woman is not pregnant. It includes the *ovarian cycle* of events in the ovaries and the *menstrual cycle* of events in the uterus.

2. *Oogenesis* is the production of eggs. Unlike spermatogenesis, it occurs in a monthly rhythm and usually produces only one gamete (egg) per month.

3. Oogenesis begins with *oogonia,* which multiply until the fifth month of a girl's fetal development. Some of these develop into *primary oocytes* and begin meiosis I before birth. Most primary oocytes undergo *atresia* during childhood, leaving about 400,000 at puberty.

4. Surviving primary oocytes undergo meiosis I to produce a small *first polar body,* which dies, and a *secondary oocyte.* The secondary oocyte progresses only as far as metaphase II if it is not fertilized. If fertilized, it completes meiosis II, producing a *second polar body,* which also dies, and an ovum that goes on to become the zygote.

5. The ovarian cycle, occurring in the absence of pregnancy, typically lasts about 28 days, with day 1 considered to be the first day of visible menstruation.

6. The *follicular phase* of the cycle extends from day 1 to day 14. The *preantral phase* of the cycle is follicular development preceding the for-

mation of an antrum; it includes *primordial* and *primary follicles*. These begin development before birth and resume in adolescence, with a cohort of follicles each month advancing to the antral phase.

7. Follicles in the *antral phase* have an antrum (internal cavity) and develop under the influence of pituitary gonadotropins. One secondary (antral) follicle outpaces the others in its development and becomes the *dominant follicle,* while the others in the cohort undergo atresia. The dominant follicle develops into a mature (graafian) follicle and secretes predominantly estrogen.

8. The mature follicle ovulates around day 14, primarily under the influence of luteinizing hormone. After releasing its oocyte, it collapses and develops into a *corpus luteum.* The corpus luteum secretes progesterone, but in the absence of pregnancy it involutes from days 22 to 26 and is an inactive scar *(corpus albicans)* after day 26.

9. The menstrual cycle is divided into proliferative, secretory, premenstrual, and menstrual phases.

10. The *proliferative phase* (days 6–14) is a period of rebuilding the lost stratum functionalis by mitosis under the influence of estrogen from the ovaries.

11. The *secretory phase* (days 15–26) is a period of thickening of the endometrium by secretion of mucus and glycogen under the influence of progesterone from the corpus luteum.

12. The *premenstrual phase* is a period triggered by involution of the corpus luteum and the resulting lack of progesterone. Spasms of the spiral arteries deprive the endometrium of blood flow, resulting in necrosis and sloughing off of the stratum functionalis. The *menstrual phase* begins when enough necrotic tissue and blood have accumulated to produce noticeable vaginal discharge of menstrual fluid.

## Female Sexual Response (p. 1085)

1. Female sexual response occurs in stages similar to that of the male, with the following major differences: In the excitement phase, the labia majora and minora, the clitoris, and the breasts become vasocongested; secretions of the greater vestibular glands lubricate the vulva; and the vagina is moistened by vaginal transudate. The inner end of the vagina dilates and its lower end constricts to form a narrow passage, the *orgasmic platform.* The uterus rises from its forward-tilted position to a nearly vertical one (the *tenting effect*).

2. In orgasm, the paraurethral glands secrete into the vulva, the orgasmic platform of the vagina constricts repeatedly, and the cervix plunges into the vaginal canal (into the semen if present).

3. In resolution, the uterus returns to its forward tilt, the orgasmic platform relaxes, the breasts become less congested, and there may be an outbreak of perspiration. Unlike men, women lack a refractory period and may experience successive orgasms.

## Pregnancy and Childbirth (p. 1087)

1. *Gestation* lasts an average of 266 days from conception to birth, but birth is predicted to occur about 280 days from the onset of the last menstrual period.

2. Fertilization occurs in the distal half of the uterine tube and the fertilized egg divides five or six times before reaching the uterus. All the products of fertilization—the embryo or fetus and the associated membranes—are called the *conceptus.*

3. The major hormones of pregnancy are *human chorionic gonadotropin (HCG), estrogens, progesterone,* and *human chorionic somatomammotropin (HCS).*

4. HCG stimulates the corpus luteum to grow and secrete estrogen and progesterone. Estrogen from the corpus luteum and later from the placenta stimulates tissue growth in the mother and fetus and softens joints of the mother's pelvic girdle in preparation for giving birth. Progesterone from the corpus luteum and placenta inhibits premature uterine contractions and stimulates the mitosis of uterine *decidual cells* that nourish the early conceptus. Estrogen and progesterone also promote mammary gland development and inhibit the secretion of FSH.

HCS from the placenta mobilizes fatty acids as fuel for the mother while it spares glucose for use by the fetus.

5. Thyroid hormone, parathyroid hormone, glucocorticoids, aldosterone, and relaxin also contribute to the developments of pregnancy (table 28.2).

6. Morning sickness, constipation, and heartburn sometimes accompany pregnancy as steroid hormones inhibit intestinal motility and the growing uterus compresses the digestive organs. The basal metabolic rate rises and one may feel hot as a result. Nutrient intake must increase moderately to meet the needs of the fetus.

7. Blood volume and cardiac output increase in pregnancy. Pressure from the uterus may cause edema, hemorrhoids, or varicose veins.

8. Breathing becomes more rapid in pregnancy as $O_2$ demand and $CO_2$ sensitivity increase, yet pressure from the uterus indirectly compresses the lungs and makes breathing shallower.

9. Glomerular filtration and urine output increase to dispose of both fetal and maternal wastes, but the capacity of the bladder is reduced by pressure from the uterus.

10. The maternal skin grows, especially on the breasts and abdomen; dermal tearing may cause *striae;* and some women exhibit melanization of the skin on the abdomen *(linea nigra)* or face *(chloasma).*

11. In the seventh month, the fetus usually turns into the vertex (headdown) position.

12. Late in pregnancy, the uterus becomes more contractile, sometimes exhibiting *Braxton Hicks contractions* weeks before the true *labor contractions* occur. A rising ratio of estrogen to progesterone may be responsible for this increased contractility late in pregnancy. True labor contractions are stimulated by uterine stretching and oxytocin (OT).

13. According to the positive feedback theory of labor, stretching of the cervix triggers reflex contraction of the uterine body, which pushes the

fetus downward and stretches the cervix still more. Cervical stretching also activates a neuroendocrine reflex that results in OT secretion, and OT stimulates more and more intense uterine contractions. The voluntary abdominal muscles also aid in giving birth.

14. The *dilation (first) stage* of labor involves dilation of the cervical canal to a diameter of 10 cm and thinning *(effacement)* of the cervical tissue. The fetal membranes typically rupture and discharge the *amniotic fluid* during this stage.

15. The *expulsion (second) stage* begins when the baby's head enters the vagina and lasts until the baby is entirely discharged. An attendant usually drains, clamps, and cuts the umbilical cord at the end of this stage.

16. The *placental (third) stage* is the discharge of the placenta, amnion, and other components of the *afterbirth.* The afterbirth is inspected to be sure

it has all been discharged and that it shows no abnormalities.

17. The *puerperium* is a period of 6 weeks postpartum marked by *involution* of the uterus and the return of other maternal anatomy and physiology to the pregravid state.

**Lactation (p. 1093)**

1. During pregnancy, estrogen, growth hormone, insulin, glucocorticoids, and prolactin stimulate growth and branching of the ducts of the mammary glands. Progesterone then stimulates the development of secretory acini at the ends of the ducts.

2. For 1 to 3 days postpartum, the mammary glands secrete a fluid called *colostrum* rather than milk. Colostrum is higher than milk in protein but lower in fat and lactose. It contains immunoglobulins that give the neonate some immunity to infection.

3. Prolactin is secreted during pregnancy but cannot stimulate milk

synthesis until after the placenta is shed. The mammary glands begin releasing milk about 2 or 3 days postpartum.

4. The nursing infant stimulates neuroendocrine reflexes in which the pituitary gland secretes OT and prolactin. OT triggers contraction of myoepithelial cells of the acini, making milk flow down the lactiferous ducts to the nipple. Prolactin stimulates synthesis of the milk that will be used for the next feeding.

5. Breast milk changes composition over the first two weeks and varies at different times of day and over the course of a single feeding. Most lactose and protein are delivered to the infant at the beginning of a feeding, and most fat at the end.

6. A woman nursing one infant eventually produces about 1.5 L of milk per day at a cost of 50 g of fat, 100 g of lactose, and 2 or 3 g of calcium phosphate. Her diet must compensate for these demands.

# Testing Your Recall

1. Of the following organs, the one(s) most comparable to the penis is (are)
   a. the clitoris.
   b. the vagina.
   c. the vestibular bulbs.
   d. the labia minora.
   e. the prepuce.

2. The ovaries secrete all of the following *except*
   a. estrogens.
   b. progesterone.
   c. androgens.
   d. follicle-stimulating hormone.
   e. inhibin.

3. The first haploid stage in oogenesis is
   a. the oogonium.
   b. the primary oocyte.
   c. the secondary oocyte.
   d. the second polar body.
   e. the zygote.

4. Plasmin is an enzyme that dissolves old blood clots, but in female reproductive physiology its function is
   a. to promote ovulation.
   b. to promote menstruation.

   c. to prevent menstrual fluid from clotting.
   d. to support the corpus luteum.
   e. to stimulate follicle development.

5. The hormone that most directly influences the secretory phase of the menstrual cycle is
   a. HCG.
   b. FSH.
   c. LH.
   d. estrogen.
   e. progesterone.

6. The ischemic phase of the uterus results from
   a. rising progesterone levels.
   b. falling progesterone levels.
   c. stimulation by oxytocin.
   d. stimulation by prostaglandins.
   e. stimulation by estrogens.

7. Before secreting milk, the mammary glands secrete
   a. prolactin.
   b. colostrum.
   c. lochia.
   d. meconium.
   e. chloasma.

8. Few women become pregnant while nursing because _____ inhibits GnRH secretion.
   a. FSH
   b. prolactin
   c. prostaglandins
   d. oxytocin
   e. HCG

9. Smooth muscle cells of the myometrium and myoepithelial cells of the mammary glands are the target cells for
   a. prostaglandins.
   b. LH.
   c. oxytocin.
   d. progesterone.
   e. FSH.

10. Which of these is *not* true of the luteal phase of the sexual cycle?
    a. Progesterone level is high.
    b. The endometrium stores glycogen.
    c. Ovulation occurs.
    d. Fertilization may occur.
    e. The endometrial glands enlarge.

11. Each egg cell develops in its own fluid-filled space called a/an

    _____.

12. The mucosa of the uterus is called _____.

13. A girl's first menstrual period is called _____.

14. A yellowish structure called the _____ secretes progesterone during the secretory phase of the sexual cycle.

15. The first layer of cells around a mature secondary oocyte is the _____.

16. A secondary follicle differs from a primary follicle in having a cavity called the _____.

17. Menopause occurs during a midlife period of changing hormone secretion called _____.

18. All the products of fertilization, including the embryo or fetus, the placenta, and the embryonic membranes, are collectively called the _____.

19. The funnel-like distal end of the uterine tube is called the _____ and has feathery processes called _____.

20. Postpartum uterine involution produces a vaginal discharge called _____.

*Answers in Appendix B*

## True or False

*Determine which five of the following statements are false, and briefly explain why.*

1. After ovulation, a follicle begins to move down the uterine tube to the uterus.

2. Human chorionic gonadotropin is secreted by the granulosa cells of the follicle.

3. An oocyte never completes meiosis II unless it is fertilized.

4. A slim girl who is active in dance and gymnastics is likely to begin menstruating at a later age than an overweight inactive girl.

5. There are more future egg cells in the ovary at puberty than there are at birth.

6. Women do not lactate while they are pregnant because prolactin is not secreted until after birth.

7. Colostrum contains more protein than milk, but less fat.

8. Several follicles develop in each ovarian cycle even though only one of them usually ovulates.

9. Progesterone inhibits uterine contractions.

10. The entire endometrium is shed in each menstrual period.

*Answers in Appendix B*

## Testing Your Comprehension

1. Would you expect puberty to create a state of positive or negative nitrogen balance? Explain. (See chapter 26 to review nitrogen balance.)

2. Aspirin and ibuprofen can inhibit the onset of labor and are sometimes used to prevent premature birth. Review your knowledge of these drugs and the mechanism of labor, and explain this effect.

3. At 6 months postpartum, a nursing mother is in an automobile accident that fractures her skull and severs the hypophyseal portal vessels. How would you expect this to affect her milk production? How would you expect it to affect her future ovarian cycles? Explain the difference.

4. If the ovaries are removed in the first 6 weeks of pregnancy, the embryo will be aborted. If they are removed later in pregnancy, the pregnancy can go to a normal full term. Explain the difference.

5. A breast-feeding woman leaves her baby at home and goes shopping. There, she hears another woman's baby crying and notices her blouse becoming wet with a little exuded milk. Explain the physiological link between hearing that sound and the ejection of milk.

*Answers at www.mhhe.com/saladin4*

# www.mhhe.com/saladin4

*The textbook website provides a wealth of interactive study materials fully organized and integrated by chapter. You will find practice quizzes, labeling exercises, and much more that will complement your learning and understanding of anatomy and physiology. The website also includes tools designed to enhance your **Anatomy & Physiology** | **REVEALED** experience.*

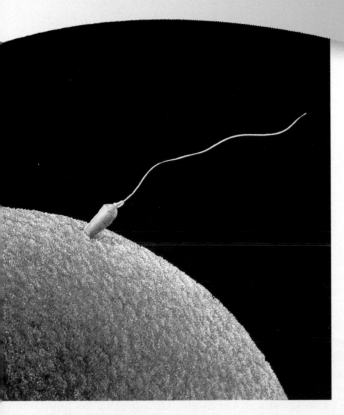

Boy meets girl: the union of sperm and egg (SEM)

# HUMAN DEVELOPMENT

## CHAPTER OUTLINE

## INSIGHTS

## Brushing Up

To understand this chapter, it is important that you understand or brush up on the following concepts:

- Structure of the spermatozoon (pp. 1053–1054)

- Anatomy and histology of the uterus (pp. 1069–1071)

- Stages of oogenesis (p. 1077)

- Structure of a mature ovarian follicle (p. 1080)

Anatomy & Physiology REVEALED®

Perhaps the most dramatic, miraculous aspect of human life is the transformation of a one-celled fertilized egg into an independent, fully developed individual. From the beginning of recorded thought, people have pondered how a baby forms in the mother's body and how two parents can produce another human being who, although unique, possesses characteristics of each. In his quest to understand prenatal development, Aristotle dissected bird embryos and established the sequence in which their organs appeared and took shape. He also speculated that the hereditary traits of a child resulted from the mixing of the male's semen with the female's menstrual blood. Misconceptions about human development persisted for many centuries. In the seventeenth century, scientists thought that all the features of the infant existed in a preformed state in the egg or the sperm, and simply unfolded and expanded as the embryo developed. Some thought that the head of the sperm had a miniature human curled up in it, while others thought that the miniature person existed in the egg, and the sperm were merely parasites in the semen.

The modern science of **embryology**—the study of prenatal development—was not born until the nineteenth century, largely because darwinism at last gave biologists a systematic framework for asking the right questions and discovering unifying themes in the development of diverse species of animals, including humans. It was in that era, too, that the human egg was first observed. Embryology is now a part of **developmental biology,** a broader science that embraces changes in form and function from fertilized egg through old age. A rapidly expanding area of developmental biology today is the genetic regulation of development. Human development from conception to death is the scope of this chapter.

# Fertilization and the Preembryonic Stage

### Objectives
When you have completed this section, you should be able to

- describe the process of sperm migration and fertilization;
- explain how an egg prevents fertilization by more than one sperm;
- describe the major events that transform a fertilized egg into an embryo; and
- describe the implantation of the preembryo in the uterine wall.

Authorities attach different meanings to the word **embryo.** Some use it to denote stages beginning with the fertilized egg or at least with the two-celled stage produced by its first cell division. Others first apply the word *embryo* to an individual 16 days old, when it consists of three **primary germ layers** called the *ectoderm, mesoderm,* and *endoderm.* The events leading up to that stage are called *embryogenesis,* and the first 16 days after fertilization are thus called the *preembryonic stage.* This is the sense in which we will use such terms in this book. We begin with the process in which a sperm locates and fertilizes the egg.

## SPERM MIGRATION

If it is to survive, an egg must be fertilized within 12 to 24 hours of ovulation; yet it takes about 72 hours for an egg to reach the uterus. Therefore, in order to fertilize an egg before it dies, sperm must encounter it somewhere in the distal one-third of the uterine tube. The vast majority of sperm never make it that far. Many are destroyed by vaginal acid or drain out of the vagina. Others fail to penetrate the mucus of the cervical canal, and those that do are often destroyed by leukocytes in the uterus. Of those that get past the uterus, half probably go up the wrong uterine tube. Finally, 2,000 to 3,000 spermatozoa reach the vicinity of the egg—not many of the 300 million that were ejaculated.

Sperm migrate mainly by means of the snakelike lashing of their tails as they crawl along the female mucosa, but they are assisted by certain aspects of female physiology. Strands of mucus guide them through the cervical canal. Although female orgasm is not required for fertilization, orgasm does involve uterine contractions that may suck semen from the vagina and spread it throughout the uterus, like hand lotion pressed between your palms. The egg itself may release a chemical that attracts sperm from a short distance; this has been demonstrated for some animals but remains unproven for humans.

## CAPACITATION

Sperm can reach the distal uterine tube within 5 to 10 minutes of ejaculation, but they cannot fertilize an egg for about 10 hours. While migrating, they must undergo a process of **capacitation** that makes it possible to penetrate an egg. In fresh sperm, the plasma membrane is toughened by cholesterol. This prevents the premature release of acrosomal enzymes while sperm are still in the male, and thus avoids wastage of sperm. It also prevents enzymatic damage to the spermatic ducts. After ejaculation, however, fluids of the female reproductive tract leach cholesterol from the plasma membrane and dilute other inhibitory factors in the semen. The membrane of the sperm head becomes more fragile and more permeable to calcium ions, which diffuse into the sperm and stimulate more powerful lashing of the tail.

Sperm remain viable for up to 6 days after ejaculation, so there is little chance of pregnancy from intercourse occurring more than a week before ovulation. Fertilization also is unlikely if intercourse takes place more than 14 hours after ovulation, because the egg would no longer be viable by the time the sperm became capacitated. For those wishing to conceive a child, the optimal "window of opportunity" is therefore from a few days before ovulation to 14 hours after. Those wishing to avoid pregnancy, however, should allow a wider margin of safety for variations in sperm and egg longevity, capacitation time, and time of ovulation—variations that make the rhythm method of contraception so unreliable.

## FERTILIZATION

When the sperm encounters an egg, it undergoes an **acrosomal reaction**—exocytosis of the acrosome, releasing the enzymes needed to penetrate the egg. But the first sperm to reach an egg is not the one to fertilize it. Sperm must first penetrate the granulosa cells and zona pellucida that surround it (fig. 29.1). It may require hundreds of sperm to clear a path for the one that penetrates the egg proper.

Two of the acrosomal enzymes are **hyaluronidase,** which digests the hyaluronic acid that binds granulosa cells together, and **acrosin,** a protease similar to the trypsin of pancreatic juice. When a path has been cleared through the granulosa cells, a sperm binds to the zona pellucida and releases its enzymes, digesting a pathway through the zona until it contacts the egg itself. The sperm head and midpiece enter the egg, but the egg destroys the sperm mitochondria and passes only maternal mitochondria on to the offspring.

Fertilization combines the haploid ($n$) set of sperm chromosomes with the haploid set of egg chromosomes, producing a diploid ($2n$) set. Fertilization by two or more sperm, called **polyspermy,** would produce a triploid ($3n$) or larger set of chromosomes and the egg would die. Thus it is important for the egg to prevent this, and it has two mechanisms for doing so: a fast block and slow block to polyspermy. In the **fast block,** binding of the sperm to the egg opens $Na^+$ channels in the egg membrane. The rapid

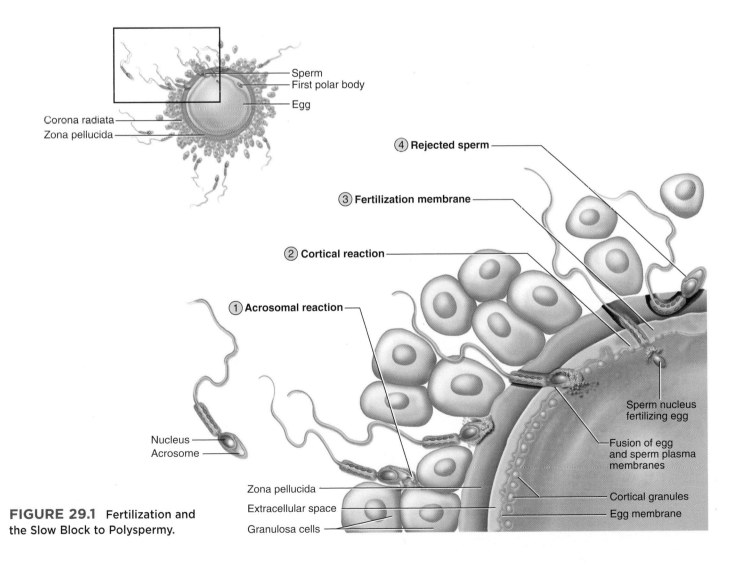

**FIGURE 29.1** Fertilization and the Slow Block to Polyspermy.

inflow of Na$^+$ depolarizes the membrane and inhibits the attachment of any more sperm. The **slow block** involves secretory vesicles called **cortical granules** just beneath the membrane. Sperm penetration releases an inflow of Ca$^{2+}$; this, in turn, stimulates a **cortical reaction** in which the cortical granules release their secretion beneath the zona pellucida. The secretion swells with water, pushes any remaining sperm away from the egg, and creates an impenetrable **fertilization membrane** between the egg and the zona pellucida.

> ### Think About It
>
> *What similarity can you see between the slow block to polyspermy and the release of acetylcholine from synaptic vesicles of a neuron? (Compare p. 465.)*

## MEIOSIS II

A secondary oocyte begins meiosis II before ovulation and completes it only if fertilized. Through the formation of a second polar body, the fertilized egg discards one chromatid from each chromosome. The sperm and egg nuclei then swell and become **pronuclei.** A mitotic spindle forms between them, each pronucleus ruptures, and the chromosomes of the two gametes mix into a single diploid set (fig. 29.2). The fertilized egg, now called a **zygote,** is ready for its first mitotic division. As in chapter 28, we will use the term *conceptus* for everything that arises from this zygote—not only the developing individual but also the placenta, umbilical cord, and membranes associated with the embryo and fetus.

## MAJOR STAGES OF PRENATAL DEVELOPMENT

Clinically, the course of a pregnancy is divided into 3-month intervals called **trimesters:**

1. The **first trimester** extends from fertilization through the first 12 weeks. This is the most precarious

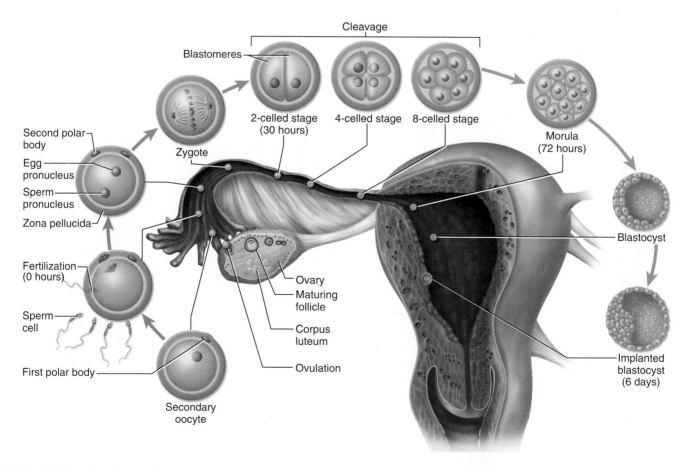

**FIGURE 29.2  Migration of the Conceptus.**   The egg is fertilized in the distal end of the uterine tube, and the preembryo begins cleavage as it migrates to the uterus.

▶ *Why can't the egg be fertilized in the uterus?*

stage, for more than half of all embryos die in the first trimester. The conceptus is most vulnerable to stress, drugs, and nutritional deficiencies during this time.

2. The **second trimester** (weeks 13 through 24) is a period in which the organs complete most of their development. It becomes possible with sonography to see good anatomical detail in the fetus. By the end of this trimester, the fetus looks distinctly human, and with intensive clinical care, infants born at the end of the second trimester have a chance of survival.

3. In the **third trimester** (weeks 25 to birth), the fetus grows rapidly and the organs achieve enough cellular differentiation to support life outside the womb. Some organs, such as the brain, liver, and kidneys, however, require further differentiation after birth to become fully functional. At 35 weeks from fertilization, the fetus typically weighs about 2.5 kg (5.5 lb). It is considered mature at this weight, and usually survives if born early. Most twins are born at about 35 weeks' gestation, and solitary infants at 40 weeks.

From a more biological than clinical standpoint, human development is divided into three stages called the *preembryonic, embryonic,* and *fetal stages.* The timetable and landmark events that distinguish these stages are outlined in table 29.1 and described in the following pages. We end this section with the preembryonic stage, in which the fertilized eggs divides into hundreds of cells, the cells organize themselves into the primary germ layers, and the conceptus becomes firmly attached to the uterine wall.

## THE PREEMBRYONIC STAGE

The **preembryonic stage** comprises the first 16 days of development, culminating in the existence of an embryo. It involves three major processes: cleavage, implantation, and embryogenesis.

### Cleavage

**Cleavage** refers to mitotic divisions that occur in the first 3 days, while the conceptus migrates down the uterine tube (see fig. 29.2). The first cleavage occurs about 30 hours after fertilization and produces the first two daughter cells, or **blastomeres.**[1] These divide simultaneously at shorter and shorter time intervals, doubling the number of blastomeres each time. By the time the conceptus arrives in the uterus, about 72 hours after ovulation, it consists of 16 or more cells and somewhat resembles a mulberry—hence it is called a **morula.**[2] The morula is no larger than the zygote; cleavage merely produces smaller and smaller blastomeres. This increases the ratio of cell surface area to volume, which favors rapid nutrient uptake and waste removal, and it produces a larger number of cells from which to form different embryonic tissues.

The morula lies free in the uterine cavity for 4 to 5 days and divides into 100 cells or so. Meanwhile, the zona

---

[1]*blast* = bud, precursor + *mer* = segment, part
[2]*mor* = mulberry + *ula* = little

| TABLE 29.1 | The Stages of Prenatal Development | |
|---|---|---|
| **Stage** | **Age*** | **Major Developments and Defining Characteristics** |
| *Preembryonic Stage* | | |
| Zygote | 0–30 hours | A single diploid cell formed by the union of egg and sperm |
| Cleavage | 30–72 hours | Mitotic division of the zygote into smaller, identical blastomeres |
| Morula | 3–4 days | A spherical stage consisting of 16 or more blastomeres |
| Blastocyst | 4–16 days | A fluid-filled, spherical stage with an outer mass of trophoblast cells and inner mass of embryoblast cells; becomes implanted in the endometrium; inner cell mass forms an embryonic disc and differentiates into the three primary germ layers |
| *Embryonic Stage* | 16 days–8 weeks | A stage in which the primary germ layers differentiate into organs and organ systems; ends when all organ systems are present |
| *Fetal Stage* | 8–40 weeks | A stage in which organs grow and mature at a cellular level to the point of being capable of supporting life independently of the mother |

*From the time of fertilization

pellucida disintegrates and releases the conceptus, which is now at a stage called the **blastocyst**—a hollow sphere with an outer layer of squamous cells called the **trophoblast**,[3] an inner cell mass called the **embryoblast,** and an internal cavity called the **blastocoel** (BLAST-oh-seal) (fig. 29.4a). The trophoblast is destined to form part of the placenta and play an important role in nourishment of the embryo, whereas the embryoblast is destined to become the embryo itself.

## Implantation

About 6 days after ovulation, the blastocyst attaches to the endometrium, usually on the fundus or the posterior wall of the uterus. The process of attachment, called **implantation,** begins when the blastocyst adheres to the endometrium. The trophoblast cells on this side separate into two layers. In the superficial layer, in contact with the endometrium, the plasma membranes break down and the trophoblast cells fuse into a multinucleate mass called the **syncytiotrophoblast**[4] (sin-SISH-ee-oh-TRO-foe-blast). (A *syncytium* is any body of protoplasm containing multiple nuclei.) The deep layer, close to the embryoblast,

is called the **cytotrophoblast** because it retains individual cells divided by membranes (fig. 29.4b).

The syncytiotrophoblast grows into the uterus like little roots, digesting endometrial cells along the way. The endometrium reacts to this injury by growing over the blastocyst and eventually covering it, so the conceptus becomes completely buried in endometrial tissue (fig. 29.4c). Implantation takes about a week and is completed about the time the next menstrual period would have occurred if the woman had not become pregnant.

Another role of the trophoblast is to secrete human chorionic gonadotropin (HCG). HCG stimulates the corpus luteum to secrete estrogen and progesterone, and progesterone suppresses menstruation. The level of HCG in the mother's blood rises until the end of the second month. During this time, the trophoblast develops into a membrane called the *chorion,* which takes over the role of the corpus luteum and makes HCG unnecessary. The ovaries then become inactive for the rest of the pregnancy, but estrogen and progesterone levels rise dramatically as they are secreted by the ever-growing chorion (see fig. 28.19, p. 1087).

---

### INSIGHT 29.1    Clinical Application

## Twins

There are two ways in which twins are produced (and, by extension, other multiple births). About two-thirds of twins are *dizygotic (DZ)*—produced when two eggs are ovulated and fertilized by separate sperm. They are no more or less genetically similar than any other siblings and may be of different sexes. Multiple ovulation can also result in triplets, quadruplets, or even greater numbers of offspring. DZ twins implant separately on the uterine wall and each forms its own placenta, although their placentas may fuse if they implant close together (fig. 29.3).

*Monozygotic (MZ) twins* are produced when a single egg is fertilized and the cell mass (embryoblast) later divides into two. MZ twins are genetically identical, or nearly so, and are therefore of the same sex and nearly identical appearance. In most cases, they share the same placenta. Identical triplets and quadruplets occasionally result from the splitting of a single embryoblast.

Reproductive biologists are beginning to question whether MZ twins are truly genetically identical. They have suggested that blastomeres may undergo mutation in the course of DNA replication, and the splitting of the embryoblast may represent an attempt of each cell mass to reject the other one as genetically different and presumably foreign.

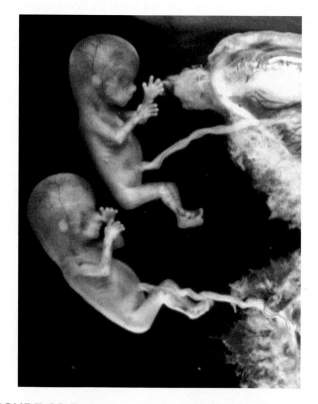

**FIGURE 29.3** Dizygotic Twins with Separate Placentas.

---

[3]*troph* = food, nourishment
[4]*syn* = together + *cyt* = cell

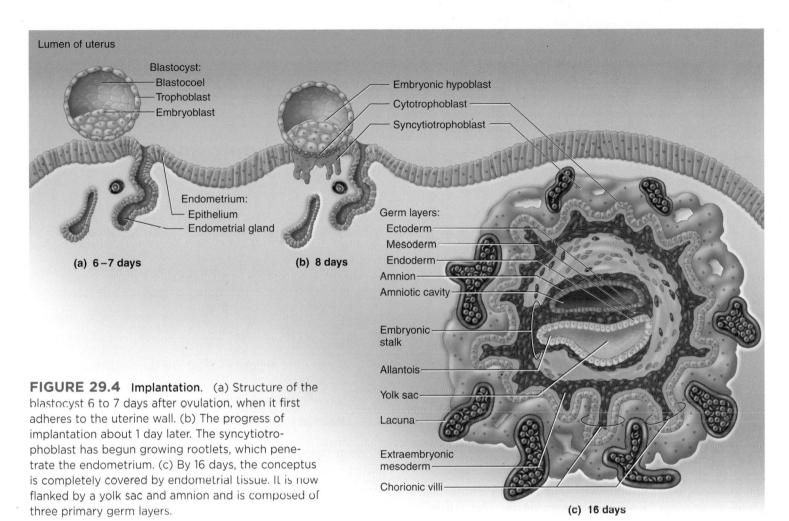

**FIGURE 29.4** **Implantation.** (a) Structure of the blastocyst 6 to 7 days after ovulation, when it first adheres to the uterine wall. (b) The progress of implantation about 1 day later. The syncytiotrophoblast has begun growing rootlets, which penetrate the endometrium. (c) By 16 days, the conceptus is completely covered by endometrial tissue. It is now flanked by a yolk sac and amnion and is composed of three primary germ layers.

---

---

[5]*ec* = outside + *top* = place

### Embryogenesis

During implantation, the embryoblast undergoes **embryogenesis**—arrangement of the blastomeres into the three primary germ layers: *ectoderm, mesoderm,* and *endoderm.* At the beginning of this phase, the embryoblast separates slightly from the trophoblast, creating a narrow space between them called the **amniotic cavity.** The embryoblast flattens into an **embryonic disc** composed initially of two layers: the *epiblast* facing the amniotic cavity and the *hypoblast* facing away. Some hypoblast cells multiply and form a membrane called the *yolk sac* enclosing the blastocoel. Now the embryonic disc is flanked by two spaces: the amniotic cavity on one side and the yolk sac on the other.

Meanwhile, the embryonic disc elongates and, around day 15, a thickened cell layer called the **primitive streak** forms along the midline of the epiblast, with a **primitive groove** running down its middle. These events make the embryo bilaterally symmetric and define its future right and left sides, dorsal and ventral surfaces, and cephalic and caudal ends.

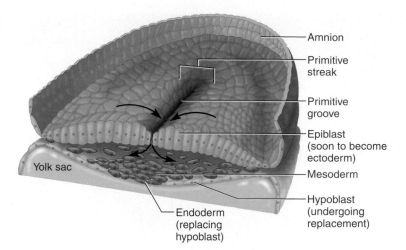

Amnion
Primitive streak
Primitive groove
Epiblast (soon to become ectoderm)
Mesoderm
Hypoblast (undergoing replacement)
Endoderm (replacing hypoblast)
Yolk sac

**FIGURE 29.5**  **Formation of the Primary Germ Layers (Gastrulation).**  Composite view of the embryonic disc at 15 to 16 days. Epiblast cells migrate over the surface and down into the primitive groove, first replacing the hypoblast cells with endoderm, then filling the space with mesoderm. Upon completion of this process, the uppermost layer is considered ectoderm.

The next step is **gastrulation**—multiplying epiblast cells migrate medially toward the primitive groove and down into it (fig. 29.5). They replace the original hypoblast with a layer now called **endoderm,** which will become the inner lining of the digestive tract, among other things. A day later, migrating epiblast cells form a third layer between the first two, called **mesoderm.** Once this is formed, the remaining epiblast is called **ectoderm.** Thus, all three primary germ layers arise from the original epiblast. Some mesoderm overflows the embryonic disc and becomes an extensive *extraembryonic mesoderm* which contributes to formation of the placenta (see fig. 29.4c).

The ectoderm and endoderm are epithelia composed of tightly joined cells, but the mesoderm is a more loosely organized tissue. It later differentiates into a loose fetal connective tissue called **mesenchyme,** which gives rise to such tissues as muscle, bone, and blood. Mesenchyme is composed of fibroblasts and fine, wispy collagen fibers embedded in a gelatinous ground substance.

Once the three primary germ layers are formed, embryogenesis is complete and the individual is considered an embryo. It is about 2 mm long and 16 days old at this point.

## Before You Go On

*Answer the following questions to test your understanding of the preceding section:*

1. *How soon can a sperm reach an egg after ejaculation? How soon can it fertilize an egg? What accounts for the difference?*

2. *Describe two ways a fertilized egg prevents the entry of excess sperm.*

3. *In the blastocyst, what are the cells called that eventually give rise to the embryo? What are the cells that carry out implantation?*

4. *What major characteristic distinguishes an embryo from a preembryo?*

# The Embryonic and Fetal Stages

### Objectives
When you have completed this section, you should be able to

- describe the formation and functions of the placenta;
- explain how the conceptus is nourished before the placenta takes over this function;
- describe the embryonic membranes and their functions;
- identify the major tissues derived from the primary germ layers;
- describe the major events of fetal development; and
- describe the fetal circulatory system.

Sixteen days after conception, the germ layers are present and the embryonic stage of development begins. Over the next 6 weeks, a placenta forms on the uterine wall and becomes the embryo's primary means of nutrition, while the germ layers differentiate into organs and organ systems. Although these organs are still far from functional, it is their presence at 8 weeks that marks the transition from the embryonic stage to the fetal stage. Here we will see how the embryo becomes a fetus, how the membranes collectively known as the "afterbirth" develop around the fetus, and how the conceptus is nourished throughout its gestation.

## EMBRYONIC FOLDING AND ORGANOGENESIS

The **embryonic stage** of development lasts from day 16 to the end of week 8. The primary developmental process to occur during this time is the formation of organs and organ systems from the three primary germ layers (table 29.2)—a process called **organogenesis.**

One of the major transformations in this stage is conversion of the flat embryonic disc of figure 29.4c into a somewhat cylindrical form. This occurs during weeks 3 to 4 as the embryo rapidly grows and folds around the yolk sac (fig. 29.6). As the cephalic and caudal ends curve around the ends of the yolk sac, the embryo becomes C-shaped, with the head and tail almost touching. At the same time, the lateral margins of the disc fold around the sides of the yolk sac

| TABLE 29.2 | Derivatives of the Three Primary Germ Layers |
|---|---|
| **Layer** | **Major Derivatives** |
| Ectoderm | Epidermis; hair follicles and piloerector muscles; cutaneous glands; nervous system; adrenal medulla; pineal and pituitary glands; lens, cornea, and intrinsic muscles of the eye; internal and external ear; salivary glands; epithelia of the nasal cavity, oral cavity, and anal canal |
| Mesoderm | Skeleton; skeletal, cardiac, and most smooth muscle; cartilage; adrenal cortex; middle ear; dermis; blood; blood and lymphatic vessels; bone marrow; lymphoid tissue; epithelium of kidneys, ureters, gonads, and genital ducts; mesothelium of ventral body cavity |
| Endoderm | Most of the mucosal epithelium of the digestive and respiratory tracts; mucosal epithelium of urinary bladder and parts of urethra; epithelial components of accessory reproductive and digestive glands (except salivary glands); thyroid and parathyroid glands; thymus |

to form the ventral surface of the embryo. This lateral folding encloses a longitudinal channel, the *primitive gut,* which later becomes the digestive tract.

As a result of embryonic folding, the entire surface is covered with ectoderm, which later produces the epidermis of the skin. In the meantime, the mesoderm splits into two layers. One of them adheres to the ectoderm and the other to the endoderm, thus opening a body cavity between them (fig. 29.6c). The body cavity becomes divided into the thoracic cavity and peritoneal cavity by a wall, the diaphragm. By the end of week 5, the thoracic cavity further subdivides into pleural and pericardial cavities.

We cannot delve at greater length into development of all the organ systems, but this description is at least enough to see how some of them begin to form. Some have also been described in earlier chapters. By the end of 8 weeks, all of the organ systems are present, the individual is about 3 cm long, and it is now considered a **fetus** (see fig. 29.11d). The bones have just begun to calcify and the skeletal muscles exhibit spontaneous contractions, although these are too weak to be felt by the mother. The heart, beating since the fourth week, now circulates blood. The heart and liver are very large and form the prominent ventral bulge seen in figure 29.11c. The head is nearly half the total body length.

⌐ **Think About It**

*List the four primary tissue types of the adult body (see chapter 5) and identify which of the three primary germ layers of the embryo predominantly gives rise to each.*

## EMBRYONIC MEMBRANES

Several accessory organs develop alongside the embryo: a *placenta, umbilical cord,* and four embryonic membranes called the *amnion, yolk sac, allantois,* and *chorion* (figs. 29.6 and 29.7). To understand these membranes, it helps to realize that all mammals evolved from egg-laying reptiles. Within the shelled, self-contained egg of a reptile, the embryo rests atop a yolk, which is enclosed in the yolk sac; it floats in a little sea of liquid contained in the amnion; it stores its toxic wastes in the allantois; and to breathe, it has a chorion permeable to gases. All of these membranes persist in mammals, including humans, but are modified in their functions.

The **amnion** is a transparent sac that develops from cells of the epiblast (see figs. 29.4c and 29.5). It grows to completely enclose the embryo and is penetrated only by the umbilical cord. The amnion fills with **amniotic fluid** (see fig. 29.11d), which protects the embryo from trauma, infection, and temperature fluctuations; allows the freedom of movement important to muscle development; enables the embryo to develop symmetrically; prevents body parts from adhering to each other, such as an arm to the trunk; and stimulates lung development as the fetus "breathes" the fluid. At first, the amniotic fluid forms by filtration of the mother's blood plasma, but beginning at 8 to 9 weeks, the fetus urinates into the amniotic cavity about once an hour and contributes substantially to the fluid volume. The volume remains stable, however, because the fetus swallows amniotic fluid at a comparable rate. At term, the amnion contains 700 to 1,000 mL of fluid.

The **yolk sac** arises from hypoblast cells opposite the amnion. It is a small sac suspended from the ventral side of the embryo. It contributes to the formation of the digestive tract and produces the first blood cells and future egg or sperm cells.

The **allantois** (ah-LON-toe-iss) begins as an outpocketing of the posterior end of the yolk sac (see fig. 29.4c), but eventually becomes an outgrowth of the caudal end of the gut. It forms the foundation for the umbilical cord and becomes part of the urinary bladder. It can be seen in cross sections cut near the proximal end of a mature umbilical cord.

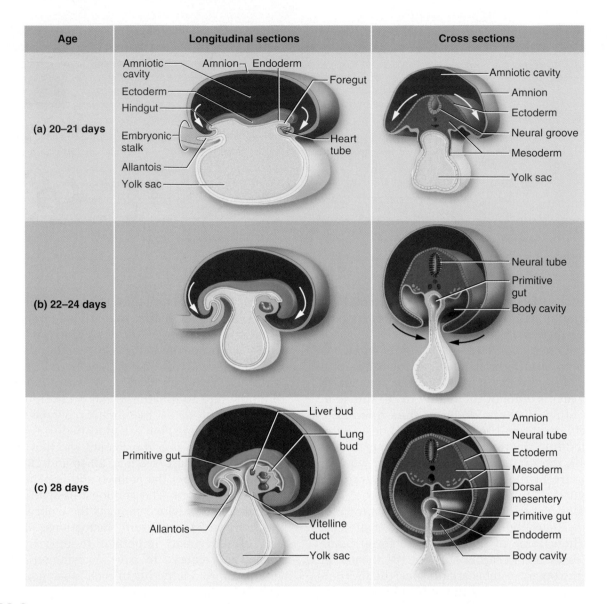

| Age | Longitudinal sections | Cross sections |
|---|---|---|
| **(a) 20–21 days** | Amniotic cavity, Amnion, Endoderm, Ectoderm, Foregut, Hindgut, Embryonic stalk, Heart tube, Allantois, Yolk sac | Amniotic cavity, Amnion, Ectoderm, Neural groove, Mesoderm, Yolk sac |
| **(b) 22–24 days** | | Neural tube, Primitive gut, Body cavity |
| **(c) 28 days** | Liver bud, Lung bud, Primitive gut, Allantois, Vitelline duct, Yolk sac | Amnion, Neural tube, Ectoderm, Mesoderm, Dorsal mesentery, Primitive gut, Endoderm, Body cavity |

**FIGURE 29.6   Embryonic Folding.**   The right-hand figures are cross sections cut about midway along the figures on the left. Part (a) corresponds to figure 29.4c at a slightly later stage of development. Note the general trend for cephalic and caudal (head and tail) ends of the embryo to curl toward each other *(left-hand figures)* until the embryo assumes a C shape, and for the flanks of the embryo to fold laterally *(right-hand figures),* converting the flat embryonic disc into a more cylindrical body and eventually enclosing a body cavity (c).

The **chorion** is the outermost membrane, enclosing all the rest of the membranes and the embryo (see fig. 29.7). Initially it has shaggy outgrowths called **chorionic villi** around its entire surface, but as the pregnancy advances, the villi of the placental region grow and branch while the rest of them degenerate. At the placental attachment the chorion is then called the *villous chorion,* and the rest is called the *smooth chorion.* The villous chorion forms the fetal portion of the placenta, discussed shortly.

## PRENATAL NUTRITION

Over the course of gestation, the conceptus is nourished in three different, overlapping ways: by *uterine milk, trophoblastic nutrition,* and *placental nutrition.*

**Uterine milk** is a glycogen-rich secretion of the uterine tubes and endometrial glands. The conceptus absorbs this fluid as it travels down the tube and lies free in the uterine cavity before implantation. The accumulating fluid forms the blastocoel in figure 29.4a.

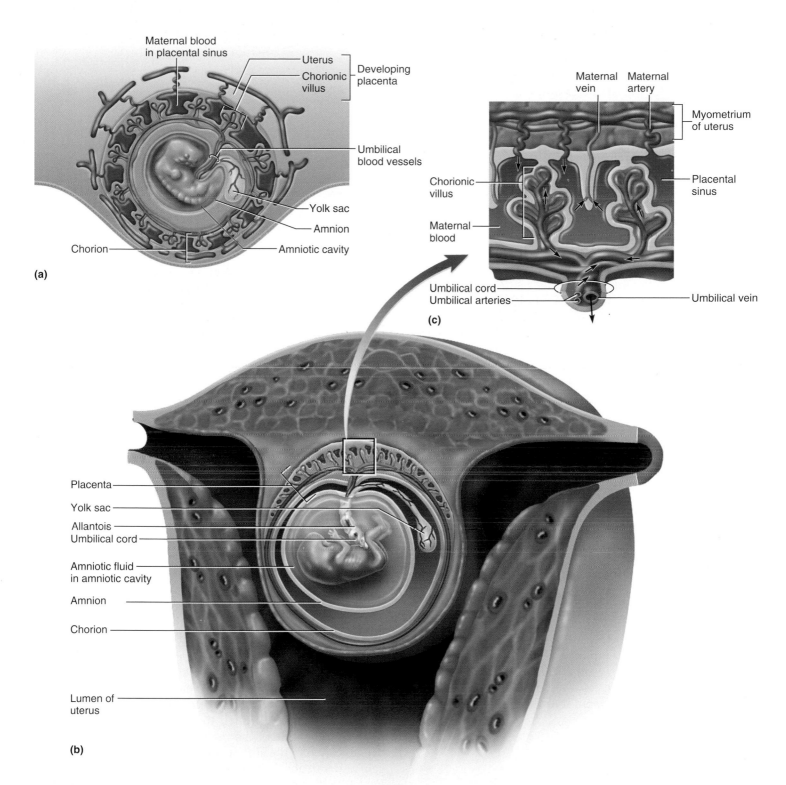

**FIGURE 29.7 The Placenta and Embryonic Membranes.** (a) Embryo at 4 weeks, enclosed in the amnion and chorion and surrounded by a developing placenta. (b) Fetus at 12 weeks. The placenta is now complete and lies on only one side of the fetus. (c) A portion of the mature placenta and umbilical cord, showing the relationship between fetal and maternal circulation.

As it implants, the conceptus makes a transition to **trophoblastic nutrition,** in which it consumes so-called **decidual**[6] **cells** of the endometrium. Progesterone from the corpus luteum stimulates these cells to proliferate and accumulate a store of glycogen, proteins, and lipids. As the conceptus burrows into the endometrium, the syncytiotrophoblast digests them and supplies the nutrients to the embryoblast. Trophoblastic nutrition is the only mode of nutrition for the first week after implantation. It remains the dominant source of nutrients through the end of week 8; the period from implantation through week 8 is therefore called the **trophoblastic phase** of the pregnancy. Trophoblastic nutrition wanes as placental nutrition takes over, and ceases entirely by the end of week 12 (fig. 29.8).

In **placental nutrition,** nutrients diffuse from the mother's blood through the placenta into the fetal blood. The **placenta**[7] is a disc-shaped organ attached to the uterine wall on one side and, on the other side, attached by way of an **umbilical cord** to the fetus (fig. 29.9). When fully developed, it is about 20 cm in diameter, 3 cm thick, and weighs about one-sixth as much as the newborn infant. The surface facing the fetus is smooth and gives rise to the umbilical cord, while the surface attached to the uterine wall is rougher. The uterine surface consists of chorionic villi embedded in the mother's endometrium.

The placenta begins to develop about 11 days after conception, becomes the dominant mode of nutrition around the beginning of week 9, and is the sole mode of nutrition from the end of week 12 until birth. The period from week 9 until birth is called the **placental phase** of the pregnancy.

Figures 29.4 and 29.7 depict the development of the placenta, or *placentation.* The process begins during implantation, as extensions of the syncytiotrophoblast penetrate more and more deeply into the endometrium, like the roots of a tree penetrating into the nourishing "soil" of the uterus. These roots are the early chorionic villi. As they penetrate uterine blood vessels, they become surrounded by *lacunae,* or endometrial spaces filled with maternal blood (fig. 29.4c). The lacunae eventually merge to form a single blood-filled cavity, the **placental sinus.** Exposure to maternal blood stimulates increasingly rapid growth of the villi, which become branched and treelike. Extraembryonic mesoderm grows

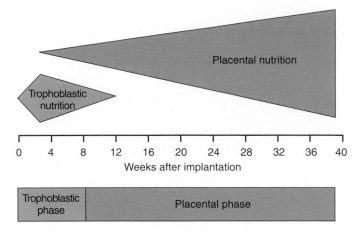

**FIGURE 29.8**  The Timetable of Trophoblastic and Placental Nutrition.  Trophoblastic nutrition peaks at 2 weeks and ends by 12 weeks. Placental nutrition begins at 2 weeks and becomes increasingly important until birth, 39 weeks after implantation. The two modes of nutrition overlap up to the eighth week, but the *trophoblast phase* is the period in which most nutrients are supplied by trophoblastic nutrition, and the *placental phase* is the period in which most (eventually all) nutrition comes from the placenta.

▶ *At what point do the two modes contribute equally to prenatal nutrition?*

into the villi and gives rise to the blood vessels that connect to the embryo by way of the umbilical cord.

The umbilical cord contains two **umbilical arteries** and one **umbilical vein.** Pumped by the fetal heart, blood flows into the placenta by way of the umbilical arteries and then returns to the fetus by way of the umbilical vein. The chorionic villi are *filled with* fetal blood and *surrounded by* maternal blood (fig. 29.7c); the two bloodstreams do not mix unless there is damage to the placental barrier. The barrier, however, is only 3.5 μm thick—half the diameter of a single red blood cell. Early in development, the villi have thick membranes that are not very permeable to nutrients and wastes, and their total surface area is relatively small. As the villi grow and branch, their surface area increases and the membranes become thinner and more permeable. Thus there is a dramatic increase in *placental conductivity,* the rate at which substances diffuse through the membrane. Materials diffuse from the side of the membrane where they are more concentrated to the side where they are less so. Thus, oxygen and nutrients pass from the maternal blood to the fetal blood, while fetal wastes pass the other way to be eliminated by the mother. Unfortunately, the placenta is

---

[6]*decid* = falling off
[7]*placenta* = flat cake

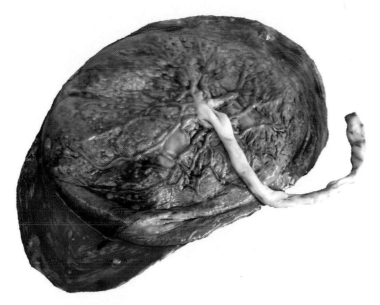

**(a) Fetal side**

**(b) Maternal (uterine) side**

**FIGURE 29.9  The Placenta and Umbilical Cord.**  (a) The fetal side, showing blood vessels, the umbilical cord, and some of the amniotic sac attached to the lower left margin. (b) The maternal (uterine) side, where chorionic villi give the placenta a rougher texture.

also permeable to nicotine, alcohol, and most other drugs that may be present in the maternal bloodstream.

Table 29.3 summarizes the nutritional, excretory, and other functions of the placenta.

## FETAL DEVELOPMENT

The fetus is the final stage of prenatal development, from the end of the eighth week until birth. The organs that formed during the embryonic stage now undergo growth

| TABLE 29.3 | Functions of the Placenta |
|---|---|
| Nutritional roles | Transports nutrients such as glucose, amino acids, fatty acids, minerals, and vitamins from the maternal blood to the fetal blood; stores nutrients such as carbohydrates, protein, iron, and calcium in early pregnancy and releases them to the fetus later, when fetal demand is greater than the mother can absorb from the diet |
| Excretory roles | Transports nitrogenous wastes such as ammonia, urea, uric acid, and creatinine from the fetal blood to the maternal blood |
| Respiratory roles | Transports $O_2$ from mother to fetus and $CO_2$ from fetus to mother |
| Endocrine roles | Secretes estrogens, progesterone, relaxin, human chorionic gonadotropin, and human chorionic somatomammotropin; allows other hormones synthesized by the conceptus to pass into the mother's blood and maternal hormones to pass into the fetal blood |
| Immune roles | Transports maternal antibodies (especially IgG) into fetal blood to confer passive immunity on fetus |

and cellular differentiation, acquiring the functional capability to support life outside the mother.

The circulatory system shows the most conspicuous anatomical changes from a prenatal state, dependent on the placenta, to the independent neonatal (newborn) state (fig. 29.10). The unique aspects of fetal circulation are the umbilical–placental circuit and the presence of three circulatory shortcuts called *shunts.* The internal iliac arteries give rise to the umbilical arteries, which pass on either side of the bladder into the umbilical cord. The blood in these arteries is low in oxygen and high in carbon dioxide and other fetal wastes; thus they are depicted in blue in figure 29.10a. The arterial blood discharges its wastes in the placenta, loads oxygen and nutrients, and returns to the fetus by way of a single umbilical vein, which leads toward the liver. The umbilical vein is depicted in red because of its well-oxygenated blood. Some of this venous blood filters through the liver to nourish it. However, the immature liver is not capable of performing many of its postpartum functions, so it does not require a great deal of perfusion before birth. Most of the venous blood therefore bypasses it by way of a shunt called the **ductus venosus,** which leads directly to the inferior vena cava.

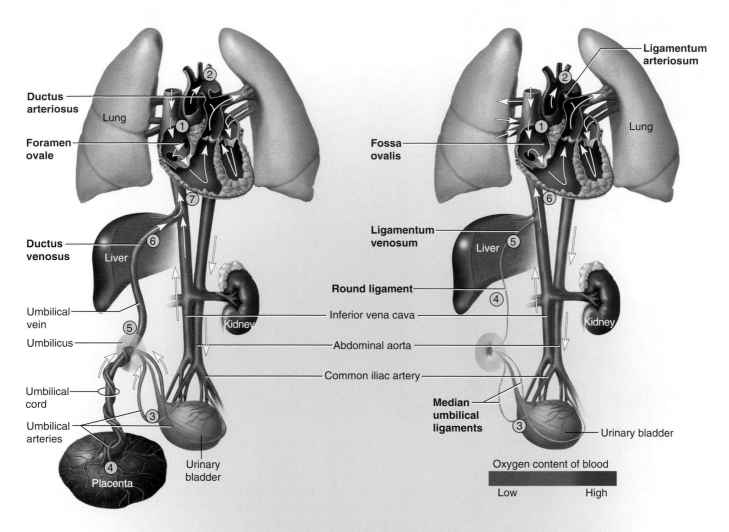

**Ductus arteriosus**

Lung

**Foramen ovale**

**Ductus venosus**

Liver

Umbilical vein

Umbilicus

Umbilical cord

Umbilical arteries

Kidney

Placenta

Urinary bladder

**Ligamentum arteriosum**

Lung

**Fossa ovalis**

**Ligamentum venosum**

Liver

**Round ligament**

Inferior vena cava

Abdominal aorta

Common iliac artery

**Median umbilical ligaments**

Kidney

Urinary bladder

Oxygen content of blood

Low                    High

**(a) Fetal circulation**

① Blood bypasses the lungs by flowing directly from the right atrium through the foramen ovale into the left atrium.

② Blood also bypasses the lungs by flowing from the pulmonary trunk through the ductus arteriosus into the aorta.

③ Oxygen-poor, waste-laden blood flows through two umbilical arteries to the placenta.

④ The placenta disposes of $CO_2$ and other wastes and reoxygenates the blood.

⑤ Oxygenated blood returns to the fetus through the umbilical vein.

⑥ Placental blood bypasses the liver by flowing through the ductus venosus into the inferior vena cava (IVC).

⑦ Placental blood from the umbilical vein mixes with fetal blood from the IVC and returns to the heart.

**(b) Neonatal circulation**

① Foramen ovale closes and becomes fossa ovalis.

② Ductus arteriosus constricts and becomes ligamentum arteriosum.

③ Umbilical arteries degenerate and become median umbilical ligaments.

④ Umbilical vein constricts and becomes round ligament of liver.

⑤ Ductus venosus degenerates and becomes ligamentum venosum of liver.

⑥ Blood returning to the heart is now oxygen-poor, systemic blood only.

**FIGURE 29.10    Blood Circulation in the Fetus and Newborn.**    Boldfaced terms in (a) indicate the three shunts in the fetal circulation, which allow most blood to bypass the liver and lungs. Boldfaced terms in (b) indicate the postpartum vestiges of fetal structures.

In the inferior vena cava, placental blood mixes with venous blood from the fetus's body and flows to the right atrium of the heart. After birth, the right ventricle pumps all of its blood into the lungs, but there is little need for this in the fetus because the lungs are not yet functional. Therefore, most fetal blood bypasses the pulmonary circuit. Some goes directly from the right atrium to the left through the **foramen ovale,** a hole in the interatrial septum. Some also goes into the right ventricle and is pumped into the pulmonary trunk, but most of this is shunted directly into the aorta by way of a short passage called the **ductus arteriosus.** This occurs because the collapsed state of the fetal lungs creates high resistance and blood pressure in the pulmonary circuit, so blood in the pulmonary trunk flows through the ductus into the aorta, where the blood pressure is lower. The lungs receive only a trickle of blood, suffi-cient to meet their metabolic needs during development. Blood leaving the left ventricle enters the general systemic circulation, and some of this returns to the placenta. This circulatory pattern changes dramatically at birth, when the neonate is cut off from the placenta and the lungs expand with air. Those changes will be described later.

Fetal growth is charted by weight and body length (table 29.4). Body length is customarily measured from the crown of the head to the curve of the buttocks in a sit-ting position *(crown-to-rump length, CRL),* thus excluding the lower limbs. Full-term fetuses have an average CRL of about 36 cm (14 in.) and average weight of about 3.0 to 3.4 kg (6.6–7.5 lb). The fetus gains about 50% of its birth weight in the last 10 weeks.

Additional aspects of embryonic and fetal develop-ment are listed in table 29.4 and depicted in figure 29.11.

| TABLE 29.4 | Major Events of Prenatal Development, with Emphasis on the Fetal Stage | |
|---|---|---|
| **End of Week** | **Crown-to-Rump Length; Weight** | **Developmental Events** |
| 4 | 0.6 cm; <1 g | Vertebral column and central nervous system begin to form; limbs represented by small limb buds; heart begins beating around day 22; no visible eyes, nose, or ears |
| 8 | 3 cm; 1 g | Eyes form, eyelids fused shut; nose flat, nostrils evident but plugged with mucus; head nearly as large as the rest of the body; brain waves detectable; bone calcification begins; limb buds form paddlelike hands and feet with ridges called **digital rays,** which then separate into distinct fingers and toes; blood cells and major blood vessels form; genitals present but sexes not yet distinguishable |
| 12 | 9 cm; 45 g | Eyes well developed, facing laterally; eyelids still fused; nose develops bridge; external ears present; limbs well formed, digits exhibit nails; fetus swallows amniotic fluid and produces urine; fetus moves, but too weakly for mother to feel it; liver is prominent and produces bile; palate is fusing; sexes can be distinguished |
| 16 | 14 cm; 200 g | Eyes face anteriorly, external ears stand out from head, face looks more distinctly human; body larger in proportion to head; skin is bright pink; scalp has hair; joints forming; lips exhibit sucking movements; kid-neys well formed; digestive glands forming and **meconium**[8] (fetal feces) accumulating in intestine; heart-beat can be heard with a stethoscope |
| 20 | 19 cm; 460 g | Body covered with fine hair called **lanugo**[9] and cheeselike sebaceous secretion called **vernix caseosa,**[10] which protects it from amniotic fluid; skin bright pink; brown fat forms and will be used for postpartum heat production; fetus is now bent forward into "fetal position" because of crowding; **quickening** occurs—mother can feel fetal movements |
| 24 | 23 cm; 820 g | Eyes partially open; skin wrinkled, pink, and translucent; lungs begin producing surfactant; rapid weight gain |
| 28 | 27 cm; 1,300 g | Eyes fully open; skin wrinkled and red; full head of hair present; eyelashes formed; fetus turns into upside-down **vertex position;** testes begin to descend into scrotum; marginally viable if born at 28 weeks |
| 32 | 30 cm; 2,100 g | Subcutaneous fat deposition gives fetus a more plump, babyish appearance, with lighter, less wrinkled skin; testes descending; twins usually born at this stage |
| 36 | 34 cm; 2,900 g | More subcutaneous fat deposited, body plump; lanugo is shed; nails extend to fingertips; limbs flexed; firm hand grip |
| 38 | 36 cm; 3,400 g | Prominent chest, protruding breasts; testes in inguinal canal or scrotum; fingernails extend beyond fingertips |

[8]*mecon* = poppy juice, opium
[9]*lan* = down, wool
[10]*vernix* = varnish + *caseo* = cheese

**FIGURE 29.11** **The Developing Human.** Parts (a) through (d) show development through the end of the embryonic stage. Parts (e) and (f) represent the fetal stage of development.

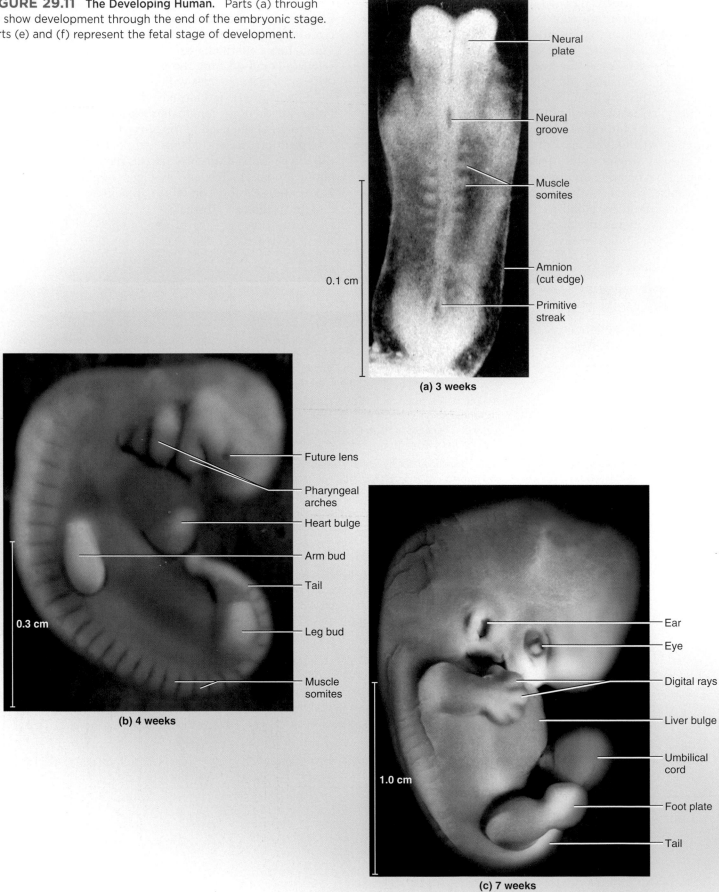

**(a) 3 weeks**

**(b) 4 weeks**

**(c) 7 weeks**

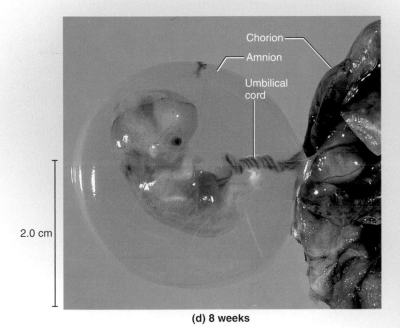

(d) 8 weeks

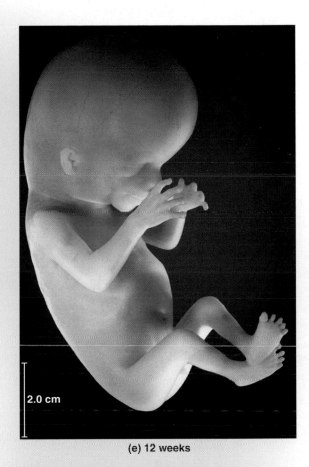

(e) 12 weeks

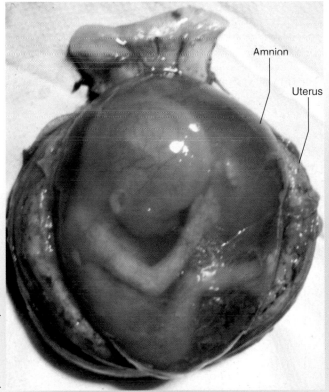

(f) 20 weeks

## Before You Go On

*Answer the following questions to test your understanding of the preceding section:*

5. *Distinguish between trophoblastic and placental nutrition.*

6. *Identify the two sources of blood to the placenta. Where do these two bloodstreams come closest to each other? What keeps them separate?*

7. *State the functions of the placenta, amnion, chorion, yolk sac, and allantois.*

8. *What developmental characteristic distinguishes a fetus from an embryo? At what gestational age is this attained?*

9. *Identify the three circulatory shunts of the fetus. Why does the blood take these "shortcuts" before birth?*

# The Neonate

### Objectives

When you have completed this section, you should be able to

- describe how and why the circulatory system changes at birth;
- explain why the first breaths of air are relatively difficult for a neonate;
- describe the major physiological problems of a premature infant; and
- discuss some common causes of birth defects.

Development is by no means complete at birth. For example, the liver and kidneys still are not fully functional, most joints are not yet ossified, and myelination of the nervous system is not completed until adolescence. Indeed, humans are born in a very immature state compared with other mammals—a fact necessitated by the narrow outlet of the female pelvis, which in turn was necessitated by the evolution of bipedal locomotion.

## THE TRANSITIONAL PERIOD

The period immediately following birth is a crisis in which the neonate suddenly must adapt to life outside the mother's body. The first 6 to 8 hours are a **transitional period** in which the heart and respiratory rates increase and the body temperature falls. Physical activity then declines and the baby sleeps for about 3 hours. In its second period of activity, the baby often gags on mucus and debris in the pharynx. The baby then sleeps again, becomes more stable, and begins a cycle of waking every 3 to 4 hours to feed. The first 6 weeks of life constitute the **neonatal period.**

## CIRCULATORY ADAPTATIONS

After the umbilical cord is clamped and cut, the umbilical arteries and vein collapse and become fibrotic. The proximal part of each umbilical artery becomes the *superior vesical artery,* which remains to supply the bladder. Other obliterated vessels become fibrous cords or ligaments: the distal parts of the umbilical arteries become the *median umbilical ligaments* of the abdominal wall; the umbilical vein becomes the *round ligament* (ligamentum teres) of the liver; and the ductus venosus (the former shunt around the liver) becomes the *ligamentum venosum* on the inferior surface of the liver (see fig. 29.10b).

When the lungs expand with air, blood pressure in the pulmonary circuit drops rapidly and pressure in the right heart falls below that in the left. Blood flows briefly from the left atrium to the right through the foramen ovale (opposite from its prenatal flow) and pushes two flaps of tissue into place to close this shunt. In most people these flaps fuse and permanently seal the foramen during the first year, leaving a depression, the *fossa ovalis,* in the interatrial septum. In about 25% of people, however, the foramen ovale remains unsealed and the flaps are held in place only by the relatively high blood pressure in the left atrium. Pressure changes in the pulmonary trunk and aorta also cause the ductus arteriosus to collapse. It closes permanently around 3 months of age and leaves a permanent cord, the *ligamentum arteriosum,* between the two vessels.

## RESPIRATORY ADAPTATIONS

It is an old misconception that a neonate must be spanked to stimulate it to breathe. During birth, $CO_2$ accumulates in the baby's blood and strongly stimulates the respiratory chemoreceptors. Unless the infant is depressed by oversedation of the mother, it normally begins breathing spontaneously. It requires a great effort, however, to take the first few breaths and inflate the collapsed alveoli. For the first 2 weeks, a baby takes about 45 breaths per minute, but subsequently stabilizes at about 12 breaths per minute.

## IMMUNOLOGICAL ADAPTATIONS

Cellular immunity begins to appear early in fetal development, but the immune responses of the neonate are still weak. Fortunately, an infant is born with a near-adult level of IgG acquired from the mother through the placenta. This maternal IgG breaks down rapidly after birth,

---

### INSIGHT 29.3    Clinical Application

#### Neonatal Assessment

A newborn infant is immediately evaluated for general appearance, vital signs (temperature, pulse, and respiratory rate), weight, length, and head circumference and other dimensions, and it is screened for congenital disorders such as phenylketonuria (PKU). At 1 minute and 5 minutes after birth, the heart rate, respiratory effort, muscle tone, reflexes, and skin color are noted and given a score of 0 (poor), 1, or 2 (excellent). The total (0–10), called the *Apgar*[11] *score,* is a good predictor of infant survival. Infants with low Apgar scores may have neurological damage and need immediate attention if they are to survive. A low score at 1 minute suggests asphyxiation and may demand assisted ventilation. A low score at 5 minutes indicates a high probability of death.

---

[11]Virginia Apgar (1909–74), American anesthesiologist

declining to about half the initial level in the first month and to essentially none by 10 months. Nevertheless, maternal IgG levels remain high enough for 6 months to protect the infant from measles, diphtheria, polio, and most other infectious diseases (but not whooping cough). By 6 months, the infant's own IgG reaches about half the typical adult level. The lowest total (maternal + infant) level of IgG exists around 5 to 6 months of age, and respiratory infections are especially common at that age. A breast-fed neonate also acquires protection from gastroenteritis from the IgA present in colostrum.

## OTHER ADAPTATIONS

Thermoregulation and fluid balance are also critical aspects of neonatal physiology. An infant has a larger ratio of surface area to volume than an adult does, so it loses heat more easily. One of its defenses against hypothermia is brown fat, a special adipose tissue deposited from weeks 17 to 20 of fetal development. The mitochondria of brown fat release all the energy of pyruvic acid as heat rather than using it to make ATP; thus, this is a heat-generating tissue. As a baby grows, its metabolic rate increases and it accumulates even more subcutaneous fat, thus producing and retaining more heat. Nevertheless, body temperature is more variable in infants and children than in adults.

The kidneys are not fully developed at birth and cannot concentrate the urine as much as a mature kidney can. Consequently, infants have a relatively high rate of water loss and require more fluid intake, relative to body weight, than adults do.

## PREMATURE INFANTS

Neonates weighing under 2.5 kg (5.5 lb) are generally considered **premature.** They have multiple difficulties in respiration, thermoregulation, excretion, digestion, and liver function. Most neonates weighing 1.5 to 2.5 kg are viable, but with difficulty. Those weighing under 500 g rarely survive.

The respiratory system is adequately developed by 7 months of gestation to support independent life. Infants born before this have a deficiency of pulmonary surfactant, causing **respiratory distress syndrome (RDS),** also called *hyaline membrane disease.* The alveoli collapse each time the infant exhales, and a great effort is needed to reinflate them. The infant becomes very fatigued by the high energy demand of breathing. RDS may be treated by ventilating the lungs with oxygen-enriched air at a positive pressure to keep the lungs inflated between breaths, and by administering surfactant as an inhalant. Nevertheless, RDS remains the most common cause of neonatal death.

A premature infant has an incompletely developed hypothalamus and therefore cannot thermoregulate effectively. Body temperature must be controlled by placing the infant in a warmer.

It is difficult for premature infants to ingest milk because of their small stomach volume and underdeveloped sucking and swallowing reflexes. Some must be fed by nasogastric or nasoduodenal tubes. Most of them, however, can tolerate human milk or formulas. Infants under 1.5 kg (3.3 lb) require nutritional supplements of calcium, phosphorus, and protein.

The liver is also poorly developed, and bearing in mind its very diverse functions (see table 26.6, p. 1021), you can probably understand why this would have several serious consequences. The liver synthesizes inadequate amounts of albumin, so the baby suffers hypoproteinemia. This upsets the balance between capillary filtration and reabsorption and leads to edema. The infant bleeds easily because of a deficiency of the clotting factors synthesized by the liver. This is true to some degree even in full-term infants, however, because the baby's intestines are not yet colonized by the bacteria that synthesize vitamin K, which is essential for the synthesis of clotting factors. Vitamin K injections are now routine (but somewhat controversial) for newborns in the United States. Jaundice is common in neonates, especially premature babies, because the liver cannot dispose of bile pigments efficiently.

## BIRTH DEFECTS

A birth defect, or **congenital anomaly,**[12] is the abnormal structure or position of an organ at birth, resulting from a defect in prenatal development. The study of birth defects is called **teratology.**[13] Birth defects are the most common cause of infant mortality in North America. Not all of them are noticeable at birth; some are detected months to years later. Thus, by the age of 2 years, 6% of children are diagnosed with congenital anomalies, and by age 5 the incidence is 8%. The following sections discuss some known causes of congenital anomalies, but in 50% to 60% of cases, the cause is unknown.

### Infectious Diseases

Infectious diseases are largely beyond the scope of this book, but it must be noted at least briefly that several microorganisms can cross the placenta and cause serious congenital anomalies, stillbirth, or neonatal death. Common viral infections of the fetus and newborn include herpes simplex, rubella, cytomegalovirus, and

---

[12]*con* = with + *gen* = born
[13]*terato* = monster + *logy* = study of

human immunodeficiency virus (HIV). Congenital bacterial infections include gonorrhea and syphilis. *Toxoplasma,* a protozoan contracted from meat, unpasteurized milk, and housecats, is another common cause of fetal deformity. Some of these pathogens have relatively mild effects on adults, but because of its immature immune system, the fetus is vulnerable to devastating effects such as blindness, hydrocephalus, cerebral palsy, seizures, and profound physical and mental retardation. These diseases are treated in greater detail in microbiology textbooks.

## Teratogens

**Teratogens**[14] are agents that cause anatomical deformities in the fetus; they include viruses, drugs, other chemicals, infectious diseases, and radiation such as X rays. The effect of a teratogen depends on the genetic susceptibility of the embryo, the dosage of the teratogen, and the time of exposure. Teratogen exposure during the first 2 weeks usually does not cause birth defects, but may cause spontaneous abortion. Teratogens can exert destructive effects at any stage of development, but the period of greatest vulnerability is weeks 3 through 8. Different organs have different critical periods. For example, limb abnormalities are most likely to result from teratogen exposure at 24 to 36 days, and brain abnormalities from exposure at 3 to 16 weeks.

Perhaps the most notorious teratogenic drug is thalidomide, a sedative first marketed in 1957. Thalidomide was taken by women in early pregnancy, often before they knew they were pregnant. It caused over 5,000 babies to be born with unformed arms or legs and often with defects of the ears, heart, and intestines. It was taken off the American market in 1961 but has recently been reintroduced for limited purposes. People still take thalidomide for leprosy and AIDS in some Third World countries, where it has resulted in an upswing in severe birth defects (fig. 29.12). A general lesson to be learned from the thalidomide tragedy and other cases is that pregnant women should avoid all sedatives, barbiturates, and opiates. Even the acne medicine isotretinoin (Accutane) has caused severe birth defects. Many teratogens produce less obvious or delayed effects, including physical or mental retardation, hyperirritability, inattention, strokes, seizures, respiratory arrest, crib death, and cancer.

Alcohol causes more birth defects than any other drug. Even one drink a day has noticeable effects on fetal and postpartum development, some of which are not noticed until a child begins school. Alcohol abuse during pregnancy can cause **fetal alcohol syndrome (FAS),** characterized by a small head, malformed facial features, cardiac and central nervous system defects, stunted growth, and behavioral signs such as hyperactivity, nervousness, and a poor attention span. Cigarette smoking also contributes to fetal and infant mortality, ectopic pregnancy,

**FIGURE 29.12    The Teratogenic Effect of Thalidomide.**    The infant in this photo was born in 2004 in Kenya to a woman believed to have used thalidomide during her pregnancy. He has no arms or legs and only rudimentary hands and feet. His birth father wanted to kill him, a common fate for deformed infants in Kenya, but he was adopted and taken to England. On the right is Mr. Freddie Astbury, president of Thalidomide UK in Liverpool. Mr. Astbury was also born with rudimentary limbs because of this teratogen.

anencephaly (failure of the cerebrum to develop), cleft lip and palate, and cardiac abnormalities. Diagnostic X rays should be avoided during pregnancy because radiation can have teratogenic effects.

## Mutagens and Genetic Anomalies

A **mutagen** is any agent that alters DNA or chromosome structure. Ionizing radiation and some chemicals have mutagenic, teratogenic, and carcinogenic effects, with extremely diverse results. Prenatal exposure to mutagens may result, for example, in stillbirths or childhood cancer.

Some of the most common genetic disorders result not from mutagens, however, but from the failure of homologous chromosomes to separate during meiosis. Recall that homologous chromosomes pair up during prophase I and normally separate from each other at anaphase I (see p. 1051). This separation, called *disjunction,* produces daughter cells with 23 chromosomes each.

In **nondisjunction,** a pair of chromosomes fails to separate. Both chromosomes then go to the same daughter cell, which receives 24 chromosomes while the other daughter cell receives 22. **Aneuploidy**[15] (AN-you-PLOY-dee), the presence of an extra chromosome or lack of one, accounts

---

[14]*terato* = monster + *gen* = producing

[15]*an* = not, without + *eu* = true, normal + *ploid,* from *diplo* = double, paired

for about 50% of spontaneous abortions. The lack of a chromosome, leaving one chromosome without a match, is called **monosomy,** whereas the presence of one extra chromosome, producing a triple set, is called **trisomy.** Aneuploidy can be detected prior to birth by *amniocentesis,* the examination of cells in a sample of amniotic fluid, or by *chorionic villus sampling,* the removal and examination of cells from the chorion.

Figure 29.13 compares normal disjunction of the X chromosomes with some effects of nondisjunction. In nondisjunction, an egg cell may receive both X chromosomes. If it is fertilized by an X-bearing sperm, the result is an XXX zygote and a set of anomalies called the **triplo-X syndrome.** Triplo-X females are sometimes infertile and sometimes have mild intellectual impairments. If an XX egg is fertilized by a Y-bearing sperm, the result is an XXY combination and **Klinefelter**[16] **syndrome.** People with

Klinefelter syndrome are sterile males, usually of average intelligence, but with undeveloped testes, sparse body hair, unusually long arms and legs, and enlarged breasts (gynecomastia). This syndrome often goes undetected until puberty, when failure to develop the secondary sex characteristics may prompt genetic testing.

The other possible outcome of X chromosome nondisjunction is that an egg cell may receive no X chromosome (both X chromosomes are discarded in the first polar body). If fertilized by a Y-bearing sperm, it dies for lack of the indispensable genes on the X chromosome. If it is fertilized by an X-bearing sperm, however, the result is **Turner**[17] **syndrome,** with an XO combination (O represents the absence of one sex chromosome). About 97% of fetuses with

[16]Harry F. Klinefelter, Jr. (1912    ), American physician
[17]Henry H. Turner (1892–1970), American endocrinologist

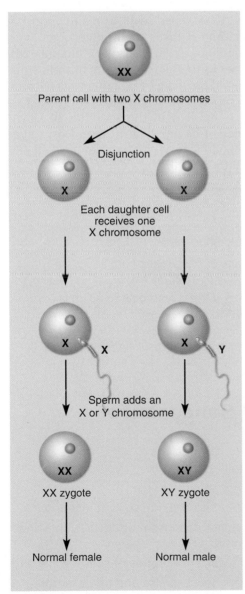

**(a) Normal disjunction of X chromosomes**

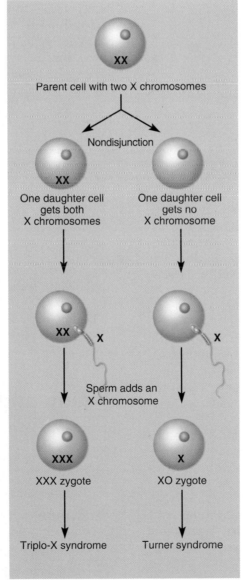

**(b) Nondisjunction of X chromosomes**

**FIGURE 29.13  Disjunction and Nondisjunction.**  (a) The outcome of normal disjunction and fertilization by X- or Y-bearing sperm. (b) Two of the possible outcomes of nondisjunction followed by fertilization with an X-bearing sperm.

▶ *In the right half of the figure, what would the two outcomes be if the sperm carried a Y chromosome instead of an X?*

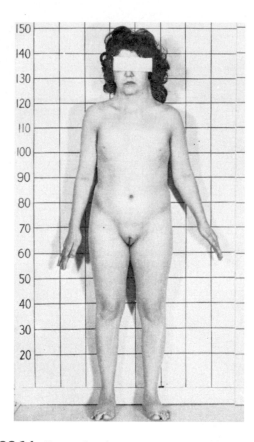

**FIGURE 29.14   Turner Syndrome.**  A 22-year-old woman with Turner syndrome, with an XO karyotype (see figure 29.13, far right). Note her short stature (about 145 cm, or 4 ft 9 in.), lack of sexual development, webbed neck, and widely spaced nipples.

Turner syndrome die before birth. Survivors show no serious impairments as children, but tend to have a webbed neck and widely spaced nipples. At puberty, the secondary sex characteristics fail to develop (fig. 29.14). The ovaries are nearly absent, the girl remains sterile, and she usually has a short stature.

The other 22 pairs of chromosomes (the autosomes) are also subject to nondisjunction. Only three autosomal trisomies are survivable: those involving chromosomes 13, 18, and 21. The reason is that these three chromosomes are relatively gene-poor. In all other autosomal cases, trisomy gives the embryo a lethal "overdose" of genes. Even the nonlethal trisomies are the leading genetic cause of mental and developmental abnormalities.

Nondisjunction of chromosomes 13 and 18 results in *Patau syndrome* (trisomy-13) and *Edward syndrome* (trisomy-18), respectively. Nearly all fetuses with these trisomies die before birth. Infants born with these syndromes are severely deformed, and fewer than 5% survive for 1 year.

The most survivable trisomy, and therefore the most common among children and adults, is **Down**[18] **syndrome** (trisomy-21). Its signs include impaired physical develop-

ment; short stature; a relatively flat face with a flat nasal bridge; low-set ears; *epicanthal folds* at the medial corners of the eyes; an enlarged, protruding tongue; stubby fingers; and a short broad hand with only one palmar crease (fig. 29.15). People with Down syndrome tend to have outgoing, affectionate personalities. Mental retardation is common and sometimes severe, but is not inevitable. Down syndrome occurs in about 1 out of 700 to 800 live births in the United States.

About 75% of the victims of trisomy-21 die before birth, and 20% of infants born with it die before the age of 10 from such causes as immune deficiency and abnormalities of the heart or kidneys. For those who survive beyond that age, modern medical care has extended life expectancy to about 60 years. After the age of 40, however, many of these people develop early-onset Alzheimer disease, linked to a gene on chromosome 21.

Aneuploidy is far more common in humans than in any other species, and 90% of cases are of maternal rather than paternal origin. These facts seem to result from the extraordinarily long time it takes for human oocytes to complete meiosis—as long as 50 years (see chapter 28). For various reasons including defects in the mitotic spindle and in chromosomal crossing over, aging eggs become less and less able to separate their chromosomes into two identical sets. This is evident in the statistics of Down syndrome: The chance of having a child with Down syndrome is about 1 in 3,000 for a woman under 30, 1 in 365 by age 35, and 1 in 9 by age 48.

## Before You Go On

*Answer the following questions to test your understanding of the preceding section:*

10. *How does inflation of the lungs at birth affect the route of blood flow through the heart?*

11. *Why is respiratory distress syndrome common in premature infants?*

12. *Define* nondisjunction *and explain how it causes aneuploidy. Name two syndromes resulting from aneuploidy.*

# Aging and Senescence

### Objectives
When you have completed this section, you should be able to

- define *senescence* and distinguish it from aging;
- describe some major changes that occur with aging in each organ system;
- summarize some current theories of senescence; and
- be able to explain how exercise and other factors can slow the rate of senescence.

---

[18]John Langdon H. Down (1828–96), British physician

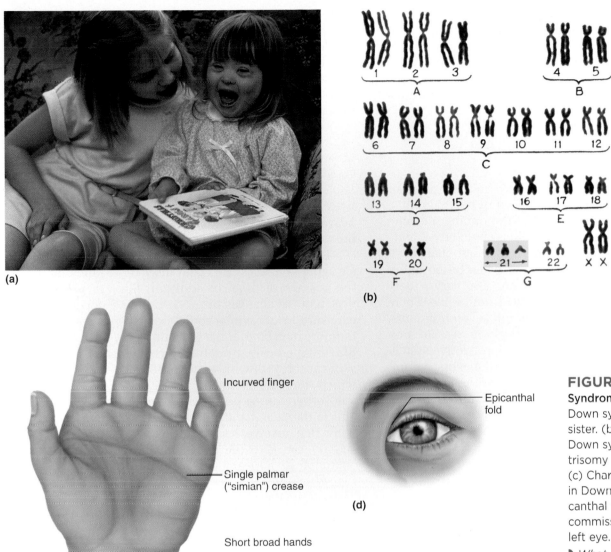

**FIGURE 29.15  Down Syndrome.** (a) A child with Down syndrome *(right)* and her sister. (b) The karyotype of Down syndrome, showing the trisomy of chromosome 21. (c) Characteristics of the hand in Down syndrome. (d) The epicanthal fold over the medial commissure (canthus) of the left eye.

▶ *What was the sex of the person from whom the karyotype in part (b) was obtained?*

Like Ponce de León searching for the legendary fountain of youth in Florida, people yearn for a way to preserve their youthful appearance and function. Our real concern, however, is not aging but senescence. The term **aging** is used in various ways but is taken here to mean all changes that occur in the body with the passage of time—including the growth, development, and increasing functional efficiency that occur from childhood to adulthood, as well as the degenerative changes that occur later in life. **Senescence** is the degeneration that occurs in an organ system after the age of peak functional efficiency. It includes a gradual loss of reserve capacities, reduced ability to repair damage and compensate for stress, and increased susceptibility to disease.

Senescence is not just a personal concern but an important issue for health-care providers. One in nine Americans is 65 or older. As the average age of the population rises, health-care professionals will find themselves increasingly occupied by the prevention and treatment of the diseases of age. The leading causes of death change markedly with age. Accidents, homicide, suicide, and AIDS figure prominently in the deaths of people 18 to 34 years old, whereas the major causes of death after age 55 are clearly related to senescence of the organ systems: heart disease, cancer, stroke, diabetes, and lung disease. The causes of senescence, however, remain as much a scientific mystery today as cancer was 50 years ago and heredity was 100 years ago.

As we survey the senescence of the organ systems, you should notice many points relevant not only to caring for an aging population but also to your personal health and fitness practices that can lessen the effects of senescence and improve the quality of life in your later years. In addition, the study of senescence calls renewed attention to the multiple interactions among organ systems. As you will see, the senescence of one organ system typically contributes to the senescence of others. Your study of this topic will bring together many concepts introduced in earlier chapters of the book. You may find the glossary helpful in refreshing your memory of concepts revisited in the following discussion.

## SENESCENCE OF THE ORGAN SYSTEMS

Organ systems do not all degenerate at the same rate. For example, from ages 30 to 80, the speed of nerve conduction declines only 10% to 15%, but the number of functional glomeruli in the kidneys declines about 60%. Some physiological functions show only moderate changes at rest but more pronounced differences when tested under exercise conditions. The organ systems also vary widely in the age at which senescence becomes noticeable. There are traces of atherosclerosis, for example, even in infants, and visual and auditory sensitivity begin to decline soon after puberty. By contrast, the female reproductive system does not show significant senescence until menopause and then its decline is relatively abrupt. Aside from these unusual examples, most physiological measures of performance peak between the late teens and age 30, then decline at a rate influenced by the level of use of the organs.

### Integumentary System

Two-thirds of people age 50 and over, and nearly all people over age 70, have medical concerns or complaints about their skin. Senescence of the integumentary system often becomes noticeable by the late 40s. The hair turns grayer and thinner as melanocytes die out, mitosis slows down, and dead hairs are not replaced. The atrophy of sebaceous glands leaves the skin and hair drier. As epidermal mitosis declines and collagen is lost from the dermis, the skin becomes almost paper-thin and translucent. It becomes looser because of a loss of elastic fibers and flattening of the dermal papillae, which normally form a stress-resistant corrugated boundary between the dermis and epidermis. If you pinch a fold of skin on the back of a child's hand, it quickly springs back when you let go; do the same on an older person and the skinfold remains longer. Because of its loss of elasticity, aged skin sags to various degrees and may hang loosely from the arm and other places.

Aged skin has fewer blood vessels than younger skin, and those that remain are more fragile. The skin can become reddened as broken vessels leak into the connective tissue. Many older people exhibit **rosacea**—patchy networks of tiny, dilated blood vessels visible especially on the nose and cheeks. Because of the fragility of the dermal blood vessels, aged skin bruises more easily. Injuries to the skin are more common and

severe in old age, partly because the cutaneous nerve endings decline by two-thirds from age 20 to 80, leaving one less aware of touch, pressure, and injurious stimuli. Injured skin heals slowly in old age because of poorer circulation and a relative scarcity of immune cells and fibroblasts. Antigen-presenting dendritic cells decline by as much as 40% in the aged epidermis, leaving the skin more susceptible to recurring infections.

Thermoregulation can be a serious problem in old age because of the atrophy of cutaneous blood vessels, sweat glands, and subcutaneous fat. Older people are more vulnerable to hypothermia in cold weather and heatstroke in hot weather. Heat waves and cold spells take an especially heavy toll among the elderly poor, who suffer from a combination of reduced homeostasis and inadequate housing.

These are all "normal" changes in the skin, or **intrinsic aging**—changes that occur more or less inevitably with the passage of time. In addition, there is **photoaging**—degenerative changes in proportion to a person's lifetime exposure to ultraviolet radiation. UV radiation accounts for more than 90% of the integumentary changes that people find medically troubling or cosmetically disagreeable: skin cancer; yellowing and mottling of the skin; age spots, which resemble enlarged freckles on the back of the hand and other sun-exposed areas; and wrinkling, which affects the face, hands, and arms more than areas of the body that receive less exposure. A lifetime of outdoor activity can give the skin a leathery, deeply wrinkled, "outdoorsy" appearance (fig. 29.16), but beneath this rugged exterior is a less happy histological appearance. The sun-damaged skin shows many malignant and premalignant cells, extensive damage to the dermal blood vessels, and dense masses of coarse, frayed elastic fibers underlying the surface wrinkles and creases.

Senescence of the skin has far-reaching effects on other organ systems. Cutaneous vitamin D production declines as much as 75% in old age. This is all the more significant because the elderly spend less time outdoors, and because of increasing lactose intolerance, they often avoid dairy products, the only dietary source of vitamin D. Consequently, the elderly are at high risk of calcium deficiency, which, in turn, contributes to bone loss, muscle weakness, and impaired glandular secretion and synaptic transmission.

### Skeletal System

After age 30, osteoblasts become less active than osteoclasts. This imbalance results in **osteopenia,** the loss of bone; when the loss is severe enough to compromise a person's physical activity and health, it is called *osteoporosis* (see p. 235). After age 40, women lose about 8% of their bone mass per decade and men about 3%. Bone loss from the jaws is a contributing factor in tooth loss.

Not only does bone density decline with age, but the bones become more brittle as the cells synthesize less pro-

## Muscular System

One of the most noticeable changes we experience with age is the replacement of lean body mass (muscle) with fat. The change is dramatically exemplified by CT scans of the thigh. In a young well-conditioned male, muscle accounts for 90% of the cross-sectional area of the midthigh, whereas in a frail 90-year-old woman, it is only 30%. Muscular strength and mass peak in the 20s; by the age of 80, most people have only half as much strength and endurance. A large percentage of people over age 75 cannot lift a 4.5 kg (10 lb) weight with their arms; such simple tasks as carrying a sack of groceries into the house may become impossible. The loss of strength is a major contributor to falls, fractures, and dependence on others for the routine activities of daily living. Fast-twitch fibers exhibit the earliest and most severe atrophy, thus increasing reaction time and reducing coordination.

There are multiple reasons for the loss of strength. Aged muscle fibers have fewer myofibrils, so they are smaller and weaker. The sarcomeres are increasingly disorganized, and muscle mitochondria are smaller and have reduced quantities of oxidative enzymes. Aged muscle has less ATP, creatine phosphate, glycogen, and myoglobin; consequently, it fatigues quickly. Muscles also exhibit more fat and fibrosis with age, which limits their movement and blood circulation. With reduced circulation, muscle injuries heal more slowly and with more scar tissue.

But the weakness and easy fatigue of aged muscle also stems from the senescence of other organ systems. There are fewer motor neurons in the spinal cord, and some muscle shrinkage may represent denervation atrophy. The remaining neurons produce less acetylcholine and show less efficient synaptic transmission, which makes the muscles slower to respond to stimulation. As muscle atrophies, motor units have fewer muscle fibers per motor neuron, and more motor units must be recruited to perform a given task. Tasks that used to be easy, such as buttoning the clothes or eating a meal, take more time and effort. The sympathetic nervous system is also less efficient in old age; consequently, blood flow to the muscles does not respond efficiently to exercise and this contributes to their rapid fatigue.

## Nervous System

The nervous system reaches its peak development around age 30. The average brain weighs 56% less at age 75 than at age 30. The cerebral gyri are narrower, the sulci are wider, the cortex is thinner, and there is more space between the brain and meninges. The remaining cortical neurons have fewer synapses, and for multiple reasons, synaptic transmission is less efficient: The neurons produce less neurotransmitter, they have fewer receptors, and the neuroglia around the synapses is more leaky and allows neurotransmitter to diffuse away. The degeneration of myelin sheaths with age also slows down signal conduction.

**FIGURE 29.16 Senescence of the Skin.** The skin exhibits both intrinsic aging and photoaging. The deep creases seen in the old woman's face result mainly from photoaging.

tein. Fractures occur more easily and heal more slowly. A fracture may impose a long period of immobility, which makes a person more vulnerable to pneumonia and other infectious diseases.

People notice more stiffness and pain in the synovial joints as they age, and degenerative joint diseases affect the lifestyle of 85% of people over age 75. Synovial fluid is less abundant and the articular cartilage is thinner or absent. Exposed bone surfaces abrade each other and cause friction, pain, and reduced mobility. *Osteoarthritis* is the most common joint disease of older people and one of the most common causes of physical disability (p. 315). Even breathing becomes more difficult and tiring in old age because expansion of the thorax is restricted by calcification of the sternocostal joints. Degeneration of the intervertebral discs causes back pain and stiffness, but herniated discs are less common in old age than in youth because the discs become more fibrous and stronger, with less nucleus pulposus.

Neurons exhibit less rough ER and Golgi complex with age, which indicates that their metabolism is slowing down. Old neurons accumulate lipofuscin pigment and show more neurofibrillary tangles—dense mats of cytoskeletal elements in their cytoplasm. In the extracellular material, plaques of fibrillar protein (amyloid) appear, especially in people with Down syndrome and Alzheimer disease (AD). AD is the most common nervous disability of old age (p. 474).

Not all functions of the central nervous system are equally affected by senescence. Motor coordination, intellectual function, and short-term memory decline more than language skills and long-term memory. Elderly people are often better at remembering things in the distant past than remembering recent events.

The sympathetic nervous system loses adrenergic receptors with age and becomes less sensitive to norepinephrine. This contributes to a decline in homeostatic control of such variables as body temperature and blood pressure. Many elderly people experience *orthostatic hypotension*—a drop in blood pressure when they stand, which sometimes results in dizziness, loss of balance, or fainting.

## Sense Organs

Some sensory functions decline shortly after adolescence. Presbyopia (loss of flexibility in the lenses) makes it more difficult for the eyes to focus on nearby objects. Visual acuity declines and often requires corrective lenses by middle age. Cataracts (cloudiness of the lenses) are more common in old age. Night vision is impaired as more and more light is needed to stimulate the retina. This has several causes: There are fewer receptor cells in the retina, the vitreous body becomes less transparent, and the pupil becomes narrower as the pupillary dilators atrophy. Dark adaptation takes longer as the enzymatic reactions of the photoreceptor cells become slower. Changes in the structure of the iris, ciliary body, or lens can block the reabsorption of aqueous humor, thereby increasing the risk of glaucoma. Having to give up reading and driving can be among the most difficult changes of lifestyle in old age.

Auditory sensitivity peaks in adolescence and declines afterward. The tympanic membrane and the joints between the auditory ossicles become stiffer, so vibrations are transferred less effectively to the inner ear, creating a degree of conductive deafness. Nerve deafness occurs as the number of cochlear hair cells and auditory nerve fibers declines. The greatest hearing loss occurs at high frequencies and in the frequency range of most conversation. The death of receptor cells in the semicircular ducts, utricle, and saccule, and of nerve fibers in the vestibular nerve and neurons in the cerebellum, results in poor balance and dizziness—another factor in falls and bone fractures.

The senses of taste and smell are blunted as taste buds, olfactory cells, and second-order neurons in the olfactory bulbs decline in number. Food may lose its appeal, and thus declining sensory function can be a factor in malnutrition.

## Endocrine System

The endocrine system degenerates less than any other organ system. The reproductive hormones drop sharply and growth hormone and thyroid hormone secretion decline steadily after adolescence, but other hormones continue to be secreted at fairly stable levels even into old age. Target cell sensitivity declines, however, so some hormones have less effect. For example, the pituitary gland is less sensitive to negative feedback inhibition by adrenal glucocorticoids; consequently, the response to stress is more prolonged than usual. Diabetes mellitus is more common in old age, largely because target cells have fewer insulin receptors. In part, this is an effect of the greater percentage of body fat in the elderly. The more fat at any age, the less sensitive other cells are to insulin. Body fat increases as the muscles atrophy, and muscle is one of the body's most significant glucose-buffering tissues. Because of the blunted insulin response, glucose levels remain elevated longer than normal after a meal.

## Circulatory System

Cardiovascular disease is a leading cause of death in old age. Senescence has multiple effects on the blood, heart, arteries, and veins. Anemia may result from nutritional deficiencies, inadequate exercise, disease, and other causes. The factors that cause anemia in older people are so complicated it is almost impossible to control them enough to determine whether aging alone causes it. Evidence suggests that there is no change in the baseline rate of erythropoiesis in old age. Hemoglobin concentration, cell counts, and other variables are about the same among healthy people in their 70s as in the 30s. However, older people do not adapt well to stress on the hemopoietic system, perhaps because of the senescence of other organ systems. As the gastric mucosa atrophies, for example, it produces less of the intrinsic factor needed for vitamin $B_{12}$ absorption. This increases the risk of pernicious anemia. As the kidneys age and the number of nephrons declines, less erythropoietin is secreted. There may also be a limit to how many times the hemopoietic stem cells can divide and continue giving rise to new blood cells. Whatever its cause, anemia limits the amount of oxygen that can be transported and thus contributes to the atrophy of tissues everywhere in the body.

### Think About It

*Draw a positive feedback loop showing how anemia and senescence of the kidneys could affect each other.*

Everyone exhibits coronary atherosclerosis with age. Consequently, myocardial cells die, angina pectoris and myocardial infarction become more common, the heart wall becomes thinner and weaker, and stroke volume, cardiac output, and cardiac reserve decline. Like other connective tissues, the fibrous skeleton of the heart becomes less elastic. This limits cardiac distension and reduces the force of systole. Degenerative changes in the nodes and conduction pathways of the heart lead to a higher incidence of cardiac arrhythmia and heart block. Physical endurance is compromised by the drop in cardiac output.

Arteries stiffened by atherosclerosis cannot expand as effectively to accommodate the pressure surges of cardiac systole. Blood pressure therefore rises steadily with age (see table 20.1, p. 764). Atherosclerosis also narrows the arteries and reduces the perfusion of most organs. The effects of reduced circulation on the skin, skeletal muscles, and brain have already been noted. The combination of atherosclerosis and hypertension also weakens the arteries and increases the risk of aneurysm and stroke.

Atherosclerotic plaques trigger thrombosis, especially in the lower extremities, where flow is relatively slow and the blood clots more easily. About 25% of people over age 50 experience venous blockage by thrombosis—especially people who do not exercise regularly.

Degenerative changes in the veins are most evident in the extremities. The valves become weaker and less able to stop the backflow of blood. Blood pools in the legs and feet, raises capillary blood pressure, and causes edema. Chronic stretching of the vessels often produces varicose veins and hemorrhoids. Support hose can reduce edema by compressing the tissues and forcing tissue fluid to return to the bloodstream, but physical activity is even more important in promoting venous return.

## Immune System

The amounts of lymphatic tissue and red bone marrow decline with age; consequently there are fewer hemopoietic stem cells, disease-fighting leukocytes, and antigen-presenting cells (APCs). Also, the lymphocytes produced in these tissues often fail to mature and become immunocompetent. Both humoral and cellular immunity depend on APCs and helper T cells, and therefore both types of immune response are blunted. As a result, an older person is less protected against cancer and infectious diseases. It becomes especially important in old age to be vaccinated against influenza and other acute seasonal infections.

## Respiratory System

Pulmonary ventilation declines steadily after the 20s and is one of several factors in the gradual loss of stamina. The costal cartilages and joints of the thoracic cage become less flexible, the lungs have less elastic tissue, and the lungs have fewer alveoli. Vital capacity, minute respiratory volume, and forced expiratory volume fall. The elderly are also less capable of clearing the lungs of irritants and pathogens and are therefore increasingly vulnerable to respiratory infections. Pneumonia causes more deaths than any other infectious disease and is often contracted in hospitals and nursing homes.

The chronic obstructive pulmonary diseases (COPDs)—emphysema and chronic bronchitis—are more common in old age since they represent the cumulative effects of a lifetime of degenerative change. They are among the leading causes of death in old age. Pulmonary obstruction also contributes to cardiovascular disease, hypoxemia, and hypoxic degeneration in all the organ systems. Respiratory health is therefore a major concern in aging.

## Urinary System

The kidneys exhibit a striking degree of atrophy with age. From ages 25 to 85, the number of nephrons declines 30% to 40% and up to a third of the remaining glomeruli become atherosclerotic, bloodless, and nonfunctional. The kidneys of a 90-year-old are 20% to 40% smaller than those of a 30-year-old and receive only half as much blood. The glomerular filtration rate is proportionally lower and the kidneys are less efficient at clearing wastes from the blood. Although baseline renal function is adequate even in old age, there is little reserve capacity; thus other diseases can lead to surprisingly rapid renal failure. Drug doses often need to be reduced in old age because the kidneys cannot clear drugs from the blood as rapidly; this is a contributing factor in overmedication among the aged.

Water balance becomes more precarious in old age because the kidneys are less responsive to antidiuretic hormone and because the sense of thirst is sharply reduced. Even when given free access to water, elderly people may not drink enough to maintain normal blood osmolarity. Dehydration is therefore common. It is often said that aged kidneys are deficient in maintaining electrolyte balance, but the evidence for this remains questionable.

Voiding and bladder control become a problem for both men and women. About 80% of men over the age of 80 are affected by benign prostatic hyperplasia. The enlarged prostate compresses the urethra and interferes with emptying of the bladder. Urine retention causes pressure to back up in the kidneys, aggravating the failure of the nephrons. Older women are subject to incontinence (leakage of urine), especially if their history of pregnancy and childbearing has weakened the pelvic muscles and urethral sphincters. Senescence of the sympathetic nervous system and nervous disorders such as stroke and Alzheimer disease can also cause incontinence.

## Digestive System and Nutrition

Less saliva is secreted in old age, making food less flavorful, swallowing more difficult, and the teeth more prone to caries. Nearly half of people over age 65 wear dentures because they have lost their teeth to caries and periodontitis. The stratified squamous epithelium of the oral cavity and esophagus is thinner and more vulnerable to abrasion.

The gastric mucosa atrophies and secretes less acid and intrinsic factor. Acid deficiency reduces the absorption of calcium, iron, zinc, and folic acid. Heartburn becomes more common as the weakening lower esophageal sphincter fails to prevent reflux into the esophagus. The most common digestive complaint of older people is constipation, which results from the reduced muscle tone and weaker peristalsis of the colon. This seems to stem from a combination of factors: atrophy of the muscularis externa, reduced sensitivity to neurotransmitters, less fiber and water in the diet, and less exercise. The liver, gallbladder, and pancreas show only slightly reduced function. Any drop in liver function, however, makes it harder to detoxify drugs and can contribute to overmedication.

Older people tend to reduce their food intake because of lower energy demand and appetite, because declining sensory functions make food less appealing, and because reduced mobility makes it more troublesome to shop and prepare meals. However, they need fewer calories than younger people because they have lower basal metabolic rates and tend to be less physically active. Protein, vitamin, and mineral requirements remain essentially unchanged, although vitamin and mineral supplements may be needed to compensate for reduced food intake and intestinal absorption. Malnutrition is common among older people and is an important factor in anemia and reduced immunity.

## Reproductive System

In men, the senescent changes in the reproductive system are relatively gradual; they include declining testosterone secretion, sperm count, and libido. By age 65, sperm count is about one-third of what it was in a man's 20s. Men remain fertile (capable of fathering a child) well into old age, but impotence (inability to maintain an erection) can occur because of atherosclerosis, hypertension, medication, and psychological reasons.

In women, the changes are more pronounced and develop more rapidly. Over the course of menopause, the ovarian follicles are used up, gametogenesis ceases, and the ovaries stop producing sex steroids. This may result in vaginal dryness, genital atrophy, and reduced libido and may make sex less enjoyable. With the loss of ovarian steroids, a postmenopausal woman has an elevated risk of osteoporosis and atherosclerosis.

## EXERCISE AND SENESCENCE

Other than the mere passage of time, senescence results from obesity and insufficient exercise more than from any other causes. Conversely, good nutrition and exercise are the best ways to slow its progress.

There is no clear evidence that exercise will prolong your life, but there is little doubt that it improves the quality of life in old age. It maintains endurance, strength, and joint mobility while it reduces the incidence and severity of hypertension, osteoporosis, obesity, and diabetes mellitus. This is especially true if you begin a program of regular physical exercise early in life and make a lasting habit of it. If you stop exercising regularly after middle age, the body rapidly becomes deconditioned, although appreciable reconditioning can be achieved even when an exercise program is begun late in life. A person in his or her 90s can increase muscle strength two- or threefold in 6 months with as little as 40 minutes of isometric exercise a week. The improvement results from a combination of muscle hypertrophy and neural efficiency.

Resistance exercises may be the most effective way of reducing accidental injuries such as bone fractures, whereas endurance exercises reduce body fat and increase cardiac output and maximum oxygen uptake. A general guideline for ideal endurance training is to have three to five periods of aerobic exercise per week, each 20 to 60 minutes long and vigorous enough to reach 60% to 90% of your maximum heart rate. The maximum is best determined by a stress test but averages about 220 beats per minute minus one's age in years.

An exercise program should ideally be preceded by a complete physical examination and stress test. Warmup and cool-down periods are especially important in avoiding soft tissue injuries. Because of their lower capacity for thermoregulation, older people must be careful not to overdo exercise, especially in hot weather. At the outset of a new exercise program, it is best to "start low and go slow."

## THEORIES OF SENESCENCE

Why do our organs wear out? Why must we die? There still is no general theory on this. The question actually comes down to two issues: (1) What are the mechanisms that cause the organs to deteriorate with age? (2) Why hasn't natural selection eliminated these and produced bodies capable of longer life?

### Mechanisms of Senescence

Numerous hypotheses have been proposed and discarded to explain why organ function degenerates with age. Some authorities maintain that senescence is an intrinsic process governed by inevitable or even preprogrammed changes in cellular function. Others attribute senescence to extrinsic (environmental) factors that progressively damage our cells over the course of a lifetime.

There is good evidence of a hereditary component to longevity. Unusually long and short lives tend to run in families. Monozygotic (identical) twins are more likely than dizygotic twins to die at a similar age. One striking genetic defect called *progeria*[19] is characterized by greatly

---

[19]*pro* = before + *ger* = old age

**FIGURE 29.17 Progeria.** This is a genetic disorder in which senescence appears to be greatly accelerated. The individuals here, from left to right, are 15, 12, and 26 years old. Few people with progeria live as long as the woman on the right.

accelerated senescence (fig. 29.17). Symptoms begin to appear by age 2. The child's growth rate declines, the muscles and skin become flaccid, most victims lose their hair, and most die in early adolescence from advanced atherosclerosis. In Werner syndrome, caused by a defective gene on chromosome 8, people show marked senescence beginning in their 20s and usually die by age 50. There is some controversy over the relevance or similarity of these syndromes to normal senescence, but they do demonstrate that many of the changes associated with old age can be brought on by a genetic anomaly.

Knowing that senescence is partially hereditary, however, does not answer the question about why tissues degenerate. Quite likely, no one theory explains all forms of senescence, but let's briefly examine some of them.

***Replicative Senescence*** Normal organ function usually depends on a rate of cell renewal that keeps pace with cell death. There is a limit, however, to how many times cells can divide. Human cells cultured in the laboratory divide 80 to 90 times if taken from a fetus, but only 20 to 30 times if taken from older people. After reaching their maximum number of divisions, cultured cells degenerate and die. This decline in mitotic potential with age is called **replicative senescence.**

Why this occurs is a subject of lively current research. Much of the evidence points to the **telomere,**[20]

_____
[20]*telo* = end + *mer* = piece

a "cap" on each end of a chromosome analogous to the plastic tip of a shoelace. In humans, it consists of a noncoding nucleotide sequence CCCTAA repeated 1,000 times or more. One of its functions may be to stabilize the chromosome and prevent it from unraveling or sticking to other chromosomes. Also, during DNA replication, DNA polymerase cannot reproduce the very ends of the DNA molecule. If there were functional genes at the end, they would not get duplicated. The telomere may therefore provide a bit of "disposable" DNA at the end, so that DNA polymerase doesn't fail to replicate genes that would otherwise be there. Every time DNA is replicated, 50 to 100 bases are lost from the telomere. In old age, the telomere may be exhausted and the polymerase may then indeed fail to replicate some of the terminal genes. Old chromosomes may therefore be more vulnerable to damage, replication errors, or both, causing old cells to be increasingly dysfunctional. The "immortality" of cancer cells results from an enzyme called *telomerase,* lacking from healthy cells, which enables cancer cells to repair telomere damage and escape the limit on number of cell divisions.

Replicative senescence is clearly not the entire answer to why organs degenerate, however. Skeletal muscles and the brain exhibit extreme senescence, yet muscle fibers and neurons are nonmitotic. Their senescence obviously is not a result of repeated mitosis and cumulative telomere damage. A full explanation of senescence must embrace additional processes and theories.

***Cross-Linking Theory*** About one fourth of the body's protein is collagen. With age, collagen molecules become cross-linked by more and more disulfide bridges, thus making the fibers less soluble and more stiff. This is thought to be a factor in several of the most noticeable changes of the aging body, including stiffening of the joints, lenses, and arteries. Similar cross-linking of DNA and enzyme molecules could progressively impair their functions as well.

***Other Protein Abnormalities*** Not only collagen but also many other proteins exhibit increasingly abnormal structure in older tissues and cells. The changes are not in amino acid sequence—therefore not attributable to DNA mutations—but lie in the way the proteins are folded and other moieties such as carbohydrates are attached to them. This is another reason that cells accumulate more dysfunctional proteins as they age.

***Free Radical Theory*** Free radicals have very destructive effects on macromolecules (see chapter 2). We have a number of antioxidants to protect us from them, but it is believed that some of these antioxidants become less abundant with age and are eventually overwhelmed by free radicals, or that some of the molecules damaged by free radicals are long-lived and accumulate in cells. Free radical damage may therefore be a contributing factor in some of the other mechanisms of senescence discussed here.

***Autoimmune Theory***   Some of the altered macromolecules described previously may be recognized as foreign antigens and stimulate lymphocytes to mount an immune response against the body's own tissues. Autoimmune diseases do, in fact, become more common in old age.

## Evolution and Senescence

If certain genes contribute to senescence, it raises an evolutionary question—Why doesn't natural selection eliminate them? In an attempt to answer this, biologists once postulated that senescence and death were for the good of the species—a way for older, worn-out individuals to make way for younger, healthier ones. We can see the importance of death for the human population by imagining that science had put an end to senescence and people died at a rate of only 1 per 1,000 per year regardless of age (the rate at which American 18-year-olds now die). If so, the median age of the population would be 163, and 13% of us would live to be 2,000 years old. The implications for world population and competition for resources would be staggering. Thus it is easy to understand why death was once interpreted as a self-sacrificing phenomenon for the good of the species.

But this hypothesis has several weaknesses. One of them is the fact that natural selection works exclusively through the effects of genes on the reproductive rates of individuals. A species evolves only because some members reproduce more than others. A gene that does not affect reproductive rate can be neither eliminated nor favored by natural selection. Genes for disorders such as Alzheimer disease have little or no effect until a person is past reproductive age. Our prehistoric and even fairly recent ancestors usually died of accidents, predation, starvation, weather, and infectious diseases at an early age. Few people lived long enough to be affected by atherosclerosis, colon cancer, or Alzheimer disease. Natural selection would have been "blind" to such death-dealing genes, which would escape the selection process and remain with us today.

## DEATH

**Life expectancy,** the average length of life in a given population, has steadily increased in industrialized countries. People born in the United States at the beginning of the twentieth century had a life expectancy of only 45 to 50 years; nearly half of them died of infectious disease. The average boy born today can expect to live 72 years and the average girl 79 years. This is due mostly to victories over infant and child mortality, not to advances at the other end of the life span. **Life span,** the maximum age attainable by humans, has not increased for many centuries and there seems to be little prospect that it ever will. There is no verifiable record of anyone living past the age of 122 years.

There is no definable instant of biological death. Some organs function for an hour or more after the heart stops beating. During this time, even if a person is declared legally dead, living organs may be removed for transplantation. For legal purposes, death was once defined as the loss of a spontaneous heartbeat and respiration. Now that cardiopulmonary functions can be artificially maintained for years, this criterion is less distinct. Clinical death is now widely defined in terms of **brain death**—a lack of cerebral activity indicated by a flat electroencephalogram for 30 minutes to 24 hours (depending on state laws), accompanied by a lack of reflexes or lack of spontaneous respiration and heartbeat.

Death usually results from the failure of a particular organ, which then has a cascading effect on other organs. Kidney failure, for example, leads to the accumulation of toxic wastes in the blood, which in turn leads to loss of consciousness, brain function, respiration, and heartbeat.

Ninety-nine percent of us will die before age 100, and there is little chance that this outlook will change within our lifetimes. We cannot presently foresee any "cure for old age" or significant extension of the human life span. The real issue is to maintain the best possible quality of life, and when the time comes to die, to do so in comfort and dignity.

### Before You Go On

*Answer the following questions to test your understanding of the preceding section:*

13. *Define* aging *and* senescence.

14. *List some tissues or organs in which changes in collagenous and elastic connective tissues lead to senescence.*

15. *Many older people have difficulty with mobility and simple self maintenance tasks such as dressing and cooking. Name some organ systems whose senescence is most relevant to these limitations.*

16. *Explain why both endurance and resistance exercises are important in old age.*

17. *Summarize five mechanisms that may be responsible for senescence.*

## INSIGHT 29.4 Clinical Application

# Reproductive Technology—Making Babies in the Laboratory

Fertile heterosexual couples who have frequent intercourse and use no contraception have an 85% chance of conceiving within 1 year. About one in six American couples, however, are *infertile*—unable to conceive. Infertility can sometimes be corrected by hormone therapy or surgery, but when this fails, parenthood may still be possible through other reproductive technologies discussed here.

## Artificial Insemination

If only the male is infertile, the oldest and simplest solution is *artificial insemination (AI)*, in which a physician introduces donor semen into or near the cervix. This was first done in the 1890s, when the donor was often the physician himself or a medical student who donated semen for payment. In 1953, a technique was developed for storing semen in glass ampules frozen in liquid nitrogen; the first commercial sperm banks opened in 1970. Most women undergoing AI use sperm from anonymous donors but are able to select from a catalog that specifies the donors' physical and intellectual traits. A man with a low sperm count can donate semen at intervals over a course of several weeks and have it pooled, concentrated, and used to artificially inseminate his partner. Men planning vasectomies sometimes donate sperm for storage as insurance against the death of a child, divorce and remarriage, or a change in family planning. Some cases of infertility are due to sperm destruction by the woman's immune system. This can sometimes be resolved by *sperm washing*—a technique in which the sperm are collected, washed to remove antigenic proteins from their surfaces, and then introduced by AI.

## Oocyte Donation

The counterpart to sperm donation is *oocyte donation*, in which fresh oocytes are obtained from one woman, fertilized, and transplanted to the uterus of another. A woman may choose this procedure for a variety of reasons: being past menopause, having had her ovaries removed, or having a hereditary disorder she does not want to pass to her children, for example. The donated oocytes are sometimes provided by a relative or may be left over from another woman's in vitro fertilization (see next paragraph). The first baby conceived by oocyte donation was born in 1984. This procedure has a success rate of 20% to 50%.

## In Vitro Fertilization

In some women the uterus is normal but the uterine tubes are scarred by pelvic inflammatory disease or other causes. *In vitro fertilization (IVF)* is an option in some of these cases. The woman is given gonadotropins to induce the "superovulation" of multiple eggs. The physician views the ovary with a laparoscope and removes eggs by suction. These are placed in a solution that mimics the chemical environment of her reproductive tract, and sperm are added to the dish. The term *in vitro*[21] *fertilization* refers to the fact that fertilization occurs in laboratory glassware; children conceived by IVF are often misleadingly called "test-tube babies." In some cases, fertilization is assisted by piercing the zona pellucida before the sperm are added (*zona drilling*) or by injecting sperm directly into the egg through a micropipet. By the day after fertilization, some of the preembryos reach the 8- to 16-celled stage. Several of these are transferred to the mother's uterus through the cervix and her blood HCG level is monitored to determine whether implantation has occurred. Excess IVF preembryos may be donated to other infertile couples or frozen and used in later attempts. In cases where a woman has lost her ovaries to disease, the oocytes may be provided by another donor, often a relative.

IVF costs up to $10,000 per attempt and succeeds only 14% of the time. A couple can easily spend $100,000 before IVF is successful, and then some attempts are "too successful." Multiple preembryos are usually introduced to the uterus as insurance against the low probability that any one of them will implant and survive. Sometimes, however, this results in multiple births—in rare cases, up to seven babies (septuplets). One advantage of IVF is that when the preembryo reaches the eight-celled stage, one or two cells can be removed and tested for genetic defects before the preembryo is introduced to the uterus.

IVF has been used in animal breeding since the 1950s, but the first child conceived this way was Louise Joy Brown (fig. 29.18), born in England in 1978. It is now estimated that worldwide, about one IVF child is born every day.

## Surrogate Mothers

IVF is an option only for women who have a functional uterus. A woman who has had a hysterectomy or is otherwise unable to become pregnant or maintain a pregnancy may contract with a *surrogate mother* who provides a "uterus for hire." Some surrogates are both genetic and gestational mothers, and others gestational only. In the former case, the surrogate is artificially inseminated by a man's sperm and agrees to give the baby to the man and his partner at birth. In the latter case, oocytes are collected from one woman's ovaries, fertilized in vitro, and the preembryos are placed in the surrogate's uterus. This is typical of cases in which a woman has functional ovaries but no functional uterus. A surrogate typically receives a fee of about $10,000 plus medical and legal costs. Several hundred babies have been produced this way in the United States. In at least one case, a woman carried the child of her infertile daughter, thus giving birth to her own granddaughter.

---

[21]*in vitro* = in glass

**FIGURE 29.18   New Beginnings Through Reproductive Technology.**  Louise Joy Brown, shown here at age 10, was the first child ever conceived by in vitro fertilization (IVF). She is holding Andrew Macheta, another IVF baby, at a 10-year anniversary celebration at the clinic near London where both were conceived.

## Gamete Intrafallopian Transfer

The low success rate of IVF led to a search for more reliable and cost-effective techniques. *Gamete intrafallopian transfer (GIFT)* was developed in the mid-1980s on the conjecture that pregnancy would be more successful if the oocyte were fertilized and began cleavage in a more natural environment. Eggs are obtained from a woman after a weeklong course of ovulation-inducing drug treatment. The most active sperm cells are isolated from the semen, and the eggs and sperm are introduced into her uterine tube proximal to any existing obstruction. GIFT is about half as expensive as IVF and succeeds about 40% of the time. In a modification called *zygote intrafallopian transfer (ZIFT)*, fertilization occurs in vitro and the preembryo is introduced into the uterine tube. Traveling down the uterine tube seems to improve the chance of implantation when the conceptus reaches the uterus.

## Embryo Adoption

*Embryo adoption* is used when a woman has malfunctioning ovaries but a normal uterus. A man's sperm are used to artificially inseminate another woman. A few days later, the preembryo is flushed from the donor's uterus before it implants and is transferred to the uterus of the woman who wishes to have a child.

## Ethical and Legal Issues

Like many other advances in medicine, reproductive technology has created its own ethical and legal dilemmas, some of which are especially confounding. Perhaps the most common problem is the surrogate mother who changes her mind. Surrogates enter into a contract to surrender the baby to a couple at birth, but after carrying a baby for 9 months and giving birth, they sometimes feel differently. This raises questions about the definition of motherhood, especially if she is the gestational but not the genetic mother.

The converse problem is illustrated by a case in which the child had hydrocephalus and neither the contracting couple nor the surrogate mother wanted it. In this case, genetic testing showed that the child actually had been fathered by the surrogate's husband, not the man who had contracted for her service. The surrogate then accepted the baby as her own. Nevertheless, the case raised the question of whether the birth of a genetically defective child constituted fulfillment of the contract and obligated the contracting couple to accept the child, or whether such a contract implies that the surrogate mother must produce a healthy child.

In another case, a wealthy couple was killed in an accident and left frozen preembryos in an IVF clinic. A lawsuit was filed on behalf of the preembryos on the grounds that they were heirs to the couple's estate and should be carried to birth by a surrogate mother so they could inherit it. The court ruled against the suit and the preembryos were allowed to die. In still another widely publicized case, a man sued his wife for custody of their frozen preembryos as part of a divorce settlement.

IVF also creates a question of what to do with the excess preembryos. Some people view their disposal as a form of abortion, even if the preembryo is only a mass of 8 to 16 undifferentiated cells. On the other hand, there are those who see such excess preembryos as a research opportunity to obtain information that could not be obtained in any other way. In 1996, an IVF clinic in England was allowed to destroy 3,300 unclaimed preembryos, but only after heated public controversy.

It is common for scientific advances to require new advances in law and ethics. The parallel development of these disciplines is necessary if we are to benefit from the developments of science and ensure that knowledge is applied in an ethical and humane manner.

## CHAPTER REVIEW

# Review of Key Concepts

### Fertilization and the Preembryonic Stage (p. 1106)

1. Sperm must travel to the distal one-third of the uterine tube if they are to encounter the egg before it dies. This traveling, or *sperm migration,* may be aided by the cervical mucus, female orgasm, and chemical attractants emitted by the egg.

2. Freshly ejaculated sperm cannot fertilize an egg. They undergo *capacitation,* becoming capable of fertilization, as they migrate.

3. When a sperm encounters an egg, it releases enzymes from its acrosome (the *acrosomal reaction*), enabling it to penetrate the cumulus oophorus, zone pellucida, and egg membrane. Hundreds of sperm may be needed to clear a path for the one that fertilizes the egg.

4. The egg has a *fast block* and a *slow block* to prevent fertilization by more than one sperm *(polyspermy).* The fast block employs a change in egg membrane voltage that inhibits the binding of additional sperm. The slow block involves exocytosis of the egg's *cortical granules* to produce an impenetrable *fertilization membrane* around the egg.

5. The fertilized egg completes meiosis II and casts off a second polar body. The sperm and egg nuclei swell and form *pronuclei.* When the pronuclei rupture and their chromosomes mingle, the egg is a diploid *zygote.*

6. Pregnancy is clinically divided into three *trimesters* of about 12 weeks each, and biologically divided into *preembryonic, embryonic,* and *fetal stages.*

7. The preembryonic stage comprises the first 16 days of development and includes cleavage, implantation, and embryogenesis, culminating in a three-layered embryo.

8. *Cleavage* is the mitotic division of the zygote into cells called *blastomeres.* The stage that arrives at the uterus is a *morula* of about 16 blastomeres. It develops into a hollow ball called the *blastocyst,* with an outer cell mass called the *trophoblast* and inner cell mass called the *embryoblast.*

9. *Implantation* is the attachment of the blastocyst to the uterine wall. The trophoblast differentiates into a cellular mass called the *cytotrophoblast* next to the embryo, and a multinucleate mass called the *syncytiotrophoblast,* which grows rootlets into the endometrium. The endometrium grows over the blastocyst and soon completely covers it.

10. The trophoblast secretes human chorionic gonadotropin, the hormone that stimulates growth and secretion by the corpus luteum.

11. During implantation, the embryoblast undergoes *gastrulation* and its cells stratify into three *primary germ layers: ectoderm, mesoderm,* and *endoderm.* This process is *embryogenesis.* When the three primary germ layers have formed, 16 days after conception, the individual is an *embryo.*

### The Embryonic and Fetal Stages (p. 1112)

1. The embryonic stage extends from 16 days through the end of week 8. It is marked by formation of the embryonic membranes, placental nutrition, and appearance of the organ systems.

2. During this stage the embryo curls longitudinally into a C shape; its lateral margins fold ventrally and convert the flat embryonic disc into a cylindrical body; a primitive gut and body cavity form; and all of the organ systems first appear. Differentiation of the organs is called *organogenesis.*

3. Four membranes are associated with the embryo and fetus: the amnion, yolk sac, allantois, and chorion.

4. The *amnion* is a translucent sac that encloses the embryo in a pool of *amniotic fluid.* This fluid protects the embryo from trauma and temperature fluctuations and allows freedom of movement and symmetric development.

5. The *yolk sac* contributes to development of the digestive tract and produces the first blood and germ cells of the embryo.

6. The *allantois* is an outgrowth of the yolk sac that forms a structural foundation for umbilical cord development and becomes part of the urinary bladder.

7. The *chorion* encloses all of the other membranes and forms the fetal part of the placenta.

8. Until it implants in the endometrium, the conceptus is nourished by a secretion called *uterine milk.*

9. After implantation, the conceptus is fed by *trophoblastic nutrition,* in which the trophoblast digests *decidual cells* of the endometrium. This is the dominant mode of nutrition for 8 weeks.

10. The *placenta* begins to form 11 days after conception as chorionic villi of the trophoblast eat into uterine blood vessels, eventually creating a blood-filled cavity called the *placental sinus.* The chorionic villi grow into branched treelike structures surrounded by the maternal blood in the sinus. Nutrients diffuse from the maternal blood into embryonic blood vessels in the villi, and embryonic wastes diffuse the other way to be disposed of by the mother. *Placental nutrition* becomes dominant at 8 weeks and continues until birth.

11. The placenta communicates with the embryo and fetus by way of two arteries and a vein contained in the *umbilical cord.*

12. Traces of all organ systems are present by the end of 8 weeks. The individual is considered a *fetus* from then until birth.

13. In the fetal stage, organs undergo growth and differentiation and become capable of functioning outside the mother's body. Major developments in the fetal stage are summarized in table 29.4.

14. The circulatory system differs most markedly from prenatal to neonatal life. In the fetus, a pair of umbilical arteries arise from the internal iliac arteries and supply the placenta. A single umbilical vein returns from the placenta and drains most of its blood into the inferior vena cava (IVC).

15. Three bypasses or shunts divert fetal blood from organs that are not very functional before birth: The *ductus venosus* bypasses the liver and carries most umbilical vein blood directly to the IVC; the foramen ovale in the interatrial septum of the heart allows blood to pass directly from the right atrium to the left atrium, bypassing the lungs; and the *ductus arteriosus* allows blood in the pulmonary trunk to pass directly into the aorta and bypass the lungs.

**The Neonate (p. 1122)**

1. The first 6 to 8 hours after birth are a *transitional period* marked by increasing heart and respiratory rates and falling temperature. The first 6 weeks of postpartum life are the *neonatal period.*

2. After severance of the umbilical cord, the proximal parts of the umbilical arteries become *vesical arteries,* which supply the urinary bladder. Other blood vessels unique to the fetus close and become fibrous cords or ligaments. The foramen ovale and ductus arteriosus close so that blood from the right heart is forced to circulate through the lungs.

3. Breathing is very difficult for the neonate as it first inflates the pulmonary alveoli.

4. Neonatal immunity depends heavily on IgG acquired through the placenta and IgA from colostrum. By 6 months, the infant produces ample IgG of its own.

5. Neonatal thermoregulation is critical because infants lose heat easily. This heat loss is compensated for to some extent by a form of heat-producing adipose tissue called *brown fat.*

6. The neonatal kidneys are not very efficient at concentrating urine, so neonates have a relatively high rate of water loss and require more fluid intake than adults do relative to their body weight.

7. Premature infants suffer especially from respiratory distress syndrome, poor thermoregulation, poor fat digestion, and multiple dysfunctions resulting from inadequate liver function.

8. *Congenital anomalies* (birth defects) can result from infectious diseases, teratogens, mutagens, and genetic disorders.

9. Some of the more common and serious infectious diseases and pathogens of the newborn are herpes simplex, cytomegalovirus, HIV, gonorrhea, and syphilis.

10. *Teratogens,* agents that cause anatomical deformities, include alcohol, nicotine, and X rays.

11. *Nondisjunction,* the failure of homologous chromosomes to separate during meiosis, can result in such congenital defects as triplo-X, Klinefelter, Turner, and Down syndromes.

**Aging and Senescence (p. 1126)**

1. *Senescence* is the degeneration that occurs in an organ system as we age. It begins at very different ages and progresses at different rates in different organ systems. Senescence of one organ system often contributes to the senescence of others.

2. Senescence of the integumentary system is marked by graying and thinning of the hair, atrophy of sebaceous glands, thinning and loss of elasticity in the skin, fragility of cutaneous blood vessels, decline in cutaneous sensory function, slower healing, and poorer thermoregulation. *Intrinsic aging* occurs inevitably with time, while *photoaging* is an added effect proportional to the amount of lifetime UV exposure. Senescence of the skin contributes to bone loss, muscle weakness, and poorer glandular secretion and synaptic transmission.

3. Senescence of the skeletal system is marked by loss of bone density (*osteopenia* or, when more severe, *osteoporosis*), increasing susceptibility to fractures, slower healing of fractures, osteoarthritis, and other degenerative joint diseases.

4. The aging muscular system exhibits muscular atrophy, loss of strength, and easy fatigue. Some loss of muscular function results from degenerative changes in the nervous system.

5. Senescence of the nervous system is marked by substantial loss of brain tissue and synapses, less efficient synaptic transmission, and declining motor coordination, intellectual function, and short-term memory, but relatively little loss of language skills and long-term memory. Senescence of the sympathetic division results in less effective homeostasis in other organ systems.

6. Visual acuity and auditory sensitivity begin to decline shortly after adolescence. Vision can be impaired by cataracts, glaucoma, and reduced dark adaptation. Declining inner-ear function can result in poor balance. Taste and smell become less sensitive.

7. The endocrine system shows relatively little senescence except for the decline in reproductive hormones. Reduced densities of hormone receptors can contribute to type II diabetes mellitus and poorer negative feedback control of the pituitary.

8. Senescence of the circulatory system is a leading cause of death. Anemia, atherosclerosis, thrombosis, varicose veins, hemorrhoids, and edema become more common in old age. Atherosclerosis contributes to weakening of the heart, myocardial infarction, aneurysm, stroke, and atrophy of all organs.

9. Senescence of the immune system makes older people more subject to cancer and infectious diseases.

10. Pulmonary functions decline as the thoracic cage becomes less flexible and the lungs have fewer alveoli. Pneumonia and chronic obstructive pulmonary diseases are major causes of death.

11. The kidneys atrophy a great deal with age, and thus elderly people are less able to maintain water balance and to clear drugs or toxins from the body. Elderly men are increasingly subject to prostatic enlargement and urine retention, and women to urinary incontinence.

12. Senescence of the digestive system includes reduced salivation, difficulty swallowing, poorer dental health, atrophy of the stomach, gastroesophageal reflux, constipation, loss of appetite, and impaired liver function.

13. Reproductive senescence is marked in men by reduced testosterone secretion, sperm count, and libido, and in women by menopause and multiple effects of the loss of estrogen secretion.

14. Exercise slows the rate of senescence and improves the quality of life in old age by maintaining strength, endurance, flexibility, and independence. It reduces the incidence and severity of hypertension, osteoporosis, obesity, and diabetes mellitus.

15. There are numerous theories of what causes senescence. *Replicative*

*senescence,* a limit on how many times cells can divide, may stem from shortening of the chromosomal *telomeres* at each cell division. Cross-linking of proteins and DNA and the misfolding and other structural defects in proteins may cause increasing cellular dysfunction. The cumulative effects of free radical damage and increased incidence of autoimmune disease may be other factors in senescence.

16. Longevity is known to be partially hereditary. Natural selection has presumably been unable to eliminate genes that cause some of the dis-

eases of old age because such genes have no effects that natural selection can act on until after the individual has reproduced.

17. *Life expectancy* has increased in modern times mostly because of our ability to reduce infant and childhood mortality. *Life span,* the maximum attainable age, has not markedly changed, however.

18. Death is usually clinically defined by an absence of brain waves, reflexes, or spontaneous respiration or heartbeat.

## Testing Your Recall

1. When a conceptus arrives in the uterus, it is at what stage of development?
   a. zygote
   b. morula
   c. blastomere
   d. blastocyst
   e. embryo

2. The entry of a sperm nucleus into an egg must be preceded by
   a. the cortical reaction.
   b. the acrosomal reaction.
   c. the fast block.
   d. implantation.
   e. cleavage.

3. The stage of the conceptus that implants in the uterine wall is
   a. a blastomere.
   b. a morula.
   c. a blastocyst.
   d. an embryo.
   e. a zygote.

4. Chorionic villi develop from
   a. the zona pellucida.
   b. the endometrium.
   c. the syncytiotrophoblast.
   d. the embryoblast.
   e. the corona radiata.

5. Which of these results from aneuploidy?
   a. Turner syndrome
   b. fetal alcohol syndrome
   c. nondisjunction
   d. progeria
   e. rubella

6. Fetal urine accumulates in the _____ and contributes to the fluid there.
   a. placental sinus
   b. yolk sac
   c. allantois
   d. chorion
   e. amnion

7. One theory of senescence is that it results from a lifetime of damage by
   a. teratogens.
   b. aneuploidy.
   c. free radicals.
   d. cytomegalovirus.
   e. nondisjunction.

8. Photoaging is a major factor in the senescence of
   a. the integumentary system.
   b. the eyes.
   c. the nervous system.
   d. the skeletal system.
   e. the cardiovascular system.

9. Which of these is *not* a common effect of senescence?
   a. reduced synthesis of vitamin D
   b. atrophy of the kidneys
   c. atrophy of the cerebral gyri
   d. increased herniation of intervertebral discs
   e. reduced pulmonary vital capacity

10. For the first 8 weeks of gestation, a conceptus is nourished mainly by
    a. the placenta.
    b. amniotic fluid.
    c. colostrum.
    d. decidual cells.
    e. yolk cytoplasm.

11. Viruses and chemicals that cause congenital anatomical deformities are called _____.

12. Anouploidy is caused by _____, the failure of two homologous chromosomes to separate in meiosis.

13. The maximum age attainable by a member of the human species is called the _____.

14. The average age attained by humans in a given population is called the _____.

15. Fetal blood flows through growths called _____, which project into the placental sinus.

16. The enzymes with which a sperm penetrates an egg are contained in an organelle called the _____.

17. Stiffening of the arteries, joints, and lenses in old age may be a result of cross-linking between _____ molecules.

18. An enlarged tongue, epicanthal folds of the eyes, and mental retardation are characteristic of a genetic anomaly called _____.

19. The fossa ovalis is a remnant of a fetal shunt called the _____.

20. A developing individual is first classified as a/an _____ when the three primary germ layers have formed.

*Answers in Appendix B*

# True or False

*Determine which five of the following statements are false, and briefly explain why.*

1. Freshly ejaculated sperm are more capable of fertilizing an egg than are sperm several hours old.

2. Fertilization normally occurs in the lumen of the uterus.

3. An egg is usually fertilized by the first sperm that contacts it.

4. By the time a conceptus reaches the uterus, it has already undergone several cell divisions and consists of 16 cells or more.

5. The conceptus is first considered a fetus when all of the organ systems are present.

6. The placenta becomes increasingly permeable as it develops.

7. The endocrine system shows less senescence in old age than most other organ systems.

8. Fetal blood bypasses the nonfunctional liver by passing through the foramen ovale.

9. Blood in the umbilical vein has a higher $P_{O_2}$ than blood in the umbilical arteries.

10. It is well established that people who exercise regularly live longer than those who do not.

*Answers in Appendix B*

# Testing Your Comprehension

1. Suppose a woman had a mutation resulting in a tough zona pellucida that did not disintegrate after the egg was fertilized. How would this affect her fertility? Why?

2. Suppose a drug were developed that could slow down the rate of collagen cross-linking with age. What diseases of old age could be made less severe with such a drug?

3. Some health-food stores market the enzyme superoxide dismutase (SOD) as an oral antioxidant to retard senescence. Explain why it would be a waste of your money to buy it.

4. In some children, the ductus arteriosus fails to close after birth—a condition that eventually requires surgery. Predict how this condition would affect (a) pulmonary blood pressure, (b) systemic diastolic pressure, and (c) the right ventricle of the heart.

5. Only one sperm is needed to fertilize an egg, yet a man who ejaculates fewer than 10 million sperm is usually infertile. Explain this apparent contradiction. Supposing 10 million sperm were ejaculated, predict how many would come within close range of the egg. How likely is it that any one of these sperm would fertilize it?

*Answers at www.mhhe.com/saladin4*

# www.mhhe.com/saladin4

*The textbook website provides a wealth of interactive study materials fully organized and integrated by chapter. You will find practice quizzes, labeling exercises, and much more that will complement your learning and understanding of anatomy and physiology. The website also includes tools designed to enhance your* **Anatomy & Physiology | REVEALED** *experience.*

# PERIODIC TABLE OF THE ELEMENTS

Nineteenth-century chemists discovered that when they arranged the known elements by atomic weight, certain properties reappeared periodically. In 1869, Russian chemist Dmitri Mendeleev published the first modern periodic table of the elements, leaving gaps for those that had not yet been discovered. He accurately predicted properties of the missing elements, which helped other chemists discover and isolate them.

Each row in the table is a *period* and each column is a *group (family)*. Each period has one electron shell more than the period above it, and as we progress from left to right within a period, each element has one more proton and electron than the one before. The dark steplike line from boron (5) to astatine (85) separates the metals to the left of it (except hydrogen) from the nonmetals to the right. Each period begins with a soft, light, highly reactive *alkali metal,* with one valence electron, in family IA. Pro-

gressing from left to right, the metallic properties of the elements become less and less pronounced. Elements in family VIIA are highly reactive gases called *halogens,* with seven valence electrons. Elements in family VIIIA, called *noble (inert) gases,* have a full valence shell of eight electrons, which makes them chemically unreactive.

Ninety-one of the elements occur naturally on earth. Physicists have created elements up to atomic number 118 in the laboratory, but the International Union of Pure and Applied Chemistry has established formal names only through element 109 to date.

The 24 elements with normal roles in human physiology are color-coded according to their relative abundance in the body (see chapter 2). Others, however, may be present as contaminants with very destructive effects (such as arsenic, lead, and radiation poisoning).

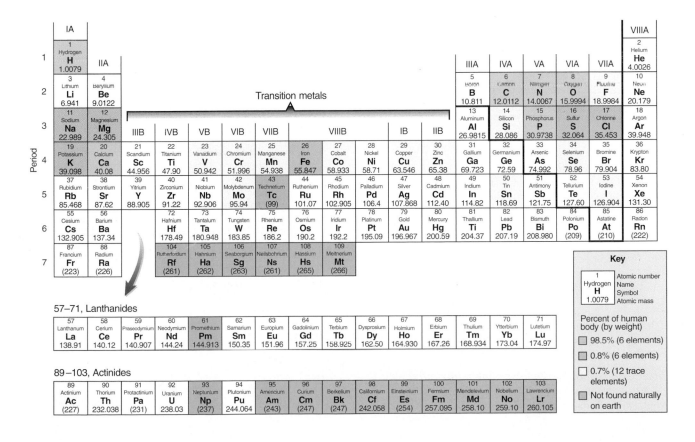

# APPENDIX B

This appendix provides answers to the end-of-chapter Testing Your Recall and True or False questions and the questions in the figure legends. In the True or False sections, all statements are true except those listed and explained here. Answers to Think About It and Testing Your Comprehension questions are posted by chapter at www.mhhe.com/saladin4.

## Chapter 1

*Testing Your Recall*

| | | |
|---|---|---|
| 1. a | 8. c | 15. homeostasis |
| 2. e | 9. d | 16. set point |
| 3. d | 10. b | 17. negative feedback |
| 4. a | 11. dissection | 18. organ |
| 5. c | 12. Hooke | 19. stereoscopic |
| 6. c | 13. deduction | 20. prehensile, |
| 7. a | 14. psychosomatic | opposable |

*True or False (explanation of the false statements only)*

3. Auscultation means listening to body sounds, not inspecting its appearance.
6. Leeuwenhoek was a textile merchant who built microscopes to examine fabric.
7. A scientific theory is founded on a large body of evidence and summarizes what is already known.
8. Both the treatment and control groups consist of volunteer patients.
10. Negative feedback is a self-corrective process with a beneficial effect on the body.

*Answers to Figure Legend Questions*

1.7   A gorilla. About 4.5 million years.
1.10  The thermostat is the sensor. The furnace is the effector.
1.11  It allows blood to circulate closer to the skin surface and lose heat through the skin.

## Atlas A

*Testing Your Recall*

| | | |
|---|---|---|
| 1. d | 8. d | 15. hand, foot |
| 2. c | 9. b | 16. meninges |
| 3. e | 10. d | 17. retroperitoneal |
| 4. d | 11. supine | 18. medial |
| 5. d | 12. parietal | 19. inferior |
| 6. a | 13. mediastinum | 20. cubital, popliteal |
| 7. a | 14. nuchal | |

*True or False (explanation of the false statements only)*

4. The diaphragm is inferior to the lungs.
5. The esophagus is in the ventral body cavity.

6. The liver is in the hypochondriac region, superior to the lateral abdominal region.
9. The peritoneum lines the outside of the stomach and intestines.
10. The sigmoid colon is in the lower left quadrant.

*Answers to Figure Legend Questions*

A.3   Median (midsagittal)
A.10  No, it lies inferior to the peritoneum.
A.14  The lungs, heart, liver, stomach, and spleen, among others
A.19  Posterior
A.21  Fat

## Chapter 2

*Testing Your Recall*

| | | |
|---|---|---|
| 1. a | 9. b | 16. -ose, -ase |
| 2. c | 10. d | 17. phospholipids |
| 3. a | 11. cation | 18. cyclic adenosine |
| 4. c | 12. free radicals | monophosphate |
| 5. a | 13. catalyst, | 19. anaerobic |
| 6. e | enzymes | fermentation |
| 7. b | 14. anabolism | 20. substrate |
| 8. c | 15. dehydration | |
| | synthesis | |

*True or False (explanation of the false statements only)*

1. The monomers of a polysaccharide are monosaccharides (simple sugars).
3. Such molecules are called isomers, not isotopes.
6. A saturated fat is one to which no more hydrogen can be added.
8. Above a certain temperature, enzymes denature and cease working.
9. These solutes have different molecular weights, so 2% solutions would not contain the same number of molecules per unit volume.

*Answers to Figure Legend Questions*

2.1   Potassium gives up an electron.
2.6   By sharing four pairs of electrons, the carbon and the two oxygens each have eight valence electrons, fulfilling the octet rule.
2.8   Since water molecules cling together, they vibrate less freely than do molecules of a nonpolar liquid, and it requires more heat to get water to boil.
2.13  Decomposition
2.26  No, the amount of energy released is the same with or without an enzyme.

## Chapter 3

*Testing Your Recall*

| | | |
|---|---|---|
| 1. e | 4. b | 7. a |
| 2. b | 5. e | 8. c |
| 3. d | 6. e | 9. d |

10. b
11. micrometers
12. second messenger
13. voltage-regulated
14. hydrostatic pressure
15. hypertonic
16. exocytosis
17. nucleus, mitochondria
18. smooth ER, peroxisomes
19. ligand regulated gate
20. cisterna

*True or False (explanation of the false statements only)*

1. Osmosis does not require ATP.
3. Second messengers activate enzymes in the cell; they are not transport proteins.
5. A channel could not move material from the outside of a cell to the inside unless it extended all the way across the membrane; it must be an integral protein.
6. The plasma membrane consists primarily of phospholipid molecules.
7. The brush border is composed of microvilli.

*Answers to Figure Legend Questions*

3.9   Adenylate cyclase is a transmembrane protein. The G protein is peripheral.
3.19  The $Na^+$–$K^+$ pump depends on ATP, whereas osmosis does not. ATP is quickly depleted after a cell dies.
3.23  Transcytosis is a combination of endocytosis and exocytosis.
3.25  Large molecules such as enzymes and RNA must pass through the nuclear pores, but not through the plasma membrane.
3.30  Like an axoneme, a centriole is a cylinder with nine bundles of microtubules. However, each bundle in a centriole has three microtubules, whereas an axoneme has only two; a centriole lacks the central pair of microtubules present in an axoneme; and a centriole is much shorter.

## Chapter 4

*Testing Your Recall*

1. a
2. o
3. c
4. c
5. e
6. b
7. a
8. d
9. d
10. a
11. cytokinesis
12. alleles
13. genetic code
14. polyribosome
15. RNA polymerase
16. genome
17. 46, 92, 92
18. ribosome
19. growth factors
20. autosomes

*True or False (explanation of the false statements only)*

1. There are no ribosomes on the Golgi complex; they are on the rough ER.
2. There are no genes for steroids, carbohydrates, or phospholipids, but only for proteins.
6. This law describes the pairing of bases between the two strands of DNA, not between mRNA and tRNA.
9. Males have only one X chromosome, but have two sex chromosomes (the X and Y).
10. Several RNA polymerase molecules at once can transcribe a gene.

*Answers to Figure Legend Questions*

4.3   The diameter would not be uniform. The double helix would bulge wherever two large purines were paired and would sink inward wherever two small pyrimidines were paired.
4.7   The ribosome would have no way of holding the partially completed peptide in place while adding the next amino acid.
4.16  It would have two X chromosomes at the bottom of the chart instead of an X and a Y.

## Chapter 5

*Testing Your Recall*

1. a
2. b
3. c
4. e
5. c
6. a
7. b
8. e
9. b
10. b
11. necrosis
12. mesothelium
13. lacunae
14. fibers
15. collagen
16. skeletal muscle
17. basement membrane
18. matrix (extracellular material)
19. multipotent
20. simple

*True or False (explanation of the false statements only)*

1. The esophageal epithelium is nonkeratinized.
5. Adipose tissue is an exception; cells constitute most of its volume.
6. Adipocytes are also found in areolar tissue, either singly or in small clusters.
7. Tight junctions serve mainly to restrict the passage of material between cells.
10. Perichondrium is lacking from fibrocartilage and from hyaline articular cartilage.

*Answers to Figure Legend Questions*

5.2   These are longitudinal sections. In the transverse plane, both the egg white and yolk would be round. In the oblique plane, the egg white would be elliptical but the yolk would be round.
5.12  The epithelia of the tongue, oral cavity, esophagus, and anal canal would look similar to this.
5.28  Gap junctions
5.30  It would be a simple sac opening directly onto an epithelial surface, without a duct.
5.31  The holocrine gland, to replace the cells that disintegrate

## Chapter 6

*Testing Your Recall*

1. d
2. c
3. d
4. b
5. a
6. e
7. c
8. a
9. a
10. d
11. insensible perspiration
12. piloerector
13. debridement
14. cyanosis
15. dermal papillae
16. earwax
17. sebaceous glands
18. anagen
19. dermal papilla
20. third-degree

*True or False (explanation of the false statements only)*

3. Keratin is the protein of the epidermis; the dermis is composed mainly of collagen.
4. Vitamin D synthesis begins in the keratinocytes.
7. The hypodermis is not considered to be a layer of the skin.
8. Different races have about the same density of melanocytes but different amounts of melanin.
9. A genetic lack of melanin causes albinism, not pallor. Pallor is a temporary, nonhereditary paleness of the skin.

*Answers to Figure Legend Questions*

6.7   Keratinocytes
6.11  Apocrine glands are associated with pubic, axillary, and beard hair, which are terminal hairs.
6.12  Asymmetry (A), an irregular border (B), and varied color (C). The photo does not provide enough information to judge diameter (D) of the lesion.

## Chapter 7

*Testing Your Recall*

1. e
2. a
3. d
4. c
5. d
6. a
7. d
8. e
9. b
10. d
11. hydroxyapatite
12. canaliculi
13. appositional
14. solubility product
15. hypocalcemia
16. osteoblasts
17. calcitriol
18. osteoporosis
19. metaphysis
20. osteomalacia

*True or False (explanation of the false statements only)*

3. The most common bone disease is osteoporosis, not fractures.
4. Bones elongate at the epiphyseal plate, not the articular cartilage.
5. Osteoclasts develop from stem cells in the bone marrow, not from osteoblasts.
7. Hydroxyapatite is the major mineral of bone; the major protein is collagen.
9. The major effect of vitamin D is bone resorption, though it also promotes deposition.

*Answers to Figure Legend Questions*

7.2   The wider epiphyses provide surface area for muscle attachment and bone articulation, whereas the narrowness of the diaphysis minimizes weight.
7.5   Spongy bone
7.7   Places where bone comes close to the skin, such as the sternum and hips
7.11  An infant's joints are still cartilaginous.

# Chapter 8

*Testing Your Recall*

| | | |
|---|---|---|
| 1. b | 9. e | 17. auricular |
| 2. e | 10. b | 18. styloid |
| 3. a | 11. fontanels | 19. pollex, |
| 4. d | 12. temporal | hallux |
| 5. a | 13. sutures | 20. medial |
| 6. e | 14. sphenoid | longitudinal |
| 7. c | 15. anulus fibrosus | |
| 8. b | 16. dens | |

*True or False (explanation of the false statements only)*

2. Each hand and foot has 14 phalanges.
3. The female pelvis is wider and shallower than the male's.
7. The lumbar vertebrae have transverse processes but no transverse costal facets.
8. The most frequently broken bone is the clavicle.
9. *Arm* refers to the region containing only the humerus; *leg* refers to the region containing the tibia and fibula.

*Answers to Figure Legend Questions*

8.10  The occipital, parietal, sphenoid, zygomatic, and palatine bones, and the mandible and maxilla
8.12  The frontal, lacrimal, and sphenoid bones, and the vomer, maxilla, and inferior concha
8.25  Vertebra C4 can be distinguished by its bifid spinous process, transverse foramina, and light body. Vertebra T4 can be distinguished by its heavier body, costal facets, and transverse costal facets. Vertebra L4 can be distinguished by its very heavy body, blunt squarish spinous process, medially directed superior articular facets, and laterally directed inferior articular facets.
8.34  The adult hand lacks epiphyseal plates, the growth zones of a child's long bones.
8.39  The tibia is a weight-bearing bone and articulates with the broad surface of the talus; the fibula bears no weight.

# Chapter 9

*Testing Your Recall*

| | | |
|---|---|---|
| 1. c | 8. d | 15. gomphosis |
| 2. b | 9. b | 16. serrate |
| 3. a | 10. d | 17. extension |
| 4. e | 11. synovial fluid | 18. range of motion |
| 5. c | 12. bursa | 19. labrum |
| 6. c | 13. pivot | 20. menisci |
| 7. a | 14. kinesiology | |

*True or False (explanation of the false statements only)*

1. Osteoarthritis occurs in almost everyone after a certain age; rheumatoid arthritis is less common.
2. A kinesiologist studies joint movements; a rheumatologist treats arthritis.
3. Synovial joints are diarthroses and amphiarthroses but never synarthroses.
7. The round ligament is somewhat slack and probably does not secure the femoral head.
9. Synovial fluid is secreted by the synovial membrane of the joint capsule and fills the bursae.

*Answers to Figure Legend Questions*

9.4   The pubic symphysis consists of the cartilaginous interpubic disc and the adjacent parts of the two pubic bones.
9.5   Interphalangeal joints are not subjected to a great deal of compression.
9.7   MA = 1.0. Shifting the fulcrum to the left would increase the MA of this lever, while the lever would remain first-class.
9.27  It is the vertical band of tissue immediately to the right of the medial meniscus.

# Chapter 10

*Testing Your Recall*

| | | |
|---|---|---|
| 1. b | 8. a | 14. hamstring |
| 2. e | 9. d | 15. flexor retinacula |
| 3. a | 10. c | 16. urogenital triangle |
| 4. c | 11. origin | 17. linea alba |
| 5. e | 12. fascicle | 18. synergist |
| 6. e | 13. prime mover | 19. bipennate |
| 7. b | (agonist) | 20. sphincter |

*True or False (explanation of the false statements only)*

3. The mastoid process is its insertion.
7. The trapezius is superficial to the scalenes.
8. Normal exhalation does not employ these muscles.
9. They result from rapid extension of the knee, not flexion.
10. They are on opposite sides of the tibia and act as antagonists.

*Answers to Figure Legend Questions*

10.21  To be answered by marking the illustration.
10.25  *Teres* refers to the round or cordlike shape of the first muscle, and *quadratus* refers to the four-sided shape of the second.
10.26  Part (c) represents a cross section cut too high on the forearm to include these muscles.
10.40  The biceps femoris, rectus femoris, fibularis longus and brevis, and tibialis posterior and anterior

# Atlas B

*Figure B.15 Muscle Test*

| | | |
|---|---|---|
| 1. f | 11. y | 21. k |
| 2. b | 12. m | 22. d |
| 3. k | 13. n | 23. f |
| 4. p | 14. e | 24. b |
| 5. h | 15. g | 25. a |
| 6. z | 16. v | 26. u |
| 7. o | 17. f | 27. j |
| 8. x | 18. c | 28. i |
| 9. c | 19. y | 29. g |
| 10. a | 20. x | 30. q |

# Chapter 11

*Testing Your Recall*

| | | |
|---|---|---|
| 1. a | 8. c | 15. acetylcholine |
| 2. d | 9. e | 16. myoglobin |
| 3. b | 10. b | 17. Z discs |
| 4. d | 11. threshold | 18. varicosities |
| 5. a | 12. complete tetanus | 19. muscle tone |
| 6. c | 13. terminal cisternae | 20. lactic acid |
| 7. e | 14. myosin | |

*True or False (explanation of the false statements only)*

1. A motor neuron may supply 1,000 or more muscle fibers; a motor unit consists of one motor neuron and all the muscle fibers it innervates.
2. Calcium binds to troponin, not to myosin.
6. Thick and thin filaments are present but not arranged in a way that produces striations.
7. Under natural conditions, a muscle seldom or never attains complete tetanus.
9. A muscle produces most of its ATP during this time by anaerobic fermentation, which generates lactic acid; it does not consume lactic acid.

*Answers to Figure Legend Questions*

11.12 ATP is needed to pump $Ca^{2+}$ back into the sarcoplasmic reticulum by active transport and to induce each myosin head to release actin so the sarcomere can relax.
11.15 The gluteus maximus and quadriceps femoris
11.16 The muscle tension line would drop gradually while the muscle length line would rise.

# Chapter 12

*Testing Your Recall*

| | | |
|---|---|---|
| 1. e | 9. d | 15. oligodendrocytes |
| 2. c | 10. b | 16. nodes of Ranvier |
| 3. d | 11. afferent | 17. axon hillock, initial segment |
| 4. a | 12. conductivity | 18. norepinephrine |
| 5. c | 13. absolute refractory period | 19. facilitated zone |
| 6. e | 14. dendrites | 20. neuromodulators |
| 7. d | | |
| 8. a | | |

*True or False (explanation of the false statements only)*

4. The $Na^+$ outflow depolarizes the neuron, and the $K^+$ inflow repolarizes it.
5. The threshold stays the same but an EPSP brings the membrane potential closer to the threshold.
7. The effect of a neurotransmitter varies from place to place depending on the type of receptor present.
8. The signals travel rapidly through the internodes and slow down at each node of Ranvier.
9. Synaptic contacts are remodeled, added, and removed throughout life.

*Answers to Figure Legend Questions*

12.11 It would become lower (more negative).
12.19 Axosomatic
12.24 One EPSP is a voltage change of only 0.5 mV or so. A change of about 15 mV is required to make a neuron fire.
12.28 The CNS interprets a stimulus as more intense if it receives signals from high-threshold sensory neurons than if it receives signals only from low-threshold neurons.
12.30 A reverberating circuit, because a neuron early in the circuit is continually restimulated

# Chapter 13

*Testing Your Recall*

| | | |
|---|---|---|
| 1. e | 8. a | 15. intrafusal fibers |
| 2. c | 9. e | 16. phrenic |
| 3. d | 10. b | 17. decussation |
| 4. d | 11. ganglia | 18. proprioception |
| 5. e | 12. ramus | 19. dorsal root |
| 6. c | 13. spinocerebellar | 20. tibial, common fibular |
| 7. c | 14. crossed extension | |

*True or False (explanation of the false statements only)*

1. The gracile fasciculus is an ascending (sensory) tract.
4. All spinal nerves are mixed nerves; none are purely sensory or motor.
5. The dura is separated from the bone by a fat-filled epidural space.
8. Dermatomes overlap each other by as much as 50%.
9. Some somatic reflexes are mediated primarily through the brainstem and cerebellum.

*Answers to Figure Legend Questions*

13.4 If it were T10, there would be no cuneate fasciculus; that exists only from T6 up.
13.9 They are in the ventral horn of the spinal cord.
13.12 They are afferent, because they arise from the dorsal root of the spinal nerve.
13.21 Motor neurons are capable only of exciting skeletal muscle (endplate potentials are always excitatory). To inhibit muscle contraction, it is necessary to inhibit the motor neuron at the CNS level (point 7).
13.22 They would show more synaptic delay, because there are more synapses in the pathway.

# Chapter 14

*Testing Your Recall*

| | | |
|---|---|---|
| 1. c | 8. d | 14. hydrocephalus |
| 2. a | 9. e | 15. choroid plexus |
| 3. e | 10. e | 16. precentral |
| 4. a | 11. corpus callosum | 17. frontal |
| 5. b | 12. ventricles, cerebrospinal | 18. association areas |
| 6. c | 13. arbor vitae | 19. categorical |
| 7. a | | 20. the Broca area |

*True or False (explanation of the false statements only)*

1. This fissure separates the cerebral hemispheres, not the cerebellar hemispheres.
2. The cerebral hemispheres do not develop from neural crest tissue.
5. The choroid plexuses produce only 30% of the CSF.
6. Hearing is a temporal lobe function; vision resides in the occipital lobe.
10. Eye movements are controlled by the oculomotor, trochlear, and abducens nerves; the optic nerve serves only to carry visual information.

*Answers to Figure Legend Questions*

14.7 The most common sites of obstruction are the interventricular foramen at label 2, the cerebral aqueduct at label 4, and the lateral and median apertures indicated by label 6.
14.9 Signals in the cuneate fasciculus ascend to the cuneate nucleus in part (c), and signals in the gracile fasciculus ascend to the nearby gracile nucleus. Both of them decussate together to the contralateral medial lemniscus in parts (b) and (a) and travel this route to the thalamus.
14.11 The reticular formation is labeled on all three parts of the figure.
14.14 Commissural tracts also cross through the anterior and posterior commissures shown in figure 14.2.
14.15 Dendrites
14.23 Regions with numerous small muscles

# Chapter 15

*Testing Your Recall*

| | | | | |
|---|---|---|---|---|
| 1. b | 8. d | 15. enteric | | |
| 2. c | 9. a | 16. norepinephrine | | |
| 3. e | 10. c | 17. sympathetic | | |
| 4. e | 11. adrenergic | 18. preganglionic, | | |
| 5. a | 12. dual innervation | postganglionic | | |
| 6. e | 13. autonomic tone | 19. cAMP | | |
| 7. d | 14. vagus | 20. vasomotor tone | | |

*True or False (explanation of the false statements only)*

1. Both systems are always simultaneously active.
3. In biofeedback and other circumstances, limited voluntary control of the ANS is possible.
4. The sympathetic division inhibits digestion.
6. Waste elimination can occur by autonomic spinal reflexes without necessarily involving the brain.
7. All parasympathetic fibers are cholinergic.

*Answers to Figure Legend Questions*

15.4  No; inhaling and exhaling are controlled by the somatic motor system and skeletal muscles.
15.5  The soma of the somatic efferent neuron is in the ventral horn, and the soma of the autonomic preganglionic neuron is in the lateral horn.
15.7  The vagus nerve

# Chapter 16

*Testing Your Recall*

| | | |
|---|---|---|
| 1. a | 8. c | 15. hair cells |
| 2. c | 9. c | 16. stapes |
| 3. b | 10. b | 17. inferior colliculi |
| 4. a | 11. fovea centralis | 18. taste hairs |
| 5. e | 12. ganglion | 19. olfactory bulb |
| 6. e | 13. Na$^+$ | 20. referred pain |
| 7. d | 14. otoliths | |

*True or False (explanation of the false statements only)*

1. These fibers end in the medulla oblongata.
3. Because of hemidecussation, the right hemisphere receives signals from both eyes.
5. The posterior chamber, the space between iris and lens, is filled with aqueous humor.
6. Descending analgesic fibers block signals that have reached the dorsal horn of the spinal cord.
10. The trochlear and abducens nerves control the superior oblique and lateral rectus muscles, respectively.

*Answers to Figure Legend Questions*

16.1  Yes; two touches are felt separately if they straddle the boundary between two separate receptive fields.
16.9  The lower margin of the blue zone ("all sound") would be higher in that range.
16.14  They tune the cochlea to improve frequency discrimination.
16.15  It would oppose the inward movement of the tympanic membrane, thus reducing the amount of vibration transferred to the inner ear.
16.28  It is the right eye. The optic disc is always medial to the fovea, so this has to be a view of the observer's left and the subject's right.
16.40  Approximately 68:20:0, and yellow
16.43  It would cause blindness in the left half of the visual field. It would not affect the visual reflexes.

# Chapter 17

*Testing Your Recall*

| | | |
|---|---|---|
| 1. b | 10. e | 17. negative |
| 2. d | 11. adenohypophysis | feedback |
| 3. a | 12. tyrosine | inhibition |
| 4. c | 13. acromegaly | 18. hypophyseal |
| 5. c | 14. cortisol | portal system |
| 6. c | 15. glucocorticoids | 19. permissive |
| 7. d | 16. granulosa, | 20. up-regulation |
| 8. c | interstitial | |
| 9. a | | |

*True or False (explanation of the false statements only)*

5. Hormones are also secreted by the heart, liver, kidneys, and other organs not generally regarded as glands.
7. The pineal gland and thymus undergo involution with age.
8. Without iodine, there is no thyroid hormone (TH); without TH, there can be no negative feedback inhibition.
9. The tissue at the center is the adrenal medulla.
10. There are also two testes, two ovaries, and four parathyroid glands.

*Answers to Figure Legend Questions*

17.4  The neurohypophysis
17.8  The thymus
17.22  Steroids enter the target cell; they do not bind to membrane receptors or activate second messengers.
17.26  Such a drug would block leukotriene synthesis and thus inhibit allergic and inflammatory responses.
17.27  She would have been a pituitary giant.

# Chapter 18

*Testing Your Recall*

| | | |
|---|---|---|
| 1. b | 8. c | 14. agglutinogens |
| 2. c | 9. d | 15. hemophilia |
| 3. c | 10. c | 16. hemostasis |
| 4. a | 11. hemopoiesis | 17. sickle cell disease |
| 5. b | 12. hematocrit or | 18. polycythemia |
| 6. d | packed cell volume | 19. vitamin B$_{12}$ |
| 7. d | 13. thromboplastin | 20. erythropoietin |

*True or False (explanation of the false statements only)*

3. Oxygen deficiency is the result of anemia, not its cause.
4. Clotting is one mechanism of hemostasis, but hemostasis includes others. Agglutination is unrelated to either of these.
6. The most abundant WBCs are neutrophils.
9. The heme is excreted; the globin is broken down into amino acids that can be reused.
10. In leukemia, there is an excess of WBCs. A WBC deficiency is leukopenia.

*Answers to Figure Legend Questions*

18.1  A nucleus
18.5  Hemoglobin consists of a noncovalent association of four protein chains. The prosthetic group is the heme moiety of each of the four chains.
18.19  Although numerous, these WBCs are immature and incapable of performing their defensive roles.
18.21  A platelet plug lacks the fibrin mesh that a blood clot has.
18.23  It would affect only the intrinsic mechanism.
18.24  In both blood clotting and the enzyme amplification mechanism of hormone action, the product of one reaction step is an enzyme that catalyzes the production of many more molecules of product at the next step. Thus there is a geometric increase in the number of product molecules at each step and ultimately, a large final result from a small beginning.

# Chapter 19

## Testing Your Recall

| | | |
|---|---|---|
| 1. d | 8. c | 14. Na$^+$ |
| 2. b | 9. a | 15. gap junctions |
| 3. d | 10. e | 16. T wave |
| 4. a | 11. systole, diastole | 17. vagus |
| 5. a | 12. systemic | 18. myocardial infarction |
| 6. d | 13. atrioventricular | 19. endocardium |
| 7. d | (coronary) sulcus | 20. cardiac output |

*True or False (explanation of the false statements only)*

1. The coronary circulation is part of the systemic circuit; the other division is the pulmonary circuit.
3. The first two-thirds of ventricular filling occurs before the atria contract. The atria add only about 31% of the blood that fills the ventricles.
6. The first heart sound occurs at the time of the QRS complex.
7. The heart has its own internal pacemaker and would continue beating; the nerves only alter the heart rate.
10. The ECG is a composite record of the electrical activity of the entire myocardium, not a record from a single myocyte. It looks much different from an action potential.

*Answers to Figure Legend Questions*

19.2  To the left
19.7  The trabeculae carneae
19.12  The right atrium
19.15  It ensures that wave summation and tetanus will not occur, thus ensuring relaxation and refilling of the heart chambers.
19.20  This is the point at which the aortic valve opens and blood is ejected into the aorta, raising its blood pressure.

# Chapter 20

*Testing Your Recall*

| | | |
|---|---|---|
| 1. c | 8. a | 14. thoracic pump |
| 2. b | 9. e | 15. oncotic pressure |
| 3. a | 10. d | 16. transcytosis |
| 4. e | 11. systolic, diastolic | 17. sympathetic |
| 5. b | 12. continuous | 18. baroreceptors |
| 6. c | capillaries | 19. the arterial circle |
| 7. e | 13. anaphylactic | 20. basilic, cephalic |

*True or False (explanation of the false statements only)*

4. Some veins have valves, but arteries do not.
5. By the formula $F \propto r^4$, the flow increases 16-fold.
8. The capillaries normally reabsorb about 85% of the fluid they filter; the rest is absorbed by the lymphatic system.
9. An aneurysm is a weak, bulging vessel that *may* rupture.
10. Anaphylactic shock is a form of venous pooling shock.

*Answers to Figure Legend Questions*

20.2  Veins are subjected to less pressure than arteries and have less need of elasticity.
20.18  Nothing would happen if he lifted his finger from point O because the valve at that point would prevent blood from flowing downward and filling the vein. If he lifted his finger from point H, blood would flow upward, fill the vein, and the vein between O and H would stand out.
20.23  Aorta → left common carotid artery → external carotid artery → superficial temporal artery
20.26  The bronchial artery
20.31  The deep and superficial palmar arches
20.35  The cephalic, basilic, and median cubital veins

# Chapter 21

*Testing Your Recall*

| | | |
|---|---|---|
| 1. b | 9. a | 15. opsonization |
| 2. c | 10. c | 16. pyrogen |
| 3. a | 11. pathogen | 17. interleukins |
| 4. a | 12. lysozyme | 18. antigen-binding |
| 5. d | 13. lymphadenitis | site, epitope |
| 6. b | 14. diapedesis | 19. clonal deletion |
| 7. e | (emigration) | 20. autoimmune |
| 8. d | | |

*True or False (explanation of the false statements only)*

1. Lysozyme is a bacteria-killing enzyme.
3. Interferons promote inflammation.
4. Helper T cells are also necessary to humoral immunity.
9. Anergy is a loss of lymphocyte activity, whereas autoimmune diseases result from misdirected activity.
10. Interferons inhibit viral replication; perforins lyse bacteria.

*Answers to Figure Legend Questions*

21.4  There would be no consistent one-way flow of lymph. Lymph and tissue fluid would accumulate, especially in the lower regions of the body.
21.16  Both of these produce a ring of proteins in the target cell plasma membrane, opening a hole in the membrane through which the cell contents escape.
21.23  All three defenses depend on the action of helper T cells, which are destroyed by HIV.
21.26  The ER is the site of antibody synthesis.
21.31  AZT targets reverse transcriptase. If this enzyme is unable to function, HIV cannot produce viral DNA and insert it into the host cell DNA, and the virus therefore cannot be replicated.

# Chapter 22

*Testing Your Recall*

| | | |
|---|---|---|
| 1. c | 9. d | 17. compliance, |
| 2. c | 10. a | elasticity |
| 3. a | 11. epiglottis | 18. inspiratory |
| 4. e | 12. bronchial tree | center |
| 5. e | 13. pulmonary surfactant | 19. ventilation– |
| 6. c | 14. atmospheric | perfusion |
| 7. b | 15. obstructive | coupling |
| 8. a | 16. anatomic dead space | 20. alkalosis, |
| | | hypocapnia |

*True or False (explanation of the false statements only)*

1. The phrenic nerves also fire during expiration and exert a braking action on the diaphragm.
4. When volume increases, pressure decreases.
5. Atelectasis can have other causes such as airway obstruction.
8. In an average 500 mL tidal volume, 350 mL reaches the alveoli.
10. Most $CO_2$ is transported as bicarbonate ion.

*Answers to Figure Legend Questions*

22.3  Bacteria can easily travel from the throat up the auditory tube to the middle ear.
22.7  The right primary bronchus is more vertical than the left, making it easier for objects to fall into the right.
22.19  $P_{O_2}$ drops from 104 to 95 mm Hg on its way out of the lungs because of some mixing with systemic blood. It drops further to 40 mm Hg when the blood gives up $O_2$ to respiring tissues and remains at this level until the blood is reoxygenated back in the lungs. $P_{CO_2}$ is 40 mm Hg leaving the lungs and rises to 46 mm Hg when $CO_2$ is picked up from respiring tissues. It remains at that level until the blood returns to the lungs and unloads $CO_2$.

22.23  About 70%

22.25  In the alveoli, $CO_2$ leaves the blood, $O_2$ enters, and all the chemical reactions are the reverse of those in figure 22.24. The blood bicarbonate concentration will be reduced following alveolar gas exchange.

22.26  A higher temperature suggests a relatively high metabolic rate, and thus an elevated demand for oxygen. Comparison of these curves shows that for a given $P_{O_2}$, hemoglobin gives up more oxygen at warmer temperatures.

## Chapter 23

*Testing Your Recall*

| | | | |
|---|---|---|---|
| 1. a | 9. c | 16. transport |
| 2. d | 10. a | maximum |
| 3. b | 11. micturition | 17. antidiuretic |
| 4. c | 12. renal autoregulation | hormone |
| 5. b | 13. trigone | 18. internal urethral |
| 6. b | 14. macula densa | 19. protein |
| 7. d | 15. podocytes | 20. arcuate |
| 8. e | | |

*True or False (explanation of the false statements only)*

1. Parathyroid hormone and angiotensin II regulate the PCT.
2. Urine contains more urea and chloride than sodium.
4. A substantial amount of tubular fluid is reabsorbed by the paracellular route, passing through leaky tight junctions.
5. Glycosuria does not occur in diabetes insipidus.
8. Urine can be as dilute as 50 mOsm/L.

*Answers to Figure Legend Questions*

23.2  Ammonia is produced by the deamination of amino acids; urea is produced from ammonia and carbon dioxide; uric acid from nucleic acids; and creatinine from creatine phosphate.

23.3  The kidney lies between the peritoneum and body wall rather than in the peritoneal cavity. The pancreas, aorta, inferior vena cava, and renal artery and vein are also retroperitoneal.

23.10 The afferent arteriole is bigger. The relatively large inlet to the glomerulus and its small outlet results in high blood pressure in the glomerulus. This is the force that drives glomerular filtration.

23.15 It lowers the urine pH because of the $Na^+$–$H^+$ antiport (see the bottom cell). The more $Na^+$ that is reabsorbed, the more $H^+$ is secreted into the tubular fluid.

23.21 The relatively short female urethra is less of an obstacle for bacteria traveling from the perineum to the urinary bladder.

## Chapter 24

*Testing Your Recall*

| | | | |
|---|---|---|---|
| 1. c | 9. d | 16. hyperkalemia |
| 2. a | 10. b | 17. hyponatremia |
| 3. a | 11. $Na^+$ | 18. respiratory |
| 4. a | 12. $K^+$ | acidosis |
| 5. d | 13. metabolic water | 19. limiting pH |
| 6. c | 14. cutaneous | 20. osmolarity |
| 7. e | transpiration | |
| 8. b | 15. fluid sequestration | |

*True or False (explanation of the false statements only)*

2. Aldosterone has only a small influence on blood pressure.
5. PTH promotes calcium absorption but phosphate excretion.
6. Protein buffers more acid than bicarbonate or phosphates do.
9. More water than salt is lost, so the body fluids become hypertonic.
10. Aquaporins are found in the distal tubule and collecting duct.

*Answers to Figure Legend Questions*

24.1  The tissue fluid
24.7  Ingestion of water
24.9  It would decrease.
24.12 Reverse both arrows to point to the left.

## Chapter 25

*Testing Your Recall*

| | | | |
|---|---|---|---|
| 1. b | 8. a | 15. vagus |
| 2. d | 9. a | 16. gastrin |
| 3. c | 10. a | 17. sinusoids |
| 4. e | 11. occlusal | 18. maltase, maltose |
| 5. a | 12. amylase, lipase | 19. chylomicrons |
| 6. c | 13. parotid | 20. iron |
| 7. a | 14. enteric | |

*True or False (explanation of the false statements only)*

1. Fat digestion begins in the stomach.
2. Most of the tooth is dentin.
3. Hepatocytes secrete bile into the bile canaliculi.
7. Intrinsic factor is involved in the absorption of vitamin $B_{12}$.
10. Water, glucose, and other nutrients pass between cells, through the tight junctions.

*Answers to Figure Legend Questions*

25.6  The first and second premolars and the third molar
25.11 Blockage of the mouth by the root of the tongue and blockage of the nose by the soft palate
25.14 It exchanges $H^+$ for $K^+$ ($H^+$–$K^+$ ATPase is an active transport pump).
25.20 From the hepatic artery and the hepatic portal vein
25.31 The autonomic nervous system controls the internal anal sphincter, and the somatic nervous system controls the external anal sphincter.

## Chapter 26

*Testing Your Recall*

| | | | |
|---|---|---|---|
| 1. a | 8. a | 15. liver |
| 2. c | 9. d | 16. insulin |
| 3. b | 10. d | 17. core temperature |
| 4. e | 11. incomplete | 18. arcuate |
| 5. b | 12. glycogenolysis | 19. cytochromes |
| 6. e | 13. gluconeogenesis | 20. ATP synthase, ATP |
| 7. c | 14. urea | |

*True or False (explanation of the false statements only)*

1. Leptin suppresses the appetite.
4. Most of the cholesterol is endogenous, not dietary.
5. Excessive protein intake can cause renal damage.
8. Gluconeogenesis is a postabsorptive phenomenon.
9. Brown fat does not generate ATP.

*Answers to Figure Legend Questions*

26.2  A high HDL:LDL ratio indicates that excess cholesterol is being transported to the liver for removal from the body. A high LDL:HDL ratio indicates a high rate of cholesterol deposition in the walls of the arteries.

26.3  It would stop at step 3, and PGAL would accumulate. Anaerobic fermentation replenishes $NAD^+$.

26.5  NADH and $FADH_2$

26.9  Acidosis (or ketoacidosis or metabolic acidosis)

26.10 From amino acids to keto acids to pyruvic acid to glucose

## Chapter 27

*Testing Your Recall*

| | | | | | |
|---|---|---|---|---|---|
| 1. | a | 9. | d | 15. | tunica albuginea |
| 2. | a | 10. | d | 16. | seminal vesicles |
| 3. | a | 11. | mesonephric | 17. | sustentacular |
| 4. | c | 12. | fructose | 18. | secondary |
| 5. | a | 13. | pampiniform | | spermatocyte |
| 6. | d | | plexus | 19. | deep |
| 7. | e | 14. | secondary | 20. | acrosome |
| 8. | c | | spermatocytes | | |

*True or False (explanation of the false statements only)*

4. Only the testes are primary sex organs.
5. Female development results from a low testosterone level, not from estrogen.
7. The pampiniform plexus prevents the testes from overheating.
8. Sperm are stored in the epididymis.
9. There is no such phenomenon as male menopause.

*Answers to Figure Legend Questions*

27.3  Both disorders result from defects in hormone receptors rather than a lack of the respective hormone.
27.5  The word *vagina* means "sheath." The tunica vaginalis ensheaths the testis.
27.11 An enlarged prostate gland compresses the urethra and interferes with emptying the bladder.
27.12 An overly engorged corpus spongiosum would compress the urethra and interfere with the expulsion of semen.
27.14 Crossing over in prophase I
27.15 The next cell stage in meiosis, the secondary spermatocyte, is genetically different from the other cells of the body and would be subject to immune attack if not isolated from the antibodies in the blood.

## Chapter 28

*Testing Your Recall*

| | | | | | |
|---|---|---|---|---|---|
| 1. | a | 8. | b | 15. | corona radiata |
| 2. | d | 9. | c | 16. | antrum |
| 3. | c | 10. | c | 17. | climacteric |
| 4. | a | 11. | follicle | 18. | conceptus |
| 5. | c | 12. | endometrium | 19. | infundibulum, |
| 6. | b | 13. | menarche | | fimbriae |
| 7. | b | 14. | corpus luteum | 20. | lochia |

*True or False (explanation of the false statements only)*

1. Only the ovum and corona radiata enter the uterine tube, not the whole follicle.
2. HCG is secreted by the placenta.
5. Many eggs and follicles undergo atresia during childhood, so their number is reduced by the age of puberty.
6. Prolactin is secreted during pregnancy but does not induce lactation then.
10. Only the superficial layer (functionalis) is shed.

*Answers to Figure Legend Questions*

28.8  The paraurethral glands
28.11 This results in a clearer image since the X rays do not have to penetrate such a thick mass of tissue.
28.19 The rising ratio of estrogen to progesterone makes the uterus more irritable.

## Chapter 29

*Testing Your Recall*

| | | | | | |
|---|---|---|---|---|---|
| 1. | b | 8. | a | 15. | chorionic villi |
| 2. | b | 9. | d | 16. | acrosome |
| 3. | c | 10. | d | 17. | collagen |
| 4. | c | 11. | teratogens | 18. | Down syndrome |
| 5. | a | 12. | nondisjunction | | (trisomy-21) |
| 6. | e | 13. | life span | 19. | foramen ovale |
| 7. | c | 14. | life expectancy | 20. | embryo |

*True or False (explanation of the false statements only)*

1. Sperm require about 10 hours to become capacitated and able to fertilize an egg.
2. Fertilization occurs in the uterine tube.
3. Several early-arriving sperm clear a path for the one that fertilizes the egg.
8. Blood bypasses the lungs via the foramen ovale.
10. Exercise improves the quality of life in old age, but has not been shown to increase life expectancy significantly.

*Answers to Figure Legend Questions*

29.2  An unfertilized egg dies long before it reaches the uterus.
29.6  Eight weeks
29.13 XXY (Klinefelter syndrome) and YO (a zygote that would not survive)
29.15 Female, as seen from the two X chromosomes at the lower right

# SYMBOLS, WEIGHTS, AND MEASURES

## UNITS OF LENGTH

| | |
|---|---|
| m | meter |
| km | kilometer ($10^3$ m) |
| cm | centimeter ($10^{-2}$ m) |
| mm | millimeter ($10^{-3}$ m) |
| μm | micrometer ($10^{-6}$ m) |
| nm | nanometer ($10^{-9}$ m) |

## UNITS OF MASS AND WEIGHT

| | |
|---|---|
| amu | atomic mass unit |
| MW | molecular weight |
| mole | MW in grams |
| g | gram |
| kg | kilograms ($10^3$ g) |
| mg | milligrams ($10^{-3}$ g) |

## UNITS OF PRESSURE

| | |
|---|---|
| atm | atmospheres (1 atm = 760 mm Hg) |
| mm Hg | millimeters of mercury |
| $P_X$ | partial pressure of gas x (as in $P_{O_2}$) |

## CONVERSION FACTORS

| | |
|---|---|
| 1 in. = 2.54 cm | 1 cm = 0.394 in. |
| 1 fl oz = 29.6 mL | 1 mL = 0.034 fl oz |
| 1 qt = 0.946 L | 1 L = 1.057 qt |
| 1 g = 0.0035 oz | 1 oz = 28.38 g |
| 1 lb = 0.45 kg | 1 kg = 2.2 lb |
| °C = (5/9)(°F − 32) | °F = (9/5)(°C) + 32 |

## UNITS OF VOLUME

| | |
|---|---|
| L | liter |
| dL | deciliter (= 100 mL) ($10^{-1}$ L) |
| mL | milliliter ($10^{-3}$ L) |
| μL | microliter (= 1 $mm^3$) ($10^{-6}$ L) |

## UNITS OF CONCENTRATION

| | |
|---|---|
| mEq/L | milliequivalents per liter |
| Osm/L | osmoles per liter |
| mOsm/L | milliosmoles per liter |
| M | molar |
| mM | millimolar |
| pH | negative log of $H^+$ molarity |

## UNITS OF HEAT

| | |
|---|---|
| cal | "small" calories |
| kcal | kilocalories (Calories; 1 kcal = 1,000 cal) |
| Cal | "large" (dietary) calories (1 Cal = 1,000 cal) |

## GREEK LETTERS

| | |
|---|---|
| α | alpha |
| β | beta |
| γ | gamma |
| Δ | delta (uppercase) |
| δ | delta (lowercase) |
| η | eta |
| θ | theta |
| μ | mu |

# BIOMEDICAL ABBREVIATIONS

| | |
|---|---|
| δ+, δ− | slight positive or negative charge |
| $2n$ | diploid |
| a., aa. | artery, arteries |
| A | adenine |
| Ab | antibody |
| ACh | acetylcholine |
| AChE | acetylcholinesterase |
| ACTH | adrenocorticotropic hormone |
| AD | Alzheimer disease |
| ADH | antidiuretic hormone |
| ADP | adenosine diphosphate |
| Ag | antigen |
| AIDS | acquired immunodeficiency syndrome |
| amu | atomic mass unit |
| ANF | atrial natriuretic factor |
| ANS | autonomic nervous system |
| APC | antigen-presenting cell |
| ATP | adenosine triphosphate |
| AV | atrioventricular |
| BBB | blood–brain barrier |
| BMR | basal metabolic rate |
| BP | blood pressure |
| bpm | beats per minute |
| C | cytosine, carbon |
| $Ca^{2+}$ | calcium ion |
| CAH | carbonic anhydrase |
| cAMP | cyclic adenosine monophosphate |
| CCK | cholecystokinin |
| CHF | congestive heart failure |
| $Cl^-$ | chloride ion |
| CNS | central nervous system |
| COP | colloid osmotic pressure |
| COPD | chronic obstructive pulmonary disease |
| CP | creatine phosphate |
| CRH | corticotropin-releasing hormone |
| c.s. | cross section |
| CSF | cerebrospinal fluid |
| DNA | deoxyribonucleic acid |
| ECF | extracellular fluid |
| ECG | electrocardiogram |
| EPSP | excitatory postsynaptic potential |
| ER | endoplasmic reticulum |
| FAD | flavin adenine dinucleotide |
| $Fe^{2+}$ | ferrous ion |
| $Fe^{3+}$ | ferric ion |
| FSH | follicle-stimulating hormone |
| G | guanine |
| GABA | gamma-aminobutyric acid |
| GAG | glycosaminoglycan |
| GFR | glomerular filtration rate |
| GH | growth hormone |
| GHRH | growth hormone–releasing hormone |
| GnRH | gonadotropin-releasing hormone |
| $H^+$ | hydrogen ion |
| Hb | hemoglobin |
| HCG | human chorionic gonadotropin |
| $HCO_3^-$ | bicarbonate ion |
| HDL | high-density lipoprotein |
| HIV | human immunodeficiency virus |
| Hz | hertz (cycles per second) |
| $I^-$ | iodide ion |
| ICF | intracellular fluid |
| IDDM | insulin-dependent diabetes mellitus |
| Ig | immunoglobulin |
| IL | interleukin |
| I.M. | intramuscular |
| IPSP | inhibitory postsynaptic potential |
| IR | infrared |
| I.V. | intravenous |
| $K^+$ | potassium ion |
| LDL | low-density lipoprotein |
| LH | luteinizing hormone |
| LM | light microscope |
| l.s. | longitudinal section |
| m., mm. | muscle, muscles |
| MHC | major histocompatibility complex |
| MI | myocardial infarction |
| mRNA | messenger ribonucleic acid |
| MS | multiple sclerosis |
| MSH | melanocyte-stimulating hormone |
| mV | millivolts |
| MW | molecular weight |
| $n$ | haploid |
| n., nn. | nerve, nerves |
| $Na^+$ | sodium ion |
| $NAD^+$ | nicotinamide adenine dinucleotide |
| NE | norepinephrine |
| NFP | net filtration pressure |
| NIDDM | noninsulin-dependent diabetes mellitus |
| $OH^-$ | hydroxyl ion |
| OT | oxytocin |
| $P_i$ | inorganic phosphate group |
| PG | prostaglandin |
| PIH | prolactin-inhibiting hormone |
| PNS | peripheral nervous system |
| PRH | prolactin-releasing hormone |
| PRL | prolactin (luteotropin) |
| PTH | parathyroid hormone (parathormone) |
| RBC | red blood cell |
| Rh | rhesus factor |
| RMP | resting membrane potential |
| RNA | ribonucleic acid |
| rRNA | ribosomal ribonucleic acid |
| SA | sinoatrial |
| SEM | scanning electron microscope |
| T | thymine |
| $T_{1/2}$ | half-life |
| $T_3$ | triiodothyronine |
| $T_4$ | thyroxine (tetraiodothyronine) |
| $T_m$ | transport maximum |
| TEM | transmission electron microscope |
| TRH | thyrotropin-releasing hormone |
| tRNA | transfer ribonucleic acid |
| TSH | thyroid-stimulating hormone |
| UV | ultraviolet |
| v., vv. | vein, veins |
| VLDL | very low-density lipoprotein |
| $V_{O_2 max}$ | maximum oxygen uptake |
| WBC | white blood cell |

# Glossary

This glossary defines approximately 1,000 terms. They are not necessarily the most important ones in the book, but they are terms that are reintroduced most often and, for lack of space, are not redefined each time they arise. The index indicates where you can find definitions or explanations of additional terms. Terms are defined only in the sense that they are used in this book. Some, however, have broader meanings, even within biology and medicine, that are beyond the scope of this text. Terms that are commonly abbreviated, such as *ATP* and *PET scan,* are defined under the full spelling. See the list of abbreviations in appendix D for complete spellings. The glossary gives pronunciation guides for many terms, with accented syllables in capital letters. A key to the pronunciation of individual syllables and letter groups can be found at the end of the glossary.

## A

**abdominal cavity** The body cavity between the diaphragm and pelvic brim. fig. A.7

**abduction** (ab-DUC-shun) Movement of a body part away from the median plane, as in raising an arm away from the side of the body. fig. 9.13

**absorption 1.** Process in which a chemical passes through a membrane or tissue surface and becomes incorporated into a body fluid or tissue. **2.** Any process in which one substance passes into another and becomes a part of it. *Compare* adsorption.

**acetate** The ionized form of acetic acid ($CH_3COO^-$). Serves as the monomer of fatty acids and the intermediate of aerobic metabolism that enters the citric acid cycle.

**acetylcholine (ACh)** (ASS-eh-till-CO-leen) A neurotransmitter released by somatic motor fibers, parasympathetic fibers, and some other neurons, composed of choline and an acetyl group. fig. 12.21

**acetylcholinesterase (AChE)** (ASS-eh-till-CO-lin-ESS-ter-ase) An enzyme that hydrolyzes acetylcholine, thus halting signal transmission at a cholinergic synapse.

**acid** A proton ($H^+$) donor; a chemical that releases protons into solution.

**acidosis** An acid–base imbalance in which the blood pH is lower than 7.35.

**acinus** (ASS-ih-nus) A sac of secretory cells at the inner end of a gland duct. fig. 5.29

**acquired immunodeficiency syndrome (AIDS)** A group of conditions that indicate severe immunosuppression related to infection with the human immunodeficiency virus (HIV); typically characterized by a very low CD4 T lymphocyte count and high susceptibility to certain forms of cancer and opportunistic infections.

**actin** A filamentous intracellular protein that provides cytoskeletal support and interacts with other proteins, especially myosin, to cause cellular movement; important in muscle contraction and membrane actions such as phagocytosis, ameboid movement, and cytokinesis.

**action** The movement produced by the contraction of a particular muscle.

**action potential** A rapid voltage change in which a plasma membrane briefly reverses electrical polarity; has a self-propagating effect that produces a traveling wave of excitation in nerve and muscle cells.

**active site** The region of a protein that binds to a ligand, such as the substrate-binding site of an enzyme or the hormone-binding site of a receptor.

**active transport** The movement of a solute through a cellular membrane, against its concentration gradient, involving a carrier protein that expends ATP.

**acute** Pertaining to a disease with abrupt onset, intense symptoms, and short duration. *Compare* chronic.

**adaptation 1.** An evolutionary process leading to the establishment of species characteristics that favor survival and reproduction. **2.** Any characteristic of anatomy, physiology, or behavior that promotes survival and reproduction. **3.** A sensory process in which a receptor adjusts its sensitivity or response to the prevailing level of stimulation, such as dark adaptation of the eye.

**adduction** (ah-DUC-shun) Movement of a body part toward the median plane, such as bringing the feet together from a spread-legged position. fig. 9.13

**adenine** (AD-eh-neen) A double-ringed nitrogenous base (purine) found in such molecules as DNA, RNA, and ATP; one of the four bases of the genetic code; complementary to thymine in the double helix of DNA. fig. 4.2

**adenosine triphosphate (ATP)** (ah-DEN-oh-seen tri-FOSS-fate) A molecule composed of adenine, ribose, and three phosphate groups that functions as a universal energy-transfer molecule; yields adenosine diphosphate (ADP) and an inorganic phosphate group ($P_i$) upon hydrolysis. fig. 2.29a

**adenylate cyclase** (ah-DEN-ih-late SY-clase) An enzyme of the plasma membrane that removes two phosphate molecules from ATP and makes cyclic adenosine monophosphate (cAMP); important in the activation of the cAMP second-messenger system.

**adipocyte** (AD-ih-po-site) A fat cell.

**adipose tissue** A connective tissue composed predominantly of adipocytes; fat.

**adrenal gland** (ah-DREE-nul) An endocrine gland on the superior pole of each kidney. fig. 17.11

**adrenergic** (AD-reh-NUR-jic) Pertaining to epinephrine (adrenaline) or norepinephrine (noradrenaline), as in adrenergic neurons that secrete one of these chemicals or adrenergic effects on a target organ.

**adrenocorticotropic hormone (ACTH)** (ah-DREE-no-COR-tih-co-TRO-pic) A hormone secreted by the anterior pituitary gland that stimulates the adrenal cortex.

**adsorption** The binding of one substance to the surface of another without becoming a part of the latter. *Compare* absorption.

**adventitia** (AD-ven-TISH-uh) Loose connective tissue forming the outermost sheath around organs such as a blood vessel or the esophagus.

**aerobic exercise** (air-OH-bic) Exercise in which oxygen is used to produce ATP; endurance exercise.

**aerobic respiration** Oxidation of organic compounds in a reaction series that requires oxygen and produces ATP.

**afferent** (AFF-uh-rent) Carrying toward, as in *afferent neurons,* which carry signals toward the central nervous system, and *afferent arterioles,* which carry blood toward a tissue. *Compare* efferent.

**afterload** The force exerted by arterial blood pressure that opposes the openings of the aortic and pulmonary valves of the heart.

**agglutination** (ah-GLUE-tih-NAY-shun) Clumping of cells or molecules by antibodies. fig. 18.13

**aging** Any changes in the body that occur with the passage of time, including growth, development, and senescence.

**agonist** See prime mover.

**agranulocyte** Either of the two leukocyte types (lymphocytes and monocytes) that lack prominent cytoplasmic granules.

**albumin** (al-BYU-min) A class of small proteins constituting about 60% of the protein fraction of the blood plasma; plays roles in blood viscosity, colloid osmotic pressure, and solute transport.

**aldosterone** (AL-doe-steh-RONE, al-DOSS-teh-rone) A steroid hormone secreted by the adrenal cortex that acts on the kidneys to promote sodium retention and potassium excretion.

**alkalosis** An acid–base imbalance in which the blood pH is higher than 7.45.

**allele** (ah-LEEL) Any of the alternative forms that one gene can take, such as dominant and recessive alleles.

**all-or-none law** The statement that a neuron either produces an action potential of maximum strength if it is depolarized to or above threshold, or produces no action potential at all if the stimulus is not strong enough to reach threshold; there are no action potentials of intermediate strength.

**alveolus** (AL-vee-OH-lus) **1.** A microscopic air sac of the lung. **2.** A gland acinus. **3.** A tooth socket. **4.** Any small anatomical space.

**Alzheimer disease (AD)** (ALTS-hy-mur) A degenerative disease of the senescent brain, typically beginning with memory lapses and progressing to severe losses of mental and motor functions and ultimately death.

**ameboid movement** (ah-ME-boyd) Movement of a cell by means of pseudopods, in a manner similar to that of an ameba; seen in leukocytes and some macrophages.

**amino acids** Small organic molecules with an amino group and a carboxyl group; the monomers of which proteins are composed.

**amino group** A functional group with the formula —$NH_2$, found in amino acids and some other organic molecules.

**ampulla** (AM-pyu-luh) A wide or saclike portion of a tubular organ such as a semicircular duct or uterine tube.

**anabolism** (ah-NAB-oh-lizm) Any metabolic reactions that consume energy and construct more complex molecules with higher free energy from less complex molecules with lower free energy; for example, the synthesis of proteins from amino acids. *Compare* catabolism.

**anaerobic fermentation** (AN-err-OH-bic) A reduction reaction independent of oxygen that converts pyruvic acid to lactic acid and enables glycolysis to continue under anaerobic conditions.

**anaphylactic shock** A severe systemic form of anaphylaxis involving bronchoconstriction, impaired breathing, vasodilation, and a rapid drop in blood pressure with a threat of circulatory failure.

**anaphylaxis** (AN-uh-fih-LAC-sis) A form of immediate hypersensitivity in which an antigen triggers the release of inflammatory chemicals, causing edema, congestion, hives, and other, usually local, signs.

**anastomosis** (ah-NASS-tih-MO-sis) An anatomical convergence, the opposite of a branch; a point where two blood vessels merge and combine their bloodstreams or where two nerves or ducts converge. fig. 20.9

**anatomical position** A reference posture that allows for standardized anatomical terminology. A subject in anatomical position is standing with the feet flat on the floor, arms down to the sides, and the palms and eyes directed forward. fig. A.1

**anatomy 1.** Structure of the body. **2.** The study of structure.

**androgen** (AN-dro-jen) Testosterone or a related steroid hormone. Stimulates somatic changes at puberty in both sexes, adult libido in both sexes, development of male anatomy in the fetus and adolescent, and spermatogenesis.

**anemia** (ah-NEE-me-uh) A deficiency of erythrocytes or hemoglobin.

**aneurysm** (AN-you-rizm) A weak, bulging point in the wall of a heart chamber or blood vessel that presents a threat of hemorrhage.

**angiogenesis** (AN-jee-oh-GEN-eh-sis) The growth of new blood vessels.

**angiotensin II** (AN-jee-oh-TEN-sin) A hormone produced from angiotensinogen (a plasma protein) by the kidneys and lungs; raises blood pressure by stimulating vasoconstriction and stimulating the adrenal cortex to secrete aldosterone.

**anion** (AN-eye-on) An ion with more electrons than protons and consequently a net negative charge.

**antagonist 1.** A muscle that opposes the agonist at a joint. **2.** Any agent, such as a hormone or drug, that opposes another.

**antebrachium** (AN-teh-BRAY-kee-um) The region from elbow to wrist; the forearm.

**anterior** Pertaining to the front (facial–abdominal aspect) of the body; ventral.

**antibody** A protein of the gamma globulin class that reacts with an antigen; found in the blood plasma, in other body fluids, and on the surfaces of certain leukocytes and their derivatives.

**anticoagulant** (AN-tee-co-AG-you-lent) A chemical agent that opposes blood clotting.

**antidiuretic hormone (ADH)** (AN-tee-DYE-you-RET-ic) A hormone released by the posterior lobe of the pituitary gland in response to low blood pressure; promotes water retention by the kidneys. Also known as *vasopressin.*

**antigen** (AN-tih-jen) Any large molecule capable of binding to an antibody and triggering an immune response.

**antigen-presenting cell (APC)** A cell that phagocytizes an antigen and displays fragments of it on its surface for recognition by other cells of the immune system; chiefly macrophages and B lymphocytes.

**antioxidant** A chemical that binds and neutralizes free radicals, minimizing their oxidative damage to a cell; for example, selenium and vitamin E.

**antiport** A cotransport protein that moves two or more solutes in opposite directions through a cellular membrane; for example, the $Na^+$–$K^+$ pump.

**aorta** A large artery that extends from the left ventricle to the lower abdominal cavity and gives rise to all other arteries of the systemic circulation. fig. 20.21

**apical surface** The uppermost surface of an epithelial cell, usually exposed to the lumen of an organ. fig. 3.5

**apocrine** Pertaining to certain sweat glands with large lumens and relatively thick, aromatic secretions and to similar glands such as the mammary gland; formerly thought to form secretions by pinching off bits of apical cytoplasm.

**apoptosis** (AP-oh-TOE-sis) Programmed cell death; the normal death of cells that have completed their function. *Compare* necrosis.

**appendicular** (AP-en-DIC-you-lur) Pertaining to the extremities and their supporting skeletal girdles. fig. 8.1

**arcuate** (AR-cue-et) Making a sharp L- or U-shaped bend (arc), as in the *arcuate arteries* of the kidneys and uterus.

**areolar tissue** (AIR-ee-OH-lur) A fibrous connective tissue with loosely organized, widely spaced fibers and cells and an abundance of fluid-filled space; found under nearly every epithelium, among other places. fig. 5.14

**arrhythmia** (ah-RITH-me-uh) An irregularity in the cardiac rhythm.

**arteriole** (ar-TEER-ee-ole) A small artery that empties into a metarteriole or capillary.

**artery** Any blood vessel that conducts blood away from the heart.

**articular cartilage** A thin layer of hyaline cartilage covering the articular surface of a bone at a synovial joint serving to reduce friction and ease joint movement. fig. 9.5

**articulation** A skeletal joint; any point at which two bones meet; may or may not be movable.

**ascorbic acid** Vitamin C; a dietary antioxidant.

**aspect** A particular view of the body or one of its structures, or a part that faces in a particular direction, such as the anterior aspect.

**atheroma** (ATH-ur-OH-muh) A fatty deposit (plaque) in a blood vessel consisting of lipid, smooth muscle, and macrophages; characteristic of atherosclerosis. fig. 19.22

**atherosclerosis** (ATH-ur-oh-skleh-ROE-sis) A degenerative disease of the blood vessels characterized by the presence of atheromas and often leading to calcification of the vessel wall.

**atrial natriuretic peptide (ANP)** (AY-tree-ul NAY-tree-you-RET-ic) A hormone secreted by the heart that lowers blood pressure by promoting sodium excretion and antagonizing aldosterone.

**atrioventricular (AV) node** (AY-tree-oh-ven-TRIC-you-lur) A group of autorhythmic cells in the interatrial septum of the heart that relays excitation from the atria to the ventricles.

**atrioventricular (AV) valves** The bicuspid (left) and tricuspid (right) valves between the atria and ventricles of the heart.

**atrophy** (AT-ro-fee) Shrinkage of a tissue due to age, disuse, or disease.

**auditory ossicles** Three small middle-ear bones that transfer vibrations from the tympanic membrane to the inner ear; the malleus, incus, and stapes.

**autoantibody** An antibody that fails to distinguish the body's own molecules from foreign molecules and thus attacks host tissues, causing autoimmune diseases.

**autoimmune disease** Any disease in which antibodies fail to distinguish between foreign and self-antigens and attack the body's own tissues; for example, systemic lupus erythematosus and rheumatoid arthritis.

**autolysis** (aw-TOLL-ih-sis) Digestion of cells by their own internal enzymes.

**autonomic nervous system (ANS)** (AW-toe-NOM-ic) A motor division of the nervous system that innervates glands, smooth muscle, and cardiac muscle; consists of sympathetic and parasympathetic divisions and functions largely without voluntary control. *Compare* somatic nervous system.

**autoregulation** The ability of a tissue to adjust its own blood supply through vasomotion or angiogenesis.

**autorhythmic** (AW-toe-RITH-mic) Pertaining to cells that spontaneously produce action potentials at regular time intervals, chiefly cardiac and smooth muscle cells.

**autosome** (AW-toe-some) Any chromosome except the sex chromosomes. Genes on the autosomes are inherited without regard to the sex of the individual.

**axial** (AC-see-ul) Pertaining to the head, neck, and trunk; the part of the body excluding the appendicular portion. fig. 8.1

**axillary** (ACK-sih-LERR-ee) Pertaining to the armpit.

**axon** A process of a neuron that transmits action potentials; also called a *nerve fiber*. There is only one axon to a neuron, and it is usually much longer and much less branched than the dendrites. fig. 12.4

**axoneme** (AC-so-neem) The core of microtubules, usually in a "9 + 2" array, at the center of a cilium or flagellum. fig. 3.11

# B

**baroreceptors** (BARE-oh-re-sep-turz) Pressure sensors located in the heart, aortic arch, and carotid sinuses that trigger autonomic reflexes in response to fluctuations in blood pressure.

**basal metabolic rate (BMR)** The rate of energy consumption of a person who is awake, relaxed, at a comfortable temperature, and has not eaten for 12 to 14 hours; usually expressed as kilocalories per square meter of body surface per hour. *Compare* metabolic rate.

**basal nuclei** Masses of deep cerebral gray matter that play a role in the coordination of posture and movement. fig. 14.16

**base 1.** A chemical that binds protons from solution; a proton acceptor. **2.** Any of the purines or pyrimidines of a nucleic acid (adenine, thymine, guanine, cytosine, or uracil), serving in part to code for protein structure. **3.** The broadest part of a tapered organ such as the uterus or the inferior aspect of an organ such as the brain.

**basement membrane** A thin layer of glycoproteins, collagen, and glycosaminoglycans beneath the deepest cells of an epithelium, serving to bind the epithelium to the underlying tissue. fig. 5.32

**base triplet** A sequence of three DNA nucleotides that codes indirectly (through mRNA) for one amino acid of a protein.

**basophil** (BAY-so-fill) A granulocyte with coarse cytoplasmic granules that produces heparin, histamine, and other chemicals involved in inflammation. table 18.7

**belly** The thick part of a skeletal muscle between its origin and insertion. fig. 10.2

**bicarbonate buffer system** An equilibrium mixture of carbonic acid, bicarbonate ions, and hydrogen ions ($H_2CO_3 \leftrightarrows HCO_3^- + H^+$) that stabilizes the pH of the body fluids.

**bicarbonate ion** An anion, $HCO_3^-$, that functions as a base in the buffering of body fluids.

**bile** A secretion produced by the liver, concentrated and stored in the gallbladder, and released into the small intestine; consists mainly of wastes such as excess cholesterol, salts, and bile pigments but also contains lecithin and bile acids, which aid in fat digestion.

**bile pigment** Strongly colored organic compound produced by the breakdown of hemoglobin; biliverdin and bilirubin.

**bilirubin** (BIL-ih-ROO-bin) A yellow to orange bile pigment produced by the breakdown of hemoglobin and excreted in the bile; causes jaundice and neurotoxic effects if present in excessive concentration.

**biogenic amines** A class of chemical messengers with neurotransmitter and hormonal functions, synthesized from amino acids and retaining an amino group; also called *monoamines*. Examples include epinephrine and thyroxine.

**bipedalism** The habit of walking on two legs; a defining characteristic of the family Hominidae that underlies many skeletal and other characteristics of humans.

**blood–brain barrier (BBB)** A barrier between the bloodstream and nervous tissue of the brain that is impermeable to many blood solutes and thus prevents them from affecting the brain tissue; formed by the tight junctions between capillary endothelial cells, the basement membrane of the endothelium, and the perivascular feet of astrocytes.

**B lymphocyte** A lymphocyte that functions as an antigen-presenting cell and, in humoral immunity, differentiates into an antibody-producing plasma cell; also called a *B cell*.

**body** 1. The entire organism. 2. Part of a cell, such as a neuron, containing the nucleus and most other organelles. 3. The largest or principal part of an organ such as the stomach or uterus; also called the *corpus*.

**bolus** A mass of matter, especially food or feces traveling through the digestive tract.

**bone** 1. A calcified connective tissue; also called *osseous tissue*. 2. An organ of the skeleton composed of osseous tissue, fibrous connective tissue, marrow, cartilage, and other tissues.

**Bowman's capsule** *See* glomerular capsule.

**brachial** (BRAY-kee-ul) Pertaining to the arm proper, the region from shoulder to elbow.

**bradykinin** (BRAD-ee-KY-nin) An oligopeptide produced in inflammation that stimulates vasodilation, increases capillary permeability, and stimulates pain receptors.

**brainstem** The stalklike lower portion of the brain, composed of all of the brain except the cerebrum and cerebellum. (Many authorities also exclude the diencephalon and regard only the medulla oblongata, pons, and midbrain as the brainstem.) fig. 14.8

**bronchiole** (BRON-kee-ole) A pulmonary air passage that is usually 1 mm or less in diameter and lacks cartilage but has relatively abundant smooth muscle, elastic tissue, and a simple cuboidal, usually ciliated epithelium.

**bronchus** (BRON-kus) A relatively large pulmonary air passage with supportive cartilage in the wall; any passage beginning with the primary bronchus at the fork in the trachea and ending with tertiary bronchi, from which air continues into the bronchioles.

**brush border** A fringe of microvilli on the apical surface of an epithelial cell, serving to enhance surface area and promote absorption. fig. 5.6

**buffer** 1. A mixture of chemicals that resists changes in pH when acid or base is added to the solution. 2. A physiological system that contributes to acid–base balance, specifically the respiratory and urinary systems.

**bursa** A sac filled with synovial fluid at a diarthrosis, serving to facilitate muscle or joint action. fig. 9.23

# C

**calcaneal tendon** (cal-CAY-nee-ul) A thick tendon at the heel that attaches the triceps surae muscles to the calcaneus; also called the *Achilles tendon*. fig. 10.38

**calcification** The hardening of a tissue due to the deposition of calcium salts.

**calcitonin** (CAL-sih-TOE-nin) A hormone secreted by C cells of the thyroid gland that promotes calcium deposition in the skeleton and lowers blood calcium concentration.

**calmodulin** An intracellular protein that binds calcium ions and mediates many of the second-messenger effects of calcium.

**calorie** The amount of thermal energy that will raise the temperature of 1 g of water by 1°C. Also called a *small calorie*.

**Calorie** *See* kilocalorie.

**calorigenic** (ca-LOR-ih-JEN-ic) Heat-producing, as in the calorigenic effect of thyroid hormone.

**calsequestrin** (CAL-see-KWES-trin) A protein found in smooth endoplasmic reticulum that reversibly binds and stores calcium ions, rendering calcium chemically unreactive until needed for such processes as muscle contraction.

**calvaria** (cal-VERR-ee-uh) The rounded bony dome that forms the roof of the cranium; the general portion of the skull superior to the eyes and ears; skullcap.

**calyx** (CAY-lix) (plural, *calices*) A cuplike structure, as in the kidneys. fig. 23.4

**canaliculus** (CAN-uh-LIC-you-lus) A microscopic canal, as in osseous tissue. fig. 7.5

**capillary** (CAP-ih-LERR-ee) The narrowest type of vessel in the cardiovascular and lymphatic systems; engages in fluid exchanges with surrounding tissues.

**capillary exchange** The process of fluid transfer between the bloodstream and tissue fluid.

**capsule** The fibrous covering of a structure such as the spleen or a diarthrosis.

**carbohydrate** A hydrophilic organic compound composed of carbon and a 2:1 ratio of hydrogen to oxygen; includes sugars, starches, glycogen, and cellulose.

**carbonic anhydrase** An enzyme found in erythrocytes and kidney tubule cells that catalyzes the decomposition of carbonic acid into carbon dioxide and water or the reverse reaction ($H_2CO_3 \leftrightarrows CO_2 + H_2O$).

**carboxyl group** (car-BOC-sil) An organic functional group with the formula —COOH, found in many organic acids such as amino acids and fatty acids.

**carcinogen** (car-SIN-oh-jen) An agent capable of causing cancer, including certain chemicals, viruses, and ionizing radiation.

**cardiac center** A nucleus in the medulla oblongata that regulates autonomic reflexes for controlling the rate and strength of the heartbeat.

**cardiac cycle** One complete cycle of cardiac systole and diastole.

**cardiac muscle** Striated involuntary muscle of the heart.

**cardiac output (CO)** The amount of blood pumped by each ventricle of the heart in 1 minute.

**cardiac reserve** The difference between maximum and resting cardiac output; determines a person's tolerance for exercise.

**cardiovascular system** An organ system consisting of the heart and blood vessels, serving for the transport of blood. *Compare* circulatory system.

**carotid body** (ca-ROT-id) A small cellular mass immediately superior to the branch in the common carotid artery, containing sensory cells that detect changes in blood pH and carbon dioxide and oxygen content. fig. 20.4

**carotid sinus** A dilation of the common carotid artery at the point where it branches into the internal and external carotids; contains baroreceptors, which monitor changes in blood pressure.

**carpal** Pertaining to the wrist (carpus).

**carrier** 1. A protein in a cellular membrane that performs carrier-mediated transport. 2. A person who is heterozygous for a recessive allele and does not exhibit the associated phenotype, but may transmit this allele to his or her children; for example, a carrier for sickle-cell disease.

**carrier-mediated transport** A process of transporting materials through a cellular membrane that involves reversible binding to a membrane protein.

**cartilage** A connective tissue with a rubbery matrix, cells (chondrocytes) contained in lacunae, and no blood vessels; covers the articular surfaces of many bones and supports organs such as the ear and larynx.

**catabolism** (ca-TAB-oh-lizm) Any metabolic reactions that release energy and break relatively complex molecules with high free energy into less complex molecules with lower free energy; for example, digestion and glycolysis. *Compare* anabolism.

**catalyst** (CAT-uh-list) Any chemical that lowers the activation energy of a chemical reaction and thus makes the reaction proceed more rapidly; a role served in cells by enzymes.

**catecholamine** (CAT-eh-COAL-uh-meen) A subclass of biogenic amines that includes epinephrine, norepinephrine, and dopamine. fig. 12.21

**cation** (CAT-eye-on) An ion with more protons than electrons and consequently a net positive charge.

**caudal** (CAW-dul) 1. Pertaining to a tail or narrow tail-like part of an organ. 2. Pertaining to the inferior part of the trunk of the body, where the tail of other animals arises. *Compare* cranial. 3. Relatively distant from the forehead, especially in reference to structures of the brain and spinal cord; for example, the medulla oblongata is caudal to the pons. *Compare* rostral.

**celiac** (SEE-lee-ac) Pertaining to the abdomen.

**cell** The smallest subdivision of a tissue considered to be alive; consists of a plasma membrane enclosing cytoplasm and, in most cases, a nucleus.

**cellular membrane** Any unit membrane enclosing a cell or organelle. *See also* unit membrane.

**central** Located relatively close to the median axis of the body, as in the central nervous system; opposite of peripheral.

**central nervous system (CNS)** The brain and spinal cord.

**centriole** (SEN-tree-ole) An organelle composed of a short cylinder of nine triplets of microtubules, usually paired with another centriole perpendicular to it; origin of the mitotic spindle; identical to the basal body of a cilium or flagellum. fig. 3.30

**cephalic** (seh-FAL-ic) Pertaining to the head.

**cerebellum** (SERR-eh-BEL-um) A large portion of the brain dorsal to the brainstem and inferior to the cerebrum, responsible for equilibrium, motor coordination, and memory of learned motor skills. fig. 14.10

**cerebrospinal fluid (CSF)** (SERR-eh-bro-SPY-nul, seh-REE-bro-SPY-nul) A liquid that fills the ventricles of the brain, the central canal of the spinal cord, and the space between the CNS and dura mater.

**cerebrovascular accident (CVA)** (SERR-eh-bro-VASS-cue-lur, seh-REE-bro-VASS-cue-lur) The loss of blood flow to any part of the brain due to obstruction or hemorrhage of an artery, leading to the necrosis of nervous tissue; also called *stroke* or *apoplexy*.

**cerebrum** (SERR-eh-brum, seh-REE-brum) The largest and most superior part of the brain, divided into two convoluted cerebral hemispheres separated by a deep longitudinal fissure.

**cervical** (SUR-vih-cul) Pertaining to the neck or any cervix.

**cervix** (SUR-vix) 1. The neck. 2. A narrow or necklike part of an organ such as the uterus and gallbladder. fig. 28.3

**channel protein** A protein in the plasma membrane that has a pore through it for the passage of materials between the cytoplasm and extracellular fluid. fig. 3.8

**chemical bond** A force that attracts one atom to another, such as their opposite charges or the sharing of electrons.

**chemical digestion** Hydrolysis reactions that occur in the digestive tract and convert dietary polymers into monomers that can be absorbed by the small intestine.

**chemical synapse** A meeting of a nerve fiber and another cell with which the neuron communicates by releasing neurotransmitters. fig. 12.20

**chemoreceptor** An organ or cell specialized to detect chemicals, as in the carotid bodies and taste buds.

**chemotaxis** (KEM-oh-TAC-sis) The movement of a cell along a chemical concentration gradient, especially the attraction of neutrophils to chemicals released by pathogens or inflamed tissues.

**chief cell** The majority type of cell in an organ or tissue such as the parathyroid glands or gastric glands.

**choanae** (co-AH-nee) Openings of the nasal cavity into the pharynx; also called *posterior nares*. fig. 22.3

**cholecystokinin (CCK)** (CO-leh-SIS-toe-KY-nin) A polypeptide employed as a hormone and neurotransmitter, secreted by some brain neurons and cells of the digestive tract. fig. 12.21

**cholesterol** (co-LESS-tur-ol) A steroid that functions as part of the plasma membrane and as a precursor for all other steroids in the body.

**cholinergic** (CO-lin-UR-jic) Pertaining to acetylcholine (ACh), as in cholinergic nerve fibers that secrete ACh, cholinergic receptors that bind it, or cholinergic effects on a target organ.

**chondrocyte** (CON-dro-site) A cartilage cell; a former chondroblast that has become enclosed in a lacuna in the cartilage matrix. fig. 5.21

**chorion** (CO-ree-on) A fetal membrane external to the amnion; forms part of the placenta and has diverse functions including fetal nutrition, waste removal, and hormone secretion. fig. 29.7

**chromatid** (CRO-muh-tid) One of two genetically identical rodlike bodies of a metaphase chromosome, joined to its sister chromatid at the centromere. fig. 4.15

**chromatin** (CRO-muh-tin) Filamentous material in the interphase nucleus, composed of DNA and associated proteins.

**chromosome** A complex of DNA and protein carrying the genetic material of a cell's nucleus. Normally there are 46 chromosomes in the nucleus of each cell except germ cells. fig. 4.15

**chronic** **1.** Long-lasting. **2.** Pertaining to a disease that progresses slowly and has a long duration. *Compare* acute.

**chronic bronchitis** A chronic obstructive pulmonary disease characterized by damaged and immobilized respiratory cilia, excessive mucus secretion, infection of the lower respiratory tract, and bronchial inflammation; caused especially by cigarette smoking. *See also* chronic obstructive pulmonary disease.

**chronic obstructive pulmonary disease (COPD)** A group of lung diseases (asthma, chronic bronchitis, and emphysema) that result in long-term obstruction of airflow and substantially reduced pulmonary ventilation; one of the leading causes of death in old age.

**chylomicron** (KY-lo-MY-cron) A protein-coated lipid droplet formed in the small intestine and found in the lymph and blood after a meal; a means of lipid transport in the bloodstream and lymph.

**chyme** (kime) A slurry of partially digested food in the stomach and small intestine.

**cilium** (SIL-ee-um) A hairlike process, with an axoneme, projecting from the apical surface of an epithelial cell; often motile and serving to propel matter across the surface of an epithelium, but sometimes nonmotile and serving sensory roles. fig. 3.11

**circulatory shock** A state of cardiac output inadequate to meet the metabolic needs of the body.

**circulatory system** An organ system consisting of the heart, blood vessels, and blood. *Compare* cardiovascular system.

**circumduction** A joint movement in which one end of an appendage remains relatively stationary and the other end is moved in a circle. fig. 9.16

**cirrhosis** (sih-RO-sis) A degenerative liver disease characterized by replacement of functional parenchyma with fibrous and adipose tissue; causes include alcohol, other poisons, and viral and bacterial inflammation.

**cisterna** (sis-TUR-nuh) A fluid-filled space or sac, such as the cisterna chyli of the lymphatic system and a cisterna of the endoplasmic reticulum or Golgi complex. fig. 3.26

**citric acid cycle** A cyclic reaction series involving several carboxylic acids in the mitochondrial matrix; oxidizes acetyl groups to carbon dioxide while reducing $NAD^+$ to NADH and FADH to $FADH_2$, making these reduced coenzymes available for ATP synthesis. Also called the *Krebs cycle* or *tricarboxylic acid (TCA) cycle*. fig. 26.4

**climacteric** A period in the lives of men and women, usually in the early 50s, marked by changes in the level of reproductive hormones, a variety of somatic and psychological effects, and in women, cessation of ovulation and menstruation (menopause).

**clone** A population of cells that are mitotically descended from the same parent cell and are identical to each other genetically or in other respects.

**coagulation** (co-AG-you-LAY-shun) The clotting of blood, lymph, tissue fluid, or semen.

**codominant** (co-DOM-ih-nent) A condition in which neither of two alleles is dominant over the other, and both are phenotypically expressed when both are present in an individual; for example, blood type alleles $I^A$ and $I^B$ produce blood type AB when inherited together.

**codon** A series of three nucleotides in mRNA that codes for one amino acid in a protein or, as a *stop* codon, signals the end of a gene.

**coenzyme** (co-EN-zime) A small organic molecule, usually derived from a vitamin, that is needed to make an enzyme catalytically active; acts by accepting electrons from an enzymatic reaction and transferring them to a different reaction chain.

**cofactor** A nonprotein such as a metal ion or coenzyme needed for an enzyme to function.

**cohesion** The clinging of identical molecules such as water to each other.

**collagen** (COLL-uh-jen) The most abundant protein in the body, forming the fibers of many connective tissues in places such as the dermis, tendons, and bones.

**colloid** An aqueous mixture of particles that are too large to pass through most selectively permeable membranes but small enough to remain evenly dispersed through the solvent by the thermal motion of solvent particles; for example, the proteins in blood plasma.

**colloid osmotic pressure (COP)** A portion of the osmotic pressure of a body fluid that is due to its protein. *Compare* oncotic pressure.

**colostrum** (co-LOS-trum) A watery, low-fat secretion of the mammary gland that nourishes and immunizes an infant for the first 2 to 3 days postpartum, until true milk is secreted.

**columnar** A cellular shape that is roughly significantly taller than it is wide. fig. 5.6

**commissure** (COM-ih-shur) **1.** A bundle of nerve fibers that crosses from one side of the brain or spinal cord to the other. fig. 14.2 **2.** A corner or angle at which the eyelids, lips, or genital labia meet; in the eye, also called the *canthus*. fig. 16.22

**complement** **1.** To complete or enhance the structure or function of something else, as in the coordinated action of two different hormones. **2.** A system of plasma proteins involved in nonspecific defense against pathogens.

**computerized tomography (CT)** A method of medical imaging that uses X rays and a computer to create an image of a thin section of the body; also called a *CT scan*.

**concentration gradient** A difference in chemical concentration from one point to another, as on two sides of a plasma membrane.

**conception** The fertilization of an egg, producing a zygote.

**conceptus** All products of conception, ranging from a fertilized egg to the full-term fetus with its embryonic membranes, placenta, and umbilical cord. *Compare* embryo, fetus, preembryo.

**condyle** (CON-dile) A rounded knob on a bone serving to produce smooth motion at a joint. fig. 8.2

**conformation** The three-dimensional structure of a protein that results from interaction among its amino acid side groups, its interactions with water, and the formation of disulfide bonds.

**congenital** Present at birth; for example, an anatomical defect, a syphilis infection, or a hereditary disease.

**conjugated** A state in which one organic compound is bound to another compound of a different class, such as a protein conjugated with a carbohydrate to form a glycoprotein.

**connective tissue** A tissue usually composed of more extracellular than cellular volume and usually with a substantial amount of extracellular fiber; forms supportive frameworks and capsules for organs, binds structures together, holds them in place, stores energy (as in adipose tissue), or transports materials (as in blood).

**contractility** 1. The ability to shorten. 2. The amount of force that a contracting muscle fiber generates for a given stimulus; may be increased by epinephrine, for example, while stimulus strength remains constant.

**contralateral** On opposite sides of the body, as in reflex arcs where the stimulus comes from one side of the body and a response is given by muscles on the other side. *Compare* ipsilateral.

**convergent** Coming together, as in a convergent muscle and a converging neuronal circuit.

**cooperative effect** Effect in which two hormones, or both divisions of the autonomic nervous system, work together to produce a single overall result.

**cornified** Having a heavy deposit of keratin, as in the stratum corneum of the epidermis.

**corona** A halo- or crownlike structure, as in the corona radiata or the coronal suture of the skull.

**coronal plane** *See* frontal plane.

**corona radiata** 1. An array of nerve tracts in the brain that arise mainly from the thalamus and fan out to different regions of the cerebral cortex. 2. The first layer of cuboidal cells immediately external to the zona pellucida around an egg cell.

**coronary circulation** A system of blood vessels that serve the wall of the heart. fig. 19.11

**corpus** Body or mass; the main part of an organ, as opposed to such regions as a head, tail, or cervix.

**corpus callosum** (COR-pus ca-LO-sum) A prominent C-shaped band of nerve tracts that connect the right and left cerebral hemispheres to each other, seen superior to the third ventricle in a median section of the brain. fig. 14.2

**corpus luteum** (LOO-tee-um) A yellowish cellular mass that forms in the ovary from a follicle that has ovulated; secretes progesterone, hormonally regulates the second half of the menstrual cycle, and is essential to sustaining the first 7 weeks of pregnancy.

**cortex** (plural, *cortices*) The outer layer of some organs such as the adrenal gland, cerebrum, lymph node, and ovary; usually covers or encloses tissue called the medulla.

**corticosteroid** (COR-tih-co-STERR-oyd) Any steroid hormone secreted by the adrenal cortex, such as aldosterone, cortisol, and sex steroids.

**costal** (COSS-tul) Pertaining to the ribs.

**costal cartilage** A bladelike plate of hyaline cartilage that attaches the distal end of a rib to the sternum.

**cotransport** A form of carrier-mediated transport in which a membrane protein transports two solutes simultaneously or within the same cycle of action; for example the sodium–glucose transport protein and the $Na^+$–$K^+$ pump.

**countercurrent** A situation in which two fluids flow side by side in opposite directions, as in the countercurrent multiplier of the kidney and the countercurrent heat exchanger of the scrotum.

**cranial** (CRAY-nee-ul) 1. Pertaining to the cranium. 2. In a position relatively close to the head or a direction toward the head. *Compare* caudal.

**cranial nerve** Any of 12 pairs of nerves connected to the base of the brain and passing through foramina of the cranium.

**creatine phosphate (CP)** (CREE-uh-teen FOSS-fate) An energy-storage molecule in muscle tissue that donates a phosphate group to ADP and thus regenerates ATP in periods of hypoxia.

**crista** A crestlike structure, such as the crista galli of the ethmoid bone or the crista of a mitochondrion.

**cross section** A cut perpendicular to the long axis of the body or an organ.

**crural** (CROO-rul) Pertaining to the leg proper or to the crus of a organ. *See* crus.

**crus** (cruss) (plural, *crura*) 1. The leg proper; the region from the knee to the ankle. 2. A leglike extension of an organ such as the penis and clitoris. figs. 27.11b, 28.8

**cuboidal** (cue-BOY-dul) A cellular shape that is roughly like a cube or in which the height and width are about equal; typically looks squarish in tissue sections. fig. 5.5

**cuneiform** (cue-NEE-ih-form) Wedge-shaped, as in the cuneiform cartilages of the larynx and cuneiform bone of the wrist.

**current** A moving stream of charged particles such as ions or electrons.

**cusp** 1. One of the flaps of a valve of the heart, veins, and lymphatic vessels. 2. A conical projection on the occlusal surface of a premolar or molar tooth.

**cutaneous** (cue-TAY-nee-us) Pertaining to the skin.

**cyanosis** (SY-uh-NO-sis) A bluish color of the skin and mucous membranes due to ischemia or hypoxemia.

**cyclic adenosine monophosphate (cAMP)** A cyclic molecule produced from ATP by the action of adenylate cyclase; serves as a second messenger in many hormone and neurotransmitter actions. fig. 2.29b

**cyclooxygenase** An enzyme that converts arachidonic acid to prostacyclin, prostaglandins, and thromboxanes.

**cytochromes** Enzymes on the mitochondrial cristae that transfer electrons in the final reaction chain of aerobic respiration.

**cytokinesis** (SY-toe-kih-NEE-sis) Division of the cytoplasm of a cell into two cells following nuclear division.

**cytology** The study of cell structure and function.

**cytolysis** (sy-TOL-ih-sis) The rupture and destruction of a cell by such agents as complement proteins and hypotonic solutions.

**cytoplasm** The contents of a cell between its plasma membrane and its nuclear envelope, consisting of cytosol, organelles, inclusions, and the cytoskeleton.

**cytosine** A single-ringed nitrogenous base (pyrimidine) found in DNA; one of the four bases of the genetic code; complementary to guanine in the double helix of DNA. fig. 4.2

**cytoskeleton** A system of protein microfilaments, intermediate filaments, and microtubules in a cell, serving in physical support, cellular movement, and the routing of molecules and organelles to their destinations within the cell. fig. 3.31

**cytosol** A clear, featureless, gelatinous colloid in which the organelles and other internal structures of a cell are embedded.

**cytotoxic T cell** A T lymphocyte that directly attacks and destroys infected body cells, cancerous cells, and the cells of transplanted tissues.

# D

**daughter cells** Cells that arise from a parent cell by mitosis or meiosis.

**deamination** (dee-AM-ih-NAY-shun) Removal of an amino group from an organic molecule; a step in the catabolism of amino acids.

**decomposition reaction** A chemical reaction in which a larger molecule is broken down into smaller ones. *Compare* synthesis reaction.

**decussation** (DEE-cuh-SAY-shun) The crossing of nerve fibers from the right side of the central nervous system to the left or vice versa, especially in the spinal cord, medulla oblongata, and optic chiasm.

**deep** Relatively far from the body surface; opposite of *superficial*. For example, the bones are deep to the skeletal muscles.

**degranulation** Exocytosis and disappearance of cytoplasmic granules, especially in platelets and granulocytes.

**dehydration synthesis** A reaction in which two chemical monomers are joined together with water produced as a by-product; also called a *condensation reaction*. *Compare* hydrolysis.

**denaturation** A change in the three-dimensional conformation of a protein that destroys its enzymatic or other functional properties, usually caused by extremes of temperature or pH.

**dendrite** Process of a neuron that receives information from other cells or from environmental stimuli and conducts signals to the soma. Dendrites are usually shorter, more branched, and more numerous than the axon and are incapable of producing action potentials. fig. 12.4

**dendritic cell** An antigen-presenting cell of the epidermis and mucous membranes. fig. 6.2a

**denervation atrophy** The shrinkage of skeletal muscle that occurs when its motor neuron dies or is severed from the muscle.

**dense connective tissue** A connective tissue with a high density of fiber, relatively little ground substance, and scanty cells; seen in tendons and the dermis, for example.

**deoxyribonucleic acid (DNA)** (dee-OCK-see-RY-bo-new-CLAY-ic) A very large nucleotide polymer that carries the genes of a cell; composed of a double helix of intertwined chains of deoxyribose and phosphate, with complementary pairs of nitrogenous bases facing each other between the helices. fig. 4.3

**depolarization** A shift in the electrical potential across a plasma membrane toward 0 mV, associated with excitation of a nerve or muscle cell. *Compare* hyperpolarization.

**dermal papilla** Bump or ridge of dermis that extends upward to interdigitate with the epidermis and create a wavy boundary that resists stress and slippage of the epidermis.

**dermis** The deeper of the two layers of the skin, underlying the epidermis and composed of fibrous connective tissue.

**desmosome** (DEZ-mo-some) A patchlike intercellular junction that mechanically links two cells together. fig. 5.28

**dextrose** An isomer of glucose; the only form of glucose with a normal role in physiology.

**diabetes** (DY-uh-BEE-teez) Any disease characterized by chronic polyuria of metabolic origin; diabetes mellitus unless otherwise specified.

**diabetes insipidus** (in-SIP-ih-dus) A form of diabetes that results from hyposecretion of antidiuretic hormone; unlike other forms, it is not characterized by hyperglycemia or glycosuria.

**diabetes mellitus (DM)** (mel-EYE-tus) A form of diabetes that results from hyposecretion of insulin or from a deficient target cell response to it; signs include hyperglycemia and glycosuria.

**dialysis** (dy-AL-ih-sis) 1. The separation of some solute particles from others by diffusion through a selectively permeable membrane. 2. Hemodialysis, the process of separating wastes from the bloodstream and sometimes adding other substances to it (such as drugs and nutrients) by circulating the blood through a machine with a selectively permeable membrane, used to treat cases of renal or hepatic insufficiency.

**diapedesis** (DY-uh-peh-DEE-sis) Migration of formed elements of the blood through a capillary or venule wall into the interstitial space. fig. 21.18

**diaphysis** (dy-AFF-ih-sis) The shaft of a long bone. fig. 7.2

**diarthrosis** (DY-ar-THRO-sis) A freely movable synovial joint such as the knuckle, elbow, shoulder, or knee.

**diastole** (dy-ASS-tuh-lee) A period in which a heart chamber relaxes and fills with blood; especially ventricular relaxation.

**diencephalon** (DY-en-SEFF-uh-lon) A portion of the brain between the midbrain and corpus callosum; composed of the thalamus, epithalamus, and hypothalamus. fig. 14.12

**differentiation** Development of a relatively unspecialized cell or tissue into one with a more specific structure and function.

**diffusion** Spontaneous net movement of particles from a place of high concentration to a place of low concentration.

**dilation** (dy-LAY-shun) Widening of an organ or passageway such as a blood vessel or the pupil of the eye.

**diploid (2n)** Pertaining to a cell or organism with chromosomes in homologous pairs.

**disaccharide** (dy-SAC-uh-ride) A carbohydrate composed of two simple sugars (monosaccharides) joined by a glycosidic bond; for example, lactose, sucrose, and maltose. fig. 2.17

**disseminated intravascular coagulation (DIC)** Widespread clotting of the blood within unbroken vessels, leading to hemorrhaging, congestion of the vessels with clotted blood, and ischemia and necrosis of organs.

**distal** Relatively distant from a point of origin or attachment; for example, the wrist is distal to the elbow. *Compare* proximal.

**disulfide bond** A covalent bond between the sulfur atoms of two cysteine residues, serving to link one polypeptide chain to another or to hold a single chain in its three-dimensional conformation.

**diuretic** (DY-you-RET-ic) A chemical that increases urine output.

**dizygotic (DZ) twins** Two individuals who developed simultaneously in one uterus but originated from separate fertilized eggs and therefore are not genetically identical.

**dominant** 1. Pertaining to a genetic allele that is phenotypically expressed in the presence of any other allele. 2. Pertaining to a trait that results from a dominant allele.

**dopamine** (DOE-puh-meen) An inhibitory catecholamine neurotransmitter of the central nervous system, especially of the basal nuclei, where it acts to suppress unwanted motor activity. fig 12.21

**dorsal** Toward the back (spinal) side of the body.

**dorsal root** A branch of a spinal nerve that enters the spinal cord on its dorsal side, composed of sensory fibers. fig. 13.2b

**dorsiflexion** (DOR-sih-FLEC-shun) A movement of the ankle that reduces the joint angle and raises the toes. fig. 9.22

**Down syndrome** *See* trisomy-21.

**duodenum** (DEW-oh-DEE-num, dew-ODD-eh-num) The first portion of the small intestine extending for about 25 cm from the pyloric valve of the stomach to a sharp bend called the duodenojejunal flexure; receives chyme from the stomach and secretions from the liver and pancreas. fig. 25.24

**dynamic equilibrium 1.** A state of continual change that is controlled within narrow limits, as in homeostasis and chemical equilibrium. **2.** The sense of motion or acceleration of the body.

**dynein** (DINE-een) A motor protein involved in the beating of cilia and flagella and in the movement of molecules and organelles within cells, as in retrograde transport in a nerve fiber.

# E

**ectoderm** The outermost of the three primary germ layers of an embryo; gives rise to the nervous system and epidermis.

**ectopic** (ec-TOP-ic) In an abnormal location; for example, ectopic pregnancy and ectopic pacemakers of the heart.

**edema** (eh-DEE-muh) Abnormal accumulation of tissue fluid resulting in swelling of the tissue.

**effector** A molecule, cell, or organ that carries out a response to a stimulus.

**efferent** (EFF-ur-ent) Carrying away or out, such as a blood vessel that carries blood away from a tissue or a nerve fiber that conducts signals away from the central nervous system. *Compare* afferent.

**eicosanoid** (eye-CO-sah-noyd) Twenty-carbon derivative of arachidonic acid that functions as an intercellular messenger; includes prostaglandins, prostacyclin, leukotrienes, and thromboxanes.

**elastic fiber** A connective tissue fiber, composed of the protein elastin, that stretches under tension and returns to its original length when released; responsible for the resilience of organs such as the skin and lungs.

**elasticity** The tendency of a stretched structure to return to its original dimensions when tension is released.

**electrical synapse** A gap junction that enables one cell to stimulate another directly, without the intermediary action of a neurotransmitter; such synapses connect the cells of cardiac muscle and single-unit smooth muscle.

**electrochemical gradient** A difference in ion concentration from one point to another (especially across a plasma membrane) resulting in a gradient of both chemical concentration and electrical charge.

**electrolyte** A salt that ionizes in water and produces a solution that conducts electricity; loosely speaking, any ion that results from the dissociation of such salts, such as sodium, potassium, calcium, chloride, and bicarbonate ions.

**elevation** A joint movement that raises a body part, as in hunching the shoulders or closing the mouth.

**embolism** (EM-bo-lizm) The obstruction of a blood vessel by an embolus.

**embolus** (EM-bo-lus) Any abnormal traveling object in the bloodstream, such as agglutinated bacteria or blood cells, a blood clot, or an air bubble.

**embryo** A developing individual from the sixteenth day of gestation when the three primary germ layers have formed, through the end of the eighth week when all of the organ systems are present. *Compare* conceptus, fetus, preembryo.

**emphysema** (EM-fih-SEE-muh) A degenerative lung disease characterized by a breakdown of alveoli and diminishing surface area available for gas exchange; occurs with aging of the lungs but is greatly accelerated by smoking or air pollution.

**emulsion** A suspension of one liquid in another, such as oil in water or fat in the lymph.

**endocrine gland** (EN-doe-crin) A ductless gland that secretes hormones into the bloodstream; for example, the thyroid and adrenal glands. *Compare* exocrine gland.

**endocytosis** (EN-doe-sy-TOE-sis) Any process in which a cell forms vesicles from its plasma membrane and takes in large particles, molecules, or droplets of extracellular fluid; for example, phagocytosis and pinocytosis.

**endoderm** The innermost of the three primary germ layers of an embryo; gives rise to the mucosae of the digestive and respiratory tracts and to their associated glands.

**endogenous** (en-DODJ-eh-nus) Originating internally, such as the endogenous cholesterol synthesized in the body in contrast to the exogenous cholesterol coming from the diet. *Compare* exogenous.

**endometrium** (EN-doe-MEE-tree-um) The mucosa of the uterus; the site of implantation and source of menstrual discharge.

**endoplasmic reticulum (ER)** (EN-doe-PLAZ-mic reh-TIC-you-lum) An extensive system of interconnected cytoplasmic tubules or channels; classified as rough ER or smooth ER depending on the presence or absence of ribosomes on its membrane. fig. 3.26

**endothelium** (EN-doe-THEEL-ee-um) A simple squamous epithelium that lines the lumens of the blood vessels, heart, and lymphatic vessels.

**endurance exercise** A form of physical exercise, such as running or swimming, that promotes cardiopulmonary efficiency and fatigue resistance more than muscular strength. *Compare* resistance exercise.

**enteric** (en-TERR-ic) Pertaining to the small intestine, as in enteric hormones.

**enzyme** A protein that functions as a catalyst.

**enzyme amplification** A series of chemical reactions in which the product of one step is an enzyme that produces an even greater number of product molecules at the next step, resulting in a rapidly increasing amount of reaction product. Seen in hormone action and blood clotting, for example.

**eosinophil** (EE-oh-SIN-oh-fill) A granulocyte with a large, often bilobed nucleus and coarse cytoplasmic granules that stain with eosin; phagocytizes antigen–antibody complexes, allergens, and inflammatory chemicals and secretes enzymes that combat parasitic infections. table 18.7

**epidermis** A stratified squamous epithelium that constitutes the superficial layer of the skin overlying the dermis. fig. 6.2

**epinephrine** (EP-ih-NEFF-rin) A catecholamine that functions as a neurotransmitter in the sympathetic nervous system and as a hormone secreted by the adrenal medulla; also called *adrenaline.* fig. 12.21

**epiphyseal plate** (EP-ih-FIZZ-ee-ul) A plate of hyaline cartilage between the epiphysis and diaphysis of a long bone in a child or adolescent, serving as a growth zone for bone elongation. figs. 7.10, 7.12.

**epiphysis** (eh-PIF-ih-sis) **1.** The head of a long bone. fig. 7.2 **2.** The pineal gland (epiphysis cerebri).

**epithelium** A type of tissue consisting of one or more layers of closely adhering cells with little intercellular material and no blood vessels; forms the coverings and linings of many organs and the parenchyma of the glands.

**erectile tissue** A tissue that functions by swelling with blood, as in the penis and clitoris and inferior concha of the nasal cavity.

**erythema** (ERR-ih-THEE-muh) Abnormal redness of the skin due to such causes as burns, inflammation, and vasodilation.

**erythrocyte** (eh-RITH-ro-site) A red blood cell.

**erythropoiesis** (eh-RITH-ro-poy-EE-sis) The production of erythrocytes.

**erythropoietin** (eh-RITH-ro-POY-eh-tin) A hormone that is secreted by the kidneys and liver in response to hypoxemia and stimulates erythropoiesis.

**estrogens** (ESS-tro-jenz) A family of steroid hormones known especially for producing female secondary sex characteristics and regulating various aspects of the menstrual cycle and pregnancy; major forms are estradiol, estriol, and estrone.

**evolution** A change in the relative frequencies of alleles in a population over a period of time; the mechanism that produces adaptations in human form and function. *See also* adaptation.

**excitability** The ability of a cell to respond to a stimulus, especially the ability of nerve and muscle cells to produce membrane voltage changes in response to stimuli; irritability.

**excitation–contraction coupling** Events that link the synaptic stimulation of a muscle cell to the onset of contraction.

**excitatory postsynaptic potential (EPSP)** A partial depolarization of a postsynaptic neuron or muscle cell in response to a neurotransmitter, making it more likely to reach threshold and produce an action potential.

**excretion** The process of eliminating metabolic waste products from a cell or from the body. *Compare* secretion.

**exocrine gland** (EC-so-crin) A gland that secretes its products into another organ or onto the body surface, usually by way of a duct; for example, salivary and gastric glands. *Compare* endocrine gland.

**exocytosis** (EC-so-sy-TOE-sis) A process in which a vesicle in the cytoplasm of a cell fuses with the plasma membrane and releases its contents from the cell; used in the elimination of cellular wastes and in the release of gland products and neurotransmitters.

**exogenous** (ec-SODJ-eh-nus) Originating externally, such as exogenous (dietary) cholesterol; extrinsic. *Compare* endogenous.

**expiration 1.** Exhaling. **2.** Dying.

**extension** Movement of a joint that increases the angle between articulating bones (straightens the joint). *Compare* flexion. fig. 9.12

**extracellular fluid (ECF)** Any body fluid that is not contained in the cells; for example, blood, lymph, and tissue fluid.

**extrinsic** (ec-STRIN-sic) **1.** Originating externally, such as extrinsic blood-clotting factors; exogenous. **2.** Not fully contained within an organ but acting on it, such as the extrinsic muscles of the hand and eye. *Compare* intrinsic.

**exude** (ec-SUDE) To seep out, such as fluid filtering from blood capillaries.

# F

**facilitated diffusion** The process of transporting a chemical through a cellular membrane, down its concentration gradient, with the aid of a carrier that does not consume ATP; enables substances to diffuse through the membrane that would do so poorly, or not at all, without a carrier.

**facilitation** Making a process more likely to occur, such as the firing of a neuron, or making it occur more easily or rapidly, as in facilitated diffusion.

**fallopian tube** See uterine tube.

**fascia** (FASH-oo-uh) A layer of connective tissue between the muscles (deep fascia) or separating the muscles from the skin (superficial fascia). fig. 10.1

**fascicle** (FASS-ih-cul) A bundle of muscle or nerve fibers ensheathed in connective tissue; multiple fascicles bound together constitute a muscle or nerve as a whole. figs. 10.1, 13.8

**fat 1.** A triglyceride molecule. **2.** Adipose tissue.

**fatty acid** An organic molecule composed of a chain of an even number of carbon atoms with a carboxyl group at one end and a methyl group at the other; one of the structural subunits of triglycerides and phospholipids.

**fenestrated** (FEN-eh-stray-ted) Perforated with holes or slits, as in fenestrated blood capillaries and the elastic sheets of large arteries. fig. 20.6

**fetus** In human development, an individual from the beginning of the ninth week when all of the organ systems are present, through the time of birth. *Compare* conceptus, embryo, preembryo.

**fibrin** (FY-brin) A sticky fibrous protein formed from fibrinogen in blood, tissue fluid, lymph, and semen; forms the matrix of a blood clot.

**fibroblast** A connective tissue cell that produces collagen fibers and ground substance; the only type of cell in tendons and ligaments.

**fibrosis** Replacement of damaged tissue with fibrous scar tissue rather than by the original tissue type; scarring. *Compare* regeneration.

**fibrous connective tissue** Any connective tissue with a preponderance of fiber, such as areolar, reticular, dense regular, and dense irregular connective tissues.

**filtrate** A fluid formed by filtration, as at the renal glomerulus and other capillaries.

**filtration** A process in which hydrostatic pressure forces a fluid through a selectively permeable membrane (especially a capillary wall).

**fire** To produce an action potential, as in nerve and muscle cells.

**fix 1.** To hold a structure in place, for example, by fixator muscles that prevent unwanted joint movements. **2.** To preserve a tissue by means of a fixative.

**fixative** A chemical that preserves tissues from decay, such as formalin.

**flagellum** (fla-JEL-um) A long, motile, usually single hairlike extension of a cell; the tail of a sperm cell is the only functional flagellum in humans. fig. 27.17

**flexion** A joint movement that, in most cases, decreases the angle between two bones. *Compare* extension. fig. 9.12

**fluid balance** *See* water balance.

**fluid compartment** Any of the major categories of fluid in the body, separated by selectively permeable membranes and differing from each other in chemical composition. Primary examples are the intracellular fluid, tissue fluid, blood, and lymph.

**fluid-mosaic model** The current theory of the structure of a plasma membrane, depicting it as a bilayer of phospholipids and cholesterol with embedded proteins, many of which are able to move about in the lipid film. fig. 3.6

**follicle** (FOLL-ih-cul) **1.** A small space, such as a hair follicle, thyroid follicle, or ovarian follicle. **2.** An aggregation of lymphocytes in a lymphatic organ or mucous membrane.

**follicle-stimulating hormone (FSH)** A hormone secreted by the anterior pituitary gland that stimulates development of the ovarian follicles and egg cells.

**foramen** (fo-RAY-men) A hole through a bone or other organ, in many cases providing passage for blood vessels and nerves.

**formed element** An erythrocyte, leukocyte, or platelet; any cellular component of blood or lymph as opposed to the extracellular fluid component.

**fossa** (FOSS-uh) A depression in an organ or tissue, such as the fossa ovalis of the heart or a cranial fossa of the skull.

**fovea** (FOE-vee-uh) A small pit, such as the fovea capitis of the femur or fovea centralis of the retina.

**free energy** The potential energy in a chemical that is available to do work.

**free radical** A particle derived from an atom or molecule, having an unpaired electron that makes it highly reactive and destructive to cells; produced by intrinsic processes such as aerobic respiration and by extrinsic agents such as chemicals and ionizing radiation.

**frontal plane** An anatomical plane that passes through the body or an organ from right to left and superior to inferior; also called a *coronal plane.* fig. A.3

**functional group** A group of atoms, such as a carboxyl or amino group, that determines the functional characteristics of an organic molecule.

**fundus** The base, the broadest part, or the part farthest from the opening of certain viscera such as the stomach and uterus.

**fusiform** (FEW-zih-form) Spindle-shaped; elongated, thick in the middle, and tapered at both ends, such as the shape of a smooth muscle cell or a muscle spindle.

# G

**gamete** (GAM-eet) An egg or sperm cell.

**gametogenesis** (GAM-eh-toe-JEN-eh-sis) The production of eggs or sperm.

**gamma-(γ-)aminobutyric acid (GABA)** (ah-MEE-no-byu-TIRR-ic) An inhibitory neurotransmitter of the central nervous system in the biogenic amine class. fig. 12.21

**gamma (γ) globulins** (GLOB-you-lins) A class of relatively large proteins found in the blood plasma and on the surfaces of immune cells, functioning as antibodies. *See also* globulin.

**ganglion** (GANG-glee-un) A cluster of nerve cell bodies in the peripheral nervous system, often resembling a knot in a string.

**gangrene** Tissue necrosis resulting from ischemia.

**gap junction** A junction between two cells consisting of a pore surrounded by a ring of proteins in the plasma membrane of each cell, allowing solutes to diffuse from the cytoplasm of one cell to the next; functions include cell-to-cell nutrient transfer in the developing embryo and electrical communication between cells of cardiac and smooth muscle. *See also* electrical synapse. fig. 5.28

**gastric** Pertaining to the stomach.

**gate** A protein channel in a cellular membrane that can open or close in response to chemical, electrical, or mechanical stimuli, thus controlling when substances are allowed to pass through the membrane.

**gene** A segment of DNA that codes for the synthesis of one protein.

**gene locus** The site on a chromosome where a given gene is located.

**generator potential** A graded, reversible rise in the local voltage across the plasma membrane of a nerve or muscle cell in response to a stimulus; triggers an action potential if it reaches threshold.

**genetic engineering** Any of several techniques that alter the genetic constitution of a cell or organism, including recombinant DNA technology and gene substitution therapy.

**genome** (JEE-nome) All the genes of one individual, estimated at 35,000 genes in humans.

**genotype** (JEE-no-type) The pair of alleles possessed by an individual at one gene locus on a pair of homologous chromosomes; strongly influences the individual's phenotype for a given trait.

**germ cell** A gamete or any precursor cell destined to become a gamete.

**germ layer** Any of first three tissue layers of an embryo: ectoderm, mesoderm, or endoderm.

**gestation** (jess-TAY-shun) Pregnancy.

**gland** Any organ specialized to produce a secretion; in some cases a single cell, such as a goblet cell.

**glaucoma** (glaw-CO-muh) A visual disease in which an excessive amount of aqueous humor accumulates and creates pressure that is transmitted through the lens and vitreous body to the retina; pressure on the blood vessels of the choroid causes ischemia, retinal necrosis, and blindness.

**globulin** (GLOB-you-lin) A globular protein such as an enzyme, antibody, or albumin; especially a family of proteins in the blood plasma that includes albumin, antibodies, fibrinogen, and prothrombin.

**glomerular capsule** (glo-MERR-you-lur) A double-walled capsule around each glomerulus of the kidney; receives glomerular filtrate and empties into the proximal convoluted tubule. Also called *Bowman's capsule.* fig. 23.7

**glomerulus** **1.** A spheroid mass of blood capillaries in the kidney that filters plasma and produces glomerular filtrate, which is further processed to form the urine. fig. 23.7 **2.** A spheroidal mass of nerve endings in the olfactory bulb where olfactory neurons from the nose synapse with mitral and dendritic cells of the bulb. fig. 16.7

**glucagon** (GLUE-ca-gon) A hormone secreted by alpha cells of the pancreatic islets in response to hypoglycemia; promotes glycogenolysis and other effects that raise blood glucose concentration.

**glucocorticoid** (GLUE-co-COR-tih-coyd) Any hormone of the adrenal cortex that affects carbohydrate, fat, and protein metabolism; chiefly cortisol and corticosterone.

**gluconeogenesis** (GLUE-co-NEE-oh-JEN-eh-sis) The synthesis of glucose from noncarbohydrates such as fats and amino acids.

**glucose** A monosaccharide ($C_6H_{12}O_6$) also known as blood sugar; glycogen, starch, cellulose, and maltose are made entirely of glucose, and glucose constitutes half of a sucrose or lactose molecule. The isomer involved in human physiology is also called *dextrose.*

**glucose-sparing effect** An effect of fats or other energy substrates in which they are used as fuel by most cells, so that those cells do not consume glucose; this makes more glucose available to cells such as neurons that cannot use alternative energy substrates.

**glycerol** (GLISS-er-ol) A viscous three-carbon alcohol that forms the structural backbone of triglyceride and phospholipid molecules; also called *glycerin.*

**glycocalyx** (GLY-co-CAY-licks) A layer of carbohydrate molecules covalently bonded to the phospholipid and protein molecules of a plasma membrane; forms a surface coat on all human cells.

**glycogen** (GLY-co-jen) A glucose polymer synthesized by liver, muscle, uterine, and vaginal cells that serves as an energy-storage polysaccharide.

**glycogenesis** (GLY-co-JEN-eh-sis) The synthesis of glycogen.

**glycogenolysis** (GLY-co-jeh-NOLL-ih-sis) The hydrolysis of glycogen, releasing glucose.

**glycolipid** (GLY-co-LIP-id) A phospholipid molecule with a carbohydrate covalently bonded to it, found in the plasma membranes of cells.

**glycolysis** (gly-COLL-ih-sis) A series of anaerobic oxidation reactions that break a glucose molecule into two molecules of pyruvic acid and produce a small amount of ATP.

**glycoprotein** (GLY-co-PRO-teen) A protein molecule with a smaller carbohydrate covalently bonded to it; found in mucus and the glycocalyx of cells, for example.

**glycosaminoglycan (GAG)** (GLY-cose-am-ih-no-GLY-can) A polysaccharide composed of modified sugars with amino groups; the major component of a proteoglycan. GAGs are largely responsible for the viscous consistency of tissue gel and the stiffness of cartilage.

**glycosuria** (GLY-co-SOOR-ee-uh) The presence of glucose in the urine, typically indicative of a kidney disease, diabetes mellitus, or other endocrine disorder.

**goblet cell** A mucus-secreting gland cell, shaped somewhat like a wineglass, found in the epithelia of many mucous membranes. fig. 5.32

**Golgi complex** (GOAL-jee) An organelle composed of several parallel cisternae, somewhat like a stack of saucers, that modifies and packages newly synthesized proteins and synthesizes carbohydrates. fig. 3.27

**Golgi vesicle** A membrane bounded vesicle pinched from the Golgi complex, containing its chemical product; may be retained in the cell as a lysosome or become a secretory vesicle that releases the product by exocytosis.

**gonad** The ovary or testis.

**gonadotropin** (go-NAD-oh-TRO-pin) A pituitary hormone that stimulates the gonads; specifically FSH and LH.

**G protein** A protein of the plasma membrane that is activated by a membrane receptor and, in turn, opens an ion channel or activates an intracellular physiological response; important in linking ligand–receptor binding to second-messenger systems.

**graded potential** A variable change in voltage across a plasma membrane, as opposed to the all-or-none quality of an action potential.

**gradient** A difference or change in any variable, such as pressure or chemical concentration, from one point in space to another; provides a basis for molecular movements such as gas exchange, osmosis, and facilitated diffusion, and for bulk movements such as blood flow and airflow.

**granulocyte** (GRAN-you-lo-site) Any of three types of leukocytes (neutrophils, eosinophils, or basophils) with prominent cytoplasmic granules.

**granulosa cells** Cells that form a stratified cuboidal epithelium lining an ovarian follicle; source of steroid sex hormones. fig. 28.13

**gray matter** A zone or layer of tissue in the central nervous system where the neuron cell bodies, dendrites, and synapses are found; forms the core of the spinal cord, nuclei of the brainstem, basal nuclei of the cerebrum, cerebral cortex, and cerebellar cortex. fig. 14.6

**gross anatomy** Bodily structure that can be observed without magnification.

**growth factor** A chemical messenger that stimulates mitosis and differentiation of target cells that have receptors for it; important in such processes as fetal development, tissue maintenance and repair, and hemopoiesis; sometimes a contributing factor in cancer.

**growth hormone (GH)** A hormone of the anterior pituitary gland with multiple effects on many tissues, generally promoting tissue growth.

**guanine** A double-ringed nitrogenous base (purine) found in DNA and RNA; one of the four bases of the genetic code; complementary to cytosine in the double helix of DNA. fig. 4.2

**gyrus** (JY-rus) A wrinkle or fold in the cortex of the cerebrum or cerebellum.

# H

**hair cell** Sensory cell of the cochlea, semicircular ducts, utricle, and saccule, with a fringe of surface microvilli that respond to the relative motion of a gelatinous membrane at their tips; responsible for the senses of hearing and equilibrium.

**hair follicle** An oblique epidermal pit that contains a hair and extends into the dermis or hypodermis.

**half-life ($T_{1/2}$)  1.** The time required for one-half of a quantity of a radioactive element to decay to a stable isotope *(physical half-life)* or to be cleared from the body through a combination of radioactive decay and physiological excretion *(biological half-life)*. **2.** The time required for one-half of a quantity of hormone to be cleared from the bloodstream.

**haploid (*n*)** In humans, having 23 unpaired chromosomes instead of the usual 46 chromosomes in homologous pairs; in any organism or cell, having half the normal diploid number of chromosomes for that species.

**helper T cell** A type of lymphocyte that performs a central coordinating role in humoral and cellular immunity; target of the human immunodeficiency virus (HIV).

**hematocrit** (he-MAT-oh-crit) The percentage of blood volume that is composed of erythrocytes.

**hematoma** (HE-muh-TOE-muh) A mass of clotted blood in the tissues; forms a bruise when visible through the skin.

**heme** (heem) The nonprotein, iron-containing prosthetic group of hemoglobin or myoglobin; oxygen binds to its ferrous ion. fig. 18.5

**hemoglobin** (HE-mo-GLO-bin) The red gas-transport pigment of an erythrocyte.

**hemolysis** (he-MOLL-ih-sis) The rupturing of erythrocytes from such causes as a hypotonic medium, parasitic infection, or a complement reaction.

**hemopoiesis** (HE-mo-poy-EE-sis) Production of any of the formed elements of blood.

**heparin** (HEP-uh-rin) A polysaccharide secreted by basophils and mast cells that inhibits blood clotting.

**hepatic** (heh-PAT-ic) Pertaining to the liver.

**hepatic portal system** A network of blood vessels that connect capillaries of the intestines to capillaries (sinusoids) of the liver, thus delivering newly absorbed nutrients directly to the liver.

**hepatitis** (HEP-uh-TY-tiss) Inflammation of the liver.

**heterozygous** (HET-er-oh-ZY-gus) Having nonidentical alleles at the same gene locus of two homologous chromosomes.

**hiatus** (hy-AY-tus) An opening or gap, such as the esophageal hiatus through the diaphragm.

**high-density lipoprotein (HDL)** A lipoprotein of the blood plasma that is about 50% lipid and 50% protein; functions to transport phospholipids and cholesterol from other organs to the liver for disposal. A high proportion of HDL to low-density lipoprotein (LDL) is desirable for cardiovascular health.

**hilum** (HY-lum) A point on the surface of an organ where blood vessels, lymphatic vessels, or nerves enter and leave, usually marked by a depression and slit; the midpoint of the concave surface of any organ that is roughly bean-shaped, such as the lymph nodes, kidneys, and lungs. Also called the *hilus*. fig. 22.9

**histamine** (HISS-ta-meen) An amino acid derivative secreted by basophils, mast cells, and some neurons; functions as a paracrine secretion and neurotransmitter to stimulate effects such as gastric secretion, bronchoconstriction, and vasodilation. fig. 12.21

**histological section** A thin slice of tissue, usually mounted on a slide and artificially stained to make its microscopic structure more visible.

**histology 1.** The microscopic structure of tissues and organs. **2.** The study of such structure.

**homeostasis** (HO-me-oh-STAY-sis) The tendency of a living body to maintain relatively stable internal conditions in spite of greater changes in its external environment.

**homologous** (ho-MOLL-oh-gus) **1.** Having the same embryonic or evolutionary origin but not necessarily the same function, such as the scrotum and labia majora. **2.** Pertaining to two chromosomes with identical structures and gene loci but not necessarily identical alleles; each member of the pair is inherited from a different parent.

**homozygous** (HO-mo-ZY-gus) Having identical alleles at the same gene locus of two homologous chromosomes.

**hormone** A chemical messenger that is secreted into the blood by an endocrine gland or isolated gland cell and triggers a physiological response in distant cells with receptors for it.

**host cell** Any cell belonging to the human body, as opposed to foreign cells introduced to it by such causes as infections and tissue transplants.

**human** Any species of primate classified in the family Hominidae, characterized by bipedal locomotion, relatively large brains, and usually articulate speech; currently represented only by *Homo sapiens* but including extinct species of *Homo* and *Australopithecus.*

**human chorionic gonadotropin (HCG)** (COR-ee-ON-ic) A hormone of pregnancy secreted by the chorion that stimulates continued growth of the corpus luteum and secretion of its hormones. HCG in urine is the basis for pregnancy testing.

**human immunodeficiency virus (HIV)** A virus that infects human helper T cells and other cells, suppresses immunity, and causes AIDS.

**hyaline cartilage** (HY-uh-lin) A form of cartilage with a relatively clear matrix and fine collagen fibers but no conspicuous elastic fibers or coarse collagen bundles as in other types of cartilage.

**hyaluronic acid** (HY-uh-loo-RON-ic) A glycosaminoglycan that is particularly abundant in connective tissues, where it becomes hydrated and forms the tissue gel.

**hydrogen bond** A weak attraction between a slightly positive hydrogen atom on one molecule and a slightly negative oxygen or nitrogen atom on another molecule, or between such atoms on different parts of the same molecule; responsible for the cohesion of water and the coiling of protein and DNA molecules, for example.

**hydrolysis** (hy-DROL-ih-sis) A chemical reaction that breaks a covalent bond in a molecule by adding an —OH group to one side of the bond and —H to the other side, thus consuming a water molecule. *Compare* dehydration synthesis.

**hydrophilic** (HY-dro-FILL-ic) Pertaining to molecules that attract water or dissolve in it because of their polar nature.

**hydrophobic** (HY-dro-FOE-bic) Pertaining to molecules that do not attract water or dissolve in it because of their nonpolar nature; such molecules tend to dissolve in lipids and other nonpolar solvents.

**hydrostatic pressure** The physical force generated by a liquid such as blood or tissue fluid, as opposed to osmotic and atmospheric pressures.

**hydroxyl group** (hy-DROCK-sil) A functional group with the formula —OH found on many organic molecules such as carbohydrates and alcohols.

**hypercalcemia** (HY-pur-cal-SEE-me-uh) An excess of calcium ions in the blood.

**hypercapnia** (HY-pur-CAP-nee-uh) An excess of carbon dioxide in the blood.

**hyperextension** A joint movement that increases the angle between two bones beyond 180°. fig. 9.12

**hyperglycemia** (HY-pur-gly-SEE-me-uh) An excess of glucose in the blood.

**hyperkalemia** (HY-pur-ka-LEE-me-uh) An excess of potassium ions in the blood.

**hypernatremia** (HY-pur-na-TREE-me-uh) An excess of sodium ions in the blood.

**hyperplasia** (HY-pur-PLAY-zhuh) The growth of a tissue through cellular multiplication, not cellular enlargement. *Compare* hypertrophy.

**hyperpolarization** A shift in the electrical potential across a plasma membrane to a value more negative than the resting membrane potential, tending to inhibit a nerve or muscle cell. *Compare* depolarization.

**hypersecretion** Excessive secretion of a hormone or other gland product; can lead to endocrine disorders such as Cushing syndrome or gigantism, for example.

**hypertension** Excessively high blood pressure; criteria vary but it is often considered to be a condition in which systolic pressure exceeds 140 mmHg or diastolic pressure exceeds 90 mmHg.

**hyperthermia** Excessively high core body temperature, as in heatstroke or fever.

**hypertonic** Having a higher osmotic pressure than human cells or some other reference solution and tending to cause osmotic shrinkage of cells.

**hypertrophy** (hy-PUR-tro-fee) The growth of a tissue through cellular enlargement, not cellular multiplication; for example, the growth of muscle under the influence of exercise. *Compare* hyperplasia.

**hypocalcemia** (HY-po-cal-SEE-me-uh) A deficiency of calcium ions in the blood.

**hypocapnia** (HY-po-CAP-nee-uh) A deficiency of carbon dioxide in the blood.

**hypodermis** (HY-po-DUR-miss) A layer of connective tissue deep to the skin; also called *superficial fascia, subcutaneous tissue,* or when it is predominantly adipose, *subcutaneous fat.*

**hypoglycemia** (HY-po-gly-SEE-me-uh) A deficiency of glucose in the blood.

**hypokalemia** (HY-po-ka-LEE-me-uh) A deficiency of potassium ions in the blood.

**hyponatremia** (HY-po-na-TREE-me-uh) A deficiency of sodium ions in the blood.

**hyposecretion** Inadequate secretion of a hormone or other gland product; can lead to endocrine disorders such as diabetes mellitus or pituitary dwarfism, for example.

**hypothalamic thermostat** (HY-po-thuh-LAM-ic) A nucleus in the hypothalamus that monitors body temperature and sends afferent signals to hypothalamic heat-promoting or heat-losing centers to maintain thermal homeostasis.

**hypothalamus** (HY-po-THAL-uh-mus) The inferior portion of the diencephalon of the brain, forming the walls and floor of the third ventricle and giving rise to the posterior pituitary gland; controls many fundamental physiological functions such as appetite, thirst, and body temperature and exerts many of its effects through the endocrine and autonomic nervous systems. fig. 14.12b

**hypothermia** (HY-po-THUR-me-uh) A state of abnormally low core body temperature.

**hypothesis** An informed conjecture that is capable of being tested and potentially falsified by experimentation or data collection.

**hypotonic** Having a lower osmotic pressure than human cells or some other reference solution and tending to cause osmotic swelling and lysis of cells.

**hypovolemic shock** (HY-po-vo-LEE-mic) Insufficient cardiac output resulting from a drop in blood volume. *See also* shock.

**hypoxemia** (HY-pock-SEE-me-uh) A deficiency of oxygen in the bloodstream.

**hypoxia** (hy-POCK-see-uh) A deficiency of oxygen in any tissue.

# I

**immune system** A population of cells, including leukocytes and macrophages, that occur in most organs of the body and protect against foreign organisms, some foreign chemicals, and cancerous or other aberrant host cells.

**immunity** The ability to ward off a specific infection or disease, usually as a result of prior exposure and the body's production of antibodies or lymphocytes against a pathogen. *Compare* resistance.

**immunoglobulin** (IM-you-no-GLOB-you-lin) *See* antibody.

**implantation** The attachment of a conceptus to the endometrium of the uterus.

**inclusion** Any visible object in the cytoplasm of a cell other than an organelle or cytoskeletal element; usually a foreign body or a stored cell product, such as a virus, dust particle, lipid droplet, glycogen granule, or pigment.

**infarction** (in-FARK-shun) 1. The sudden death of tissue from a lack of blood perfusion; also called an *infarct*. 2. An area of necrotic tissue produced by this process.

**inferior** Lower than another structure or point of reference from the perspective of anatomical position; for example, the stomach is inferior to the diaphragm.

**inflammation** (IN-fluh-MAY-shun) A complex of tissue responses to trauma or infection serving to ward off a pathogen and promote tissue repair; recognized by the cardinal signs of redness, heat, swelling, and pain.

**infundibulum** (IN-fun-DIB-you-lum) Any funnel-shaped passage or structure, such as the distal portion of the uterine tube and the stalk that attaches the pituitary gland to the hypothalamus.

**inguinal** (IN-gwih-nul) Pertaining to the groin.

**inhibin** A hormone produced by the testes and ovaries that inhibits the secretion of FSH.

**inhibitory postsynaptic potential (IPSP)** Hyperpolarization of a postsynaptic neuron in response to a neurotransmitter, making it less likely to reach threshold and fire.

**innervation** (IN-ur-VAY-shun) The nerve supply to an organ.

**insertion** The point at which a muscle attaches to another tissue (usually a bone) and produces movement, opposite from its stationary origin; the origin and insertion of a given muscle sometimes depend on what muscle action is being considered. *Compare* origin. fig. 10.2

**inspiration** Inhaling.

**insulin** (IN-suh-lin) A hormone produced by beta cells of the pancreatic islets in response to a rise in blood glucose concentration; accelerates glucose uptake and metabolism by most cells of the body, thus lowering blood glucose concentration.

**integration** A process in which a neuron receives input from multiple sources and their combined effects determine its output; the cellular basis of information processing by the nervous system.

**integumentary system** (in-TEG-you-MEN-tah-ree) An organ system consisting of the skin, cutaneous glands, hair, and nails.

**interatrial septum** (IN-tur-AY-tree-ul) The wall between the atria of the heart.

**intercalated disc** (in-TUR-kuh-LAY-ted) A complex of fascia adherens, gap junctions, and desmosomes that join two cardiac muscle cells end to end, microscopically visible as a dark line which helps to histologically distinguish this muscle type; functions as a mechanical and electrical link between cells. fig. 19.13

**intercellular** Between cells.

**intercostal** (IN-tur-COSS-tul) Between the ribs, as in the intercostal muscles, arteries, veins, and nerves.

**interdigitate** (IN-tur-DIDJ-ih-tate) To fit together like the fingers of two folded hands; for example, at the dermal–epidermal boundary, intercalated discs of the heart, and pedicels of the podocytes in the kidney. fig. 23.10b, c

**interleukin** (IN-tur-LOO-kin) A hormone-like chemical messenger from one leukocyte to another, serving as a means of communication and coordination during immune responses.

**interneuron** (IN-tur-NEW-ron) A neuron that is contained entirely in the central nervous system and, in the path of signal conduction, lies anywhere between an afferent pathway and an efferent pathway.

**interosseous membrane** (IN-tur-OSS-ee-us) A fibrous membrane that connects the radius to the ulna and the tibia to the fibula along most of the shaft of each bone. fig. 8.33

**interphase** That part of the cell cycle between one mitotic phase and the next, from the end of cytokinesis to the beginning of the next prophase.

**interstitial** (IN-tur-STISH-ul) 1. Pertaining to the extracellular spaces in a tissue. 2. Located between other structures, as in the interstitial cells of the testis.

**interstitial fluid** Fluid in the interstitial spaces of a tissue, also called *tissue fluid*.

**intervertebral disc** A cartilaginous pad between the bodies of two adjacent vertebrae.

**intracellular** Within a cell.

**intracellular fluid (ICF)** The fluid contained in the cells; one of the major fluid compartments.

**intravenous (I.V.)** 1. Present or occurring within a vein, such as an intravenous blood clot. 2. Introduced directly into a vein, such as an intravenous injection or I.V. drip.

**intrinsic** (in-TRIN-sic) 1. Arising from within, such as intrinsic blood-clotting factors; endogenous. 2. Fully contained within an organ, such as the intrinsic muscles of the hand and eye. *Compare* extrinsic.

**intrinsic factor** A secretion of the gastric glands required for the intestinal absorption of vitamin $B_{12}$. Hyposecretion of intrinsic factor leads to pernicious anemia.

**involuntary** Not under conscious control, including tissues such as smooth and cardiac muscle and events such as reflexes.

**involution** (IN-vo-LOO-shun) Shrinkage of a tissue or organ by autolysis, such as involution of the thymus after childhood and of the uterus after pregnancy.

**ion** A chemical particle with unequal numbers of electrons or protons and consequently a net negative or positive charge; it may have a single atomic nucleus as in a sodium ion or a few atoms as in a bicarbonate ion, or it may be a large molecule such as a protein.

**ionic bond** The force that binds a cation to an anion.

**ionizing radiation** High-energy electromagnetic rays that eject electrons from atoms or molecules and convert them to ions, frequently causing cellular damage; for example, X rays and gamma rays.

**ipsilateral** (IP-sih-LAT-ur-ul) On the same side of the body, as in reflex arcs in which a muscular response occurs on the same side of the body as the stimulus. *Compare* contralateral.

**ischemia** (iss-KEE-me-uh) Insufficient blood flow to a tissue, typically resulting in metabolite accumulation and sometimes tissue death.

**isometric contraction** A muscle contraction in which internal tension rises but the muscle does not shorten.

**isotonic** Having the same osmotic pressure as human cells or some other reference solution.

**isotonic contraction** A muscle contraction in which the muscle shortens and moves a load while its internal tension remains constant.

# J

**jaundice** (JAWN-diss) A yellowish color of the skin, corneas, mucous membranes, and body fluids due to an excessive concentration of bilirubin; usually indicative of a liver disease, obstructed bile secretion, or hemolytic disease.

# K

**ketone** (KEE-tone) Any organic compound with a carbonyl (C=O) group covalently bonded to two other carbons.

**ketone bodies** Certain ketones (acetone, acetoacetic acid, and β-hydroxybutyric acid) produced by the incomplete oxidation of fats, especially when fats are being rapidly catabolized. *See also* ketosis.

**ketonuria** (KEE-toe-NEW-ree-uh) The abnormal presence of ketones in the urine as an effect of ketosis.

**ketosis** (kee-TOE-sis) An abnormally high concentration of ketone bodies in the blood, occurring in pregnancy, starvation, diabetes mellitus, and other conditions; tends to cause acidosis and to depress the nervous system.

**kilocalorie** The amount of heat energy needed to raise the temperature of 1 kg of water by 1°C; 1,000 calories. Also called a *Calorie* or *large calorie.*

**kinase** Any enzyme that adds an inorganic phosphate ($P_i$) group to another organic molecule. Also called a *phosphokinase.*

# L

**labium** (LAY-bee-um) A lip, such as those of the mouth and the labia majora and minora of the vulva.

**lactation** The secretion of milk.

**lactic acid** A small organic acid produced as an end product of the anaerobic fermentation of pyruvic acid; a contributing factor in muscle fatigue.

**lacuna** (la-CUE-nuh) A small cavity or depression in a tissue such as bone, cartilage, and the erectile tissues.

**lamella** (la-MELL-uh) A little plate, such as the lamellae of bone. fig. 7.5

**lamina** (LAM-ih-nuh) A thin layer, such as the lamina of a vertebra or the lamina propria of a mucous membrane. fig. 8.22

**lamina propria** (PRO-pree-uh) A thin layer of areolar tissue immediately deep to the epithelium of a mucous membrane. fig. 5.32

**larynx** (LAIR-inks) A cartilaginous chamber in the neck containing the vocal cords; the voicebox.

**latent period** The interval between a stimulus and response, especially in the action of nerve and muscle cells.

**lateral** Away from the midline of an organ or median plane of the body; toward the side. *Compare* medial.

**law** A verbal or mathematical description of a predictable natural phenomenon or of the relationships between variables; for example, Boyle's law and the second law of thermodynamics.

**leader sequence** A sequence of bases in mRNA that is not translated to protein but serves as a binding site for a ribosome.

**length–tension relationship** A law that relates the tension generated by muscle contraction to the length of the muscle fiber prior to stimulation; it shows that the greatest tension is generated when the fiber exhibits an intermediate degree of stretch before stimulation.

**lesion** A circumscribed zone of tissue injury, such as a skin abrasion or myocardial infarction.

**leukocyte** (LOO-co-site) A white blood cell.

**leukotriene** (LOO-co-TRY-een) Eicosanoid that promotes allergic and inflammatory responses such as vasodilation and neutrophil chemotaxis; secreted by basophils, mast cells, and damaged tissues.

**libido** (lih-BEE-do) Sex drive.

**ligament** A cord or band of tough collagenous tissue binding one organ to another, especially one bone to another, and serving to hold organs in place; for example, the cruciate ligaments of the knee, broad ligament of the uterus, and falciform ligament of the liver.

**ligand** (LIG-and, LY-gand) A chemical that binds reversibly to a receptor site on a protein, such as a neurotransmitter that binds to a membrane receptor or a substrate that binds to an enzyme.

**ligand-regulated gate** A channel protein in a plasma membrane that opens or closes when a ligand binds to it, enabling the ligand to determine when substances can enter or leave the cell.

**light microscope (LM)** A microscope that produces images with visible light.

**linea** (LIN-ee-uh) An anatomical line, such as the linea alba.

**lingual** (LING-gwul) Pertaining to the tongue, as in lingual papillae.

**lipase** (LY-pace) An enzyme that hydrolyzes a triglyceride into fatty acids and glycerol.

**lipid** A hydrophobic organic compound composed mainly of carbon and a high ratio of hydrogen to oxygen; includes fatty acids, fats, phospholipids, steroids, and prostaglandins.

**lipoprotein** (LIP-oh-PRO-teen) A protein-coated lipid droplet in the blood plasma or lymph, serving as a means of lipid transport; for example, chylomicrons and the high- and low-density lipoproteins.

**load 1.** To pick up a gas for transport in the bloodstream. **2.** The resistance acted upon by a muscle.

**lobe 1.** A structural subdivision of an organ such as a gland, a lung, or the brain, bounded by a visible landmark such as a fissure or septum. **2.** The inferior, noncartilaginous, often pendant part of the ear pinna; the earlobe.

**lobule** (LOB-yool) A small subdivision of an organ or of a lobe of an organ, especially of a gland.

**locus** *See* gene locus.

**long bone** A bone such as the femur or humerus that is markedly longer than wide and that generally serves as a lever.

**longitudinal** Oriented along the longest dimension of the body or of an organ.

**loose connective tissue** *See* areolar tissue.

**low-density lipoprotein (LDL)** A blood-borne droplet of about 20% protein and

80% lipid (mainly cholesterol) that transports cholesterol from the liver to other tissues.

**lower limb** The appendage that arises from the hip, consisting of the thigh from hip to knee; the crural region from knee to ankle; the ankle; and the foot. Loosely called the leg, although that term properly refers only to the crural region.

**lumbar** Pertaining to the lower back and sides, between the thoracic cage and pelvis.

**lumen** (LOO-men) The internal space of a hollow organ such as a blood vessel or the esophagus, or a space surrounded by cells as in a gland acinus.

**luteinizing hormone (LH)** (LOO-tee-in-eye-zing) A hormone of the anterior pituitary gland that stimulates ovulation in females and testosterone secretion in males.

**lymph** The fluid contained in lymphatic vessels and lymph nodes, produced by the absorption of tissue fluid.

**lymphatic system** (lim-FAT-ic) An organ system consisting of lymphatic vessels, lymph nodes, the tonsils, spleen, and thymus; functions include tissue fluid recovery and immunity.

**lymph node** A small organ found along the course of a lymphatic vessel that filters the lymph and contains lymphocytes and macrophages, which respond to antigens in the lymph. fig. 21.12

**lymphocyte** (LIM-foe-site) A relatively small agranulocyte with numerous types and roles in nonspecific defense, humoral immunity, and cellular immunity. table 18.7

**lymphokine** Any interleukin secreted by a lymphocyte.

**lysosome** (LY-so-some) A membrane-bounded organelle containing a mixture of enzymes with a variety of intracellular and extracellular roles in digesting foreign matter, pathogens, and expired organelles.

**lysozyme** (LY-so-zime) An enzyme found in tears, milk, saliva, mucus, and other body fluids that destroys bacteria by digesting their cell walls. Also called *muramidase.*

# M

**macromolecule** Any molecule of large size and high molecular weight, such as a protein, nucleic acid, polysaccharide, or triglyceride.

**macrophage** (MAC-ro-faje) Any cell of the body, other than a leukocyte, that is specialized for phagocytosis; usually derived from blood monocytes and often functioning as antigen-presenting cells.

**macula** (MAC-you-luh) A patch or spot, such as the *macula lutea* of the retina.

**malignant** (muh-LIG-nent) Pertaining to a cell or tumor that is cancerous; capable of metastasis.

**maltose** A disaccharide composed of two glucose monomers.

**mammary gland** The milk-secreting gland that develops within the breast in pregnancy and lactation; only minimally developed in the breast of a nonpregnant or nonlactating woman.

**mast cell** A connective tissue cell, similar to a basophil, that secretes histamine, heparin, and other chemicals involved in inflammation; often concentrated along the course of blood capillaries.

**matrix** 1. The extracellular material of a tissue. 2. The fluid within a mitochondrion containing enzymes of the citric acid cycle. 3. The substance or framework within which other structures are embedded, such as the fibrous matrix of a blood clot. 4. A mass of epidermal cells from which a hair root or nail root develops.

**mechanoreceptor** A sensory nerve ending or organ specialized to detect mechanical stimuli such as touch, pressure, stretch, or vibration.

**medial** Toward the midline of an organ or median plane of the body. *Compare* lateral.

**median plane** The sagittal plane that divides the body or an organ into equal right and left halves; also called *midsagittal plane.*

**mediastinum** (ME-dee-ah-STY-num) The thick median partition of the thoracic cavity that separates one pleural cavity from the other and contains the heart, great blood vessels, and thymus. fig. A.7

**medulla** (meh-DUE-luh, meh-DULL-uh) Tissue deep to the cortex of certain two-layered organs such as the adrenal glands, lymph nodes, hairs, and kidneys.

**medulla oblongata** (OB-long-GAH-ta) The most caudal part of the brainstem, immediately superior to the foramen magnum of the skull, connecting the spinal cord to the rest of the brain. fig. 14.8

**meiosis** (my-OH-sis) A form of cell division in which a diploid cell divides twice and produces four haploid daughter cells; occurs only in gametogenesis.

**melanocyte** A cell of the stratum basale of the epidermis that synthesizes melanin and transfers it to the keratinocytes.

**meninges** (meh-NIN-jeez) (singular, *meninx*) Three fibrous membranes between the central nervous system and surrounding bone: the dura mater, arachnoid mater, and pia mater. fig. 14.5

**menopause** Cessation of the menstrual cycles, occurring during female climacteric.

**merocrine** (MERR-oh-crin) Pertaining to gland cells that release their product by exocytosis; also called *eccrine.*

**mesenchyme** (MEZ-en-kime) A gelatinous embryonic connective tissue derived from the mesoderm; differentiates into all permanent connective tissues and most muscle.

**mesentery** (MEZ-en-tare-ee) A serous membrane that binds the intestines together and suspends them from the abdominal wall; the visceral continuation of the peritoneum. fig. 25.3

**mesoderm** (MEZ-oh-durm) The middle layer of the three primary germ layers of an embryo; gives rise to muscle and connective tissue.

**mesothelium** (MEZ-oh-THEEL-ee-um) A simple squamous epithelium that covers the serous membranes.

**metabolic pathway** A series of linked chemical reactions, most of which are catalyzed by a separate enzyme; glycolysis, for example.

**metabolic rate** The overall rate of the body's metabolic reactions at any given time, which determines the rates of nutrient and oxygen consumption; often measured from the rate of oxygen consumption or heat production. *Compare* basal metabolic rate.

**metabolic waste** A product of metabolism that is not useful to the body but is potentially toxic and must be excreted.

**metabolism** (meh-TAB-oh-lizm) The sum of all chemical reactions in the body.

**metabolite** (meh-TAB-oh-lite) Any chemical produced by metabolism.

**metaplasia** Transformation of one mature tissue type into another; for example, a change from pseudostratified to stratified squamous epithelium in an over-ventilated nasal cavity.

**metastasis** (meh-TASS-tuh-sis) The spread of cancer cells from the original tumor to a new location, where they seed the development of a new tumor.

**microtubule** An intracellular cylinder composed of the protein tubulin, forming centrioles, the axonemes of cilia and flagella, and part of the cytoskeleton.

**microvillus** An outgrowth of the plasma membrane that increases the surface area of a cell and functions in absorption and some sensory processes; distinguished from cilia and flagella by its smaller size and lack of an axoneme.

**milliequivalent** One-thousandth of an equivalent, which is the amount of an electrolyte that would neutralize 1 mole of $H^+$ or $OH^-$. Electrolyte concentrations are commonly expressed in milliequivalents per liter (mEq/L).

**mineralocorticoid** (MIN-ur-uh-lo-COR-tih-coyd) A steroid hormone, chiefly aldosterone, that is secreted by the adrenal cortex and acts to regulate electrolyte balance.

**mitochondrion** (MY-toe-CON-dree-un) An organelle specialized to synthesize ATP, enclosed in a double unit membrane with infoldings of the inner membrane called cristae.

**mitosis** (my-TOE-sis) A form of cell division in which a cell divides once and produces two genetically identical daughter cells; sometimes used to refer only to the division of the genetic material or nucleus and not to include cytokinesis, the subsequent division of the cytoplasm.

**moiety** (MOY-eh-tee) A chemically distinct subunit of a macromolecule, such as the heme and globin moieties of hemoglobin or the lipid and carbohydrate moieties of a glycolipid.

**molarity** A measure of chemical concentration expressed as moles of solute per liter of solution.

**mole** The mass of a chemical equal to its molecular weight in grams, containing 6.023 3 $10^{23}$ molecules.

**monocyte** An agranulocyte specialized to migrate into the tissues and transform into a macrophage. table 18.7

**monokine** Any interleukin secreted by a monocyte or macrophage.

**monomer** (MON-oh-mur) **1.** One of the identical or similar subunits of a larger molecule in the dimer to polymer range; for example, the glucose monomers of starch, the amino acids of a protein, or the nucleotides of DNA. **2.** One subunit of an antibody molecule, composed of four polypeptides.

**monosaccharide** (MON-oh-SAC-uh-ride) A simple sugar, or sugar monomer; chiefly glucose, fructose, and galactose.

**monozygotic (MZ) twins** Two individuals who developed from the same fertilized egg and are therefore genetically identical.

**motor end plate** A depression in a muscle fiber where it has synaptic contact with a nerve fiber and has a high density of neurotransmitter receptors.

**motor neuron** A neuron that transmits signals from the central nervous system to any effector (muscle or gland cell); its axon is an efferent nerve fiber.

**motor protein** Any protein that produces movements of a cell or its components owing to its ability to undergo quick repetitive changes in conformation and to bind reversibly to other molecules; for example, myosin, dynein, and kinesin.

**motor unit** One motor neuron and all the skeletal muscle fibers innervated by it.

**mucosa** (mew-CO-suh) A tissue layer that forms the inner lining of an anatomical tract that is open to the exterior (the respiratory, digestive, urinary, and reproductive tracts). Composed of epithelium, connective tissue (lamina propria), and often smooth muscle (muscularis mucosae). fig. 5.32

**mucous membrane** A mucosa.

**mucus** A viscous, slimy or sticky secretion produced by mucous cells and mucous membranes and consisting of a hydrated glycoprotein, mucin; serves to bind particles together, such as bits of masticated food, and to protect the mucous membranes from infection and abrasion.

**multipotent** Pertaining to a stem cell that has the potential to develop into two or more types of fully differentiated, functional cells, but not into an unlimited variety of cell types.

**muscle fiber** One skeletal muscle cell.

**muscle tone** A state of continual, partial contraction of resting skeletal or smooth muscle.

**muscularis externa** The external muscular wall of certain viscera such as the esophagus and small intestine. fig. 25.2

**muscularis mucosae** (MUSS-cue-LERR-iss mew-CO-see) A layer of smooth muscle immediately deep to the lamina propria of a mucosa. fig. 5.32

**muscular system** An organ system composed of the skeletal muscles, specialized mainly for maintaining postural support and producing movements of the bones.

**muscular tissue** A tissue composed of elongated, electrically excitable cells specialized for contraction; the three types are skeletal, cardiac, and smooth muscle.

**mutagen** (MEW-tuh-jen) Any agent that causes a mutation, including viruses, chemicals, and ionizing radiation.

**mutation** Any change in the structure of a chromosome or a DNA molecule, often resulting in a change of organismal structure or function.

**myelin** (MY-eh-lin) A lipid sheath around a nerve fiber, formed from closely spaced spiral layers of the plasma membrane of a Schwann cell or oligodendrocyte. fig. 12.7

**myocardium** (MY-oh-CAR-dee-um) The middle, muscular layer of the heart.

**myocyte** A muscle cell, especially a cell of cardiac or smooth muscle.

**myoepithelial cell** An epithelial cell that has become specialized to contract like a muscle cell; important in dilation of the pupil and ejection of secretions from gland acini.

**myofibril** (MY-oh-FY-bril) A bundle of myofilaments forming an internal subdivision of a cardiac or skeletal muscle cell. fig. 11.2

**myofilament** A protein microfilament responsible for the contraction of a muscle cell, composed mainly of myosin or actin. fig. 11.3

**myoglobin** (MY-oh-GLO-bin) A red oxygen-storage pigment of muscle; supplements hemoglobin in providing oxygen for aerobic muscle metabolism.

**myosin** A motor protein that constitutes the thick myofilaments of muscle and has globular, mobile heads of ATPase that bind to actin molecules.

# N

**necrosis** (neh-CRO-sis) Pathological tissue death due to such causes as infection, trauma, or hypoxia. *Compare* apoptosis.

**negative feedback** A self-corrective mechanism that underlies most homeostasis, in which a bodily change is detected and responses are activated that reverse the change and restore stability and preserve normal body function.

**negative feedback inhibition** A mechanism for limiting the secretion of a pituitary tropic hormone. The tropic hormone stimulates another endocrine gland to secrete its own hormone, and that hormone inhibits further release of the tropic hormone.

**neonate** (NEE-oh-nate) An infant up to 6 weeks old.

**neoplasia** (NEE-oh-PLAY-zee-uh) Abnormal growth of new tissue, such as a tumor, with no useful function.

**nephron** One of approximately 1 million blood-filtering, urine-producing units in each kidney; consists of a glomerulus, glomerular capsule, proximal convoluted tubule, nephron loop, and distal convoluted tubule. fig. 23.8

**nerve** A cordlike organ of the peripheral nervous system composed of multiple nerve fibers ensheathed in connective tissue.

**nerve fiber** The axon of a single neuron.

**nerve impulse** A wave of self-propagating action potentials traveling along a nerve fiber.

**nervous system** An organ system composed of the brain, spinal cord, nerves, and ganglia, specialized for rapid communication of information.

**nervous tissue** A tissue composed of neurons and neuroglia.

**net filtration pressure** A net force favoring filtration of fluid from a capillary or venule when all the hydrostatic and osmotic pressures of the blood and tissue fluids are taken into account.

**neural tube** A dorsal hollow tube in the embryo that develops into the central nervous system. fig. 14.3

**neuroglia** (noo-ROG-lee-uh) All cells of nervous tissue except neurons; cells that perform various supportive and protective roles for the neurons.

**neuromuscular junction** A synapse between a nerve fiber and a muscle fiber. fig. 11.6

**neuron** (NOOR-on) A nerve cell; an electrically excitable cell specialized for producing and transmitting action potentials and secreting chemicals that stimulate adjacent cells.

**neuronal pool** (noor-OH-nul) A group of interconnected neurons of the central nervous system that perform a single collective function; for example, the vasomotor center of the brainstem and speech centers of the cerebral cortex.

**neuropeptide** A peptide secreted by a neuron, often serving to modify the action of a neurotransmitter; for example, endorphins, enkephalin, and cholecystokinin. fig. 12.21

**neurotransmitter** A chemical released at the distal end of an axon that stimulates an adjacent cell; for example, acetylcholine, norepinephrine, or serotonin.

**neutral fat** A triglyceride.

**neutrophil** (NOO-tro-fill) A granulocyte, usually with a multilobed nucleus, that serves especially to destroy bacteria by means of phagocytosis, intracellular digestion, and secretion of bactericidal chemicals. table 18.7

**nitrogenous base** (ny-TRODJ-eh-nus) An organic molecule with a single or double carbon-nitrogen ring that forms one of the building blocks of ATP, other nucleotides, and nucleic acids; the basis of the genetic code. fig. 4.2

**nitrogenous waste** Any nitrogen-containing substance produced as a metabolic waste and excreted in the urine; chiefly ammonia, urea, uric acid, and creatinine.

**nociceptor** (NO-sih-SEP-tur) A nerve ending specialized to detect tissue damage and produce a sensation of pain; pain receptor.

**norepinephrine** (nor-EP-ih-NEF-rin) A catecholamine that functions as a neurotransmitter and adrenal hormone, especially in the sympathetic nervous system. fig. 12.21

**nuclear envelope** (NEW-clee-ur) A pair of unit membranes enclosing the nucleus of a cell, with prominent pores allowing traffic of molecules between the nucleoplasm and cytoplasm. fig. 3.25

**nucleic acid** (new-CLAY-ic) An acidic polymer of nucleotides found or produced in the nucleus, functioning in heredity and protein synthesis; of two types, DNA and RNA.

**nucleotide** (NEW-clee-oh-tide) An organic molecule composed of a nitrogenous base, a monosaccharide, and a phosphate group; the monomer of a nucleic acid.

**nucleus** (NEW-clee-us) 1. A cell organelle containing DNA and surrounded by a double unit membrane. 2. A mass of neurons (gray matter) surrounded by white matter of the brain, including the basal nuclei and brainstem nuclei. 3. The positively charged core of an atom, consisting of protons and neutrons. 4. A central structure, such as the nucleus pulposus of an intervertebral disc.

**nucleus pulposus** The gelatinous center of an intervertebral disc.

# O

**olfaction** (ole-FAC-shun) The sense of smell.

**oncotic pressure** (ong-COT-ic) The difference between the colloid osmotic pressure of the blood and that of the tissue fluid, usually favoring fluid absorption by the blood capillaries. *Compare* colloid osmotic pressure.

**oocyte** (OH-oh-site) In the development of an egg cell, any haploid stage between meiosis I and fertilization.

**oogenesis** (OH-oh-JEN-eh-sis) The production of a fertilizable egg cell through a series of mitotic and meiotic cell divisions; female gametogenesis.

**ophthalmic** (off-THAL-mic) Pertaining to the eye or vision; optic.

**opposition** A movement of the thumb in which it touches any fingertip of the same hand.

**optic** Pertaining to the eye or vision.

**orbit** The eye socket of the skull.

**organ** Any anatomical structure that is composed of at least two different tissue types, has recognizable structural boundaries, and has a discrete function different from the structures around it. Many organs are microscopic and many organs contain smaller organs, such as the skin containing numerous microscopic sense organs.

**organelle** Any structure within a cell that carries out one of its metabolic roles, such as mitochondria, centrioles, endoplasmic reticulum, and the nucleus; an intracellular structure other than the cytoskeleton and inclusions.

**organic** Pertaining to compounds of carbon.

**origin** The relatively stationary attachment of a skeletal muscle. *Compare* insertion. fig. 10.2

**osmolality** (OZ-mo-LAL-ih-tee) The molar concentration of dissolved particles in 1 kg of water.

**osmolarity** (OZ-mo-LERR-ih-tee) The molar concentration of dissolved particles in 1 L of solution.

**osmoreceptor** (OZ-mo-re-SEP-tur) A neuron of the hypothalamus that responds to changes in the osmolarity of the extracellular fluid.

**osmosis** (oz-MO-sis) The net diffusion of water through a selectively permeable membrane.

**osmotic diuresis** (oz-MOT-ic DY-you-REE-sis) Increased urine output due to an increase in the concentration of osmotically active particles in the tubular fluid.

**osmotic pressure** The amount of pressure that would have to be applied to one side of a selectively permeable membrane to stop osmosis; proportional to the concentration of nonpermeating solutes on that side and therefore serving as an indicator of solute concentration.

**osseous** (OSS-ee-us) Pertaining to bone.

**ossification** (OSS-ih-fih-CAY-shun) Bone formation.

**osteoarthritis (OA)** A chronic degenerative joint disease characterized by loss of articular cartilage, growth of bone spurs, and impaired movement; occurs to various degrees in almost all people with age.

**osteoblast** Bone-forming cell that arises from an osteogenic cell, deposits bone matrix, and eventually becomes an osteocyte.

**osteoclast** Macrophage of the bone surface that dissolves the matrix and returns minerals to the extracellular fluid.

**osteocyte** A mature bone cell formed when an osteoblast becomes surrounded by its own matrix and entrapped in a lacuna.

**osteon** A structural unit of compact bone consisting of a central canal surrounded by concentric cylindrical lamellae of matrix. fig. 7.5

**osteoporosis** (OSS-tee-oh-pore-OH-sis) A degenerative bone disease characterized by a loss of bone mass, increasing susceptibility to spontaneous fractures, and sometimes deformity of the vertebral column; causes include aging, estrogen hyposecretion, and insufficient resistance exercise.

**ovary** The female gonad; produces eggs, estrogen, and progesterone.

**ovulation** (OV-you-LAY-shun) The release of a mature oocyte by the bursting of an ovarian follicle.

**ovum** Any stage of the female gamete from the conclusion of meiosis I until fertilization; a primary or secondary oocyte; an egg.

**oxidation** A chemical reaction in which one or more electrons are removed from a molecule, lowering its free energy content; opposite of reduction and always linked to a reduction reaction.

**oxytocin (OT)** (OCK-see-TOE-sin) A hormone released by the posterior pituitary gland that stimulates labor contractions and milk release.

# P

**pancreas** (PAN-cree-us) A gland of the upper abdominal cavity, near the stomach, that secretes digestive enzymes and sodium bicarbonate into the duodenum and secretes hormones into the blood.

**pancreatic islet** (PAN-cree-AT-ic EYE-let) A small cluster of endocrine cells in the pancreas that secretes insulin, glucagon, somatostatin, and other intercellular messengers; also called *islet of Langerhans.* fig. 17.12

**papilla** (pa-PILL-uh) A conical or nipple-like structure, such as a lingual papilla of the tongue or the papilla of a hair bulb.

**papillary** (PAP-ih-lerr-ee) **1.** Pertaining to or shaped like a nipple, such as the papillary muscles of the heart. **2.** Having papillae, such as the papillary layer of the dermis.

**paracrine** (PERR-uh-crin) **1.** A chemical messenger similar to a hormone whose effects are restricted to the immediate vicinity of the cells that secrete it; sometimes called a local hormone. **2.** Pertaining to such a secretion, as opposed to *endocrine.*

**parasympathetic nervous system** (PERR-uh-SIM-pa-THET-ic) A division of the autonomic nervous system that issues efferent fibers through the cranial and sacral nerves and exerts cholinergic effects on its target organs.

**parathyroid glands** (PERR-uh-THY-royd) Small endocrine glands, usually four in number, adhering to the posterior side of the thyroid gland. fig. 17.10

**parathyroid hormone (PTH)** A hormone secreted by the parathyroid glands that raises blood calcium concentration by stimulating bone resorption by osteoclasts, promoting intestinal absorption of calcium, and inhibiting urinary excretion of calcium.

**parenchyma** (pa-REN-kih-muh) The tissue that performs the main physiological functions of an organ, especially a gland, as opposed to the tissues (stroma) that mainly provide structural support.

**parietal** (pa-RY-eh-tul) **1.** Pertaining to a wall, as in the parietal cells of the gastric glands and parietal bone of the skull. **2.** The outer or more superficial layer of a two-layered membrane such as the pleura, pericardium, or glomerular capsule. *Compare* visceral. fig. A.8

**pathogen** Any disease-causing organism or chemical.

**pedicle** (PED-ih-cul) A small footlike process, as in the vertebrae and the renal podocytes; also called a *pedicel.*

**pelvis** A basinlike structure such as the pelvic girdle of the skeleton or the urine-collecting space near the hilum of the kidney. figs. 8.35, 23.4

**peptide** Any chain of two or more amino acids. *See also* polypeptide, protein.

**peptide bond** A group of four covalently bonded atoms (a —C=O group bonded to an —NH group) that links two amino acids in a protein or other peptide. fig. 2.23b

**perfusion** The amount of blood supplied to a given mass of tissue in a given period of time.

**perichondrium** (PERR-ih-CON-dree-um) A layer of fibrous connective tissue covering the surface of hyaline or elastic cartilage.

**perineum** (PERR-ih-NEE-um) The region between the thighs bordered by the coccyx, pubic symphysis, and ischial tuberosities; contains the orifices of the urinary, reproductive, and digestive systems. figs. 27.6, 28.8

**periosteum** (PERR-ee-OSS-tee-um) A layer of fibrous connective tissue covering the surface of a bone. fig. 7.2

**peripheral** (peh-RIF-eh-rul) Away from the center of the body or of an organ, as in peripheral vision and peripheral blood vessels.

**peripheral nervous system (PNS)** A subdivision of the nervous system composed of all nerves and ganglia; all of the nervous system except the central nervous system.

**peristalsis** (PERR-ih-STAL-sis) A wave of constriction traveling along a tubular organ such as the esophagus or ureter, serving to propel its contents.

**peritoneum** (PERR-ih-toe-NEE-um) A serous membrane that lines the peritoneal cavity of the abdomen and covers the mesenteries and viscera.

**perivascular** (PERR-ih-VASS-cue-lur) Pertaining to the region surrounding a blood vessel.

**pernicious anemia** A deficiency of hemoglobin synthesis resulting from inadequate vitamin $B_{12}$ ingestion or absorption.

**pH** A measure of the acidity or alkalinity of a solution; the negative logarithm of hydrogen ion molarity ($pH = 1/\log [H^+]$). A pH of 7.0 is neutral, a pH <7 is acidic, and a pH > 7 is basic (alkaline).

**phagocytosis** (FAG-oh-sy-TOE-sis) A form of endocytosis in which a cell surrounds a foreign particle with pseudopods and engulfs it, enclosing it in a cytoplasmic vesicle called a phagosome.

**pharynx** (FAIR-inks) A muscular passage in the throat at which the respiratory and digestive tracts cross.

**phosphorylation** Addition of an inorganic phosphate ($P_i$) group to an organic molecule.

**physiology** **1.** The functional processes of the body. **2.** The study of such function.

**piloerector** A bundle of smooth muscle cells associated with a hair follicle, responsible for erection of the hair; also called *arrector pili.* fig. 6.7

**pineal gland** (PIN-ee-ul) A small conical endocrine gland arising from the roof of the third ventricle of the brain; produces melatonin and serotonin and may be involved in timing the onset of puberty. fig. 14.2

**pinocytosis** (PIN-oh-sy-TOE-sis) A form of endocytosis in which the plasma membrane sinks inward and imbibes droplets of extracellular fluid.

**pituitary gland** (pih-TOO-ih-terr-ee) An endocrine gland suspended from the hypothalamus and housed in the sella turcica of the sphenoid bone; secretes numerous hormones, most of which regulate the activities of other glands. fig. 17.4

**placenta** (pla-SEN-tuh) A thick discoid organ on the wall of the pregnant uterus, composed of a combination of maternal and fetal tissues, serving multiple functions in pregnancy including gas, nutrient, and waste exchange between mother and fetus. fig. 29.9

**plantar** (PLAN-tur) Pertaining to the sole of the foot.

**plaque** A small scale or plate of matter, such as dental plaque, the fatty plaques of atherosclerosis, and the amyloid plaques of Alzheimer disease.

**plasma** The noncellular portion of the blood.

**plasma membrane** The unit membrane that encloses a cell and controls the traffic of molecules in and out of the cell. fig. 3.6

**platelet** A formed element of the blood derived from the peripheral cytoplasm of a megakaryocyte, known especially for its role in stopping bleeding but also serves in dissolving blood clots, stimulating inflammation, promoting tissue growth, and destroying bacteria.

**pleura** (PLOOR-uh) A double-walled serous membrane that encloses each lung.

**plexus** A network of blood vessels, lymphatic vessels, or nerves, such as a choroid plexus of the brain or brachial plexus of nerves.

**pluripotent stem cell (PPSC)** **1.** A cell of the inner cell mass of a blastocyst that is capable of developing into any type of embryonic cell, but not into cells of the accessory organs of pregnancy. **2.** A cell of the red bone marrow that can give rise, through a series of intermediate cells, to leukocytes, erythrocytes, platelets, and various kinds of macrophages.

**polymer** A molecule that consists of a long chain of identical or similar subunits, such as protein, DNA, or starch.

**polypeptide** Any chain of more than 10 or 15 amino acids.

**polysaccharide** (POL-ee-SAC-uh-ride) A polymer of simple sugars; for example, glycogen, starch, and cellulose.

**polyuria** (POL-ee-YOU-ree-uh) Excessive output of urine.

**popliteal** (po-LIT-ee-ul) Pertaining to the posterior aspect of the knee.

**positron emission tomography (PET)** A method of producing a computerized image of the physiological state of a tissue using injected radioisotopes that emit positrons.

**posterior** Near or pertaining to the back or spinal side of the body; dorsal.

**postganglionic** (POST-gang-glee-ON-ic) Pertaining to a neuron that transmits signals from a ganglion to a more distal target organ.

**postsynaptic** (POST-sih-NAP-tic) Pertaining to a neuron or other cell that receives signals from the presynaptic neuron at a synapse. fig. 12.18

**potential** A difference in electrical charge from one point to another, especially on opposite sides of a plasma membrane; usually measured in millivolts.

**potential space** An anatomical space that is usually obliterated by contact between two membranes but opens up if air, fluid, or other matter comes between the membranes. Examples include the pleural cavity and the lumen of the uterus.

**preembryo** A developing individual up to 16 days of gestation, prior to the existence of the three primary germ layers. *Compare* conceptus, embryo, fetus.

**preganglionic** (PRE-gang-glee-ON-ic) Pertaining to a neuron that transmits signals from the central nervous system to a ganglion.

**presynaptic** (PRE-sih-NAP-tic) Pertaining to a neuron that transmits signals to a synapse. fig. 12.18

**prime mover** The muscle primarily responsible for a given joint action; agonist.

**programmed cell death** *See* apoptosis.

**prolactin (PRL)** A pituitary hormone that promotes milk synthesis.

**pronation** (pro-NAY-shun) A rotational movement of the forearm that turns the palm downward or posteriorly. fig. 9.18

**proprioception** (PRO-pree-oh-SEP-shun) The nonvisual perception, usually subconscious, of the position and move-

ments of the body, resulting from input from proprioceptors and the vestibular apparatus of the inner ear.

**proprioceptor** (PRO-pree-oh-SEP-tur) A sensory receptor of the muscles, tendons, and joint capsules that detects muscle contractions and joint movements.

**prostaglandin** (PROSS-ta-GLAN-din) An eicosanoid with a five-sided carbon ring in the middle of a hydrocarbon chain, playing a variety of roles in inflammation, neurotransmission, vasomotion, reproduction, and metabolism. fig. 2.21

**prostate gland** (PROSS-tate) A male reproductive gland that encircles the urethra immediately inferior to the bladder and contributes to the semen. fig. 27.11

**protein** A large polypeptide; while criteria for a protein are somewhat subjective and variable, polypeptides over 100 amino acids long are generally classified as proteins.

**proteoglycan** (PRO-tee-oh-GLY-can) A large molecule composed of a bristle-like arrangement of glycosaminoglycans surrounding a protein core in a shape resembling a bottle brush. Binds cells to extracellular materials and gives the tissue fluid a gelatinous consistency.

**proximal** Relatively near a point of origin or attachment; for example, the shoulder is proximal to the elbow. *Compare* distal.

**pseudopod** (SOO-doe-pod) A temporary cytoplasmic extension of a cell used for locomotion (ameboid movement) and phagocytosis.

**pseudostratified columnar** A type of epithelium with tall columnar cells reaching the free surface and shorter basal cells that do not reach the surface, but with all cells resting on the basement membrane; creates a false appearance of stratification. fig. 5.7

**pulmonary** Pertaining to the lungs.

**pulmonary circuit** A route of blood flow that supplies blood to the pulmonary alveoli for gas exchange and then returns it to the heart; all blood vessels between the right ventricle and the left atrium of the heart.

**pyrogen** (PY-ro-jen) A fever-producing agent.

**pyruvic acid** The three-carbon end product of glycolysis; occurs at the branch point between glycolysis, anaerobic fermentation, and aerobic respiration and is thus an important metabolic intermediate linking these pathways to each other.

# R

**ramus** (RAY-mus) An anatomical branch, as in a nerve or in the pubis.

**receptor** **1.** A cell or organ specialized to detect a stimulus, such as a taste cell or the eye. **2.** A protein molecule that binds and responds to a chemical such as a hormone, neurotransmitter, or odor molecule.

**receptor-mediated endocytosis** A process in which certain molecules in the extracellular fluid bind to receptors in the plasma membrane, these receptors gather together, the membrane sinks inward at that point, and the molecules become incorporated into vesicles in the cytoplasm.

**receptor potential** A variable change in membrane voltage produced by a stimulus acting on a receptor cell; generates an action potential if it reaches threshold.

**reduction** **1.** A chemical reaction in which one or more electrons are added to a molecule, raising its free energy content; opposite of oxidation and always linked to an oxidation reaction. **2.** Treatment of a fracture by restoring the broken parts of a bone to their proper alignment.

**reference man** A healthy male 22 years old, weighing 70 kg, living at a mean ambient temperature of 20°C, engaging in light physical activity, and consuming 2,800 kcal/day. A standard of reference for typical adult male physiological values.

**reference woman** A healthy female 22 years old, weighing 58 kg, living at a mean ambient temperature of 20°C, engaging in light physical activity, and consuming 2,000 kcal/day. A standard of reference for typical adult female physiological values.

**reflex** A stereotyped, automatic, involuntary response to a stimulus; includes somatic reflexes, in which the effectors are skeletal muscles, and visceral (autonomic) reflexes, in which the effectors are usually visceral muscle, cardiac muscle, or glands.

**reflex arc** A simple neural pathway that mediates a reflex; involves a receptor, an afferent nerve fiber, sometimes one or more interneurons, an efferent nerve fiber, and an effector.

**reflux** A backward flow, such as the movement of stomach contents back into the esophagus.

**refractory period** **1.** A period of time after a nerve or muscle cell has responded to a stimulus in which it cannot be reexcited by a threshold stimulus. **2.** A period of time after male orgasm when it is not possible to reattain erection or ejaculation.

**regeneration** Replacement of damaged tissue with new tissue of the original type. *Compare* fibrosis.

**renal** (REE-nul) Pertaining to the kidney.

**renin** (REE-nin) An enzyme secreted by the kidneys in response to hypotension; converts the plasma protein angiotensinogen to angiotensin I, leading indirectly to a rise in blood pressure.

**repolarization** Reattainment of the resting membrane potential after a nerve or muscle cell has depolarized.

**reproductive system** An organ system specialized for the production of offspring.

**resistance** **1.** A nonspecific ability to ward off infection or disease regardless of whether the body has been previously exposed to it. *Compare* immunity. **2.** A force that opposes the flow of a fluid such as air or blood. **3.** A force, or load, that opposes the action of a muscle or lever.

**resistance exercise** A physical exercise such as weight lifting that promotes muscle strength more than it promotes cardiopulmonary efficiency, endurance, or fatigue resistance. *Compare* endurance exercise.

**respiratory system** An organ system specialized for the intake of air and exchange of gases with the blood, consisting of the lungs and the air passages from the nose to the bronchi.

**resting membrane potential (RMP)** A stable voltage across the plasma membrane of an unstimulated cell.

**reticular cell** (reh-TIC-you-lur) A delicate, branching macrophage found in the reticular connective tissue of the lymphatic organs.

**reticular fiber** A fine, branching collagen fiber coated with glycoprotein, found in the stroma of lymphatic organs and some other tissues and organs.

**reticular tissue** A connective tissue composed of reticular cells and reticular fibers, found in bone marrow, lymphatic organs, and in lesser amounts elsewhere.

**ribonucleic acid** (RY-bo-new-CLAY-ic) Any of three types of nucleotide polymers smaller than DNA that play various roles in protein synthesis. Composed of ribose, phosphate, adenine, uracil, cytosine, and guanine forming a single nucleotide chain.

**ribosome** A granule found free in the cytoplasm or attached to the rough endoplasmic reticulum, composed of ribosomal RNA and enzymes; specialized to read the nucleotide sequence of messenger RNA and assemble a corresponding sequence of amino acids to make a protein.

**risk factor** Any environmental factor or characteristic of an individual that increases one's chance of developing a particular disease; includes such intrinsic factors as age, sex, and race and such extrinsic factors as diet, smoking, and occupation.

**rostral** Relatively close to the forehead, especially in reference to structures of the brain and spinal cord; for example, the frontal lobe is rostral to the parietal lobe. *Compare* caudal.

**ruga** (ROO-ga) **1.** An internal fold or wrinkle in the mucosa of a hollow organ such as the stomach and urinary bladder; typically present when the organ is empty and relaxed but not when the organ is full and stretched. **2.** Tissue ridges in such locations as the hard palate and vagina. fig. 25.12

# S

**saccule** (SAC-yule) A saclike receptor in the inner ear with a vertical patch of hair cells, the macula sacculi; senses the orientation of the head and responds to vertical acceleration, as when riding in an elevator or standing up. fig. 16.19

**sagittal plane** (SADJ-ih-tul) Any plane that extends from ventral to dorsal and cephalic to caudal and divides the body into right and left portions. *Compare* median plane.

**sarcomere** (SAR-co-meer) In skeletal and cardiac muscle, the portion of a myofibril from one Z disc to the next, constituting one contractile unit. fig. 11.4

**sarcoplasmic reticulum (SR)** The smooth endoplasmic reticulum of a muscle cell, serving as a calcium reservoir. fig. 11.2

**scanning electron microscope (SEM)** A microscope that uses an electron beam in place of light to form high-resolution, three-dimensional images of the surfaces of objects; capable of much higher magnifications than a light microscope.

**sclerosis** (scleh-RO-sis) Hardening or stiffening of a tissue, as in multiple sclerosis of the central nervous system or atherosclerosis of the blood vessels.

**sebum** (SEE-bum) An oily secretion of the sebaceous glands that keeps the skin and hair pliable.

**secondary active transport** A mechanism in which solutes are moved through a plasma membrane by a carrier that does not itself use ATP but depends on a concentration gradient established by an active transport pump elsewhere in the cell.

**secondary sex characteristic** Any feature that develops at puberty, further distinguishes the sexes from each other, and promotes attraction between the sexes; examples include the distribution of subcutaneous fat, pitch of the voice, female breasts, male facial hair, and apocrine scent glands.

**secondary sex organ** An organ other than the ovaries and testes that is essential to reproduction, such as the external genitalia, internal genital ducts, and accessory reproductive glands.

**second messenger** A chemical that is produced within a cell (such as cAMP) or that enters a cell (such as calcium ions) in response to the binding of a messenger to a membrane receptor, and that triggers a metabolic reaction in the cell.

**secretion** 1. A chemical released by a cell to serve a physiological function, such as a hormone or digestive enzyme. 2. The process of releasing such a chemical, often by exocytosis. *Compare* excretion.

**section** *See* histological section.

**selectively permeable membrane** A membrane that allows some substances to pass through while excluding others; for example, the plasma membrane and dialysis membranes.

**semen** (SEE-men) The fluid ejaculated by a male, including spermatozoa and the secretions of the prostate and seminal vesicles.

**semicircular ducts** Three ring-shaped, fluid-filled tubes of the inner ear that detect angular accelerations of the head; each is enclosed in a bony passage called the semicircular canal. fig. 16.20

**semilunar valve** A valve that consists of crescent-shaped cusps, including the aortic and pulmonary valves of the heart and valves of the veins and lymphatic vessels. fig. 19.8

**semipermeable membrane** See selectively permeable membrane.

**senescence** (seh-NESS-ense) Degenerative changes that occur with age.

**sensation** Conscious perception of a stimulus; pain, taste, and color, for example, are not stimuli but sensations resulting from stimuli.

**sensory nerve fiber** An axon that conducts information from a receptor to the central nervous system; an afferent nerve fiber.

**serosa** (seer-OH-sa) See serous membrane.

**serous fluid** (SEER-us) A watery, low-protein fluid similar to blood serum, formed as a filtrate of the blood or tissue fluid or as a secretion of serous gland cells; moistens the serous membranes.

**serous membrane** A membrane such as the peritoneum, pleura, or pericardium that lines a body cavity or covers the external surfaces of the viscera; composed of a simple squamous mesothelium and a thin layer of areolar connective tissue.

**serum** 1. The fluid that remains after blood has clotted and the solids have been removed; essentially the same as blood plasma except for a lack of fibrinogen. Used as a vehicle for vaccines. 2. Serous fluid.

**sex chromosomes** The X and Y chromosomes, which determine the sex of an individual.

**shock** 1. Circulatory shock, a state of cardiac output that is insufficient to meet the body's physiological needs, with consequences ranging from fainting to death. 2. Insulin shock, a state of severe hypoglycemia caused by administration of insulin. 3. Spinal shock, a state of depressed or lost reflex activity inferior to a point of spinal cord injury. 4. Electrical shock, the effect of a current of electricity passing through the body, often causing muscular spasm and cardiac arrhythmia or arrest.

**simple epithelium** An epithelium in which all cells rest directly on the basement membrane; includes simple squamous, cuboidal, and columnar types, and pseudostratified columnar. fig. 5.3

**sinus** 1. An air-filled space in the cranium. 2. A modified, relatively dilated vein that lacks smooth muscle and is incapable of vasomotion, such as the dural sinuses of the cerebral circulation and coronary sinus of the heart. 3. A small fluid-filled space in an organ such as the spleen and lymph nodes. 4. Pertaining to the sino-atrial node of the heart, as in *sinus rhythm.*

**skeletal muscle** Striated voluntary muscle, almost all of which is attached to the bones.

**skeletal system** An organ system consisting of the bones, ligaments, bone marrow, periosteum, articular cartilages, and other tissues associated with the bones.

**smooth muscle** Nonstriated involuntary muscle found in the walls of the blood vessels, many of the viscera, and other places.

**sodium–glucose transport protein (SGLT)** A symport that simultaneously transports $Na^+$ and glucose into a cell.

**somatic** 1. Pertaining to the body as a whole. 2. Pertaining to the skin, bones, and skeletal muscles as opposed to the viscera. 3. Pertaining to cells other than germ cells.

**somatic nervous system** A division of the nervous system that includes efferent fibers mainly from the skin, muscles, and skeleton and afferent fibers to the skeletal muscles. *Compare* autonomic nervous system.

**somesthetic** 1. Pertaining to widely distributed *general senses* in the skin, muscles, tendons, joint capsules, and viscera, as opposed to the *special senses* found in the head only; also called *somatosensory*. 2. Pertaining to the cerebral cortex of the postcentral gyrus, which receives input from such receptors.

**sperm** 1. The fluid ejaculated by the male; semen. Contains spermatozoa and glandular secretions. 2. A spermatozoon.

**spermatogenesis** (SPUR-ma-toe-JEN-eh-sis) The production of sperm cells through a series of mitotic and meiotic cell divisions; male gametogenesis.

**spermatozoon** (SPUR-ma-toe-ZOE-on) A sperm cell.

**sphincter** (SFINK-tur) A ring of muscle that opens or closes an opening or passageway; found, for example, in the eyelids, around the urinary orifice, and at the beginning of a blood capillary.

**spinal column** See vertebral column.

**spinal cord** The nerve cord that passes through the vertebral column and constitutes all of the central nervous system except the brain.

**spinal nerve** Any of the 31 pairs of nerves that arise from the spinal cord and pass through the intervertebral foramina.

**spindle** 1. An elongated structure that is thick in the middle and tapered at the ends (fusiform). 2. A football-shaped complex of microtubules that guide the movement of chromosomes in mitosis and meiosis. fig. 4.14 3. A stretch receptor in the skeletal muscles. fig. 13.20

**spine** 1. The vertebral column. 2. A pointed process or sharp ridge on a bone, such as the styloid process of the cranium and spine of the scapula.

**splanchnic** (SPLANK-nic) Pertaining to the digestive tract.

**squamous** (SKWAY-mus) Having a flat, scaly shape; pertains especially to a class of epithelial cells. figs. 5.4, 5.12

**stem cell** Any undifferentiated cell that can divide and differentiate into more functionally specific cell types such as blood cells and germ cells.

**stenosis** (steh-NO-sis) The narrowing of a passageway such as a heart valve or uterine tube; a permanent, pathological constriction as opposed to physiological constriction of a passageway.

**stereocilium** An unusually long, sometimes branched microvillus lacking the axoneme and motility of a true cilium; serves such roles as absorption in the epididymis and sensory transduction in the inner ear.

**steroid** (STERR-oyd, STEER-oyd) A lipid molecule that consists of four interconnected carbon rings; cholesterol and several of its derivatives.

**stimulus** A chemical or physical agent in a cell's surroundings that is capable of creating a physiological response in the cell; especially agents detected by sensory cells, such as chemicals, light, and pressure.

**strain** The extent to which a body, such as a bone, is deformed when subjected to stress. *Compare* stress.

**stratified epithelium** A type of epithelium in which some cells rest on top of others instead of on the basement membrane; inclues stratified squamous, cuboidal, and columnar types, and transitional epithelium. fig. 5.3

**stress 1.** A mechanical force applied to any part of the body; important in stimulating bone growth, for example. *Compare* strain. **2.** A condition in which any environmental influence disturbs the homeostatic equilibrium of the body and stimulates a physiological response, especially involving the increased secretion of hormones of the pituitary–adrenal axis.

**stroke** *See* cerebrovascular accident.

**stroke volume** The volume of blood ejected by one ventricle of the heart in one contraction.

**stroma** The connective tissue framework of a gland, lymphatic organ, or certain other viscera, as opposed to the tissue (parenchyma) that performs the physiological functions of the organ.

**subcutaneous** (SUB-cue-TAY-nee-us) Beneath the skin.

**substrate 1.** A chemical that is acted upon and changed by an enzyme. **2.** A chemical used as a source of energy, such as glucose and fatty acids.

**substrate specificity** The ability of an enzyme to bind only one substrate or a limited range of related substrates.

**sulcus** (SUL-cuss) A groove in the surface of an organ, as in the cerebrum or heart.

**summation 1.** A phenomenon in which multiple stimuli combine their effects on a cell to produce a response; seen especially in nerve and muscle cells. **2.** A phenomenon in which multiple muscle twitches occur so closely together that a muscle fiber cannot fully relax between twitches but develops more tension than a single twitch produces. fig. 11.14

**superficial** Relatively close to the surface; opposite of deep. For example, the ribs are superficial to the lungs.

**superior** Higher than another structure or point of reference from the perspective of anatomical position; for example, the lungs are superior to the diaphragm.

**supination** (SOO-pih-NAY-shun) A rotational movement of the forearm that turns the palm so that it faces upward or forward. fig. 9.18

**surfactant** (sur-FAC-tent) A chemical that reduces the surface tension of water and enables it to penetrate other substances more effectively. Examples include pulmonary surfactant and bile acids.

**sympathetic nervous system** A division of the autonomic nervous system that issues efferent fibers through the thoracic and lumbar nerves and usually exerts adrenergic effects on its target organs; includes a chain of paravertebral ganglia adjacent to the vertebral column, and the adrenal medulla.

**symphysis** (SIM-fih-sis) A joint in which two bones are held together by fibrocartilage; for example, between bodies of the vertebrae and between the right and left pubic bones.

**symport** A cotransport protein that moves two solutes simultaneously through a plasma membrane in the same direction, such as the sodium–glucose transport protein.

**synapse** (SIN-aps) **1.** A junction at the end of an axon where it stimulates another cell. **2.** A gap junction between two cardiac or smooth muscle cells at which one cell electrically stimulates the other; called an *electrical synapse.*

**synaptic cleft** (sih-NAP-tic) A narrow space between the synaptic knob of an

axon and the adjacent cell, across which a neurotransmitter diffuses. fig. 12.20

**synaptic knob** The swollen tip at the distal end of an axon; the site of synaptic vesicles and neurotransmitter release. fig. 11.6

**synaptic vesicle** A spheroid organelle in a synaptic knob containing neurotransmitter.

**synergist** (SIN-ur-jist) A muscle that works with the agonist to contribute to the same overall action at a joint.

**synergistic** An effect in which two agents working together (such as two hormones) exert an effect that is greater than the sum of their separate effects. For example, neither follicle-stimulating hormone nor testosterone alone stimulates significant sperm production, but the two of them together stimulate production of vast numbers of sperm.

**synovial fluid** (sih-NO-vee-ul) A lubricating fluid similar to egg white in consistency, found in the synovial joint cavities and bursae.

**synovial joint** A point where two bones are separated by a narrow, encapsulated space filled with lubricating synovial fluid; most such joints are relatively mobile.

**synthesis reaction** A chemical reaction in which smaller molecules combine to form a larger one. *Compare* decomposition reaction.

**systemic** (sis-TEM-ic) Widespread or pertaining to the body as a whole, as in the systemic circulation.

**systemic circuit** All blood vessels that convey blood from the left ventricle to all organs of the body and back to the right atrium of the heart; all of the cardiovascular system except the heart and pulmonary circuit.

**systole** (SIS-toe-lee) The contraction of any heart chamber; ventricular contraction unless otherwise specified.

**systolic pressure** (sis-TOLL-ic) The peak arterial blood pressure measured during ventricular systole.

# T

**target cell** A cell acted upon by a nerve fiber, hormone, or other chemical messenger.

**tarsal** Pertaining to the ankle (tarsus).

**T cell** A type of lymphocyte involved in nonspecific defense, humoral immunity, and cellular immunity; occurs in several forms including helper, cytotoxic, and suppressor T cells and natural killer cells.

**tendon** A collagenous band or cord associated with a muscle, usually attaching it to a bone and transferring muscular tension to it.

**testis** The male gonad; produces spermatozoa and testosterone.

**tetanus 1.** A state of sustained muscle contraction produced by temporal summation as a normal part of contraction; also called *tetany*. **2.** Spastic muscle paralysis produced by the toxin of the bacterium *Clostridium tetani.*

**tetraiodothyronine** (TET-ra-EYE-oh-doe-TIIY-ro-neen) See thyroxine.

**thalamus** (THAL-uh-muss) The largest part of the diencephalon, located immediately inferior to the corpus callosum and bulging into each lateral ventricle; a point of synaptic relay of nearly all signals passing from lower levels of the CNS to the cerebrum. fig. 14.12a

**theory** An explanatory statement, or set of statements, that concisely summarizes the state of knowledge on a phenomenon and provides direction for further study; for example, the fluid mosaic theory of the plasma membrane and the sliding filament theory of muscle contraction.

**thermogenesis** The production of heat, for example, by shivering or by the action of thyroid hormones.

**thermoreceptor** A neuron specialized to respond to heat or cold, found in the skin and mucous membranes, for example.

**thermoregulation** Homeostatic regulation of the body temperature within a narrow range by adjustments of heat-promoting and heat-losing mechanisms.

**thorax** A region of the trunk between the neck and the diaphragm; the chest.

**threshold 1.** The minimum voltage to which the plasma membrane of a nerve or muscle cell must be depolarized before it produces an action potential. **2.** The minimum combination of stimulus intensity and duration needed to generate an afferent signal from a sensory receptor.

**thrombosis** (throm-BO-sis) The formation or presence of a thrombus.

**thrombus** A clot that forms in a blood vessel or heart chamber; may break free and travel in the bloodstream as a thromboembolus.

**thymine** A single-ringed nitrogenous base (pyrimidine) found in DNA, complementary to adenine in the double helix of DNA. fig. 4.2

**thymus** A lymphatic organ in the mediastinum superior to the heart; the site where T lymphocytes differentiate and become immunocompetent. fig. 21.10

**thyroid gland** An endocrine gland in the neck, partially encircling the trachea immediately inferior to the larynx. fig. 17.9

**thyroid hormone** Either of two similar hormones, thyroxine and triiodothyronine, synthesized from iodine and tyrosine.

**thyroid-stimulating hormone (TSH)** A hormone of the anterior pituitary gland that stimulates the thyroid gland; also called *thyrotropin.*

**thyroxine ($T_4$)** (thy-ROCK-seen) The thyroid hormone secreted in greatest quantity, with four iodine atoms; also called *tetraiodothyronine.* fig. 17.19

**tight junction** A zipperlike junction between epithelial cells that limits the passage of substances between them. fig. 5.28

**tissue** An aggregation of cells and extracellular materials, usually forming part of an organ and performing some discrete function for it; the four primary classes are epithelial, connective, muscular, and nervous tissue.

**tissue gel** The viscous colloid that forms the ground substance of many tissues; gets its consistency from hyaluronic acid or other glycosaminoglycans.

**totipotent** Pertaining to a stem cell of the early preembryo, prior to development of a blastocyst, that has the potential to develop into any type of embryonic or adult cell.

**trabecula** (tra-BEC-you-la) A thin plate or layer of tissue, such as the calcified trabeculae of spongy bone or the fibrous trabeculae that subdivide a gland. fig. 7.5

**trachea** (TRAY-kee-uh) A cartilage-supported tube from the inferior end of the larynx to the origin of the primary bronchi; conveys air to and from the lungs; the "windpipe."

**transcription** The process of enzymatically reading the nucleotide sequence of a gene and synthesizing a pre-mRNA molecule with a complementary sequence.

**transducer** Any device that converts one form of energy to another, such as a sense organ, which converts a stimulus into an encoded pattern of action potentials.

**translation** The process of enzymatically reading an mRNA molecule and synthesizing the protein encoded in its nucleotide sequence.

**transmembrane protein** A protein that extends through a plasma membrane and contacts both the extracellular and intracellular fluid. fig. 3.7

**transmission electron microscope (TEM)** A microscope that uses an electron

beam in place of light to form high-resolution, two-dimensional images of ultrathin slices of cells or tissues; capable of extremely high magnification.

**triglyceride** (try-GLISS-ur-ide) A lipid composed of three fatty acids joined to a glycerol; also called a *triacylglycerol* or *neutral fat.* fig. 2.19

**triiodothyronine ($T_3$)** (try-EYE-oh-doe-THY-ro-neen) A thyroid hormone with three iodine atoms, secreted in much lesser quantities than thyroxine. fig. 17.19

**trisomy-21** The presence of three copies of chromosome 21 instead of the usual two; causes variable degrees of mental retardation, a shortened life expectancy, and structural anomalies of the face and hands.

**tropic hormone** (TROPE-ic) A hormone of the anterior pituitary gland that stimulates secretion by another endocrine gland. The four tropic hormones are FSH, LH, TSH, and ACTH.

**trunk 1.** That part of the body excluding the head, neck, and appendages. **2.** A major blood vessel, lymphatic vessel, or nerve that gives rise to smaller branches; for example, the pulmonary trunk and spinal nerve trunks.

**T tubule** A tubular extension of the plasma membrane of a muscle cell that conducts action potentials into the sarcoplasm and excites the sarcoplasmic reticulum. fig. 11.2

**tunic** (TOO-nic) A layer that encircles or encloses an organ, such as the tunics of a blood vessel or eyeball.

**tympanic membrane** The eardrum.

# U

**ultraviolet radiation** Invisible, ionizing, electromagnetic radiation with shorter wavelength and higher energy than violet light; causes skin cancer and photoaging of the skin but is required in moderate amounts for the synthesis of vitamin D.

**umbilical** (um-BIL-ih-cul) **1.** Pertaining to the cord that connects a fetus to the placenta. **2.** Pertaining to the navel (umbilicus).

**unipotent** Pertaining to a stem cell that has the potential to develop into only one type of fully differentiated, functional cell, such as an epidermal cell that can become only a keratinocyte.

**unit membrane** Any cellular membrane composed of a bilayer of phospholipids and embedded proteins. A single unit membrane forms the plasma membrane and encloses many organelles of a cell, whereas double unit membranes enclose the nucleus and mitochondria.

**unmyelinated** (un-MY-eh-lih-nay-ted) Lacking a myelin sheath. fig. 12.7

**upper limb** The appendage that arises from the shoulder, consisting of the brachium from shoulder to elbow, the antebrachium from elbow to wrist, the wrist, and the hand; loosely called the *arm,* but that term properly refers only to the brachium.

**uracil** A single-ringed nitrogenous base (pyrimidine) found in RNA; one of the four bases of the genetic code; occupies the place in RNA that thymine does in DNA. fig. 4.2

**urea** (you-REE-uh) A nitrogenous waste produced from two ammonia molecules and carbon dioxide; the most abundant nitrogenous waste in the blood and urine. fig. 23.2

**urinary system** An organ system specialized to filter the blood plasma, excrete waste products from it, and regulate the body's water, acid–base, and electrolyte balance.

**uterine tube** A duct that extends from the ovary to the uterus and conveys an egg or conceptus to the uterus; also called *fallopian tube* or *oviduct.*

**utricle** (YOU-trih-cul) A saclike receptor in the inner ear with a horizontal patch of hair cells, the macula utriculi; senses the orientation of the head and responds to horizontal acceleration, as when riding in a car that starts and stops. fig. 16.19

# V

**van der Waals force** A weak attraction between two atoms occurring when a brief fluctuation in the electron cloud density of one atom induces polarization of an adjacent atom; important in association of lipids with each other, protein folding, and protein–ligand binding.

**varicose vein** A vein that has become permanently distended and convoluted due to a loss of competence of the venous valves; especially common in the lower extremity, esophagus, and anal canal (where they are called hemorrhoids).

**vas** (vass) (plural, *vasa*) A vessel or duct.

**vascular** Pertaining to blood vessels.

**vasoconstriction** (VAY-zo-con-STRIC-shun) The narrowing of a blood vessel due to muscular constriction of its tunica media.

**vasodilation** (VAY-zo-dy-LAY-shun) The widening of a blood vessel due to relaxation of the muscle of its tunica media and the outward pressure of the blood exerted against the wall.

**vasomotion** (VAY-zo-MO-shun) Collective term for vasoconstriction and vasodilation.

**vasomotor center** A nucleus in the medulla oblongata that transmits efferent signals to the blood vessels and regulates vasomotion.

**vein** Any blood vessel that carries blood toward either atrium of the heart.

**ventral** Pertaining to the front of the body, the regions of the chest and abdomen; anterior.

**ventral (anterior) root** The branch of a spinal nerve that emerges from the anterior side of the spinal cord and carries efferent (motor) nerve fibers.

**ventricle** (VEN-trih-cul) A fluid-filled chamber of the brain or heart.

**venule** (VEN-yool) The smallest type of vein, receiving drainage from capillaries.

**vertebra** (VUR-teh-bra) One of the bones of the vertebral column.

**vertebral column** (VUR-teh-brul) A dorsal series of usually 33 vertebrae; encloses the spinal cord, supports the skull and thoracic cage, and provides attachment for the limbs and postural muscles. Also called *spine* or *spinal column.*

**vesicle** (VESS-ih-cul) A fluid-filled tissue sac or an organelle such as a synaptic or secretory vesicle.

**vesicular transport** The movement of particles or fluid droplets through the plasma membrane by the process of endocytosis or exocytosis.

**viscera** (VISS-er-uh) (singular, *viscus*) The organs contained in the dorsal and ventral body cavities, such as the brain, heart, lungs, stomach, intestines, and kidneys.

**visceral** (VISS-er-ul) **1.** Pertaining to the viscera. **2.** The inner or deeper layer of a two-layered membrane such as the pleura, pericardium, or glomerular capsule. *Compare* parietal. fig. A.8

**visceral muscle** Single-unit smooth muscle found in the walls of blood vessels and the digestive, respiratory, urinary, and reproductive tracts.

**viscosity** The resistance of a fluid to flow; the thickness or stickiness of a fluid.

**vitamin** A small organic nutrient that is absorbed undigested and serves a purpose other than being oxidized for energy; often serves as a coenzyme. Most vitamins cannot be synthesized by the body and are therefore a dietary necessity.

**vitreous body** (VIT-ree-us) A transparent, gelatinous mass that fills the space between the lens and retina of the eye.

**voluntary muscle** Muscle that is usually under conscious control; skeletal muscle.

**vulva** The female external genitalia; the mons, labia majora, and all superficial structures between the labia majora.

# W

**water balance** An equilibrium between fluid intake and output or between the amounts of fluid contained in the body's different fluid compartments.

**white matter** White myelinated nervous tissue deep to the cortex of the cerebrum and cerebellum and superficial to the gray matter of the spinal cord. fig. 14.6

# X

**X chromosome** The larger of the two sex chromosomes; males have one X chromosome and females have two in each somatic cell.

**xiphoid process** (ZIFF-oyd, ZYE-foyd) A small pointed cartilaginous or bony process at the inferior end of the sternum.

**X ray 1.** A high-energy, penetrating electromagnetic ray with wavelengths in the range of 0.1 to 10 nm; used in diagnosis and therapy. **2.** A photograph made with X rays; radiograph.

# Y

**Y chromosome** Smaller of the two sex chromosomes, found only in males and having little if any genetic function except development of the testis.

**yolk sac** An embryonic membrane that encloses the yolk in vertebrates that lay eggs and serves in humans as the origin of the first blood and germ cells.

# Z

**zygomatic arch** An arch of bone anterior to the ear, formed by the zygomatic processes of the temporal, frontal, and zygomatic bones; origin of the masseter muscle.

**zygote** A single-celled, fertilized egg.

Pronounce letter sequences in the pronunciation guides as follows:

| | |
|---|---|
| ah | as in father |
| al | as in pal |
| ay | as in day |
| bry | as in bribe |
| byu | as in bureau |
| c | as in calculus |
| cue | as in ridiculous |
| cuh | as in cousin |
| cul | as in bicycle |
| cus | as in custard |
| dew | as in dual |
| eez | as in ease |
| eh | as in feather |
| err | as in merry |
| fal | as in fallacy |
| few | as in fuse |
| ih | as in fit |
| iss | as in sister |
| lerr | as in lair |
| lur | as in learn |
| ma | as in man |
| mah | as in mama |
| me | as in meat |
| merr | as in merry |
| mew | as in music |
| muh | as in mother |
| na | as in corona |
| nerr | as in nary |
| new | as in news |
| nuh | as in nothing |
| odj | as in dodger |
| oe | as in go |
| oh | as in home |
| ol | as in alcohol |
| oll | as in doll |
| ose | as in gross |
| oss | as in floss |
| perr | as in pair |
| pew | as in pewter |
| ruh | as in rugby |
| serr | as in serration |
| sterr | as in stereo |
| sy | as in siren |
| terr | as in terrain |
| thee | as in theme |
| tirr | as in tyranny |
| uh | as in mother |
| ul | as in bicycle |
| verr | as in very |
| y | as in why |
| zh | as in measure |
| zy | as in enzyme |

# Credits

## Photographs

### Chapter 1

Opener: © Dr. Yorgos Nikas / SPL / Photo Researchers, Inc.; 1.1: National Library of Medicine / Peter Arnold, Inc.; 1.2: Art Resource; 1.3: © SPL / Photo Researchers, Inc.; 1.4a: Courtesy of the Armed Forces Institute of Pathology; 1.4b: © Bettmann / Corbis; 1.6: © Tim Davis / Photo Researchers, Inc.; 1.13a: © U.H.B. Trust / Tony Stone Images / Getty Images; 1.13b: Custom Medical Stock Photo, Inc.; 1.13c: © CNR / Phototake; 1.13d: © Tony Stone Images / Getty Images; 1.13e: © Monte S. Buchsbaum, Mt. Sinai School of Medicine, New York, NY; 1.14: © Alexander Tsiaras / Photo Researchers, Inc.; 1.14 (inset): © Scott Camazine / Sue Trainor / Photo Researchers, Inc.

### Atlas A

A.1: © McGraw-Hill Companies / Joe DeGrandis, photographer

### Chapter 2

Opener: Alfred Pasieka / Science Photo Library / Photo Researchers, Inc.; 2.3: © American Institute of Physics / Emilio Segre Visuals Archives, W.F. Meggers Collection; 2.10: © Ken Saladin

### Chapter 3

Opener: © P.M. Motta & T. Naguro / Photo Researchers, Inc.; 3.3: © K.G. Muri / Visuals Unlimited; 3.4a,b: From Cell Ultrastructure by William A. Jenson and Roderick B. Park © 1967 by Wadsworth Publishing Co., Inc. Reprinted by permission of the publisher; 3.6a: © Don Fawcett / Photo Researchers, Inc.; 3.10a: © Ed Reschke; 3.10b: Biophoto Associates / Photo Researchers, Inc.; 3.11a: Custom Medical Stock Photo, Inc.; 3.11c: © Biophoto Associates / Photo Researchers, Inc.; 3.16(a–c): © Dr. David M. Phillips / Visuals Unlimited; 3.22(all): Company of Biologists, Ltd.; 3.23: © Don Fawcett / Photo Researchers, Inc.; 3.24b: Courtesy of Dr. Birgit Satir, Albert Einstein College of Medicine; 3.25a: © Richard Chao; 3.25b: © E.G. Pollock; 3.26a,b: © Don Fawcett / Photo Researchers, Inc.; 3.27a: Visuals Unlimited; 3.28a,b: © Don Fawcett / Photo Researchers, Inc; 3.29a: Dr. Donald Fawcett & Dr. Porter / Visuals Unlimited; 3.30a: From: Manley McGill, D.P. Highfield, T.M. Monahan, and B.R. Brinkley "Effects of Nucleic Acid Specific Dyes on Centrioles of Mammalian Cells", published in the *Journal of Ultrastructure Research* 57, 43–53 (1976), pg. 48, fig. 6 © Academic Press; 3.31a1: © K.G. Murti / Visuals Unlimited; 3.31a2: © Biology Media / Photo Researchers, Inc.; 3.31b: © K.G. Murti / Visuals Unlimited

### Chapter 4

Opener: © Gopal Murti / Phototake NYC; 4.1a: © P. Motta & T. Naguro / SPL / Photo Researchers, Inc.; 4.4a: From "The Double Helix" by James D. Watson, 1968, Aentheneam Press, NY. Courtesy of Cold Springs Harbor Laboratory; 4.4b: Courtesy of King's College, London; 4.4c: © Bettman / Corbis; 4.9: © E.V. Kiseleva; 4.14(1–4): © Ed Reschke; 4.15b: © Biophoto Associates / Science Source / Photo Researchers, Inc.; 4.16: Peter Arnold, Inc.; 4.17a(both): The McGraw-Hill Companies, Inc. / Joe DeGrandis, photographer; 4.18(top): Comstock / Getty Images; 4.18(middle): Photodisc Red / Getty Images; 4.18(bottom): Photodisc Green / Getty Images; 4.19(top): From G. Pierrard, A. Nikkels. April 5, 2001, "A Medical Mystery." *New England Journal of Medicine,* 344: p. 1057. © 2001 Massachusetts Medical Society. All rights reserved; 4.19(middle): British Journal of Ophthalmology 1999;83:680 © by BMJ Publishing Group Ltd / http://www.bjophthalmol.com; 4.19(bottom): From G. Pierrard, A. Nikkels. April 5, 2001, "A Medical Mystery." *New England Journal of Medicine,* 344: p. 1057. © 2001 Massachusetts Medical Society. All rights reserved.

### Chapter 5

Opener: © Dr. Andrejs Liepins / Photo Researchers, Inc.; 5.4a, 5.5a: © The McGraw-Hill Companies, Inc. / Dennis Strete, photographer; 5.6a: © Lester V. Bergman / CORBIS; 5.7a: © The McGraw-Hill Companies, Inc. / Dennis Strete, photographer; 5.8a: © The McGraw-Hill Companies, Inc. / Joe DeGrandis, photographer; 5.9a: © Ed Reschke; 5.10a: Visuals Unlimited; 5.11a: © The McGraw-Hill Companies, Inc. / Dennis Strete, photographer; 5.13: © The McGraw-Hill Companies, Inc. / Rebecca Gray, photographer / Don Kincaid, dissections; 5.14a–5.20a: © The McGraw-Hill Companies, Inc. / Dennis Strete, photographer; 5.21a: © Ed Reschke / Peter Arnold, Inc.; 5.22a: © The McGraw-Hill Companies, Inc. / Dennis Strete, photographer; 5.23a–5.26a: © Ed Reschke; 5.27a: © The McGraw-Hill Companies, Inc. / Dennis Strete, photographer; 5.34: AP Wide World Image

### Chapter 6

Opener: © Ed Reschke / Peter Arnold, Inc.; 6.2b: © The McGraw-Hill Companies, Inc. / Dennis Strete, photographer; 6.3(left): © Tom McHugh / Photo Researchers, Inc.; 6.3(right): © The McGraw-Hill Companies, Inc. / Joe DeGrandis, photographer; 6.4: © Meckes / Ottawa / Photo Researchers, Inc.; 6.5a:

© The McGraw-Hill Companies, Inc. / Dennis Strete, photographer; 6.6a,b: © The McGraw-Hill Companies, Inc. / Dennis Strete, photographer; 6.7b: © CBS / Phototake; 6.7c: © P.M. Motta / SPL / Custom Medical Stock Photo, Inc.; 6.8 (all) & 6.10 (a–c): © The McGraw-Hill Companies, Inc. / Joe DeGrandis, photographer; 6.11a © NMSB / Custom Medical Stock Photo, Inc.; 6.11b: © Biophoto Associates / Photo Researchers, Inc.; 6.11c: © James Stevenson / SPL / Photo Researchers, Inc.; 6.12a: © SPL / Custom Medical Stock Photo, Inc.; 6.13b,c: © John Radcliffe / Photo Researchers, Inc.

## Chapter 7

Opener: © Cabiso / Visuals Unlimited; 7.5a&c: © D.W. Fawcett / Visuals Unlimited; 7.5d: Visuals Unlimited; 7.6: © Robert Calentine / Visuals Unlimited; 7.9 © Ken Saladin; 7.11: © Biophoto Associates / Photo Researchers, Inc.; 7.12: Courtesy of Utah Valley Regional Medical Center, Department of Radiology; 7.13: Victor Eroschenko; 7.14: © The McGraw-Hill Companies, Inc. / Joe DeGrandis, photographer; 7.21a: © Doug Sizemore / Visuals Unlimited; 7.21b: © SIU / Visuals Unlimited; 7.22a: © Michael Klein / Peter Arnold, Inc.; 7.22b: © Dr. P. Marazzi / Photo Researchers, Inc.; 7.22c: © Yoav Levy / Phototake

## Chapter 8

Opener: © SIU / Visuals Unlimited; 8.20: © The McGraw-Hill Companies, Inc. / Bob Coyle, photographer; 8.34c: © NHS Trust / Tony Stone Images / Getty Images; 8.42b: © Walter Reiter / Phototake

## Chapter 9

Opener: © Simon Fraser / SPL / Photo Researchers, Inc.; 9.1: © Gerard Vandystadt / Photo Researchers, Inc.; 9.12a–9.23a: © The McGraw-Hill Companies, Inc. / Timothy L. Vacula, photographer; 9.25a: © McGraw-Hill Companies / Rebecca Gray, pho-

tographer / Don Kincaid, dissections; 9.26: © Richard Anderson; 9.27: © The McGraw-Hill Companies, Inc. / Rebecca Gray, photographer / Don Kincaid, dissections; 9.30a: © Richard Anderson; 9.30b: CNRI / Science Photo Library / Photo Researchers, Inc.; 9.31a: © SIU / Visuals Unlimited; 9.31b: © Ron Mensching / Phototake; 9.31c: © SIU / Peter Arnold, Inc.; 9.31d: © Mehau Kulyk / SPL / Photo Researchers, Inc.

## Chapter 10

Opener: Custom Medical Stock Photo, Inc.; 10.1c: Victor Eroschenko; 10.5: © The McGraw-Hill Companies, Inc. / Rebecca Gray, photographer / Don Kincaid, dissections; 10.6 & 10.11: The McGraw-Hill Companies, Inc. / Joe DeGrandis, photographer; 10.12: © McGraw-Hill Companies / Rebecca Gray, photographer; / Don Kincaid, dissections; 10.16: © imaging-body.com; 10.19: © McGraw-Hill Companies / Rebecca Gray, photographer / Don Kincaid, dissections; 10.23a,b: © McGraw-Hill Companies / Rebecca Gray, photographer / Don Kincaid, dissections; 10.34: © The McGraw-Hill Companies, Inc. / Rebecca Gray, photographer / Don Kincaid, dissections; 10.36a,b: © The McGraw-Hill Companies, Inc./Photo and Dissection by Christine Eckel

## Atlas B

B.1a–B.15b: © The McGraw-Hill Companies, Inc. / Joe DeGrandis, photographer

## Chapter 11

Opener: © Don W. Fawcett / Photo Researchers, Inc.; 11.1: © Ed Reschke; 11.4a: Visuals Unlimited; 11.6a: Victor B. Eichler; 11.24: Shady Awwad, MD.

## Chapter 12

Opener: © SPL / Photo Researchers, Inc.; 12.7c: © The McGraw-Hill Companies, Inc. / Dr. Dennis Emery, Dept. of Zoology and Genetics, Iowa State University, photographer;

12.10: © Arici / Grazia Neri / Sygma / Corbis; 12.19: © Omikron / Science Source / Photo Researchers, Inc.; 12.31a: Custom Medical Stock Photo, Inc.; 12.31b: © Simon Fraser / Photo Researchers, Inc.

## Chapter 13

Opener: © Ed Reschke; 13.3: Biophoto Associates / Photo Researchers, Inc.; 13.7: MMP / Campix; 13.12: © From "A Stereoscopic Atlas of Anatomy" by David L. Bassett. Courtesy of Dr. Robert A. Chase, MD; 13.16: © The McGraw-Hill Companies, Inc. / Photo and Dissection by Christine Eckel.

## Chapter 14

Opener: © CNRI / Photo Researchers, Inc.; 14.1c: © The McGraw-Hill Companies, Inc. / Rebecca Gray, photographer / Don Kincaid, dissections; 14.2b: © The McGraw-Hill Companies, Inc. / Dennis Strete, photographer; 14.6c: © The McGraw-Hill Companies, Inc. / Rebecca Gray, photographer / Don Kincaid, dissections; 14.18a: © The McGraw-Hill Companies, Inc. / Bob Coyle, photographer; 14.27b: © The McGraw-Hill Companies, Inc. / Rebecca Gray, photographer / Don Kincaid, dissections; 14.34c: © The McGraw-Hill Companies, Inc. / Joe De Grandis, photographer; 14.40: © Marcus E. Raichle, MD, Washington University School of Medicine, St. Louis, Missouri

## Chapter 15

Opener & 15.3: © From "A Stereoscopic Atlas of Anatomy" by David L. Bassett. Courtesy of Dr. Robert A. Chase, MD

## Chapter 16

Opener: Wellcome Dept. of Cognitive Neurology / Science Photo Library / Photo Researchers, Inc.; 16.6c: © Ed Reschke; 16.10: © The McGraw-Hill Companies, Inc. / Joe DeGrandis, photographer; 16.14: Quest / Science Photo Library / Photo Researchers, Inc. Inc.; 16.22: © The McGraw-Hill Companies, Inc., Joe DeGrandis,

photographer; 16.27: © Ralph C. Eagle / MD / Photo Researchers; 16.28a: © Lisa Klancher; 16.34a: © The McGraw-Hill Companies, Inc. / Joe DeGrandis, photographer; 16.35a: Courtesy of Beckman Vision Center at UCSF School of Medicine / D. Copenhagen, S. Miltman, and M. Maglio

## Chapter 17

Opener: © P. Motta / SPL / Photo Researchers, Inc.; 17.5a: © Dr. John D. Cunningham / Visuals Unlimited; 17.5b: © Science VU / Visuals Unlimited; 17.9b: © Robert Calentine / Visuals Unlimited; 17.12c: © Ed Reschke; 17.13a: © Manfred Kage / Peter Arnold, Inc.; 17.13b: © Ed Reschke; 17.27(all): From "Clinical Pathological Conference Acromegaly, Diabetes, Hypermetabolism, Protein Use and Heart Failure" in American Journal of Medicine, 20:133, 1986. Copyright © 1986 by Excerpta Media, Inc.; 17.28: © CNR / Phototake; 17.29a,b: From "Atlas of Pediatric Physical Diagnosis", 3/e, by Zitelli & Davis, fig 9–17 1997. Mosby-Wolfe Europe Limited, London, UK; 17.30: Dr. John Money

## Chapter 18

Opener: © Juergen Berger, Max-Planck Institute / Photo Researchers, Inc.; 18.3: © Ton Koene / Peter Arnold, Inc.; 18.4b: © Richard J. Poole / Polaroid International Photomicrography Competition; 18.4c: © Don W. Fawcett / Visuals Unlimited; 18.10: © Meckes / Ottawa / Photo Researchers, Inc.; 18.11: Courtesy Schomburg Center for Research in Black Culture, The New York Public Library; 18.14: © Claude Revy / Phototake; 18.17: © SIU / Photo Researchers, Inc.; Tb18.7(1–3): © Ed Reschke; Tb 18.7(4&5): © Michael Ross / Photo Researchers, Inc; 18.19a: © Ed Reschke; 19.19b: © SIU / Photo Researchers, Inc.; 18.20a: NIBSC / Science photo Library /

**a-** no, not, without (atom, agranulocyte)

**ab-** away (abducens, abduction)

**acetabulo-** small cup (acetabulum)

**acro-** tip, extremity, peak (acromion, acromegaly)

**ad-** to, toward, near (adsorption, adrenal)

**adeno-** gland (lymphadenitis, adenohypophysis)

**aero-** air, oxygen (aerobic, anaerobe, aerophagy)

**af-** toward (afferent)

**ag-** together (agglutination)

**-al** pertaining to (parietal, pharyngeal, temporal)

**ala-** wing (ala nasi)

**albi-** white (albicans, linea alba, albino)

**algi-** pain (analgesic, myalgia)

**aliment-** nourishment (alimentary)

**allo-** other, different (allele, allograft)

**amphi-** both, either (amphiphilic, amphiarthrosis)

**an-** without (anaerobic, anemic)

**ana- 1.** up, build up (anabolic, anaphylaxis). **2.** apart (anaphase, anatomy). **3.** back (anastomosis)

**andro-** male (androgen)

**angi-** vessel (angiogram, angioplasty, hemangioma)

**ante-** before, in front (antebrachium)

**antero-** forward (anterior, anterograde)

**anti-** against (antidiuretic, antibody, antagonist)

**apo-** from, off, away, above (apocrine, aponeurosis)

**arbor-** tree (arboreal, arborization)

**artic- 1.** joint (articulation). **2.** speech (articulate)

**-ary** pertaining to (axillary, coronary)

**-ase** enzyme (polymerase, kinase, amylase)

**ast-, astro-** star (aster, astrocyte)

**-ata, -ate 1.** possessing (hamate, corniculate). **2.** plural of -a (stomata, carcinomata)

**athero-** fat (atheroma, atherosclerosis)

**atrio-** entryway (atrium, atrioventricular)

**auri-** ear (auricle, binaural)

**auto-** self (autolysis, autoimmune)

**axi-** axis, straight line (axial, axoneme, axon)

**baro-** pressure (baroreceptor, hyperbaric)

**bene-** good, well (benign, beneficial)

**bi-** two (bipedal, biceps, bifid)

**bili-** bile (biliary, bilirubin)

**bio-** life, living (biology, biopsy, microbial)

**blasto-** precursor, bud, producer (fibroblast, osteoblast, blastomere)

**brachi-** arm (brachium, brachialis, antebrachium)

**brady-** slow (bradycardia, bradypnea)

**bucco-** cheek (buccal, buccinator)

**burso-** purse (bursa, bursitis)

**calc-** calcium, stone (calcaneus, hypocalcemia)

**callo-** thick (callus, callosum)

**calori-** heat (calorie, calorimetry, calorigenic)

**calv-, calvari-** bald, skull (calvaria)

**calyx** cup, vessel, chalice (glycocalyx, renal calyx)

**capito-** head (capitis, capitate, capitulum)

**capni-** smoke, carbon dioxide (hypocapnia)

**carcino-** cancer (carcinogen, carcinoma)

**cardi-** heart (cardiac, cardiology, pericardium)

**carot- 1.** carrot (carotene). **2.** stupor (carotid)

**carpo-** wrist (carpus, metacarpal)

**case-** cheese (caseosa, casein)

**cata-** down, break down (catabolism)

**cauda-** tail (cauda equina, caudate nucleus)

**-cel** little (pedicel)

**celi-** belly, abdomen (celiac)

**centri-** center, middle (centromere, centriole)

**cephalo-** head (cephalic, encephalitis)

**cervi-** neck, narrow part (cervix, cervical)

**chiasm-** cross, X (optic chiasm)

**choano-** funnel (choana)

**chole-** bile (cholecystokinin, cholelithotripsy)

**chondro- 1.** grain (mitochondria). **2.** cartilage, gristle (chondrocyte, perichondrium)

**chromo-** color (dichromat, chromatin, cytochrome)

**chrono-** time (chronotropic, chronic)

**cili-** eyelash (cilium, superciliary)

**circ-** about, around (circadian, circumduction)

**cis-** cut (incision, incisor)

**cisterna** reservoir (cisterna chyli)

**clast-** break down, destroy (osteoclast)

**clavi-** hammer, club, key (clavicle, supraclavicular)

**-cle** little (tubercle, corpuscle)

**cleido-** clavicle (sternocleidomastoid)

**cnemo-** lower leg (gastrocnemius)

**co-** together (coenzyme, cotransport)

**collo- 1.** hill (colliculus). **2.** glue (colloid, collagen)

**contra-** opposite (contralateral)

**corni-** horn (cornified, corniculate, cornu)

**corono-** crown (coronary, corona, coronal)

**corpo-** body (corpus luteum, corpora quadrigemina)

**corti-** bark, rind (cortex, cortical)

**costa-** rib (intercostal, subcostal)

**coxa-** hip (os coxae, coxal)

**crani-** helmet (cranium, epicranius)

**cribri-** sieve, strainer (cribriform, area cribrosa)

**crino-** separate, secrete (holocrine, endocrinology)

**crista-** crest (crista galli, mitochondrial crista)

**crito-** to separate (hematocrit)

**cruci-** cross (cruciate ligament)

**-cule, -culus** small (canaliculus, trabecula, auricular)

**cune-** wedge (cuneiform, cuneatus)

**cutane-, cuti-** skin (subcutaneous, cuticle)

**cysto-** bladder (cystitis, cholecystectomy)

**cyto-** cell (cytology, cytokinesis, monocyte)

**de-** down (defecate, deglutition, dehydration)

**demi-** half (demifacet, demilune)

**den-, denti-** tooth (dentition, dens, dental)

**dendro-** tree, branch (dendrite, oligodendrocyte)

**derma-, dermato-** skin (dermatology, hypodermic)

**desmo-** band, bond, ligament (desmosome, syndesmosis)

**dia- 1.** across, through, separate (diaphragm, dialysis). **2.** day (circadian)

**dis- 1.** apart (dissect, dissociate). **2.** opposite, absence (disinfect, disability)

**diure-** pass through, urinate (diuretic, diuresis)

**dorsi-** back (dorsal, dorsum, latissimus dorsi)

**duc-** to carry (duct, adduction, abducens)

**dys-** bad, abnormal, painful (dyspnea, dystrophy)

**e-** out (ejaculate, eversion)

**-eal** pertaining to (hypophyseal, arboreal)

**ec-, ecto-** outside, out of, external (ectopic, ectoderm, splenectomy)

**ef-** out of (efferent, effusion)

**-el, -elle** small (fontanel, organelle, micelle)

**electro-** electricity (electrocardiogram, electrolyte)

**em-** in, within (embolism, embedded)

**emesi-, emeti-** vomiting (emetic, hyperemesis)

**-emia** blood condition (anemia, hypoxemia)

**en-** in, into (enzyme, parenchyma)

**encephalo-** brain (encephalitis, telencephalon)

**enchymo-** poured in (mesenchyme, parenchyma)

**endo-** within, into, internal (endocrine, endocytosis)

**entero-** gut, intestine (mesentery, myenteric)

**epi-** upon, above (epidermis, epiphysis, epididymis)

**ergo-** work, energy, action (allergy, adrenergic)

**eryth-, erythro-** red (erythema, erythrocyte)

**esthesio-** sensation, feeling (anesthesia, somesthetic)

**eu-** good, true, normal, easy (eupnea, aneuploidy)

**exo-** out (exopeptidase, exocytosis, exocrine)

**facili-** easy (facilitated)

**fasci-** band, bundle (fascia, fascicle)

**fenestr-** window (fenestrated)

**fer-** to carry (efferent, uriniferous)

**ferri-** iron (ferritin, transferrin)

**fibro-** fiber (fibroblast, fibrosis)

**fili-** thread (myofilament, filiform)

**flagello-** whip (flagellum)

**foli-** leaf (folic acid, folia)

**-form** shape (cuneiform, fusiform)

**fove-** pit, depression (fovea)

**funiculo-** little rope, cord (funiculus)

**fusi-** 1. spindle (fusiform). 2. pour out (perfusion)

**gamo-** marriage, union (monogamy, gamete)

**gastro-** belly, stomach (gastrointestinal, digastric)

**-gen, -genic, -genesis** producing, giving rise to (pathogen, carcinogenic, glycogenesis)

**genio-** chin (geniohyoid, genioglossus)

**germi-** 1. sprout, bud (germinal, germinativum). 2. microbe (germicide)

**gero-** old age (progeria, geriatrics, gerontology)

**gesto-** 1. to bear, carry (ingest). 2. pregnancy (gestation, progesterone)

**glia-** glue (neuroglia, microglia)

**globu-** ball, sphere (globulin, hemoglobin)

**glom-** ball (glomerulus)

**glosso-** tongue (glossopharyngeal, hypoglossal)

**glyco-** sugar (glycogen, glycolysis, hypoglycemia)

**gono-** 1. angle, corner (trigone). 2. seed, sex cell, generation (gonad, oogonium, gonorrhea)

**gradi-** walk, step (retrograde, gradient)

**-gram** recording of (electrocardiogram, sonogram)

**-graph** recording instrument (sonograph, electrocardiograph)

**-graphy** recording process (sonography, radiography)

**gravi-** severe, heavy (gravid, myasthenia gravis)

**gyro-** turn, twist (gyrus)

**hallu-** great toe (hallux, hallucis)

**hemi-** half (hemidesmosome, hemisphere)

**-hemia** blood condition (polycythemia)

**hemo-** blood (hemophilia, hemoglobin, hematology)

**hetero-** different, other, various (heterozygous)

**histo-** tissue, web (histology, histone)

**holo-** whole, entire (holistic, holocrine)

**homeo-** constant, unchanging, uniform (homeostasis, homeothermic)

**homo-** same, alike (homologous, homozygous)

**hyalo-** clear, glassy (hyaline, hyaluronic acid)

**hydro-** water (dehydration, hydrolysis, hydrophobic)

**hyper-** above, above normal, excessive (hyperkalemia, hypertonic)

**hypo-** below, below normal, deficient (hypogastric, hyponatremia, hypophysis)

**-ia** condition (anemia, hypocalcemia, osteomalacia)

**-ic** pertaining to (isotonic, hemolytic, antigenic)

**-icle, -icul** small (ossicle, canaliculus, reticular)

**ilia-** flank, loin (ilium, iliac)

**-illa, -illus** little (bacillus)

**-in** protein (trypsin, fibrin, globulin)

**infra-** below (infraspinous, infrared)

**ino-** fiber (inotropic, inositol)

**insulo-** island (insula, insulin)

**inter-** between (intercellular, intervertebral)

**intra-** within (intracellular, intraocular)

**iono-** ion (ionotropic, cationic)

**ischi-** to hold back (ischium, ischemia)

**-ism** 1. process, state, condition (metabolism, rheumatism). 2. doctrine, belief, theory (holism, reductionism, naturalism)

**iso-** same, equal (isometric, isotonic, isomer)

**-issimus** most, greatest (latissimus, longissimus)

**-ite** little (dendrite, somite)

**-itis** inflammation (dermatitis, gingivitis)

**jug-** to join (conjugated, jugular)

**juxta-** next to (juxtamedullary, juxtaglomerular)

**kali-** potassium (hypokalemia)

**karyo-** seed, nucleus (megakaryocyte, karyotype)

**kerato-** horn (keratin, keratinocyte)

**kine-** motion, action (kinetic, kinase, cytokinesis)

**labi-** lip (labium, levator labii)

**lacera-** torn, cut (foramen lacerum, laceration)

**lacrimo-** tear, cry (lacrimal gland, nasolacrimal)

**lacto-** milk (lactose, lactation, prolactin)

**lamina-** layer (lamina propria, laminar flow)

**latero-** side (bilateral, ipsilateral)

**lati-** broad (fascia lata, latissimus dorsi)

**-lemma** husk (sarcolemma, neurilemma)

**lenti-** lens (lentiform)

**-let** small (platelet)

**leuko-** white (leukocyte, leukemia)

**levato-** to raise (levator labii, elevation)

**ligo-** to bind (ligand, ligament)

**line-** line (linea alba, linea nigra)

**litho-** stone (otolith, lithotripsy)

**-logy** study of (histology, physiology, hematology)

**lucid-** light, clear (stratum lucidum, zona pellucida)

**lun-** moon, crescent (lunate, lunule, semilunar)

**lute-** yellow (macula lutea, corpus luteum)

**lyso-, lyto-** split apart, break down (lysosome, hydrolysis, electrolyte, hemolytic)

**macro-** large (macromolecule, macrophage)

**macula-** spot (macula lutea, macula densa)

**mali-** bad (malignant, malocclusion, malformed)

**malle-** hammer (malleus, malleolus)

**mammo-** breast (mammary, mammillary)

**mano-** hand (manus, manipulate)

**manubri-** handle (manubrium)

**masto-** breast (mastoid, gynecomastia)

**medi-** middle (medial, mediastinum, intermediate)

**medullo-** marrow, pith (medulla)

**mega-** large (megakaryocyte, hepatomegaly)

**melano-** black (melanin, melanocyte, melancholy)

**meno-** month (menstruation, menopause)

**mento-** chin (mental, mentalis)

**mero-** part, segment (isomer, centromere, merocrine)

**meso-** in the middle (mesoderm, mesentery)

**meta-** beyond, next in a series (metaphase, metacarpal)

**metabol-** change (metabolism, metabolite)

**-meter** measuring device (calorimeter, spirometer)

**metri-** 1. length, measure (isometric, emmetropic). 2. uterus (endometrium)

**micro-** small (microscopic, microcytic, microglia)

**mito-** thread, filament, grain (mitochondria, mitosis)

**mono-** one (monocyte, monogamy, mononucleosis)

**morpho-** form, shape, structure (morphology, amorphous)

**muta-** change (mutagen, mutation)

**myelo-** 1. spinal cord (poliomyelitis, myelin). 2. bone marrow (myeloid, myelocytic)

**myo-, mysi-** muscle (myoglobin, myosin, epimysium)

**natri-** sodium (hyponatremia, natriuretic)

**neo-** new (neonatal, gluconeogenesis)

**nephro-** kidney (nephron, hydronephrosis)

**neuro-** nerve (aponeurosis, neurosoma, neurology)

**nucleo-** nucleus, kernel (nucleolus, nucleic acid)

**oo-** egg (oogenesis, oocyte)